VR
88
011
142239

TURNER PUBLISHING COMPANY

Turner Publishing Company
412 Broadway • P.O. Box 3101
Paducah, Kentucky 42002-3101
(502) 443-0121

Turner Publishing Company Staff:
Publishing Consultant: Douglas W. Sikes
Project Coordinator: John Mark Jackson
Designer: Shelley R. Davidson

ISBN: 1-56311-381-3
Library of Congress Catalog
Card Number: 98-88601

Printed in the United States of America. Additional copies may be purchased directly from the publisher. Limited Edition.

Table of Contents

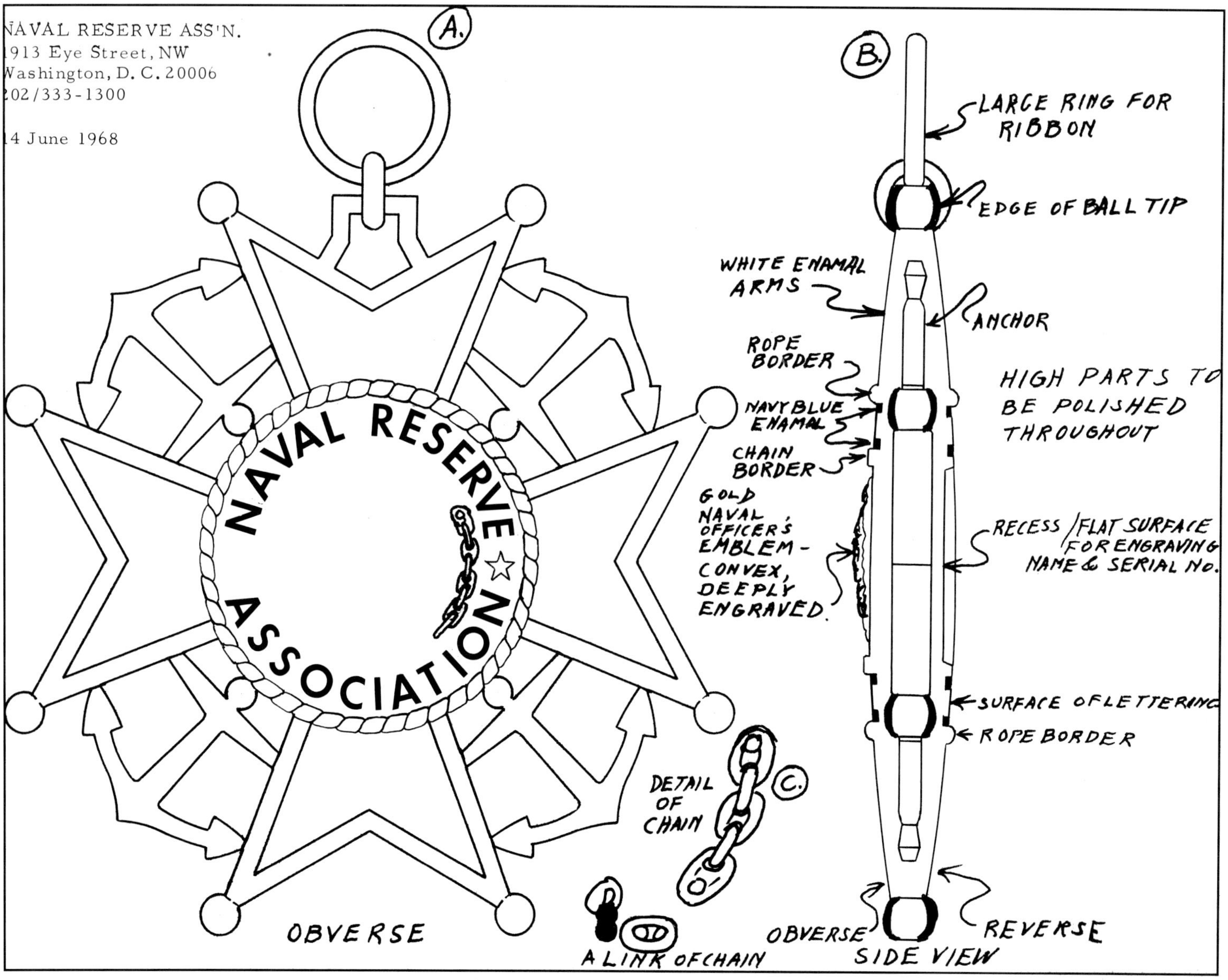

Original working drawing of the obverse.

Some of the Stages in the Development of the Badge of the Naval Reserve Association

Pictures of the first badges. From left: Standard Badge, Service Ribbon, and Miniature Badge. The Gold Star is worn by Life Members.

At left: *Artist's Rendering of Design, by LTCDR Richard Biddle. Subsequent Changes including color changes: subsequent changes include use of dolphins for the medal hanger, reducing the size of the center device, using blue arms instead of white, and changing the ribbon to Navy blue, gold, and white.*

INTRODUCTION

This is the story of the Naval Reserve Association; its origins, and its history; from the earliest attempts to form the association, to the magnificent people that had the vision to make it what it is today.

The Naval Reserve Association owes its origin and purpose to 43 Naval Reserve officers who, on 18 September 1954, had the courage, stamina and foresight to found and organize the Naval Reserve Association and the fortitude and persistence to ensure the association's early and continued growth and credibility.

We salute our founders for their dedication and their generous sacrifice of time, talent and resources in establishing the Naval Reserve Association.

It is a story of visionary pioneers and glorious triumph, as they developed, tried, refined and perfected our association.

This is their story, told often in their words, as seen through their eyes, as you read the story, page through the book and try to visualize those who have gone before us, those who gave many hours for the defense of freedom and our way of life, those whose legacy to today's force is a tradition of dedication to professional excellence and to uncompromising honor.

The Naval Reserve Association is pleased to present to you this history of the Naval Reserve Association. We believe that you will indeed enjoy the magnificent first portion about our founders, in which it presents the history during the lifetimes of those whose biographies are the cornerstone of this book. Similarly, we believe that all will be delighted by the remarkable presentation of the early years and the following years of development.

In the other sections, corporate histories of the Naval Reserve and the assistance provided by the Naval Reserve are discussed, bringing additional perspectives to our beloved profession.

We wish to thank the many members of the Naval Reserve, heroes all, in taking time to contribute and make this effort possible.

Arthur C. Monson
Captain, USNR (Ret)
National President

Publisher's Message

In a world filled with minute by minute sound bites of continual political and military unrest somewhere at all times, bringing to mind an era of "wars and rumors of wars," one can rest more assuredly knowing that preparedness is war's greatest deterrent. And for preparedness, nothing exceeds the United States Reserves in the provision that strength remains present, accessible and ever more importantly, affordable. And none exhibit that necessary, far-reaching presence as well as the Naval Reserve. This clear vision of expandable national defense is nowhere better supported and promoted than by this organization. Therein is the story of the Naval Reserve Association.

This important chronicle is the product of many and their labor of love for their Naval Reserve Association. Some of those that we would like to recognize for their contribution are RADM James Carey, Past President; RADM J. E. Forrest, former Executive Director; RADM Thomas F. Hall, current Executive Director; CAPT John Rice, Jr., Past President; CAPT Arthur C. Monson, current National President; and especially for the persistent efforts of Naval Reserve Association Headquarters Staff who helped make it all come together into the book you now hold.

Turner Publishing Company would like to thank the Naval Reserve Association for allowing us to help tell your noble tale and wish you continued success in your purpose.

Dave Turner

Dave Turner, President
Turner Publishing Company

Naval Reserve Association History

History Of The Naval Reserve Association

Predecessors of the NRA

The Naval Reserve Association is the premier association supporting the US Naval Reserve, but like the Reserve Force itself, its beginnings go back earlier than just the founding of the organization.

The Naval Reserve was founded in 1915, as the nation prepared for World War I. It grew out of the various naval militias organized by several different states and out of the need to have a larger navy for the coming conflict. More than 300,000 personnel (including 11,275) served in the Naval reserve during World War I. By November 1918 the reserve's strength stood at 20,000 officers and 270,000 enlisted men.

Many of the returning veterans wanted to find some way to retain a connection both with the Navy and with their shipmates. Towards this end, they formed the Naval Reserve Officers Association (NROA). The new organization was comprised of Navy and Marine Corps officers and received the support of the Assistant Secretary of the Navy, Franklin D. Roosevelt.

The new organization had its work cut out for it. national apathy and the idea that the nation had just fought "The War to End All Wars," meant there was little public support for maintaining a large Navy. In September 1921 nearly the entire Naval Reserve Force was disenrolled. The only members allowed to remain were former regular officers and enlisted members of the USN, and members of the reserve who agreed to affiliate in a non-pay non-drill status. While eviscerating the reserves, the agreement to keep the former regulars and the on-pay personnel meant a small core of committed personnel was retained for the future.

Matters did not stay this way for long, and the Navy agreed in 1922 to a reserve structure of 107 drilling units comprised of 2,000 officers and 14,000 enlisted men. These men drilled both at shore stations and aboard 56 Eagle patrol boats and gunboats assigned for their specific use.

The growth of the reserves led, in October 1922, to the formation of the Reserve Officers Association (ROA) of the United States. The ROA in early 1923 formed a Naval legislative Committee to deal specifically with naval issues.

The Naval Reserve maintained a lean existence during the 1902s and 1930s. While never completely disestablished by Congress or Executive branch cutters, the reserves did suffer reduced pay during the Depression. The lack of funding for the Reserves also was created, in part, by the desire of the regular Navy to retain as much funding for itself as it could.

The Naval Reserve Force, and the nation, awoke out of the doldrums during the late 1930s, as war clouds loomed over the horizon. On September 8, 1939, now President Franklin D. Roosevelt declared a Limited National Emergency. This was followed in 1940 by the Selective Service Act, which drafted men for service in the military and the organization of the V-7 program. Men who entered the V-7 program received 90 days training and a commission as officers in the Naval Reserve. The "90-day Wonders" provided the bulk of naval officers during World War II, while the program served as the progenitor of Officer Candidate School.

Despite the growing fears of war, the Navy had no concrete plan for recalling reservists as late as the summer of 1940. A scattered recall had been going on, and by the autumn of that year 816 officers and 1,136 enlisted men were on active duty. this situation changed in 1941. On May 27, President Roosevelt declared an Unlimited National Emergency, and by July 8, 1941 all 30,000 naval reservists had been recalled to indefinite active duty. The Naval Reserve at this point lost its separate organizational and administrative identity and all personnel were considered part of the regular Navy. As part of the move to war, the ROA and the NROA suspended their operations for the duration.

By the end of World War II, more than 3,000,000 men and women had served in the Naval Reserve. They formed more than 80 percent of the Navy's manpower and in the process won 13 Medal of Honor and 200 Navy Crosses. They commanded more than 85 percent of the patrol craft, minesweepers, and submarine chasers that proved so vital to winning the war effort. With the end of the war, most of these men returned back to their civilian careers, but once again, as in World War I, many wanted to retain a connection with the Navy.

As the war ended, some reservists in Washington DC organized the US Reserve Officers of the Naval Services (RONS). Both they and the ROA were allowed to recruit at separation centers, leading to suspicions that the Navy Department did not want a strong NROA confronting it for resources after the war.

NROA was reactivated in early 1946 and held its first post-war convention in October 1946. Most of the members of NROA were relatively senior officers. To take advantage of the vast pool of younger officers, the NROA began discussions with the RONS about a possible merger. These discussions continued through 1946 and 1947 and culminated in a merger in December 1947 as the RONS-NROA. The national officers contained many future members of the NRA, including CDR John Bracken, CAPT Jesse Draper, CAPT Alvin Fisher, LCDR Fred Kempe, CDR Paul Abbott, CDR Jim Crowder, LCDR Guy Stillman, LCDR Ross Currier, CAPT Mack Young, and CAPT Jim Ross.

That same year, the Air Force became a separate service, growing out of the former Army Air Corps. This decision was to have a major effect on the RONS-NROA. ROA revised its constitution to provide for an Air Force section, and began discussions with RONS-NROA about a potential merger. This merger took place in the autumn of 1948. The new organization agreed to respect service autonomy and rotate the Presidency among the three services.

World events meant the Naval Reserve was continually in the news. More than 2,000 naval reservists volunteered for active duty during the Berlin Airlift of 1948/49. Less than a year later, the Korean War found more than 250,000 Naval Reservists recalled for active duty. They formed almost 30 percent of the active-duty Navy and flew 75 percent of the combat air sorties during the war.

The end of the Korean War was different from the period after any other war. The Soviet Union presented a growing threat, and the briefing paper National Security Council (NSC) 68 recommended a strategy of meeting this threat by a military buildup and support for allied nations. Part of the new military buildup was a revision of reserve force structure. Studies for a new force structure had begun during the Korean War, as part of the National Security Training Commission's deliberations, and the Armed Forces Reserve Act, passed on July 9, 1952 mandated the creation of separate Ready and Standby Reserves, within the overall reserve force structure. The Ready Reserves could be called for active duty during a war declared by Congress or a Presidential-declared national emergency. The Standby Reserve could only be used during a war declared by Congress.

Reserve manpower policies also underwent a change during the mid-1950s. The 1954 report of the Reserve Forces Policy Board had reported serious deficiencies with the reserves, including authorized strength not geared to mobilization requirements, an unready Ready Reserve with very few personnel receiving serious or practical training, and a requirement that Congress determine the required numbers of reservists, rather than the President.

To remedy these shortcomings, President Eisenhower asked Secretary of Defense Charles Wilson and the Joint Chiefs of Staff for a reserve manpower plan in December 1954. Eisenhower wanted the reserves:

(1.) To be organized into units "in an immediate mobilization state of readiness as opposed to merely "ready" in the sense of having legal liability for immediate mobilization.

Early 1950s, left to right: Warren Cougenour, Kinsey Owens, Charles Buscher, and Vaughn Coale; drilling unit "W-1."

(2.) Men between 17 and 19 are allowed to volunteer directly into the Reserves and receive six months of training before returning to their units for nine and a half years' drilling obligation and, if insufficient quantities volunteered, the authority to draft men into the Reserve was requested.

(3.) Incentives are made to encourage men to enlist in National Guard components and receive training in the active forces.

(4.) Personnel who refused to satisfy drilling requirements be involuntarily "recalled to active duty in order to maintain or restore competencies." In addition, that the granting of honorable discharges be contingent upon individuals satisfying Reserve obligations as well as performance on active duty. Personnel should be screened to determine availability for service.

(5.) States is authorized to establish local defense forces to be available when the National Guard is federalized.

The Navy created its own force structure in line with the Armed Forces Reserve Act.

Category	Group	Regular Drills	AT
A	Ready	48	2 weeks required
B	Ready	24	2 weeks required
C	Ready	12	2 weeks required
D	Ready	0	2 weeks required
E	Standby	0	2 weeks permitted

As part of their force structure revision, the Naval Reserve also had to involuntarily separate 2,000 officers in 1954, because the service had more officers than authorized. A Naval Reserve officer Review Board was established in April 1954 to screen the records of 8,000 officers to find an additional 1,000 candidates for discharge in 1955. Officers who had not trained in two years were also discharged under the congressional mandates. When the cuts were completed, 45,000 officers had been removed in 1954 and 16,500 officers were let go in 1955.

The Navy also wanted to see what missions the Naval Reserve could perform effectively, and which it should shed. A Naval Reserve Evaluation Board, under retired Adm. Robert Blandy began looking into these questions in October 1953. The board interviewed 426 witnesses, including admirals, seamen, drilling reservists, and disgruntled former reservists. It also consulted with the senior reserve officers of the other armed services and Britain's Royal Navy.

The board's report, submitted in August 1954, recommended 105 changes. It said training had taken on too much of a "night school" atmosphere and needed to become more "hands-on," to produce "men-o-wars men." It called for giving an aircraft carrier to the Naval Air Reserve and more ships to the surface reserves. Communications needed improvement, because many veterans were angry about their being recalled to service during the Korean War. Reservists had to be told of their obligations to drill and face mobilization. In February 1955 Assistant Secretary of the Navy Albert Pratt recommended adopting 90 of the minor changes. But the board's controversial suggestions such as having a Deputy Chief of Naval Operations for Naval Reserve, using punitive measures to compel drill attendance, and increasing afloat training were shelved.

Several other changes were made to the Naval Reserve force structure during the mid-1950s, to improve its capabilities. The Naval Reserve Training Command was established in Omaha, NE, on July 1, 1956. The command was responsible for non-aviation Naval reservists and reported to the Chief of Naval Operations through the Deputy Chief of Naval Operations for Personnel. It took over the duties of the Assistant Chief of Naval Operations for Reserves. In addition, 53 Fleet divisions were established to augment the active duty crews of destroyers. On Feb. 13, 1958 the Chief of Naval Operations established the Selected Reserve as a separate category within the Ready Reserve. The Selected Reserve would consist of aircraft squadrons and surface ship units with an anti-submarine warfare mission. They had a higher priority than other reserve units, to ensure they could meet their wartime mission.

In 1954, at the suggestion of Capt. J.L. Counihan, a reserve squadron, VP-741, deployed to the Mediterranean for its two-week annual training. Additional squadrons deployed to the Mediterranean and Western Pacific in 1955 and 1956. Overseas annual training was also authorized for surface reserve units, following the success of these missions. In 1958 the mission of the Naval Air Reserve also changed, with an emphasis on antisubmarine warfare. Attack squadrons were decommissioned, and ASW squadrons added to reflect the new mission.

But while the Department of Defense and the US Navy were preparing for any eventualities including war, and many reservists were celebrating their return from the Korean War, another war was brewing on the homefront, between the reserve components of the different services.

When the Naval Affairs Committee of the ROA met in January 1953, chairman CDR Merrill Hammond reported that the senior leadership of the ROA was ignoring many of the recommendations of the Naval Affairs Committee. Differences between the Army and the Air Force on one side and the Navy had been growing since the merger of ROA with RONS-NROA, but the concern with the war in Korea had kept these issues from coming to a boil. During the 1953 Congressional budget hearings, the views of the Naval Affairs Committee were ignored by ROA leadership, if those views conflicted with those of the Army/Air Force delegations. Many members of the Naval Reserve began to consider forming/reforming their own organization to press for naval reserve issues, while remaining members of the ROA to keep their hands in the "big picture" organization.

Matters came to a head in April 1954 when LCDR Hank Doyle was fired as Assistant Director for Naval Affairs of the ROA headquarters staff. The termination was taken without consulting with the Navy Section's representatives, a violation of the ROA constitution. The move also violated the recommendations of the Naval Section, which had recently commended LCDR Doyle. The Naval Affairs Committee met on April 24, but the members decided to postpone any action until the national convention in June in Omaha. But when they arrived in Omaha, the Naval Section realized that no solution was possible, because of the vast gulf. The result was the decision to organize a separate Naval Reserve Association, the NRA.

Foundations of the NRA and Early Growth

The NRA was formally established on June 18, 1954 at the Hotel Fontenel, in Omaha, Nebraska. The 43 founding members elected CDR Blaney C. Turner as President, LCDR Tom Fortson as Vice-President, and LT Frank Pohanka as Secretary-Treasurer. They agreed to hold their first national convention in September in Washington DC.

The first annual national convention was convened on September 18, 1954 at the Washington Hotel. CAPT Mack Young was elected President, RADM Jim Ross was elected First Vice President, LCDR George Moorman was elected Second Vice President, LT Frank Pohanka was elected Secretary and CDR Len Bailey was elected Treasurer. CDR Merrill Hammond was appointed to chair the Nominations Committee, with CAPT Alvin Fisher and LCDR Ed Cairns serving as well. LT Al Tiedemann chaired the Resolutions Committee, with CDR Ham Lokey and LT Bill Waylett, Sr. as his assistants. The First Vice President handled Budget and Finance, while the Second Vice President handled Legislation.

Founder, first Secretary-Treasurer, holder of nine other national offices, and a true "sparkplug" during the early years of NRA – CDR Frank S. Pohanka, Jr. Frank also served actively on the National Advisory Committee until his untimely passing in 1973. Photo circa 1970.

The conferees agreed that officers of the Marine Corps and Coast Guard would be eligible for associate membership, while aviation cadets and reserve midshipmen were eligible for active memberships. The convention was addressed by VADM Felix Johnson, who discussed the Naval Reserve Evaluation Board and RADM Kenmore McManes who led a forum on the status of the Naval Reserve. The conferees agreed to set annual dues at $5.00, with the NRA year running from 1 November through 31 October. No provision was made for long term memberships at the convention because the conferees had fears that the new organization might founder as its predecessors had.

The conferees adopted several resolutions. The first read that "Be it resolved by the Naval reserve Association that this organization, in its very inception, adopts the philosophy that an association is not necessary to protect the Naval reserve from the Naval Establishment but that on the other hand an association is vitally needed to cooperate with the Naval Establishment in the solution of the many and complex problems which of necessity arise in the administration of a Navy composed of both regular career and reserve personnel, and be it further resolved that this association pledges its best efforts to such a program of cooperation and dedicates itself to the high purpose of working for those things which will improve the Navy and make it better able to perform its role in the National Defense of our country by presenting the viewpoint of the Naval Reservist, and be it further resolved that this association shall seek ever to measure its actions by the yardstick of patriotism rather than that of self-interest."

Another resolution adopted during the first annual convention called for more afloat training for Naval Reservists. This brought forth a reply from the Chief of naval Operations, ADM Robert Carney, who urged Fleet Commanders, in a letter, to remember that "Personnel in Fleet ships must bear in mind that the Reservist they are helping to train today may be needed to relieve them after a long mid-watch in time of war."

The infant organization immediately adopted a very active program, of advice and comment. The members began to solicit membership by mail in late September 1954 for their views on the National Naval Reserve Policy Board. A report was prepared based on the results, and in late 1954 LCDR Doyle and LT Pohanka briefed the National Naval Reserve Policy Board on the results of the poll and the purposes and goals of the NRA. The end of 1954 also saw the nascent organization hire the former Assistant Chief of Naval Personnel for the Naval Reserve, RADM John H. Shultz as its office manager.

The dawn of the new year saw the NRA off to a good start. LCDR Bill Waylett, Sr. was designated Chairman of the NRA naval Reserve Committee, and LCDR Hank Doyle became editor of the newsletter, NRA NEWS. During the congressional budget hearings, CDR Burwell, LCDR Doyle, and LT Pohanka testified before the House Subcommittee on reserve matters concerning the National reserve Plan, HR.2967, which called for a Presidential authority to call up the Ready reserve and a provision to draft men for six months followed by 9-1/2 years compulsory drilling in reserve units, including the Naval Reserve. The NRA opposed the latter proposal, because it would shatter the voluntary character of the reserves. The NRA scored a major victory when the subcommittee threw out the legislation, despite support from the Army and Air Force reserve organizations.

Most of the remainder of the bill was passed, and became law. It established the Reserve Officer Personnel Act of 1954 (ROPMA) which established a promotion system for reservists similar to that for regular officers. Other legislation passed by Congress that year as Public Law 305 authorized the President to call up to active duty up to 1,000,000 reservists without Congressional approval during an emergency. The law also established six years as the total obligated services for reservists.

The NRA also fell afoul of the anti-Communist hysteria then sweeping the nation. The February 1955 issue of NRA NEWS had been printed in red ink as an attention grabber. But a copy was sent to the Federal Bureau of Investigation (FBI) with a note that any organization using red ink was subversive. The FBI took no action.

The NRA's influence was acknowledged during the Second National Conference, in Chicago, when UnderSecretary of the Navy Thomas S. Gates said that the organization's effectiveness during its first year had amazed the Navy Department. The Second Conference also elected CWO3 Sidney Fields as National President, while establishing a Third Vice President for handling Naval aviation issues. CDR Dick West was elected to this post.

By the middle of the year, the NRA had 500 members with chapters in six cities, Atlanta (CDR Charles Lokey, President); Chicago (LCDR Cole Keyes and LCDR George O'Connell, Presidents); Columbia, SC (CDR Tom Fortson); New York (CAPT Howard Stevens); Philadelphia (CAPT Len Bailey); and Washington DC (CAPT John O'Pray). Finances were still rather parlous, and as of the end of the year, the NRA had a bank balance of $287.78. LCDR Russ Miller, Treasurer, reported that the organization had survived only through personal contributions from some members and that the NRA should look into buying its own headquarters building as a way to cut rental costs.

The NRA saw another example of its growing power and respect in 1956, when CDR Hank Doyle testified for the NRA before the Reserve Officer Promotion Evaluation Board, which had been established under the directorship of VADM Felix Johnson to evaluate the Naval reserve promotion system. The NRA had been calling for such a board practically since the organization was founded. CDR Doyle called for nearly two dozen changes in the current promotion system, to make it more equitable for reservists.

NRA members also testified before Congress in 1956, in this case, the Senate Armed Services Committee. Senior NRA officials said that retired (with pay) reservists should have the same rights and benefits as retired members of the regular Navy. The bill had not said anything about the health and care of reserve retirees and dependents. The interest in promotions was mirrored by a service offered by the NRA. Members interested in their promotion status or other individual matters could call the national headquarters and get an answer. This was the beginning of one of the NRA's longest-standing and most valuable services, and a major reason for new members joining.

The 1956 Executive Committee meeting saw one suggestion made that never came to fruition, the establishment of a US Naval Reserve Academy, to train candidates for Officer Candidate School (OCS). OCS was established in 1952 to provide the large numbers of naval officers required by the burgeoning fleet. The school's curriculum was modeled on that of the World War II V-7 program that turned out fresh officers in 90 days. OCS had a 19-week curriculum. Graduates of OCS received a commission in the Naval Reserve and were required to serve three (later four) years on active duty. The proposal for a Reserve Academy never came to pass.

The NRA had more success with another suggestion in 1956, when the Navy decided to establish a separate unified command for the surface reserve force, mirroring the Naval Air Reserve command structure. The new Naval Reserve Training command, at Fort Omaha, included four primary responsibilities, Fleet programs, Fleet Support Activities, Shore Establishment Augmentation, and Schools.

NRA saw several other milestones in 1956. Membership passed the 1,000 point, and the organization moved to a regular office building, the National Press Building at 14th and F Streets in downtown Washington. For its first two years of existence the NRA had leased a small office in Pohanka Service Inc. Oldsmobile Sales and Service, an auto dealers known informally as "Pohanka's Garage" on 20th Street NW in Washington DC. In addition, two new Vice Presidencies, Membership and WAVES, were established at the Third National Conference in Cleveland. The conference also elected CDR Richard K. West as National President.

Word of a good thing gets around fast, and so it was with the NRA, which doubled in size in 1957, passing the 2,000 mark by the middle of the year. Word of the NRA's successes had another result that year, when, on March 25, the Naval Enlisted Reserve Association was formed in Washington, DC. The new organization opened its membership to all enlisted members of the Navy and Marine Corps reserve. Recognizing fellow members of the sea services, the NRA immediately pledged its full assistance and cooperation.

The NRA continued to make itself active on the legislative front, opposing proposals to transfer the Navy's fleet logistic air wings to the Military Air Transport Service. The year also brought changes in the NRA's hierarchy. RADM John Schultz was relieved as NRA's Executive Officer by RADM Chauncey Camp, USNR (Ret).

When the NRA held its annual convention in Palm Beach on Dec. 6-7,1957, the members could look back over a very satisfactory year, with many accomplishments. The convention elected CDR A. Winfield Chapin as National President, and outgoing President CDR Dick West as National VP-Legislation. Sadly, Dick West did not live to assume his new duties. One week later, on Friday, December 13, his light plane crashed as he was flying from Rochester to NAS Glenview.

NRA continued its active lobbying efforts on several fronts during 1958. In February President Chapin denounced the fact that reservists were ignored in discussions planning the use of the reserves and potential reorganizations of the Department of defense. He called on other reserve organizations, including AFA, AUSA, FRA, MCROA, NERA, NGA, and ROA to hold joint deliberations to work out jointly-approved positions on the use of the reserves.

Naval Reserve capabilities were further enhanced in 1958 when six destroyers and 30 destroyer escorts were assigned to Naval Districts as Naval Reserve training ships, with mixed active-duty/selected-reserve crews. The Selected Reserve expanded further in 1960 when it was assigned 11 minesweepers and new mine warfare duties. Reserve missions and training also changed during this time. Additional reservists served in the Active Fleet Augmentation Component, which would become a fourth watch section aboard active-force ships, upon mobilization. These reservists comprised 65 percent of the Selected Reserve and drilled, for the most part, in Reserve Centers, although a few drilled aboard reserve training ships.

The NRA also mounted a strong opposition to the DoD reorganization Act of 1958, which, according to VP-Legislation CDR Clayton Burwell's testimony to the Senate Armed Services Committee would (1) transfer authority over national defense from Congress to the Executive Branch, (2) establish a "single chief of staff" concept, and (3) limit, if not deny access to Congress for individual Service Secretaries and members of the Joint Chiefs of

The Naval Reservist's Creed

I serve voluntarily in the service of my choice... the United States Navy

I serve willingly because it is my priviledge as well as my duty to serve my country

I serve with pride because of the heritage passed down to me by our long line of proud Naval volunteers

I serve with dignity because I know that to my associates among the civilian population I am the Navy

I serve with honor because the Navy's honor is mine to cherish and guard

I serve with confidence because I am certain of the ability of my Naval leaders and the future of the Navy

I serve with fellowship because I am secure in the unique and splendid comradeship of the Navy

I ask only that I be given by my country and my fellow citizens the opportunity to better prepare myself to serve in the defense of our beloved nation

Originated and adopted by the Naval Reserve Association Third National Conference, Cleveland, Ohio, September 1956

Staff. While in Washington, Chapin, Burwell, and LCDR Cole Keyes, VP for Membership were briefed by the Assistant Chief of Naval Operations for Naval Reserve and the Assistant Chief of Naval Personnel for Naval Reserve on the status of different reserve programs, particularly the Selected Reserve Program.

The NRA gained greater visibility during 1958 when two members, Alvan Fisher, past National VP for the 1st Naval District and John McElroy, National VP for the 12th Naval District, were selected for promotion to Rear Admiral. RADM McElroy was frocked and recalled to active duty as Deputy Commander for NAVRESTRACOMD.

When the Executive Committee met in Washington in September, they approved the establishment of a NRA Distinguished Service Award, meritorious Service Award and a National Award of Merit, and membership awards for 100 percent NRA units.

The year also a continued steady growth. By December 1958, membership had passed 3,000 and was still growing, and 11 new chapters had been formed, including Cleveland; Birmingham; St. Louis; Indianapolis; Anderson, SC; Springfield, MO; Boston; Des Moines; Salt Lake City; Sioux City; and Bakersfield.

Because of this growth, the NRA Executive Committee decided it required a Headquarters Executive Officer to manage the main office. Frank Pohanka, Bill Waylett Sr., Merrill Hammond, and Hank Doyle formed a committee to find the right person. Two highly qualified candidates emerged, both retired Rear Admirals, one a regular, and the other a reservist. National President Chapin recommended choosing the reservist, because the NRA was a reserve organization. But other board members decided to let both flag officers serve, with RADM Hilliard B. Holbrook, USNR (Ret), as Executive Officer, and RADM John S. Lewis USN (Ret) as Deputy.

RADM Holbrook was an enlisted man in World War I, commissioned an Ensign USNRF in 1922, and served in the Caribbean and the Marianas during World War II. His last duty was in the Naval Reserve Training Program, before he worked for two years as the ROA's Assistant Director for Naval Affairs, while RADM Lewis served 25 years on active duty, including duty on ADM R.K. Turner's staff in World War II, before retiring in 1957.

The membership drive found a new tool in 1959, when a '58 gray Opel stationwagon was donated to the organization. RADM's Hilliard and Holbrook immediately fitted the auto with NRA stickers on the port and starboard doors and began sending it around the country to help recruit new members, organize new chapters, and spread the word about the NRA. Use of the wagon helped get chapters up and steaming on Long Island and Waterloo-Cedar Falls, pushing chapter numbers to 21 by mid-1959, with a resultant membership of 5,000, according to National VP-Membership LCDR George O'Connell.

Rear Admiral Jack Lewis, USN (Ret.), past Executive Director of the Naval Reserve Association.

The nearly doubled increase in membership was reflected by a decision to move the NRA's headquarters to a larger building at 1905 K St., NW from the cramped quarters at the National Press Building. In a similar move, NRA News announced it would suspend publishing KUDOS and KING KUDOS for those members signing up five or more new members a month. So many members were qualifying, that the four-page newsletter was in danger of becoming just a list of KUDO qualifiers.

As with past years, NRA continued to focus much of its attention on legislation affecting Naval reservists and the Navy. During the 1959 legislative season, NRA members testified before the National Naval Reserve Policy Board. While they expressed their opinions on all the matters at hand, the NRA staffers expressed particular concerns over three points. They emphasized their support for below-the-zone consideration for outstanding Naval Reserve officers and for restoration of the post of ACNO for Naval Reserve, with a direct line relationship to the CNO. They also strongly opposed the proposal that the post of Assistant Secretary of the Navy for Personnel and Reserve Forces be abolished and his functions scattered.

In a follow-up to their testimony before the Policy Board, VP-Legislation Bob Burke testified before a subcommittee of the House Armed Services Committee, urging parity between the Naval Reserve and the Regular Navy in the matter of below-the-zone promotion. In other personnel-related issues, NRA lobbied strongly in favor of retention of the combat-citation retirement advancement law. (More commonly known as the "tombstone promotion" act, the law gave one-grade promotions upon retirement to veterans of combat.) Although President Dwight Eisenhower supported extending the law's cut-off date, the authority for "tombstone" promotions expired on 1 November 1959.

At the NRA's fifth anniversary convention in Atlanta, members had much to be proud of. The organization had become a force to be reckoned with on Capitol Hill and in the halls of the Pentagon. Senior Regular Navy leadership was taking the organization's viewpoint into consideration on matters affecting the Naval Reserve. With 5,000 members, the organization had a solid membership cadre, ready to spread the word and work at recruiting new members. The anniversary was covered in a Special NRA Conference Section in NAVY TIMES, courtesy of Editor Bill Kreh, a member of the NRA. Members received a no-holds-barred discussion of the Naval Reserve and where it was headed, from RADM "Dutch" Deutermann, COMNAVRESTRACOMD Omaha; RADM R.T. Keith, ACNO-Naval Reserve; RADM Allen Smith, CNARESTRA; and ADM James Russell, VCNO.

The NRA also adopted its own official prayer, composed by National Chaplain CDR Sam Hill Ray, Chaplain Corps. "Almighty and merciful God, we humbly ask Thy blessing upon our Association. Grant us, we humbly petition, the knowledge, the strength, and the courage to accomplish the high purpose for which we are united, to promote the serious purpose of our great Navy and the difficult tasks of our national defense and national security. By mutual assistance may we promote unfailingly the sincere interest of our Naval Reservists in the Navy and the sincere interest of our Navy in its Reservists.. Amen."

The 1960s, NRA's Era of Major Growth

The start of the new decade saw NRA offer two services that have become major recruiting tools for the organization and outstanding membership services. The first was the Record Review Service, whereby senior reservists in the association would review a member's service record and inform him/her what items were missing and how the record could be improved and made more competitive before the selection board. The second service, while offered by other fraternal organizations, was still a strong inducement for new members, a Group Accidental Death and Dismemberment insurance plan. By July the new Record Review Service had become so popular, and the record review so intensive, that the EXCOM decided to institute a charge for it. Fortunately, the charge could be treated as a tax-free contribution.

The major news of the year was, without a doubt, the NRA's stunning growth. At the beginning of 1960, RADM John McGregor, VP for the 12th Naval District, announced plans to establish 12 new chapters on the West Coast. These would join San Francisco's Chester Nimitz Chapter, which was inaugurated by FADM Nimitz himself. Among the new chapters were Lexington, Kansas City, Rockford-Beloit, Santa Barbara, Pomona-San Gabriel, Orange County, San Diego, Los Alamitos, Los Angeles, Pasadena, Stockton, Peninsula Area, Monterey, Santa Cruz, Sacramento, San Jose, and Idaho Falls. The St. Croix chapter was in the process of being formed.

By the time the National Conference convened in late 1960, the West Coast had chartered 15 new chapters, including one at Pearl Harbor. At the meeting, National VP-membership, George O'Connell reported that membership exceeded 5,000 with 55 chapters formed or being formed. The good news for membership growth was echoed by the news that as of July 1960 membership had almost doubled in the previous 18 months and tripled in the preceding 30 months, while the number of chapters had more than doubled.

The growing number of members, and the large number of members who traveled on business, meant that districts and chapters had to revise their meeting policies. Committees in the 11th and 12th Naval Districts took the lead in implementing these new policies. In the 11th, LCDR Russ Miller, formerly the National Treasurer and then the National VP for the 11th ND, began a Council of Chapter Officers, who held District wide meetings. The new policy allowed a greater exchange of information and better coordination on lobbying efforts and membership initiatives. In the 12th ND, several members proposed that all Chapters schedule regular luncheon meetings on the third Friday of each month. That way, out-of-town Association members would always know when the local chapter was meeting and could make arrangements to attend.

As the Association grew competition between chapters and districts heated up, to see whom could have the first 100 percent NRA membership in the area. In 1960, several new records were set. In Atlanta, Naval Reserve Law Company 6-6 became the first in the 6th Naval District to achieve 100 percent NRA membership, under the leadership of CDR Julius Lunsford, then also serving as National Judge Advocate. That same year, NROS 9-3, Glenview Extension became the first NROS unit to make the 100 percent goal. The unit CO was CDR Herbert Rawlins.

Chapter growth meant the NRA needed a larger home, and new quarters were found at 1737 H St. NW. Along with a new home came new staff members. Tony Prockelo, a Navy veteran, became Membership Assistant, working alongside YNC Ernie Royall USN (Ret). Prockelo replaced PNC Dick Durren, who had been the membership Assistant at National Headquarters as a part-time job. Durren had to leave when he received a promotion to Warrant Officer and orders to the USS Forrestal.

Other Association staffers also received major accolades and promotions in 1960. National President CDR A. Winfield Chapin was on the year's CAPT-selectee list. This made him the second National President to be promoted, following in the footsteps of Sidney Fields, who had been promoted to Warrant Officer-4 while serving as National President. The flag list also saw strong NRA representation. Among the flag selectees for the year were CAPTs Len Bailey, Bob Copeland, Ed Reeder, and Jim Hardy. CAPT Bailey was an NRA founder, while CAPTs Copeland and Reeder were past national officers. In addition, CAPT Bill Hughes was selected as the first TAR Admiral in history.

The NRA continued its lobbying efforts during 1960, and this work bore fruit in 1961, when the Secretary of the Navy approved several items the NRA had focused its efforts on. Among the items that had been high on the NRA's agenda, and which were approved were the continuation of multiple drills for NROS, uniform allowances for Reservists, continuation of annual "incentive" cruises to the Far East and the Mediterranean for outstanding reservists, make-up drills in advance of planned absences, and additional financing for the Reserve Anti-Submarine Warfare program.

When CAPT Bob Burke, National-VP for Legislation, addressed the National Naval Reserve Policy Board, he made several additional recommendations. Particularly high on his list was the creation of a new Naval Reserve Evaluation Board, fashioned on the "Johnson Board" of 1953-54. According to CAPT Burke, NRA chapters would be happy to supply "grass roots" points of view to such a board. By the middle of 1961, such a Board was in fact formed.

Almost as soon as the new Naval Reserve Evaluation Board was formed, CDR Jim Mann, 11th Naval District District President and CDR Mark Ogden, from the same district, appeared before it and recommended that the Navy establish in each community a Naval Career Consultant/Committee to assist and advise officers just released from active duty.

Simultaneously, the NRA's national board was considering changes of its own to help reservists in the association. CAPT Chip Chapin, National President, organized a special national committee to develop an NRA Chapter Manual. The NRA's explosive growth, which had reached 6,000 by mid-1961, meant that not all the chapters were conducting business in the same manner. the board's members included CAPT Bob Bolt, CAPT Les McDowell, CAPT Don Osborne, CAPT John Sewell, CDR Ken Kalmbach, CDR Tom Keane, CDR Horace Lyon, CDR Edgar Martin, CDR Mark Ogden, LCDR Harold Mitchell, LCDR Art Foss, LT Herman Javson, LT Frank Vignola, and CWO4 Sid Fields.

One of the issues the new manual addressed was how to handle having several Chapters in the same geographic area. Partly as a result, three chapters in the San Francisco Bay area, NIMITZ, SAN FRANCISCO, and EAST BAY, merged to form the FADM CHESTER W. NIMITZ-GOLDEN GATE Chapter with LCDR George Concannon as the first president.

Additional changes were made to the NRA's Constitution and By-Laws at the National Convention in Philadelphia. Among the changes were the creation of two additional Vice -President billets: National VP for Naval Affairs and National-VP for Education & Activities; changing the title of Vice President Naval District to District President, and changing the National Executive Officer's title to National Executive Director. Other changes included modifying the NRA's fiscal year to run from 1 October through 30 September, creating a National Advisory Board of five members, and changing the dues structure to four dollars per stripe effective 1 January 1962.

RADM Jack Lewis receives from C.N.O. SEC NAV Distinguished Public Service Award. Pictured are (from left): CAPT Clarence Williams, CAPT Bud Wulff, RADM Jack Lewis, CAPT Bob Bolt, ADM Thomas Moorer (C.N.O.), and LCDR Tom Curran.

The National Convention welcomed members from eight new chapters as well, including York, PA; Willow Grove; Tampa; Mobile; New Orleans/ Acadian; Albuquerque; and Gary, IN.

The Convention also recognized individual achievements by several members. CDR Art Solberg of Glenview received he Membership Achievement Award for breaking his own record and signing up 70 new members in one month. CAPT James Wilson took over the editorship of NRA News from CDR Charlie Gilchrest, who had been handling it as a collateral duty since 1959. RADM Holbrook was commended for his efforts as National Executive Officer when he stepped down to become a part-time assistant to RADM Jack Lewis, who succeeded him.

The year also saw several changes in Navy policy towards reservists. Officers who failed to earn at least 12 retirement points in FY62 or beyond would be transferred to Standby status or retired or discharged. These changes came at a time when the Selected Reserve force was again undergoing modification. The new revisions divided the Selected reserve community into two categories. Category A was authorized 135,000 men who would be available for immediate activation. They would be responsible for anti-submarine warfare, mine warfare, and augmenting active fleet, direct fleet support and key shore establishment units. Category B was authorized to have 117,000 men available 30 days after mobilization for the reactivation of reserve fleet ships and additional fleet and shore establishment units.

The new reserve force structure faced its first test in 1961 during the early years of President John F. Kennedy's administration. In response to Soviet threats over the status of West Berlin, Kennedy asked for a recall of reservists. On Aug. 25, 1961, the Navy recalled 6,400 reservists to help augment fleet units. Because of the lack of time, reserve ships were not reactivated. Instead, Naval Surface Force reservists were sent to augment the crews of 13 destroyers, 27 destroyer escorts, and Naval Air Reservists helped man five patrol (VP) and 13 antisubmarine warfare (VS) squadrons. An additional 1,957 Naval Reservists were recalled on Sept. 13, 1961. All the reservists remained on duty until August 1962.

Only Selected Reserve personnel were recalled involuntarily, and they were sent to units in which they trained, where possible. Recall policies for the aviation personnel called for using volunteers to the maximum extent possible and not recalling veterans of World War II or the Korean War if at all possible. Despite these policies, many personnel were not satisfied and left the reserves as soon as possible. Most of the 7,980 Naval Reservists mobilized for the Berlin Crisis left the reserves at the earliest chance, in fact only 50 percent of the enlisted personnel and 75 percent of the officers mobilized stayed on. Because of the complaints, the Navy recalled very few reservists in October 1962 for the Cuban Missile Crisis, preferring to extend the enlistments of active duty personnel.

While the Berlin Crisis was going on, the Navy Department formed a Naval Reserve Evaluation Board to suggest improvements to the Naval Reserve. In May 1962 the board issued a report with 132 recommendations. Secretary of the Navy Paul Nitze approved 118 of the suggestions. In November 1962, while these suggestions were being implemented, Congress asked the General Accounting Office (GAO) to study the Naval Reserve. The GAO completed its study in June 1963 and issued a critical report in September 1963. Most of the GAO's criticisms focused on the various enlisted programs and did not comment on officer programs.

If the preceding years had seen some stability, 1962 could be called the year of the wandering headquarters. The building where NRA was headquartered was sold in May, forcing the Association to move to new quarters at 1700 Eye St. NW. But in November that building was torn down, causing the NRA to end the year at 1205 Connecticut Ave NW.

But building changes weren't the only differences that had to be contended with. RADM Holbrook retired, and CDR Charlie Gibson had to leave his post as National Public Affairs officer. CAPT James Wilson, NRA News editor, became dual-hatted, assuming the post as National PAO while keeping editorship of the newsletter. The first National Advisory Committee also was appointed, consisting of CAPT Bob Burke, CDR Jim Conway, CDR George Dutton, CDR Mark Ogden, and CDR Sam Hill Ray. Their duties also included serving as members of the National Executive Committee.

The NRA also made a hesitant entrance into the computer age in July 1962, when it transferred some personnel records to IBM cards. The evolution was rather painful, with missing and mutilated cards causing some record snafus. Nonetheless, it marked an acknowledgment that business would be handled differently in the future, and that technology, like time, constantly marches on.

One of the decisions made by the EXCOM that year was to officially approve the Association's logo. While the NRA had long had an unofficial logo, used on its letterhead, consisting of an officer's cap device surrounded by fancywork and line, it had never officially been adapted. LCDR Dick Biddle received the credit for coming up with the design. The committee also copyrighted the Naval Reserve March, composed by SSgt John R Bourgeois of the Marine Corps Band, with lyrics by CAPT Winfield Chapin. The piece was premiered at The Watergate on 12 August at a concert attended by 10,000 people.

The organization's growth and the interest in its services also led to changes in NRA policy. The Record Reviews had been handled basically by CAPT Robert Cofer, since its inception. Because of the growing popularity of the service, RADM Holbrook agreed to provide assistance, along with CAPT Floyd Myhre, who joined the staff as a consultant. The NRA also decided to add an additional service, the Washington Jacket Check. The Washington Jacket Check was not designed as a replacement for the Review. Rather, it would tell an officer just entering a promotion zone whether his service jacket had all the documents required, and ensure their accuracy.

Another change began making itself felt in the NRA in 1962. Prior to that year, the organization had been heavy with World War II veterans as it membership. But CDR Smokey Stivers of the Wardroom chapter reported that more than 23 percent of his members were junior officers, in the grades of Lieutenant and below. Although this was far above the national average, it marked a trend that would grow as the years went on. Both senior and junior officers made up the ranks of the seven new chapters formed in 1962, in Erie; Winter Haven, FL/Ridge; Fort Lauderdale; Winston-Salem/Ole North State; Duluth; Glenview; and Puget Sound.

Senior officers continued to predominate in the Association, and when the CDR -selectee list appeared, there were 88 members on it. Founding members who got picked up for 0-5 included LCDRs Frank Pohanka (National Treasurer), Stephen Shry, Al Tiedemann, and Bill Waylett. In addition, Art Foss (Eighth ND president), Mike Kapnas (President of Gary/Calumet Chapter), and Joe Neblett (Tampa Chapter President) also got the nod.

Other senior officers were in the news, including CDR George O'Connell, Jr., who was elected national president at the 9th National Conference, relieving Winfield Chapin. Retiring President Chapin received the Secretary of the Navy's Meritorious Public Service Citation from principal speaker, CNO ADM George Anderson. Chip did not ride off into the sunset, however, but became chairman of the National Advisory Committee.

The year ended on a down note for many Naval Reservists. RADM Chauncey Camp, a stalwart of the NRA and its Executive Director from February 1957-February 1958 died in Hattiesburg, MS. And in October and November 1962, several thousand naval reservists had to leave hearth and home as they were recalled to active duty during the Cuban Missile Crisis.

The new year, 1963, dawned on a promising note. Some reservists were already being released from their active duty recalls, and more reservists expressed an interest in the NRA. The Association once again showed its legislative clout when it successfully fought a proposal in Congress that called for a military pay raise for voluntary reservists, but not for those personnel drilling under a Reserve obligation.

The growing size of the NRA meant that the newsletter, NRA News could finally be expanded from a four-pager to an eight pager. Along with the expansion came a change in personnel. Bill Kreh, long active in the Association, left his job as Managing Editor at Navy Times to become the newsletter's new editor. And with the expansion in size came a decision to reinstate the KUDOs list, honoring outstanding recruiters.

One expression of the growth was the completion of the 1,000th Record Review since the service's inception. With the growing popularity of the service, CAPT Hal Burch, USNR (Ret) and CAPT David McLaughlin (USNR) signed on as additional consultants for the service.

While the review drew in new members, the NRA took on another mission that had the potential for drawing in new members. It offered to set up counseling boards at Naval Reserve Training Centers and Naval Air reserve activities for personnel returning to the civilian community from active duty. The CNO gratefully accepted the offer, which was established as "Operation Affiliation."

As part of the preparations for Operation Affiliation, BUPERS asked the Washington DC/Wardroom Chapter to perform an advance study of how to develop a Professional Development Plan for Naval Reserve officers. The Chapter's study group recommended creating formal Reserve Counseling Boards at each reserve site. Much of the work was based on a study undertaken in 1961 by CDRs Jim Mann and Mark Ogden for a Naval Reserve Career Consultant program.

Manpower was not the only critical need addressed by the NRA during 1963. At the National Naval Reserve Policy Board, NRA President George O'Connell called for upgrading facilities and equipment at reserve centers as well as replacing the aircraft used by Naval Air Reserve transport squadrons.

In a bow to the electronic age, the Navy established an electronic-computer-supported activity called the Naval Reserve Manpower Center (NRMC) at Bainbridge, MD. The new facility would centralize record keeping for Reserve and retired personnel not participating in a Naval Reserve training program. The first commanding officer at NRMC was CAPT Robert Feiten, a TAR aviator and NRA member.

Other NRA members kept busy in different parts of the country. CAPT Bernard Johnson, Thirteenth ND President, established a Construction Battalion Training Facility as Sand Point Naval Air Station (Seattle) for Seabees in the district to drill and perform their ACDUTRA. While at the site, the Seabees also performed community service, including upgrading local softball fields.

As the year ended, and the NRA held its 10th National Conference in Kansas City, it made several changes to the Constitution. Dues were changed to $10 per year or $27 for three years. In addition, the National Advisory Committee (NAC) members were to be non-voting members of the National EXCOM, rather than the voting members they had previously been. In addition, the Convention decided that no member could serve concurrently in a National and District elective office, no national officer would be eligible for election to the same office for more than two consecutive terms, and it established a counterpart "District-officer structure under each district president. The final resolution adopted called for two new standing national committees, a Resolutions Committee and an Awards Committee.

Perhaps the highlight of the conference was the "George meets George" awards ceremony. NRA President CDR George A. O'Connell Jr. USNR-R, received the National Award of Merit from CAPT George A. O'Connell Jr. USN for his service as Assistant Chief of Naval Personnel (Reserve Affairs). Neither officer was related to the other, to their knowledge. The Conference also welcomed members from the new Chapters at Springfield, MA; New Jersey/FADM Halsey; New Jersey/Neptune; St. Petersburg; Miami; Memphis; Houston; and Colorado Springs.

Some of the other resolutions put forth at the 1963 Convention bore fruit, but not until 1964. In that year, the Navy decided to change its plans to scrap eight PCE-type Naval reserve patrol craft, scrapping only three vice the eight originally planned. In addition, the Navy received 27 C-118 transport aircraft after requesting 35 such planes. Both the request and the reduced scrapping schedule came about after strong NRA lobbying.

But if it won some battles, the NRA also lost a few others, when DoD imposed a five-year ceiling of 126,000 drill-pay strength. The NRA supported the position of Congressman F. Edward Hebert, who had recommended an end strength of 166,000.

The NRA continued its lobbying activities on other fronts. National VP-naval Affairs Frank Pohanka testified before the National naval Reserve Policy Board urging replacement of additional outmoded Naval Reserve transport aircraft, fiscal-year accounting of Civil Service military leave, vice the calendar-year policy then in effect, and TAR appointments for enlisted TEMAC recruiters.

The Navy formally adopted another proposal, by CDRs Jim Mann and Mark Ogden of the Eleventh Naval District, for Naval Reserve Counseling Boards. The first Boards were officially established in early 1964. The Boards grew out of the study the two NRA members performed and the Association's Operation Affiliation.

In keeping with what had become a tradition approaching that of the Flying Dutchman, the NRA again had to switch its headquarters location in 1964, moving a couple of blocks to the sixth floor of 1913 Eye St., NW. Once established at the new location, Th Association began offering a new

This 1958 EXCOM meeting attracted none other than ADM Arleigh Burke (then CNO). Other attendees included: Merrill Hammond, Ralph Hellstrom, Frank Pohanka, Bob Burke, Hugh Howell, Toni Robinson, Chip Chapin, George O'Connell, and others.

Ladies at the Naval Reserve Association Conference visit the Truman Library in Independence.

Speech Service, with a library and background material for speeches about every aspect of the Navy, naval operations, seapower, and the service's role in national defense.

Members, or the public, interested in the Navy, could also turn their gaze towards the efforts of two other Association members, CAPT Gordon Keating and LCDR Dick Biddle, who were running the US Naval Historical Display Center, more popularly known as the Navy Museum, at the Washington Navy Yard.

When the 11th Annual Convention met in Tampa, it elected CAPT Donald Osborne as National President. CDR O'Connell, the outgoing President, received the Secretary of the Navy's Distinguished Public Service Award Conferees also welcomed members from the new chapters at Newport, RI; Hartford; Akron; Montgomery County, MD; Jackson, MS/Magnolia State; Topeka; Wichita, KS; Denver; Great Lakes; Reno; and Spokane.

The delegates to the Conference also adopted resolutions calling for an increase in drill pay strength to meet mobilization requirements, legislation to provide full protection to Naval Reservists injured while in a travel or duty status in connection with drills; below-zone selection of Reserve flag officers, and establishment of an Indian Ocean fleet.

The 50th Anniversary of the Naval Reserve, March 3, 1965, found the Naval Reserve Association moving full steam ahead in a variety of ways. That year, the NRA began offering members a Group Life Insurance Plan, providing $10,000 of level term insurance up to age 70. During the initial period, more than 20 percent of the Association's membership signed up for the program. The NRA also began offering life memberships. Interestingly enough, two of the first four NRA Life Members were women, with the first four including LCDR Marie Frauens, LCDR Ruth Kasky, CDR Frank Pohanka, and CAPT Armistead Clay.

The NRA continued its vigorous legislative program, and showed that it worked beyond parochial Navy issues by opposing a plan by Secretary of Defense McNamara to abolish the Army and Air Force Reserve and merge them with the Army and Air National Guard programs.

In its dealings with the Regular and Reserve components of the Navy, the NRA was scoring more "attaboys," too in 1965. RADM R.E.M. West, COMNAVRESTRACOMD (and NRA member), commended the NRA, noting that early feedback showed that the NRA-conceived Naval Reserve Counseling Boards were producing many dividends, providing valuable information and "Gouge" to Naval Reservists.

To keep chapter officers informed on how to retain the interest of their members, CAPT Les McDowell, National VP-Education and Activities, wrote a booklet, "CONTACT" describing various possible chapter and district activities. Copies of the booklet also were sent to Regular Navy Distric Commandants and flag officers who regularly dealt with the reserve.

When the 12th National Conference was held in Memphis, it elected CAPT Robert Barto as National President. Barto had previously served as National VP-Membership and District President of the 13th Naval District. Outgoing President CAPT Don Osborne received the Secretary of the Navy's Distinguished Public Service Award.

The Conference also welcomed the newest Chapters to the NRA, including Buffalo; Alexandria, VA/Matthew Fontaine Maury; Augusta, GA; Little Rock; Austin; Fort Worth; Dallas; San Antonio; Milwaukee; Minneapolis-St Paul; Detroit; Tucson; Fresno; and Portland, OR. Another NRA'er honored that year was CDR Ken Kalmbach, lauded by the Washington/Wardroom chapter for his services to the NRA.

The convention did not pass as many resolutions as some previous gatherings had, but one resolution that was passed was a sign of the NRA's growing concern for the surrounding community. Outgoing President Don Osborne pushed for the development of a set of Ocean Science Study Kits for high school students and teachers.

NRA chapters provided another sign of their civic involvement in 1965 when Hurricane Betsy smashed through the New Orleans area. To help with the reconstruction efforts, the Acadia chapter in the city undertook Operation BETSY Restoration. The chapter helped hundreds of Navy victims find repair services and organized repair parties to help provide additional services. The chapter also held a major drive to collect furniture, clothing, appliances, and money.

In 1965 and 1966, the NRA donated two awards to the Navy, to be presented annually to honor outstanding retention by Naval Reservists in the aviation community and performance in the field of anti-submarine warfare.

The first award, donated in 1965, was the Richard K. West trophy, commemorating CDR Dick West, an organization founder and former President who died in a plane crash in 1957. The award was to be presented annually to the Naval Air Training Unit (NARTU) or Naval Air Station (NAS) with the best retention record. NAS Willow Grove was the first station to win the award. The second award, donated by the NRA in 1966 for premier presentation in 1967, was the A. Winfield Chapin Trophy, commemorating CAPT Winfield "Chip" Chapin, another past National President. The Chapin Trophy was to be given annually to the ASW Reserve Crew with the best retention record.

The NRA's legislative tracking team was kept quite busy in 1966, because of numerous initiatives put forth by the Navy, with some being improvements, but many not. Among the improvements was the decision to do away with the promotion-point requirement for officer promotions. This decision, praised by many, did away with a bureaucratic requirement for promotions. The decision retained the requirement that officers had to earn at least 12 points annually, other than gratuitous points.

Some of the other decisions concerned the size and shape of the reserve establishment. Plans unveiled by Secretary of Defense Robert McNamara called for downsizing or closing 40 USNR training activities by 1970, with most to close by 1968. The plans called for 25 Naval Reserve Training Centers to close, three centers to be reduced by Naval Reserve Training Facilities, and three Facilities to close. Along with the reductions in infrastructure, the Defense Department again called for setting drill-pay strength at 126,000. The decision went against congressional direction, which had called on several occasions for Naval Reserve drill-pay strength to be set at 132,000. The Defense Department also proposed putting the entire Naval Reserve on four-hour drill periods. National President Bob Barto immediately questioned this in a letter to Secretary McNamara. In response to Barto's letter, which pointed out the flexibility in the current system, DoD reevaluated its plans.

But while the Navy was cutting back on portions of the reserve infrastructure, it also was asking USNR officers on active duty to apply for Training and Administration of the Reserves (TAR) billets as soon as they finished their active duty period, to relieve a shortage of TAR officers.

In other actions on Capitol Hill, Rep. F. Edward Hebert and Sen. John Stennis began hearings on a "Reserve Bill of Rights." The Bill passed the House of Representatives 333 to 6, but the Senate took no action. The hearings bore some fruit later in 1966, when a bill sponsored by Rep. Hebert was signed into law. While the new bill was not the "Reserve Bill of Rights," it did contain a key guarantee. Once a reservist was officially told he had 20 satisfactory years of service and was eligible to receive retired pay at age 60, the subsequent discovery of any clerical error could not be cause to have the pay revoked.

Other NRA efforts during 1966 included the publication of the Ocean Science Study Kits for high school students and teachers first proposed in 1964. NRA'er CDR Clarence Lindquist provided key material assistance in developing the kits, which were originally proposed by then National President CAPT Don Osborne. The first kit was presented by Oceanographer of the Navy RADM Odale Waters to National President CAPT Bob Barto, with a Bravo Zulu and thanks to the NRA for coming up with the idea and designing the kits. Also in the scientific realm, the first mountain-climbing expedition to Antarctica, which scaled the Vinson Massif, planted the NRA burgee at the peak, in appreciation for the help extended by the NRA.

Another NRA-sponsored initiative in 1966 was Navy Sabbath. Proposed by CDR (Chaplain) Gerry Kuhn and CDR Al Tiedemann, the idea quickly gained the support of prominent clergy of many faiths, Secretary of the Navy Paul Nitze, and the Chief of Chaplains. Navy Sabbath was to take place in October coinciding with the Navy's birthday.

When the National Conference met in Denver, the Association heard keynote speaker CDR Scott Carpenter, USN, describe his experiences as both an astronaut and an aquanaut in Project Mercury and Project SeaLab. The attendees heard a presentation from the new National Public Affairs Officer LT Dave Woods. They also saw Chief of Information RADM "Hank" Miller deliver Secretary of the Navy Distinguished Public Service Awards to National President CAPT Bob Barto and past National President-Naval Air CAPT Jim Conway for their service to the Navy.

On the growth front, attendees heard that membership had passed the 7,000 mark, with particular help from top recruiters CAPT Herb Tuttle (Newport), CDR John Shanklin (Albuquerque), CAPT Bud Wulff (Chicago), LCDR Robert Baumbach (Lafayette, LA), and LCDR John Ellis (Portland, OR). Part of the growth also came from 11 new chapters, including Shreveport; Amarillo; Baltimore; Lubbock; Norfolk; Roanoke; Oklahoma City; Lafayette, LA; Forest Park, IL; Fort Omaha; and Phoenix. The Baltimore chapter held its first meeting aboard the sailing frigate Constellation, moored in Baltimore Harbor.

Several NRA members received naval honors during the year, including Past national President Don Osborne, commissioned an Admiral in the Georgia Navy and ADM J.J. "Jocko" Clark of the New York Navy chapter, elected Commander General of the Naval Order. ADM Clark succeeded CAPT Jerry O'Shea of the Nimitz-Golden Gate chapter.

The Vietnam Years

The growth continued in 1967, when membership passed the 8,000 mark, thanks in large part of the efforts of recruiters CAPT Herb Tuttle (Newport), CDR John Shanklin (Albuquerque), CAPT Bud Wulff (Chicago), CDR Bob Olson (Amarillo), and CAPT Spencer Church (San Diego). Also contributing to the 1967 growth was the establisment of new chapters. Liberty Bell in Willow Grove, PA, chartered itself with a record 250 members several years after being founded. Other new chapters included San Juan; Corpus Christi; Madison, WI; Salem, OR; and Eugene, OR.

The NRA entered 1967 sporting a new look—specifically for the NRA News. The paper, previously printed in blue ink on gold paper (and referred to, by some wags, as "Yellow Journalism" or the "Yellow Rag") was turned into a standard black ink on white paper newsletter. This was accomplished at the same time as the conn was relieved, with Allen Scott taking over editor's duties from Bill Kreh, who began work on a history of the U.S. Naval Reserve.

In another sign of its growth, the NRA joined seven similar organizations to form a Navy-Marine Corps Council in 1967. The other organizations included the Navy League, Fleet Reserve Association (FRA), Naval Enlisted Reserve Association (NERA), Marine Corps League, Marine Corps Reserve Officers Association (MCROA), Navy Mothers Clubs of America, and the Navy Wives Clubs of America. The council would coordinate on their response to legislation before Congress and coordinate other activities benefiting the Navy/Marine Corps reserve team.

When the National Executive Committee met in March 1967 in Kansas City, they decided to concentrate their lobbying efforts on four key measures—portal-to-portal coverage for reservists, the Reserve Bill of Rights, dependents' benefits, and deposits of savings for personnel Missing in Action (MIA)s. The EXCOM also voted to adopt an incremental-payment plan for life members. While these were the primary issues, the NRA also got caught up in many other issues on Capitol Hill and in the Pentagon.

Shortly after the EXCOM met, the NRA held its 14th National Conference in Chicago. The conference elected CAPT Robert B. Bolt, National Vice President-Budget & Finance, as National President. Several resolutions were also passed, expressing support for the Reserve Bill of Rights, for portal-to-portal coverage for reservists, for nuclear propulsion for all new construction major surface ships, and for increased congressional funding for oceanographic research.

One of President Bolt's first actions was to protest to the Secretary of the Navy the recent decision to reduce funding for Naval Air reserve flight time as cost savings. CAPT Bolt pointed out that reduced flight time could easily lead to reduced safety through unfamiliarity with aircraft.

One potential issue, that fortunately was stopped before if became DoD policy, was a proposal by Secretary of Defense Cyrus Vance to have the reserves lower their minimum standards to match those of the regular Navy. This would have required the reserves to accept some Mental Group IV recruits—the category which provided many of the sailors who couldn't adjust to service life or learn one of the skills upon which the increasingly technological Navy was coming to rely. Another idea that didn't make it past the starting blocks was a proposal to eliminate four overseas Naval Districts and three continental United States Naval Districts by merging them. The House of Representatives passed legislation defeating this proposal.

Fortunately, some other ideas did come to fruition, particularly the Reserve Bill of Rights, which was reintroduced after Congress reconvened in early 1967. In a classic case of second time lucky, the bill passed both houses. It gave each service an Assistant Secretary for Manpower and Reserve Affairs, mandated an annual manpower floor under each component, required the Secretary of Defense to provide more equipment, established Selected-reserve components within each service, and blocked future attempts at service mergers. This last issue had little relevance to the Navy, but was critical to the Army and Air Force, which had seen proposals to merge their reserve components and their National Guard components in 1965.

Unfortunately, the portal-to-portal coverage issue, to provide medical care for reservists from the time they left home to when they returned for weekend drills, also failed. The issue had been a major sticking point with the Naval Reserve since the 1930s. Unfortunately, despite five bills in recent Congresses, the bill failed again in 1967.

With its growing strength, some members began tackling personnel issues that were bedeviling the reserve force. LCDR Gordon Brown of Tucson's Desert Sailor chapter, published an article in Naval Institute Proceedings analyzing naval reserve forces and manpower and suggesting 19 changes to make the Naval Reserve more ready. National PAO LT Dave Woods also published an article in Proceedings dealing with the surface community and the problems of officers assigned ashore for their active-duty tour. Woods suggested that either a separate career track be established for these officers, or updating seagoing qualifications be given paramount priority for training duty.

As 1968 began, more and more reservists found themselves supporting active duty forces in the growing Vietnam conflict. Such support included flying patrol aircraft in the South China Sea, supplementing active duty crews aboard warships, working with the Construction Battalions (SEABEES) and a whole host of other duties. The war also led to vastly expanded Officer Candidate School and Aviation Officer Candidate School programs. Many of the graduates of these programs went directly to duty in Vietnam with the Riverine Force or in the Sea Air Land (SEAL) teams operating in the swamps of Vietnam. Other flew off carriers on Yankee Station or operated with the Tonkin Gulf Yacht Club, aboard destroyers on the gun line or logistics ships supplying the task forces. And many of these officers joined the NRA, providing the key leadership cadre that would run the organization during the 1980s and 1990s.

Vietnam might have provided an opportunity for the Naval Reserve to display its capabilities, but while individual Naval Reservists served and displayed the naval reserve at its best, there was little organized call up. The primary reason for this was

CAPT Bob Bolt congratulates CDR John Shanklin and awards him the National Meritorius Service Award.

President Bob Bolt addresses the National Executive Committee in Dallas on March 29, 1969, following a moment of silence in memory of General of the Army Dwight David Eisenhower.

Awards luncheon in the Crown Room at the Hotel del Coronado, October 4, 1969.

Senator Richard Russell receiving a Distinguished Service Award for 1967 from President Bob Bolt.

political, with President Lyndon Johnson not wanting to risk setting off the political firestorm that would have accompanied calling up as many as 235,000 reservists, as some military figures wanted.

Between 1965 and 1968 the US government considered mobilizing the reserves for service in Vietnam. Arguments in favor of mobilizing the reserves included the belief that it would show Hanoi and its supporters in the Soviet union and China of the resolve of the United States and that it would provide more troops for service in Vietnam, helping to hasten the end of hostilities. These arguments were countered by arguments that Hanoi might view the call-up as a sign of US desperation, that it could cost at least $10 billion per year, and that it would lead to an extremely divisive national debate. Arguments against a massive reserve call-up won the day, and the reserves did not see widespread use during the conflict.

As part of the debate about reserve mobilization, in early 1967 the Joint Chiefs of Staff Navy wanted to activate 15 reserve destroyers to help relieve the pressures on fleet destroyers, which were finding themselves in increasing use as gunfire support platforms off Vietnam. The JCS also considered mobilizing the reserve 4th Marine Division, which would have required mobilizing supporting Naval Reserve units. Other JCS proposals called for activating additional "Reserve aircraft, equipment, and reservists. These steps were all rejected for the same reasons that a general reserve mobilization was not called. These destroyers would have been replaced in the Naval Reserve Force with destroyers taken from the mothball fleet.

But while there was no large scale organized call-up, reservists served in many capacities throughout the Vietnam conflict. Four reserve destroyer escorts that were recalled during the Berlin Crisis, Edmonds, Walton, McGinty, and Whitehurst served in the South China Sea in 1962 and 1963. In May 1965, Adm. Thomas Moorer, Commander-in-Chief Pacific Fleet, asked Chief of Naval Operations Adm. David McDonald for Naval Reserve fleet tactical air support to help shuttle supplies to Vietnam. Volunteer Naval Air Reservists serving on extended active duty for training flew C-54s between Naval Air Station (NAS) Alameda, CA and NAS Barber's Point, HI. Other reservists, flying C-118 cargo planes, shuttled cargo onwards to Cubi Point NAS, Philippines and Saigon and Da Nang in Vietnam.

In January 1968, the United States did find a requirement to mobilize some reservists. North Korea's seizure of the intelligence-gathering ship Pueblo (AGER 2) and the shooting down of an EC-121 reconnaissance aircraft led to fears of a growing global conflict. On January 26, 1968, six Naval Reserve tactical aircraft squadrons, including 593 men and 72 aircraft, were mobilized. The mobilization revealed major problems with the three fighter and three attack squadrons.

Their A-4 and F-8 aircraft were not carrier compatible. And while the squadrons had an authorized strength of 1,115, only 593 personnel were assigned. The shortfall was made up with 206 volunteers, 229 TARs, and 107 assigned from other units. But none of the squadrons deployed, and eight months after they were mobilized, they had completed less than 75 percent of their training syllabus. As a result, the crews and planes were released from active duty on Nov. 1, 1968. As a result of the shortcomings the call up revealed, Chief of Naval Operations Adm. Thomas Moorer ordered a study of Naval Air Reserve capabilities. The study revealed that most aircraft in the reserves lacked the avionics required for current carrier operations. Faced with these problems, the Navy ordered the replacement of all reserve aircraft with up-to-date planes. 1971 accomplished this.

While the Pueblo activation was underway, the Joint Chiefs recommended the activation of 600 Naval reservists, including medical, dental, and chaplain corps personnel, to support the 4th Marine Division. This recommendation was not approved, but in May 1968 1,000 men in Naval Reserve Construction Batallions RMCB 12 and RMCB 22 were recalled to active duty for service in Vietnam. The two units went to Vietnam in August and September respectively, for eight-month deployments. They were deactivated within 30 days of their return to the United States in 1969.

As part of the conflict, in January 1968 six Naval Air Reserve squadrons were called up, VA-776, VA-831, VA-873, VF-661, VF-703, and VF-931. National president Bolt called upon NRA chapters to provide all assistance possible to the families of those reservists called to active duty. The six squadrons transferred from their normal reserve sites to operational Naval Air Stations at Lemoore and Cecil Field. They served at these stations until the fall of 1968 when they were released from active duty. The recall of the reserve squadrons had a beneficial effect on the Naval Air Reserve program, with the Navy voting to restore Naval Air reserve flight funds. (Sixty percent had already been restored following the strong protests lodged by the NRA. The recall caused the restoration of the remaining funding.) In other war-related actions, a freeze was placed on drill-pay Reserve releases. But some reservists were forced out, mainly those commanders with 26 years of service and two pass-overs.

While the war affected those NRA'ers in some of the aviation, surface, and SEABEE communities, for many other reservists, life continued as before. The Naval reserve saw a change in its hierarchy, as CPT D.W. Davis, Jr., USNR retired as Deputy Commander of NAVRESTRACOMD at Omaha. Following his retirement, he came to the NRA as Deputy Executive Director. Having been in the TAR program since 1957, he brought a vast wealth of knowledge to the NRA about this program. The war also led to CDR Bob Olson, Eight District President, going back to active duty to take command of the Birmingham Naval and Marine Corps Reserve Center. LCDR John Dodson was appointed to fill out the remainder of CDR Olson's term.

In other association news, CAPT Ray Ackerman, a National Advisory Committeeman from Oklahoma was appointed Special Assistant to the National VP-Membership, with the mission of building up NRA membership in the Naval Air Reserve. Naval Air membership might also have grown because of a new policy in response to the recalls, the NRA announced that any member recalled to active duty would have his membership extended free for a year. This might have caused some reservists to sign up. Other reservists signed up after discovering that association members heard from the NRA when they had been selected for promotion faster than some units sent out notifications. Along with a new hand on board, headquarters also announced plans to begin fully automating membership records.

Awards were also a point of order during 1968. The EXCOM authorized a Twice a Citizen Award, to be awarded once a year per District as the highest district-level award. An NRA badge also was designed, for production as a large medal, miniature medal, and service ribbon. The award will be worn on the appropriate occasions, behind all other US medals but ahead of any foreign awards.

When the 15th National Conference was held in Kansas City, the NRA had a lot to congratulate itself on. Membership had passed the 9,000 mark at the end of July, spurred by the efforts of CAPT

Herb Tuttle, CDR John Clayton, CAPT Ray Ackerman, CDR John Shanklin, and CAPT Bob Clark. New chapters had formed in China Lake; Billings, MT; Louisville; Watertown, NY; St Joseph, MO; Trenton; Beaumont-Port Arthur; Santa Fe-Los Alamos; Waco; El Paso; Lincoln, NB; San Fernando Valley; Tacoma; and Springfield, MO, which was rechartered.

During Conference, resolutions were adapted calling for an extension of SGLI coverage to include inactive duty Reservists, and the establishment of two or more carrier airwings or groups in the Naval Air Reserve Force. CAPT John Brasel of the New York Navy chapter argued that the National organization should collect local dues jointly with national dues and rebate the local's shares to the local chapters. The Conference heard presentations from CAPT Phil Saylor, USN, the acting Deputy Assistant Secretary of Defense for Reserve Affairs, who told the conferees that the Defense Department was studying the reserve programs of all the services and analyzing their missions and requirements. He added that there would be a greater emphasis on unit training towards the goal of performing particular mobilization tasks.

Several NRAers were honored during the Conference as well. Chief of Naval Operations ADM Thomas Moorer presented the Secretary of the Navy's Distinguished Public Service Award to National President CAPT Bob Bolt and National Executive Director RADM Jack Lewis. RADM Lewis became only the second retired USN officer and first admiral to receive the Navy's top civilian award.

The good news as 1968 ended came as a major boost into 1969. But problems loomed. Portal-to-portal coverage was again defeated in Congress, despite the best efforts of Rep. F. Edward Hebert and Sen. Henry Jackson. The problems on Capitol Hill were matched by problems in the Pentagon, where Deputy Assistant Secretary of Defense for Reserve Affairs Ernest Massad resigned, criticizing his superiors for a lack of access and clout. He told the Senate Armed Services committee the post should be upgraded to an assistant secretaryship. Reserve matters were high on the Pentagon's action list in 1969, as the long-planned Naval Reserve Force Study Group initiated its hearings and investigations. RADM Julian Burke, USN, the study group leader, brought key members of his study group to Atlanta to brief participants in the NRA's Naval Affairs Meeting. The Naval Affairs Meeting was an annual national meeting held alternately on the East or West Coast. RADM Burke's panelists got invaluable feedback from those with a direct concern in what happened to the Naval Reserve.

In another, but unrelated matter, the Pentagon announced plans to close 28 Naval Reserve training activities, along with Naval Air Stations at Olathe and Twin Cities. In another cost savings measure, the Navy halted the practice of providing per diem to reservists who commuted to and from their homes during annual active duty.

Problems also cropped up within the NRA, when all membership records were finally computerized. Some Districts found they had no membership rosters—or no members. Other districts and chapters had undecipherable printouts or ghost members. After much hair pulling and teeth gnashing by the districts, chapters, headquarters, and the contractor, the computers were squared away and order prevailed.

But if the computerization of membership records was the downside of the coin, there was an upside. Membership passed the 10,000 mark with the accession of LT Robert Ayers of Rhode Island. His recruiter was the consistent high scorer, CAPT Herb Tuttle, who was followed in 1969 by CDR "Daisy" Wilhoit, CDR John Shanklin, CAPT Ray Ackerman, and CAPT William Lang. Along with the new members, new chapters were formed at Bangor, ME; Boise; Poughkeepsie; Wilmington, DE; Kingsport, TN; Midland-Odessa, TX; Tulsa; Olympia; Bremerton; Bellingham; Pullman; Everett; and Boston, which was rechartered. In another sing of the NRA's growing reputation, when the first three women Captains in the Naval Reserve were selected, all three were active NRA members. Bravo Zulu to CAPTs Irene Wolensky, a past National VP-Women; Rilla Lovell, and Eleanor Kehoe.

Headquarters saw some good news and some grief that year as well. Bill Kreh returned to the editorship of NRA News, being relived at navy Times by current NRA News editor Allen Scott. Bill Kreh's return came after the completion of his NRA-sponsored history of the Naval Reserve, Citizen Sailors—The U.S. Naval Reserve in War and Peace. Another important work was published by headquarters in 1969, We're Telling it Like It Is—Straightforward Answers to the 20 Questions Most Often Asked by Naval Reservists Being Released to Inactive Duty. The publication received immediate praise in the fleet, particularly among the thousands of junior officers leaving active duty as their tours came to an end.

But the joy at Headquarters at the publication of these works, and the phenomenal growth in the association was tempered by the death in 1969 of RADM John H. "Tex" Schultz (USN) Ret, NRA's first National Executive Officer, from 1954-1957 on March 8 at Bethesda Naval Hospital.

When the EXCOM meeting was held in Memphis in the summer, the chief topic on conversation was whether to raise dues because of rising costs and expenses. The keynote speaker was a Marine Corps Reservist, MAJ Walter Cunningham, who discussed his role as an astronaut on the flight of Apollo 7 in 1968. Shortly after the EXCOM, the National Conference was held in Coronado, CA, with 575 delegates in attendance. The delegates voted down the dues raise. CAPT Ray Ackerman, National VP-Naval Air, was elected National President. A past President, CAPT Bob Burke, was elected Commander General of the Naval Order.

Bill Kreh supplied the comedy at the EXCOM. While Eleventh District President CAPT Jim Mann was conducting an auction in the hospitality suite to raise funds, Bill innocently walked in on some business, with his hand over his head. "Sold," yelled Jim, who proceeded to relieve Bill of $25.00 while handing him a cactus plant.

In late 1969, the Washington DC chapter changed its name from Wardroom to Kalmbach-Wardroom. The name change honored the late CDR Ken Kalmbach, one of the chapter's more devoted members. Ken had served for many years as Fifth District President and as National VP for Education and Activities.

The dawn of 1970 saw bad news for the Naval Reserve. As part of the cutbacks in non-Vietnam related defense spending, Secretary of Defense announced the closing of a number of bases, including three Naval Air Stations at Seattle, Los Alamitos, and New York and 21 Naval Reserve Training Centers and Facilities. The cuts also included the decommissioning of more than 40 warships, including 10 Naval Reserve training ships. The ships included eight on the Great Lakes, one in Baltimore, and one in Tacoma. The announcement indicated that additional decommissionings were highly likely, as proved to be the case. Not all the cutback news was bad. After vigorous protests by National President CAPT Ray Ackerman and others, funding was restored to the Coast Guard Reserve. The planned budget cuts would have had the result of wiping out the Coast Guard Reserve.

Two personnel moves affected the NRA in mid-1970. Fourth District President CAPT Gordon Gardow was transferred out of his District, necessitating the appointment of CAPT Sam Cummins as the new District President. In a move affecting headquarters and the entire association, National President Ackerman found it necessary to have the assistance of a Parliamentarian. LCDR John Rice (New Orleans-Acadia Chapter) became the first parliamentarian.

Interestingly enough, the largest Naval Reserve recall during the Vietnam era was for domestic duty, when 3,900 Naval Reservists in the New York City area were recalled from Mar. 23-26, 1970 during a postal workers strike. Most reservists did not get a chance to deliver the mail because of the short duration of what was the only nonmilitary recall of the Naval Reserve. The mail strike callup also featured mailmen who had gone on strike having to sort and carry mail in their capacity as reservists.

The end of the draft and scaling down of US commitment in Vietnam saw 1,200 first-tour junior officers discharged from active duty as an economy move. In other budget cuts, the Navy continued closing Naval reserve facilities and decommissioning ships, leaving ship strength by mid year at 659, a considerable drop from the nearly 1,000 in service just three years previously. On the bright side, many of the ships being decommissioned were growing long in the tooth and lacked the modern systems required to cope with the threats of the 1970s.

The cutbacks in endstrength had a side benefit for the NRA. Following the success of the experimental complimentary membership program, which produced a bumper crop of new paid members, the NRA voted to begin a full-scale complimentary membership program for junior officers being released to civvy side. As another tool towards gathering new

November, 1971, NRA EXCOM, 18th National Conference.

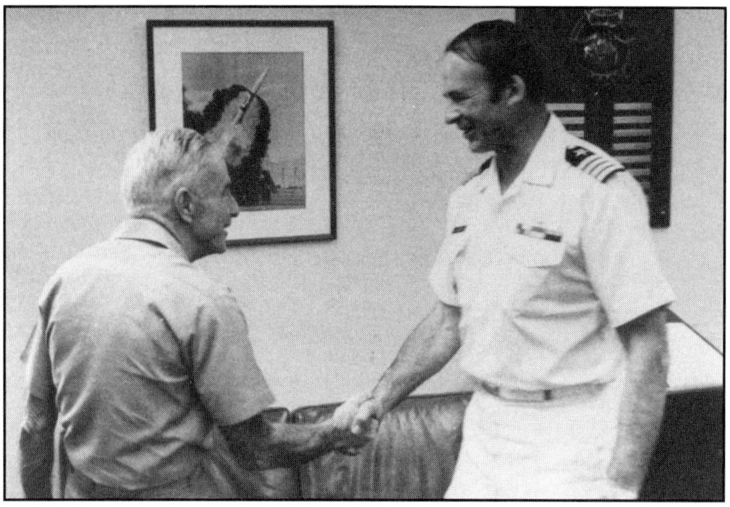

February 1972, HQ CINCPAC, Honolulu. ADM John McCain thanks CAPT Ray Schuman for special studies on personnel training and public relations performed during ACDUTRA.

From left: ENC Joe Wasson, SecNav James Johnson, CNO ADM Zumwalt, SecNav John Warner, President Les Willig, Executive Director Nev Shaffer.

members, the NRA released Twenty Questions Revisited, a revised and updated version of the We're Telling It Like It Is booklet of 1969, which already had seen 150,000 copies printed and distributed.

The scaling down of the force and the implementation of the Naval Reserve Force Structure Study meant new concepts began to be looked at. One was a combined Regular-Reserve riverine unit. The NRA had lobbied hard for the new unit, pointing out that the vast majority of the riverine force officers in Vietnam had been reservists.

The entire concepts of naval reserve training ships also was revised, in line with the Gaining Command concept, whereby the command a reserve unit would report to in time of hostilities would be the one to exercise control over the units. As a result, Naval Reserve Force training ships were transferred from the operational control of Commander Naval Reserve Training to control of the Fleet Commanders-in-Chief. Further proposals were floated to have the title Commander Naval Reserve Training changed to Commander Naval Reserve Force and that Naval Reserve Training Centers become simply Naval reserve Centers. The Pentagon turned down these proposals by Commander Naval Reserve Training, RADM Edward Rosenberg. Another proposal, the Fleet Expansion Unit, whereby specific USNR units would have certain rates, ratings, designators, and grades and be assigned to specific ships was also turned down.

The Navy was not the only service facing problems. The Department of Transportation again questioned the need for the Coast Guard Reserve. But vigorous lobbying by the NRA and action by Congress prevented these ideas from being implemented, and instead, Congress authorized a 15,000 man endstrength.

The NRA saw another year of strong growth. Once again, CAPT Herb Tuttle led the recruiters, followed by LCDR Donald Clark, CDR Robert Wilkins, CDR Doyle Quisenberry, and CAPT Hugh Murphy. The new recruits helped boost the number of chapters to over 150, with the newest sites being Staten Island; Wilmington, DE; Allentown; Williamsport; Newport News; Huntington, WV; Raleigh; Jacksonville; Knoxville; Nashville; Charleston, SC; baton Rouge; Decatur, IL; Pueblo; Benton Harbor; Cape Girardeau; and Anchorage.

To help the new sites (as well as the old ones) aid their constituents and operate smoothly, Headquarters publishes the NRA Operations Manual. A strong Bravo Zulu goes to the committee that produced the work, with CAPT Les McDowell and LCDR Marie Frauens heading a team comprised of CAPT Bill Gilmore, LCDR Dave Woods, and LCDR Lee Douglas. The new looseleaf notebook is distributed to all national, district, and chapter officers. At the same time, Headquarters publishes a leaflet Introducing NRA, for use in membership recruiting and distribution to new members.

Lee Douglas – erstwhile National Membership VP, since 1970 national retention membership statistician, and since 1979 National Historian – who among many other contributions dreamed up the "Log."

Receiving the Naval Reserve Association's National Distinguished Service Award is the Honorable Strom Thurmond. The award was presented on October 7, 1972 by CAPT Leslie A. Willig.

With these triumphs, it was a festive occasion when the EXCOM met in the spring, prior to the National Conference in Dallas. The EXCOM voted to support deep-selection for USNR officers, a recording activity for enlisted reservists, widow's equity legislation, restored and continued funding and support for the Coast Guard Reserve. It also went on record opposing plans to transfer the procurement and operation of commercial shipping from the Military Sealift Command to the Army. The EXCOM also voted to establish NJROTC Libraries in 14 secondary schools that had NJROTC programs. At the time of the vote, 120 high schools had such programs and another 150 were planned.

When the National Conference took place, delegates elected CAPT Leslie Willig as National President. Outgoing President CAPT Ray Ackerman received the Secretary of the Navy's Distinguished Public Service Award. Members also heard the good news that CAPT Mal Munger was elected Commander General of the Naval Order. The Conference saw 130 more members transition to life member status, thanks to the efforts of CAPT "Big Bad John" Clayton, special assistant for long term recruiting. Conference delegates also elected to adopt a sliding scale for Life Membership dues, which would decrease by $10 per year from $200 at age 50 to $100 at age 60.

Members heard the Navy announce plans (which never came to fruition) to give the reserve an attack aircraft carrier, a mixture of other ships, and put operational Naval Reserve ships back on the Great Lakes. Other plans called for giving the Naval Air Reserve several used, modern air transports. The Naval Air reserve also announced the formation, in 1971, of two Reserve Attack Carrier Air Wings and the logging of more than 3,321 accident-free landings.

The Navy continued its slide in officer strength in 1972, as 1,100 more reserve officers received early outs. These officers were some of the last to have file numbers, because starting in 1972 all military personnel will use their Social Security numbers vice having officer file numbers or enlisted service numbers. In other changes in the active side of the house, women are cleared to command shore activities, and seven women reservists are granted permission to remain in Reserve program after they become mothers.

Other changes see COMNAVRESTRACOM's title changed to Commander Naval Surface Reserve and CNARETRA's title changed to Commander Naval Air Reserve/Commander Naval Air Reserve Force. Rumors have it that the two officers will abandon their respective headquarters of Omaha, NB and Glenview, IL and consolidate in New Orleans, LA. Moving to the Big Easy doesn't come

Top honoree at the 1972 Norfolk conference, Senator Strom Thurmond, shared a few remarks at the Awards Luncheon.

Coronado, California, 1973.

One hundred percent membership desk name plate is presented to CAPT Bob Beskind, Commander, Readiness Command Region 13, by CDR (Sel) James Carey, NRA's National Vice President for Public Relations. From left are Region 13 (and NRA) staff members LCDR Roger Coats, CDR Jack Morehead, CDR Everett Lockridge, CDR Ellis Vestal and LCDR George Kouba.

as fast as some would like it to. But by the end of the year, the rumors become true, when Secretary of the Navy John Warner announces the establishment of the Naval Reserve Command in New Orleans, across the Mississippi River from the Algiers Naval Station.

The Naval Reserve Command will be supported by a stronger NRA, which passes the 14,000 member mark in 1972, paced by recruiters CAPT "Big Bad John" Clayton, CDR Doyle Quisenberry, CDR John Shanklin, RADM Roman Ziolkowski, and CDR Julian Benjamin. Along with the new members come a record 23 new chaptersin Manchester, NH; Worcester; Syracuse; Albany; Rochester; Jamestown; Jersey City; Cumberland, MD; Pensacola; Wilmington, NC; Huntsville, AL; Tuscaloosa; Orlando; Asheville; Gainesville; Abilene; Fargo; Grand Rapids; Terminal Island/Long Beach; Channel Islands/Point Mugu/Hueneme; Santa Maria; San Bernardino; and Missoula.

The Post-Vietnam Era and Drawdown

The dawn of 1973 saw further interest in the Readiness Command concept, as the Deputy Chief of Naval Operations for Manpower outlined plans for the new commands, at least for the Surface Reserve Force. The first Readiness Commands were formed in the spring of 1973. Baltimore, Houston, and San Francisco were the three sites chosen, and stalwart NRA members RADM Jud Smith, RADM Phil Koelsch, and RADM Tony Braccia respectively were chosen to honcho the commands. In a slightly related move, towards consolidating the reserves, RADM James Ramage USN was appointed the first Commander Naval Reserve on 1 February, with his headquarters at the new command in New Orleans. On 12 April, VADM Damon Cooper USN, who became the first permanent Chief of the Naval Reserve, relieved RADM Ramage.

In the aftermath of the Vietnam War, retention in the armed forces might have been a problem, but the situation was not matched in the NRA, which was experiencing a 90 percent rate, according to LCDR Lee Douglas, National Retention Officer. Other helping the rate included top recruiters CAPT Bill Cooper, CAPT "Daisy" Wilhoit, CAPT Julian Benjamin, CAPT Hugh Murphy, and CDR Gene Musselman. Other factors helping growth were the new chapters formed during the year in Portland, ME; Providence; New Haven; Perth Amboy; Bridgeport; Lancaster, PA; South Charleston, WV; Lexington, KY; Charlottesville-Staunton; Pascagoula; Des Moines; Bakersfield; Pasadena; Santa Monica; Oakland; Helena; and Butte.

When the National Conference met in Coronado, CAPT Mort Leavitt was elected National President, while, in what was becoming a tradition, outgoing President CAPT Les Willig received the Navy's Distinguished Public Service Award. CAPT Leavitt was succeeded as Chairman of the National Nominating Committee by CAPT Fred Carl. Eleventh District president CAPT "Daisy" Wilhoit received the KEN KALMBACH award for best district communication. The attendees were heartened to hear Secretary of the Navy John Warner tell them what many knew all too well, that "As the Navy goes down in active strength, the importance of the Reserve goes up."

The Conferees approved several resolutions. The key items included support for Naval Air reserve acquisition of jet-powered transports, opposing making TAR billets civilian, full exchange privileges for Reservists, support for earliest implementation of the Reserve Merchant Marine program, and urging that contributory support ACDUTRA not be charged against CAT D allocations. The Merchant Marine resolution came just as the Navy was considering reinstating the program, and with the NRA's whole hearted support, the program was reinstated in 1974.

Several other association stalwarts garnered kudos during the summer of 1974. CAPT Hugh Howell, who had served as Sixth District president and five terms as National VP-Legislation was selected as the first reserve Rear Admiral in the Judge Advocates General Corps. Past National President George O'Connell received the Legion of Merit for his service with the Commander Naval Air Reserve Flag Support Component. The Presidency of the Naval Order passed from CAPT Cal Munger to RADM Alban "Stormy" Weather of the Glenview Chapter.

Headquarters also saw changes. NRA Business Manager John Fudge retired to Los Angeles. His successor came from the Army's Quartermaster Corps, COL David DeVille. Not all the news that year was cheerful, however. Long time headquarters staffer RADM Jack Lewis died in late June. In his honor, the new award for outstanding NRA chapter was renamed the RADM John S. Lewis Award. CAPT Spencer Church designed the award, which was presented later that summer to Long Beach's Terminal Island chapter.

Thanks to hard work done by CDR Jalmer Rolfson of the Kalmbach-Wardroom chapter, the IRS issued a ruling bringing the tax provisions of the military Survivor Benefit Plan (SBP) onto a basis similar to that of the Retired Serviceman's Family Protection Plan. CDR Rolfson had proposed the idea and supplied much of the steam power needed

A selection of the 1974-75 national officers includes: Larry Schuler, Roy Caves, Mort Leavitt (national president), Kay Brice, Julian Benjamin, Emile Bonnot, Ed Grandin, and Joe Reppert.

to get Uncle Sam to issue this ruling. But Uncle Sam giveth and Uncle Sam taketh away. That same year, while drilling reservists were told they had unlimited Exchange privileges on their drill day, non-pay drillers were told they were not counted as drilling reservists for purposes of the benefit.

The new year 1974 opened on a happy note for National President CAPT Mort Leavitt, when he married Mrs. J.N. Rassler on 12 January in the Naval Academy chapel. The couple then had a working honeymoon, visiting NRA districts and chapters on the West Coast. Looking after the health of the districts and chapters was important, because with the post-Vietnam drawdown definitely underway, the Department of Defense initiated a number of moves to restructure the reserves in 1974. The Chief of Naval Operations was first out of the gate, announcing plans to make the Naval Surface Reserve a Contingency Response Force, with mobilizable units in 11 force programs. These units would provide the "most crucially needed, first reaction capabilities required by the active forces."

The Secretary of Defense's office in turn, called for a study on "Guard and Reserve in the Total Force," to be performed by Dr. Ted Marrs, Deputy Assistant Secretary of Defense for Reserve Affairs. The study would concentrate on availability, force mix, limitations, and potential capabilities of the Selected Reserve in any national emergency.

If the one numbers game was not enough, many reservists found themselves wondering what their status was. More than 25 units were disestablished during the drawdown, and many other reservists found themselves cut from 48 to 24 drills. Along with the reduction in drills, many reservists received found inaccurate summaries of their service as the Navy computerized its records. Although it took a year, eventually all the records were sorted out. (Although many reservists to this day believe either gremlins lurk in the computers, or the program never was installed properly.) Another change in the same vein was the decision by the services to remove the 60-point annual ceiling on creditable Reserve retirement points.

With the restructuring underway, the NRA's Washington Conference had some major chores ahead of it. Meeting in the nations's capitol, it was an ideal chance to lobby lawmakers.

The conferees could look at a year well spent. NRA membership had passed 15,000 and retention was more than 90 percent. In addition, new chapters were organized in Scranton; Lakehurst; Patuxent River; Annapolis; Greenville-Spartanburg, Davenport; Sioux Falls, SD; and Hutchinson, KS. On the personal side, past National President Ray Ackerman was one of the new RADM-selectees—the first former National President to achieve this honor, but hardly the first NRAer. Three former Chapter Presidents of the Nimitz-Golden Gate, RADMs Robert Colwell, Ray Hemming, and Ed Wilson had achieved the honor, as had RADM John McElroy from the 12th District.

In early 1975, a Navy study, conducted by Op-605, recommended a strength of 102,000, but the Pentagon's FY76 budget called for only 92,000, representing a cut of 15,000. Congress challenged these figures, with the House of Representatives calling for 112,000 and the Senate going along with the Pentagon. The Conference Committee settled on 106,000, but the eventual budget was for 92,000. National President CAPT Leavitt became a regular on Capitol Hill, testifying that Pentagon officials were calling for more use of the reserves, but that the changing numbers showed little idea of what was required.

The continuing debates between Capitol Hill and the Pentagon over the size and shape of the Reserve Force required the national officers and staff to become regular commuters up to the Hill to testify. The national officers were able to point to a growing body of man (and woman) power behind them. By mid-1975 the NRA had more than 16,000 members thanks to the efforts of many, but particularly CDR Dick McCann, CDR Gene Musselman, and CAPT "Daisy" Wilhoit. The membership boost also was helped by a new policy allowing junior officers on active duty to become members, without having to wait for their release from active duty. Another sign of the membership growth came with the chartering of chapters in Pelham; Toldeo; and Dayton, bringing the number of chapters to more than 200.

With its ever-growing clout, the Association's leaders felt it was time to take on their service and its indecisiveness over how to use the reserves. With this in mind, the NRA choose the 1975 Spring EXCOM in New Orleans to issue a policy statement/position paper criticizing the Navy and its failure to define the roles and missions of the Naval Reserves. The paper called on the Navy to:

1.) Define and identify the Reserve's roles and missions; 2.)Use the Naval Reserve as a cost-effective means of maintaining a ready-forces posture; 3.)Grant the Chief of the Naval Reserve the policy authority to establish and operate a viable Naval reserve program, similar to the authority granted to the Army and the Air Force; 4.) Cease further reduction in Naval Reserve personnel until the Force's roles and missions were defined.

Shortly after the Spring EXCOM, a major change took place at NRA headquarters. CAPT M.F. "Fridge" Lovern retired as Assistant executive Director. CAPT Philip G. Saylor USNR relieved Fridge.

The spring EXCOM was followed several months later by a summer EXCOM in Buffalo. This meeting featured two key guests, CNAVRES VADM Pierre Charbonnet, and RADM Frank Guest, com-

President Mort Leavitt and Secretary of the Navy Middendorf.

President Leavitt calls on CNO ADM James L. Holloway III.

ADM Elmo Zumwalt was a featured speaker at the 1976 chartering of the Blue Ridge Chapter at Afton Mountain, Virginia. Also pictured are RADM Bernard Nash, CAPT Sam Cummins, and CDR Bob Finney.

CAPT and Mrs. Robin Goodenough and RADM George R. Muse, USN, Commander, Naval Reserve Training Command.

mander of the Third Naval District. Following up on their proposals from the Spring EXCOM, the EXCOM said they wanted to hold more frequent meetings with CNAVRES and that the Association should develop an organizational development plan, and that the NRA help update Navy school curricula to teach the active duty Navy more about the Reserves. The guests also saw that NRA'ers of all grades were active, when ENS Phil Seghers, the spring EXCOM's Registration Chairman, was able to announce his recent election as President of the New Orleans/Acadia chapter.

Just when the EXCOM was meeting, the Pentagon released its Total Force Study, prepared by Dr. Ted Marrs, which looked at the Roles and Missions of all the services reserve elements. While most elements of the military took some hits, the Navy and the Naval Reserve took a heavy one. It recommended cutting Naval Reserve paid drill strength to 92,000 and cutting Reserve Seabee battalions from 17 to eight. The primary thrust of the report was that the Navy was least able to use its Selected Reserve units effectively, so it should move towards relying on individual-skill augmentation. The NRA came out in strong opposition to the report's conclusions. In some ways the NRA's opposition had an effect, as the Navy has made the least use of individual mobilization augmentee billets in its reserve force structure.

The "Total Force Study" was followed by the Defense Manpower Commission's interim report. This report questioned the viability of the All-Volunteer Force and claimed the Navy utilized its Reserves to a much lesser degree than did the Army or the Air Force.

With the one-two punch of anti-Reserve sentiment, the 22nd National Committee meeting, in Kansas City, felt it had to make a strong statement, and fast. The Conference adopted four key resolutions in this regard. The first called for establishing a separate Assistant secretary of Defense for Reserve Affairs. The second called for a two-year moratorium on any structural changes to the US Naval Reserve. The other two called for the NRA to become more involved in legislation affecting its members, while the last called for more coordination between the NRA and other professional support associations, particularly the Naval Enlisted Reserve Association (NERA) and the ROA. Other resolutions called for "portal-to-portal" coverage for drills and other inactive duty support functions and making active Navy training facilities available to the Naval Reserve for weekend use, thus avoiding the segregation of too many reserve units to Reserve Centers and Facilities.

The conference agreed with the Department of Defense on one issue. In regards to the calls by some non-service lobbyists for a serviceman's union, a committee comprised of RADM Jim Forrest and Ray Ackerman; CAPTs Jim Austin, Bill Gilmore, George O'Connell, and Phil Smith; and CDR Dave Woods was appointed to consider the issue. It took little time for the committee to issue a report stating that "on the whole unions have served a constructive purpose. However unionization is deemed inappropriate to meet the unique needs of Armed Forces personnel. Professional associations representing members of Armed Forces provide a much more suitable form of representation."

Conference elections saw CAPT Robin Goodenough elected National President. Members gaining recognition in other ways included Past National President CAPT Bob Bolt, who was elected Commander General of the Naval Order, the Sixth District under CAPT Bill Gilmore, which won the Ken Kalmbach Award, and the Nimitz-Golden Gate chapter, which won the Jack Lewis Award for outstanding chapter. The District membership-performance competition saw the Fourth, led by CAPT Tom Nabors, District VP-Membership and CDR Russ Olsen, President, narrowly edge out the Fifth District. The Conference also recognized National VP-Naval Affairs CAPT Fred Carl, who had received the Navy Commendation Medal for dynamic leadership of the Indianapolis Recruiting District Assistance Council in 1972-1974.

During the Conference, CAPT Chip Chapin announced he was presenting the yacht "Reservist, Twice a Citizen," to the Navy. Acting Secretary of the Navy David S. Potter accepted the gift.

The association also began a Job Placement Service under LCDR Jack Westerfield, using the NRA News as the medium. Shortly before this, the NRA News contained an item mystifying many members. Among the units listed as 100 percent NRA units was Investigative Operations 2214 Detroit. But no commanding officer was listed, unlike all the other units, which had the CO's name displayed.

Early in the Bicentennial Year 1976, the NRA once again had to weigh anchor and move its headquarters, this time to leased space in the Barr Building, on Farragut Square, just across from the Army-Navy Club.

The Naval Reserve Association had much to congratulate itself about in early 1978. The year marked the association's Silver Anniversary, and the spring saw membership passing over 18,000, with Life Members hitting the 2,000 mark when CAPT Con MacLane, immediate past District President of the Thirteenth NRA District paid his dues. Hitting the 2,000 mark meant Life Membership had doubled in slightly over three years—and moved up ten-fold in less than eight years. More than 200 chapters had already been formed, and Columbus, OH;

Left to right: Jim Forrest, Don Osborne, Robin Goodenough, and CNO ADM Holloway.

Melbourne, FL; Gainseville, FL; and a new group in Pittsburgh were the newest branches to enter the fold. The association also had chapters overseas, although none had received official recognition as yet. With this new strength, NRA found itself well prepared to fight the budget cutters, who had already bared their teeth in January.

When the Carter Administration released the FY79 budget at the beginning of 1978, there was new bad news for the Naval Reserve. The new budget called for drill-pay strength of 51,400, with the concurrent loss of eight of the 16 Readiness Commands (REDCOMs), between 70 and 90 Reserve Centers, nine NARDETs, nine of 17 Seabee battalions, and four helicopter squadrons. The budget also proposed transferring all airlift functions to the Military Airlift Command and not replacing the Naval Reserve's C-118s with newer aircraft. Control of military airlift was an annual problem, and Secretaries of Defense had been proposing a consolidation under MAC since the late 1950s.

But not all was bad news. Shortly after the budget was released, RADM Richard Altmann, director of the Reserve Compensation System Study, recommended that Survivor Benefit Plan coverage be extended to retired reservists under the age of 60. Later in the year, this proposal was adopted in a modified form, with Survivors' Benefit Protection being offered to Reservists who had retired but not turned 60. Two options were offered either SBP payable when the deceased Reservist would have reached 60 or SBP payable immediately upon a reservists death regardless of the Reservist's age.

In a somewhat similar vein, the NRA's Life Insurance Plan coverage was increased from $50,000 to $100,000 and the highest eligibility age increased from 70 to 100. At the same time, NRA announced a life insurance plan for members' children and an automobile lease-purchase plan.

CDR John Shanklin (left) and CAPT Fred Carl.

President Fred Carl congratulates Senator John Tower upon presenting him NRA's Distinguished Service Award at the 25th National Conference in New Orleans, 1978.

National President CAPT Fred Carl told members to go to General Quarters and set Condition 1, only to find most already on station. When it came to the annual budget proposal and fighting with the Defense Department, most members hadn't relaxed their readiness condition for the past five years—it was merely whether all hands were closed up for action or on modified GQ.

To counter the budget cutters, NRA headquarters organized an ad-hoc committee under past Third District President CAPT Bud Holman. CAPT Holman's team was charged with putting together a legislative package proposing proper stabilization, funding, and organization of the Naval Reserve. The team would allow the NRA to issue its own proposed legislation, instead of waiting until the summer/fall National Conference to issue resolutions.

Playing on the Silver Anniversary theme, the NRA launched a Silver Anniversary Membership contest, with points awarded for recruiting 1-year, 3-year, and Life Members prior to the 25th National Conference in New Orleans. Prizes would be awarded to all qualifiers. The contest, initially designed only for the national organization, was modified at the suggestion of Sixth District President LCDR Phil Pauze' to run concurrently with District contests. The closing date for all contests was 31 August.

One of the major plagues of the late 1970s, inflation, hit the NRA in its NJROTC Library program. The cost of each library rose from $350 to $375, requiring chapters and the national organization to each pay $187.50 rather than the previous $175. The program had already attracted much favorable press, with 75 schools that had NJROTC units having the collections of books.

The spring also saw several NRA members garnering kudos for their work. National Executive VP CAPT Phil Smith received a Meritorious Service Medal for outstanding service during a 1976-1977 active duty as senior officer of a Navy organic airlift analytical project and service as Director Navy Air Logistics for the Chief of Naval Reserve. This work helped CAPT Smith pick up his first star. This made him the third national officer in a row to receive this signal honor, following his predecessor as national Executive VP, RADM Sam Cummins and RADM Bill Gilmore, who made flag while serving as Sixth District President.

The spring EXCOM in Pensacola was relatively quiet. The EXCOM attendees voiced their opposition to turning the Panama Canal to Panamanian control and approved plans for an annual NRA reception for members of Congress. Attendees also approved two proposals for increasing membership, by using recruiting posters, and placing newspaper advertisements on Reserve issues.

The members also opposed two proposals by Navy and Pentagon officials that would have affected personnel. The first was that Naval Aviators who resigned their Regular commissions be denied participation in the Naval Reserve. The proposal was made to stem the hemorrhage of active duty aviators that was hurting readiness. The second proposal called for placing CNAVRES under the Deputy CNO for Manpower. NRA opposed this proposal, pointing out that it would lower the status of the Naval Reserve and make it less effective.

Much of the NRA's awareness of what was going on in Washington came from headquarters and the devoted staff. But some of the awareness came from members who had their ears to the ground in key places, such as Rep. Bill Chappell (R-FL) and his special assistant Berton Otto. Both men were stalwart NRA members and CAPTs in the Naval Reserve. In thanks for their strong support, NRA and NERA threw a luncheon in June for them. More than 200 attended.

Another group of members in key places was the NRA members on the National Naval Reserve Policy Board. Of the senior board membership, both RADMs, all four CAPTs, and eight of the 11 CDRs were NRA members. In addition, one of the four Chiefs was an associate member. Therefore, when the NRA presented its recommendations to the board, shortly after the spring EXCOM, they received a favorable hearing. The NRA made 11 recommendations to the board in 1978, including

- Appointing a USNR officer as CNAVRES, and giving him planning/programming/budgeting authority.
- Making CNAVRES directly responsible to the CNO.
- Providing formal instruction within the Navy training and educational system about the reserves.
- Having all USNR-related issues and proposals referred to the Naval Reserve Policy Board for comment.
- Having Selected Reserve officers appointed to command Reserve Air Squadrons.
- Restoring PCO and PXO billets in NRF ships and providing more billets for career USNR petty officers.
- Returning drilling REDCOM commanders to Cat A pay status.
- Restoring all Cat B billets (24-drill) to Cat A status (48-drill).
- Having all USNR flag officers present at all annual flag-officer briefings, and having a discussion of Naval Reserve issues at these briefings.
- Transferring NROTC funding out of the Naval Reserve budget.
- Publishing the Policy Board's report upon completion, rather than waiting for the Secretary of the Navy's comments and actions.

Shortly after the recommendations were published, the summer EXCOM was held in Phoenix. The meeting was attended by 137 NRA members and 38 guests—a five-year record, and the first held at a non-military location. EXCOM attendees voiced support for the House of Representatives plans to strengthen the Reserve Bill of Rights, particularly the provisions making CNAVRES a three-star flag officer and upgrading the Deputy Assistant Secretary of Defense for Reserve Affairs to an Assistant Secretary of Defense.

The attendees also changed the dues structure, raising dues to $15 for one year, $36 for three years, and tacking on a $5 service charge for those life members paying on the installment plan. But a plan to eliminate the Complimentary Memberships was tabled. In other action, "minimum performance standards" for chapters were established and published; and a separate District fund of $.50 per member was established. CDR Jim Carrico was appointed General Chairman for the first congressional reception, to be held in 1979.

Scarcely was the EXCOM meeting over than Capitol Hill boiled over, as President Carter vetoed the defense appropriations bill, which contained funding for a nuclear-powered aircraft carrier. Secretary of Defense Harold Brown followed with a letter to Sen. John Stennis, chairman of the Senate Armed Services Committee, calling for the committee to accept the President's cuts to Naval Reserve strength. The NRA promptly responded to the challenge, calling for the reinstatement o Naval Reserve transport aircraft and Selected Reserve strength. When the smoke cleared and the joint Conference resolution was reported out, the bill funded 87,000 Selected Reserve billets, a new C-9B aircraft, the acquisition of three C-131H aircraft from the Air National Guard, and the extension of the C-118 program.

Left to right: VADM Pierre Charbonnet, CDR John Shanklin, CAPT Jim Foster, and CAPT Lester Bob Smith.

VADM Pierre Charbonnet addresses an NRA EXCOM as the Chief of Naval Reserve.

At the same time, the Senate asked the Pentagon to study alternative force structures to replace the all-volunteer force, which was seen as failing. While DoD was gearing up for this study, the Review of the Guard and Reserve (ROGAR) II came out. ROGAR II was a follow-on to the original study, which pointed out that the Naval Reserve Force destroyers then in service were too old to be of great use during a conflict. The study added that the new ships just entering the fleet were too sophisticated for use by the reserves and would require full crews aboard at all times.

The Naval Reserve suffered another loss when the decision was made to have CNAVRES remain a two-star billet. In accordance with this decision, RADM Fred Palmer USN relieved VADM Pierre Charbonnet. RADM Palmer brought a wide-ranging experience to his post, having been an enlisted Naval Reservist and surface line officer for 10 years before becoming a naval aviator and augmenting into the regular Navy.

Other events also kept the NRA busy during 1978. The NRA Operations Manual was completed and the new edition distributed to District and Chapter leaders. In a side issue, the dates of Navy Sabbath were changed from the weekend of the Navy's birthday in October to Veterans Day weekend in November. The association also decided that the Ken Kalmbach Award would not be awarded in 1978, but the Jack Lewis Award for outstanding chapter would go to the Sutter Chapter in Sacramento, led by CAPT Robert Harris. In a tradition for leading officers of the NRA to move on to a sister organization, NRA Past National President CAPT George O'Donnell became president of the Naval Order. In a similar move, Past National President George O'Connell was appointed a Rear Admiral in the Illinois Naval Militia.

When the Silver Anniversary National Conference was held in New Orleans, members could look back on a very good year. CAPT Stu Jacobs of Miami's Sidney Fields-Sunshine Chapter won the Silver Anniversary Membership Contest. CAPTs John Wilhoit (San Diego) and Lee Isaacson (Polaris-San Jose) took second place. The Twelfth District qualified for a cash prize in the District Competition, while CAPTs Stu Jacobs, Jack Oglesbee (Atlanta) and Phil Campbell (Pensacola) received honors as individual top recruiters for the year. Thanks to their efforts and those of untold hundreds, paid retention for the entire association was over 91 percent and every district achieved an over 90 percent rate.

Because of the success, the association voted to have the contest continued for 1978-1979 in a somewhat modified form, with prizes for individual efforts. New categories were established for 1-year members, 3-year members, Life Members, and total gain. In addition, the A. Winfield Chapin Trophy would be awarded to the District achieving the highest membership renewal rates during the year.

The Conference approved the dues increase, extended eligibility for dues-paying associate memberships to spouses, widows, and widowers, and created four new national standing committees, Site Selection, Missions and Mobilization, Retired Officer Affairs, and Junior Officer Affairs. Only three resolutions were adopted, calling on Congress and the Administration to observe vital security factors in negotiating and approving a SALT II treaty, calling for an immediate shipbuilding program, and calling for a National Patriotism Week. This last item had earlier been suggested at the Phoenix EXCOM by Lori Cox, recipient of the NRA's Meritorious Service Award.

The highlight of the National Conference was an address by Chief of Naval Operations ADM. Thomas Hayward. Calling for action to implement the One Navy goal, ADM Hayward laid out seven tasks for meeting the goal. These included

"Revitalizing the Surface Reserve;"

"Stabilizing the authorized numbers of Selected Reserve personnel;"

"Ensuring that all roles are meaningful and complement the active forces...and addressing the Naval reserve requirements concurrently with the active force requirements in the programming and budgeting system."

"Providing the necessary training support for the Reserve forces, ensuring the highest practicable mobilization readiness;"

"Improving communications between the active and reserve forces;"

"..getting a greater awareness of needs;"

"...and ensuring that the Naval Reserves get their fair share of top quality officers to run the organization."

The National Conference also saw some levity, as four life members received life membership numbers matching their designators. CAPT Sig Seigel (Philadelphia) received number 2205, followed by CDR David Strickland, who received 1105, RADM Martin Andrew with 1635 and LT Phil Kazanjian with 1655. The Conference also saw the publication of a Silver Anniversary Cruisebook, NRA LIGHTS THE WAY, produced by LCDR Kleinschmidt, CWO4 Shirley Barlow, CDR Rick Allen, CAPT Spencer Church, LCDR Lee Douglas, and CDR Latka. A separate companion brochure with no illustrations, TIDBITS FROM 25 YEARS, accompanied the cruisebook.

Shortly before the National Conference, the NRA's calls for more Navy airlift capability were highlighted during a worldwide mobilization exercise, Nifty Nugget. The major complaints from the joint Commanders-in-Chief were the lack of tactical airlift capability. Because of the lack of airlift, among other problems, the best review the exercise received was "fairly good."

In a separate active duty shortly after the Conference, CAPT Dick Reade (New York/Long Island) went on active duty to help develop the Navy Manpower Mobilization System (NAMMOS). NAMMOS was to meet the requirement for system showing actual total force requirements, while providing a capability for rapid response to changes in scenario, mission, function, and task requirements.

If 1978 was an interesting year, 1979 promised to live up to the old curse calling for a person to "live in interesting times." Early in the year, National President CAPT Fred Carl warned of an increasing anti-reserve bias in the Pentagon, as Reserve supporters left the Department of Defense. He also answered critics in the organization, who said that the NRA should remain focused as a professional and social organization. He pointed out that if the NRA did not support the Naval Reserve and keep the heat on; the Naval Reserve might drastically shrivel in size, leaving nothing to support.

If CAPT Carl wanted support for his views, he couldn't have asked for better allies than the Carter Administration's Defense Department. The ROGAR II study recommended a minimum Selected Reserve strength of 100,000. But shortly after ROGAR II was issued, the FY80 Budget Request came out. The request called for a 48,700-man Naval Reserve. Cutting back to this level would have seen the loss of 20 NRF destroyers, four VR squadrons, three C-131 aircraft, four ASW helicopter squadrons, two light attack helicopter squadrons, to VC squadrons, 50 percent of the Ship Intermediate Maintenance Activities (SIMAs), one-third of the cargo handling battalions, all cryptologic communications personnel, all Naval Beach Group augmentation units, one-third of the intelligence units/personnel, all Auxiliaries Activities programs, and all medical, management, logistics, and training support activities.

To cap off the bad news, the Navy shortly thereafter announced plans to close 36 additional Naval Reserve Centers and Facilities. These plans would have resulted in the transfer of more than 2,300 reservists. This was followed by news of the closing of five Naval Air Reserve Detachments (NARDET)s. The facilities being closed included those at Brooklyn, Lakehurst, Miami, Niagara Falls, and Seattle.

Responding immediately, the NRA, along with every other service-support organization, immediately protested the drastic cuts. Congress heard the outcry and supported the organizations, commencing hearings that seriously questioned Administration premises of only very short wars or nuclear holocausts. Many Congressmen gained a favorable view of the NRA during the March 27 Congressional Reception at the Rayburn House Office Building. CDR Jim Carrico and the Congressional Reception committee asked all NRA members to purchase "sponsorship invitations" to send personally to whichever Congressman they wished. The idea behind the invitations was to help defray costs, which were quite high.

Nearly 1,000 people attended the reception, including 83 Congressmen, four Senators, many senior Congressional staff, and senior Navy and Defense officials, including Secretary of the Navy Graham Claytor, CNAVRES RADM Fred Palmer. More Senators would have attended, except for a roll call vote that took place during the reception. NRA National Presidential Awards were presented to Senator Sam Nunn (D-GA), Representative Bob Wilson (R-CA), and John Ford, director of the House Armed Services Committee staff. Attendance was so large the reception paid for itself and cost the Association not a penny. In a strong show of support for the Association, many attendees were from Chapters well outside the Washington DC area.

Congress wasn't the only group getting a good view of the Naval Reserve. In the spring of 1979, Reserve Air Wing 20 conducted a special one-week active duty aboard the aircraft carrier Independence to train junior ships' company in cyclic operations. The Wing won the highest praise from observers for their pride and professionalism. Two of the senior observers aboard for the exercise were CAPT Bob Lowry, National VP-Naval Air and CAPT John Bell, a National Headquarters Consultant.

Continuing its campaign against the Pentagon's planned budget cuts, the Spring EXCOM, which met in San Antonio, drafted an official policy statement opposing the decommissioning of the 20 NRF destroyers, without a simultaneous replacement of the capability by newer warships. National President CAPT Fred Carl thanked the members for their attention to the issues and asked for them to keep the cards and letters coming in to their congressmen. During the EXCOM, CAPT Carl went public with his criticisms, telling a press conference that President Carter's actions "seriously degraded the ability of the US Navy to respond to a major crisis, in order to appease a very vocal but very narrow-minded minority of his constituency."

Not all the news that spring was bad. In March, CNAVRES announced that all enlisted Cat B (24-drill) billets had been converted to Cat A (48-drill) billets retroactive to March 1. That same month, the House Armed Services Committee voted for a naval Reserve Selected Reserve strength of 87,000 and called for the funding to come from DoD agencies rather than just Navy funds. And when National President CAPT Carl testified before the House Armed Services Committee and Senate Armed Services Committee in April, he met a very favorable reception.

During these and other hearings, Rep (and CAPT, USNR) Bill Chappell (D-FL) pointed out that approximately 16 of the NRF destroyers the Administration wanted to decommission could be repaired and upgraded, giving them several more years of service. When the Senate Armed Services Committee met, Sen. Sam Nunn (D-GA) suggested that if manning was a problem, the Navy should look at recalling reservists to active duty. In other action on Capitol Hill, more than 25 Senators sent a letter to Senate Armed Services Committee chairman Sen John Stennis (D-MI) expressing concern over the Carter Administration's continued attempts to gut and eviscerate the Naval Reserve.

Congressional support for keeping the reserves at their current strength could hardly have come at a more critical time. The fleet was so short of medical officers that a special call went out for volunteers to support deploying ships and Naval Hospitals. In response to the call, 280 Naval Reserve medical officers volunteered to serve 3- to 20-week periods on active duty. The recall paid off in two ways. A critical need was filled, and the Navy received invaluable training in how to conduct a mobilization. But as RADM Fred Palmer told attendees at the Ninth Districts annual District Conference, until the annual budget battle halted, manning stability would be a myth. In another step to alleviate the manning problems, reenlistment bonuses for Selected Reserve personnel in 13 critical ratings was approved. And in another healthy step, the TAR program began advertising for recruits again. Manning was not a problem across the board, because Selected Reserve strength stood at 86,618 in June 1979—a gain of 5,000 on the strength at the beginning of the calendar year.

In between working Capitol Hill and NRA headquarters, Executive Director RADM Jim Forrest noticed that while the Reserve flag officers were all NRA members and attended the conference and briefing, they had never had an informal get together. Resolving to remedy this shortcoming, he organized a reception at the Army-Navy Country Club in Arlington on May 24. Nearly 50 Reserve admirals attended the reception, along with more than 100 other NRA members. The chance to bat ideas around in an informal setting proved invaluable to all participants.

Attendees also had a chance to comment on a recent letter to the editor that appeared in Navy Times from Past National President Chip Chapin. Angry at the unisexing of the Navy terminology, he blasted the Navy's plans to "crew" vessels rather than "man" them, call ships "it" instead of "she," and "other asininities," as he put it. Chip wondered when the Navy would do away with "sideboys," "quartermasters," "the old man," "man o' war," and other traditions of the sea services.

The summer saw at least one benefit coming to Reservists, when the Navy began mailing Survivor Benefit Program information and applications to reservists. Reservists had to return their decision by September 30. In a similar, but unrelated move, in the autumn, President Carter proposed to Congress that the military retirement system be changed to provide two levels of retirement. One was reduced benefits prior to age 60 then full benefits. The other called for retirement benefits at age 60 for personnel who served on active duty for between 10 and 20 years.

When the Summer EXCOM met in Alameda in June 1979, the attendees heard a request from the Navy to rescind the NRA's position statement in favor of retaining the 20 NRF destroyers. After examining the issue and considering both the condition of the ships and their utility to the Reserves, the attendees decided not to rescind the statement. On the financial front, the attendees voted to provide a subsidy of $100 as seed money for new Chapters and increase the stipend for reactivated chapters from $50 to $100. They also voted to increase the District Presidents' fund for membership purposes from $.50 to $.60.

The Summer EXCOM also considered and recommended several changes to the Constitution and Bylaws of the NRA. These changes included

- The period within which National Conferences could be scheduled was broadened to be 15 August to 15 November. CAPT Curt Coleman recommended the change, noting that it allowed the National Conference to take advantage of off-season rates.
- Associate members became eligible to hold Chapter and/or District office, but remained ineligible to hold National office.
- The title National Judge Advocate was changed to National VP-Legal Affairs, and a standing legal committee was to be established.
- The Missions and Mobilization Committee was to be disestablished.
- The complimentary membership program for newly commissioned active duty junior officers was eliminated. The major problem with the program was the constant deployments so familiar to Navy life, with the consequent lateness of mail and inability to participate actively in local Chapters. But the complimentary membership program for junior officers recently released from active duty was retained. In 10 years of operation, the program showed considerable success and profit. Once the released junior officers settled down and affiliated with a local reserve unit and put down stakes, they became the core of the future NRA.

While the EXCOM was in session, the NRA Policy Committee met and issued a statement calling for immediately reactivating Selective Service registration and classification. The draft was not to be reactivated unless critical manpower shortages developed in the active duty forces.

Summer was also the time for many NRA awards to be given out, and 1979 was no different from any other year. The Eleventh District, led by CAPT Kay Brice, won the Ken Kalmbach award for best District communications program, while the Southern New England-Providence Chapter, led by LCDR Harry Snyder won the Jack Lewis award for outstanding chapter. The Nimitz-Golden Gate Chapter, directed by CAPT Charles Carlsen received special honorable mention as a close runner up for the prize.

Kudos for the recruiters went to CAPT Jim McNallen for top individual effort. Jim tied with CDR William Kerr for attracting new 1-year members and with CAPT Ed Grandin for recruiting new 3-year members. Honors for life member recruitment were shared by CAPT "Daisy" Wilhoit, CDR Dave Woods, CAPT Lucille Mellish, and CAPT Winfield Chapin. These recruiting efforts helped lead to 1979s' outscoring eight of the previous 10 years for recruiting, setting an all-time retention record for the third year in a row, registered the second-highest gain in a single decade, and passing the 45 percent mark in long-term membership. Moreover, all this came during a year in which dues were raised.

Two other NRA members to win praise and/or recognition during 1979 were CAPT Bob Beskind, who won the first John Paul Jones Award from the Bluejackets Association. Bob's honor came after his efforts to foster and implement the idea of One Navy while serving as the active duty Readiness Commander at Great Lakes and President of the Great Lakes NRA chapter. That same summer, LCDR Phil Kazanjian, President of Rose Bowl-Pasadena Chapter, played up the importance of the CNO Sea Power Presentation Program in an article in NRA News. Phil suggested that NRA Chapters make the program an official activity, coordinating their efforts with local Reserve Readiness Command.

Other events during the year included a 10th reunion by Mobile Construction Battalion 22, on the anniversary of their release from active duty, which included an eight-month tour in South Vietnam. The unit gained great recognition for its efforts, and its then commanding officer, CDR Roy Dunlop, appeared at the reunion as RADM-select

Dunlop of the Waco Chapter. In another memory of the Vietnam War, CAPT Ruth Kasky visited ex-POW CAPT John S. McCain III, head of the Navy Liaison Office in the US Senate. During the war, when numerous members of the NRA wore POW bracelets with the name of a POW engraved, then LCDR Kasky had a bracelet with the name of then LCDR McCain. During her visit, CAPT Kasky presented the bracelet to CAPT McCain.

In signs of the changing times and computer age, eight Naval Reserve Centers received ShipBoard Simulators (SBS). The SBS, comprising an integrated complex of bridge, CIC, damage control, engineering, and communications modules, can generate a full-scale battle problem. Plans call for 20 more to be installed by the end of FY80. The Naval Air Reserve also saw changes, with the seven remaining NARDETS in Columbus, Denver, Minneapolis, Miramar, Moffett Field, Olathe, and Patuxent River becoming Naval Air Reserve Centers. Upgrading the status of the centers meant the officers-in-charge became commanding officers.

While the Air Reserve Centers were being upgraded, other parts of the Naval Air Reserve also were assuming greater importance, with the Fleet Commanders and active duty aviation planners requesting more support from P-3 reserve squadrons to meet the growing need for air ASW assets. These calls came at a time when the Defense Department was claiming no funding was in the pot for any upgrades to the overworked Reserve P-3s.

The capabilities of the Reserve P-3 community were highlighted in 1979 when Naval Reserve Patrol Squadron 60, under the command of CDR Howard Lynse of the Glenview Chapter received two coveted honors. First came the Noel Davis Trophy for best squadron of its kind; second came the Lockheed Golden Wrench Award for best maintenance of squadron aircraft in the P-3 community. Much of the credit also went to the squadron's TAR coordinator, LCDR Howard Redmon, also of the Glenview Chapter.

The highest award for the year, the NRA's Distinguished Service Award, went to CAPT Winfield "Chip" Chapin. One of NRA's most tireless stalwarts, Chip barely missed being a founding member. He more than made up for this by serving as national VP-Budget and Finance in 1956-1957, General Chairman of the Fourth National Conference, and five terms as National President, from 1957-1962. He followed this by serving from 1962-1964 as first Chairman of the National Advisory Committee. After leaving national office, he continued lending financial, political, and moral support to the association, always being there when there was a need. The late 1970s saw him leading the informal Past Presidents' Group and continuing to speak out on issues.

One other kudo during the year came as the Naval Order held its Biennial Congress and elected NRA founder CDR Fred Kempe as its next Commander General. CDR Kempe's honor came shortly before the NRA held its National Conference in Buffalo. During the Conference RADM Phillip W. Smith Jr. from Florida was elected National President, becoming the first flag officer elected President.

Conferees adopted several resolutions, which were inserted into the Congressional Record on 30 October by Sen Strom Thurmond (R-SC). The resolutions included

- A recommendation to the Senate to reject the SALT II agreement as being unacceptable in its present form.
- A recommendation for the Secretary of the Navy and the Chief of Naval Operations to reject any proposal to scale down the Reserve Inshore Undersea Warfare (IUW) community, particularly because the active duty IUW community was scheduled for disestablishment in 1980.
- A recommendation to consider a sea-based MX missile system as being safer and more cost-effective than a land-based system.
- A recommendation that the Department of Defense and Congress push for program improvements for all Reserve components, particularly in their hardware, training facilities, and budgets.

Conferees also heard Deputy Assistant Secretary of the Navy for Manpower Mary Snavely Dixon review the results of the NAMMOS Study. The study had been completed in the spring and reviewed by the Fleet Commanders and senior Navy Department and Defense Department hierarchy over the summer. It called for a Naval Reserve personnel strength of 106,000—more than double the figure the Defense Department called for at the beginning of the year.

At the same time, a study of the total reserve force structure by the General Accounting Office (GAO), initiated in 1978, came out with two major reports. One report concluded that manpower shortages would restrict the use of the Reserve and National Guard forces in any future conflict. The report said better pay and benefits were urgently needed to attract and retain people. The other report said that while training deficiencies had been rectified to some extent, much remained to be done. Particularly important was the need for improved platforms—long a point the NRA had been trying to make.

In a move designed to boost the numbers of reservists, the Navy established a Naval reserve Indoctrination Team for the Naval District Washington similar to teams already existing in Norfolk and San Diego. But the one-hour briefing was not given automatically. Military members had to know about the briefing and then specifically request it.

When 1979 came to a close, the NRA, the Naval Reserve, and the Navy had cause for hope and despair. On the down side, ships were unable to get underway and planes were unable to fly for lack of spare parts. The services were hemorrhaging people, particularly mid-grade officers and petty officers. Procurement and force structure policies were fluctuating wildly and zigzagged radically from one year to the next. But on the positive side, the Iranian hostage situation and the Soviet invasion of Afghanistan had awakened many government officials to the fact that there were threats beyond our shores. And there were comforting signs that policies might change in 1980 as fear of these threats mounted.

For the NRA, the year ended on an uphill note. Congress kept Selected Reserve strength at 87,000, rather than the Administration's 48,700. Congress kept the force structure of 12 air squadrons and nine Seabee battalions, rather than the massive cuts favored by the Administration. Congress also voted to keep 12 of the 20 NRF destroyers in service and provide the Navy with the authority to call 500 reservists to active duty to help with their manning. Congress also voted to keep most of the administrative support units and procure two C-9B air transports in FY80. The year also saw CNAVRES reprogram money earlier in the fiscal year to avoid the perennial problem of making September drills non-pay and having to cancel summer active duty orders because there was no money available.

The year 1980 dawned with the US military facing growing threats and the Carter Administra-

End of an Era – RADM Philip W. Smith, USNR (Ret.) was the guest speaker for the Decommissioning Ceremony for the C-118 at NAS Atlanta.

NRA and CNO.

tion beginning to see that the military might require more funding. But it was only beginning to see the light, and when the Joint Chiefs of Staff decried the low state of military pay, the President said that "pay was not a major consideration" when he was in the Navy between 1947 and 1953. Although the Joint Chiefs did not continue complaining, others did, and National President Phil Smith fired off a letter to the White House.

Facing the existing problems with funding and capabilities, the NRA released their legislative proposals early in the year. They called for a supplemental FY80 appropriation to allow an USNR Selected Reserve strength of 89,000 with the ability to build to 92,000 in FY81. They also called for putting all USNR drilling personnel in Cat A (48 drills) and authorizing the recall of 1,000 volunteer reservists in critically short skills. Other proposals included the procurement of at least six C-9B aircraft, phasing in the P-3A/B tactical navigation modification upgrades. On the personnel side there were proposals to enact DOPMA and develop a Reserve Officer Personnel Management Act (ROPMA) for Reservists. Finally, the NRA called for a budget supplement to relieve the $27 million USNR-budget shortfall anticipated for the remainder of FY80.

The budget proposals were not the only area where the NRA made a contribution. NRA members began proposing the possibility of transferring some FF-1052 Knox class frigates or FFG-7 Perry class frigates, or building a Coast Guard cutter-like vessel to take the place of the NRF destroyers, which were obsolescent and would reach the end of their effectiveness by the mid-1980s at the latest. A GAO study highlighted this, pointing out the cost needed for required overhauls.

In one way, the Naval Reserve was ahead of the active force in training capabilities. The ShipBoard Simulators installed in 17 Reserve Centers were so much more capable and realistic than the systems available to the active fleet that the Naval Reserve was asked to make these systems available to the Regular Navy, as was practicable.

Another point of congratulation for the NRA was the selection of Past National VP-Legislation Tom Stansbury, Past Eighth District President Lester Smith, and National Parliamentarian Julian Benjamin for flag rank. One NRA member who had already achieved his flag, RADM Robert Spiro, from the Jacksonville Chapter, moved to another service in early 1980, when he was nominated and confirmed as Secretary of the Army.

Other members of the Association were distinguishing themselves in different ways. The Louisville Chapter, led by CDR Ray Wireman, sponsored four more Navy Libraries for NJROTC units, bringing the total number of such facilities to 82—of which Louisville contributed six. Libraries help foster writers, and two NRA members won national prizes for their literary efforts. CDR Victor Meyer (ADM Byrd-Richmond Chapter) won the Freedom Foundation's annual "Defender of Freedom" award for the Foundation's letter writing contest. The same contest saw CAPT Jim Johnston (Everglades-Fort Lauderdale Chapter) winning the George Washington Honor Medal. To spur a similar effort within the NRA, Past National President Chip Chapin sponsored an NRA Loyalty essay Contest, with three awards of $500, $300, and $200. The winners would be announced at the National Conference in San Francisco.

Two other NRA members contributed even more to the society. LCDR Ken Goodsell (San Diego Chapter) and LCDR Mike Scuro (Arrowhead-San Bernardino Chapter) saved the life of a San Diego fisherman whose boat sank more than 10 miles from shore. The two performed their life saving mission while returning from an ASW exercise as pilot and co-pilot of a helicopter from Helicopter Antisubmarine Squadron 84.

With the recent literary and lifesaving accomplishments of its members, the NRA had much to congratulate itself about on 18 March, when it held its Second Annual Congressional Reception. With no votes or roll calls scheduled this time, 160 Senators and Representatives, amounting to 30 percent of the US Congress showed up to meet and speak with more than 1,000 NRA members and guests. NRA National Presidental Awards were presented to Sen. John Stennis (D-MS), Sen. Jake Garn (R-UT), and Rep Bill Chappell (D-FL) for their roles in forging a stronger and more stable Naval Reserve.

Shortly after the congressional reception, the NRA held its spring EXCOM in Kansas City on 28 March. One of the EXCOM's first orders of business was to make the Congressional Reception an annual event, because of the success of the first two gatherings. Other positions adopted by the spring EXCOM included calls for increased competition for active duty personnel, supplemental appropriations to adequately fund travel and training in FY80, and building a 650 ship Navy centered around 16 carriers and 17 air wings and the procurement of at least 400 aircraft per year. (The current naval strength was 529 ships, of which only 453 were active.)

The spring EXCOM could look back with satisfaction at certain policies that had been established earlier in the year. One called for TAR officers to have a career pattern similar to that of their active duty counterparts, including greater opportunities for command. The TAR program underwent another major change in 1980, when it was opened to women Naval Reserve officers. Another called for Selected Reservists assigned to units, whether pay or non-pay, to be able to fly Space-A at any time, and not solely in connection with drill or active duty.

The question of surface ships for the Naval Reserve was another issue tackled by the spring EXCOM when it met in Kansas City. The EXCOM opted for the FFG-7 or some other fully mission-capable ship rather than the proposed FFX corvette. But the conferees called for retaining the current destroyers until mission-capable platforms were available. The conferees knew all too well from past experience that assets lost were not easily regained, particularly in the tight budgets of the late 1970s. Another reasons for keeping the ships was because of the growing manpower of the Naval Reserve.

In the middle of the spring of 1980, the Naval Reserve saw its recruiters meeting 124 percent of their goals, far ahead of the 89 percent of goal achieved in the first half of FY79. The retention of veteran Reservists also was up six percent over the year before. Part of this was due to a decision to let Naval Reserve medical officers (2105) who had exceeded the age-in-grade limits remain on an active status until age 64 if they met all other requirements. Some of these recruits of course ended up in the NRA, where they found a new Introducing NRA brochure, updated and expanded to include new material about the Association and the advantages and benefits of joining.

To cope with the growing numbers of reservists, 20 officers were recalled to active duty to serve in "265" program billets managing different aspects of the Naval Reserve. The assigned officers were sent to commands in Washington, New Orleans, and at the Naval Reserve Readiness Commands. One of the first items these officers (and others already on active duty) had to cope with was a message from the Department of Defense ordering the services to clean up the has existing in the Standby Reserve. The 25,000 in the Navy (USNR–S1 and –S2) and the 75,000 in other services, were in more than a few cases unmobilizable, because of their jobs or other factors. Most were reassigned to the Ready Reserve, particularly the Individual Ready Reserve.

The growing number of sailors also needed homeports, and in the summer of 1980, Long Beach was upgraded from a Support Facility, which it had become in 1973, to its previous status. To help upgrade the facility required the services of many active duty and Reserve sailors. The beachhead of these teams was a 250-person Naval Reserve unit YR-85, manning a barge of that name. Commanded by CAPT Charles Adolf (South Bay Chapter), the unit swiftly established their repair barge as a floating intermediate repair facility for the 30 ships that soon called Long Beach home.

Not all the reservists were getting paid for their efforts. A bumper sticker :reading "Naval Reservists Do It For Nothing," circulated by Chicago reservists CDR Fran Nielsen and CDR Jim Carey turned out to be particularly applicable to the situation in 1980. Only eight percent of USNR admirals were drawing drill pay, compared to 79 percent in the Army reserve and 100 percent in the Army and Air National Guards, Air Force Reserve and Marine Corps Reserve. Only the Coast Guard fared worse, with neither of its two reserve admirals drawing drill pay. The situation was somewhat different in the O-6 ranks. Here, 29 percent of Navy captains received drill pay, compared to 27 percent for colonels in the Army Reserve, 30 percent for Marine Corps Reserve, 35 percent of Coast Guard Reserve captains, 75 percent of Air Force Reserve colonels, and 99 percent and 100 percent for Army National Guard and Air National Guard respectively.

The pay problem filtered to other ranks in June/July, when pay for one weekend had to be cancelled to help make up a $14.6 million shortfall for the Navy's travel budget. But to achieve the savings, the drill weekend savings of $6.6 million had to combined with a $6.9 million savings from reprogrammed Naval Reserve programs and $1 million in cuts to other programs.

Some Reservists found a way around this by paying for their travel out of pocket. Seven officers and three chiefs from Naval Control of Shipping (NCSO) 113 paid $900 apiece to travel to their mobilization site in Pusan, Korea. As unit commanding officer CDR Francis Stokes (Trident-Chicago Chapter) saw it "With the world situation in Afghanistan and Iran as unpredictable as it is, we decided that this was no time to stay in Chicago for training."

If the NRA leadership wanted a perfect example of the funding shortfall, it came during National President RADM Phil Smith's visit to the Pearl Harbor Chapter. While in Hawaii, he saw an operational C-118 transport, VC-118 number 128424, which he had instructed in 1951 and 1952 during the Korean War. With 29 years under the hood, the plane was still serving faithfully.

Despite the age of their aircraft, the Naval Air Reserve could take great pride in the announcement by Chief of Naval Operations ADM Thomas Hayward, that they had passed 100,000 accident free hours of flight time, making them "...the standard bearer of excellence in Naval Aviation." The CNO found one tangible way to express these feelings during the summer of 1980, when he visited Naval Air Station South Weymouth, MA and presented Helicopter Antisubmarine Squadron 74 with its sixth CNO Safety Award. After visiting several other reserve units at South Weymouth and Great Lakes IL, he was interviewed by CBS newsman Peter Hackes (also known as CAPT Peter Hackes, Kalmbach-Wardroom Chapter). During the interview, ADM Hayward stated "The Joint Chiefs of Staff make an assumption about the Reserve forces—we assume they're ready to go...The Naval Reserve is doing a fine job today and I'm really pleased and proud of them."

In addition to the CNO, some members of Congress also were worried by the situation. Rep. Paul Trible (R-VA) told the Fifth District Conference in Alexandria that he and Rep Charles Bennett (D-FL) would introduce a National Defense Compensation Act. The act would call for additional compensation for the active duty and Reserve forces because "We can no longer maintain a first rate defense with a third rate system of compensation." At the same time, Trible also called for adding an additional carrier battle group to the Navy and increasing the size and capability of the Merchant Marine.

Another Congressional supporter of the Naval Reserve was Sen. Charles "Mac" Mathias (R-MD) who led a long, tough, but ultimately successfully battle to get drill pay and ACDUTRA reimbursement reinstated following their cancellation. But the hearings only opened Pandora's Box when it became apparent that the Reserve Personnel Navy budget for FY81 was already short by $10 to $17 million.

While members could be heartened by these words, other events in the spring and summer of 1980 caused grief within the ranks of the NRA, as four early leaders made their final voyage. The first to die was LT Lou Hurst, who had served as Assistant to the Executive Director from 1959 to 1970 and done much to expand member services and the Association's outreach programs. One of the Association's founders, LT Paul Sidebotham, who had been District President for the Fifteenth District from 1954 to 1960. The third leader to go was CAPT Bob Barto, who had left a major legacy to the Association as Thirteenth District President in 1964, National VP-Membership 1964-1966 and National President from 1965 to 1967. The final key figure to leave in 1980 was CAPT Leo Bilger, who served as National VP-Budget and Finance from 1969-1971 and Chairman of the Constitution and By-Laws Committee from 1971-1977.

The Twin Cities Chapter hosted the summer EXCOM and the Naval Affairs Meeting in Minneapolis from 25-27 July. The first item of business was hearing a report from the NRA Memorial Building Fund Committee, led by CAPT Pat Lucci (Western New York Chapter). The committee had been meeting for the previous month to determine how and when to construct a permanent headquarters building and end the Association's peripatetic wanderings. Pat's report called for establishing a NRA Memorial Building Fund, with the express purpose of raising $2 million by mid-1982. Each member of the Association would be asked to contribute approximately $100. The fund raising effort got off to a good start, raising $5,000 in pledges and contributions right on the spot.

In other business, the EXCOM approved proposals to urge Congress to increase the number of reservists the President could call to active duty in the absence of a national emergency from 50,000 to 100,000. The attendees also called for enacting military-retirement legislation to "grandfather" all existing pay and benefit systems. The EXCOM attendees also considered how to reduce the cost many members faced of having to attend two EXCOMs, the National Conference, and the Congressional Reception. National Site Committee Chairman Bob Brice recommended combining the spring EXCOM with the Congressional Reception. The Committee also said it would continue trying to rotate the Summer EXCOM and the National Conference, given that the Spring EXCOM would become, by default, a Washington, DC event.

Summer EXCOM attendees also approved a resolution urging the CNO and CNAVRES to find interim substitute warships to replace the FRAM 1 NRF destroyers. Many of the NRF destroyers were about 35 years old and showing their age. To go along with new ships for the NRF, the EXCOM also called for new fleet ships and aircraft, noting that the Soviet threat was growing and becoming more dangerous.

CAPT J.L. Sobien, USNR, Commanding Officer NARTU Andrews, receives the Richard K. West Trophy from RADM W.S. Guest, USN, Chief of Naval Air Reserve Training.

The Summer EXCOM also called on the NRA to consider reinstating the presentation of awards to Naval Reserve units. The practice had fallen into abeyance during the 1970s because of the continual reorganization of the Naval Reserves, which meant that units that were in service one year might not be around in time to collect their award. Following discussions with CNVARES, the decision was made to have the NRA revive two existing awards and establish two new ones, all for presentation starting in 1981.

The four awards were the Richard K. West Trophy, the Sidney Fields Trophy, the Robert I. Barto Trophy and the Leo V. Bilger Trophy. The Richard K. West Trophy, previously awarded to the unit with the best NARS/NARU retention, would in the future be presented to the Naval Reserve Center with 250 or more Naval Reservists assigned. The Sidney Fields Trophy, previously awarded to the best Large Surface Division, was revised for presentation to the best Naval Reserve Center with fewer than 250 Naval Reservists assigned. The Robert I. Barto Trophy would be awarded to the best reinforcing Naval Air Reserve unit, while the Leo V. Bilger Trophy would be for the best reinforcing Naval Surface Reserve unit.

Naval Reservists have always been called "Twice the Citizen," and the activities of many members of the NRA in 1980 only bore this out. In the fall, Former National Treasurer CAPT Irene Wirtschafer served as deputy tour leader of a special delegation to China. During the tour, Irene attended the opening of the US Consulate in Shanghai, met with Vice Chairman Madame Soong Ching-Ling and was granted an interview with then Vice Premier Deng Xiaoping. Upon returning, Irene reported that Deng opposed the Soviet invasion of Afghanistan, saw the Vietnamese as the Cubans of East Asia, and wanted increased US trade an a US-South Korea-North Korea summit to reunify Korea peacefully.

Closer to home, LT Doug Guinard (New Orleans/Acadia Chapter), but a resident in New York City, helped catch a suspected purse snatcher, holding the man until police arrived. Running was easy for Doug, whose hobby was running marathons. Another Association member, CAPT Jim Johnston, (Fort Lauderdale/Everglades Chapter), who had already won the George Washington Honor Medal from the Freedoms Foundation, won first prize in the NRA's Loyalty Essay Contest sponsored by the Past Presidents' Group. Second prize went to CAPT Lowell Howard (Columbus Chapter), while LCDR Marshall Colt (San Fernando Valley Chapter) won the third prize.

In another case of one NRA member receiving a Bravo Zulu from another member, National President RADM Phil Smith presented the certificate of permanent appointment as a Master Chief Hull Technician to Ray Vaughan. Both are members of NARU Norfolk, and Ray also is National President of our sister organization, the Naval Enlisted Reserve Association (NERA).

All was not hard work within the Naval Reserve Association. During the summer, the Western New York chapter co-sponsored the first Great Lakes International Challenge Cup Race at the Buffalo Yacht Club. Unfortunately, the Buffalo team was defeated by the Naval Academy Team in straight hearts, with both sides using Tartan 10-meter single-masted racing class boats. Thirsting for revenge, the Western New York NRA'ers announced plans for a rematch one year later in August 1981.

The fall of 1980 saw several major changes coming to the Navy and the Naval Reserve. Effective 1 October, the Naval Districts were officially disestablished. The NRA retained the traditional Naval-District structure for its own purposes, while considering what other regional structural format it might change to.

As the new fiscal year began, RADM Fred Palmer, CNAVRES, listed the improvements that had taken place since 1978. These included providing MIUW units with their own AN/TQS-108 vans and multiple rolling stock; assigning four assault craft to ACU units, with six LCM-8 boats in process of being assigned. HCU units had received diving equipment. The Shipboard Simulator program was moving along very nicely. Seventeen

Simulators had been installed, 12 were being ordered, and plans called for an eventual total of 44. The system had also expanded to include a Damage Control Trainer module, with seven installed and nine under construction. Six active Shore Intermediate Maintenance Activities (SIMA) had been established and four Reserve Yard Repair (YR) barges to work with the SIMAs. The SIMA/YR program would require 120 units with 5,000 reservists. Sixty Naval Control of Shipping Organization (NCSORG) units would be established in FY81 with 6,400 pay billets.

CNAVRES also noted that the Seabees had 80 percent of the heavy equipment required to perform their duties. Fleet training schools provided more than 9,000 quotas to the Naval Reserve for NEC/NOBC training; and monthly airlift had been approved for the fourth drill weekend to support WET training. Unfortunately, the airlift program was hindered by a lack of available aircraft and inadequate travel funding. On the waterfront, by the end of FY80, the Naval Reserve Force had three amphibious cargo ships (LKAs), two ammunition ships (AEs), one salvage ship (ARS), one fleet tug (ATF). Two tank landing ships (LSTs) and an additional LKA were due in FY81. In addition, the CNO had approved a pilot program to have a Maritime Administration Ready Reserve Fleet cargo ship at the Philadelphia Naval Base to serve as a training platform for Reserve Cargo Handling Battalions. To ensure that Reserve issues received an adequate hearing, a Surface Warfare Naval Reserve Advisory Committee (SWNRAC) to assist the Deputy CNO for Surface Warfare and keep reserve issues high on the list of concerns.

The early fall also saw the NRA offering new services to members and changing some of its policies on other services. The high inflation of the late 1970s had forced the Association to raise the price of the Record Review Service. The new prices ranged from $45 for Ensigns and Lieutenants (JG) to $85 for Captains. The Washington Jacket Check remained $15 regardless of rank. The Association also began offering a money management service, The Fund for Ready Income. The fund offered several investment plans and invests in US Government securities.

Although these accomplishments were not to be looked down on, association members attending the 27th National Conference in San Francisco from 24-27 September realized they still had a ways to go to see all their goals met. RADM Fred Palmer, CNAVRES, presented the major address to the convention. After listing the accomplishments the Naval Reserve had made during the previous year, he gave the attendees the good news/bad news. Good news was that recruiting was excellent and fleet commanders were increasingly asking for Reserve support. Bad news was that retention was low, and that as with the active duty, pay and benefits were not in good shape. This was reinforced by Deputy Assistant Secretary of the Navy for Manpower Mary Snavely Dixon. In her address she said that because of the hemorrhage of petty officers from the Navy, the Navy was considering various ways to have the Naval Reserve relieve the shortage.

The Conferees adopted six resolutions in San Francisco. These called for rebuilding the nation's naval forces, accepting no arms control agreement that jeopardized national security, and having a national strategy of peace through strength along with a strong economy. Other resolutions called for having all youths register for the draft and reinstating the draft and raising military pay and benefits to a comparable level with the civilian sector. The sixth resolution called for making educational benefits weighted in favor of military rather than civilian service and in favor of those who enlisted rather than draftees.

The Association presented Rep. Larry McDonald (R, GA) with the Distinguished Service Award. In his acceptance address, Rep. McDonald told the gathering that America had been lied to about the situation its armed forces were in, and that the United States was in danger of taking second-place status if matters were not corrected swiftly.

Other awards presented at the Conference included the Ken Kalmbach Award for best District communication, which went to the Twelfth District. The credit for the program here went to District President, CDR Ken Rice and District Vice President for Public Relations CDR John Chase. The Forest Park Chapter, led by Chapter President LCDR Lou Hoornbeck took the Jack Lewis Award as outstanding chapter.

In other good news, Headquarters Business Manager COL Dave Deville announced that interest paid to the General Fund from the separate Life Member Fund represented a contribution of $13.13 per life member, marking the first time that this was greater than the average dues contribution of one- and three-year members, which stood at $13.12. When Dave made his report on August 31, life membership stood at 2,784.

In the area of recruiting, Fifth District President CAPT Jim McNallen won honors as outstanding recruiter for bringing in the most new one-year members. LT Randy Corgan (Cincinnati Chapter won the award for most three-year recruits. The Chapin Award for membership renewal went to the Third District under District President CAPT Ed Grandin. Other awardees included First District President CWO4 Ray Gauvin for largest Life Member Increase, Sixth District President CAPT Claude Caviness for three-year membership increase, and Thirteenth District President CAPT Chuck Bryant for One-Year member increase. Recruiting was helped by the formation of two new chapters in 1980, at LaCrosse , WI; (CAPT Peter Berg, President) and NAS Miramar (CWO2 Mike Fulgham, President).

As the fiscal year drew to a close, the Navy took several steps changing its personnel policies. The age ceiling for commissioning in the Naval Reserve was raised from 27-1/2 to 35 for warfare officers and many of the restricted line communities, including the Supply Corps, Nurse Corps, Civil Engineer Corps, and Medical Service Corps. For the Medical and Judge Advocates General Corps the age limit was raised to 42. The revisions caused a change in the classes at Officer Candidate School and Officer Indoctrination School, which found many officers entering service from other careers in the civilian world. The Navy also established a Limited Duty Officer (LDO) Aviator Program, which would accept 35 enlisted sailors from the active and Reserve fleet, to begin training in April 1981.

While all this was going on, NRA members continued to make the news. LT John Roach (Matthew F. Maury Chapter) completed a 15x52 ft painting of the battleship Arizona on a commission from the Pearl Harbor Survivors Association. John, the Navy's preeminent combat artist, took the painting to the National Park Service's Arizona Memorial Visitors Center aboard a transport from VR-57. National President CAPT Phil Smith and National VP-Public Affairs LCDR Sherry Kleinschmidt accompanied John to the dedication.

A former National VP-Public Affairs, CDR Jim Carey (Great Lakes Chapter), was elected National President of the Bluejackets Association. He succeeded CDR Fran Nielsen (Chicago-Trident Chapter) who had served two terms at the helm of the 8,000-member association.

National President CAPT Smith also was principal speaker at the dedication of the Navy Memorial Park at the former Olathe Naval Air Station. NAS Olathe KS, trained more than 10,000 naval aviation cadets between 1942 and 1945 before becoming a reserve naval air facility. The facility closed in 1970.

Some of personnel news that fall was somber. CAPT J. Mack Young died in New York on 17 November. Mack was a founder of NRA and second National President. A stalwart member of the old NROA, he helped lead the merger of NROA and RONS in the early 1950s, prior to the establishment of the NRA.

The end of 1980 also saw Congress take action on several measures affecting the Naval Reserve. The first was the passage on 15 December of the Defense Officer Personnel Management Act (DOPMA) regulating and making uniform the laws affecting the appointment, promotion, separation, and retirement of regular officers from all services. The law took 15 years to craft, leading many to wonder how long a rumored ROPMA for reservists might take to design. One of DOPMA's provisions, changed after only a few years, was the creation of a one-star Commodore Admiral. The new paygrade soon reverted to the traditional Commodore.

The Defense Appropriations Bill authorized a Naval Reserve end strength of 92,000 and the retention of six helicopter squadrons. The bill also provided $17 million for P-3 Tac/Nav Modifications and an additional $10 million for drill pay and travel pay for Naval Reservists. Shortly after the bill was passed, Public Law 96-584, was signed on 23 December 1980, increasing the number of reservists the President could recall to active duty without a national emergency declaration from 50,000 to 100,000.

But the most important news out of Congress was a statement of policy for the Naval Reserve Surface Force. The statement declared maritime convoy escort to be a wartime mission of the Naval Reserve. The NRF ships responsible for this and other duties would have an evenly mixed active/Reserve crew, and TAR personnel would replace the active personnel as soon as was feasible. The NRF would retain at least eight ships, including all with less than two years since their most recent overhaul, until newer ships entered the inventory. At least four new ships would join the NRF by FY82, with a further 12 being accessed between FY82 and FY87 for an eventual strength of 24 ships. During peacetime, these ships would perform crew training under the operational and funding control of CNAVRES. The TARs would be paid out of Reserve Personnel accounts rather than active duty Military Personnel budgets. To ensure the policy was implemented, the Navy had to present Congress with a plan of implementation by 1 April 1981.

And finally, in a bit of holiday cheer, Naval Reserve G-7 Station Hospital Headquarters 106, commanded by CDR Perry Banda (Stephen Decatur Chapter) helped their active duty counterparts in a major evolution. The reservists, all medical personnel in their civilian lives, helped move 160 patients, including some in critical and ambulatory condition, from the old Bethesda Naval Hospital to more modern facilities at the upgraded Bethesda National Naval Medical Center.

The NRA, along with the other reserve associations and the nations's armed forces received another piece of good news at year's end. The incoming Reagan Administration had pledged to end the decline in military strength and pay.

The 1980s, Years of Resurgence

One of the first priorities of the new Administration was to reverse the downhill slide suffered by the armed forces. Although outgoing Carter Administration called for some increases in defense programs in its the final, proposed FY82 budget,

Left to right: the Honorable E.C. Grayson, Jr., Deputy Assistant Secretary of the Navy (Manpower); Captain Curtin R. Coleman, II, USNR, Executive Vice President, Naval Reserve Association; the Honorable John F. Lehman, Jr., Secretary of the Navy; and the Honorable J. Ronald Denny, Principal Deputy Assistant Secretary of the Navy for Manpower and Reserve Affairs. All of the above officials are drilling reservists and are members of the Naval Reserve Association.

not all was rosy. The budget called for reducing USNR end-strength to 87,600—a cut of about 100 in the officer/enlisted TAR community and commissioning 1,300 NROTC and 155 ROC officers. It also proposed increasing travel and transportation funding by six percent, and boosting reserve operations and maintenance funding by approximately the same amount, which was less than the rate of inflation.

It didn't take the NRA long to counter these calls with its own proposals for the FY82 budget. NRA's recommendations called for a Selected Reserve end-strength of 97,000. To help fund this increase, NRA called for the return of $12.4 million authorized by Congress to increase manpower to 92,000 and $5 million authorized to expand the TAR program by 1,000 personnel, but which were both reprogrammed by the Secretary of Defense to other programs. The NRA also called for rejecting the inclusion of TAR personnel in Selected Reserve end strength. Other personnel requests included providing "portal-to-portal" injury coverage for drilling reservists and changing DOPMA so that mobilized reservists would be placed on a separate lineal promotion list for the duration plus six months, rather than be placed on the same list as regular active duty officers.

The NRA made two recommendations for equipment changes. One called for a supplemental appropriation for procuring four C-9Bs in FY81, rather than six C-9Bs in FY82. The association also recommended a coherent multi-year aviation procurement plan be enacted. The other equipment recommendation called for modernizing 36 P-3A/B aircraft with the Tac/Nav modification, vice the 15 P-3s called for in the budget submission.

February saw several signs that the NRA's proposals would receive a favorable hearing on Capitol Hill and in the Pentagon as five members of the Association joined the Reagan Administration in key Defense Department positions overseeing the Navy and Reserve matters.

LCDR John Lehman (MF Maury Chapter) was nominated and confirmed as Secretary of the Navy. A reserve Naval Flight Officer and President of the Abingdon Corp., a defense consulting firm, Lehman had called for a 600-ship Navy centered around 15 carrier battle groups and four battleship battle groups. A longtime student of naval history and national security policy, SECNAV Lehman promised to change much about the Navy.

CDR Larry Korb (Southern New England Chapter) was named Assistant Secretary of Defense for Manpower, Reserve Affairs, and Logistics. CDR Korb had previously been director of defense studies at the American Enterprise Institute, a Washington DC-based think tank, and prior to that a professor of management at the Naval War College. Assisting CAPT Korb as Deputy Assistant Secretary of Defense for Reserve Affairs was CDR Edward Philbin.

The fourth NRA member filling a key position within the Administration was CAPT J. Ronald Denney (Philadelphia chapter) who became Principal Deputy Assistant Secretary of the Navy for Manpower and Reserve Affairs. CAPT Denney's assistant, and the fifth NRA member, was MAJ Fred Davidson III, USMCR, a Vietnam veteran, who became Deputy Assistant Secretary of the Navy for Reserve Affairs.

Other Additional members also moved into senior positions in the spring of 1981. RADM Ted Levy (Atlanta Chapter) was recalled to active duty as Deputy Director of the Naval Reserve in Washington, relieving RADM "Cricket" Johnson (MF Maury Chapter). RADM Johnson in turn went to Philadelphia as Commander Naval Base to relieve RADM Carl Brettschneider (Philadelphia Chapter) who was retiring. Yet another member of the MF Maury Chapter, RADM Bill Daniels, became the Deputy Chief of the Naval Reserve in New Orleans. Finally, RADM Bob Garrick (Pasadena-Rose Bowl Chapter), the Naval Reserve's first Public Affairs flag officer, was named Deputy Counselor to the President by President Reagan.

With so many NRA members coming into positions of authority, it was no surprise that one idea the Association advocated, National Patriotism Week, was signed into law. Originally proposed by the winner of the NRA's 1978 Meritorious Service Award, Lori Cox, and supported by a resolution at the NRA's 25th National Conference, the week would begin on the third Monday in February.

The proliferation of NRA members and senior reservists in the government caused the Defense Department to consider changing the policy requiring all Ready Reservists to be capable of reporting for duty within 24 hours of a call-up. Current policy called for anyone not meeting this standard to be put into the Standby Reserve—to which many of the new senior federal appointees had to transfer. DoD now considered expanding the policy to cover Reservists in key state and local government positions and, possibly, key individuals in certain industries, particularly the defense industry.

While these changes were going on in the Pentagon, across the Potomac on Capitol Hill, CAPT (Rep) Bill Chappell (D-FL and Gainesville Chapter) introduced a resolution calling for the creation of a new post, Assistant Secretary of Defense for Reserve Affairs. The new position would come about by splitting the functions of the Assistant Secretary of Defense for Manpower, Reserve Affairs, and Logistics office.

Congress considered other NRA-related items during the spring of 1981. In February 1981, Rep Sonny Montgomery (D-MS) announced plans to offer an amendment to the GI bill legislation providing a mini-GI Bill for Reservists and the National Guard. The legislation allowed Reservists to receive up to $140 per month in education allowances for 36 months of satisfactory service.

February 1981 also saw the formation of a new professional association for Naval officers. Several Regular and Reserve Hispanic Naval and Marine Corps officers formed the Association of Naval Services Officers (ANSO) to boost the number of Hispanics in the Navy, Marine Corps, and Coast Guard officer corps.

Many of the new Congress and Administration joined with other NRA members at the third NRA Congressional Reception, which was combined with the Spring EXCOM. The EXCOM was held in Arlington and drew 180 registrants, setting an eight-year record, and an all-time record for an event held at a non-military facility. The Congressional Reception drew considerably more attendees—over 1,100, including more than 100 Senators and Representatives. Four Congressmen were honored for their contributions to the Naval Reserve, Sen. Ted Stevens (R-AK), Sen. Henry "Scoop" Jackson (D-WA), Rep. G. William Whitehurst (R-VA), and Rep. Jack Kemp (R-VA).

Attendees also heard a message from the Vice Chief of Naval Operations, ADM James Watkins. ADM Watkins stated that the Navy could not have performed many missions during the late 1970s without the help of the Naval Reserve. Many units and individual members performed key duties upon request, allowing the Active Duty Navy to concentrate on other critical tasks.

Some Naval Reservists almost played a key role in an unexpected manner, somewhat reminiscent of the 1970 postal workers strike. When the air traffic controllers went on strike in the spring of

1981, there were plans to bring in military Reservists to replace the strikers. FAA controllers were key federal employees, and therefore not supposed to be in the Selected Reserve. But because of different circumstances and loopholes, the Navy reported that some air traffic controllers were members of the Selected Reserve. While no Reservists were recalled, once again, the nation almost saw striking federal employee reservists coming close to performing their civilian duties by way of a military call-up.

With many proposals from the past being high on the new Reagan Administration's plans for modernizing the military, the Spring EXCOM did not have to lobby for many new issues. The EXCOM did call for a GI Bill that would provide for both active duty and reserve personnel. It also approved funding for an engineering and financial feasibility study to help find a new headquarters building for the association. To save on other costs, it recommended that the Congressional reception be held every other year, following the national general elections. The final action endorsed a proposal to extend complementary membership to newly commissioned Naval Reserve warrant and limited duty officers. This was viewed as important because these officers, lacking a further active-duty commitment, would be ineligible for any other complementary membership.

The issue of a new home for the NRA became more important during the spring of 1981 when the landlord at Farragut Square slammed the Association with a 120 percent rent increase. Some wags suggested that either the after-hours meetings of some national committees were becoming too noisy, while others suggested this was the downside of the improving economy. But the increase was too steep to bear, so on June 1, the Association moved across the Potomac to temporary headquarters at Suite 307, 110 North Royal Street in Alexandria, VA. The move marked a milestone, being the first time the Association moved out of the 20006 ZIP Code and downtown Washington, DC.

Such moves were all too familiar to the NRA, which had had to move on an average of once every three years since its founding in 1954. (The June 1981 move was the Association's ninth move.) The 1981 move was not the first pier-head jump either. Old timers claimed that during the 1962 move, Jack Lewis and other staffers nearly got buried beneath the wreckage of the building as it got torn down.

One member who had seen just about every headquarters the Association had been in was CWO4 Vaughn Coale, known both as the Crusty Old Gunner and a 27-year member of the NRA. In March, Vaughn decided that rather than renewing his membership in one-year and three-year increments, he would become a Life Member, with Number 3250. Vaughn's record of long-term one-year renewals beat the previous record, held by CAPT Al Tiedemann. Shortly after Vaughn made his big switch, his wife, Naomi, also converted to a Life Membership.

While the defense budget was winding its way through the halls of Congress, the Navy made an announcement that warmed many hearts in the NRA and the Reserves. Beginning in FY82, four FF-1052 Knox class frigates would be transferred from the active force to the Naval Reserve Force. Two would be homeported in Newport, RI and two in San Diego. Starting the following year, in FY83, two more Knox class frigates would be added each year, until 12 frigates were in the NRF. The plans also called for NRF to have 42 ships in FY82. This good news was followed scuttlebutt that FFG-7 Perry class guided missile frigates, the state of the art surface combatants, would start entering the NRF in the mid 1980s.

The frigates would replace the Gearing/Carpenter-class FRAM II destroyers that were showing the age but soldiering on faithfully. The announcement said three old tin cans, Owens, Vogelgesang, and Steinaker, would be decommissioned in 1982. The three ships, had an average age of 36 years, making them among the oldest in the Navy. All the remaining World War II-era tin cans would leave service by 1984.

Members of the NRA, while applauding the new plan, also noticed some major problems. The plan did not specify that the frigates would be under the operational control of CNAVRES. A close reading of the Navy's plans led to the conclusion that the fleet commanders-in-chief (CinCs) would have the all-important OpCon. Nonetheless, the provision of modern ships was like manna from heaven to the NRF.

The Department of Defense and the Navy Department, revitalized with the promise of greater support from the Administration, also began changing several policies. The first, and probably most critical to NRA members, was considering the possibility of a Reserve Officer Personnel Management Act (ROPMA), similar to the DOPMA legislation which would go into effect on 15 September 1998. Pentagon officials wanted the new legislation to go into effect in 1983.

While DoD discussed ROPMA, the National Naval Reserve Policy Board prepared a report with 27 proposals. The key proposals called for establishing a Physician Assistant Warrant officer program in the Naval Reserve, moving the office of CNAVRES to Washington from New Orleans, and adopting an advancement-multiple system for enlisted Naval Reservists similar to that used by the Active Navy.

Bringing in more physicians was critical, because the Reserves had only 41 percent of their required strength in this field. The Naval Reserve had 55 percent of its required end strength, less than the Air Force Reserve's 75 percent, but far more than the Army Reserve's 27 percent. To entice more doctors, the DoD proposal called for paying bonuses to Reserve physicians and recruiting them at higher ranks and pay grades.

Looking at other ways to help the Reserves, Deputy AsstSecNav (Reserve Affairs) Maj Fred Davidson III suggested allowing drillers to remain in the Naval Reserve until age 60, portal-to-portal insurance coverage, and one-day-a-month commissary privileges for drilling Reservists.

While Congress debated and the Pentagon planned, NRA members continued, as usual, to personify the expression "Twice the Citizen" and perform outstanding service to their communities and the Navy. CDR Dave Woods (Kalmbach-Wardroom Chapter) former National PAO and National VP-Membership, wrote and edited the introduction to two volumes of Arno Press's series Historical Studies in Telecommunications, specifically, Signalling and Communication at Sea. Dave had already written another book, A History of Tactical Communication Techniques in 1975, which was published by Arno Press.

Another NRA author was CDR Dan Englehardt, another Kalmbach-Wardroom member, received the Navy Commendation Medal for heading the team that researched and rewrote the Command and Control Manual (NWP-8). Dan's team received credit for being the first reservists to entirely write an NWP during drill and ACDUTRA periods.

CAPT Chuck O'Malley (Pres-San Bernardino/Arrowhead Chapter) developed an ion exchange resin regenerator system for demineralizing the water used for boiler feed water in steam powered ships. The Naval Sea Systems Command, recipient of the system, estimated it could save $1 million annually through the end of the century.

Yet another example of NRA members gaining accolades for their service to the fleet occurred in March and April, when members of FLTTRAGRU WestPac 118, commanded by CAPT Samuel Anderson (Mile High Chapter) spent their 13 days ACDUTRA conducting the complete refresher training for the USS Denver (LPD-9). No Navy officials could recall a previous instance of a Reserve FLTRAGRU administering RefTra for an active force ship.

While some units served the sea-going navy, other units and individuals provided ace support to the shore-side Navy. NOSIC 0166 and FOSIC 0166, two intelligence units, received the Meritorious Unit Citation for supporting the Navy Field Operational Intelligence Office at Ft Meade, MD. NOSIC 0166 was led by CAPT Andrew Biache (Kalmbach-Wardroom), while CDR David Mercer (Matthew F Maury) commanded FOSIC 0166.

Further south, RADM Bill Gilmore (Ole North State) received the Legion of Merit for outstanding service and introducing numerous innovations while serving as commander REDCOM Seven in Charleston, SC between 1977 and 1980.

Other members of the Association devised more popular and prosaic pieces of hardware. CDR Olen Stewart (Admiral Byrd-Richmond) came up with the idea of "Naval Reserve" license plates. Olen claimed the idea had three purposes; enhancing pride and professionalism, aiding recruiting and retention, and facilitating access to mobilization sites. Thanks to Olen's drive, Virginia became the first state to make such licenses available.

Other NRA members had different ideas for drumming up recruits. CAPT Joseph Azzolina (Perth Amboy) and president of the Foodtown Supermarket chain put a nautical décor into two of his Foodtown Supermarkets in New Jersey. The Woodbridge Township store had a small Navy recruiting booth, with brochures describing the different careers available in the Navy. The Oakhurst store, meanwhile, featured a customer-service booth shaped like the bow of a ship, a large photograph of the battleship USS New Jersey, and walls covered with portholes and lithographs of clipper ships and schooners.

While some NRA members were members of Congressmen, and others were Capitol Hill staffers, others found different ways to aid the Hill. CAPT Irene Wirtschafter, Past National Treasurer, participated as a Senior Citizen intern for US Sen Paula Hawkins (R-FL) in a week-long program in Washington.

Another NRA'er, CAPT William Head (Kingsport-Upper East Tennessee) was elected International President of the International Society for Philosophical Enquiry (also known as "The Thousand").

NRA members also continued their literary efforts. CAPT Bruce Pruitt (Peninsula) wrote a column in Naval Institute Proceedings recommending that all four BB-61 Iowa class battleships be returned to active duty, with two on each coast, as reserve training platforms. Bruce said this would ease manning problems, since their major use would be during major conflicts. During peacetime, Reserve crews would sail the ships for one weekend a month and two weeks per year—the standard Reserve duty schedule. As things turned out, CDR/SecNav John Lehman was pushing to bring the four battlewagons back as members of the active fleet.

The opposite tack was taken by another NRA member, LCDR Roger Crossland (Bridgeport-Isaac Chauncey), who wrote an article for Proceedings warning against relying on the big gun ships. Roger, a SEAL, called for maritime guerilla operations, using naval saboteurs, mini-submarines, and "Q-ship" warships disguised as merchant ships. These assets would perform hit and run attacks, harassing an opponent's forces.

The same issue of Proceedings saw CAPT David Kenney (M.F. Maury) author an article describing a "Counter-Terrorism Program." Dave noted that terrorists often act rationally, but that accepting terrorism as a fact of life would wreck international negotiations. He suggested that the United States and other concerned nations take five steps to deal with terrorism. The major Western powers must convince the world that terrorism backfires on those nations supporting it; exchange data about counter-terrorism; reduce the media publicity given to terrorists; suspend the protection of law for terrorists; and develop counter-terrorist doctrines, weapons, and tactics as a full-time professional discipline.

Other NRA members active in the media included the Public Affairs specialists, and one PAO unit, Internal Relations Activity Detachment 319, led by LCDR Phil Kazanjian, President of the Pasadena-Rose Bowl Chapter, won the National Mobilization Training Award for the third year in a row. As the only unit to achieve this distinction, the unit received the award from RADM Bruce Newell, Chief of Information.

Awards got given, articles got written, and Capitol Hill continued to stir. Sen Jeremiah Denton (R-AL) and a retired active duty Captain and Vietnam POW, introduced a bill expanding Presidential recall in non-national emergencies. The Denton bill would expand the President's power from calling up 100,000 Reservists for 90 days to recalling 250,000 Reservists for 180 days. This would amount to more than 25 percent of total (all services) Reserve end strength.

Another Naval Aviator making the news in 1981 was the NRA's National President, RADM Phil Smith, who became the Navy's "Silver Falcon," or drilling reservist with longest service since being designated as a Naval Aviator. When he achieved this designation, Phil had 11 months more service than the "Gray Eagle," or active duty aviator with the greatest service. Phil thus became bull aviator of the Navy. But he declined to pose for the "And If Called, We Reservists Will Be Happy to Serve" poster.

With all the activities the Reserves had been engaged in, the NRA could look back at an interesting time as it met in Albuquerque, NM for its 28th National Conference. CAPT Curtin R. Coleman II of Florida was elected National President. At that Conference, two new national offices were created—National VP-Retired Personnel, with CWO4 Vaughn Coale (Stephen Decatur) as National VP and National VP-Junior Officers, with LCDR Mileva Hartman (Corpus Christi) filling the billet.

RADM Bill Daniels (left) and RADM Fred Palmer.

Left to right: ADM Hayward, CAPT Coleman, and RADM Palmer.

The Conference honored several individuals for various services over the years. Chief among the awardees was Navy Times Editor Bill Kreh and his wife Beverly. Bill had attended 21 National Conferences and done much to help the Association, as editor of Navy Times, editor of NRA News, and author of the history of the Naval Reserve Citizen Sailors: The US Naval Reserve in War and Peace. Many old timers regard Bill as the unsung recruiter extroadinaire because of the many members who heard about the NRA through his articles.

Other members were honored for more tangible recruiting efforts. The District membership competition was won by the Eighth (CAPT Dick Dolan, District President, CAPT Doug Lier VP for Membership) in the three-year and total net gain categories. The winner for one-year District recruitment went to the Eleventh (CAPT John Leahy, LCDR Matt Quinlan). The Life membership honors went to the First District (CAPT Paul Garber, CDR Bill McGurk) for the third year in a row. The Ninth District took honors for highest member renewal rate.

In the overall NRA recruitment competition, CAPT James White (VP-Membership, Spirit of St Louis Chapter) and CAPT Roy Willians (President, San Diego Chapter) took top honors. The Providence-Southern New England Chapter also won an award for recruiting, when RADM Floyd Miller, Commander Navy Recruiting Command presented it with CRUITCOM's Sustained Superior Performance Award. CWO4 Ray Gauvin accepted the award on behalf of chapter President CDR Patricia Burke.

Other NRA members were honored in different ways at the Conference. The Twelfth District (CDR Ken Rice, District president; LCDR Win Hayward district, VP-Communications; LTJG Alan McKean, Newsletter Editor) won the Ken Kalmbach Award for a second time and the Polaris Chapter (CDR Robert Francis, President) won the Jack Lewis Award.

The keynote speaker for the Conference was CNO ADM Thomas Hayward. He told the Conference that budget cuts in the Reserve accounts would force the decommissioning of two NRF destroyers and four minesweepers in 1982, although these would be somewhat offset by the addition of two Knox class frigates to the NRF roster. He also said plans called for a USNR SelRes end strength of 98,700 by 1986 rather than the NAMMOS target of 111,000 to 114,000.

The Conference made several changes to the NRA's Constitution. The term Naval District was changed to District, although the same geographic boundaries were retained. Complimentary membership for new direct-commission officers in the USNR was approved, eliminating the requirement for having served on active duty. In addition, the National Award of Merit was restricted to five awards per year, vice the previous unlimited quantity. Two Golden Helm awards also were established, to go to the best Naval Reserve Echelon 3 commands.

The Conference adopted a number of resolutions. The major ones called for support for a GI Bill for both active-duty and Ready Reserve personnel to aid in recruiting and retention. Equipment-related resolutions called for reactivating four battleships and two Essex class aircraft carriers, transferring 24 P-3 type aircraft to the VR program as an interim measure to maintain an airlift capability, and developing a sea-based MX Missile system rather than a new land- or air-based system. The attendees called for supporting the CNO's implementation of a plan to modernize the management of Naval Reserve surface forces by CNAVRES. Another personnel-related resolution supported the CNO's plans to raise SelRes end strength to NAMMOS-recommended levels gradually rather than immediately. The final resolution supported the Defense Department's interpretation of postal regulations, allowing, with probable cause, the inspection and search of military mail for contraband.

Shortly after the NRA's National Conference, our sister organization, the Naval Order held its Biennial Congress in San Diego. During the Congress, LCDR Ray Cross (Chicago-Trident) was elected Commander General, succeeding NRA Founder CDR Fred Kempe. RADM Win Weese (Acadian), past NRA National Surgeon and incoming Eighth District President, was elected Vice Commander General.

Autumn 1981 saw RADM Bill Daniels, member of the Acadian Chapter and serving in a TAR billet as Deputy CNAVRES, presented with the Legion of Merit for his outstanding service as Director Naval Reserve Personnel Management Department at NMPC and as Special Assistant to the DCNO for Reserve Affairs.

The waning days of the year also saw ADM Arleigh Burke (Kalmbach-Wardroom) designated as Honorary Naval Aviator 14. As Chief of Naval Operations from 1955-1961 the noted surface warrior did much to expand naval aviation's roles in ASW and electronic warfare.

While members were getting honored, Congress and the Navy Department were making changes affecting thousands of other Reservists. Dependents of drilling Reservists were given permission to shop unaccompanied by the sponsor at Exchanges. All they need do is show their sponsor's privilege-authorizing document and proof of relationship.

CNAVRES announced plans to limit enlisted SelRes to a 30-year career in order to make room for more promotions to E-8/E-9. Waivers would be allowed for critical billets. The new policy would be introduced in increments starting with a 36 year limit on 1 January 1983 and dropping to the 30-year rule by 1 January 1986.

The last two months of 1981 saw the beginning of a turn-around in fleet strength, and not a moment too soon. On December 7, 1981, the Navy released figures showing that while the fleet had 587 combatant warships and 323 auxiliaries on December 7, 1941; the corresponding December 7, 1981 figures were 421 combatants and 129 auxiliaries. The totals were, therefore, 910 ships in 1941 and 550 ships in 1981. Reflecting certain changes in Naval force structure, the aircraft total had increased, with 4,000 aircraft prior to Pearl Harbor and 5,765 aircraft (including Naval Reserve, Marine Corps, and Marine Corps Reserve) in 1981.

NRA Conference in New Orleans.

Emile Bonnot (center) keeping an eye on things.

The newest ship coming into service in 1981 was the Oliver Hazard Perray (FFG-7) class frigate. In a nice move for the Naval Reserve, CNO ADM Hayward told CAPT Curt Coleman, NRA National President, that the Naval Reserve would receive FFG-7s following the introduction of the FF-1052s that were slated to enter the NRF in 1982.

The news about the FFG-7s was only part of the good news at year's end. The Senate/House joint conference reported out on 15 December. Among the highlights of the FY82 appropriations bill was the decision to fund a Naval Reserve endstrength of 94,000, retain the two NRF destroyers and four NRF minesweepers slated for decommissioning, and provide added funds for modernizing the Naval Air Reserve force.

All was not totally well in Congress. Deputy Assistant Secretary of Defense for Reserve Affairs (and NRA member) Edward Philbin told the Senate Armed Services Committee that the Reserves required at least $17.8 billion, particularly in medical equipment, to meet their requirements. Also in government news, President Reagan proposed capping COLA raises for military retirees, and round down annuity checks to the next lowest dollar.

While Congress was providing additional funding, the Navy was changing some policies to allow it to use the funding. One change was increasing the age at which enlisted Naval Reservists could be recalled to active duty. The new regs only require a person to complete 20 years duty by age 60, vice 55. This will allow more reservists to be eligible for voluntary recall to man the battleships New Jersey and Iowa, now being prepared for reactivation.

The end of 1981 also saw various members in the news. CDR Jim Carey, past National Vice President-Public Relations, was confirmed by the Senate and sworn in as Commissioner of the Federal Maritime Commission.

Jim is followed to Washington in early 1982 by RADM Phil Smith, NRA's immediate past President, who becomes Jim's Confidential Assistant. Phil and Jim confer and wonder how many other cases have seen a Reserve Admiral reporting to a Reserve Commander. They admit it has provided a shock value to some of the Washington crowd.

Another NRA member receiving kudos was RADM Sam Cummins, past National Executive VP, who received the Legion of Merit upon his retirement as Commander REDCOM Four. Presenting the award was NRA member and Deputy CNAVRES RADM Bill Daniels.

The new year 1982 started off with a major boost to the Naval Reserve Force, when the Knox (FF-1052) class frigates MILLER (FF-1091) and LANG (FF-1060) were assigned to Newport and Long Beach respectively. This marked the first time the NRF had state-of-the-art warships in its inventory. Plans called for the frigates Valdez and Gray to join the NRF in July.

Shortly after the two frigates are turned over, CNO ADM Thomas Hayward and Secretary of the Navy John Lehman (both NRA members) testify on Capitol Hill that the Naval Reserve has made "impressive improvements" since 1978. Major changes include the addition of new ships and new aircraft, horizontal integration of new aircraft, and a projected growth to 94,000 personnel in FY82. (Horizontal integration means new equipment is phased in to the active duty and Reserve sides of the house simultaneously.) In a further sign of the growing integration, the Navy announces that two slots in each future class of the Senior Enlisted Academy will go to TAR or SelRes enlisted. But as National President CAPT Curt Coleman points out in a letter to President Reagan, much more can be done to build up the Reserves.

SecNav Lehman has a suggestion for increasing Reserve readiness. He would require Reservists recalled for active duty in a national emergency to serve for up to two years, even if they are already at or near age 60 and thus eligible for retirement.

Deputy Assistant Secretary of Defense for Reserve Affairs (and NRA member) Edward Philbin strikes a different note when he testifies before Congress. Ed notes that while all the Reserve components will reach the endstrength required to carry out their assigned missions by 1987, they could lack the modern equipment needed to carry out those duties. This is particularly vital since so much Reserve equipment is facing block obsolescence.

His office later notes another problem. Calendar year 1981 saw 2,898 Naval Reservists accept voluntary recall to active duty, meaning Reserve Recruiters had a difficult time reaching their goal of 7,000 accessions for the year. One way to solve this problem could be a Pentagon proposal to authorize bonuses of $1,000 for entry into the Individual Ready Reserve and $900 for re-enlistment. The plan also calls for extending minimum service obligations from six to eight years.

One way the Navy will increase Reserve participation is by easing the regulations on Navy veterans. Now a vet has up to four years to affiliate in the Reserves without losing his rate. Anyone out more than four years can keep the paygrade provisionally, with three years to re-qualify. In another move, promotion zones for USNR officers will go by designator, following the active duty policy as per DOPMA. Some officers will do better, some worse, under the new scheme, but with the running mate system there is little alternative that makes any more sense.

NRA received support from its fraternal partners, particularly the Naval Enlisted Reserve Association. NERA officials testified before the Senate Armed Services Committee that CNAVRES should have a four-year tour in the billet, while Reserve flag officers should also be the deputies for air and surface warfare.

A tradition-bound organization such as the Navy often displays traces of the past in the present. In this vein, the Navy announced it was reinstating the grade of Commodore. The one-star paygrade had not been used since World War II. The rank had been a courtesy title for captains commanding squadrons until 1862 when it became a permanent rank. Abolished in 1899, it was revived in 1939 as a temporary rank for 147 officers, before being retired (along with the officers) in 1949. DOPMA plans called for 45 USN officers receiving the rank, along with a proportionate number of Reservists. (Tradition has it that the title was originated by the Dutch during the wars with England in 1652, when senior officers were needed, but there was opposition to giving too many the rank and pay of an admiral.)

Two months later, when the selection board results appear, four NRA stalwarts are among the first Staff Corps USNR Commodores. The selectees are Donald St.Angelo (SC), Charles Smith (CEC), Jim Austin (MC), and Will Yamanouchi (DC). Jim and Will are very well known to NRA members, both having been past National Surgeons/VPs—Health Services.

Unfortunately, the Line Commodore board does not go so well. The first board had to be dissolved amid charges of tampering and outside interference. Not until late spring does the Naval Reserve announce the lucky captains. Among them is LeRoy Collins, Jr. (Florida West Coast Chapter) who holds his wetting down aboard the USS Requin, a decommissioned World War II submarine serving as a tourist attraction in Tampa, FL. LeRoy manages to get 250 friends and attendees below decks—possibly a record for number of people aboard an old boat.

The Pentagon did not have to look for advice on using the Naval Reserve. The National Naval Reserve Policy Board, headed by RADM Ted Levy (Atlanta Chapter) issues a report to the SecNav. Ted's report notes that the Navy's growth to 15 carrier battle groups can only be accomplished by extensive use of USNR mobilization assets. He calls for using NRF ships, giving USNR aviation squadrons fleet compatible aircraft, and establishing a clearly understandable way of measuring the Naval Reserve's contribution to Naval readiness.

The Board's report goes on to call for expanding the number of fulltime USNR career counselors and expanding minority recruiting. It recommends providing pension vesting for Reservists who are physically disqualified after 10 years service, establishing overseas USNR units, issuing ID cards to Reserve dependents, increasing commissary privileges, and providing portal-to-portal injury coverage for drilling Reservists.

In another tie with tradition, RADM Chip Taylor (Houston-Space Center Chapter) returned to his alma mater, Northwestern University, to help dedicate a memorial honoring the school's 23 NROTC midshipmen who lost their lives during World War II. Chip played a key role in establishing and maintaining the memorial and associated monument.

Despite all the seriousness, there was still time for some lighthearted levity. Two NRA members got their names and faces in the newspapers when Navy Times ran a feature, with photographs, showing Twelfth District President CDR Ken Rice and Kingsport Chapter's RADM Frank Anderson with their custom license plates of NAVRES (California) and NAVY-1 (Tennessee). Still not sure who thought of the idea, but they could be great recruiting tools.

Not all was wine and roses for the NRA in 1982. Sadly, CAPT McKenzie Shannon of the Pearl Harbor Chapter crossed the bar in the early part of the year. McKenzie, a longtime Fourteenth District rep to the National Executive Committee, was known more for his often pungent opinions in NRA News about the state of the Navy and the Naval Reserve. These opinions brought him a legion of followers who regarded him as the Sage of Portlock Road.

The Spring EXCOM, held from 19-20 March in San Diego, featured symposia for Public Affairs and Health Services NRA'ers. Several military and civilian leaders briefed the attendees about these fields. The capstone of the event was the San Diego Naval Reserve Officers' Ball with guest of honor Secretary of the Navy CDR John Lehman (MF

NRA Headquarters building – before.

Maury Chapter). John said the Naval Reserve was finally receiving state-of-the-art ships and aircraft, vice "cheap talk."

Attendees also heard that the Golden Helm Awards for top Readiness Command/Naval Air Station retention went to REDCOM Six in Washington DC and NAS Willow Grove in Philadelphia. Individuals also heard that the ROA's Navy Section had chosen LCDR Phil Kazanjian to receive its Outstanding Junior Officer of the Year Award. Phil is the past President of the NRA's Pasadena Rose Bowl Chapter.

Other NRA members received various awards later in the year. National Treasurer CAPT Bob Burkhart was named Man of the Year by the Cleveland Chapter, while past National President CAPT Mort Leavitt received the Man of the Decade award.

Another member of the Cleveland Chapter received a different sort of honor, when Shirley Leavitt was appointed first chairman of a panel to consider the establishment of an auxiliary organization of associate members which would become the Anchors.

Two NRA districts presented the Association's Twice a Citizen Awards to Naval Reservists who contributed to Navy or NRA activities, LCDR Robert McCorry (Southern New England Chapter) and CWO4 Lorna Bridenstine (Kalmbach-Wardroom Chapter).

National President CAPT Curt Coleman was invited by SecNav to speak at the Naval War College's Current Strategy Forum. At the forum, attended by many senior military, political, and civilian leaders, Curt spoke on lessons of the Falklands War and how it proved the need for a Navy based on 15 carrier battle groups.

In another sign of Naval War College recognition of Naval Reservists, two USNR quotas have been approved for 17-week in-residence courses of Strategy and Policy and Selection and Application of Naval Forces. Later in the year, approval is given for Reserve Captains and Commanders to apply for a seat in the Naval Warfare course, while Reserve Lieutenant Commanders may apply for a seat in the Command and Staff Course.

Other recognition of the Naval Reserve comes when Deputy Assistant SecDef for Reserve Affairs (and NRA member) Fred Davidson announces that the SelRes will have first shot at billets aboard the recommissioned battleship IOWA. The move makes up for the Navy's having reneged on plans to give SelRes first shot at billets aboard her sister ship the NEW JERSEY.

Donations also were on the mind of some NRA members in 1982. The NRA extended its Navy Library program, which had previously covered only NJROTC high schools, to all NROTC colleges and universities. More than 100 NRA libraries already exist across the country.

In a different vein, past National President Winfield (Chip) Chapin donated the large guesthouse next to his Palm beach residence, to the Association. It becomes the NRA Camelia Mary Chapin Memorial Building.

Faced with a shortage of doctors, dentists, and nurses, the Navy is looking at how to increase their numbers in the Reserves by allowing them to participate without attending regular drills. Many medical personnel have had to quit because of scheduling conflicts.

In a sign of shortages being redressed, the Navy is redressing its shortage of skilled petty officers. When the repair ship HECTOR issued a voluntary recall for Reservists for a West Pac deployment, more than 42 percent of the volunteers liked it so much they shipped over as TARs or Regular Navy.

Personnel questions also hit Capitol Hill, where Congressmen and their staffers are asked who are "key employees" and could not be mobilized. More than 1,600 key federal employees have had to leave the Ready Reserve for this reason, and now Congress finds itself in the same situation.

The Summer EXCOM in Phoenix was a happy occasion. Fun was partly on the agenda, with the wetting down of COMO Wil Yamanouchi, but there was also business. The EXCOM proposed that the Navy establish a Naval Reserve Unit Command insignia, similar to the active duty command at sea/command ashore insignia. It also established a Long-Range Planning Committee to study the NRA's organization and operations and propose possible changes.

Members also heard that the new CNAVRES would be RADM Robert F. Dunn, an active duty aviator and graduate of the Naval Academy class of 1951. In addition to having surface combat experience in Korea and air combat experience in Vietnam, RADM Dunn is a member of the NRA.

Almost concurrently with the announcement of RADM Dunn's selection came news that CNAVRES would keep his primary headquarters in Washington, with the Deputy CNAVRES, RADM Bill Daniels (also an NRA member in New Orleans). Members also heard that the old proposal to make CNAVRES a three-star billet was once again being kicked around.

NRA member RADM Bob Garrick (San Diego Chapter), spoke to Oakland's Jack London Chapter about his recent tour as Deputy Counselor to the President (Public Affairs) and the attendant problems dealing with the White House press.

The spring of 1982 brought sad news too. In June CAPT Hank Doyle passed away. Hank was one of the NRA's founders, father of the NROS program, early editor of NRA News, early National VP-Legislation (which often brought him up to Congress) and a bundle of energy in the early days. Some credit Hank's being fired by the ROA headquarters staff with being the aegis behind the founding of NRA. He will be sorely missed.

Shortly after, one of the NRA's oldest chapters, Chicago's Trident Chapter goes out of business as its 215 members cross over to the Great Lakes Chapter.

The spring also saw the National Naval Reserve Policy Board report its findings. It calls for portal-to-portal injury coverage, the use of Exchanges by "pink-card retirees" (if they are eligible for pay at age 60) and that CNAVRES be made a three-star billet!

Some NRA members found themselves farther afield during the summer of 1982. CMACH (W-4) Stephen Drabouski (Denver-Mile High Chapter) took special active duty aboard the Yokosuka, Japan-based carrier Midway for a month. During his tour he used his computer skills and knowledge of damage control to create a Computerized Moment Stability System (CMSS) that generated a stability solution much faster than the time previously required. Following Bravo Zulu's from Midway's damage control assistant and repair department, the CMSS system is being studied in Washington for application aboard other Navy ships.

Another NRA member in West Pac is RADM Herb Bridge. He fills in on short notice as Commander MSC Far East in Yokohama. His tour finds him inspecting ships of the Near term Prepositioned Force and shore facilities in Japan, Korea, the Phillipines, and Diego Garcia.

Autumn soon rolled around, and with it, the 29th National Conference in Detroit. The opening address was by the new CNO, ADM James Watkins who stated that "One Navy" was not just a concept, but had become an operational concept. The NRA Distinguished Service Award was presented to RADM Bobby Inman.

Some NRA members had particular reasons to remember the occasion, among them Bill and Lucille Mellish. Their neice, Pamela Mellish, a soprano who performs with the Los Angeles Opera Theater, sang the National Anthem.

Other members also would remember the Conference well. CAPT Bob Shaffer (Kalmbach-Wardroom Chapter) received a Special Presidential Award for his role in the DC-9 acquisition program. CDR Charlie Watkins (Santa Cruz-Cabrillo Chapter), NRA's Special Assistant for Long-Term Membership, was cited for his efforts.

Two awards were made by RADM James Williams, Commander Navy Recruiting Command. The Sustained Superior Performance Award went to NRA's National Headquarters Staff while the Outstanding Recruiting Support Award went to Oakland's Jack London Chapter, represented by CAPT Ian Thompson, Chapter President.

The traditional awards also were presented at the Conference. The Ken Kalmbach Award went to the Eleventh District led by President CAPT John Howe and District VP-PR CDR Sherry Kleinschmidt. The Jack Lewis Award for Outstanding Chapter went to the South Bay Chapter led by President LT Joe Edone.

New NRA Headquarters – after.

The awards for Membership and Recruiting represented a broad swath of NRA chapters and personnel. The Eighth District, led by president RADM Win Weese and CAPT Earl Rowe, District VP-Membership had the greatest net gain. The Fourth District (CAPT Will Shelly, DP; LT Randy Corgan, DVP-M) and the Twelfth District (CDR Otis Hutchinson, P and CAPT Richard Miles DVP-M) won for Life Membership increase and one-year-member increase respectively. The Chapin Trophy for highest renewal rate award went to the Third District (CDR George Sinnott and CDR Jim O'Reilly DVPs-M and CAPT Ron Partnoy DP).

Individual membership-recruiting awards went to CWO4 Paul Howard for most points and Life conversions, CDR Mileva Hartman for most complimentary members and runner up for total points and to CDR Dale Kletke for most one-year members. High point awards also went to CDR Bill Waylett, Jr. and CDR James Lange.

Shortly after the Conference, the first Naval Reserve Center in Puerto Rico was opened on 23 October in under the command of LCDR Forrest Curran, San Juan Chapter.

Also shortly after the Conference, HMCM Manny Ratner (New York Navy Chapter) was elected National President of the NERA at their 25th Anniversary Conference. NRA member and Deputy Assistant Secretary of the Navy Fred Davidson told the NERA Conference that additional non-hardware units were being established to support Naval Air Wind augmentation. He also said two 250-bed fleet hospital units would be stood up by 1985.

The Navy's second Varsity Player exercise, honing mobilization skills, took place in San Diego during the autumn of 1982. Regular and Reserve ASW units took part, with CAPT John Riess as Battle Group Commander and CDR Gil Grigsby as ASW Force Commander. Both John and Gil are stalwarts of the San Diego Chapter.

Another NRA member found himself leaving the West Coast, as RADM Howard Roop (Balboa Chapter) left his post as Commander REDCOM-19 to return to active duty as Deputy Director of the Naval Reserve in Washington, D.C.

As the fiscal year ended and the Navy released the results of its safety records, the Naval Air Reserve Force came in with an accident free record for its 162,000 flying hours. The force includes 53 squadrons flying 53 types of aircraft.

Other records in the past fiscal year included eight NJROTC libraries established, with three coming from the Long Island Chapter.

NRA members continued to serve their communities. LCDR Dennis Conte (CHC) (New Hampshire Alan Shepard Chapter) is cited for heroism by the cities of Revere and Everett, MA. While responding to a fire, in his civilian role as a volunteer Chaplain for the Revere Fire Department, Dennis is told by the Rabbi of Temple Tefereth Israel in Everett that the Temple's priceless Torah scrolls were still in the sanctuary. Dennis fearlessly led several others inside to rescue the scrolls. After being hauled out to safety and treated for smoke inhalation, Dennis went back inside three more times (twice alone) and retrieved the rest of the scrolls and various other religious items.

Few NRA members could equal the heroism of a Dennis Conte, but serving the community is something all could aspire to. CDR Al Eucare (Kalmbach-Wardroom Chapter) is named as the General Chairman of the 1983 Congressional reception, which will henceforth take place every other year at the start of the new Congress. One of the new members Al will meet in 1983 will be CDR Charles Britt (Ole North State Chapter) who is now Rep. Charles Britt (D-NC) from the Sixth District. Charles will be assigned to the House Armed Services Committee.

Other NRA members serving their communities in 1982 include CDR James Celebrezze (Cleveland Chapter) who is elected Associate Justice of the Ohio Supreme Court. CDR Donald Fossedal (MF Maury Chapter) was featured in a Washington Post article for his work as Director of Marketing for the Government Printing Office. Don promotes the sales of thousands of publications including the Cross-Cut Saw Manual, Infant Care, and the Manual of Patent Examining Procedure, to name just a few.

NRA members also played their role in the public affairs arena. The Command Information Bureau established for the recommissioning of the battleship NEW JERSEY features CAPT Donald Wylie (San Diego Chapter), CDR David Henley (Balboa Chapter), LCDR Carl Benscheidt (South Bay Chapter) and LCDR Tom Pinard (Arrowhead Chapter). Countless other NRA members worked the public affairs arena by helping present many of the ore than 2,000 Seapower presentations that Naval Reservists presented to audiences of 1,900,000 people throughout the nation in 1982.

Perhaps the final sign of the year's closing was the retirement of past National President Phil Smith. Phil's retirement means the title of Silver Falcon, or Senior Reserve Naval Aviator, goes to CAPT Billie Spencer (San Diego Chapter) and a member of the NARU at North Island.

January 1983 started off badly for the NRA, as another Founding member and stalwart crossed the bar on 14 January. RADM Bob Burke of the New York Naval Militia (CAPT USNR) had served as National VP-Legislation from 1958-1961 and 1970-1972, National President of the ROA, Commander General of the Naval Order, and National Director of the Navy League. Many throughout the Navy and Reserve communities will miss him.

On the good news/bad news banner, the Selected Reserve is over end-strength, meaning people can't perform active duty until the new fiscal year. The SelRes also have to find a way to cut 3.8 percent of their people out of pay billets.

NRA members continued to display their literary efforts, as CWO4 T.J. Smith (Terminal Island Chapter) and CDR Jacob Wolfsberger (Akron Chapter) send commentaries to Naval Institute Proceedings discussing mine warfare and close-support bombing respectively.

Another NRA member with a stronger connection to the Naval Institute is CDR Paul Stilwell (USNA Chapter) who directs the Institute's Oral History program. Paul is preparing a history of the battleship NEW JERSEY, aboard which he served during the Vietnam era.

A very well known author with NRA ties is LT Herman Wouk, lifetime member of the Kalmbach-Wardroom Chapter and the author of the Caine Mutiny, Winds of War, and War and Remembrance—all novels with a Navy theme. The televised version of Winds of War runs from 6-13 February and draws a record audience, naturally including many NRA members.

In other California news, three San Fernando Valley Chapter members, CAPT Roger Nichols, LCDR Jeffrey Nielsen, and CDR Jim Renton join CDR Gil Knipe (South Bay Chapter) to hold a three-day Navy/Marine Corps Reserve Judge Advocate seminar at San Diego. About 240 JAG Corps officers attend, with more than 70 percent being active duty types.

On the other side of the country, CAPT Bill Baker (ADM Byrd Chapter) becomes Aide-de-Camp to Virginia Governor Charles Robb.

Further good news—all 11 Reserve Commodore selectees are NRA members, including past National VP-JAG/VP-Legal Affairs CAPT Bob Wiss (Great Lakes Chapter).

The Spring EXCOM in Arlington hears CNAVRES RADM Robert Dunn discuss long range Naval Reserve plans. He feels CNAVRES should not get OPCON of NRF ships and that a three-star CNAVRES would make no difference is the person is a Reservist—since a Reservist would not be able to hold his own in the beltway infighting for funds and resources. (This view is later echoed by RADM James Ramage, a former CNAVRES and San Diego Chapter life member, in a letter to NRA News.) The EXCOM also hears about the ARAPAHO project, which converts merchant ships to helicopter carriers, and the proposed ROPMA legislation.

Although the defense budget has been rising substantially the past three years, the NRA still believes it can be improved. It recommendations for FY84 include having 105,000 USNR SelRes plus 12,942 TARs. Additional P-3C Update III aircraft and P-3A/B TAC/NAVMODs should also be procured. The post of CNAVRES should go to a Reservist

With mobilization always on a Reservist's mind, it's high time a manual for the subject was produced. CAPT Fred Farmer (S. Decatur Chapter) and CDR David Skeen (MF Maury Chapter), produce the Fleet Staff Training Manual, a 200-page comprehensive manual for personnel on duty with major shore and fleet staffs. Fred is the skipper of CINCLANTFLT Det-206, while Dave is the project leader for the task.

Fred and his team complete their task in the autumn and garner a Meritorious Unit Commendation for their efforts. The award is presented by Deputy Assistant SecNav Ellison Grayson. (Ellison better known as E.C. has another role—captain in the Naval Reserve and member of the MF Maury Chapter.) Also attending the ceremony is Fred's predecessor as unit CO, CAPT Jim Abert (Kalmbach-Wardroom Chapter).

Another focus of the Navy is the Bear. CDR Robert Suggs (MF Maury Chapter) produces an article for the April 1983 issue of Naval Institute Proceedings about the pressures facing Soviet Navy leader ADM Sergei Gorshkov. In the same issue, CAPT Henry Beam (South Central Michigan Chapter) describes the Project Seafarer/ELF Communications System being installed Michigan's Upper Peninsula.

Two months later, CAPT David Kenney (MF Maury Chapter) writes an article for Naval Institute Proceedings analyzing the Falklands Conflict. Major conclusion—high tech wars are wars of attrition, meaning a high technology combat environment requires great logistical depth.

Mobile Diving & Salvage Unit One Det 220, led by CDR James Spagnole (Sutter Chapter) and CDR Jack Scholl (Peninsula Chapter) wins the Meritorious Unit Commendation Award. The unit is commended for its salvage services to active commands between 1980 and 1982.

Commendations of another sort come to YN1 Janis Broad, on TEMAC at NMPC, when she is commissioned LTJG in the Medical Service Corps and converts from associate to a regular NRA membership. Past National President RADM Phil Smith performs the commissioning honors.

Congratulations also go to LCDR William Grannan (Constitution Chapter), chosen as ROA's Outstanding Naval Reserve Junior Officer of the Year. Future NRA members might receive the NROTC Award being established by the San Diego Chapter in honor of the late CAPT Ben Fisher (MC) USNR. The award will go to the outstanding midshipman from the joint University of San Diego/ University of California (San Diego) NROTC unit.

Life membership passes the 5,000 mark when Renee Beskind signs on the dotted line. Renee is wife of NRA Deputy Executive Director CAPT Bob Beskind and a strong recruiter in her own right. At the same time, membership statistician LCDR Lee Douglas reports that more than 50 percent of the

NRA's membership are Life or three-year members, with Life passing at the 25 percent mark.

The NRA Founders Group, comprised of the 43 officers who founded the Association, elect CDR Fred Kempe as their spokesman.

A spokesman of another sort is RADM Matt Backer (Spirit of St.Louis Chapter) who fires off a letter to Navy Times rebutting an editorial that opposed transferring Active-duty missions to the Reserves. RADM Backer claims such transfers would be a prudent use of the taxpayers' money.

His letter is seconded a month later, when CAPT Bob Beskind sends in a full page OpEd piece discussing readiness, age, and skills and saying that a "Naval Reserve unit, provided with like hardware, can equal or outperform the active force any time."

Not every member of the NRA writes an article or OpEd for a newspaper. Some members are featured in profiles, or even special sections, if you're CAPT Guillermo (Willie) Zuniga (Alamo Chapter). When Will puts in his papers, the lifetime resident of Laredo finds the Laredo Morning Times publishing a special eight-page Sunday supplement, devoted to his life and times, both as a Naval officer and civilian.

CAPT Pat Lucci (Western New York Chapter) and National VP-Public Relations is promoted to the rank of Rear Admiral in the New York Naval Militia.

Another New Yorker being promoted is William Seidenstein (Long Island Chapter). He becomes the first Reservist to be commissioned an LDO Ensign in Physical Security (Designator 6945) Bill, an NYPD detective lieutenant in civilian life, is commissioned by his CO, CAPT Raphael Mur, National VP-Legal Affairs.

The Long Island Chapter also sees 17 members retiring in 1983. Chapter president CDR John Patafio presents them with certificates and notes that three have service predating Pearl Harbor. Five served in World War II, four in Korea, one in Vietnam. Three have WWII and Korean service, while one has service in all three wars. One was a POW of both the Germans and the Japanese. Between them, they have more than 500 years of Navy service.

Retirement is also a subject familiar to CAPT Rolland Hamilton, two-term President of the Twelfth District, when he is elected President of the California State Federation of the National Association of Retired Federal Employees (NARFE).

Another NRA stalwart taking a leadership position outside the NRA is CDR Scott Miller (Santa Fe Trail Chapter) is elected National President of the U.S. Naval Sea Cadet Corps. More than 8,000 youths are in the Corps, which is sponsored by the Navy League.

Also being congratulated is CDR Betty Lawson (NC) (Jacksonville Chapter) who is invited by the University of Tokyo Hospital to be their guest at the 20th anniversary of their Physical Medicine and Rehabilitation Center. Betty, who taught in several hospitals in Japan, helped develop the Center.

LCDR Jo Ellen Oslund also saves lives, but in a different manner. The San Diego Chapter member is copilot of a helicopter flying a training mission out of North Island NAS. Hearing a distress call her aircraft proceeds to rescue a downed Marine Reserve A-4 pilot whose aircraft developed mechanical difficulties.

Another CDR also sees signs of his past, when CDR George Woodward (Milwaukee Chapter) performs his active duty aboard the battleship Iowa, which is being prepared for recommissioning. Going to the aerology office, he sees the Navy Weather Service symbol painted on the bulkhead by then-ENS George Woodward in 1956.

The first Naval Reserve unit to receive an E-2C hawkeye is VAW-78 of Norfolk, commanded by CDR Charles Fisher (Hampton Roads Chapter). The E-2C is one more sign that current equipment is finally entering the Reserve inventory.

The Summer EXCOM in Kansas City makes a momentous decision. Tender an offer for a building at 1615-1619 King Street in Alexandria, VA, to become permanent NRA Headquarters. One member of the committee that will do the tendering is LCDR Russ Miller, who, as National Treasurer in 1955 recommended that a permanent site be bought because real estate prices tend to rise faster than expected.

RADM Russ Gorman (Polaris Chapter) had an interesting two months tour on special active duty. He served as Deputy Commander of MSC, but two weeks saw him filling in as acting Commander, when the Commander takes annual leave.

The SecNav approves some recommendations of the National Naval Reserve Policy Board's FY82 report, but not all. He approves of having a three-star Director of Naval Reserve, but not necessarily a Reserve flag officer in the billet. The USNR will see two new commands established at New Orleans, Naval Surface Reserve and Naval Air Reserve. Deputy CNAVRES RADM Bill Daniels and Deputy Chief of Staff for Air Readiness RADM Tommie Rinard will head the two commands. Both are long-time NRA stalwarts.

Women Reservists will be considered for flag billets regardless of Congressional proposals to write this into law. In another move, the Naval Reserve will consider different ways of acquiring drill credit for specialized Reservists with difficult work schedules—principally members of the medical communities.

Chapters also find time to party. The Santa Fe Trail Chapter holds a dining out. Chapter President LCDR Sam Migliazzo hosts, while RADM James Mantel is President of the Mess and LT Michael Pandzik is Mr Vice.

On a happy note, the FY84 Defense Authorization Bill approves portal-to-portal medical and dental coverage for drilling Reservists.

Unfortunately, the 30th National Conference begins on a sad note. As members are arriving in Phoenix, they learn that Korean Air Lines Flight 007 has been short down with more than 200 passengers, one of whom is CAPT Larry McDonald (Kalmbach-Wardroom Chapter). Larry, a Georgia Congressman, won the NRA's Distinguished Service Award in 1980.

The major award during the Conference goes to outgoing National President CAPT Curt Coleman, who receives the Navy's Distinguished Public Service Award. Deputy Assistant Secretary of the Navy (Reserve Affairs) Fred Davidson presents the award. (Fred, it should be mentioned is also a Marine Reserve lieutenant colonel and a member of the NRA).

CAPT John C. Rice Jr. of Louisiana is elected National President. The position of National Secretary, abolished in 1958, is reestablished, to be automatically filled by the NRA Executive Director, currently RADM Jim Forrest. In another election, a son inherits what had been his father's mantel. CDR Bill Waylett Jr. is elected National VP-Surface/Subsurface. The post, previously National VP-Naval Affairs, was held by CDR Bill Waylett Sr. during the 1950s.

No chapter wins the Ken Kalmbach Award, but the Jack Lewis Award for Outstanding Chapter goes to the M.F. Maury Chapter, led by CDR Walt Steiner. An honorable mention goes to the San Diego Chapter under CAPT Charles Wilhoit. This was the first time two large chapters competed for the award.

Membership performance awards go to the Fifth (CDR Dick Morgan DVP-Membership, CAPT Bill Dunbar President) for highest increase in Life Members and the Eleventh (CAPT Marv Duncan, DVP-Membership, CAPT Charles Wilhoit, President) for 3-year member gains and total net increase. The best renewal rate goes to the First District (CDR Bill McGurk DVP-M, CAPT-sel John Wheeler DP). Awards for individual top recruiters go to CDR Mileva Hartman, CAPT Goerge Linton, CAPT Bill Chisum, LCDR Joel Frank, CAPT Ed Grandin, and CDR Bill Waylett, Jr. CAPT Bob Beskind should have earned an award, but as a Headquarters staffer he is ineligible.

With the defense budget in good shape, the Conference adopts relatively few resolutions. One commemorates Congressman McDonald and Sen Henry "Scoop" Jackson (D-WA), who won NRA's 1970 Distinguished Service Award. Other resolutions call for full ACDUTRA funding for Category-D personnel and urge the CNO to ensure 100 percent manning of USN billets at all Naval Reserve activities.

The Conference decides to establish a NRA Alcohol and Drug Abuse Referral Program for NRA members. Several members have been pushing for such a program, including CDR Gere Bunn (San Diego Chapter), who has performed active duty

Cigar Store Indians (left to right) John Rice, Jim Forrest, Win Weese, Don Osborne.

From left: CAPT Rice, RADM Forrest, Mayor Moran and SECNAV Warner at Headquarters dedication.

National Past Presidents' Meeting at Headquarters.

The Conference Room at Headquarters.

Corridor outside of the Conference Room at Headquarters.

tours since 1974 in Navy alcohol/drug abuse activities, including a recent tour as CO, Naval Alcohol Rehabilitation Center San Diego.

The major news at the Conference is that the Executive Committee in a special meeting immediately before the Conference site has selected a permanent headquarters, as the Association commits itself to the King Street site in Alexandria. The Building Site Acquisition Committee reconstitutes itself as the Building Renovations Committee under RADM Bill Gilmore and LCDR Russ Miller. Planned occupation date is 1 July 1985.

One of the brighter, but sadder pieces of news heard by the Conferees concerns the Surface Force. The good news is that the Navy will place the first OLIVER H. PERRY (FFG-7) class guided missile frigate into the Naval Reserve Force in early 1984. USS DUNCAN (FFG-10) will be stationed at Long beach Naval Station. But the Navy is also decommissioning the last two World War II-era Gearing class NRF destroyers, WILLIAM C. LAWE (DD-763) and USS HAROLD J. ELLISON (DD-864). While the NRA had long urged the replacement of these ships, there are some sad eyes as these last representatives of their class, which gave valiant service, leave service.

Shortly after the Conference, the Naval Order holds its Biennial Congress and elects RADM Win Weese Commander General. RADM Weese is the outgoing President of the NRA's Eighth District. The Naval Order's Commander General-Elect is CDR Stan Majka (Jack London Chapter).

When the Long Island Chapter holds its Navy Birthday Ball, President CDR John Patafio welcomes CNO ADM James Watkins and Mrs. Watkins. The $3,300 netted by the Chapter is donated to three local charities.

For those officers wanting advice about selection boards, members are urged to read the September issue of Naval Institute Proceedings, where CAPT Torrey Sylvester (Constitution Chapter) gives the complete gouge on current plans and procedures.

If anyone doubted the ability of Naval Reservists to fill in for active duty officers, it was disproved in the summer of 1983, when personnel changes left a gap in the USN watch captains in the Navy Command Center. Navy Command Center 106, commanded by CAPT Donald Knuth (MF Maury Chapter) provides a list of five USNR captains qualified for the duty. Filling the watch bill over a three-month period are CAPTs Harry Boggs (Cumberland), Henry Gill (Kalmbach-Wardroom), Robert Hoch (Constellation), Maurice Lynch (Hampton Roads), and Jack Smith (Atlanta).

In a lighter vein, in Navy Times, CAPT Joseph Sheridan (Grand Rapids Chapter) describes the origin of the term gedunk. Turns out it came from a cartoon character in the 1930s who liked "gedunk sundaes," just when the Navy's battleships and carriers were establishing soda fountains.

And in a practical vein, CAPT Gordon Brown (Desert Sailor Chapter) writes an article in Proceedings recommending ways to solve the problem of too many senior officers in units and qualifications. Among his recommendations are delaying the reserve promotion flow points to LCDR and CDR. He also calls for annual professional tests for Surface Officers wishing to remain in drill pay and improving the ability of Reservists to qualify as Surface Warfare Officers. He also recommends making coastal REDCOM Commanders non-deploying NRF Surface Group Flotilla Commanders.

On a more serious publishing note, three members of the Triangle Chapter, LCDR James Galloway, LCDR Michael Zapata, and LCDR James Smith, are completely revising and rewriting NWP 19-3: Deep Ocean Search, Inspection, and Recovery. The two-year project will see an updated Tactical Warfare Publication reflecting many changes in the fleet, particularly the use of computers, in deep sea operations.

The scientific and technological bent of some Naval Reservists is reflected in the November Proceedings. CAPT George Hill (Neptune Chapter) and a professor of surgery at the University of Medicine and Dentistry in New Jersey, analyzes cold-weather requirements for personnel. He points out that contingencies might require US forces to deploy to frigid areas, but little has been done in the way of training and research in treating cold weather casualties.

While some NRA members could ponder and write, others had been in the thick of things in October 1983. When a suicide bomber blew up the Marine BLT 1/8 quarters at Beirut Airport, a Reserve C-9B from VR-56 was on active duty in Rota Spain. Squadron CO CDR Russell Kuhl and plane commander LCDR Mack Simons, both NRA members, quickly reconfigured their aircraft for Medevac duty and made flew four missions to Beirut to bring in aid and evacuate casualties.

The plight of the Marines in Beirut was much on the mind of newly elected National President John Rice who in his first testimony before a Congressional Committee spoke of saving of combat search & rescue aircraft which were in the inventory. The outcome would have been different here and in Vietnam. He described existing equipment & efforts as Kamikaze missions. He won the NAVY's first two SAR helicopters.

Also serving off Lebanon is CAPT Joseph Azzolina (Perth Amboy Chapter), who leaves his job as a Navy planning officer to deploy to Central America and Lebanon aboard the battleship USS NEW JERSEY. He completes his Surface Warfare Officer qualification and provides major assistance to the ship's diplomatic and public affairs efforts.

That same month, US forces launched a rescue mission to the island nation of Grenada in the Caribbean. Five Reserve Air Crews support this operation, including members of VR-56 from NAS Norfolk and VR-58 from NAS Jacksonville. Among the crews are NRA stalwarts CDR CDR Russell Kuhl, CO of VR-56; CDR Jim Horan, CO of VR-58; CDR "Skip" Wilbur and CDR Gerry Sheppard. Russell, who flies out the day after re-

Executive Director RADM Jim Forrest.

Presentation of the RADM John Lewis trophy.

turning from Beirut, might entitle his tour "Half-way Around the World in Just Over Eight Days." Other NRA members help man up the Navy Command Center and the CINCLANTFLT HQ Operational Control Center.

Closer to home, there is sad news as CAPT Bill Davis, long-time NRA Executive Director, passes away in Orlando.

December's Proceedings features two NRA authors. CDR David Grover (Vallejo Chapter) and a former dean at the California Maritime Academy, warns that the US may lose a generation of Merchant Mariners because less than 20 percent of Merchant Marine Academy graduates finds a seagoing job. He suggests the Coast Guard, Maritime Administration, and Naval Reserve work together to solve the problem. The same issue sees CAPT Henry Beam (Battle Creek Chapter) analyzing four leadership approaches towards Navy enlisted. He says as the Navy moves towards a participatory system of leadership, it will increase its effectiveness because of higher retention rates among skilled personnel.

Retention is high on the mind of RADM Joe Miller (MC) (Memphis Chapter). As the newly appointed Deputy Director of Naval Medicine/ Deputy Surgeon General for Reserve Affairs, he must solve a shortfall of 800 physicians and more than 10,000 enlisted medical personnel to meet current Navy medical requirements.

RADM Miller might talk to RADM Cecil Kempf, USN, (M.F. Maury Chapter) the new Chief of Naval Reserve. An Academy graduate and pilot, he has commanded squadrons and ships and been Assistant DCNO for Air Warfare. RADM Kempf relieves RADM Robert Dunn who will become Commander Naval Air Force Atlantic.

The Pentagon's reserve hierarchy could have a new face. CAPT Larry Korb (MF Maury Chapter) and Assistant Secretary of Defense (Manpower, Reserve Affairs, Logistics) tells NRA members that the position of Assistant Secretary of Defense for Reserve Affairs has finally been approved. Unfortunately it will take nearly five months to find a candidate to fill it the position.

In Washington, CAPT George Scaboo (MF Maury Chapter) Deputy Archivist at the National Archives, has completed a four-year task of assembling the private papers of Richard Nixon and preparing them for public release. His 12-man team had to arrange and sort 628 cubic feet of documents.

The first news out of NRA Headquarters in 1984 was good news. Effective 1 February, every active and associate member would automatically receive $1,000 worth of accidental death and dismemberment insurance coverage. Fully paid-up Life Members would receive an additional $1,000 worth of insurance. Hartford Accident and Indemnity Coverage provided the insurance. In addition, NRA planned to offer up to $100,000 more in life insurance under a group program being developed. The same announcement said that premiums for NRA's term life insurance would be cut 10 percent across the board and maximum coverage increased to $150,000.

Congratulations to USNR Unit Intelligence Center Pacific (IPAC) 0171, which is awarded the first Joint Meritorious Unit Award. Unit CO CDR James Zurcher (Mile High Chapter) accepts the award and simultaneously receives the Joint Service Commendation Medal. IPAC 0171 is a 100 percent NRA unit.

Sad news marks the passing of ADM James L. Holloway Jr. As Chief of Naval Personnel following World War II, Holloway developed the "Holloway Plan" of Regular NROTC scholarships, which became the college education and commissioning source for many NRA members.

Navy Times Associate Editor Ted Bush points out the downside of Naval Reserve contributory support. Much of the funding comes out of funding for mobilization training, but the duties performed aren't always related to mobilization. An example is Reserve ASW P-3 squadrons flying anti-drug missions in the Caribbean. They're performing valuable service, but it isn't part of their mobilization training plan. Either some mobilization training plans should be changed or funding categories revised.

RADM Robert Spiro (SC) (M.F. Maury Chapter) and former Under Secretary of the Army, is named the Executive Director of the ROA. Also out of ROA comes the news that LCDR William Ratner (San Fernando Valley Chapter) is the Outstanding Navy Junior Officer of the Year. Will commands Fleet Training Group Det 219.

LCDR Tracy Connors (MF Maury Chapter) submits a Professional Note to the January 1994 Proceedings describing the recent Global War Game conducted at the Naval War College's Center for War Gaming. The annual exercise took three weeks and involved more than 350 people. The February Proceedings finds CDR James Browning (MF Maury Chapter) describing a similar 11-day war game at the National Defense University's War Gaming & Simulation Center.

The February Proceedings also finds RADM Penrose "Whitey" Albright (Kalmbach-Wardroom Chapter) commenting on the parlous state of the US Merchant Marine. He recommends tax credits, so that it will become financially advantageous to ship cargoes in US hulls.

March is the International Navies Issue of Proceedings, and LCDR Brad Hahn (Orlando Chapter) provides an incisive analysis of the Chinese Marine Corps. It has a strength of 120,000 men, with 30,000 trained Reservists. Brad notes that the Chinese Marine Corps dates its heritage back 1,745 years, when it conducted its first overseas amphibious operation.

Writing in a more querulous vein to Navy Times, CDR Louis Zurcher (Austin-Capital Chapter) decries the confusion currently shown in the naming of warships. He blames it on "mindless political expediency." Navy Times also sees a letter from CAPT Ted Jakaboski (El Paso Chapter) questioning negative assumptions about the capabilities of Naval Reservists to perform active-duty assignments and missions.

In another side of the literary world, the increasing cost of books means NRA NJROTC Libraries now cost almost twice what they did. To keep the program intact, local chapters will see their cost increasing by about seven percent to $200, while the number of books will drop from about 40 to less than 25. (When the program began in 1972, the number of books per package was 50.)

The Spring EXCOM in March in Charleston approves National President CAPT John Rice's suggestion that a space be set aside in the new Headquarters Building as a Naval Reserve Historical Research Center. The EXCOM also votes to discontinue the Congressional Reception, disestablish the National Publications Committee and reschedule the Summer EXCOM from July in Las Vegas to August In Chicago.

May finds the Wilmington, Delaware Naval Reserve Center celebrating its 25th anniversary, with many First State Chapter members on hand, including chapter member LCDR Mary Ann Rowe, Center CO.

In the May issue of Proceedings finds CAPT John Foster (Monterey Chapter) recommends that the Naval Reserve form a coastal patrol force of fast patrol boats integrated with the Mobile Inshore Undersea Warfare units (MIUW) and the light Attack Helicopter Squadrons.

Finally in May James H. Webb Jr. is appointed as Assistant Secretary of Defense (Reserve Affairs). He is a 1968 Naval Academy graduate and won the Navy Cross for valor in Vietnam. He also wrote the critically acclaimed novel Fields of Fire.

CAPT John Francis (Polaris Chapter) makes his mark when he qualifies, nearly simultaneously, as a Navy Parachutist and as a Surface Warfare Officer.

Washington finds quite a bit of goings on concerning both the NRA and the Naval Reserve. Past National President CAPT Fred Carl, who also is Executive Director of the Armed Services YMCA, signs an agreement with Secretary of Defense Casper Weinberger. The agreement formalizes the relationship between the YMCA and the military, which has existed since 1861.

Also in Washington news comes word that RADM Neale Smith (Space Center Chapter) will relieve RADM Howard Roop (Balboa and MF Maury Chapters) as Deputy Director of Naval Reserve. And finally, CNAVRES RADM Cecil Kempf announces that career guides will be coming around 1 October for most specialties in the Unrestricted Line specialduty line, and Staff Corps communities.

Washington also is the scene of a major Naval Reserve policy study under the direction of RADM Bruce Newell, former Chief of Information and currently Special Assistant to the CNO. RADM Newell's study recommends expanding the missions of the Naval Reserve by making them responsible for maritime coastal defense and protection of Caribbean sea lines of communication. He says Reserves should be responsible for airborne and seaborne mine countermeasures, and man some Air Cushion Landing Craft (LCACs), and Oilers and Supply Ships. The Naval Reserve also should man two Repair ships. The air side should be loaned some P-3Cs for training and be completely responsible for the 15th and 16th Carrier Air Wings. Other air Reserve missions should include carrier on-board delivery and land-based aerial tanking. The Medical Reserve should augment the active duty personnel aboard the two hospital ships being built.

RADM Newell's study is considered, some steps are taken to implement some of his conclusions, but overall, little action is taken. Some of the recommendations of the 1983 National Naval Reserve Policy Board are implemented. These include a Reserve Unrestricted-Line direct commission program, increased manning of regular billets at Reserve activities, incentive pay for Reservists in high-priority units, and crediting all retirement points earned through inactive duty training.

One NRA stalwart who will be coming to Washington more frequently is CAPT Irene Wirtschafter (Melbourne Chapter). SecNav has just appointed her to the Committee on Retired Personnel.

Inland NRA members haven't been inactive either. When massive flooding hits Missouri, Pony Express Chapter President CAPT George Ruwwe class tells his membership to man their craft, in this case a 12-ft flatbottom Jonboat. Chapter members assist the National Guard and Coast Guard in ferrying property owners through the area and performing patrol duties.

Other NRA members continued to grace the pages of Naval Institute Proceedings. The August issue saw four members published, with LCDR Steve Schady (Hartford Chapter) discussing the press blackout during the Grenada operation and LT Tom McCaffery (Mile High Chapter) commenting on officer manning difficulties in the Merchant Marine. CAPT Henry Beam (Battle Creek) had an article "Help Yourself" giving tips to polishing one's fitness reports and CAPT John Ahlstrom (Wilmington Chapter) providing his reminiscences of the battle of Leyte Gulf.

Farther afield, CAPT Larry Korb (MF Maury Chapter) in his role as Assistant Secretary of Defense for Manpower, Installations, and Logistics, dedicates the first Naval Reserve Readiness Center at Naval Reserve Center Houston. The center will be one of five mega centers in REDCOM 10 that will consolidate training resources. If all goes well, this pilot program will be expanded to other REDCOMs.

Another revised Center is the Naval Construction Battalion Center in Davisville, RI. Reserve Seabees of the 21st Reserve Naval Construction Force Regiment, commanded by CAPT Fred Anibal (Philadelphia Chapter) will train here.

Some training facilities are afloat, however. The first shipboard Naval Reserve Center is dedicated as USNRC Manhattan aboard the former aircraft carrier INTREPID, a World War II and Vietnam veteran, now a museum ship on the city's West Side. RADM Joseph Callo (New York Navy Chapter) leads the dedication ceremonies. The center occupies about 20,000 sq. feet and will be home to 250 Naval Reservists. As is pointed out in the dedication, Naval Reservists train aboard ships, but there has never been an entire floating Reserve Center until now.

A different form of floating meeting takes place across the continent in Seattle, when the Puget Sound Chapter and the Thirteenth District board the 123-ft ship El Primero to hold their annual election meetings. The fast cruise also sees CDR David Satterlee relieve CAPT Dennis Vaughan as Thirteenth District President.

The National Conference, in San Antonio, sees awards going to several chapters and individuals, although all are members of the San Diego Chapter, presided over by CAPT Charles (Daisy) Wilhoit (who also leads the Eleventh District). The San Diego Chapter the Jack Lewis Award for Outstanding Chapter, while the Eleventh District receives the Ken Kalmbach Award for best District communication (along with District VP-PR CDR Sherry Kleinschmidt). Other honorees from the San Diego Chapter include LT "JR" Knight who takes the paid-recruitment honors, while CAPT Bill Chisum leads in the recruitment of complimentary members.

The Conferees decide to back off from the previous policy of advocating that CNAVRESFOR control NRF ships. They decide to disestablish the National Publications Committee and increase the number of Golden Helm Naval Reserve Retention awards from two to five categories. They decide to increase the number of NRA Outstanding Naval Reserve Junior Officer awards to two annually, with one going to a SelRes and the other to a TAR or officer on recall to active duty.

Conferees adopt resolutions asking the Navy to study the possibility of having NRF crews be located remotely from their home port and modernize all light Attack Helicopter Squadrons. Other resolutions call for legislation allowing concurrent receipt of military retired pay and VA disability compensation in particular cases and providing USNR affiliation incentives for former enlisted personnel with critical skills.

During the Conference, CNAVRES RADM Cecil Kempf tells Conferees that the Navy is considering adding a new DCNO billet, DCNO Naval Reserve. Having such a billet would allow the Naval Reserve to compete for resources earlier in the budgetary process, rather than having him beg for resources from the other DCNOs and ACNOs, as is currently the situation.

As 1984 winds down, the Headquarters Building Fund stands at $1 million—about 40 percent of the goal. The downside is that only five percent of NRA members have contributed to it.

The year's end issue of Proceedings also sees LCDR Paul Hendrickson (Liberty Bell Chapter) comment on the different types of commanding officers one might encounter during a career. Paul's comment: "Ask yourself how I would like to have me as a boss?"

The dawn of 1985 coincided with the beginning of the second Reagan Administration and the defense budget reaching the highest point it would reach in real dollars during the 1980s buildup. With firm support from the Congress and the Administration, it was no surprise that SecNav Lehman approved 21 of the 1984 National Naval Reserve Policy Board's recommendations.

The recommendations include increasing SelRes participation in the management of the Surface Reserve program, reducing admin reporting requirements for SelRes units by 35 percent and notifying SelRes unit COs well before the implementation date of any proposed unit personnel changes resulting from the NAMMOS adjustments. Other recommendations call for establishing a commissioning program for senior enlisted TAR personnel. Other key recommendations called for a system to monitor professional licenses, with procedures to discharge personnel who lost their licenses for professional misconduct. The Board also called for having CWO-4s sit on all Warrant Officer selection boards and developing procedures to pay BOQ/BEQ service charges for Reservists not in a per-diem status.

In a similar vein, although not part of the Naval Reserve Policy Board's recommendations, the SecNav called on the Navy to return to its nautical linguistic roots and leave the generic language of the 1980s behind. Specifically, dining halls would revert to being called mess decks or wardrooms. Stairs, ceilings, floors, walls, and hallways vanished as if by magic, and ladders, overheads, decks, bulkheads, and passageways returned. And the Unaccompanied Officer Personnel Housing (UOPH) and Unaccompanied Enlisted Personnel Housing (UEPH) reverted to Bachelor Officers Quarters (BOQs) and Bachelor Enlisted Quarters (BEQs) or barracks.

The beginning of the new year also saw one of the NRA's stalwarts retire as the indefatigable Bill Kreh (Stephen Decatur) retire from the editorship of Navy Times on February 15. When Bill, who had also been editor of NRA News for many years, stepped down after 30 years with Navy Times, he received the Secretary of the Navy's Distinguished Public Service Award, the highest award the Navy can present to a civilian. No editor of Navy Times before or since achieved this recognition. Bill's retirement plans—to write a column for Navy Times and contribute to other sea-service oriented publications.

Fleeet Commander's brief.

Keynote Luncheon at the National Conference.

Yet another Association stalwart to gain recognition was CAPT Bennett Sparks USCGR, (M.F. Maury), who was selected to Commodore in the Coast Guard Reserve. Bennett, a past President of the ROA (and the first Coast Guard officer to hold the post) also has the distinction of serving in every rank from Seaman Recruit through Chief Petty Officer and Ensign through Commodore.

On the other side of the nation, LCDR Frederick Curlin, Medical Affairs VP, San Bernardino-Arrowhead Chapter, received the Long Beach Police Department's Community Service Award for rendering emergency on-scene treatment to an injured police officer.

Another member of the NRA gaining kudos was LT Byron Coleman (M.F. Maury), who received the ROA's Naval Reserve Outstanding Junior Officer of the Year Award in early March 1985. A committee from COMNAVRESFOR selected Byron, after it received nominations from Readiness Commands across the country. The presenter was also an NRAer, Deputy Asst. SecNav for Reserve Affairs Fred Davidson III, also a Lt Col in the Marine Corps Reserve and a member of the Kalmbach-Wardroom Chapter.

On March 31, NRA members heard sadder news, as word came of the passing of CWO-4 Vaughn Coale, the Crusty Old Gunner. Vaughn had gained quite a reputation within the Association extending from when he joined the newly-founded NRA in 1954. A founder and first president of the Stephen Decatur Chapter in Maryland, Vaughan also served three times as Fifth District President, on many national committees, and most recently, in 1981, as first National VP-Retired Affairs. Vaughan was sorely missed.

People can be missed, and so, sometimes can aircraft. April 1985 saw the decommissioning of the last C-118 transport at NAS Alameda. The guest speaker was past National President RADM Phil Smith, who, in an younger life, had flown and instructed in the first C-118. Very few of the C-118's civilian equivalents (DC-6) were left either. The decommissioning, coming two years after the last two Gearing class NRF destroyers were decommissioned, marked the end of an era.

In another case of déjà vu all over again, CAPT (CHC) Aaron Landes, NRA's immediate past National Chaplain, dedicated an Ark and a Torah scroll aboard the aircraft carrier Forrestal (CV-59) following her Service Life Extension Program (SLEP) at the Philadelphia Naval Shipyard. Thirty years previously, in 1955, the first Jewish service aboard the newly commissioned Forrestal was conducted by then LT (CHC) Aaron Landes.

Reserve aircraft and personnel also made the news in other ways in 1985. Calendar year 1984 marked the third year in a row that the Naval Air Reserve Force had the best aviation mishap-free rate in the Navy, with only two mishaps. Neither resulted from pilot or maintenance error. The low rate of 1.15 Class A mishap rate gave the air community something to strive to maintain.

NRA members also provided much assistance to foreign navies during 1985. In April, the People's Liberation Army in Taiwan asked CAPT Larry Korb, Asst SecDef for Manpower, Installations, and Logistics (Southern new England) for assistance. Seems the PLA wanted to improve their spare parts supply system. CAPT Korb's quick answer—use computers. In an editorial, Navy Times suggested that the abacus might be better, given the Pentagon's problems with computers. But the Chinese were impressed enough to follow up with another question—how the US military compensation worked. This, Larry had to admit, was much harder to answer.

Several months after helping the Chinese, another group of NRA members found themselves helping the Peruvian Navy in June. After a storm damaged one of their bases, the Peruvians asked CNO ADM James Watkins for assistance. In response, Seabees from RNMCB-16 (Los Alamitos) and RNMCB-17 (Port Hueneme) spent two weeks in Callao, Peru cleaning and repairing the sewer system, restoring damaged communications networks, and repairing buildings. Commander First Construction Brigade RADM Thomas Maddock (South Bay) commended his troops for their work, saying it marked a first for cooperation between Reserve Seabees and the Peruvian Navy.

Other NRA members traveled north of the border for more festive reasons. June saw 50 NRA members going to Halifax, Nova Scotia, to attend the 40th annual National Conference of the Naval Officers Association of Canada and help celebrate the Canadian Navy's 75th anniversary. The lucky 50, led by National Exec. VP CAPT Pat Lucci had a delightful time with their north of the border brethren.

Public Affairs also played a key role in the NRA's 1985 activities. CAPT Dave Woods (Kalmbach-Wardroom) was elected National President of our sister organization, the ROA in May. Dave had already seen legion service as NRA's National VP-Membership, National PAO, and Fifth District President. Dave became the first Naval officer since 1975 to serve as ROA President. Aside from bringing his legendary public affairs skills to his new post, Dave also brought his reputation as an auctioneer and prankster.

Another NRA public affairs type, CDR James Myers (Space Center) and the NR Office of Information Det 310, which he commanded, developed a major upgrade to the Navy's Fleet Home Town News Center (FHTNC) in Norfolk. Jim's project will allow FHTNC to provide news releases on Reservists' activities to their hometown newspapers, the same way the active duty system functions. All Surface and Air Reserve PAOs received the kits. Hopefully, the new system will increase the awareness of the Naval Reserve among the civilian community.

Not all the proposals made by the NRA or its members were greeted so successfully. The medical community, particularly the Nurse Corps and the Medical Service Corps proposed adding Reserve Flag billets in their specialties, because of the quadrupling and doubling of billets in the two fields. While Flag billets exist on the active side, there are none for the Reserves, who will play a key role in any conflict. It would be approximately 10 years before these proposals were accepted.

The summer EXCOM meeting, in Denver, was relatively quiet. The major proposal that was approved established a Naval Reserve Fund, with NRA as the sole member, thereby making NRA tax exempt as an educational and charitable organization as well as a veterans organization. The fund handled all headquarters-building fund assets for the new headquarters building, then nearing completion. The fund would simplify accounting procedures and personal tax procedures for persons contributing to the building fund. The EXCOM also supported an initiative of Sen William Roth (R-DE) establishing a medal and ribbon for POWs, endorsed a proposed national resolution to support the current military retired-pay system, and authorized preparation of an amendment to the NRA constitution establishing a National VP-Capital Development.

Without a doubt, the major news of 1985 was the occupation of New Headquarters. August 22 saw the dedication of the NRA's own home, at 1619 King Street, in Old Town Alexandria, VA. No fewer than three ribbon cuttings were held that day. The first floor ribbon was cut by Sen John Warner (R-VA), Mayor Jim Moran of Alexandria, and National President CAPT John Rice. Second floor honors went to CAPT Bob Shaffer (M.F. Maury) and third floor ribbon was cut by Conference Room sponsors CDR William Mellish and CAPT Lucille Mellish (Twelfth District).

The move into the building marked the Association's 10th move in 30 years and the first time it didn't have to worry about fleeing ahead of doubled rents or a wrecker's ball. It was also a long way from "Pohanka's Garage," the backroom of the family auto dealership of the late Founder LT Frank Pohanka. The new headquarters is a renovated three-story 1920s mercantile building. During the two-year overhaul preceding the move, the façade was ripped away and replaced with a replica of the original and a large window with small panes installed on the first floor. The interior was gutted and totally rebuilt to the NRA's requirements. The NRA took possession of the second and third floors, renting out the first floor to another firm.

Two events in the Navy hierarchy in mid-1985 affected various NRA members. In June the Naval Materiel Command (NAVMAT) was disestablished and its functions redistributed between the five Systems Commands under its control. The move was made to streamline the decision-making and acquisition management processes. The move caused some Reserve units to be disestablished and other reorganized to handle new responsibilities. The middle of the year also saw Congress turn down the Navy's proposal to create a seventh Deputy CNO for Reserve Affairs. But the Navy promoted Chief of the Naval Reserve RADM Cecil Kempf, an NRA life member, to Vice Admiral, thereby making the CNR billet a three-star slot—equal to the fleet type commanders and warfare sponsors.

Any philatelists among the NRA had an interesting year in 1985. On February 22, the Postal

Some of the nearly 500 junior officers attending the JO symposium listen to the remarks of Naval Reserve flag officers who comprised one of the symposium's many presentations.

Service issued a $.50 postage stamp honoring the late FADM Chester Nimitz, who had been a member of the Golden Gate Chapter (since renamed Nimitz-Golden Gate). The Postal Service issued a stamp honoring Korean War veterans on July 26. CAPT Bob Lunney (New York Navy) represented the NRA at the dedication in Washington DC at the express invitation of the Postmaster General. Bob, National VP-Legal Affairs, is a Korean War veteran of the Chosin Reservoir/Hungnam campaign.

While most of the remainder of 1985 might have seemed anticlimactic for NRA members, following the dedication of the new headquarters, many events occurred and many members showed the Naval Reservist at his/her best, as twice the citizen. In August a new identifying logotype was unveiled for the Surface Navy. The logotype, showed a destroyer, bow-on, superimposed upon a crossed cutlass and sword, with three arrows pointing up, out, and down to represent the surface fleet's operations against air, surface, and subsurface threats. The unveiling ceremony at the Navy Memorial Museum at the Washington Navy Yard, honored former CNO ADM Arleigh Burke (Kalmbach Wardroom). The ceremony also featured the first display of the Surface Warfare sweater, a blue pullover sweater.

Some NRA members have found a way to make the VTU an interesting duty, if not a paying proposition. CAPT Joe Davenport (Houston-Space Center Chapter) began a program in 1983, Operation Opportunity, to recruit former active duty officers into the Naval Air Reserve program. Contact is made with officers as they come up on their end of service date or shortly thereafter. Cooperation has been good, and by 1985 a number of officers have joined the Reserves because of the program. Joe credits a number of other NRA members with support, including CDR Ron Grooters (Jack London Chapter), CAPT Charles Orr (Pascagoula Chapter), LCDR Ray Gibbons (Western New York Chapter), CAPT William Sample (Denver Chapter), CAPT George Dresser (Waco Chapter), and CAPT Jerry Donevant (South Bay Chapter). With such a strong contingent on the air side, they propose pitching the idea to the Naval Surface Reserve.

To keep its members up to date, the Eleventh District held a one-day seminar "The State of the Naval Reserve" More than 150 people gathered to hear NRA members RADM Howard Roop, former Deputy Director Naval Reserve; CAPT John Rice, National President; and COMO Bruce Beran, Commandant Eleventh Coast Guard District. Other speakers included Rep. Daniel Lungren; CAPT Frank McGirr, area Readiness Commander; CAPT James Curry, CO NAR North Island; and CAPT James Callan, area Inspector General.

The attendees also took time to recognize the achievements of various NRA stalwarts. District President CAPT Larry Danielson presented Twice a Citizen Awards to CAPT John Howe and LCDR Bob Brice, District Junior Officer Awards to LCDR Michael Nemechek and LCDR Bryan Lucas, and the CAPT Paul Hummel Award for lifetime service to CAPT Charles (Daisy) Wilhoit.

The following month, other NRA stalwarts were recognized at the National Conference, held in Norfolk. Outgoing National President CAPT John Rice received the Secretary of the Navy's Distinguished Public Service Award from the Chief of the Naval Reserve, RADM Cecil Kempf.

The Ken Kalmbach Award for outstanding District Communication went to the Sixth District, led by CAPT Melvin Klein, the NRA's Deputy VP-Public Relations. CDR Broadus (Jack) Dempsey, TAR CO of the Roanoke Reserve Center, won the individual membership recruiting award.

The Conference elected CAPT Pasquale (Pat) Lucci as National President. It adopted several resolutions. These called for a Korean War memorial, strict acounting for Vietnam POWs and MIAs, and procuring additional ASW trainers for NRF FFG-7 and FF-1052 class frigates. Other resolutions called for recruiting additional medical personnel, and increasing the number of USNR Flag billets from 48 to 50 to accommodate one Medical Service Corps and one Nurse Corps Flag officer.

The air side saw another gain in late 1985, when VFA-303 became the first Naval Reserve squadron to convert to the F/A-18 Hornet. Many active duty units have not yet received the new strike fighter, so the changeover marks a major success for the Reserve community. The commanding officer of the squadron, CDR Bob Greathouse, is a TAR and NRA life member.

Other NRA members won kudos in late 1985 as well. CDR Michael Matt, a TAR, won the Stephen B. Luce Award as the top graduate in his Naval War College class. CDR Stan Majka (Jack London Chapter) was elected Commander General of the Naval Order, succeeding RADM Win Weese (New Orleans-Acadia Chapter).

The winners of the NRA's Outstanding Junior Naval Reserve Officer Awards went to LT Mike Winkler (Houston Space Center Chapter) for the SelRes category and LT Don Roberts (Terminal Island Chapter) for the TAR-Active Duty category.

LCDR Ken Nichols (Shreveport) pointed out in a letter to Navy Times that while most elements of the Navy were doing better on funding than they had, this did not hold true for the small arms program, both in the active duty and, particularly Reserve, components. He noted the lack of funding for Navy personnel wishing to compete as Navy personnel at interservice and other rifle/pistol matches. Ken also decried the lack of training ammunition, saying one REDCOM had 4,000 rounds for 4,000 Reservists.

The NRA and its "sister" organizations could look back on many accomplishments as 1986 dawned. The Navy was well on its way to reaching the 600-ship goal, new frigates and aircraft were coming into the Naval Reserve, and morale among the officers and enlisted had never been higher. But there were signs of trouble ahead.

In a column in Navy Times early in the year, NRA Life member Bill Kreh said he saw signs that "lip service" from the administration was on its way back. Specifically, Bill asked why the Reagan Administration was not making a firm statement about Reservist use of commissaries.

Scarcely had the column come out than the Gramm-Rudman-Hollings deficit reduction act denied cost of living increases (COLA)s to military retirees and SBP widows. These were two of the few groups not getting COLAs. Social Security recipients and Veterans with disability pensions did receive the COLAs.

To fight these trends,the NRA joined with 15 "sister" organizations to form the Military Coalition. The Coalition had a combined membership of 1.3 million active, Reserve, and retired military personnel. The organizations comprising the Coalition included the Air Force Sergeants Association, the Association of Military Surgeons, Fleet Reserve Association, Marine Corps Reserve Association, Army Warrant Officers, Coast Guard Chief Petty Officers Association, Coast Guard Chief Warrant Officer and Warrant Officer Association, Reserve Officers Association, Naval Enlisted Reserve Association, Commissioned Officers Association, Noncommissioned Officers Association, and the National Military Family Association.

Nonetheless, when the 1986 budget was approved, it could have been called the NRA's wish list. Among the items that had been cut during the year but were restored following NRA lobbying were nine P-3C aircraft, with six for the Naval Reserve; full funding to develop an onboard ASW system trainer for NRF frigates, full funding for the AN/AQA-7 signal processor for P-3A and P-3B aircraft and four more Combat Search and Rescue helicopters.

While the major action was on Capitol Hill, out beyond the beltway NRA members continued to shine. CAPT Dane Canete (Pearl Harbor) served as the cardiologist on an open-heart surgery team that flew to China to teach coronary artery by-pass grafting at a hospital in Hanzhou. In Philadelphia, LT Joseph Picogna became the first Naval reserve Supply Corps officer to win the Surface Warfare Supply Corps Officer device. And in San Francisco, LT Antonio Reyes (Jack London) was at a local restaurant during having lunch on a drill weekend. While eating, he saw a man fall to the floor choking. Jumping up, Reyes went over and performed the Heimlich Maneuver and CPR. After saving the man's life, he returned to his meal, and then back to drill.

Another member of the NRA received his award as a capstone to his career and his life. March 6 saw Albuquerque proclaim "CDR John Shanklin Day," in honor of the Commander. Many civic and military groups presented him with awards. Most fitting was the Secretary of the Navy's Superior Public Service Award, which was presented by Past National President Phil Smith. Sadly, two days later, CDR Shanklin passed away.

Another NRA stalwart received an honor which would give him a chance to influence both the Naval Reserve and the Navy. CAPT Bill Chappell (Kalmbach-Wardroom Chapter) (also known as Rep Chappell D-FL) became chairman of the House Defense Appropriations Subcommittee.

One reason members were able to accomplish their varied successes was because they had to contend with less paperwork. A major example was the Naval Reserve Unit Inspection Guide, which had been cut from 50 pages to 23 and from 307 items to 161. The cuts came after an edict from Secretary of the Navy John Lehman that Naval Reserve paperwork be cut by 30 percent.

With the FY86 bill coming in a success, and members more active than ever, the NRA was ready to turn its sights on the FY87 budget. The legislative recommendations for FY87 included continued P-3C procurement, continued Towed Array procurement for NRF frigates, 18 SH-2 LAMPS helicopters, and initiation of a weapon system improvement program for SH-2s already in service. Other initiatives included the procurement of 14 Mine Countermeasures ships (MCM)s and 17 Mine Sweeperhunters (MSH)s, with all to become NRF assets after a year with the active fleet. The NRA wanted a force structure of two Combat Search and Rescue helicopter squadrons and provision of suitable helicopters for light helicopter attack squadrons HAL-4 and HAL-5.

Manning all the new systems would not a problem, either. DoD's FY87 Manpower Requirements Report postulated that by FY91 Selected Reserve strength would be 170,700, compared with FY86's 155,700 level. The Marine Corps Reserve would show similar growth, from 44,400 to 48,000. The projected growth rates would be four times as great as those of the active duty forces.

While this good news was being digested, several hundred Naval Reservists were preparing for an event that showcased the Naval Reserve's capabilities and allowed the service to show itself at its best—the 1986 Liberty Weekend in New York City. The celebration included the relighting of the Statue of Liberty with an International Naval Review of more than 40 warships from 20 nations and 20 square-rigged tall ships. More than 800 Naval Reservists played a key role as boat crews, Shore Patrol detachments, Public Affairs specialists, mess cooks, administrative staff, escort officers and foreign naval liaisons, and other duties. Supervising the entire naval side of the event was RADM Dick Young (Denver-Mile High Chapter).

While many Naval Reservists helped prepare for the big event, and thousands more watched in person or on television, one found a unique way to get to and see the events. CDR Richard McCormick (Western New York Chapter) and his wife packed a tent and supplies in their canoe, "Seneca Chief II" and paddled for three weeks and six days from Cazenovia Creek near Buffalo to Caven Point in New York harbor. The pair retraced the route of the canal packet 'Seneca Chief' took when it opened the Eire Canal in 1825. But in a sign of the times, Richard and his wife took much longer than the original nine days journey required by Gov. DeWitt Clinton 161 years before Liberty Weekend.

As the Liberty Weekend festivities were winding down, on 7 July, the Supreme Court proclaimed the automatic sequestration provision of the Gramm-Rudman-Hollings bill to be unconstitutional. The court gave Congress a 60-day window of opportunity to modify the act. This would have reversed the denial of COLAs to military retirees and SBP widows, but Congress did not vote to restore COLAs.

The middle of the summer, normally a quiet time in the nation, saw two Lieutenant Commanders from the NRA making the news. LCDR Leonard McRoskey (NoVA-Maury Chapter) was named Deputy Assistant Secretary of the Navy for Reserve Affairs, filling a post previously held by Marine Corps Reservist Lt Col Fred Davidson III (Kalmbach-Wardroom Chapter). Also on the East Coast, retired ADM Stansfield Turner, a past President of the Naval War College, presented the Naval War College Past President Prize for the best professional essay, to LCDR Christopher Yates (Stephen Decatur Chapter).

Further south along the Atlantic Coast, the summer also saw CDR Steve Israel (Orlando Chapter) save the Navy $400,000 at Naval Weapons Station Charleston. Steve led a special team that dismantled a marine railway and drydock dating back to World War II, while preparing the area for a dock expansion.

The onset of fall brought tears to the eyes of many NRA stalwarts, when word was received of the passing of CDR Fred Kempe, who died in September after a long illness. One of the Founders of the Association, Fred had led several committees as well as his Chapter and District. But far more importantly, Fred's out-of-pocket contributions were one of the key reasons NRA remained afloat during its early years. In addition to many other honors and posts, Fred also was a past Chairman of the NRA Founders Group and a past Commander General of the Naval Order.

With the growing number of reservists in the Navy, Congress and the Administration saw the need to make several changes in personnel policy in the FY87 Defense Authorization bill. In a sign of the changing times and growing numbers of women in the Navy, Congress voted to allow female Reservists to be promoted to Flag rank. The bill also called for men and women with the same designator to be on the same lineal list for promotion, rather than separate lists. This change made Reserve personnel policies similar to those of the active Navy, which had combined lineal lists in 1981. The FY87 bill also authorized the President to call as many as 200,000 Reservists from all services to active duty for 180 days during a crisis. This was a major expansion over the previous authority, which was limited to 100,000 Reservists for only 90 days.

The act also allowed single parents to enlist in the Reserves in certain cases and expanded medical, dental, and pay benefits for Reservists on active duty for less than 30 days. While not as generous as the benefits provided to active duty forces, the benefits nearly matched them, providing Reservists with far greater protection than they had previously. In a final acknowledgement of the contributions of Reservists and the spirit of one military, the groundwork was laid to allow Reservists to use commissaries for a number of visits equal to the number of days they spent on active duty during a year.

Nonetheless, the same Congress that giveth also taketh away. The 1986 revisions to the Tax Code hit some Reservists in their pockets. Expenses related to drilling, including transportation, lodging, and meals, could still be deducted, but they could be considered "business-related" only if they exceeded two percent of the Reservist's adjusted gross income.

With membership in the Navy and the Naval Reserve blooming, it was only fitting that membership statistician LCDR Lee Douglas (Acadia Chapter) reported that membership at the end of FY86 stood at 24,000. Thanks to the efforts of National Retention Officer LCDR Mileva Hartman (Polaris Chapter) the Association was averaging a 94 percent retention rate—nearly 10 percent higher than the average for professional associations. Keeping the members coming in required ace recruiters, and the NRA had more than its share of such recruiters. Leading the way was CDR Shannon Jones (Fresno Chapter President). Shannon was closely followed by CAPT Steve Scott (Columbus Chapter), CDR Mileva Hartman (NoVA-Maury Chapter), CDR Bill Tewelow (Atlanta Chapter), and CAPT William Chisum (San Diego Chapter).

This good news was announced at the National Conference, which was held in Reno, NV at the Bally's Casino. While most members flew in, several dozen from the West Coast took a specially chartered train from Oakland, CA that ran through the Sacramento Valley and the High Sierras.

One of the first items the National Conference took care of was the ever thorny question about dues, which hadn't been raised since 1978. Annual dues were raised to $20 per year, with three-year dues raised to $48. Life member dues remained at $200 for those under 51 years of age, with the sliding scale moving upwards to age 60, when life member dues became $100.

The Conference also passed a number of resolutions. These included a call to maintain the current Reserve retirement system, and opposition to limiting inflation-based COLAs on retired pay. Conferees also called for maintaining the benefit structure of the SBP program. In a move of particular concern to Reservists on active duty, the Conference recommended rescinding the "fair market value" charges being levied by BOQs and BEQs and returning to the nominal cost-of-service charges. The conferees recommended that no particular segment of society absorb more than its fair share of the cuts being recommended as deficit reduction measures. They also called for a priority accounting for MIA and POW personnel.

Several resolutions dealt with fleet matters. Conferees voted to support the Strategic Homeporting Plan for basing ships in non-Navy ports. They also called for providing command-at-sea opportunities for qualified SelRes Surface Warfare Officers. Other resolutions dealing with ships and aircraft called for procuring enough SH-2 LAMPS 1 helicopters to meet the full mission capability of the FFG-7 NRF frigates, procuring the most mission capable Towed Array Sonars for NRF FF-1052 and NRF FFG-7 frigates. In line with these last two resolutions, conferees also called for a complete modernization of the Naval Reserve's VP squadrons by the middle 1990s.

Two NRA members were singled out for honors at the National Conference. CAPT Dave Woods (Kalmbach-Wardroom Chapter) received the Secretary of the Navy's Superior Public Service Award for his service as president of the ROA. After relinquishing the presidency of the NRA's sister organization, Dave began organizing a Reserve VTU devoted to Naval History, which would report to the Director of Naval History and support the needs of the Naval Historical Center. The Secretary of the Navy also presented CAPT Wells McCurdy (Puget Sound Chapter) with the Distinguished Public Service Award for his service as special assistant to the Deputy Assistant SecNav for Logistics. During his tour, Wells saved the Navy more than $30 million by resolving uncertainties concerning land-leases in the Northern Marianas.

Shortly after the National Conference, in an unrelated move, the Navy commissioned its first Reserve Helicopter Mine-Countermeasures unit, HM-18, on October 4. CDR Walt Steiner (M.F. Maury) became the first commanding officer at a ceremony presided over by Commander Naval Air Reserve Force RADM Tommie Rinard (Acadian). The squadron boasted 20 pilots, 200 SelRes enlisted, 120 TAR personnel, and of course, five RH-53D Sea Stallion helicopters. October also saw SelRes being asked to volunteer for a two-year recall to active duty to help recommission the battleship WISCONSIN. Plans called for the first group to report to Norfolk in November 1987.

Another Naval Reserve first in October 1986 was the inauguration of the first Naval Reserve Readiness Unit. Commanded by CAPT John McKinley (Kingsport), the new unit, based in New Orleans, was to be the first of 18 units designed to provide instructors and specialized training equipment to the four Readiness Centers. These centers would, in turn, aid the regional Reserve Centers. Commander Naval Reserve Surface Force, RADM Al Rieder, noted the new system would alleviate much of the paperwork currently burdening surface units.

The new initiative was part of the Surface Program to Upgrade Readiness (SPUR). SPUR was a set of initiatives and policies to better integrate the Surface Reserve with the Active fleet and ensure smooth sailing all around. NRA officers played a key role in developing SPUR and met several times with the Director of Naval Reserve (OP-09B) to discuss the new program and its implementation and impact on the Surface fleet. In a move related to SPUR, the Commander Naval Reserve Force established a "hot line" for recommendations concerning paperwork reduction in the Naval Reserves.

Two NRA members made the news in separate ways in November 1986. CAPT Andy O'Rourke (New York Navy) ran for Governor of New York State on the Republican ticket. While losing, he did very well against an extremely popular Democratic incumbent. In the Mid-West, CAPT Lou Hoos (Grand Rapids) provided legal assistance in establishing the first grave for an Unknown Soldier at Fort Custer National Cemetery in Battle Creek, MI. Lou provided the legal assistance in having the remains of a "colored unknown soldier 102 US" (102 Infantry, Michigan) moved from an unkempt corner of a local cemetery. The ceremony included an Honor Guard, Army Band, and Air National Guard flyover.

In the young follow their elders department, ENS Thomas Forrest, USN, serving aboard USS STARK (FFG-31) became Life Member 8000. Tom's father is NRA Executive Director RADM Jim Forrest.

NRA members were working smoothly, but the same couldn't always be said of Congress. In a rather typical saga, the defense appropriations had to be contained in a "continuing resolution" because Congress failed to meet its deadlines. The FY87 continuing resolution called for a USNR SelRes end

NRA's Flag Reception at the Navy Museum.

strength of 149,486 (the same as the previous year), plus 21,476 active duty support personnel (mostly TAR). Hardware procurement included nine P-3C Orions, with four going to the Naval Reserve; 84 F/A-18 Hornet aircraft, with an unknown number for the Naval Reserve; six SH-2F LAMPS Mk1 Sea Sprite helicopters; and four Combat-SAR helicopters.

The Navy and the Naval Reserve Association could take great pride in themselves at the beginning of 1987. Both the Gallup and Harris polls showed the military as the most trusted institution in the United States, coming out ahead of organized religion, higher education, the medical profession, and the Supreme Court. It was nice to be in a situation different from that of the 1970s, when the military had been ranked considerably lower.

If the first half of 1987 belonged to any members of the NRA, it was members and units residing/drilling in California, which saw a host of activities and events highlighting their accomplishments.

The first event, at the beginning of 1987, saw a newly constructed 930-ft pier dedicated at Naval Station Treasure Island on January 6. The pier, funded by the Naval Reserve, was designed to accommodate eight NRF ships, three of which arrived on January 10, the Knox class frigates ROARK (FF-1053), GRAY (FF-1054), and LANG (FF-1060). The dignitaries gathered at the ceremonies included NRA members RADM Vince Anzilotti, Commander Naval Readiness Region 20, NRA Executive Vice President CAPT Ken Welch, and LCDR Leonard McRoskey in his civilian role as Deputy AsstSecNav for Reserve Affairs. Other attendees included RADM Robert Toney, Commander Naval Base San Francisco and San Francisco Mayor Dianne Feinstein.

California NRA members showed another side in late January during change of command ceremonies at Naval Reserve Center Fresno, when commanders of all eight units received 100 percent NRA Membership Certificates. Both new and old ResCen Cos LCDR Martin Kauchak and LCDR Judith Castle were members, along with CAPT Armand Dollinger, CDR Gary Watkins, CDR John Seil, CDR Masamitsu Mori, CDR John Tenney, LCDR Michael Bussey, LCDR Lee Gholz, and LCDR James Makofske. Other honored guests and NRA members included RADM Russ Gorman, RADM Carlos Baker, CDR Shannon Jones (President of the proud Fresno Chapter), and Twelfth District President CDR Dave Foote.

Other California Naval Reservists made the news when doctors, nurses, corpsmen, and other medical personnel volunteered to deploy on a five-month mission to the Philippines aboard the Oakland-based hospital ship USNS MERCY. Many of the medical Reservists making the deployment were attached to Reserve units reporting to Naval Hospital Oakland.

California NRA-member CDR Howard Montgomery, President of the San Jose-Polaris chapter, received the FY-86 Exceptional Individual Performance Award from Command Naval Recruiting Command for being the most outstanding of the 2,000 nation-wide Sea Power Presentation Program speakers. The competition was monitored by the CNO's office. Further up the coast, in Pasadena, RADM Bob Garrick was named one of Pasadena's 10 most courteous and well-mannered citizens. Bob headed a PR firm in Pasadena, after leaving his previous job as White House public relations counselor.

In another California-related item, San Francisco's Jack London Chapter, led by CDR Mo Thorson and the Twelfth District hosted a Career development Symposium at Treasure Island, which attracted more than 200 officers of all ranks. The highlight of the event was a mock selection board held by two flag officers and five captains, all of whom had served on boards.

Appearances to the contrary, California was not the only site of NRA or Naval Reserve activity in early 1987. In Cleveland, the Navy Finance Center agreed to begin processing Reserve payrolls twice a month rather than once a month, meaning some Reservists would no longer have to wait up to six weeks for their pay check. And in Washington, the Congressional Budget Office issued a warning about future problems for the Reserves. With recruitment and retention running so well, the CBO said the large numbers of prior service officers and enlisted could lead to fewer promotion opportunities and greater manpower costs. They also warned that many Reservists had risen to levels in their civilian jobs where they would have a difficult time being mobilized. The warning came out while the Sixth Quadrennial Review of Military Compensation was meeting.

Two other events in Washington impacted on Naval Reservists—one being food for morale and the other just plain food. Two new service ribbons were authorized, the Navy/Marine Corps Overseas Service Ribbon for 12 months (cumulative or consecutive) of active duty service outside the United States service and the Naval Reserve Sea Service Ribbon for 36 months satisfactory service in an NRF unit. The DoD also formally approved expanded use of commissaries by Reservists. Reservists could now use commissaries on days of their choice, with the number of days equaling the number of days spent on active duty.

NRA officials, meanwhile, were debating the merits of the SPUR program. The Spring EXCOM meeting featured a panel with officers from the Director of Naval Reserve, COMNAVSURFRESFOR, and CNET leading the discussion. The debate failed to change any opinions, and the NRA went on record as continuing to hold the same reservations about SPUR that had already been listed in a February letter from the National President to the Director of Naval Reserve.

Washington also saw two NRA members and former Chiefs of Naval Operations being honored. ADM Arleigh Burke (Kalmbach-Wardroom) and ADM Thomas Moorer (M.F. Maury) each received a numbered print of the painting "Vietnam Patrol" depicting the reactivated NRF destroyer escort USS VAMMEN (DE-644) off South Vietnam during a tour in 1962 when she carried a mostly Reserve crew. CAPT Charlie Nelson (Raleigh-Triangle) presented the paintings in a ceremony at NRA headquarters. Charlie, who served under then-CAPT Burke in DESRON 23, the famed "Eager Beaver" squadron, commanded the VAMMEN in 1962 and later commissioned the painting from the former official artist of the Navy, Arthur Beaumont.

Washington also saw two interesting twists on NRA membership and the senior Navy hierarchy in mid-1987. An officer in the Regular Navy and an NRA member, VADM James Zimble, (M.F. Maury) named Surgeon General and Director of Navy Medicine.

During the summer, a Naval Reserve officer, RADM Francis Neale Smith, (Houston-Space Center) was named Director of the Naval Reserve and Commander Naval Reserve Force. Neale, who had been serving on recall as Deputy Director when he was named, was the first SelRes officer to take the conn of the Naval Reserve. This was a case of the Navy taking a step that the Army and Air Force had taken long before—making a Reservist the chief of the Reserve component.

One of Rear Admiral Smith's first acts as CNAVRES was to attend the summer EXCOM meeting in Phoenix. He conducted a special briefing and Q-and-A session for attendees, during which he addressed the broad range of Naval Reserve issues, from equipment to personnel policies. Of particular concern to most members and to ADM Smith was the SPUR program and other initiatives to improve Naval Reserve readiness.

Washington also saw the NRA get involved in battles it was hoping to have forgotten—Budget Wars. The coming of the late 1980s, and worries about the budget deficit, lead to various suggestions about how and what programs to cut. One of the suggestions was to have DoD submit a two-year defense budget. In a case of good news/bad news, this meant that some amount of stability would come to the budgeting process, but that those programs that got short shrift would suffer for two years. In addition, the two-year cycle only applied to the authorization budget—not the all-critical appropriations.

The NRA's legislative lobbyists saw many problems in the FY88/89 budget submitted in February 1987. The NRA immediately countered with its own proposals. These called for continuing procurement of the P-3C Orion ASW aircraft, rather than terminating production. The NRA also called for adding $18 million in FY88 for upgrade kits for the SH-2 LAMPS Mk 1 helicopters and $26 million for upgrade kits in FY89. A third aircraft-related proposal called for adding $2.5 million for F/A-18 modification kits, with an additional $2.5 million for aviation support equipment.

NRA proposals called for the Surface Force included a call for funding to procure and install FFISTS electronic tactical systems aboard all FF-1052 Knox class frigates at a cost of approximately $24 million. The NRA also advocated that the Mine Hunter Coastal (MHC) program be placed on a fast acquisition track. Another proposal called for adding $13 million per year to the FY88 and FY89 budgets to develop the Arapaho/Nomad System and procure the first Nomad (Naval Operations/Maintenance Aviation Deck) Facility. Arapaho is the system that allows Vertical/Short Take-Off and Landing (VSTOL) aircraft to be carried aboard merchant ships.

Other NRA proposals called for increasing the funding for the Military Construction Naval Reserve budget and the Operations & Maintenance Naval Reserve budget accounts. A final proposal called for providing full funding to support personnel growth in the Naval Reserve, to allow the Reserves to take account of new systems coming on-line.

While pressing ahead on the budget front, NRA members also took time to commemorate several members who passed over the bar in spring and early summer 1987. CAPT Jack Hessey (Fifth District President 1973-1975 and Chairman, National Nominating Committee, 1977-1985) and CAPT George O'Connell, National President, 1962-1964 and President of Chicago's Trident Chapter during the early years of the NRA. George was also one of the key players in doubling NRA's national membership between 1958 and 1961. Even after his active days were over, George remained as a consultant, always willing to provide sage advice.

Diseases of other sorts were on the mind of Navy planners in the spring of 1987, when they issued NavOp-28, requiring all Reservists assigned to drilling status or wishing to be so affiliated, to be screened for HIV, the AIDS-associated virus. Anyone screening positively would be denied a pay billet. The NavOp also called for annual screening to ensure that Reservists remained in good health. On another health and fitness related note, the Navy required that service-jacket photographs were to be full-length in summer khakis, rather than the previous head-and-shoulder photographs. The change was implemented to help catch those Reservists (and active duty personnel) carrying beer bellies.

While some Reservists might have been out of shape, just the opposite held true for LT Robert "Espy" Espeseth, Jr. (Willow Grove-Liberty Bell Chapter). Robert, one of several father-and-son teams in the NRA, was named 1986 Male Athlete of the Year by the U.S. Rowing Association for his performance as a member of the 1976, 1980, and 1984 U.S. Olympic Rowing Team.

Other junior officers won renown for their work in other parts of the country. LCDR Lisa Benscheidt (San Fernando Valley Chapter) devised a school earthquake-preparedness program for Pacific Palisades, CA. Other schools in the area picked up on Lisa's program and used it to create their own plans. The City of Los Angeles honored Lisa for her work in this critical area.

The NRA also honored junior officers, in particular, two life members. LT Melody Mock (Billings Chapter), CO of the Billings Naval Reserve Center; and LT Rita Szymanski (Glenview Chapter) received Outstanding Junior Officer Awards. Individual chapters and districts had their own awards, and LT Ken Denman (Columbus Chapter) received the Fourth District's Junior Officer of the Year Award.

But junior officers were not the only NRA members being honored. CDR Doyle Quisenberry (Hampton Roads Chapter) received the Navy Meritorious Civilian Service Award for his civilian work as Atlantic Fleet Logistics Plans Officer.

The summer of 1987 saw two NRA Past National Presidents elected to the two top offices of the Naval Order at its Congress in Galveston. CAPT Fred Carl was elected Commander General of the Order for a two-year term, while CAPT John Rice was named Commander General-Elect.

Shortly after the Naval Order held its Congress, the NRA held its 34th National Conference in Houston. CAPT Kenneth Welch (Maryland) was elected National President. The Conference had a busy agenda, with a large number of resolutions being passed by the attendees. Among the resolutions, the NRA called on the Administration to increase the number of full-time support personnel dedicated to the Reserve.

Another NRA stalwart who had been serving the Navy as a civilian was CDR John Lehman (MF Maury). In mid-1987 John decided to leave his job as Secretary of the Navy for greener pastures on Wall Street. John's successor, while not an NRA member, was a highly decorated and renowned veteran of the Marine Corps, James Webb. Webb, currently serving as Assistant Secretary of Defense (Reserve Affairs) won the Navy Cross for his Vietnam service and is a noted author.

Taking Webb's place as Assistant Secretary of Defense (Reserve Affairs) was an NRA member, CAPT Stephen Duncan, who in the process changed his membership from the Denver Chapter to the Northern Virginia/MF Maury Chapter.

The summer of 1987 also saw the Navy catch up with the Annual Retirement Point Reports, which had been running two years behind. Some civilian employees of NRPC had dumped all the data relating to USNR officers drill attendance, with no one catching the error. The foul-up meant all drill attendance data had to be entered by hand. In other critical matters affecting all Naval Reservists, in the autumn of 1987, DoD sent the Reserve Officer Personnel Management Act (ROPMA) to Congress. Six years in the offing, ROPMA was supposed to codify all the different Reserve personnel policies that had developed through the years, and parallel the Defense Officer Personnel Management Act (DOPMA) of the active duty side. A proposed ROPMA had been sent to Congress once, in November 1981, but no action had been taken on it at the time.

While Congress debated issues, some NRA'ers made names for themselves on the front lines of the Navy. LCDR Christopher Schur (Hartford Chapter) took the skills learned in a Convoy Commodore unit and volunteered as a Naval Liaison officer aboard the "reflagged tankers" the Navy was escorting in the Persian Gulf. Chris did double duty in this capacity, doing his first two-month tour in the summer/autumn of 1987 before coming back in late 1987 for a second two-month tour.

With the Navy making a name for itself in so many quarters, it was time for a unit to document the Navy's operations. Accordingly, in October 1987, at the beginning of FY88, the Navy stood up Naval Reserve History VTU 0615. The unit was dedicated to historical research and assisting the Naval Historical Center, which was perpetually short-handed. CAPT Bill McClintock (Hampton Roads) was named the first commanding officer.

Sadness also touched the NRA in the autumn of 1987 as ENC Joseph Wasson passed on. Wasson was one of the founders of the Naval Enlisted Reserve Association in 1957 and served 25 years as its Executive Director. He joined the NRA as a Life Associate Member as soon as that option became available in 1976, after having received the NRA's Meritorious Service Award in 1972 for his services and contribution to the Naval Reserve.

During the autumn of 1987, several changes also took place at NRA Headquarters. Mrs. Helen Duncan retired from the staff as Membership Assistant, to be succeeded by Mark DeVille. In the Headquarters building, a "Wall of Honor" was dedicated to memorialize NRA members' service in support of the NRA, the Naval Reserve, and the Navy. These members would be remembered through bronze plaques with inscriptions.

The NRA also issued its first credit cards in late 1987. The Naval Classic MasterCard worked like any normal Master Card, with one difference. The NRA received a slight rebate on all purchases charged.

Perhaps the closing event of 1987 was the most uplifting yet surprising. When San Antonio's Alamo Chapter announced they were the first NRA Chapter to sponsor a Sea Cadet Division, they got stares of surprise. But after checking the records, it turned out they were right—no other NRA Chapter had sponsored a Sea Cadet unit. CAPT-sel Scott Miller, (Kansas City Chapter) national board chairman of the Naval Sea Cadet Corps, presented the commissioning warrant to Alamo Chapter Committee Chairman CAPT Jim Franklin.

The new year 1988 found several senior NRA members busier than ever with new duties and plans for the coming year. CAPT John Eric Lindell (Beaumont Chapter), who wrote the NRA News feature "Tax Tips for Naval Reservists" found himself having to revise much of his article and many tips because of the changes that took place in the 1986 tax act.

CAPT Tony Rodriguez, Past President of the New Orleans-Acadia Chapter, was appointed by the Secretary of Transportation to the 21-member National Rules of the Road Advisory Council, to recommend changes to the maritime rules.

VADM Robert Dunn (Kalmbach-Wardroom Chapter) was appointed chairman of a SecNav Advisory Panel to review the SPUR program. Admiral Dunn, a previous CNAVRES, was serving as Assistant Chief of Naval Operations for Air Warfare when he received his additional duties.

RADM Howard Roop (South Bay Chapter) also received new duties, being named Commandant of the Defense Intelligence College at Bolling Air Force Base. The new job came only four years after Howard left his post as Deputy Director of the Naval Reserve.

Yet another NRA member facing new duties in 1988 was CDR Michael Simpson (Syracuse Chapter), who was appointed President of Utica College in New York State.

Some NRA members were honored in their chosen careers and professions. LTJG John Maniscalco received the Outstanding Junior Officer award from the San Diego Chapter. John is a 27-year Navy veteran who rose through the ranks to Warrant Officer in 1983 and Supply Corps LDO in 1987.

Opening Session at the Conference.

Another NRA stalwart honored in early 1988 was CDR Margaret Balacki (Western New York Chapter) who received the first Nielubowicz Award for Navy Nursing. The award, named for the retired director of Navy Nursing RADM Mary Nielubowicz, recognizes original research contributing to and enhancing the Navy Nurse Corps. The award was presented by RADM Nielubowicz and Navy Surgeon General VADM James Zimble (Northern VA-MF Maury Chapter) in ceremonies at Las Vegas.

The other major news at the beginning of 1988 was the completion of the FY88 Defense budget appropriations bill, which was actually completed days before Christmas. Reflecting strong bipartisan support for the Naval Reserve on Capitol Hill, the bill provided the Naval Reserve with 130,609 billets for drilling reservists and 21,991 TAR billets for full-time support. Congress added $258.8 million for unrequested equipment, including AN/SQR-17A acoustic processors for MIUW units and $194 million for six P-3C aircraft. The previous year (1987) the Naval Reserve received its first P-3C Update III aircraft, marking the first time that state of the art equipment was provided to Reserve Air ASW forces. The bill also provided nine HH-60 combat SAR helicopters at $103 million, rather than the requested three helos for $42 million. Other add-ons included $55 million for SH-2 LAMPS Mk 1 helicopter upgrades, rather than the requested $19 million.

While Naval Reserve strength was growing, the Navy was downsizing, with the April 1988 announcement of plans to decommission 10 GARCIA (FF-1040) class frigates and six Brooke (FFG-1) class guided missile frigates. Secretary of the Navy James Webb announced his disagreement with the decision and resigned in protest. His replacement was a Naval Reserve officer, William L. Ball III, who was serving on the White House staff as Assistant to the President for Legislative Affairs.

The Naval Reserve's growing manpower was boosted by the commissioning of the 66th currently-active NROTC unit, this one at Carnegie Mellon University. But another program designed to supply manpower, the Sea College Program, which recruited people for two years active duty and six years in the Selected Reserve, was terminated by Congress because it did not fill specific shortages.

And yet another example of on the one hand...and on the other hand occurred in the Coast Guard. The Coast Guard stood up two new Selected Air Reserve Training Units, in Sacramento, CA and Elizabeth City, NJ. These two units, plus a Ground Support unit in Clearwater, FL being converted to an Air Support unit, would help meet the Coast Guard's mobilization requirements to support the Navy. But just a couple of months later, the Coast Guard had to drop 1,200 Selected Reservists from its rolls in mid-1988 because Congress once again cut its funding while simultaneously increasing its responsibilities.

While all this was happening, Congress created 250,00 veterans overnight. Where did they come from—the World War II Merchant Mariners, who were finally being granted veterans status for all laws and programs administered by the Veterans' Administration. These men had a higher casualty rate than the Army or Navy, and the step was applauded by many.

An NRA member, who almost dated his service back to World War II was RADM Vince Anzilotti, who became the Naval Air Reserve's Silver Falcon in mid-1988. The Silver Falcon designation went to the Naval Air Reservist with the earliest date of designation as an aviator or flight officer—which in Vince's case was 1952.

Another NRA veteran with service dating back to the Korean era was CAPT Bob Lunney (New York Navy Chapter) who was a featured interviewee in a television special "Korean War – The Untold Story," which aired in May. The show, watched by many NRA members, was hailed it for its use of historic film, graphics, and captured enemy footage. Some members kidded Bob that he had hit the big time, especially with a show hosted by the actress Loretta Swit from the television show "M*A*S*H."

Not everyone in the Naval Reserve were set to stick around as long as Vince had, according to a survey conducted of 12,000 officers and 52,000 enlisted from all Reserve components in the spring of 1986. According to the survey, released in the early 1988, only 54 percent of the officers and 40 percent of the enlisted said morale in their units was "high" or "very high." The Navy and the Marine Corps had the highest number of satisfied personnel. The greatest number of gripes concerned the lack of opportunity for education and training, and the lack of individual skill training. The survey showed Reservists had an average age of 29.5 years for enlisted and 37.9 years for officers. The average enlisted had seven years Reserve service, while the average officer had 10.5 years Reserve time.

If anyone might be in a position to correct the results of the survey, we could hope it would be CDR Martin Lancaster (Raleigh Chapter) in his role as Rep Lancaster D-NC, who got elected by the House Democratic Caucus to a seat on the House Armed Services Committee.

The middle of 1988 also saw another NRA-Congressman announce his retirement as CAPT Sam Stratton, D-NY announced his retirement after 30 years in the House of Representatives. Sam did not retire from the Albany Chapter, however.

Some Naval Reservists might have gotten satisfaction from the medical community. The Navy announced in the summer of 1988 that six more civilian-run NavCare centers had opened to care for dependents and retirees in Charleston, Norfolk, San Diego, Oceanside, Long Beach, and Oakland. In other medical news, Naval Reserve personnel were told they would be assigned some billets drilling aboard the two new hospital ships USNS MERCY (T-AH-19) moored in Oakland and USNS COMFORT (T-AH-20) moored in Baltimore. Many other Reservists, assigned to units mobilizing to Bethesda Naval Hospital and Oakland Naval Hospital, were told they would fill in for the active duty personnel who would man the ships during any emergency.

Awards Luncheon.

Yet another form of hospital became available to Reservists in 1988, when 10 former Public Health Service hospitals were converted to Uniformed Services Treatment Facilities and made available to eligible military beneficiaries. The 10 are located in Portland (ME), Boston, New York, Baltimore, Cleveland, Port Arthur, Galveston, Nassau Bay (TX), and Seattle.

Sadly, the medical community saw another side of the NRA in 1988 as two major contributors to the Association and the Naval Reserve passed away. CDR Bill Waylett, Sr., was one of the few founders who retained an active role in the Association. After serving as the first Naval Reserve Affairs Committee Chairman, he served as an unpaid volunteer on the Headquarters staff, Fifth District President, and on national committees too numerous to mention.

A couple of months later, CAPT Winfield (Chip) Chapin crossed the bar. While Chip was not a Founder, coming on several months after the Association was founded, he served as National President from 1957-1962 and proved a major supporter, both financially and personally, to the Association. More than 100 members attended his funeral with full military honors at Arlington National Cemetery.

For many other members of the NRA, life continued at a fast pace, particularly those on deployment in the Middle East. In one such case, CAPT John Spratt (Louisville Chapter) while on the carrier CORAL SEA (CV-43) on active duty for training, ran into his son, LT Robert Spratt (Hampton Roads Chapter), an A-6E pilot with VA-55, which was part of CORAL SEA's air wing.

A more senior aviator member of the NRA was enshrined in the Naval Aviation Hall of Fame ADM Tom Moorer, former CNO and JCS chairman, and a member of the Northern Virginia Maury Chapter.

The Washington NRA staff also found itself busy with aviation matters, when it had to go to Capitol Hill to get the House and Senate Armed Services Committees to restore funding for six P-3C aircraft, which, after being funded in the FY88 bill, were being rescinded in the FY89 bill.

Aviation was not the only fight the NRA was involved with on Capitol Hill. Once the Reserve Officer Personnel Management Act (ROPMA) reached the Hill, it became fair game for the different Reserve organizations to comment on. The NRA's legislative analysts had to ensure that changes applicable to other services were not forced on the Navy.

Not every member of the NRA could get to the Middle East or Capitol Hill—two continuing battle zones. Some members performed more prosaic duties closer to home. LCDR Greg Slavonic, representing the Central Oklahoma chapter, presented the Mayor of Oklahoma City with a $100 check as a contribution to the Oklahoma City Fund. The fund helps defray miscellaneous costs associated with the commissioning of the nuclear submarine OKLAHOMA CITY. Ties between the submarine and the city are further cemented by the presence of several NRA members on the Mayor's commissioning committee.

Closer to home, a Bravo Zulu goes to 14 officers from Naval Security and Investigative Command 0166, commanded by CAPT-sel Dave Fitzgerald (Richmond Chapter) who completed a difficult 12-month qualification security and

Naval Affairs Symposium.

counter-terrorism course. Completion of the course means they can stand watch in the Navy's Anti-Terrorist Alert Center (NAVATAC).

Another NRA member receiving a Bravo Zulu is LCDR Ray Quinto (San Bernardino Chapter) who received the Congressional Excalibur Award. The prestigious award goes to the "outstanding of the outstanding" among civilian federal employees. Kudos also went LCDR Linda Lou Borges-Dubois (Providence Chapter) who received the First District's "Twice a Citizen Award."

The final NRA member being recognized for his/her achievements in 1988 is RADM Dick Young (Denver Chapter) who received the Navy League's John Paul Jones Award for inspirational leadership. Dick is the first Reservist to receive the award, and many say its truly well deserved for his leadership of the International Naval Review aspects of the July 4, 1986 festivities in New York City.

The personnel draw down began to hit the Naval Reserve in mid 1988. The Navy noticed that the Naval Reserve Captain (O-6) community had exceeded its mandated ceiling, with captains remaining in an active status administratively, although they no longer drilled. To solve the problem, the Navy will start continuation boards in FY89, with those officers weeded out being told to retire or move to the Standby Reserve.

The beginning of 1989 saw the Naval Reserve looking at a changing world. Plans were announced for Soviet warships to visit Norfolk in April and US warships to make a reciprocal port call in Vladivostok. With these changes, some observers began questioning how long the Navy would continue aiming towards a 600-ship fleet, and whether the Navy might not start downsizing. Among those leading the questioning was former SecNav CAPT John Lehman (Northern VA Maury Chapter), who called for putting more ships into the Naval Reserve Force rather than retiring them, and having Naval Reservists performing their drills once or twice a year for two- to-three-week periods. The Pentagon did not immediately act on John's suggestion, but the budget request for FY90, and plans for decommissioning some of the older Charles Adams (DDG-2) and Coontz (DDG-37) class destroyers in 1990 did indicate a fleet dropping in size.

But even if the fleet was dropping in size, the NRA wasn't. In January 1989, CAPT Lou Anciaux (Raleigh Chapter) became Life Member 10,000. Lou was a past president of the San Diego Chapter and currently awaiting a voluntary recall to active duty to serve at the Portsmouth Naval Shipyard. At this point, of the 10,000 NRA'ers who took Life Membership 9,800 are still on deck and kicking, including Number One, LCDR Marie Frauens.

Some senior NRA members passed on during the early part of 1989, including CAPT Blaney Turner, NRA's first National President. Blaney, a Vice President of Aetna Life & Casualty Company in Hartford, had not been active in the Association for some years, but he was one of the stalwart few in 1954 who founded the NRA.

Also departing on his final voyage was RADM Dave Griggs (Houston-Space Center Chapter) who was killed in the crash of a World War II-vintage aircraft he was flying over Arkansas. Dave flew aboard the space shuttle Discovery in 1985 and was in training to be the pilot of an upcoming space shuttle mission.

With all of its support, we can hope the NRA will be last as long as the Sovereign Military Order of Malta, the oldest military order in the world, dating back to approximately 1095. In late January, CAPT Bob Lunney (New York Navy Chapter) is invested a knight of Malta in a ceremony held at St Patrick's Cathedral in New York. The presiding official is retired RADM John Cardinal O'Connor, former Navy Chief of Chaplains.

The early months of 1989 saw many other NRA members making contributions or being recognized by their communities and other organizations. CAPT Tom Grassey (Monterey Chapter) received $750 and a Bronze Medal from the US Naval Institute for his essay "Justifying the Cost of Seapower," which he submitted for the Arleigh Burke Essay Contest. The contest, only recently named in honor of the former CNO and late member of the NRA's Kalmbach-Wardroom Chapter, was previously known as the USNI Prize Essay Contest.

CAPT Tom Wachtel, immediate past National VP—Health Affairs, was elected President and Chairman of the Board of Trustees of the American Burn Association. The 3,500 member association helps research in the prevention and treatment of burns. It also provides training in these subjects and sponsors publications in the field.

Another health care professional in the NRA, CDR Thomas Dembinski, stationed at Naval Dental Clinic Yokosuka, achieves a major professional goal when he becomes Board certified in Orthodontics. Normally this might not be a feat, but board regulations call for a dentist to work with and document the progress of 10 individual patients for six years—and not many Navy dentists remain in the same location or have the same patients moving with them from base to base. At least CDR Dembinski's wife can understand the difficulty. LT Elsa Dembinski also is a NRA member stationed at Naval Dental Clinic Yokosuka.

They and other Navy medical professionals will soon report to a new command, as the service announces plans to revive the old Bureau of Medicine and Surgery (BUMED), which was replaced by the Naval Medical Command in the late 1970s. Under the new policy, the Navy Surgeon General will both make and execute navy medical policy.

Another government servant making his name was CAPT Andy O'Rourke (New York Navy), who received the Pro Patria Award from Secretary of Defense Richard Cheney. Andy received the grateful thanks from SecDef for what all he had done to support all Reservists from all the services, while he was County Executive of Westchester County, New York.

As a brief aside, the Navy reported that 74 percent of eligible Navy people voted in the November 1988 elections. This was higher than the 50 percent of the eligible general public who voted. Although the survey didn't break down the mix between Reserve and active duty folk, it has long been known that Reservists tend to vote more than their active duty counterparts.

In another announcement concerning government employees, the US Public Health Service and the Department of Defense signed an agreement making several hundred officers of the USPHS Commissioned Corps available to replace active duty and Reserve military health specialists sent overseas during an emergency. To help implement these plans, the Bush Administration is preparing legislation to create a USPHS Reserve capable of mobilizing in emergencies short of war. The USPHS Reserve would have to perform monthly training drills and annual active duty. Initial plans call for a Reserve membership of approximately 150 medical and health professionals. The legislation also calls for a USPHS officer to be assigned to the Joint Staff to assist with planning.

The USPHS proposals fortunately do better than do calls from a Merchant Marine Panel, which has prepared a report stating that the nation's merchant fleet is fallen from hard times to dire straits, and that it would require a $13 billion program to rectify the problem. The Administration takes note of the report and files it away.

Another area of maritime deficiency is inshore mine countermeasures. Former Congressman (D-FL) and NRA member CAPT Bill Chappell (Northern VA-Maury Chapter) writes an article for NRA News, "Inshore Mine Countermeasures: How Naval Reservists Can Save the Day," noting that close-in mine sweeping would require losing ships at the rate of one ship per mine. He recommends procuring YP-683 class patrol craft for the Super COOP (Craft of Opportunity) Program. The craft would be located in ports throughout CONUS, Alaska, and Hawaii and manned by Naval Reservists. Unfortunately, Bill dies of cancer shortly thereafter. But other NRA members and Congressmen take up his call.

The funds for such vessels might come from the savings the Pentagon expects to receive from closing military bases. The House of Representatives approved plans to close 86 bases in the United States by 1994. Five other bases would be reduced in size and scope and 54 other bases would have to receive or lose some of their activities. Rumors abound that some of the bases being considered include some of the larger navy complexes.

Carrying out any directives regarding the Naval Reserve would be two NRA members, RADM Wallace Guthrie (Orlando Chapter), who became Deputy Director Naval Reserve on 1 July 1989. He will report to RADM James E. Taylor USN, who relieved RADM Neale Smith as Chief of Naval Reserve and Commander Naval Reserve Force two months later.

One of the first headaches they will have to deal with is a decision by the 3rd US Circuit Court of Appeals, that "inconvenience to the employer" can be considered in determining a Reservist's right to take leave for military duty. No federal court has taken this step before, and fears are that others might use the 3rd District's ruling as a precedent. Job requirements and employer pressure have traditionally been one of the major causes for Reservists leaving service—accounting for at least 33 percent of such cases according to a 1981 Pentagon survey.

NRA chapters across the nation showed their spirit in various ways in late 1989. Miami' Sidney Fields/Sunshine Chapter donated two large plaques to the Miami Naval Reserve Center, one for the Active Duty Sailor of the Quarter and another for the SelRes Sailor of the Quarter. The Fresno Chapter, along with other local civic organizations, helped sponsor the Fresno Water Tower to the memory of the late RADM George Bowman, the city's first electrical engineer and general manager of the water system. CAPT Shannon Jones, Chapter President, presided over the ceremonies.

In the good news department, the Navy announced that 24 FF-1052 class frigates would join the NRF during FY90-FY93. Sister ships have been in the NRF since 1982, and many Reservists are familiar with their systems.

The 1989 Summer EXCOM meeting at Maine Maritime Academy in Castine, ME heard two pieces of good news. The first, from CAPT John Rice, Chairman of the Building Fund Committee, was that the fund had reached 85 percent of its $2.5 million goal. The second, from CAPT George Beck was that a Navy officer's sword would be presented by NRA to an outstanding graduate of the Maine Maritime Academy's NROTC unit.

...Outgoing Chief of Naval Reserve RADM Neale Smith poses this question in an interview with NAVY TIMES: What good is a modern, well-trained Reserve if you can't use it? His reference is to operations in the Persian Gulf, where the Naval Reserve's minesweeping capability definitely was needed but no involuntary recall of personnel was made — because the President feared that a call-up would be seen as obligating the President to abide by the War Powers Act. He also points out that the trend toward shifting more forces to Reserve responsibility will only aggravate the problem of accessibility unless the political will to use call-up authority can be found.

Perhaps because of the questions of employer support and political will, the Chief of Naval Reserve decided to open the Individual mobilization Assignment (IMA) to the entire Naval Reserve as Training Category B. Selected officers and enlisted personnel who required little training and had trouble meeting the regular drill requirements, would perform 12 days annual training but be excused from the normal pattern of weekend drills. The naval Reserve medical community uses this program, but Naval use has lagged far behind that of the other services, which widely use the IMA approach.

Even if the political will wasn't always there, the Reserves were. When senior Pentagon officials flew out to the carrier EISENHOWER to observe the Naval Air Reserve's Carrier Air Wing 20, they were highly impressed with the proficiency and professionalism of the Wing. Noting the capabilities the Wing displayed, the officials said more and more senior officials were coming round to the idea that more burdens could be placed on the Reserves.

This news cheered the attendees at the 36th National Conference in Los Angeles. The conferees elected RADM Lester (Bob) Smith as national president. The South Bay (Los Angeles) Chapter, under CAPT Dick Abele, won the Jack Lewis Award for Outstanding Chapter, while the Ken Kalmbach Award for outstanding communications went to the Eleventh District under CAPT Gil Grigsby. CAPT Dick McCann and CAPT Ed Grandin received the award for individual membership recruits, while CDR Mary Jo Majors won the award for most new one-year members recruited.

The Conferees adopted several resolutions. These called for supporting the Mobile Inshore Undersea Warfare program upgrades, procuring sufficient ASW systems to bring NRF ships and aircraft up to modern standards, granting commissary privileges to VTU drillers on the same basis as pay drillers, concurrent receipt of retired pay based on years of service and VA disability compensation, and development and production of the F-14D and the V-22.

In another sign of the times they are a changing, the Pentagon decreed the standardization of terms describing training. Annual Active Duty for Training (ACDUTRA) is out, Annual Training (AT) is in. Special Active Duty for Training (SPECACDUTRA) becomes Active Duty Training (ADT). Regular drills become Inactive Duty Training (IDT). The change is mandated so all services speak with the same tongue.

Some NRA members saw proud moments in the personal lives. National President and Mrs. Ken Welch pin the shoulder boards on ENS Patrick Welch, as he is commissioned from the University of Missouri's NROTC unit.

In another family note, LT Robert Fitzpatrick (Constitution Chapter) commissions his wife, LT (NC) Genevieve Fitzpatrick, aboard the USS CONSTITUTION, Old Ironsides.

Past National President CAPT John Rice was elected Commander General of the Naval Order, relieving Past National President CAPT Fred Carl.

CDR Doyle Quisenberry, in his civilian role as CINCLANTFLT Logistics Plans Officer, receives an award for his suggestion to centralize computation of war reserve requirements utilizing the JDS/TPFDD data base. The suggestion is being adapted Navy-wide and might go service-wide.

Although no one knew it when the year began, 1990 would prove to be a major year in the history of the Naval Reserve. The year began in a spirit of celebration, with 3 March marking the 75th anniversary of the Naval Reserve.

To mark the anniversary, the US Naval Institute sponsored an essay contest. Entrants had to answer how the Naval Reserve should respond to budget cuts and what its roles and missions should be in a changing world.

One sign of a changing world was the fate of personnel. Enlisted Reservists facing high-year tenure cuts found out they would not be moved from the Selres to the IRR if they were using their Mongomery GI Bill benefits.

Those personnel in the Naval Reserve faced a growth in certain responsibilities, particularly medical support and anti-drug operations. In a March interview with Navy Times, Chief of Naval Reserve RADM James Taylor (MF Maury Chapter) said the Naval Reserve provided 22 percent of the Navy's flight hours and 25 percent of the steaming days in the drug war. Most of this work went on in the Caribbean and the Gulf of Mexico. RADM Taylor also said that physician recruiting was improving and would increase, as would support man-days.

Medical and sea-going personnel were not the only Naval Reservists or members of the NRA busy in early 1990. CAPT Mileva Hartman (MF Maury Chapter) was recalled to active duty as Staff Director for the Secretary of Defense Reserve Policy Forces Board.

Across the country, CAPT Bob Greathouse was elected first President of the new Whidbey Island chapter. Executive Director RADM Jim Forrest presented the chapter with its charter and noted how the chapter was being stood up when the common talk was of base closings and ship decommissionings.

Some members of the NRA could look at all the talk and put it in a historical perspective. One such West Coast officer was ADM James Russell USN (Ret) (Tacoma Chapter). ADM Russell was inducted into the National Museum of Naval Aviation's Hall of Fame for his contributions, which included winning the Collier Trophy for helping develop the F-8 Crusader fighter and a lifetime of aviation experience, dating back to winning his wings aboard the Navy's first carrier, LANGLEY.

As the Naval Reserve marked its 75th anniversary, the Spring EXCOM met in Washington from 1-3 March to hear the CNO, ADM Carlisle Trost; Deputy Director of the Naval Reserve, RADM Wallace Guthrie; CAPT Larry Korb of the Brookings Institution; Director of the Strategy/Plans/Policy Division RADM Phillip Smith; and Chief of Naval Personnel VADM Mike Boorda. For RADM Guthrie and CAPT Korb it was a homecoming of sorts, as both are NRA stalwarts.

The EXCOM also enthusiastically accepts the offer by LT Peg O'Connell (Glenview Chapter) to sponsor an annual award to the best Naval Reserve Intelligence Unit. The award will be named for her late husband, Past National President CAPT George O'Connell.

The first day of the EXCOM coincided with a reception and luncheon at the Hart Senate Office Building to commemorate the 75th Anniversary. Members heard a Sense of the Congress Resolution read formally recognizing the 75th anniversary. Further ceremonies were held at the Navy Memorial on 3 March with a formal ceremony presided over by newsman CAPT Peter Hackes (Kalmbach-Wardroom Chapter). These events were capped off that evening with a formal dinner and ball at Union Station.

During these events, members paid homage to the 15 Naval Reservists who were awarded the Congressional Medal of Honor. The 15 included ENS Augustus Joseph Sullivan, ENS Charles Hammann, LT (MC) Orlando Henderson Petty, and LCDR James Jonas Madison during World War I; LTJG William E. Hall, PH 2/c William David Halyburton, WT 1/c Elmer Charles Bigelow, LTJG Rufus Herring, ENS Francis Flaherty, PH 3/c Jack Williams, ENS Herbert Charpoit, LT Arthur Murray Preston, and ENS John Joseph Parle during World War II; and LT (CHC) Vincent R. Capodanno and LTJG Robert Kerrey during the Vietnam Conflict. Six of the men, Halyburton, Flaherty, Williams, Jones, Parle, and Capodanno, were awarded posthumous medals.

Capitol Hill also saw an appearance by CAPT John Lehman, former Secnav and active NRA member. John pushes the idea that the fleet should perform shorter deployments, concentrate on duties closer to home, and mothball older warships. Many of the ships remaining should be manned by crews with a 50-50 mix of active/Reserve personnel, with Reservists filling in for active duty personnel in as many other areas as possible.

John might get an argument on some of his points from CAPT Jay Avella (MF Maury Chapter) who is serving on an extended tour with the Office of the Assistant Secretary of Defense Reserve Affairs. Based on his work, Jay writes a letter to the NRA, with copies to DoD officials, pointing out problems with Title 10 US Code 673b, the law governing Presidential Recall Authority over the Reserves.

Jay points out that use of the recall authority is both a military and a political decision with domestic and international ramifications. Domestically it could affect relations between the White House and Congress about the use of military force, while internationally it could send "the wrong signal."

This, in turn, led to a question of the contradiction between recalls that were both politically practical and militarily justifiable. According to CAPT Avella, while the Persian Gulf was militarily sound, it was not politically viable. Conversely, Panama was politically sound, but not politically viable.

Swearing in a new team.

Another problem Jay discussed was that the active duty force might not want to see a successful recall. With falling budgets and personnel cutbacks, a successful recall might spur further cutbacks in the active force.

Jay also noted that the law is restrictive, as it extends to units and individuals not in units, but not to portions of units, as with a Ship Augment Team in a Readiness Unit. He noted that during the recent Panama Crisis, individuals in units could not be called unless their entire unit was recalled. This sometimes led to mobilizing people and leaving them with nothing to do.

To ameliorate this problem, the paper recommended changing the law's language to allow the recall of up to 200,000 Selected Reservists whatever their assignment or location, and deleting the language specifying "units and individuals not assigned to units."

The paper recommended that the Presidential Recall Authority should not be used as a blunt club, with massive recalls, but rather as a rapier, with recalls of as many or as few personnel as required.

While few could see it at the time, these suggestions, and the entire Presidential Recall Authority, would be highlighted within four months.

Several senior officers found themselves receiving promotions in early 1990. CAPT W. David Sullins (MSC) (Atlanta Chapter) and CAPT Maryanne Ibach (NC) (MF Maury Chapter) became the first Reserve officers selected for Flag rank in the Medical Service Corps and Nurse Corps respectively.

The spring also saw RADM Jim Carey (MF Maury Chapter) and Commander REDCOM Six, knighted in the Military and Hospitaller Order of St. Lazarus of Jerusalem. The knighting ceremony took place in New Orleans.

Shortly afterwards, CAPT Charlie Buesner was piped over the side at Headquarters, where he had been the Legislative Assistant for many years. His relief was well acquainted with the Washington scene, being none other than past National President RADM Phil Smith.

Two Northeastern NRA members found themselves honored for their work benefitting all servicemen and women, active and Reserve. CAPT Emile Bonnot (New York Navy Chapter) and LCDR Myron Smith (Constitution Chapter) are elected to the Emiritus Board of the Armed Services YMCA. The Board's members have all performed major services during past years for the Armed Services YMCA.

June 1990 saw CAPT Ted Mason, National Awards Chairman and a member of the Burke Chapter, presenting an award of a different sort. He commissioned his son Andrew Mason. Andy immediately became a member of the Association as well.

The summer EXCOM, was held earlier than usual, on 29-30 June in San Francisco. The early date and location was in consideration of those members of the NRA who would be attending the Naval Order's Special Centennial Congress on 1-4 July. While our sister organization normally holds its Congress in the fall, this year's was timed to coincide with the Centennial, of the Naval Order's founding on 4 July 1890. In charge was CAPT John C. Rice, Jr., NRA Past National President & current Commander General of the Naval Order.

The EXCOM agrees to recommend a dues increase, and upon specific recommendations to the National Conference including creation of a National Vice president for Supply Corps Affairs.

Some EXCOM members headed back to Buffalo that same month to dedicate a compartment to the Naval Reserve, in honor of the late past National President CAPT Winfield (Chip) Chapin aboard the former guided missile cruiser USS LITTLE ROCK, currently a museum ship in Buffalo, NY. The compartment, set up like an office, will display much of the memorabilia Chip acquired during his service to the Navy and the NRA.

Other honors that same month went to IPAC 0319, an Intelligence Unit that drills at Nellis AFB. The unit, commanded by CDR Gerry Turner (South bay Chapter) receives the "Best Reserve Production Unit" Award from RADM Gene Dickey, also an NRA member, and Commander Naval Reserve Intelligence Command.

Early August 1990 found most Naval Reservists on vacation or suffering through the dog days of summer. But on 2 August Iraq invaded the nation to its south, Kuwait, and matters changed. Three days later Saudi Arabia, fearing a similar attack, asked for United States assistance, and Operation Desert Shield was announced. Two Carrier Battle Groups were moved to the Gulf of Oman, while elements of the 82nd Airborne Division and the 1st Marine Expeditionary Force were airlifted and sealifted to Saudi Arabia.

But these forces were insufficient for stopping a full-scale invasion, and on 22 August, President Bush exercises his recall authority to call-up Selected Reservists. The first call-up authorized the recall of 6,243 Naval Reservists, 3,000 Marine Corps Reservists, and 1,250 Coast Guard Reservists exchange their civilian clothing for uniforms. But hundreds of other Reservists volunteered their services.

The actual call-up, on 24 August, saw over 3,000 Naval Reservists recalled. These included 2,400 medical personnel, 400 Mobile In-Shore Undersea Warfare specialists, 60 Naval Control of Shipping Reservists, and 50 minesweeper crew members. Some personnel immediately were sent overseas, while many others, particularly in the medical communities, found themselves taking the place of active duty medical staff at Bethesda and Oakland Naval Hospitals who were being deployed aboard the hospital ships MERCY and COMFORT.

By early September another 3,000 Naval Reservists had been recalled, including approximately 600 cargo handlers, 360 intelligence personnel, 40 Combat Search and Rescue specialists, 1,700 logistics support personnel, and 300 staff personnel.

Some Naval Reservists manned 24-hour action centers at Military Sealift Command headquarters in Washington, D.C. Across town, other Naval Reservists established a crisis action center at the Navy Annex to handle any personnel requirements.

While many Reservists were being recalled, or waiting for a call-up notice, life continued as it had for many others. NR COMSEVENTHFLT won the Meritorious Unit Commendation, a rare honor for a reserve unit. The citation recognized the more than 6,000 man-days of ADT fleet support in 1989 and its contributions to PACEX '89 when it worked with Japanese, Korean, and Thai military personnel and aboard the USS Blue Ridge, COMSEVENTHFLT's flagship. The award was accepted by the unit's commanding officer, CAPT Kenneth Fisher (Dallas Big-D Chapter).

Also on the West Coast, CAPT Patrick Buttner received his Convoy Commodore designation. Pat has been a merchant mariner for more than 22 years and a San Francisco bar pilot since 1978.

One NRA function on the West Coast is rather sparsely attended. The San Diego Chapter, led by CDR Elaine Allen, holds a "Diamond in the Rough" Award ceremony to honor 75 of the area's most promising junior officers. Although many are unable to attend because of other commitments, those who do receive an originally commissioned, framed, and personalized certificate personally presented by CNAVRES RADM James Taylor. Those who didn't attend will have to wait for "Mail Call" to get their certificates.

Back in Washington, LCDR Kathryn Sullivan (MF Maury Chapter) presents a miniature Naval Reserve 75th Anniversary poster to NRA headquarters. The poster's claim to fame—it flew aboard the space shuttle Discovery earlier in 1990 when Kathryn, an astronaut, made a space flight. Kathryn is also the first American woman to walk in space.

Operation Desert Shield did not deter the NRA from holding its 37th National Conference in Dallas on 6 October. Speakers included the CNO ADM Frank Kelso, RADM James Carey (in his civilian role as Chairman of the Federal Maritime Commission), CNAVRES RADM James Taylor, Deputy Assistant SecNav for Reserve Affairs Donald Morency, Director of Naval Intelligence RADM Thomas Brooks, Deputy Director Naval Reserve RADM Wallace Guthrie, and Mrs. Cindy McCain, standing in for her husband Sen John McCain (R-AZ).

The Conferees decided to increase dues to $25 for annual members and $60 for three-year memberships. No change was voted for Associate Member or Life Member dues. Conferees vote down the proposal for a National VP for Supply Corps Affairs and a proposal to shorten the 3-months lead time used to determine membership strength, and thus, Chapter and District voting strength at the Conference, however this will be enacted later.

The Conferees commended the President for his action in Operation Desert Shield. They also called for maintaining a substantial number of FF-1052 class frigates in the NRF, transferring active duty HSL squadron helicopters to the Naval Reserve. They recommended basing frigates at other than traditional homeports, establishing a third carrier air wing in the Naval Air reserve if two active wings were decommissioned, and transferring the entire mine countermeasures mission to the Reserve.

The Conferees also called for strengthening the US Merchant Marine, and establishing a Harbor Mine Hunter program using small, low-cost, shallow-draft vessels. Other resolutions called for treating the retirement system as a good faith contract and not reducing COLA adjustments, and funding upgrades for LAMPS Mk 1 helicopters and FF-1052 class frigates.

It would be hard to find a member of the US armed forces reserves who didn't remember where they were at the beginning of 1991. As the year opened, more than 100,000 Reservists from all the services had been recalled for Operation Desert Shield. On 16 January Operation Desert Storm commenced. More than 100 warships were operating in the Red Sea/Persian Gulf region, including six aircraft carriers and two battleships. Aboard the ships were 75,000 sailors.

More than 20,000 of these were Naval Reservists, with the majority coming from the Medical, Seabee, Minesweeping, and Military Sealift Command communities, but practically every community was affected, and every Naval Reservist knew someone who had been recalled.

Some members of the Naval Reserve and NRA found pressing matters on the home front. CDR Kip Nicely (Rio Grande Chapter) was elected to the New Mexico State House of Representatives.

Other NRA members also found themselves performing major civic functions. In Seattle, CAPT Mike Montgomery, past District President of the Thirteenth District is named SEAFAIR Volunteer of the Month by SEAFAIR, an annual civic festival with more than 50 events.

NRA members also spearheaded the drive by Navy and Marine Corps Reserve space community personnel to establish a memorial to NRA member and astronaut RADM Dave Griggs. The memorial is a scholarship fund to send needy youngsters to Space Camp. Space Camp, located at the Space & Rocket Center in Huntsville, AL and at the Kennedy Space Center, FL, encourages youngsters to prepare for careers in science, high technology, and space.

Meanwhile, in Nebraska, CWO4 Dale Andersen (Fort Omaha Chapter) is elected President of the University of Nebraska College of Arts & Sciences Alumni Board.

CDR Joseph deVito (Commodore Barry Chapter) also had pressing duties on the home front. He commissioned his son ENS Joseph Peter DeVito into the JAG Corps USNR and the NRA.

And some NRA father-son pairs serve together. LT George Pearson is a retired volunteer serving under non-pay/non-credit permissive orders in NCSO Korea in Forest Park IL. The unit's Admin officer happens to be CDR Robert Pearson. George has Navy and Merchant Marine experience dating back to World War II.

Bad News. The Pentagon Inspector General slams the Navy's management of the Selected Reserve and says it wastes millions on dollars while providing substandard combat readiness. Apparently, some Naval Reservists are in drill-pay status but lack mobilization billets, and are not counted against unit readiness, and will not deploy with the units. The IG recommends terminating drill pay for these people, establishing procedures to preclude the commencement or continuation of drill pay for such Reservists, and placing a moratorium on SelRes billet structure changes until all personnel either have a mobilization billet or have been removed from the SelRes.

On February 24, 1991 Iraq surrenders and Operation Desert Storm ends. As of that date, more than 21,000 Naval Reservists had been recalled to active duty, with 6,856 being deployed to the Middle East. The Marine Corps saw 30,586 Reservists recalled and 13,500 sent overseas.

The end of the war finds many Naval Reservists preparing to return to civilian life. One who is returning to civilian life (but was not mobilized specially for the Persian Gulf crisis) is RADM Al Rieder. He leaves active duty as Commander Naval Base Philadelphia and past Surface Reserve Force and joins the NRA Headquarters Staff to organize and lead an NRA program to assist recalled Reservists being demobilized.

RADM Rieder briefs the plan to the Spring EXCOM in San Diego in late March. The EXCOM also votes to provide a special certificate or some other recognition to all Naval Reserve officers who served in Operations Desert Shield/Desert Storm and provide complimentary memberships to those officers not already enrolled.

Late spring sees the NRA Special Certificates of Appreciation completed and endorsed by the Chief of Naval Reserve. All officers recalled for Desert Shield/Desert Storm will receive certificates, as will other NRA members recalled to active duty. The certificates will be distributed by local NRA Chapters, in coordination with drill site commanding officers.

One of the ways returning Reservists are being helped is through the Reserve Transition and Recognition (RESTAR) program. The program will be directed by NRA member RADM James Taylor, in his role as CNAVRES. Phase 1 program development project leader was CAPT Mary Jo Majors, working under the direction of CAPT Ernest Milner, both NRA stalwarts. Phase II will see four RESTAR Transition Managers chosen to implement Phase III, where returning reservists will be welcomed and assisted as necessary.

LCDR Arthur Loeben (Philadelphia Chapter) receives a letter of Commendation from the U.S. Attorney for the Eastern District of Pennsylvania. The letter thanks Arthur for his work as a Special Assistant U.S. Attorney from 1987 through 1990 handling cases involving personnel at local military facilities.

Another member of both NRA and the legal profession is RADM Bob Wiss (Great Lakes Chapter) and a past National JAG/VP-Legislative Affairs. He has been nominated and confirmed as a Judge on the U.S. Military Court of Appeals. The seven-year assignment will require him to move to Washington, D.C.

Yet two more NRA JAG types are busy this spring. CAPT Norman Miller (Phoenix-Saguro Chapter) and CDR Duane Benton (Spirit of St. Louis Chapter) are two of five USNR JAG officers chosen to research the laws impacting recalled Reservists. One of their findings is that federal law does not provide for staying or suspending state proceedings in which a service member is involved. They draft the legislation to fix this glitch.

Headquarters itself receives recognition, when NERA executive Director HMCM Manny Ratner and NERA President MMC Larry DiPasquale present NRA with an Award of Appreciation in recognition of its many years of assistance and cooperation. NRA Executive Director RADM Jim Forrest accepts the bronze plaque in behalf of the Association.

The FY92 Defense budget, now being prepared, calls for an FFT-1052 Training Frigate program. Eight Knox class frigates will be used to train both their own SelRes augment crews and four other crews of 10 officers and 55 enlisted. Each of these crews will provide the core mobilization crew for an inactive FF-1052 class frigate. These frigates will be maintained in a status permitting their reactivation within 180 days.

The FY92 budget contains many cuts for the active duty and Reserve forces. In light of this, NRA has a long list of recommendations. Key proposals include retaining Squadron Augmentation Units, retaining the current NRF structure, and maintaining SelRes strength at 153,400. Other recommendations call for considering reserve basing when making Base Closure decisions, increasing the appropriation for SH-2G helicopters, maintaining 13 P-3 Naval Air Reserve squadrons, and increasing funding for ASW upgrades for NRF ships. NR also wants the COOP program retained and a new class of Harbor Mine Hunters funded.

CAPT Marianne Drew (M.F. Maury Chapter) and Deputy Assistant Secretary of the Navy (Force Support and Families) is the guest speaker in April for the Seventh Annual Awards Luncheon of the Westchester County/Avon Women's Hall of Fame in Rye Brook, NY. Welcoming her to the new York suburbs is Westchester County Executive CAPT Andy O'Rourke and LCDR Phyllis Zagano, both of the New York Navy Chapter.

CAPT Jim Hannagan from Michigan is elected ROA's National President.

Other honors go to CDR Richard Bradley (Hampton Roads Chapter), who receives the Defense Logistics Agency's first Outstanding Navy Reserve Individual Mobilization Augmentee Award. DLA Director LGEN Charles McClaudland, USAF, presents the award, along with a Joint Service Commendation Medal. Richard is being recognized for designing and implementing a fuel consumption projection system. The system, already being implemented by CINCLANTFLT, CINCPACFLT, and CINCUSNAVEUR, is being considered by the Joint Staff for use by all the services.

In other award news, CAPT Thomas Baer (Knoxville Chapter) receives the Navy League's Stephen B. Luce Award at the Naval War College. The award goes to the outstanding student at the College of Naval Warfare and is based on academic performance, participation in Naval War College and community activities, and promotion of the Navy and the Marine Corps.

Florida sees a migration of Naval Reservists in the autumn. They first come to Fort Lauderdale to the Naval Order's Annual Congress. Here, Naval Order Commander General (and past NRA National President) CAPT John Rice heads a panel discussing Operations Desert Shield/Desert Storm. Many senior officers from the sea services are present to discuss what occurred.

Shortly afterwards, the NRA holds its 38th National Conference in Miami Beach. CAPT Robert Lunney (New York Navy Chapter) is elected National President. One of his first items of business is to report that the building fund has received $2,617,242.

The 1991 Conference revises four resolutions from the 1990 conference. These call for maintaining a large number of FF-1052 class frigates in operational service, but with more Reservists in the crews; and transferring decommissioned active duty light ASW helicopters to the reserve. Other resolutions call for basing some of the FF-1052 and FFG-7 NRF ships at the strategic homeports and on the Great Lakes and for transferring the entire mine countermeasures mission to the Naval Reserve, along with all Surface and Air Mine Countermeasure assets.

Conferees also approved a number of new resolutions. They called on the navy to consider Reserve manning and readiness before closing any Reserve facilities and maintaining the current NRF force structure until Congress analyzed the implications of such moves on Total Force structure. The Conferees wanted the Navy to use the new Training Carrier, USS FORRESTAL, as a Reserve Training platform, as long as that did not interfere with training naval aviators. Equipment-related resolutions called for continued funding for ASW AQH-11 combat mission recorders, SQQ-T1 sonar trainers, and SQR-17A signal processors. The Conferees recommended maintaining at least five T-AGOS class warships in the NRF. Another aviation-related resolution called for establishing two Naval Reserve VS squadrons with S-3B aircraft and kits.

Autumn 1991 saw the Pentagon announce that the Navy was down to approximately 450 ships and would go down to 414 ships by 1997. On the brighter side, RADM Wallace Guthrie, Deputy Director of the Naval Reserve and an NRA member, announced that of 21,000 Naval Reservists recalled for duty during Operations Desert Shield/Desert Storm, only three failed to show up and only six required exemptions. These totals did not take into account the hundreds of volunteers who requested and received orders without going through mobilization.

The year 1991 rings down on several major changes across the world that affect the U.S. Naval Reserve. The Soviet Union formally dissolves, marking the end of the Cold War. In the Philippines, Mt. Pinatubo's explosion leaves Naval Base Subic Bay badly damaged and causing the Navy to decide to leave the base as soon as possible. And in the United States, the FY92 Defense Authorization Act decrees that starting with the class of 1996, all US Naval Academy graduates will receive Reserve commissions.

TAR problems were one source of friction between the active and Reserve sides of the house in 1992. A Navy working paper recommended eliminating the TAR program, giving control of Naval Reserve personnel and units to the Gaining Commands and letting the Gaining Commands have control of the appropriate funding. This suggestion had been raised and batted down in 1977, and it has found supporters and opponents in both the Active and SelRes communities.

The Navy Memorial between 7th and 9th Streets on Pennsylvania Ave. in Washington, D.C. is completed after a three-year construction program. Dedication ceremonies find a large NRA turnout.

At the Spring EXCOM in Albuquerque, members decided to increase the amount of financial assistance available to a Chapter being newly established of renovated, raising it from $100 to $150. The EXCOM also votes to approve the mergers of the Fort Worth Chapter with the Dallas Big-D Chapter and the Long Beach Terminal Island Chapter with the South Bay Chapter.

Mine Countermeasures again hits the headlines when CAPT Frank Avigone (Charleston Chapter) testifies before the House Armed Services' Committee's Seapower Subcommittee. Frank, who has commanded a COOP squadron and been in other MCM units, tells the committee the COOP program must be retained because of the possibility of terrorist mining and the proven threat mines hold for surface ships. He recommends that the YPs being used be replaced by 75-ft fishing boats with current combat suites.

Despite the controversy about mine countermeasures, many members note the decommissioning of the battleship MISSOURI, the last battleship on active duty.

While battleships depart and mine countermeasures face turmoil, the carrier fleet steams on. In May, the commanding officer of the KITTY HAWK, CAPT James Maslowski, visits Minneapolis to brief his augment unit, NR CV-63 Kitty Hawk 0178. CDR Dave Niedermaier (Twin Cities Chapter), CO of the unit, meets CAPT Maslowski, his training officer, and command master chief. According to CDR Jim Thurmond (Twin Cities Chapter) and CO of NAVAIRESCEN Minneapolis, no other carrier CO has ever made the trip to brief his Reserve unit—its always the other way around.

Yet another transition sees RADM John Natter (Birmingham Chapter) relieve RADM Wallace Guthrie as Deputy Director of Naval Reserve. RADM Natter enters office having to analyze the NRA's FY93 legislative recommendations. These call for retaining a USNR end strength of 141,545, maintaining 13 P-3 Reserve squadrons, retaining the COOP program, instituting a construction program for inshore MCM vessels, and replacing the NRF MSOs with more modern MHC or MCM craft. Conferees also recommend retaining augment units for active ships and squadrons, passing ROPMA, using the carrier FORRESTAL to train Reservists, and reactivating the FFT-1052 Training Frigate program. The conferees also recommend funding many items of air and surface equipment that were not funded in the President's budget.

Once again the commendations come forth. Naval Reserve Special Warfare Unit Two, in Kearney, NJ, receives the Meritorious Unit Commendation for its work during Operations Desert Shield/Desert Storm. Accepting the award on behalf of the unit is commanding officer CDR James Ayres (Neptune Chapter).

Members of New Jersey's Halsey Chapter are active in another way, when CDR Robert Herman (CEC) commissions his son, ENS William Herman (CEC) at Arlington National Cemetery and enrolls him in NRA.

When the Summer EXCOM meets in Buffalo, it recommends a dues increase for the National Conference. It also votes to recommend that the Puerto Rico and Virgin Islands be incorporated into the Sixth District.

NRA members and Naval Reservists can be found wherever the Navy is active. In that spirit, CAPT Alexander Monroe (ADM Byrd Chapter) of the Naval History VTU deploys to Guantanamo Bay, Cuba to document the Haitian boat lift operation. The Detachment gathers oral history interviews, takes photographs, and collects artifacts from participants. Also working on the project at CINCLANT is CAPT Bill McClintock (Hampton Roads Chapter). Bill has a double interest in the project because he serves in the unit and his civilian job is CINCLANT Command Historian. Other unit members assist the project at the Naval Historical Center in Washington, D.C.

In the spring of 1992, NRA Sixth District President CAPT Bill Waylett Jr. is transferred out of the District. His relief is CAPT Myles Eastwood (Atlanta Chapter) who had been National VP-Legal Affairs.

Also in Washington, CAPT Arnold Pock (MF Maury Chapter) reports that the condition of records he saw during a recent selection board was atrocious. Missing fitness reports, old photographs, no photographs, missing orders, missing AQQs, awards reported in fitness reports with no supporting documentation. But all is not the fault of the officers involved. Some reporting seniors write narratives greatly at odds with the grades they give. Seems some folks aren't taking advantage of the NRA Jacket Check or Record Review.

Almost at the same time, the M.F. Maury Chapter holds a farewell reception for Rear Admirals James Taylor, Wallace Guthrie, and Jim Carey. The three Maury Chapter stalwarts are leaving their posts as Chief of Naval Reserve, Deputy Director of Naval Reserve, and Commander REDCOM Six respectively.

The three JO Award winners.

Another retirement is that of CWO4 Fred Burkle (Raphael Semmes Chapter). Fred, with 37 years' service, is the Senior Warrant Officer in the Navy, and one of the longest-serving individuals at this date. He has been in charge of maintenance for VP-94 since the squadron was stood up in 1970. Other significant service includes tours on E7/E8/E9 selection boards and the National Naval Reserve Policy Board.

Summer also sees the 39th National Conference in Chandler, AZ. The Conference agrees to sponsor the annual Bea Ratner Award for the outstanding Reserve Junior Nurse Corps Officer on inactive duty. Other key decisions see Puerto Rico and the Virgin Islands incorporated into the Sixth District rather than being at large and dues raised from $25 to $30 for one year and $60 to $75 for three years.

The Conferees also vote to make major constitutional changes. Chapters can no longer predesignate delegates to the Conference. Each regular member who is registered for the Conference and in attendance will be a voting delegate. Proxies will not be allowed for Chapters or individuals. Chapters that are not represented will have their voting strength distributed amongst the other delegates from the Chapter's District. Each chartered Chapter's voting strength will be based on one vote for every 50 regular members. The final change called for voting procedures at District Conferences to be the established in the same manner as at the National Conference.

The Conferees also adopted a number of resolutions. These called for maintaining the current Reserve Force strength and structure, including the COOP program, P-3 force levels, Light Helicopter Squadrons and SelRes end strength of 141,454. Before closing Reserve bases, Congress should carefully analyze the NR manning and readiness. Conferees called for the formation of a Joint Reserve/Guard Drug Demand/Reduction Task Group to conduct planning and coordination of these operations. The JROTC program should be supported. The ban on homosexuals in the military should be continued. Material upgrades should include the LAMPS Mk I helicopter upgrade, AQH-11 ASW mission recorders for SH-2Gs, AQS-14 trainers for NRF operators, AQA-7B(V) 18/19 IPADS equipment for P-3B squadrons, and AQS-14 and Magic Lantern mine detection sets for active duty and reserve forces. C-2 and C-130 aircraft procurement should continue.

Other resolutions called for some form of managed care, such as CHAMPUS Prime, for military retirees, and a self-funded dental care plan. Members called on DOD not to reduce the age of 65 for termination of CHAMPUS benefits, and provide all military retirees, including those over 65, with voluntary enrollment a selection of plans. This would include making CHAMPUS a second payer for military retirees over age 65 who had to convert to Medicare.

Immediately following the Conference, an EXCOM is held. The EXCOM votes to make several Chapter changes. In New York State, Staten Island is merged into New York Navy Genessee Valley-Rochester is deactivated and its members reassigned to Central New York-Syracuse or Western New York-Buffalo. Further south, the Cumberland Chapter which was never chartered, is deactivated and members reassigned as appropriate. The US Naval Academy-Annapolis Chapter sees a similar fate—disestablishment and reassignment of members. Other Chapters being deactivated/merged included Everglades-Fort Lauderdale, which was being merged with Sidney Fields/Sunshine-Miami. On the West Coast, the three San Francisco Bay area chapters, FADM Nimitz/Golden Gate, Polaris, and Jack London, were to be combined into one FADM Nimitz Chapter.

Two Naval Reservists from NR NAVSPACECOM 0166 flew the space shuttle Discovery in 1992. Unit commanding officer CDR Steve Oswald (Space Center Chapter) served as pilot while Chief of Staff CDR William Readdy served as Flight Engineer.

The aviation community also sees the Silver Falcon statue passed down , as RADM Jack Smith (Atlanta Chapter) retires and turns over the title to RADM Sam McWilliams (Nimitz Chapter).

With the growing concern over drug abuse among school children, it was only a matter of time before the Naval Reserve became involved. Autumn 1992 sees RADM Scott Logan (Commodore Isaac Chauncey Chapter) appointed as the Naval Reserve's Campaign Drug Free Flag Advocate and program sponsor. During Campaign Drug Free, members of the Naval Reserve use video tapes and live presentations at elementary, middle-, and high-schools to warn youths of the danger of illegal drugs.

For the NRA, 1992 ends on a happy note when CDR Odette Willis (Nimitz Chapter) and current National VP-Health Affairs becomes Life Member 11,000.

Another year, another fight. As 1993 opens, the NRA and other Reserve organizations lobby against a report from the Rand Corporation, Assessing the Structure and Mix of the Future Active and Reserve Forces. The report states that only 65,000 Naval Reservists are required to fight two major regional contingencies, with a Navy of only 350-450 ships. No USNR personnel took part in the Naval analysis, which was conducted by the Center for Naval Analysis. The report adds a caveat that "having more Reservists than the estimated minimum would be a hedge against uncertainty" and that the support Reservists provide in peacetime is another argument for larger Reserve forces.

One person who might have argued the Naval Reserve's case unfortunately is no longer around. January also saw retired TAR CAPT Charlie Buesener, NRA's Director of legislation from 1984-1990, cross the bar.

But the ROA's Naval Affairs Committee, led by NRA stalwarts CDR Bill Legg and RADM Russ Gorman, is hiring retired specialists to conduct an independent study paralleling and challenging the Rand figures. The NRA also is assembling a panel to prepare a White Paper, From the Sea: The Role of the Naval Reserve. The NRA's panel, comprised of retired USN and USNR officers from a variety of designators and experiences, will present their own report of what the future Naval Reserve should look like and how it will be able to support the Navy.

All is not number crunching and worry. Beauty must also exist, to sooth the stormy waves. Towards this end, CAPT Torrey Sylvester (Constitution Chapter) donates an oil painting by the late LCDR Edward Stevenson USNR (Ret) to NRA. The painting displayed in the Headquarter's Conference Room, depicts a guided missile cruiser design of the late 1950s.

To better serve the Fleet, four NRA stalwarts have developed a deployable Naval Reserve Logistic Task Force Atlantic (LTFLANT). The LTFLANT consists of a headquarters staff along with an Advanced Logistic Support Site/Forward Logistics Site (ALSS/FLS). These will be used to augment an existing US facility or operate in an area where there is no US presence. The four officers who developed the plan are RADM Jack Smith, CAPT Doug Jeffords, CAPT John Totushek, and retired TAR CDR Doyle Quisenberry, who handles these issues as a civilian as CINCLANTFLT's Logistics Plans Officer. The LTFLANT will be led by two NRA members, RADM Stephen Yusem (Philadelphia Chapter) and RADM Bob Crates (Chattanooga Chapter) as Commander and Deputy Commander respectively.

When the 1993 Base Closure Commission presents its recommendations, the list includes 30 Naval Reserve Centers/Facilities, and another eight major bases hosting Naval Reserve activities. The Commission also recommends closing and consolidating three REDCOMs, Eighteen (Olathe), Two (Scotia), and Five (Ravenna).

The NRA White Paper is prepared rapidly, and briefed at the Spring 1993 EXCOM at Fort McGuire by VADM Robert Dunn (M.F. Maury). The conferees approve it and vote to have it rapidly published and promulgated. Other presentations come from RADM Al Rieder from Headquarters, describing the NRA College Scholarship Program and CAPT Tom Parke on the Navy's Drug Demand Reduction Program.

The White Paper, From the Sea: The Role of the Naval Reserve, makes several recommendations. These include:

- Decouple Reserve strength from active force levels so that lock-step proportional or share-the-pain force cuts do not preclude the most effective utilization of the Naval Reserve in the Total Force.
- Develop U.S. Naval Reserve initiatives that further integrate Naval Reserve personnel into day-to-day Navy operations.
- Review proposed Reserve installation closures in terms of demographics nd the effect upon the readiness of the Naval Reserve.
- Test Fleet CINC Operational Control of a limited number of Reserve units.
- Continue ongoing tests of gaining commands' direct control of Reservists on a limited basis until evaluated.
- Continue the selection of the Chief of Naval Reserve from the ranks of the best qualified flag-officer community.
- Continue the full time support TAR program.
- Undertake a formal and wide-ranging review of the Naval Reserve organization and administration.

The spring term at the Naval War College sees a new professor. CAPT George Waller (Palmetto Chapter) takes leave from his duties at Wake Forest University School of Law to serve as the Charles H. Stockton Professor of International Law. He will teach international law and maritime law and conduct research.

Another NRA member has been conducting research of a different sort, as CAPT Steve Smith, President of the Columbus, GA Chapter prepares a career Development program for Naval Reserve officer career guidance and planning. Steve's effort contains three major parts. The first uses outlines, program notes, handouts, and viewgraphs to describe the selection board system, managing records, and career planning. Part 2 provides an expanded version of NRF pamphlet P-1040.1 to describe different career paths, educational information, and general guidance. The third section contains an interview checklist and interview guide for officer career planning. The program is being evaluated by the Assistant Chief of Naval Personnel for Reserve Affairs.

The Navy Memorial has plaques describing the different facets and activities of the Navy, so it is fitting that NRA sponsored and funded a Twice a Citizen bronze plaque honoring Naval Reservists. More than 300 guests, including 70 USN and USNR flag officers gather to dedicate the plaque on 14 April 1993. The principal speaker is Chief of Naval Reserve RADM Thomas Hall. Also speaking is the guest of honor, NRA National President CAPT Bob Lunney. The dedication is held in conjunction with the NRA's annual Reserve Flag Officer Reception. The plaque, honoring all Naval Reservists, depicts a Naval Reservist boarding a frigate.

Library reception in Headquarters Conference Room.

To help ensure that all new recruits are dentally ready when they join the fleet, the Navy is sending 260 Dental Reservists to NTC Orlando for their two-week AT to help the active duty Dental Corps personnel at the base. CAPT Craig Marcello (DC) (Acadian Chapter) says it is the largest peacetime operation the Naval Reserve Dental Corps has ever faced.

Florida also sees a Naval Reservist and NRA member become the first woman selected to serve in a combat air squadron. LCDR Kathryn Hire (Melbourne Chapter) is assigned to VP-62, a P-3 squadron based at NAS Jacksonville. A 1981 Naval Academy graduate, she was designated an NFO in 1982 and transferred to the Naval Reserve in 1989. She works as a space shuttle test project engineer for Lockheed in her civilian career.

The Summer EXCOM in Seattle marks the first time an NRA national meeting has come to the Pacific Northwest, which is a growing Navy area with major bases at Puget Sound/Bremerton, Everett, and Bangor. Attendees hear three presentations, from CAPT Walt Steiger on the Campaign Drug Free program, from RADM Phil Smith on the From the Sea: The Role of the Naval Reserve White Paper, and from CAPT Jack Ringelberg Chairman of the Ad-Hoc Committee on National Meeting Scheduling. Jack recommends that there be two national meetings annually, a three-day event between 1 February and 30 April and a two-day event between 1 July and 30 September.

The summer also finds National President CAPT Bob Lunney testifying before the Senate Armed Services Committee's Subcommittee on Force Requirements and Personnel. Bob testifies that closing 38 Reserve installations has made it difficult for many Reservists to continue serving. The closure of many naval air facilities in the Northeast and Midwest adds to the problem, with many reservists having to travel hundreds of miles to a facility or stop participating in the Naval Reserve.

Other summer events see RADM James Miller (SC) USN (Ret) relieve RADM Bill Thompson USN (Ret) as President of the U.S. Navy Memorial Foundation. RADM Thompson has headed the Foundation since 1978. He now becomes Vice Chairman of the Foundation's Board of Directors. Both officers are members of the M.F. Maury Chapter.

In other Washington area personnel changes, RADM Al Rieder, who has been a special consultant at Headquarters, relieve CAPT Bob Beskind as Deputy Executive Director on 31 August. Bob and Renee Beskind retire to St. Augustine Beach FL and join the Jacksonville Chapter.

While some members retire, others join. CAPT Bill Mauck National VP-Legal Affairs and CAPT Sam Mauck (both of the ADM Claude V. Ricketts Chapter) commission ENS Victoria Mauck, a student at the University of Missouri (Kansas City) Law School. ENS Mauck immediately affiliates with the Santa Fe Trail Chapter.

For other NRA members interested in continuing their education, the Commandant of the Joint Military Intelligence College (formerly the Defense Intelligence College) briefed the NRA on the establishment of a weekend-format Postgraduate Intelligence Program for reserves. Students will be able to earn a master's degree at the end of a two-year part-time course of study. Plans call for the first class to begin in October.

The 40th National Conference elects CAPT Robert Bell of Florida as National President. The conferees adopt a plan calling for two conferences to be held each year. A National Naval Affairs Conference will be held between 15 January and 15 May to deal with external matters, particularly legislative concerns, while an Association management Affairs Conference will be held between 15 July and 31 October to deal with internal matters, particularly the election of officers. Only two EXCOM meetings will be regularly scheduled, either immediately before or after the two major conferences.

The end of the Cold War saw the US military holding exercises with former opponents of the Warsaw Pact, and NRA members have been active in this program. LCDR Eric Germain, with CINCNAVEUR Det 130 in London is participating the Mil-to-Mil Program, assisting former Warsaw Pact nations in transitioning their militaries to civilian control. Eric spends several weeks in Bulgaria before returning to London and preparing a report with lessons learned and recommendations.

One problem Eric and other members have faced is the lack of a Chapter in London. This is rectified in late 1993 when the London Chapter holds its inaugural session. CAPT Nick Von Baillou becomes the first President. Most of the members come from CINCNAVEUR Det 130, VTU 3001, and COMNAVACTS UK Det 230.

Another NRA officer providing lessons learned is CAPT James Myers (Space Center Chapter). After giving a seapower speech, Jim realized the presentation was out of date. After acquiring updated materials from OPNAV, he began giving the revised seapower speech to any audience he could find. The senior CNAVRES PAO heard about Jim's successes and requested he prepare a two-day six-hour mini-course on how to deliver speeches. The first course, before a group of senior Air Reserve officers, was enthusiastically received, and Jim has been teaching both seapower and speech power.

The end of 1993 finds CDR Doyle Quisenberry receiving his second Navy Civilian Superior Service Medal. This one, presented by CINCLANTFLT ADM Henry Mauz, is for Doyle's work and leadership in developing the two Naval Reserve Logistic Task Force units, one each for CINCLANTFLT and CINCPACFLT, which will organize Navy logistics at overseas locations when required. Doyle then receives a Logistics Task Force Atlantic plaque from commanding officer RADM Steve Yusem (Philadelphia Chapter).

Another NRA officer exercising leadership is LT Rich Kolko (Atlanta Chapter). But people are surprised when they learn how Rich exercises his leadership—as coach of the Puerto Rico Bobsled team in the XVII Winter Olympic Games in Lillehammer, Norway in February 1994. Rick is a previous Olympian, having raced with the US Luge team and won the U.S. national championship in 1985. Rick gave up active competition when he was accepted for P-3 pilot training in 1986, but in 1988 and 1992 he was asked to, and coached, the Puerto Rico Luge Team.

RADM James Taylor (MF Maury Chapter) and former CNAVRES, finds himself in a new leadership role when he becomes the Executive Director of the Navy/Marine Corps/Coast Guard Residence Foundation, in McLean, VA. The Foundation operates Vinson Hall, a retirement for military retirees that is open to retired Reservists, their wives or widows and their parents.

March 1994 sees the indefatigable CAPT Bob Lunney, past National President receive the Navy Distinguished Public Service Award from SecNav John Dalton.

NRA also holds its first Spring National Conference in March, in San Francisco. A two-day Junior Officer Symposium cosponsored by REDCOM 20 attracts more than 200 junior officers from the region, in addition to those JOs attending the Conference.

The Conference adopts resolutions calling for retaining all airborne mine countermeasures forces in their present configuration. If these forces had to be reduced, at least two squadrons with new MH-53E helicopters were to be retained. New state-of-the art aircraft were required to replace the C-9 fleet. JTFR Article U7150 had to be revised so that all Reserve personnel had TDY/TAD orders for inactive duty training requiring travel from a member's residence to duty station and back.

While some NRA chapters were eliminated in the 1990s, others were being formed. CAPT Dave Woods, Fifth District President and CAPT Harry Smith, Fifth District VP announced the formation of a new Chapter at NAF Andrews. Most of the cadre for the new Chapter come out of Reserve Intelligence Area 19.

On a good note, the Navy announces that VP-68, which flies the P-3C, became the first squadron to achieve 100,000 accident free hours. The major milestone occurred during a flight flown by VP-68's CO, CDR Byron Fisher (Kalmbach Wardroom Chapter). Byron is accompanied by Deputy Assistant Secretary of the Navy for Reserve Affairs (CAPT USNR) Wade Sanders, of the San Diego Chapter

.... NRA has been selected by the Department of Defense as a **World War II Commemorative Community** in recognition of its special programs honoring veterans of that war. For the next two years, the NRA will post the distinctive **WWII Commemoration Flag** at Headquarters, and sponsor activities throughout the United States honoring WWII veterans. National Historian CAPT John Wheeler is coordinating a **preservation program** for all NRA members with WWII memories.

Asst. SecDef (Reserve Affairs) Deborah Lee releases her long-awaited report, **THE FUTURE NAVAL RESERVE ROLES AND MISSIONS, SIZE AND SHAPE**. The report advocates that the Naval Reserve should become more fully integrated in all aspects of the Navy's roles and missions within the constraints of forward-presence requirements. More reliance upon the Naval Reserve is advocated. The study regrettably, confirms the Bottom-Up Review's projected Naval Reserve strength of about 100,000 SelRes, but states that the implementation of its recommendations could add another 10,000. Those recommendations are:

New Missions and Forces:

Transfer **two repair tenders** to the NRF from the 12 tenders being decommissioned.

If **Airborne Electronic Warfare squadrons** are redeployed to CONUS, the Naval Reserve could assume a role in the mission.

Provide **Seabees** with equipment to decontaminate ports, facilities, and airfield after a **chemical warfare** attack or **biological warfare** attack.

Expanding Current Missions and Forces:

Expand USNR involvement with **Integrated Undersea Surveillance System (IUSS)**.

Same for **Combat Search & Rescue (CSAR)** and **Naval Special Warfare**.

Reconfigure the Naval Reserve **frigate force** to a smaller, less expensive force and provide training for **augmentees**, with additional **Reserve crew members** assigned.

Sustaining Current Missions and Forces:

Increase the **ratio of Reserve** to total force fleet hospitals from 40% to 60%.

Airwing mix of 9 active and **2 Reserve**.

Adversary and electronic warfare functions **fully Reserve**.

Maritime patrol squadron mix of 12 active and **10 Reserve**.

Enhance **intra-theater airlift** capability.

Increase number of Reservists assigned to **Naval Reserve intelligence units**.

Allocate about **half** of the **mine countermeasures mission** to the Naval Reserve. (The program currently is scheduled to be consolidated at Ingleside, TX; this creates demographic and other challenges the Naval Reserve will need to overcome to make this a successful program.)

Peacetime Support to Navy Operations:

Navy should consider providing **additional funding** for peacetime support, and ensure that this funding not be transferred to other programs.

Navy should actively encourage and support innovations demonstrated by Naval Reservists: closer gaining command relationship, flexible drilling, streamlined command structure.

Reference is made in the January 1995 issue of the NRA NEWS in which the establishment of a library/research facility dedicated to Naval Reserve and Naval Reserve Association History was announced. Construction began in early February and NRA President CAPT Bob Bell dedicated the facility on 20 April 1995.

Immediately afterward, work began on cataloging books on hand. These came from a number of members most however from NRA Past President CAPT 'Chip' Chapin and especially NRA Founder CDR Frank S. Pohanka, Jr.

This was barely underway when Ambassador and former SECNAV J. William Middendorf, II visited Headquarters. Impressed by what he saw, the Ambassador donated 237 volumes from his personal library. Most were first editions, containing personal notes to him signed by the author. Shortly afterwards, he followed up with an authorization to purchase 100 books of our own choosing to improve the collection. The shelves were filling up fast, so fast in fact, that no general solicitation to send materials was made in the NEWS.

In short order two bonanza donations occurred. The first was the personal library of NRA Member LT Hugh Clenaghan. Hugh, a Merchant Mariner had left his collection to NRA Member CDR Lawrence D. Hawkinson who donated then to the Library some 52 boxes of books and periodicals relating to Maritime interests worldwide.

This was immediately followed by the donation of the personal library of naval historian and U.S. Naval Academy professor CAPT Michael T. Isenberg some 600 books from his office and 1200 more from his residence. We were given 1st option to select the volumes we wanted.

The upshot was we were chock-a-block in the library. At this juncture seven sections each with seven shelves were added in the adjacent corridor. These spaces too were all full.

Most of our members do not have vast book collections but they do have other items of Naval Reserve historical import. Here's a sample of what was brought in:

- The original film CHINFO prepared for Navy Sabbath which we originated.
- Histories of several Naval Reserve Centers
- Photos of original Lighter Team (LTA) units exercising
- Amphibious and Construction Battalion Units at work
- ASW operations flown by our Naval Reservists
- Corpsmen in action
- Photos & logbooks from WWI, WW II, Korea, Vietnam & the Persian Gulf
- Slides with narrative of Riverine craft in Vietnam

Two exceptional items are:

- A piece of the forward boathouse of USS ARIZONA. The Navy authorized sale of these to raise funds for the Arizona monument. One of our members bid at auction to get this for us.
- And a one of a kind item-the only Kamikaze aircraft propeller existing today. It struck the USS Hanley and was salvaged by then Ensign A.D. 'Gus' Caire, an NRA Life Member.

Our Founders too responded with their holdings. CAPTs John P. Bracken, Al Tiedemann supplemented their earlier donations adding to those of RADM Bob Burke, CDR Don Burnett, CAPT Hank Doyle, RADM Alvin Fisher, CAPT Tom Fortson, CDR Fred Kempe, CDR Charlie Stone, CDR Bill Waylett Sr. and most recently LCDR Russ Miller.

The Founders papers predate NRA since many were active in the Reserve groups which preceded NRA. These are primary source documents of the origin & establishment of the Naval Reserve itself as well as NRA's own beginning.

Concurrent with these Library efforts was the expansion and upgrading of our communication capabilities (computer upgrades, worldwide web, e-mail, in-house roster preparation, and other rapid response techniques).

One major area of continued growth is the jacket check and record review service. A corollary to this, is space for individual counseling. Our membership services will provide direct or referral capabilities for career patterns, job placement, spousal benefit assistance, pre-retirement briefing and how to, where to, info throughout.

Individual Naval Reserve Centers had Naval Reserve volunteers who gave their time to help those caught in the Berlin Call up and Cuban Crisis. NRA's unique referral office as manned by RADM Al Rieder at the time of Persian Gulf War was tremendously effective and without comparable offerings in any service active or Reserve.

The Naval Reserve Library at NRA Headquarters.

"Parade the Colors"

NAVAL RESERVE SPECIAL STORIES

A-6 Diary

by Lt. Allen H. Smith III

It has been over 20 years (two decades) and during that time, I have rarely spoken about my two air tours in Vietnam. I have never opened up to anyone about my very personal experiences in combat.

In a way, I guess I am a silent, non-medical victim of that very unpopular war. I came back on crutches, both physically and emotionally. The Navy and my buds thought I was a hero. But they were the only ones. I never really had a chance to discuss the war with the people I was "directly" fighting the war for or to receive their gratitude. After returning to Philadelphia, I soon learned I couldn't even vaguely discuss where I had been, let alone what I had done. When I did try, it ended in a serious shouting match, being berated or humiliated and sometimes even attacked.

I quickly disciplined myself to forget the three years on active duty and finally managed to survive by separating myself from who I had been and what I had seen and done.

So why am I now writing an article about my combat tours and my experiences as an A-6 pilot? I had been offered many opportunities to speak to veteran, Navy and educational groups, but I had always turned them down. So why now? Because this time, the circumstances are different.

I now have one last A-6 mission to perform: to honor and pay tribute to the plane that saved my life, that of several of my B/Ns (bomber/navigators) and thousands of other A-6 crews. I am now helplessly witnessing the horrible and inappropriate destruction of the A-6 community. It is being done all too quietly and without the tribute this plane deserves.

When asked to contribute my thoughts on the A-6, I seized the chance as one pilot's opportunity to glorify the A-6 with a first-hand account of what a truly remarkable aircraft it is. This diary is also intended to be a tribute to the men and women at Grumman who built her, the crews that flew the A-6, the personnel who maintained her and all of us who loved and were passionate about the Intruder.

The most mystifying part of any flight-training program is exactly how do you take your average B+student, an immature, beer-drinking, Saturday night party kind of guy/gal, and transform that person into a highly skilled tactical warrior? As soon as the transformation begins, they are immediately prepared to engage in mortal combat at a second's notice and never acknowledge the dangers. How does this happen?

The answer is discipline, total dedication and focus. These were the most distinguishable characteristics that my training as a naval aviator imparted to me: discipline, and the ability to focus my entire essence on the job at hand.

Starting in 1967, my freshman year, my transformation was rapid and comprehensive. I spent that summer as an AVROC II cadet in Pensacola, FL, and then came back home, broke my two-year engagement to my high school sweetheart, changed major from pre-med to aerospace engineering and prepared myself for a life dedicated to being a high-tech warrior shot off flat-top boats.

In the 30 years since that first summer, I have never lost that disciplined manner of thinking. But instead of using it to embrace Navy life and combat, I have used it to isolate myself from it and some of the memories it generated.

When I graduated from Penn State, I had already completed my two summers at Pensacola with the AVROC program. Since there were shortages of pilots, I was commissioned as an ensign within hours of graduation; then, three days later, off to VT-1 at Whiting Field in Milton, FL, to start primary flight training.

This was the most exciting time of my life, and I was so caught up in the anticipation of "catting" off a carrier that I never paid much attention to the moral or ethical aspects of modern warfare. It wasn't Navy doctrine to ignore that part of the job, and neither did they avoid blunt contact with the issue. But, I think, in retrospect, that it was just a human trait to go into denial.

Obviously, the idea of my going to Vietnam was a very delicate issue with my mother, compounded by the fact that my dad had passed away when I was just 15. Therefore, as the only son, at the age of 17, I already had an exemption from military service so I had to ask her to sign my enlistment papers. To get her to do that, I had to promise her that I would never see combat. Of course, the Navy never promised that, and it's a good thing she never asked me how I planned to deliver on that promise.

As I went through the Navy flight-training program, everything became increasingly more difficult and demanding - mentally. The hardest physical parts of my career were the weeks in the INDOC Battalion going through the P-T obstacle course and the hours we spent in the pool with the Dilbert Dunker and, later, in the Gulf of Mexico for survival training. After that, I found the physical work easy, and the brain work became the key to almost everything.

There were no classes to psychologically prepare us to drop bombs and napalm on people; we were trained to be technicians, manipulate and aircraft, to fly a profile and drop ordnance with surgical precision. But the consequences of these acts were almost incidental and left for us to sort out. Somehow, I managed to put all of it in perspective and compartmentalize and deny it all in the name of God and country.

The normal 20-month training program between VT-1 and my wings took only 15, because there was a shortage of attack pilots and we were put through a "must pump" program. After only seven weeks and 65 hours of flying, I had to choose a pipeline for final aircraft selection. I had gotten above-average grades in training, so I got to choose which pipeline I wanted. I chose Strike because I had studied all the aircraft in the Navy and wanted the A-6 over fighters, props or helicopters. I liked the A-6 because it was the only truly all-weather aircraft that could launch and recover without the pilot ever having to look outside. It carried a devastating bomb load of 18,000 pounds more than the legendary B-17 Flying Fortress, and enough fuel for an incredible loiter and mission profile of almost 2.8 hours. Above all, I liked the idea of a two-man crew with specialized duties and the side-by-side seating for crew coordination. However, I didn't like the idea that it was subsonic and had no permanent guns for self-defense. Although the A-6 could be armed with Mk 4 gun pods and heat-seeking sidewinder AIM-9 missiles, it was rarely done, and the Navy played to the A-6 strong suit, stealth bombing in the worst weather imaginable.

After being winged and designated a naval aviator, I got only a scant two-day leave to savor with my classmates our mutual accomplishments. Forty-eight hours later, it was on to the RAG (fleet replacement) Squadron VA-42 at NAS Oceana for A-6 transition training.

Because I had some civilian flight and aerobatic training, a degree in aerospace engineering and had gone through the slow integration of the AVROC program, I was prepared for all the book work and long hours required for Navy flight training. I did well and was considered by my classmates and instructors to be a good technical and intuitive pilot.

When doing carrier landings, I knew the technical reasons why the A-6 would and wouldn't land on the rocking boat - 5% luck and 95% energy management. Instead of just seeing the ball on final, I visualized an upper and a lower energy envelope that I had to keep the plane inside of. Too much energy and I would miss the three-wire and hopefully catch the #4, or I would miss them altogether and would bolter, aborting the landing and going around to make another attempt.

The worst case was too little energy, and I would be below the envelope and unable to recover from a ramp strike or a full-blown collision with the stern of the boat. If you were lucky, you could punch out; if not, the Navy would lose an eight-million-dollar plane, a half million in training and incur significant damage to the boat. Every time I briefed, preflighted or flew, I was thinking of the plane and how it would respond and perform under the anticipated conditions.

Each time I made an attitude or power change, I was able to visualize the movement in the center of lift under the wings and the three-dimensional energy vectors. The A-6 has one of the best engine response times of all the jets in the fleet, but even so, you had to anticipate the lag time between throttle input to the pair of J-52-P-8 turbojet engines and the time it took the engines to redirect the flight path of a 30-ton aircraft in the direction you wanted it to go.

I had already completed my first combat tour with VA-165, the Boomers, on the USS *Constellation*. Throughout that entire cruise of nine months and almost 170 days on line with combat Ops, only two men were lost. Several planes and crews were downed by AAA, SAMS and accidents, and all but two survived to return home.

Unfortunately, very early in the cruise, at the end of only the second on-line period, it happened that we lost one of our best pilots, LCDR Frank Holmes. He had been an instructor Stateside, a great stick and well-respected by everyone - and we still lost him. Despite the tragic loss of one of the best, I finished that cruise feeling immortal and almost charmed. Maybe it was that VA-96s aces, Cunningham and Drisscoll, were kicking butt, the effectiveness of our mining of Haiphong Harbor, or whatever. But I felt great in the plane, comfortable in combat and, literally, bulletproof.

These were all closely guarded personal secrets, because if divulged, there was a good chance that you'd be perceived as cocky, which equals carelessness, which would have translated into no one wanting to fly with you. It's possible that others felt the same way, but as long as no one lost it or got careless, we talked tactics and survival and we kept the heavy personal stuff inside.

Having completed that first tour, never losing an entire airplane, just some large pieces of one or two, and never losing a B/N or wingman, I felt I wanted to do it all over again. So, I requested a second tour and got assigned to the VA-35, the Black Panthers, who were coming on line in a few weeks on the USS *America*.

Not only was I getting what I wanted, but there was a little extra bonus. I was now considered a "Salty Dog" - a seasoned combat veteran who had survived aerial combat, beaten AAA and SAMS and was expected to pass along valuable first-hand information to a new squadron.

Unlike night traps, I loved cat shots, especially at night—black and rainy. It makes no sense, but it was a time to just trust the airplane—which I did.

There is only so much you can learn in training; beyond that you needed the real front-line version of how to fight the war and manage the all-too-often limiting Rules of Engagement (ROGs). This really suited my ego and made me feel all the more like I would live forever and nothing would happen to me, at least not while on Yankee Station.

I spent part of the intervening weeks at Subic on some R&R and the rest on TDY (temporary duty) on the *Coral Sea* as a Salty Dog with a Marine squadron that was starting its first A-6 combat tour with TF-77 in several years. Just before July 4, 1972, I caught the COD (carrier-on-board-delivery) from Cubi Point out to the *America* and checked in with my new squadron.

From the get-go, it was hard to act like the new guy on the block and be humble and unassuming in my openly acknowledged role as an "expert." It's tough enough being an anal-retentive, left-brain, engineer with a Navy pilot's ego and credentials, but then to be put into a position of suggesting or telling 30 accomplished Navy pilots and B/Ns how to fly their missions was asking a lot.

For the first time, I fully realized the enormity of what I had taken on; this was not going to be just one of those long week-ends in Hong Kong when you swap tales and big lies with A-6 drivers from other squadrons. I was now in a formal position of shaping squadron doctrine and SOP (standard operating procedure), any or all of which will be tested without delay by VA-35 in combat.

How was I going to feel if I was directly responsible for someone making a decision or issuing an order based on my VA-165 doctrine, and people died. It was a sobering thought. I had been able to manage and keep my own mortality in perspective, but now I had 30 more souls that I was responsible for, and I could now imagine how any CO felt. We had less than two weeks before our first line period, and we had a lot of work to do. We flew almost every day, since the weather was never a problem for the A-6, and we all handled the task like professionals - as long as I kept my ego in check.

It helped that I was a living testimonial that what I was talking about works, and that I wasn't going to bolt out of there. I was there for the duration and was prepared to go into combat based on the doctrine I was expounding. But I was destined to fly with senior officers almost all the time - in particular, our CO, CDR Verne Donnelly. Understandably, he would ultimately be responsible for the squadron, and he wanted to personally get and evaluate the input, test it, and then pass it along as squadron doctrine through senior officers and not pilot-to-pilot.

The bonus for me was that the *America* arrived on station with the new A-6C, "Charlie," and I got a chance to fly it and get qualified in it before we headed up North. The A-6C provided both the B/N and the pilot with a TV-like presentation of the enemy target area, and this proved to be highly effective in the night-attack and surveillance profile.

Our first day on line at Yankee Station was July 14, and we flew an excellent mission up north and all planes recovered and there were no losses or injuries. So it went for the next two days while we flew cyclic Ops, 12 hours on and then 12 off. The up time was your briefing, planning, preflighting, flight/combat time. The other 12 were supposed to be for sleeping, maintenance, training and paperwork.

For those next two days, we had successful missions with no casualties, and I was beginning to feel like the charm was still working and that what I had passed on was being validated.

A few key points of the "new doctrine" included:

1) Only one pass at the target per plane; never go back to the same target twice on the same mission.

2) Go in on the target flying a wagon-wheel pattern. Don't all go in like ducks in a row from east to west; each plane/element comes in like a spoke of a wheel.

3) Because of their incredible bomb loads and accuracy, the A-6s go in last; they usually obliterate the target for other aircraft visually sighting the target and put a lot of debris up in the air.

4) Stay above 3,000 feet, out of small-arms fire range.

5) Vary your altitudes randomly for AAA sighting.

6) And no matter what happens, if you're in trouble, head 120 degrees; it will get you someplace safe.

July 17 started out with a decidedly upbeat team mood of acceptance, trust and proven success throughout the squadron spaces. Unfortunately, up north, the weather had deteriorated to the point that the only things that might fly that day were A-6s; no problem, that was our stock in trade, our hallmark, and we loved it when everyone else stood down and we were the only aircraft that could launch.

The Air Plan was approved, and we were briefed for a mini-alpha strike of hard bombing up north on the Bac Giang powerplant, 40k NW of Hanoi. We were scheduled for a 20:00 launch with three two-plane A-6 elements in the group. My lead would be the CO CDR Verne Donnelly, and I was already in a hard crew with LCDR Ken Buell as my B/N. Our call signs were Panthers 507 and 512, respectively.

Everything seemed normal, at least, as it ever gets, except that an unusual number of little problems were cropping up. The weather forecast for up north kept changing from bad to worse - possibly worsening on Yankee Station. Even the A-6 has limit for coping with sea swells and winds, mostly for recovery. Then maintenance found unusual debris in the fuel samples on 512 (my plane).

The A-6 had a history of debriding small pieces of fuel-cell liners that could clog the fuel strainers and flame out one or both engines even though they had largely independent systems.

Also, Strike Ops kept generating a little confusion about the specific bomb loads. They kept changing between saying we'd have a full hard load of 30 bombs, or an average load of 24 bombs, or a composite load of Mk-82 500-pound Snakeye retarded bombs, CBU-24s, Rockeye cluster bombs - even possibly a mixed load with 1,000-pound Mk-84s. The CBU and clusters were anti-personnel types with a frag component in the warhead. The question, I guess, was the number of defenders at this location that could be taken out by frag as opposed to a 500-pound detonation.

But everything seemed to settle down, yet it still gave me a spooked feeling about the mission.

Crew members of the 2nd Division, Shipley Bay (CVE-85).

Crew members of VF(N)-53.

The weather continued to push back the launch cycle several times until it ended up at 22:30. Now the queasy feeling really started. Nobody liked night traps, and even after 370 of them (over 120 at night) I still didn't like them. But at midnight, in horrible weather, after an exhausting 2.5 hour mission, we would be at the end of our cycle and have to perform at our very best with the greatest fatigue. But that was our job: just refocus and get on with it.

At around 20:00, the word came down that we were going to take a mixed load of hard bombs and ARMSs (anti-radiation/radar missile) missiles with us. Since the SAM sites would be up and have only us to play with, why not take advantage of it.

The ARM missile was able to lock on to the radar signal from the SAM site and follow it right back down to the trailer and launchers and deposit a warhead on target using the site's own telemetry as guidance. The only way to defeat an ARM was to shut down the SAM site, but that left their SAM missiles without guidance, and they would go ballistic. But, if their SAM was locked on to your A-6, this was just what you wanted. Further, if they didn't shut down the site fast enough, the ARM would already be tracking its deadly trajectory and would impact pretty close to its guided target anyway, therefore accomplishing 60 to 100 percent of the maximum destruction.

Due to weather and short notice, it was a very difficult task to re-bomb all six planes, but as always, the guys in the red shirts got their jobs done consummately well, and we launched on time. Unlike night traps, I loved cat shots, especially at night - black and rainey. It makes no sense, but it was a time to just trust the airplane (which I did) and if you trimmed it right, it would just fly off the deck, sink a little and drop the stick right back into your carefully positioned waiting hand with a positive climb.

We poked above the gloom and murk at about 10,000, and it was a bright moonlit night. So far, it looked like we had a good weather brief and the target would be partially clear. This was not an unusual night environment for us, but it always created an incredible paradox in almost everyone's mind.

The scene was beautiful, almost romantic, and we could have been on a commercial or general-aviation flight anywhere back home. Yet, in this almost ideal setting, we were about to drop 60 tons of high explosives with the obvious, and not so romantic results, the only romance was that we loved doing it.

On nights like this, it was always hard to keep focused. The guys who had girlfriends or wives back home, I thought, had a particularly difficult time of it. Despite the tranquillity, all the pilots' heart rates went up because we knew that we would soon have to poke our planes back down through the goo and set down a 20-ton plane on what amounted to a matchstick floating in an Olympic swimming pool.

The group rendezvoused as briefed with a 360 TACAN form-up, and all checks were normal. It was coming together fast, and we headed northwest at 15,000 feet to the coast in-point north of Haiphong. Once across the coast, we went up to 21,000 feet and followed the briefed route to the target; it would circumnavigate known SAM and AAA sites, as per the Intel briefing.

All was quiet, but I still felt odd, despite the exhilaration of the cat shot. Usually, I chatted a lot and did double-checks on the checklist, etc. No kids, dogs, wife, car stuff - just A-6 stuff and the mission. But I was quiet, so much so that Ken, my B/N, asked me what was bothering me. I gave him the anticipated and humorous answer; I said "the trap." Even though we were about to go into combat and still had to face the rigors of the bombing mission, 85mm and possibly 102mm AAA, SAMs, etc., pilots would worry about a night trap that was almost one and a half hours away. This was a recurring phenomenon with Navy pilots, but it made Ken laugh, and it got my mind working again.

The weather was still crystal clear, and the formation lights shone with unusual intensity. The CO called for the formation to open up because he knew that we would soon be near a former AAA site and would soon be starting our pre-bombing checklist, and attention would be drawn inside the cockpit. So he smartly added a measure of safety to our spacing.

Right in the middle of maneuvering, we all of a sudden picked up the tone in our helmets of the Firecan radar detection warning, and the visual threat-panel light illuminated. This alert tone indicated that the NVC (Northern Viet Cong, aka CNV or Communist North Vietnamese) AAA gun-control radar was searching the skies for something to lock on to and use as a target for its 85mm (3.5 inch diameter) shells.

With no radio chatter, we went to our briefed AAA spread of staggered altitudes and separations because we were about to get pelted with a very heavy barrage of shells. Any one of the 85mm shells could easily take out a single plane, so why make the target any easier?

The tone went from low to high warble and the threat-panel light started flashing, which meant that the NVC had a good lock, and the shells were already out of the muzzles heading for us.

AAA was the single biggest threat to us, so the pucker factor instantly went up, and controlled panic set in while we tried to be oblivious. We could see the explosions in the moonlight sky and it was kind of like July 4th, but we were on the wrong end of the display. Fortunately, for the most part, they were off by as much as 500 feet, but the planes still bounced and rolled with the shock waves and concussion.

Triple-A was a total crapshoot, and there was no doctrine on how to jink (maneuver) around it like there was for SAMs. You stood as much chance of diverting smack into a round as of getting hit where you were. So we maintained course and tried to keep control of the aircraft, and we were through the area in just a few minutes.

The check-in verified that Panther 503 had been hit, so he and his lead broke off and returned to the boat, and our remaining four planes headed for the target. Unfortunately, when 503 returned, they could confirm that the safe/abandoned AAA site was, in fact, HOT!

We'd been feet-dry for about 10 minutes, and we were wired (consummately had our act together) as Ken started his pre-bombing checks. Thank goodness we had great maintenance and both the search and track radars were up. Having both wasn't critical, but it made the bombing easier and more accurate.

The A-6 was the smartest plane in the fleet, but you had to manually define mission parameters by setting drum switches relating to your requirements for weapons selection and delivery. That's why the A-6 had a two-man crew, a pilot and a B/N. Without the B/N doing his job just right, there wouldn't be a mission.

My job was to get the plane on and off the boat, fly the plane and to drop the weapons when cleared by the B/N. There was a ton of data to confirm and switches to set, and Ken was busy confirming all the settings prior to the IP (initial point), where the profile for the bombing run starts. (Our nickname for the B/Ns was "MATOs - Master Arms Turner-Onner").

We had briefed the wagon-wheel run-in, so we broke up and took varied speeds and headings to our separate IPs so that we would arrive at them at staggered times and altitudes. This put each element over the target at a different time so we wouldn't be dropping bombs on each other, making it harder for AAA to get a fix and would be out of the majority of the frag pattern from the previous plane's bombs. Tonight, these A-6s were going to leave 500-foot-long strings of explosions that had the potential of fragging the following aircraft.

I was now 100 percent on the instruments and flying the plane manually while studying the TV-like presentation of flight data in front of me. We would have to stick our heads in and out of the clouds several times, so there soon was going to be a total black-out, and I wanted to be ready.

The timing was right; I got the boss's "Arm Weapons" call on schedule and then his in-bound call exactly two minutes before my roll-in. That would put us a textbook 90 seconds behind him over the target. We were number two on the target, so I focused on the radar-generated pipper on my display, validated what I thought the presentations should be and got ready to maneuver the plane into a bomb-run profile. The computer would actually

From left: LT Jim Ozbirn, LCDR Bill Evans, and LT Ed Guilbert.

The procession to receive commission as Ensign, 1931.

release the bombs from the racks, based on a myriad of speed, wind and other data. But first, I had to verify all the data that I could; only then would I authorize and "commit" the computer to take control of the bombing from the buttons on my control stick.

The CO called off the target just as we entered the clouds, and all I was thinking about was the job, the technical aspects of the bomb run. My constant and ever-present fear was altitude; did we have the right settings? Was I going to fly this plane right in to the target? It's one thing to be shot out of the sky with AAA or smoked by a MiG, but to die due to pilot error was the worse fate.

I triple-checked everything that I had control over and trusted that Ken had done his job. He had, and I got "Commit" indications; now we were just along for the ride. My heart was pounding like a trip hammer, and I felt the slightest nausea from the nerves. Two deep gulps of oxygen and a final check of the instruments, and then we felt the reassuring shudder of the aircraft as the first bomb load peeled away from the plane; then another; and then there was an uncomfortable delay until we felt the third bomb release—and it was my plane again. I didn't like being down low, so I called, "Clear target 512," pushed the throttles forward and climbed the hell out there so the second element led by Steve wouldn't run right up my tailpipes.

Without the drag of the hard bombs, the plane flew superbly, and felt like a rocket by comparison. We still had a SHRIKE standard ARM missile AGM 45A on our #2 station, but were still relatively clean. It was time to be a hot dog, since we had to jink to avoid small-arms gunfire, AAA and any possible SAM setups.

There were no rules; you just flew the plane in as unpredictable a manner as possible—reckless almost. This is where a lot of B/Ns lost their lunches and, inevitably, the B/N would bounce his helmet off the canopy more than a few times.

Because of the sudden release from the pressures of doing the job right, I always felt like doing a victory roll or something outrageous at this point, but we were too close to the ground, less than 3,000 feet, and it was against doctrine!

We could feel the concussions from our bombs and all readings were normal, so the A-6 and *Diane* (the onboard computer systems) had done their job again. We knew even without photos that if the Intel was right, our bombs had taken down that power station 100 percent. We all climbed, formed up on Verne and headed on a somewhat twisted route to the southeast coast and out toward the boat.

At this point, I realized that I was exhausted and wished Ken could fly the plane back while I took a nap, but no such luck. B/Ns had a stick between their legs all right that kind of looked like the one pilots had, but it wasn't attached to any flight-control surfaces. It was used to slew data and input information to the A-6 systems.

I fought the groggy feeling for a while, and then I got my wake-up call. My helmet reverberated with the sound of the Fan Song tone, which was different from the "Fire Can" sound and indicated that Charlie had his SAM radar looking for us. The threat panel lit up, and we went to our briefed separation for SAM attacks. The briefed positions not only reduced the threat of a SAM hit, but also reduced the likelihood of a midair with our own planes while we evaded the SAMS. I shouldn't admit it, but if SAM attacks were inevitable, I loved them.

It was the ultimate application of my technical flying skills as we put the A-6 through every conceivable maneuver and condition—most of which the factory pilots never dreamed of doing.

It was an energy-management game, and the best way to beat a SAM was to beat its energy. I had been successful with this approach during the first tour, and most others who understood what I was saying succeeded as well.

The way to beat a SAM was to literally play chicken with it: angle toward it on what ultimately would be a collision course, and then, at the last possible moment, make a sudden break sharply toward it in a turn that it couldn't match because of its almost Mach 3 speed. The faster something is going, the larger its turning radius; the slower, the tighter the radius.

In this game, I was the tortoise, and it was my job to make the hare overshoot. There was no way to outrun the SAM. Despite the terror in your heart knowing that there was a Mach 3, telephone-pole-size missile with a 300-pound warhead headed right for you, you absolutely had to stay there in the snake pit with these damn things and beat the snake. If you ran, you were dead meat; it would close in on you in seconds and turn you into a cinder.

Almost everything in this game was visual, and you couldn't afford glances at your airspeed and altitude indicators; if you were concerned, you asked the B/N to squawk out the data. In a three-dimensional chess game in the dark, it's easy to get FUBARed (fouled up beyond all recognition) and think up is down, etc., and fly right into the ground.

Within seconds, the alert tone changed, and we had one, then two flashes of the telltale solid-fuel booster that launched the Russian S-2 SAMs that the NVC used. Good thing Strike Ops had the ARMS loaded because Bob, my lead's B/N, had a lock on the radar site right away, and Verne pickled his ARM. The SAM site should have been able to detect the launch of an ARM missile heading in their direction and then shut down. If they did, the SAM would be flying dumb and just go ballistic.

The second element of A-6s were not in the game; they were out of the firing zone and kept a safe distance. Verne and I were the closest to the two flashes off our left wings, and we were the targets. At this point, we had two SAMs coming up and one ARM going down; it was simply a matter of timing: who got there first and how close the missiles would be to their targets.

The CO called the north SAM and that left me the one to the east, and I turned toward the flash of the launched SAM to play chicken with it. We both had good initial positions to beat these mothers; the trick now was heads-up awareness of where they were at all times and choosing when and where to break to make them overshoot.

Ken understood the tactics and was great at calling out directions, range and altitude of the bandit/SAM as well as our own. At this point, I was flying the plane by the seat of my pants. I didn't even have to look at the airspeed or the attitude indicator; my eyes were glued outside, scanning for the SAM.

With the SAMs going 2,000 mph, it meant a closure rate of almost Mach 4, so my feel for the plane was good enough; the important part was early detection of the bandit and split-second timing.

After what seemed like an eternity, I gave up on the ARM missile getting there first and making things a little easier. Verne was a great stick and beat the SAM by a mile and flew out of the zone to circle around and clear the area for me. I was getting closer to mine and was in a steep climb, planning to pull the missile up with me, and then split-S out (doing the second half a loop from the inverted position).

The SAM was closing incredibly fast and I didn't have time to get all of my altitude for a clean split-S down and away. I rolled the plane inverted at 320 knots and pulled the stick back into my gut and split-S. While hanging upside-down, there was a tremendous concussion, the plane shuddered and shook and felt as if it was being pushed 500 feet straight down.

Trusty Shellbacks Ducharme and Punahoale, in background far right is Johnny Johnson.

Obviously, the SAM had exploded, but how? It wasn't a direct hit, or we would have been in postage-stamp-size pieces all over North Vietnam. Either the proximity fuse detected our near midair, or this bird was controlled from a second radar site, and the NVC on the ground did a ground-controlled detonation. Who knows?

What I did know was that I was out of altitude—like now! Still semi-inverted, I pushed the stick forward with both hands and a giant grunt and flew out the bottom of the split-S inverted and in the opposite direction to that intended.

Once stable, I slowly rolled the airplane to the left to the upright position. I don't know how we made it out of that position, but I do know that we lost the starboard engine. I confirmed it was the right engine and not the left, them I pulled back the right throttle to shut it down prior to an air restart.

We climbed on the left engine, up to almost 5,000 feet while I reset the ignition CBs (circuit breakers), checked my master engine and fuel switches for "On" and attempted an air restart. It took two shots, but it did start, even though I couldn't get full power out of it. Since there was no fire, even if the engine failed totally, there was a chance that we would get back to the boat; single-engine traps are possible—but there are no bolters.

Normally, the A-6 handles stiffly and you have to use a lot of stick forces, but it has superb spool-up time and the engines reacted to throttle inputs almost instantaneously. Now 512 was flying like a dump truck; I literally needed two hands to fly against heavy control forces, and the throttle reaction time had deteriorated dramatically.

The little voice in my head kept saying "Steer 120 degrees," which was the safe way out of the north. At times like this, even the little things like the way home had to be committed to memory in advance. I was already turning the plane for the coast outpoint and resetting circuit breakers when I heard Verne on the radio saying that he had two more SAMs in the air and he wanted some help.

Without asking, Ken barked the vectors into the ICS (intercom) and I radioed that we were on the way.

I didn't have any idea what I was doing at this point; I had no plan. All I knew was that I had a bud in trouble, the plane was still flying, and the CO's time may be measured in seconds. For the first time

ever since I had been flying in Vietnam, I felt my chest to make sure that my 9mm gun and extra 14 shot clips were there. There was still no fire, but I was concerned about the dropping hydraulic pressure.

I traded hands on the stick and felt down with my right hand to confirm the location of the switch for the back-up hydraulic pump, should I need it. On the Fan Song radar, we had the tone scanning then locking. Damn it! There were two more SAM sites—not really surprising since there were more than 300 in North Vietnam since the last build up during our stand down and our unfathomable Rules of Engagement.

I told Ken to ready the ARM on #2 station "NOW!" so that we'd get the jump on knocking out the SAM site. After Ken had set the switches and targeted the SAM site, he looked up at me as I happened to glance over. I could tell from his expression that he wasn't sure if this was such a good idea. I felt bad that I was subjecting him to all this; I didn't ask his opinion, and I didn't stop long enough even to think about it. At this point, he was a captive passenger, and despite his senior rank, I was the pilot in command and making all the decisions. B/Ns place blind trust in their pilots, and even though we had hard crews and flew a lot together, you never knew what to really expect until there was a crisis.

I could only see one of the SAMs coming up, and I wasn't really sure where the second one was; Ken didn't have any Intel for me. He gave me the "Good to go" on #2, so I pickled the ARM, and just at that moment, I picked up the glare of the second SAM's engine. Turning toward it, I called "Panther 512, missile away and I've got SAM #2 to the south." As we approached the SAM as close as I dared get with a damaged plane, I did a hard left roll and dished out in the bottom so that I would be diving toward the ground. Just about the time I realized that the radar altimeter doesn't work in this altitude, we lost the tone on the SAM site, and either our ARM found its target, or they shut down; either way, as I rolled out, the SAM on my tail went ballistic and darted over our heads, missing us by less than 300 feet.

Verne had similar good luck; the missile came from behind him and passed off his right wing—not a good spot to have been in. I knew I had made the right decision when he said he owed me one and was going to name one of his kids after me, or at least my call sign, "Attila."

With 85 percent power on the right engine, we limped back to find the CO. As we climbed through about 3,000, we started to get additional engine alarm lights, and the plane felt even rougher. As I headed for the coast, I told Verne my situation and asked him to form up on me to inspect for damage. We were still 50 miles from the coast, and at this speed, it would be 10 minutes or more before we got there. I slowly traded some of our very precious airspeed for altitude and tried to get above 4,000 so that I could maximize any flight/glide path that I might have out to the water.

We started to pick up some light AAA and small-arms fire. What a great target we were—low and slow. The AAA didn't last long, but it did manage to do in what was left of the right engine, and I had to shut it down. As long as we didn't have a fire, we could stay with the plane, get out over the water and try to make it home for a single-engine trap. No one ever practiced single-engine traps, and I had never met anyone who had done one. But the books that I trusted so much said it could be done, so I was game to try.

But we had apparently also taken some damage to the good left engine. I gave up any hope of getting back to the boat and acknowledged dying as an ever greater possibility. At this point, I felt that the least I could do for Ken was to get him a water landing and possibly avoid being tortured or being a POW for the next four years. I told Verne to wave off. "We're critical and going in. Just make sure Sandy and the rotor guys know our location."

I always thought there would be more to it if I ever had to say those three words, "We're going in." Now, I was prepared to die, and I was so matter-of-fact about it that I couldn't believe it. I had worked hard, done everything right, and the system still beat me. *Diane* was off line, and I was flying by stick and rudder, steering 120 degrees and trying to remember the best glide speed, clean with one engine wind-milling. I was sure it was around 210 knots, but how good was my ASI (airspeed indicator). At OP (optimum) glide, I might get a 12:1 ratio. That meant if I got up to 7,000 feet, I'd glide about 14 miles—not bad, but I wasn't climbing anywhere without major loss of precious airspeed. I was flying now just to get my passenger as safe and close to home as I could.

As we approached the coast, I became more optimistic because we'd at least have a water landing and not have to confront the nightmare of landing in the jungle at night and being totally out of our element, with only a peashooter for self-defense.

Yet I might soon find out the answer to the question that all pilots asked: "Where do I put the last bullet? In the bad guy or in myself?"

Now all we had to worry about was hypothermia and sharks. We went through the checklist for ejection, and I briefed Ken that he would go first, through the canopy—the best way to get out of an A-6, unfortunately.

I fixed my eyes on the left engine gauges; as soon as the engine exceeded its limit, we were punching out. The CO took a high position over the water and started to coordinate the rescue. I didn't have to worry about that part; if we got far enough out to sea, I knew we had a good chance of being rescued.

At this point, I started to feel negligent and ashamed that I wasn't bringing my plane back. I had brought them back broken and damaged before, but I always brought them back. I told Ken that the CAG (officer in charge of the Air Wing) would be angry that I lost a plane. He replied that at least we got out of making the night trap I was worried about. We both kind of laughed.

I was hoping to get at least 20 miles out and wait for the helos from the America and Red Crown to find us. But that wasn't to be our luck; the left engine temperature had just spiked. I gestured to Ken that he was to eject immediately.

I was not prepared for how violent the ejection would be. When Ken pulled the face curtain, the noise was worse than the SAM explosion, and the light was blinding. I thought, "Oh, Man! They could take pictures of this from downtown Hanoi. With that much light, there will be boats out looking for us before we hit the water."

We had been briefed on the locations and numbers of small boats patrolling the waters waiting for such an occurrence, and we were right in the middle of the "NVN Fleet." The only good news was that most of these boats were NVN regulars, and there was less chance of being killed outright by them than by enraged villagers in the jungle. The NVN wanted all Yankee air pirates alive—at least initially, so they could be tortured and made to confess their "crimes" publicly.

If I wasn't already convinced that I was going to die that night, watching Ken go through the canopy with all that noise and the rocket tail convinced me that I wasn't going to survive this one last ordeal.

I had to count five seconds before I ejected so that Ken and I wouldn't become entangled with our chutes.

After he punched out, I lost all track of time and thought to myself: "Five seconds must have passed, but if I'm wrong, I'll drive Ken's ankle up through his esophagus. That would be the ultimate irony. As if I haven't screwed up enough already!"

I told my mom I loved her; there wasn't anyone else. I asked God to thank the guys at Grumman who designed this incredible plane that got us this far, and at the top of my lungs I yelled, "Ho Chi Minh, you son of a bitch" as I pulled the face curtain on my ejection seat.

Right on cue, I was out of there. The acceleration was incredible; if I lived, I would definitely be two inches shorter. It was the wildest ride of my life. I thought the rocket would never cut off and I hardly noticed as the top portion of my seat poked a hole in the inch thick Plexiglas canopy for me. We were so low that the barostat that controlled man/seat separation activated almost immediately and I was tumbling through the air.

I caught just a brief glimpse of #512 as she headed toward home without us. I could see the navigation lights faintly. Damn it! I should have turned them off to conceal our position; it just wasn't my night.

Wham bang! There was a substantial tug on my torso harness and I was under a fully deployed canopy. No kidding; this stuff really works.

I looked down at the water and saw dozens of boat lights all around—I mean 360 degrees all around me. At this rate, I wouldn't have to worry about sharks or a water landing. I'd land right on the deck of one of these NVN river boats.

We were surrounded; could there be that many boats around us that fast? But they weren't just the usual red and green running lights; they were pinks and blues. Something's wrong here; reality check time. I squinted my eyes and the lights went away for the most part. "I must be seeing spots from the flash of Ken's rocket motor. That's it; the old flashbulb syndrome."

I listened for a moment, and I couldn't hear anything through my helmet so I just discounted the lights as mirages. It was a medium-moonlit, partially overcast night, and it didn't seem like I was coming down at all; I felt suspended. To see if I could find Ken, I twisted around as much as I could and worked the risers to go left and right, but no sign of him.

Time for a mental checklist for water landing; feet crossed, raft and survival gear dangling underneath me, and release koch fittings just before touchdown, about 10 feet or less. This way, if the chute does get dragged under or caught in a current. I won't go along for the ride. But I could only guess where the water was and how fast it was coming up; the lighting and the soundproofing of the helmet were very disorienting—kind of neat, but also very disconcerting.

I was worried about the plane. I really felt bad about losing it—major big-time FUBAR. But I guess that is part of the job, and at least, so far, all four of us were healthy, and if it hadn't been my plane, it would most likely have been the CO's and we would have lost both Verne and Bob his B/N. I guess it was a trade-off; I hope the Navy thought so. "What are they going to do - court-martial me? After what the NVC have tried to do to me for the past eight months, I could survive that. But we're not out of here yet."

I started thinking about currents and drifting the last 80 miles home, when, out of the corner of my left eye, I saw something that startled the hell out of me. I turned quickly to catch a glimpse of a fast moving target. It was Ken, several hundred yards off to my left, and I guess his chute passed in front of the moon and cast a shadow over me for a second.

I still wasn't sure if we were totally alone, so I didn't yell, but it looked like he had made it all right. He had tried to steer closer to me so we'd stay together, but I guess with my heavier body, I was falling a little faster.

I was trying to keep sight of Ken when out of nowhere, I was in the water, coughing, gagging and sinking. "Mother, now I'm going to drown." I managed to inflate half my life vest and to get one good gulp of air between the gulps of sea water. I struggled to release my koch fittings to free myself from the chute entanglement. It was while trying to manipulate the fittings that I noticed something odd about my right index finger—like it wasn't there.

I was still wearing my Nomex gloves and I could just tell that the right one was shredded around the index finger area and the glove finger was hanging limply down with something stuffed inside. I thought about looking at it with my flashlight, but I really didn't want to see what might be or not be there, so I tried to ignore it. I did a quick check of other body parts and everything else seemed to be in place.

I ditched my beloved helmet and sorted out the rest of my gear. From all the ready-room discussions, we had a prepared doctrine about this kind of situation. You wanted only the essentials and to stay as submerged as you could; to make it harder to see you from an approaching enemy boat, don't go bobbing around in your raft. I got my flares, survival radio and flashlight and kept my survival vest on, with my gun and three extra clips. I had decided to decline the Navy issue .38 caliber weapon in favor of a Browning 9mm Parabellum with a 13-shot clip and one in the chamber.

The water was bone-chilling, but refreshing in a way, and it gave me a new shot of energy to go on with what might be a long night, or swim, or both. After a while, I forgot about the water; either my body was numb, or I was burning up so much adrenaline that it kept me warm. I started to swim toward where I thought Ken would be, but the breast stroke in flight shoes wasn't going to work, so I treaded water as well as I could, got rid of them and started swimming again.

Then it dawned on me: "Here I am, swimming back toward Hanoi—great! Why don't I just make it really easy for them to find me? I'll swim and meet them halfway." Fortunately, Ken had the same idea, and after about 30 minutes, we established flashlight contact. Shortly after that, we joined up in almost absolute silence. While swimming, we both had heard the sound of tiny outboard motors chugging through the water, and were now starting to see searchlights on the horizon.

Floating in the water like this, your view is about 38 miles to the horizon. But looking back toward North Vietnam, it seemed as if we were 100 yards off the beach instead of what I calculated roughly to be closer to 12 miles.

In whispers, we developed our game plan; if a single small boat approached, with my gun in my hand, I would lie on my right side as if I'd been severely injured, and I'd allow the boat to approach me from the left. At the same time Ken would hang on to me and act disoriented while turning me so I would be in position. As the boat approached, Ken would choose the moment to open fire and I would spring into action and empty my clip into the boat.

If there were two boats or more than 10 guns in the one boat, Ken would do nothing, and when I figured the gig was up, I would just drop my gun to the bottom of the ocean. If it was our intent to surrender, why provoke a fire fight against odds that Ken had already figured were bad? We also joked that if it was more than 10 guns, we would just shoot each other.

For at least an hour, we listened while the engines noises got louder, the lights got bigger, and our hearts beat like crazy. If we tried to swim, we would attract not only the NVN, but also sharks, so we had to stay where we were and wait.

Our lack of movement made the water feel colder and colder, and we both started to shiver—especially Ken. I also worried about the performance of this beautiful gun; would it work after being submerged all this time? Could I use my middle finger accurately? No time or place to find out.

We had to be absolutely passive like a couple of Florida alligators sitting in the swamp. Without going negative on the buoyancy we just tried to keep our noses and mouths out of the water with our heads tilted back. Every once in a while, we'd get some sea spray and start to cough, but we had to bite the bullet and not react.

With our ears submerged, we started to hear the engines even louder under water. There was no doubt that they had seen our ejection, and these weren't just night-fishing boats. This was the worst part: we couldn't use our survival radios for fear the enemy were eavesdropping; we couldn't swim toward safety; and no way could we fire off our flares. We would use our flares only when the helos got close—but, so far, nothing.

"Oh, God, what I wouldn't do for that beautiful thump, thump, thump, of a Navy helicopter. If a Sea King showed up right now, I would never again say a nasty or demeaning thing about a rotor pilot ever again."

We couldn't figure out any pattern to the boat movements. One second, we'd think a particular boat was bearing right down on us and we'd submerge; then it would turn and head off on a weird heading. Thank God, they didn't seem to have a pattern; it was free-for-all, like tossing candy and peanuts in the air at a kids' picnic. This time, though, Ken and I were the prizes, two Yankee air pirates, and our captors would be rewarded with a lifetime supply of rice and ammo.

Almost an hour and 45 minutes after we had ejected, we heard the first faint sound of rotor blades. It was an incredible pure adrenaline boost, and all of a sudden, we weren't cold anymore. If we used flares, maybe the helo would see them and project our location, but the NVN would definitely get to us first. Before we popped the flares, we had to bide our time and, with almost uncontrollable impatience, wait until we agreed that the helo was low enough and headed right for us.

"Now!" I'm sure that we blinded the helo pilot, but the crew was on the ball; searchlights lit up the sky. We popped smoke, and I used the radio to steer them right to us. Within minutes they were over us, and there were swimmers in the water. I sent Ken up first. While the swimmer and I waited for the harness, we couldn't see or hear the fast-approaching NVC boats because of the rotor wash.

We were still in the water when the helo started taking small-arms fire, which they couldn't return because they didn't know where to aim. The gunfire just was coming out of the darkness, and the Sea King wasn't equipped to just lay down suppresser fire without the chance of hitting us.

We were barely out of the water when the helo started to head for home, and we were skimming the water. The pilot killed all the lights, and we were in total darkness. The interminable winch ride back to the helo in total darkness gave me a really eerie feeling—like being tractor-beamed into an invisible alien ship.

By the time we got onboard, we were out of small-arms range. While we guzzled down hot, black coffee and cookies and huddled under a pile

USS LST-451 at the Leyte Assault, Philippines, WWII.

of blankets, the air crew pointed out at least eight holes in the helo that came from the approaching NVC boats.

Back on the America, we landed to a hero's welcome and were rushed down to sick bay. Ken was in great shape and would return to full flight status as soon as the medicinal brandy wore off. But I had managed to traumatically amputate my right index finger and to sever about three-quarters of the Achilles tendon in my right ankle. The finger injury most likely came from the ejection through the Plexiglas, and the ankle wound, from an AAA fragment. I was just glad to be alive and get all the "attaboys" from the squadron, Ken, Verne and the CAG.

Every once in a while, I really felt bad about losing the plane, but not a single person mentioned it, so I wasn't going to worry about it. The sick bay put my pieces back in approximately the right positions, soaked them in sterile saline and bandaged me for a rapid medi-evac on the COD to Cubi Point.

By the time I had recovered, Stateside, the war was winding down and the Navy was looking to attrit pilots for the smallest medical or visual imperfection. So, even though I appealed the decision twice, I lost my medical rating in tactical jets, and I didn't want to fly helos, big jets or props. I wanted a tailhook, or I wanted out. Having been one of Nixon's finest hired warriors, and decorated for my efforts, I felt this was extremely unfair, and the Navy and I parted ways on some very unsavory terms.

I will always love the big ugly A-6. While completing its mission, it saved my life and the lives of at least three others that night, but it gave up its own life in the effort.

Today, #512 lies in pieces on the bottom of the Gulf of Tonkin. If it had been Navy personnel, doing what it did that night, it would have gotten the Congressional Medal of Honor, for unselfishly giving up its life to save those of others. I want this article to be its testimonial and a permanent epitaph to the A-6 Intruder, the finest aircraft ever to wear Navy colors.

Today, I own and fly a Navy T-34B, painted in Navy colors, just like the very first Navy plane I ever flew. Frequently, I wish that I had stayed in the Navy flying helos (off carriers).

I'm No Hero

by CDR Joel K. VanderWaal, USNR (RET)

Let me first say, I'm no hero. These pages are filled with officers who are more deserving to have their stories told. I just wish to say if it wasn't for the Navy, I'd still be de-tasseling corn back in central Illinois. Early on, I learned of the NRA when past NRA President, CAPT Robert Bolt honored me in 1973, as a midshipman, with the Naval Order of the US CDR Horner award. Then in 1993, CAPT Bolt again honored me at a NRA conference with a lifetime membership award.

In 1977, after winging, I was assigned to VR-24 in Sigonella. At first I was disappointed because it was not the way I wanted my career to start. Soon the challenges of flying in a non-radar environment (like Europe and the MED in the 1970s) combined with the ASCOMED tasking caught my interest. I became a TAC (Transport Aircraft Commander) while still a LTJG. With only about 500-700 total flight hours, I found myself flying medivacs, senators, Miss America, and various dignitaries all over Europe and the Middle East. I flew in many VIPs into Cairo for the Carter peace talks. I was most pleased to command the flights of ADM James Watkins (later CNO), COMSIXFLT at that time. He was a class act.

The point of the story concerns stranding ADM Watkins in Nice, on July 4, 1979. I had flown him in for an event at the embassy, was sitting, incognito, on the beach enjoying the view, when someone from the embassy approached me (how he found me I'll never figure out).

"Are you LTJG VanderWaal?"

"Yes sir."

"You have been reassigned to fly to RAF NORTHOLT, at 0700 tomorrow."

I said, "What about ADM Watkins?"

He said "Code 2"

Whatever it was, I hated to leave ADM Watkins stranded. But a Code 2 meant that I had no choice.

We arrived at the Queens Airfield, RAF NORTHOLT on time. Northolt is a famed battle of Britain field, and still currently used by British royalty. When I, a lowly LTJG, walked into ops, unknown why I was there, the duty lieutenant, a grizzled RAF veteran said to me "Thats OK, sony, you can send in the aircraft commander now."

I said, "I am the A/C."

He said, "Oh no, this will never do, Lord Louis never flies with any officer under the rank of major."

I said, "Well I'm all you've got."

Puzzled, I got the paperwork and returned to the aircraft. A red carpet was rolled out, and soon a resplendent British Admiral wearing whites and a lot of fruit salad appeared, nonchalantly poked his head in the cockpit, ignoring my single silver bar and said:

"Hello, I'm Louis Mountbatten."

Trying to emulate his nonchalance, I introduced myself and my co-pilot.

"How's the weather in Naples?"

To shorten the story, my VIP turned out to be Lord Louis Mountbatten, Uncle of Prince Charles, Viceroy of India, and CINC of Southwest Asia during WWII! We flew him to Naples to award the Battenburg Cup to a US Sixth Fleet ship for seamanship!

For good reason the Navy did not tell me who the lift was. He was a model and gracious passenger and we got him to Naples exactly on time.

A week later, Lord Mountbatten was killed in an IRA terrorist attack. What a tragedy.

From Pearl Harbor To Attu To The South Pacific As A Navy Pharmacist

by Ray L. Wilson

As told to Pacific Drug Review by Ray L. Wilson, Pharmacist, USNR, Warrant Officer of the Hospital Cops, USN.

I am very happy to describe for readers of the *Pacific Drug Review*, the duties of a pharmacist in the Navy and include therewith some of my experiences while on active duty in several war zones.

First as to the duties: The rank of pharmacist, a warrant officer of the hospital corps, is a designated rank to correspond to the rank and pay of boatswain of the line. The duties mostly are of an administrative nature in hospitals, large dispensaries, aboard hospital ships or with mobile hospital units and with the Marines. Ordinarily this rank is the ultimate goal of the Navy's hospital corpsman. Here are some of the professional subjects, aside from the military duties, a hospital corpsman will have to know as he progresses and advances upward: clerical forms and procedure; dosage of ordinary drugs; dosages and restrictions of dangerous drugs; management of common diseases and injuries; first-aid treatment in emergencies; preventive-medicine measures: vaccination, antityphoid and venereal prophylaxis; practical work in nursing, pharmacy, urinalysis and microscopic technique.

The pharmacist usually administers the hospital corps personnel as the personnel officer, or he may be the commissary officer, the record officer, the property and accounting officer or the handling and the expediting of medical stores from a naval medical supply depot.

The duties of the hospital corpsman in the field with the Marines or aboard one of our warships is an emphasis of all his training put into real practical use, such as: stopping the flow of blood, giving life-saving blood plasma, first-aid and minor surgery, bedside care and other such duties; also the routine activities such as camp building, field hygiene and sanitation, dressing-room duties and transportation of the sick and injured. In these activities the corpsman, regardless of rate, is frequently placed in a situation where he is on "independent duty," in which he must make fast decisions when the occasion demands, such as administering emergency treatment and providing transportation for the injured during and after battles, and following bombing attacks. The corpsman is well-respected and is commonly referred to as "Doc."

No experience stands out in my mind so vividly as the unorthodox Japanese attack upon Pearl Harbor. I was visiting in Pololo Valley that fateful Sunday morning with Mr. Harlan Giles, the Hawaiian manager for the Winthrop Chemical Company. It was just prior to 0800 in the morning when we noticed the neighbors (some were Japanese) excitedly running into the street and gazing skyward. We followed their gaze and I counted 11 Japanese bombers flying overhead. Fortunately for us they were saving their bomb loads for vital installations. They were heading over the "Pali," no doubt aiming for Kaneohe air station on the northern, or windward side of the island. At that moment the radio was announcing that Pearl Harbor was being bombed and we could distinctly hear the distant booming from the increasing number of bomb explosions. The radio was also ordering all civilians to stay off the streets. So after a little discussion, Harlan agreed to drive me to my station, the old naval dispensary on the waterfront of downtown Honolulu. As we passed through Waikiki we noticed evidence of bomb bursts and wrecked cars. The streets were ominously quiet and deserted. Upon arrival at my station, I was immediately dispatched with two doctors to Pearl Harbor to give all aid possible to the casualties. Believe me, I shall never forget that ride; it was about the longest eight miles that I have ever ridden. Cars before us had just been strafed and lay wrecked along the highway; Army trucks were crazily hurrying here and there. I just knew, with US Navy stenciled over our official car, that we were the target of the whole Japanese air force. As we progressed we could see great columns of black smoke rising skyward. Upon seeing this, my first thought was that the oil storage tanks had been hit, but it proved to be some of our mightiest warships were afire—the battleships. The most fiercely burning was the *Arizona*. We finally arrived at the Navy yard dispensary, only after passing quite a few very angry and determined Marines. This was about 0855 in the morning and events were happening so fast that it was difficult to keep pace with them.

I received orders to take one doctor to the officers' club near Merry Point in the harbor which was quickly being converted into a first-aid station. The other doctor and I then proceeded to the recreation arena, near the Army's Hickam Field, to help set up hospital cots, provide surgical instruments and medical supplies. Enemy planes were coming over us now at housetop level, heading for Hickam Field. This continued intermittently until about noon, when all raids seemed to have stopped.

I worked until about sundown and was then ordered back to my regular station in Honolulu. The town itself had taken on a new appearance; it was

Troop ship APR-47 (auxiliary personnel barracks) USS Kingman, departed from Evansville, Indiana, and departed from New Orleans to Nagasaki, Japan, 1944 or 1945.

making ready for any eventuality; barbed-wire barricades, extra guards and patrols and pillboxes were being erected along the streets. This was quite a change from the Honolulu I had known but a few hours before.

I was ordered aboard ship on March 12, 1942, and served through the Midway battle, the Coral Sea battle, the occupation of Guadalcanal and the occupation of Attu Island before being transferred to my present duty at the USN Training Station, San Diego, CA.

My ship is a fleet oil tanker and its primary function is to keep the fleet fueled and ready for battle. This necessary function is one of great importance, especially since modern naval warfare covers such long supply routes. We usually tag right along with the task forces; that is, unless a battle seems imminent. These in some cases have taken place awfully close. We know that if ever hit we would be nothing but an extra large ocean-going incinerator. We are commonly called "Tanker Stiffs," which is a very apt name for a tanker man. Yet it is a great ship and if called back to sea I hope to be aboard the same vessel.

Our sick bay is well equipped and well arranged. It has the maximum facilities in a minimum of space; operating room ward, dressing room, sterilizing room, pharmacy and office. We performed two appendectomies while at sea. The appendices were successfully removed by the ship's doctor, LCDR R.E. Olsen, a former pathology instructor at Wayne University. Two corpsmen and myself were assisting, giving the anesthetic, conveying the instruments, handling the sterilized supplies, etc.

One time two injured fliers were transferred aboard. They had been shot down on the initial landing at Guadalcanal. One flier had quite an experience, as he had to bail out of his burning plane behind the Japanese lines. He spent three days dodging snipers, but finally hobbled into some friendly natives who cared for him until he managed to reach his own lines. We extracted quite a few pieces of shrapnel from his swollen and painful legs and ankles.

The other flier was more seriously injured. He had been hit by a 20mm explosive shell. It went right through his left buttock and came out through the interior portion of his thigh, leaving a great gaping hole. Luckily no bones were hit, otherwise the missile might have exploded. These men both praised the Medical Department for the kind and attentive care they received. They gratefully thanked us all before departing.

While in the Coral Sea one morning about 0200, our battle alarm sounded and we all hustled to our battle stations on the double, not knowing just what was happening. A lookout had spotted two small boats bobbing up and down in the heavy swell just to our port beam - how he ever saw them I will never know, as the night was jet black. Upon investigation they were found to contain 35 survivors of a recently-sunk merchant vessel. This merchantman had been torpedoed just the night before. We did not find out what happened to the remainder of the crew, but I do know that we had a happy and grateful bunch of men. That look-out was properly commended.

I hope I have answered all of your questions to satisfaction and also given you something of interest. I will end this with the Navy Medical Department's slogan: "To keep as many men at as many guns as many days as possible."

The Importance Of The Naval Reserve Military Sealift Command Program

by CAPT Irwin D. Nathanson, USNR (Ret)
Former XO and CO of various NR MSC Units

Strategic Sealift is one of the four primary missions of the entire USN. Yet whereas an officer in the Army or Air Force can progress from 2nd lieutenant to general while specializing, respectively, in land and air transportation, there is no similar career path in the active USN.

For example, as a commander, I performed three weeks at the US Transportation Command (TRANSCOM) at Scott AFB. It didn't take me long to realize that the typical Army or Air Force officer I met there knew infinitely more about Sealift than the naval officers assigned to TRANSCOM. The Army and Air Force officers had been specializing in transportation for their entire careers. Although experts in land and air transportation, they had learned something about ocean transportation as well. Their Navy counterparts, on the other hand, were for the most part surface warfare officers and aviators with no interest in, nor prior experience with, Sealift. They were at TRANSCOM merely to get their "Purple Suit" tickets stamped before moving on to other assignments unrelated to Sealift.

So wherein does the Navy's Sealift expertise reside? This knowledge and related skills resides with the civilian employees of the Military Sealift Command and with the officer and enlisted Naval Reservists drilling with MSC Naval Reserve Units! At least in the past, these personnel, like the author, have tended to "homestead." And this is to the advantage of the Navy as a whole!

Case in point: The most massive sealift in recent years was in support of the US and Allied involvement in the Gulf War. Official Navy videos covering the Navy's activities in this conflict devote the majority of footage to spectacular views of Tomahawk Missiles flying off the forecastles of BBs, and F-14s taking off from carriers. Only a very few moments are devoted to the massive sealift effort which made the entire progress of the war possible. Yet, in reality, the vase majority of the Navy's contribution to the success of the Gulf War was in surge and re-supply sealift. By comparison, while much more glamorous, the Navy had a very minor role as a combatant. To fulfill its sealift mission, MSC recalled approximately 80% of its Naval Reservists. Without them, the Navy could not have carried out its strategic sealift mission in support of the Gulf War.

Someday, the Navy may decide to treat Sealift as a distinct specialty in its own right, with its own warfare designator and separate career path. Until then, the only military personnel who really know something about sealift will be MSC Naval Reservists. Plus some Army and Air Force personnel who specialize in transportation.

From The Frying Pan Into The Fire

by Elwood W. Epps. CDR, CHC, USNR (Ret)

On August 5, 1965, Chaplain Elwood W. Epps departed the Continental United States from Travis AFB in California to serve one year with the Marine Corps in Vietnam. On August 6, 1965, he reported to 3rd Marine Division (Rear) in Okinawa. On August 11, 1965, he reported to 3rd Marine Division (Forward) in Da Nang, Vietnam. He was assigned to C Co., 3rd Med. Bn. as the Protestant Field Hospital Chaplain. The chaplain was to find that going from pastoring a church in the USA to pastoring Marines in the combat zone was like going from "frying pan into the fire."

C Co., 3rd Med. Bn. was a few miles inland from Da Nang Harbor at the base of Hill 327. C Co., also called C Med, was made up of tents with a helicopter landing pad for incoming casualties; the wounded were usually transported directly from the

Chaplain Epps in front of his quarters/office, "C" Company, 3rd Medical Battalion.

3/7 Chapel, Chu Lai.

combat area to the field hospital. In one 24 hour period over 120 casualties were received by the hospital. The medical officers, dental officers, corpsmen, dental technicians and others were involved without stop at such times.

Visitation in the wards was one of the ministries of the chaplains. One Marine infantryman said that he was on patrol when he was wounded by a booby trap. He lost one leg below the knee, part of his other foot, part of one arm and he had shrapnel in the abdominal area. His surgery took over four hours. His heart stopped in the midst of surgery and he had to be resuscitated. When asked on the next day as to how he thought he would deal with his injury; he said, "Chaplain, I'll be fine." He was looking forward to seeing his family; he was 17 years old.

Chapel services at C Med. were held under a fly tent. It was an impressive view with only a canopy overhead with rice paddies to the North and Hill 327 to the South as two or three dozen men sang and prayed together. One of the corpsmen at C Med. was helped in his adoption of a seven-year-old Vietnamese girl. The C Med Dental Unit provided dental care for the children at the Worldwide Evangelization Crusade's Home for Children.

After three months duty at C Med., Chaplain Epps was assigned on November 13, 1965, to 3rd Bn., 7th Marines (an infantry battalion) in Chu Lai, Vietnam. Chu Lai was described by Bob Hope in his Christmas show there in 1965 as "the biggest catbox in the world." It had sand stretching several miles inland from the south China Sea. The Marines had metal mesh on top of the sand to provide a runway for air operations. Several battalions were defending the miles of perimeter around the airfield; 3rd Bn., 7th Marines (3/7) was one of those battalions.

A few days after the assignment of Chaplain Epps to 3/7, an Army of the Republic of Vietnam (ARVN) Unit in Thack Tru was overrun by Vietcong and 3/7 was called on to go to their support. It was traumatic and tragic to see the dozens of ARVN troops that died. The Vietcong left the area and 3/7 returned to the defense of the Chu Lai Airstrip until called upon for other operations such as: Mallard (1/66), Texas (3/66), Hot Springs (4/66), Yuma (5/66) and Morgan (5/66).

Chapel services in the battalion were held in the Command Post in the thatch and bamboo chapel. On the perimeter six services were out of doors. Several services were conducted each week at different locations and services were arranged for other faith groups. There were many counseling opportunities dealing with such concerns as fear, family problems, the unpopularity of the war, discouragement and finances. Fear and foxhole religion was more easily understood with incoming mortar and rifle rounds. It made a person face up to their mortality and reach out to a power beyond themselves. Memorial services were held often to honor the deceased.

In one operation the chaplain was talking to the commander of L Co. and there were seven incoming mortar rounds. They scrambled for cover and checked out the situation. One Marine had an injured hand, a corpsman bandaged his hand and he stayed in the field. Later that night the chaplain transferred to a different company. An incident occurred during the transfer that showed the courage, resourcefulness and character of American troops in Vietnam. A Marine infantryman was accompanying the chaplain on the transfer. As they walked they heard some explosions. The Marine instantaneously pushed the chaplain down and fell on top of him. It was heartening to see this kind of behavior repeatedly not only on operations but also in the monotony of everyday living on the isolated perimeter of the Chu Lai Airstrip. That night there was one incoming mortar round and one man had to be taken to the field hospital due to head injuries. As the wounded man and the chaplain waited for the evacuation helicopter to arrive they shared their faith in God's providential watch and care over them and the rest of the battalion despite the anxieties at the time.

It is difficult to summarize briefly an experience such as this. In the first year that 3/7 was in Vietnam, 46 men from the battalion gave their lives and over 400 men from the battalion were wounded. It is appropriate to honor them by concluding this review with a quote from a letter from one their mothers, for it also expresses their courage and faith as well as hers.

"...I received the picture of my son yesterday.
I was so glad to get it.
As you know he was killed over there...
That has taken all the joy out of me.
He was my only son.
I loved him so very dearly...
We are praying for you fellows over there,
and we love you all.
Again, I will say you are brave and true."

This letter reminds one of the Father who gave His only Son that the world might have life and peace on earth as it is in heaven. On August 1, 1966, Chaplain Epps was detached from 3/7 for rotation to the Continental United States.

Good Friends

by Charles Franklin LaRue, CHC, LCDR, USNR (Ret)

My seven years of active duty with the USN, and 16 years as a Ready Reservist, were the finest years during my working career. Without a doubt, the largest single factor was the outstanding friendships I made among officers and enlisted personnel of the Navy and of the Marine Corps. Many of those friendships continue after some 30 years. My first ship was USNS *GEN M.M. Patrick*, manned by Civil Service and the Military Department of which I was a part. The Master was Olie Svehaug, tall and straight, having gone to sea at age 12, he was now 69. Our doctor was LT William Lukash and our nurse was LT Lyda Frank.

Miss Frank was a full-blooded Navajo and became a LVN at Ganada Presbyterian Mission, and she became a RN at Los Angeles Presbyterian Hospital. Dr. Lukash once referred to Miss Frank as the most competent nurse he had yet to meet. She retired with 20 years and the rank of LCDR Dr. Lukash remained in the USN for 30 years, with rank of RADM. President Nixon brought him into the White House, President Ford kept him in that position, as did President Carter during his four years in office. Possibly no other doctor spent so many years as White House physician.

LanShip Ron Five consisted of eight LSTs. I relieved an outstanding chaplain, LT Everett Nelson, in 1958. It was great to follow such a man. There were twin brothers on different ships of the squadron. Both assisted me as Roman Catholic lay-leaders. Both David and George Ungerer became helpful friends; David the more so, as he was an officer on the command ship. Chaplain Henry Exum Austin had set records with the Marines on Iwo Jima and in Korea before I met him at Moffett Field. Chaplain Austin and his whole family of wife and four children became my family. Casualty calls were difficult, but made much easier with the support of the other chaplains and the wives of Navy pilots.

The final names of outstanding Naval Personnel I have served with are LCOL Martin J. Sexton, CO of 3rd Bn., 3rd Div. and Grant Randall, the chaplain's clerk. COL Sexton was the most remarkable of all leaders and CPL Randall left little for the chaplain to do but preach. Good friends are a benefit of naval service.

Vestige Of An Epic Sea Battle

by Joseph M. Mathias

The SS *Kansan*, a freighter on which I was the officer in charge of the Navy Armed Guard crew, steamed into the harbor of Montevideo, Uruguay, on August 23, 1943. That was 43 days after we had discharged our cargo of war supplies for Russia in the Persian Gulf. Our voyage from there had been unescorted and mostly in heavy seas and icy gales.

As I stood on deck enjoying the sight of land once again, there came into view about a mile to starboard the upper hull of a sunken fighting ship. All at once, I was transported back to the early days of the war, before our Nation got into it, to an epic sea battle that grabbed the headlines world-wide. I said to the chief mate standing beside me, "That must be the *Graf Spee*."

"Sure is," he replied.

Ranging the South Atlantic in late 1939, the *Spee*, a German pocket battleship, had sunk 50,000 tons of merchant shipping carrying food and other necessities to Britain. At daybreak on December 13, 1939, a British hunting squadron of three cruisers intercepted the *Spee* as she headed for the mouth of the Rio de la Plata in search of other merchant prey.

The British ships were the *Ajax*, flagship of Commodore Henry Harwood, *Achilles* and *Exeter*. The *Ajax* and the *Achilles* mounted 6-inch guns and the *Exeter* 8-inch guns. None of them could match the firepower of the *Spee* which was armed with 11-inch guns.

A furious 90-minute battle ensued. Commodore Harwood's tactic was to use the superior speed of his squadron to divide the *Spee's* fire. At the outset, the *Spee* concentrated on the *Exeter*. The 11-inch shells demolished the *Exeter's* bridge and put her gun turrets out of action. But when the *Spee* began to close for the kill, the *Ajax* and *Achilles* raced in with guns blazing, enabling the *Exeter* to escape.

When two of the *Ajax's* gun turrets were knocked out, the British light cruisers withdrew from the action, but shadowed the *Spee* as she headed for Montevideo to repair her damage. The *Exeter*, with 64 of her officers and men dead,

Chaplain LaRue giving Communion to Marines at Camp White Beach, Okinawa, Christmas Eve, 1961.

headed for the Falkland Island naval base. The *Spee* had 36 officers and men killed. One of the *Exeter's* eight shells tore a six-foot hole in the *Spee's* hull above the water line and her superstructure was badly battered.

The *Spee* made the harbor at Montevideo, safe for the time being from her pursuers. There followed several days of high-powered diplomatic negotiations over the right of Uruguay, a neutral nation, to allow a combatant nation's warship to remain in her harbor. Ultimately, the *Spee's* captain, Hans Langsdorff, was given 72 hours to take his ship out. Meanwhile, the *Ajax* and *Achilles*, which had been joined by the British heavy cruiser *Cumberland*, waited farther down the estuary, waiting for the battle that seemed inevitable.

At 6:15 p.m. on December 17, the *Spee* got underway with her battle ensigns flying. Montevideons crowded the waterfront. They had front-row seats for an historic naval engagement. But abruptly the *Spee* came to a stop. Her crew was transferred to a German freighter that had come alongside. Then there was an explosion. Flames shot skyward and the German raider, scuttled by her captain's order, began slowly to settle on the river bed. And that's where I saw her three years and eight months later.

We did not tarry long in Montevideo, but steamed up the La Plata to Buenos Aires in Argentina. And that's where the final scene of the drama of the *Graf Spee* was played out. Captain Langsdorff had been given asylum by Argentina, but the loss of his ship was more than he could take. He wrapped himself in the Imperial Navy Ensign under which he had fought at Jutland in WWI and put a pistol to his head. He left a note that read in part, "...for a captain with a sense of honor, it goes without saying that his personal fate cannot be separated from that of his ship."

Prelude To Third Savo

by Roy E. Fors

The many sea battles in Iron Bottom Sound in 1942 were usually preceded by attempts to replenish troops and supplies by the Japanese navy and the USN. The troop transport, USS Zeil*in*, was on such a mission. On November 9, 1942, Task Group 62.4, under the command of RADM Norman Scott on board the USS *Atlanta* (CL-51) left Espiritu Santo, New Hebrides, on their mission to Guadalcanal. This task group was made up by the USS *Atlanta* (CL-51), troop and cargo carrier, USS *Zeilin* (PA-9, later to become APA-3), cargo transports USS *Betelguese* (AK-28) and USS *Libra* (AK-53). The destroyer screen consisted of USS *Aaron Ward* (DD-483), USS *Fletcher* (DD-445), USS *McCalla* (DD-488) and USS *Lardner* (DD-487).

This task group hoped to elude long-range air detection by passing north of San Cristobal Island, but the Japanese had sent several submarines to the lower Solomon's to establish a picket line. It was a I-class submarine carrying a seaplane that sighted the force on November 10 and managed to stay beyond anti-aircraft range. Our sighting was reported and the element of surprise was gone.

Our task group passed Lunga Point and arrived off Kukum early on the morning of November 11 and began discharging cargo and units of the Army's Americal Division to join their buddies of the 164th Inf. Regt. Condition Red sounded at 0830 hours, and from the deck of the *Zeilin* you could spot aircraft from the Cactus Air Force taking off from Henderson Field and circling for altitude waiting to intercept. At 0930 hours nine "Val" bombers from the Japanese carrier Hiyo appeared overhead escorted by 18 Zekes. Our ships were underway and violently zigzagging to present an ever-moving target as the Vals peeled off on the transports. There were many near misses straddling the ships as they fired back with their 3-inch anti-aircraft guns and 20mm guns. The danger of collision was great, as was the possibility of casualties from friendly fire as the guns followed the planes all the way down. The "Vals" were also firing their 7.7mm cowl guns at us.

VF(N)-53 Detachment, 1944.

From the fantail of the *Zeilin*, where I was gun captain of our 5-inch 50-surface gun, we watched helplessly as we tried to elevate this gun above its 16 degrees maximum. Our 3-inch guns on the tubs above us were firing away rapidly as five of these planes had singled us out. Watching the fixed landing gear of these planes and the bombs angling down on us had us transfixed until the bridge passed the word to our talker to take cover. We had two near misses, one astern and one off our port quarter that lifted our fantail out of the water. A third bomb glanced off our starboard side between #7 and #8 cargo holds and exploded under water. As a result of these blows, the *Zeilin* shipped a considerable amount of water and suffered cracked plates a broken propeller shaft. In retrospect we considered ourselves lucky. Had the bomb landed on deck or in the cargo holds, we could have been drawing flight pay to this day since one hold still had barrels of high-octane gas in it and the other 155mm ammunition. The attack was over quickly (although it seemed like forever) and all the "Vals" were reported down by the combined fire power of all the ships in our group. Condition Red continued and at 1030 hours a second air attack occurred by 25 "Betty" bombers and a like amount of "Zeroes." The focus of this attack was against dock and storage area.

The ship continued unloading troops and cargo as our stern had settled several feet and we had a distinct starboard list. With the addition of several pumps from other ships and the hard work of our damage control and working parties, we remained afloat. We had disembarked all our troops, but not all of our cargo when we were detached from Task Group 62.4 and ordered in the late afternoon to depart the area with the USS *Lardner* (DD-487) as our escort and protector. At this point the forces were building up for the third sea battle for Guadalcanal. As we slipped out of Sealark Channel to the east, somewhere during the night, we passed by the trained and ready guns of RADM Richmond Kelly Turner's Task Force 67 of troop and cargo carriers, and the fighting ships that were to participate in the coming sea battle.

Our progress was very slow on our trip back to Espiritu Santo and the feeling of dread was very strong since we knew that there were several Japanese subs in the area and we were a fat, crippled sitting duck. We probably would have been more scared if we knew what was taking place back in Iron Bottom Sound. Perhaps the Japanese subs were drawn to that action because we limped back into Segond Channel, Espiritu Santo, safely a few days later. A short time after that, the heavy cruiser, USS *San Francisco*, joined us there. We were in awe of it and its ability to survive in its shell-riddled condition. After temporary repairs we left the area to return to the States by way of Samoa. After three separate landings at Guadalcanal since August 7, 1942, we were happy to return to the Terminal Island Naval Base, Long Beach, CA, for repairs and overhaul.

The USS *Zeilin* was repaired and updated and returned to the Pacific War for several more invasions from the Solomon's to the Aleutians. Between the Aleutian Campaigns in 1943, it had the honor of having the shipboard scenes of *Guadalcanal Diary* filmed on board it. It was a very lucky ship having survived a kamikaze attack in the Philippines with heavy damage and loss of life and ultimately surviving the war. It was decommissioned in 1946 after earning eight Battle Stars and a Naval Unit Commendation.

The Love Day Invasion of Okinawa

by Robert N. Colwell

Dawn on April 1 saw our main invasion force standing off the Hagushi Beaches lowering invasion craft, and deploying the rocket boats that were to augment ADM Blandy's ever-continuing bombardment. As each fully loaded invasion craft was hoisted over the side of one of the transports, it sped a short distance to the nearest circle of sister craft, where the circle expanded to accommodate it. Then with almost unbelievable ease and rapidity, and still without enemy opposition, the last few amphibious craft of the assault force had entered the water and joined their respective circles. The time was H-hour (0830), the date was April 1, 1945, and the day was not only our long-awaited Love Day, it also was both Easter Sunday and, of course, April Fools' Day.

By now it was a truly impressive site that I was privileged to view as I stood, with other members of GEN Buckner's staff, on the flagship's main deck. For a stretch of about five miles to our north and five miles to our south, the entire offshore area was churning with the circles of troop-laden boats. Probably at no time in the entire history of warfare has there been a smoother coordination of vast forces than that which was to unfold before my very eyes. The frogmen of the Underwater Demolition Teams had now completed their removal of the hedgehogs and other underwater obstacles off the beaches.

And the rocket boats were moving into place, each capable of unleashing firepower that, for a few moments, would be greater than that of the mightiest battleship.

Even so, our off-lying battleships and other elements of ADM Blandy's Bombardment Support Force were blasting away, to their fullest, truly awesome capability.

According to my notes, at that particular time the actual US bombardment by ships alone was coming from 10 battleships, nine cruisers, 23 destroyers and 177 gunboats. And although our aircraft carriers were many miles away, their aircraft roared through the skies around us, ready to strafe and bomb the beaches in support of our invasion craft as they made their perilous runs shoreward. I noticed that GEN Buckner was making heavy use of his field glasses to observe this critical phase of the operation, as he would a few minutes later when the first wave of our assault force actually hit the beach.

Would the weather hold, at least through the main assault landing? The answer to that question had been very uncertain even a few hours earlier, but now the answer was "Yes! A storm may strike later today, but now H-hour of Love Day, the weather is okay with only moderate offshore waves and gentle surf conditions. Yes, the 'sea state' is favorable. And, of course, the tide is high, all according to a plan we had made many weeks earlier."

But will the landing craft clear the shallow, wide, nearly flat reef as we had predicted from our aerial photo measurements of water depths (given these weather and time conditions) or will they hang up perhaps hundreds of yards offshore as had so disastrously happened at Tarawa, a few amphibious landings earlier? We soon would know.

On schedule to the minute, those 10 miles of circling invasion craft were breaking from the circles; and now the whole armada of them, forming a line nearly 10 miles long was streaking landward, each craft toward its respective beach segment. Talk about "the sound and the fury." A scene such as this probably will never again be duplicated in warfare. Midst the cheers from those few of us who were lining the rails of the ships - cheers that certainly were drowned out by the roar of it all - the armada skimmed landward. There! See the ones heading toward the beaches nearest to us? They have already cleared the outer fringe of the reef, now they're over the middle of it, now they're within 100 yards of dry land.

Now they've run aground, but very near to the shoreline, just as we predicted. Look at those boat ramps splash down; look at those troops wade and scramble; wade and scramble. And look to the north, and to the south, almost as far as you can see. They have all cleared the reef, just as had been predicted from the tide stage and from aerial photo-based measurements of water depths that we had made nearly one month earlier! Soon we could even see the troops at one of the nearest assigned beaches as they deployed their ladders, quickly scaled the seawalls and scrambled inland the first few yards.

But there was one major surprise: Thus far our landing was not being opposed by the enemy!

COL Ely had estimated, mindful of experiences that the Allied forces had had elsewhere (especially at Tarawa and Iwo Jima), that 4,000 troops would die in the landing phase - perhaps the most dramatic few minutes in amphibious warfare. The official report given the next day to GEN Buckner stated that, not 4,000 men, but one man had died during the assault landing. Combat-laden he had stepped into a deep hole in the coral reef and in the confusion of things, had drowned before his buddies noted his plight.

"Why" I wondered, "had the Japanese decided not to oppose the landing?" (Surely our pre-invasion bombardment, heavy though it was for seven or eight days, had not been so effective as to wipe them out.) Perhaps such vital information was known to us but tightly held at higher levels of security than had been entrusted to me, even though I felt sure that I was being entrusted with all information, regardless of its security classification. Did GEN Buckner and ADM Turner know? I soon got my answer from GEN Buckner, who had been pacing the deck of our amphibious command ship while watching the assault landings. I heard him say to those of us on his staff who were there lining the rail with him.

"I don't know what the Japanese are up to by not opposing our landing, but so far it suits me just fine!"

I can only conclude that our pre-invasion intelligence was lacking in this vital respect. But it fits perfectly with the Japanese plan, as we later confirmed it, of letting virtually all of our troops get ashore but with only limited supplies. Then, with their 100,000 man army that was already stationed there at Okinawa when we landed, they intended to annihilate us at their leisure, aided by their suicide boats, their kamikaze aircraft and their mighty, already-approaching surface ships.

Near Miss

by CDR Paul J. Lundbeck

As a LTJG and ASW officer, I became a plankowner of the USS George E. Davis (DE-357), commissioned at Orange, TX, in August 1944. Our eight ship division of DEs departed Boston, MA, after six weeks operational training in Bermuda, en route to the Leyte Gulf invasion, Philippines; via Panama Canal, Galapagos Islands, Bora Bora, Tahiti and Hollandia, New Guinea.

Upon entering Leyte Gulf in November 1944, after a month plus at sea, we were detached to refuel alongside an almost empty Dutch oil tanker. While we were refueling, the cruiser, USS Minneapolis arrived, as SOPA, waving us off for their refueling. We hover to off their starboard waiting to resume refueling. Half-way through their refueling we observed a Japanese Kamikaze plane appear over the Samar Mountains, heading for the Minneapolis. The kamikaze aimed for the bridge of the cruiser in a steep dive. Fortunately it missed, crashing instead through the bridge of the tanker, then into the empty #2 oil hold, penetrating the outer hull about three feet above the water-line. The bridge was destroyed, killing or wounding many personnel.

We were at General Quarters, firing all guns. The cruiser fueling detail, meantime had chopped all the 8" manila hawser lines with azes to free the ship. The entire zoot fueling detail jumped overboard. We spent the remainder of our first day in the gulf picking up sailors out of the water.

Vietnam Reminiscences

by CAPT Edmund P. Willis, USNR (Ret)

When President Kennedy called up the Reserves in 1961, I was a lieutenant with the Reserve Crew of Alvin C. Cockrell (DE-366) at Alameda, CA. Cockrell took station off Vietnam with a patrol to interdict supplies from the north to the Viet Cong. While there I was ordered to temporary duty as naval liaison officer in Da Nang and advisor to Vietnam Navy Lieutenant Nguyen Ba Trang, commander of the small naval base.

One day a team of Navy Seals arrived to begin training Vietnamese sailors. They wished to rent a place to live. I introduced the officer in charge to Madam Trung, whom I called the "Dragon Lady." She had the contract to build a 10,000 foot runway at the air base and had all the right connections. She put us in touch with an owner of a house for rent. He showed us through the house, but wouldn't let us enter one room. Someone had recently died there and it was the custom not to enter the room until the spirit of the deceased had departed, normally in 30 days.

On another occasion I met a Canadian missionary, the Rev. Gordon Smith, who was mentioned in Robin Moore's book, *The Green Berets*." Rev. Smith did evangelical work with the Montagnards, or mountain people such as the Hmong tribe. Rev. Smith's organization, the Worldwide Evangelization Crusade, also operated an orphanage and a leprosarium in Da Nang.

In those years the International Control Commission patrolled north and South Vietnam, hoping to prevent the introduction of additional arms into either zone. The ICC posted teams of three officers in various locations, including Da Nang. An Indian major lead each team; the other two being Canadian and Polish. One day an American ammunition ship arrived and anchored in the harbor. The Polish officer wanted the ICC team to inspect it. MAJ Nair pulled out his rule book and read that the team could inspect a ship only if it was alongside dock. When

The USSC-432, 1937. Reserve training subchaser, Great Lakes.

The officers and crew of the USSC-432, 1937, Ludington, MI.

Jeep Carrier CVE-57 Anzio, Pacific Ocean, spring 1945. Composite Squadron 13, VC-13, all the airmen.

Jeep Carrier CVE-57 Anzio, Pacific Ocean, spring 1945. Composite Squadron 13, VC-13, all the pilots.

it came time to unload the cargo of bombs for the South Vietnamese air force, a South Vietnamese army unit escorted the ICC team out of town on an inspection tour. MAJ Nair said the North Vietnamese air force, a South Vietnamese army unit escorted the ICC team out of town on an inspection tour. MAJ Nair said the North Vietnamese did the same, preventing them from inspecting trains from the USSR, arriving via China.

LT Trang made a lasting impression on me as a patriotic and dedicated representative of South Vietnam. Until drafted he had been a high school history teacher. He expected to be in uniform for the duration of the conflict. After I left the country, we corresponded. In 1966 he invited me to Philadelphia where he and his crew took command of a US patrol craft being transferred to the South Vietnamese navy. At the re-commissioning ceremony they renamed it *Ngoc Hoi*, commemorating a military victory in ancient times against the Chinese. Trang later advanced to commander and led a VNN force in the Mekong Delta. The US awarded him the Bronze and Silver Stars.

Composite Squadron Thirteen

by John J. Sharer

After winning the Navy Unit Commendation for outstanding performance of duty in combat with Air Force Atlantic Fleet in 1943 and 1944, VC-13 was re-deployed to the Pacific in December 1944. After 10 months of duty with Air Force Pacific, the squadron returned to the US for leave and reassignment.

In 30 months of continuous sea duty from Casablanca to the Third Fleet's strikes on the Japanese Home Islands this two ocean squadron's submarine sinking specialists hunted down and attacked 18 Nazi and Japanese subs. Eight of these were sunk, six were damaged and as one pilot expressed it, "the rest got the hell scared out of them." In addition to the submarine sinking specialty, the squadron participated in the invasion and occupation of Okinawa accounting for 11 Japanese planes destroyed in the air and many land and sea targets.

In making this record the squadron's Wildcat Fighters and Avenger Torpedo bombers flew 14,546 hours of sorties, 2,005 hours being at night from escort aircraft carriers in the North and South Atlantic and Western Pacific. Serving on five "Jeep" aircraft carriers the squadron cruised 140,684 nautical miles or almost six times around the world tracking down the enemy underseas craft. Twelve officers and 16 aircrewmen lost their lives during the total tour of duty with the Atlantic and Pacific Fleets.

The Squadron flew from the USS Core, Guadalcanal, Tripoli, *Kassan Bay* and *Anzio*. Foreign ports of call included Casablanca; Bermuda; Recife, Brazil; Argentina, Newfoundland; Pearl Harbor and Hilo, T.H.; Guam; Pelelieu; Ulithi; Leyte and Samar, Philippine Islands; and Kerma Retto, Okinawa. On April 3, 1944, the squadron became a "shellback" Squadron as the Equator was crossed aboard the USS Tripo*li*. On February 28, 1945, they became Golden Dragons as the International Date Line was crossed aboard the transport CVE USS *Long Island*.

In March and April 1945, taking a "vacation" from the tedious, unrewarding and dangerous day and night anti-submarine warfare, the squadron flew in 33 days at Okinawa 884 sorties, 61 of these at night.

During the period from May 23 to June 15, 1945, 24 hour flight operations were conducted continuously with only three interruptions - once for zero flying weather and twice for ship refueling.

From June 3 at 1030 to June 10 at 1820, planes were kept in the air continuously for 175 hours and 50 minutes which is believed to be a record for continuous carrier operations.

On April 6 one division led by LT Douglas R. Hagood of Fort Worth, TX, while flying combat air patrol over Okinawa encountered the squadron's first Japanese planes and, by way of baptism, shot down 10 of the Japanese who were attempting suicide dives on two DDs below. No damage was inflicted on the division which included in addition to LT Hagood who accounted for two of the Japanese; LT William E. Davis, Little Rock, AR, who shot down three; LTJG Eugene D. Pargh, Nashville, TN, who accounted for three and LTJG Thomas N. Blanks who shot down two.

When two Zekes approached the carrier on April 16, 1945, Division leader LT Hyde and his wingman ENS John Sharer, Section Leader LT Davis and his wingman, LTJG John W. Alexander, were vectored out to intercept them. The division split into two sections to keep the Zekes between them. LT Davis and LTJG Alexander spotted the Zekes first. The squadron's eleventh Japanese plane was shot down by LTJG John W. Alexander, Lake Charles, LA. LTJG Alexander shared this kill with his section leader, LT Davis. The other Zeke ran away so fast neither pilot could overtake it.

In addition to the 11 planes destroyed in the air, two were knocked out on the ground, one Japanese tanker was damaged and probably sunk, five suicide boats were destroyed and 25 were damaged; gun emplacements were attacked and damaged on 10 missions; trucks and tanks on four; a radio station was destroyed; one ammunition dump was bombed; enemy airfields at Yontan and Naha were bombed and strafed on four strikes; buildings were hit on 22 missions; and on 10, caves and trenches were attacked.

...contrary it was lucky and as a deliberate challenge to fate they selected for the squadron insignia the Black Cat superimposed on the supposedly equally unlucky 13. The insignia was designed for the squadron by Walt Disney in 1942.

The insignia proved unlucky for the Germans and Japanese only, as on July 13, 1943, the present squadron commander, LCDR Robert P. Williams, formerly of Snoqualmie, WA, now of Cardiff, CA, sighted and sunk the squadron's first Nazi sub. At that time the squadron was flying from the USS Core (CVE-13). The squadron boarded the USS *Anzio*, its permanent base in the Pacific on March 13, 1945. The squadron returned to the US with 13 Japanese flags (11 for planes destroyed in air and two for subs sunk) and just to clinch the argument for the non-superstitious arrival in San Francisco was on September 13, 1945.

Outstanding among the squadron's Atlantic actions was the one beginning July 13 and continuing to August 23 during which LCDR Williams, then squadron operations officer, now its commander, sighted four Nazi subs and sunk three of them. Thirty-three prisoners were taken from the first sunk on July 13; three from the July 16 sinking and on August 23, 36 prisoners were taken including the U-boat's captain. For this outstanding individual accomplishment, LCDR Williams received the Navy Cross, the Distinguished Flying Cross and two Gold Stars in lieu of second and third DFCs. His radioman at this time, Morris G. Grinstead, ARM1/c of Letts, IA, who is now at Mercer Field, Trenton, NJ, received the Distinguished Flying Cross and promotion to chief petty officer. His gunner Melvin H. Paden, AMM2/c of Salinas, CA, now on duty at NAS San Pedro, was promoted to AMM1/c and awarded the Air Medal. Several of Grinstead's pictures taken during these attacks proved to be the outstanding anti-submarine action shots of the war and appeared in papers and magazines throughout the country. One national news magazine (*Time*) used one shot for a cover in one of its few covers not of an individual. LCDR Williams in a previous tour of duty in the Pacific in 1941-42 with VB-2 aboard the *Old Lex* won another Navy Cross for his attack on a Japanese aircraft in the battle of the Coral Sea.

LCDR Williams' first attack on July 13 followed strafing runs by three of the squadron fighter pilots. LTJG Earl H. Steiger of Buffalo, NY, who was shot down by the sub and went down with his plane; CDR Charles W. Brewer, USN, of Tulsa, OK, then squadron commander and LT Julius R. Brownstein of Corpus Christi, TX. LTJG Steiger received the Navy Cross posthumously for his attack. CDR Brewer and LT Brownstein received DFCs. CDR Brewer was missing in action after winning a Navy Cross for Japanese planes shot down over Japanese held Roi. Another TBM piloted by LTJG James F. Schoby of Bode, IA, assisted LCDR Williams in the attack. LTJG Schoby was later lost at sea aboard USS Guadalcanal when, while attempting a landing in a rough sea, his plane went over the side. Assisting LCDR Williams in the August attack was LT Martin G. O'Neill of Jersey City, NJ, fighter pilot who received the Distinguished Flying Cross and who is now on duty with VOC-1 LT W.W. Felter of Port Huron, MI, now on duty with ComFair West Coast sank a Nazi U-boat on August 24, 1943, only five hours after LCDR Williams sank his third U-boat.

On January 16, 1944, LT William M. McLane of Jacksonville, FL, and LTJG Bert Hudson of Coeur d'Alene, ID, now on duty at NAS Westerly, RI, sighted a large Nazi U-boat refueling another on the surface. Hudson made a run, firing rockets and dropping depth charges between the U-boats. The smaller sank slowly by the stern while some 40 men abandoned the larger. McLane then came in for a similar rocket and depth charge attack resulting in an explosion which blew men off the deck and caused the larger sub to sink stern first. Although LT McLane dropped life rafts and vests no survivors could be found when DEs arrived. The weather was foul, the sea rough and it is believed the Germans drowned in the heavy sea. Returning to the tiny carrier, Hudson found it corkscrewing in the heavy seas. The plane ahead of him crashed on the pitching deck. Night was falling and it was impossible to clear the deck. In attempting to land after dark Hudson crashed over the side and he and his crew were picked up by a DE. For this attack LT McLane and LTJG Hudson received Distinguished Flying Crosses. LT McLane's crew: K.G. Dodd AOM1/c of Pasco, WA, and A.C. Torres, ARM1/c of Los Angeles, CA; LTJG Hudson's crew: R.J. Horacek, ARM2/c of Lincoln, NE, and Peter Schwahn, AOM1/c of Oregon, all received Air Medals.

On August 23, 1943, LT McLane sighted, attacked and damaged another Nazi U-boat. In this attack he was assisted by LT Robert E. Bomar of Tullahoma, TN, fighter pilot who swept the deck clear of gun crews before LT McLane's bombing attack. On July 16, 1945, while flying anti-submarine coverage for the Third Fleet strikes on Japanese Home Islands, LT McLane sighted a fully surfaced Japanese submarine. Closing the target, he dived through a small cloud, strafed the conning tower and fired his rockets at almost point blank range with positive hits. The sub was badly damaged and could not escape. It submerged and was finished off by a DE and another squadron plane piloted by LT Rex W. Nelson of Port Orchard, WA. Evidence of the kill was afforded by a large oil slick, debris which littered a square mile area and included much deck planking, a Japanese magazine, packages of personal letters in Japanese, a photograph of a Japanese family and Japanese money. Dodd and Torres served as his crewmen in all three of these actions.

On October 20, 1943, LT Robert Hayman of Red Wing, MN, flying from the USS *Core* sighted and sank a U-boat with depth charges. He was awarded a Distinguished Flying Cross for this successful attack. His crew each of whom received the Commendation Ribbon comprised Charles L. Roades, ARM1/c, Radioman of South Bend, IN, and William J. DeWildt, AMM2/c of Grand Rapids, MI, now with CASU-21. LT Hayman was assisted by LCDR Brewer who strafed the surfaced sub.

In May 1944 the Allied anti-submarine campaign was so successful that the Nazis surfaced only at night. The squadron received special training in night attacks at Norfolk and Quonset Point, RI, and on August 19, 1944, LT Carl E. Lair of RFD #2, Celina TX, at the time on loan to another squadron, made the first successful night attack using a searchlight to spot the sub. These new tactics broke the already ailing back of the Nazi underseas craft for now it was not even safe to surface at night to charge batteries and refresh the crew. His crew of Henry W. Iwanicki ARM1/c of Lowell, MA and Charles Paris of Lynchburg, VA, now an aviation cadet, received Air Medals.

All during August and September of 1944, the squadron operated day and night in extremely bad weather off Newfoundland. The climax came August 27 when a heavy Grand Banks fog closed the ship in four Avenger torpedo bombers and five Wildcat fighters were forced to ditch in the water. All personnel were rescued by escorting DEs.

The squadron flying in unusually bad weather in the North Atlantic and North Pacific has the dubious and "damp" distinction of having what is believed to be more water landings than any other CVE based squadron. Aside from the squadron's fatal loses, 17 Torpedo bombers and eight fighters either had to be landed in the water due to impossible landing conditions on deck or went over the side as a result of rough seas. Only seven of the water landings were due to engine failure.

There was such an epidemic of water landings that the squadron yeoman ran out of application forms for the "Sea Squatters Club" as with their crew of three the 17 TBM landings made 51 members of the squadron eligible in addition to the fighter pilots. The Yeoman mimeographed application forms and the secretary of the club noting the unorthodox application blanks pouring in wrote the squadron offering to send more applications. The letter was received the day the squadron started back t the States and squadron commander replied that he didn't think they would be needing any more.

In November 1944 the squadron participated in what might have been an outstanding story of the war but fortunately for the East Coast did not materialize. On November 6 reports were received that the Nazis were planing to launch buzz bombs against New York, Boston and other East Coast cities. At the time the squadron was at Norfolk preparing for its seventh anti-submarine cruise in the Atlantic. On two hours notice the entire squadron was moved to Quonset Point, RI, so the squadron's sub busting pilots would be available if the Germans showed up. The movement was made so rapidly and secretly that many of the pilots had to phone their wives from Quonset that night to tell them they would not be home for diner. Day and night patrols were flown in search of the buzz bomb launching subs, but none appeared and the East Coast was freed from the strongest possibility of attack during the war.

In the squadron's second successful sub kill in the Pacific, LTJG Sam L. Stovall of San Diego, while on a night anti-submarine search on May 31, 1945, in the sea lanes between Okinawa and Guam sighted a fully surfaced Japanese sub. An apparently accurate rocket run was made before the sub dove. Shortly after the first attack a second was launched. The success of the latter was evidenced by an oil slick which covered a 2 square mile area and debris which included splintered deck planking, cork, a report form printed in Japanese, flashlight bulbs stamped with Japanese characters and bits of human flesh. LTJG Stovall's crew in this attack was William D. Rodrigues, AOM1/c gunner, San Leandro, CA and Adam Green ARM2/c, Baltimore, MD. Radioman, LTJG Stovall has recommended for the Distinguished Flying Cross for this attack and Rodrigues and Green the Air Medal.

Three officers and one aircrewman of the squadron returned to sons and daughters whom they had never seen. The fathers are LT Nelson, LT Brownstein, LTJG T.N. Blanks of Meridian, MS, and K.G. Dodd, AOM1/c of LT McLane's three sub threesome.

One member of the squadron, Joseph Gill AMM1/c of Ozone Park, L.I., returned a year older but minus a birthday as his birthday occurred the day the squadron crossed the International Date Line east of Guam and the day was skipped.

VC-13 was placed in commission at NAS Seattle, WA, August 5, 1942, with CDR Thomas Ashcraft, USN, of Coronado, CA, as commanding officer. CDR Ashcraft was succeeded by LCDR Charles W. Brewer, USN, Tulsa, OK. The squadron's third commanding officer was LCDR Adrian Perry, USN, of Pasadena, CA, who became air officer of the USS Tulagi. The present commanding officer, LCDR Williams assumed command on June 22, 1944, at Fentress, VA. In addition to LCDR, Brewer who succeeded to command the following two officers served as executive officer, LCDR Victor L. Zimmermann of New York City now on duty with Air Group 93 and the present executive officer, LCDR Edmund B. Jackson of Boston, MA.

The squadron first sailed against the Germans on March 24, 1943, aboard the USS Core and against the Japanese on March 21, 1945, aboard the USS Anzio.

USS George F. Elliott

by Roy E. Fors

Bombed and sunk during the Guadalcanal invasion August 8, 1942, 11 killed and 13 wounded.

Killed: Samuel T. Broomall CMP; Gordon II. Craig LT; Thaddeus J. Durski, MM2/c; John M. Delony, ENS; Henry H. Glorch, MM1/c; Francis E. Kelly, F2/c; William Landsburg, S2/c; Stanley II. Roof, F2/c; Robert E. Schenk, F2/c; Marion G. Schleich, MM1/c; Stanley Sizemore, WO.

The *George F. Elliott* (AP-13) had the misfortune of being the first large ship sunk at Guadalcanal, Aug. 8, 1942. She was the only large troop transport that was lost during the first six months of the Guadalcanal Campaign.

George F. Elliott (AP-13), formerly the *City of Los Angeles*, was built in 1918 (one of five sister ships) by the Bethlehem Steel Company in Alameda, CA. She was acquired October 30, 1940, and put in commission January 10, 1941, at Maryland Drydock Company, Baltimore, MD. Almost half of her complement was made up of the 21st and 30th Naval Reserve Divisions from Chicago and Peoria, IL.

USS G.F. Elliott (AP-13) and the USS McCawley (APA-27) in Guantanamo Bay, 1941.

She spent 1941 carrying units of the 1st Marine Brigade to the Caribbean for training exercises, and operated out of Norfolk before departing New York February 19, 1942, with over 1,100 men bound for Europe. After joining a convoy off Halifax, Nova Scotia, she reached Belfast, Ireland, March 3 to debark her passengers. She took on passengers at Guerock, Scotland, and subsequently returned to New York, March 25.

After embarking 1,229 fighting men, the ship got underway April 9 with a convoy bound for Tongatabu, arriving one month later and debarking her troops. She sailed May 19 (with some Lexington survivors aboard) and arrived San Francisco June 5 for repairs.

Soon ready for sea, she embarked 1,300 men of the 2nd Battalion, 1st Marines, and stood out under the Golden Gate June 22 in convoy reaching Wellington, New Zealand, July 11 where combat gear and stores were loaded. As part of Task Force 62 she departed July 22 for the 1st Marine Division's amphibious assault on Guadalcanal. After conducting landing maneuvers in the Fiji Islands she proceeded to Guadalcanal.

Closing Lunga Point on D-day, August 7, 1942, *George F. Elliott* sent her boats away at 0733 and simultaneously began discharging cargo. Despite enemy air attacks, she continued to work far into the night, ceasing unloading only when the beach head became too congested. The next day, August 8, she got underway at 1056 to avoid an imminent air attack and at 1159 opened fire on Japanese twin-engine bombers coming in very low and fast over Florida Island. Her anti-aircraft guns made repeated hits on a plane approaching the starboard beam only 30 feet off the water; suddenly it swerved and crashed into the ship amidships, spreading raging flames and rupturing the water mains. In spite of the crew's valiant efforts, the fires continued out of control. As salvage was impossible, she was abandoned, and the gutted ship was later sunk.

With the help of the USS *Hull* (DD-350), most of the survivors were transferred to the transport USS *Hunter Liggett* (AP-27) and transported back to Noumea, New Caledonia for further assignment.

With only a relatively small turnover of personnel, the crew of *Elliott* served together for almost 18 months. They had attained a high rating of efficiency in cargo and troop handling and gunnery during that period. Because of that, it had been the hopes and prayers of the survivors that they would be sent back to the States as a unit for leave and new construction. It didn't happen that way as they were transferred in small groups to other ships and stations in the South Pacific.

USS *George F. Elliott* was struck from the Navy list October 2, 1942. She was awarded one Battle Star for WWII service.

Japanese Occupation

by L. Vernon Frye

I was commissioned an ensign on July 9, 1945. For service in the Pacific operational area and occupation assignments, I received the American Campaign, Asiatic-Pacific Campaign, WWII Victory and Navy Occupation Service.

I was a senior in high school and living with my parents on December 7, 1941. I had been away from the house and returned in the evening to find everyone gathered around the radio. They were listening to a news broadcast about the Japanese surprise attack on Pearl Harbor. Like most everyone, I had been following the events of the war in Europe, but there was no concern expressed regarding a possible warring situation with Japan. I clearly remember the shock and compassion felt for those servicemen involved. It was beyond comprehension that the reports we were hearing could be true. It was just unbelievable at the time.

My 18th birthday occurred on March 11, 1942, which necessitated registering for the draft, although 18-year-olds were not drafted prior to November 13, 1942. Upon graduation from high school in June 1942, a close friend and classmate, Frank R. Watkins, and I drove to Pittsburgh, PA, to enlist in the Marine Corps. I was informed that as a draft registrant, I could not enlist. Frank was only 17 and had passed the physical examination, but refused to enlist without me. This occurrence was most fortunate for both of us. Frank, who incidentally was a descendent of Heinrich Frey through Benjamin, Samuel and the latter's son Abraham, enlisted in the Air Force and became a pilot serving in the China-Burma-India Theater. I returned to my job, as a bulldozer operator, with Republic Steel Corporation at their coal stripping mine near Charleroi, PA. This work was classified as essential to the war effort, which resulted in draft deferment status.

In early 1943, the armed forces became concerned about the impeding shortage of officers and undertook to supplement those being trained at the military academies and certain colleges in the Army's ROTC and Navy's NROTC programs. A nationwide examination was scheduled for April 2, 1943, to select qualifying applicants for these new training programs. I applied to take the examination for the designated Navy V-12 Program. Nearly 316,000 applicants took the written examination and 125,000 were enrolled in the V-12 program of which nearly 60,000 successfully completed. I was inducted into the Navy at Pittsburgh on July 1, 1943, and placed on inactive duty until October 1. After reporting for active duty, I completed four semesters, or two years, of college and related Navy training before graduating from the USNR Midshipman School at Notre Dame on July 9, 1945.

I had requested shipboard duty on a destroyer, which resulted in my being assigned to the Navy Destroyer School at Miami, FL, for advanced training. Pending the start of the next class, I was ordered to Fort Schuyler, NY, for temporary duty. While at this station, the surrender of Japan occurred and the war was over. This resulted in an abrupt change of plans. Instead of advance training for combat, the Navy's objective centered on replacing veterans who had earned enough points for release. I reported to Miami on schedule, but was advised that the school had been closed and new orders were issued for me to report aboard the USS English (DD-696) on station in the Pacific.

Transportation was by air and sea following a circuitous route that resulted in reporting aboard the ship some 49 days after leaving Miami. The ship was anchored in Tokyo Bay at the former Japanese Naval Base, Yokosuka, Japan.

The last leg of the journey was from Guam by air to Tokyo, Japan, and by bus to the Yokosuka Naval Base. I arrived before noon and went to the Port Director's Office only to be told that the ship was not berthed at that site. I asked about lodging and was directed to the Officer's Club. It was located nearly a mile distant, which I walked, only to be told the facilities were filled to capacity. After at least five round trips, toting all of my luggage, between the two facilities and getting the same response from different personnel, I arrived back at the club about 10 p.m., thoroughly exhausted. I pleaded with the petty officer on duty for a place to lay down. He took pity and offered me a cot and blanket that he kept in a garage, or barn located a short distance from the club. Grateful, I sought out my berthing for the night, but upon entering the dark building with a flashlight, I became apprehensive concerning my safety. There were no doors or window panes and the building abounded with junk and debris. It was totally accessible to the public and many Japanese people were roaming the street. However, the night was uneventful and I awoke rested, but with a lingering memory of the event. Reporting again to the Port Director's Office, they still insisted the *English* was not in the bay. In desperation, I walked down to the waterfront and noted a motor whale boat displaying the numerals "696." I asked if they were from the *English* and the response was "yes." I later learned the ship had been anchored in the same berth for several weeks. So much for "snafus" that were very common in that post war period.

The air view of Tokyo gave an appearance of scattered wooden and masonry debris laying in flat incongruous patterns. Although it gave a lifeless appearance, there were undoubtedly people living in sheltered areas of their former homes. The only section of Yokosuka that I saw was near the former naval base and those structures appeared to have not been damaged. Our task of inspecting and disarming naval vessels also took us into other port cities such as Sasebo and Nagasaki. We were able to take a jeep tour of the latter. Roads had been bulldozed through the rubble consisting of bent and twisted molten steel and glazed masonry. The destruction resulting from conventional bombing of Tokyo was extensive leaving little that was salvageable, but in comparison Nagasaki's damage was near total. It was an unforgettable experience. Although the route that we toured had been declared safe, I have often wondered if it would have been safe by today's criteria and testing measures. I have talked with a number of people from the *English* and other ships, some of which toured the area a month before I did, but none have reported any problems. In carrying out our inspection and disarmament obligations, we found the Japanese to be most courteous and very cooperative, The same could be said for the ordinary citizens who we encountered.

In early January of 1946, we received orders to return to the US, and with exception of the captain, nearly all officers had been replaced by new inexperienced officers. We arrived in San Francisco in late January and by way of the Panama Canal arrived in Baltimore in mid-February. The *English* had traveled over 150,000 miles since leaving the east coast 18 months earlier. The *English* carried out designated Atlantic Fleet assignments and visited east coast ports until placed in the reserve fleet and berthed in Boston in mid-1946. I was released from active duty on July 1, 1946, and returned to inactive duty until recalled for the Korean War in 1951-53.

World War II

by L.A. Basinger

In WWII when Pearl Harbor was attacked, we as teachers at Willard High School, and as families, at first were stunned. Then we were angry. No country could get by with a sneak attack on our United States and kill thousands of our boys. We all joined in helping the war effort in every way we could, even when such efforts seemed to be ever so small in results, we felt it all helped. I took my agriculture classes to the field to do hand work, when what we accomplished seemed to be only a drop in the bucket—pulling carrots out in the marsh. We had to come up with faster methods, but that is another story. I made a food drier and put it on display in the local bank, so people could see how they might put up food, which is sure to become short in wartime. Everybody did things like that. To save money and to contribute to the war effort, we all tried to buy "war stamps" and war bonds.

A "Civil Defense" unit was established in most communities. We felt this to be especially needed in our railroad town. We knew that the enemy might

INSTRUMENT OF SURRENDER

We, acting by command of and in behalf of the Emperor of Japan, the Japanese Government and the Japanese Imperial General Headquarters, hereby accept the provisions set forth in the declaration issued by the heads of the Governments of the United States, China and Great Britain on 26 July 1945, of Potsdam, and subsequently adhered to by the Union of Soviet Socialist Republics, which four powers are hereafter referred to as the Allied Powers.

We hereby proclaim the unconditional surrender to the Allied Powers of the Japanese Imperial General Headquarters and of all Japanese armed forces and all armed forces under Japanese control wherever situated.

We hereby command all Japanese forces wherever situated and the Japanese people to cease hostilities forthwith, to preserve and save from damage all ships, aircraft, and military and civil property and to comply with all requirements which may be imposed by the Supreme Commander for the Allied Powers or by agencies of the Japanese Government of his direction.

We hereby command the Japanese Imperial General Headquarters to issue at once orders to the Commanders of all Japanese forces and all forces under Japanese control wherever situated to surrender unconditionally themselves and all forces under their control.

We hereby command all civil, military and naval officials to obey and enforce all proclamations, orders and directives deemed by the Supreme Commander for the Allied Powers to be proper to effectuate this surrender and issued by him or under his authority and we direct all such officials to remain at their posts and to continue to perform their non-combatant duties unless specifically relieved by him or under his authority.

We hereby undertake for the Emperor, the Japanese Government and their successors to carry out the provisions of the Potsdam Declaration in good faith, and to issue whatever orders and take whatever action may be required by the Supreme Commander for the Allied Powers or by any other designated representative of the Allied Powers for the purpose of giving effect to that Declaration.

We hereby command the Japanese Imperial Government and the Japanese Imperial General Headquarters at once to liberate all allied prisoners of war and civilian internees now under Japanese control and to provide for their protection, care, maintenance and immediate transportation to places as directed.

The authority of the Emperor and the Japanese Government to rule the state shall be subject to the Supreme Commander for the Allied Powers who will take such steps as he deems proper to effectuate these terms of surrender.

Signed at TOKYO BAY, JAPAN *at* ______ *on the* SECOND *day of* SEPTEMBER*, 1945*

By Command and in behalf of the Emperor of Japan and the Japanese Government.

By Command and in behalf of the Japanese Imperial General Headquarters.

Accepted at TOKYO BAY, JAPAN *at* ______ *on the* SECOND *day of* SEPTEMBER*, 1945*

for the United States, Republic of China, United Kingdom and the Union of Soviet Socialist Republics, and in the interests of the other United Nations at war with Japan.

Supreme Commander for the Allied Powers

United States Representative — *Republic of China Representative*

United Kingdom Representative — *Union of Soviet Socialist Republics Representative*

Commonwealth of Australia Representative — *Dominion of Canada Representative*

Provisional Government of the French Republic Representative — *Kingdom of the Netherlands Representative*

Dominion of New Zealand Representative

Japanese Articles of Surrender, WWII, signed September 2, 1945.

be coming over the Arctic Circle to try to put our B&O Railroad out of commission by dropping a bomb in the center of the "Hump," where trains were made up. One machine gun placed in the center of the town of Willard could completely immobilize us. The whole town could be paralyzed, and since we were the division line for the B&O RR, the effect would be very far reaching for passenger transportation as well as for freight. We had plans for rerouting transportation should we be wiped out. Some important industries set up "fake factories' about two miles away from the real factories, and lit up the dummy plants at night, so if the enemy did try to bomb us out, they very likely would hit the lit areas and miss the real targets which, of course were blacked out.

We had short courses in first-aid, so civilians could help the wounded if we were attacked. We were scared, but we remained calm.

In the high school at Willard, OH, I taught an adult class. The main subject was taught in the farm shop, with units in preventive maintenance of farm machinery and equipment. We taught such things as how to get double use out of hammermill screens, and how to harden plow shares using the electric welder to lay a bead along the under side; anything to save money and steel, which was becoming short in supply. We needed plenty of steel for war equipment, ships, planes, tanks and guns.

At first this course, for adult farmers, was funded with state and federal monies. A woman supervisor drove a Packard automobile from Columbus to Willard, a total of 120 miles or more, to check on why I had not attached an invoice for $1.25 repair bill I had claimed in one of the reports I had to send to her office. Using her gas stamps, we were rationed in the use of gasoline, to find out if I was cheating, made me furious. I took her to the man who had repaired an electric motor for the school shop, and let him tell her that he just was too busy to make out a bill at the time, and that he had never thought she would drive to Willard to check on such a thing. I took this story to the HS Superintendent, and he was as furious as I was. He told me to never use these state and federal funds again, our school board would approve anything we needed to continue this program. We did not need the above kind of efficiency, but people did things like that. Give them an ounce of authority and they surely could find a way to waste time and money.

The first former FFA boy killed in action, at Salerno, tore me up inside. This kid did not deserve to die. Then I lost another, in the South Pacific. My superintendent at New Haven had volunteered and had taken a direct commission in the Navy. At Willard the athletic coach and the chemistry teacher both volunteered, taking commissions in the Navy. Some of these guys were older than me. They wrote and told me that if I volunteered for the Navy, to ask for Destroyer duty, and not to get too good a grade in my navigation or I might be placed on a bomber headed for the far Pacific. "You would only make one mistake." Then they wrote telling me not to ask for destroyer duty, it was extremely rough when the seas got to rolling.

In the meantime, I did assistant coaching of football, as our staff at the schools became smaller and smaller, older retired teachers came to "fill in." Some couldn't handle the discipline. Almost every man under 29 years of age was in the service of his country. Some people thought I should be in the armed forces, they harassed my parents with remarks like, "We don't see why your sons don't have to go." "Is it true that you have talked to the draft-boards so your sons don't have to go to service?" That did it. With a wife and three children, I felt I must enter the armed forces. I'd taken enough of this crap, even though I knew it was thoughtless, and certainly not true. Hopefully, such people will get their just rewards in one way or another.

I decided that if I went to war, I would get all I could, in such a way as to take care of my family before, during, and after my service. I planned so my family would have a place in which to live. My wife understood, as she had three brothers in the armed forces, two of them reached the rank of colonel. I went to Detroit to apply for a commission in the USN. I had no idea, for sure, if I could qualify.

The Detroit Office of Naval Procurement was in the Cadillac Building. I got into line and picked up papers-filled them out, and waited my turn to talk to the interviewing officer. As I sat there just outside this officer's door, I couldn't help but overhear the questions being asked the candidates. The thought occurred to me that if the candidate just ahead of me could make it, maybe I might also, all at once I heard this: In a whiny voice, "I feel that maybe I ought to be in service, so I came in to see what the Navy has to offer." The interviewing officer was a Marine Captain, with a whole chest-full of "fruit salad" (ribbons and medals) and his answer to this wimpy guy came out with a roar, "Did you come in here to go shopping? Get out of here."

I took my cue from the above, and when the officer invited me into his office, I went in and waited for him to tell me to be seated. He looked over my paper work, then asked, "What can I do for you?" My answer was, "I came to apply for a commission in the Navy, Sir." His reply was, "You are an agriculture teacher. Don't you know you are needed at home to help produce food?" My reply to this question was, "Who is kidding who?" His answer was, "You might do. Let's see if you can pass the physical and mental tests."

That day they gave me a very thorough physical examination and was pronounced "fit for service." I passed my mental examinations that afternoon, and was told that if everything was okay in my security clearance, I would be sworn in. This last would take a few weeks. They gave me a "date of rank" and a commission as ensign in the USNR. I was 29 years old. I had to go to Cleveland for my active duty physical examination and for swearing in. At this point I discovered that once I became a commissioned officer of the USNR, I would be an officer until I die. Protecting our country would always be a part of me, from this day on. (In a few weeks this last would be drilled into me, so much so, that if I ever saw anything at all that suggested action or even leaning toward the overthrow or injury to our great country, I would be obligated to report what I knew to the USN Department.)

In the Cleveland Office of Naval Officer Procurement, everything was about the same as in the Detroit office. Naked men walking around from one station to another carrying sheets of paper. One guy lost a paper for a few minutes, and the entire office panicked, maybe the paper flew out the open window, we were on fourth floor. Then they found it, a little Wave had it, over at her typing table, where she was completing some of the requirements for the man carrying this paper. She sat over in a corner, so busy with her work she never noticed the men walking around her, or the search being made for the papers she was working on. In fact, no one had time to be embarrassed by human frailties. The man who had lost his paper was so badly needed, he was sent directly to service, without uniforms, and without ever going home. I never found out what his specialty was, but they wasted no time getting him sworn in.

I bought uniforms in Cleveland, actually a bit over-fitted, I found out later that one set of dress blues would have gotten me through, but while one suit was in the cleaners, it was nice to have another.

My first "duty" was to report on August 24, 1944, to the Naval Training Station at Princeton, University, Princeton, NJ. I had my uniform shoes too tight, which I found out in 24 hours with all the marching we had to do and had to get Navy issue shoes.

I was assigned to Patton Hall (billeted) and assignment to a unit came next. Men who were from all walks of life made up the companies. A recent full-back from a large university, a quarterback, former secret serviceman, former FBI men, former IRS men, attorneys, teachers, salesmen, chemists, clergy, farmers, professors, all were in my outfit. Some could attend opera and symphonies and know exactly what was coming next. Again I felt very humble, but proud to be a part of this group - all volunteers.

We soon learned to make up our bunks, and that an "Irish pennant" which is a corner of a sheet showing under the bed, or a bit of dust picked up by the inspecting officer on his white gloves, likely would result in your being restricted for the week-end. In other words, no liberty; you "don't go off the base this week-end, sir." It does not take long to learn that the Navy means exactly what it says. Exactly. We learned the five S's (s..., shave, shower, shampoo and shine) to be performed in a few minutes. To be "out of uniform" was a cardinal sin, and could cost you a "week-end" or worse.

It soon became apparent we could eat about double the amount of food eaten in civilian life, and still lose weight. We became strong, solid men in a short time. No fat around the middle. The first day in physical training class, the instructor, a chief petty officer who had been a track star at one of the eastern universities the year before, told us we would run around Carnegie Lake. I asked him how far it is around Carnegie Lake. His reply, "Three miles." You could imagine the groans from the guys. Some of us had to walk that first day, and for several days, for short laps of the three miles, but with this guy's encouragement, in a few days we could trot all the way around Carnegie Lake and not even puff. We could swim for long distances, and take care of ourselves in the water. We could do calisthenics almost forever. Military drill was "duck soup" after ROTC at Ohio State in 1933 and 1934.

At Princeton we learned naval courtesies, orientation in conduct of ourselves under almost any circumstance, socially and in the battle conditions, small arms care and firing at targets, communications by Morse Code, flag hoists, and semaphore. We had considerable training in military discipline. Dead reckoning navigation, reading charts, and we studied during every waking hour. Celestial navigation would come later. We toughened up our bodies, and our attitudes regarding facing the enemy. Our spiritual life was not neglected, there are no atheists in foxholes, someone has said, and there are none at sea.

At .45 caliber pistol range, one day, we had a new officer who put all his shots "in the black." We thought he had missed the target altogether. Upon examination, his holes were all nicely grouped in the center. The instructor asked him where he had learned to shoot, and his answer was, "With the FBI, sir. When he first came to the Naval Training Center, this man had looked so ineffective, but boy, how he could shoot. And a .45 is one of the most difficult, to get a good score. This gun is unwieldy, heavy, and has quite a recoil. It is hard to hold it steady, but the advantages of the .45 are that the muzzle velocity is slow, compared to most guns, about 800 feet per second, and since it fires a large heavy bullet, it is excellent to stop a man, or to spin him around when he is hit, where a bullet fired from a higher muzzle velocity and smaller bullet, the projectile will possibly pass right through an enemy's body, allowing him to continue to charge. If he is moving toward us with bayonet, knife, or his own gun, and he is not stopped in his tracks, it can be extremely dangerous to us.

We worked every waking minute. Flash cards, study devices were used all the time. We tested each other as we sat around in the evenings. We had semaphore to learn, flag hoists that we would be seeing hung from nearby ships to tell us something without breaking radio silence. We studied Morse Code, the coding and decoding of messages. In the first class every morning, our instructor would give us a briefing of the news of the world, because we were completely cut off from newspapers and newscasts. This officer would always end his news report with something like, "Little Orphan Annie is having a hell of a time." We were becoming "90 day wonders."

One day my wife wired me to tell me our oldest daughter, who was about seven years of age, had had an emergency appendectomy a couple of days before. Rebecca didn't want to worry me with it, so she waited until Kay was "out of the woods," and on the mend.

At Princeton we had to take our shots, and be vaccinated against a number of diseases we might come in contact with. I remember how the yellow fever shot burned. We got it in both arms, and we soon were so sore we could hardly let anybody touch us where it hurt. That is when my wife, bless her, decided to come to see me. I got a room in Philadelphia, and we should have had a second honeymoon but I was so sore from those damned shots I could only lie on my back or my belly. Becky didn't seem to mind, in fact she thought it was funny.

Former World Boxing Champion Gene Tunny was the head of our physical training program. Once in a while we would see him. One time he was riding a bicycle with one arm loaded with groceries and we pulled up with a salute (he was a lieutenant commander) and he couldn't salute back, then he would say, with a big grin on his face, "Darn you guys."

In swimming, we had instruction in the various strokes, diving, and abandon ship. To abandon ship, before jumping into the sea, you grab your left shoulder with your right hand, grab your nose with your left hand, cross your legs, then jump. All this to prevent one from being knocked out by any debris floating in the water, or to be wounded in the groin should you fall astride of something. It makes sense. Life-saving came later.

We graduated in the beautiful chapel at Princeton, receiving a nice certificate. Our assignment for further service came next. I was sent to General Line School at Hollywwod, FL. I had been in the Navy two months.

At Hollywood we were quartered on the sixth deck of the Hollywood Beach Hotel. We had to muster the first thing every morning. We "messed" and went to PT (physical training). We now had to learn celestial navigation. We had more training in leadership, communications, including coding and sending/receiving radio messages. More sophisticated stuff, and more skills and also how to use classified information in convoy.

Physical training at Hollywood was rougher and tougher than at Princeton. Living on 6th floor (now called a "deck") was one method to toughen up our legs. We had to make at least six trips up and down every day, with no use of the elevator, except when going on liberty and dressed in whites. Swimming was tougher. Water polo taught us to take care of ourselves in the water, the referee would say, "No holds barred, no fouls unless you draw blood, and I'm color-blind." Then he would toss the ball up for grabs. We got bumped, skinned, and held under water until we thought our lungs would burst, but nobody drowned.

Our swimming instructor gave us a lesson in life-saving. That is saving of another person, it went like this, "Now, when you go to save a guy from drowning, it would be stupid for you to lose your own life in the process. If, when you try to help a drowning man he should get a strangle hold around your neck, and if they are frightened, and they will be, to break his strangle hold, let yourself go deep into the water. Get a good breath of air first. Usually they will let go of you, for going deep in the water is the last thing this guy wants to do.

One day, while on the beach, we saw a dive bomber, an SPD Dauntless, fail to come out of a dive and go directly into the drink about two miles out at sea. We felt helpless, there were two men in that airplane, and we couldn't get to them to help. The next morning a diving crew recovered one wheel from the wreckage. That is all.

In celestial navigation we were taught to use the sextant at twilight of stars, which is that time on a clear evening when just before dark, the stars and planets appear nice and clear. We were taught to use the Rude Star Finder to determine what heavenly body we had shot. We took three star sights, at wide angles, so we could set up a triangle on the chart, showing where we were. From now on, any map is a "chart." Each star-sight had to be timed exactly, so we used a rated man with a stop-watch to record the exact time, and to record the "angle of declination." This latter term means the angle of height above the horizon. By using Dutton's "Navigation" and charts of the positions of all heavenly bodies commonly seen from our position at a given time of the year, at any precise second we could plot our position. We had to be able to come up with our position to within two miles of our known position. "This is where we are." This whole procedure probably did more to convince me there is a God and a precise creation of the universe. It was, and still is, awe inspiring.

We learned to use the book, *Tides and Currents*, to find out the height of tides and the direction of currents for anyplace in the world, and at any time. This was absolutely necessary, to know the depth of water under a ship when going into or out of a harbor. Currents can work with you, or against you, and we had to learn where these are likely to be, and so we would go with a current to make a good time, and save fuel.

We were required to learn how to use the *International Light List*. This book had every light of any kind on all sea coasts in the world, and gave the position of the light, the kind, the color, and the flashing frequency. No two lights are the same in any one area along a coast. By timing the flashes of a light, we can tell where we are by looking up the light that we are viewing by description in this book. It may be a flashing light, one flash every three seconds. It may be a light with flashes at variable speeds, like one flash, three seconds later two quick flashes.

Channels in harbors are marked by red and black buoys by day, and by red and green lights at night. In coming into a port, we keep red lights to our right side, "Red right returning."

The rules of the road will tell what lights ships are to "show" at night. Foremost lights are a few feet lower than the light on the main mast. Thus, by looking at a ship's lights, we can tell which direction he is headed. The starboard "running light" which is on right side of ship is always green. The port running light is always red. A stern light, near the main deck of the vessel will be white. If a ship is pulling a tow, the tow will show appropriate lights.

Aboard USN ships, most have radar, which allows us to follow the course and speed of another ship long before we can see his running lights at night, or before we can see and identify him in daylight. For this reason, during war time, all ships are considered to be "targets" until he identifies himself by answering our message, "Identify yourself, friend or foe." If he is our friend, he should give the correct pre-arranged reply. If he does not, we may hear our captain give the order, "Commence firing," we have been ordered to battle stations as soon as the vessel is picked up on radar, if we don't know who is out there.

In time of war, only hospital ships run under lights. All the rest of our ships under darkened ship conditions. No flashlights are permitted topside, and no cigarettes. A lighted cigarette can be seen for a mile on a clear night, especially with binoculars.

It takes Combat Information Center on a ship only a few moments to plot the course and speed of a target, and they can tell us exactly when we will be making our closest approach to this vessel, if neither vessel changes course or speed. We can track a ship for hours without being detected. We can also tell if a submarine is under us, anywhere close by the use of our sonar gear.

A paragraph or two back I told you something about lights used in navigation at night. When we go into a harbor, there will be a "Range," of probably two red lights, to get lined up in our path ahead, to mark the approach to the channel. As a rule, a ship must take a trained pilot aboard when entering a harbor. No matter how experienced the captain of a ship may be, the pilot knows that particular channel better than the captain.

The ship's whistle sounds mean something, the length of the blast, the time between blasts have a meaning. Before any voyage, the whistle must be tested. Usually there is a race to pull the whistle cord among officers on the bridge, when the captain gives the order, "Test the whistle." One long blast, from a vessel at anchor or moored also lets crews of other ships know that some movement by the ship sounding the blast is about to come about. Two short blasts signal that we are about to back down. After we have backed out into the harbor, one long blast indicates we are about to get under way. By observations from their own decks, and by the use of navigational aids, other ships can tell what we are doing, and their knowledge of the harbor or slip locations and conditions, they can very well predict where we are headed. In peace time, radio communications will also tell our intentions.

The ship's bell is used to let the crew know what time it is. The day starts at 0:00 to 04:00. The second watch is from 04:00 to 08:00. The third watch is from 08:00 to 12:00. The fourth watch is from 12:00 to 16:00. The next watch is called the "first dog watch," and is from 16:00 to 18:00. The second dog watch is from 18:00 to 20:00. The next watch is from 20:00 to 24:00 (midnight). The purpose of the two hour dog watches is to move the duty times of the crew, so they do not stand the same watches all the time. One thing it really means is that one night a crewman will get six and one-half hours sleep, the next night he will get four and one-half hours sleep, and the third night he will get two and one-half hours of sleep. After three days, the schedule starts all over again, and will go on until the voyage is ended by arrival in the next port. It takes a lot of black coffee for the crew under way. This is in peace-time. In war, under battle conditions, the crew may do entirely without sleep for several days at a time. This is the reason why every Navy person must pass a very rigid physical examination upon entering service, and the reason the Navy chow is nutritious, plentiful and well prepared. Pity those ships at sea with water so rough the crew cannot get a hot meal for days and days. That is when they eat "sea rations" that are planned and preserved to be as nutritious and as tasty as possible, even though they come out of cans, or eat dry bars of chocolate and fruit.

I neglected to mention earlier, the ship's bell is used in times of breakdown, when a ship is limping along or dead in the water. Also in time of dense fog. Even though radar may be working, the ship's bell is sounded at regular intervals during periods of fog. Any nearby ships will hear our bell, for they will be moving slowly under extreme fog conditions, as a rule, and every ship will have a man on watch forward to report by sound-powered telephone, or regular phone, to the bridge at the very moment he hears or sees another ship. Other watches may be posted around the sides and stern of the ship, also in the "crow's nest" at the top of the masthead. Sometimes these posts are relieved more often than every four hours, when it is imperative that the crewmen on these assignments be extremely alert. You may be sure they are supplied with plenty of coffee, and officers keep in direct contact with them, to be sure they are awake and sharp. It should be kept in mind that when near any shore line, these crewmen on watch may from their vantage points be the first to sound the alarm if we are getting too close to shore. With a fathometer and all electronic equipment in operation, and with all visual look-outs on duty, it is possible, and sometimes happens, a ship can run aground causing possible serious damage to the ship, loss of valuable time, and additional expense to the taxpayer.

Emergency measures are practiced by the crew at times of drill. Sometimes the drills are unannounced, and no one but the captain and the executive officer, and the officer of the deck may know if the alarm is sounding a real emergency or a drill. This keeps the entire crew alert and ready for whatever happens. Every crewman, when not on regular watch, has an assignment, a place where he is to be under any emergency condition, and when the alarm is sounded, he had better be at his post as fast as he can get there. Sometimes we find crewmen getting to an emergency station in all stages of dress or undress; it can happen at any time of day or night.

All seamen are trained for certain primary duties. Their emergency assignments, as much as possible, correspond to their training for the rank or rate they hold. A roster, called the "Watch, Quarter, and Station Bill," posted at several places on the ship lists the assignments of all personnel at all times and under all conditions. As promotions take place, and crewmen advance in rate or rank, as they gain competence, their assignments also change to more responsible duties, both regular and emergency.

Seamanship includes everything from a knowledge of the ship's equipment, navigational aids, such as radar, compasses, sonar, radio, intercom, charts and maneuvering board, the anchor and mooring gear, navigational charts and books, life-boats, and life-rafts, how to use all the above, and when. Terminology and skills have to be learned, and speed in using this information must be developed. Lives of the crew depend on it. What to do in case of fire, man over-board, and collision are all a part of what crewmen, officers and men must know how to do, and what individual part he is responsible for in every circumstance.

Crewmen, sailors, may be trained in one or several competencies but officers of the line must be knowledgeable in all phases of seamanship. Officers in specialty fields, such as medicine, finance, or law are not expected to be as informed as line officers, and only line officers give commands in the operation of the ship. Officers not of the line are expected to take command of any situation should all line officers be wounded or killed.

How to handle men is also a skill needed by an officer. To get performance from men (or women) of the crew is the number one priority. To be respected by the crew usually gets the best performance. If an officer is not liked he can be in deep trouble. There are many ways in which a crewman can make an officer look bad, while the crewman appears to be just plain innocent. For an officer to use force to get orders carried out should be the last resort. To use force is sometimes necessary, and if a situation is that bad, the one receiving the order usually fully understands the need for a command or order to be carried out, and he will respect an officer who is fair, and not afraid to do his own part, and to take the necessary action, which protects the entire crew, when eminent danger occurs. By example: I was asked by a chief petty officer, who in civilian life was a close friend of mine, and who could have been a commissioned officer himself, but he preferred to remain an enlisted man, "Why do you carry a 45 on your hip?" I replied, "When I have the duty, as I do now, if I ask you to do something, you are going to do it." He told me, "You wouldn't use a gun to get me to do something." I told him, "Just don't ever try me." In wartime, I might have to use a gun to get an order carried out. Only commissioned officers can understand this last, and this is why they must be so carefully screened, when the lives of an entire crew, or of a larger number of a fighting force may depend upon the performance of one single man at one precise moment.

We were taught to prepare our eyes so we would have good night vision by wearing red goggles for about 20 minutes before going on watch, on the bridge. "Night vision" is when your eyes are prepared to see by the use of the rods in your eyes instead of the bars. One can see in almost total darkness, when your eyes are adapted. On the bridge, there will be absolutely no lights, and especially when operating under "darkened ship" conditions. There is to be no light topside, a flashlight could give our position away to the enemy, a lighted cigarette can be seen a mile away, especially if the viewer has binoculars. It may seem unreal, but true, present day binoculars and gun sights allow us to see in almost total darkness.

Time at Hollywood, FL, went very fast. We were permitted to go on liberty and usually took it in Miami. The married men usually did not stay off base over night. I visited friends and saw a show or two. But we had to study most of the time. In chapel, I never will forget the Navy Hymn, and the reciting of the Lord's Prayer by 500 Navy officers. These men did not know where they were going to be sent, or if they would be coming back! They boomed it

USS Evans (DD-552) in camouflage.

out. These men were offering themselves for what they believed, regardless of the sacrifices, whatever the outcome might be.

At the end of our training period at Hollywood, our next assignments came like this: we were called in by the officer in charge of placement. We were told that we had three choices for our next duty: "Amphibious forces, or we could take amphib, or a third choice, we could have amphib." I told him I would take amphib. My next duty was at Fort Pierce, FL.

Fort Pierce, FL, was considered by many in the Navy to be the worst place in the world. It was under the command of a four-striper who had goofed up by allowing an entire flotilla of five ships to secure their boilers while moored off the coast of Africa. A German submarine came up in the midst of this flotilla and sank all five ships. This sea captain was banished to commanding the Fort Pierce training base for the duration of the war, and probably to face a court-martial after the war. (I hope so. Any man in his right mind would not have allowed ships in a battle zone to permit their ships to secure boilers). Scuttlebut (rumor to you)had it that when you saw his Annapolis ring, you were to shudder. If we saw a green automobile, be it Chrysler, Dodge, DeSoto, or Plymouth anywhere on the base even at distances of 100 yards or less, we had better come up to a salute, because the old man would have a paddy wagon following who would put you on report, and you would have to face captain's mast for not saluting. Captain's mast is like traffic court in civilian life, held every morning by the captain, and he can mete out just about any punishment his little wooden heart might desire. He seemed to be mad all the time. Only one person, again according to scuttlebutt, and probably true, as it sometimes is, really put the old man in his place. This was the head nurse at the hospital in Fort Pierce. It seems this sea captain ventured into the hospital to visit a sick friend (we didn't know he had any friends) and this nurse stopped him, and asked where he thought he was going. The old man said he was going to visit his friend. The nurse told him, "No one is visiting that man." The captain asked, "Do you know who I am?" The nurse's reply was, "I don't care if you are the good Lord Himself, you are not going into that room. Now get the hell out" The captain left the hospital.

At Fort Pierce we learned to handle the 36 foot attack boats. Some were "hard noses," without a ramp, and some were LCVPs with ramps. Also, we handled 50 foot attack boats with ramps, that could haul trucks, jeeps, as well as personnel. LCVPs had a single 225 horse-power Grey Marine Diesel engine. The LCM's, 50 footers, had two Grey Marine Diesel engines. These engines operated on diesel fuel, had water-cooled engines, but the water cooling the engines circulated through the block was cooled by running it through a "heat exchanger" which was cooled in turn by sea water pumped from the ocean. The sea water had to filtered, and the device called a "sand trap" had to be watched continually by the engineer of the crew, and the filter had to be changed periodically as the need arose. This meant two filters were a part of the assembly.

A special device aided in starting this engine in cold climates and in my total experience with attack boats, in the Atlantic and in the Pacific, for about a year and a half, we never had a problem in starting these diesels.

Both the 36 footer LCVPs and the 50 foot LCMs had crews of four men: a coxswain, engineer, signalman, and bowhook, all having been especially trained for the duty they were assigned. An officer was assigned to six attack boats when "hitting a beach." Communications used included semaphore, blinking lights, and hand-carried "walkie-talkies." The signalman could communicate with the mother ship, as well as with other boats, and especially the boat carrying the boat-group commander (the boat officer). Orders for any maneuvers, whether in a drill, or in an attack, all came from the Combat Information Center aboard the mother ship. CIC orders were from the captain of the ship, sometimes through the executive officer, to CIC and to the OOD, all receiving the same information at the same time. The captain may or may not have received his orders by coded message from the admiral of the fleet or a subordinate. (No one but CIC, then the captain was privy to any coded messages, as these might be highly classified in nature).

Both the small and larger attack boats were equipped with "Mouse Traps" (rocket launchers which could fire a volley at the proper distance from the beach being attacked). These rockets could only be used on the first wave of each attack group. If used on successive groups hitting a beach, very likely our own fighting men would be killed by our own rockets, those men we had put on the beach in the first wave and from all waves thereafter.

Boat groups would, after loading men from the mother ship, circle in a broad enough area to accommodate all six boats in the group, as they had to load off the ship one at a time and as soon as all six were in the circle, and the command came from the ship to hit the beach, the boats would fall into a straight line formation and head for the beach, usually at top speed. When a group would hit the beach, and all the beaches were marked by the underwater demolition teams who had gone in earlier so we knew we were to hit the correct beach. While we were on our way to strike a beach, planes of the Navy would fly smoke screens over our heads, and close to the beach to protect us from they eyes of the enemy. Sometimes we had to rely entirely on our compasses, and on orders from the CIC of the mother ship, who was watching our every move on their radar. In this way, we could land hundreds of men carrying armament, with jeeps and trucks, loads of ammunition, food, medical supplies, in one fell swoop! We appreciated the smoke-screen cover, you can bet. Think of the practice it required to learn to coordinate an operation, such as I have just described. It required that every person involved do his job and be alert. Mistakes could be costly in time and in loss of life. Some mistakes are bound to happen, due to human error, or due to equipment failure. The latter almost never happened if equipment were properly handled.

At Fort Pierce we practiced landings until we were adept at backing off into the surf, turning, and heading back to the mother ship, in smooth and in rough water, without broaching (taking on excessive amounts of water or upsetting). Life-jackets were a must at all times when hitting the beach, and we practiced at night as well as in daytime. Discipline was a must at all times.

One night, when maneuvers were over, and each boat-group was to assemble and prepare to go back into the base harbor, after the boat group commander satisfied himself that all boats had checked in and ready to go in, we had one boat officer who got tired of waiting, and without orders from the commander, took his six boats and crews into the base. After checking and re-checking all boat groups, and commander finally accepted the idea that this officer had taken his group in, we all received the order to follow the commander's boat into the base landing. When we arrived in at the landing, about one hour late, the commander and all the rest of us found the "lost" boat all neatly tied up in its berth, and its crew nowhere around. Upon proceeding to quarters, we found the officer and his 24 men all in the sack, they simply had come in and gone to bed. The commander searched out the quarters until he found the officer who had left the flotilla and come in early, I have heard a few cases where officers were chewed out by a superior, and times when men had to be dressed down. But this officer got what he really deserved. He had caused us to lose one hour of our night's sleep, and a lot of fuel wasted while we were trying to find out where he was, out there at sea. We all were aware that this officer might at any time disobey an order, and this cannot be countenanced under any circumstance. I hoped this officer learned his lesson! We knew that if he could not follow orders, his men under him could very well get the same habit and the damage could snowball; in fact, we considered him to be dangerous to have around. He might be expected to bolt discipline when circumstances became tense, and he could endanger lives of many people around him. There are very few times when it is advisable to disobey orders.

At Fort Pierce I stood watch, had the duty on a regular basis, looking out for the safety of some 300 men. We had them billeted in large quarters, with bunks three high. One night we had a man become deathly ill, and we transferred him to sick-bay, where he died in a few hours. We did not let the rest of the men know he had died. He had been diagnosed as having spinal meningitis. My job was to keep this from spreading to other personnel. We had a couple of other men get sick and we sent them to sick bay at once. We changed bunking arrangements so that the men slept head to foot with others. This reduced breathing each other's breath. We prayed we wouldn't lose anymore men, and we didn't.

Most officers got a case of catarrhal fever (we called it "cat" fever). Many went to sick bay (hospital), but I refused to go. I had no fever, but stuffy nose and a cough. It was extremely cold in the barracks, especially for Florida, with no heat; the room temperature got to around 55 degrees F. These barracks were tents mounted on solid wood floors. We had folding cots with thin mattresses. I found a spare mattress, and covered myself with my overcoat over my blanket at night. I slept in long-johns under pajamas, and lay there shivering. I recovered from the cat fever, and didn't get another cold while in service, that must have more or less immunized me.

At Fort Pierce I learned to "mid-night requisition." Some would call it stealing, but what the heck, everything belonged to the taxpayers anyway. The commanding officer sent me for materials to build a gant line, clothes line to you landlubbers. The supply officer refused to give me what I required when I went through channels. When I told this to the CO, he swore a bit, then told me, "You go and get that stuff, and don't you come back until you have it." So, I found three sailors, took them with me, and we grabbed the first truck we saw (with no one's permission) and started out to drive through the Navy base. When we saw something we could use, like posts, we said, "There's our posts." and took them. We picked up some wire the same way. Every time we saw anything we needed, we got it. We loaded it all on that truck and headed back to Camp 4, where we were billeted. When we drove into our camp, the CO was standing outside and the only remark he made was, "That's more like it."

One evening I got some special duty - *Life* magazine had visited the base, and asked to photograph some attack boats "hitting the beach." In trying to please the photographers, the boat crews chosen to make this effort sank an LCVP by diving it onto an underwater railroad rail obstruction placed there to keep out the enemy. We had heard that the Germans were wearing our uniforms, landing on our beaches, and attending our movies at night. Anyway, a small boat was sunk off Coon Island. The CO grabbed me, told me to get eight men, and go out to Coon Island and keep watch over the sunken boat. At low tide, we were to try to repair the bottom of the boat and get her back to base. Out on that beach, the sand fleas were so bad we built a fire and roasted one side of our bodies while the sand fleas ate the other side. When the tide went out, we found a hole in the bottom of the boat about 15 feet long and two feet wide. No way could we fix that so we could float her out. The next day we had a jeheemie, an A-frame, large enough to straddle the boat, hoist her up, and hauled her out and back to the base. Those sand-fleas were one of the reasons Fort Pierce was sometimes called the "worst place in the world." That night was the most miserable night I ever spent, before or since.

When it came near time for us to "ship out, the CO told me to go to the supply officer and get three sea chests: one for the CO, one for his executive officer and one for me. I went to see the supply officer, a commander, and told him I was to get three sea chests. He refused three, but would okay two. I took the paper approving two sea chests to the shop where they were made, and I was in a quandary. If I went back to Camp 4 with only two sea chests, the CO would chew me out. So I asked a yeoman sitting behind a desk if he could change a two into a three. I showed him where, on the paper, and he made the change. I took the order to the lieutenant commander in charge and told him I had an order for three sea chests. He took one look at the paper and asked, "Who changed that two into a three? Did you?" I told him, "No Sir! which was the truth. He said, "Someone did, but I'll give you three sea chests." I still have that sea chest, and when I got home with it, my wife immediately took it over for a blanket storage and sewing materials.

At Fort Pierce several times a couple of us talked the CO into letting us take a small boat out to sea to check to see if any boats were out there in trouble. He knew we only wanted to take a boat ride, but he let us do it. We hit waves big enough to knock us to our knees and got soaking wet when we took on water, but I still say it taught us more about boat handling than some of the regulation stuff. It was fun, and we did not actually take any bad chances. We loved it and were not afraid.

One time while waiting for the other boats in our group to join us, we tried to lasso a shark, a hammerhead about seven feet long. He liked to stay in the shadow under the bow of our boat. We had a sash-cord about 12 feet long which we made into a slip-knot and tried to slip it over this shark's head. I don't know what we would have done had we been successful in getting that noose over his head.

We had to learn to make flotation gear out of our pants. While swimming, take off your trousers, hold them so the seat is in your face, then throw the waist out and scoop air, bringing the waist down at arms length. You will find you have scooped enough air to hold you up in the sea water for several minutes. The wet cloth will stay inflated until it leaks air, then you simply repeat the process. This knowhow has saved many lives. The buoyancy you get is a great help, and in salt water, it does not take much to keep you afloat. Any garment that can be made to balloon up, when wet can be used in the same way. A jumper top, or a plain blue shirt will serve.

At Fort Pierce I experienced for the first, and actually the only time in my entire Navy career, when an enlisted man, in any way, refused my com-

Only this text in English is authoritative

ACT OF MILITARY SURRENDER

1. We the undersigned, acting by authority of the German High Command, hereby surrender unconditionally to the Supreme Commander, Allied Expeditionary Force and simultaneously to the Soviet High Command all forces on land, sea, and in the air who are at this date under German control.

2. The German High Command will at once issue orders to all German military, naval and air authorities and to all forces under German control to cease active operations at hours Central European time on and to remain in the positions occupied at that time. No ship, vessel, or aircraft is to be scuttled, or any damage done to their hull, machinery or equipment.

3. The German High Command will at once issue to the appropriate commanders, and ensure the carrying out of any further orders issued by the Supreme Commander, Allied Expeditionary Force and by the Soviet High Command.

4. This act of military surrender is without prejudice to, and will be superseded by any general instrument of surrender imposed by, or on behalf of the United Nations and applicable to GERMANY and the German armed forces as a whole.

5. In the event of the German High Command or any of the forces under their control failing to act in accordance with this Act of Surrender, the Supreme Commander, Allied Expeditionary Force and the Soviet High Command will take such punitive or other action as they deem appropriate.

Signed at Rheims France on the 7th day of May, 1945.

On behalf of the German High Command.

IN THE PRESENCE OF

On behalf of the Supreme Commander, Allied Expeditionary Force,

On behalf of the Soviet High Command,

- 2 -

Major General, French Army (Witness)

German Articles of Surrender, WWII, signed May 7, 1945.

mand and about half cussed me out. If it had not been so funny and had he been wrong, I would have placed him on "report" and made him appear at a Captain's Mast (like a police court to you). So in good humor I let him get by with it. He was a chief petty officer, and under the circumstances, I didn't have the heart to come down on him. We had a ship's party on Friday night, and when I brought him a bunch of dirty dishes Saturday morning just before the galley was to undergo weekly captain's inspection (where an officer inspected the whole place for cleanliness and neatness and this chief would have been found guilty of not keeping his kitchen in "shipshape" condition), I couldn't blame him for exploding. I simply took the dirty kitchenware and hid them where no inspecting officer would look, so I didn't mess up the galley at 0900 that day.

By this time, we were ready to take part in an invasion. We were about to go to a ship and "go over." We were told to select our crews for own "lead" boat. I chose a big coxswain and an engineer who had both been in the landing at Salerno, Italy. I firmly believed, and still do, that they were the best men in the flotilla. I chose a bow hook and an engineer. All at once I sensed trouble. My coxswain came to me and told me, "If Taylor goes, I don't want to go." My signalman told me the same thing. Neither man would tell me why. In fact, they told me none of the boat crew would tell me, but there was a man in another flotilla who would tell me what was the matter with Taylor. I took this man's name and called him into my office. He told me Taylor was homosexual. I called Taylor's name over the bull horn and told him to report to my office. Upon questioning him, he denied it all, saying that he had put on a show in the barracks that "just looked like it." I told Taylor that I would give him a choice, either he go with me to see a psychiatrist at once, or he would get a transfer immediately. He said, hanging his head, "I'll take a transfer." Then, after I replaced this man on our team, my crew was happy.

At about this point I received word that I was to lead one of the first waves into Tokyo. There would, of course, be several beaches like, "red beach, yellow beach," etc. I had a battle plan for Iwo Jima three days after we had hit that beach, and I knew that of the first wave, all boats and all men made it in to that beach without casualty but that none, and I repeat, none of the boats nor a man of the second wave made it to the beach. The boats of the first wave all had mouse-traps (rockets) but none of the waves after that could use rockets, because of the danger to our own men. At Iwo we paid a terrible price. Securing this island was very costly. This plan was very highly classified and I had to sign for it. When I found out I could go home to see my family for three days before shipping out for overseas, I tried to get rid of that classified paper. No one on the base would sign for it and take it over. So, I took it home to Ohio with me. When I got home I found my brother-in-law, COL John Tapp, also home on leave. He had been with GEN Patton in Europe. I told John about the paper I was carrying and told him I wanted to burn it and have him witness it, as I stirred the ashes. He agreed, as both knew this was the proper procedure to get rid of "classified matter."

Our training period at Fort Pierce was over. We received our orders to proceed to Brooklyn Navy Yard to put our ship, the USS Oglethorpe, into commission. We had our list of officers and men, and we entrained as a troop movement.

While on this trip to New York, I had to carry a classified portfolio to 90 Church Street, a Navy Department Office. I did not know what was in the folder, but I was instructed to never allow this folder out of my sight. In other words, for this duty I was a courier. When I got to my destination with the portfolio, I have never before nor since seen so much "gold braid," admiral and general officers all over the place. I was pumping my right arm all the time (saluting).

We made up our ship, the USS *Oglethorpe* (AKA-100), and christened it with full dress. She was named after a former governor of Georgia. Formal engraved invitations were sent to the families of the crew, and we had a grand ceremony. My wife, Rebecca, came to visit me and to attend the commissioning ceremony. We stayed in the St. Regis Hotel (special rates for servicemen). After one week at this fine hotel, we had to move out, as they had a policy of rotation. We then moved into the Beaux Arts, which was nice, but nothing to compare with the St. Regis, where we had maid service with the bed turned down every evening, etc.

While in New York we could get free tickets to opera, stage plays and all types of entertainment (given as an honor to servicemen). All we had to do was go to "A" deck at the Commodore Hotel and ask for tickets to the shows we wished to see.

I reported aboard ship every morning at 0800 to muster my division and to carry out my orders for certain work to be done. Getting familiar with a large ship, 486 feet long and 68 feet beam, and at the same time being assigned certain duties is a great experience. I started out as a junior officer, with my first duties to be in charge of attack boats and their crews. I was responsible for the housekeeping of the entire mid-section of the entire ship. I found I was "father confessor" to all my men, and responsible for their well-being, not only physically, but mentally as well. I had to make personal inspection of all "mess cooks" to be sure they were clean and well-groomed. I sampled the crew's mess to be sure it was appetizing and sufficient. I was in charge of examinations for men seeking ratings in the field care, pharmacist's mate, and the like. Procedures and materials were well prepared and easy to supervise.

Junior officers stand regular junior deck watches in port, and at sea. Shortly after we went to sea I was one of six, out of 36 officers, to pass an examination to serve as senior deck officer at sea, as well as in port. This last was quite an honor, but with it came very heavy responsibility. A senior deck officer, while he "has the deck" is responsible for all the activities going on aboard the entire ship, and for the carrying out of all orders of the captain, including the course and speed underway, and in fact for the safety of the entire ship and all its crew. An officer of the deck answers only to the captain of the ship and his executive officer. Your commands to the navigator, the gunnery officer, the engineering department, engine room, Combat Information Center, must all be carried out. You had better not make a mistake. You have rated enlisted men as well as junior officers who can carry out your orders, who are under the same strict obligation that you are.

We had to learn to give the correct commands for all emergency situations. The crew had to be trained in their duties in all situations, as well. We conducted drills for the entire ship to ensure personnel would function as they should to preserve the safety of the ship and her crew. I really screwed up the first time I received orders from the captain to sound the proper alarm (I didn't know it was only a drill), and I did everything wrong. The captain called me in and told me he would give me about 30 minutes to learn the correct procedures in all drills. He then tried me again, and this time I did it properly. We had drills for "fire," "Man overboard," "abandon ship," "battle stations" and to "prepare to attack." Every man had a specific duty in each of these conditions, and his duty was posted on the "Watch, quarter and station bill" where he might look and remind himself of his post at any time.

We all had to learn "damage control," what to do when our ship took a torpedo, had a collision, fire, or heaven forbid, an explosion of any kind aboard ship. Damage control involves shoring up of bulkheads (walls to landlubbers) to prevent further breakage when strains have been caused by enemy armament, or by storm, or whatever, and the pumping of water from the holds to prevent the sinking of the ship. All jobs must be performed and there may be little time to prevent the order, "abandon ship," from coming over the loud speaker (if it is still working). If in the drills the crew learned how to load and launch life-boats and life-rafts in drills, many lives may be saved. Duties required in the above evolutions can only be carried out if the officers and men assigned know what they are to do and have the necessary skills. Not everyone knows how to operate a handy billy, a portable pump, or firefighting gear, or the boat davits. There may be no time for fumbling when emergencies occur.

One of my regular routine duties was to inspect the life boats and life rafts once each week, to be sure the proper amount of medical supplies was in place, emergency rations, and that lines for boat handling were in good shape. Every time I made my inspections, I found the morphine surrets missing, someone was stealing them. We posted watches, but never caught the culprits.

Every ship has to take a "shakedown" trip to find out how the crew functions, how the ship responds to needs for speed, maneuvering, sudden stops and the like. Our ship required a full nautical mile to get stopped dead in the water from full speed ahead, by stopping all engines and running her full astern. We had to learn that all vessels with only one propeller all back to port, they do not back in a straight line. With full way on, it took about 20 seconds for the ship to "answer the rudder" (start to turn) so we were constantly reminded to think ahead.

On the bridge, my watch station while officer of the deck, we were 55 feet above the water. This for the AKA-100. Many other ships are different. While in the Chesapeake Bay, on shakedown, we rode out a storm while at anchor. We had both anchors down and the engine turning ahead one-third speed, and we still drug anchors. We measured the wind to be over 90 knots above that we could not measure. I had to report to the captain what our position was every few minutes, getting my information by watching the radar PPI scope. Once I reported to the captain that we had closed our position relative to the aircraft carrier behind us to 1,000 yards (one half mile). The captain's answer was, "What are you trying to do? Scare the hell out of me?" When I went off duty at midnight, I immediately hit the sack, as I was dead tired. I had every confidence in our crew. I did not know until the messenger awakened me in the morning that pitch and roll of a ship will give one a monstrous headache.

In the ward room (officer's mess) our tables had little fences that could be snapped up in position to keep tableware from sliding off the tablecloth, in times of rough seas. Our chairs had chains to the deck to keep us from sliding around when it got rough. Our ship was big enough that the greatest roll we ever had was 23 degrees, which actually was not very much compared to Destroyers and smaller vessels, whose crews never had a hot meal for days at a time when sailing under stormy weather. We never missed a hot meal.

Some of the officers and men got seasick in rough weather. I know of some who get seasick immediately after their ship casts off from a dock or mooring buoy. For me, gentle ground swells would make me feel slightly queasy in the stomach, but I never vomited. While around those who are seasick, we were advised, and rightly so, to never sympathize with them, or you will become sick yourself. So, when I saw a person about to heave his

LCDR William F. Badcock at launching of the USS Selinur.

cookies, I told him, "That's the way! Throw it as far as you can, old boy!" The bad part of it was that not all got to the lee side of the ship, but tried to throw up into the wind, then their vomit would come back onto the ship, and if they were on one of the upper decks, it would be brought back onto the ship by the wind. There was a rule, hard to enforce, that every man had to clean up his own mess. Sometimes the men got so sick they couldn't.

I remember two men in my attack boat who got seasick every time we went out on training runs. I reported this to my CO and he said he would take them out for a ride and see for himself. They could be "goofing off" to avoid sea duty. When he came back after about a half hour with these two sailors at sea in a small boat, he declared them to be not fit for sea duty and transferred them to permanent shore duty (through no fault of their own).

My assigned steward's mate seemed always to get a dose of Gonorrhea every time he went on liberty. He would have to go to sick bay and get his shots and would be quarantined for several days before he could serve in the wardroom, or perform any other duty for that matter. I would visit him in sick bay, while he was laid up and ask him what was the matter. He told me he had gotten into some "bad meat." I asked him if he couldn't tell by the smell. He said, "No, sir, all de same."

One officer I went with on training cruises after WWII got sick as soon as we cast off. He would carry a gallon bucket with him everywhere he went, on duty, everywhere. He was a fine officer and did his duties well. I used to get a sandwich (liverwurst) and walk over where he was standing watch, just to tantalize him, he would be feeling a bit green around the gills, and he would yell at me, "Get the hell out of here with that!" He was one heck of a nice guy.

After the storm in the Chesapeake Bay, we were sent to Newport, RI, as a training ship. This was not good duty. The trainee sailors broke up our attack boats, when they hit the beach in practice runs, and we always had a lot of repair work to do after they took our boats out. I had nine "miss-fires" while these green horns were firing our 20mm guns. When we have a misfire, this means a live round is stuck in the barrel, and when the barrel has been in use for a lot of firing, it is very hot, and we could have a "cook-off" in about one minute, which could injure or kill any personnel near-by. The only way to remove the live round from the barrel is to take a ramrod and gently push the round out of the hot barrel. If you should ram the live round too hard, you could detonate the nose-fuse and the shell would explode, doing the same damage as a cook-off. So every time I had one of these misfires, I detected it because the barrel of the gun did not come back into "battery" after firing; gunners mates should see these things. That is what they have been trained for, but these guys didn't, it left little old me to do the dirty work. I would say a little prayer, like, "Please God, don't let it happen this time." Then I would go after that live round, push it out, and throw it into the sea. I had eight of these guns to watch at the same time, all placed amidships, each with a gunner, fastened in the harness, a gunners mate, and an ammunition handler. I still can't fathom why none of my gunners mates caught those misfires. I believe that was the closest I ever came to death while in service. Only a direct hit from enemy shelling might have been worse than a cook-off. And this was only in training duty. The good Lord must have heard me.

While at Newport, RI, we junior officers had to make liberty runs into the War College. A liberty boat, an LCVP, would make the run every hour, leaving the ship which was anchored out in the harbor. Round trip took almost an hour. The crew on such a run was the coxswain, engineer, bow hook, signalman and an officer. It was foggy much of the time, and the run was several miles from our anchorage to the beach. The ship's Combat Information Center would watch our progress using radar, and they would call us on our boat using walkie talkies, to guide us through the fog. CIC would warn of traffic near us and their course and speeds. CIC would also warn of us of our approach to an island in the harbor and the like. One time on this run, I realized that CIC was steering us right onto Rose Island. The tide was in and about a foot of water covered the tips of the island, which CIC couldn't see. I realized what was happening, and ordered our coxswain to slow down so we could just keep enough way on to have steerage, then I called CIC on the ship and asked for course 150 degrees true, which would take us around the tip of the island, it wasn't too late already. CIC came back, with "Who's doing this, anyway? You or are we? Remain-on course." Good old Smitty, the officer on CIC. He refused to give us the course that I knew would get us around the island. In a few minutes we ran aground. At such a slow speed we wouldn't damage the boat much. But with all the rocks on that shoreline, damage to the propeller was something else. After letting the coxswain try for about 10 minutes, I took over. After shifting all the men aboard, crew and passengers to the stern of the boat, I took the helm and worked her loose by using full power in reverse and ahead. I finally got her off the rocks, and we headed back to the ship. Our boat vibrated terribly, because I had damaged the prop, worth $1,100, beyond repair. We limped in and all the while kept contact with the ship. As soon as we came alongside, and tied up at the gangplank, the word came over the loud speaker for Mr. Basinger to report to the captain's stateroom. Mr. Basinger reported to the captain. He asked me a lot of questions and I tried to protect Smitty, the CIC officer, without making him look bad. All at once the captain said to me, "Aw, why don't you admit it? Smitty gave you a bum steer."

It was here at Newport that our sailors started to rob lobster traps in the bay. This was an offense that if caught, carried a $50 fine for the first offense. When the skipper found this out, he ordered lobster traps made by our carpenter shop, and all ship's company had lobster once in a while, all nice and legal.

One night when I had the 2000 to 2400 watch, my friend Bill Axtman came to me and told me he saw three men, my own boat crew, the coxswain, engineer and bow hook in an LCVP belonging to our ship, and heading away from the ship in the dark. As OOD, I quickly organized a shore patrol party, got orders typed up by the ship's secretary, and sent them after the boat and crew. We knew they had to be heading for Portsmouth, across the bay. This was a serious matter, since they had stolen a boat, besides being AWOL. The shore patrol, led by LTJG Haier, our navigator, wearing dress uniforms, an carrying .45 caliber pistols and arm bands had police authority and more. I told them, before they shoved off to capture those men, bring them back to the ship, and to recover the boat. Then I placed the three men "on report," meaning they would go to the next "Captains Mast," which is a police court aboard ship convened by the captain of the ship. The SP party wasn't back yet when I went off duty, so I hit the sack. The next morning I asked the duty officer who had relieved me if anything happened during his watch from midnight to 0400. He said the only thing was that a little white rowboat was adrift near the ship at about 0300, and he had the messenger and boatswain mate of the watch go out and pick up this boat and secure it. At muster, all the missing men were present. I had placed them on report, so, as their division officer, I had placed the charge, and I also had to defend them. The captain, in charge of "Captain's Mast," as judge and jury, read the charges against these men, then asked them how did they plead. "Not guilty," all three said. Then the captain asked me, "Mr. Basinger, did you cause a search of the ship to be made?" My reply was, "Yes, sir!" The captain's next question, "And you did not find them?" "No, sir," I replied. "What do you have to say for yourselves?" he asked the men. "We were sleeping in #7 boat sir," was their reply. The old man then turned to me, and asked, "Did you cause a search to be made in the #7 boat?" I knew if I said, "Yes," he would really throw the book at those guys. If I said, "No, Sir," which was the actual truth, he'd probably hang me. So, I answered, "No, sir." The captain grumbled a bit, then roared, "I know you are guilty as hell, but I can't prove it. Dismissed." My expected chewing out never came. In fact, no one mentioned it until months later when on a beer bust on one of the islands in the Pacific, I asked the three men what happened that night back in Newport. They told me they had stolen the boat and just got tied up in Portsmouth, when they heard my shore-patrol party coming after them. They hung by their fingers under the dock, while my SPs walked right over their heads, into town. These men then followed the SPs at a safe distance so they were not detected until the SPs went back to the dock, took their own boat, retrieved the stolen boat and went back to the ship. Then the three had to figure out a way to get back to the KA-100. They stole the little white rowboat rowed out to the ship, and climbed the anchor-chain to get back aboard. It was easy to get past the deck watch, on the bridge, and the hawsepipes being at the very bow of the ship. They then went to bed, and in the morning reported to muster.

One of the pleasures of sea duty was that while at Newport at anchor out in the harbor, every third night we drew liberty. We would go ashore, visit the city, then as we went back to the ship, we would stop at a little restaurant and have steamed clams with a bottle of beer. This place was located on the waterfront.

The same three men, mentioned above, my attack boat crew, had been at Salerno, Italy, had made that landing and didn't wish to see any more action, out in the Pacific. I didn't really blame them, but they had a hitch to finish out, so they had to go with us. They went "over the hill" (deserted) for 29 days. If they had stayed away for 30 days, they would have been court-martialed, and probably shot for desertion in time of war. This way, by coming back before the 30 days, they were placed on bread and water for 30 days with a full meal every third day—a terrible punishment, but they survived, and came back to the ship looking pale and wan, and like skeletons. They went with us, of course when we started overseas.

We were all relieved when we were taken off training duty at Newport, RI, and sent out to the Pacific. We were loaded out for the Pacific, with food, medicines, ammunition, belly-tanks for aircraft, parts, and one hold full of beer. We were finally "going over."

We arrived at Colon, Panama at about 2000. It was rough going around Cape Hatteras, and we were told we were never out of the sight of a wreck as we went down the east coast of the United States, even though we could not always see one. This is how devastating the German submarines were in WWII. We passed close to Cuba on the way to Panama.

The night we arrived in Panama, the skipper allowed liberty at once, since we would only be at Colon that one night, and I drew shore patrol duty. I had to work the red light district armed with a loaded .45 pistol wearing the arm band, and all. This was a real eye-opener to me! I never realized how some people live, I guess. The girls had their beds just inside the door of their little houses, and even though I was an officer, with arm-band on and carrying my pistol. They would grab me by the arm and try to coax me in. When we got back to the ship at 2400, one fellow officer took off his bursard and .45, and went back to the red-light district. I told him this would be a mistake, but the pressure was too great.

The next morning I stood the senior deck watch taking our beloved USS *Oglethorpe* through the Panama Canal. We would go through locks where we would have to be still in the water while they closed the locks, floated us to the desired depth to go through. After some time, I noticed our helmsman was standing at the wheel with his shoes off. I asked him why he was not wearing his shoes. His reply was, "I've been standing here for six and one-half hours, sir." I knew we had only two men "qualified" to stand helmsman duty in the canal, the other one was a chief petty officer and he had let this man stand for six and a half hours without relief. I called that chief quartermaster to the quarter deck and gave him a chewing out for all in ear-shot to hear; I reamed him for allowing this man, a first class quartermaster, to stand for so long without relief, when both he and the chief should have relieved each other every half hour. As I write this, I still seethe at that chief. I bet he never tried that again, at least not with me on duty.

As we left Panama, one of my men received word that his father was near death. He asked for emergency leave, so he could see his father one more

Part of 432 crew on liberty in Manistee, MI, 1938, reserve training.

From left: CAPT Gary W. House, his daughters Lael and Julie, and their uncle CAPT Karl T. House (Gary's brother).

time. Our captain refused to grant this man leave. I worked through the Red Cross and got the man permission for a 10 day leave plus travel time.

Once we were out of the Panama Canal, in the Pacific, headed for Pearl Harbor, we were in convoy. Deck officers had to maintain space and interval in relation to the positions of the other ships by the use of instruments that measured the height of mastheads of near-by ships, next to us in formation. All ships turned together on zig-zag courses as determined by a plan agreed upon for each day and laid down in a highly classified publication. Before an officer took over the deck as senior OOD, he had to copy off his plan for the coming four hour period of duty and have another senior deck officer figure the same courses and speed for that watch and initial it to be correct. This to eliminate errors. To make a mistake could cause a collision, and there is nothing worse than a collision at sea. It was pretty serious stuff. Maintaining position in convoy required a fine touch. Small adjustments were better than radical changes in speed or direction. The engineroom did not appreciate it when I called down and gave an order to put on one more turn of the propeller per minute, or "take off one turn." This was not easy for them to do, I knew it, but it kept us from lagging, or crowding our neighboring ships.

We had a submarine on our tail at one time, soon after leaving the Panama Canal, while we were forming up, and not up to our cruising speed. (We could only go as fast in convoy as the top speed of our slowest ship.) Our sonarmen picked this guy up on their instruments, he never surfaced, or we would have got him. We had no trouble out-running a submarine at that point.

Our men were at "battle stations" several times, though, and we deck officers had to warn the guys to "knock off the chatter" over the battle phones. Some of the men got to sharing their homesickness over the sound-powered phones.

One sailor on duty during the mid-watch one night forgot where he was, and who he was talking to. He was so homesick to see his wife, he told me all about how he loved her and how pretty and lovely she was. Kind of pathetic, but I could empathize and his secrets were safe with me.

While in New York Brooklyn Navy Yard, I had a fine rated man, a first class carpenter's mate receive a "Dear John" letter from home. He tried to commit suicide three times by jumping over the side. I had the boatswain mate of the watch and the messenger to put him in the brig for the night. If I got to close to him and he cussed me out, I would have to place him on report and he would appear before captain's mast. The next morning he was sobered up and promised he was over it, so I let him go. It was a shock for him to find out his wife was deserting him, but he was able to adjust.

We had one hold full of beer on our way to the South Pacific. Other cargo included food, medical supplies, clothing, ammunition, trucks, and a variety of military needs. Since we had cargo nets stowed in the upper "tween decks," we could not lock and seal off this hold full of beer. These cargo nets were oily, and could cause a spontaneous combustion and a serious fire. Therefore, we had to post a watch to guard the beer round the clock, day and night. The JOOD's had to check on the sentries every half hour, day and night. The sentries each had a loaded gun, and were instructed not to allow anyone to get into the beer. All went well until the captain made a surprise inspection. He found 15 empty beer bottles. He was furious. He called a meeting of all watch officers and announced, "Somebody's been in the beer. I am going to find out who it was, and I'm going to hang him, or them, and I'm going to hang the officer who let them do it!" Then he asked if any of us had caught a sentry asleep. None admitted to this. Then the old man called us into his stateroom and grilled us one at a time, quizzing us to no end. None of the officers had any knowledge that would solve the "who stole the beer." About this time we were getting close to Pearl Harbor and the old man announced over the PA system that there would be no liberty for anybody until he caught whomever had broken into the beer. Pretty tough on a crew that had been at sea for a considerable time, and most of whom were coming into Hawaii for the first time. We had a communications officer, a lieutenant who was on the ball. He went to the captain and asked for permission to go ashore to pick up the secret publications needed for the next voyage. This was his job, and no one knew exactly when we would be sent out to sea again, a requirement, and the skipper knew it. In fact, he couldn't legally keep the communications officer from doing this. Upon hitting the beach, this officer went directly to the base chaplain at Pearl Harbor, who was a rear-admiral, and gave them the story that our captain was refusing liberty for the crew because of 15 empty beer bottles. The chaplain must have done something, for it was announced at once, "port and starboard liberty," for everybody and we saw the captain of our ship go ashore in a hurry. This meant half the crew could go ashore, for one liberty, and then the other half could go. We figure the old man got a "dressing down." Now, the truth of the matter was, when we finally got to comparing notes, every officer who had checked the sentries on the beer watch, had found one man, one of the sentries guarding the beer, and this man was nicknamed "vitamins" because he was a skinny, puny fellow and every one of the six watch officers had found him asleep on watch. When I caught him taking a nap, I woke him and warned him that it was a court-martial offense to sleep on watch during time of war and he could be executed. But like the rest of these officers, I felt it to be a pretty cruel punishment, to be executed for falling asleep while guarding beer, so I took the chance on it somehow working out okay. All six officers had lied to the captain, when he asked his questions.

The neck of the entrance to Pearl Harbor is very narrow. Only one large vessel can come or go at a time. This is why our Navy was ravaged by the Japanese on December 7, 1941. As we approached the entrance to Pearl, we got lined up to enter the channel, all at once we started receiving blinker from another approaching ship, several miles away. It read, "We are coming in. Please do not interfere." It was the *Sara* (*Saratoga*, an aircraft carrier) with an admiral aboard. Our skipper had to stop and back down and let the flagship go in ahead of us. I have seen some angry men in my time, but this captain of ours was something else. He could swear for 15 minutes never using the same word twice. This time he literally blistered the deck. No one stayed on the bridge who did not have the duty. I was standing senior deck watches by this time and everyone not working got off the bridge, even if we were coming into Pearl Harbor for the first time.

The "A" bombs had just been dropped on Japan, and everything was in limbo to see what the outcome might be. The fighting suddenly came to an end. No one knew what his next assignment would be. Was it all over? Could we go home? Might it start all over again? Nothing as huge as our operations could come to an abrupt halt. It would take months to bring everything to a dead stop, to grind to a halt. The entire operation was too big. Even if Japan surrendered, the ships, and planes, and bases could not be closed down or left stranded. People had to be processed out gradually.

So, we went to Quadulain, then Eniwetok, stopping briefly to unload supplies. Most danger was over, as we were operating under peacetime conditions, now. The remainder of the trip was sort of a lark, except that our at home jobs were askew, and we continued to be homesick. We now were getting to see a little more of the world.

At Quadulain Atoll, one officer became angry because he, with five others, was pushed into the drink. We had been at the officers' club, and some were feeling no pain. Some got into a shoving match, while standing on the pier, waiting for our liberty boat to take us back to the ship. It all started when the ship's secretary leaned too far over the edge of the dock, and a commander, the ship's doctor, in fact, gave this guy a shove. Into the drink he went, his hat floating away, never to be recovered. This officer splashed around and got his bearings, swam back to the dock where the rest of us were standing, and reached up for help to get back up with the rest of us, and the good doctor reached down, took his hand, and the man in the water gave a quick pull and the doctor was in the water. More pushing and shoving followed until five officers were in the drink. The only person to get mad was a lieutenant from South Carolina, and when we got back to the ship, he turned the rest of us in to the skipper who just laughed and didn't do anything about it.

At Eniwetok another officer, an ensign, went ashore to see one of the programs put on by the USO. Then he stole a jeep and drove it to the far end of the island, where he parked it, and entered the sick-bay. He was picked up by the shore patrol trying to get into the sick-bay's medicinal alcohol. For this he was placed on report by the base admiral. Our skipper placed him "in hack" for one week, confined to his state-room, and this meant the rest of us had to do his duty, what ever it was. He ended up with all of us hating his guts, he never got us paid back. This was the same officer who, back in Newport, RI, had given me a bum steer that ran me on the rocks with the liberty boat. A bright enough man, but no discipline.

When in port, we had port and starboard liberty, as I have mentioned before. Half the ship's company could go ashore at a time, leaving enough personnel to operate the necessary procedures of the ship, for safety and comfort of the entire crew.

As we approached Guadalcanal, we found we had an off-shore current, and the captain had to back off and make a second attempt to get close enough to the dock to have lines thrown over to moor. The second time he was successful, he should have been, he brought her in like a destroyer. We wondered if he would hit the dock, but he just about had to do it that way. After we got tied up at the dock we were instructed not to wander off the marked trails in the jungle, as the jungles were still mined. The Japanese had a mine that we called the "emasculator" that if tripped would jump up about 30 inches before it detonated and would catch a man just about the right spot to clear off his genitals. Many of our men were permanently incapacitated sexually by these land mines, if not killed outright.

Since I was in charge of maintaining the hull of the ship, one day while tied up at Guadalcanal, the captain called me to his stateroom and told me, "We're going back to the States, maybe to New York. I want this ship's hull painted from stem to stern as soon as possible. In fact, there will be no liberty until we have this ship painted." That last order was all the incentive I needed to get myself in gear. We'd get that old tub painted pretty soon or I'd be catching heck from the entire crew. Since we were tied up along the dock, the first thing I did was to "commandeer" a tow-motor (finger lift). Then we got a pallet and an operator. I had a man in the paint locker mixing paint. We had several men lifting five-gallon buckets of paint to the main deck via man-lifts. Our compressed air equipment would easily handle two paint sprayers. With the tow-motor, we could reach the main part of the hull-leaving only the fantail and the bow to be reached from the water. I located a punt (flat boat low in the water) and put a couple of men down there with

USS Pioneer (MCM-9), Sturgeon Bay, WI, 1992.

the paint sprayer nozzles, and we had the entire one side of the ship painted in one day. We did not bother to do any chipping, and from the thickness of the paint on the sides of the ship, I don't think any chipping was done the time before. We didn't have much trouble with barnacles, those little shell crustaceans that attach themselves to the hull of ships. Then I asked the captain for permission to turn the ship around-this would have to be done by himself, or the executive in charge. The captain couldn't believe that I had one side already painted, so I invited him to go ashore and inspect, which he did, pronouncing it a good job, well done. We turned the Ogle*thorpe* around and next day we repeated the process and painted the port side.

About this time, an officer took a shower, in his stateroom head, and forgot to stow four Longines Amphibious wrist-watches in his safe. When he got out of his shower, the watches were gone. They had been stolen. He immediately reported the loss to the captain, who caused a search to be made of the entire ship-every crew's quarters, all officer staterooms, except the captain's. I searched the quarters of my entire enlisted crew members, and found two-fifths of whiskey, which I confiscated and stowed in my safe. In the executive officer's stateroom I found 17 fifths of whiskey, which he had brought with him to sell (it would bring $75 a fifth out in the islands). We never found the watches. When we got back to Pearl Harbor, I gave my men their two bottles of whiskey. They were surprised and never expected to see it again. We did nothing about the exec's whiskey, we liked him, and the old man probably knew nothing about this incident.

We had to keep tight security at all times, even at night, which was the worst time for depredations to happen. I remember one night when we were tied up in down-town Honolulu, I had the deck and about 3:00 in the morning, a stranger sneaked up the gangplank. How he got past the sentry on the dock I don't know. When he stepped onto the quarterdeck, I tripped him, and stuck a loaded, cocked .45 caliber pistol in his ribs, telling him, "One move and you're dead." He did not move. Then I backed off, one should never stand too close to a culprit with a gun in his ribs, he might whirl and knock the gun out of your hand. I then turned him over to the boatswain mate of the watch, and gave this petty officer my gun and went to get the exec. This was the time I caught the exec in what might have been an embarrassing condition. When the exec came out, by this time the man we were holding regained some of his composure and identified himself as a dock foreman who simply wanted to open hatch covers early so his crew might go to work earlier in the morning. He had papers authorizing him to do this work-and why he didn't produce those papers and come aboard properly, I will never know! From then on, I suspect he did it correctly, because if I had been trigger happy, he could have been a dead man.

By this time, we had a new captain. He seemed competent enough, but he lacked some of the personality of the former skipper. While we were at Guadalcanal, we found out that the officials here had only 30 days before turning the island over to the British. Much good equipment was "deep sixed" because we had not enough time to load it out and reclaim it for the USA. We saw heavy equipment like bull-dozers loaded on barges, and the next day those barges were without cargo. They had been unloaded during the night. They had 25,000 .45 caliber pistols to get rid of. All we had to do to get a .45 was go ashore to the warehouse and sign our name beside the serial number of the pistol we had picked out, and it was ours to take home, a beautiful souvenir. When we got back to the ship after picking up our 45s, the new captain asked us, "Where's mine?" Didn't you bring me one?" We informed him he would have to get his own, and that settled that.

We were told that it would cost more to carton it up, load it and take it home than the stuff they threw away would be worth. Besides, my own thought, the taxpayers will never know.

Our next stop was Noumea, New Caladonia. Here, when we went ashore we could drink no water, because these natives fertilized their vegetables with human excrement, and drained their bathroom stools down into the curbs along the streets. On a warm day, the town smelled of raw sewage. In fact I only ate when I could get back to the ship. We did see the great cathedral in Noumea, but that is about all. The cathedral was over a hundred years old. Nothing in French New Caledonia for us, except that we did refuel there.

We visited an island in the Ellis Group, called Nuku Fe Tau-we. We couldn't anchor at this island because of the too shallow reefs, so we just kept the ship in position by slowly running the engines and turning the propeller enough to hold us where we needed to be. Our job here was to float in a water evaporator and a truck to the garrison stationed at this island, about 17 or 18 people, enough to keep a radio station to communicate with our forces. I had not pulled duty on this day, so the old man asked me to take a small boat to the island and try to pick up a load of coconuts. So, I took my regular attack boat crew, and went to the island in an LCVP. We were met at the shore line by natives, jabbering and letting on like they spoke no English. The men wore a kind of loin-cloth; the women wore what I call "see-through" dresses like were in style in the States around 1920, with nothing under them. We tried to tell these people that we wanted coconuts, but they seemed to be unable to understand us. By sign language we managed to talk them out of about one wheelbarrow full, so after about an hour of seemingly unsuccessful bargaining, I told my crew we should lift anchor and head back to the ship. When the natives saw that we were going to leave, one of them said, plain as day, "You going back to the ship now?" We had been had. They understood our every word. When we got back to the ship we found that the chief of the island tribe had come aboard and offered gifts for every member of our crew, if we would stay overnight. Our captain told him we had to be underway at 1700, and thanks, but not this time. I neglected to tell you, we had traded four bags of buggy flour for the coconuts we did get on that trip to the island. A side story: When the cooks found bugs in our flour, they tried to hide them by using raisins in the bread. When we discovered the bugs, we could not eat the bread for a while. Later, we managed to just ignore the bugs and eat the bread. Then we said we looked for bugs to put into the bread (real tasty). This last was not actually true.

As we were preparing to leave Nuke Fe Tau at 1700, we somehow got the frapping line caught in one of the davit hooks. (A frapping line is a wire cable holding the two dait hooks from swinging far apart while hoisting or lowering a boat). Each of these davit hooks weighs several hundred pounds, and swinging loosely could badly injure or kill a man or knock him overboard. Before we could hoist our LCVP, we had to free those davits. The skipper was getting quite concerned. I called the carpenter's shack for a bolt cutter and was told it wouldn't work. I then remembered that in our classes on damage control, we had been taught that we had a damage control locker and what was in it. I needed a cable cutter, and not knowing what one looked like, I had to go myself. I found this tool in the damage control locker right where it was supposed to be, but it was in two parts and covered with cosmolene (a heavy thick grease to prevent rust. On my way back up the boat deck, I wiped off the grease and assembled the tool. My training and experience as a farm shop teacher saved my old nanny. As I approached the boat deck, the men and captain were all standing around wondering what was about to happen. "Now that you found the darn thing, you won't know how to use it." This last from the captain. I ignored him and handed the tool to a man perched so he could cut the cable, and told him to slip the cutting jaws over the cable, use the fast crank to take up the slack, then the more powerful crank to do the cutting. That cable was cut like hot butter. I turned to the captain who had made the insulting remark saluted and told him, "We'll be ready to shove off in five minutes, captain." As soon as we were underway and in safe waters, the captain called me to his stateroom. He didn't apologize, but he did ask me what that tool was that I had used, and how I knew it was in that tool chest. I told him I had been taught that it should be there and it was. He then asked me to go and get that tool, he wanted to get a closer look at it. After examining it, he asked what the copper cable (which was about six feet long) was for. I told him it was to ground the tool if we were to cut an electrically charged wire. He said, "Well, I'll be darned." As I passed the stateroom of the communications officer on the way back to the tool box, he wanted to look at and handle the tool and asked the same questions the skipper had asked.

Most Navy chow is good. We ate officers' mess for a long time, in the ward room, paying around $35 to $45 per month. Commissioned officers have to pay for their meals. Then we found out we were eating hamburgers while the crew, ship's company, were eating steaks. Our wardroom officer was buying too many fancy condiments and sauces (he was some kind of "fancy Dan type"), so we called a meeting and elected to go on crew's mess which would cost us only $21 per month.

Aboard our ship, most of the time we had a good water supply. For the crew of our size, 36 officers and 360 men, we had a rather large water evaporator. Very seldom did we have "water hours" when we had to limit the use of fresh water. When this did occur, we would have to take salt water showers, which we did not like. The salt water healed every little scratch, but it left a gritty feeling on the skin and itched like the dickens.

One night at sea, I had the mid-watch. Everybody was sleeping but my bridge crew, the engine room crew, special watches, crow's nest, steering aft, and lookouts. All these men at their posts are connected to the bridge by telephones and report to the OOD at regular times to be sure they are awake and alert. We always kept plenty of coffee on hand, it helps to keep personnel awake and alert. "Tis said that the grounds of Navy coffee are never thrown out." Anyway, while on this particular watch, I saw a falling star, starting its fall from high in the zenith, like high overhead, and some of my watch crew saw it, too. Since it was so high in the sky, I knew it could not be a flare, it was simply a meteorite, so I didn't bother to wake the exec or skipper. But wouldn't you know, one of the men reported to the skipper that I had failed to report a flare while on midwatch. The skipper called me in to see him, and he asked me about it. I told him for a flare to be started that high, it would have to had come from our own vessel so concluded it was a meteorite. He agreed and that ended that.

Traveling under darkened ship, we watched other vessels with our radar for hours. CIC computed when we would be making our closest approach, the direction and speed of the vessels sighted, and the like. One night we watched three small vessels for hours. When we got close enough to talk to them by blinker, we gave them the IFF (identify yourself, friend or foe). They gave the right answer, so we talked to them by blinker. They were lost and wanted an estimated position. These little LCIs had not had a starsight in three days from where they were. We offered them some ice cream if they could come alongside—which they did.

One night during my mid-watch, we were traveling under darkened ship. We had adapted our eyes by the use of red goggles for about 20 minutes before going on watch. Absolutely no lights were permitted topside after dark. A lighted cigarette could be seen for at least a mile with field glasses. On this particular night, all at once I saw a flashlight being used on deck. I was furious as we were still in war. I yelled at the person with the light, "Put out that light," then went to see who it was. It was the captain, and I immediately apologized. I expected a chewing out, but he said, "Its okay, I forgot myself. I've been sounding our fuel tanks." He would not trust his crew to tell him how much fuel we had. Maybe he wasn't as bad of a person as I had thought him to be.

One other time I saw a light in the darkness, while running under "darkened ship conditions" it turned out to be a lighted cigarette. I have said before a lighted cigarette can be seen a mile away when field glasses are used, and when I saw this lighted cigarette, I ran after him and yelled "Douse that cigarette." I never caught the man because he took off over a laced ladder we had painted, and how he got over all that lacing in the dark, I can't guess. I know I scared the heck out of him, and he is probably still smarting from the cussing I gave him. The life of every man on that ship could have been placed in danger by that one lighted cigarette.

Every so often an enlisted man would ask to be tested in his rating for advancement. I would take him to a private area where I could give him the test and not be bothered by routine activity. These tests are prescribed by the Navy, for each rate and most require a lot of study and preparation. Not all sailors pass these tests the first time. They have been known to study for weeks and sometimes months. When they pass, these men can wear the newly authorized rating badge, which tells what their duty now is and what skills they now have. A man receives a higher pay rate when promoted, and this is reflected by a higher pension if he remains in service for 20 years or longer. Such men are called "career men." It was my privilege to promote many men when they deserved advancement.

As a division officer, I listened to men's personal problems, and these are private matters which shall remain private. If I could help them over a rough spot in their lives, this made me happy. If I felt someone with more training and ability than I had was needed, I would not hesitate to refer these men to chaplain or to someone else qualified. I did accumulate the reputation of being a good listener. I guess the word got around that I had a "short fuse" and could explode on short notice, but never unless it was a just cause. I did, and still do not countenance laziness, stupidity, don't care attitudes or unnecessary excuses. If I give an order to be carried out, it had better be done, or the person given the command had better have a good reason. I can be reasonable and if something of sufficient importance comes up making it wise to follow another direction than I have given, and if the person going over my head gives me a reasonable explanation, I will go along with it. Often in war time, there is not enough time for any person under any officer's command to be given reasons for such orders. When there is time, you may be sure, all hands with a direct need to know are briefed on anything coming up. I had been known to have been wrong once, but only once; this is a no-no, but human.

One time when I was out in the Pacific about 10,000 miles from home, a man asked me where I was from. When I told him I was from north central Ohio, he asked if I knew where Kenton is. I answered in the affirmative. He said it is a great place and they have beautiful girls. Small world. I moved to Kenton several years later.

When I did come to Kenton, my senior agent, Boyce Bradford, who had served as an Army officer in WWI, told me about the Belmont Hotel, another "cat-house" that flourished in Kenton when he came to town, many years earlier. Brad said he did not know about the Belmont Hotel when he first came to Kenton and he stopped in that place to buy a package of cigarettes. He told that when he entered the place, a pretty girl came tripping down the open stairway and asked, "May I help you?" Brad said he told he wanted to buy a pack of cigarettes. She told him, "We don't sell cigarettes here."

When we crossed the equator, a very special ceremony took place. All "pollywogs" (those who had never crossed the equator) had to be initiated by the "shellbacks" who had gone through this event before. For over a week, shellbacks kept a little note book, making entries every time a pollywog failed to offer them butter, bread or for that matter any food before they served themselves. An entry was also made if we failed to open doors for them, or if we sat down before they did or anything they could cook up all in good fun. We were approached by a man dressed as a priest, and asked to bend over and kiss the cross he was carrying on a long staff. As we bent over, another shellback hit us over the rear-end with a huge paddle, and these were not love taps, they stung. Each of us was placed in a barber's chair fastened to the deck of the ship and we were clipped to the bone. After our heads were clipped, we were painted on top of the head with oil, then tipped over into a huge tank full of water, held under for a bit, then taken out and slid down a chute and flailed with a stick. Following this we were guided into a huge pipe which had a fire hose turned on us from both directions before we were allowed to crawl out. We were now shellbacks and were given a card (presented by the captain)so stating and a nice certificate. We were told to never be without that card when we went to sea, or, if we should have to cross the equator again, we would have to be initiated again. I kept that card in a place where I could find it.

Crossing the International Date Line involved no ceremony, but they did present each of us a card, denoting the time and date of crossing. The International Date Line is that spot where each day begins; you can be crossing one way on one day and be a few minutes later at a spot on the day before.

On one occasion at Pearl Harbor, we were moving the ship to Honolulu, about 10 miles from our mooring. The captain called me into his stateroom and told me that due to the time it would take to load the attack boats and the fact we would have to unload in Honolulu, that immediately after getting there, we should just follow the ship with the boats. Since I was the boat group commander, I would be with the men in the boats. When the skipper ordered the mooring buoy cast-off, just before we backed the ship around to head out of the harbor and on to Honolulu, the first division (forward most division) let both ends of the wire-preventer loose when they should have held on to one end. This is a wire and hemp rope-like cable 1,000 feet long valued at over $1,000 and I knew we only had two aboard the KA-100. I also knew that if the skipper ever found out the mistake, the officer in charge of first division allowed to happen, he would get a chewing out he'd never forget, or worse. (He is the same officer who had failed to warm up radar while the ship was underway in a heavy fog.) I figured that just by chance, the "old man" (captain) might not have seen what happened, so I let him take the ship about one mile ahead, then I told the men in the attack boats, using hand held walkie talkies, that we would retrieve that wire preventer, stowing it in the #1 boat, which when hoisted aboard would be closest to where the wire preventer had to go. After we got that thing loaded, we took after the ship, going at "flank speed," which is top speed, and the men all liked this. The old man never need be told what happened. And I don't think he ever found out. When we got to Honolulu, he called me in and asked me why we had waited so long to follow and I told him, "Oh, you know how those sailors always like to travel at flank speed, and by hanging back, I could let them do it and catch up with the ship at just about the right time." You know, he bought it.

One time while at Pearl, we helped a sailor catch his Destroyer, which was leaving the harbor.

USS Paducah, reserve training ship on the great Lakes, pre-WWII.

Why he wasn't aboard that tin can, we'll never know, but he came running along the dock with his sea bag yelling, "There goes my ship, and they are going home." We picked him up with our LCVP and went after that destroyer. The tin can was not going very fast, so we easily caught up with her and pulled alongside with both vessels having way on. The men aboard the tin can saw what we were doing, so they lowered a rope ladder, and the sailor scrambled aboard. We simply turned about and headed back to our ship. A good turn done.

After our stops at Quadulain, Eniwetok, New Caledonia and the Ellis Islands, I received orders to be detached and to head for home, to be discharged from active duty. However, the old man insisted that I remain aboard, since we were all headed for Seattle anyway. I stayed aboard and remained as ship's company, and continued to stand deck watches all the way back through the South and North Pacific. Going into the Straits of Juan de Fuca, I drew the mid-watch (from midnight to 0400). A lieutenant commander, who was actually a passenger, came to me and said, "Bass, I know the entrance to Juan de Fuca. I will volunteer to guide you into the harbor." I gladly accepted his offer, and we came in full speed. Besides it was December 20, 1945, and believe me when I say we came in fast. I was relieved at 0400, so I got two and one-half hours of sleep. Before we got off the ship, we had to be inspected by a customs officer. He called those of the officers who were leaving the ship into the wardroom and asked us, "Do you have anything in your luggage that you shouldn't have? Say, 'no.'" So we all said, "No." That was our passing through customs when we got back to the States. From customs inspection we left the ship for the last time, and went to the railroad station. USN buses were making regular runs to the RR station. When we got there we found over 1,200 service men and women who had arrived at the station the night before. By train time there would be a thousand more. I proceeded to buy my ticket to Chicago, and sat down to wait for 12 hours. I had the promise of a Red Cap that he would get me on the train, The Great Northern, at 2015 that night, so I felt I had it made. But I only thought so. About 15 minutes before boarding time, that Red Cap came to me and said, "I go off duty now, sir." So there I was back at square one. I surely was glad I hadn't paid him any money. I visited a small office call "Military Escort Bureau." A Navy chief petty officer was in charge and I showed him my "radio dispatch orders," and told him that if I didn't get on that train that night, I would have to go out to the base and check in with the base admiral. He agreed and asked me, "Will you go on with a dead body?" I told him, "Try me." So when it came time to load that passenger train, the chief ordered the crowd to "give way" for the color guard, the casket, and old man Basinger. Once aboard the train, I was even glad to sit on my luggage. I was on my way home.

At Glendive, MT, it was 20 degrees below zero. We didn't notice this until we got off the train. Most of the guys ran for sandwiches; we were all hungry. But I chose to go to the ticket office, and asked the agent if there just might be a sleeper bunk open between Glendive and Chicago, and lucky me, there was an upper without a body in it, so I gladly took it, got off the train I was on and waited a few minutes for the sleeper section. This way I got about six or seven hours of sleep, getting into Chicago early in the morning.

When I got to the Great Lakes Training Station, I was examined physically, and cashiered out from active duty. I found out that I could fill out my paper work by myself, instead of following the reader line by line, so I was ready to go, paid off, signed, and everything, then on my way home. I got into Lima Station at 0400 in the morning of December 25, 1945, and called my brother Norman who came and picked me up. He took me to Bellefontaine, OH, where Rebecca and the kids were living with Rebecca's folks. This was a very Merry Christmas.

I don't think I should leave this out of the story. While waiting in the Lima RR station for Norman to come for me, there was quite a little group, waiting for family, or for the next train. Among the people there were two blacks, still in uniform. One of the two men was very profane in his language and used all the little four letter words and there were ladies present. Since I was the only "rank" present, I felt it was my undeclared duty to somehow shut this guy off. So I motioned the other man to come to me, and he did. I told him to tell his buddy to take his fifth of whisky outside, where we wouldn't have to listen to him, and he could kill that entire bottle if he wanted to. I knew that as cold as it was outside, he probably wouldn't feel a thing, and maybe it might sober him up a little. The sober one went over to his friend, and I don't know what he told the inebriated one, but the drinking guy came over to me and said, "I'm sorry sir, if I have offended you." I told him to take his bottle outside and he did. But not the end of the story. About a month after I had come home, Rebecca and I attended a concert in Columbus Grove, the Harmoneers from Radio Station WLW, Cincinnati, OH. Now, get this. The main first tenor, standing in the front row was the man I ordered outside of the RR station in Lima about one month before. I told Rebecca, "I won't dare to let him see me. I have no rank to pull now and he might just clean my clock."

I am now a civilian again with a permanent rank of LT, USNR (Ret) and carry an identification card showing name, rank, serial number and marked "indefinite." This card has opened doors for me, especially when visiting USN establishments on the mainland and in Hawaii.

It has been my privilege to visit Pearl Harbor on two occasions since WWII. It makes me sad when I visit the USS Arizona where over a thousand men remain interred where they died on December 7, 1941. We have also visited the Vietnam Memorial in Washington, DC, where my son Dick's name is on the "Wall." He gave his life over there like the rest of those heroes whose names are with his. Dick wore my "Rail-road tracks" that I wore in WWII. He was a helicopter pilot.

In Dick's last letters to me, he said, "Don't worry about me, Pop. I'm doing what I want to do." Later he said he was scared. He told me once that he could never trust man, woman, or child, as they would hold a hand grenade, pull the pin and roll it under a person. Dick was decorated with two Purple Hearts, Legion of Merit, 17 Air Medals, Navy Commendation medal and some high awards from the country of South Vietnam.

A Magnolia tree grows at the Hardin County Mary Lou Johnson Library, in Kenton, in Dick's memory. A huge chime clock stands in Presser Hall at Ohio Northern University in his memory, both of the above presented by Dick's parents. The magnolia at the library was planted by Dr. Calvin Jackson, a retired colonel in WWII, who survived the Death March in the Philippines and was a prisoner of war for over three years. A 25 acre living memorial given by his parents to the USDA Soil and Water Conservation Service to honor Dick is being developed as a place where youth and adults may be taught the wonders of nature, soils studies, forestry, studies in water quality and wild-life management. This last is called The Silver Creek Center for Environmental Studies, and has been visited by over 1,500 people each year. Many school days find at least one school bus at the Center, and sometimes as many as three school buses. The SCS has assigned one member of their staff to work with the center.

I have often wondered what actually motivated Dick to get into the service of his country. I know that I went to the Naval Reserve meetings once a week for years, after I came home from WWII, and one time when the USS *Oglethorpe* (KA-100) came to Toledo, I took my family to see her. When I approached the quarterdeck and registered in, the Officer of the Deck wrote in the log, "LT Lorain A. Basinger returned to the ship." Perhaps these things caused Dick to enroll in the Marine Corps immediately after graduating from Ohio Northern University. He had been an enlisted man in the Naval Reserve before a couple of "Jirenes" got to him.

Many are the times when I felt honored to have served my country. The last time I visited Pearl Harbor, with Patty, my second wife, we found the USS Miss*ouri*, battleship tied up along one of the piers. I got out my ID card and asked permission to visit the ship. The sentry looked at my card, and said, "Yes Sir." So, we went up the gangplank. When we got to the quarterdeck, I saluted the colors and the OOD then asked permission to come aboard. The quartermaster logged us aboard, after I presented my ID card. The OOD gave us a third class petty officer as a guide, and we visited the bridge, wheelhouse, gun turrets, saw the ground tackle, the whole works, top-side, except one area which was classified and not allowed visitors. It was a great experience. I had visited several battle wagons and aircraft carriers, but Patty had not. When we left the *Big Mo*, I turned to salute the flag, just by habit, and I saw a smile on Pat's face—the OOD and men on the quarterdeck were saluting me.

They Sure Tasted Good!

by LCDR Gale S. Hanks, USNR-RET

I was aboard a gun boat LCI(G) during the invasion of Leyte in the Philippines during WWII. After the invasion we proceeded up the coast with about 20 guerrillas aboard to liberate a small village.

We helped to run the Japanese out and to put the former mayor and other elders in charge. The people of the village were so thankful that they gave the ship's crew a big party in an old barn. We had lots of chicken and roast pork and other goodies I didn't know the name of. They served a drink which was the same as they used for fuel in their old cars.

One dish was popcorn, something like our caramel corn and was very good. Since there were Japanese snipers in the neighborhood, we only had candles for a light so we couldn't see too well as to what we were eating. Along with the caramel corn were hard boiled eggs with a spicy flavor. The eggs were very good but now and then a little piece of something got caught in your teeth.

I took some of the popcorn and eggs back to the ship with me. A few days later one of the crew asked me how I liked the caramel corn and hard boiled eggs. I replied, "They were sure good!" Better take a closer look at them," he replied.

I checked the popcorn in the daylight and found that each kernel contained a tiny grasshopper. I peeled one of the eggs and found inside a baby chicken. The pin feathers were what got stuck in my throat.

It was kind of a realization that when you are hungry and in the dark, you will eat most anything, but those popcorn kernels and hard-boiled eggs sure tasted good!

1969 Naval Reserve Association National Officers. Front row: (left to right) CAPT John Shanklin, CAPT A. W. Tiedemann, CAPT Robert Bolt, LCDR Ruth Kasky, CAPT Morton Leavitt, CDR Thomas Stansbury. Back row: LCDR David Woods, CAPT Ray Ackerman, LT John Wheeler, CAPT L.L. McDowell, CAPT F.J. Ballinger, LCDR Robert Shirley, CAPT Albert Sobel. Not pictured: CAPT Hugh Howell.

Naval Reserve Veterans

EDGAR GLYNN ABEL, CDR, born June 29, 1913, Carroll County, MI. He received BS and MS degrees.

Joined the service May 2, 1942. Abel's stations include: New Orleans, Houston, San Francisco and the South Pacific.

He served with USS *Dash,* Hunter's Point Naval Station. In the South Pacific he escorted ships. Achieved the rank of commander and was discharged October 1945.

His memorable experiences include: typhoons and the South Pacific scenery.

Abel married Leah Jane "Dukie" and they have a son, Daniel G. Abel. Employed as assistant to mayor and Dean of Men. Retired in 1973.

RABBI REUBEN ISRAEL ABRAHAM, LCDR, (Previous name, Robin Scott Jones), born Nov. 4, 1951 in Rockford, IL. He received a BA, Rockford College; M.H.L, Reconstuctionist Rabbinical College.

Joined the service Aug. 6, 1976. Abraham's stations and units served with include: OCS, Newport, RI; NSCS, Athens, GA; USS *Coral Sea* (CV-43); NAS Alameda; COMNAVAIRPAC, NAS North Island, San Diego, CA; OIC COMSYSTO, NAS Bermuda.

He received the Navy Achievement Medal, Jewish Lay Leader, COMPATWINGSLANT. Achieved the rank of lieutenant commander, SC, USNR and was discharged Oct. 1, 1987.

His memorable experiences include: running in Bermuda International 10K road race and Bermuda marathon while stationed at NAS Bermuda as OIC COMSYSTO Bermuda.

He married Jennifer Stamm and they have son, Aaron David Jones; stepson, Dani Gabriel; stepdaughter, Adeena Gabriel.

Abraham is Rabbi, Adath Tikuah-Montefiore, Philadelphia, PA and Jewish chaplain, Holy Redeemer Health System, Meadowbrook, PA.

ALLAN B. "BLAKE or ANIMAL" ADAMS, JR., CAPT, born Sept. 23, 1952 in New Orleans, LA. He received BA, Political Science, University of Richmond.

Joined the service May 19, 1972. Stationed in AVROC P'Lola, NAS Whiting, Corpus Christi, Moffett Field, Misawa Japan and New Orleans.

Served with VT5, VT3, VT31, VP31, VP50, VT2, VP94, VP5682, VR54, Fairmed 1082.

Adams was awarded MSM, Coast Guard MUC, Battle E, National Defense, Sea Service, Overseas Deployment, Coast Guard Spec. Ops, Naval Res w/Hour Glass. He achieved the rank of captain 06.

His memorable experience includes: XO/CO of VR54. The Navy's first C130 VR Sqdn. with four C130 T aircraft. Numerous actual prosecutions of Soviet subs in the Sea of Japan, Pacific, Atlantic and men with VP50 and VP94.

Adams is married to Eileen and they have children, Brigette and Katy. He is a pilot with Fed Ex, 727 Flight Standards Department.

BILLY J. ADAMS, CAPT, joined the USNR as a seaman recruit in 1948 when he started college at the University of Oklahoma. In 1950 he joined the ROC program and upon graduation from OU in 1952, he received his commission and reported to NSCS, Bayonne. Upon graduation he reported aboard the USS *Henderson* (DD-785) as supply and disbursing officer. After two tours in WESPAC he was released to inactive duty in September 1954, subsequently joining a supply company in Los Angeles.

In 1958 he joined the USS *Vammen* (DE-644) SR ship where he served until his recall to active duty as a result of the Berlin Crisis in October 1961. He reported aboard the USS *Vammen* as supply and disbursing officer, assisted in its recommissioning and subsequent tour of West Pacific where the ship participated in the landing of US Marines in Thailand during the 1962 Laos Crisis, as well as working with the South Vietnam navy in the Gulf of Siam.

In October 1962, after assisting in the decommissioning of the *Vammen,* he was released to inactive duty where he continued serving aboard the *Vammen* and COMDESDIV staff until his promotion to commander in 1966. Additional units served with were CO, SSOD11-1; AIRTYCOM 219, Point Mugu; senior supply coordinator, ASDARP; CO, USNR, DCASR 1019, Los Angeles, and CO, VTU Supply 1904, Long Beach. He retired from the Naval Reserve in July 1983, after serving 35 years.

He completed numerous ACDUTRA's at sea, as well as performing ACDURTA's at Management Transportation School, Oakland; National Security Seminar, Lake Charles; Counter Insurgency, Coronado; CAPT/CDR Selection Board, Washington; COM 14, Pearl Harbor; COMNAVAIRPAC, North Island; Naval War College, Newport; DCASR HQ, Los Angeles; President LT/LCDR Selection Board; NAVSUP and Office of CNO, Washington, DC.

His civilian occupation was director, Manufacturing Administration and Production Control, Rocketdyne Div., Rockwell, supporting the Space Shuttle, Atlas, Delta and Peacekeeper Programs. He retired from Rockwell in January 1996.

CAPT Adams is a member of the State Bar of California, as well as previous co-owner with his wife, Paula, of Adams' Specialty Shop, a mastectomy boutique in Northridge. His wife, Paula, retired in November 1995 after selling the business.

He received a MA degree from USC in 1961 and Juris Doctor of Law degree in 1969 and is a member of the California State Bar.

He and his wife have three married daughters, 10 grandchildren and now spend their retirement traveling (cruises and RV) or residing at their beach home at the Baja Beach & Tennis Club, on the Todos Santos Bay, Ensenada, Mexico, or at their primary residence in Westhills, CA.

DAVID EARL "DAVE" ADAMS JR., CAPT, born Aug. 23, 1946, Wichita, KS. Earned his BS degree at USNA in 1968, MSA from George Washington University in 1977 and his MBA from Wichita State University in 1988.

Joined the service Aug. 23, 1963, and stationed in San Diego, Long Beach, Naples, Vietnam and Norfolk, VA. Served with VS-41, USS *Decatur,* USS *Lester* (DE-1022), NWTGLANT, USNS WACCAMAW, COMFWG/COMNAV.

Discharged with rank of CAPT. His medals include the Navy Achievement, Vietnam Service, Armed Forces Reserve Medal, National Defense and Small Arms Marksman.

He and wife Mary have two children, David and Tara (McDaniel), and three grandchildren: Christian McDaniel, Cody and Courtney Adams. Still working as systems engineer and structural engineer for Boeing.

DENIS L. ADAMS, CAPT, born July 8, 1940, Missoula, MT. He was reared in Winnett, MT and received a BS degree, University of Montana, 1962.

Joined the service October, 1962. Commissioned ensign, Supply Corps, USNR, March 1963. His stations include: OCS, Newport, RI; Navy Supply Corps School, Athens, GA; USS *Finch* (DER-328); Naval Area Audit Service, San Francisco. Released from active duty March 1967.

Adams served with following Reserve units: Medium Surface Unit, Billings, MT; USS *Prairie* (AD-15) Det. Great Falls, MT; SACLANT DET. Spokane, WA; NSC Puget Det. Boise, ID; VTU, Billings, MT. He achieved the rank of captain, Supply Corps, USNR and retired Sept. 1, 1993.

He was employed as CPA, 16 years public accounting; vice-president of mining company; director, Montana Department of Revenue; tax commissioner, Crow Tribe of Indians.

FRANK E. ADAMS JR., LCDR, born Feb. 8, 1921, Minneapolis, MN. He received a BS degree, University of Minnesota, August 1943.

Joined the service June 1942 and Sept. 12, 1950. His military stations include: USS *Rhea* (AMC-58) Pacific, USS *Witter* (DE-636), Pacific; USS *Kalmia* (ATO-23), ATL-PAC, USNTC-San Diego. He achieved the rank of lieutenant commander, USNR and was discharged Dec. 12, 1946 and March 11, 1953. Retired USNR, Feb. 8, 1981.

Adams married Joanne Cronin and they have children: Catherine, Bruce, Annette and David and grandchildren: Russell, Clayton, Gloria, Paula, Bonnie, Andrew, Brian, Ethan, Sarah Caitlin.

He was employed as district attorney investigator, junior college instructor of Police Science, retiring Sept. 29, 1978.

GREGORY S. "GOMEZ" ADAMS, LCDR, born May 27, 1962, in Abington, PA, originally from Feasterville, PA. Graduated with a BS degree and was commissioned from the USNA in May 1985 and received M.Ed., Holy Family College. After flight school in Pensacola, FL, he went on to his first fleet tour with HS-3 flying the SH-3 Sea King helicopter out of Jacksonville, FL and deploying aboard USS *Saratoga* (CV-60).

He then served at the Bureau of Naval Personnel, Washington, DC, from January 1990 until his separation from active duty in September 1992. For this tour he was awarded the Navy Commendation Medal.

Adams immediately returned to Pennsylvania for affiliation with HSL-94 as a drilling reservist at NAS Willow Grove, PA, flying the SH-2 Sea Sprite helicopter. While deployed with HSL-94 during Operation Restore Democracy off the coast of Haiti, his crew rescued 10 Venezuelan merchant marines whose ship had sunk. He was awarded his second Navy Commendation Medal for this action.

LCDR Adams again lives in Pennsylvania with his wife, Bonnie and two children, Cody and Emily. As a civilian, he teaches in Neshaminy School District in the very school he himself attended.

JAMES MILLER ADAMS, LCDR, born in Cleveland, OH, Sept. 9, 1924. Entered Case, September 1942, enlisted, June 1943, V-12, November 1943. Received BS (ChE), April 1945. Commissioned ensign, Columbia Midshipmen's School, August 1945.

Reported USS *Casablanca* (CVE-55), (assistant navigator). He made three voyages: to Maui, Noumea and Yokosuka. Transferred USS *Whythe* (APB-41), (gunnery officer) Tsingtao. Made one voyage to Inchon, then "stateside" for "moth-balling" at Galveston. Released from active duty, Great Lakes, August 1946. Transferred to Retired Reserve, as LCDR, July 1971. Awarded Naval Reserve Medal and Armed Services Medal.

Attended Virginia Polytechnic Institute. Received BS (Biology), 1948, MS (ChE), 1949 and PhD (Biochemistry), 1954. Following various research positions, most of career spent with US Customs Service; Regional Director, Baltimore laboratory, 1969-82; Chief, Research Branch, New York Laboratory, 1982-86. Author or co-author of 40 or more scientific papers or publications.

Currently resides in Joppa, MD with wife, Lorraine (Waryold). He has seven children and eight grandchildren, so far.

RONALD E. "RON" ADAMS, CAPT, born June 11, 1947, Willimantic, CT. He received BSCE 1970; MBA 1976 University of Kansas, Lawrence, KS.

Appointed midshipman, August 1965. Commissioned June 2, 1970.

His stations include: NROTC, University of Kansas, Lawrence, KS, Aug. 28, 1965-June 2, 1970, midshipman.

Active duty assignments: USS *Intrepid* (CVS-11), Quonset Point, RI, June 03, 1970-Feb. 4, 1972, assistant damage control assistant/repair division officer; USS *Josephus Daniels* (DLG-27), Norfolk, VA, Feb. 5, 1972-June 28, 1974, damage control assistant.

Reserve duty assignments: PHIBGRU STAFF DET 217 (ORU), Kansas City, MO, Aug. 24-Dec. 26, 1974, enlisted training officer; ACLANT DET 317 (ORU), Kansas City, MO, Dec. 27, 1974-Feb. 28, 1977, training officer; NR COMPHIBGRU 1 DET 118, Kansas City, MO, March 1, 1977-Sept. 30, 1978, combat cargo officer; NR AR-6 AJAX DET 218, Kansas City, MO, Oct. 1, 1978-Sept. 30, 1979, commanding officer; NR YR-84 HQ DET 122, Tacoma, WA, Oct. 1, 1979-Oct. 31, 1981, accounts, analysis and reports officer; NR YR-86 HQ DET 122, Tacoma, WA, Nov. 1, 1981-Nov. 30, 1982, accounts, analysis and reports officer; NRMTF PUGET SOUND HQ DET 122, Tacoma, WA, Dec. 1, 1982-Sept. 30, 1983, accounts, analysis and reports officer; NRMTF PUGET SOUND DET 322, Tacoma, WA, Oct. 1, 1983-Oct. 31, 1984, officer-in-charge; NRMTF GREAT LAKES HQ DET 113, Great Lakes, IL, Nov. 1, 1984-Sept. 30, 1989, executive officer; REDCOM 13 STAFF Great Lakes, IL, Oct. 1, 1989-Sept. 30, 1992, training coordinator; USNR VTU 1318, Madison, WI, Oct. 1, 1992-Sept. 30, 1993, executive officer, USNR READINESS UNIT 1652-1, LaCrosse, WI, Oct. 1, 1993-Sept. 30, 1994, commanding officer, USNR VTU 1318, Madison, WI, Oct. 1, 1994-Jan. 31, 1997, executive officer. Achieved the rank of CAPT (0-6) and transferred to Retired Reserve without pay February 1997.

Adams was awarded the Navy Commendation Medal, Meritorious Unit Commendation, National Defense Medal w/Bronze Star, Armed Forces Reserve Medal w/HG.

He was employed with Kansas Department of Transportation, June 1976-July 1979 as state rail engineer; August 1979-April 1984, Washington State Department of Transportation, rail program manager; Wisconsin Department of Transportation, May 1984-present, chief, rail project management section. Adams married Joan and they have children, Penelope L. Adams and Melanie E. Adams.

PATRICK STEPHEN AGNEW, LCDR, born Oct. 28, 1958, Washington, DC. He received Doctor of Podiatric Medicine, Pennsylvania College of Podiatric Medicine.

Agnew joined the service, Nov. 6, 1986. His military stations and locations include: NRC Philadelphia, PA; N&MCRRC Norfolk, VA; N&MCRC Washington, DC; Fleet Hospital 20, 4th Marine Div, 4AAVBN DET A, OPNAV 093-106, PHIB CB2 0206.

His awards include: Navy Commendation, Navy Achievement, National Defense, Military Outstanding Volunteer Service Medals and Fleet Marine Force Ribbon. Agnew achieved the rank of LCDR.

Memorable experiences include volunteering for service in the Persian Gulf War, but every time the orders arrived, his name was absent. So when another podiatrist was called from out of town to cover Portsmouth Naval Hospital, he volunteered to take his place, without pay. He was able to serve, keep his private practice and received a medal!

He married Heidemarie and they have children, Megan Brittany and Michael Patrick. Presently employed in private practice, and vice-president Am. Col. of Foot and Ankle Pediatrics.

JOSEPH O. AGRA III, LCDR, born in Pila, Laguna, Philippines, June 4, 1954. He enlisted in the USN at Subic Naval Base, Philippines, Dec. 4, 1972. LCDR (then ADJ-3) Agra served as aviation machinist mate (ADJ) for over six years active duty with light attack (A-4E, A-7E), Medium Attack (A-6A), and fighter squadrons (F-4S, F-14A), along the West Coast and Cubi Point, Philippines. He participated and supported the Operation New Life and Eagle Pull (Evacuation of S. Vietnam, April 1975). After over six years of active duty, then (ADJ-2) LCDR Agra got out in March 1979 to complete his college degree.

He was commissioned 2LT in the USAR and the California NG via the California Military Academy on July 22, 1983. He attended and completed the Air Defense Arty. and Chaparral Vulcan Officer Course in Fort Bliss, TX (Jan-May 1984). He served as heavy weapons platoon leader, an executive officer of an infantry company and evaluator in California from 1983-87. Feeling homesick, he then transferred back to the USNR in June 1987. He graduated with BBA and MBA from National University.

Newly commissioned LTJG, he immediately reported to Flt. Maint. Fac. Det. 0280 as avionics division officer. In October 1987, he reported to NASC Pensacola, FL for the ten week Aircraft Maintenance Officer Course. After two successful deployments to NAF Misawa, he was recalled back to active duty in October 1989 for 90 days SPECACT as maintenance/material control officer of newly commissioned HM-19 at NAS Alameda. In October 1991, then LT Agra reported to VF-302 as AAMO and participated in several squadron deployments.

In July 1994, he reported to CINCUSNAVEUR, London, UK as FLTAVNREDOFF for the N43 staff. Under the Peacetime Support Program of the late ADM J. Boora, he augmented HC-4, NAS Sigonella, Italy several times in support of Operations Deny Flight, Deliberate Force, Joint Endeavor (Bosnia), Southern Watch (Iraq) and Assured Response (Liberia). He is currently serving as a USN representative to the US Military Observer Gp assigned to the UN Peacekeeping Mission to Western Sahara for 179 days (Jan. 15-July 12, 1997).

In civilian life, he is a career program integrator (AV-8BII+, C-17A & MV-22A [UK] for the Defense Contract Management Command at London, United Kingdom. He married Cynthia and they have two sons, Joshua and Joseph IV. They currently reside outside London suburb, calling Laguna Hills, CA as home. He enjoys coin collecting with numerous specimens of Greeks, Romans, Byzantine and Ottoman Empire coins.

JOHN D. AHLSTROM, CAPT, born Dec. 26, 1917, New York City. He received BA, Williams College; MA Duke University.

His military stations include Navy Department Washington, VC-76; Pacific USS *Petrof Bay* (CVE-80); Naval Intelligence Grad School (Korean Service). Battles and campaigns include the Pacific, Palau, Leyte, Iwo Philippines.

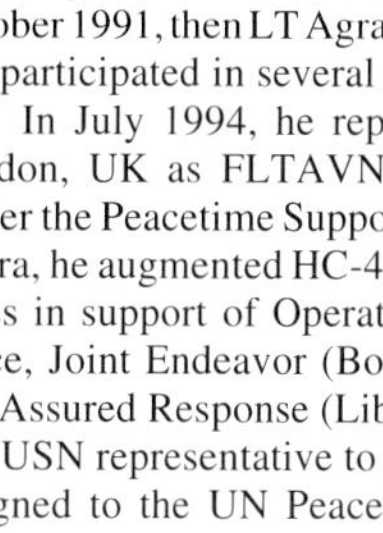

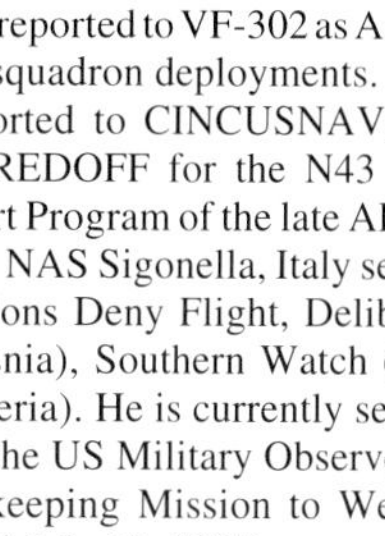

Ahlstrom received the Presidential Unit Citation, Philippine Liberation and Pacific Campaign. He retired 1965. His memorable experiences include the Battle of Leyte Gulf.

He married Doris and has three stepsons, two daughters: Austin, David, Paul, Debby and Susan; and grandchildren: Louise, Meg, Michael and Alexandra. He was employed in marketing and sales, retiring from DuPont Co. in 1984.

DONALD S. ALBRIGHT JR., RADM, born in Alliance, OH. He was educated in public schools in Sebring and Alliance, OH and attended Mount Union College in Alliance, and Ohio State University in Columbus before entering the US Naval Academy at Annapolis, MD in 1946. Upon graduation as an ensign from the academy in 1950, RADM Albright was assigned duties as gunnery division officer aboard the USS *Cabot* (CVL-28).

During 1951 and 1952, he attended flight training at Pensacola, FL and advanced training at Corpus Christi, TX. Graduated from flight training as a naval aviator in May of 1952 and was assigned to squadron duties at Airship Sqdn. One in Weeksville, NC and then as Anti-Submarine Warfare Tactics instructor at the Fleet Airborne Electronics Training Unit at Norfolk, VA.

He was ordered to the US Naval Post Graduate School in Monterey, CA and graduated in 1958 with a BS degree in electronics engineering. He served as program manager of avionics systems in carrier based AEW aircraft at the Bureau of Aeronautics, and as materiel officer at the Norfolk Detachment of Operational Test and Evaluation Forces.

Following active duty, he immediately affiliated with the USNR in 1962. Assignments in the USNR included commanding officer of HS-872, which twice won the Noel Davis Trophy as the outstanding helicopter squadron in the USNR, and commanding officer of VTU-5120 and commanding officer of VTU 8787.

Upon promotion to RADM in 1977, his assignment was as the commanding officer of the USNR Readiness Command Region 20 with headquarters in San Francisco. Albright was also assigned to the billet of commander of the naval base of San Francisco (upon mobilization) until his retirement from the USNR in 1984.

He entered the securities business in 1962 with Schwabacher & Co. Monterey, CA. During the following years he was employed as manager for DuPont Walston & Co. and later as vice-president for Paine Webber Inc. where he managed their two offices in Monterey and in Salinas. Retiring from Paine Webber, Inc. in 1988, he then established an office for Planned Investments, Inc. in Salinas, where he does financial planning for individuals, as well as for corporate clients. He is a registered principal with the National Association of Securities Dealers.

Active in community affairs, he has served on the Board of Directors of the American Cancer Society in Monterey, and the Monterey Chapter of the Navy League. He is a Past Master of Monterey Masonic Lodge F&AM #217, a member of Scottish Rite Bodies, Islam Temple Shrine, BPOE Elks, and the American Legion. He holds the Legion of Merit, Meritorious Service Medal, WWII Victory Medal, National Defense Service Medal and the Armed Forces Reserve Medal.

LANCE R. ALDERMAN, CDR, graduated from the NROTC unit at Jacksonville University on April 21, 1979. He reported to his first ship, USS *Darter* (SS-576) at its homeport of Sasebo, Japan. His second ship the USS *Daniel Webster* (SSBN-626) conducted patrols out of Holy Loch, Scotland. He was discharged from active duty Dec. 30, 1983.

As a reservist, CDR Alerman has served in an NCSO, NCSLO and CNO N87 reserve units. Notable experiences were exercises in Iceland, Germany, Netherlands, and Norway including OIC of a Reserve Convoy Unit to Narvik in Exercise Battle Griffin.

At the Pentagon, he has supported Submarine Live Fire Test (OSD) and the Joint Warrior Interoperability Demonstration (JWID).

CDR Alderman as a civilian is a division head in a Submarine Electronics Program office, Program Executive Office (PEO) Submarines, Arlington, VA.

DONALD ALF, CAPT, born Sept. 14, 1944, is a native of Berlin, WI. He attended the University of Wisconsin on a NROTC scholarship and graduated with a BSEE and a BS in Naval Science in June 1967. He did graduate work at Stanford University, receiving a MSEE, before performing active duty on the staff of ADM

Rickover as a nuclear propulsion engineer in the Naval Nuclear Propulsion Program. He resigned from active duty in September 1973 and accepted a civilian position in the Naval Nuclear Propulsion Directorate.

He affiliated with the USNR in June 1974 with the Office of Naval Research (ONR) Reserve Program. His duty assignments in the USNR have been Research Co 5-8; ONR 106; ONR Tactics, Development & Evaluation 306, ONR Technology Mobilization Management Team 206, program manager Sea Theater Nuclear Warfare 106, and ONR Science & Technology HQ 106. He was the commanding officer of ONR Science & Technology HQ, 106 as well as the director of ONR's Science & Technology Reserve Program.

He retired with 30 years of commissioned service on July 1, 1997. He significant Navy awards consist of the Legion of Merit, Meritorious Service Medal (w/Gold Star), Navy Unit Commendation, National Defense Medal (w/Bronze Star), and Armed Forces Reserve Medal (w/ Silver HG).

Alf is currently the branch head for nuclear safety and reactor containment in the Naval Nuclear Propulsion Directorate. He resides in Alexandria, VA with his wife Susan.

FREDERICK S. ALLEN JR., CAPT, born in Chicago, Oct. 15, 1938. The son of a naval officer, his family moved to San Francisco in 1941. He was graduated from the University of Pennsylvania with an AB/Honors in 1960. Commissioned ensign through the NROTC program he was assigned to the staff of commander Amphibious Sqdn. THREE where he served through 1962. Upon release from active duty, he continued with his reserve career with the Amphibious Beach Div. in San Francisco, and then in Los Angeles with the Telecommunications Censorship Div., the Wartime Information Security Program, USNR Commander Naval Forces Japan (commanding officer), and the Voluntary Training Unit (commanding officer) in Santa Barbara, CA. He attended classes at the US Naval Academy, the Naval War College and the National Defense University. He retired from the USNR in 1985 and is a lifetime member of the NRA.

In civilian life Allen had a 30 year career with Wells Fargo Bank, served as president of the local Rotary Club and served on the boards of a number of non-profit agencies. He has four children and resides in Santa Barbara, CA with his wife Suzanne and their two younger children. He is the president of The Frederick Allen Studio, specializing in International and Sports Photography.

JANA S. ALLEN, LCDR, born Sept. 12, 1962, Atlanta, GA (native of Dundee, IL). Joined NROTC May 20, 1985 and received BA degree, University of Illinois (Champaign-Urbana).

Military locations and stations include USS *Hector* (AR-7), auxiliary and electronic officer, 1986-87; USS *Acadia* (AD-42), MPA, 1987-89; engineering instructor Surface Warfare Off School SWOSCOLCOMPAC commanding officer NCTAMS EASTPAC DET 319, 1989-91; Reserves CINCPACFLT DET 719, 1991-96.

Discharged Sept. 20, 1991 and still in Reserves. Awards and medals include Navy Achievement Medal (4), Battle E, Sea Service (3), Meritorious Unit Commendation (2), Armed Forces Expeditionary and National Defense Service.

Memorable Experiences: serving as MPA while USS *Acadia* spent 30 days repairing USS *Stark* in 1987 after it was hit by two exocet missiles; three six month WESTPAC Persian Gulf/IO deployments; and serving as CO of NCTAMS EASTPAC DET 319.

Married LCDR Russell C. Allen, USNR, father is CAPT John D. Lindstrom, USNR (Ret). Civilian employment as manager, engineering data management department, Solar Turbines, Inc.

RICHARD E. "RICH" ALLEN, CAPT, born Sept. 15, 1939, DuQuoin, IL. A naval reservist and teacher at Hawthorne High School, retired from the Navy on Sept. 1, 1991, after 30 years of military service. Allen's last command position was as the commanding officer of reserve unit USS *Independence* (CV-62) 0176 at Naval Air Reserve Point Mugu, CA.

He has flown on two combat A4 deployments with Attack Sqdn. 55 on the USS *Ticonderoga* and the USS *Ranger,* earning the Distinguished Flying Cross for heroism in Vietnam.

Allen has resided in Westchester, CA for 25 years, moving there in September 1966 after his release from the active duty Navy.

He earned his M.Ed. at Loyola University in 1971 while teaching and coaching baseball at St. Bernard High School. Allen has been an English teacher at Hawthorne since that time and has also served as a master teacher for the LMU Education Department. He is well known in scholastic journalism for his perennial state champion yearbook staff at Hawthorne High.

CAPT Allen has served as president of the South Bay chapter of the NRA and has also served locally as chairman of the board at University Christian Church, Westchester Chamber of Commerce Education Committee, Westchester High School Academic Booster Club, Rotary Club, Westchester Veterans Committee and as a Westchester Park baseball coach.

A graduate of Northwestern University in 1961, Allen has served his alma mater as director of the Alumni Admissions Council and as president of the Northwestern Club of Los Angeles. He received the 1994 Northwestern U. National Service Award and 1996 National Paddle Tennis Champion.

Allen's wife, Lynne, is a life-long resident of Westchester and a first grade teacher at Westchester Neighborhood School, where she was recently named "Teacher of the Year." Their children are Rick, Aimee and Lindsay. His parents are Gene (USMC veteran) and Charlene Allen of Colorado Springs.

RUSSELL E. "RUSTY" ALLEN, LCDR, born Aug. 10, 1960, Decatur, IL. Earned BS in biology, University of Texas (Austin). Joined the service Dec. 20, 1984; served with VT-27, HT-18 (Milton), HC-3, HC-8, HC-3 (instructor duty), HC3-5 drilling reservist.

Stations include Navy Flight School, Pensacola, Corpus; FRS (H-460) NASNI, NAS Norfolk, NASNI, Pt. Munga. Memorable experiences include flying night VERTREP during three med deployments, NVG SEAL missions w/HCS-5 in Thailand Cobra Gold 94 and in Hawaii PUMPAC 96. Total 2300 hours including H-46D and HH-60H, winged in 6186.

Married Jana, no children. Civilian employment flying Metrolines III aircraft (1000 hours) for Conquest Airlines.

STEPHEN R. "STEVE" ALLEN, CAPT, born Oct. 19, 1952, Minneapolis, MN. Joined the NROTC 1970-74; USN August 1974-February 1981; USNR February 1981-December 1996. His military stations and locations include: Penn State NROTC, NAS Pensacola, NAS Chase Field, NAS Oceana.

Allen served with the following units: VT-5, VT-26, VT-25, VA-42, VA-176, NISCPROD D102, VP-1123, VP-2644, VP-64, PATWING 0593. He was awarded the Battle E. Achieved the rank of CAPT 0-6.

Employed as airline pilot for People Express, Continental and United, retiring Dec. 1, 1996. He is married to Glenda and they have sons Gregory and Richard.

WILLIAM HAND "BILL" ALLEN, CAPT, born April 2, 1952, Sharon, PA. He received BS EE, Georgia Tech; MS Applied Physics, Johns Hopkins (1985); MA Business, National University (1981).

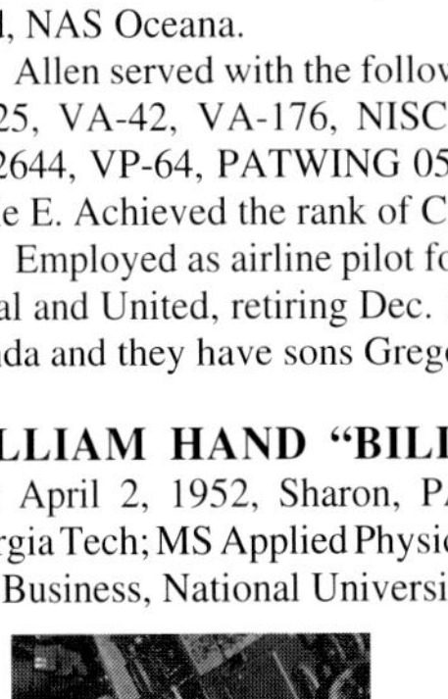

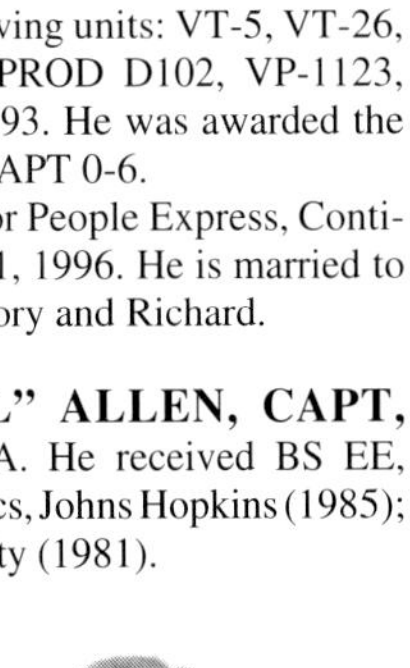

Allen entered service June 6, 1975. His military stations and locations include: USS *Haddo* (SSN-604), 1976-80 DCA, WEPS, MPA, RCO, EO, Submarine Training Facility, 1980-81, Engineering Deputy Director. He participated in two Western Pacific deployments and was awarded two Navy Achievement Medals, Navy Commendation Medal, Navy Expeditionary Medal, National Defense Medal, Armed Forces Reserve Medal.

His memorable experiences include: Nuclear Power School, Bainbridge; Nuclear Power Training Unit, Ballston Spa, NY; Sub School, New London. Reserve duties, Norfolk Naval Shipyard, USS *Simon Lake,* NAVSEA, SPAWAR, VTU FY 92 & 93, CO SUPSHIP Newport News; FY 94 & 95, CO SUPSHIP New York.

Married Deana Bradstock and they have children, William Yazvac and Daniel Burker. He was employed with Allen & Shariff Corporation - principal. He is presently in Reserves, VTU-603.

FRANK J. ALLSTON, RADM 0-8, retired June 1, 1985 after serving more than 34 years in the Navy and USNR, including active duty in Washington during the Korean War. At the time he retired, he was the senior drilling Naval Reservist, serving as the director, Naval Reserve Aviation Supply Development and Readiness Project. Upon retirement, the Chief of Supply Corps asked RADM Allston to research, write and arrange publication of the Bicentennial history of the USN Supply Corps. His book, *Ready For Sea,* was published by the US Naval Institute Press in 1995.

He was born on Oct. 2, 1930 at New Bern, NC, the son of Frank J. Allston Sr., a Navy enlisted man in WWI, and Louise Moore Allston. Both parents were career school teachers and high school athletic coaches. He attended public schools in New Bern and Durham, NC, and Arlington, VA. He was graduated from the University of North Carolina, Chapel Hill, in 1952 with an AB degree in journalism and political science. In 1988, 36 years after receiving his bachelor's degree, he received a master's degree in business administration (MBA), with a major in general management, from the Graduate School of Business, The University of Chicago.

He enlisted in the USNR in February 1951 and was commissioned an ensign, Supply Corps, through the NROC (ROC) program at the Naval Station, Long Beach, CA, in August 1952. He was ordered to active duty at the Navy Supply Corps School, Bayonne, NJ, in September 1952. He next reported to the Bureau of Supplies and Accounts, Navy Department, WA, where he served as associate editor and acting editor, *Navy Supply Corps Monthly* newsletter, from February 1953 through August 1954.

Following release from active Navy duty in August 1954, Allston served in management positions with General Electric Company, Bunker Ramo Corporation, and IC Industries, Inc. At the time of his retirement from the corporate business community in March 1989, he was vice-president, Corporate Affairs, Illinois Central Railroad. He is presently a business communications consultant. In 1992, he wrote and published *Con-glom-er-ate,* the story of the diversification of the original Illinois Central Railroad into a diversified multi-national conglomerate, from 1966 to 1987.

When he first returned to civilian life in 1954, LTJG Allston immediately affiliated with the Ready Reserve. He was commanding officer of three Naval Reserve units - Ships Supply Officer Div. 3-3, New York City (1966-67); Supply Company 6-8, Greenville, SC (1968); and Ships Parts Control Center Mechanicsburg 113, Chicago (1974-77). He founded the Recruiting District Assistance Council Program (RDAC) for the Navy Recruiting Command, and served as the nation's first RDAC chairman, 1972-73 in Chicago. He attended the National War College, Washington; and the Naval War College, Newport, RI; and participated in Logex 1968, Fort Lee, VA; and the Senior Reserve Officer Orientation, US Atlantic Fleet. He served as president (1978) of both the Naval Supply Corps Association and the Chicago Council, Navy League; and in 1978 as chairman, Armed Forces Week in Chicago. He is a life member of the Navy League of the US, the NRA, and the US Naval Institute. He is also a member of the Naval Order of the United States and The Retired Officers Association.

He was awarded the Navy Meritorious Service Medal in 1977 for "outstanding meritorious service for support of recruiting, Naval command accomplishments and for community affairs activities" and the Legion of Merit in 1985 for "outstanding performance" as a flag officer. He was also awarded the National Defense Service Medal and the Armed Forces Reserve Medal w/HG Devices signifying second and third awards. He earned two Centurion Awards for making 215 presentations to civilian and military audiences in 12 states and Canada as a charter member of the Chief of Naval Operations Sea Power Presentation Team, from 1968 to 1980.

He is an elder of the United Presbyterian Church, and a member of the Rotary Club of Naperville. He served as chairman of the city of Naperville Transportation Advisory Board (1977); and general chairman, Naperville Bicentennial Commission (1981). He is currently a commissioner, Riverwalk Commission, city of Naperville, a director of the Chicago Council of the Navy League, and has been a visiting lecturer at North Central College, Naperville.

He and his wife, the former Barbara French Brown, of Woodstock, VA, and a classmate at the University of North Carolina, were married in September 1955. They reside in Naperville, IL. The Allstons enjoy gardening, bridge, music, and world travel in addition to community and church activities. Their two adult sons are employed in the insurance industry; James Frank of Tinley Park, IL, a corporate systems planner; and John Pierson of Carol Stream, IL, an appraiser. They also have a grandson, Pierson John of Carol Stream, and grandchildren, Christine Loraine and Nathan Lenard, of Tinley Park.

PAUL R. ALWINE JR., CAPT, a native of Binghamton, NY and Limestone, ME, is a 1962 NROTC graduate of the University of Southern California and an MBA graduate of California State University, Fullerton. Upon commissioning he attended the Navy Supply Corps School in Athens, GA, then served in USS *Valley Forge* (LPH-8). He was in WestPac during the Gulf of Tonkin incident and served in Vietnam operations during 1964. NAS Point Mugu was his next duty station and he was released from active duty in July 1966, when he joined the USNR. He served in various Supply Corps units in southern California, including four command tours, until his retirement in 1992 after 30 years.

A strong supporter of Naval Aviation, he proudly wears the "Wings of Gold" of a Naval Aviation Supply Officer. He also holds commercial pilot and flight instructor ratings. He was awarded the Defense Meritorious Service Medal along with other personal commendations and service medals.

In civilian life he is a sales executive with Digital Equipment Corp. and resides in Irvine, CA with his wife Binnie. They love traveling the world. He treasures the many friends with whom he served and made his Navy experience so great!

RICHARD HILLER "DICK" AMBERG JR., LTJG, CAPT, born Oct. 26, 1942, Philadelphia, PA. He received a BA in 1964 from Harvard University. He joined the service July 1, 1964.

Amberg's military stations and locations include: USS *Yancey* (LKA-93) as communications officer. He was awarded the Meritorious Service Medal and Navy Commendation Medal. He achieved the rank of LTJG (active duty), CAPT (Reserves). He was discharged July 1, 1966 (active), July 1, 1994 (Reserves).

His memorable experiences: CO of five reserve units; attended Naval War College, National Defense University, Armed Forces Staff College; served on various selection and policy boards.

He married Beverley Sharp and they have a daughter Elizabeth. He is a publisher and advertiser, Montgomery, AL.

OLIVER F. AMES, CAPT, born Dec. 13, 1920, Boston, MA, received AB from Harvard College, 1943 and MBA, 1951, Harvard Business School.

He joined the service September 1939. His military locations and stations include: Harvard NROTC, USS *Crosby* (APD-17), USS *Little* (DD-803), USS *Fiske* (DD-842). He fought campaigns in Solomon Islands, New Britain, New Guinea, Iwo Jima and Okinawa.

Ames was awarded the Bronze Star Medal w/Combat V, Navy Unit Commendation, Asiatic Pacific w/6 stars. He achieved the rank of CAPT USNR and retired in 1964.

His memorable experiences include witnessing the first flag raising on Mt. Suribachi, Iwo Jima. Survived sinking of USS *Little* (DD-803) by kamikaze aircraft, May 3, 1945, on radar picket station #10 off Okinawa. He retired in 1964.

He married Esther D. and they have children: Oliver F. Jr., Abby, Samuel and Mary; and grandchildren, Benjamin and Caroline. He held positions as professor, trustee director, state senator, Massachusetts Legislature 1962-70.

HAROLD "HOPPY" ANDERSEN, LCDR, born Nov. 20, 1917, Detroit, MI, received BS, Industrial Engineering, General Motors Institute.

He joined the service Aug. 16, 1943. His military locations and stations include: Norfolk Naval Shipyard, Portsmouth, VA, serving with the Bureau of Ships. He was awarded the American Campaign, WWII Victory Medal, Naval Reserve, and Armed Forces Reserve.

Anderson achieved the rank of LCDR and retired in 1977. He married Marguerite E. and they have children: Roger Lars, Gary Lee, David Christian and Richard Kirk; grandchildren are: Jeanette, Hillary, Elizabeth and Kristen. He was employed with the GM Corporation, national truck manager, Chevy Div. He retired in 1980.

A.C. "TED" ANDERSON, CAPT, born May 7, 1930, Chicago, IL, received BS, geology, Indiana University and Beloit College.

He joined the service July 1952. His military locations and stations include: COM-12 (San Francisco) and COM-17 (Kodiak, AK). He attended Naval Intelligence School and served at various duty stations.

Anderson achieved the rank of CAPT and was discharged from active duty November 1955. He later retired from the Reserves in May 1990.

His memorable experiences include receiving his commercial pilot's rating for single engine land and sea aircraft and private pilot's rating for glider aircraft.

He has held positions as manager Energy Liaison, Shell Oil Company, retiring in November 1991. Presently manages Discovery Bay Wilderness Resort.

BLANE K. ANDERSON, CDR, born June 8, 1958 in Barstow, CA, received BA in journalism from the University of Oklahoma in 1980 and MA in public administration, Oklahoma State University in 1986. He was commissioned May 10, 1980 via Naval ROTC, University of Oklahoma.

His military locations and stations include: USS *Inchon* (LPH-12), legal officer/boilers officer, 1DIV officer, 1980-83; inactive duty: NR USS *Dixon* (AS-37) DET, Houston, TX, 1983-84; Naval Reserve Center, Stillwater, OK, commanding officer, 1984-86; USS *New Jersey* (BB-62), 16-inch turret officer, 1986-87; Naval Reserve Readiness Command, Region TWO, Scotia, NY: Ashore program manager 1987-90; Commander, Naval Surface Reserve Force, New Orleans, LA, deputy chief of staff for Information Systems Management and Reserve Billet Training Requirements Officer, 1991-94; chief of Naval operations staff, director of Naval Reserve (N095), Washington, DC, assistant for Surface Training and Head, Support Programs Branch, 1994-present.

Battles and campaigns: Multi-National Peacekeeping Force, Beirut, Lebanon, 1982-83. His awards and medals include: Navy Commendation Medal, Navy Achievement Medal (2), Navy Unit Commendation Award, Meritorious Unit Commendation, Battle Efficiency "E", Navy Expeditionary Medal, National Defense Service Medal, Sea Service Deployment Ribbon (2), Armed Forces Reserve Medal. Anderson achieved the rank of commander (0-5) and is presently still active.

His memorable experiences include: deploying in 1986 with the first battleship battle group (BBBG) aboard USS *New Jersey* (BB-62) since the Korean War.

He married Terry Thorpe and they have children: Kelcey, Ashlyn, Kevin and Cameron.

BURTON M. "ANDY" ANDERSON, CAPT, born Jan. 10, 1915, Jamestown, NY, attended school two years at the University of Cincinnati.

He joined the service December 1942. His military locations and stations include: NAS Corpus Christi; NAS Bunker Hill; NAS Floyd Bennett; CASU F-20, VA-859 Niagara Falls, CO VP-859; CO USNRO 3-68 (2). Served on staff CAG-8, 1951-52.

He received awards for Asiatic-Pacific, Korean, American Campaign, including the WWII Victory Medal and Armed Forces Reserve Medal.

Anderson achieved the rank of CAPT and left the service January 1975.

His memorable experiences include being flight instructor, ferry pilot, test pilot. He qualified as carrier pilot in AD and was qualified in many types including: FM-2, F6F, F4U, SNB, SNJ, NH-1, P2V, AD.

He married Elinor and they have children: Barbara Lyons, Nancy Belschner and David Anderson. Anderson has held positions as partner-manager for Anderson Cleaners, Jamestown, NY, retiring in 1976.

CHARLES CLYDE ANDERSON, MD, CAPT (MC), born Oct. 10, 1928, Moline, IL, received BA from the University of Iowa in 1950 and an MD from University of Illinois in 1954.

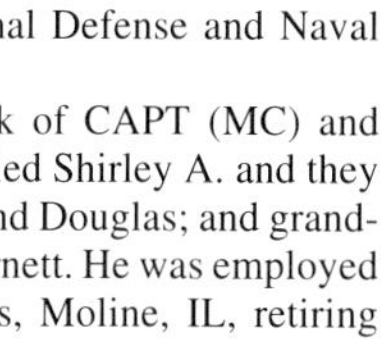

He joined the service Jan. 10, 1955. His military locations and stations include: USS *Telfair* (APA-210), May 1956-August 1957; NAV STA San Diego, August 1957-April 1958; NRTC Forrest Park, IL, 1960; NRG September 1968; Moline, IL, 1961-69; T&S NRTC, Davenport, IA, 1969-77; VTN 1603, 1977-82.

He was awarded the National Defense and Naval Reserve Medals.

Anderson achieved the rank of CAPT (MC) and discharged Oct. 1, 1982. He married Shirley A. and they have children: Elizabeth, James and Douglas; and grandchildren, Payton and Madison Burnett. He was employed with the Medical Arts Associates, Moline, IL, retiring July 1, 1991.

DORRANCE I. ANDERSON, CAPT, born 1917 in Minnesota. Earned BS degree, Iowa State; enlisted December 1941, V-7 and commissioned ENS, August 1942, Notre Dame, USS *Prairie State* (also, Vance, Wouk); August 1942, USS *Kankakee*, South Pacific; August 1943, Noumea, passed flight physical; September 1943, flight training, Dallas.

Designated naval aviator, June 1944, Pensacola; 1944-45, PV's Lake City, PB4Y's, Chincoteague, Key West; 1945, Norfolk, USS *Albemarle* to Bristol, PB4Y *Dunkeswelle*, Wales, Iceland, Labrador, Westover, Philadelphia, Still Water Grave Yard, in soup over city day B-25 into Empire State Building; 1945, NAS Seattle, in prospect, PB4Y2s, Whidbey, Attu, Kamchatka Run; 1945 two big ones, thankful; November 1945, inactive.

1949, DDS, *Minnesota*, commission in Regular Corps, USPHS; 1953, Santa Barbara; 1955, re-joined USNR; 1940, Champion, Collegiate (USA, Canada, Mexico), dairy products judging, Convention Hall, Atlantic City. Marketing Specialist (USDA), Los Angeles to Pearl Harbor disaster. Reported Market News (National Consumption). Graded (top quality) butter, cheese, eggs, dressed poultry sold US Military, Southern California, permanent deferment, Armed Forces.

In 1959 started national organization (useful occupations, short term volunteers) to assist third world. Gave rise to Peace Corps. Anderson family donated 36 times to present; 1960, MADAG as CAR, pioneer, established clinic (TSIRANINA); 1961 NEPAL, pioneer, established clinic (Fabulous Flemings); 1962 Amazonia (Uncle Cam); 1963, Gabon (Schweitzer); Sierra Leone (Birch); Guinea (SS Hope, Walsh); Borneo (Baughman); Haiti (Mellon); New Guinea (Braun); the Gambia, Sumatra, etc. Elected Explorers Club, 1965, Emeritus, 1985; Alumni Merit Award, Iowa State, 1969; Governor, Rotary International District 524, 1972-73.

Married Katherine Erickson (Iowa) and has two sons, Michael and Thomas.

Mike, BA, Yale, 1969, NROTC Outstanding Midshipman Sword (Brewster). Rickover Interview, Nuclear Power School (submarines); 1978 MD (cum laude), Yale, USNMC, San Diego; 1990 retired, chief oncology, USNMC, Bethesda. Sheryl (attorney), daughter, admiral Paul Lacy, COMSUBPAC, Thomas, Charles.

Tom, BS, USNA, 1974; MS, Monterey; MS, ICAF, associate dean, ICAF; MS, George Washington University; Downes, Sunbird, Bronstein, Ft. Fisher, Denver (XO), Ft. McHenry (CO), Wasp (CO, 1997). Patricia (nurse), daughter, CAPT Thomas Powers, USNR, Julie, Tyler, Joseph.

EDDIE G. ANDERSON, CDR, born Jan. 23, 1955, Moultrie, GA. He is a graduate of Valdosta State College (GA) and enlisted in the USN in April 1978. After boot camp and DP "A" School, was selected for OCS and commissioned ENS, USNR in September 1979.

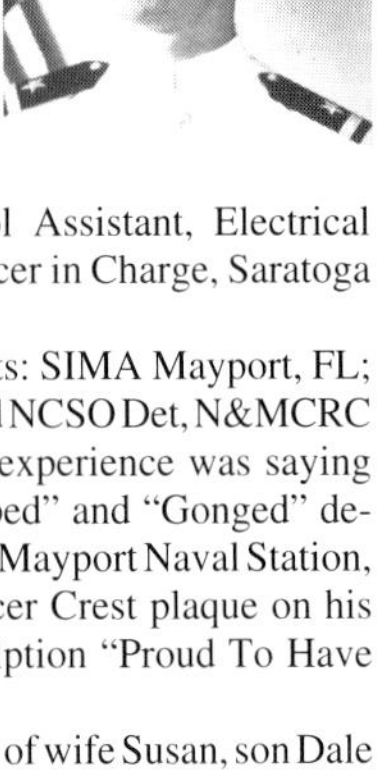

Sea duty assignments aboard USS Holland (AS-32), USS W.S. Sims (FF-1059), USS Dale (CG-19), USS Saratoga (CV-60) in several diverse assignments: Communications Officer (Afloat), CMS Custodian, Damage Control Assistant, Electrical Offier, E Division Officer and Officer in Charge, Saratoga Shore Det. RAD in March 1984.

Inactive SELRES assignments: SIMA Mayport, FL; SIMA-NRMF Philadelphia, PA; and NCSO Det, N&MCRC Wilmington, DE. His memorable experience was saying farewell to the officers; being "Piped" and "Gonged" departing his favorite ship, USS *Dale*, Mayport Naval Station, in September 1982. A Naval Officer Crest plaque on his office wall has this personal inscription "Proud To Have Served."

The Anderson family consists of wife Susan, son Dale and daughter Audrey. They reside in Drexel Hill, PA and he is employed as a resource management analyst with the Defense Logistics Agency in Philadelphia, PA.

ERIC D. ANDERSON, LCDR, born Sept. 22, 1947, Sarasota, FL, received BBA from Georgia State University and MS, Naval Postgraduate School.

He joined the USMC, December 7, 1966 and the USN, Jan. 12, 1981. His military locations and stations include USS *Oldendorf* (DD-922); USS *Leahy* (CG-16); USS *Constant* (MSO-427); USS *Esteem* (MSO-438). He served with 2nd Bn., 12th Marines, 3rd Bn., 8th Marines.

Battles and campaigns: USMC: RVN October 1967-October 1968; USN: Earnest Will, December 1987-April 1988. Anderson received the Purple Heart, Navy Commendation Medal, Vietnamese Cross of Gallantry. He achieved the rank of SGT, USMC and LCDR in the USN. Discharged from the USMC Sept. 18, 1969.

His memorable experiences include: surviving TET 1968; mine clearance OPS in Arabian Gulf during Earnest Will.

He is married to Parrish and they have children: Jason, Candace, Sarah, Jordan, Lauren and Samantha. Anderson is employed as eastern regional manager, American Red Cross.

GARY L. "TUGBOAT" ANDERSON, CAPT, born in Republic, WA on Aug. 1, 1941 and entered the USNR after graduating from the US Merchant Marine Academy, Kings Point, NY in 1963.

He is a licensed CG engineer and received his reserve commission in December 1963.

Anderson is a true naval reservist, having never served on active naval duty. He has sailed extensively to foreign countries with the Merchant Marine, and more recently saw foreign service as part of annual reserve training exercises, with NATO in Europe and Team Spirit in Korea. He is a Naval Academy Blue Gold Officer, and US Merchant Marine Academy information officer. He has also had the pleasure of performing two weddings; one Marine and one AF.

Anderson has been a member of Reserve units in New Jersey, Virginia, and Washington State. He has served in many units and commands, including Merchant Marine, Military Sealift Command, Mobilization Team, Surface Ship Intermediate Maintenance Activity and Readiness Unit. He has attended the Naval War College, National Security Seminar, and Naval Academy Blue Gold Info Program. He has served as commanding officer of PhiBase Coronado 322, USS *Jason* (AR-8), Sima Pearl 1022 and Readiness Unit 2217.

CAPT Anderson's awards include an Overseas Service Ribbon, Expert Rifle, Expert Pistol, Armed Forces Reserve Medal w/HG. Retired November 1991, CAPT 0-6.

He is married to the former Pamela Sue Cornelius of Republic. They have three children, Michael Anne, David Gary and Scott Charles, who are all college graduates. He is vice-president and CEO of Anderson's Grocery and Flour Mill, Inc.

LANE S. ANDERSON III, CAPT, born in Charleston, WV, Dec. 17, 1944. An NROTC Scholar at the University of North Carolina-Chapel Hill, he was selected as the number one midshipman three out of four years. Commissioned in June 1967, Lane served aboard the USS *Little Rock* (CLG-4) which was the flagship of the Sixth Fleet homeported in Italy. He was operations division officer, EWO, AIC officer (F-4 Phantoms), assistant CIC officer, and a fleet qualified officer of the deck. After completing Amphibious Warfare School at Quantico, he became the staff intelligence officer for COMPHIBRON 12 based in Norfolk with deployments to the Caribbean Sea.

In 1971, LT Anderson joined the Naval Reserve Intelligence Program. He worked as a Soviet analyst in support of Fleet Intelligence Center-Europe and Atlantic, the Defense Intelligence Agency and the CIA. In addition he had three assignments to the Naval War College. He served in two XO billets before being selected as commanding officer of the Naval and Marine Corps Reserve Intelligence Training Center in Dam Neck, VA. Selected for captain in 1986, he eventually became director, Naval Reserve Intelligence Program.

In 1991, CAPT Anderson retired, but remains active as director of the Area 4 Blue/Gold program which screens candidates for the US Naval Academy. Lane received his doctorate in educational administration and is a school principal, as well as an educational consultant.

Lane's wife of over 30 years is the former Joanne Creasy of Lynchburg, VA and their daughter, Julie, is an interior design graduate of Salem College. Joanne and Lane live on three wooded areas in Summerfield, NC with their two dogs and five cats.

ROBERT L. ANDERSON, LTJG, born on Aug. 21, 1912 in Baker, OR. Enlisted in the service, Oct. 7, 1942. Attended boot camp at San Diego, CA, then optical school at San Diego, and then transferred to Treasure Island to await transfer to the USS *Pennsylvania* in January 1943. The next 34 months on the ship were spent in the North and South Pacific. They were engaged in 13 amphibious attack operations starting with Attu (May 11-19, 1943). It was followed by Kiska, Makin, Kwajalein, Eniwetok, Saipan, Tinian, Guam, Peleliu, Anguar, Leyte, Luzon, Lingayen Gulf, and the surface engagement with Japanese fleet in the Surigao Straits. The ship was then ordered to Buckner Bay, Okinawa, when it was hit by an aerial torpedo.

The ship returned to Bremerton, WA, Oct. 24, 1945. He left the ship Oct. 25, 1945, a well decorated sailor and was honorably discharged Oct. 28, 1945.

He re-enlisted in the Naval Reserve (inactive) as FCOC2c on Dec. 13, 1946 and was honorably discharged April 28, 1950. The years between 1945-52 he obtained a BS degree and a DMD degree.

On Feb. 24, 1952, he was appointed a LTJG in the Dental Corps. He continued his dental practice and his Naval Reserve activities until he retired from the Navy on Aug. 21, 1984, as a senior officer. He retired from his practice on Aug. 21, 1986. He presently resides in the Palm Springs area with his wife Karen.

RUSSELL F. ANDERSON, CAPT, served as a reservist for 33 years, having been recruited by Naval Intelligence in the 1940s when he was a foreign correspondent for International News Service. Among other assignments he covered the Spanish Civil War, the Russo-Finnish War, and the evacuation of the British at Dunkirk.

He was called to active duty on Dec. 7, 1941, as an ensign. After a tour of duty with ONI, he was assigned to CINPAC as a liaison officer with correspondents covering the Pacific.

Anderson returned to inactive status September 1945 and changed his 1600 designator (Intelligence) to 1665 (Public Relations). He served as CEO of reserve units in New York City and London. He retired in June 1974, with the rank of captain. He received the usual campaign and service medals, including the Legion of Merit.

SAMUEL ARVID "SAM" ANDERSON, CAPT, born Oct. 23, 1928, Denver, CO. Earned AA, BA and JD degrees from Graceland College and Wyoming University. Joined the USN in May 1950 and was commissioned in July 1953.

Stationed at Pensacola, San Diego, Long Beach, WestPac, Cheyenne, Des Moines, Denver, Great Lakes, Brooklyn, St. Petersburg, Alameda, ASW/SEACON Ops, Newport, Dam Neck, Mazatlan, Washington DC, New Orleans, Norfolk, St. Louis, San Francisco, Pearl Harbor, Yokosuka and Subic Bay.

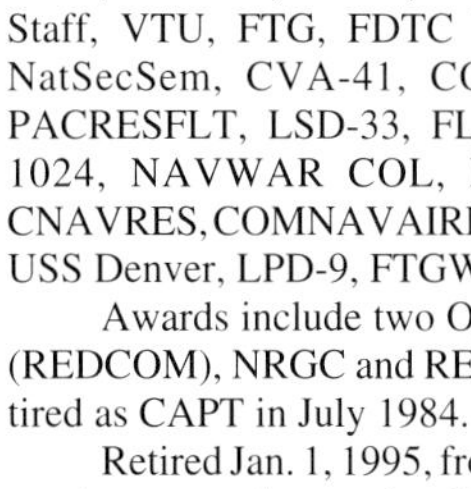

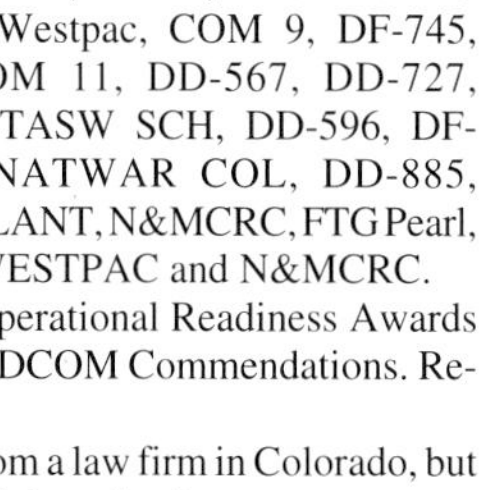

Commands: NAVCAD, Flt. Photo, Res. Off. Candidate School, Naval Justice Firefighting and Crypto/Command, CVA-21, NRSD, NRGC, NRFD, CRUDESPAC Staff, VTU, FTG, FDTC Westpac, COM 9, DF-745, NatSecSem, CVA-41, COM 11, DD-567, DD-727, PACRESFLT, LSD-33, FLTASW SCH, DD-596, DF-1024, NAVWAR COL, NATWAR COL, DD-885, CNAVRES, COMNAVAIRLANT, N&MCRC, FTG Pearl, USS Denver, LPD-9, FTGWESTPAC and N&MCRC.

Awards include two Operational Readiness Awards (REDCOM), NRGC and REDCOM Commendations. Retired as CAPT in July 1984.

Retired Jan. 1, 1995, from a law firm in Colorado, but continues to solo practice. He is active in numerous community and civic affiliations, including being mayor of town of Columbine Valley, CO for four years. Has been member of Sertoma International since 1961 and served as Distinguished Governor, President, Chief Executive and Administrative Officer.

Married Mary Etta and they have four children: Angela (Curtis), Suzanne (Bremer), Anthony and John;

five grandchildren: Scott Curtis, Jennifer Murphy, Patrick, Cole and Alec Miller; and one great-grandchild.

DR. JOHN R. ANDRADE, LCDR, born Aug. 23, 1944, Providence, RI. He received BS and MS degrees from the College of the Holy Cross, 1966/1967. PHD Brown University 1979.

He joined the Naval Reserve Association in 1973. Commissioned as ensign USN, June 8, 1966.

His military locations and stations include NGMS, Dam Neck, VA (July 1967-October 1967); USNA, Annapolis (Feb. 1971-June 1972); USS *Dahlgren* (DLG-12) October 1967-August 1969; DE-1028 (August 1969-December 1970).

Andrade was awarded the Natonal Defense Medal. He achieved the rank of LCDR and retired from active duty, June 30, 1972 and Reserves, Aug. 1, 1986. His memorable experiences include: crossing the equator at 00 latitude and 00 longitude on Jan. 31, 1969.

He is married to Paula and they have children: Mary, Judith, Lisa, Christa, John and Joseph. He has held positions as associate director, Pharmaceuticals Dev. Wyeth-Ayerst.

MARIO IACOBUCCI ANDREA, LCDR, born May 21, 1917. Andrea joined the service February 1944 and was called to active duty May 1, 1941. His military locations and stations include: US Navy Yard Boston; US Amphibious Adv. Base, Plymouth, England. He served engineering duty (BUSHIPS); Naval Architect and Marine Eng. Participated in shore duty to build, repair and maintain ships. Awarded medals for Pre-Pearl Harbor, WWII Victory Medal, European, Naval Reserve.

A former nuclear engineer, naval architect and marine engineer, owns and operates The Gem Tree jewelry store in Bethesda, MD.

Dr. Andrea holds a BS degree in Naval Architecture and Marine Engineering from the Webb Institute of Naval Architecture, Glen Cove, NY; an MS degree in engineering (nuclear) from the Catholic University of America, Washington, DC and a PhD degree in Engineering Science from Pacific Western University, Encino, CA. He also graduated from the Oak Ridge School of Reactor Technology and completed graduate courses in advanced mathematics and quantum physics at the University of Maryland, College Park and advanced nuclear engineering at the University of California, Los Angeles. He is a licensed professional engineer in Maryland.

In 1977 Dr. Andrea received his graduate gemologist diploma from the Gemological Institute of America (GIA), Santa Monica, CA. He is a charter member of the GIA Alumni Association, a charter member of the National Association of Jewelry Appraisers and a member of the International Society of Appraisers.

He retired from the USNR with the rank of LCDR in 1961. He retired as a general engineer from the Maritime Administration, US Department of Commerce, in the early seventies. Earlier he was the project engineer for the construction of the nuclear power reactor for the NS *Savannah,* first nuclear powered merchant ship in the world.

He has published articles on science and engineering, including *Economic Considerations for the Application of Nuclear Power to Merchant Ships"* and *"How the Crystal Structure of Gems Affect Their Color."* His doctoral dissertation was on the *Helical Ship,* a new hull design for which he now holds the patent.

A native of Haverhill, MA, received the Shattuck Prize for the highest average in science when he graduated from Haverhill High School in 1935. His educational and professional background in science, engineering, and gemology provides a unique approach to his present pursuits as gemologist and appraiser. He is a recognized forensic gemologist and has been called to testify as an expert witness in court. He serves also as president of the National Institute of Gemology, Inc. He was selected for inclusion in the twenty-first edition (1986-87) of Marquis' *Who's Who in the East.*

His memorable experiences include: first launch LCT from LST; assemble LCT in one day in England before D-Day; nuclear reactor engr. for NS *Savannah;* tech supervisor naval architect (MSTSPACAREA, Japan).

He is married to Elma Williams and they have children: Gail, Patricia, Marjorie, Lucia and Janet; and grandchildren: Valerie, Elizabeth, Robert Jr., Ricky, Jimmy and Jessica. He was employed with GE Co., MSTS (Navy), Martime Administration, retiring May 1961 after 20 years.

ROBERT D. ANDREWS JR.. CAPT, born in Muskegon, MI, Dec. 20, 1939. He is a graduate of Muskegon High School (1957) (their class produced three known Navy captains), Albion College (AB-1961) and the University of Michigan Law School (LLB-1964).

Andrews was commissioned LTJG with OCS Class 507 in March 1965. Following Naval Justice School he served as claims and investigations officer for the commandant, 8th Naval District, New Orleans, LA. Thereafter he served as legal officer, NAS, Meridian, MS, from March 1966 until release from active duty February 1968. Reserve service includes duty as XO, NR NAMARA (govt.) Det 103, Chicago and as CO, NR NAMARA (Govt) Det 104, Detroit.

Following RAD he was employed as assistant prosecuting attorney for Kent County (Grand Rapids), MI, including a period of martial law in Grand Rapids. Thereafter Andrews served as prosecuting attorney of Mason County (Ludington), MI. Since 1973 he has been in private law practice, emphasizing real property, probate & estate planning in Western Michigan.

He is married to Mary Anne Harris, a Navy Junior. They have three children: Stacy, high school Spanish and social studies teacher; Robert III, a mechanical engineer, and Mark, an aspiring actor and director.

CAPT Andrews holds the National Defense Service Medal and the Armed Forces Reserve Medal. Memorable experiences include: Hurricane Betsie in New Orleans September 1965, and thereafter, as ASDO, coordinating initial recovery plans and logistics and ship damage investigations; advising a CO who ordered motorcycles barred from base; reviewing a claim investigation report for loss of a milk cow. His first military trial was as assigned defense counsel in the GCM of a career officer (acquitted).

THOMAS HAW ANDREWS, CDR, born September 1918, Muskegon, MI. Following graduation from Albion College in 1940, he enrolled in Northwestern University Law School. Enlisted USNR June 1941. Ordered to active duty January 1942.

Commissioned ensign, USNR, May 1942. Graduated Bomb Disposal School, Navy Shipyard, Washington July 1942. Gunnery officer, NAS Palmyra July 1942-May 1943. Bomb disposal officer, Central Pacific Region, July 1942-May 1943. Bomb Disposal School staff, American University, Washington, DC, May 1943-October 1944. Base ordinance officer, CB Base, Camp Parks, CA, October 1944 to RAD November 1945. Organized NRSD 9-102L Muskegon and served as commanding officer from May 1946 to December 1948. Recalled January 1949. Assistant ordinance officer, 3rd Naval District, New York City, and Reserve Ordinance Program officer, 3rd Naval District, January 1949-June 1950. Admin. officer, Naval Mine Depot, Yorktown, VA, June 1950-October 1952. XO, Naval Ammunition Depot, Hingham, MA, October 1952-August 1954. RAD August 1954. CO NRSD 9-103 October 1954-October 1957. Admin officer, Brigade Command, NTC Chicago, October 1966-October 1969.

Awards and Medals: Pearl Harbor, American Theater, Asiatic-Pacific Theater, Navy Reserve Service, Armed Forces Reserve Service, Expert Rifleman.

Civilian employment: Holds lifetime secondary education teaching certificate; Michigan Licensed Real Estate Broker 1955-91. Realtor Chippewa County, MI 1980-91; Administrative officer, SBA Central Region, Chicago, IL 1965-69; SBA EDC Development and Organization Director, Detroit 1969; Community Planning and Development Representative, HUD, Detroit 1969-77; Chippewa County EDC vice-president, president, 1977-80. Retired from US Civil Service 1980.

His awards and civic service: Mayor, North Muskegon, two years. President, Michigan Education Association, North Muskegon Chapter, Muskegon County Chapter, West Michigan District. Served on MEA National Committee on Ethics; president, EUP Realtors Board 1983-84; Director Chippewa County Chamber of Commerce; United Way Board; Chippewa County Republican Chair three years; Kinross Township Trustee 1988-92.

He was married to Maxine Huether for over 50 years. Their son Tom Jr. is a lawyer in Muskegon. He is presently married to Corrine.

RICHARD ANDRIANO-MOORE, CDR, a native of Petaluma, CA, received his commission as ensign, USNR, in 1957, upon graduation from the Officer Candidate School, Newport, RI. He received his BA degree from San Jose State University, his MBA from Pepperdine University and attended several Navy schools, including the Naval War College at Newport.

After commissioning he served aboard ships in both the Atlantic and Pacific Fleets. Shore assignments included command of several Naval and Marine Corps Reserve Centers; duty on the staff, 11th Naval District. Director of Administration and Operations Officer, National Committee for Employer Support of the Guard and Reserve, and as commander, Naval Reserve Recruiting Region One. CDR Andriano-Moore completed his Naval career as chief of staff, Naval Reserve Readiness Command Region Twenty. Retired after over 25 years of active Naval service, Aug. 31, 1985. He is a member of The Naval Reserve Association, The Retired Officers Association, the Naval Order of the United States, and served as commander of the California Commandery, Military Order of the Loyal Legion of the US.

His awards include: Defense Meritorious Service Medal, Navy Commendation Medal (two awards), Armed Forces Reserve Medal, Navy Expert Pistol Shot Medal w/ Silver "E," National Defense Service Medal, Command Ashore Insignia, Surface Warfare Insignia, two Gold Wreaths for Recruiting Excellence, and the Office of the Secretary of Defense Identification Badge for service within the Office of the Secretary of Defense.

After retirement CDR Adriano-Moore resided in Petaluma, CA where he served as scoutmaster of Troop 18, at the Petaluma Coast Guard Training Center. He was awarded the Scoutmaster Award of Merit and served as a District Scout Executive and currently serves as a substitute teacher for Tomales High School, Tomales, CA. He currently resides in Santa Rosa, CA and has a son Stephen and daughter Erika.

CHARLES WILLIAM "BILL" ANESHANSLEY, LCDR, born July 11, 1920, Lancaster, OH, Fairfield County. He graduated Naval Air Station, Pensacola, FL. Joined V-5 Program, Oct. 15, 1942.

His military stations and locations include: NAS Clinton, OK; NAS Floyd Bennett, NY; NAS Dallas, TX. Served with VRS-1, VR-1, Ferry Command, NAS Norfolk, VA. Served six years active duty and 17 years Reserve duty, retiring as LCDR, July 11, 1980.

His entire service time, both active and reserve was memorable. Being a member of the US Naval Air Program from beginning to retirement. It was a privilege to serve his country with all the different groups and people that he was associated with during his service time.

He checked out in 20 plus types of aircraft, military and civilian types and holds a commercial pilot license - single and multi-engine instrument rated. Accumulated 5,000 military and 6,000 civilian flight hours.

His civilian time was spent in developing a distributorship for wholesale and retail oil and gasoline products that is now - Anoilco, Inc. Totally owned and still operated by his family, representing a major nationally known oil and gasoline supplier servicing a six county area with 18 company owned units, serving also as a director of one of the area's largest banks for approximately 30 years.

Married 51 years to Frances S. and they have children: Michael I., James Alan, Janice Carol and William Bradley; grandchildren are Kristi Lynn, William Scott, William Brian, Jill Renea, Autumn Nicole, Leigh Kashay, William Baerett. Stepgrandson, Matt Charlies Harris; and great-granddaughter Kelisie Layne.

RAYMOND L. "ANDY" ANTONCZYK, CDR, born July 2, 1923, Chicago, IL. He attended Wright Jr. College, 1949 and Illinois Institute of Technology, receiving a BSME 1957.

Enlisted V-5, Nov 27, 1942. Activated March 3, 1943. Commissioned ensign Nov. 22, 1944. Designated naval aviator Nov. 22, 1944 at Corpus Christi.

Military locations and stations include SB2c Operational Cecil Field, FL; CarQual USS *Guadalcanal* April 7, 1945. He served with VB-98 Los Alamitos, CA. ReQual day and night USS *Ranger* Aug. 3, 1945.

Antonczyk was awarded American Theater, WWII Victory, Naval Reserve and Armed Services Medal. He achieved the rank of CDR Sept. 1, 1961.

His memorable experiences include: serving as naval aviator with VB713, VA57A, VA721, VA727, V5723, VP725, NARS72, NARMU 722, and NARMU 725; XO of VS723 and VP725; NARMU 722 and 725. Released to inactive duty, Nov. 19, 1945. Transferred to inactive service list and retired July 2, 1983.

He was married to Patricia Mary Woods. They were divorced in 1983. Their children are Andrew, Mark and Mary Pat. He was employed with Skil Corp in 1942; 1969, director of engr., 1969-76 chief engineer. Presently owner manager of Framing and Art Gallery (Chicago). Semi-retired consulting in construction industry.

THOMAS A. ANZALONE, CDR, born July 17, 1943, in St. Louis, MO. After graduating from St. Louis University in 1965, he went to Pensacola, FL, for flight training. After flight training, he was assigned to HA(L)-3 and served in Vietnam from June 1967, until May 1968, flying 309 combat missions in the UH-1B gunship. Among his combat awards are 33 Air Medals and a Purple Heart.

In June 1968, CDR Anzalone was assigned to VRF-32. As a ferry pilot, he flew the A-4, A-7, F-4, F-9, H-1, H-34 and S-2. In October 1969, he left active duty for the USNR where he flew the A-4L with VA-204, NAS Memphis. In 1978, he was assigned to USNR NAS Oceana (located at NAF Washington, DC) and served two years as XO and two years as CO of the unit. In 1985, he was transferred to the Standby Reserve. In civilian life, he was a B-727 pilot for 13 years with Eastern Air Lines until it went bankrupt in 1989. In 1990, he joined the FDIC where he is currently an information systems analyst.

FRANK E. APOSTOLICO, LT, born May 2, 1946, New York, NY. He received BS, Penn State, 1968; MBA University of Scranton, 1982.

He joined the NROTC, July 1966. Commissioned ensign June 1968. His military locations and stations include USS *Nantahala* (AO-60) July 1968-July 1969; USS *Brule* (AKL-28), July 1969-June 1970. Served with Navy Recruiting District, Philadelphia, July 1970-March 1973.

Served in-country Vietnam, Rivers of Mekong Delta. Awarded Vietnam Service Medal, Vietnam Campaign Medal, National Defense Service Medal, Armed Forces Reserve Medal. Apostolico achieved the rank of LT in active duty and CDR in Reserves. He left active duty July 31, 1973 and retired from USNR, July 1, 1994.

His memorable experiences include action in the rivers of the Mekong Delta, July 1969-June 1970 aboard the USS *Brule* (AKL-28) an in-country tour. Recruited OCS candidates on college campus, July 1970-March 1973.

He is single and has held positions as president, American Assets Inc., excutive vice-president, WEA Inc. Div. of Time-Warner.

ROBERT E. "BOB" ARCHIBALD JR., CDR, born Nov. 15, 1936, Los Angeles, CA. Earned BS degree in forest management, 1959, Humboldt State University. Joined the service Feb. 3, 1959 (OCS, Class 43).

Stationed at OCS, COMNAV Marianas, NAVCOMM Unit Seattle. Ships and shore stations ACDUTRA: NRSD 12-32(S), VTU 1908, NRSD 11-50(M), ALLAS CO. Memorable experience was being only ensign on admiral's staff at COMNAV Marianas, first duty station. He routinely delivered classified messages to the admiral(office and quarters).

Left service March 1, 1983, as CDR, USNR (Ret). Received National Defense and Naval Reserves (Hour Glass) medals.

Married 36 plus years to former Margaret Nichols. They have two children, Hope and Brian. Employed as land manager, Bur. Land Mgt., USDI (33 years) until retirement Oct. 1, 1995 (36 1/2 years federal service). He is now semi-retired as a private mortgage broker.

RICHARD K.F. "DICK" ARCOLEO, LCDR, born July 31, 1932, Bronx, NY. He received bachelor of mechanical engineers, Clarkson College, Potsdam, NY.

He joined the service May 9, 1956, ONOP Macon, GA. His military locations and stations include: OCS Newport, AFSWP Albuquerque, SWUPAC San Diego, NSA Naples. He served with USS *Ticonderoga* (CVA-14) (1957-58), USS *Ranger* (CVA-61), (1958-59).

Arcoleo participated in battles in Indonesian Alert (1958), Berlin Crisis (1961). He received the National Defense, Armed Forces Reserve, CT Naval Militia, 20 years. Achieved the rank of LCDR, discharged RAD Aug. 28, 1962.

His memorable experiences: around the Horn (Norfolk, VA-Alameda, CA) June-August 1958 on board USS *Ranger* (CVA-61) and explosion below duty station; special weapons magazine spaces on board USS *Ranger* (CVA-61) - winter 1959.

Arcoleo married Jan. 3, 1962 to M. Isabel Peschiera. They have children: Frederick, Cynthia Ines, and Kenneth Anthony; grandchildren are Erin, Anthony, Amanda, Adrian and Casey. His civilian employment and positions held include PWA-Experimental Engr. & Hamilton Std. Quality Engr. Retired July 1, 1977, with pay Aug. 1, 1992.

ROBERT B. ARDIS, CAPT, born Bay City, MI, 1925. Entered Navy V-12 program at the University of Michigan in 1944. Commissioned upon graduation in 1946. Received BSEE and MSEE degrees from University of Michigan in 1946 and 1947 and JD degree from New York University in 1951.

Served in USNR billets in New York City area 1951-68 and in Chicago 1969-1974. Commanding officer, Surface Div. 3-79(L), Brooklyn, NY, 1963-66. Chief staff officer, Group Command 3-22, Whitestone, NY, 1967-68. Attended Reserve Officer Command and Staff Course at Naval War College in 1963 and Senior Reserve Officer Course there in 1966.

Joined Bell Telephone Laboratories, Inc. in 1947 as member of its patent attorney staff. Retired in 1988 from AT&T Bell Laboratories as General Attorney (Intellectual Property Matters). Currently legislative officer and newsletter editor for Northern New Jersey chapter of The Retired Officers Association. Bob and his wife, Catherine, live in Mountainside, NJ.

EDWARD BRADFORD ARMSTRONG JR., CAPT, oral and maxillofacial surgeon, naval officer. He was born Teaneck, NJ, Sept. 24, 1928, the son of Edward Bradford and Ruth Elizabeth (Fippinger) A. He received AB degree from University of Pennsylvania, 1950; DDS, NYU, 1954. Married Dusanka Vladimirovna Jakovljevic, Nov. 5, 1960. They have children: Edward Bradford, III, James B., Hugh B.

He served as LTJG, USN, 1954, advancing through the grades to CAPT, 1971. Intern in oral surgery, Roosevelt Hospital, NYC, 1958, assoc. attending oral surgery, 1959, attending oral surgeon outpatient department, 1959; chairman, moderator Oral Surgery Staff Confs., 1963-70; resident Carle Hospital, Urbana, IL, 1959; assoc. attending oral surgeon Flower and Fifth Ave., hosps., NYC, 1960-68; asst. attending oral surgeon Hackensack (NJ) Hospital, 1963-65; adminstrative officer NR Dental Co. 3-2, 1965-68.

Exec. officer, 1968-71, commanding officer, 1971-73; comdt.'s rep. 3rd Naval District, Naval Academy, 1972-78, 3rd Naval District for Dentistry, 1973-75; group staff officer for dentistry and medicine, 1973-75; Ready Res. Unit 502, 1975-77, VTU 0207, 1977-79, ret., 1979; assoc. clin. prof. oral surgery New York Medical College, 1963; adj. assoc. clin. prof. oral surgery Columbia University School Dentistry, 1973-89; chairman board E & R, Armstrong, Inc., Albany, NY, 1966-77; pres. Edward B. Armstrong, P.C., NYC, 1979-90; direc. Songtime, Inc., Boston; director, mem. exec. com. PGP Internat. Corps, Inc. Board of directors, trustee Christian Mission Farms of Paraguay, Inc., 1974-84.

President, trustee Central Bible Chapel, Palisades Park, NJ; area rep., ann. giving University Pennsylvania, 1960-68; Blue and Gold Officer Naval Academy. Admissions Com.; sec. bd. dirs., trustee Boys' Club of New York Health Services, Inc. Diplomate Am. Bd. Oral Surgery Fellow NY Acad. Dentistry (sec. dir., pres. 1979-80), Am., Internat. colls. dentist, American College Oral and Maxillofacial Surgeons (founding).

American Dental Society Anesthesiology (hon. life), member American Society Oral Surgeons (NJ rep. H of Dels. 1963-65), New York Soc. Oral Surgeons (chairman audit and budget com 1972-79), ADA, First District, NY, Brooklyn, Yokosuka (hon.) dental society Association Mil. Surgeons US Military Order of World Wars, NRA (life) Union League (chairman art com. 1973-76, board of governors (1974-77, 1982-84, v.p. 1977, 1980, 1985-88), Met. Club, University of Pennsylvania Club, University Pennsylvania Club of Met. NJ (direc. 1982-) Acacia, XI Psi Phi, Psi Omega (hon.), Delta Sigma Delta. Member Plymouth Brethren Church. Resides in Leonia, NJ.

LLOYD ARMSTRONG, CAPT, received a BSME. Joined the USN, June 6, 1956 and was stationed on USS *Bristol* (DD-857). He achieved the rank of CAPT, USNR, retiring from active duty in 1959 and from the Reserve in 1985.

He married Susan Bainbridge. Armstrong held the position of president of American Medical Systems, retiring April 1995.

RICHARD S. "RICH" ARNOLD, CDR, born in Montgomery, AL, received a BA from the University of South Alabama; MA from University of Victoria and PHD from the University of Alabama.

He was commissioned OCS, September 1980. His current station: CO, ATGPAC DET 322, Seattle, WA. He has served with USS *England* (CG-22), USS *Pledge* (MSO-492), MARDEZ (Seattle, WA).

Arnold was awarded the Sea Service w/3 stars, NAM w/star, Meritorious Unit Commendation, National Defense Medal w/star. He is currently commander, USNR (SE/Res).

His memorable experiences include: standing OOD underway as a Reserve LCDR on a minesweeper in transit from Seattle to San Diego. He realized it was still possible to get seasick.

He is married to (neé) Robin Marrion and they have children, William and Camilla. His civilian employment includes Malaspina University-College, Nanaimo BC (professor of English).

ARTHUR JOSEPH "ART" ARSENEAULT JR., LCDR, born Oct. 29, 1924, Somerville, MA. Graduated Massachusetts Maritime Academy, September 1944. He was commissioned Sept. 24, 1944. Reported for active duty, Oct. 8, 1944.

His military locations and stations include USS *Griggs* (APA-110), 1st Div. off., 1944-45; USS *Lejeune* (AP-74), R Div. off., 1945-46; US Naval School EOD, Indianhead, MD, student, 1946-47; Staff ComPhibLant, USS *Taconic* (AGC-17) EOD off., 1947; Underwater Demolition Team FOUR EOD off/XO/CO, 1947-49; USS *Mississippi* (EAG-128), 5th Div. Off., 1949-50; Scripps Institution of Oceanography, student, 1949-50; USS *San Pablo* (AGS-30), assistant hydrographic officer, 1950-51; Northwestern University, student, 1952-54; General Line School,

Monterey, student, 1954; USS *Haynsworth* (DD-700), operations off., 1954-57; Staff ComMinLant, Charleston, SC and OinC EODU-TWO, 1957-59; USS *Denebola* (AF-56), XO, 1959-61; USN Explosive Ordnance Disposal Facility, RDT&E Dept., 1961-63; 2nd Air and Naval Gunfire Liaison Co., FMFLant Camp Lejeune, operations off.

He was awarded the American Area Campaign Medal, European Area Campaign Medal, WWII Victory Medal, National Defense Service Medal, Naval Marine Expeditionary Medal w/FMF Insignia, Master EOD Badge, Naval Parachutist Badge. Arseneault achieved the rank of LCDR, USN and retired July 1, 1965. Transferred from USNR status to USN status in 1946.

He married Willidean Blazier LTJG NC USN in 1952. They have children: Carl Joseph is a retired US Army captain; Daniel Scott is a retired USN LCDR; Nanette Marie is a commander, RN, USPHS, Suzanne Terese married a retired USN BT1, David Wayne is a MMCS on active duty, Stephen Paul and Michelle Grace is married to an ex-corporal, USMC. They have eight grandsons and three granddaughters.

His civilian employment includes: Boy Scouts of America, Atlanta Area Council, District SE, 1965-66; Health Department, State of Georgia, RN Recruitment Program, 1966-67; GA Science and Technology Commission, Research Associate, 1967-71; GA State Crime Commission, police planner, 1971-72; Cobb County Georgia Police Department, administrative assistant, 1973-87. He retired April 1, 1987.

PAUL K. ARTHUR, RADM, born in Kansas City, MO, Jan. 14, 1931. He enlisted in the Navy, July 1949 following one year at Taylor University. After boot camp, he was in the deck force of the USS *Oglethorpe* (AKA-100) and then attended Electronic Technicians' School. Following ET School, he was ship's company in USS *Leary* (DDR-879) and later in USS *Braine* (DD-630) as an ETN3. He was discharged from USN in November 1953 and reenlisted in USNR. Advanced to ET1 and was pending DOR to CPO when he was commissioned LTJG in 1962 as an engineering duty officer (EDO).

His assignments included CO of five USNR Units: SURFDIV 8-54, SIMA 6610, WEAPSTA 411, Pearl Harbor NSY A211, and NAVSEA HQ 811. Other positions were inspector general and REDCOM Eleven Staff EDO. His last USNR position was the director of Reserve programs for Naval Sea Systems Command and Space and Naval Warfare Systems Command. His mobilization assignment was DOD coordinator of ship conversion and repair. He retired in November 1987 after over 38 years naval service.

ADM Arthur received a BSEE from Purdue University in June 1956 and completed graduate work at New Mexico State University. In his civilian career, he has been employed by the Department of the Army at White Sands Missile Range (WSMR), NM, since November 1956, and is presently the director of the Materiel Test Directorate with responsibility for weapons systems test and evaluation.

He resides in Las Cruces, NM with his wife, Joy. They have two adult children, Greg and Lia, both are engineers. He enjoys backpacking, gardening, karate and running.

SAMUEL W. ASBURY, LT, born in Salinas, CA, May 5, 1965. He graduated with High Scholarship from Oregon State University and completed the University's Honor Program. Upon graduation, he entered active duty in the USN and served four years, during which time he qualified as surface warfare officer aboard the USS *Nimitz* (CVN-68).

After leaving active duty, he entered Gonzaga University School of Law and joined Mobile Inshore Undersea Warfare Unit (MIUWU) 102 as a drilling reservist. Asbury was awarded the Navy and Marine Corps Achievement Medal while serving at MIUWU-102. He graduated from law school with a J.D. cum laude and with experience as a law review editor.

He started his own law practice in Portland and transferred to MIUWU-110. His campaign and service medals to date are: National Defense Service Medal, Southwest Asia Service Medal w/Bronze Star and Kuwait Liberation Medal.

He performed an annual training period at the US Naval War College and helped draft the Navy's Total Force Policy. He has completed the Naval War College course on International Law with distinction, and he is currently enrolled in the Naval War College Command and Staff graduate diploma curriculum.

Asbury and his wife Melinda (neé Gaskin) live in Fairview, OR, which is near Portland. He is a life member of the Naval Reserve Association, the Reserve Officers Association, and the National Eagle Scout Association.

CHARLES IVAN ASHBAUGH, CAPT, born Sept. 13, 1935, Orleans, NE, received BS Aerospace Engineering, University of Texas, 1960.

He joined the service June 4, 1960. His military locations and stations include: NPTU Idaho Falls; NPTU West Milton, NY; Yokosuka Japan; Long Beach CA. He served with DLG-15, DLG/CG-24, DLGN/CGN-35, DD-528, DD-731, APA-248, FTG-819, VN-1904. He served in Vietnam and received the Navy Unit Citation, Armed Forces Reserve Medal, Vietnam Service w/3 stars, RVN Campaign (60-), Armed Forces Expedition (Korea). Ashbaugh achieved the rank of CAPT, USNR (ret). Discharged to Retired Reserve, Oct. 1, 1989.

His memorable experiences include: interview with ADM Rickover; playing "Tag" with a Russian nuclear submarine between Long Beach, CA and Pearl Harbor, HI; selected and trained (prize crew engineering officer) to "rescue" USS *Pueblo* from North Koreans.

He is married to Pat and they have children: Chuck and Robin. He has stepchildren: James, Richard and Kathrine. Their grandchildren are: Megan, Jedediah, Melisa and Emily. He has held positions with the city of Los Angeles, Department of Water and Power, as generating station manager, now retired. Presently working for wife in balloon decorating business. Retired from LADWP, Nov. 1, 1995 and USN with pay, Sept. 13, 1995. Ashbaugh is a life member of the NRA #11039.

DAN D. ASHCRAFT, LCDR, born Jan. 2, 1949, Los Angeles, CA, received MS, Industrial Engineering and Management, Oklahoma State University in 1980.

He joined the service June 14, 1967 and his military station was the USS *Oklahoma City* (CLG-5). He served with SIMA SOGO 2611, SIMA SDGO 2718, DESRON 35, and WPNSTA Concord EOT 14, fighting in the Vietnam campaign.

Ashcraft received the National Defense Service Medal, Armed Forces Expeditionary Medal (Korea), Vietnam Service Medal w/3 Bronze Stars, Vietnam Campaign Medal w/device (1960), Naval Reserve Meritorious Service Medal (1st, 2nd, 3rd) and Armed Forces Reserve Medal (1st, 2nd).

His memorable experiences: ran a ground off the coast of Vietnam while serving as a 3rd class machinery repairman on board the USS *Oklahoma City* (CLG-5) during gunfire support operations with the Seventh Fleet. It took two fleet tugs from Da Nang harbor to pull them off the sand bar after several unsuccessful attempts at backing out.

He was the only chief machinery repairman advanced in the USNR in 1980. Received direct commission in 1982 as an engineering duty officer (EDO) after completing an industrial engineering masters degree. Completed EDO qualifictations in San Diego, in 1987. Served as commanding officer of four USNR unit detachments from October 1989 to September 1997.

Ashcraft has achieved the rank of LCDR and is married to Jaime Latisha. They have children Michael and Brian. His civilian employment and positions include: Chance Operations Inc., engineering projects coordinator. Presently drilling Reservist.

CECIL S. "SPAN" ASHDOWN, LCDR, born May 14, 1923, New York, received BS in engineering from Princeton in 1944 and an MBA from Harvard in 1948.

He enlisted as apprentice seaman, Jan. 8, 1944. His military locations and stations include: Attended USNR Midshipman's School, Annapolis, MD (V-7 Program); USS *Wrangell* (AE-12), 1944-46, as electrical engineering officer. First ship to transfer ammunition underway at sea. He participated in Philippine Liberation, Iwo Jima, Okinawa and the Bombardment of Japan. Achieved the rank of LCDR and separated from active duty, Oct. 5, 1946, and retired from the USN, July 1, 1973.

His memorable experiences: CO, USNR MSTS/MSC Co. 3-24, New York City, 1966-67. Serve on three non-profit boards of directs. Active in four maritime history organizations.

Married Marie M. and they have children: Stephen, Vivian Marie, Cecil III and Charles. Their grandchildren are Mary Grace and Helen Ashdown.

He joined Alcoa Steamship Co., Inc., New York City in 1948. Later transferred to Mobile, Trinidad, Suriname Pittsburgh, New Orleans. Finally returned to New York as procurement manager in 1959, retiring Jan. 1, 1988.

AARON R. ASRAEL, CAPT, born Nov. 14, 1940 in Baltimore, MD. He entered the Navy through Officer Candidate School, Newport, RI, where he was commissioned an ensign, Nov. 22, 1963. Following his commissioning, he reported to the Staff, Commander US Second Fleet, where he served aboard the flagships USS *Newport News* (CA-148) and USS *Little Rock* (CLG-4) until he left active duty in 1966.

Immediately upon leaving active duty, he affiliated with a Naval Reserve unit in Baltimore, MD. Upon disestablishment of that unit in 1974, he affiliated with the Naval Reserve Intelligence Program and was assigned to Naval Reserve Unit OSD/JCS 106. When the Naval Reserve Intelligence Program was restructured in 1976, he was assigned to Naval Reserve Unit NISHQ 0166 where he served as project officer of the unit's Criminal Fraud Project and in 1984, he became the unit's executive officer. In 1986, Asrael was selected to serve on the RIAC staff where he was assistant deputy RIAC for Administration, and then as Deputy RIAC/Inspector General. Asrael assumed command of NISCOM 0166 on Oct. 1, 1988.

Asrael received a bachelor of arts degree from the University of Maryland and a Juris Doctor degree from the University of Maryland School of Law. He is a member of the Bars of the State of Maryland and the District of Columbia. In civilian life, he is a senior manager for the National Science Foundation where he has been employed since 1969.

Asrael and his wife Gloria reside in Columbia, MD, and have one daughter, Jill who is married to Jeff Borten.

GEORGE H. ASTER, CAPT, born April 22, 1932 in Detroit and raised in Grand Rapids, MI. He enlisted in the USNR as a seaman recruit May 1, 1951. By the time of his graduation from the University of Michigan in 1955 with a bachelor of architecture degree, he had advanced to the rate of CT3.

Commissioned as ensign USNR through the ROC program at Newport, RI in August 1955, he was ordered to US Naval Station, Argentia, Newfoundland, where he served successively as communications watch officer, personnel officer, and Special Services officer. Released from active duty in August 1957, he migrated to San

Francisco, where he worked many years as an architect. In San Francisco he affiliated with the Naval Reserve Intelligence Program, and during the following 23 years advanced to the rank of CAPT.

He retired from the USNR, June 1, 1981 and from his civilian employment with Pacific Gas & Electric Co., May 1, 1992. Aster and his wife Pat reside in Larkspur, CA. They have two sons and, as of 1996, five grandchildren.

OLIVER T. "O.T." ATKINSON JR., CAPT, born Dec. 14, 1940, Norfolk, VA, received BS VA Tech (1962); MSFS American College, 1979.

He joined the service Aug. 8, 1963. His military locations and stations: chief engr., USS *Hammerberg* (DE-1015); XO USS *Darby* (DE-218). Served with US Central Command, US Special Operations Command US Readiness Command, CO, US Central Command NR 108.

Atkinson was awarded the Navy Reserve Medal, National Defense Medal and Expert Rifle. He achieved the rank of captain and was discharged July 31, 1967.

His memorable experiences: Elint patrols off Cuba; NATO Matchmaker Cruise; Serving in HQ of three CINC commands; US Readiness Command; US Central Command and US Special Operations Command.

He is married to Loraine and they have children: Oliver III, Paul and Andrew; and grandchildren, Olivia and Benjamin. Atkinson was president-CEO Atkinson, Gower Risk Management, Inc., 1972-present. Retired Jan. 12, 1990 from USNR.

JAMES J. AUKOFER, CWO-4, born in Minneapolis, MN, Oct. 8, 1933, graduated from St. Thomas Academy in 1951 and from the University of Minnesota with a BEE degree in 1962.

He enlisted on April 15, 1953 and completed boot camp at Great Lakes before assignments to USS *Worcester* (CL-144). Four years of active duty included two Mediterranean cruises and a WestPac cruise as well as electronics technician class "A" and "C" schools. Released from active duty on April 4, 1957 as electronics technician second class (ET2).

He affiliated with the Naval Reserve Security Group in Minneapolis and changed rating to communications technician. Served with units in Davenport, IA; Santa Monica, CA and Atlanta, GA as well as surface, shipboard and communications units in Cedar Rapids, IA and Wichita, KS.

He was advanced to chief communications technician (CTMC) on March 16, 1963 and selected for the Warrant Officer Program (cryptologic technician) on Dec. 15, 1968. Advanced to CWO-4 on Dec. 15, 1978. He retired on Oct. 8, 1993 with over 40 years of continuous naval service.

CWO-4 Aukofer has been awarded the Navy Achievement, Good Conduct, Naval Reserve Meritorious Service (2), Navy Occupation Service (E), National Defense Service (2) and Armed Forces Reserve (3) medals and the Overseas Service Ribbon (2).

He married Joy Ann Schultz and they have children: Anthony, Andrea, Michael and Lori. He held position as area manager, Rockwell-Collins Service Centers and retired Oct. 8, 1993.

ERSKINE POND "PETE" AUSBROOKS JR. CDR, born Nov. 1, 1929, Nashville, TN, received BA, MA, MALD, US NavPostGrad School, Naval Justice School.

Joined USNR, 1946; 1947 (Aviation Midshipman NACP); ensign USNR, 1950; 1954, USN. Military locations and stations: NROTC Vanderbilt; Pensacola, FL; Norfolk, VA; San Diego, CA. He served with CVS-39, CVS-20, ZP-2, VW-1, NATTC Memphis, NAVCRUITSTA Philadelphia; Clarksville, Base, TN.

Ausbrooks participated in Cuban Missile Crisis, MATSU-Quemoy, Vietnam. He was awarded the American Defense Medal, Navy Expeditionary Medal, Vietnam, Navy Commendation. He achieved the rank of CDR, USN and was discharged Dec. 31, 1976.

His memorable experiences: carrier duty (weapons & XO); typhoon tracking; nuclear weapons; losing two engines while transpacing from Japan; CO NAVCRUITSTA Philadelphia, flight training and operations, shipbuilding in Newport News, VA.

He married Helen Virginia Russel and they have children: Erskine III, Jon R., Daniel P., Judith Ann; and grandchild Erin Goodwin. He held positions with Tennessee Department of Education; city manager and mayor of Hendersonville. He retired December 1976 (30 years honorable service). He is a life member of the NRA. His hobbies include: old cars, family history and carpentry.

STANLEY E. "STAN" AUSLANDER, LCDR, born Nov. 7, 1917, New York, NY, received BA Hofstra College, Hempstead, NY, 1939.

He joined the service, January 1941, Floyd Bennett Field, Brooklyn, NY, "E" Base Training. His military locations and stations include: Jacksonville, FL, flight training; Brazil; England. Served with VP-94, VPB-105, and Project "Cast."

Auslander participated in the American and European Theaters, receiving the Distinguished Flying Cross and two Air Medals. He achieved the rank of LCDR and was discharged October 1945.

His memorable experiences: as patrol plane commander in PBYs with VP-94 in Brazil dive bombed a German U-boat, reached a speed of 200 knots, pulled out of dive 10 feet from water's surface - result submarine was sunk (date July 19, 1943). Flew B-24 Liberator in England with UPB-105 from Dunkeswell before and after D-Day, patrolling the Bay of Biscay and English Channel with many flights during the invasion. Attacked one U-boat and homed in HMS Fame to the location. Worked and flew (many types of planes) with radar researchers from MIT and Harvard, at Project "Cast" in Bedford, MA. Many of these people went on to work for NASA.

He married Jean Hamilton in 1944. They have children, Gail A. Ginnetty and Dean H. Auslander. He was employed as president of chemical company, retiring in 1982.

EDWARD U. "ED" AUSTIN, RADM, born Oct. 15, 1922, Raleigh, NC, attended University of North Carolina, CH, Emory University and University of Pennsylvania.

He joined the service in 1942. He was awarded the WWII Victory Medal and Korean Commendation. Austin achieved the rank of RADM and was discharged Oct. 15, 1982.

He married Marjorie B. Austin and they have children: Robert, Steve and Don; and grandchildren: Thad, Emily, Pace, Gregory and Erin. He was employed as an oral surgeon, retiring October 1983.

JAMES ALBERT "JIM" AUSTIN, RADM, born Sept. 23, 1931, Phoenix, AZ, received his BS from University of Southern California; MD, George Washington; MBA, Pepperdine University.

He joined the service June 28, 1956 and was stationed in Bethesda N. Hospital 1956-60; Rota S, 1961-64; San Diego Naval Hospital, 1964-66 Guam N. Hospital 1979-81. Reserve units served with: RNCB, 4th Eng Bn USMC, USS *Jason*.

Austin married Sharon M. Reichle and they have children: Cynthia Ludgin, Lauri Fuller and Wendy Rea. He held position as CEO, Ultra Link, Inc.; VP Care Continuum; medical director, Southern California, Prudential Health Care.

TOM A. AUSTIN, CDR, born Nov. 11, 1925, San Francisco, CA, received BA and MA from San Diego State University.

He served in the military from May 18, 1943 to August 1985 ranks from SA to CDR. His military locations and stations include: Washington, DC; Hawaii; San Diego; Charleston, SC. Served with USS *Bernadou* (DD-153) in the North Atlantic in Bizerti and sub encounters et al. He was awarded the Asiatic and Pacific (Euro w/2 stars).

His memorable experiences: using lifeline on destroyer in North Atlantic. Nothing to stand on, just holding life line. On KP duty at the time. OCS Newport, RI. Something like *Officer and Gentleman* movie.

He married Lois A. and they have children: Doug, and Gary. Their grandchildren are Elizabeth, Michael and Dillon. Austin held position as teacher and principal of SD Unified School District 34 years.

FRANK T. AVIGNONE III, CAPT, a native of New York, enlisted in the USN during the Korean Conflict and served aboard the USS *Hawkins* (DDR-873).

Upon completion of active duty he entered Georgia Tech in 1957 where he received his BS, MS and PHD degrees in physics.

He was recalled to active duty as CPO in 1961 during the Berlin Crisis and was commissioned an ensign in 1962.

Avignone joined Naval Reserve Destroyer Sqdn-34 in 1966 and served aboard the USS *Soley* (DD-707). He later served aboard the USS *Gainard* (DD-706), the USS *Hardwood* (DD-861), the USS *Strong* (DD-758), the DESLANT Staff (DET-607), the USS *Cone* (DD-866) as prospective commanding officer and commanded Mine Warfare Command Detach 107. In 1973 he volunteered for active duty during which he graduated from the US Naval Destroyer School Command Course.

Upon selection for afloat command he served from 1978-80 as commander, Mine Div-126 composed of: USS *Exultant* (MSO-441), USS *Fearless* (MSO-442), USS *Engage* (MSO-433), and USS *Impervious* (MSO-449). In 1980 Avignone lead in the establishment of the Techology Mobilization Program, a nationwide organization under the Chief of Naval Research and later served on the Central Management Team as fleet liaison officer, field officer, senior field officer, executive officer and from October 1986 until October 1988, as the officer-in-charge. From October 1988 until October 1989 he served as an assistant to the director of Maintenance and Modernization (OP-43) on the staff of the Chief of Naval Operations until he assumed his duties as comander, Craft of Opportunity Mine Sqdn. TWO TWO in Charleston, SC. He was promoted to CAPT in 1983 and served continuously in the Navy and on active and inactive duty from 1952 until his retirement in 1991.

Avignone was awarded the Meritorious Service Medal w/star, the Navy Commendation Medal, the Battle-E Ribbon, the Navy Good Conduct Medal, the China Service Medal, the Naval Occupation Medal, and the National Defense Medal w/star, the Naval Reserve Sea Service Ribbon and the Armed Forces Reserve Medal-third award. He holds an Endowed Chair Professorship and has served and the chairman of Department of Physics and Astronomy at the University of South Carolina since 1979.

He is married to the former Norma Llorraine Novia of Cambridge, MA. They have two children Mrs. Michelle Inere of Pensacola, FL and Frank IV of Plano, TX who served five years in the USN.

DAVID B. AVISON, CDR, a native Oregonian, attended Willamette University and Lewis and Clark College, obtaining his BA, MPA and JD degrees. He practices law in the Portland area, is married to the former Jean Bell and is the proud dad of James, Joshua, Jessica and Krista.

His naval career began in 1968. He served three years active duty: first in the pre-commissioning detail for USS *Juneau* and then as a naval gunfire liaison officer with the USMC in Vietnam where he was awarded the Naval Commendation Medal ("V") as well as Unit Citation, Campaign Stars and Marksmanship Awards.

Avison served 21 years in the Ready Reserve, commanding three units: ARD-30, LSD-28 and Fleet Support Training Unit. His active training included two sessions at the Naval War College. He retired in 1992 and is a life member of the USNI and NRA.

JOSE AYALA II, LT, born April 15, 1961, Chicago, IL, received a BSE and MBA degree.

He joined the service Aug. 13, 1988 and his stations include: NRC Denver; NAS Alameda; NAF Washington. Served with ASP 1471, ASP 1187, NADEP 0187 and ASP 0266.

Ayala received the National Service Medal and NAM, achieving the rank of LT. He married Karen and they have a daughter Katherine. He is employed as a senior staff engineer/project leader with Lockheed Martin Corp.

CHARLES O. "CHUCK" AYARS, CAPT 06, born Oct. 27, 1924, San Diego, CA. His father Fred C. Ayars, was one of the builders of *Spirit of St. Louis.* He received a BA (math) from UCLA in 1949.

Joined the service Nov. 28, 1942. Ayars was V-5 Naval Aviation Cadet, stationed Marianas Island, Los Alamitos, North Island and CINCLANT. He served with VJ 12A and VJ-3. He fought in WWII, receiving the Meritorious Service Award and Service Commendation. Ayars achieved the rank of CAPT 06 and retired after 39+ years, Jan. 1, 1982.

His memorable experiences: 1946-VJ-3, flying and controlling radar control drones (TD2C's); 1961-62, VS-771, requalifying day and night carrier landing after 17 years from previous car/quals; 1977-80, CO, CINCLANT Flt 106, battle staff supervisor.

He married Janet in 1946 and they have children: Deborah, Charles Jr. and Thomas; and grandchildren: Jeremy, Rachel, Caitlin, Jana, Jordan and Jonathan. Employed as banking assistant vice-president, (Civil Service GM-15, USAF Realty officer (Washington, DC, 1975-89). Ayars is a life member NRA, ROA, ANA and TROA.

WILLIAM FOSTER "BILL" BABCOCK, CAPT, son of R. Fayerweather Babcock, well-known Chicago artist, and Anita Joy Biggs Babcock, a granddaughter of Civil War General Herman Biggs, US Army, was born June 29, 1918 and raised in Evanston, IL, graduating with honors from Northwestern University in 1939 with a BS in Commerce. He was in the first group of Clarion DeWitt Hardy Scholars and a member of Phi Eta Sigma, Beta Gamma Sigma and Phi Gamma Delta. An NROTC graduate, he was commissioned ensign, USNR (on inactive duty) and was a trainee with General Mills, Inc. from Aug. 1, 1939 to Feb. 28, 1941, when he was ordered to active duty at Grand Rapids, MI, helping to get Naval Reserve Div 9-13 ready for sea, and with it to Newport News, VA, where an ESSO tanker, renamed USS *Salamonie* (AO-26), was being converted (including a destroyer's armament), and on board after commissioning.

Salamonie was in the convoy occupying Iceland in the summer of 1941 which found a lone lifeboat about 200 miles south of Greenland containing crew members of a ship torpedoed 12 days earlier. *Salamonie* remained in Iceland awhile to refuel USN ships, then participated in various Atlantic operations, visiting Scotland, Ireland, England, Algeria and Morocco among others.

Babcock rose through the ranks to LCDR, serving as assistant communications officer, communications officer, and navigator. On Feb. 12, 1943 *Salamonie* was in a collision 800 miles east of Bermuda with SS *Uruguay* while returning to base course in a zigzagging convoy of 35 ships plus 13 destroyers headed for Casablanca. *Salamonie* returned to Bermuda with a destroyer escort, where a temporary bow was installed, then to Maryland for permanent repairs. While there he met his future wife, Margaret Burke Allen, daughter of Herschel Heathcote Allen, a WWI major and prominent civil engineer who designed airports, bridges and turnpikes, including among others the Pennsylvania, Ohio and Indiana turnpikes and the Chesapeake Bay Bridge.

Salamonie continued operating in the Atlantic, and he was transferred in December 1943 to EO USS *Marias* (AO-57) under construction near Baltimore, and he was married in January 1944. After shakedown, *Marias* operated through the Pacific.

Early in 1945 he was ordered to Providence, RI as commanding officer of USS *Selinur* (AKA-41) under construction. *Selinur* passed through the Panama Canal June 2, 1945 and carried Marines from Hawaii to Majuro and Kwajalein and returned to Hawaii. Then, carrying a Marine Corps hospital unit, it joined a flotilla of 27 transports and seven escorts, carrying the 5th Marine Div and the 5th Amphibious Corps for the occupation of Japan, setting up a Marine hospital at Sasebo. *Selinur* then took troops from the Philippines to Japan and then at Okinawa took on six officers and 298 men for the long trip back home. *Selinur* returned to the Philippines for another load of soldiers for discharge in San Francisco, then went to Philadelphia, where it was transferred to Pennsylvania State Maritime Academy for a school ship.

Babcock then made one trip on USS *Elokomin* (AO-55) from Brownsville, TX to Boston before reporting to NROTC at University of Colorado as assistant professor of Naval Science in the fall of 1946. While there, he resigned from USN (to which he had transferred earlier under the option offered NROTC graduates) and took a USNR commission again.

Returning to Evanston, he joined Naval Reserve Div. 9-21, and later Naval Reserve Officers School 9-3, where he was promoted to CAPT before retiring as CDR with over 30 years of federal service. On Dec. 29, 1946 he joined Chicago's Continental Bank, retiring in 1990 as a vice-president. He and his wife have five children: Margaret Anita, Barbara Allen, Martha Cecelia, W.F. Jr. and Richard F. II, and 12 grandchildren. They live in Wilmette, IL. His only brother, Richard F. Babcock Jr., attended Northwestern two years, then graduated from Annapolis in 1945 (with the Class of 1946) and served in USS *Maddox* (DD-731) 1945-47. He was also in the Naval Reserve in Evanston and was recalled to active duty, serving in USS *Doyle* (DMS-34) 1951-53 and transferring as LCDR to Retired Reserve in 1968.

JOHN MATTHEW BACKER, CAPT, born Aug. 15, 1929, Baltimore, MD, attended Baltimore Polytechnic Institute; Illinois Institute of Technology, BS electrical engineering 1954; Massachusetts Institute of Technology, graduate study, Urban & Regional Planning (1958-60); University of Baltimore, graduate study, Management Information Systems.

He joined the service May 13, 1954. Commissioned Officer Candidate School (Class 18), Dec. 16, 1954.

His military locations and stations: US Naval School, Explosive Ordnance Disposal, US Naval Powder Factory, Indian Head, MD, graduated August 1955; USS *Observer* (MSO-461)*; Atlantic Fleet and Sixth Fleet 1955-56; USS *Swerve* (MSO-495)*; Atlantic Fleet; 1956-57; temporary duty: USS *Turkey* (MSC(O)-49) May-June 1957. *NOTE: primary duty assigned as explosive ordnance disposal (EOD), but concurrently held operations officer billet and others as well.

Naval Reserve Units (listed chronologically); NR Research Co. 1-1, Boston, MA; NR Research Co. 4-1, Baltimore, MD, CO (July 1967-June 1969); Inshore Undersea Warfare Div. 5-4, Baltimore, MD; NR Group 5-1(L), Baltimore, MD; SURFWEPCEN DET 206, Baltimore, MD, CO (July 1975-February 1977); EOD FAC IND 106, (EOD Technology Div.), Washington, DC, CO (March 1977-September 1979); CINCLANTFLT DET 106, (CINCLANTFLT Command Center), Norfolk, VA; NR General Volunteer Training Unit 0603 (NR GVTU 0603), Baltimore, MD. Assigned: NR IUW Group TWO (inspector general); NR GVTU 0603, Baltimore, MD, CO (October 1984-September 1985).

Backer was awarded the National Defense Service Medal, Armed Forces Reserve Medal, Surface Warfare Officer Insignia, Senior Explosive Ordnance Disposal Badge. He achieved the rank of CAPT. Discharged from active duty Dec. 2, 1957.

His memorable experiences: No billets requiring EOD qualifications existed until "restructuring" of the Reserve in 1977. Upon restructuring of the Reserve, selected to command of one of the two EOD units established - the detachment supporting the EOD Technology Div., Indian Head, MD. He had the satisfaction of demonstrating to the active duty counterpart the pay-off from utilization of reserve manpower.

Sixth Fleet and NATO operations (1956) aboard the USS *Observer* as part of the first MSO division to successfully complete a six-month deployment with a full complement of five minesweepers.

Selection for and completion of the Senior Reserve Officer Course at the Naval War College, Newport, RI.

Navigating the USS *Turkey* from Charleston, SC via the Old St. Lawrence Seaway system and the Great Lakes to Toledo, OH for delivery to its Naval Reserve Center (1957).

Mobilization assigment at EOD FAC IND 106 was to relieve the commanding officer of the Naval EOD Facility (now Naval EOD Technology Div.) Final billet with CINCLANTFLT Det 106 was assistant battle staff supervisor. Qualified as surface warfare officer, senior explosive ordnance disposal officer. Life member of NRA, and has been a member since 1959. Member, Board of Trustees, US Navy Memorial Foundation, Washington, DC 1994-present.

He married Mary Eaton and they have a daughter, Mary Patricia. He has held position of assistant electrical engineer with Underwriters' Laboratories, Inc., Chicago, IL (1952-54); Baltimore Regional Planning Council, Baltimore, MD, regional planner (1963-70); Department of Planning, Baltimore, MD, city planner (capital programs), (1960-63); Chief, Research & Info Systems, assistant director (1970-83); Maryland Mass Transit Administration, manager, Capital Programs and Grants, manager, Strategic Planning (1984-92). He is also life member of American Planning Association (APA) and American Institute of Certified Planners (AICP). Retired and transferred to Retired Reserve, Oct. 1, 1985.

MATTHIAS H. "MATT" BACKER JR., MD, RDMU MC (0-8), born Dec. 19, 1926, St. Louis, MO, pre-Med and Medical at St. Louis University.

He joined the service Sept. 12, 1944. His military locations and stations: GLKS, NH San Diego, University of Colorado, NH Bethesda also BUMED & Pentagon + 26 NavHosps. Served with the following units: CO 9-1; CO COs 9-22 and 2812; CRUITCOM.

Backer was awarded the Defense Sup Serv, Legion of Merit, Navy Achievement, National Defense, AFRES, NAVRES, WWII Victory. He achieved the rank of RDMU MC (0-8) and retired Sept. 12, 1984.

His memorable experiences: recruited 400 med students into Navy; taught and served in 26 hospitals, IN and OCONUS; Dir Nav Medical Recruiting; asst dep surgeon general; Navy member of Reserve Components Medical Council (DOD).

Backer is a widower (wife Laverne died Oct. 16, 1992) and has 13 children: Kay, Matt, Lynn, Lou, Don, Bob, Ed, Susan, Carol, Pat, Joe, Brian and Denis; and 23 grandchildren. He was employed Head Medical Group (Corp), 1954-84; St. Louis University, 1985-92, prof/chm ob/gyn.

ROBERT E. "BOB" BAIER, CAPT, born Oct. 31, 1939 in Buffalo, NY, received BES (physics), 1962; PhD (biophysics) 1966.

He joined the service, April 1, 1966. His military locations and stations include NRID 5-1, Washington Navy

Yard; DEFATTACHE 0-166 Andrews. Served with NISC 103, NRL 105, Rochester, NY.

Baier was awarded the Navy Commendation Medal and achieved the rank of CAPT. He retired Oct. 1, 1996.

His memorable experiences: Plankowner NISC 103; CO NRL 105; at-sea cruises aboard USNS *Haynes,* USNS Gillis, R/V New Horizon; providing intel briefs to crew and wardroom of USS *LaSalle* prior to deployment as COMMIDEASTFOR.

He married Corinne May Bongiovanni and they have children: Valerie Ann (b. 1962) and Ann Marie (b. 1972). He has held the position as principal scientist, Calspan Advanced Technology Center 1968-84; professor at SUNY Buffalo.

JAMES A. "JIM" BAIRD, CAPT, born in San Antonio, TX, Nov. 15, 1947. He enlisted in the USN, June 1, 1966 and attended boot camp in San Diego, CA. He served aboard the USS *Klondike* (AR-22) and his last year of active duty was at the Puget Sound Naval Shipyard in Bremerton, WA. While attending college at Central Washington University in Ellensburg, WA (EdM degree), he remained in the USNR. In 1972 he was advanced to first class petty officer and in 1974 received a direct commission as ensign in the USNR.

Most recently, he served as XO of Reserve Intelligence Unit Joint Intelligence Center, Pacific One Six Eight Nine (JICPAC 1689) from Oct. 1, 1992 to Sept. 30, 1993. He then served as the CO of Reserve Intelligence Unit Joint Intelligence Center, Pacific One Four Eight Nine (JICPAC 1489) from Oct. 1, 1993 to Sept. 30, 1995. Noteable awards include a Meritorious Unit Commendation and a Navy Commendation Medal.

Baird retired from the USNR, June 1, 1996. He has taught secondary education in Olympia, WA for the past 21 years. His family consists of wife Sharon and son James "Lee".

DONALD L. "DON" BAKER, CDR, born July 19, 1927, Philadelphia, PA. Education includes New York State Maritime Academy graduate; Amherst College, AB; Harvard Business School, MBA; New York Univ., GBA, MBA plus credits for PhD in business administration; the New School for Social Research, MA, international economics plus PhD credits.

Joined the service July 8, 1945 (midshipman). During active duty he served in USNR, MMR; USS Rockwall (APA-230) during Korean War; sailed in Merchant Marine as licensed deck officer.

Inactive duty included National War College and numerous service schools; commanding USNR Surface Divs.; USMCR Intelligence Unit; Served on Group Command Staffs and other line units; annually served on Afloat ACDUTRA. Retired in October 1974 as CDR.

He is NRA life member, has held both chapter and district offices, been appointed to and served on national committees. He is founder and commander, Philadelphia/Delaware Valley Commander, Naval Order of US; LCOL, CAP, group director, Legislative Affairs; CAPT, New York Naval Milita; past post commander, VFW and American Legion; The Army and Navy Club, Washington, DC; Admiral Farragut Academy Class Trustee; Fernwood Lodge #543, F&AM; Valley of Philadelphia, AASR, 32*; National Sojourners, Benjamin Franklin Ch. #16; Royal Arch Chapter #348; TROA, ROA, VP Navy; Navy League of the US.

President, Donald L. Baker & Associates; President, MBA Sales/Marketing Services; business consultant and held corporate top management positions. Divorced, he has five children: Wendy, Jeffrey, Donald II, Sarah and David; grandchildren are Halley and Graham Griffin and Jane Baker.

JOSEPH W. "JOE" BAKER, CAPT, born Oct. 11, 1925, Fitzwilliam, Yorkshire England, received a BA and MBA from Dartmouth College, Hanover, NH.

He joined the service May 15, 1943. He was stationed at various V-12 and V-5 training locations, serving with USS *Fargo* (CL-106) Fire Control Div.

Baker was awarded the Naval Reserve Medal, Armed Forces Reserve Medal, Good Conduct Medal, European Occupation Medal, WWII Victory Medal, Atlantic Theatre Medal. He achieved the rank of captain and was discharged Aug. 15, 1946.

His memorable experiences: Trieste Italy 1946; sending ship's Marines ashore to keep peace between Italians and Yugoslavs; the Graf-Spee scuttled in the harbor of Montevideo, Uruguay; three day South Atlantic storm.

His wife Frances died in April 1990. He has children: Susan, Joan, Joseph W. Jr. and Carolyn; and six grandchildren. He was employed in sales and marketing with General Electric Company, retiring April 1981.

LYNN D. BAKER, CAPT, born in Tucson, AZ, Aug. 27, 1952. He was a graduate of Purdue University with a BCHE degree and later completed his MBA at the University of Phoenix. Baker graduated in 1974 and was commissioned in May from the NROTC program of Purdue. During his first year in the USN he went through the Naval Nuclear Power Program, graduating in May 1975.

During his active duty, Baker was first assigned to the USS *Bainbridge* (CGN-25) as ASW officer (1975-77) then to the USS *Long Beach* (CGN-9) (as nuclear EOOW and B-2 division officer (1977-79). After leaving active duty in June 1979, he joined the USNR where he has served as operations officer, training officer, XO, and CO with various units.

Baker was president of the NRA, Southern Arizona from 1988-91. He has had a direct membership since 1981. While serving with the USNR he has been employed with IBM Corporation.

He was awarded the Navy Commendation Medal (2nd Gold Star), and the Navy Achievement Medal. His campaign and service medals include: Battle Efficiency "E"; Sea Service Ribbon, Overseas Ribbon, National Defense, Recruiting Ribbon, Naval Reserve, Armed Service, Expert Rifle, Expert Pistol, Surface Warfare Officer Pin and Polaris Patrol Pin.

Baker is married to Dianne and they have a son Andrew O. and daughter Kathleen D.

MICHAEL S. BAKER, CAPT, is a general, vascular and trauma surgeon practicing in Walnut Creek, CA and Director of Surgery for NR Fleet Hospital 21. He has served as staff medical officer for Special Boat Unit 11, 4th MAW MED MAG 42, and COMUSNAVCENT DET 220. Recalled for Operations Desert Shield/Storm with the 1st Med. Bn., 1st FSSG he performed the only successful battlefield vascular repair, and later served with the 4th Civil Affairs Gp. in Kuwait. In 1996 he was MEDCAP commander for Exercise Cobra Gold in Thailand.

Baker has published over 40 articles on military medicine, wound management, warm water diving and medical readiness training. He wears the Navy and Marine Corps Commendation (w/Gold Star), Army Commendation, Air Force Commendation, Navy Achievement, Fleet Marine Force and Combat Action Ribbons, Meritorious Unit Citation, Southwest Asia Service w/Marine Corps Globe and anchor and three campaign stars, and the Combat Craft OIC Warfare Device.

He is married to Marian and they have children Philip and Jacob. His civilian employment and positions include geneal, vascular and trauma surgery. He is a drilling reservist.

MARGARET FOOTE "MAGGIE" BALACKI, CAPT, born April 24, 1950, Scranton, PA, received BS from George Mason University, 1979, and MS SUNY at Buffalo, 1984.

She joined the service June 17, 1970. Her military locations and stations include: St. Albans NH, NRMC Jax; NH Roosevelt Roads PR NNMC Bethesda; G7 Sta Hosp; VTU; 4th FSSG Det C Hosp Co.; PTSMTH 305, 205, FH 7 & 18; Bethesda 106.

She was awarded two National Defense Medals, Meritorious Unit Citation, NAM; Naval Commendation Medal, Recruiting Medal. Balacki achieved the rank of CAPT and was discharged from active duty Aug. 18, 1977.

Her memorable experiences: first recipient of the Admiral Nielobowicz Award for Navy Nursing; first Director of Nursing of Fleet Hospital 18. Her most memorable experience: the people she's met in the USNR.

She was married to CAPT John A. Balacki, MC, USNR and they are now divorced. Their children are Katherine (attending Georgia Tech on NROTC scholarship) and Alexander. Balacki is employed as nurse practitioner and psychotherapist. Still a drilling Reservist NR NNMC Bethesda 106 Acelphi, MD.

WILLIAM H. "BALDY" BALDEN, LT, born Oct. 13, 1921, Mercer County, KY. Earned BS degree in business administration. Joined the USN in July 1942 and was stationed at Pensacola, FL; Miami; Seattle; Maui; USS *Enterprise.* Served with VT-10 and VT(N)90.

Participated in action at Marshalls, Truk, Hollandia, Guam, Saipan, Okinawa and Luzon. Memorable experience was first night carrier attack on ships in Truk Harbor - every night flight from USS *Enterprise* while in night Air Group 90.

Discharged in December 1945 as LT. Received numerous Air and Distinguished Flying Crosses.

Married the former Corinna Cook and they have one daughter, Madge Balden Adams, and grandson, William Balden Adams. Retired as manager of Southern Bell Telephone Co. - farmer and tobacco warehouse business.

ROBERT O. "BOB" BALL, CAPT, born Sept. 20, 1927, Rockville Centre, New York, received a BA from St. John's University, Brooklyn, NY.

He joined active duty Aug. 1, 1945. Served in Reserve 1946 to July 1, 1976. He was stationed in China and various Reserve commands in New York. Ball achieved the rank of captain, USNR, Commodore (New York State), discharged July 1, 1976.

He married Irene Lennon and they have children: Bob Jr., David, Tom, Chris, Mary Anne, Jon, Suzanne, Matt, Lauren and Jenny. Their grandchildren are: Mariel, Brian, Devin, Francesca, Maggie, Matt, Elizabeth, Catherine, Marian, Caitlin, Jonathan and Erin. He was employed as VP, Marsh & McLennan, New York, retiring June 1994.

JERALD TYLER "TYLER or JERRY" BALLANCE, CDR, born Nov. 23, 1955, Norfolk, VA, received (BSA) Christopher Newport University, Naval Post Graduate School (Fin. Mgmt), City University of Seattle, MBA.

He joined the service May 18, 1979. Completed OCS August 1979. Commissioned by his brother, Master Chief Bobby Ballance. Served in USS *Blandy* (DD-943), boilers officer; USS *Long Beach* (CGN-9), terrier missile officer; USS *Kitty Hawk* (CV-63) one propulsion group officer; USS *Bradley* (FF-1041), weapons officer; USS *Racine* (LST-1191) 1LT; COMNAVSURFRESFOR (Dir. Curriculum Development); commissioned USS *Pioneer* (MCM-9), as CO.

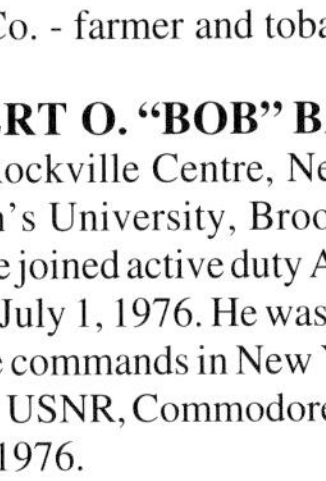 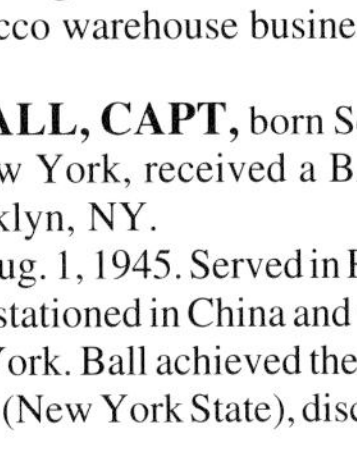

His memorable experiences: sailing USS *Pioneer* (MCM-9) with the commissioning crew from Sturgeon Bay through the Great Lakes to her new home port in Ingleside, TX. The entire crew of *Pioneer* marching in the city of Portland, TX Veterans Parade to cheers of the town people.

Served in three deployment campaigns: Persian Gulf, Alaska, RIM PAC and was awarded the Navy Achievement Medal, Navy Commendation, Meritorious Service Medal.

Ballance is single and is president of EnviroImpact Inc., an environmental research company in Virginia.

ROBERT E. "BOB" BALLOW, CAPT, born Feb. 27, 1933, Fawnesdale, AL, received a BA in math from Vanderbilt University. He also attended East Tennessee State, Graduate School, economics.

He joined the service June 4, 1954. His military locations and stations include: USS *Olmested* (APA-188); USS *Tweedy* (DE-214), Reserve Ship; USNR, Kingsport, TX; Alexandria, VA; Yonkers, NY; Brooklyn, NY; and Atlanta, GA.

Ballow achieved the rank of CAPT and was discharged July 1984. His memorable experiences: CO NRSD 3-64, CNET, VTV.

He and his wife Jean have children: Stephen, Cynthia, Lindsay and Michael, and grandchildren: Cristine, Cathrine, Thomas, Stephen, Barry and Benjamin. He was employed with IBM for 26 years.

ADELBERT A. "DEL" BALUNEK, CDR, born Feb. 5, 1933, Cleveland, OH, received AB from Miami University; and JD degree Cleveland Marshall Law School.

He was commissioned ensign Miami University NROTC Unit 1955. His military locations and stations: Navy Supply Corps School, Athens, GA. Served on active duty USS *Yorktown* (CVA-10), 1955-57.

Balunek achieved the rank of LTJG and was discharged 1957. His memorable experiences: active duty carrier takeoffs and landings in right seat of AD-5 aircraft; Reserve engineering officer and XO USS *Hollidaysburg* (PCS-1385) on Lake Erie, Fleet Div. 4-7D. He retired in 1981 as CDR, USNR. He is a charter member #955 of Association of Naval Aviation and was NRA chairman, National Missions and Mobilization Committee, 1978-79.

He and his wife Klara have children Klara Helen, Konrad and Victoria. He has been employed as an attorney.

EUGENE MILLER "GENE" BANE JR., CDR, born March 23, 1940, Grundy, VA, received a BA from Emory and Henry College; BS and OD from Southern College of Optometry.

He joined the service July 27, 1965. His military locations and stations include: NNMC, Bethesda, MD; NHosp, San Diego, CA. Served with NROS, Roanoke, VA; MCRU-706 Roanoke, VA; 2nd MARDIV B-306 4 FSSG-Atlanta.

Bane received the National Defense Medal, Fleet Marine Ribbon, Armed Forces Reserve Medal, Rifle and Pistol Qual. He achieved the rank of CDR, MSC, USNR and was discharged Oct. 1, 1987.

His memorable experience: after 15 days of a resphiblex exercise at Camp Lejeune, everyone else was ready to go home, but he had been so well indoctrinated he wanted to stay for another two weeks.

Bane married Rebecca F. and they have children: Katherine, Eugene III and Melissa. He held the position of doctor of optometry and retired Oct. 1, 1987.

JOHN PETER BANKERD, CAPT, born April 25, 1942, Long Beach, CA, and reared in Whittier, CA. Earned BA degree, University of California, Santa Barbara (1965), MBA degree, California State University, Long Beach (1970).

Enlisted in the Naval Reserve in February 1964 and commissioned ENS via OCS November 1965. Completed Communications Officer School and reported aboard USS *Kearsarge* (CVS-33). During his three year tour, he served as CR division officer, communications watch officer, CICWO, and anti-submarine air controller.

Two deployments to Vietnam, included three major campaigns including the Tet Offensive of 1968. He participated in operations in the Sea of Japan after capture of the USS *Pueblo*.

From 1969-93 he held various billets in a variety of Naval Reserve units, including command tours as CO Shore Intermediate Maintenance Activity/Naval Reserve Maintenance Facility San Francisco DET 220 and CO SIMA/NRMF SF HQ DET 120.

Received the Navy Commendation Medal, Meritorious Unit Commendation, National Defense, Armed Forces Expeditionary, Vietnam Service, Overseas Service, Armed Forces Reserve, Vietnam Meritorious Unit Commendation and RVN Campaign. Retired from the NR March 1, 1994.

Memorable experiences: swearing in his son Chris when he enlisted in the NR; being guest speaker at his son's CM "A" School graduation; being guest speaker at the retirement of an active duty CPO Command; and the outstanding shipmates that are too numerous to list.

He and his wife Ginny reside in Concord, CA. Their family includes Christopher (CM3, USNR), Jennifer, Ryan and Debbie. Civilian employment as purchasing manager for Robert Mondavi Corp.

CHARLES OLD "PAT" BARCLAY JR., MD, LCDR, born March 17, 1918, Portsmouth, VA, received a BS from the College of William & Mary in 1939 and MD the University of Virginia in 1950.

He joined the service July 17, 1941. His military locations and stations: USS *Prairie State;* USS *PC 549;* Submarine Bases in Pearl Harbor, Midway Island; Brisbane, Australia; Subic Bay, Philippines, medical officer, USNH Ports, VA; medical officer for local unit to 1970.

He was awarded the American Defense Medal, three area campaign medals, WWII Victory, Philippine Liberation, Philippine Victory, Naval Reserve Medal, Department of Defense.

Barclay achieved the rank of LCDR, CDR (MC). Discharged Feb. 5, 1946 and Jan. 7, 1970.

His memorable experience: sea and foreign duty for the entire WWII, medical officer 1950-51 in Korean Conflict. It would take a volume to enumerate them all. He would not take a million dollars tax free for them. He was and is a real "feather merchant".

He married Norma and they have children, Rebecca O. and Charles Old III. Employed as MD for the past 45 years. Retired from Reserve on July 1, 1970. Still a practicing MD.

ROBERT S. BARGAR, LCDR, born in Jamestown, NY, Aug. 8, 1919. He graduated from the Wharton School of Finance and Commerce, University of Pennsylvania in 1941.

In February 1942 he was commissioned as an ensign in the USNR. Beginning in May 1942 he served as signal officer and assistant communications officer aboard the submarine tender, USS *Fulton.*

He then served as a communications officer, with the service forces, Southwest Pacific from January 1943-May 1944. From June 1944 until April 1945, he was communications officer of US Naval Base, Sydney, Australia.

Switching to the Supply Corps, he was waterfront administrative officer of the Naval Supply Annex, Stockton, CA and finally special assistant to the executive officer at that facility. He retired from active duty in April 1946.

In civilian life has been an officer and director of the SM Flickinger Co. of Jamestown, NY, a wholesale food distributor, retiring in 1982 as president of the Jamestown concern.

During active business life he was director of the Empire Worsted Mills, Dahlstrom Manufacturing Co., Jamestown Star Publishing Co., the Jamesway Corporation, Debell Vending Co., Jamestown Electroplating Co., the Lakeview Cemetery Association, Clark Laboratories, and the Town Club.

He was president of Jamestown Civic Auto Ramps, the Southern Chautauqua County Chapter of the American Red Cross and the Jamestown Area Chamber of Commerce. He also served as president and director of the Sportsman's Club and a trustee of Chautauqua Institution and the Chautauqua Foundation.

Bargar is currently chairman of the board of the Fluvanna Realty Co. and partner in the JBC Co. He was selected as "Man of the Year" by the Jamestown Chapter of the National Conference of Christians and Jews.

He and his wife Je'Anne live in Chautauqua, NY in the summer and Naples, FL in the winter. They have four children and six grandchildren.

CHARLES D. "CHARLIE" BARHAM JR., CAPT, born July 7, 1930, received BS and JD degrees from Wake Forest University.

He joined the service Oct. 15, 1954. For active duty he was stationed Washington, DC (Pentagon); OPNAV 922Y and OPNAV 922N (1955-58); Active Reserve stations include Raleigh and Durham, NC, NRIU 6-1-8 (1958-77).

Barham received the rank of CAPT and retired Jan. 31, 1977.

He married Margaret C. and they have children, Margaret D. and Charles D. III. He has held the position as executive vice-president, Carolina Power & Light Co. and is now retired.

THOMAS C. "TOM" BARKER, CAPT, born May 25, 1931, Cedar Rapids, IA, received BS in 1954 from State University of Iowa, Iowa City; MA, 1960; PhD 1963.

Enlisted as seaman recruit July 1949. Selected for Reserve Officer Candidate (ROC) Program. Advanced to BMG 3. Returned to university while continuing NR participation. Commissioned ensign June 1954, returned to active duty, attended school in Treasure Island, CA; assigned to USS *Etlah* (AN-79), homeported in Yokosuka, Japan. Released from active duty after having served as XO and acting CO of *Etlah* in 1956. Served in the active Reserve in multiple units and assignments until retirement in 1985. Served as CO of six units. Last assignment CO, Commander in Chief, Atlantic, Detachment 206, Norfolk, VA.

Barker was awarded Joint Services Commendation Medal, Navy Commendation Medal, China Service Medal, National Defense Service Medal, Armed Forces Reserve Medal, Navy Meritorious Unit Award, Air Force Meritorious Unit Award, Surface Warfare Badge. He achieved the rank of CAPT and retired July 1, 1985.

His memorable experiences: Never a dull moment. He enjoyed it all. Particularly the surface warfare and command and control.

He married Mary Irene and she died Sept. 2, 1996. His children are Cheryl, Thomas C. Jr. (deceased), David E. and Laura Ann; and grandchildren, Thomas C. III and Jeffrey Sean.

Barker retired as Dean at the university, July 1, 1996. Now serving as Dean Emeritus and Professor Emeritus of Health Administration, Gerontology and Community and Public Health of the Medical College of Virginia, Health Sciences Division of Virginia Commonwealth University. Retired from USNR July 1, 1985.

CHARLES B. "CHARLIE" BARNETT, CAPT, born April 3, 1926, Greer, SC. Attended Furman University, BA, 1948; Medical College of Virginia, DDS, 1956. In November 1943 he enlisted in V-5, V-12 programs and took boot camp at Great Lakes, February-March 1945; Lake Union, Seattle, Fire Control School, March-September 1945.

He served in part of the Occupation Forces in the early

days after the surrender on V-J Day Aug. 14, 1945. Served in the USS *Southerland* (DD-743) as part of a four destroyer squadron with small aircraft carrier USS *Hancock*. In December 1945 they sailed down to Guam, Tinian and Saipan.

On Tinian he saw the large fleet of B-29 bombers that had leveled Japan. Their ship was disabled with engine trouble so was detached from the squadron. They set sail for Pearl Harbor via Kwajalein. From Pearl they sailed to San Diego and he was later discharged in New Orleans.

Re-enlisted in the USNR in June 1948, discharged in July 1952 and joined the USNR Dental Corps in 1963 as LCDR. He rose to rank of CAPt. Started up RedCom 07 in Charleston, 1979, as dental staff officer. Served as CO of Reserve Unit for NDC1, Charleston; NDC1 Camp Lejeune; NDC1 Parris Island. He attended all the Naval Reserve Dental Seminars, post graduate training seminars at Naval Dental School, Bethsada, MD. He was selected to sit on two Promotion Selection Boards. Retired April 1986 after 32 years of satisfactory service.

Received the Naval Reserve, Armed Service, Occupation, South Pacific and WWII Victory Medal.

Married Elizabeth Simpson Feb. 1, 1950. They have two children, Elizabeth Jane (Blackstone) and William Charles, and one grandchild, Hannah E. Blackstone. Retired August 1994 from private practice of general dentistry.

WILLIAM S. "BILL" BARRACK JR., CDR, born July 26, 1929, Pittsburgh, PA, received BS in engineering, University of Pittsburgh, Graduate work, Dartmouth.

Direct commission Dec. 28, 1950. His military locations and stations: Postgraduate School, Monterey, CA. He served with the following units: Naval Post Graduate School, Monterey, CA; Reserve Fleet, Charleston, SC; Com DesLant, Newport, RI; US Atlantic Fleet; US 6th Fleet; Naval Fuel Supply Department, Washington, DC; COMNAV Europe, London, England; US European Command, West Germany; US Naval War College, Newport, RI; MSTS Washington, DC.

Barrack was awarded the National Defense Medal, European Occupation, Armed Forces Reserve and Naval Reserve Association. He achieved the rank of CDR and was discharged June 10, 1953. His memorable experiences: first NATO Naval exercises North Sea/Arctic October 1952; RCM Operations, Baltic Sea, 1952.

He married Evelyn Irene (Ball) and they have children: William P. and Elizabeth Irene (Howells IV), and grandchildren: William W., John N. and Jennifer W. Barrack; Alexandra I. Howells.

His civilian employment includes: Chairman and CEO, Texaco LTD; Sr. vice-president and director, Caltex Petroleum Corp., Texaco, Inc.; director, Consolidated Natural Gas Co.; director, Standard Commercial Corp.; director, International Executive Service Corp.; director, Foreign Policy Association; director, The Mary Rose Society, England; Board of Visitors, University of Pittsburgh. Retired July 26, 1989.

EDWARD L. "NED" BARRETT JR., CAPT, born Aug. 18, 1941, Lynchburg, VA, received BS degree from Virginia Tech.

He joined the service November 1962 and was stationed in San Diego and Long Beach, CA, serving with USS *Tulare* (AKA-112); USS *Reaper* (MSO-467). He is a Vietnam Veteran.

Barrett was awarded the National Defense, A/F Expeditionary Medal, Vietnam Service Medal, AF Reserve Medal, Vietnam Campaign Medal and Rifle Expert Pistol.

He achieved the rank of CAPT and was discharged June 1967. His memorable experiences: 3 WESTPAC cruises.

He is now now divorced and has three children: Edward III, William Daniel and Mary Melissa and three grandchildren: Austin III, Daniel and Laura Rose. He was employed at BP Oil Co. and retired Sept. 30, 1992.

STEPHEN D. BARRETT, CAPT, was raised in Dallas, TX. He received a BS degree in zoology from the University of Oklahoma in 1968.

Commissioned at OCS, Newport, RI in May 1969, he served as 1LT on USS *Eversole* (DD-789), home ported in Long Beach, CA and ASW officer on USS *Worden* (CG-18), then (DLG-18), forwarded deployed from Yokosuka, Japan.

Released from active duty in December 1972, he attended Texas A&M University. He was subsequently employed by the Texas Parks and Wildlife Department, serving as a superintendent and law enforcement officer. He affiliated with the Selected Reserve serving as XO of an ARS augmentation unit and CO of a DE augmentation unit.

He was recalled to active duty in July 1976 in the Training and Administration of Reserve (TAR) program. He attended Surface Warfare Officer Department Head School and in 1977 was assigned as weapons officer on USS *Joseph Hewes* (FF-1078).

He next commanded, from 1981-83, Naval and Marine Corps Reserve Center Peoria and three other Naval Reserve Centers in Illinois and Iowa, followed by a two year tour as current operation officer on the staff of commander Second Fleet. His next assignment was XO of USS *Coronado* (AFG-11), Middle East Force flagship and later Commander Amphibious Gp Two flagships.

Transferred in 1985 to Minneapolis, MN, he served for one year as manpower officer for Naval Reserve Readiness Command Region Sixteen and for an additional year as chief staff officer. He was subsequently assigned in London from 1987-90 as Reserve Plans coordinator on the staff of Commander in Chief, US Naval Force Europe, where he was selected to his present rank.

He served as Atlantic Representative, Commander Naval Surface Reserve Force from June 1990-September 1993. In October 1993 he returned to Minneapolis and relieved as commander, Naval Reserve Readiness Command Region 16. He reported to the Bureau of Naval Personnel as Surface TAR Officer Detailer in October 1995. He assumed command of Naval and Marine Corps Reserve Center, WA on Sept. 28, 1996. His awards include the Legion of Merit, Meritorious Service Medal (four awards), the Navy Commendation Medal and other service awards.

He is married to the former Joyce Sachse of Seminole, OK. They have a son David and two daughters, Erin and Carrie, who live in Austin, TX.

CAMERON A. BARRON, CAPT, born May 20, 1935, in Jackson, MS. After graduating from Lanier High School, he matriculated at Tennessee State University. He majored in chemistry and received his BS degree in 1956.

In September of 1956, he took his military oath in

Memphis and entered the Aviation Officer Candidate Program in Pensacola, FL, where he received his commission. In 1958 he received his wings of gold and was off to see the world.

Stationed at NAS Agana, Guam; NAS Pax River; NAF Washington, DC, and served with VW-2, VW-3; AEWTUIANT, VP-68, NISHQ. In 1972, he switched from the aviation community into the Intelligence community and was assigned a 1635 designator. For eight years, he served as an information officer for the USNA. After completing many Naval seminars, schools, symposiums and graduate work at the University of Maryland, he retired from Naval service in 1991.

His career included assignment overseas and CONUS. Significant assignments included: XO of the Naval Investigative Service HQ; CO NIS Unit, Baltimore; and CO Naval Intelligence Processing Unit, Suitland, MD.

He married Betty Rivers and they have children: Yvonne E., Adrienne D., Cameron II, Kathleen A. and Erica A.; and a grandchild, Taylor A. Barron. He was employed as special agent with the Federal Government, retiring Aug. 31, 1991. He is a member of many professional and civic organizations, including the Omega Psi Phi fraternity, ROA, NRA, NNOA, a certified lay speaker and a 33 degree mason.

GARY R. BARRON, LT, born March 30, 1958 in San Pedro, CA. He received an AAS in aeronautics from Long Beach City College in 1980, a BS in business computer methods from California State University, Long Beach in 1982 and a MBA from University of Phoenix in 1985. In June 1988 he received a direct commission as an ensign in the USNR and was designated an intelligence officer.

Upon commissioning, Barron affiliated with Reserve Intelligence Area FOUR (RIA4) at NAVAIRES San Diego and was assigned to Fleet Intelligence Rapid Support Team Pacific 0919 (FIRSTPAC 0919) as an intelligence analyst in geo-political and special warfare studies. In November of 1990, he was selected to become the first air intelligence officer for Tactical Air Control Sqdn 1294 (TACRON 1294). In November 1993 he joined COMNAVSURFPAC 0994 where he is currently drilling as an amphibious intelligence analyst, assistant training officer, and OIC of the COMPHIBRON 3 DET.

Attending the Direct Commission Officer Course at NAS Pensacola, he graduated top in his class and received an outstanding student award from the Naval Aviation Schools Command. He also attended Basic Intelligence Training Course (BITC) and Image Interpretation Course (IIC) at FITCPAC San Diego, graduating top in his class with honors for both courses. In July 1994 LT Barron graduated with honors from the Basic Naval Operational Intelligence Course (BNOC) at NMITC in Dam Neck, VA. He has participated in Yama Sakura XXIII 1993 in Sendai Japan, Cobra Gold 95 in Korat Thailand, FLEETEX 92-1M and PAC JTFEX 96-1 aboard the USS *Tarawa* off the coast of Southern California.

In January 1989, he joined the Naval Reserve Association (NRA), as a life member. He is currently president of the 11th District and past; National VP of Junior Officer Affairs, National VP Education and Chapter Activities, president of the South Bay Chapter, VP Junior Officer Affairs for the NRA 11th District, and past VP Budget & Finance/Treasurer of the South Bay Chapter. In 1990 he was the recipient of the 11th District Outstanding Junior Officer Award, and in 1992 he was selected as the Naval Reserve Association's Naval Reserve Junior Officer of the Year (Selected Reserve). He is also a life member of the Naval Order of the US and CDR of the Long Beach Commandry, life member of the US Naval Institute, Navy league of the US, Naval Intelligence Professionals, Reserve Officer Association, and a member of the USS Arizona Reunion Association.

His decorations include the Navy Achievement Medal, National Defense Service Medal, Expert Rifleman Medal and the Expert Pistol Shot Medal.

He is married to the former Pamela J. Baune (a chiropractor) of San Pedro, where they live with their nine year old daughter Jessica and their six year old son Jonathan. He is presently employed as a senior manager at McDonnell Douglas in Long Beach, CA, working as the C-9D program manager leading the McDonnell Douglas efforts to replace the USN C-9B Fleet Logistic Support Aircraft.

JOHN G. BARSNESS, CDR, born Dec. 21, 1923, Black Earth, Dane County, WI, received a BS degree and Dr. of Law degree from University of Wisconsin. He joined the service Dec. 7, 1942. His military locations and stations include: Pensacola, FL; Jacksonville, FL; Glenview, IL NAS. Served with naval aviation TRA Commands and various VP-VR-VS Sqdns.

He was designated naval aviator May 10, 1945. He achieved the rank of CDR and transferred to Reserve status, Jan. 18, 1946.

Barsness is married to Nancy and they have children: Gregory (naval aviator), Christopher, Zoe and Amy. His grandchildren are Greer, Booth, Jackson and Thor. He was employed as lawyer, Barsness Law Offices since 1951 and retired Dec. 21, 1983.

JOHN DONNINGTON BARTLETT, CAPT, born in Bozeman, MT, Jan. 24, 1932, the son of LTC John D. Bartlett, USAF and Rea Anceney Bartlett. His career in the USN began after high school (1951) during the Korean War by enlisting and serving on the USS *Consolation* (AH-15) for three years as a striker operating room technician. After four years he completed his education at the University of Michigan receiving a BS degree in 1959, an MD degree in 1963 and surgical residency in 1968. He then served on the teaching staff at the University of Michigan becoming an assistant professor of surgery.

His research interests included catechol amine physiology and pharmocology, and he set up and directed the Catechol Amine Laboratory studying patients with Pheochromocytoma. Throughout his medical career, he has published 23 surgical papers and co-authored one chapter in a surgical text. He then entered private practice as assistant chairman of surgery at the Lovelace Clinic, Albuquerque, New Mexico, and was instrumental in establishing surgical residency affiliation with the University of New Mexico. His practice centered around general surgery, endocrine surgery, and peripheral vascular surgery. While there he was appointed clinical associate professor of surgery at the University of New Mexico. After seven years at Lovelace Clinic, he then entered solo private practice for seven years in Butte, MT.

His interest in teaching prompted reentry into the USN June 1984 as a CDR. He since has been stationed at the Naval Hospital, Oakland, rising to become chairman, Department of Surgery, program director, director of surgical services and specialty advisor to the surgeon general in general surgery. Bartlett has been instrumental in establishing full integration of his residency program with Dr. Claude H. Organ Jr., University of California, Davis East-Bay as of July 1991. He was appointed clinical associate professor of surgery, USUHS and clinical professor of surgery, University of California, Davis.

As a military surgeon, he served seven months as the head of general surgery on the USNS *Mercy* (TAH 19) during operation Desert Shield and Storm. He was voluntarily released from active duty January 1993 and reentered the active Naval Reserve. He is a fellow of the American College of Surgeons, a member of the Frederick A. Coller Surgical Society, and the Western Surgical Association, the Association of Military Surgeons of the US and the San Francisco Surgical Society.

His military awards include the Navy Commendation, Navy Unit Commendation, Meritorious Unit Commendation, Good Conduct, Southwestern Asia (two stars), National Defense (one star), Korean Service (three stars), Humanitarian Service, United Nations, Kuwait Liberation w/Palm, Republic of Korea Presidential and Vietnam Presidential Unit Citation.

He and his wife, Joanne, have been married 39 years and have one son John and three daughters, Deborah, Jennifer and Wendy.

ROBERT I. BARTO, CAPT, born Feb. 2, 1917, in Seattle. He was a graduate of the Univ. of Washinton and graduated as a naval aviator at NAS Corpus Christi in 1942. He was assigned to the Aircraft Delivery Unit, NAS, Terminal Island, CA, as an acceptance, test and ferry pilot.

In 1943 he went to the Texas American Airlines (Navy Contract) School and then for two years was administrative pilot, ferry pilot, test pilot, and flight safety officer for VRF-3 at NAS Terminal Island, CA. From 1945-58 he served as a naval aviator at NAS Seattle and had various billets in squadrons, including CO of FasRon 896.

In 1958-59 he was a unit instructor at NAS Seattle and then became assistant air wing commander (AW-89M), administrative. Selected CAPT, and assigned as a member of the CAPT's Reserve Advisory Board.

CAPT Barto has about 4,000 hours of Navy flight time and 500 civilian flight hours. He has flown more than 80 types of military and civilian aircraft and has a commercial pilot's license with three ratings.

Delegates to NRA's National Conference unanimously elected Barto as National President. CAPT Barto had been NRA president for Membership the previous year and before that was president of the 13th Naval District. He has been a NRA member since 1955 and active in Seattle's Puget Sound Chapter.

He was a partner with his brother, a retired USNR LCDR in a general finance firm. In addition, he was an officer in three local finance corporations and a partner in two small insurance agencies.

The Barto family consists of wife Meg, son Robert J. "Bart" (now in the USN), daughter Bonnie and her husband Brian. They reside in Seattle, WA. He enjoys model railroading and stamp collecting.

STEPHEN C. "STEVE" BARTO, CDR, born Nov. 24, 1958 in Cincinnati, OH, received BA in accounting, 1980 Michigan State University, received MBA, 1989, Darden Graduate School of Business, University of Virginia.

Joined May 1, 1981. Commissioned an ensign Oct. 2, 1981, Aviation Officer Candidate School, Pensacola, FL. Graduated from Armed Forces Air Intelligence Training Center, Denver, CO, March 26, 1982. Assigned to USS *John F. Kennedy* (CV-67), May 1982 to January 1985; multi-sensor interpretation officer in the Carrier Intelligence Center (CVIC); Ship's MWR Fund Custodian. Managed tactical reconnaissance exploitation and reporting of Syrian and other deployed militias in Lebanon, in support of Multi-National Peacekeeping Forces in Beirut including USMC.

Assigned to Defense Intelligence Agency (DIA), January 1985 to July 1987; National Photographic Interpretation Center (NPIC); Iran/Iraq War analyst and Arabian Peninsula Country Military analyst.

Recalled to active duty Operation Desert Shield/ Storm, DIA September 1990 to August 1991; imagery/ Intelligence dissemination policy from the Pentagon.

Barto was awarded the Joint Service Commendation Medal, Navy Commendation Medal, Joint Service Achievement Medal. Selected unit's NRA Junior Officer of the Year 1995. Selected for the rank of commander, June 1996. Date of rank April 1, 1997.

He is married to the former Judith Marie Norman (Judy). They have no children. He is employed as financial consultant, Merrill Lynch in Alexandria, VA. Former president Board of Directors of the Clarendon Alliance, a local community development organization in Arlington, VA. Treasurer, Beirut Veterans of America.

His memorable experiences: managing exploitation of TARPS imagery of Lebanon in support of peacekeeping forces and US Marines in Beirut in 1983-84, and providing support for retaliation air strike was exhilirating highlight of his sea duty.

EDWARD ALBERT BASDEKIAN, CAPT, born Nov. 21, 1928, Georgetown, Washington, DC. Graduated Florida Southern College, University Florida, Central Michigan University. Military Education: Air War College (2070th USAR School); NavWarCol (nonres. seminars); ICAF; NROS. November 3, 1949 received intersvc trans. from USAR (enl. USAR, Aug. 25, 1947).

His military locations and stations: NOB, NORVA, NAAS Oceana, VA; NAS Jax, FL; R.R., P.R., NAS ROTA, SP; RVN. Served with FASRON-3; VF-42; 1st 2nd 3rd MARDIV; VP742; VR60F1; AWS74; COMNAVFORV (MAAG). His battles and campaigns: Rep. Vietnam (MAAG) April 1972 camp.

He was awarded the MSM, NCM, Good Conduct, Combat Ribbon, Navy Unit Commendation, NavResMeritSvc, Navy Occupation (Europe), NDSM, RVN SVC ArmedForRes, RVN Gallentry Cross w/palm; CNO CentAward.

Basdekian achieved the rank of CAPT. His memorable experience: recom. of VF42; advisor, USAID/LAOS; FLTSUPOFF, Athens, GR; 6th FLT-USS *Saipan* (CVL-48); USS *Newport News* (CA-148) ProjOff, USS *JFK* (CVA-67); Dep. SpecAsst to SGN for Reserve Affairs and Dir. TemAcDuProgMedical 1979-85.

He is married to Alice Louise Thompson, and he is an assistant dean and professor of pharmacy (emeritus).

L.A. "BASE" BASINGER, LT, born April 18, 1915, Putnam County, OH, received BS in education; MSC in education, OSU; graduate work, Colorado State; USN Indoc. Princeton General Line School, Hollywood, FL.

He began active duty Aug. 15, 1944. His military location and stations: USN NAV TR Station, Princeton, General Line School, FL.

Basinger served with USS *Oglethorpe* after amphibious training, Ft. Pierce, FL. He fought battles in the American Theater and Asiatic-Pacific Theatre. Awarded the American Theater Medal, Asiatic-Pacific Medal and the WWII Victory Medal, he achieved the rank of LT and was discharged January 1946.

His memorable experiences: Helping put ship in commission, NYC; training ship Newport, RI; visiting Pearl Harbor, Kwajalein, Eniwetok, Guadalcanal, New Caledonia and Ellis Islands.

He married Patty Ann and they have children: Mrs. Kay Evans, Mrs. Ray Merry and Mrs. Thomas Oates. Basinger was Vo-Ag teacher for 16 years. County agent for Ohio State University for six years, rank professor. Number three man in a grain marketing facility, and five farm equipment lines. In present location he has been president of local Chamber of Commerce, president of Rotary Club, president of Colorado Cancer Society, chair of a River Conservancy District, president of Colorado Auto Club, Bible teacher; past commander of American Legion Post 198, Department of Ohio; Judge Advocate of Richard L. Basinger Amvets Post 1994 (named after his son, killed in Vietnam May 12, 1967). He has been semi-retired since approximately 1980.

JAMES A. BATEMAN, CDR, born July 16, 1946 in Baltimore, MD. He graduated from Georgia Institute of Technology in June 1968 with a bachelor of aerospace engineering degree and received his commission from the Georgia Tech NROTC unit. He completed his MS in Aerospace Systems Engineering in 1969 and received his aviator wings in January 1971. His first duty assignment was VX-1 at NAS Key West, FL where he was an analyst

and system test pilot for several ASW programs such as LAMPS MK-1, S2-G and P3-C. Trained as a helicopter pilot and qualified as aircraft commander in the SH-3, CDR. Bateman transferred to HS-15 at NAS Lakehurst, NJ and flew from the USS *Guam* (LPH) and the USS *Wainwright* (DLG) in development of the Sea Control concept.

In 1976 he was designated USNR-TAR and was assigned to NARU Alameda, CA, as the O-in-C of the ASW helo augment units HS-128 and 287. He developed with his units and HS-85 a resume of San Francisco Bay Area Helo Landing Sites as well as shipboard landing training for the area's reserve forces. He deployed with HS-85 aboard the HMCS Provider. While on RIMPAC 1978 exercises, he would have had to ditch his helicopter had the HMNZS Waikato not come to the rescue with a landing deck, albeit, too small for the proper landing of an SH-3. He executed a precarious rearward diagonal landing in very rough seas at dusk and received a commendation from COMHELWINGRES. A photograph of his helo on the Waikato made the New Zealand newspapers the next day and was published in *Naval Aviation News* with the caption "Elephant on a Matchbox."

In 1977 he augmented to the regular Navy and was transferred to HS-9 at NAS Jacksonville, FL for his department head tour. He made deployments aboard USS *America* and USS *Saratoga* (CVAs) and USS *Nimitz* (CVN) as well as detachments on the USS *Albany* (CG). In 1982 he was stationed at COMNAVAIRLANT as the helo training OPS and safety desks. His next tour of duty was with NAVAIRSYSCOM where he was division head of all air ASW and helo training systems program managers (S-3, P-3, H-3, H-2, V-22, H-1, H-46, H-53) and also was responsible for the procurement of the SH-60F training system. He also received a designation of Material Professional (MP) and moved to the Office of the ASST SECNAV Shipbuilding and Logistics, as deputy for the Director of Reliability, Maintainability, and Quality Assurance. While with ASN he was program manager for the Government Industry Data Exchange Program (GIDEP) where he was instrumental in upgrading all the Department of Defense's contractual data collection/dissemination from manual/microfilm/fiche to computer.

During his 22 years active duty he was honored twice as the East Coast Helo Pilot of The Year and twice as the Naval Reserve Helo Pilot of the Year. He also received the Meritorious Service Medal (2), Navy Achievement Medal, Navy Commendation Medal, Expert Rifle and Pistol, National Service and Armed Forces (Navy) Reserve Medal.

Bateman retired June 1990 to Fairvax, VA with his wife Elizabeth and two sons, Brandon and John. He has since started his own company, Bateman Custom Construction, engaged in residential and light commercial construction and remodeling.

JOHN WILLIAM BATEMAN, LCDR, born May 31, 1919, Baltimore, MD, received BE of engineering from Johns Hopkins University.

He joined the service January 1942. His military locations and stations: Anacostia, Kanehoe, SoPac, Alameda, Moffett, Floyd Bennett. He served with the following units: USNRAB, VP24, VP14, HEDRON2, HEDRON8, VR5, VR834.

Bateman fought battles and campaigns in Solomon Islands, Rabaul, NG, SoPac (1943-44) and was awarded American, Pacific Theaters Medals, WWII Victory Medal and Naval Reserve Medal. He achieved the rank of LCDR and retired May 31, 1979.

He married Patricia Albers and they have children: James, Douglas and Diana B. Lacke; and have grandchildren: Brandon, John, Michael, Stephanie and Gregory. Employed as draftsman with Martin Aircraft and Pan Am as captain, retired May 31, 1979.

THOMAS R. BATES, MD, born Dec. 19, 1935, in Mansfield, OH. Graduated from the Ohio State University in 1957 and 1960 with a BA and MD degree. He began his Navy career in 1962. After completing aviation medicine in

Class 100, was flight surgeon for VMF 335 El Toro, CA. He spent one year as the medical officer in Antartica, at McMurdo Sound, where he was involved in the first mid-winter rescue mission, subsequently, medical officer Christchurch, New Zealand.

He began residency in ophthalmology, Philadelphia Naval Hospital July 1965. July 1, 1968, organized the eye clinic at newly commissioned Naval hospital, Orlando, FL. Discharged from active duty, June 30, 1971. Remained in the Winter Park/Orlando area. Remained in the USNR and is currently on high year tenure.

Overseas Reserve duty tours include: Eluthera, Bahamas; Yokosuka, Japan; Naples, Italy; Antiqua, West Indies; Roosevelt Roads, and Okinawa, Japan. Served as a Blue Gold Officer for the Naval Academy. He taught residents and performed patient care on many occasions at the National Naval Medical Center in Bethesda, MD and at the Tripler Army Hospital. Representing the Reserve Forces, he served with the chiefs of ophthalmology of the Navy, Army, and AF, as part of a blue ribbon advisory panel, advising the surgeon general on equipment needs of field medical systems.

Currently active with the Bethesda 506 Support Unit spending much of his reserve time teaching at Bethesda.

Dr. Bates' private practice is in Orlando, FL, and he is an associate clinical professor of ophthalmology at University of South Florida.

He recently received the James Clower, MD Community Service Award, given by the Florida Society of Ophthalmology to a member who makes extraordinary contributions to his community and to patient care. The award was given him with the following notation:

During his nearly 30 years practicing ophthalmology, this doctor has been a leader wherever he has been. He helped found both the Central Florida Affiliate of the National Society for the Prevention of Blindness and the Medical Eye Bank of Central Florida. He has been chairman of the ophthalmology departments of two major central Florida hospitals, and was Mid-Florida Regional Chairman of the USF Eye Institute campaign for funds in 1984. He has been a president of the Florida Society of Ophthalmology and the Central Florida Society of Ophthalmology, and a member of the American Academy of Ophthalmology's Board of Councillors. Yet it is for his community service that he earns this award.

His church services include serving on the Vestry of his local Episcopal church on numerous occasions and he has been senior warden on three occasions. He is a sub deacon and Lay Eucharistic Minister at the Cathedral Church of St. Luke, and he has been a chorister for 27 years. At the Diocesan level, he has been on the Diocesan Board for two, three-year terms and is currently completing the third year of his second term as a member of the Standing Committee. He was a member of the Bishop's Search Committee and also has served as a member of a Rector's Search Committee. On the national church level, he serves on the Commission on Health and is the treasurer. He will be a deputy to the general convention of the Episcopal Church in 1997, having also served as a deputy in 1988, 1991, and 1994. During the convention, he was on the Social (Medical) and Urban Affairs Committee.

Currently, he is serving on the medical board of the Orlando Coalition of the Homeless. He is vice-president of SHARE - The Self Help and Resource Exchange. He also serves on the Board of Trustees of Vision Health International, which annually sends volunteers to poor countries to provide eye care in under-served areas. He has made five trips to Costa Rica and two trips to Poland.

He currently resides in Orlando, FL, with his wife Helen. They have six children and 10 grandchildren.

WAYNE L. BATES, CAPT, born Oct. 14, 1947, Rome, NY, received BA (history), St. Lawrence University, Canton, NY, 1969.

He entered OCS July 19, 1969. His military locations and stations include NOB, Norfolk; USNAB LCRK; NAVSTA Key West. Units served with: UDT-22, UDT-21, Naval School Underwater Swimmers and Seal 2.

Bates was awarded the Naval Parachutist Medal, Seal Insignia, Expert Rifle and Pistol. He achieved the rank of CAPT, USNR-R (present rank), leaving active duty Aug. 30, 1974.

His memorable experiences: completed underwater demolition team replacement training at USNAB Little Creek, Dec. 18, 1970; competed on military reserve officer pentathlon team in 1975.

He is married to Dianna and they have children, Jennifer and Douglas. Bates has been employed as special agent with the US Treasury Department, Orlando, FL, since 1977.

DR. JAMES W. BATTEN, a native of Micro, NC attended the University of North Carolina, earning the AB, MA and Doctorate. He began his Naval career in 1941, attended Midshipman's School and was commissioned ensign in 1942.

During WWII, he served with the 3rd Inf. Div. as Naval gunfire liaison officer for the North African Campaign in November 1942. Subsequently, he was assigned to the construction of a new fleet oiler, the USS *Schuylkill* (AO-76) and assembled the crew. From April 1943 to March 1946, he served on board as division officer, educational officer, gunnery officer, ship's service officer, navigator, and executive officer.

The USS *Schuylkill* crossed the equator 32 times and the 180th meridian six times. She participated in the campaigns for the Aleutians, Gilbert Islands, Marshall Islands, Marcus, Wake, Marianas, Western Carolines, Formosa, Luzon, Lingayen Gulf and Iwo Jima. She was a part of the 3rd Fleet that entered Tokyo Bay, Sept. 5, 1945. The *Schuylkill* had considerable armament for antiaircraft, surface ships and submarines and had special equipment for fueling ships and transferring mail, passengers, and freight at sea. During WWII 706 ships were fueled during the many Pacific campaigns. Dr. Batten was authorized to wear the service medals for the American, European and Pacific Theatres, WWII Medal, Naval Reserve, Philippine Liberation and his ship received Commendation from the Commander of the Third Fleet, Admiral W.F. Halsey. Dr. Batten remained in the USNR until he was approved for transfer to the Retired Reserve.

He is married to Sara S. and is professor emeritus, East Carolina University.

KENNETH H. "KEN" BAUER, CAPT, born Feb. 20, 1938, Corder, MO, received BA physics; math; MS physics; MBA.

He joined the service March 9, 1961. His military locations and stations: San Diego, Norfolk, Vietnam, Pt. Mugu, Washington, DC. Served with USS *Alamo,* USS *Boxer,* River Patrol Flotillas, Air Systems Command.

Bauer was awarded Vietnam Campaign, Vietnam Meritorious Unit Citation, Meritorious Service Medal, Navy Commendation Medal, and Presidential Unit Citation.

He achieved the rank of CAPT, retiring Aug. 31, 1994. His memorable experiences: US Naval Post Graduate School; chief explosive safety inspector for Naval Sea Systems Command; inspector general support for Naval Air Systems Command.

He married Jo Ann and they have a child Chris and grandchildren: Nicole, Crystal and Kenny.

Employed with the US Department of Energy, program manager for Weapons and Technology; Naval technical representative for A-GE Tram at Hughes Aircraft Company.

DALE N. BAUMANN, LT, born Sept. 4, 1941, St. Louis, MO. He is a high school graduate. Joined the USAR, February 1959; USNR, March 1980.

His military locations and stations: Fort Leonard Wood, NRNAS Memphis. Served with the following units: 102nd Inf. Div., VR-53, VR-60, MMFG, NR NAS Cubi Point, NR NAF Misawa.

Baumann was awarded the Army Commendation, Navy Achievement, Naval Reserve, Meritorious Service, National Defense, Armed Force Reserve (2).

He is presently LT, USNR-R and was discharged from the Army February 1967 and the Navy March 1989. His memorable experiences: 1982 Rota Det with VR-53 on C-118 as an aircrewman on a VR-60DC-9, his selection as an LDO, and standing up the contributory support program at NAR/NRRC Memphis.

Married to Joyce and they have three children: Michael, Timothy and Mark. He was employed with Union Electric Co., as maintenance and construction supervisor.

HUGH W. BAUMGARDNER, CAPT, born in Akron, OH and spent his youth in Chillicothe, OH. Following graduation from Marietta College in Ohio, he attended Aviation Officer Candidate training in Pensacola, FL and was commissioned an ensign in 1963.

His first tour was with the Navy hurricane hunter squadron, VW-4, based in Puerto Rico where he was designated as a naval flight officer and collateral duty intelligence officer. In 1967 he was assigned to RCVW-4 and later to LATWING ONE located in Jacksonville, FL before leaving active duty. He affiliated with the USNR in 1970 and was redesignated as a special duty (intelligence) officer. Subsequently, he served as an air intelligence officer with RTU-62, VP-1407, and VP-62 until 1977 when he was assigned to DIAHQ 0408. In 1979 he was recalled to active duty at the Naval Intelligence Command in Washington, DC and, upon completion of active duty, was assigned to DEFATTACHE 0166. In 1981 he was assigned as XO and later, in 1983, as CO of the National Military Intelligence Center Naval Reserve Unit, DIACURRENTINTEL 0166, in the Pentagon. In 1986 he was assigned as CO of the USNR headquarters unit of the Naval Investigative Service, NISHQ 0166, and in 1989 as XO of the Defense Technology Security Administration Naval Reserve Unit, OSD Tech Transfer 0166, within the office of the Secretary of Defense. In 1990 he was recalled to active duty during Operation Desert Shield/Storm and served as deputy assistant to the Director of Naval Reserve for Legislation, Liaison and Information.

He retired in 1993 with 30 years of commissioned service and is authorized to wear the Meritorious Service Medal, Joint Commendation Medal, Navy Commendation Medal, Joint Unit Commendation (second award), National Defense Medal (second award), Armed Forces Reserve Medal (second award), Navy Overseas Service Ribbon, and Battle Efficiency Ribbon.

Baumgardner was awarded a JD degree from George Mason University School of Law in 1980 and is a member of the Virginia Bar. He lives with his wife, the former Molly Allison Rahe, a CDR in the USNR, and son Mark in Arlington, VA.

LOREN L. "DUTCH" BAUMGARDNER, LT, born Aug. 23, 1922, Springfield, OH, received BSC and BA from the Ohio State University. He joined the service Nov. 4, 1942. His military locations and stations include: naval aviation cadet, commissioned Nov. 1, 1944, Pensacola, FL, serving with VF-44.

Baumgardner achieved the rank of LT USNR and is retired. His memorable experiences: training as a naval aviation cadet; flying an F6F hellcat with VF44 and as a "weekend warrior" at Port Columbus, OH.

He is divorced and has children: Dane, Holly and Lori, and grandchildren: Danica, Paige, Naomi, Joseph, Brent, Carlee, Layne and Taryn and great-grandchild Brionna. He was employed as real estate broker and certified property manager. Retired from the Navy Aug. 23, 1982 and property management, 1994.

MANUEL F. BAUTISTA, CAPT, an honors student and valedictorian was a 1965 graduate of the College of Medicine, University of Santo Tomas. He was the vice-president of the UST-IMSA and president of his interns class at the Veterans Hospital. He joined Booth Hospital, Flushing, NY, July 1968 and later Essex County Hospital Center, Cedar Grove, NJ. He established the first Behaviour Modification Program and became the coordinator for hospital's CO-ED program. He joined the USN as a senior medical officer at NTTC Pensacola, FL and NAVCAMSTA, Guam, MI. He enjoyed the closeness of the community second only to the E&ML enabling him to go around Asia and the Philippines.

He left active duty (May 1977-November 1982) and plunged into his psychiatry residency with SIU, Springfield, IL and WMHI OshKosh, WI. He moved to Lakewood, WA and became consultant with COMPHEALTH, Salt Lake, UT and VCMHC, Auburn, WA. He's presently chief of the medical staff, Western State Hospital and staff psychiatrist WA DSHS. He has privileges with hospitals of the Franciscan Health Systems Northwest. He is attached with FH CBTZ NINE where he was formerly head Department of Psychiatry, but is now the SMO of FH NINE DET K in Tacoma, WA.

The Bautista family includes wife Jazmin, two sons, Giancarlo, a combat systems engineer with USAF, Kirtland AFB, Albuquerque, NM; and Robin, his 14 year old tonic of a man. He is a member of the Elks Club, Knights of Columbus and former Board Member, GLMHC. In a hectic but enjoyable professional life, he maintains a practice in psychiatry and his ties to the community as guest speaker if not adviser to various functions of the Filam League, PNIA and UIPS, (local cultural organizations) both in his capacity as parent/physician and above all a US Naval medical officer.

JON W. BAYLESS, CDR, is a native of Dallas, TX. He was commissioned in May 1979 from the USNA where he received a BS degree in systems engineering. Following commissioning and while awaiting flight training, he served six months at the Naval Plant Representative Office (NAVPRO) for Douglas Aircraft in Long Beach, CA where he had the opportunity to watch the last A-4 Sky Hawk and the first KC-10 roll off the assembly line. He was designated a Naval aviator in April of 1981 and completed fleet replacement training at Patrol Sqdn. 30 (VP-30) in November of that same year.

From December 1981 to December 1984, he was assigned to Patrol Sqdn. 16 (VP-16) in Jacksonville, FL. During this tour he was designated a patrol plane commander and mission commander making deployments to NAS Bermuda, NS Rota, Spain and NAS Keflavik, Iceland. The "War Eagles" were recognized with the "Battle E", the "Golden Wrench" and three Meritorious Unit Commendations during this period. In January of 1985, he reported to Training Sqdn. 27 (VT-27) in Corpus Christi, TX as an instructor pilot in the T-34C. Serving as a flight leader, he instructed basic familiarization, basic instruments, radio navigation, night navigation, precision acrobatics, and was an initial safe for solo check pilot.

Upon separation from active duty in July of 1986, Bayless affiliated with the USNR. After brief assignments with Patrol Sqdn. 91 (VP-91) and FIRSTPAC 0571, he reported to the "Executioners" of Patrol Sqdn. 93 (VP-93) in November of 1989. During this tour, he served as department head for Administration, Operations, and Maintenance. With the disestablishment of VP-93, he received orders to Patrol Sqdn. 68 (VP-68) in the spring of 1994. He served as an executive assistant and as the XO of the "Blackhawks" before reporting to Patrol Sqdn. 66 (VP-66).

He proudly serves as a life member of VP International, Reserve Officer's Association, Naval Reserve Association, Association of Naval Aviation, Naval Academy Alumni Association, and the Navy League of the US. He is employed as an Airbus 320 First Officer with Northwest Airlines at the Detroit Metropolitan Airport. He is married to the former Denise Michele Weatherford of Farmington Hills, MI. They reside with their daughters Amanda, Brooke and Shannon in Commerce Township, Michigan.

KENNETH C. "KEN" BEACHLER, CAPT, born Oct. 11, 1935, Battle Creek, MI, received a BA with high honors, Michigan State University.

He joined USAR, Jan. 23, 1957-Jan. 1959; USNR commissioned Aug. 21, 1968. His military locations and stations include: USAREUR, Karlsruhe, Germany. He served with USNR: NRPAC 9-6 Detroit, OI DET 813, SACLANT 113, NAVINFO 113.

Beachler was awarded the Army Good Conduct USNR 20 years, Navy Commendation. Achieved the rank of CAPT and was discharged Oct. 11, 1995.

His memorable experiences: duty as PIO at SACLANT 113 (1987-90) was high point, especially joint exercises at NATO HQ, Brussels and CINCEASTLANT, Northwood, UK.

He is single and retired as executive director, Wharton Center for Performing Arts, Michigan State University.

JOSEPH J. "JOE" BEARD, CAPT, born Aug. 27, 1933, Winthrop, MA, received BS (electrical engineering) Tufts, J.D., Suffolk Law School; MBA Babson College; LLM (tax) Boston University; SJD Harvard.

He joined the service Aug. 8, 1951, NR Sub Div. 1-7, stationed USS *Colcaterra* (DER-390); USS *Harder* (SS-508), NROTC Brown.

Beard was awarded the National Defense Medal and Reserve Medal. He achieved the rank of CAPT. His memorable experience: under ice OPS in North Atlantic while in *Harder* (LANTSUBICE EX 59).

He is divorced and has children: Ariel, Christopher and Holly and grandchildren: Douglas, Curtis, Wendy and Andrew. Employed as professor of law, St. Johns University, School of Law. Retired December 1981.

EDWARD L. BEARDMAN, CDR, born in Sharon, PA, Feb. 16, 1907. A 1931 graduate of Carnegie Inst. of Technology, Pittsburgh and attended graduate school at Carnegie and at Case Inst. of Technology. After 10 years in the steel industry he joined the Navy Department (BuShips) as a metallurgical engineer.

He joined the USNR on Aug. 15, 1942, and was called to active duty, August 29. He was chief of the Miscellaneous Metals Section of the Office of Procurement and Material from Aug. 29, 1942 to Aug. 25, 1945. He aided the war production board and in the Navy in developing substitutes for materials in short supply through the modification of existing procedures thus shortening delivery time and opening new fields of supply.

He was chief of the Metals and Minerals Unit from Aug. 25, 1945 to July 10, 1946 and was released from active duty on Oct. 26, 1946. He was retired from Navy on March 1, 1967.

GEORGE A. BECK, CAPT, a longtime resident of Westport, CT, attended the University of Virginia, receiving his BA degree.

Ordered to the YMS 183, he transited the Panama Canal to the South Pacific, receiving three Battle Stars.

He then attended the Naval Mine Warfare School, serving as operations officer and instructor afloat.

He became executive director of Pi Lambda Phi fraternity, retiring in 1992. He was awarded the National Interfraternity Conference Gold Medal for Distinguished Service to Youth. Beck served NRA as chapter and district president, then as chairman of site selection for the past 15 years.

LEMUEL EDWARD BECK SR., CWO4, born March 28, 1938, Hannibal, MO. He joined the service June 12, 1956. His military locations and stations include: USS *Ticonderoga* (CVA-14); NAS Olathe, KS; NAS Grosse Ile, MI. Served with ships company fighting in the Formosa Straits, 1958.

Beck was awarded the Good Conduct Medal, Expeditionary Medal and National Defense. He achieved the rank of CWO4.

His memorable experiences: crossing the equator and going around the horn (South America).

He married Mary Evalyn and they have children, Lemuel Jr. and RA Renee Francis; and grandchildren: Lemuel III and Joshua, Celeste Amy and A. Moyers. Retired Jan. 1, 1989 as assistant supervisor.

PIERRE ROBERT BECKER, CDR, born April 20, 1922, Woodhaven, NY. Joined the service as a cadet in April 1941 and stationed at Ft. Schuyler, NY State Maritime Academy. Units served: SS *Executive*, USMMA, SS *William C.C. Claiborne*, SS *Bald Eagle*, SS *Marine Raven* and USS *General John Pope* (AP-110).

Duty was in in the North and South Atlantic and Pacific oceans and Indian Ocean plus numerous seas and bays within.

Memorable experience was trip to and from LeHavre, France, aboard the TransAtlantic French liner SS *DeGrasse* for a summer visit with his grandparents in La Chaux-de-Fonds Switzerland in 1928. Also memorable was making thre dives aboard a WWI USS submarine (training) in 1942; serving as chief engineer and CO afloat, plus having served on two capital ships of the USN; and serving as the only merchant marine officer aboard the CL-144 with a wardroom of 66 officers in 1952.

Married Mildred Annette Bruns Oct. 17, 1943. They have two daughters, five grandchildren and three great-grandchildren.

WALTER H. BECKHAM JR., CAPT, born April 18, 1920 in Albany, GA, received an AB, 1941, Emory University; LLB cum laude 1948, Harvard Law School; JD, 1969. He joined June 1941. His military locations and stations include: NAS, Miami, FL; Naval Supply Corps at the Harvard Graduate School of Business Administration; USS *Portland* (CA-33), Pacific Theater; USS *Sperry* (AS-12), Pacific Theatre.

Beckham served with various Pacific Fleet Units and Pacific Submarine Fleet. His battles and campaigns include Midway, June 3-6, 1942; Guadalcanal-Tulagi Landings, Aug. 7-9, 1942; Eastern Solomons (Stewart Islands), Aug. 23-25, 1942; Santa Cruz Islands, Oct. 26, 1942; capture and defense of Guadalcanal, Aug. 10, 1942-Feb. 8, 1943; Guadalcanal (Third Savo), Nov. 12-15, 1942.

He was awarded the American Defense, American Theater, Asiatic-Pacific (six Battle Stars), WWII Victory Medal, Navy Unit Commendation, Naval Reserve, Armed Forces Reserve. He achieved the rank of CAPT, Supply Corps, USNR and is now retired.

His memorable experiences include serving on Portland as USS *Yorktown* escort in Battle of Midway. Participating in third battle of Savo Island, November 1942.

He married Ethel Koger and they have children: Barbara DeLeo, Walter Beckham III and James K. Beckham; and grandchildren: Jonathan and Rebecca Rice and Justin and Blake Elizabeth Beckham.

Civilian employment and positions held: 1968-Present Of Counsel, Podhurst, Orseck, Josefsberg, Eaton, Meadow, Olin & Perwin, P.A.; 1967-82 Professor of Law, University of Miami Professor Emeritus - 1982; 1950-67, partner in law firm of Nichols, Gaither, Beckham, et al; 1948-49 Associate Professor of Law, University of Miami.

GERALD P. "JERRY" BECKNEL, CAPT, born Feb. 28, 1937, New Orleans, LA, received a BA in marketing. He joined the service June 23, 1955. His military locations and stations include: USS *Kenneth Whiting* (AV-14).

Served active duty 1956-58; CO AD319, Bakersfield, CO 919 ORD San Bernardino, and others. His memorable experiences include getting ensign commission through integration program on June 30, 1965 from rate of YN1.

Becknel was awarded the Armed Forces Medal (3 awds) Award and Navy Commendation Medal (2 awds). He achieved the rank of CAPT and retired Jan. 1, 1992.

He is married to Rae and they have a son, Gerald P. Becknel Jr. and grandchildren: Jayson, Zachary and Elyse. He will retire Feb. 17, 1999 as sales executive from Equifax.

PETER H. "PETE" BECKWITH, RADM, born Sept. 8, 1939, Battle Creek, MI, received AB and ThD (honoris causa) Hillsdale College; M. Div University of the South; S.T.M. & L.H.D. (honoris causa) Nashotah House.

He entered the service Sept. 1, 1972. His military locations and stations: NAVRESCEN, Saginaw; NAVRESCEN, Fort Custer; NMCRC, Columbus. Served with FRSA Log 113; 4 MAF REL 405; COMNAVSURFLANT Det 505; OPNAV O1H 806; NAS Oceana 0281; 4th MAW MED; RNCF; REDCOM 5; 4th FSSG; & CNRF (N1F).

Beckwith was awarded the MSM, NCM, NAM, Navy Fleet Marine Force Service Ribbon, National Defense Medal w/Bronze Star, Reserve Medal w/HG, Expert Rifle and Expert Pistol.

He married Melinda J. Foulke and they have children Peter II and Michael J. He will retire as Bishop, Diocese of Springfield, IL on Sept. 30, 1999.

ROBERT Y. BEESBURG, CAPT, born in Princeton, NJ and grew up in the Trenton, NJ area. In October 1951 he was recalled to active duty and spent two years in an anti-submarine squadron at Quonset Point, RI as an aviation electrician's mate and radar crewman. During this two year tour, he completed two carrier deployments.

After release from active duty, he attended Bucknell University, graduating in June 1958 with a BS degree in electrical engineering.

In September 1959, Beesburg received a direct commission as an aeronautical engineering duty officer (AEDO) and was assigned to the NAR at Willow Grove, PA. Over the next 22 years, he affiliated with various NAVAIR system reserve units at Willow Grove becoming the XO of Weapons Systems Unit-203 in 1980.

Upon his promotion to CAPT, in 1981 he transferred to the Naval Air Systems Command in Washington, DC to become the XO of the NAVAIR Command Center Unit (0266) and later its CO.

In April 1987 he became the CO of NAVAIR's Vendor Awareness Program Unit - 1366. In April of 1989, he was assigned to NAVSEA INDUST MOB DET 204 at the Philadelphia Navy Yard.

His awards and service medals include the Navy Commendation Medal, National Defense Medal and Naval Reserve Medal. He retired in April 1993 with 37 years of service.

In civilian life, he worked at the Naval Air Development Center, Warminster, PA over 32 years in various supervisory and managerial positions developing airborne/shipboard weapons systems. He retired from Civil Service in July 1990 and is presently employed by Navmar Applied Sciences Corporation, Warminster, PA.

He is married to the former Sylvia C. Arena of Trenton, NY. They have two children, April and Robert Jr. The Beesburgs currently reside in Yardley, PA.

JAMES W. BELCHER, LCDR, born May 28, 1924, Salisbury, MO. He attended Salisbury High School and one year at Kansas City Jr. College. He joined the service March 3, 1943 and served with VPB-17, VPB-212, BARTV 701, NASRU 701.

He was awarded the Naval Reserve Medal, American Campaign, Asiatic-Pacific Medal, Philippine Liberation, WWII Victory Medal and Armed Forces Reserve Medal.

Belcher achieved the rank of LCDR and was discharged May 19, 1946. Retired LCDR May 28, 1984, USNR. His memorable experiences: designated aviator, Oct. 31, 1944 at Pensacola; and flying Black Cat missions in PBMs in VPB-17.

He married Martha L. and they have children, Barbara J. Gilbert and Jacqueline A. Belcher; and grandson, Craig R. Gilbert. He was employed with Bell Helicopter 29 years.

EDWARD J. "SKI" BELINSKI, CAPT, born June 23, 1927, Parish, NY, received BS, Syracuse University; DDS, Temple University, intern VA Hospital, Philadelphia, PA.

Entered the service in September 1951. His military locations and stations include: Pensacola, Key West and Atlantic City. He served with: FAWTU LANT, VC 33, Naval Reserve Dental Corps.

Belinski achieved the rank of CAPT September 1957. His memorable experiences: first carrier landing; first night car qual in a serious storm on 600 foot flight deck; nite talk landing by LSO in severe rain squall; mirror was not discernible. Lost AcDu Tra aboard *Nimitz.*

Married to Marion and they have children, David and Karen. Retired from dentistry in 1980.

GARLAND E. BELL, CDR, born in Durham, NC, May 4, 1919. Attended University of North Carolina 1938-1940. Joined Navy Aug. 19, 1940, from Washington, NC. Assigned to 7th Naval District Intelligence Office, Miami, FL as a Y-3rd Class. Started flight training at Pensacola, FL and received Navy Wings in October 1943.

Joined VS-45, Inshore Patrol Sqdn. at NAS Trinidad, BWI. Reassigned to CA-37, heavy cruiser USS *Tuscaloosa* in 1944 as a scout observation pilot. Awarded Distinguished Flying Cross for missions over Iwo Jima and Okinawa.

Completed BS degree from Rollings College, Winter Park, FL in 1974. Retired as CDR USNR 1979 and as chief of the FAA Flight Inspection District Office, Orlando, FL in May 1975.

He married Caroline and they have children, William E. and Carol Ruth; and grandchildren: Patricia, Joshua, Leslie and Stanley.

R. VICTOR BELL, CAPT, born Sept. 6, 1935, Lynchburg, VA. Received a BA in economics. Bell earned his aviator wings in 1960, served on active duty as a regular naval officer for nine years and then in active USNR for 15 years.

Logged over 4,100 pilot hours and made 172 carrier landing on 13 different aircraft carriers from the straight deck of the USS *Lake Champlain* (CVS-39), to the nuclear USS *Enterprise* (CVA-65). Flew 141 flights across the Atlantic Ocean as a member of the North Atlantic Early Warning Wing.

Qualified as CIC officer and served CO of three Reserve aircraft squadrons and as reserve wing commander for MATWING One at NAS Oceana.

In civilian life, he is a CEO of Radco Industries, a provider of industrial equipment and resides in Charlotte, NC. Married Anne and they have three children: Greg, Michael and Ivan; and one grandchild Eric.

WILLIAM F. BELL, CAPT, a native of New York City, enlisted in the Naval service on Nov. 23, 1953. He entered flight training at Pensacola, FL in September of 1954 and was commissioned and designated a naval aviator in May of 1956.

In June of 1956 Bell was assigned to Patrol Sqdn. NINE at NAS Alameda, CA. There he participated in deployments to Iwakuni, Japan and Kodiak, AK. In August of 1960 he was transferred to Training Sqdn. 28 at NAS Corpus Christi, TX. He served as flight instructor for three years in both advanced instrument training and aircraft carrier qualifications.

He was assigned to Fleet Tactical Support Sqdn. 21 (Det) Japan in September 1963. He saw duty in support of 7th Fleet operations in the Philippines and Vietnam. Bell was Anti-Submarine Sqdns. program manager at the Naval Air Reserve Training Unit in Alameda, CA from 1966 through 1970. He then served three years at the Naval Air Reserve Unit, Washington, DC as training officer and manpower management officer.

In May of 1973, he was transferred to the staff of the Director of Naval Reserve in the Office of the Chief of Naval Operations, and was subsequently assigned to the Chief of Naval Reserve Staff in New Orleans, LA as aviation plans officer. He assumed command of the Naval Reserve Center, Long Beach, CA in June of 1977.

He retired in September of 1980. Married Brenda and they have children, Bill and Michele. Retired as purchasing manager, United Speaker Systems, Oct. 1, 1980.

TOWNSEND M. BELSER JR., CAPT, born April 30, 1937. Commissioned 1958, Georgia Tech ROTC (Navy). Served 1958-1962 ADM Richover's Eng. Staff. He also served with Atomic Energy Com and Code 1500 BuShips, CO of various Reserve Units, including company and Naval Legal Service Unit, Charleston Naval Base.

Belser is a partner in the firm of Pollock, Vande Sande & Priddy, R.L.L.P. and has specialized in chemical and nuclear processes, mechanical equipment and devices, and related patent matters since 1966. He earned a BCHE degree with honors from the Georgia Institute of Technology in 1958, a certificate of reactor engineering from Bettis Reactor Engineering School of Westinghouse Electric Company in 1960, and a JD degree from George Washington University in 1965. Among his academic honors were the Tau Beta Phi and Phi Kappa Phi honorary societies.

He was registered to practice before the US Patent and Trademark Office in 1966 and was admitted to the South Carolina Bar in 1965 and the District of Columbia Bar in 1974. He is admitted to practice before the US Court of Appeals for the Fourth Circuit, the US Court of Appeals for the Federal Circuit, the US Court of Military Appeals and the US Supreme Court. Before specializing primarily in Intellectual Property Law in 1978, Belser's practice included general litigation before state and federal courts and administrative agencies.

He is a member of the American Bar Association, the American Intellectual Property Law Association, the Carolina Patent, Trademark & Copyright Law Association, the Washington International Trade Association, the Technology Transfer Society, and the Licensing Executives Society. He holds the rank of CAPT in the Judge Advocate General Corps of the Naval Reserves and served on active duty as a naval nuclear propulsion engineer with the Atomic Energy Commission from 1958 to 1962.

Belser has lectured in seminars on intellectual property law as a visiting law instructor at the University of South Carolina, and is the author of the paper entitled "Something About Patent Law for the General Practitioner," published by *The South Carolina Lawyer* (January-February 1990).

He is now divorced and has children: William Barney Belser and Elizabeth Davis Belser Kistler; and a grandchild Lauren Davis Kistler.

WILLIAM R. "BILL" BELTZ, CAPT, born Oct. 28, 1945, Brawley, CA, received BA from Auburn University. He joined the service June 1968. His military locations and stations include: Pensacola, Vietnam, San Diego, Seattle, Alameda and Jacksonville.

Beltz served with HAL-3; HS-4, NROTC University of Washington; HS-85; NAR JAX. He fought in Vietnam. Received the Silver Star, Distinguished Flying Cross (2), Single Action Air Medal (2), Air Medals (22), Navy Commendation.

He was discharged as CAPT. Transfered to Reserves June 1976. His memorable experiences: his year in Vietnam as an attack helicopter pilot with the HA(L)-3 Seawolves; command of 14S-85 at Alameda.

He is married to Nancy and retired as chemical abuse counselor, September 1993.

WALTER R. BENAVITZ JR., LCDR, born Feb. 14, 1944, in Honolulu. He is a graduate of the University of Hawaii and enlisted in the USN in February 1961. Attended the Naval Communication Technician School in Pensacola, FL, and served as a communications tech at San Miguel, Philippines on board the flagship *Oklahoma City* (CL-G5) and aboard the USS *Liberty* (AGTR-5).

During the Vietnam War on board the USS *Epperson* (DD-719) between October 1965-October 1969, he served in the Tet Counteroffensive and in the Vietnamese Counteroffensive Phase II, IV and V, and in the Tet Counter Offensive.

Benavitz joined the USNR in 1969 and was promoted to chief radioman in 1980 and to chief warrant officer in 1981. He was the commanding officer of NR Mobile Mine Assembly Group Det 2720 and of the NR Competent (AFDM-6) between 1985-87.

He was awarded the Navy Commendation Medal and his campaign and service medals include the Navy Unit Commendation, RVN Gallantry Cross w/palm, Naval Combat Action Ribbon, Armed Forces Expeditionary Medal (Vietnam).

The Benavitz family consists of wife Kathy; sons: Walter, Jeremy, Jon, Micah; and daughters, Emily and Evelyn. They reside on Oahu, Hawaii.

CHARLES A. "CHUCK" BENCIK, LCDR, born March 23, 1938, Chicago, IL, received BS, Physics, Illinois Institute of Technology; MA, Human Resource Management, Pepperdine University.

Commissioned through NROTC (Illinois Institute of Technology) July 1961-63; USS *Castor* (AKS-1); NR Surface Div 11-41L, 1963-66. TAR assignments: N&MCRC Phoenix, 1967-71. ComNine 1971-75; NRC Altoona, PA 1975-78. REDCOM TEN 1979-82. N&MCRC San Bernardino 1982-85. He commanded Reserve Centers in Altoona, Johnstown, Curwensville, PA; San Bernardino.

Bencik received the Navy Commendation, Humanitarian Service, Gold Wreath for recruiting w/star.

He achieved the rank of LCDR and was discharged July 1985. His memorable experiences: Johnstown, PA flood recovery; Meritorious Citation, 1977; when San Bernardino Reserve Ctr. won best Medium Size Reserve Center trophy in 1984; writing a naval history of the Great Lakes, also publishing history of San Diego Naval Militia organization.

He married Helga in 1962 in Yokosuka, Japan. Their children are Martin C. (EO3), Thomas E. and Victoria M. He retired as historian from San Diego Maritime Museum. He was also an actor and singer. At present he is governor, Pacific District, Flying Scot Sailing Assoc.

JULIAN R. BENJAMIN, RADM, born Aug. 24, 1926 in Chicago, IL. He attended University of Miami School of Law and received JD, 1949 cum laude.

Joined the service March 20, 1944. His military locations and stations: Navy Service School (EE & RM); NOB 230 Adak, AK; WWII ET2/c. Reserve: NRLC 6-3, Miami, NROS 6-3, Miami, director, Naval Reserve Law Program, 1980-84.

He was awarded the American Defense Medal, WWII Victory Medal, Armed Forces Reserve w/HG, Naval Reserve Medal, and Legion of Merit.

Benjamin achieved the rank of RADM (08) JAGC, USNR. His memorable experiences: Past National Judge Advocate, NRA; Past National Parliamentarian, NRA; Past Chairman, NBC, NRA; Pi Lambda Phi, Council Key.

He is married to June and they have children, James and Jeffrey; and grandchildren: Kaitlyn, Courtney, Rachel and Hailey. His civilian employment: attorney at law and president Judge Advocates Association (1978). He retired June 1, 1984.

WILLIAM B. BENNER, CDR, born in Pittsburgh, PA, July 25, 1944, and raised in Winter Park, FL. Upon graduation from high school he entered the USN as a seaman apprentice, completed Aviation Electronics "A" School and served as a C-121 radio operator in AEWTULANT, Patuxent River, MD.

In 1965, he entered naval flight officer training as an officer candidate airman (OCAN) receiving his wings and commission as ensign in 1966. He was assigned to RA-5C's in RVAH-4 and RVAH-6, and was deployed aboard the USS *Ranger* in the Tonkin Gulf. He was then assigned as a RATTC watch officer at Patuxent River.

Upon his release from active duty in 1969, LT Benner entered the Ready Reserve and requested a change of designator to the restricted line as an aeronautical engineering duty officer (1525). He served as maintenance officer in Intermediate Maintenance and Fleet Logistics Support Units such as VR2608 (C118s) and VR58 (C9s), retiring in 1984 at the rank of CDR.

He has been awarded the National Defense, Armed Forces Expeditionary, Vietnam Service and Armed Forces Reserve Medals and a Meritorious Unit Commendation.

Benner has a BS in Secondary Education and a PhD in management/economics. He is married to the former Patricia Barrineau and they have three children and four grandchildren and reside in Huntsville, AL.

EARL CHARLES BENNETT III, LDCR, born Jan. 12, 1945, Fresno, CA. He received a BA from the University of Washington, 1971, Near East Languages and Literature.

Joined the USAF Feb. 16, 1965, serving four years. Entered USNR, May 30, 1980.

His military locations and stations include: Iraklion Air Station, Crete, Greece. (Khobar Towers, Dhahran, Saudia Arabia). He served with 6931st Security Sqdn., 202nd MI BN, participating in Operation Desert Storm.

Bennett received the BSM, JMUA (3), AMUA, AFOUA (2), NDSM (2), SWASM, KLMK, SSDR, OSR, AFRM, NPMM(E).

He has achieved the rank of LCDR. His memorable experiences: Desert Storm Document Exploitation, UN Special Commission Nuclear Inspection Team #16; Baghdad, Iraq, October 1991, and missile inspection team #45, October 1992.

Bennett married Barneata Jean and they have children, Kristina Lee Hall and Roy Lee Bennett. Their grandchildren: Kody Lee Bennett, Mindy Marie and Steven Anthony Knight. He is presently employed with Army Intelligence and Security Command as intelligence operations specialist.

LINCOLN J. "LINK" BENNETT, LCDR, born Sept. 8, 1922, Portland, OR, received a BSE in physics from the University of Michigan in 1947.

He joined the service Oct. 28, 1942. His military locations and stations include: V-12, Central College, MO and University of Michigan. He served with APA-163 and AP-173.

Bennett was awarded the American Area Medal, Asiatic-Pacific Medal and Naval Reserve Medals. He achieved the rank of LCDR and was discharged July 25, 1946.

Married Anne G. and they have a daughter, Marjorie L. Warner, and grandchildren: Beth, Amy, Laura and Steven. He retired as manager, Product Group, BF Goodrich, Sept. 8, 1982.

ROBERT L. BENNETT, CDR, born June 29, 1921, Traverse City, MI, received a MA from Colorado State College. He joined the service June 22, 1942.

His military locations and stations: Iowa Pre-flight, AF Navigation, Hollywood Beach, Pensacola, FL, serving with VB-109 (July 1943-August 1944). He fought campaigns in Gilberts, Marshalls and Marianas Islands, Pacific Theatre.

Bennett was awarded the Admiral Letter Commendation and Presidential Unit Citation. He achieved the rank of LTJG plus active reserve to CDR, retiring in 1981.

He married Marjorie M. in 1945 and they have children: William M., James R. and David L.; and a granddaughter, Elizabeth A. He has been employed as a teacher and administrator, Alpena, MI, retiring June 30, 1977.

WILLIAM D. "DAVE" BENNETT, LTJG, born July 13, 1948, Hartford, CT, attended State University of NY, Morrisville NY, receiving a AAS degree in 1968. He also attended Northampton Community College, Bethlehem, PA, receiving an AA degree, 1983.

Enlisted Oct. 30, 1968, and was commissioned CWO2 Jan. 1, 1989 and LTJG, Oct. 1, 1992. His military locations and stations: NAS Memphis (NATTC); NAS Lakehurst (AIMD); NAS Willowgrove; NAS Roosevelt Roads; NAS Pensacola; NAS Bermuda; NAS Gitmo Bay, VR-52.

Bennett was awarded the Armed Forces Reserve Medal, Meritorious Unit Commendation, Overseas Service Medal, Battle E, Navy Achievement, Navy Good Conduct, Reserve Meritorious Service and National Defense.

He achieved the rank of LDO LT/03E and is still presently active. His memorable experiences: being initiated October 1976 as a CPO; being commissioned as a CWO2; being commissioned as a LDO LTJG; serving with the men and women of VR-52.

He married Johnette and they have children, Justen and Alexis. He has been employed with IBM as customer engineer and system services representative since June 1979.

MICHAEL J. BENSON, CAPT, born in Brooklyn, NY, Sept. 18, 1942, graduated from Rutgers University in 1963 and was designated a naval flight officer in 1964.

Assigned to Patrol Sqdn. 26 at NAS Brunswick, ME, he participated in the Cuban surveillance during 1964 and 1965, flying P2V Neptunes from bases in Guantanamo Bay and Key West.

During 1966 and 1967, he served in Keflavik, Iceland and Rota Spain, tracking Soviet submarines, during which time he flew P-3B Orions.

Upon release from active duty in 1967, he affiliated with the USNR at Floyd Bennett Field, NY and later at NAS Willow Grove, PA. In addition to his flying duties, he served in numerous squadron billets, including schedules officer and operations officer in VP-834, VP-62 and VP-66. In 1981 he was selected as CO of PATWING 593, and later that year assumed command of Patrol Sqdn. 2644 at NAS Willow Grove, PA.

In 1985 Benson transferred to the inactive reserve, but continued serving the USN as an adjunct faculty member at the National Defense University in Washington, DC, until his retirement in 1989.

He and his wife Motria reside in Simsbury, CT. They enjoy sailing their sloop, *Serendipity,* out of Niantic, CT. They have three children: Christine, Theresa and Michael Jr.

GEOFFREY D. "GEOFF" BENTLEY, CAPT, born Sept. 30, 1950, Grand Forks, ND, received a BA degree from Concordia College in 1972; BS degree from University of Minnesota in 1974 and DDS degree from University of Minnesota in 1975.

He joined the service Aug. 19, 1974. His military locations and stations: NRDC San Diego, CA; Branch Dental Clinic NAS Miramar. His unit, Branch Dental Clinic NAS Miramar. He was awarded the National Defense Medal, Sharpshooter, Rifle and Pistol Medal and Armed Forces Reserve Medal.

Bentley achieved the rank of CAPT and was discharged March 1978. His memorable experiences: squadron dental officer VF-154; providing actual support to this excellent group of aviation personnel.

He married Pamela and they have three children: Erik, Sarah and Reid. He is employed in private dental practice. Bentley is also an avid pilot. He holds a private pilot certificate with instrument and multi-engine ratings. He has over 1800 hours of flight time.

W. DUANE BENTON, CAPT, has served as staff JAG to the Reserve Construction Force (Gulfport, Washington and Norfolk) and the Readiness Command Region Eight (Jacksonville). Previously, he was CO for NRLSO

Memphis and NRLFS San Diego. On active duty, he served in Vietnam, New London, Great Lakes and Memphis.

While an NROTC midshipman, he was battalion commander, a distinguished naval graduate, and a delegate to the Naval Academy Foreign Affairs Conference.

In civilian life Benton is a judge on the Missouri Supreme Court, the only CPA on any supreme court in America. Previously, he was State Revenue Director, a private attorney and Chief of Staff to a US Congressman.

Graduate of Northwestern University (BA), Yale Law School (JD), Memphis State University (MBA), Virginia Law School (LLM), and Central Missouri State University (honorary LLD).

Married to Sandra Snyder, formerly LT, NC, USN, and they are the proud parents of Megan (April 18, 1985) and Grant (May 28, 1990).

GORDON JEROME "GORDY" BENTSON, CAPT, born Dec. 12, 1937, Dawson, MN, received a BA in economics from the University of Minnesota.

He enlisted October 1956, seaman recruit. Commissioned ensign SC June 1962. His military locations and stations: Great Lakes, Green Cove Springs, FL and Athens, GA. Units served with: AWS 81, VP 810, NARDIV V6, NARDIV V5, NAS 1416, ASB LG HQ 216, REDCOM16, NSD S/B HQ A116, VTO 1618.

Bentson achieved the rank of CAPT and was discharged June 1966. His memorable experiences include taking 45 degree rolls on USS *Lloyd Thomas* during midshipman summer cruise; great HCTDUTRAs to Subic Bay, Rota Spain, Oakland, NAS Lemoore; Pt. Mugu, San Diego, the great camaraderie of the USNR.

He is married to Dolly and they have five children: Steven, Mark, Jeff, Paul and Ken; and a grandchild, Eli. Employed as buyer, Target, SuperAmerica, Shopko, GMM, Sibley Catalog Showrooms, Thrifty White Drug, retiring in December 1992.

ALAN E. BENTZ, CAPT, born in San Antonio, TX, Jan. 1, 1933. He enlisted in the USN in July 1950, served overseas, both ashore and aboard ships, notably USS *Bellatrix* (AKA-3), (Attack Cargo Ship), as an assault boat coxswain and craft master during his 30 month Korean West-Pac deployments.

In January 1958, he was promoted to CPO (E7), BMC. He then attended the University of Texas at Austin, where he graduated in January 1961 with a BS degree in civil engineering and in June 1961 he was commissioned ensign. In April 1982, he was promoted to CAPT.

Among his numerous assignments, Bentz attended the Naval War College and various additional navy schools. He had nine Command Tours, which included: Surface units, amphibious units, navy recruiting, navy liaison officer (National Security matters) to the State of Texas, and most recently served as navy liaison officer to Commander Fifth US Army. He has served the San Antonio Recruiting District since 1991. During this period, Capt. Bentz was selected as Recruiting Area Seven RDAC chairman of the year, for three consecutive years. To date he has earned 22 Recruiting Wreaths for recruiting excellence.

During his 41 years of service he has served aboard almost every type of Navy ship and vessel. His service awards and military decorations include: Distinguished Service Medal, Navy Meritorious Service Medal, and US Army Meritorious Service Medal.

Bentz retired from naval service, Oct. 1, 1991. At this time, he and his wife, Yvonne (Bugs), reside and operate their Anchor B Ranch in Poteet, TX. They have a daughter, Alisha Seiders, who lives in Austin, TX, and a grandchild, Jessie.

DONALD M. BERG, CDR, a native San Diegan, was born in 1921, and graduated from San Diego State in 1943. Commissioned ensign, USNR at Northwestern Naval Reserve Midshipman School August 1943. LCVP training Little Creek, VA and Ft. Pierce, FL.

Assigned to LST 289 January 1944. April 1944 participated in Tiger Operation, a D-Day rehearsal. LSTs 289, 507 and 531 were torpedoed by German E-Boats in the

English Channel, with a loss of over 700 men. LST 289 was saved by LCVPs, Berg was the small boat officer.

In the Pacific Theater he served as XO LST 245, CO LST 452, and amphibious rep. Philippine Off Shore Patrol Training, 1945-46. Separated from active duty at Subic June 1946, then was master LST 883 Army Transport Command for one year. Served with Submarine Div. 11-43, NROS Program 11-2, and Mobilization Team Div. 11-4 from 1947-70.

Married Theresa, LT USN Nurse Corp., in 1948, and they have three sons: Donald, Douglas and Kenneth. He was an educator with San Diego City Schools from 1948-83. Since retiring he has played senior softball. His tournament team, the San Diego Silver Hawks, have won five-Senior Softball World Series rings from 1989-96. CDR Berg is NRA Life Member #3719.

ARTHUR W. "ART" BERGER, born Dec. 12, 1925, Maywood, IL, received a BS in marketing, University of Illinois; Chartered life Underwriter and Chartered Financial Consultant from the American College, Bryn Mawr, PA.

His active service began June 20, 1944. Commissioned ensign, SC USNR, Dec. 20, 1948. His military locations and stations: Boot camp, Great Lakes; ATB NORVA; USS *LST-967;* USS *Ulysses* (ARB-9). He served with NavResSupp Co 9-7, Chicago; NavResSupp Co 9-1, Forest Park. Participated in battles and campaigns: American Theatre, Asiatic-Pacific Theatre and China Service.

Berger was awarded the WWII Victory Medal, China Service Medal and Good Conduct Medal. He achieved the rank of LCDR and was discharged July 1, 1975.

His memorable experiences: promoted to YN3/c on USS *Ulysses;* member of initial crew of the *Ulysses* at commissioning. Sailed through Panama Canal to San Francisco to Pearl Harbor, Eniwetok, Saipan, Tinian, Okinawa and Shanghai; USS *Ulysses* was decommissioned at Mayport, FL in 1946.

He married Jacquelyn D. (Lynch) and they have two children, Kevin J. and Terrence A.; and three grandchildren: Kelsey, Kara and Kent Berger. He has been employed as agent and broker with the Equitable Life Assurance Society for 39 years. Currently in partnership with son, Terrence.

CARL RUDOLF "RUDY" BERGHULT, CDR, born April 15, 1905, W. Duluth, MN. Graduated from Denfeld High School, Duluth, MN, 1923. Received AB degree Gustavus Adolphus, St. Peter, 1926; MA Teachers College, Columbus University NYC, 1928; completed course and language requirements for PhD Columbia University, 1932; Attended Institute of Public Admin. NYC 1931-32; Geneva Schools of International Studies, Switzerland, July-August 1932; Naval Indoctrination School, Princeton, NJ, October-December 1942; Subchaser Training Center, Miami, December-February 1942-45; Civil Affairs Administration, Roseneath, Scotland, February-April 1945.

His military locations and stations: February 1943-May 1944, CO, Subchaser USS 1291, Convoy Control Center, Key West, FL; one of five escorts for 84 ship convoy, UK-France, June 5-6, 1942; control ship Normandy Beaches Easy Red & Fox Green; convoy and escort duties, June 7-November; Deputy Chief (Assist Head), US Naval Component, SHAEF Mission to Norway, April-November 1945; assistant to Naval Attache and Naval Attache for Air, US Embassy, Oslo; Admin. Asst to CO, USNRS, Seattle, WA, January-September 1946; HQ 13th Naval District, PIO, OinC and Special Events, Staff of Commandant, September 1946; 1949, July-August: Coordinator for CO, San Diego Naval Station, liaison with ROCS; training officer, CO, Naval Reserve Unit 6-69, PAFB, FL (1950s).

Berghult was awarded the American Area Campaign Medal, EAME w/star, WWII Victory Medal, Bronze Star Medal, Haakon VII Freedom Medal (Norway).

His wife, Eva Louise (Drewes), passed away Jan. 23, 1995. He has two children, Sondra Louise Dietert and Carl Robert; and six grandchildren: Eric, Amy and Paul Dietert; Christina Parks, Carla and Richard. His great-grandsons are Joshua R. and Aaron M. Parks. He is chaplain of the Mountain Chapter of NARFE.

DR. SAMUEL I. BERMAN, CAPT, born July 17, 1938, Minneapolis, MN. Earned his MD degree from University of Minnesota in 1964. Joined the service in January 1966.

Stations include NAS Pensacola, FL; NAS Agana Guam; Persian Gulf; Naval Aerospace Medical Institute.

Discharged in March 1991. Memorable experience was flying into a typhoon in the Pacific.

Married Mary Ellen and they have two children. Presently staff physician at Minneapolis VA Medical Center.

WILLIAM N. "BILL" BERNARD, CDR, born Nov. 18, 1928, Hastings, MN, received bachelor of aeronautical engineering degree at University of Minnesota; LLB and JD from William Mitchell College of Law.

He joined the service Aug. 20, 1948. His military locations and stations: Pensacola, Corpus Christi, San Diego, Barbers Point, Midway Island, Seattle, Minneapolis. He served with VU-1, VU-7, FAWTU PAC, VF-812, VF-813, VA-891, HU-813, HS-10.

Bernard achieved the rank of CDR and was discharged Dec. 8, 1952. His memorable experiences: carrier qualifications, flew single and multi-engine, land and sea conventional and jet, seaplanes and helicopters; 2800 military hours; 1500 civilian hours, flew 20 military types, six civilian types; stationed Midway Island at start of Korean War flying PBY(old) restricted to airport landings only in Air-Sea Rescue mission on meeting air evac planes from Tokyo Japan, one made a water landing and they couldn't land in the water.

He has been married to Catherine "Katie" Hablos of Fargo, ND, since 1955. They have three children: Tim, Kathleen and Mary Ann; and a grandchild Shannon Bernard. Employed with Honeywell, engineer (1955-59); Sperry Rand Univac Eng.-lawyer, (1959-64); started private law practice in 1964 and is still practicing.

JOSEPH BURRILL "DICK" BERNHEIM, LT, born June 4, 1922, Gulfport, MS, attended University of Mississippi and Loyola University.

Stationed at NavSta, Green Springs, FL, and served in USS *Albermarle* as dental officer. Participated in campaigns in San Diego, CA and the Korean Conflict. Bernheim achieved the rank of LT.

He has retired from his dental practice in Gulfport, MS. He is divorced.

LOUIE BERTOLUCCI, LCDR, born Aug. 6, 1918, Stockton, CA and attended college for two years. He joined the service Jan. 2, 1942. His military stations and locations: Armed Guard, Treasure Island, San Francisco, USS *Wileman.* He served with NSD, Clearfield, UT.

Bertolucci was awarded the WWII Victory Medal, American Campaign Medal, Asiatic-Pacific Ribbon. He achieved the rank of LCDR and was discharged June 27, 1946.

He married Regina Anita and they have three children: Kathleen Marie, Ronald Louis and Marcia Jean; and seven grandchildren: Justin, Ahmon, Shawna, Monica, Christine, Regina and Michayla. Employed as clerk, Assoc. Oil Co. (salesman Acme Beer Co. and Schlitz Beer), retiring Aug. 6, 1984.

JOHN R. "BERT" BERTRAND, CDR, born Aug. 3, 1914, Gray County, TX. Earned BS in agricultural education and MS in agricultural economics, Texas Tech Univ. and PHD in rural sociology from Cornell Univ. Joined the service Feb. 3, 1942, and was plankowner through six war patrols in USS *Bowfin* (SS-287) in the SW Pacific and USS *Sea Leopard* (SS-483), both launched and commissioned at Portsmouth Navy Yard in Maine.

During WWII he served as submarine officer and, along with several other awards, twice received the Silver Star for action aboard the USS *Bowfin.* He stayed in Ready Reserve until December 1959 when he accepted honorary retirement as CDR, USNR (Ret). His awards include the PUC, NUC, Silver Star, Gold Star in lieu of second Silver Star and Submarine Combat Insignia with stars for six patrols.

He served as president of Berry College and Berry Academy, 1956-80. Currently, he is President Emeritus. He came to Berry from the University of Nevada where he was professor of agriculture and dean and director of the Max C. Fleischmann College of Agriculture. He had served as dean of the basic division and earlier as assistant dean of agriculture at Texas A&M University where he was also associate professor of rural sociology. Previously he served as dean of men at Sam Houston State University in Texas where he was associate professor of sociology.

Bertrand is a native of White Deer, TX, where he graduated from high school in 1932. He began his college study in 1933-34 at West Texas State University. He enrolled in Texas Tech in 1937 after farming for five years in Carson and Deaf Smith Counties. In 1941-42 prior to reporting for active duty in the USNR, he was enrolled in the University of Missouri as a doctoral candidate in rural sociology.

By Presidential appointment served from 1970-74 as a member of the board of visitors of the USNA. He was one of 20 college and university presidents in the US selected in 1971 to receive the Danforth Foundation short-term leave renewal grant. Texas Tech recognized him as a "Distinguished Alumnus" in 1971 and the West Texas Chamber of Commerce honored him as a "Top Displaced West Texan" in 1972. In 1974, Oglethorpe University awarded him the honorary degree of Doctor of Humane Letters.

Bertrand has been married since 1942 to the former

Annabel Lee Hodges of Shreveport, LA, who, before their marriage received her BS and MA degrees in art from Texas Woman's University and was instructor in art at Sam Houston State University. They are members of the United Methodist Church and have four children: J. Thomas, Diana Bertrand Williams, Karen Bertrand Wilson and J'May Bertrand Rivara; grandchildren are Maglyn Brenna Bertrand, Maggie and Matthew Rivara, Anne Elizabeth and Nathan B. Williams.

FRED S. "FRITZ" BERTSCH JR., CAPT, born Oct. 8, 1920, Chicago, IL, received AB from Hope College in 1941 and MA Western Michigan University in 1960.

Enlisted USNR, July 15, 1940. Cruised in USS *Wyoming*. Commissioned Sept. 12, 1941. Served in USS *Case* (DD-370) 41-3. Actions: Pearl Harbor attack; Aleutians; Guadalcanal; various carrier actions with USS *Enterprise*, USS *Victorious*, USS *Monssen* (DD-798). Actions: Saipan; 1st Philippine Sea Battle; Pelileu. Instructor, Damage Control School, Philadelphia, 1945-46; General Line School, 1946-47; XO, USS *Henry W. Tucker* (DD-875) 1947-48. China station; participated in 1948 Eniwetok atom bomb tests.

Flight training, 1948-49. VP42, 1949-51; Korean War patrols based at Iwakuni, Japan. Staff, CFAW-14, San Diego, 1951-53. Assistant PNS, University of Minnesota, 1953-55. VP-47 1955. Resigned regular commission and accepted USNR. Various USNR assignments. Retired as CAPT June 30, 1973.

He was awarded the Distinguished Flying Cross, three Air Medals, American Defense Service (fleet) Asiatic-Pacific (6), American Area, Victory, Occupation (Japan), China Service, National Defense Service, Korean (4), United Nations, Naval Reserve Medal, Korean Presidential Unit Citation.

Bertsch married Lorraine and they have four children: COL Susan Christoph, USA (Ret); Martha Ann Neiboer; CAPT Fred S. Bertsch III, USN; M.D., Bertsch, MD. He was employed as principal at Holland (MI) High School, 1964-78.

EDWARD L. "ED" BETHEL, CDR, born April 16, 1930, Sublette, KS, received BSEE from Kansas State University.

He enlisted in the service Oct. 4, 1948. His military locations and stations: NTC San Diego, NAS Miramar, NAS North Island. Served with VF-781, VF-121, FASRON 7, participating in Korean War in USS *Oriskany*.

His awards and medals include the Navy Commendation (2), Air Medal (3), and Korean Service (3). He achieved the rank of CDR and was discharged April 8, 1954. His memorable experiences: two ejections on take off: F9F-6 at NAS Olathe and A4 at NAS Los Alamitos.

Bethel married Joyce M. and they have children, Christine and David; and grandchildren: Gabriel, Amber and Jared. Employed in electrical engineering and retired December 1994.

WALTER F. "WALT" BETHUNE, CAPT, born in Glenmora, LA in 1926. He enlisted in the Navy in Seattle, while attending the University of Idaho in lieu of his fourth year of high school in Klamath Falls, OR. He enlisted in November 1944 in the Air Crew Program (Aerial Gunner and Aviation Machinist Mate).

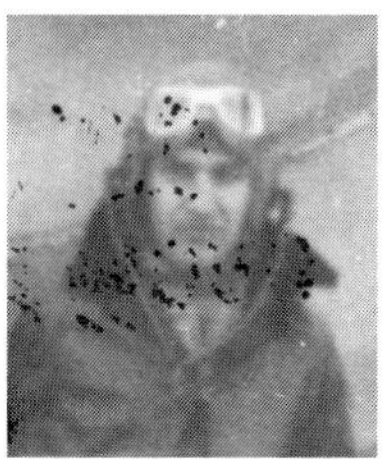

The V-5 pilot training program re-opened, so he transferred with three years of high school and one summer of college. Duty was at Memphis, TN; Norman, OH; Marshall, MO; Miami University, Oxford, OH; Oregan State University; Stearman solo at NAS, Los Alamitos, Long Beach, CA and Pre-flight, Ottumwa, IA.

Resigned from the USN (March 1947) and attended Pre-med Dental at OSU and Dental School (I of O) in Portland, OR. After graduating (DMD) he joined the Naval Reserve Dental Company in Portland, OR working up to command. ACDUTRA included The Naval War College, *Kitty Hawk* (CVA-63), sub tenders and many other activities.

He practiced dentistry in Lake Oswego, OR (Portland) for 34 years and was very active in his local Dental Society and local Academy of General Dentistry, achieving president in both.

He retired from the USNR in December 1986. His wife Patty is deceased. He has five children and seven grandchildren: Keith, Miranda, Heather, Spencer, Kooper, Alex and Alysha.

ROBERT H. BEYER, CAPT, born April 4, 1933, NYC, NY. He enlisted in the USNR while still in high school He volunteered for recall and was activated in October 1951. Upon completion of AT School, he was assigned to VP-46 where he served as an aircrewman aboard PBM and P5M aircraft. Upon release from active duty, he attended NYU while attached to VP-831 at NAS New York.

Graduating with a degree in electrical engineering, he received a direct commission as ensign USNR in 1962. In 1963 he was transferred to Yokosuka, Japan as a NAESU engineer on the DASH helicopter. After receiving a MA degree from Sophia University in Tokyo, he returned Stateside in 1970 and affiliated with the NAVAIRSYSCOM Reserve Program. While in WESTPAC, he had numerous tours of duty aboard ships of the 7th Fleet.

In 1976, he was selected to escort the Japanese government representative during the Bi-Centennial Fleet Review in New York. Promoted to captain in 1982, he subsequently commanded four NAVAIR reserve units. Upon his retirement in 1993, he was the director of the NAVAIR Rapid Industrial Mobilization Program.

He is a life member of NRA. He currently resides in Fairfax, VA with his wife Helen. His four children are Ken (CDR USNR), Karen (LCDR USPHS), Kevin, Kim; and 10 grandchildren compete for time with his hobby of model railroading.

SILVID JOHN "BIANC or DR. B" BIANCALANA, CAPT, born July 19, 1911, Madera, CA, received DDS degree. He joined the service Sept. 14, 1942. His military locations and stations include Treasure Island and South Pacific. He served with CG Hospital Unit in the South Pacific.

Biancalana achieved the rank of CAPT and was discharged Jan. 1, 1946. His memorable experience: when the war was over.

He is married to Madeline D. and they have a son Timothy and grandson William. He is employed as dentist in his own practice.

JOHN W. "JOHN" BIDDULPH, CAPT, born Nov. 4, 1946, Phoenix, AZ, received a DDS degree in 1972 from the University of Southern California School of Dentistry.

He joined the service June 1968 and was stationed NAS Lemoore and NRMC Oakland with 4th FSSG, CPEND 119.

Biddulph received the rank of CAPT 06 and was discharged January 1, 1993. His memorable experiences: POW repatriation at Oakland, CA; exercises at 29 Palms.

He is single and has a child Jamie Lou Biddulph. Civilian employment as president, Biddulph and Associates Ltd. General Dentistry.

ROBERT P. BIELKA, CDR, born in Berkeley, CA, Feb. 19, 1925. He is the son of a career naval officer, LCDR Rudolph Paul Bielka, killed in action at Pearl Harbor, Dec. 7, 1941 on the USS *Utah*. His family lived in many places, including the Far East, Hawaii and the USA.

During WWII, his service included the University of Washington V-12 program for one year and the US Navy Radio Materiel Program in Chicago, Monterey and Treasure Island. He was commissioned an ensign in the Reserve in August, 1948. During the Korean Conflict (1951-53), he was assigned to the USS *Bon Homme Richard* (CVA-31) first as assistant electronics repair officer, then as electronics repair officer. He returned to Seattle where he joined the Boeing Company as research engineer until 1956 when he joined the Crosley Division of Avco Mfg. Co. in Cincinnati, OH, returning to Boeing in 1958. He retired from Boeing in 1986 after assignments in Renton, Everett (WA), Toronto and New York in the fields of Reliability and Maintainability. He co-authored a study of four USAF aircraft, F111A, F4D, C141, and B52G, as related to their accidents and causes, with recommendations for ways to avoid them in the future. As a result of this publication, he was nominated to conduct a safety study of the hydraulic system of the Space Shuttle, although his Boeing assignments precluded that effort.

Bielka entered the Naval Intelligence Program in 1957 while in Cincinnati and continued until his retirement in 1975. Prior to his retirement, he was transferred to the Naval Attache branch of the program, partly as a result in his ability to speak German and Italian languages. In addition to these two, he has studied Spanish, Japanese, Chinese and Russian.

The family of five includes wife Maxine, son Robert Bruce, New York; daughter Janice Dressler, Renton, WA; daughter Barbara Wans, Slidell, LA; and four grandchildren: Aaron and Anna Dressler; Alyssa and Lacey Wans. Retired from Boeing in August 1986. Hobbies include model railroading, electronics and photography.

GLADE S. BIGLER, CAPT, born April 21, 1928, Brigham City, UT, received a BS, LLB and JD degree.

He was commissioned June 12, 1950, NROTC, University of Utah. His military locations and stations: USS *Valley Forge* (CVA-45); USS *Bexar* (APA-237). He served as CO of NRSG Div. 12-9 Salt Lake City and CO NRSG Command, San Francisco. He served three years active duty in Korean War, 27 years in Reserve.

Bigler was awarded the Japanese Occupation Medal, Korean Service Medal, United Nations Medal and Unit Commendation Medal. He achieved the rank of CAPT and retired March 1, 1981.

He married Lois A. and they have five children: Cathy, Nadine, Elaine, Thad and Patricia; and 17 grandchildren. Retired as chief attorney and district counsel, US Department of Veterans Affairs, Dec. 30, 1995.

ARTHUR JAY BILENKER, LCDR, born in Brooklyn, NY, Aug. 6, 1956, attended Rutgers College and Farleigh Dickinson Dental School.

In 1925 he entered the USNDC program while in school and received initial training at NAEC Lakehurst, NJ. Upon graduation in 1982 he served at NRDC Philadelphia

and later aboard USS *Forrestal* (CV-59) and NSA Brooklyn, NY. Following his release from active duty in 1986 he served for six years with the 14th Dental Co., 4th FSSG, 4th MARDIV.

Bilenker was awarded the Navy Achievement Medal, Navy Reserve Medal, the FMF Ribbon and a Meritorious Unit Commendation.

In 1990 he opened a private dental practice in Cranford, NJ and retains his reserve affiliation in IRR Status. He lives in Cranford with his wife Teri and his children, Daniel and Carrie.

RODNEY B. BINDON, CAPT, born in Seattle, WA, Nov. 1, 1941. Graduated from University of Puget Sound, 1963. Attended Navy OCS Newport, RI, September 1963-January 1964. Navy Supply Corps School, Athens, GA, February-August 1964.

Reported to USS *Constellation* (CVA-64) at Yankee Station September 1964. Completed two deployments to Vietnam as assistant sales, wardroom and sales officer. Released from active duty January 1967.

Affiliated with USNR 1968 at NRRC Seattle as a member of SSOD 13-1. Later was a member of TTTMD 13-1, NSP-1922, NSC Puget 1022, and served as XO of NSD YOKO 122 and CO of NSC Puget Sound 122. On October 1987 was assigned to USCINCPAC DET 322 and later VTU 2208 as POL plans officer and in 1990 as USCINCPAC Deputy Foreign Policy Advisor. As acting Foreign Policy Advisor was on the CINC's battle staff during Desert Shield/Desert Storm.

He was the project officer for US State Dept. Chiefs of Mission Conf. in 1991 and 1994; escort to all US Ambassadors to 50th Anniversary of Pearl Harbor Commemoration Ceremonies and meeting with President Bush, Secretary Cheney and Chairman Powell; USCINCPAC coordinator for the removal of the remains of Ferdinard Marcos to the Philippine Islands.

Simulated the role of President, Secretary of State and Chairman JCS during numerous CINCPAC exercises and wargames. Served as direct liaison between USCINCPAC staff and the 54 US State Department Embassies and diplomatic posts in the PACOM AOR. Authored: *Proposal for a Combined Naval Task Force for Humanitarian Aid, Disaster Assistance and Sea Control in the WESTPAC, SEA Op Area.* Co-authored: *Arms Control, Anti-nuclear Sentiment and NCND; An Alternative.* Co-authored: *The Strategic Importance of Hong Kong to the United States Pacific Command.*

He was awarded the Defense Superior Service Medal, Navy Commendation Medal, National Defense Medal (2), Armed Forces Expeditionary Medal, Vietnam Service Medal (4) Naval Reserve Medal, Vietnamese Cross of Gallantry and Republic of Vietnam Campaign Medal. Bindon retired from the USNR, July 1, 1994, as CAPT SC.

He lives with his wife Carole in Medina, WA where they enjoy boating, traveling and golf. They have two children, Julie and Angela. Retired July 1, 1996 from civilian employment as president, Bindon & Assoc. Inc., Commercial Real Estate.

JON A. BIRD, CAPT, born in Ypsilanti, MI, Aug. 26, 1939. He attended the University of Michigan where he obtained AB and MBA degrees and was commissioned there as an ensign in the Supply Corps in 1961 through the Regular ROTC program. He served on active duty until 1965 during which time he was twice awarded the Armed Forces Expeditionary Medal for service during the Berlin Crisis and Cuban Blockade as disbursing officer of the USS *Pawcatuck* (AO-108).

Following a tour at Wright Patterson AFB as assistant controller of BWFRR-CEN, he accepted a USNR commission and served an additional 23 years with reserve units located in Michigan, Illinois and Ohio including tours as logistics officer for Commander Service Group Two (1982-83), CO of NR DCASR Chicago 613 (1984-86) and NR DESC Dayton 105 (1986-88). He has been a NRA member since 1970 and served as treasurer of the NRA National Conference which was held in Detroit in 1982.

He retired Oct. 1, 1988 and continues to reside in Birmingham, MI with his wife, the former Jan Brucker Eberly of Mansfield, OH. They have one daughter, Karen, who lives in Morristown, NJ. He is employed as vice-president, Comerica Bank Detriot.

JOHN WILBUR BIRDSALL, LT, born May 24, 1917, Iowa City, IA, received BS in CE Georgia Tech, June 1940, Civilian Pilot Training Program for private certificate.

He enlisted October 1935. His military location and stations: Atlanta, GA (October 1935-October 1939); NTC Williamsburg, VA (April 1943-June 1943); Camp Endicott Davisville, RI (July 1943-August 1943); Londonderry Northern Ireland (September 1943-November 1943), London England (November 1943-October 1944), Plymouth England (October 1944-March 1945); Camp Endicott Davisville, RI (April 1945-August 1945); Guam (September 1945-June 1946); Raleigh, NC (April 1956-December 1959).

He served with the following units: Atlanta BN USNR, 67th CB, 97th CB, COMNAVU, 13th Construction Regt., Commander Amphibious Bases UK, 5th Construction Bde, Civil Engineer Co. 6-5.

Birdsall achieved the rank of apprentice seaman, Sm2c, Sm1c, CM1c, ensign, LTJG and LT, retiring Oct. 1, 1961.

His memorable experiences: The bombing of London by the Germans, that included fire bombs, V1 (bussbombs) and V2 bombs. He had temporary quarters in the Palace Hotel and was returning to the hotel by bus one day about noon. He heard a V1 flying over, and heard the engine quit. A few second later he heard the explosion that was quite close. When he arrived at the hotel he found that the bussbomb had hit a building across the street from the hotel. It had damaged the whole side of the building, killed an employee of the hotel and hurt many people. His room was on the other side of the building and the window was blown out.

He married Barbara M. Bolding and has one son, LTC Ian A. Birdsall, USAF (RET); and two grandchildren, Caroline and Heather. He has held positions with the Georgia Highway Department as inspector; USDA (surveyor); USA Corp of Engineers, (civil engineer); Birdsall & Pattillo (surveyor); Piston Ring & Parts Co. (bookkeeper); North Carolina Highway Comm. (design engineer). Retired December 1979.

JAMES BENSON "BEN" BIRINDELLI, CAPT, born Jan. 12, 1939, Richmond, VA. Earned BS, USNA, 1962, and MBA, Old Dominion University, 1971, business management. Joined the USNR May 6, 1956, and stationed at NSC Oakland and NATO HQ SACLANT.

Served in the USS *Ranger* (CV-61) and USS *Decatur* (DD-31) and participated in Tonkin Gulf and Vietnam (1964). Memorable experiences: CO Defense Fuel Supply Ctr. Unit during Desert Shield/Desert Storm; had three volunteers in Saudi Arabia and six at Defense Fuel Region, Houston.

Retired Sept. 1, 1992, as CAPT, SC, USNR.

Married Nancy and has three children: Tyler Anne (Kolenboski), Michael and David; and grandchild Joey Kolenboski. Employed with Ingalls Shipbuilding as manager and plans to retire in 2001.

JOHNNY M. BITTICK, CAPT, born Jan. 17, 1940, Ryan, OK. He joined the USN as an aviation cadet and received his commission and designation as a naval aviator in August 1963.

After graduation from Maintenance Officers School, ENS Bittick was assigned to fly the A4 Skyhawk at NAS Oceana. Prior to his release from active duty, he made a NATO cruise and a WESTPAC deployment with VA-72, and a MED cruise with VA-83.

After affiliating with the Naval Reserve at NAS Dallas and transitioning to the F8 Crusader, LT Bittick was recalled to active duty with VF-703 in response to the capture of the Pueblo by the North Koreans. The squadron was released from active duty in October 1968.

LCDR Bittick returned to active duty in 1971. He deployed with VA-164, serving as maintenance officer and operations officer during a 10-month WESTPAC deployment. He returned to civilian life having flown 320 combat missions in the Vietnam conflict. His decorations include the Distinguished Flying Cross, Meritorious Service Medal, Air Medal (7 individual, 25 strike/flight), Navy Commendation Medal (6 awards), Navy Achievement Medal, Vietnamese Cross of Gallantry and various other campaign and unit awards.

LCDR Bittick re-affiliated with VF-202 at NAS Dallas and served as safety officer while flying the F8 and F4. He has a total of 3,000 hours in the A4, F8 and F4 aircraft and 600 carrier landings.

CAPT Bittick was selected to help stand up NR Carrier Group 770 in February 1979, serving as operations officer, XO, and CO. He also commanded the Voluntary Training Unit and NR NAS North Island before retiring in 1989.

In civilian life CAPT Bittick is a pilot with US Airways, accumulating over 20,000 hours flying time. He is married to the former Jane Castle of Davison, MI. They reside in Argyle, TX, and have three children and six grandchildren.

JAMES BENJAMIN "BLACKIE, JIM or BUD" BLACK, CAPT, born Oct. 10, 1924, Charleston, SC, received a BS in business administration from the University of South Carolina, Columbia, SC.

He enlisted Dec. 7, 1942 as apprentice seaman V-1; July 1, 1943 program modified, apprentice seaman V-12; NROTC Student USC, Columbia, SC, 1942-44, commissioned ensign Feb. 27, 1944; communications student, Harvard University, Cambridge, MA; amphibious training, Solomons, MD; Communications Watch Officer, Staff (temporary) Com LCI (L) Flotilla 27 aboard USS *LCI-780;* communications watch officer-staff (temporary); Com Transport Div five aboard USS *Olmstead* (APA-188); 1944-45, communications Watch Officer-staff (temporary), Com Task Unit 78.3.9. Aboard USS *Gilliam* and then ashore at Toloso, Leyte, Philippine Islands; USS LST-464 - Pacific, communications watch officer (two months), navigator (11 months), XO (four months). Returned ship to US and decommissioned it at Beaumont, TX, CO (one month). He was released from active duty June 16, 1946.

Recalled to active duty for Korean Conflict, Oct. 30, 1950; served as inspector, instructor-navigation (CIC); PhibOpTra Element Atlantic, Little Creek, Norfolk, VA; released from active duty Oct. 29, 1952.

Reserve Duty: 1947-50 rate group officer and personnel officer, Organized Surface Div. 6-38 Georgetown, SC; 1953, rate group officer, NR Surface Div. 6-23 Charleston, SC; 1953-68 progressively rate group, training, administrative and CO officer, NR Surface Div. 6-38, Georgetown, SC; 1968-71 assistant officer in charge (Inactive duty) (3 years) NR Training Facility, Georgetown, SC; 1971-73 group commander (3 years) NR Group Cmd. 6-4(M), Charleston, SC. All NR Units in the Charleston area came under the command of the Group Commander. There were eight such units: Surface Div. 6-33, Charlestown, SC; Naval Reserve Training Facility, Georgetown, SC; Surface Div., Inshore Undersea Warfare Div., Security Gp., Naval Reserve Shore Patrol Div., Submarine Support Div., Intelligence Unit, Ship Activation, Maintenance and Repair Div., Officer School.

His Group Command received a "Commendation" letter, Nov. 6, 1973, from the Commandant 6th Nav. Dist. for achieving the First Place standing for Fiscal Year 1973 in the 6th Naval District's Performance Rating System.

The most significant of the many training duties were: 1965-one of two officer teams which evaluated the Status of Training and Technical proficiency. of the USS *Leahy* (DLG-16); 1968-Operations and Logistics Staff Training at 6th Nav. Dist. HQ, Naval Base Charleston, SC; 1970-On job training, operations, CinCland Flt. Staff, Norfolk, VA; 1971-war game instruction, NWC, Newport, RI; 1973-Senior Reserve Officer Crs., NWC.

He assisted task group in organizing NRA Charleston Chapter #147, November 1971. The chapter consisted of 76 members. He became its first elected president March 23, 1972. Black retired July 1, 1974 and transferred to Retired Reserve with the grade of CAPT but without pay. Placed on Retired Roll with pay Oct. 10, 1984. He was awarded the ATM, AP, WWIIVM, NOSM w/Asia Clasp, NRM, NDSM, AFRM, HGD in lieu of 2nd AFRM and the PLM.

His memorable experiences: the USS LST-464 is the only known ship in the Pacific to be converted to an armed First Aid Ship. Instead of carrying tanks it carried hospital beds in a sick bay alongside of operating rooms. It was a floating hospital in a combat zone with 10 medical officers and 50 hospital corpsmen. A chaplain was also assigned. This ship was painted green (not the white of a hospital ship). Therefore it did not enjoy the immunity that a hospital ship (which stays some 20 miles outside of the combat zone) does. This LST served as a station hospital, medical supply, and evacuation ship. It evacuated patients to hospital ships or to air strips to be carried to rear area hospitals.

Most significant civilian positions were: 1978-82, manager of Finance Services in the Corporate Office's Logistics Department located in New York City; 1982-85, financial controller of construction for the reconfiguration of two large paper mills. Before retirement he was a CPA in Connecticut, Arkansas and South Carolina.

1972 State President of "Confederation of South Carolina Local Historic Societies; 1989 chairman, Georgetown County Historic Commission; 1987-89 President of Georgetown Myrtle Beach Chapter #341 of National Assoc. of Accountants; 1989-92 President of the Navy Club of Georgetown, SC; 1996-97 adjutant for local camp of Sons of Confederate Veterans.

Married Carolyn Haigler and they have two children, Bryan Anthony and Steven Douglas; and two grandchildren, Allison Marie and Hunter Benjamin. He retired from International Paper Co. Dec. 1, 1985.

GEOFFREY A.L. "BLACKIE" BLACKMAN, LCDR, born June 14, 1920 in Canada, received an AA degree. He joined the service June 1941.

His military locations and stations: Pensacola, naval aviator #10012, served with VP-23, NAS Corpus Christi 18, VPB 71. He fought in Midway and Guadalcanal. He was awarded the Distinguished Flying Cross and three Air Medals.

Blackman achieved the rank of LCDR and left the service December 1976. His memorable experiences are Midway and Guadalcanal invasion Aug. 7, 1942.

He married Dorothy and they have two children, Marjorie and Thomas; and two grandchildren, Kristia and Erik. He is a partner in large insurance brokerage firm and retired in 1962.

JAMES F. "JIM" BLACKSTOCK, CAPT, born Sept. 19, 1947 in Los Angeles, CA, received BA in English from the University of Southern California in 1969 and a JD degree in 1976.

Served NROTC, 1966-69 and active duty July 1, 1969-July 12, 1973. Units served with include: DESRON 23; ASW GRU ONE; Seventh Fleet. His campaigns include: Vietnam Campaign 1969-73.

Blackstock was awarded the Navy Commendation, Navy Achievement Award (2), Armed Forces Expeditionary Medal, Vietnam Service. He achieved the rank of CAPT (Reserves); LT (active duty), discharged from active duty July 12, 1969.

His memorable experiences: ASW in WestPac, carrier screens and plane guarding in the Tonkin Gulf; gun fire support off the coast of Vietnam; Haiphong Harbor in July 1993 during minesweeping operations; circumnavigation of South America aboard USS *Cook* spring 1972.

He married Kathleen Weigand and they have two children, Kristin and James Jr. A lawyer and law firm partner, he is currently serving in the Selected Reserve.

THOMAS M. "MIKE" BLAIR, CAPT, born April 1, 1946, Baltimore, MD. He is a graduate of Towson State University (BA, political science) and the University of Baltimore Law School (JD). He served as an enlisted cryptologic "O" Branch Tech in the USNR from April 1, 1963-April 1, 1971, including an 18-month tour of active duty with the Naval Security Group Activity, Kamiseya, Japan, 1969-71.

In April 1976 he was commissioned as a cryptologic special duty officer and served in various billets within Reserve Naval Security Group Units in Baltimore, MD; Adelphi, MD and Atlanta, GA, including tours as XO (1982-86) and CO (1986-88). In 1990 CDR Blair entered the Naval Reserve Intelligence Program on an exchange tour and served as an administrative officer or production officer of several units supporting the Office of Naval Intelligence and other National Intelligence and Law Enforcement Agencies within the US and Allied governments where he continues to serve. CDR Blair served at the Navy International Programs Office in Washington, DC during the latter stages of Operation Desert Storm in 1991.

His service medals include the National Defense, Naval Reserve Meritorious Service, Armed Forces Reserve, Navy Unit Commendation, Meritorious Unit Commendation, Naval Reserve Sea Service and Naval Reserve Overseas Service. He has been a Naval Reserve Assoc. member since 1976.

Since 1978 CDR Blair has practiced as an attorney at law in both Pennsylvania and Washington, DC with an international business practice specialty and he is president and cofounder of several international consulting companies including Business Ventures International and Global Tracking Systems, Inc. The Blair family consists of wife, Brenda, and daughters, Lauren and Shannon. They reside in Baltimore County, MD.

ROBERT WALLACE BLAKE, LCDR, born May 24, 1921, Quantico, VA, received BS in aero engineering from MIT 1941. Graduate study at MIT Columbia Graduate School of Business and US Navy War College.

He joined the service April 7, 1944. His military locations and stations: USNPFS St. Mary's, NAS Memphis, Glenview, Corpus Christi, Pensacola, New York, Quonset Point, Oceana, Cecil, Rota (Spain), Seattle, CINCUSNAVEUR (London). Units: VA 832, 3; BARTU 832; WEPTU 891.

He was awarded the Naval Reserve Medal, American Theater, WWII Victory Medal, Armed Service Reserve. Blake achieved the rank of LCDR and was discharged Jan. 27, 1947.

His memorable experiences: Logging the events of the Seven Days War while serving on ACDUTRA at CINCUSNAVEUR in June 1967; sea duty in USS *Saipan* January 1947; USS *Gilbert Islands* in June 1954 and USS *Lloyd Thomas* in June 1964.

His wife Ruth died in 1989. He was employed for 41 years with Pan American World Airways 1941-82. Retired from Reserve July 1, 1971; with pay, June 1, 1981; from Pan Am, May 1, 1982. After retirement served as a consultant, writer and lecturer. Active throughout his career in industry and professional society committees and working groups of AIAA, SAE, ATA, IATA, ICAO and RTCA. Articles are published in SAE *Transactions*, Seattle *Times*, Seattle *P-I*, NWC *Review*, Naval Institute *Proceedings* and Washington *Spectator*.

RUSSELL S. "RUSS" BLAKE, CDR, born Feb. 15, 1924, Kansas City, MO, received BSBA University of Missouri, Columbia, MO June 1948.

Joined the service July 6, 1943, and served as naval aviator with USNR, VF-97 (Air Gp. 97). Achieved the rank of ENS and was discharged Nov. 29, 1945.

Memorable experiences: continued in the Selected Reserve, reaching rank of CDR and served as CO of Reserve All Weather Fighter Sqdn. VF-881.

He married Ruth Richards and they have three children: Rhayma Ann, Roger T. and Brian R.; and four grandchildren: Kristen, Trisha, Danny and Adam Blake. Employed with FAA as air traffic control specialist/automation staff specialist/manager, retiring May 3, 1994.

BILLY J. BLANKENSHIP, CAPT, born Longview, TX, Feb. 13, 1928. His mother was a school teacher and father a railroad engineer who fought with Texas/Oklahoma, 90th Inf. Div., WWI, recommended for Distinguished Service Cross and was wounded. They lived in Mineola, TX; lettered football, track when 14, and recommended for Carnegie Medal for Heroism (rescued elderly woman from burning home).

Graduated Polytechnic High, Ft. Worth, 1944 age 16; Army ROTC North Texas Agriculture College; Branch Texas A&M, Arlington. At 17 volunteered for Merchant Marine; Maritime School Catalina Island, CA. Served aboard troop transport during Okinawa invasion, ashore for short period of land combat and later, tankers and freighters, Atlantic/Mediterranean/North Sea.

When 18 volunteered for Regular Navy and served two years in Pacific aboard destroyers; merchant ships worldwide for another three years. Was in China during fall of Tientsin and ended career as licensed third officer.

He earned BS, BA, MED, degrees with honors, Stephen F. Austin State, Nacogdoches. A 1958 honor graduate, University Texas Dental Branch, DDS degree, OKU Honor Society. Two years USPHS Post-Doctoral Fellowship Pursuing PHD commissioned LT DC USNR 1958. Organized, commissioned, commanded USNR Dental Company 8-3. Battalion Dental Officer 1/23, 4th MARDIVUSMCR.

A 1963 honor graduate, University Texas Medical Branch, MD degree; AOA Honor Society; Surgery Internship/Residency Training for board certified oral/maxillofacial surgeon. Assistant professor surgery and chief, Division Oral/Maxillofacial Surgery 1967-69. Battalion Surgeon 1/23, and Delta Company, 4th AMTRAC BN, USMCR. Attended Army Parachute School, Ft. Benning.

Volunteered for Vietnam with US State Department and USNR as surgeon and with US/Vietnamese Special Forces, 1966. Received several Commendations from Government South Vietnam and Humanitarian Service Award from American Medical Association. Promoted CDR, Dental Corps, USNR, 1968; private practice, Harlingen, TX, 1969-71 .

Professor of Surgery/Pharmacology, University Texas Health Science Center, San Antonio. Major, USAFR flight surgeon and hospital commander; 500 hours C-130 aircraft, 433rd Tactical Airlift Wing; graduated USAF Air War College, Air Command & Staff College, and Aerospace Medicine, 1971.

Re-entered Navy in 1975 as commander, MC, active duty; Submarine/Diving School and Naval Special Warfare; two years Western Pacific operating from Subic Bay, Philippines in USS *Grayback* as submarine/diving medical officer for naval special warfare (UDT/SEALS) and 7th Fleet; SPECOP with Seals.

Promoted to CAPT and force medical officer, Submarine Force, Atlantic Fleet, Headquartered CINCLANTFLT, Norfolk. Served TAD on several nuclear fast attack and fleet ballistic missile submarines, submarine tenders and diving vessels, 1977.

Returned to Okinawa after 35 years, as division surgeon, 3rd MARDIV, and force surgeon, III Marine Amphibious Force, headquartered Okinawa. Japan, Korea, Philippines, Thailand, 1980-81.

XO/director Clinical Services, Naval Hospital Corpus Christi, 1982-83; left active duty in 1983 to command USNR Combat Support Hospital 10-3 and to practice oral/maxillofacial surgery, hyperbaric, occupational, and emergency medicine. Medical director, South Texas Nuclear Project three years. Special Boat Unit 22 (SPECWAR), New Orleans and flight surgeon, Naval Hospital, Corpus Christi.

Retired from Navy in 1988 as medical/dental officer, flight surgeon, parachutist, deep sea diver, submarine/diving medical officer. Also drug medical review officer/

aviation medical examiner. Commercial pilot, instrument, multi-engine rated. Traveled 90 foreign countries, and 50 US States. Served briefly WWII, China Civil War, Korean and Vietnam Wars.

Boards in oral/maxillofacial surgery, family practice, hyperbaric medicine. Belonged to 70 scientific societies, published/presented 150 scientific articles and been on staffs at 30 hospitals.

He was awarded the Legion of Merit, Navy and Marine Corps Medal, Purple Heart, Navy Meritorious, Navy Commendation (2), Combat Action, Navy Unit Commendation, Meritorious Unit Commendation, Air Force Outstanding Unit, Organized Marine Corps Reserve (2), Armed Forces Victory, Navy Occupation, China Service, National Defense, Vietnam Service, Humanitarian Service, Sea Service (2), Navy and Marine Corps Overseas, Fleet Marine Force Ribbon, Air Force Longevity, Armed Forces Reserve Expert, Merchant Marine Victory, USMM Pacific War Zone, USMM Atlantic War Zone, USMM Mediterranean War Zone, Vietnam Meritorious Unit Citation Gallantry Cross, Republic Vietnam, China War Memorial, Association Military Surgeons, Letter of Commendation (5), Letter of Appreciation (4), Parachutist Wings, Basic, Flight Surgeon Wings, Navy Parachutist Wings, Submarine Dolphins, Diving Medical Officer Badge, Command Ashore Badge, USMM WWII Badge, Vietnamese and Philippine Parachutist Wings.

Married Carolyn Keeling Jan. 26, 1952 and they have daughters: Roberta, national sales rep for watch company, Chicago; Jimmie, handicapped at home; LT Jean Blankenship, combat rated patrol plane commander presently flying Arabian Gulf. He has Gold Naval Aviator Wings and Army Parachute Wings.

C. DENNIS BLANZ, CDR, born Feb. 20, 1935, Chicago, IL, received BA, De Paul University of Chicago; MBA, Loyola University of Chicago.

Joined the service June 16, 1958. Military locations and stations: USS *Taluga* (AO-62), Naval Reserve Intelligence. Served with 7th Fleet, Pacific, engineer officer; Unit 9-1 Glenview, IL, executive officer.

He was awarded the National Defense, Armed Forces Reserve Pistol Marksman. Blanz achieved the rank of CDR, retiring Feb. 20, 1995.

His memorable experiences: Goodwill tour of Japanese Ports; Plebe at the US Naval Academy, Annapolis, MD, (1956-1957).

He married Marjorie and they have three children: Barbara, Jennifer and Michelle; and five grandchildren: Lindsey, Rachel, Mary Caitlin, Rachel Elizabeth and Joseph John.

Blanz is senior vice-president, Bank of America Trust Co. of Florida, Baca Raton, FL.

NORMAN L. "NORM" BLEMASTER, CAPT, born July 26, 1940, St. Johns, MI, received a BA degree from Michigan State University and MS degree from George Washington University.

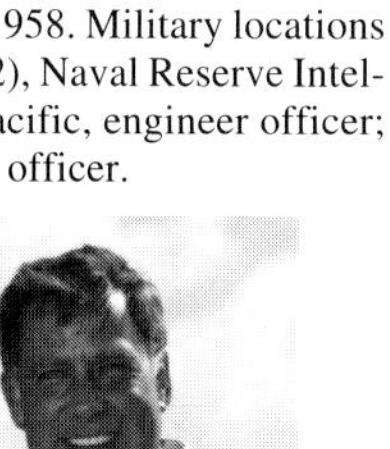

He joined the service July 1962. His military locations and stations: Vietnam; Norfolk, VA; OCS Newport, RI; Dallas, TX. Served with Intelligence units in Dallas, S. Weymouth, Memphis, New York and Minneapolis.

Blemaster was awarded the Expert Pistol Medal, Navy Unit Commendation, Armed Forces Expeditionary, Meritorious Service Medal, and Navy Commendation, Vietnam Service Medal. He achieved the rank of captain.

His memorable experiences: two years service as CDR, Reserve Intelligence Area Ten and a tour in Vietnam.

He married Lou and they have two children, Kimberly and Robert. He retired as vice-president of NOVUS Inc. in July 1995.

ROGER D. BLEVINS, LCDR, born March 12, 1947, Harlan, KY and reared in Tampa, FL. He attended Florida State University, earning a BS degree in physical science (1974) and MS in land use planning in 1978. He earned an additional MPA degree from the University of North Florida in 1982. An associate of science in aerospace technology was completed in 1972 with a commercial pilot certificate, with instrument and multi-engine ratings.

He began his military career in 1966 with an enlistment in the USAF. He completed electronic technical training at Kessler AFB, MS in 1967; served on active duty until 1970; following two years as an intelligence specialist second class, he was selected for direct commission in 1980. His awards include the NDSM w/star, MUC, AF Good Conduct, Joint Services Commendation and Defense of Kuwait.

LCDR Blevins was recalled to active duty September 1990 and served until June 1971. He served on the US Central Command, CINC J-2 staff in Riyadh and other locations in Saudi Arabia. His memorable experience include Scud attack in Riyadh and Dhalran.

LCDR Blevins is currently the program manager for AF Comprehensive Planning, at the AF Center for Environmental Excellence. He is responsible for executing comprehensive planning for AF bases world-wide.

The Blevins family consists of the former Elizabeth Beaty of Indianapolis, IN, and their three children: Robert, Matthew and Emmalyn. They reside in San Antonio, TX.

CHARLES W. "CHARLY" BLOOD, LCDR, born Aug. 27, 1924, St. Louis, MO, received BSEE from the University of Illinois, 1948; JD, St. Louis University, 1975. Commissioned Ens. Civil Engineering Corps. Registered Missouri Bar Association; registered engineer in Missouri.

Joined the service March 27, 1943. His military locations and stations: Farragut, ID; San Diego, CA (Sonar School); USS *Weyandotte* (PC-1141); University of Notre Dame V-12; Indiana State V-12; University of Illinois V-12; NROTC; Opa Loca Marine Corps Air Station, Opa Loca, FL 1952-1954; Several volunteer reserve units in St. Louis; WWII, antisubmarine convoy duty in Caribbean; Korean Campaign, Opa Loca, FL.

He was awarded the Good Conduct Medal, American Campaign Medal. Achieved the rank of SOM2/c enlisted; LCDR (CEC) commissioned. Retired in 1973 w/o pay. Discharged as SOM2/c in 1946, but immediately re-enlisted in NROTC.

The whole Navy, from boot camp in Idaho to convoy duty in the Caribbean to V-12 college to active duty with CEC in Opa Locka, were all memorable.

He married Patricia Mary Muesenfechter June 24, 1950. They have six sons: Charles W. Jr., Gerard M., John R., Richard E., Joseph B. and Daniel B.; and 18 grandchildren. Retired as manager from Telecommunications Services, Union Electric Co., St. Louis, MO in 1969.

FELIX EDWARD "TURBO or PHIL" BLUM, CDR, born Sept. 26, 1928, Garden City, NY, attended college for three years. He joined the service July 8, 1949. His military locations and stations: Pensacola North Island; Barbers Point, HI. Served with VC-3, VF-884, VF-44, NAS Moffett, Pax River. Fought in Korea 1952-53, USS *Kearsarge* and USS *Boxer*.

Blum received Distinguished Flying Cross and five Air Medals. He flew 72 missions, achieved the rank of CDR and was discharged June 6, 1975. He served seven years active duty and 18 years in USNR.

Memorable experiences include being a VC-3 night fighter pilot F4U Corsair in USS *Kearsarge* and USS *Boxer;* Korea 1952-53 and graduate US Navy Test Pilot School, 1955-56.

Blum has worked 41 years as an experimental test pilot for McDonnell Douglas is still working. He married Genevieve "Gen" and they have three children: Phillip, James and Lynne; and a granddaughter, Sabrina Blum.

ARTHUR W. "ART" BLUME III, CDR, born Oct. 27, 1936, in Springfield, MO, received BA (1958); MED (1970); from Drury College, Springfield, MO.

He joined the service Jan. 7, 1954. His military locations and stations: USS *King County* (AG-157) (Pacific) ACU-1 Coronado CA (Pacific). Units served with: USS *King County* (AG-157) and ACU-1.

His awards: Armed Service and Naval Reserve Medals. He achieved the rank of CDR and retired April 1, 1977. His memorable experiences: early involvement in guided missile test program at Pt. Mugu and holding five USNR Commands.

He married Vikki D. and they have five children: Art IV, De De, Steve, Robyn and Julie; and seven grandchildren: Nicole, Amanda, Rachael, Keirston, Dalton, Nathan and McKenzie. He is owner of Exec-U-Travel, Inc.; charter president of Claud V. Rickett Chapter Naval Reserve Association; active member of Rotary International, Blue Lodge, York Rite (KT), Scottish Rite 32 degree, Shriners, Fleet Reserve Association, Retired Officers Association and American Society Travel Agents.

EDWARD A. BOBRICK, CAPT, born in Chicago, IL, June 30, 1933. Graduated from the University of Illinois in January 1958, received a BS degree in marketing and graduated from DePaul University Law School with a JD degree, June 1964.

He began his naval career on March 3, 1958 at OCS, Newport, RI and was commissioned an ensign in the USNR on July 1, 1958. He was assigned to the USS *Cromwell* (DE-1014), where he qualified as an officer of the deck, (fleet operations), and was selected as Weapons Department Head. He was selected for temporary assignment to Commander Destroyer Atlantic Forces as part of a group of destroyer officers tasked with creating and bringing into existence a professional school for officers aspiring to become a Department Head aboard a fleet destroyer, this school initially was named Destroyer Officers School, later renamed Surface Warfare Officers School.

When released from active duty in June 1961, he joined the USNR holding many assignments which included, XO of Naval Reserve Surface Div. 9-3(M), and 9-5(L); CO of Naval Reserve Fleet Div. 9-20(D), Naval Reserve Surface Div. 9-4 (M), and Naval Reserve PERSMOBTM 1813. He was an original member of Naval Reserve COMSECFLT Staff 113. He was selected as plans and mobilization officer for Naval Reserve Readiness Command Region 13. He ended his naval career as CO of Naval Reserve Voluntary Training Unit 1314 (G).

Bobrick distinguished himself while on active duty for training (ACDUTRA), re-qualifying as an officer of the deck (Fleet Operations) aboard the destroyers USS *Crow* (DE-252), USS *Henley* (DD-762), USS *Bristol* (DD-857), USS *Ault* (DD-698) and USS *Huntington* (DD-851). He also qualified as an XO aboard the USS *Dyess* (DD-880). While performing ACDUTRA with CDR SECOND FLEET, in underway exercise "Agate Punch," he was the first selected USNR officer to qualify as a flag staff watch officer with that command.

He participated in NATO exercises "Establish Contact" 76 and 78 at the US Naval War College, and was one of the first reserve officers assigned duties regularly handled by COMSECFLT active duty members. He performed ACDUTRA with the CDR Naval Reserve Forces, New Orleans, LA, Deputy Chief of Naval Operations for Logistics, (DCNO OP-40) and commander in chief of the Atlantic Forces. While with DCNO OP-40, his duties involved correcting a difficult problem in Navy logistic planning, and his work was noted to have far reaching positive effects for the Navy in the future. He completed two courses of study at the US Naval War College, Newport, RI.

He qualified as a surface war officer and was authorized to wear the Surface Warfare Officer Breast Insignia. He was awarded the Navy Commendation Medal, the National Defense Service Medal, the Armed Forces Reserve Medal and the Reserve Sea Service Ribbon. Bobrick completed 30 years of commissioned service as a surface warfare officer, retiring July 1, 1988.

The Bobrick family resides in Skokie, IL and consists of his wife Jane and their children: Benjamin, Rachel and Nathaniel. He is a US Magistrate Judge in the US District Court for the Northern District of Illinois, and has served in that position since 1990.

S. DAVID BOGART, CAPT, born Sept. 3, 1937, NYC, USA, received a BA and MD degree from Dalhousie University, Halifax, Nova Scotia.

He joined the service May 1962; stationed in the USS *Raleigh* (LPD-1) and USS *Spiegel Grove* (LSD-32); and served at the San Diego Naval Hospital. He fought in Dominican Republic Civil War (1965).

Bogart received the Commendation Medal in 1966. He achieved the rank of CAPT, MC, USNR, retiring Aug. 1, 1967.

He married Arlene and they have two children, Adam and Laura. He has held the position of radiologist, Director of Department, Coler Memorial Hospital, New York City, retiring October 1991.

MARION WILLIAM BOGGS, LCDR, born May 17, 1909, Mexico, MO, received a BA in 1932; MA 1933 University of Missouri, PhD 1940, University of Chicago.

He joined the service Feb. 3, 1943, NTS, Dartmouth 1943-44, Navy School of Military Government, Columbia University 1943-45, AVP 21 1944-45, HQ 12th Fleet, London 1943-46 (assistant to naval advisor, European Advisory Commission (EAC) 1944-45, deputy secretary-General EAC 1945).

Boggs was awarded the Commendation Ribbon and European Theater Ribbon. He achieved the rank of LCDR. Ordered to inactive duty, Sept. 23, 1945.

His wife Ida Elizabeth Boggs is deceased. His children are Charles William Boggs, attorney Missoula, MT and Elizabeth MacLeish (nee Elizabeth Baird Boggs) professor Trinity College, Hartford, CT.

He has held positions as foreign affairs analyst, Dept. of State, 1942-43, 1945-47; staff, National Security Council, 1947-59, deputy executive secretary, National Security Council, 1959-61; foreign service reserve officer, Department of State, 1961-70. Retired from the USNR, Jan. 1, 1954.

ROBERT W. "ITCH" BOHN, CAPT, born Oct. 13, 1932, Cleveland, OH, received a BS in metallurgical engineering. He joined the service in 1956, was stationed at Pensacola for flight training; at NAS Whidbey Island; MCAF Iwakuni Japan; NAS Glenview, IL, and served with VP-50 (active) VP-90 and many reserve units at NAS Glenview, IL.

Bohn was awarded the Naval Reserve Medal. He achieved the rank of CAPT and retired in 1987. His memorable experiences include when he soloed, becoming designated patrol plane commander, being CO of VP-90 and having command of three other Reserve aviation units at NAS Glenview, IL.

He married Esther Jane and they have two children, Sandra and Cynthia. Employed as manager of operations in metals industry.

LELAND E. "LEE" BOLT, CAPT, born Aug. 30, 1928, Payette, ID, received BS, USNA, MS (systems management), USC. He joined the service June 1, 1951. His active duty military locations and stations: USS *Lenawee* (APA-195); USS *Princeton* (CV-37); Fleet Gunnery School San Diego; USS *Uhlmann* (DD-687); USS *Wiseman* (DE-677); Staff CINCPACFLT and OASD M&RA Total Force Study. Ready Reserve: CO USS *Wiseman* (DE-667), Staff COMRESDESDIV 11ND, Staff COMCRUDESPAC, and CO IUW Unit 11-1.

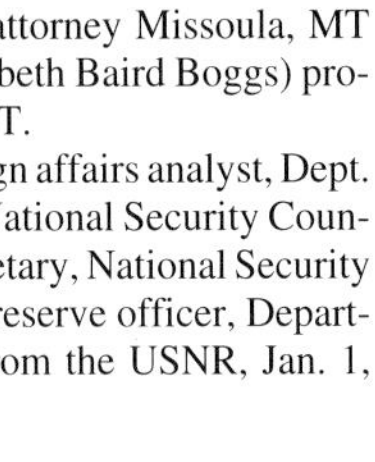

Battles and Campaigns: Korea and Vietnam. Bolt was awarded the National Security Industrial Association Award of Merit and the Navy Commendation Medal. He achieved the rank of captain; separated from active duty Feb. 19, 1974.

His memorable experiences include command of USS *Wiseman* (DE-667); active duty as CO of his Systems Analysis Unit and RADM Woods (COMCRUDESPAC) inviting him to lunch or other functions for discussions. Fortunately, home was on the way to the flag ship, so he could change into and out of uniforms to work both jobs. Also memorable was the day all his USS *Princeton* (CV-3) boiler gang manned their fueling stations in black shirts.

He is married to Jacqueline and they have three children: Leland Jr. (veterinarian in Asheville, NC); Jon (project engineer, Hewlett Packard, Boise) and Clay (clinical pharmacist in San Diego). They have three grandchildren: Ryan, Jordan and Taylor.

Bolt has been employed with General Dynamics, Convair Div., as research engineer, project engineer, Tomahawk Support Systems manager and manager of technology development. He retired from USNR Oct. 1, 1977 and General Dynamics Sept. 1, 1992.

ROBERT BASHFORD "BOB" BOLT, CAPT, born March 20, 1917, Berkely, CA, graduated from Stanford University in 1940 and attended USNA PG School, 1945-46. He joined the service Sept. 6, 1940, USN Midshipmens School Abbott Hall, Chicago, 1940-41. 1946 (following Post Graduate Course at US Naval Academy). Reported to USS *Duluth* (CL-87) in Pacific as head of department. Transferred to Reserve Forces 1947 had command of division, battalion, group command (2).

His military locations and stations: USS *Indianapolis,* 1941; Naval Armed Guard, 1942. Served with Destroyer Escorts (Atlantic) 1943-45, Cort Ron 26-1945-46, USNA; Indian Ocean April 1942-North Africa 1942; PG School. Served in all areas of the world during WWII, American, European, Pacific, China and Indian Ocean.

He was awarded the Silver Star, Navy Commendation (4), Armed Forces Medal, Reserve Medal. Bolt achieved the rank of CAPT (0-6). His memorable experiences: Korea 1950-52, PACFLTTRNGCMD (gunnery and camera party); national president, NRA 1967-69; Commander General, Naval Order of the United States, 1975-77.

Employed 1947-50 and 1952-58 Travelers Insurance Co.; 1958-70, Life Insurance Co. of North America; 1970-82, Old Republic International. In each company served as manager, training officer, vice-president involved with agent development, training, corporate development. Held All Lines Insurance Brokers License, Held Securities License (NASD). Licensed in 12 states as non-resident broker.

He married Margaret Kingsland Hewitt and they have one child, Leslie Topliff Dennis; and grandchildren, Jeffrey Mercer and Brian Bolt Dennis.

EMILE L. BONNOT, CAPT, born Orange, NJ, May 11, 1907, BA Rutgers University in 1929. He was a plank owner, of USS *Fremont* (APA/RAGC 44) where he served in various billets including XO.

During his extensive tour in the Pacific, *Fremont* was flagship for Admirals Blandy, Fechteler & Loomis and Commanding Generals of Army and Marine Div. in the invasions of Saipan, Pelelieu-Anguar, Leyte, Lingayen, Iwo Jima and Okinawa.

Fremont as commanders headquarters afloat stayed from the initial landings until the Islands were declared secured. Aboard *Fremont* Bonnot survived shelling, strafing, bombing, kamikaze attacks, underwater swimmers carrying mines, typhoons and an engine room fire. While on watch, through his binoculars he witnessed the two flag raisings at Iwo Jima.

In the Naval Reserve, Bonnot commanded Bn. 3-10, founded at MSTS Co. He commanded MSTS Div. 3, NCSO Div. 3-4 and served on Group Command Staff.

A companion and life member of the Naval Order of the US, he is past commander of the New York Commander, past Historian General and now Historian General Emeritus. He received the general Commandery Certificate of Appreciation in 1985 and 1989. He was honored at the Centennial Congress in 1990 for his work as historian. He joined the Naval Reserve Association in 1954 and is life member #46. He was president of New York Navy Chapter; 3rd District President; National Treasurer & National V-P Budget and Finance. He was instrumental in establishing the insurance program.

He and his wife Dorothy (De) were plankowners and major donors to the NRA Building Fund. In her memory he has established the De Bonnot Anchors Award to honor annually a member of the anchors group.

WILKERSON "BILL" BOONE, CAPT, born Dec. 22, 1921 in Philadephia, PA, received BS engineer, Lafayette College and MBA Temple University.

He enlisted Feb. 4, 1939-Dec. 21, 1949 and was commissioned Dec. 22, 1949-Oct. 31, 1977. His military locations and stations: NAS WilGro PA; Guam, West Pac, Japan. Unit served with: Arou#4, VS-931, USS *Sicily,* USS *Badoeng Strait.*

Battles and campaigns: supported operations in Korea. He was awarded two Good Conduct Medals. He achieved WWII-ACMM, retired as CAPT, retiring Oct. 3, 1977.

His memorable experiences: everything was a great experience, from enlisting in WWII to serving in a commissioned status.

Boone married Isabelle in 1948. His daughter Linda was born in 1952 and Sandra in 1964. He retired from Philadephia Gas Works, as manager of transportation in February 1982.

RICHARD C. "DICK" BORDEN, CAPT, born Aug. 11, 1936, Minneapolis, MN; attended USMA West Point; received a BA in 1962 and MS in 1968 from University of Minnesota.

Joined the service Jan. 15, 1954. His military locations and stations: NAS Twin Cities, NARC Twin Cities, NAS Pt. Atugu, NAS Dallas. Served with NRIU 9-1-9, VP-65, NIAC 0278, RIA ID, COAQNAVRESINTCOM.

Borden was awarded the Joint Meritorious Unit Award, National Defense Service Medal, Armed Forces Reserve Medal (2), Expert Pistol Medal. He achieved the rank of CPT, USNR and retired Sept. 30, 1992. His memorable experiences: CO, NIAC 0278, 1984-85; Reserve Intelligence Area commander Area Ten, 1988-90.

He married Joyce M. and they have children, Eric and Christopher. He retired as president of Signal Affiliates, LTD, Sept. 30, 1992.

GLENN EDWARD BORGARD, CAPT, born Dec. 17, 1932, St. Louis, MO, received a BS and MS in engineering from the University of Missouri, Rolla. He joined the service October 1954. His military locations and stations: Newport, RI; San Diego, Great Lakes, IL, NS San Diego. Served with LST-836, Pacific 1955-58; NAS Glenview, NAS Memphis. He is a Korea veteran.

Borgard was awarded the National Defense Medal, Armed Forces Reserve Medal and Naval Reserve Medal. He achieved the rank of CAPT and retired December 1992 after 31 years active and reserve years.

His memorable experiences: XO LST-836; three years operations in Pacific; CO of WEPTU NAS Glenview; CO of NASRU NAS Memphis including training of pilots going to Vietnam; command duty officer NS St. Louis, MO.

He married Shirley D. and they have four children: Kevin, Michael, Steven and Gregory; and four grandchildren: Nicole, Nathan, Austin and Tylor. He is vice-president Volz Engineering, St. Louis, MO.

LOREN C. BORGWARDT, CAPT, born Feb. 4, 1929, Alma, Buffalo County, WI. Earned BS degree in electrical engineering and BS in naval science, University of Wisconsin. Joined the military Sept. 26, 1946, Naval

Reserve Officers Training Corps, Regular Program.

Stationed from 1946-53 in USS *Albany* (CA-123); *Boxer* (CV-21); *Toledo* (CA-133); *Los Angeles* (CA-135). From 1953-77 at various facilities and units: NR Electronics Facility, Eau Claire, WI; NR Training Center, La Crosse, WI; HQ 9th Naval District, Great Lakes, IL; *Ticonderoga* (CVA-14); *Rowan* (DD-782); NR Naval *Security Group Div. 9-7; Newport News* (CA-148); Skaggs Island, Sonoma, CA; *Neosho* (AO-143); Fleet Anti-Submarine Warfare School, San Diego, CA; Naval and Marine Corps Reserve Training Center, Twin Cities, MN; USNA.

Participated in Korean War. USS *Los Angeles* made three deployments to WesPac during his active duty assignment. CAPT Borgwardt's final retirement was Feb. 4, 1989. Received the WWII Victory Medal, National Defense Service Medal, Korean Service w/5 Battle Stars, UN Service Medal, Korean Presidential Unit Citation Award, Armed Forces Reserve medal w/HGD, four formal commendations and many informal commendations.

Married Jean and they have three children: Steven, Joan (Ludvigson) and David; and two grandchildren, Brian and Cara Ludvigson. Employed with Northern States Power Company (Wisconsin) and retired as director of hydroelectric projects Sept. 1, 1988.

WILLIAM B. "BILL" BORRMANN, CDR, born June 28, 1920, Buffalo, NY, received BS (aero engr.), MBA and BS (math). He joined the service in 1944 and was stationed in NAS JAX, NAS Grossell, NATTC Memphis. Borrmann achieved the rank of CDR. His wife Georgia is deceased. He was employed at Ford Motor Co. and was department manager at time of retirement.

JAMES A. "JIM" BOULDIN, CAPT, born Feb. 7, 1930 in Tennessee, received a BS in engineering from Tennessee Tech.

He joined the service July 9, 1952. His military locations and stations: 1952-56, OCS Newport, RI; Naval Gun Factory, Washington, DC. Served with units: Taiwan, Formosa-Nationalist China Navy, served on PC-1168 and PGM-31 as advisor, XO Naval Comm Sta. Key West, FL, 1956-79 served as CO of Surface Divs. 6-75, 6-76 and Group Command 6-22, Chattanooga, TN. He achieved the rank of captain.

Married to Jane and they have two children, Margaret and Meredith. He is president of Cannon Equipment, Chattanooga, TN. Retired in 1979.

ROBERT SHEPARD "BOB" BOURCIER, born Feb. 10, 1938, Milwaukee, WI, received a BS and MA in business and is certified in public administration. He joined the service in June 1955. His military locations and stations include USS *Yorktown*, fleet work study group, PAC, San Diego.

Bourcier achieved the rank of CDR and was released July 1976. Memorable Experiences: the Cold War; 4 years active duty; 17 years Reserve duty; 1961 when a Russian Badger flew over their ship.

He married Nancy Kalbfell and they have three children: David, Suzanne and Diana; and four grandchildren: Garrett, Reneé, Kevin and Tracy. He has held positions in city government including city manager for a total of 18 years; vice-president of electronic manufacturing, manufacturing amplifiers for USN. He will retire in 1998.

ALLEN "SNOOKS" BOUSKA, CDR, born Sept. 18, 1945 in Vallejo, CA, received a BS, Chico State, CA. He joined the service March 1967 and designated naval aviator, March 1969. His military locations and stations: Pensacola Plow Back VT-2 formation instructor. He served with VP-9, Moffett NAS, VIP Pilot NAVAIRPAC.

Bouska participated in Vietnam patrols from Subic, Da Nang and Thailand; was CO, Concord Naval WPN STA DET; CO Naval Communications, DET Wahiwa, HI and was discharged in September 1974. Retired as CDR, March 1992. His memorable experience: He did not spill the Admirals coffee.

He married Vickie Jett and they have two daughters, Lori and Allison. He was owner of Aerial Application, 1975-85, presently owner of Dairy Queen.

BRADFORD T. "BRAD" BOWEN JR., CDR, born July 16, 1922, Providence, RI, received a BS in agriculture. He joined the the USN as an aviation cadet July 1942, but was not called to active duty until May 1943.

In May of 1944 was in primary flight training at Norman, OK. When word of the purge was passed, he went to Midshipman School at Ft. Shuyler, NY and was commissioned ensign in March of 1945. In July of 1945 he was assigned to VR-12 and sent to Johnston Island.

Discharged in March 1946, graduated from Rhode Island State College in September 1946, moved to Miami and joined the Air Reserve. He served with UR-801 until Opa Locka closed then transferred to UR-742 in Jax FL. Served in various squadrons and retired in 1977 with 35 good years as a commander.

Married Dorothy and they have one son Brad III and a grandson Brad IV. He was owner of Bowen & Son Roofing Inc., Sebring, FL until his retirement in 1984.

R. RONALD "RON" BOWERS, CAPT, born July 8, 1933, born Newark, NJ, received a BS from the University of Maryland. He joined the service June 25, 1951, enlisting as HSSR.

His military location and stations: NTC Great Lakes; Hospital Corps School Portsmouth, VA; USNH, Quantico, VA; Naval Dispensary Pax River, MD; NNMC, Bethesda, MD RAD May 7, 1954. Served in 18 different Reserve units over a 39 year period including three tours as CO of Medical Units and three XO tours.

Memorable Experiences: appointed chief hospital corpsman July 1, 1961; appointed ENS MSC USNR, Oct. 23, 1962; served as director, Medical Services, REDCOMSIX, 1990-1992; recalled and served during Desert Storm as director, Public Safety, NNMC, Bethesda, MD.

Discharged as CAPT MSC USNR, July 8, 1993. He was awarded Meritorious Service Medal, MUC (Desert Storm), National Defense (two awards), Naval Meritorious Service Medal and Armed Forces Reserve Medal.

He married Charlotte L. and they have four children: Richard Jr., Elaine, Brian and Christopher. Employed as district commander, Montgomery County, Maryland Police Department, retiring Feb. 29, 1980.

DAVID R. BOWES, CAPT, born Aug. 31, 1948, Corning, NY. Earned BBA in marketing from New Mexico State University in 1970. He joined the service June 1, 1972.

Stationed at NAF, COMFAIRMED, Naples, Italy, AOCS and Flight School, San Diego, CA and Atsugi, Japan. Served with HC-1 and HC-9. Capt. Bowes was discharged in June 1978 as LT and retired from USNR in November 1995.

Memorable experiences include long cruise to MED in CV-42; last cruise in USS FDR; saving a fellow airman whose plane went down at sea; long cruise in CV-41, USS *Midway*, Atsugi, Japan; shellback; Blue Nose above Arctic Circle Norway; 2,000 flight hours (C-1, S-2, C-131, H-1, H-2, H-3, H-46).

Married Judy Steinkerchner and they have two children, Lauren and Leigh. He was east area government sales mgr. for Herman Miller until retirement in November 1995.

JOHN B. BOYD JR., CDR, born April 1, 1941, Monroe, NC, graduated from East Carolina University in 1965 with a degree in geography.

He joined the service June 1965. His military locations and stations: Pensacola, FL; Corpus Christi, TX; Jacksonville, FL, VP-45. He served with VP-45 and was instructor VT-27. Boyd did tour in Vietnam 1968-69, Mediterranean tour 1969-70. He achieved the rank of LT and was released from active duty in 1971.

Served 20 years in the USNR at NAS Dallas, TX serving in numerous units, two XO tours, retiring in 1991 with the rank of CDR.

Married Lucy and they live in Corpus Christi, TX. He retired as assistant area manager for Texas Workforce Com. in August 1996 with 33 years service. Currently employed with Lockheed Martin at NAS CO, TX as a flight simulator instructor.

FREDRIC MARTIN "FRED" BOYDEN, CAPT, born Aug. 12, 1935, Schuyler, NB, received a BA and MD during his radiology residency. He joined the service fall 1953.

His military locations and stations include NAS Lincoln, NE; NSAT Hospital Taiwan; LeMoore Naval Hospital, LeMoore, CA. Served with Reserve units and three years active duty. Participated in Desert Shield and Desert Storm at Lemoore, CA. He received the usual medals 1955-1995. Achieved the rank of CAPT, USNR (MC).

Memorable experiences were two years in Taipei where their two oldest children were born; eight months Desert Storm duty in Lemoore ACDUTRA; teaching radiology residents at Oakland and San Diego.

Married to Jeanne Charvat and they have four children: Eric, Jenny, John and Scott; and four grandchildren: Zack, Emma, Ted and Abby. Employed as diagnostic radiologist for 32 years, retiring Aug. 12, 1995.

JACK I. "JB" BOYLAND, CAPT, born Sept. 11, 1929, Newark, NJ. Education: Seton Hall University, BS (chemistry/English;pre-medical); Philadelphia College of Pharmacy and Science, BS (pharmacy); USN War College and Fairleigh Dickinson University, Graduate School of Business, MBA cum laude, marketing.

Served in the USS *Holder* (DDE-819) 1951-53 and with reserve divisions at Newark, Perth Amboy, Bayonne and Clifton (all in New Jersey). CAPT Boyland retired from the USNR in September 1989. His medals include the European Occupation w/Mediterranean Clasp, Navy Reserve Medal, Armed Forces Medal and Expert Pistol and Rifle.

Registered pharmacist in New Jersey, Pennsylvania and Delaware; management in pharmaceutical industry for over 30 years; director product communications, professional services. Retired in March 1995.He is member of NRA, VFW and several professional societies

Married Audrey Jane and they have two children, Susan Catherine and Wendy June; and two grandchildren, Jack Ryan and Jordon Quinn.

HARRY VOSS "PERRY" BOYLES, JR., CAPT, born February 1921 in Kentucky, received BBA from the University of Cincti; NYU (even.) Notre Dam. Harvard B. (military).

Joined the service Oct. 6, 1942. His military locations and stations: ND University, SD, NO. LA Harvard U. Supply, ABSTU Norfolk, VA; Palawain, PI and Truk, Atoll. Units served with: ABSTU, Norfolk, VA; ACORN 52; Occu. for Truk and Atoll and Central Caroline Island.

Boyles was awarded the Asiatic-Pacific, Philippine Liberation, Naval Reserve, Armed Forces, American Defense and WWII Victory Medals. He achieved the rank of CAPT USNR (ret) LTJG in WWII and was discharged July 31, 1946.

His memorable experiences: CO of SSQD1-1 Boston, MA and CO TTHD, Boston, MA. Supply officer on staff BGEN Robert Blake Truk and Central Caribbean Island in 1945-46.

Married Helen Reed and they have a son Craig Stuart and granddaughters, Julia Audrey and Kathryn

Elizabeth. He was employed for GE Co., Frank Brown Co. and Mrs. Days Ideal Baby Shoe, retiring in 1975.

MATTHEW P. "MATT" BRAATZ, CDR, born Sept. 9, 1942, Wausau, WI, graduated from Tulane University, June 1966 and received a BA in history. He joined NROTC, commissioned Sept. 22, 1966, USNR. His military locations and stations: Norfolk, VA, USS *America* (CVA-66); CIC watch officer, assistant legal officer and division deck officer.

Braatz fought in Vietnam. He was awarded National Defense Medal, Navy Unit Commendation, Vietnam Gallery Cross, Vietnam Campaign Medal w/ 2 stars, NAVRES 10 Year Medal.

Reserve Billets: 1979-91: CO, XO, training officer PERSMOBTM 302; November 1977-December 1978, training officer, VTU 0219; May 1977-November 1977, XO MACG 0213. Attended MIL Justice School, Newport, RI, October 1966-November 1966. Retired from the USNR in 1991.

He achieved the rank of CDR and was discharged September 1969 from active duty. His memorable experience was crossing the Equator and becoming a "Shellback."

He married Sandra and they have four children: Robert K., Matthew P. Jr., Jennifer M. and Krissy (Nielsen); and grandchild Morgan Nielsen. Employed with Korter Management Co. (commercial real estate) as vice-president.

PAUL S. BRADY, CAPT, born in 1931 in Johnstown, NY and graduated from the University of Rochester in 1953. He did graduate work in English Law at Oxford University in England and received a JD from Harvard University in 1961.

He was commissioned an ensign in 1953 from OCS and served four years active duty at Naval Security Group (NAVSECGRU) activities in Hawaii and Philippines, and at the National Security Agency (NSA), Washington, in operations, communications and planning assignments.

As a Naval Reservist, Brady served for 31 years in Boston, New York City, Baltimore, Adelphi, Washington and Alexandria, including almost 10 years in command of six different NAVSECGRU units. He was also the first cryptologist assigned to a Readiness Command Staff, and served 8 1/2 years as the NAVSECGRU program manager of two Readiness Commands and a group command. His training assignments at national headquarters and field activities included NSA, NAVSECGRU Command, National War College, CINCPACFLT, CINCUSNAVEUR, NMPC for five reserve selection boards, and NAVSECGRU activities in Homestead, Northwest and Winter Harbor.

An attorney in private life, Brady was a member of the DOD General Counsel's staff, and retired in 1995 as the Principal Deputy General Counsel of NSA.

KENNETH N. BRAGG, LT, born Portland, OR, Aug. 21, 1923. Lived in Rochester, WA and attended St. Martin's College, Lacey, WA and Pacific University, Forest Grove, OR. Entered active duty at Navy V-12 unit, Washburn Mun. University, Topeka, KS, June 28, 1943.

Commissioned ensign June 1, 1944. Assigned to Submarine Chaser Training Center, Miami, FL, June 7, 1944 and then to the USS *Register* (APD-92) upon its commissioning at Charleston, SC, Jan. 11, 1945. Ensign Bragg was assistant communications officer and OD underway. At Okinawa "kamikaze" attacks disabled the *Register* after 3 "zeros" were destroyed by *Register* gunnery.

Following repairs the *Register* picked up survivors from the USS *Indianapolis* sunk by Japanese torpedoes after delivery of the atom bomb to Tinian in July 1945. Upon the surrender of Japan in September 1945, Bragg was assigned to the USNS, Seattle, following emergency leave to visit his gravely ill brother Harry, USNR.

He was promoted to LTJG Jan. 1, 1946 and released from active duty on Sept. 2, 1946. He had reserve cruise duty in 1947 on the USS *Swenson* (DD-729) and in 1948 on the USS *Buck* (DD-761).

He has medals for Philippine Liberation, WWII Victory and Asiatic-Pacific w/star. Following his military service LTJG Bragg received a BA in public administration and an MA in international relations from the U of W, Seattle. He then was staff director for legislature in Oregon, MD and New Jersey as well as a legislative analyst for the US Senate Appropriations Committee.

On Jan. 6, 1946 he married Margaret Bosworth. They have three children Janet Wainwright, Geoffrey and Janell Bragg; and two grandchildren, Shawn and Dana Miller.

ARTHUR W. "ART" BRAMBLE, CAPT, born Oct. 13, 1938 in Hammond, IN, received BS degree in Purdue University.

He joined the service March 1961. His military locations and stations: Pensacola NAS; Glynco NAS; USS *Kitty Hawk;* Glenview NAS; VP-722, VP-90, VTU-7272.

Married Martha J. and they have a son Bradley W. Employed with Northern Indiana Public Service Co., district manager, retiring Dec. 31, 1993.

JOHN H. "BIG JOHN" BRAME, CDR, born Dec. 25, 1929, Los Angeles, CA. Enlisted June 23, 1948 in Marine Corps Reserve. Served in A Co., 13th Inf. Bn., 1st Marine Div. based at Santa Monica, CA.

Served as enlisted man in the Naval Reserve from December 1950 in Patrol Sqdn. Seven Seven Three homeported at NAS, Los Alamitos rising to personnelman third class.

Commissioned an ensign in the USNR on Aug. 11, 1956. Inactive duty assignments included Auxiliary Aviation Unit Seven Seven Three, Air Antisubmarine Sqdn. Seven Seven Six, and Attack Sqdn. Seven Seven Six all of which were homeported at NAS, Los Alamitos, CA; and NARD NAS, Los Alamitos, CA. Served as administrative/personnel officer in above squadrons and as division officer in Naval Air Reserve Div.

Performed active duty Naval Air Stations, Lemoore and Los Alamitos, CA from January 1968 to November 1969 in support of USS *Pueblo* (AGER-2) crisis. From February 1971 to December 1976 served on extended active duty in the Navy Recruiting Div., Bureau of Naval Personnel and the Navy Recruiting Command when that division was redesignated as the Navy Recruiting Command in April 1971. Served as Naval Academy liaison officer 1990-1996.

He was awarded the Navy Commendation Medal, National Defense Service Medal, Armed Forces Reserve Medal. Brame achieved the rank of CDR and was discharged Dec. 30, 1990.

His memorable experiences: plankowner, Navy Recruiting Command, division officer, Navy Recruiting Command, Education Liaison Div.

He is divorced and has two children, Roderic Ian and Christine Laurel; and five grandchildren: Michael, Rita, Gabriel, Hannah and Nathanael. Brame is a high school teacher/coach and real estate sales associate.

JOE E. BRANT, CAPT, born Nov. 3, 1916, in Crescent, OK and graduated from Oklahoma State University in 1938. Entered Naval Aviation Cadet Training in November 1940, and was commissioned an ensign and naval aviator in September 1941, at NAS Jacksonville. He was assigned to VP-22 (PBY) in Pearl Harbor on Dec. 7, 1941.

Other active duty assignments: Patrol Wing 10 in the Dutch East Indies, Patrol Wing 4 in the Aleutian Islands, Sqdn. XO at NAS Corpus Christi, and Pilot VR-2 at NAS Alameda. He was released from active duty in December 1945.

Naval Reserve Training assignments: NAS Dallas, Denver, CO VR-711, St. Louis, Los Alamitos; CO FASRON 772, 14th Naval District Composite Co.; Senior member Naval Reserve Policy Board, NAS Olathe. Retired in 1972.

Civilian employment: Co-pilot Braniff International Airways 1946-1947; Federal Aviation Administration 1948-1974, retiring as a Flight Standards branch chief.

Brant and his wife, Esther, now reside in Leawood, KS.

STEVEN E. BRAUN, CDR, born April 13, 1952, New York, NY, received AB, Rutgers College, and JD Southwestern University. He was commissioned March 5, 1975. His military locations and stations: NLSO Norfolk, VA (1978-1981); Naval Justice School, Newport (1978). Served with USS *Sierra* (AD-18) (1978-1979).

Braun was awarded the Navy Achievement Medal, Meritorious Unit Commendation, National Service Defense Medal, and Armed Forces Reserve Medal. He achieved the rank of CDR, JAGC, USNR. His memorable experiences include Mediterranean cruise on USS *Sierra* as a legal officer, defense attorney at NLSO Norfolk.

He married Lynda and they have a son Jordan. Employed as senior assistant prosecutor, Passaic County Prosecutor's Office.

MARK K. BRAZIER, CDR, born Dec. 23, 1954, Anchorage, AK, received a BS from computer science and MS, industrial engineering, AM Texas & A&M University.

Joined the service May 6, 1978, and stationed at NASNI, San Diego, CA; NAS Dallas; NAF NASH; NAS Jax, FL. Served with: VS-37, FACSFAC, NAVAIR 770, NAVPRO 470, NAVAIR. Brazier achieved the rank of CDR.

He is married to Lauri and they have two children, Alexis and Garrett.

DUANE L. BREAUX, LT, born Sept. 20, 1955, Baton Rouge, LA, received a BA from St. Joseph Seminary, St. Benedict, LA; MDIV from Notre Dame Seminary, New Orleans, LA.

Joined the service August 1973. His military locations and stations: (Enlisted) Naval Base, Charleston, SC; NAS, Norfolk, VA; USS *Thomas C. Hart* (FF-1092); USS *Ainworth* (FF-1090); Naval Reserve Center, Baton Rouge, LA; (Officer) Naval Reserve Center, New Orleans, LA; Belle Chase Air Station, New Orleans, LA; Naval Reserve Center, Baton Rouge, LA.

Units Served With: (Enlisted) Mine Counter Measures Unit Bravo; Ground Electronics Div., NAS Norfolk; USS *Thomas C. Hart;* USS *Ainsworth;* NRC, Baton Rouge. (Officer) NRC, New Orleans; 4th Marine Air Wing, Belle Chase Air Station, New Orleans, NRC, Baton Rouge.

He was awarded the Navy Achievement Medal, Good Conduct (two awards), Overseas Ribbon, Sec Nav Unit Citation. Achieved the rank of LT and was discharged Oct. 1, 1994.

His memorable experiences: clearing the Suez Canal of mines; celebrating a Roman Catholic Mass at the Recruit Chapel in Orlando, FL, where he had gone to "boot camp" almost 20 years earlier.

Breaux is a Catholic priest for the Diocese of Baton Rouge-pastor of St. Joseph the Worker, Pierre Part, LA; Chaplain for Scouting, Diocese of Baton Rouge. He retired Oct. 1, 1994.

WILLIAM RICHARD "BILL" BREMER, COL, born in San Francisco, Jan. 5, 1930. He received his BS (Business Administration) from Menlo College in 1952 and headed for Quantico, VA as a Marine Private assigned to the Fourth OCC. When Commissioned, he completed the

14th SBC and Air Control School, and was ordered to Korea as a GCI air controller, MGCIS-1, Kunsan, 1st MAW.

He was assigned as an exchange controller to the 608th AC&W Sqdn., USAF, Chodo Island, and recorded eight enemy aircraft intercepts. Released from active duty in 1954, he served in the Marine Corps Reserve holding a variety of billets and commands through February 1982, whereupon he retired. He holds the Meritorious Service Medal.

He received his JD, University of San Francisco School of Law in 1958, and was admitted to the California Bar. He has been in the private practice of law since. Bremer serves as a Judge Pro Tem for the County of Marin and the County of San Francisco and as an Arbitrator for both Marin and San Francisco Courts.

He served on the board and was president of the Marines' Memorial Club, San Francisco; is a life member of the Navy League and was San Francisco Council President, Northern California State President and a National Director for 12 years. He is presently a National Deputy Judge Advocate; a life member of the Naval Order of the United States, has served as San Francisco Commandery Commander, Judge Advocate General and Commander General (1993-95), the first Marine to hold that office. He was elected a Councilman, Town of Tiburon and served as Mayor. He has served as an officer and director of the Marshall Hale Memorial Hospital Board; San Francisco Children's Hospital Board; The Bridgeway Plan for Health; and as a director for Bay Area USO and for the Center of the Pacific Rim, University of San Francisco. He is a life member of MCROA, the Naval War College Foundation, in which he serves as a Regional Vice-President, the Association of Naval Aviation, ROA, TROA and Naval Reserve Association. He is a past Commodore of the Corinthian Yacht Club, a past President of the Montgomery Street Motorcycle Club.

He is married to the former Margaret (Peggy) Herrington, and they have two children, Karen and William Jr. Peggy and Bill Jr. are life members of NOUS.

STEPHEN A. "STEVE" BREWER, LCDR, born July 26, 1945, Louisville, KY, received a BS degree in environmental engineering, from University of Louisville in 1973 and is a registered professional engineer. He served in the USAF August 1963-August 1967. Enlisted in the USNR May 1975 and commissioned in 1978.

His military locations and stations: USAF, Glasgow AFB, MT; USNR, Louisville, Little Creek, VA. He served with RMNCB 23, RNMCB 20, RNCR 2, OICC MED, NSWC Crane, 13th FIS.

Brewer achieved the rank of LCDR and retired November 1994. He was awarded the Good Conduct (USAF), MUC (2), Battle "E" (3), NDM, Expert Rifle, SS Pistol.

He married Marilyn F. Shemwell and they have two children, Stephen W. an electrical engineer, and Stuart A. a musician. Employed at the Naval Ordnance Station, Louisville as environmental engineer; LANTNAVFACENGCOM, Env. Engr.; NOS Louisville, Supv. Civil Engr.

KAROLLA "KAY" BRICE, CAPT, born April 3, 1920. Graduated from nurses training, Feb. 5, 1942, receiving a RN CRNA, BS. She joined the USNR Jan. 28, 1943. Stayed on active duyt for seven years, seven months. Augmented to the Regular Navy in 1946.

During WWII she had many duty stations and one tour on a hospital ship. She met her husband on the ship and they were married in 1947. In 1947 she resigned her commission and returned to civilian life. In April 1948 the Bureau of Naval Personnel asked her to re-enlist in the USNR and she went back on active duty from January 1949-January 1951. Released from active duty and returned to a Drilling Reservist for the next 29 years. Her rank progressed from ensign to captain. She retired from all naval duty on April 30, 1980. Brice received the Asian-Pacific Campaign Medal w/Battle Star, Philippine Liberation Medal, American Theater Medal, WWII Victory Medal and the Reserve Medal.

She has no children and has been a certified registered nurse anesthetist for 35 years. Her nursing career spans 50 years.

HERB BRIDGE, RADM, born in Seattle, March 14, 1925 and is a graduate of the University of Washington. Enlisting in 1942 shortly after his high school graduation and 17th birthday, he was an enlisted radio technician in the Pacific before selection for V-12 from the fleet. Following commissioning at Columbia Midshipman School, he served aboard the USS *Breton* (CVE-23), as a fighter director and finally as its navigator.

Recalled to Korea, he helped commission Tactical Air Control Squadron 5, serving in Korea and aboard the USS *Mt. McKinley* and USS *Estes.* Among his many commands in the USNR was the Readiness Command for the five northwestern states. In 1982 he was the interim active duty commander of Military Sealift Command Far East with 45 ships and responsibilities that extended from the Pacific to the Persian Gulf.

Decorations include the Legion of Merit w/Gold Star in lieu of a second award, the Naval Unit Commendation, Korean War awards w/3 Bronze Stars, as well as WWII overseas medals.

Bridge is the co-chairman of the family business, Ben Bridge Jeweler, a 51 store company extending into nine states. A certified gemologist, he is chair of the American Gem Society Trustees. He is past chair of the Greater Seattle Chamber, the Downtown Seattle Association, past president of the Washington Athletic Club, as well as having numerous board responsibilities that include the Washington Mutual Bank, the Naval Academy Foundation, the Naval Undersea Museum, the USO, and officership in the United Way.

The Admiral and his wife of 49 years, Shirley, live in downtown Seattle. They have two married sons, CAPT Jon Bridge, USNR, and Rabbi Daniel Bridge, as well as four grandchildren.

JONATHAN J. BRIDGE, CAPT, JAGC, born in Seattle, March 19, 1950. He is a magna cum laude, Phi Beta Kappa graduate of the University of Washington, where he also graduated law school.

He was assigned to the USS *Perry* (DD-844) in June 1972, as its gunnery and ECM officer immediately upon being commissioned USN from NROTC.

Following the Vietnam War he was given leave for law school, and with a change of designator to JAGC, was stationed at Puget Sound Naval Station and Whidbey Island NAS. Following nine years active duty in the USN, he transferred to USNR where he is in drilling status. He is chair of the NRD Recruiting District Assistance Council and is the initiator of the reserve liaison program with the University of Washington NROTC Unit, where he serves as the unit's Navy law instructor.

His awards include the Naval Commendation Medal, the Joint Service Commendation Medal, the Combat Action Ribbon and Vietnam Service Medal.

He is the Co-CEO and co-owner of Ben Bridge Jeweler Inc., a 51 store, nine-state company that does in excess of $90 million annually. He is past president, U of W Law School Alumni, past president also of the Seattle Urban League, the State's American Jewish Committee, and of the Pacific NW Jewelers Association. He is a member of Rotary, and on the Boards of the Greater Seattle Chamber of Commerce, Seattle Alliance for Education, KUOW-FM radio, Evergreen Children's Association, and many other civic and charitable boards.

He is married to Superior Court Judge Bobbe Bridge, and they have two children, Donald and Rebecca.

H. EDWARD "FAST EDDIE" BRIDGEMAN, CAPT, born Nov. 20, 1942, Texarkana, TX. Earned BS degree in 1964, Baylor University and DDS in 1967, Baylor Dental College. Joined the service in May 1967 and was stationed at Great Lakes, IL, NTC, 1967-68, HSA Tacpei, ROC 1968-70, NAS Corpus Christi, 1970.

Served with REDCOM 11 SDO 4th Dental Co. 4FSSG Co. Retired Sept. 1, 1993, as CAPT. His awards include the Navy Commendation, National Defense, Fleet Marine Reserve Service Medal, Expert Pistol and Rifle w/ C Device.

His two years in Taiwan was a great learning experience; he served with some of the finest men and women in the Navy Reserve for 23 1/2 years.

Married, he and Suzanne have four children: Jeb, Brett, Jay and Lee. Civilian employment: H.E. Bridgeman, DDS Inc.

JOSEPH A. BRODER, CDR, born Hartford, CT, Jan. 19, 1939, received BA from Trinity College, CT in political science, 1963 and JD degree from Harvard Law School in 1963.

Enlisted Connecticut Army NG, July 1963. Transferred to USNR, commissioned LTJG, June 1965; Naval Reserve Law Co. 3-12, 1965-77; (CO, 1974-76); PERSMOBTEAM 101, 1977-78; VTU LAW 0107, 1978-89; ACDUTRA assignments include NAS Brunswick, NAS Jacksonville, NAS Key West, NS Treasure Island, NS Long Beach, NS New London and NAS Keflavik.

Senior partner - Broder & Butts, Attorneys, Colchester, CT; Connecticut Superior Court Attorney Trial Referee; Member-Glastonbury Town Council. Admitted in state/ federal courts Connecticut and New York, US Supreme Court. Member, Connecticut and American Bar Associations, American Trial Lawyers Association, National Association of Elder Law Attorneys. Past President - Colchester Rotary Club. Life member, NRA, former member Connecticut State Legislature.

Resides in Glastonbury, CT, with wife Andrea and son Michael. He retired Jan. 1, 1989.

TERRENCE P. BRONSON, CAPT, born Oct. 12, 1949, Toledo, OH, received BA from Villanova University and a JD degree from Cooley Law School. He joined the service May 16, 1971. His military locations and stations: Active duty, 1971-73, USS *Vulcan* (AR-5), legal officer, assistant navigator.

Reserve duty: training officer and XO, NR USS *Piedmont* (AD-17) Lansing, MI, 1975-80; CO, NR USS *Piedmont* (AD-17), Lansing, MI, 1980-81; CO, NR ABFC (E8), Southfield, MI, 1981-82; CO, NR USS *Barney* (DDG-6), Southfield, MI, 1982-84; AO, NR CONVOYCOM Guam/Pearl, Detroit, MI, 1984-85; XO, NR FTG, Guantanamo Bay, Cuba, Southfield, MI, 1985-87; CO, Surface Readiness Unit (four surface ship units, approx. 200 personnel assigned), Southfield, MI, 1987-89; training officer, CINCUSNAVEUR, Southfield, MI, 1989-91; CO, VTU, Selfridge, MI, 1991-94.

Bronson achieved the rank of CAPT (0-6). His memorable experiences/accomplishments: Performed an investigation regarding the misappropriation of nearly $100,000 of disbursing funds, USS *Piedmont* (AD-17), 1980; qualified as a surface warfare officer, USS *Barney* (DDG-6), 1984; qualified as a ship handling and navigation instructor, Fleet Training Gp., Guantanamo Bay, Cuba, 1986; qualified OOD (Underway) USS *Racine* (LST-1191).

He also became a shellback, crossing the Equator for the first time as a full commander, 1988; qualified as a watch officer, US Seventh Fleet on board USS *Blue Ridge* (LCC 19). Visited Thailand and Hong Kong, 1989; port liaison officer, Boston, MA, 1990; assigned to USEUCOM HQ, Stuttgart, Germany. He worked on the naval aspects of joint training exercises and NATO mobilization plans, 1991; operations officer for Naval Control of Shipping Exercise, CINCUSNAVEUR, London, 1991; DISTAFF director, Naval Control of Shipping Exercise, CINCUSNAVEUR, London. Included in this exercise were elements of Military Sealift Command and Allied Merchant Fleets, 1993. Since 1993 he has been a facilitator for Command Excellence seminars. He also helped coordinate the move of NRC Southfield to Selfridge AB in 1992-93.

He was awarded the Defense Service (2), USNR Sea Service (4), Battle E, Overseas (2), USCG Special Operations, Navy Unit Commendation, Meritorious Reserve, Surface Warfare Device.

He is married to Loretta (Lorie) and they have five children: Kathleen, Michael, Patrick, Philip and Daniel. Bronson has been employed as Chief District Judge, Monroe, MI.

ERNEST DOUGLAS "DOUG" BROOKS, LTSG, born May 4, 1920, Oklahoma City, OK, received a MA degree. He joined the service May 1941. His military locations and stations include: St. Louis, MO; Corpus Christi, Norfolk, VA; USS *Ranger,* USS *Changngo.* Served with Fighting Sqdn. 9 and 28, officer-in-charge Ground School, Lee Field, FL.

Participated in campaigns in North Africa, assisted in sinking French battleship *Jean Barte.* Brooks received ribbons for North Africa, Good Conduct, USS *Swanee,* USS *Changngo.* He achieved the rank of LTSG and was medically retired. All his time spent in the military was memorable.

His wife Irma is deceased. He has four children: Noel, Lynette, Kerry and Steve, and 11 grandchildren: Bobby, Marie, Lara, Tahlia, Margi, Doug, Greg, McKenzie, Evelyn, James and Amanda. He was employed as chairman of the Art Department, Phoenix College.

ROBERT "BOB" BROW, LCDR, born Jan. 13, 1937, Philadelphia, PA, received a BA from Penn State University and MS California State, Northridge. He joined the service June 1960 and stationed at NAS Point Mugu, NAS Los Alamitos, NAS Quonset Point, McGuire AFB.

Served with VX-6, VR-3, VP-65. Brow was awarded the Antarctic Service Medal, National Defense and Vietnam Service Medal. He achieved the rank of LCDR and was discharged October 1982. His memorable experience: navigator on record flight for C-130, McMurto Sound to Christ Church, New Zealand, 1963.

Brow married Gail and they have two children, James and Beatrice. Employed as administrative manager for Kaiser Permanente, retiring January 1997.

DAVID A. BROWER, CAPT, born Dec. 25, 1952, Providence, RI, received a BS from the US Naval Academy in math and an MS degree from the US Naval Postgraduate ASW.

He joined the service June 29, 1970 and was stationed at Pensacola NAS, Mayport NAF, NAS South Weymouth, NWC Newport. Served with VT-2, 6 HT 8; HSL-30, 36, 40, 42, 74, USS *Saratoga* (CV-60).

Brower was awarded the Meritorious Service Medal, Unit Citation, Navy and Coast Guard and Humanitarian Service Expeditionary.

Discharged as CAPT in October 1986. His memorable experiences include flying ASW missions from small deck frigates and destroyers.

He married Phyllis Anderson, Dec. 1, 1979 and they have three children: Jonathan, Paul and Geoffrey. He is CFO S.P. Harris, East Providence, RI.

ALLEN K. "AL or BROWNIE" BROWN, CAPT, born Aug. 12, 1932, Seattle, WA, received a BA, BS and DDS (Doctor of Dental Surgery). He joined the service May 21, 1950.

His military locations and stations: USNH Camp Pendleton, USS *Bryce Canyon,* Bremerton and Bangor Sub Base. He served with USS *Bryce Canyon* (AD-36). Participated in campaign in Laos.

Brown was awarded the Naval Reserve and National Defense Medal. He achieved the rank of CAPT and was discharged Sept. 1, 1986.

His memorable experience: Internship Naval Hospital, Camp Pendleton and CO Sand Point Dental Clinic; boot camp at Sand Point as a seaman recruit and FASRON 896 with Bob Bartas as skipper.

He married Janice and they have three children: Troy, Daryn and Gregory. Employed as associate professor University of Washington, State Board of Dental Examiners, private practice of dentistry.

JOHN BARDON BROWN, CAPT, born April 22, 1927 in Chicago, IL, attended Loras College, Loyola Dental School and received a DDS degree. He joined the service October 1953. His military locations and stations: Great Lakes, NS; Camp Pendleton, Forest Park, IL NRS. Served with FMF Pacific, Nav. Reg. Dental Center Great Lakes, Indianapolis NRS.

Brown was awarded the WWII Victory Medal, Asiatic-Pacific Theater Medal (1945-1946 enlisted man). He achieved the rank of captain, USNR and discharged to Retired Reserve April 22, 1987.

His memorable experiences: CO of Naval Regional Dental Center (Naval Reserve Unit at Forest Park, IL Reserve Center).

He is married to Mary Catherine and they have five children: James, Daniel, Thomas, John and Catherine; and two grandchildren: Amanda and Ryan Brown. Employed in general dentistry practice 1957 to present.

JOHN NOBLE "MOHAWK or MO" BROWN, LTJG, born Sept. 12, 1923 in Boston, MA, attended V-5 USNR, Michigan State for two years. He joined the service December 1942. His military locations and stations: NAS Guinan Philippine Islands. Served with AVU-4, AVU-11, VIP 11, VP-732 (three years), Grosse Ile, MI.

Brown received the Asiatic-Pacific Medal and WWII Victory Medal. He achieved the rank of LTJG and was discharged March 6, 1946. Memorable experience was passing under Golden Gate Bridge on return from overseas aboard USS *Macyntyre .*

He married Jane and they have a daughter Nancy Armstrong. He is employed with Brown Bros. Harriman Co. and Gray Marine Motor Co.

PERRY E. BROWN, LTJG, born Dec. 28, 1964, West Palm Beach, FL. Earned BS degree at Palm Beach Atlantic College and MBA at Baylor University. He joined the military service Nov. 12, 1993.

Stationed at NAS Dallas, TX; NAS Jacksonville, FL; NAVCENT 108 and served with ABFC SSU411, FISC EAST 108, NR COMUSNAVCENT 108, Tampa, FL.

Awards/Medals include the Joint Unit Meritorious Service Award, Expert Rifle and Pistol. He is still serving with rank LTJG.

Married Debra Denise and they have two children, Jordan and Ethan. He is director, Corporate Relations and Career Services, University of Florida MBA Programs.

PETER R. "PETE" BROWN, RADM, born May 10, 1928, Newton, MA, received BS, MS-CE and MBA degrees. He joined the service April 1, 1951, achieving the rank of RADM and was discharged Oct. 1, 1981.

Brown married Marina and they have two children, Peter and Carolyn; and three grandchildren. Employed as chairman, Peter R. Brown Construction, Inc.

RICHARD R. BROWN, CAPT, born Oct. 30, 1924, Wymore, NE. He joined the service November 1942.

He was awarded the USMS Combat Bar, American, European Pacific, China Service, WWII Victory Medal, Philippine Liberation, Korea, Vietnam, USNR. He achieved the rank of CAPT and retired November 1982.

Married Jacqueline and they have two children, Steven and David.

VICTOR H. BROWN, CDR, born in Clearfield, IA, Sept. 28, 1917. He received a BS degree from Tarkio College in 1939 and an MS degree from Iowa State University in 1954. On April 5, 1942, he was called to active duty as an SK3, USNR. On March 12, 1943 he graduated from Reserve Midshipman School, Columbia University and was commissioned an ensign, USNR-R.

September 1943, he reported to the USS *SC-729* an escort, patrol and control ship, first in the South Pacific Theatre and later in the Southwest Pacific. August 1944 through March 1945, he was CO. During this period, the ship participated in the Mindoro amphibious landing and the reinforcement to Lingayen Gulf Reinforcement Operation. From June-September, he was XO of the USS *PC-1211* in the Panama Sea Frontier, on Oct. 14, 1945, he was released from active duty.

September 1950, he returned to active duty; XO of the USS *Bisbee* (PF-46) under COMNAVFE and COMNAVPHIL, NAVSTA Washington, DC, October 1951-September 1952 and finally as XO NROTC, University of Kansas from October 1952-January 1954. He was released from active duty, Jan. 28, 1954.

Inactive duty assignments included commanding officer, Surface Div. 9-26(M) and several years as instructor in NROS 5-6 and 5-7 (ASW and Oceanography) 1957-68. Highest rank achieved was CDR, USNR-R. Tansferred to the USNR Retired List without pay in March 1969 and with pay on Oct. 5, 1977.

From December 1956-May 1977, he was employed by the Navy as an education specialist, about 11 years in the Bureau of Naval Personnel as program manager for ASW Training Aids, Devices and Trainers, four years as educational advisor to COMNAVSURFRES Omaha, NE; two years as educational advisor to COMNAVCRUITAREA SIX, Omaha NE; two years as technical director of the career training analysis group (CTAG) a select group of about 30 military and 30 civilians who were assembled by CDR, Navy Education and Training in Pensacola, FL and tasked with developing a computer based model that incorporated job/task/skill analysis into a system for planning, programming and evaluating Navy training.

Brown is authorized to wear the Command-at-Sea Insignia and the following ribbons: American Theater, Asiatic-Pacific (two stars) and the Philippine Liberation (Star). Also the National Defense, Korean Service, UN Service and Naval Reserve Medals.

Early on, Brown was a high school teacher and coach. Since retirement in 1977, he has stayed busy with part-time and project work at Iowa State University.

He and Margaret have been married for 55 years. Margaret is an honorary Navy Recruiter. They have three children: William K. (active duty Naval Reserve Officer during the Vietnam War and a career hospital administrator), Annette M. (McCarthy) and Pamela S. Brown, both elementary school teachers.

WARD W. BROWN, CAPT, born in Geneva, OH, July 23, 1919 and attended Cleveland College. In 1940, he entered the V-7 Program at Tower Hall in Chicago, received his commission as an ensign, and was assigned to the USS *Conyingham* (DD-371) at Pearl Harbor.

In 1943, he was reassigned to the USS *Hoel* (DD-533) in the Pacific Fleet, and then in December to the USS *Chase* (DE-158) in the Atlantic. The *Chase* was converted to APD-54 in time for the Okinawa invasion and was severely damaged by a kamikaze. In September 1945, he was released from active duty in Honolulu and joined the surface unit at Pearl Harbor, where he served as a CO in the division and squadron units.

Recalled in May 1951 to active service as XO, on the USS *Fletcher* (DD-445) in the Pacific. His campaign and service ribbons include nine stars on the Pacific, two on the Atlantic and two on the Korean. He resides in Honolulu with his wife, two sons, one daughter and three grandchildren.

BERNARD S. BROWNING, RADM SC, enlisted in the USNR in 1943 and was assigned to the V-12 Program at the University of Richmond. He was commissioned in the Supply Corps Nov. 1, 1944 and served in the Atlantic Fleet aboard the USS *Mount Vernon* (AP-7) and in the Pacific Fleet aboard the USS *Lavaca* (APA-22). He was selected for graduate school training and earned an MBA from Harvard University in 1947. Subsequent duty stations were NAS, San Diego, Navy Ships Store Office, Brooklyn, CINCNELM Staff, London and BUSANDA, Washington, DC.

Resigning from the regular Navy in September 1954 as a LT, he was deep selected for flag rank in 1972. His business career has been devoted to small business and franchising. He founded General Business Services, Inc. in 1962 pioneering the concept of small business consulting on a mass basis. He served as CEO until his retirement in 1986 when GBS had 1100 franchised offices in 50 states plus Guam and the Virgin Islands. Inducted into the Franchising Hall of Fame as a pioneer in service franchising, he also served two terms as chairman of the National Small Business Association. He earned the Meritorious Service Award for his year as chairman, National Advisory Council, Small Business Administration.

His career has brought many other honors but he is proudest of his four children: Frances Elaine, Virginia Diane, John Scott and Lawrence Rogers Browning. He and Adeline Rogers from South Carolina were married in August 1955.

DAVID B. BRUBAKER, CAPT, joined the USNR at Akron, OH in 1950 while in high school. He served as plane captain and flight engineer on various reserve patrol aircraft.

He accepted a direct commission as an aeronautical engineering duty officer in 1959. He affiliated with Naval Air Systems Command reserve units at NAS Grosse Ile and NAS Alameda. He served with Alameda Air Systems Program (ASP) Units from 1973 until his retirement in 1991. At Alameda, he was CO of three ASP units.

In civilian life he was employed as the Navy Air Systems Command deputy program manager for various Navy P-3 and A-3 type aircraft. He retired from this position in 1994, and since has participated as a management consultant and freelance technical writer.

He holds a BS degree in aerospace technology from Kent State University and a MA degree in journalism from Ohio State University.

He is married to the former Marian Schott of Akron, OH. They have three daughters and one grandson. He now resides in Lafayette, CA.

STEPHAN R. "STEVE" BRUCE, CAPT born July 14, 1949 in Seattle, WA, received a BSAE, USNA and MBA from the University of Virginia. He joined the service June 6, 1967. His military locations and stations: Pensacola, Pax River, Jacksonville, Long Beach carriers. Served with VT-6, HS-7, VT-2, USS *Camden,* USS *Saratoga,* USS *Waldo County.*

Bruce participated in the Vietnam War. He achieved the rank of CAPT, USNR. Still in the Reserves. His memorable experiences: received wings 1972; carrier qualification; chief maint. test pilot; flight instructor for Iranians; Aeronautical Engineering duty officer; CO for two Reserve Units.

He married Conni and they have a daughter Stephanie. He is presently supervisor and chief marketing officer for CNA Insurance.

WARREN "HAIRY" BRUCE, CAPT, born April 26, 1930 in St. Louis, MO, received a BS and MS degree in chemical engineering from Princeton, 1952 and a PhD from the University of New Mexico, 1962.

He joined the service June 9, 1953. His military locations and stations: OCS Newport; Pensacola, FL; Patuxent River; San Ford, FL. He served with VC-7, VAH-7, Res Res 8-9, Res. Intelligence Unit D.C.

Bruce received the National Defense Medal, European Occupation Medal and Presidential Unit Citation. He achieved the rank of CAPT and was discharged from the USNR June 1, 1980.

His memorable experiences: flying on operation "Teapot"; nuclear test series; flying with VAH-7 off Lake Champlain during Suez crisis; CO Reserve Intelligence Unit, Washington, DC.

He married Kathy and they have four children: Warren IV, Cynthia, Thomas and David; and seven grandchildren: Elizabeth, Rachael, Sarah Joy, Thomas, Colleen, John David and Jacqueline. Fellow, Los Alamos Scientific Lab; Research manager DuPont, retiring Aug. 1, 1992.

Member of the American Chemical Society, American Institute of Chemical Engineers, Naval Reserve Officers Association, Reserve Officers Association, Naval Intelligence Professionals, National Rifle Association, Tailhook Association, Aircraft Owners and Pilots Association, Nuesse Sailing Association/Oriental Sailing Club and the Navy League. Elected County Commissioner Pamlico County, NC (1996).

WILLIAM L. BRUCKART, CDR, born Dec. 10, 1919, Washington, DC. Graduated from Riverside Military Academy, Gainesville, GA in 1937 and joined the USNR in July 1937. Attended USNA, Class of 1942; BS degree in metallurgical engineering, University of Kentucky, June 1942; Dartmouth College, July 15, 1942; Fire Control School, Washington Navy Yard; Naval Training School (radar, MIT); Bowdoin College; Fleet Service Schools, Destroyer Base, San Diego; CINCPAC for Radar Maintenance School.

Volunteered for submarines and ordered to USS *Sailfish,* Oct. 6, 1943; USS *Pampanito,* Feb. 23, 1944; SubDiv 201 Relief Crew, August 1944; USS *Kingfish,* September 1944; new construction at Portsmouth Navy Yard; USS *Odax,* April 1945; USS *Plunger* as XO; inactive duty Nov. 15, 1945.

Served as engineer for Southern Liquid Gas Co., Dothan, AL; Graduate School at Ohio State University and earned MS degree, 1953; development engineer and later manager of sales, Refractomet Div., Universal-Cyclops Steel Co., Bridgeville, PA.

Promoted to CDR, USNR, Jan. 1, 1956; set up as an independent technical and marketing consultant in 1960 and later affiliated with Aerojet-General Corp., as manager, Metals and Refractories Dept., Azusa, CA. Retired from the USNR in 1968.

Was assistant professor, Industrial Technology, School of Engineering, California Polytechnic College, retired as professor in December 1984.

Married Jessie Hocker Francis, June 1942, and they have two daughters and three sons. Divorced and married Angela Mary Heath in December 1963.

DONALD J. "JOE" BRUNDAGE JR., LCDR, born March 8, 1950 in Atlanta, enlisted Jan. 21, 1970. Achieved CPO (AMHC) in 1980.

Active duty: VF-201, NAS Dallas, VP-67, NAS Memphis; flight crew plane captain on the SP-2H. Selres: RTU-54 Det Atlanta; qualified C-118B flight engineer. Active duty: VP-68, NAS Patuxent River; qualified P-3 flight engineer; VP-62, NAS Jacksonville; completed BS degree.

Commissioned Sept. 26, 1980, assigned to VT-10 for NFO training, disqualified for sinus problems, became an aircraft maintenance duty officer. Next VP-45, NAS Jacksonville; USS *Carl Vinson* (CVN-70) and VX-5, China Lake, completed MS degree. Next, O-I-C Medium Attack Weapons Det, NAF El Centro, and maintenance officer, Fleet Logistics Support Wing, NAS Dallas; retired from active duty July 31, 1995.

Highest award of 10 medals, three ribbons and Meritorious Service Medal.

Brundage and his wife Sheryl and children, John and Rachel, reside in Colorado Springs, CO.

DANIEL P. "BRUNER" BRUNSVOLD, CDR, born April 14, 1948, Fergus Falls, MN, received a BBA from Morehead State University. He joined the service Jan. 11, 1973. His military locations and stations: VP-48, NAS Moffett Field, NAS Glenview, NARDET Minneapolis, VP-48, VP-90, NRD Minneapolis, VP-0622, VP-0448, VP-0122. He served mostly in the Officer's Club with Don Porterfield.

Brunsvold was awarded the Battle E, Unit Commendation, Sea Service, Meritorious Service Medal. He achieved the rank of CDR and discharged from USNR, July 1, 1996.

His memorable experiences: Great friends: Don Porterfield, Frank Rose, Gary Bakken, Mike Byrnes and Greg Metcalf.

He is married to Elizabeth "Liz" and they have a daughter Aimee. Employed with 3M as corporate pilot since 1980.

DOUGLAS E. BRYAN, CDR, born April 8, 1917, Auckland, New Zealand. Attended University of California (Berkeley) and received an AB in history in 1942; BS (Entomology) 1948; PhD (Entomology, Plant Pathology, Ecology) 1952.

He joined the service Feb. 5, 1943. His military locations and stations: boot camp, Farragut, ID; USS *Azimech* (AK-124) 1943-44 QM3c; officer training, Tucson, AZ. Commissioned Sept. 5, 1944; assigned USS *Buckingham* (APA-141) ensign (1105); boat officer, assistant navigator, 1944-45.

Joined Naval Reserve Research Co. 8-13 Stillwater, OK 1953; assumed command 1962. Transferred to NRR Co. 11-7 Tucson, AZ 1964; assumed command 1967. Transferred to Retired Reserve 1970; retired from Navy 1977. ACDUTRA locations: Disease Vector Control Center-NAS Jacksonville, FL; DVCC-NAS Alameda, CA DVCC-USMC Camp Lejeune, NC.

He was awarded the American Theater Medal, Asiatic-Pacific Medal, Victory Medal and Naval Reserve Medal. Achieved the rank of CDR, MSC.

Married Marilyn Wilson in 1941. They have four daughters: Mrs. Deborah Flanagan, Mrs. Judith Miller, Mrs. Wendy Borchgrevink and Kathleen Bryan; and four grandsons: Bryan, Benjamin and John Flanagan and Matthew Borchgrevink.

Professor and Research Scientist at Oklahoma State University, Stillwater, OK 1952-1963. Research scientist at U.S. Department of Agriculture Insect Biological Control Laboratory, Tucson, AZ 1964-74; Laboratory leader 1974-76. USDA National program staff scientist for Biological Control of Insects 1976-79. Retired 1979.

WILLIAM JAMES BUCKINGHAM, born Aug. 25, 1955, Ann Arbor, MI, graduated from Ferris State University, College of Optometry, May 1979, with a doctor of optometry degree (OD). Received Armed Forces Health Scholarship in April 1977 while attending Ferris State University, College of Optometry. Commissioned in June 1979 as a lieutenant in the USNR.

Served on active duty as an optometrist at the following locations: Naval Regional Medical, Corpus Christi, TX, June 1979-June 1982; Branch Clinic, Brunswick, ME-June 1982-June 1985; released from active duty June 1985, affiliated with the Reserves 1985-87 NH Glakes 513, Battle Creek, MI, asst. TO; 1987-89 NH Glakes 513, Battle Creek, MI, TO; 1988-89 NH Glakes 813, Lansing, MI, TO; 1989-92 NH Glakes 813, Lansing, MI, CO; 1992-94 NH GLakes 913, Saginaw, MI, CO; 1994-96 Fleet Hospital 23, Detachment Kilo, Saginaw, MI, OIC; 1996-present NH Glakes 813, Saginaw, MI, CO.

Buckingham was awarded the National Defense Medal, Naval Reserve Medal. Promoted to CDR, Nov. 1, 1991. Still affiliated with the USNR.

His memorable experiences: 1983, Corpus Christi, TX, landed and cat shot of a carrier; active duty for training and ADSW's; Adak, AK (three times); Pearl Harbor, HI (two times); FHOTC, Camp Pendleton, CA; Naval Hospital Great Lakes (four times); NNMC, Bethesda, MD; NAS Glenview, IL.

Married Janis Fleming and they have three children: Alice, Scott and Kathryn. He has been employed at Ferris State University, College of Optometry, Big Rapids, MI, associate professor; September 1988-to present; Duane Waters Hospital, Jackson, MI, optometrist, September 1988-to date; William J. Buckingham, O.D., P.C., Jackson, MI, optometric practice, April 1990-date.

BYRON "BY" BUELL, CAPT, born April 8, 1937, Highland Park, MI, received a BME from Rensselaer Polytechnic Institute, 1959; MBA Rochester Institute of Technology in 1973.

He joined the service June 5, 1959. His military locations and stations: USS *Samuel B. Roberts* (DD-823) 1959-60; USS *McCaffery* (DDE-860) 1961-62. Served with NR NRL 105, Rochester, NY ultimately as CO.

His memorable experience: serving as chief engineer on the *McCaffery* during a Frame II overhaul and shakedown in GTMO with major failures in a modified evaporator. He was awarded the Reserve Service Ribbon. Achieved the rank of CAPT, USNR and retired Jan. 1, 1984.

He married Barbara and they have a daughter Catherine. Employed in manufacturing management, Eastman Kodak Co. for 30 years, retiring Dec. 1, 1991.

CARL A. "BUFFY" BUFFLAP, LT, born Feb. 10, 1922, York, PA, attended Shippensburg University, Shippensburg, PA, 1940-42, Drexel University, Philadelphia, PA 1946-48 BSME degree.

He joined the service July 12, 1942, V5 Naval Aviation Cadet, received wings September 1943. His military locations and stations: NAS Anacostia, Pensacola, Jacksonville, Pasco, WA, Willow Grove. Served with Composite Sqd. VC-94, VF-58E USS *Shamrock Bay* (CVE-84).

He fought campaigns in Lingayen Gulf, Iwo Jima and Okinawa. Awarded the Distinguished Flying Cross, Air Medal w/3 stars, American Campaign, Asiatic-Pacific Campaign Medal. Bufflap achieved the rank of LT and was discharged December 1945, USNR. Retired from the Reserve Program April 1961.

His memorable experiences: first catapult shot off USS *Saratoga* (CV-3) 1944; having tail shot up at Lingayen Gulf and making emergency landing January 1945. Checking out in the FH-1 Phantom Jet in the Reserve at Willow Grove 1952.

He is married to Elizabeth "Betty" and they have two children, Cathy and Nancy; and three grandchildren: Jake, Hannah and Jamie Peerless. Employed in management with SKF Industries and Taylor Instruments, retiring August 1985.

MARK K. BUNKER, CDR, was raised in Santa Clara, CA. Attending Oregon State University on an NROTC scholarship, he graduated with a degree in computer science and was commissioned an ensign in June 1977. After Flight School, CDR Bunker received his Naval Flight Officer Wings in January 1979. He received the Markey Award for graduating number one in his flight school class.

CDR Bunker reported to Patrol Squadron Nine, then home-ported at NAS Moffett Field, CA, and deployed to WestPac and Atlantic bases during the height of the Cold War. In 1982 he returned to Mather AFB as a navigation instructor and obtained an MBA in telecommunications management from National University. He transferred to the Naval Reserves in early 1986 and served in two patrol squadrons, VP-93 and VP-67. He reported to NR COMSEVENTHFLT 111 at NAS Dallas in 1994. He has been employed with EDS (Electronic Data Systems) since 1985.

CDR Bunker has over 2,500 hours in P-3 and T-43 aircraft, qualifying as navigator, tactical coordinator and mission commander. As a private pilot, CDR Bunker has flown both production and experimental aircraft, his favorite being his father's Breezy.

The Bunker family includes his wife, Karen; sons, Ryan and Matthew; and daughter, Catherine.

F. GREGG "COUNT" BURGER, CDR, born in Staten Island, NY, received a BS in business administration from Washington Lee University.

He joined the service in 1941. His military locations and stations: Lakehurst, Glynco, Elizabeth City and Weeksville. Served Port Lyauty, ZP-12, ZP-14.

Burger achieved the rank of CDR and was discharged in 1945. His memorable experiences: North Africa, Italy and France.

Married to Martha and employed as vice-president, Director of Sales, L.A. Dreyfus Co., retiring in 1980.

GARY L. BURGUND, born May 23, 1952, Cleveland, OH, received an MBA from Cleveland State University and a BBB from Kent State University.

He joined the service May 1, 1970 and was commissioned February 1984. His military locations and stations: Hakata AFB, Japan; Naval Hospital Portsmouth, Naval Reserve Recruiting Command (Detachments Two and Five).

Burgund achieved the rank of LCDR. He was awarded the Unit Commendation and Naval Reserve Recruiting (x2). His memorable experiences: Buffalo inspection with LCDR Haynes and LT Petrush; Portsmouth Naval Hospital - ADCUTRA's 1984-87; Camp Pendleton, Fleet Hospital Training.

Married Sept. 25, 1976 to Karen and they have four children: Nicole, Gregory, Jessica and Colleen. He is employed as assistant administrator - Human Resources.

PAUL ANDREW BURKE, LCDR, born Feb. 5, 1964, Pittsburgh, PA and was raised in Portage, MI. He received a BS in electrical engineering at Western Michigan University, graduating April 1986.

He joined the service May 25, 1984 (NUC Propulsion Officer Candidate Program, Sept. 26, 1986 (Officer Candidate School. His military locations and stations: NPS Orlando; NPTU Balston SPA (MARF S7G), SOBC Groton, CT. Served with USS *Key West* (SSN-722), Norfolk, VA, 1988-91.

Served in campaigns in Desert Shield and Desert Storm. Burke received the NAM, NUC, Navy Expeditionary Medal, Sea Service (2), Arctic Service. He achieved the rank of LT (active) and LCDR Select - FY97 (Reserves) and was discharged March 31, 1991.

His memorable experiences: two submarine deployments; North Atlantic (1989); Mediterranean (1990), extended to six months due to Desert Shield; as DCA ship's diving officer and docking officer, docked USS *Key West* (SSN-722) in Trident Defit facility Kings Bay dry dock for emergency repairs; first time L.A. class docked in the TRF Kings Bay.

Married July 27, 1991 to Melanie J. Sanic. They have no children but are the proud owners of two Labrador Retrievers, Nickie and Sandie. Employed with Mobil Oil Corp., Torrence, CA, reliability engineer (1991-93); Consumers Power Co., Palisades Nuclear Plant, system engineer, (1993-present).

Presently still active SELRES, XO of NR Subase Pearl, DET 113 (Greatl Lakes NRRC).

MICHAEL T. "MIKE" BURKHARDT, CAPT, born Dec. 10, 1945, Bethlehem, PA, received a BA West Chester University and MBA from William and Mary University.

He joined the service May 16, 1967. His military locations and stations: Yokosuka, Japan, USS *Terrell County,* CNFJ; Little Creek Amphibious Base, Virginia Beach, VA SNEP FIT.

Burkhardt served on two deployments to Vietnam, February 1968-June 1970. He was awarded the National Defense Medal, Vietnamese Cross of Gallantry, Vietnam Service, Vietnam Campaign (7), Meritorious Unit Citation (3), Presidential Unit Citation, Navy Commendation w/ Combat V, and Gold Star in lieu of second award, Armed Forces Reserve Medal.

He achieved the rank of CAPT, USNR. His memorable experiences: Best: aide to ADM Arleigh Burke for a month on the 20th anniversary of Panmunjon; most memorable: River clearing operation on the Son Bo De River, Vietnam, in preparation for seafloat establishment.

He is married to Rita McGann and they have two children, Catherine and John. Employed as vice-president, sales and marketing, Gardner Asphalt Corp., retiring Sept. 1, 1992.

FREDERICK W. BURKLE, CWO4, a native of West New York, NJ, was born in Elizabeth, NJ, Jan. 29, 1932. He enlisted in the Naval Air Reserve Jan. 6, 1951 at NAS New York as a weekend warrior. After one year in a drilling status, he requested active duty and was assigned to CQTU-4 at NAAS Barin Field, Foley, AL. In March 1952, while attached to CQTU-4, he was a plane captain for the following aircraft: SNJ, F6F, F8F, TBM and AD, and worked in the structural sheetmetal repair shop. With the end of the Korean Conflict, July 1953, he was released from active duty in December 1953 and completed his enlistment in the stand-by reserve.

July 26, 1958 CWO Burkle re-enlisted in the Naval Air Reserve as an AMS3 at NAS New Orleans assigned to VA-822 and qualified as a plane captain on the F9F (Cougar). May 1960 he was transferred to VR-822 and qualified as a flight crew plane captain on the R4D-8 (DC-3) Skytrain and May 1962 the C-54 (DC-4) Skymaster. He was promoted to CPO May 1965 and December 1967 promoted to W-1 assigned as maintenance officer. July 1968 he was transferred to VP-61X1 as the maintenance officer, his introduction to the Antisubmarine Warfare Community and SP-2H "Neptune" aircraft.

November 1970 saw the reorganization of the Navy Reserve with the establishment of RESFORON's, one Navy concept, reserve squadrons manned as their regular Navy counterpart and the commissioning of VP-94 with 38 Reserve Officers. He was assigned as maintenance officer until 1976 when the squadron transitioned to the P-3A Orion. The complement was increased and he was assigned as the maintenance/material control officer. December 1976 he received an early promotion to his present grade and served as W-1 for over 24 years. He was the most senior warrant (USN and USNR) until his retirement. He was the last officer plankowner.

His military decorations include the Navy Enlisted Air Crew Wings, Navy Commendation Medal, Navy Achievement Medal, Coast Guard Meritorious Unit Commendation w/Distinguishing Device and Gold Star, Navy Battle Efficiency "E" Ribbon, Navy and Marine Corps Overseas Deployment Service Ribbon (8) Awards, and the Armed Forces Reserve Medal w/2 HGs.

He is a retired supervisor engineering technician Civil Service (Naval Aviation Depot-PNS) and Civic Center manager, City of Foley. He retired with 38 years of Naval service in January 1992 while assigned to Patron Nine Four, NAS New Orleans.

He is married to the former Carol Barnes and they reside in Foley, AL. He has four children: Frederick C., Merrilee Slota, Roderick, Derrick; and five grandchildren.

RICHARD H. "RICK" BURN, RADM, born March 9, 1945, Chicago, IL, received an MBA from Utah State University and a BS from University of Wisconsin-Madison.

He joined the service Aug. 14, 1967. His military locations and stations: USS *Columbus* (CG-12); Hill AFB. Served with NSC Norfolk 207; RCHB Four; SPEC 107; DESC.

Burn achieved the rank of CAPT, SC, USNR and was discharged Oct. 4, 1971. His memorable experiences: working with RADM Allston in SPEC 107 and managing the Norfolk Air Terminal.

He is married to Linda L. and they have two children, Laura L. and Richard. He is employed as director of transportation Logistics, Michigan Sugar Co., and will retire September 1997.

GARY G. BURNETT, CDR, born Nov. 6, 1932 in Roseburg, OR, received a BS in 1960 and a DMD in 1962 from the University of Oregon.

He joined the service Aug. 20, 1952 and served until Oct. 18, 1955 in the US Army. He then served from Dec. 19, 1972-March 31, 1990 in the USN.

His military locations and stations include the Philippines and NR Portland, OR. He served with the dental unit in NR Portland.

Burnett achieved the rank of SGT in the Army and CDR in the Navy. Discharged April 1, 1990.

He is married to Marilyn and they have three children: Bruce, Brenda and James; and eight grandchildren: Johnathan, Matthew, Reese, Joe, Dusty, Andrew, Barry and Taylor. He retired as a dentist in June 1992.

DAVID S. "DAVE" BURR, CAPT, born Jan. 6, 1943 in Honolulu, HI, received a BS in business administration from Oregon State University in 1964 and a MS in management from the University of Arkansas in 1973.

He joined the service June 6, 1964. His military locations and stations: USS *Cochrane* (DDG-21); DESRON 23 Staff; CINCUSNAVEUR. Served on various deployments to and missions off of Vietnam 1965-71. Achieved the rank of LT during active duty and CAPT in Reserves. Discharged from the Reserves Sept. 30, 1991.

Burr received the Navy Achievement w/Combat V, Meritorious Unit Commendation, National Defense Service Medal, Vietnam Campaign Medal w/Silver Star, Vietnam Service Medal, Naval Reserve Medal, Vietnam Gallantry Cross w/OLC, Expert Pistol and Expert Rifle.

He is married to Cheri and they have two children, Elizabeth and Emily. He is presently employed with Houston Cellular Telephone Co., as a direct sales representative. He has held other positions in sales marketing, operations management and management consulting with various firms which include: American Telco, Inc., Coast Chemical Co., Industrial Handlign Engineers, Inc., T.H. Hill Associates, Keith-Stevens, Inc. and Phillips Petroleum Co. (and subsidiaries)

DONALD A. "DON" BURRISTON, CDR, born April 13, 1930 in Oakland, CA, received a BS from the University of California.

He joined the USMCR July 1948 and NAVCAD Pensacola in 1951. His military locations and stations: Pensacola, FL; Oakland, CA; North Island; Los Alamitos. He served with VS-871, VS-37. Commissioned and designated naval aviator, Jan. 21, 1953. He received awards for China Service (extended) and Tachen Evacuation.

Burriston achieved the rank of CDR and was discharged in 1971. His memorable experiences: Far East Cruise in *Princeton* (CVS-37), 1954; four years active, remainder in Ready Reserves. Total years 20 years, 11 months and 2 days.

He is married to Janet Walton and they have two children, Donna (Price) and Brian; and grandchildren, Jordan and Ragleigh Price and Goleta and Nathan Burriston. Employed as assistant county assessor, Contra Costa County, CA, retiring April 1990.

ROBERT D. "BOB" BURRUS, CDR, born June 6, 1954 in Jersey City, NJ, received a BS in civil engineering and MS in manufacturing engineering.

He joined the service July 30, 1980. His military locations and stations: NAS Pensacola, NAS North Island, NAS Whiting Field. He served with Schools Command, VT-2, HT-8, HT-18, HSC-31, HT-8, HSC-84.

Burrus achieved the rank of CDR and still remains active in the Reserves. He was awarded the Battle E and Navy Achievement Medal.

He married Cecilia and they have three children: Robert, Carly and Madison. He is employed as quality assurance manager with Applied Superconetics.

ROBERT CHAPMAN BURTON, LT, born Dec. 16, 1959 in Oklahoma City, OK, received a BS in industrial engineering and management from Oklahoma State University in 1984.

He joined the service May 2, 1990. His military locations and stations: NAS Dallas, TX. Served with Naval Mobile Construction Bn. 22.

Burton is currently ranked LT 0-3. He has received a Battle E, NAM, NDSM, Expert Pistol and Rifle. His memorable experiences: first military exposure as a direct commission SELRES at age 30. The camaraderie and pride of serving our country.

He is married to Charlotte and they have three sons: Austin, Jacob, Daniel, and a daughter Kelly. He is employed with the city of Oklahoma, Public Works as a civil engineer.

RICHARD S. "RICK" BUSH, CDR, born Aug. 12, 1948 in Bethesda Naval Hospital, Bethesda, MD. He received an AB from Brown University 1970 and a JD from the University of Mississippi Law School in 1975.

He joined the service in 1968. His military locations and stations: Washington DC OLA 1970-72. He served with Naval Reserve Legal Service Office, 209 NAS Memphis, TN.

Bush achieved the rank of CDR USNR JAGC 2505. He has received the National Defense Medal. His memorable experiences: retired in 1995; being XO and CO of NRLSO 209 NAS Memphis, TN.

He is divorced and has three children: Jason R., Lisa C. and Matthew E.

WILLIAM C. BUSHNELL, CAPT, enlisted in a Minnesota Reserve unit Sept. 30, 1963. Seaman recruit Bushnell had no idea he would one day return to command that same unit...helping it earn recognition as the top Naval Reserve Security Group unit in the nation.

Completing boot camp at NTC Great Lakes in December, he rose to CTR3 before being commissioned an ensign at OCS, Newport, RI Aug. 18, 1967.

Following a 47-week course in Russian at the Defense Language Institute, Washington, DC, active duty assignments included Direct Support Division Officer and Assistant Net Control Officer for Atlantic HFDF net (during the early years of the Bullseye System).

CAPT Bushnell served as CO of NRSECGRUACT LANT 316, Lacrosse, WI (two tours), NRSECGRUACT HANZA 616, Oshkosh, WI and NRSECGRU 216, St. Paul, MN. During his tenure, this last unit earned the James D. Delaney Memorial Award as the outstanding NRSECGRU unit in the country.

He represented REDCOM Region 16 at the COMNAVSECGRU conference that established the first hands-on training program for Security Group reserve units. He also served as OIC of one of the first programs involving Navy, Marine and Air Force personnel drilling at an Army regional training facility.

CAPT Bushnell was special advisor for cryptology for REDCOM Region 18, Assistant Regional Cryptologic Area Coordinator for RCA Central and served on the national staff of the Naval Reserve Security Group Program.

DAVID H. BUSWELL, CAPT, born in Chicago in 1934 and graduated from Northern Illinois University in 1957. After active military service from 1957-59, he moved to Washington, DC. He was commissioned in the USNR in 1966 as a public affairs officer and retired in 1994.

During his career he was CO of several reserve public affairs units and served as public affairs officer for the Reserve Naval Construction Force. He was recalled to active duty at the Pentagon during Operation Desert Storm. Among his awards are three Navy Commendation Medals.

His civilian positions include public information officer for the National Academy of Sciences, Director of Public Information for the Federal Trade Commission and vice-president for Government Relations of the Fiat Corporation.

He lives with his wife, Yoko Ogata Buswell at their home "Fair Winds" in Heathsville, VA. They have two children, Janice and Scott.

JAMES H. BUTCHER, CDR, born March 21, 1928 in Kent, OH, received a BS from Kent State and a JD degree from the University of Akron. He joined the service May 2, 1945. His military locations and stations: RT School at Monterey; USS *Coral Sea* (CVB-43).

Butcher achieved the rank of CDR. He received the Victory Medal, National Defense and Armed Forces Reserve Medal. His memorable experiences: all three Naval Reserve Commands, especially NRSD 4-68.

He is married to Claire and they have two children, James Jeffrey and Elaine Brook (Moorefield); and five grandchildren: Aaron, Rebekah, James Bennet, Seth and Laura Beth. Employed as senior tech. manager, General Counsel Colington Harbour. He retired Jan. 1, 1977.

JOHN VERNON BUTZ, CAPT, born March 26, 1935, Baltimore, MD, attended the US Merchant Marine Academy.

He joined the service December 1954. His military locations and stations: Norfolk, Charleston, Newport, New Orleans, Naples, Washington, DC. Served with Military Sealift Command, Sixth Fleet, N-4, CINCLANT, USS *El Paso,* USS *Shenandoah,* Pensacola.

Butz participated in campaigns in Lebanon 1958; Congo Emergency 1961; Vietnam 1965-67; Desert Shield/Storm, 1990.

He achieved the rank of CAPT and retired June 15, 1996. He was awarded the Good Conduct Medal, National Defense Medal, Antarctic, Vietnam, Desert Storm, AM Exp. Medals. His memorable experiences: Too many! Washed overboard and rescued in Antarctic Sept. 7, 1963.

He is married to Shirley Ann and they have a daughter Susan Alice. Employed with Military Sealift Command, Master-UNREP Ships, USNS *Norwalk, Pawcatuck, Marias, Missinena, Rigel, Wallamaw, Wright.*

JAMES W. "JAMIE" BYERS, CDR, born Dec. 29, 1953 in Ft. Worth, TX, received a BS in Marine Transportation from Texas A&M University and MBA from Regent University.

He joined the service July 20, 1977. His military locations and stations: Sailed merchant ships as part of MMR program. He served with NCS Panama, MSCO Red Sea, MSCO N. Japan and MMROCH Washington.

Byers participated in Operation Desert Storm. He received NAM (3), NCM, NUC, SWA, Kuwait, Armed Forces Reserve. He achieved the rank of CDR and is presently active in the Reserves.

He is married to Cynthia K. Freiley and they have four children: Amy, Casey, Joanna and Joshua. He is employed in Marine Operations as data processing supervisor.

JOHN E. BYERS, CDR, born in Cleveland, June 27, 1949. He was a graduate of John Carroll University in 1971. Enlisted for a time, completed RTC at Great Lakes and PN "A" School at NTC Orlando. He attended Navy OCS at Newport and was commissioned February 1972.

As ensign and LTJG, he served with J-2 USS SOUTHCOM in the Panama Canal Zone. He then served as CO aboard the USS *Hoel* (DDG-13), making a Westpac deployment out of San Diego.

As a reservist, his assignments included ACU-2, Little Creek; the USS *R.H. McCard* (DD-822), Tampa; NAVSTA Rodman, Panama, CZ; SIMA Mayport, FL where he served as CO from 1984-86. He then served with the Joint Deployment Agency, MacDill AFB; Military Sealift Command, Southwest Asia during Desert Shield/Storm and then completed his 23-year career with the Special Operations Command at MacDill AFB.

His medals and awards include Surface Warfare Officer, Joint Meritorious Unit, National Service, Armed Forces Reserve, Sea Service, Naval Reserve and Navy Overseas Service.

Commander Byers' family consists of wife Barbara and son John Patrick. They reside in Tampa, FL. He is a self-employed business owner.

ALBERT D. "GUS" CAIRE, CAPT, attended Tulane University NROTC from 1939-43 and was commissioned ensign upon graduation.

Active duty began as gunnery officer aboard the USS *Evans* (DD-552) during operations and engagements in the Marshall Islands; Marianas operations capture and occupation of Saipan, Tinian and Guam; Battle of the Philippine Sea (Marianas Turkey Shoot); 2nd Bonin Raid; the West Caroline Island Operation; including capture and occupation of the Palau Islands Peleliu and Anguar; assault on Philippine Islands, on Yap; the Leyte Operations; the Luzon attacks; assault and occupation of Iwo Jima, followed by Okinawa Gunta.

Caire and his gunner's mates bagged 26 Japanese planes while *Evans* was hit by four kamikaze aircraft crashes and 2,500 lb. bombs. One of those kamikaze propellers, thanks to him, is on display in the lobby of NRA HQ Building. The *Evans* received a Presidential Unit Citation for her gallantry.

His SelRes career would remain focused on destroyers. ACDUTRA included USS *English* (DD-696), USS *W.L. Lind* (DD-763), and update enhancement training at Flt. Gunnery School in Washington, DC; damage control in SanFran; ASW in Key West; Advanced Weapons in Key West and Charleston; and Senior Officer Tactics AAW at Dam Neck.

He was awarded the outstanding reserve officer of the center plaque for his leadership and performance. He served successively as Admin/Pers Off, XO and CO of SURFDIV 8-29(L). 8-29 won the 8th Naval District (five state) Commandant's Cup three year's running and captured the Forrestal Trophy (1st among 600 SURFDIVs). Tours followed as Battalion 8-8 Chief of Staff, SURFDIV 8-12 XO and MobTeam CO.

Caire shaped the careers of most of the Naval Reserve officers at N&MCRTC New Orleans as senior instructor in the NROS course program and as chairman of the Counselling Board helping junior officers chart their active and inactive duty assignments. He capped his exceptional career as Naval Reserve Group Command 8-1(L) CO.

He was married to the late Beth Glass Caire. Their children are Jennifer Caire Wein, mother of Ashley and Collins; A. David Caire Jr., father of James and Jason; and Miles Hand Caire.

NICHOLAS J. "NICK" CALISE, born Sept. 15, 1941, New York, NY, received AB degree from Middlebury College, 1962; MBA Columbia Graduate School of Business, 1965; LLB from Columbia Law School 1965.

He joined the service Nov. 4, 1965. His military locations and stations: Newport, RI; Washington, DC. Served with office of the Judge Advocate General; Navy Council of Personnel Boards. He achieved the rank of CAPT and was discharged November 1968 RAD. He retired as CAPT, JAGC, USNR April 1, 1996.

He is married to Mary and they have two children, James and Lori. Employed with BF Goodrich Co., vice-president, associate general council and secretary.

SYDNEY R. "SID" CALISH, LCDR, born May 15, 1920, Boston, MA, graduated from University of California, Berkeley, CA, May 1941.

Signed up for V-7 program in July 1941; commissioned ensign USNR in May 1942. His military locations and stations: USNA V-7 Program, January 1942-May 1942; Diesel School, Cornell University-May 1942-September 1942; engineer officer, USS *Dominant,* AMc 95, September 1942-February 1943; engineer officer, USS *Nuthatch* (AM-60), February-April 1943; Sub-Chaser Training School, Miami, FL (with TAD at GE Turbo-Electric School and Naval Boiler School June-August 1943) April 1943-September 1943; engineer officer, USS *Whitehurst* (DE-634), (September 1943-August 1945); US Naval Advanced Base Personnel Depot, Tanforan, CA, September-November 1945. Released from active duty in December 1945.

Active in various USNR units until retirement. Battles and campaigns: minesweeping and convoy duty in Atlantic Theatre, New Guinea Campaign, Philippine Campaign and Battle of Leyte Gulf, Battle of Okinawa.

Awards and medals include Atlantic Theatre, Pacific Theatre, Philippine Campaign and Naval Reserve. Achieved the rank of LCDR, officially retired on May 15, 1960.

He has many memorable moments from his naval career most of which are associated with his shipmates and many of whom were the very best of their kind that he has ever had. The most dramatic, thrilling and sad experience occurred aboard the USS *Whitehurst* through the first three weeks of the Okinawa Campaign. They had to sail from Ulithi to Eniwetok in the Marshalls to act as escort for the old battleship USS *Texas* which was to join the bombardment group off Okinawa. The *Whitehurst* went at flank speed (24 knots) for over 36 hours to make the rendezvous and he was proud of his engineering division. After arriving at Okinawa they witnessed the bombardment and acted as part of the anti-sub screen for the operation until April 12, 1945 when they were assigned to radar picket station, a suicidal mission at that time. Sure enough, a batch of Japanese suicide planes were picked up on their radar and five of them began to circle their position. One after another they made a run for their ship. They responded with all the AA fire they could muster, and for the first time their four barrel 1.1 inch gun got off 476 rounds without a hang fire or misfire. Four of the five planes were shot down before hitting the ship, but the fifth, although hit, managed to slam into the bridge at the conning level. They lost 43 dead and 24 wounded out of about 210 aboard. They managed to sail the ship into Kerama Retto to accomplish repairs, remove the dead and get help for the wounded but their fighting days at Okinawa were over. They sailed back to Pearl Harbor via Saipan and Kwajalein for complete overhaul and he was transferred after the repairs were completed in August 1945.

He married Judith and they have two children, Steven A. and Randall F. He has held positions at Chevron Research Corporation (now Chevron Research and Technology) as research engineer, senior research engineer, research associate; Chevron Corporation, Product Engineering Department as staff product engineer; Chevron USA Inc. as product engineering and senior staff product engineer, retiring December 1981; S.R. Calish Consultants, January 1982 until retirement in December 1990.

HARRY R. CALLISON, born Sept. 11, 1918 in Beard, WV, attended college and received a BS degree in agriculture.

He joined the service Feb. 18, 1942. Served with USS *Chinquapin* (AN-17); USS *Yazoo* (AN-97); USS *Marias* (AO-57). His battles and campaigns: Marianas Campaign (Kwajalein-Eniwetok, Saipan-Guam, Okinawa).

Callison was awarded the Pacific Campaign Medal w/3 Stars. He achieved the rank of CDR. His memorable experiences: kamikaze attacks at Okinawa. One occurred after Japan said they were giving up.

He married Dolores and they have two children, Richard and Peggy; and four grandchildren: Katie, David, Amy and Nicole. Employed as diesel power plant shift supervisor for the city of Los Angeles, retiring February 1981.

JOHN EDWARD "JOHNNY" CAMERON, born March 4, 1932 in Des Moines, IA, received a BBA from the University of Hawaii in 1959; graduate school, University of Hawaii 1959; and Drake in 1960.

He joined the service May 11, 1951, Des Moines, IA. His military locations and stations: San Diego, Monterey, Hawaii, Des Moines; Topeka, Kansas City, Oklahoma City, USNG Gen. Line School; NAVPGSCH; USS *Greenlet;* Reserve Units 14-1, 9-14, 9-62, LPH-717; AACT 117, COMPHIBGRP1-118, NAVMTO 311.

Cameron was awarded the Good Conduct Medal, National Defense Medal, Armed Forces Reserve Medal, Naval Reserve Meritorious Service. He achieved the rank of CWO4 and retired Oct. 1, 1986.

His memorable experiences: swearing in his two sons to Navy active duty; during construction of NAV PG SCOL acted as liaison between department heads, professors and Frisco naval purchasing office to guarantee proper specifications were issued to purchase equipment to furnish completed school.

Married Phyllis J. Kennedy and they have three children, Kevan Charles, Lori Angeline and John James Kennedy; and five grandchildren: Whitney, Andrew, Ashley, Danielle and Mason. Employed at Goodyear Tire and Rubber factory as accounting manager. He expects to retire in late summer 1997.

EDWIN P. "COMBAT QUACK" CAMIEL, MD, born June 24, 1950, Philadelphia, PA. He joined the service April 12, 1979. His military locations and stations: NAS Willow Grove, AFRC Folsom, Ft. Dix. Served with 4th MAW MED MAG-49; Fleet Hospital 15, FH20, FH9. His battles and campaigns: Southwest Asia.

Camiel received the Navy Commendation Medal, Navy Achievement, Pistol Expert, MUC, JMUC, Armed Forces Reserve, SWASIA Campaign, Kuwait Liberation, Overseas Service, FMF Ribbon, Sea Service Deployment and Rifle Marksman.

He married Judith Anne Gervasi and they have a child, Kristen. He is employed as a physician.

ARMAND V. CAMPILLO, born July 16, 1930 in San Diego, CA, received a BS in economics and business administration, St. Mary's College, CA in 1952.

Enlisted in the USN Aug. 7, 1952. Commissioned ensign July 1953. His stations: various schools then USS *Philippine Sea* (CVA-47), 2 1/2 years. His battles and campaigns: various Korea Westpac.

Campillo achieved CDR, USN, RAD May 1956. His memorable experiences: After RAD, joined NAR, served in BARTU, WEPTU, NASRU units. Retired 1976. Retired w/pay 1990, total service 24+ years.

He married Norma Louise "Ghio" and has four children: John G., Paul A., Mark J. and Margo; and a granddaughter, Ashley Nicole Campillo. He has held various management positions with the City of San Diego, spanning 31+ years. Retired from civil employment July 1988. Retired from Navy Reserves October 1976.

FRED GROVER CAMPILLO, LCDR, born Dec. 24, 1956 in Inglewood, CA, received an AA in psychology from El Camino College, CA 1978; BA psychology from California State University, Long Beach, 1980; MS financial management, Naval Post Graduate School, 1987; currently enrolled in Texas Christian University, proposed graduate date 1998. He graduated from OCS, Jan. 15, 1981 and commissioned in May 1981. Joined NRA, Jan. 7, 1991.

His military locations and stations: Long Beach, CA; Manama, Bahrain; Monterey, CA; Newport, RI; Pearl

Harbor, HI; Agana, Guam; Decatur, IL; Norfolk, VA; Dallas, TX.

Units served with: USS *Racine* (LST-1191); USS *LaSalle* (AGF 3); USS *Badger* (FF-1071); USS *Niagara Falls* (AFS 3); NAVRESCEN Decatur, IL; USS *George Washington* (CVN-73); REDCOM 11.

Battles and campaigns: Operation Earnest Will (Iran-Iraq War); Operation Desert Shield (Gulf War); Operation Desert Storm (Gulf War); Operation Southern Watch (Gulf War); Operation Deny Flight (Bosnia).

Campillo was awarded the Meritorious Service Medal, Navy Commendation Medal, Combat Action Ribbon, Navy Unit Commendation Ribbon w/Bronze Service Star, Meritorious Unit Commendation w/Bronze Service Star, Navy Combat Efficiency "Battle E" Ribbon; National Defense Service Medal; Armed Forces Expeditionary Medal, Southwest Asia Service Medal w/3 Bronze Campaign Stars, Armed Forces Service Medal, Humanitarian Service Medal, Sea Service/Deployment Ribbon w/ 3 Bronze Service Stars, Reserve Sea Service Ribbon, Armed Forces Reserve Medal, Kuwait Liberation Medal (Saudi Arabia), Kuwait Liberation Medal (Kuwait), Rifle Marksmanship Ribbon, and Expert Pistol Shot Medal. He achieved the rank of LCDR.

His memorable experiences: sailing on a frigate (single ship) nose-to-nose with a Russian battlegroup across the Sea of Okhotsk. Rescuing dying Vietnamese refugees adrift in the South China Sea. Being on TAO watch in the Persian (Arabian) Gulf when the call went out for the cruise missile launches that marked the transition from Desert Shield to Desert Storm; giving the keynote address before a capacity crowd and the local news cameras as part of the 1991 Memorial Day commemoration in Decatur, IL; and many, many nights on watch at sea.

He married Chris Andrea Cherches and they have a daughter Madeline Elaine.

JESSE D. CANNON JR., born Oct. 11, 1947, Portsmouth, VA, received a bachelors in architecture, 1972 and master of urban and regional planning in 1991.

He joined the serivce Oct. 11, 1964. His military locations and stations: NRC, Laurel, MS; NRC, Little Creek, VA; NRC, Baton Rouge, LA; Naval Reserve Readiness Center, New Orleans, LA; Reserve Support Site, Shreveport, LA; Construction Bn. Center, Port Hueneme, CA; Construction Bn. Center, Gulfport, MS.

Billets and Units served with: electrician's mate E-1, Reserve Unit 6-72M, Laurel, MS, October 1964-September 1965; electrician's mate and engineering aide, (E-2) Reserve Unit and Seabee Detachment, Little Creek, VA October 1965-May 1967; engineering aide (E-3), Naval Schools Construction, Port Hueneme, CA, November 1966-March 1967, (active duty); engineering aide (E-3 through E-5), Seabee Detachment 8-22, New Orleans, LA, March 1967-May 1970; engineering aide (E-5), Reserve Naval Mobile Construction, Bn. 22, Gulfport, MS, May 1968-March 1969 (active duty); engineering aide (E-6), Seabee Detachment 8-28, Baton Rouge, LA, June 1970-December 1972; engineering office and assistant company commander (ENS through LT), Reserve Naval Mobile Construction Bn. 28, Shreveport, LA, January 1973-February 1979 (commissioned ENS, USNR, Civil Engr. Corps, November 1973).

Inactive Ready Reserve, March 1979-August 1983; Co. CDR (LCDR), Reserve Naval Mobile Construction Bn. 28, Shreveport, LA, September 1983-September 1984; FSO (LCDR), officer-in-charge of construction, EFD Guam, September 1984-September 1987; equipment officer (LCDR), Navy Cargo Handling Bn. 13, Gulfport, MS, October 1987-September 1990; facilities officer coordinator (CDR), Engineering Field Div., Pacific, Houston, TX, October 1990-September 1993; Unit-Officer-in-Charge (CDR), Environmental Engineering Unit, NAVFAC, Pensacola, FL, October 1993-September 1995 (Unit OIC); Deputy Branch Manager (CDR), Naval Reserve Contingency Engineering Program, European Command West, Dallas, TX, October 1995-current.

Cannon achieved the rank of CDR and received the Navy Achievement Medal and Ribbon w/Gold Star, Combat Action Ribbon, MUCR, RVNCM and Ribbon, RVN Civil Actions Unit Citation, RVN Gallantry Cross w/Palm Unit Citation, NDSM and Ribbon, VSM and Ribbon, AFRM and Ribbon and the Navy Battle Efficiency "E" Ribbon. Authorized to wear insignia of a Seabee Combat Warfare Specialist.

He married Mary and they have two children, Jessika and Mattie. He held the position as Fifth Circuit Court of Appeals, architect.

JOSEPH E."JOE" CANTWELL, CWO3, born June 27, 1918, Cincinnati, OH, graduated high school and Naval Electrical School. He joined the USN, Aug. 13, 1936. His military locations and stations: Norfolk, VA Destroyer Base San Diego,; APA Pacific commissioning schools, Seattle, WA. Served with USS *Indianapolis,* USS *Maryland,* USS *Antaeus,* USS *Frederick Fonston,* USS *Lenawee,* NRTC, CIN O.

Cantwell participated in battles and campaigns in Sicily, Salerno, Mediterranean, Philippines, Iwo Jima and Okinawa, Japan. He was awarded the Good Conduct Medal, American, Pre-War, European, Pacific, WWII Victory, Philippines. Achieved the rank of CWO3, chief electrician and retired in 1978.

His memorable experiences: three years, 1937-1940, USS *Indy* Com. Scout Force; all battle experiences; flag raising on Iwo; all our power, Tokyo Bay, Sept. 2, 1945.

Married Noma and they have three children: Jerry, Tim and Margaret Ann; and six grandchildren: Brian, Brad, Brent, Jeremy, Chris and Amy. Employed in electrical construction in all positions including supervisor. Retired from the USNR, June 27, 1978 and from civilian employment Dec. 31, 1981.

JAY S. CAPUTO, LT, born Nov. 25, 1949 in Brooklyn, NY, received a BA degree from Adelphi University, Garden City, NY. Graduate of FBI National Academy, 148th Session.

He enlisted in service June 24, 1970. Served as an air controlman and rose to second class petty officer, NAS Millington, TN; NAS Glynco, GA; NAS NORVA W4, NAS NORVADET 1602. Commissioned Sept. 14, 1981.

Military locations and stations: Reserve Intelligence Area 17, Freeport, NY; Reserve Intelligence Area 16, Willow Grove, PA. Served with FIRSTEURLANT 0302, NISRO 0602, MARDEZLANT 0102, DEFOPS 0102, NIS 0893, IVTU 0104, ONI 0102.

He was awarded the Navy Achievement Medal, Naval Reserve Meritorious Medal (three awards), Armed Forces Reserve Medal (two awards), National Defense Service Medal, Navy Overseas Service Ribbon (two awards), Navy Pistol Expert Medal, Navy Rifle Expert Medal. Caputo achieved the rank of CDR.

His memorable experiences: Operation Liberty 1986. Counter-Narcotics operations with the Spanish National Police and Royal Gibraltar Police. Achieving command as the CO of ONI 0102.

Married Kathleen Koubek and they have two children, Janet and Dana. Employed with Nassau County Police Department, LT, Office of Chief of Operations.

FRED D. CARL, CAPT, born in Washington, DC, June 16, 1920, and graduated from American University in June 1947. He enlisted in the USN in March 1942, served for 18 months at the Washington Navy Yard Communications Office, advancing to the rate of yeoman 2/c. He was commissioned an ENS in December 1943 upon completion of Northwestern University Midshipmen School in Chicago.

His first sea duty was aboard the USS *Eldridge* (DE-172), operating in the Atlantic and Mediterranean. Following training at the Sub Chaser Training Center in Miami, he reported aboard the USS *Waters* (APD-8) (former DD-115), which carried Underwater Demolition Teams, and soon became the CO. During the year and a half he was aboard the *Waters* was engaged in combat throughout the Iwo Jima and Okinawa campaigns. Following WWII, he served in Reserve Surface Warfare assignments, including that of group commander. He retired in 1976 after 34 years of service, with the rank of CAPT.

His professional career was with the YMCA in Washington, Indianapolis, New York City and nationally. He retired 1986 as the national executive director, Armed Services YMCA of the USA.

He was the national president of the NRA, 1977-79. He previously served as national vice-president, Surface Program, national treasurer and president of the Stephen Decatur Chapter. He was the commander general of the Naval Order of the US, 1987-89, after first being the Commander General-Elect and Treasurer General, and has served for 13 years as the managing editor of its newsletter. Between 1993-95, was the national chairman, US Naval Sea Cadet Corps. He was a national vice-president of the Navy League of the US, 1990-95, and earlier was the president of the Chicago Council. He currently serves as a national director, and as chairman of its Youth Program Committee.

Carl's decorations include the following medals: Navy Commendation, Navy Superior Public Service, Navy Meritorious Public Service, Coast Guard Meritorious Public Service, American Pacific (with two engagements stars) and European Theaters; WWII Victory; National Service; and the Naval Reserve.

Carl and the former Janet A. MacIntosh were married in 1996 and reside in Indialantic, FL. He has two married daughters, Cheryl C. Hughes and Sandra C. Barber, by his first wife of 45 years, Murial H. Carl (deceased).

JAMES W. CARLSON, CDR, born Nov. 27, 1952, Olongapo, Philippines. He grew up in Southern California, attended Marina High School and eventually received an appointment to the USAF Academy in 1971 and graduated in 1975 with BS degrees in history and international affairs and a commission in the Air Force. He was also president of his graduating class and received his master of science in 1971 in human resource management from Gonzaga University in Spokane, WA.

CDR Carlson has over 1500 hours of Air Force and Navy flight time in a variety of aircraft. In 1976 he received his navigator wings from Mather AFB in California and was assigned to Fairchild AFB in Washington where he flew B-52Gs with the 325th BS 92nd BW. In 1981 he earned the wing bombing navigation prize and competed with the "best in SAC" for that year's bomb/nav trophy. In 1982 he resigned from the Air Force to pursue opportunities in computer programming and design. Three years later, he affiliated with the Naval Reserves as an Intelligence Officer and his unit assignments were with FIRSTPAC 0919 (1985-86), VFA-305 (1986-89), VA-304 (1989-90) as an A-6E bomb/nav, JICPAC 1287 (1990-91), NSWC 0287 (1991-94), DIAHQ 0878 (1994) and JICPAC 0178 (1994-97).

As a reserve intelligence officer, CDR Carlson served as an instructor in the Navy's Strike Warfare Center in Fallon, NV; a watch chief in Saudi Arabia for Operation Southern Watch; the daily Yugoslavia briefer for CINC AFSOUTH in Naples, Italy; and as the Joint RECON Officer for HQ EUCOM in Stuttgart, Germany for Operation Joint Endeavor in the Balkans and Central Africa.

He is a life member of the NRA, an active member of the Academy Assoc. of Graduates, and currently pursuing a law degree through the Univ. of Minnesota Law School in Minneapolis. His daughter Jennifer is enrolled at Norwich Univ. in Vermont and serves in the Corp of Cadets in a leadership position.

CDR Carlson has been awarded the JSCM, Naval Achievement Medal, SWA Service Medal, Joint Meritorious Unit Award, NDSM w/Bronze Star, CRM, HSM, AFOUA, Navy and Marine Corps OSR, AFRM w/M Device, NATO Medal, Navy Pistol Expert Award and Navy Rifle Expert Award.

RICHARD A. CARLSON, CDR, born in Canton, OH, June 28, 1924. He began his naval aviation career in 1943 and was commissioned ensign in 1944. During WWII he served in the Atlantic Campaign in various air antisubmarine squadrons. Released to inactive duty in 1945 he joined the active Reserves at NAS Akron and attended Mt. Union College receiving a BA degree.

Recalled to active duty in 1951. During the Korean War he served on various carriers during a tour with Air Antisubmarine Sqdn. 21. He became a TAR (Training/ Administration Reserves) officer in 1954 and remained thus until retirement in 1972. During the TAR career he served in various billets at NAS Columbus, NARTU Jacksonville, NAS New Orleans, and NAS Los Alomitos. Interspersed with AS duties were two tours of duty for five years at the Bureau of Naval Personnel. Flew approximately 6,000 hours in single, multi-engine and jets.

Spent 15 years Civil Service with chief of USNR. His awards and medals include American Defense, Naval Reserve, WWII Victory, Korean Service, United Nations, Armed Forces Reserve, National Defense Service and Navy Commendation Medal.

ROBERT O. "RED-SWEDE" CARLSON SR., born Nov. 23, 1919, Albert Lea, MN. He joined the service Jan. 9, 1939. His military locations and stations: USNTC Great Lakes; USS *Oklahoma* (BB-37); VC-80. Served with VC-80; HEDRON 14; FASRON 0123; NAS Oceana.

Carlson fought in Pearl Harbor, Dec. 7, 1941. He achieved the rank of LCDR and was discharged May 1, 1964. His memorable experiences: survivor of USS *Oklahoma,* Dec. 7, 1941.

He has been employed as clerk with Juvenile and Domestic Relations Courts, State of Virginia.

WM. H. "BILL" CARLYON, CAPT, born June 29, 1924, Salt Lake City, UT, received a BS and BA from the University of Washington.

He joined the service December 1942. His military locations and stations: USS *Batfish* (SSB10), USS *C.F. Hughes* (DD-428), WWII, Korea, Naval War College, National War College, AFS College. Units served with: CO, Sub Div. 13-17; CO Sub Div. 13-6; group commander 13-4; Deputy Naval Reserve Readiness Command Region 22; CO CINC PAC 322.

Carlyon was qualified SS and SWO. He received Letters of Commendation. Achieved the rank of CAPT and discharged June 1984. His memorable experiences: designated CDR TG 133.3 (OTC Joint Reserves 73).

Married Phyllis and they have two children, William and Caron; and two grandchildren, Melanie and Alyson. Employed as Reg. director, Aircraft Contracts, Boeing, retiring May 1990.

JAMES E. "JIM" CARNEY, CDR, born Feb. 21, 1925 in Kearny, NJ, received a BS in marketing from Rutgers University, 1951. He joined the service Feb. 21, 1943. His military locations and stations: NAS Pensacola, NAS Lakehurst, NJ. Units served with: NAAS Ellyson, LP751, LP753, AW575, HU752, CO of NARDIV 751.

Carney achieved the rank of CDR. He was awarded the American Campaign Medal, WWII Victory Medal, Reserve Medal and Armed Forces Reserve Medal. He retired Feb. 27, 1985.

His memorable experiences: commissioned naval aviator, April 1946. Qualified LTA airship pilot May 1953; CO NARDIV 751 Lakehurst NAS September 1967; served in recruiting and Blue and Gold officer for US Naval Academy. Retired June 30, 1990 from USNR w/pay.

He married Estelle and they have four children: Ed, Jim, Rich and Kathy; and grandchildren, Danielle and James. He contracts with Kearfost Guidance and Navigation Corp. as a contract specialist.

JOSCELYN "JOSH" CARON, LT, is a native of Montreal, Canada. He immigrated to the US with his family in 1971. He completed high school in Maine, undergraduate studies with the University of Maine system and the University of the State of New York.

In 1990 he obtained a MS-nurse anesthesia from the University of New England College of Arts and Science. He is presently employed by Waterville Anesthesia Associate P.A. in central Maine, a CRNA/MDA group.

Caron has been affiliated with the USNR since 1989 when he was first commissioned as an ENS being attached to NRC Augusta, ME. Branch clinic Brunswick NAS.

His next billet was with Combat Zone Fleet Hospital #20, Primus Unit where he remained until last year returning to NRC Augusta, ME, now holding a billet with National Naval Medical Center Bethesda, MD. He was on active duty during the Gulf War from January to April 1991, providing anesthesia care with Fleet Hospital #15 in Al Jubayl, Saudi Arabia. LTJG Caron voluntarily took first call, first case on OR schedule to prove his professional skills in anesthesia administration.

Service ribbons and medals include Meritorious Unit Commendation, National Defense Service Medal, Southwest Asia Service Medal, Navy and Marine Corps Overseas Service Ribbon, Liberation of Kuwait Medal (Kuwait), Kuwait Liberation Medal (Kingdom of Saudia Arabia).

The Caron family consists of mom, Donna, a pediatric registered nurse; son, Eric; and daughter, Taylor. They reside in Belgrade, ME. His hobbies are mostly outdoor activities performed with the family such as skiing and swimming.

DONALD BLODGETT "DON or CARP" CARPENTER, LCDR, born Aug. 20, 1916, in New Haven, CT. His youth was spent in Hanover, NH and Burlington, VT. He graduated from Burlington High School in 1934 and the University of Vermont with a PhB degree in 1938.

He entered the USNR on Sept. 8, 1942, and completed indoctrination training at NTS Newport, RI. He was assistant biography editor on the staff of the *Golden Boot*, the class "Cruise" book of the Officers' Indoctrination School. Carpenter was also managing editor of the *Scuttlebutt,"* the school's daily newspaper. He was commissioned an ENS in November 1942. While the members of this graduating class were often referred to as "90-day Wonders", they actually finished their intensive training in two months.

He was assigned duty as a CWO at staff headquarters of the Western Sea Frontier, in San Francisco, CA. From there he served on active duty as a CLO for the staffs of Convoy Commodores in the Asiatic-Pacific Theater of Operation. When WWII ended, on Sept. 25, 1945, he was serving aboard the flagship, as a CLO and beachmaster, for the "CULA" Expedition in the Arctic waters off Point Barrow, AK. Following separation from the service on Feb. 26, 1946, he remained in the USNR, ending up as XO, and then, CO of a NR MSTS Unit, located at Treasure Island, CA. On June 30, 1968, he retired with the rank of LCDR, USNR.

His military awards, medals and ribbons include the Commendation Ribbon, Letter of Commendation, WWII Victory Medal, Naval Reserve Medal, American Area Campaign Medal, the Asiatic-Pacific Area Campaign Medal, and Reserve Officers Association Medal.

He is a life member of the American Legion, NRA, Naval Order of the US, ROA, TROA, Navy League of the US, MOWW, National Association for Uniformed Services, and US Naval Institute.

Since his two brothers are also US Veterans of WWII, three branchs of the services were represented. John T. Carpenter was in the Marines in the Asiatic-Pacific Theater of operation, and Torrey C. Carpenter was in the Army in the European area campaign.

Single and employed as teacher and coach; real estate appraiser; REI. He retired Dec. 31, 1988, and lives in Mendocino, CA.

CHARLES REED CARR, LCDR, born June 14, 1937, Berkeley, CA, received a BS and MS from Brigham Young University, 1972-73.

He enlisted in the service June 10, 1955. His military locations and stations: boot camp. Started flight training, February 1956. Commissioned July 3, 1957, age 20 years three weeks. Units served with: HS-4, VRF-32, HT8, NAVSTA Trinidad, HS-85.

Carr received the Navy Commendation Medal. Achieved the rank of LT ACDU, LCDR USNR, and was discharged June 1967 active duty and September 1975 Reserves. His memorable experiences: helping to rescue 10 people off sinking fishing boat, Pensacola, FL, 1964.

He married Donna and their children are Jennifer, Michelle, Teri, Lisa, Reed Jr., David and Rachel; and so far, 19 grandchildren. Presently teaching at Rice University; sales of securities.

MICHAEL F. "MIKE" CARR, LCDR, born April 7, 1935, Syracuse, NY, received a bachelor of business administration, 1956 from the University of Notre Dame and master of business administration, 1963 from New York University. He joined the service Aug. 7, 1956, Officer Candidate Seaman apprentice, OCS Class 29, Newport, RI. Commissioned Dec. 13, 1956.

Active duty: DO, PO, Atlantic Fleet Phototriangulation Group, Guantanamo Bay, Cuba Detachment. Permanent Shore Patrol as collateral duty, 1957-58.

Officer-in-charge Fleet Camera Party, Newport Detachment (name changed from Atlantic Fleet Phototriangulation Group, and now called Combat Camera Group). Port control duty officer as collateral duty, 1959. He achieved the rank of LTJG on active duty. RAD Dec. 13, 1959.

Served as student and instructor with Naval Reserve Officer School 3-2 New York City; GVTU, Brooklyn, NY; division officer, XO and CO of Convoy Commodore Command, New York/Naples. Retired 1980 as LCDR, USNR and CDR, New York Naval Militia.

Civilian: Securities Analyst, Institutional Securities salesman and investment portfolio manager since RAD. Presently president of Carr & Associates, Inc. a registered investment advisory firm in Lodi, NJ.

Past president of Third District of Naval Reserve Association. Past Commander of the New York Commandery of the Naval Order of the US.

HELEN PAULINE CARTER, LT, born Oct. 14, 1913, Attala County, MS. Earned BA, Blue Mtn. College; MA, U. of Alabama; MED, U. of Mississippi.

Joined the service May 24, 1943; stationed at USNPG,

Dahlgren, VA; served with Bureau of Ordnance; ordnance officer, Dahlgren, VA; NRSD6-71.

LT. Carter was discharged May 1, 1946; Naval Reserve Retired List, Oct. 14, 1973. She received the American Campaign Ribbon.

Single, she retired July 1, 1979 from civilian employment as a high school counselor at Jackson Public Schools.

WILLIAM L. "NICK" CARTER, LT, born Nov. 21, 1916 in Rye, NH, received an AB in economics from Brown University in 1939. He joined the service Jan. 15, 1942. His military locations and stations: V-7 USNR; USS *Prairie State;* Torpedo Station, Newport. Served with: USS *Parker* (DD-604); USS *Beatty* (DD-756).

Carter participated in battles in campaigns in North Africa, Sicily, Italian Coast (Anzio Southern France). He achieved the rank of LT and was discharged Feb. 22, 1946. Received six Battle Stars.

He married Mary Brauer and they have five children: Thomas, Anthony, Joan, Daniel and Peter; and 10 grandchildren: Thomas, John, Elizabeth, Kate, Anne, Nicolas, Haley, Regina, Leal and Geoff. Employed at First National Bank Boston as loan officer, Harnischfeger Corp. as executive vice-president, CFO, director, retiring Jan. 1, 1981.

PETER "PETE" CARTY, LT, born Sept. 29, 1965 in Portland, OR, received a BSME from Virginia Military Institute (1987); MBA from the University of Southern California (1995).

He joined the service May 15, 1987. His military locations and stations: USS *Jarrett* (FF6-33), 1988-91, MIUWU 105, 1991-96; NRACU5, 1996-present.

Carty participated in Desert Storm and has received the Combat Action, Battle E, MUC, SWA Service and NAM. He achieved the rank of LT. He is married to Blair and is employed with Parker Hamilton Corp. as territory manager.

RICHARD F. CASWELL, CAPT, graduated from the University of Minnesota and was commissioned as an ENS on June 25, 1958. After two years of active duty with the District Public Works Office, Great Lakes, IL he returned to civilian life and joined Naval Mobile Construction Bn. 25 where he served in various billets including the CO of Seabee Reserve Div. 9-46 Racine, WI.

From 1969-71, he served on active duty with the Antarctic Support Activities as the PWO with two deployments to the Antarctic during Deep Freeze 70 and 71. In 1971 he transferred to the Naval Construction Bn. Center, Davisville, RI as the TO.

Caswell returned to civilian life in June 1973 and reaffiliated with the Seabee Reserve Program. He subsequently commanded Naval Mobile Construction Bn. 25, the 30th Reserve Naval Construction Regt., the 2nd Reserve Naval Construction Regt. and served as chief staff officer of the 6th Reserve Naval Construction Regt. He also served on the staff of the First Reserve Naval Construction Brigade as operations officer, facilities planing officer and assistant chief of staff for plans and training.

He has earned the Command Ashore Insignia and been awarded the MSM w/Gold Star in lieu of a second award, the NCM w/Gold Star in lieu of a second award, the NDSM, the Antarctic Service Medal and the AFRM. Ribbons include the MUC, Pistol Sharpshooter and Rifle Marksman.

He and his wife, the former Darlene Kay Langseth reside in Greendale, WI. They have three grown daughters: Lori, Judy and Mary. He retired from the USNR on Sept. 30, 1985.

THOMAS J. CAULFIELD, CAPT, born March 14, 1932 in Buffalo, NY, received a BS, in 1954 and MS, 1963 from Canisius College; Ed.D, 1969 from Suny at Buffalo, NY.

He joined the service in 1955. His military locations and stations: Pensacola, FL; Hutchinson, KS; Brunswick, ME; Willow Grove, PA. Units served with: VP-23, VP932-37; NAR DET, Niagara Falls, NY.

Caulfield achieved the rank of CAPT and was discharged March 14, 1992. His memorable experiences: Iceland deployment during Lebanon crisis 1959 with VP-23; CO NARDET Niagara Falls 1974-78; patrol plane commander (PPC) in P2V-7 aircraft; Chair, counselor education at Canisius College 1972 to 1997.

He married Camille and their children are Rena Mount and Jennifer Van Petten; and grandchildren are Kathryn, Benjamin, Thomas, Timothy, Emily and David. He has held positions as school teacher, school counselor and college professor.

WILLIAM ROBERT CAUTHEN, CDR, born July 14, 1931 in Winston County, AL, received a BS in electrical engineering from the University of Alabama.

He joined the USNR as seaman recruit in August 1948 for minority at age 17. Called to active duty as EMFN in September 1950, served two years aboard USS *Seer* (AM-112) and released to inactive duty in September 1952. Attained rate of EMC (E-7) and served 15 years Reserve time as enlisted.

Received commission as LTJG engineering duty, in 1965. Retired from Reserve Nov. 1, 1990 as CDR, EDO, with total 40 years satisfactory enlisted and commissioned service.

Served as CO of three different reserve units supporting various shipyards and naval ship engineering in Washington, DC.

He married Joyce S. and their children are LeAnn Cauthen, Jane Hughey, Beth Rawlson, Robert H. Cauthen and Beverly Payne; and they have 10 grandchildren. He retired from Tennessee Valley Authority after 30 years service.

CESAR E. CAVANNA, born Dec. 1, 1932 in Modesto, CA, received a BS (mechanical engineering) from the University of California, Berkeley, 1956.

Drafted into the Navy in 1956. His military locations and stations: active from 1956-59, repair officer, heavy attack aircraft carrier. Responsible for the pipe shop, CO2 shop, carpenter shop, and damage control to the ship.

USN retired. Retired from active reserve status in 1978 with rank of CDR. Assignments included CO of Research Reserve Unit to CO of Military Training Div., along with various sea duty training exercises. His memorable experiences: aboard CVA-31, ship gave air show to GEN Chiang Kai-shek of Formosa.

He married Arleen and their children are Cynthia born 1963, Catherine born 1966, Christine born 1967, Cheryl born 1969, and Colette born 1972. A granddaughter Chelsea was born in 1993.

He was employed from 1960-86 with Lox Equip. Co., as executive vice-president; 1987-93, consulting engineer, Boster Kobayashi Assoc. Presently president of Cesar Cavanna Consulting Engineer, an engineering firm specializing in the mechanical engineering aspects of accident reconstruction. He is listed in Who's *Who in California* (1994), *Automotive Consultants Directory* (SAE) (1987) and *Director of Railroad Safety Consultants* (FELA Reporter) 1989. His name is listed on four cyrogenic equipment patents.

KENNETH C. CAVANAUGH, CAPT, born April 30, 1916 in Fremont, MI, received a BS degree from Michigan State University. He was commissioned ENS in USNR, April 22, 1943. Active service in WWII May 31, 1943-Dec. 31, 1945.

His military locations and stations: Cornell University, Ithaca, NY; Norfolk Navy Station, Norfolk, VA. At sea: USS *Glennon* (DD-620) and USS *Tolman* (DM-28). Battles and campaigns: Normandy and Okinawa; American, European, Asiatic-Pacific.

His awards and medals include Purple Heart (combat wound-Okinawa); American, European, Asiatic-Pacific Campaign Medals; Bronze Stars for Normandy and Okinawa invasions; WWII Victory Medal; Naval Reserve Medal; Armed Forces Reserve Medal.

Cavanaugh achieved the rank of CAPT and was transferred to Retired Reserve April 1, 1973. His memorable experiences: Aboard USS *Glennon* (DD-620) when it was sunk by a German-placed mine on June 8, 1944 during Normandy Invasion of France during WWII. Aboard USS *Tolman* (DM-28) when it was hit by Japanese torpedoes during Okinawa invasion, Apirl 1, 1945. He was wounded and received the Purple Heart.

Employed with the Federal Civil Service with Federal Housing Agencies from 1945-73. Retired as director of Housing Programs, Dept. of Housing and Urban Development (HUD); San Francisco Housing Authority, controller and deputy director, 1973-78. Self-employed as housing consultant until Dec. 31, 1985.

He married Blythe and their children are James Lee, John Thomas and Patricia Ann Arnett; grandchildren are Marie Loomis, Julie Barnard, Kelly Xitco and Ashley Cavanaugh; and great-grandchildren are Ashley Marie Loomis, Jordan Leigh Barnard and Andrew Mark Xitco.

ROY D. CAVES, CAPT, born on April 30, 1933 at Waynesboro, MS. He earned a BS degree from Mississippi State University and a MBA degree from Mississippi College. At age 17 he joined the National Guard and served there until graduating from college and being accepted for Navy OCS where he was commissioned ENS in January 1956.

Caves' active duty included amphibious orientation at Coronado, CA and 30 months aboard the USS *DeKalb County* (LST-715), including two tours of duty in the Western Pacific. Aboard DeKalb County he served as stores officer, 1LT and gunnery officer, and chief engineer. He completed active duty with the COM 13 Military Assistance Team and was released to inactive duty in January 1959. His reserve billets included assignments as TO, XO, CO of three units and group commander.

He served on the 6th Naval District Policy Board, Secretary of the Navy Policy Board, a Selection Board, and the staff of CINCPAC. On June 1, 1981 he transferred to the Retired Reserve with just over 30 years of service.

As organizer of the Magnolia Chapter of NRA, served as it's first president. He later served NRA as 6ND president, as national vice-president, Naval Affairs, and the first ever national vice-president, Surface/Subsurface. He has been a life member of the NRA since 1972.

He married the former Ann White and they have one daughter, Mrs. Christopher (Bronwyn) Burford, and one granddaughter, Elizabeth Faulkner Burford. He is the founder and CEO of BAC Management Services, Inc., an association management firm in Jackson, MS.

LAWRENCE L. "LARRY" CECIL, JR., born March 6, 1923, Comus, MD, received a BS degree from Georgetown University School of Foreign Service.

Joined the service May 11, 1943 as air aviation cadet

and separated Aug. 17, 1946. Recalled Jan. 15, 1953 to VS-22 and separated July 1, 1954. His military locations and stations: Corpus Christi, Boca Chica, Barbers Point, Guam, Willow Grove, Lakehurst. He served with: VTN-53, VJ-NINE, VS-22, VS-932, VS-752 and VTN-90. Flew 17 aircraft types (including TBF, F4U, F6F, S2-F). Logged 3300 hours, 106 CVE carrier landings.

Cecil was awarded the Naval Reserve Medal, American Theater Medal, WWII Victory Medal, Asiatic-Pacific Medal. He achieved the rank of LCDR.

His memorable experiences: As co-pilot, ditched a B-26 Marauder off the coast of Guam. Rescued by native fishermen in a dugout canoe and became a life member of the sea squatters club.

He married Margret "Peggy" in 1953 and they have three sons: Lawrence "Skip," Chris, Peter; a daughter Robin; and grandchildren: Simon, Noah, Wilson, Noelle, Aaron and Daniel. Cecil was employed as international market manager for Agchem. He retired July 1, 1982 from Diamond Shamrock.

WALTER T. CEDERHOLM, CAPT, born in New York City on Nov. 8, 1942, and was raised in Huntington, NY. He attended Brown University on an NROTC scholarship, and graduated with an ScB in mechanical engineering in 1965. Following commissioning, he completed training at the US Naval Submarine School, Groton, CT, and Naval Nuclear Power School, Mare Island, CA and Idaho Falls, ID. He was then assigned to the USS *Bang* (SS-385), where he served as assistant engineer, supply officer, and weapons officer. Cederholm next completed four patrols on USS *Will Rogers* (SSBN-659) Gold, and finished his active duty as the assistant engineer, Submarine Sqdn. TEN.

Shortly after leaving active duty, he joined the USNR, where he served continuously from 1971-92. He served in 12 Reserve units, six with the Naval Control of shipping from program, four submarine Reserve units, and NAVSEA and SPAWAR units. Included were command tours of COMSUBLANT Battle Gp. Support 201, AS-11 *Fulton* Detachment 201, AS-18 ORION Detachment 101, NAVSHIPRESCEN 101, and VTU MSC 0115. His final assignment was director COMSUBLANT Battle Group Support Program. He has been awarded the Navy Commendation Medal and the Meritorious Service Medal.

When he left active duty, he joined the ITT Corporation as a sales engineer for the Nesbitt Div., where he stayed for five years. Following a two year period as a self-employed sales management consultant, he joined the General Electric Corp., where he held several marketing positions and was the quality manager of an operating department. In 1992, he joined Asea Brown Boveri, Ltd., and currently (1996) serves as the director Total Quality for its Combustion Engineering Nuclear Systems Div. in Windsor, CT.

He has been active in several community and professional organizations. He is a senator in Jaycees International, has served as the president of a Federal Credit Union, and has been a member of the Board of Examiners for the Malcom Baldridge National Quality Award. He is a life member of the NRA, and a member of the Naval Submarine League and the US Naval Institute.

Cederholm is an avid golfer and a certified wine judge and an accomplished amateur winemaker. He has two children, Matthew (1966) and Laura (1968). He resides in Granby, CT.

JAMES J. CERDA, RADM, born Dec. 29, 1930, Brooklyn, NY. Enlisted in the USN, 1951. Served aboard the USS *Steinaker* (DD-863), USS *Robert L. Wilson* (DDE-847) and on staff of Commander Escort Destroyer Div. 42, 1951-54. Honorably discharged from the USN, 1954 as personnelman 2/c. Commissioned ENS, Medical Corps USNR, 1959.

Military locations and stations: USS *Lake Champlain* (CVS-39); medical officer, University of Maryland, School of Medicine, USS *Derby* (DE-218) and USS *Snowden* (DE-246); district medical officer, Destroyer Sqdn. 30; medical school programs officer, University of Pennsylvania; Associate Professor of Medicine, University of Florida; medical officer, USS *Waldron* (DD-699), and USS *Meredith* (DD-890); squadron medical officer, Destroyer Sqdn. 34; speciality advisor in Gastroenterology to the Commander Naval Medical Command.

Promoted to Flag rank, April 1, 1985 and RADM (upper half), March 1987. His last assignment was as staff medical officer, Commander in Chief US Naval Forces Europe. Cerda is decorated with the Legion of Merit, Navy Unit Commander Medal, Navy Occupation Medal, Navy Good Conduct Medal, National Defense Service Medal and the Naval Reserve Medal.

In 1989 made both an "Honorary Chief Petty Officer" and an "Honorary Lieutenant" USN Nurse Corps. Life member of the NRA, ROA and the US Naval Institute.

He serves as professor, Department of Medicine, Division of Gastroenterology, Hepatology & Nutrition, University of Florida College of Medicine. He married Mariana Cosimi and they have three children: James J. Jr., Deborah and Carla. He retired in October 1991 after 36 years of active and reserve duty.

RICHARD E. "RICK" CHAMBERS, LCDR, born Aug. 1, 1952 in Tacoma, WA, received BSEE from U.T. El Paso, 1982, MSEE Standford 1983 and MBA, University of Dallas, 1994.

Enlisted and service active duty May 1971-June 1978 electronics technician (Communications). He has served in the Reserves from September 1986 to present, aerospace engineering duty officer.

His military locations and stations: NAVCOMMSTA Philippines (1972-74); NRRF Kamiseya/NAVCOMMSTA Yokosuka, 1976-78); USS *Excel* (MSO-439), 1974-76; NAVAIR Air Systems Program Units at NAS Pt. Mugu and NAS Dallas, 1986-present.

Chambers was awarded the Navy Achievement Medal, Joint Meritorious Unit Award, Good Conduct, Naval Reserve Sea Service, National Defense SM (2), Overseas Service.

He achieved the rank of ETN2 (E-5) active duty; LCDR (0-4) Reserves and was discharged from active duty June 1978. His memorable experiences: airlifted via Japanese sea plane from NAS Atsugi to JSDF *Aki Gumo* (Japanese destroyer) at sea off coast of Guam to participate in joint US/Japan exercise - SOAP (Soviet Out of Area Patrol) 1977; Responsible for establishing/maintaining communications systems for interoperability.

Married Ann and they have two children, Garrett and Travis. Employed with Raytheon E-Systems as engineering manager.

KATHRYN "KATE" BRUTON CHANDLER, CAPT, born on the Army post at Ft. Knox, KY, she went on to earn a BS degree from The Ohio State University College of Education. After receiving her commission in October 1969 upon graduation from Naval Women Officers School, Newport, RI, she held successive junior officer administrative billets within the jet training community (VT-26 and VF-126). It was during these tours that she logged backseat flight time in the TF-9J and the TA-4 aircraft. Her subsequent active duty billets were at NAB Coronado and later at NAS Miramar, San Diego, CA.

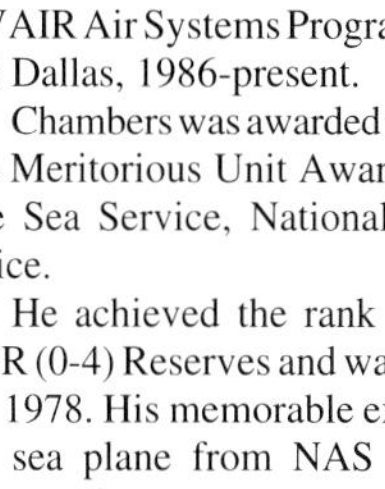

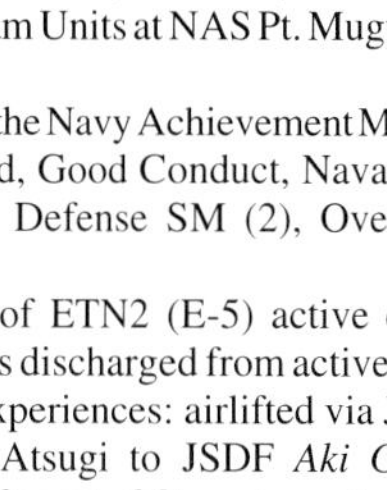

As a new Ready Reservist and junior LCDR, she joined CDR, Naval Inshore Undersea Warfare Gp. One for four years working in administration and training. Upon selection to CDR, she served a memorable tour as CO, Personnel Mobilization Team 2619. Following a department head tour with CDR, Military Sealift Command, Far East Detachment 219, she was promoted to captain while serving on the headquarters staff of CDR, Naval Reserve Readiness Command, Region 19.

Retired from the USNR since May 1, 1994, she remains highly involved with the navy training community through civil service employment as an instructional systems specialist. Completing six years working directly with "A" and "C" technical training schools at Service School Command, Naval Training Center, San Diego, she was awarded the Meritorious Civilian Service Medal in August 1996. She is currently employed with Navy Environmental and Preventive Medicine Unit Five, Naval Station, San Diego, as training department head. Travel and outdoor activities continue to provide enjoyment and interest.

JOHN C. "JOHN" CHAPMAN, CAPT, born in Damariscotta, ME April 10, 1940. He was commissioned Dec. 21, 1962 via OCS after graduating that year from Colby College. Following completion the flight officer program he was assigned to Airborne Early Warning Sqdn. 11 as a CIC officer with collateral duties as assistant squadron legal officer. He logged over 1450 hours in the EC121(k). During one Christmas Eve G-I-UK barrier patrol he cooperated with the plane commander in reporting the transit of an ECHO satellite as a mysterious flying object bound for every child-occupied home on the planet.

In 1965 he was assigned USS *Wasp* (CVS-18), homeported in South Boston, as air controller, tactical briefing officer, and operations administrative assistant. During the 14 month tour he participated as a member of primary recovery ships company during the NASA Gemini Space Program.

In 1966 he became an insurance agent in a family business in Maine, and organized a successful real estate brokerage in the Pemaquid Region of Maine. He held many volunteer positions and is a past president of the Lincoln County Multiple Listing Service, Lincoln County Association of Realtors, Lincoln Academy Board of Trustees, Damariscotta-Newcastle Rotary Club and Damariscotta Baptist Church Trustees.

Although a designated naval flight officer, he successfully completed transition to surface line while serving in various positions in ship maintenance and repair reserve units located at NRC, Portland, ME. He was a plank owner of the Casco Bay Chapter, NRA, and a recipient of the Citizen Sailor Award granted in commemoration of the 200th birthday of the Navy.

Following the USNR reorganization under the Total Force concept, he was selected to command the ship maintenance facility reserve detachment in Portland, ME, and subsequently the Military Sealift Command HQ Mediterranean unit in Bangor, ME.

Following service as a US Naval Academy information officer assigned to Maine mid-coast communities, he retired from the USNR in 1990. He and his wife Allison live in Topsham, ME. Their five grown children reside in Maine, Massachusetts, New York, California and Germany.

BRUCE HIRSCHL CHARNOV, CAPT, born Nov. 16, 1946 in Grand Rapids, MI, received a BA, MA, Rabbi, MBA, PhD and JD degree.

He joined the service July 9, 1970. His military locations and stations: NTCSD; 3rd MARDIV; 2nd MARDIV; SUBLANT; REDCOM 2, 11. Served with NASNURVA; PERSMOBTM; SUBLANT; 3/12 3rd MARDIV.

Charnov was awarded the National Defense (2), NAM, NCM, Reserve Service (2), FMF, BUC. His memorable experiences: 3/12 3rd MARDIV; COMFAIRMED just after Desert Storm; CNFJ Yokosuka; NSA Rota Spain.

He is single with three children: Miryam Esther, Aharon "Roni" and Jessica Lauren. Employed as associate professor, Hofstra University. His anticipated retirement date is August 1998.

RICHARD "DICK" HOVEY CHASE, born June 17, 1912 in Boston, graduated Newton High School and received a JD degree from Suffolk University 1942.

He joined the service and his military locations and stations include NAS Squantum; NAS So. Weymouth, Staff Command Oakland, CA. Served with VR-12, Honolulu, Samar, Guam.

Chase received the Asiatic-Pacific and American Ribbons. He achieved the rank of CAPT and retired June 7, 1972.

He was married to Barbara Week and is now a widower. His children are Lawrence A. Chase, Sandra R. Keach and Patricia A. Wage; and grandchildren are Susan, Barbara and Joyce Chase; Jennifer and Amanda Keach and Colleen and Nicholas Wade. Retired from New

England Mutual Life after 30 years as underwriter, Massachusetts General life underwriting vice-president.

NORMAN V. "NORM" CHATFIELD, LCDR, born Feb. 23, 1921, Salt Lake City, UT, graduated from Weber State University, (formerly Weber College), Ogden, UT.

He enlisted in the service Sept. 29, 1942. His military locations and stations: USNS Treasure Island, CA, 1942; USNS Chicago Armory, Chicago, IL, 1942-1943. Enlisted Utah State University, Logan, UT (Navy School), 1943; NAS Ware Island, Corpus Christi, TX, 1943; NAS Sqd. 8-A, Pensacola, FL, 1943-45; Navy Res Training Center, Ogden, UT, 1946-47; USN Gen. Line School, Monterey, CA, 1947-49; USN Post Grad School, Monterey, CA; 1949-52; Boston Navy Shipyard, Charleston, MA, 1956-58; Potomac and Ser. River Navy Command, Naval Warfare Plant, Washington, DC, 1958-60.

Chatfield was awarded the Good Conduct Medal and Air Crew Wings. He achieved the rank of LCDR and retired Feb. 23, 1981.

He has been married for 56 years to Catherine deMik and they have four children: Tamara, Michael, David and Leilani; 19 grandchildren; and 10 great-grandchildren. Employed with the US Civil Service and electronic engineer GS14, retiring June 1975, McClellan AFB, Sacramento, CA.

DALTON K. "CHAT" CHATTERTON, CDR, born Aug. 31, 1917, in Blandensville, IL. He attended the University of Minnesota 1938 and 1939 studying aeronautical engineer.

In 1939 he formed the Twin Cities Flying Club and soloed May 6, 1940. He was called to active duty Nov. 6, 1940 as a member of a reserve unit. He graduated from NAS Pensacola in September 1942. He was assigned to the test and ferry squadron VR-1 at NAS Floyd Bennett, NY. There he flew the first FM-2 No. 15954 from the factory at Trenton, to the Navy Test Center at Patuxent River, MD. In 1944 he went to Roanoke, VA to Penn Central Multi-Engine School, a Navy contract R5D school. In July 1944 he was assigned to the staff of ADM A.S. Carpender, commandant of the Ninth Naval District as a staff pilot. He served there until he was discharged from active duty, December 1945.

From 1946-77 he flew for Northwest Airlines until he retired at age 60. During his airline career, he spent 20 years in the USNR, Minneapolis, MN, where he was CO of VR-813, which received a commendation from ADM Brown, the Commandant of the 6th Fleet, Naples, Italy. He was qualified in 20 different Navy aircraft and has 3,500 Navy hours and over 25,000 airline and civilian hours.

He married Delores and they have two children, Dennis and Darlene; and grandchildren: Dawn, Andrew, Mia, Tiffany and Beau. The Chatterton family has retired in Boca Raton, FL.

RICHARD M. "DICK" CHAUNCEY, LCDR, born July 11, 1910, Albion, MI, attended Albion College and graduated Wharton School of Finance and Commerce, University of Pennsylvania.

He joined the service in April 1942. His military locations and stations: USS *Agressor,* mine sweeper; USS *Dauntless,* seagoing gunship, Flagship of ADM Ernest J. King, Chief of Naval Operations, WWII, gunnery officer; asst. personnel director, Philadelphia Navy Yard, Ship Building Dept.; gunnery officer, Missionary Ridge, Armed Guard.

Chauncey was awarded the Atlantic Fleet, Pacific Fleet and American Theater medals. He achieved the rank of LCDR and was transferred to Navy Retired Reserve, Feb. 1, 1955.

His memorable experiences: while he was officer of deck, ADM King introduced him to GEN George C. Marshall, US Army Chief of Staff and promoter of Marshall plan. He was also introduced to GEN Wap Arnold, air commander. He was also introduced to Averill Harriman, American Ambassador to Soviet Union and he met other dignitaries including all of ADM King's staff. ADM King and most of his staff lived aboard the Flag ship.

He married the late Carolyn K. and is now married to Esther B. His children are: Ann C. Masternak, Mary C. Avery and Margaret C. Cramer; and grandchildren are Becky, Jeff and Mary Masternak; Caroline and Kevin Avery; Molly and Nicholas Cramer. Employed with Mutual Benefit Life Insurance Co. as agent.

MILTON "MILT" CHAZEN, CAPT, born April 17, 1914 in Danbury, CT, attended CEC Renselaer Polytechnic Institute. He joined the service April 1941. His military locations and stations: Argentia, New Foundland, Camp Perry, Camp Pendleton, Maui, Samar, North China. Served with 17th Naval Const. Bn. 30th Naval Const. Bn.

Chazen was awarded all campaign medals for all campaigns and Korea. He achieved the rank of CAPT, USN and was discharged in 1952. His memorable experiences: VJ-Day in Pacific.

He married Fay Scheer and they have two children, Cathy Stone and Richard; their grandchildren are Levi, Caitlin and Colin Chazen, and Mary and Andrea Stone. Employed as high school teacher, SEC Chazen Companies.

ALAN CHERTOK, LCDR, born in Philadelphia, April 5, 1953. He completed undergraduate studies at the Salisbury State College (MD), and holds master's degree from the University of Baltimore and Central Michigan University.

In 1996 he served as the 11th William J. Fulbright Fellow for UK Police Studies. Chertok received a direct commission in 1987 and served through September 1991 with the Defense Attaché Unit, Arlington, VA. There, he gained the 9635 NOBC. He served a number of reserve active duty tours in Latin America, mostly in support of Operations Units. He was also a member of the USAFR East Coast Spanish Language Emerson Team. Chertok's current assignments is with AIC1986, Norfolk, VA.

Chertok, a career police officer, currently serves as a major within the Newport News (VA) Police Department. He has 400 civilian flight hours and is a civilian commercial pilot. He pilots his own Cessna 182.

MICHAEL B. CHESSON, CAPT, born in Richmond, Sept. 5, 1947. Graduated from William and Mary with high honors in history in 1969, he enlisted in the USNR. Assigned to the Naval Security Group, he graduated from boot camp at Great Lakes in 1970 as RPOC of the color company. He attended "A" School at NTTC Corry Field, Pensacola, 1970-71, and graduated second in his class as a CTRSN. Sent to NCS Rota, Spain, he served there a year, returning to his Richmond unit and advancing to CTR3.

Beginning graduate school at Johns Hopkins, he drilled with the Fort McHenry unit, 1972-73. Transferring to Harvard the following year, he drilled in Boston. He was commissioned an ENS in February 1975 and promoted to CAPT in 1996. He received the PhD in history from Harvard in 1978 and accepted a position as an assistant professor at the University of Massachusetts-Boston, winning early promotion to associate professor in 1982, and to full professor and department chairman in 1996.

Chesson has served with units at Lawrence (where he was XO when the unit won the Delaney Award), New Haven, Great Lakes, South Weymouth, and Devens. He has held five commands, in Quincy, Worcester, Providence, and Lawrence. He is currently national editor and assistant DCOS for Admin. NR Security Group HQ, Dallas. He has been awarded the Meritorious Unit Commendation, and service medals for Meritorious Service, National Defense, Overseas Service, and Armed Forces Reserve.

He lives with his wife Jane Sherwin and their children, Mark and Virginia, in Belmont, MA.

THOMAS W. CHEWNING, LCDR, born Oct. 4, 1922 in Stevensville, VA, received a BAS in computer systems.

He joined the service May 25, 1944. His military locations and stations: USS *Fargo* (CH-106); NAS Anacostia, JAX, Kingsville, Willow Grove, Dallas, Seafley. Served with VF-916, 83.

Chewning received the rank of LCDR and retired Dec. 31, 1959. His memorable experiences: 1st org. reserve from NAS Anacostia to NAVCAD 1949.

He married Barbara and they have three children: Thomas W. III, Steven E. and John W.; and a grandson Michael. Employed as machinist, programmer/analyst, retiring Dec. 31, 1969.

JOHN NORRIS "NORRIS" CHILDS, III, CAPT, born Sept. 21, 1946 in Abington, PA, received an AB degree from Harvard College and an MD from Johns Hopkins School of Medicine.

He joined the service Sept. 7, 1989. His military locations and stations: Al Jubail S.A. Operation Desert Storm. Served with: Fleet Hospital 20/15, Nav. Mob. Construction Bn. 21. His battles and campaigns: ODS with Fleet Hospital 15.

Childs was awarded the Navy Commendation Medal. He achieved the rank of CAPT (sel).

He is married to Usha Mathew, MD, and they have three children: John Norris IV, Ashok Mathew and Katharine Ann. Employed in private practice, general surgery.

JAMES ALTON CHINN, CDR, born July 29, 1917 at Lemoore, Kings County, CA. He joined the USN in November 1941 and was issued orders to start aviation training in a V-5 program scheduled to begin Feb. 2, 1942 at the Oakland Municipal Airport which was a Navy Reserve Training Base.

Prior to enlisting and while attending Fresno State University, he enrolled in the Civilian Pilot Training Program developed by the Dept. of Commerce. During this course the Japanese bombed Pearl Harbor and the Fresno group of pilots moved to Bishop, CA to complete the course.

Upon completion of the training at Oakland he reported to Corpus Christi as a member of the Class 5B-42-C(C). He graduated as an ENS, A-V (N) USNR on Oct. 8, 1942. Orders were received to report to Norfolk, VA to the air group, VB-17, and to the USS *Bunker Hill* which was being built in Massachusetts. There was further training at Opa Locka and Jacksonville for carrier tactics. On July 11, 1943 the bomber, fighter, and torpedo pilots and planes landed on the carrier for the first time. After the shakedown cruise to Trinidad they sailed to San Diego where they took on additional personnel, supplies, and military equipment and sailed to Hawaii. At NAS Kaneohe, they had training in night tactics.

The USS *Bunker Hill* had her first taste of battle at Rabul, followed by action at Tarawa where the pilots made their first carrier take offs at night. The ship saw action in December, January, and February at Nauru, Kavieng and Eniwetok which often called for 300 mile searches. On February 16 and 17 they bombed Truk and made reconnaissance flights. The squadron returned to Alameda for reorganization and to Fallon, NV for further training and then back to Hilo and into the Pacific.

In February 1945 the orders were to report aboard the USS *Hornet.* They bombed islands on the way to Tokyo which included Iwo Jima. They bombed Tokyo on February 25, Okinawa and Bonin Islands from March to April. May 13-22 they bombed mainland Japan. Following the hurricane the *Hornet* returned to Alameda for repairs and the squadrons were sent to Jacksonville for reassignment. Because of the end of the war, Chinn accepted an assignment into the Reserves. He continued as a weekend warrior flying at Oakland Reserve Base, Fresno, and Los Alamitos. He retired July 29, 1977 as CDR USNR.

For his action, he received ribbons for the Pacific and Asiatic Theater of war aboard the USS *Bunker Hill* and the USS *Hornet.* He received two Flying Crosses and the Air Medal w/5 stars.

One memorable occasion was in 1973. He and his wife and several squadron members met at Coronado for the decommissioning of the USS *Bunker Hill* knowing she was to be towed to Oregon and cut up for scrap.

After the war he married Barbara Deibert and they raised three children. He attended USC graduating as a Doctor of Pharmacy and they eventually owned the La Ondra Drugs in La Mirada. Barbara was a teacher and administrator. They now enjoy retirement at Bishop, CA where they are active in the community, especially Civil Air Patrol, and enjoy equestrian activities. Their travels have taken them to Mexico, Alaska, Italy, Jamaica, Europe, Russia and Africa, and throughout the USA. In 1995 they celebrated their Golden Wedding anniversary.

KATHERINE M. CHINQUE, LCDR, born Feb. 25, 1909, in Elyria, OH, received a BS from Wayne State University, 1947; master's MPH, University of Michigan, 1950.

She joined the service February 1944 and served in the Navy Nurse Corps. Her military locations and stations include Great Lakes, IL; Miami Beach, FL, Portsmouth, VA.

She achieved the rank of LCDR and was discharged in 1969. Chinque received the American Campaign and WWII Victory Medal. Single, she is employed with the University of Washington, Seattle, WA, educator; consultant Michigan Dept. of Public Health, retiring in 1974.

MELVIN H. CHIOGIOJI, RADM, a native of Hawaii, graduated in 1961 from Purdue University with a BS degree in electrical engineering, received a MBA degree from the University of Hawaii in 1968, and a Doctor's of business administration degree, majoring in Operations Research, from George Washington University in 1972.

He was commissioned an ENS in the USN in 1961 following completion of the Naval Reserve Officers Training Corps program at Purdue University. He spent five years on active duty at various assignments. Selected as RADM in 1987. He retired from the USNR in 1993. His military decorations include the Legion of Merit w/Gold Star, Meritorious Service Medal w/2 Gold Stars, and the Navy Commendation Medal w/Gold Star.

In his civilian career, he spent 27 years as a federal civil servant. He served as deputy assistant secretary of the US Dept. of Energy from 1980-85, and completed his federal career, as deputy director and construction manager for the New Production Reactor Program at the US Department of Energy, a senior executive service, level 5 position in January 1993. His professional experience includes 32 years of government management experience (19 years as a GS-16, GS-17, SES-4 and SES-5) in project and program management in the engineering and research fields. He has broad experience in the areas of energy efficiency programs, energy economics and financing, international energy programs, industrial waste utilization and minimization, and construction management of fossil fueled and nuclear power plants. He also has experience in engineering and evaluation of naval weapons and weapons systems.

He has been the owner of an auto repair business for the past 15 years and is presently assisting his wife, who owns MELE Associates, Inc., a engineering and environmental services firm, as director of marketing. He is a registered professional engineer, a senior member of the Institute of Electrical and Electronics Engineers, and a life member of the Naval Reserve Association and the Society of American Military Engineers. He is listed in *Who's Who in America, Who's Who in World, Who's Who in Technology,* and *American Men and Women of Science*. He and his wife, the former Eleanor Oura of Wailuku, Maui, HI, live in the Washington, DC metropolitan area. They have two children, Wendy and Alan; and four grandchildren.

GREY CHISHOLM, LCDR, born in California, raised in Louisiana, New Mexico, Michigan, and Colorado, is a man of many ports. He acquired his love of the sea when he commercial fished in Alaska in 1972. He enlisted in the CG in 1973 then served in Alaska and San Diego. He made chief quartermaster in the Coast Guard Reserve while attending California Maritime Academy. Upon his graduation in 1983 he accepted a commission in the USNR, Merchant Marine Program.

The next four years, he spent as chief mate, captain, and pilot on tug boats. He operated off the US West Coast from San Diego to Seward, AK, the Gulf of Mexico, and then Panama Canal. Chisholm was active in the USNR serving between tug voyages on the USS *Hoel,* USS *William H. Stanley,* various schools. He activated his commission in 1987 and joined the USS *Sacramento* in October as combat systems officer. He became the assistant navigator just before WestPac 1988-89 and highly enjoyed his job until returning to civilian life in 1990.

Eleven days after re-entering the job market he was conning a 200,000 ton super tanker through the Straits of Hormuz. He left that ship as second mate and became actively involved with the USS *Pledge* in Seattle. He earned a chief mate's license after sailing on the product tanker USNS *Sealift China Sea.*

He spliced the line with Michele on Twin Peaks in San Francisco May 1992. To try to stay home with his new bride, he worked as a civilian rigger on the USS *Abraham Lincoln,* then served his Reserve time on the carrier when it came out of availability. His last vessel was the break bulk ship SS *Pembina.*

He has temporarily swallowed the anchor and is working as a computer graphic artist in Albuquerque, NM. He has been enjoying teaching midshipmen at Annapolis during summer AT and in 1996 spent four months at the academy running the MMIRRG augmentation program placing reserve officers on the yard patrol craft during the midshipmen summer cruise.

TIMOTHY KIRKPATRICK "PAT" CHISM JR., born in Tompkinsville, KY, March 24, 1954. He received his BA in mass communications from Western Kentucky University in 1976, and his JD from University of Louisville School of Law in 1980.

He was sworn into the USNR, Dec. 13, 1979, in Louisville, KY, in the JAG Corps Student Program, while in law school. He attended Officer Indoctrination School (OIS), and Naval Justice School at NETC, Newport, RI, in 1980.

Chism served at Naval Legal Service Office Detachment, NAVSTA Mayport, FL, 1981-82; and as deputy SJA for COMNAVBASE Pearl Harbor, HI and COMNAVLOGPAC, 1982-84. He was released from active duty August 1984.

He has served in the Selected Reserve for NR SUBRON SEVEN 620, N&MCRC HONOLULU, 1984-85; NR LSO Pearl 320, 1985-88; NR SIXTHFLT 109, N&MCRC Louisville, 1988-90; NR LSO MEMPHIS 209 (as executive office), NRC Jackson, MS, 1990-93; and NR VOLTRAUNIT 0913, NRC Nashville, TN, 1993-present.

He is entitled to wear the Naval Commendation Medal, and the Navy Achievement Medal, the Armed Forces Service Medal and the National Defense Medal.

Chism has been in private practice since his release from active duty and a prosecutor for the Commonwealth of Kentucky since 1988. He was admitted to the Florida Bar in 1982 and the Hawaii Bar in 1984. He married Dorothea Highland and they have one son, Timothy Major Chism.

STEPHEN R. "STEVE or CHIZ" CHISMARICH, CAPT, born March 16, 1942 in Washington, DC, received Predental, UCSB, no degree (1960-64); DDS, St. Louis University (1964-68).

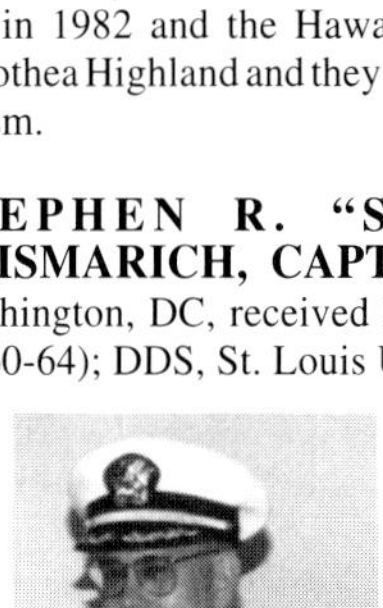

He joined the service Feb. 1, 1964. MCRD, San Diego (1968-69); 9th MAB, Okinawa (1969); 1st MAR WING (REAR) Iwakuni, Japan, 1969-70; US NAV Hosp., San Diego 1970-71 (end active duty). All dental departments Reserve Units: Bakersfield, CA 1971-77; Hannibal, MO 1977-80; NORVA 218, St. Louis (CO 87-89); Ima, Olathe, KS (1990-91); NORVA 119, Omaha (CO 1991-92) Dental Department, Fleet Hospital Olathe, KS (1992-94).

He achieved the rank of CAPT, (DC) USNR and was discharged June 30, 1994. He was awarded the National Defense (2nd) and Naval Reserve (2nd).

His memorable experiences: Hard work and wonderful social experiences of NORVA 218 at gaining command (1984-90), Norfolk-Little Creek, VA; NORVA 119-17 days Hawaii June 1942; Paraguay (SOUTHCOM) Health Mission trip at Fleet Hospital 22, February 1993.

He married Mignon Francis Purcell and their children are Stephen Patrick, b. Nov. 21, 1971, and Michael Donald Sean, b. Feb. 2, 1976. Employed as general dentist at Delano, CA, July 1971-October 1977; Bowling Green, MO, November 1977-present; retired June 30, 1994.

WILLIAM R. "BILL" CHISUM, CAPT, born Feb. 6, 1937, Little Rock, AR, received an AB from San Francisco State University. He also did graduate work at San Diego State University.

He enlisted in the service Feb. 25, 1954. Commissioned via ROC program June 1958. His military locations: USS *Prairie* (AD-15) 1958-60; personnel off, legal off, asst Ops off. Served with Recruit Training Command, San Diego 1960-62; Bn. CDR Command Admin Off.; CO of NR Distribution Team 11-2 and also NR Processing Team 2821.

Chisum achieved the rank of CAPT (06) and was awarded National Defense Service Medal, Recruiting Gold Wreath Award, Armed Forces Res. Medal w/2 Hourglass Devices.

His memorable experiences: being sea detail JOOD entering Yokosuka harbor and having a collision with a Japanese merchant ship. Also, Christmas 1959 when *Prairie* adopted an orphanage in Yokosuka and the many hours they spent there repairing it and assisting the children to have a great Christmas.

He married Isobel Mary and they have two children, William Thomas and Steven Robert; and granddaughter, Elizabeth Marie. Employed as educator with San Diego Unified School District for 35 years, retiring June 1996. Chisum is a life member of the NRA.

DAVID S. CHRISTIAN, born April 28, 1934, Barnes (Warren County) PA, attending Clarion State Teachers College, Clarion, PA and Washington University, St. Louis, MO.

He joined the service June 1958. His military locations and stations: OCS, Newport, RI, 1958; USS *Observation Island* (EAG-154), Norfolk, VA and Port Canaveral, FL, 1958-60; USS *Whippoorwill* (MSC-207), Sasebo, Japan, 1961-62; Naval Reserve Centers, St. Louis, MO and Clifton, NJ, 1963-80.

Christian retired as LCDR, April 1994. His memorable experiences: participated in the initial launches of the A-1 Polaris from USS *Observation Island* (EAG-154), USS *George Washington* (SSBN-598), and USS *Patrick Henry* (SSBN-599), 1959-1960.

He married Bonnie Jill Matthews Politte and they have four children: Paula, Amy, Connie and James. Employed as teacher at Wampum, PA and Smethport, PA; data processing at Monsanto Co., St. Louis, MO, Monsanto Flavor Essence Company, Montvale, NJ; Adjunct Faculty St. Louis Community College, St. Louis, MO; and retail sales at Radio Shack in St. Louis, MO.

ALLISON LEWIS "CHRIS" CHRISTOPHER, CAPT, born Sept. 21, 1932, Gadsden, AL, received a bachelor of mechanical engineering, Alabama Polytechnic Institute (Auburn University).

He joined the service June 5, 1954. His military locations and stations: USS *Calvert*, NAVRESCEN's Long Beach and San Diego. Served with USS *Calvert Samar* 11-2, SAMAR 11-3, SMR 2819, NAVSEA DETS 519 and 419 and NAVAIRES San Diego. Battles and campaigns: Passage to Freedom (1954).

He was awarded the Navy Commendation, National Defense, Armed Forces Reserve, Marksmanship, RVN PUC. Christopher achieved the rank of CAPT and retired July 1, 1987.

He married Marlene Young and they have three children: Susan, Kelly Ann and Lynn Marie; their grandchildren are Nicole, Brian and Alicia Velo; Lemuel and Madelyn Zeigler; and Patrick Rowland. He was employed as Naval Ordnance Center Pacific Div., executive director, retiring July 1, 1987.

PHILIP D. "PHIL or "CHRIS" CHRISTOPHER, LTJG, born Aug. 18, 1926 in Los Angeles, CA, graduated San Francisco State University, June 11, 1955 and AB Business Administration with major in accounting, enlisted Oct. 26, 1944.

He received direct commission as LTJG March 25, 1957. He has served 52 years active and inactive. His military locations and stations: various active duty and USNR (Ready Reserve and Standby).

He achieved the rank of LT (03); enlisted CPO (YNC). Still a member NAVRES (Ret). His memorable experiences: was aboard USS *General Sheridan* (PA-51), en route to Leyte/Samar (Philippines) when the cruiser *Indianapolis* (CA-35) was sunk by a Japanese submarine approximately 100 miles off their course (July 1945).

Christopher was awarded the Good Conduct Medal, National Defense Medal, American Area Campaign, Asiatic-Pacific Area Campaign, Philippine Liberation Medal, Naval Reserve Medal, WWII Victory Medal, Armed Forces Reserve Medal (Navy).

He is married to Marianne L. Casperson. Self-employed, enrolled agent, enrolled to practice before the Internal Revenue Service. Transferred to the Retired List with pay Aug. 18, 1986.

DAVE "DAVE" CHUBB, LCCR, born Sept. 23, 1933 in Philadelphia, PA, received a MBA from the University of Nevada, Reno.

He joined the service March 9, 1956. His military locations and stations: Karamursfel Turkey, NCS Guam MI; USEUCOM Paris, France; NMCRC Chattanooga, TN. Served with NAV SECGRU, NAV & MARCORP Reserve Cent.

Chubb received a Joint Service Commendation and the Navy Unit Citation. He achieved the rank of LCDR and was discharged June 30, 1976. His memorable experiences: primary, intelligence officer for EUCOM during six-day Arab Israeli War.

He married Shirley and they have four children: Kara, Lori, Mitchell and Jason; and a grandchild, Katie. Employed as financial manager and retired Sept. 2, 1995.

PHILIP R. "PHIL" CIAFFA, LT, born Aug. 22, 1953, in New York, NY, received a BA in Communications Arts, Fordham University, Bronx, NY (1975); Certificate of Naval Studies, Naval Cert. Basic Qualification Course Supply Officer afloat Navy Supply Corps School, Athens, GA (1981), Officer Candidate School, NETC, Newport, RI (1981);

He joined the service Sept. 24, 1980. His military locations and stations: Naval OCS, Naval Education and Training Center, Naval Station Newport, RI (November 1980-March 1981); Navy Supply Corps School, Athens, GA (April-October 1981); USS *England* (CG-22) Homeport: Naval Station San Diego, CA (October 1981-September 1982); Staff: CDR, Naval Surface Force, US Pacific Fleet (Force Supply Office) Naval Amphibious Base Coronado, San Diego, CA (September 1982-April 1983); Naval Reserve Force Inactive Reserve Status (April-September 1983).

Drilling Reserve Unit CAT A: NR Defense Contract Administrative Services Region (DCASR) New York 202, Naval & Marine Corps Reserve Center, Fort Schuyler, Bronx, NY (September 1983-November 1983); Drilling Reserve Unit CAT A: NR Naval Supply Center (NSC) San Diego HQ 119, NAS North Island Annex, San Diego, CA (December 1983-May 1984); Drilling Reserve Unit CAT A; NR Cargo Handling Bn. (CHB) SIX Detachment C 602, Naval and Marine Corps Reserve Center, New Rochelle, NY (June 1984-September 1985); Drilling Reserve Unit CAT A: NR Military Traffic Management Command (MTMC) 202, MTMC Eastern Area HQ, Military Ocean Terminal, Bayonne, NJ (October 1985 to August 1987).

Drilling Reserve Unit CAT A: USS *Boulder* (LST-1190) Naval Reserve Force Primary Crew, NR LST-1190 *Boulder* Alternate Crew 9002, Naval and Marine Corps Reserve Readiness Center, Floyd Bennett Field, Brooklyn, NY (August 1987-October 1988); Drilling Reserve Unit CAT A: NR LHA-2 Saipan USS *Saipan* (LHA-2), Detachment 202, Naval and Marine Corps Readiness Center, Floyd Bennett Field, Brooklyn, NY (October 1988 to October 1989); Drilling Reserve Unit CAT A: NR Ships' Part Control Center (SPCC) Mechanicsburg, PA 202, Ships' Parts Control Center, Mechanicsburg, PA (October 1989-April 1992); Drilling Reserve Unit CAT A: NR Defense Personnel Support Center (DPSC) Philadelphia, PA 202, Naval and Marine Corps Reserve Center, Fort Schuyler, Bronx, NY (April 1992-March 1993); Retired Reserve Status (April 1993-present).

Ciaffa was awarded the Armed Forces Reserve Medal and National Defense Service Medal. He achieved the rank of LT (0-3), Supply Corps, USNR, retired March 31, 1993.

His memorable experiences: WESTPAC deployment onboard USS *England* (CG-22), USS *Ranger* (CV-61) Battle Gp. April to September 1982.

Divorced, he remarried March 8, 1997 to Priscilla Hart. Employed as business unit manager, Life Cycle Operations Division, General Electric Information Technology Solutions, Inc. (formerly Ameridata, Inc.).

DONNA M. CIGGIA, born Sept. 18, 1956 in Buffalo, NY, received a BSN D'Youville College, Buffalo, NY 1980 and MS from San Diego State University 1993.

Commissioned May 1980 and was stationed at Camp Lejeune Naval Hospital, 1980-83; Charleston Naval Hospital, 1983-86, currently Med Unit P1942A out of NTMCRRC San Diego.

Ciggia achieved the rank of CDR and is presently an active Reservist. Memorable experience: Navy camaraderie in general.

JOHN J. "JACK" CISTRIANO, CAPT, is a native of Philadelphia, PA. He attended the University of Southern Mississippi, receiving both a BS and a MS degree. Additionally, he received a PhD from the University of Maryland. He was commissioned as an ENS after graduating from OCS in February of 1962.

He served on active duty after graduating from CIC and Air Control School in Glynco, GA, and became a plankowner of the USS *Semmes* (DDG-18). While aboard *Semmes*, he served as first lieutenant, CIC officer, and operations officer. Following *Semmes* he was assigned for a four-year tour at the US Naval Academy as an instructor and assistant baseball and basketball coach.

In August of 1969, he transferred to the Naval Reserve Program and has served in a variety of Surface Warfare Units. He has been CO of NRSD 5-1(M), AB 506, Amphibious Bn. 3806, Assault Craft Unit 306, ACU 2 LCM 8, Naval Beach Group Two - DET 206, GVTU 0609, ACNO Surface Warfare DET 106, OP-07 DET 106 and Naval Historical Center DET 206. He served as amphibious program officer in OPNAV 03 and as director of Readiness and Training for Navy Command Center 106. He is one of the select group of Naval Reserve officers qualified to stand the Navy Department duty captain watch in the Navy Command Center.

He was awarded the Navy Commendation Medal for his performance as CO of Naval Beach Group Two, the Meritorious Service Medal and Meritorious Service Medal Gold Star for his performance as CO of ACNO Surface Warfare DET 206 and OP-07 Det 106. He has performed a vareity of ACDUTRA's on board USS *Albany* (CG-10), USS *Austin* (LPD-4), USS *Sumter* (LST-1181), USS *Borie* (DD-707), JCS Exercises, Solid Shield 81, CACEX 82, and has attended the Senior Officer's Surface Warfare School.

In his civilian profession, he is the special assistant to the president of the Chesapeake College in Wye Mills, MD. He is actively involved in activities of the US Olympic Committee, regional director for the National Junior College Athletic Association, a USNA information officer, vice-president of the Academy Chapter of the Naval Reserve Association, Naval Order, Naval Reserve Association and the Surface Warfare Association.

He is married to the former Judy Marie DiGuglielmo of Conshohocken, PA. They have one daughter, Gina, and reside in Annapolis Landing, Riva, MD.

ETHEL BARBARA CLARK, CAPT, born Sept. 26, 1924 in Philadelphia, PA, received a BA from the University of Pennsylvania June 1951.

Joined the service May 22, 1945 as seaman recruit and was stationed at NAS Willow Grove ONOP Philadelphia and Pittsburgh Hunter College.

Clark achieved the rank of CAPT, USNR, retired September 1984. Civilian employment: vice-president of Methodist Hospital, Philadelphia, PA. Clark is single.

JAMES DONALD "DON" CLARK, LCDR, born Jan. 1, 1918 in Plainville, IN, received a BS (1947) NMIMT; MS (1949) Oklahoma University naval aviator.

S2c (March 28, 1941); AvCad V-5 (1941). Commissioned ENS April 10, 1942, Jax, FL. His military locations and stations: NAS Pearl Harbor, 1942; South Pacific 1943-44, NAS San Diego, Western Pac. 45-46. He served with Utility Sqdn. 189 COM U and RONS, Fund Area, Guam.

Clark participated in battles and campaigns: Solomon Islands, New Britian, Manus and the Philippines. He was awarded the Air Medal, American Area, American Defense, Asiatic-Pacific w/2 stars, Philippine Liberation and WWII Victory.

He achieved the rank of LCDR, sep. inactive Reserves, Sept. 11, 1946. His memorable experiences: Flew land and amphibious planes, single and multi-engines, became patrol plane commander in January 1943. Numerous air-sea rescue missions 1943-44. Operated from many of the Western Pacific Islands.

He married Lura Alice Bennington, May 28, 1944. They have four children: Richard Douglas, Ronald Dean, Donna Kay and James Jeffrey; their grandchildren are Kimberly, David, Michael, Christopher, Jamie and Tiffany. Employed as petroleum engineer, chief reservior engr. UNOCAL, petroleum consulting engineer. He retired from the Retired Reserves March 1, 1954.

PHILIP C. "PHIL" CLARK, LT, born June 4, 1926 in Hutchinson, KS, received a BS from Northwestern University in 1946; BS, Lawrence College, 1948, and attended two years University of Colorado, 1954-55, civil engineering.

He enlisted in the service Dec. 6, 1943. Released June 24, 1946 and recalled Dec. 3, 1948. His military locations and stations: Little Creek, VA and Washington, DC, serving with V-12; NROTC; Nav. Beach GR. 2; UDT-4; CIA.

Clark was awarded the Good Conduct, American Campaign, WWII Victory, National Defense Service and Naval Reserve Medals and Parachutist Wings.

He achieved the rank of LT and was released Jan. 5, 1954. He retired from the USN Sept. 1, 1966. His memorable experiences: 1949, resupply of arctic weather stations-Beachmaster; 1950-51, O-in-C, Submersible Operations Unit, Underwater Demolition Unit-2; 1952-53, UDT-Type training of foreign nationals - CIA; 1952, qualified parachutist, Ft. Benning, GA.

He married Meredith K. Plier, June 24, 1950, and they have two children, James W. and Meriweather K.; their grandchildren are Desireé, Joseph, Jonah, James, Danae, Nathanael, Susanna, Noah and Isaiah. Employed as land surveyor, insurance agent, Title and Trust Co., real property appraiser; senator and vice-president of Ninth Legislature of the US Virgin Islands, retiring June 1988.

SCOTT M. CLARK, LT, born Nov. 4, 1968, Atlanta, GA, received a BJ from the University of Missouri, post-grad. business study Virginia Commonwealth University.

He joined the service Jan. 17, 1992, AOCS and VT-10 at Pensacola NAS; NR-OI DET 206 at Washington, DC. Served with VT-10, NR-OI DET 206, FICPAC NR 0176.

Clark was awarded the National Defense Medal. He achieved the rank of LT and was discharged June 18, 1993. His memorable experiences: AOCS, flew F/A-18D *Hornet* at McDonnel-Douglas. Flew across Antarctica to South Pole with Naval Support Force Antarctica and VXE-6 "ice pirates" in support of the Navy's 40th anniversary.

He is single and employed with an advertising agency as account executive.

JAMES C. CLAUGHTON, CAPT, born in Denver, CO, received primary and secondary education in Amarillo, TX. Enlisting in Class V-1 Naval Reserve in July 1942, he completed the V-12 program, receiving a BS in Electrical Engineering from the University of Oklahoma in 1944. After completing Reserve Midshipman training at the US Naval Academy, he was commissioned ENS, April 1945.

After a brief tour aboard the USS *Santa Fe* (CL-60), he was released from active duty. Recalled to active duty in November 1950, and served aboard the tanker USS *Noxubee* (AOG-56) until November 1952 while deployed in the Mediterranean.

Claughton served in USNR units in Amarillo: 1956-60, CO Nav Sec Gru 8-20; 1960-64, CO NAV RES CB Div. 8-15 (received SEC NAV Commendation, winning J.J. Manning Trophy 1962); 1965-66, CO NAV RES Surf Div. 8-75; 1969-72, CDR NAV RES GRU 8-2; promoted to CAPT 1966 and retired June 30, 1975. Retired from SPS, regional electric utility as manager, Planning and Engineering Services, 1986. He is married, has two children and is a life member of the NRA.

CHARLES F. CLIFFORD, CAPT, born in McDonald, OH, Oct. 7, 1925. He is a graduate of Bucknell University and the LIMI at Purdue University. He began his naval career in June 1943 as an aviation cadet and was commissioned an ENS in 1945.

He served in the USS *Lassen* (AE-3) as DO in the Deck Department. In 1946 he served in the USS *Cumberland Sound* (AV-17) as fire control officer during Operations Crossroads at Bikini Atoll. During 1950-52, the Korean War, he served in the USS *Salisbury Sound* AV-13 as gunnery officer. He is a qualified surface warfare officer.

From 1953-78 he served in various billets in the surface forces and held commands in Surface Div. 4-115 Youngstown, OH; Naval Reserve Gp. CDR 4-32, Columbus, OH; Naval Reserve Gp. CDR, 9-66, Lincoln, NE; CO, NRVTU 517; and CO NR Com Cru Des Gro5. His campaign and service medals are: American Campaign, Asiatic-Pacific Theatres; WWII Victory, Occupation, National Defense, China Service, Korean Service, United Nations, Korean Presidential, Naval Reserve Medal, Armed Forces Reserve Medal. He was selected for CAPT in 1968 and retired June 30, 1978.

In his civilian occupation CAPT Clifford was a district agent, agency manager, Director of Education and Training and agency director for the State Farm Insurance Companies. He is a Charter Life Underwriter (CLU) and a Chartered Financial Consultant (ChFC). He is a member and past president of the Lincoln Chapter American Society of CLU and ChFC.

Clifford is a life member of the NRA, US Naval Institute, Reserve Officers Association of the US, VFW, and American Legion.

His family consists of his wife, Ann; son, Charles F. Clifford III; daughters, Marianne Clifford Upton, Julianne C. Taylor and Carol Anne Jenkins. Their grandchildren are Chris, Corey, Carissa and Connor Clifford; Emily and Carolyn Taylor; Dan, Nick and Scott Jenkins.

GORDON H. CLOW, CAPT, born Jan. 5, 1942, Washington, DC, received a BS from the USNA 1964 and MBA from the American University 1975.

He joined the service June 3, 1964. His military locations and stations: NAS Patuxent River, MD; NAVAIRSYSCOM Washington, DC. He served with: VP-8 (1965-68) NAVAIRSYSCOM 1968-69; VP-68 (Reserve) NCC-106 (Reserve). His battles and campaigns: Vietnam missions June-December 1966.

Clow was awarded the Air Medal (2) and Navy Unit Commendation. He achieved the rank of CAPT and was discharged ACTUDRA Oct. 31, 1969. Retired from Naval Reserve Nov. 1, 1985.

He is married to Holly M. and they have two children, Lori Ann and Michael G. He is co-founder and executive vice-president, Vanguard Technologies. "Retired" 1988.

DAVID M. CODERRE, born and grew up in Manchester, NH. After graduating from the University of New Hampshire and OCS, he completed Navy Supply Corps School in Athens, GA.

On active duty, he served aboard the USS *Emory S. Land* (AS-39), Defense Contract Management Area, Bridgeport and the USS *Stephen W. Groves* (FFG-29). Highlights included an Indian Ocean deployment on *Emory S. Land* and a 12 port NATO cruise on *Stephen W. Groves.*

On reserve duty, he has served with two SIMA units, an NSC, FISC, DDRE and NAVMTO unit. Highlights included, selection under his command (NAVMTO) as the Region Readiness Command (REDCOM 8) unit of the year and recruitment of five new NRA members.

Coderre currently resides in Jacksonville, FL with his wife, Kiki. He is a lifetime member of the NRA.

JOSEPH P. "JOE" COFER JR., CAPT, born Sept. 11, 1923, Bluefield, WV, received a BS, BSME and MSMETE degree.

He joined the service June 23, 1943. Served six years in DDs, two years in minesweeps, two years in cruisers. Served four tours with 6th Fleet 1951-58.

Cofer received nine awards and medals, including a Navy Commendation. He achieved the rank of CAPT. He retired Jan. 1, 1980. CLU, CHFC, NSNA Annapolis Class 1947.

He is married to Randy and employed with General Agent Life.

RICHARD "RICK" W. COFFEY, LCDR, was born into a military family on June 4, 1947. He is at least a sixth generation career military officer. He enlisted in the USN in July 1967 and received his commission in February 1974. Both his undergraduate and graduate work were in business administration. The University of Washington and Seattle University, respectively.

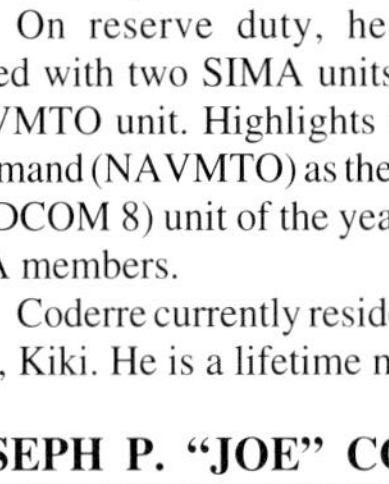

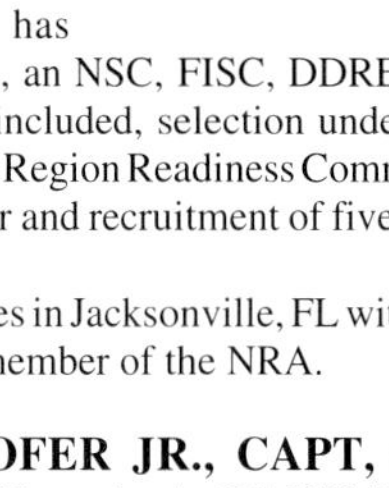

In 1968 (then) RMSN Coffey graduated from Radioman Class A School in San Diego, DA. From 1968-70 he worked in satellite communications at NAVCOMSTA Honolulu. After completing his undergraduate degree and while serving with the Naval Security Group (Unit 13-2, Seattle), he was promoted from RM2/c to ENS (1615 cryptology).

LCDR Coffee had many exciting tours of duty, both in the US and Asia, in his 27th year naval career. He was a graduate of the senior officer cryptologic course and received many unit and personal awards, including the Navy Achievement Medal. He served as executive officer of Naval Reserve Security Group Units in Honolulu, Portland and Seattle. His last duty assignment was as commanding officer of NR Security Group, Seattle. LCDR Coffey retired in July of 1994.

LCDR Coffey is married and has three grown daughters, Roxanne, Kimberlee and Erica; and a son Kyle. The Coffey family lives in Woodinville, WA. Currently, LCDR Coffey is vice-president of PC Northwest, Inc., the Pacific Northwest's largest independent human resource management consulting firm.

HOWARD CHARLES COHEN, CAPT, born April 28, 1943 in St. Louis, MO, received a BA in Princeton University 1965; JD, UCLA School of Law 1972; CCE, Naval War College 1989.

He originally joined the service Sept. 3, 1965. He was honorably discharged in 1976. Recommissioned Dec. 17, 1980. Served with USS *Robert A. Owens* (DD-827), 1965-67; USS *John Paul Jones* (DDG-32), 1967-68; USS *Constellation* (CV-64), 1968-69; VTU (LAW) 1911, 1980-1984, 1988, 1990-1993, 1996; NR LSO 119, 1984-87, 1988-90; NR NAMARA 110, 1993-95; NR CINCPAC 120, 1996-present.

He was awarded the National Defense Medal w/ Bronze Star, Vietnam Service Medal w/Bronze Star, Outstanding Military Volunteer Service Medal, Armed Forces Reserve Medal, RVN Gallantry Cross and Rifle and Pistol Sharpshooter Ribbons.

He achieved the rank of CAPT, JAGC, USNR and is an active reservist. Married Avalee Nicholls and their children are Alicia Avalee and Jessica Louise. Employed as a research attorney, Hon. William Enright (S.D. Cal.) 1972-74; Defenders Inc, 1974-79; Law offices of Ben A. Borevitz, 1979-84; Appellate Defenders, Inc, 1984-present.

MARTIN F. "MARTY" COHEN, born Dec. 24, 1934 in Easton, PA, received a BSBA degree, Lafayette College, 1956; accounting, Florida International University, received a CPA Certificate 1991.

He joined the service Sept. 19, 1956; active duty (OCS) Oct. 15, 1956. His military locations and stations: OCS Newport, NSCS Athens; NAF OPPAMA, Japan. Served with units: OCS-Class 30.

Cohen was awarded the Armed Forces Reserve Medal. He achieved the rank of LTJG on active duty; CDR, SC in Reserves. Transferred to standby Reserve, Aug. 1, 1977.

His memorable experiences: spending 28 months (October 1957-February 1960) in Japan; serving as disbursing and fiscal officer with military and Japanese civilian personnel; enjoying group friendships at OCS, NSCS and a small naval air facility.

He is married to Barbara and they have two children, Lisa Newman and Barry Cohen. Their grandchildren include a granddaughter Taylor and a grandson is expected July 4, 1997. He has been employed as a savings and loan bank accounting officer from 1960 until present.

VINCENT J. "VINCE" COLAN, CAPT, born June 13, 1914, Astoria, Long Island, NY. He enlisted in the USNR and NYNM at New York City on his 18th birthday, as an apprentice seaman. Annual active duty for training was performed aboard the USS *Ellis* (D-154), USS *Overton* (D-239) and USS *Arkansas.*

Holding the rate of signalman 3/c at the time, he enlisted in the USN on Oct. 14, 1935. It being peacetime, he had to revert to apprentice seaman again. After boot camp at Newport, RI he attended communication school at Norfolk Naval Base, and upon completion, sent to the USS *San Francisco* (CA-38) on the West Coast in September 1936 as a signal striker. From 1938-39, RADM Husband Kimmel, as ComCruDiv 7, used the signal bridge as his command post, where he practically became a member of the signal gang.

In 1939, when he again achieved the rating of signalman third class, he was transferred to the USS *Fairfax* (D-93) and then to the USS *Manley* (D-74). When his enlistment expired on Oct. 13, 1939, he left the Navy and went to work for the US Government in Washington, DC, where he again reenlisted in the Naval Reserve. In the summer of 1940, his reserve division spent its two-week annual training period getting the USS *Satterlee* (D-190) ready for transfer to the British as part of the 50 destroyer lend-lease deal.

On Jan. 1, 1941, his Reserve Div. was called to active duty and set to the Panama Canal to take over the USS *J. Fred Talbott* (D-156). In October 1942, as a signalman first class, he was recommended for commission as ensign and became an ensign in February 1943. Several months later, he was ordered to the USS *David W. Taylor* (DD-551). He was designated as first lieutenant and damage control officer. Upon commissioning, the ship was ordered to the South Pacific.

When the war ended in August 1945, he left the Navy, and returned to his US Government employment in Washington, DC and again associated with the USNR. In 1951, he was called to active duty for the Korean War. After attending CIC School at Glenview, IL, he was ordered to the USS *Johnston* (DD-821). He became CAPT, USNR in 1963. Colan retired from US Government employment, Oct. 31, 1969. After 41 years of Navy and USNR service, he was transferred to the Retired Reserve on July 1, 1973. On his 60th birthday, June 13, 1974, he was retired from the USNR with pay.

Awarded the Silver Star Medal, Good Conduct Medal w/2 Stars, American Defense Medal w/Fleet Clasp, American Campaign Medal, Asiatic-Pacific Medal w/8 Battle Stars, Navy Occupation Medal w/A&E Clasps; Philippine Liberation Medal and Presidential Unit Citation.

His memorable experiences: 1938-39 served with RADM Husband Kimmel aboard *San Francisco.* February 1943, as a signalman first class, received commission as ENS. January 5, 1945, USS *D.W. Taylor* struck a mine while bombarding ChiChi Jima. As DCO, directed damage control parties in keeping ship afloat for safe return to Saipan.

He married Jane and they have two children, Susan and Peter; their grandchildren are Heather and Sarah Colan. He is assistant director for Technical Services, US Department of Labor.

GAGE N. "JAKE" COLBY, born June 28, 1923 in St. Paul, MN, graduated from the University of Minnesota with a BA degree in 1944 and a DDS in 1946.

He joined the service Feb. 18, 1942. His military locations and stations: NAS Quonset Point, RI. He served with FAS RON 2, USS *Oriskany* (CV-34), USS *Providence* (CL-82).

Colby achieved the rank of CAPT DC USN and was discharged July 1, 1970. He was awarded the WWII Victory Medal, Naval Occupation (Europe), American Theatre and National Service Award.

His memorable experiences: D Div. officer USS *Oriskany* (CV-34) on Sept. 25, 1950; Commissioning Day. Brooklyn, Naval Yard, NY, NY. Stationed at NAS Port Lyautey 1956, Ft. Morocca at their independence.

He married Virginia and they have four children: Christopher, Gregory, Katharine and Rebecca; grandchildren are Jessica Lorraine, Jessica Nocale, Gretchen, Colen, Nicolas, Madeline, Ryan and Drew. Employed in his private practice of dentistry, 1970-95. He retired July 1, 1995.

JOHN G. "JACK" COLGAN, CAPT, native of Philadelphia, PA, St. Josephs University, BS 1952.

Assignments: 1953-53, flight training. 1954-56, VS-23 flew TBM and S2F aircraft; two WESTPAC deployments aboard USS *Princeton* and USS *Boxer.* 1957-58, FAWTUPAC instructor in SNB, T-28B, TV-2 and F9F-8; 1958-60 GA Tech NROTC instructor; 1960-62 VT-21 Kingsville, TX, instructor in F9F-8; 1962-64, USS *Ranger* (CVA-61), flight deck officer. 1964-65, US Naval War College. 1965-68, VF-101, F-4 training; VF-103, operations officer; two Mediterranean deployments aboard USS *Saratoga* (CVA-60). 1968, VF-121 XO. 1968-69, CVW-11, Air Wing operations officer.

Combat cruise to SE Asia aboard USS *Kitty Hawk* (CVA-63); flew F-4s with VF-114. 1970-71, COMNAVAIRLANT staff, NATOPS coordinator. 1971-72, CO of VA-43 at NAS Oceana. 1972-74, navigator of USS *Forrestal* (CVA-59), two Mediterranean deployments. 1975-79, BUPERS, special assistant to CNP for POW/MIA Affairs. 1979-81, CO of Naval Air Reserve Units, Norfolk, VA. 1981-85, US Embassy, Ottawa, Canada, Naval Attache. August 1985, retired from USN after 33 years. Colgan accumulated 5,000 flight hours after 750 carrier landings.

He was awarded the Defense Superior Service Medal, Meritorious Service Medal, Air Medal (3), DOD Joint Commendation Medal, Navy Commendation Medal (2) (1 with Combat V), Navy Unit Commendation, Meritorious Unit Commendation, China Service Medal, National Defense Service Medal w/Bronze Star, Expeditionary Medal, Vietnam Service Medal w/3 Bronze Stars, RVN Unit Commendation and Civil Action Medals and the Vietnam Campaign Medal.

The Colgan family includes wife Catherine, son James and stepdaughters Sara, Martha and Rebecca.

Civilian employment: 1986-88, Center for Space and Advanced Technology, director, International Marketing. 1989-92, "Colgan Communications" consulting firm - partner.

Associations: Association of Naval Aviation (ANA) - National Trustee, Chairman of 1992 National Convention, former CO of Hampton Roads Sqdn., US Naval Institute; US Naval Reserve Association; Naval Officers Association of Canada, Honorary Life Member; Enlisted Naval Reserve Association; The Retired Officers Association.

Political Experience: Bush/Quayle 1988 - Deputy National Chairman of Veterans Coalition; 1993-97 - Legislative Assistant to State Senator Ken Stolle; 1994-95 - member, Governor's Commission on Base Retention and Defense Adjustment; Republican Party of Virginia Beach, VA, Member of City Committee.

THOMAS E. COLLINS, CAPT, born in Superior, WI on Jan. 9, 1938. He graduated from Marquette University in Milwaukee, WI in June 1960 with a BS degree in business administration.

Collins attended Naval OCS, Newport, RI and was commissioned ENS in May 1961. He served three years aboard USS *Eldorado* (AGC-11) homeported in San Diego, CA. During this period he made two WESTPAC deployments and deployed to the Caribbean during the Cuban Crisis on October 1962.

From 1967-84 served as a drilling reservist at Naval and Marine Corps Reserve Center, Twin Cities in various billets. Among them was CO of SERVRON 516 and CO of PERSMOBTM 116. He was selected for CAPT in January 1983.

His service medals are as follows: Armed Forces Expeditionary Medal, National Defense, and Naval Reserve Medals.

Collins family consists of his wife Winnie and three daughters: Katie and husband Dan Walsh, Maureen and husband Jim Englund, and Molly. They all reside in Minneapolis, MN along with four grandchildren: Brian, Anne and Laura Walsh and Cullen Englund.

He is employed in sales with Van Waters and Rogers, Inc. in St. Paul, MN. Collins retired from the USNR in October 1989.

ROBERT NEIL "BOB" COLWELL, RADM, born in Star, ID, Feb. 4, 1918. He attended the University of California, Berkeley, where he received a bachelor of science degree in forestry in 1938 and a PhD in Plant Physiology in 1942.

Early in WWII, he was commissioned as an ENS in the USNR. He was an air combat intelligence officer for the invasion and capture of Guadalcanal; then chief instructor of the Navy's Photo Intelligence School in Washington, DC; and finally, Chief of Photo Intelligence for the planning and execution of the Okinawa Campaign. At the end of the war he was placed in charge of the Navy's training programs in photo interpretation and photogrammetry.

During the Korean War he was recalled to active duty for two years and placed in charge of the Navy's research programs in photo interpretation and photogrammetry and also headed up the Navy's Air Objective Folder Program in support Allied Forces in Korea. In the Vietnam War he was the CO of a Special Studies Unit comprised of 25 Navy captains and directed activities of that unit as requested by CNO, CINCPAC, CNAVRES, and others.

In 1974 Colwell was selected for Flag rank and concurrently became the first director, Naval Reserve Intelligence Program, a position he held until October of 1977. Among his military awards are the Legion of Merit, the Bronze Star, the Navy Commendation Medal, two Presidential Unit Citations, WWII Asiatic-Pacific ribbon with numerous Battle Stars, and Korean Service.

In 1947 he became an assistant professor of Silviculture/Remote Sensing on the Berkeley campus of the University of California, where he remained for nearly four decades.

He is the author of more than 400 scientific and professional articles, almost all of which relate either to remote sensing or the use of radioactive isotopes in plant physiological research.

Colwell is married to the former Betty Louise Larson of Oakland, CA. They have four children: Arthur, John, Nancy and Robert.

JOHN R. COMINS, CAPT, born in Clayton, NY on June 18, 1946. Comins is a 1969 graduate (ME) of Rochester Institute of Technology, Rochester, NY. He entered Naval Officers Candidate School, Newport, RI in April 1970.

After commissioning in August 1970, ENS Comins reported for training at Damage Control Officers School, Treasure Island and Combat Information Center Watch Officers School, San Diego. Completing training in December 1970, he reported to the USS *Eversole* (DD-789), stationed in Long Beach, for duty as the Damage Control Assistant and "R" Div. Off. He served on the *Eversole,* until September 1972, making two WESTPAC cruises during the Vietnam War. Since 1977, he held command of and XO billets in NR LST-1194 Lamoure 9405, NR NCSO Erie 305 and NR NRL Science of Technology 105. He also served as XO of Mobile Inshore Undersea Warfare (MIUW) 214, Buffalo, NY and as an assistant training officer in NR SACLANT DET 205, in Buffalo, NY. He was promoted to CAPT in May 1992. In addition to his Navy Reserve participation, he also served the State of New York as a member of the federally recognized New York Naval Militia (NYNM). Appointed in October 1982 as CO, Bn. 5-3 for the Rochester, NY district, he served in that position until retirement in May 1996. After retirement, he continued voluntarily service to the Naval Militia and was promoted to the rank of RADM (L/H) NYNM on Sept. 27, 1996.

Comins' medals and awards include the Navy Commendation Medal, Combat Action Ribbon, Navy Unit Commendation Ribbon (2nd award), National Defense Service Medal (2nd award), Armed Forces Reserve Medal (10 yr.), Vietnam Service and Campaign Medals, Coast Guard Special Operations Ribbon, Naval Reserve Overseas Ribbon and the Navy Expert Pistol Medal. In addition he holds the following New York State awards: Military Commendation Medal (2nd award), Conspicuous Service Medal and the New York Naval Militia Long and Faithful Service Medal (15 years) and the Militia Recruiting Medal.

He is married to Jane E. Harrison of Lima, NY. They have a daughter, Jamie Erica. Since 1969, he has been employed in various engineering and managerial positions by the Eastman Kodak Co., Rochester, NY and is currently a project manager in the Image Acquisition Systems Div. He and his family reside on Orchard Creek Lane in Rochester.

EDWARD J. "JOE" COMPTON, CDR, a Michigan native, graduated as a naval aviator at NAS Pensacola in 1945. He attended the University of Michigan, obtaining AB and MA degrees. He drilled with Reserve Sqdn. (AVUA #4), NAS Grosse Ile, 1950-51.

Recalled during Korean War, he served as CIC and Air Intelligence officer with Composite Sqdn. Three (VC-3), NAS Moffett Field. He deployed with VC-3 detachment "Charlie" (VF-112, aboard the USS *Kearsarge* (CVA-33) as LOW-Visibility Intercept (LVI) controller and AIO in 1953.

He remained in the Reserve Air Intelligence Program at NAS Los Alamitos and NAS Seattle 1954-57. He was CO, NAIRU 891, NAS Seattle 1963-64. He retired June 1, 1979.

In civilian life, Compton spent more than 30 years in various marketing and management positions with IBM. The Compton family consists of wife Rita, daughter Cathleen and granddaughters, Dana and Robin. They reside in Boca Raton, FL.

DON WRIGHT CONLEY, CAPT, born in Chandler, OK, May 9, 1933. He graduated from Oklahoma A&M in 1956 with a BS in mechanical engineering. After graduating from OCS at Newport, he attended AGO school at NAS Jacksonville, FL. He was then assigned to the maintenance department of VF(AW)4 at NAS Atlantic City from January 1957 to September 1959. He went to sea with squadron detachments on the USS *Valley Forge* and USS *Tarawa*. Upon closing of Atlantic City NAS, he moved with his squadron to NAS Quonset Point, RI.

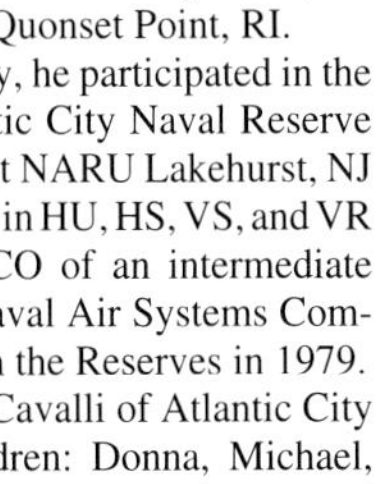

Upon release from active duty, he participated in the reserves for one year at the Atlantic City Naval Reserve Center and an additional 20 years at NARU Lakehurst, NJ where he was a maintenance officer in HU, HS, VS, and VR squadrons. He also served as a CO of an intermediate maintenance support unit and a Naval Air Systems Command support unit. He retired from the Reserves in 1979.

Married the former Barbara Cavalli of Atlantic City in 1958 and they have five children: Donna, Michael, Karen, John and Lori. After release from active duty in 1959, he worked for the Federal Aviation Agency, Atlantic City, NJ as an aerospace engineer performing research related to in-flight and crash fire protection of commercial transports. He was the project manager of several major programs with the FAA.

He took a promotion assignment with Dept. of Army at Aberdeen Proving Ground, MD in 1966 where he performed test and evaluation of Army aircraft and related systems. He learned to fly at APG and eventually earned his commercial certificate and flight instructor ratings and was a flight instructor with the APG flying club. He retired from government service in 1988 as a temporary GS-15. He then worked an additional five years with a commercial firm providing technical support to the Army's binary weapons development and biological defense effort.

MARTIN H. "MARTY" CONLIN, LT, born Jan. 15, 1923, Scranton, PA, received a BA degree from the University of Oregon, 1947.

He joined the service May 5, 1942. His military locations and stations: V-12, Los Angeles, CA; USNRMS, New York, NY, ATB, Coronado, CA; Amphibious Communications, Waipio, AOB, HI; Waianae ATB, Navy #36, TH; SS *Sea Runner,* XAP; Waianae, ATB, Navy #36, T.H.; SS *Cape Georgia* XAK, Waianae ATB, Navy 36, T.H., PerSepCen, Bremerton, WA.

Served with: NRED 13-22, Klamath Falls, OR, XO and CO. His battles and campaigns: Peleliu, Iwo Jima and Okinawa.

His awards and medals: American Area, Asiatic-Pacific Area (3 stars), WWII Victory Medal, Naval Reserve Medal and Armed Forces Reserve Medal. Conlin achieved the rank of LT. His memorable experiences: landing on all three islands.

He married Barbara L., LTJG, Nurse Corps, WWII in Samaar, and they have five children: David, Laurie, Marty Ann, Janelle and James (RET USN STSC - 20 years). Grandchildren are Chad, Jamie, Michele, Lindsey, Derek, Sara and Scott. Employed as accountant, corporate controller. He retired Jan. 15, 1983.

ROBERT D. "BOB" CONN, CAPT, a native of Osborne, KS attended Central Missouri State University, Kansas University, and Oklahoma State University. He obtained a BS and MS degrees in education and Ed.D in educational administration at OSU.

He began his naval career in 1952 by enlisting in the USN. During his enlisted service he was deep sea diver and served aboard the USS *Aeolus* (ARC-3). Upon receiving his direct commission in 1960 he embarked upon an exciting career in the Naval Reserve as well as a career in education administration.

Served as CO, 9-130-M St. Joseph Mo., CO, NR USS *Jouiett* DG, CO, NR USS *Roanoke* AOR, CO, NR VTU, Joplin, MO. Other noteworthy service includes three years 1972-75 on the 9th Naval District Policy Board and National Naval Reserve Policy Board. In 1986 Captain Conn moved to Massachusetts to be superintendent of the state's largest regional school district, Wachusett Regional High School/School Union #64.

He joined the Newport, RI, VTU at the Naval War College in 1986. He actively participated in the Global War Games 1986-1991. In 1988 he completed the Naval Reserve courses at the War College and was assigned instructor in the operations department at NWC until his retirement in 1991.

Conn is a life member of the NRA. He served as 9ND NRA VP Education and Chapter Activities and 9ND NRA president from 1970-75. He was awarded the Good Conduct Medal, Naval Reserve, United Nations.

He completed 33 years of service to his country when he retired in 1991. In 1997 he retired from public education completing 32 years of service to the youth of America.

Resides on his farm in Cherryvale, KS with his wife Mary. They have two daughters, Sara of Newton, NJ and Jennifer of Newton, KS. He also has a stepdaughter, Susan, of Wellington, KS. Grandchildren are Joshua, Aubrey, Logan and Savannah.

JAMES G. CONNELL JR., CAPT, born Sept. 21, 1939, in Adel, GA, a descendant of Scots who moved to south Georgia from South Carolina in 1799. He graduated from the USNA in the Class of 1961, then served four years as a surface warfare officer in the USS *Farragut* (DDG-37) and USS *Lawrence* (DDG-4). Transferring to the Naval Reserve, he earned an MA in comparative literature at the University of Georgia, and MA and PHD degrees in Slavic Languages and Literature at Ohio State University. He continued to drill with Surface Naval Reserve units in Athens, GA and Columbus, OH.

From 1971-81, Connell taught Russian at Valdosta State University in Valdosta, GA, and was XO of Fleet Training Center Mayport and Commander, South Atlantic Naval Force units in Jacksonville, FL. Returning to active duty, he served as director of active duty for training services for the Commander, Naval Reserve Force, in New Orleans for four years. He was instrumental in introducing the Reserve Financial Management System (RESFMS) to the Naval Reserve with an order format still in use as of 1997.

Back in the Ready Reserve, Connell served as commanding officer of the Naval Reserve Merchant Marine Headquarters unit in New Orleans. Returning to active duty a third time in 1988, he used his Russian language skills with Intermediate Nuclear Forces (INF) Treaty implementation with the On-Site Inspection Agency in Ulan-Ude in the Lake Baikal area of the Soviet Union, with the Navy Liaison Units in Munich and Washington, and, for a year, with the American Embassy in Rome. Connell retired in 1992 after 31 years in the Regular Navy and Naval Reserve. He holds the Meritorious Service Medal and the Navy Commendation Medal.

In 1992 he went to American Embassy Moscow to work with the US-Russia Joint Commission on Prisoners of War/Missing in Action, created by Presidents Bush and Yeltsin in March of 1992, to investigate allegations that American POWs may have been transferred to the former Soviet Union in connection with WWII, the wars in Korea and Vietnam, and the 38 American planes shot down during the Cold War, and to help account for Soviet MIAs in Afghanistan and other countries. In connection with this work, by summer 1997, Connell had interviewed hundreds of citizens of the former Soviet Union, visited every former republic except Armenia, and traveled throughout Russia and Kazakstan to visit prisons, psychiatric hospitals and former Gulag prison camp sites. He became chief of the office in 1996.

In Moscow, Connell served as warden and trumpeter of St. Andrew's Anglican (Episcopal) Church, established in 1825, confiscated in 1920, and open again for services since 1991. He worked in an active campaign to have the church restored and returned to the full use of the congregation from Melodia Recording Firm, which had sole occupancy of the building between 1960 and 1991. He took advantage of the rich cultural life of Moscow: ballet, opera, symphony. In Moscow he had the opportunity to meet a large number of famous people including President Clinton, Queen Elizabeth II, former President Nixon, and Russian Orthodox Patriarch Alexy II.

Divorced, he has three children in America: son James III, a lawyer with the federal government; daughter Ashley, a graphics artist who teaches World Wide Web homepage design at the University of Georgia; and Thomas, in elementary school in New Orleans.

JAMES FRANCIS "JIM" CONNOLLY, CWO4, born March 3, 1919 in Butte, MT, attended Apprentice School at Mare Island.

Connolly joined the service in 1944. His military locations and stations: Port Hueneme and Okinawa. He served with 36th Special CB's. His battles and campaigns: Okinawa and Guam.

He achieved the rank of CWO4 and retired March 22, 1979. Connolly was married to Beth for 52 years. Their children are Timothy, Terri and Thomas; and grandchildren are Armand, Nicole, James, Buffie, John, Janni, Katie and Jake. Employed as leadingman rigger and real estate agent, retiring in 1977.

G. CHRISTOPHER "CHRIS" CONNOR, LT (EDO), born Nov. 28, 1940 in Norfolk, VA, received a BS in engineering, Widener University, Exec. Management Course, Yale University.

Joining the service February 1962, his military locations and stations include USS *Ethan Allen* (SSBN-608) NavRes Philadelphia, PA. Served with: USS *Ethan Allen,* Sub Div. 4-37 Sub RVN 6, Sub support. Battles and campaigns: 8 Detachment patrols N. Atlantic and Mediterranean FAC New London. He was awarded the Good Conduct Medal, National Defense, MUC, Naval Reserve Meritorious Service Medal, Armed Forces Reserve.

He achieved the rank of LT (EDO) and retired November 1983. His memorable experiences: Tours in Holy Loch Scotland and Rota Spain; Missile Launch Cape Canaveral, FL; making chief, CWO, and LDO in Reserves: ACDUTRA in New London as repair officer.

Connor is married to Joan and they have children, Wendy and Amy. He held positions of general manager, Nuclear Support and Services, Hope Creek/Salem Nuclear Stations, retiring May 1996.

ROGER A. "ROG" CONNOR, CDR, born Jan. 21, 1920 in Allenstown, NH, received a BA from St. Anselm College and MD from Laval Univerity of Medicine.

He joined the USN in February 1942, ENS AVP. Appointed LTJG in 1947. Recalled to active duty November 1950. Reported to Camp Lejeune, NC. Transferred to Camp Pendleton, Oceanside, CA, 1951. Attached to Co. A, 3D Amphibian Tractor Bn. Force Troops, FMF Pacific. Released from active duty in 1953. Drilled as participating member of the Marine Unit, Lynn, MA 1953-56. Assumed duties as medical officer, 7th Naval Reserve Training Center in Pittsfield, MA, 1956-75.

Connor practiced dermatology in Pittsfield, MA, 1956-85. Volunteer in the practice of dermatology in a clinic for indigent patients, Naples, FL, 1986 to date. He and Mary M. Monahan celebrated their 50th wedding anniversary in 1997. They have six children: Michael, Kathleen, Marybeth, Timothy, Christopher and Matthew; and six grandchildren: Meghan, Patrick, Timothy, Christopher, Sean and Griffin.

RONALD D. CONRAD, CDR, born in Norfolk, VA, in January 1954. He is a graduate of Old Dominion University, enlisted in the USN in 1977 and attained the rating of SK2. He served aboard USS *David R. Ray* and USS *Saginaw.*

After commissioning in 1981, he served as a supply officer in various reserve billets including XO of NSC Norfolk B506. He received his MS degree from Florida Tech in 1993. He served as OIC/ACO during Operation Uphold Democracy (Haiti) and as WO during Operation Joint Endeavor (Bosnia).

He received two NCMA Outstanding Academic Achievement Awards and scholarships totaling $4,500 for graduate work. For 1995, DLA presented him and "Reservist of the Year Award (Navy Senior Officer)." Military awards include the Defense Meritorious Service, Joint Service Commendation, Army Achievement, and two Navy Achievement Medals, and numerous other awards.

His civil service career began in 1981 and he is currently a level III contracting officer with the Customs Service.

Conrad is a lifetime member of NRA. He also holds memberships with NCMA and PTA. He lives in Fredericksburg, VA with his wife, Holly; son, Jeremiah; and daughters Rebekah and Amanda.

JOHN P. CONROY, CAPT, born July 22, 1936 in Pittsburgh, PA, received a BS degree from Duquesne University and MA from Xavier University.

He served June 23, 1954-July 3, 1957; November 1965-July 22, 1996. He was awarded the Meritorious Service Medal, National Defense (three awards), MUC (3), PUC, Overseas Ribbon, Naval Reserve (three awards). Conroy achieved the rank of CAPT and retired July 22, 1996.

His memorable experiences: becoming Reserve Intelligence Area CDR, Area 12, NARCEN Columbus, OH; responsible for naval intelligence Reservists in three states.

He married Barbara and they have two children, Mary and Laura. Employed as vice-president, investments, Prudential Securities.

JAMES CONWAY, CAPT, born Feb. 21, 1921 in Chicago, IL, attended Marquette University and Columbia College. He joined the service September 1942. His military locations and stations: NAS Minneapolis, NAS Pensacola, NAS Glenview. Served with: VN 9-7; FASRON 56; VS 722; ACORN 9-7.

Conway served in American Theatre Campaign. He was awarded the SecNav Distinguished Public Service Award. He achieved the rank of CAPT USNR, was discharged September 1972.

His memorable experiences: producing Navy film in Morocco and Mediterranean area. Producing mock air attack by NAS MPLS aircraft on Chicago with NAS Glenview defending. CO VS 722 CO ACORN 9-2.

Conway married Audrey and they have four children: James, Victoria, Michael and Richard; and 11 grandchildren. He was employed as a radio and TV commentator and producer, retiring in 1986.

JOHN F. COOK, CAPT, a native of Green Bay, WI, attended Marquette University, obtained PhB and JD degrees. He began his naval career in 1942 as a Naval Reserve Officers Training Corps cadet at Marquette University. He was placed on active duty in July 1943 and was commissioned ENS in February 1945, upon graduation.

After communications and destroyer schools, he served as communications officer of the USS *Bordelon* (DD-881), first involved in occupation of Japan and then as part of Task Force 77 in the Pacific. After return from active duty in September 1946, he served in Naval Reserve Units in Milwaukee, Oshkosh, WI; Alexandria, VA; Washington, DC and Sacramento, CA continuously until 1983.

During this period he was commanding officer of NAVSECGRUDIV 9-13 (1956-59) NAVSECGRUDIV 9-8 (1968-69) and NAVSECGRUDIV 5-15 (1973-74). His campaign and service medals are: American Theatre, WWII Victory Medal and the Occupation (Japan), China Service, Naval Reserve and Armed Forces Reserve Medals.

While in the USNR, he was assistant city attorney for Milwaukee for 21 years, followed by service with the US Department of Justice. He then served as Federal Administrative Law Judge for 21 years, retiring as Chief Judge of the United States Department of Education in 1994.

SCOTT LOUIS COOPER, CDR, born Feb. 4, 1956 in Los Angeles, received a BS degree in business management.

He joined the service May 1979. His military locations and stations: ROTC University of Idaho; Surface Warfare Officer School, Coronado, CA; Communications Officer School Fleet Training Center San Diego. Schools: Fleet Guarded Missile School Dam Neck, VA; Fleet Intelligence Officer School, Fleet Intelligence Training Center Pacific, San Diego, CA.

Commands: USS *Halsey* (C6-21), CO; USS *Prairie* (AD-15), assistant operations officer, NavSta San Diego; Communications Officer School, classified material systems school department head, Fleet Training Center, San Diego, CA.

He was awarded the West Pac Deployment Ribbon, nominated Readiness Command 19, Junior Naval Reserve Officer of the Year. Cooper achieved the rank of CDR and was discharged September 1983.

His memorable experiences: shore patrol office Subic Bay, Philippines, Military exchange officer, Korean and Japanese Navy, created first cold weather operations school NavSta San Diego.

He was Top 10 Agent out of 15,000 agents nationwide with Farmers Insurance Gp. and KNX Los Angeles Citizen of the Week. Retired from his civilian employment June 1, 1996. He is married to Rariba and they have children, Chaniel and Sean Anthony.

TOMMY A. COOPER JR., CDR, born Sept. 30, 1917 in Frederick, OK. He attended two years of college and joined the service Aug. 25, 1941.

His military locations and stations: E-Base Hensley Field, Grand Prairie, TX, advance training, Corpus Christie, TX. Land plane and sea plane, Kingsville, TX, Sec. Officer in 10th Cadet Bn.; Corpus Christi, TX; flight instructor, Atlanta, GA; flight instructor, Lake City, FL; PPC-PVI Beaufort, SC; Nav and Gunnery Attu-Aleutian Islands VPB-131 Pensacola, FL; flight instructor USS *Gen. J.C. Breckenridge,* transportation and cargo officer; VS-21 ASW Sqdn., North Island, San Diego, CA, air phase officer, in CIT School, Glenview, IL.

Served with VPB-199; VPB-131; VS-21. Battles and campaigns: Inchon invasion; South Korea; Hungnam Evac.; South Korea.

Cooper achieved the rank of CDR, USNR (Ret) and was awarded the Air Medal, American Defense, American Theatre, Asiatic-Pacific Naval Reserve; National Defense Service. His memorable experience: getting checked out and flying a new A.F. plane.

He married Muriel and they have a daughter, Lisa Jane Cooper Berarducci. He was employed as air traffic control specialist and retired Sept. 30, 1977.

BRUCE D. CORB, CDR, is a native of Scarborough, ME. After receiving a BS degree in Marine Engineering, a third assistant engineer's license, and a USNR commission from Maine Maritime Academy, he commenced flight training at NAS Pensacola, FL in 1969.

Following designation as a naval aviator in 1970, he reported to the Naval Research Laboratory, Washington, DC. After transitioning to the EC-121K aircraft, he completed numerous research and development detachments to the Indian Ocean, Alaska, Hawaii and Europe. While at the Naval Research Lab, he was directly involved in the development of the AWG-9 Phoenix Missile System. He served as schedules, logs and records and scientific liaison officer to electronic warfare, as well as plane commander in the EC-121K and S-2D research aircraft.

In 1973, Corb reported to Patrol Sqdn. 23 homeported at NAS Brunswick, ME. He completed operational deployment to NAS Sigonella, Sicily, NS Rota, Spain, NAF Lajes, Azores, and NAS Bermuda. He served as nuclear weapons training and aircraft division officer, plane commander and instructor pilot. He received a letter of commendation from Commander Patrol Wings Atlantic for superior performance with the Squadron's Nuclear Technical Proficiency Inspection (NTPI).

Following his release from active duty in 1976, he accepted employment as a pilot with TransAmerica Airlines flying the L-188C, L-382G and DC-8-63/73 series aircraft until the airline ceased operations in 1986. That same year he accepted a position piloting BE-200 and CE-550 series aircraft for Federal-Mogul Corporation in Detroit, MI.

His RESFORON experience includes VP-0810, NAS South Weymouth, MA as flight officer and instructor pilot for six months: VP-2644 & VP-93, NAF Detroit, MI as schedules and assistant operations officer, plane commander and instructor pilot for two and one half years; VP-0919 and VP-91, NAS Moffet, CA as electronic warfare and assistant maintenance officer while completing transition to the TACNAVMOD update aircraft for one year, and VP-92, NAS South Weymouth, MA as assistant operations officer and plane commander for one year. His final RESFORON assignment was to VP-93, NAF Detroit, MI in 1982 as Gold Wing administration officer, plane commander and instructor pilot. He completed his tour as CO of VP-2473, NAF Detroit, MI in 1989. During his tenure, readiness increased from R-4 to R-2, with three detachments to NAF Keflavik, Iceland. Corb retired in 1989 with over 20 years of service.

In his civilian occupation, he has been a property manager for Chateau Properties since 1990, managing a portfolio of 1,000 units with assets exceeding $20,000,000. He resides with his wife, Debbie, in Clinton Township, MI.

WALTER CORNELIUS, CAPT, born March 8, 1916, in Scranton, PA obtained an aeronautical engineering degree in 1938 and recognition as a Distinguished Engineering Alumnus in 1973 from the University of Cincinnati. After being commissioned an ENS in November 1942, he served as the Bu-AER resident representative at the Turbo Engineering Corporation, Trenton, NJ, 1943-44, and engineering officer of AROU-2 at NAS Admirality Islands and NAS Samar, Philippine Islands, 1944-46.

After being released from active duty in June 1946, he became active in the Naval Organized Reserve Program in March 1948 serving in aircraft maintenance and administrative capacities in WF, VA and VR squadrons and Wing Staff at NAS Grosse Ile and Wing Staff at NAF Detroit, retiring as CAPT on July 1, 1973.

As an executive research engineer at the General Motors Research Laboratories, 1937-81, he was involved in the research and development of conventional and unconventional automotive reciprocating engines, automotive regenerative and aircraft gas turbines and a stationary gas turbine. He has been the co-editor of two technical volumes, granted four engine patents and has authorized 24 technical papers with 1975 listing in the Library of Congress was recognized in *American Men of Science* and *Who's Who in Middle West.*

Cornelius and wife, Mary, reside part-time in Troy, MI and north Ft. Myers, FL. They have two children, Lisa and Paul. He is a past president of Michigan Retired Officers Association, life member of the NRA and ROA Associations and Member of the Navy League.

RICHARD W. COST, CAPT, born June 23, 1942. He enlisted May 29, 1962 Liverpool, NY in ROC program while attending Syracuse University. Commissioned June 1964. Reported USS *Zellars* (DD-777) as main battery assistant, fire control and DASH officer. Crossed Arctic Circle and Equator. Reported 1967 to COMSERVLANT as aide and flag lieutenant to RADM Ben W. Sarver. Released from active duty Oct. 2, 1969. Served in USS *Willard Keith* (DD-775) and USS *Huntington* (DD-781), Surface Warfare designation. USS *Gilmore* (AS-16), USS *Orion* (AS-18), Det 304, USNA Blue and Gold Officer.

CO Weapons Station Earle Detachment 1104 and WEPSTA Earle, HQ 104. Named to NAVSEASYCOM and SPAWAR Reserve Policy Steering Gp. CO VTU 0411. 1988, Staff of Readiness Commander Region Four. Attended Reserve Advanced Management Seminar, Leesburg. Co-Chair Recruiting District Assistant Council, North Jersey.

1990, CO Readiness Unit, Lehigh Valley, PA. Attended Naval War College, National Security Decision-Making Course. Taught Command Excellence and Equal Opportunity Seminars. CO Naval Activities United Kindgom, Det 504 Harrisburg. Military Academy Selection Committee and Veterans Committee of Congressman Paul McHale. Retired July 1, 1994. Awarded the Meritorious Service Medal, National Defense Service Medal (2), Armed Forces Reserve Medal (2).

Currently vice-president for Institutional Advancement, Bridgewater State College and executive director of Bridgewater State College Foundation. MBA Old Dominion University, Ed.D Rutgers University, Institute for Educational Management, Harvard University. Director Fuller Museum of Art, Director Area Chamber of Commerce. Resides Middlesboro Mass with wife Ellen a higher education consultant. They spend as much time as possible in a daysailer on Narragansett Bay. E-mail rcost@bridgew.edu

JOHN G. "BALLS" COTTON, CAPT, born Feb. 28, 1951 in Norfolk, VA. Earned BS degree in aerospace engineering, USNA, 1973 and aviator wings in 1974 (third generation military pilot). Deployed from Japan with VA-93, USS *Midway* (CV-41), flying A-7. Served on Staff, CDR Light Attack Wing, Pacific Fleet, and released from active duty 1980.

Joined Reserves NAF Washington, flew A-7 with VA-2082, NAS New Orleans, then RF-8G with VFP-6366, VFP-306, VFP-206, NAF Washington. Returned NOLA 1987 as VA-2082 Ops, CO, then VA-204 Ops, Training, F/A-18 transition and VFA-204 XO, CO.

While CO, VFA-204 received Safety "S", Battle "E", Meritorious Unit Commendation and F. Trubee Davison Awards as best reserve tailhook squadron. CO NAS, Keflavik 1066 at NAF Washington in 1994 and CO, Navy Command Center 106 at Pentagon in 1996. Awarded Meritorious Service (2) Navy Commendation and Navy Achievement Medals.

He is married to Cindy and they have children: Jennifer, Charles and Christine. Employed as pilot with American Airlines, based in DC.

JOHN T. COUNTS, CDR, hailing from East Greenwich, RI, born in Taipei, Taiwan, son of LCDR and Mrs. William D. Counts. Graduating from the USNA, he received his BS degree and commission in May of 1979. Nine months were spent assigned to VS-37 at NAS North Island, CA, flying the S-3A while awaiting flight school. After completing flight training at NAS Whiting Field, FL, he was designated a naval aviator in March of 1981.

Counts' first fleet assignment was with HC-3 (sea duty), NAS North Island, flying the venerable CH-46D and serving as FRAMP division officer and avionics division officer. He deployed onboard USS *San Jose* (AFS-7), USS *White Plains* (AFS-3) and USNS *Kilauea* (TAE-26) in various billets including Det Operations and Assistant OIC.

In October 1984, he reported to Naval Recruiting District (NRD) Louisville, KY as an office programs officer flying the T-34B. He received the NRD Louisville Officer Recruiter of the Year award for 1986.

He returned to Southern California in August 1987 as the air operations officer for CDR, Amphibious Sqdn. 7 based at NS San Diego. He flew the UH-1N twin Huey off of USS *Okinawa* (LPH-3) and USS *Tarawa* (LHA-1).

His next assignment took him overseas in January 1990 to HC-5 at NAS Agana, Guam. Again flying the CH-46D, he served as Training, Safety and Operations Department head. Deploying onboard USNS *Kilauea* (TAE-26) as Det OIC, his detachment served with the USS *Independence* Battle Gp., Amphibious Readiness Gp., and the USS *Theodore Roosevelt* Battle Gp., from the beginning of the Operation Desert Shield through the completion of Desert Storm.

In August 1992 he affiliated with the USNR and joined the Golden Gates of HS 85 at NAS Alameda, CA flying the SH-3D. He has served as the administrative and operations department head through the squadron's redesignation as HC-85. In August 1996 he became the XO.

Counts' personal awards include the Navy Commendation Medal w/2 Gold Stars and the Navy Achievement Medal w/Gold Star. After leaving active duty in July 1992, he accepted a position in the Marine Operations Dept. of Sea-Land Services, in Long Beach, CA and presently holds the title of vessel planner.

He is married to Lisa Hunt and they live with their two children, Ian and Mia, in Rancho Palos Verdes, CA.

JAMES HUSTON "JAMIE" COWAN JR., LT (USN), CDR (USNR-R), born Aug. 25, 1943 in Corvallis, OR, received a BSE (EE) from Princeton University in 1965 and a MBA from the University of Miami in 1974. He joined the service Sept. 3, 1965 and was commissioned after four years Navy ROTC at Princeton.

Military locations and stations: USS *Skate* (SSN-578), New London, CT; USS *Ulysses S. Grant* (SSBN-631). Served with (Gold), Charleston, SC.

Cowan achieved the rank of LT (USN), CDR (USNR-R) and was discharged May 1971. His memorable experiences: steaming under the polar ice cap on USS *Skate* and trailing Russian warships at the height of the Cold War, followed by 14 years of good fun and friendships in the USNR in Miami, FL.

Cowan is divorced and has a child, Ashley Atherton Cowan. Employed as information systems consultant in the healthcare industry.

GORDON L. COX, CAPT, born Nov. 11, 1931 in Mound Valley, KS, attended two years college at Coffeyville College and Parsons Junior College, KS.

He joined the service Jan. 4, 1952. NAS Olathe, KS, reported flight training April 1952; NAS Pensacola as NAVCAD. Designated and commissioned Nov. 12, 1953, NAS Hutchinson, KS. His military locations and stations: NAS Olathe, KS, VS-38 NAS North Island, VS-33 NAS North Island (plank owner), Hawaiian Sea Frontier Operations Control Center, VW-2 NAS Patuxent River Naval Air Technical Training Unit.

He served with VR-882 and VR 21, K28 VR1K3 (XO), NAS Olathe, NARS MI VR52-2, VR-2512 (CO), VR-53 DET "OinC" NAS Memphis.

Cox achieved the rank of CAPT and was discharged from active duty June 30, 1964 and Reserve May 30, 1984. His memorable experiences: about 200 traps aboard USS *Bennington* (CVS-20). About 7,000 hours pilot time (CIV and MIL). Flew: SNJ, PB4Y2, PB-1W, WV-2, R7V, SNB, T-34, TY-2, S2F, C-54, C-118, Constellation and Super Constellation, B-727, DC-9, Falcon, Fanjet. He has airline transport rating.

He was awarded the National Defense Service Medal, Expert Pistol Medal, Vietnam Medal, Reserve Medal, Navy "E" for ASW, Isabell Trophy, and 4 CNO Safety Awards.

Cox is single and has three children: Laura, Gordon Jr. and Lisa; grandchildren are David and Christopher Dawson, Daniel and Riley Smith, and Paul Freeman. Employed as TWA first officer, owner/operator trash service company, retiring Nov. 11, 1991.

MACK COX, CAPT, born June 18, 1937 in Dallas, TX, received a BS in geology from Baylor University. He joined the service Sept. 9, 1959. His military locations and stations: OCS (Newport, RI); Aviation Ground Officers School (Jacksonville, FL); Naval Intelligence School (Washington, DC); VFP-62 (Cecil Field, FL).

Cox received the Meritorious Service Medal, Navy Unit Commendation, three Navy "E" Medals, Naval Expeditionary Medal, National Defense Service Medal, Armed Forces Expeditionary Medal. He achieved the rank of CAPT and was discharged from active duty Nov. 19, 1963 and from the USNR, March 1, 1988.

His memorable experiences: squadron flew low level missions over Cuba; recalled to active duty-operation homecoming; Vietnam, 1973; Debriefed POW LT Ed Miller; worked on Damage Assessment Task Force, Walker Spy Case; CDR Reserve Intelligence Area 6 (was RIAC).

He is married to Linda and has daughter Heather; step-daughter, Aimee Kerrigan; and step-son, Brad Kerrigan. He is employed with Engineered Magnetics, Inc. in marketing.

CHARLES L. CRAGIN, CAPT, a Portland, ME native enlisted in 1961. He was separated in 1964 as a communications technician second class. In 1968 he was commissioned an ENS (Public Affairs) and currently serves as the NEPLO to the 1st US Army. He has commanded USACOM 506 and NAVINFO 101 and served as Reserve special assistant to the Chief of Naval Operations.

Cragin earned a BS and JD degree from the University of Maine. He was the 1982 Republican nominee for governor for Maine and later served as chairman of the Republican National Committee's Budget Committee. He was sergeant-at-arms of the 1988 Republican National Convention.

He practiced law in Maine for 20 years prior to his appointed by President Bush and Senate confirmation in February 1991 as chairman, Board of Veterans' Appeals. He married LCDR Maureen P. Ford.

NAYLOR CRAGIN, LCDR, born in Crosby, MS, July 20, 1921. Enlisted in USN, Dec. 8, 1941. Boot camp at San Diego. Assigned to USS *Arcadia* (AK-41). Delivered bombs and aviation gas to South Pacific areas. In 1943 as SK1c reverted to AS and sent to V-12 at LaTech. After six months, assigned to Harvard Graduate School for one year. Became ENS in 1944 and assigned to USS *Arcadia* (AD-23). Released to inactive duty August 1946.

Graduated LSU in 1950 BS. Recalled to active duty for Korean War April 1951. Served two years as assistant for Disbursing Naval Station, Algiers. Remained in Reserve until completion of 20 years.

Retired July 20, 1981. Campaigns and service medals: American and Pacific Theatre, two Battle Stars, WWII Victory Medal, Korean Service and Naval Reserve.

He is married to Flo Guerin and children are Michael, Patricia, Timothy and Kelly; and grandchildren are Stacey and Kerry Joe Cragin; Holly and Matthew Rudisel; Allison, Patrick and Sally Rose Cragin. Retired as director and executive vice-president Capital Savings in 1986.

WESLEY P. CRAIG, CAPT, born Dec. 9, 1918, Oakland, CA. Signed up in Navy 1937 UC Berkeley NROTC. Received degree and commissioned ENS. Called to active duty summer 1941 aboard USS *Cassin* (DD-372).

He was OOD Dec. 7, 1941. Sounded GQ at 0759 in Drydock #1 Pearl Harbor. Ship blew up, burned. Last officer off ship.

Made secret voyage South Pacific, Spring 1942 with RADM Richard E. Byrd for seven future Navy bases, finally landing New Zealand. He was on destroyers and amphibious LST's. Made two assaults D-Day, H Hour landings as senior beachmaster Lingayen Gulf and Okinawa. Credited with eight Battle Stars, Commendation Medals from Philippines and Chinese Governments, Pearl Harbor Congressional Commemorative Medal, Naval Unit Commendation, etc.

Separated from active duty November 1945 as LCDR. In 1949 joined Active Naval Reserve, MSTS Unit. Served seven years XO, three years as CO. President Reserve Officers Association, Sonoma County; 30 years as officer; retired as Navy CAPT.

In civilian work, joined Dean Witter and CO. financial firm 1945-1979. Retired vice-president, Santa Rosa, CA.

He has been married for 53 years to Ruth R. Roemer and they have three daughters: Claudia, Lynnea, and Diane; grandchildren are Leah, Cindy, Steven, Michael, Blas, Tom; and two great-granddaughters, Emily and Chrissa.

RICHARD S. CRAMER, CDR, born Aug. 2, 1943, Newark, NJ, and started his naval career after graduating from Teaneck High School, Teaneck, NJ. He enlisted in the USN during the Berlin Crisis in July 1961. After boot camp at Great Lakes, he was assigned to Sonar School in Key West, FL. He graduated first in his class of then a 33 week "A" School. He was then assigned to the USS *Ault* (DD-698) then in FRAM II overhaul. He was sent to "C" School back in Key West and just completed the course as the Cuban Blockade started. He spent his active duty time on the *Ault*, doing refresher training at GITMO, a Great Lakes Cruise, various East Coast and Caribbean exercises and an extended Mediterranean deployment.

Upon release from active duty he affiliated with the Naval Reserve Program on board the NRF DEs in the New York area. At the rate of STCS(SW) he was selected for the Chief Warrant Officer Program in 1976. As CWO 2&3, Operations Specialist, he served aboard the FRAM II DDs in the New York area as Weapons Officer. He was selected for the LDO Program in 1980 and as LTJG, Deck LDO, was SWO qualified on board the USS *Vogelgasang* (DD-889).

After 20 years of destroyers and a support group, NR Fleet Training Group, he was assigned to the amphibious surface Navy on the USS *Bolder* (LST-1179). He served as weapons officer, gunnery officer, assistant first lieutenant and SELRES Administrator CDR Cramer upon becoming Surface Warfare Qualified (SWO) requested and had approved a designator change to Unrestricted Line Officer, 1115.

He was then assigned to NR Fleet Training Group DET 102 as executive officer then commanding officer of the unit. During his tenure the unit accomplished many firsts then unheard of for Naval Reserve Units. He has been SELRES Coordinator on board the USS *Estocin* FFG-15 and been with a PERSMOBTEAM Unit, a NR Fleet Support Training Unit and currently drilling with the newly commissioned Naval Coastal Warfare Group Two.

JAMES BRAXON CRAVEN III, LCDR, born Dec. 8, 1942 in Portsmouth, VA, attended USNA, UNC-Chapel Hill (AB 1964) and Duke University (JD 1967 M.Div. 1981).

He joined the service July 5, 1960. Commissioned April 1974, LTJG, JAGC, USNR). His military locations and stations: USNA-Annapolis. Served with JAGC CO 0708, 1974-82; Naval Academy information officer (1976-present).

Craven was awarded the NAM, National Defense Service w/Star, Armed Forces Reserve w/Silver Hour Glass, Navy Expert (rifle and pistol) and Navy Recruiting Ribbon. His memorable experiences: Helping approx. 100 young men and women get into USNA and become officers in the Navy and Marine Corps.

He married Sarah Harris Aug. 22, 1964. They have three children: Jamie, Joe and Will. He has been employed as a lawyer since 1967 and was ordained an Episcopal Priest in 1985.

CHARLES W. CRAWFORD, CAPT, born Feb. 17, 1931 in Spindale, NC, received a BEE degree from North Carolina State University and MSEE from Clemson University.

He joined the service June 30, 1948. Enlisted and commissioned ENS USNR July 1957. His military locations and stations: USS *Floyd B. Parks* (DD-884) and Charleston Naval Shipyard. Served with the Fire Control Gang on Parks & Repair Div. at shipyard. Fought battle and campaigns: Korean War and siege of Wonsan Harbor.

Crawford was awarded the Navy Commendation, Meritorious Unit Commendation, Good Conduct, NAV OCC, China Service, National Defense, Korean, NAV RES, UN.

He achieved the rank of CAPT (officer), discharged June 1952 and POZ (enlisted), October 1989. His memorable experiences: serving as MK 42 rangefinder operator on the Parks in Wonsan Harbor in 1951 and serving as ship superintendent at CHAS NAV SHPYD in 1965 and 1966 on ship conversion (sea plane tender *Albermarle* to helicopter repair ship).

He is married to Betty and they have three children: Debra Crisco, Gregory and Kent Crawford; grandchildren are Michael, Rembert and Brooke Crawford and Tiffany Thomas. He is owner of MacGuire and Crawford, Inc., mfg. rep, industrial instruments, Charlotte, NC.

LOUIS R. CRIEZ, CAPT, was born in Seattle, WA, Sept. 30, 1929. He is a graduate of Washington State University. He enlisted in the USNR in 1948 and served as an RDSN on a minesweeper attached to the Seattle Reserve Training Center until selected for Reserve Officers Candidate School in 1950. Upon being commissioned as an ENS in 1952, he was called to active duty in the Korean War during which time he served as 1LT, engineer officer and XO of the USS *Flamingo* (AMS-11) and as engineer officer of the USS *Impervious* (AM-449).

Upon release from active duty in 1954 he held various billets in reserve units in the Puget Sound area until 1960 when the USS *Ruff* (MSC(O) 54) was placed in reserve force status. He was assigned to the *Ruff* where he served as engineer officer, XO and CO. While CO he received a Secretary of the Navy Commendation. Following completion of command of the USS *Ruff* he was assigned to the 13th ND Group CDR as staff assistant and subsequently CO of NRSD 13-13(L). In 1970 he participated in the establishment of Mine Div. 53(NRF) on Puget Sound and was appointed Div. CDR. Upon completion of the Division Command and promotion to CAPT, he held various shore billets in the Seattle area.

Criez retired from the Boeing Co. in 1986 and as a C.G. Licensed 100T Master has since been engaged in boat deliveries and enjoying recreational sailing on his boat.

A. JAY "AJ" CRISTOL, CAPT, born Sept. 25, 1929 in Fountain Hill, PA, received a BA from the University of Miami in 1958 and a JD degree in 1959.

He joined the service Nov. 4, 1951. His military locations and stations: NAS Los Alamitos, CVS-37, NAS North Island, MCAS Miami, NAS Jacksonville, VS-37, FAWTUPAC, VR-801, VR-802, VR-805, VR-804, VTU Law Miami, LSO 108. He fought in Korea and Vietnam campaigns.

Cristol achieved the rank of CAPT and his awards

and medals include the Meritorious Service Medal, Navy Commendation Medal, Navy Achievement Medal.

His memorable experiences: Day and night carrier qualified; four engine plane commander, honorary professor at Naval Justice School, Wise Old Owl of the Jag Corp., commanding officer of VTU Law Miami and LSO 108.

He married Eleanor "Elly" and their children are Stephen and David. They also have a granddaughter, Samantha. He held the position of Chief Judge, US Bankruptcy Judge, Southern, FL.

HOWARD E. CROMER, CWO-4, born Aug. 14, 1939, St. Johnsville, NY, received an AA degree from Long Beach City College, Long Beach, CA.

He joined the service Aug. 29, 1956. His military locations and stations: USS *Ranger* (CV-61), USS *Kearsarge* (CVS-33). Served with NAS Meridian, MS; NAS Los Alamitos, CA; VA-305, Pt. Mugu, CA. He served in Vietnam in 1964.

Cromer was awarded the MUC, Good Conduct, Armed Forces Expeditionary, and Naval Reserve Medal. He achieved the rank of CWO W-4 and retired June 1, 1988. His memorable experiences: involved with MA-9 recovery-Mercury program of Astronaut Gordon Cooper and Faith-7 space capsule on USS *Kearsarge* May 11-18, 1963.

He married Eve and they have two children, Howard Steven, ex-LT USN, and Kenneth Earl. Employed at Long Beach Naval Shipyard as eng. tech., retiring Aug. 3, 1988.

CORNELIUS LEO "NEIL" CRONIN, CDR, born May 11, 1922 in Pawtucket, RI, attended Rhode Island State College.

He joined the service Oct. 1, 1942. His military locations and stations: St. Louis, MO; Pensacola, FL; South Weymouth, MA. He served with VPB-128, VP-912 and HS-912. Participated in WWII Pacific Campaign November 1944-September 1945.

Cronin was awarded the Distinguished Flying Cross, Air Medal, American and Pacific Theater, NUC, NR and WWII Victory Medal. He achieved the rank of CDR 05 and was discharged April 1946/June 1970.

He married Jean Louise and they have two children, Marshall Michae and Ann Jane; their grandchildren are Ian, Clara, Tracey, Kelly and Kate. He held position of chief aero/inspector State of Rhode Island; president East Prov. Can Co., 1946-77.

RALPH C. CROPP, LT, SG, born May 6, 1923 in Chicago, IL, received a BS, AOE.

He joined the service March 4, 1943. He was stationed at various NAS, served with VT-98, VT-23, VA-75A and VA-921.

Cropp achieved the rank of LT SG and retired May 6, 1983. His memorable experiences: designated naval aviator Oct. 31, 1944. NRA life member #3020.

He married Ruth B. and they have four children: Kenton, Cherie, Lance and Joyce. Employed with US Civil Service, GS-14, retiring Jan. 3, 1985.

JAMES KERR "TERRI" CROSBIE III, LCDR, born Jan. 27, 1934 in Stoneham, MA, received a AS in business administration in 1974, from the University of ME, Portland; a BS in voc. educ. teacher, University of ME, Portland, Gorham; 1986, MS in education admin., University of Southern ME, Gorham.

He enlisted July 21, 1953 and May 20, 1967. Commissioned April 1, 1981. His military locations and stations: NTC Bainbridge, MD (boot camp); NATTC Norman, OK; (airman primary training) - NATTC Lakehurst, NJ (aerology 'A' School); NAS Oceana, VA (weather office); SAU #1, London, England (Green door weather); NAS South Weymouth, MA (forecaster, leading chief, forecast duty officer, division officer, XO); NAS Brunswick, ME (forecaster, section leader, forecast duty officer, OIC Reserve Weather Detachment); CV-67 USS *John F. Kennedy* (communications watch officer, OIC Reserve Detachment on Board, Dept. Head Reserves on Board Underway).

Crosbie was awarded the Good Conduct Medal, Naval Reserve Meritorious Service (2), Armed Forces Medal, Battle E Ribbon, Meritorious Unit Citation, National Defense (2). He achieved the rank of chief (enlisted), LCDR (commissioned) and was discharged July 20, 1961/April 1, 1981. Retired as LCDR, USNR, Jan. 1, 1993.

His memorable experiences: shock test aboard (CV-67), his commissioning as line officer (LDO); the many wonderful people he lived, worked and relaxed with throughout his naval career; the night he followed 'his buddy' ACAN Arnold (in his capacity as night watchman-a moonlighting position) on an inspection tour of a local VFW Hall, where upon opening the only door he was shot and Crosbie was bowled over by a fleeing, panicky burglar, as he made his apparent escape. A temporary situtation.

He married Barbara Ann Peters on April 12, 1959; she passed away Nov. 29, 1996. He has one daughter, Kimberly Anne White, and two grandchildren, Joshua Matthew and Danielle Nicole White. He has been employed in the automotive industry for 30 years as mechanic/technician, parts manager, service manager, general manager, factor representative and small business owner. He was also a teacher for 15 years.

WAYNE H. CROW, CDR, born April 4, 1954 in Philadelphia, PA, received a BA from Duke University in 1936; MBA, Wheaton School, University of Pennsylvania (1982).

He joined the service NROTC September 1972-May 1976. Commissioned May 1976. Reserves March 1981. His military locations and stations: NSCS Athens, GA (1976); USS *LA Moore County* (LST-1194) (1976-78); USS *Leader* (MSO-490) (1979), supply officer, NISMD Philadelphia, PA (1979-81).

His battles and campaigns: Northern Wedding, Undercover Ambush. He was awarded the Naval Achievement Medal, Meritorious Unit Award, Navy Unit Commendation, Meritorious Unit Commendation, Battle E, National Defense Medal, Arctic Service Pistol Ribbon.

He achieved CDR, SC. His memorable experiences: Reserve Unit: COMSERVGRU TWO DET 104 (Philadelphia, PA) (1981-82); Suppo NASCAJES AZ 0473 (1982-83); XO, NAS Whiting Field 0143 (1983-84); At NAF Detroit; XO (1984-85) and CO (1985-87) ASB (SM) 313 at NRC South Field MI; S-4A RN MCB 26 (1987-88); ASO Phil HQ 104 (1988-91); DCMDM Phila. 304 (1991-94); NAVICP Philadelphia 104 (1994-)

He married Deborah and they have three children: Kevin, Gary and Nathan. Last employment was with New Holland North America 1989 to present.

ALAN CRUIKSHANK, CAPT, CEC, entered the USN in 1945 as an apprentice seaman. Following boot camp in Bainbridge, MD assigned to the Office of Naval Research in Washington, DC and Port Washington, Long Island, NY.

Discharged in 1946 as a MMs3/c. Immediately enlisted in the USNR. Subsequent to receiving a mechanical engineering degree from Princeton University in 1948, directly commissioned ENS, CEC. CB Div. 3-7, Brooklyn, NY. CB Div. 3-13, Rochester, NY, culminating as CO. Ten years with RMCB-19: H Co. CDR 1 year, Ops 3 years, XO three years, CO three years. RNCR-4. CO, FAC 202, Clifton, NJ. Area coordinator, Naval Academy Information Program. CDR, RNCR-7, Davisville, RI.

Training duty at Davisville, RI; CECOS, Port Hueneme, CA; MCB-1, Guantanamo Bay, Cuba; Little Creek, VA; Camp Lejeune, NC; Goshen, VA; Gulfport, MS; NavFac, Washington, DC; NorthDiv, Philadelphia, PA; US Naval Academy, Annapolis, MD; Quantico, VA.

MBA New York City University. JD University at Buffalo, NY. Assistant District Attorney, Rochester, NY. Wife-Fran. Children: Jill (deceased), Janet (deceased), Andrew. Retired as CAPT, CEC, USNR, July 1, 1984.

JOHN ARTHUR "ARCHIE" CULVER, CAPT, born Feb. 8, 1927, Somerville, MA, received a BS degree in Marine Engineering from the Massachusetts Maritime Academy degree credits from George Washington and American Universities. Professional engr. (PE) in Massachusetts, New Hampshire and California.

He joined the service February 1945, Midshipman USNR, Mass Maritime Aca. His military locations and stations: 1948-1952, USS *Weiss* (APD-135), USS *Tumult* (AM-127); (USS *Linnet* (AMS-24); USS *Gull* (AMS-16), USS *Harry E. Hubbard* (DD-748); 1969 (2 months), Naval Ship Systems Command (Minesweep Program manager); 1981-1983, Naval Sea Systems Command, program director, Navy Vendor Motivation Program; 1985 (two months), Space and Naval Warfare Systems Command, special assignment for implementing Vendor Motivation Program.

Served with: Naval Reserve Surface Div. (May 1937); Bureau of Ships Co. (11-16); Naval Reserve Officers School (1-7), Attack Sqdn. (VAJ 911); Naval Reserve Officers School, CO; Group Command (1-1), Boston; Engineering Co., (1-1), CO; Ship Maintenance and Repair Unit 4001, CO; Readiness Command, Region One, Advanced Ship Repair Base (YR-26), CO; Navy Liaison, Mass Army NG.

Battles and campaigns: Combat minesweeping operations in Korean waters 1951-1952 aboard the USS *Gull* (AMS-16). Shore bombardment operations in Korean waters 1952 aboard USS *Harry E. Hubbard* (DD-748).

He was awarded the Meritorious Service Medal, Navy Commendation Medal, China Service Medal, American Theater Medal (WWII), Victory Medal (WWII), National Defense Medal, Korean Service Medal (three Bronze Stars), Naval Reserve Medal (two Hour Glasses) and the United Nations Medal.

He retired July 1, 1985 as CAPT, engineering duty, USNR. His memorable experience: the first day he reported to a Navy ship, the USS *Weiss* (APD-135), was an experience since he had been an engineer on merchant ships prior to reporting. The difference in operation was a memorable experience. The only common ground was the machinery. Establishing and implementing, and being the first CO of the "Naval Reserve Engineering Co. (1-1)" at the Boston Naval Shipyard, which was the beginning of the Naval Shipyard Reserve Units and the "Advanced Ship Repair Base (YR-26)" at Newport, RI.

Retired from USNR after 38 years of service July 1, 1985. He married Fern and they have three children: Joel, Amy and Julie; their grandchildren are John, Andrew, Natalie, Ingrid, Rebecca, Wesley, Elizabeth, Bridget and Brianna. Following active Navy duty during the Korean conflict he was employed by the Navy Department for 30 years, retiring in 1983.

CURTISS "CURT" CUMMINGS, CAPT, born Jan. 6, 1916, Brookline, MA. Education: HS, 1935; AB degree, Princeton, 1939; MD degree, Columbia University College of Physicians and Surgeons, 1943; National War College, Georgetown.

Joined the service in 1942; stationed at USNH; ATB Solomons, BUMED; served with LCI(M) Group 17; admin. officer NAVMED Co. 3-4; CO NR Hosp. Corps. Div. 3-1; director, NAVAS Div. BUMED.

Received campaign ribbons and medals for Leyte, Lingayen, Iwo Jima, Okinawa. His memorable experiences include seeing the flag raised on Mt. Suribachi, Iwo Jima and being appointed admiral in great Navy state of Nebraska (Omaha). Retired in 1970 (60% disabled) as CAPT MC USNR.

He lives with his wife Norma in Rockville, MD.

FRANCIS "FRANK" CUMMINGS, CDR, born July 23, 1914 in Balto, MD, received a BA from Loyola College, Balto. He joined the Naval Reserves, 1932-36; active 1940-46; N.R. 1947-74.

His military locations and stations: DD-240, DD-663, DE-424 SCTC, Miami, Staff. Served with N. Atlantic Convoy 1942-42. His battles and campaigns: African invasion 1942; various actions in Philippines, Okinawa, S. Pacific. He was awarded various theater medals.

Cummings achieved the rank of CDR and retired June

23, 1974. His memorable experiences: DD-240 sunk Gulf Mexico April 1942; various actions; Philippines 1945; Lingayen Gulf; Lubangi.

He is married to Marian P. and they have two children, Carl F. (CDR Chaplain) and Marga C. Hill. Their grandchildren are Jennifer, Brian, Patrick and Melissa Hill.

JAMES SYLVESTER "JIM" CUNNINGHAM, CPO, born May 4, 1925 in Ashland, MO, is a graduate of Ashland High School, Ashland, MO. He joined the service September 1943, discharged in November 1958 from USNR.

Recruit training at US NTS, Farragut, ID and also attended Class A Storekeeper School. Disbursing School at US Naval Base, Bremerton, WA; US NRS, Treasure Island, San Francisco, CA; USS *Uvalde* (AKA-88), plank owner in August 1944; US Naval Mobile Construction Bn. No. 6, plank owner in May 1951.

Served with USS *Uvalde* (AKA-88) and Mobile Construction Bn. 6. Fought campaigns in the Philippines and Okinawa in WWII. He was awarded the American Campaign Medal, National Defense, WWII Victory, Asiatic-Pacific w/2 Battle Stars, Philippine Liberation w/star, China War Memorial Medal and Navy Reserve.

He achieved the rank of CPO - DKC, discharged May 1946, WWII, August 1952, Korean War. His memorable experiences: present at the USS *Mount Hood* AE-11 explosion on Nov. 10, 1944 in Seeadler Harbor, Naval Base, Manus, Admiralty Islands in which 350 Navy men were killed.

He married Alva Jane Sappington and has daughter, Laurel Jane Cunningham Fuemmeler; and grandchild, Vivaca Jane Fuemmeler. Employed as vice-president and secretary treasurer of MFA Oil Co., retiring Oct. 1, 1990.

JOHN F. CURRAN, CAPT, born in Honesdale, PA, a small town in northeastern Pennsylvania, on June 2, 1926. He was class president in the 9th, 10th, and 11 grades, and president of the student council in 12th grade. He won varsity letters in soccer and basketball.

He began his naval career in July 1944 in the combat aircrew program and qualified as an aviation radioman, soon winning his wings as an aircrewman. Subsequently, he entered the Navy V-5 aviation cadet program. In 1947 Curran was released from active duty to attend the University of Pennsylvania where he received a BS in economics and was commissioned an ENS USNR in 1949. He then pursued studies in mechanical engineering at Drexel University.

After qualifying as an air intelligence officer at NAS Willow Grove, PA, he had a number of assignments: as department head in training, support and plans. He served as a component director, Air Wing Staff 936; XO and CO of Naval Air Intelligence Reserve Unit 931; CO Area Analysis Unit 504; Attack System Unit 204, and Intelligence Volunteer Unit 104.

Assuming command as Reserve Intelligence Area CDR of four reserve intelligence units, Area 16 USNR Intelligence Program. He retired from the USNR in December of 1980 after 33 years of service. His service medals are WWII Victory, American Campaign, USNR and Armed Forces Reserve. Memorable experiences: selection to captain, appointment as RIAC 16 NRIP, establishment of the NRIP in 1974.

The Curran family consists of his wife Adalyn, daughter Hollis, son John and four grandchildren. He is a member of NRA, TROA, and he is active in the naval retirees activities program. His hobbies include woodworking and model building. He is active in church and community affairs.

DALE CURTISS, born Feb. 23, 1954 in Del Rio, TX and has been the CO of Reserve units serving the Naval Research Laboratory and a Shore Intermediate Maintenance Activity. Other assignment included Commander-in-Chief, Atlantic Fleet; Norfolk Naval Shipyard; and the USS *Perserver* (ARS8).

He graduated from the US Merchant Marine Academy in 1977 with two BS degrees and has a masters of engineering administration with a concentration in managing of research and development from George Washington University.

His civilian experience includes sailing on merchant ships as an engineering officer and as a deck officer. He has 18 years of shipbuilding experience at Newport News Shipbuilding. He has supervised all propulsion plant and electrical plant testing of an aircraft carrier during its overhaul. He is certified by NAVSEA as a combat systems production manager.

He is married to the former Carole Searcy of Long Island, NY and they have three children. His main priority is serving God.

DEAN A. CURTISS, CAPT, born Jan. 30, 1934, in Syracuse, NY, received a BA in physics, MSEE and MBA in marketing. He was drafted March 23, 1956. Boot camp, OCS later in 1956.

His military locations and stations: NAESU, Washington, DC, active duty. Served with various Naval Air Systems Command Reserve Units, BARTUS, WEPTUS, four years active duty and 24 years Reserve duty.

He achieved the CAPT and retired in 1984 after 28 years. His memorable experiences: called to do (2) separate ACDUTRAS to Turkey some 40 days over a two year period and helped to establish first Turkish naval aircraft maintenance program for their first naval aircraft - S2Fs.

He has been married to Theresa H. for 40 years and they have children David, William and Deanna and grandchildren Jennifer, Megan and Matthew Curtiss. Employed as field engineer, engineer, program manager, marketing manager. Retired after 32 years with Bendix and Allied Signal in 1992.

THEODORE D. "TED" DABAGH, born Aug. 8, 1905 in Berkeley, CA, received a BA degree from the University of California at Berkeley, 1931.

He joined the service September 1927, midshipman, University of California, Naval ROTC, graduated June 1931, ENS USNR. He was stationed mostly in the Central Pacific area. Served with the local defense forces; NAVAIR STA, Kaneohe, French Frigate Shoals, HWN Sea Frontier. He fought in the Pearl Harbor attack.

Dabagh was awarded the Asiatic-Pacific Medal, American Defense, WWII Victory, Pearl Harbor Survivor Commendation Medal and the Naval Reserve Medal.

Debagh achieved the rank of CAPT. His memorable experiences: Pearl Harbor attack; OinC at French Shoals; meeting with ADM Nemitz after 21 years and at his request in April 1944; National Naval Reserve Policy Board, 1954, 1957, 1958, 1959.

He is married to Eleanor Jean Lynch and is employed in managerial and administrative positions. Presently on pay status: USNR.

GERALD G. "GERRY" DACKERT, CAPT, born April 8, 1932 in Bryn Mawr, PA, received a BS in economics from Villanova in 1953. He joined the NROTC in 1949 and was commissioned in June 1953. He was stationed aboard USS *Shannon* (DM-25).

Dackert achieved the rank of CAPT and retired June 1955. He was awarded the National Defense Medal and Naval Reserve Medal.

He is married to Patricia "Trish" M. and they have three children: Kristin, Jennifer and Marshall. He retired October 1977 as president and owner of Marshall Associates, Inc.

R. JACK "BLACK JACK" DALBEY, CDR, born Dec. 25, 1931 in Rochester, NY, received a BA from Williams College in 1954, Williamstown, MA.

He joined the service August 1954. His military locations and stations: NAS Mirarmar; NAS Floyd Bennet; NAS Los Alamitos; NS Coronado; USS *Constellation;* USS *Ranger.* He served with VA-126; VA-113; VA-773; VA-831.

Dalbey was awarded the Meritorious Unit Commendation, National Defense Medal, Naval Reserve, Sharpshooter Ribbon. He achieved the rank of CDR and was discharged May 1959.

His memorable experiences: 2100 jet flight hours, F7U-3 Cutlass; F9F-8 Cougar; F2H Banshee; A4B Sky Hawk; RDAC Council Navy Recruiting, LA; TAC briefing off. Coronado, CA; member VA-126 Air Ground Champs - 1958 Air Weapons meet - ELC, CA.

He married Nancy Jean in March 1956. Their children are Darralyn, Robyn, John, James and Jeffrey; and grandchildren are Robert, Matt, Conor, Jordan, Jackson and Blake. He held the position of vice-president E.F. Hutton and Co., Newport Beach, CA, retiring June 1982.

RICK "WHIZZER" DALE, CDR, born Nov. 13, 1955, Nashville, TN. Earned BS degree (cum laude) in 1977 at Texas A&M University. Joined the military via OCS, Sept. 7, 1977, in Quantico, VA, and was commissioned Nov. 11, 1977, 2LT, USMC.

Active duty from Sept. 6, 1977-Aug. 15, 1985. He received interservice transfer to USNR in August 1987. Duty stations include 1977-78 Quantico, VA; 1978-79 NAS Pensacola and NAS Whiting for primary T-34 training with VT-2; 1979-80 NAS Kingsville, TX as student naval aviator in VT-23 and VT-21.

Designated naval aviator Sept. 5, 1980; September 1980-August 1982 A-4 instructor pilot and accumulated over 1000 A-4 hours; August 1982-August 1985 MCAS El Toro, CA, member of 1st Fleet F-18 Sqdn.; 1987-88 VFC-13 NAS Miramar A-4E/F advisory pilot; 1988-1990 VF-201 NAS Dallas, F-14A pilot; 1991-94 NAS Key West 0170; 1994-present NR COMSEVENTHFLT 111.

USMC career was from 1977-85 and received honorable discharge in 1985. Served in USNR from 1987 to present. His awards and medals include two Navy Commendation Medals and one Navy Achievement Medal. His current rank is commander.

Memorable experiences: Member of first fleet F-18 Sqdn.; participated in three Red Flag exercises; 39th aviator to trap aboard a carrier in the Hornet. USS Vinson (CVN-70); Adversary pilot qualified in A-4 E/F at VFC-13, NAS Miramar; transitioned to the F-14 at VF-201 NAS Dallas.

Married Kate Ahern in September 1992, and they have one daughter, Caroline. Civilian employment: 747-400 First Office at Northwest Airlines (1985-present).

PAUL PATRICK DALEY, CAPT, born July 10, 1941 in Boston, MA, received an AB in Boston College in 1963; MBA and JD, Harvard Business School and Harvard Law School 1973, is a senior partner at Boston's Hale and

Dorr, where he specializes in commercial, bankruptcy and corporate law. He is admitted to practice before the Supreme Court, the Massachusetts and New York Courts, the First and Fifth Circuit Courts of Appeals. He is an inaugural fellow of the American College of Bankruptcy, an active member of the Business Bankruptcy Committees of the American, Massachusetts and Boston Bar Associations, a former director of the American Sail Training Association, Chairman of the Board of Trustees of the Massachusetts School of Professional Psychology and a Trustee of the Naval War College Foundation.

Between college and law school, he was attached to Fighter Sqdn. 96 aboard USS *Enterprise,* where he flew 212 combat missions in F4Bs mostly over North Vietnam for which he received the Distinguished Flying Cross, 16 Air Medals, the Navy Commendation Medal and the Vietnamese Air Gallantry Cross. He then taught Naval History at Yale University as assistant professor.

He continued his service at NAS South Weymouth, where he served in a number of ASW squadrons and units. He attended the Junior and Senior Reserve Officer Courses at the Naval War College and the Defense Strategy Course at the National War College. He has commanded the FASO Training Gp., NAS Oceana, NR-USS *John F. Kennedy* (CV-67), NS Rota and VTU-9191. He earned his parachutist wings after completing the Army Abn. course at Ft. Benning in 1981. He retired from the USN in 1995 after 31 years of service.

He and his wife, Dr. Barbara Daley, and their son, Patrick, live in Waban, MA.

CHARLES W. DALY, CDR, born July 19, 1927 in Lowell, MA, attended Boston College and Boston College High. He joined the service May 9, 1945, serving with VB-4, VPB-114, VP-911 and HS-911.

Daly served in the American Theatre, WWII, Atlantic Patrol. His awards include the American Theatre Medal, WWII Victory Medal, National Service Medal, Naval Reserve Medal and Armed Service Medal.

He achieved the rank of CDR and was discharged July 20, 1973. His memorable experiences: plane captain to ADM C.T. Joy's aircraft. Aircrewman PBY's, PB4Y-2s; Navy Mustang officer.

Married Ruth and they have 10 children and 12 grandchildren, retired from advertising sales July 19, 1973.

JAMES E. "JIM" DALY, CEC, born Dec. 18, 1919 in Duluth, MN, received a BS degree from the University of Minnesota. Attended Graduate School, UMD, San Diego State; University of Texas, Austin.

He joined the service Sept. 14, 1937. His military locations and stations: USS *Wright* (AV1) 1938-41; USS *Vulcan* (AR-5) 1941-43; USS *Eurale* (AS-27) 1943-44; USS *Grundy* (APA-111) 1944-46.

His campaigns: New Guinea, Philippines, Okinawa. He was awarded the Good Conduct, American Defense (Fleet) European, Pacific, Philippine Defense, China Service, WWII Victory Medal. He achieved the rank of CWO-4 (carpenter). Later changed to CEC, discharged Feb. 12, 1946.

His memorable experiences: aboard *Vulcan,* Iceland, with USS *Grundy* (for repair) when Pearl Harbor was hit. Relieved USS *Holland* in New Guinea. Made chief warrant on Eurale. Commissioned *Grundy.* Built sub rest camp on Manus. Returned to US from China, South Korea as "Pox Ship"

His wife, the former Margaret Louise Larsen, is deceased. His children are Roger James (1944) and Judith Lorene (1947); and grandchildren are Rick and David Daly (in Tucson) and Ronald Dent (in California). Employed as teacher, principal, superintendent; foreman, superintendent and construction planner and estimator (Miramar/professor), retiring Dec. 17, 1979.

ARNO EDWIN "ARNIE" DAMEROW, CDR, born Sept. 1, 1926 in Austin, MN, received a BS from Franklin and Marshall College and DDS from the University of Minnesota.

He joined the service July 1, 1944. His military locations and stations: NTC San Diego, CA; Camp Pendleton, Oceanside, CA.

Damerow achieved the rank of CDR and was discharged June 30, 1973. His memorable experiences: CO, Naval Reserve Dental Co. 9-17, Minneapolis, MN, 1971-73.

He married Nancy Bryant in 1949. They have four children: Judy, Martha, Bryan and Mary; and 10 grandchildren: Rachel, Brent, Michael, Carl, Christopher, Marin, David, Victoria, Elizabeth and Jordan. Retired from dental practice in 1994.

PETER WHITON "PETE" DAMISCH, CDR, born Feb. 27, 1953, Chicago, IL. Earned BS degree, aerospace engineering, USNA, Annapolis, MD; MSE, nuclear engineering, USN Nuclear Power School, Orlando, FL; USN Nuclear Prototype Training Unit, West Milton, NY; MM/MBA, Kellogg Graduate School of Management, Northwestern University, Evanston, IL.

Joined the service June 30, 1971, and served with JICPAC 1187, Dublin, CA; JICPAC 0620, Ogden, UT; JICPAC 0620, Draper UT; DEFINTERROG 0118, Omaha, NE; DIAHQ 0878, Minneapolis, MN; NISCPROD 0820, Monterey, CA; USS *Texas* (CGN-39), Norfolk, VA; USS *Nimitz* (CVN-68) Norfolk, VA; Navy Congressional Aide, Office of Legislative Affairs, Washington, DC (1975).

Achieved rank of CDR. His awards and medals include the Navy Achievement Medal, Meritorious Unit Commendation, National Defense Service Medal w/star, Sea Service Deployment Ribbon w/star, Joint Meritorious Unit Award, Navy Expeditionary Medal, Rifle and Pistol Expert.

He lives with his wife, Kathleen in Del Rey, CA. Civilian employment with FMC Corp., 1982-91; Arrow Dynamics, Inc., 1991-94; Pearce Systems International, Inc., 1994-96; Golf One, Inc., 1995-96; Compass Rose, Inc., Senior Management Consulting, owner, founder, president and CEO, 1995-present, Marina Del Rey, CA.

EDWARD W. D'ANGELO, CWO4, born May 17, 1930 in New York City, NY. He joined the service April 9, 1947 and served with VA-86E TBM Sqdn., stationed at Sub Base New London, CT, 1953-54. He also served with USS ARD7, NMCB 19, 13-1954.

D'Angelo achieved the rank of CWO4 and received the National Defense Medal w/Star, Armed Forces Reserve w/Hour Glass, Naval Reserve Meritorious, Expert Rifle w/ Silver "E" and Expert Pistol w/ Silver E. His memorable experiences: CO NMCB19 Det. 3-15, 1968-72; commissioning his son Edward upon graduation from college in 1986 and pinning on his Naval Aviator Wings in 1992.

He is married to Dolores and they have children: Michael, Catherine, Dolores, Edward and James; and grandchildren: Robert, Michael, Philip, Maria, Bryon, Christina, Nicholas and Kelli. He retired as general foreman, May 1, 1990, from various elect.. construction jobs in New York City, IBEW Local #3.

KIRK J. DANIELS, born May 29, 1953 in Tipton, IN, received a BS in oceanography from the US Naval Academy in 1971. He joined the service June 1971 and served with HSL-30, HSL-32, HS-94, HS-84, HSL-84, NAVAIRSYSCOM.

Daniels achieved the rank of LCDR and retired September 1994. His memorable experiences: wardroom friendships.

Civilian employment: vice-president, Planned Inv. Co. Inc.; asst. vice-president, new business Kaman Corp.; test pilot Kaman Corp. He married Wendy J. and they have a daughter Alexandra.

THOMAS "DAN" DANIELS, CAPT, a native of Clarksville, TX, is a graduate of Paris Junior College and Texas A&I at Corpus Christi obtaining a BA and BS degree. He began his naval career in 1963 as a member of the USNR Electronics Div. 8-12 which was commanded by CAPT Paul H. Daniels his father.

During the Vietnam War in 1965-66 he served on board the USS *Gunston Hall* (LSD-5) and the USS *Magoffin* (APA-199) obtaining the rank of YN3. Following active duty he reaffiliated with USNR Surface Div. 8-12, and was commissioned ENS in 1975 as a 1655 public affairs officer, assigned to the Office of Information Detachment 310 Houston, TX. He rose to the rank of CDR and XO when recalled to Desert Shield/Desert Storm and served as the director of electronic internal information in Reyahd S.A. Upon return to the USNR he was promoted to CAPT in 1996 and became the CO of VTU 1004 Corpus Christi, TX.

Daniels has been awarded the Joint Service Commendation, the Navy Commendation, plus numerous other medals.

He is the owner of Cobra Broadcasting and Production Co. The Daniels family consists of son LT Rodney Daniels (3rd generation naval officer) serving on board the USS *Yorktown* (CG-52) and son BMSN Bradly Daniels of Chicago, IL.

LARRY R. DANIELSON, CAPT, born in North Bend, OR, he grew up in Los Angeles and received an NROTC commission upon graduation from UCLA in 1962. He served on active duty in USS *Samuel N. Moore* (DD-747) and was with the *Kitty Hawk* task group that established "Yankee Station" in the Gulf of Tonkin in May 1964.

Danielson's USNR assignments included service in seven NRT/NRF ships and in numerous staff billets including Fleet Training Gp., COMNAVSURFPAC and CINCPACFLT units. He was promoted to CAPT in 1984 and retired in 1994.

He received his law degree from the University of La Verne in 1969 and has practiced law for the California Department of Transportation since then, currently as assistant chief counsel in the Los Angeles regional legal office.

An NRA member since 1975 and a life member since 1983, Danielson has served as a chapter president and as 11th District President and on numerous national committees. He resides with his wife, Brenda, of Stamford, CT, a life member of Anchors, in Seal Beach, CA. They have three grown children and two grandchildren.

F. ROBERT "BOB" DANNI, CAPT, born Oct. 26, 1939 in Pittsburgh, PA. Attended the US Merchant Marine Academy and received a BS in engineering, US Merchant Marine Officers License, 3rd Asst. Engr. Licensed PE, New York State; Licensed Chief Stationary Engr., City of Buffalo, NY.

He was commissioned July 31, 1961. His military locations and stations: US Merchant Marine, American Export Lines, SS *Export Ambassador,* SS *Export Banner.* His units served with: NR Surface Div. 3 - 60 (M); NR Gp. Command 3-17; NR FLETREGRU 2803; NR Sima Chasn, Buffalo, NY; NR WPNSTA Earle, Buffalo,

NY; NR VTU-0503; NR Readiness Unit 2, Buffalo, NY; NY Naval Militia, Region III CDR.

Danni was awarded the Navy Commendation, National Defense Medal, Armed Forces Reserve Medal, New York State Conspicuous Service Medal, New York State Military Commendation and New York State Long and Faithful Service Medal.

He achieved the rank of CAPT and his memorable experiences: crossing North Atlantic in January; grain shipments to Yugoslavia; docking in Beirut, LEB while under fire.

Married Kathleen and children are Mark, Traci, Scott, Todd and Adam; and stepchildren: Kristen, Kari and Derek Setter. He is employed as assistant commissioner of Building in Amherst, NY. He retired from the USNR Oct. 1, 1991.

GEORGE A. DANSKIN, CDR, born in Atlanta, GA in 1938 and grew up Cranford, NJ. He graduated from the University of Montana in 1964 and was commissioned at Pensacola in January 1965. Thereafter, he served three years in USS *Franklin D. Roosevelt* (CVA-42) during which he completed a deployment to southeast Asia in 1966-67.

Duties during this period included DO, tactical signals officer, 6th Fleet courier coordinator, and cryptological security. One of the more memorable events of this assignment was the port call at Cape Town, South Africa where shore leave was canceled as a means of expressing the US disagreement with the racial policies of the South African government. As an alternative, the officers and men of the USS *Franklin D. Roosevelt* hosted 150,000 visitors of every race, creed, and color during the three day stopover.

From 1968-88 he held billets in a number of USNR Intelligence Units including FIRSTEURLANT at Floyd Bennett Field, NY and MARDEZLANT located on Governors Island, NY. Billets included training, geo-political analysis, XO and deputy area commander.

Danskin had a 28 year career with the New York State Department of Environmental Conservation holding a number of management positions at both regional and headquarters levels. In 1982, selected as one of New York State's "Top 20" managers, he attended a nine-month graduate level development program conducted at the Urban Academy for Public Management in New York City.

Prior to retirement in 1996 he was the chief permit administrator for the Department of Environmental Conservation. Since then, he has affiliated with a real estate firm in New Palz, NY where he lives with his wife, Victoria. He has two grown daughters, Susan and Melissa. He passionately pursues fly fishing for trout throughout the East and in Montana.

LEE E. DARLEY, LCDR, born Aug. 23, 1925 in Ft. Worth, TX, received a BA from North Texas State Teachers College. He joined the service July 1943.

His military locations and stations: Ft. Schyler, NY; Hollywood Beach, FL; Philippine Islands. He served with LCT-245, LCT-922, APC-44, VA-703, VF-210, LCT FLOT 8, APC-43, NAS Dallas, TX.

Darley participated in the Philippine Liberation. He achieved the rank of LCDR and retired Aug. 1, 1964. His memorable experiences: Subic Bay fire works (starshells, tracer 20mm and 40mm) upon notice of Japan surrender;

He is married to Wanda and they have children, Nancy Kessler and Sandra Hardin, and grandchildren: Dan and Amy Hardin, Jeff and Eric Kessler. Employed as purchasing manager with Texas Instruments.

JOSEPH D. DAVENPORT, CAPT, born on a Nebraska farm, he was commissioned ENS and designated a naval aviator in November 1954 following graduation from the University of Nebraska.

Active duty included 6th Fleet deployments in VF-32 aboard USS *Ticonderoga* (CVA-14) and in VAH-5 aboard USS *Forrestal* (CVA-59).

Reserve service includes command of VP-RTU-94, NAVAIRPAC-509 and VTU-8282 with consulting tours at major staffs, BUPERS and in the Pentagon. Davenport wears the Meritorious Service and the Navy Commendation Medals.

His life insurance career as agent, training director and general agent for major mutual life insurance companies culminated in formation of his own firm in Houston. Holder of the FLMI, CLU and ChFC designations and graduate business degrees from New York University (MBA) and The American College (MS), he is past Regional Vice-President of the American CLU/ChFC and past national president of the UN-Kearney Alumni Association.

A life member of NRA, he has been president of the Space Center Chapter and chairman of the National Nominations Committee.

He and his wife Elizabeth reside in Houston near their daughters Lisa and Michele. He has been self-employed in life insurance sales, sales management and consulting since 1959.

FRED DAVIDSON III, COL, life member and a native of Indianapolis, attended Central State University, Howard University School of Law and received the LHD degree from the National University in San Diego. He is also an associate alumni of the US Naval Academy and a life member of the New York Naval Militia Association.

He began his naval career in 1963 at the USMC Officer's Candidate School in Quantico. Upon his commissioning, he reported to Camp Lejeune for training with continuing orders to WESTPAC.

While in Vietnam, he was assigned to the First Marine Aircraft Wing. After the completion of his various logistical and legal assignments at the group level, he was again assigned to Quantico. Davidson was released from active duty in December 1969 and immediately affiliated with various OMCR units. Reserve assignments included: XO, 4th Communications Support Co. (FMF), USMC advisor NAVRESRESCOM Region 13, MTU HQMC, training officer, Congressional Reserve Unit, and served as the State Emergency Preparedness Officer, USMC representative, Office of the Governor/State Adjutant General, State of Indiana. In 1991, he was recalled to active duty for Operation Desert Storm and served as XO of RTB-7. In July 1993, he was retired from the USMC.

His military education includes various staff schools with both LFTC (LANT) and (PAC), the Amphibious Warfare School, Command and Staff College and the Armed Forces Staff College.

As a civilian, Secretary Davidson held several senior positions in private industry, was elected to public office and served in the Administration of Presidents Kennedy and Johnson as a White House intern and President Reagan as the Deputy Assistant Navy Secretary (RA). In 1985, he was granted knighthood in Oslo, Norway. Additionally, he was awarded the Distinguished Civilian Service Award, SECNAV, Meritorious Public Service Award USCG, and received a letter of appreciation, Office of the Secretary of Defense RFPB. Finally, Secretary Davidson was elected as trustee of the US Naval Academy Foundation.

CHARLES G. "CHARLIE" DAVIS, born Jan. 16, 1914 in Albuquerque, NM, attended the University of NM and received in BS in mechanical engineering. He also attended Navy Diesel School.

He joined the service Dec. 24, 1943. His military locations and stations: University of Arizona, Brooklyn Navy Yard and Charleston Navy Yard. He served with LCI Flotilla 7, South Pacific. Fought in the American and Philippine campaigns.

Davis achieved the rank of LCDR, USNR and was discharged Jan. 16, 1974. His memorable experience was after the action in the South Pacific and while on the long trip returning to the States. In the accompaniment of other LCIs they encountered stormy seas and one of them nearly broke in two. There was a quick transfer of a welder and after a night of welding, the ship was saved.

He is a widower and has children: Lynn L., Bruce G. and Ford G., and grandchildren: Desiree, Brian and Art. He is self-employed.

DR. HOWLAND S. DAVIS, CDR, born Nov. 21, 1913 in Baltimore, MD, attended three years at the University of Virginia. He joined the service Aug. 15, 1939 at NAS Anacostia, DC.

His military locations and stations: AvCad, Pensacola; Navy Wings 1940; Iceland through Brazil. Served with VP-55, VP-74; NAS Banana River; Arctic-Sub N&S Atlantic. Participated in some antisub action all in PBM type.

Davis received the Distinguished Flying Cross for submarine sinking action in S. Atlantic. He achieved the rank of CDR, USNR and was discharged in 1960. He flew for a total of 18,000 hours military and civilian.

He married Frances and they have two children, Frances and Catharine; and grandchildren: Eric, Ashley and Jason. Employed as corp. pilot for 16 years, then Bendix/NASA for 15 years.

JOSEPH R. "JOE" DAVIS, JR., CDR, born Jan. 24, 1922 in Dolores (extinct) Webb County, TX, attended St. Edwards University and the University of Texas (both in Austin, TX).

He joined the service on this 20th birthday, Jan. 24, 1942 and joined the USNR after war. His military locations and stations: USS *Kalinin Bay* (CVE-68), ComAirPac, NAS, Daytona Beach, FL and NAS Banana River, FL.

He was awarded the American Campaign, Asiatic-Pacific Campaign w/4 stars, Philippine Liberation w/star, Purple Heart, Presidential Unit Citation, and WWII Victory Medal.

Davis achieved the rank of CDR and was discharged from active duty March 6, 1945 and from the USNR, July 1, 1964. His memorable experiences: Battle of Leyte Gulf, Palau, Marshal and Marianas Campaigns.

He married Margaret "Meg" June 4, 1948. They have children: Megan Porter, Linnean Gorelick, Joseph R. III (deceased) and Andrew; and grandchildren, Darwin and Daniel Porter. Employed with Ford Motor International, resident marketing manager in Cuba, Peru, Kenya, Spain and Puerto Rico, retiring November 1981.

L. DOW DAVIS, CAPT, Naval Reserve Duty: 1966-67, Navigator, Military Airlift Cmd.; 1967-69, Fuels and Air Admin. officer, USS *Ticonderoga* (CVA-14); 1969-70, navigator, NAF Washington, DC; 1970-72, division officer for Line Div., Maint. Dept., VR-52; 1972-76, aircrew training officer, VR-52; 1976-78, personnel officer, VR-52; 1970-80, legal/admin. officer, VR-52, Detachment, Washington, DC; 1980-83, legal officer, NRS, Keflavik. In 1982 graduated with honors from Navy Reserve Officer's Justice School; 1982-86, executive administrator for Secretary of Navy's National Naval Reserve Police Board and received Meritorious Service Medal; 1983-85, special asst. for Mobilization and Personnel to Deputy Asst. Secretary of the Navy for Reserve Affairs.

Commanding and XO, 1985-88, of Joint Staff Mobilization Unit which advised the asst. Secretary of Defense for Reserve Affairs on mobilization and crisis management; 1988-91, XO of NR DCNO 306; 1992-93, Navy Representative at Federal Emergency Management Agency.

Civilian employment as general counsel to Senator Don Nickles of Oklahoma, 1985-87; senior trial attorney, Dept. of Energy, Office of General Counsel, 1987-present.

NORMAN C. "NORM" DAVIS, CDR, born Jan. 4, 1941 in Schnectady, NY, received a bachelor of architecture from the University of Illinois in 1965.

He joined the service Sept. 1, 1966. Active duty stations: Vietnam, Long Beach, CA; Reserve: SF Bay area; Seattle, WA. Active duty units: NMCB 74, NSC Long Beach; Reserve: NARU, EFD, RNMCB 18, REDCOM 22.

Davis fought in Vietnam from June 1968-February 1969 and August 1969-January 1970. He was awarded the National Defense Medal, Vietnam Campaign Medal, Vietnam Armed Forces Unit Citation, Armed Forces Reserve Medal, Navy Commendation w/Combat "V", Combat Action Ribbon, Navy Unit Commendation, Navy Unit Citation, Vietnam Services, Meritorious Unit Commendation Unit "E" Award Best of Type.

He achieved the rank of CDR, CEC and retired Jan. 1, 1990. His memorable experiences: Bravo Co. CDR; Seabees; NMCB74; Operations officer; S-3 RNMCB18.

He is married to Birgit and they have a son Christopher. Employed with Civil Service, USN and Public Health Services as project manager and eng. director, retiring Oct. 1, 1996.

CHARLES POPE "CHARLIE" DAY JR., CDR, born 1932 in NYC and commissioned through Princeton NROTC in 1954, he served aboard USS *Dortch* (DD-670) 1954-56 and received his MA from Vanderbilt in 1959.

Active in USNR Composite Co. 3-6, he transferred to Naval Reserve Intelligence in 1964, where he served in NYC Units until his retirement in July 1, 1980 and retired list with pay Sept. 7, 1992.

Service medals include: WWII European Occupation, National Defense, Navy Expert Rifle and Pistol, and Armed Forces Reserve w/Hourglass.

A life member of the NRA, TROA, ROA, Association, the Naval Order of the United States, the Military Order of Foreign Wars, and the National Rifle Association, he is also a member of Naval Intelligence Professionals, and the Army and Navy Club in Washington, DC.

Day is a writer and lobbyist for pro-gun causes. His uncle, CAPT Bernard Pope Day was Founder of the Naval Reserve Association. Divorced, he has one daughter, Margaret Genevieve Mary Day. Civilian employment as vice-president and director of Joseph P. Day, Inc., Real Estate auctioneers. He is presently a writer and lobbyist for pro-gun issues.

JAMES SPURGEON "JIM" DAY JR., CAPT, born Oct. 4, 1940 in Greenville, NC, received a BS in business administration from Virginia Tech in 1962. He joined the service April 1962. Attended Officer Candidate School, Newport, RI and was commissioned December 1962.

His military locations and stations: USS *Stoddard*

(DD-566); USS *Interdictor* (AGR-13); USS *Mt. Katmai* (AE-16). Participated in TET, 1965 and Vietnam.

Day received the Meritorious Service Medal and Vietnam Service Medal. He achieved the rank of CAPT and discharged to inactive duty January 1966.

He married Betty Forrester and their children are James Michael, Robert Neil and Stuart Morrison. He retired as president and CEO DeJarnette & Beale Insurance Agency, September 1992.

JACK O. DAYTON, LCDR, born Oct. 1, 1954 in Memphis, TN, received an associate of arts, general studies and is working on a BA in business.

He joined the service Oct. 2, 1972. His military locations and stations: Boot camp, RTC, NTC, San Diego; PN Class "A" School, NTC San Diego; USS *Schenectady* (LST-1185); NAS Kingsville, TX; NAVFAC Eleuthera Bahamas; VA-35 NAS Oceana, VA.

Dayton served with NR SUBASE Pearl Det 618, NR PERSMOBTEAM Det 2318, NR WEPSYSENGSTA Port Hueneme Det 618, NR NUCWEPTRAGRUPAC Det 218, NR FF-1062 WHIPPLE Det 6218, NR AE-35 KISKA Det 3519, NR SIMA SDGO Det 2218, NR AD-42 ACADIA Det. 17.

He was awarded the Navy Commendation Medal, Navy Achievement Medal (three awards), Navy Good Conduct (two awards), Meritorious Unit Commendation, Expert Pistol Shot, Navy and Marine Corps Overseas Service Ribbon, Armed Forces Reserve Medal, National Defense (two awards), Battle "E" Award and Navy Expeditionary Medal.

Enlisted as SA, advanced to PN1. Received commission through the LDO program. Currently a LCDR. Still active in reserve component.

He married Mary G. and their children are Amanda, Kenny and Steve. Employed as a management analysis specialist II, Missouri Division of Energy.

REGINA L. LAUDI-DAYTON, LCDR, born April 27, 1952 in Cleveland, OH. Received a BGSU and BA/MA from Heidelberg College. Joined the service Aug. 4, 1980, and served with AIC 1081 Columbus, OH.

Dayton was awarded Armed Forces Reserve Medal and National Defense Medal, Joint Meritorious Unit Award and Navy Marine Corps Achievement Medal and achieved the rank of LCDR. Memorable experience: AT Pearl Harbor, HI.

She married Timothy John Dayton, CDR, USNR, and is employed as social studies teacher, secondary. She retired Nov. 1, 1996.

TIMOTHY JOHN DAYTON, LCDR, born Aug. 8, 1954 in Port Clinton, OH. Attended USMMA/CSU and BS/MS degrees. He joined the service June 2, 1976, and served with CNSRF New Orleans, LA.

Dayton was awarded the Armed Forces Reserve Medal and National Defense Medal. He achieved the rank of CDR. His memorable experiences include AT Pearl Harbor, HI.

He married Regina L. Laudi, LCDR, USNR. Civilian employment as Merchant Marine officer, master.

FLORENCE E. "FLORRIE" DEANER-MCCARTHY, CDR, born Nov. 30, 1944 in Minneapolis, MN, received a BSN, 1967 from the University of Minnesota, MSN, 1985 from the University of Wisconsin, Milwaukee.

She joined the service July 15, 1989. Her military locations and stations: NRC Tampa, FL. Served with NAV Hosp. JAX.

Deaner received the Marksman Medal and National Defense Medal and achieved the rank of CDR. Her memorable experiences: two weeks active duty at Naval Hospital, Naples, Italy, 1995 and FHOTC at Camp Pendleton, 1992 and TNCC at Orlando Regional Medical Center.

She is married to the Rev. Dennis F. McCarthy and is employed as RN, Sr. Community Health Nursing Supervisor.

ANDREW H. "AHDY" DEARTH, CAPT, born March 2, 1937 in Medford, MA, graduated from Malden, MA High School. Attended Springfield College and received a MA and BS degree in physical education. He was certified in physical therapy at Boston University and received an AL and MS in management from Troy State University.

He joined the service May 1, 1961. His military locations and stations: OCS, Newport 1961; Sandia Base-NM; USS Robert H. McCard (DD-822); USS *Albany* (C6-10) Pre-comm; USS *Decatur* (DD-936). Dearth achieved the rank of CAPT. He was discharged from the Drilling Reserve, Oct. 1, 1989 and retired from the USN March 1997.

His memorable experiences: Returning from a Mediterranean cruise in 1964 and riding out a hurricane in the Atlantic; serving as SELRES coordinator for USS *Robert A. Owens* (DD-827) when assigned to Pensacola, FL 1977-79.

He married Alma Ruth and they have four children: Pamela (m. Kirk Nichols), Craig (m. Sandra), Jeffrey and Cindi (m. Scott Crumbo); their grandchildren are Toni Crumbo, Shahara J. Dearth and Shannon Nichols. Employed as hospital administrator with MCA and Quorum Health Resources.

MARY V. DECICCO-WALLMARK, CDR, born Dec. 3, 1953, in Scranton, PA, received DMD (dental degree); UMDNJ, 1979; BS biochemistry from University of Scranton June 1976.

She joined the service in 1976 with the Health Scholarship Program. Her military locations and stations: NDC Orlando, FL 1979-82; USS *Holland* (AS-32), 1982-83; BDC New London, CT 1983-88; NAS JRB Willow Grove, NRBR MED/DEN 0193 1988-present.

DeCicco achieved the rank of CDR. Her memorable experiences: possibly was the first women "Eucharistic Minister" deployed on any ship; USS *Holland*, Indian Ocean, deployment six months April-October 1983; CO, Branch Medical 0193, 1993-95.

She is married to John Wallmark and started her own dental practice in 1994.

PETER W. DEFORTH, CAPT, born Nov. 4, 1941 in Glendive, MT, received a AB degree from Dartmouth College 1963; MA Michigan State University (geography/Russian Studies). Graduated from Naval War College (non-resident command and staff program) 1981.

He joined the NROTC and Dartmouth 1959-63. Commissioned June 1963. His military locations and stations: USS *Essex* (CVS-9) June 1963-June 1965; USS *Agerhold* (DD-826) July 1965-June 1966; USS *Cromwell* (DD-1014) July 1966-August 1967.

Deforth served with NRSD9-102(L) 1967-71; CINCLANTFLT 206 1977-81; PMS TNW 106 1981-89; REDCOMSIX 1989-91 (PLANS). He participated in the Vietnam Campaign (Coastal Patrol and SAR 1966). Awarded the MSM, Navy Commendation (2). He achieved the rank of CAPT (06) in 1984 and was discharged Aug. 27, 1967. His memorable experiences: Service as plans director on REDCOMSIX Staff; CO of NR PMS TNW 106 (Theater Nuclear Warfare Unit supporting PMS-423).

Family includes wife Carol; child, Leigh; stepchildren: Brian, Leon and Austin. He has held various positions with the Federal Government for the past 24 years as intelligence professional. He retired from the Reserves Sept. 1, 1993.

RODMAN DRAKE "ROD" de KAY JR., LT, born June 22, 1923 in New York City, NY, received a MS in industrial engineering from Columbia University, 1952 and a BE mechanical from Yale University, 1947.

Joined the US Army Nov. 19, 1942 and February 1949. His military locations and stations: Hollandia, New Guinea, Manila, PI. Served with US Army Repair Ship *Fitch* (Transportation Corps).

He participated in campaigns in SW Pacific and received Battle Star, Presidential Unit Citation, Good Conduct. Achieved the rank of T-5 Army Seaman; LT limited engineering duty. Discharged from the Army in February 1943 and from the Navy in September 1968. Retired from the US Army Feb. 26, 1946 and from the USNR July 1, 1969.

His memorable experiences: infantry replacement training, Navy Diving School; underwater riveting, driving landing craft, putting pitch in propellers with sledge hammers, guarding drug store after Port Chicago, CA blast.

Married Margaret and has two children, Dr. Joseph Rodman Drake and George Gilbert Timpson de Kay; and grandchildren: Nicholas, James and David de Kay. Employed as associate budget examiner, NYS Div. Budget-Higher Education, retiring March 28, 1991.

CHARLES F. "CHUCK" DELEOT III, CAPT, born Sept. 13, 1945 in New York, NY, received a AB degree from Duke University; MA (2) from Central Michigan University and MS from the University of Southern California.

He joined the service Aug. 11, 1967. He was stationed at Pearl Harbor and served with FICPAC, CINCPACFLT, FOSICPAC 120, CINCPACFLT/GT 120.

Deleot achieved the rank of CAPT (06) and was discharged June 30, 1972. He was awarded the Navy Commendation Medal (2), Navy Unit Commendation and Expert Pistol. Retired Nov. 1, 1990.

His memorable experiences: SERE School 1970 Warner Springs. On duty in FOSIC CINCPACFLT when order given to mine Haiphong Harbor; two tours as unit commanding officer; five years as chief of staff to Reserve Intelligence Area Three.

He is married to Laurie and is currently tech director for C4I, CINCPACFLT N6B.

JAY A. DELOACH, CDR, was born in San Diego, CA on Dec. 9, 1955 and became the third generation to serve in the USN. He graduated from the US Naval Academy in 1978 earning a bachelor's degree in Marine Engineering. He attended the Naval Nuclear Power School and Prototype and completed the Submarine Officers Basic Course.

In 1980, he was assigned to the USS *Patrick Henry* (SSBN 599) Gold Crew, then stationed in Pearl Harbor and advanced based out of Guam where he earned his Submarine Gold Dolphins during three strategic patrols. It should be noted that CDR DeLoach is the great-great-great-great-grandson of the famed Patrick Henry of Virginia. The USS *Patrick Henry* eventually offloaded its ballistic missiles, converted to a slow attack submarine, and changed home port to Bangor, WA while awaiting its turn in the decommissioning schedule. During this time, CDR DeLoach passed his engineer's exam at Naval Reactors and was immediately fleeted up to engineer officer of the USS *Patrick Henry*—one and half months before an Operational Safeguard Exam.

In 1983, he attended the Submarine Officers Advanced Course in Groton, CT before being assigned as the engineer officer of the USS *Lewis and Clark* (SSBN 644) Gold. In 1987, he was assigned to the Naval Sea Systems Command in Washington, DC where he served as project manager for submarine periscopes and later as the branch chief for sonar transducers. He left active duty in 1989 and affiliated with USNR where he was assigned to Naval Maritime Intelligence Center at Suitland, MD conducting analyses of foreign submarine operations. DeLoach has also served in reserve units associated with the CNO Submarine Warfare Office (N87), the staff of CDR, 7th Fleet, Naval Recruiting District San Antonio, and the Inspector General office of the Naval Sea Systems Command. He is currently the CO of the Submarine Sqdn. Six reserve unit out of Reading, PA.

In the civilian sector, he became the Washington area marketing director for a California-based company, Ocean Technologies, Inc. In 1990, he joined an independent executive government agency, the Defense Nuclear Facilities Safety Board, as a senior engineer and program manager where he conducts technical safety evaluations of the Department of Energy's nuclear weapons production facilities. He has also earned a MA in business from Central Michigan University and a Masters of Engineering in Nuclear Engineering from the University of Virginia.

DeLoach has been awarded the Navy Commendation Medal w/Gold Star, Navy Achievement Medal w/2 Gold Stars, Navy Unit Citation, Meritorious Unit Citation w/Bronze Star, Battle E Ribbon w/3 Es, National Defense Service Medal w/Bronze Star, Sea Service Deployment Ribbon w/Bronze Star, Foreign Overseas Deployment Ribbon, Strategic Deterrent Patrol pin w/10 stars and four Gold Wreaths for Recruiting Excellence.

The DeLoach family consists of wife, Jodi; and daughters, Jessica and Jaclyn. They currently reside in Herndon, VA.

CHARLES WILLIAM "PIZZA" del PIZZO JR., born Aug. 18, 1943 in Philadelphia. He joined the service June 4, 1965. His military locations and stations: training command; NAS JAX; NAS Cecil; NAS ATL CV-11. Served with VA-44, VA-66, VA-203, NRCG-0667, NRUSS (CV-60) 0167.

His battles and campaigns include Vietnam, July 1968-December 1968; 85 missions VA-66. He achieved the rank of CAPT and was discharged Oct. 31, 1969. Decorations include the seven Air Medals and 2 NavCom (one w/Combat V).

He was designated a naval aviator in 1967 and flew 85 missions with VA-66 from USS *Intrepid* in 1968. In 1969 he affiliated with VA-2F1 which became VA-203. Lt. del Pizzo suggested the "Blue Dolphin" as the squadron's symbol and designed the squadron patch and plaque. In 1978 he joined the NRCG-0667. In 1984 he became the first USNR to qualify as a CARGRU battle watch captain. During Fleetex 1986 his all Reserve Battle Watch team was on watch when CNO visited COMCARGRU Six's completely integrated staff. Commanding NRCG-0667 in 1987 he awarded 13 Enlisted Surface Warfare Pins to Sailors completing a five year, one of a kind program he instituted. He commanded NRCV-60-0167 through 1991 and was Atlanta Chapter President to his 1993 retirement.

His memorable experiences: plankowner VA-203; designated patch, plaque and mascot; CO of NRCG -0667 when first eight enlisted earned ESWAS pins; first NR designated a carrier group battle watch captain, Atlanta chapter president 1991-93. Retired Sept. 1, 1993.

He is married to Jan and has children: Charles III (Tre) and Lauren; and stepchildren, Mark and Michelle. Employed Eastern Air Lines 1971-89; American International Airways 1995-present.

JUANITO DEL ROSARIO, LCDR, a native of the Philippines, was a graduate of San Diego State University (BSEE), National University (MBA) and West Coast University (MSEE). He began his naval career in 1968 as a steward's mate and was discharged in 1980 as CPO.

He joined the USNR in 1980 and was commissioned ENS in 1983. He has been serving as engineering duty officer.

After COMINELANT, he served in USS *Ticonderoga* (CVS-14), USS *Dixon* (AS-17), USS *Dixie* (AD-14), USS *Racine* and at SSC, NTC, San Diego as an electronics instructor.

Memorable experiences: Two WestPac deployments to Vietnam (1971-1973), and rescue and recovery of Apollo 15, Apollo 16 and Skylab.

Del Rosario, a registered professional engineer, is an electronics engineer at the Naval Research and Development, San Diego. He works in the areas of EMI/EMC, GPS and HF communications systems. He lives in San Diego with wife, Fe, and daughters, Hilda and Eve.

WILLIAM A. "FROG" DEMERS, CDR, born Jan. 21, 1953 in Derry, NH, received a BS in math from WPI and MS in computer science from NPS.

He joined the service Nov. 2, 1976. His military locations and stations: Norfolk, New Orleans and Monterey, CA.

Demers served with VAW-123, VAW-120, VAW-78, NAR NORVA, CNARF, NPS, CNRF, NRISO, NALO. He achieved the rank of CDR and was awarded the NCM (2), NAM (3), MUC (4), Expeditionary Medal, Armed Service, Reserve Service (2). His memorable experiences: working with SELRES at RESFORON. Flying 6-2C Hawkeye (about 2,500 hours).

He is married to Cheryl G. and they have children, Mark and Aimee.

EDWARD J. DEMPSEY, CAPT, received his AB degree from Boston College in 1964 and was commissioned upon completion of Officer Candidate School on Dec. 18, 1964. For the next three years he served aboard *USS Waller* (DD-466), homported at Norfolk, VA. During 1965 and 1966, *Waller* was assigned duties supporting various Atlantic Fleet operations. In late 1986 *Waller* deployed for duty with the 6th Fleet and an independent assignment to CDR, Middle East Force in the Persian Gulf. During this time his primary assignments were as electronics material officer and CO, with the collateral duty of legal officer.

Upon release from active duty in September 1967, CAPT Dempsey joined his wife, Eileen (a Navy nurse stationed at Bethesda, MD), in Washington, DC area and entered the Columbus School of Law at The Catholic University of America from which he received his JD degree in 1970. CAPT Dempsey, a qualified surface warfare officer, spent his first five years of inactive duty aboard ships of the USNR Force at Baltimore, MD, first as CIC officer aboard USS *Mills* (DER-383), then operations officer USS *Allen M. Summer* (DD-692) and USS *Steinaker* (DD-863). In 1975 he became a "plankowner" of OPNAV-04 Component 106, a unit supporting DCNO (Logistics). Service with this unit included several major Command Post Exercises at the Pentagon and the Alternate National Military Commander Center.

In 1978 CAPT Dempsey returned to *Steinaker* as XO of the Selected Reserve Crew (SELRES). He became CO (SELRES coordinator) of that unit in 1980, serving until *Steinaker's* decommissioning in 1982. About that same time, CAPT Dempsey and his family moved to Connecticut where he served in a detachment of VTU 0111 supporting the Naval Submarine School at New London. In 1985 CAPT Dempsey transferred to VTU 0119 (NWC), a unit supporting the War Gaming Department of the Naval War College. He specialized in logistics and participated in various war games, including NATO's "Open Road" and the Global War Game. He was Logistics Department head for the unit at the time of his retirement on Sept. 30, 1991.

In civilian life, CAPT Dempsey, a member of the bar of the District of Columbia and Connecticut, is director, Industrial Relations and Labor Counsel for United Technologies Corporation. He and his wife, Eileen, live in Avon, CT. They have four children: Kristen, Katie, Shelagh and Jim.

DOUGLAS RICHARD "DOUG" DENNY, born Feb. 11, 1947, Cleveland, OH, received a BS from the United States Merchant Marine Academy (1969) JD, The Ohio State University College of Law (1973).

He joined the service Aug. 22, 1965 as midshipman USNR, commissioned June 4, 1969. His military locations and stations: deck officer aboard various US Flag merchant ships. Denny served with NR ACU-2 LCM 6 DET 2605;

RedCom 5 Staff; NR Law Co. 4-20; NR DDG-39; NR PHIB CB-2 DET 605; NR NCSO SW Pacific 105; NR VTU 0505 G.

He achieved the rank of CAPT, USNR. His awards include the National Defense, Armed Forces Reserve and the Vietnam Service Medal. His memorable experiences: Naval War College and National Defense University; Duty at NAB Little Creek, VA.

Married to Barbara and they have two children, Christopher and Matthew. Employed as attorney-partner, Ray Robinson, Carle, Davies and Snyder of Cleveland, OH and Chicago, IL.

JOHN F. DERR, CAPT, born in Chicago, IL, Aug. 23, 1936. He was a graduate of Purdue University in 1958. He attended OCS where he was commissioned an 1105 ENS in May of 1959. He first served on the USS *Lofberg* (DD-759) out of San Diego as 1LT and NAV and later as ASWO *Lofberg* was part of a Hunter Killer Gp. He wrote an enemy SS *Hold Down* Tact Note that was used in the Pacific for two years. He earned the National Defense and Expeditionary Medals. In 1961 LTJG Derr was transferred to the USS *Dewey* (DLG-14) out of Norfolk where he was ASWO, CICO, and assistant OP's officer. He wrote the first doctrine coordinating conventional ASW weapons with nuclear weapons to be approved by the AEC. He left active duty after five years in 1963.

In 1964 Derr joined the Selected Naval Reserve and began and 18 year tour with Naval Reserve Destroyers including Staff Operations officer on 3rd ND Sqdn. Staff, Reserve XO on the USS *Dyess* (DD-880), Reserve CO on *Dyess* and Reserve CO on USS *William R. Rush* (DD-714). He earned the Sea Medal. In 1976 CAPT Derr began his non-sea duty as CO of the Yorktown Naval Weapons Stations Reserve Unit in Great Lakes, CO of the Santa Barbara VTU, and then six years with REDCOM 19 in San Diego where he developed a senior officer think tank, the first Reserve CO School, and REDCOM 19 Strategic Plan.

He completed his active USNR career in 1988 as CODE 3 of REDCOM 19. During the six summers from 1983 until 1988 he served his active duty for training at The Pentagon working for the assistant secretary of Defense for Health Affairs developing strategic plans. He was qualified to stand duty in the OSD Crisis Center. Upon retirement in 1988 after 30 years of USNR duty, he received the Meritorious Service Medal for his work at the Pentagon and for REDCOM 19. In 1996 he was appointed to the NRA Health Affairs Committee.

Derr has been the president of five healthcare companies and held high executive level positions with four healthcare Fortune 500 Corporations. His wife Polly; daughters, Dr. Debbie Derr and Mrs. Jennifer Neel, live in the Westlake Village, CA area.

PAUL R.G. DES JARDINS, CAPT, born in Schenectady, NY in March 1917, is a 1938 graduate of MIT and worked as an engineer with Worthington Pump & Machinery Corp. Commissioned an ensign June 25, 1941, he served as assistant engineering member of the Board of Inspection and Survey from September 1941 to February 1946 conducting operational trials and inspections of 334 vessels of types from PT boats to battleships and carriers.

In January 1950 he became CO of Buships Unit 5-4 Charleston, WV; and was recalled to active duty June 1951 to help establish and operate the USN shipbuilding scheduling activity as material planning superintendent involved with shipbuilding, conversion and repair in the US and NATO countries.

Released from active duty in April 1953, he joined Composite Co. 3-6 Chatham, NJ and served terms as planning officer, XO and CO before requesting placement on the honorary retired list October 1964 until official retirement March 1977.

Training duty included Nuclear Science at Brookhaven National Laboratories 1956; Industrial College of the Armed Forces 1958; the Progress Analysis Group of the Office of the Chief of Naval Operations 1959; and the Naval War College 1961. He made CAPT July 1, 1960.

When not on active duty, he was with Worthington in sales, marketing, engineering, planning, and general management positions until retirement in March 1982. Since 1990 he has lived with his wife Madelyn in McLean, VA at Vinson Hall, a retirement residence founded by the Navy Marine Coast Guard Foundations.

JOSEPH A. DESPRES, CAPT, came to the Naval Officer Candidate School, Newport, in February 1964, with a bachelor's degree in philosophy from Cathedral College, Brooklyn, NY. Upon commissioning as ENS in June, he reported to USS *Interdictor* (AGR-13), Treasure Island, San Francisco, as damage control assistant. A year later, he transferred to USS *Point Defiance* (LSD-31), Long Beach, as repair officer and ballast officer. Ship's duties included ferrying Marines from Camp Pendleton, to various WESTPAC locations during the Vietnam conflict, as well as conducting amphibious landing operations with numerous battle action groups.

Upon release from active duty in June 1967, he joined MTD 3-20, Naval Reserve Center, Huntington, NY. He served with several surface divisions, assuming increasing responsibility in each one. His first command was PERSMOBTEAM 202, at NRC, Huntington; then a tour at NAVRESREDCOMTWO, Scotia, NY, as assistant SAM coordinator led to command of USS *Puget Sound* (AD-38). Promoted to CAPT, he capped his career as CO of Readiness Unit 1002, NRC, Amityville, NY.

Despres' ATs included many shipboard tours, involving special assignments, inspections and operational duties. The years with PERSMOBTEAM saw duty at NAVPERSCEN, Washington, DC. During one AT, he prepared a paper title *Expanding the Peacetime Mission of PERSMOBTEAMS"*, which significantly improved the role of those units in the task of Reserve recall. A tour of Special Active Duty at Naval Inspector General, WA, dealt with a study of the sustainability of the Navy in war. Additional tours included National Security Seminars, as well as multiple ATs the Naval War College, Newport, RI. Despres retired in September 1991 after 27 years of service.

The Despres family consists of wife, the former Kathleen Dowling, and two sons, Michael and David. In civilian life he has been a high school guidance counselor, and adjunct professor at Long Island University. He holds a doctor of education degree from Teachers College, Columbia University, and is the author of two books, *Major Decisions: A Guide to College Majors, and Tech Decisions: 140 Hottest Tech Prep Careers.* He is a life member of the NRA.

VINCENT NICHOLAS DEVITO, CAPT, born June 3, 1930 and died April 23, 1993. He was a past president of the Long Island Chapter of the NRA.

He was raised in Manhattan, NYC and graduated from Brooklyn Technical High School. A 1951 graduate of the US Merchant Marine Academy, he then sailed with Grace Lines as a Marine engineer. In 1953, he requested active duty with the USN and was assigned to the USS *Johnnie Hutchins* as a 1LT and from 1953-55 as its' gunnery officer.

He was employed by Grumman Aerospace corporation for 36 years as a manufacturing engineer. During his career he worked on the Lunar Module and Hydrofoil programs, alignment of the Space Shuttle wings as well as the A6-E upgrade and A6-F programs.

DeVito is survived by his wife of 37 years, Francesca Corso DeVito; son, Steven of Bethpage, NY; daughter, Linda; and three grandchildren: Nicholas, Maria and Vincent Kuchenbuch, all of Floyd, VA.

RONALD P. "RON" DEWALT, CDR, born Sept. 2, 1932 in Marion, IN, received a BSEE from Purdue University in 1954; MEES, Penn State, 1972; NROTC Purdue, 1950-54.

He joined the service May 1954. His military locations and stations: USS *Toledo* (CA-133), 1954-57) repair officer. Served with NR Units - Erie, PA (1958-80), 26 years SAT Service.

Dewalt fought in the Korean War. He was awarded the China Service Medal. Achieved the rank of CDR, USNR (Ret). Retired in 1980 with 26 1/2 years service. His memorable experiences: three cruises to West Pacific (six months each), registered professional engineer (PA) and surveyor (PA) received IEEE Centennial Award (1984) and Erie Outstanding Engineer Award.

He married Adelle and they have children: Paul, James, Janet and John; and grandchildren: Daniella, Tim, Dave, Amanda, Daniel, Ian, Aaron and Chris. Employed with Pennsylvania Electric Co. (31 1/2 years) Div. Engr. manager.

MALACHI DEWS JR., CWO4, born Aug. 31, 1943 in Smith County, Whitehouse, TX, attended Prairie View A&M University, Tyler Junior College and the University of Texas at Tyler, TX.

He joined the service Oct. 25, 1962. His military locations and stations: USS *Chevalier,* USS *Cree;* USS *Preble;* USS *Stanley;* USS *Cook;* USS *Hyman;* USS *Putnam;* USS *Vancouver;* USS *Chowanac;* USS *Carpenter;* USS *McKean;* USS *Quapaw;* USS *Ajax* and USS *Cape Cod* A-4-3.

Dews served in Vietnam and was awarded the Navy Achievement Medal, Good Conduct Medal, Navy Reserve Meritorious Service Medal, Republic of Vietnam Campaign Medal, Armed Forces Expeditionary Medal and the National Defense Service Medal.

He achieved the rank of CWO4 and was discharged Nov. 30, 1966. His memorable experiences: In 1978 while aboard USS *Cree* was accidentally hit by three 500-lb bombs. Two bombs exploded and one did not explode. The bomb squad disarmed the live bomb. The bomb was removed from the ship.

He married Sharon and their children are Selwyn, Katrina and Malachi III. Employed at Kelly-Springfield Tire Company, Tyler, TX. He retired June 1, 1994.

JOHN J. DEYAK, CDR, a native Pennsylvanian, was born Jan. 14, 1927. A Navy combat veteran who saw active service during WWII and the Korean Conflict and a long time Naval Reservist.

He is a graduate of Penn State Class of 1951 and is married to the former Merlyn Lorraine Aiken.

Deyak is life member No. 584 of the NRA.

CANIO J. "JOE" DI CAIRANO, born Sept. 23, 1913 in Staten Island, New York City, received a BA and took four graduate courses.

He joined the service Sept. 28, 1943. His military locations and stations: Indoctrination NYC; officer Commo.

Served with COMSERVICEPAC, PEARL; East Sea Frontier HQ.

Di Cairano was awarded the Victory Medal, Asiatic-Pacific and Naval Reserve Medal. He achieved the rank of CDR, USNR and was discharged June 1946.

His memorable experience: simultaneous French to English for NHTO Conferences, plus consecutive for second conference, 1948 and 1955; special affairs officer, USS *Constellation* (CWA-64).

He married Virginia Nilan and they are now divorced. His children are Gerard and Margaret and grandchildren are Timothy and Meredith. Employed as linguist, editor, US Government. He retired Sept. 23, 1973 (OCT Card).

GENE P. DICKEY, RADM, born Aug. 25, 1933 in Stantonville, TN, received a BS from the University of Tennessee in 1954.

He joined the service Feb. 23, 1955. His military locations and stations: NAS Glynco, Argentia Newfoundland, Yucca Flats, NE, Denver, CO, Taipai, Taiwan. Served with ZP-2, VW11, Taiwan Defense Command, Atomic Energy Commission.

He was awarded the Legion of Merit, Joint Service Commendation Medal, Navy Commendation Medal, National Defense Medal, Armed Forces Reserve. Dickey achieved the rank of RADM. His memorable experiences: establishing the USNR program as an Echelon IV Command.

Married Fanna Loma and they have three children: Gene Jr., Kevin and David; and two grandchildren, Ashlie and Brandon.

Employed as associate administrator, US Department of Agriculture, Regional Administration, US Department of Agriculture CEO PMI-McLaughlin. He retired June 27, 1992.

GLEN L. DICKEY, born Dec. 19, 1936 in Pomona, CA, received a BA from the University of California and a MA from the University of Oregon.

He joined the service June 1959 and served with USS *Rochester,* COMCRUDES FLOT 11. He achieved the rank of CDR and was discharged from the Retired Reserves.

His memorable experiences: USC NROTC Unit, WESTPAC Cruises, ACDUTRA afloat.

Dickey is married to Betty and they have children, David and Darcy. He retired from teaching Dec. 19, 1996.

GARY W. DICKSON, CAPT 06, born May 17, 1949 in Pittsburgh, PA, received a PhD in Zoology (1980); MS Biological Sciences (1976); BS Biological Sciences (1971).

He joined the service July 8, 1971. His military locations and stations: Navy OCS; Norfolk, VA- ASWFORLANT, COMSECONDFLT, CINCLANTFLT. Served with CINCLANTFLT 307, NRL 307 and CNO IAC 206.

Dickson was awarded the Navy Commendation, Navy Unit Commendation, National Defense Service (2), Armed Forces Reserve, Navy Pistol Marksmanship.

Married Barbara J. and they have two children, Ryan and Chris. He is employed by CIBA Crop Protection as head of environmental safety.

HIRAM B. DICKSON, CDR, born Nov. 8, 1920 in Admire, KS, received a BS in Kansas State University and MS Emporia, Kansas University.

He joined the service June 12, 1942. His military locations and stations: Underwater Demolition Team #9, CASU #22, Fleet Marine. Served with Force-Atlantic, Seabees - Camp Endicott, USS *Belknap,* USS *Brooks,* USS *Sands.* His battle and campaigns: Leyte and Lingayen.

Dickson was awarded the Silver Star and others. He achieved the rank of CDR and was retired June 30, 1970. His memorable experiences: VDT #9 boat sunk by Japanese at Leyte Oct. 18, 1994 (where MacArthur landed Oct. 20, 1994); kamikaze hit ship at Lingayen Gulf, January 1945).

He married Maxine E. Marx on Sept. 7, 1945, and their children are Richard (MD), Kenneth (attorney) and Diane Mittenmeyer. Their grandchildren are Inger, Ashley, Erica, Kristin and Douglas (USAF) Dickson; and Kindra and Mark Mittenmayer. Employed with US Civil Service commissioned, investigator-in-charge. Retired June 30, 1970. He is life member of NRA.

JAMES M. DIEHL, CAPT, born in Phoenix Aug. 16, 1944, enlisted in the USNR August 1961. As an ATR3 assigned to VS-661 at NAF Andrews, he received a BS (EE) from George Washington University in 1967, was married to Patricia Kay Dryden from Lexington, KY, and commissioned ENS (by his father, NRA life member CDR Arthur C.V. Diehl (Ret). He was electronics and communications officer on the USS *Warrington* (DD-843) in the Atlantic.

Rejoining the Reserves in 1970, he served on the USS *Dickson* (DD-753) in Boston, USS *Limkin* (MSC-195) and as XO on USS *Direct* and Exultant (MSO 430 and 441), the latter three in Perth Amboy, NJ. Moving to Upper Marlboro, MD, in 1972, he served in staff and command positions in Washington, DC and Yorktown, VA, including command of three units. In 1979, he earned a Juris Doctor with honor from University of Maryland. In 1989, he became Navy liaison officer to Maryland, served on the OPNAV staff, and retired in March 1997, after a tour at Office of Secretary of Defense and 35 1/2 years service. He graduated with Distinction from the US Naval War College in 1988 and was nominated as Outstanding Graduate.

After 20 years as an engineer in communications and aviation, he began a solo law practice and teaches college part-time. His older son Bill (William J, 3-14-69) is a LT, USN (Duke 1991) and his younger son John M. (May 3, 1972) graduated from Frostburg State University in 1996. His grandson, Joseph M. Diehl, born Jan. 30, 1991, lives in Chesapeake, VA. He is an airline transport pilot, flight instructor, and aircraft mechanic (A&P). He was awarded the Joint Service Commendation Medal for planning the June 1991, New York City parade recognizing returning Desert Storm Veterans and the Defense Superior Service Medal for his work while at OSD.

GERARD L. DIETRICH, LCDR, born May 15, 1934 in Philadelphia, PA, he received a BS in physics from St. Joseph's University with graduate studies in nuclear engineering at Pennsylvania State University.

He began his career as an airman recruit 1952 serving with AWS 93 and VS-932 achieving the rate of AE2 and qualifying as an ASW aircrewman. After college graduation in 1961 he left the Reserves until 1978 due to worldwide assignments: Exploration of the North Sea, Gulf of Alaska, Sulu Sea/South China Sea, Algeria, Egypt, Iraq and the Amazon Basin (Brazil and Venezuela), Aurora Studies (Alaska), Volcano Studies (Hawaii), and assignment with the Egyptian Air Force (Russian Aircraft Modifications).

Re-affiliating in 1978 as an AT1 he served with FASOTRAGRULANT and in 1980 was commissioned an ENS through the LDO program. He served with NR NAS Bermuda in various unit roles and later as detachment OIC. Subsequent assignments included: NAS Willow Grove, PMTC Point Magu, NADC, Naval Space Command, and US Space Command. Later assignment to HS-75 included a tour on the USS *Eisenhower,* and then to HSL-94 serving in various squadron roles and as assistant maintenance officer. He was selected as LCDR in 1991 and retired in 1994.

His awards include the Navy Achievement Medal, Naval Reserve Meritorious Service Medal, National Defense Service Medal, Armed Forces Reserve Medal, Naval Reserve Medal and the Battle E.

The Dietrich family includes his wife Jane; daughters: MaryEllen, Anita, Rosanne and Michelle; and granddaughter, Kaitlin. They reside in Centre Square, PA. He is a project manager with Lockheed Martin M&DS in Valley Forge, PA.

RONALD L. DIETRICH, CAPT, born in Kiel, WI, Dec. 5, 1949, graduated from the University of Wisconsin, Madison with a BS degree in physics December 1971. After completing OCS May 1972, he served four years as a instructor at the Naval Nuclear Power School, Mare Island, CA.

From 1976-94 he remained in active reserve, holding various billets including OIC PHIBBASE CORONADO, XO COMNAVFORCARIB, CO NCSO San Juan and XO NCSO Korea.

He lives in the Chicago area with his wife Judy and three children: Jennifer, Jonathan (cadet at AF Academy) and Jessica. He received his MBA from the University of Chicago and is currently the plant manager for Sanford Corporation, a division of Newell.

Dietrich is a life member of NRA, relished his service time and is most grateful for the opportunities to travel and serve his country in Korea, the Indian Ocean and the Pacific.

JOHN DIGIUSEPPE, CAPT, born July 21, 1942 in Plymouth, PA. A Wyoming Valley native, was "piped ashore" for the last time in a military retirement ceremony capping 30 years of Naval Reserve service at the Naval Air Station at Willow Grove.

Son of the late Mr. and Mrs. Orazio DiGiuseppe, of Hunlock Creek, CAPT DiGiuseppe entered active service upon graduation from Wilkes College. He was assigned to "Operation Deep Freeze," the Navy's International Geophysical Year research project in Antarctica. He then earned his wings as a Navy pilot after training in Pensacola, FL.

His active duty career included service in Naples and Athens. In addition to flying with the 6th Fleet in the Mediterranean, he helped to establish the Fleet Support Office in Athens, networking with Greek officials, the American Embassy and various military services for which he was awarded the Navy Commendation Medal.

In 1972, after being separated from active service, CAPT DiGiuseppe joined the USNR at Willow Grove. His assignments, in addition to flying, included Celestial Navigation Instructor, Deputy Inspector General and XO of a Naval Reserve unit which augmented the staff of the Aircraft Carrier *America* during open-seas flight operations.

With two other offices, he established the Keflavik Iceland Reserve Unit that brought the first Willow Grove detachments to that country. His special duty assignments included war games roles with the Naval War College and the Air Force Air University.

He also participated in a National Defense University seminar, served in the Congressional Support Office of the Pentagon and on a national Navy Promotion Board, ad was on the CO Executive Staff at the naval base in Norfolk, VA.

Most recently, he served as XO, then as CO of the Senior Reserve Command at the Willow Grove NAS. In Harrisburg, he is employed by the State Department of Labor and Industry. He has also been involved with various civic, Christian-educational and professional activities. Never married, he lives in Harrisburg, PA. Employed as personnel supervisor with Pennsylvania Dept. of Labor. Retired Sept. 1, 1993.

J.A. "AL" DILLAN JR., born July 3, 1929 in Duluth, MN, received a BS from the University of Minnesota; MM Northwestern University. He has attained: CLU CPCU LUTCF degrees (professional insurance degrees).

He joined the USNR January 1948 and was commissioned in 1951. His military locations and stations: USS *Abbot* (DD-629), 1951-53 (active duty). Served with COMNAVRESMINDIV 13; CO USS *Ruff* MSC(0) 54; IUW and MSC Units. Member SECNAV NR Policy Board 1975-78. Attended Naval War College and CINCLANT Senior Officer College.

Dillan achieved the rank of CAPT, retiring in 1984. His memorable experiences: life member NRA (past president Puget Sound Chapter), ROA, Navy League; past president/director Emeritus of Bremerton Symphony Association; member Masonic Lodge, Scottish Rite, Shrine, Bremerton Area Chamber of Commerce; member and past president Olympic Peninsula Life Underwriters Association.

He is divorced and has four children and 11 grandchildren. Owner/president of Dillan Insurance Agency, Inc.

GREGORY "GREG" BRETT DILLON, CDR, born Nov. 3, 1955 in Joplin, MO, received a BS in business administration from Southwest Missouri State University.

He joined the service March 13, 1978 and was commissioned in 1979 through the Officer Candidate School in Newport, RI. On active duty he attended Surface Warfare Officer School in Coronado, CA and Main Propulsion Assistant School in Newport, RI before being assigned as gunnery officer aboard USS *Henry B. Wilson* (DDG-7). After leaving active duty, he was assigned to NR Mobile Inshore Undersea Warfare Unit 13 as 1LT before transferring to MIUWU 1319 to serve as training officer.

Dillan then served as Mine Countermeasures Officer and Selected Reserve coordinator aboard the minesweeper USS *Pluck* (MSO-464). He next served as chief staff officer of Mine Div. Five-Four, San Diego. He was then assigned to MIUWU 106 where he served as Maintenance/Logistics Department head, Operations Department head and XO. After his tour as XO of MIUWU 106, he was assigned to NR Special Warfare Gp. One DET 119 as administration officer and chief staff officer. Assumed command of Mobile Inshore Undersea Warfare Unit 106 San Diego, Oct. 26, 1996.

Dillon's awards include the Navy Commendation Medal, two Navy Achievement Medals, the Humanitarian Service Medal and several other personal and unit decorations. He resides in Chula Vista with his wife Terri and their three children: Brenna, Brett and Kelcey. In his civilian career, he is the owner and manager of the Prudential California Realty Chula Vista-Eastlake Office.

KEVIN L. "THE DILL" DILLON, received a BA in economics from St. Bernard College in 1964, graduate studies George Washington University 1974-1976.

He joined the service July 7, 1964. His military locations and stations: NAS Norfolk; NAS Patuxent River; Washington, DC. He served with Naval Flight Training Sqdns: VP-56, VP-68, NCC 106, CINCLANTFLT DET-206.

He achieved the rank of CAPT and retired July 7, 1990. His children are Michelle C. (b. 1971) and Kevin E. (b. in 1974). He is employed as CAPT with United Airlines.

JAMES J. DI MAIO, CDR, born March 7, 1918 in Providence, RI, received a BS in 1949 from the University of RI, (mechanical engineering option aeronautics).

He joined the USNR in 1942. His military locations and stations: VR 12, NAS Alameda, VR-5 Seattle; Kodiak, AK. Served with AROU Navy 140 and VR5 & VR12. His battles and campaigns: American Theatre and Asiatic-Pacific and awarded the USNR Medal.

Di Maio achieved the rank of senior CDR, and retired in 1972. His memorable experiences: weekend warrior (USNR) duties as aircraft maintenance officers in VP-912 Sqdn. and FASRON NAS NY.

He is married to Angela and they have a son, James J. Jr. and daughter-in-law, Francesca. Employed DOD (AROU), NAS Quonset Point, senior value engr. Retired from USNR and DOD in 1972.

ROBERT H. DINEGAR, CDR, born Dec. 18, 1921 in New York City, received an AB degree (Cornell University) and AM and PhD from Columbia University.

He joined the service Sept. 1, 1942. His military locations and stations: Aircraft 7th Fleet, Fleet Air Wing 17, VP-ML 63. He served with NRRC 8-9, NAS Banana River, SSC, San Diego.

His awards and medals include the American Theatre, Asiatic-Pacific, WWII Victory, Vietnam Era Service, National Defense and Naval Reserve Medals.

Dinegar achieved the rank of CDR and retired June 30, 1975. His memorable experience: service with 7th Fleet in SW Pacific Theater of War, 1943-45; NESEP Program, San Diego, 1971-74.

He married Elaine L. Coleman and they have children: Janice, Barbara and Robert; and grandchildren: Nicolas, Anthony, Robert and Amber. Employed at Los Alamos National Laboratory 1950-87, retiring May 1, 1987. Episcopal Priest, 1959 to present.

GINO DiPACE, CAPT, a native of Detroit, MI, was born Jan. 30, 1924. He completed his education at Denby HS, Detroit, MI, January 1942. His naval career began December 1942 with his enlistment in the USN V-5, aviation cadet program.

In July 1943 he reported for active duty to Wooster, OH for the first phase of the naval aviation cadet training. Additional training was received at the following bases: Lewiston, ID; St. Mary's College, CA; Norman, OK; and NAS Corpus Christi, TX. Completion of this sequence of bases resulted in his designations as naval aviator, and he received his commission as ENS May 1, 1945.

His first assignment was Primary Flight Instructor School, and upon graduation he was transferred to NAS Ottumwa, IA as a flight instructor. His last transfer to NAS, Kingsville, TX from which he was released to inactive duty, March 1946.

DiPace, returned to Detroit, MI. Contacted NAS Grosse Ile, MI, and joined the "Weekend Warriors". Following Warriors fixed-wing assignments, the transition to helicopters in 1955 at NAS Ellison Field, FL and graduating as a qualified helicopter pilot, he returned to NAS Grosse Ile, Helicopter Sqdn. H.S. 731, where after a period of service, he became "CO" of H.S. 731 from 1963-66. After the decommissioning of NAS, Grosse Ile, Naval Air Facility established in 1969 at Selfridge Air National Guard Base, Mount Clemens, MI.

He joined VTU-7373, "Weekenders" as XO of the unit and retired as CAPT in 1976, with about 3800 hours of flight time. He is a member of the ROA, ANA, TROA, Tailhook Association, Naval League, VFW, American Legion, NRA and Quiet Birdmen.

The DiPace family resides in Clinton Twp., MI and includes wife Ilene; two sons, Donald and Robert; and granddaughter, Jessica.

MICHAEL T. "MIKE" DiPUPPO, CAPT, born Sept. 9, 1941 in Philadelphia, PA, received a BS in business administration, marketing major Penn State University, graduating in August 1963.

Military locations and stations: USN Officer Candidate School, NPT RI, Aug. 12, 1964 commissioned ENS USNR; USS *Mathews* (AKA-96) Long Beach, CA January 1964-May 1965, deck officer; Explosive Ordnance Disposal School, Indian Head, MD June 1966-March 1967 student; USS *Holland* (AS-32) Charleston, SC, March 1967-June 1968 EOD officer; USN School Diving and Salvage, Washington, DC, June 1968-October 1968, student; Communications School, NPT RI November 1968-April 1969, student; USS *Tringa* (ASR-16) NLON CT - Keywest, FL May 1969-November 1970 LT, operations officer HeO2 diving officer.

Served with Reserve Training USNR Training Center Folsom, PA; Fleet Processing Team, NAVBASE, Phila. Administrative officer; Mobile Diving and Salvage Unit 2 DET 304 NAVBASE, Phila, PA; two tours XO and one tour CO; volunteer training unit and inspector generals staff - inspecting units and centers in REDCOM Region 4 Phila. area. Returned to Mobile Diving and Salvage Unit 2 DET 304 for second tour as CO and diving officer.

His battles and campaigns: Conducted two West Pac cruises as 1st and 2nd division officer aboard USS *Mathews* (AKA-96), ferried Korean Tiger Div. for service in Vietnam.

DiPuppo received the National Defense Service Medal, Armed Forces Expeditionary Medal, Vietnam Service Medal (2 Stars), Republic of Vietnam Campaign Medal.

He achieved the rank of CAPT USNR and was discharged from active duty Nov. 31, 1970 to Ready Reserve. Retired from USNR Sept. 30, 1991.

Memorable Experiences: 1965-66 as deck officer USS *Mathews* (AKA-96) Beach Liaison Officer with Marines for off-loading supplies and equipment in Chu-Lai; - ferried Korean Tiger Division from Korea for service in South Vietnam; - ferried US Marines and equipment between Okinawa and South Vietnam; - as EOD officer USS *Holland* (AS-32) 1967-68, assisted in clearing unexploded Civil War cannon balls and ordnance from Charleston, SC undeveloped areas; - safety officer nuclear weapons transfer to/from FBM submarines; - trained gunnery crews for reactivated 3"-50 gun mounts; - as ops officer USS *Tringa* (ASR-16); - 1969 escorted ADM Rickover and crew on shakedown cruise of NR-1 research submarine; - as CO MDSU2 DET 304; - 1974 plankowner; - 1977 senior officer present during de-mudding and salvage operations to raise sunken barge *Elmar II* in Mississippi River, New Orleans. Conducted dives to de-mud barge.

Married to Patricia and they have children: Michael, Julia and Kathryn. Employed as senior vice-president, Branch Manager and Financial Planner: Devon Financial Group, Ltd., King of Prussia, PA.

RICHARD A. DIRKS, CAPT, born 1933, Indianapolis, IN, received a BA from Amherst College in 1955. Commissioned March 1956. Active duty with Naval Security Gp. Detachment, CINCLANTFLT, 1956-59. Member of the Air Intelligence Reserve Program, 1959-74, and the consolidated Naval Reserve Intelligence Program (NRIP), 1974-87, primarily at NAVAIRES Alameda.

Billets included: CO, NAIRU G1, 1973-73; Special projects officer for the director, Naval Reserve Intelligence Program (DNRIP), 1974-77; Deputy Reserve Intelligence Area Coordinator, Area 2, 1977-83; and DNRIP Chief of Staff/CO, NICHQ 0187, 1983-87. Also served as secretary,

NRIP Training Advisory Board, 1981-87. Member of three promotion selection boards.

Following retirement, he was recalled to active duty 1987-88 as Director of Naval Intelligence project officer to establish the Naval Reserve Intelligence Command. Subsequently recalled for various periods, 1988-95, as BUPERS project officer for Navy fitness report and enlisted evaluation policies and instructions.

Civilian: investment manager and orchardist. Awards include the Legion of Merit, Meritorious Service Medal, Navy Commendation Medal (2) and Meritorious Unit Commendation.

NEWTON B. "BRAD" DISMUKES, JR., CAPT, born Sept. 25, 1936 in Houston, TX, received a BA in history from Rice University in 1959 and PhD studies, American University.

He joined the service February 1959. His military locations and stations: Pearl Harbor, Jacksonville, FL, Washington, DC. Served with FICPAC, FICEUR, DIA.

Dismukes achieved the rank of CAPT and received the Joint Service Commendation and Defense Meritorious Service Medal and was discharged Sept. 25, 1980.

He married Jake Martin and is employed with Center for Naval Analyses as program director, retiring Sept. 25, 1996.

DAVID ALLEN "DITTY" DITTMEIER, CDR, born March 28, 1946 in Louisville, KY, received a BA degree in 1968 from Marquette University and MS degree in 1982 from George Washington University.

He joined the service Dec. 21, 1968. Commissioned NROTC Marquette University. His military locations and stations: USS *Kearsarge* (CVS-33) 1968-70; USS *Kitty Hawk* (CVA-63) 1970-71.

Dittmeier served in Vietnam and received the Vietnam Service and National Defense Medals. He achieved the rank of CDR, retiring in 1984. His memorable experiences: The J.O. Bunkroom USS *Kearsarge.*

He is married to Jami and retired as US Government Security Specialist, Sept. 30, 1995.

ERNEST P. "ERNIE" DOCLAR, CDR, born Oct. 9, 1930 in New Orleans, LA, received a BA degree from Loyola University, New Orleans, May 1951.

He enlisted in the service July 13, 1951. Commissioned ENS VS NR July 24, 1952. His military locations and stations: Rec.Tra, NTC, San Diego; OCS, Newport, RI; NAS Atsugi, USS *Lake Champlain* (CVA-39) NRS, New Orleans; USS *Huse* Rescrew; USS *Woodson* Rescrew (DE-359), XO, 1961-64; 1964-76 NR Center, Perth Amboy, NJ.

Doclar received the National Service Medal, Korean Service Medal, UN Medal, Occupation Medal, Armed Forces Medal and USNR Service Medal. He achieved the rank of CDR and RAD July 24, 1954 and August 1963. Retired from USNR April 1, 1977.

He married Jane May and their children are Paula Dart, Mary Doclar and Laura Parsons Muller; their granddaughter is Madeleine Dart. Employed with the Boy Scouts of America, 1956-94, editor of *Scouting* magazine. He retired Nov. 1, 1994.

ELLIOTT G. "ACE" DODGE JR., CDR, born July 1, 1915 in Narberth, Philadelphia, PA, received AB Grad. Union College, Schenectady, NY, 1938; graduate work at Northwestern University, Chicago IL, Case Institute of Technology, Cleveland, OH, and University of Pittsburgh.

He joined the service Sept. 9, 1940, V-7 USNR. Commissioned June 12, 1941. His military locations and stations: USS *New York* (BB-34) Atlantic, USS *Vulcan* (AR-5), Caribbean, Iceland, Oran, Algeria; USS *Dutchess* (APA-98) navigator-Pacific.

Served with: Atlantic-COMSERVLANT; Pacific-COMPHIBPAC. Commands held: Founder and CO USNTC Lorain, OH 1946-49, Ninth Naval District.

Battles and campaigns: The USN April 1, 1945 landing on Okinawa proved to be the last and greatest air battle in the Pacific. The Japanese in their deeply dug, honey combed underground defenses fully intended to make this fight another Dunkirk. By May 1 when his ship, the USS *Dutchess* (APA-98), arrived at Orange Beach to off load 2,000 Marines, 1MARDIV, they were subject to constant attack by hundreds of Japanese kamikazes and suicide dive bombers for eight days. By June 30th after fierce fighting their victory was assured in the Pacific.

He was awarded the American Defense with Fleet A, American Campaign, European African Middle Eastern Campaign, Asiatic-Pacific Campaign, WWII Victory, Navy Occupation Service, National Defense Service, Armed Forces Reserve, Naval Reserve and Philippine Liberation. Dodge achieved the rank of CDR and was discharged July 16, 1966. He retired July 16, 1975.

His memorable experiences: 1) In 9-41 Vulcan entering Hvalfjordur, Iceland from the vast Atlantic, to establish a base as Senior Officer Present Afloat (SOPA) for next two years to assist their fighting ships and convoys in war against German U-boats; landing in Algiers, that alabaster city of minarets (mid-1943) prior to establishing their base at Oran to support Allied invasion of Anzio, Italy from North Africa. 2) 4-44 as naval courier his flight from Oran, Algeria to Cairo, Egypt via RAF and American Transport Command (ATC), flew over battle fields where British 8th Army fought Rommel's German Panzer Divs. to stand still in the North African Desert campaign. Their DC-4 set down at their Benghasi AB in the Sahara for the night then proceeded to Cairo arriving Easter morning. Just one year before Roosevelt, Churchill, Stalin and Chiang-Kai-Chek held their historic conference in the Mena House across from the Sphinx and Cheops Pyramid. At dinner they greeted the British and Aussies at the world famous Sheppard Hotel and toured the ancient City of the Dead. En route to Algiers they spent the night at the Translantique Hotel Tunis which just months before was German HQ. 3) His safe arrival home after 61 months of foreign sea duty. He wrote *Five Years at Sea Around the World* in 1996, published in *WWII Reminiscences* by John H. Roush, Jr., USA Ret., sponsored by the Reserve Officers Assn. of the US Department of California, memorializing the history of WWII 50 years ago.

NRA activities: Life member; as president, Fort Omaha Chapter NRA, presented the Most Active Chapter Award of the 9th Naval District at Dallas, TX October 1971. Served two years as President 9th Naval District covering 11 states and for many years served on NRA's National Advisory Committee since Nov. 4, 1975. In California he is active in the FADM Nimitz-Golden Gate Chapter, San Francisco, CA.

Other military organizations: Naval Reserve Association, life member; Navy League of the US, life member; Retired Officers Association, life member; Naval Order of the US, life member; Association of the Naval Aviators and Naval Officers Association of Canada. His wife is a member of the NRA and NOUS.

Married Anita Weymouth Vieth, June 6, 1941, and she passed away April 15, 1977. He married Winifred O'Leal on Jan. 7, 1981. His children are Elliott G. III, Randall D., Nancy C., Jeffrey N (naval architect, Pearl Harbor); stepchildren: Jo Anne, Leland D., Linda Joy, Mark D. and Gail C. O'Leal.

He retired March 1, 1976 from a full 30-year career as an executive in operations, sales and engineering department of the US Steel Corp.

LLOYD P. DODGE, CAPT, born March 22, 1915 in Bellmore, NY, received an AB degree from Amherst and a JD from Yale Law School. He joined the service June 4, 1941. He served as supply officer, USS *Holland* and CANWA, Philadelphia Navy Yard.

His battles and campaigns: Manila, Philippines, WWII, Asiatic Station for three years. Dodge achieved the rank of CAPT and was discharged November 1945. His memorable experiences: between the USS *Pecos* and USS *Langley* February 1942 when they moved to protection of Christmas Island to help destroyer when *Langley* and *Pecos* sunk.

He is married to Lorayne and they have three children: Sybil, Lloyd P. Jr. and Henley; and grandchildren: Lloyd P. Dodge III, Michael, Mark, Matthew and Margaret.

JOHN R. DODSON, CDR, a native of East Texas, he joined the USN in April 1944, went to service in the V-12 program and was commissioned an ENS. He served aboard DE-701, LSM-442 as deck officer in the Chinese Naval Training Command, and XO of LCI(L)-1054, involved in evacuation of Marine troops at Tsingtao, China. He mustered out in mid-1947 at age 20 and returned to finish college.

He then spent three years in Brazil for oil exploration on the Amazon River as a water-born surveyor, returned for US, joined the Texas Highway Dept. in 1952, worked as a design and construction engineer in East Texas, expressway engineer in Corpus Christi and Railroad Coordinating Eng. for Texas in Austin until his retirement in 1987.

During his engineering career, he served in the USNR-R as a "weekend warrior" and a lifetime member of the NRA. Included in his billets were CO of Seabee Unit 8-72 at NAS Corpus Christi, Public Works officer at the NRTC in Austin, then Reserve Public Works officer at NAS Dallas until his retirement from USNR-R in 1978. During the late 1960s, he was NRA 8th Naval District President for three years, instituted the District Area Vice-President system for 8ND and other districts, and assisted by the late LCDR Lee Douglas, developed the current executive vice-president National NRA office. He received the National Award of Merit for outstanding service to the Navy in 1970. His service medals include the American and Pacific Theatres, WWII Victory Medal and China Service Medal.

He is presently a consulting engineer in Rail-Highway matters, residing in Austin, TX. Married to Dawn and they have a son Larry and five grandchildren.

FRANCIS J. "FRANK" DOHERTY, LCDR, born in Everett, MA, Nov. 19, 1916. Enlisted Naval Communications Reserve as a radioman third class, April 4, 1937.

Volunteered active duty, Nov. 10, 1939. Served until April 1942 when honorably discharged to accept appointment as warrant radio electrician.

Served with local defense forces, Eastern Sea Frontier; twice ordered Port of Boston closed against U-boat attacks while Navy Harbor Defense watch officer.

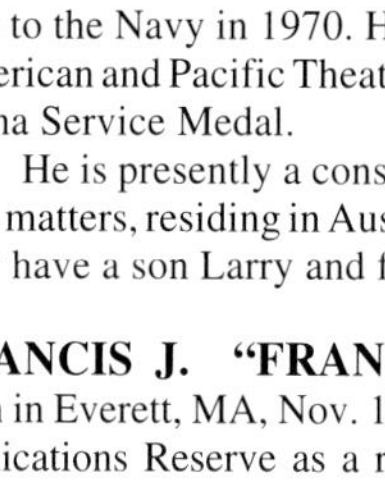

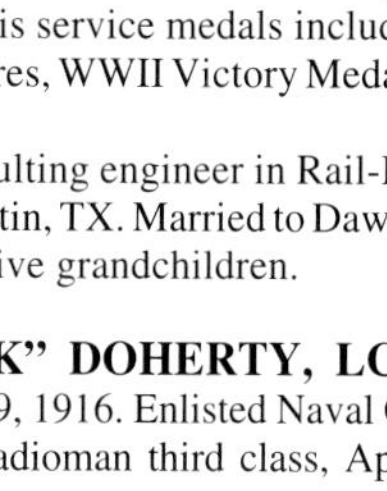

Detached Local Defense Forces December 1943, ordered USS *Altamaha* (CVE-18) as radio officer. Participated in Battle of Leyte Gulf aboard USS *Altamaha* (CVE-18) and Philippine Liberation. Detached USS *Altamaha* to attend Warrant Officer School, NRL, Anacostia, DC. Released from active duty Nov. 13, 1945 with a Certificate of Satisfactory Service signed by James Forestal, Secretary of the Navy. Promoted to chief radio electrician, Oct. 1, 1945.

Served as assistant CO and air maintenance officer, Air Wing Staff 91 and ACORN Div. 1 (L) from June 1949 to June 1968 in the Active Naval Reserve both AIR and SURFACE. Recalled to active duty during Korean Conflict.

Selected Naval Academy Information Officer, June 1968; served to June 1984. Awarded Commendation Certificate for work for Naval Academy. Graduated Tufts University AB 1948; EDM 1949.

Educator/counselor with Revere, MA school system. Retired December 1979 with 31 year of service.

Married Thelma King of Revere, MA, April 5, 1942. He is the father of four sons and five daughters.

Transferred to USN Retired List with pay and allowances after 35 years active and inactive service in the USN, November 1976.

He is a member of the following associations: NRA (life member), Retired Officers Association (life member), The Reserve Officers Association (life member), Fleet Reserve Association.

His military decorations include the American Campaign, American Defense, Asiatic Pacific Campaign Medal w/2 Bronze Stars, WWII Victory Medal, Philippines Liberation Medal, Armed Forces Reserve Medal, United States Naval Reserve Medal w/star, Philippine Independence Ribbon.

MACARIO L."MARC" DOMAGUING JR., MD, CAPT, born May 4, 1939 in Quezon City, Philippines, and received his doctor of medicine degree.

Joined the service July 6, 1976, and was stationed at NRMC Camp Lejeune, NC; NRMC Jacksonville, FL. Served with NHLB 219; NHLB 319; NH Pendleton 319. He fought in Desert Shield/Desert Storm in 1991.

He achieved the rank of CAPT 06 MC USNR and was discharged November 1981 from active duty USNR. He received the Navy Commendation, Armed Forces, National Defense Meritorious Unit, Sea Deployment Medals.

His memorable experiences: Sqdn. medical officer 6th Fleet 1981; Reserve Unit, as CO; winning the Best Unit 1992 Region 19; winning the Golden Anchor Award 1992 Region 19.

Married Lou and they have two children, Michael Lyndon and Marilou Lyn. Employed in solo practice of family practice and surgery.

WILLIAM B. "BILL" DONLEY, CAPT, born June 22, 1921 in Marcus, WA, received a BA and LLB. He joined the Naval ROTC, September 1939 and active duty June 1943. His military locations and stations: USS *San Diego* Pacific Ocean and USS *Sirago* (SS-485), Atlantic Ocean.

The USS *San Diego* participated in the following actions while Donley served aboard prior to being transferred to the submarine *Cachalot* in January of 1945: Buka-Bonins Strike, Rabual Strike, Gilbert Islands Occupation, Kwajalein-Wotje Raid; Kwajalein and Majuro Occupation; Joluit Atoll attack; Truk attack; Saipan-Pagon attacks; Iwo Jima attacks; Bonins Raid; Saipan Occupation; Guam Occupation; Tinian Occupation; Philippine Sea Battle; Southern Palau Islands Raid; Philippine Islands Assault; Okinawa attack; Northern Luzon, Formosa attacks; Luzon attacks; Visayos attacks.

He achieved the rank of CAPT and was discharged June 1946. Retired from the Reserves in 1974.

Married Barbara Coleman and their children are Colleen D. Flower, Kathleen D. Burnett and Judy D. Weller; and grandchildren: Patrick, Matthew, Eric, Amanda, Billie and Jamie. Employed as vice-president, Merrill Lynch & Co., retiring in 1984.

CECIL C. DONOHO, CAPT, born Batesville, AR, Jan. 31, 1920. At Ft. Smith, AR 1935, transferred from premed training at Quachita College to aviation training at Arkansas State Teachers College and received private pilot's license in 1942 and commercial pilots license and flight instructors rating the same year. Taught in War Training Service (Army then all Navy cadets). Joined the USN 1943 as SR assigned to same duties (WTS). Called to active duty, 1944 as Aviation Pilot First Class NAS Dallas (eight weeks primary training) to NAS New Orleans (eight weeks Instructors School). Commissioned January 1945, assigned to NAS Glenview, released to inactive duty November 1945 after soloing 109 cadets.

In 1950 rejoined the active reserve in NRSD 8-14 Fort Smith, AR XO administration officer. Back to aviation at NSA Memphis with VR-791 to VR-792. Assigned to NARS as inspector general for 2nd component when promoted to commander. 1969 transferred to NAS Dallas as senior IG and Component director for NARTV and senior advisor to CO and IG to NAS. Promoted to CAPT in 1972.

His medals are the Naval Reserve Medal, Armed Forces Reserve, National Defense Service, WWII Victory Medal, American Campaign, American Defense Service and Navy Commendation Medal.

He was president and CEO of a mortgage corporation, insurance agency and title agency. He acquired two mortgage companies and three insurance agencies before retiring in 1980.

His family consists of wife Martha Ann, daughter Charlotte Cecille McGowan, son Robert R. (COL USA); and six grandchildren: Amy, Annabeth, Matthew McGowan; Drew, Katherine and Charlotte Donoho.

MICHAEL RICHARD "MIKE" DONOVAN, CAPT, born May 12, 1952 in Bay Village, OH, received a BS in naval architecture, USNA 1974, MS in naval architecture and Marine engineering, MIT 1975; naval nuclear power training Vallejo, CA and Idaho Falls, ID.

He graduated as ENS, USN from USNA June 5, 1974. His military locations: San Diego, USS *Bainbridge* (CVN-25) and USS *Long Beach* (CGN-9) and Washington, DC at Naval Sea Systems Command. He served with NAVSEA DET 419 SUPSHIP 919, NAVSEA in dust MOB 319, REDCOM 19 technical training advisor for program 26 and 29 units.

Donovan was awarded the Navy Commendation (3), Unit Meritorious Service, Battle E, National Defense, Humanitarian Service, Sea Service, Expert Pistol and Armed Forces Reserve. He achieved the rank of CAPT (06) USN discharged from the USNR November 1983.

His memorable experiences: qualified nuclear engineering officer and surface warfare officer.

Married Christine Simony and their children are Peter Michael (b. Feb. 10, 1983) and Elizabeth Anne (b. Feb. 2, 1985). Employed with Solar Turbines Incorporated as project manager, international operations.

DAVID J. DORAN, CAPT, born in Connecticut, and a graduate of Fairfield University, was commissioned at OCS in 1959. He completed Aviation Ground and Electronic Officer Schools and served as VA-42s Avionic Div. officer and assistant maintenance officer. As OinC of numerous CARQUAL Maintenance Detachments, he served on nine East Coast carriers. He also assisted with the A6 Intruder Fleet introduction. His active duty was extended to accommodate the Cuban Missile Crisis.

He was attached to Air Reserve Units at NAS South Weymouth: VS914; ACORN 1-1; RSAND Z-1; Patwing 101 and several NAVAIRSYSCOM. Sqdn. ACDUTRAs took him to Cuba and Halifax supporting fleet operations. Designated an aerospace engineering duty officer in 1968, his career with diverse NAVAIRSYSCOM assignments culminated in command of AIRSYS-0891, AIRSYS-1391 and AIRSYS-0591.

Doran retired from the USNR on board the USS *Constitution* in 1992 and from GTE in 1993 as the business segment manager for Navy Systems. His program management responsibilities included Voice and Data Security Systems, Tactical Communications Systems and most notably the USN Elf Communications System for submarines.

He resides with his wife, Linda, in Framingham, MA. They have four sons, three of whom earned their Wings of Gold as naval aviators.

BRENT A. DORMAN, CDR, a native of Randallstown, MD, was commissioned May 16, 1981 through the Naval Reserve Officer Training Corps program. He graduated from the University of North Carolina at Chapel Hill with a BS degree.

CDR Dorman entered aviation indoctrination at Pensacola, FL and was designated a naval aviator in May 1982. After completing training at Helicopter Anti-Submarine Sqdn. ONE (HS 1), he reported to Helicopter Anti-Submarine Sqdn. THREE (HS 3) at NAS Jacksonville, FL.

While attached to HS 3, he served as nuclear safety officer, materials control officer and personnel officer; and deployed aboard USS *Saratoga* (CV-60). He transferred to Helicopter Training Sqdn. EIGHT (HT 8) where he flew 1,300 hours as an instructor and completed his active duty as a maintenance functional check flight pilot.

August 15, 1988 he relocated to Port St. Lucie, FL and instructed at Flight Safety International's Training Academy located at Vero Beach Municipal Airport. Besides acquiring his civilian flight instructor rating, CDR Dorman has achieved airline transport pilot ratings in both rotorcraft and multi-engine airplanes.

In December 1988, Dorman affiliated with Helicopter Submarine Sqdn. SEVEN FIVE (HS 75) and has served as administrative, operations, aircraft maintenance, and safety department head. He has accumulated 4,000 military flight hours. Two of his major achievements include being selected as 1990 NAVAIRES Jacksonville Aviator of the Year and receiving the Navy Commendation Medal for the only night ocean rescue during Operation Desert Storm.

He is presently employed by PEMCO World Air Service as a functional test pilot for reworked H-3 helicopters. He is a member of the NRA, US Naval Institute, UNC and UNC NROTC Alumni Association, and the National Honor Society.

Dorman and his wife, the former Linda M. Coulter of Corpus Christi, TX reside in Orange Park, FL. Their son, Airman 1st Class Brian C. Coulter is a USAF Pararescueman stationed at Moody AFB, Valdosta, GA.

MARSHALL L. DORMAN JR., CDR, born Oct. 25, 1931 in Everett, WA, received a BSME from the University of Washington, June 1959.

He joined the service June 1949 and served with Surface Div., USS *Edmonds,* MSC Div. and NavSea Div. Dorman achieved the rank of CDR and was discharged September 1986.

His wife, Agnes Denise, is deceased; his children are Marshall M. Dorman, Maryann Anderson, Catherine Roe and James A. Dorman. Employed as principal engineer, Boeing and RCA, systems design.

PIERRE E. DORRATCAGUE, CAPT, born in New York City, 1920. He was a graduate of Columbia University, School of Engineering in 1942 and was commissioned ENS, Aviation, in 1943.

During WWII he served with ARGUS 8, a Pacific Island Radar Early warning unit as radar officer and during his second Pacific duty as a technical observer and Radar Combat crew member in aerial flights aboard the lead Torpedo Bomber of VT-12 attached to aircraft carrier USS *Randolph;* participated in the first carrier Tokyo campaign initiating the use of electronic and passive countermeasure and in flight support of the Iwo Jima and Okinawa campaigns.

From 1947-1963 he served in various electronics and administrative billets with VA-51A, Air Gp. CVG-51 in Anacostia, DC and at NAS Seattle with FAS-896, 891, WEPTU 893 and XO 891. He was selected CAPT in 1963. He served in the following years as member of the Captain's Reserve Advisory Board of AW 89(M) and ACOS of NARS(M)T1. He retired in 1973.

He was awarded the Air Medal w/2 stars, Campaign, Theatre and Naval Reserve Medals.

Employed by the Boeing Company in the capacity of a senior engineering research specialist in the design, development of an advanced radar systems and the integration of electronics countermeasure system associated with missile and aircraft projects. Retired January 1985.

KEN RICHARD DOUGHERTY, CDR, born Nov. 3, 1946 in Lexington, KY, received a BS in accounting, University of Kentucky, 1971.

Enlisted USN, Dec. 1, 1965. His military locations and stations: USNRS, Ft. Allen, P.R. (September 1966-October 1969). Separated Oct. 23, 1969 as RM1 (radioman first class). Served in the Reserves in various units as XO and department head and Deputy Reserve Intelligence area CDR (Intelligence and Security), Area Twelve, 1992-93. Reserves Oct. 24, 1969-present (31 years service so far).

Commissioned June 2, 1975 as ENS, special duty (Intelligence) (1635)

His medals and ribbons include the Joint Meritorious Unit Award, Meritorious Unit Commendation (3), Good Conduct Medal (Navy), Naval Reserve Meritorious Service, National Defense Service (2), Naval Reserve Sea Service, Naval and Marine Corp Overseas Service, Armed Forces Reserve (two Silver Hourglass), Expert Rifle Shot and Expert Pistol Shot.

He married June 10, 1967 to Janis E. Durkin and they have three children: Kimberly A., Edward H. and Andrea E.. Employed with Brown and Dougherty, PSC, CPAs-partner and vice-president.

GREG "GREG" DOWBAK, CDR, born Nov. 26, 1947 in Brooklyn, NY, received a MD from SUNY Downstate; general surgery, vascular surgery, plastic surgery, Hand Fellowship, Parkland Memorial Hosp.

Joined the service October 1994. He is stationed 5/14 9th MARDIV. Achieved the rank of CDR. His memorable experiences: Volunteer surgeon, UNHCR disaster, assistance relief team (DART), Sarajevo, Bosnia 1994-two trips. Employed as hand surgeon, Santa Barbara Cottage Hospital.

GEORGE E. DOWLING, CAPT, born in Arkansas in 1925. After a year at University of Arkansas, qualified for Navy V-12 training at Central Missouri State University in 1944. Commissioned ENS at Fort Schuyler, NY, midshipman school.

Reported for duty in USS PCS-1430 in Pearl Harbor. After WWII, became CO - brought the ship back to San Francisco (COM12). In civilian life, became a civil engineer. In 1951 (Korean War), recalled to active duty as gunnery officer in USS *Borie* (DD-704). Qualified for command at sea in destroyer types, and participated in "Operation Mainbrace," followed by tour with Sixth Fleet. Reported to USS *Cronin* (DEC-704) for mothball operations. Continued active reserve affiliation, going to sea each year to maintain sea qualifications. Held several commands and served as group commander, Little Rock, with broad command of all drilling reservists in Arkansas.

During retirement ceremonies, June 30, 1978, was awarded the Navy Commendation Medal by Chief Naval Reserve. Other medals include Navy Occupation, Naval Reserve, Armed Forces Reserve, WWII Victory Medal, National Defense, American Defense, and Asiatic Pacific Theatre.

Dowling and wife Marilyn reside in Dallas. Son Mike resides in Little Rock, AR, and daughter Sharon in Portland, OR.

THOMAS E. DOYLE, CAPT, born May 6, 1941, in Detroit. He is a graduate of Brown and Harding Universities and holds a master's degree. CAPT Doyle was commissioned ENS in 1964, OCS, Newport. Initial assignment was to Naval Communications Unit, Key West. Subsequent active duty assignments were to COMCARDIV 7 communications embarked on USS Midway (CVA-41) for operations off Vietnam, and to USS *Warbler* (MSC-207) as operations officer, deploying from Sasebo, Japan, for anti-infiltration operations, Vietnam.

After serving as a drilling reservist in various surface divisions, he was assigned to a Personnel Mobilization Team in Memphis. As training officer, he participated in JCS exercise Nifty Nugget which tested the nation's mobilization capability. As a result of that exercise, and upon request of the Mobilization Plans Division, Chief of Naval Reserve, he returned to active duty to assist in developing NR mobilization policy, procedures and systems.

Subsequently he was ordered to the Joint Deployment Agency, MacDill AFB, FL, as evaluator for the Joint Deployment System. Detached from active duty in 1985, he was promoted to CAPT in the same year while assigned to NR JDA108, a staff unit which augmented the Joint Deployment Agency.

Thereafter, CAPT Doyle was selected as CO, NR 118, US Transportation Command, Scott AFB, IL. The unit was mobilized to augment crisis action deployment operations for the Persian Gulf War. During that assignment, Capt. Doyle was instrumental in establishing the military's first joint reserve unit which included members of all the services.

CAPT Doyle is married to the former Janice McClurg of Verona, MO, whom he met as an English teacher, King High School, Sasebo, Japan. CAPT Doyle's awards include various personal Navy and Joint Commendations. He was four times commanding officer. He retired in 1992 after 28 years as a Reservist.

DANIEL JAMES "DAN" DREW, LT, born July 23, 1958 in Portland, OR, received a BS in aviation administration.

He enlisted Aug. 3, 1976 and was commissioned May 4, 1987. His military locations and stations: active duty: 1980-82 US Naval Communications Station, Japan; 1978-80, US Naval Mobile Construction Bn. 4; 1976-78. USS *John R. Craig* (DD-855).

Reserve duty: NR ABFC FTU D109, Indianapolis, IN, 1994-present; NR DDNV, Indianapolis, IN, 1993-94; MIUW 213, Milwaukee, WI, 1991-93; NR ABFC FTU 210 HQ, Houston, TX, 1989-91; NR NSC Oakland, Sacramento, CA, 1986-89; NR NSC San Diego, San Bernardino, CA; 1984-86; Naval Weapons Station, Charleston, Orlando, FL; 1982-84.

Drew participated in Operations Desert Shield/Desert Storm. His awards and medals include the Navy Achievement, Joint Services Commendation, Navy Meritorious Unit Commendation, Enlisted Good Conduct, Southwest Asia Campaign Medal w/2 stars, Navy Overseas Deployment, Navy Overseas Shore Assignment, Kuwaiti Liberation Medal (Saudi Arabia), Kuwaiti Liberation Medal (Kuwait), Expert Marksman (Rifle).

He achieved the rank of LT, Supply Corps. His memorable experiences: COP initiation; commissioning as a naval officer; recall to active duty in the Middle East during the Gulf War.

Married to Lori Ann and they have children: Brady, Adam and Cameron. Employed as WESCO Distribution, Incorporated.

Employed as Midwest District manager, Elkhart, IN, 1996-present; branch manager, Elkhart, IN; 1993-96, account representative, Madison, WI; 1991-93, account representative, Houston, TX, 1989-91; account representative, Houston, TX; account representative, Sacramento, CA; 1986-89, account representative, Sacramento, CA; 1984-86, sales/management trainee, Riverside, CA.

JOSEPH A. DRISCOLL, CDR, enlisted in the USNR in 1957 and attended Officer Candidate School under the ROC program while attending St. Peters College. He was commissioned an ENS in June 1961 upon graduation.

Driscoll served aboard USS *Thor* (ARC-4) as damage control assistant, USS *Halsey Powell* (DD-686) as weapons officer, and USS *Thaddeus Parker* (DE-369) as officer-in-charge from June 1961 through November 1967.

Upon release from active duty, he affiliated with the NRF Destroyer Program serving aboard USS *Robert K. Huntington* (DD-781), USS *W.B. Rush* (DD-714) and Fleet Training Group 202 in various billets including navigator, weapons officer and XO.

He attended a number of service schools including Destroyer School, Naval Justice School and ASW School. He is designated surface warfare officer.

Driscoll is a CPA and recently retired as a partner in the accounting firm of KPMG Peat Marwick. He currently sits on the Board of Directors of a number of companies and is an independent financial consultant.

He resides in Manhasset, NY with his wife Corrine. They have two married daughters. He is a life member of the NRA.

DONALD DAVID DRONE, CAPT, born March 8, 1942 in Evansville, IN, received a BS/MS Forest Products, Michigan State University, East Lansing, MI.

He joined the service July 1966, active duty-January 1967-December 1969. His military locations and stations: USS *Endurance* (MSO-435) operations officer, XO. Served with operations with Pacific Fleets. Reserve affiliation July 1997 to present: CO, NR FF-1044 Brumby October 1978-90; NR NAVMAG GUAM Det 113, October 1980-January 1981; NR WPNSTA CONCORD 2513, February 1991-September 1982; NR WPNSTA CONCORD 2513 October 1986-September 1989; NR Readiness Unit 1310 October 1989-December 1991; NR NH Glakes 1813 October 1989-December 1991; NR FF 1080 Paul October 1989-December 1991; NR VTU 1305 December 1991-May 1992; NR VTU 1307 June 1992-September 1996.

He was awarded the Republic of Vietnam Campaign Medal w/Device (1960-), National Defense Service Medal w/star, Navy Commendation Medal, Vietnam Service Medal w/3 Campaign Stars, Armed Forces Medal w/2 HGs.

Married Cynthia Kay and they have son, David Todd. Civilian employment: vice-president Carmi Lumber, president Carmi Area Industrial Sites, Director White County Bank. Retired June 1997.

STANLEY T. "STAN" DROSKI, LCDR, born Dec. 5, 1923 in Grand Rapids, MI. He was drafted into the service December 1942 and allowed to finish the semester Grand Rapids Jr. College.

Served with Midshipmen School, Ft. Schuyler, NY; USS *OBrien* (DD-725) and the war was almost over when he went on board the *O'Brien*. He served in the Korean War August 1950-August 1952.

Droski was awarded the American Defense, Victory Ribbon, Korean Medal, Korean Victory Medal and the Reserve Medal. He achieved the rank of LCDR and retired in 1983.

His memorable experiences: served in Communication Center Bldg. 1, Treasure Island (Korean War); Cryptographic Section for two years—fascinating. He feels he owes the Navy far more than the Navy owes him.

He married Betty and they have two children, Betsy Quinlan and Kevin D. Their grandchildren are John and Casey Quinlin. Employed as insurance business owner and also farms. He retired in 1983.

DANIEL N. DUBE, CDR, born Feb. 5, 1958 in Hartford, CT, received a BA in 1980; MBA 88, Cornell University; MS in 1997 Creighton University.

He joined the service May 26, 1980. His military locations and stations: Pearl Harbor, HI, USS *Cochrane* (DDG-21), 1981; Key West, FL, USS Gemini (PHM-6) 1984; Omaha, NE: NR NEAT 218 (AS CO) 1991-93; 1986-91, various units in the Northeastern US; NR PhibCB-1 218 (AS CO) 1993-95; NR SRU 1362 (AS CO) 1995-now.

He was awarded the Surface Warfare Pin, and nine other decorations including Vietnam w/star. He is currently ranked as CDR; left active duty as LT. He was discharged Feb. 13, 1986 from active duty and still in Reserves.

Dube married Dorothy and they have three children:

Derrick, Ryan and Paul. He is a market research manager, CONAGRA a "Fortune 20" food manufacturer.

THEODORE B. "TED" DUBS, CAPT, born in Canton, OH on March 9, 1940. He graduated from the US Naval Academy in 1962 and was commissioned an ENS. He graduated as a naval aviator at New Iberia, LA in 1963 and was assigned to VS-34 at NAS Quonset Point, RI. In 1967 he was assigned to Replacement Air Gp. (RAG) VS-30 at NAS Boca Chica, Key West, FL.

In 1968 he left active duty and went to Pan American World Airways as a pilot and was based in San Francisco. At that time he also affiliated with the Naval Reserve at NAS Alameda, CA and continued his naval career culminating with his promotion to CAPT in 1986 and retirement in 1988.

In 1970 he was furloughed from Pan American and began a career with AT&T initially as a plant engineer and then as an executive pilot with destination around the world.

Dubs has 10,000 hours of flight time and has 200 carrier landings. His experience has taken him to all parts of the globe in a dozen types of aircraft.

His family consists of wife Mary Jane; sons, Paul and Billy; daughters, Sissy and Tonya; and seven grandchildren: Natasha, Mina, Elizabeth, William, Joey, Ryan and Tiffany

JAMES D. "DUBIE" DUBUAR, born Oct. 25, 1920 in Seattle, WA, received a BA and JD degree from the University of Washington. He joined the service in 1993 NAVAL ROTC. Commissioned ENS, June 1943. His military locations: SCTC Miami, Fleet Sonar School, Key West. Served with USS *Daniel* (DE-335), USS *Gilligan* (DE-508).

His battles and campaigns: North Atlantic convoys and killer groups-Pacific-Aleutians, Mediterranean Caribbean. He was awarded the American, European Middle Eastern Theater; WWII Victory and Korean War Victory medals. Also wore Merchant Marine WWII Victory Ribbon, Merchant Marine Pacific Area Ribbon, Naval Reserve Ribbon and Armed Services Reserve Ribbon.

Dubuar achieved the rank of CAPT. Discharged from WWII, June 10, 1946 and from Korean War, Dec. 4, 1952.

His memorable experiences include intercepting a Portuguese schooner (blue and red sails with part yellow hull in the golden glow of the setting sun off of the Azores).

Married to Elaine H. Dubuar and has children, Janeane and Melanie. He is employed as lawyer in Dubuar Liraus and Engel, 44 years. Total service on active duty in USNR, 44 years.

JOHN R. DUCK JR., born March 10, 1943, Yeadon, PA. He entered USNA from Olympia High, Greece, NY, graduating in 1965. Trained in USS Piper, NUPOWSCOL Bainbridge, NPTU West Milton, USS *Dogfish*, SUBSCOL and NGMS Dam Neck. He completed submarine qualification, record-breaking overhaul and three MED and WESTPAC FBM Deterrent Patrols as MPA/ANAV/AENG/OOD/EDO on USS *John Adams* at Charleston, Bremerton and Honolulu.

Joining the USNR in 1971, he earned a MBA from Harvard. He has worked in real estate investment, insurance brokerage, nuclear consulting, ASME standards' development, life insurance and business brokerage. He volunteers as a high school science fair judge. Currently, he is with Liberty Business Exchange in Dallas, TX, where he resides with his wife, the former Mary Clare Peebles of New Iberia, LA.

Reserve units included SURFDIV 3-2M, SUBDIV 1-8, AS-2201 FULTON, SUBSUPPFAC-402, FFG-3 SCHOFIELD, CARGRU-370 and Aviation VTU-7070. First non-aviator Carrier Group XO with CARGRU-770. Appointed COM7THFLEET Pol-Mil/LR Plans (N61), Exercise ACOS, CTU-71.6.1, and Flagship NTCS-A Battle Watch (BW) Captain while with C7F-111. Mobilized during Desert Shield initial phases as COMUSNAVCENT air ops/C41 Liaison to Royal Saudi Air Force, Riyadh. As NAVCENT BW Captain, he managed "Multinational Maritime Intercept Force" enforcing UN Arabian Theater blockade of Iraq. SECNAV awarded Navy Commendation Medal. Promoted to CAPT in 1990. As CO he organized VTU-1104 from scratch.

He executed 50 AcDuTras, including 24 WESTPAC/IO operations, rating the Overseas Service Ribbon with a Silver and four Bronze Stars. Fleet Surface BW during PACEX-89, the largest combined surface force operation since WWII. Senior Watch Chief at Japanese MSO (Rappongi), CinCSDFLT (Funakoshi) and JASDF (Fuchu) Headquarters. Oversaw Fleet interactions in Korean exercises as CinC UNC/CFC Seoul's "Joint Gaming Center" Senior BW Captain, Fleet Controller and Navy Game Director. Acted as Crisis Action Team XO for USCinCPAC.

As USCinCPAC-111 Chief of Staff, presented with MSM by the President for earning RADM J.T. Natter Award as best medium unit with unprecedented 95% readiness.. Other Reservist awards: DMSM (USCinCPAC), NUC (USS Blue Ridge), two MUCs (C7F-111), second NDM, SWA Service Medal w/Bronze Star, Navy Sea Service Ribbon (C7F), AF Reserve Medal w/Bronze M and Silver HG, Kuwait Liberation Medal. Achieved the rank of captain 0-6; discharged Feb. 25, 1971 and retired Nov. 30, 1990.

Memorable Experiences: during ADSW w/COM 7 Flt. in Japan, did "Sistership Cruise" w/JDS Shirane; climbed Mount Fuji and mobilized in P-3 to Persian Gulf in August 1990.

Married Mary Clare Peebles. Employed as associate, Liberty Business Exchange, Dallas, TX. Retired June 30, 1995, at Broken Arrow Joint Reserve Center, OK.

DONALD JOHN DUDZIAK, CAPT, born Jan. 6, 1935 in Alden, NY, received a BS in Marine engineering, USMMA, 1956; MS Radiological Physics, University of Rochester 1957; PhD, mathematics, University of Pittsburgh, 1963.

He joined the service Aug. 2, 1952 and achieved the rank of CAPT, USNR.

Married Judith and they have three children: Alan Joseph, Matthew John and Karin Marie. Employed as professor and department head, Nuclear Energy, North Carolina State University. Since 1990 he has been Fellow, Los Alamos National Lab. He retired July 1, 1987.

ROBERT F. DUGUAY, CDR, born July 2, 1931, Waterbury, CT, received a BS Commerce, Washington and Lee University, and a MBA in banking, NYU Graduate School of Business Administration.

He joined the service as Naval Reserve Officer Candidate (ROC Program) October 1951. Served with 6th Fleet (Mediterranean).

Duguay achieved the rank of CDR and discharged July 1988. His children are Robert S. (USNA, Class 1989) and Douglas R. (Penn State 1989). He is senior vice-president and trust officer, Penn Security Bank and Trust Co., Scranton, PA.

WILLIAM M. DUNBAR, CAPT, born and raised in Baltimore, MD, graduated from the Johns Hopkins University with a BS in industrial management in 1953. He was commissioned ENS in April 1954 at OCS in Newport, RI.

He was subsequently assigned to the Naval Station,

Adak, AK and rounded out his active duty at the Naval War College. He completed 30 years in the active reserves, holding billets as commanding officer of two surface divisions, chief staff officer to the Baltimore Gp., and a plankowner on the staff of then Readiness Command-Baltimore. He completed 26 training duties and retired as CO of the local volunteer training unit on July 1, 1988.

In civilian life, Dunbar worked for the C&P Telephone Co. for 30 years before transferring to AT&T in 1983. Most of his career was in External Affairs including stints at community relations manager and legislative affairs manager with C&P. For AT&T, he was liaison with the Maryland Public Service Commission and retired as director of Government Affairs.

RANDY E. DUNCAN, CDR, born April 7, 1957 in Kalamazoo, MI, received a BS in Newberry College. He joined the service June 25, 1981. His military locations and stations: NAS Pensacola, NAS Norfolk, US Naval Academy, NS Mayport, NAS Atlanta. Served with: VT-10, VAW-22 (deployed USS *Independence* and USS *Forrestal*), 33rd company officer USNA, USS *Forrestal,* COMPHIBGRU 2 DET 0867, AIC 0967, RIA 14.

He fought campaigns in Grenada and Lebanon. He was awarded the Navy Commendation Medal, Navy Achievement Medal (2), Joint Meritorious Unit Commendation, Navy Unit Commendation, Meritorious Unit Commendation, Battle "E", Navy Expeditionary Medal, National Defense Service Medal, Armed Forces Expeditionary Medal, Sea Service Ribbon (3), and Armed Forces Reserve Medal.

He achieved the rank of CDR (sel) and is presently an active reservist. His memorable experiences: Combat cruises to Grenada and Lebanon, achieving over 350 traps; being the first E-2C NFO designated as a copilot, and being a company officer at the Naval Academy.

Duncan married Lori Pujazon in 1982. They have two children, Evan and Christina. Employed at IBM/TSS as a national customer service manager.

WILLIAM F. "BILL" DUNCAN, CDR 05, born March 11, 1943 in Summit, NJ, received a BS in elementary education 1965; MS education. Enlisted Sept. 26, 1962, Naval Reserve Electronics Facility, Emporia, KS. His military locations and stations: Naval Reserve Facility, Emporia, KS (inactive duty September 1962-May 1965). June-August 1963 and June-August 1964 ACDUTRA as ROC, Newport, RI).

Commissioned upon graduation May 26, 1965. USN Deep Sea Diving School, Washington, DC, student, June-October 1965; USS *Takelma* (ATF-113), Diving Deck, Supply, OPS, OOD, CDO, November 1965-August 1967; US Naval Justice School, Newport, RI, student, September-November 1967; USS *Amphion* (AR-13), legal, diving, weapons OOD, CDO, November 1967-June 1969; US Naval and Army Reserve Center, Hutchinson, KS, CO and Naval Reserve Electronics Facility, Emporia, KS, OinC, August 1969-November 1972; USS *Opportune* (ARS-41) XO, November 1972-July 1975; Naval Reserve Readiness Command, Region 7, Afloat Programs Mgr., July 1975-June 1978; Naval Surface Gp., Western Pacific and Commander TF-73-scheduler, OPS, July 1978-July 1980; Naval Reserve Center, Oklahoma City, OK, CO and Naval Reserve Center, Stillwater, OK, CO, September 1980-July 1984; Atlantic representative, Commander Naval Surface Reserve Force, staff, July 1984-September 1986.

He was awarded MSM (2), NCM, MUC (2), National Defense, Sea Service Deployment, Armed Forces Reserve (2) and Expert Pistol.

His memorable experiences: Salvage projects with OPPORTUNE; Iranian crisis with CTF-73; all the super people at each NAVRESCEN.

Employed as fifth grade teacher, Norfolk City Public School, Norfolk, VA. Presently education officer, Jamestown-Yorktown Foundation, Williamsburg, VA. He is married to Leah Will and they have children, Kathy and LTR. Bruce Duncan, USN; and a grandson, Devon Gardner.

H. LEE DUNN JR., CAPT, born Aug. 7, 1945 in Akron, OH, received the BBA from Kent State University, 1967; Stonier Graduate School of Banking Rutgers University, 1981.

He joined the service Jan. 24, 1968. His military locations and stations: USS *Wasp* (CVS-18); Quonset Point, RI, 1968-71. Served September 1975-February 1977, NAVSHOREACT 7005; March 1977-September 1978, USS *Yarnell* October 1978-February 1979, ANTE DEFCOM 105 March 1979-September 1981 NREDCOM 5; October 1981-September 1990 COMIDEASTFOR 105; October 1990-September 1992 RU-2 Dayton, OH.

He was awarded the Navy Unit Commendation, Meritorious Unit Commendation, Overseas Expeditionary Medal, National Defense Service Medal. Achieved the rank of CAPT and was discharged Sept. 1, 1994. His memorable experiences include Operation Snaptrap and eight annual training exercises in the Middle East.

He is married to Terri and they have children, David and Ryan. Employed as the president, Charter One Bank FSB Western Ohio Division.

EZRA L. DUNNAVANT, LCDR, born May 19, 1924 in Lunenburg Co., VA, received a BA from Lynchburg College, 1950; MDiv Wesley Theological Seminary, 1957; D. Min Lexington Theological Seminary, 1974. He served June 23, 1944-Jan. 31, 1946 in Army Infantry in European Theater.

Joined the USNR Oct. 13, 1958; stationed at NRTF Hagerstown, MD; NRTC Alexandria, VA; Washington Navy Yard; NRTC Cumberland, MD; NRTF Fairmont, WV; NRTC Parkersburg, WV.

ACDUTRA: Norfolk Navy Base 1959; Bainbridge, MD, 1960; Carrier Antietam 1961; Chaplain School, summers of 1963 and 64; Little Amphib Base 1965; NTC Great Lakes 1966; Camp Lejeune RMCB 1967; Wavan Weapons Station, Yorktown, VA 1968; USCGRTC Yorktown, VA, 1969; Naval Hospital Camp Lejuene 1970; Naval Station Annapolis, MD 1971; Cherry Point Marine Air Station 1972; Marine Corps Dev. & Ed. Command Quantico, VA 1975; NNMC Bethesda, MD, 1976.

He received the Armed Forces Reserve Medal and achieved the rank of LCDR. He was discharged July 1, 1978.

Married Edith Higgins and they have four children: Nellie Durham, Kay Slonaker, Terry A. and Anthony L.; grandchildren are Lofton, Amanda and Jonathan Durham; Erin, Bridget, Caitlin and Tara Dunnavant; John, Lena, Anna and Michael Slonaker.

Employed as pastor with the Christian Church Disciples of Christ. Served congregations in Virginia, Maryland and West Virginia. Served in many capacities, most notable position was Moderator of the Christian Church Disciples of Christ in the region of Virginia, which is the highest elected position of that region, retiring August 1990.

FRANK (n) DURBIN III, CWO-4, born in Salem, OR, July 31, 1928. Enlisted in the USN Sept. 3, 1943, 34 days after his 15th birthday. Boot camp at San Diego, CA. Graduated ARM and Radar School, NATTC Memphis, TN; Aerial Gunnery School, NAGS, Yellow Water, FL; and SBD Dauntless Operational Training School in NAS Deland, FL.

Transitioned to SB2C Helldivers in August 1944 and in December 1944 proceeded to the Central Pacific Theater of War, flying numerous patrols and missions as an SB2C dive bomber radio/gunner in the Marianas Campaign and the invasion of Okinawa. Received an Air Medal for service and two Battle Stars. Discharged from the Navy in April 1945.

Affiliated with the Naval Air Reserve in January 1947 serving in Helldiver, Catalina, and Ventura squadrons as radio operator. Abruptly in March 1951 VPML-57 was recalled to active duty for the Korean War, redesignated V-871, and transitioned into PB4Y-2 Privateer aircraft, and transferred to NAS Sand Point, WA. He was assigned as first radioman and right waist turret gunner on the PB4Y-2. Suddenly his crew and one other were dispatched to NAS Atsugi, Japan, joining Detachment Able of VP-28. While with VP-28 he flew over 30 combat missions on "Operation Lamplighter" out of K1 near Pusan, and later out of K14 near Seoul, Korea.

The night missions entailed lighting up North Korean targets for destruction by Marine Night Fighter Sqdn. VMF(N) 513 F7F's and F4U's. On one mission his crew broke the record in one fire fight in which 25 trucks were destroyed at one road block with 30 secondary explosions before the accompanying aircraft was shot down. He was awarded three Korean War Battle Stars. Received Air Medal for service. Discharged 1952, returning to University of California (Berkeley) where he received his BS in engineering.

Re-enlisted as AT1 in the USNR in 1958, serving in Banshee, Tracker, Rotary Wing, and C-118 Transport Sqdn. When the Cuban and Berlin Crisis arose in quick succession he returned to active duty as a USN recruiter for a total of two years, this time receiving advancement to CPO. In summer of 1965 as training chief of an 85 day boot camp, he led them to Number One of 18 commands, and later received commendations for transporting priority cargo to Vietnam as a C-118 crewman. In 1965 he received the RADM Richard D. Fowler Outstanding Reservist Award and shortly thereafter, was promoted to warrant officer.

In January 1968 he was individually recalled to active duty for the Vietnam War and assigned as avionics/weapons officer for the USS *Hancock* (CVA-19). While on *Hancock* he completed two full Vietnam deployments spanning almost three years, taking part in "Operation Rolling Thunder," the bombardment of North Vietnam, Laos, and Cambodia earning his four Battle Stars which included the Tet Offensive. Under his supervision of the DECM shop, NO HANCOCK AIRCRAFT were downed by SAM missiles over both deployments. After a one year tour at Barber's Point as NAESU Team Leader he returned to civilian life in February 1972.

Upon his separation he joined VP-91, an Orion Squadron, where he remained for 11 years as avionics/weapons officer. During his period he served 106 days in the Western Pacific, 35 days in the Aleutians, and 16 days in Spain, earning numerous citations and his Navy Achievement Medal. In 1983 he was assigned as training officer in "Mobile Vans Unit" MMV-0280 and later Diego Garcia AIMD 0180. He retired with the rank of CWO-4 on his 60th birthday, July 31, 1988 with naval service spanning of 45 years. He had the honor to be: the last Non-Flag Rank WWII veteran on the active rolls, the only five time recalled reservist in the US, and one out of 102 men who fought in WWII, the Korean and Vietnam.

He received the following awards: Combat Aircrew Wings (3 Combat Stars), Aircrew Wings (gold), Air Medal (one star), Navy Achievement Medal, Navy Unit Commendation, Navy Meritorious Unit Commendation, Navy Good Conduct Medal, Navy Reserve Meritorious Service Medal, American Campaign Medal, Asiatic-Pacific Campaign Medal (two Battle Stars), WWII Victory Medal, Navy Occupation Service Medal, National Defense Service Medal (one Star), Korean Service Medal (three Battle Stars), Armed Forces Expeditionary Medal, Vietnam Service Medal (four Battle Stars), Navy and Marine Corps Overseas Service Medal (Silver Star), Armed Forces Reserve Medal (two Hourglasses), Korean Presidential Unit Commendation, Republic of Vietnam Cross of Gallantry, United Nations Service Medal, Republic of Vietnam Campaign Medal, Republic of China War Memorial Medal, Navy Expert Rifleman, Navy Expert Pistol (Sharpshooter Award), Navy E (Certificate Only), VUMS Medal (Organizational award from Veterans of Underage Military Service).

Retired science and mathematics high school teacher (San Francisco School District). Life member, NRA. Resides in Daly City, CA. Family includes wife of 34 years, Marion Diana Ferreira (also ex-Navy) and daughter Erin Meredith Tiras.

GEORGE C. DUVALL, LTJG, born July 14, 1909 in Davenport, IA. He graduated from high school in 1927.

He joined as ENS Sept. 5, 1935, and achieved the rank of LTJG in 1960. Duvan was an airline captain and business owner. He retired July 14, 1971. He is a widower and has a daughter Debora J.

BERNARD J. "BERNIE" DVORSCAK, CAPT, born Nov. 20, 1927 in McAdoo, PA, received bachelor's degree in aero engineering, Georgia Institute of Technology, 1956, Class XXI USN Test Pilot School PAX RVR MD, 1959.

He joined the service November 1945. His military locations and stations: NAS JAX VF-117 (Banshees) flying midshipman. Released from active duty August 1953. Served through in USNR through 1975.

Dvorscak achieved the rank of CAPT and was discharged July 1978. His memorable experiences: Carrier qualified in F2H-1 June 28, 1950; MACH 1-05 in F4D-1 Sept. 8, 1954; CO VR-54, July 1975.

He married Carolyn Rose Davie and their children are Michael, Holly and Joan; grandchildren are Gene, Timothy, Luce, Jennifer and Holly. Employed as engineering test pilot, Lockheed, GA, retiring March 1991.

HUGH E. EAGLETON, CAPT, born in Washington, DC and is the son of Sterling and Ruth Eagleton. He is married to the former Jean Hayes Teeter of Oakboro, NC who resides with their three children in Bethesda, MD. Upon his graduation from the University of North Carolina, Chapel Hill in July 1966, he was commissioned an ENS, USNR and subsequently advanced to CAPT to date from Dec. 1, 1987. A Vietnam veteran he retired with 30+ years on Nov. 1, 1996.

Recalled in December 1991 to six months active duty, served as the watch captain, in the Navy crisis action center during Desert Shield/Desert Storm/provide comfort. Previously for three years he qualified and stood the Navy Department duty captain watch in the Navy Command Center, Pentagon. Other reserve assignments included the Readiness Command Staff and command of surface reserve units at Indian Head, MD; Washington, DC; Cumberland, MD and Andrews AFB. He has developed special expertise in JCS/CPX and war games.

His awards include Meritorious Service (2) and Navy Commendation Medal. Post graduate education includes MBA (Finance and Operations Research) from George Washington University, graduate of the American Institute of Banking and the Naval War College. He has lectured at numerous universities and trade groups. An ex-commercial banker, he is in the Office of the Chairman, Federal Deposit Insurance Corporation. Previously, he served with the Comptroller of the Currency, US Treasury Department. Currently, he is completely his DA at George Mason University.

Eagleton's volunteer activities include USNA Sailing Coach; Ruling Elder and Trustee, Chevy Chase Presbyterian Church; director and treasurer, Children's Hospital National Medical Center, and district vice chairman finance, Boy Scouts.

ELLIS R. "ROPE" EASTERLING, CAPT, born Jan. 19, 1928, Hattiesburg, MS. Joined the service in April 1941; stationed at VP-91 and in several ships. He participated in battles and campaigns in the Midway and Solomons.

His memorable experience was being struck by lightning. He achieved rank of CAPT in 1962 and received two Air Medals and Navy Unit Citation.

Married with six children: Mary, Lyn, Liz, Bob, Mickey and Charlene. Retired Jan. 28, 1972 as director of marketing.

GLENN H. "DUKE" EASTON JR., LIFE MEMBER, began his military career as an enlisted machine gunner with Btry. H, 240th Coast Arty. in 1936.

After two years he transferred to the USN and served on active duty from 1938-40, 1943-46, 1950-54 and 1970. He participated in the WWII landings at Leyte, Luzon, Iwo Jima and Okinawa. Later he became CO, NCSO, Seward, AK.

Among others his Naval Reserve billets included CO of a Surface Div., Bn. CO and Gp. Co. After his USNR retirement in 1979 he assisted Secretary Ray Donovan in the US Department of Labor as a political appointee. He is currently president of Glenn Easton and Associates, a management and insurance consulting firm.

Easton is married with four grown children and lives in Reston, VA. For the past 20 years he has been listed in *Who's Who in America* and *Who's Who in the World.*

WILLIAM E. "BILL" EATON, LCDR, born Dec. 11, 1915 in Baltimore, MD. He was commissioned Sept. 27, 1939. Units served with: Navy Yard, Pearl Harbor and Seattle, Boston and New York.

His memorable experience was duty in Navy Yard at Pearl Harbor, Dec. 7, 1941. He achieved the rank of LCDR, USNR.

He is married to Marjorie and they have a son William III. Released to inactive duty June 24, 1946 and retired 1975.

FREDERICK W. EBEL, CAPT, born and raised in Minnesota graduated from the University of Montana with a BS degree in forestry. He attended OCS and was commissioned ENS in 1960. Active duty assignments include the USS *Thetis Bay* (LPH-6) and COMPHIBGRU 3 staff.

As a reservist, he served in Naval Reserve Surface Div., SIMA, VTU and NAVSEA DET units including six years as CO. He attended the Naval War College in 1979 and National Defense University in 1982. Promoted to CAPT in 1981, he served as assistant I.G. and the mobilization plans officer for REDCOM 22. Prior to his retirement in 1990, he was the naval liaison officer to the adjutant general of the Oregon National Guard.

He was awarded the Navy Commendation, Armed Forces Expeditionary, Naval Reserve and National Defense Service Medals plus the Oregon National Guard's Exceptional Service Medal.

He is president of a natural resources consulting company and serves on the board of several business, civic and professional organizations. He and his wife Nancy reside in Colbert, WA. They have two daughters.

GEORGE J. EBERLY III, LT, born in Portland, OR on June 21, 1963. Enlisted in the Delayed Entry Program on Dec. 30, 1980, and entered active duty, Aug. 21, 1981. Graduated from basic training at RTC San Diego in October, and Machinist Mate "A" School in December. Attended Nuclear Power School in Orlando, FL, graduating in March 1983. Completed nuclear power training at NPTU Idaho in September 1983 and was selected as a staff instructor at the S5G prototype.

Transferred to the USS *Georgia* (SSBN-729 Blue) in October 1985 and made four strategic deterrent patrols, qualifying in submarines and receiving a COs letter of commendation for qualifying in less than three weeks into his first patrol. Left active duty in December of 1987 as an MM1(SS). Received NAM, GC, MUC and NDSM.

Joined the USNR in January 1988 and was promoted to CPO in September 1989, while with SUBASE Pearl Harbor DET 322. Applied for commission under the LDO program in the Nuclear Power Field while working as a licensed reactor operator at the Trojan Nuclear Plant. Selected and commissioned in December 1991 as an ENS.

Served as an assistant to the force nuclear officer at COMSUBPAC N4, and as a department head with COMSUBPAC Det 122. Currently serving as XO of SUBASE Pearl Harbor Det 322.

Eberly currently works for Intel Corporation in Aloha, OR. He has a wife, Rinda; a son, George IV; and a daughter, Genifer.

WILLIAM GRAY EBERLY, CAPT, enlisting in 1943 at age 17, he served on active duty during WWII. Qualified as an aerial gunner and aviation radioman, he was awarded Combat Aircrewman Wings. At age 20, having completed primary pilot training, and naval air navigator training, he was commissioned ENS and awarded Aerial Navigator Wings.

Following wartime active duty, he joined the Ready Reserve and affiliated with the last Navy LTA "blimp" squadron. In 1957 LTJG Eberly entered the Air Intelligence branch of the Navy, serving in a succession of intelligence billets until 1983, including three as CO. In 1973, during the US involvement in Vietnam, was recalled to active service for OPERATION HOMECOMING, to aid in the return of American prisoners-of-war from Southeast Asia. He was subsequently awarded the Navy Commendation Medal for his service. He retired from 40 years of active reserve status in 1983.

THOMAS M. "TOM" EDEN JR., born Sept. 6, 1927, Miami, FL, received a BS, Auburn University, Auburn, AL, AU professor, television talent 13 years.

He joined the service August 1945. His military locations and stations: Norfolk, San Diego, USS *Bottineau* (APA-235). Served with Amphibious Forces, NAS, Birmingham, Atlanta. His battles and campaigns: WWII, Korean War.

Eden received medals for Pacific, Atlantic, North Atlantic and Caribbean actions. He achieved the rank of Navy CDR and was discharged November 1952. He retired from the USNR Sept. 6, 1988. His memorable experiences: carrying 2,000 Marines all over the world; landing craft asst. BGC, flying in Europe, CO Air Wing Staff, Atlanta.

His wife Carolyn, died April 1, 1996. Their children are Tommy Eden III, Eve Golden and Steve A. Eden; grandchildren are Stephanie, Sherry, Phillip, Drew, Luke, Ashley. He is the owner of Garden of Eden Services, Inc.

VICKI L. REININGHAUS-EDGAR, *See her bio in R Section*

JOHN S. EDINGER, CAPT, a native of Wilmington, DE, obtained a BA degree from Amherst College and an MBA from Harvard Business School. After completing OCS, he was commissioned ENS in January 1954.

After completing CIC School at NAS Glenview, IL, he served 2 1/2 years aboard the USS *Coral Sea* (CVA-43) on three cruises with the 6th Fleet in the Mediterranean. On Coral Sea, he was a CIC officer, air control and OI Div. officers.

Beginning 1957, inactive duty assignments included CO Naval Reserve Officers School, Gp. CDR Wilmington, DE, ACOS Naval Reserve Readiness Command Region 4 and Naval Representative Military Support Delaware.

Service medals include: Navy Occupation, National Defense Service, Armed Forces Reserve and Conspicuous Service Cross Delaware.

He is a retired department director from Zeneca Inc. His wife of 35 years died in 1990. He has one son, John Jr. and two grandchildren, Lindsey and John III.

PAUL B. EDWARDS, LCDR, born in Ridge Spring, SC, Nov. 12, 1915 and graduated with a BS degree in chemistry from the University of Tampa in 1937. In 1958, he earned a master's degree from Harvard University and, in 1972 a doctor's degree from the George Washington University.

He reported for active duty as an ENS April 16, 1942. His subsequent training included indoctrination at Fort Schuyler, aviation ordnance and gunnery at NAS Jacksonville, Norden bombsight operation and maintenance at Dahlgren Naval Proving Ground, and fire control and flight instruments at Sperry Gyroscope Corporation. He then instructed at Dahlgren, NAS Jacksonville and NATTC Jacksonville, with duties involving flying as a technical observer. He was assistant OinC of Fire Control School (successor to bombsight school) when returned to inactive duty January 13, 1946.

After the war, his participation in Naval Reserve activities included: an aviation unit (AVUA #1, Miami, FL); NRSD 6-50, Tampa; and NROS 6-12 (as instructor). He also performed active duty for training in areas ranging from guided missiles to the National War College. A member of the Ready Reserve, he carried orders to report as CO of an ACORN in the event of mobilization. He completed a number of correspondence courses and was retired with pay on his 60th birthday.

As a civilian, he operated a wholesale and retail hobby supply doing business in the southern US, Cuba and Venezuela. He then served eight years in the public schools of Hillsborough County, FL as a mathematics teacher and Head of Department. Under fellowships awarded by DuPont and the National Science Foundation, he did graduate work at the University of North Carolina, Harvard, George Peabody, and Wesleyan University (Connecticut). In 1963 he joined the Johns Hopkins University's Applied Physics Laboratory where he supervised education and training for a staff of over 2,000 employees and directed a graduate school offering master's degrees in six scientific and engineering areas. He was a member of the principal professional staff when he retired in 1981.

In April 1997 he will have been married for 54 years to the former Sarah Dee Barnes. Their daughter, Susan Dee Edwards, married James David von Suskil (now CAPT USN, retired), and they have two children David Bryan and Julie Elisabeth.

MICHAEL M. EGNETOVICH, CAPT, born Sept. 14, 1944 in Carbondale, PA, received a BS USNA 1967; BBS New Hampshire College, 1971; MS Iowa State University, 1974; MBA Temple University 1984.

He joined the service June 7, 1967. His military locations and stations: NAS Brunswick ME; NROTC Iowa State University, Ames, IA; NAS Willow Grove, PA. He served with units: VR-11, VP-64, ASO 104 HQ (XO), NAVAIRSYS 0366 (CO), NADOC 1O6 (CO), NAVORD.

Awarded the NCM. Achieved the rank of CAPT and retired July 1, 1997. His memorable experiences: mission commander of first reserve maritime patrol aircrew to participate in joint maritime operational exercise with NATO active duty participants in Scotland, 1978.

He is married to Stephanie and they have two children, Michael Steven and Jonathan Mark. Employed as director Software Productivity Consortium; vice-president Business System's Integration and Research, Inc., retiring July 1, 1997.

LANCE D. EHMCKE, CAPT, a native of Cleghorn, IA, and graduated from Iowa State University and the NROTC program in 1972. His Vietnam service was in USS *Midway* (CV-41) and USS *Welch* (PG-93). He served as Stores and ADP Officer in USS *Jason* (AR-8). All ships where CAPT Ehmcke served as a supply officer were awarded the Supply Blue "E." Frequently deployed overseas, he is a Blue Nose, a Shellback and a Golden Dragon.

CAPT Ehmcke served on two NATO/JCS strategic exercises from response centers at CINCUSNAVEUR, London, UK. His reserve specialty is supply planning and integrated logistic support. He was the first CO of NR-NSD Subic Bay B-816, Sioux City, IA, 1984-87, and HQ Commander at NR-NSD Subic Bay HQ B-516, Des Moines, IA, 1988-90. He is a life member of the NRA.

CAPT Ehmcke is a private practice, business attorney in Sioux City, IA. His family includes his wife Kathy and daughter Elizabeth.

RICHARD L. "DICK" EHR, CAPT, born Aug. 24, 1932 in Muskegon, MI, received BSAE Notre Dame; MSEE Marquette University.

He joined the service June 1954, flight training Pensacola, September 1954-July 1955. Designated naval aviator December 1955, Corpus Christi July-December 1955.

His military locations and stations: VA-145, NAS Miramar, FASRON 4, NAS North Islands: Reserves: VS-722, VS-41V1, VP-RTU, AU NAS Glenview.

Ehr achieved the rank of CAPT. Retired from active duty June 1959 and Reserves June 1975. His memorable experiences: flew AD-6 off USS *Hornet* (WESTPAC January-July 1957) flew T-28 AD-5/6, F9F-8 F2H-1, F3D-2, C-1, S-2 and AWD P-2 Aircraft CO of OPCON UMT at NAS Glenview 1971-72.

He is married to Luanne and they have four children: Deborah, Stephen, Andrew and Amy; and grandchildren: Joan Slayton and Alex, Kelsy and Kiersten Ehr. Employed as CAPT with American Airlines (1965-92).

JOHN D. EICHER, CAPT, born Feb. 10, 1918 in Seattle, attended the University of Washington, business and NROTC.

He joined the service in July 1940 and served in the USS *Pennsylvania* (BB-38). Served with the USN and USNR and received 13 Battle Stars.

Eicher achieved the rank of CAPT and retired in 1960. His memorable experiences: was officer-of-deck Dec. 7, 1941, 0400-0800 watch (BB-38) Pearl Harbor.

He is married to Betty J. and they have three children: John G., James N. and Shirley; and seven grandchildren. He is co-owner of Eicher-Richards Co. Seattle, (25 years).

GUINN L. EILAND, CAPT, born July 6, 1934 in Mermentau, LA, received a BBA in engineering route to business at University of Texas.

He joined the service Sept. 12, 1956 (Officer Candidate School, Newport, RI); Aviation Ground Officers School, Jacksonville, FL) Naval War College, Newport, RI. His military locations and stations: USS *Independence,* USS *F.D. Roosevelt.* Served with: VF-43, VF-11, VR-703, VR21D3, NRS 70D1, NROS8-4(M), NARDIV D2, RTU-52-1, KRU 5111, VTU 510, VT4 1007.

Eiland was awarded the Armed Forces Reserve Medal. Achieved the rank of CAPT. Released from active duty Feb. 29, 1960, transferred Jan. 1, 1979 Retired Reserve without pay.

His memorable experiences: as personnel officer achieved the highest re-enlistment rate in their units (active and reserve). This was accomplished by an aggressive program of providing information on rights, benefits and opportunities. Treating each person as though they were special-even to assisting in preparing resumes to get a job if they left the Navy.

He is married to Beverly Ann and they have three children: Gary, Gerald, Julie Futrell; and grandchildren: Geremy, Erin and Grayson Eiland; Hayley Futrell. Employed as cost analyst w/U.S. Gypsum-owner/operator dry cleaners, shipping specialist w/ Exxon Chemicals Co., retiring July 6, 1994.

JOHN A. EKLEM, LCDR, native of Logansport, IN. Enlisted at Indianapolis, IN in March 1943, and attended boot camp and Machinist's Mate School at Great Lakes. Assigned to USS *Kenmore* (APA-162/AK-221) November 1943. Received four Battle Stars in the Pacific Theatre. Discharged as MML1/c in April 1946. Re-enlisted in the Reserves in 1948 and commissioned ENS after graduating from Purdue University in 1952. Recalled for Korean Conflict and assigned to USS *Frontier* AD-25. Released to inactive duty January 1954.

Employed at Delco Electronics for 27 years in Kokomo, IN as general supervisor in semi-conductor production. Participated in various reserve units attaining rank of LCDR. Retired in 1976 with 31 years active and reserve service. Wife Betty, and three children: Julie, John and Jennifer. Activities are Blue and Gold Officer, church treasurer, state coordinator for AARP Tax-aide Program and member of AARP State Leadership Council.

NORMAN O. ELDRED, LTJG, was commissioned ENS, USNR, line engineering officer in June of 1943, following USN V-7 Training aboard the USS *Prairie State* (former Battleship *Illinois*), 137th St. and Hudson River, New York City.

He was assigned directly to the destroyer USS *Henley* (391), DesRon 5, 7th Fleet based at Buna, New Guinea, Asiatic-Pacific waters. On Oct. 3, 1943 while engaged in violent anti-submarine action, the destroyer USS *Henley* was torpedoed and sunk by a Japanese submarine, off Morobe New Guinea.

Ensign Eldred's next assignment was at the US Navy Graduate School of Electronics (Radar) at Harvard University, The Massachusetts Institute of Technology, Boston, MA and the Bell Telephone Labs for War Training, New York City.

As a LTJG SE3T, USNR he was then assigned to CDR Service Force, Pacific Fleet, Pearl Harbor, HI. His awards include the Purple Heart, Bronze Star, Asiatic-Pacific, American Theatre and Victory Medals.

Received his BSE in chemical engineering in June 1938 at the University of Michigan, Ann Arbor, MI. Masters work in Ch.E. during 1946-47. National Honorary Scholastic Society memberships include Phi Lambda Upsilon, Phi Kappa Phi, Sigma Rho Tau and Phi Eta Sigma. His professional accomplishments have been cited in biographical listings including *Engineers of Distinction, FIDEC-International Federation of Consulting Engineers, Who's Who in Engineering, Who's Who in Atoms* and *Chemicals Whos Who.*

He is a member of AICh.E, ASME, IEEE, ASTM, NSPE, ISPE and the US Naval Institute of Annapolis, MD. Eldred is a registered professional engineer (State of Illinois) and is engaged in consulting engineering in nuclear reactor design and petroleum-refineries.

Single, he is a member of the Chicago Chapter of the MOWW, a life member of the ROA, VFW and the NRA. A PUFL member of the American Legion. Other memberships include the American Defense Preparedness Association (Michigan Chapter) and the USN Institute, Annapolis, MD.

RICHARD G. "DICK" ELLIOTT, born July 28, 1929 in San Francisco, CA, received a BS and ME in civil engineering. He joined the service June 6, 1952. His military locations and stations: Sea duty (AKA 104), USS *Seminole* and various Naval Reserve units.

Elliott served in the Korean campaign and received the National Defense, United Nations and Korean Service w/2 stars. He achieved the rank of CAPT and was discharged March 1, 1973.

He is married to Sandra and they have four children: Catherine, Patricia, Susan and Karen. Employed as deputy district director Operations CALTRANS, retiring Dec. 31, 1991.

EUGENE D. "GENE" ELLIS, born Aug. 10, 1932 in Kosciusko, MS. Earned BBA with distinction from the University of Mississippi. Joined the service as midshipman in 1950-54 and was commissioned as ENS June 1954.

His military locations and stations: Naval flight training, Pensacola, FL and Kingsville, TX; NAS North Island, NAS Moffett Field; NAS Alameda, USS *Missouri;* USS *Pittsburg;* CVS-47; USS *Philippine Sea;* NAS Los Alamitos; NAS Calendar Fields, New Orleans.

He served with VS-38, VS-37, VS-821, CORTU-81, CO OPCON NAS Moffett, Chief of Staff, Safety NAS Alameda VPU-5120. Ellis was awarded the Navy Expeditionary Forces Medal, National Defense Medal and Navy Reserve Medal. Achieved the rank of CAPT. His memorable experiences: Served 11 months on West Pac cruise, CVS-47 (VS-37) 215 day and night straight deck LSO landings.

Recalled in Cuban Crisis served in VS-821/eight months GITMO and NAS Key West flying anti-shipping patrols off Cuba in the Devils Triangle - asst. ops officer. Briefing Admiral daily in GITMO. XO-VS NAS Los Alamitos; CO, RTU 81 NAS North Island; CO OPCON NAS Moffett Field. It was a great experience and a real challenge working with the best officers and men.

He is married to Marianne and has two children: Eugene David and Martin Daniel; and five grandchildren: Daniel Gene, Amber, Alisha, Stephanie and Ryan. Retired Sept. 2, 1992. Employed at Holiday Inn, El Paso, manager, United Air Line, CAPT. Flew B727, DC-8, DC-10.

JOHN R. ELLIS, CDR, born in Panhandle, TX, Dec. 5, 1932. Finished two years at North Texas State College in 1951. Enlisted as E-1 in the Naval Air Reserve January 1951. Active duty as NAVCAD in November 1951. Commissioned ENS and naval aviator June 1953 at age 20.

Joined VF-174 "Hell Razors" of Air Gp. 1 at JAX October 1953 and cruised aboard USS *Midway* CVA-41 round-the-world December 1954 to July 1955. Midway joined 7th Fleet off Formosa (now Taiwan) for show of force to Red Chinese who were shelling Formosa daily.

Released from active duty January 1956 and went

into Ready Reserve at NAS Dallas. Earned BS in 1957. Affiliated with VR-704 etc. and ended Ready Reserve at NAS Dallas. Affiliated with VR-704 etc. and ended Ready Reserve career in 1971 as CO of NARDIV D-1, Dallas.

Semi-retired from insurance-finance career with wife, Mary. They have five children: Rick, Becky, Ann, Cecile and Janene; and eight grandchildren. Officially retired December 1992.

RICHARD M. "RICK" ELSTER, born March 5, 1941 in Brooklyn, NY. Attended high school Bangor, ME, 1959, and the University of Pittsburgh Class of 1974. Enlisted in USNR 1958, shipped over to USN 1959, discharged from active duty November 1966 and enlisted in USNR June 1984 as an SK1 under the APG program.

Active duty stations: Texas Reserve Fleet, Orange, TX; USS *Waller* (DDE-466), Norfolk, Sub School, Groton, USS *Redfish* (AGSS-395), San Diego, USS *Kamehameha* (SSBN-642) Guam. Memorable experiences include: 1962, awarded Silver Dolphins; 1986, becoming chief petty officer; 1989, commissioned ENS-through the LDO program.

He was awarded two Navy Achievement Medals, two Good Conduct Medals, Naval Reserve Meritorious Service, Naval Expeditionary Medal, two National Defense Service Medals, Armed Forces Expeditionary Medal, Vietnam Service Medal, Armed Forces Reserve Medal, Military Outstanding Volunteer Service Medal. Achieved the rank of LT.

He is married to Judith Ann "Judy" and they have two children, Bonnie and Sherri; and granddaughter, Sydney,. He is employed with Veryfine Products, Inc.

VERN EMBERY, CDR, attended SDSU San Diego, CA. Began his naval career in 1942 as an aviation cadet, was commissioned ENS and designated naval aviator #1638 4-44 at Corpus Christi, TX. Progressed through civilian pilot training at Sky Harbor Tempe, AZ; pre-flight at Del Monte, CA; "E" base flight training at Livermore, CA, and subsequent graduation at Corpus Christi, TX.

After commissioning additional instructor training was completed at Instrument Flight Instructors School NAS Atlanta, GA. Assignment was to basic and instrument flight instructor duties with Sqdn. 13-C at Bellville and Cudihy Fields, TX.

In 1946 active duty was terminated and CDR Embery continued to serve in the USNR at NAS North Island and Los Alamitos Air Stations until retirement in 1966. During the Reserve period he served in the following squadrons: FASRON 68, VF778, VF-79E, VA-771, HU-771, HS-771 and AWS77L. Aircraft: SNJ, SNB, F6F, F4U, TBM, AD, TV and FJ-1. Served in many billets and as XO in AWS. Active in recruiting and cadet selection boards.

He has 3,000 hours in 16 different single and multi-engine military aircraft plus 1,500 hours in 40 different varieties of civilian aircraft. Many of these civilian hours were accumulated on training and search missions as a LTC in Sqdn. 57 of the CAP. He served CAP as a Command, mission and check pilot for 20 years, was active in the San Diego Chamber of Commerce, Mission Bay Committee, past president of the San Diego Exchange Club and life member of TROA .

The Naval Reserve was a most memorable experience because its personnel contained such diverse expertise in military and civilian accomplishments: Ted Williams, baseball great and Marine aviator; Zeke Cormier, cheerleader, naval aviator, Blue Angels CO; ENS Bray, Chicago Bears football star athletic, instructor, pre-flight Del Monte; ENS Aerroge, ski instructor; Lana Turner's husband, an "E" base flight instructor; Tyrone Power, movie star, Marine instrument flight instructor. There were many others. Where else could there be the opportunity to mingle and learn from and enjoy the comaraderie of such individuals other than national military reserve service. It was and is an unparalleled experience.

Embery married Shirley Irene Johnson and they have three children: Linda, Joan and Janet; and two grandchildren, Jennifer Cole and Lisa Williams. He is employed in management with Packard Bell as staff accountant, traffic superintendent, retiring in January 1979.

EMERSON EMORY, CAPT, born Jan. 29, 1925 in Dallas, TX. Attended Prairie View A&M University. He received a BA cum laude from Lincoln University, PA and a MD from Meharry Medical College.

He joined the US Army in 1943 and the USN in 1949. His military locations and stations: Camp Wheeler, GA; Ft. Jackson, SC; Ft. Benning, GA; England; France; Germany and Philippines. He served with 2515th ASTU, 3617 QMCTK Co., 3533, QMC, NAS Dallas, NAS Los Alamitos. His battles and campaigns: ETO, AP.

Emory was awarded: ETO, American Defense; AP, WWII Victory, Philippine Liberation, Naval Reserve w/2 HGs. He achieved the rank of CPL, US Army; CAPT, USN. Discharged from Army April 1946 and Navy, 1979.

His memorable experiences: volunteer physician, Vietnam 1966. Served as a member of Sect NAV; Advisory Board for Retired Personnel. Served as area coordinator; US Naval Academy, Blue & Gold Program. Founding president, National Naval Officers Association.

He is employed as a physician in private practive, Psychiatry and Int. Medicine.

ALLEN H. EMRICH, LT, born Dec. 13, 1927 in Columbus, OH, received a BFA, BBA and MBA. He joined the service September 1946 NROTC Ohio State University (1st Alternate Anapolis). His military locations: USS *Tripoli* (TCVE-64).

Awarded the National Defense Service Medal and the Korean Service Medal. Achieved the rank of LT (permanent) 493803/105.

Memorable experiences include active duty December 1951-December 1953; Naval Research Reserve CO; atomic defense officer. He is a life member NRA.

Married to Rachel K. and they have two children, Matthew and Todd; and two grandchildren, Alexander and Zachery. Employed as ceramic engineer; accountant, retiring in 1993.

WEYLIN G. ENG, CAPT, a fifth generation Chinese American native of Oakland, CA, was born on Nov. 22, 1942. The Eng family has a proud tradition in the USN. His father, Dr. Raymond L. Eng a Navy League Member, was the industrial eye safety consultant to the Naval Supply Center Oakland. CAPT Eng's uncle, Chief Petty Officer Timothy Eng Lenneau, was among the first Chinese American to serve on active duty in the USN during WWII and completed over 35 years in the USNR. CAPT Eng's brother, Wellington R.L. Eng, is also a naval reserve captain, Dental Corps.

CAPT Eng received his BS and doctors of Optometry from University of California Berkeley. He was commissioned as ENS, Medical Service Corps in 1965 prior to his first active duty station at the National Naval Medical Center, Bethesda, MD, and was promoted to CAPT in 1984. His 30 years naval career included assignments to Naval Hospital Oakland, Naval Aerospace Regional Medical Center, Pensacola, Bureau of Medicine, Washington, DC and as "Ship's Optometrist" on the USS Carl Vinson and USS Abraham Lincoln. He conducted clinical research for the Navy in environmental medicine and industrial eye safety and pioneered vision research with high altitude wearing of contact lenses. His works has been published in US Navy Medicine, Journal of Aerospace Medical Association, Reader's Digest and Encyclopedia Britannia.

CAPT Eng was commanding officer of Naval Hospital Oakland 320, First Mar. Div. Med. 120 and National Disaster Medical System 120. In 1991 CAPT Eng was recalled for active duty during Operation Desert Storm. One of the most challenging "job opportunity" in his naval career was to organize, implement, and to coordinate the first Joint Service Military Reserves (Navy, Army, Air Force) Fleet Eye Care Program for aircraft carriers berthed at NAS Alameda, CA. Currently he is the Navy Professional School Liaison Officer to University of California Berkeley.

CAPT Eng's civilian activities included: President, Alameda and Contra Counties Optometric Society and Director of Clinics of the University of California Berkeley School of Optometry. Currently he is appointed as associate clinical professor and director of professional affairs at University of California Berkeley. He also maintains private practices in Oakland and Orinda, CA. He is active in the Orinda Rotary Club and Orinda Masonic Lodge. CAPT Eng is a founding member of the Armed Forces Optometric Society and Association of Medical Service Corps Officers Reserves. His decorations include Meritorious Service Medal, Navy Commendation Medal and the Meritorious Unit Commendation Medal. CAPT Eng is married to the former Roselyn Yagi of Honolulu, HI. They have four daughters: Kristine, Tiffany, Shannon and Aimee.

HAROLD G. "ACE" ENGLE JR., CDR, born Jan. 20, 1930 in Hershey, PA. He joined the service April 29, 1952. After graduating from Officers Candidate School he was assigned as maintenance officer to the US Air Antisubmarine Sqdn. VS-22 at NAS Norfolk, VA.

During his last 18 1/2 years of active service he was stationed at Naval Reserve Air Station Willow Grove, PA. There he was assigned as maintenance officer to squadron VS-932, 934 and VP-64.

Engle retired Dec. 31, 1974, from active naval service with 21 1/2 years and had achieved the rank of CDR. Memorable experiences include while serving as maintenance officer for US Air Antisubmarine Sqdn. VS-934 at NAS Willow Grove, PA their squadron was awarded The Noel Davis Trophy for the year 1966. He is married to Doris H. Engle.

PAUL M. ENGLE, LCDR, born Jan. 20, 1922 in Dayton, OH, received a BChE 1943, University of Dayton; MS Che 1954 Ohio State University.

He joined the V-7 Program in 1942. Active duty May

6, 1943 at USNA. His military locations and stations: USNA, Harvard, MIT, Norfolk Navy Yard. Served with USS *Prince William* (CVE-31). Served during battles and campaigns in Atlantic-Morocco, Pacific, Kwajalein, Ulithi, Guam, Philippines. He continued in reserve units. He has Reserve retirement.

Engle achieved the rank of LCDR and was discharged Feb. 13, 1946. His memorable experience was the crash of F4U into ships bridge structure and the subsequent fire.

Married to Emma May and they have five children: Joan, William, David, John and Thomas. Employed as plant and division manager, Carrier Corp. He retired September 1982.

LEON JAMES ENTAS, CAPT, born Oct. 26, 1936 in Boston, MA, received his BS from Boston University. Graduate of ONI Basic Agents School; Amphibious Warfare; Industrial College of the Armed Forces; various ASW schools; Armed Forces Air Intelligence Training Center; CI School; NILO: SERE; Vietnamese Language; and Defense Intelligence School.

Served with Units: (1953-63) VA-911, S. Weymouth NAS, MA; NAS Denver, CO; NAS New York, NY; VS-911 NAS S. Wey; VR-911 NAS S. Wey; (1963-72) VP-911 NAS S. Wey; NAIRU-ZI NAS S. Wey; AFAITC, Denver, CO; Amphib Base, Coronado, Camp Pendleton, CA; Vietnam: NILO Quang Tri; COMNAVFOR; MAC ADV Team 19; NAVADVGRP, Team 143, Intel Officer (TAR); Military Region One; NAVSUPACT Da Nang; (1973-88) DIA, NMIC; JCS, NMCC; JCS, NEACP; NAVINVSERVHQ; NAF Washington, DC.

Awards and decorations include the Legion of Merit, AF Aircrew Insignia, Bronze Star w/Combat V, Meritorious Service w/Gold Star, Air Medal (3 Awards), Combat Action Ribbon, Joint Service Commendation, Joint Meritorious Unit (DIA), Navy Unit Commendation (COMNAVFORV, NAVSUPACT, DaNang), Meritorous Units (Navy Advisory Gp., Team 143, NAVINVSERV), AF Combat Readiness, Vietnamese Honor Medal, 1st Class, Vietnamese Gallantry Cross and Civic Action Rifle Expert, Navy Good Conduct, Naval Reserve Meritorious Service, National Defense, Armed Forces Reserve (three awards), Vietnamese Service (four stars) and Vietnamese Campaign.

He is married and has three children: James Frederick, Jeffrey Pierce and John Michael. Employed as Defense Intelligence Agency, senior intelligence officer.

THOMAS E. "TOM" ENTENMANN, CDR, born Oct. 15, 1922, born in Philadelphia, PA. Attended FKD High School, Philadelphia; BS Millersvill University; MS NYU; Port MS Temple University.

He joined the service Nov. 20, 1942 in Philadelphia, PA. His military locations and stations: Midshipman School, Northwestern University; European Theater; Korean Conflict. Served with LST 77; LST 551; European Theater, invasion of Southern France; ARL121-Korea. Battles and campaigns: Invasion of Southern France on LST77; flew 1952 (Carregian) on ARL 21 (USS *Proserpine).*

Entermann was awarded the American Defense, EAME w/star; Naval Reserve Medal and the Victory Medal. He achieved the rank of CDR, five years. Transferred to Retired Reserve on April 1, 1973.

His memorable experiences: his five years of active duty on amphibious type ships and rounding out his 20 years in Reserves as CO of two surface divisions; one fleet division; one mobilization team and finally as a USNA information officer. Also served as chief-of-staff of a Naval Reserve Surface Bn.

He married Margaret Louise "Peg" and they have four children: Richard and David (twins), Thomas Jr. and Martha L. Covert. Their grandchildren are Christine, Scott, Thomas, John, Todd, Derek, Lindsay, William and Emily. Employed with the school district of Philadelphia. Teacher of Industrial Arts; supervisor of I.A.; assistant director of Career Education. Retired in 1979 (35 years service).

DONALD C. EPP, LCDR, born Nov. 29, 1924 in Kansas City, MO, attended college for four years. He joined the service Dec. 3, 1942. His military locations and stations: NAS Corpus Christi, Cabiness 12C; NAS Olathe, KS; VF 88-2, 3, 4, 5, 6.

He served with the National Treasurer 1967-68 and is life member #440, NRA.

Epp was awarded the American Theater Medal, WWII Victory Medal, Navy Reserve Medal, Armed Forces Medal and achieved the rank of LCDR, Naval Aviator, 1945. Never discharged from the service. Retired Nov. 29, 1984.

His wife, Martha A. is deceased. He has a daughter Carolyn and grandchildren, Christina and Lily. He retired Nov. 29, 1984.

ELWOOD W. EPPS, CDR, born in Carter, OK, May 14, 1931. He graduated from Biola University, Wheaton College (IL) and Eastern Baptist Theological Seminary. He pastored three American Baptist Churches in Clifton, NJ; Ventura, CA and Taft, CA.

Epps was in the USNR from 1962 until 1986 including active duty from 1965-70. His active duty was with 3MARDIV, Vietnam; NTC, San Diego; and the USS *Estes.* Reserve duty was with various units at several Navy and Marine Corps Reserve Training Centers: Los Angeles, Santa Barbara, District of Columbia, Austin, San Antonio and Phoenix.

His campaign and service medals are: Republic of Vietnam Campaign Medal and Device, Vietnam Service Medal, National Defense Service Medal, Navy Unit Commendation, Presidential Unit Citation, Combat Action Ribbon, Naval Reserve Medal and Armed Forces Reserve Medal.

Since completing a residency for clergy at the National Institute of Mental Health he has worked in behavioral health services for Texas Rehabilitation Commission, Mesa Community College, and Arizona Department of Economic Security. He is now retired and enjoys part time work, jogging, reading, traveling and his family.

His family consists of wife Patricia, son Steven, son Wade (a graduate of the Naval Academy), daughter Vernie Keim, son-in-law Howard Keim Jr., grandson Jeffrey and granddaughter Taylor. They now live in Arizona.

RICHARD A. "RICH" ERHARDT, CDR, born Jan. 24, 1947 in New York City. Received a BS degree in electrical engineering, Renssalaer Polytechnic Institute, Troy, NY, 1969; JD Golden Gate University 1977, San Francisco, CA.

Joined the service June 12, 1969. His military locations and stations: flight training school, Pensacola, FL. Active duty, Jacksonville, FL. Reserves at Pt. Mugu, Moffett Field and Alameda, CA. Erhardt served with VP-5, VP-65, VP-91, Naval Air Systems Command Units. He achieved the rank of CDR 05 and was discharged July 1, 1994.

Received the Overseas Ribbon w/star, Armed Forces Reserve Medal, National Defense Medal, Navy Pistol Marksmanship Ribbon w/E, Humanitarian Service Medal, Squadron Excellence "E" Ribbon, Meritorious Unit Commendation.

He is single and has two children, Kimberly and Christopher. Employed as captain, United Airlines, flying Boeing 727.

CHARLES A. "CHUCK" ERICKSON, CAPT, born Dec. 9, 1949 in Seattle, WA, received a BS degree in commerce from the University of Virginia; and MBA from Pepperdine University.

He joined the service June 5, 1971. His military stations: NASNI; NAWS Pt. Mugu, CA. Served with HSL-31, HSL-33, HAL-5, HCS-5. His battles and campaigns include Desert Storm.

Erickson was awarded the Bronze Star, Meritorious Service Medal and the Air Medal. He achieved the rank of CAPT and was discharged March 31, 1978. His memorable experiences: CO HCS-5.

Married to Penny and they have two children, Nikki and Lindsay. Employed as a senior marketing manager with Litton Aero Products.

HILMER "LEIF" ERICKSON, CWO4, born June 9, 1932 in Minneapolis, MN, received an AS in criminology from Miami Dade and Naval War College, 1964; BA, anthropology, 1965 and master education, 1970; executive cert. public admin., 1993.

He joined the service January 1951. His military locations and stations: MCAS, Opa Locka, FL; USS *Norton Sound* AVM. Erickson served with: NRSD 6-44, NRSD 6-39, FASRON 805, IRU, Duluth NRTC, MAS. Participated in Korean Conflict. He received the National Defense, Armed Forces Reserve, Navy Achievement, Meritorious Service Medals and was inducted into the Alumni Hall of Fame, Florida Atlantic University, Nov. 12, 1990. Achieved the rank of CWO4 boatswain and was discharged June 1992.

Memorable Experiences: mobilization assignment with CIA. Worked subversives 1965-68 in conjunction with government civilian job—Top Secret Clearance.

Life member of ROA, American Legion, NRA, USN Institute, National Rifle Assoc., US Army Warrant Officers Assoc. and others. He married Gloria A. Olesen and was employed as detective sergeant, senior counselor, Department of Labor, retiring in June 1992.

CLIFFORD E. ERNST, CDR, accepted commission as ENS while in medical school at Indiana University. Appointed May 15, 1943 V-12(3) USNR. Discharged from H-VP Unit, Indiana University School of Medicine, Aug. 23, 1943. Appointed assistant surgeon with rank of LTJG MC-V(g) USNR, Aug. 22, 1943.

Served a nine month internship Indianapolis Methodist Hospital, Indianapolis, IN, August 1943; nine month residency in pathology, Indianapolis Methodist Hospital, May 1944, ordered to active duty, March 23, 1945, reporting to USNH, Portsmouth, VA; reported aboard USS *General Tasker H. Bliss* (AP-131) as junior medical officer, senior medical officer as of July 1946. Separated from active duty through Great Lakes Naval Station, 9th Naval District, Aug. 17, 1946. Appropriate duty with pay, USNR Training Center, Indianapolis, IN, Jan. 27, 1953.

Ernest practiced medicine in Phoenix, AZ from 1954-93. Attached USNR Naval Training Center, Phoenix, AZ, June 1960. Promoted to CDR in 1973. Retired from the USNR in 1978.

He was awarded the WWII Victory, American Theater, European-African and Asiatic-Pacific Ribbons.

Married Mary Frances Stansel and they have three children: Paul, James and Judith Ann; and grandchildren: Jason, Tyler, Michael and Stephanie. Retired from medical practice in 1993.

EDWARD E. ERPELDING, CAPT, born in Ft. Wayne on July 30, 1940. He attended the Athenaeum of Ohio, Xavier University, St. Francis College (IN), Ball State University and Catholic University of America, obtaining a BA, MA in Ed, and MA in Th (honors program). He was a hospital chaplain, parish curate and licensed social studies teacher and principal in Ft. Wayne, Huntington, Mishawaka and South Bend.

Commissioned a LT, CHC, USNR-R, Dec. 7, 1970. He served units of N&MCRTC Ft. Wayne and AFRC, So. Bend (70-74). Recaled to active duty he was assigned with: the Coast Guard at Petaluma (1974-77) and Kodiak (1977-78); Navy civilian graduate studies (1978-79). He served tours: as pastor, instructor and hospital chaplain at Newport (1979-82); regimental chaplain, 4th Marines, Okinawa (1982-83) and USCG, Gusahi; pastor, hospital chaplian NAS Corpus Christi, ADDU to USCG (1983-86); ship's chaplain USS *Guam* (LPH-9), USS *Guadalcanal* (LPH-7) (1986-88); director of Credo, San Diego (1988-91); command chaplain NSB New London/ SUBGRU TWO (1991-93) and USS *Dwight D. Eisenhower* (CVN-69)/IKEBATGRU (1993-96).

Erpelding participated in operations: Earnest Will, Nimble Archer, Uphold Democracy, Vigilant Warrior, Southern Watch, Deny Flight, Provide Comfort and Sharp Guard.

He was awarded the LOM, MSM w/Gold Star, NCM, NUC, NMUC, Navy E, NDSM w/Bronze Star, AFEM w/ Bronze Star, SWASM w/Bronze Star, AFSM, SSDR 2/ Bronze Stars, OSSR, NATO Medal.

Retired July 1, 1997 and now pastors St. Martin de Porres Parish, Syracuse, IN. Life member of Naval Reserve Association #1392.

ROBERT D. "BOB" ESPESETH, CAPT, born July 11, 1930 in Cameron, WI, received a BS, 1952 and MS 1956 from the University of Wisconsin-Madison.

He joined the NROTC, September 1948 and was commissioned June 1952. His military locations and stations: USS *Purdy* (DD-734), June 1952-June 1954. Served with NRSD 9-212, Madison, WI; SSS Units Madison, WI, Detroit, MI and Springfield, IL

His battles and campaigns include Korean War April 1953-August 1953. Espeseth was awarded the Navy MSM, Commendation Medal, SSS Exceptional Medal and the Army Civilian Service. Achieved the rank of CAPT, USNR and was discharged June 1982.

Memorable Experiences: while aboard the USS *Purdy* they made an "Around the World Cruise," a Med cruise and three Caribbean cruises. They visited 23 foreign countries and had travel opportunities he never dreamed of before the Navy.

Married 44 years to Mary Ann and they have five children: Robert Jr., Michael, Steven, David and Karen; and five grandchildren: Brett, Abby, Hannah, Michael and Gretchen. Employed as chief park planner, WI DNR; professor, University of Illinois, retiring September 1995.

ROBERT C. "BOB" ESTERBROOKS, born April 29, 1924 in Glendale, CA, received a BS in civil engineering from the University of Texas and a masters in engineering from UCLA.

He joined the service Dec. 4, 1942. His military locations and stations: V-12, University of Texas; 35th Naval Construction Bn. Philippines, Naval Torpedo Station, Keyport, WN; NCB Ctr. Port Hueneme, CA. Battles and campaigns: Pacific Theatre, WWII, Philippine Theatre, WWII.

Esterbrooks was awarded the Legion of Merit, Area Medals and Victory Medal. He achieved the rank of RADM, CFC and was discharged May 1, 1984.

He is married to Viola and has three children: Linda, James and Sandra; and five grandchildren: Alice, Ann, Brendon and J.D. Castile, and Brooklyn Esterbrooks. Employed as assistant city engineer, San Diego; city engineer, Phoenix, assistant county manager, Phoenix. He retired Nov. 1, 1996.

ALBERT CHARLES "ACE" EVANS, CAPT, born Oct. 28, 1929 in Chicago, IL, Cook County, received a MBA from DePaul University.

He joined the service June 6, 1952. His military stations and locations: USS *Northampton* (CLC-1); USS *Henrico* (APA-45). Served with PHIBRON 1. He was awarded the China Service Medal and achieved the rank of CAPT. Discharged Aug. 5, 1957 and retired from USNR, June 30, 1985.

He is married to Lucy Pumilia and they have two children: Christopher Charles and Kathleen Frances; and a granddaughter, Rebecca Lynn Evans. Employed with IBM in marketing, retiring June 30, 1987.

JOHN TEMPLAR EVANS, CAPT, born Feb. 13, 1915 in Columbus, OH, received AB from Yale University in 1937; MS from Brown University 1940; MA Harvard University and PhD 1950.

Commissioned NROTC at Yale, June 22, 1937. Division officer, Organized Reserve, Pawtucket, RI 1938-40. Active duty June 15, 1941. *Rodman* (DD-456) 1941-43; *Braine* (DD-630) 1943-45; CO *Gridley* (DD-380) 1945. Areas served included North Atlantic, Central Atlantic, Middle Pacific to South Pacific to South China Sea.

He was awarded the Bronze Star w/Combat V, Naval Reserve Medal, American Defense, European-Africa Theatre w/2 stars, Asiatic-Pacific Theater w/6 stars, Occupation Medal, WWII Victory Medal, Armed Forces Reserve w/HG for second award, Philippine Liberation w/2 stars, Philippine Presidential Unit Citation, and Russian Commemorative Medal.

He left active duty March 1, 1946 as CDR. September 1946 assumed command of Quincy-Hingham Bn. 1-6 (to March 1947). August 1949 joined the Naval Reserve Medical Co. 1-1, Boston, MA (His line commission was changed at his request to Medical Service Corps Oct. 6, 1948). Promoted to captain, MSC, USNR, Sept. 9, 1957. July 1, 1969 assumed command of Medical Co. 1-1 (until June 30, 1972).

Retired with pay having reached the statutory age, Feb. 13, 1975. Married April 8, 1942 to Mary Alma Sjobeck (deceased March 25, 1995). He has two children, Albert Ellis (1950) and Pamela Ann (1952); and five grandchildren: Brian Mathew, Kristine, Jennifer, Brian A. and Joseph. Employment included 1946-49 Mental Hygiene Clinic, VA Boston, MA. Naval Hospital Boston, MA 1949-73. As clinical psychologist Jan. 1, 1974. Retired from government service (although was not retired from USNR until Feb. 13, 1975).

MILES J. "EVIE" EVANS, born Oct. 18, 1915, Pleasantville, IA. Attended Knoxville, IA High School 1933. Attended University of Iowa, 1933-34 and again starting in 1939; USN Hospital Corps School 1st Class to Recomm. April 1935; San Diego, CA, Class "A" Operating Room Tec., 1935; College of Mortuary Science, St. Louis, MO, 1946; mortician 1947.

Enlisted December 1934, USRTC San Diego, CA; USNH Corps School 1935; USNH Mare Island, CA, 1935-36; USNH, Guam, M.I. 1936-38; USS *Gold Star* 1937; USNH Pensacola, FL, 1938.

Re-enlistment June 1942, USNH Mare Island, CA; NAS Moffett Field, CA 1942; South Pac. Acorn #8, 1942-43. Island "X" New Georgia; Fleet Air Wing 1, New Georgia and New Hebries, 1944; US Fleet Hosp., Guadalcanal, 1945; USS *LST* (H) 951, 1945 (Okinawa Invasion Easter Sunday morning 1945); NAS Bunker Hill, IN, 1945.

Relieved from active duty as a warrant 1 on Oct. 1, 1945. USNR Center in Des Moines, IA and USNH Great Lakes qualified his 36 1/2 years for retirement. Medical Service Corp W-4 designator 8179. Retirement Oct. 1, 1975. Probably one of the few remaining.

He is married to Frances Howard and they have three children: Mark J., John Scott and Mrs. Joy Patterson. Their grandchildren are Anna, Stacia, Callie, Taylor, Chad and Scott. Retired from civilian employment as a funeral director. Presently still a recruiter for the Navy.

DAVID J. EVANS, LT, born May 21, 1953, Houston, TX, Received BME from Texas Tech University, 1981. Joined the service Oct. 19, 1970 as enlisted man and on May 22, 1981 as officer.

Stations/Units: San Diego NavSta 1971-73; Pensacola, FL, 1981-83; NAS Oceana 1983-89; USS *Jovett* (DLG-29) Ops. Specialist, 1971-73; NATC 1981-83; VA-42, 1982-83; VA-75, 1983-86; Vietnam August 1972-March 1973; Lebanon December 1983 and Desert Storm January-February 1991.

Discharged Jan. 1, 1992, as lieutenant. Accumulated 1,000 flight hours, designated NFO April 1982, primary aircraft A6 (BN) and received various medals.

Married Teresa Kim Henderson and has one child, Lee. Civilian position with US Postal Service. Retired May 30, 1994.

NORMAN S. EVANS, LCDR, born July 14, 1920 in Houston, TX, received a BA in sociology/English from Baylor University, Waco, TX.

He joined the service April 18, 1942 and was discharged Nov. 30, 1953. His military locations and stations: Staff-COMDESPAC, Pearl Harbor 1943-45, communications; USS *Gansevoort* (DD-608) Marshall Island 1944 asst. CO; 8th Naval Destroyer HQ, 1946 Com. Center; Military Govt. Unit-Truk Atoll, Central Carolines, Pacific 1946-47; Rehabilitation of natives; STAFF-ComNavFe, Tokyo, Japan, 1952, Communications; Recruit Training Command, Camp Elliot, San Diego, CA, 1953; Special Services officer, Naval Training Center, San Diego, CA, 1953. In charge of all recreational activities on base.

Units served: Destroyer Task Force, Marshall Islands. 1944, patrolling Japanese occupied islands in Marshall chain.

Memorable experience: in 1946 he was assigned duty with the Military Government Unit at Truk Atoll in the central Caroline Islands of the Pacific. Prior to WWI the Pacific Islands were controlled by Germany. German missionaries had visited the Islands, helping to educate the natives and teach them English. Following Germany's defeat in WWI the islands of Micronesia were mandated to Japan by the United Nations, thereby coming under Japanese rule and occupation. Truk had been bypassed as invasion site during the war since the atoll (consisting of several large islands and many scattered smaller ones) was used as a harbor for the Japanese fleet and a communication center. Only bombing raids were carried out on Truk.

When our small contingent of officers arrived on the main island of Moen at Truk, Japanese who had occupied the islands were still being held in a stockade guarded by Marines. After all the Japanese had been returned to their homeland, they were assigned quarters (Quonset huts), and their families were cleared for transportation to join us on the island.

Field trips were made to outlying islands from their headquarters on Moen. The purpose of these trips was to assist in rehabilitation of the natives who had been under Japanese rule for so many years. On one field trip he made, the port of call was the small island of Kapingamarangi. The skipper of the vessel making the trip had agreed to allow his wife Nancy and their little dog Ozzie to accompany him as he had ample accommodation aboard.

As there was no harbor at Kapingi access to the island was via outrigger canoe navigated through the surf by the natives. When they arrived on the island they were greeted most graciously by King David. The island had always had a ruler and King David looked every inch the part. He had handsome features, a most imposing figure, standing seven feet tall and weighing 300 lbs. Since the background of many of the natives on the islands of the Pacific are shrouded in mystery, it is significant to note the natives of Kapingi are extremely tall as opposed to the short stature of the natives on the other Pacific Islands. On Kapingi the natives have their own language and dialect that is spoken nowhere else in the world. It is also true the natives on Truk speak Trukese which is like no other language.

They spent the day in King David's company exploring the entire island. They were given the royal treatment with a banquet of many native foods prepared in their honor. His wife Nancy was the first American woman to visit Kapingi and the natives were in awe. Also, the children of the island had never seen a dog. They would run ahead of them shrieking and hiding in the bushes as Ozzie trotted along the pathway.

King David plied them with many gifts. Evans favorite was a coconut stool King David presented to him, which today serves as a hassock in his den. These stools are carved out of sturdy wood. The women (who perform all the menial tasks) straddle these saddle shaped stools, which have a sharp protruding blade on one end. They ream out the meat from the coconuts which the young boys of the island have cut down from the surrounding trees. The coconut meat falls into a basket which is placed under the sharp blade of the stool (the stool in his den has the sharp blade removed). From this bounty, copra and coconut oil can be made, which is a great export item for the islanders. While they were ashore many of the native export items were loaded on their ship. In return they brought ashore ample supplies the natives could use in their daily lives.

As they departed the island amid fond farewells, he had doubts they would make it through the high surf from their outrigger. Laden with gifts, he was holding the coconut stool in one hand and Ozzie in the other. He could just picture the outrigger capsizing and all their gifts and their little dog being consumed by the high waves. However, with their experience the natives skillfully navigated the rough surf and returned them safely to their ship anchored off shore. This was a day that will long live in his memory as one of the highlights of his career in the USNR.

Employed with Department of Motor Vehicles, Hawthorne, CA, supervisor, driver's license section, 10 years; California Race Tracks, CA; three major race tracks (thoroughbred racing); terminal operator, computers, 20 years. Retired July 14, 1980.

He is married to Gladys and they have four children: Robert, Steven and Jane Evans, Catherine Garcia; three grandchildren: Jennifer Detherage, Alexandra Tovar, Dale Maylor; and three great-grandchildren.

WILLIAM M. "BILL" EVANS, LT, born in Sparta, WI, March 11, 1920. He is a graduate of the University of Wisconsin under an accelerated USN V-7 Program, trained at Columbia University, 15th Class Midshipmen's School. Graduated Georgetown Law School, LLB, JD and attended Post Graduate, one semester University of Wisconsin Law School.

He was a member of the Wisconsin National Guard, 1937-39, USNR, March 4, 1942-60, when he was retired. Served aboard LST-285 at Normandy and Southern France and in the Pacific. Served in cruiser, *Albany* (CA-123), with the 6th Fleet and cruiser, *Manchester* (CL-83), Korea. Employed mainly interdiction and bombardment with over 29,000 rounds (five and six inch) expended and escort carrier, *Badoeng Strait* (CVE-116), Korea. The heroic missions of Marine Pilots were mainly close air support.

His memorable experiences: The Normandy invasion, H-hour plus one with several trips from England carrying men and machinery of war in and wounded back. After the campaign the large number of fine young corpsmen were transferred for duty in Iwo Jima. Word was received that one-half had been killed, including their chief. LST-496 and LST 523 were sunk close aboard with few survivors. At Southern France LST-282 became a huge torch when sunk at dusk with her oil catching fire.

When their LST went through the canal to the Islands of the Pacific they went over a year without seeing a white woman. Evans is single and has no children. Self-employed at Evans Lumber Co. and is over 99% retired, no phone-he travels "A great deal."

WINTHROP S. "WIN" EVANS, LCDR, born June 21, 1939 in Santa Monica, CA, received an AA from Fullerton College, BA from Cal State University and AD from Western State.

He joined the service April 1957. His military stations: NAS Pt. Mugu, NAS North Island. Served with GMU-41, VU-7 and VRF32.

Evans achieved the rank of LCDR and was discharged Aug. 31, 1965. Retired from USNR Jan. 1, 1976.

He is married to Carlene and they have four children: Theresa, Shalene, Shanna, and Michelle; and two grandchildren, Anthony and Gregson. Employed with American Airlines, captain; attorney, retiring Feb. 1, 1997.

TERRELL A. EVERETT, CDR, born March 21, 1943 is a native of Fort Wayne, IN. He graduated from California State University at Los Angeles, in 1965 and received his commission from Officer's Candidate School in Newport, RI in 1966. Everett completed Explosive Ordnance Disposal School in 1967 and was assigned as officer in charge, Explosive Ordnance Disposal Mobile Unit, Detachment Adak, AK.

In 1969, he earned a master of public administration degree from the University of Southern California. On inactive USNR duty he was assigned to the Staffs of Naval Reserve Intelligence Units in Southern California. In 1977 he was assigned as intelligence officer for Underwater Demolition Unit 119. Naval Reserve Explosive Ordnance Disposal Group One was formed in 1979 and was assigned as XO.

Recalled to active duty in 1986, his assignments include: recruiting officer for Naval Reserve Readiness Command Region Two; director, CDR Naval Reserve Force Recruiting coordinator Region VI; officer-in-charge, Naval Reserve Recruiting Command Detachment SIX, before retiring in 1993.

Everett is authorized to wear the National Defense Medal, Expert Rifle Medal, the Expert Pistol Medal, Recruiting Service Ribbon (2) Reserve Meritorious Service Medal (2) and Navy Unit Commendation.

He is married to the former Laura Ann Wesley of Glendale, CA. They have three children Meghan, Seth and Caitlin.

FRANCIS J. FABRIZIO, RADM, is a native of Washington, DC, graduated from Georgetown University Dental School in 1935. He was commissioned a LTJG in the USNR in January 1936 and was assigned to the 5th Bn., USMCR where he served until early 1941 at which time he was promoted to LT.

In 1943 he reported to active duty at NATTC, Jacksonville, FL. In 1944 he reported for duty at NRS, Anacostia, DC, for which duty he received his first Letter of Commendation. Upon release from active duty in 1946 he returned to private practice and teaching at Georgetown University Dental School following postgraduate training in Periodontics.

Active participation in the USNR continued and included the organization of the first Naval Dental Reserve Unit in a dental school. Following the implementation of this program he was directed to organize the naval residency program in Periodontics. For these efforts he was awarded further commendations.

Continuing his interest in the Naval Dental Corps Reserve and active duty assignments he was selected for promotion to RADM July 1, 1966. He retired July 7, 1971.

His wife is deceased; his children are Francis J. Fabrizio Jr. and Carol Ann Knoll. He also has five grandchildren and one great-grandchild.

RICHARD FAHY JR., a native of New Hampshire, attended Tulane University where he graduated in 1985 with a degree in International Relations. Commissioned from Naval ROTC, he served on USS *Frederick* (LST-1184) as first division officer and CIC officer for three years. During this time, the ship participated in two deployments to the Western Pacific.

His follow-on tour was to the staff of Amphibious Sqdn. Five as the plans and operations officer (N31) where a ten month deployment include the Gulf War. In 1992, he spent 15 months as an instructor and division officer of the Amphibious Warfare Team Trainer and Naval Warfare and Tactics Department.

Fahy entered the USNR in 1993 to pursue a career in business and education. An avid traveler, he recently returned from a two month visit to Africa, his fifth contingent. He has also completed climbs of Mt. Aconcagua (23,000 ft. summit in Argentina) and Klimanjaro (Tanzania). In 1996 he graduated from the University of New Hampshire with a masters in business administration. He has been awarded the Navy Commendation Medal (2) and Combat Action Ribbon.

JOSEPH "JOE" FAIRBROTHER, JR., CDR, born Sept. 30, 1944 in Brooklyn, NY, received a BBA in Finance from St. Bonaventure University in 1967.

He joined the service Jan. 29, 1968. Commissioned July 19, 1968 at OCS in Newport, RI. Attended Communications Officer School, San Diego, CA, July 1968-September 1968. Assigned to Commander Landing Ship Sqdn. 3 Staff (headquartered on Guam), serving as CO, operations officer and RPS custodian. While on the staff, he sailed on squadron ships (542 LSTs) to the Philippines, Japan, Taiwan and Vietnam, September 1968-January 1970; assigned to USS *Fresno* (LST-1182) as CIC officer, electronics material officer and RPS custodian. Deployed to WESTPAC (Vietnam) most of tour, February 1970-July 1971; attended Defense Intelligence School in Washington, DC as a part of the Naval Post Graduate degree program; August 1972-January 1971, assigned to USS *Point Defiance* (LSD-32) as CO and "fleeted up" to 1LT. Deployed to WESTPAC (Vietnam) most of tour.

Released from active duty Jan. 5, 1974. Reported for Reserve Duty to YR-64 (Freeport Detachment) in January 1978 and has been with YR-64, NRMTF NY HQ 102, SIMA/NRMF and SSFLON SIMA DET ever since.

At YR-64/NRMTF/SIMA Served in the following billets; admin. officer, training officer, quality assurance officer, planning and estimating officer, assistant officer-in-charge, officer-in-charge, repair officer, XO.

At YR-64/NRMTF/SIMA performed active duty at: Mayport, FL; Newport, RI; Norfolk, VA; Long Beach, CA; Dam Neck, VA; Naval War College and Brooklyn, NY (6 times).

His memorable experiences: met commodore Tran Van Chon CNO of Vietnamese navy; served with Heisman Trophy winner CAPT Joe Belino on "REPAIREX '79."

He is married to Ellen and they have three children: Joy, Renee and Brian. Employed as resource manager for Northrop Grumman Corp. He retired Oct. 1, 1994.

FRANKLIN E. "FRANKIE" FAIRHURST, CAPT, born Aug. 3, 1922 in Freetown, MA, received a BS in engineering and business administration.

He joined the service in 1942. His stations include: Pensacola, CASU-5, VC-88 (VT-84), Quonset NAS. Served with Squantum, South Weymouth. His battles and campaigns: Lea, New Guinea, Philippine Sea, Luzon, Lingayen Gulf, Iwo Jima, Western Islands, Okinawa, Miyako, Ishigaki.

Fairhurst was awarded the Distinguished Flying Cross and Air Medals. He achieved the rank of CAPT and retired in August 1982. His memorable experiences: pick up at sea, shot down, USS *Kendell C. Cambelle* (DE-443).

He married Jean M. and their children are Sybil, Franklin Jr., Paula, Margaret, Kathleen and Raymond; and grandchildren are Dana, Ronald, Heide, William, Walter, Eric, Sara and Anna. Civilian Employment: Shell Oil Co., engineering.

JOHN P. FAJNOR, LCDR, born April 26, 1929 in Chicago, IL, received a BE (civil engineering), 1950 Yale University. He joined the NROTC Unit, June 1950 Yale University, ENS, USN. His military locations and stations: Norfolk, VA and Pearl Harbor, TH.

Units served with: USS *Missouri* (BB-63); US Naval Comm Sta. Pearl. Participated in action in Korea 1950-52 aboard USS *Missouri.* He achieved the rank of LTJG USNR; LT USNR; LT CEC USNR, LCDR CEC, USNR. Discharged from active duty in 1954 and retired from USNR 1979.

His memorable experiences: Service aboard USS *Missouri* during Korean War, OIC Fleet Post Office, Pearl 1953-54; XO NRCB Unit, White Oak, MD 1978-79.

He married Carolyn and has two children, Craig and Lisa; and one grandchild, Rebecca Giacosie. Employed as engineer/program manager, IBM, 1966-91.

HARRY C. FALL, CAPT, born Feb. 28, 1931 in Edmonton, Alberta, Canada, received a BS from Lawrence University. He joined the service Sept. 11, 1952. His military locations and stations: NAS Barbers Point; NAS Glenview; NAS Grosse Isle.

Served with VP-28, VP-723, VS-721, VS-722, VS-724, VP-661. Fall's awards and medals include the National Service Medal, Naval Reserve Medal, Korean Service Medal and the UN Medal. He achieved the rank of CAPT.

He married Arden and they have a daughter Pamela and two grandchildren, McKenna and Jackson. Employed as insurance broker, vice-president, Marsh and McLennan, Inc. He retired Aug. 1, 1993.

TIMOTHY O. FANNING JR., ADM, born in 1943 in Hartford, CT. He graduated from Hamilton College in 1965 and was commissioned an ENS upon completion of Officer Candidate School. In 1971, he received a master of business administration degree from the Wharton Graduate Div. of the University of Pennsylvania.

On active duty, ADM Fanning completed Deep Sea Officer School and then served aboard the USS *Seneca* (ATF-91), including deployments to the Caribbean and Mediterranean. He subsequently served as executive officer on the USS *Mataco* (ATF-86) during two deployments to Vietnam.

ADM Fanning affiliated with the USNR in 1971 and has served in a variety of command and staff billets. He has commanded the following Naval Reserve units: *Puget Sound* (AD-38) Det. 404, CDR Naval Surface Force Atlantic Det. 404, Shore Intermediate Maintenance Activity Philadelphia HQ Det. 104; and Naval Base Philadelphia. His staff assignments have included tours as director of plans and assistant deputy for readiness, Naval Reserve Readiness Command Region Six, Washington.

His active duty assignments include three years on the Secretary of the Navy's National Naval Reserve Policy Board; assistant to CDR, 7th Fleet; and assistant to CDR in Chief, US Atlantic Fleet. His professional education includes the CDR in Chief Atlantic Fleet Senior Reserve Officer Course; the Defense Economics Course, at the Naval War College; and the National Security Course, at National Defense University.

In civilian capacity, ADM Fanning is managing partner of Fitzpatrick-Fanning, a real estate management and development firm in Downington, PA. He is a former national chairman of the board of directors of the US Naval Sea Cadets Corps and past national vice-president of Youth of the Navy League. He is a life member of the Navy League, Naval Reserve Association and the Surface Warfare Association.

His decorations include two awards of the Meritorious Service Medal, the Navy Commendation Medal, the Meritorious Unit Commendation and several campaign and service medals.

Fanning is married to the former Nancy Orr of Bennington, VT. They have four children: Heather, Tom, Brock and David.

His Reserve mobilization billet is chief of staff at US Atlantic Command Headquarters in Norfolk, VA.

JAMES FOSTER FANSEEN, CAPT, attended Baltimore Polytechnic Institute, 1946; University of North Carolina, AB, 1950; University of Maryland School of Law, JD, 1954; Michigan State University, 1958.

Commissioned as 2LT, USAFR, June 1950 and served as reserve judge advocate general officer through rank of major and transferred to USNR in 1969. Past CO, OJAG Litigation 206, USNR; legal advisor for Space Affairs, Naval Space Cmd. Reserve Unit, Dalgren, VA, rank of CAPT, retired in 1988.

From 1954-67 was partner in law firm of Fanseen & Fanseen, Baltimore, MD; 1967-71, commissioner on Federal Maritime Commission; 1971-81, partner in law firm of Fanseen & Fanseen, World Trade Center, Baltimore, MD; 1981-90, National Aeronautics and Space Administration as special assistant and consultant to the administrator; 1990-present, partner, Fanseen & Fansee, Linthicum, MD.

Membership in USNA Officers and Faculty Club and numerous other professional affiliations.

ALAN JAMES "AL" FARLING, LCDR, born Feb. 26, 1946 in Bryan, OH. Graduated high school in 1964; Wayland, MI, bachelor of science in engineering, 1973; Old Dominion University, Norfolk, VA, master of engineering, 1976, Old Dominion University, Norfolk, VA.

He enlisted in the service Oct. 11, 1965. His military locations: N&MC Reserve Center, Grand Rapids, MI, seaman recruit to seaman, 1965-66; USS *Canopus* (AS-34) (Rota Spain), seaman to draftsman second class, 1966-68; Naval Reconnaissance and Technical Support Center, draftsman second class, 1968-69; NAS, Norfolk, VA, draftsman second class, 1969-70; Naval Air Intelligence Reserve Unit-R1, draftsman first class, 1970-73; Area Analysis Unit 706, ENS to LTJG, 1973-77; Fleet Ocean Surveillance Intelligence Facility Rota 0186, LTJG to LT, 1977-79; CINCLANTFLT DET 106, LT, 1979-81; Helicopter Attack Sqdn. Light 4, LT to LCDR, 1981-84; CDR Naval Air Forces Atlantic 01-LCDR, 1984-86; Naval Operational Intelligence Center 0566, LCDR, 1986-89.

Farling was awarded the Meritorious Unit Commendation, Good Conduct Medal, National Defense Service Medal, Armed Forces Reserve Medal w/HG, Armed Forces Reserve Medal and Expert Pistol Shot Medal. He achieved the rank of LCDR (special duty intelligence). Transferred to the Retired Reserve, Nov. 1, 1989.

His memorable experiences: serving as the liaison officer between HAL-4 and SPECWARGRP-2, planning and participating in operations where HAL-4 provided air support for SEAL teams; and being on watch in the CINCLANT OPCON Center when the hostages were taken in Iran.

Farling has been married 29 years to Carrie Ellen O'Neill and they have two children, Monica Michelle and Alex Jon. Employed with Henningson Durham & Richardson as project engineer, 1973; City of Portsmouth, VA, Public Works Engineer, 1976; York County, VA, director of Public Works, 1978; city of Dover, DE, director of public works, 1982; Tilcon, DE, construction superintendent, 1984; Natural Resources and Environmental Control, state of Delaware, Environmental Program administrator for Groundwater Protection, 1986. In 1983 he joined the Robbins Hose Company No. 1, Inc. (Dover, DE volunteer fire company) and is currently, first assistant fire chief. Life member of the Naval Reserve Association.

RONALD ANDREW "RON" FARMER, LT, born Dec. 15, 1962, Aurora, CO. Earned BS in mechanical engineering, New Mexico State University, 1985. Joined the military Oct. 17, 1984 and stationed at OCS, NETC, Newport, RI; NPS Orlando, FL; S7G Prototype, Ballston Spa, New York; SOBC, New London, CT; Charleston, SC; Kings Bay, GA.

Units served with include USS *Woodrow Wilson* (SSBN-624), USS *James Madison* (SSBN-627), USS *Francis Scot Key* (SSBN-657). Participated in action in Desert Shield and Desert Storm.

LT Farmer was discharged May 31, 1991. His awards and medals include the Navy Achievement Medal, Defense Service Medal, Humanitarian Service Medal, Commander, Submarine Group, and six Letters of Commendation.

Memorable experiences include completing submarine warfare qualification; OOD mid-watch on the surface in the middle of the Atlantic Ocean; new chiefs' initiation ceremony; his last underway with the best officers and enlisted men in the submarine service aboard the USS *Francis Scott Key* (SSBN-657).

Employed at Sandia National Laboratories as reactor supervisor at ACRR; promoted to LCDR in December 1996; currently serving as XO of LKA support, DET C in Albuquerque, NM; served four years in the Pearl Harbor Naval Shipyard Engineering Support Unit in Albuquerque 1992-96, and served as training officer of CONVCOM Unit in Albuquerque, NM, 1991-92.

Married Laura Hillman and they have two children, Joshua and Amanda.

RICHARD L. FARRELL JR., CAPT, a native of Pittsburgh, PA, attended Clarion State College and the University of Notre Dame obtaining BA and MBA degrees. Commissioned through AOCS, he was designated a NFO in January 1970. Reporting aboard VW-1 at NAS Agana, Guam he flew typhoon penetrations and made detachments to Da Nang, South Vietnam flying reconnaissance for CTF-77.

In March, 1971 he reported aboard TACRON 13 qualified as Tactical Air Controller Airborne and later deployed aboard USS *Dubuque* as aviation plans officer. In October 1982 he joined NR VP-0122 at NAS Glenview, IL flying P-3B MOD aircraft as tactical navigator. Aboard NR PATWING 0472 he qualified as NATO WO and served as XO and acting CO. Subsequently he joined NR COMNAVAIRLANT 0172 as information systems officer, with further assignments as ships C3 systems and tactics D&E officers.

In September 1993 he was chosen CO of NR NAS Adak, AK 0172. For his performance he received the Meritorious Service Medal. In October 1995 was selected project support officer for NR SPAWARS HQ 213 and later mobilization officer for CONAVRESREDCOMREG 13. He attended the National Defense University and the Naval War College.

Farrell is a CPM and held procurement positions with Westinghouse Electric Corporation and Notre Dame. Currently, he is general manager of a Midwest distribution company and resides in South Bend, IN.

VINCENT D. FARRELL, CAPT, joined the USN June 30, 1947 and served on active duty for nine years, which included the USS *Midway* from 1951-55, making two Med cruises and a world cruise, USS *Coral Sea,* USS *Boxer* during Inchon invasion in Korea, and various stations in the US. He played on the Bainbridge and Lakehurst baseball teams. In the USNR he served as CO of 6 NR units and as SJA. He retired Aug. 12, 1989.

His awards are: Navy Commendation (2), China Service, Good Conduct (2), Korean Service, PUC, Korean PUC, NUC, MUC, UN, European and Japanese Occupation, Armed Forces Reserve, Expert Pistol and Rifle.

He graduated from St. Mary's College in Kentucky and from St. John's University School of Law. He is an attorney with Farrell, Farrell & Burke, which includes his son Ed and daughter Patty Burke. He is married to Margaret O'Brien, a professor at Nassau Community College. They have two other children, David and Maureen, and two grandchildren, Kevin Edmond and Conor John Burke.

WILLIAM "BILL" FAWX, LCDR, born Sept. 17, 1924 in San Francisco, CA and enlisted in the USN in September of 1942. He received NROTC training at the University of California at Los Angeles and amphibious training at Camp Bradford in Norfolk, VA. Fawx attended several gunnery and firefighter schools.

He served on two LSTs in two different wars, the LST 1045 in WWII and the LST 1080 during the Korean conflict. He was the gunnery officer and chief engineer on the 1045 and served as the engineering officer, supply and stores officer, and senior watch officer on the 1080.

During WWII he and the LST 1045 served in the Pacific and Asiatic areas. They traveled to Panama, Saipan, Hawaii, Eniwetok, Guam, Philippines, Japan, Korea and Peleliu. During the Korean conflict he again visited Saipan, Guam, Japan and Korea in addition to Hawaii and Kojedo.

Between service in WWII and Korea, Fawx was an active member of a naval reserve division unit in Sacramento, CA. After service in Korea, he was again affiliated with several naval reserve units, serving the last 15 years of such service with a mobilization team. He accumulated a total of 26 1/2 years of Naval Reserve Service.

He was awarded the Naval Reserve Medal, American Area Campaign Medal, WWII Victory Medal, Navy Occupation Medal, Asiatic-Pacific Campaign Medal, National Defense Medal, Korean Service Medal (three Battle Stars), Presidential Unit Citation and Armed Forces Reserve Medal.

As a civilian Fawx was employed by the state of California for 37 years, retiring in 1985. He married Emma E. on June 20, 1953 and they have four daughters and two granddaughters. At the time of this writing he resides in Newcastle, CA.

JAMES W. "PUTTER" FEE, CAPT, born June 13, 1932 in Hutchinson, KS, received a BA from the University of Kansas. He joined the service Dec. 10, 1950. His military locations and stations: USS *Albany* (CA-123). He was awarded the National Defense Service Medal, European Occupation Medal.

Fee achieved the rank of CAPT USNR and was discharged Sept. 10, 1956. His memorable experiences: two Mediterranean cruises; two midshipmen cruises to Northern Europe.

He is married to Martha J. and they have three children: James W. Fee Jr., Stacy J. Shaw and Allen K. Fee; and grandchildren, Robert T. and Daniel J. Fee. Employed as president, Fee Insurance Group, retiring June 13, 1992.

HARRY J. "JERRY" FEENEY III, CDR, born Feb. 15, 1940 in Tampa, FL, received a BS from the USNA C-62, BS EE USNPGS 10-67, MS SCIMGT - USIU 6-71.

He joined the service June 30, 1958, USNA, and was commissioned ENS, June 6, 1962. Locations and Stations: USNA (June 1958-June 1962); Mayport (June 1962-May 1964); NORVA (May 1964-December 1965); USNPGS (December 1965-October 1967); Vietnam (January-December 1968); USNA (6-5-6-62); USS *Zellar* (DD-777) (June 1962-May 1964); USS *D.H. Fox* (DD-779) (May 1964); USNPGS December1965-October 1967; CTF111 January 1968-December 1968; CDR RIV Div. 513 in Vietnam; (TET Spring Offensive and three others).

His battles and campaigns: Cuban Missile Crisis, 1962; Cyprus Crisis (1964), Ind. Pak. (1965), Vietnam (1968). He was awarded the Navy Commendation (V), PUC (2), NUC, National Defense, AFEXP, Vietnam Service (five stars) and Vietnam Gallantry.

Feeney achieved the rank of CDR USNR and retired May 19, 1989. His memorable experiences: Cuban crisis, 30 days with V3 crew, Vietnam; TET 68 and Spring offensive.

He married Consuelo Aug. 24, 1991. Employed as Naval Ocean Systems Center (September 1971-June 1989). Retired (to Fleet Reserve) May 19, 1989.

JOHN PAUL "JACK" FEERICK, CAPT, born Aug. 15, 1950 in New York, NY, received a BS (1972); MD (1978), Georgetown University, Washington, DC. Joined April 1979.

His military locations and stations: NAF, Washington, DC; MCALF-15 (Bogue Field), Cherry Point; NAS, Pensacola, FL (NAMI) NS Keflavik, Iceland, London, UK, BUMED, Washington, DC. Units: 2nd Marine AIRWING; MEDCRU 404; Fleet Hosp. 250 (Combat Zone) 7D; NMCRRC (Columbus, OH). Currently serving as Code 25 operational medicine and fleet support BUMED, Washington, DC.

His memorable experiences include flight training at Pensacola (AEROSPACEMEDICINE), service in Iceland and flights to the North Pole and flying from carriers *Lexington* and *Nimitz* during cruises.

His decorations include the Navy Commendation Medal, Meritorious Unit Commendation, National Defense Service Medal and Armed Forces Reserve Medal. Feerick is a companion of the Naval Order of the US. Also NRA/ROA (chapter officer), USNI, AMSUS, National Maritime Historical Society.

He is a neurologist and stroke researcher, editor-in-chief *Neurorehabilitation News*. (NOUS ancestor, Albion Burnham, commander of USS *Carnation* 1865 So. Atlantic Blockade Sqdn.)

Married to Patricia and they have four children: John, Meaghan, Catherine, Thomas.

JAMES C. "CHRIS" FELCHLIN, CDR, born March 27, 1944 in Yuma, AZ, received a BA in business and economics, University of California, Santa Barbara, 1966; MBA in Finance, University of California, Berkeley 1966.

He joined the service Oct. 24, 1964. His military locations and stations: Long Beach 1967; West Pac Deployment 1968; Pearl Harbor 1969. Served with USS *Force* (MSO-445) 1967-69, operations officer.

Achieved the rank of CDR and was discharged Jan. 1, 1990. His memorable experiences: drilling reservist 1970-85. CO of SIMA SF DET 420, CDO of SIMA SF DET 120, XO of MSC PAC DET 220.

Married to Vera Freeman and they have two children, Christopher A. and Sarah. Employed as a private banker with the Citibank Private Bank, San Francisco.

JOHN J. "BIG JOHN or JJ" FELKER, CAPT, born April 8, 1929 in New York City, received a BBA from Manhatten College and a MS from Cornell University. He joined the service Sept. 18, 1952. His military locations and stations: OCS, Newport, RI; NSCS, Bayonne, NJ; NSD Mechanicsburg, PA.

Felker was awarded the National Defense Service Medal and Armed Forces Reserve Medal. He achieved the rank of CAPT and was discharged Dec. 15, 1955. His memorable experiences: Served as CO of DCSC 105, Columbus, OH; logistics officer of FACS - 4105, Pittsburgh, PA; CO, XO, training officer and administrative officer of NRSC 4-2, Pittsburgh, PA.

While serving at NSO Mechanicsburg, he was responsible for selling over 100 million dollars of surplus naval material to the general public.

He is married to Mary Rose. His first wife Jean died in 1991. His children are Cynthia A. Leszunov and Bonnie M. Shields and grandchildren are Jeanne, Tiffany, Anthony and Michelle Leszunov. Employed with LTV Steel Co., as director, arbitration, manager ind. relations, retiring July 1, 1983.

HARRY DAVID "FEETS or BLUEFISH" FELSENTHAL, JR., CDR, born Dec. 31, 1915 in Piru, CA, received a BA (physics), Pomono College, 1937; MA (physics), Dartmouth College 1939; MS (Aero. Engr. 1957).

He joined the service Oct. 1, 1941. Served active duty in V-7 Program, beginning Jan. 9, 1942 at USNA. His military locations and stations: Reserve Midshipman School, Jan. 5-May 5, 1942; Radar School Harvard & MIT, June 1-Nov. 30, 1942; Radar Center 128, Pearl Harbor; USS *Gurnard* (SS-254); Sub Div. 121; USS *Bluefish* (SS-222); Dec. 25, 1942-Nov. 15, 1945; Second and Third War Patrol; *Gurnard;* Fourth through Ninth war patrols, *Bluefish.*

He was awarded the Bronze Star Medal, two Letters of Commendation, Submarine Combat Pin (five successful war patrols). Felsenthal achieved the rank of LT when released from active duty; CDR when retired from USNR, 1968. He was discharged Nov. 15, 1945.

His memorable experiences: his first war patrol; first sea duty; introduction to attacks on enemy shipping and subsequent depth charging; typhoon on return to Pearl; other attacks on ship and sweating out attacks; rescuing aviators; becoming qualified in submarines.

He married Shirley Hartford Oct. 12, 1944 during the overhaul of *Bluefish.* They have a son Charles Louis Felsenthal II, named after his uncle, a B-24 pilot, who was lost with his entire crew in May 1944. Employed as weather observer, Sept. 1939-December 1941. Design engineer, Raytheon Co., January 1946-September 1948; electronic engineer, US Naval Missile Center, September 1948-May 1972.

MARSHALL E. "GATOR" FELTHAM, CDR, born Oct. 1, 1935 in Atlanta, GA, received a BAAS-SWTSU; JD, John Marshall Island.

He joined the service Feb. 16, 1956. His military locations: Oceana/Jax-Cecil, FL; NORVA; Olathe; Atlanta/Washington, DC. Served with VA-72; VA-44; VRF-31; VA-205. His battles and campaign Lebanon 58. Awards/Medals include the Armed Forces Reserve w/ HG, National Defense and Armed Forces Expeditionary. Feltham achieved the rank of CDR and retired Nov. 1, 1976.

His memorable experiences: 20,000 LDG, CVA-62; flew last TET out of NAS Olathe; night eject from A-4 and rescued by DD *Wood;* logging 2500+ A-4 hours; OinC VA-205.

Married Mary Lacey, COL USA (Ret) and they have two children, Valerie and Marshall II; and four grandchildren: Tiffany, Holly, Chase and Victoria. Member Georgia Bar Association 1979; FAA Flight Examiner 1985.

KENNETH D. "KEN" FELTON, CDR, born May 26, 1930 in Ft. Worth, TX, received a BS from the University of Miami. Attended Graduate School, University of Texas; ADVD MGMT Program, University of Southern California.

He enlisted in the service Sept. 9, 1948. His military locations and stations: USS *Helena* (CA-75); USS *Everglades* (AD-24); USS *Union* (AKA-106). Active in USNR until retirement in 1976. His last position was CO of Surface Reserve Unit in Bakersfield, CA. Felton achieved the rank of CDR and was discharged March 1960.

He married Janis and they have four children: Mark, Steven, Daniel and David. Employed as vice-president Hexcel Corp; president Urethane Plastics, Inc.; president, Urethane Associates. He is still working.

ROBERT LEE "BOB" FERGUSON, CAPT, born Jan. 1, 1931 in Tulsa, OK, received a BS in chemistry from the University of Tulsa in 1952. He joined the service November 1954 and was stationed in USS *Point Cruz* and at NAVCOMMFAC London.

Ferguson is a Korean veteran and achieved the rank of CAPT USNR. Retired from active duty February 1958 and from Reserve duty in 1982.

He is married to Mary Ann and they have five children: David, Sheryl, Michael, Linda and Robert; and six grandchildren: Amanda, Sean, Hanna, Emily, Brittany and Danielle. He is a general partner with R/W Acquisition Associates.

DEWITT T. "DEWEY or JACK" FERRELL JR., CDR, born Jan. 14, 1930 in Dallas, TX. Attended one year in Oklahoma A&M and three years in Central State College.

He joined the service Aug. 28, 1950. His military locations and stations: Flight training, Pensacola; NAS Alameda; USPG, Line School NAAS Baren. He served with: VF-194; BTG-5; CAG-11 and VA-212. His battles and campaigns: Korean War. He was awarded three Air Medals and the Letter of Commendation "V". He achieved the rank of CDR. Discharged June 1964.

His memorable experiences: flying with his section leader LTJG Sam Catterlin - the best lead in the USN.

He is married to Sue and they have four children: Mary, Michael, Melissa and Megan; and seven grandchildren: Katie, Matthew, Ryan, Tyler, Jimmy, Morgan and Allison. Employed with the United Air Lines as CAPT, retiring Jan. 14, 1990.

RODNEY J. FINK, CDR, born April 10, 1934 in Oregon, MO. He earned BS, MS and PhD degrees. Joined the service Oct. 5, 1952, and served with MIPULANT and CINCLANTFLT.

Served active duty 1956-60 and retired from the Reserves in 1981. He achieved the rank of CDR, USNR and received the Good Conduct Medal.

He married Bertha and they have four children: Ross, Lori, Brenda Fink Lincoln and Lisa Fink Moore. Their grandchildren are Rachel and Rhett Fink. Employed as Dean, College of Applied Sciences, Western Illinois University, retiring in 1993.

H. RAY FISHER, CAPT, a native of Salt Lake City, UT, graduated from the University of Utah and entered the USN through the Aviation Officer Candidate program in 1968. He received his wings in 1970 after completing advanced jet training at NAS Kinsgville, TX. He was first assigned to NAS North Island at NATOPS officer in the CRT (Combat Readiness Training) program, flying F-4, U-11 and T-28 aircraft.

He reported in December of 1972, to URC-50 at NAS Cubi Point, Philippines. He served with numerous detachments in Japan and South East Asia, including Atsugi, Saigon and Da Nang. He was next ordered to VT-21 at NAS Kingsville, TX where he qualified as a weapons instructor and standardization pilot in the A-4. He later served in VR-1 and URC-40 at Norfolk, VA until leaving active duty in June 1979.

In October of 1979, he affiliated with VR-52 at NAS Willow Grove, PA flying the C-9 aircraft and in 1990 became the CO of that unit. Currently serving as chief of staff for Army Joint Service Unit 6396. At Ft. Douglas, UT. Fisher has been awarded the Navy Commendation Medal w/Gold Star, Navy Achievement Medal (two), Meritorious Unit Citation (two) and various other unit and service medals.

Fisher is a pilot for Northwest Airlines and resides in Sandy, UT with his wife Vickie. They have five children: Marc, Melinda, Daniel, Jeremy and Christopher.

JACK G. FISHER, CAPT, born March 30, 1925 in Olathe, CO, received a BS from Colorado A&M, 1950 civil engineering.

He joined the service June 1, 1943. His military locations and stations: Colorado Springs, CO; Farragut, ID; Memphis, TN; Purcell, OK; Jax, FL; Seattle, WA; Whidbey Island, WA. Served with VPB-2, VPM2-75; RNCB-18, RNCB-28; RNCR-5.

Fisher was awarded the American Theater Medal and Victory Medal. Achieved the rank of CAPT and was retired July 1, 1978.

He married Karen and they have four children: Judith, John, Jaclyn and Janet; and three grandchildren: Kirsten, John and David Lindsay. Employed as chief engineer, Idaho Fish and Game, consulting engineer.

RICHARD "DICK" FISHER, LCDR, after one year at Millersville, STC, entered the Navy V-12 Program and attended the University of Pennsylvania for three semesters. This was followed by US Naval Midshipmen School at the University of Notre Dame where he was commissioned an ensign in the USNR.

He then attended Destroyer School at Norfolk, VA, and subsequently Line Officer's School in Miami, FL. After graduation he was assigned to the USS *Borie* (DD-704), which was at Hunter's Point, San Francisco, having been repaired after being hit by a kamikaze plane in Okinawa. The ship rejoined its squadron as the war ended and was assigned to the inactive fleet in Charleston.

Because of its renovated condition, the *Borie* was designated as plane guard for new carriers coming off the line for shakedown in the Caribbean (e.g., the *FDR* and the *Tarawa*).

In 1946 he was released from active duty, returned to Millersville STC, and completed the requirements for his degree in 1948. He was in the USNR for 22 years, attaining the rank of lieutenant commander. Having served in many units, including five summers on active duty as a staff instructor, he participated in a variety of training duties. In 1964 he earned a MS in education from Penn.

In 1958 he married Alta Acheson, a fellow Millersville graduate. They have one daughter, Judith, who was born in 1964. Judy is a graduate of the University of Pennsylvania, with a BS in computer science engineering and a BA in mathematics.

He was an industrial arts teacher in the Philadelphia School System for 10 years and the Radnor Township School District for 26, retiring in 1984.

Since retirement, he and Alta have traveled extensively in the US, often by travel trailer. In addition, their foreign junkets include the Caribbean, Mexico, Canada, much of Europe, and major locations in the Far East.

EDWARD M. "ED" FITCH JR., CAPT, born Aug. 26, 1920 in Phoenix, AZ, received a BS degree from USC. He joined the service March 14, 1942. His military locations and stations: NAS Corpus Christi; Chemical Warfare School. Served with VTN 42 (1953-56). His battles and campaigns: Iwo Jima, Japan Homeland.

Fitch achieved the rank of CAPT USNR(R) and was discharged Dec. 31, 1947. His memorable experiences: Survival of the kamikaze attack on CV-3 USS *Saratoga* at Iwo Jima.

He is married to Beverly and they have three children: Ed II, Donald and Wanelle; and seven grandchildren: Derek, Danielle, Diana, Darren, Alix, Marie and Luke. Employed as insurance broker, president, retiring Dec. 31, 1986.

MARSHA A. FITZGERALD-WILLIAMSON, CAPT, born in Waterbury, CT, received a BSN and MSN. She joined the service in 1970. Her military location and stations: Newport, RI; NAVHOSP Boston, RedCom 13, 8, 9.

Units served with include Naval Hospital and Fleet Hospitals.

CAPT Fitzgerald was awarded the MSM, Navy Commendation, Meritorious Unit, National Defense and star and 10 Year Reserve.

She is married to CAPT Joseph Williamson, SC, USNR, and is employed as a nurse paralegal and litigation consultant.

HUGH M. FLANAGAN, CAPT, born May 22, 1939 in Indianapolis, IN, earned BSME in 1961, Purdue University; MSEE in 1966, Michigan State and JD in 1970, Loyola University, Los Angeles.

He joined the service Aug. 16, 1961 via NROTC Purdue University. Reported to staff US Naval Nuclear Power School, New London, CT; graduated from Basic Submarine School, qualified in submarines; USS *Tench* (SS-417).

Released from active duty and affiliated with the Naval Reserve. Changed designator to the Judge Advocate General Corps in 1971 and served in various billets including CO VTU LAW 2011; CO LSO 220 (San Francisco); SJA, REDCOM 20; Sr. Res. JA, CINCPACFLT.

Flanagan was qualified in Submarines. He received a MSM, AFRES (3), National Defense (2), Pistol Expert Rifle (Mark). Instrument rated private pilot and life member of NRA. He achieved the rank of CAPT and was discharged in August 1964. Memorable experiences include duty in Rota Spain; Rome, Italy; Honolulu.

He is married to Norma and their children are Mike, Mark, Matt and Tim; grandchildren are Andrew, Morgan and Allison. Employed as senior member, Flanagan, Mason, Robbins et. al, Attorneys, retiring Aug. 31, 1992.

FRANCIS M. "FRANK" FLEMING, CDR, born Oct. 4, 1916 in Cathlamet, WA. Attended two years at the University of Portland, Portland, OR. He joined the service Nov. 21, 1941 at NAS Sand Point, Seattle, WA. His military locations and stations: Served on three aircraft carriers, test pilot at NATC Patuxent River. His units included VF-16 (AG-16) on USS *Lexington,* USS *Bon Homme Richard* and USS *Randolph.*

His battles and campaigns: Tarawa, Marshall Islands, Marianas Islands, New Guinea, Truk, first Battle Philippine Sea and Marianas Turkey Shoot.

Fleming was awarded the Silver Star, two Distinguished Flying Crosses, five Air Medals, Presidential Unit Citation, Area Ribbons.

He achieved the rank of CDR. Separated from the service July 27, 1946. His memorable experiences: Graduated as naval aviator in 1946 out of Pensacola and Opa Locka, shot down first enemy aircraft in Tarawa campaign November 1943; torpedoed on *Lexington* in December 1943.

He married Margarete and they have two sons, Kieran and Stephen; and grandchildren: Ian, Matthew, Heather, Nicole and Katherine. Employed at General Motors Insurance Div. Senior adjuster and underwriter.

Retired Nov. 1, 1972 after nearly 21 years service in USNR.

FREDERICK F. FLETCHER, CDR, during 1965, graduated from San Jose State College, received a commission at the Naval OCS, Newport, RI and completed Combat Information Center School in Glynco, GA. He was then assigned to the Operations Department of the anti-submarine warfare (ASW) support aircraft carrier USS *Bennington* (CVS-20).

During 30 months on board the CVS, he served as a CIC watch officer, electronic warfare officer and anti-air warfare officer, completing a deployment to the Western Pacific, recovering the Apollo IV spacecraft during mid-Pacific operations and transiting the Sea of Japan. In 1968, after completion of Defense Information School, Ft. Benjamin Harrison, IN, he reported for duty with the Commander First Fleet staff as an assistant for public affairs.

Following completion of the Commander First Fleet tour, he assumed duties as operations officer of the Pacific Fleet ammunition ship USS *Chara* (AE-31). During 1969-71, through two Vietnam deployments. Fletcher authored the publication *Skeet for the Fleet* in the Naval Engineer's Journal. During this assignment he added a deck qualification to his record. In 1972 he transferred to the Active Reserve. Serving in a variety of billets within a broad range of units, Fletcher retired from the USNR in 1983 as a CDR.

Returning to graduate school after leaving active duty, he was awarded a masters degree in industrial safety during 1982. Fletcher has been employed in a variety of professional safety management and engineering positions with the National Safety Council, the county of San Mateo, the city of Santa Rosa and the state of California.

Married and divorced, he continues to live and work somewhere in the great Central Valley of California.

ROY W. FLETCHER JR., CDR, graduated from the CPT (Civilian Pilot Training) program at Northwestern University in June 1940. Enlisting in the USN pilot training program at Glenview, IL in April 1941, he was commissioned ENS, A-V(N), USNR, naval aviator #9680, at NAS Corpus Christi (Class 4a41c) on Nov. 28, 1941. He remained there as a flight instructor until ordered to Sqdn. VS-6-D14(VS-56) for inshore patrol duty in the Aleutian Islands during the campaign to retake Attu and Kiska.

Ordered to ComFair West Coast, Nov. 24, 1943, he trained with Sqdn. VF-20 at NAS North Island, and Hawaii until Air Gp. Twenty boarded USS *Enterprise* (CV-6) on Aug. 4, 1944. In November 1944, Air Gp. 20 transferred to USS *Lexington* (CV-16) at Ulithi Atoll. The Air Gp. returned to the States in February 1945, and CDR Fletcher was reassigned to the Tactical Development Unit at North Island, until leaving active duty in September 1945. He continued service in Organized Reserve Sqdn. VF-71A, based at NAS Los Alamitos, retiring as squadron executive officer on Sept. 6, 1948. His civilian occupations included petroleum and mining geology and business executive.

He participated in actions against the Japanese at the Bonin Islands, Palau, Formosa, Philippines, Hong Kong, Iwo Jima, Okinawa, including Battle for Leyte Gulf attacks on the Japanese fleet, in the Sibuan Sea and off Cape Engano. His campaign and service citations include the American and Pacific Theatres, Silver Star, Distinguished Flying Cross, Air Medal w/eight stars, and Presidential Unit Citations awarded to *Enterprise* and *Lexington.* He became a life member of the NRA on Jan. 7, 1985.

JOHN W. FLUKE, CDR, born Feb. 27, 1944 in Columbus, MS, received a BS in mechanical engineering. He joined the service June 1, 1966.

His military locations and stations: USS *Seminole* (AKA-104); Basic Submarine Officer's Course. He served with USS *Bonefish* (SS-582), Pearl Harbor. Fluke achieved the rank of CDR and was discharged Oct. 21, 1971.

He is married to Lorraine and they have a son, Andrew. Employed as manager, Predictive Maintenance, HSB Reliability Technologies, Alexandria, VA. He retired June 1, 1995.

HENRY M. "HANK" FLYNN JR., born Oct. 30, 1935 in Bangor, ME, received a bachelor of architecture from the University of Virginia. He was commissioned June 1960, USN, ROTC regular. Served with USS *Mt. Katmai* (DE-16); USS *Denebola* (AF-56).

His battles and campaigns: Cuban Missile Crisis, Blockade Line. Awarded the Navy Commendation, Expeditionary (Cuba) and National Defense. Flynn achieved the rank of CAPT and was discharged June 1964.

His memorable experiences: seven gunnery "E's" in two days as gunnery officer; Senior Course, Navy War College, three inter-American War games at NWC CO of Units at 04, 05, 06 levels, Cuban Missile Crisis; great friendships.

He married Katherine Harrell and they have three children: Katherine (Kit), Henry III and Harrell. He retired as president, CEO, Henry M. Flynn, architect, P.C. in 1987.

JOHN EDWARD FLYNN, CAPT, born Oct. 25, 1953 in Brooklyn, NY.

He joined the service Dec. 27, 1974. His military locations and stations: USS *Forrestal* (CV-59); USS *Coontz* (DD6-40).

Flynn was awarded the Naval and Marine Corps Commendation Medal. He achieved the rank of CAPT, USNR.

His memorable experiences: served aboard USS *Forrestal* (CV-59) which hosted the president during the Nation's Bi-Centennial Celebration, July 4, 1976.

He is married to Kathryn J. and they have children John C. and Thomas H. Employed as USN, AEGIS program office, general engineer.

BRUCE SOMERNDIKE "BRUCE" FOERSTER, CAPT, born March 22, 1941 in Ann Arbor, MI, received an AB from Haverford College, 1963 and MBA University of Pennsylvania, 1971.

He joined the service April 30, 1963. He achieved the rank of CAPT and was discharged in 1989. His memorable experiences: having the ships CO fall overboard during an Equator "Crossing the Line" ceremony and successfully recovering him.

He married Gail Homer Beckham and they have three children: Samantha, Otto and Lucy.

WILLIAM KERN FOGERTY, CAPT, an educator with both BA and MA degrees, was commissioned through OCS in December 1966. Active duty assignments included two WESTPAC deployments on USS *Berkeley* DDG-15 as DCA, a Vietnam tour as a patrol officer for IUWG-1 in Nha Trang, RVN, and recall during Desert Storm as CO Military Sealift Command Office, Jacksonville, FL.

Reserve billets include five tours as CO. His awards include Navy Commendation Medal (2), Combat Action Ribbon and 17 other service medals and ribbons.

From 1992-96, he taught the Naval War College Joint Military Operations Course, participated in 10 joint, multinational exercises, and lead a military mission to Poland. In 1993, he developed an exceptional Campaign Drug Free Program in Atlanta.

Fogerty lives in Altanta with his wife, the former Ann Friemel of St. Louis. He is self-employed, the owner of a marketing and distribution company.

CARL R. "RAY" FOLEEN, CDR, born Oct. 21, 1929 in Portland, OR, received a BS from the University of Portland 1952.

He joined the service September 1952, Officer Candidate School (1952-53). His military locations and stations: USS *Boxer* (CVA-21) (1953-55) Lake Mead Weapons Base (1955-57); TAD-Albuquerque NM-1955, Key West, FL, 1954. Served with Task Force 77 - West Pac. His battles and campaigns: Korea 1953.

Foleen was awarded the Korean Service Medal, United Nations Medal and National Service. He achieved the rank of LT (active duty) and CDR (Reserves). Discharged November 1957 from active duty and December 1974 from Reserves. His memorable experiences: At GQ during final negotiations of Korean War truce July 1953.

He is married to Harlene and they have four children: Carol, Carla, Lori and Linda; and four grandchildren: Chris, Megan, Garret and Kaley. Employed with Bonneville Power Admin. (Deputy Administrator) 1958-81; power consulting (own business) 1981-88; director of utility (1988-94).

ROBERT B. "BOB" FOLSOM JR., LCDR, born May 24, 1939 in Huntington, NY, received BA from the University of South Sewanee, TN; JD Tulane University, School of Law New Orleans, LA.

He joined the service June 1960. His military locations and stations: USS *Independence* (CVA-62). He achieved the rank of LCDR and was discharged June 1981.

Folsom is married to Hervey and they have a daughter Margaret. Employed as vice-president and general counselor, Bank of Wedowee, Wedowee, AL.

WILLIAM E. "BILL" FORD, LCDR, born April 6, 1921, El Paso, TX. Earned BS degree in finance, University of Notre Dame; MBA, Golden Gate College, San Francisco. Joined the service Dec. 31, 1942.

After graduating from Midshipman School, he took Gunnery School at San Diego then sent to San Francisco with orders to proceed to Espiritu Santo where the *Bennett* was getting ready for the Bougainville operation. They made the 30 day trip aboard the Liberty Ship, SS *Samuel Lancaster,* in 30 days without an escort.

Boarded the *Bennett* the day IT left for Bougainville and helped escort a number of transports filled with Marines. The first night at Bougainville they watched a classic naval battle between ADM Merrill's cruisers: Cleveland, Columbia, Montpelier and Denver, and ADM Burke's destroyers and a Japanese cruiser task force coming down out of Rabaul to try and destroy our destroyers and transports. Merrill, Burke and company saved our necks.

Memorable experience when hit by kamikaze off Okinawa while on radar picket station. The bomb from kamikaze killed several of his shipmates. The ship went back to Bremerton for repairs and the war ended before they returned to the Pacific. Retired as LCDR, USNR.

Later, as a civilian he served as a docent on the SS *Jeremiah O'Brien* in the 1980s. Ford, his wife, daughter and two granddaughters were in Normandy for 50th anniversary. The *O'Brien* was the only US ship there. Ford and his wife took a 26 day cruise aboard the *Royal Viking Star* to the South Pacific.

Married Patricia Ann Fagan and they have two children, Sherry and Michael; two grandchildren, Amy and Megan Miller. Employed with Chevron Oil Co. until retirement in July 1985.

JAMES E. FORREST, RADM, SC, USN, a native of Palms, CA, enlisted in the USN in 1940, serving aboard the battleships *Tennessee* and *Wyoming* and the fleet oiler *Cuyama* before receiving a Secretary of the Navy appointment to the Naval Academy in 1942, graduating with the accelerated class of 1946 on June 6, 1945.

He commanded three minesweepers before transferring to the Supply Corps in 1948. He later served aboard USS *Saratoga* (CVA-60) and USS *Hyman* (DD-732).

Forrest was selected for flag rank in 1971. Immediately prior to becoming national executive director of the Naval Reserve Association in June 1975, RADM Forrest and was auditor General of the Navy for three years during the period July 1972-June 1975. His former assignments included Deputy Director and Director, Simulation and Computer Directorate, Industrial College of the Armed Forces; CO and XO, Navy Fleet Material Support Office, Mechanicsburg, PA; assistant fleet and force supply officer, Pacific Fleet Service Force and Deputy Commander, Military Traffic Management Command and Terminal Service.

RADM Forrest has been awarded the Secretary of the Navy Distinguished Public Service Award, the Legion of Merit w/3 Gold Stars in lieu of second, third and fourth awards, and the Navy Commendation Medal. His campaign and service medals include American Defense, American Campaign, Asiatic-Pacific Campaign, WWII Medal, Occupation Medal and the National Defense Medal.

He holds a master's degree in business administration from Stanford University and has attended the Navy Supply Corps School; Navy Postgraduate School, Monterey, CA; the Industrial College of the Armed Forces and the Executive Management Program at the University of Pittsburgh.

He was appointed XO of the Naval Reserve Association in June 1975, a position he held for over 21 years, retiring in November 1996. During his term as XO he personally raised over three million dollars for the purpose of constructing an office building for the association including a Naval Reserve Library. Part of these funds were also used to finance a Naval Reserve Assistance Program for those Naval Reservists who participated in the Persian Gulf Campaign.

He is married to the former Elisibeth Hanson of East Lansing, MI. They have five children: Ann, James, Jean, Ellen and Tom.

ROY E. FORS, CWO4, born in Chicago, IL, July 25, 1915. Enlisted in the Illinois Naval Militia on April 9, 1935. Division federalized into active service, November 1940. Served aboard the USS *George F. Elliott* (AP-13) in North Atlantic duty, and, until she was sunk Aug. 8, 1942 at Guadalcanal. Transferred to the USS *Zeilin* (APA-3) for further activity at Guadalcanal until she was badly damaged on Armistice Day, Nov. 11, 1942. *Zeilin* was sent back to Terminal Island for major repairs.

Remained aboard *Zeilin* for the Attu, Kiska, Tarawa, and Kwajalein landings. Promoted to warrant bos'n January 1944. While at Guadalcanal, April 1944, was transferred back to the States and assigned to Harbor Defense Training. Later took part in the Iwo Jima invasion and remained there until the war ended. Awarded eight Pacific Battle Stars and two Naval Unit Commendations.

Graduated with a BA degree from USC, then married Derith. They have one son Greg and now reside in San Francisco. He remained associated with the USN until July 1975 attaining the permanent rank of CWO4, USNR.

JAMES WHEELER FORSYTH, LTJG, born May 17, 1923 in Staten Island, NY, received a LLB St. Johns University.

He joined the service Nov. 13, 1942. His military locations and stations: European Theatre of Operation, Pacific Theatre of Operations. Served with USS *Chateau Thierry;* LCT (5) Flotilla Ten; USS *LC (FF) 571.* His battles and campaigns: WWII, Anzio, Sicily, North Africa, Italy, Philippines, Occupation of Japan.

Forsyth was awarded EAME Campaign Medal w/4 Battle Stars, WWII Victory Medal, Asiatic-Pacific Campaign, American Theatre Campaign Medal, Navy Occupation Service Medal w/Asia Clasp; Navy Unit Commendation Medal, Philippine Liberation Ribbon, Philippine Republic Presidential Unit Citation. He achieved the rank of LTJG, was discharged March 23, 1946, and retired from the USNR May 1, 1962.

Married to the former Janis A. Hall and they had four children: J. Michael, John Patrick, Timothy and Nancy; and four grandchildren: Deirdre, Katherine, Caitlin and Kimberly. Proctor in Admiralty and Merchant Marine officer. Forsyth died Jan. 30, 1987.

TIMOTHY FORSYTH, CDR, born Feb. 3, 1959 in Staten Island, NY, received his BS from SUNY Maritime College at Ft. Schuyler (1981); J.D. Brooklyn Law School (1987).

He joined the service Aug. 16, 1981. His military locations and stations: New York, NY; San Diego, CA; Newport, RI; Little Creek, VA. Served with COMUSMARDEZLANT. His battles and campaigns: Operation Uphold Democracy.

Forsyth was awarded the Navy Achievement Medal, Armed Forces Reserve Medal w/Bronze "M" Device, National Defense Service Medal, Navy Expert Rifleman, Navy Expert Pistol Shot.

He achieved the rank of CDR. His memorable experiences: Exercise Fuertas Detensas 95 in Panama; Exercise Allegiant Sentry 95 in Cape May, NJ.

Forsyth is married is Susan M. Murphy, MD. Employed as attorney at law and Merchant Marine officer (still serving in USNR).

JOSEPH R. FORTIER, MCPO, born May 29, 1922, Portland, OR. He joined the USNR while a junior in high school, Jan. 22, 1940. He drilled one night a week in Tacoma, WA at the Perkins Building and one week-end a month on the Eagle Boat (57). His first cruise was on board USS *Crosby* (DD-164), a four pipe destroyer.

He was ordered to active duty Nov. 29, 1940, before graduation, as a F3/c at $36.00 per month, to serve on board the USS *Pawtucket* (YT-7), a coal burner. During WWII, besides the *Pawtucket,* he also served onboard the *Black Douglas* (PYC-45) and *Wyman* (DE-38). The nickname "one punch" *Wyman* was credited sinking of two Japanese submarines and was overseas 21 months attaining 150,000 sea miles.

He rejoined the USNR January 1946 in Tacoma, WA; was a drill instructor, enlisted classifier, chief master at arms and served on many ships and stations. The ships: *Brannon* (DE-446), *Gilligan* (DE-508), *Rombach* (DE-364), *Shangra-La,* (CVA-38); *Constant* (MSO-427); Watts (DD-567); *Salisbury Sound* (AV-13); *Navarro* (APA-215); *Delta* (AR-9); *Twinning* (DD-540), *Klondike* (AR-22), *Mullany* (DD-528); *Buck* (DD-761) and served 7 1/2 years as master chief of the command, onboard *Orleck* (DD-886). Retiring with 42 plus years.

He was awarded the following medals: Navy Achievement, Good Conduct, Naval Reserve Meritorious Service, American Defense, American Campaign, Asiatic-Pacific Campaign, WWII Victory, Armed Forces Reserve and United States Naval Reserve.

His wife, former E. Marie Miller, has been his partner since, Aug. 17, 1943. Daughters are Cherie Marie Murchie, Judith Elizabeth Fortier; granddaughter is Patricia Marie Mitchell; grandson is Philip Matthew Mitchell; and great-granddaughters are Carson Alexandria Johnson and Melaina Marie Johnson.

FREEMAN P. FOUNTAIN, CAPT, born in Hawley, MN on April 16, 1921. He is a graduate of Stanford University (BA) and the University of Colorado (MS, MD).

During WWII he was an FMF Corpsman in the Pacific Campaign. He was attached to the 2MARDIV on Saipan and in the Occupation of Japan. During the Korean Conflict, he served as an Army medical officer at Fitzsimons Army Hospital. Later, as a Navy medical officer, he treated Vietnam casualties at the San Diego Naval Hospital.

He is a life member of the NRA, VFW and TROA. He is active in the Military Order of the World War and served as their surgeon general for four years.

The Fountain family consists of his wife Jeri, three children: Diana, David and Daniel; and six grandchildren: Sarah, Rabekah, Angela, Julie, Andrew, and David. Doctor and Mrs. Fountain reside in Scottsdale, AZ.

EUGENE F. FOWLER, CAPT, enlisted in the USN at Dallas, TX in 1954 and began his naval career as a seaman recruit at Naval Recruit Training Center, San Diego, CA. Completion of Class "A" School in Electronics at Naval Training Center, Treasure Island, San Francisco, CA, led to sea duty in WestPac, the Far East and Alaska aboard USS *Mathews* (AKA-96). He completed active duty in 1958 at NAS Corpus Christi, TX as electronics technician first class. When he returned to his home in Denton, TX, he affiliated with a Naval Reserve Security Gp. Unit. Upon graduation in 1962 with a BS degree in physics from North Texas State University, he was commissioned ENS (cryptology). In 1963 he graduated with MS degree in nuclear physics.

He served between 1971-91 as CO of four Naval Reserve Security Gp. units in the Dallas, TX area. In a national role as software technician director for the Reserve Security Gp. from 1984-86, he led a team in the development of mission-critical software that was distributed nationwide for use in the Naval Reserve Security Gp. program. Assigned 1986-89 to Naval Reserve Readiness Command Region 11, Dallas, TX as special assistant for cryptology, he was responsible for four NRSG units within TX, OK and NM. In early 1993, while assigned on staff at the Cyrptologic Division Officer School, Naval Security Group HQ, Washington, DC, he created advanced training courses on new technology weapon system. During his naval career he was awarded the Meritorious Service Medal, two awards, the Navy Commendation Medal and the Enlisted Good Conduct Medal.

In civilian life, Fowler was technical staff member and program manager at Texas Instruments, Inc., Dallas, TX where he held positions from 1962 through 1991 in Research and Development of Electro-Optical missile guidance systems for the US Government. In civic activities he served as a director of a private elementary school and actively participated in local community programs.

He resides in Richardson, TX where he is a consultant in intellectual property and telecommunications matter. He and wife Kay have two adult daughters, Liesl and Julie.

JAMES R. FOWLER, RADM, born on Feb. 4, 1937 in Douglasville, GA where he completed his primary and secondary education. He was commissioned an ENS in 1959, and graduated from Emory University School of Medicine in 1964. Following an internship at Harborview Medical Center, University of Washington, Seattle, he reported to NAMI, Pensacola, FL in 1965 as a student naval flight surgeon. Upon designation he was assigned to the 3MAR Aircraft

Wing at El Toro. His active service in the States included VMT-2, El Toro and VMT-103 USMCAS, Yuma as a naval flight surgeon until 1969.

While in-country he served with HMM-163, MAG-16, 1st MAW from 1966-67.

Fowler's reserve career includes service with the 4th FSSG, Atlanta, GA; commanding officer Surgical Team 220, Salt Lake City, UT and 4th MAWMED MAG-42, Alameda and director of Health Services, REDCOM 20 San Francisco. He was the CO, Fleet Hospital CBTX-500, No. 9, REDCOM 22, Seattle, when selected for Flag rank in 1990.

Dr. Fowler is in the practice of surgery of the hand and upper extremity in Salt Lake City, UT. He is certified by the American Board of Plastic and Reconstructive Surgery and has a certificate of added qualification in surgery of the hand. He is assistant chief, Bureau of Medicine and Surgery and Surgeon General for Reserve Affairs, Washington, DC and was selected for his second star in 1993.

CECIL C. FOX, CAPT, native of Canton, MS received BBA degree from the University of Mississippi. Enlisted May 3, 1944. Reported for active duty June 6, 1944. Completed boot training Camp Peary, VA, reported aboard USS *Arkansas* (BB-33) September 1944. Participated in the invasions of Iwo Jima and Okinawa as a communication yeoman in the Radio Div.

Discharged June 1946. Accepted direct commission as ENS SC on October 1949. Recalled Sept. 5, 1950, reporting to USN Supply Corps School, Bayonne, NJ. Graduated in January 1951 and reported to the USS *Tarawa* (CV-40), served as disbursing officer January-November 1951.

Detached and released from active duty. While serving in various units rose to the rank of CAPT. Units served included: NRS Div 6-71 Jackson, MS; AWS-79 NAS, Memphis, TN; TT&TMD 8-1, New Orleans, LA; NR Support Unit 8-37, New Orleans, LA. Transferred to Retired Reserve Jan. 1, 1982.

He is married to Anne and they have three children: Cecil Jr., William and Linda. Employed as wholesale food distributor, chairman of the board, retiring June 9, 1986.

EDWARD MICHAEL "ED" FOX, CDR, born Feb. 2, 1924 in Dallas, TX, received a BA from Drake University; BBA from the University of Texas, Austin.

Enlisted in the service Dec. 15, 1942. Commissioned March 1947 as ENS AV-L. His military locations and stations: NAS Whidby Island; Washington NAS Alameda. Units served with include Hedron 6 AS/R; NAS Alameda; NAS Dallas FASRON; NAS Glenview; Naval Air Test Center, Paxtuxent River. He participated in the Cuban Missile Crisis;

Fox was awarded the Good Conduct Medal, American Theater Medal, WWII Victory, Unit Commendation (2). He achieved the rank of CDR and retired June 30, 1970.

He is married to Alyce and they have two children, Quinn and Emily; grandchildren are Jonathan and Christina. Employed with Meredith Corp./Better Homes and Gardens for 29 years, as branch manager, Chicago and Dallas.

NORA J. FRAIN-HEYER - *See her bio in the H section.*

PETER J. "JOHN" FRANCE, LCDR, born July 20, 1959 in Albuquerque, NM. He received a BSBA 1984 Rockhurst College, Kansas City, MO. He did graduate work, 15 hours Madonna University, Livonia, MI.

Joined in the delayed entry program Oct. 14, 1976, and entered active duty April 10, 1977. His military locations and stations: RTC Orlando (April 1977-June 1977); AFAITC, Lowry AFB, CO ("A" School) (June 1977-September 1977); USS *Independence* (CV-2) (October 1977-March 1981); DEFDOC Kansas City, MO 0218 (April 1981-September 1984); FIRSTEURLANT 1673, NAF Detroit, MI (September 1984-April 1987); AIC 1773, NAF Detroit, MI (April 1987-May 1992); ONI 0218, Scott AFB, IL (May 1992-September 1996); JICTRANS 0182, USTRANSCOM, Scott AFB, IL (September 1996-present).

France served in Persian Gulf Hostage Crisis, 1979-80; Support of Desert Shield/Desert Storm, 1980-90. His memorable experiences: in 1986 being selected, then initiated as a CPO, then within the year being commissioned an ENS, limited duty officer-Intelligence.

He was awarded the Navy and Marine Corps Achievement Medal (2), Navy Good Conduct Medal, Navy Expeditionary Medal, National Defense Medal, Armed Forces Reserve Medal, Joint Meritorious Unit Commendation, Meritorious Unit Commendation (2), Sea Service Ribbon (2), Marksman Pistol Ribbon, Sharpshooter Rifle Ribbon. France achieved the rank of LCDR.

He married Mary Frances and has a son Christopher Peter and a daughter Allison Marie. Employed as market manager, Technique Products, Sherwood-Davis & Geck, St. Louis, MO.

RALPH H. FRANCE II, CAPT, born July 9, 1941 in Baltimore, MD. He joined March 13, 1966.

His military locations and stations: Naval Supply Depot, Seattle; NAVSTA ADAK, NNMC Bethesda, MD. France achieved the rank of CAPT, JAGC, USNR.

He married Carol and has three children: Ralph Hayward III, Robert B. and Jill Elizabeth. Employed as Judge, District Court II, Washington Co., MD.

HENRY "BLAND" FRANKLIN JR., CAPT, born May 14, 1934 in Raleigh, NC, received a BS in animal husbandry and MS in sociology/economics, North Carolina State University.

He joined the service May 1956. His military locations and stations: OCS Newport, RI; USS *Merrimac* (TAO-47); USS *Truckee* (AO-147); Service Sqdn. Four embarked on USS *Vulcan* (AR-5).

Franklin was awarded the Valley Forge Essay Freedom Award in 1972; Outstanding Sea Power Presentation Award.

He achieved the rank of CAPT and was discharged August 1984. His memorable experiences: Going through Suez Canal, tracking ICMB's in South Atlantic launched from Cape Canaveral, FL. Breaking refueling record at sea for most ships and most fuel transferred in 24 hours period while serving on USS *Truckee.*

He is married to the former Mary Frances Fulp and their children are Henry III, Benjamin and Shelly; grandchildren are Daniel, Grabriel and Jadon. Employed with NC Agricultural Extension Service, Polytechnic Institute; faculty, Virginia State University. He retired November 1992.

JOHN A. FREEMAN, CAPT, supports the NRA as a lifetime member. Starting life in El Reno, OK, on March 23, 1944, and graduating from Oklahoma State University with a master's degree in Solid State Physics in 1969. He enlisted in the USN and experienced boot camp at NTC, San Diego, where his company was selected as the most outstanding academically. He completed OCS in Newport, RI, and was commissioned an ENS in 1970. His first assignment was as assistant administrative officer and intelligence officer aboard USS *Meridith* (DD-890) homeported in Mayport, FL, and toured the Mediterranean Sea in the winter and spring of 1971 while participating in surveillance duties resulting from the Jordanian

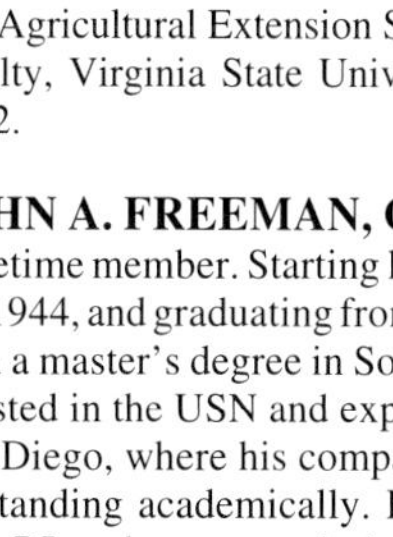

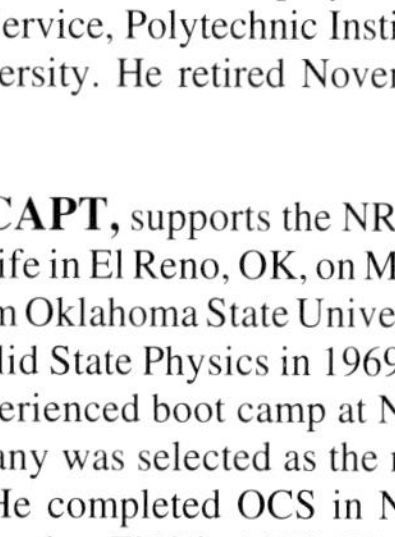

Crisis. CAPT Freeman's first shore assignment took him to the Pentagon and the Defense Intelligence Agency where he managed multi-service surveillance requirements.

Affiliating with the Reserves in 1975, as an intelligence officer in Naval Reserve Intelligence Area 16, Willow Grove Joint Reserve Base, PA, he served in various units. The first of these was with Naval Attache Unit 0106 followed by Area Analysis Unit 504. Subsequently, he served as Air Intelligence Officer with Patrol Wings 504 and 593 and with Anti-submarine Patrol Sqdn. 66 where he trained squadron personnel in ship recognition. While on duty in Rota, Spain and the Azores, he reported on Russian troop ships en route to the uprising in the Congo. In 1981 CAPT Freeman moved to Fleet Intelligence Rapid Support Team, Europe and Atlantic 0693. In 1985 he served with Intelligence Volunteer Training Unit 0104. In 1986 he was a member of a unique fleet deployable unit, COMSERVGRU2DET204 at the Lehigh Valley Reserve Center, Allentown, PA, performing the duties of training officer and intelligence officer.

CAPT Freeman's most memorable tour was as plankowner and CO, Navy Operational Intelligence Center 0293 from October 1987-September 1989. This unit received the ADM Cronin Award as the most outstanding unit in Reserve Intelligence Area 16. During this period he received the Navy Commendation Medal recognizing his leadership achievements. After another year with the Intelligence Volunteer Training Unit 0104, he became Commanding Officer, Atlantic Intelligence Command 0593 in 1990. In 1991 he served with the Intelligence Volunteer Training Unit 0104 as inspector general for three years. Before leaving Area 16, he was commanding officer, Defense Intelligence Agency Unit 0593, 1994-96.

CAPT Freeman wears the Navy Commendation Medal, Meritorious Unit Commendation, Joint Service, National Defense, Armed Forces Reserve Medals and Overseas Ribbon.

As an electronics engineer at the Naval Air Warfare Center, Patuxent River, MD, he designs acoustic signal processing software used aboard the Navy's P-3 Orion anti-submarine patrol aircraft. CAPT Freeman enjoys wood turning and resides with his wife, Margaret, in Lexington Park, MD. He maintains a second residence in Langhorne, PA.

DONALD RALPH "DON" FREESE, CAPT, born July 8, 1941, Duck Island, IL, received a BS, USNA 1963; MS in 1970 and MBA 1976 Binghamton University; JD 1994 Chase College of Law. He joined the service July 7, 1959. He had a diverse 26 year career in the USNR following graduation from the US Naval Academy and four years active duty aboard destroyers and cruisers.

He is surface warfare qualified and believes he is one of the few who have qualified at sea in each of three decades, 1964 USS *New* (DD-818); 1979, USS *Virginia* (CGN-38) and 1988 USS *Wainwright* (CG-28). He had eight commands, including two Naval Research Units, the USS *Virginia* (CGN-3802), a PMT, two GVTUs, and two Readiness Units. His first command, the USS *Virginia* Unit, won the 1979 Readiness Command Two Outstanding Unit Award and the NRL 105 Unit won the 1987 Readiness Command Five Competition.

Freese received National Defense Medal, 20 year Reserve and Battle E. He achieved the rank of CAPT and retired Sept. 1, 1993. His memorable experiences: Mediterranean and Middle East deployments 1965-66; Gemini recovery 1965.

He married Sherry G. and has two children 1LT Marc D. Freese, USMC, and Michelle L. Freese. Employed as attorney in private practice.

HOWARD FREIBERGER, CAPT, born April 2, 1921 in New York, NY, received THHS 1937; BS CCNY 1941; MA Columbia University 1949.

He joined the service June 28, 1944. His military locations and stations: Great Lakes, Chicago, Gulfport, Yokosuka and Sasebo. He served with COMFLTACT, Yokosuka, COMPHIBPAC, Mine Countermeasures Sta.

His battle and campaigns included the Occupation of Japan and Korean War. Freiberger was awarded the WWII Victory Medal, Korean Presidential Unit Citation, UN Medal, USNR Medal.

He achieved the rank of CAPT and retired April 2, 1981. His memorable experiences: teletypewriter repairman Occupation of Japan, Staff COMPHIBPAC 1950, Mediterranean Cruise USS *Leary* 1948; CO NR Communications Div. 3-1 1961-64; CO NR Officers School 3-2 1969-72.

Employed as chief, Research and Development Div., VA Prosthetic and Sensory Aids Service, retiring in 1982.

FRED R. FRENCH JR., LCDR, born June 2, 1933 in Brockton, MA, received a BA and MA.

He joined the service September 1951 Midshipman, NROTC Brown University; commissioned June 1955. His military locations and stations: USS *CADMUS* (AR-14). He served with NR Composite Co. 1-25, Boston; NR Support Unit 1-44 and VTU -114, Providence.

French achieved the rank of LCDR and retired May 1, 1980. His memorable experiences: as member of the NR Composite Co. 1-25, Boston, was asked in 1960 to establish a program of rehabilitation for confinees in NAVSTA Brig. Program became a prototype throughout the Navy. In 1973 established similar program at Newport, RI Brig.

He is single and employed as president, North River Foundation, Inc.; psychotherapist.

ERIC B. FRETZ, LT, born June 6, 1967 in Cheverly, MD, received a BA from the University of Michigan; BS Southern Illinois University.

He joined the NROTC 1985-89. Commissioned in 1989. His military locations and stations: Precommissioning DET for USS *Chosin,* USS *Chosin* (CG-65) also Naval Training Center Great Lakes, IL and Selective Service System.

His battles and campaigns: Desert Shield/Storm. His awards and medals include NCM, NAM, MOVSM, Battle E (2), Sea Service, SW Asia (2), National Defense, Master Training Specialist.

Fretz achieved the rank of LT and left active duty March 1996 and joined the USNR. His memorable experiences: Tour in Gulf, visiting Australia, qualifying as OOD & EOOW, passing OPPE.

He is married to Jennifer and they have a son Mason. He is employed as teacher and computer consultant.

JOHN J. FRIEL, CAPT, born Nov. 22, 1944 in Riverside, CA, received a BA from the University of Pennsylvania NROTC (1996); MA Temple University (1972); PhD University of Pennsylvania (1975).

He joined the service September 1966. His military locations and stations: NOB Norfolk, Navbase Guam, Navbase Philadelphia (NRF). Served with USS *Taconic* (LCC-17); USS *Iredell County* (LST-839). Selres-USS *Douglas H. Fox* (DD-779); USS *Robert L. Wilson*(DD-847); USS *Harold J. Ellison* (DD-864); USS *Rich* (DD-820); USS *Hawkins* (DD-873); Fleet training Gp. (FTG) DET 104 as CO; USS *Clark* (FFG-11) as reserve coordinator; Office of Naval Research (ONR) DET 104 as CO.

Friel was awarded PUC, NUC, National Defense (2) NR Sea Service, Vietnam Campaign, AF Res. He achieved the rank of CAPT and discharged Sept. 18, 1969.

He married Patricia Murphy and their children are Joseph, Kathleen, Christine and David. Employed as technical director, Princeton Gamma-Tech and retired Oct. 31, 1991 from Selres.

WAYNE R. FRITZ, CAPT, born Nov. 3, 1934 in South Haven, MI. Graduated from high school. Attended US Naval Academy Class of 1957, receiving a BS in engineering and US Naval Postgraduate School and BS in physics.

He joined the service June 1953 as a midshipman, US Naval Academy. His military locations and stations: USS *Lyman K. Swenson* (DD-729), based in Long Beach, CA; Submarine School, Groton, CT; USS *Menhaden* (SS-377), based in San Diego, CA; Naval Postgraduate School, Monterey, CA; Guided Missile School, Dam Neck, VA; USS *Lafayette* (Gold) SSBN 616, based in Rota, Spain (families lived in Charleston, SC); Staff COMSUBGRU SIX, based in Charleston, SC; USS *Bluegill* (SS-242), based in San Diego, CA; instructor, Nuclear Weapons Training Center, North Island, CA.

Naval Reserve Duty: chief of staff, Naval Reserve Group Command (3-25S), Liverpool, NY; CO, Naval Reserve Surface Div. (3-95L), Liverpool, NY; XO, NavRes Submarine Reserve Unit (AS2603), Liverpool, NY; CO, NavResSubUnit AS34 Canopus (3402), Liverpool, NY; Staff CINCLANTFLT Det. 206, Washington, DC; Staff COMNAVRES Readiness Command (Region SIX), Washington, DC; Also served as USNA Blue & Gold Officer (10 and 1/2 years).

Fritz' battle and campaigns and awards and medals include: National Defense Service Medal w/Bronze Star; Naval Expeditionary Medal (Quemoy/Matsu Chinese conflict 1958); Vietnam Service Medal (w/2 Bronze Stars, Tet I & Tet II); Naval Reserve Service Medal; Naval Reserve Association Medal; Letter of Commendation COMSUBGRU SIX for outstanding performance as operations and navigator of USS *Lafayette* during the Cold War for readiness and accuracy of Polaris II missile tests 1966.

He achieved the rank of CAPT USNR, retiring Jan. 1, 1980. His memorable experiences: Greatest satisfaction was turning around a defunct and listless Naval Reserve Organization in Liverpool, NY as CO of NRSD 3-95(L). Through the obtaining of funds, restructured training at real Naval Schools in Damage Control, Fleet Operations, and other training which had a great effect on sailor motivation. Including the families in the overall program significantly raised the morale of the division and had an outstanding effect on their readiness to perform their assigned duties.

He married Joan and their children are Michael, Scott and William. Employed as application engineer Shipboard Electronics and Combat Systems, GE Aerospace Syracuse, NY; manager Undersea Electronics Programs and Business Development, GE Aerospace, Washington, DC; director Business Development Submarine Electronics and Combat Systems, Martin Marietta, Corp., Washington, DC. He retired Jan. 1, 1995. Presently working for the Naval Academy Visitors Center as tour guide and military reunion coordinator.

WILBERT FRITZ, CAPT, born in Sheboygan, WI, Nov. 30, 1914, of immigrant parents. They spent six years in Sheboygan then moved to Fresno, CA for two years, then to Portland, OR for six years and Ridgefield, WN. He graduated from Ridgefield High School in 1936 and lettered in baseball four years and in football three years.

He attended Washington State University in 1936, graduating in 1940 with a BS degree. He attended for a fifth year and earned a B.Ed. in 1946. In 1963 received a master degree in education administration from Gonzaga University.

He taught school at Central High for two years. February 6, 1943 he entered recruit training at Farragut, ID. Graduated as E-4, as honor man, Co. 88 and recruit COP. Transferred to Bainbridge, MD. Graduated as CPO and transferred to Saufley Field, Pensacola, FL

Commissioned as ENS as Camp Shoemaker, CA. Officer Indoctrination School, Ft. Schuyler, NY, 1944; Amphibious training, Coronado, CA, 1944; Assigned to USS *Audrain* (APA-59), served as boat officer, deck officer, 1DIV officer, 1944.

Initial landings: New Guinea, Mindoro, Luzon, Nagasaki and Okinawa. Fritz was released to inactive duty in 1946. He joined the USNR, 1947 served as traffic officer, XO, of Surf. Div. 13-6, 13-19, 13-21, Tra. off Mobil Team, CO Mpb Team, Group CDR, Naval Res. Tra. Ctr.

He was recalled for the Korean War in 1950. Attached to Rec. Tra. Cmd. as assistant legal officer, then legal officer, 1953, June, released to inactive duty. Assigned to NR Surface Divs. 1953. Assigned to NR Counseling Board, 1966. Promoted to CAPT, permanent in 1966. Fritz retired Nov. 30, 1974. He thoroughly enjoyed the USN and the many friendships, really an outstanding family.

He married Gladys and is employed as Central Valley School District Teacher, Administrator, retiring May 1974.

JAMES H. "JIM" FROST, CAPT, born Aug. 25, 1946 in Seattle, WA, received a BA in business administration from the University of Washington.

He joined the service May 12, 1966. His military locations and stations: US Naval Base Yokosuka, Japan; Amphibious Base Coronado, CA. He served with units: USS *Windham County* (LST-1170); Coastal River Sqdn. 1. Participated in battles and campaigns in Vietnam and Desert Storm.

Frost was awarded the Navy Commendation Medal w/Combat V (2) and Combat Action Ribbon. He achieved the rank of CAPT and was discharged Sept. 30, 1993.

His memorable experiences: duties as an officer-in-charge Mobile Support Team. Two in support of air operations with Naval Special Forces Vietnam, LST-1170, various Reserve Commands.

He married Sheryl Leo and they have two children, Aaron and Shelby. Employed as insurance broker, executive vice-president.

LAURANCE A. "LARRY" FROST SR., CWO4, born April 26, 1926 in Malden, MA, attended Franklin Technical Institute for four years.

He joined military service May 6, 1943. His military locations and stations: Newport, RI, Mare Island, CA; Pearl Harbor, HI, Midway and Saipan Islands, San Diego, CA; New London, CT and Key West, FL.

Frost was awarded the Navy Good Conduct Medal, American Defense Service Medal, Asiatic-Pacific Campaign, WWII Victory Medal, National Defense Service Medal, Navy Reserve Medal and Armed Forces Reserve Medal. He achieved the rank of CWO4 and was discharged April 26, 1947, USN. Transferred to USNR on the same day.

His memorable experiences: stationed at the Mare Island shipyard he was assigned to work on the ill-fated USS *Indianapolis* (CA-35) to replenish ammunition that was expended in the invasion of Tarawa. The *Indianapolis* was the last ship lost in WWII. She was torpedoed by a Japanese submarine on July 30, 1945 at 12:05 a.m. and sank within 15 minutes. Of the 1,196 men on board, 880 officers and sailors died.

In late 1943-early 1944 the *Indianapolis* arrived at the Pearl Harbor Submarine Base. Frost was transferred to the Aiea receiving barracks. The next assignment was to work on the battleship *Oklahoma* (BB-37), one of the first ships lost in WWII. The 16" guns and deck houses were all removed. Frost and another seaman were assigned hip boots, weather gear, helmets, battle lanterns and a bushel basket. As the compartments on the *Oklahoma,* now in dry dock, were vented of poison gases, he went below to search for the remains of the warrant officers in the state rooms.

Frost's first ship was the submarine tender USS *Futon* (AS-11) started on June 25, 1944 and ended April 1986 when he was "Piped Over The side" on board *Fulton.* During those years his Reserve Division supported *Fulton.*

Widowed in 1973 he remarried in 1975 on the Ceremonial Deck of *Fulton* by the ship's chaplain. On Sept. 9, 1991, Mr. and Mrs. Frost attended the 50th reunion and decommissioning of the USS *Fulton in* New London, CT.

VANCE H. FRY, RADM, is a native of Chattanooga, TN, and graduated from the US Naval Academy in 1960 where he received a BS degree and a commission as an ENS in the USN. Upon graduation from the Navy Supply Corps School in Athens, GA, he served as supply officer of the USS *Maddox* (DD-73) in the Pacific.

After three years with International Business Machines Corporation in Atlanta, GA, Fry returned to Chattanooga and founded the Electronic Computer Programming Institute where he served as president for 12 years. Upon graduation of Tennessee School of Law he established a law office which serves as a focal point for managing business and property development interests.

During Operations Desert Shield/Desert Storm, eight of the Navy's 14 Cargo Handling Bns. were deployed to SW Asia, Europe and the Pacific. As CDR, Navy Cargo Handling Force, Fry was selected in June 1991 to lead the Naval Reserve Component in the Desert Storm Washington, DC victory parade.

He is past-president of the National Navy Supply Corps Association, Inc. Other Navy involvement's include: ROA, NRA, Navy League and US Naval Academy Alumni Association.

Fry holds the Legion of Merit, the Joint Services Commendation and Armed Forces Reserve Medal w/HG.

He and his wife Charlynne live in Chattanooga, TN. They have three children: Vance Jr., Randal and Melodye. Retired Sept. 1, 1995.

DAVID B. FRYE, CAPT, born Jan. 15, 1937 in Peterborough, NH, received a BA in 1958 from the University of New Hampshire and a MBA in 1978 from Bryant College.

He joined the USN, Oct. 8, 1958. His duty stations: OCS NWPT RI, CIC School Brunswick, GA; Air Control School Brunswick, GA; *W.C. Law* (DD-763); *MC Fox* (DDR-829); OODF, CIC, Communications, Operations, Flight School NAS Pensacola, FL; various VT squadrons; HS-1, HS-5, CO HS-74; CO NAS Brunswick, ME.

Reserves: XO Naval War College Reserves, National Defense University, NASA Recovery Force, Mercury Project, Gemini 5 Recovery (Conrad and Cooper), Apollo Program CVS-39; Essex, CVA-59, 62, 63, 66, 69 and others, CNAVES Advisory Board, and Wardroom Club, Boston, MA.

He achieved the rank of CAPT and was discharged from active duty June 1969 with 4,400 total flight hours, 4000 H3's. His memorable experiences: Gemini 5, Apollo Recoveries; training VMX1 pilots for Presidential pilot duty in H-3's Quantico and Key West. Command at Sea HS-74 (five safety awards 60,000+ accident free flight hours) and OODF at sea on destroyers September 1959 through June 1962.

Frye is married to the former Cherie Bodine and they have a daughter, Karin McClure Frye Jacoboski, Southington, CT. She received BS/MS in finance, Bryant College and University of Hartford. She is employed with United Technologies; daughter, Julie Collins Frye, BS/MS Physical Therapy, Simmons Residency, Hartford, CT; son, Nathan Howard Frye, UNH, student, Criminal Justice, completed Marine Corps Aviation Officer Candidate School Junior Class, Quantico, 1996, Army Jump School, Army ROTC, Ft. Benning, GA, 1995.

They have one granddaughter, Emily M. Jacoboski born Aug. 7, 1996. He is employed as vice-president BankBoston, president/CEO Mutual Advisory Corp.; president and CEO Eastern Portfolio, Inc.; president/CEO/owner, Frye Investment Associates, LLC. Retired Jan. 31, 1992.

LUTHER VERNON "LU" FRYE, LCDR, born March 11, 1924 in Charleroi, PA, received a BS in Civil Engineering in 1950 from the University of Pittsburgh.

He was accepted into the Navy V-12 Program June 26, 1943, Purdue University. His military locations and stations include: Purdue University Nov. 1, 1943; DePauw University, July 1, 1944; Notre Dame, March 11, 1945.

Frye served with units: USS *English* (DD-696), Oct. 17, 1945-June 30, 1946; USS *John R. Pierce* (DD-753), March 1951-March 1953.

He participated in the Occupation of Japan and the Korean War. He received the Naval Reserve Medal, American, Asiatic-Pacific, WWII Victory, Navy Occupation, Armed Forces Reserve, Korean Medal w/star, UN National Defense. Frye achieved the rank of LCDR and retired March 11, 1984.

His memorable experiences: standing on top of main battery director, unable to get section watch officer to clear hatch to get to battle station, as gunnery officer, while Pierce was taking seven hits off coast of Korea while guarding mine sweeper. Fortunately, no loss of life.

He married Alice H. and they have two children, Wesley and Nancy; grandchildren are Shannon, William, and Christopher Herron; and great-granddaughter is Sydney Taylor. Employed with Municipal, Water and Wastewater Engineering and Management, retiring May 1, 1992.

BOYD L. "THE FACE" FUGAL, LCDR, born Oct. 22, 1921 to Niels and Viola Young Fugal, in Pleasant Grove, UT. Graduated from Pleasant Grove High School and Timpanogos Stake Seminary. Attended Utah State University 1939-41; BYU, 1946.

He served in the USNR, 1942-46. Achieved the rank of LCDR. Pilot Navy combat fighter plane aboard aircraft carrier CV-18 Wasp in the Pacific Theater.

His memorable experiences: serving as naval aviator piloting a F4-U Corsair, pressing home strafing and bombing attack; low altitude in the face of intense and accurate anti-aircraft fire from Japanese strongholds.

He married Venice Kathleen Jacob of Provo, UT, March 15, 1944, Sanford, FL. Sealed in Salt Lake Temple April 20, 1944. They have six daughters, three sons, 43 grandchildren and four great-grandchildren. Employed as heavy equipment operator; president Niels Fugal Sons Construction Company, retiring Dec. 14, 1959.

JOSEPH B. "JOE" FULLER, CDR, born May 9, 1921 in Augusta, GA, received a BS in 1943 from Presbyterian College, Clinton, SC.

He joined the V-7 Program June 1, 1942. Commissioned Jan. 20, 1944; USNRMS, Notre Dame University. Units served: (WWII) OinC LCT 1077, Flot 31, Gp. 91; (Korea) BMUI, NAVBCHGp1.

Fuller achieved the rank of CDR and was awarded the Navy Commendation Medal, Presidential Unit Citation, American Theatre, Asiatic-Pacific Theatre, WWII Victory Medal, Occupation Service, National Defense, Korean, Naval Reserve Medal and Philippine Liberation Medal.

He is married to Margaret Wake and they have two children, Russell and Pete; grandchildren: Christy, Barry, Andy and Michael. Employed with Merry Brothers Brick and Tile Co., retiring Jan. 1, 1985.

LEE M. FULLER, JR., CAPT, born March 14, 1938 in Mt. Vernon, NY, attended Princeton University in 1960 and Columbia University School of Law in 1965.

He joined the service in 1960. His military locations and stations: USS *Essex* CVS-9, Quonset Point, RI, 1960-62. Served with various USNR Intelligence & JAGC units in New York, 1963-92.

His awards and medals include the Cuban Expeditionary Medal, National Defense Medal and the Naval Reserve Medal. He retired in 1992.

He has three children: Scott Fuller, Lisa ten Grotenhuis and Kristina Fuller; grandchildren are Nicolaas ten Grotenhuis and Erin Fuller. He is an attorney, partner in law offices of Lee M. Fuller Jr.

WALTER F. "WALLY" FUNK, LTSG, born March 30, 1920 in Fargo, ND, received a BA from Carleton College, Northfield, MN.

He joined the service Jan. 6, 1942. His military locations and stations: San Diego Sound School; Pearl Harbor Decoding Unit. Served with units: USS *Mugford* (DD-389), assistant communications officer.

Funk served in nine campaigns: Solomons, New Guinea to Okinawa and Japan. Received nine stars for the Pacific Theater and also American Theater. He achieved the rank of LTSG and retired in 1962. His memorable experiences: hearing Tokyo Rose declare US sunk off Morobe Harbor, New Guinea in 1943. After the late landings; CAPT Corey hanging onto the direction finder in front of the bridge. He dodged all the Betty's bombs.

He is single and was employed as accountant, public and private-Lakedale Telephone Co., retiring June 30, 1987.

EDWARD JOSEPH FURDAK, CAPT, born in Detroit, MI, Jan. 24, 1934 and attended secondary schools in Dearborn, MI. A graduate of the University of Michigan he holds a BS degree in civil engineering. He was commissioned an ENS in 1955 through the Aviation Officer Candidate Program, and was designated a naval aviator in 1957 after completion of flight training at Pensacola, FL and Corpus Christi, TX.

After receiving his wings, Furdak reported to Airborne Early Warning Sqdn. Twelve (VAW-12) based at Quonset Point, RI. During his tour he piloted the Navy's A1 (Skyraider) aircraft and flew numerous carrier missions. As part of the Carrier Air Group Three (CAG-3) aboard the USS *Saratoga,* he participated in the Lebanese crisis. His duties in VAW-12 included ordnance officer and 1LT.

In March 1978, he was assigned to VTU-7373. Concurrently, he affiliated with the US Naval Academy Information Officer Program and subsequently became Director for the state of Michigan. Retired from the USNR in June 1985 with 30 years of service.

In civilian life he worked for Ann Arbor Construction, International Business Machines and Ford Motor where he retired in January 1996 after 31 years of services.

Furdak and his wife reside in Plymouth and East Point, MI. They have two children, Susan Baldwin and Robert Furdak. Susan, a first grade teacher, lives in Northville, MI with her husband Scott and sons, Michael and Andrew. Robert, managing director, State Street Global Advisors, resides with his wife Dr. Kirsten Eckland, in Wellesley, MA.

LOUIS A. "LOU" GABRIELLI, LCDR, born June 26, 1916 in West Hoboken, NJ. He graduated from Emerson High School, Union City, NJ.

Enlisted in USNR, 29th Fleet Div., Hoboken, NJ as apprentice seaman, Jan. 22, 1935. First trip to sea, USNR cruise, USS *Badger* (DD-126). First trip to sea US Merchant Marine, deck cadet, SS *Black Osprey,* Black Diamond Lines, Jan. 15, 1936. He attended US Maritime Service School, Hoffman Island Training Station, NY. Trained on *Joseph Conrad*, full rigged ship. Attended Kweder's Navigation School, New York, NY.

Received a ENS commission USNR, Feb. 14, 1942; Participated in Casablanca Invasion November 1942 while serving as second mate on the troop ship SS *Santa Elena.* He was under air attack at Oran, Algeria, Naples, Italy, Palermo, Sicily, Bizerte, Tunisia. First command at age 27 on May 26, 1944, SS Falcon, Grace Lines. Other commands: SS city of *Savannah*, SS *Samuel F. Dewing*, SS *Grenville M. Dodge*, SS *George L. Farley*. LT commission, USNR, Jan. 1, 1946. Commissioned LCDR, USNR, April 1, 1952.

Served in US Merchant Marine from 1936-51, afloat with Black Diamond Lines, United Fruit Co., Morgan Lines and Grace Lines; 1951-70, served with Grace Lines, ashore; 1970-80, served with Sea-Land Service, Inc., ashore; Retired Oct. 1, 1980. Gabrielli received the Atlantic War Zone Bar, Pacific War Zone Bar, The Merchant Marine Combat Bar, Mediterranean Middle East War Zone Bar, WWII Victory Medal Merchant Marine, USNR 10 Year Medal, New York State Conspicuous Service Cross. He is a member of 17 military organizations.

He is married to Victoria and they have a daughter Elaine and grandchildren, Emily and Erica Wilts. Employed as steamship company executive, retiring October 1, 1980.

LEONARD H. "LEN" GAFFGA, LCDR, born Aug. 11, 1942 in New York City, NY, received a BA economics (Yale University); MBA International Business (Wharton Business School).

He joined the service Oct. 24, 1964. His military locations: OCS-Newport, RI; Guam, MI; Seventh Fleet, Sixth Fleet. Served with: Naval Communications Station (Guam); USS *F.D. Roosevelt* (CVA-42). His battles and campaigns: Vietnam and Yankee Station.

Gaffga was awarded the National Defense Service Medal, Vietnam Service Medal, Armed Forces Reserve Medal and Republic of Vietnam Campaign Medal.

He achieved the rank of LCDR and was discharged Aug. 12, 1968. He retired from the USNR Oct. 1, 1987. His memorable experience: crossing the Equator for the first time; typhoons, Pamela and Nancy; dry dock in Japan; transiting the Straits of Gibraltar at night; being called up in support of the Lydian strike.

He is married to Catherine and they have children: Christopher, Nicholas and Timothy. Employed with IBM (retired); Dekalb College (professor).

OSCAR "OZZIE" GALENO, LTJG, born Nov. 1, 1923 in San Francisco, CA, received a BS from the University of San Francisco and a MA San Francisco State University.

He joined the service August 1942. His military locations and stations: Sand Point, WA. Served with VPB-13; PB2Y Sqdn. His battles and campaigns: Philippines, Iwo, Okinawa. Galeno was awarded the Distinguished Flying Cross and Air Medal. He achieved the rank of LTJG and was discharged Jan. 1946 to inactive service.

His memorable experiences: 12 hour patrols and missions up both sides of Japan. Seaplane tender (home ship) being hit by kamikaze. Intensive and excellent training provided; flight, navigation, engines, physical leadership.

He has been married 49 years to Jane Ann and they have children: Richard, Janet, Lynn and Joan; and grandchildren, Camille and Merlin. Employed as school principal with San Francisco Public Schools. He retired in September 1979.

FRANCIS X. "FRANK" GALLAGHER, LCDR, born July 29, 1925 in Baltimore, MD, received a BA from the University of North Carolina (NROTC); JD University of Baltimore.

He joined the service May 7, 1943. His military locations and stations: Navy Supply School, NAS-Jacksonville. Units served with: Submarine Batt 5-1, Acorn 5-2.

Gallagher received the American Theater, WWII Victory, Korean Service and USNR. He achieved the rank of LCDR and retired in 1988.

His memorable experiences: practicing trial attorney, Baltimore City Council 1959-87; serving four terms as vice-president. Retired as president. Chair of the Board of Estimates; director of various educational and charitable boards.

He is married to Eileen B. and has children: Michael J. Gallagher and Katherine A. Numbers. Employed as public service, founder and managing partner of law firm, city of Baltimore 1987; law 1993.

F. TREVOR "TRIG" GAMBLE, CAPT, a Vermont native, he grew up in New York State where he attended Colgate University under the Aviation Midshipman Program. He was designated a naval aviator in 1950 following flight training in SNJs and ADs.

Assigned to VS-26 flying TBM Avengers, he had deployments aboard the carriers *Siboney, Mindoro, Saipan, Kula Gulf, Palau* and *Salerno Bay;* he completed All-Weather Flight School in 1952. Returning to shore duty in 1953, he attended aviation electronics (O) and Guided Missile Schools and in 1954 returned to the fleet as guided missile officer aboard the USS *Shangri La* (CVA-38). During this tour and a 1956 Pacific deployment, he qualified as an OOD underway.

A Naval Air Reservist from 1957-81 he flew FJ, F2H, S2F, P2V and SNB aircraft at Niagara Falls, South Weymouth, Gross Ile and Detroit. He commanded VS-64Y1 from 1969-71 as well as an NAS augmentation unit and a VTU.

Gamble earned a PhD in physics from Connecticut and served on the faculty of Denison University from 1963-96.

He is married to Carolyn and has three children: F. Trevor Jr., Rebecca Pyle and Melinda Lovern; and six grandchildren: Sarah, Jameson, Lauren Pyle, Brittany, Brodee Gamble and Madison Keelyn Lovern. Employed as professor, retiring July 10, 1988. Denison University. Retired from the USNR Jan. 1, 1982.

PAUL WILLIAM GARBER, CAPT, born Nov. 16, 1934, Harvard College AB (mcl) 1956, commissioned ENS contract NROTC. Harvard Law School, 1961; US Naval War College School of Naval Warfare, 1970. Served on active duty aboard USS *Crow* (MSC (O)-7) 1956-57; thereafter spent 28 years in Naval Control of Shipping Organization and Military Sealift Command Reserve Units, including five commands.

Retired 1986 as CAPT. He received the Navy Achievement Medal in 1970. Past president Constitution Chapter and First Region NRA and Navy Chapter 5, ROA; Judge Advocate, Massachusetts Bay Council, NLUS; Board of Overseers, USS *Constitution* Museum; former member of Board of Visitors, Massachusetts Maritime Academy; director, West End House Boys and Girls Club, Boston.

Honorary Consul of Chile, Boston 1974; New England (less CT) 1995; awarded Order of Bernardo O'Higgins in grade of Knight Commander 1979. Practices law with his twin brother Philip, COL, USAR (ret).

DAVID L. "DAVE" GARDEA JR., born July 8, 1941 in El Paso, TX, received a BBA from the University of Texas, El Paso in 1965 and M.Ed in 1983.

He joined the service July 16, 1958 (Enlisted) Commissioned Oct. 22, 1965. His military locations and stations: (active duty). Enlisted Submarine School, New London, CT, July 1960-October 1960; USS *Sea Devil* (AGSS-400), San Diego, CA, October 1960-June 1962; Officer Candidate School, Newport, RI; June 1965-October 1965; Boat Group Officer Training, PHIBASE Coronado, CA, October 1965-December 1965; USS *Blanco City* (LST-344), 1LT, combat duty, RVN, January 1966-September 1968; CMDR SVC Force, Commo/Ops/Plans, Pearl Harbor, HI, December 1968-January 1972; DD School, Department HQ, Newport, RI, January 1972-July 1972; USS *Knox* (DE-1052), OPS/SWO, Pearl Harbor, HI, August 1972-November 1973; Com14thNavDistr, assist. Ops, Pearl Harbor, HI, November 1973-June 1974; Naval Station PERL Hbr Ops, Pearl Harbor, HI, June 1974-June 1975.

Inactive duty-USNR: October 1975-September 1976, NAVORD 1611, training officer, ELP, TX; October 1976-September 1978, TCSPHIL 511, CO, ELP, TX; October 1978-November 1979, PHIBRON 111, XO, ELP, TX; November 1979-September 1980, PHIBASECOROSD 211, CO, ELP, TX; October 1980-September 1982, FLTRAGRU SDGO 211, CO, ELP, TX; October 1982-September 1984, SIMA SDGO 1511, CO, ELP, TX; October 1984-September 1985, CGN39 TEXAS, CO, ELP, TX; October 1985-September 1987, COM7THFLT, ASST OPS Dept Head, NAS, Dallas, TX; November 1987-October 1988, VTU-1105, training officer, El Paso, TX; November 1988-September 1989, VTU-1105, CO, ELP, TX; SURFREDUNIT, CO, ELP, TX.

He was awarded the Surface Warfare Officers Pin, Silver Dolphins, Navy Achievement Medal, Combat Action Ribbon, MUC, National Defense Medal, Naval Reserve Meritorious Service Ribbon, RVN Campaign Medal w/60-, RVN Gallantry Cross w/Palm, RVN Civil Action Medal w/Palm, Navy and Marine Corps Overseas Service Ribbon w/Star, Armed Forces Reserve Medal w/HG, RVN Service Medal w/3 stars. He achieved the rank of CAPT 06 and retired in Oct. 1, 1991. His memorable experiences: combat duty in Vietnam waters serving on LST-344 on various duties with the Mobile Riverine Force in the rivers of South Vietnam, Market Time Operations along coastal patrol areas, and logistical life/resupply to coastal ports.

Married Delia and they have three children: Gina Marie and David Joseph. Employed with US Army Air Defense Artillery School, Ft. Bliss, TX; foreign military sales training coordinator/Special Projects, Operations.

JOSEPH JOHN "JOE" GAREFFA, CAPT, born Feb. 12, 1941 in Allentown, PA, received a BA in psychology.

He joined the service September 1959 NROTC University of Utah, June 1963, active duty. His military locations and stations: USS *Roosevelt* June 1963-October 1963; VP-46 Moffett Field, CA; VP-64 NAS Willow Grove; VP-46, Okinawa; Sangley Point, PI; Sattahi P. Thailand; Adak, AK; VP-64 NAS Bermuda, NAS Cajex Azores, Rota, Spain.

His battles and campaigns: Vietnam, (North Korean Elections, 1966). He was awarded the Air Force Expeditionary Medal, National Defense, Vietnam Service, Vietnam Campaign, Armed Forces Reserve Medal, NUC and MUC.

Gareffa achieved the rank of CAPT (06) and was discharged from active duty June 1969. His memorable experiences: operations from Philippines to Yankee Station (1965-66-67); first P-3s to Thailand (Sattahip), 1966-67; Adak deployment (1969).

He is married to Marily Louise and they have children, Anthony James and Gregory Neil. Employed with United Airlines, 767 captain (1969-present), furloughed seven years (1971-78); Citicorp Leasing, senior account executive (1973-78). He retired Oct. 1, 1985.

BRUCE M. GASCOIGNE, CDR, born Nov. 8, 1921 in Oak Park, IL, received a BA from the University of Washington in 1949.

He joined the service June 19, 1942, USMC; USN, Aug. 15, 1944. His military locations and stations: Seattle, New Jersey, New York, Illinois; USS LST-853 and 947, Sand Point NAS, Seattle, Air Wing ACOS ADM.

Gasciogne's battles and campaigns: Eniwetok, Saipan, Tinian, Lingayen Gulf, Okinawa, Occupation of Japan. He was awarded the Asiatic-Pacific Medal, Philippine Liberation Medal, Naval Reserve Medal, Armed Forces Medal. He achieved the rank of CDR and retired June 30, 1970.

His memorable experiences: torpedo attack; zero attack and hit; D-Day; Okinawa; typhoon 2; watching signing of surrender on the *Missouri.*

He is married to Dorothea J. and they have four children: Robert B. II, Bruce M., Victoria A. and James E.; and five grandchildren: Angela, Alicia, David, Danielle and J. McMillan. Employed as owner-Gasciogne Labor Co., Seattle, WA, as vice-president and president Feely Lawlor Lumber Co., Seattle, WA.

ALEXANDER "SANDY" GASTON, QM3, born Oct. 26, 1946 in New York City, NY, graduated high school and attended college. Joined the service November 1963. Served with USS *Princeton* (LPH-5); USS *Wexford County* (LST-1168). His battles and campaigns include TET offensive and various combat operations.

Gaston received the National Defense Medal, Vietnam Service Medal, RVN Expeditionary, Meritorious Unit Commendation and Combat Action Ribbon. He achieved the rank of QM3 and was discharged November 1969.

His memorable experience: his entire 21 months of active duty, from January 1967 to October 1968 were all memorable. He is single and self-employed in investments.

ROLAND E. GAUDETTE, CAPT, born in Worcester, MA, June 22, 1929 and received a BA from Assumption College, an OD from the New England College of Optometry and a MBA from Nichols College.

Attending OCS in Newport, RI he was commissioned ENS (1105) in 1953. Active duty assignments were with SACLANT and as training officer, USS *Macon* (CA-132). Reserve assignments were in the surface program with Command of two Reserve Surface Divisions, a Naval Reserve Officers School and a volunteer training unit. He was promoted to CAPT on July 1, 1973 and retired from the Reserve Program July 1, 1978.

In civilian life Gaudette was an optometrist for 31 years in Spencer, MA and then assoc. dean for clinical education at Nova Southeastern College of Optometry, north Miami Beach, FL for four years, retiring to Naples, FL in August 1995. His family includes his wife Therese, six children and 10 grandchildren.

JOSEPH A. "JOE" GAUDIO, CAPT, born Sept. 30, 1931 in New York, NY, attended Columbia University, chemical engineer (1954); OCS Newport, RI.

Joined the service March 5, 1959. Commissioned ENS July 1959. His military locations and stations: VQ-2 Sqdn., Rota Spain and USS *Springfield.* (Homeported: VilleFranche-Sur-Mer-France. Served with VQ-2 and USS *Springfield* (CLG-7), Sixth Fleet flagship. Served in active duty July 1959-November 1965.

Gaudio received the Sharpshooter (Pistol), Unit Meritorious Award, Reserve Medal. He achieved the rank of CAPT (06), USNR and retired March 1, 1990.

His memorable experiences: people-to-people officer in Rota, Spain; organized a "Caseta" for USN participation in the famous "Feria" of Sevilla; foreign claims officer of Sixth Fleet investigated massive oil spill at Cannes, France by carrier *FDR.*

He is married to Josephine and they have a daughter Marcy Maggio. Employed with Johnson & Higgins, Global Insurance Broker; International vice-president in charge of Latin America, retiring Sept. 30, 1996.

JAMES A. "JIM or PAY" GAYLE, CDR, born Oct. 7, 1921 in Dubois, ID, received a BS/1 year Graduate School MBA.

He joined the service July 6, 1942 as apprentice seaman. His military locations and stations: various ships/stations-Pacific Theater. Served with 11 & 12 Naval Dist., 6th Fleet, Sub Base, PHTH, COMSERVPAC. Participated in various campaigns in South Pacific, 1942-43.

Gayle received two Battle Stars, Asiatic-Pacific, etc. He achieved the rank of CDR, USNR Supply Corps, discharged RID. He retired from the military Oct. 7, 1981.

He is married to Phyllis and they have two children, Robert S. and Janis E.; and four grandchildren: James, Robin, Scott and Craig. Employed as president nationwide truck line, retiring August 1989.

RONALD G. GELINEAU, CAPT, born in Duluth, MN, July 25, 1931, he attended the University of Minnesota, Duluth (UMD) for two years and U of M, Minneapolis for fall quarter 1951. He enlisted as a naval aviation cadet in November 1951. He graduated as naval aviator and was commissioned an ENS at Pensacola in June 1953. After All Weather School at Corpus Christi, he reported COMNAVAIRPAC and was assigned to VS-23, NAS North Island, the first West Coast squadron to receive the Grumman S2F-1, twin engine antisubmarine, carrier based aircraft. After a deployment to WestPac aboard the USS *Princeton* (CVS-37), he reported to VU-7, NAAS Chula Vista, where he flew fleet support missions in the JD-1 and F9F-5 aircraft.

In April 1956, LTJG Gelineau was ordered to NAS Seattle as a TAR and assumed the duties as assistant flight training officer (VS). During the next two years, he introduced and transitioned four Naval Air Reserve squadrons from single engine Grumman AF and Douglas AD aircraft to the S2F Tracker.

Released to inactive duty in December 1957, LT Gelineau remained in the NAR and began a civilian flying career with West Coast Airlines, Boeing Field. He was recalled to active duty by President Kennedy on Oct. 1, 1961 as a member of the VS-891 during the Berlin Wall Crisis and released to inactive duty in August 1962. He subsequently served in VS-892, VS-893, VS-894, and as CO of VS-35T3 and VS-68T2. CDR Gelineau served on two BUPERS Reserve Officer Selection Boards and at the close of NAS Seattle, he was assigned to the Naval Air Reserve Staff, NAS Whidbey Island, and served as mobilization officer and staff recruiting officer.

Gelineau is married to the former Shannon Lee Richmond of Burbank, CA. He is a life member of the Naval Reserve Association and retired from the USN in November 1977. Awards: China Service, National Defense, Armed Forces Reserve (w/HG).

As a naval aviator, he had the privilege to fly various types of aircraft including the SNJ, TBM, S2F, JD, F9F, AF, AD, SNB, T-34, accumulating 3,550 flight hours.

As an airline pilot, he flew as a co-pilot and then CAPT on Douglas DC-3, Fairchild F-27 and Douglas DC-9 aircraft accumulating 24,000 flight hours in his 34 years with first, West Coast Airlines and through the ensuing mergers and corporate buy-outs, Air West, Hughes Air West, Republic, and Northwest Airlines. A long time member of the Air Line Pilots Association, he participated as local council co-pilot representative, master executive council chairman, member of the National Board of Directors, local grievance chairman, member of Seniority Merger and three contract negotiating committees.

CAPT Gelineau retired from the Northwest Airlines at the mandatory retirement age of 60 on July 25, 1991.

JACOB "JOCK" GELLER, born May 30, 1930 in Houston, TX, received a BA (Rice); MEd (University of Houston); DDS, Baylor Dental.

Enlisted in the AF while a senior at Dental School (Baylor) in 1956. Commissioned June 1956, 1LT USAF Res. By August he was in England at RAF Chelveston (3914th ACRON), which started as a SAC base and changed to TAC (10th TAC Hospital).

He left in 1960 and returned to Houston to practice. Signed on in USAF about 1961 into a Reserve Unit at Ellington Field. Promoted to MAJ. In 1967 transferred to USNR as LCDR. Served in USNR until 1988.

His most memorable experiences: 48 active duty tours and others - but most at non-pay station. Retired with over 3,500 retirement points in 1988, as CAPT, USNR. He received the Navy Commendation in 1988.

He is married to Martha and employed as associate professor, 1967-86.

LAWRENCE E. "LARRY/DUKES" GEMMA, CAPT, a native of Worcester, MA, he entered the Naval Aviation Cadet Program from the University of Massachusetts in 1965 after four years in the enlisted Naval Reserves. He joined HS-3 stationed in NORVA, and deployed aboard the USS *Randolph* and USS *Yorktown* flying ASW missions from 1967 to 1970. He was then stationed at NAS Pensacola for Search and Rescue duty and attended the University of West Florida from 1970-72. Immediately joining the NAR at South Weymouth, MA, he flew for 15 years with HS-74 and HSL-74, ending his flying career as CO of HSL-74 from 1985-87. While attached to the VTU at NAVSTA Newport, RI, Gemma volunteered for Desert Storm in 1991, and became deputy operations officer for COMUSNAVCENT/COMIDEASTFOR stationed aboard the USS *Lasalle* in the Arabian Gulf from May 1991 to May 1992.

He was awarded the Meritorious Service Medal, Navy Commendation Medal, Navy Meritorious Unit Commendation, Navy "E" Ribbon, Naval Reserve Meritorious Service Medal, National Defense Service Medal w/Bronze Star, Southwest Asia Service Medal W/Bronze Star, Sea Service Deployment Ribbon, Naval Reserve Sea Service Ribbon, Navy and Marine Corps Overseas Service Ribbon, Armed Forces Reserve Medal w/two HGs, Saudi Liberation of Kuwait Medal, Kuwait Liberation of Kuwait Medal and Pistol Expert Medal.

He is co-owner of Mystic's Army Navy Store in Mystic, CT, his home of 25 years. He has two daughters, Michelle and Maria.

JOHN GENERAL, CAPT, born July 26, 1943 in Dunfermline Scotland, received a BS with distinction USNA; MS OPS Analysis NPGS.

He joined the service June 6, 1965. His military locations and stations: USNA; flight training; NAS NORIS; NAS Moffett; NPGS. Served with VP-50; COMFIRSTFLT; COMTHIRDFLT; CINCPACFLT; BUPERS.

General served in Vietnam. He received the MSM, Air Medal, NAM, MUC, Expert Pistol. He achieved the rank of CAPT and was discharged September 1994.

His memorable experiences: CamRahn Bay; 1st carrier landing T-28 *Lexington;* 1st son Pensacola; 2nd son Ft. Ord; management assistance CINCPACFLT IG, 1978-91.

He is married to Ellen and they have two children, Matt and Sean; grandchildren are Cece, Jenny, Stewart and Marshall. Employed as CEO Delfin Systems; executive vice-president St. Research; SRI; ESL, retiring September 1994.

LEON S. GEOFFREY, LCDR, born Aug. 31, 1917 in New Orleans, LA, received a BSS from the College of City of New York. Graduate work, American University, Washington, DC.

He joined the service Aug. 9, 1943. His military locations and stations: Naval Training School (radio) Bedford, PA; Navy Supply Corps School, Harvard University; Naval Supply Operational Training, Bayonne, NJ; GROPAC 15; USS *William P. Biddle* (APA-8).

Geoffrey achieved the rank of LCDR. He received the American Theater, Asiatic-Pacific Theater, WWII Victory Medal; Naval Reserve (10 years) and Armed Forces Reserve (10 years).

His memorable experiences: crossing the equator (Neptune ceremonies); training for invasion of Japan; 1993, 1994 and 1996 reunions with crew of USS *Biddle*.

His wife Rose is deceased. His children are Joel and Alan and grandchildren are Laura and Ryder. Employed with Census Bureau as statistician; War Productions Board (economist); US Mutual Security Mission to China (public administrator and advisor); patent office (manager, library and search center); Virginia Professional Standards Review Foundations (executive director), retiring 1978.

THOMAS W. GEORGE, CAPT, a native of Baltimore, MD, attended the University of Maryland and received a BS in aeronautical engineering, 1961. Attended Florida Institute of Technology, earned MS degree in space technology in 1968. Commissioned in 1971 as an aeronautical engineering duty officer in the USNR.

Enlisted in the USNR in 1952; served with SURDIV 5-4, Ft. McHenry. Enlisted in the USMC in 1953, served with VMA-333, Opa Locka and deployed onboard USS *Saipan*. Served in Korea with H&HS1, Pohang; Japan with MARS 17, Iwakuni; HMR (M) 461, New River; released from active duty in 1957. Served in the Reserves, 1958-60, with VMA-321, Anacostia.

Since 1971 served continuously with Jacksonville Naval Air Systems Command Reserve Unit and is currently CO of NADEP 0474. Recalled to active duty for Operation Desert Storm January 1991 and served with Naval Aviation Depot Ops Ctr. European Repair and Rework Activity, Naples, Detachments Bahrain and Saudia Arabia as the senior officer-in-charge of aircraft battle damage repair activities for Navy and Marine units in the Persian Gulf Theater.

He is an Associate Fellow of the American Institute of Aeronautics and Astronautics. Awards include the NCM, MUC, Marine Corps GCM, NDM, SWA Campaign Medal, AFRM, Kuwait Liberation Medal and the Navy Overseas Ribbon.

Employed by NASA as a technical manager for Space Shuttle operations at the John F. Kennedy Space Center, retiring Nov. 1, 1994. He and his wife Shirley have three children: Kathryn Riley, Elizibeth and Andrew; two granddaughters, Mary Kathryn and Patricia.

WILLIAM ARTHUR "ART" GEORGE, CAPT, born Jan. 29, 1910 in Pittsburgh, PA, received a BS from the University of Pittsburgh and DMD in 1932. Member ODK; OKU.

He joined the service Oct. 14, 1942. His military locations and stations: Amphibious Force Training Base, Solomons, MD; Mactan Philippines Islands. Units Amphibious Training Base; Acorn 51; Reserve Units after active duty: 4-2 and 4-10. Retired from active duty Feb. 20, 1946.

Joined Reserve Unit 2-2 in 1948 and was released Aug. 30, 1953 to be CO of Unit 4-10 University of Pittsburgh Reserve Dental Co. for recruiting, training and assigning of dentists for active duty. From 1956 through 1959 30 officers were assigned each year. The maximum quota was then reduced to 11 per year and was decreased in subsequent years. During these years and until his university retirement in 1988 when he became a professor emeritus of prosthodontics and associate dean emeritus. He retired from the Reserves April 1974.

He is married to Wilma Marie and they have three children: William A. Jr., Ruth Ann and Richard A.; grandchildren are William A. II, Kevin A., Richard T. George; Peter A. and Lynn Ann Stoyak, Rachel E., Benjamin T. and Mara C. George; great-grandchildren: Belinda J., William A. IV; Siera N. Victoria George.

JOHN D. GERELL, CDR, born Aug. 27, 1920, Newport News, VA. Educated at the University of Missouri, 1939-41. Joined the service in March 1942.

Stations and Units: NAS Corpus Christi, Kingsville, instructor, Atlantic City NAS, VFN-53, VC-4, USS *Midway* in Mediterranean area, NAS Minneapolis (TAR), NAS St. Louis (TAR).

Participated in action in both Pacific and Atlantic areas. His memorable experiences include flying over 60 different models and types of aircraft during his career; as a special weapons F2H Banshee pilot aboard the USS *Midway* in 1953-54.

Discharged from active duty in February 1958. Awards include Asiatic-Pacific Campaign, Naval Reserve Medal, Navy Occupation Medal and others.

Married Lillian (deceased 1993) and has two children, Robert and Lynn; and two grandchildren, Robyn and Keola. Married Nancy and they live in Honolulu. Retired Aug. 27, 1980 as FAA air carrier inspector - DC-9 specialist.

CARL CHRISTIAN GERLACH JR., LCDR, a native of Glenside, PA, he attended Penn State in the ROTC program for two years. Reported there in July 1943 in the Navy V-12 program. Transferred March 1944 to USN Midshipman School, Camp Macdonough, Plattsburgh, NY. Graduated and commissioned June 1944 as ENS. Attended Amphibious Training School, Camp Bradford, Norfolk, VA. Assigned to LST 630 in Seneca, IL August 1944. Sailed down Mississippi River, through Panama Canal, to the South Pacific. Made many amphibious landings in the Philippines, Okinawa, Japan, China and French Indo-China.

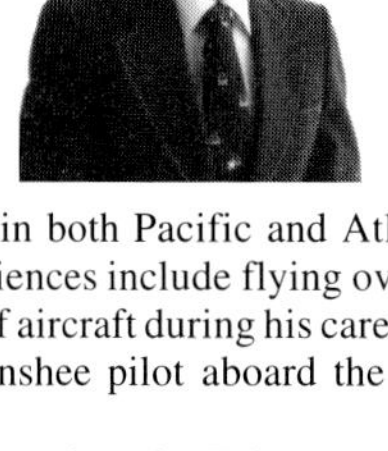

Served as gunnery officer, assistant first lieutenant, and loading officer. Separated from USN in Philadelphia July 1946.

Graduated from Penn State in 1948 with BS in psychology. He was a personnel officer with the State of Pennsylvania.

Gerlach became active in Reserves at NAS Willow Grove, PA in 1955. He was administrative/personnel officer in squadrons VR933, VS932 and VR52. Retired in 1982 with the rank of LCDR.

GERALD E. GILBERT, RADM, commissioned OCS, 1956. Active duty, 1956-59. Served on USS *Iron County* (LST-840), Pearl Harbor and in Communications Office, Great Lakes.

He transferred in 1962 as designator to JAG Corps, USNR. Selected to flag rank in 1987, received second star in 1990. Served as USNR senior judge advocate until retirement in 1992; received numerous awards including Legion of Merit.

Received BA Denison University (Distinguished Alumni Award 1984), 1955; JD University of Virginia, 1962. Following 1962 admission to BAR, served as assistant US Attorney, Washington, DC; General Counsel's Office, US Department of Commerce, and In-house Counsel, Nonprescription Drug Manufacturers Association.

Joined Hogan & Hartson in 1968; made partner 1970; twice elected to Executive Committee; presently a senior partner and managing partner of Virginia office. Concentrates on international arbitration and trade association representation.

Past National President of Federal Bar Association (received highest award, Earl Kintner Award in 1991) and Past National President of Judge Advocates Association. General Counsel to Direct Selling Association (inducted into Hall of Fame 1992). General Counsel to International Hardwood Products Association (received IHPA's first Distinguished Service Award in 1996). Past member, Board of Directors, Navy Memorial Foundation; present board member of Navy Marine Coast Guard Residence Foundation.

ROGER GILBERTSON, RADM, was commissioned at the Officers Candidate School (OCS) at Newport, RI, in July of 1961. His active duty was with the Naval Tactical Data System (NTDS) - initially in the technical office at Minneapolis and then at the Bureau of Ships in Washington, DC. Upon leaving active duty in 1964 he immediately affiliated with a Reserve unit.

The highlights of his Navy career include: Commissioned as a USNR (ED). Earned the Presentation Sword as first among the 300 who started in the OCS. He was a plank owner of NTDS, the first transistorized computer system. Commanded a Reserve Unit as a LT. Commanded two reserve units as a CAPT. Selected as a RADM while his mother was in her 92nd year. Served on a dozen Officer Selection Boards. He was the first USNR flag officer to be a Capstone Fellow at the National Defense University. Became the senior engineering duty officer responsible for 1,000 ED Reservists in 200 Reserve Units. Gave hundreds of motivating and energizing speeches. Promoted to two-star RADM.

He is a life member of the Naval Reserve Association, the Reserve Officers Association, the US Naval Institute, the American Society of Naval Engineers, the Navy League, and The Retired Officers Association.

Gilbertson will have served for 37 years of commissioned service in an ED and will be close to 62 years old upon Navy retirement. His intent is to speak, write, teach and consult - to fill newly discovered time.

In his civilian career with the US Department of Commerce he has been a program manager for basic research. He has administered $75 million in industry-centered basic research. His second civilian job has been on the adjunct faculty at Shenandoah University; teaching MBA students.

His undergraduate degree in Industrial Engineering is from Georgia Tech in 1958. He is also a graduate of Purdue University, the Harvard Graduate School of Business, the Industrial College of the Armed Forces (ICAF), and the University of Southern California (USC). He received a doctorate from USC in 1988, based on a dissertation, *The Search for Wisdom Systems*.

He is the father of grown children, a daughter, Gambi

and a son, Erik. Proud to be an American and honored to have had the privilege of wearing the naval uniform of the US of America for 37 years. Fair winds and following seas to all!

GERALD W. "JERRY" GILL, CAPT, born May 22, 1932 in Ohio/USA, received a BS in Ohio State University.

He joined the USNR, 1950 MIDN. Commissioned ENS in 1954. Achieved the rank of CAPT USNR. Naval aviator 1956. Life member Naval Reserve Association.

He is married to Katherine and they have two children, John and Steven; grandchildren are Kevin and Rebecca. Employed as captain with American Airlines and retired in 1992.

WILLIAM F. "BILL" GILLEN, LCDR, born June 15, 1918 in Philadelphia, PA, graduated from Girard College, St. Joseph's University, Philadelphia with a BS degree in 1959.

AVCAD USNR Sept. 3, 1942-Aug. 6, 1943. Commissioned Oct. 12, 1946 to June 20, 1967. His military locations and stations: US Maritime Service September 1943-September 1946, LTJG. Served with USS *Murrelet* (AM-372), Nov. 24, 1951-Nov. 14, 1953, Korean War.

Gillen was awarded the Bronze Star, Combat V, SECNAV Commendation, Korea Service w/3 stars, Great Patriotic War Medal, Russian Government, 1944; Murmansk-Archangel.

He achieved the rank of LCDR and was discharged June 20, 1967. His memorable experiences: (1944) SS *Benjamin Schlessinger;* Murmansk-Archangel Bro'T Russian officers and men to Greenock to man captured Italian ships. Commanded whaleboat crew, May 1952. Capturing Sampans and enemy troops off Korea.

His wife Alyce died Sept. 16, 1994 and they had children: Constance Loux and Louise Gillen Holland. His grandson is Brett. Employed as secretary, Girard College Alumni Association, executive vice-president, Philadelphia, Del. Valley Restaurant Association.

DR. FRANK P. "DOC" GILLEY, CDR, born April 25, 1922 in South West Harbor, ME, attended University of Maine, 1940-42; University of Maryland Dental School, 1942-45; North Western University Dental School, 1946-47.

He joined the service May 1942, Army Reserve, 2nd LT, 1943-44, PFC ASTP. His military locations and stations: 1945, LTJG USNR Dental Corps; Navy Base, Newport, RI, 1945-46; Boston Naval Shipyard, 1953-54.

Gilley was awarded the Army Good Conduct, American Theatre, Naval Reserve, Korean Conflict. He achieved the rank of CDR, DC, USNR, discharged April 25, 1982. His memorable experiences: enjoyed his work helping with the dental health of Navy men during WWII and Korean Conflict, USN is like a fraternity, men helping men.

He married Mary Ellen and their children are Ellen, Sue, Judy, Francis and Cindy; grandchildren are Greg, Chris, Andy, Kristi, Matt, Ryan, Todd, Libby and Jeff (Jeff Parola was a Maine State Trooper and killed in line of duty Nov. 12, 1994). Employed as dentistry-orthodontics, Bangor, ME, 41 years, retiring August 1988.

CHARLES H. GILLILAND JR., CAPT, born Dec. 4, 1942, Alton, IL, received a BA, MA, PhD from the University of Florida.

He joined the service August 1965 (enlisted in Ready Reserve). Commissioned September 1966. His military locations and stations: USS *Douglas H. Fox* (DD-779); COMDESDIV 601 Staff (Key West), May 1966-September 1969. Served with BUPERS, Fleet Training Center Norfolk; January 1982-August 1985, US Naval Academy (English Dept.); January 1991-July 1991: BUPERS in support of Operation Desert Storm.

Gilliland was awarded the Navy Commendation w/2 Gold Stars and the Navy Achievement, NUC. He achieved the rank of CAPT.

He married Carol and their children are Anne-Marie, Alexandra, Elizabeth and Alice. Employed as an English professor at VMI and since 1986 at US Naval Academy, retiring Nov. 1, 1996.

Major experiences: Command of NR Fleet Training Center Norfolk, 1991-93.

DAVID S. "DAVE" GILMER, CAPT, born Dec. 1, 1937 in Manila, Philippines; BS-USNA 1959; MS-University Michigan 1967; PhD,University of Minnesota, 1971.

He joined the service USNA 1959; USN 1959-64; USNR 1964-89. His military locations: USN-submarine; USNR-ONR assignments. Served with USS *W.R. Rush* (DDR-714); USS *Croaker* (SS-214); USS *Dogfish* .

Gilmer achieved the rank of CAPT and left the service in August 1989. He is married to Diane F. and they have two children, Dave Jr. and Sarah. Employed as research biologist, USDI, retiring December 1997.

GERALD L. GLAHN, CAPT, born Nov. 13, 1932 in Enid, OK, received a regular commission in 1955 at University of Oklahoma and gunnery orders to *Coral Sea* (CVA-43). Admin. and engineering billets followed aboard *Valley Forge* (CVS-45). He and his wife, Nancy, have four children and five grandchildren. She teaches gifted and talented classes; he is a registered professional engineer. He holds BS, MChE and MBA degrees. His civilian specialties were Math/Computing/Systems engineering for Phillips Petroleum, Humble Oil and Exxon Chemical. Professional activities included AIChE, ASNE, NRA and USS Kidd Memorial.

After release from active duty, he reported to Naval Research Co. 8-8, Bartlesville, and became CO of NRRCo 8-4, Houston, in 1967. With transfer to Essochem Europe in 1971, he joined NavEur-4 NROS in Brussels, Belgium. Upon return to CONUS, he attended NROS 8-11, Baton Rouge, until appointment as CO of NavOrd 3910, Houston, in 1974. He has been certified in engineering management by LSU, in professional development by AIChE, and as ship engineering specialist by Naval Sea Systems Command. Glahn achieved NOBCs as CO shore activity, Supervisor of Shipbuilding, Conversion & Repair, and Mathematics Research Officer (Computers). Other command billets were SupShip NOLA HQ 110 and NavRepFac SDGO 310 in 1977-80. Subsequent assignments with SIMA SDGO 110 included collateral billet as engineering duty qualifying officer for NavSea.

At retirement in July 1988, he was "bull" Captain in reserve ED program. His naval honors are led by Navy Commendation Medal for meritorious engineering achievement and RedCom 10 commendation for superior command performance. His Armed Forces Reserve Medal includes 2nd and 3rd awards.

FRED STEPHEN "STEVE" GLASS, RADM, born Oct. 17, 1940 in Asheboro, NC, received a BA from Wake Forest University in 1963 and a JD degree from Wake Forest University School of Law in 1966.

He joined the service June 19, 1967. His military locations and stations: active duty from October 1967 to February 1971; NAS, Miramar, CA; staff judge advocate, trial counsel, defense counsel, legal assistance officer, division officer; 11th Naval District: Special Courts-Martial Military Judge.

Units served with: Staff Judge Advocate, NR Readiness Command Region SIX, Washington, DC, (1982-92); Force Judge Advocate, Naval Reserve Construction Force (1985-89); CO, NR Volunteer Training Unit 0708 (1984-85); CO, NR Legal Service Office Norfolk 107 (1978-79); CO, NR Officers' School 0607 (1975). He retired on Nov. 1, 1996 after completing his four year (1992-96) tour of duty as the Naval Reserve Judge Advocate General.

His awards include the Navy Distinguished Service Medal, Meritorious Service Medal, (Second Award), Navy Commendation Medal (Second Award), Meritorious Unit Commendation, Military Outstanding Volunteer Medal, Recruiting Service Ribbon, National Defense Service Medal (Second Award) and the Seabee Combat Warfare Specialist designation. He is the recipient of the Naval Reserve Association's 1995 National Meritorious Service Award.

He is married to the former Martha G. Daughtry and has two children, Elizabeth, Lynne and John. He is a partner in the Cary, NC law firm of Brooks, Stevens & Pope, P.A. Retired Nov. 1, 1996.

HARRY J. GLASS II, born Jan. 23, 1927 in New Orleans, LA, received a LL.B, Tulane University College of Law.

Entered NROTC September 1942 at Tulane University, enlisted January 1944 into V-12, he was commissioned ENS June 1945. After serving aboard USS *Impeccable* (AM-320) as CO, he earned his LLB at Tulane in 1984.

Selected Reserve billets in Naval Air units spanned the years 1949-73, including fighter director officer, CIC officer, OPCONCEN watch officer, and OPS/plans officer, with CVG-85, AWS71, AWS88, and NARDIV 882, as well as administrative/personnel officer with VA-881.

Upon commissioning of NARSD2 at NARDET Denver, CDR Glass served as administrative officer, CSO, and staff commander, transferring to retired July 1973.

Active in Navy-related and civic affairs, including NRA (life member), Navy League, Kiwanis and other organizations and committees, he was president/CEO of Harry J. Glass and Associates, manufacturers' representative in Denver. With his wife, the former Shirlee Nelson of Denver, he retired to southwest Alabama. Four children and 10 grandchildren complete the family.

KENNETH M. GLASS, CAPT, born on a farm near Dundee, IA, Feb. 26, 1922; received his early education in a small community school and attended Upper Iowa University from 1939 to 1941. Following active naval service he received a BS and a master of education from Wayne State University in Detroit. In 1963, he earned the Doctor of Education from Michigan State University.

He started his naval career at Wold Chamberlain Field in Minneapolis, MN on June 4, 1942. He earned his wings and was commissioned as an ENS on Jan. 1, 1943 at Corpus Christi, TX. He joined Air Gp. Two at Quonset

Point, RI in June 1943 and subsequently served aboard *Hornet* CV-12 as a pilot with Torpedo Sqdn. Two from March 1944, until late September 1944. During this period, he completed 33 combat missions, earning four Battle Stars, a Presidential Unit Citation, seven Air Medals and a Distinguished Flying Cross.

Following carrier duty he concluded his active duty as the XO of VT-100 Barbers Point, HI. Leaving active duty in June 1946, he continued as an active USNR pilot flying out of Grosse Ile, MI, until his retirement in 1972 with the rank of CAPT.

Highlights of USNR assignments include: XO of an F4U Sqdn.; CO of VA-731 (a USNR Douglas Skyraider squadron). In 1952, his squadron was awarded the prestigious Noel Davis trophy, denoting it as the most proficient attack squadron in the US Naval Command for that year. He also served as an Air Wing Staff Commander for four years.

He married Rose Pytroczko in 1950 and they have five sons: Kenneth, Richard, Robert, Philip and Paul, all graduates of Miami University. He and his wife Rose now live in Oxford, OH where he was an administrator and professor of education at Miami University for 25 years. In 1985, he and his wife were named Parents of the Year by students at Miami University.

A major and significant achievement has been completion of *The Hornets and Their Heroic Men* with co-author Harold L. Buell. The overwhelming positive responses from shipmates are a high tribute to CAPT Glass and CDR Buell. Departed shipmates would have been proud to share in this effort.

His memorable experiences: 35 combat missions as carrier pilot; winner of the Noel Davis Trophy as CO of VA-731 Grosse Ile, MI 1952; Elected to Ohio Veterans Hall of Fame 1996; Hornet Man of the Year 1993.

PARK R. GLASS JR., CDR, a native of Connellsville, PA, born June 23, 1922. He attended Youngstown College, Westminster College, PA and the University of Pittsburgh, obtaining a BA and masters degree in education. He began his naval career in 1942 in the V-7(G) program and was commissioned ENS in 1944.

During WWII he served aboard the USS LSM-158 as gunnery, communications officer, XO and CO.

His campaigns and service medals are the American-Pacific Theaters, Philippine Campaign, Japan Occupation, WWII Victory Medal, Korean Service and the USNR.

He married Mintie Eberhart July 27, 1948. They have one son, two daughters and two grandchildren. Glass taught in secondary education and coached football from 1947-84. He now volunteers in Pastoral Care Unit, NNMC Bethesda, MD where he serves as president of the Protestant Advisory Council.

DAVID RICHARD "GLASSY" GLASSMIRE, LT, born Feb. 6, 1961 in Lackawanna, NY, received a BS degree in business administration, Chapman College and Master of Divinity Christ the King Seminary, E. Aurora, NY.

He joined the service Dec. 2, 1979. His military locations and stations: RTC Great Lakes, IL; NAEC Lakehurst, NJ; USS *Ranger* (CV-61); San Diego, CA; NWTGP; NAS North Island; DIA Washington, DC; DEFATT 0166, DIAHQ 0705, 4th MD, 3rd BATT, 25th Marines Co. I.

Glassmire received the Enlisted Aviation Warfare Specialist Award, Good Conduct Medal, National Defense Service Medal, Navy Expeditionary Medal, Humanitarian Service Medal, Sea Service Deployment Ribbon and Rifle and Pistol Marksman Ribbon.

He achieved the rank of LT and was discharged as selected reservist. His memorable experiences: meeting Queen Elizabeth, Ronald Reagan and George Bush; commission, ordination as an R.C. Priest; commission in the USN Chaplain Corp. Glassmire is single and employed with PRC Corporation, project analyst; Roman Catholic Diocese of Buffalo, Parochial Vicar.

JORDAN ALAN "JORDAN" GMACH, LCDR, born June 22, 1941 in Sheboygan, WI, attended the University of Wisconsin, September 1960-January 1963 and January 1969-January 1973, receiving his BS in physics and math and worked on MSNE.

He served active duty Feb. 26, 1963-Jan. 3, 1969; USNR February 1973-present. His military locations and stations: active Great Lakes, Vallejo MINSY, NPTU Idaho, DLGN 35; NY Ship New Jersey; Long Beach, CA; Bremerton and Tonkin Gulf. He served with USS *Trutun* DLG(N)-35; active: 18 units to date; USNR: CPG2 ROAN; FF 1079 TL; ARS41 ROAN; AS36 BLT, MD; Sima PSMTH; CINCLANTFLT; VTU0607 Norfolk; NAVBASE NFK-LEPS. Served in Vietnam February-November 1968.

He was awarded the National Defense Medal, Armed Forces Expeditionary, Expert Rifle, Expert Pistol, Vietnam Service, Vietnam Campaign w/Device, Good Conduct (four), MUC, Naval Reserve Meritorious Service.

Achieved the rank of BT1 in active duty and LCDR, USNR. Discharged from active duty Feb. 25, 1969 and is still currently in the USNR.

His memorable experiences: just clearing the harbor of Sasebo, Japan after their last R&R before heading to the Tonkin Gulf, Vietnam, they (and the *Enterprise* and *Halsey*), received orders to proceed flank speed to North Korea; he was on board and salvage phase II and also a member of the ships rifle team (which had just taken the 3rd and 4th Naval District championship) when word came to change direction at flank speed they immediately went into planning and drills target to take back the *Pueblo* by force if necessary. This was his first gut wrenching and soul searching about "defending with his life" It will always be with him. They arrived on station early the next a.m. Koreans would not talk unless *Enterprise* left. They both did by late morning.

He married Charlaine Joy and their children are Jon Tankersley, Tara, Matthew, Melodi and Jordan Jerid; grandchildren are Don Korona and Joshua Gmach. Employed at nuclear power plant, service engineer (SG), Sr. SE, Sr. PTNC, SE, supervisor and unit manager field services, analysis and diagnostics instructor, engineering recruitor.

ROBERT B. GODFREY, CAPT, born June 15, 1914, Berkeley, CA. Earned AB degree in economics, Michigan State College, 1940. Volunteered for the Naval Reserve Officers Trng. program in 1940; was apprentice seaman aboard the USS *Arkansas*; attended USNA and was commissioned May 15, 1941. He reported for active duty and went to Gunnery School at Little Creek, VA.

Served as CO aboard SS *Exchange* and SS *John D. Archbold.* Entered Ordnance School, Washington Naval Yard, the Submarine Chaser Trng. Center, Miami and the Fleet Sound School at Key West. Served aboard the USS *Dale W. Peterson*, transferred to ATB, Little Creek, VA, then duty in USS LST-881 assigned to Task Group 13.13.6.

Sailed from Pearl Harbor bound for Okinawa, Eniwetok, Saipan, Guam. Transported personnel and supplies to Iwo Jima; departed for Leyte Gulf to transport occupation troops to Japan; transferred to inactive duty Dec. 20, 1945.

Employed by Michigan Adjustment Bureau, then Duplex Truck Co. before entering the general insurance business in 1949.

Volunteered for active duty in the Korean War and was assigned to the USS *Rockwall* as Ops officer in 1951. Assigned as inspector-instructor of Naval and Marine Corps Reserve Trng. Ctr., Little Rock, AR, 1952, and released to inactive duty Jan. 31, 1953.

Served as CO of Organized Surface Bn. 9-25 in Lansing, 1953-55; promoted to CPT in 1960 and retired April 1, 1963. Awards include ADSM, ACM, EAME Campaign Medal, Asiatic-Pacific Campaign Medal, Navy Occupation Service Medal w/clasp, NDSM, WWIIVM, Navy Reserve Medal and Command at Sea Insignia.

Returned to insurance business and retired as vice-president and treasurer, Hastings Mutual Ins. Co. in 1979.

Married Reva Neller July 6, 1941, and they have six children: Linda (Lincolnhol), Lisa Ann, Laurel (Kennedy), David (deceased), Lucy (Banks) and Lisbeth (Gierman), and nine grandchildren.

DOUGLAS R. "DOUG" GODSHALL, CAPT, born in 1935 and resided in Collegeville, PA until attending Brown University under an NROTC scholarship, graduating with an engineering degree in 1959. His initial active duty tour was "A" Div. officer on the USS *Okanogan* (APA-220), followed by chief engineer on the USS *Tom Green City* (LST-1159) in Japan until August 1962.

After graduate studies in material science at the University of Washington, he was employed at Mare Island Naval Shipyard on the nuclear submarine program before marrying Suzanne Lyford of Richmond, CA in 1965. Subsequent employment with the AF in Sunnyvale, CA included QA for the Global Network of Satellite Tracking Stations and Lockheed development of the space shuttle thermal protection shield.

From 1982 until retirement in 1995 Godshall has worked for Lockheed, primarily as a reliability manager for development of the MILSTAR communications satellite. He presently resides in Pennsylvania as a consultant engineer.

JOHN T. GODWIN, CDR, born Dec. 2, 1917 in Social Circle, GA, received a BS and MD degrees.

He joined the service June 1942. His military locations and stations: Pensacola, North Africa, Italy, France and Okinawa. Served with 144 Construction Bn., ARLI (SS *Archedous.*

Godwin participated in battles and campaigns in North Africa, Italy, Southern France, Okinawa and the Philippines. He was awarded the Unit Commendation. He achieved the rank of CDR. He retired July 1969 to Retired List, Dec. 2, 1977. His memorable experiences: returning home in good condition.

He married Sara M. and their children are Elizabeth, Thomas and Patricia; and grandchildren are Jennifer, Bradley, Catherine and James. Employed as pathologist.

J.B. GOEBEL, CDR, born Nov. 12, 1948 in Chicago, received a BA, MA and PhD in psychology.

He served in the Army Reserve May 1970-April 1976 and has served in the USNR from September 1986 until present. His military locations and stations: Mesquite, TX; NRC Forest Park, IL. He served with Hospital Unit, Great Lakes, Office of Naval Research.

Goebel received the National Defense Ribbon w/star, Armed Forces Reserve Medal, and Navy Unit Commendation Ribbon. He achieved the rank of CDR June 1996. His memorable experiences: two weeks AT 1991 aboard USS *Eisenhower.*

He married Amy Andereck and they have two children, Alison and Lauren. Employed as clinical psychologist in private practice.

KENNETH C. GOEBEL, LTJG, born Aug. 16, 1922 in Houston, TX, attended the University of Texas Naval V-5 Program; US Merchant Marine Cadet Corp USNR #275006-1109; USMMR #43-10654, (was a cadet officer) also attended numerous naval officers schools in fire control, gunnery, military justice, oceanography, etc.

He joined the service Sept. 26, 1942 under J.P. Goebel as aviation cadet. He served at numerous locations and stations, South and SW Pacific 1943-46, with Navy and Army Transport Service, Troopships, ammunition ships, etc.

Participated in American and Pacific Theater, USNR and others. He achieved the rank of LTJG, USNR. His memorable experiences: being sunk in Manila Harbor, ramming and sinking a Jap sub with the troop ship USAT *Yarmouth,* barely missing floating mines; barely clearing *Texas City* aboard the naval tanker *Tomahawk* (in 1948 during the Berlin Airlift); when *Texas City* blew up behind them killing over 500 people. They had loaded high octane close to the *Van Camp* which blew up, they were bound for Hamburg, Germany.

Goebel is divorced and has a grandson Jeremy Paul Johnson. He is a retired member of Master Mates and Pilots, licensed also in insurance and real estate, etc. Retired Feb. 1, 1972 USNR #275006-1109, 8th Naval District, NROS-8, Houston, TX.

ROBERT A. GOFF, CAPT, a native of Long Beach, CA, enlisted after graduating from Compton High School. Served as an electrician's mate onboard USS *Ticonderoga* (CVA-14). After release from active duty, graduated from San Diego State University in 1965. Returned to active duty from OCS, Newport, RI. Subsequently augmented to TAR.

This surface warfare officer's continuous career includes service in amphibious assault ships, destroyers, cruisers and frigates in both the Atlantic and Pacific Fleets. Sea service includes combat operations in Vietnam, command of the frigate USS *Lang* and Commodore of Surface (Destroyer) Squadron ONE.

His service ashore is highlighted by major staff assignments and command of Naval Reserve Readiness Center Columbia, while concurrently OinC, Naval Facility Florence, SC. A sub-specialist in Manpower, Personnel and Training, and Anti-submarine Warfare Management, he held many key assignments, primarily in Washington, DC, in such areas as personnel and fiscal resource management and coordination of Naval Reserve Forces.

He was CDR, Naval Reserve Readiness Command Region 19 when he retired from active duty in July 1996 having completed 37 years of distinguished service.

Goff's awards and decorations include the Legion of Merit Medal (3), Meritorious Service Medal (4), Navy Commendation Medal, Navy Achievement Medal w/Combat V, Joint Service Meritorious Unit Award, Coast Guard Meritorious Unit Commendation (Operations), Battle Efficiency "E" Ribbon, Navy Good Conduct Medal, National Defense Service Medal (2), Armed Forces Expeditionary Medal (3), Vietnam Service Medal w/Silver and Bronze Stars, the Military Outstanding Volunteer Service Medal, Sea Service Deployment Ribbon (3), Naval Reserve Sea Service Ribbon (4), Coast Guard Special Operations Ribbon, Armed Forces Reserve Medal (3), Vietnamese Unit Cross of Gallantry w/Palm, Republic of Vietnam Campaign Medal and the Navy Marksmanship Medal (pistol).

The Goff family consists of wife Connie, daughter, Stephanie and her husband Craig Orona residing in San Diego; sons Robert and Thomas of Los Angeles. CAPT Goff has been a member of NRA since 1970.

THOMAS "TOM" GOLDENBERG, LCDR, born Jan. 18, 1953 in Los Angeles, CA, received a MD from USC in 1975.

Joined the service Jan. 4, 1993. Served with Primus P2063 (NH Oakland); NRFH500-CBT2-9; DETP2263A.

Goldenberg was awarded the Navy Commendation Medal and achieved the rank of LCDR.

Married Tabitha and their children are Daniel, William, Benjamin and Elizabeth. Employed as physician with Tahoe Womens Care.

JOHN MADISON GOLDING JR., CWO4, born in Conroe, TX, July 15, 1938. During high school he enlisted in the NAR at NAS Dallas, July 23, 1955. During his civilian employment and NAR service he earned BS and MEd degrees at Sam Houston State University, Huntsville, TX.

He made aviation ordnance CPO, August 1969 and was selected as WO1 in December 1970. He served as aviation division officer, safety officer and executive officer as a reservist during the Cuban and Vietnam conflicts. He retired Oct. 1, 1984 as CWO4 in charge of training of NAVWEPENGSTA DET510, Pt. Hueneme, CA attached to NRTC Houston, TX.

Golding's other duty stations were VA-704, AAU-708, VR-942, VS-823, NARDIV 02, NARDIVX2, NAVORD 3910, and VF-301 at NAS Dallas, Ellington AFB, NAS NOLA and NAS Miramar.

During his naval career he was awarded and authorized to wear the Naval Reserve Meritorious Service Medal (three), National Defense Service Medal, Armed Forces Reserve Medal (two) and the Navy Expert Pistol Medal.

In civilian life he worked as a lab analyst, businessman, school teacher and school administrator from which he retired July 31, 1993.

He treasures his experiences, duties, service, and friends made during his 29 years, two months, eight days total service in the Naval Air Reserve. He now enjoys investing, hunting and gun collecting and continues to reside in Conroe, TX.

The Golding family consists of wife JoAnn; daughters, Danni Jo Golding, Penni Ann Biehl and her husband Geoffrey; grandchildren are Madison Golding, Brady and Zachary Biehl.

DONALD CHAD GOLDY, CAPT, born in Annapolis, MD, attended high school in Lockport, NY. He attended the University of Denver (on a basketball scholarship) receiving a BS/BA (accounting) in December 1952.

He was commissioned at Newport, RI, September 1953 and served on staff at COMAIRLANT until assigned to USS *Roanoke* (CL-145) as assistant navigator in February 1955. He served in the Mediterranean and the Orient. He achieved the rank of CAPT and was discharged Sept. 4, 1956.

In September 1956 he entered law school at the University of Denver, receiving LLB and JD degrees. He served as an assistant US attorney for Colorado 1972-74, and has been in private practice since 1974.

While serving in the USNR, Goldy attended the Naval War College and the National War College. He is a life member of the NRA. The Goldy family resides in Denver, CO where his sons, Eric Chad and Darin Michael, also reside. Grandchildren are Michael and Megan. He retired March 25, 1991.

ROBIN WINCHESTER GOODENOUGH, CAPT, served in WWII and in Korea as a Navy Seal. He was a combat wounded veteran with distinguished combat awards. He also served as a special assistant in the office of the Secretary of Defense during the Vietnam War, contacting military assistance groups around the world.

He served as president of the NRA from 1975 through 1977. He was group commander of all Reserve units in the San Diego NROS. He served as mayor of Coronado, CA in 1960 and 1970 and was in elective SD County Office for 17 years.

He has a BA from Yale; an MA from SD State University; a JD from USD and a PhD and diplomate in clinical psychology. He is a professor of Law and Psychiatry and Law and Medicine at the State of Maryland's University of Baltimore.

A graduate of the Naval Justice School holder of the Joint Service Commendation Medal and SECNAV's Distinguished Public Service Medal. His wife is Marina and daughters are Joy, Merry and Gay—the three cheers.

PETER R. GOODLATTE, CDR, is a native of East Longmeadow, MA. After graduation from Claremont M. College in California and completion of OCS in Newport, RI, he was commissioned ENS USNR in November 1970. While on active duty through 1972, he served onboard the USS *Mars* (AFS-1), which included underway bridge watches and three campaigns in the combat zone waters adjacent to Vietnam when his ship was forward deployed from Alameda to Subic Bay, PI.

In 1986 Goodlatte affiliated with the Naval Reserve Intelligence Program at the Naval Air Facility, Andrews AFB and drilled with Defense Attache 0166 in Arlington, VA through 1995. For annual active duty tours in the Reserves, he served as the assistant naval attache to the USDAO in Lisbon Portugal five times and twice to Rio. From 1993 to 1995, he was selected to fulfill the Defense Attache role in three consecutive joint exercises to test the Embassy Country Team function and interaction during major military operations involving coalition power projection overseas.

After receiving an MS in 1977 from San Diego State University and an MBA in International Business from George Washington University in 1983, he graduated from the Naval War College Seminar Program in 1993. In 1996, he was recalled to active duty as part of the coordination team preparing for the Worldwide Defense Attache Conference and was awarded the Joint Service Commendation Medal for his efforts. Based on a temporary gap in 1997, he volunteered to serve as Naval Attache in Cairo Egypt.

He resides in Silver Springs, MD with his wife Maria, son James and daughter Theresa.

VERNON LEE "TOOTIE" GOODMAN, ENS, born March 19, 1923 in Pratt, KS, received a bachelor's degree in business administration from the University of Kansas.

He joined the service Oct. 28, 1942. His active duty military locations and stations: Flight School, Iowa City, IA, Dec. 1, 1942-Feb. 15, 1944; assigned to Corpus Christi NAB for flight training, 1944; flight duty #XVF200 on Aug. 28, 1945, ENS A-1; assigned to Daytona Beach, FL in November 1945 as assistant gunner officer; assigned to Air Gp. 58 aboard aircraft carrier USS *Saipan;* stationed in Norfolk, VA on March 18, 1946 with Fighting Sqdn. 58. Completed active duty Nov. 11, 1946.

Goodman achieved the rank of ENS and was discharged Nov. 11, 1946. NAR military locations and stations: Joined the NAR in Olathe, KS in 1947; served in Sqdn. VF-884; transferred from VF-884 to Sqdn. VF-70A in 1951; received rank of LT in 1952 and rank of LCDR with Sqdn. VF-883, Sept. 1, 1957; transferred to Sqdn. VA-882 in 1958; promoted to rank of CDR, July 1, 1971; transferred to Sqdn. VR-70K1 at NAS in Memphis, TX upon closure of Olathe AS; served as CO of NAR Div. Mike I, a training squadron of Reserves organized to augment the RESFORON Groups. Promoted to rank of CAPT July 1, 1972. Retired from USNR, Jan. 7, 1973 after 29 years, eight months and five days of qualified service.

His memorable experiences: he hitchhiked from Lawrence, KS to NAS in Olathe, KS, one day in 1947 to look into joining the NAR. When he arrived, he was the only person at the base who knew how to fly the few planes assigned to the air station - early model Corsairs. The CO signed him up on the spot.

Married Mary Harmon Hiatt, Aug. 18, 1948 in Pratt, KS and they have a son, Philip; two daughters, Lee and Linn; granddaughters, Allison, Alicia and Erin; and grandson, Brian.

Early career included jobs in insurance and real estate fields. Joined Mortgage Guaranty Insurance Corporation as Kansas sales representative in 1968. Retired from MGIC in 1989.

WILLIAM ALBERT "BILL" GOODWIN, born Feb. 13, 1936 in Des Moines, IA, received a BS (industrial engineering); Iowa State University, June 1959; MBA (finance), Stanford University, June 1965.

He was commissioned ENS May 29, 1959 from NROTC, Iowa State University. His military locations and stations: USS *Bearss* (DD-654), Homeport Norfolk, VA 1959-62; N&MCRC, Des Moines, IA, 1962-63; N&MCRC, San Mateo, CA, 1963-65; N&MCRC, Des Moines, IA, 1965-82.

Units served with: on ACDU: DesRon 28; ResDesRon 34; ResDesRon 4; Task Group Alfa. On INACDU: various. Commanded a surface division; communications station unit; mobilization team and VTU (volunteer training unit).

Goodwin received service ribbons and achieved the rank of CAPT (DOR April 1980). Discharged USNR (ret) July 1982.

His memorable experiences: serving as a capsule recovery officer in the Mercury Space Program, nearly recovering COL John Glenn; running before Hurricane Donna and making port in Boston just in time; as part of Task Group Alfa, sitting on a Soviet sub for several hours and almost forcing him up, which would have won their wardroom a case of Jack Daniel's from COMCARIBSEAFRON, RADM Dan Gallery.

He is married to Elizabeth Snider and they have children: Andrea Goodwin Barton (b. June 1962); Geoffrey L. (b. March 1975); Jocelyn E. (b. February 1978); grandchildren are Nicholas Goodwin Barton (b. February 1993) and Lauren Elizabeth Goodwin (b. December 1994).

Employed with Iowa Clay Pipe Co., 1962-63, production engineer; Roll, Inc., masonry products, 1965-69, administrative assistant and project engineer in plane design; Can-Tex Industries, 1969-73, regional vice-president for masonry products; Iowa Concrete Block and Material Co., 1973-96, president and subsequently owner); Cerro Gordo Land Co. (residential land development), president, 1995 to the present.

Retired from USNR as above. In civlian life still active as of this writing October 1996. Member of NRA since 1962.

DELORIS JANE "LOIS or GORDIE" GORDON-WIEMERS, ENS, born Sept. 5, 1923 in Brainerd, MN, attended St. Mary's School of Nursing, Minneapolis, MN.

She joined the service Feb. 24, 1945. Her military locations and stations: USNH, Great Lakes, IL; USNH Sampson, NY. Served with McIntire Unit, Great Lakes, IL.

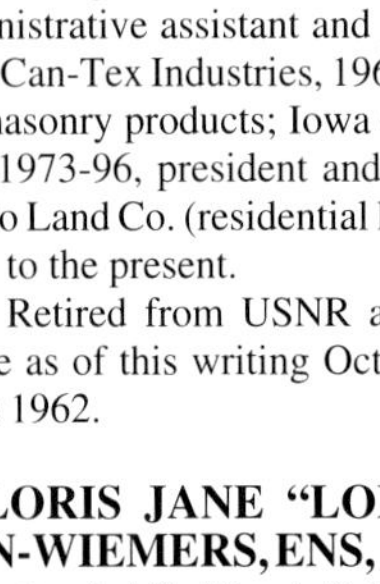
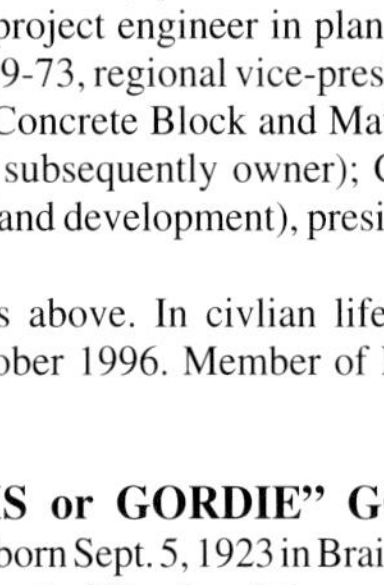

Gordon received the WWII Victory Medal and achieved the rank of ENS, Nurse Corps, USNR, discharged Oct. 31, 1946. Her memorable experiences: being in Chicago on V-E Day; being in Milwaukee on V-J Day; and day she married.

Married Eugene H. Wiemers June 14, 1946, and they have children: Michael, Jeanne, Mary, Theresa, Joseph, Constance, Joanne, Annette, Christine and John and grandchildren: Melissa Theresa, Blaine Mathew and Carolyn Kay Wiemers; Lynne Therese Van Volkenburg; Rebecca Jane and Stephanie Marie Green; Albert W. III and Daniel Eugene Kelly; Jerome Douglas Dolan; Aaron Neal, Jennifer Elaine and Kaitlin Jean Panowicz; Alexandra V. and Andrew John Gonzales; Michelle Marie Stump; Justin James and Forrest Gordon Eldred.

ENS D. Jane Wiemers died July 14, 1996 and is buried at Ft. McPherson National Cemetery, Maxwell, NE.

JAMES W. "JIM" GORDON, CAPT, born March 1951 in Pittsburgh, PA, graduated USNA in 1973.

He joined the service June 1969. His military locations and stations: VP-48, Moffett Field, CA. Served with units VP-48, VT-28 and VP-62. He was awarded the Battle "E", NAVCOM.

Gordon achieved the rank of CAPT and was discharged February 1980. His memorable experiences: CO, HASW-0174. Single and employed as United Airline pilot.

FREDDIE A. "FRED" GOYNES, LCDR, born in Cavite, Philippines December 1921. Upon graduation from high school in Manila, he shipped out as a "workaway" aboard the USS *President Taft* bound for San Francisco, leaving behind his father and five siblings. However, upon arrival in Honolulu he was offered a job at the Pearl Harbor Naval Shipyard which he accepted and was present during the December 7 bombing.

Inspired by the fact that his family was captured and interned by the Japanese, Goynes immediately entered the US Merchant Marine Academy, Kings Point, NY.

Upon graduation in 1943 as an engineering officer, he was commissioned ENS, USNR and ordered to active duty aboard the USS *Taylor* (DD-468) which participated in numerous actions from Guadalcanal to Tokyo Bay.

Three memorable experiences occurred during this tour of duty, namely: returning to the Philippines at the same time as GEN MacArthur at Leyte; making an impromptu flight from Subic Bay into Manila and finding his entire family alive and together at the Santo Tomas Internment Camp (3). The *Taylor* escorting the USS *Missouri* into Tokyo Bay for the official surrender of the Japanese.

Goynes was released to inactive duty in 1946 and recalled 1950 for duty aboard the USS *Sproston* (DDE-577). Next assignment in 1951 was as engineer of the USS *Sausalito* (PF-4) which was designated station ship at Cavite, Philippines.

Under CDR, Philippine Sea Frontier, the *Sausalito* made a good will tour of Southeast Asia, visiting Saigon, Bangkok, Singapore and the island of Penang.

Active duty was concluded in 1954 where Goynes was the schools quota officer for the Fleet Underway Training Group, San Diego, CA.

Retired in 1972 while a member of the Naval Reserve Seabees at San Diego and now living with his wife Caroline and two sons, Michael and Lawrence, in San Antonio, TX after 10 years of overseas civilian construction work in Saudi Arabia and Indonesia.

CHAPLAIN WILLIAM O. "BILLY" GRAHAM, LCDR, born June 14, 1923 in New Liberty, KY and grew up in Covington, KY, received a BA, Mississippi, 1950; BD, MRE, ThM, New Orleans Baptist Theological Seminary, 1950-56.

His military locations stations WWII June 30, 1943-

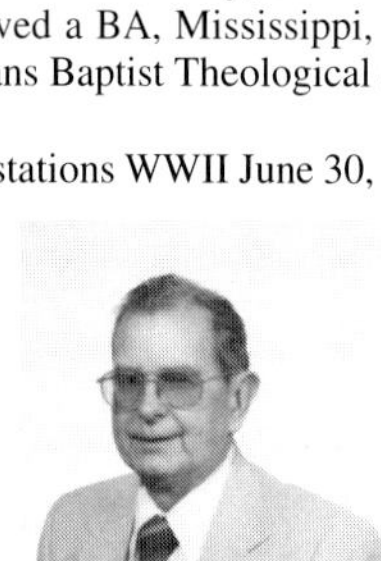

March 8, 1946: S1C gunner's mate aboard liberty ships in Pacific Theatre; Armed Guard Center, New Orleans. Battles and campaigns: Nov. 23, 1944, survived torpedoing of *Gus W. Darnell* in Leyte Harbor. Gasoline and ammunition exploded. Hellzapoppin would be a good name for what happened. After the CG pulled the partially burned and crippled ship to an upright position in the harbor. The crew went back aboard to man the guns for 52 days to fight off more than 200 air raids and alerts.

He was awarded the American Area Ribbon, Asiatic-Pacific Ribbon w/3 stars, Philippine Liberation Ribbon w/ star, WWII Victory Medal and Presidential Citation.

Military service March 30, 1957-June 30, 1962: chaplain service, USNR-ret, June 14, 1983; Military Sea Transport Service, Brooklyn Army Terminal, 1957-59; Naval Hospital, Memphis, TN, 1959-61; Marine Bn., Camp Lejeune, NC 1961-62.

Married Jean Lyon, March 23, 1947, BRE New Orleans Seminary. They have children: Billy Harold, Marsha BeAndra and Franklin Ray and grandchildren: Brent, Brad, Brandon, Elizabeth and David Graham and Michael and Alisha Van Ham.

Employed 1962-86, Chaplain Service: Veterans Administration Medical Center, Wilmington, DE, 1962-65; St. Petersburg, FL (Bay Pines), 1965-73; Montgomery, AL, 1973-87. Retired July 3, 1987.

VIC H. "VIC" GRAMOUNT JR., LT received an AB from Emory University, Atlanta, GA (1963); teacher certification (TN), University of Tennessee, Chattanooga June 1976; 1992 certified in special education; graduate from SW Oklahoma State University. Worked his way through both prep school and college. Much of college part-time. Listed on the Dean's List three times during freshman and sophomore years.

Served for nine years as aircrewman (E6-avionics) while attached to VP Sqdn. at NAS Atlanta (P2V-5Fs). One of four squadrons called up to serve during Cuban Crisis, but "missed" them. First assigned to ECM-Radar-MAD seat; then, volunteered to be radioman, over five years in days when sending/receiving Morse Code still commonplace. They helped doing photo recon which finally led to their spotting the Soviet missiles shipped to Cuba.

Commissioned as LTJG in 1966, after graduating from Emory University, Atlanta, GA. Served as "the" PAO at USN and MCRTC, Chattanooga, TN nearly five years. Helped commission carrier *John F. Kennedy* in 1968; involved hand-addressing 10,000 personal invitations to guests, including Jackie and Caroline. While attached to *Kennedy*, "early upped" to full LT.

Commission retired after eight years, but chose to revert to old rating of AT1/c and as drilling reservist until retired in January 1986. He got back his commission which almost doubled his retirement checks!

Gramount is married to Sarah and they have three children: Diana, Sheryl and Stuart; grandchildren: Corey and Jessica. Employed as GED/Adult Education instructor at local prison (presently); public private school teacher, over 15 years; contract instructor for USN.

EDWARD S. GRANDIN IV, born June 13, 1914 in New York City. He joined the service Sept. 10, 1942. His military locations and stations: Armed Guard Centers, New York and New Orleans. Served as gunnery officer on three merchant ships, Eastern Sea Frontier. His battles and campaigns: Casablanca, Convoy UG6, Normandy landings.

Achieved the rank of LT USNR and was discharged May 9, 1946. His memorable experiences: UGS6 convoy, March 15, 1943 attacked by submarines for seven days. Many ships sunk, his ship hit by dud torpedo.

He was honored for the 50 years of service he has given to make the Brooklyn Memorial Day Parade a success. He is the Naval Reserve Association's 1000th life member.

The CAPT, a member of the NRA New York Navy Chapter, has served as vice-president of many NRA offices. He has also chaired many national committees.

Grandin is a CLU of the American College of Life

Underwriters and holds a bachelor's degree from Columbia University and a masters from New York University.

He served on active duty from 1942-46 and has been associated with the NR surface program since 1947. He has commanded a division, two battalions, and a Reserve Group.

The Memorial Day Parade Committee of Kings County honored Grandin, chairman of the group for the past 10 years, as its Man of the Year.

During WWII, he was a gunner on merchant ships and then was involved in antisubmarine patrols. In 1973, he retired from the USNR as a CAPT, but under special orders continued to train once a month at Floyd Bennett Field and holds the title of ADM in the New York Navy Militia.

HAROLD GRANLUND, LT, born Aug. 22, 1921 in Troy, ID, BS in business, University of Idaho; V-12 University of Redlands, Northwestern University; Midshipman Harvard University (communications).

He served Oct. 7, 1942-46 WWII; Korea 1951-54. His military locations and stations: S. Pacific, Japan, Korea. Served with CTD-35, CTF-95 various ships.

His battles and campaigns: Philippines, Wonsan-Korea. He was awarded the American Theatre, Asiatic-Pacific, WWII Victory Medal, Korean Medal, Philippine Liberation.

Granlund achieved the rank of LT and was discharged in 1946 and 1954. His memorable experiences: trip up Hai Ho River in China to Tienstin and viewing dead bodies floating by; Wonsan, Korea having shrapnel hitting ship.

Granlund is single. Employed banking 15 years; Spokane County Assessor Office, etc. Retired August 1984.

JAMES V. "JIM" GREALISH, RADM, born in San Francisco, Aug. 8, 1920, graduating from the University of San Francisco in 1942 with a degree in accounting.

He enlisted in the USN in 1942, reporting to USNR Midshipman's School, from which he was commissioned an ENS on March 5, 1943. His WWII service included duty aboard the USS *North Carolina* (BB-55), USS *Waters* (APD-8) and as CO of USS *Cronin* (DE-704).

In 1951, he was recalled to active duty for the Korean Conflict and served as XO of the USS *Nawman* (DE-416) and USS *Cushing* (DD-797).

Remaining active in the USNR, he held shore and afloat command billets, including CDR, Naval Reserve Readiness Command, Region 20. He retired July 1, 1979, after 37 years of service, with the rank of RADM.

Among his awards and decorations are the Legion of Merit, Navy Meritorious Service Medal, Asiatic-Pacific Theater Ribbon w/Seven Battle Stars, Philippine Defense, Korean Service, Navy League Distinguished Service Award and the Order of Malta Cross of Merit.

Since again returning to civilian life in 1953, he has served in financial management positions and presently is chairman of Protected Investors of America, a full service investment planning firm founded in 1934.

Married Jean B. and has three children: Kent M. (CAPT, SC USNR, RET); Kathleen Ciardella and Susan Flanigan; grandchildren: Tony, Todd, Jon, Nicole, Danielle, Meaghan, Kaitlin and Kelly. Employed as chairman Protected Investors of America, retired July 1, 1979.

JOSEPH B. GREEN, LCDR, is a native of Panama City, Republic of Panama. The son of a career army soldier, he traveled throughout the US and Panama during his youth.

He entered military service in 1971. From 1971-75

Green served as a personnelman aboard the USS *Blue Ridge* (LCC-19), during the Vietnam War.

Upon completion of his enlistment he graduated from Gonzaga University in 1978 and Gonzaga University School of Law in 1981 obtaining his BA and JD degrees. He was then commissioned LTJG in the JAG Corps from 1983-87. He served as Defense Counsel and LAO at NLSO, NTC, Great Lakes, IL and then as trial counsel, defense counsel, claims officer, LAO and assistant OIC of the NLSO Det, NAS Whidbey Island, Oak Harbor, WA.

At the end of his JAG tour, he affiliated with the USNR from 1988 to present. His first assignment was recruiting and retention officer for DIA Unit 0522, Spokane, WA where he later served as security and awards officer. In 1990 he was recalled to active duty during Desert Shield where he served as operations officer on the IRAQI Intelligence Task Force. Green changed his designator from JAG officer to Special Duty Intelligence in 1992. Since then he has served as department head in various departments such as security, recruiting, retention and awards, training, and administration.

He has been awarded the Joint Service Achievement Medal, Navy Achievement Medal and Letter of Commendation. His service medals are: Good Conduct Medal, Combat Action Ribbon, Joint Meritorious Unit Commendation, National Service Medal, Vietnam Service Medal, Armed Services Medal, Sea Service Ribbon and Rifle Expert. He was selected Junior Officer of the Year of his unit in 1991 and 1995. He was selected as Junior Officer of the Year for Reserve Intelligence Area One in 1991.

He is currently the XO of Fleet Detachment Whidbey 1489, NAS Whidbey, WA. Green maintains a full time law practice in Spokane, WA where he is active in Boy Scouting, church and community affairs.

The Green family consists of wife, Kari; sons Jacob A. and Lucas T., and daughter Alizabeth K.

RAYMOND B. "RAY" GREEN, CDR, born July 12, 1929 in Hartford, CT, received a BA from Yale College in 1951. Graduated Yale Law School in 1954.

He joined the service January 1955. His military locations and stations: OCS and NAVIUS School, Newport, RI; Seventh Fleet WestPac and PacFlt; Subic Bay, Philippines. Served with USS *Philippine Sea* (CVA-47); US NAVSTA Subic Bay; NROS and NRLC Hartford, CT.

He achieved the CDR (SA at OCS, ENS and JG on active duty. LCDR and CDR in active reserves NRDS and NRLC Hartford), retiring July 1, 1979.

His memorable experiences: Westpac sea duty line officer USS *Philippine Sea* (CVA and CVS-47; legal officer NAVSTA Subic Bay; especially sea duty defense at Taiwan in Formosa Strait and antisubmarine in North Pacific during Suez Crisis.

He is married to Barbara Miller and their children are Elizabeth and William and grandchildren are Anne, Kendra, Tyler and Claire. Employed as attorney (partner) Day, Berry and Howard 1958-present. Part-time judge of probate, Canton, CT, 1966-present.

ROBERT W. "BOB" GREEN, CDR, born March 27, 1935 in Warroad, MN, received a BS degree. He enlisted in the service May 28, 1958 and was commissioned Dec. 18, 1958.

His military locations and stations: OPNAVCOMMUNICATIONS-Pentagon, CWO, 1959-60; Naval Communications Station, div. officer, RPS custodian, circuit control officer, Londonderry, North Ireland, 1960-63; USS *Taconic* (AGC-17), Norfolk, VA, CO OOD (fleet), 1963-65; NRC Burlington, IA, CO, additional temporary duties: CO NRCs Peoria, IL and Davenport, IA, 1965-68; USS *Fort Snelling* (LSD-30), Norfolk, VA, operations officer, senior watch officer, acting XO (three months), 1968-70; Fourth Naval District, Philadelphia, PA, surface programs officer, district recruiting officer, 1970-73; NRC, Washington, DC, Plans and Policies Department, recruiting support, fleet liaison officer, 1973-77; Naval Reserve Center, St. Petersburg, FL, CO, 1977-79.

He was awarded the SWOS, NCM, National Defense, Armed Forces Reserve w/HG. He achieved the rank of CDR (0-5) and discharged March 1, 1979 (ret). His memorable experiences: Mediterranean and Caribbean deployments; assignment to North Ireland; holding command and significant staff positions; making lasting friendships.

He is married to Peg and they have children: Patricia, David, Brian and Sara; and grandchildren: Andrew, Robert, Michael, Amanda, Garrick and Jamie. Employed with UNISYS Corp. (formerly Sperry) as senior field engineer, retiring Nov. 24, 1994.

EDWARD J. GREENAN, CAPT, born April 20, 1939 in Woonsocket, RI, received a BA, STB, MA and PhD degree.

He joined the service Aug. 19, 1963. His military locations and stations: Naval Security Gp., Newport, Rota, Ft. Mead, Washington HQ. He served with USS *Saratoga,* USS *America,* COMSUBFLOTS, 2&4, USS *Clifton Sprague.*

Greenan received the Meritorious Service (two awards), Navy Expeditionary Medal, National Defense (2), and Air Force Reserve. He achieved the rank of CAPT, USNR (ACG LCDR SEL), discharged June 30, 1972. His memorable experiences: served as Russian translator to CO, USS *Clifton Sprague* during first joint US Russian Naval Operations after end of Cold War. Retired June 30, 1972.

He is married to Margery Hamilton and they have children: Beverley, Anne, Cheryl, Christopher, David and Jennifer; grandchildren are Matthew, Cheryl, Wesley, Emma and Caleb. Employed as chair, Department of Foreign Languages Department, North Kingstown School, Department and Adjunct Professor, Community College of Rhode Island. Retired April 1, 1995.

STEPHAN M. "GREENIE" GREENBERG, MD, CAPT, born March 4, 1941 in Far Rockaway, NY, received a BA Alfred University, 1962; M.D., New York Medical College 1966.

He joined the USPHS 1966-69; USNR 1988-present. Served as battalion surgeon with 8th Tanks, 4th Marines, Fleet Hospitals 7, 22 Director Surgery.

Greenberg was awarded the Navy Commendation, National Defense, Meritorious Unit, Vol. Service Medal. He achieved the rank of CAPT MC, USNR.

His memorable experiences: Operation Arctic Care 96; Northwest Alaska; attended Strategic Medical Readiness Contingency Course and senior Navy Tendership Course, Washington, DC, lecture Ocular trauma AMSUS 1995.

He is married to Nurit and they have three children: Lior, Eran and Galit. Employed as opthalmologist, solo practice.

ELVIN GREENE, LCDR, born April 19, 1932 in Newark, NJ, received a BS in chemical engineering, New Jersey Institute of Technology (1952); MS chemical engineering University of Tennessee (Knoxville) 1961.

He joined the June 3, 1954. His military locations and stations: USNAV Clarksville Base, Clarksville, TN.

Greene was awarded the National Defense and achieved the rank of LCDR. He was discharged from active duty May 27, 1957. His memorable experiences: witnessed the detonation of an air-dropped nuclear weapon during exercises at Desert Rock, NE in 1957.

He is married to Maye and they have children, Beth and Laura; grandchildren are Jonathan, Sara, David and Aron. Employed at Union Carbide/Viskase Corp., district operations manager, Research and Development, project manager, retiring Nov. 1, 1993. Retired from the USNR July 1, 1974.

ALAN E. GREENSPAN, CAPT, born Sept. 10, 1931 in Philadelphia, PA, received a BS in business administration, Temple University, 1953.

He joined the service June 9, 1953. His military locations and stations: USS *Grapple* (ARS-7); Pearl Harbor; USS *Seneca* (ATF-91), Norfolk, VA. Served with NRSD 9-23, NRSD 9-26 CINCPAC DET 216 CSR-4 DET 116, NRGC 9-68S.

Participated in Korea and Vietnam (Passage to Freedom); Greenspan was awarded the China Service Medal, NR Medal (three awards). He achieved the rank of CAPT 06 and was discharged July 1, 1985.

His memorable experiences: Navy Salvage and Diving School, Bayonne, NJ; typhoon Philippine Sea September 1954; salvage OPS during storm at Cape Henry, VA, January 1956; member, Line Commander Selection Board (1984).

Greenspan is married to Dolores (Dolly) and they have a daughter, Dr. Carrie E. Greenspan-Puller, MD (OB-GYN); grandchildren are Benjamin, Daniel and Zachary. Employed in automotive sales management. Retired from the USNR, Sept. 10, 1991.

V. WILLIAM H. GREENWAY, born April 14, 1924, in France. He is an American citizen and married. Education: Perry School 1942, magna cum laude, National Honor Society; BA degree in history, Yale University, 1948, Dean's List; University of Maryland, 1951-52, international relations; St. Johns College, Oxford, NATO Senior Officers Course, 1968; Universidad Complutense de Madrid, Curso de Verano, 1992.

Military Service: WWII, Pacific Theater, command at young age; 1950-52, USN Rhine River Patrol, NATO operational command: liaison with Royal Navy Rhine Sqdn., French Navy, Belgian Army Engineers and Royal Dutch Engineers; command of the German Labor Service Unit; NATO International Military Staff, Public Affairs Advisor to Chairman Military Committee, ADM Sir Nigel Henderson, RN; PIO to GEN Spivy; liaison with Dept. of Defense, Pentagon. Had Cosmic Top Secret clearance; NATO (SHAPE) Advanced Weapons Course, Oberammergau, Germany.

Awards include JSCM; USNRM; Atlantic Campaign Medal; Pacific Campaign Medal; WWIIVM; NDSM; German Occupation Medal; Philippine Liberation, PUC and Independence Medals; AFRM, China Victory Medal, French L'Ordre National du Merite, NRA Medal, Joint Chiefs of Staff Badge and NATO Badge.

Languages include English, French, Italian and German. He is active in top-level international consulting and is a seasoned negotiator. Greenway is a member of the NRA and many other clubs, affiliations and professional associations.

HOWARD E. GREER, VADM, is a native Texan who graduated from the USNA in June 1943. He served in the USS *Princeton* through the island-hopping campaigns from Tarawa to the Philippines, departing for flight training in October 1944. Wings were awarded in November 1945. He served in fighter squadrons and fleet staffs deploying to the 6th Fleet five times and to the 7th Fleet seven times. He commanded VF-64, CVG-10, USS *Ponchatoula,* USS *Hancock,* the Naval Air Reserve Force; Carrier Group Three and Naval Air Forces, Atlantic. Mixed in with these tours he attended the University of Southern California, the Naval War College, and the Industrial College of the Armed Services and George Washington University (MBA).

He held four positions with the 7th Fleet during Vietnam including ACOS (operations) 1964-66, CO, USS *Ponchatoula* 1966-67; CO, USS *Hancock* 1968-69 and CDR Carrier Group Three, 1972. While serving as Yankee Station CDR he directed all naval air operations during Linebacker I and the mining of the port of Haiphong.

Decorations include two Distinguished Service Medals and four Legions of Merit. Two highlights of his career on shore duty were: serving as a writer for ADM Arthur Radford as he prepared for congressional hearings in 1948-49 during the "Revolt of the Admirals" and reorganizing the Naval Air Reserve in 1970-71 to bring reserve squadrons into alignment with fleet squadrons in manning levels in equipment.

Greer is married to the former Dale Price, daughter of ADM John Dale Price and they live in LaJolla, CA.

FREDERICK PARKER GREGG, LCDR, born Jan. 14, 1944 in Houston, TX, received a BA from the University of Texas (Austin); MD, University Texas Medical Branch (Galveston, TX).

Joined the service in 1967. His military locations and stations: NRMC, Camp Lejeune, NC. Gregg achieved the rank of LCDR, MC, USNR and was discharged in 1980.

He is married to Cecilie M. Pienny and they have two children, Shannon Marie and James Parker.

H. JACKSON "JACK" GREGORY, CAPT, born Jan. 11, 1942 in Lancaster, SC, received a BA in 1963 from Furman University; JD degree 1968, USC Law School.

He joined the service Dec. 17, 1964 SCANG and June 1, 1970, USNR. His military locations and stations: SC Air National Guard, McEntire ANG Base, SC. Served with 169th Combat Support Sqdn., NR 6-20 Law Co., NR LSO 107 and NR VTU Law 0701.

Gregory was awarded the National Defense Service Medal and Armed Forces Reserve Medal. Achieved the rank of SSGT/SCANG and CAPT/NAVRES, and at present, still serving.

His memorable experiences: Gooney Bird flight from South Carolina to Texas in 1965 for active duty basic training in USAF.

He is married to Pat Clyburn and has a daughter Alexis Elizabeth. Employment: Gregory & Gregory, attorneys-owner.

ROBERT S. "ROB" GREGORY, LCDR, born Jan. 6, 1960 in Nice, France, received a BS in computer science, Auburn University, 1986.

He joined the service June 2, 1980. Enlisted Nuclear Power Program, NROTC Scholarship. Commissioned March 16, 1986. His military locations and stations: NAVSTA Rota, Spain; NROTC Unit Southern University, Baton Rouge, LA. His battles and campaigns: Beirut, Lebanon 1983-85/Liberia 1989; Desert Shield.

Gregory achieved the rank of LCDR. His memorable experiences: discharged from active duty May 1992; serving with USNR 1992-present at NAS Atlanta, GA; NR NS Rota 0167.

He is married to Karen and they have a child Jordan born in 1991. Employed with Andersen Consulting, analyst.

EDSON GRISWOLD, CAPT, born Sept. 20, 1943 in Oakland, CA, received a BS in (geography); MF (Forest Management).

He joined the service June 1965. His military locations and stations: USS *Renshaw* (DD-499); COMPHIBRON ONE; COMCOSSURUFOR, RVN.

His battles and campaigns: Vietnam War. He was awarded the Navy Commendation Medal (2) and Service Medals and Ribbons. He achieved the rank of CAPT, USNR and retired July 1989.

He is married to Virginia C. and has two children, Karyn Elizabeth and William Edwin. Employed with Lockheed Martin Technical Operations Co., senior staff engineer.

WILLIAM "BILL" GROGG JR., enlisted May 12, 1942, Kansas City, MO as EM2c, then to Great Lakes Training Center. Assigned to USS PC-1119 at commissioning in New Orleans. To SW Pacific and Australia. Ordered back to US for ENS commission, and assigned to USS LST-866 under construction and commissioning at New Orleans, December 1944.

On to Western Pacific, Guam, Saigan, Tinian, others. Left LST in Western Pacific and returned to US and was released to inactive duty at Great Lakes on Nov. 24, 1945.

Returned to St. Joseph, MO, and Quaker Oats. He was assigned to USNR Surf Div. 9-130 and recalled to active duty Nov. 21, 1950. Assigned to Underway Training Element, San Diego. Released to inactive duty, Aug. 27, 1952. May 1953-June 1961 Naval Surface Reserve Units, Akron and Canton, OH. Promoted to LCDR Jan. 1, 1955.

September 1961, assigned to NROS 9-3 Evanston, IL. Transferred to USNR Retired List July 18, 1974. He had the satisfying experience of never getting seasick, no matter how bad the weather.

He is married to Thelma H. and has children, William R. and Martin H. Employed with Quaker Oats Co. Engineer-Design and Safety, retired Jan. 1, 1981 (46 years, including Navy times).

JOHN C. GROOM, QM3C, born July 8, 1947 in Bangor, ME, received a BS in accounting in 1969 from the University of South Carolina; MBA, 1981 from the University of South Carolina.

He joined the service Aug. 3, 1964. His military and locations and stations: Navy Supply Corps School, Athens, GA; CDR, US Naval Forces Philippines; Aviation Supply Office, Naval Audit Site; USS *Mount Baker* (AE-34).

Units served with: prior enlisted service was spent in various Surface and Submarine Support Reserve units. Accepted into the Reserve Officer Candidate Program in July 1966. OCS training was split between the Naval Academy and Newport, RI. Prior to commissioning in October 1969, attained rank of QM3c.

Commissioned service: NR, Naval Surface Forces Atlantic 107 (supply officer); NR, Naval Warfare Command 107 (plans and logistic officer); NR, Naval Supply Center HQ 107 (admin., training and XO); NR, Naval Air Terminal Norfolk 107 (CO); Naval Reserve Readiness Command Region Seven (Command Evaluation Officer); Naval Reserve Readiness Center, Charleston (command evaluation officer).

He was awarded the Naval Unit Commendation, Meritorious Unit Commendation w/Bronze Star, Navy Battle E Ribbon, National Defense Service Medal, Naval Reserve Meritorious Medal, Sea Service Ribbon, Armed Forces Reserve Medal w/HG and Philippine Presidential Unit Citation.

Groom achieved the rank of CAPT, Supply Corps. His memorable experiences: while deployed with the Sixth Fleet, the USS *Belknap* had a collision at sea with the USS *Kennedy* and was escorted to Augusta Bay Sicily by the USS *Mount Baker.* As the head "chop" of the largest ship present, the *Mount Baker* assumed a primary role in the logistics support. They provided over 1,500 Thanksgiving dinners for the *Belknap* crew and other ships participating in the rescue and salvage operation. Observing the survivors of the *Belknap* made everyone appreciate their lot in life and made their recent cruise extension to be a small inconvenience in comparison.

He is married to the former Julie Ann Staats and they have a daughter, Dawn Cherie, and grandson, Tyler Mathew. Employed as team leader, chief financial officer audit team, deputy comptroller, Naval Aviation Depot, Cherry Point.

WILLIAM SAMUEL "BILL" GRUBB, CDR, born March 1, 1934 in Miami, OK, received a BS from the

University of Oklahoma. Graduate studies Rice University, University of Houston.

He joined the service February 1957. His military locations and stations: USS *Irwin* (DD-794); USN and MCRTC, Houston, TX. Served with NR MOB Team, MSTS and MSC Res Units, NCSO Units, New Orleans, LA.

Grubb achieved the rank of CDR, USNR (Ret), and discharged February 1959. Retired from USNR in 1981.

He is married to Bette P. and their children are Deborah G. Hall, John W. Grubb, Steve Thompson, Carol T. Crawford and Samantha Crawford; grandchildren are Jessica, Brian and Heather Hall; Eric, Logan and Brian Thompson; and Samantha Crawford. Employed with Amoco Corp., (17 years); George N. May and Association (18 years), consultant owner.

WILLIAM ALLEN "BILL" GRUN, LCDR, born April 29, 1920 in Philadelphia, received a BS from Penn State; MED, Penn State. He did graduate work at Temple Princeton Seminary.

He joined the service Sept. 21, 1941. His military locations and stations: Navy Base Philadelphia; SEC BASE Cape May, NJ; USS *Canary* (AMC-25). Served with USS *Wm Seiverling* (DE-441); NAS Atlantic City, NJ. His battles and campaigns: Philippines, Lingayen Gulf, Iwo Jima and Okinawa.

Grun was awarded the Good Conduct, Atlantic, Pacific, Philippine Liberation and Naval Reserve Medals. He achieved the rank of LCDR and was discharged April 1992. His memorable experiences: swearing in daughter, Margaret; kamikaze attacks; exploding German mine; Atlantic.

Divorced, he has two children, Marylou and Margaret; grandchildren, Jonathon and Christopher (Tangye). Employed Chair Abington High School, Industrial Arts teacher, retiring in 1980 after 37 years in education.

WILLIAM GUDERIAN, CAPT, entered the USN via the AOC program at NAS Willow Grove following graduation from Pennsylvania Military College in June 1957. He received his commission as ENS, USNR in November 1957.

Following training at NAS Pensacola, Key West, and Albuquerque. He was assigned as OinC of AUW Unit 0405 at Barbers Point, HI. In May 1960, he was assigned to NATTC Lakehurst where he was training officer of the Parachute Rigger School. Guderian transferred to the Naval Air Reserve in May 1962 and was assigned to VS-752. Later assignments included NAVIRDIV N-1, WEPSYS 793, and as CO, WEPSYS 793 at NAS Willow Grove. He was XO of NAVSEA 1403 and Readiness Unit B at NWS Earle. His last command was CO, GVTU 0404 at NARRC, Kearney, NJ. He retired from the USN in September 1994 having served 37 years.

He served both in the Naval Air Reserve and Naval Surface Reserve in various capacities. He was designated a naval aerospace duty officer and has performed ACDUTRA at NAVAIR, NAVSEA, CNM, OPNAV, COMNAVFORCARIB, and COMLOGRON II. He earned a MA in management from Pepperdine University in 1976.

Guderian is married to Catherine "Kippy" and has two children, Mary and William A., and one grandchild, James.

JAMES W. "JIM" GUERIN, CDR, born Feb. 14, 1923 in Chicago, received a bachelor of naval science, Notre Dame, 1945; bachelor of science, philosophy, University of San Francisco, 1947.

Joined the USNR May 27, 1942. His military locations and stations: John Carroll University, Cleveland, as V-12. University of Notre Dame, as NROTC. Commissioned 2-45; USS PC 797, as gunnery and ASW officer; USS *Prinz Eugen* (IX 300) former German heavy cruiser; then radiological officer of Operations Crossroads, the post-war experiment in nuclear weapons.

After active duty: plank owner in Reserve Submarine Div. in San Francisco. First officer to qualify in submarine (SS) without previous active duty in submarines.

He was awarded the WWII Service, Expert Pistol Shot, American Campaign, Asiatic-Pacific, Armed Forces Reserve and Naval Reserve.

Guerin achieved the rank of CDR and retired Dec. 28, 1982. His memorable experience: won for his weight in a V-12 boxing tournament. Being in Operation Crossroad in Bikini; also the pleasure of being OOD and 1st Div. officer of that PRINZ EUGEN; and being radiological officer in flagship. Possibly he was the closest person to see Bomb "B." No problem from anyone despite the later Communist propaganda.

As a civilian, he has been a municipal bond consultant and personnel officer. He has never married. He has done much volunteer work, including work at TELICOM, the organ of the best IQ groups.

MAX JOHN "JACK" GUHL, CAPT, born July 24, 1932 in Madison, WI, received a BBA from the University of Wisconsin-Madison, 1955; MBA University of Wisconsin-Madison, 1959.

Commissioned ENS USN from Regular NROTC Program. His military locations and stations: USS *Okanogan* (APA-220); USS *DeKalb County* (LST-715). Served Naval Reserve Units in Madison, WI; Oshkosh, WI. His battles and campaigns: Appleton, WI; Sheboygan, WI and Horseheads, NY.

He was awarded the Naval Reserve Medal w/two HG Devices. Guhl achieved the rank of CAPT and released from active duty Jan. 23, 1958. His memorable experiences: overall satisfaction of over 33 years of commissioned service to his country. Participation in the USN Academy Blue and Gold Program.

He married Mary Jane Ebling and their children are John E., James C. and Linda Ann Guhl Wolski; grandchildren are Rick W. and Brian J. Wolski; and Kavia J. Guhl. Employed as Human Resources manager, Oshkosh Coil Spring Co., retired from the USNR July 1, 1988.

EDWARD H. "ED" GUILBERT JR., CDR, born June 26, 1943, Miami, FL, but grew up in Pensacola. Joined the service Sept. 16, 1965, and had active duty at VFP-63, NAS Miramar (RF8G), VT-19, NAS Meridian (T2BC), NAS Pensacola (AIMD QA); Reserve duty: NR NAS Cecil Field 0667, NAS Atlanta (trng. off.), instructor.

Battles and campaigns: Vietnam 1968, USS *Hancock* (CVA-19), 55 combat missions. Awards & Medals: five strike/flight Air Medals, Navy Commendation Medal w/combat V, and others. Retired July 1, 1995, from USN.

Memorable experience was the thrill and elation of walking across the flight deck after landing—having met the ultimate challenge yet again. Has a total of 255 carrier landings and has been a proud member of NRA since 1979.

Married Sara, a registered nurse, and they have two children, Ellen and Eddie. Guilbert retired from Bell South as a manager-engineering, September 1995. He has not found the professionalism and camaraderie of naval aviation anywhere in the civilian world, and if he were young enough, he'd do it all over again.

RAYMOND GODWIN "RAY" GUNN, LCDR, born Nov. 28, 1925 in Walnut (Tippay County) MS, received a BS in accounting, Mississippi State University in 1948.

He accepted commission as ENS, Navy Supply Corps, Dec. 19, 1949. Appointed Nov. 23, 1949. Placed on USNR Retired List, Nov. 28, 1985. Attended Supply Corps School, Bayonne, NJ; supply officer of USS *Roberts* (DE-749); Naval Supply Center, Norfolk, VA.

WWII - US Army Signal Corps (274th Signal Heavy Construction Co., New Guinea, South Philippines, Luzon, occupation duty in Japan, Feb. 24, 1944-Feb. 16, 1946, Naval Station, Rodman, Canal Zone Staff, Commander Western Sea Frontier (Treasure Island, San Francisco, USS *Arcadia* (AD-23).

His medals include: Asiatic-Pacific Campaign, WWII Victory Medal, Occupation Service Medal (Japan), National Defense Service Medal, Armed Forces Reserve Medal, Philippine Liberation Ribbon w/Bronze Star and Army Good Conduct Medal. He achieved the rank of LCDR and placed on Naval Reserve Retired List Nov. 28, 1985.

Memorable Experiences: riding out a typhoon on board an LST between the Philippines and Japan in October 1945.

Married Betty E. Robinson and their children are Raymond Godwin Gunn II (b. Jan. 26, 1949) and Alan Lee Gunn (b. June 15, 1958). Grandchildren are Christopher M., Laura and Matthew Gunn. Employed as Resource Tech., US Forest Service, Jackson, MS, retired April 30, 1975.

M.L. GUNTER, LT, born Nov. 9, 1915 in a jail in Wheeler, TX (his father was a sheriff). Went to Texas A&M in Troop D Cav. two years, dropped out and became a rural school principal in 1935. Resigned in 1941, joined the USN and arrived at San Diego, CA, on Pearl Harbor Day. As they got off the train they were given 1903 Springfield rifles and shells and told to "hit the beach."

He was assigned to the USS *Zeilin* (APA-3) and went by Pearl Harbor, Fuji and arrived in the Solomons. Saw several ships sunk and damaged. They were in two battles landing Marines on Tulagi acrosss from Guadalcanal. Once, 24 Japanese torpedo planes came at their group and 12 were shot down. The *Zeilin* was assigned to take Japanese POWs to Wellington, New Zealand.

Became ENS and assigned to USS *Patokai* (AO-90) and spent 22 months hauling oil off the east coast of South America, from the Amazaon to Uruguay and was in Rio eight times. On the *Patokai* they survived three typhoons off Okinawa and Japan. The *Patokai* was the only Navy ship to have had a mooring mast for dirigibles.

Gunter achieved the rank of LT first class and was discharged in 1946. Decorations include the WWII Victory, Asiatic-Pacific Campaign, American Campaign and Occupation Service.

His memorable experiences: Landing Marines on Solomons; taking Japanese prisoner to Wellington, NZ aboard USS *Zeilin* (APA-3); USS *Patoka,* hauled oil along S.A. coast, was refitted to minesweep tender, went to Japan.

He and Lois have been 51 years and have four children: Pam, Mala, Terrell and Lee III; grandchilcren, Jennifer, Ian, Chris, Valentina, Francisco and two others. Employed as a teacher in Texas for 20 years and New Jersey for 20 years, retiring June 1978. A few years ago, they moved from New Jersey to Wheat Ridge, CO, where three of their children live. Life has been good. Gunter bought his wife a lottery ticket for Mother's Day and she won 1.2 million.

ARTHUR L. "DIESEL GEEZEL" GUNZEL, LT, born March 21, 1916 in Bronx, NY is a high school graduate.

He joined the service Nov. 4, 1935. His military locations and stations: West Coast USA and South America.

Units served with: USS *Minneapolis,* USS *Henley,* USS *Bowditch* (LST-123).

Gunzel was awarded the Good Conduct, American Defense, American Campaign, Asiatic-Pacific, WWII Victory and Philippine Liberty. Achieved the rank of LT and discharged March 21, 1976. His memorable experiences: "The whole bit."

He is married to Ann C. and they have children: Frederick, Ann and Daniel; grandchildren: Lucia, Frederick, Kathleen, Kristopher, Jacob, Jillian and Jonathan. Employed with New York Telephone Co. as garage mechanic, lineman and installer, retired June 23, 1975.

MICHAEL N. GUSSIS, GMM3, born Oct. 1, 1951 in Kearny, NJ, received a BS in construction management, Kean College of NJ (1976). He joined the service Nov. 11, 1969. His military locations and stations: USS *Somers* (DDG-34); USS *Worden* (DLG-18).

Gussis participated in four Vietnam Campaigns, 1970-73. His awards and medals include the Navy Achievement, Commendation, Combat Action, Good Conduct, Vietnam Service.

Achieved the rank of GMM3 and was discharged Aug. 15, 1973. His memorable experiences: becoming commissioned in 1980; selected to CDR 1996 (USNR); gunfire support 1970 as gun captain on 5"54; April 14, 1972, hit by missile while on USS *Worden;* June 1973 operation endsweep Haiphong Harbor North Vietnam.

Married Rosemary and has children: Brian, Glenn and Staci. Employed as president, Matthew A. Segall and Co., Inc. Presently commander, Civil Engineering Corp., USNR-R.

DONALD ROY GUSTAVSON, CAPT, originally of Evanston, IL, attended Northwestern University (BS 1946) and University of Washington (MBA 1963). At Northwestern he completed the NROTC basic course, then attended Midshipman School.

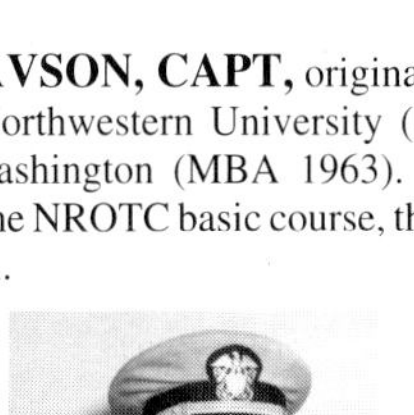

Commissioned as ENS Sept. 12, 1941, he immediately went on active duty. As CO of USS SC-739 he took this subchaser from Miami, FL to Freemantle, Australia. Then he became XO and navigator of USS *Brattleboro* (PCER-852), which transported German prisoners of war (from U-505 capture) from Bermuda to Norfolk, VA, and then participated in Leyte, Lingayen, and Okinawa invasions.

In October 1945 he was promoted to LCDR and became CO of USS *Brattleboro,* taking her from Hawaii to Philadelphia. He was CO, USN and MCR Training Center, Boise, ID 1947-48. He received 10 decorations during his 5 1/2 years of active duty. He was promoted to CAPT Jan. 1, 1962 and retired July 1, 1972. He currently resides in Richland, WA.

GENE E. "GUS" GUTHRIE, born Feb. 16, 1923 in Wenatchee, WA, received a BA Air Transportation, University of Washington.

He joined the service Nov. 17, 1942. Military locations and stations: CPT-Wenatchee, WA; Del Monte Preflight; NAS Hutchinson, KS; Corpus Christi; NAS Jacksonville; NAS Whidbey; Attu; Adak; Dutch Harbor; NAS Naha; Okinawa.

Served with VPB-43; VP-ML-62; VP-891; VP-892; VP-4. He was awarded the Naval Reserve Medal, Armed Forces Medal, American Theater, Asiatic-Pacific, WWII Victory, China Theater, Korean War. Achieved the rank of CAPT, USNR and discharged Feb. 16, 1983.

His memorable experiences: losing an engine in PBY5-A in the fog north of Dutch Harbor on patrol. Landed safely at Ft. Glenn for an engine change. A GCA controlled orientation, approach and landing in a snow storm at night, landing in below minimum flying conditions at Attu in November 1944. Having his own crew as PPC and flying the P2V-5 aircraft in the Far East in VP-4 during the Korean War. Commanding officer of VP-892 at NAS Seattle for three years.

He has been married 41+ years to Nancy and they have two children, Lee Elizabeth Cummings and Mary Louise Moris; three grandchildren: Kenneth Christopher, McCauley Jean and Shannon Cummings; Colin and Cady Moris. Employed as insurance broker, V.P. Talbott Carroll Co., Inc. and owner of his own insurance brokerage before retirement, retiring July 1, 1984.

WALLACE N. "WALLY" GUTHRIE JR., LTJG, born Feb. 22, 1939 in New York City, received a BS, MS, EdS. He joined the service July 29, 1956. His military locations and stations: Washington, DC; New London, CT; San Diego, CA. Served with USSTRATCOM, CNO (NO95), *Abraham Lincoln* (SSB-602).

Guthrie was awarded the DSSM, LOM, DMSM, MSM, NDSM, AFRM. He achieved the rank of RADM and retired Sept. 1, 1994.

Memorable Experiences: serving as navigation officer aboard USS *Abraham Lincoln* (SSBN-602) as a LTJG; recall to full-time active duty to serve as deputy director of USNR.

Married Virginia and they have three children: Wally, Gail and Ginger. Employed as superintendent of schools.

ROBERT H. "BOB" GWYNN, CDR, born April 17, 1940 in New York, NY, received a BA in economics from Monmouth College, Monmouth, IL. Joined the service May 27, 1963; stationed at OCS, Newport, RI; NSCS Athens, GA. Served with USS *Vance* (DER-387).

His battles and campaigns: Vietnam Operation Market Time. Gwynn was awarded the Combat Action Medal, Vietnam Service and Campaign Medal, National Defense and Naval Reserve Medal. He achieved the rank of CDR, SC, USNR (retiring May 1987). Retired from active duty Feb. 7, 1967.

His memorable experiences: qualified as OOD-F; Northern Pacific Barrier Patrols; served as supply and disbursing officer and USS *Vance* (DER-387) September 1964-January 1967. Served under LCDR Marcus Anrelius Arnheiter for his 99 day command.

Gwynn is single and has two children: Lori Gwynn Palumbo and David Scott Gwynn; and a grandchild Taylor Ellyse Palumbo. Employed with J.D. Edwards & Co. as senior management consultant, Lockheed Martin, MRP II consultant.

CHRISTOPHER P. "CHRIS" HAAKON, CAPT, born June 23, 1939 in New York, NY, received a BA from Columbia University.

He received a direct commission, ENS USNR in July 1966. Upon graduation from Columbia University in 1962, he joined the USNR where he achieved the rate of second class photo interpreter.

His 33 years of Reserve affiliation includes two XO tours and four CO assignments culminating as the Reserve Intelligence Area CDR RIA-19, Washington, DC. As RIAC he commanded 1,004 officers and enlisted, comprising 20 units. Haakon holds the record of the most command tours in the reserve intelligence program. He retired July 1, 1995.

His military decorations include the Meritorious Service Medal, three Navy Commendation Medals, Navy Reserve Medal, three Meritorious Unit Awards and an Overseas Ribbon.

Haakon is executive vice-president on the Board of Directors of a high technology research and development company, Autometric, Inc. He resides in Fairfax, VA, with his wife Lorraine and they have three grown children: Constance Spadafore, Michael and Peter Haakon; and granddaughter Emma Spadafore.

DALE V. HAFER, CDR, born May 11, 1954 in Detroit, MI, received a MBA in finance and marketing.

He joined the service Aug. 22, 1978. His military locations and stations: USS *Lapon* (SSN-661), Norfolk, VA. Served with ONR, CNO N091, COMSUBGRU TEN, NSWC.

Hafer was awarded the MSM, NCM, NAM w/Gold Star. He achieved the rank of CDR and discharged USNR (SELRES).

His memorable experiences: OOD; Cat and Mouse games in the Arctic.

He is married to Mel and they have children, Michael and Kevin. Employed as vice-president, SFA, Inc.

S. CHARLES "CHARLIE" HAFER, CDR, born Dec. 13, 1951, Gettysburg, PA, graduated from Pennsylvania State University with a BS degree in business economics. Completed Aviation OCS in Pensacola, FL; commissioned ENS, March 1978; flight training, Corpus Christi, TX (T-28 and T-44) and designated a naval aviator in 1979.

First fleet assignment was with VAW-125 as an E-2 pilot deployed on board the USS *Forrestal* (CV-59) in the Mediterranean Theater. Reported in March 1982 to Tactical Support Wing 1 in Norfolk, VA. Transitioned to Training and Administration of Reserves Program and involved in the consolidation of airlift scheduling to the Navy Air Logistics Office.

Reported in 1985 to Naval Air Facility, Washington, DC as a reserve program manager and C-12 officer, maintenance officer, administrative officer and operations officer.

Released from active duty in July 1990 and affiliated as a Selected Reservist with VR-56 until transfer in November 1991 to VR-46 in Atlanta, GA. After assignment as safety officer and XO, he was selected as the CO of VR-46, serving from March 1995-October 1996. He is presently the CO of NR NAS Pensacola 0867, drilling at NAS Atlanta.

Personal and unit awards include the MSM, Navy Achievement Medal, Navy Meritorious Unit Commendation Ribbon, Navy "E" Ribbon, SW Asia Service Medal, NDSM, Navy Sea Service Deployment Ribbon, Navy & Marine Corps Overseas Service Ribbon, AFRM and Navy Expert Pistol Shot Medal.

Civilian employment as pilot with Delta Air Lines, currently based in Atlanta, GA. Married Bonnie J. Ellis of Salisbury, MD. and they reside in Marietta, GA.

THOMAS BAILEY HAGEN, CAPT, of Erie, PA is a recognized business executive, community leader, and public official. His first association with the USN was as a youngster through his grandfather, Frank J. Bailey (1874-1961), CAPT NYNM and LCDR USNR, a charter member of New York Naval Militia and USNR who served in both World Wars.

CAPT Hagen joined the inactive Naval Reserves in January 1957 while in college. He was called to active duty in June 1958 following graduation, as an OCSA in Class 34 of the Navy Officer Candidate School, Newport. He was commissioned as ENS, Supply Corps, USNR, Nov. 1, 1957, taking the oath of office from his grandfather who presented him the sword given him when he was commissioned in 1909. CAPT Hagen carried that sword throughout his 30 year USNR career.

He attended Navy Supply Corps School, Athens, GA, graduating May 1958. He was ordered to independent duty as supply and disbursing officer of USS *Harwood* (DDE-861) homeported in Newport with Destroyer Sqdn. 24 assigned as ASW Task Gp. Bravo. In January 1960, he reported to Navy Supply Depot, Newport as issue control and fleet liaison officer. Upon detachment from active duty as a LTJG in October 1960, he affiliated with the Naval Reserve Center, Erie, PA, until his retirement on Dec. 1, 1986. He was promoted to CAPT on June 1, 1985. His USNR assignments included: logistics officer of Mobilization Team 4-10, assistant joint petroleum officer of CINCLANT Detachment 105 and senior officer of a Voluntary Training Unit. Active duty for training included: Navy Finance Center, Cleveland; various mobilization exercises; Navy Supply Center, Norfolk; Naval Activities, UK and CINCLANT logistics staff.

A life member of the NRA and Naval Order of the United States, CAPT Hagen was born in Buffalo, NY, Sept. 19, 1935. He moved to Erie, PA in 1943 where he attended public schools and Penn State-Erie, The Behrend College, 1953-55. He is past chairman of the College Council of Fellows, and in 1988, received the prestigious and lifetime Penn State Alumni Fellow designation. He graduated from The Ohio State University in 1957 with a BS degree in commerce. He was a Griffith Scholar, member of Phi Kappa Psi and received the Alumni Citizenship Award. In 1996, Edinboro University of Pennsylvania conferred the honorary degree of doctor of public service.

Hagen had a distinguished business career spanning more than 40 years with Erie Insurance Group where he rose from part-time file clerk to chairman and CEO. He continues as a member of the board of directors. He is also a board member of: GPU, Inc., Pennsylvania Housing Finance Agency, Team Pennsylvania, and chairman of The Team Pennsylvania Foundation. Additionally, Hagen is a board member of The Athenaeum of Philadelphia, board member and past president of The Pennsylvania Society, honorary director of Erie Conference on Community Development, honorary director and past president of Erie Philharmonic, past chairman of Insurance Federation of Pennsylvania, trustee emeritus of Griffith Foundation at Ohio State, and past president of Erie Council of the Navy League.

He is a former board member of: Pennsylvania Business Roundtable, Pennsylvania Chamber of Business and Industry, First National Bank of Pennsylvania, Union Bank and Trust Company, Chautauqua Hotel Company, Erie Western Pennsylvania Port Authority and Preservation Pennsylvania.

From 1995-97, he served in the Cabinet of Governor Tom Ridge of Pennsylvania, as Secretary of Commerce and then as first Secretary of Community and Economic Development. He was a member of the Governor's Executive Board and past chairman of: Pennsylvania Industrial Development Authority, Pennsylvania Economic Development Finance Authority, and Ben Franklin/Industrial Resource Centers Partnership.

Long identified with community revitalization, CAPT Hagen earned the "Pennsylvania Preservationist of the Year Award" and the "Honor Award" of the Pennsylvania Society of Architects for contributions to the profession by a non-architect. He also received the "Outstanding Community Service Award" in Erie and was named a "Distinguished Pennsylvanian" by both Greater Philadelphia Chamber of Commerce and William Penn Society of Gannon University, Erie. He is listed in *Who's Who in America* and *Who's Who in the World*.

DENNIS LLOYD HAGER, CAPT, born July 23, 1942 in Duluth, MN. His naval service was spent primarily in Southeast Asia with assignment as 1LT and WO aboard the USS *Jerome County* (LST-848). He continued to serve in the USNR in a variety of assignments, including five as CO. Subsequent to his final command tour as Commodore of NR BEACHGROUP TWO, he served as a Navy Department duty captain in the Navy Command Center, a team chief in the OSD Crisis Coordination Center (CCC) and as a combat documentation team leader for the Naval Historical Center.

Currently chief of the USCG Office of Intelligence and also serves as the principle intelligence advisor to the assistant commandant for operations. Prior to his present position, Hager was the first technical director of the USCG's Intelligence Coordination Center. In that capacity he was instrumental in creating and organizing the Coast Guard's first national level intelligence activity. He was also employed by the Drug Enforcement Administration (DEA) and predecessor organizations of the Office of Naval Intelligence (ONI). He retired June 1995.

He received his BA in history from UCLA in 1964 and his MA in Latin American studies from San Diego State in 1969. In addition to the Defense Superior Service Medal (DSSM) and Meritorious Service Medal (MSM), he has received a variety of other military and civilian awards including numerous outstanding performance awards from the USN, DEA and the CG. As supervisor of the Maritime Intelligence Unit at DEA's El Paso Intelligence Center (EPIC) he received the DEA's top Unit Award. As officer-in-charge of the OSD CCC during the Gulf War his unit received the Joint Meritorious Unit Award.

Hager is married to the former Bernadette A. Francis of Lockhart, TX. They have three children and reside in Huntingtown, MD. Lloyd and Andrew attend the University of Maryland, while their daughter Katrina is a cadet at the USCG Academy. Lloyd is also a USMC reservist attached to VMFA-321 at Andrews AFB.

WALTER R. "WALT" HAIDET, CAPT, born in Alliance, OH, holds BA and MBA degrees respectively from Mount Union College and the University of Pennsylvania (Wharton School).

Following OCS at Newport, RI he was commissioned ENS Supply Corps in October 1952. After Supply Corps School (Bayonne, NJ) he was assigned to USS *Wasp* (CVA-18) in June, 1953 and transferred to USS *Waldron* (DD-699) at Sasebo, Japan in February 1954 as supply and disbursing officer. In this assignment he raised an "unsatisfactory" Supply Department to "outstanding" before his release to inactive duty in October 1955.

Upon RAD, he joined the accounting staff of Aluminum Company of America and retired 36 years later at company headquarters, Pittsburgh, PA, from a position of functional manager of plant and facilities accounting worldwide.

Haidet remained active in the USNR until retiring in 1981 with 32 years service. He held several command billets and was promoted to CAPT, SC in 1973.

He and his wife of 41 years, Carmela, reside in Pittsburgh, PA. They have three children: Keith, Ellen and Paul, all practicing physicians, who are married and have produced three grandchildren: Jared, Victoria Rose, Christina Marie.

F. LAMBERT HALEY, CDR, born on Oct. 14, 1929 in Buffalo, NY. He was graduated from Canisius College in 1950 and awarded a JD degree from the State University of New York and Buffalo Law School in June, 1953.

After completing Naval Officer School in Newport, RI in October 1953, he attended Naval Justice School before reporting to the USS *Yorktown* (CVA-10) in San Francisco. He served as the ship's legal officer and also performed unrestricted line officer duties for three years, during which there were two cruises to the Far East. He was promoted to LTJG in April 1955 and released from active duty in November 1956.

CDR Haley joined the USNR Surface Div. 3-59 in Buffalo in 1957 and served as its XO and later the CO, being promoted to the rank of CDR, USNR on July 1, 1969. He retired from the USNR in October 1989. He is a life member of the NRA.

For 25 years, he was a trial attorney in private practice in Western New York State. He was appointed a US Administrative Law Judge with the Social Security Administration in 1980 and serves in the Buffalo office as well as many other states and Puerto Rico.

Judge Haley and his wife Lucy (Curley), reside in Buffalo. They have one son, F. Lambert Jr., of Newport Beach, CA.

JOHN DAVID "JACK" HALEY, LCDR, born March 16, 1924 in Denver, CO, graduated as professional engineer, Colorado School of Mines in 1948.

He joined ERC, Nov. 12, 1942; active duty, Aug. 19, 1943-Dec. 4, 1945 and March 1, 1952-March 6, 1954, serving with VF-54, *Valley Forge* (CV-45).

Haley achieved the rank of LCDR USNR (R) and discharged Dec. 4, 1945 and March 6, 1954.

He married Laurie Breeden and their children are Laura, Patricia, Brian, Sharon, Norine and Kathleen; grandchildren are Travis, Alia, Corina, Matthew, Chelsea, Michael, Kathleen and David. He is president of Haley Engineering, Inc.

DANIEL B. HALL JR., CDR, born in Frostproof, FL, Sept. 30, 1933. He graduated in 1956 from Stetson University, DeLand, FL. He completed Navy Officer Candidate School, Newport, RI, in December 1956 and reported to the USS *Surfbird* (ADG-383) homeported in Sasebo, Japan. He later served in Florida with a Reserve Fleet Command, and in the public affairs office of the 8th Naval District, New Orleans. He was released from active duty in June 1961 and worked a year as a management trainee for a newspaper chain in Louisiana.

Returned to active duty in June 1962 as assistant officer in charge, Navy Exhibit Center, Washington, and later served in the Office of Information, Navy Department. He then spent 18 months in the helicopter carrier USS *Boxer* (LPH-4) homeported in Norfolk. He returned to the Office of Information as a civilian, but was voluntarily recalled to active duty in September 1967 during the Navy's expansion to support commitments in Southeast Asia. He was assistant public affairs officer, Bureau of Naval Personnel, for a year, then reported to the joint Army-Navy Mobile Riverine Force operating in the Mekong Delta of South Vietnam. Later assignments include public affairs officer for the ASW Force, Atlantic, and the US Second Fleet in Norfolk; and Director, Program Analysis Div., Navy Internal Relations Activity, WA.

Released from active duty in June 1973 and participated in Reserve units in the Washington area until his transfer to the Retired Reserve Jan. 1, 1982. From October 1979-September 1981 he served as CO of Office of Information Detachment 206. Hall's military decorations include the Navy Achievement Medal w/Combat V, Combat Action Ribbon, Vietnam Service Medal, Vietnam Campaign Medal w/4 stars, National Defense Service Medal, Armed Forces Reserve Medal, Presidential Unit Citation, and Navy Unit Commendation.

Following release from active duty he pursued a graduate program in public relations at the American University, Washington, and was awarded the MS degree in August 1975. In January 1976 he joined the public affairs staff of the Military Traffic Management Command, a jointly staff major Army command, as an editor and later as the senior civilian public affairs officer. In 1981 he moved to the American Forces Information Service, an element of the office of the Assistant Secretary of Defense (Public Affairs), as a policy analyst for print media policy.

Retired from there in April 1992, and resides in Annandale, VA. He is a life member of the NRA, ROA, and TROA. He is also a member of the American Legion and the Vietnam Veterans of America.

DAVID E. HALL, LCDR, born in Fort Worth on Aug. 7, 1957. He is a graduate of Saint Thomas, receiving a BA in meteorology and his commission from the Naval ROTC Unit at Rice University in 1980.

His first assignment was with Naval Aviation Schools Command for flight training. Later active duty assignments include: CO, NCS Iceland; resources and financial liaison officer, COMNAVTELCOM Washington; assistant electronics material officer, USS *America* (CV-66); maintenance/material control officer, AIMD, NAS New Orleans; assistant maintenance officer, VA-85. Reserve assignments: Maintenance/material control officer, VR-56; maintenance officer, USS *Kitty Hawk* Det 0170; and training officer, Naval Air Warfare Center Det 0170, NAS Dallas. Active training includes assignments with: NAWC HQ Washington and COMNAVAIRSYSCOM IG (AIR 8.0G).

Service schools: NCO; Aerospace Maintenance Duty Officer; Defense Systems Management College and Defense Security Institute.

In August 1990, Hall was selected in the first group of 36 officers for the new TAR Aerospace Maintenance Duty Officer (1527) designator. In 1994, he was selected for the Naval Air System Command, Air Systems Program (1525) designator. He is a life member of the NRA, USNI and Naval Order of the US.

LCDR Hall has been awarded the Navy Achievement Medal. Campaign and Service Medals are: Naval Unit Commendation, Meritorious Unit Commendation, National Defense Service Medal, Armed Forces Expeditionary Medal, Sea Service Deployment, Navy and Marine Corps Overseas Assignment, Naval Reserve Medal and Navy Expert Pistol Shot Medal.

In civilian life, he is a federal law enforcement officer with the Bureau of Engraving and Printing, US Department of Treasury in Fort Worth. He resides in North Richland Hills, TX.

LAWRENCE J. "LARRY" HALL, CAPT, born Nov. 14, 1935 in Vallejo, CA, received a BA from Stanford University.

He joined the service Aug. 25, 1958 and was commissioned March 19, 1959. His military locations and stations: VAH-3, JAH-7, JAH-1, RVAH-3, RVAH-5, NAS Sadford, FL; VA-125, VA-10; NAS Leemore, CA; JP-G7, JA-2323; NAS Memphis; SACLANT-118; NRTC Kansas City MO.

His battle and campaigns: Vietnam (USS *Kitty Hawk);* Israeli-Arab War (USS *America).* Hall was awarded the National Defense Service Medal, Armed Forces Expeditionary Medal, RVN Campaign Ribbon. He achieved the rank of CAPT and retired July 1, 1987.

His memorable experiences: four carrier aviator cruises; three that were nine months in duration (USS *Independence,* USS *Kitty Hawk* and two on board USS *America);* over 500 carrier landings, deployments to West Pac and Atlantic (Med Cruise).

He has been married to Nancy Ann for 39 years and they have two children, Randy and Karen. Employed as TWA pilot, 1968-97 as CAPT, retiring Nov. 1, 1997.

RALPH NORMAN HALL, CDR, born July 12, 1931, Kansas City, MO. Attended Kansas City Junior College and Northwestern University, Evanston, IL, earning BS degree in business. Joined the USNR Feb. 12, 1951, attended OCS Class 54-42 at Newport, RI.

Active assignment 1953-56 was aboard the USS *Cavalier,* a division and later a squadron flagship. Reserve assignments as enlisted: NRSD 9-10 and 9-14 in downtown Chicago; NRSD 9-21, Evanston, IL.

Reserve duty as officer: NRED 9-29S, Evanston; NRSD0-75M Benton Harbor, MI; NRSD 9-46M, Michigan City, IN; NRGRPCOMD 9-75S, Gary, IN; NRSD 9-37L, Gary; Naval Academy Information Officer, Annapolis, MD, for 20 years; CO: NRED 9-29S, NRSD 9-37L, 0-75M; group commander: NRGRPCOMD 9-65S, Forest Park, IL.

Released from active duty in Japan and before going home visited Hong Kong, Calcutta, Beirut, Cairo, Athens, Rome and a few others. In Europe he traveled through Italy, Switzerland, Austria, Germany, Denmark, Sweden, Norway, England, Holland and Germany. Awards and Medals: Korean, UN, National Defense and Navy AFRM.

Married Arlette Marie and they have three children: Lance, Pennie and Heidi; and four grandchildren: Erik and Hazel Hall, Christopher Walter and Joslyn Haldeman. Worked 39 years for Sears Roebuck and retired in 1987.

STEPHEN FARRELLY HALL, LT, born in Manhattan and attained his undergraduate degree in government and philosophy from Boston University and graduate degree in international management from Thunderbird-The American Graduate School of International Management.

Enlisted in the US Regular Army, February 1961. Graduated July 1961, first in his class from the advance infantry course at Ft. Dix, NJ and assigned to Co. D, 1st Battle Gp., 20th Inf. (Sykes Regulars) at Ft. Kobbe in the Republic of Panama. Assignments included special training/operations in Puerto Rico and Costa Rica while deployed aboard USS *Middlesex County.*

Affiliated with the USAR as SGT (E-5) and was assigned chief harborcraft operator with the 1170th US Army Terminal Unit at the Boston Army Base, March 1964.

Honor graduate from the 94th Command HQ Non-Commissioned Officers Academy at Ft. Drum, NY, July 1966. He entered the standby reserve in April 1969.

Joined the USNR as an intelligence specialist first class with Fleet Intelligence Rapid Support Team 0291, NAS S. Weymouth, MA, September 1979. Advanced to chief intelligence specialist, September 1982.

March 1986, commissioned a reserve intelligence officer and transferred to NR Atlantic Intelligence Command 0191. Naval Reserve assignments included underway watches under instruction as: JOOD, primary flight control officer, ship signals exploitation system watch, Combat Information Center watch and engineering watch aboard USS *Wasp* (LHD-1). In addition, served as WO with CDR 7th Fleet, CDR US Forces Korea, HQ US Special Operations Command and HQ US Southern Command. Also served as analyst on the Secretary of the Navy's Espionage Damage Assessment Task Force, and was a geopolitical analyst with the Operation Desert Shield/Desert Storm Intelligence Support for Strike and Amphibious Forces Cell at the Atlantic Intelligence Command. He retired May 1996.

LT Hall is managing director of Murro Consulting, Inc., a management consulting firm specializing in advancing organizational performance. He lives in Tucson, AZ with his wife, the former Patricia J. Teagle, and two sons, James T.F. "Tully" and Robert C.T. "Colin" Hall.

ARTHUR L. "ART" HALSTEAD III, 0-6, born Nov. 5, 1945 in Norfolk, VA, received a BA from the University of Virginia; DDS, VCU, Medical College of Virginia.

He joined the service July 28, 1971. His military locations and stations: Norfolk Naval Shipyard, Portsmouth, VA; NR NDCL NORVA 106, NR NAVMEDCOM 606, NR OPNAV 093-106, BUMED 106.

Halstead received two NCMs and achieved the rank of 0-6. His memorable experiences: two weeks AT to BDC London UK; three week AT to SMRCC at Bethesda, MD; one week selection board duty; selection as Vaughn dental officer 1988; Vaughn Unit 1994. He is divorced and self-employed.

KENT W. HAMLIN, CAPT, born in Sharon, PA, Nov. 15, 1948. Graduated from the USN Academy in 1970 with a BS and served on nuclear submarines USS *Tinosa* (SSN-606) and USS *Francis Scott Key* (SSBN-657) on active duty. Naval Reserve duty included a variety of billets in submarine and engineering duty communities and CO of two units, NR US COMEASTLANT SUPSHIPS 108, and NR NAVSEA Industrial Mobilization 611.

Hamlin worked for Combustion Engineering, American Nuclear Insurers, Institute of Nuclear Operations, and World Association of Nuclear Operations, to improve the safety and reliability of nuclear plants world-wide. Additional education includes MBA and Harvard University's Advanced Management Program.

His awards include Navy Commendation Medal, Navy Achievement Medal (2), and a variety of unit and service medals.

His family consists of his wife Teresa and sons, Kent Jr., Derek and Reid. He enjoys scuba diving, tennis, softball and computers.

CRAIG A. "MEL" HAMNER, LCDR, born Jan. 9, 1962 in Niagara Falls, NY, received a BS from the USNA; master public administration, George Washington University.

Joined the service May 23, 1984 (graduation and commissioning from USNA). His military locations and stations: USNA Annapolis, MD; NAS Pensacola, FL; NAS Oceana, VA; Washington, DC; NAS Atlanta, GA; Monterey, CA; Selfridge ANGB, MI.

Served with units: VT-10, VT-86, VA-42, VA-85, NSGA Ft. Meade, BUPERS, NAS Atlanta Naval PG School, NAVAIRESACT Selfridge.

Hamner was awarded the Navy Commendation Medal, Navy Achievement (2), Armed Forces Expeditionary, MUC, NDSM, Expert Pistol. He achieved the rank of LCDR 0-4 and presently is still on active duty (TAR).

He married Susan M. Smith and they have two children, Adam and Ryan.

W. CHRIS HAMPTON, LCDR, is a native of San Antonio, TX. He was commissioned through the NROTC program at the University of Oklahoma in December 1983, where he received a BS degree in computer science. He earned his masters degree in computer science from Chapman University, Orange, CA in 1989.

He is a surface warfare officer and holds a sub-specialty code in Computer Science. His first assignment was main propulsion assistant and then CIC officer on USS *Reasoner* (FF-1063). His next tour was at Fleet Combat Training Center, Pacific (FCTCPAC) where he was assigned as the division head, for the Computer Aided Training Div. His last active duty assignment was operations officer aboard USS *Kinkaid* (DD-965). His operational experience includes deployments to the Western Pacific, Persian Gulf and Mediterranean Sea.

LCDR Hampton left the active naval service in 1992, and entered the USNR. His first assignment in the USNR was to the Deputy Chief of Naval Operations for Surface Warfare (NR DCNO N86 106) where he provided contributory support to the ASW and Surveillance Branch (N863E). From October 1994 to October 1996 he served as CO of NR Personnel Support Detachment, CINCLANTFLT 106. He is now assigned to NR COMSEVETHFLT 111.

In his civilian career, Hampton is a management and information technology consultant specializing in data communications and network engineering. He resides in Highland Village, TX and is married to the former Elizabeth D. Salmon of Nutborne, West Sussex, England.

BRUCE H. HANDLER, CAPT, born Oct. 19, 1935 in Chicago, IL. Graduate of Princeton University 1958.

Active service: OCS Class 40 (1958); Patrol Squadron 40, USNS Sangley Point, R.P. (1959-61); NAO, OinC, CIO, CDO, SDO, CBDO, customs officer, Armed Forces Courier, Special Escort/Interpreter CIN-CPAC, TS control officer, survival officer and narcotics custodian; (1962) COMNAVDEFEAST-

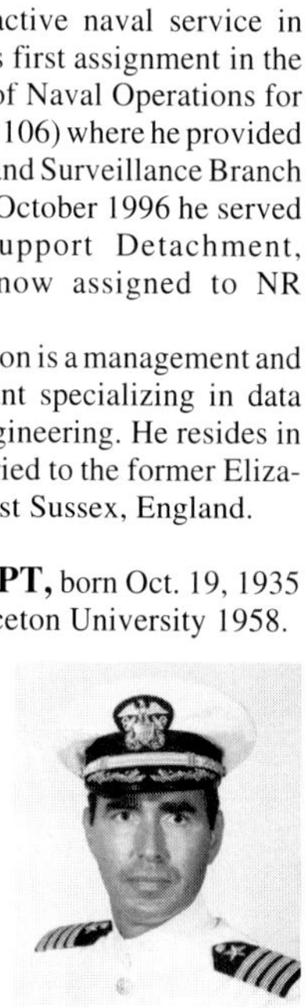

PAC/COMWEST-SEAFRON/CINCPACRESFLT: assist. N-2/ASSO/Flag Briefing officer/Command Rep.

Reserve Service: Joined non-pay Politico-Military experimental unit 1963; CO 1979-85 and 1987-89. AT and SPECAT: COMPHIPAC, COMNAVDEFEASTPAC, OPNAV (OP-06), OASD/ISA, OSD/ISA, US Department of State (PM/MC, PM/ISO, et al.) & Policy Planning Staff, US/OAS, US Department of Commerce (BXA), US MBFR DEL Vienna, USCINCEUR/EUCOM, SHAPE, NATO/IMS, CSCE and National Security Council (White House), among other assignments. Representational duties for SECDEF, SECNAV and other agencies. Acting chief of mission (ambassador) US/EC. Special projects and studies for various Federal Departments CONUS/OUTCONUS. Coordinator, PBM Recovery Project.

Former naval flight officer (NFO). Awarded French Knighthood (Order of National Merit); Italian Knighthood (Order of Merit of the Italian Republic); US Legion of Merit, National Defense, Armed Forces Expeditionary and Naval Reserve Medals among other decorations. Badges: US-JCS and NATO-IMS.

Former founder, chairman, CEO, general partner or principal of 10 entities; pro bono projects for US and foreign institutions; president, Politico-Military Council.

GALE S. "WINDY" HANKS, LCDR, born Aug. 15, 1917, grew up in Crawford, NE and earned his degree in metallurgical engineering at SDSM&T. Following graduation, he became an assistant metallurgist at the Frankford Arsenal in Philadelphia, PA. He entered the USN in 1943. He left the USN as a LCDR in 1946 and went to work at the Kennecott Copper Corporation in McGill, NV. In 1949 He began a 27-year career with the Los Alamos National Laboratory in Los Alamos, New Mexico. He pioneered the development of metallurgy and fabrication methods for uranium and other metals used in nuclear energy.

Hanks has published over 100 papers related to the metals used in atomic energy. He retired in 1976. He and his wife, Edith, live in Albuquerque, NM. They have two children, Kenneth and Jerry Hanks; and grandchildren: Tracey, Chris, Ron, Brian, Pat, Christina and Ben.

CODY A. HANNA, CDR, born Aug. 13, 1944 in Pittsburg, KS. He graduated from Buffalo, MO High School in 1962 and enlisted March 3, 1966. After boot camp at San Diego he was assigned to Defense Language Institute at Monterey, CA for training as a Russian linguist. He reported to Naval Security Group Activity Kamiseya, Japan, September 1967, and served until December 1969. During this period he earned Aircrewman wings with VQ-1 and spent a six-month deployment aboard USS *John R. Perry* (CT-12). He immediately affiliated with the USNR in Springfield, MO.

On June 3, 1971, he was commissioned an ENS, USNR. He remained an active reservist until March 1994, attaining the rank of CDR. During this period he formed the NAVRESSECGRU unit in Springfield, MO and served as CO of NAVSECGRU Reserve units in Springfield, MO; Lincoln, NE and St. Louis, MO. He served as OIC of a Bn. 15 Seabee detachment for three years and had duty with NAVSECGRU aboard USS *Belleau Wood* (LHA-3) and battleship USS *Missouri* (BB-63). He graduated from the University of Missouri School of Law in 1974 and was an attorney in private practice until 1988 when he was appointed a judge of the 30th Judicial Circuit of Missouri. He married Judith Sayer in 1970 and they reside in Buffalo, MO where continues to serve as judge.

PETER C. HANSEN, CAPT, born Jan. 10, 1921 in Riverhead, LI, NY, received a MS in business administration, Akron University.

He enlisted in the USN Sept. 26, 1941, Hospital Corps. His military locations and stations: USNH, Norfolk, VA; USMC New River, NC; USMC Camp Pendleton, CA; Served with USNH, Pearl Harbor; CASU 8, San Nicholas Island.

Hansen achieved the rank of CAPT, USNR Med Ser. Corp and received the Good Conduct, American Theatre Medal, Asiatic-Pacific Theater, Victory Medal and Naval Reserve Medal.

His memorable experiences: receiving commission as ENS Med. Serv. Corp USNR. Qualified as Marine parachutist 1942.

He is married to Ellennae C. and they have a daughter Judith Ellen and grandchild Christian James Hansen. Employed as sales and marketing branch manager, Garret Corp., aircraft industry. He retired Jan. 10, 1981.

RALPH J. "R.J." HANSEN, LCDR, born Jan. 18, 1951 in Rome, NY, enlisted in Submarine Reserve while still in high school. After graduation, three years active duty in TM2(SS), qualified USS *Trigger* (SS-564). He remained in Sub Reserve until it was disbanded.

Switched to naval aviation (VP-65) while earning his college degree and began his civilian career in law enforcement. In 1977 he changed services to the Coast Guard Reserve as a port securityman and served as credentialed special agent, CG Intelligence and Legal from 1978-87.

Highlights included undercover narcotics operations and dignitary protection for the yachting venue, 1984 Olympics. In 1987, received his direct commission, Naval Intelligence, USNR. Since 1988, has served as a credentialed agent for the Naval Criminal Investigative Service.

Currently employed as the hazardous materials investigator for the Irvine, CA, police department. He is married to the former Rosemary Rae Hill of Los Angeles.

CHARLES P. "CHUCK" HANSON, CAPT, born Jan. 30, 1946 in Minneapolis, MN, received a BS in accounting from the College of St. Thomas, St. Paul, MN.

He joined the service May 30, 1968, USNR. His military locations and stations: Pensacola, Meridian, Kingsville, Miramar, Whidbey, Alameda, Norfolk. He served with: VT-1, VT-7, VT-9, VT-4, VT-22, VF-126, VAH-123, VAQ-130, VAQ-135, VAQ-33, VAK-208, VAK-308, VTU-3004.

Hanson participated in Vietnam 1970-73. He was awarded Air Medals (7), MUC, NUC, etc. Achieved the rank of CAPT 06 and was discharged Sept. 1, 1979.

His memorable experiences: Armed Forces Staff College; Naval War College; National Defense University; USC Safety School; Naval Postgraduate School, Monterey.

He is single and employed as CAPT, Southwest Airline, Boeing 737, retiring Sept. 30, 1994 after 26 years.

MARTIN PHILIP "MARTY" HANSON, CAPT, born in Watseka, IL, Feb. 4, 1937. Graduate of the US Naval Academy. Served at sea in the USS *Bigelow* (DD-942), USS *Sampson* (DDG-10), USS *Lawrence* (DDG-4) and USS *Enterprise* (CVAN-65). Instructor at US Naval Destroyer School, Newport, RI. Resigned from active duty in 1969. In 1971 he obtained a USNR commission in the Civil Engineer Corps.

As a reserve officer served on the staffs of Naval Facilities Engineering Command, SEVENTH Reserve Naval Construction Regt., Naval Reserve Readiness Command Region FOUR, Chief of Naval Construction Regiment, Naval Reserve Readiness Command Region, Chief of Naval Operations (OP-44), and Naval Reserve Readiness Command Region THIRTEEN. Commanded Reserve Naval Construction Bn. TWENTY-SEVEN, Brunswick, ME, and TWENTY-FIRST Reserve Naval Construction Regt., Davisville, RI.

Participated in Beirut, 1958; Cuban Missile Crisis, 1962; and Vietnam, 1968-69. He was part of team that extinguished the disastrous fire at sea aboard USS *Enterprise* (CVAN-65) Jan. 14, 1969 and part of the team that restored operations when the USS *Enterprise* came too close to bottom of Pearl Harbor and ingested mud into all power plant cooling systems (1969).

Awards and medals include the Navy Unit Commendation and Navy Commendation Medal (2). Discharged July 6, 1969 and retired Sept. 1, 1992.

In his civilian career he was a project engineer, Consumers Power Company, Jackson, MI, project engineering manager, United Engineers and Constructors, Inc., Philadelphia, PA, Configuration Management Specialist, Systematic Management Services, Inc., Argonne, Il, project engineer, MAC Technical Services, Co., Argonne, Il, and senior project management specialist, Aguirre Engineers, Inc. Argonne, Il.

Married Virginia and they reside in Darien, IL. They have two sons, Martin P. Jr. and Adam Gunnar, and grandchildren, Melissa Paige and Josephine Pauline.

PAUL GERHARD HANSON III, born April 19, 1953 in Des Moines, IA, received a BA in government at University of Notre Dame (May 1975); JD (law) at University of San Diego (May 1982).

Appointed midshipman, Sept. 8, 1971, USNR-R while at NROTC at the University of Notre Dame. Commissioned ENS, USN, after graduation from University of Notre Dame, May 17, 1975.

His military locations and stations: University of Notre Dame NROTC Unit, (midshipman USNR), September 1971-May 1975; Surface Warfare Office School (SWOS), Newport, RI (ENS student), June 1975-August 1975; USS *Berkeley* (DDG-15), San Diego, CA and Bremerton, WA (navigator, gunnery assistant). WESTPAC Cruise in 1976, September 1975-July 1979.

Individual Ready Reserve IRR (inactive reservists), August 1979-April 1985; NR MARDEZESECAK at Naval Reserve Center, Everett, WA (Staff Mine Warfare Officer), May 1985-September 1989; NR USS *Conquest* (MSO-488), Seattle, WA, (SelRes coordinator, October 1989-August 1991; VTU 2205G at Naval Reserve Center, Everett, WA (MCM officer); September 1991-September 1992; NR MARDEZSECPACNW at Naval Reserve Center Everett, WA (Staff Mine Warfare Officer), October 1942-September 1994; VTU 2205G at Naval Reserve Center, Everett, WA (XO), October 1994-present.

He was awarded the Military Outstanding Volunteer Service Medal, National Defense Service Medal, Armed Forces Reserve Medal, Naval War College Graduate Diploma (with distinction),

Hanson achieved the rank of CDR, USNR as of Jan. 31, 1997. His memorable experiences: As SelRes coordinator for USS *Conquest* (MSO-488), it was commanding his reserve crew unit; preparing for imminent mobilization during Operation Desert Shield/Desert Storm (1990-91). Also participating in the Alaska "BrimFrost" coastal defense fleet training exercises in January 1987 and January 1989 with MARDEZSECAK. And attending the Naval War College extension program in Seattle (1987-90). Requalifying and served as OOD Underway on the USS *Conquest* (MSO-488) (1989-91) (after more than 10 years from shipyard duty on the USS *Berkeley* (DDG-15).

He is divorced and has two children, Paul Ryan Hanson and Dianna Kimberly Hanson. Employed as Lynnwood, WA City Prosecutor (1983-87); Mountlake Terrace, WA Public Defender (1987-88); trial lawyer at Paul Hanson, Attorney at Law, Lynnwood, WA (1989-present).

STEPHEN L. "STEVE" HANSON, MD, CAPT 06, born July 23, 1940 in Fargo, ND, received a MD degree from the University of Minnesota in 1966.

Joined the Army July 5, 1967 and Naval Reserve Medical Corps July 1, 1983. His military locations and stations: AFEES, Ft. Jackson, SC. Served with MED CRU 316, NR Corpus Christi, NR Keflavik, Med Dent 0178.

He was awarded the National Defense (2), Naval Reserve, Army Commendation Medal, Meritorious Unit Commendation, Fleet Marine Force.

Hanson achieved the rank of CAPT 03 (Army); CAPT (06) Navy and discharged July 4, 1969 from the Army. His memorable experiences: Air combat element (ACE) surgeon CAX8-88; CO 4th Marine Air Wing, medical (MAW MED) October 1986-October 1989; Naval Hospital Keflavik, Iceland; Pensacola, FL.

He is married to Oronah and their children are Bradley, Brian, Joel, Erica, Sara and Kirsta; and their grandson is Andrew. Employed with Prudential, regional medical director; 11 gallon blood donor; family physician in rural Minnesota (underserved area).

HENRY L. HARDER, CAPT, born Oct. 8, 1936 in Van Buren, AR, received a MA from the University of Arkansas; PhD University of Maryland.

He joined the service Aug. 10, 1960 (AOC). Commissioned ENS December 1960. His military locations and stations: VR-21, Barbers Point, Hawaii, 1962-65; US Naval Academy 1965-69. Naval aviator, 1961.

Harder served campaigns in Vietnam. He was awarded the Navy Commendations (2), Vietnam (3). Achieved the rank of CAPT and retired from active duty July 1969 and retired from USNR September 1991.

His memorable experiences: Reserves: CO, NUWPNTRAGRUPAC 218; CO, COMNAVMARIANAS 318; assistant deputy for Readiness, NAVRESREDCOM Region 18.

He is married to Ramona F. Johnson and their children are Karen, Monica, Kenneth, Stephen and David; and grandchildren are Julie, Andrew, Austin, Edward, Caitlin, John Henry and Megan. Employed as professor, Missouri Southern State College.

HOWARD O. HARDY, CDR, born Feb. 4, 1915 in Seattle, WA, received a BA and M.Ed from WWU. He joined Washington NG April 1932; US Army October 1937; USNR January 1942.

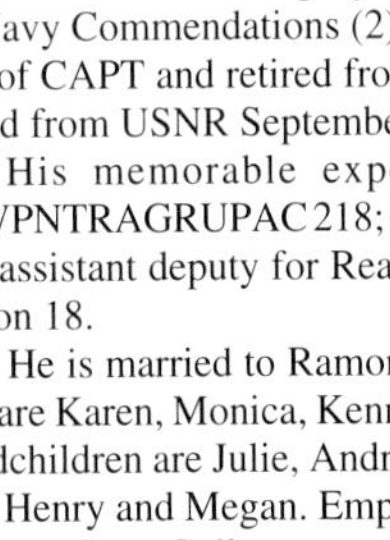
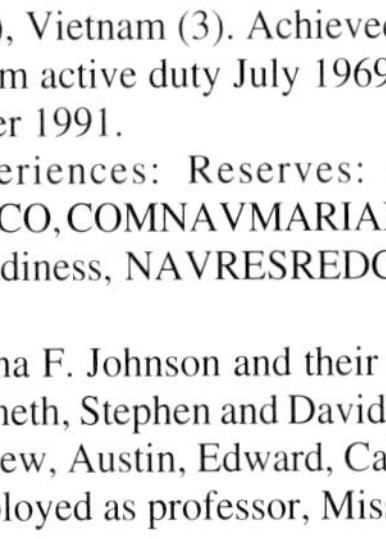

Military locations and stations: Ft. Shafter, HI; Pearl Harbor, Guam, Japan; PI. Served with 146th FA; 64th CA; USS SC-539; CVE-65; CHB-2; CHB-8. His battles and campaigns: WWII and Korean Conflict. Hardy received 11 awards and achieved the rank of CDR, SC, USNR.

Married ENS Lillian B. Campbell and they have four children: Laura, Virginia, Keith and Gloria; and five grandchildren: Sean and Nathan Greene, Tony Ali; and Brandon Conley. Employed as teacher, Seattle Schools. Retired Reserve October 1970; USN, Ret., Feb. 4, 1975.

RICHARD E. "RICH" HARKINS, CAPT, born Nov. 18, 1932 in Tulsa, OK, received a BA and MA from the University of Oklahoma.

He joined the service Jan. 6, 1954. His military locations and stations: USS *Badoeng Strait* (CVE-116), West Coast and WESTPAC. Awarded the China Service, American Defense and Reserve Armed Forces Medal. He achieved the rank of CAPT and retired from the Reserves July 1, 1987.

FLOYD HARLIN, LCDR, born April 23, 1924 in Lovington, IL, received a BS and EE degree from the University of Illinois.

He joined the service Jan. 26, 1943, 875-36-52. His military locations and stations: Great Lakes, Solomons, MD; V-12, April 1, 1945-June 23, 1946 Princeton University, John Carroll University, University of Louisville. Served with USS *LST-47,* USS *LST-688.*

His battles and campaigns: Palau, Leyte Gulf (two times) and Lingayen Gulf. Harlin was awarded Good Conduct, Group Letter of Commendation. He achieved the rank of QM1/c and discharged June 23, 1946.

His memorable experiences: June 30, 1949, appointed ENS, CECR, USNR. Served various active reserve assignments including CO CB Bn. 4-26. Retired LCDR April 23, 1984.

He has been married for 50+ years to Welda Jean and they have a son, Michael Kent. Employed with General Electric Co., engineer/manager for 37 years, large and small aircraft engines, large stationary gas turbines and missile and space re-entry vehicles.

NORMAN W. HARPER, CAPT, born April 13, 1931 in Kansas, attended Phillips University, Enid, OK (1949-51), received a BS from the US Naval Academy (1955) and an MS from Johns Hopkins University (1973). Active duty assignments were USS *Abbott* (DD-629); USS *Sea Robin* (SS-407); USS *Crevalle* (SS-291); Staff, Basic Officers Submarine School, New London, CT; and USS *Casimir Pulaski* (SSBN-633).

Reserve staff assignments in addition to CO billets were in: Naval Reserve Submarine Div. 5-7 Baltimore; Naval Reserve Gp. 5-1(L), Baltimore; Naval Reserve COMSUBLANT DET 106, Alexandria, VA and NR VOLTRUANUT 0614, Washington, DC. An NRA member since 1969 Harper became a life member in 1993.

During his tour as CO of NR SUBDIV 5-7, Baltimore (1972-74) the unit received a Secretary of the Navy Commendation and was the first Submarine Reserve Unit to prove their repair capability aboard a submarine tender. The unit played a significant role in transforming the submarine reserve program from a mobilization only role to an effective contributory role. The submarine reserve was among the first programs to make this shift, a forerunner of the one Navy concept.

While CO of NR COMSUBLANT DET 106 in 1977-79 the unit conducted many studies and projects for Commander Submarine Forces Atlantic Fleet. In addition the unit pioneered the use of qualified submarine reserve officers to stand duty officer watches at the COMSUBLANT Operations Command Center in Norfolk, VA, from where all submarines in the Atlantic Fleet are controlled. During this tour unit members became active participants in both war games and fleet exercises.

Norm and his wife Jeanie reside in Silver Springs, MD and their three children and two grandchildren live in the Washington Metropolitan area.

JAMES J. HARRINGTON, CAPT, comes from Oklahoma City, OK. He graduated from Oklahoma State University with a bachelor in fine arts.

Entering the USN in 1962, Harrington was commissioned ENS in 1963 after completion of Aviation Officer Candidate School in Pensacola, FL. He followed his aviation training with Combat Information Center/Air Intercept Control School in Georgia. His other tours have included duty aboard the USS *Investigator* (AGR-9); USS *Cadmus* (AR-14); at Naval Investigative Service Office, New York and Field Operations Support Group Detachment, NY; aboard the USS *Aludra* (AF-55); CO, Naval Reserve Center, Galveston, TX.

He has also had orders to Commandant 11th Naval District. Reserve Supplement, San Diego, CA as OinC of Fleet Ships Assistance Program in Long Beach, CA; CDR, Mine Sqdn. Five (NRF), Long Beach; CDR, Naval Reserve Readiness Command Region 11, Dallas, TX as director of training and program management; CO, Naval and Marine Corps Reserve Center, Salt Lake City, UT; Naval Reserve Readiness Command Region Eight, Jacksonville, FL as chief of staff; Naval Reserve Readiness Command Region Ten, New Orleans, Louisiana as deputy commander and as CO, Naval Reserve Readiness Center, Dallas, TX.

His decorations consist of the Meritorious Service Medal (Second Award), Navy Commendation Medal (Second Award), Navy Achievement Medal, Armed Forces Expeditionary Medal, Vietnam Service Medal (two Bronze Stars), RVN Campaign Medal, Armed Forces Reserve Medal (HG Device) and National Defense Medal.

Post retirement activity: Past president, Eighth District Naval Reserve Association; National VP-Retired Personnel, Naval Reserve Association; life member of the following: NRA; Navy League of the United States; Retired Officers Association; Disabled American Veterans and American Association of Retired Persons.

Harrington is married to the former Nancy Wilson, also of Oklahoma City. They have two grown children: a daughter, Kimberly and son, Philip.

DAVID J. "DAVE" HARRIS SR., CAPT, born March 23, 1928 in Atlanta, GA, received a BA from Emory University, Atlanta, GA, and joined the service Aug. 1, 1951.

Military locations and stations: Camp Pickett, VA (US Army); CNO Washington, DC; USS *Randolph* (CVA-15). Basic training (Army), SecNav liaison radio/radio division officer USS *Randolph* (CVA-15).

He achieved the rank of LT, retiring from USNR March 23, 1988. His memorable experiences: two Mediterranean tours; Hurricane Edna, 1955, at sea aboard USS *Randolph* (CVA-15) and Hurricane Hazel, 1955, in dry dock at Portsmouth.

Harry is married to Polly Anna and they have three children: David Jr., (CAPT USNR/MC), Frank P., and Paula Ann; and four grandchildren: David III, Andrew, Lauren, Madeleine Harris. Employed as a factory rep, Thermo Products, retiring Jan. 1, 1990.

HORACE HARRISON, CAPT, born in Bourne, MA, Jan. 1, 1932. Graduated from the Massachusetts Maritime Academy in 1954 with a BS in nautical science and later earned a BS in Marine Transportation. Upon graduation he received his commission as an ENS, USNR, and a USCG Merchant Marine License as third mate.

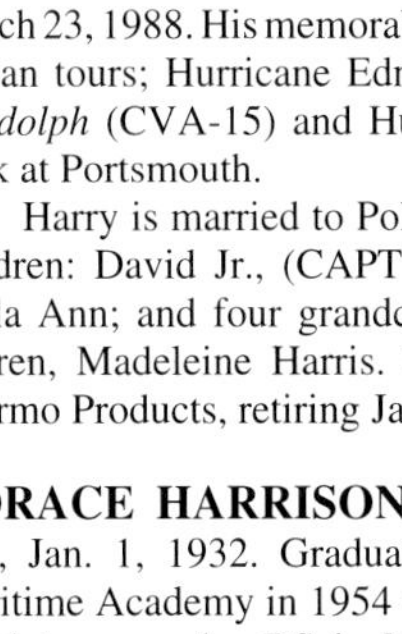
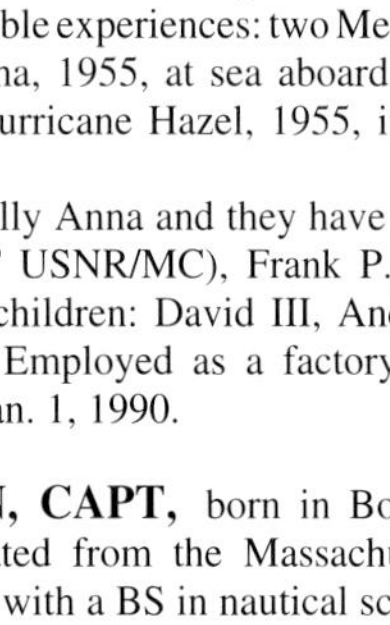

From 1954-56 and 1958-60 was employed in the US Merchant Marine as a deck officer/navigator sailing the US

East and West Coasts and Central and South America. During 1956 and 1957 served on active duty in the USN as XO/navigator aboard the Minehunter USS *Sanderling* (MHC-49). Served a total of 31 years in Military Sealift Command units and retired with the rank of CAPT.

From 1960-present he has been employed in the US Space Program at Kennedy Space Center, FL in a variety of management and engineering jobs.

REGINALD WOODROW "WOODY" HARRISON JR., CDR, born Dec. 12, 1942 in Washington, DC, received a BA in 1964; JD 1967 from the University of North Carolina at Chapel Hill.

He joined the service June 1967. His military locations and stations: Pensacola, FL; Lowry AFB; Virginia Beach; Newport, RI; Coronado, CA. Served with: VF-11, COMNAVAIRPAC, CINCLANTFLT Reserve Units. His battles and campaigns: Airpac Intelligence briefer during Vietnam conflict.

Harrison achieved the rank of CDR and was discharged July 1990. His memorable experiences: Served as OinC for Operation Home Coming for the Navy on the West Coast as COMNAVAIRPAC Rep.; had 55 returnee debriefs released under his name and coordinated all Intelligence OPS in the debriefs at San Diego, Oakland and Whidbey.

Harrison is a practicing attorney and is married to Nancy Rogers Harrison. They have two children, Elizabeth Harrison, a lawyer in Raleigh, NC, and son Reggie Harrison of Wilson, NC.

CHARLES A. HART, CDR, born April 5, 1917 Green Cove, VA. Received a BS from Wake Forest College 1942 and immediately began working for the Bureau of Ships. Commissioned ENS July 4, 1944, he served aboard LSM-314, AKA-20, APA-40 and AGC-16 during WWII and Korea as deck officer, navigator, electronics officer and radiological defense officer.

Released from active duty in 1946, CDR began a 25 year continuation of his Navy career. Returning to active duty in BUPERS, he was responsible for setting up the organization of the Surface and Submarine Program for the training of USNR personnel and received numerous Letters of Commendation and Appreciation. In 1951-52 served aboard the Flagship USS *Pocono* and upon release attended George Washington University, completing an MA in education. Continued studies at American University with 108 hours toward PhD in Public Administration. Graduate studies completed at night.

During the periods of inactive duty, CDR Hart was associated with a local USNR unit at the Washington Navy Yard and Alexandria, VA, serving on the USNR Counseling Board and as a recruiter officer for NROTC and NAVCAD information officer. During this time he was employed by the Navy's David Taylor Model Basin and later by BUPERS as education specialist. Transferred to Retired Reserve in 1971. Member of the Naval Reserve Association since 1960 and a life member since 1972, serving as vice-president for membership for the 5th Naval District in 1965-66 and writing several articles for the NRA News in 1967-68.

His medals include: Asia Pacific Campaign Ribbon, WWII Victory Medal, Atlantic Campaign Ribbon, Naval Reserve Medal. Hart and his wife, Mildred, reside in Hendersonville, NC. They have children: Maxine Lynn and Brian and grandchildren: Joshua, Paul, Thomas and Mara Lang and Mandi and Daniel Hart.

CHARLES B. HART, CAPT, born Dec. 10, 1934, Long Prairie, MN. Earned BA in zoology at USC and DDS at Case Western Reserve. Served active duty 1961-63 and Reserve duty 1963-88. Life member of NRA and USN Institute; member ROA and Assoc. of Military Surgeons.

Served with Dental Co. 12-1, Dental Co. 11-2, NRDC 3720 and 21, REDCOM 20, NRDC HQ YOKO 120, NRSE Seapower REDCOM 20, NRDC SF 320.

Stationed at West Coast Navy Reserve; Periodontic, Letterman Hospital; National Security Seminar, Sacramento; Oak Knoll Hospital; Dental Clinic, T.I.; Reserve Seminar BUMED; Moffett Field; Selection Board BUPERS; Naval Regional Dental Ctr., Yokosuka, Japan; NAS Moffett Field; Phoenix Reserve Seminar; NRA Seminar Nuclear Medicine; ADA Naval Symposium; NRA Detroit, MI.

Married Wanda and they have one child, Marc. He has been in private dental practice for 35 years.

ROYCE NELSON HART, LCDR, born Jan. 27, 1924 in Ft. Worth, TX. Attended some college classes. He joined the service Feb. 27, 1941. His military locations and stations include the Pacific Coast and Region, Panama and Philadelphia, PA.

Served with units: YP-37, USS *Tippecanoe*; USS *Lexington* (CVZ), USS PC-579, USS *Laws* (DD-558), USS *Wateree* (ATF-117), USS *Sappho* (AKA-38), USS *Whetstone* (LSD), USS *ATA-196*; USS *Pasadena*; USS *Boston* (CA-69-CAG1), USS *Lexington* (CV-16), USS *Yorktown*; USS Gilligan, USS *Charles E. Brannon*; USS *McGinty*; USS *McKean*; USS *Point Defiance,* USS *Dubuque,* USS *Pickaway.*

His battles and campaigns include Salamoa, Lae, Coral Sea, Tinian, Saipan, Guam and Okinawa. Hart achieved the rank of LCDR and was awarded the Armed Forces Reserve Medal, National Defense Medal, WWII Victory Medal, Asiatic-Pacific Theater Campaign, American Theater Campaign, American Defense, USNR and Navy Good Conduct.

His memorable experiences: saw LT "Butch" O'Hare shot down the Japanese planes to become an "Ace"; survived sinking of *Lexington* in Coral Sea; served Christmas dinner on two successing days in 1941; survived sinking of *Wateree;* played chess with Japanese POW; first CPO assigned to CAG1.

Married Catharina and they have three children: Janet, Chris and Sally; and seven grandchildren: Marty and Michael Niehaus, Jill Messenger, Jaime Niehaus-Parker, Kelley and Carlye Caveney, and Jason White. As of this writing there are four great-grandchildren and counting.

Hired out to Northwestern Electric Co. as a hole digger. Progressed to groundman. Northwestern taken over by Pacific Power & Light Co. Pacific was taken over by Clark County Public Utility District. Progressed to apprentice lineman, lineman, working foreman, safety superintendent, technical superintendnet, and superintendnet of buildings and grounds. He hired out on strictly a temporary job; nobody ever told him any different. He finally retired over 34 years later.

PHILIP J. "PHIL" HARTER, LCDR, born March 31, 1932 in Watertown, NY, received an AB degree from Hamilton College in 1954 and a MBA from Harvard University in 1956.

He joined the service Aug. 3, 1956. His military locations and stations: Newport, RI; Norfolk, VA; Damneck, VA. Served with OCS, August 1956-December 1956; USS *Des Moines,* January 1957-June 1958; USS *Salem,* June 1958-December 1958; COMCRUDIV 2, December 1958-September 1958; AAWTC, September 1959-December 1960.

He achieved the rank of LT (SG), active duty, retiring Dec. 10, 1960. His memorable experiences: Naval Reserve Fleet Div. 3-1 (January 1961-March 1964); NR TELECOM Unit 3-1, March 1964-March 1973; NR (Select) March 1973-September 1977.

He married Joanna Trayser Dec. 22, 1956. They have children: Lisa, Laurie, Jeffrey J. and Jennifer; and grandchildren: Miranda, Eian, Zoe and Christian. Employed with Eastman Kodak Company. Retired from the USNR as LCDR, September 1977 and EKCO 1986; Navy w/pay 1992.

MILEVA M. HARTMAN, CAPT, born in Lubeck, Germany. She and her family immigrated to the US in 1951, settling in Piketon, OH. Commissioned in 1968, she served two years on the staff of Commander, Service Force Atlantic. Upon release from active duty in 1970, she affiliated with the USNR in Scotia, NY.

As a Selected Reservist, she has served in numerous staff and command positions including CO, Naval Weapons Station Concord 1010. Since 1984, Hartman has served numerous tours of active duty on the SECDEF and OPNAV staff. She currently serves as special assistant to the CNO.

She is a graduate of Ohio University and holds an MS from Corpus Christi State University. Her personal awards include the Defense Superior Service Medal, Defense Meritorious Service Medal (two awards), the Navy Meritorious Service Medal (two awards) and the Navy Commendation Medal.

In civilian life, she has held positions in advertising and personnel management. She is married to Richard Hartman. Their daughter, Laura, attends the University of Virginia.

WILLIAM L. "BILL" HARTMAN JR., CDR, born Feb. 13, 1923 in New York, NY, attended Yale University.

He joined the service Dec. 15, 1942. His military locations and stations: NAS Ft. Lauderdale; NAAS Ream Field; NAS NY; Akron; Detroit. Served with: VA-53B; VF653; VR-651; VR-734; VR-731; VR-733; VR-62Y2-VR-6241.

His battles and campaigns: RTU-52; NARS Y-1; Korea. He was awarded the Air Medal (4); Presidential Unit Citation, NUC, Navy Reserve, American Theater, WWII Vicotry, Navy Occupation, National Defense, Korea (2) Reserve w/HG, Korean Presidential Unit Citation, United Nations. Achieved the rank of CDR. Retired without pay July 1, 1977 and with pay Feb. 13, 1983.

Hartman is single. He is a major stockholder, Laurence, Inc., electronics sales/service.

RICHARD P. HARTMANN, CAPT, born in Terre Haute, IN, Sept. 4, 1934. Graduated from Capital University, Columbus, OH in 1956. Graduated from OCS in December 1956 and was assigned to the USS *Windlass* (ARSD-4) in Norfolk, VA. During 1957 he graduated from second class divers school and served as assistant diving officer in addition to his duties as operations officer, chief engineer, and XO (18 months). Qualified for command and received the surface warfare officer designation.

In 1958 and 1959 *Windlass* divers were called upon to lay early warning detection cable for submarines along the east coast in areas from Barbados, BWI to Argentia, Newfoundland. Retrieved crashed aircraft from the Chesapeake Bay and Atlantic Ocean.

From 1960-87 he served as Seabee Div. XO, CO of two surface divisions, and group commander of the Naval Reserve Center, Dayton, OH. He also commanded a PMD in Cincinnati, OH, was the Commandants representative in Columbus, OH and served on the staff of REDCOM 5 in Ravenna, OH. From 1969-97 was SW OH area coordinator for USNAs information program. Performed ACDUTRA at Naval War College, USNA, the Pentagon and NSG 4 in Newport, RI. Awarded the Navy Commendation Medal. Life member of the NRA. He retired June 30, 1987.

He is a CPA and operates his own firm in Kettering, OH where he was on the city council for 12 years and served as mayor for eight years from 1989-97. He was married to Marcia in 1956 and they have three sons: Richard P. Jr., Jeffrey B. and Roger A.

KENNETH W. "KEN" HARTWELL, CAPT, born June 30, 1910 in Detroit, MI, received a BS in Aero Engineering from the University of Michigan.

He joined the service July 1940. His military locations and stations: NAS: Grosse Ile; Ottuma; Whidbey Island, Katchan Honko Okinawa. Served with aircraft repair units at each station. Participated in Okinawa campaign.

Hartwell achieved the rank of CAPT and was discharged July 1945. His memorable experiences: Outfitting PBY airplanes for the Russian air force while at Whidbey Island, Oak Harbor Seaplane Base; assisting in final outfitting of four convert LSTs into 2 ARVA and 2 ARVE and serving on them on Okinawa.

He married Joyce in 1950 and they have three children: Kathleen, Sherry and Douglas; and grandchildren: Gayle and Bryant Hartwell, Eric Brian, and Matthew Levens. Employed as industrial purchaser and material managers, retiring in 1962.

RAY W. "EAGLE EYE" HARVEY JR., CAPT, born in Baton Rouge, LA. Graduated Vanderbilt (BA NROTC) and Tennessee (JD).

Active duty: USS *Clarion River* (LFR-409) as Weapons Dept. head. *Clarion River* fired 130,000 rounds of 5" rocket in Vietnam. Commissioning USS *Portland* (LSD-37) as gunnery, CIC officer.

Reservist: Commanded ABFC 1409 and DDG-10 Sampson: assistant IG, REDCOM Nine; to of COMNAVFORCARIB Staff.

Active duty: four years COMNAVSURFRESFOR Staff, New Orleans, Train-The-Trainer program. Developed, implemented (Command Excellence Seminar), nationwide; trained 800 instructors.

Gulf War: six months no-pay as legal officer and deputy director of Family Assistance Operations Center for Guardsmen, Reservists and family members. Cofounder of TNNG Youth Camp, and mentor of NJROTC.

Decorations: Army Commendation, Navy Commendation, National Defense, Armed Forces Reserve, Vietnam Service, Campaign and RVN MUC Medals, Rifle and Pistol Expert Medals and Naval and Marine Corps Reserve Overseas Service Ribbon, MUC and Navy E. Resides in Nashville with wife Mary and daughter Mary-Grace.

ALLEN LEE HASLUP, CAPT, born Nov. 4, 1926 in Manila, PI, received an AA from Princeton in 1946; MD George Washington University in 1950.

He joined the USMCR Nov. 3, 1944. Transferred Nov. 20, 1950 to MC USNR. His military locations and stations: Naval Mine Depot Yorktown, VA. Served with 2nd MARDIV, Camp Lejeune; Norfolk Naval Hospital, Bethesda Naval Hospital, 4th Assault Bn., Tampa, FL.

Haslup is a veteran of WWII and Korean War. He achieved the rank of CAPT (MC) USNR and was discharged Nov. 4, 1986.

His memorable experiences: did reserve duty at MAG in Taiwan prior to its decommissioning. His father was a career Marine. A daughter is a MAJ in the USAF Medical Corps.

He married Eva Copley Haslop and their children are Allen L. Haslup II, Elizabeth Haslup Lowe, Forrest C. Haslup and Mary Jo Olsen. Employed in private practice, medical missionary.

CARMRID G. HATHAWAY, CAPT, joined the military Feb. 14, 1943. Served with VC-98, 1944-45; Reserve Sqdn. VA65-E and VS-661, 1947-51; VS-21, 1951-53; VC-35, 1953-54.

Stations include NARTU Anacostia; NAS Atlanta, 1954-57; NARTU Anacostia, 1957-60; OPNAV, 1960-64; NARTU Andrews AFB, 1964-66; NAS New York, 1966-68; NAS Grosse Ile, 1968-70; OPNAV, 1970-72; Naval and Marine Corps Reserve Center, Washington, DC, 1972-75 as CO.

Memorable experience was flying a single-engine Gruman Avenger and his engine was knocked out by AA during a bomb run. The forward canopy was covered with engine oil making it almost impossible to see. He was going to make a water landing when he heard a US Marine tower operator announcing that they had just captured the Yunton AF and he was able to make it there for a dead-stick landing.

Retired from active duty June 30, 1975. Awards include ADSM, AP w/3 stars, KSM w/star, UNSM, NDSM w/star, Navy Occupation Svc. Med., Philippine Lib. Rib., MSM, AM, PUC w/2 stars, Naval Res. Med., WWIIVM, AFRM, and Philippine Lib. Med.

Married Margaret Deese (d. 1988) and has two children, Pletcha and Carmrid III (CDR, USN); and two grandchildren, Sandra and Carmrid IV. He married Klara Dadmanesh Sept. 30, 1994. From 1980 to present he has been president of Harness Racing, Inc.

MICHAEL J. "MIKE or MJ" HAUBNER, CDR, born Nov. 11, 1953 in Evergreen Park, IL, received a BS from Maryville University, St. Louis; JD, John Marshall Law School, Chicago.

He joined the service in 1978-79. His military locations: NavBase-Philly; NETC-NPT; NAVSTA San Diego; Camp Pendleton, CA. Served with NSLO, USS *Ajax* COMNAVBASE - SD, NISRA, Camp Pendleton.

Haubner was awarded the Navy Commendation, Navy Achievement (3), Sea Service, National Defense, Rifle and Pistol Expert, Reserve Medal, Military Outstanding Volunteer Medal, JMUA. He achieved the rank of CDR, USN and retired September 1985 and May 1991.

His memorable experiences: 1982 WESTPAC Cruise; defended US sailors before British magistrate in Diego Garcia; in 1984-85 he was on "loan" to US Attorney in San Diego.

He is married to Louise and they have children Susanna and Joseph. Employed as US Administrative Law Judge, Office of Hearings and Appeals. Presently still in the Reserves.

ROBERT RICHARDS "BOB" HAUPT, CAPT, born Jan. 20, 1934 in Roanoke, VA, enlisted in the USNR as a construction recruit in 1952. Upon graduation from high school he served two years on active duty where he was promoted to construction mechanic third class. His active duty time was served with MCB6 in the Bahamas building a naval facility and ACB TWO in Little Creek, VA.

Upon release from active duty he attended college at the Virginia Polytechnic Institute, Blacksburg, VA and received a BS in mechanical engineering. During this time he remained active in the USNR, Seabee Div. in Roanoke and was promoted to construction mechanic first class.

Transferred to Portsmouth, VA when he accepted employment at the Naval Weapons Station, Yorktown, VA and was commissioned by direct appointment to ENS, Civil Engineering Corps in October 1962. He is a plank owner in Reserve Naval Mobile Construction 23 (RNMCB23) upon its commissioning in 1962. During his tour with RNMCB23 he served as company commander for three companies, plans and training officer, operations officer, XO and CO.

Served as the 6th Reserve Naval Construction Regt. (RNCR) plans and training officer, chief of staff with Reserve CBLANT and the 8th RNCR, CO ACB2 Detachment 106, contingency plans officer for Reserve Div. Naval Facilities Engineering Command, CO EFD CHES/LANT, CO 8RNCR and Staff Civil Engr. for the Naval Reserve Readiness Command, Region Six.

He is a registered professional engineer in the state of Virginia. He is a retired civil servant of 31 years of service with the Department of Navy, a member of the NRA, Naval Memorial, Seabee Scholarship Association, Seabee/CEC Foundation and Society of American Military Engineers.

Decorations include the National Defense Medal, Navy Commendation Medal, Armed Froces Reserve Medal w/HG, Naval Meritorious Medal, Navy Unit Citation and Navy E.

Haupt married the former Bettye Gresham of Salem, VA and they have two children, Teri Byers and Robert Jr. He and Bettye are proud grandparents of four girls: Sheri and Heather Haupt and Amy and Allison Byers.

ALBERT CHARLES HAUSMANN, CDR, born Jan. 26, 1932 in Lakewood, OH, received a BA from Capital University in psychology and MFA, Bowling Green State University (Ohio) sculpture.

Joined the service February 1955. His military locations and stations: OCS Newport; AGO Jacksonville; Washington, DC; NAS Pensacola. Served with VA-New Orleans; VS-Grosse Ile, VA and VR Memphis.

Hausmann achieved the rank of CDR and was discharged January 1979. His memorable experiences: baseball coach NAS Pensacola. Played Cleveland Indians and in Wrigley Field, Chicago.

He married Barbara Ann Bunge Hausmann, July 10, 1955 and their children are Teri Michelle, Kimberly Sue and Robert Shawn; grandchildren are Chris, David, Stephen and Lesley. Owner, Advertising Consultants Pensacola; Owner, Milo Lanes, Toledo; professor "Ole Miss," Oxford and North Alabama, Florence.

ROBERT C. "BOB" HAVEN JR., born June 25, 1907 in Wilkinsburg, PA, received the BSE (CE) in 1932 from the University of Michigan and a CE in 1947.

He joined the service Nov. 11, 1943 as LTJG Line USNR. His military locations and stations: BUSHIPS (OinC) Okinawa, Guam, CNO, Bay City, MI. Served with Ad. Base Sec; Ship Repair Base Baten Ko; Op40; Res Sup ship. His battles and campaign: WWII.

Haven was awarded Citation JLPC Logistic Plans Floating and Shore Based Ship Repair Reg. Forward Area. He achieved the rank of CDR and was discharged June 6, 1946 and July 1, 1957.

His memorable experiences: JLPO logistic planning and meetings for six Pacific OPS. Photography and hist. adv. base construction and ship and base destruction by Okinawa typhoon Oct. 9, 1944, Buckner Bay. Construction and launchings of XMSB23 (Plastic): DE-1021, DE-1022 Defoe Shipbuilding, Bay City, MI.

He married Jane Carlton and their children are Ruth Jean, Anna Elizabeth and Merrily Margaret; grandchildren are Robert, David, Charles, Kathi, Jeni, Carolyn, Haley and Kristin. Employed as USBUR Reclamation Asst. Engr. Colorado, California, Utah; Consulting Firms (4) Engr., California and Michigan. He retired July 1, 1967.

WILLIAM FRANKLIN "FRANK" HAWKINS, CDR, born Aug. 23, 1933 in Greenville, SC, received a BSEE, Clemson University, SC. He joined the service January 1953 at Greenville, SC; boot training at San Diego, CA ATTC, Norman, OK; NATTC, Memphis, TN.

His military locations and stations: NAS Oceana, VA; NAS Norfolk; HAL for Malta, Charleston, SC. Units served with: NAS Oceana; NAS Norfolk; FASRON (Special) 201; VP-674; NAS Atlanta; USS *Soley* (DD-707); Mine Warfare Command 107.

Hawkins participated in the Korean War. He was

awarded the Good Conduct, National Defense, Korean War, Aircrewman (radio/radar) P2V-5F. Achieved the rank of CDR and was discharged Dec. 1, 1983. His memorable experiences: Enlisted man January 1953-February 1962 (AT1); Commissioned February 1962 until retired 1983 (CDR).

He married Frankie Eugenia Whitmire; they later divorced. Their children are Krista F. Hawkins Hancock and Mark Edward Hawkins. Grandchildren are Audrey Whitmire Lake and Garret Daniel Hawkins. Employed with Bell Telephone Labs; Western Electric-Design Eng.; TVA supervisor industrial safety and fire protection, retiring Aug. 1, 1988.

STUART A. HAYES, CAPT, born on Jan. 10, 1955 in Kodiak, AK, and was raised in Jacksonville, FL. He attended Jacksonville University on a NROTC scholarship, where he also played varsity baseball and basketball. He received BS degree in sociology and was commissioned in April 1977.

Following his designation as a naval aviator in December 1978, Hayes reported to the "Skinny Dragons" of Patrol Sqdn. 4, (VP-4) at NAS Barbers Point, HI in August 1979. He qualified as patrol plane commander, mission commander, instructor pilot and maintenance check pilot. During his three year tour he deployed twice to NAS Cubi Point, Republic of the Philippines and detachment operations to the Indian Ocean, Guam, and Adak, AK. Additionally, he served as the assistant aircraft division officer, public affairs officer, schedules officer, and flight officer.

He transferred to Patrol Sqdn. 31 (VP-31) at NAS Moffett Field, CA in August 1982. He served as a fleet replacement instructor pilot, facilities manager, pilot familiarization track manager, and pilot courseware officer.

In May 1985, Hayes joined the "Broadarrows" of Patrol Sqdn. 62 (VP-62), NAS Jacksonville, FL. He served as patrol plane commander, mission commander, schedules officer, pilot NATOPS officer, flight officer, flight training officer, training department head, operations department head, executive assistant, and XO. On Nov. 18, 1995 he became the 17th CO of Patrol Sqdn. 62 until his tour was over on March 22, 1997. His squadron won the 1996 Battle Efficiency "E" and the 1996 Retention Excellence Award, the 1996 Administrative Excellence Award, and the 1996 Top Bloodhound Award for torpedo delivery excellence.

He is authorized to wear the Meritorious Service Medal, Navy Achievement Medal, Meritorious Unit Commendation, Coast Guard Meritorious Unit Commendation w/Operation Distinguishing Device, Battle Efficiency "E" Ribbon (two awards), Navy Expeditionary Medal (two awards), National Defense Service Medal, Armed Forces Service Medal, Humanitarian Service Medal, Sea Service Deployment Ribbon (two awards), Navy and Marine Corps Overseas Service Ribbon, Coast Guard Special Operations Service Ribbon (two awards), and the Armed Forces Reserve Medal.

After 3,750 flight hours as a pilot flying the T-28, T-44, and the P3B and P3C, he has been selected to be the CO of NR NAS Cecil Field, FL in October 1997.

Hayes is married to the former Gwen E. Bohanan of Jacksonville, FL. They have two daughters, Kristen and Melissa, and reside in Miami, FL where he is a pilot for American Airlines.

JOHN P. HAZELRIG, CAPT, born in Atlanta, April 30, 1947, began his naval career at the US Naval Academy.

Designated a naval aviator, April 1971, he reported to Lemoore, CA for A-7E Corsair transition. He made two Southeast Asia deployments on USS *Coral Sea* and was an A-7 instructor pilot.

In 1978 he began his career in the USNR. Assignments included squadron OIC; COM-NAVAIRESFOR TACAIR program manager; CO VFA-305 and CDR Carrier Air Wing 30 (CAG).

From 1992-96, he served on the CNO staff, for Director Air Warfare and Chief of Naval Reserve. He is currently the CO of Naval Air Reserve Santa Clara, Moffett Field, CA.

Hazelrig has BS and MBA degrees. His medals include two Legion of Merit, two Meritorious Service, two Navy Commendation and other campaign/service awards. He and his wife Susan have three sons.

DOROTHY BEAN HAZELWOOD, born July 31, 1939 in Medford, OR, received a BS degree, School of Nursing, Brigham Young University.

She joined the service Sept. 26, 1968, USNNC (Reserve). Her military locations and stations: Naval Hospital, Annapolis, MD; NAVSTA Midway Island. Served 24 years in the USNR NAVREACEN Ft. Lauderdale, FL and NAVRESCEN, Albuquerque, NM.

Awarded the Vietnam Defense Medal, Naval Reserve and Sharpshooter Medal. She achieved the rank of CDR and retired Oct. 1, 1994.

Her memorable experiences: In December 1969, a distress call was received from the Russian fishing trawler, Pogranichnic Strelinkov, by the Midway Island Naval Station. The medical facility at Midway responded to a request for help from the Soviet vessel to transfer a female crew member, suffering a tubal pregnancy, for emergency care.

The Soviet woman required treatment beyond the Midway Island clinic capability. Transported by USAF C-141 aircraft, LT Hazelwood, NC, USNR, accompanied the patient during the transport from Midway on to Tripler Army Hospital, Honolulu. During the flight the patients condition was critical and on arrival at the Hickam AFB there were immigration difficulties with the entry of a Soviet citizen in to the US—a new experience for the Navy Nurse, LT (Bean) Hazelwood.

She is married to Ernest R. Hazelwood and employed with Latter Day Saints (LDS) Hospital, SLC, UT June 1961-September 1968.

GEORGE J. HAZLEY, CAPT, born Aug. 28, 1934 in New York City, received a BS in math from Manhattan College and Grad. Bus., Fordham University.

He joined the service March 1957, Officer Candidate School, Newport, RI. His military locations and stations: Norfolk, VA, qualified surface warfare officer. Served with USS *Tutuila* (ARG-4) 1957-60, operations officer.

Hazley received the National Defense Medal, Naval Reserve Medal and Marksmanship Medal. He achieved the rank of LT (active duty) and CAPT (Reserve), discharged in 1960.

His memorable experiences: 25 years USNR duty; member of Naval Control of Shipping Org. (NCSORG 3-4, Elizabeth, NJ) XO and CO of CONVOYCOM 3-4, Bayonne, NJ, 1978-85.

He is married to Barbara and they have two children, George Jr. and Gregory. Retired IBM marketing executive (1960-92). Presently marketing consultant.

PETER S. "PETE" HEAD, LCDR, born June 21, 1936 in San Diego, CA, received a BIE from the Georgia Institute of Technology.

Joined the service June 21, 1953 as a Reserve AT NAS Atlanta until flight training Pensacola and Corpus Christi. Active duty at NAS Rota Spain; Reserve at NAS NY and NAS Willow Grove. Active duty VQ-2. Reserve, VP-66.

Head participated in Cuban crisis. He achieved the rank of LCDR. Retired from active duty Dec. 24, 1964 and USNR June 21, 1973.

He is married to Sheilagh and they have children, Meredith and Patrick. Employed as pilot with TWA for 32 years. Captain on L1049G, B-707, B-720, B-727, B-757, B-767, B-747, L1011.

WILLIAM I. "BILL" HEAD, CDR, born April 4, 1925 in Tallapoosa, GA, received a BS, Georgia Tech, 1949; PhD, Columbia Pacific University, 1988.

Joined the service June 7, 1943 (voluntary inductee). His military locations and stations: NAS JAX and De Land, FL (1943-44); NAS Alameda (1944); Ford I. VS-53, (1944); NAS Kahulu I, Maui (1944-46). Participated in Atlantic and Pacific Theaters (no combat stars).

Received direct commission 1949; volunteered for Korea era but never received orders; promoted to CDR, Supply Corps 1965; attached to Selective Service 1969; passed over for CAPT twice (Supply); switched to line and passed over twice more (he had convinced Chief PERS VADM Bagley that boards were not "looking at" Selective Service Officer); promoted to CAPT 1974.

He was awarded the Navy Commendation Medal in 1980. He achieved the rank of CDR (supply); CAPT (1974-83), discharged July 1, 1983.

Currently listed in *Who's Who in America* and *Who's Who in World*, among others (Finance & Industry, Science & Engineering).

EDWARD R. "ED" HEALY, CAPT, born April 18, 1946 in New York City, received a BS, BBA and MBA. Joined the service October 1972. His military locations: Jax, Key West, Tampa, England (UK). Served with AO57/ MSD/ SIMA, CINCEAROPE, WPC.

Received the Vietnam Service, Battle E Award and Navy Commendation Medal. He achieved the rank of CAPT 06 IRR.

Married Shirley and has stepchildren: Todd, Greg, Sharlene and Richard Pitts; grandchildren are Jillian, Danny, Christopher, Amanda, Candace, Lexi, Ricky and Bob.

ALAN F. HEARL, CAPT, born and raised in Sacramento, CA and graduated from the University of California, Davis, 1962 (BS) and Fuller Theological Seminary, Pasadena, CA, 1965 (MDIV). Ordained in 1967 in the Evangelical Convenant Church, he has pastored churches

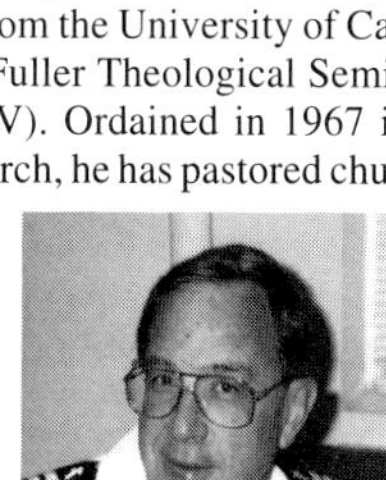

in Pennsylvania, Minnesota, California and Connecticut.

Hearl volunteered for Naval Service in 1971. After Chaplain's School he served with 3rd MEFREL 116, Minneapolis, MN; 1st MEFREL 220, Treasure Island, CA (CO); Special Boat Unit 11, Mare Island, CA; group chaplain MAG-42, NAS Alameda, CA; staff chaplain, NRREDCOMREG 20, TI, CA; and staff chaplain, Icelandic Defense Force, Providence, RI.

He completed Navy Survival School (SERE) and qualified to fly backseat in A-4 Skyhawk's and F-18s. He was was awarded the National Defense Ribbon w/star and the Naval Reserve Medal w/HG.

Hearl is married to the former Karen VanSteenwyk. They reside in Manchester, CT and have two sons, David and Timothy; and grandchild, Kirsten.

GARY LEE HEARNE, CDR, born Wilmington, DE, March 20, 1948; graduated John Dickinson High School 1966 and the University of Delaware; BS in agricultural engineering 1971. Sworn into Navy Reserves, Nov. 7, 1970 at NAS Willow Grove, PA. Commenced active duty Sept. 21, 1971 at NAS Pensacola, FL Aviation OCS.

Primary flight in T-34A, February 4, 1972, NAS Saufley Field, VT-1. Commissioned Feb. 18, 1972. Transitioned to T-28, NAS Whiting Field, Milton, FL at VT-6 and VT-3; and VT-5 for carrier training. January 1973, NAS Corpus Christi, VT-28 for multi-engined training in TS-2A and US-2B. Designated naval aviator June 8, 1973.

June 28, 1973-Sept. 5, 1973, VT-3, NAS Whiting Field as flight instructor and ground training officer. Received Chief of Naval Air Training Letter of Commendation for 740 accident free instructor flight hours from RADM B.H. Shepherd.

Sept. 5, 1975-June 29, 1978, VC-6, NAS Norfolk, VA, as remote control flight officer and instructor for the Northrop MQM-74C Chukar, Detachment Delta OinC. Attended Pacific Missile Test Center, Pt. Mugu, CA. Served with Fleet Target Gp., Royal Naval Air Station, Portland, Dorset, England and RAE Aberporth, Wales providing instruction and integration of target services; TemDu to VRF-31 as ferry pilot in 18 different aircraft. Received Letter of Commendation, Fleet Tactical Support Wing One-Commodore G.R. Olson; Detachment of the Quarter Letter of Commendation 1976, and Letter of Appreciation-Commodore G.E. Jacobssen Jr., Service Gp. TWO. Joined Order of the Ditch aboard USS *Vreeland,* Aug. 17, 1977, and Ancient Order of Shellbacks Aug. 21, 1977. Received Achievement Award as Chukar Controller by Northrop Corporation, Ventura Div. on Nov. 25, 1976.

June 29, 1978-March 31, 1983, VRC-40, NAS Norfolk, as carrier on-board delivery pilot, aircraft commander, flight instructor, detachment OinC, and Administrative Department Head. Received Battle "E" Award and CNO Safety Award in 1980. Received Letter of Commendation 1978, and Letter of Commendation from Fleet Tactical Support Wing-one Commodore Jack P. Smith. Logged over 200 traps, and and 2,400 flight hours and given Achievement Award for over 1,000 accident free hours in the C-1A by Grumman on Feb. 28, 1982. Left active duty Navy 31, March 1983 and joined active Reserves on May 21, 1983.

May 21, 1983-Sept. 30, 1989, NRNS Keflavix 1066, NAF Andrews, MD as training and Admin. Department Heads and acting executive and CO; mobilized five times overseas to NS Keflavik, Iceland as OinC. Received Navy Commendation Medal from RADM W.F. Flagg, Letter of Appreciation from RADM R.K. Maughlin, Unit Letter of Commendation from SECNAV Ball, and joined Realm of the Arctic Circle on Aug. 22, 1985.

Oct. 1, 1989-Sept. 30, 1992, Readiness Command Region Six, Washington Navy Yard, as Readiness Analysis officer; frequently acted as assistant deputy for readiness. Awarded Admiral Neale Smith Readiness Trophy in 1990 and Navy Commendation Medal by RADM M.J. Bresnahan, Jr.

Sept. 3, 1991-Nov. 27, 1991, requested by ADT by deputy assistant secretary of the Navy (Reserve Affairs), Honorable Donald C. Morency; resolved Information Resources Management Pay and Personnel problems from Operation Desert Shield/Storm. Actions directly resulted in Navy's first IRM Policy signed by SECNAV March 1992.

Oct. 1, 1992-Sept. 30, 1995, NR Navy Command Center 106, Washington, DC as CNO Operations Briefer and Assistant Navy Department Duty Captain. Awarded Best Unit of the Year Award 1994 in REDCOMSIX and Unit Letter of Commendation on Sept. 22, 1994 from ADM J.M. Boorda. Completed retraining in the T-34C turbo and initial training in the HT-57B/C helicopter 1995.

Oct. 1, 1995-Sept. 30, 1996, NR Carrier Group 0466, NAF Andrews as Training Department Head; served aboard USS *Enterprise* (CVN-65) as Commander Carrier Group Four Staff.

GVTGU-6666, NAF Andrews, from which requested assignment to the Individual Ready Reserve, Oct. 1, 1996 to Dec 26, 1996.

He wears the Navy Commendation Medal, Navy Achievement Medal, Battle "E" Ribbon, National Defense Medal (2nd Award), Overseas Ribbon (2nd Award), Sea Service Ribbon (2nd Award), Armed Forces Reserve Medal w/Silver HG and Bronze "M" Device and the Expert Pistol Medal.

Hearne is married to the former Gloria Jean Riner of Washington, DC and has five children: Michael, Colene, Deborah, Alison and Kathryn and one grandchild, Brennan.

Currently employed with RMS Information Systems Inc., of Lanham, MD as a senior principle systems engineer. As subcontractor to TRW, Inc., supports the FAA's EnRoute Programs Office managing the installation of the $2.5 Billion Voice Switching and Control System (VSCS), VSCS Emergency Access Radio System (VEARS), the VSCS Training and Backup System (VTABS), and the $2 Billion Display System Replacement (DSR), which changes the way the FAA controls aircraft in the 21 air traffic control centers throughout the US.

JERRY L. HEATH, CDR, a native of Carrollton, GA, attended West Georgia College and the University of Georgia where he graduated with a BSA degree in 1968. He began his naval career in 1968 and was commissioned ENS in 1969. After completing multi-engine flight school in 1970 at NAS Corpus Christi, TX, he was designated as a naval aviator.

In 1970, he was assigned to VAW-121 at NAS Norfolk, VA, where he flew as a fleet pilot flying the E-1B, an airborne early warning aircraft. During this time, he served with Detachment 18 on the USS *Wasp* and Detachment 11 on the USS *Intrepid* as an aircraft carrier pilot.

Upon release from active duty in 1972, he affiliated with the USNR and served until 1979 flying the E-1B and the E-2B with VAW-207 and VAW-78 at NAS Norfolk, VA. He was a designated plane commander, carrier aircraft plane commander and a maintenance test pilot. He held numerous billets in the squadrons and was a flight safety instructor for the Naval Air Reserve. He later served four years with TACRON 23 until his promotion to CDR in January of 1984. After his promotion, he served three years as a senior staff officer with the CDR Naval Air Forces Atlantic Command. CDR Heath has been awarded the following medals: Battle "E", Armed Forces Reserve, National Defense and Meritorious Unit.

In civilian life, he is the owner of of a real estate rental and management business. He is a life member of NRA and the Retired Officers Association.

STANLEY P. HEATH, CWO-4, born in Spokane in 1925 and left high school to enlist in June 1943. He was released from active duty in 1946 after serving in the Pacific on the carrier USS *Boxer.* He affiliated with NAS Spokane and was named as the first "sailor of the year" in 1961. Several squadron changes led to plank ownership in Spokane's VR-901. Base closures took him to Seattle and Whidbey Island mostly in the VR community. He was one of two known E-9s selected for Warrant 1.

He used Naval and civilian transports to commute to activities for the next 25 years. Without missing a scheduled reserve activity, he earned a doctorate degree at Washington State University in 1978. The SecNav granted him waivers from mandatory retirement until he had 39 years of satisfactory service and was discharged as CWO-4 in1982.

He retired to Olympia, WA close to his two children, Jay and Jennifer.

F.J. HEFESTAY JR., LCDR, enlisted in the USNR in 1978 and completed cross-rate training to IS1 in 1981. Commissioned in 1982, he was assigned to NR FIRSTPAC 0571 as an imagery analyst. During this tour he was designated as RIA FIVE's Officer Intelligence Analyst of the Year for 1986.

He also served on COMNAVRESINTCOM's HQ Staff (1991-94) as PAO and national director for Basic Reserve Intelligence Training. In 1996 he performed six months active duty supporting DECISIVE ENDEAVOR at the Combined Air Operations Center, Vicenza, Italy. Currently, he is assigned to NR ONI 1666 in RIA NINETEEN and is a FY 1997 select for CDR.

His decorations include the Joint Service Commendation Medal, Navy Commendation Medal, three Navy Achievement Medals, Joint Meritorious Unit Award, Meritorious Unit Commendation, Armed Forces Service Medal and the NATO Medal.

He is a graduate of the University of Colorado and is employed by Lockheed Martin.

ERNEST A. HEGI, CAPT, born Sept. 16, 1939, Rahway, NJ. He graduated from the SUNY, Maritime College with a BS and the University of Hartford with an MBA.

He enlisted in September 1956. After completing boot camp at NTC Bainbridge, MD he was assigned to several reserve destroyers. Over the next six years he made ship visits to 17 nations and territories in Europe and the Mediterranean.

In March 1964, he received a direct commission to ENS and reported for duty to USS *Catamount* (LSD-17) in Sasebo, Japan where he served as ships boatswain, combat cargo officer and later navigator. During this tour he participated in the initial movement of US Marines to Vietnam plus other amphibious and river operations. LT Hegi was then assigned as material officer for Naval Base New York and the Third Naval District. As collateral duty he was assigned as the liaison officer to the training ships, *Juan Sebastian De Elcano, Sagres* and *Custodio De Mello.*

His various USNR assignment and command tours include several Surface Units, Naval Reserve Officers Schools, USS *Fulton,* SIMA, NPT, MSC Central Europe, Readiness Command Staff and NLO to NH. He served additional tours of extended active duty during OpSail 1976, OpSail 1980 and for SIMA training program.

He has been awarded personal military decorations for duty in Vietnam and other naval operations. He has two sons, Eric and Adam.

THOMAS C. "TOM" HEIDRICK, LCDR, born Aug. 23, 1934 in Poplar Bluff, MO, received a BS in business administration from the University of Nebraska in 1957.

He joined the service April 1957. His military locations and stations: Alameda, CA, serving with USS *Midway* (CV-41), September 1957 to September 1960.

Heidrick received the Naval Reserve Medal and achieved the rank of LCDR. He was discharged Sept. 1, 1960. He spent about 3 1/2 years on active duty and about 17 years in active reserves.

Married Marilyn L. and they have children: Suzanne E., John T., Margaret L., Carolyn M. and James W., and seven grandchildren. Employed as auditor and taxpayer service for IRS, retiring Sept. 19, 1993.

WALTER F. "WALT" HEISEY, CAPT, born April 10, 1919 in Chowchilla, CA, received a BA degree from Fresno State and did graduate work at UCLA.

He joined the service Jan. 26, 1943. His military locations and stations: CO, USS SC-989; CO, US PC-569. Spent the war escorting merchant convoys, Caribbean, Gulf of Mexico and Central Pacific Ocean area.

Heisey received the Theatre Medals and Victory Medal. He achieved the rank of CAPT USNR and discharged April 10, 1979.

He married Catherine C. Nov. 2, 1946. They have children: William, Thomas, Kathleen, Sally, Rebecca and Patricia; and 17 grandchildren. Employed as president, Ace Financial Corp. and president, Bakersfield Sandstone Brick Co.

HENRY S. "HANK" HELLWEG, CDR 05, born Nov. 15, 1922 in Houghton, IA, received a BS degree and degree in Naval Science, Northwestern University, Evanston, IL.

He enlisted as apprentice seaman in regular Navy on March 3, 1941 for six years. His military locations and stations: Naval Training Station, Great Lakes, Navy Storekeeper School, Toledo, OH; USS *Wyoming;* Flag Unit, CDR Air Force Atlantic Fleet; V-12 Pre-Midshipman School, Asbury Park, NJ; St. Ambrose College, Northwestern University. Commissioned as ENS in June 1947, then inactive reserve.

Hellweg achieved the rank of CDR (0-5) and retired from USNR June 1964.

Recalled to Korean War September 1952, Navy Supply Corps School, Bayonne NJ, then supply officer at the Marine Corps Air Facility; Santa Ana, CA; Weekend Warrior from 1954-64.

He is married to Marge. His first wife Inez died, July 25, 1987. He has children: Joan E. Hellweg-Barrick and Dean V. Hellweg and grandchildren, Jason E. and Christopher P. Ross (enlisted in Navy December 1996). Employed as life insurance agent (CLU) and securities sales.

JOHN FREDERICK HELPAP, CDR, born May 1, 1940 in Saginaw, MI, received a BA from Albion College, Albion, MI.

He joined the service June 20, 1962, Recruit Training Command, Great Lakes. Active duty USS *Cogswell* (DD-651). Served with USS *Cogswell* (DD-651); Recruit Training Command; Great Lakes.

Helpar served in Vietnam and received the National Service Medal and Naval Reserve Medal. He achieved the rank of CDR and was retired July 1, 1983.

He is married to Susan and they have two children, Allyn and Beth; and a grandson, Michael John Olson. Employed with Baxter Healthcare, vice-president, retiring December 1990.

VICTOR H. HEMMY JR., CAPT, SC, USNR (Ret.) born April 19, 1944, New York City, NY. He joined the service Jan. 22, 1967 and was commissioned from Regular NROTC Program at the University of Oklahoma. Obtained master's degree in business administration from the University of Hawaii.

Reported to Supply Corps School, Athens, GA, 1968 and graduated in 1969. Various assignments as asst. supply officer, USS *Skabit* (LKA-105); ship's stores officer USS *Proteus* (AS-19), Guam, 1972; Navy exchange officer, Garden City and Brooklyn, NY, 1974; supply officer, USS *Hassayampa* (AO-145), 1977; deputy and exchange officer, Pearl Harbor, HI, 1977.

Transferred to active Navy Reserve in 1977 and served in various Reserve billets at COMNAVLOGPAC, COMSUBPAC, COMSUBRON Seven, Naval Supply Center, Pearl, CINCPACFLT, CINCPAC and Logistics Task Force Pacific, and COMUSNAVCENT. Received the Navy Commendation Medal w/star and Joint Services Commendation Medal.

Highlights: selected as Commanding Officer of NR Naval Supply Center, Pearl DET 220 in 1983. The following year recalled to active duty to serve as comptroller of COMUSNAVCENT in Pearl Harbor. In 1995 recalled to AD to serve as the deputy director of the Naval Coordinational Committee at CINCPACFLT to commemorate the 50th anniversary of the end of WWII. Recalled to AD to serve as the CINCPAC Deployment Process Special Action Group as Deputy Director.

Married Candace Pratt and they have three children: Malia, Victor III and Mina. Retired Feb. 1, 1997. Currently an analyst for Cubic Applications, Inc. and principal in Hemmy & Associates, a real estatebrokerage and consulting firm.

RICHARD W. HENDEL, CAPT, born in Buffalo, NY in 1931 and graduated from the State University of New York, Brockport in 1953 with a BS in education. Due to the Korean War, he immediately went to NAS Pensacola for flight training and received his commission and wings at NAS Corpus Christi in November 1954.

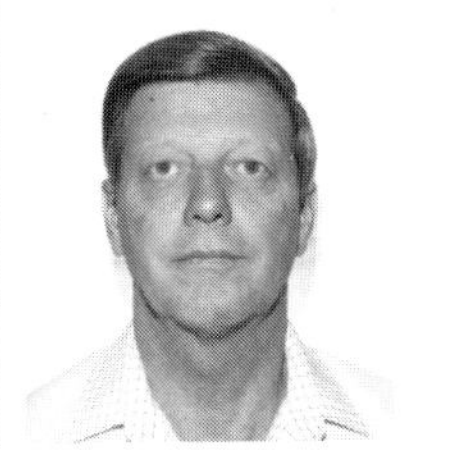

His first active duty was in VA-104 (CAG-10) at NAS Jacksonville, flying the AD "Skyraider" and deploying on a Mediterranean cruise aboard the CVA *Coral Sea.* Released from active duty in 1957, he became a pilot for Capital Airlines which later merged into United Airlines.

In 1963, Hendel returned to active duty ordered to VR-1 NAS Patuxent River. Here he was assigned to the VIP Div., flying two years on CNOs crew and qualified as aircraft commander on the C-118 and C-130. He then transferred to VX-6 at NAS Quonset Point where he deployed for six months to Antarctica flying the C-130 with skis.

Returning to the airline, he affiliated with the USNR in the VR community at NAS Los Alamitos and became CO of VR-733. When Los Al closed, he transferred to NAS Whidbey Island and became CO of VR-51 and upon relief, became CO of the Reserves first C-9 jet transport squadron, VR-55, at NAS Alameda. After that tour he was selected as CO of NAS 0194 at NAS North Island and in 1984, retired as CO of VTU-9494 at NAS Point Mugu after 31 years service.

In 1991 he retired as a B-747 captain for United Airlines with 34 years of service and total flight time over 21,000 hours. Hendel resides in Westlake Village, CA with his wife Carol and has three adult children, Susan and twins Linda and Rick.

BARBARA J. "BARBI" HENDERSON-MORRILL, CDR, born May 11, 1958 in Bryn Mawr, PA, received a BS in economics from the Wharton School of the University of Pennsylvania.

She was commissioned May 19, 1980. Served with VP-17, FICPAC, NOIC 0293, FIRSTEURLANT 0. Participated in the Desert Storm/Desert Shield campaigns.

Henderson was awarded the Navy Achievement Medal, Navy Meritorious Unit Commendation, National Defense Service Medal, Sea Service Deployment Ribbon, Overseas Service Ribbon. She achieved the rank of CDR, currently XO of ONI 0102.

She is married to William C. Morrill and is marketing communications manager at General Signal Networks, Mt. Laurel, NJ.

NELSON HENDRICKS, LCDR, born Sept. 4, 1961 in Honolulu, HI, received a BA from UCLA, geography; MA Chapman College, education.

He joined the service August 1983. His military locations and stations: San Diego: VS-33, Fleet ASW Training Center, NAR SO. Units: Whidbey: VAQ-309; NAF Andrews: VAQ-209.

Hendricks achieved the rank of LCDR and is married to Pam.

F. SCOTT "SCOTT" HENDRICKSON, LT, born Jan. 3, 1941 in Alton, IL, is a dental graduate.

He joined the service June 1966. His military locations and stations: NAS Norfolk; USS *Sierra* (AD-18). Served with USS *Sierra.*

Hendrickson achieved the rank of LT, retiring in 1968. His memorable experiences: He loved it. He is a self-employed dentist.

JAMES R. HENRY, CAPT, born in Green Bay, WI, March 9, 1934. He enlisted in the USN as an E-1 in Minneapolis, MN, June 4, 1952, and completed basic training at Great Lakes, followed by Airman Preparatory School at Norman, OK and Aviation Electronics Technician School at Memphis, TN.

He was selected for the naval aviation cadet program in 1954, while stationed at NATC Patuxent River. After completing basic flight training at Pensacola, and advanced flight training at Hutchinson, KS, he was commissioned in April 1956 and received orders to Utility Sqdn. FIVE at Atsugi, Japan where he flew the JD-1 and F9F-5.

Henry served as a naval aviator in VU-5, JTTU, VFAW-3, VAH-16, VAH-123 and NAAS Meridian, MS. Subsequently he completed Air Intelligence School and served as the Air Wing intelligence officer for CVW-19, deploying to Vietnam aboard USS *Bon Homme Richard* (CVA-31).

After leaving active duty, he completed the BS and MS and the Doctor of Education degrees at the University of Kansas. He served as CO of several USNR intelligence units and as the Reserve Intelligence area coordinator for Area Ten.

Henry is married to the former Kay Swafford of Hutchinson, KS and they have two daughters, Barbara and Jennifer, and two grandsons.

He retired July 1, 1985 in Minneapolis, MN, the same city in which he enlisted 33 years earlier.

THOMAS EDWARD HENRY, CAPT, born April 7, 1927 in Bay Shore, NY, attended Saint Paul's School, Garden City, NY; Massachusetts Maritime Academy 1951. He joined the service March 1945 and served with USS *PCS 1444*

Henry was awarded the American Theater Medal and WWII Victory. He achieved the rank of GM3/c and was discharged July 15, 1946.

His memorable experiences: Massachusetts Maritime Academy 1951; Merchant Marine Korean Service Medal. Had the great privilege of carrying military cargo (120,000 tons) ammo and general military, for six years to Vietnam. Rockets in DaNang, Vietnam Service Medal-Merchant Marine.

He is married to Barbara E. Henry. A Master Mariner, U.S. Lines Inc., he retired in March 1984.

LARRY DONALD "L.P." or "DOC" HENSLEY, CAPT, born Oct. 6, 1944 in Radford, VA, received a BS degree from Hampden-Stoney College in 1964 and DDS from the Medical College of Virginia in 1968.

He joined the USN June 25, 1968; USNR, Aug. 21, 1971. No unbroken service.

His military locations and stations: Portsmouth Naval Hospital, Norfolk Naval Station; Camp Lejeune Naval Base, Little Creek, VA; NAS, Norfolk; USS *Vulcan*; VTU-106; FMF LANT; CMCHS; NDMS; USS *Vulcan* (AR-5); NARS, Norfolk; MATWING 106 Norfolk, VA Dental Fellowship Academy of General Dentistry, 1976; President SDS 1987-88.

He was awarded two National Defense Medals and the Naval Reserve w/HG. Achieved the rank of CAPT and retired Nov. 1, 1993.

His memorable experiences: Sea duty USS *Vulcan* (AR-5), 1969-71; Field Medical School (Camp Lejeune), 1979; Internship (US Naval Hospital, Portsmouth, VA), 1968-69; US Naval Academy Inf. Officer/Candidate Guidance (US Naval Academy) 1975-present; CO, NR CMCHS/NDMS 106 Norfolk, 1983-87; FMF LANT DENT A-206 (XO), 1977-80.

He is married to Anne and they have a daughter Mary Lynnette. A grandchild is expected in August 1997. He is self-employed - family general dental practice. He is a life member of NRA.

JOHN G. "JACK" HERBEIN, CAPT, born April 13, 1938 in Macedon, NY. He received a BS from the USNA, 1960, and attended various technical schools, including Navy Nuclear Power School, 1964.

He joined the service in 1956 as midshipman, US Naval Academy. His military locations and stations: Destroyers, *Beatty* (756); *Purdy* (734); *Moale* (693); *Gainard* (706). Served with Navy Destroyer School (1962); Nuclear Power School 1964-65, ASW Key West, 1961.

His battles and campaigns: Cuban Missile Crisis, Santo Domingo Crisis. Awards and campaigns: Federal Expeditionary, Navy Expeditionary and Navy Achievement Medal. He achieved the rank of CAPT, USNR and retired September 1990.

His memorable experiences: brigade, boxing champion USNA 1957, 1958, 1959. Winning engineering E on USS *Moale* in 1965; Arleigh Burke, teaching first class of Destroyer School (1962); five Med cruises; serving as inspector general for REDCOM 4 and the eight tours as a Reserve Unit CO in Pennsylvania (the most memorable as CO of the *Williamsport* "JL Brown" Unit.

He is married to Angela and they have children: Michael, Christine, Jeffrey and Steve; and grandchildren: Scott and Sarah. VP Generation-GPU GENCO, retiring December 1995.

B. DEANE HERBERT, LCDR, born Dec. 31, 1921 in Bangor, ME, received a BS in Marine Engineering, Marine Maritime Academy.

He joined as midshipman, October 1941. Commissioned as ENS, May 29, 1943. His military locations and stations: South and Western Pacific; WWII, Atlantic and Korea. Served with USS *Ballard* (AVD-10); USS *Laning* (APD-55) in WWII; USS *Cabot* (CVL-28) in Korean conflict.

Awarded the Atlantic, Pacific w/2 stars, and WWII Victory Medal and Korean Medal.

He achieved the rank of LCDR and was discharged in February 1952. Memorable experiences: meeting people from all over the US, friendship is still in tact after 55 years in many cases, attend ships reunions each year.

Married Avis Burns. He is employed with Thiokol Corp. Rocket Research and Development, test eng., program manager, assistant general manager, retiring Jan. 1, 1985.

ROBERT STEVEN HERBERT, CAPT, born Nov. 1, 1945 in Washington, DC, received bachelor of city planning, University of Virginia 1968; masters administration, GWU, 1979.

He joined the service June 1968. His military locations and stations: Vietnam, Guam, NYC, Anapolis (USNA), Norfolk. Served with USS *Canon* (PG-90), USS *John King* (DDG-3), USS *Josephus Daniels* (CG-27).

Served in Vietnam and awarded Navy Cross, Meritorious Service Medal, Purple Heart and Navy Commendation.

His memorable experiences: WEPS officer, USS *Canan,* USS *John King,* USS *Josephus Daniels,* Aide to commandant Third Naval District, EA-Director of Professional Development, USNA, CO-FCDSSA DET 106, CO-SURFLANT DET 606, CO-TACTRAGRULANT 106.

Married to Sally and they have children, Jack and Peter. Employed as assistant manager, Suffolk, VA. Currently XO CINCLANTFLT DET 206.

WILLIAM G. HERBERT, CAPT, enlisted Aug. 31, 1955. After serving three years, including assignment to FASRON NINE and advancement to AT2, he was selected from the fleet for the regular NROTC program. He graduated from Vanderbilt University with a bachelor's degree in electrical engineering and was commissioned in 1962.

He attended several training commands before serving as missile systems officer in USS *Semmes* (DDG-18), Tartar Missile Systems Battery Officer in USS *Albany* (CG-10), Terrier Fire Control Officer in USS *Bainbridge* (DLGN-25) and Combat Systems Training Officer at MISTRAULANT.

Herbert joined the USNR in 1971. Major assignments as a drilling reservist included: XO Inshore Undersea Warfare Div. 11-5, "Plankowner" CO Inshore Undersea Warfare Group ONE Detachment 119, CO Personnel Mobilization Team 2519, Assistant Inspector General Naval Reserve Readiness Command Region NINETEEN and CO CINCPACFLT Detachment 219. Designated a surface warfare officer, and served over 32 years of naval service upon his retirement April 1, 1992.

His awards included the Navy Commendation Medal, Navy Unit Commendation, Navy Good Conduct Medal, National Defense Service Medal (2), Armed Forces Expeditionary Medal, Vietnam Service Medal (4), Armed Forces Reserve Medal (2), RVN Campaign Medal and Navy Expert Pistol Shot Medal.

Traditional Navy certificates accorded CAPT Herbert include *Shellback, The Realm of the Arctic Circle,* and *The Domain of the Golden Dragon.* In his civilian profession, he was employed for over 25 years as a supervisory electronics engineer at Port Hueneme Div., Naval Surface Warfare Center ("Nemesis") providing in-service engineering support for the HARPOON Weapon System program.

ROBERT EARL "EARL" HEREFORD JR., CAPT, born March 11, 1942 in Ft. Benning, GA, received a BS degree in metallurgical engineering from the University of Pennsylvania in 1964; MBA UT Chattanooga in 1985.

He joined the service May 1964. His military locations and stations: Nuclear Power School, Vallejo, CA and West Milton, NY, Sub School. Served with USS *Flasher* (SSN-613); USS *Daniel Webster* (SSBN-626); various reserve units. He was CO of NR Naval Shipyard, Portsmouth for a total of five years; five Polaris patrols on the *Webster* and two WESTPAC deployments on the *Flasher*.

Hereford received the Presidential Unit Citation, Navy Commendation Medal, Vietnam, Korean Expeditionary. He achieved the rank of CAPT USNR and was discharged in July 1970.

Married since 1976 to Harriette, they have three children: Kathryne, Rob and Anne. Employed with Tennessee Valley Authority, 24 years; various engineering and management positions. Retired from TVA in October 1994. Employed with American Express Financial Advisors as personal financial advisor since May 1995.

ROBERT L. "BOB" HERMAN, CDR, born July 16, 1927 in New York, NY, received a AB from Columbia University and BS in civil engineering, Columbia University School of Engineering.

He joined the service June 1, 1949. His military locations and stations: CECOS Port Hueneme, Saipan, MI; Guam, MI; NSD Bayonne, NJ. Served with CBD 1504, 103rd NCB. He achieved the rank of CDR, CEC, USNR.

Married Susan Marie and they have two children: Candia Ruth Herman-Flaum and William Neal Herman (LT, CEC, USNR); and grandchildren: Chloe Phyllis Herman and Cameron Pierce Flaum. Employed as senior chairman, Global Technology Systems, retiring July 16, 1987.

HERBERT R. HERN, CAPT, born in New York City, March 22, 1925. Commissioned ENS, USNR in 1944 upon graduation from New York State Maritime Academy. Served as engineering officer on merchant ships in North

Atlantic convoys and in Mediterranean and Western Pacific areas until 1946. Graduated from Cornell University with BS degree.

Called to active duty in 1950 and served as XO aboard USS LST-883 involved in Inchon landing, Wonsan landing and Hungnam evacuation. Served as material officer, Recruit Training Command, San Diego and then ordered in 1955 to USS *Calhoun County* (LST-519) as CO. Calhoun County engaged in deep water disposal of condemned ammunition and radioactive waste on the eastern seaboard. Next served as CO of USN & MC Reserve Training Center, Binghampton, NY from 1957 through 1959.

Released from active duty at his own request in 1959, he served in various reserve units in Ithaca, NY, Scranton, PA and the San Francisco Bay area. Promoted to CAPT, USNR, July 1, 1966. Assigned group commander, USNRTC Treasure Island, CA, 1968-71. Member of the National Naval Reserve Policy Board, 1967-70. Served as deputy commander, Naval Reserve Readiness Command, San Francisco from 1973 until transferred to the Retired Reserves on July 1, 1975. Participation as a plankowner of a Readiness Command in the realignment of the USNR was a very memorable experience. Awarded the Meritorious Service Medal.

Married to the former Betty Criswell and they have four children: Larry (a former naval aviator), Robert a (former naval officer), Michael and Kathleen. Employed since 1960 as a management representative in the labor relations arena, he retired from his position as president of the Industrial Employers and Distributors Association in 1991. The Herns reside in Niceville, FL.

ROBERT A. HERRING III, CDR, was commissioned ENS, USN, upon graduation from the University of Mississippi NROTC in 1969. Sea tours were gunnery officer aboard USS *Robert L. Wilson* (DD-847) (including Mediterranean deployment, and engineering officer with Assault Craft Unit One (including WESTPAC deployment to Subic Bay). Shore tours were Director, Counseling and Assistance Center, NAS Lemoore, CA; and treatment unit officer, Naval Drug Rehabilitation Center, NAS Miramar, CA.

He left active duty in 1979, affiliating with the Ready Reserve. He earned his MBA and PhD in management at Florida State University. After joining the Selected Reserve in 1981, tours of duty included SIMA 1008, PERSMOBTEAM 1208 (XO and CO), PERSMOBTEAM 3106 (CO), USCINCLANT 206 (navy operations officer), and FLTSUPTRAUNIT 1206 (CO). He retired in 1995.

Herring earned the Joint Service Commendation Medal, as well as qualifications as surface warfare officer, basic parachutist, EOD swimmer and assault craftmaster.

He is currently employed as the Assistant Director of the Division of Business and Economics at Winston-Salem State University. He resides in Greensboro, NC, with his wife Gail. They have two adult children, Aimee and David.

WILLIAM E. "BILL" HERRON, RADM, born in Philadelphia, PA in November 1945. He enlisted in the USNR in November 1965. He entered the Aviation Reserve Officer Candidate Program while attending La Salle University. Upon commissioning in 1967, he was assigned as a maintenance officer with VC-3 at NAS North Island, San Diego, CA. While assigned to VC-3, he became a qualified missile operations conductor at the Pacific Missile Range, Pt. Mugu NAS. He was subsequently assigned in 1969 as OinC of VC-3 DET, Yuma, AZ, where he ran the air-to-ground missile range.

Upon his release from active duty in 1970, he was assigned as a maintenance officer with the Aircraft Intermediate Maintenance Department at NAS Willow Grove. In 1976 he was transferred to VP-64 Maintenance Department where he deployed with the squadron to NS, Rota, Spain, Naval Air Facility Lajes, Azores, and to NAS Bermuda. In 1981 he was appointed as the maintenance officer for VP-64. In 1982 he was transferred to training officer to NADC Warminster 0193. In February, 1984 he was transferred as XO to NATC Patuxent River 0193 and became the CO in September 1989. In 1991 he transferred to NAF Washington and was selected as CO of NAVAIRSYSCOM 2366. In 1992 he became CO of NR-NADOC 0166. He assumed his next assignment as director, Naval Reserve Air Systems Program in November 1994.

He has been awarded the Navy Commendation Medal, Meritorious Unit Commendation Medal, Armed Forces Reserve Medal, National Defense Medal and the Reserve Overseas Service Medal.

In civilian life, as a CPA, he is a general partner in the international public accounting and consulting firm of Arthur Andersen & Co. specializing in the real estate and financial services industries. He is very active in civic and community affairs. He is past president of the Philadelphia Historic Preservation Corp., former treasurer of the Rudolphy Home for the Blind and the Willow Grove Historical Aircraft Association. He is currently on the Armed Services Committee of the Union League of Philadelphia. He is a Knight of Malta and a member of the Patron's Foundation. He is a life member of the NRA and the Association of Naval Aviation.

He resides in Bethesda, MD with his wife, the former Regina E. Moffatt of Phildelphia. They have five children: Jennifer, Timothy, Liam, Jessica and Rebecca.

JERRE F. HERSH JR., LTJG, born Dec. 19, 1924 in Kansas City, MO, received a BS at Colorado State and JD at University of Kansas.

He joined the service June 14, 1943. His military locations and stations: Cecil Field, FL and Olathe NAS, KS. Served with Sqdn. VF-883.

Hersh achieved the rank of LTJG and was discharged in 1953. He was awarded the Victory Medal, WWII Victory and American Campaign Medal.

He is married to Mary Louise Clough. He was a real estate attorney, retiring July 15, 1984.

ALBERT M. "AL" HERZEL, CWO 4, born April 15, 1916 in Phoenixville, PA, attended Norfolk Div. College of Wm. and Mary; Wharton Evening School, University of Pennsylvania (USN Sup School.)

He joined the USMC Nov. 26, 1935, CSK (AA); USNR Jan. 10, 1942. His military locations and stations: Paris Island, SC; Rec Sta Norfolk, VA; Athens GA, LST Flotilla; MAR Corps Schools DET; Disbursing R/S NOR/VA, GSK RS NORVA.

Herzel campaigns: securing Saipan. Established small craft pay office USS *Baham* (AG-71). He was awarded the Good Conduct USMC, USN, Atlantic, Pacific Theater, USNR. Achieved the rank of CWO 4, USNR and discharged June 16, 1947.

His memorable experiences: w/money bag climbing aboard to hold pay day in mess compartment. He is married to Virginia E. and employed Supply Dept. Naval Shipyard, USAECOM GS-12, retiring June 30, 1978.

WILLIAM CURTIS "CURT" HESS, born Nov. 24, 1948, Williamsport, PA and began his naval career as an enlisted naval reservist in 1967. In early 1968, he transferred to USN and was assigned to Recruit Training Center, San Diego, CA. After completing Interior Communications "A" School, he was assigned to the Repair Dept. of USS *Nereus* (AS-17). In June 1970, after completing nuclear power training, he was accepted into the Naval Enlisted Scientific Education Program (NESEP). In 1973 he graduated with highest honors from North Carolina State University receiving a ES in computer science. After attending Officer Candidate School in Newport, RI, he was commissioned an ENS in September 1973.

Following OCS, Hess served on *Nimitz* pre-commissioning crew, and deployed to the Mediterranean as additional duty on USS *Forrestal*. After commissioning, he served as CIC watch officer on USS *Nimitz* (CVAN-68) where he was the first officer to earn surface warfare officer (1110) designation. In 1977, he was assigned as DD-963 System Integration Team officer at Fleet Combat Direction Systems Support Activity, Damneck, VA.

After being designated an engineering duty officer in 1978, Hess attended Naval Postgraduate School, Monterey, CA, where he earned an MS-electronic engineering in digital systems. In 1980, his first ED assignment was to the staff of CDR, Naval Electronic Systems Command, Washington, DC, in the Software Engineering Support Office where he served as the micro-computer standards expert. He completed engineering duty qualification (1440) in 1981. In 1982, he was transferred to the Naval Sea Systems Command to the ED Plans and Policies Office (SEA OOZ). In 1984, he resigned from active duty to manage full time his own computer system integration business. He accepted a commission in the USNR, and joined his first unit in 1987.

As a Naval Reservist, Hess has served in the Office of Naval Research TAC 106 unit as training officer. After serving as XO and CO of SURFWEPCEN DET 106, he was selected to command ONR Science and Technology Unit 406. His next assignment as CO, NAVSEA PMS Aegis 106. Divorced, he has two children, Will and Kate. He resides in Raleigh, NC where he is employed in the Internet industry.

DEAN J. HEWITT, CAPT, a native of Milwaukee and attended University of Missouri, obtaining bachelor and master of journalism degrees.

He began naval career in 1949 as NROTC midshipman and was commissioned ENS in 1953.

He served in USS *Menard* (APA-201) 1953-55 as communications officer in Western Pacific and California. In 1955-56 he was photographic intelligence officer on commissioning staff, Fleet Intelligence Center Pacific, Pearl Harbor.

From 1957-84 he served in USNR commands in six states. He was CO of Surface Divisions in Appleton and Green Bay, WI; Area Command Staff, Southfield, MI; Office of Information Detachment, Oklahoma City, and was officer in charge, Naval Reserve Facility, Appleton. He was deputy inspector general, Readiness Command Thirteen, Great Lakes, and mobilization plans officer, Readiness Command Eleven, Dallas.

He received Korean Service, United Nations, National Defense, Armed Forces Reserve Medals and Vietnam Presidential Unit Citation.

His memorable experiences: participation in Operation Passage to Freedom August 1954 in which 293,000 Vietnam refugees sealifted from Haiphong to Saigon; photo-intelligence officer on commissioning staff of Flt Intel Center Pacific, HI; CO 4 Nav Res units, OinC NR Facility, Appleton, WI; Deputy Ins. General REDCOM 13, Mobilization Palns of REDCON 11.

He married Alice Arntzen and they have three children: Thomas, Jeffrey and Dana Jean; and grandchildren: Ann, Alexis, Kelly and Clare. Employed VP marketing, Purolator Products; Public Relations; director, Kimberly-Clark Corp, Bendix Corp. He was discharged May 24, 1956 and retired March 1, 1984.

NELSON O. HEYER, RADM, first flag officer to head Reserve Cryptologic Force of the USN. He was born in Rochester, NY, on Feb. 10, 1925. Joined the Navy V-12 Program, June 1943, at Union College where he was commissioned an ensign.

Served during WWII on USS *McCrea* (APA-149) (troop transport) and as assistant chief engineer, USS *Cadmus* (AR-14), a repair ship. Returned to inactive status and completed a BSS in electrical engineering, (1947) Union College and a MS in engineering (1951) Stevens Institute of Technology. Joined the Reserve Naval Security Group to use computer engineering experience with the IBM Corporation.

From 1947-83 commanded many Naval Security Group divisions and staff positions. Selected 1979 as first reserve flag officer to head the RESNAVSECGRU. Responsible for overall program direction, and coordination with the active force, NAVSECGRU Command, WA. Mobilization billet NSA Ft. Meade. Executive, IBM Corporation (1947-86) design engineer, manufacturing and personnel management, retired as director of Human Resource Planning. Established, HEYER Associates, human resource consulting firm as president. Fellow, Boston University Human Resources Policy Institute, Vice Commander Development, Naval Order of US. Life member NRA and CDR, Scarsdale American Legion.

Married LTJG Nora Frain USNR, formerly cryptanalyst assigned to NSG, 3801 Nebraska Avenue, during WWII. They have three daughters: Jane, Anne, Susan; one son, Thomas; and five grandchildren.

NORA J. FRAIN-HEYER, LTJG, born on Sept. 19, 1919 in Manhattan, NY. She enlisted in the USN Waves (W), USNR in March 1943 and after attending Yeoman School in Cedar Falls, IA was rated as yeoman petty officer third class and assigned to Communications Annex, 3801 Nebraska Ave (now Naval Security Group Command) as a crypt analyst assistant. After attaining the rating of Specialist Q, 1st class she was selected for OCS and was commissioned an ENS and reassigned to the Naval Communications Annex as a cryptanalyst. She was awarded a Unit Commendation from the Secretary of the Navy.

She believes the highest honor and privilege of her life was to serve during WWII in the "Greatest Navy in the World" and the continued affiliation with the Navy through the USNR and the Association.

Following return to inactive reserve status she attended the University of Chicago. Business assignments included administrative assistant at Huron Milling Headquarters in New York City. She married RADM Nelson O. Heyer and they have three daughters: Jane Price, Anne Dremluk, Susan Brunner; one son, Thomas Nelson Heyer; and five grandchildren: Matthew, Michael, Laura, Douglas and Taylor.

She is a life member and active in the NRA. She is active in the Naval Order of the United States, holding several commandery positions and Scarsdale Post 52, American Legion.

CURT W. "BILL" HIBBARD JR., CDR, a native of Jefferson, WI, after graduating from high school there, enlisted in the USN on Aug. 14, 1945, making him one of the youngest WWII veterans. About two hours after he enlisted, the Japanese surrendered to end the fighting.

After attending boot camp and receiving some electronic technician training, Hibbard was honorably discharged. Whereupon he signed up for the Naval Reserve program and eventually was promoted to ET3.

After earning a journalism degree from the University of Wisconsin-Madison, he received a LTJG commission with a 1635 designator. He became a member of the Milwaukee Naval Reserve Intelligence Unit 9-1-10, where he served in every billet, including officer-in-charge twice.

CDR Hibbard also served on the Milwaukee Naval Reserve Group 9-55 (L) Command Staff as liaison with the Intelligence, Security Group and Law units, and on the 9th Naval District Reserve Intelligence Div. 9-1 Command Staff as public information and publication officer.

He retired after 30 years of active and reserve participation. He is married to the former Edith Molinaro and they have two children, Curt W. III and Mary Ellen. All live in Milwaukee, where CDR Hibbard spent 38 years as a reporter, feature writer and editor at *The Milwaukee Journal*, receiving many regional and national awards for his work.

CDR Hibbard was travel editor for *The Journal* nearly 20 years, spent six years as auto editor and covered the 1964 Winter Olympics at Innsbruck, Austria. In spare time, he was the Milwaukee correspondent for the US Information Agency in the 1950s, taught writing in night school at Marquette University in the late 50s and early 60s, served as Midwest editor for *Ski Magazine* in the 70s and freelanced stories and photos to many other newspapers and magazines. In 1978 he served as national president of the Society of American Travel Writers.

Retired from the paper, Hibbard and his wife now write freelance travel articles, which he illustrates with his own photography. He is a life member of the NRA and of the Reserve Officers Association.

JOHN P. HICKEY, CAPT, born Feb. 7, 1930 in Holyoke, MA, received a BA, American International

College, 1951 and a MA from the University of Massachusetts in 1953.

He joined OCS November 1952. Graduated Navy Supply Corps School August 1953. His military locations and stations: USS *Bataan* (CVL-29); USS *Stockham* (DD-653). Served with USS *Franklin D. Roosevelt* (CVA); USS *Concord;* USS *Lindenwald.* His battles and campaigns: Korean War (3.5 years active duty); 30 years active Reserve. He was awarded the Defense Meritorious Service Medal, United Nations Service Medal.

Hickey achieved the rank of CAPT 06 and was discharged July 1, 1986. His memorable experiences: Armed Forces Medal, Korean Service Medal, National Defense Medal. Two weeks active duty on *Franklin D. Roosevelt* CVA under war time conditions, fully operational.

He married Constance M. "Connie" and their children are John P. II, Janellen, David D. and Annemarie; and grandchildren: Brian and Sara Calkins and Ashley Hickey. Employed as manager Cignare (group reinsurance), retired from Cigna Corp., May 29, 1992.

FRED M. HICKS, CDR, a member of the Navy's Merchant Marine Reserves, he received a BS in nautical science from the California Maritime Academy in 1973.

Since graduation, he has performed memorable ACDUTRA as a ship landing instructor, at Naval Amphibious School, Little Creek, VA and a convoy commodore of naval, civilian Maritime Administration vessels and Army rolling stock assets in JLOTS.

His civilian shipboard and shoreside employment covers position as an Alaskan and Marianas Islands (Saipan) harbor pilot, ships chief mate (XO), port captain damage control officer and tug master among others.

Married to Bridget June Riley (USN 1989-93), also a USCG licensed master (CAPT) the Hicks' enjoy traveling. Light aircraft flying and sports fishing when not working together as a "husband and wife" Master and Mate licensed officer team aboard merchant vessels. A son, Tyler Boulton Hicks, born Nov. 1, 1980, lives in California.

GERALD F. "JERRY" HICKS, CAPT, born May 15, 1940 in Fargo, ND, received a BBA from the University of Wisconsin (Madison). He joined the service June 4, 1962. His military locations and stations: MCAS Iwakuni; NAS North Island; NAS Corpus Christi; NAS Alameda. He served with VP-50, VT-27, Reserves: VS-82, VAQ-308.

Hicks participated in Vietnam, Reserve Readiness Test USS *Ranger* (1977). CO of VAK-308 at NAS Alameda. He achieved the rank of CAPT and retired January 1984.

His memorable experiences: operation in Cam Rahn Bay, Vietnam, before opening of Cam Rahn AB; being only Navy Sqdn. at Marine Air Station Iwakuni, Japan; splashes and dashes in P-5M; his first night operations on USS *Ranger* in 1977.

He married Kathleen Morey and they have two children, Brian and Kristen. Employed as CAPT, Delta Airlines; president, Innovations Inc., retiring January 1984.

REBECCA "BECKY" HIGHFIELD-MINTON, *See M Section.*

DAVID E. "DAVE" HILL, CWO-4, born Nov. 24, 1933 in Monmouth, OR, attended one year of college.

He joined the service Jan. 8, 1952. His military locations and stations: Boot camp NTC San Diego; School Command T.I. San Francisco. Served with USS PCE (C) 896; USS *Seminole* (AKA-104) numerous reserve billets.

Hill served in Korea and was awarded the UN Service Medal, Korean Medal, National Defense Medal, Good Conduct Medal, Reserve Meritorious Service, Armed Forces Reserve, Expert Rifle and Expert Pistol. He achieved the rank of PO 1 in active duty and CWO 4 in Reserves. Retired from active duty April 6, 1956 and USNR, Dec. 1, 1990. Retired with pay from USNR Nov. 30, 1993.

He married Jean and they have three children: Jeanette Bernards, Kathy Olsen and David A. Hill; and five grandchildren: Tricia, Marc, Curt, Jenna Bernards and Kassy Olsen. Employed as insurance investigator, retiring Nov. 30, 1993.

RALPH A. HILL, CWO4, born in St. Joseph, MO, July 26, 1934. Graduate of Northwest Missouri State University. Joined the USN, Feb. 4, 1952 at St. Joseph. Went to boot camp at Great Lakes, IL. After boot camp reported on board the USS *Coral Sea* until 1956. Discharged in 1956 and served in USNR with units NRSD-9-180 St. Joseph, MO and USS *Ajax* DET 218 Kansas City, MO until September, 1986.

Recalled to active duty September 1986 to February 1992. Served on active duty at NAS North Island, San Diego, CA as an officer recruiter.

His memorable experiences include his retirement from teaching in Missouri and being recalled to active duty September 1986. He married Carol Jean Koelling. They have a son, daughter, and three grandchildren.

Retired from USNR July 26, 1994, and presently living in Gulf Breeze, FL.

CHESTER I. "CHET" HILLS, CDR, born June 4, 1934 in Chicago, IL, received a BA from St. Olaf College, 1956 and MS Chicago State University, 1969.

He joined the service August 1958 and commissioned August 1958 (OCS). His military locations and stations: USS *Chewaucan* (AO6-50), 1LT, DCA, ENG, XO. Served with NRSD 9-5L; NRSD 9-4M XO, CO.

Participated in Armed Forces Exp. to Lebanon, 1958. He was awarded the Armed Forces Expeditionary Medal, National Defense, Naval Reserve Medal. He achieved the rank of CDR and retired USN July 1985.

His memorable experiences: SOLANT Amity cruise to Africa March-August 1961; CO NRSD 9-4M.

Hills married Sandra and they have a daughter Sarah and granddaughters, Ashley and Stephanie. He retired from public school education in June 1994.

THOMAS G. "TOM" HILT, CAPT, born Aug. 11, 1947 in Newport, RI, received a BA in psychology, Stanford University, CA, 1969; MA, economics, San Jose State University, CA, 1979.

He joined AOCS, June 1969, Pensacola, FL. Commissioned Oct. 17, 1969, NAS Pensacola, 1635 (Intel) Designator. Reserves: February 1977, RIA-2, NAS Alameda, CA.

His military locations and stations: AOCS and VT-10, NAS Pensacola, FL, 1969-70; Armed Forces Air Intelligence Training Center, Lowry AFB, CO, 1970; VF-151, NAS Miramar, CA and USS *Midway* (CVA-41), 1970-73; Naval Investigative Service, MCAS Iwakuni, Japan, 1973-76.

Primary Reserve Units served with: Navy Field Operational Intelligence Office 0305, Treasure Island, CA; CNO Intelligence Plot 0166, Pentagon; DIA Current Intelligence 0166, Pentagon; Naval Intelligence Command 0166, Suitland, MD (XO); NISRO 1066, Washington Navy Yard (CO); ONI 0466, Suitland, MD (CO); Intelligence VTU 0105, NAF Washington, DC (CO).

Hilt served in Vietnam from May 1971-March 1973. He was awarded the Navy Commendation Medal and Navy Achievement Medal. He achieved the rank of CAPT and is still presently active in the USNR.

He is married to Juliana "Julie" and is employed CIA, Directorate of Intelligence.

CLIFFORD G. HINES JR., LT, born May 29, 1924 in St. Thomas, US Virgin Islands. Enlisted Oct. 24, 1942, Philadelphia, PA NROTC University of Pennsylvania. Commissioned ENS Jan. 24, 1945.

Served in USS *Azimech* (AK-124) Asiatic-Pacific Theater Campaign; to Baltimore, MD to decommission Dec. 11, 1945.

To division officer USS *Navasota* (AO-106) Sun Shipyard, Camden, NJ, Feb. 27, 1945. Operations as fleet oiler in Atlantic.

Released from active duty May 24, 1946. Campaign and service medals: American and Asiatic Pacific Theater, WWII Victory Campaign and Service Medal, Naval Reserve and Armed Forces Reserve, Expert Rifleman.

Hines graduated with bachelor of architecture, University of Pennsylvania, June 1947. Member of Naval Reserve Composite Co. 5-36 (Washington Navy Yard), July 27, 1951, administrative officer July 1, 1963, retiring June 30, 1966.

NRA member since 1954. Plankowner Wardroom Chapter (Washington Navy Yard), now Northern Virginia Matthew Maury Chapter.

He has been an architect in private practice 1956-96. He is a member of American Institute of Architects, various architect, building and construction boards and panels; member American Arbitration Association. Registered in DC, MD and VA. Family includes his wife Pat; sons, Mark and Christopher; daughter Nancy; five grandsons; and one granddaughter.

Memorable experiences: traveling half the globe, duty watch officer, navigation, gunnery, 1LT etc. before 22nd birthday.

HUBERT O."BUD" HINES, CDR, born Aug. 10, 1942 in Phoenix, AZ, received a MA and BA in sociology from San Diego State University. He joined the service July 31, 1960 (as E-1). Promoted to SK1, finished college and received a direct commission as ENS 1635.

His military locations and stations: USS *Sperry* (AS-12); USS *Lynn McCormick* (DDG-8); USS *Shields* (DD-596); HS-84; USS *Matthew* (AKA-96); USS *Vamen* (DE-644); USS *Marsh* (DE-699); USS *Sommers* (DDG-34); USS *Samuel Gompers* (AD-37). Units served with: NRSD 11-35, NRSD 11-41, NRI 11-1-2, DEF Interro 0-0166, DEF Attache 0166, CNO Intel PCOT-0166, IVTU-0106.

Hines was awarded the National Defense, Armed Forces Expeditionary, Armed Forces Reserve, RVN Campaign, Expert Pistol, Navy Achievement, Naval Reserve, Meritorious Unit, Joint Meritorious Unit, Overseas Service. He achieved the rank of CDR and was released from active duty June 18, 1965. Retirement date Oct. 17, 2000.

Memorable experiences include first duty aboard ship; being petty officer in charge of Color Guard and raising flag at Morning Colors and lowering the flag at Evening Colors; and his last duty aboard ship.

He is married to Hsio-Ying Hines and employed with IRS as computer specialist, union chapter president, and revenue officer. He plans to retire from civilian position in 1997.

DAVID O. HINTON, CAPT, from Gibsonville, NC, joined the USN in 1960 as an aviation electronic technician. He attended North Carolina State University under NESEP, earning a BS degree in electrical engineering.

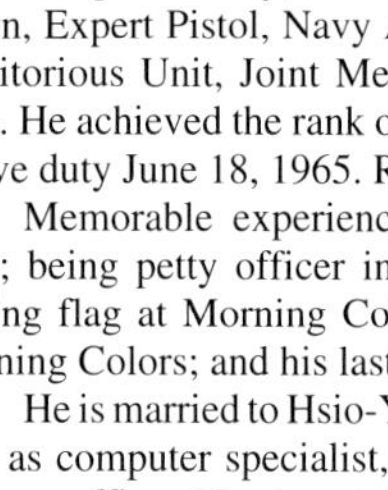
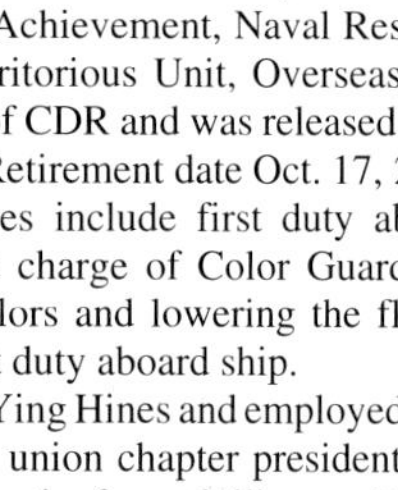

Receiving a commission in 1965, Hinton became the electronics Division officer aboard USS *Sampson* (DDG-10). In 1968 he completed Naval Flight School.

As a tactical coordinator, he was assigned to ASW

Sqdn., VP-23, Brunswick, ME. His next assignment was maintenance officer for the Naval Research Laboratory's Flight Support Det., Patuxent River, MD. In 1977 CAPT Hinton transferred to the US Public Health Service as an engineering officer assigned to EPA's National Environmental Research Center, Research Triangle Park, NC.

He served as project officer, quality assurance officer and deputy director, Human Exposure Research Division. CAPT Hinton retired in 1995 as director, Atmospheric Measurements Research Division. CAPT Hinton retired in 1995 as director, Atmospheric Measurements Research Division.

His awards include the Navy Enlisted Good Conduct Medal, USPHS Commendation Medal, EPA Bronze Medal and Naval Reserve Medal. He is a life member of NRA.

The Hinton family consists of wife Marie and son David Scott. They reside in Clayton, NC.

KENNETH G. "HIP" HIPPE, CAPT, born Jan. 2, 1918 in Burlington, IA, received a BS from Bradley University.

He joined the service Dec. 6, 1939. His military locations and stations: USS *Geo. F. Elliott,* Pensacola, FL. He served two years as primary instructor, VC-3 and VF-66. His battles and campaigns: Saipan, Palau and Leyte Gulf Battle.

Hippe received the Navy Cross and Distinguished Flying Cross. He achieved the rank of CAPT and was discharged Sept. 7, 1945. Most memorable experience was shooting down five Japanese Lillies in about 10 minutes. It was the most exciting day of his life.

He is married to Nell Louise and they have four children: Jane Reilly, Barbara Mansfield, Pattie White and Pam Henderson; and seven grandchildren: Jennifer, Lindsay, Jody, Morgan, Kate, Clayton and Lil. He was employed for 32 years as a TWA pilot, retiring Jan. 2, 1978.

EDWARD J. "ED" HLOPAK, CAPT, a New Jersey native, graduated from the US Naval Academy in 1965. Upon commissioning, he completed submarine school, nuclear power school and nuclear power prototype training before reporting to the USS *Henry L. Stimson* (SSBN-655) (Blue) in April 1967. He made six FBM deterrent patrols and was designated "Qualified in Submarines."

He left active service in 1970, accepted a commission in the USNR, and then served in: NRSSD 1-11; NSSF 101; NSSF NLON DET 301; Readiness Command Region One and Ten Staffs; VTU G 1007; NRMTF HQ 110; and the DSW N87 unit at the Pentagon. He also commanded NSSF NLON DET 210, AS37 Dixon DET 3710, CSS 1 DET 110, Readiness Unit "Delta" Houston, and Readiness Unit "Alpha" San Antonio, and was a Navy Blue and Gold Officer for 10 years during his 25 years in the USNR from which he retired on June 30, 1995.

He married the former Dorothy Marie Atkinson of Gloucester, NJ and they currently reside in Houston, TX. They have two children: Edward II, a West Point graduate and David, an alumnus of Texas Tech University, a daughter-in-law, Holly, and a granddaughter, Hannah.

His civilian experience includes more than 19 years, and positions, in engineering, business development and management at two major architect/engineer/constructors, Stone & Webster Engineering Corp. and Ebasco Services Incorporated and for four plus years with Destec Energy, Inc., a leading independent power project developer.

He is currently president of Visitech, Inc., a small company which designs and manufactures high technology electronic products for the traffic industry. He has an MBA from Babson College and is a registered professional engineer.

FRANK M. HOAK III, a native of San Pedro, CA, has been CO of Fleet Intelligence Rapid Support Team 0593, NAS Willow Grove, since October 1980.

He entered the USNR as a Seaman Recruit in 1955 at Surface Div. 12-35 San Jose, CA. After graduating from San Jose State University in 1959 with a major in journalism, he was commissioned at OCS, Newport, RI in April 1960. Upon completing the Naval Intelligence School, Washington, DC in 1961, he reported to Carrier Air Gp. TWENTY-ONE (CVG-21) aboard the USS *Lexington* (CVA-16) as assistant air intelligence officer. In 1962 he transferred to Attack Sqdn. 216 as air intelligence officer and made two deployments to Southeast Asia aboard the USS *Hancock* (CVA-19).

After separation from active duty in 1963 Hoak affiliated with Naval Air Intelligence Reserve Unit G-1 (NAIRU G-1) at NAS Alameda, CA. In subsequent years he was affiliated with NAIRU H-1 at NAS North Island, CA; NAIRU X-1 at NAS New Orleans, LA and NAIRU N-1 (later area analysis Unit 404 and FIRSTEURLANT 0575) at NAS Lakehurst, New Jersey.

Recalled to active duty for Operation Homecoming in 1973, he was assigned to Naval Hospital Portsmouth, VA assisting with the debriefing of returning US POWs from North Vietnam.

In October 1978, Hoak was named XO of Naval Investigation Service Office Unit 0893 at NAS Willow Grove; in October 1979, XO of Fleet Intelligence Rapid Support Team 0593 at NAS Willow Grove and CO, Fleet Intelligence Rapid Support Team 0593 in October 1980. Under his command, the unit received numerous letters of commendation from its Gaining Command, Fleet Intelligence Center, Europe and Atlantic in Norfolk, VA, for Mobilization Readiness excellence.

ANDREW HOBOKAN, CAPT, born in Phillipsburg, NJ, Jan. 25, 1923. Enlisted in the USN in 1941, and attached to USS *Matagorda* (AVP-22) on Dec. 8, 1941. Convoyed across the North Atlantic and conducted searches for German submarines in the North and South Atlantic, becoming a chief radioman in 1943. Later assigned to NRL, COMDEVFORLANT aboard USS *Cadmus* (AR-13). Employed at the Naval Air Test Center in 1947; managed many flight projects on airborne radar, navigation, ELINT, and countermeasures equipment. Graduated with a BSEE from the University of Maryland and received a direct commission to LTJG in the unrestricted line.

Joined the Space Task Force in 1960 at McDonnell Aircraft; managed major segments of Mercury and Gemini spacecraft programs. He was NASA launch preparation manager for Gemini mission GT-4; became NASA resident manager at McDonnell for Apollo Soyuz and Skylab spacecraft components.

He became NAS resident manager at Grumman Corporation in 1967 for the Apollo Lunar Modules for landing men on the moon and launching them for return to Earth. After the successful moon landings and returns, became manager, manufacturing and test, for the Space Shutt Orbiter. He was appointed a charter member of the Senior Executive Service by President Carter. Retired in 1979 and became a consultant to McDonnell Douglas on the Space Station Program.

During his military career, he served on numerous ships as senior watch officer underway, as administration officer, XO of many units, and held two reserve commands ashore. Held all chairs including chapter and department president in ROA.

His memorable experiences: convoying across North Atlantic in winter at 10 knots; being high over PaxRiv chasing target planes; being president manager for "Eagle" "Aquarius" and all other Lunar modules and a charter member of the Senior Executive Service.

CAPT Hobokan was discharged Oct. 10, 1947. He received the Navy Good Conduct, American Theater, European African Theater, Victory Medal and the AF Reserve Medal.

Hobokan married Antoinette Contrada and they have four married daughters: Anne Roy, Ellen Hill, Nance Allen and Margaret Wallace; and two grandchildren, Rebecca Jean Hill and Catherine Elizabeth Wallace.

ROBERT L. HOBSON, CAPT, a native of Crown Point, IN, graduated from Wabash College in June 1957. He entered OCS at Newport and was commissioned an ENS in March 1958.

After attending Fleet Sonar School, Key West, he was assigned to the US Naval Facility Coos Bay, OR, and then to the US Naval Facility Big Sur, CA. He was reassigned to Coos Bay as the CO before being released from active duty.

Hobson graduated from the University of California, Hastings College of the Law in June 1964. During that time and thereafter he affiliated with several USNR units in the San Francisco Bay area, serving as CO of seven reserve units and as assistant chief of staff for the Naval Reserve Readiness Command San Francisco and in other assignments with the Readiness Command Region 20. He was assigned for three years to the National Naval Reserve Policy Board by the Secretary of the Navy.

He currently practices law in San Francisco in tax and estate planning. His family consists of wife Mary, a school counselor; and two sons, Steve and Mike, both graduates from the University of California.

FRED H. "HANK" HODELLA, CDR, born March 14, 1919 in Yonkers, NY, received a BBA from the College of William and Mary, Norfolk, VA.

He joined the service July 17, 1942. His military locations and stations: Newport, RI; NAS Sand Point, Seattle. He served with USS *Tripoli* (CVE-64). His battle and campaigns: ASW, North, South, Atlantic, AFRO-European.

Hodella received the Good Conduct, American Theater, AFRO-European Area and Victory Medal. He achieved the rank of CDR and was discharged March 14, 1979.

His memorable experiences: submarine detection; escort Allied ships through German submarine area.

He married Laura Starrfield; she is now deceased. Employed as equipment specialist GS-11. He retired July 3, 1972.

GEORGE LOWRANCE HODGE, CAPT, born in Waldo, AR, Nov. 11, 1934. After two years of college he entered flight training in Pensacola as a NAVCAD. He was commissioned and designated as a naval aviator at NAS Corpus Christi in 1956. He served his active duty as a pilot in Patrol Sqdn. (VP-46) at NAS North Island with overseas deployment to the Far East.

Upon release from active duty in 1959 he affiliated with the Naval Air Reserve and entered the University of Arkansas where he graduated with a BS degree in industrial engineering. After two years in industry as a practicing engineer he returned to the campus at Oklahoma State University and earned a master of science degree in industrial engineering/management.

From 1961-66 he served as naval aviator in several reserve ASW (VP) and Transport (VR) squadrons in NAS Memphis, NAS New Orleans and NAS Dallas. In 1967-68 he was an instructor in the Naval Reserve Officer's School at NTC Baton Rouge, LA. From 1969-75 he held command positions in several training units at NAS Dallas, TX, including CO of NARDIV D-1 and NAS 5111.

In 1977 he was selected for CAPT and completed his 32 years of service in Volunteer Training Units with special assignments to the CO, NAS Dallas. He retired in 1987.

In civilian life he was a director in management services for Arthur Young & Company and later the senior vice-president of manufacturing operations and engineering for May Kay, Inc.

CAPT Hodge's family consists of his wife of 40 years, Marilyn; daughters: Kathy, Julie and Rosemary; one son, Jerry, who graduated from the Naval Academy in the Class of 1986; and five grandchildren: James, Brandon, Emma, Jared and Carolyn. They all reside in Dallas area. He enjoys reading, fishing and traveling.

NORMAN "NORM" HODGES, CAPT, born Aug. 5, 1925 in Silver City, NM, received a BA, BS, LLB, JD from the University of New Mexico.

He joined the service Dec. 7, 1942. His military locations and stations: Philippine Islands; Tokyo Bay; Shanghai, China. Served in USS *Mt. Olympus* (AGC-8). His battles and campaigns: Philippine Islands (Leyte, Manila).

Awarded the Philippine Liberation Medal, Japan, China Occupation, Pacific. He achieved the rank of CAPT, USNR (R) and retired Aug. 5, 1985.

His memorable experiences: On *Mt. Olympus* when it collided with a US oil tanker between Guam and Leyte, Philippine Islands; on *Mt. Olympus* at Tokyo Bay, Sept. 2, 1945 for Japanese surrender.

He is married to Tress and has a stepson William Weiland. Employed as lawyer, district attorney and district judge, retiring Jan. 1, 1987.

A. ALAN "AL" HOELTING, CAPT, born April 20, 1919 in Roundup, MT, received a BA from the University of Washington (plus P.G.)

Enlisted August 1940 as apprentice seaman on USS *Tuscaloosa* in Atlantic, then to Midshipman School, Northwestern University. Commissioned ENS 1941. Served with USS *Tuscana* (AKN-3); USS *Holly* (AN-19) as CO.

Hoelting participated in neutrality patrol Atlantic; Eniwetok; Kwajalein; Saipan; Iwo Jima (Pacific). He received only the usual area ribbons plus one Commendation for Iwo Jima.

He achieved the rank of LT, USNR (WWII); CAPT USNR (Ready Reserve), discharged 1945 and 1952.

His memorable experiences: recalled to BUPERS, Washington, DC (Korean War); served as mobilization planning officer for officer and enlisted training (Navy); traveling to all naval districts to organize (plan) training facilities or universities and trade schools.

He has been married 52 years to Patricia and they have two daughters, Leslie and Dana; sons, Kimball and Kurt; and nine grandchildren. Employed as Marine Ins. underwriter and real estate developer, marketer, retiring in 1982.

JOHN PAUL HOFER, LCDR, born in Chicago, graduating from St. Rita High School in 1969, volunteered for the draft in 1970, served in the US Army with Co. B, 5/46, 198th Brigade of the Americal Div. as infantry combat medic earning the Army Combat Medical Badge west of Chu Lai in Ky Tra, and Dragon Valley in I Corp.

Upon honorable discharge, graduated from University of Illinois, Champaign-Urbana. Obtained the rank of staff sergeant in the US Army Reserves and while attending dental school at Southern Illinois University he was gratefully shanghaied by an ex-Marine, LCDR Jim Parkhill into the USNR serving as DT-1 until graduation, and accepting a direct commission into the Dental Corps he has had the assignments with NDCL NORVA 218, FLT HOSP COMMZ HQ 13, HQ 14 DENCO, 4th FSSG, 4th MARDIV, NMCB-20 and currently with NMCB-15. He has had memorable duties normally not assigned to a dental officer such as H Co. commander with NMCB-20 while on MILEX 94 at Camp Lejeune, NC.

Awards include the Bronze Star Medal, Navy Achievement Medal and Seabee Combat Warfare Specialist designator. He is a staff dentist at VA Hospital Chillicothe, OH, serves as dnetal implant coordinator placing and restoring implants and is clinical instructor at Ohio State University's Dental School. Active with veterans organizations and the father of nine year old triplets, but he still finds time to help coach baseball and basketball in Grove City, OH.

EDDIE W. HOGAN, CAPT, born April 11, 1915 in Waco, TX, received a BS from Texas A&M in marketing.

He joined the service September 1941. His military locations and stations: USA, European, Asia Pacific. Served with minesweeper, East Coast, Japan.

Awards and medals: before Pearl Harbor, Europe-Asia, Pacific. He achieved the rank of CAPT and retired April 11, 1975, USNR-R.

He married Barbara Ann Roddey; she is now deceased. He has four children: Walter L., Roddey D., Eddie M. and Suzie; and five grandchildren: Eddie, Jimmie, Randy, Matthew and Camaron L. Employed with the state of Texas.

RAYMOND L. "RED" HOGDEN, CAPT, born March 7, 1917 in Ettrick, WI, received a BA in 1939 from Luther College, Decorah, IA.

He joined the service Nov. 27, 1940. His military locations and stations: Corpus Christi, VC-75, Commander AF, Pacific Fleet. Served with VC-75. His battles and campaigns: Palau, Leyte, Battle of Leyte Gulf, Mindoro and Lingayen Gulf.

Hodgen was awarded the Navy Cross and Air Medal. He achieved the rank of CAPT and discharged June 30, 1973. His memorable experiences: sinking a Japanese heavy cruiser Oct. 25, 1944; survived sinking of CVE-79 USS *Ommaney Bay,* Jan. 4, 1945; CDR Naval Air Reserve Staff E1 July 1, 1967-June 30, 1969.

His wife Eleanor is deceased. He has three children: Anola Leisso, Marc and Linda Hogden; and two grandchildren, Jeffrey Urbanek and William Hogden. Employed as general contractor, partner and owner, retiring Aug. 31, 1977.

EIKE H. HOHENDAHL, CDR, born in Garmisch, Germany, Aug. 20, 1941, he immigrated to the USA, Nov. 16, 1956. He enlisted in the USN, Sept. 14, 1960. After recruit training he was assigned to USS *Esteem* (MSO-438) deploying twice to the western Pacific, and participating in operation Market "Time" as the first major naval unit to take part in the Vietnam conflict.

He left active duty on Sept. 15, 1964, and immediately affiliated with the USNR, advancing to chief within the next five years, serving as Command Chief of NRSD 11-21(L) for 105 enlisted personnel and 12 officers. He enrolled in college under the newly established GI bill in the fall of 1966.

Upon graduation with a degree in physics in 1970, he received a direct appointment as ENS (special duty-intelligence), advancing to CDR before retiring Oct. 1, 1995. In January 1988 he accepted a one year appointment as an exchange scientist with the Australian Navy, after having been selected by the office of The Chief of Naval Operations through his employment within the Naval Sea Systems Command.

During his career, Hohenadl served on nine ships including USS *Esteem* (MSO-438); USS *Kitty Hawk* (CV-63); USS *Halsey Powell* (DD-686); USS *Vammen* (DE-699), USS *Whetstone* (LSD-27); USS *Shields* (DD-596); USS *Mount Whitney* (tone-20), two tours; USS *Bealleau Woods* (LHA-3) and USS *Peleliu* (LHA-5). He participated in 10 fleet exercises, including exercise "Northern Wedding" 1979 - the largest exercise ever conducted by NATO. Served with NRSD 11-21(L); NRIU 11-1-3; AAU-2519; FICPAC 0915; FICPAC 0919; PHIBGRU-TWO; PHIBGRU-3; IVTU-0121.

At age 28 he became one of the youngest CPO in the peacetime USNR, and at age 50, he was the oldest sailor ever to graduate from Ft. Benning Army Airborne School. At age 51 he qualifed for and was awarded the USN/USMC Gold Jump Wings.

His decorations include Vietnam Service, Coast Guard Achievement, Armed Forces Expeditionary, Navy Meritorious Service, Naval Reserve Meritorious Service, Republic of South Vietnam Service, Armed Forces Reserve (three awards), National Defense Service (two awards), Expert Pistol and Expert Rifle Shot Medals, Sea Service and Overseas Service Ribbons, US Army Parachute Wings, US Navy/USMC Parachute Insignia.

Memorable experiences: part of first major naval unit in Vietnam, establishing "Operation Market Time," Saigon River transit; participated in NATO history's largest operation: Exercise Northern Wedding 1978; damage control assistant USS *New Jersey* (BB-62); Translator/analyst-naval liaison unit Munich.

He was married to the former Joyce Christoff on June 6, 1994, and they have one son, Eike Allan, born on June 2, 1966. Joyce has been a strong supporter of his naval career, also a Naval Reservist she is the Command Chief of PERSMOBTEAM 0219. Together they are sometimes called the "Commander and Chief".

He is an avid skydiver with 1,750 jumps (as of Jan. 12, 1997), and a runner, having completed the 1994, and 1995 Los Angeles marathons. In his retirement he plans to spend more time on the drop-zone, stay physically active and travel with his wife Joyce. He is grateful for having had the opportunity to serve with the finest men and women in the finest Navy in the world.

WADE G. HOLCOMB, LCDR, enlisted the USNR in 1938, as a RM3/c (age 17). In 1940 he was ordered to the USS *Crosby* (DD-164) as a RM2/c. He was duty radioman on the *Crosby,* tied up at the DESABASE San Diego, on Sunday Dec. 7, 1941 when he copied the message on the Fox Schedule Frequency "Japs Attack Pearl Harbor X This Is No Drill," an experience he says he will never forget!

He was transferred to Naval Research Laboratory Electronics' Material School, Washington, DC in December 1943. Following graduation he served as an instructor in ET schools in Illinois, California, and Michigan. He was promoted to ENS USNR in May 1945. Following his release he attended CalTech and UCLA (1945-51).

Recalled to active duty in 1951 and assigned to the Officer's Electronics School, TI. He served as OinC, Officer's Electronic School in 1953. Returning to civilian life in 1954 he performed engineering work in underwater acoustics at Bell Telephone Labs, NJ (1954-59). From 1959-62 he worked for Airborne Instruments as a team leader on antenna pattern measurements of the Nike Zeus ABM systems at White Sands, NM and Kwajalein, MI. From (1963-66) he helped design an instrument to read microscope slides (CYDAC) into a computer memory at the University of Pennsylvania. He went to Yale School of Medicine in 1966 (faculty appointment) to support a RF cardiac pacemaker circuit development. In 1982 he joined Yale's first Magnetic Resonance Instrument (MRI) group as an RF engineer.

Retired from Yale in 1988. Eight of his almost 43 years in the Naval Reserve was spent on active duty, in electronics' or communications, he transferred to the Retired Reserve in 1981. His decorations included Good Conduct, Asiatic-Pacific, American Theater, American Defense w/star) and the National Service Defense Medal.

Holcomb is a life member of The Institute of Electrical and Electronics Engineers, NRA, Reserve Officers Association and US Naval Institute. He is also a member of Sigma Xi, Four Stack APD Veterans, Quarter Century Wireless Association and American Radio Relay League. he has been a radio amateur since 1938 (W1GHU).

Married Mary "Molly" Louise Boynton in January 1945 and they have three children: Anne, Julia and Derek; and four grandchildren: John, Victori and Elizabeth Jarvis in Fairfax, VA and Sarah Paradise in Griffith, IN.

His father (ENS Wade A. Holcomb) served in the USN from 1915 to 1921 and survived the submarine sinking of the USS *President Lincoln* in May 1918 in the North Atlantic.

WILLIAM H. HOLDEN JR., CAPT, is a native of Chicago, IL. While in high school he enlisted in the USNR, affiliating with a submarine division.

He received an appointment as a midshipman in the NROTC program at the University of Illinois, where he received a BS degree in journalism and was commissioned ENS, USN.

Reporting to USS *Norfolk* (DL-1), he served as 1DIV officer, torpedo officer, assistant ASW officer and assistant CIC officer. His next assignment was to USS *Merrick* (AKA-97) as 1LT. In 1960, he reported to the Naval Preparatory School, Bainbridge, MD as assistant academic director and assistant OinC.

He attended the University of Maryland, receiving a JD degree, with honors, in 1965, and subsequently engaged in the practice of law in Baltimore and Bethesda, MD.

From 1965 until his retirement in 1982, he served in NR Surface Div. 5-2 (L), where he was appointed CO, and Navy Command Center 106 in the Pentagon, where he was assistant navy department duty captain.

MELANIE A. HOLMES, CAPT, born May 31, 1922 in Aspinwall, PA. Unable to afford college after graduation from high school, she went to work as a stenographer. Attracted by the GI Bill promise of college, she enlisted in the Navy WAVES in August 1944.

Basic training was at Hunter College, NY, followed by Yeoman School, Cedar Falls, IA. Assigned to the Pilotless A/C Div. of BuAer, she was introduced to the fascinating world of science. She was yeoman to the board which selected Pt. Mugu as the Naval Missile Test Center. She recorded interviews with German missile scientists after WWII. Discharged from active duty, September 1946, she joined the USNR. She was inactive in the Reserves until while studying for an advanced degree in physics, she was commissioned LTJG.

From June 1950 to February 1957 while employed as an assistant research physicist at Carnegie Mellon Univeristy, she designed, executed and analyzed shaped charge firing using a Kerr cell and an exploding wire light source to provide nanosecond photography.

In 1956 she wrote an article published in the *Pittsburgh Press* and quoted in papers across the country refuting a UP article, *Unready Ready Reserves.*

In February 1957 she joined Rand Corp., an AF sponsored "think tank," to assist in building the world's first large scale real-time computer system for air traffic control.

Holmes considers herself very fortunate to have programmed the world's first large scale computer and designed and implemented computer programs for the AF/CIA first reconnaissance satellite in 1971. She continued as a project manager developing computer programs for reconnaissance satellites until her retirement in 1989.

As CO of RSAND U-1, she was the first woman to command a Reserve Unit at NAS, Point Mugu, CA.

ROBERT W.P. HOLSTROM, CAPT, was commissioned an ENS, USNR July 9, 1945 at the USN Midshipmen's School, Notre Dame University. He served briefly aboard ship during the latter months of WWII. In 1952, he graduated from law school with a JD degree and was later admitted to practice in California and Florida, as well as before several administrative tribunals and the Court of Military Appeals. Soon thereafter, he was recalled to active duty and reported to the Air Intelligence School, Naval Observatory, Washington, DC.

In April 1952, he was assigned to Fighter Sqdn. 12 and was on board the USS *Wasp* when the carrier was involved in a collision at sea with the destroyer *Hobson.* His next assignment was in the Pentagon as a legislative and legal liaison officer in the office of the chief of naval operations.

Then in February 1956 he became an analyst at the Fleet Intelligence Center, Port Lyautey, Morocco. Upon returning to inactive duty in 1958, he joined the Active Reserve Program at NAS, Alameda and later commanded Naval Air Intelligence Reserve Unit 871. During his command, the unit won the prestigious Noel Davis Trophy. He was promoted to the rank of CAPT in 1969. Holstrom was soon after selected to command the Fleet Management Assistance Unit 11-2, Treasure Island, under the supervision of the inspector general, CDR, Pacific Fleet.

In the early 1970s, was president of the Nimitz Chapter of the NRA, at which time the Chapter received top honors as the most outstanding in the nation. He was also a member of The Navy League and the Naval Order in San Francisco.

In civilian life, as an attorney, he is active in trust and estate planning, and involved in various investment areas, as well as, president of an investment advisory firm in San Francisco. In addition, he is a member of Rotary International in San Francisco and various other civic and professional organizations.

GEORGE W. HOLYFIELD, CAPT, a native of Kansas City, KS, currently residing in Houston, TX, attended the Kansas City, KS Junior College and the University of Kansas. Appointed a midshipman in the Naval Reserve Training Corps, he graduated in 1955 with a degree in Civil Engineering and a Naval Commission.

Following flight training at NAS Pensacola, FL and NAS Hutchinson, KS, he was designated a naval aviator in 1956. Holyfield was ordered to AEW Sqdn. 14 at NAS Barbers Point, HI, flying EC121 aircraft.

Released from active duty in 1958, he joined Reserve Transport Sqdn. 883 at NAS Olathe, KS. He was designated as Transport Aircraft CDR in both C-54 and C-118 aircraft at that station.

Employed by Chevron, USA, Holyfied was transferred to Dallas, TX in 1967 where he joined Transport Sqdn. 704 at NAS Dallas. He served as XO and CO of several flying units at NAS Dallas and in 1976 assumed command of Fleet Logistic Support Sqdn. 53 at NAS Memphis and NAS Dallas. During his tenure as CO, VR-53 was awarded the Noel David Trophy for excellence. He has been awarded the National Defense, Armed Forces Reserve, Vietnam Campaign, Navy Commendation and Meritorious Service Medals.

A life member of the NRA, he has served two terms as chapter president, Space Center Chapter. He was a member of the 34th National Conference Committee and has served on the National Advisory Committee. Holyfield served as president, NRA Eight District in 1990 and 1991.

JAMES A. "JIM" HONER, CDR, born Aug. 13, 1930 in Richland Center, WI, received a BS in agricultural education, University of Wisconsin-Platteville, WI.

He joined the service July 29, 1952. Commissioned Dec. 21, 1952. His military locations: 1953, USS *Franklin D. Roosevelt* (CVA-42); 1954-55 USS *Altair* (AKS-32).

He was awarded the National Defense Service Ribbon, and Navy Occupation Medal (European). He achieved the rank of CDR and was discharged Aug. 13, 1990.

His memorable experiences: 6th Fleet carrier duty. Underway replenishment operations with the Sixth Fleet in the Mediterranean.

He is married to Virginia and they have children: Mark A. Honer and Mari E. (Honer) Werther. He has a grandchild Chase Dagny Werther. Employed as worked for the USDA-Farmers Home Administration, as rural housing loan chief. Retired April 3, 1986. Presently a WI certified general appraiser. Active in the Madison Chapter of the National Association of Retired Federal Employees (NARFE) and a National Council member of the Org. of Prof. Employees of USDA - (OPEDA).

LOUIS ARTHUR "LOU" HOORNBEEK, LCDR, born in Ellenville, NY on Feb. 11, 1930. He is a graduate of Union College (BA) and Cornell University (MBA). Soon after the start of the Korean War he enlisted as seaman recruit. Subsequently, he was selected for the Reserve Officer Candidate (ROC) program.

He was commissioned as an ENS in 1952 after completing the ROC School, Long Beach. In August 1952 he reported on board the USS *Corregidor* TCVE 58 on which he served as CIC officer. He participated in five cruises to the European Theater and two cruises to Japan. He was honorably discharged from active duty in 1954 at Treasure Island, San Francisco.

His reserve assignments included surface and NCSO division as well as ONR and NROS units. In 1969 the Surface Div. 11-15 (S) for which he was training officer, was rated by the Commandant 11th Naval District as its best small surface division.

In 1970 he completed the Command and Staff Course at the Naval War College. In 1973 he was assigned as CO of the Naval Reserve Officer's School 9-19(S), Joliet, IL.

In 1974 he served with the Office of Naval Research Branch 313, ORU, Chicago. There he participated in the unit's feasibility study of the Reserve Merchant Ship Defense System (RMSDS) for COMNAVAIR/ATLANTIC. Although he was transferred to Retired Reserve status in 1976, he continued to support this unit under orders on a voluntary basis.

He has been an active life member of the NRA since 1964. He served as president of the Forest Park, IL Chapter from 1978 through 1980. This chapter was selected as the Outstanding Chapter of the Year 1980 by NRA.

In his civilian career he was a subcontract and contract administrator and manager for various defense contractors, Argonne National Laboratory and ITT Research Institute.

The Hoornbeek family consists of his wife, the former Lynda Couch, and three married sons, John (wife, Laurie), David (wife, Sue) and Mark (wife, Darcy) and three grandchildren: Allison, Daniel and Hannah. He enjoys playing tenor saxaphone with a Dixieland band and traveling.

His campaign and service medals are the National Defense Service, Korean Service, Navy Occupation, European Clasp, Naval Reserve and Armed Services Reserve Medal.

E. LOU HOOS, CAPT, born in Muskegon, MI. He is a graduate of Michigan State University and obtained a Juris Doctorate Degree from Wayne State University Law School. He entered the USN via USN Officer Candidate School, Newport, RI, and was commissioned an ENS in June 1962. Hoos first served aboard the USS *Kitty Hawk* (CVA-63) as an assistant navigator, air transportation officer and officer of the deck (fleet operations). He served two tours in Vietnam and participated in the conflict between communist mainland China and the Taiwanese nationalists involving the shelling of Matsu and Quemoy Islands. During Desert Storm he served with CDR, Military Sealift Command, Far East, Yokohama, Japan.

He served as CO of the following Naval Reserve Commands: Naval Reserve Surface Div. (9-96M); USS *Pharris* (FF-1094), naval liaison officer to the State of Michigan and the Michigan National Guard Adjutant General; and Military Sealift Command Office Middle Pacific 113.

During his career Hoos has served in the following commands: USS *Havre* (PC-877); USS *Compton* (DD-705); USS *Hyman* (DD-732); USS *Vulcan* (AR-5); United States Naval Justice School, Newport, RI; USS *Shreveport* (LPD-12); USS *America* (CVA-66); USS *Lexington* (CVT-

16); United States Naval War College; Industrial College of the Armed Forces; USS *Vogelsong* (DD-862); Naval Base, Norfolk, VA; USS *Meredith* (DD-890); Shore Intermediate Maintenance Activity, Newport, RI; Readiness Command Region 13, NTC Great Lakes, IL; US Naval Training Center, Great Lakes, IL; Lowry AFB, Denver, CO; Naval Construction Bn. HQ, Gulfport, MS; Robbins AFB, Macon, GA; CINCLANTFLT, Norfolk, VA; CDR Military Sealift Command Pacific, Oakland, CA; Michigan National Guard Camp Grayling; CINCPACFLT, Pearl Harbor, HI; Military Sealift Command Office, Pearl Harbor, HI; Commander Military Sealift Command Far East, Yokohoma, Japan.

CAPT Hoos was designated as a surface warfare officer and awarded the Meritorious Service Medal, Army Commendation Medal, Navy Unit Commendation Ribbon, Navy Meritorious Unit Commendation, Navy "E" Ribbon, National Defense Service Medal, Armed Forces Expeditionary Medal, Vietnam Service Medal, Armed Forces Reserve Medal, RVN Campaign Medal, Navy Rifle Expert Medal and Navy Pistol Expert Medal.

He is a life member of the NRA, American Legion, Veterans of Foreign Wars and Vietnam Veterans of America, and is president of the NRA Admiral VanKammen Chapter in Grand Rapids, MI. He is a practicing attorney in Grand Rapids, MI and a member of the American Bar Association, State Bar of Michigan and the Grand Rapids Bar Association. He and his wife, Marty, have three children: Jennifer, Stephanie and M. Bradley.

WILIAM H.P. "BILL" HOPKINS, CDR, born Aug. 6, 1926 in Bangor, ME, graduated from US Naval Academy, 1950.

He enlisted in the service April 1944. USNA graduate, officer June 2, 1950. His military locations and stations: various enlisted, Atlantic WWII; USS *Borie* (DD-704); USS *Plouer* (AMS-33), Korea. Served with various units in WWII and Korea. Served five years sea duty, officer; ROTC, Holy Cross College. His battles and campaigns: Korea-Inchon, Wonson, Hungnam evacuation.

Hopkins was awarded the American Campaign, National Defense, Korean Service w/4 stars, United Nations, and Korean President Unit Citation. He achieved the rank of CDR and was discharged in 1972. Memorable experiences: attending Storekeepers Class A, USNA "50," Naval Gunfire Support, Mine Warfare, assist prof. Naval Science, Holy Cross College.

He is married to Anne Grey and they have six children: Jane, Bill, Mary, Joan, Patricia and John; and, so far, eight grandchildren. Employed with GM-EC&WL Hopkins Inc., SIS, Corman Bag Co.

IFFATH ABBASI HOSKINS, CAPT, is a member of the USNR Medical Corps. After obtaining her MD degree from Dow Medical College in Karachi Pakistan she did an internship and residency in psychiatry at St. Elizabeth's Hospital and George Washington University in Washington, DC. In 1979, she joined the USN on active duty and did an ObGyn residency at the National Naval Medical Center (NNMC) in Bethesda, MD. Following this, she worked as an attending in ObGyn at NNMC for one year.

This was followed by a two year fellowship to obtain special competence in high risk obstetrics at the Uniformed Services of the Health Sciences in Bethesda. This involved her working for one year at NNMC and one year at Walter Reed Army Medical Center (WRAMC) in Washington, DC. This was followed by two years as an attending in the High Risk Obstetrics Division of the Department of ObGyn at NNMC. During this time, she created a Center for Obstetric Genetics at NNMC, in which patients from CONUS and overseas could seek counseling and treatment for various genetic problems for themselves or their babies. She, also, co-directed the Human Genetics Fellowship at the National Institutes of Health.

In 1987, she left NNMC and active duty to move to New York City with her family. She joined the High Risk Obstetrics Division of the Department of ObGyn at New York University (NYU) as Director of Research and Infectious Disease.

She is currently an associate professor at NYU and chief of obstetrics, and director of the Antepartum Testing Center at Bellevue Hospital (the oldest public hospital in the U.S.) and the flagship hospital of the Health and Hospitals Corporation, the largest public hospitals system in the US. She also has a private practice in High Risk Obstetrics at NYU. Hoskins is a member of the American College of ObGyn and of the American College of Surgeons.

Iffath is also a member of several other national and international clinical and research organizations in ObGyn. She has served on numerous local, national and international committees that impact on the health care of pregnant women. She is also active at the university serving as a member of the Medical School Admissions Committee, and being co-director of a Faculty Advisory College that mentors medical students.

In addition, she is past president of the NY State Perinatal Association (the largest health care lobbying organization in New York State); chairman of the American College of Obstetrics and Gynecology District II Committee on Underserved Women; member of its Committee on Obstetric Practice (which sets practice policy in Obstetrics) and a member of the PROLOG Task Force on Obstetrics (a teaching guide provided by ACOG). She is currently a member of the medical board of Bellevue Hospital.

She has authorized over 60 articles and chapters on many clinical and basic science topics in obstetrics. She is currently editor of a text book in preparation on high risk obstetrics and an obgyn manual for residents.

Hoskins has been active in the USNR since 1987. She currently serves as OIC of a PRIMUS Unit at Ft. Schuyler, Bronx. In addition, she is a member of the Centralized Credentialling and Priveliging Authority based in Jacksonville, FL. She has also served on numerous Reserve Selection Boards.

She is married to William John Hoskins (a USN veteran who served 20 years on active duty). He is currently deputy physician in chief and chief of gynecology oncology at Memorial Sloan Kettering Cancer Center in New York. She has four children: Tonya, a retailer who lives in Knoxville, TN; Bill Jr. a restaurant chef who lives in Washington, DC; Jamie a pre-law student at University of Vermont; Mariya an Art History student at the American University of Paris.

RICHARD C. HOSKINS JR., CDR, born May 26, 1946. Received BS in Business Administration, Elmhurst College.

He joined the service February 1967. His military locations and stations: NAS Pensacola, Glenview, Corpus Christi, Norfolk, Pax River, Jacksonville, FL. Served with VT-10, VT-29, VP-30, VP-45, VP-90, NAS Glenview

Awarded the PPTC in SP-2H and P-3A & B aircraft models - PPMC P-38. He achieved the rank of CDR, USNR, retiring April 1, 1989.

His memorable experiences: many overseas duty stations, including: NAF Sigonella, Souda Bay Crete, NAS Rota Lajes, Bermuda Roosevelt Roads, Barbers Point, Agana Guam, Atsugi, Misawa, Cubi Point, Utaphao and the many DETs from these bases.

FREDERICK L. "FRED" HOTES, CAPT, born July 23, 1918 in Portland, OR, received a BS in civil engineering, University of Washington, 1940, and MS degree from the State University of Iowa, 1941.

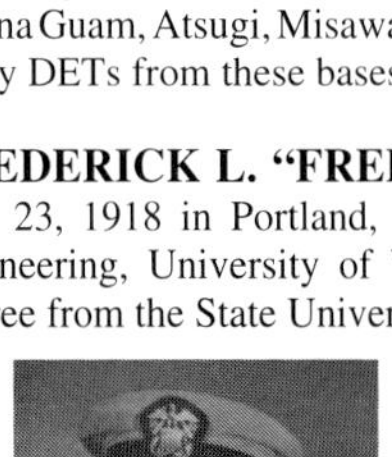

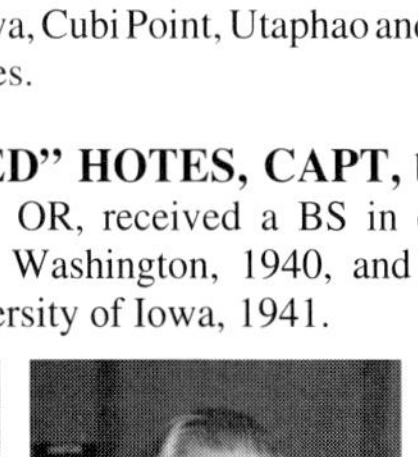

He joined the service Feb. 26, 1942. His military locations and stations: Noumea, New Caledonia; Camp Parks, CA; Okinawa; Washington, DC; San Francisco Bay area.

Served with Third Naval Construction Bn. (1942-44); Officer Training Dept., Camp Parks, CA (1944-45); Commander Construction Troops, Okinawa (1945); Bureau of Yards and Docks and Bureau of Naval Personnel, Washington, DC (1950-52); various Naval Reserve Units in San Francisco Bay area (1947-72), including Seabee units, instructor in Naval Reserve Officer School, Politico-Military Affairs Unit, and last as CDR, Fifth Reserve Naval Construction Regt. with honorary rank of "Commodore," commanding Reserve Seabee Units in Northern California, Nevada, Utah, Oregon, Idaho, Washington, Montana, Hawaii and Alaska.

Hotes participated in the invasion of Okinawa, April 1945. He was awarded the Asiatic-Pacific w/Battle Star, American Theater, WWII Victory, WWII Occupation, National Defense, Naval Reserve and Armed Forces Reserve w/star.

He achieved the rank of CAPT, Civil Engineer Corps, USNR. Active duty: 1942-45 and 1950-52 and total active and reserve duty of 30 years before retirement.

Memorable experiences: Commanding a 90mm AA Btry. while with the Third Seabees in New Caledonia 1943. Invasion of Okinawa 1945. Training duty on CinCPac Staff 1964. Commandant's Local Representative 1967-68. Training duty with Naval Office of Legislative Affairs 1969, with desk in Congressional House of Representatives Office Building. President of CEC/Capt. CDR Selection Board in BuPers 1970.

Married to Ruth E. (former WWII Navy Nurse), since Dec. 26, 1944. They have children: Sunwolf, F. Michael, Barbara Ann (now Rocchi) and Bruce B. and grandchildren: Ian, Marina, Vanna, Kyra (all Rocchi). Employed with Tennessee Valley Authority, International Engineering, Faculty University of California, Berkeley, Development and Resources Corp. Iran, Uniconsult, World Bank. Consultant to many companies and government agencies. Elected 1987 to Honorary Member Status in American Society of Civil Engineers. Retired July 23, 1978 from military and September 1983 from World Bank.

MARVIN F. "MARV" HOUGH, CAPT, is a native of Bon Secour, AL. He graduated from Livingston University in 1967 with a BS degree in business administration and later received his master of science degree in management from Troy State University in 1979.

He joined the USN in October 1967 and after receiving his commission in February 1968, served 10 years as an aeronautical maintenance engineering duty officer. His active duty included tours at the Naval Air Test Center, Patuxent River, MD, USS *Oriskany* (CVA-34), Training Sqdn. 5, NAS Saufley Field, and Aircraft Intermediate Maintenance Dept., NAS Pensacola, FL. While serving on board the carrier *Oriskany* from April 1971 to July 1974. CAPT Hough made three Western Pacific cruises during the Vietnam Conflict. Since transferring to the USNR in July 1978, he has served with NADEP Pensacola 0582 as CO VTU-1087, NAS Atlanta as XO, NADEP Norfolk/Cherry Point 0386 as XO and CO since October 1994.

As a civilian, Hough has served as assistant program manager for Logistics (APML) for the Navy's T-2B/C, T-39D, and T-34B aircraft with program manager AIR 225. He is presently assigned to the office of program manager AIR 222, Jacksonville, FL, where he is APML for the Fleet Electronic Warfare Support Group (FEWSG).

He wears the following decorations: National Defense Medal, Meritorious Unit Commendation, Naval Reserve Service Medal, Vietnam Service Medal (three campaigns), and RVN Campaign Medal.

Hough is married to the former Mary Nell Walker of Orange Beach, AL. They have two children, a daughter, Kelly Anne and a son, Jonathan Andrew.

EDMUND "ED" HOULIHAN, CDR, born Sept. 10, 1939 in Newport, RI, received a BS from Providence College 1962; MBA from Manhattan College, 1990 and PMC from Iowa College, 1992.

He joined the service April 1963. His military locations and stations: USS *Kepper* (DD-765) (1963-65); USS *Barry* (C-1966 DD-933).

Houlihan served in Vietnam and achieved the rank of CDR. He was discharged in 1984. His memorable experiences: Around the world cruise Sept. 29, 1965-April 8, 1966. Five months at sea service in Vietnam as communications officer for COMDESRON 24.

He is married to Josephine B. and they have two children, Edmund M. Jr. and Jennifer Ann. Employed as chief operating officer Cairns Clifton, NJ.

GARY W. HOUSE, CAPT, born June 8, 1941 in Seattle, WA. Graduate of University of Washington, with BS in Political Science in 1964. Graduated from UPS Law School in Tacoma in 1974 went through two years and was a member of Law Review.

He reported to Newport, RI for OCS December 1964. Graduated from OCS in May 1965. Attended Military Justice Course, seven weeks in Newport following OCS graduation. Reported to USS *Belle Grove* (LSD-2) in July 1965. Spent three years aboard this ship during which they made three WestPac tours (Vietnam). He was their communications and legal officer. Also RPS custodian. Spent 20 years in the USNR following his separation from active duty in May 1968.

Significant units as follows: drilled in units in California, Oregon and Washington. Executive officer and then CO of SDB-36, Turnwater, WA; CO of BCT 1422 ORU, Eugene, OR; CO of TELCOM (ABFC) C347 422, Eugene, OR; XO then CO of CONVCOM Sub/Man 222, Eugene, OR; CO NCSO Alaska 422, Everett, WA; last unit before retiring, XO of USCINCPAC DET 322, Seattle, WA, (his Acdutras were with USCINPAC staff as Crisis Action Team Chief).

Employed as assistant prosecutor for five years, Lane Co., Eugene, OR. Practiced Maritime Law, five years. Defense counsel for major insurance company, Seattle, WA.

His organizations and memberships: Washington and Oregon Bar, American Bar, Washington State Trial Lawyers Association, NRA and ROA, Maritime Law Association of United States, Propeller Club of the United States, Association of Marine Underwriters.

He was awarded the National Defense Service Medal, Vietnam Campaign Medal, Vietnam Service Medal and Armed Forces Reserve Medal.

Married to Cheryl Diane for 27 years. They had two daughters: LT Lael L. House, separated from USN, January 1997. She was the Director of Protocol for ADM Charles Larson, USNA and received the Navy Achievement Medal, National Defense Medal, Navy Commendation Medal. She will continue in the USNR as did her father. LT Julie A. House is on active duty as assistant intelligence officer on the USS *Essex,* currently returning from the Persian Gulf Deployment. She received the Navy Achievement Medal, Navy Commendation and National Defense Medal.

CAPT House is deceased. His wife states that "This is a USN family through and through and one in which he would be proud to know his girls are carrying on in his footsteps and serving their country proudly."

RAYMOND J. "RAY" HOUSER, CWO4, born May 4, 1917 in Fort Wayne, IN, attended two years college at Purdue, ME Major.

He joined the regular Navy, Jan. 17, 1939. Entered V-5 AVCAD February 1944 (Reserve Entry). His military locations and stations: All East Coast bases. Three years Minesweeping Base, Trinidad. Served with USS *Ranger,* USS *Denebola,* USS *Benson,* USS *Chachalaca* and eight other ships.

His battles and campaigns: North Atlantic patrol convoy, Minesweeping Caribbean Sub Rescue. Houser received campaign medals (none officially awarded). He achieved the rank of CWO4 machinist and was retired Aug. 1, 1974. His memorable experiences: Navy Comradeship. Attempted rescue of German sub. crew off Long Island while serving aboard the sub rescue vessel ATA 125.

Married 39+ years to Mary Helen. They have a daughter, June Theresa DeMarco; son, Lynn Raymond Houser; and grandchildren: Joseph DeMarco and Scott Raymond DeMarco. Employed with Dana Corp., plant engineer and construction project manager, retiring April 1982.

ANDREW CLARK "ANDY" HOWARD, LCDR, born March 27, 1953 in Carterville, GA, received a BSEE from the US Naval Academy.

He entered USNA, June 30, 1971. His stations: Surface Warfare Officer School 1977, Newport, RI. Served in USS *John F. Kennedy* (CV-67), 1977-80 various billets in ship's company. Naval Reserve in Houston. Subtender Unit NCSO, 1980-84; IRR, 1984-96.

Howard received the Naval Reserve and Armed Service Medal. He achieved the rank of LCDR and retired July 31, 1996. Memorable experiences: Represented 6th Fleet at International Sailing Regatta in Alexandria, Egypt in 1978; NRA life member; USNA Alumni Association life member.

He is married to Tina Hickam, a nurse practitioner. Exxon Computing Service Co., 1980-present, telecommunications engineer; loan assignment to ARAMCO in Dhahran Saudi Arabia, 1984-89.

WILLIAM LEE "BILL" HOWARD, CAPT, born June 5, 1924 in Atlanta, GA, received electrical/electronic degree from Georgia Tech.

Enlisted in the USN, Sept. 28, 1941. Served on the USS *Yorktown* (CV-5). Participated in the Coral Sea battle and the Battle of Midway Island where the *Yorktown* was sunk. Served on the USS *Altamaha* (CVE-18) which steamed the South Pacific, Western Pacific, and the Indian Oceans. Port visited Manus, Ulithi, Ellice, Philippines, Hawaiian and Fiji Islands; Perth, Melbourne, and Brisbane Australia; Panama Canal; Karachi, India; Norfolk, VA; San Diego, San Francisco and Bremerton, WA. Duties were seaman, plane captain, responsible for readiness of a TBF (torpedo, bomber, fighter), plane used in ASW patrols against the Japanese submarines. Duties as a weather observer/forecaster for ASW patrols in the western Pacific and NAS Mayport/Jacksonville, FL.

He was awarded the American Defense Service Medal, American Campaign Medal, Asiatic-Pacific Campaign Medal w/7 Battle Stars, WWII Victory Medal, Naval Reserve Medal, Good Conduct Medal, Armed Forces Service Medal, Philippine Presidential Unit Citation Badge, Philippine Liberation Ribbon and Navy Commendation Medal. Achieved the rank of AERM 3/C-USN CAPT, USNR, retiring from the USN, 1945 and USNR, 1984.

His memorable experiences: swimming from Battle of Midway, 1942; USS *Yorktown* (CV-5) sunk June 1942; ABMA/NASA Control System-Redstone, Jupiter, Saturn-Man to the Moon, Space Lab.

Howard's wife Anne is deceased. He has children: Andrew Clark, Elizabeth Lee and Carolyn Sue; and grandchildren: Lillis Taylor and Scott Perry Youngblood. Employed with the US Army Corps Engr, ABMA NAS, retiring December 1980.

JOHN WILLIAM "COLONEL" HOWE, CAPT, born Feb. 1, 1925 in Milwaukee, WI, received an AA degree from Santa Monica College and a BS degree from UCLA. He did graduate work at USC.

He joined the service Jan. 27, 1943. His military locations and stations: Corpus Christi, TX; NAS North Island; Los Alamitos; Pt. Mugu; Sand Point. Served with TRARON 18-B, VF-771, VF-772, VS-771, VS-774, VS-775, VS-776.

Howe acquired 6,000 hours flight time. His battles and campaigns: WWII, Korea, Berlin build up. His awards include the WWII, American Theater and NR. Achieved the rank of CAPT and retired Feb. 1, 1985.

His memorable experiences: member of first Tarmac Unit in USN. Aircraft flown: PBY, SBD, F6F, F4U, AF and S2F. CO of VS-771, Air Group 77. CO VTU-7676. Charter life member #91 Tailhook Association. Life member NRA, Combat Pilots Association, Naval Order of the United States.

He is married to Norma Jean "Beanie" Howe and they have three children: Pamela, John Jr. and Susan; and grandchildren: Anthia Hawkins and John Jr. II. Employed as superintendent Beaches and Harbors; Captain Ocean Lifeguards, LA County Lifeguard and Firefighter City of Santa Monica.

JOHN G. "JOHN" HOWELL, CAPT, born Sept. 9, 1906 in Berkeley, CA. Attended UC Berkeley for four years; MIT one year and received BS and EE degrees.

He joined the service March 1931, USNR F3/c. His military locations and stations: Philadelphia USNR 1931, ENS, 1933. Served with USNR, Oakland, CA, 1934. Battles and Campaigns: Mine Force, San Francisco Bay, 1940-45 and Pearl Harbor.

He received a commendation from ADM Nimitz. Achieved the rank of CAPT and discharged in 1945. His memorable experiences: making the minesweeping gear work and developing repair procedures for the magnetic sweeping cable. Prepared watch quarter and station bills for their sweepers.

He married Katharine Hartt; she is now deceased. His children are Ann and Jane; and grandchildren are Laura and Katherine Marceau. Employed as chief electrical and mechanical engineer for Owens-Illinois Glass Co., Pacific Coast Div., retiring in 1971.

FRED J. HOWLETT, CAPT, born April 21, 1939 in Crete, NE, received a BS and EE degree from University of Nebraska; MBA from the University of San Francisco and JD from Seton Hall Law; MS from Pace University.

He joined the service in 1956 as midshipman. Commissioned in 1961. His military locations and stations: Submarine School and Nuclear Power School. Served with USS *Bugara* (SS-331) and USS *Sculain* (SSN-590).

Howlett received the Presidential Unit Citation and Navy Expeditionary Medal. He achieved the rank of CAPT USNR and retired from active duty in 1966 and Reserve in 1983.

He is widowed and has three children: Anne Halstater, Joan Kaiser and Fred M. Howlett. Employed with AT&T as international director, retiring Sept. 1, 1995.

SUSAN D. "SUE" HOYNE, CDR, born May 5, 1917 in Chicago, IL, received a BA from the University of Chicago, plus work toward an MA.

She joined the service in 1943. Her military locations and stations: Midshipman School, Northampton, MA. Served with Staff, COMFAIR Seattle.

Hoyne was awarded the American Theater and Victory Medals. She achieved the rank of CDR and retired May 1977. Her memorable experiences: working with the best group of people, both enlisted and officer.

She is single. Employed as assistant agency secretary, Bankers Life and Casualty, retiring April 1994.

GEORGE T. HUBBARD, CWO4, enlisted at age 17 after graduating from West High School in Minneapolis, MN. He received his boot training at Great Lakes NTC, was a member of the Drum & Bugle Corps and was the regimental bugler. He attended EM School at Purdue and was assigned to the USS *Biloxi* (CL-80).

The *Biloxi* was preparing for the assault on Japan when the A-Bomb ended the war. The *Biloxi* entered Nagasaki Harbor and evacuated English, Australian and New Zealand POWs from a nearby prison camp.

Hubbard was recalled for active duty during the Korean conflict, serving aboard the USS *Chikaskia* (AO-54). While aboard, he received the Navy and Marine Corps Medal for attempting to save his EMC from drowning after both were washed overboard in heavy seas.

He retired in 1986 after 42 years of active and reserve service. He served as ICC for 10 years, was selected for WO1 in 1965 and CWO4 in 1980. He has served on 31 ships varying in size from Swift boats and MSOs to battleships and carriers.

His decorations, awards and campaign medals include the Navy and Marine Corps Medal, Navy Commendation, Naval Reserve Medal, USNR Meritorious Service, Good Conduct, American, South Pacific, China Service, Japanese Occupation, WWII Victory, Korean Service, National Defense, UN Korean Service, Korean PUC, Armed Forces Reserve w/3 Devices, Expert Pistol and Battle E.

Hubbard is an Eagle Scout, has served as Scoutmaster, is active in his church and worked for Sears for 33 years.

His wife of 48 years, LaDonna always said "We were in the Navy together." They have a son Rick (Eagle Scout), daughter Linda and three grandchildren: Sarah, Christopher (Eagle soon) and Emily.

JOHN F. HUBERT, CAPT, born on a dairy farm near Erie, PA, where he attended St. Boniface and Cathedral Preparatory Schools. He graduated from Rensselaer Polytechnic Institute with a bachelor of metallurgical engineering degree and from Lehigh University with a master of metallurgy and material science degree.

He was commissioned an ENS in 1955 after graduating from OCS Newport, RI Class 21. His naval training included Aviation Ground Officers School at NAS Jacksonville and Aviation Electronic Officers School at NATTC Memphis.

He joined Carrier Air Gp. 4 as avionics and armaments officer at NAS Jacksonville to conduct shakedown of USS *Saratoga* (CVA-60), and deploy in USS *Randolph* (CVA-15) to the 6th Fleet.

On release to inactive duty he joined VP-933 at NAS Willow Grove in 1958, involved in ASW with P2V-5Fs. Hubert was recalled to active duty during the Berlin Crisis in 1961. Deploying to NAS Guantanamo where he was designated NAO. In 1965 he was designated a naval flight officer, serving as navigator/tactical coordinator. He subsequently transitioned to the P3A and was assigned XO RTU 64 and then CO VP 1604. He retired after drilling 20 years at NAS Willow Grove, during his 23 years Naval Reserve career.

In his civilian career he has worked on several aerospace programs including the Space Shuttle and currently on the X33 Venture Star single stage to orbit prototype.

The Hubert family consists of wife Joy, and daughters: Karen and family, Kathleen and Kristen. John competes as a masters runner and swimmer, relaxes as a gardener.

JAMES ALLEN "JIM" HUFF, CAPT, born Nov. 6, 1936 in Texarkana, USA, received a BS and MS from the University of Houston; MHA Baylor University, 1973.

Enlisted USMC (R) 1954-62. Commissioned USNR (Intelligence) 1968. His military locations and stations: USMC - West Coast; eight stations: USNR 1968-94. Served with 14 different USNR Intelligence units. Participated in the Gulf War.

He was awarded the Defense Superior Service Medal in 1991. He achieved the rank of CAPT 06 and was discharged September 1994.

Huff's memorable experiences: senior debriefer of returned Navy POWS, St. Albans Naval Hospital 1973; "Shadow" attache to Australia 1984-88; OinC of Special Intelligence Team, Gulf War 1990-91.

He is married to Elizabeth Young and they have children, Allen Huff and Chris Huff; and a granddaughter Nicole Young. Employed as senior executive service - VA, retiring May 1994.

HERBERT H. HUGHES, CAPT, born in New Mexico, graduating from the University of New Mexico. He has an MS and PhD from Florida State University where he also coached football.

He joined the USNR in 1947, but from 1951-55 served in the Regular AF in both enlisted and officer status. Later as a USNR officer he served in Surface and Research units and as CO of medical units.

During the Korean War, he flew in the "Fighting 4th" all-weather fighter squadron in the Far East. He had later collateral assignments as a 29th Air Div. administrator.

He has been an active NRA member since 1966 and an elected Albuquerque city counselor. He has been a college professor, businessman/banker and public administrator.

The Hughes family includes wife Nancy, daughter Katherine and son Bryn. He and his wife reside in Albuquerque, NM.

STEVE R. HUJARSKI, CDR, born Jan. 18, 1947, is a native of Cleveland, OH, and has received an associates in computer sciences, bachelors in business, and is a graduate of the Naval War College.

On Dec. 7, 1965 he enlisted in the USN as an E-1. After spending two years on active duty at NAVSECSTA from January 1967-October 1968 in support of Vietnam operations, he reaffiliated with a Cleveland Reserve Unit as a DP3. In November 1957 he achieved the rank of chief and was assigned to REDCOM-5. In April of 1979, he was commissioned a CWO-2 (DP) and assigned to reserve unit FIC 1081. On August 1981, he was commissioned a LTJG 1635 intelligence officer. In 1987, he was assigned to COMMIDEASTFOR Reserve billet and deployed twice to the Persian Gulf aboard the USS *LaSalle.* As a LCDR, he was recalled to active duty for Desert Storm and was attached to AIC in the ISSAF (Intelligence Support for Strike and Amphibious Forces) cell.

He was selected for CDR in October of 1996, and is currently the deputy RIAC for Manpower in Area-17. Hurarski is a NRA life member and serves on the Cleveland Chapter Executive Board. He has over 30 years of data processing experience, and is currently the director of data processing for the Summit County Auditor in Akron, OH.

Married to Cynde and they have four children: Karen, Ellen, Kristen and Stephen.

CHARLES S. HUMPHRIES, CAPT, is a native of Coral Gables, FL, graduated from the University of Florida and was designated a naval flight officer June 1964.

He reported to Patrol Sqdn. 18, Roosevelt Roads, Puerto Rico, serving as tactics officer, nuclear weapons loading officer, NATOPS tactical coordinator instructor and OinC of VP-18 Detachment Seven, Naval Station Guantanamo Bay, Cuba.

Humphries affiliated with the USNR in March 1969. He became a plankowner in the newly commissioned VP-62 and transitioned from the SP-2H Neptune to the P-3A Orion aircraft, qualifying as a patrol plane tactical coordinator and mission commander.

His military decorations include the Battle E Ribbon, National Defense Service Medal, Armed Forces Expeditionary Medal, Navy Overseas Service Ribbon and the Armed Forces Reserve Medal.

Humphries is married to the former Eleanor Harriett Broome and resides on Lake Santa Fe near Gainesville, FL. He retired Oct. 1, 1991.

RONALD E. "RON" HUNTER, CAPT, although born in McMinnville, TN, was raised in Minneapolis, MN. He graduated from Minneapolis North High School and received his BA in mathematics from St. Olaf College in 1970. Eight years later he received his JD degree from the University of Minnesota. His naval career began in 1971 when he reported to OCS in Newport, RI.

Following his commissioning, then ESN Hunter attended Damage Control School in Philadelphia and Serv Pac Petroleum School in San Pedro. From 1972-75, he served aboard the USS *Tolovana* (AO-64) as DCA and liquid cargo officer, and was awarded the Meritorious Unit Commendation, Vietnam Service Medal and Vietnam Campaign Medal. Subsequent reserve detachment assignments included: various billets, including XO, on USS *Norton Sound* (AVM-1); XO of Fleet Training Gp., San Diego; XO of USS *Cape Cod* (AD-43); XO of CINCPACFLT; VTU Gen 1618 component director and CO of COMNAVFORJAPAN.

Hunter served the State of Minnesota as an assistant state public defender and a special assistant attorney before becoming a senior attorney in the Law Department of Cargill, Inc. He resides in a Minneapolis suburb with his wife Renee and their daughter Lindsey.

JOHN W. HURD, LTJG, born July 8, 1961 in Long Beach, CA, received a BA in Media Management/Psychology, Marietta College, OH.

Enlisted in active duty Oct. 25, 1984. Russian Language, DLI-FLC Monterrey, CA. Reserve: commissioned March 7, 1994.

His military locations and stations active duty: Monterrey, CA; Goodfellow AFB, TX; Misawa, Japan; Ft. Meade, MD. Served with Naval Security Group. Reserves: NR Security Group, Washington, DC.

Hurd was awarded the Naval Expeditionary Medal, Good Conduct Medal, Meritorious Unit Commendation and Pistol Expert. He achieved the rank of CTI2/E-5, active and LTJG Reserves. Discharged from active duty Feb. 22, 1990 and is still serving in the Reserves.

He is married to Kathleen Coolahan and employed as analyst, Department of Defense since August 1996.

WILLIAM WISEMAN "BILL" HUSS, CAPT, born Sept. 9, 1932 in Gastonia, NC, graduated from the University of Southern California of Business in 1962 and its Law School in 1965, receiving a BS and JD degree. He had been on active duty from 1951 to 1954 as a CT working in cryptology. He went to school at Imperial Beach and boxed as a welterweight. He served as a Reservist until 1989, after obtaining a commission in 1965. He was CO of Law Co. 11-2 in Los Angeles and became a specialist in Public International Law and the Law of War. He served in USS *Siboney,* USS *Cascade,* USS *Ranger* and USS *Kitty Hawk.*

In 1953 he served TAD at Plymouth Navy Base in Plymouth UK with the British and Canadian navy in Operation Mariner. He retired in September 1989 and accepted a commission as a COL in the California Military Reserve.

His memorable experiences: welterweight champion, USNAVRAD STA Imperial Beach.

He is married to CAPT Marlene C. Huss, MC, USNR and they have children: Matthew, Charlyn and Damon; and grandchildren: Helen, Nathaniel and Joshua. Employed as judge, Los Angeles Superior Court.

RALPH W. "HUTCH" HUTCHINSON, LCDR, born Dec. 30, 1918 in Pittsfield, MA. He joined the USN, Construction Bn., July 6, 1944 with the rank of first class petty officer. He received his basic training at Camp Perry, VA, Davisville, RI and Camp Parks in California. He was

shipped overseas to the Philippines in 1945 and served there as a CPO. He was an instructor for heavy equipment for the Seabees and also served in Japan, and China.

Hutchinson was discharged in Boston, MA, January 1946. He joined the USNR early in 1947. In 1950 he was promoted to LTJG. He made commander of the Naval Base Reserves CB 115 in Pittsfield, MA and served there many years. His last promotion was in 1961, when he was made LCDR. He was proud to serve the USN and his country for a total of 32 years.

He married Ruth in 1940 and they have children: Raymond, Renee and a son Michael, who died in 1990. Their grandchildren are Christine, Elizabeth and Matthew. He owned a realty company, sand and gravel company and water company, retiring in 1978.

GEORGE W. HYDE, LCDR, July 4, 1912 in Jersey City, NJ, attended NYU and received a BA and LLB degree.

He joined the service April 7, 1942. His military locations and stations: Local Defense Force; State Pier, New London, CT; Mine Warfare, USS *Acme* (AMC-61); USS *Sentilal* (AM-113); Naval Mine Warfare School, Yorktown, VA; Chief of Naval Operations, Mine Warfare Section, Minesweeping Desk, Washington, DC; Minecraft Training Center, Little Creek, VA; USS *Capable* (AM-155), Mine Div. 29, COMPAC.

Battles and campaigns: Minesweeping east coast from Newport through New York; stationed at Tompkinsville, Staten Island, NY; convoy duty North Atlantic and Mediterranean; Sicilian Invasion; Convoy duty Pacific Saipan to Ulithi to Seattle to Alaska; Defence Council at General Court Martial, San Diego, CA.

Hyde received spot promotion to LT from LTJG after Sicilian campaign. Achieved the rank of LCDR and discharged Dec. 26, 1945. To Naval Research Reserve Unit 3-1 July 31, 1949. Achieved the rank of CAPT, retired July 4, 1972. His memorable experiences include the sinking of the USS *Sentinal* (AM-113) off Lacata, Sicily during the invasion of that island.

He is married to Anne Shropshire and employed in individual private practice of law. Partner law firm Pendry, Schneider & Hyde; Garfunkel & Hyde.

RONALD "RON" HYMAN, LCDR, enlisted in the Naval Air Reserve 4x10 program April 19, 1969 while still in high school. He attended recruit training in Memphis, TN and Aviation Electrician's Mate Class "A" School at NATTC, NAS, Jacksonville, FL, graduating on March 15, 1970. Upon graduation, he started drilling at NARTU, Norfolk, VA, assigned to Unit NARDIV S-1. He has completed over 28 years of drilling with no active duty time rising through the enlisted ranks from E-1 to E-8 and the LDO grades of 0-1 to 04.

The summer of 1970, he was assigned to unit VAW-207 as an "AE" working on the old E1B Tracer aircraft. He stayed with VAW-207 until it was disestablished, transferring to unit VAW-78, where he stayed advancing through the rate of senior chief petty officer. He applied for a LDO commission in 1982 and was commissioned an ENS in April 1983.

LCDR Hyman's assignments since 1983 have been: VAW-1086, NAS-2186, VRC-4086 and his current unit COMNAVAIREASTLANT 0286, where he is the communications and security officer. He is still an active reservist completing over 40 days of "AT" each year reporting to COMFAIRMED for assignments throughout the Mediterranean.

LCDR Hyman's awards include the Navy Achievement Medal, Meritorious Unit Commendation Medal, Navy E Ribbon, Navy Reserve Meritorious Service Medal, National Defense Service Medal, Armed Forces Service Medal, Navy Overseas Service Ribbon, Armed Forces Reserve Medal, Military Outstanding Volunteer Service Medal, Navy Expert Rifle and Navy Expert Pistol.

He is a life member of the Naval Reserve Association and the Reserve Officer Association; he is also an active member of the Virginia Committee for the Employers Support of the Guard and Reserve Program. Active in his community, he is an auxiliary police detective on the Portsmouth Police Department. He is employed by Bell Atlantic where he is a LAN administrator/assistant manager in Virginia Beach, VA. He holds a BS degree with a major in business administration. He holds professional certifications in LAN and personal computer applications, hardware and software.

LCDR Hyman is married to the former Marea Ann Stone and is the father of six children: Angela, Barry, Carrie, Danny, Eddie and Zachary. His three grandsons are: Alexander, Jacob and Tristan. He can be contacted on-line at: rmhyman@erols.com or at work at ronald.b.hyman@bell-atl.com. He and his family reside Virginia Beach, VA.

STEPHEN D. IHRIG, CAPT, born Oct. 18, 1946, Detroit, MI, graduated from Michigan State University in 1969 with a BS degree in psychology, from Pepperdine University in 1976 with a MBA, and Cooley Law School in 1980 with a JD.

CAPT Ihrig was commissioned in May 1972 and was designated a naval flight officer in November 1973. After completing P-3 training at VP-31, NAS Moffett Field, CA, he reported to VP-22 at NAS Barbers Point, HI. CAPT Ihrig participated in operational deployments to NAS Cubi Point in the Philippines and Kadena AFB in Okinawa, for which the Blue Geese won the Battle "E." In November 1976, CAPT Ihrig was released from active duty and affiliated with the Executioners of VP-93 in February 1977. CAPT Ihrig was instrumental in transitioning naval flight officers from various aviation communities to the P-3 aircraft. As NATOPS instructor for both tactical coordinators and navigators, CAPT Ihrig has contributed significantly to the transition of VP-93 to an operational patrol squadron. CAPT Ihrig became executive officer of VP-93 Sept. 12, 1987. From Feb. 3, 1990-Aug. 10, 1991, CAPT Ihrig served as CO of VP-93. On Aug. 18, 1991, he assumed the duties of CO NR CVN 69 DET 0373 and served as CO until Oct. 18, 1993. CAPT Ihrig participated in VTU 7373G until May 1995.

CAPT Ihrig wears the Coast Guard Meritorious Commendation, Battle "E," National Defense Service Medal, Armed Forces Expeditionary Medal, Sea Service Deployment Ribbon, Coast Guard Special Operations Ribbon, Navy and Marine Corps Overseas Service Ribbon, Armed Forces Reserve Medal, and Meritorious Service Medal.

He is employed by EPI Printers, Inc. as vice-president and corporate counsel. CAPT Ihrig resides with his wife, the former Patricia Schaefer, and their two children, Steve and Carrie, in Richland, MI.

JOHN M. "MICKEY" IMEL, CAPT, born Aug. 4, 1932, Cushing, OK. He graduated from the University of Oklahoma, receiving his commission through the NROTC program, was assigned in 1954 to the USS *Colonial* LSD-18 stationed out of San Diego, CA, and served in the Seventh Fleet in the Pacific Area. While aboard the USS *Colonial*, 1954-56, he served as gunnery officer and served as defense attorney under the Uniform Code of Military Justice. Upon return from the USN, he enrolled in the University of Oklahoma College of Law and graduated in 1959, was appointed municipal judge in Tulsa, OK, in 1960, and later appointed US attorney in 1961 by President Kennedy and reappointed by President Johnson in 1965. He resigned in 1967 when he joined the law firm of Moyers, Martin, Santee, Imel & Tetrick of which he is a partner and active member.

After graduation from law school, he continued with his USNR career, transferring to naval intelligence, and served in various capacities. He was CO for three tours of Naval Reserve Intelligence 8-1-7 in Tulsa. He was selected as CAPT in 1975 and retired from the program in 1977. As an attorney, he was selected as a fellow of the American College of Trial Lawyers, fellow of the American Bar Foundation, specializes in corporate defense litigation, and is listed in the *Best Lawyers in America* under the heading of Business Litigation. He was appointed by the governor and served on the University of Oklahoma Board of Regents (1981-88, as chairman 1987-88).

He is widowed and his family consists of his four daughters: Blythe, Katy, Dixie and Sally, all married, and have made him a grandfather of six. He still resides in Tulsa and for hobbies he enjoys golf, tennis and swimming.

SALVATORE J. "SAL" INDIVIGLIA, CDR, born Nov. 16, 1919, New York City, NY, attended Pratt Institute where he received his BA and MFA degrees, various art schools and private instructors. He served apprenticeship with Alfred D. Crimi in mural and fresco painting and completed 17 NROS professional courses.

He joined the US Army/USAAF in June 1942 and the USN where he received a direct commission as LCDR, USNR, PIO & official USN combat artist, Oct. 29, 1962.

His military locations and stations: Submarine Sqdn. 12, Key West, FL; Pt. Mugu Pacific Missile Range, Oxnard & Port Hueneme, CA; Navy Electronics Laboratory (Bathscape Trieste) San Diego, CA; Hawaii and Midway Island; Amphibious Base, Little Creek, VA; and USNA, Annapolis, MD.

The units he served with were: Vietnam, Riverine Force, River Assault Gp. 21; Seabees; Navy Support Activities and Junk Force; 1967 Vietnam: Formed the first Navy writer and artist combat team covering the Navy

medics in Rach Gia, Seal Team TWO, Market Time & Game Warden Operations with CF 115, 116 & 117, Sea Wolf Operations, COMSEVENTHFLT (Det. Charlie), Navy Support Gps. Sea Dragon Operations with cruiser *St. Paul* and destroyers; Seabees & Marines in Dong Ha on the DMZ; carriers USS *Forestal*, USS *Intrepid*, USS *Bonhomme Richard*, USS *Blue*, USS *Mispillion*, LST USS *Garrett County* and HMAS *Hobart*, Australian navy. CONUS: PAC 3-1, NROS 3-5, VTU 602 and member of the New York Naval Militia.

Battles and campaigns he participated in were Italy 1943-45, Naples-Foggia; Rome-Arno; Vietnam 1964 Advisory Effort, Riverine Force, NAVFORV, Market Time Operations; Vietnam 1967 Game Warden Operations; Sea Dragon Operations and Seawolves Bassac River and Rung Sat areas.

Indiviglia was awarded the Navy Commendation Medal, Combat Action Ribbon, Presidential Unit Citation, Navy Unit Commendation Medal, Meritorious Unit Commendation Medal, WWII Victory Medal, Armed Forces Expeditionary Medal, (Italian) Croce Al Merito Di Guerra: Meritorious War Cross, RVN Gallantry Cross w/Palm (Unit Citation), USNA Superintendent's Award (MVP) Most Valuable Player as coordinator of Blue and Gold Program, USNA Alumni Association Awards and Director's Awards, Nassau County Citation, Midshipmen's Parents Club of L.I. "Grateful Appreciation Citation" Army Good Conduct.

His memorable experiences: 1967 senior officer on board a STAB (Seal Tactical Assault Boat) as one of three to extract Seal Team II Alpha Squad from mangroves on the Rung Sat river, when they were discovered and overrun by VC. Engaged VC in a firefight within small arms range, until "Mike Boat" with heavier weapons covered their extraction and withdrawal. In 1962 was designated official USN combat artist. 1964 Operations Palette: Exhibited 12 paintings of combat operations in Vietnam at Smithsonian Institute, Washington, DC. Represented in USN Combat Art Collection USN Museum, Washington, DC; Amphibious Museum, Little Creek, VA; and Seabee Museum in Port Hueneme, CA. He achieved the rank of 05 CDR and retired Nov. 16, 1979.

He married Quinta (deceased Jan. 9, 1986); his children are Joseph, Barbara Ann (deceased), Lawrence USNA 1977, and Dianne. He has two grandchildren, Joseph and Lauren. Civilian employment as art director, vice-president, account executive, art instructor and art lecturer. Indiviglia retired June 14, 1984.

JACK D. INGOLD, CAPT, born Dec. 2, 1933, Dewey, IL, received his BS degree from the University of Illinois in 1956 and his MBA degree from Adelphi University in 1966.

He joined the service April 25, 1956. His military locations and stations: Newport, RI; Glencoe, GA; Glenview, IL; Barbers Point, HI; and Patukent River, MD. Served with VW-2 and VW-14.

His memorable experiences: Participating in establishing of the "Dew" line in Pacific as C/C officer and airborne air controller, operating with fleet in Mediterranean and qualifying as patrol plane tactical coordinator in Reserves. He achieved the rank of CAPT and was discharged July 3, 1961.

Ingold married Jerilyn and they have one son, Darrell Paul. He served as director of military marketing with Grumman Aerospace and retired June 30, 1995.

CHARLES MARSHALL INGRAM, CAPT, born July 4, 1948, Goldsboro, NC, received his AB degree from the University of North Carolina, Chapel Hill and his JD degree from Cumberland School of Law, Samford University, Birmingham, AL.

He joined the service June 1, 1970. His military locations and stations: USS *Strong* (DD-758), Charleston, SC and served with NR OPNAV 09B; NR BUPERS (06); and NR COMNAVBASE NORVA. He was recalled during Operation Desert Storm.

Ingram was awarded the Meritorious Service Medal and Navy Commendation Medal. He was discharged with the rank of CAPT.

Ingram is married to Carolyn and is an attorney.

BUSTER LEE IRVING, QM2C, born July 17, 1925, Anderson, IN, attended four years of high school, four years of college and four years of Medical School.

He enlisted in the USN Air Corp. March 10, 1943. His military locations and stations: Navy V-12 University of Michigan, USS LST 34, ATA 187, and PCS 1430. Served with Service School Command, Great Lakes, CDR Seventh Fleet.

Irving was awarded the Philippine Liberation Ribbon, Victory Ribbon, Asiatic-Pacific Campaign Medal, American Area, and Good Conduct Medal. He achieved the rank of QM2C and was discharged May 1946.

His memorable experiences: Re-enlisted March 6, 1981; achieved rank as CAPT; served as CO MED CRU 608; Marine Support Camp Lejeune; discharged July 1, 1985.

Ingold married Juanita and they have five children: Walter Lee, Michael, Susan, Tom and Toni. Their grandchildren are: Bill, Michael, Todd, Douglas, Brad, Mary, Jaimie, Cortney, Clayton, Jesica and Shawn.

His civilian employment included 23 years in private medical practice and 11 years as a VA physician. He retired in May 1989.

SHAHIDUL "SHAHID" ISLAM, CAPT, born Oct. 6, 1946, in India. He graduated from Medical School there in 1969, came to the US in 1970, took all his medical training including a post-doctoral fellowship in Psychosomatic (Mind-Body) Medicine from Harvard and Vanderbilt Schools of Medicine.

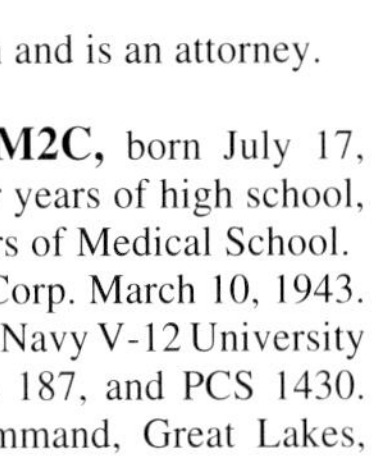

He joined the faculty at Louisiana State University School of Medicine in 1978. In 1981 he joined the USNR Medical Corp. as a LCDR and was promoted to CDR in 1983. In 1992 he was promoted to the rank of CAPT.

CAPT Islam has served overseas in active duty in Italy, Japan and Iceland in time of need. He enjoys working for the USNR and looks forward to future challenges of service. He was awarded the medal for 10 years of service with the USNR.

Islam is married to Hosneara "Jolly" and they have one daughter, Shefa "Tara." He is an assistant professor at Louisiana State University School of Medicine.

DAVID A. ISQUITH, CAPT, born Jan. 11, 1935, Brooklyn, NY, received his BS degree from the USNA in June 1957 and his MS in administration from George Washington University Feb. 21, 1972.

He joined the service Jan. 11, 1952, as SR while at Adm. Farragut Academy to qualify to take USNA entrance exams. Designated as honor man of NRTU during NTC Bainbridge, MD boot camp and advanced to SA. Re-entered USNR, from active duty, in July 1967. Isquith retired in June 1982 with the rank of CAPT USNR.

His assignments: 2nd Div. officer, USS *Bremerton* (CA-130); gunnery officer USS *Blue* (DD-744); Submarine School, New London; OPS and navigation officer, USS *Meredith* (DD-890); staff officer, COMOCEANSYSLANT, Norfolk; OPS and navigation officer, USS *Ashland* (LSD-1); assistant base OPS officer, NAVPHIB Base, Little Creek, VA. With the USNR he served as training officer, Surface Div. 5-37(L); OPS/PXO ROBERTS (DE-749); XO, MTD 5-7; CO, Surface Div. 5-4(L); assistant chief of staff, Gp. 5-5(M); CO, HQ BUPERS 3006, PM, NAVSEADET 106; training officer, VTU 102; XO, PMS Port Hueneme Det 106, CO, VTU 102.

Isquith is married to Clare and they have four children: Jeffrey, Aileen, Robin and Hillary; and six grandchildren: Ariana, Olivia, Zachary, Max, Sydney and Rachel. Civilian employment as branch head, NAVSEASYSCOM Command & Control/Combat Systems Design & Arrangement (July 1967-June 1988); technical consultant, Wheeler Industries (March 1989-March 1991); Rolling Airframe Missile Program, manager, EG&G (March 1991-October 1996).

STEPHEN S. ISRAEL, CAPT, born in 1944, Baltimore, MD. He graduated from the USNA in 1967 and served aboard USS *Borie* (DD-704) from 1967-69 as Drone Anti-Submarine Helicopter (DASH) and gunnery officer. CAPT Israel deployed to Vietnam where *Borie* served as plane guard and gun ship. Other sea duty included COMDESDIV 182, and later COMDESDIV 362 as operations officer, serving until 1971. CAPT Israel received a master's degree in weapon systems acquisition from the Naval Post Graduate School in 1973. His last active duty through 1977 was with the Production Div., DCASR Atlanta, GA.

USNR duty since 1977 included four commanding officer tours within REDCOM 8 in Florida: SUPSHIP Pascagoula 608, NSY Philadelphia 208, Weapons Station Charleston 808, and NAVSEA 1308. Recent tours include the Pentagon, N43 department head for supportability, maintenance and modernization, and NAVSEA 1606 Command Center.

CAPT Israel, an NRA member since 1984, has served as president of the Metroplex Chapter in Dallas-Fort Worth and the national vice-president for education and chapter activities.

Awards include the Meritorious Service, Joint Service, Navy Commendation Medals (3), Volunteer Service, Vietnam Service Medals, the Navy Unit Commendation, the Recruiting Ribbon and the Recruiting Gold Wreath.

He and his wife, Patricia, have three grown children and two grandsons. They currently reside in Fort Worth, TX, where CAPT Israel serves as director of marketing for EFW, Inc., a defense electronics company.

CHARLES U. "BUD" JACKSON, CDR, born June 27, 1921, San Diego, CA, graduated California Maritime Academy in 1943.

He joined the service Dec. 18, 1942. His military locations and stations: Naval Amphibious Base, Coronado. Units served with included First, Second, Third, Fifth, Sixth and Seventh Fleets.

He achieved the rank of CDR and was discharged Feb. 2, 1958.

Jackson was awarded the Asiatic-Pacific Campaign Medal, National Defense Service Medal, WWII Victory Medal, China Service Medal, and United Nations Service Medal w/2 stars (Korea).

He married Dorothy Mary and they have three children: Martin V., James Michael and Anita Marie. Civilian employment was in architectural planning and design. Jackson retired June 30, 1965.

ROBERT E. "BOB" JACKSON, CDR, born Dec. 21, 1931, Flushing, NY, received his BA degree from Tulane University (1960) and a JD degree from Tulane Law School (1963).

He joined the service Nov. 2, 1952. His military locations and stations: Pensacola Flight School, Class 5-53; CO of Law Co. 3-6, New York. He served with VP-44 Norfolk, VA; VP-821 New Orleans, and VS-New Orleans. He achieved the rank of CDR.

His memorable experiences: Missile tracking flights, Key West, FL and ASW tracking, VP-44, NAS Norfolk, VA.

Jackson married Patricia Ann and they have two children, Robert E. Jr. and Lisa Jackson Hawk. Civilian employment was as an attorney with the National Labor Relations Board in New Orleans. He retired Dec. 21, 1991.

DAVID W. JACOBS, CAPT, born May 14, 1945, New York City, NY, received a BS degree from the University of Rhode Island, a MS degree from Columbia University and a MS degree from Polytechnic Institute of New York.

He joined the service May 10, 1968. His military locations and stations: Dam Neck, VA. Jacobs served at the Fleet Computer Programming Center. He achieved the rank of CAPT and was discharged in June 1971.

Jacobs is married to Marilyn. Civilian employment was as a senior construction engineer for Metro-North Railroad Co. He retired in September 1993.

FRANCIS T. "TED" JACOBS, CAPT, USNR, born June 13, 1945, Harrisburg, PA, received a BS degree in education (math, physics and psychology) from the University of Missouri. He received a NROTC scholarship from the University of Missouri in 1963; was commissioned in August 1967; and earned his wings Dec. 14, 1968.

He was awarded the Meritorious Unit Commendation, Noel Davis Award (2), National Defense Service Medal, Naval and Marine Corps Overseas Service, Naval Reserve Medal and Expert Pistol. Jacobs achieved the rank of CAPT and retired Sept. 30, 1993.

His memorable experiences: Completed Naval Reserve component courses at the Naval War College, Providence, RI and National Defense University, Washington, DC. Served on officer and warrant officer promotion boards, Washington, DC.

Jacobs married Pollyann Mills and they have three children: Ted, Laura and Tara. He is currently a pilot with Southwest Airlines. Previous employment as pilot, Morris Air, Ryan International and Braniff; and owner, Wells Abstract Co.

KAY L. "JAKE" JACOBS, CAPT, born Feb. 25, 1924, Pleasant Grove, UT, received a BS degree from Brigham Young University, Provo, UT.

He joined the service in August 1942. Military locations and stations: Naval Aviation Cadet Training, VP Sqdn. Lake City, FL; Beaufort, SC; VF Sqdn. NAS Denver, CO; AAU Salt Lake City, UT; NRED Provo, UT; NRSD Salt Lake City, UT; and a number of ACDUTRA ships and stations.

His memorable experiences: Flight time in the PV 1 & 2 Ventura, F8F BearCat, F2H Banshee and F9F Cougar. He achieved the rank of CAPT and retired Feb. 25, 1984, with 26 years of service.

Jacobs married Nedra H. and has four children: Sheila Forsyth, Harley W., Shellie Gray and John H. (by his first marriage to Joyce H.). He also has 17 grandchildren. Civilian employment was as President CEO of Deseret Ban Corp. He retired in January 1990.

STUART K. "STU" JACOBS, CAPT, born Dec. 21, 1929, Cleveland, OH, received an AB degree. He joined the service April 4, 1951. Military locations and stations: USS *Knapp* (DD-653), USS *Kenneth M. Willett* (DE-354). Jacobs served with the Mobilization Team 6-10; MSCO Carib. 108; and Gp. Command 6-15.

He achieved the rank of CAPT. Jacobs was awarded the Navy Occupation Service Medal, National Defense Service Medal, Korean Service Medal, United Nations Service Medal, and the Armed Forces Reserve Medal.

He married June and they have three children: Douglas, Sue and Stanley; and five grandchildren: Dennis, Bryan, Spenser, Stephanie and Sydni. Civilian employment as a chartered life underwriter. He retired Sept. 31, 1979.

ARTHUR W. JAHNS, CAPT, born Jan. 23, 1929, Milwaukee, WI, received a BS degree from the University of Wisconsin-Milwaukee and an EDM degree from Marquette University.

He enlisted in the service July 25, 1953. His military locations and stations: OCS, Newport, RI; USS *Oriskany* (CVA-34); COMTRAPAC, San Diego; USS *Rupertus* (DD-851); Far East; and China Patrol.

Jahns was awarded the American Defense Service Medal, China Service Medal, China Service Extended, Vietnam Campaign Medal, Battle Stars (4), Commendation Medals (4), Naval Reserve Medal, and various command awards. He achieved the rank of CAPT and was discharged Dec. 7, 1959. He retired from the USNR in July 1976.

His memorable experiences: NR transfer to retired reserve 1976; *Oriskany* Show the Flag Tour 1953-55 from Alaska to Indonesia; Battle of Quemoy 1957-58; first Navy ship to Guam since WWII; Brisbane; Australia; visited Antarctica and all major Asian ports; Battle Award for best gunnery and torpedo warfare.

Jahns is single and civilian employment was as a junior high and high school principal. He retired in July 1987.

THEODORE P. JAKABOSKI, CAPT, born March 17, 1931, Middletown, CT, graduated from Yale College in 1958, Magna Cum, and from Yale Law School in 1961.

He enlisted in the service in October 1947 as a SR, Div. 34-M, Cromwell, CT. His military locations and stations were AcDuTra at most major naval stations on all three coasts. The units he served with included USS PC-1182, USS CG-22 and LST-1192. He retired with the rank of CAPT (O-6) March 17, 1991.

In his 66th year, as he looks back at his life he can discern two reliable, and key, principles affecting his life. They are, first, that his association with the Navy and Naval Reserve have caused him to be a better human being and citizen. And second, through the storms in the life of an American male born in 1931, the Navy and Naval Reserve have always been there, in ways he never imagined; showing up as a more constant factor than the university, several marriages, siblings, clubs, and even his legal profession. The Navy has been his first family—pretty much the same way that the French Foreign Legionnaire has "La Legion" as his family.

A Legionnaire is required to spend Christmas with the unit - which reminds him of the time he attended Christmas Eve Mass aboard USS *Orion*, tied up in Charleston in late 1976. His military service took place in each of the last six decades.

His military experience started in 1947 as a 16-year-old whose father helped get him into the USNR although underage. In 1950 with the Korean War starting, and having finished high school, he entered the USAF after the USN said they didn't need him for a while. Four years later he was pleasantly surprised to be accepted at Yale. Then came a scholarship to Yale Law School, and off into the legal profession. A highlight was a stint as a federal immigration judge in El Paso, TX. The last 15 years have been in the private practice of immigration law.

His main contribution to the Naval Reserve Assoc. was an article appearing in the *NRA News* of July 1989, entitled "Looking Beyond Perestroika," a future oriented alert to many of the issues stemming from a serious reduction in the resources available to the Navy.

In 1982 on AcDuTra at Subic Bay, he wrote a manual on the immigration aspects of naval personnel marrying Filipino women. Ships served on AcDuTra were the PC-1182, USS *England* CG-22, and the USS *Racine*, LST-1192.

WILLIAM L. "BILL" JAMES, CDR, born April 26, 1927, Chicago, IL, received BSME from Purdue University and graduate studies at the University of Chicago.

He enlisted in the service in July 1945. He was stationed at various military locations and stations. The units he served with were VF-726, VP-723 and NAS Glenview. He achieved the rank of CDR and was discharged in 1980.

James married Dolores M. and they have two sons, William L. III and Robert A. Civilian employment was as president of Fennell Corp. He retired in 1991. Presently he is president of James & Associates Engineers.

DION S. JANETOS, LCDR, born May 26, 1917, Dover, NH, attended the University of New Hampshire and Tufts University Dental School receiving his DMD degree in June 1941.

He enlisted in the service in December 1941. His military locations and stations: Naval Training Center, Newport, RI, January-April 1942. USS *Beaver* (submarine tender), April 1942-January 1944. They sailed to Europe in October 1942 where their subs participated in the North African Invasion. Then to Alaska in 1943 for the Aleutian Campaign. USNTC, Great Lakes, IL from February 1944-January 1945 where he supervised a dental clinic. NATTC, Memphis, TN from January-December 1945 where he supervised operative dentistry section. PSC, Lido Beach in December 1945.

His memorable experiences: The October 1942 Atlantic crossing in a convoy of 38 ships. Only 13 survived the German torpedoes. As the squadron dental officer he had to treat "trench mouth" that was rampant among the submarine crews and the personnel aboard the Beaver. His findings necessitated a publication in the Navy medical bulletin in 1944 entitled "Trench Mouth Aboard a US Naval Auxiliary Vessel."

Janetos was awarded three Area Campaign Ribbons and is very proud to have served in the USN. He was discharged July 22, 1946, with the rank of LCDR.

His civilian employment was in private dental practice until his retirement in February 1963.

BRUCE JANIGIAN, CAPT, born Oct. 21, 1950, San Francisco, CA, attended the University of California and George Washington University receiving his AB, JD and LLM degrees.

He enlisted in the service Feb. 1, 1968. His military locations and stations: Rota, Spain and Washington, D.C. Served with Naval Legal Branch Office; USN and USMC Court of Military Review. Janigian served during the Vietnam era.

His memorable experiences: Served as commissioner of the Navy-Marine Corps Court of Military Review; later special legal assistant to CINCUSNAVEUR and assistant fleet judge advocate to CINCPACFLT.

Janigian was awarded the Navy Achievement Medal, National Defense Medal, Reserve Medal and the Unit Commendation Medal w/star. He achieved the rank of CAPT (O-6) and is still serving in the USNR.

He married Susan Elizabeth and they have two children, Alan Michael and Alison Elizabeth. Civilian employment as vice-president and director, The Salzburg Seminar; chairman, California Agricultural Labor Relations Board; professor of international law, University of Salzburg; and Fulbright Scholar.

HILTON JAYNE, CDR, born Aug. 27, 1910, Elmira, NY, attended Cornell University and received his BA degree.

He was called to active duty as yeoman 3/c April 15, 1941, and assigned to USS *Howard* (DD-179) for duty in the North Atlantic. Commissioned ENS, (IVS) June 15, 1992, and assigned to the District Intelligence Office, 5th Naval Dist. From December 1992-September 1993 he served on staff of Advanced Naval Intelligence School, New York, NY. In October 1993 he was assigned to Naval Gp. China and proceeded to Chungking. In May 1994 he was appointed CO INTEL ONE and proceeded to HQ of General MA Chia Chiao, CO of LPA Guerrilla Army. Returned to ONI, Washington in October 1945.

Achieved the rank of CDR and was discharged from active duty March 1, 1946, and retired from USNR May 31, 1961. From September 1949-May 1961 he served with Public Relations 1-1. Jayne was awarded the Order of Cloud and Banner w/border of red #262 by Chinese National Government.

His memorable experiences: The five evacuations of his naval intelligence unit in the Ti Mo Shan Mts., southwest Shanghai, by Japanese forces.

Jayne married Frances S. and they have two children, Parker H. and Sarah J. Everdell. Their grandchildren are: Toby and Alice Jayne and Nicholas and Molly Everdell. His civilian employment was with Liberty Mutual Insurance and Central Intelligence Agency. He retired May 31, 1971.

WILLIAM E. JENNINGS, CDR, born Jan. 15, 1943, Cleveland, OH, graduated from Kent State University, obtaining a BS and MS degree. He began his naval career in 1964 as a reserve officer candidate and was commissioned ENS in 1968. He served with commander, Amphibious Training Command followed by service with Naval Control of Shipping and Military Sealift Command.

While serving in a variety of USNR Surface Units, a major portion of his career was devoted to the Military Sealift Command. CDR Jennings served in every unit officer capacity including executive officer and

CO. Recalled during Operation Desert Storm, CDR Jennings served in Saudi Arabia as operations officer, Military Sealift Command Office, Dammam, and chief staff officer, CDR Military Sealift Command Southwest Asia.

CDR Jennings is a public school principal in Norfolk VA. He resides in Virginia Beach with his wife, the former Myra Wilkinson, and they have three children: Elizabeth, Myra Briana and William Jr.

ARTHUR S. JENSEN, CAPT, born Dec. 24, 1917, Trenton, NJ, received from the University of Pennsylvania the degrees of BS (physics) 1938, MS (physics) 1939 and PhD (physics) 1941. He married Lillian Elizabeth Reed Aug. 9, 1941; they had twin sons in 1942 and a daughter in 1959.

On Aug. 1, 1941, he reported for active duty at the USNA to teach physics and physics of aviation in the Dept. of Electrical Engineering until March 1946. At RCA Laboratories Dr. Jensen developed the first practical high-density random access memory, used in second generation computers, serving a part in the DEW-line radar, and enabling the first all electronic telephone central office, over-the-horizon radar, and airborne moving target radar. He served as training officer in USNR Organized Surface Bn. 4-22. From 1957-94 at Westinghouse Defense and Space Center Dr. Jensen invented a high brightness display storage tube that was used for radar display in high altitude aircraft, an infrared camera tube that aided design of the SR-71 aircraft, and low noise integrated circuits. He contributed to the Defense Meteorological Satellite and other reconnaissance programs. Dr. Jensen's inventions resulted in 25 US patents, election to Life Fellow of the Institute of Electrical and Electronic Engineers and inclusion in *Who's Who in America*. His service as the electrical engineer on the Maryland State Board for Professional Engineers earned him the Governor's Citation in 1986. CAPT Jensen served as CO of Naval Reserve Research Co. 5-4.

Within 22 years of satisfactory service he earned the American Defense Service Medal, American Campaign Medal, WWII Victory Medal, Naval Reserve Medal, and Armed Forces Reserve Medal. He retired in December 1977.

CHRISTIAN EDWARD JENSEN, CAPT, born July 7, 1933, Newark, NJ, earned his BS and MA degrees from Rutgers; MD degree from Duke; and MPH from the Medical College of Wisconsin. He also graduated from the Naval War College in 1984 "with distinction."

He joined the service June 16, 1955. His military locations and stations: OCS, Newport; NSCS, Athens; USS *Cambria* (APA-36); USS *Rizzi* (DE-537); along with many other stations and reserve units; Fleet Hospital 15; and Naval Medical Clinic, Annapolis. Participated in Operation Desert Storm (Fleet Hospital 15, Al Jubail, Saudi Arabia).

His memorable experiences: CO of four naval reserve units. Jensen was awarded the Naval Commendation Medal, Kuwait and Saudi Liberation Medals, Southwest Asia Campaign Medal and numerous others. He achieved the rank of CAPT and is still on full-time active duty, Navy Medical Clinic, USNA, Annapolis, MD.

He married Gail Baxter and they have two children, Christian and Wendy, and four grandchildren: Christian, Jonathan, Jacob and Andrew. Civilian employment as medical director for DuPont Co., Delmarva Foundation for Medical Care, plus private practice.

JOHN THORVOLD "TORVOL" JENSEN, CAPT, born June 5, 1919, Duluth, MN, earned his BBA degree from the University of Minnesota.

Joined the service July 24, 1943. Basic training at Northwestern University, Chicago, IL. Overseas deployment: European Theater; England; Battle of Normandy. Brought in wave of small boats (LSVPs) to Omaha (Red Beach) with US Marines in second wave. His boat hit underwater mine so he spent the afternoon and evening on the beach.

Had 30 day leave in the States and was then sent to the Pacific, which included three landings in the Philippines, followed by numerous landings on most islands from the Philippines all the way up to Japan. He had a great small boat crew of five men who he loved dearly. He was sent back to the States in December 1945 and was discharged in February 1946. He achieved the rank of CAPT, USNR.

His memorable experiences: D-day, Battle of Normandy, numerous islands in the Pacific, including the falling of Japan.

Jensen is presently married to Lilian and has four children and seven grandchildren. Civilian employment in digital computer sales. He retired July 1, 1969.

KATHLEEN "KATHI" JENSEN, born May 24, 1953, earned her BA degree in language arts. She initially joined as a SK 3 in the APG (Advanced Paygrade) program using her civilian experience to be accepted into this program. Her first unit was NSC Yokosuka (now FISC) and was highly supported in her applying for a direct commission. She was commissioned Jan. 10, 1981, and spent the next three years working her way through Supply Corps School.

She joined the NRMTF (Naval Reserve Maintenance Training Facility) immediately after graduation (1984-87). She then joined NSD Puget (1987-90) until women were accepted into the Seabees. Jensen spent the next six years moving from the regiment, brigade and battalion staff. During this time she earned her Seabee Combat Warfare designator, Overseas Duty Ribbon and a NAM for performance during one of several field exercises. She is currently with the COMNAVSURFPACNW, but will move into the REDCOM 22 N4A position at the new fiscal year. She is also co-chairing the 1998 Supply Community Workshop to be held in Seattle, WA in May.

Jensen is a single mother, taking care of a great 3-year-old son, while attending the University of Washington-Bothell, earning an additional degree in computer software and systems.

ROBERT G. JENSEN, CDR, born Dec. 29, 1937, Winner, SD, earned his BA degree from the University of Colorado in 1961; his MPA degree in 1966; and his JD degree from the University of Denver in 1974.

He enlisted in the service in July 1961 and was commissioned in November 1961. His military locations and stations: LST-1126 and on the staff of COMPHIBPAC. He served with Recomm. JAE Corps, 1982, CIVLAWSUPACT 115, Denver.

Jensen was awarded the Navy Commendation Medal, Navy Reserve Medal, National Defense Service Medal, Navy Expert Rifleman, and Expert Pistol. He achieved the rank of CDR and retired in 1994.

His memorable experiences: active duty 1961-65; missile range, South Pacific, 1961-63; and witnessing two atom bomb tests.

Jensen married Carol and they have two children, Kalyn Bowers and Glenn; and two grandchildren, Chris and Jaron Bowers. In civilian employment he served as director of administration, Denver Water Board, 1966-92 and House Legal Counsel, United Power, 1993-present.

EDWARD P. "ED" JEPSEN, LCDR, born March 17, 1917, Alameda, CA, earned his BS degree in forestry from the University of California, Berkeley in 1939.

He enlisted in the service in December 1943 and served with the Motor Torpedo Boat Sqdn. 17, Naval Reserve, Alameda, CA 1946-66. He participated in the Philippine Liberation.

Jensen was awarded the American Theatre Ribbon, Pacific Theatre Ribbon, Philippine Liberation Medal, Naval Reserve Medal, and WWII Victory Medal. He achieved the rank of LCDR and was discharged in 1966.

He was married first to Josephine, who passed away in 1983, and is now married to Karen. Jensen has two children, Margaret J. Bowles and Helen J. Wall; and four grandchildren: John and Megan Bowles and Stephen and JoAnna Wall. Civilian employment was with Wells Fargo Bank as vice-president. He retired from the bank in August 1979.

WILLIAM B. "BILL," "JEP" JEPSON, CAPT, born July 28, 1928, Sioux City, IA, earned his BS degree from the University of Iowa.

He enlisted in the service June 8, 1948. His military locations and stations: San Francisco, Moffett Field, Monterey, Subic Bay, Alameda, New York City, Port Lyautey, and Alexandria, VA. He served with the office of naval intelligence/naval investigation service.

His memorable experiences: The opportunity to have lateral careers in the USNR and as a special agent provided fascinating opportunities domestically and abroad. He achieved the rank of CAPT and retired in June 1981.

Jepson was married first to Catherine (deceased) and is presently married to Patricia. He has three children: Frederica Jeannine, Christine Suzanne and Alexandra Marie; and six grandchildren: Avery, Willem, Morgan Leigh, Chris, Eric and Carey. He was employed as a special agent.

JASPER TRUETT "TRUETT" JERNIGAN, CAPT, born Jan. 26, 1945, Lillington, NC, attended the University of North Carolina at Chapel Hill and various naval schools. He earned a degree in business administration in 1967.

He enlisted in the USN Feb. 8, 1968, Raleigh, NC and was sent to boot camp at Great Lakes Naval Training Center. While in boot camp he received orders to report to OCS in April 1968, graduating as part of Co. B-6809. Out of OCS ENS Jernigan was ordered to NTDS School, San Diego, CA; then AIC School at GLYNCO, Brunswick, GA in preparation for duty in the operation department and combat information center aboard USS *Forrestal* (CVA-59) January 1969-July 1971 (including three Mediterranean cruises and a rare visit to Istanbul, Turkey). Following sea duty LT Jernigan was ordered to instructor duty at NROTC Unit, Jacksonville University, Jacksonville, FL as the first naval science instructor, instructing the nation's "first Women Officer Candidates in the NROTC Program" August 1971-August 1973 (under the command of CAPT "Mo" Muncie).

After more than five years on active duty Jernigan advanced his career through various billets/commands at the Naval & Marine Corps Reserve Center, Jacksonville, FL from August 1971-February 1994, retiring with 24 years of service as CAPT, USNR, Feb. 8, 1994. His unit received the "first" CAPT Jones Award for Command Excellence.

His memorable experiences: (Early 1970s)-served aboard USS *Forrestal* (CVA-59), the last carrier to visit Istanbul, Turkey; and "first" naval science instructor at NROTC Unit-JU; (1980s)-the first training officer to certify Naval Reserve Enlisted and Officers to be fire fighting instructors at FTC, Mayport Naval Station, FL; and (1990s) serving as naval liaison officer to US Embassy in Rio, Brazil and Quito, Ecuador during UNITAS 32 & 33, respectively, while serving with CSOLANTFOR.

Jernigan married Carlotta McGarvey in July 1971 and they have three children: Carlotta Cannon, Tessa Powell and Michael Meehan; and four grandchildren: Austin and Hunter Cannon and Brooke and Christian Powell. His civilian employment has included estate planning and brokerage management; Florida real estate broker; relocation specialist, HUD; and presently, a human services planner, Planning & Development Dept., Jacksonville, FL.

DR. ROY JOHANSEN, LCDR, born Sept. 12, 1939, Melvern, KS, earned his BS degree in education (1961) and his MS degree in education (1967) from Kansas State Teachers College; and his PhD in education in 1975 from Kansas State University.

Enlisted in the USN June 7, 1961. His military locations and stations: USS *Requisite* (ACS-18), USS *Mt. Baker* (AE-4) and USS *Surfbird* (ACS-383). He lived in Japan for two years.

Johansen was awarded the National Defense Service

Medal and Vietnam Service Medal. He achieved the rank of LT and was discharged in July 1966.

He and his wife, Judy Irene, have two children, Susan and Eric. He is employed as assistant superintendent at Pleasant Hill R-III.

CLARENCE SPAETH "CLINK" JOHNSON JR., LCDR, born Dec. 4, 1925, Newark, NJ, attended East Orange High; earned an AA degree from Princeton; BA and MA degrees from Montclair; and EdD degree from Rutgers.

He enlisted in the service Feb. 17, 1943. His military locations and stations: Columbia midshipman and GOOS, Washington, D.C. Units served with include USS *Cape Gloucester* (CVE-109), YMS-179 and USS *Des Moines* (CA-134). He participated in WWII and the Korean War (Mediterranean) with the Sixth Fleet.

Johnson was awarded the Asiatic-Pacific Campaign Medal, WWII Victory Medal and the Mediterranean Service Medal. He achieved the rank of LCDR and retired after six years active duty and 20 years of federal service in January 1972.

His memorable experiences: Atlantic command executive (USS *Kearsarge*, 1949); USS *Beak*; LST Terrebourne Parish; USNS Geigerts Breuerhaven; USS *Lake Chaplain* CV; and USS *Columbus* (T duty).

He and his wife, Dorothy R., have four children: Laurel, Valerie, Dorna and Julianne; and six grandchildren: Todd, Emily, Eliot, Reed, Francis and Alyssa. He was employed as an elementary school teacher for five years and as associate professor, NJIT, for 32 years. Johnson retired from civilian employment in June 1987.

CLAYTON H. JOHNSON III, CAPT, born June 5, 1944, Cos Cob, CT. His family moved to Columbia, MO in 1947 and he attended Hickman High School and the University of Missouri-Columbia, graduating with a BA in June 1966. He received a DDS from Washington University in St. Louis, MO in 1970.

He joined the USNR in September 1963 as an airman recruit, advanced to dentalman and drilled at NAS Olathe, KS until 1966 when he was commissioned an ENS in the 1925 program following his acceptance into Dental School. He participated in dental officer indoctrination at Newport, RI during the Summer of 1968 and completed an eight week clerkship at Great Lakes, IL during the Summer of 1969. Following graduation from Washington University School of Dentistry in June 1970, he reported for active duty aboard the USS *Hancock* (CVA-19), where he served until August 1972 as assistant dental officer and division officer.

His service included two WESPAC cruises, mostly off the coast of North Vietnam. During the interval between the cruises, he completed a 30-day anesthesia course at the Navy Hospital in Oakland, CA. During the second cruise he provided general, regional or spinal anesthesia for 45 cases. After his discharge in August 1972, he returned to Missouri where he affiliated with the USNR in Kansas City, MO. He served in a variety of positions in a variety of units in Kansas City and Springfield, MO from 1972-85 when he was selected as OIC of the First Detachment of the 14th Dental Co., 4th Dental Bn., 4th FSSG. He served in that post until September 1990. He served as an instructor in NR GEN IMA (D) 1318 and NR VTU GEN 1808 until his transfer to IRR. CAPT Johnson retired in June 1995.

Decorations include National Defense Service Medal, Vietnam Service Medal, Vietnam Campaign Medal, Meritorious Unit Commendation, Rifle Expert, Pistol Expert and Armed Forces Reserve.

CAPT Johnson established a private practice in August 1973 in California, MO after working parttime for Dr. Hadley Stonner in Tipton, MO. He also worked part-time as a dental officer at Missouri State Penitentiary and the Boonville (MO) Correctional Center from 1974-86. He is a member of the American Dental Assoc. and active in the Missouri Dental Assoc., the Central District Dental Society and the Jefferson City (MO) Dental Society. He has held a variety of positions in these organizations. He is also active in a variety of local civic and social organizations.

CAPT Johnson married the former Eleanor Jane Gallivan of New Madrid, MO Sept. 9, 1972. They have three children: Emily, Virginia and Letitia. The family lives near California, MO on a small acreage.

CLINTON B. JOHNSON, CAPT, is a native of Detroit, MI. He began military service in the 398th Military Police Bn. while at Wayne University. He is a 1960 graduate of the USNA. Designation as a naval aviator was November 1961. His first assignment was to VU-1, NAS Barbers Point, HI. Two Vietnam combat tours followed in 1965 and 1966 with VA-25 where he was credited with a MiG-17 kill. He left active duty after RAG instructor duties with VA-127, NAS Lrmoore.

CAPT Johnson joined Delta Airlines in July 1968. While with Delta he flew the DC-9, B-727 and L-1011. For seven years he was a B-727 line check airman. In 1975 he was the ALPA pilot contract negotiator. He retired from Delta in March 1997.

While with Delta he served in the USNR in the following squadrons: VF14B1, VA8V2, VA-209, VA5Y1, VA-1008 and VA-1074. He commanded VA-1008, VA-1074, NRNS Midway, NRNAS Agana. His last reserve assignment was naval liaison officer to the state of Wisconsin, 1988-90. As a naval aviator he flew T34, T28, A1, A4, A7, T33, F9, S2, C45, C12, P2, H25, 01 and 06 aircraft and remote controlled Regulus I missiles. CAPT Johnson completed three courses at the Naval War College and instructed one. He was a member of eight reserve promotion boards.

His awards include the Silver Star, Meritorious Service Medal, 11 Air Medals, Navy Commendation Medal, Combat Action Medal, Navy Unit Commendation Medal, National Defense Service Medal, Vietnam Service Medal, Overseas Service Ribbon, Armed Force Reserve Medal, Vietnam Air Gallantry, Gallantry and Civil Actions Medals, Vietnam Campaign Medal and RVNAF Senior Pilot Wings.

DONALD L. JOHNSON, CAPT, received his BA of education (1961) and MA of education in school administration (1971) degrees from Southwest Missouri State University, and the EdD in educational administration (1977) from East Texas State University. He graduated from Navy OCS in May 1961. Subsequently, from 1961-68 he served on active duty as a communications officer at the Naval Air Facility in Naha, Okinawa; attended the Defense Language Institute (Russian) in Monterey, CA; and served as an operations watch officer at the National Security Agency, Ft. Meade, MD.

From 1969-86 he served in a variety of naval reserve billets in Springfield, MO; Dallas, TX; and Little Rock, AR. A life member, he joined the National Reserve Assoc. in 1969.

CAPT Johnson presently resides in Russellville, AR with his wife, Dinah. He has served since 1978 as dean of graduate studies and continuing education at Arkansas Tech University.

ELLSWORTH LEE "ELLSWORTH" JOHNSON, LCDR, born Sept. 8, 1932, Norfolk, VA, received his BS degree from Maine Maritime Academy and his BBS degree from Old Dominican University.

He joined the service in April 1954. Johnson achieved the rank of LCDR and was discharged in May 1956.

He and his wife, Shirley D., have three children: Susan, Beth and Ellsworth Jr.; and one granddaughter, Catherine Elizabeth Baskerville. He was employed with NAVMASSO in the shipboard division until his retirement Jan. 3, 1966.

JAMES R. JOHNSON, LCDR, born Aug. 2, 1956, Sandusky, OH, earned his BS degree in environmental health in 1980.

He joined the service Feb. 9, 1983. His military locations and stations: Naval Reserve Center, Toledo and Naval Hospital, Cherry Point. He served with NR NH Portsmouth 1105, NR NH Great Lakes 1613, NR FH 250 CBT2 7, and FH 500 COMMZ II. He served in Operation Desert Storm at the Cherry Point Naval Hospital.

Johnson was awarded the Navy Commendation Medal, Navy Achievement Medal and Outstanding Volunteer Service Medal. He achieved the rank of LCDR and is still serving.

His memorable experiences: Recalled to active duty in support of Desert Storm. Selected for LCDR in September 1992.

Johnson married the former Nancy Jo Miller March 26, 1983. They have two children, Stephen James and Melanie Kathryn. He is employed as assistant superintendent of operations, city of Fremont.

JORDAN EDWARD "JOHNNY" JOHNSON, CDR, born Sept. 9, 1921, Don Pedro Dam, CA, graduated Narbonne High School, Lomita, CA and El Camino College, Gardena, CA.

He joined the service Sept. 4, 1942. His military locations and stations: Pre-flight, St. Mary's College, Oakland, CA (17th Bn.); NAS, Livermore, CA (Primary Flight School); NAS, Corpus Christi, TX (NATC); designated naval aviator Dec. 21, 1943; NAS, Ft. Lauderdale, FL (VTP type training); NAS, Glenview, IL (CARQUAL); VC 99 aboard USS *Hoggatt* CVE-75 WWII; NAS, Terminal Island, San Pedro, CA; Sept. 17, 1946, separated from active duty, NAS, Los Alamitos, CA, VA 72A.

Recalled to active duty July 3, 1948, NAS, Sand Point, Seattle, WA, NAAS Saufley Field, FL (flight instructor). He was discharged from active duty Aug. 2, 1949.

Units served with were VC-99, VA-72A, VA-214, VC-772, VS-772 and HS-771. Battles and campaigns: Operations against the Japanese forces in the vicinity of Okinawa Gunto, Sakishima Gunto and Ryukyu Island, Philippine Liberation and Occupation of Northern Hunshu.

Johnson was awarded the Air Medal w/2 Gold Stars, American Area Campaign Medal, Navy Occupation Medal (Japan), Asiatic-Pacific Campaign Medal, Navy Unit Commendation (CVE 75 and attached squadrons), WWII Victory Medal, and Naval Reserve Medal. He achieved the rank of CDR, USNR, and retired July 1, 1970.

His memorable experiences: On their way to the Philippine Islands the USS *Hoggatt Bay* was caught between two major typhoons, they were lucky no major damage. The first kamikaze attack against their carrier was quite an experience.

He and his wife, Elizabeth Jean, have two daughters, Barbara Jean and Marilyn Elizabeth; and four grandchildren: Jack Ernest III, Jason, Justin and Betsy. He was employed with Hughes A/C Co. as supervisor of Falcon Missile Planning Group and TRW assistant program manager. Johnson retired from TRW Oct. 1, 1986.

PIERCE J. "PIERCE" JOHNSON, RADM, born July 3, 1944, Columbus, NE, received his BS degree from the USNA in 1966 and his MS degree from the Naval Postgraduate School in 1974.

He enlisted in the service June 27, 1962. His military locations and stations: CNO (OPS-96/01), NPS, Military Sealift Command. He served with USS *Collett* (DD-730), USS *O'Bannon* (DD-450), USS *Everett F. Larson* (DD-830) and USS *Seattle* (AOE-3). He participated in the Vietnam War and Operation Desert Shield/Storm.

Johnson was awarded two Legions of Merit, Bronze Star Medal, Navy Achievement Medal w/Combat "V," Meritorious Service Medal (2 Awds.), Navy Commendation Medal, Combat Action Ribbon and various sea service awards.

He has three sons: Benjamin, Stephen and Samuel; and one grandson, Chase. RADM Johnson recently retired as head, Studies and Resources Administration Branch for the director, Assessment Division (N81) in the office of the Chief of Naval Operations.

DR. RONALD E. "RON" JOHNSON, CDR, born Oct. 14, 1939, Portland, OR, received his BS and MS degrees in civil engineering and his PhD in physical oceanography.

He received his direct commission as LT Feb. 5, 1980. He served with the Naval Meteorology and Oceanography Reserve Activity 2186.

Johnson was awarded the Navy Commendation Medal, Navy Unit Commendation Ribbon and Expert Rifle/Pistol. He achieved the rank of CDR and is still serving.

His memorable experiences: Completed tours as CO in October 1996, now in GUTU.

He and his wife, Roberta, have two sons, LT Robert and Ronald R. He is employed as a faculty member, Dept. of Oceanography, Old Dominican University.

STANWOOD W. JOHNSON, CAPT, a native of Plaistow, NH, graduated from Boston University in 1938, and enlisted as yeoman 2nd class in Boston Dec. 27, 1941. He served at NAVCRUITSTA Boston until he was commissioned the following year. Following commissioning he was assigned to COMONE for communications duty, remaining there until ordered to the Mediterranean.

From May 1943 until the end of WWII he served with RADM John L. Hall Jr., USN, in the latter's successive flag commands that planned and executed the amphibious assaults on Sicily, Salerno, Normandy (Omaha Beach) and Okinawa. As commander Tokyo Force, RADM Hall led his ships, loaded with occupation troops, into Tokyo Bay while the surrender ceremony was underway aboard the USS *Missouri.*

Between WWII and the Korean Conflict, CAPT Johnson served at COMONE as a traveling recruiting officer, and at COMINLANT in Charleston, SC, as communications officer. During the Korean Conflict he was assigned to NRS&ONOP Detroit until ordered to COMPHIBGRU 3 as aide and flag secretary.

From January 1955-July 1973, he served in various Ready Reserve assignments as commanding officer of several surface divisions, a surface battalion and a naval reserve officers school. He was a member of a selection board at BUPERS, and for several years a naval academy information officer (Blue and Gold Program). He was transferred to the Naval Reserve Retired List effective November 1976.

CAPT Johnson has been awarded the Navy Commendation Medal w/Combat "V". His campaign and service medals are: Naval Reserve Medal, American Campaign Medal, EAME Campaign Medal, Asiatic-Pacific Campaign Medal, WWII Victory Medal, Navy Occupation Service Medal, National Defense Service Medal, Korean Service Medal, Armed Forces Reserve Medal, United Nations Service Medal and Philippine Liberation Medal.

He and his wife, Gretta, live in Carlsbad, CA. His extended family includes daughter, Judy, and husband, Jack, who reside in New York with their grown children, Tristen and Brian; son, Stanwood, an attorney who lives in San Diego; son, Douglas, and wife Linda, who reside in Anchorage; and son, Sherman, and wife Susan, and their children: Mike, Matt and Jamie Ann, who live in Escondido.

THOMAS T. "TOM" JOHNSON, CDR, born April 25, 1935, Lincoln, NE, earned his BS degree from Syracuse University and JPA from March University.

He enlisted in the service Oct. 21, 1954, Burlington, VT. His military locations and stations: Norfolk, VA; Pt. Hueneme, CA; Davisville, RI; Brunswick, ME; and Newport, RI. He served with NMCB-21, NMCB-27, 7th NCR and 21st NCR.

Johnson was awarded the usual "reserve" plus pistol "sharpshooter" and Navy Commendation Medal. He achieved the rank of CDR and retired April 25, 1995.

His memorable experiences: Director, RCECOS; CO, NMCB-27; CDR, 7th NCR.

He and his wife, Sheri, have two sons, Michael and Peter. He has been self-employed as an architect since 1970.

WILLIAM HARDING "JOHNNY" JOHNSON, CDR, born March 6, 1921, Smithwick, SD, earned his BS degree in industrial education and his MS degree in trade and industry.

He joined the service Aug. 8, 1939; promoted through enlisted ratings to machinist mate first class; commissioned Jan. 12, 1946, as ENS; and promoted to CDR, July 1, 1962. Retired to Retired Reserve April 1, 1973. Served on USS *Concord* during WWII; NAVRADIOSTA (Navy 220), Londonderry, Northern Ireland; and on USS *Gilligan* during Korean War.

Johnson was awarded the American Defense Service Medal w/Fleet Clasp and one star, American Campaign Medal, Asiatic-Pacific Campaign Medal, National Defense Service Medal, WWII Victory Medal, Armed Forces Reserve Medal, UN Service Medal, Korean Service Medal, Navy Reserve Medal and Good Conduct Medal w/Bronze Star.

His memorable experiences: Becoming a Trusty Shellback the first time crossing the Equator March 13, 1941. Crossing the International Date Line in March 1941.

He and his wife, Edith Kelley, have three children: Larry William, Barbara Kay and Edward Alan; and eight grandchildren: Richard, Susan, John, David, Daniel, Ellen, Kathy and David. He was employed as a teacher-machine tool technology.

WINTON H. "WINT" JOHNSON, CAPT, born April 19, 1910, College Pt., Long Island, NY, earned his BA degree in 1934 from Columbia College and his MD degree in 1938 from the College of Physicians and Surgeons, Columbia.

He enlisted in the service Aug. 4, 1941. His military locations and stations: USN Hospital, Norfolk, VA; USS *Roper* in Atlantic; *Louisville* CA-28, Pacific 22 months. He served with the Pacific Seventh Fleet. Battles and campaigns participated in include North Atlantic Convoys, Pacific from Aleutians to liberation of Philippines.

Johnson was awarded the Bronze Star Medal (1945), Atlantic and Pacific Campaign Medals and 11 Battle Stars. He achieved the rank of CAPT and retired April 19, 1970.

His memorable experiences: Rescue of baby born in lifeboat on Atlantic; sinking of the first German submarine on Atlantic coast; battle of Lingayen Gulf; surviving two kamikaze attacks.

He married the former Louisa Tate and they have four children: Roland E., Emily J. Ference, Frederick L. and Winton Paul (deceased); and three grandchildren: Jennifer, Ian and Keri Lynn. He was employed by Hackensack Hospital Surgical Staff and in private medical practice 1946-85.

ERVIN F. JOHNSTON, CAPT, born Jan. 4, 1931, St. Louis, MO, earned his BS degree in civil engineering in 1952 and his JD degree in 1958 from the University of Arizona.

He enlisted in the service in June 1952. His military locations and stations: OCS, Newport (1952); USS *Ridgemont* AD-17 (1952-54). He served with COMSERVPAC (1954-55); IUW (1960-71); and group command, Long Beach (1971-73).

Johnston was awarded the Korean Service Medal and United Nations Service Medal. He achieved the rank of CAPT and was discharged from active duty in August 1955.

His memorable experiences: As commanding officer of IUW Div. 11-2 at Long Beach was instrumental in obtaining first equipment for any IUW Div. nationwide.

He was married June 6, 1953, to Ella Louise, and they have two children, Debbie and Paul, and four grandchildren, Paula, Timothy, Megan and Matthew. He served on patent council and general council of Naval Ocean Systems Center, San Diego, CA. He has been in private practice as an attorney specializing in patents since 1993 in San Diego.

ROBERT GRIFFIN JOHNSTON, CAPT, born Oct. 11, 1939, Memphis, TN, earned his BA degree in 1961 from Mississippi State University and his LLB degree in 1963 from Vanderbilt University.

He enlisted in the service Aug. 18, 1963, and served with NMCG-21. Johnston was awarded the Armed Forces Reserve Medal.

His memorable experiences: The judge advocate general of the Navy administered his oath as commander.

He married the former Elizabeth Boone Knight and they have two children, Arthur S. III and Amy Nell. Civilian employment as an attorney-at-law. He retired Sept. 30, 1993.

WILLIAM R. JOHNSTON, CAPT, born June 17, 1933, Pittsburgh, PA, received his BSIE degree from Pennsylvania State University.

He was drafted and volunteered for naval service at call up March 2, 1956. His military locations and stations: USNAD, Bangor and Bremerton, WA. He served with various units in Wheeling, WV; Pittsburgh, PA; Hazelton, PA; Rochester, NY; Springfield, MO; and Akron, OH.

Johnston was awarded the Naval Reserve Medal. He achieved the rank of CAPT and retired Sept. 30, 1984. His memorable experiences: Boot camp at Great Lakes, IL, March 2, 1956; commissioned out of OCS, Newport, RI, Oct. 6, 1956.

He and his wife, Margaret M., have two children, Paul and Mark, and two grandchildren, Devon and Eryn.

CHARLIE C. JONES, CAPT, born Oct. 13, 1927, Samson, AL, received his BA degree in 1954 from Baylor University, Waco, TX.

He enlisted in 1947 as SR USNR, 1947-54 various reserve assignments Pensacola, FL and Waco, TX. Commissioned ENS in 1954 via ROC Program. Active duty 1954-65: NAS, Atsugi, Japan; STRIKFORSOUTH, Naples, Italy; USS *Marias* (AO-57); USS *Northampton* (CC-1); CO, NRTC Waterbury, CT. 1966-85: Held various USNR billets in Corpus Christi, Ft. Worth and Dallas, TX.

Achieved the rank of CAPT and retired from the USNR in 1985. Jones was awarded the Naval Reserve Medal, National Defense Service Medal and Armed Forces Reserve Medal.

His memorable experiences: OPNAV liaison officer aboard USS N*orthampton*, President Kennedy's National Emergency Command Post Afloat.

He married the former Helen E. Hollingshead, and they have three children: Denise, Mark and Drew; and two grandchildren, Brianna and Camille. Jones retired as vice-president, United Services Planning Assoc. Inc., in January 1991. He now owns and operates Century Oaks, a bed and breakfast in Waco, TX.

GEORGE BRYAN JONES JR., CDR, born Jan. 8, 1947, Scotia, CA, received his BS degree in pre-medicine, 1969; MA degree in community psychology, 1976; MEd in rehabilitation counseling, 1986; and PhD in counseling psychology, 1987.

He joined the USNR in June 1969. His military locations and stations: Site one, Holy Loch, Scotland and served with USS *Hank* (DD-702), 1969-70 and USS *Canopus* (AS-34), 1972-73.

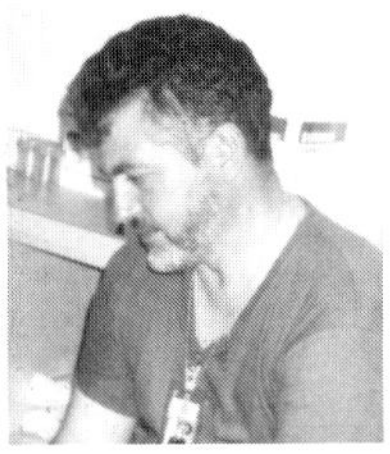

Jones achieved the rank of CDR and was discharged in April 1990. Jones was awarded the Unit Commendation (2 awards), Hazardous Duty Service Ribbon, Crisis Response Service Award, Commendation Medal, Achievement Medal, National Emergency Preparedness Service Ribbon, National Defense Service Medal, and Armed Forces Reserve Medal w/2nd award.

His memorable experiences: Graduating first in his class at Naval Hospital Corps School, 1970; Yom Kippur War, October 1973; being selected for CAPT, USNR after leaving naval service June 1991.

He married the former Cynthia L. Chester and they have two children, Carrie Beth and Christian Bryan. Civilian employment includes part and full-time faculty member; licensed psychologist; and certified professional counselor. He is now on continuous active duty in the US Public Health Service.

M. ADDISON JONES JR., CAPT, a native of Ashland, VA, graduated from the USNA in 1969 and was designated an unrestricted naval aviator in April 1971. His fist assignment was Helicopter Antisubmarine Sqdn. 5 (HS-5) at Quonset Point, RI. HS-5 deployed on the *Wasp* CVS-18, *Intrepid* CVS-11 and *Independence* CV-62 (CVW-7).

From 1974-75 he was the aviation program manager for Naval Recruiting District Boston. He recruited two of seven of the Navy's second group of female naval aviators.

CAPT Jones received a reserve commission in July 1975 and joined HS-75 in Lakehurst, NJ. When HS-75 moved to Willow Grove he was assigned to the COMNAVAIRLANT 0293.

In 1984 CAPT Jones joined the MARDEZLANT unit at NOB Norfolk, VA as the training officer. In 1986 he was selected as commanding officer for HM-1486 an augmentation unit for the Active Duty Helicopter Mine Sweeper Sqdn., HM-14.

In 1988 CAPT Jones joined the OP-03 unit at Anacostia supporting OPNAV in the Pentagon. In 1991 he joined OP-05 (DAW N88 0166) as the CV helicopter requirements officer. He attended all three Naval War College courses, the Reserve Components National Security Course at Ft. McNair and the Mine Warfare School in Charleston, SC.

CAPT Jones received the following medals: Meritorious Unit Commendation, National Defense Ribbon w/ star, Naval Reserve Ribbon w/HG, and Expert Pistol.

He and his brother established The Supply Room Companies, Virginia's largest independent office and furniture supply business. They operate 13 distribution centers throughout Virginia and Baltimore, MD.

He is married to the former Jessica Sanders and has three sons and a daughter (Kemper USNA '97, Taylor, University of South Carolina '00, Ben, Adelina). They reside in Richmond, VA.

THOMAS L. "TOM" JONES, CAPT, born March 1, 1950, is a native San Franciscan and 1972 USNA graduate. Following a tour on USS *Voge* (DE-1047) and NFO training, he reported to VP-48 flying the P-3C Orion.

After teaching NROTC at Miami University (Ohio), he became a TAR and served in a variety of billets in VP-69, VP-65 (OINC), and VP-66 (XO/CO). He also served on commander, Reserve Patrol Wing Atlantic's staff as chief staff officer and CNO's staff as director for Legislation and Information Management with the chief/director of Naval Reserve's staff.

In September 1996 he assumed command of Naval Air Reserve Whidbey Island. He is a life member of NRA, ROA, NERA, ANA, and TROA and currently serving a three-year term on the National Naval Reserve Policy Board. His highest earned award is the Legion of Merit.

CAPT Jones is married to the former Jean Kirkpatrick of Seattle and has two children, Jennifer and Matt.

WILLIAM N. JONES, LTJG, born Dec. 25, 1919, Hackensack, NJ, attended public schools 1926-33, Berkshire School 1933-38, Amherst College 1938-41, USMCR 1941-45, University of Rochester Medical School 1946-50. Internship: Rhode Island Hospital 1950-52. Orthopedic Residency: Children's Hospital, Boston, MA 1952-53; Massachusetts General Hospital, 1953-54; Worcester City Hospital, 1954-55; and VA Hospital, New Haven, CT, 1955.

He joined the USMCR in 1940 and transferred to USNR in 1952. His military locations and stations: Quantico, VA; Philadelphia Navy Yard; Camp Lejeune, NC; Southwest Pacific War Theater, 1942-44; USN Flight Training, Dallas, TX and Corpus Christie 1945. He served with the 11th Marine Regt., 1st Mar. Div., FMF 1942-44. Battles and campaigns participated in were Guadalcanal, Cape Gloucester and Peleliu.

He achieved the rank of CAPT in the USMCR (1944) and transferred to LTJG, USNR (1953).

His memorable experiences: Japanese air raid on transport fleet Aug. 8, 1942 and naval battle Aug. 9, 1942; Jap naval bombardment Oct. 13, 1942; band fighting at Tenaru River; landing at Cape Gloucester; landing on Peleliu crossing coral reef against hostile fire; warm welcome home stateside in California November 1944.

Married to the former Rosalind Loutrel and they have four children: Margarett, Rosalind, Amy and Natalie; and nine grandchildren: Cyrus, Clancy, Grey, Nina, Sarah, Shepherd, Lydia, Jackson and Rosalind. He is employed as an orthopedic surgeon in New London, CT.

RICHARD E. "DICK" JORGENSEN JR., CDR, born June 23, 1938, Los Angeles, CA, attended the University of California at Davis, receiving a BS degree in June 1960 and later attended the University of California Hastings College of Law. He enlisted in the USNR in 1961 and received his commission from Officer's Candidate School in 1962. Served aboard USS *Cavalier* APA-37 at Boat Group Commander in the Pacific. This tour was followed on board USS *Glennon* DD-870 as assistant CIC officer and navigator. His final two years on active duty were as instructor of military law at the Officer Candidate School, Newport, RI.

On inactive Naval Reserve duty, he served in various surface units in Ohio, Minnesota and California. In 1971 he affiliated with Naval Air Reserve Center, Moffett Field and served in various anti-submarine warfare units, being designated a naval aviation officer in 1979.

In 1984 he assumed the duties as executive officer of ASWOC 880 remaining at Moffett until 1986 when he transferred to NUWES DET 220, Keyport located at Reno, NV. In 1988 he was selected as commanding officer of NRVTU 2003, Pacific Grove, CA remaining in this assignment until his retirement in 1989. Additionally he served as a "Blue and Gold" officer on the Naval Academy information team from 1988-90.

CDR Jorgensen resides in Oakdale, CA having retired from Eli Lilly & Co. He currently is a regional manager for Agrimerica, Inc.

IVAN L. JOSLIN, CAPT, born Sept. 9, 1935, Fayetteville, NC, earned his BS in education with English and biology as his majors.

He enlisted in the service Jan. 24, 1954. His military locations and stations: DDs, 1958-60; submarines, 1960-71; and England, 1985-92. He served aboard the USS *Hartley* (DE-1029), USS *Sea Poacher* (SS-406) and USS *U.S. Grant*.

He achieved the rank of CAPT and retired from CINCLANTFLEET July 31, 1992.

His memorable experiences: Broken service—Active duty 1958-70; active reserves 1970-85; active duty 1985-92. Was PO2 when commissioned June 10, 1958.

He married the former Marjorie B. Fitchett and they have five children: Stephen, Beverly, Lessie, Brian and Michael; and 10 grandchildren: Carleen, Cole, Daniel, Wil, Caleb, Megan, Benjamin, Christine, Michele and Brenda.

GEORGE WILLIAM DOUGLAS "DOUG" JOYNER, CDR, born July 6, 1922, Franklin County, Louisburg, NC, earned his AA degree from Louisburg Jr. College; AB degree from Atlantic Christian College; and MA degree from East Carolina University.

He enlisted in the service Dec. 2, 1942. His military locations and stations: Pensacola, FL; Ellyson Field; Bronson Field; Atlanta, GA; Olathe, KS; Camp Shelton Naval Separation Center, Little Creek, VA; Norfolk, VA; Long Beach, CA; and Glenview, IL.

Joyner was awarded the American Theatre Medal, WWII Victory Medal, Naval Reserve Medal and Armed Services Medal. He achieved the rank of CDR and retired July 1, 1970.

His memorable experiences: Was in Pre-Flight School, University of North Carolina, Chapel Hill, NC the summer of 1943 with Ted Williams, Johnny Pesky and Johnny Sain, notable major league baseball players, and his physical education teacher was future president, LTJG Gerald Ford. Summer baseball games with local teams was exciting and interesting to say the least.

Married the former Edna Moye and they have one daughter, Debra Lynn Joyner Sawyer, and one grandson, Kenneth Douglas "Kenny" Plummer. He was employed as director, Audiovisual Education and director, Library Media Services at Wayne County School, Goldsboro. He retired July 1, 1988.

THOMAS J. "TOM" JUDGE, CAPT, born Feb. 21, 1935, Kansas City, MO, earned his BA degree from St. Mary's College, Winona, MN in 1956.

He enlisted in the service Sept. 14, 1956, and served on active duty October 1956-August 1961. His military locations and stations: OCS, Newport, RI and Electronics Officers School, Great Lakes, 1957. He served aboard the USS *Savage* (DER-386), Pearl Harbor, August 1957-February 1960 and USS *Henderson* (DD-785), San Diego, March 1960-August 1961.

Judge was awarded the Surface Warfare Officer and Naval Reserve Medal w/2 10-year HG Awards. He achieved the rank of CAPT and retired April 1, 1987.

His memorable experiences: CO, NRSD 11-15(S), 1968-69; CO NRSD 11-22(L), 1971-72; CO, VTU 1904, 1979-80.

He and his wife, Mary T., have three children: Robert, Cathryn and Richard; and two grandchildren, Taylor and Olivia. He was employed as manager,

Computinh, North American Aircraft Rockwell International. He retired Feb. 24, 1995.

CARROLL MICHAEL "MIKE" JUDICE, CDR, born May 25, 1947, Jeanerette, LA, graduated from the University of Southwestern Louisiana in 1969 and entered Aviation Officers Candidate School at NAS Pensacola, FL in May 1970 and was commissioned in September 1970. Following training at VT-29 in Corpus Christi, TX, he was designated a naval flight officer and received his "Wings of Gold" in March 1971.

After completing training with VP-30, Patuxent River, MD, in the P-3C, he reported to VP-49 at NAS Jacksonville, FL, where he served as acoustics officer, aviation equipment branch officer and assistant CMCS officer. He deployed to NS Keflavik, Iceland, three times with the "Woodpeckers" of VP-49.

CDR Judice was released from active duty in August 1974 and affiliated immediately with a squadron augment unit at NAS Jacksonville, FL. In August 1976 CDR Judice affiliated with Patrol Sqdn. 94. While attached to VP-94, CDR Judice served as airframes officer, assistant NFO training officer, assistant administrative officer, assistant maintenance officer, maintenance officer and executive assistant to the commanding officer. As a "Crawfisher," he completed detachments to NS Rota, Spain; NAF Lajes, Azores; NS Sigonella; and NS Bermuda.

His military decorations include the Navy Achievement Medal, Coast Guard Meritorious Unit Commendation Ribbon, Navy Battle Efficiency "E", National Defense Service Medal, Sea Service Deployment Ribbon, Armed Forces Reserve Medal (2 awards), Navy Expert Rifleman and Expert Pistol Medals.

CDR Judice is employed by the Iberia Parish School Board and is currently a principal in the school system. He is married to the former Pam Steen of Metairie, LA and they reside in New Iberia with their six children: Kevin, Brett, Brad, Kimberly, Chad and Ashley.

ALBINAS K. "AL" JURKONIS, CDR, born July 21, 1922, Springfield, IL. Enlisted in the NavCad Program Aug. 22, 1942. Graduated at Pensacola as naval aviator and commissioned an ENS. Assigned to Instrument Flight Instructor School. Upon completion, returned to Pensacola to instruct cadets the art of instrument flight. Also served with the Air/Sea Rescue Unit based at Cory Field.

Qualified as PPC in PBY, PBY-5A, and PBM-5 when sent to the Pacific Theater to VP-53 based on Saipan. On arrival, ordered to Majuro Island in the Marshall Islands to establish a solo Air/Sea Rescue Unit. They were airborne and on assigned station during Operation Crossroads as one of their assignments. Months later, VP-53 received orders to the States and based at NAS Whidbey Island. On arrival received orders to VP-62 based on Adak Island in the Aleutians where he served until released from active duty in January 1947.

CDR Jurkonis became very active in the Naval Ready Reserve Program to maintain his skills. Recalled to active duty in 1953 during the Korean War. Assigned to Hurricane Hunter Sqdns. VJ-2 and VW-4 based at NAS Jacksonville, FL. Designated PPC in the PB-4Y and P2V3W. Upon release from active duty he continued to be active in the Naval Ready Reserve Program.

CDR Jurkonis has over 4,000 hours in military aircraft. Besides aircraft already mentioned, a variety of single and multi-engine trainers, the R4D and a bi-wing single float relic J2F6. Has a commercial pilot license for single and multi-engine aircraft with instrument rating.

Awarded the National Defense Service Medal, Naval Reserve Medal, American Area Campaign Medal, WWII Victory Medal and Asiatic-Pacific Campaign Medal. He achieved the rank of CDR and retired from the USN March 1, 1970.

CDR Jurkonis is a life member of the Naval Reserve Assoc. He played a major part in organizing the Houston Space Center Chapter and served as the chapter's charter president.

He and his wife, Mildred A., have one son, Paul M. and one daughter, Aldonna K. Kroeker and two grandchildren, Paul M. II and Shauna R. Kroeker. He was employed in a number of positions as industrial engineer and plant manager until his retirement in August 1984.

JOHN FRANCIS PATRICK "GUNNER" KAHOE, CWO3, born Sept. 15, 1944, Cheltenham, England, joined the service Dec. 28, 1962. His military locations and stations: ET "A" School, San Francisco, CA; NAVSTA Midway Island; Naval Ammunition Depot, Oahu, HI. He served aboard the USS *Interceptor* (AGR-8), USS *Canopus* (AS-34), Momag Det. 1711, and Naval Reserve Weapons Station Det. 1311.

He was discharged from active duty Dec. 21, 1970, with the rank of PO1 and achieved the rank of CWO3 at time of retirement Aug. 1, 1985. Kahoe was awarded the Navy Commendation Medal, Navy Achievement Medal, Navy Good Conduct Medal, Naval Reserve Meritorious Service Medal, Naval Reserve Medal and National Defense Medal.

His memorable experiences: Appointment as CWO, chief's initiation, crossing the Equator, three year tour in Hawaii, tours in Spain and Scotland, 11 months-two weeks-five days on Midway Island, refresher training in Guantanamo Bay, Cuba, first visit to the Vietnam Memorial, CDR W.H. "Harry" Munyon.

He and his wife, Jan, were married Oct. 2, 1966, and they have two sons, Sean and Todd. He was employed by General Motors Corp. as maintenance supervisor and body shop technical manager (18 years). Journeyman electrician with International Brotherhood of Electrical Workers Local Union 1141.

DAVID W. KALEEL, LCDR, born April 25, 1960, Jacksonville, FL. He began his naval career as a nuclear officer candidate at the University of Florida receiving a BSEE degree with honors in December 1982. After nuclear submarine training, he was assigned to USS *Minneapolis-St. Paul* in Norfolk, VA from January 1985-June 1988 completing three overseas deployments. His next assignment was operations officer at FBM OTSU-II aboard USNS *Range Sentinel* in Port Canaveral, FL during the Trident-II D5 missile test program. He left active duty in October 1988 and has since been a program manager at Harris Corporation in Melbourne, FL.

After release from active duty, he received his MBA from the University of Central Florida and has remained active in the reserves. He is currently assigned to COMSUBGRU-8 at Orlando Reserve Center regularly performing annual training at the gaining command in Naples Italy.

LCDR Kaleel has been awarded the Navy Commendation Medal, Navy Achievement Medal w/2 stars, Joint Meritorious Service Citation, Navy Unit Citation, Meritorious Unit Citation, Expeditionary Medal and various other service ribbons.

The Kaleel family consists of his wife, the former Lesa Ochab, and daughter, China.

JOHN L. KALTENBACH, CAPT, born Nov. 11, 1930, Midwest, WY, earned his BS degree in geology at the University of Wyoming in 1953.

He joined the service Nov. 3, 1948, as SR, USNR. His military locations and stations: NR Training Center, Billings, MT; Navy OCS, Newport, RI; NAS Norfolk, VA; NAS Denver, CO; NR Training Center, Denver, CO; NAS Dallas, TX; and NAS New Orleans, LA. He served with TACRON 21, ORGSURDIV 13-25, AIRTU 701, NAIRU 701, INTEL Div. 9-1, NAIRU 821/XI (CO), INTEL VTU 109, INVEST OPS 1510 (CO), COM 8 INTEL 110 (CO), NR NICHQ 0118, RES INTEL Area 7 (coordinator/CDR).

He achieved the rank of CAPT, USNR and retired July 1, 1987. Kaltenbach was awarded the Navy Meritorious Service Medal, Navy Occupation Service Medal (Europe), National Defense Service Medal, Armed Forces Reserve Medal w/2 Hourglass Devices.

His memorable experiences: Recalled to active duty in 1973 during Operation Homecoming to serve as senior debriefing officer to USN POWs. Prepared special project reports for the director of Naval Intelligence on earth observations oceanography and meteorology features and phenomena observed from NASA orbiting manned spacecraft. Graduate of the Senior Reserve Officer's Course and the Command and Staff Course at the Naval War College, the Defense Strategy Course at the National Defense University, and the Senior Naval Reserve Officer Orientation Program at CINCLANTFLT. Served on a naval reserve line selection board.

He and his wife, Carolyn, have one daughter, Julie Ann, and one grandchild, John "Josh" Garrett. He was employed by NASA, Johnson Space Center; manager, Space Shuttle Earth Observations Project Office. Kaltenbach retired Jan. 5, 1985.

BEAU K. KANSTEINER, LCDR, born March 15, 1934, St. Charles, MO, earned his BS degree in engineering from the University of Kansas.

He joined the service in September 1952. His military locations and stations: University of Kansas and USS *Gregory* (DD-802). He served with DesRon 17 and 21 and participated in battles and campaigns at Quemoy and Matsu.

He achieved the rank of LCDR and retired March 15, 1994. Kansteiner was awarded the National Defense Service Medal, Air Force Expeditionary (Taiwan, Quemoy, Matsu) and Naval Reserve Bronze Star.

His memorable experiences: Pac deployments: Yokosuka, Sasebo, Kobe, Subic, Kao Shung, Hong Kong, Guam and Pearl Harbor.

He was employed as an engineer with Boeing and as manager, Leavenworth Water Dept.

STEVEN B. KANTROWITZ, CAPT, born Dec. 25, 1953, Brooklyn, NY, earned his BS degree at Cornell University; JD degree at Villanova University; and LLM degree at George Washington University.

He joined the service in September 1974 with the rank of ENS, USNR. His military locations and stations: Office of the Judge Advocate General. He served with NRLSO 204, Philadelphia; VTU Law 0413; NAVJAG 106; and SJA-Reserve Readiness Command Region Four.

CAPT Kantrowitz was awarded the Meritorious Service Medal w/Gold Star, Navy and Marine Corps Commendation Medal w/Gold Star and Navy and Marine Corps Achievement Medal.

He and his wife, Christine, have two children, Lisa and David. He is a managing partner of McBreen, McBreen and Kopko Law Firm in Philadelphia, PA.

MICHAEL GEORGE "MIKE" KARAGIANS, CAPT, born Aug. 6, 1934, Chicago, IL, has earned his BS, BSD and DDS degrees.

He joined the service in July 1960. His military locations and stations: MCRD San Diego, CA and GLNTC, IL. He served in the dental department, Reserve 9-4, Forest Park, IL.

CAPT Karagianis was discharged in August 1962. He was awarded the National Defense Service Medal and

Expert Rifle and Pistol. He is a life member of the Naval Reserve Assoc.

His memorable experiences: USS F. Marion 16 day cruise; port mishap; surface vessel collision with six casualties; emergency sick bay; reserve relief; Naval Hospital, Naples Italy; and medical evacuation.

He and his wife, Vicki, have three children: George, Lydia and Leland; and two grandchildren, Kyleigh and Alexandra. He was employed as a dentist in private practice for 37 years. Karagianis retired in October 1988.

ROBERT E. "BOB" KARAS, CDR, born May 25, 1934, Chicago, IL, earned his BSCE degree at Marquette University, advanced science, and attended US Postgraduate School Monterey.

He enlisted in the service in June 1956. His military locations and stations: 1956-58, Harbor Defense Units, San Francisco and Yokosuka Japan; 1959-61, USS *Laffey* (DD-724); 1961-62, Postgraduate School, Monterey, CA and Steam Engineering School, San Diego, CA; 1962-64, USS *Wasp* (CVS-18); 1964-65, chief of staff and material officer commander Destroyer Sqdn. Two; 1965-67, engineering head, staff of commander training, Command Atlantic Flt.; 1967-69, executive officer, USS *Noa* (DD-841); 1969-71, staff analysis and project officer, Center for Naval Analysis; 1971-73, commanding officer, USS *Henderson* (DD-785); 1973-76, staff OP-03, Washington, D.C.; 1976-80, commanding officer Naval Reserve Center, Chicago and Forest Park, IL.

He participated in the 1962 Cuban Crises aboard the USS *Wasp*; the Vietnam Cruise aboard the USS *Noa*, 1968-69; and Vietnam Cruise aboard the USS *Henderson*, 1972-73.

CDR Karas retired in June 1980. He was awarded the Navy Commendation Medal, Second Award w/Combat V, Combat Action Ribbon, Navy Unit Commendation, National Defense Service Medal, Vietnam Service Medal w/3 stars, Armed Forces Expeditionary Medal, Vietnam Service Medal w/3 stars, RVN Gallantry Cross Unit Citation and RVN Campaign Medal. Authorized to wear: Surface Warfare Device, Command at Sea Device and Major Command a Shore Device.

He married the former Barbara E. Falbe. He was employed as manager, technical market, Everpure, Inc. until his retirement in July 1996.

DANIEL K. KAST, LT, born Nov. 15, 1919, earned his BS degree in 1942 at New York University; attended the New York University Graduate School of Business Administration, 1942-April 1943; earned his JD degree at New York Law School in 1953; and his MBA (taxation) degree at Pace University in 1970.

Military Service: April 1942, A.S. "boot training," Sampson, NY. May 1943-September 1944, yeoman 3/c, "satisfactory active service, released under honorable conditions." September 1944, appointed ENS USNR, permanent commission USNR. April 1946 promoted to LTJG and released to inactive service. January 1954 assigned to Naval Reserve Law Co. 3-3. Designator changed from line to (1105) law specialist. Promoted to LT in 1961. Retired to inactive reserve in 1966.

Service schools included Yeoman School, Newport, RI; NTS Tucson, AZ (University of Arizona); ATS (Amphibious Training Base) Coronado, San Diego. He served aboard the USS LST 611, the USS *Oneida* APA-221 and the USS ATR-72. Military locations and stations: China, Korea, Okinawa, New Guinea, Kwajalein, Ulithi, and the Philippines.

Kast was awarded the American Theater Medal, WWII Victory Medal, Navy Occupation w/Asia-China Clasp, China Service Medal and Naval Reserve Medal.

He and his wife, Leonore S., have two children, Robert A., M.D. and Caren Sue. He was employed as an attorney and counsellor at law in New York; a certified public accountant in New York; and professor of accounting, law and taxes at Iona College (1961-90). He retired in 1990 as professor emeritus (accounting and law).

WILLIAM J. "BILL" KASTNER, CAPT, born May 27, 1928, Jersey City, NJ, earned his BA degree in mechanical engineering (cum laude) at Catholic University in 1951 and his MA degree in public administration at USC in 1988.

He joined the service Aug. 11, 1952. His military locations and stations: Navy Mine Defense Lab, Panama City, FL and NAVSEA, Washington, D.C. He served with NAVSEA Reserve Units and Buships Co. 3-1.

CAPT Kastner retired July 1, 1985. He was awarded the Legion of Merit, Meritorious Service Medal (3 awards) and the Navy Commendation Medal.

His memorable experiences: Aid to Igor Sikorsky on visit to base; taking a WWII PT boat on training trip in Gulf of Mexico (with his wife to be) in 1955.

He and his wife, Nelda, have six children: Joseph, Margaret, Ann Marie, Timothy, David and Jack; and nine grandchildren: Jimmy, Christie, Mick, Zach, Julia, Ronnie, Matt, Craig and Weston. In civilian life: Production, sales and marketing engineer and regional sales manager for two production line packaging machinery companies; production manager of a brewery; corporate engineering manager of a national soft drink manufacturing corporation; vice-president and general manager of a building supply manufacturing company; vice-president of marketing for two packaging machinery companies; and sales manager for a national gas turbine manufacturer. Presently director of the Naval Sea Systems Command, Naval Reserve Office, Navy Dept., Washington, D.C.

HOWARD TODD KAUDERER, CDR, earned his BSIE degree in industrial engineering at Technological Institute, Northwestern University in 1976; his MM degree in management at Kellogg School, Northwestern University in 1986; and his MS degree in strategic intelligence at Defense Intelligence College in 1991. (He was the first NR officer called to active duty for one year to attend program.) From 1994-present he attended the University of Southern California where he earned his DPA in public administration (RD&A).

He was commissioned in June 1976, NROTC Program at Northwestern University; 1976-79, USN assistant CIC officer and firewatch officer, USS *Independence* (CV-62); 1979-80, USN assistant chief engineer, main propulsion, USS *Deyo* (DD-989); 1980-84, USN intelligence officer, Navy Operational Intelligence Center; 1984-present, USNR, National Military Joint Intelligence Center.

Kauderer was awarded the Defense Meritorious Service Medal, Navy Commendation Medal w/Gold Star, Joint Achievement Medal and Navy Achievement Medal.

He married the former Nancy Susan Norton of River Forest, IL and they have three children: Amanda M., Robert G. and Heather A. From 1984-90 he was employed as director of operations and senior program manager, SATELLEASE, Inc. Presently senior professional staff, Joint Warfare Analysis Dept., Applied Physics Laboratory, The Johns Hopkins University.

KIRK K. KAZARIAN, CAPT, born June 11, 1930, New York, NY, graduated from Bucknell University with a BS in electrical engineering. He attended OCS Newport and commissioned ENS in 1953. During the Korean War he was an engineering duty officer (electronics). After release from active duty he attended medical school and continued post graduate surgical training at New York Medical College and served in the Medical Corps as MSLO and RDAC, NRD New York.

CAPT Kazarian retired as chief of surgery at Stamford Hospital, Connecticut in 1990 and was recalled to active duty in June 1990 at the Naval Hospital and Naval Medical Research Institute, Bethesda in the Septic Shock Research Program.

He is professor of surgery at New York Medical College and research professor at the Uniform Services University of Health Sciences, Bethesda. He and his wife, Micheline, enjoy retirement. His son, Karl-Antoine graduated from the USNA and is a surface warfare officer stationed in San Diego.

REV. PIUS F. KEATING, CDR, born Dec. 7, 1916, Plymouth, PA, graduated from St. John's Franciscan College, New York and attended the Catholic University of America, Washington, D.C. from 1940-46 where he was ordained to the priesthood Feb. 9, 1946.

He joined the service Feb. 15, 1951. His military locations and stations: USS *Valley Forge* CVA-45. He participated in the Korean War for two years and earned four stars.

Keating was awarded the Armed Forces Reserve Medal, National Defense Service Medal w/star and United Nations Service Medal. He achieved the rank of CDR and retired March 1, 1971.

He was an ordained Catholic priest for nearly 50 years.

EARL EUGENE KEELE, LCDR, attended William Jewell College, Liberty, MO, obtaining the degree of BA in history in 1956; Midwestern Baptist Theological Seminary obtaining a THM degree in 1961. He began his active duty naval career June 29, 1962.

After attending Chaplains Orientation School in Newport, RI, he served two years attached to the Marine Corps Base Camp Lejeune, NC followed by a tour on Adak, AK with NS, Adak, NAV COM Station and Coast Guard Loran Station, Adak. After a two year tour with NAS MFS, Millington, TN he was assigned to the Fleet Marine Force Pacific and sent to the First Marine Aircraft Wing, Da Nang, Vietnam, in December 1967, serving with Marine Aircraft Gp. 11 at Da Nang, and Marine Aircraft Gp. 36 stationed at Phu Bai near Hue City. His last tour before being released from active duty was with the Naval Disciplinary Command, Portsmouth, NH with ADDU PORTSMOUTHNAVSHIPYARD, and Portsmouth Naval Hospital February 1969-June 26, 1971.

Chaplain Keele returned to Missouri and entered Southeast Missouri State University, Cape Girardeau, where he obtained a BSc in secondary education, with a lifetime teaching certificate in 1974. Chaplain Keele and his family moved to Grand Junction, TN where he taught science three and one-half terms to junior high students, attending Memphis State University, Memphis, TN for seven hours of science courses.

In 1979 Chaplain Keele was hired by the Dept. of Corrections of the state of Tennessee as Chaplain I and retired from that position as Chaplain II March 24, 1989.

Chaplain Keele was retired from USNR, while attached to ABFC-HQA 0179 NAS MFS Millington, TN July 1, 1981, and retired from the USN June 12, 1988.

Prior to his naval career, Chaplain Keele also served in the USAAC, which became the USAF, July 1947-July 1950. His overseas tour of duty was on the island of Guam with a B-29 unit working as an aircraft structural mechanic.

Chaplain Keele currently enjoys retirement with his wife, Naomi, in Greenville, MO. They have three children and three grandsons. He has been on a 30 day archaeological dig in Israel and enjoys reading about, and following, the course of Near Eastern archeology, landscaping their yard, woodwork, and adding to their home. Their oldest daughter lives in Wales, and the youngest daughter lives in Tennessee. One child, their son, is deceased. He is currently a member of the Wayne County Historical Society and spent most of the spring, summer and fall in 1995 helping tear down, move and restore a historical log house.

Chaplain Keele is proud of his medals and military plaques, but especially proud of a set of rocking chair wings presented to him at MAG-36.

RICHARD R. "DICK" KEISTER, CAPT, born April 25, 1939, Pomona, CA, attended the University of Southern California where he received his BS degree in marketing management in 1962.

He enlisted in the service in February 1962. His military locations and stations: Navy Supply Corps School, Athens, GA; Weapons Supply School, PACFLTRACOM, San Diego, CA; and US Naval Construction Battalion Center, Port Hueneme, CA.

Keister was awarded the National Defense Service Medal, Antarctica Service Medal and Naval Reserve Service Medal (20 years). He achieved the rank of CAPT and transferred to the Retired Reserve March 1, 1986.

His memorable experiences: Upon release from active duty in February 1966, affiliated with Naval Reserve Supply Co. 11-10, North Hollywood, CA and became XO in 1970. In 1971, affiliated with the Naval Air Reserve, Pt. Mugu, CA and was a plank owner in newly commissioned NARDIV-U1. Served as supply officer in nine different reserve units and Component A director of VTU 7676, Pt. Mugu, CA.

He and his wife, Donna, have two children, Bryon and Cheryl; and six grandchildren: Jake, Nicole, Vanessa, Marc, Joshua and Andrew. His civilian employment was with Port Hueneme Div., Naval Surface Warfare Center Head, Management and Organization Support Office, Code 5A04.

PHILIP A. "PHIL" KEITH, CAPT, born Aug. 24, 1946, Springfield, MA, earned his BA degree in 1968 at Harvard; a MS degree in 1975 at San Jose State and attended the Naval War College.

He joined the service Sept. 1, 1964. He served as naval aviator and naval intelligence officer with VP-6, CINCPAC, VP-31, aboard the USS *Constellation*, and ONI. He participated in the Korean and Vietnam Wars.

Keith was awarded the Presidential Unit Citation, Air Medal, Navy Commendation Medal, VC, National Defense Service Medal, Armed Forces Expeditionary Medal, Vietnam Service Medal, Expert Rifle, VCG and Vietnam Campaign Medal. He achieved the rank of CAPT and retired in 1987.

ARTHUR V. KELLER JR., LCDR, born Dec. 13, 1942, Elyria, OH, graduated from Bowling Green State University, Bowling Green, OH.

LCDR Keller enlisted in the USN Aug. 8, 1960. He completed Interior Communications School at Great Lakes and was assigned to the USS *Independence* CV-62. While on board he participated in the Cuban Blockade. He was transferred to the USS *Fulton* AS-11 in 1963 and was discharged as a PO2 Dec. 9, 1963.

On July 27, 1977, he enlisted in the Naval Reserve under the APG program as a 1C boiler technician and was selected for CPO in 1981. He was commissioned ENS under the LDO program in 1985 and promoted to his present rank of LCDR in 1995.

LCDR Keller's awards include Enlisted Surface Warfare Specialist, National Defense Service Medal, Armed Forces Expeditionary Medal, Naval Reserve Meritorious Service Medal, Good Conduct Medal, Armed Forces Reserve Medal and the Naval Expeditionary Medal.

The Keller family consists of wife, Linda; sons, Arthur and Louis; daughters, Michelle and Kathleen; and grandson, Dale Louis. They reside in Avon Lake, OH.

WILLIAM L. KELLICK, CAPT, born July 24, 1923, Niagara Falls, NY, graduated from Holy Cross College, was member of NROTC and commissioned ENS in February 1944.

During WWII he served on board LST-707 (July 1944-June 1946) in Asiatic-Pacific Campaign, participating in invasions at Leyte, Luzon and Mindanao; landings at Iwo Jima and Okinawa; and was awarded five Battle Stars.

After discharge June 28, 1946, he participated in Naval Reserve for 30 years, filling various billets at NAS Niagara Falls, NAS New York and NAS Willow Grove until retirement July 1, 1974. He was promoted to CAPT June 1, 1965.

Campaign and service medals: American and Pacific Theatre, Philippine Liberation Ribbon, Occupation (Japan) Medal, WWII Victory Medal and Naval Reserve Medal.

He obtained a JD degree from Fordham Law School, practiced law in Niagara Falls and served for 30 years as family court judge until retirement Dec. 31, 1993. In 1986 Niagara University awarded him honorary degree, Doctor of Laws.

Married in 1951 to the former Joan Quinn, they are parents of five children and have seven grandchildren.

His memorable experiences: (a) Each of the many kamikaze attacks; (b) at time of Japanese surrender in September 1945 their ship joined hundreds of Allied warships at anchor in Tokyo Bay—a most impressive sight; and (c) bringing their LST-707 "home" from the war—sailing under the Golden Gate Bridge in January 1946.

HUSTON GLENN KELLOGG, CAPT, born April 6, 1924, Los Angeles, CA, earned his BS degree at Yale University and his MD degree at Washington University Medical School.

He joined the service Oct. 27, 1942. His military locations and stations: V-12 Yale and Washington University Medical School. He served with MSTS Pacific, USNH, San Diego, 4th Tank Bn., Medical Co. and NRMC 5621, San Diego.

CAPT Kellogg was discharged April 6, 1984. He was awarded the Good Conduct Medal, American Campaign Medal, WWII Victory Medal, Japanese Occupation Medal, National Defense Service Medal, Naval Reserve Medal and NRA National Award of Merit (1984).

He and his wife, Eleanor, have two children, Jacob and Paul, and one grandson, Derek. He was employed as a pediatrician until his retirement May 25, 1995.

JOHN S. "JACK" KELLOGG, CAPT, born Jan. 26, 1924, Downers Grove, IL, enlisted in Naval Reserve in October 1942, at age 18, in V-12 program, reporting for active duty July 1, 1943, at Central Michigan College, Mt. Pleasant. Next was Midshipmen's School, Plattsburg, NY. Commissioned ENS June 27, 1944. Following Tactical Radar School, Hollywood, FL, assigned as CIC officer, USS *Lenawee* (APA-195). Participated in troop landings at Iwo Jima and Okinawa. Completed active duty at Naval Base, Tinian.

After graduating from Law School, University of Michigan, (GI Bill) in 1950 came to Denver, retired from law practice in 1993. Retired from Naval Reserve, 1982. Reserve "career" in legal units, mostly non-pay, Denver. Served as CO of both types.

He married the former Margaret Pietz Dec. 27, 1952, and they had three children. She died in 1991. On June 26, 1993, he married Sondra Jackson.

ALBERT B. "AL" KELLY, LCDR, born Oct. 30, 1922, Yonkers, NY. He joined the service June 20, 1942. His military locations and stations: USMMA Kings Point, Long Island, NY and served on board liberty ships. He participated in the Battle of the Bulge, Antwerp.

He achieved the rank of LCDR and was discharged Sept. 28, 1982.

His memorable experiences: Christmas of 1942 in the Persian Gulf on ship in Antwerp during Battle of the Bulge.

He married the former Myrene Bailey and they have four children: Frank J., Albert, Mark and Myrene; and seven grandchildren: Robin, Erin, Stephanie, Krista, Joe, Bill and Julia. He was employed as a plant engineer until his retirement Nov. 1, 1987.

KENNETH D. "KEN" KELLY, LCDR, born Aug. 21, 1925, Coopersville, MI, earned his BS degree in civil engineering at Michigan State University.

He joined the service Aug. 3, 1943. His military locations and stations: Camp Perry, VA; Gulfport, MS; and Pacific Theatre WWII. He served with the 127th USNCB, 78th USNCB, 82nd USNCB, USNR Div. 9-103, and USNR SEC GRU 9-5. Battles and campaigns participated in include Admiralty Islands, 1944 and Okinawa, 1945.

LCDR Kelly was discharged from active service March 18, 1946 and retired from USNR July 1, 1969. He was awarded the American Theatre Medal, Pacific Theatre Medal, WWII Victory Medal and Good Conduct Medal.

His memorable experiences: Serving as enlisted man in Construction Bn. 78 in Pacific Theatre in 1944-45. Return to Okinawa after 50 years in June 1995 - a mountain top experience.

He and his wife, Ruth, have been married for 50 years. They have three children: Roger, Richard and Scott; and seven grandchildren: Gina, Asa, Michael, Nicole, Carly, Kenneth and Leanna. He was owner of Kelly Sales and Engineering for 33 years. Now works part time as CEO and engineer, Kelly Sales and Engineering.

GORDON L. "POWDER RIVER" KENNEDY, CDR, born May 27, 1919, Toas Indian Pueblo, NM, earned his BS degree in agricultural education.

He joined the USN June 15, 1941. His military locations and stations: Cadet at NAS Jacksonville and Opa Locka, FL. Assigned to USS *Saratoga* from October 1942-July 1943 and to USS *Yorktown* from August 1944-February 1945. Participated in battle of Solomon Islands, carrier strikes on Philippine Islands, French Indo China Sea, Formosa and invasion of Iwo Jima. Stationed VT-3 Torpedo Sqdn. 3.

Kennedy was awarded the Distinguished Flying Cross, five Air Medals and Navy Presidential Unit Citation. He achieved the rank of CDR, USNR, and retired Jan. 6, 1946.

He and his wife, Gloria M., have three children: Jeannine M., Marlene L. and Christine L.; and two grandchildren, Leanna and Micheal G. He was employed as an

in the CIC School at NAS Glenview and then assigned to VS-23 as a naval aviation observer (controller) at NAS North Island. From 1951-54 he served as an airborne air controller in TBM-3Ws for VS-23 which included a WESPAC tour and being assigned alternately to NAS Agana, USS *Point Cruz* (CVE-119), and NAS Atsugi. The squadron was on board during the Point Cruz visit to Shimoda, Japan for the Black Ship Festival to commemorate the centennial of Commodore Perry's visit with the Black Ship Sqdn.

Transferred to inactive duty in 1954. Assigned to the Ready Reserve at NAS Los Alamitos from 1955-71, he served in VS-773, VR-772, VR-70L1 and NARS-L1 as NAO (controller), NAO (navigator), personnel officer and administration officer, respectively. He was transferred to the Retired Reserve in 1971 at NAS Point Mugu and to the Naval Reserve Retired List in 1984. His campaign and service medals are the American Theatre Medal, WWII Victory Medal, National Defense Service Medal, Korean Service Medal, United Nations Service Medal and the Naval Reserve Medal.

Between tours of active duty, 1946-51, he was a student at Southwestern State (Oklahoma) and Oklahoma A&M and then employed as a seismic observer in seismograph field crews. Following his second duty tour, he attended UCLA. From 1955-86 he was employed in the field of human factors by the same company as it progressed from a division of RAND to SDC to a Burrough's Subsidiary. His assignments evolved from the training of personnel in the application of systems software/hardware to defining the requirements for system applications software. He retired in 1986 as a systems requirements analyst.

The Kisor family includes wife, Vivian; daughter, Shari, her husband, Philip Petruso, and their son, Anthony; and daughter, Michelle, her husband, Steven Smith, and their son, Taylor.

JOSEPH T. "TED" KISS, CAPT, born Nov. 16, 1937, Willimantic, CT, earned a BS degree at the University of Connecticut and a BS degree at the Naval Postgraduate School, Monterey, CA.

He joined the service Nov. 10, 1961. His military locations and stations: Seattle, WA; El Paso, TX; New Orleans, LA; Newport, RI; Monterey, CA; and Presidio, RI. He served on the USS *Providence* (CLG-6), USS *Widdes* (DD-489), USS *Lowery* (DD-770), USS *Perkins* (DD-870) and USS *Vega* (AF-59). He participated in the Vietnam War.

Kiss was awarded the Meritorious Service Medal w/ Gold Star, Navy Commendation Medal w/Combat V and Navy Achievement Medal. He achieved the rank of CAPT and retired Aug. 1, 1988.

He and his wife, Barbara E., have two children, Diana L. and Eric T.; and two grandchildren, Julie N. and Jenny L.

GEORGE M. KISSINGER, CAPT, born Oct. 20, 1946, Minneapolis, MN, graduated from the University of Minnesota in 1969 with a BA degree and received his commission through the NROTC Program at the University of Minnesota.

CAPT Kissinger is a Vietnam veteran, having served a tour as communications and operations officer in USS *Pitkin County* (LST-1082), conducting combat logistics support operations primarily out of the port of Da Nang. He is a "plankowner" of USS *Lockwood* (FF-1064) and deployed for a second time to Vietnam during the naval blockade, serving as operations officer for Destroyer Sqdn. 23, "The Little Beavers."

He affiliated with the Naval Reserve Intelligence Program in 1976 and served as executive officer of NR FIRSTPAC 0278 and as commanding officer of both NR FIRSTPAC 0178 and NR DIAHQ 0878. He served as reserve intelligence area commander, Area 10, from Oct. 1, 1994-Sept. 30, 1996. CAPT Kissinger retired from naval service Oct. 1, 1996.

CAPT Kissinger wears the Meritorious Service Medal, Navy Commendation Medal (2), Joint Meritorious Service Award, Meritorious Unit Commendation, National Defense Medal w/star, Naval Reserve Service Medal, Vietnam Campaign Medal and Vietnam Service Medal.

He is married to the former Diane Leslie Brodin of Minneapolis, MN. They have one son, Karl.

JOHN SCOTT KISTLER, CAPT, is a naval aviator and helicopter pilot who joined the service in 1968, after graduating from Principia College in Elsah, IL. Born in 1945 in Miami, FL, Kistler has served on six aircraft carriers, and while twice aboard USS *LaSalle* (AGF-3) he was officer in charge of HS-1, Det. ONE. During his second tour in *LaSalle*, in 1979, his men originated the now world-famous Desert Ducks.

CAPT Kistler's most recent tour was as deputy commander, Naval Air Reserve Force. Presently in command of Naval Air Reserve, Jacksonville, he is authorized to wear the Legion of Merit (2) and other unit and personal awards. He has been married to the former Karen Andersen of Chicago since 1968 and their children are: Jay, Scott, Laura and her husband, Mark.

GEORGE F. KLEIN JR., CDR, born Sept. 15, 1923, Minneapolis, MN, earned his BA degree in architecture at the University of Minnesota in 1949.

He joined the service in November 1942, V-5 naval aviation pilot training 1943-45. He received Wings of Gold June 15, 1945, Corpus Christi. Served as instrument instructor in operational type A/C SB2C, SNJ at NAS Daytona Beach, Cecil Field.

He achieved the rank of CDR, USNR, squadron commanding officer 1965-67. Klein retired June 1, 1967.

His memorable experiences: After college graduation in March 1949 he "re-upped" and joined naval reserve squadrons VA-811 and VA-813; two attack squadrons (formerly VF); flew SB2C, F4U, F9F, A1E, SNB and A4B aircraft for 17 years.

He was married to Helen L. (deceased) and they have four children: Katherine L., Margaret J., Jennifer C. and Heidi J.; and two grandchildren, Sara and Tara Rose. Civilian employment as architect and founder of "Klein McCarthy & Co. Architects." He retired from business in 1991.

SHARON KAY "SHERRY" KLEINSCHMIDT, born Dec. 29, 1936, farm house, Todd County, MN, earned her BS degree at the University of Minnesota.

She joined the service Oct. 2, 1959. Active duty, Pacific Missile Range, Oxnard, CA. She was editor of 25th NRA *Guiding Light* history, San Diego Chapter *Honker* newsletter editor and serving as national public affair officer NRA and naval reserve ball chairman.

She achieved the rank of CDR and retired from reserves in December 1996 with 36 years of duty.

Her memorable experiences: Active duty, Yokosuka Japan; flying in a F4 jet; CO of the Wave Recruit Div., San Diego.

She is a home economics instructor.

FRANK MARK KLICAR, CDR, born Dec. 10, 1941, Chicago, IL, earned his BS degree in Zoology at Roosevelt University, Chicago.

He joined the service in December 1963. His military locations and stations: USS *Fremont* (APA-44); NAVSUPPACT, Da Nang Vietnam, COM 4 staff, Philadelphia. He achieved the rank of CDR.

Klicar was awarded the National Defense Service Medal, Vietnam Service Medal, RVN Campaign Medal, RVN Gallantry Cross and Air Force Reserve Medal.

His memorable experiences: Engagement in small arms action with Vietcong infiltrators during security patrol in Da Nang in 1966.

He is married to the former Ann Cruse. Klicar is president, Megamark Productions/video producer.

FRANK KLING, CDR, born Dec. 2, 1916, New York, NY, enlisted in the service April 10, 1942. His military locations and stations: WWII - Eastern Sea Frontier HQ; Aug. 30, 1943, attached to staff CINCLANT aboard USS *Vixen* (PG-53); Sept. 15, 1945, administrative assistant to deputy commander Western Sea Frontier; Aug. 3, 1946, separated. Korean War - Sept. 11, 1950, attached to staff Commandant Third Naval Dist.; Sept. 10, 1952, separated.

Kling was awarded a commendation from commander-in-chief US Atlantic Fleet. He achieved the rank of CDR.

He and his wife, Consuelo A., have two children, Frank W. and Michael A.; and four grandchildren: Andrew, Caroline, Margaret and Michael. He was employed in investment management. Kling retired May 10, 1989.

MATTHEW W. "MATT" KLOCEK, CAPT, born Jan. 1, 1939, Glassport, PA, earned his BA degree in history at the University of Rochester, NY. Commissioned ENS with orders to flight training at NAS Pensacola, FL. After completing multi-engine flight training at NAS Corpus Christi, TX he was designated naval aviator in May 1963.

Ordered to VP-6 at NAS Barbers Pt., HI; attended the Naval Justice School, Newport, RI; and VP-31 at NAS North Island, CA. In August 1964 he was deployed with VP-6 to WESTPAC in response to the Gulf of Tonkin Resolution. At the end of his deployment he had flown out of NAF Naha, Okinawa; NAS Cubi Pt. and NAS Sangley Pt.; Philippines; Tinian; Taiwan; Hong Kong; MCAS Iwakuni and Misawa AFB, Japan; and NAS Agana, Guam as well as many hours over Yankee Station. Immediately following WESTPAC, he deployed with VP-6 to Johnston Island in support of a weapons testing program. In 1965 he completed P-3 Orion flight training at VP-31, NAS Moffett Field, CA and in 1965 deployed to NA Adak, AK. While assigned to VP-6 he served as squadron legal officer, aircraft line officer, avionics/weapons officer, as well as P-2 navigator and co-pilot, and P-3 plane commander.

He resigned his regular commission and was released from active duty in 1967 and affiliated immediately with the Naval Air Reserve Unit drilling at COMPATWING TWO, NAS Barbers Pt. He was the first reservist to qualify and serve as PATWING TWO watch officer on a drill basis.

Following a short stint with a VTU which he drilled non-pay with COMTHIRDFLT Det 120 at Ford Island, he was assigned to CINCPACFLT Det 120 where he served as department head for reserve projects, senior watch officer, and assistant battle watch captain in the CINCPACFLT Command Center.

Upon promotion to CAPT he was again assigned to a VTU but continued to drill with the CINCPACFLT cadre which was establishing the US Naval Forces Central Command. When NAVCENT was commissioned and its reserve augmentation unit Det 120 was established, CAPT Klocek screened for command and was named the first executive officer, to augment as ACOS for operations. In 1984 he deployed as the acting commander to Vandenberg AFB for "Gallant Eagle 84."

CAPT Klocek transferred to the IRR in 1985 and retired in 1989. He was awarded the National Defense Service Medal, Armed Forces Expeditionary Medal (Vietnam), Navy Unit Commendation and the Armed Forces Reserve Medal.

In civilian life he has served as management analyst for manpower and budget on the staff of CINCPACFLT and in a career spanning more than 30 years, including 20 years as CAPT with Aloha Airlines, he has logged more than 10,000 hours combined military and civilian time in Viscounts and Boeing 737s.

He lives with his wife, Priscilla, in Kaneohe, HI where he is active in boating, fishing and yacht club activities. His grown daughters; Sharon (by former wife Anne) is married and lives in Michigan and adopted daughter, Amy, lives in California.

PAUL J. KLOTE, CAPT, born Jan. 4, 1946, Knox County, MO, graduated from Northeast Missouri State University. Commissioned August 1968 OCS Newport and assigned to the USS *Coates* (DE-685) as the weapons and operations officer. January 1970 assigned to the USS

Shangri-La (CVS-38) as the 5th Div. officer. December 1971 affiliated with the reserves. He served in several units, commanding six.

Klote was awarded the Navy Commendation Medal, Meritorious Unit Commendation, National Defense Medal, Navy and Marine Corps Overseas Service Medal, Vietnam Campaign Medal, Vietnam Service Medal and the Armed Forces Reserve Medal. He was promoted to CAPT in February 1990.

He is the chief, real estate appraiser for the Kansas City District, US Army Corps of Engineers. He is a professionally designated member (MAI) of the Appraisal Institute, a licensed auctioneer and commercial pilot. Klote has been a NRA member since 1971.

His wife, LuAnna, is an accounting and finance manager. They reside in Merriam, KS.

WILLIAM ALONZO KLUTTS, LCDR, born June 26, 1928, Ripley, TN, earned his BA degree at the University of Chicago in 1947; graduate study there (History Dept.; Committee on Social Thought).

Editor/publisher, the weekly *Lauderdale County Enterprise*, Ripley, 1949—. Winner, record number Tennessee Press Assoc.-University of Tennessee State Press Awards; Pulitzer Prize nominee, 1954. National Guard; Army (Armor) 1950-52 (SFC); Reserve, 1952-64 (CAPT); USNR (1955), retiring 1977.

Ex-president and Paul Harris Fellow, currently assistant permanent secretary-treasurer, Ripley Rotary Club; ex-president, Ripley Chamber of Commerce. Scouting Silver Beaver. Former chairman, deacons/trustees, First Baptist Church, Ripley; trustee, Union University, Jackson, TN; vice-president, West Tennessee Historical Society; member, Tennessee Civil War Centennial Commission. Antiquarian bookman; US Membership Secretary, Private Libraries Association; member, Typophiles, Gutenberg-Geschellschaft. Philatelist, collects American art pottery, especially Roseville. Unmarried. Destination: Maplewood Cemetery, Ripley.

FRED B. KNIGHT, CDR, born Dec. 12, 1925, Waterville, ME, earned his BS degree in Forestry at the University of Maine and his Master's and Doctor's degrees at Duke University.

He joined the service April 26, 1943, and served in active duty July 1, 1943-June 18, 1946. His military locations and stations: Navy V-12 (Harvard University); Midshipman's School, Ft. Snyder, NY; commissioned ENS March 6, 1945. He served aboard the USS *Satterlee* (DD-626) and USS *Massey* (DD-778), 1946. Knight participated in the Asiatic-Pacific convoys as sonar officer.

He was awarded the American Theatre Medal, Asiatic-Pacific Theatre Medal, Occupational Medal, Victory Medal, Naval Reserve Medal and Armed Forces Medal. He achieved the rank of CDR and retired from the military Dec. 12, 1985.

His memorable experiences: Riding out a typhoon near Iwo Jima in 1945; news of war's end (in Okinawa harbor) in August 1945; and his civilian career as a forester and educator.

He married the former Jane Wooster and they have three children: Mary Jane, Susan and James; and three grandchildren: Heather, David and Caitlin. He held many civilian positions with his last being Dean at the College of Forest Resources and professor of Forest Resources at the University of Maine.

PHILIP D. KNOWLES, CDR, born Feb. 15, 1930, Brooklyn, NY, attended college and earned his MBA degree.

He joined the service in 1951. His military locations and stations: OCS, Newport; USS *Monterey* (CVL-26); USS *Leyte* (CVS-32); USS *Toro* (SS-422); and other boats as reservist.

Knowles was awarded the Naval Reserve Medal. He achieved the rank of CDR and was discharged Dec. 17, 1954.

His memorable experiences: Guantanamo training, Philadelphia shipyard, hurricane Carol, qualifying OOD, EOW and qualified in submarines as reservist. Went to Leyte after catapult explosion which killed 34—replaced A Div. officer.

He and his wife, Eileen, have four children: James, Sally, Gregory and Jeffery; and six grandchildren. Grant, Meredith, Scott, David, Hannah and Madeline. Knowles retired in 1971.

THEODORE EMMETT "TED" KNUDSON, CDR, born April 21, 1917, Marshalltown, IA, earned his BS degree in business administration, at the University of Southern California.

He joined the service in June 1942. His military locations and stations: Supply Corps School, Boston and aboard the USS *Colohan* (DD-658) August 1943-February 1945.

Knudson believes he is entitled to eight Battle Stars. He achieved the rank of CDR and was discharged in January 1946. He was recalled and served from June 1951-June 1953, NAS, St. Louis, MO.

He and his wife, Naomi, have three children: Richard T. (CAPT, SC, USNR), Mary and Linda; and three grandchildren. He was employed as agency manager with Northwestern Mutual Life Ins. Co. Knudson retired in 1979.

GEORGE J. "PORKY" KOEHLER, LCDR, born Sept. 2, 1920, Paulsboro, NJ, earned his BS degree in air transportation at Purdue University.

He joined the service July 29, 1942. His military locations and stations: V-5 cadet, Squantum, MA; Pensacola, FL; Hollywood, FL; and Norfolk, VA. He served with FAW-7 at Azores Islands and Dunkeswell England. Participated in anti-sub warfare, Atlantic Ocean (VPB-114 Searchlight Sqdn.).

Koehler was awarded the Presidential Citation, European Theatre Medal, Atlantic Theatre Medal, WWII Victory Medal, Naval Reserve Medal and American Defense Medal. He achieved the rank of LCDR and retired from the USN in 1969.

His memorable experiences: Was plane CDR in PB4Y-1 and PB4Y-2. Also flew twin Beech, SNJ and AD4 in reserve training.

He and his wife, Alice, have two sons, Richard (retired USN CAPT) and Robert; and two grandchildren, Clay and Madison. He was employed as president, Koehler Bros. Landscaping and Garden Center. Koehler retired in 1984.

RICHARD J. "DICK" KOFSKY, LCDR, born Dec. 22, 1931, Chicago, IL, earned his BSEE degree at Valaparaiso University in 1954.

He joined the service June 18, 1954. His military locations and stations: Great Lakes; Newport, RI; Pensacola; Patuxent River; Argentia, Newfoundland; and Glenview. He served with VW-15 and VR-722. Kofsky participated in the Dew Line; Cuban blockade.

He was awarded the National Defense Service Medal, Naval Reserve Medal, Vietnam Service Medal and Cuban Blockade Medal. He achieved the rank of LCRD.

He and his wife, Anne (deceased), have three children: Richard Jr., Kristine and Jeffery; and three grandchildren: Stephanie, Jeremy and Lindsey. He was employed as manager, Ace Hardware, Mundelein, IL and as owner, Ace Hardware, Saugatuck, MI. Kofsky retired in June 1976.

ALBERT J. "AL" KOLLAR, CWO4, born Nov. 3, 1919, Richmond, CA, graduated from high school and took college extension and correspondence courses involved with mechanical engineering.

He joined the service Oct. 28, 1940 (V-3, V-6 USNR). His military locations and stations: American and Asiatic-Pacific Theater of Operations. He served with Sect. Base, Treasure Island, CA; MINEFOR (Coastal); Amphibs. USS LST-451; DD; NAVRES SAT Div. 12-2; SMR-720; T&S Unit 11-76; and VTU 1220. He participated in the assaults of Saipan, Tinian, Leyte, Luzon and Okinawa.

Kollar was awarded the Navy Unit Commendation, Good Conduct Medal, Naval Reserves Medal, American Defense Service Medal, American Campaign Medal, Asiatic-Pacific Campaign Medal w/5 Battle Stars, WWII Victory Medal, Armed Forces Reserve Medal, Philippine Liberation Ribbon, Philippine Republic Presidential Unit Commendation, Philippine Defense Ribbon and the Philippine Independent Ribbon. He achieved the rank of CWO4 and retired with pay Nov. 3, 1979.

His memorable experiences: Conducting minesweeping operations in extremely heavy weather and standing the long watches to maintain the formations. Serving aboard USS LST-451 during their five major combat assaults and beachings. The premature launching of the LCT during heavy weather from the deck of LST-451 with loss of life. He and several ship's company nearly getting shot up on Saipan through an unauthorized act. Serving with pride and professionalism in their great Navy.

He and his wife, Faye, have two daughters, Linda K. Heath and Dianne K. Caron; and three grandchildren, Kristen, Justin and Kayla. He was employed at Mare Island NSYD in the planning department of Ordnance Design Section as a mechanical engineering technician. Kollar retired July 1, 1972.

CHARLES J. "CHUCK" KONDRACK, CAPT, born June 16, 1953, Glendale, CA, earned his BS degree in chemistry at the USNA and his MS degree in engineering at California Polytechnic University, Pomona, CA in 1987.

He joined the service June 4, 1975 with his first assignment at Puget Sound Naval Shipyard, Bremerton, VA. He completed Submarine Officer School and reported aboard the USS *Scamp* (SSN-588) in early 1977. He resigned from active duty in June 1980 and affiliated directly with the USNR.

With AS-19 PROTEUS Detachment 219 he served in various positions aboard the USS *Dixon* (AS-41). As a LCDR he was selected for command of NR NWS Seal Beach Detachment 519. After his command tour, he joined COMSUBPAC Detachment 219. He became officer-in-charge, Submarine Sqdn. 11 Reserve Unit.

His next assignment was to Naval Sea Systems Command, Washington, D.C. He attended the Naval War College and has recently been assigned as the CO, NCCOSC RDT&E Div. 119, San Diego, CA.

CAPT Kondrack is currently a senior project manager for TRW Strategic Systems Div. (SSD), San Bernardino, CA. He provided program and technical support to all the USAF strategic ballistic missile programs. He led TRW's initial environmental systems efforts working with the Dept. of Energy (DOE) and DOE contractors on nuclear waste cleanup at the DOE Hanford Reservation where he managed the local TRW office in Richland, WA.

Kondrack was awarded the Navy Commendation Medal.

He married the former Terry Heinzman of Aberdeen, SD and they have one daughter, Robyn, and one son, Douglas. The Kondracks presently reside in Redlands, CA. Currently, he is managing government and commercial engineering integration projects for information and healthcare systems.

GEORGE N. KOVCH, O-5/CDR, born Dec. 19, 1949, Saskatoon, Saskatchewan, Canada, earned his BA degree in political science and history at Wayne State University, Detroit, MI in 1972; attended various USN/USNR technical and managerial courses and seminars, 1974-91; and DoD contract management and management support courses at Defense Systems Management College, Ft. Belvoir, VA, 1982-90.

He joined the service in August 1973 (active duty) and entered the USNR in February 1980. His military locations and stations: Pensacola, FL; Norfolk, VA; Charleston, SC; Souda Bay, Crete, GR during active service. While with the reserves he was stationed at Earle, NJ; Charleston, SC; Bayonne, NJ; Oakland, CA; London, UK; Rotterdam, NE; Vainhingen, GE; Washington, D.C.; and Naples, IT. During active duty he served with the Aviation Officer Candidate School; aboard the USS *San Diego* (AFS-6); COMSERVGRU TWO; COMOMAG; MOMAG DET SIX. Reserve duty units included: NR SURFWARCEN DET 106; NR MOMAG DET 906; NR COMSC HQ WASH DET 206; NR COMSCEUR DET 206; NR DCNO OP-01 (N1).

Kovch was awarded the National Defense Service Medal, Sea Service Medal, Marksman Medal, "E" Ribbon, Meritorious Unit Commendation, Overseas Deployment Medal, Armed Forces Reserve Medal (2) and Navy Commendation Medal. He achieved the rank of O-5/CDR and was discharged from active duty in September 1978 and from the USNR in May 1995.

His memorable experiences: Tours as OIC of (active and reserve) MOMAG units, 1977, 1980-82. XO of Naval Reserve Military Sealift Command, Europe detachment, 1990-93. Letter of Commendation (from RADM J.J. Carey, USNR), 1990. Received two Navy Commendation Medals, 1991 and 1993.

He and his wife, Mary Ann, have two children, Andrew and Kristina. Civilian employment includes Advanced Marine Enterprises, Inc. as systems analyst, 1979-83; Wheeler Industries, Inc. as systems engineer, 1983-87; and Integrated Systems Analysts, Inc. as senior principal systems analyst, May 1987-present.

JOHN TRUBERT KRATTIGER, CAPT, born Aug. 30, 1916, Denison, TX, earned his BA degree at Austin College, Sherman, TX; his MA degree in mathematics at Southern Methodist University, Dallas, TX; and his EdD at the University of Oklahoma, Norman, OK.

In 1944 he joined the USNR as an ENS. He completed his indoctrination training Nov. 22, 1944, at Hollywood, FL with Class 3-45. He served with Plt. 15, Co. E in Japan with duty assignment in Nagasaki.

They boarded the troop ship APR-47 (Auxiliary Personnel Barracks), USS *Kingman* at Evansville, IN and departed from New Orleans. Only four ships of this kind were built.

He served on active duty from Sept. 22, 1944-June 30, 1949. He retired from the USNR with the rank of CAPT Aug. 30, 1976.

Krattiger served 36 years on the faculty at Southeastern Oklahoma State University where he taught mathematics; he coached tennis four years. He retired June 30, 1984, with the honorary title of emeritus vice-president for student services and professor of education. He served as chairman of the Oklahoma-Arkansas Section of the Mathematical Assoc. of America and president of the Southwest Assoc. of Student Personnel Administrators.

He is married to the former Ruth Whitlock Lance and they reside in Durant, OK. His son by a previous marriage, John Leroy, and his wife, Lou Ann, have three children: Jake, Robbie and Maggie. They reside in Norman, OK.

WILLIAM JOHN "BILL" "TARZ" KRAYER, LCDR, born Feb. 6, 1923, Erie, PA, earned his BS degree in forestry at Purdue University.

He joined the cadet program at Moravian College. His military locations and stations: University of Pennsylvania; University of North Carolina; NAS Gross Isle; NAS Corpus Christi; NAS Sanford; NAS Livermore; NAS Los Alamitos; NAB Roi Namur (Kwajalein); NAB Tanapag; NAS Glenview; and NAS Seattle. He served with VBF-98.

Achieved the rank of LCDR and was discharged in July 1947 (active duty) and October 1962 (USNR). His memorable experiences: CO at Roi Namur Naval AB, Kwajalein Atoll.

He married the former Frankie S. Sullivan and they have five children: Lynn Dee, William Brad, Lisa, Lisete and Lori; and nine grandchildren: Megan, John, Silvia, Emily, Monica, Desiree, Alexandria, Conner and Marianna. He was employed with the US Forest Service as a forester. Krayer retired in July 1978.

FRANK A. KREIDLER, CDR, born Jan. 20, 1947, Cleveland, OH, earned his AA degree at Palm Beach Junior College, 1968; his BS degree at Florida State University, 1970; and his JD degree at Florida State University, College of Law, 1973.

Joined the USMCR in 1965, AMTRAC crew chief, commissioned in the USNR JAG Corps in 1976. He served with JAGC Unit, Miami, FL, 1976-94 and as USNA recruiter 1981-present.

He and his wife, Mary Ann, have four children: Catherine, James, Kristine and Kimberly; and one grandson, Andrew. He has been an attorney for 23 years; president of the Palm Beach Chapter of the Naval Reserve Assoc., 1990-present; and adjunct professor, Florida Atlantic University, 1982-present.

EARL MICHAEL KRIEG, CAPT, born April 6, 1940, Detroit, MI, earned his AB, MD, JD degrees; naval flight surgeon, FACS, FACOS.

He joined the service July 1, 1962. His military locations and stations: Pensacola; Cecil Field; CVW-8 staff; Jacksonville, FL; NARA Milwaukee; and NARA Detroit. He served with CVW-8 staff; VA-82, USS *America*; and VA-86, USS *Shangri-La*. He participated in Vietnam, Desert Shield and Desert Storm.

His memorable experiences: Flight surgeon, Vietnam; orthopaedic surgeon, Desert Shield and Desert Storm; and many others from Sydney to Wellington. He achieved the rank of CAPT and was discharged in April 1973.

He and his wife, Elizabeth, have three children: Michael, Matthew and Carolton. He is employed as chief, orthopaedic surgery, Detroit Veterans Hospital.

DARLA LOUISE KRUPSKI, LT, born Oct. 24, 1956, New Haven, CT, attended Salve Regina College and the University of the State of New York, obtaining a BS degree in sociology. She began her naval career in 1980 under the delayed entry program and entered active service in April 1981 as a dental technician and hygienist. She was commissioned in 1987.

During her active enlisted service she served as the LPO at dental clinics in Norfolk, VA; Newport, RI; and Yorktown, VA. She began her reserve career as assistant LPO at NRNDC, Little Creek, VA in December 1986.

After her direct commission she was assigned to NRNMPC, Norfolk, VA as the financial/administrative officer. In 1988 she transferred to NRCINCLANTFLT, Norfolk, VA as the assistant training officer, standing duty as a battle staff watch officer and admiral's aide.

From 1992-96 she was assigned as the training department head at NRMSC, Norfolk, VA. Her mobilization assignment was as port services officer. In October 1996 she was selected as CO of NRPSDSOEUR, Greensboro, NC.

LT Krupski was recently nominated and elected as one of the finalists for the ROA Junior Officer of the Year for 1996. She has been awarded the Navy Achievement Medal (four awards), Good Conduct Medal, and National Defense Service Medal. She also received the Navy League Award upon graduation from recruit training in recognition of her selection as honor recruit.

LT Krupski is married to retired LCDR Paul Krupski. They have a son, Alexander Joseph, and reside in Williamsburg, VA. LT Krupski is an avid gardener and enjoys her work as a teacher's assistant.

DEAN H. KRUSE, CAPT, a native of Illinois, was commissioned through the NROTC program at Iowa State University in 1969. Active duty assignments included weapons officer, USS *Mills* (DER-383) and Naval Advisory Gp. Vietnam.

Completing active duty, Kruse earned a MS degree in biology at San Diego State University. Concurrently he began his reserve career specializing in the Inshore Undersea Warfare and Maritime Defense Zone communities. His awards include the Naval Commendation Medal w/Combat V as well as the Coast Guard Achievement Medal. A life-time member of the NRA, he retired in 1993.

CAPT Kruse is the dean of science and technology at Portland Community College. He will complete his PhD in environmental sciences and resources in 1997. Married to the former Deborah Mueller, he is the father of daughter and son, Kelly and Marc.

FREDERICK W. KUHL, LCDR, enlisted in the USN in June 1942. Entered V-12 program at Park College (MO) in July 1943. Commissioned ENS at Camp McDonough, Plattsburg, NJ in June 1944 and assigned to subchaser training schools Naval Base Charleston (SC) and Naval Training Center, Miami.

Served aboard barracks ship USS APL-9 in Admiralty Islands, attached to USS ABSD-2 and USS ABSD-4, Service Force, Seventh Flt., providing berthing, messing and medical services to crews performing battle damage repairs to various Pacific Fleet vessels. Changed designator to intelligence during Navy's 1950s reserve recruitment efforts. Retired in 1975 after nearly 34 years active and reserve service. NRA life member #5301.

Graduated from the University of Oregon with BS degree (1947) and MS degree (1961) in journalism, with minors in English and history; awarded membership in Kappa Tau Alpha, national journalism scholastic honorary. Earned DEd in 1981 from Nova University. Listed in *Who's Who in California* and *Who's Who in the West*.

Ten years in corporate public relations, advertising and newspapers in Washington and Oregon before career in education. Professor emeritus at American River College, Sacramento, CA. After retiring from USNR, served more than 10 years in California State Military Reserve as LTC, public affairs, headquarters unit and as advisor to the National Guard's *Grizzly* magazine, winning numerous area and national honors for "journalistic excellence." Awarded Military Dept.'s Meritorious Service Award, two Medals of Merit and the Commendation Medal.

ANTHONY T. "TONY" KURAS, CDR, born July 15, 1925, Greenpoint, Brooklyn, NY, earned his liberal arts degree in education at Fordham University in 1971. He joined the Navy Seabees in September 1943 and attended boot training at Camp Peary, VA. Later served at Camp Endicott, RI; Camp Parks and Port Hueneme, CA. Pacific bound, he served in Hawaii, the Marshalls, Carolines, Palaus, Philippines, and finally in the invasion of Okinawa in April 1945. His stint with the Seabees in the Pacific during WWII ended in April 1946 when he transferred into the Surface Navy.

During his 35 years of service in the USN, at sea and shore billets, he advanced in rate from SA to CPO, to warrant and chief warrant officer and also from ENS to CDR Line, having served in WWII, the Korean War, the Cuban Missile crisis, and the Vietnam War. He qualified as OOD(F) aboard *Wasp* and during the Cuban Missile Crisis, served as interpreter, utilizing his state department level in the Polish language.

Kuras and his wife, Diane, have two daughters, Linn and Laura, and three grandsons: Tony, Patrick and Seth. He was employed as Charlotte County Veterans Service officer from 1981-92. Kuras retired Oct. 31, 1978.

SAM KUPRESIN, CAPT, native of East Moline, IL, graduated in June 1974 from the USNA with a BS degree in operations analysis.

After receiving his naval aviator wings in October 1975 CAPT Kupresin joined his first command, Patrol Sqdn. 46, at NAS Moffett Field, CA. During that tour, he was designated a patrol plane commander/mission commander and instructor pilot in the P-3C aircraft and participated in operational deployments to Kadena, Okinawa; Adak, AK; and Misawa, Japan.

Reporting to VP-90 in July 1979, CAPT Kupresin spent his last year on active duty and the next 12 years as a selected reserve in this squadron. During his 13 years in VP-90 CAPT Kupresin made 14 WestPac deployments and participated in one UNITAS exercise in South America. CAPT Kupresin ended his tour in VP-90 as the CO of the squadron. Upon leaving the "Lions" of VP-90, CAPT Kupresin served consecutively as the CO of NR NAF Midway Island 0972; the counterpart to N7, Strategy and Tactical Development, on the staff of commander, Naval Air Forces US Atlantic Fleet (COMNAVAIRLANT 0172); CO of deputy chief of Naval Operations (DCNO) N3/5 Det 113; and CO of COMNAVAIRLANT 0172.

CAPT Kupresin's awards include the Meritorious Service Medal, Navy Commendation Medal, Navy Achievement Medal, Meritorious Unit Commendation, Navy Battle "E", Navy Expeditionary Medal, National Defense Medal, Sea Service Deployment Ribbon, Reserve Sea Service Deployment Ribbon, and the Reserve Officers Medal.

In civilian life, CAPT Kupresin is a captain for Delta Air Lines. He is married to the former Patricia Chambers, also from East Moline, IL. They reside in Palatine, IL with their two children, Sam Patrick and Kara Dale.

GEORGE KURUZOVICH, CDR, born April 16, 1933, Cleveland, OH, earned his BSBA degree from Bowling Green State University. CDR Kuruzovich was commissioned through the Naval Preflight School, Pensacola, FL, Sept. 1, 1961.

Operational tours included: naval aviation officer, VW-1, NAS Agana, Guam and VAW-13, NAS Alameda, CA; assistant air intelligence officer, USS *Hancock* (CVA-19), WestPac; and, first coastal zone intelligence officer, Da Nang, RVN.

Staff assignments included: air intelligence program officer, NAS Seattle, WA; reserve intelligence program officer, Area 13, NARU Jacksonville, FL; and, management analyst/intelligence officer, DIA, Washington, D.C. He also had prior enlisted service with the USAF from 1951-55.

His personal military decorations include the Bronze Star w/Combat V, A/S Air Medal, Joint Commendation Medal, Navy Achievement Medal, Good Conduct Medal, and RVN Staff Honor Medal, First Class.

After retiring from the USN, he was employed by the Naval Intelligence Command as a management analyst from 1980-85. He also served as a senior intelligence analyst from 1985-95 with the USMC at Quantico, VA, supporting combat development and RD&A.

CDR Kuruzovich is currently retired in northern Virginia after 38 years of federal service. His wife's name is Loretta "Lori."

EUGENE W. KUTHY, CAPT, born Dec. 17, 1930, Dearborn, MI, graduated with a BA degree from the University of Michigan in 1954, with a JD degree from the Detroit College of Law and is a member of the State Bar of Michigan. He entered the USNR in November 1954 as an officer candidate SA and was commissioned ENS, USNR in May 1955 and retired as CAPT USNR in July 1988.

The US Naval Security Station, Washington, D.C. was his first active duty assignment where he served as a department head in the Naval Security Group's worldwide distribution system for classified and registered publications and served on the Joint Communications-Electronics Committee, Joint Staff, Joint Chiefs of Staff. He finished his active duty tour on the Staff of the Chief of Naval Operations at the Pentagon as aide and administrative assistant to RADM H.C. Bruton, director of Naval Communications.

CAPT Kuthy joined the Naval Security Group Reserve in Detroit, MI in 1958. He successively served as assistant training officer, training officer, administrative officer, EO and CO of NRSG 9-1, Detroit, also serving on the staff of the Detroit Naval Reserve Group Command 9-19(L). Subsequently he served as CO of NRSG 9-12, Southfield, MI; NRSG 9-16, Indianapolis, IN; National Security Group 314, Southfield, MI; VTU 1325, Southfield, MI; and DIRNSA FMDE 204, Philadelphia, PA.

Under his command, DIRNSA FMDE 204 was judged to be the unit that had made the most outstanding contribution to the overall mobilization readiness of the Naval Reserve Security Group for the previous fiscal year. It was the first reserve unit to be so chosen, besting over 100 NRSG units nationwide in criteria that included training/readiness, operations, administration, management improvement, and special projects.

He has also served as the staff cryptologic officer for Readiness Command Region FOUR, Philadelphia. His final NR assignment was in the same capacity, in Region FIVE, Ravenna, OH.

CAPT Kuthy has served on active duty for training at the National Security Agency, Defense Intelligence Agency, Naval Security Group Headquarters Command, Office of the Chief of Naval Operations, and the Armed Forces Courier System, as well as virtually all the Naval Security Group US field activities. He has been awarded the Navy Commendation Medal and the Armed Forces Medal and Chief Naval Reserve Letter of Commendation.

As a civilian, CAPT Kuthy has had over 30 years of executive management experience at major corporations, and accounting and consulting firms. This includes his most recent corporate assignment as chairman of The Chamberlain Organization, Ltd., a firm providing counsel to financial institutions and governments.

His government experience includes service from 1983-90 as commissioner of financial institutions for the state of Michigan, as a gubernatorial appointee, senate confirmed, regulating state-chartered banks, savings and loans, credit unions, mortgage companies and finance companies. During his term he was elected national chairman of the Conference of State Bank Supervisors where he championed numerous financial industry issues before the US Congress and federal agencies.

In the non-profit field he was vice-president of the American Academy of Transportation from 1966-72. He served for over 25 years as an adjunct associate professor in the School of Management, University of Michigan, Dearborn Campus and was the first appointed leader in residence at Walsh College, Troy, MI in 1991.

CAPT Kuthy is married to the former Tara G. Robinson; has three married sons: Douglas, James and Thomas; a married daughter, Robin Kuthy Richter; and four grandchildren.

GERALD P. KYNETT JR., CDR, born Dec. 10, 1925, Media, PA, earned his BS degree in business at the University of Richmond, Richmond, VA.

He joined the service Dec. 8, 1943. His military locations and stations: Sampson, NY; Newport, RI; Gulfport, MS; Camp Bradford; NOB Norfolk. He served aboard the USS LST-867 and USS *Linnet* MSC(O)-24. He participated in battle at Okinawa, Philippines and Japanese occupation.

Kynett was awarded the Asiatic-Pacific Campaign Medal w/1 star and the Japanese Occupation Naval Reserve Medal. He achieved the rank of CDR and was discharged in 1965.

His memorable experiences: ORI, Little Creek, Chesapeake Bay as CO USS *Linnet* MSC(O)-24.

He married the former Virginia Smith and they have three children: Linda K. Harris, Gerald P. III and Michael S.; and seven grandchildren: Elizabeth, Samantha, David, Gerald IV, Jessica, Tate and Trey. Civilian employment as a partner of NYSE Investment Counsel Firm.

SUSAN COOPER "CHRISSY" LABHARD, LCDR, born Oct. 20, 1955 in Baltimore, MD, received a MSN 1993; BSN 1982; BS, HEC (Concentration Child Dev. Family Rel.) 1977.

She joined the service Nov. 15, 1983. Her military locations and stations: NAVRESCENs; Rochester, NY; Butte, MT; Portland, OR; Salem, OR. Served Programs 32 and 36 (NH Bremerton, WA and NRFH 500 CBTZ 9 DETS G, PRIMUS, H.

Her battles and campaigns: Team Spirit-Korea 1989 (She has always been in Reserve).

Awarded the 1993 Sigma Theta Tau Outstanding Masters Student Award; First recipient to receive FY91, NRFH 22 "Nurse Corps Officer of the Year"; 1st recipient for 1993 NRA Outstanding Naval Reserve Jr. Nurse Officer, Beatrice M. Ratner Award.

Labhard was awarded 1989 Meritorious Unit Commendation Ribbon, 1991 National Defense Ribbon, 1994 Armed Forces Reserve Medal, 1997 NAM.

Lubhard's memorable experiences: Mediterranean assist on helicopter in Korea; two week at sea as RN aboard USS *Lexington* (AV-16); ambulatory care RN-NH Sigonella, Italy; two weeks as Mediterranean TO aboard USS *Constellation* (AV-64). Carrier qualified at sea; CO/OIC of Med Units; receiving first NRA/Bea Ratner Award.

She is married to Michael Edison Labhard, MSEE. They have four dogs (Amsel, Wolfie, Deschutes and Rex); one cat (Lily); 1 Conure (Chuckles). Her organizations: lifetime NRA; AMSUS; Lifetime ACCH; NAON; Sigma Theta Tau Honor Society. Nurse manager outpatient clinic-Shriners Hospital for Children. Enjoys Wildlife Conservation; volunteering for Humane Society; Advocate for Physically/Mentally Challenged children; skiing and travel.

HOWARD WAYNE LACEY, CAPT, born Sept. 1, 1936, born in Houston Harris County, TX, received a BA in Valparaiso University; MBA University Houston, Naval War College, Industrial College of Armed Forces and National Defense University.

He joined the service Aug. 14, 1958 in Athens, Norfolk, VA. His military locations and stations: OCS Newport, NSCS, USS *Ault* (DD-698), Texas Group-Orange, TX. He achieved the rank of CAPT 06 and discharged May 31, 1962 and retired Sept. 1, 1989.

His memorable experiences: first armed US warship to enter Black Sea since WWII.

Lacey is married to Jeanette Marie and they have children: Stephen, Micheal, David and Sharon and grandchildren: Katie, Rebecca, John and Benjamin. Employed as professor, Dean of School of Business Concordia University, Austin, TX.

WILLIAM M. "BILL" LAMB, CAPT, born Jan. 26, 1935 in Ft. Worth, TX was a 1958 engineering graduate of the University of Texas and commissioned the same year through OCS. He served five years active duty in submarines and Navy nuclear weapons programs and had Reserve Commands in Albuquerque, New Mexico; Denver, CO; and Washington, DC where he retired in 1989 after 31 years of naval service. Lamb is a registered professional nuclear engineer, a 1974 graduate of the Industrial College of the Armed Forces and a 1979 graduate of the National War College.

He is also retired from the U.S. Department of Energy's Senior Executive Service after having served over 25 years in this nation's nuclear weapons and nation defense programs. He then founded Lamb Associates, Inc., and Maryland based, small business that grew to over 100 scientists and engineers. He and Janet, his wife, served as its first and only corporate officers and he as Chairman of the Board. The company will be merging with a California based engineering/services company where he will also serve on the board of directors.

He and Janet were married in 1955, and they have three grown children: Barbara, Dana and Jon; and two grandchildren, Samantha and Taylor. The Lambs reside in Rockville, MD, where they enjoy the two grandchildren and he sneaks in a little golf.

WILLIAM "JON" LAMBDEN, CAPT, born Oct. 1, 1937 in Durant, OK, attended UC Berkeley; University of Melbourne; USN War College, USF (EdD).

He joined the service Dec. 29, 1960. His military locations and stations: US Seventh Fleet, White Beach, Okinawa, Atsugi Japan, DAO Canberra, Santiago Caracas. Served with: COMTAIWANPAT for CTF-72, VQ-1, DEFATTACHE 0166.

His battles and campaigns: Rolling Thunder/Market Time/UNITAS XXVII. He achieved the rank of CAPT and was discharged July 1, 1995.

He was awarded the Defense Joint Meritorious and Navy Unit Commendation, Overseas Service, National Defense, Vietnam Service, AF Service and Expert Pistol and Rifle. Lambden achieved the rank of CAPT and retired July 1, 1995.

His memorable experiences: flying unarmed and unescorted reconnaissance sorties from USS *Coral Sea* (CVA-43) to and around Handi; retirement aboard USS *Hornet* (CV-12), a ship he had visited with his father and brother in 1942-43. Currently serving as a volunteer to keep her as a national historic monument.

He is married to Connie Pamela (SLEAP) Lambden. He is married to Katherine and Erinne. Employed as businessman, consultant, educator, professor.

HERMAN L. LANCASTER, CDR, born Jan. 22, 1933 in Wellston, OK, received a BBA from the University of Oklahoma in 1959.

He joined the US Army April 13, 1953 and the USNR, April 20, 1964. His military locations and stations: Wonsu, Korea and Sendai, Japan. Served with the 52nd Transportation Bn.

Lancaster was awarded the UN Service Medal, National Defense Service Medal, Naval Reserve Medal and Korean Service Medal. He achieved the rank of CDR-USNR and discharged March 24, 1955. Retired from USNR April 1, 1983.

His memorable experiences: service in Korea; ACDUTRA aboard USS *Constellation* (CV-64).

He is married to Christine and they have children: René Horn and Craig Lancaster and grandchildren: Sarah Horn, Stephanie Horn and Alyssa Lancaster. Employed with Amoco Corp., as senior staff accounting analyst, retiring May 31, 1990.

JEFFREY J. LANCE, LT, born July 11, 1958 and enlisted as a seaman in 1976 after high school in Scranton, PA. He was assigned to Naval Communications Center, Orlando, FL following one year of electronics technician schooling in Great Lakes, IL and Fort Gordon, GA. This was followed by additional training in radar systems repair, then assignment to USS *Koelsch* (FF-1049) in March 1980.

During the next 31 months he made several deployments, including an extended 130 day tour of the Persian Gulf during the Iran-Iraq War. During this time, he was awarded USS Koelsch Sailor of the Year honors (1981), and received the first of his four Navy Achievement Medals for "Outstanding performance and maintenance of the ship's communications and radar equipment." Shortly following his advancement to electronics technician first class in 1982, he accepted employment with Raytheon Company, a world leader in military radar systems.

Affiliating with the USNR in 1983, assignments included NR *Fulton* (AS-11) 301 and USS *Duncan* (FFG-10), during which he received his commission as a limited duty officer (electronics). He has since served with two units at SIMA Newport and as CO of NR SIMA Portsmouth 501, where he was selected as Readiness Command Region One's 1993 Junior Officer of the Year. LT Lance is currently assigned as training officer for NR Space and Naval Warfare Systems Command HQ 101. He has been active on committees for both the NRA and the Navy League.

LT Lance holds a BS degree from Northeastern University, and is employed by Raytheon as a senior engineering specialist supporting the Navy's Aegis Radar Program. He is married to the former Wendy Stapleton to Smithtown, NY. They reside in Framingham, MA with their daughter Kylie.

LEEROY (NMN) "LEE" LANCE JR., LCDR, born Dec. 27, 1950 in Corpus Christi, TX, received a BS and MS in Industrial Education from Texas A&M University.

He enlisted in the service, Feb. 2, 1972. Received direct commission, ENS 1635, Jan. 19, 1981. His military locations and stations: NRTC Great Lakes; RM "A" Bainbridge, MD; NTCC LCRK (ACDU Enlisted). Served with NR NISRO's 2210, 2310, 2010 and 2182; recalled in support of Desert Storm; NISRA NTC San Diego (Desert Storm Recall); JTF 160 GITMO (Staff J2A/Intell Ops officer); JTF 6 RoosRds PR (OIC, Intell Analyst); Officer, RES AcDuTra); NISRA Subic Bay RP; NISRA London UK; JTF-4 Key West, FL; NISRA Roos Rds PR (Op Ocean Venture 93).

Lance was awarded the NCM, NAM, JMUC, NDSM (2), HumSvc, MilOutVolSvc, NMCOvrSeasSvc (3), Expert Rifle and Pistol, Navy Meritorious Unit Commendation. Received the rank of RM1 (enlisted) and LCDR (commissioned).

His memorable experiences: Oct. 13, 1975, his son, David was first born to active duty personnel on Navy's 200th birthday (however, not at naval hospital); commissioned his sister, Jeanette as ENS, 1635; Security detail for Statue of Liberty re-dedication ceremonies; being VertRep'd aboard USS *Harlan County* during OP Ocean Venture.

He has been married 25 years to the former Linda Lee Miller and they have two children, David Ray and Michael Lee. Employed as shop teacher for 12 years; police officer six years. Owner/president of medical claims processing (one year). He has been NCIS reserve officer agent since 1985. recruit petty officer chief, CO, 907 Great Lakes, 1972; Honor Guard RM "A" School; NRA life member.

JOSEPH W. "JOE" LANDAU, CDR, born May 23, 1930 in Buffalo, NY, received a BA from Cornell University 1951 and MD Cornell University 1955.

He joined the service in 1956. His military locations and stations: NAS Pensacola, School AVMED, MAW FMR El Toro, MAW, FMK Oppama, Japan. He served as a flight surgeon.

Landau achieved the rank of CDR, MC USNR and is now retired. His memorable experiences: active duty, January 1957-June 1959; Reserve July 1959-76 (NAS So. Weymoth, Los Alamitos and Pt. Mugu.

He is divorced and has children: Brenda, Kenneth and Jason. Employed on the staff on UCLA Medical Center and has a private practice in dermatology.

LEE E. LANDES, RADM, born NYC March 17, 1924. Enlisted Dec. 12, 1942; V-12 on July 1, 1943 Union College; AB degree there February 1944. Commissioned ENS Aug. 16, 1944. WWII and Korea: NAS Pensacola; LCI Group 61 Pacific; Fleet activities Sasebo; NAS New York and Squantum; Com 3 and Com 9. Harvard MBA 1949. RADM November 1972; retired November 1979. Legion of Merit. Commendations: SecNav, Com 9, DLA, and Michigan Legislature (four times).

Retired: Ford Motor Company, February 1980, as world-wide technician planning manager; also worked in finance and product planning. Resides in Livonia, MI with wife, Suzanne. Son, Stephen; his wife, Cornelia and their children, Benjamin and Gwenyth, live in Ann Arbor. Younger son, George, killed by drunk driver in October 1981; founded MADD in Michigan in 1982. Both sons graduated from University of Michigan. Volunteer: NLUS, Sea Cadets, Navy recruiting, ROTC, victims support, and reading program in Detroit.

Life member: NRA, ROA, NLUS. Member of American Legion, Air Force Association, Association of Naval Aviation, Fleet Reserve Association, Society of Automotive Engineers and Naval Officers Association of Canada (honorary).

ROBERT E. "BOB" LANDO, CDR, born Sept. 18, 1920, in San Diego, CA. He earned BS degree at San Diego State and MA degree, education and teaching certificate.

Joined the service April 7, 1942 and was stationed at Navy Dept., Washington, DC; Ocracoke, NC; Anacostia; Jacksonville, FL, NAS Opa-Locke, FL, COMNAVFE.

Served with Co. H; Beach Jumper Units 6, 7, 8, 9; Photographic School, Intelligence Cmd., Jacksonville, Advance Trng. School, COMNAVFE USS *Helena*. Participated in action in Palau, Mindoro, Luzon and Korea.

CDR Lando was discharged in 1946. His awards include the American Theater, Asiatic-Pacific w/star, WWII Victory Medal, Korean Victory, Vietnam Service Medal, Reserve Medal, and others.

Some of his memorable experiences include New Guinea and Philippines, Korea, Intelligence G2 Tokyo, Photo Intelligence COMNAVFE and USS *Helena*.

Civilian employment: photogrammetrist, USGS and US Forest Service, elementary classroom. He retired in September 1976. Married Phyllis Ann Burr; they are now divorced.

JAMES A. "BARTMITH" LANDRITH JR., SGT E-5, born Sept. 16, 1970 in Peoria, IL, attended college.

He joined the service Jan. 16, 1989. His military locations and stations: MCRD San Diego, Camp Johnson, NC, Camp Lejeune and Henderson Hall. He served with 2nd Force Service Support Group, H&S Bn. adjutant; MC Legis. Asst to CMC.

Landrith participated in Desert Shield and Desert Storm. He received the Navy Achievement Medal, Navy

Commendation Medal, Good Conduct Medal. He achieved the rank of SGT E5. He is still active in the USNR.

Married Cheryl Gamble and they have two children, Ryan and Charles. Employed as executive assistant to Hospital Trade Association and officer on Board of Directors to Push Literacy Action Now.

LEO R. LANDRY, CAPT, a Louisiana native, was graduate of the University of S.W. Louisiana. He enlisted in the USNR in 1950 and was commissioned ENS in 1954.

He served two active duty tours, with USNAVCOMMSTA Washington, DC (1954-56), and with COMNAVBASE GTMO and the National Security Agency (1961-65).

Landry was affiliated with a number of Naval Reserve Units in a variety of assignments. He commanded three Security Group Units. One unit directed special operations performed by over 70 units located throughout the US. His medals are the National Defense Service Medal, Armed Forces Expeditionary Medal and the Armed Forces Reserve Medal.

He held a variety of management positions within the COMSEC and R&D Directorates of the National Security Agency from 1965-93.

He married E. Kay and they reside in Crofton, MD. They have five children and three grandchildren.

ROBERT C. "BOB" LANE JR., CDR, born Sept. 15, 1937 in Miami Beach, FL, received BBA and JD degrees from the University of Miami.

He joined the service Sept. 1, 1959. His military locations and stations: USS *Boston* (CAG-1). Served with USS *Lexington,* USS *Beatty,* USS *English,* USS *Thrush,* CO NR SURF DIV 6-44; CO MSC; VTU 0808; MSC 204.

Lane received the Cuban Expeditionary Medal and achieved the rank of CDR and retired Jan. 1, 1982.

His memorable experiences: standing off of the Bay of Pigs unable to assist the invasion.

He married Phyllis M. and they have two children, Elizabeth Lawhorn and Robert C. Lane III; and three grandchildren: Sarah Lane, Zach Lane and Jack Lawhorn. Employed as attorney.

ROLLAND A. "LANDY" LANGLEY, CAPT, born Aug. 22, 1931 in San Francisco, CA, received a BS, MS and MBA degree.

He joined the service June 6, 1953. His military locations and stations: USS *Bekar* (APA-237), Naval Air Training Command. Served with VS-861, VS-873.

His battles and campaigns: Korea and Cuba. Langley was awarded the Armed Forces, UN, Korea, Armed Forces Expeditionary Reserve. He achieved the rank of CAPT and discharged August 1957 and September 1962.

His memorable experiences: Berlin recall of his reserve squadron VS-861 and subsequent service at Guantanamo Bay, Cuba, where they made the first sitting of Russian missiles going into Cuba.

He married Chiara and they have three children: Owen C., Cynthia and James R. Employed as president and CEO, BNFL, Inc. He retired in July 1982.

JAY C. LANGNESS, CAPT, a native of Fargo, ND, enlisted in the USNR in 1969, through the Reserve Office Candidate Program, while a freshman at North Dakota State University. He rose to the rank of QM2 prior to graduation in 1971, and then attended OCS in Newport and was commissioned in September 1971.

He served as the tanker operations assistant to CDR, Military Sealift Command, Washington, DC. After three years with MSC, he attended Surface Warfare Officer School, Newport and weapons related courses at Dam Neck. He served at assistant fire control officer and missile officer aboard USS *Dale* CG-19, homeported in Mayport, FL. In 1977, he was appointed to the TAR program and assumed command of Naval Reserve Center, Burlington, VT.

In 1979 he was released from active duty and moved to Wisconsin to work in the transportation industry. He commanded the following reserve units: ARS-41/USS *Opportune;* Port services, Norfolk; Convoy Commodore Staff, Key West, and Readiness Unit 1665, all in Wisconsin. He served as chief staff officer for convoy exercises Rainbow Gulf/Diego Garcia and Rainbow Reef/Guam.

He earned a master of science in management degree from Cardinal Stritch College, Milwaukee, WI. After promotion to CAPT, he relocated to Georgia and served as the Naval Emergency Preparedness Liaison officer to FEMA, Region 4 in Atlanta, working several hurricanes. He accepted a job transfer to Pennsylvania in 1995 where he is vice-president of operations for an environmental services company. He is currently assigned to the Individual Ready Reserve.

CAPT Langness and his wife RaNae have three daughters: Kristen, an archaeologist with the Army Corps of Engineers; Kari, a junior at the University of Georgia; and Jana, a junior in high school.

MILTON C. "MILT" LAPP, CAPT, born Feb. 8, 1927 in Buffalo, NY, attended the University of Buffalo, Undergraduate and Medical School, MD degree in 1952.

He joined the service Jan. 25, 1942, hospital corpsman, served to pharmacist mate second class. After discharge he returned to college and remained in the Reserves. On entering medical school in 1948 was commissioned as an ENS in the Hospital Corps and as LTJG Medical Corps upon graduation June 1952. Continuous reserve until discharge from active duty February 1987. Active duty from December 1982 to February 1987.

His military and stations: Sampson Naval Training Center, Hospital Corps School, San Diego, CA; Sampson Naval Hospital, NY; PSC Lido Beach, NY; Naval Hospital, St. Albans, NY; Submarine Medical School, New London, CT; Experimental Diving Unit, Washington, DC; Medical Research Facility, Bethesda, MD; Naval Hospital, Orlando, FL; Reserve Naval Construction Force: Battalions 17, 27, Regiments 4, 7 Brigade Staff as force medical officer 1980-82; chief of Naval Reserve staff as force medical officer and medical programs director-New Orleans, LA; Naval Medical Command Staff, special assistant to Director of Physical Standards and Readiness Div., Washington, DC.

He was awarded the Meritorious Service Medal w/ Gold Star (second award), Navy Commendation Medal, Naval Research Meritorious Service Medal, WWII Victory Medal, American Campaign Medal, Armed Forces Reserve Medal w/2 HGs and the Expert Pistol Medal. He achieved the rank of 0-6 and was discharged Feb. 20, 1987.

Memorable experiences: all of his experiences in regards to the USN were memorable but the time he spent working for the USNR were the most memorable as it was during the period of expansion of the Reserve Medical Programs in the period 1982-85. The subsequent use of the expanded medical force during the Persian Gulf encounter proved invaluable to the Naval Medical Department.

He married Margaret Munzesheimer and they have four children: Susan Phelps of Burke, VA; Dr. Josef Lapp of Corning, NY; Gretchen Hamlin of Skaneateles, NY; Julia Lapp of Fairbanks, AK. Their grandchildren: Malia, Emaleigh and Josef Phelps of Burke, VA; Olivia and Virginia Hamlin of Skaneateles, NY.

As a civilian he was in private practice as a surgeon holding positions as hospital staff chief of surgery, president etc. county and state medical society offices. Retired December 1982 to enter active duty as force medical officer for the Chief of Naval Reserve.

ALEXANDER E. LARDIS, CAPT, born in 1941, raised in Warren, OH and graduated from Hiram College with a BA in 1964. He was commissioned November 1964 at OCS, Newport, RI and assigned as communications/ electronics officer, Mobile Inshore Undersea Warfare Unit 23. In 1966, he served as detection, security and patrol boat officer, Harbor Entrance Control Post, Qui Nhon, Vietnam. Lardis then served as OIC of Mobile Unit 23 until assignment as cyrpto security officer, RPS custodian and traffic officer, NAVCOMMSTA Greece from 1968-70.

In 1970 Lardis attended Kent State University and received his MS in Aquatic Ecology in 1973. He joined the USNR in 1972 and served with Medium and Large Surface Divisions, NROS, NAVWEPSTA and COMNAVIUWG-2. His final assignment was CDR, Ashore Mobile Contingency Communications Command in 1991. He retired in 1993.

He received numerous campaign and service awards including the Navy Commendation Medal, Navy Achievement Medal, Combat Action Ribbon, Joint Meritorious Unit Award and Meritorious Unit Citation.

Lardis is a senior research scientist and technical manager for the Navy Environmental Quality Applied Research Program with Naval Surface Warfare Center. Lardis is a life member of NRA, and married to the former Victoria M. Nicholas of Norfolk, VA. They have one son Mike, and reside in Annapolis, MD.

CHRIS LARGER, CDR, born Aug. 12, 1952 in New York, NY. He received a BA, Miami of Ohio, 1975, and MBA, University of Texas, San Antonio, TX, 1989.

He joined the service, June 8, 1975. His military locations and stations: USS *Virginia* (CGN-38), 1975-78; USS *OCA Peterson* 00969, 1978-80; Saudi/Naval Expansion Program; USNR 1982-96; PG511, Navy liaison officer, 1980-82.

Larger received the Battle "E" for USS *Virginia* and USS *Peterson.* He achieved the rank of CDR, USNR and retired May 1, 1996.

His memorable experiences: commissioning crew, USS *Virginia* April 15, 1982. Discharged CGN-38 September 1976. Also taking PG-511 HMS AS SIDDIQ RSNF through the Great Lakes, St. Lawrence seaway to Little Creek NAB as a Navy liaison officer for RSNF.

Larger is single and employed USAA Life Insurance Company, 1986-present as account sales representative.

E. ARTHUR "ART" LARSON JR., CDR, born Aug. 17, 1923, born in Astoria, OR, received a BA from Pacific Lutheran University; B.Ed, M.Ed, University of Puget Sound.

He joined the service Oct. 6, 1942. His military locations and stations: V-12 Dickinson College, Dickinson, ND, Pre-Midshipman's School, Asburey Park, NJ; Midshipman's School, Columbia University, New York, NY; US Naval Training Center, Terminal Island, (San Pedro), CA; USS *Moosehead* (IX-98); USS *Gardiners Day,* (AVP-39)l USNS *Archer T. Gammon* (AK-243); USNS Pvt. Sadao S. Munemori (T-AP-190); MSTS NORPAC, Seattle, WA.

Served with Naval Reserve Surface Div. 13-23; Naval Reserve Surface Div. 13-16; Naval Reserve Surface Div. 13-17; Mobilization Team Div. 13-6; Naval Academy information officer, (All in Tacoma, WA).

Battles and campaigns: Okinawa, Kerama Rhetto, Inchon and Pusan.

Larson was awarded the USN Reserve Medal, WWII American Campaign Medal, Asiatic-Pacific Campaign Medal w/2 Battle Stars, WWII Medal, Occupation Medal

(Japan), China Service Medal, National Defense Medal, Korean Service Medal, Armed Forces Reserve Medal and UN Korean Service Medal.

He achieved the rank of CDR and was discharged July 1, 1971. His memorable experiences: surviving three hurricanes; living through a number of kamikaze attacks at Kerama Rhetto; entering Tokyo Bay and establishing a seaplane base prior to the surrender ceremonies on Sept. 2, 1945.

He married P. Lorraine Akehurst and they have three children: Curtis Thayne, Susan Carol, Darold Wayne (wife, Elizabeth); and grandchildren: Bryan Arthur and Hannah Kristine Larson. Employed as teacher, coach, counselor, vice-principal, and principal on the high school level for the Puyallup School District. Retired July 30, 1979; Washington State Employment Officer Puallup. Retired Oct. 30, 1986; Pacific Lutheran University Adjunct Professor in Education. Retired June 30, 1992.

REUBEN R. "REUB" LARSON, CAPT, born Aug. 9, 1932 in Sioux Falls, SD. He received a BA degree (accounting) at Carleton College in 1954 and attended University of Minnesota.

He joined the service November 1954. His military locations and stations: Pax River, MD; Argentia, Newfoundland. Served AEWRON 11 (VW-11).

Larson achieved the rank of CAPT and was discharged in 1958. He married Betsy and they have children Richard, Mark and Kari, and a granddaughter Sara. He is owner, R.R. Larson Assoc., retiring in 1981.

CHARLES FRANKLIN LARUE, LCDR, born April 1932 in Kempner, TX. Public schools, Dallas, graduating 1949. BA Maryville College (Tenn) 1953; M.Div. McCormick Theo. Seminary, Chicago, 1956. M.Ed. North Texas State University 1968.

Took Navy oath and reported at Newport in 1956; 1956-57 Bainbridge NTC; 1957-58, MSTS, NorPac; 1958-59 LanShipRon Five; 1959-61, Moffett Field NAS; 1961-62, 3rd MARDIV, Okinawa. Summer augmentation: NAB, Little Creek 1964; Camp Lejeune 1965; NTC San Diego, 1966; Camp Pendleton 1967. Ready Reserve: MSTS Of. Co.; Seabee Bn. 28; Mar CHC Co. Retired July 1979.

LOYT L. LATHROP, LCDR, born Oct. 20, 1910 in Mound Valley, KS, received a BS from Kansas State in 1934.

He joined the service in 1942. His military locations: Boston. Served with airborne radar. He achieved the rank of LCDR and was discharged in 1946.

Lathrop is married to Jennie S. and they have children, Kenneth and Winifred L. (Manzo), and grandchildren: David, Alisa, Kristen, Joe and Terry Lathrop, and Jason and Nathaniel Manzo. Employed at Sandia Labs, Albuquerque, as section supervisor. He retired in 1974.

MITCHELL L. LATHROP, CAPT, born in Los Angeles on Dec. 15, 1937. He attended the Thatcher School and graduated with his BS degree from the US Naval Academy. He was assigned to the USS *Thetis Bay* (LPH-6), one of the first US units involved in the Vietnam conflict.

He was next assigned as XO of USS *Inflict* (MSO-456) and returned to Southeast Asia for Operation Market Time, the operation to prevent the resupply by sea of the Viet Cong in South Vietnam. He left active service to attend law school at the University of Southern California, graduating with a JD degree. He transferred to the Judge Advocate General's Corps in 1966.

Lathrop is a senior partner of the national law firm of Luce, Forward, Hamilton & Scripps LLP and divides his practice between the firm's San Diego and New York offices. He is married to Denice.

REGINA L. LAUDI-DAYTON, *See her bio in D Section.*

JAMES H. "HOWARD" LAUGHON, CAPT, born Jan. 6, 1932 in San Diego, received a BS in civil engineering.

He joined the service Dec. 17, 1950. His military locations and stations: Pensacola, Corpus Christi, NAS North Island. Served with: VC-35, VS-771, VAW-88. His battles and campaigns: Tachen Island evacuation. He was awarded the National Defense, China Service, Navy Commendation.

Laughon achieved the rank of CAPT and was discharged in 1975. His memorable experiences: night landings on straight deck carriers. CO VS-771 and VAW-88 (plank owner CO).

He is married to Marlene and they have children: David, Jana, Greg, Carolee, and has grandchildren: Chelsea, Geoff, Aaron, David, Kevin, Nathan and Katelyn. Employed as El Cajon city engineer, American Airlines as captain, retiring Jan. 6, 1992.

DOUGLAS P. "DOUG" LAWLESS, CDR, born July 25, 1946 in Indianapolis, IN, received a BA, June 1968 from the University of Florida and a JD degree December 1973 from Florida State University.

He joined the service August 1968. His military locations and stations: AOCS NAS Pensacola, Little Creek, NE. He served with VT1, VT2, COMPHIBLANT, NIRA 108, PACEN 108, USREDCOM, USCETCOM.

Lawless was awarded the National Defense Medal and achieved the rank of CDR. His memorable experiences: CO NIRA 108, CO PACEN 108, OIC JIB Bold Eagle 1984; PAO for 1st CVWR-20 deployment; USS *J.F. Kennedy* (CVA-67).

He is married to Cindy and they have children: Sara L. Klossner, Sharon Lawless and Ken A. Lawless and a grandson Jacob. He retired as attorney, Oct. 1, 1990.

F. PARKER LAWRENCE, CAPT, born Nov. 24, 1944 in Gainesville, FL and graduated from the University of Florida in 1967 with a bachelors degree in electrical engineering. On July 18, 1962 at age 17, he enlisted in the Naval Reserve Security Gp. Div. 6-2, Gainesville, FL as a seaman recruit (E-1). He was commissioned ENS, special duty, cryptology at Naval Officers Candidate School, Newport, RI on Dec. 15, 1967.

He served active duty tours at NAVCOMMSTA HONO, Wahaiwa, HI and as an assistant operations officer at NAVSECGRUDET, CINCPACFLT. Upon release from active duty in 1970 he re-joined his former reserve unit and attended law school at the University of Florida, where he graduated in 1974 and joined the Florida Bar in 1975.

From 1974 until 1991 he served tours as CO of Naval Reserve Group units in Atlanta, Jacksonville, Gainesville, FL, Orlando and Tampa. During this time he also served two separate tours as a Naval Security Gp. staff representative with REDCOMREGEIGHT, Jacksonville, FL. He retired in the permanent grade of CAPT on Dec. 31, 1991 from NRVTU Aldelphia, MD.

In civilian life, he served as General Counsel for the University of Florida Foundation from 1980 to 1991, entering private practice in 1991 and he continues to practice law. He enjoys his family which includes wife Mala and daughter Mala Kay. They enjoy traveling and occasional golf. CAPT Lawrence has served as president of both the Rotary Club of Gainesville and the Kiwanis Club of Gainesville. He also serves as a member of the Alachua County Health Facilities Authority.

JAMES TARLTON LAZAR JR., CAPT, born Oct. 24, 1922 in Whiteville, NC. Reared and attended schools in Florence, SC. He received a BS from Clemson College, Clemson, SC; MS, Cornell University, Ithaca, NY; PhD, NC State University, Raleigh, NC.

Joined the USN as a naval aviation cadet in December 1942. Called to active duty in July 1943 and attended Floyd Prep War Training Service, Pre Flights and flight training. Transferred to boot camp at Great Lake, IL. He was commissioned an ENS in August 1944 and attended Officer Training School at Ft. Schuyler, NY. Assigned to Amphibious Forces and trained at Fort Pierce, FL. Assigned to USS *Montague* (AKA-98) as small boat officer. After placing this ship in commission, he served in the Pacific Theater of Operations and participated in the battle of Okinawa and the occupation of China. Returned from the Pacific and placed the USS *Montague* in mothballs in the York River, VA. He was discharged from active duty in July 1946. Later served as the commanding officer of Naval Reserve Units in Anderson, SC, Greenville, SC and Asheville, NC.

Served as Commandants' representative in Sixth Naval District and as Red Com 6 representative for the Greenville and Spartanburg, SC, Savannah and Augusta, GA and Asheville, NC Naval Reserve units. Retired from USNR in 1982 with the rank of CAPT.

Some of his memorable experiences: serving as small boat officer and later as communications officer for the USS *Montague.*

He is married to Mary D. Lazan. He has served as director of Official Production Testing (dairy cattle) for the South Carolina Experiment Station located at Clemson, SC. Later he served as research assistant at Cornell University, Ithaca, NY and as assistant professor, associate professor and professor at Clemson University, Clemson, SC. Retired as professor emeritus Clemson University in July 1982.

LEO M. "LEE" LAZO, CAPT, born April 2, 1927 in Boston, MA, attended Boston College High, Northeastern University, receiving a BS in Business Administration. He also received a JD degree from Boston University Law School.

He joined the service March 12, 1945. His military locations and stations: USNTS Sampson, NY; NTS, San Diego, CA; Camp Elliott, CA. Served with: ATB Coronado; USS *Argonne* (AG-31); USS *Brownson* (DD-868) and USS *Portsmouth* (CL-83).

His battles and campaigns include the American and Pacific Theaters, WWII.

Lazo was awarded the Meritorious Unit Commendation, Pacific Theater Medal, American Theater Medal, WWII Victory, Japanese Occupation Medal, Navy Reserve Medal, Armed Forces Reserve Medal, Recruiting Ribbon, National Defense Medal and Expert Pistol Medal. He achieved the rank of CAPT, USNR. Enlisted BM2/c. Retired April 2, 1987.

He married Jane and they have children: Valerie, Michael and Christina. Employed as attorney-at-law, Badger, Sullivan, Kelley and Cole, retiring April 2, 1987.

EUGENE HARVEY LEAKE, CDR, born Aug. 21, 1927 in Sperryville, VA, attended Maritime School, Brooklyn, NY as marine engineer.

He joined the Merchant Marine, August 1944. Commissioned in USN, May 1950. His military locations and stations: 5th Naval District, operations officer, October 1954-October 1956. He served with the Merchant Marine in the Pacific Theater.

Leake was awarded the Pacific War Zone Bar, issued by the War Shipping Administration. He achieved the rank of CDR, USNR. Discharged from active duty October 1956. Retired USNR, August 1976. Served as CO of MSC Div. 5-3, July 1969-July 1971, Washington, DC.

His memorable experiences: Pacific sea duty; the storms; and 26 years USNR service.

He married Mary V. and they have a daughter Deborah Leake Harrington. Employed with the US Army Engineers as mechanical engineer, retiring May 1983.

MORTON LEAVITT, CAPT, born Nov. 27, 1911 in Cleveland, OH. Graduated Carnegie Institute of Technology in 1936, as bachelor of architecture. In 1943 received a direct commission as LTJG in USN with orientation at Princeton University.

Served active duty as Armed Guard officer in command of gun crew in Atlantic and Caribbean areas. Reassigned as CO of a rocket firing ships (LCI-704(R)) with duty in South Pacific areas, in support of 3rd Div., USMC assault forces for invasion support fire at Okinawa. Upon conclusion for hostilities, reassigned to the China Sea area on patrol of Taku River and based at Tienstin, China.

Returned to Cleveland in May 1946 and immediately joined Selected Reserve Force in August 1946. Became active in NAR. Was chairman of National Nominating Committee, National Surface/sub-surface Affairs and National Executive V.P. In 1973 was nominated and elected National President of Naval Reserve Association. Served in two years term.

He is authorized to wear the Bronze Star Medal w/V, Navy Commendation Medal w/V Atlantic and Pacific Medals, Chinese Theatre Medal, Philippine Liberation and Victory Medal and Public Service Medal.

Married to the former Shirley Rassler of Elyria, OH who is also a life member of NRA and past chairperson of NRA "Anchors". He has two sons, Andrew and Jeffrey, by former marriage and three grandchildren.

DONALD H. "DON" LEDIN, CDR, born July 13, 1914 in Lindstrom, MN. He graduated from Forestry School University of Georgia.

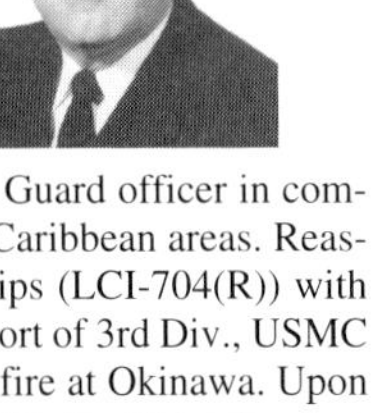

Enlisted USNR 1937 and commissioned ENS 1943. His military locations and stations: VR-6 Norfolk, VA; VR-2 and Alameda, CA. His total military time, 38 years.

He achieved the rank of CDR USNR and discharged June 1968. His memorable experiences: pilot on Mars flying boat.

He is married to Margaret Winkle and their children are Donald, Daryll and Debra. Employed as trails coordinator, pilot, assistant superintendent of Game Section, Minnesota Department of National Resources. He retired in September 1981.

BEN F. LEE, CWO2, married Chrystelle H. Lee. Ben Lee passed away Dec. 21, 1995.

RONALD P.E. "RONALDO or NALDO" LEEDS, born Nov. 14, 1939, in New York City. He received BSE degree, Wharton School (U. of P.) and advanced graduate studies at A.U.B.

He joined the service in 1960. His military locations and stations: Charlestown, Norfolk, Little Creek, Quonset, "Gitmo". Units served with: USS *Notable* (MSO-460); USS *Donner* (LSD-20); HS-1; HS-5; Naval Attache US Embassies of Jordan, Cyprus, Syria, and Lebanon. His battles and campaigns: Bay of Pigs; Cuban Blockade; Dominican Republic; "6 Day War".

He was awarded the Special Citation, Department of State-Beirut 1967 "6 Day War". Leeds achieved the rank of LT and discharged active duty 1968-69; Reserves 1976.

Memorable experiences: Gemini and Apollo Recovery Pilot; events leading up to, during, and after the 6-Day War in the Middle East.

Married Darcy and they have a daughter Natalie. Civilian employment as Exerv Corporation S.A., president; KLM Group, Chairman; Jesup and Lamont (member NYSE) chairman; Republicans Abroad (the International Arm of the Republican National Committee), co-chairman,; Cancer Research Fund of the Damon Runyon-Walter Winchell Foundation, director.

ROBERT WALLACE "WALLY" LEES, LTJG, born Oct. 16, 1911 in Oakland, CA. He attended college.

He was commissioned as ENS DM USNR 1939. Called to active duty March 1942. Assigned to USS *Charles Carrol,* an APA, fitting out at Bethlehem Steel, Sparrow Point Shipyard, Baltimore, MD, and on board when placed in full commission.

Due to lack of naval offices and facilities at Bethlehem Steel, he was temporarily assigned to Maryland Drydock and Repair, industrial manager's office and assisted in the conversion and repair of the various vessels.

Detached June 1942 from industrial manager, Maryland Drydock and assigned to USS *Rochambeau* (AP-63) fitting out in San Francisco area and on board when placed in full commission.

The *Rochambeau* was assigned to Commander Western Sea Frontier, San Francisco and operated between West Coast of USA and Western Pacific, Southwest Pacific and South Pacific.

Upon full commissioning he was assigned as navigator by CO while still only LTJG. In 1944 detached from *Rochambeau* and was assigned to USS *Herald* of the Morning A.P. 173 at 1LT and subsequently XO. He has been promoted to LT and LCDR while still serving on board the *Rochambeau.* USS *Herald of the Morning* was also assigned to Commander Western Sea Frontier and immediately reassigned to Commander Amphibious Forces, Pacific.

Participated in landings on Saipan, Anguar Island, Ulithi, Leyte Gulf, Mindoro, Lingayen Gulf and Iwo Jima. Reinforced landed elements in Philippines.

Awarded battle ribbons: Western Pacific, Southwest Pacific w/Silver Star, Philippine Liberation and Occupation, two ribbons each w/Bronze Star. Achieved the rank of LCDR while on active duty. Participated in USNR Training Program and Correspondence courses and was promoted to CDR in 1951. Included in Ready Reserve. He was released to inactive duty Sept. 30, 1945, and retired from USNR, Oct. 1, 1959.

A Widower, he was married to Florence Renfrew. His children are Nancy Odegard and Wallis Ahlstrom, and grandchildren are Marina Behla, Cleave Odegard, R.W. and Brooke Ahlstrom.

Held various assignments as deck merchant marine officer. Seven issues of unlimited masters license. Principal surveyor, Pacific District, US Salvage Association Inc., retiring Oct. 1, 1976.

ALEX LUCERO "ALEX" LEGASPI, LT, born Oct. 14, 1951 in Makato Aklan, Phillipines, philosophy major, bachelor in Sacred Theology.

He joined the service Feb. 8, 1991. Commissioned May 8, 1991. Military locations and stations: Naval Station, Treasure Island, SFCA; NAVMARCORESCEN, Alameda, CA. Served with I Marine Expeditionary Force Rel. 220; 1st FSSG-FMF PAC Camp Pendleton, CA.

He was awarded the National Defense Ribbon and Fleet Marine Force Ribbon. Achieved the rank of LT. His memorable experiences: rappelling with 18 year old Marines in Bridgeport, CA and freezing in the middle of the cliff (8,000 feet). That's when he regretted joining the Marine unit.

Legaspi is single and the associate pastor of St. Emydius Catholic Parish in San Francisco. Still active in the USNR.

ROBERT LEE "BOB" LEJEUNE, CDR, born Sept. 5, 1938 in Iota, LA, received a BA from Maryknoll College; MED Northwestern; MBA Lamar University.

He joined the service October-September 1962. His military locations and stations: Hampton Roads AADCP, Norfolk; NAV Shipyard Boston; NAVSTA San Diego; NAVSTA LBeach; NAV Base Yokosuka, Japan. Served with AADCP, Norfolk; USS *H.W. Tucker* (DD-875); Reserve Centers Lafayette, LA; Houston; Beaumont, TX; Orange, TX; San Antonio, TX; Waco, TX.

His battles and campaigns include the Tonkin Gulf and Vietnam. He was awarded the Meritorious Unit Citation, AF Expeditionary, Vietnam Service w/star, AF Reserve, NR Medal, Expert Rifle, Expert Pistol, National Defense, RVN Campaign. He achieved the rank of CDR and retired October 1987.

His memorable experiences: onboard USS *HW Tucker* as weapons officer; first ship to shore bombard North Vietnam; many excellent two weeks training duties in San Diego, Washington, DC, Portland, OR; LBeach; New Orleans; Concord, CA; the Orlando, FL and others.

He married Doris Marie LeBlanc and their children are Simone, Suzanne, André, Marcel and Yvette, and their grandchild is Celeste Marie Triggart. Employed with Texas Department Human Services, University of Houston, Lamar University, USL Lafayette, LA. He will fully retire in September 1998.

J.W. LEMAY JR., CAPT, enlisted in 1942, and was released from active duty in 1945. His active duty service was on board USS APC-37 (enlisted) and USS *Rushmore* (LSD-14) (commissioned) in the amphibious forces, where he participated in landings on numerous Japanese Occupied Islands. His Navy campaign and service medals include: American Theater, Asiatic Pacific w/4 stars, Philippine Liberation w/star, Victory Medal w/star, consolidations New Guinea, Philippines.

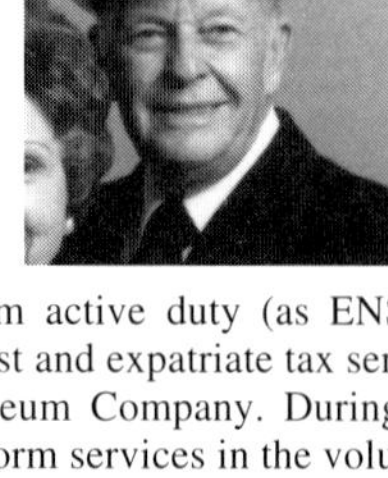

Upon his release from active duty (as ENS) he became foreign tax specialist and expatriate tax services director for Phillips Petroleum Company. During this period he continued to perform services in the volunteer USNR and advanced in rank to CAPT before being transferred to the Retired Reserve after completion of 30 years of continuous commissioned service.

He and his wife Katherine reside in Bartlesville, OK. Their daughters: Karen Meyer and Pam Snelson and their families lives nearby, while their son Bill Lemay and family lives in Kansas City.

DONALD R. LENDERINK, LCDR, a native of Grand Rapids, MI and graduate of Hope College and Western Theological Seminary in Holland, MI, and became a chaplain near the end of the Korean Conflict in 1954 following graduation. He served aboard MSTS troop transports in the Atlantic and Mediterranean for which he was awarded the National Defense and Navy Occupation (Europe) Service Medals.

He left active duty in 1956 after which he ministered to Reservists at Clifton, NJ, while serving a pastorate in Little Falls, NJ and at Sioux City, IA, while serving in Orange City. He held many positions of leadership in the Reformed Church in America, and was awarded an honorary doctorate from Northwestern College in Orange City, IA. He began his ministry sailing out of Brooklyn, NY, and concluded it as a pastor in Brooklyn, OH, retiring in December 1949. His family consists of wife Ruth, four married children and eight grandchildren.

RONALD M. "RON" LEONARD, LCDR, born Dec. 21, 1935, New York City, NY. Earned BS and MBA degrees in business management. Joined the military service Feb. 5, 1954.

Stationed at NASNY, NAS NORVA, N&MCRRC Floyd Bennett Field, Brooklyn, NY. Served with VF-84, VS-835, Telecom Censors 3-1-1, NARDIV R-1, NARMU-14R1, NR Recruiting Cmd.

Awards and medals include N&MC Commendation Medals, N&MC Achievement Medals, NR Meritorious Medal, four AF Service Medals, MUC, two National Defense Service Medal, three Recruiting Ribbons and Enlisted Air Crew Wings. He retired Dec. 20, 1995.

Married Esther and they have six children: Desiree, Ronald Jr., Kathleen, Mena, Todd and Antoinette; three grandchildren: Trina, Christopher and Serena. Civilian employment with Exxon Intl. Co., New York Telephone, city of New York.

JOHN J. LEPORE, CDR, born in Everett, MA, March 19, 1931, was educated at Boston College and St. John's Seminary. In August 1966 he was commissioned a Navy chaplain and served in Gitmo, Cuba, Vietnam, Pax River, Maryland, Yokosuka, Japan and NSGA, Winter Harbor, ME. In Vietnam, Chaplain Lepore received the Purple Heart, Combat Action Medal, Presidential Unit Citation among other awards.

After leaving active duty in 1974, Lepore served the Reserve Navy as a public affairs officer. He affiliated with Navy Office of Information Boston/New England and retired as XO of the unit in 1990.

In his civilian capacity spent 20 years in the Malden, Massachusetts schools as a counselor, psychologist, and coach, retiring in 1994.

He now resides in Coronado, CA with his wife, Martha. Lepore is an avid tennis/racquetball player and sailor. He is a life member of the NRA.

MALCOLM FRANCIS F. LEQUEUX, CDR, born in Henry, LA, Jan. 14, 1919, received a BA, University of Southwestern Louisiana, 1940; MEd, University of Houston, 1956; Advanced Counseling Institute, The University of Texas at Austin, 1963-64; Advanced Certification: Educational Diagnostician, 1974.

Midshipman training: Midshipman training, Northwestern University, Chicago, September-December 1941. Commissioned ENS, Jan. 16, 1942. His military assignments: WWII, February 1942-December 1946, Asiatic-Pacific arena, on the following Cargo and Troop Transport ships: USS *Talamanca,* USS *Rutilicus,* USS *Griggs* and the flagship USS *Estes* (homeport: Tsingtao, China).

Korean Conflict: returned to active duty. Assigned as inspector-instructor/CO of the USNR Training Center, Austin, TX, January 1951-February 1953. XO, USS *Waccamaw* (homeport: Newport, RI), 1953-54.

Naval Reservist Activities: When not on active military duty, remained active as a member of the Naval Reserve Surface Div. 8-97, Port Arthur, TX. Served as unit CO, 1950-51.

Summer active duty: instructor, Reserve Officer Candidate (ROC) Program, US Naval Station, San Diego, CA, 1949; 1950; acting CO, US Naval and Marine Corps Reserve Training Center, Albuquerque, New Mexico, summer 1957.

Married to Lou for 52 years and they have a son Blayne and three daughters: Diane, Lydia and Julie. They have eight grandchildren: Karen, David, Suzanne, Tracey, Katherine, Elizabeth, John, Jennifer.

Employed by the Port Arthur, TX, ISD in various capacities. Teacher, activities director, counselor, principal and Pupil Personnel Services until retirement, June-1979. He retired January 1979 as CDR, USNR (Ret).

RUSSELL TED LERCH, CAPT, born Dec. 26, 1937, Freeport, NY. Earned BS in ind. eng. and MBA in finance and economics. Joined the service in June 1959 and was stationed at Great Lakes, IL, Norfolk, VA, Washington, DC, and Willow Grove, PA.

Duty as management engineering officer, assistant ops and plans officer, special project consulting and special projects with NAVMASSO, Norfolk, VA. He was in charge of NAVAIR Command and Control Ctr. operations, managed and reorganized the 83 Navy POM.

Organized, directed and completed a study of P-3 part problems, completed NATO portion of 81 Navy POM report for the Navy at OPNAV-09, Washington, DC, and completed evaluation of the electronic cal-lab system for aircraft components.

Evaluated analog/digital test systems for avionics blackboxes; made management survey of NARF at Norfolk; was special mgt. analysis of NAVAIR Systems Cmd.; completed a CASS research project study on aircraft carrier fires; Mob assignments in Surface Missiles Div. and ASW and management study of Naval Air Engineering LAB (SI) Philadelphia. Retired July 1, 1994, as CAPT.

Employed with GE, IBM, then AMP. Married Rowena and they have three children: David, Patti and Gregory, and one grandson Jack.

JULIAN LERNER, CAPT 06, born July 15, 1940 in Brooklyn, NY, received a BS in economics, Wharton School of University of Pennsylvania.

He joined the service May 1962. His military locations and stations: NAS NY, NAS Norfolk. Achieved the rank of 0-6 and discharged May 1991.

Lerner is married to Nancy and they have children Scott and Jaymie. Employed as president, Carolina Rugmakers, Inc.

JOHN W. LESTER, CAPT, born in Binghamton, NY on Sept. 1, 1930. He was graduated from Clarkson University with a bachelors degree in mechanical engineering in 1952. He was employed with General Motors Research Laboratories in Detroit, MI from 1952 to 1954.

In October 1954, he entered the USN through Officer Candidate School in Newport, RI. Upon commissioning in March 1955, he was assigned to USS *Dunn County,* homeported in San Diego. During his two years aboard, he served as the chief engineer and was deployed during two summers to the Arctic regions where he gained considerable experience in the ice seamanship. After a brief tour at NAB, Coronado, he was assigned as an engineering instructor at the US Naval Academy. Following this two year tour, he was released from active duty in June 1959.

He earned a master's degree in mechanical engineering from the University of Minnesota in 1961. He then took a job with the Garrett Corporation (now AlliedSignal Aerospace Co.) in Phoenix, AZ. He was primarily involved with the design and development of gas turbine engines for the aerospace industry. He retired from AlliedSignal in 1990.

Lester was active in the USNR, and was subsequently designated an engineering duty officer. He held two commands, reporting to the Naval Sea Systems Command. He has served as a Naval Academy Blue and Gold Officer since 1972. He was retired as a CAPT in 1987, but has continued to serve voluntarily in the Blue and Gold Program. He is currently the Arizona state coordinator for that program, coordinating the activities of 13 officers.

He is a member of the NRA, US Naval Institute, the Society of Automotive Engineers, and is a registered professional engineer in Arizona.

Lester is married to Shirley and has two children and two grandchildren. He resides in Tempe, AZ.

ROYAL A. LETT JR., CDR, born June 23, 1920, in Miami, FL.

Joined the service September 1940. His military locations and stations: V-7 Midshipmen's Program. (Class #3) at USS *Prairie State* at New York City, NY, graduating June 6, 1941.

His military locations and stations: NAS Sand Point, Seattle, WA, VP-42; NAS Kodiak, AK, VP-42; NAS Dutch Harbor, AK, VP-42; Advance Base, Umnak Island, AK, PATWING 4; NAS Moffett Field, CA, LTA Flight Train ing; NAS Lakehurst, NJ, LTA Advanced Training; NAS Tillamook, OR, ZP-33 lighter than air command pilot; NTS, New Orleans, LA, refresher training for sea duty as XO.

Served with VP-42, HEDRON, PAT WING 4, training off Z P-33, PXO AK201 (Fleet Cargo Ship) USS *Pemiscot.*

His battles and campaigns: Aleutian Campaign, Unit Citation, Pre-War. Post war: XO, 2P-952, NAS Tustin, CA, 1950-55 LTA. He achieved the rank of CDR 1955 and retired July 1, 1966.

Married to June and they have children: Vickie, Sandy, Carolyn and Alex and grandchildren: Vikki, Scott, Marge, Robyn, Steven, Andrea, Jay, Leigh Ann and Amy and his great-grandchildren: Emelia, Carolyn, Olivia.

Employed as owner, photo/TV company, Long Beach, CA, 1945-55; EX VP, AV Electronics, Inc. (Public Corp), Fresno, CA; President, Rolaben Corp. (Private Corp), Clovis, CA, 1971-83. Retired in 1983 and now living east of Sanger, CA in the foothills of the Sierras.

ROBERT J. LEY, LCDR, born Sept. 29, 1918 in Joliet, IL, received an AB in 1939 from the University of Illinois; JD in 1946 from DePaul University of College of Law.

He joined January 1942. His military locations and stations: Midshipman NYC, USS LST 220. Served with Central, South and Southwest Pacific Amphibious Forces. His battle and campaigns: Solomon Islands, Kwajalein, New Guinea, Marianas, Leyte, Luzon, PI.

He was awarded campaign ribbons. Achieved the rank of LCDR, USNR and discharged November 1945. Serving under J. Warren Wright, CO, LST 220, an inspired leader.

He married Janet Musselman in July 1943. They have children: Ellen L. Satenstein, Caroline L. Kennedy, George M. Ley, Catherine L. Lawler, and grandchildren: Aaron, Charles and Ryan Satenstein, Peter and Clare Kennedy, Madeline and Grant Ley, Patrick and Anna Lawler. Employed as attorney in private practice, Chicago and Buffalo Grove, IL.

FREDERIC C. "FRED" LIBBY, CDR, born Sept. 7, 1920 in South Portland, ME, received a BS in engineering physics from the University of Maine, Orono, ME.

He joined the service Aug. 10, 1942. His military locations and stations: boot camp, Newport, RI; NATTC Jax, FL; NAS Puunene, TH. Served with CASU 44; NAS Corvallis, OR; NAS Alameda.

Libby achieved the rank of CDR and retired June 1974. His memorable experiences: Many; foremost of which was that he was never shot at.

He married Margaret Marley April 19, 1947. After her death, he married Joan R. Case, April 29, 1978. His children include Frederic C. Libby Jr., Lorna Libby Barrett Johnson and grandchildren: Carl Libby, Erik Libby, Scott Barrett and Katharine Barrett. Employed as photographic engineer, supervisor apprentice training, Eastman Kodak Co., Rochester, NY, retiring April 1981.

MARC E. LIEBMAN, ENS, was commissioned in 1968 through the AVROC program and designated as naval aviator in 1969. He flew combat support and special operations missions in UH/HH-2C/D helicopters during deployments to the Mediterranean, North Atlantic and Vietnam in HC-2.

Liebman transferred to HS-75 and its training units flying the SH-3A/D/G in 1972. In 1978, he moved to Dallas, TX and was promoted to LCDR and selected CO of the USS *Ranger* augment unit. Promoted to CDR in 1982, he joined COMSEVENTHFLT 111 until selection to CAPT and CO of COMCARGRU 770.

COMCARGRU 770 was recalled for Operation Desert Shield/Storm in support of its gaining command, CARGRU 7 embarked on the USS *Ranger.* COMCARGRU 770 was the only Navy carrier group staff recalled and provided, among other major taskings, a fully combat qualified flag watch team. CAPT Liebman was the assistant chief of staff for Operations and Plans for the Persian Battle Force Anti-Surface Warfare commander. He was selected as CO of COMSEVENTHFLT 111 before retiring in July 1992.

As a Reservist, CAPT Liebman was awarded three Meritorious Service Medals and four Navy Commendation Medals plus awards for heroism and campaign medals from the US, Vietnam, Kuwait and Saudi Arabia.

Liebman is the principal shareholder in ADI International, Inc. which repackages and ruggedizes electronic equipment. He is responsible for business development and finance. Earned a BS in Applied Mathematics and an MBA in marketing. He has owned a full service advertising and marketing consulting company. Frequently contributes as a freelance writer to publications on defense and foreign affairs. CAPT Liebman is married to Betty Kay Yanowitz and has two children, Michele and Seth.

PETER R. LINDELL, CAPT, born March 3, 1940 in Red Wing, MN, received a BS in physics, Gustavus Adolphus College, St. Peter, MN.

He joined the service Sept. 11, 1962. His military locations and stations: Navy flight training, NAS Pensacola, FL and NAS Corpus Christi, TX; USNS Argentia Newfoundland, Canada 1964-65; USNS Keflavik, Iceland, 1964-65; NAS Moffett Field, CA, 1966-67; MCAS Iwakuni, Japan, 1966-67; NAS Twin Cities, MN, 1967-69; NAS Glenview, IL, 1969-79; Naval Air Reserve Center Twin Cities, 1979-83.

Units served with/aircraft flown: VW-11/EC-121P Super Connie; VP-19/P3A Orion; VP-60/SP-2H, P3B Orion; VP-0472/P3B; NR PMTC-0278/None.

He was awarded the Armed Forces Reserve Medal, Vietnam Service Medal, National Defense Medal, Unit Citation and Unit "E" Award. Lindell achieved the rank of CAPT and was discharged from duty, September 1967.

His memorable experiences: May of 1967 when VP-19 was assigned to detach from NAS Moffett to NAS Agana Guam to monitor a Soviet ECHO Class nuclear submarine off the FBM repair facility in Guam. SOCUS had tracked two ECHO subs from Vladivostok. One had gone into the Tonkin Gulf and the other to Guam. They had a new passive acoustic sonobuoy which allowed them to fly over him around the clock for several days. One of their destroyers put a depth charge in the water quite some distance away and the ECHO boat made a sudden high speed departure. The resulting "knee curve" acoustics were a welcome addition into the Navy's Soviet submarine data base. Their squadron was awarded the Unit Citation for this six week deployment.

Maintenance officer of Patrol Sqdn. VP-60 at NAS Glenview, IL from June 1975 to September 1978; CO of Augment Unit VP-0472 at NAS Glenview, IL from October 1978 to March 1980; CO of Naval Reserve Pacific Missile Test Center Unit PMTC-0278 at Naval Air Reserve Center Twin Cities from September 1982-October 1983.

He married Judy Anderson in 1963. They have children: Steve, NROTC Northwestern University, class of 1989. Has had three deployments with VP-26 (NAS Brunswick, ME) to Sigonella and is instructing multi-engine students at VT-31 in NAS Corpus Christi, TX; Kari, national accounts manager, with Northwest Airlines; Kim, secondary English teacher.

Employed by Honeywell Flight Operations for 30 years, flying company executives and customers worldwide. Captain on Gulfstream G159, G1159, GIV and Citation C650. He has accumulated over 16,000 flight hours of which 3,000 are USN. A past soccer referee commissioner for communtiy soccer. Enjoys sailing in the Apostle Islands of Lake Superior, computer geneaology, hunting and fishing. He retired Oct. 1, 1989.

LORI LINDHOLM, CDR, born Aug. 10, 1967 in Gloversville, NY, received a BA, Theater Arts SUNY College of Oswego, 1979; MS Systems Management, University of Southern California, 1988.

She joined the service January 1980. Her military locations and stations: San Diego, NAS North Island; NAS Cubi Point; NAS Pensacola; NavSta Anacostia. Served with VRC-30; VC-5; HC-1; USS *Lexington,* NODAC; NCC106; OUSD/PCCC.

Lindholm achieved the rank of CDR, USNR, Drilling Reservist and received the MUC, USCG Spec Ops, Overseas, Marksman, Navy Achievement Medal (2), and Navy Commendation. She is presently USNR Drilling Reserve.

She is married to Ross and is employed as program manager, Strategic Resources, Inc.

JOHNATHAN W. "JOHN" LINDSAY, CAPT, born Aug. 27, 1921 in Gainesville, TX, received BA and MA degrees.

He joined the service Sept. 2, 1942. His military locations and stations: USS *Stoddard,* USRS & ONOP Los Angeles, Navy Prep School, San Diego. Served with CO, NRSD 11-48; CO NRSD 11-32; COMOR NRCG 11-9.

Lindsay was awarded the American Theater, Pacific Theater, two Battle Stars, Naval Reserve Medal, and Armed Forces Reserve Medal. He achieved the rank of CAPT and retired July 1, 1975.

His memorable experiences: shore bombardment, Kurile Island; kamikaze attacks, Okinawa; plane guard duty (on DD) outside Tokyo Bay at the signing of Japanese surrender on BB *Missouri;* held three commands of Naval Reserve Units.

He married Kathleen O. and they have children: Kathe D. and Sally E. Lindsay, and grandchildren: Lindsay D., Ryan J. and Rayna E. Gasik. Employed at Texas Tech, UCLA, Cal State University, Northridge, professor and associate dean, retiring Dec. 31, 1989.

JOHN D. LINDSTROM, CAPT, born July 30, 1934 in Minneapolis, MN, received a BS degree in 1956 and a MS degree in 1961 from the University of Minnesota.

He joined the NROTC from the University of Minnesota, 1952. His military locations and stations: NAS Moffett, NAS MPLS, NAS Atlanta, Norfolk, Glenview. Served with VR-7, VR-813, VR-724, VR-56, VR-51, NR Agana, Guam.

Lindstrom served in Vietnam. He was awarded the Navy Commendation Medal, National Defense Medal, Vietnam, Naval Reserve, Unit Commendation. He achieved the rank of CAPT and retired in 1986 from the USNR. His memorable experiences: Noel Davis Trophy VR-51 1980, BARTO Trophy 1984 NR Unit Agana Guam.

He is married to Jan and they have children: Jana Allen, Jill Lindstrom and Jodie Frerichs and a grandson Zachary Frerichs. He achieved the rank of CAPT MD-11, L-1011, B-767/757/727, DC-9, 7, 6; chief pilot, Chicago Delta Airlines 1983-88, retiring July 30, 1994.

MURT "K" LININGER, CAPT, born Sept. 17, 1932 in Bowman, ND, attended California State University at San Luis Obispo and University of California at Davis obtaining a BS degree in engineering and a certificate in program management.

He joined the service USMC, Sept. 24, 1952; USNR commissioned ENS, May 28, 1968. His military locations and stations: K-3 Korea, Taipeh Formosa; NAS Alameda. Served with VMJ-1, NRIU: 12-1-3, FICPAC, DNRIP.

Awarded the USMC Good Conduct, China Service, National Defense, Korean Service, Armed Forces Reserve, United Nations, Navy Expert Pistol Shot and Secretary of the Navy Letter of Commendation.

Lininger achieved the rank of SGT (E-4); CAPT (06) USNR retiring Nov. 1, 1991. His memorable experiences: It was an honor and a privilege for him to serve his country while in uniform with the USMC and the USN.

He married Marie Ann and they have children: Glen Dale Lininger, Deborah Lynn Wolfbrandt and Lori Ellen Mote; and grandchildren: Lisa, Amy, Timothy, Kenneth, Micah, Shari and Kali. Employed with the state of California as supervising engineer and program manager, retiring May 2, 1995.

EDWARD M. "NED" LINKER, LT 03, born May 29, 1926 in Chapel Hill, NC, received a BSME from Duke University.

He joined the service July 1, 1944. His military locations and stations: Duke University V-12/NROTC Served with Naval Reserve Composite Companies 5-20, 5-5-9; ONR VF861 and VS861; NARTU NAS Norfolk.

Linder was awarded the American Campaign and WWII Victory Medals. He achieved the rank of LT 03 and was discharged June 25, 1946. Transferred to Retired Reserve Aug. 1, 1973 and then transferred to Naval Reserve Retired List effective, May 29, 1986.

He married Jean Stone and they have children: Edward M. Jr., Jean Jane, Mary Frances and Walter Stone; their grandchildren are Lucy L. Bernard and Edward M. III. Employed as engineer with Du Pont for 11 years and Tultex Corp. for 26 years, retiring Dec. 31, 1984.

JOHN W. "JACK" LINKER, LCDR, born Dec. 25, 1937 in Takoma Park, MD, received a MPA from Florida Atlantic Univesity and PhD from University North Texas.

He joined the service August 1960. His military locations and stations: Norfolk and Miami. He served with USS *Diamond Head;* COMSERVLANT; OCS. Participated in the Cuban Expedition. Linker was awarded the Cuban Expeditionary Medal. He achieved the rank of LCDR.

His memorable experiences: visit to Morovia Liberia by *Diamond Head* on mission "Brother's Brother" to provide aid.

Linker is divorced. Retired as director, Technical Training, USDA in 1994.

DAVID R. LIPINSKI, CAPT, a native of Chicago, IL, received his commission upon graduating from the US Naval Academy in June 1974. His Trident Scholar dissertation, *Toward a Cosmic Ray Astronomy*, was published by the Naval Academy in 1974. He holds a BS in physics from the Naval Academy and a MBA from the University of New Haven. His military education includes courses at the Naval War College, Newport, RI. Lipinski is a registered professional engineer and is a chartered financial analyst.

After commissioning, Lipinski entered the nuclear power program. Following Submarine School at Groton, CT, he reported to USS *Sand Lance* (SSN660). Aboard *Sand Lance,* he served as a division officer and department head. With *Sand Lance,* he cruised in the Mediterranean, North Atlantic and Caribbean. Lipinski then returned to Submarine School as a tactics instructor and the Sonar Div. director. He left active duty and accepted a Reserve commission in 1981.

Since 1981, his reserve activities have largely involved the intelligence community. Reserve assignments have included intelligence officer with VP-92 an VP-90 maritime patrol squadrons and staff assignments within OPNAV as well as XO and CO positions in intelligence production units. Annual training duty has included squadron deployments to Bermuda, the Azores, and Japan; staff assignments in Washington, Norfolk and the UK; and a variety of exercises in European and Far Eastern Commands.

Lipinski has received the Navy Commendation Medal, National Defense Medal, Naval Reserve Medal, Joint Meritorious Unit Award, Sea Service Ribbon, Reserve Overseas Service Ribbon and the Expert Pistol Medal.

CAPT Lipinski resides in Lake Forest, IL. He is married to the former Karen L. Marino and is the father of two children, Ashley and Eric.

ROBERT D. LITTEN, CWO4, born in Bedford, IN, June 1, 1926. He moved to Miami, FL with his family in 1934. Upon graduating from Miami Senior High School in 1944, he was immediately called to active duty. After boot training at Camp Perry, VA and engineering schools in Gulfport, MS reported aboard USS *Alchiba* (AKA-6). During the last year of WWII, served in the western Pacific and in 1946 was transferred to the USS *Strickland* (DE-333) for mothballing. Discharged and then re-enlisted in the USNR became a plankowner in Surface Div. 7-8 and later plankowner in Surface Div. 6-45.

Called to active duty in 1950, served in Boat Unit 1-Naval Beach Gp. 1 until released to inactive duty the latter part of 1952. Reported back aboard Naval Reserve Center, Miami and reassigned to Div. 6-45. Appointed to warrant machinist in 1957. In 1964, assigned to SAMAR Div. 6-4 as training officer and later administrative officer. In 1968 promoted to CWO4. Became a plankowner in Military Sealift Div. 130 in 1976.

He was redesignated CWO, Civil Engineer Corps in 1977 and assigned to CBC Gulfport as training and operations officer. Later assignments followed in RNCB 14, RNCB 24, 3rd RNCR and CB Hospital Unit, Atlanta.

His campaign and service medals are American and Asiatic Theatre, WWII Victory, National Defense, Naval Reserve, Armed Forces Reserve w/2 HGs, Korean Service, United Nations, 3-E Awards and the Korean Presidential Unit Citation.

Retired from the USNR in 1986 after 42 years on pay status. The experiences of traveling, schools, cruises and the many different billets over the years has given CWO Litten a broad understanding of naval life that few experience in their lifetime.

He has been a member of the NRA since 1957 and a life member since life membership became available. In 1973, he attended classes at Miami Dade Community College and earned a degree in Business Administration and one in Supervision of Management.

He has been working in the electrical field since 1948 and has been the owner and manager of his own electrical contracting business since 1980.

He and his wife, Gwen, have one son, Joe, and grandsons Jeff and Jameson. They have resided in Franklin, NC since 1978.

His hobbies include coin and stamp collecting, woodworking and playing country and Bluegrass on the mandolin. He has been active in the Boy Scouts since joining in 1938.

GERALD V. "JERRY" LITTIG, CDR, born Nov. 13, 1911 in Coeur d'Alene, ID, attended the University of Virginia, engineering, 1932.

He joined the service April 1941. His military locations and stations: USNOB Guantanamo, Cuba; US naval liaison officer, Cartagena Columbia; assistant naval attache, US Embassy, Monte Video, Uruguay.

Littig was awarded the Defense and Victory Medals for Antisubmarine Campaign-Caribbean. He achieved the rank of CDR, USNR, and retired February 1946.

His memorable experiences: matching wits with pro Nazi underground in Cuba and Columbia (attached to ONI, naval intelligence officer).

He is married to Rosa Menéndez and they have children, David and Marina (Mrs. Kip Harris); and grandchildren, Mark and Melissa. Employed as executive director, Long Range Strategic Planning, UpJohn Company, retiring in 1981.

PALMER O. "SUNNY" LITTLE, LT, born July 6, 1927, Boxford, MA. Earned BA degree from San Diego State University; Certificate in Aerospace Mgmt., Univ. of California; Certificate in business mgmt., Univ. of Southern California. He served honorably for 27 years in enlisted and officer billets of the USNR. He served in many classes of ships and submarines and was involved with administrating and training of Naval Reserve units.

Station/Units/Ships: (1945-58) USS *Robinson, Tillman, Ulhman, Steelhead, Ocella, Fort Marion, Flicker, Warbler, Koiner;* Pacific 5th and 7th Fleets, Atlantic 2nd Fleet, Minsqrd 5, Minflot 1, USNS Long Beach, CA; USN Minewarfare Staff Officer School, Naval Post Graduate School in Undersea Ordnance; Naval Liaison Officer, American Air Defense Command (NORAD).

Memorable Experiences: steering ship through Panama Canal; flight operations support in the Atlantic; riding out a hurricane on the Tillman; rescuing a pilot after he crashed in the ocean; , working with Japan's military self defense forces; and many others.

Served in the USNR 1959-80: USN Electronics Unit 11-2, Santa Marie, CA; USNR Officer School, Santa Maria and Santa Barbara, CA; USNR Surface and Composite Unit, Tustin, CA.

Married Betty and they have two children, Christopher and Heather; grandchildren, Matthew and Samantha Little, Katlyn Gilchrist, Joshua and Benjamin Wise.

Civilian employment with Convair, Martin Marietta Co. and McDonnell Douglas Aerospace. Retired Dec. 22, 1994.

JAMES "LANE" LITTRELL, CAPT, born Nov. 12, 1951 in Oklahoma City, received business degree with transportation major from the University of Texas at Austin 1975.

He joined the service Oct. 15, 1975 through Aviation Officer Candidate School (AOCS). His military locations and stations: VQ-1, Guam; VAQ-33, Key West; Naval Air Reserve Center, Minneapolis; Pentagon, Washington, DC.

His awards and medals include the Navy Commendation, Navy Expeditionary Medal, Navy Unit Commendation and Expert Pistol. He achieved the rank of CAPT 06 and is presently still serving.

His memorable experiences: flying A-3s (Whales) aboard the ship at night.

He is married to Nancy and they have children: Alison, Austin, Abby, Alayne and Alexis. Employed as airline pilot, Northwest Airlines, St. Paul, MN.

ARNO R. LIVINGSTON, CDR, born June 5, 1949, is a native of Arlington, VA and graduated from Virginia Polytechnic Institute and State University. He took his masters from the University of Southern California. He began his military career in 1971, enlisting in the Army Reserves. He served in HQ 2174, Salem, VA and the 955th ENG, Ft. Belvoir, VA.

In 1975 he was commissioned in the USNR. He served in Naval Air Systems Command Units 0166 and 0266 from 1971 to 1987. His final assignment was with NADOC-0166, from which he retired in 1994.

Livingston is presently employed by The Allergy Center of Huntsville, and also trades stocks and options for his own account. He previously was president of the Western Pacific Corporation, director of Huntsville Operations for Joint Research Group, and project manager for D.K. Dixon and Company.

Awards include Defense Service Medal w/star, Expert Rifle, Sharpshooter Pistol, Armed Forces Reserve Medal and Navy Commendation Medal.

He is married to Linda and they have children: Ryan, Mark, Erin and Yadi.

LEO O. "RENEGADE" LOB, LT, born June 1, 1924, Baton Rouge, LA. Earned BBA at LSC and MBA at Tulane Univ. Joined the service in 1941 and served at NAS Corpus Christi, NAS Pensacola and in USS *Hubbard*, and *Intrepid*.

Participated in battles and campaigns in the Philippines, Okinawa and Korea. Memorable experience was four kamikaze attacks; and being the first US Military at Manchuria two days prior to Russian army's arrival. Discharged in 1954 as LT. Received the Purple Heart, Distinguished Flying Cross, Air Medal and Philippine Liberation Medal.

Married Kaye and they have four children: Chuck, Jackie, Davy and Nancy. Still working as broker, rep., insurance and securities.

JOHN JAMES LOCKWOOD, LCDR, born April 24, 1947, Waukegan, IL. Earned BS degree in business administration. Joined the service in March 1981, and served in the USS *Bigelow*, 1981-92 and USS Guadalcanal, 1982-85.

He attended numerous military schools and training

tours. From 1986-91 he served as Commander, Military Sealift Cmd., Southern United Kingdom Det 308, Orlando, FL; 1991-93; Naval Reserve AD-19, Yosemite Det 1908, Orlando; 1993-present as commander, Military Sealift Cmd., SWA Det 208, Orlando as assistant training/AT coordinator.

Awards include Navy Expeditionary Medal, NDM (2), AFRM, SWA Service Medal; Navy Unit Comm. Rib. Navy Battle E Rib., Sea Svc. Deployment Rib., Navy/Marine Corps Overseas Rib. and Marksman Rib.

Married Marilyn Ann and they live in Orlando. Presently a defense contractor.

JOHN F. LOFTIS, CDR, born at Stuart, OK. He was first released from active duty in 1955 after four years duty with Hospital Corps in naval hospitals, the USS *Glynn* (APA-239) and USS *Navarro* (APA-215). Transferring to reserve status, he drilled at Norman, OK while attending the University of Oklahoma, graduating in 1960 with a degree in pharmacy.

Commissioned an ENS, MSC in 1966 he drilled with several units throughout his career, serving as CO of medical units at Everett and Bremerton, WA; Eugene, OR and XO of NR Fleet Hospital 9, Seattle, WA.

In 1989 he returned to active duty at the Medical Clinic, Naval Submarine Base, Bangor, WA from which he retired in August 1992 after more than 40 years of naval service.

A life member of NRA, he is married to the former Shirley Price of Calgary, Canada. The parents of five children, they now reside at Boring, OR.

WILLIAM J. LOGAN, RADM, a native of Chicago, IL, graduated from Purdue University in January 1970 with a BS degree in Industrial Management, and an associates degree in General Aviation Flight Technology. He entered flight training at NAS Pensacola, FL, graduated as "flight honorman", and was designated a naval flight officer in December 1970.

After undergoing fighter transition training in the F-4J Phantom II at VF-121, RADM Logan reported to VF-21, where he subsequently deployed with the "Freelancers" aboard USS *Ranger* (CVA-61) for a combat cruise to Southeast Asia in 1972-73, and again in 1974.

In July of 1974, he reported for duty as a fleet replacement squadron flight instructor with VF-124, flying the F-14A Tomcat. RADM Logan concurrently attended school during this tour and in July of 1977, graduated from the University of Southern California with a master of science degree in Systems Management.

In May 1978, he reported to VF-111 as initial cadre during the "Sundowners" transition to the F-14A Tomcat.

In May of 1979, he left active duty and joined the Naval Air Reserve. While a member of the USNR, he has served in VF-1285, VF-302 (ultimately serving as CO), VTU-9494 (Inspector General Staff), NAVAIRPAC 1094 (War Plans and Exercises), NAS North Island 0194 (CO) and Naval Air Pacific 1094 (CO). He was selected for promotion to the rank of RADM (Lower Half) and is currently serving as deputy commander Naval Air Forces US Pacific Fleet. In June of 1996, Logan was recalled to active duty and served as deputy commander Joint Task Force Southwest Asia, Riyadh, Kingdom of Saudi Arabia. His professional military education includes completion of the Naval War College curriculum, and the National Defense University's CAPSTONE curriculum.

RADM Logan's civilian career includes positions as chief pilot with Flight International Inc., flight manager with Pacific Express, Checkairman with Jet America Airlines and is currently employed as a captain with Alaska Airlines.

During his military career, he has flown over 2,700 hours in fighter aircraft. He has over 350 carrier landings and over 750 combat missions. RADM Logan's personal decorations include the Defense Meritorious Service Medal, two Meritorious Service Medals, three Strike/Flight Air Medals, the Navy Commendation Medal w/Combat "V" and Gold Star, three Battle Efficiency Ribbons, the Armed Forces Expeditionary Medal, the Vietnam Service Medal and the Vietnam Campaign Ribbon.

RADM Logan resides in Encinitas, CA with his wife Suzanne and their son Brice.

THOMAS D. "TOM" LOMACCHIO JR., CAPT, born Aug. 4, 1944, Brooklyn, NY. He earned BS degree at General Science and Engineering USNA, 1966. He was commissioned June 8, 1966.

He served in USS *Okinawa* (LPH-3) and USS *John King* (DDG-3); stationed at NATTC Georgia, CIC and AIC Schools. Lomacchio participated in two Vietnam campaigns, 1967 and 1968-69.

His awards and decorations include the National Defense, NUC, Vietnam Service w/2 Bronze Stars, Vietnam MUC Gallantry w/Palm. He achieved the rank of captain 0-6. He transferred to Retired Reserves May 1, 1944.

Memorable experiences include Apollo VI Recovery while serving as OOD and assistant CICO aboard LPH-3 in 1968; receiving NYS Naval Military Award for Long and Faithful Service (15 years); Vietnam service, especially as OOD on 15 minute medevac.

Married Barbara and they have three children: Lisa, Dana and Tom. Civilian employment as president, High Mark Enterprises, Inc.

DOUGLAS S. LONG JR., CDR, was with Naval Aviation Pre-Flight School in Iowa City, IA as his first assignment. From there he went to Williamsburg, VA, Camp Perry, the 98th Seabees Bn., then on to ABD Quonset Point, RI. Their battalion was shipped by train to embarkation point of Gulf Port, MS. They were due in to North Africa but Rommel capitulated before they arrived so back to ABD Quonset Point then down to Camp Lejeune, NC to go through Marine Survival School after which they were sent to Norfolk, VA.

He boarded the USS *Ernst* for a trip down the canal and out to Hawaii and placed in the Marine Transient Center at Pearl Harbor. Next aboard ship bound for Guadalcanal to join III Phib Corps 155mm Howitzer Bn. Then up to Guam to liberate them, and on to Okinawa to liberate them. Because of wounds Long suffered, he was sent back to base hospital on *Espiritu Santo* for recovery then to San Francisco for further treatment. From there he went to San Diego Naval Training Center where he was officer in charge of South Clinics. When the war ended he transferred to Reserve status and returned to Kansas City where he now resides.

He has six daughters: Karen, Pamela, Kathy, Diane, Daphne and Kimberly; two sons, David and Douglas; 17 grandchildren and seven great-grandchildren. He is now retired and spends time chartering his yacht in the Virgin Islands.

JAMES T. LONG, CAPT, born Jan. 24, 1933 in Dyer County, TN, received a BS and MS degrees at the University of Tennessee and PhD at Texas A&M University.

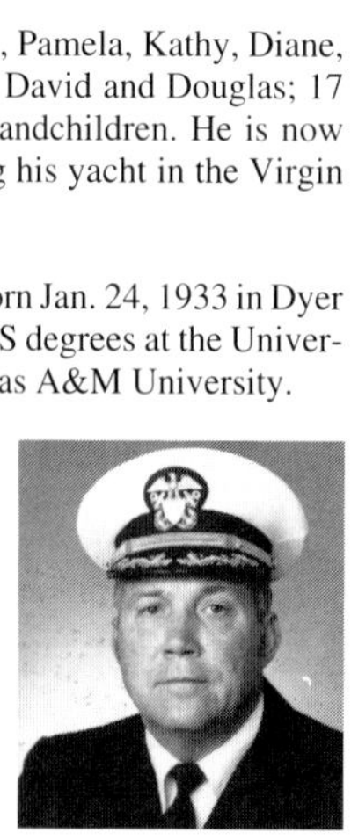

Enlisted USN 1952, E-1 G AM2 in 1956/direct appt. to 0-1 1962. His military locations and stations: VX-1 Key West, FL and USS *Coral Sea* (CVA-43). Served in USNR VP&VR units at NAS Memphis and NAS Dallas.

He received Unit Citation with "E", Clasps for Second Award, Navy Good Conduct, National Defense Ribbon, Occupation Medal and NR Service Medal. Long achieved the rank of CAPT USNR (1555) and was discharged in May 1956. He had many memorable experiences.

He married Peggy Joy Douglass and they have children: Tim, Melinda and Amy; and grandchildren: Meagan Pember and Kelley, Kasey and Kyle Long. Employed as professor with Texas A&M University and Murray State University. Retired from Murray State University in July 1995 and from the USNR, Jan. 1, 1993.

LOUIS J. "LOU" LONG, LCDR, born Sept. 16, 1923 in Westville, NJ, attended Wharton School of U of P and received a BA in accounting and finance.

He joined the service March 1942 as V-5 Cadet, Philadelphia, PA. His military locations and stations: V5 Aviation Cadet Training, Class 8A44P; Flight Preparatory, Murray, KY; WTS, Purdue University, West Lafayette, IN; Preflight, Iowa City, IA; Primary, Grosse Isle, MI; Advanced, Pensacola, FL.

Commissioned an ENS, designated naval aviator, April 24, 1945. Aviation primary training command: Flight Instructors School, class 5D45, NAS NOLA; NAS Bunker Hill, IN and NAS Memphis, TN. Detached from active service 1946.

Reserve service: NAS Willow Grove, PA 1947-49; Naval Reserve Officers School, Clifton, NJ, 1955-60; NAS NY, NY WEPTU 836, NAIRU 836 (1961-68); NAS Grosse Ile, MI, NAIRU Y1 (1969); NAS Detroit, MI (Selfridge AFB, NAIRU Y1 (1970-71).

Security clearance: Secret NAC March 23, 1959. Qualified: air intelligence officer (Antisubmarine Warfare). Achieved the rank of LCDR and retired Sept. 16, 1983.

His most memorable experience: ACDUTRA aboard USS *Randolph* (CVS-15), January 1960. Submarine hunting with TaskGroupALFA.

He is married to Shirley Patrick Long and they have children: Candice Long Davis and Stephen Patrick Long and grandchildren: Rebecca Elysse Long and Amanda Patrick Long. Employed in accounting and financial management with Otis Elevator; internal audit manager, UNILEVER US. He retired Sept. 16, 1988.

NEVILLE S. LONG, CDR, born Sept. 19, 1923 in Christchurch, New Zealand, received a BS and MS in Civil Engineering, California Institute of Technology.

He joined the service Nov. 13, 1942. His military locations and stations: V-12 CalTech, Davisville, RI, Guam. He served with 76th and 23rd Naval Construction Bns. His battles: Pacific. Awarded the Pacific, American, WWII Victory, Naval Reserve, Armed Forces Reserve and Expert Pistol. Long achieved the rank of CDR and was discharged June 19, 1946. His memorable experiences: continued in USNR, retiring Oct. 1, 1968.

He is married to Anne and they have children: Donald, Arthur Dennis and Margaret Fisher. They also have nine grandchildren. Employed as coast engineer, project manager, vice-president, Bechtel International.

DANIEL P. "PETE" LOUCKS, CAPT, after obtaining his commission from Naval OCS in Newport, RI, 'Pete' began his flight training in December 1955. He received his aviator's wings in February, 1957, and served in VW-11 flying WV-2s and SNBs until February 1959.

His reserve duty included VR-911 at NAS South Weymouth, VR-704 at NAS Dallas, VR-833 and VR-60R2 at NAS New York and as CO of RTU-52, CO of VR-1604 and OinC and CO of VR-52 at NAS Willow Grove, flying C-54 and C-118 aircraft.

As CO of VR-52, he commanded the then largest Air Transport squadron in the USN, with detachments at NAF Andrews and Detroit. For services involving operations in the US, Europe, Southeast Asia and South America, Loucks was awarded the Navy Commendation Medal in January 1981.

Professor Loucks consults, teaches and conducts research in engineering at Cornell University. He is married to Marjorie Grant and they have two daughters, Jennifer and Susan.

ROBERT J. LOVE JR., LCDR, born in East Lansdowne, PA, Aug. 2, 1950. After graduating from Ocean County College he enlisted at Lakehurst NAS in 1970. He served in Photo Intelligence in VS-71. He was transferred as an intelligence specialist to NAIRU after completing active duty.

In 1973 he attended Colorado State University, graduating in 1976 with a BS in Forestry. A plank owner in FIRSTPAC 0571 where he finished his Reserve enlistment and became one of the first direct commission officers after the Vietnam War. He continued to serve in such units such as NORAD, IPAC, CNFK Operational Intelligence after receiving his MS in forest economics in 1978.

During this interval he was a professional forester leaving the US Forest Service at the Black Hills in 1986 when he became a registered consulting forester in South Dakota. He was a plankowner contributing to the effective integration of NSIU 0171 into the Joint Aerospace Reserve Program, based from Colorado Springs, CO. In 1994 pioneering targeting solutions for this Joint Command, he was awarded the Joint Service Achievement Medal. The same year he was recalled to active duty. He established an analysis and dissemination infrastructure at Operation Provide Promise during the Balkans Conflict.

Love was awarded the Joint Service Commendation Medal for successfully implementing leading edge information management. He is currently the CO of VTU-2204G at the Naval & Marine Corps Reserve Center Eugene, OR.

He currently resides in Corvallis, OR working as a forest products extension agent at Oregon State University. The Love family consists of his spouse, Gem; daughter, Natalie; and son, Adrian. He enjoys the outdoors, travel and reading.

ROBERT W. "BOB" LOVE JR., CAPT, born June 28, 1929 in Springfield, MO, postgraduate MD.

He joined the service July 15, 1959. His military locations and stations: USNH MFS TN.

Love was awarded the Naval Reserve and National Defense Medals. He achieved the rank of CAPT and was discharged July 15, 1961 and retired Oct. 1, 1988.

He is married to Barbara and they have children: Robert III, Carry and Rebecca and grandchildren: Robert IV, Aubrey, Elizabeth, Mark and Catherine. Employed with Veterans Administration, retiring Oct. 1, 1990.

ROBERT D. "BOB" LOVISONE, LT, born June 19, 1939 in Denver, CO, received an associates degree from the University of Denver, Red Rocks Community College.

He enlisted in the service July 27, 1958. His military locations and stations: 1958 NTC San Diego; NATTC Norman, OK, 1959; NATTC Memphis, TN; 1959-61 NATTU Olathe, KS, 1961-62; VU-3 NASNI, CA.

Enlisted USNR March 1982, NARCEN Denver, CO. Served with USNR, ABFC; NR CV-41 0171; NR CV-61 0171 and still serving with NR CV-64.

Lovisone was awarded the MUC, Good Conduct, Meritorious Service two awards, USNR, National Defense two awards and Reserve Service. He achieved the rank of LT.

His memorable experiences: Commissioning ceremony at NARCEN Denver Nov. 1, 1961.

He has been married 36 years to Barbara and they have two children, Dee Anne and Darrel. Retired after 32 years State of Colorado was chief engineer telecommunications; senior engineer, microwave systems, project engineer, systems development. Retired May 1994 State of Colorado. Presently serving in the USNR.

JERRY W. LUFFMAN, LCDR, born Nov. 9, 1949 in Elkin, NC, received bachelor applied science, Electronic Facilities Management, Troy State University, Troy, AL.

He joined the service Nov. 20, 1968. His military locations and stations: Enlisted boot camp, Orlando, FL, 1968-69; USS *Perry* (DD-844) Mayport, FL, 1969-72; RNMCB-24 Red Stone Arsenal, AL, 1972-75; Commissary Store, Keflavik, Iceland, 1975-76; USS *Cochrane* (DDG-21) Pearl Harbor, HI, 1975-76; Electronic Warfare School, Pensacola, FL, 1976-77; USS *Saratoga* CV-60, Mayport, FL, 1977-79; Officer Candidate School, Newport, RI, 1979-80; Navy Supply Corps School, Athens, GA, 1980; USS *Frederick* (LST-1184), San Diego, 1980-83; Food Service Officer Pearl Harbor, HI, 1983-85; Reserve Naval Construction Force, Greensboro, NC, 1985-87; Naval Reserve Center, Greensboro, NC, 1987-91; Naval Marine Corps Reserve Center, Raleigh, NC, 1991-96.

He was awarded the National Defense Service Medal, Meritorious Unit Commendation, Vietnam Service Medal, Armed Forces Expeditionary Medal (Korea), Good Conduct Medal, Navy Sea Service Ribbon, Navy Marine Corps Expeditionary Medal, Naval Reserve Medal and Navy Achievement Medal. Luffman achieved the rank of LCDR, Supply Corps and left active service September 1987.

Participated in Vietnam service aboard USS *Perry* (DD-844), 1969. Selected for Officer Candidate School onboard USS *Saratoga* (CV-60) while serving as electronic warfare supervisor, 1979.

Completed Officer Candidate School and Navy Supply Corps School in 1980. He was supply officer, USS *Frederick* (LST-1184), 1980-83; food service officer, Naval Station Pearl Harbor, 1983-85; staff supply officer/ logistics officer commander Reserve Naval Construction Force (active duty) 1985-87; commanding officer Naval Air Terminal Unit 307 NAS Norfolk, VA served during Desert Shield/Desert Storm, 1990-91.

He is single and a self-employed computer consultant, retiring May 1996.

RANDALL DAN LUKE, CAPT, born June 4, 1935 in New Castle, PA, received a BA with honors in economics and JD in law from University of Pennsylvania.

He joined the service Oct. 2, 1960. His military locations and stations: OCS; Naval Station Long Beach, CA 1961-64. Active duty 1960-64; Reserve duty 1964-81. He achieved the rank of CAPT, USNR and was discharged March 1, 1964 and retired July 1, 1981.

He married Saralee K. and they have children: Lisa E., Stephanie and Randall S.; and grandchildren: Skye, Oliana, Danielle and Adelyn. Employed in law practice 1964-80; vice-president Assoc. Gen. Counsel and Secretary, The Firestone Tire and Rubber Co., 1981-91; Of Counsel Hahn Loeser & Parks LLP 1991-present.

DONALD L. LUKINBEAL, CAPT, born in Salem, OR in 1937. In 1954 he joined the Ready Reserve as an airman. Commissioned in 1961 at OCS, he subsequently became a TAR (1967). USNR tours: XO N&MCRTC T.I>; COMUSMARDEZPAC; CNO (OP-094); CO N&MCRTC Alameda; Deputy NAVREREDCOMREG 20; and CDR NAVRESREDCOMREG 5. USN tours: NAVCOMSTA Japan; USS *Austin* (LPD-4); USS *New Orleans* (LPH-11); and USS *Blue Ridge* (LCC-19). During his 36 year career he received 10 service and four personal awards. He has a BSed from West Oregon State College, and an MAed from The George Washington University. Married in 1960, he and Kay have three children. They live in Browns Valley, CA.

CLAUDE C. LUMPKIN III, CAPT, born Oct. 3, 1943 in Little Rock, AR, received a BS degree from the US Naval Academy, 1965 and MBA, University of Louisville, 1976.

He joined the service June 9, 1965, commissioned ENS, USN. His military locations and stations: Norfolk, VA (SURFLANT); San Diego, CA (SURFPAC); Guam (PHIBPAC). Served with USS *H.E. Yarnell* (CG-17); USS *Hull* (DD-945); USS *Sedgwick County* (LST-1123).

His awards and medals include the Navy Commendation Medal, Navy Achievement Medal w/Combat "V", National Defense Service Medal, Combat Action Ribbon, RVN Campaign Medal w/3 stars, Vietnam Service Medal, Presidential Unit Citation, Navy Unit Commendation, Meritorious Unit Commendation and COMSEVENTHFLT Unit Commendation. He achieved the rank of CAPT, USNR and retired Jan. 1, 1995.

His memorable experiences: served in USS *Hull* (DD-945) during 1968 TET offensive. Fired guns continuously until the last mount was out of commission. The crew was heroic, forming a human chain to strike powder charges and shells from the magazines to the gun mount. Achieved exceptional results on behalf of Allied forces ashore.

Lumpkin is divorced and has children: Claude F. and Mary C., and a granddaughter, Emma J. Lumpkin. Employed as project engineer, Lucent Technologies (1984-present).

CAROL WALSH LUND, CDR, born April 21, 1957 in Binghamton, VA, attended Villanova University and the University of Virginia.

She joined the service May 11, 1979. Her military locations and stations: NRMC San Diego, Parris Island, SC. Served with NR MED DEN 0166 and NNMC Bethesda, MD; NR NNMC 206; NAVHOSP Pax River, NMU Annapolis.

Lund received the Navy Achievement Medal and three Meritorious Unit Citations. She achieved the rank of CDR.

She is married to Daniel and they have three children: Thomas, Peter and Gregory. She is employed as a nurse.

PAUL J. LUNDBECK, CDR, born Oct. 9, 1918, San Francisco, CA. Earned BA degree in 1940 at Stanford University. He was commissioned ENS at Cornell University in 1942.

Stations/Units: 1943, Panama Canal Zone, USS *Dortch*; 1944, JCTC, Miami, FL, commissioned (plankowner) USS *Geo. E. Davis* (DE-357), ASW officer; 1944-45, 7th Fleet, Leyte Gulf Invasion, Philippines, carrier escorts, ASW Patrols, Convoying, etc.

Participated in liberation of Manila, Peleliu, Okinawa invasion and escorted first troopship to Sasebo, Japan after surrender. Stationed 1946-50, at various Reserve billets; 1950-53, recalled to active duty for Korean War, flew to Sasebo, Japan, joined the 7th Fleet, TF-77 off Korea, boarding USS *Higbee* (DDR-806) as ops officer and later as XO. 1953-68, RTC, T.I., San Francisco, NR Composite Co. 12-46, CO 3 years, politico militaffairs Co. 12-1, annual op. 62 (Navy Dept.) project leader.

Retired in 1968 after 27 years. Awards and Medals include NDM, American Campaign, Asiatic-Pacific Campaign, Philippine Liberation, China Service, Asia Occupation Service, WWII Victory, Naval Reserve, Armed Forces Reserve, Korean Service, UN Korean Medal, Korean Govt. and War Veterans Medal and Bronze Stars for Asiatic-Pacific, Philippine Liberation and Korean Service.

He is a widower, no children. Civilian employment in sales, contract, marketing, executive and wholesale distribution until retirement in 1975.

JON C. LUNDBERG, LCDR, born June 26, 1961, Royal Oak, MI, received a BS from the University of Southern Colorado.

He joined the service in 1988. His military locations and stations: NAS Atlanta and NAS Norfolk. Served with NAVINFOSE, PSYOPS, AVU-0286, IMAGU 0186.

Lundberg was awarded the NAM (2), MUC, Reserve SS, Overseas Deployment, Expert Rifle and Pistol. He achieved the rank of LCDR.

He is married to Lisa Marie and they have children, Samantha Leigh and Nelson Taylor. He is president/CEO, The Corporate Image, Inc.

STANLEY H. LUNDGAARD, CAPT, born in Minneapolis, MN. He enlisted in the USNR in July 1956 and was attached to Transport Sqdn. 811 (VR-811) NAS Minneapolis. Upon receiving a bachelors degree from the University of Minnesota in June 1962 he reported to Aviation Officers Candidate School in Pensacola, FL. He was commissioned an ENS in December 1962, and was designated a naval flight officer in June 1963.

While on active duty, Lundgaard served with Patrol Sqdn. Nineteen (VP-19), home based at NAS Moffett Field, and deployed with the squadron throughout the Pacific. He later served as an instructor in VP-31 at NAS Moffett.

Upon release from active duty he reported to a Reserve P3 Sqdn. forming at NAS Moffett Field, which subsequently became the present VP-91. While in VP-91 his positions included readiness officer and maintenance officer. He served three years as a special projects officer in a volunteer training unit (VTU) prior to reporting to Coastal Defense Coordination Component 320 at Treasure Island. CDCC 320 was redesignated Maritime Defense Zone Pacific Northern California (MARDEZPAC).

In November 1986, Lundgaard reported to Space and Naval Warfare Systems Command Unit (SPAWAR 2001) in San Jose where he held positions as readiness officer, XO and CO. He retired from the USNR in July 1992 having served 36 years duty for his country.

Lundgaard has served as president of the Polaris Chapter and 12th District President, NRA. He holds an MBA degree from the University of Santa Clara and is employed by Lockheed Martin Missile and Space Corporation.

He resides in Sunnyvale, CA with his wife, the former Beverly Cross of Seattle, WA. They have two daughters Sara and Sonja.

J. ROBERT LUNNEY, CAPT, born Dec. 15, 1927 in New York City. He resides with his wife, Joan and his son, Alexander in Bronxville, NY. He served in the USNR from May 1, 1945 to Dec. 15, 1987.

Enlisting in the USNR at 17, he served on occupation duty with the Naval Amphibious Forces, Pacific (Eniwetok; Kwajalein; Saipan; Guam; Iwo Jima) 1945-46, as a QM3/c. During the Korean War he served in Korea, 1950-51, with the US Merchant Marine as a staff officer (Inchon Landing, Hungnam Evacuation). His awards include the Navy Commendation Medal, Asiatic-Pacific Campaign Medal, Korean Service Medal (two stars), Merchant Marine Gallant Ship Citation and the Korean Presidential Unit Citation. His Gallant Ship citation commends him for courage and resourcefulness in participating in "one of the greatest marine rescues in the history of the world" at Hungnam, North Korea during the Chosin Reservoir campaign, December 1950.

In February 1994, Hon. John H. Dalton, Secretary of the Navy, decorated him with the Department of the Navy Distinguished Public Service Award. The citation commends Lunney for exceptionally outstanding service to the Navy for over four decades and especially for his dedication and leadership in improving the strength and readiness of the Navy and the capabilities of the Department of Defense to carry out essential wartime missions.

He received his BA degree (1950) from Alfred University, Alfred, NY and his Doctor of Law degree (1954) from Cornell Law School, Ithaca, NY. By appointment of the US Attorney General he served five years with the Department of Justice as an assistant US Attorney in New York (1955-59). In 1968, after eight years with the New York firm of Shearman & Sterling, he founded his own law firm, now Lunney & Murtagh, LLC specializing in litigation.

CAPT Lunney served as the 22nd National President (1991-93) of the NRA. His other civic and charitable activities include the Sovereign Military Order of Malta (American Association); Sons of the Revolution in the State of New York (past president); Lavelle School for the Blind; NYC (Bd. Directors); Navy League of US, NY Council (Board of Directors); Naval Order of US (past commander, NY Commandery); NY Naval Militia (RADM); The Chosin Few (past National Director) and Military Order of World Wars (vice commander, NY Chapter).

MARTIN ALMS LUNT, LCDR, born May 14, 1934, Miami, AZ. In May 1951 he joined an electronics unit, Miami, AZ, as a junior in high school; 1954, Recruit Trng. Ctr., San Diego, CA, assigned to attend Airman "P" School, Norman, OK, graduating Dec. 23, 1954; January-March 1955, NATTU, Norman, OK; March-August 1955, Naval Photographic School, Class A, Pensacola, FL; September 1955-September 1956, attached to Utility Sqdn. One, VU1, Photo Lab, NAS, Barbers Point, HI; October 1956-March 1957, attached to Naval Station, Honolulu; April 1957-August 1958, attached to Naval Photographic Unit, NAS, El Centro, CA.

Discharged Aug. 8, 1958 and worked for Southern California Gas Company until January 1961. Matriculated at Imperial Valley College January 1961 and graduated in June 1962. Matriculated at San Diego State College, Imperial Valley Campus, September 1962, majoring in elementary education and minor in English and social studies. Graduated June 1964 with BA and teaching credential.

Re-enlisted in June 1964 in the USN to attend OCS and was commissioned ENS, October 1964; November 1964-March 1965, attached to HQ Field Cmd., DASA Sandia Base, Albuquerque, NM; April 1965-September 1967, USS *Yorktown* (CVS-10); October 1967-May 1969, USS *Truxtun* (DLGN-35); June 1969-October 1972, NRTC, Pueblo, CO as CO; November 1972-October 1974, USS Durham (LKA-114); November 1974-February 1975, HQ, 11th Naval District, Reserve Supplement, San Diego, CA; March 1975-January 1977, Coastal River Sqdn. One, Naval Amphibious Base, Coronado, CA; February 1977-August 1980, USN and USMCR Ctr., Boise, ID as CO.

Retired from active service Sept. 1, 1980, as LCDR. Medals and awards include the Good Conduct, Vietnam Service Medal w/stars, National Defense Medal, Armed Forces Expeditionary Medal (Korea), RVN Campaign Ribbon, RVN Armed Forces Meritorious Unit Citation (Gallantry Cross Medal) and the Navy Commendation Medal.

Matriculated at Boise State University, earning master's degree in reading instruction and curriculum in May 1983. Taught at Boise State 1983-87. From 1988-92, worked at Hewlett Packard Electronics. He is now retired. Married Shirley Lee Walker and they have two children, Crystal and Aaron.

NEAL A. LUTH, LCDR, born April 8, 1939 in Brooklyn, NY, received a BA in Business Administration from Penn State University.

He joined the service Sept. 15, 1961. His military locations and stations: US Naval Station, Washington, DC. Served with Naval Intelligence.

Luth was awarded the Armed Forces Reserve Medal and achieved the rank of LCDR. He retired from active duty Sept. 12, 1964. His memorable experiences: Military Sealift Command; Miami Reserve duty.

He married Mary and they had sons, Todd and Scott. He retired from Inactive Reserves March 1, 1983. Luth died March 13, 1996.

LARRY G. LUTZ, CAPT, a native Oklahoman, born Jan. 30, 1938, despite living in southern Maryland, still calls Edmond, "home". He received his BS from the University of Oklahoma (1961) and MA from Central Michigan University (1982). He began his naval career as a regular NROTC student in 1956.

After graduation he reported to pre-flight school in Pensacola, began flight training at NAAS Saufley, basic training at NAAS Whiting and advanced training at NAS Corpus Christi. He reported to VR-21 DET Japan for WestPac COD duty in April 1963. After three years fleet combat support on "Yankee Station", "Dixie Station", and in-country RVN, he returned to VT-28 as an advanced flight training instructor. His second Vietnam tour was as AOINC, VAW-111 DET 31, deployed aboard USS *Oriskany*, May-November 1970. He ended his first active duty tour with 2,500 combat support flight hours out of a total of 4,400 plus total hours.

Lutz enjoyed a period as a "drilling reservist" from 1971-78 with intermediate maintenance units at NAS Dallas. He returned to active duty as a "265", voluntary recall, with the CDR, Naval Air Reserve Force staff in New Orleans, where he was selected to the TAR program. He served as a TAR officer until his retirement in September 1992, serving as the Naval Air Logistics Center (later Naval Air Depot Operations Center), Patuxent River, and as the first TAR IG for Commander-in-Chief, US Naval Forces, Europe.

Lutz earned four Meritorious Service Medals, three Air Medals, the Navy Commendation Medal, two Armed Forces Defense Medals, five RVN Campaign Medals, the Armed Forces Reserve Medal w/HG, RVN Medal, Expert Pistol Medal and Navy Unit Commendation, two Meritorious Unit Commendations and Overseas Service Ribbon w/ 2 stars.

He is married to the former Darlene K. (Tammy) Thomas of Canon City, CO, and they have five daughters: Lori, Niko, Merry, Gabrielle and Rachael and four grandchildren by Lori: Lee, Michelle, Joshua and Hannah, to bless their lives.

CAPT Lutz is currently working as senior logistics consultant with The Prometheus Co., Inc. of Alexandria, VA.

ROBERT HENRY "BOB" LYDON, LT, born Nov. 12, 1922 in Denver, CO, received a BS from the University of Colorado and an MS degree from Indiana University.

He joined the service Oct. 26, 1943. Enlistment, November 1942. His military locations and stations: NAS San Luis Obispo, CA; Boulder, CO; Recruit Training USNTS, Great Lakes, IL; USNAMS, Northwestern, Chicago, IL; Pacific Theater USS LCI-549; amphibious forces.

He participated in American Theater, Pacific Campaign, Solomon Islands, Okinawa and the Occupation of Japan. Achieved the rank of LT and was discharged from inactive duty July 14, 1946 and Retired List in 1968.

His memorable experiences: making smoke, Okinawa, Buckner Bay, Occupation Kyushu.

His wife Marjorie Jean is deceased. His children are Doris, Richard, Kathleen, Mary and Robert M.; and grandchildren are Jeffry D'Amico; Timothy, Jason and Theresa Herbig; Jeremy, Melinda, Melissa, Joshua and Justin Lydon; Robert, Jacqueline and Marcus Vietor; Jennifer and Kristina Gibbar.

He was employed USGS cartographer, USAF Defense Mapping Agency.

CHARLES G. "CHARLIE" LYNCH JR., LCDR, born Feb. 23, 1925 in Henderson County, NC, received a bachelor degree in Business Administration. He also did some graduate work.

He joined the service April 19, 1943. Served with US Naval Training Station, Bainbridge, MD, boot camp; CICGTC, Little Creek, VA, 1943; USS *Devosa* AKA, 1944-45; Princeton University, 1945; Howard College, Birmingham, AL, 1945; University of South Carolina, 1945-46; NAS, Patuxent River, MD, 1946; US Naval Reserve Training Center, Asheville, NC, 1946-51; USS *Tills* (DE-765), 1951-52; Naval Reserve Surf. Div., Miami, FL, 1955-58; Naval Reserve Intelligence Unit, Miami, FL, 1958-75.

He achieved the rank of LCDR USN. His memorable experiences: progressed from SA to LCDR, step by step.

He married Ramona S. and they have children, Sandra Diane, Shirley (deceased), Ronald B. and Sheila; and grandchildren: Allen and Rusty Johnson and Caron Lynch. Employed as administrative officer, Metro Dade County, FL. He retired June 30, 1975.

FRANK H. LYONS, CDR, born May 12, 1945 in Hot Springs, SD, received a BS in mechanical engineering from Colorado State University in 1968.

He joined the USN, Feb. 24, 1969. Reported to NAVOCS, Newport, RI, March 13, 1970. Commissioned an ENS, USNR, July 17, 1970.

Ordered to USS *Maddox* (DD-731) at Long Beach, CA on commissioning, with 10 week training at DATC, San Diego in transit. Reported on board Oct. 19, 1970. Served as main propulsion assistant and chief engineer while on board. Decommissioned USS *Maddox* on July 6, 1972 and was released from active duty on the same day. While on board, he served as chief engineer of the transfer crew for USS *Maddox*, as it was being transferred from the US to Nationalist China at decommissioning.

Affiliated with USNR at Denver, CO on Sept. 11, 1972. Transferred from drilling status to IRR status on Feb. 24, 1992. Transferred to retired status as a CDR on April 1, 1994. Served with the following USNR units, all at Naval and Marine Corps Reserve Center, Denver, CO: NROS 9-2; NRSD 9-235(L); CGN-9 (SRU); USS *Denver* (LPD-9); NR DDG12 Robison; NCSO-FE and NR-VTU 1804 Denver.

He was awarded the National Defense Medal, Naval Reserve Medal w/Hourglass and Naval Expert Rifle (permanent).

Married Judy M. Letoumeau on Oct. 18, 1968. They have two children: a daughter, Heather, and son, Christopher.

Currently employed with the National Park Service as a project supervisor for the Denver Service Center. He is presently stationed at Mammoth Cave, KY having recently transferred from Cumberland Gap, TN.

Prior employers include: Natkin & Co., mechanical contractors; Honeywell, Inc., temp control contractors; Jefferson County Schools, maintenance division and the DEH for the US Army at Fort Carson, CO.

Current membershps include Registered Professional Engineer in Colorado, NRA, plank owner, USN Memorial, Society of American Military Engineers, American Society of Mechanical Engineers, life member in the CSU Alumni Association.

RICHARD J. LYSTER, CAPT, born in Hollywood, and grew up in Santa Monica. His BS and MBA degrees are from UCLA. Commissioned ENS in 1969, Lyster has had assignments with CHINFO offices in Washington, Chicago and Los Angeles; Naval Base San Francisco; Patrol Wings Atlantic; MSC San Francisco and the Far East; the USDAO in Paris; CINPACFLT; and Desert Shield.

He served as PAO for JTF5, the Pacific's anti-drug force. As part of Politico-Military Affairs 2011, he saw frequent duty at NATO HQ in Brussel. While PAO for REDCOM 20 he was the spokeperson at AFSOUTH for NATO's first ever military action in Bosnia, and later began managing public information for NATO exercises.

He has commanded Public Affairs Center 220, and Office of Information 1018.

In civilian life, Lyster is a foreign service officer, currently assigned to Dhahran, Saudi Arabia, where he lives with his wife Diana, son Todd and daughter Alice.

ALANSON H. "LANCE" MABRY, born March 31, 1941 in Akron, OH, received a BA from DePauw University in 1963.

He joined the service June 5, 1963. Graduated OCS Newport, RI, Nov. 22, 1963. His military locations and stations: *Principally* September 1964-June 1967. EOD Tech Center, Indian Head, MD. Served as EOD officer, foreign weapon-intelligence officer, Tech Center.

Achieved the rank of 04, eventually 1145 officer, discharged June 13, 1967. His memorable experiences: Joined the USNR, Akron, OH September 1967 and retired Oct. 1, 1991 with the rank of CAPT.

He is married to Jacqueline S. Mabry and they have two children, Tiffany and Mabry. Employed as president and owner, E.A. Mabry Inc., and Akron Jewelers Supply Inc., since 1970.

EDNA M. MACEWICZ, CAPT, born Jan. 16, 1920 in Worcester, MA. She served in WWII Oct. 15, 1943-August 27, 1946; June 5, 1951-Jan. 25, 1952. Served with USNH Chelsea, Portsmouth Naval Dispensary; NH USNHS San Diego.

She achieved the rank of CAPT and retired Jan. 16, 1980. Macewicz enjoyed all of his assignments. Employed as director Public Health Nurses, Worcester Department of Health.

L. FRANK "FRANK" MACH, CAPT, born Aug. 2, 1941 in Chicago, IL, received a BS degree from the University of Illinois in 1964. Graduated from the Naval War College in 1990 and attended National Defense University, Washington, DC and Naval Transportation School in Oakland, CA.

He joined the service July 27, 1964. Commissioned Feb. 5, 1965, OCS, Newport, RI. His military locations and stations: USS *Robert L. Wilson* (DD-847) to OPS department head December 1967. Reserve: CO, USS *Leahy* (CG-16). Served with Support Unit; CO MilSealiftCmd NorVA; CO, MSC Southwest Asia.

His battles and campaigns: Dominican Republic, CO of unit deployed to Desert Storm, 1990-91.

Mach achieved the rank of CAPT and retired June 1, 1995. His memorable experiences: officer in Tactical Command, Commodore CTU 48.1.16 Joint Logistics Over The Shore (JLOTS 1988) and Commodore, OTC CTU 48.1.18 Solid Shield JLOTS 1989.

Married Connie Kight Dec. 9, 1967, and they have three children: Vallie Lynn, Mandy Anne and Hunter Lewis. Employed with US DOT, Maritime Administration, South Atlantic Region. Projected retirement to 20 years government service, October 2005.

HENRY MATTHEW MACHINA, LT, born May 14, 1936 in Chicago, IL. Joined the USNR program at NAS Glenview in June 1954. He served with VF723 and later transferred to VF722 where he earned his air crew wings. After eight years of service he was honorably discharged as an AMS3 in June 1962.

In March 1979 LT Machina re-enlisted in the USNR and affiliated with FIRSTEURLANT 1472 as an intelligence specialist third class. He served with Patrol Sqdn. VP-90 from October 1982 through September 1984 and returned to First 1742 as an ISI.

LT Machina applied for an LDO Commission in April 1985 and was appointed as a LDO law enforcement/physical security officer in May 1986 and began his commissioned career with NR NCISRO 1972. He remained with this "great little" investigative unit until its decommissioning on Sept. 9, 1995. While with NR NCISRO 1972 he earned permanent NCIS agent credentials and functioned as an investigator, instructor, analyst, production officer and various collateral duties in support of the agent force.

In October 1995 he was assigned to DIAHQ 0972 where he functioned as an analyst until his retirement on May 14, 1996.

He was awarded the Naval Reserve Meritorious Service Medal, National Defense Service Medal, Expert Pistol, Expert Rifle and Armed Forces Reserve Medal, retiring in May 1996

His civilian credentials include an associate of arts and science degree in law enforcement from Chicago City College. He has been Chicago police officer since February 1961 and with (CID) Criminal Investigations Div. for over 25 years, holding the rank of detective first class. Machina retired from the Chicago Police Department Feb. 15, 1996 after 35 years of dedicated service.

He lives on the northwest side of Chicago with his wife Carole, a retired Coldwell Banker real estate agent/broker. Immediate family includes three daughters: Gail Marie Cabala, a licensed architect; Jill Marie Machina a Chicago Public School teacher with a masters in school administration and Karen Ann Machina a Niles Township special education high school teacher. Grandchildren include: Nicholas Michael and Matthew Henry Henkle.

STANLEY R. "STAN" MACKAY, CDR, born Aug. 18, 1927 in Boston, MA, received a BS in chemistry from Tufts University.

Enlisted as third class electricians mate, October 1945. Served with USS *Tarawa* (CV-40) and discharged November 1946. He was recalled to active duty October 1950 and served in USS *D.J. Buckley* (DDR-808). Attended Officers Candidate School, Newport, May 1951-September 1951. Served on the USS *Beale* (DDE-471) as ENS, electrical officer, legal officer and discharged November 1952.

Mackay joined the USNR December 1952 and served with the Surface Div. in Salem, Sperryville and Lawrence. He retired to the Retired Reserves September 1977. Retired from the USN, Aug. 18, 1987, as CDR.

He was awarded the American Campaign Medal, Asiatic-Pacific Medal, Victory Medal, Korean Service Medal, National Defense Service Medal, Naval Reserve Medal and Armed Forces Reserve Medal.

His memorable experiences: Transit Panama Canal on aircraft carrier.

He married Joanne C. and they have three children: S. Rae III, Anne and Jane; their grandchildren are Ryan, Emily, Matthew and Elizabeth. Employed with GTE Products Corporation since 1979 as manager strategic planning, precision materials group.

STEPHEN M. MACNISH, CDR, born in Troy, NY, June 9, 1947, graduated from Clarkson College of Technology in June 1968, with a BS in industrial management.

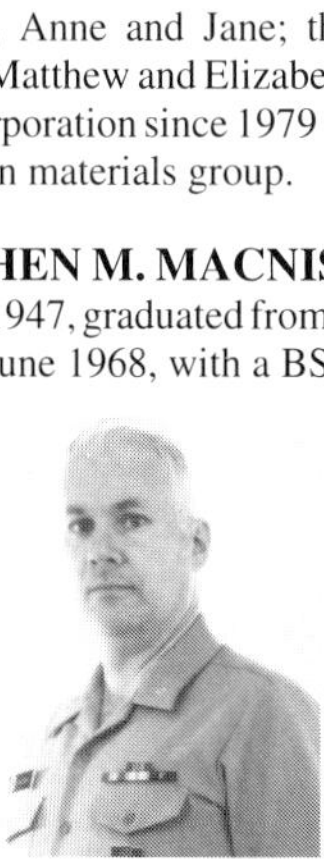
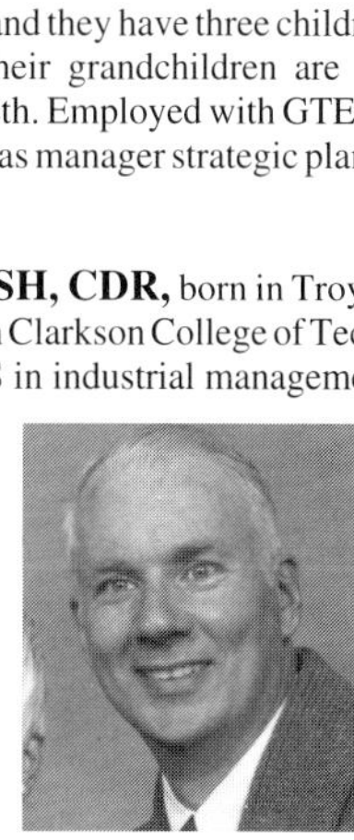

Commissioned from OCS in Newport in 1971. He served in the Deck/Weapons Dept., USS *Guam* (LPH-9) until his early release in 1973. Affiliating with the Reserve Intelligence Program, he held numerous assignments in Poughkeepsie, NY intel units supporting the NavIntSupCen. In 1985, he became XO of NavOpIntCen 0102 in New London, CT. He returned to Poughkeepsie's Intel VTU-102, and in 1988, became assistant intelligence officer, ComStrikFltLant 0102, Albany, NY. Upon completion, he returned to Poughkeepsie's Intel VTU 0101, retiring in 1991.

After 25 years with IBM, primary as IBM

Poughkeepsie Site Energy Coordinator, he retired in 1993 and earned his paralegal certification. He works in energy law for Couch White in Albany, NY.

He and his wife, Sue, live in Pleasant Valley, NY. Their son and daughter are in college. Steve serves as treasurer of the Pleasant Valley Free Library.

DANIEL MICHAEL MADDEN, LT, born Oct. 30, 1916 in New York, NY, received a BA CCNY 1939.

Commissioned an ENS May 15, 1942. His military locations and stations: Communications School Harvard (teacher). Served with carriers *Intrepid* (CV-8), *Ticonderoga* (CV-14). Participated in several battles in South Pacific up to Japan's surrender.

Madden was awarded the area and battle ribbons. He achieved the rank of LT and retired in 1962. His memorable experience: When *Ticonderoga* took attacks by two kamikazes when task force exiting South China Sea Jan. 21, 1945.

His wife Huan Wilson is deceased. Madden is a writer-author.

LOUIS MARION MADUELL, ENS, born Nov. 18, 1919 in New Orleans, LA. Attended Jesuit High School, Louisiana State University for 2 1/2 years and Naval Aviation School for seven months.

He joined the service May 25, 1942. Called to active duty July 30, 1942. Commissioned ENS, USNR and naval aviator March 1, 1943. His military locations and stations: NAS Corpus Christi; Trinidad, BWI; Aleutian Campaigns; Port Lyautey North Africa. Served with VP-53, Trinidad; VPB-62, Amchitka-Attu (Aleutians); VPB-114 Lyautey

Maduell participated in the American Theater Campaign, Asiatic-Pacific (Aleutians), Port Lyautey, France, Morocco. He received the European Medal, Naval Reserve Medal, American Defense and Victory Medal. He achieved the rank of LCDR, USNR and retired Nov. 18, 1979.

His memorable experiences: Convoy patrol and ASW (Atlantic); Aleutian patrols; Bering Sea and North Pacific patrols; ferry of a/c TransAtlantic; XO of Naval Air Reserve transport VR-822, operations officer "Old" NAS New Orleans and Grosse Ile, MI.

He is married to Audrey and has five children: Mariana, Louis Jr., Roger, Mildred and Grace; and seven grandchildren: Stephen, Melanie, Michelle, Jude, Michael, Daniel and Amanda. Employed as foreman, shipyard (pre-war); FAA General Aviation inspector GADO. He retired from the USNR in 1963 and FAA in 1976.

HENRY B. MADURO, CAPT, earned a bachelor's degree in business adminstration from the University of New York at Albany in 1972, and on graduation entered Officer Candidate School. He received his commission from the Navy Supply Corps School in June 1973.

He served on active duty as assistant material officer at USNS Roosevelt Roads, Puerto Rico. Afloat, he served as assistant supply officer of the USS *Cook* (FF-1083) homeported in San Diego, CA.

He entered the USNR in 1978, and served as administrative officer for Military Sealift Command Pacific, Detachment 120, Oakland, CA; Logistics officer for Aviation Supply Office 202, Philadelphia, PA; training officer for Defense Personnel Support Center 202, Philadelphia, PA; CO, Naval Activities United Kingdom 402; CO, Naval Supply Center Norfolk 402; and CO, Defense Logistics Agency (DLA) Contingency Support, Materiel Management Team (DCST-MMT), Philadelphia, PA.

Madura was recalled to active duty in January 1991 for Operation Desert Storm and served as assistant to Chief of Subsistence at the Defense Personnel Support Center, Philadelphia, PA; in October 1994 for Operation Uphold Democracy serving as Chief of Staff for the Joint Logistics Support Command, and CDR for the DLA Contingency Support Team (DCST); and in March 1995 for the UN Mission in Haiti where he served as DCST CDR.

His decorations include the Defense Meritorious Service Medal, the Joint Service Commendation Medal w/OLC, the Joint Service Achievement Medal, the UN Medal, and various unit and service awards.

He is married to the former Denise Buete. They have one son and two daughters: Josias, Celestina and Anellia.

THOMAS H. MAGEE, CAPT, a native of New Hartford, NY, received BSME and JD degrees from the University of Rochester (1965) and Syracuse University College of Law (1973).

After being commissioned ENS USN in June 1965 (NROTC Regular), he reported aboard USS *Easton* (DD-510) as damage control assistant upon completion of Damage Control School in Philadelphia, PA. He served aboard *Eaton* in Norfolk, VA until June 1967 when he was assigned management analyst for fleet work study group Atlantic, leading working sampling teams afloat in support of the Shipboard Manning Document program until February 1969.

After completing predeployment training at Naval Amphibious Base Coronado, CA, he reported to CDR Naval Forces Vietnam Staff in Saigon, Republic of Vietnam (RVN) where he served as communications operations officer, being awarded the Navy Commendation Medal w/Combat "V," Navy Unit Commendation, RVN Meritorious Unit Citation, Vietnam Service and Vietnam Campaign Medals. Upon release from active duty in July 1970, he affiliated with the USNR.

After serving as department head, training officer and XO in various USNR (NR) Surface Divisions in Liverpool, NY and Trenton, NJ, he assumed duty in 1975 as CO, NR TELCOMBCT NAS JAX 104 at Reading, PA which was cited by the Pentagon for outstanding performance during its 1976 Unit ACDUTRA. He served as CO, NR TELCOM CAMSLANT 104 in Folsom, PA from 1977 to 1979 when he became training officer for Naval Special Warfare Gp. Alpha 104 in Elizabeth, NJ. While serving in NR MILPLANS 304 as FEMA REG HQ 3 PHILA in 1985, he was selected for promotion to CAPT. After serving as senior inspector for REDCOM 4 IG Staff, in 1987 he reported as CO, Readiness Unit (A), NRRC Kearny, NJ. In 1990 he reported as CO, NR VTU 0413 (G), NAVRESREDCEN, Philadelphia, PA.

Magee is a graduate of the Naval War College RO Strategy & Policy Course at Newport, RI, the National Defense University RCNS at Ft. McNair, Washington, DC and the Naval Postgraduate School DRMEC 91-1 Course at Monterey, CA. He is a life time member of the NRA, US Naval Institute and the Navy League. He resides with his wife, the former Judith Stone, and son Michael in Wilmington, DE where he is senior counsel for E.I. DuPont de Nemous and Company.

ENAR V. "MAGGIE" MAGNUSON, CAPT, born Oct. 16, 1920 in Sweden, Province of Dalarna, received a BS in forestry from the University of Washington in 1951.

He joined the service October 1942. Commissioned Dec. 15, 1943. His military locations and stations: NAS Olathe, KS and NAS Seattle, WA; NAS Kodiak. Qualified plane commander, PBY, R4D, R5D. Served with VR-3 and VR-5. Life member of NRA and ROA.

Magnuson achieved the rank of CAPT, USNR, retired January 1946. He received the routine theater medals. His memorable experiences: flying the Aleutian chain out of NAS Kodiak, AK; also to Fairbanks and Pt. Barrow.

He married Helen E. and they have four children: Linda, Don, Paul, Deann; and five grandchildren. Employed as sales rep., Simpson Timber Co. Retired from Naval Air Reserve in 1968. Certificate of retirement Oct. 16, 1980.

JAY A. MAHER, CAPT, born March 30, 1947 in Chicago, received a BA in Carthage College. He joined the service September 1965. His military locations and stations: USS *Forrestal* (CVA-59). Served with VF-74 and VT-22.

His awards and medals include the National Defense Medal and Naval Reserve Medal (25 years). He achieved the rank of CAPT and discharged October 1991.

He married Susan and they have three children: Christopher, Ryan and Patrick. Employed as president, Central Specilaities Ltd.

PAUL V. MAIER, CAPT, born Nov. 6, 1947 in Seattle, WA, received an MBA from Harvard Business School and a BS from Pennsylvania State University.

He was commissioned an ENS, SC, USN, June 14, 1969 (NROTC Penn State, Sept. 27, 1965). His military locations and stations: USS *Westchester County* LST-1167), Navy Supply Center, Oakland, CA. Served with Riverine Force/HAL-3/ Seafloat. His battles and campaigns: RVN Campaigns XII, XIII, XIV, XV. Awarded the Joint Service Commendation Medal and Navy Commendation Medal (2).

Maier achieved the rank of CAPT, SC, USNR. Retired from active duty, Aug. 15, 1973, and USNR, Dec. 3, 1995. His memorable experiences: CO, Reserve Units: Navy Regional Contract Center Long Beach; AIRPAC Supply San Diego, Fleet & Industrial Supply Center, San Diego.

He is married to Shirley and is a senior vice-president & CFO, Ligand Pharmaceuticals, Inc.

RICHARD L."DICK" MAIER, CDR, born Sept. 9, 1935 in Chicago, IL, received a BS in management, DePaul University, Chicago, IL.

He joined the service November 1957. His military locations and stations: NAS Pensacola; Kingsville; Los Alamitos; North Island; USS *Hornet.* Served with VS-37, VS-35, VS-722, VS-66 and CINCPAC.

He was awarded the Armed Forces Expeditionary Medal and achieved the rank of CDR and discharged January 1963.

His memorable experiences: off Okinawa a recovery on a stormy night with low visibility, a carrier controlled approach to the ship, windshield wipers quit, the windscreen blurred. He was thankful for a ready deck, cut and a wire.

He married Patricia and they have two children, Michael and Christine; and two grandchildren, Amanda and Samantha. Employed with IBM in sales and with TWA as captain. He retired in April 1981.

NANCY A. MAIHOFF, CDR, born in Blue Island, IL, Nov. 14, 1939. She is a graduate of Southern Illinois University (BS); San Francisco State University (MA); and Michigan State University (PhD). She was assigned to the

Naval Communications Station-Stockton, CA after completing OCS.

She joined the USNR after release from active duty, and her last assignment was as CO of Military Sealift Command Philadelphia 104. She received a Navy Commendation Medal for service while commanding MSCO Phil 104. She has earned two awards of the National Defense Service Medal and the Armed Services Reserve Medal. She retired on Nov. 1, 1990 aboard the historic USS *Olympia* in Philadelphia.

She is the director of Teacher Education at the University of Charleston, in Charleston, WV, and spends her free time hiking and regularly traveling to New York, Philadelphia and/or the San Francisco Bay area.

GARY D. MAILANDER, CAPT, born in Santa Monica, CA, Sept. 20, 1938. He graduated from Tufts University School of Dental Medicine in 1963 and was first commissioned on May 8, 1961 as an ENS. He was appointed LT in the Dental Corps in June of 1963 and served on active duty at the US Naval Dental Clinic, Yokosuka, Japan from June 1963 to July 1966.

Following active duty, he did graduate training in orthodontics at Tufts University School of Dental Medicine and entered private practice in Mountain View, CA. Affiliating with the USNR, he served in various dental/medical units in REDCOM TWENTY. He served as CO of Yoko HQ 120, NavMedCom Northwest Det520 and Naval Hospital, Lemoore 0190. He also served as staff dental officer for REDCOM TWENTY. He served aboard USS *Midway* (CVA-41) and USS *Hector* (AR-7) as acting head, Dental Department. He retired from the USNR, Sept. 1, 1989.

His awards include the Navy Commendation Medal, National Defense Service Medal and Navy Unit Commendation. He married Kathryn and they have four children: Mark, Eric, Darcy and Kirsten. He resides in Truckee, CA.

STANLEY J. "STAN" MAJKA, CDR, born Jan. 28, 1919 in Three Rivers, MA, graduated from Wocester Polytechnic Institute with BS degrees in mechanical and aeronautical engineering.

Joined the service April 19, 1941. Navy service includes Massachusetts Institute of Technology, Bureau of Aeronautics, USS *Monterey* (CVL-26), Aircraft Sqdn., VC-30 (VF and VT aircraft), US Naval Attaché at US Embassy, Belgrade, Yugoslavia and Naval War College.

Engaged in nine major battles aboard CVL-26 in Pacific Ocean, starting with Makin and Tarawa through the battle of the Philippines.

His awards include the Navy Commendation Medal, Pacific Theater (with nine stars), European Theater, Philippine Liberation Medal, Naval Reserve and Naval Order of the United States Medal (he is a past commander general of the Naval Order of the US). Achieved the rank of CDR and was discharged June 30, 1967.

His most memorable experience occurred on Dec. 18, 1944 when their carrier, CVL-26, was struck with a typhoon, lasting 36 hours. They lost all planes off the flight deck due to winds of unbelievable strength followed by a hangar deck fire which disabled the carrier's forward elevator. The maximum roll was recorded at 38 degrees. They were part of ADM "Bull" Halsey's Task Force.

He married Peggy June McCollum, and they have two sons, Paul and Stanley Jr. He is employed by Chevron, USA for 32 years and is a certified financial planner with 27 years experience (still active).

DENNIS C. MAKARAINEN, CAPT, born Sept. 29, 1941 in Painesville, OH, received a BS degree from Cornell University; MBA in management/marketing from Syracuse University; MS in accounting/logistics, Syracuse University.

He joined the service March 4, 1964. His military locations and stations during active duty: Great Lakes NTC; NAS Oceana; OCS Newport; Supply Corps School, Athens, GA; USS *Frank E. Evans* (DD-754), Long Beach, CA; Navy Exchange School, Brooklyn; Military Sealift Command, Oakland, CA; Defense Personnel Support Center, Philadelphia, PA.

Reserve: supply officer for units in Utica, Rochester, Liverpool and Syracuse, NY; XO, Naval Activities Europe: Supply officer, USS *Canopus* (AS-34); Supply officer, RNMCB 13, Peekskill, NY; CO, NSC Norfolk 402, Syracuse, NY; Supply officer, 7th RNCR and RNCBC, Davisville, RI; XO and CO, Defense Personnel Support Center 202, Philadelphia, PA; XO, Navy Exchange Command 102 Norfolk, VA.

Makarainen was awarded the Defense Meritorious Service Medal, Joint Service Commendation, Navy Commendation, Navy Achievement, Joint Meritorious Unit Award w/Gold Frame, Meritorious Unit Commendation w/star, Navy "E" w/eff. ltr "E," National Defense w/star, Armed Forces Expeditionary, Vietnam Service w/3 Bronze Stars, Armed Forces Reserve w/"M," Vietnam Meritorious Unit Citation w/Gold Frame w/Palm, Vietnam Campaign w/Date Bar.

He achieved the rank of CAPT, Supply Corps, USNR. His memorable experiences: to knowingly help to serve the US for freedom; the comraderie of military personnel; chance to travel and see parts of the world; while CO of DPSC 202, the unit was mobilized for ODS providing the opportunity to demonstrate that the unit was prepared and could fully integrate into the daily operations of the active command and provide on-going support of the mission.

He married Jeanette and they have two children, Michael and Lorna. Employed as data processing business manager, County of Onondaga, Syracuse, NY. He retired from the Navy, Oct. 1, 1996.

BERNARD A. "BERNIE" MAKI, LCDR, born Feb. 15, 1932 in Eben Junction, MI, received a BA degree in economics and MA degree in education.

He served November 1950 to February 1954. Enlisted March 1958-1982. His military locations and stations: USS *Palau* (CVE-122); USS *Wasp* (CVA-18); USS *Leyte* (CVS-32). Served with USS *Interdictor* (AGR-13); CIC officer 1958-60; operations officer, 1960-61.

Maki was awarded the Good Conduct Medal, National Defense Medal, Armed Forces Reserve Medal. He achieved the rank of LCDR and was discharged February 1982.

His memorable experiences: all his time in USN and USNR was memorable. Most significant assignment was OIC of GVTU, Reserve Center, San Diego, 1978-80.

He married Eeva K. and they have three children: James, Elizabeth and Anna; and five grandchildren: Amanda, Rachel, Noah, Timothy and Adam. Self-employed with Maki and Associates, Inc. as president.

VERNON M. "VERN" MALEC, CAPT, born May 19, 1939 in Chicago, IL. Undergrad in New Mexico State University. Graduated from San Diego State University.

He joined OCS Aug. 3, 1962. Commissioned Dec. 21, 1962. His military locations and stations: USS *Bridget* (DE-1024); FLT AAWTRACEN, San Diego. Served with: USS *Shields* (DD-596); USS *Buck* (DD-761); USS *Southerland* (DD-743); USS *Enhanse* (MSO-437); CO, SIMA 419; CO NAVSEA SYSCOM 519; REDCOM 19; TACTRACGRU PAC, CO.

Malec achieved the rank of CAPT 06 and was awarded the National Defense Medal, Reserve Service Medal and Marksman (45).

His memorable experiences: served on last fletcher class destroyer on active duty; USS *Shields* (DD-596) before it was retired in 1971. He was XO of Reserve crew.

He married Judith Basquez and they have two children, Paul Edward and Kris Margret. Employed at Navy Personnel R&D Center; and project manager, retiring Oct. 30, 1992.

CHESTER W. MALINS, CAPT, born July 20, 1920 in Chicago, IL, received a BA from Millikin University; MS, from Texas A&M, Kingsville; CLU, American College of Life Underwriters, Bryn Mawr, PA.

He joined the service April 16, 1942. His military locations and stations: Alaskan waters and Asiatic-Pacific Theater. Served with Anti-submarine Warfare and 7th Fleet Amphibious Forces.

His battles and campaigns: Philippine Liberation Campaign and Southeast Asia Campaign. He was awarded Naval Reserve Medal, American Theater, Asiatic-Pacific w/3 Bronze Stars, WWII Victory Medal, China Service, Armed Forces Reserve, Philippine Liberation, Philippine Presidential Citation, and China Service Occupation.

He achieved the rank of CAPT 06 and retired July 1, 1973. His memorable experiences: July 1945 participated in the last amphibious invasion of WWII in Borneo.

He married H. Toni Placek and they have six children: Kris, Beth, Jo, Bo, Bob and James; and 10 grandchildren: Sarah, Rachael, Jeff, Rick, James, Lara, Steve, Joe, Kevin and Lisa. Employed with New York Life Ins. Co., chartered life underwriter. He retired Jan. 1, 1982.

NORMAN R. "NORM" MALMBERG, CAPT, born Jan. 26, 1938 in Red Oak, IA, received a BS in math, ISU; business, SD State, UC Davis, USC; Law, USD.

He joined the service September 1956 as midshipman; Nov. 23, 1960. Commissioned ENS.

His commands: Reserve Officers Training Corps, Ames, IA, summer cruise 1957, USS *Wisconsin* (BB-64); Summer cruise 1959, USS *Valley Forge,* 1957-60; USS *Topeka* (CLG-8) in various billets, 1960-63; Fleet Training Center, San Diego, OOD School Instructor, 1964-67; USS *Marsh* (DE-699) Long Beach, CA, senior watch officer, 1967; Military Sea Transportation Service, San Diego, CA, 1967-68; Military Training Div. 12-6, San Francisco at Treasure Island; USS *Kearsarge* (CVS-33), assistant navigator, 1969; XO, Naval Reserve Center, Seattle, WA, 1970-72; Program Sponsor's Rep., Naval Reserve Officer Schools, Chief of Navy Technical Training, 1972-74; TSU 13-70, training officer, 1974-75; SD 13-31S, 1975; Naval Reserve Center, Medford and Central Point, OR. Serving in various billets - approximately 15 years as a CO of VTU and the senior officer present at Naval Reserve Center.

He was awarded the National Defense Service Ribbon, American Vietnam Campaign, Vietnam Gallantry Cross, Vietnamese Campaign, Naval Reserve Meritorious Service Medal, Naval and Marine Corps Overseas Service Medal (two awards).

He achieved the rank of CAPT in 1981 and retired December 1990. His memorable experiences: the 30 years were one great adventure from his first midshipman cruise on the *Wiscy* to Cuba, Panama and Chili to conning the USS *Topeka* in a typhoon with 60' seas to leading the greatest reserves in the nation at Medford, OR for the last 16 years.

He married Pamela Mae "Pam" and they have two children, daughter Anne married to Gary Lundgren and son Ryan.

Employed as owner, Norm Malmberg, Ranch and Land Broker and Associates, 22 years, retired from the Navy Dec. 1, 1990.

RICHARD D. "DICK" MALOIAN, LCDR, born in 1925 in Detroit, and received a BS in chemistry from Wayne University.

He joined the service July 1, 1943. His military locations and stations: V-12 unit Illinois Inst. of Tech.

Served with NSC Guam, Naval Air Reserve Fasron 735 Grosse Isle, MI.

Maloian was awarded the Pacific Area Medal and achieved the rank of LCDR.

He married Rose and they have two children, George and Anne; and a grandson Matthew. Employed as superintendent Cadillac Motor Car Div. He retired in December 1984.

WILLIAM E. "BILL" MALONEY, LCDR, born Sept. 26, 1918 in Golden Gate, IL, received a BA, LLB, JD.

He joined the service Nov. 18, 1936. Commissioned Sept. 15, 1943. His military locations and stations: USS *Pennsylvania* 1937-40; MTB commissioning detail, New Orleans 1942-44; USS *Maryland* 1945, HQS Command, Yokosuka, Japan 1950-52.

Maloney achieved the rank of LCDR and was discharged June 29, 1962. He married Winifred and they they have a daughter Janice Mainland. Their grandchildren are Mark, Jason and Grant Mainland. Employed Bank of America, public relations; Farmers Merchants Bank, vice-president trust officer; attorney at law, California State Bar. He retired Sept. 26, 1978.

JAMES W. "JIM" MANGIMELI, CDR, born Jan. 13, 1945, a native of Omaha, NE, attended Marquette University on an NROTC scholarship and was commissioned in June 1967. He reported aboard USS *Thomas J. Gary* (DER-326) in Key West, FL for three years, serving as electronics officer, CIC officer and operations officer. During this period *Gary* received the Naval Unit Commendation and CDR Mangimeli received the Navy Achievement Medal for surveillance and intelligence collection missions in the North Atlantic, Mediterranean and Caribbean.

After a tour at NROTC Unit, University of Minnesota, he transferred to the USNR, serving in the Naval Reserve Intelligence Program for 18 years in various units in Minneapolis, Glenview, IL and Lincoln, NE.

CDR Mangimeli, a high school teacher and administrator, received his MA at the University of Minnesota and has done additional post-graduate work at Creighton University. He and his wife Carol reside near Waterloo, NE with their children.

HORRIS G. "JOHNNY" MANISCALCO, LCDR, born Aug. 24, 1934, in Biloxi, MS. He along with his twin brother, Morris, joined the USN in 1951. He now resides in National City, CA. He was honorably discharged after 10 years as a dental technician second.

He went into the Reserves in 1964 and 1978. After various training he changed his rate to aviation storekeeper first. Advanced to chief and later promoted to CWO2. In 1987 he was promoted to LTJG and graduated from Supply Corps School, Athens, GA. After his promotion to LCDR her retired with over 34 years of service.

He is authorized to wear: Good Conduct (3), National Defense, Korea and China Service, Korea, Marksman, Armed Forces and Naval Reserve Medals.

Maniscalco has been married for 40 years to Audelia Miranda. They have two adult children, Thomas and Alison; and two grandchildren, J. Anthony and Benjamin.

He retired with over 43 years in Civil Service. Enjoys honey dos, grandchildren, performing on stage and in film.

MARK WILLIAM MANLOVE, CAPT, born Aug. 10, 1947 in Panama Canal Zone, received a BS in accounting, San Diego State University and MS degree, USC.

He joined the service Nov. 24, 1968. His military locations and stations: NAUCS Pensacola, VA-128 Whidbey, VR-30 Alameda. Served with Reserve Intelligence Area Four/FIRSTPACS.

Manlove achieved the rank of CAPT 06 and was discharged Jan. 30, 1972. His memorable experiences: San Diego Intell Community/FICPAC ACTDUTRAS.

He is married to Elizabeth Green and they have children: Todd and Elise. Employed as supervisory auditor, Defense Contract Audit Agency. He retired June 30, 1996.

KENNETH PAUL "KEN" MANNING, RADM, born Jan. 18, 1942 in New York City, received a bachelor mechanical engineer, Rensselaer Poly Inst. and a MBA from the American University.

He joined the service Feb. 1, 1963. His military locations and stations: NASC Washington, DC; APD-107; APD-90. Served as director NR Air Systems Program.

He was awarded the Legion of Merit and achieved the rank of RADM and retired Nov. 19, 1967.

Manning married Maureen and they have six children: Kenneth J, LT John J., M.D. Paul, Elise, Carolyn and Jacqueline. Employed as president and chief executive officer, Univ. Food Corp. He retired Jan. 1, 1995 from the Reserves.

GENE A. MAREK, CAPT, born in Goose Creek, TX, was raised in nearby Crosby. He received a technology degree from Lee College in 1964 and a bachelor of science degree from the University of Houston in 1966.

After commissioning through the AOC program, he received his wings at NAS Corpus Christi in April 1968. Reporting to his first squadron (VS-32), he flew the S-2E during three North Atlantic/Mediterranean deployments onboard USS *Wasp*. After a tour a officer programs officer at NRD Houston, he reported on USS *Coral Sea* as CCA officer, C-1A pilot and ATO. Most significant West Pac cruise was during the Cambodia/Vietnam Evacuation and SS *Mayaquez* recovery, where he also provided air logistics assistance for CTF-77.

Joining the USNR in 1977, he began C-118 flying duties with VF-21 DET Nola, VR-54, and VR-53/59. Following RESFORON duties, he began the first four tours with CG-0282 at NAS New Orleans where he served as XO and ultimately CO. During his CO tour, Marek attended the National Defense University Course, represented COMCARGRU Two at two planning conferences with the Egyptian military, and provided assistance onboard USS *John F. Kennedy* during their MED 93-1 deployment off Bosnia. His duties prior to his March 1997 retirement were that of command inspector for NAS JRB New Orleans. His 15 awards include the Navy Commendation Medal.

He is employed as a corporate pilot and staff aviation safety representative for Shell Oil Company in Houston, TX. He is married to the former Glenna Ann Poland Marek. Their family includes two sons, Dale (wife April) and Jeff; and grandson, Tyler.

ETHEL MARIAM, CAPT, enlisted in USNR June 16, 1943 (20th birthday), discharged Feb. 1, 1946 as a SKD2c; reenlisted on July 10, 1947 as a DK2, discharged Oct. 28, 1949; commissioned ENS Supply Corps, USNR Oct. 28, 1949; retired June 16, 1983 as CAPT, SC USNR.

Boot camp NTS (WR) Hunter College, NY. Duty stations and schools: Storekeeper School NTS Indiana University, Bloomington, IN; NTS Bainbridge, MD (recruit and stewards mate disbursing offices); Demobolization School, NTS Great Lakes, IL; Separation Center, NATTC Memphis, TN, on staff opening this activity until discharge February 1946. Inactive Reserve 1947-49 in NR Units, surface divisions 138 and 139 Reserve Training Center, St. Louis, MO and mobilization officer and classification officer; Defense Contract Administration Services Region, St. Louis, CO DCASR 718 from 1977-78; also attended Naval Reserve Officer School and served on selection board for CDRs and CAPTs Supply Corp.

She was awarded the American Theater, WWII Victory Medal, Naval Reserve and Armed Forces Medals.

Civilian career includes staff member of St. Louis accounting firms and administrative assistants for fiscal operations for the Jewish Federation of St. Louis and Consolidated Neighborhood Services, Inc.

Retired in 1987. Member of Jewish War Veterans Post 644 and life member of NRA; Reserve Officer Association; and Retired Officer Association. Volunteer at USO and member of Eagles Nest.

EUGENE C. "MARKIE" MARKIEWICZ, LT, born in Pennsylvania, June 1, 1919. Graduated 1940 in accounting, St. Vincent College, Latiola, PA.

Employed by Stupakoff Manufacturing Co. of Latiola, PA as a cost accountant to June 1944. Drafted by US Army serving in Camp Hood, TX. For the convenience of the government transferred in August 1944 to the USN as an ENS. Attended OCS US Naval School in Hollywood, FL. Assigned to US Naval Bomb Disposal School, American University in Washington, DC. In October 1944 to April 1946 served on USS *Lunga Point* (CVE-94) in gunnery department. Returned from active duty May 8, 1946. Recalled during Korean Conflict, Dec. 14, 1950, to Maryland. Served as head of surface ordnance disposal instruction until discharged July 1952.

He received the American Theatre, WWII Victory, Asiatic-Pacific w/2 stars, Philippine Liberation, Presidential Unit Citation for the Battle at Leyte Gulf.

Self-employed from 1946 except for Korean War. His wife Jean is deceased. His children are John, Francis Patrick, Catherine Edwards and Nancy Shaffer; and grandchildren are Pat, Andy, Steve, Scott, Tracy, Beth, Adam and Marky.

JAMES T. "JIM" MARKLEY, LCDR, born Jan. 10, 1933 in Altoona, PA, received a BS and MS in geography from Florida State University.

He joined the service May 21, 1952, USN, YN2; 7-59, USNR. His military locations and stations: USS *Saga City* (MSO-469); USS *Grackle* MSC(0) 13. He served with Reserve Intelligence Units-Miami, Charlottesville, Norfolk, VA.

His awards and medals include the Navy Good Conduct, National Defense (2), Expert Pistol, Armed Forces Reserve (2). He achieved the rank of LCDR, Reserve, YN2, active USN. Retired July 1981.

Memorable experiences: active, commissioning crew, USS *Saga City* (MSO-469); USNR, Intelligence Debriefer Operation, homecoming Vietnam POW February 1973-March 1973; TEMAC FOSIF Rota, Spain June 1970-July 1970.

He married Mary and they have five children: Timothy Earl, Daniel Philip, Cindy, Beverly, and James Michael; grandchildren are Thomas, Brent, Bryan, Megan, Brock, Samuel and Kyle. Retired in May 1993 from college as geography professor.

ELVIN A. "AL" MARNE, CDR, born Aug. 1, 1940 in E. Meadow, NY, received a BA degree, Towson State University, 1967.

Joined the service in Baltimore, MD April 1959. His military locations and stations: USN Prep School, April 1959-August 1960; VA-163, August 1960-January 1963; Naval Special Warfare Center (NAVSPECWARCEN), July 1986-April 1987; CDR, Naval Special Warfare Command (COMNAVSPECWARCOM) June 1987-October 1987.

His Reserve: Various Naval Air Intelligence Reserve Units (NAIRU) in Washington, DC and Jacksonville, FL; Various Fleet Intelligence Rapid Support Teams, (FIRSTPAC) Pacific in Los Alamitos and San Diego, CA; XO Fleet Intelligence Training Center, Pacific (FITCPAC) 0194 October 1982-October 1984; XO CDR Naval Air Forces, Pacific, Operational Intelligence Unit (CNAPOPINTEL) 0194 October 1987-October 1989; Intelligence Volunteer Training Unit (IVTU) 121 October 1989-January 1993.

He was awarded the Navy Commendation Medal, Good Conduct Medal, National Defense Medal, Naval Reserve Longevity Medal, Navy Pistol and Expert Medal. Marne was discharged April 12, 1965.

His memorable experiences: serving with the Navy Seal community during the Persian Gulf Crisis, July 1986-October 1987. Established an intelligence base for the newly created Strategy and Tactics Dept. (NAVSPECWARCEN); becoming a plankowner of the newly created COMNAVSPECWARCOM, and serving as special security officer (SSO), from June 1987-October 1987.

He is married to Elizabeth (Liz) and was employed with Los Angeles Unified School District as teacher, retiring Oct. 1, 1994.

DANIEL E. "DOC or DAN" MARSALEK, CAPT, born April 10, 1931 in Cleveland, OH, received a BS degree at University of Notre Dame; DDS (Case-Western Reserve University).

Marsalek joined the service November 1955. His military locations and stations: USNH Great Lakes; USS *Essex* (CVA-9), USNR-R Cleveland.

He was awarded the Armed Forces Expeditionary and the Naval Reserve. Achieved the rank of CAPT. His memorable experiences: line crossing, covered Lebanese landings, 1957 on *Essex;* internship (dental).

Married Jan Alice and they have two children, Daniel and David. Self-employed, Bay Village, OH. He retired 1983.

MYRLE FREDERICK "FRED" MARSH, CAPT, born July 5, 1929 in Plymouth, Marshall County, IN, received a BS in 1951 from Indiana University and a MD in 1954.

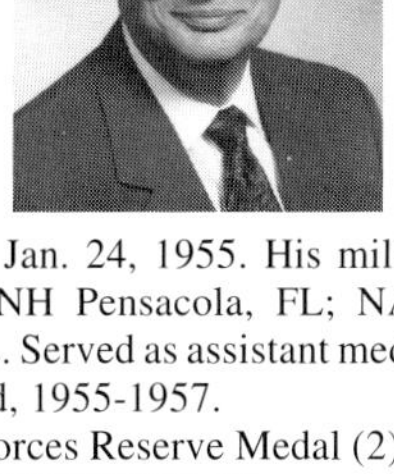

He joined the service Jan. 24, 1955. His military locations and stations: USNH Pensacola, FL; NAAS Saufley Field, Pensacola, FL. Served as assistant medical officer, NAAS Saufley Field, 1955-1957.

Awarded the Armed Forces Reserve Medal (2). He achieved the rank of CAPT (MC) USNR. His memorable experiences: NR Medical Co. 9-3; NAVRES MOB Team Div. 9-17 NRGG 9-37(M) training and support 9-64 (medical); CO NAVREGMEDCEN 8517 ORU; NR Lafayette River Brch 118; NR FLT Hospital 18-B NRCOMPHIBRONS-518.

He married Marilyn and they have two children, Theresa J. Marsh Holloway and Steven F. Marsh; and granddaughter, Ashley E. Marsh. Employed as MD/Ob-Gyn, private practice. He retired July 5, 1989.

ARTHUR H. "ARI" MARTIN, born Feb. 3, 1928 in Ames, IA, received BSCE, Iowa State University.

His military locations and stations: MCAS, Cherry Point, NC; USN Construction Base. Served with USNRCB-17; USNC Regiment 9; USNREC 1 Port Hueneme, CA.

His memorable experiences: winner of Manning Trophy, CO USNRMCB-17 CO USNRCR-1; CO USNRCR-2; CO USNRMCDIV 11-6.

He married Virginia and they have two children, Andrew Kris and Mariah Lori Martin Dewitt; and two grandchildren: Seth Andrew and Jeremy Christian. Employed as president RGA, Engineering Corps, Principle Engr. CDM Corps. He retired Jan. 30, 1990.

CHARLES W. MARTIN, CDR, the day WWI began, April 3, 1914, was born in Philadelphia, PA. He graduated from Temple University in 1936 with a bachelor's degree in education. On Oct. 30, 1942 he enlisted in the USCG in New York, NY. From seaman second class he advanced to yeoman first class. He served more than three years which included duty on the USS *Key West.* He was honorably discharged on March 29, 1946.

He joined the USNR on March 17, 1947 in Philadelphia where he had an assignment at the Cramp Shipyard and later transferred to the DIO at the 4ND in Philadelphia. Charles was recalled to active duty (Korean Conflict) summer 1950. While at the Pentagon in Washington, DC he received his commission on Aug. 15, 1950 as LT, Supply Corps. He was honorably discharged as yeoman first class and transferred to the Supply Corps School in Bayonne, NJ as a student. Later he served on the USS *Cabot* (CVL-28), the USS *Tappahannock* (AO-43) before returning as an instructor at the Supply Corps School. He was separated on Jan. 15, 1954 from active duty, but he continued to serve in several USNR units. During this time, he successfully completed 14 correspondence courses (1955-66). He was commissioned as LCDR Supply Corps (Jan. 1, 1955) and as CDR, Supply Corps (Dec. 1, 1959). As of April 3, 1974 he was transferred to the USNR Retired List.

His civilian employment was as a secondary school teacher where he taught German and Business Education. He earned a masters in education 1948 and a doctorate in education 1962. Later he had his own practice as a public accountant.

Shortly after Charles joined the USNR (1947), he married Althea who was born at the time WWI ended. They have a son Charles W. Martin, Jr. (nuclear engineer) and a daughter Althea (computer specialist). The son has two children, Charles W. Martin III and Nicole Elizabeth; and daughter has three children: Gwyneth Rose, Lee Anne and James Martin Whieldon.

Genealogical research revealed that Charles has seven ancestors who participated in the Revolutionary War. In addition to this search Charles and Althea's retirement is being spent enjoying their home and family.

He was authorized to receive the following medals: Asiatic-Pacific Area, EAME Area, American Area, WWII Victory, USCG Conduct and National Service Defense.

DAVID P. MARTIN, CAPT, a native of Auburn, NY, attended State University of New York and Syracuse University obtaining BA and MS degrees. He began his naval career as seaman recruit in 1955, later yeoman first class, and was commissioned an ENS (public affairs) in the USNR in 1965.

During his active duty career he served as a yeoman with Fleet Air Wing THREE; NAS Quonset Point, RI and Brunswick, ME; and NAS Guantanamo Bay, Cuba.

After separation from active duty in 1959, he served in the reserve as public affairs officer, NRC Syracuse; XO, CDG8; PAO, REDCOMTWO; CO, OI DET 102 and NAVINFO EAST 102. During both CO tours, each unit received CHINFO Mobilization Awards for Fleet Week celebrations in NY Harbor.

Martin has been awarded the Good Conduct Medal and Navy Commendation Medal (two awards).

The Martin family consists of wife Gretchen, an associate professor; daughters: Lizabeth, Pamela and Kimberly, who are pursuing careers in law and counseling.

He continues college teaching while finding more time for hiking, skijoring, fishing and increasing numbers of grandchildren.

JAMES MILTON MARTIN, CAPT, born May 15, 1914, Waxahachie, TX. He attended elementary, high school, Trinity University, Waxahachie. He earned BS degree in engineering physics at University of Oklahoma, 1938; USN Postgraduate School, Annapolis, 1945; MS degree, physics, Sigma Xi, Renssalear Polytechnic Institute, Troy, NY, 1946; postgraduate study University of Maryland, 1947-50; Columbia University, 1949, exploration geophysics six years; USN Ordnance Laboratory, mine countermeasures, January 1942.

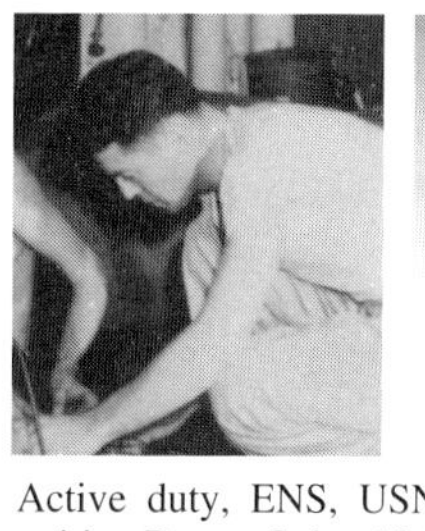

Active duty, ENS, USNR, April 6, 1942; Naval Ammunition Depot, Oahu, 1942-43, submarine mine test, modification, evaluation, assembly, operational planning for COMSUBPAC; COMAIRPAC training exercises for visual low altitude mine laying tactics. Commander Minecraft, US Pacific Fleet, 1943-45, aerial mine modification, preparation, and use against Palau, March 1944 and Truk, April 1944; mining officer USS *Lexington* (CV-16) for TF-58 operations against Palau; mining and rocket safety officer, USS *Yorktown* (CV-10) during First Battle Philippine Sea; participated in plans for mining Japanese homeland with Army Air Force B-29s from Marianas Islands, one of three naval officers assigned XXI Bomber Command January 1945; responsible with B-29 crew, 313th BW, Tinian, for developing radar tactics, location methods, and potential laying accuracy for mining Japan (Operation Starvation). Postgraduate School, 1945-46; released to active duty Oct. 5, 1946. Command billets NR weapons units; attended Naval and National War Colleges; assigned duty OPTEVFOR, USS *Coral Sea* (CVB-43), USS *Canberra* (CA-70), promoted to CAPT, USNR, April 1962; member US Mine and Mine Countermeasures Task Team, Korean War, 1950; retired Naval Reserve, May 15, 1974.

Civilian: USN Ordnance Laboratory, 1947-74; Department Chief, Underwater Evaluation, 1957, technical evaluation of weapon systems; organized and sponsored 16 annual conferences on the naval minefield, 1958-73; CNO representative in formulating, selecting sites for two-year measurement programs in important water areas with Turkey, Denmark and Japan; and for USN-Japan Weapon Development Program; project engineer for evaluation and selection of Gould Inc. Torped Mk 48 Mod 1; project manager for RDT&E, production, and fleet introdution; retired from NOL, 1974. Engaged in post WWII research, seamines in wars and minor hostilities; papers published resulting from this research include, "Seamines and the USN" Naval history, USN Institute, Summer and Fall editions, 1989; "Seamines in Nicaragua;" proceedings, USN Institute, September 1990; "We

Still Haven't Learned," Persian Gulf War, proceedings, USN Institute, July 1991.

Received USN Bronze Star for WWII service, Navy Superior Civilian Service Award, Meritorious Civilian Service, Sustained Superior Performance, US Commendations and foreign defense establishments.

Married Margaret McLemore of Sulphur Springs, TX, in June 1941. They have three sons: James, Lawrence, Thomas; two grandchildren and two step-grandchildren.

He is member of the USN Institute and The Assoc. of Minemen; life member of TROA, ROA, NRA and The Naval Minewarfare Assoc.

RICHARD G. MARTIN, CAPT, born and bred in Michigan, he won a NROTC scholarship to University of Michigan. He was battalion commander of the class of 1952, swam on the varsity team, and got orders to Pensacola upon graduation and commissioning. He was designated a naval aviator in September of 1953 which he considers the most important event in his life.

Ensign Martin's active duty career began with VA-155 which deployed aboard *Yorktown* in 1954, flying AD-6 aircraft. After a Westpac tour of nine months he returned to Moffett where he served with FASRON 10 until release from active duty in October of 1955. He joined the USNR at Los Alamitos. There he flew the AD, F9F8, S2F, and finally the P2V. About the time that the Reserve was converting to P3V, then CDR Martin was promoted to CAPT and the flying days were over. He continued his Reserve career in various units and finally retired with 34 years total service.

Martin had no wars and no medals other than service ribbons. His major accomplishments included tours as CO of VP65L1 from 1968 to 1971, CO of NARS L1, CO of ATC 219, and CO of VTU 7676.

In civilian life Martin owns and insurance agency and has served as founding director of Western Bank for 23 years. He and his wife Ginny have two children, Richard and Sarah; and five grandchildren: Jason, Robert, Ryan, Katie and Sarah. They enjoy golf at Los Angeles Country Club. His major hobby is flying a fully restored NAVION.

RICHARD GEORGE "SPIKE" MARTIN, CDR, born Jan. 23, 1953 in Brockton, MA, received the bachelor of business administration from the University of Texas at Austin, 1979.

Commissioned Dec. 21, 1979. His military locations and stations: NAS Dallas; NAS New Orleans; Pensacola; San Diego; Jacksonville, FL. Served with VT-25, 26, 27; VT-10, VS-41, VS-28 "World Famous Gamblers;" COM7thFLT 111. Supported Multi-National Forces, Beirut, Lebanon 1983-84, USS *Independence.*

He was awarded the Navy Commendation Medal, Navy Achievement (2), Navy Expeditionary Medal and National Defense (2). He achieved the rank of CDR and discharged SelRes.

His memorable experiences: completed two deployments on board USS *Independence* (CV-62) 1983-85 to the North Atlantic, Mediterranean Sea, Indian Ocean and N. Arabian Sea. Exercises in Korea and Japan. Completed 180 carrier arrested landings in the T-2, TA-4J and S-3A Viking aircraft.

Married Monica Sue Sweeten and they have a son Andrew. Employed with American Airlines, pilot. Retired active reserve, COMSEVENTHFLT III, Dallas, TX.

EDUARDO VIDAL "EDDIE" MARTINEZ, LCDR, born Sept. 27, 1955 at Travis AFB, CA, received a BA from the University of Texas, Govt & Intel Relations (1978); MA, Antioch School of Law, Center for Legal Studies (1983); JD, Mission College, 1990.

Joined the service 1983 OCS, Newport, RI; Commissioned June 3, 1983. His military locations and stations: Surface Warfare Officer School, Coronado, CA; San Diego, CA. Served with USS *Samuel Gomper* (AD-37); USS *Vancouver* (LPD-2); Naval Control of Shipping; USS *Papago.*

Martinez was awarded the Sea Service Deployment Medal, Reserve Overseas Service Ribbon, Expert Rifle, and Pistol Medals, National Defense, Armed Forces Service, NAM and Air Force Commendation. He achieved the rank of LCDR and was discharged July 1, 1987.

His memorable experiences: aboard amphibious squadron standing offshore during Marcos overthrow in Philippines in event of evacuation of Americans was necessary.

He married Mary Kim and they have a son Anthony Michael. Employed as program analyst, US Department of State, Special Assistant, Attorney General.

GILBERT HERRERAS MARTINEZ, CAPT, born March 26, 1943 in Tucson, AZ to Delia H. and Gilbert E. Martinez. He and several buddies joined the USNR while seniors at Tucson High School.

He continued with his studies at the University of Arizona and upon graduation began his two years of active duty as a hospitalman second class on the USS *Hassayampa* (AO-145). During this time he served two tours of duty in Vietnam.

After release from active duty, he continued drilling and returned to the University of Arizona to work on his masters degree. Subsequently, he applied for and received a direct commission as a LTJG in the Medical Service Corps.

Highlights during this time include his selection to command several units: a Marine medical unit, medical logistics unit and a hospital unit. After a short recall during Desert Shield/Desert Storm, he returned to serve as CO of the fighting VTU.

Martinez is married to Paula and they have two sons, Timothy and Daniel. They reside in Tucson, AZ.

JOHN R. MARTINEZ, LCDR, born Oct. 13, 1923 and was sworn into the USN after graduating from high school and successfully passing the qualifying physical and written examinations for the Navy's V-5 Naval Air Cadet Training Program on July 1, 1942. He completed civilian pilot training in Dayton, OH, March 1943; Navy Preflight School University of Iowa, June 1943; elementary flight training, Minneapolis, September 1943.

Commissioned ENS and was designated naval aviator, February 1944. He served with Patrol Sqdns. VP-52 and VP-53 in New Guinea and Philippine Liberation Combat Operations. He was discharged from active duty Dec. 15, 1945 and remained in the Naval Air Reserve with various squadrons until finally earning retirement status in October 1983.

Martinez was awarded two Air Medals. His memorable experiences: rescued three survivors from downed B-25 aircraft, June 29, 1944 in McClure Gulf New Guinea. Designated patrol plane commander pilot PBY type aircraft February 1945.

He presently resides in Southern California with his wife, the former Tana M. Romo of Detroit, MI, whom he met in high school. They have five children: Mary Louise, Judith Ann, John, Annette and James; and 16 grandchildren who live nearby and they enjoy immensely. He enjoys playing golf and woodworking.

THEODORE T. MASON, CAPT, a native of New York City, attended State University of New York Maritime College and Columbia University Graduate School of Business obtaining a BME and MBA-finance degree. Commissioned ENS in 1957, he served in the Pacific aboard USS *Guide* (MSO-447) as engineer officer. CAPT Mason is a surface warfare officer.

From 1960-88 he served in over 15 Naval Reserve organizations, including Surface, Mobile Inshore Undersea Warfare, Intelligence, Coastal Defense, Systems Analysis, Readiness and Maritime Defense units or commands. He was CO of IUD 3-2, MIUW 504, HDCC 606, and CCDL 106. He was a major force in standing up Navy Coastal Defense Command, Atlantic and it's successor US Maritime Defense Zone, Atlantic. CAPT Mason received four awards of the Navy Commendation Medal: As CO HDCC 606, Norfolk; as advisor to CDR Maritime Defense Zone Atlantic; as chief of staff to Commodore Smith and Senior Naval Liaison Officer to Commander, US Maritime Defense Zone, Atlantic, concerning re-emergence of the coastal defense mission; and, for this role in execution of the 1986 International Naval Review and the Statue of Liberty Centennial.

Naval Reserve Association participation included co-founding the J.F. Burke Chapter, President NY Navy Chapter, VP Surface/Subsurface NRA 3rd District, Chairman, National Resolutions Committee, National VP-Budget and Finance, National VP Surface/Subsurface, chairman; National Awards Committee and Chairman, The National Advisory Council. For accomplishments in Naval Inshore Warfare he received NRA's National Award of Merit.

Mason's industry positions included: Pratt & Whitney Aircraft; KPMG Peat Marwick; Ted Mason Venture Associates; vice-president, Penntech Paper/ Willamette Industries; and managing director, EastWind Power Partners, Ltd., a power plant developer. He is vice chairman or trustee of eight mutual funds, overseeing 1.5 billion dollars.

His family includes his wife Carol and two sons, Andrew Mason, LT, USNR (R1) an intelligence officer aboard USS *Abraham Lincoln* (CVN-72) during Operation Desert Storm and is now an investment banker, and Robinson Mason, a multi-media software engineer.

ROBERT G. "GLENN" MASSENGALE, CDR, born in Stewart, AL, June 4, 1915. At an early age he moved to Birmingham, AL where he attended the public schools and Birmingham-Southern College, graduating in 1935 with the AB degree. Subsequent degrees included the BD (1939) and the PhD (1950) from Yale.

On Dec. 27, 1943 he was appointed a LTJG in the Chaplain Corps of the USNR and reported for duty on Jan. 15, 1944 to the chaplain school (Williamsburg, VA). From March 18, 1944 until Feb. 1, 1945, he served as the first chaplain of the Naval Air Facility (Columbus, OH) serving there until he was assigned to the Bremerton, WA Naval Yard for a pre-commissioning school in connection with the USS *Siboney* (CVE-112), being built in Tacoma, WA. After the commissioning in Tacoma on May 14, 1945, the ship set sail to join the forces destined for the invasion of Japan. Massengale recalls that one of his most memorable experiences occurred when he was one day out of Pearl Harbor, HI. The ship captain's voice came over the ship's speaker system loud and clear announcing the unconditional surrender of the Japanese. After about three months of touring the Japanese and other islands round about, Massengale returned to San Diego, CA where he was ordered to NAS New Orleans, LA for release from active duty with the rank of LT, CHC, USNR, effective April 11, 1946. His campaign medals included the American Theater of War, the Asiatic-Pacific Area, and the WWII Victory.

He remained active in the USNR through correspondence, training duty, and various other assignments and was transferred to the Retired Reserve in 1969 with the rank of CDR. While his duty assignments were numerous and various, aboard ships and on stations, he recalls the following with special pride: chaplain assigned to the USS *New Jersey* (BB-62) for the two months midshipmen's cruise to Europe (1956), chaplain for US Naval and Marine Corps Reserve training centers in Montgomery, AL (1951-1961) and Birmingham, AL (1961-69).

His family consists of his wife Elizabeth; daughter Lois M. Knowles; two granddaughters, Beth K. Brinley and Kris K. Crawford; and two great-granddaughters, Ashton Brinley and Kaci Crawford.

HAROLD J. "HAL" MAST, CAPT, born Feb. 8, 1944 in Volga, SD, received a BA in economics; MA in Health Care Administration.

He joined the service July 31, 1968. His military locations and stations: Naval Hospital, Oakland, CA from October 1968 to September 1991. Served as XO and CO of numerous medical units in Michigan. Most recently as medical staff officer of NR BUMED 106 in Washington.

Mast was awarded the National Defense Medal, several letters of commendation and appreciation. He achieved the rank of CAPT, USNR.

His memorable experiences: serving as first medical reserve officer as part of the BUMED inspector general's staff, and participating in three NAVHOSP inspections. Three years of active duty in Oakland Naval Hospital.

He has been married to Angie K. Mast for 30 years and they have four children: Christopher, Kerianne, Jacki Lynn and Erick; and a grandson, Jordan Scott. Employed as director of residential services at Pine Rest Christian Mental Health Servies in Grand Rapids, MI. He is still active in VTU.

JOSEPH M. MATHIAS, CDR, born Jan. 23, 1914 in Frankfort, KY, he grew up in Washington, DC. He is a 1935 graduate of the Univesity of Maryland and a 1942 graduate of the Southeastern University Law School.

In 1943-44 he served as OinC of Armed Guard crews on three merchantmen. Most noteworthy was his first trip, a 211-day around the world voyage on a freighter that carried Russian war supplies to Bandar Shahpur in the Persian Gulf. The last year of the war he spent as a deck officer and assistant navigator on the USS *Cleburne* (APA-73) in the Pacific.

CDR Mathias practiced law in Maryland from 1946 until he was named to the bench in 1965, and after his retirement in 1981, for nine years as counsel to a law firm. Before entering the practice of law he was an assistant city editor at the *Washington Post.*

He married Ruth Wellington and their children are Mark, Marcia and Marilyn; and grandchilren are Theo and Olivia. He resides in Kensington, MD, and is a retired judge of Maryland's Sixth Judicial Circuit, consisting of Montgomery and Frederick counties.

TYLER R. MATTHEW, CAPT, born Aug. 18, 1918 in Berkeley, CA, received a BSEE from Cal Tech in 1939.

He joined the service Jan. 9, 1941. His military locations and stations: Boston, MA; Green Cove Springs, FL; Tontouta, New Caledonia; Vella la Vella; Guadalcanal;

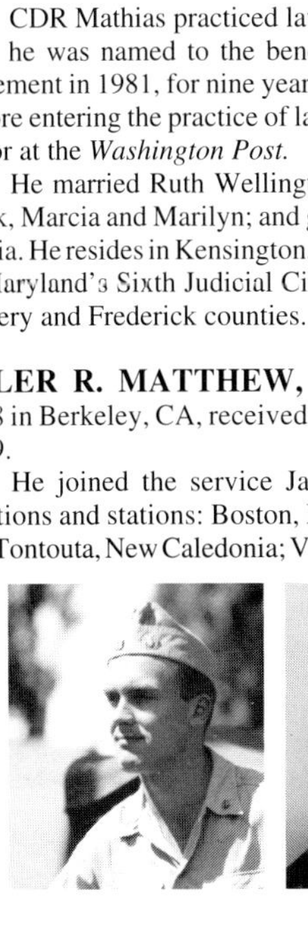

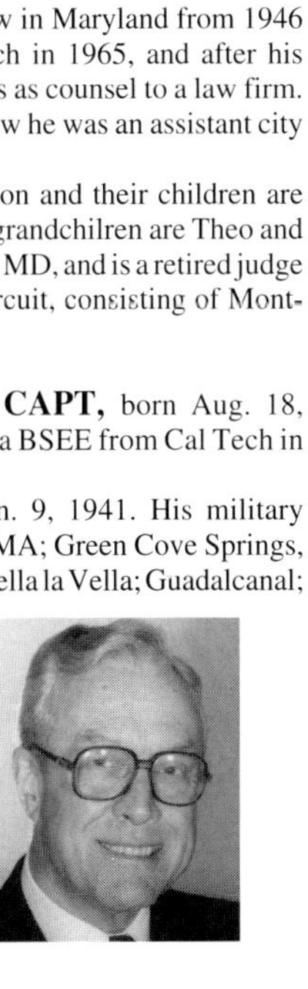

Pasco, WA; Orote, Guam; Cherry Point, NC; Yellow Sea, Philadelphia, PA; Seattle, WA.

Units served with: Navy Supply Corps School; NAAS, Lee Field; CASU-3; ASA Navy 145; NAS Pasco, WA; Com FAir WesPac; MCAS Cherry Point; USS *Sicily* (CVE-118); ASO; CO, SSOD 13-3 and TTTMD 13-1.

His battles and campaigns: American Theater; Asiatic-Pacific Campaign; Korean War.

Matthew was awarded the Navy Unit Commendation Ribbon, American Defense, American Theater, Asiatic-Pacific Theater, WWII Victory, National Defense, Korean War Service w/3 stars, Navy Occupation Service (Asia Clasp); Armed Forces Reserve w/2 HGs, UN Service (Korea Clasp), Korean Presidential Unit Citation. He achieved the rank of CAPT, SC USNR and was discharged from the USN, Jan. 28, 1955. Retired from the USNR, Aug. 14, 1978.

His memorable experiences: as supply officer, CASU 3 on Vella la Vella, provided supply support to the "Black Sheep" Sqdn., VMF-214, MAJ Gregory "Pappy" Boyington USMC, CO, 1943-44; as supply officer, USS *Sicily* (CVE-118), again supported VMF-214, providing close air support over the Korean Peninsula, 1951-52.

His wife Virginia Knight Matthew is deceased. Their children are Ann Louise Wood and Eleanor Lee Matthew.

Employed with Southern California Edison Co., 1939-41, Jr. power salesman; Northrop Aircraft Co., Junior stress engineer, March-September 1941; Lockheed Aircraft Service International, section supervisor, materiel, 1955-56; The Boeing Co., Sr. Spares engineer, Sr. Spares supervisor, 1956-70; American Mail Line, inventory control manager, 1971-72; logistics engineer, GS-12, NSWC, Dahlgren, VA, 1972-80. Retired Feb. 29, 1980 from NSWC, Dahlgren, VA.

THOMAS C. MATTHEWS, CDR, born May 28, 1910 in Monroe, WA, received a BS in physics from the University of Washington.

He joined the service in 1954. His military locations and stations: BuOrd, Washington, DC and Boston, MA. Served with BuOrd Exp Unit, Polaroid Corp., BuOrd, Washington, DC.

Matthews achieved the rank of CDR, USNR and discharged Nov. 16, 1946. His memorable experiences: celebration at end of WWII.

He married Ruth and they have a son Thomas Matthew Jr. and two grandchildren, Aaron and Tien Marie. Employed as spectroscopist, state of Oregon. Also employed as electrical engineer in Fairbanks, AK. He retired May 28, 1970.

LANCE MAURO, LCDR, born March 1, 1958, in Buffalo, NY. Earned BA degree in economics, May 1985, and BS degree in management/finance, December 1984, both at State University of New York, Buffalo. Earned his AS in business administration, Erie County Community College, Buffalo, NY, January 1982.

He joined the service June 7, 1977 and served in USS *Kilauea* (AE-26) and USS *Nashville* (LPD-13). He was a supply officer with NMCB-13; a one man operation at the Navy Transportation Equipment Management Center Detachment at Ramstein AB, Germany; supply officer, Naval Mobile Construction Bn., Peekskill, NY, 1989-93; led special projects during military operations in Guam, Panama, Japan, New York, North Carolina and California, 1992-93.

Received the Navy Commendation, Navy Achievement and Humanitarian Service medals.

Married Cheryl and they have one daughter Tivi. Civilian employment: president and owner of Rental Yard, LTD, Buffalo, NY and real estate broker.

KENNETH A. "MAX" MAXWELL, CDR, born May 12, 1939 in Columbus, OH, received a BS from Tufts University.

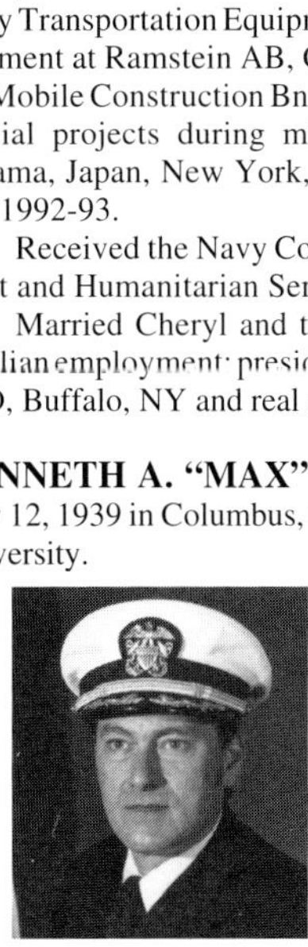

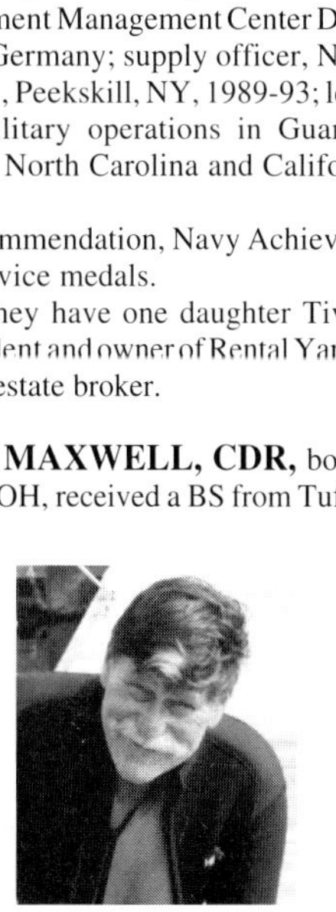

He joined the service September 1953, NROTC Midshipman Tufts University. His military locations and stations: HS-5 Key West and Quonset Point; Ellyson Field, PNS; NAS Keflavik, Iceland. Reserve stations: Floyd Bennett Field and NAS NY; NAS Lakehurst.

Maxwell achieved the rank of CDR and retired July 1979. His memorable experiences: lots of fun and good people.

He married Arlene and they have two children, Allen Maxwell and Wendy Maxwell Koke; and three grandchildren: Kenneth Daniel Maxwell, Rebecca and Benjamin Koke. Employed with American Airlines as pilot; sitting, retired June 1986.

JOHN D. MAY, CDR, born May 5, 1945 and is a fifth generation Virginian. He graduated from Roanoke College in 1968 and enlisted in the Navy as an ET. He received a commission in 1971 and was assigned to the USS *Nashville.* Following a tour as the CO of the Naval Reserve Center, Cape Girardeau, MO, he served department head tours on the USS *Dyess* and the USS *LaMoure City.*

He then received his MS from the Naval Postgraduate School with a follow on tour as the XO of the USS *Nashville.* He was then assigned as the director of Naval Reserve Officer Promotion Branch in Washington, DC (BUPERS). His next assignment was chief of staff for REDCOM SIX. CDR May retired in 1991 and now works as a record review consultant for the NRA.

GEORGE DOUGLAS MAYO, CAPT, a product of the Navy V-7 Program, began his indoctrination at the University of Notre Dame, April 15, 1942, followed by Naval Reserve Midshipman training aboard USS *Prairie State* in New York City.

Commissioned Aug. 5, 1942, he married Mildred Dulcie Harris, and was assigned to the Amphibious Force, US Atlantic Fleet. On Sept. 15, 1943 he placed USS LCI (L) 468 in commission, serving as CO. The ship was assigned to the Pacific Fleet and converted to close-in fire support ship USS LCI (G) 468 before being torpedoed and sunk in the Marianas Operation, June 17, 1944.

Later active duty assignments included CO, USS LCI (L) 476, and assistant professor at the NROTC Unit, Rice University. Inactive duty assignments included CO, Naval Reserve Officers School 6-13, Memphis, TN. Completing the PhD degree after WWII, civilian work included personnel and training research and college teaching.

STEPHEN B. "DOC" MAZER, CAPT, born Jan. 15, 1939 in Bronx, NY. After interning in New Brunswick, Canada, he began his naval career as a LT in the Medical Corps in 1974. Stationed in NAS, Jacksonville as associate with the Minor Acute Illness Clinic. He was in charge of enlisted sick calls and brig medical officer. He served at the dispensery training facility in Athens, GA. He founded his private medical practice in Clearwater, FL in 1977 in which he currently maintains to date. In 1995 he was selected for CAPT and is currently in active USNR and stationed in NRC, Clearwater, St. Pete.

He was awarded the National Defense Service Medal, Armed Forces Reserve Medal w/Bronze Star, Navy and Marine Corps Overseas Service Ribbon, Rifle Sharpshooter, and Pistol Sharpshooter.

His family consists of wife Marilyn and four daughters, Anne, Bari Lynn, Michelle Mari and Jennifer Lee and grandchildren: Ryland and Joshua Israel. Affiliated with the Academy of Military Surgeons, American Medical Association, Florida Medical Association, American Academy of Family Practice. His personal interests include coin and gem collecting, shooting and foreign travel.

RUSSELL A. "RUSS" MAZZEO, born May 16, 1937 in Waterbury, CT, received a BS from Central CT State University, 1960; MS, Syracuse University, 1963.

Enlisted June 1, 1954 in the USNR, Seabee Reserve Div. 3-22, as a CR. His military locations and stations: USNRTC, Bainbridge, MD; CBC, Davisville, RI; CBC, Gulfport, MS; NMCB Det. CTMO, Cuba; EFD, Philadelphia, PA; EFD MED, Naples. Served with: CB Div. 3-22; NMCB-13; 21st RNCR; 7th RNCR; NREFD North; NRFHCBTZ 2D.

He was awarded the Seabee Combat Warfare Officer, Meritorious Unit Commendation Medal, Res. Spec. Commendation Medal, Armed Forces Res. Medal, 3rd Award; Expert Rifle Medal.

Achieved the rank of CDR retired to retired list, July 1, 1994. His memorable experiences: received direct commission to LTJG on Oct. 1, 1968; mobilization officer for 7th RNCR during Mobilization Exercise in which NMCB-12, 13 and 27 were successfully "mobilized" to and from Camp Lejeune, NC during their annual two weeks AT.

Mazzeo has never married. Employed with Uniroyal Chemical Co., Inc. as senior research scientist.

LEE F. MCALLISTER, LT, born April 15, 1958 in Kalamazoo, MI, received a BSBA from the University of Michigan in 1981.

He joined the service, Feb. 11, 1983. His military locations and stations: OCS Newport, RI; USS *Durham* (LKA-114). Served with USNR PHIBGRU III November 1987-May 1989.

McAllister achieved the rank of LT, USNR and discharged July 27, 1987. All of his experiences were memorable.

He is married to Jill and employed as zone manager, Monroe Auto Equipment Co.

ROBERT MCBRIDE, LCDR, born March 9, 1960 in Woonsocket, RI, received a BA in psychology from Providence College 1982; MBA Bryant College 1994.

He joined the service June 1985. His military locations and stations: USS *Paul* (FF-1080) Mayport, FL; NETC Newport, RI, N&MCRC Worcester, MA; N&MCRRC Providence, RI. Served with: USS *Paul* (FF-1080); NETC, SWOSCOLCOM; Reserves: MSCO 201, NCSO 201, FTG 201.

McBride was awarded the Navy Achievement Medal (2x), Expeditionary Medal, National Defense, Coast Guard Special OPS (2x), Sea Deployment (2x), 10 year Reserve Ribbon. He achieved the rank of LCDR and is presently still in the USNR. His memorable experiences: several Mediterannean cruises; seeing parts of Europe, Egypt and Israel.

Mcbride is single and employed at Hoechst Marion Roussel, Inc. as account rep for a pharmaceutical company.

JOHN J. MCCARTHY, LCDR, born May 26, 1937, in Philadelphia. He graduated from Saint Joseph's College (Univ.) in 1959 and received his MBA degree from New Hampshire College in 1982. In between, he spent 20 plus years on active duty, being commissioned an ensign with OCS Class 47 at Newport, RI, April 8, 1960.

Duty assignments included NAS Guantanamo Bay, Cuba (1960-62), USS *Lake Champlain* (CVS-39), 1962-64; VF-101, NAS Key West, 1964-67; commissioning crew of the USS John F. Kennedy (CVA-67), 1968-70; Naval Intelligence Cmd., 1970-73 and Defense Intelligence Agency, 1973-75, both in Washington, DC. He retired from FASOTRAGRULANT DET (NAS) Brunswick, ME, 1975-80, on Aug. 31, 1980. During the course of his active duty, he served as a 1355, 1635 and 1637.

Awards include the Armed Forces Commendation Medal, Navy Expedition, Armed Forces Expedition, National Defense medals. After retirement he worked tax seasons for H&R Block, 1981-84, in Maine and Rhode Island. He accepted an auditor's position with the Rhode Island Department of Business Regulation, Division of Banking.

Married to Judy since 1962, they are the proud parents of Christy, Carolyn McDaniel and Patricia DeFrancesco, all graduates of Rhode Island College.

STEPHEN FRANCIS MCCARTNEY, CAPT, born Nov. 12, 1948 in Santa Monica, CA. He graduated from Reseda High School in 1996 and entered the USAF where he was stationed in Texas. After discharge he served in the California Air National Guard in the 146th Aeromedical Evacuation Sqdn. where he attained the rank of airman 2nd class upon discharge in 1970. He entered college in 1969 and graduated from Loyola University in 1973 with a bachelor's in biology. He entered medical school afterwards and graduated with an MD from the University of Oklahoma in 1978. He then trained in general surgery at Harbor UCLA Medical Center in Torrance, CA.

He volunteered for the USN in 1984 after completing chief resident duty. He was assigned to Camp Pendleton Naval Hospital where he was staff surgeon until 1984 when he transferred to Naval Hospital Orlando in Florida. He performed general surgery there until 1987. During that time he was TAD to Keflavik, Iceland, Rota, Spain and to the carriers USS *Nimitz* (CVN-68) and USS *Constellation.*

He was selected for Naval Postgraduate Fellowship and spent the next year at McGill University in Montreal Canada at the Royal Victoria Hospital as fellow in peripheral vascular surgery. He returned to duty as chief of Vascular Surgery Division at the US Naval Regional Medical Center at Oakland, CA from 1988-90. During this time he became double board certified in general surgery and vascular surgery by the American Board of Surgery. He was inducted into the American College of Surgeons and Society for Clinical Vascular Surgery.

He left active duty after six years and entered private practice in Santa Monica, CA. He was recalled three months later for Operation Desert Storm where he remained until May 1991. At that time he entered the USNR at the rank of CDR which was attained on active duty in 1989. He was assigned as the battalion surgeon for 2nd BN/23rd Marines, 4th Marine Div. (FMF) which he served until 1996.

McCartney was awarded the Navy Commendation Medal for service to the battalion during combined arms exercises at Marine Corps air ground combat training base at 29 Palms, CA. During that time a Marine severed his arm nearly completely and he performed emergency surgery with his assistant battalion surgeon with limb being saved. He was promoted to CAPT in 1995 and then assigned as regimental surgeon to the 23rd Marines, 4th Marine Div. FMF. He was recommended for Meritorious Service Medal at that time. Other notable achievements were attained of Fleet Marine Services designator, Navy Reserve Medal, Naval Humanitarian Medal and Naval Expert Pistol Ribbon.

The McCartney family consists of wife Jane; sons, Jeremy Jon and Stephen Daniel; a grown daughter, Andrea Beth; and a granddaughter, Jennaleigh. CAPT McCartney has been an NRA member since 1994.

JOHN J. "DUTCH" MCCARTNY, CDR, enlisted August 1942 in Philadelphia. He attended Villanova V-12 and Northwestern Univ. Midshipman School and was commissioned ENS, March 17, 1944. Served aboard the USS LCI(L) and (G) 46 as 4th officer, XO and CO for 26 months.

Service includes American-European (Italian Campaign) and picket duty in the Pacific Campaign. He was awarded the Navy Asia Occupational Medal. He was asst. personnel officer, Bainbridge Trng. Station, MD. Released from active duty in July 1946 as LTJG.

Memorable experiences include relating to evacuating refugees from the Anzio Beachhead and transporting and landing French African and Brazilian troops on the Southern France Beachheads.

McCarthy worked for Mobil Oil Corp. for 37 years in various managerial positions. He retired in 1983 as manager of light and heavy products, Valley Forge, PA. During the same period he continued association with the NR (in spite of 10 corporation moves and three being out of state). He retired as CDR with 26 years of service in September 1981.

He served as training officer and CO of surface divisions in Eddystone, Folsom and Pittsburgh, PA areas. Married 50 years to Mary and they have six daughters, 10 grandchildren and two greats. Presently he and Mary run a tree farm (PA Stewardship program). As a world class master runner, he has participated in five world veteran games and five senior Olympics. His events have included zoom up to the marathon as well as the javelin and discus.

JOE C. MCCASLIN, CAPT, born July 31, 1932 in Maiden, NC, received a MS and ME from Purdue University and a BS and ME from North Carolina State University.

He joined the service September 1950 and his military locations and stations: USS *Yellowstone* (AD-27). Served with units: NRMCR Z4 and NCR 3 and Res. Div. NAVFAC.

McCaslin received the Naval Reserve Medal, Good Conduct, National Defense and European Occupation Medal. He achieved the rank of CAPT CEC.

Married to Vel and they have children: Faith McCaslin Shipps and Mitzi McCaslin Aspiri and grandchildren Kramer, Amber, Madison and Alyssa. Employed as engineer and engineer mgt., engr. manager, L&S Cryogenics.

GERALD S. MCCAUGHEY, CDR, born in Montreal, Oct. 4, 1925. Educated post-WWII at McGill University, receiving a BA Honours English, 1951; MA University of Washington, 1958. Taught at universities in Canada and US, mainly at University of Alberta, retired 1989.

His military service: RCAF as pilot trainee, 1943-44; RNVR Fleet Air Arm as pilot, 1944-46; RCAF Auxiliary (#401 Sqdn.) 1948-52; RCNR (VC 922 Sqdn.) 1954-60; CO UNTD, HMCS Malahat (1960-63); NCSO officer, HMCS Nonsuch (1978-79); training commander, NRTC, HMCS Naden (1980-83). Final rank CDR.

He flew 40+ types of aircraft. Memorable experiences: vampire jet came apart after take-off crash landed (St. Hubert, 1952); His #2, TBM (taking off) collided with a Banshee (landing) (HMCS Shearwater, 1958), both pilots died; landing a C-45 with radios out, heavy snow, fire in stbd engine, low ceiling, and visibility (Vancouver, 1959) no problem.

Married to Jennifer Ang, barrister-at-law, June 1989. He has five children and six grandchildren from a previous marriage.

JAMES J. "JIM" MCCLURE JR., LT, born Sept. 23, 1920 in Oak Park, IL, received a BA in 1942 from the University of Chicago and JD in 1949 from the University of Chicago Law School.

He joined the service Aug. 1, 1942. His military locations and stations: Midshipman School, Chicago, Subchaser Training Center, Miami; USS SC 1025, Gunnery School, Washington, DC. Served with USS SC 1025, CO; USS *Weber* (APD-75), exec.

McClure was awarded the Asiatic Theatre Medal. He achieved the rank of LT and was discharged April 6, 1946. His memorable experiences: typhoon of Buckner Bay 1945; first American ships into Nagasaki after war ended; reunions with *Weber* crew.

He has been married to Carolyn Phelps for 47 years and nine months; they have children: Jack, Julie and Don and grandchildren: Andrew, Laura, Jamison, Corey, Christine and Margaret. Employed as partner, Gardner, Carton and Douglas (law firm). He retired Dec. 31, 1991.

DENNIS NEILL "DENNY" MCCOMB, CAPT, born March 11, 1943 in Van Nuys, CA, received a BS from the US Naval Academy June 1965; and MBA from Western New England, February 1973.

He joined the US Naval Academy July 1, 1961. His military locations and stations: Naval Academy Annapolis; Submarine School, New London; two Submarines in Charleston, SC and New London. He served with USS *Odax* (SS-484) and USS *Cobsler* (SS-344) and various reserve units. Participated in Northern European and Carribean campaigns.

McComb achieved the rank of CAPT and was discharged May 1988. His memorable experiences: submarine qualification officer of the deck, duty officer, diving officer, commanding officer, Submarine Reserve Unit; qualified in Submarines in 1967.

He has been married for 31 years to Karen Rayner McComb and they have children: Dennis Jr., Christopher and David. Employed in real estate, manufacturing. Now owns a real estate company, Keller Williams Region Southwest.

RICHARD K. "DICK" MCCORKLE, CAPT, born July 20, 1923 in Portland, OR, received a BS degree from Seattle University and a ME from the University of Washington.

He joined the service Dec. 5, 1942. His military locations and stations: North Island, Whidbey, Washington, DC; Agana Guam; Japan. He served with units: VP-199, VU-78, VU-9, BuAer, VP-891, 892, 893, Line School. Participated in the Korean Theatre.

McCorkle was awarded the WWII Victory Medal, Naval Reserve Medal, Japan Occupation Medal and the American Defense Medal. He achieved the rank of CAPT and was discharged October 1978. His memorable experiences: VP CO, BuAer, Korean Theater, and his years in Reserve.

He is married to Genie Kilcup McCorkle and they have children: Susan Gayle and Richard Alexander McCorkle and grandchildren: Jason Lee Achberger, Kristin and Janelle McCorkle.

Employed with Boeing Service, Airworthiness, Mil. Sales, retiring June 1987.

CHARLES L. "MAC" MCCOY, CDR, born July 27, 1920 in Collingdale, PA, received a BS from Millersville STC, Millersville, PA.

McCoy joined the service March 1942. His military locations and stations: CH-NC, Glenview, Corpus, South Pac V565, V5933, VP936. Served with units, V565, VS933, VP936 (PPC). He achieved the rank of CDR (05) and was discharged July 1969.

He married Vivian Kreider and they have two children, Vivian and Margie; and five grandchildren: Charles, Joseph, Jennifer, Jason and Justin. Employed with Industrial Arts Secondary School in Pennsylvania, retired June 1980.

JOHN H. MCCOY, CAPT, born in San Francisco on Sept. 29, 1945. He grew up in Marin County, just north of San Francisco. He joined the USN as a seaman recruit in 1963 before finishing high school. While on deferment, he graduated from Sir Francis Drake High School and began attending junior college.

In 1964 he began working for the Pacific Telephone and Telegraph Company as an engineering aide. While working as a frameman in San Rafael in 1966, his deferment was canceled. He entered active duty as an enlisted man, attaining the rate of data systems first (DS1) while making two tours to Vietnam (1966-1968) aboard the USS *Enterprise* (CVA(N)-65). The most significant event that occurred while he was aboard the *Enterprise* was the North Korean's capture of the USS *Pueblo*. The *Enterprise* was the first ship on the scene. DS2 McCoy had no way of knowing at the time that the mission of the *Pueblo* and his future as an officer with the USN would be so intertwined.

Following active duty, he returned to work for PT&T, transferring to San Luis Obispo in 1969. He attended California State Polytechnic College while working part-time. It was in San Luis Obispo, in 1969, that he met and married the former Judy Burns of Tehachapi, CA.

In 1971, CAPT McCoy graduated with a degree in Computer Science/Statistics. He then applied and was accepted into the Direct Commission Program. He was commissioned ENS, USNR, Cryptology, January 1972.

The then ENS McCoy joined NRSG 12-1 at Treasure Island in 1973. From 1974 through 1985 he drilled at the San Bruno, CA, Surface Reserve Center, where he completed one XO and four CO tours. He then returned to the Treasure Island unit to head up project Classic Buoyant for the next five years. This was followed by three years in the Treasure Island SECGRU VTU. In 1993 he left TI for another CO tour, this time in Salt Lake City. While in Utah, he was promoted to CAPT, USNR, in June 1995.

CAPT McCoy accepted the position of Deputy Chief of Staff for Security on the staff of the Director, Naval Reserve Security Gp., November 1995. In October 1996, he accepted the position of DCOS for Operations. He is currently very active in integrating USNR Cryptologic resources into the DODs Total Force Policy.

McCoy retired from Pacific Bell after 31 years of service in 1996. The McCoy family resides in San Rafael, CA. His son, Scott, is living in Santa Rosa, CA, and son, Sean, lives at home while attending college. The family enjoys tinkering with PCs, and discovering their Scottish/Irish heritage.

RAYMOND MCCOY JR., LT, born July 17, 1924 in Breckenridge, TX. Attended the University of Minnesota. Served April 1943-August 1948. Organized Reserve October 1948-April 1951. Active duty April 1951-July 1953.

His military locations and stations: Naval Air Facility April 1946-August 1948. At NAF Majuro in Athens, he was field officer and operations and communication officer. Served with VF-14, VF-813 (Minneapolis), VF-921, VF-84.

His campaigns include the American Defense, Asiatic-Pacific, Pacific and WWII Victory. McCoy was awarded the American Defense Medal, Asiatic-Pacific Medal and WWII Victory Medal and Air Medal. He flew over 30 military and civilian aircraft. He achieved the rank of LT and was separated July 1953.

Divorced he has four children: Cynthia, Charles, Mary and Molly. Employed as stockbroker for 37 years, (Paine Webber, 21 years).

ROBERT U. MCCURLEY JR., born May 11, 1929 in Jasper, Jasper Co., TX, received a BS from Sam Houston State University.

He joined the service Feb. 21, 1951. His military locations and stations: Pacific and Far East. Served with USS *Fletcher* (DDE-445). His battles and campaigns: Korean and Winter 1952, Korean Summer, Fall 1953.

He was awarded the China Service, American Defense, Korean Service, UN Medal, Korean Presidential Unit Citation. McCurley achieved the rank of CDR. His memorable experiences: CO NR Fleet Div. 8-2d, CO NR Surface Unit 8-80(L), CO NR Research Co. 8-12, Chief of Staff Naval Reserve Gp. CDR 8-21(L).

He married Dolores G. and they have two children, Lori-Ann McCurley Burns and Laura Susan McCurley Anderson and grandchildren: Forrest Nicole Anderson, Algeria Lee Anderson, J. Robert McCurley Burns, John Rocco McCurley Burns. Employed as Texas Instrument, member group technical staff.

RICHARD L. MCFARLAND, CAPT, born July 9, 1929. Grew up on a farm near Fredericksburg, IA. Eldest son of Donald (deceased) and Dorothy McFarland.

Attended elementary and secondary schools, Fredericksburg, IA; graduated 1947 (salutatorian). First year of college at University of Northern Iowa (ISTC); transferred to University of Iowa; received BA degree in Political Science, February 1951.

Attended University of Pennsylvania (Wharton School) following discharge from USN; received MBA in January 1957. Completed hours for PhD (economics) at University of Pittsburgh, transferring by ALOCA before dissertation; degree not finished. Evening school instructor University of Pittsburgh and University of Minnesota; courses in transportation.

Enlisted in the USN January 1951. Served in Western Pacific (Korea/Japan). Promoted to GM3, selected of OCS, Newport, RI. Commissioned an ENS July 1952. Assigned to various 2nd MARDIV battalions, in and out of Camp Lejeune, NC for next two years. Deployed to 6th Fleet for two Mediterranean cruises plus numerous missions in Caribbean. Then assigned to amphibious forces; USS *Rushmore* (LSD-14) and USS *W.B. Cobb* (APD-106) during final year of active duty. Released from active duty July 1955, having completed 4 1/2 years service. Affiliated with USNR units in Pittsburgh, St. Paul and St. Louis. Held various billets; most recently CO of NAVBASEGUAM and Military Sealift Command/Japan reserve units.

Retired July 1985 at the rank of CAPT with 34 1/2 years. He was awarded the Navy Occupation Medal, American Defense, Korean w/2 stars, Armed Forces Reserve w/2 HGs, South Korean Ribbon and UN Ribbon.

Employment: 1947-50, (summers), bank teller, 1st State Bank, Fredericksburg; 1957-60 Alcoa, Pittsburgh & Sidney OH in traffic department; 1960-63, 3M St. Paul, MN, distribution analyst; 1963-90, Monsanto Chemical Co., St. Louis, MO, Logistics and production management in agriculture, industrial chemical and plasticizer divisions of Monsanto. Took early retirement in April 1990.

He married Ramona J. Staack of Sumner, IA in August 1955. They have three sons and seven grandchildren, all of whom continue to live in the St. Louis area.

His community and professional associations: YMCA Indian Guides program and both Cub and Boy Scouts over the years when sons were active; Cub Master for four years; United Way; Red Cross blood drives;

Junior Achievement advisor; NRA; Navy League; certified member of the American Society of Transportation and Logistics; member of Interstate Commerce Commission as practitioner.

They are members of Webster Groves Methodist Church. Currently serving on stewardship committee and ushering coordination.

His hobbies include 35mm photography; classic automobile restoration, primarily 1955 Chevys (owns six of them). He is currently restoring his 1942 Willys military jeep.

GREGORY J. MCGIFFNEY, LCDR, born in Camden, NJ, June 11, 1963. After attending East Brunswick High School in New Jersey, he received an appointment to the US Naval Academy in 1981. Upon graduation from the Naval Academy in May 1985, he received his commission as an ENS, USN directly from the hand of President Ronald Reagan.

After graduation, and initial warfare training, and also a brief tour as a Naval Science instructor at the University of Pennsylvania, LCDR McGiffney served in USS *Berkeley* (DDG-15) in a variety of division officer billets within both engineering and weapons departments. It is here that he received both his surface warfare officer designation and engineering officer of the watch (1200 psig steam) qualification. After serving on major Western Pacific and Indian Ocean fleet deployments, LCDR McGiffney resigned his regular Navy commission in June 1990.

In July of 1990, LCDR McGiffney affiliated with the USNR. He served in the NRF fleet in a variety of positions until June 1995. His initial assignment was an engineer officer of the USS *Lockwood* (FF-1064) SELRES Pre-Crew. He then served in USS *Duncan* (FFG-10) as auxiliaries officer and ASW officer. He next served in USS *Copeland* (FFG-25) as operations officer.

In July of 1995 LCDR McGiffney was promoted to his present rank. He is currently serving as assistant training officer, unit training coordinator, and USMC liaison officer for NR COMNAVFORKOREA DET 4. He also concurrently serves as a US Naval Academy "Blue and Gold" officer.

LCDR McGiffney's awards include the Navy Achievement Medal (two awards), Armed Forces Expeditionary Medal, and Battle "E." He was also selected as the 11th Naval District NAVSURFRESFOR Junior Officer of the Year in 1993. In 1996 LCDR McGiffney was elected as a NRA national vice-president, as well as being the Bakersfield Chapter President since 1995.

In his civilian career, LCDR McGiffney is an account specialist with BetzDearborn, Inc., where he has been employed since leaving active duty in 1990. He is also a member in good standing of many other professional organizations, and is active in many of those having local chapters within his community.

LCDR McGiffney resides with his family in Bakersfield, CA. He is married to the former Holly Logan of Kalispell, MT. They have two children, a son Kyle (4) and a baby daughter, Kelly. They are active members of All Saints Episcopal Church in Bakersfield.

NEIL F. MCGINN, CDR, born Aug. 19, 1959 in Worcester, MA, received a BA in economics from Holy Cross College.

He joined the service May 27, 1981. His military locations and stations: NAS Pensacola; NTTC Corry Field; NAS Whidbey Island; Camp H.M. Smith, HI. Served with units: VT-10, VT-86, VAQ-130, USCINCPAC.

McGinn was awarded the Defense Meritorious Service Medal. He achieved the rank of CDR.

His memorable experiences: 2 WestPac deployments aboard USS *Kitty Hawk* (CV-63); naval flight officer EA-6B Prowler. Received wings Sept. 3, 1982.

He is married to Wendy and they have a daughter, Erica.

Employed as mission planner at USACOM.

ROBERT S. MCGINNIS, JR., CAPT, born in Owensboro, KY, March 3, 1934, McGinnis graduated from his hometown Kentucky Wesleyan College (1955) and received master degree from Southern Seminary (1958), University of the South (1965), Tulane University (1971), University of New Orleans (1980) as well as a doctorate earned at Vanderbilt University (1971).

After serving as an Episcopal priest he entered the Liberal Catholic Church in 1973, and received the episcopacy in 1979. At that time he became Bishop for the Armed Forces for the American Province of this church and remains so today.

Chaplain McGinnis, commissioned LT in 1969, served continually as a USNR chaplain until his retirement in 1994. First assigned to NRSU 837, New Orleans, he then became battalion chaplain, 4th MPBn, ForTrps, FMF, USMCR and later 3/23, 4THMARDIV, FMF, USMCR. He was also CO, 4THMAFREL (510), and chaplain, NR MEDCLIN (110). Moreover, he was district chaplain, Eighth Coast Guard District and had three tours as staff chaplain NRREDCOMREGTEN, which was his last command.

The chaplain served aboard USS *Putnam* (DD-737) and USS *Lexington* (AVT-16) as well as duty at NAS Signoella, (Sicily) Italy; USCGSUPCEN, Kodiak, AK; USCG Island, Alameda, CA and Governor's Island, NY; NAVSTA, Norfolk, VA and San Diego, CA; NAS, Alameda, CA; Miramar, CA; Norfolk, VA; Memphis, TN; Pensacola, FL and Whidbey Island, WA; USNAVSUB Base, King's Bay, GA; NAVHOS, Charleston, SC and Oakland, CA; NTC, San Diego, CA and Great Lakes, IL; NETC, Newport, RI; COMNAVRESFOR, New Orleans, LA; Navy Chief of Chaplain's Office and HQ, Commandant, US Coast Guard, Washington, DC; USMCAS Cherry Point, NC and Camp Pendleton, CA; USMC Base, Camp Lejeune, NC and Camp Pendleton, CA; and USMCMWTC, Bridgeport, CA.

He was awarded the Meritorious Service Medal, Navy Achievement Medal, Coast Guard Achievement, National Defense Medal, Humanitarian Service Medal and Armed Forces Reserve Medal w/HG and other decorations.

Bishop McGinnis currently heads the Philosophy and Religion Area at Dillard University, New Orleans, LA as well as being chairperson of the Division of the Humanities. He is married to the former Laura Mendoza of New Orleans. They have eight children and 14 grandchildren.

E. RICHARD MCGONAGLE, CDR, born Sept. 26, 1920 in Royalton, MN, received a BBA from the University of Minnesota.

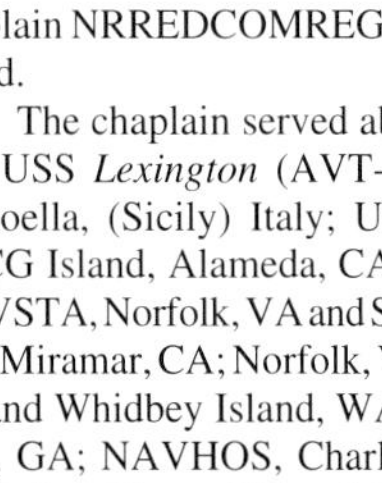

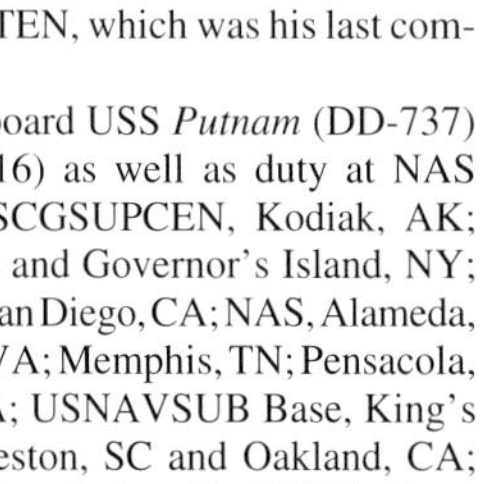

Enlisted as seaman second class, March 12, 1942. Stationed NAS Minneapolis. Aviation cadet August 1942, Corpus Christi. Commissioned ENS and designated naval aviator, Feb. 5, 1943. Miami (F2A); Jacksonville (F4F). LSO School Jacksonville June 4, 1943; NAS Norfolk and USS *Charger* (CVE-30, October 1943; USS *Croatan* (CVE-25) from Jan. 7, 1944 to Sept. 20, 1945. Duties: LSO; V-1 Div.; flight deck officer.

Released from active duty Jan. 8, 1946. Organized Reserve: CO VA-814. Promoted LTJG March 1944; LT August 1945; LCDR February 1952; CDR May 1957. Retired December 1964 with 20 years satisfactory service.

Joined the NRA, November, 1955, life member. Employed 1946-64, Northwest Airlines; flight dispatcher, Minneapolis, MN; Anchorage and Shemya, AK. Obtained CPA license and formed own firm in Sauk Rapids, MN. Retired July 1984. Married to Mildred, they have no children.

DOONAN DWIGHT MCGRAW, CAPT, born in Andalusia, AL, May 30, 1925. Attended Auburn, then Georgia Tech, receiving a bachelor of mechanical engineering degree from Tech in 1948; JD degree with honors, from George Washington University Law School in 1956.

Enlisted 1943, commissioned June 1945. Mine Warfare School, then swept their own magnetic mines on the YMS-337 in the Inland Sea of Japan. RAD June 1946. Recalled in 1951. On USS *Instill* (AM-252) and USS *Herald* (AM-101). RAD January 1953.

Participated in Surface Reserve Programs in Indianapolis, Washington, Detroit, and Dayton. Commands: Fleet Div., Large Surface Div., Group Command, and a Military Sealift Command Unit in Detroit. AcDuTras were on DD/DE and MSC, several officer selection boards. Retired July 1977. A life member of the NRA and the Reserve Officers Association.

Campaign and service medals: American, Asiatic-Pacific w/Battle Star, WWII Victory, Occupation Service, National Defense, Naval Reserve and Armed Forces Reserve.

He worked for GM for many years, first as an engineer, then as a patent attorney becoming a recognized expert in vehicle brake patents, retired to Naples, FL in 1988, and continues occasional patent prosecution. He and his wife, Phyllis, have three children: Robbie, Nancy Katharyn and Bruce; and eight grandchildren: Allison, Christopher, Katie, Sarah, Becky, Pat, Sean and Bridget. His father and two sons served in the USN.

GEORGE SCOTT MCGUIRK, LCDR, born Dec. 2, 1950 in Bronxville, NY, received a BA from Holy Cross and MBA from New York University, JD from St. John's University.

He joined the service Nov. 16, 1985 (Direct Commission, Intelligence) 1635). His military locations and stations: Floyd Bennett Field, Brooklyn; NASJRB Willow Grove, PA. Served with Naval Operational Intelligence Center 0402; Atlantic Intelligence Command 0393.

McGuirk was awarded the Meritorious Unit (2), Naval Reserve, National Service, Expert Pistol. He achieved the rank of LCDR. His memorable experiences: N-2 on USS *Shreveport* for exercise Agile Provider 1994.

He is married to Ellen and they have children: Jack, Rosemary and William. Employed with Smith-Cairno Ford, Inc. as general manager.

THOMAS J. MCGURR, CWO4, received a BA in math from Fairleigh Dickinson University, Teaneck, NJ and a MBA FDU. He joined the service Sept. 20, 1948. Achieved the rank of CWO4 and discharged March 14, 1991.

He is married to Claire and is employed as systems analyst/self-employed electrical contractor.

JON WILLIAM "TUCKER" MCHUGH, CDR, born July 29, 1944 in St. Paul, MN, received a BA from Oklahoma State University and a MBA from the University of Oklahoma.

He joined the service Nov. 27, 1967. His military locations and stations: Pearl Harbor Naval Base, NAVFAC Pt. Sur. Served with FTG Gitmo, MSCFE, Flt Hosp 21. He participated in campaigns in the Vietnam era.

McHugh was awarded the NCM, National Defense (2), Naval Reserve, Meritorious Unit (2), Expert Rifle and Pistol. He achieved the rank of CDR and was discharged July 21, 1971.

His memorable experiences: P-3 flight from NAS Adak to track Russian subs in Petropavlovsk OP area. Flight shadowed by Russian migs.

He is married to Susan and they have children: Kelly, Courtney, Carrie and Matthew. He retired Dec. 4, 1994.

MALCOLM A. "MAC" MCINNIS, CAPT, born June 25, 1929 in Seattle, WA, received a BA degree from the University of Washington.

He joined the service Aug. 22, 1950. His military locations and stations: NAS North Island, NAS Seattle, NAS Alameda, NAS Pensacola, and NAS Sangley Point. Served with VP-40 during the Korean War and Vietnam War.

McInnis was awarded the Air Medal (3), China Service, Vietnam (2) and Korean (2). He achieved the rank of CAPT and retired Aug. 14, 1978. His memorable experiences: greatest experience and education a person could ever have.

He is married to Jane and they have children: Malcolm Jr., Mike, Mitzi and Marshal; and grandchildren: Kelley, Tyson, Megan, Scott, Tucker, Stuart, Flint and Madeline. Employed as owner, General Contractor Corp., retiring from contracting in 1996.

MONI MCINTYRE, CDR, born in Detroit, MI and holds five degrees including a PhD in theology (Toronto).

In 1979 she enlisted in the advanced pay grade program. YN2 McIntyre served with the Personnel Mobilization Team at the reserve center in Southfield, MI. In 1981 after her direct commission, ENS McIntyre joined Service Gp. Two at Southfield.

In 1985, McIntyre joined the Naval Weapons Station Earle unit in Buffalo and became their gaining command liaison officer until 1990. Assigned to the Surface Group Mediterranean unit in Pittsburgh (1990-93), she later became the CO of the Mobile Mine Assembly Group (1993-95) there. She is currently assigned to the Military Sealift Command North Persian Gulf unit in Pittsburgh.

CDR McIntyre is assistant professor of theology at Duquesne University in Pittsburgh where she teaches ethics. She is a member of the Sisters, Servants of the Immaculate Heart of Mary, Monroe, MI.

ROBERT TODD "BOB" MCKINLAY JR., CAPT, born Dec. 11, 1938 and received his BA degree in psychology, Yale University, 1960; MD in medicine, Univ. of Pennsylvania School of Medicine, May 1964; Naval Aerospace Medical Institute, Pensacola, FL, naval flight surgeon, 1966; Naval Hospital, Bethesada, MD, ophthalmology residency, 1969-72.

Military locations and stations include NH Pensacola, NAMI, CAW-9 (VF-96), CVA(N)65, MCRD, San Diego, Station Hospital, NAVSUPPACT, Da Nang, RVN, NH Bethseda, NH Boston. He participated in five Vietnam campaigns.

His awards and medals include the National Defense Medal (2nd Awd.), Vietnam Service Medals (5 campaigns), Navy and Marine Corps Overseas Service Ribbon, Armed Forces Reserve Medal, RVN Meritorious Unit Citation (Galantry Cross Medal Color w/Palm), RVN Civil Actions Unit Citation w/Palm, RVN Campaign Medal, Navy Expert Rifle and Pistol Medals. He achieved the rank of CAPT and was discharged in June 1974.

Memorable experiences: CAT shot off CVA(N)65 in an F4 Phantom, 1966; A-3 stick time off coast of North Vietnam; drafting SOP for Ophthalmology Dept. USNS Mer. TAH-19; Research/Recommendations to SG on Reserve Flag Medical Mobilization Billets; meeting ADM Jeremy Boorda when CH NAUPERS; writing chapter for eye injuries in 1975.

Married Helen Elizabeth Ann Griffin and they have five children: Kathleen Miller, Lisa Salyers, Allyson Mitzel, Elizabeth Smith and Stephen; and two grandchildren, Jennifer and Jay Thomas Miller.

President and CEO, Comprehensive Eye Care of Central Ohio, president and CEO, Ohio Eye Care, LTD and clinical associate professor of ophthalmology, Ohio State University. He plans to retire in 1998. Member of Society of Military Ophthalmologists, Assoc. of Military Surgeons of the US and a life member of Navy League of Columbus and the NRA.

DAVID STUART MCLEAN, CAPT, born July 20, 1930 in Norwood, MA, received an AB degree from Tufts University and an MBA degree from Northwestern University.

He joined the service May 25, 1953 in Boston, MA. His military locations and stations: Newport and Providence, RI; Philadelphia; NORVA; Orange, TX; Little Creek; Portland, OR; Chicago; San Francisco Bay area. Served with various ships and stations in above locations (active and reserve).

McLean received service awards and performance medals, no decorations. Achieved the rank of CAPT 06. His memorable experiences: commissioning from OCS; surface warfare designation; all duty as a CO; promotion to CAPT and administering the oath of a naval officer this son.

He married Brenda Williams McLean and they have children: LCDR David S. McLean, Jr. (aviator); Heather McLean Keller and grandchildren: Jonathan Hastings McLean and Abigail Williams Keller. Employed in packaging industry, sales and sales management. Retired from the Navy June 30, 1984 and from civilian employment Oct. 6, 1994.

WALTON J. "WALT" MCLEOD III, CAPT, born June 30, 1937 in Walterboro, SC, received a BA degree from Yale University in 1959 and a LL.B degree from the University of South Carolina in 1964.

He was commissioned June 6, 1959. His military locations and stations: Yale NROTC; Newport, RI; San Diego; Washington, DC; Naples, Italy. Served active duty with units: USS *Ticonderoga* (CVA-14), 1959-61; BuPers, (1963); Naval Support Activity, Naples, 1990. Participated in Desert Shield, 1990.

He was awarded the Meritorious Service Medal, Navy Commendation, National Defense Service, Navy Marine Corps Overseas Ribbon, Armed Forces Reserve, Rifle Ribbon, Expert Pistol Medal, 1994 Hugh Howell Award, East Coast as outstanding senior Naval Reserve JAG Officer. McLeod achieved the rank of CAPT, JAG, USNR, and retired from USNR Sept. 1, 1992.

His memorable experiences: Two WestPac cruises on board USS *Ticonderoga* 1960-61; AT, Naval War College, 1988; CO, Oklahoma City NAMARA Defense Unit, 1989-91; 32 days active duty at NSA Naples during Operation Desert Shield, 1990; National President, Judge Advocates Association, 1991-92; State President ROA, 1981-82; President, SC Naval Reserve Lawyers's Association, 1993-present.

He is married to Julie Hamiter McLeod and they have a son Walton James McLeod. Employed as attorney, State Representative, 1996-present.

THOMAS J. "TOM" MCMAHON, CAPT, born June 17, 1929 in Syracuse, NY, received an AB from Holy Cross College in 1951; JD from Georgetown University Law Center, 1957.

He joined the service Sept. 4, 1951. His military locations and stations: OCS Newport. Served with USS PC-572, F.P.O. NY (later named USS *Tooele*.

McMahon was awarded the National Defense Medal and Armed Forces Reserve Medal w/2 repeats. Achieved the rank of LTJG on active duty, CAPT in USNR. Retired July 1, 1988. Released to inactive duty on Dec. 18, 1954.

He is married to Alcida Rose (Levesque) McMahon and they have children: Sharon (Sawyer), Alcida (Litchfield), Michelle (Sawan), Thomas McMahon and grandchildren: Kathryn, Aaron, Abigail, Daniel (Sawyer) and Joseph (Sawan).

Employed as attorney (law firms and Major Corps). Professor Suffolk University Law School in Boston full time for 22 years at Suffolk University Law School. Now professor emeritus. Retired (professor) Aug. 1, 1996.

ARTHUR T. "ART" MCMANUS, CAPT, born Dec. 21, 1937 in Queens, NY, received a BS from Holy Cross College (1960); MBA Wharton Graduate School, University of Pennsylvania (1968).

He joined the service Aug. 8, 1960. His military locations and stations: August 1960, USS *Fiske* (DDR-842), NS Mayport, FL; September 1960, NAS Pensacola, FL, student naval aviator; August 1961, NAS New Iberia, LA, student naval aviator; April 1962, VS-37, NAS North Island, CA and USS *Hornet* (CUS-12), (S2 Aircraft); July 1965, NROTC Unit, University of Pennsylvania, Philadelphia, PA, assistant professor of Naval Science; December 1968, VP-64, NAS Willow Grove, PA (P2 & P3 aircraft); VP-1123, NAS Willow Grove, PA, CO (P3 aircraft), October 1976; ASO-104, Aviation Supply Office, Philadelphia, PA, XO, December 1978; June 1982, Inactive Reserve.

He was awarded the National Defense Service Medal, Armed Forces Expeditionary Medal, Armed Forces Reserve Medal.

McManus achieved the rank of CAPT and was discharged from active duty June 28, 1968. His memorable experiences: NROTC Program, Holy Cross College. Designated naval aviator March 1962.

He is single and has children: Arthur T. III; LT David E., USN. Employed as president, PRM Corporation and retired Jan. 3, 1993.

RONALD D. "MAC" MCMASTERS, LCDR, born March 5, 1924 in Upland, CA, received an AB and MS degree Fresno State University.

He joined the service September 1943. His military locations and stations: Pensacola, FL; Quonset Point, RI; North Island, El Centro. Served with VA7A, instructor Pensacola, VA923, FAGU, Reserve Squad.

Participated in Korean campaign. Awarded the Air Medal (2) and Distinguished Flying Cross. He achieved the rank of LCDR and retired in 1965. His memorable experiences: primary instructor. Carrier Squadrons VA7A and VA923. Instructor (VA) Fleet Air Gunnery Unit (FAGU).

He is married to Jessie and they have two children, Mike and Barbara; grandchildren: Kelley, Jennifer, Carrie, Sarah; and great-grandson, Kevin. Employed at Bakersfield College as dean of students. Retired in 1984.

JAMES P. "JIM" MCMENAMIN, CDR, born May 21, 1922 in Philadelphia, PA, received a BS from LaSalle University, Philadelphia, PA.

He joined the service Aug. 26, 1941. His military locations and stations: V-12, Villanova University, Pennsylvania; Midshipman Harvard University Massachusetts. Commissioned ENS May 30, 1944. He served with VR-4 Oakland, CA; Aviation Supply Office, Philadelphia, PA; NAS Willow Grove, PA; NAS Lakehurst, NJ; USS *Shelikoff* (AVP-52).

McMenamin achieved the rank of CDR. His memorable experiences: taking the USS *Shelikoff* (AVP-52) out of moth balls at the Philadelphia Naval Base.

He married Miriam Jane and they have a daughter, Patricia Jean Joseph; and three grandchildren: Stephanie Joseph and David Joseph. Employed as customer service, Union Carbide; banking officer, Pennsylvania Lottery, retiring July 1, 1982.

EARL D. MCMICHAEL JR., CAPT, born March 14, 1925 in Youngstown, OH. Joined the service July 1943. His military locations and stations: NTC, Great Lakes, HCS Great Lakes, NAVHOSP Sampson, NAVHOSP Brooklyn, Fox V29 LST 284, Pre-Invasion Training England, Normandy (Overlord) June 6, 1944, Resupply and Hosp. Shp Runs-Omaha/South Hampton, Transit to AATB Bizerte.

Rammed en route west of Gibraltar; repair/rehab/training, Naples, Sardinia, Southern France invasion at St. Raphael D-Day August 1944, NAVHOSP/Disp Oran, Stateside July 1945; Deploy through Shoemaker and TI to Pearl, Independent duty PHM3 12 LST 868; Philippines, Marianas, Japan, Hondis, PTS - March 1946. In active reserve until April 1947. NAS COLS 1947-50. Commissioned ENS (direct procurement) Feb. 2, 1950.

Served with AMPHIB WWII, Aviation Units in Reserve NAS/COLS, Grosse Ile, NAF Detroit. Participated in battles and campaigns: Normandy, Southern France, Independent duty PHM3/2c LST 868 in Western Pacific, ASW Aug GRP 3Y2 as CO at ComIce ASW, Keflavik, Iceland; COMFAIRKEF.

McMichael was awarded the Navy Unit Commendation, Atlantic, Mediterannean, Asiatic-Pacific, China Service, WWII Victory, NUC and NRA Medals. He achieved the rank of CAPT and was discharged March 1946 and April 1985.

Married Lois Rawlins McMichael and has three children: Earl D III, Karen Marie and Kevin Charles; and five grandchildren: Heather Ann, Earl D. IV, Christopher John, Daniel James and Benjamin Earl. Employed as teacher (English, Spanish, Journalism), secondary administrator, Franklin County Public Schools, Avon Lake, OH. Retired from the military in 1985 and education 1981/85.

J. RICHARD "RICK" MCMICHAEL, CDR, born March 9, 1943, Berkeley, CA, received a BA in 1964 and JD in 1969 from the University of California, Berkeley.

He joined the service June 1962. Enlisted reservist until commissioned in 1964. Served on USS *Maddox* and USS *Hamner*. His battles and campaigns: Vietnam.

Awarded the Vietnam Service Medal, Armed Forces Service Medal. He achieved the rank of CDR and discharged from active duty 1966. His memorable experiences: service in USS *Hamner* (DD-718). First ship to fire on both North Vietnam and South Vietnam. Retired from Reserves in 1982.

He is married to Karen McMichael and has children Richard and Eric. Employed as vice-president Weyerhaeuser Real Estate Company.

CHARLES M. MCMILLAN, LCDR, born Jan. 16, 1932, in New York City. He graduated from Stuyvesant High School in 1950, Marietta College in 1954, OCS in Newport, commissioned in March 1955.

Attended EOD School, Indian Head, MD; USS

Bunting (MHC-45), Key West, 1956-57; EODTC Indian Head, MD, 1957-58; UC Berkeley Law School, JD degree, 1961. For 18 years he was attached to reserve units at Treasure Island (MSTS, naval control of shipping and political affairs companies.

Memorable events: 18 years of views going to Treasure Island and disposal of 5,000 lb. warhead.

Married to Evelyn with children: David, Cindy, Jay, Denise and April. Grandchildren are Andrew, Emily, Megan and Jonathan. He was managing partner in a law firm until retiring. Retired from USN in 1978 and from his law practice in 1991. Two weeks later the house burned down in the Oakland Hills Fire and all records and photos were destroyed

HARLEY I. MCPEEK JR., CDR born on Feb. 3, 1943 in Parkersburg, WV. He attended West Virginia University before entering the Navy as a NAVCAD in June 1964. CDR McPeek earned his naval aviator wings and commission on Feb. 3, 1966, at NAS Corpus Christi.

Assigned to VS-28 and VS-22 at NAS Quonset Point, CDR McPeek flew S-2 aircraft from USS *Wasp* and USS *Essex*. Reserve assignments included flying the S-2 (VS-70Z1) at NAS South Weymouth, the SH-3 (HS-74) at NAS Quonset Point and NAS South Weymouth and the P-3 (VP-5056) at NAS New Orleans. Other billets included XIO NAS 3682, XO FAIRMED 1082 and member of VTU-8282 at NAS New Orleans.

McPeek obtained degrees from the University of Rhode Island (BA 1/72) and the University of South Alabama (MBA 6/80). He is a senior manufacturing engineer, with Lockheed Martin, manufacturing the Space Shuttle External Fuel Tank.

He married Sandra Joan Collins and they have one daughter, Jennifer Ann Mutz, and grandson, Timothy Lee Mutz II. They reside in Slidell, AL.

WAYNE H. "MAC" MCQUERRY, CDR, a native of Missouri, was born in St. Joseph on Sept. 28, 1922 and raised on a small farm near Blythedale. He graduated from the local country and high schools and, in 1949, following four years in the Navy, from Northwest Missouri State University, Maryville, MO, with majors in physics and mathematics.

Navy career began in 1943 as a Tarmac at NAS, Hutchison, KS. He became a Naval Aviation Cadet and was designated naval aviator and commissioned ENS, USNR, in December 1945, at NAS, Corpus Christi, TX. Subsequent duty assignments included VB-82 (flew SB2C Helldivers off the USS *Randolph*), and VRF-1, NAS, Norfolk, VA. He was released from active duty in November 1947.

He was recalled to active duty in November 1951. He was assigned to VF-884 and flew 47 combat missions in F4U Corsairs off the USS *Kearsarge* operating in Korean waters. Following release from active duty in December, 1953, Mac served in numerous USNR billets, flying and non-flying, at NAS, Los Alamitos, CA, and NAS, Grosse Ile, MI. He was CO of WEPTUs 772 and 776 prior to retiring in May 1970.

Operational aircraft flew included the SBD, SB2C, F6F, TBM, F4U, F8F, AD, and F9F. Awards include the Air Medal w/Bronze Star, Naval Reserve Medal, American Theater and WWII Victory Medals, China Service Medal, Korean Service (two stars) and UN Medals.

Civilian occupation was design and development engineer, aerospace radio communications and radar equipment. Most of his career was with North American Aviation, Inc., later Rockwell International, on the Apollo Lunar Landing and Space Shuttle Programs. He was a member of the NAA/Rockwell Support Teams at Johnson Space Center, Houston, TX, for all Apollo Lunar Missions and the Appollo-Soyuz Mission.

He married Ruth Collins, of Westboro, MO in December 1945. They had four children: LTCOL Thomas Owen McQuerry, USMC, (deceased), Patricia Ann, Samuel Allen and Bruce Dean. They have seven grandchildren and four great-grandchild. Wayne and Ruth reside in Orange, CA.

DONALD JAMES "MICK" MCRAE, LCDR, born May 5, 1926 in Kewanee, IL, received a BS from Northwestern University 1948; JD Northwestern University School of Law 1951.

He joined the service July 1, 1944. His military locations and stations: NW Mo. State College; Olathe NAS; Iowa City Preflight; Ottumwa pre-flight; and Corpus Christi NAS; naval aviation cadet; various reserve unit Chicago, IL; Galesburg, IL; Moline, IL and Davenport, IA.

He was awarded the WWII Victory Medal and the American Theater. He achieved the rank LCDR and was discharged May 8, 1946.

McCrae is single. Self-employed attorney retired, May 5, 1986.

ROBERT HENRY "MAC" MCSWEENY, CAPT, born Jan. 5, 1933 in Detroit, MI, received a BA from the University of Michigan in 1955 and a JD degree in 1966.

His military service: Midshipman, Regular NROTC program, University of Michigan, 1952-55; USS *Chilton* (APA-38), Boat Gp. CDR, CIC officer, Combat Air Controller, operations officer (ENS/LTJG), 1955-57; USS *Parle* (DE-708), gunnery officer, EX (LTJG), 1957-59; Aide and flag LT to chief, Navy Gp., JUSMMAT, Ankara Turkey (LT), 1959-62; USS *Tombigee* (AOG-11), CO (LT/LCDR) (Resigned to attend law school), 1962-64; Commissioned LTJG, Judge Advocate General Corps, USNR-R, 1968; Served in various posts including command of Portland JAG reserve units; ACDUTRA in JAG offices in Seattle, San Diego, JAG HQs, Washington, DC, served on selection boards for LCDR, CDR and CAPT; selected for 0-6 1978, 1968-86; Staff Judge Advocate, REDCOM 22, Seattle, 1983-85; Retired, CAPT, JAGC, USNR-R, 1986.

He was awarded the Antarctica Service Medal and Reserve Service Ribbon.

His memorable experiences: commissioning; command at sea; sailing to Antarctica, with visits to New Zealand and Australia; flying home from Havana on Eastern to beat the stork for their first son, Rob; three years in Turkey with the family (their youngest son born there); Service on several CDR and CAPT selection boards; Retirement ceremony at Portland Naval Reserve Center.

Married Dec. 21, 1955 on a nine day leave from USS *Chilton* (APA-38) to Jean Alexander of Shaker Heights, OH. Two sons: Robert H., Jr. who is a nurse living and working in Seattle, and Michael, an executive chef who lives in Orlando, FL. Two grandsons: Alexander and Rowan of Seattle, WA.

Civilian employment: 25 years as a practicing attorney in Oregon; retired in 1991; currently serving as an arbitrator for the courts of three counties; started Nonprofit Resource Group in 1993 to help nonprofit organizations get started and now doing extensive volunteer work as a nonprofit board member, fund raiser and coach.

Served as president of Portland, OR chapter of Navy League of the United States; member, Board of Trustees, Episcopal Diocese of Oregon; member of Willamette Valley Development Officers, Oregon State Bar and NRA.

FRANCIS BAIN MCVEY, CDR, born on Feb. 19, 1945. He grew up in Tucson, AZ and joined the USNR in 1964 while still in high school. He graduated with honors from the University of Arizona in 1968 and entered the Naval Flight Office Program, receiving his wings in 1969.

CDR McVey was assigned to VP-50 at NAS Moffett Field, CA, completing two overseas deployments between

1970 and 1972. The first deployment was split between Sangley Point, Republic of the Philippines and Cam Rahn Bay, Vietnam while the second was to Adak, AK, Misawa, Japan and Agana, Guam. In addition to being a tactical coordinator/mission commander in the P-3 Orion, CDR McVey was the squadron ASW/Air Intelligence officer.

Upon release from active duty, CDR McVey affiliated with VP-91 at Moffett continuing as a flight crew mission commander. He held various positions in the training, scheduling, and operations departments during his time with VP-91.

In 1984, having completed the flying portion of his career with over 3,000 hours, CDR McVey joined ASWOC 1080 and quickly earned his watch officer designation. Between 1984 and 1990 he held department head positions in training, mobilization, readiness and operations. Each year he completed active duty for training at Misawa, Japan or Kadena, Okinawa. From 1990-92 CDR McVey served as XO of Reserve Patrol Wing Pacific 0180.

He has been awarded the following medals and decorations: Air Medal, Navy Commendation, Navy Meritorious Unit Commendation, Coast Guard Meritorious Unit Commendation, Navy "E" Ribbon, Navy Reserve Meritorious Service, National Defense Medal, Armed Forces Expeditionary Medal, Vietnam Service, Navy Overseas Service, Navy Reserve Overseas Service, Armed Forces Reserve Medal, Vietnam Gallantry Cross, Vietnam Civil Action, Vietnam Campaign and Navy Marksman.

He served as vice-president of the Polaris Chapter of the NRA from 1988 to 1990. He has continued his education in the graduate level Urban Planning Program at San Jose State University and, more recently, completed a Masters of Arts in Organizational Management from the University of Phoenix.

He is employed by the City of San Jose, CA and has held positions as city planner, development officer, housing administrator, assistant director of Neighborhood Preservation and currently deputy director of the Department of General Services.

KEITH D. "MEAKS" MEAKINS, LCDR, born Oct. 23, 1931 in Sterling, IL, received a BS in accounting from Colorado University, 1959.

Enlisted in the service June 2, 1952. Received direct commission ENS, Supply Corps, Feb. 9, 1960. Served as enlisted machine accountant, COMSERV PAC. Served with PAMI and 4th ND PAMI. He was awarded the Good Conduct, National Service, Naval Reserve w/HG. Reserve assignments: ASB MED HQ 218; DSA Depot 118; NR ship supply officer Div. 9-2. Discharged as LCDR, July 1, 1980.

He is married to Doris Hendrix Meakins and they have a son Bruce. Employed 35 years with Public Service Co. of Colorado, as data processing operations supervisor, retiring April 1, 1994.

JOHN JAMES "JOHN" MEILERT, CAPT, born March 12, 1932 in Wichita, KS, received a BS degree in business administration, accounting major, ECON minor, University of Wichita (now Wichita State University).

His military locations and stations: USNR Div. 9-64 Wichita, KS, October 1949-April 1950, enlisted as seaman; May 1950-May 1951, SA, attended ROC School at Treasure Island 1951 (San Francisco) Terminal Island, Long Beach - five-six months supply school; SN, May 1951-September 1954; NSCS, Athens, GA, October 1954-April 1955; USS *Philippine Sea,* aviation stores officer, May 1955-May 1956; US NAS Twin Cities, assistant supply officer included serving as part-time supply officer, June 1956-November 1958; MOB Team, Wichita, KS, supply officer, December 1958; USNR Surf. Div. 9-64, assistant training officer and administrative officer, April 1965; US NAS, Grosse Ile, MI, supply/comptroller officer, May 1965-June 1968; NAF, Washington, DC, supply officer, July 1968-June 1971; NAVSUPSYSCOM, Washington, DC director, Reserve Div. (SUP OP2), June 1971-June 1975; DSA IG Field Office, Northeast, Mechanicsburg, PA, Chief, DSA IG FO, July 1975-June 1977; CNAVRES, New Orleans, LA, deputy chief of staff for supply, July 1977-August 1980. Retired from the Navy and went into the restaurant business, September 1980-present.

He was awarded the Legion of Merit, Meritorious Service Medal, Armed Forces Reserve Medal (3), National Defense Service Medal. He achieved the rank of CAPT and discharged Aug. 31, 1980.

His memorable experiences: while serving as NAVSUP-0P2. He was fortunate to be heavily involved in the first meaningful effort to restructure the Reserve Program, particularly for the Supply Corps, in 25 years many of the programs they developed are alive and well and have been expanded. Such as Res. Cargo Handling Bns.

He married Margaret J. Copernoll and they have three children: Cary Meilert, Stephanie Holder and Suzanne Lubin; and three grandchildren: Danielle Lubin, Shayla Holder and Blake Holder. After retirement he was owner and operator of five Popeye chicken and biscuits restaurants in Northern Virginia.

LUCILLE W. (MEEKS) MELLISH, CAPT, joined the USN Nurse Corps as a LTJG in April 1957 in the direct commissioning program at the age of 36. Born Oct. 8, 1920 in Gardner, MA, she graduated from Brattleboro (VT) High School in 1938. In August 1945 she graduated from the Memorial Hospital School of Nursing, Wilmington, DE, where she had been a student nurse since August 1942. While a student, she had in October 1943 also become a member of the Cadet Nurse Corps.

Her nursing career included service as an operating room nurse, in Veterans Administration Hospitals in Albuquerque, NM and Palo Alto, CA from August 1951 until she retired from nursing on Aug. 15, 1977.

She started her naval career on June 7, 1957 at St. Albans Naval Hospital, Long Island, by attending a six week orientation course before reporting for duty at Oaknoll Naval Hospital in Oakland, CA, on July 25, 1957. Following a tour of duty at the station hospital, US Naval Ammunition Depot in Hawthorne, NV, from Jan. 10, to Sept. 2, 1958, she returned to Oaknoll, completed her tour, and was released from active duty in Oakland on June 5, 1959. Lucille remained active in the Ready Reserve until retirement on July 1, 1979 as a CAPT with over 22 years service.

In August 1965 she joined the NRA, became a national officer and served on the Budget and Finance Committee from 1966 to 1995. She was instrumental in the decision to buy and finance the building that is now the national headquarters in Alexandria, VA. Lucille married CDR William Mellish, Dec. 23, 1967. They reside in Palo Alto, CA.

WILLIAM J. MELLISH, CDR, born Feb. 26, 1922 west of the Allegheny River in Bradys Bend Township, Armstrong County, PA, and graduated from East Brady High School in June 1940. On Sept. 16, 1942 while attending Grove City (PA) College he enlisted in the USNR as an apprentice seaman. In June 1943 the Navy placed him in the V-12 program at Brown University where he continued until November 1944 when he was transferred to the 22nd class of Midshipmen at Columbia University. Upon commissioning as a line officer (1105) ENS, March 8, 1945 he was assigned to the heavy cruiser USS *Columbus* (CA-74) in the engineering department and immediately sailed to the Pacific Theater. Following duty in Japan, after the surrender, Columbus spent the winter of 1945 in Tsingtao and the spring of 1946 in Shanghai, China. Upon release from active duty in September 1946 he returned to Brown University and graduated in June 1947 with a BS degree in engineering. As a Ready Reservist he was recalled for Korean duty aboard USS *Missouri* (BB-63). When not in Korea or Japan during this two-year tour of sea duty the battleship visited England, Norway, France, Cuba, Haiti and Hawaii.

In 1955 he changed his designator from line to Civil Engineer Corps (5105) to match his USNR service more appropriately with his civilian occupation as a highway construction engineer with the State of California. As a Selected Reservist from July 1964 to July 1967 he was CO of Reserve Mobile Construction Bn. 28. Personnel for this battalion came from the northern halves of California and Nevada plus Utah. When a shortage of officers developed in July 1969 he volunteered to return for a third two-year tour of active duty as head of the Seabee Reserve Program for the 12th Naval District located at Treasure Island, San Francisco, CA.

His campaign and service medals are: China Service, American Campaign Service, Asiatic-Pacific Campaign, WWII Victory, American Occupation Service, National Defense Service, Korean Service w/3 Bronze Stars, Armed Forces Reserve, Naval Reserve, Korean Presidential Unit Citation, UN Service and Navy Rifle Marksmanship.

He retired from the USN in July 1971 after 29 years of honorable service. He retired from state highway construction in September 1987 after 40 years. As a life member of the NRA, he has given many hours of service promoting the association's programs. He is also a life member of the Naval Order of the US, the Association of Naval Aviation and the Navy League.

He will always be outranked by his wife, CAPT Lucille Mellish, NC, USNR (Ret). The family consists of his son Markham and his wife, Roslyn, and their children, Alan and Alison.

BURTON G. "STORMY" MENDELSON, LT, born May 5, 1925 in Chicago, IL, received a BS in engineering from Cal Tech in 1945 and MBA in marketing from the University of Chicago in 1954.

He joined the service in 1942. His military locations and stations: Pensacola NAS; Port Lyautey French Morocco; Pax River (Korean War).

Mendelson was awarded the American Theater, WWII Victory, Occupation, National Service, Korean Service and Naval Reserve. He achieved the rank of LT.

He married Grace and they have three children: Cyrile Ann, Keith Joel and Alan Mitchell; and grandchildren: Michele, David, Benjamin, Daniel, Hannah, Lydia, Joey and Emily. Employed as VP Motorola Inst. and Control; vice-president Western Union Corp., president E&M Group Inc. He retired in July 1986.

CURTIS H.P. MENEFEE, CAPT, born March 2, 1938 in Denver, CO, received BA from the University of Colorado 1960; JD University of Denver, 1965.

Enlisted in the USNR June 11, 1955. Commissioned July 1960, NROTC University of Colorado.

His military locations and stations: USS *Hancock* (CVA-19); V-6 Div. Asst. AIO. Served with FASRON 715; MOBTM 904; MARDIV K-2; CIVLAW SUPPACT 118 (CO); SJA RED COM 18; UTU Law 1804. His battles and campaigns LAOS/VIETNAM 1961-62.

He achieved the rank of CAPT, judge advocate general's comp. and retired Dec. 1, 1994.

Menefee is married to Kathryn A. House and has children: Tim EC and C. Christopher P.

Employed with the Department of the Interior, as assistant regional solicitor Rocky Mountain Region.

MARTIN MENEZ, CAPT, attended elementary and high school primarily in Chicago, IL, except for short periods in the 1960s when he lived with his family in Lima, Peru; and Rome, Italy.

Following graduation from Marquette University in 1972 with a BS degree in theoretical mathematics, he was commissioned and served for over seven years on active duty as a surface warfare office. He held nine leadership positions in the operations, engineering, and weapons department of USS *Bausell* (DD-845), USS *Ouellet* (FF-1077), and three shore stations. His varied tours were all outside the continental US, based in Yokosuka, Japan; London, England and Pearl Harbor, HI.

CAPT Menez left active service in 1979 to attend Northwestern University's Kellogg Graduate School of Management full-time as an Exxon fellow, earning his MM (MBA) degree. In 1981 he joined AT&T Long Lines where he had assignments in operations, engineering, and marketing. He also managed Strategic Air Command long distance communications out of Offutt AFB, NE.

In 1986 CAPT Menez established AT&T Communications España in Madrid, Spain and served as its first managing director. He directed all of AT&T's long distance business between Spain and America. In 1991 CAPT Menez repatriated to a position as international programs director with AT&T Federal Systems International in Greensboro, NC, where his groups managed complex multimillion dollar telecommunications projects for allied foreign governments. Over two thirds of his personnel were assigned in the Middle East. In 1994 CAPT Menez again went on foreign assignment, this time to Mexico City as project management director for AT&T de México, where he formed an organization dedicated to post-sale implementation of AT&T's major equipment contracts in Mexico. CAPT Menez is currently program director with AT&T Communications (UK) Ltd in London, leading its services integration with the pan-European telecommunications joint venture, Uniworld.

CAPT Menez has had four reserve commands. CO tours were of ship augmentation units (LPH-9, DDG-41), a NAVACTUK unit in London, England, and an MSC unit. His AT's have included courses at the Naval War and Armed Forces Staff Colleges, membership on promotion and CO/GCLO selection boards, and as liaison officer to the Mexican and Spanish forces. His most recent prior tour was as mobilization officer for Readiness Command 13 in Charleston, SC.

CAPT Menez was recalled to active duty in February 1991 at CINCUSNAVEUR, London, in direct support of Operation Desert Storm. His decorations include the Navy Commendation Medal, Combat Action Ribbon, Meritorious Unit Commendation, Armed Forces Expeditionary Medal (two), Vietnam Service Medal, Humanitarian Service Medal, Sea Service Deployment Ribbon, Overseas Service Ribbon (three), and Armed Forces Reserve Medal (with mobilization "M"). He is an active member of the US Naval Institute, Navy League (1995 men's council president in Mexico City), and the following associations: Surface Navy, Naval Reserve and Reserve Officers.

PETER F. "PETE" MERENDA, CAPT, born July 18, 1922 in Everett, MA, received a BS and MA from Tufts University; CAS from Harvard; PhD from Wisconsin.

He joined NROTC-Tufts in 1941. Commissioned ENS 1944. His military locations and stations: Amphib Tra. Base 1944-45; Atl. Flt. Tra. Com 1945-46. Served with NRSD 5-12; NRSD 9-210, NRRC 9-7; NRGC 1-2; NRRC 1-2, VTU 1-2, GC 1-2. He was awarded the WWII Medal, NR, ND, AT.

Merenda achieved the rank of CAPT. His memorable experiences: one of the Navy's first line psychologist 1665. In 1974, his 30th year of commissioned service, since he was the sole 0-6 in the designator and therefore could not be passed over for 0-7, he was not attrited and remained on active status in the Naval Reserve for six more years.

He has been married 50 years to Rose and they have children: Anne, Rosemary and Pamela; and grandchildren: J. Peter, Lauren, Kathryn and Pamela. Employed as professor emeritus of psychology and retired April 1980.

EDWIN W. MERGELE III, CDR, born in San Antonio, TX, Jan. 10, 1952, he graduated from the University of Texas in 1975. After commissioning he reported to USS *Albany* (CG-10), in Norfolk and later forward-deployed to Italy.

Transferring to the USNR in 1979, he served in several units in San Antonio and Houston, TX. In 1984 he earned his MBA at UTSA. In 1985 he worked for COMNAVRESFOR in New Orleans. In 1991 he mobilized with Mobile Inshore Undersea Warfare Unit 108 for Operation Desert Storm, serving in Saudi Arabia and Kuwait. From 1992 to 1994 he was CO of MIUWU 108. He is currently assigned to COMNAVRESREDCOM REG 11.

Mergele is a surface warfare officer and a life member of the NRA, Space Center Chapter, Houston and the Naval Institute. He is a senior architect for the city of Houston.

HARDY L. "HARDY" MERRITT, CAPT, born Aug. 15, 1943 in Poplar Bluff, MO, received a BA from Southeast Missouri State University; MPA and PhD from the University of South Carolina.

He joined the service Dec. 18, 1965. His military locations and stations: NAS Jacksonville, NAS Glenview, NAS Moffett, Pentagon. Units served with: VP-16, VP-45, PAT WING 11, VP-90, REDCOM 7, OP-095, RESPATWING PAC.

Merritt was awarded the LOM (2), MSM, DMSM, NCM, NAM, Joint Service Achievement. He achieved the rank of CAPT, USNR and was discharged Oct. 31, 1992. His memorable experiences: as a wing commander (COMRESPATWINGPAC) watching his squadron personnel taking off for Desert Storm; and then welcoming them all back.

He is married to Billie and has stepsons, Glenn and Jim Harris; son, Owen Bray Merritt; and granddaughter, Amanda Harris. He is employed as management analyst for the State of South Carolina.

SIDNEY M. "SID" METCALF, LTSG, born July 4, 1912 in Vineta, OK, received a BS in business from the University of Oklahoma.

He joined the service March 11, 1942. His military locations and stations: recruiting stations: Ft. Smith, AR; Los Angeles, CA; Argus 28, Lion 6, NOB Guam.

Served with 4th Echelon during the invasion of Guam. He was awarded the Asiatic-Pacific Campaign, American Campaign, WWII Medal, Commendation Is. Comdr. Metcalf achieved the rank of LT(sg) and discharged February 1946. His memorable experiences: sub alert and air alert on way to Guam. Air alerts while on Guam.

His wife died in 1991. He has a daughter, Karen Ann Kleager; and two grandchildren, Jon Kleager, Gering, NE and Sarah Dunkum, Richmond, VA. He operated and owned theatres in Texas, Oklahoma, Nebraska and Arkansas. Worked for retail Cr. Co. and Lone Star Ord. Co. as welfare director.

PHILIP J. "PHIL" METRES JR., CAPT, born Nov. 11, 1942 in Brooklyn, NY, received a BA from Holy Cross College; MA, social psychology and PhD, professional psychology.

He joined the service June 10, 1964. His military locations and stations: USS *Purdy* (DD-734); USS *Gainard* (DD-706); Naval Advisory Group; Vietnam; Naval Amphibious School, Coronado (USN); USNR-RESCREW, USS *Bauer* (DE-0125) and USS *Shields* (DD-596); DCNO, OP-06 113.

His battles and campaigns: TET Offensive, Republic of Vietnam 1968; three Vietnam Campaigns.

He was awarded the Bronze Star Medal w/Combat V, Meritorious Service Medal, Combat Action Ribbon, Vietnam Staff Service honor Medal, 1st Class. He achieved the rank of CAPT and was discharged Oct. 1, 1992.

His memorable experiences: CO VTU 1314 G and DCNO OP-06 113 at NRC G.L.; tremendous camaraderie with officers and men of Naval Advisory Group, Vietnam, especially during TET Offensive 68.

He married Katherine "Kay" and they have three children: Philip J. III, Katherine M. and David M. Employed as clinical psychologist, North Shore Center for Counseling and Therapy, Northbrook, IL.

DOROTHY M. DABBERT-MEYER, LTJG, born July 14, 1932 in Geneva, IL, received a diploma, St. Joseph's Mercy School of Nursing, Aurora, IL.

She joined the service May 27, 1955. Military locations and stations: USNH, St. Albans, NY; USNH, Oakland, CA.

She achieved the rank of LTJG, NC, USNR and was discharged May 31, 1957. Her memorable experiences: working in orthopedics with Korean vets at USNH Oakland.

She is married to CAPT Jack Meyer, USNR, (Ret). Employed as RN, Meadowbrook Care Center, retiring June 1, 1993.

JACK A. MEYER, CAPT, born April 14, 1932 in Sacramento, CA, received a BSIE from the University of California, Berkeley, CA, June 1954.

He joined active duty June 25, 1954. His military locations and stations: USS *Onslow* (AYP-48); USS *Firedrake* (AE-14); USNH Oakland, CA. Served with numerous USNR units in California, Illinois and Ohio.

He was awarded the Defense Meritorious Service Medal w/ OLC in lieu of second award. Meyer achieved the rank of CAPT, Supply Corps. Retired from active duty June 25, 1957. Retired from USNR July 1, 1987.

His memorable experiences: The outstanding people with whom he served throughout his career; aboard *Onslow* in Kaomsiung, Formosa (1955) when Chinese ammo barge exploded nearby; on *Firedrake* in WestPac during November 1956 Hi Alert due to that Suez Crisis.

He married to Dorothy Dabbert and they have children: Patrice, Scott and David; and grandchildren: Nicole, Matt and Ryan Martinez; and Rachel, Chloe and Tess Meyer. Employed with Proctor & Gamble, numerous plant and corporate positions. His final position, business analysis manager, World Wide Product Supply Finance. He retired Sept. 1, 1993 after 36+ years.

JOHN F. MICHAELS JR., CAPT, born in Boston, Nov. 29, 1941, graduated from Boston College and St. Louis University School of Medicine. After internship and psychiatry residency, he entered the USN through the Berry Plan in July 1972, serving as head, Psychiatric Branch, NAMI, PNCLA FLA until July 1974.

Returning to civilian life, he affiliated as a SELRES in March 1975 at N&MCRTC STL. In April 1980 he relocated to Johnstown, PA, drilling with N&MCRESCEN. He was selected for CAPT in 1985. From 1987-94 he was senior medical officer at NAVRESCEN Macon, GA. He was twice selected to Staff Officer Selection Boards, and completed the C-4 Course at Ft. Sam Houston in 1990.

He has been a NRA member since 1976 and AMSUS since 1983. He retired from the USNR in May 1994. Certified by the American Board of Psychiatry, he is medical director, Satilla Park Hospital, Waycross, GA.

STEPHEN CARL "STEVE" MIDDENDORFF, CAPT, born July 25, 1943 in Sacramento, CA, received a BS in Science from Oregon State University; BS (pharmacy) OSU; Navy OCS, Damage Control School, Petroleum School, convoy commodore, NCSO, OCA.

He joined the service Jan. 30, 1964. His military locations and stations: USS *Cacapon* (AO-52); Joint Task Force Middle East (Bahrain); MSC Far East; MSC Atlantic. Served with USS *Cacapon* (AO-52), NRSD 13-19(s); AO-921 SRU; AO-922 SRU, Convoy LOM Subic; Manila

122; RU 2204; NCSO 222, NCSO 122.

He participated in battles and campaigns: Vietnam (five campaigns); Operation Earnest Will; Kuwaiti Tanker Escort; Operation Sea Signal; Haitian Immigrants.

Middendorff was awarded two Meritorious Service Medals, Navy Commendation Medal, Navy Achievement Medal and 10 others. He achieved the rank of CAPT and is still presently active in the Reserve.

His memorable experiences: acted as senior naval liaison officer aboard Kuwaiti reflagged tankers; CO of convoycom Subic/Manila 122; CO of Readiness Unit 2204; CO of NCSO Portland; CO of NCSO Northwest Pacific; Active: CO of Military Sealift Command Office (MSCO) Kingston, Jamaica; CO of MSCO Guantanamo Bay; Commander Task Unit 48.1.8.

He married Carol and they have five children: Michael, Heidi, John, Janie and Marshall; and grandchildren, Kay-Lee and Michael. Employed as pharmacist. President of Aslin Enterprises (construction/retail).

JOEL A. MIELE, CAPT, born in Jersey City, NJ, May 28, 1934 and raised in New York City. He graduated (BCE) from the Polytechnic Institute of Brooklyn in 1955. Enlisted in the USNR in 1957 and was commissioned ENS, CEC, through OCS that same year.

Served over three years on active duty and 29 years on inactive duty (retired 1988) with NMCB1, 5ND staff, NMCB13, 4NCR and 7NCR, commanding both NMCB13 and 7NCR. Served as Director of Logistics and Operations, International Naval Review 1986 and Liberty Weekend.

Personal Decorations include the Navy Commendation Medal, Navy Unit Commendation, Expert Rifleman and Expert Pistol Medals (w/Silver "E" in lieu of second and third award). Service awards include the National Defense Service Medal w/Bronze Star and the Armed Forces Reserve Medal w/2 HGs.

In civilian life, CAPT Miele was a consulting professional engineer in private practice, and upon retirement therefrom, served New York City as commissioner of city planning (1990-94); Commissioner, Department of Buildings (1994-96); and presently as commissioner of the Department of Environmental Protection.

He is past president of the New York State Society of Professional Engineers and the Queens Borough Public Library. He presently serves as chairman of the boards of Peninsula Hospital Center, Peninsula General Nursing Home and Creedmoor State Hospital and as vice-president and executive board member of the American Parkinson Disease Association.

Married in 1959 to the former Josephine Ann Cottone and they have a son, Joel Arthur Jr., and two daughters, Vita Marie and Janet Ann. Joel Jr. is married to the former Eileen Daab and they have three children: Joel Arthur III, John Christopher and James Nicholas. Vita Marie is married to Gary Chu and has a daughter, Caroline Ann. Janet Ann is married to William Powers.

PAUL P. MIESZCZANSKI, CDR, born June 25, 1954 in Bridgeport, CT, received MS, in mechanical engineering from Naval Post Graduate School, BS in physics from Fairfield University.

He entered OCS, Feb. 19, 1977. His military locations and stations: OCS Newport, RI; SWOS, San Diego; NAS Alameda, Bremerton, NSY. Active duty: USS *Coral Sea* (CV-43), SUPSHIP Brooklyn, Philadelphia Naval Shipyard. Participated in Iranian Hostage Crisis Expedition.

Decorations include the Sea Service Ribbon, Armed Forces Medal, Navy Achievement Medal, Expeditionary Medal, Meritorious Unit Citation, Reserve Medal. He achieved the rank of CDR, USNR and discharged from active duty, Jan. 1, 1988 still a drilling reservist.

His memorable experiences: Western Pacific deployment; operational overhaul duties aboard (CV-43), ship superintendent for CG-19, project officer for FF-1079, AOR-4 and LPD-15.

He married Maria Trinidad Mieszczanski and is employed in Naval Sea Systems Command, project engineer for foreign military sales and CV/CVN Systems Engineer.

JACK H. "JACK" MIHELICH, CAPT, born May 6, 1933 in Paw Paw, MI, received a BS degree, Western Michigan University, aviation management.

He joined the service Nov. 29, 1955. His military locations and stations: NAS Brunswick, NAS Rota Spain. Served VP-10, VT-3, VR-52, VP-734, VA-434, VR-1413, NAAS Whiting.

His medals accident free operation 1958-59, Atlantic Fleet Battle Excellence, 1958-59. He achieved the rank of CAPT (0-6) and retired July 1, 1979.

His memorable experiences: charter member Association of Naval Aviation, member of Naval Reserve Association, member of the Retired Officers Association, member of Reserve Officers Association, member of the Disabled American Veterans Association.

Married Esther De La Torre and has two children, John P. Mihelich and Timothy J. Mihelich. Achieved the rank of CAPT with United Airlines, Inc., retiring July 1, 1979.

FRED R. MIKRUT, CDR, born Chicago, IL, Aug. 20, 1941. Graduated from Wisconsin State University-River Falls, May 1963. Entered preflight at NAS Pensacola, FL, September 1963. Commissioned an ENS, January 1964. Received wings and was transferred in July of 1964 to VR-21 at Barbers Point, HI, as navigator on C-118 and C-130 aircraft.

After release from active duty in December of 1967, he joined a C-54 Sqdn. located in Minneapolis, MN, January 1968. After serving there a number of years, he was transferred to VR-51 based in Glenview, IL. Joined Retired Reserves in March of 1984 after serving the USN over 20 years.

On active duty with VR-21 he helped support the personnel in Vietnam by navigating aircraft carrying people and cargo to that destination. He traveled to Wake, Guam, the Philippines, Okinawa, Japan, Hong Kong, Bangkok, and various NAS on the west coast of the US in the Reserves he spent most of his 16 day ACDUTRAS in support of the forces in EASTPAC while based in Rota, Spain. During active and reserve time, he accumulated over 5,100 hours of flight time.

The Mikrut family consists of wife Judy, daughter Monique and sons, James and Brian.

ERNEST P. "E.P." MILES JR., LCDR, born March 3, 1919 in Birmingham, AL, received a BS (Birmingham Southern), MA, PhD (Duke University).

He joined the service July 15, 1941. His military locations and stations: NAS Jacksonville, FL; NSCS Boston, MA (Harvard); NS New Orleans, LA; NASTC Corpus Christi, TX; ATD Port Hueneme, CA; ACOM 18, ACOM South Pacific, ARSD-60; ASA-145; USN Hospital, Bethesda, MD.

Miles achieved the rank of LCDR and was discharged Nov. 1, 1946. He received a Pacific Theatre Ribbon, American Theatre Ribbon, WWII Victory Medal.

Married Audrey Vicknair and they have two children, Austin Miles and Lauren Allison Miles. Employed as mathematics professor and computer scientist, Auburn University and Florida State University, retiring Nov. 1, 1946.

PAUL A. MILES, CDR, born in San Bernardino, CA, Aug. 24, 1937. Enlisted in the USN, as a "kiddy cruiser" (under 18 years old), in September 1954. Accomplished recruit training (boot camp) in Company 290 San Diego Naval Training Center graduating as an airman apprentice (E2) in November 1954.

After airman preparatory training in Norman, OK, was assigned to the USS *Coral Sea* (CVA-43) (a straight deck then) homeported in Norfolk, VA, in March 1955 and departed for a Mediterranean deployment that same month. He was assigned as a "plane pusher" (blue shirt) aviation boatswain's mate working between the catapults in "crew one."

He was reassigned in September 1955 to the USS *Forrestal* (CVA-59) and became a "plank owner" when the ship was commissioned in October 1955. Reassigned in November 1956 to the Naval Air Facility, Naval Ordnance Test Station, Inyokern, CA (later named Naval Weapons Center, China Lake) in crash and salvage/fuel farm/visiting aircraft line/Security. He was released from active duty in July 1958 and discharged from the USNR in February 1962.

After serving a machinist apprenticeship at the Naval Weapons Center he rejoined the USNR as a steelworker 2nd in January 1965 assigned to Reserve Naval Mobile Construction Bn. (RNMCB) 17 with Permanent Drill Site at Port Hueneme, CA.

He was advanced to machinery repairman chief (E7) in August 1970. After graduating from Cal Poly State University, Pomona, CA, he received a "direct appointment" and was commissioned a LTJG (02) engineering duty officer in June 1973 and supported the USS *Constellation* (CVA-64) in North Island/San Diego.

His designator was changed from 1405 to 5105 (Civil Engineer Corp) in 1974 and was assigned first as officer-in-charge (OIC) of a Construction Bn. Center, Port Hueneme, augment unit and then to RNMCB-17 as HQ Co. CDR and personnel officer and, after promotion to LT, Company Commander of a general construction company. He was also OIC of the China Lake, CA Seabee Detachment.

Transferred to Washington, DC in 1978 and was assigned to RNMCB-23 (Fort Belvoir, VA) as a B Co. CDR and Det. OIC of their Baltimore, MD (Fort McHenry) unit.

After promotion to LCDR (04) 1981, he was assigned to the 8th Reserve Naval Construction Regt. at Fort Mead, MD as their personnel officer and was reassigned to RNMCB-23 as the operations officer in 1982. During this period Bn. 23 completed numerous construction projects, conducted military training and a Green Stinger air detachment deployment and was recognized as the "Best of Type" of all USNR Construction Bn.

Transferred in 1984 to Engineering Field Div. Chesapeake/Atlantic drilling at the Washington, DC Naval Shipyard as personnel officer and facilities support officer (FSO). In 1985 he transferred to Albuquerque, NM and was reassigned to the 9th RNCR NAS Dallas, TX, as assistant IG and in 1986, transferred to officer in charge of construction (OICC) Guam, also drilling in Dallas, as Lead FSO for Reserve facilities in Amarillo, San Antonio, Laredo, El Paso, Midland, Lubbock, Abilene, TX, Albuquerque, NM, Oklahoma City and Broken Arrow, OK. During this period he supported the Evaluation of Mobilization Requirements for naval activities in Guam, MI and Yokosuka, Japan. After promotion to CDR (05) in 1986, he was assigned to OICC Mid-Pacific as a senior FSO and became a Seabee

Combat Warfare Officer (SCWO), after that warfare designation was authorized, in January 1993.

In January 1994, after 38 years of service (16 years enlisted and 22 years commissioned), he was transferred to the Retired Reserve. He is a life member of the NRA (#3770).

Awards receive include the Meritorious Unit Commendation, Naval "E" Ribbon, Navy Occupation (European), Good Conduct, Naval Reserve Meritorious Service, National Defense, Armed Forces Reserve w/Bronze HG, Navy Expert Rifleman w/Sharpshooter E, and the Navy Expert Pistol Shot Medal.

ALLEN L. MILLER, LCDR, born Oct. 13, 1932 Greeley, CO, received a BS in Civil Engineering from the University of Colorado, Boulder, CO.

He joined the service March 3, 1956; Buckley Field NAS Denver, CO. His military locations and stations: NAS Pensacola, FL and vicinity. Served with AOC Class 13-56 Naval Air Training Command, Pensacola.

Achieved the rank of LCDR in USNR and was discharged July 13, 1959. His memorable experiences: Reserve service with research company 9-21, Denver, CO from February 1962 through April 1978.

He married Dale P. Sanchez in April 1957 and has three children: Janice C., Ann Leigh and Allen L. Jr.; and grandchildren: James, Andrea and Dawn Schmidt; and Holly, Lisa and Andrew Pritchett. Employed at Bureau of Reclamation, Department Interior, GS-13. Retired from USNR in April 1978; Bureau of Reclamation in August 1987.

GEORGE HENRY "SI" MILLER, LCDR, born Jan. 5, 1915 in Manville, Jefferson County, IN. Attended two years Hanover College; BJ School of Journalism, University of Missouri; MSJ School of Journalism, University of Illinois. Graduate work at the University of Florida.

He joined the service Feb. 29, 1944. His military locations and stations: Pacific area, Shoemaker Separation Center, Great Lakes Separation Center, others.

Served with Armed Guard, then PR work and writing "Training of the Atlantic Fleet". His battles and campaigns: Cape Britain, Lingayen Gulf, and Zamboanga, all Pacific invasions.

He was awarded many medals. He achieved the rank of LCDR, USNR, and discharged to Reserves October 1946.

His memorable experiences: gunnery officer, Armed Guard, aboard SS *George Taylor,* delivering troops to three invasions, later in Reserve duty being the Navy PR officer for the opening of the St. Lawrence Seaway where they USN escorted the *Queen* to Chicago, 1959.

He is married to June and they have son Jacob Lee Miller. He teaches journalism at several universities. He retired from the USNR 1972, with rank of LCDR.

JOE MILLER, CAPT, is a Northwestern University/ NROTC graduate of 1970. He began flight training in Pensacola that same year and was designated a naval aviator at NAS Corpus Christi on St. Patty's Day 1972. Active duty assignments included NAF Naha, Okinawa, VT-6, NAS Pensacola, CNET Flag Pilot, and NAF Atsugi, Japan. Miller flew the T-28, C-117, P-2, US-2, C-131, CT-39, and C-12.

Miller is a Cubi Dog Centurion and completed over 100 missions at Mamma Lee's in Taegu, Korea. As a Reservists, he was assigned to COMASWWINGPAC, San Diego; XO of USS *Carl Vinson,* CVN-70 0183 Olathe, KS; and CO of Surface Readinesss Unit, Omaha, NE.

He is a former CAPT for Braniff International and now flys for America West Airlines out of Phoenix. He resides in Kansas City with his wife, Caryn, and daughter, Robyn.

NORMAN L. "NORM" MILLER, CAPT (06), born Jan. 31, 1936 in Detroit, MI, received an AB in 1957 from the University of Michigan and a JD in 1960 from the University of Michigan Law School.

He joined the service October 1973 and was stationed in numerous locations. Served with NAVMAR Trial Judiciary.

His battles and campaigns: Meritorious Unit, Battle E, National Defense, Sea Service Deploy, Armed Forces Reserve, Meritorious Service, Navy Commendation, Outstanding Volunteer Service. He achieved the rank of CAPT (06) and was discharged Jan. 31, 1996.

His memorable experiences: starting International Mil. Education Program for USN-JAG, four years as GCM Military Judge, 22 years of great camaraderie.

Married Carol and they have four children: Norman L., Jr., Kristina, Katherine and Thomas. He is a founding partner, Miller LaSota, PLC, retired Jan. 31, 1996.

ROBERT LEMUEL "LEM" MILLER, CAPT, born Oct. 19, 1936 in Ft. Worth, TX, received a MA in mathematics.

He joined the service April 1959. His military locations and stations: Heavy Attack Sqdn. 6, Whidbey Island/ USS *Ranger.* Served with four Reserve commands-DIA, N/C Security, NIPSSA (2).

He achieved the rank of CAPT, USNR, RAD September 1962.

His memorable experiences: overflight of Chinese Islands, NATO exercise from Keflavic, ASW operations Mediterranean, Intelligence support for operations in WesPac.

Miller is single and has children: Troy Randall Miller and Elaine Ann Miller Ball; and grandchildren: Austin Becket Miller. Employed as senior systems engineer, SAIC, retired Oct. 19, 1966.

RUSSELL B. "RUSS" MILLER SR., LCDR, born April 14, 1912 in Woodland, MI, received a BS from Western Michigan University.

Joined the service Aug. 1, 1943, USNR. Joined the USNR May 1943. Began training at Princeton University with advanced training at Harvard University. First duty station, naval personnel office, Washington, DC as a classification officer writing standards for interviewing officer programs. Notre Dame as assistant interviewing officer, Navy Manpower Survey Board, HQs Third Naval District, interviewing officer for officer billet assignments.

Released to inactive duty February 1946 and assigned to Staff Bn. No. 9-22, Detroit, MI. Continued active duty training in personnel until retirement from Naval Reserve April 1972. Assisted in initial organization of Naval Reserve Brigade, Detroit, MI. Volunteer recruiting officer 9th Naval District, State President for Navy, Reserve Officers Association, State of Michigan, Secretary-Treasurer, ROA, District of Columbia, National Treasurer NRA, vice-president 11th Naval District NRA, Natural Historian.

His wife Hulda died. His children are Russell B. Miller Jr. and Eugene F. Miller; and grandchildren are Steven R., Melissa J. and Brandon Miller Lindsay.

ROBERT SCOTT "BOB" MILNE, LCDR, born Sept. 3, 1917 in Dighton, MA, received a BA from Clark University, Worcester, MA. He retired from the USN, 1966.

He joined the service March 1942. His military locations and stations: Pearl Harbor, Combat Intelligence. Served with FRUPAC (Pacific Fleet Radio Unit), Naval Attache; Athens. He achieved the rank of LCDR and discharged April 1946.

He married Gaby Milne and they had two children, Scott (decease) and Fay Nalani Myers, and grandchildren, Bryce and Kristi. Employed as independent travel writer, editor and publisher of Travelwriter Marketletter.

ANDREW HAMMILL "DREW" MILSTEAD, CDR, born March 28, 1934 in Statesville, NC, received a BS (physics), University of North Carolina 1956; MS (physics), Ohio State University, 1959; MA (astronomy), UCLA, 1964; PhD Engineering, UCLA, 1968.

He joined the service Jan. 9, 1974 in APG program as E-7 (EWC). Direct commission as LCDR (1515) in December 1975.

His military locations and stations: N&MCRC, Encino, CA; Naval Weapons Center, (now NAWC, Weapon Div.), China Lake, CA; PMTC, Point Mugu. Served with NWC 0276.

He was awarded the Expert Pistol, Expert Rifle, Armed Forces Reserve Medal w/HG, National Defense Medal. He achieved the rank of CDR (05) and retired w/pay March 28, 1994.

His memorable experiences: Special active duties for training, 1987 on 7th Fleet Staff embarked aboard USS *Bluebird;* 1989, Taegue and Seoul, Korea, both times supporting simulated war exercises.

Married Suzane "Suzy" Milstead. He is currently employed as senior systems engineer, TRW Defense Systems Div. Retired from position of chief scientist, Hughes Aircraft Co., Space and Communications Gp., December 1984. Retired from USNR, March 28, 1994 on his 60th birthday.

He holds airline transport pilot license (airplane single and multi-engine), private helicopter license, certified flight instructor license (airplane-single multi-engines, instrument; ground instructor certificate, 4400 hours flight time (all civilian).

JAMES RAYMOND "JIM" MIMS JR., CAPT, born Dec. 2, 1911 in Oklahoma City, OK, received a BS C.E. Tristole University.

He joined the USMR, May 1934; USNR May 1943. His military locations and stations: Camp Perry, VA; Davis Hill, RI; Gulf Port, MS; Port Hueneme, Pearl Harbor, Saipan, Okinawa. Served with 134th CB/CBMU 600; 130 CB.

His battles and campaigns: Saipan, Okinawa landing April 1, 1945. He was awarded five or six medals. He achieved the rank of CAPT, CEC, USN, discharged Jan. 26, 1946 but stayed in Reserves until retirement, Dec. 1, 1971. His memorable experiences: Armistice Day 1918; Saipan clean up; FDR, McArthur and Nimitz at Pearl Harbor; landing at Okinawa.

He is widowed and has children: Margaret Hamilton, Ellen Stuart, J.R. Mims III; two stepchildren; six stepgrandchildren; seven grandchildren and four great-grandchildren. Employed as senior vice-president of a construction company, retiring Nov. 1, 1982.

REBECCA "BECKY" (HIGHFIELD) MINTON, CDR, born April 11, 1957 in Dayton, OH, received a BA in English, magna cum laude, University of Florida.

She joined the service June 9, 1979. Served with DPSCPAC Pearl Harbor, NAVCRUITCOM, PSD Miami, MIUW U208; NRMARDEZ SEC 4, NR NAVSTA Panama 108, Harbor Defense Command 207.

She was awarded the NCM, NAM, Overseas Service Ribbon. She achieved the rank of CDR and is presently a drilling reservist. Her memorable experiences: NROTC Scholarship.

She is married to Dr. Kenneth Lee Minton and they have two children: Sara Elizabeth Minton and Brandon Ross Minton.

ROBERT EDWIN "MIKE" MISKOSKY, CDR, born Aug. 26, 1922 in E. Patchogue, NY. Attended Navy Academic Refresher Unit at Cornell College, IA, USN flight training; USN School of Music, Washington, DC, MU3/c.

He joined the service Nov. 18, 1941. His military locations and stations: NAS Quonset Point; Naval Base Norfolk, NAS San Diego area; NAS New York; USS *Mawry;* Washington, DC. Served with VU-4, VU-5, VU-7, VF-833, V-832, VR-833. Participated in WWII and Korea active duty (12 years). Awarded Pre-Pearl Harbor Medal, American Defense, WWII Victory, Korean Victory and Armed Services Reserve. Miskosky achieved the rank of CDR and discharged from active duty March 1, 1953. Retired w/o pay Aug. 267, 1967 and w/pay Aug. 26, 1982.

His memorable experiences: After three years as a Navy musician to become a naval aviator with almost 20 years of active reserve flying; best experience, a tour on the USS *Maury* (AG-16) in the Persian Gulf at Arabia and Kuwait; flying off the Desert (1950) before modernization of countries.

He married Wilhelmina E. and they a daughter, Linda D. Bellofatto, and a grandson, Michael. Employed in the family florist business.

JOHN T. MISUREK JR., CAPT, born May 5, 1946 in Milwaukee, WI, received a BA degree from the College of St. Thomas (now St. Thomas, University of).

He joined the service June 30, 1968 in Great Lakes, IL. His military locations and stations: NAS Moffett Field, NAS Glenview, NAS Pt. Mugu, NAB Coronado. Served with VP-47, VP-90, VP-65, COMNAVSURFPAC DET 219. His battles and campaigns: Vietnam War. His awards and medals include: Vietnam Service Medal, Armed Forces Expeditionary Medal, Korea.

He achieved the rank of CAPT, USNR. His memorable experiences: VP-47 deployment to Adak, AK as first P-3C Sqdn. in Pacific Fleet including Arctic ice patrols PARPROS, Midway Island detachments and participation on crew chosen to fly AEC-configured P-3C in support of Project Cannikin, a nuclear bomb test explosion on Amchitka Island, AK.

He is married to JoAnn and they have two children, Valerie and Brian. Employed with TD Technologies, Inc., Western Regional sales manager. He retired November 1991.

JOHN PATRICK MITCHELL, CAPT, born July 31, 1944, Owensboro, KY. Attended Brescia College, Owensboro for his BS degree in business administration, 1966, and earned 6 units toward his MBA from University of Santa Clark, CA and was commissioned in August 1966.

Units/Stations from 1971-79: USS *Putnam, Inchon, Holder, Fairfax, Ponce, Plymouth, Fidelity*; War College, New Port, RI; COMPHIBGRU 213, Mt. Whitney; COMSURFLANT, Norfolk, VA; 1980-92, N&MCRD, Jacksonville, FL; REDCOM 8, Jacksonville; Naval War College, Newport; Chemical SC, Ft. McClellen, AL; COMSECOND FLT, Mt. Whitney; NAS Pensacola, FL; COMSECOND FLT, Mt. Whitney; CINCLANTFLT DESRON, Norfolk, VA; COMSECOND FLT, Mt. Whitney; NAVOP-07, Pentagon and CINCLANTFLT, Norfolk, VA; and EX USS *Shadwell*, Mobile, AL, 1992.

His decorations include the Navy Achievement Medal w/Combat V, Navy Unit Commendation, RVN Campaign Medal w/Device, Vietnam Service Medal w/4 Bronze Stars, National Defense Service Medal, and Meritorious Unit Citation Gallantry Cross. Achieved the rank of CAPT, July 1, 1989.

Civilian employment in finance/automotive industry as South Regional Sales Manager. He retired Sept. 2, 1992.

PAUL M. MITCHELL, CDR, born in Norfolk, VA, Sept. 9, 1922. Enlisted in the V-5 Aviation Cadet Program in August 1942, commissioned ENS and designated naval aviator in February 1944 at NAS Corpus Christi, TX. Following WWII was a civilian flight instructor and attended William and Mary-VPI, Norfolk division (later renamed Old Dominion University).

In October 1950 was recalled to extended activity duty during the Korean conflict and completed tours as XO USS LST 503; CO USS LST 503; CO US Naval and Marine Corps Reserve Training Center, Owensboro, KY; Staff Amphibious Operational Training Unit, Little Creek, Staff CDR, Amphibious Force Atlantic; XO, USS *Donner* LSD-20; and CO, USNR Training Center, Norfolk, (St. Helena Annex).

Released from active duty in 1963 and completed inactive tours as CO, USNR Surface Div. 5-19 (L), October 1963-June 1964 (Class National Winner, 1965) and CO, ship activation maintenance and Repair Div. 5-1(M) July 1966-October 1967. Returned to active duty October 1967 as CO, USNR Training Center, State Island, NY, completed 20 years active service (28 years Naval Service) and retired from active service on April 1, 1971. Subsequently completed 20 years of federal civilian employment and retired from civil service in September.

He is married to Hazel Robinson-Mitchell and they have children, Michael and Robert; and grandchildren, Damian, Tina, Tabitha and Robert.

WILLIAM FRENCH "BILL" MITCHELL, LCDR, born April 30, 1918 in Melrose, MA, received a BS from Bowdoin College; MA Middlebury College; Diplomé, University of AIX-France Marseille.

He joined the service September 1940. His military locations and stations: Norfolk, Key West; at sea. Served with VP-83; USS SC-676; USS PCS 1492; USS *Acree* (DE-167). His battles and campaigns: Gulf of Mexico; Southwest Pacific; Atlantic-Mediterranean Sea.

Mitchell was awarded all war areas and the Navy Cross. Achieved the rank of LCDR, USNR and was discharged September 1945. His memorable experiences: many; most unbelievable.

He is married to Norma Dean Mitchell and they had a daughter, Elizabeth M. Bell (deceased). Employed as professor of French and the Humanities.

VINCENT F. MIZERAS, CDR, entered the USN in July 1965 after graduating from Catawba College with a BA degree in Business Administration. He attended Officer Candidate School, Newport, RI and was commissioned an ENS. He attended Naval Justice School, Newport, RI and Gunnery School, Dam Neck, VA prior to reporting to the USS *Boxer* (LPH-4).

While on USS *Boxer*, he served as legal officer, X Div. officer, CWO, OOD Underway, OOD in port, registered publications officer, assistant navigator and N Div. officer. During the time he was onboard USS *Boxer,* the ship participated as part of the recovery effort for the Gemini Space Program and a tour of Vietnam. Mizeras departed the USS *Boxer* in November, 1968.

He affiliated with the Naval Reserve Center, Lancaster in 1969 where he remained until 1977. During this period, CDR Mizeras served as training officer of Naval Reserve Surface Div. 4-19; CO of Naval Reserve Officer School 4-14; Administrative Assistant of Individual Reserve Unit 804 and Volunteer Training Unit 0409.

In October 1977, he transferred to the Armed Force Reserve Center, Harrisburg, where he became training officer for Supply Detachment 104. In February 1980, he transferred to Naval Reserve Center, Avoca as a training officer for Antilles Defense Command Detachment 104.

He transferred to Naval-Marine Corps Reserve Center, Reading in February 1981 as assistant training officer of Volunteer Training Unit 0414. In February 1982 he associated with US Naval Academy Information RC-4 as a Blue and Gold Information Officer, where he remained until his retirement. During this period, he counseled and interviewed 174 high school seniors and had 23 attend and graduate from the Naval Academy.

Mizeras participated on afloat active duty for training periods on board USS *Lexington,* USS *Miles C. Fox* and USS *Portland.* Ashore periods included 4th Naval District Administrative Course, Inactive Ship Maintenance Facility, National Security Seminar, Naval Reserve Center, Lancaster, Naval War College, Naval Aviation School and four tours at the US Naval Academy.

He is designated a surface warfare officer. He had been awarded the National Defense Medal, Armed Forces Reserve Medal w/HG, Vietnam Service Medal w/Bronze Star, and the Meritorious Unit Citation Medal.

He is employed by High Industries, as human resources director. He is married to the former Carole Stokes. They have two sons: Bradly who is a pilot with the US Army and Craig who lives in Lancaster, PA. His grandchildren are Samantha Paige Mizeras and Sydney Nicole Mizeras.

KERMIT T. MOCK, CAPT, born Aug. 28, 1933 in Millen, GA, received a BS from Berry College, Mt. Berry, GA.

He joined the service Dec. 14, 1956. His military locations and stations: USS *FDR* (CVA-42); NAVBASE, CHASN, SC.

Mock was awarded the Naval Reserve Medal w/2 HGs. He achieved the rank of CAPT and was discharged July 1, 1987.

He is married to Joyce and they have two children, Gregory Alan and Teresa Lynn; and grandchildren: Nathan, Payton and Justin. Employed at Cannon Equipment, prod. const. manager.

ARTHUR P. MOEBIUS, LTSG, born July 29, 1912, in New York City. He graduated Dartmouth College, Class 1934 and joined the service in November 1942.

Primary flight instructor, NAS Olathe, KS and St. Louis, MO. Served with USS CVE-103, 3rd Fleet, June-November 1945. His memorable experience was suppling pilots and planes for 3rd Fleet. He was discharged in November 1945.

Married Ruth, there are no children. Employed 40 years as sales rep. in Ohio and Indiana until he retired July 29, 1977.

RICHARD E. "DICK or DOC" MOFFITT, CAPT, born July 10, 1928, Portland, OR. He joined the USN July 24, 1945. Boot camp was at San Diego, USNH Long Beach, NAS Seattle, USNH Chelsea, MA.

Served in USS O'Brien (DD-725), USS Hollister (DD-788), USS Hawk (AMS-17), NRRC-13 MASS Corvalls, OR. Memorable experience was commendation for service to critically injured personnel on USS Hollister during explosion in Mount 3 in 1949.

Discharged July 1, 1979 to Retired Reserve. His awards and medals include the Good Conduct w/2nd award, American, WWII Victory, Naval Reserve Medal, National Defense, and Asian Occupation.

Married Helene and they have children: Susan, Michele, Kathleen; grandchildren: Jimmy, Mindy, Jessica, Ben, Nichole, Laura and Ginny; he also has four step-children, eight step-grandchildren and two step great-grandchildren. A self-employed dentist, he retired Sept. 1, 1988.

VIRGINIA IRIS "MO" MOJICA, LT, born July 6, 1961 in Rio Piedras, Puerto Rico, has a master of education, May 31, 1997, National University, San Diego, CA; San Diego State University, Dec. 31, 1991; BA San Diego Community College, Aug. 16, 1988.

She joined active duty June 2, 1980. Joined the Reserves May 17, 1989. Her military locations and stations: (AD) USS *Ajax* (AR-6) Naval Station San Diego, January 1981-February 1984; February 1984-November 1987, Fleet Training Center, NAVSTA San Diego.

Served with USNR NRCINCPACFLT 219 May 1989-September 1993; NRDICCPAC September 1993-June 1997; NR NCSO PAC 219, July 1997-present, all in San Diego.

She was awarded the Navy Achievement Medal (2), Battle E, Navy Good Conduct, Naval Reserve Meritorious, National Defense, Sea Service, Rifle and Pistol Expert Medals. Mojica achieved the rank of LT (03E), retired Sept. 1, 1997, USNR YNZ (AD), Nov. 16, 1987 (AD).

Her memorable experiences: first and only WESTPAC. When they visited Berbera, Africa they showed some African boys (children). How to play frisbee. Although they didn't speak English, they all enjoyed playing. Prior to leaving the beach, they gave the kids their frisbee and they smiled!

She is single and has three godchildren: Samantha Iris, Kaitlyn and Krystoffer; and adopted Mick Dalton Boyce. Employed with Central Elementary, Imperial Beach, CA, as sixth grade bilingual teacher, South Bay Union School District.

RICHARD W. "DICK" MOLTHOP, LCDR, born in Chicago, IL, May 18, 1920. He graduated from the University of Michigan. He also had schooling at the University of Chicago and Argon National Laboratory.

He entered the USN in 1942 and he was trained as a flight director, and radar man. He was the head radar man of the USS *Franklin* when it was bombed. He was then sent to Jamestown, RI as the officer in charge of SX radar prototype.

Was discharged from active service in early 1946. In the USNR he was at Glenview NAS, Glenview, IL. During the Korean War he was the liaison to the CIA for Argon National Laboratory for four years, and later did research for the USN. After 30 years he retired from the USN.

He is divorced and has two daughters, Susan and Julia; and four grandchildren. He had a stroke in 1972. He a computer fan. He resides in Denver.

ROBERT E. MONAHAN JR., CAPT, a native of Americus, GA, entered Georgia Institute of Technology in September 1946 as a NROTC regular student. He received his ENS USN commission upon graduation in 1950 and was ordered to the USS *Hugh Purvis* (DD-709) at Newport. Deployments included a seven month tour with the Sixth Fleet in the Mediterranean.

In March 1952 he was transferred to the five-month Naval Combat Information Center Officer's School, Glenview, IL as a student. Upon completion he was ordered to the USS *Radford* (DDE-446) at Pearl Harbor. Deployments included: Operation IVY, the detonation of the first hydrogen bomb at Eniwetok Atoll, Marshall Islands; and a seven-month tour with the Seventh Fleet during the Korean conflict.

Upon release from active duty in December 1953 he re-entered Georgia Institute of Technology and completed his second undergraduate degree. He accepted employment with US Steel Corporation in April 1955 at their plant operations in Fairfield, AL. Assignments included: operations, safety, production planning and systems development area. He retired in August 1986 after nearly 32 years.

He began his USNR participation in 1954 and continued participation until retirement in July 1985, completing almost 32 years of service. He held CO billets in the following Naval Reserve Units: Underway Training Unit 423, Convoy Commodore Galveston 109 and Naval Control of Shipping Office Mobile 109.

His military decorations are: China Service Medal (Extended), National Defense Service Medal, Korean Service Medal w/star, Armed Forces Reserve Medal w/ HG device, UN Service Medal, Korean Presidential Unit Citation and Surface Warfare Officer Insignia.

CAPT Monahan is married to the former Tony Shariett of Birmingham. Their children are Robert E. III, Nan R. and Ann C. They reside in Hoover (Birmingham), AL.

THOMAS P. MONAHAN JR., CAPT, born Jan. 11, 1937 in Staten Island, NY. He enlisted in the USNR, March 3, 1954 and received a discharge from the USNR to enlist in the USN on Nov. 10, 1954. Attended 12 weeks of boot camp at Bainbridge, MD, 10 weeks at Boilerman Class "A" School in Philadelphia, followed by assignment to the USS *Rigel* (AF-58).

In April of 1957 he was transferred to the USS *Sabine* (T-AO 25) and subsequently discharged in November 1957. Upon discharge he was employed as a marine stoker (boilerman) for the Staten Island ferries. While working and studying his way toward an engineers license he rejoined the USNR serving as an engineering instructor in Naval Reserve Surface Div. 3-98(M) on Staten Island.

Upon receiving his first asst. engineers license he applied for a commission through the "Seaman to Admiral" program and was commissioned ENS in June 1965. He served in various surface reserve units as training officer, XO and CO, before assignment as repair officer on NRMTF NY HQ 102 the forerunner of SIMA Staten Island. Subsequently was assigned as XO, and finished his career as CO of the SIMA S.I. Reserve Unit.

During his active duty for training he served tours on various destroyers, APAs, destroyer tenders, an aircraft carrier, (USS *Intrepid),* the Naval War College, and the Armed Forces Industrial College. In his civilian life CAPT Monahan retired from the NYC Ferry System as chief engineer and director of construction and repair.

His personal awards and decorations include: Navy Commendation Medal, Good Conduct, National Defense Service Medal w/two Bronze Stars, Armed Forces Reserve Medal w/HG, and the Navy Expert Rifle and Pistol Medals. He is a graduate of St. Mary's High School, Perth Amboy, NJ, and Strauss School of Engineering in New York.

He and his lovely wife for the past 38 years, Patricia, reside on Staten Island, NY. They have seven children: Thomas, Theresa, Kevin, Scott, Russell, Craig and Melissa. They also have eight grandchildren: Kaitlyn, Kevin, Thomas, Christina, Joseph, Russell, Scott and Catherine (Catie).

MARY JANE (ESTES) MONROE, CDR, born Sept. 19, 1924 in Memphis, TN, received a BS degree.

She joined the service April 1956. His military locations and stations: Bethesda, Portsmouth, Puerto Rico, Memphis and Pensacola. Served with mostly NAS Naval Hospitals.

She achieved the rank of CDR and was discharged September 1978. His memorable experiences: privilege of serving with the best people in the world.

Employed at Memphis VA Hospital, staff and supervisor. She retired Oct. 1, 1984. Her husband Clyde is deceased.

EDWARD "EDDIE" MONSOUR, CAPT, born Jan. 24, 1918 in Meridian, MS, received an MS and BS in geology, Louisiana State University.

He joined the service March 14, 1942. His military locations and stations: Interior of China, Asiatic-Pacific, French Morocco, USN, USAF, USMC, Navy Group China, TAD with 14th USAAF China, USS *Franklin* (CV-13), TAD with Marine Bombing Sqdn. 612, Photographic Interpretation Center Washington, DC, and COMNAVACTS French Morocco.

His battles and campaigns: Asiatic-Pacific Theater, China Campaign, India and Burma Campaign, Okinawa Campaign, Third Fleet Operations Against Japan Campaign and American Area Campaign.

He was awarded the Bronze Star Medal w/Combat Star, Presidential Unit Citation w/Combat Star, Navy Unit Citation, Republic of China War Memorial Medal, Republic of China Sino-American Cooperative Organization Medal, American Area Campaign Medal, Asiatic-Pacific Area Campaign Medal w/4 Battle Stars and Marine Corps Device, WWII Victory Medal, Korean Service Medal, Navy and Defense Reserve Medals.

He achieved the rank of CAPT and was discharged Jan. 24, 1978. His memorable experiences: Active duty WWII and Korean Conflict. Helped train Siamese college students attending US Schools (Free Thai Army) and embarked with them aboard liberty ship SS *Abraham* taking 89 days at sea from Baltimore to New York to Panama Canal escorted and unescorted from the Canal, south of New Zealand and Australia to Bombay, India.

Served as naval observer in interior of China and TAD with MAJ GEN Claire L. Chennault of Flying Tigers fame. Survivor of ill-fated carrier USS *Franklin* (CV-13) which was heavily bombed and damaged while attacking Japan. Served with Marine Bomb Sqdn. on Iwo Jima and Okinawa. During Korean Conflict served on staff of COMNAVACTS Morocco while the colony was attempting to gain independence from France.

Married Barbara M. and they have one daughter, Annette M. Monsour, LCDR, USNR (Ret). Employed as supervisory analyst w/DOD and Navy and geologist w/ Army Engineers.

MARION W. "MOO" MOOBERRY, LCDR, born Feb. 22, 1920 in Wichita, Sedgwick Co., KS, graduated high school and attended college.

He joined the service May 1942. His military locations and stations: Primary flight training, Kansas City, KS; NAS Pensacola, and flight instructor. Served with Air Gp. 14, AG-26, Korea AG-9, VF-91.

He achieved the rank of LCDR. Married Helen M. in

September 1943; she died in 1972, and he married Dora Mae October 1978. His children are Bruce A., Brian W. and Susan K. Self-employed and postmaster, Cheney, KS.

FRANK E. MOONEY, CAPT, born July 2, 1919 in McAllen, TX, received a masters degree in administrative education.

He joined the service April 15, 1942. His military locations and stations: Indoctrination, Notre Dame University; Midshipman School (Ship), Prairie State, New York, NY; Bomb Disposal School, Georgetown University, Washington, DC; Naval Sea Plane Base, Florida Island, Solomon Islands; Naval Station, New Georgia, Solomon Islands; Naval Air Base, Guadalcanal, Solomon Islands; Office of Strategic Services, Washington, DC; Office of Strategic Services, London, England.

Awarded the Asiatic-Pacific Campaign, American Campaign Medal, EAME Campaign, WWII Victory Medal, Naval Reserve, Faithful Service. Achieved the rank of LT(sg), active duty, retiring Dec. 20, 1945; CAPT, USNR, 1963, San Antonio, TX Naval Reserve Training Center, retiring 1966.

Married Ruth Hall and they have two children, Barbara and Frank Robert (Bob); and grandchildren: Casey, Laurel and Retta. Employed for 39 years as Industrial Arts teacher, at San Antonio, and San Marcos, TX, retiring May 1996.

CHARLES E. MOORE, CDR, born Oct. 20, 1930, Oskaloosa, IA. Earned BA degree at University of Missouri, 1953; graduate study, 1954. He joined Regular Navy NRSTC, August 1954; served with amphibious forces Atlantic, Norfolk, VA, 1954-57; Naval Reserve, 1957-70 at Buffalo, NY and 1970-74 at St. Louis, MO. Served in USS *Thuban* (AKA-19); Staff ComTran PhibRon II; CO, NRSD3-60 and CO NRSD 3-57, Buffalo, NY, 1968-69.

Left active duty and transferred from Regular Navy to Reserves in 1957; transferred to Inactive Reserve in 1974, CDR, USNR. His awards and decorations include the Korean Service, Good Conduct and others.

Memorable Experiences: Resupply mission Pac. line construction Artic, July-September 1955; Sixth Fleet during Suez crisis, 1956, June-November; and duty as commanding officer.

Divorced, he is the father of six children: Kathleen, Charles, Susan, Thomas, John and Steven; and has 10 grandchildren: Joan, Michael, Anne, Catherine, Megan, Tommy, Rebecca, Stephen, Mathew and Kevin. Employed 28 years with the Pillsbury Company, he was VP of export grain operations at time of retirement in 1986. Since then, has been self-employed as investor.

ERNEST VICTOR "SKIP" MOORE JR., CDR, born Dec. 16, 1926 in Memphis, TN, attended two years college and received an AA from Christian Brothers University.

He joined the service May 2, 1944. His military locations and stations: boots at Camp Wallace, TX; Gunnery Officers Ordnance School, Anacostia, DC. Served with USS *Neshanic* (AO-71), USS *Tolovana* (AO-64); USS *Kaskaskia* (AO-27); VRS Aviation Sqdn.

Served as an enlisted man in WWII. Present at Iwo Jima, Okinawa, 3rd Fleet strikes on Home Islands and at Tokyo Bay at surrender signing. Rated coxswain at Naval Mine Depot Yorktown, VA after the war.

Commissioned ENS (DL) after college in 1948. Recalled to active duty for Korean Conflict, serving as gunnery officer aboard USS *Tolovana* and operations officer aboard USS *Kaskaskia.* Served three tours in the Korean area and two tours as station tanker at Keelung and Kaohsiung, Formosa.

Joined Naval Air Reserves at NAS Memphis after release from ACDUT and served as staff photographic officer, various duties with various VR and VP squadrons and as wing staff officer after promotion to CDR in 1970.

Retired April 1, 1973 at NAS Dallas, TX. He was awarded the Asiatic-Pacific, Korean China Service (2), Naval Reserve Medal (2).

Married Dorothy James Hayes and they have children: Ernest III, Robert Douglas, USMC and David Jame, USMC; and grandchildren: Elizabeth, Andrew, Rebekah and Benjamin. Employed with Giton Fire and Marine Underwriters for 39 years, as senior underwriter, retiring Sept. 1, 1987.

FORREST G. MOORE, CAPT, born Feb. 9, 1916, Wadena, IA. Earned BA degree, 1939; MA in 1945 and PHD in 1953. He joined the service July 14, 1946.

Military Locations and Stations: (1941-42) USNR AB, Dallas, TX; USNAS, Corpus Christi, TX; USNAS, Norfolk, VA; (1942-44) TT Sqdn. Atlantic, Banana River, FL; USNAS, Norfolk, VA; Harvey Point Air Group, Harvey Point, HQ Sqdn. 5, Herford, NC. (1944-58) VPB-21 for overseas assignment to Marshal Islands, Palau, Saipan and Okinawa. From 1949-69 he served in a variety of Reserve units, primarily Air Intelligence as officer and an instructor and from 1969-74, NARS E-1, Sea Cadets Program.

Memorable experience was carrying torpedoes on PBM to target between Korea and Japan; flying back from the China Coast after losing one engine to anti-aircraft fire.

Awards and medals include Philippine and Palau Service Stars, American Campaign Medal, Asiatic-Pacific Campaign Medal, WWII Victory Medal, Air Medal w/4 stars, Distinguished Flying Cross and Naval Reserve Medal. Discharged Sept. 15, 1945 and retired Feb. 1, 1974.

Employed as HS teacher and assistant to full professor until retiring from the University of Minnesota in 1980. Married the former Margaret Hall; she passed away in January 1994. His children are Eleanor Beth, Marilyn and Shirley Moore, Jacqueline Eddy and Gloria Moore Clow; grandchildren are Christopher and Katie Eddy.

HENRY "HANK" MOORE, JR., CAPT, born April 29, 1916 in Los Angeles, CA, received a BS in mechanical engineering, UC Berkeley, CA 1940.

He joined the service April 10, 1941. His military locations and stations: LB, CA; NAS Jacksonville, FL; NAS CC, TX; NAS Ford Island, Pearl Harbor; NAS San Diego; NAS Norfolk; NAS Los Alamitos, CA.

Served with VJ-1, VJ-7; VJ-9; ATTDTU NAS Norfolk, VA; VP FASRON 775; VP-776; VP-774. His battles and campaigns: Submarine search patrols, Pacific Theater; Target towing for PAC Fleet.

He achieved the rank of CAPT, USNR naval aviator #11274 and retired from active duty October 1945.

His memorable experiences: runaway propeller on P2V-5; in a violent storm lost all radio units on new JM-2; hit by flak while target towing over Coronado Island; the great experience of flying missions in 25 different types of Navy aircraft.

Moore is married to Agnes Holman Moore and they have three children: Mary Ellen, William Henry and Timothy Holman; and grandchild, Rachel Anna Moore. Employed as vice-president of California Structures, Inc.; Aeropace Engineer Bureau of Naval Weapons Rep., Burbank, CA, retired from May 1, 1976 from USNR.

MOORE MOORE JR., RADM, born March 9, 1909 in Memphis, TN, USA, received an MD from the University of Tennessee.

He joined the service March 1934. His military locations and stations: USNH Charleston, SC and New Orleans, LA; Panama Canal Zone. His battles and campaigns: North Atlantic Invasion.

He was awarded Personal Commendation w/Combat V. Moore achieved the rank of RADM (MC), USNR and discharged February 1946.

His memorable experiences: senior medical officer of USS *Eden Rutledge* sunk by torpedoes at FEDALA North Africa on Nov. 12, 1942.

He married Houston Westbrook Veller. Employed as physician (orthopedic surgeon) (now retired).

GEORGE R. MOORMAN, CAPT, born Nov. 17, 1919 in Ft. Worth, TX, a founder of NRA, and its first vice-president for legislation, he was born in Ft. Worth, TX on Nov. 17, 1919, attended NTAC Junior College and Southern Methodist University (BBA 1939, JD 1941). A licensed attorney since 1941, post WWII, he founded the firm of Moorman, Tate, Moorman, Urquhart & Haley attorneys in Brenham, TX, also founded South Central Savings. Retiring in 1985, he remains active in Mediation and Arbitration activities for the 33rd District Court in the five counties surrounding Horseshoe Bay, TX, his residence.

Prior to WWII he received an appointment as an FBI special agent in Washington, DC. Enlisted in the V-7 program, he was called to active duty on Jan. 21, 1942, completing Midshipman's School in Chicago, as an ENS, assigned duty aboard a seaplane tender, USS *Sandpiper* (AVP-9), performing convoy escort operations out of Greenland in the North Atlantic during 1942-1943; then escort and supply duties in the South Atlantic. In early 1944 he was appointed CO. In 1945, he reported as XO to the USS *Fortune* (AVS-2); made CO until decommissioning, discharged from active duty in November 1945.

His reserve career started in Lubbock, TX establishing a Surface Div. Moving to Waco, TX, he established and was CO of a surface division, responsible for obtaining a training center. He received a Letter of Commendation for assistance in the Waco Tornado. The unit received 8ND honors for the five state area. Before founding NRA, he was Texas Department President of the Reserve Officers Association. Commissioned CAPT in 1962, he remained active until retirement in 1984.

Married Billie Scoggin and they have two sons. Hal is married to Lucy and has three children and Rob is married to Becky and has five children; both live in Brenham. He is active in golf, with three Robert Trent Jones golf courses in the resort community, Horseshoe Bay on Lake LBJ, an avid water skier, and boating enthusiast.

JAMES M. "JIM" MOOSE JR., CAPT, born Sept. 23, 1923 in Pinebluff, AR, received a BS from the University of California, Berkeley 1948; LLB University of California, Berkeley 1957.

He joined the service April 1, 1942. Served with USS *Oak Hill* (LSD-7); LCT Flotilla 16; USS LSM 432. His battles and campaigns: Leyte, Lingayen and Okinawa. He achieved the rank of LTJG active duty; CAPT in USNR, discharged August 1946.

He married Virginia and they have four children: Catherine, Anne, James and Elizabeth. Employed as attorney, administrative law judge, State of California. He retired in August 1995.

BASILIO MORAN, LT, born July 2, 1931 in Los Angeles, CA, East Los Angeles College, AA business, 1956; Los Angeles State College; BS Business, 1958; California State University, Fullerton, MPA, Public Administration, 1975.

He joined the USAF, Nov. 1, 1948 to Oct. 31, 1951; US Army, Nov. 1, 1951 to Sept. 1, 1954; USAFR, Aug. 17, 1955 to Aug. 16, 1958; USNR, Dec. 7, 1960 to Dec. 6, 1964; USNR, April 5, 1967 to Dec. 31, 1988. Retired Jan. 1, 1989.

His military locations and stations and units served with: USAF; Shepard AFB, Wichita Falls, AFB; Wichita Falls, TX, basic training; Smyrna AFB, Nashville, TX; HQ

& HQ Sqdn. 3750 BT Wing & 314th AB Gp.; Ft. Warren AFB, WY; 3468th Training Sqdn., 3450th Tech. Training Wing; Hamilton AFB, San Rafael, CA; 43rd Comm. Sqdn. Comd. & 542 AC&W Gp.; Ft. McArthur, San Pedro, CA, 669th AC&W Sqdn.

US Army, Camp Roberts, CA, basic training; Ft. Monmouth, NJ, Co. C, OC Bn.; Pusan, Korea, 8226 AU and HQ Det 8063; Sasebo, Japan, HQ Det. 8023AU; Camp Kilmer, NJ, 6215ASU;

USAFR, unassigned. USNR, NRSD 11-8(L)-N&MCRC Los Angeles, CA; NRSD 11-5(M), Los Alamitos NAS, Los Alamitos; NRGC 11-1(M), Los Alamitos NAS, Los Alamitos, CA; CBC 919-AFRC, Los Alamitos, CA; 1st RNCR-AFRC, Los Alamitos, CA; NRMCB16-AFRC, Los Alamitos, CA; NRMCB17, Las Vegas, NV; RESCOMCBPAC, Pt. Hueneme, CA.

He was awarded the National Defense Service Medal, UN Service Medal, Korean Service Medal, AF Good Conduct Medal, Armed Forces Reserve Medal (three awards) and Navy Achievement Medal. He achieved the rank of CPL, USAF; CPL, US Army; A/1c, USAFR and LT, USNR.

Married Norma Jean English. He had 37 years of public service, not including military time. Served primarily in the Human Resources area, working in one state college, five school districts in California and Nevada and worked for the State Division of Highways in California. Now an adjunct professor of business for the Community College of Southern Nevada. Retired from the USNR, Jan. 1, 1989 and from civilian employment, July 31, 1995.

WILLIAM T. MORAN, CAPT, native of San Francisco, CA, enlisted in USNR January 1951 while in high school. Economics major at Stanford University. Completed Reserve Officer Candidate (ROC) Program and commissioned an ENS upon graduation from Stanford in 1955.

Ordered to the Naval Air Technical Training Center, Memphis, TN. Graduated with distinction Naval Justice School, Newport, RI.

Returned home in October 1958 to NRSD 12-20. Recalled to active duty as a lieutenant in the Training Administration Reserve (TAR) Program, 1959.

Served in the USS *San Joaquin County* (LST-1122) as 1LT/NAV and for second year as XO/NAV. In November 1962 assigned as CO of the Naval Reserve Center, Lake Charles, LA.

Transferred to the Naval Reserve Manpower Center, Bainbridge, MD as head of the mobilization branch for shore activities in November 1965. In July 1966 was fleeted up to head of the Processing Department. March 1968 transferred to the USS *Hermitage* (LSD-34) as operations officer.

Served in the Bureau of Naval Personnel as head of the Surface/Subsurface Programs Branch for the USNR and also the Special and General Programs Branch, 1970-74. In 1973 became director of mobilization for all aviation, surface and subsurface programs of the USNR.

Graduated with distinction from the Senior Course, Naval War College, Newport, RI, 1975. Served in the Pentagon under the director of Naval Reserve as head of the Surface Plans, Policy and Programs Branch, 1975-76.

Served one day as CDR, Cleveland Readiness Command and two years as CDR, Naval Reserve Readiness Command Region Five. Retired October 1981 after three years as special assistant for Reserve Affairs to the ASN(M&RA) and recorder for the Secretary of the Navy's National Naval Reserve Policy Board. Has the following awards: Legion of Merit, Meritorious Service Medal, Naval Reserve Meritorious Service Medal, Naval Reserve Medal (3), National Defense Medal (2).

He was NJROTC department head and instructor, Northwestern High School, Adelphi, MD 1981-88. Executive director, the Master Builders' Association, Inc. and executive director, Constructor Contractors Council, 1988-95. Retired 1995 to serve as consultant for Labor Relations; also sits as a board member on a number of Union Trust Funds.

Married to the former Mitzi Carole Brummel of Union City, TN, the Morans live in Fairfax Station, VA and have two daughters, Tina, married to Richard A. Buckingham, and Shari, married to Joseph A. Trotter III. They also have three grandchildren, Joshua and Amy Buckingham and Michael Trotter.

DANIEL JOHN MORAVEC, LCDR, born Feb. 14, 1944 in Waco (McLennan), TX, received a AAS, micro computer application/data processing, May 1991; business administration/management, May 1983; management/business, May 1978.

He joined the service June 18, 1964. His military locations and stations: USS *Henry County* (LST-824), naval support activity, USS *John W. Weeks* (DD-701), combat support 3rd Marines. Served with NR-Surface Support, USS *Texas,* USS *Kinkaid,* USS *Gompers,* Battle Group 1. His battles and campaigns: Vietnam 1965, 1966.

Moravec was awarded the Vietnam Campaign, Vietnam Service, Good Conduct, Meritorious Award, Expert Rifle. He achieved the rank of LCDR 04 and was discharged from the active reserve.

He married Carol A. and they have two children, Rachel A. and John D. Employed at Diesel Power Supply Company, Waco, TX, engine and sub assembly, overhaul supervisor.

JAMES R. MORELAND, CDR, born on NAS Pensacola, Aug. 27, 1950. He was a graduate of Embry-Riddle Aeronautical University, FL and Troy State University, AL, earning a BS, MS and an MBA. He completed AOCS and was designated a naval flight officer in 1977.

From 1978-86, he served active duty assignments with VAQ-129, VAQ-137 at NAS Whidbey Island, WA, deploying to the Western Pacific and the Caribbean. He served as aviation fuels officer on USS *Saratoga* (CV-60), deploying to the Mediterranean and Caribbean, and a SLEP period at the Philadelphia Naval Shipyard. His shore assignment was to NAS Pensacola to air operations and AIMD.

Leaving active duty in 1986, he reported to VP-62 where he served in the aircraft maintenance department. Promoting to CDR in 1994, he then drilled with Naval Air Warfare Center, Training Systems Div., Orlando, FL. Currently he serves as XO, PATWING 1174, NAS Jacksonville.

CDR Moreland is a boating accident/fatality investigator with the Florida Game and Freshwater Fish Commission, Division of Law Enforcement. He lives with his wife Shelley and son Spenser in Sanford, FL.

DONALD C. MORENCY, CAPT, a native of Salem, MA, received his BA degree in economics from Boston College in 1960. Upon completion of Officer Candidate School at Newport, RI, he was commissioned an ENS in the USNR on July 14, 1961.

On active duty he served as 1LT/gunnery officer aboard the USS *Summit County* (LST-1146), and participated in Operation Dominic in 1962, the last nuclear tests in the Pacific before the Nuclear Test Ban Treaty with the Soviets. He was then assigned as flag lieutenant/aide to commander, amphibious training command, Pacific until his release from active duty in July 1964.

CAPT Morency affiliated with the Naval Air Reserve at Los Alamitos, CA as personnel officer with VR-774, then with VR-21L4. In 1969, he transferred to Washington, DC and became associated with VR-62A1 at the Naval Air Facility.

In 1972, he was selected to assist in establishing VR-52 under the RESFORON concept; in 1974, following the RASP Project, NAS 1006; and in 1977, under Project Readiness, NR Naval Station Keflavik 1066 to which he was assigned as XO.

CAPT Morency was promoted to his present rank in May 1983, and assigned to VTU-6666 as the Selected Reserve Recruiting/Retention Officer for NAF Washington. He then served as CO of NR CV-62 *Independence* 0166 between 1983 and 1985; as CO of VTU-6666 at NAF Washington in 1986; as senior inspector on the IG Staff of REDCOM SIX; and as CO of NR DCNO OP-06 106 between 1987 and 1989. Upon being appointed as deputy assistant secretary of the Navy for Reserve Affairs in 1990, he was placed in an inactive status and retired in 1993.

Additional duty assignments have included the 1974 CNAVRES Policy Board, Naval War College, National Defense University, CINCLANTFLT, SACLANT, OPNOV, selection boards, Navy Legislative Liaison, Navy Comptroller's Office, READEX 1-83/1/84/2-85, and chairman of the Recruiting District Advisory Council for the Naval Recruiting District Washington from 1978-1982.

In civilian life, CAPT Morency held executive positions in the aerospace industry and since 1971 has been a management consultant in Washington specializing in government relations. He also serves on the Board of Advisors of American Defense International, Inc. Over the last 30 years, he has been active in various civic and service related organizations. He is a life member of the NRA and was president of its Washington Chapter.

Among his military and civilian decorations are the Meritorious Service Medal w/Gold Star, the Navy Commendation Medal, the Meritorious Unit Commendation w/2 Bronze Stars, the Navy "E" Award, the National Defense Service Medal, the Navy & Marine Corps Overseas Award, the Armed Forces Reserve Medal, the Navy's Distinguished Public Service Award, the Navy's Superior Public Service Award, and the Navy League's highest award and the Distinguished Service Award.

ARTHUR FORT MORGAN, CAPT, born April 19, 1929 in Vienna, GA, received a BBA from Georgia State University and a MS degree from Georgia Tech.

He served from September 1948-February 1955. His military locations and stations: NAS Pensacola, Corpus Christi, Alameda, ComAir PAC. He served active duty: PATRON 19, Reserve Patron 671, NARS B-1, OPCON, 5B2, VIU 5180.

He participated in Korean Campaign and received the National Defense Medal, Korean Defense Medal, UN and Armed Forces Reserve Medal. He achieved the rank of CAPT and retired in August 1977.

His memorable experiences: V-5 Program flight training-carrier qual. Patrol in Sea of Japan and North China Sea involving ASW, ECM, shipping surveillance and night cover for Task Force.

He married Diane and they have three children: Michael, Nancy and Gregory; and grandchildren: Candace, Ginger, Sam, Austin and Sarah. Employed at National Bank of Georgia, senior vice-president, corporate personnel officer. Employed in real estate sales from January 1985 to present.

RICHARD S. MORGAN, LCDR, is a native of Miami, FL. He entered the USN as a seaman recruit in 1963. Graduating from basic submarine school he reported to the USS *Scorpion* (SSN-589) COMSUBRON SIX, Norfolk, VA. As a 'mustang' torpedoman, he earned his submarine qualifications and advanced to TM2(SS) at his end of tour in 1966.

TM2 (SS) Morgan served with NRSD-644 Submarine Div. in Miami, FL in support of COMSUBRON-12, Key West. Duties as a torpedoeman on board the USS *Quillback* (SS-424); USS *Amberjack* (SS-522); USS *Barracuda* (SST-30); and the USS *Sea Cat* (SS-399). Instructed basic submarine principles, fire fighting, damage control, basic electricity and electronics.

In 1976 Chief Torpedoman Morgan reported to NAS Jacksonville as the assistant weapons officer with NAR-644, NAS 508, and NAS-0374 NAS JAX, FL. Professional career development included various weapons courses and NUCWEPS Training Gp., Norfolk; Polaris Missile Facility, Charleston, SC; and Pensacola, FL. In 1982 Chief Morgan was commissioned as a chief warrant officer (CWO2) underwater ordnance technician.

LTJG Morgan was commissioned as an Aviation Ordnance LDO in 1984 at NAS Jacksonville where he served as maintenance officer with MMF-0174. He later transferred to HASW-0174 as weapons officer in support of COMHSWINGLANT. His assignments have included administration officer; training officer, XO, and in October 1995 to his current assignment as CO. LCDR Morgan completed his undergraduate degree from Rollins College with honors and is currently advanced to candidacy at Webster University in pursuit of an MBA in Computer Resources.

LCDR's memorable experiences were his time on board the USS *Scorpion* and the continual memories of his shipmates and friends who were tragically taken on May 22, 1968.

LCDR Morgan is employed as program manager, Lucent Technologies, CIO Business Communications Systems Div. He married the former Lynda F. Long of Miami, FL and they have two children, Renee' (Morgan) Harris and Scott Morgan.

RUSSELL E. "RUSS" MORGAN, LCDR, born Feb. 14, 1909 in Minersville, PA. Attended Minersville High School, Lebanon Valley College, BS; Temple Medical School, MD, VAFP, MS surgery.

He joined the service July 30, 1944. His military locations and stations: South Pacific/Sea duty on USS *Chourre*. Served with USS *Chourre,* Navy Hospital, MANUS. He achieved the rank of LCDR and was discharged Aug. 13, 1946.

Morgan's wife Elizabeth is deceased. He has three children: Russell, Gwendolyn and Martha; and grandchildren: Emily Morgan, Jesse and Marg White. Employed in private surgical practice, retiring June 1944.

STEPHEN LEE MORGAN, LCDR, born in Orlando, FL in 1959. He enlisted in the AF in 1976, assigned to Grand Forks AFB, ND.

Enlisting in the USNR in 1980, he was commissioned in May 1981. He served with Naval Intelligence Space and Weapons Technology Unit, and was later assigned to CDR, South Atlantic Force, supporting the UNITAS XXIV and XXV exercise in Chile and Peru. Recalled to active duty in 1991 for Operation Desert Storm, he served with Naval Space Command. In 1994, he was selected as assistant deputy chief of staff for intelligence support, CDR, Naval Reserve Intelligence Command.

His decorations and service awards include: Navy and Marine Corps Commendation Medal, National Defense Service Medal, Armed Forces Reserve Service Medal w/M Device, and Navy & Marine Corps Overseas Service Ribbon as well as four Navy and Air Force unit commendations. He was on the founding committee of the NRA's Theodorus Mason Chapter (NAF Washington) in 1994.

Morgan is now executive manager of the National Capital Office for Scientific-Atlanta, Inc.

JOHN F. "JACK" MORIEN, CDR, born Oct. 24, 1943 in Evergreen Park, IL, attended Marquette University, Master in Business (MBA) 1968; bachelor of science, business administration (accounting), 1966.

He joined the service June 28, 1968. His military locations and stations: August 1968-December 1968, Navy OCS, Newport, RI; January 1969-July 1969, Navy Supply Corps School, Athens, GA; July 1969-July 1970, Naval Area Audit Service, Norfolk, VA; July 1970-July 1972, USS *Meyerkord* (DE-1058), San Diego, CA.

Served with: NR NSD Subic Bay DET 519, San Diego, CA; October 1977-July 1980; NR MIUW 1319, San Diego, August 1983-September 1983; NR NSC SD HQ B419, San Diego, CA, October 1983-September 1985; NR NSC SD DET B619, Pomona, CA, October 1985-September 1987; NR NAVSEA DET 419, San Diego, CA, October 1987-September 1989; NR NRCC SD 519, San Diego, CA, October 1989-September 1993; NR DCMD LA A919, Los Angeles, CA, October 1993-September 1996; NR VTU Supply SD 1915, San Diego, CA, October 1996-June 1997.

He was awarded the National Defense, Vietnam Service, Meritorious Unit Commendation, Reserve Defense, Joint Meritorious Unit and Joint Service Commendation.

Morien achieved the rank of CDR, USNR. His memorable experiences: wonderful people he met and challenging jobs he was assigned.

He married Patricia A. and they have three children: Kathleen, Christopher and Anthony. Employed as senior accountant, county of San Diego, auditor and controller department, retiring July 1, 1997.

GARRY G. MORISSETTE, CDR, born Jan. 11, 1956, Lawrence, MA. He received his commission from Aviation OCS, Pensacola, FL in May 1978 after graduating second in his class. In May 1980 he was awarded his Naval Aviator Wings at Beeville, TX and assigned as a SERGRAD providing basic jet flight instruction in the T-2C aircraft. He completed A-7E transition training in Lemoore, CA, with VA-122 and was then assigned as a fleet aviator with the Mighty Shrikes of Attack Sqdn. 94.

Morissette completed two WESTPAC cruises on board the USS Enterprise, accumulating over 250 carrier landings. After leaving active duty in 1985, he was assigned to Naval Reserve Mobile Maintenance Facility B-0191 at NAS South Weymouth, where he performed as training officer and acting Reserve Space and Warfare Command HQ 101 supporting the JOTS Program Office at SPAWAR HQ in Washington, DC.

He received his promotion to commander June 1, 1993 and was assigned in October 1993 to MARDEZLANT Sector 1A, now renamed Harbor Defense Command Unit 201. CDR Morissette participated as a member of the evaluation staff during Allegiant Sentry '94, an HDC exercise held in Norfolk, VA. He was mobilized and assigned as the HDC Operations Officer/N-3 during Operation Uphold Democracy in Port Au Prince, Haiti. CDR Morissette is currently assigned as the Readiness officer/N-7 with Naval Coastal Warfare Group Two in Portsmouth, VA.

His awards include a Navy Commendation Medal, Meritorious Unit Commendation, Coast Guard Meritorious Team Commendation, Armed Forces Expeditionary Medal, Humanitarian Service Medal, Battle "E" Ribbon, National Defense Service Medal, Sea Service Deployment Ribbon (second) and Expert Pistol Medal.

CDR Morissette received a BS degree in mathematics from the University of Lowell and a master's degree in business administration from the Univ. of New Hampshire. He is a marketing manager with Dynamics Research Corp., a government contractor based north of Boston. He lives in Andover, MA, with his wife Sarah; sons, Gregory and Nicholas; and daughter, Laura.

JAMES M. MORRELL, CAPT, is CDR, Destroyer Sqdn. ONE. A TAR officer from Tacoma, WA, he enlisted in the USN in 1967 as a seaman recruit and was commissioned in December 1971 through the ROC program. He has a BA from Western Washington State College and an MBA from the University of Puget Sound.

He served on seven ships in the Pacific Fleet and was CO, USS *Fresno* (LST-1182) from 1989 to 1991. He assumed command of Destroyer Sqdn. ONE in March of 1996.

He also commanded Naval Reserve Center, Bremerton, WA; was chief of staff, CDR, Naval Surface Reserve Force and was the Surface TAR Detailer.

Morrell wears the Legion of Merit, Meritorious Service Medal (two awards), Navy Commendation Medal (three awards), the Navy Achievement Medal and various campaigns and service awards. His wife, Rennie, was a TAR LCDR who retired in 1995 and is presently working on her MBA at the University of San Diego.

VERENA "RENNIE" HART-MORRELL, LCDR, is from Anderson, SC. She enlisted in the USN in 1973, advanced to AZ1, received an associate's degree from Embry Riddle Aeronautical University in 1980 and was selected for the Enlisted Commissioning Program. Upon graduation from the University of South Carolina and OCS, she was commissioned in 1983.

She was the assistant maintenance training director at ASW Wing One at Cecil Field, FL; protocol officer at COMFLEACT, Sasebo, Japan after which she became a TAR officer and served at REDCOM 22, REDCEN Long Beach.

She wears the Meritorious Service Medal, the Navy Commendation Medal (three awards), the Navy Achievement Medal (two awards) and other service awards. She is working on her MBA degree at the University of San Diego and is married to CAPT Jim Morrell.

MARILYN "LYN" WILLIAMS-MORREY, LCDR, born Feb. 27, 1948 in Camden, SC, received a bachelors in nursing from UT/Austin and nursing diploma, Presbyterian Hospital, Charlotte, NC.

She served active duty, March 15, 1969 to March 15, 1972; Reserve, Sept. 30, 1988 to present. Her military locations and stations: Beaufort Naval Hospital, Beaufort, SC and Harold E. Holt, Exmouth Western Australia. Served with the Navy Nurse Corps. Her battles and campaigns: Vietnam era; Persian Gulf War aboard hospital ship *Mercy.*

Morrey received the Armed Forces Reserve Medal, Liberation of Kuwait, Saudi Arabia, National Defense, SW Asia. Presently LCDR, USNR. Discharged from active duty March 15, 1972.

Her memorable experiences: living in house in Australia, 750 miles from the newest city of Perth, Western Australia. Flying doctor trips across the obstacles with patients in small planes, dirt landing strips and taking off at dusk with car headlights as runway guide lights.

She married John Morrey and is employed in intensive care nursing and also as a legal nurse consultant.

CHERI CARSON-MORRILL, LCDR, born Feb. 17, 1961 in Pensacola, FL, received a bachelor of science from the University of Florida.

She joined the service Aug. 4, 1984. Her military locations and stations: Naval Ocean Processing Facility, Ford Island, HI, 1985-86; CDR Oceanographic System Pacific, Pearl Harbor, HI, 1986-89; US Naval Facility, Guam, 1989-90.

Served with NR US Naval Facility, Keflavik Iceland, November 1992-February 1996; XO/training officer; NR CDR US Naval Forces Central Command DET 108, February 1996-present, security manager.

She was awarded the Navy Commendation, Navy Achievement Medal w/Gold Star in lieu of second award, Defense Service Medal and Armed Forces Reserve Medal. She achieved the rank of LCDR.

Her memorable experiences: working with ASW communities during exercise and real time operations, especially when their mission was achieved.

She married Philip H. Morrill and has three children: Alyssa, Samantha and Jennifer. Full-time mother with the exception being her naval reserve duties. Her hobbies include camping, fishing, and canoeing with the family.

WILLIAM C. "BILL" MORRILL, CDR, born April 23, 1958 in Philadelphia, received BS in industrial engineering from Pennsylvania State University.

He was commissioned Nov. 29, 1980. Served with VRC-50, VQ-3, VR-52. His battles and campaigns: Desert

Storm/Desert Shield. He was awarded the Navy Achievement (two awards), SW Asia Medal. He achieved the rank of CDR currently XO of VR-52.

He married Barbara J. Henderson and is employed as pilot with TWA.

HENRY EARL MORRIS JR., LCDR, son of Henry Earl and Eva Heatley Morris. Born Jan. 14, 1921, in Pueblo, CA. Received BA in Botany, 1944 in University of Denver.

Joined the US Army, 1942-46; Camp Crowder Signal Corps Camp, 1942-43; University of Pennsylvania, Chinese Studies 1943-44; Chief Clerk, China Theatre Replacement Service, 1945-46; Discharged as technical sergeant, Jan. 10, 1946 w/Bronze Star Medal.

Assistant placement advisor, Lowry AFB 1946; Iliff School Theology, Masters of Theology, 1949; Colorado Annual Conference, Methodist Church, 1947-49; California-Nevada Annual Conference, United Methodist Church, 1949-present. Served 14 churches, retiring July 1, 1986.

Served USNR, 1955-73, LCDR, Chaplains Corps. Chaplain, USNRTC Fresno 1955-58; Reno, NV, 1958-60; Fresno, 1960-67; Burlingame, 1967-69, Sacramento, 1969-73.

Member of MCA 1956-present (life member 1971); president Golden Gate Chapter 1965-69; area vice-president, 1969-73; member TROA (life member); member The NRA (life member); plankowner, USN Memorial; member Auburn (CA) Faith Community Hospital.

WILLIAM L. "BILL or WAR EAGLE" MORRIS, CAPT, born Feb. 27, 1932 in Monroe, LA, received a BS in aero engineering, Auburn University, June 1954.

He joined the ROTC Midshipman September 1950, commissioned June 1954. His military locations and stations: flight instructor, NAS Whiting, LT Jet attack pilot, NAS Miramar. Served with VA-146; CAG-14; USS *Ranger*; active Res SQDS, NAS New Orleans, Seattle Memphis and Whidbey Island. Flew FJ4B; F9F-8; S2F; A-4 A,B,C, TV-2; T2V. His awards and medals: National Defense, Naval Reserve, Rifle and Pistol Expert Marksman.

He achieved the rank of CAPT. Discharged from active duty September 1959 and from active Reserve, October 1976.

His memorable experiences: 1958 "Round the Horn" cruise, West Pac cruise on USS *Ranger* in 1959 and Tiger cruise in *Ranger* (Hawaii-SDA) in 1989 with son, LT John Morris of VAW-116, 30 years later, 3,500 flight hours.

Married Mavis Stewart and they have two children, Randy, MAJ, USAF, and John, LT USNR; and two grandchildren, William Thomas and Philip John.

Employed with the Boeing Co. for 33 years, missile test OPS MGMR, 10 years, E-6, flight test director, nine years, retiring Aug. 1, 1992.

W. WALLACE MORTON JR., CDR, born in Fredericksburg, VA, on Nov. 1, 1936. He graduated as an aeronautical engineer from the University of Virginia in June 1959 and was commissioned through the NROTC program as an ENS. From 1959 to 1963 he served aboard the USS *Oriskany* (CVA-34), where he was Air Department administrative assistant, Second Div. officer, and qualified as officer of the deck underway.

After release from active duty, CDR Morton joined a Naval Air Reserve Unit at Clinton County AFB, OH. After several years, he returned to Virginia and transferred to The Naval Air Systems Program (ASP) at Andrews AFB. As this program matured, he supported the P3 program manager at NAVAIR HQ. In several ASP units, he served as administrative, training, and XO. He retired in 1983.

CDR Morton has worked at the Naval Surface Warfare Center, Dahlgren Div., Dahlgren, VA, for the past 29 years. He is an expert in weapons vulnerability and lethality.

The Morton family consists of wife, Michaele; daughter Fenton, a pharmaceutical sales representative in San Antonio, TX; and son William IV, a writer for CNN Financial News in New York City. CDR and Mrs. Morton live in King George, VA. He enjoys Boy Scouting, skiing and photography.

GERALD P. "GERRY" MOTL, CAPT, born July 31, 1946 in Milwaukee, WI, received a BS from the US Naval Academy, 1968; MBA, University of Maryland, 1978; Wharton School, Advanced Management Program 1993.

He is commissioned June 5, 1968. His military locations: Pearl Harbor, Idaho Falls, ID. Served with USS *James Monroe* SSBN 662 (Gold).

Motl achieved the rank of CAPT, USNR, discharged from active duty June 30, 1973. Retired from USNR Sept. 1, 1992.

His memorable experiences: Started 1967 Army-Navy Football Game, 1967; Rickover interview, Panama Canal Passage, MIDN cruise in Mediterranean during 7-day war, 1967.

He is married to Roxanne and they have two children, Christine and Brian. Employed as vice-president, Science Applications International Corp.

EDWARD M. MULHERN, CAPT, born in Brooklyn, NY in 1947. He graduated from the US Naval Academy at Annapolis in 1968. He served an initial tour as MPA aboard USS *Harry E. Hubbard* (DD-748) which deployed to WESTPAC and Vietnam. He qualified as SWO and received orders to USS *Capricornus* (LKA-57) in Little Creek, VA.

Following this tour, CAPT Mulhern volunteered for and received orders to the Naval School of Deep Sea Diving and Salvage in the Washington Navy Yard. He proceeded to Pearl Harbor, HI to serve aboard the diving and salvage ship USS *Bolster* (ARS-38) which made two WESTPAC deployments including service in Vietnam. CAPT Mulhern served in various assignments including diving and salvage officer, OPS and XO. In 1972 he was ordered to Harbor Clearance Unit One (HCU-1) as XO and transferred into Reserves status following this tour.

As a reserve officer he maintained his diving qualifications and qualified as a special operations officer. Among other billets, he has held three commands including HCU-1, LongBeach, Naval Weapons Station HQ at Seal Beach, CA and the Mobile Diving and Salvage Unit (MDSU) in San Diego. He was the Region 19 diving officer in 1980; he has taken courses at the Naval War College and the National Defense University. He retired from the Reserves in 1996.

CAPT Mulhern completed an MS in oceanography at Florida Tech in 1976. He has worked at a number of aerospace companies and in the Defense, Environmental and Health Care industries. He is now an engineer-manager with SAIC in San Diego, CA. He is married to Maria Candida Valdez, a native of Baguio, RPI.

RICHARD F. MULLER JR., CAPT, born in New Orleans, LA, Dec. 29, 1923. He graduated from Kings Point, US Merchant Marine Academy in 1946 after spending nine months as a midshipman aboard an Army transport operating between USA and England during the Battle of the Atlantic.

He returned to Tulane University of LA and graduated in mechanical engineering in 1947. From 1947 to 1952 he served in various billets in the Surface Naval Reserve in N.O.

Recalled to active duty in 1952 and served as engineering officer on the USS *Lewis* (DE-535) from 1952 to 1954 actively engaged on patrol in enemy waters in the vicinity of Wanson, Korea. During one patrol in 1953 the *Lewis* was hit by enemy shore batteries. Muller received a Navy Commendation Medal w/Enemy Action "V" for his part in the action.

He served on NR Staffs and as CO of various NR units from 1954 to 1983 when retired as CAPT USNR.

In civilian life he designated and built Petro Chemical Plants for Kaiser Aluminum and Chemical Corp. His family consists of wife Ann, three children and six grandchildren. He and Ann live in Metairie, LA.

Muller's mobilization billet for many years was CO of the Charleston Navy Yard.

RICHARD W. "RICK" MUNSELL, CAPT, born Feb. 7, 1949, received BMed 1971 University of Southern Mississippi, DVM 1984 Mississippi State University.

He joined the service July 15, 1971. His military locations and stations: CV-41, NAS Norfolk, NAS Memphis, NAF Washington, DC. Served with VAW-115, VAW-120, VP-67, CVN-72, NAS Keflavik.

Munsell achieved the rank of CAPT and his memorable experiences: watching a South Vietnamese Cessna land on the Midway.

He married Jane and they have three children: John, Scott and Megan. Employed as North Shelby Animal Hospital, chief veterinarian.

ALEXANDER "LEX" J. MURPHY, JR., LCDR, born Uniontown, PA, on Aug. 17, 1952. Graduate of Villanova University, BS in physics. Commissioned via the Villanova NROTC in May 1974.

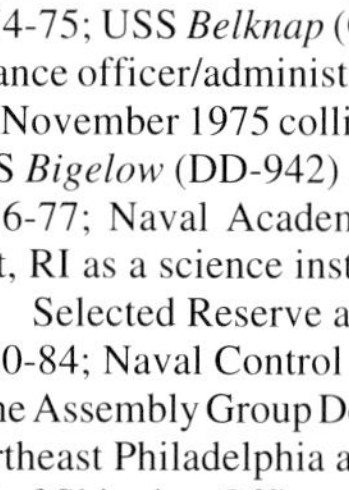
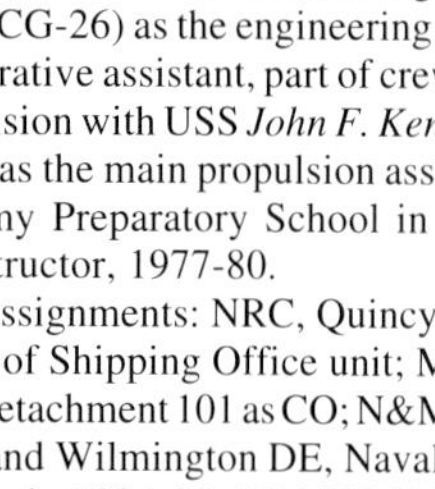

Active duty assignments: USS *Newport News* (CA-148), as M Div. officer, part of decommissioning crew, 1974-75; USS *Belknap* (CG-26) as the engineering maintenance officer/administrative assistant, part of crew during November 1975 collision with USS *John F. Kennedy;* USS *Bigelow* (DD-942) as the main propulsion assistant, 1976-77; Naval Academy Preparatory School in Newport, RI as a science instructor, 1977-80.

Selected Reserve assignments: NRC, Quincy, MA, 1980-84; Naval Control of Shipping Office unit; Mobile Mine Assembly Group Detachment 101 as CO; N&MCRC Northeast Philadelphia and Wilmington DE, Naval Control of Shipping Office unit, 1984-88; N&MCRC, Reading, PA, Readiness Unit as XO, 1989-94, retired as LCDR, June 1994.

Currently, product assurance manager for Lockheed Martin Missiles and Space, Valley Forge.

The Murphy family consists of wife, Anne; son, Brian; and daughter, Gwendolyn. They reside in Reading, PA.

DENISE ELLEN ACKLEY-MURPHY, CDR, born Sept. 9, 1953 in Spokane, WA, the daughter of a career USAF officer. She is a 1975 graduate of Arizona State University.

She entered Aviation Officers Candidate School (AOCS) in October 1976 with the first class to admit women. Upon graduation in February 1976, she was commissioned an ENS in the USNR and was designated an aviation intelligence officer.

Upon completion of the Naval Intelligence Officer Course at Lowry AFB, she reported to Patrol Sqdn ONE, based at NAS Barbers Point, HI. Assigned from 1977 to 1981, she completed two deployments to NAS Cubi Point, Philippines and a detachment to NAS Agana, Guam.

CDR Murphy left active duty in 1980 and affiliated with the USNR. She drilled with NRFIRSTEURLNT 0191 at NAS South Weymouth, MA; CLFINTELL 186 at NOB Norfolk, VA; FOSICPAC RU 120 at Honolulu, HI; NISCPROD 0820 at NPS Monterey, CA; and FIRSTEURLNT 0774 at NAS Jacksonville, FL. In 1990 CDR Murphy joined the Individual Ready Reserve. She is a distinguished graduate of the Naval War College.

Murphy is married to Leo Murphy, USN. They currently reside at NAS Keflavik, Iceland with their four children: Kelli, Ben, Joel and Colin.

NED D. "TED" MURPHY, LCDR, born Jan. 9, 1912 in Grantville, KS, received a BS in business administration, Baker University, 1939.

He joined the service Jan. 1, 1943. His military locations and stations: Brooklyn, NY; served with CO Naval Armed Guard. Served in no organized battles or campaigns.

He received the American-European Theatre, Mediterranean Theater and Asiatic Theater Medals. Achieved the rank of LCDR and discharged Jan. 9, 1972.

His memorable experiences: while traveling in convoy to Italy, as they approached Gibralter, they received a flag message from the convoy commodore, that enemy subs were in their vicinity. At that moment, his ship broke down and they had to drop out of convoy. About an hour was spent on repairs, then full speed ahead to rejoin convoy.

He married Dolores and they have two children, Rodney and Donna; and grandchildren: Rick, Sean, Stephanie, Heather, Larry, Tim, Christy, Lori and Jennifer. Employed as public school teacher, instructor and supervisor in Federal Bureau of Prisons.

His final 15 years in the USN was spent in the Naval Reserve Officers School in Kansas City, MO as follows: three years as student; ten years as instructor and two years as XO.

ANTHONY H. "TONY" MURRAY JR., RADM, born June 7, 1926, Bethlehem, PA. Entered Pennsylvania Maritime Academy, Feb. 10, 1944, graduated and commissioned ENS Aug. 10, 1945. He sailed in Merchant Marine, 1945-46; USS *Mississippi*, 1946-47 and released to inactive duty in July 1947.

Attached in 1949 to Submarine Reserve Div. 4-37, Philadelphia. Recalled to active duty Oct. 14, 1951 and assigned to USS *Threadfin* (SS-410); assigned to USS *Icefish* and transferred to Royal Netherlands navy in February 1953 to ComSubLant Staff as liaison officer. In June 1953 transferred to USS Sarda and released to inactive duty in August 1953.

Joined the Naval Reserve Air Wing Staff 75, October 1953, Lakehurst, NJ; transferred September 1954 to VS Sqdn. 751; April 1955 to AAU 931 Willow Grove; Jan. 1, 1956, VP-931; Submarine Reserve Div. 4-37, July 1, 1957, taking command in July 1958 and was XO, 1959-62.

Attended NWC, Newport, RI, 1961; CO, Submarine Reserve Div. 4-37, 1962-65; NWC, Newport, 1963; chief of staff of Naval Reserve Group Cmd. 4-9(L), 1965-67; NWC, Washington, DC, 1966; served as chairman and senior member of the Fourth Naval District Reserve Policy Board; group commander, NR Group Command 4-9(L), 1970-71; member of the Reserve Captain Selection Board in BUPERS; 1971; chairman of the Counseling Board, NR Group Cmd. 4-10(M), 1971-72; naval representative to Pennsylvania Military Support Plans Officer, 1972-73; Commandant's local rep. as OinC of the Philadelphia Navy Department Bicentennial Group, 1973-75; Readiness Commander, Philadelphia Readiness Command, 1975-77; director NR Merchant Marine Program, 1977-81; Commander Military Sealift Command, Atlantic, April 1981 until his retirement in August 1981.

Awards and medals include Legion of Merit w/Gold Star, Meritorious Service Medal, Navy Commendation Medal, (Secretary of the Navy) Distinguished Public Service Award, Pennsylvania Commendation Medal, American Theater WWII, Pacific Theater WWII, WWII Victory Medal, Armed Forces Reserve Medal w/HG, Naval Reserve Medal, Atlantic War Zone WWII Merchant Marine, Pacific War Zone WWII Merchant Marine, and Victory Medal WWII Merchant Marine.

Married Wilma A. Murray and their children are Susan E. Kozub, CDR Anthony Murray III, LCDR Kenneth Murray, Virginia Franklyn and LCDR Michael L. Murray. They also have seven grandchildren.

Graduated, St. Joseph's College, Philadelphia, PA, June 1950, with BS in business administration and from Temple University, School of Law, Philadelphia with JD degree in 1957 and admitted to practice law in Pennsylvania. From 1974 to present has been in private law practice and Real Estate Broker.

JAMES W. "JIM" MURRAY, CAPT, born March 9, 1929 in Oakland, CA, received a BSCE from the University of California, Berkeley, January 1951; master of engineering, January 1956.

Midshipman NROTC September 1947-August 1949, UCLA; University of California, Berkeley, September 1949-January 1951.

Active duty: 3 1/2 years commencing Jan. 26, 1951 as ENS, USN. His military locations and stations: USS *John R. Craig* (DD-885), Korean waters with Task Force 77 and 95, April 1951-August 1951; CEC Officers School, Port Hueneme, CA. Transferred from Line to Civil Engineer Corps, September 1951-December 1951; 103rd NCB & MCB-10, Guam of which three months were with Detachment A on ChiChi Jima, Bonin Islands, January 1952-April 1943; Operational Training Unit and Naval Construction Battalion Center, Port Hueneme, CA, May 1953-July 1974. Released from active duty as LTJG, CEC, USNR, July 24, 1954.

In active duty in the USNR. From 1954-73, the major portion of his inactive USNR service was in various assignments with Seabee units with headquarters in the San Francisco Bay area. This duty included one year as CO of Seebee Div. 12-2; two years as XO of MCB-2, and two years as chief staff officer of the Fifth Reserve Naval Construction Regt. He served on two CEC Selection Boards, one in May and one in August 1971. He was promoted to CAPT, Civil Engineers Corps in 1971.

From 1973 to 1983 he served with the Captain's Management Assistance Unit at NAS Alameda and at NAS Moffett Field, CA. CAPT Murray retired from the USNR on Feb. 1, 1983.

His military awards include the Navy Commendation Medal, Korean Service Medal w/3 stars, Korean Presidential Unit Citation, National Defense Service Medal, Armed Forces Reserve Medal (HG Device), and UN Service Medal.

He was registered in California as a civil engineer in 1958 and as a structural engineer in 1962. He worked in the area of structural design, plan review, and construction of buildings for nearly 40 years commencing in 1956, with the last 34 years involved almost exclusively with schools and hospitals. He retired from civilian employment in November 1995. He married Alice in 1987 and has four children, three stepchildren and eight grandchildren.

MICHAEL J. MURRAY, CDR, born April 23, 1951 in Providence, RI, attended De La Salle Academy, Newport, RI; received an AB from the College of Holy Cross, 1973; JD, 1980, Franklin Pierce Law Center.

He joined the service June 1, 1973 and his military locations and stations: legal officer, USS *Lexington;* CO, MSCO London, UK; Department head, War Games and Analysis Dept., Naval War College. He was awarded the Navy Commendation Medal; Surface Warfare Office.

Murray achieved the rank of CDR. His memorable experiences: USN, 1973-77; USNR, 1977-93.

He married to Teresa T. and they have three children: Mellissa, Rebecca and Alison. Employed as partner, Murray & Cutcliffe, Attorneys At Law, retiring June 30, 1993.

JAMES C. "JIM" MYERS, CAPT, born Aug. 26, 1941 in Chicago, IL, received a BA from Washington University in 1964. He joined the service Dec. 18, 1962. His military locations and stations: NAVCOMSTA Honolulu, HI. Served with NROIDET 310.

His awards and medals: five Navy Commendations, four Meritorious Unit Medals, Recruiting and Overseas. He achieved the rank of CAPT. His memorable experiences: created 75th anniversary poster-USNR; revised fleet hometown news form to include civilian info; 50th anniversary Pearl Harbor; 100th anniversary Statute of Liberty Celebration; campaign drug free.

Married to Judy Schnitzer and they have children: Jeffrey, USN and Jeremy and grandchildren: Daniel and Rebecca. Employed as president, Motivators, Inc., retiring Nov. 1, 1996.

CHAPLAIN JIMMY LAIRD MYERS, LCDR, born March 11, 1953 in Winston-Salem, NC to Tildren R. and Mattie Lois Laird Myers. He received a BA cum laude in history from Wake Forest University in 1975, and an MDiv magna cum laude from Emory University in 1978. He was ordained an Elder in the United Methodist Church on June 8, 1979, and he served as pastor of Greer's Chapel and Mt. Carmel United Methodist Churches of Lexington, NC from June 1978 until September 1985.

He was commissioned a LTJG in the USNR on Sept. 19, 1985, and he graduated with class B86001 at the Naval Chaplains School in Newport, RI in November 1985.

He reported for active duty with the 2d Marine Div., and was initially assigned to HQ Bn. as assistant battalion chaplain in December 1985, but in January 1986, he was assigned to 1st Bn., 10th Marine Regt. as battalion chaplain. He participated with his unit in NATO exercises 200 miles north of the Arctic Circle in Norway during February and March 1986 when it endured extreme cold weather conditions reaching a wind chill factor of -68 degrees F. He was promoted to LT on Nov. 1, 1986, and he received the Fleet Marine Force Ribbon for his service with the USMC in December 1986. In June 1986, Chaplain Myers was assigned to 5th Bn., 10th Marine Regt., as battalion chaplain.

Chaplain Myers left active duty and entered the US Naval Ready Reserve in October 1988, and reported for duty to Camp Lejeune Naval Hospital Detachment 207 in Greensboro, NC in January 1989. He was given additional duty as Unit Chaplain with Communications Co., HQ and Service Bn., 4th Force Service Support Gp., 4th Marine Div., USMCR. While serving with this unit, he received the National Defense Medal and a Unit Commendation Ribbon for stateside service during the Persian Gulf War.

Chaplain Myers was transferred to Personnel Mobilization Team 1007 of the US Naval Ready Reserve on Oct. 1, 1992. He served with this unit at three Reserve centers: Greensboro, NC until April 1, 1993; Charleston, SC until March 1, 1996; and Wilmington, NC. He was promoted to LCDR on Aug. 1, 1993. He completed the Advanced Reserve Chaplain's Course at the Naval Chaplain's School in Newport, RI in July 1994.

Chaplain Myers was transferred to Naval Reserve Naval Hospital Beaufort and cross assigned into Naval

Reserve Naval Hospital Charleston Detachment 607 on Nov. 1, 1996. He drills with that unit in Greensboro, NC.

In May 1991, he received a JD degree from the School of Law of the University of North Carolina at Chapel Hill. He practiced law with Mr. Garry W. Frank of Lexington, NC from September 1991 to October 1992 when he opened his own law offices in Advance and Lexington, NC. During this time, he held a special appointment beyond the local church as a legal mediator in the Western NC Conference of the United Methodist Church. He also served on the Ethics Committee of the Lexington Memorial Hospital. In November 1994, he was elected District Court Judge in the 22nd Judicial District of North Carolina. He took office in December 1994.

Chaplain Myers lives in Advance, NC and maintains a teaching and preaching ministry in addition to his duties as a judge and Naval Reserve Chaplain. He is also very active in local civic organizations.

JOHN J. "JACK" MYHRE, CDR, born April 15, 1924 in St. Paul, MN and attended three years of college.

He joined the service July 18, 1942. His military locations and stations: V-1 then V-5 Program received wings and commissioned Corpus Christi, TX, April 25, 1945. Served with utility unit NAS Banana River, FL; NAS Kobler Field, Saipan.

Served in the American Theater Area; Victory; and Asiatic-Pacific Campaigns. Myhre achieved the rank of CDR and discharged from the service Dec. 2, 1945 and retired May 3, 1947.

His memorable experiences: just being able to fly in the Navy was great.

He is married to Elsie R. and they have children: Mary Margaret, Steven Jackson and John Jay and a granddaughter, Ashley Fidler. Employed with the US Department of Agriculture as inspector and pilot. Retired with no pay, Feb. 1, 1976; w/pay April 15, 1984.

DAVID GEORGE MYRICK, CAPT, a native of Montgomery, AL, he graduated from Huntingdon College in 1967 and was commissioned ENS USNR. He served on active duty as a CIC watch officer onboard USS *Coral Sea* (CVA-43) and made two WESPAC deployments in support of US operations in Vietnam.

In 1971 after release from active duty, he affiliated with the USNR and served in various billets for the next 24 years. His significant commands included NR LST 509, Troy, AL; NR FFG 4 TALBOT, Montgomery, AL; MIUW 209, Mobile, AL; and Readiness Unit, San Antonio, TX. In addition he served as maintenance officer on the staff of COMNAVIUWGRU TWO, Williamsburg, VA and as Director of Plans for Naval Reserve Readiness Command Region Ten, NOLA.

His personal decorations include the Meritorious Service Medal and Naval Commendation Medal (gold star in lieu of 2nd award). Unit awards include Navy Unit Commendation and Meritorious Unit Citation. Myrick retired on Oct. 1, 1994.

He serves as a district manager for the Alfa Insurance Companies and resides in Mobile, AL.

JOHN C. NATASI, LT born Jan. 2, 1942 in Brooklyn, NY, graduated from RCA Institute, NY, 1964, with associate electronics technology degree; graduated Polytechnic Institute, NY, BSEE, 1975; graduated Adelphi University Garden City, NY, MBA, 1979.

Enlisted USNR Aug. 16, 1961 (2x6 program) seaman recruit. His military locations and stations: two years active duty, USNAVCOMSTA, ASMARA Ethiopia from 1964 to 1966. Served in Brooklyn, NY, 3-49(L), 3-53(L)9-COM DIV 3-1, 1961-64; Brooklyn, NY, 3-49(L), 1967-73; Staten Island, NY, Repair Unit AD3502, 1974-76; Whitestone, NY, Fleet Training Gp. (DET 102), 1977-79; Staten Island, NY, NRMTF, NY (DET 102), 1980-93.

His awards and medals include the National Defense, Naval Reserve Meritorious, Armed Forces Reserve. Achieved the rank of CWO2, 1980; LTJG, 1984; LT, 1986 and discharged August 1993.

His memorable experiences: during his two year active duty in Ethiopa, North East Africa he was lucky to visit the Holy Land (during leave) in 1966 (one year before the Israel/Arab War.

He married Loretta Ciccolallo and they have two children, Joanne and John. Employed as electronic tech, R&D (1967-70) and NYNEX, associate director, 1970-96. He retired Oct. 31, 1996.

IRWIN D. NATHANSON, CAPT, a native of Brooklyn, NY, graduated (first in his class) from the State University of New York Maritime College, Fort Schuyler June 1969. After sailing on his deck license in the Merchant Marine, he came ashore and started working in foreign military sales with the (then) Naval Ordnance Systems Command in Washington. At the same time, he started actively drilling in the USNR.

His first unit was Military Sea Transportation Service (MSTS) Unit COMHQ DET 106 (Shortly thereafter, MSTS became MSC - the Military Sealift Command.) Thus began a 27 year affiliation with MSC and related USNR programs.

As his civilian career progressed, so did Nathanson's career in the Reserves. He served in units in Adelphi, MD; Washington, DC; Houston and Orange, TX; Binghampton and Syracuse, NY and finally Baltimore, MD. As a CDR, he also served nine months active duty at MSC HQ.

As a CDR he served as XO for four years and as a CO for two. He then served two consecutive CO tours as an 0-6 before retiring in September 1995. At the time of his retirement, CAPT Nathanson was the last military person (officer or enlisted) to have served in MSTS. He was also the most experiences naval officer (active duty or reserve) in military sealift.

He is currently vice-president, International of DENRO, Gaithersburg, MD. He is responsible for all foreign sales of DENRO's Voice Communications Systems used for civil and military Air Traffic Control and for military Command, Control and Communications.

His sons, Joshua and Matthew, attend college in New York, respectively at the School of Visual Arts and the Culinary Institute of America. Julie, his wife of 27 years, is a data base manager for Bell Atlantic's Tele-TV Div.

His medals include Meritorious Service and Navy Commendation. He holds an MS in International Commerce, from George Washington University. He has been a member of the NRA since he was an ENS.

JAMES A. "JIM" NATTER, CAPT, born Aug. 17, 1950 in Birmingham, AL, received BS from US Naval Academy; MS, Georgia State University; PsyD, Adler School of Professional Psychology.

He joined the service June 1968 at US Naval Academy. His military locations and stations: 5W05, Newport: NR USS *Orion* (AS-18) DET 101; NR FLTSUPTRA; DET 1108; NR COMSECONDFLT; CO DET 113; NR CINCUSNAVEUR DET 513; CO MIUW 213; USS *Milwaukee* (AOR-2); USS *Josephus Daniels* (CG-27); USS *Valdez* (FF-1097).

He was awarded the MSM, NCM, National Defense, Vietnam Service and NR Sea Service. He achieved the rank of CAPT. Discharged from active duty December 1979. Still in Reserves.

His memorable experiences: Serving off Vietnam in 72-73 replenishing four carriers; tracking the Kiev in the Mediterranean; traveling to Bermuda and Honduras on *Valdez;* traveling to Mid-Canal Panama in 1995 and withstanding seal attacks during fleet exercise in 1996 while CO of MIUW 213.

Natter is single with four children: Daniel, Joseph, Scott and Tiffany. Employed with Central Dupage Pastoral Counseling Center, director of counseling services.

KEVIN S. NAUGHTEN, ENS born April 19, 1964. His military locations and stations: NAS North Island, NAB Coronado. Units served with: USS *Kitty Hawk* (CV-63) (1986-87), Seal Team Three (1987-90). His battles and campaigns: Operation Earnest Will, 1987-88; Operation Praying Manits (1988).

He was awarded the Joint Chiefs of Staff crest, Small Craft Insignia, Basic Parachutist, Navy Achievement Medal, Army Achievement Medal, Joint Meritorious Unit Commendation (3x), Meritorious Unit Commendation (2x), Battle "E", Good Conduct Medal, Naval Reserve Meritorious Service Medal, Navy Expeditionary Medal, Armed Forces Expeditionary Medal, Sea Service Awards (2x), Armed Forces Reserve Medal, Pistol Marksmanship Medal w/Expert and Rifle Marksmanship Medal w/"Expert."

He received direct commission in August 1995. Currently ENS (0-1E).

His memorable experiences: attacking the Iranian oil rigs on his birthday April 19, 1988 during Operation Praying Mantis in the Southern Persian Gulf. Receiving his direct commission from MAJ GEN Patrick Hughes, US Army, Joint Chiefs of Staff J-2 aboard the USS *Barry* (DD-933) at Washington Navy Yard with his (Mustang) CO, father, wife, son and family present.

He married Kelley and they have a son, Conner H. Employed with US Army Technical Testing and Training Agency, as operations officer.

EDWARD J. "ED" NAVA, LCDR, born Nov. 27, 1952 in Wiesbaden, Germany, received a master of science, EE, Stanford and BS, EE University of New Mexico.

He joined the service October 1985. His military locations and stations: Pearl Harbor, Long Beach, Puget Sound. Served with Naval Shipyards Long Beach and Pearl Harbor, SPAWAR.

His awards and medals: Military Outstanding Volunteer, MUC, Recruiting, NAT Defense Service, Rifle Expert, Pistol SS. He achieved the rank of LCDR.

His memorable experiences: completed engineering duty officer qualification program 1989, NRA national vice-president 1994-96.

He is divorced and has three children: Mark, Lorenzo and Rebecca. Employed at Sandia National Labs, as technical department manager.

TINA L. NAWROCKI, LCDR, born Sept. 17, 1960 in Lewiston, ME, received a MSN (masters in nursing) from Catholic University, Washington, DC.

She joined the Reserves August 1989; active duty January 1, 1991. She is still on active duty.

Her military locations and stations: New Orleans, LA; Jacksonville, FL; Boston, MA; Washington, DC.

Served with BRCL El Centro, NavHos JAX; NRD New England Catholic University (master's student). She was awarded the Sea Service Deployment Ribbon, Recruiting Ribbon, five Navy Achievement Medals, Humanitarian Service Medal, National Defense and Joint Meritorious Unit Commendation.

She achieved the rank of LCDR. Her memorable experiences: Humanitarian mission; Haitian Relief Crisis, Guantanamo Bay, Cuba, December 1991-March 1992.

Naworcki is divorced and has two children, Kristin and Catelyn.

DOROTHY M. NAZARENUS, CAPT, a native of Denver, CO, graduated from the University of Denver in 1961 with a bachelor of music degree. She received her USNR commission at Newport, RI on Dec. 18, 1964 and reported to NPC in Washington, DC for her first tour. After augmenting to USN, she had tours of duty in Kunia, HI; Boston, Charleston and Virginia Beach.

During her four year tour at the naval station in Charleston, she earned her master of business administration degree at The Citadel and also acquired a commercial pilot license with an instrument rating.

She left active duty in May 1978, accepted a USNR commission and began drilling with the NCSO Far East 118 unit in Denver. She had three commands, which are NAVCOMMSTA Philippines (Denver), SIMA San Diego (Wichita, KS) and NCSO 118 (Denver). She served three tours on two major staffs - as deputy for readiness and retention officer for REDCOM18 (Olathe, KS); and as mobilization officer for REDCOM20 (San Francisco).

JOHN S. "JACK" NEAL JR., CAPT, born May 23, 1917, in Lincoln, NE. Earned AB degree from Franklin & Marshall College and LLB from Temple University. Joined the service in September 1940V-7 Program.

Military Locations and Stations: 1940-43 Prairie State, USS *Illinois*, NY Naval Militia, Amphibious Force, Atlantic Fleet, Norfolk, VA. 1943-44 Little Creek, VA Amphib. Force; OIC, Joint Air Amphib. signal Communications School. 1944-45, USNA Post Graduate School, Annapolis, MD and Cambridge, MA Harvard University.

From 1946-51, NR Officer School, Chester, PA; 1951-53, Staff Potomac River Naval Command; 1953-62, NR, Officer School, Princeton University as instructor of military justice and strategy and tactics.

Memorable Experiences: North Atlantic winter crossings; North African landing amid combat - in Morocco, landing in Sicily, Italy; duty in Washington DC moving US Constitution to new archives building.

Married the former Barbara Cade Stratton and they have four children: Judith, Nancy, George and William; 11 grandchildren: Adam, Matthew, Eleanor, Megan, Wendy, Kurt, Jeanne, Benjamin, Margaret, Meredith and Elizabeth; and two great-grandchildren, Jennifer and Amanda.

ROBERT P. NEAL, LT, served 1957-61; honorable discharged an ET2-P2. Reenlisted 1978 under APG program as AT1 with VR-51 at Glenview NAS. Flew more than 1000 hours as flight communications operator on the C-118 and more than 1000 hours as flight attendant on the C-9, all with VR-51 while serving as work center supervisor and division chief.

Earned the rank of ATC authorized to wear Air Warfare Wings and subsequently (through the LDO program) was commissioned ENS in 1987. Promoted to LTJG in 1989 and full LT in 1991, he served as division officer and then maintenance/material control officer with many collateral duties including PAO and weight and balance officer. He was awarded a Navy Achievement Medal in 1994.

From VR-51's decommissioning date September 1994 until June 1996, was attached to VTU1314G, tad to NAVINFO Midwest and VTU7272G. He was then assigned to NR Adak, AK as division officer/PAO and when the unit was changed to NR NSA Souda Bay, Crete, he was assigned as training officer and public affairs officer where he continues to serve.

He continues to reside in Wadsworth, IL with his wife Pat. They have five children and four grandchildren. He is a life member of NERA, NRA, Naval Order, USN Public Affairs Alumni Association, Navy League (serving on chapter board), Bluejackets Association (serving as national president), and ROA (serving as legislative chairman for Illinois).

WILLIAM S. "BILL" NEAL, ESQ., LCDR, born July 14, 1948, Darby, PA. Earned AB degree, Dartmouth College; AA, Bucks Co. Community College; graduate work, Dickinson, Law and Temple University Business. Joined the service July 1968, NROTC, Purdue University, West Lafayette, IN; NROTC Dartmouth College, Hanover, NH.

Stationed at Naval Base, Long Beach, CA; Fleet Anti-Air Warfare Training Center, Dam Neck, Virginia Beach, VA; Naval Shipyard, Charleston, SC; Naval Base, Newport, RI; Naval Base, Norfolk, VA; USNR Center, Trenton, NY and Harrisburg, PA; USNR/USMCR Center, NE Philadelphia; USNR Center, Philadelphia; USNR/USMCR Center, New York at Naval Militia Armory, New Rochelle, NY; USNR Readiness Center, Kearney, NJ.

Units served with include USS *Fechteller* (DD-870), *Trippe* (DE-1075), *Paul* (DE-1080), *Sierra* (AD-18), *Connole* (DE-1078), (DD-863), *Tripoli* (LPH-10), *Guam* (LPH-8), NROTC Unit Purdue University and NROTC Unit Dartmouth College.

Memorable experiences include officer training, underway in destroyers, duty in Turkey, CO tour as reservist, joint duty, and conning a helicopter carrier in the Pacific.

Discharged Jan. 29, 1972. Awards and medals include National Defense, Naval Reserve, Rifle Marksmanship (S), Pistol Marksmanship (E). Naval Reserve Association, Reserve Officer Association (past Navy VP, Dept. of PA) and New York Naval Militia (five years).

Married Ellen Michelle Westerman and they have two children, Adam and Matthew; and two grandchildren, Jennifer and Amanda. Civilian employment as lawyer, civic and professional posts. He retired Sept. 18, 1994.

NORMAN W. NELSEN, CAPT, born in Seattle, WA, Sept. 30, 1926. Raised in Tacoma, WA, after the age of four.

A graduate of the University of Washington, he began his naval career in July of 1944 in the V-12 Unit at the U of W. The V-12 concluded in 1946, he continued his college education in the peacetime NROTC. Graduated June 1948 with bachelor of science in mechanical engineering plus commission as ENS, USNR.

Active duty from early 1953 to 1955 as engineering officer, USS *Leray Wilson* (DE-414).

Attached to Fleet Div. 13-2 in Tacoma from 1952 to 1961 aboard *Brannon* (DE-446), *Robin* (MSCO 53) and *Watts* (DD-567). Fleet Divs. had become selected res crews. Watts a part of "Berlin" call up (October 1961-August 1962).

Subsequent assignments: XO, USS *Watts* Res Crew; MOB Team 13-6; CO USS *Marshall* (DD-676) Res Crew October 1965-January 1968; Group Command 13-4 (s) various billets ending as chief of staff (13-4). Retired as captain Sept. 30, 1986.

Nelsen's civilian employment consisted of four years as a professional mechanical engineer with a consulting firm plus 27 years with an industrial mechanical contracting firm. He assisted in the start-up of same in 1962, eventually becoming principal stockholder and retiring as president of the firm in 1989.

The Nelsen family consists of wife Constance, three daughters: Leslie, Erica and Victoria; and two grandchildren, Lauren and Kaitlyn.

CHARLIE S. "CHAZZ" NELSON, CAPT, born July 23, 1920, Pelzer, SC. Earned AB degree in journalism, UNC, Chapel Hill, NC.

Military Locations and Stations: USNR Midshipman School, Columbia Univ., NY,NY; NROTC Univ. Idaho, Asst. Prof. Nav. SCI; NROTC, Vanderbilt Univ.; Operational Development Force, Norfolk, VA, Tactical Development Off.; Naval and Marine Corps Reserve Training Center, Seattle, WA, XO/CO; NR Group, Los Alamitos, CA Command; NR Fleet Mgmt. Asst. Unit, Pacific Fleet, Honolulu, HI, CO; CincPacFleet Command Operations Control Center.

Units Served In: USS *Claxton* (DD-571), DESRON 23; USS New York (BB-34); USS *Consolation* (AH-15), OPS/NAV; USS *Edisto* (AGB-2); USS *Basilone* (DDE-624) XO; USS *Vammen* (DE-644) CO; Reserve Destroyer Div. 271, Commander (two tours); Commander USS *Vietnam* Task Element, Gulf of Siam.

Battles: Empress Augusta Bay, Cape St. George, Leyte Gulf, Samar; Invasions: Leyte Island, Negros, Cebu, Palawan Island, Lingayan Gulf, Corregidor, Nasugbu; and South Vietnam Campaign off shore patrols.

Memorable Experiences: damage control officer of DD when rear end blew off twice; all battles with 31 Knot Burke; battles of Surigao Strait and off Samar, Philippines; discovery and return of ADM Peary's documents proving North Pole discovery; first transit of NW Passage from east to Arctic Ocean at Melville Island; first transit of Foxe Basin, NW Terr. from Fury and Hecla Strait, south to Hudson Strait; OinC CINCPAC Fleet Command and Control Center during up-temp bombings using "smart bombs during S. Vietnam Campaign. Retired, CAPT, USNR, March 1974.

Married Margaret Alexander and has two children, two step-children, five grandsons and two great-grandsons. Civilian employment with United Services Life Ins. Co., Arlington, VA

DEAN C. NELSON, CAPT, CAPT, raised on a farm near Viborg, SD, graduated from South Dakota State University with a degree in mechanical engineering. He entered the Navy's NAVCAD program in 1953, receiving his wings and commission in March 1955. Following LTA training, he served the remainder of his active duty (1955-57) with Airship Sqdn. TWO (ZP-2). As a drilling reservist he held several billets in Naval Surface Units, including that of CO, and VTUs. Later he served in the Naval Engineering Branch.

He and his wife Beverly, also from Viborg, now enjoy life in Port Ludlow, WA, after having lived in Idaho, NM and Washington where he held several engineering and project manager positions in the field of nuclear waste management. He retired from Los Alamos National Laboratory in 1993.

DONALD L. NELSON, CAPT, born Oct. 15, 1942 in Duluth, MN, received a BA from the University of Minnesota in 1964; JD from the University of Wisconsin Law School, 1971.

He joined the service July 17, 1968. Achieved the rank of CAPT and is still serving in the USNR.

Nelson is divorced and has custody of two children, Donald L. Jr. and Katherine Anne. He is an attorney.

GEORGE G. NELSON JR., LCDR, born in Haughton, LA, July 7, 1931. He attended Louisiana Tech University and Louisiana State University, graduating in 1952 with a BS degree in chemical engineering. He was accepted into Officer Candidate School at Newport, RI, graduating sixth in his class, December 1952. After six months at the Naval Intelligence School, he was assigned as an intelligence analyst and security officer with the Scientific and Technical Intelligence Unit, Office of Naval Intelligence, Washington, DC.

Released from active duty in December 1955, he

continued as an active member of the USNR, retiring as a LCDR after 20 years service. Reserve duties included training officer for a medium surface division, recruiting and instructor Naval Reserve Officers School, Mobile, AL. Active duty for training included intelligence assignments, two-weeks at the Foreign Service Institute in Washington, DC and two weeks aboard DDG-2 (Charles F. Adams). He is a life member of the NRA and a plankowner, US Navy Memorial.

Nelson retired from International Paper in June 1992, with 36 years service including chief-pulp research, Southern Kraft Div.; manager technical and environmental services and manager employee training, Natchez mill. Later, as a training consultant at International Paper's mill in Selma, AL, he developed and supervised implementation of an operator training certification program from a major mill expansion project.

The Nelson family includes his wife, Dona, who was a Navy Department secretary, 1955-56, and two married daughters. The Nelsons are active in their church and community and in model railroading. They live in Fairhope, AL. *Submitted by George G. Nelson Jr.*

HOMER B. NELSON, LT, born Feb. 21, 1916 in Osterville, MA, received a BA in mechanical engineering, plus 20 hours aero engineering.

He joined the service June 1942. His military stations and locations: NAS Jacksonville, FL. Served with BUAER, Navy Department, Washington, DC. Nelson achieved the rank of LT, USN and was discharged Dec. 1, 1945.

Married Mildred and they have a son, Brent A. Nelson. Employed as senior analytical engineer, Cooper Ind. Superior Div. He retired Feb. 21, 1981.

RONALD DUANE "RON" NELSON, CDR, is the only son of Veteran P-47 fighter pilot 1LT Lincoln W. Nelson, USAAF, deceased, and his wife Rose Marie. He was born in Santa Monica, CA, Aug. 30, 1945, and earned his BA in political science/international relations at Whittier College, CA, February 1968. He has an MBA in management from California State University, Carson, CA.

In June 1966 the University of Southern California AFROTC cadet captain received the outstanding achievement award from the Sons of the American Revolution. The AF program also let him earn a private pilot's license. Switching to the Sea Service he attended Aviation Officer Candidate School in Pensacola, FL and was commissioned an ENS, Oct. 11, 1968. He completed VT-1 (T-34B Mentor), VT-7 (T-2A Buckeye jet), VT-9 (T-2B), and VT-4 (T-2B) where he completed aerial gunnery and on Sept. 11, 1969 carrier qualified on the USS *Wasp* (CVS-18) in the twin engine-centerline T-2B Buckeye jet. Following 59 hours of advanced jet training in the TF-9J Cougar at NAS Kingsville, TX, LTJG Nelson was transferred to USNS Midway Island as RPS custodian. The last five of his 17 months on Midway he was the communications department officer, a LCDR billet.

Leaving active duty as an aviation ground officer (1355), LT Nelson affiliated with the drilling reserves as an 1105 (unrestricted line, surface) at USNS Long Beach serving as department officer in: NRSD-11(L), AD7019, LPA-219, LPA-248 USS *Paul Revere* and CNSP 319 (the first reserve unit on the west coast to write and execute a multi-service Fleet Amphibious Assault Exercise "Silver Strand"). He was XO of ACU-1 1819(LCM 8). At NS Philadelphia he was attached to the GVTU and PMT 504. Moving to the Milwaukee area of Wisconsin, in addition to time in the VTU: NAVELEX 313, SPAWAR HQ313, and CO of Personnel Mobilization Team 1813 in nearby Great Lakes, IL.

Professionally challenging annual training sessions: afloat with DESRON 37 twice, and DESRON 27/SURFRON 1 five times; ACU-1 twice, COMNAVSURFLANT, NRPC three times, Space and Naval Warfare Systems Command four times and the Naval War College's Center For War Gaming.

With two National Defense, Naval Reserve and a Sea Service Awards, he retired as CDR July 1, 1995. Divorced he has two sons Duane (October 1979) and Brian (May 1982). CDR Nelson is a life member of both the USNR and Air Force Associations. A former marketing management associate and district manager with Merck Shar & Dohme, he has been self-employed since 1991, entertainment productions.

JOSEPH J. NEMANIC, CAPT, born Aug. 31, 1920 in Colorado, received an AA in forestry and a BS in Wildlife Management.

He joined the service May 1942 in San Francisco, CA. His military locations and stations: Pacific (DE-402) and 408) radar and XO. Served with Escort Div. 68; CVE Task Group. His battles and campaigns: Morotai, Leyte, Luzon, Iwo Jima, Okinawa.

Awarded the American-Pacific w/5 stars, Philippines. Achieved the rank of CAPT and discharged in 1973.

His memorable experiences: Philippine Sea Battle; kamikaze attacks. Rescuing survivors from *Swanee, Shelton* and ditched airmen. His first ship *Zaca* (Errol Flynn's yacht) a sail ship armed with 50 cal. patroling northern water is suspected Jap activity. Hit major storm, lost main boom, fresh water and almost ship. His life in balance first swinging boom missed his head by one-inch. Second volunteered to go over side and cut boom free. The boom sprung up hitting him in face. He blacked out, his feet dangled, his arms held.

He married Geneve and they have two children, Edward and John. Self-employed real estate, farming. Retired from the USN in 1973 and civilian employment, 1985.

MICHAEL P. "MIKE" NEMCHICK, RADM, born July 6, 1927, Eckley, PA. He enlisted in the USN in 1945 and was discharged in 1947. Using the GI Bill, he enrolled in King's College, Wilkes-Barre, PA and received his bachelor's degree in 1951, graduating magna cum laude.

Military Service: 1945-47 active duty; 1947-50 selected reserve duty; commissioned ENS, USNR, 1950, officer member of NR Surface Div. 4-57, Kingston, PA; 1951 officer member of NR Surface 5-40, Wash., DC; 1965 CO, NR Surface Div. 5-40; 1966 CO Recruit Trng. Unit, Wash., DC; 1969 chief staff officer NR Gp. 5-12, Wash., DC; 1972 CDR, NR Gp. 5-5(M), Adelphi, MD; 1975 CO, Dir. Naval Education & Training HQ Unit 106; 1976 Deputy Readiness Commander, Baltimore, MD; 1978 Deputy Readiness Commander, Naval Reserve Readiness Command, Region Six, Washington, DC.

In August 1979, promoted to RADM, USNR; 1976-80 Inspector General, NR Readiness Command, Region Six, Wash., DC; 1980-83, Reserve Flag Officer Unit, Chief of NR New Orleans. Retired with pay July 1987, RADM 0-8.

Civilian employment: teacher, principal, owner of personal income tax service and agent for Nationwide Ins. Co. He passed away in December 1993. He is survived by wife Gerri, four children: Michael, Barbara, Donna and Joan; and four grandchildren: Julia, Rebecca, Michael Jr. and Anthony Vallone.

EDWIN C. "ED" NEUMAN, CAPT, born Dec. 4, 1931 in Philadelphia, PA, received a BS from Drexel University.

He joined the service Sept. 28, 1954. His military locations and stations: NAS Jacksonville; NAS Glenview; NAS Willow Grove. Served with VP-5; VP-725; VP-933; VP-66, CO VP-64.

Neuman received the US Atlantic Fleet Battle Readiness Excellence Pennant "E", Armed Forces Reserve, National Defense and US Navy Expeditions. He achieved the rank of CAPT and discharged December 1991.

He is married to Alice M. Neuman and they have a daughter, Nancy Matthews. Employed with Allis-Chalmers; ITT A-C pump, general manager, retiring December 1993.

GEORGE NEVEN JR., CWO4, born in New York City on April 30, 1929. He enlisted in the USN in July 1947 at Albany, NY. Attended three Naval Photo Schools in Pensacola, FL. He served at NAS Seattle, VU-7, VU-1, Salisbury Sound AV-13, Combat Camera Gp. and MSTS San Francisco as an aerial photographer. He was honorably discharged in July 1956 as a PH1.

Employed as a photographic specialist for 35 years at Cape Canaveral and the Kennedy Space Center, where he was directly involved with missile launches including: Atlas, Titan, Thor Delta, Saturn IV & V, Polaris, Trident, Minuteman, Pershing and Manned Space Launches; Mercury, Gemini, Apollo, Space Shuttle.

He joined the USNR in Jacksonville, FL in May 1972 as PH1 and was promoted to CWO2 in December 1977. He served as an intelligence specialist with a unit at PAFB, FL until he retired as a CWO4 in April 1989.

Married 48 years, the Neven family consists of wife Sharon; sons, Christopher and Lance; daughters, Candace M. Adamick and Dr. Lisa G. Neven-Godwin. Son Chris is a retired ATC (AW) from the TARS and daughter Lisa is a research scientist with the USDA in Yakima, WA. They have four grandchildren: Aaron J. and Alexis H. Adamick and Holly M. and Joy E. Neven.

They reside in Mims, FL and he has retired after 35 years in the space program but remains very active in the Gateway Masonic Lodge as a past master and secretary of the lodge. He is also active with the Scottish Rite in Orlando, and the AZAN Shrine in Melbourne, FL. They have a RV and do extensive traveling of the US.

ROBERT T. "BOB" NEWELL, CAPT, born Nov. 13, 1946 in Whittier, CA, received a BS, PhD-physics; MS-Aeronautical Systems.

He joined the service July 1968. His military locations and stations: NAS Miramar, USS *Kitty Hawk* (CV-63); N&MCRL Albuquerque. Served with VF-114, VF-124. Reserve units: N. Ship Yard Pearl Harbor 211; N Ship Yard Long Beach 311.

Newell participated in Vietnam campaign. He was awarded the Individual Air Medal, 10 S/F Air Medals, 2 NCM w/Combat V and Vietnam Gallantry Cross. He achieved the rank of CAPT and was discharged Aug. 1, 1994.

His memorable experiences: the time they shot LTJG Hal (Wierd Harold) Crites through the closed doors of the NAS Cubi Point O'Club basement into an empty swimming pool. Hal survived, but the "Cubi Flyer" sustained strike damage.

Married Betty and they have two children: Darcy and Jenna. Employed at Applied Research Associates, Inc. as senior scientist.

ERNEST BURCHIE "ERNIE" NEWSOM, CAPT, born Oct. 17, 1937, Memphis, TN. Earned BS and MS degrees in psychology, Tennessee State Univ.; MDiv in theology, Louisville Presby. Seminary; DMin, Chicago Theological Seminary.

Joined the service in 1978 with Basic and Advanced Chaplain School at Newport, RI; 1980-81, VMGR-234 El Toro MACS, CA; 1982, Marine Indoc Camp Pendleton, CA; 1983, RNMCB-26, Quantico; 1984, Base Chapel, NAS, Memphis, TN; 1985, USS *Lexington*, Philadelphia, PA; 1986, MCAS, Cherry Point, NC; 1987, CNO Washington, DC; 1988, Relig. Support, Parris Island, SC; 1989-90, Command and Staff College, Quantico; 1990, NETC Newport, RI; 1992, Cherry Pt. MCAS, NC; 1992, NNMC Bethesda, MD; 1994, Operation AGILE Provider, Camp Lejeune; 1995, Battle Griffin '95 Norway; 1996, USS *LaSalle*, Gaeta, Italy.

Memorable experience was being selected as USN Chaplain of the Year in 1993 by the Military Chaplain Assoc. He was the second Afro-American USN Chaplain (Reserve) selected, CAPT 0-6 in the history of our nation.

Married Nolia Brentice and they have two children, Jacqueline Polk and Kelvin; and two grandchildren, Michael Polk Jr. and Britanny Polk. January 1984, Chief Chaplain Service, Alvin C. York Veteran Med. Ctr., Murfreesboro, TN; 1977-84, staff chaplain, North Chicago, Veteran Med. Ctr., North Chicago, IL; 1970-77, pastor, Fellowship Presbyterian Church, Huntsville, AL, Interim Pastor, Mt. Tabor-Salem Presbyterian Church, Columbia, TN.

ROBERT R. "BOB or NICK" NICHOLAS, CAPT, born Aug. 5, 1908 in Appleton, WI, received BS in EE.

He was commissioned LTJG, CEC, USNR, July 17, 1942. After indoctrination at Norfolk, VA he was assigned to Navy Yard Marine Island, CA and served as transportation superintendent until August 1945. Spent a brief period then at Providence, RI in preparation for assignment to a construction battalion. No new battalions were being sent out at that time due to the surrender of Japan so he was returned to Mare Island and assigned to the position of officer in charge of waterfront construction. A major contract administered was construction of piers for berthing submarines and tenders of the mothball fleet. He was released from active duty May 23, 1946 with rank of LCDR.

He remained active in the USNR by taking correspondence courses, going to two weeks training duty annually, and participating in weekly drill at the Naval and Marine Corps Reserve Training Center, Denver, CO. In August 1954 he was appointed CO, Naval Reserve Construction Bn. 9-29 in Denver. This duty was interrupted for a period of about two years while he served as public works officer for the training center. He was promoted to CAPT, CEC, USNR July 1, 1961 and retired Sept. 1, 1968. During the period 1961-68 served at intervals as senior evaluation officer for CB Bns. on two weeks training duty.

Nicholas is deceased. Children are David and Richard Nicholas; and grandchildren are Lori, Tami, Lisa, Robin, Heather and Linda; and seven great-grandchildren. Employed with head estimates section, Chief Engineer's Office, US Bureau of Reclamation. He retired from USBR 1969; USN, 1968.

WILLIAM B."BILL" NICHOLS, CDR, born July 6, 1921, Kennett, MO. He earned associate of arts degree at Flint Jr. College, Flint, MI.

Military Duty: 1942-47 at NAS Pensacola, FL; NAS New Orleans, LA; NAS Livermore, CA; NAS Columbus,

OH; NAS Alameda, CA; VR-11 NAS Honolulu, HI. Inactive duty from 1947-52 associated with NR NAS Oakland, CA; co-pilot navigator for Pan American and worked as salesman in food brokerage business.

Duty from 1952-55 with VR-3 NAS Moffett Field, CA; 1955-60, VR-22 NAS Norfolk, VA; 1960-64, VR-21 NAS Barbers Point, HI; 1964-67, was chief of Aircraft Alloc. & Tech Services Div., HQ MAC Scott AFB, IL.

Awards and Medals include Asiatic-Pacific, American Area, WWII Victory Medal, Naval Reserve Medal, National Defense Service Medal (2nd Awd.).

Memorable experiences include flying around the world in 1954; flying scintists from Washington, DC, to Christ Church, New Zealand for South Pole Project; flying through a typhoon on high priority mission, and others.

Civilian employment from 1967-94 as flight and simulator instructor, was also director of Sabrelmen training, 1974-94. He was designated FAA examiner for Sabrelmen type ratings and completed 497 type rating check rides. Retired from flight safety after 27 years.

He and wife Edith have four children: Cheryl, Gregory, Melinda Slack and Deborah; grandchildren are Ryan and Kacie Slack

WALLY R. "NICK" NICKOLI, CDR, born Feb. 8, 1952 in Aberdeen, SD, received a BS in secondary education, Northern State University and masters in educational leadership, Northern Arizona University. He began his naval career at OCS in July 1979 and completed explosive ordnance disposal, surface warfare officer and parachute training and served as special operations officer with Explosive Ordnance Disposal Mobile Unit ONE where he was OIC, Detachment 19 (CV-41) *Midway*, Mine Counter Measures.

A Selective Reservist, he served as CO for NR Mobile Technical Unit FIVE Detachment 419, Mobile Inshore Undersea Warfare Unit 105 Detachment 1, NR LST-1195 Barbour County, NR Amphibious Construction Bn. ONE DET 219, Mobile Diving and Salvage Unit ONE DET 319; and as XO, NR Seal Beach Weapon Station DET 716, and NR LST-1195 Barbour County.

Recalled for Operation Desert Shield/Desert Storm, September 1990-June 1991, he served on Port Security Harbor Defense Group III staff. There he wore many hats and served as: intelligence officer, communications officer, Joint Emergency Operations Center watch officer, SecMat officer, and Saudi Royal Frontier Forces liaison officer.

He is married to Susan and they have a son Ryan. Employed as corporal, Pima College Police.

DOUGLAS C. NICKSON, CDR, born Dec. 22, 1945, Pueblo, CO, received a BA in education, BA (business), MBA, MSCLS (master of science in Commerce and Legal Studies).

He joined the service Nov. 4, 1967. His military locations and stations: NAS Lemoore, NAS Whidbey Island. Units served: Lemoore RATCC, OPCON 8T2 (Noel Davis Trophy); XO of NASL 3490, MATWING 289, VTU 8989 (chief of staff).

His battles and campaigns: Vietnam Era; Desert Storm Era. He was awarded the National Defense, Naval Reserve, Pistol Sharpshooter, Overseas Service.

He achieved the rank of CDR and discharged from active duty Jan. 30, 1970. His memorable experiences: (Reserves) 1989 winner of the Northwest Naval Aviation Rendezvous Award, ACDUTRA (Seattle-New York-Rome-Somalia-Tailhook on the USS *Carl Vinson,* Hong Kong-Seattle in 17 days, Navy War College, Armed Forces Staff College, Pentagon, Cubi Point, flying a P-2 out of Midway from the bow, Korean War games, having CAPT Ed Nixon (the late President's brother) as his CO during Watergate with his last name pronounced the same as CAPT John Thompson from Marlin, TX as CO at NAS Lemoore.

Nickson is divorced and retired December 1991. Currently director of sales and marketing for Cresmont Inc.; past president of Pacific Marketing International, Northwest District manager for Exide Battery; president of Basic Petroleum (CA).

JOHN "F" NIENSTEDT, CAPT, born in Emporia, KS, July 21, 1938 and is a graduate of Kansas University. He completed OCS in 1961 and was assigned to the USS *Weiss* (APD-135) as communications and CIC officer. While serving on the *Weiss* in 1963 he was involved in extensive reconnaissance missions in Vietnam as a boat commander for Recon Marines.

In 1972, while serving as XO of Surface Unit 123 in Kansas City, LCDR Nienstedt was mobilized to command the Navy's honor company for President Harry S. Truman's funeral in Independence. The Navy was the only service to call on reservist to fulfill that mission.

His command assignments include: CO TELCOMMSTA MOROCCO, CO COMNAVPHIPGRU-1 Det. and CO COMNAVSURPAC Det. 219. He also served extensively on the staffs of REDCOM's 18 and 19. He achieved the rank of CAPT.

He resides in Phoenix and is a professional speaker and author on leadership and self motivation. His first book *The Outstanding Man, A Guide to Personal and Professional Success* was published by Harbinger House in 1992. He has two children, John Brook and Barrie Sue.

JOHN E. NIESSE, CAPT, born Nov. 30, 1927 in Indianapolis, IN, received a BS in naval science and MS and ScD metallurgy.

He joined the service June 2, 1950. His military locations and stations: USS *Warrington* (DD-843); USS Trout (SS-566), USS *Argonaut* (SS-475); USS *Harlan R. Dickson* (DD-708) Reserve Crew.

He achieved the rank of CAPT, USNR and retired July 1980. Niesse is married to Elaine and they have two children, Ann L. Kaeding and John A.; and grandchildren, Elizabeth Niesse and Jasper J. Kaeding. Employed with the Monsanto Co., group consultant, materials engineering.

ROBERT L. NIFONG, CAPT, born in Pittsburg, KS in 1929. He enlisted in the USNR as a seaman recruit in August 1947 and advanced to radarman third class. He was selected for and attended Reserve Officer Candidate School at San Diego in 1949 and 1950. He received a BS in architectural engineering from the University of Kansas in 1952, was commissioned as an ENS in the Civil Engineer Corps in June 1952, and ordered to active duty at the US Naval Construction Bn. Center at Port Hueneme, CA.

After completion of active duty he returned to Kansas and worked as a structural engineer in Topeka. In 1960 he accepted a position as a structural engineer in the USN in Port Hueneme. In 1961 he transferred to the Pacific Missile Range at Point Mugu, CA. He remained at Point Mugu for 28 years advancing through various positions including senior engineering manager, division head, and program manager in the Range and Electronic Warfare Directorates. He retired from federal service in 1989 and then worked as a technical advisor for INFOTEC in Camarillo, CA for approximately three years.

During his reserve career as an enlisted man was attached to Organized Surface Divs. in Joplin, MO; Topeka, KS and Gary, IN. As a seabee officer he was with Seebee Divs. in Topeka and Port Hueneme; Naval Air Reserve, Point Mugu; Facilities Engineering Program, Los Angeles; Readiness Command Region 19, San Diego,

VTU 1913; Santa Barbara, and served as CO of Research Co 11-8, Point Mugu; NAVFAC 519, Los Angeles; and RNCBPAC, Port Hueneme. He transferred to the Retired List in 1985.

Extensive travel as a staff officer and CO during the last eight years of his reserve duty required him to visit and units in the US and the Pacific, including Adak, Hawaii, Guam, Philippines and Japan to coordinate training and project work with the active duty and reserve Seabees.

Nifong is a licensed professional engineer in Kansas and received an FAA private pilot license in 1982. He was also qualified aircrewman on P-3 aircraft.

He married Elizabeth in 1950 and they reside in Oxnard, CA. Their six children are Cathleen, Teresa, Chris, Angela, Monica and Aimee; and seven grandchildren are Melissa, Caree, Lucas, Morgan, Rhiannon, Robert and Nina, reside in various cities in California and England.

FRANK W. NIKISCHER, LCDR, born Sept. 26, 1953. His military locations and stations: USS *Olmstead* (APA-188), Norfolk, VA and various other combatant ships and shore stations.

Served with amphibious force, Atlantic Fleet. Awarded the National Defense Service Medal and European Occupation Medal. Achieved the rank of LCDR and discharged July 4, 1976.

His memorable experiences: graduate of Penn State NROTC; commissioned in Regular Navy 1953. Served on active duty during Korean conflict.

Married to Judith and they have three children: Frank Jr., David and Wendy; and grandchildren: Kristel, Michael, Frank III, Gabrielle (all Nikischer); Caitlin Keim and Oliver Winston Scofield. Employed as president and CEO of Double N., Inc. He retired Sept. 1, 1996.

JAMES J. "J.J." NIXON, CDR, born Jan. 12, 1927 in Woburn, MA, received an AB from Monmouth College and JD, Suffolk University Law School.

He joined the service Sept. 15, 1944. His military locations and stations: NATTC Memphis, NAVRECSTA Algiers, LA, Ottumwa, IA; Corpus Christi, TX. Served with Arm School, Preflight and Primary CCTX, VRF-1 Brooklyn, NY. He was awarded the Naval Reserve 2x, Victory and American Theatre Medals.

Nixon achieved the rank of CDR and was retired from USN, July 1981. His memorable experiences: two years, active duty, 35 years in Reserve units.

He married Margaret P. Treanor and they have a daughter, Patricia M. Hinds, and grandchildren, Joseph and James Hinds. He is a retired superior court judge, now practicing lawyer. Retired from the court November 1992.

W. BARRY NIXON, CAPT, born on March 18, 1930, a native of Woodstown, NJ. He was a graduate of Yale University (BE) and Princeton University (MSE) majoring in aeronautical engineering. He began his naval career in 1955 at OCS Newport, RI and was designated a naval aviator at NAS Hutchinson, KS in 1956.

Assigned to VW-16 at NAS Barber's Point, HI he trained in the WV-2 before transferring to VU-1 there flying the JD, F6F, and F9F. Released from active duty he joined the Ready Reserve at NAS Lakehurst, NJ in 1959 flying the S2F and was recalled to active duty in VS-751 during the 1961 Berlin Crisis, which included deployment to NAS Guantanamo, Cuba. Released in 1962 he returned to Lakehurst to eventually become CO of ResForRon VS-73 and, subsequently, two augmentation units. Selected for CAPT in 1975 he remained active in the Reserve until retiring in 1985 with the Naval Expeditionary, National Service and USNR medals. He became a life member of the NRA, Tailhook Association and ANA.

In his civilian career he was a research pilot/aero engineer/lecturer at Princeton University's Flight Research Laboratory for 25 years and also flew several corporations and a commuter airline. Lately he flys a hot air balloon appropriately named "Wings of Gold" around Lawrenceville, NJ aided by crew chief/wife Marian and copilot/son Todd.

LASALLE L. NOLIN, LCDR, born March 18, 1927 in Cohoes, NY and received a bachelors in Marine Engineering, USMMA.

He joined the service August 1952, XO on USS LSMR 520. He achieved the rank of LCDR. Single and has a son, Peter M. Nolin and grandchildren Katy Bea and Margaret Sue. Employed as port engineer Military Sea Lift Command. He retired in January 1993.

ELLIOTT WYNN NORRIS, CDR, born at Cedar Bayou, TX on May 31, 1935. He is a graduate of The University of Texas. Following completion of Naval Officer Candidate School, he began his commissioned service as gunnery officer aboard USS *James E. Kyes* (DD-787) in 1958.

In 1961, as gunnery officer aboard USS *Hornet* (CVS-12), he was selected for augmentation into the Regular Navy. Subsequent duty assignments included: 1LT, USS *Windham County* (LST-1170); staff operations officer, WESPACDET Naval Beach Group ONE; Destroyer School; weapons officer, USS *Rogers* (DD-876); and fleet team director, Manpower Validation Office, Pacific. He received medals for service in the Taiwan Straits and the Vietnam War.

Upon discharge from the USN in 1969, after 11 years active duty, CDR Norris entered the USNR. All his reserve assignments were at Orange, TX, and included a command of a ships activation, maintenance and repair unit. He retired in 1980.

His civilian employment has been primarily in insurance, risk management and safety. He was recently promoted to manager of Community Relations for the Coast Corporation on the Caribbean Island of Aruba.

Norris and his wife, the former Majorie Rose McMaster, reside on Aruba and have been married 37 years. Their family includes daughter and son-in-law, Sarah E., and Terry Wick, and their three children, of Poteet, TX; son, Stephen M. Norris, of Ft. Worth, TX; and son and daughter-in-law, Todd Q. and Marcia Norris, also of Fort Worth.

EILIV M. "PILOT" NORVIK, CDR, born Oct. 21, 1911, Bergen, Norway. Attended St. John's Univ. for one year; joined the service May 16, 1939 with active duty beginning July 18, 1940.

Stationed at Receiving Station, Casco Bay, ME, OinC. Units served with include Solace; Seattle Trng. Ctr.; Denebola (flagship) Commander Destroyer Atlantic Fleet; Celeno (AK-76); Argonne (AG-31) Flagship Commander South Pacific; Caney (A0-95) as CO and Pawcatuck (AO-108) as XO.

Battles and campaigns include Pearl Harbor - Midway; consolidation of Southern Solomons; assault and occupation of Okinawa Gunto; and Third Fleet Operation against Japan.

Discharged Sept. 21, 1946. His decorations include Letter of Commendation Medal w/Ribbon (Military Merit), American Defense, American Campaign, Asiatic-Pacific w/3 stars, WWII Victory Medal, Naval Reserve, 26 years pre-Pearl Harbor, and NY State Conspicuous Service.

Memorable experiences: Guadalcanal on June 16, 1943 and Celeno was attacked by a swarm of Japanese bombers; his volunteer duty as harbor pilot in addition to his duties as navigation officer on the Argonne; while serving as navigation officer on the Argonne, ADM Halsey ordered him to teach "practical navigation" to the officers and he wrote a book calling it Navigator's Guide. The book was distributed throughout the Fleet.

Military Sea Transportation Service, master. He retired July 31, 1970. He and wife Margaret have two children, Anne Altieri and Susan Arsenault; grandchildren are Thomas, Russell, David and Alan.

LOUIS "LOU" NUSINSON, CAPT, born Aug. 16, 1929 in Menno, SD, received BS Northwestern University, Evanston, IL.

He joined the service June 24, 1948 SR, USNR; ENS June 1951. His military locations and stations: USS *Whitley* (AKA-91), 1951-53. Served with CO, NRSD 9-5(L), Chicago 1970-73.

He achieved the rank of CAPT and retired 1980. He married Deborah H. Stone and they have a daughter Shari D. Conrad. He is employed as associate, Neil Enterprises, Inc.

ANDREW "ANDY" TODD NUTT, CAPT, born Jan. 3, 1934 to Andrew and Lurie Todd Nutt in Gary, TX. Graduated Gary School 1951; BBA from Stephen F. Austin University-1954; MEd from SFA University-1965; PhD from University of Texas at Austin, 1974.

Joined 49th Armd. Engineering Div., Texas National Guard, 1953. Transferred to USN, Officer Candidate School, 1955. Commissioned September 1955; Air Ground School, NAS JAX, 195; CIC School-NAS Glenview, 1956; Air Control School, NAS Glenco, GA, 1956; VC-11, NAS North Island, 1956; VAW-11, NAS N. Island, 1957-58; VAW-11, DET Echo, USS *Yorktown* (CVA-10), 1956-57; VAW-11, DET Foxtrot, USS *Ranger* (CVA-61), 1958.

USNR Air Intelligence Command, NAS Dallas 1958-83; XO AAU-1111, NAS Dallas, 1975-76; CO FIRSTPAC-370, NAS Dallas, 1977-79; CO IVTU-111, NAS Dallas, 1980.

He was awarded the Sharpshooter-M1 Rifle; Naval Air Observer (controller) wings; Armed Forces Reserve Medal w/HG; Navy Unit Commendation w/E and cluster.

Memorable experiences include: first landing on USS Yorktow*n*, first catapult shot at 4 a.m.; crossing the equator on USS *Ranger* 1958. Winning FICPAC commendation as first CO of FIRSTPAC 370 which also his first command. Retired as CAPT, USNR, 1994.

He married June Weaver on Christmas Day 1958. Their two children are Rebecca Nutt Austin and Cristi Nutt Ferguson; grandchildren are Amy and Kimberley Ferguson.

Civilian employment includes: accountant, teacher, elementary school principal, county school superintendent; division director of three different divisions and administrative assistant to the commissioner for General Education of the Texas Education Agency (a branch of state government); superintendent of Beckville Schools; and adjunct professor, graduate school, SFA University.

He retired from public schools in 1991, from the USNR in 1994, and sometime in the future he plans to retire from other employment with SFA University and Consulting.

JOHN M. NUTTER, CAPT, born July 21, 1924 in Liberty, MO, joined the USNR on Dec. 14, 1942. Allowed to complete one semester at the University of Arizona before going on active duty in the V-12 Unit at the University of New Mexico. Graduated with a BS in civil engineering in June of 1945. Assigned to Midshipman's School at Davisville, RI and was commissioned an ENS, CEC on Sept. 8, 1945.

Went overseas to Okinawa and was assigned to Naval Construction Bn. 86 as a construction officer on the construction of a naval hospital. When the battalion was deactivated in early 1946, he went to Pearl Harbor and was reassigned to Construction Bn. Detachment 1106, a part of Material Salvage Unit 3. Went by LST to the island of Manus where they loaded ships with surplus material. Returned to the states and was released to inactive duty June 1946.

From April 1948 until April 1952 he was a member of Construction Bn. Det. 11-5 in Phoenix. Recalled to active duty in April 1952 and assigned to Naval Ordnance Test Station, China Lake, CA until 1954. Was in charge of Contracts Div. of the Public Works Department, responsible for the inspection and testing of construction projects.

From April 1954 until August 1956, he was assistant resident officer in charge of construction contracts, Pacific, San Francisco, CA. September 1956 until November 1958 was public works officer, Naval Air Facility, Oppama, Japan. December 1958 until June 1959 he was assistant PWO at naval amphibious base, Coronado, CA. Released from duty June 30, 1959.

Inactive duty assignments: PWO, Naval Reserve Training Center, Tucson, AZ; CO, CB Det. 11-5, USNRTC, Phoenix, AZ; CO, CB Det. 11-19, USNRTC Phoenix, AZ; chief of staff, First Reserve Naval Construction Regt., CO, Reserve Naval Construction Bn. 17; CDR, first Naval Construction Regt., assistant chief of staff, First Reserve Naval Construction Brigade. Retired June 30, 1980.

Civilian occupation was in road construction with a number of contractors starting as an estimator, then a superintendent, general superintendent, vice-president and executive vice-president. Also spent seven years with Maricopa County, AZ, retiring as assistant county engineer in September 1989.

He married Doris and they have three children: Kathryn, Linda and Frank; and grandchildren, Brandon and Kevin Leeds, Danielle and Christopher Brower, and Beckilyn Nutter.

ANN BRITT O'BRIEN, CAPT, born Oct. 9, 1929, Brooklyn, NY, attended the St. Vincent's Hospital School of Nursing, NYC 1950; BS Nursing, St. John's University, Brooklyn, NY, 1959; MA Delivery of Nursing Care, NYU 1978.

ENS Britt reported for active duty to the US Naval Hospital, Newport, RI Island on March 29, 1951. Thereafter, she was stationed at the Naval Hospital, Bainbridge, MD; Tripler Army Hospital, HI Midway Island and the Naval Hospital, St. Albans, NY. She served in various nursing positions and services. She was released from active duty in January 1956. She affiliated with the Selected Reserve and served in various positions including CO of a Medical Co. She attained the rank of CAPT. She retired in 1989.

O'Brien was employed as a registered nurse in various positions and services. Her longest employment was with the Veteran's Administration. She retired from the VA in October 1990 having held the position of Associate Chief of Nursing Service, certified in Nursing Administration Advanced.

She was married to Henry J. O'Brien in 1959.

GEORGE A. O'CONNELL JR., CAPT, born in Chicago and graduated from the University of Illinois in 1937. Did graduate work at Columbia University, Northwestern and University of Chicago.

Enlisted in the USN at the start of WWII and was stationed in the Pacific as a photo intelligence officer in Guadalcanal. In 1944 he was reassigned to Washington, DC where he met and married a WAVE officer Peg Young.

Upon returning to Chicago he continued his Reserve career at NAS Glenview. Through the following years he was the national president of the NRA 1962-64. national director of the Navy League; president of the Ninth Region Navy League; national vice-president of the US Sea Cadets; vice commander of the Veterans Assistance Commission of Cook County; member of the Naval Order; Veterans of Foreign Wars; American Legion; American Veterans of WWII; commander of the Chicago Chapter of the Military Order of the World Wars. At Glenview he commanded Sqdn. NAIRU - V2 in 1968-70.

In 1951 he was called back for the Korean War and served two years. He was also called to special duty during the Vietnam Conflict.

He received the Distinguished Public Service Award and the Legion of Merit. He was made an honorary Naval Aviator (one of eight to receive it in the country).

His outside activities consisted of sports: An official of the US Cycling Federation, Official for the Pan American Games, Official of the US Boxing Federation, Member of the US Olympic Committee from 1952 and an active game official from 1952 on. He was a member of the US International Skating Association. International Board of Directors and a licensed world official of the U.S. International Union.

He participated in 10 Olympic games and world meets as a speed skating official and was elected to the Skating Hall of Fame for over 50 years of participating in that sport.

O'Connell passed away July 27, 1987. He is survived by his wife Peg; son George A. III; daughters, Shannon and Erin; and two grandchildren, Elena and Devon O'Connell.

JOHN A. "OAKIE" O'CONNELL, CWO3, born June 26, 1922 in Fall River, MA, attended college for two years.

He joined the service Nov. 20, 1942. His military locations and stations: Newport, RI; School-Boston, USN Comp. Lab, Dayton, Washington, DC; Pearl, Guam.

He was awarded the Presidential Citation, Atlantic and Pacific Theater. He achieved the rank of CWO3 and discharged Nov. 30, 1945.

His memorable experiences: getting shot at on mail runs in Guam.

He is married to Mary and they have children: John Jr. and Carol Ann and grandchildren: Dan, Mick, Matt, Shawn and Mike. Employed with IBM as custom engineer. He retired June 1, 1984.

MARGARET O'CONNELL, LT, born Aug. 8, 1919 in Brooklyn, NY, received a BA, Notre Dame College, Staten Island, NY in 1941; MA, 1943 Fordham University, NY.

She joined the service February 1943 and commissioned ENS in June 1943. Military locations and stations: Mt. Holyoke Midshipman School, 1st Naval District Boston, MA. Served with Naval Research Lab and BuOrd, Washington, DC.

She was awarded the WWII Victory Medal and achieved the rank of full LT. She was discharged in June 1946.

Her memorable experiences: first assignment after being commissioned was as shore patrol officer at the 1st Naval District in Boston in charge of the WAVES, SPARS, and Women Marines. Women were new in those branches of the service and the Navy wanted to know how they were behaving and to have an officer in charge.

She worked out of the police station with an S.P. band on her arm flanked by two stalwart S.P. Sailors. Being only five feet tall, she looked like a period between two exclamation points.

They toured at night into all the clubs and bars from the sleaze places to the best hotels to see if the women were behaving. Meeting all the managers and Maitre D's in charge, they would call her if their were any problems. She was a very popular date, for the clubs would always find a table for her, even on the most crowded nights, and they never gave her escort a tab. The best part was that the service women came through with flying colors.

Her husband, George A. O'Connell (former President of NRA) is deceased. She has three children: George A. III, Shannon and Erin; and grandchildren: Elena and Devon O'Connell. Employed in real estates sales since 1963. She is still going strong.

ALF T. ODEGAARD, CDR, born June 12, 1916 in Waterville, IA, received BS and MS in agriculture, Iowa State College, Ames, IA.

He joined the service April 24, 1942. His military locations and stations: Mid'n School, NT; Fire Control School, San Diego. Served with USS *Woodworth,* PG School, Harvard University; USS *Oregon City.*

His battles and campaigns: SoPac, Georgia, Bougainville, Bismark Archapalego, Philippines, Okinawa. American Theater, Asiatic-Pacific, Philippine Defense, USNR.

Achieved the rank of CDR and retired June 12, 1976. His memorable experiences: CO of USS *Woodworth* was (then) LCDR W.P. Mack who later was superintendent of US Naval Academy when their son was a midshipman there. After joining reserve units, had 13 tours of active duty for training.

He married Barbara I. John and they have four children: Judith Heslop, Susan Schmidt, Louise Odegaard and CAPT John Odegaard; and grandchildren: Jennifer and Katie Schmidt, Jacquelin Heslop, Katherine and Adam and Lauren Odegaard. Employed as agriculture investment consultant, Metropolitan Life Insurance Co.

RONALD "RON" ODEGARD, CAPT, born June 8, 1947 in Jacksonville, FL, received a BA in history, University of Kansas; MBA, USIU, San Diego.

He joined the NROTC, July 1, 1969. His military locations and stations: NAS North Island. Served with HEL SUPPRON Three. His battles and campaigns: Vietnam.

Odegard achieved the rank of CAPT and discharged June 30, 1977. He was awarded the Vietnam Service, Pistol (E2, Vietnam Campaign MUC).

Married Myra Beth (also USNR CAPT) and they have children, Erik and Chris. Employed as pilot, American Airlines.

GEORGE E. O'DONNELL, LT, born March 3, 1929 in Brooklyn, NY, received a BS in marketing, New York University; others, Polytechnic Institute, Washington SQ College, City College of New York.

Joined the service April 16, 1947. His military locations and stations: NAS Floyd Bennett Field, Glenview and South Weymouth. Served with many including CV-67 Ken*nedy* and CVN 68 *Nimitz.*

He was awarded the USNR, Good Conduct, Expert Pistol. He achieved the rank of LT and retired in 1989. His memorable experiences: TWA hostage crisis while aboard *Nimitz* and Libya, shoot down aboard *Kennedy.*

O'Donnell is married to Gloria Ann and has children: Eileen, George, Gerard, Steven and Mary Ann; and grandchildren: Jennifer, Ashley, Craig, Luke and John. Employed as instructor Aviation Mechanics International marketing and distribution mgr. for various Fortune 500 companies.

JACK L. OGLESEBEE, CAPT, a native of Savannah, GA, received a BA degree from Lenoir Rhyne College (Hickory, NC) in 1952. He enlisted in the USNR in 1947. In 1954 Oglesbee was designated a naval aviator and was assigned to VP-49 at NOB Bermuda. Upon his release from active duty in January 1957, he affiliated with the Naval Air Reserve and served at NAS Willow Grove and NAS Atlanta. CAPT Oglesbee's significant assignments was CO of VP-674, VP-67B2, NAS Jacksonville DET 167, and VTU-6767.

He served as president of the Atlanta Chapter of the NRA and was retired from the Naval Air Reserve in 1988 after 40 years of enlisted and commissioned service. Oglesbee resides on Wilmington Island near Savannah, GA. He is divorced and has a son, John D. Employed in various sales management positions.

FLOYD H. OKADA, born Aug. 8, 1935 in Honolulu, HI, received a BS in Stanford, 1957; MD, Loyola University, Chicago, 1961. He joined the service Jan. 24, 1967. His military locations and stations: Camp Schwab, Okinawa, USNH Camp Pendleton, and St. Albans. He served with 3rd Med. Bn., 3rd Marine Div., FMF, NSA Hospital. He participated in RVN Campaign and Vietnam Service.

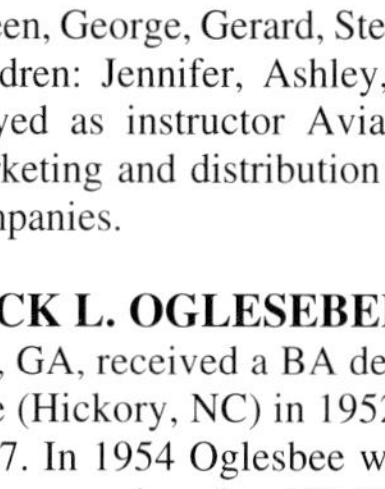

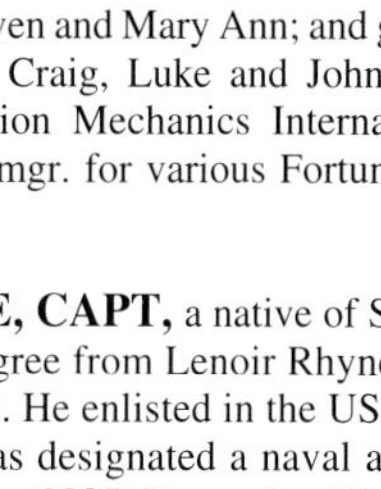

Okada was awarded the National Defense and Presidential Unit Citation. He achieved the rank of LCDR and discharged June 14, 1969. His memorable experiences: 1968 TET offensive, Vietnam, NSA Hospital, Da Nang Third Med. Bn., Phu Bai, RVN, US Naval Hospital, St. Albans, NY.

He married Janet and they have children: Kelly and Daniel. Employed as staff surgeon, Santa Clara Kaiser Hospital, retiring July 31, 1996.

HAROLD E. "HAL" OLD JR., CAPT, born Sept. 11, 1939, Mansfield, OH. Earned BS degree in business administration, Marquette Univ., 1961; MA in political science, Western Michigan Univ., 1970; PhD, political science, Michigan State Univ., 1973; post-PhD economics and finance, Michigan State Univ., 1978-79; Naval War College, 1983.

Joined the military in September 1957, midshipman, NROTC Regular, Marquette Univ. and commissioned ENS in June 1961. Locations, stations and unit served with: 1961-63, USS Union (AKA-106) San Diego, CA boat group commander OOD(F)/CDO; 1963-65, USS Monticello (LSD-35), San Diego, CA, CICO/EMO. OOD(F)/CDO/ CDO afloat; 1965-67, amphibious operational training unit PAC San Diego, CA operations officer; 1967-69, commissioned USNR MED Surf DIV 9-100M, XO/trng. officer, Kalamazoo, MI; 1969-72, large surf DIV 9-102L, XO, Lansing, MI; 1972-73, group commander staff 01-7S, Springfield, MA; 1973-74, MED SURF DIV 9-100M, XO, Battle Creek, MI; 1974-75, SURF READ Unit LST-1414, CO, Battle Creek, MI; 1975-77, SACLANT Unit 214, Staff officer, OIC IRU, XO, NMCRC, Battle Creek, MI; 1977-78, Consolidated training department supervisor, Battle Creek, MI; 1978-80, Consolidated training dept. supervisor, Lansing, MI; 1980-82, VTU(G)1317, CO, Lansing, MI; 1982-83, student, Naval War College, Newport, RI; 1983-85, SACLANT Unit 214, CO, Battle Creek, MI; 1985-87, VTU(G)1307, asst. inspector general, REDCOM13, senior advisor, CO, NMCRC, Detroit, MI; 1987-89, CINCUSNAVEUR Unit 413, CO, Southfield, MI, CINCUSNAVEUR CONUS Rep. for naval control of shipping, London, UK; 1989-91, Readiness Unit 1315(B), CO, Southfield, MI, CINCUSNAVEUR CONUS rep. for naval control of shipping, London, UK.

Battles and Campaigns: 1961 Berlin Crisis; 1962 Cuban missle crisis; 1964-65, Vietnam Tonkin Gulf attack.

Awards and medals include Meritorious Service Medal (2nd Awd.), Armed Forces Expeditionary Medal, Vietnam Campaign Medal (2nd Awd.), Armed Forces Reserve Medal (2nd Awd.), Overseas Service Ribbon and Expert Pistol Ribbon. Retired Sept. 1, 1991 and achieved the rank of CAPT.

Memorable experiences include out of area operations and deployments for 42 months, 1961-65; first line reserve officer selected to attend Naval War college full-time in 1982; 40 AT/ADTs performed, 1967-91; exercise director, 2 large scale EASTLANT/MED NCS exercises; four years as commaner in chief, USN Forces Europe CONUS rep. for naval control of shipping; member two selection boards; and command of seven units with over 12 years in Command.

Married Donna and they have two children, Eric and Philip (LCDR, USNR). Civilian employment: professor, American Military University; indepenent consultant and professional trainer; log home marketing and sales, licensed builder; human services administrator, strategic planner, public policy analyst and program developer, researcher and evaluator and educator-trainer of human service professionals/research manager and editor of labor market analysis and monographs; state of Michigan. He retired in 1997.

JACK A. "OLE" OLSEN, USNR, CAPT, born Sept. 29, 1943 in Williston, ND, is from Flaxville, MT, attended Concordia College in Moorhead, MN and earned a BA degree in chemistry and mathematics. He began his career in 1965 as an aviation cadet and was commissioned an ENS in 1965. He received his wings in February 1967.

He joined VA-75 in August 1967 and deployed to Vietnam aboard the USS *Kitty Hawk.* He served three years with the Sunday Punchers flying the A-6 Intruder and performing LSO duties. A Mediterranean cruise was completed in 1970 aboard the USS *Saratoga.* Shore duty followed with VA-42 as a RAG instructor until released from active duty in February 1972.

CAPT Olsen was awarded the Navy Flying Cross, Navy Achievement Medal, Air Medal, Vietnam Service Medal, Navy Commendation Medal, Presidential Unit Citation, Vietnam Campaign and Air Defense Medal.

As a reservist, Ole flew the F-8 Crusader with RTU-201 at NAS Dallas in 1972 and 1973 and the A-7 Corsair with VA-2267 at NAS Atlanta from 1977 to 1980. With VA-2267 he served as operations officer and XO. He finished out his career with VTU-6767. CAPT Olsen retired in October 1991.

Jack was hired by Delta Air Lines in November 1972 and has flown the L-100, DC-9, L-1011, MD-88 and B-727. He is presently flying CAPT on the B-757 and B-767. Jack and his wife, Fran, make their home in Bostwick, GA.

DAVID O. "DAVE" OLSON JR., CAPT, born Jan. 11, 1925 in Staten Island, NY, graduated from US Merchant Marine Academy, February 1945.

He joined the service March 1943 as cadet/midshipman USMMA. His military locations and stations: various merchant ships, APA-195 and LSMR-403, Korea. Served with various convoys WWI; USN Amphibious Forces, Pacific, Korea. His battles and campaigns: Murmansk Run, Winter 1943; Atlantic, Mediterranean, Asian/Pacific Theatres. Awarded Victory, Naval Reserve, Korean Service, United Nations Medal, two Bronze Stars, Korea.

Achieved the rank of CAPT, USNR and was discharged from active duty December 1952. Retired from USNR, Jan. 11, 1985. His memorable experiences: "Murmansk Run", winter 1943, WWII (very) close inshore fire support, Korea.

Married to Eileen A. and they have children: Joanne E., William D., Eric P. and Alan R.; and grandchildren: Andrew and Erin Olson, Rebecca Weeks, Tyler Gowans and Lee Olson. Employed as manager, building design and construction, Southern Bell, Ft. Lauderdale, FL.

ROBERT D. "BOB" OLSON, CDR, born Feb. 14, 1924 in Duluth, MN, received a BA from Dakota Wesleyan University. He joined the service in 1943. Served with UDT 30; TAR Program. Olson achieved the rank of CDR and was discharged in 1974. He is married to Janis.

CHARLES A. O'MALLEY, CAPT, a native of Ubly, MI, enlisted in the USNR, June 1945; Storekeeper School, *Arthur Middleton* (APA-25); inactive, August 1946. University of Detroit, direct commission, ENS, Specialist Ordnance, April 1950; B. Ch.E., June 1950.

Korean War: September 1950-November 1951, *Bexar* (APA-237); Deck Div., assault boat officer. Memorable duty: Ashore, with three others, during Chinese breakthrough, December 1950; north of Inchon, persuading Koreans to send rice south. November 1951-September 1952; ENS/LTJG; operations and communications officer, LST-883; even shore bombardment, UN Task Force, 95 with 40mm/20mm Inactive; September 1952.

M.S.Ch.E., Wayne State University, January 1958; registered professional engineer, California, July 1962. Supervised preliminary design of the Apollo Transtage Engine, which took man from earth orbit to moon orbit, and return.

Commanded 3 Weapons Units; helped select fire-control *Ticonderoga* (CG-47); with three reservists outlined her training program. NRA; 70s: president/Arrowhead Chapter. Retired August 1987.

Married Patricia McDermitt of Detroit and they have nine children. Research engineer, iron/steel; project engineer, program manager, aerospace, consulting engineer, San Bernardino, CA.

RAY C. OMAN, CDR, prior to being in the USNR, served in the USMC as a disbursing officer in Vietnam and with the 2nd Marine Div. at Camp Lejeune, NC and was discharged a CAPT in 1967.

Oman taught high school for two years and then went to Penn State and received his MS in 1971. He entered the federal government and while pursuing his career took part-time graduate studies at George Washington University and received a MPA degree in 1976 and a DPA in 1983.

In 1984 he was commissioned a LT in the Supply Corps Reserve Direct Commission Program. As a naval reservist, he has been assigned to several units in the Washington, DC area, including NSC HQ B306, NRCC WASH 206, DCNO (Logistics) N4, and NAVSUP HQ 106. In 1992 he was selected to be CO of the ABFC H14K Fuels Unit 706. While CO, he trained his unit at Roosevelt Roads, PR, naval base at Rota, Spain, and at air stations in PAX River, Oceana, Norfolk, and Cherry Point, NC. Now in a joint service billet, in 1994 he was assigned to the DLA FSC 706 Unit where he has served as training and as administration officer.

MARY F. O'NEILL, CAPT, entered the USN at Great Lakes in 1951, then went to NNMC Betheseda, MD until 1954. She spent the following years in Reserve and Selected Reserve (eight more years) at Chicago Naval Armory. She retired in 1982.

DENNIS G. OPACKI, CDR, began his career with the USN when he enlisted in 1958 in Detroit shortly after graduating from high school. He completed boot camp and electronics technician "A" School at Great Lakes. Afterwards he was assigned to the USS *Independence* (CVA-62).

Before leaving the USN in 1962, he served two complete Mediterranean cruises aboard the *Independence,* earned the Good Conduct Awards, the National Defense Service Medal, the Navy Expeditionary Medal (Cuba) and rose to the rate of electronics technician second class. After his discharge, he went to college. He earned a bachelor of arts degree from the University of Michigan in 1966 then a JD from the University of Illinois. While practicing with a law firm in Seattle, he learned of the formation of a USN JAG Reserve Unit and decided to join.

In 1973, he was commissioned LTJG in the JAG Corps and was promoted to LT later that year. Most of his Reserves career was with the Volunteer Training Unit, Seattle with the exception of a two-year assignment with the unit that drilled at Naval Legal Service Office—Bremerton. His active duty for training assignments as a JAG reservists included Treasure Island, Pearl Harbor, San Diego, Sand Point and Military Justice School Reserve Lawyer Refresher Course. He earned the rank of CDR in 1984 and the Naval Reserve Service Medal before transferring to the Retired Reserves in 1991.

While practicing law in Seattle, CDR Opacki was house counsel for the Chicago, Milwaukee, St. Paul and Pacific Railroad (the Milwaukee Road) where he developed litigation experience before becoming house counsel in 1979 for PACCAR Inc. where he is now senior counsel. CDR Opacki specializes in environmental law and products liability litigation.

He married the former Judith Lee Herman in 1966 and they live in Seattle. His son Dennis, born in 1970, currently serves aboard the *Henry M. Jackson* as a nuclear qualified machinist's mate first class.

OLIVER K. "O.K." OSBORNE, CAPT, born July 5, 1918 in Rockford, IL, received a BA in economics, Beloit College. He joined the service Jan. 22, 1942, USNR Midshipmans School, Chicago, IL. Commissioned ENS, USNR, May 5, 1942.

His military locations and stations: Torpedo School, Naval Torpedo Station, Keyport, WA May-July 1942, gunnery officer; USS *Parrott* (DD-218) July 1942-June 1946; Gunnery School, Navy Yard Washington, DC, July 1-Oct. 1, 1944; Gunnery officer, XO, USS *Murphy* (DD-603), Oct. 23, 1944-March 9, 1946. Units served with: CO USNR Surface Div. 9-2 USNR Armory Chicago, IL 1953-55; CO USNR, Surface Div. 9-9 USNR Armory, Forest Park, IL 1955-56; Recruiting officer, USNR, Armory Forest Park, IL, 1957; CO, USNR Mobilization Team Div. 9-2, USNR, Armory Joliet, IL 1958-61; Training officer, 1961-63, CO, 1963-64; USNR, Surface Div., 9-22 USNR, Armory Joliet, IL.

His battles and campaigns: USS *Parrott* conducted anti-submarine escort for convoys and anti-submarine aircraft carriers Atlantic Ocean. USS *Murphy* anti-submarine escort Atlantic, secret mission to Jidda, Saudi Arabia. March 1945 orders to Pacific. Eniwetok on V-J Day. Orders proceed to Okinawa, Japanese ports Nagoya, Nagasaki, Yokosuka. Return to Charleston, SC Naval Base December 1945.

Awarded the American Campaign, EAME, Asiatic-Pacific Campaign, Victory Medal, USNR Medal and Armed Forces Reserve Medal.

His memorable experiences: voyage of USS *Murphy* across Atlantic Ocean through Mediterranean Sea, through Suez Canal, Red Sea to Jidda, Saudi Arabia to take aboard King Abdul Aziz Ibn Saud 49 member retinue. Then return to the Suez Canal Great Bitter Lake and have the *Murphy* tie up alongside cruiser USS *Quincy* (CA-71) so the King could go aboard the cruiser to meet with President Franklin Roosevelt. Released from active duty May 16, 1946.

He achieved the rank of CAPT on July 1, 1963. Transferred to inactive USNR 1964. He retired from the USNR, July 5, 1978.

He married Marilyn and they have children: Oliver, Barry, Randy and Loren; grandchildren: Milan, Cami, Clinton, Shayne and Erin; and four great-grandchildren. Employed as purchasing agent, Divisions of General Motors Corp., 1946-74.

EDWARD J. "ED" O'SHEA, born Dec. 4, 1931 in New York, attended Notre Dame 1954; Georgetown Dental, DDS, 1956; University of Buffalo, 1960; MSD Orthodontics.

He joined the service in 1954. His military locations and stations: active duty, two years, NOB, Norfolk. Served with composite units, dental units, staff REDCOM, COMNAVRES FOR.

He was awarded the Navy Commendation Medal and Legion of Merit. Achieved the rank of RADM, upper half. His memorable experiences: reviewing officer, Great Lakes, member, medical flag council.

Married Judy and has six children: Nancy, Kevin, Megan, Maura, Kathleen and Christine. Employed as president, CEO Eastern Long Island Orthodontic Association founder, retiring 1995.

JOHN G. OSTHAUS, CAPT, born in Toledo, OH, Sept. 5, 1941. He is a graduate of Kalamazoo College and University of Michigan Law School. He was commissioned at Officers Candidate School, March 10, 1967.

He served as communications officer/RPS custodian at Naval Communications Station Honolulu. His most memorable experience was inventorying communications equipment, kept in CINCPAC's bedroom, on his hands and knees.

In January 1970 he joined Naval Reserve Security Group. He served tours in every department and four command tours with NRSG units in Washington, DC. He also served as a naval liaison officer with Federal Emergency Management Agency and as a mobilization plans officer for the assistant secretary of the Defense for Reserve Affairs. He retired April 1, 1997.

His awards include: Defense Superior Service Medal, Defense Meritorious Service Medal, Navy Unit Commendation, National Defense Service Medal and Armed Forces Reserve Medal.

He is Federal Legislation director for the Pan American Development Foundation, Washington, DC. The Osthaus family consists of wife, Donna, and sons, John and Christopher. They reside in Springfield, VA.

JOHN W. OWEN, SR., CDR, born May 24, 1944, Ft. Payne, AL, received BS in pharmacy, University of South Carolina (1973). Enlisted in the USN in 1967, stationed at Great Lakes, IL for both boot camp and Hospital Corp School.

First duty station was Naval Hospital, Portsmouth, VA, followed by a tour aboard USS *Ticonderoga* and USS *Constellation.* He received his commission as an ENS, Medical Service Corp., Columbia, SC on Aug. 1, 1979. Instrumental in seeing CDR Owen apply for a commission and who followed the course of the application was then LT Douglas Johnson (now RADM Johnson, USNR MSC Corp Chief), the CO of the unit to which Owen was assigned.

Upon receiving his commission, he applied for membership in NRA as a life member and served for a number of years as District 13 vice-president of membership. Transferring to the USNR Readiness Center, Charleston, SC, in 1986, Owen continued to serve in various unit leadership positions including command of the NDMS Coordinating Team. He was with the unit from its establishment, including its recall to ODS while serving as its CO, until its disestablishment. During ODS, CDR Owen was stationed with Naval Hospital, Charleston where he was assigned as assistant director of patient administration and in charge of the NDMS program. Incidentally, when the NDMS unit was recalled to active duty for ODS, the unit consisted of five pharmacists. The Naval Hospital, Charleston, also gained from their assignment for the pharmacy officers were able to practice their specialty by supplementing the busy department's staff.

Promoted to rank of CDR in 1995, Owen is the owner of an independent pharmacy practice in Columbia, SC, as he continues to drill, currently as a PRIMUS unit member in Augusta, GA, the unit where he also tenured as CO. He is married to Lynda Trotter, Owen, a school teacher and they have three children: John Jr., (a minister), a daughter Tara (a graduating senior in pharmacy, Auburn University) and Richard, also at Auburn University. CDR Owen serves as an adjunct clinical instructor with the College of Pharmacy, University of SC (his alma mater), as a preceptor for pharmacy students in his pharmacy practice and serves as a consultant pharmacists at three other medical organizations. Active in local and state pharmacy associations and his local church, CDR Owen has integrated his Reserve career with his civilian and family life to be both the citizen and sailor professional. The Owens family resides in Columbia, SC.

LEWIS A. "LOU" OXFORD, CDR, born Jan. 31, 1927 in Cambridge, MA, received an AB in economics and M.Ed from Tufts University.

He joined the service Nov. 4, 1944, USCGR; July 1947, USNR. His military locations and stations: Coast Guard Academy (enlisted); *Gen G.M. Randall* (AP-115). Served with USCGR, stewards mate, storekeepers; USNR, RES Intell. Program.

Oxford was awarded the Area Defense and WWII Victory Medal. Achieved the rank of CDR and retired in 1987.

His memorable experiences: As an 18-year old sailor repatriating Japanese Americans to Yokohama, Japan on the USS *Gen G.M. Randall* (AP-115); also, his years in the Naval Intelligence Program were stimulating.

He married Marjorie and has three children: Lewis Jr., Susan and Judith; and grandchildren: Kyle and Cory Craig and Shawn M. Montgomery. Employed as high school counselor, director of guidance, tax preparer.

THEODORE "TED" PACALA, CAPT, born April 26, 1928, Indianapolis, IN, earned his BA degree. He attended Franklin College and Indiana University.

He joined the service as a SR on Jan. 24, 1949, commissioned ENS June 5, 1953. His military locations and stations: September 1953-November 1954, new construction ship (AM-427), Fulton Shipyards, Antioc, CA and November 1954-October 1955, electronics officer, PacResFlt, L Beach Gp. From 1956-84 he served in various billets at the Naval Reserve Training Center, Indianapolis, IN. Units and billets worthy of note follow: NR Surface Div. 9-42(M), executive officer 1963-65, CO 1965-68; NR Officers School 9-13(M), executive officer 1970-71, CO 1971-74; NR Gp. Command 9-11(M), chief staff officer, 1974-76; and NR Voluntary Training Unit 1315, executive officer 1980-83, CO 1983-84.

Pacala was awarded the National Defense Service Medal and Armed Forces Service Medal. He achieved the rank of CAPT in August 1975 and was discharged July 1, 1984.

He and his wife, Sharon, have three children: Susan, Terri and Jenifer; and five grandchildren: Andrew, Phillip, James, Patrick and Nakia. Civilian employment with Allison Gas Turbine, a division of General Motors, as supervisor-component development, materials engineering. He retired May 1, 1988. His hobbies include gardening, hunting, fishing and handcrafting 18th century black powder rifles and accoutrements.

JERRY A. PACILIO, CAPT, born July 6, 1925, Paterson, NJ, earned his BA degree in chemistry and BS degree in naval science, Vanderbilt University, 1949.

He joined the service June 23, 1943. His military locations and stations: Armed guard, cruisers, carriers, mine sweeper, submarines. He served aboard the USS *Little Rock, Des Moines, Brambling, Tench, Argonant,* ComCarDiv 4, *Forrester.* Participated in the battle of Manus Island in 1944.

Pacilio was awarded the American Campaign Medal, Asiatic-Pacific Campaign Medal, European Theater, Good Conduct Medal, Navy Occupation Service Medal and Meritorious Service Medal. He achieved the rank of CAPT USNR and retired June 30, 1985.

He and his wife, Elizabeth, have three children: Jerry Jr., Betsy Jane and Beth Ann and three grandchildren: Jerry III, Erin and Hunter. Civilian employment as service systems analyst, IBM Corp.; computer specialist, Atlantic Intel. Command. Civilian retirement Oct. 31, 1996.

JOHN F. PADDOCK JR., RADM, born in Syracuse, NY. Following his graduation from Syracuse University, he was commissioned an ENS in 1966 and commenced active duty, serving aboard five combatant ships, a fleet oiler, a minesweeper, a destroyer, a guided missile destroyer, and a frigate. Ashore he served as strategic analyst for the chief of naval operations executive panel while in Washington, DC. He received his Master of Science in Operations Research from the Naval Postgraduate School.

Leaving active duty in 1980, RADM Paddock commanded various USNR units, including those for a frigate, striking fleet Atlantic staff, and the Readiness Unit in Albany, NY. He has also been training support advisor and mobilization plans officer for the staff of Naval Reserve Readiness Command Region Two located in Scotia, NY. His active duty periods included service in the Office of the Chief of Naval Operations and with the staffs of the Commander Naval Reserve Force, Commander-in-Chief US Naval Forces Europe, Commander Naval Surface Force Atlantic, Commander US Second Fleet, Commander Striking Fleet Atlantic, Commander-in-Chief Eastern Atlantic, Naval War College and Commander US Naval Forces Central Command/Commander Fifth Fleet in the Arabian Gulf area. His most recent active duty was in the European Theater Command Center, Stuttgart, Germany, as director, Crisis Action Team for Commander-in-Chief, US European Command, in which he was involved with Operations Joint Endeavor, Assured Response, Quick Response and Guardian Assistance.

RADM Paddock's initial flag assignment was deputy commander, Combat Logistics Group Two. He then served as deputy for Total Force Readiness, Surface & Expeditionary Warfare, OPNAV N86TF, in the Pentagon. He was also CDR, Naval Reserve Readiness Command Region TWO. RADM Paddock serves as one of three flag members of the Force Executive Steering Committee for CDR Naval Surface Reserve Force, a member of the Joint Military Intelligence College Senior Officer Advisory Panel, and chairman, Secretary of the Navy National Naval Reserve Policy Board. His current Naval Reserve billet is deputy for Total Force, DJ3RC, Headquarters US European Command, Stuttgart, Germany, which supports US Armed Forces activity in most of Europe, Africa and the Mediterranean Sea.

His awards and decorations include the Legion of Merit w/Gold Star, the Defense Meritorious Service Medal, the Joint Service Commendation Medal, the Navy Achievement Medal and the Vietnam Service Medal. He has been selected for promotion to two stars, RADM (Upper Half).

In his civilian career, RADM Paddock manages various ship combat system projects for Lockheed Martin Ocean, Radar & Sensor Systems, Syracuse, NY.

He is the son of Mrs. Carolyn Doran Paddock of Syracuse and is married to the former Sharon Lawton of Jordan, NY. They reside in Liverpool, NY and has three adult children, John, of Middletown, CT, also a Naval Reservist, and Michael and Rebecca, both of Boston, MA.

KEITH I. "KEG" PALMER, LT, born Oct. 5, 1923, Newark, NY, graduated high school and took 15 years of correspondence courses.

He joined the service in September 1942. His military locations and stations: Chappel Hill, Pensacola and Olathe Kanhon. He served with V-5 and instrument instructor for VR-11. Battles and campaigns participated in include the Pacific.

Palmer was awarded the American Theater Medal, Pacific Theater Medal and WWII Victory Medal. He achieved the rank of LT and retired Oct. 5, 1983.

His memorable experiences: Was married on base in 1944 and the birth of their baby on base.

He and his wife, Jane, have one son, David, and one grandson, Craig. Civilian employment as auto service manager and armed bank guard. He retired in 1989.

RANDALL W. PALMER, LCDR, born Oct. 26, 1919, Adrian, MI, earned his BS degree in biology at Adrian College in 1942.

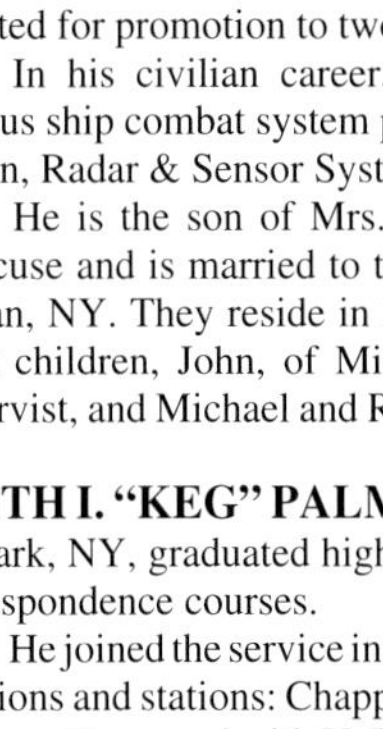

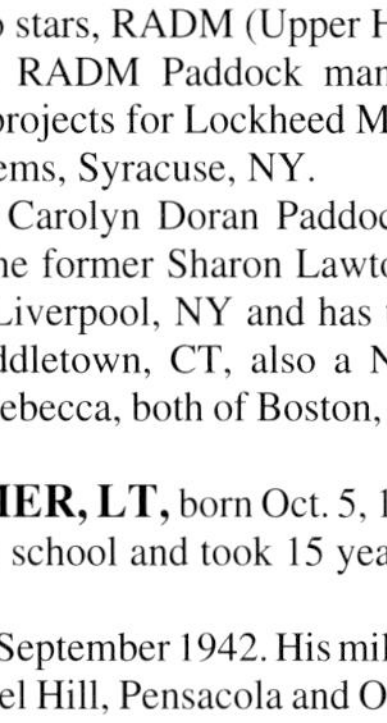

He joined the service July 27, 1942. His military locations and stations: NAS Banana River, NAS San Diego and NAS Corpus Christi. He served with VH-1 (air-sea rescue flying PBM "Mariner").

Palmer was awarded the American Theater Medal, Asiatic-Pacific Campaign Medal and WWII Victory Medal. He achieved the rank of LCDR and retired July 1, 1970.

He and his wife, Kathryn M., have one son, James R., and two grandchildren, Shayna and Trevor. Civilian employment as chemist at the University of Florida, Gainesville, FL. Palmer retired Nov. 26, 1984.

DONALD L. PANHORST, CDR, born March 11, 1932, St. Clair, MO, graduated from Central College, Fayette, MO. CDR Panhorst was commissioned at USNAVOCS in January 1954. His active duty stations include: the Reserve Officer Briefing Board at BUPERS, the USS *Otterstetter* (DER-244), the USNAVFAC San Salvadore, BWI and the USNAVOCS. After release from active duty in January 1958 his reserve assignments were: instructor at USNAVOCS for four summers, CO of two surface divisions, chief of staff and subsequently CDR of a reserve group command.

CDR Panhorst earned master and doctoral degrees from the Eastman School of Music of the University of Rochester. He had a 32-year career as a teacher, university professor, dean and vice-president. In 1973 he authored the book *How Ordinary People Can Become Extraordinary Leaders*. He and his wife, Dorothy, reside in Gulf Breeze, FL, and they have two married children and two grandchildren.

DONALD C. PANTHEN, ENS, born May 21, 1963, New York City, NY, graduated from Edgemont High School in June 1981; earned an AAS in accounting, September 1988, and a BBS in finance, May 1989, both at Pace University.

He joined the USN DEP program Sept. 23, 1980. His military locations and stations: Boot camp, Great Lakes, IL, Sept. 22-November 1981; Builder 'A' School, Port Hueneme, CA, November 1981-February 1982; NMCB 1, Gulfport, MS from November 1982-July 1986 (Guam from April 1982-January 1983; Sigonella Sicily from July 1983-February 1984; Okinawa, Japan from October 1984-June 1985; Puerto Rico from January-July 1986); NMCB 13, Operations Dept., Camp Smith, Peekskill, NY from September 1986-September 1990; 7th Naval Construction Regt., Operations Dept., Newport, RI from October 1990-September 1996 (Iceland, Operation Northern Viking, July 1995); NMCB 133 Det, New York, AOIC from October 1996-present (JO1 School, July 1996 and JO2 School, March 1997).

Panthen was awarded the Navy Achievement Medal (2), Navy Reserve Meritorious Service Medal (2), Navy "E" Ribbon (2), Naval Reserve Medal, National Defense Service Medal, Sea Service Deployment Ribbon (4), Expert Rifleman Medal and Sharpshooter Pistol Shot Ribbon.

He married the former Carolyn Marie Ferri Oct. 13, 1990, and they have two children, Kyle Craigie and Jordan Henry. Civilian employment as network manager, Pfizer Inc., New York, NY.

SALVATORE J. "SAM" PAPPALARDO, CAPT, born Jan. 8, 1937, Brooklyn, NY, earned his BA degree at the University of Rochester in 1958; his MS degree at State University College, Oswego, NY in 1966; and his EdD at SUNY, Albany, NY in 1971.

He joined the service in July 1958. His military locations and stations: OCS 1958; USS *Graham County* (LST-1176), Amphibious Operational Training Unit L/Creek. He served with NRSD 3-86(M), NRSD 3-25(M), HS Unit 3-84, NRGC 3-25(S), FCS-103, SacLant Det 205 and NRUTO 0503.

Pappalardo was awarded the National Defense Service Medal and Armed Forces Reserve Medal. He achieved the rank of CAPT and retired Oct. 1, 1985.

His memorable experiences: His service as CO, NRSD 3-93(M) Scotia, NY, and CO, SacLant Det 205 (Buffalo, NY).

He and his wife, Mary Anne, have two children, Christine and Joan; and three grandchildren: Angela, Elizabeth and Alexandra. Civilian employment as professor/chairman, Dept. of Education, Niagara University, New York.

LOUIS PARHAM, CAPT, born July 10, 1942, Lagrange, GA, earned his BS degree at Tuskegee Institute in 1968 (electronic technology) and his MSBA at the University of Massachusetts in 1974.

He joined the service Aug. 12, 1960. His military locations and stations: Enlisted - VA-64 (USS *Enterprise* CVA(N)-65); VP-672 (NAS Atlanta); VP-68AI/NAF Washington. Officer - OSD/JCS PI-106 (Baltimore, MD); NIC Task Gp; DIA HQ 0266; NOIC 566; NOIC 666; NIS HQ NAVMIC 466; IUTU 106; OSDTT 0166 (NAF Washington). He participated in the Cuban blockade in October 1962.

Parham was awarded the Good Conduct Medal, Navy and Marine Corp Expeditionary Medal, Armed Forces Expeditionary Medal, Meritorious Unit Commendation, Joint Meritorious Unit Award, National Defense Service Medal and Armed Forces Reserve Medal.

His memorable experiences: NATO cruise on USS *Wasp* in 1962; "Med" cruises on USS *Enterprise* in 1963 and 1964.

He and his wife, Dorothy Marie, have three children: Jacqueline Marie, Verice Lennette and Louis; and two grandchildren, Joseph Lamar and Gregory Ro. Civilian employment as senior engineer, Schindler Elevator Corp., Gettysburg, PA.

JAMES J. "J.J." PARKER, LCDR, born April 20, 1952, New Haven, CT, earned his BS degree in marketing at Central Connecticut State College and his MS degree in management at Renselaer Polytechnic Institute.

He joined the service Dec. 27, 1971. His military locations and stations: USS *Sam Rayburn* SSBN-635(B) and Nuclear Power School. He served with NCIS 0201 and 0501.

Parker was awarded the Navy Achievement Medal, Battle "E" (2), Good Conduct Medal, National Defense Service Medal and Unit Commendation.

His memorable experiences: Travels to Holy Loch and Rota during active service; experiences in Bahrain in the reserves; meeting COL Oliver North and President George Bush.

He is divorced and has two children, Alyson and James Jr. Civilian employment as territory manager, Johnson & McGill.

JAMES W. "JIM" PARKER, CAPT, born May 24, 1927, Burlington, VT, earned his BA degree at the University of Vermont and his MS degree at North Carolina State.

He joined the service in April 1945. Parker was awarded the Meritorious Unit Commendation, American Campaign Medal, Asiatic-Pacific Campaign Medal, WWII Victory Medal, National Defense Service Medal, Naval Reserve Medal and Armed Forces Reserve Medal w/HG. He achieved the rank of CAPT and retired July 1, 1982.

He and his wife, Andrea Thelin, have two children, Alysson and Kirsten. Civilian employment at Naval Submarine Medical Research Laboratory; director, Submarine Rescue and Diving Research. He retired July 1, 1989.

WAYNE A. PARKER, CAPT, born April 9, 1908, in Oklahoma, earned his BS and EE degrees. He joined the service Feb. 12, 1942. His military locations and stations were in the Atlantic and Pacific.

He served with DD-500, USS *Ringgold* and *Harry E. Hubbard*. Battles and campaigns: assault and landing, Tarawa Atoll; bombing, Marcus Island; assault and landing, Kwajalein; bombardment, Wake Island; assault and landing, Saipan; first Philippine Sea battle; assault and landing, Guam; assault and landing, Eniwetok; assault and landing, Hollandia; assault and landing, Leyte Gulf; and assault and landing, Okinawa.

Parker was awarded the Navy Cross. He achieved the rank of CAPT and was discharged May 1, 1968.

His memorable experiences: Assault and landing on Tarawa Atoll, assault and landing on Guam, first Philippine Sea Battle, assault and landing on Okinawa, and evacuation of GEN Wainwright from Manchuria.

He and his wife, Deborah (deceased), have two children, Deboral and Andy; and five grandchildren: Laura, Parker, LeAnne, Jennifer and Neal. Civilian employment as foreman, superintendent, vice-president and president of OG&E, an electric utility company. He retired May 1, 1973.

JAMES W. "PARKY" PARKINSON, CDR, earned his BA degree in psychology at College of Idaho in June 1965.

He joined the service in September 1965. His military locations and stations: NAS Pensacola, FL, basic training; NAS Pensacola, FL, commissioned ENS (January 1966); NAS Glynco, GA, radar intercept training; NAS Glynco, GA, designated naval flight officer (August 1966); NAS Miramar, training in F-4 Phantom II aircraft; NAS Miramar, assigned to VF-154 Black Knights; USS *Ranger* (CVA-61), October 1967-May 1968, Vietnam cruise; NAS Miramar, training in F-4J (May-October 1968); USS *Ranger* (CVA-61), October 1968-May 1969, Vietnam cruise; NAS Glynco, GA, NATTC as Rio instructor; and NAS Glynco, GA. Left active duty in September 1970. Joined the reserves in 1971 at NAS Jacksonville, FL where he served in several reserve units.

Parkinson was awarded the Air Medal (9), Vietnam Service Medal, Vietnam Campaign Medal and Unit Commendation. He achieved the rank of CDR (O-5) and retired in 1994.

He married the former June Gulley and they have one daughter, Paige Nicole. Civilian employment at Southeast Communications (1971-73), owner and operator, telephone answering equipment; and Hercules Inc. (1973-present), presently serving as purchaser.

RONALD ALLEN "RON" PARTNOY, CAPT, born Dec. 23, 1933, Norwalk, CT, graduated from Yale University with a BA degree and received his law degree from Harvard Law School. He was commissioned ENS in 1956 and served as a deck and watch officer aboard the aircraft carriers *Coral Sea* and *Ranger*. He then was affiliated with USNR surface divisions in Connecticut and rose to the rank of CAPT. Prior to his retirement in 1984 he was active in Navy recruiting.

In civilian life CAPT Partnoy has been general counsel of Remington Arms Co. and senior counsel of the DuPont Co. He has served as NRA 3rd Dist. president, Navy League national director and Connecticut state president. He is listed in *Who's Who in America, Who's Who in Finance and Industry,* and *Who's Who in American Law*. He and his wife, Diane, reside in Kennett Square, PA.

JOHN J. PATAFIO JR., CDR, a native of Long Island, NY received a BS in business administration from Columbia University.

At 17 years old in 1943, he began his naval career in the Duke University V-12 program and was commissioned July 3, 1945, from the Ft. Schuyler, NY Midshipman School. He served as fire control officer aboard the USS *Davison* (DMS-37) in the Pacific Theater until released to inactive duty in 1946.

Recalled in 1951 to active duty for two years during the Korean War, he served aboard the USS *Melvin* (DD-680) as gunnery and senior watch officer. The *Melvin* engaged in two deployments with the Sixth Fleet.

LT Patafio then affiliated with Surface Div. 3-36, Freeport, NY. In 1956 he became XO of Fleet Div. 3-2 the predecessor to the Gp. II NRF Destroyer Escort program. In 1958 LCDR Patafio continued on as XO of the USS *Albert T. Harris* (DE-447) reserve crew.

In October 1961 the *Harris* along with her reserve crew was recalled to active duty during the Vietnam conflict. He served as XO and navigator and was promoted to CDR. Upon returning to ready reserve status, he continued as XO to rebuild the *Harris* reserve crew. For that effort, the *Harris* received the 1963 England trophy as best Atlantic Fleet NRF DE.

He then served as CO of the following units: Surface Div. 3-48 Large, USS *Bristol* (DD-857) reserve crew and SAMAR Div. 3-4(S). He retired with 25 years service. He later retired as CAPT in the New York State Naval Militia.

He earned the following medals: Navy Expeditionary, China Service, American Campaign, Asiatic-Pacific Campaign w/star, WWII Victory, Navy Occupation w/ Europe & Asia Clasps, National Defense Service w/star, Armed Forces Reserve w/HG and Naval Reserve. Additionally, he was awarded the following ribbons: Navy Meritorious Unit Commendation, NR Sea Service and Navy Recruiting Service.

CDR Patafio joined the NRA in 1968 and served three terms as president of the Long Island chapter.

His civilian career was in the advertising business. He retired as vice-president marketing for Ambassador Mail Advertising Co. owned along with his two brothers. He is married to the former Margaret "Margo" S. McAulay of Newtom Merns, Scotland.

VINCENT "VINCE or VINNY" PATTAVINA, CAPT, born May 14, 1923, Quincy, MA, graduated high school, attended three and one-half years college and four years medical school.

He joined the service in November 1942. His military locations and stations: USNTS Sampson, V-12 Unit Syracuse College Medicine. He served with USNH Newport, USN MSTS Staten Island Transports, USNAS So. Waymouth, NAS Sigonella, and USN sub base Holy Loch Scotland.

Pattavina was awarded the WWII Victory Medal, Naval Reserve Medal, Meritorious Unit Commendation, etc. He achieved the rank of CAPT and was discharged Aug. 5, 1983.

His memorable experiences: Sea duty with MSTS TAD in Scotland, Holy Loch and NAS Sigonella learning to fly airplanes. Now private pilot SEL and instrument rated.

He and his wife, Frances, have four children: Vincent J., Dr. Charles, Elizabeth and Thomas B; and several grandchildren. Civilian employment in private practice for 20 years. He retired in 1985.

WILLIAM B. "BILL" PATTON, CAPT, born Feb. 12, 1939, Mason City, IA, earned his BA degree at Grinnell College in 1961.

He joined the service Feb. 12, 1962, Des Moines, IA. His military locations and stations: USS *Bexar* APA-237 and Naval and Marine Corps Reserve Center. He served aboard the USS *Bexar* as communications officer and CO, NR TelComSta Greece, NR VTU-1605 and NR MH Plane. He participated in the Cuban Missile Crisis and two tours in Vietnam with the Amphibious Ready Task Group.

Patton was awarded the Armed Forces Expeditionary Medal (2), RVN Campaign Medal, National Defense Service Medal and Armed Forces Reserve Medal (2). He achieved the rank of CAPT and was separated from active duty June 15, 1965.

His memorable experiences: As CO, NR TelComSta Greece 116 his unit was instrumental in staging and implementing the first naval reserve unit to take active duty for training at their active duty counterpart.

He and his wife, Sally R., have four children: Scott, David, Shawn and Heidi; and one grandchild, Tyler. Civilian employment as vice-president/shareholder, Stitzell Electric Supply Co. He retired Feb. 1, 1989.

R. JOHN "JP" PAUL, CDR, born June 24, 1951, North Kingston, RI, earned his BA degree at UCLA.

Joined the service June 16, 1974. Initially assigned as an assistant maintenance office, VF-24, NAS Miramar, F-8 Sqdn. Trained as a naval aviator, VT-24, NAS Chase Field, Beeville, TX. First cruise completed aboard USS *Enterprise* with VS-38, NAS North Island, flying S3A Vikings. Returned to VT-24 as flight instructor until October 1980.

Joined the USNR in 1980 and continued as a flight instructor, VT-24 through 1982. Other assignments include: NAS North Island, ASWWINGPAC, HS readiness officer; NAS Alameda, VAK-1087, training officer; VAK-208, pilot; NR CVN-65, XO; NR CVN-72, CO; EUCOM Stuttgart, Germany, NR Det. 130, Plans, Policy, and Manpower; London, UK, CINCUSNAVEUR Det. 130, XO assigned to N43 Aviation Readiness.

CDR Paul was awarded the Navy Commendation Medal, Meritorious Unit Commendation (3), National Service Medal (2) and Naval Reserve Medal.

He and his wife, the former Maria T. Brennan, have two sons, Christopher and Conor. Civilian employment as pilot with PAN AM and now employed overseas with Kuwait Airways.

RICHARD L. "RICK" PAUL, CAPT, born July 14, 1929, Silver Creek, NY, earned his BS degree at Westminster College and his MSED degree at Alfred University.

He joined the service July 13, 1951. His military locations and stations: Whidbey Island, Niagara Falls, New York City and Willow Grove. He served with USS *Corregidor* (VP-17), FAS-112 VP-832, VP-837, VP-18 and VP-62.

Paul was awarded the Korean Service Medal, UN Medal and National Defense Service Medal. He achieved the rank of CAPT (O-6) and retired March 1, 1974.

His memorable experiences: Flying VF-1 Arctic patrol building Dew Line in 1955 stationed at Pt. Barrow; losing all instruments on ASW flight in Bermuda; flew SNJ P2V, SP2H, VF1, PB4Y2 and SNB.

He and his wife, the former Christine Klinger, have two children, Karen Lynn and Eric; and three grandchildren: Matthew, Courtney and Devin. Civilian employment as college professor, division chairman, science-math. Retirement from Retired Reserve July 13, 1989.

GRANVILLE E. PAULES III, CAPT, born March 22, 1937, Columbia, PA. Grew up in Ft. Worth, TX and attended the University of Texas in Austin on a full NROTC scholarship. He graduated with a degree in electrical engineering and later obtained an MBA from the University of Maryland.

On graduation in 1960 he reported to the USS *Coontz* (DLG-9, now DDG-40) in San Diego as a member of the commissioning crew. He completed four years of active duty on the *Coontz* advancing to communications officer and then to missile officer and participated in two WestPac tours. He was qualified as DLG at-sea GQ officer of the deck and import command duty officer.

In 1964 he left active duty and began a government career serving with NASA, then the Dept. of Transportation, and returning to NASA in several R&D program management positions. Initially he worked for NASA in Mission Operations at what was to become the Johnson Space Center. He was a major participant as a Mission Control Center Apollo guidance officer for all missions through Apollo 14 receiving several awards including the Presidential Medal of Freedom, a group award for Apollo 13 rescue efforts.

On leaving active duty he transferred to the USNR as a LT (1105) serving subsequently with: ship activation, maintenance, and repair 8-2(M), Galveston, TX; NRSD 1-5(L), Boston, MA; NAVMAT HQ 106, Washington, D.C.; and with NAVSEA DET 1506 (inspector general) where he retired as CAPT (1115) in 1990. At all commands his major focus was on maximizing integration of reserve and active force missions and activities.

He lives with wife, Diane, in Rockville, MD. They have a son, Skip, and daughter, Allison.

ALBERT G. "AL" PAULSEN, RADM, periodontist, educator and practitioner, Business Data: A full service dentistry which emphasizes the treatment of referred patients with periodontal disease. The practice was founded in 1958.

Associations and Civic Backgrounds: Licensed to practice in Virginia since 1954 and the District of Columbia since 1945. Retired naval officer; RADM (retired) Dental Corps, USNR; retired clinical professor at Georgetown University School of Dentistry; diplomate, American Board of Periodontology; life member, American Dental Assoc.; life member, District of Columbia Dental Society; life member, Omicron Kappa Upsilon; American Academy of Periodontology; life member, Virginia Dental Assoc.; Fairfax Dental Society; American Equilibration Society; life member, American College of Dentists; International College of Dentists; held various consulting positions since 1958; published various articles and received numerous awards and honors.

Medals: China Service Medal, American Campaign Medal, Armed Forces Reserve Medal, WWII Victory Medal and Naval Reserve Medal.

Educational History: Graduate, New Haven High School, New Haven, CT, 1939; Loyola College, Baltimore, MD, major: predental, 1939-42; Georgetown University, DDS, 1942-45; Rotating Dental Fellowship, School of Dentistry-Georgetown University, 1945-46; Northwestern University, MSD, 1948-50; attended numerous postgraduate courses: University of Michigan, Naval Graduate Dental School, University of Illinois, Georgetown University, Walter Reed Army Medical Center, Medical College of Virginia, also took yearly correspondence courses at the Naval Graduate Dental School, 1957-74.

Personal data: Paulsen is proud to be a flag officer, in the Navy, dentist and educator. His mother and uncle are his inspiration to succeed. Paulsen was born Aug. 21, 1919, New Haven, CT. He is married to Sara J. They have one daughter, Beverley Susan, one son, Jeffrey Jordan and three grandsons: Christopher, Nicholas and Gregory.

CLEMENTS E. PAUSA, CAPT, a native of South Gate, CA. He attended the University of California at Los Angeles (UCLA) and earned a BS with honors in engineering, a MS in metallurgy, and a graduate certificate in business management from the University of California at Berkeley. He began his naval career in 1950 as a regular NROTC midshipman at the University of California, Berkeley, and was commissioned ENS USN in 1954.

He served on active duty in the Western Pacific (WestPac) aboard the USS *Montrose* (APA-212) as CIC, communications, administration, OOD Underway, and aboard the USS *Gull* (MHC-46) as engineering officer.

He transferred to the USNR in 1958 and had been a member of various surface, MSTS, and MSC divisions in assignments from instructor to CO, completing qualifications for commanding officer naval shore activity. He joined CincPacFlt Det 420 in 1972 after completing three years as CO of MSC Div. 12-1. He then spent the remainder of his naval reserve career providing management assistance to various Navy commands and activities. As a naval reserve officer, CAPT Pausa's active duties included eight management assistant studies, three as OIC; plus duty at sea, at various MSTS, MSC and NCSO courses and activities; with ComNavPhil and CTF35; and the Naval War College and Armed Forces Staff College. CAPT Pausa retired from the USNR in 1982.

He is a broadly experienced corporate officer and manufacturing executive with a consistent record of successes accumulated over 37 years in high tech. industries.

Pausa is a vice-president of the Alphatec Gp., a group of affiliated, but separately owned and operated high tech. companies which share common management and strategic direction. He is also a director of Coopers & Lybrand Consulting. He began consulting in October 1990 after his retirement from National Semiconductor Corp., where he had served as corporate vice-president of International Manufacturing and Services. His responsibilities included manufacturing operations in Hong Kong, Indonesia, Japan, Malaysia, the Philippines, Singapore, and Thailand; the DynaCraft Packaging Material Subsidiary; Package Engineering; Materials and Services; and Automotive Operations.

Pausa has more than 37 years of broad experience in semiconductor manufacturing with direct profit and loss responsibilities and has directed extensive multinational operations. During his career, he has directed 33 plants and subsidiary companies in 18 foreign countries, and 11 plants in six US states. He has served as a director on the boards of 10 foreign subsidiary companies in Australia, China, Haiti, Hong Kong, Indonesia, Malaysia, Puerto Rico, Thailand and Singapore. He is a member of the American Management Assoc., ROA and NRA.

CAPT Pausa resides in Atherton, CA with his wife, Janice; son, Geoffrey in Phoenix, AZ with his wife, Shey; and son, Ronald in San Carlos, CA with his wife, Kristi.

MARK J. PAWLAK, CDR, a native of New Jersey, graduated Miami University (Ohio) May 1979 with baccalaureate degrees in aeronautics and aero-math and commissioned an ENS. In February 1980 he was designated a naval flight officer and assigned A-6s out of NAS Oceana. CDR Pawlak completed two deployments with Attack Sqdn. 34 during 1980-83 and served as a flight instructor for VT-86 from 1984-86.

Upon release from active duty in May 1986, CDR Pawlak affiliated with the USNR and has served in numerous aviation and command staff units. He was activated by COMUSNAVCENT for Desert Shield/Storm in Riyadh, Saudi Arabia and responsible for the coordination of all naval air assets. Currently he is the CO for NR NAS Miramar 0170, Dallas.

CDR Pawlak is employed as a pilot for Continental Airlines and is married to the former Kathleen Elizabeth Miller of Virginia Beach. They have two children, Joseph and Danielle.

CHARLES A. "CHUCK" PAYNE, CAPT, born Feb. 23, 1932, Louisville, KY, attended medical school at Georgetown University, Washington, D.C., and five years post graduate school.

He joined the service in 1957. Military locations and stations: Washington, D.C.; Bethesda, MD; Portsmouth, VA; Yokosuka, Japan; USS *Independence*; and USS *Eisenhower*. He served with multiple reserve units, Fleet Hospital 14 and Fleet Hospital 6. Participated in Desert Storm, Bahrain, Fleet Hospital 6.

Payne was awarded the Naval Commendation Medal, Meritorious Unit Commendation, Air Force Reserve Medal, National Defense Service Medal, Overseas Service Medal, Saudi and Kuwait Medals. He achieved the rank of CAPT (O-6) and retired Feb. 23, 1992.

Memorable experiences: Service in multiple naval hospitals and on two carriers and in Bahrain during Desert Storm, all as a reservist.

He married the former Sylvia Negron (also an O-6 MC USNR, retired) and they have six adult children and 14 grandchildren. Civilian service as neurologist in private practice in San Juan. Still in active practice of medicine.

H.W. "DUB" PEACE II, CAPT, born May 21, 1935, Clinton, OK, earned his BS degree in geology in 1959 and his MS in geology in 1964, both at the University of Oklahoma.

He joined the service in April 1959. Military locations and stations: Newport, RI; NAS Jacksonville, NAS Brunswick, GA; Pt. Loma; Barbers; NAS Dallas; NAS New Orleans. He served with OCS, AGO, CIC School, FARWTC, AEWBARRON PAC, VP Dallas, VR-702, VR-54.

Peace was awarded naval flight officer (NFO) 1325. He achieved the rank of CAPT (O-6) and retired from Ready Reserve in May 1983 and the Navy retired list May 21, 1995.

His memorable experiences: Designated air intercept controller at AEW BARRON PAC - TransPac inflite refuel control for F-8 and A-4 Marine squadrons from El Toro to Iwakuni. Also controlled numerous aircraft for Johnston Island A bomb tests with Thor missiles in 1961-62. CO, VR-54 at retirement as first NFO CO in VR.

He and his wife, Norma, have two children, Hugh William and Susannah Lee; and one grandchild, Austen Michael. Civilian employment as president, CEO and director of Panhandle Royalty Co.

ROBERT W. "BOB" PEARCE, CAPT, born Oct. 12, 1936, Los Angeles, CA, attended Santa Monica City College and Northrop Institute majoring in aeronautics/mechanical engineering. He joined the USNR in 1955 at NAS Los Alamitos, and entered the NAVCAD program in October 1960. He received his naval wings and commission in 1962 at NAS New Iberia, LA, and reported to VAW-11 at NAS North Island.

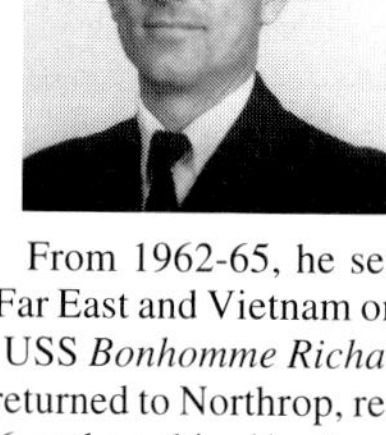

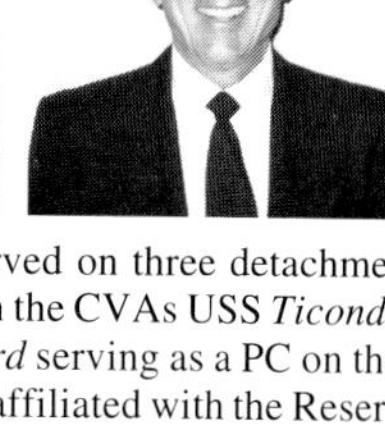

From 1962-65, he served on three detachments to the Far East and Vietnam on the CVAs USS *Ticonderoga* and USS *Bonhomme Richard* serving as a PC on the E-1. He returned to Northrop, reaffiliated with the Reserves in 1966, and was hired by American Airlines and assigned to Los Angeles in April 1966, as a flight engineer on the 707.

His USNR billets include VS-774, VS-81, Tactical Support Center 192, NARU NORIS, and as executive administrative officer of NAS 0194. He has served on selection boards and attended Naval War College. Awards include the Air Medal, Navy Commendation with V, National Defense Service, Armed Forces Expeditionary, Naval Reserve and Vietnam Service Medals.

CAPT Pearce has accumulated over 24,000 hours flying the E-1, E-2, S-2, 707, MD-80, DC-10, and other civilian aircraft. Retiring in October 1996 he formed a GPS sales company and is active in civic activities. He is a life member of the NRA. The family resides in Redlands, CA and consists of his wife, Robynn; sons, Ken and Scott; and daughters, Linda and Christine.

FRANK V. PEASE, CDR, born Feb. 14, 1927, Savannah, GA, earned his BSEE degree at Auburn University. He joined the service June 22, 1945. His military locations and stations: USHRSD 6-40 Gadsden, AL and HRSD 6-9 Columbus, GA.

Pease was awarded letters of commendation from the Secretary of Navy, chief of naval operations and commanders, Naval Reserve Training Command for the achievement of NRSD 6-40(M) moving from close to bottom in nation to #2 in the nation. They had a fine team of men working together.

He and his wife, Virginia C., have three children: Shirley J., Leila and Frances. Civilian employment as district manager, Georgia Power Co. He retired July 1, 1978. Pease established a Navy League and a Naval Reserve Assoc. chapter in Columbus, GA.

MICHAEL DAVID "MIKE" PECK, LCDR, born Sept. 26, 1951, Houston, TX. Earned BA degree in 1973 at Univ. of Texas at Austin and JD degree in 1976, Univ. of Texas Law School.

Joined the service May 24, 1985 and was stationed at NAS Belle Chase; NAS Corpus Christi; and NAS Dallas. Units served with include First 1182, NISRO 2310, NISRO 2010, NISRO 2182, SIMA Ingleside DET 1, NR COMINEWAROM 107 and AIRPAC 0270.

Achieved the rank LCDR, USNR, Active. His awards include the National Defense, Overseas, and Expert Pistol. He and wife Debbi have one child, Trevor. He is a self-employed businessman, attorney and college instructor and has no plans for retirement.

JAMES J. PENNE, CDR, born Dec. 6, 1914, Bonesteel, SD, earned his BA magna cum laude, MA in history, JD and LLM degrees.

He joined the service Feb. 28, 1942. His military locations and stations: DCNO, Washington, D.C.; NAS New Orleans, NAS Minneapolis, NAS Memphis, NAS Glenview, IL, JAG, and Bauer.

Penne was awarded the American Theatre Medal, Victory Ribbon end Expert Pistol Medal. He achieved the rank of LT (active duty) and was discharged Oct. 2, 1947; and the rank of CDR (inactive duty) from which he retired Dec. 4, 1974.

He and his wife, Elizabeth, have one child, Ellen F. and two grandchildren, James and Erica. Civilian employment as assistant counsel, Off. Gen. Counsel, Navy Dept.; assistant corp. counsel, Garrett Corp.; and US magistrate judge, Los Angeles, CA. He retired as judge, Jan. 19, 1988.

DENTON LOUIS "LOU" PEOPLES, CAPT, born Dec. 3, 1940, Kansas City, MO, earned his BS with distinction at Stanford University in 1963 and his MBA with honors at Harvard Business School in 1972.

He joined the Naval ROTC Regular at Stanford June 15, 1963. His military locations and stations: Submarine School, Nuclear Power School and Prototype. He served with the USS *George Bancroft* (SSBN-643); AIW Nuclear Prototype; and USS *Tinosa* (SSN-606).

Peoples qualified in submarines and was awarded the FBM Patrol Medal w/1 star, National Defense Service Medal, Armed Forces Reserve Navy and Naval Reserve Assoc. He achieved the rank of CAPT, USNR and retired in 1995.

His memorable experiences: Active duty - June 1963-February 1970; USNR active drilling February 1970-July 1985; retired in 1995. Commissioning crew of USS *George Bancroft* new construction at electric boat in Groton, CT; sea trial tests to validate naval engineering of 640 class of FBMs; and command of Naval Reserve unit.

He married the former Mary Ann Zannon and they have three children: John, Anne and James. Civilian employment as vice chairman and CEO, Orange and Rockland Utilities, Inc.

LOUIS N. "SWABBIE COLONEL LOU" PERNOKAS, CAPT, born Dec. 21, 1926, Boston, MA, earned his BA degree at Dartmouth College and his MD degree at Harvard Medical School.

He joined the service in April 1944 and was assigned to Navy V-12 program, Dartmouth College 1944-45. Discharged at Fargo Bldg., 1st Naval Dist. in 1945. He was commissioned ENS in 1950. Recalled in 1951 and served with 2nd Bn., 1st Marines in Korea, 1951-52; served USNH NPT R-1, 1953-54; served with several Marine Reserve units of 4th Div. Retired as CAPT, USNR.

Pernokas was awarded the National Service Medal, WWII Victory Medal, Korean Service Medal w/3 stars, Korean Presidential Unit Citation, American Presidential Unit Citation, Navy Unit Citation, UN Medal, Commendation Medal w/Combat V (Marines).

His memorable experiences: His year as a battalion surgeon with a Marine infantry battalion in combat. Performing an appendectomy on a troop transport during a storm on the Pacific.

He and his wife, the former Charlotte Johnson, have one son, Nick. Civilian employment as vascular surgeon on staff of several Boston and Brookline, MA hospitals; taught students and residents on Harvard surgical services; chief of surgery, Kent Hospital, Warwick, RI; Vascular Surgeon of the Year for Rhode Island in 1993. He retired in 1989.

ALDEN TAYLOR PERRY, CDR, born Aug. 9, 1943, Charleston, SC, earned his BSEE degree at the University of Houston, TX in 1972; graduate studies, engineering management, University of Tennessee-Chattanooga in 1991.

He joined the service (active) May 28, 1964, and the Reserves Oct. 18, 1975. His military locations and stations: Active - NTC San Diego, NATTC Memphis, NAS Cecil Field, NAS Oceana; Reserve - NAS Norfolk, Redstone Arsenal, NAV Base Charleston, NRC Atlanta, NAV Base Norfolk, NAVFAC HQ, and CINCUSNAVEUR HQ. He served with the following units: Active - VF-14 (June 1965-May 1968); Reserve - VR-56 (October 1975-December 1978), NMCB-24 (November 1980-September 1986), EFD South (October 1986-September 1991), OICC Southern Europe (October 1991-September 1993), OICC LANT (October 1993-September 1994), EFD LANT (October 1994-September 1995), NAVFAC Augment (October 1995-September 1997), Mobilized - LANT DIV NAVFAC 09A Emergency Operations Center (September-November 1994), CINCUSNAVEUR N7 DCOS Facilities (February-November 1997).

While on active duty he participated in Task Force 77 and Tonkin Gulf (June 1966-March 1967). Reserve duty included: JTFs 180 and 190, (Mobilized) Haitian Operation UPHOLD DEMOCRACY (September-November 1994), Bosnian Operations (February-November 1997).

Awards earned include: Active - the National Defense Service Medal, Vietnam Service Medal, Vietnam Campaign Medal; Reserve - Navy Commendation Medal, Meritorious Unit Commendation (2), Navy Battle Efficiency "E", Naval Reserve Meritorious Service Medal, National Defense Service Medal, Armed Forces Reserve Medal (2), Mobilized (2), Navy Expert Rifle Medal and Navy Expert Pistol Medal.

His memorable experiences include: Active - Vietnam Campaign, USS *FDR* (CVA-42); Reserve - plank owner and C-98 qualified aircrewman, VR-56 (1976), direct commission LTJG, CEC (1980), Battle "E" and Operation "Greenstinger" NMCB-24 (1984), plank owner and training officer, OICC LANT (1994), director EFD LANT 09A-EOC for Haitian Operation UPHOLD DEMOCRACY (1994), presidential mobilization to CINC US Naval Forces, Europe-N72 for Bosnian operations (1997).

He and his wife, Carolene, have been married for 27-1/2 years and they have three children: Alicia, Karen and Betsey. Civilian employment with TVA Services and Sequoyah Nuclear Plant, project engineer.

GILBERT B. PERRY, CAPT, a native of Watsonville, CA, joined the USN as an enlisted man in June 1955 at age 17. He went through boot camp in San Diego and then was assigned as a station keeper at NAS Willow Grove, PA. From there he attended the PN "A" School at Bainbridge, MD. During this period, he was promoted to PN3. While at Bainbridge, he was selected to attend the USNA Preparatory School also located there.

Upon successful completion of the Prep School, he was admitted as a midshipman to the USNA in July 1957. Upon completion of four years at the Academy, he was commissioned an ENS, USN in June 1961 and assigned as main propulsion officer on USS *Lowry* (DD-770) at Norfolk, VA.

Successive active duty assignments included: chief engineer officer, USS *Soley* (DD-707); assistant training officer, Fleet Training Gp., Guantanamo Bay, Cuba; executive officer, USS *Windham County* (LST-1170) Vietnam; and Hydrofoil Program officer, Naval Ship Research and Development Center, Washington, D.C. During this period he qualified as officer of deck (Task Force Operations, Independent Steaming, In Port), and Combat Information Center watch officer among others and was designated as a surface warfare officer.

Left active duty in June 1971 and was commissioned USNR. Reserve unit duty included three executive officer and four CO assignments among others. These involved Weapons Station, Destroyer, Repair Ship and Ship Systems Command HQ units. During the period, he qualified for and was designated engineering duty officer (ship engineering specialist).

After 37 years of both enlisted and commissioned officer service, CAPT Perry retired in July 1992. During his career he was awarded the Navy Unit Commendation Medal, Meritorious Unit Commendation, Navy "E" Ribbon, National Defense Service Medal (2), Armed Forces Expeditionary Medal (Cuba), Vietnam Service Medal w/ Bronze Star, Armed Forces Reserve Medal w/HG, RVN Meritorious Unit Citation (Gallantry Cross), RVN Unit Citation (Civil Actions Medal) and the Expert Pistol Shot Medal.

CAPT Perry is a life member of NRA and served as president of the Saguaro Chapter during the period 1982-90, general chairman of the 1978 EXCOM Phoenix Meeting, general chairman of the 30th National Conference, Phoenix and was awarded the association's "Award of Merit."

ROBERT EDWIN "BOB" PERRY, ASO-4, born Dec. 30, 1935, Teaneck, NJ, earned his BA degree in 1958 at Whittier College with post graduate studies at Long Beach State.

He joined the service in September 1953. His military locations and stations: NAS Los Alamitos, NAVSTA Newport, RI (2) and OCS. He served aboard USS *Fiske* (DDR-842), 1959-60 and USS *Boston* (CAG-1), 1960-62. He participated in the smuggler disruption operations in the Mediterranean, late 1950s and the Bay of Pigs, April 1961.

Perry was awarded the Air Wing Staff "E", National Defense Service Medal, Navy Expeditionary Medal, and Armed Forces Reserve Medal. He achieved the rank of ASO-4 and retired in September 1977.

His memorable experiences: One of few to earn DTG-3 in Ready Reserves since WWII. Working with CAPT G.D. Swisher, D.C. in NAS Los Alamitos Dental Dept. Caribbean activities. Huge NATO fleet exercise above Arctic circle - got Russian's attention; numerous Mediterranean cruises.

He married the former Magdelyn Christine "Lyn" Ferrari and they have one daughter, Beth Anne. Civilian employment as self-employed business owner, financial services since 1964.

GERALD E. "JERRY" PETER, CDR, born Feb. 22, 1915, Trenton, IL, earned his BS degree at the University of Illinois.

He joined the service in June 1942. His military

locations and stations: USS *Defiance* (AMC-73) and USS *Eager* (AM-224). He participated in battle at Okinawa.

Peter earned the Unit Citation. He achieved the rank of CDR and retired in 1960.

His memorable experiences: post-war minesweeping, Japan.

He and his wife, Margeria, have two children, Susan and Nancy; and three grandchildren: Bret, Claudia and Jerimie. Civilian employment as president, Durrie Sales Co. He retired in 1970.

CLARENCE A. "PETE" PETERS JR., LT, born April 24, 1919, Cincinnati, OH, attended Duke University, University of Cincinnati, Ohio Wesleyan, Gonzaga University, Spokane, WA and St. Mary's College in California.

He enlisted in the service Aug. 24, 1942, and entered active service Jan. 5, 1943. His military locations and stations: NAS Corpus Christi, NAS San Diego, and NAS Kaneohe, HI. He served with VPB-21, based on tender (seaplane) AV-10, and USS *Chandeleur*.

Peters was awarded the American Theater Medal, Pacific Theater Medal w/2 stars and Philippine Liberation Medal. He was commissioned ENS and received Gold Wings June 28, 1944, NAS Corpus Christi. He achieved the rank of LT, USNR (Ret.) and was discharged Oct. 21, 1945.

His memorable experiences: During Okinawa Campaign crashed in Yellow Sea between Korea and coast of China. Lost five crewmen. Fortunately spent only two days in life raft.

His wife's name is Ruth. He has one adopted son, Eric and one stepson, Ronald. Civilian employment as supervisor, NATORP three garden stores. Peters retired in January 1983.

DEAN LEHMER "PETE" PETERSEN, CAPT, born Nov. 13, 1926, York, NE, enlisted as SA, V-12 Jan. 29, 1944, and assigned to Doane College, Crete, NE; Western Michigan College, Kalamazoo, MI; and Northwestern University, Evanston, IL; being honorably discharged June 20, 1946. He enrolled in the NROTC Program, University of Nebraska, graduating and commissioned ENS June 6, 1947.

On June 6, 1949 was assigned to NRIP 9-1, Unit 8, Omaha, NE. Volunteered for active duty July 13, 1950, and assigned as student of Class V, Naval Intelligence School, Anacostia, VA. After graduation he was assigned as O-in-C, Investigation Section, DIO-6ND, Charleston, SC and Intelligence Office, COMNAVFE, Yokosuka, Japan, supervising investigations in Japan and Korea.

Returned to civilian life in July 1954 and was assigned to NRED 9-7, St. Cloud, MN and then NAVRES, SURFDIV 9-143, Lincoln, NE. In January 1960 he reactivated Naval Reserve Intelligence Div. 9-1, Unit 8, Omaha, NE, serving as O-in-C until June 1964.

During his years in the Reserve he was assigned various billets including the NRMSPO, attached to AG, state of Nebraska from 1970-74.

Then assigned as IO, Invest C&C 417. Next assignment as CO, Contingency "A" 878. In October 1977 was assigned CO of FIRSPAC 278, NAVAIRESCEN, Minneapolis, MN followed by CO, IVTU 116.

CAPT Petersen was the designated Reserve Intelligence area coordinator, Area 10, NICHQ 0118 Oct. 1, 1980. After two years he was then assigned to IVTU 0116 until transferred to the Naval Reserve Retired List Nov. 13, 1986, having completed 42 years, nine months and 14 days service.

CAPT Petersen was awarded the Naval Commendation Medal. His campaign and service medals are: American Theatre, WWII Victory Medal, National Defense Service Medal, Korean Service Medal, Armed Forces w/ 2 HGs, Naval Reserve Medal, UN Service Medal and Pistol (Expert) Medal.

He and his wife, Darlene M., have seven children and 10 grandchildren. Civilian employment as president and CEO, Nebraska Bulk Transport & Fitch Trucking, Inc.

ROBERT A. "PETE" PETERSEN, CAPT, born May 1, 1938, New York, NY, graduated from Ursinus College, Collegeville, PA in 1960 and enlisted in the USN in the rate of OCSA (officer candidate seaman apprentice). Upon completion of OCS he was assigned to the USS *Boxer* (LPH-4), as O-I Div. officer.

In 1962 CAPT Petersen volunteered for UDT/SEAL training, graduated August 1962, and was assigned to UDT-21 where he served as platoon CDR, air concept officer and training officer. He was also officer-in-charge of the Atlantic recovery team for the Mercury 7 Space Project. He also worked in the test and evaluation of recovery gear for Projects Gemini and Apollo.

In 1965 he was assigned as navigator aboard the USS *Henrico* (APA-45). During an extended WestPac deployment he also assumed the duties of Boat Gp. CDR and operations officer.

In 1967 CAPT Petersen left active duty and affiliated with the Selected Reserve. He spent his reserve career in the Naval Special Warfare Community, commanding four different Specwar units. As a LCDR he commanded the first Reserve Seal team, UDT/SEAL Lant 106. As a CDR he had command of NavSpecWarGru Two, Det 106. His commands as a CAPT include NavSpecWarGru Two, Det 402 and a second tour as CO of NavSpecWarGru Two, Det 106.

CAPT Petersen established the NavSpecWar Reserve Recruiting and Screening program for applicants to BUD/S (Basic Underwater Demolition/SEAL training) and formed the group that briefed the NROTC colleges on Special Warfare capabilities.

He spent most of his civilian career in sales and sales management in the computer hardware/software industry. Currently he and his spouse, Maureen, own Bayside Gift Basket Emporium, Naples, FL. CAPT Petersen's children: Dawn, Denise, Robert Jr. and Danielle, are located in California, Virginia, Colorado and Florida, respectively.

DAVID I. "DAVE" PETERSON, CAPT, born May 14, 1945, Nyack, NY, earned his BA in economics at St. Michael's College.

He joined the service in August 1967. His military locations and stations: New Port, RI; Washington, D.C.; Pearl Harbor, HI; Subic Bay, RPI, Vietnam; and Philadelphia, PA. Units served with include: Active duty - Officer Candidate School, USS *Grapple* (ARS-7) and Harbor Clearance Unit ONE. Reserves - Naval Control of Shipping, Harbor Clearance Unit TWO Det 104, Mobile Diving and Salvage Unit TWO Det 304 and Naval Sea Systems Command Det 1006. He participated in Vietnam in 1968, 1969 and 1970.

Peterson was awarded the Meritorious Service Medal, Navy Commendation Medal w/"V", Vietnamese Campaign Medal w/6 stars, Vietnam Service Medal, Vietnamese Honor Medal Second Class, Naval Reserve Medal and National Defense Service Medal. He achieved the rank of CAPT and retired Dec. 1, 1994.

He and his wife, Susan W., have three children: David I. Jr., Julie A. and Cathleen M.; and one grandchild, Kayla M. Civilian employment at US Bureau of the Census, 1971-96 as survey statistician and National Agricultural Statistics Service, 1997 as agricultural survey statistician.

VERNOR H. "PETE" PETRIE, LCDR, born June 12, 1916, Fairview, KY, graduated high school, attended two years of college, and aviation cadet at Corpus Christi.

He joined the service Dec. 18, 1941. His military locations and stations: AVCAD Corpus Christi, NAS Kaneohe, FAW-17, Alameda-NAS, Guam, Honolulu and NAS Opa Luka. He served with VP-24, VP-101, VH-3, VE-2, VR-11, VR-3 and VR-7. Battles and campaigns participated in were South WestPac (New Guinea) and CenPac (Iwo Jima).

Petrie was awarded the Air Medal and Unit Commendation Ribbon. He achieved the rank of LCDR and was discharged in September 1945.

His memorable experiences: Black Cat operations, New Guinea area. Air evacuation of Iwo Jima searching for downed air crews off New Guinea and in Marianas. Crossing Owens-Stanley Mountains (New Guinea) in PBY to support Aussie commandoes on Sepik River.

He and his wife, Gwendolyn, have one child, Robin. Civilian employment as CAPT, United Air Lines. Petrie retired June 12, 1976.

RICHARD A. "DICK" PFEIL SR, CAPT, born March 9, 1930, West Point, NY, earned his BS degree at US Merchant Marine Academy and his graduate certificate at American University.

He joined the service Jan. 28, 1954. His military locations and stations: Fleet Sonar School and Mine Warfare School. He served with USS *Direct* (MSO-430) homeport, Charleston, SC (engineering officer, New York state). He served in the Korean Campaign.

Pfeil was awarded the National Defense Service Medal, Korean Service Medal, UN Service Medal, Navy Commendation Medal (2) and Conspicuous Service Cross. He achieved the rank of CAPT (O-6) and was discharged Jan. 4, 1956.

His memorable experiences: Entire USN shipboard experience. First CO, Naval EXSYSCOM Reserve Unit (Washington, D.C.) and CO, Norfolk Naval Shipyard Reserve Unit (NORVA).

He and his wife, Betty, have four children: Richard Jr., John, Deborah and Robert Mark; and six grandchildren: Richard Jr., Jack Cervilla, Eric Michael, Zachary, Mark and Kevin. Civilian employment as mechanical and general engineer. He retired from USNR March 9, 1990 and NAVSEASYSCOM April 3, 1995.

JOHN PFRIMMER, CAPT, born Jan. 23, 1950, Schenectady, NY, earned his BSEE degree at Auburn University in 1973 and his MBA degree at the University of West Florida in 1979.

He joined the service June 5, 1973. Following flight officer training served with Attack Sqdn. 75 Sunday Punchers flying the A-6 Intruder for two deployments aboard USS *Saratoga* 1975-77. Served 1977-80 as an NFO instructor with VT-86 and on the staff of CTW-6 while completing an MBA. Transferred to the Reserves in 1980 and served with the following reserve units: USS *Inchon*, US HQ European Command, Commander Fleet Air Mediterranean and COMCARGRU TWO. Served as CO of reserve units USS *Kitty Hawk* and currently Carrier Battle Gp. Two. Promoted to CAPT in October 1994.

Pfrimmer was awarded the Navy Commendation Medal, Armed Forces Reserve Medal, Overseas Reserve Medal, Sea Service Active and Reserve Medals, National Defense Service Medal w/Bronze Star, Expert Pistol Medal and Meritorious Unit Service Medal.

He and his wife, Carolyn, have two children, Gray and Fran. Civilian employment as district manager for engineering services, General Electric.

EDWIN R. "FAST EDDIE" PHELPS III, CAPT, born June 13, 1941, Murphysboro, IL, attended Washburn University and graduated with a BBA degree in June 1963. He entered the USN through the AOCS program at Pensacola, FL and was commissioned an ENS in November 1963. CDR Phelps received his wings in February 1965 and reported to VA-56 where he completed two combat cruises to Southeast Asia. He was then assigned to VA-127 at NAS Lemoore, CA, as an instrument and jet transition flight instructor.

Following release from active duty in July 1968, CDR Phelps joined the Naval Air Reserve and served with VA-3K1 at NAS Olathe, KS before joining VA-204 in Memphis. He has been with the squadron since its inception in 1970 and has served as CO since October 1981. During his career CDR Phelps has been awarded the Distinguished Flying Cross, Air Medal (21), Navy Commendation Medal (3), Vietnamese Cross of Gallantry and various campaign ribbons.

CDR Phelps is employed as a pilot with US Air and resides in Leawood, KS with his wife, Sara, and three daughters: Julie, Jennifer and Martha.

GEORGE A. PHILACTOS, LCDR, born Oct. 6, 1918, New York, NY, graduated high school, took courses at NROS and received teacher license at New York State Private Trade School.

He joined the service July 18, 1938, and served on active duty WWII from May 11, 1941-Nov. 20, 1945. His military locations and stations: USS *Delta*, Radio Materiel School, Treasure Island, CA and USS *Biscayne*. He served with the Atlantic Fleet, 8th Amphib. NW African Waters and Pacific Fleet. Participated in the invasions of Sicily, Salerno, Anzio, Southern France and Okinawa.

Philactos was awarded the Navy Unit Ribbon, Good Conduct Medal, Naval Reserve Medal, American Defense Campaign Medal, China War Medal and WWII Victory Medal. He achieved the rank of LCDR and was discharged Nov. 20, 1945.

His memorable experiences: Witnessed sinking of HMS *Sparton* after being hit by Nazi radio controlled rocket bomb at Anzio; also damage to USS *Savannah* by same bomb (radio directed bomb).

He and his wife, Genevieve, have four children: Angela, Gregory, Judy and George E.; and four grandchildren: Christian, Kate, Avery and Benjamin. Civilian employment as senior engineer for Western Electric from 1941-77. Retired from Western Electric Co. (AT&T) Feb. 28, 1977.

CHARLES J. "CHICK" PHILLIPPS, CWO4, born Sept. 12, 1936, Wheeling, WV, earned his BS degree in electrical engineering.

He joined the USNR Nov. 29, 1954. His military locations and stations: Recruit Training Command, Great Lakes, IL; USS *Horace A. Bass* (APD-124) (two tours); USS *William T. Powell* (DE-213); Fleet Training Center, Newport, RI (two tours); USN and MCRTC, Steubenville, OH; USS *Chukawan* (AC-100); USS *F.D. Roosevelt* (CVA-42); USS *San Pablo* (AGS-30); 4th Naval Dist. Schools, Philadelphia, PA; USS *Mountrail* (APA-213); Fleet Training Center, Norfolk, VA; COMNAV Marianas, Guam; USS *Gilmore* (As-16); USS *Vulcan* (AR-5); Fleet Sonar School, Key West, FL; USS *Manitowac* (LST-1180); Naval Amphib. School, San Diego, CA; Naval Amphib. Base, Little Creek, VA (two tours); Naval Ship Engineering Center, Norfolk, VA (five tours); Philadelphia Naval Shipyard, Philadelphia, PA; Fleet and Mine Warfare Training Center, Charleston, SC; MOMAG Det 2, Machrehanish UK; and Naval Weapons Station, Seal Beach, CA. The units he served with were 4-108, 4-81, 4-86, ACU-205, DDG-39, Naval Ship Engineering Center, AR-5, Naval Shipyard San Francisco, NR MOMAG Det 605.

Phillipps was awarded the Order of the Founders and Patriots of America Award of Excellence and Outstanding Unit in REDCOM 5 as CO of MOMAG Det 605. He earned the Navy Achievement Medal w/Bronze Star, Armed Forces Reserve Medal w/2 HGs and Navy Sharpshooter Pistol Shot Ribbon. He achieved the rank of CWO4 and retired Oct. 1, 1986.

His memorable experiences: Outstanding active duty for training assignments, especially those out of CONUS. Award for having the Best Reserve Unit in REDCOM 5.

He and his wife, Theresa "Tipi" have five children: Dianne, Patricia, David, Mary Anne and Charles Jr.; and three grandchildren: Jessica, Amy and Aryelle. Civilian employment with LTV Steel Corp. as general supervisor electrical services. He retired Dec. 31, 1996.

BRADFORD G. PHILLIPS, CDR, born in Washington, D.C. and raised in Amityville, Long Island, NY. He earned a BA of business administration degree from Temple University, Philadelphia, PA in 1969. After attending Navy Officer Candidate School in Newport, RI, he was commissioned an ENS in the Supply Corps in 1970.

CDR Phillips' military education includes the Navy Supply Corps School in Athens, GA and numerous transportation and mobilization courses at the Navy Transportation Management School, Oakland, CA and with the US Army at Ft. Sam Houston, San Antonio, TX.

Subsequent to Supply Corps School, CDR Phillips was assigned as disbursing and assistant supply officer in the pre-commissioning crew for the USS *Luce* (DLG-7), the DESRON 24 flagship, homeported in Newport, RI. Subsequent to *Luce's* re-commissioning at the Philadelphia Naval Shipyard in 1971, he participated in various Caribbean training exercises and a six month deployment to the Mediterranean Sea.

From 1974 through retirement in 1995, CDR Phillips had many reserve assignments of increasing responsibility: CO and executive officer, Military Traffic Management Command 320, Oakland, CA; supply officer, Navy Electronics and Systems Engineering Center 220, San Jose, CA; staff, Navy Supply Center A120, Oakland, CA; staff, Aviation Supply Office 202, Ft. Schuyler, NY; staff, Military Traffic Management Command 202, Brooklyn, NY; supply officer, Naval Surface Force, Atlantic, Philadelphia, PA; and supply officer, Auxiliary Repair Unit 104, Philadelphia, PA.

CDR Phillips' decorations include the Joint Services Commendation Medal, awarded by the Military Traffic Management Command for service during Operation Desert Shield and Desert Storm, the Navy Meritorious Unit Citation (2) for Aviation Supply Office, Ft. Schuyler, NY and Philadelphia, PA, the National Defense Service Medal w/ Bronze Star, the Armed Forces Reserve Medal w/HG Device, and the 45 caliber pistol Marksman Medal.

In his civilian career, CDR Phillips is a certified public accountant and has held a variety of financial and operational staff and management positions. He took the position of chief financial officer for SIMCO Electronics, Santa Clara, CA after several years as vice-president-commercial operations. He and his wife, Lynne, both East Coast natives, have lived in Menlo Park, CA since 1982.

HARRY R. PHILLIPS, CAPT CEC, born Dec. 8, 1922, Washington, D.C., earned his bachelor's degree in civil engineering at Rensselaer Polytechnic Institute.

He joined the service in December 1942. His military locations and stations: Camp Perry, Gulfport, Guam, and V-7 Notre Dame. He served with MCB-41 & 83 - 56. He achieved the rank of CAPT CEC, USNR, retired and was discharged from active duty May 19, 1946.

His memorable experiences: First CO at formation of Reserve US Naval Construction Bn. 24.

He and his wife, Grace, have three children: Harry R. III, T. Martin and Grace; and six grandchildren. Civilian employment as president and CEO, Piedmon Steel Inc.

REED M. PHILLIPS, CAPT, born Jan. 5, 1938, Hyrum, UT. CAPT Phillips began his naval career as an NROTC student at the University of Washington. Upon graduation in 1960 he was commissioned ENS and graduated as a naval aviator at NAS Corpus Christi in 1962.

CAPT Phillips' squadrons included VP-17, VAH-21, VT-27 and VR-52. He received the Distinguished Flying Cross and Air Medal (5) while serving in VAH-21 at Cam Ranh Bay, Vietnam in 1968-69.

CAPT Phillips' assignments included US Naval Postgraduate School, WST Patuxent River, US Naval Forces Korea, JUSMAG Korea and Naval Reserve San Diego. CAPT Phillips was O-in-C of CV-64 DET San Diego and inspector general of Naval Reserve San Diego. CAPT Phillips has about 5,000 hours of Navy flight time.

As pilot for Trans World Airlines, CAPT Phillips qualified on the B-707, 727 and L-1011 aircrafts. He serves as an instructor check airman on the L-1011 aircraft.

The Phillips family consists of wife, Sue; son, Matt; and daughters, Page, Renn and Layne (who resides with her husband, Adam Steelhammer at Ft. Campbell, KY). CAPT Phillips is a life member of the Naval Reserve Assoc. He and his wife enjoy traveling and outdoor activities.

WILSON HAYWOOD "W. HIP" PHILLIPS, LT, born May 8, 1919, Buies Creek, NC, earned his BS and BA degrees in accounting in Washington.

He joined the service in March 1943. His military locations and stations: Harvard Supply School, Cambridge, MA. He served at the Naval Air Station, Coco Solo, CZ.

He achieved the rank of LT, USNR, Supply Corps and was discharged in February 1946.

His memorable experiences: As paymaster-NAS with 5,000 military and about same civilians on payroll, finding safe jammed shut on arrival for pay day. All to be paid in cash.

He and his wife, Nancy, have three children: Elaine, Woody and Susan; and five grandchildren: Nan, John, Emily, Megan and Todd. Civilian employment with CPA firm from 1946-48. Opened his own CPA firm in 1948. He partially retired in 1994.

JAMES B. PHILPITT, CDR, born in Charleston, SC, raised in Longwood, FL, received his BS from Florida State University and commission as an ENS in 1977. A former Regular Navy officer, he completed flight training and served with VP-24, VP-30, USS *Kitty Hawk* and NASC/TRAWING SIX staffs. He holds an MPA from the University of West Florida and is a graduate of the Naval War College.

Transferring to the USNR in 1988 he reported to VP-62 at NAS Jacksonville, requalifying as a P-3C Update III Mission CDR and NATOPS evaluator. He served as maintenance officer, operations officer, executive assistant and detachment officer in charge, leading squadron detachments in support of Operations SHARP GUARD and SUPPORT/UPHOLD DEMOCRACY, and accumulating over 3,500 flight hours.

In 1994 he reported to NAWCTSD Orlando as assistant P-3 program manager. Graduating from the USAF Air War College in 1996 he is currently assigned to the COMPATWING ELEVEN staff at NAS Jacksonville.

CDR Philpitt has been awarded the Navy Achievement Medal, numerous expeditionary and service medals, and various unit awards. He is married to the former Victoria Quinlan of Lakehurst, NJ.

ROBERT WILLIAM "BOB" PIAZZA, CAPT, born March 28, 1944, Rochester, NY, earned his BS degree in marine engineering at California Maritime Academy in 1965.

He enlisted in the service May 17, 1961, and was commissioned July 24, 1965. His military locations and stations: CO AOG-202, O-in-C YR-85, CO NAVSEA INDUST MOB DET 402. He served with the USS *Finch* (DER-328). Participated in the Vietnam conflict from 1965-67.

Piazza was awarded the National Defense Service Medal, Vietnam Campaign Medal, Naval Reserve Medal w/HG, Vietnam Service Medal, Merchant Marine Vietnam Service w/2 stars. He achieved the rank of CAPT (O-6), USNR and retired June 1, 1991.

His memorable experiences: Two years (1977-79) attached to Command Services Reserve Unit, US Embassy, Paris, France.

He and his wife, Penny, have been married for 32 years and have two children, Valerie and Danielle; and one grandchild, Stephanie. Civilian employment as vice-president, sales for Dresser Pump Div. and president and CEO, Price Pump Co.

DON B. "PIPE" PIEPER, LCDR, born Sept. 26, 1919, Milwaukee, WI, earned his BS degree at Marquette University in 1947.

He joined the service Dec. 16, 1941. He served with VC-16, VT-18, VBF-19 and VR-11. Battles and campaigns participated in include: Palau, Philippines, Okinawa, Iwo Jima, Formosa and Japanese Fleet.

Awards and medals include the Silver Star, Air Medal, American Theater, Asiatic-Pacific w/4 stars, WWII Victory Medal and Philippine Liberation w/2 stars. He achieved the rank of LCDR, USNR and was discharged Oct. 3, 1946.

His memorable experiences: Forced parachute jump, five days on D.O. in typhoon, saw 16" shell pass over wing Jap fleet, torpedo run, Kinron Harbor, Formosa.

He is divorced and has one daughter, Karen Marie. Civilian employment in sales. Pieper retired Sept. 26, 1979.

KEITH A. PIERCE, CDR, born April 21, 1930, Ellendale, ND, earned his BA degree in 1952 at Wheaton College, Illinois; his MED degree in 1958 and his EDD degree in 1965 at the University of Oregon.

He joined the service Oct. 12, 1952. His military locations and stations: Communication Station Pearl Harbor and USS *Wasp* CVA-18. Also Naval Reserve: Surface in Portland, OR; NavSeaGru in Denver; NROS in Spokane, WA and Reno, NV. Pierce participated in the Korean War.

He achieved the rank of CDR and retired from the reserves April 21, 1990.

He married the former Doris E. Green March 13, 1953. They have three children: Mark, Jann and Susan; and three grandchildren: Sarah, Dana and Lisa. Civilian employment as teacher/counselor in Portland, OR; counselor educator in four colleges/universities. He retired June 30, 1994, as professor of Counseling and Education, Psychology, University of Nevada, Reno.

VERNON LYNN PIERCE, CAPT, born Jan. 24, 1940, Pauls Valley, OK, earned his BA degree at Central State College, Edmond, OK and his MED degree at Oklahoma University, Norman, OK. He entered Naval OCS at Newport, RI in January 1963. After commissioning he was assigned to Naval Supply Corps School, Athens, GA, completing this assignment in November 1963.

ENS Pierce was next assigned to the USS *Valley Forge* (LPH-8), serving two years in various Supply Corps billets and as officer of the deck (in port) and as gun mount officer. During this period the ship made four WestPac cruises. His next assignment was as assistant supply officer, Long Beach Naval Station, Long Beach, CA.

Released from active duty in June 1966, LTJG Pierce affiliated with the USNR in Oklahoma City. He remained as a selected reservist until his Navy retirement in September 1993. During his reserve duties he held a variety of unit and command positions. These include assignment to NRSD 8-46(M), CO, ASB(SM)-211, CO, NAVMTO 311, NROS, CincPac 111, NAVRESRED Com. Reg. 11.

His campaign and service ribbons include the Meritorious Unit Commendation, National Defense Medal (2), Vietnam Service Medal w/3 Campaign Stars, Armed Forces Reserve Medal (3), RVN Meritorious Unit Citation, Gallantry Cross w/Palm, RVN Campaign Medal and other awards.

In civilian life CAPT Pierce is a consultant and field representative for Rose State College, and a coordinator at Tri-City Youth and Family Center, both located in Oklahoma City. His master's degree and post-master's certificate were completed at the University of Oklahoma.

CAPT Pierce is married to the former Linda Farino and has two children: Sarah L. and Aaron F.

WALTER ELDRIDGE "PETE" PIERCE JR., YNCM, born Aug. 8, 1922, Burlington, NC, graduated from high school in May 1939; attended Guilford College, Greensboro, NC, 1948-49; earned his BBA degree, magna cum laude, at Emory University, Atlanta, GA, August 1953; and his MBA degree at East Tennessee State University, Johnson City, TN, August 1976.

His service dates include: Oct. 3, 1942-March 6, 1946; March 14, 1946-March 13, 1950; March 14, 1950-March 13, 1954; July 8, 1958-March 13, 1966; July 8, 1966-July 7, 1976; and July 8, 1976-Aug. 8, 1982. Military locations and stations: NTS San Diego, CA; Sub. Div. 41; USS S-40; USS S-41; Sub Base P.H., T.H.; Pre-Mid School, Asbury Park, NJ; USS *Daniel T. Griffen*, USS *Barber* (APD-57); PERSEPCEN, NORVA; Com. 6, Charleston, SC; USNRTC, Greensboro, NC; NAS Atlanta, GA; USNRS, Charleston, SC; USNRS NORVA; USS *Robert L. Wilson* (DD-847); USNRC, Brooklyn, NY; various drilling reserve units, Atlanta, GA and Kingsport, TN. He participated in the Battle of Okinawa, May-July 1945, USS *Barber* (APD-57).

Pierce was awarded the Navy Good Conduct Medal, Naval Reserve Meritorious Service Medal (6), American Campaign Medal, Asiatic-Pacific Campaign Medal w/ Bronze Star for Okinawa, WWII Victory Medal, Navy Occupation Medal (Korean Service), National Defense Medal, Armed Forces Reserve Medal w/Bronze HG, Naval Reserve Medal (10 year service) and Philippine Liberation Medal.

His memorable experiences: Standing watch on officers confined to quarters on USS *Olympia* while being used as officer's brig alongside Pier 92, New York, NY. USS *Olympia* was ADM Dewey's flagship during Spanish-American War.

He married the former Vanira T. Walker May 10, 1952, Berlin, NJ. They have three children: Nancy Lee, CAPT Linda Sue (USAR, Nurse Corps) and LTC Walter Eldridge III (US Army); and nine grandchildren: Sarah Elizabeth, Norie Anne, Laura Elizabeth, Rebecca Erin, Maria Ellen, John Tyler Jr., Jennifer Marie, Walter Eldridge IV and Philip Joseph.

Civilian employment as newspaper carrier-salesman for three years as a boy, assistant to Head Dyer (hosiery) 1939-42, Comptrollers Div. of Tennessee Eastman Co., manufacturing division of Eastman Kodak Co., Kingsport, TN (cost accountant, internal auditor, insurance and tax accountant) 1953-88.

GEORGE M. PIERSON, CDR, born Nov. 17, 1912, Des Moines, IA, earned his BCS degree at Drake University, Des Moines in 1935.

He joined the NAVCAD USN in 1939; first active duty VP-14, North Island PBYs. His military locations and stations: San Diego, Kaneohe Bay, NAS Lake City, FL - USMC Cherry Point, NC, NAS Dallas, NAS Minneapolis - 8th ND, NAS Los Alamitos, USS *Tangiers* (CV-40) and INS MAT Louisiana. He served with VP-14.

His memorable experiences: Pearl Harbor Vet. (PBY-Kaneohe), South Pacific, Coral Sea, Solomons. Co-pilot - one of three PBYs (out of Kaneohe) before December 7 attack on Pearl Harbor.

He and his wife, Virginia, have one son, Gregory T. and two grandchildren, Douglas and Krista Rae. Civilian employment as agent, Equitable Life Assurance New York. He retired in 1962.

SIMON PATRICK "SAM" PIGNATARO LT, born Nov. 6, 1950, Newark, NJ. He is a graduate of Rutgers University, Kean College and Seton Hall University. He is a registered professional nurse and a licensed marriage and family counselor and is certified in psychiatric and mental health nursing and gerontology. He holds a master's and post-master's degree (EDS).

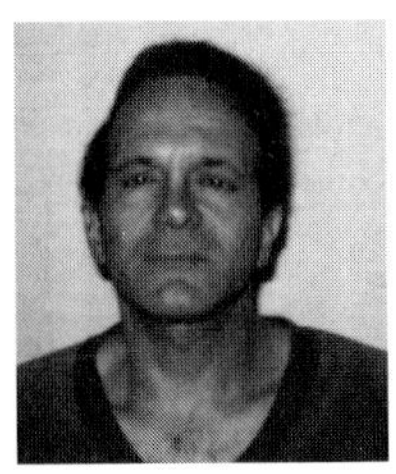

In 1983 he came into the USN through the Advanced Pay Grade (APG) program as an E-5 and was commissioned as an ENS in the Navy Nurse Corps in 1988 just prior to taking his chief's exam. In 1990 he graduated from Officer Candidate School (OCS) in Pensacola.

Among his duty stations include: Bethesda Naval Hospital, Portsmouth Naval Hospital, Philadelphia Naval Hospital, Willow Grove Air Station, the Lakehearst Dispensary, Walson Army Hospital and the Naval Hospital in Sigonella, Sicily.

During the Persian Gulf War his Fleet Hospital Naval Reserve Unit was mobilized to Al Jubail, Saudi Arabia and served in medical support operations through Desert Shield/ Storm from Jan. 16-May 15, 1991. There he was assigned to the Intensive Care Unit and later to the Acute Care (medical-surgical) Unit and Psychiatric Unit.

The military courses which he completed include: emergency medical technician course, medical entomology, command excellence, leadership management and educational training (LMET), the tri-service combat casualty care course (C-4) at Camp Bullis, TX.

His hobbies include traveling, acting, theatre, bicycle and horseback riding and free-lance writing. He is presently single and self-employed, pursuing a degree as an advanced geriatric nurse practitioner and serves as AOIC of his reserve unit in Kearny, NJ.

THOMAS C. PINARD, CAPT, native Californian, graduate of CSU San Bernardino, was commissioned at 34, 13 years after discharge from USAF. Owning community newspapers from 1961-90, CAPT Pinard's civilian experience dovetailed with the USN. He served in every chief of information unit, then in COMNAVFOR JAPAN, CINCPACFLT, COMPHIBGRU THREE and LOGISTICS TASK FORCE PACIFIC.

Along with major exercises throughout the Pacific, CAPT Pinard was OPERATION REMEMBRANCE '92 Public Affairs officer, commemorating Guadalcanal and South Pacific operations. Two Washington active duty assignments included six months directing the publication of the new *Sea Legs*, a guidebook for the Navy family. In 1996 and 1997 he was tasked by Navy Recruiting Command as GREAT LAKES CRUISE Public Affairs officer.

Since 1993, CAPT Pinard has taught journalism, computers and history at a rural school near his home of 35 years in Wrightwood, CA. A community activist, he served as an elected high school, then college trustee for 13 years and continues on various civic and taxpayer association boards.

JOSEPH CHARLES "CHARLIE" PLUMB JR., CAPT, born Nov. 3, 1942, Gary, IN, earned his BS degree at the USNA in 1964.

He joined the service in July 1960, Annapolis, MD. His military locations and stations: Pensacola, San Diego, *Kitty Hawk*, and Hanoi Hilton. He served with VF-121, VF-114, VA-2123, VP-4748, REDCOM 18, CV-43, and LATWINGPAC. He participated in 74 missions over Vietnam.

Plumb was awarded the Silver Star, Legion of Merit, Bronze Star, Purple Heart (2), POW Medal, Air Medal (3), Navy Commendation Medal (2), Combat Action, Navy Unit Citation, National Defense Service Medal, Vietnam Service Medal, Armed Forces Reserve Medal, Vietnam Merit Unit Citation, and Vietnam Campaign Medal. He achieved the rank of CAPT and retired in October 1991.

His memorable experiences: At age 24 LTJG Plumb (code name "Plumber") was shot down on his 74th combat mission over Vietnam. Nearly six years later, he emerged with the other POWs from Vietnam with prayers of thanks to God and expressions of gratitude to this nation for the opportunity to serve.

"They thought we'd been brain washed," says Charlie, "in reality we had just discovered some true meaning to life."

Plumb went on to serve in the Reserves with four different commands. As a civilian, he wrote a best-selling autobiography book *I'm No Hero* now in its 26th printing. He has made over 4,000 speeches in every state and several foreign countries, and was voted one of America's top 10 speakers by *Successful Meetings* Magazine. Plumb has authored several books and audio and video tapes and continues to speak 100 times each year. He lives with his wife, the former Catherine Cash, and two children, Joseph Charles III and Evelyn Ashley, in Santa Barbara, CA where he flies a Beechcraft Bonanza and a WWII open cockpit antique airplane.

CARL A. PLUNKETT, LT, born May 28, 1918, Elsie, MI, attended two years of college.

He joined the service in September 1942. His military locations and stations: Hawaii, Samoa, Mindoro and Leyte Gulf. He served with Naval Air Transport Service, VR-13.

Plunkett says that all of his experiences were memorable. He achieved the rank of LT and was discharged in April 1946.

He married the former Dorothy Rottz and they have one son, Edward, and one grandchild, Brandy. Civilian employment as self-employed contractor and in trucking. Plunkett retired May 1988.

ARNOLD "ARNY" POCK, CAPT, born July 19, 1934, Chicago, IL, graduated from the Scholl College of Podiatric Medicine in June 1956. He joined the USNR (active duty) June 27, 1956, under a two year enlisted program, inasmuch as commissions for podiatrists were closed. Discharged in 1958, he entered private practice and joined a Naval Research Unit at Northwestern University.

One year later, he received a commission in the Medical Service Corps as ENS, USNR. Coupling his private practice in Podiatry with Naval Reserve assignments, he excelled in both and rose through the ranks, delivering foot care at Navy and Marine Corps Recruit Training Commands throughout the country. In 1989 Dr. Pock acquired a master's degree in management/human resources and propelled himself even farther in areas of administration and command responsibilities. He served as CO of NR Naval Hospital Great Lakes, Unit 313, and multiple tours as staff administrative officer and staff medical officer, Readiness Command, Region 13, Great Lakes. He then served four years at NRREDCOMREG13GLAKES as director of Health Services.

In August 1991 CAPT Pock was recalled to active duty (BUMED) for four years. He served as special assistant for Reserve Integration, assigned to the dental division (MED-06) and double hatted in Reserve Matters (MED-07). Dr. Pock also conducted a foot clinic at BUPERS which he equipped and organized independently. His tour at BUMED was eventful. He orchestrated important changes in Chapters 6 and 15 of the Manual of the Medical Department which paved the way for reservists to secure complete physicals and four year dental examinations. NPQ status for dental_neglect was a paramount achievement.

CAPT Pock was awarded the Legion of Merit in July 1994 for his tour at BUMED. He also received the Meritorious Service Medal, Navy Achievement Medal, Expert Pistol and Rifle Medals, National Defense Medal and Gold Wreath Award for Medical Department Recruiting during his career. Dr. Pock is married and has two children. His daughter, Arnyce, is a physician (LTC) in the USAF, currently assigned to the Surgeon General's staff at Bolling AFB, Washington, D.C. His son, David, is a social worker in Chicago, IL.

CAPT and Mrs. Pock are retired and reside in Pensacola, FL.

ANDREW J. PONTA JR., CAPT, born May 2, 1933, Sacramento, CA, earned his BA degree.

He joined the service Aug. 2, 1956. His military locations and stations: COMSERVRON ONE, San Diego, CA and USS *Cree* (ATF-84). He served with the Reserve Surface Divs., Treasure Island, CA and Sacramento, CA.

He achieved the rank of CAPT and retired Oct. 1, 1980.

His memorable experiences: Participating in "Operation Hardtack" at Eniwetok during the summer of 1958 and the salvage of five Republic of Korea navy ships that were "beached" during Typhoon Sarah (near Chinhae, South Korea) in the fall of 1959.

He and his wife, Patricia Ann (of Indianapolis, IN), have two children, Andrew Jeffery and Suellen Marie; and two grandchildren, Kathleen M. and Anna Marie. Civilian employment as sub-contract administrator, Aerojet Gen. Corp. Ponta retired in April 1989.

DANTE CARL "DON" PONZELINE, CAPT, born May 21, 1927, Des Moines, IA, graduated from high school in 1945; University of Georgia, business management, 1962; University of Pittsburgh, engineering, 1968-69; and Penn State, mining engineer, 1972.

His military service includes: US Army, July 1945-December 1949; USN, March 1, 1950, active and reserve until retirement in July 1986. Military locations and stations: Army and Navy all sea duty aboard public vessels of USA. Then NAS Glenview, Norva and San Diego. He served with Trans. Corps, US Army 1945-49, Navy MSTS, CTF-43, HS-861, ASDARP and Atlantic Region director 1950-86.

Ponzeline was awarded the WWII Victory Medal, ETO Campaign Medal, Army Good Conduct Medal, American Defense Service Medal, Antarctic Service Medal (2 years), Navy Commendation Medal and Armed Forces Reserve Medal (3). He achieved the rank of CAPT, USN (O-6) DOR July 1, 1976, until retirement in July 1986.

His memorable experiences: Interpreter aboard First Troop Transport returning POWs to Europe in 1945, same duty on first War Bride Transports Europe to USA 1945-46, repatriation of European Displaced Persons 1946-49, Arctic construction and resupply 1950-55, Antarctic CTF-43 Operation Deepfreeze I and II 1955-57, charter member ASDARP 1973 and Atlantic director 1979-84.

He and his wife, Elvira E., have one son, Michael; three daughters: Denise, Deborah and Dina; and four grandchildren: Michael, Cameron, Brittany and Sentia. Civilian employment with US Rubber Co., H.K. Porter Co., National Mine Service and founder and president since July 1983 of Arizona Mine Service Co.

NEIL A. POPPENSIEK, CDR, born June 28, 1945, Teaneck, NJ, earned his BA degree in 1967 and his EDM degree in 1989.

He enlisted in the service in May 1967, Philadelphia, PA. His military locations and stations: Newport (NAVSCOLSCOM); NAVBASE Charleston (active duty). He served with Mobile Mine Assembly Team 0327, Charleston, SC; NR DEFDOCUMENT 0102, Syracuse; and NR NTIC 0305, Rochester.

Poppensiek was awarded the National Defense Service Medal, Unit Commendation Ribbon (2), Expert Rifle and Pistol Medals, and Armed Forces Reserve Medal.

His memorable experiences: Transition day—enlisted to officer, May 9, 1975; giving his first intelligence briefing; and ADMAT inspection while unit CO.

He and his wife, Danilee, have two children, Michelle and Megan. Civilian employment as insurance manager, Cornell University, teacher, writer. Poppensiek retired Oct. 1, 1993.

TERRANCE MICHAEL "MIKE" PORTER, CAPT, born Nov. 16, 1945, Austin, TX, earned his BSME degree in 1968 and his EDM degree in 1992, both at Oregon State University.

He joined the service June 2, 1968. His military locations and stations: VT-22, NAS Kingsville; VS-83, NAR Whidbey; and ASWOC 389. He served with VT-22, VS-83, PATWING 389, ASWOC 389 and VTU-8989.

Porter was awarded the National Defense Service Medal, Armed Forces Reserve Medal and Overseas Service Ribbon.

He achieved the rank of CAPT (O-6) and was discharged in September 1971.

His memorable experiences: Instructor pilot in TF-9J and TA-4J aircraft in VT-22, 1969-71. CO, ASWOC 389 at NAR Whidbey, and 19 years of airlifts from Portland, OR to NAS Whidbey for drill weekends.

He and his wife, Barbara N., have three children: Brian, Sharon and Jeffrey. Civilian employment as construction project manager, Intel Corp. Still employed with Intel. Retired from Reserves in 1995.

VENTZEL JACOB "VENTZ" POTOCHNIK, CDR, born Nov. 10, 1945, Monroeville, PA, earned his MA degree in management and supervision at Central Michigan; BSE (EE) at the University of Michigan; and his associate in science degree at Macomb C.C.C.

He joined the USAF (active duty) in August 1963 and the USNR in March 1978. His military locations and stations: Lackland AFB, TX; Keesler AFB, MS; NAF Detroit, MI; and N&MCRRC Buffalo, NY. He served with 754th Radar Sqdn.; NR FICEURLANT 1673; VP-93; NR DIAHQ 0705; and NR ONI 1805. He is a Vietnam era veteran.

Potochnik was awarded the Navy Commendation Medal, Navy and Marine Corp Achievement Medal, Joint Meritorious Unit Citation, USAF Good Conduct Medal, National Defense Service Medal and Armed Forces Reserve Medal. He achieved the rank of CDR (O-5) and was discharged from active duty in August 1967.

His memorable experiences: Air intelligence officer, VP-93; XO, NR ONI 1805; direct commission, August 1980; USAF enlisted service, August 1963-August 1967.

His wife's name is Susan M. Civilian employment in engineering at Chrysler Corp. Potochnik plans to retire in November 2005.

SCOTT M. POTTINGER, CDR, born Aug. 11, 1958, Quincy, MA, earned his BS, business administration (dual major: finance, real estate) at Ohio State University, 1980; MBA (finance concentration) at Xavier University, 1987; and MPA (management concentration) at Indiana University, 1994.

He completed Naval ROTC at Ohio State University and commissioned by his father, CDR Ian Geoffrey Pottinger, SC, USN (retired), Dec. 12, 1980. Began active duty Feb. 14, 1981. His military locations and stations: Active - Navy Supply Corps School, Athens, GA, February-September 1981; USS *Carl Vinson* (CVN-70), October 1981-January 1984, disbursing officer 12 months, sales officer 16 months; Defense Supply Center, Columbus, OH (DSCC), formerly Defense Construction Supply Center (DCSC), Columbus, February 1984-March 1986, branch chief, Stock Control Div. Reserve - NR DCSC Columbus 105, various billets, April 1986-September 1989; NR NAVORDSTA Louisville 109, executive officer, October 1989-September 1991; NR NAVAIRTERM Norfolk 609, CO, October 1991-September 1993; NR DFAS Columbus 209, assistant training officer, administrative officer, training officer, currently executive officer, October 1993-present.

Pottinger was awarded the Joint Service Commendation Medal (2), Navy Commendation Medal, Navy Achievement Medal, Armed Forces Reserve Medal w/Bronze HG, National Defense Service Medal, Expert Pistol Medal, and Sea Service Deployment Ribbon. He achieved the rank of CDR, SC, USNR and is still serving as a drilling reservist.

His memorable experiences: Father, brother and he graduated from Ohio State University and its NROTC program - all became Navy Supply Corps officers - father CDR, SC, USN (retired after 20 years of active duty serving from 1954-74) (now deceased); brother is active duty Supply Corps CDR - supply officer aboard the USS *Inchon* at Corpus Christi, TX. All three of them, at different times, visited Pottinger Street in Hong Kong - thought to be ancestor. He is a plank owner of the USS *Carl Vinson* (CVN-70) and was aboard on her "around the world" cruise - circumnavigating the globe (order of magellan) and passed the equator at 0 longitude, 0 latitude Royal Diamond Shellback.

Life member of the Naval Reserve Assoc., member of the Louisville Armed Forces Committee, committee chairperson - Ft. Knox Chapter of the American Society of Military Comptrollers.

He is married to the former Dr. Paula Hogan, of Louisville, KY and they have one child, Brandy. Civilian employment: April 1986-January 1988, numerous positions with Defense Logistic Agency; February 1988-May 1996, Naval Ordnance Station, Louisville, KY including budget director from 1988-92; director Planning and Resource Allocation 1993-94; master production scheduler 1995; director, planning and administration, Industrial

Operations 1996; June 1996-present director of Resource Management and Logistics, US Army Third Recruiting Bde., Ft. Knox, KY.

RICHARD R. "SUNSHINE" POWELL, CAPT, born Dec. 10, 1918, Salt Lake City, UT, earned his AB degree at the University of Tampa in 1939; pre-dental studies at Duke University, 1951-52; and his DDS degree at Loyola University School of Dentistry, 1952-56.

He was commissioned ENS in September 1940 and ordered to active duty in June 1941. His military locations and stations: NSCS, Harvard Business School, June-September 1941; ASO & SO, USS *Laramie* (AO-16), 1941-43; NSCS staff, Harvard, 1943-45; commissioning SO, USS *Amphion* (AR-13), Tampa, 1945-March 1946; OPT, BuSandA, Washington, D.C., 1946-Dec. 24, 1946 (separation); SO, USNR Organized Surface Bn. 6-16, 1947-51; organized reserve units until 1964, with various training duty orders, including NATO, Norfolk, VA, 1955.

He served with USS *Laramie* in the "Battle of the Atlantic," 1941-43, supplying AAF bases in Greenland and USCG. "Polar Bear" certificate for crossing the Arctic Circle.

Powell was awarded the American Defense Medal w/ "A" Fleet Clasp, American Campaign Medal, European-African Campaign Medal, WWII Victory Medal, and Naval Reserve Medal. He achieved the rank of CAPT, SC, USNR (1964) and retired Dec. 10, 1978.

His memorable experiences: USS *Laramie* torpedoed in Belle Isle Straits, August 1942, with 40' square hole in port bow. Returned to Boston for repairs and back in service in 1942. Leaving Boston for Maine, rammed by US Mine Sweep, with little damage, in deep fog. In 1943 rammed by "runaway" LST, resulting in hole in starboard bow and destroying the officers' stewards and cooks quarters, while waiting for convoy departure the next day. Navy welders patched with outer surface steel plates and they sailed with a 49 ship convoy on schedule. On that convoy, 21 ships were torpedoed and lost by U-boats on three consecutive nights. They were the middle ship on the next front line of seven vessels, but the U-boats did not return.

In 1943 he married the former Evanell Klintworth (deceased) and they had three sons: Winston, Bruce and William. He married Jane Oliver in 1976 and they have two daughters, Olivia and Rachel. He has four grandchildren: Davis R., Jonathan, Rebecca and Rachel. Civilian employment as broker, real and mortgage loans, Tampa, FL, 1947-51; general dentistry, private practice, 1957-88. Sold his practice and moved to Marianna, FL. 1990-present, senior dentist, Florida Dept. of Juvenile Justice, Dozier School for Boys, Marianna, FL.

JOHN B. "JACK" PRATT SR., CDR, born Dec. 17, 1922, White Hall, IL, graduated White Hall schools, 1940; BA degree at Illinois College, 1948; and JD in law at University of Louisville, Kentucky, 1971.

He joined the service May 1, 1942-September 1946; Reserve Unit, St. Louis, 1946-48; active duty, July 1948-October 1968. His military locations and stations: NAVCAD - Murray, KY; Terre Haute, IN; Iowa City, IA; NAS Minneapolis; NAS Corpus Christi, TX; commissioned ENS, USNR, NAS Miami, FL; NAS Glenview, IL; NASS Kingsville, TX; NAS Vero Beach, FL; NAS Key West, FL; NAS St. Louis, MO; (reserve) NAS Norfolk; NAS Key West, FL; NAS Atlantic City; NAS Cecil Field, FL; NAS Port Columbus, OH; NAS Dallas, TX; NAS Olathe, KS; USN Reserve Training Center, Forest Park, IL, CO. He served with VTN-90 (new 1946); VC-4, 1948-52; ATV-200, 1952-54; USS *Intrepid* CV-11, June 1954-March 1956, CIC officer; VF-14, March 1956-December 1957 (Air Gp. One, USS *Forrestal*). He served in Vietnam.

Pratt was awarded the National Defense Service Medal, American Campaign Medal, European Occupation Medal, Naval Reserve Medal, WWII Victory Medal and Vietnam Service Medal. He achieved the rank of CDR, USNR and retired Oct. 1, 1968.

His memorable experiences: Cold cat shot on Coral Sea in AD-1Q in 1949. Cat shot engine lost on two occasions in F2H-2N Tarawa and FDR 1951. (Never put a scratch on himself or his aircraft.)

He married the former Kathleen Ruth Turnbaugh (deceased Jan. 16, 1994) and they have six children: John Jr., Kevin, Alan, Robert, Timothy and James; and six grandchildren: Shane, Shannon, Dawn, Cara, Lisa and Christine. Civilian employment as attorney-at-law, self-employed.

FINIS H. PRENDERGAST JR., CAPT, born June 30, 1940, Shreveport, LA, earned his BS in forestry at Louisiana State University in 1963 and his master's degree in forestry at Stephen F. Austin State University in 1969.

He joined the service in September 1963, Navy Officer Candidate School, and was commissioned in February 1964. His military locations and stations: NAS Alameda, CA; South China Sea; Quang Ngia, South Vietnam; and DASA, Albuquerque, NM. He served with (active) USS *Coral Sea* (CVA-43); Naval Intelligence liaison officer, Quang Ngia, South Vietnam; instructor, Defense Atomic Support Agency; (reserve) CO, executive officer and training officer, Assault Craft Unit One, NR Det 110; CO, SIMA San Diego, NR Det 210; executive officer and training officer, SIMA San Diego, NR Det 110; and naval liaison officer, Fifth US Army. He participated in the Vietnam War from February-May 1965 and June 1965-June 1966.

Prendergast was awarded the Meritorious Service Medal, Air Medal (4), Navy Commendation Medal, Navy Achievement Medal, Naval Unit Citation (2), Presidential Unit Citation, Armed Forces Reserve Medal, Vietnamese Campaign Medal (4), Vietnamese Medal of Honor, Vietnamese Cross of Gallantry, Vietnamese Civil Action Medal and Vietnamese Campaign Medal. He achieved the rank of CAPT (0-6) and was discharged in February 1967 (active) and retired in May 1990 (reserve).

His memorable experiences: As a young ENS in the field collecting intelligence in Vietnam; commissioning four officers from enlisted ranks as CO of ACU Det 110; as naval liaison officer at Fifth Army, he was involved in the development of plans which were used in the Desert Shield/Desert Storm deployment. The satisfaction of watching, from retirement, as the plans he assisted in developing took life and worked.

He and his wife, Amy M., have one son, F. Homer "Trey" III, and one daughter, Amy Elizabeth "Libby." Civilian employment with Louisiana-Pacific Corp., 1972-92, procurement and land manager, export woodchip manager; Mannington Wood Floors, 1992-94, fiber transport and sales manager; F.H. Prendergast Consulting, 1994-present, president and owner.

GEORGE DANIEL PRETTYMAN, CAPT, born Aug. 3, 1940, Santa Monica, CA. He is a graduate of California State University at Los Angeles, obtaining BA and MA degrees. He began his Naval career as an enlisted Naval Reservist in 1963. In February 1966 he graduated from Officer Candidate School, Newport, RI and was commissioned ENS, USNR. He then reported to Amphibious Base, Coronado, CA for amphibious warfare training.

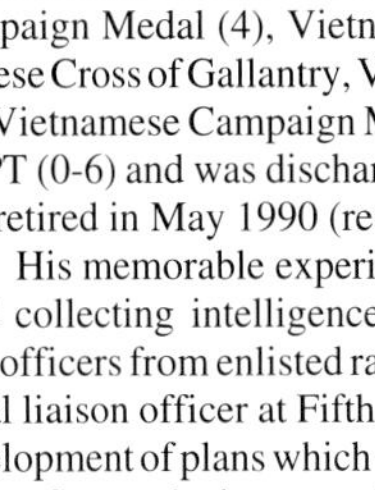

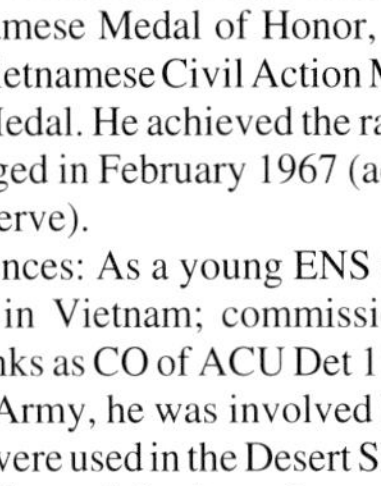

CAPT Prettyman served as gunnery officer and assistant first LT in USS *Sumner County* (LST-1148), during the period April-November 1966, operating in the I Corps region of Vietnam. He also served as a Deck Div. officer and later as navigator in USS *Duluth* (LPD-6), making two deployments to Vietnam during the period November 1966-January 1969.

CAPT Prettyman has served in various Naval Reserve assignments from 1969-94. He has held command of: Support Gp. Charleston, DET 309; Surface Force Atlantic, DET 110; Military Sealift Command, Europe, DET 110; Military Sealift Command Office, Middle East, DET 110; and Voluntary Training Unit 1021.

CAPT Prettyman has been awarded the Navy Unit Commendation and Meritorious Unit Commendation. His campaign and service awards are: Navy Achievement Medal, National Defense Service Medal w/Bronze Star, Vietnam Service Medal w/3 Bronze Stars, Navy and Marine Corps Overseas Service Ribbon, Armed Forces Reserve Medal w/2 Bronze HGs, RVN Gallantry Cross Unit Citation w/Bronze Palm, and RVN Campaign Medal.

CAPT Prettyman is married to the former Fredrica Lea Kersh and they reside in Pensacola, FL. They have two sons, James Michael of New York, NY and George Stephen of Arlington, VA.

RONALD WILLIAM PREVI, O-6, born Jan. 20, 1939, New York City, NY, earned his BA degree at Columbia University and his MS degree at LaVerne University.

He joined the service in July 1960. His military locations and stations: Norfolk, VA; Cornwall England; Brunswick, GA; San Diego, CA; and Washington, D.C. He served with VAW-12, 849 Sqdn. (RN), Ryan-110.

Previ was awarded the Meritorious Service Medal (2) and Air Medal. He achieved the rank of O-6 and retired in July 1991.

He and his wife, Marie, have four children: Dianne, Susan, Lynn and David. Civilian employment with The Kauffman Group as field superintendent.

SILVIN "SIL" PRIBYL, LCDR, born Feb. 2, 1922, Steele County, MN. He joined the service in September 1942. He was assigned to many training stations. He served with Navy Sqdn. VPB-135.

He achieved the rank of LCDR and was discharged Nov. 18, 1945.

His memorable experiences: Flying in the Aleutian Islands, specialty Attu Island (being a Navy pilot). Total of 32 years - active duty and reserve time.

He and his wife, Joyce M., have two children, Marie and Paul; and six grandchildren: Stacy, Kelly, Chris, Erin, Andy and Alie. Civilian employment as county zoning administrator and farmer. Pribyl retired Jan. 31, 1987.

JERRY L. "FISH" PRICKETT, CAPT, born Sept. 24, 1935, Wyandotte, MI, earned his BS degree in electrical engineering at Tri-State University, Angola, IN; attended the National Defense University JCS/AFSC; and is a registered professional engineer in California.

He entered the USN in June 1953 at NAS Gross Ile, MI; was promoted to CPO (AEC) in May 1962; and was later commissioned to LTJG in May 1967 at NAS Los Alamitos. He was assigned to VP-22L as a NAO/navigator and later qualified as an NFO/PPTN at NAS North Island with VP-28.

During his 38 year naval career he has been assigned to over 10 VF, VA, VS and VP squadrons, accumulated over 2800 aircrew flight hours, and has qualified in over eight aircrew positions in seven aircraft. He held numerous positions in seven air system program units, including CO. From 1976-87 he was project, then program manager for USN P3 FIAT and other support projects, and has completed numerous assignments at NAVAIR SYSCOM HQ and the Pentagon prior to his retirement in September 1991 at PMTC Pt. Mugu.

During the Lebanese crisis of 1958 he flew missions in the Mediterranean in support of the 6th Fleet.

CAPT Prickett has held several technical management positions in his civilian career while employed by

Aerojet-General and TRW Systems, including manager/task force leader for numerous Nuclear Special Projects, supervisor of Satellite Testing (USAF), consultant/resident engineer for NASA at Redstone Arsenal, Huntsville, AL and Mississippi Test Facility on the Apollo program.

Two memorable cruises were aboard the USS *Yorktown* in 1954 and the USS *Nimitz* in May 1996 with his son, Jason, who is an ES3A pilot in VQ-5.

CAPT Prickett is married to the former Marian Ma, daughter of GEN and Mrs. T.C. Ma, and resides in Diamond Bar, CA. They have two children, Andrea and Andrew. He also has two children from a previous marriage, Jason and Julianna, and one grandchild, Jacob.

ROGER L. PRIGGE, CDR, born Jan. 18, 1935, Howard Lake, MN. He joined the service Sept. 27, 1955. His military locations and stations: NAS Norfolk, VA and NAS Twin Cities. He served with VP-24 and VP-813.

He achieved the rank of CDR and was discharged Oct. 1, 1980.

He and his wife, Barbara, have four children: Candace, Kevin, Brendan and Celeste; and two grandchildren, Weston and Michala. Civilian employment at Norwest Corp. as vice-president. He retired Dec. 31, 1993.

ALWIN F. "BUD" PRIPP, CAPT, born Sept. 30, 1916, Seattle, WA, earned his BS and MS degrees at the University of Washington.

He joined the service in January 1943. His military locations and stations: V-5 pre-flight, North Carolina; Del Monte; St. Marys; NATTC Chicago; Ottumwa Islands; Iowa pre-flight; Sand Pt.; and NAS Whidbey. He served with VF-894, VR-895 and Air Wing Staff, Whidbey.

He achieved the rank of CAPT, USNR and was discharged in 1946 and then retired in 1973. His memorable experiences: The whole Navy experience. Meeting great Naval people. Tour in Morocco.

He and his wife, Margaret A., have one daughter, Mary Louise, and two grandchildren, Trisha and Christopher. Civilian employment as high school and college baseball coach. Pripp retired in 1976.

ROBERT HUNTINGTON "BOB" PROCTOR JR., O-6, born July 16, 1909, Brooklyn, NY, earned his BA degree in 1931 at the University of North Carolina, Chapel Hill, NC.

He joined the service in 1935. His military locations and stations: Flight instructor, Pensacola, FL and VR-4, Pearl Harbor. He served with VR-4 and participated in WWII.

Proctor was awarded the Naval Reserve Medal w/ Bronze Star, American Defense Service Medal, American Campaign Medal, Asiatic-Pacific Campaign Medal, WWII Victory Medal, Armed Forces Reserve Medal, and US Navy Expert Pistol Shot. He achieved the rank of O-6 and was discharged July 16, 1969.

He and his wife, the former Ella Mae Daniel, have four children: Daniel, Carol, Bruce and Alan; and 10 grandchildren: Ben, Andrew, Nathan, Mark, Laura, Joe, Allison, Elise, Devin and Tommy. Civilian employment with FAA as flight operations specialist.

ROBERT WHITAKER "BOB" PROCTOR, LT, born June 1, 1923, Boston, MA, earned his AB degree at Harvard College.

He joined the service Jan. 6, 1943, and served with USN Submarine Service "R-20" and "Hardhead."

Proctor was awarded the WWII Victory Medal, American Theatre Medal, Good Conduct Medal and Naval Reserve Medal. He achieved the rank of LT and was discharged June 6, 1946.

He and his wife, Sheilah, have three children: Sarah, Nancy and Charles; and one grandchild, Taylor Wilson. Civilian employment as financial executive-treasurer and vice-president finance. He retired July 1, 1971, with 20 years active and reserve service as LT.

DALE LORENZ PUHRMANN, CAPT, born Sept. 17, 1950, Lemars, IA, earned his BS degree at the USNA in 1973.

He was commissioned into the service June 6, 1973. His military locations and stations: (active) 1973-76, USS *Pharris* (FF-1094); 1977-78, USS *America* (CV-66); (reserve) 1979-80, NR ONR TAC 306; 1980, NR NCC 106; 1980-81, NR COMSERVRON 4; 1981-82, NR USS *Fox* (CG-33); 1982-83, NR USS *Paul F. Foster* (DD-964); 1983-84, NR USS *California* (CG-36); 1984-88, NR NCSO Far East 118; 1988-89, NR IMA GEN 1318; 1989-90, Readiness Unit Cheyenne; 1990-92, REDCOM 18; 1992-93, NR JTRU 118; 1993-94, NR NCSO North Atlantic 106; 1994-96, NR COMSCFE 118; 1996-present, NR NCSO Atlantic 113.

Puhrmann was awarded the Navy and Marine Corps Commendation Medal, Joint Service Achievement Medal, Joint Meritorious Unit Commendation, National Defense Medal (2), Armed Forces Reserve Medal w/"M" and Bronze HG, Overseas Deployment Ribbon, Sea Service Deployment Ribbon, Pistol, Sharpshooter, Rifle, Expert Marksman. He achieved the rank of CAPT and is currently inactive (drilling) reserve.

His memorable experiences: Voluntarily recalled for Operation Restore Hope (Somalian Relief) serving as a J5 watch team analyst.

He and his wife, the former Denise Law, have three children: Aaron, Janelle and Ryan. Civilian employment: 1978-80, Commonwealth Personnel Services, owner; 1980-96, Caterpillar Inc., various positions; 1996-present, Deere Power Systems Group, Waterloo, IA, On-Highway Product Support Manager.

HOLLIS T. "PUTT" PUTNAM, CAPT, born June 13, 1943, St. Cloud, MN, earned his BA degree in business management and economics at St. Cloud University, St. Cloud, MN.

He joined the service Nov. 18, 1967. His military locations and stations: NAS Pensacola, FL; NAS Kingsville, TX; NAS Glynco, GA; and NAS Miramar, CA. He served with VF-121, VF-96, VF-143 and various Naval Reserve Units. Reserve units from 1974-94: NAS Whidbey Island, WA; Amphib. Base Coronado, North Island, CA; SIMA Pearl Harbor, HI; and NCSO South Korea, Japan, Diego Garcia and Guam.

Putnam was awarded the Overseas Service Medal, Meritorious Service Medal, Expert Pistol and Rifle, and National Defense Service Medal. He achieved the rank of CAPT and was discharged in April 1971.

His memorable experiences: Receiving wings at NAS Glynco, GA; Fighter Weapons School (Top Gun) at NAS Miramar; and Carrier Quals.

He and his wife, Barbara Ann, have two daughters, Nicole Ann and Ann Katherine, and one grandchild, Brody. Civilian employment as CEO and owner of Metroquip, Inc., Boise, ID. Retired from the Naval Reserve in October 1994.

KENNETH F. PUTT JR., CAPT, born July 12, 1941, Harrisburg, PA, graduated from Rider University, Lawrenceville, NJ with a BS in business administration.

Beginning his naval career in the AOCS program, he was commissioned an ENS in February 1964. Following the NAO training pipeline, his first duty station was VC-8 (1964-67) NS Roosevelt Roads; followed by instructor duty (1967-69) at NATTC Glynco, GA. From 1969-73 LT Putt served as an electronic warfare evaluator with VQ-2 NS Rota, Spain flying the EA-3B (105 CVA landings), EC-121M and EP-3E.

Released from active duty in June 1973 he joined the RESFORRON and flew with VP-66 (1975-81) at NAS Willow Grove, PA. At NAVMARCORESCEN Washington, D.C. (1982-91) CDR Putt drilled with DCNO OP-06 (chief of Naval Operations, OP-611) and the Navy Command Center. USNR accomplishments include: (1) P-3A ASW TAC/NAV, (2) won the ASW "Liberty Bell" Trophy, (3) CNO OP-06 action officer, and (4) stood Navy Dept. "Duty Captain" watch. CAPT Putt transferred to Retired Reserve list Oct. 1, 1991.

Married to the former Karen Jacobsen of Plainsboro NJ, they have one son, Jonathan A. CAPT Putt was employed by the Office of Naval Intelligence as a research analyst from 1973 until March 1997. Life member (#3018) of the Naval Reserve Assoc.

D.I. "IKE" PUZON, CAPT, born May 16, 1949, Wilmington, NC, earned his BS degree in political science at East Carolina University; his MS degree in management at Pepperdine University; and his MS in national security strategy at the National War College.

He joined the service Aug. 21, 1970. Commissioned through the Aviation Officer Candidate program in 1972. First fleet assignment was with Patrol Sqdn. 56, Jacksonville, FL, supporting Sixth and Second Fleet operations in the Mediterranean and Atlantic, 1973-76. Active duty with Patrol Sqdn. 60, NAS Glenview, IL. In 1980, LCDR Puzon was detailed to Patrol Sqdn. 90, NAS Glenview, in support of Third and Seventh Fleet operations in the Pacific, 1976-81. Reassigned to the Reserve Anti-Submarine Warfare Training Center, NAS Willow Grove, responsible for evaluations of 13 Reserve Sqdns. and 12 Sqdn. augment units. In 1984 he was transferred to Patrol Sqdn. 94 at NAS New Orleans. He assumed the command of Patrol Sqdn. 68 in 1990.

As a member of the US Assistance negotiating team for the former Soviet Union, he participated in negotiations with all former Soviet Union Republics, NATO, the EC, Israel and Turkey.

Puzon was awarded the Defense Superior Service Award, Legion of Merit Medal, Meritorious Service Medal (2), Navy Commendation Medal (3), Navy Achievement Medal (3), Navy Battle Efficiency Award (5), National Defense Medal (2), Humanitarian Service Medal (Vietnam, Cuba, Russia). In 1979 he was chosen as the Junior Officer of the Year for Commander Naval Reserve Force.

He and his wife, Diann, have two children, Courtney and Chase, and one grandson, Tristen.

HENRY PAUL "HANK" PYKOSH, CAPT, born Aug. 11, 1941, Donora, PA, earned his BS degree from California University of Pennsylvania, California, PA and his MA degree from American University, Washington, D.C.

He enlisted in the Naval Reserve May 25, 1961, and was commissioned via the ROC program in August 1963. Served on active duty from August 1963-August 1969. Re-joined the Naval Reserve in 1969 and retired in 1991. His military locations and stations: USS *Pocono* (LCC-16), Basic Propulsion Engineering School, Service School Command, Great Lakes, IL, and the Naval Command Systems Support Activity, Washington, D.C. He served with Naval Reserve Surface Div. 4-21(M), Naval Reserve Systems Analysis Div. 5-1, NAVSHIPS Planning Programming & Financial Management 1006, OPNAV 03 (Surface Warfare) Component 106, and MSC HQ Washington Det 106.

Pykosh was awarded the Navy Commendation Medal, Navy Unit Commendation, Meritorious Unit Commendation, National Defense Service Medal and Armed Forces Reserve Medal.

His memorable experiences: Participated in large Amphibious Exercise STEEL PIKE I while on active duty aboard the USS *Pocono*. Spent much of his Naval Reserve career supporting the OPNAV 03 (Surface Warfare) Program at the Pentagon where he was awarded the Navy Commendation Medal. Life member of the Naval Reserve Assoc.

He and his wife, Patricia, have three children: Paul, David and Anthony. Civilian employment with the US Food and Drug Administration, Center for Veterinary Medicine, deputy director, Office of Management. Pykosh retired Oct. 1, 1991.

STEPHEN T. "STEVE" QUIGLEY, RADM, born in Wabasha, MN, Oct. 24, 1920. He graduated from the College of St. Thomas with a BS degree in 1942. He received a MS degree in 1950 from the University of Detroit, and a PhD degree from the University of Maryland in 1959. He was elected to the National Honor Societies, Phi Kappa Phi, Sigma Xi, Alpha Chi Sigma, and the Cosmos Club. As an ENS he received a professional certificate in meteorology from UCLA in 1944.

His 48-year career in the USNR reflects broad experience including command of both air and surface units, an extensive background in systems analysis, program planning and management, and research and operational involvement in aviation operational development. He was commissioned an ENS in the USNR with a date of rank of July 8, 1942. He was sworn in as a RADM in the USNR by ADM James L. Holloway III, Chief of Naval Operations, Aug. 29, 1974.

RADM Quigley served in the Pacific Theater during WWII and in the Far East during the Korean Conflict, having a total of 9 1/2 years active service. From 1946-48 he served as the operational representative of the Head, Naval Aerology and the Office of Naval Research, with duty involving flying as a technical observer, with three different Joint Army-Navy-USAF and US Weather Bureau Research Projects.

In his last command assignment, he served as CDR, Naval Reserve Readiness Command Region SIX, from Oct. 16, 1977 to Oct. 14, 1979. From 1972-90, he also served as a special assistant to five Chiefs of Naval Operations, Admirals Carlisle Trost, James D. Watkins, Thomas B. Hayward, James L. Holloway III and Elmo R. Zumwalt Jr., and as CDR Naval Air Reserve Force area representative for the Chief of Naval Reserve, Vice Admiral Pierre N. Charbonnet, Jr. RADM Quigley's last mobilization billet before retirement was as deputy director, Command and Control (OP-094) in the Office of the Chief of Naval Operations.

Prior to these assignments, he served as CO of Reserve Systems Analysis Div.-A2 in the Office of the Deputy Chief of Naval Operations, Air Warfare (OP-05), 1970-72; as assistant chief of staff for administration, Naval Air Reserve Staff-A1, NARTU, Washington, DC, 1967-70; and as senior meteorologist, Naval Air Wing Staff-81(1), Naval Air Station, Twin Cities, Minnesota, 1960-66.

He served as a member of the Secretary of the Defense Reserve Forces Policy Board (statutory 3-term, 1977-80). He also served as a charter member of the Chief of Naval Operations' Reserve Affairs Advisory Board (1975-78), and as member of the Secretary of the Navy's National Naval Reserve Policy Board (1976). He attended the Naval War College Senior Reserve Officers' Course and the Current Strategy Forum, the Commander-in-Chief, US Atlantic Fleet Senior Reserve Officer Orientation Course and has completed Senior Reserve Officer Courses at the US Naval Postgraduate School. In 1976, RADM Quigley was the first Navy flag officer to attend the Senior Reserve Component Officer Course at the US Army War College.

In his civilian profession, Dr. Quigley is a science/public policy and management consultant. He was the director of chemistry and public affairs for the American Chemical Society, 1966-78. From 1963-66, he served two terms and commissioner of administration of the State Government of Minnesota, having been appointed to that position by Governor Karl F. Rolvagg. During this period, he also was a member of the Board of Managers of the National Council of State Governments. In 1965, he was appointed by the Secretary of Commerce to a five year term on the US Department of Commerce Technical Advisory Board. He was a National Cancer Institute Research Fellow in 1955-57, and from 1957-63 he was senior research chemist and research supervisor in the Central Research Laboratories of the Goodyear Tire & Rubber Co. and the 3M Co.

RADM Quigley's decorations include the Legion of Merit, presented personally by the Chief of Naval Operations, ADM Thomas B. Hayward, in 1979 for his performance of duty as CDR, Naval Reserve Readiness Command Region SIX, and the Meritorious Service Medal awarded for his performance of duty as special assistant to the Chief of Naval Operations, ADM Elmo R. Zumwalt, Jr., during the period 1973-74. In 1989, he received the Secretary of Navy Distinguished Public Service Award, having been nominated for it by the Chief of Naval Operations, ADM Carlisle Trost. His other awards include the Naval Reserve Medal and the Armed Forces Reserve Medal w/ Hourglass device in lieu of second award. His campaign and service medals are the American and the Asiatic-Pacific Campaign Medals, Victory Medal, Navy Occupation Service Medal (Asian Clasp), National Defense Service Medal, Korean Service Medal, and the United Nations Service Medal.

RADM Quigley is married to Carol L. Rogers (retained her name) and they have three sons: CAPT Stephen T. Quigley, Jr., USN (RET); Kevin C. Quigley (Barbara); and John B. Quigley; and three daughters: Mrs. R.L. Wicks (Catherine Mary); Mrs. J.H. Stevens (Mary Carol); and Eileen A. Quigley. They have 13 grandchildren and one great-grandson.

RADM Admiral Quigley's statutory retirement date was Aug. 30, 1980, and he retired administratively from the Office of the Chief of Naval Operations on March 1, 1990.

ALTON "DOYLE" QUISENBERRY, SWO, qualified for command at sea, resides in Virginia Beach, VA native of Amarillo, TX, graduated from Amarillo College (AAS), West Texas A&M (BS), Pepperdine University (MA). Married Edna "Frances" Hudson in 1950 and they have two sons, Bob and Kerry.

Enlisted USNR 1947, called to active duty in 1950 during Korean conflict, served in USS *Skagit* (AKA 105) as boatcoxswain, released from active duty (RAD) in 1951, completed Reserve Officer Candidate program. Returned ACDU 1953 and served as XO and chief engineer, USS *Teaberry* (AN-34) during Korean conflict. Left Navy, taught Industrial Arts, various Amarillo junior high schools 1956-61.

Temporary active Navy duty at Service School Command, San Diego summers of 1957-59, 60. Returned active duty 1961 and selected for TAR program, was chief engineer, USS *George Clymer* (APA-27) off coast of Vietnam in 1964 shortly after Tonkin Gulf incident and during severe typhoon in Buckner Bay, Okinawa.

Logistics officer, staff, Commander Amphibious Group Two, embarked in flagship USS *Pocono* (LCC-16), and in USS *Francis Marion* (APA-249), USS *Austin* (LPD-4) and later in USS *Guam* (LPH-9) during 1970 Jordanian crisis. Commanding officer, Naval and Marine Corps Training Center, Amarillo, 1964-65, staff duty (Naval Reserve Administration) at COMFIVE, COMSIX, COMEIGHT, COMTHIRTEEN, ADDU and instrumental in development of first USNR Readiness Command, Baltimore (now Washington, DC).

Upon retirement in 1996 after 20 years active duty, was employed BUPERS, transferred to Staff Commander-in-Chief, US Atlantic Fleet in 1978, retiring in 1993 as a GM-13. He was CINCLANTFLT Senior Logistics Planner and Fleet Logistics Plans Officer (civilianized captain USN billet). Additional duty to CINC Western Atlantic (NATO) as staff logistics officer.

Significant accomplishments: Architect-Naval Reserve Logistic Task Forces, Atlantic and Pacific (LTFLANT/LTFPAC) 1992 and Naval Reserve Cargo Rig Teams (CART) 1991. Key role in development of wartime maintenance support at bare bases overseas for deployed NATO and US Maritime Patrol Aircraft (P3 Orion and Nimrod) (Seven 40 van mobile maintenance support system (MM/SS) complexes at a cost of $75M each). Instrumental development of mobilization readiness measurement of NR Inshore Undersea Warfare Units and funding to equip them in 1970, also allocation and proper outfitting of selected merchant ships to augment Navy Combat Logistic Forces following operation Desert Shield/Desert Storm.

Life member number 72 and past national officer of the NRA. Received the NRA "Twice-a-Citizen" and national membership awards. Reserve Officer Association of U.S. (ROA) Silver Minuteman, life member and past president of ROA Dept. of Virginia. Member, Naval Order of the US Eagle Scout w/Bronze, Gold and Silver palms and Honor Camper at BSC Camp KIO-A-WAH (one of 98 campers).

Formal awards include: Navy Civilian Superior Service Award and Medal (two awards), Navy Civilian Meritorious Service Award and Medal, Department of Navy Desert Shield/Desert Storm Award, Navy Commendation Medal (two awards), Meritorious Unit Commendation, National Defense Service Medal (two awards), Armed Forces Expeditionary Medal, Vietnam Service Medal, RVN Campaign Medal w/device, Naval Reserve Medal, Armed Forces Reserve Medal. Authorized to wear the Surface Warfare and Command of Shore Activity Insignia.

JAMES D. "JIM" RAACK, CAPT, born Nov. 20, 1937 in St. Louis, MO, received a BS degree in Oregon State University and MS from UCLA.

He enlisted April 1956. His military locations and stations: USS *Brown* (DD-546), USS *Coontz* (DL-69). Served with XO MIUW 1519; CO MIUW 1719; XO HDCC 119. He was awarded the Naval Reserve Medal (2) and achieved the rank of CAPT. He was discharged May 1, 1988.

His memorable experiences: First WESTPAC trip 1962; apartment sharing with junior officers of USS *Coontz* 1962-63; being fortunate enough to be CO of MIOW 1719 with a great group of officer and men.

He married Carmela and they have two children, Thomas and Steven. Employed as athletic director, WEST Los Angeles College.

ROBERT J. "BOB" RADFORD, CAPT, born in Fawn River Township, MI, in 1930 and graduated from Bronson High School (MI) in 1947. He attended Albion (MI) College prior to earning a BSE (EE) from the University of Michigan in 1952, and was commissioned as an EDO ENS in OCS Class 10, Newport, RI, in 1953.

Radford served two years as ship superintendent (later senior ship sup.) at Naval Repair Facility, San Diego, and one year at ComPacResFleet, 12ND, before his release from a active duty in 1956. He then worked in the petroleum industry for 30 years; half with Chevron Corporation in the San Francisco Bay area, and the other half with either Aramco Overseas Co., The Hague, The Netherlands, (on three "temporary" assignments) or with Aramco Services Co., Houston, TX (one "temporary" assignment).

He concurrently maintained his involvement in the USNR (being promoted to CAPT in 1973), and served in USNR Units in San Francisco, Kansas City, and Brussels, including four years as CO of the Brussels NROS/VTU Unit. His Brussels Unit was the first to receive CNO approval to conduct joint training exercises with a foreign Naval Reserve Unit (Inshore Minesweeper Div. 3001) in 1977. He attended the National Congresses of The Association of Naval Reserve Officers of Belgium each year from 1977 to 1986, and was a member of The Royal Association of Naval Reserve Officers of The Netherlands from 1980 until retirement.

CAPT Radford's voluntary AcDuTra assignment (virtually all "at no cost to the Government") were, of course, predominately engineering billets, ranging from US Naval Shipyards (Puget Sound; Mare Island; Pearl Harbor; Hunters Point (three times) to staff work (industrial manager, 12th ND); supervisor of shipbuilding, 12th ND; ComCruDesPac (AR-7), San Diego; BuShips HQ, Washington, DC; ComServForSix, Naples, Italy; Ship repair office, Athens, Greece; NavSta (and ARDM-1), Rota, Spain; NATO HQ, Brussels; USN CoastalSystemCenter, Panama City, FL (four times); CinCUSNavEur, London (four times). CAPT Radford assisted in the design of SeaLab II in February 1965 and the conversion of SeaLab II to SeaLab III in 1971. He was involuntarily transferred to the inactive Retired List in 1985.

Subsequent to retirement from his civilian career at

AOC, Leiden, The Netherlands, in 1986, CAPT Radford and his wife, Wilma, established their primary residence in Clearwater, FL, and also established/maintained their summer home in Earls Colne, England, in order to spend "quality time" with Wilma's sons and their families in Essex, East Anglia. CAPT Bob and Wilma also visit her brother and sisters (and respective families) in Germany almost every year. CAPT Radford and Wilma's two sons (their father was an RAF B-24 pilot in) are founder members of the American Air Museum in Great Britian, at Duxford, Cambridgeshire, near the Earls Colne summer home.

JOHN R. "RAGS" RAGLAND, CAPT, born April 5, 1917, Olney, TX. He earned BS degree in electrical engineering, Texas Tech College in June 1939. He joined the service Aug. 28, 1936 as RM3/c in the Naval Communications Reserve of the USN; transferred to Pearl Harbor in the Radio Communications Staff of the USS *Maury* then to Naval Research Laboratory, Anacostia, DC.

The *Maury* was very active in the Pacific Theater of Operations, including the sea battle of Midway. Returned to Pearl Habor in June 1942 and commissioned ENS with assignment to the Navy Dept., Bureau of Ships, Washington, DC. Assigned in January 1944 to the USS *Mt. McKinley* (AGC-7), Philadelphia, PA. After its commissioning in June 1944, the ship sailed for Pearl Harbor to begin its service as a Flag and General Officer Command Vessel in the amphibious landings in the Pacific Theater of Operations.

Shortly after the Battle of Okinawa, he was released from active duty Jan. 25, 1946. Decorations include the Presidential Unit Citation Ribbon w/star, Asiatic-Pacific Theater Ribbon w/2 Battle Stars (for ops by *Maury*), Navy Unit Commendation Ribbon, Asiatic-Pacific Theater Ribbon w/5 Battle Stars (for ops by *Mt. McKinley*), Philippine Liberation Ribbon w/star, American Defense Ribbon w/ star, American Area Ribbon and Victory Ribbon.

He remained active in the NR until July 1969. During the 20 year span from 1949-68, he was allowed to spend 17 of such two-week training duty periods in the Electronic Warfare Design Section of BuShips. He achieved the rank of CAPT in June 1961.

Married Nov. 22, 1943, to Norma Deane Jerman; they have three children: Donna Sue Smith, Glynn and Bryan; grandchildren are Stephanie and Heather Smith, Michael, Christopher and Timothy Ragland. He is a retired professional engineer (registered in 28 states). Retired Jan. 30, 1984 from Hicks & Ragland Engineering Co., Inc., P.E.'s.

ROBERT R. RALSTON, CWO-4, born July 14, 1922 in Duluth, MN. Graduated Roosevelt High School, Virginia, MN, 1940. Enlisted USN, July 23, 1940, Minneapolis, MN. Attended boot camp Great Lakes, CO 54-40. Transferred to USS *Colorado* (BB-45), Oct. 9, 1940. Operated in the Pacific out of Pearl Harbor until December 1941.

At PSNY on Dec. 7, 1941. Operated West Coast until transferred to USS *Indiana* (BB-58). Commissioned the USS *Indiana* on April 30, 1942 (plank owner) at Newport News, VA. Transferred to USS *Hamul* (AD-20), Dec. 28, 1942. Operated in the Atlantic until June 13, 1944. March 1, 1944, advanced to CMM (AA), March 1, 1945 to CMM (PA). June 13, 1944, transferred to Destroyers PSNY Bremerton, WA. Commissioned the USS *Rooks* (DD-804) in Seattle. Served on destroyers until July 1946. Served in Southwest Pacific including Iwo Jima and Okinawa. Returned to Bremerton December 1945.

Married Marie Johnson of Virginia, MN, Dec. 29, 1945 and they have four children and nine grandchildren. He advanced to WO1 Mach - June 30, 1957. Retired as CWO mach., June 30, 1976. Employed 30 years as power plant engineer and superintendent. Presently self-employed.

Active in VFW, American Legion, 40 & 8 Fleet Reserve Association, The Retired Officers Association and Masonic Lodge. Member of Naval Institute, USS *Colorado* and USS Hamul Almuni Association. Enjoy hunting and fishing in Northern Minnesota and Canada.

SELWYN P. "SEL" RAMSAY, CAPT, born in 1930 in the state of California. He joined the service in 1952 and was the most senior Navy Captain at time of his retirement in 1991, serving 18 years in grade. An electrical engineering graduate of the University of Southern California working on the analog computer at the Naval Ordnance Test Station, China Lake, he entered the Navy in 1952, attending OCS as a pre-designated aeronautical engineering duty officer.

Following AGO and AEO training, he served as electronics officer and aircraft maintenance officer for the carrier USS *Randolph* (CVA-15). Designated NAO/NFO in 1955, he was transferred to the Naval Air Experimental Station at NAMC Philadelphia for duty as project instrumentation engineer for test flying the XZPG-1 (XN-1) Lighter-Than-Airship out of Lakehurst, NJ.

He joined the Bureau of Aeronautics Reserve in 1959, and served in the BARTU-WEPTU-NASRU Program throughout his Reserve career at NAS Los Alamitos, NAS Pt. Mugu, NWC China Lake, and NAS North Island, San Diego, including two tours as commanding officer. In addition to regular Reserve Unit participation, CAPT Ramsay was a dual-affiliate for 13 years with Naval Reserve Officers School, served as president of the South Bay Chapter of the NRA, and sat on the Captains Selection Board.

CAPT Ramsay is a registered professional electrical engineer and is president of the Ramsay Engineering Company, which he formed in 1959 for designing and manufacturing oceanographic research instruments. He also works in technology transfer involving aviation and computing institutes in the former Soviet Union.

ALFRED A. "AL" RAPPUHN JR., born Nov. 17, 1925 in Flushing, LI, NY, attended Tulane University and USN PG School, Monterey, CA.

He joined the service Aug. 23, 1943. His military locations and stations: Navy V-5 Program; VBF-153, V2 16A, VF-53, USS *Boxer* CV-21. His battles and campaigns: and Korea. His awards and medals: Joint Service Commendation, Navy Occupation, China Service. He achieved the rank of CAPT and was discharged in 1985.

He is married to Dotty and they have children: Richard, Cynthia, Diane and Douglas and grandchildren: Steven, Mark, Natalie and Katy. Employed as director of special events, Naval Aviation Museum Foundation and retired in 1985.

RICHARD ALLEN RASMUSSEN, CAPT, born on a farm, June 4, 1913 in Lakeview, Montcalm County, MI. Graduate of Greenville High School 1931; Olivet College (AB) 1935; University of Chicago School of Medicine (MD) 1938, intern Blodgett Memorial Hospital 1938-39, surgical graduate training University of Chicago 1939-October 1942.

Entered USN, September 1942. Initial duty, Great Lakes NTC to December 1942. Assigned to 66th USN Construction Bn., as junior medical officer to January 1945, Aleutian Island (Adak); reassigned to Great Lakes Naval Hospital as staff medical officer, charge Thoracic Surgical Service to discharge November 1946; Ready (Fleet) Reserve status; Grand Rapids Navy and Marine Corps T.C. January 1947 to retirement June 4, 1973.

He married June 30, 1936 and has two sons (MDs), Eric and Richard; daughters, Linda M. Rasmussen and Karen R. Schaefer; grandchildren: Kirsten McCall, Leif and Alaric Rasmussen; Marcia L. Maloney, Deborah L. MacLean, Robert and William Lowry; Charlotte, Theodore, Rebecca and Caleb Schaefer; Andrew, William and Richard Rasmussen.

Pioneered practice of thoracic and cardiovascular surgery in West Michigan at Grand Rapids, MI, December 1946 to retirement 1983, leaving five associate surgeons.

RICHARD L. RAUDABAUGH, CAPT, entered the USNR May 1955 while a student in high school. He advanced through petty officer second class (Hospitalman) prior to earning a commission as an ENS in the Reserve Officer Candidate (ROC) Program. He then reported to the Naval Security Group HQ, Washington, DC where he served as assistant security officer and storage officer in the Navy Registered Publications System Headquarters.

Upon release from active duty in 1963, CAPT Raudabaugh immediately affiliated with Naval Reserve Intelligence Unit 4-1-7 in Harrisburg, PA, and during the next several years held a variety of assignments including unit officer-in-charge. From 1976-80 CAPT Raudabaugh was a member of Intelligence Volunteer Training Unit 1104 and performed drills with Contingency Group 604 and Naval Investigative Service Office Unit 1104 in Harrisburg, PA and with Fleet Intelligence Rapid Support Team, Europe and Atlantic 0575 in Lakehurst, NJ.

He was assigned as XO, FIRSTEURLNT 0693 in October 1980, and as CO FIRSTEURLNT 0693 from Oct. 1, 1981-Sept. 30, 1982. In his subsequent assignment to IVTU 0104, CAPT Raudabaugh served as the first Area Sixteen IS "A" (BRIT) director and as Deputy Reserve Intelligence area commander for training and readiness. In 1984, he became XO of IVTU 0104 and in 1986 the unit commanding officer. He was the Reserve Intelligence CDR of Area Sixteen from 1988-90. CAPT Raudabaugh received a BA degree from Lycoming College, Williamsport, PA in 1960 with majors in BA and psychology. In 1964 he received a MBA in marketing from the American University, Washington, DC.

He is authorized to wear the Navy Commendation, National Defense, Armed Forces Reserve, Naval Reserve, Coast Guard Meritorious Commendation, Coast Guard Unit Commendation w/Operational "O," and the Navy Pistol Sharpshooter Medal.

In his civilian occupation, CAPT Raudabaugh is the founder and owner of the Fabric Center in Carlisle, PA and of Draperyland and a decorating and drapery fabric business in Lemoyne, PA. He and his wife Juanita and their family reside in Mechanicsburg, PA.

JAMES MICHAEL "JIM" RAUSA, CDR, born Jan. 3, 1947 in Syracuse, NY, received a BS from SUNY in forestry.

He joined the service November 1969. His military locations and stations: Alameda, CA; Lakehurst, NJ; Willow Grove, PA. He served with VAQ 135; VAQ 130, VP-64; VP-2429 NAVAIRLANT DET 275.

His battles and campaigns include Vietnam, Cold War and volunteered for Desert Storm. He was awarded the Air Medal (3), Vietnam Service Campaign, MUC, Good Conduct, Reserve (20 years). Achieved the rank of CDR and was discharged January 1992.

Memorable experiences include Da Nang AFB, Alameda, Lakehurst, Willow Grove, Operation Wagon Wheel, Lajes, Rota, Sigonella, Keflavik, Scotland, VP-64 Admin. Award, Bermuda.

He is married to Phyllis Ann and they have children: Jeffery and Lisa Reneé; and grandchildren, Briana, Mike and Tom. Employed as journeyman electrician, Local 400 IBEW.

TRACY LEE RAY, ENS, born June 6, 1965 in Columbus, OH, received a bachelor of business administration.

He joined the service Jan. 29, 1985. Served with USS

Voge (FF-1047), NCSO, USS *Gates* (CG-51), USS *Fahrion,* NAVMTO DET 108. He was awarded the National Defense Medal, four Meritorious Service, Reserve Armed Forces.

Achieved the rank of E-1 to 0-1 he is still active. His memorable experiences include receiving his direct commission on May 10, 1996. Being on the USS *Fahrion* during UNITAS XXXVI.

He is married to Darlene and they have two children, Austin and Christian. Employed with the US Attorney's Office as contracting officer.

RALPH M. RAYNER, LT, born Dec. 19, 1921 in Racine, WI, received a BS in mechanical engineering, from the University of Illinois, 1943.

He joined V7 April 1942. Commissioned ENS USNR at Anapolis, Aug. 31, 1943. Served with USS LCI (L) 944 May 1944-January 1946, Philadelphia Atomic Defense Instr. 1951-53.

Rayner was awarded the American Theater, Asiatic-Pacific Theater, Philippine Liberation, American Campaign and National Defense Medal. He achieved the rank of LT and was discharged 1946 and retired in 1953.

His memorable experiences: typhoon at Naha Okinawa; kamikaze at Ulithi Atoll; Shooting floating mines in Inland Sea of Japan; tour Hiroshima following the atomic bombing.

Married Eva Crawford and they have three children: Anita D. Litteer, Kerry A. Rayner and Judy A. Sluyter; and grandchildren: Steven and Kristopher Rayner; Sean and Melissa McIlwain; Amanda Litteer; Stephen and Jason Sluyter.

Employed as research engineer at Young Radiator; Dev. engineer at Goodyear Aircraft and Sandia National Laboratory, retiring in 1983.

FREDERICK W. "FRED" READ JR., born July 30, 1908 in Providence, RI, received an AB from Columbia College and LLB from Columbia Law School.

He joined the service June 1942. His military locations and stations: US Naval Air Station, Banana River, Cocoa, FL. Served as legal officer at above stations.

Read achieved the rank of CAPT. His memorable experiences: active duty, June 1942-December 1945; active Ready Reserve, 1945-73; CO USNR Law CO 3-3; (see *Who's Who in America* for further biographical details).

He married Evelyn A. and they have two children, Frederick III and Cynthia; and grandchild, Matthew. He is vice-president and general counsel, Home Life Insurance Co. He was placed on retired list, Aug. 1, 1973.

RICHARD S. READE JR., CAPT, enlisted 1942. Member first Tarmac Bn., Olathe, KS, 1943; wings Pensacola, 1945; search team for Flt. 19 (Bermuda Triangle) Ft. Lauderdale OPTrg; Torpedo 4, WestPac (USS *Tarawa);* joined Reserves 1947; over 3,000 hours fixed and rotary wing (TBF, F4U, F6F, AD, AM, HUP, SH-34 and SH-3) shipboard and at Reserve Stations Floyd Bennet, Dallas, Akron, Grosse Ile, Willow Grove and Lakehurst; CO of HS-752 (Noel Davis Trophy), NWS-704, NAL-275, AWS-75; retired 1981.

US School of Naval Justice; Naval War College (2x), tours in NAVAIR and Pentagon; and a Sikorsky Rescue Award. Helped conceive and inagurate MOBS Program tying Reservists to their MoBillets.

Civilian career as engineer, project/program manager of aerospace programs at Bell, Sikorsky, Goodyear, Gyrodyne, and Grumman (helo development; DASH; F-14; E-2/C-2) after MME from Cornell University 1952. Life member of NRA. He now resides in Old Field, LI, NY with his wife of 43 years, nee Thomasine Hocart of Montclair, NJ.

THOMAS G. "TOM" REALE, LCDR, born Oct. 4, 1955 in New Britain, CT, received a MS in Purdue University.

He joined the service April 20, 1983. His military locations and stations: Great Lakes, Norfolk, Monterey, Ft. Meade, Alameda. Served with USS *Mississippi,* DLI, NSGA Ft. Meade, USS *Enterprise* (active). His battles and campaigns: Earnest Will; Reserve: NRSG Indianapolis, NRSG Great Lakes.

He was awarded the Navy Achievement Medal (2), MUC, Battle E, Navy Expeditionary Medal. Achieved the rank of LT (active) and LCDR (Reserve). Discharged Sept. 30, 1990 (active duty). Still in Reserves.

Reale is married to Pauline B. and they have a daughter Joan. He is employed as high school principal.

JAMES B. "JIM" REAP, RADM, born in Mt. Vernon, NY, received a BA from Wesleyan University and a JD from Harvard Law School.

He served from Feb. 14, 1951 to Oct. 1, 1990, as CDR NAVRESRFD COM REG II; CDR NAVACTS UK 14 years of command.

Reap was awarded two Presidential Legions of Merit; Meritorious Service. Achieved the rank of RADM. His memorable experiences: selected Who's Who in America 1997; Chief Judge City Court City White Plains, NY; Commanding seven different residential units; 14 years command.

He is married to Nancy Leigh and they have children: J.J., Michael and Jessica; and grandchildren: Michael II and Alexandra Leigh. Employed as US Administrative Law Judge, retiring Oct. 1, 1990.

ALVA L. "RED" REDFIELD, LCDR, born Aug. 29, 1921 in Canandaigua, NY, attended high school, Canandaigua Academy, RAMI (now RIT) graduate 1941.

Sworn in July 10, 1942. Preflight Chapel Hill, NC, "E" Base. His military locations and stations: Squantum, MA; Pensacola wings September 1943. Served with Espiritu Santo, Guadalcanal, VC-7 invasion New Guinea, Pilot pool, combat team training. VC-41, December 1944. West Coast, Hilo, HI; USS *Makin Island.*

Recalled Feb. 1, 1951, seven months aboard USS *Tarawa;* six months USNDC, Johnsville.

Awarded Presidential Unit Citation VC-7, USS *Manila Bay.* Retired July 1, 1969 as LCDR. His memorable experiences: riding out typhoon in Wakayama harbor Japan September 1945 while on board the USS *Makin Island.*

He married Beatrice G. and they have three children: B.J. (Barbara Jeanne), Nancy and Stuart; and grandchildren: Robbie, Chris, Susi, Darius, Stephanie (the Dutchman), Jeffery, Brian, Heather, Timothy and Cassy (Casandra). Employed at EKCO Kodak Park, 42 1/2 years. He retired Feb. 1, 1982.

GARLAND E. REICHLE, LCDR, born Jan. 7, 1935 in Murphy, NC, received a BS in engineering physics, University of Chattanooga, 1958.

He joined the service March 3, 1953, Tennessee National Guard, 1960; USN, Reserve, Ret., 1977. His military locations and stations: US NAD, Oahu, HI. Served with IUWD1-2, Newport, RI; NROS 6-14 Orlando, FL, NASRU F-1, JAX, FL.

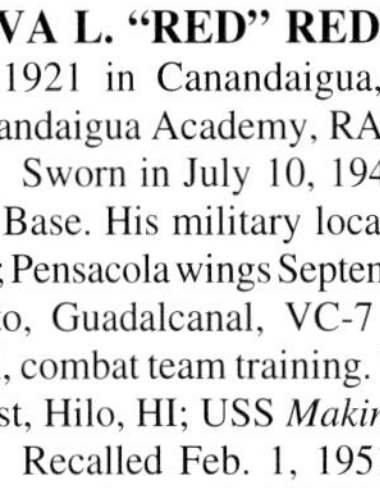
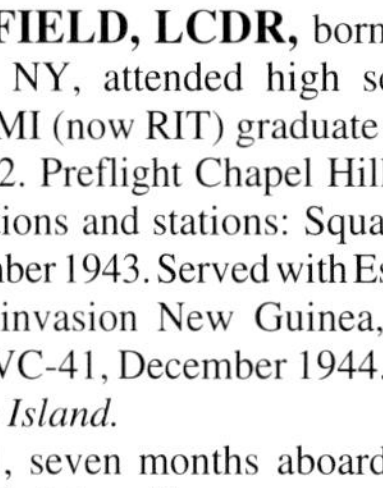

Awarded the American Defense Reserve w/HG, NRA. Achieved the rank of LCDR. His memorable experiences: OCS Class 49, SEC NAV LTR/Appointment to aeronautical engineering duty; active duty experiences at WestLoch; NAD Oahu, HI, spoke with ADM Rickover, Newport News Shipyard when working on reactor installation, USS *Enterprise.*

He is married to Margaret (Peggy) Duarte and has children, Heidi and Erich; and grandchildren, Alex and Aaron. Employed as engineer, combustion engineer, Kaiser Aluminum, NASA, Kennedy Space Center, division chief, GSE Design, 30 years service.

WILLIAM S. "BILL" REID, CAPT, born June 23, 1920 in Camden, NJ. Attended Shippensburg University in Shippensburg, PA, 1939-42. Entered the V-5 program (naval aviation cadet) at the Philadelphia Navy Yard, March 1942. Received "Wings of Gold" and commissioned ENS USNR at NAS Pensacola, FL, Nov. 16, 1942.

During the war years was assigned to several photographic squadrons flying the USN version of the B-24, (The PB4Y-1), in the South Pacific (Guadalcanal/Guam). Also during the war was a flight instructor on the PB4Y-1 and PB4Y-2 at NAS Hutchinson, KS. He was released to inactive duty in November of 1946.

After WWII and intertwined with his civilian activities, he stayed in the Reserve program. While serving in the Active Reserve he was assigned to a number of patrol and transport squadrons (VP/VR) at NAS Opa-Locka and later NAS Jacksonville, FL. During this time he held the usual squadron assignments, including XO and CO of these squadrons. His last tour of Active Reserve duty was as assistant wing staff commander NAS Jacksonville. (After being promoted to CAPT).

His civilian experience is best described chronologically so as to give it some continuity: 1946-61, airline pilot with Pan American Grace Airways (PANAGRA) based in Lima, Peru; 1961-65, Air Carrier Operations inspector (DC-8 specialist) with the Federal Aviation Agency (FAA) based in New York and Miami; 1965-67, Engineering/Acceptance Test Pilot with Douglas Aircraft Co. on the DC-8 and DC-9 programs, based in Long Beach, CA; 1967-75, in charge of DC-8 flight/simulator training program for Braniff International Airways, based in Miami, FL; 1975-80, returned to line flying for Braniff on DC-8 and B-727 equip. (served as check pilot on both equipment). On June 23, 1980 (age 60 rule) retired from both Braniff and the USN as a CAPT. (CAPT Braniff (Ret), CAPT USNR RET).

Married to Fay and they have children: George and Keith.

ROBERT P. "BOB" REIDENBACH, LTJG, born July 15, 1925 in Canton, OH, graduated high school and attended three years college.

He joined the service July 1, 1943. His military locations and stations: V-12 John Carroll University, Northwestern Midshipmens School. Served with USS *Staunch* (AM-307). His battles and campaigns: mine operations, China Sea, Tuschima Straits.

Reidenbach was awarded the Atlantic Theater, Asiatic-Pacific Theater, and Victory Medal. He achieved the rank of LTJG, USNR and was discharged July 1946. His memorable experiences: mine operations, Mine Sqdn. 12.

He married Carolyn and has three children: Susan E. Morey, Patricia C. Supon, SM2c David P., USN; and grandchildren: John P., Joshua P., Jessica and Jennifer Morey and Aaron Supon. Employed as manager, production planning, Steel Div., Timken Co.

VICKI L. REININGHAUS-EDGAR, LCDR, born Jan. 7, 1962 in Toms River, NJ, attended the Ann May School of Nursing in 1985 and Graceland College, Lamoni, IA in 1993. She obtained a diploma in nursing then BSN.

She joined the service May 13, 1986. Her military locations and stations: Naval Hospital San Diego, 1986-89; NH Bethesda 2306, July 1989-September 1989; FH

500 COMMZ 14, DET D, Jacksonville, FL, September 1989-September 1993; FH 500 COMMZ 11, DET A, Memphis, TN; September 1993-February 1996; Recalled to active duty, stationed at Naval Hospital, Okinawa, Japan; February 1996-to present.

January 1991 recalled from Jacksonville, FL to serve with Fleet Hospital 15 during Operation Desert Storm in Al Jubail, Saudi Arabia.

She was awarded the Navy Achievement Medal 1996, Navy Arctic Service 1992, National Defense 1991, Sharpshooter Pistol 1987, Kuwait Liberation 1991, Saudi Arabia 1991, Southwest Asia Medal 1991, Navy Sea Service and Meritorious Unit Citation.

She achieved the rank of LCDR. She is married to Robert A. Edgar, LCDR, MSC, USN (active). They have no children.

RONALD O. "RON" REIT, CAPT, born in 1921 in Philadelphia, PA. Living in South Jersey all of his life. He attended Temple University, Rutgers University (BA) and Glassboro State College (MA). He began his naval service as a S2c-AMM2/c and on to aviation cadet (V-5) program, graduating a naval aviator/ENS early in 1945.

After operational training in SBD & TBF was assigned to VB-17 USS *Hornet* in the Pacific Campaign. During his naval career he has served on 13 carriers in VB, VT and VF/VA squadrons and VX-3 flying propeller and many jet aircraft (Korean Conflict). CAPT Ron Reit served as XO, VS-935 in the Berlin/Cuban at "GITMO' and the squadron commander, VS-932 (1963-66) and then AWS-93, 1966-69 and then being assigned as a member of the CAPTAINs advisory board NAS WILGRO.

In civilian life, CAPT Ron Reit operated a home construction business in and about Ocean City, NJ. He also taught construction technology to high school (11th and 12th) and post secondary students (20 years) in addition to being a FAA certified flight instructor CFIT (1946-86), still flying his own twin-engine aircraft.

His wife Nita is deceased. He has two daughters, Karen and Deborah.

GARY M. REITER, CDR, born in Hanover, PA and received a BBA in 1975 from the University of Miami, FL.

He joined the service in 1976. His military locations and stations: NSCS Athens, USS *Prairie* (AD-15). Served with DCASR Boston, COMLOGGRU TWO, SPCC MECH. Participated in Atlantic Resolve 94 and achieved the rank of CDR.

Married Marie and they have two children, Ben and Alice. He is employed as information systems director.

FRANK J. REPP, LCDR, born Nov. 30, 1920 in Waukegan, IL. He attended college and joined the service Oct. 12, 1940. His military locations and stations: NAS Glenview, NAS Moffett, USS *Belleau Wood,* USS *Oriskany.* Served with VF-87, VF-723, VF-31, VF-193.

His battles and campaigns: assault of Japan. He was awarded the Air Medal and Presidential Unit Citation. He achieved the rank of LCDR and was discharged Jan. 26, 1946.

Repp is married to Patricia A. He was employed in sales, retiring Dec. 24, 1980.

PHILIP NICHOLAS "PHIL" RETSON, CDR, born Nov. 25, 1919, Appleton, WI. Earned BA in chemistry at Lawrence University, June 1940 and was a chemist from 1940-42. Served in the Wisconsin NG, 1935-39 and in the USNR with active duty, 1942-46 and ready reserve, 1946-79. Post Graduate School from 1942-43, USNA, Annapolis, MD, and graduated with aerological engineer degree.

Assignments: August 1943, Fleet Air Wing 15, Port Lyautey, Navy 214; 1944-45, Cuers, France with BLINDRON 14 DET; April-May 1945, NAS Grosse Ile, MI; May 1945-June 1946, USS *Reina Mercedes*, USNA, Aerolog. Off., VN8 Sqdn. and USNA, instructor in aerology for USNA; June 1946-November 1979, Ready Reserve.

Married Catherine in 1946 and they have four children: COL Nicholas P. Retson, AUS; Nancy Kay; Thomas N. and Barbara Kay; and five grandchildren: Gregory, Anastasia, Philip, Katie and Nicky. Civilian employment as restaurant owner and chemist.

ROBERT "REX" REXFORD, LTJG, born June 3, 1934 in Meadville, PA, received a BS, Northwestern University, business administration.

He joined the service June 1956. His military locations and stations: supply officer, USS *Marshall* (DD-676). Served with San Diego, material officer. His battles and campaigns: Service School CDR, NTC, San Diego.

Rexford achieved the rank of LTJG (active duty); LCDR (Reserves) and discharged June 1959. His memorable experiences: Pursued a career in enlistment ranking in NYC and Houston, also retired from USNR after 22 years service.

He is married to Margaret Sue and they have two children, Robert Stanley and Thomas Steele; and grandchildren, Robert Steele and Reagan Allexis. After retirement from Oppenheimer and Co. in 1996, he is pursuing personnel investments.

DAVID ALBEE "DAVE" REYNOLDS, CAPT, born Nov. 1, 1929, Binghamton, NY. Earned AB degree at Colgate Univ. and MBA at Rutgers Univ.

He joined the service Jan. 29, 1952 and served in USS *Des Moines* (CA-134); USS *Northampton* (CLC-1). Reserve Surface Units: Perth Amroy, NJ; Yonkers, NY; USMSTS Ft. Schuyler, NJ; Group Commander, Rochester, NY; MSTS, Liverpool, NY. He was discharged May 22, 1955.

Memorable experiences include commissioning Northampton, Reserve tour at Naval War College, National War College and Group Commander of Rochester, NY, Center.

Married Phyllis and has three children: Steve, Lawrence and Sara; grandchildren are Robert, Allison, Alex, Carrie, Julia, Kath and Elise. Civilian employment as assistant acct., NJ Bell; project mgr., IBM Corp. He retired Dec. 1, 1980.

FRANK P. "DR. RADICAL" REYNOLDS, MD, born Oct. 10, 1954 in Naval Hospital, Portsmouth, VA, received a BA in chemistry, Kalamazoo College (1975); MD from University of Michigan (1979).

He joined the inactive reserve in June 1975; active duty, June 1979; selected reserve, August 1987. His military locations and stations: Naval Hospitals; Jacksonville, Bremerton; Branch Hospital Adak. Served with USS Eisenhow*er* (CVN-69); (General Medical Officer Tour).

Reynolds was awarded the Battle "E" Sea Service, Meritorious Unit Commendation, Expert Pistol Shot, Navy and Marine Corps Expeditionary Medal. He achieved the rank of CAPT and is still a Reservist.

His memorable experiences: experiences garnered while general medical officer on USS *Eisenhower,* including catapult launch and deck landings. Two week "A.T." activities in Yokosuka, Ft. Sill, FHOTC, combat casualty care course.

He is married to Jo Ellen Reynolds, CDR, NC, USNR. Employed in occupational medicine; part-time medical consultant to National Hot Rod Association. Still in SelRes with GAS 2A 4th FSS6.

WILLIAM H. "BILL" REYNOLDS, CAPT, born Jan. 18, 1944 in Minot, ND, received a MS degree.

He joined the service Oct. 3, 1961. His military locations and stations: NRTC San Diego, NTC Lakehurst, NJ; NTC Bainbridge, MD. Served with NAF Naples, Italy; NNMC Bethesda, MD; NRMC Memphis, TN, NRMC; San Diego, CA; DET 118 EPMUS Lincoln, NE; NNMC Bethesda; Lincoln, NE FH-22 DET H&I Lincoln/Omaha.

He was awarded the National Defense, Southeast Asia, Kuwait, Pistol, Rifle, Mertiorious Unit Citation and achieved the rank of CAPT, MSC, USNR.

He has children: Christopher, Chad and Corrie; and a grandchild, Brady. Employed as a physical therapist.

C. ERWIN RICE, LCDR, born Oct. 5, 1921 in Syracuse, NY, received a BS from Suny College of Environmental Science and Forestry and also attended graduate courses.

Rice joined the service October 1942. His military locations and stations: NAS Squantum, NAS Pensacola, NAS Honolulu, North Island. Flight instructor Saufley and Branson Fields; VR-11 and VC-11.

He achieved the rank of LCDR and was discharged April 1946 and May 1954. His memorable experiences include the tour of duty on USS *Oriskany* off Korea.

Rice is married to Gwenn I. (Taylor) and they have children: June Bigler, David Rice and James Rice and grandchildren: Amanda Bigler, Benjamin Bigler and Jacqueline Taylor Rice. Employed as soil scientist USDA Agriculture. His wife and daughter are also USN veterans.

EDWARD W. "EDDIE" RICE, CDR, born July 24, 1911 in Great Falls, MT and received a BS in economics and LLB and JD from the University of Kansas.

He joined the service September 1943. His military locations and stations: University of Arizona, Tucson; Ft. Schuyler, NY; Amphib, San Diego. Served with USS *Haskele* (APA-117); USS *Faribault*(AK-179), XO and CO. His battles and campaigns: Lingayen, Subic Bay and Okinawa invasions.

Rice was awarded the Philippine Liberation PTO, American Theater and Command at Sea. He achieved the rank of CDR and retired from the USNR in 1976. His memorable experiences: great friendships throughout his Navy duty.

He married Patrica Murray Arnold (deceased in 1991) and has children: Barbara Beth and Catherine Marie, and grandchildren: Jennifer, Bryan and Melanie Green; Anne and Patrick Westcott. Employed as medical management consultant, Idaho Legislature.

EMORY C. RICE JR., CAPT, born Dec. 8, 1930 in Baltimore, MD and attended college in Georgetown University, Washington, DC and received a BS in physics.

Attended Officers Candidate School, Newport, RI and was commissioned in January 1954. Served four years' active duty, three years aboard the USS *Payette County* (LST-1079) and one year as assistant sailing officer, US Naval Academy, Naval Academy Sailing Sqdn. (NASS). Transferred from active military service Jan. 25, 1957. Continued to participate in the Inactive Naval Reserve Program until Jan. 1, 1982. During this period held four commands: Naval Control of Shipping Organization Unit Hagerstown, MD (three years); WEPLAB 206 Solomons, MD (two years); GVT Unit Washington, DC (one year); HQ Military Sealift Command Adephi, MD (two years).

He achieved the rank of CAPT, USNR and discharged Jan. 1, 1982. His memorable experiences: the training he received from his first skipper, CAPT Joe Y. Christian, outstanding leader, sailor, and superb ship handler, while on board USS *Payette County* (LST-1079).

Participated in the 1956 Newport-Bermuda Race on the USNA Yacht Swift as port watch captain. Skipper was William R. Maclay LTJG - USNR. Awarded the NASS Stephen W. Gerber Trophy and the Dan A. Kimball Trophy and set a record that was to last 25 years.

He is married to Ann and has children: Josephine Herring, Ann-Bowie Bleinberger and grandchildren: Benjamin, Tilghman, Sarah and Naisby Herring. Retired from civilian life in December 1990.

JOHN C. RICE JR., CAPT, born July 25, 1929, New Orleans. Earned AA, BS and ME degrees. Upon graduation from Loyola University of New Orleans, CAPT Rice was commissioned through OCS in 1953. ACDU followed at NAVCOMMSTA Boston and USS *Taconic* (AGC-17) in communication, mobilization, planning billets.

After RAD he held several billets in Advanced Base Planning Unit 8-3; Group CMD 8-1(L) Admin; SURDIV 8-21(M) CO; Group CMD War Gaming; Fleet Processing (Mobilization) Team CO; Force Area Commander Staff; REDCOM Team Retention IG; Merchant Marine DIV CO; and Navy Staff Liaison to state adjutant general followed.

Rice joined NRA in 1968 and has attended every EXCOM & Conference since. He has been chapter secretary, ex-VP and president, District VP and President, National Parliamentarian, VP E&CA, VP Surface/Subsurface, Executive VP & National President. He holds the Association Award of Merit and two National President Awards.

He was NRA's first National Parliamentarian: wrote the oath of office and authored numerous changes to the Constitution and By-laws; as E&CA he introduced chapter standards; as XVP he organized and chaired the first Long Range Planning Committee. The HQ Building was purchased and converted during his presidency. He has chaired the Building Fund Committee and more recently the Library Development Project.

He wears the Navy Department's highest civilian award, The Distinguished Public Service Award, presented him by Secretary of the Navy John F. Lehman for his services to the USN.

Married Claire and they live in New Orleans, LA. Employed as secondary school teacher and administrator, associate superintendent and exec. director of private non-profit corporation.

HOWARD L. RICH, JR., CAPT, born in Cambridge, MA on Feb. 18, 1915. He attended Ohio State University, majoring in mechanical and industrial engineering in the Class of 1938. He joined Lynd-Farquhar Co. in Boston, MA as a sales engineer remaining with that company until 1996 when he retired after selling the company.

His naval career started on Feb. 10, 1940 when he was commissioned an ENS D-V-S USNR and last for 35 years. He was assigned to USS *Thatcher* (DD-168) on patrol Western Atlantic. He was assigned to USS *Thatcher* (DD-168) on patrol Western Atlantic. On November 1940 he reported to the LDF School at Charlestown Naval Shipyard as navigation instructor. He was selected for P.G. School at Penn State College to be diesel design engineer graduating in 1941.

He was ordered to Manitowoc, WI submarine plant until Pearl Harbor Day when he requested sea duty. On being ordered to USS YMS-38 as XO on escort of convoy from Halifax to Galveston for a year. Next came orders to USS STAFF AM 114 as XO. After a year he was named PCO USS *Adopt* (AM-137). After fitting out he was selected to join the staff of COMSERVFORLANT in charge of underway training of new construction vessels. In late 1944 he was chosen as CO of USS *Towhee* (AM-388) one of the most advanced ASW ships of that era. In September 1945 he joined the Ready Reserve for 29 years serving as CDR MSTS, later as CDR Naval Control of Shipping. On March 1, 1960 became CAPT (DEVG). Recommended for command of USNR DD class vessels by RADM Wm Ball USNR-R. The last years involved in his Mobilization Billet as COS and deputy CDR NAVBASE Newport.

In 1975 he retired and returned to business full time. He is married to the former Ruth Rosenkrans of Needham, MA. They have two daughters Pamela Marston and Penelope Nuse and four grandchildren. CAPT and Mrs. Rich reside in Harvard, MA.

RICHARD L. RICHARDS, CAPT, born in Chicago on June 4, 1933. He was commissioned ENS USNR at OCS in 1956. After designation as NAO (C), he was assigned to VAW-12 at NAS Quonset Point and aboard USS *Randolph.*

In 1958 LTJG Richards became one of the first RIO Candidates for the F4H. He was designated NFO (I) and augmented to USN in 1960. He served with VF-41 and VF-84 embarked in USS *Independence;* the VF-84 tour included a deployment to WESTPAC and Vietnam. He left the USN in 1966 to become a flight test RIO with McDonnell Aircraft Company.

After returning to the USNR, LCDR Richards reported to VP-724. In 1970 he joined NASRU M-1 in St. Louis. During this assignment he was promoted to CDR and served as the unit's CO. He was promoted to CAPT in 1980 and in 1981 he commissioned VTU-0872 and served two tours as CO.

His decorations include, among others, the Distinguished Flying Cross, eight Air Medals, two Navy Commendation Medals w/Combat "V", the Navy Unit Citation and the Vietnam Service Medal.

CAPT Richards holds a BA degree from Beloit College and an MBA from St. Louis University. He flew with McDonnell Douglas for 13 years; he was later a contracts manager and is now retired from MDC.

Richards and his wife, Nancy, live in St. Louis. They have a daughter, Wendy, and a son, Richard Jr.

JOHN M. "JACK" RICHARDSON JR., LCDR, born in Plainfield, NJ, Nov. 6, 1942. Enlisted USN, July 3, 1962. Assigned BUPERS, Dec. 26, 1962. Released from active duty, Aug. 2, 1966 as YN2. Affiliated with NRSUBDIV 5-8, September 1966. Affiliated with Group Command Staff 5-4(M), 1968.

Discharged Sept. 3, 1970 as YN1 to accept commission as ENS (Intelligence) - date of rank Aug. 2, 1970. Ordered to NAVRES Intelligence Div. 5-2. Subsequently served in units supporting Naval Intelligence Command HQ, Naval Investigative Service HQ, and OPNAV. Navy civilian employee, 1968-79, in BUPERS, NAVSEA, and OPNAV. Navy contractor/consultant 1979-88.

Retired USNR, Aug. 2, 1983. Graduate, 1993, Grace Theological Seminar, Winona Lake, IN (MA Christian School Administration). Listed in Marquis' *Who's Who in Religion,* (1993 edition), and 1996/1997 editions of *Who's Who in the South and Southwest.* Member ROA, Rotary International, American Legion, life member of NRA.

He is married and has three children and two grandchildren. Employed as superintendent with Blue Ridge Christian School, Bridgewater, VA.

LYLE L. RICHMOND, LTJG, born April 10, 1930, in Chicago, IL, received a BA in 1952 from Wesleyan University, Middletown, CT; LLB, 1955 Yale Law School, New Haven, CT.

He joined the USNR June 6, 1955. His military locations and stations: OCS, NS Newport, RI; CIC School, NAS Glenview, IL; Airborne CIC School, NAAS Glynco, Brunswick, GA; NAS North Island, San Diego, CA.

Served with: NAO (C) Naval Aviation Observer (controller) aka Airborne Air Controller with VAW-11, NAS North Island, San Diego, CA. He was awarded the Armed Forces Reserve Medal. Achieved the rank of LTJG and discharged Nov. 8, 1958.

His memorable experiences: WESTPAC, USS *Hornet* (CVA-12), VAW-11 DET "F", January 21-July 25, 1957.

Ready Reserve Inactive duty in USNR. Beginning date, Nov. 8, 1958; retirement date, July 1, 1984. He retired as CAPT, JAGC, USNR.

Married to Violet P. and they have children: Jon S. Richmond, Pepele V. Faaiu, Leslee R. Armstrong and David E. Richmond; grandchildren are Fouina Rhine-Lee Faaiu, Aristarchus David Faaiu and Justus Lyle Faaiu.

He has been employed as associate justice, High Court of American Samoa, 1991-present.

ROBERT GEORGE "BOB" RIFFEL, CWO 4, born March 14, 1922 in Orange, NJ. Joined the USN, April 3, 1904 at age 18 years of age. Obtained recruit training at Newport, NTC, Newport, RI. Quartermaster School in Naval Base Norfolk, VA.

Reported to USS New York (BB-34) for duty in the North Atlantic with stops in the Caribbean, Nova Scotia, Newfoundland, and Iceland. While in dry-dock in Portsmouth Navy Yard, Japan attacked Pearl Harbor. As petty officer second class, transferred to USS *Miser* (AF-12) in December of 1941. Cruised the North Atlantic from Caribbean to Iceland, Newfoundland, Nova Scotia, Boston, New York, and Norfolk. In May 1942, went through Panama Canal to Pacific Theater to New Zealand, Sydney, Australia, Brisbane, Albany, Melbourne, Townsville, Cairns and New Guinea, Bora Bora and seven times to San Francisco supplying allied ships and bases.

In February 1943 transferred to the new USS *Hornet* (CV-12) as chief quartermaster, age of 21. The *Hornet* was in every battle in the Pacific from Eniwetok until the end of the war. After a tour of recruiting and duty at Portsmouth, NH Naval Base, was discharged in 1947. During the war years he received two Unit Citations, Good Conduct Medal, American Defense w/Fleet Clasp, American, Canadian, European and African Medals, WWII, Navy Occupation of Asia w/Clasp, Armed Forces Reserve, Philippine Liberation, and Philippine Victory.

Joined the Naval Reserve and made CWO boatswain, serving until March 14, 1982 and retired as CWO 4 boatswain.

Personal story and account: Although having gone through several hurricanes in the North Atlantic and several typhoons in the Pacific Oceans, one typhoon in the Pacific Ocean was the worst. Typhoon Viper really did damage to the third fleet. They lost three destroyers that were capsized. The cruiser *Pittsburg* lost it's bow back to the #1 turret.

He was on the bridge of the *Hornet* and watched in amazement as 5" guns, radio antenna and other gear was torn loose by the waves. The flight deck broke loose and flopped with the waves, rising several feet above the number 1 elevator. Finally the forward flight deck crumbled down on the bow of the ship.

Typhoon Viper did more damage to the ship than the Japnese could. After the storm, ADM Clark ordered the ship to launch aircraft. This had to be done while the ship backed into the wind and aircraft were launched over the stern. To his knowledge aside from the *Saratoga,* the *Hornet* is the only other carrier to launch aircraft over the stern and the only carrier to launch over the stern in time of war.

CARL O. RIGGS JR., CDR, born Dec. 4, 1918 in Norfolk, VA, received a BS in naval architecture and marine engineering from Webb Institute, 1940 and MS from Drexel University, 1956.

He was commissioned an ENS, CC-V(S), USNR with date of rank, June 25, 1940. His military locations and stations: Norfolk, Baltimore, Britain, Germany, 1940-46. He served at naval shipyard, industrial manager, landing craft repair bases, Navy Dept. (Bureau of Ships), Naval Station Kodiak, Naval Shipyard Philadelphia, Navy Department (Office of Naval Material), 1950-61.

His battles and campaigns: support of invasion of Normandy; German shipyard management. He was awarded the Naval Reserve Medal, Defense Service, American Theatre and European Theatre.

He achieved the rank of CDR July 1, 1951 and retired Dec. 4, 1978. He married Evelyn Hall and they had sons, Lee and Stephen, and daughters, Anne and Susan. He is now married to Amy Palmer. His grandchildren are: Raymond Herring, Mark Riggs (son of Stephen), Paul Herring (deceased in 1993), Joshua Riggs (son of Stephen, USNA, 1971, resigned as CAPT USMC. Josh is USNA 1996, 2LT, USMC), Julie Riggs (daughter of Stephen), George Herring, Sarah Miller, Leighton Anne Herring, Lindsey Miller and Claire Miller.

Riggs was employed as aerospace engineer with Chrysler from 1963-69; engineer, Panama Canal, 1969-79.

JOHN J. RIORDAN, CAPT, born in New York City in 1934, he attended Manhattan College and Pace University, receiving BBA and MBA degrees. He is a CPA.

After two years enlisted service in the Marine Corps, he received a direct commission as an ENS in the USNR in 1959. He served with TTT Mgt. Div. 3-1 from 1959-74 and was assigned as OIC, CHB Hatch Team Det 3602 until 1976. He served as DCASR NY in 1978 and from 1980 to 1983 and as CO of the following units: NR SPCC MECH HQ 202 (1978-80), DLA Wash 302 (1983-85), NR DPSC PHIL 202 (1985-87) and NAVRESSO HQ (1987-88). He retired from the Naval Service in 1989.

As a civilian, CAPT Riordan joined Deloitte & Touche, CPAs after college and subsequently completed a career in corporate tax management as senior tax officer of The Nestle Co. (1960-69), Ciba-Geigy Corp. (1969-73) and Combustion Engineering (1973-90). He was active in the Tax Executive's Institute as chapter president of both the Connecticut Valley and Westchester-Fairfield Chapters and as New England regional vice-president.

He and his wife, Joan, have six children and five grandchildren. They reside in Stamford, CT.

GEORGE A. RIPSOM, CAPT, born Dec. 1, 1933, Alton, IL. He is an Illinois native but spent his early years in South Texas and attended St. Mary's Univ. for three years before enlisting in the USN in 1953. Following A School in Washington, DC, he served aboard USS *Capterton* as FT1/c and leading PO before reporting to USS *Yosemite* as repair team leader.

Entering the NR as an enlisted instructor with Surface Div. 4-17(L), he received BS degree in EE from Pennsylvania State Univ. in 1960 and was commissioned ENS Aug. 16, 1960. Assigned to Ship Activation, SAMAR 4-1(L) Maintenance and Repair Div. 4-1(L). He earned MS degree in EE from U of Pennsylvania in 1962 and reported to Surface Division 1017(L), Lawrence, MA as training officer. Served as XO of Ship Activation, Maintenance and Repair Div. 1-1 before reporting to First Naval Dist. Fleet Ship Support Project as ship superintendent.

He was plankowner of Naval Electronic Systems Cmd. HQ 101 and twice commanded Portsmouth Naval Shipyard HQ 101. Served with Naval Sea Systems Cmd. Det. 901 and as staff engineer duty program officer for Commander, Naval Reserve Readiness Cmd. Region One. A graduate of the Naval War College Sr. Off. Crs. he was recalled to active duty in 1986 with Space and Naval Warfare Systems Cmd.

CAPT Ripsom is past-president and state committeeman of the ROA; VP of Massachusetts Bay Chapter of Navy League of the US; National Committee, American Society of Naval Engineers; member of Navy Recruiting District Assistance Council, Boston; director, Surface Navy Assoc.; life member, NRA; member, Naval Enlisted Reserve Assoc., Constitution Chapter and the Boston Wardroom Club.

He wears the Meritorious Service Medal, Meritorious Unit Commendation (4th Awd.) Good Conduct Medal w/ Bronze Star, National Defense Service Medal, WWII Occupation Medal (Europe), Recruiting Silver Wreath (2nd Awd) and other service awards.

Married Elizabeth and they reside in Chelmsford, MA. Their four children are CAPT Margaret Bleau, USAFR; DP1 Lissa Ripsom, USNR; Cheryl Ripsom Gardiner and George Junior. CAPT Ripsom is a principal staff engineer with the Product Assurance Laboratory of the Raytheon Co. Missile Systems Laboratories. He retired Sept. 1, 1992.

JACK F. RITTER JR., CAPT, born in Austin, TX, Dec. 19, 1933. He graduated from the University of Texas at Austin in June 1955 with a business degree in management. Commissioned ENS, USNR on June 3, 1955, after completing the NROTC program at Austin, he was assigned active duty on the 7th Fleet Destroyer, USS *Colahan* (DD-658). During his tour aboard ship he served as the gunnery officer, ASW officer, OOD-Underway, and department head. While gunnery officer, his department received an "E" for the ship as the result of shore gunfire support accuracy.

Released from active duty in 1957, he returned to the University of Texas at Austin and obtained a law degree in 1960. He established a career in the practice of law in Austin, TX.

CAPT Ritter has held billets in Reserve Intelligence Units in Austin, Houston, and New Orleans. He served as officer-in-charge NRIU 8-1-13 from July 1970 through June 1972 and July 1974 through September 1974. Later, he served as CO of Naval Intelligence Command and Control Unit 310 (ORU), Austin, TX, until that unit was relocated in 1976. In March he became XO of COM 8 Intelligence Unit 110, Houston, TX. In July 1979 he became the CO, IVTU 0109, NAS, New Orleans, LA, continuing through September 1981.

In October 1982, CAPT Ritter became Deputy Reserve Intelligence area coordinator (later commander), Area 7, Recruiting and Retention. During four successive years, he managed the Direct Commission Officer Program in Area 7, in close cooperation with the Reserve Intelligence program officer, Area 7, and with the Naval Intelligence Command, Washington, DC. He retired in July 1987.

CAPT Ritter has been awarded the Navy Commendation Medal, the China Service Medal and the Armed Forces Reserve Medal.

He is presently the chief executive officer of a Central Texas petroleum marketer coroporation, selling Conoco gasoline and other petroleum products through retail locations within 100 miles of Austin, TX. He has served in this position since October 1982.

CAPT Ritter is married to the former Mary Hamilton Carroll, of El Dorado, AR. They have two married daughters, Susan and Valerie, and two unmarried daughters, Diane and Cheryl.

MANUEL E. "MANNY" RIVERA-ALSINA, CAPT, born July 21, 1949 in Agvadilla, Puerto Rico, received MD and MBA degrees. He joined the service in 1984. He was stationed at two naval medical centers.

He was awarded the Overseas Service Humanitarian, National Defense and MUC and he achieved the rank of CAPT.

He is married to Christine C. and they have four children: Manuel, Javier, Nicole and Gabriella. Employed as assistant professor, University of Texas and University of Puerto Rico.

CECIL A. ROARK, LCDR, active duty began May 12, 1941 at NAS Dallas for preflight training in the first class at this newly commissioned base. Subsequently, he received appointment as a Naval Aviation Cadet 112699, 5A41C at Corpus Christi, TX, and designated a naval aviator 11461, Feb. 13, 1942. He was assigned to VP-24 Pearl Harbor. Other duty stations include USS *Curtis*, a sea plane tender in the Southwest Pacific; NAS Oak Harbor Washington; NAS Kaneohe, HI; NAS Pago Pago, Samoa.

Upon separation from active duty he participated in Ready Reserve units AUU-702 Tulsa, OK and Composite Co. 8-11. He has logged approximately 3700 hours of flight time during his tenure as a Navy pilot. Billets include squadron executive, CO Composite Co. 8-11, and flight training officer. He will long remember soloing Friday June 13, 1941, in a "Yellow Pearl" #13.

RICHARD R. ROBBINS, CAPT, born April 8, 1935 in Omaha, NE. He joined the service Jan. 26, 1957.

His military locations: NAS Pensacola, FL; NAS New Orleans, LA; NAS Alameda, CA. Served with Naval Air Training Pensacola; VA-822; VA-11x1; VP-45x2; NARDIV-x1; NAS-3609; VF-821; VA-821; VSF11x1; FAIRMED-1082; NASNOLA-3682; VTU-8282; RS-RSD-1182; CARGRU-0282.

He was awarded the Armed Forces Reserve Medal. Achieved the rank of CAPT and retired Feb. 1, 1986. His memorable experiences: while member of VF-822 and assigned as administrative/personnel officer unit awarded Noe Davis Trophy and designated the finest of its type in the Naval Air Reserve Training Command.

He married Ann Marie and they have children: Richard R. Jr., Randall R., Robert S., Russell M. and Ryan H.; and grandchildren: Daniel J., Michael R., Austin C., Melissa A., Amanda L., Nicole M. and Aryanne D. Self-employed as printing broker.

CHARLES H. ROBERTS, CDR, born May 5, 1954 in Peoria, IL, received a BA in mathematics from Duke University in 1977 and a MSSM, Systems Management from USC in 1987.

He joined the service May 25, 1977. His military locations and stations: June 1977, AOCS Pensacola, FL; October 1977, Naval Intelligence School, Lowery AFB, CO; March 1978, VA-82, Cecil Field, Jacksonville, FL embarked aboard USS *Nimitz;* July 1980, RDJTF/USCENTCOM MacDill AFB, Tampa, FL, July 1983 NISC Suitland, MD; May 1985, FOSIC Europe London England; March 1987 NAVRESCRUITCOM NAF Washington, DC; October 1993 ONI 0866 Suitland, MD; October 1994 IVTU RIA TWO NAS/FAF Moffett Field, CA.

He was awarded the Joint Service Commendation Medal, Navy Commendation Medal and Navy Achievement Medal. Roberts achieved the rank of CDR 1635.

His memorable experiences: his final 10 1/2 month cruise aboard the USS *Nimitz* with the port calls and Iranian hostage situation. Deployment to the desert and the Iran Iraq War while assigned to the RDJTF. The world of Washington Intelligence while assigned to NISC. London! Being a production recruiter for 6 1/2 years and meeting the recruits. Being involuntarily separated from active duty the month after making CDR. Served as DRIAC for inspections RIA TWO and making final preparations for his retirement.

Employed as physical/layout engineer for Exponential Technology, a computer chip design company. He retired in October 1997.

DEREK G. ROBERTS, LCDR, born Nov. 14, 1961 in Fort Ord, CA, received a BSME, USNA in 1983.

He joined the service July 1979. His military locations and stations: NPS, Orlando, FL; NPTU, Idaho Falls, ID; USS *Truxton* CGN-35. He served with USS *Truxtun* (CGN-35). Awarded the NAM (2), MUC, Navy Expeditionary, Armed Forces Expeditionary and the National Defense.

He achieved the rank of LT in active duty and is still in the Reserves as LCDR. Discharged from active duty in June 1988. He is still drilling.

Roberts is married to Melissa and they have two children, Sarah and Rebekah. Employed with NALCO Chemical Co., as field engineer.

JAMES E. ROBERTS JR., CAPT, a West Virginia native (born in Charleston, raised in Vienna/Parkersburg), and 1956 graduate of Greenbier Military School, Lewisburg, CAPT Roberts achieved a BS degree from Florence State College (now U. of N. Alabama), and an MBA from The Ohio State University.

CAPT Roberts was commissioned in March 1961, at OCS, Newport, RI. After some pilot training in Pensacola, he transferred to sea duty aboard the USS *Waldo County* (LST-1163), Little Creek, VA, and served with amphibious forces in the Caribbean and Mediterranean.

After active duty, he served for 21 years in ten reserve units in five reserve centers in Ohio and Pennsylvania. Billets included several tours as XO and CO. He retired in 1984.

Roberts employers included The Goodyear Tire & Rubber Co., AMOCO Chemicals Corporation, and his own business, Jim Roberts Plastics. He has lived in Canton, OH, since 1968, and has volunteered at Central-Hower High School, Akron, OH, since 1986, assisting his wheelchair-bound wife, Mary Jane, teach mathematics and computer science. They have a daughter, Mary Kathryn, of Columbus, OH.

TIM H. ROBERTS, CAPT, born July 1935 in Asheville, NC. He earned a BS degree from the University of Tennessee in March 1957. In April he joined the Navy, attended OCS followed by Minecountermeasures School and assignments to USS *Pivot* (MSO 463).

Released from active duty in 1960, he affiliated with the Reserves in Asheville. After the Cuban missile crisis, he returned to active duty, attended ASW School and began a career in destroyers. He served in four destroyers, twice as CO. He had two tours in Destroyer Tenders; one as XO and one as CO. He served 14 years afloat, experiencing deployments in the Atlantic, Mediterranean, Caribbean and Pacific. Awards include the Meritorious Service Medal and two Legions of Merit.

He served in six shore commands including one as CO. Roberts retired Oct. 1, 1985. He lives with his wife Gloria on their Gowensville, SC farm. They have a daughter, Sheridan, and a son, Danny.

WILBUR E. "BILL" ROBERTS, RADM, born April 17, 1915 in Detroit, MI, received a BS in aeronautical engineering from the University of Michigan in 1938 and a masters degree in engineering, executive program, UCLA in 1961.

He joined the service July 1938. His military locations and stations: Pensacola as student; later as a flight instructor, Pearl Harbor. Served with Bombing Sqdn. 6 on USS *Enterprise.*

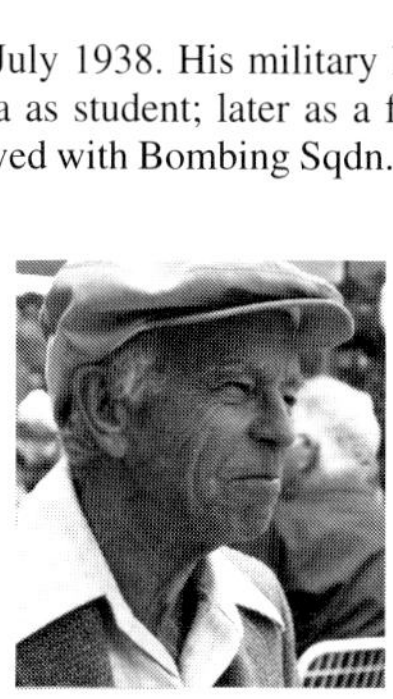

His battles and campaigns: Pearl Harbor, Marshall Island, Wake Island, Marcus Island and Midway Island. He was awarded the Navy Cross, Air Medal, Unit Citation for the USS *Enterprise.* He achieved the rank of RADM (honorary promotion) and retired from USNR, Sept. 1, 1959.

His memorable experiences: Shot down by own aircraft at Pearl Harbor, Dec. 7, 1941. Bombed Japanese aircraft carrier *Acagi* at Battle of Midway. Abandoned aircraft carrier *Yorktown* which was sinking during Battle of Midway.

He is married to Lois and they have children: Doir, Kate, Marianchild, Judith and Dave and grandchildren: Jaxon and Thomas. Employed at Hughes Aircraft Co., as designer, manager of flight operations; manager of aero. engineering department; manager of Hughes Flight Test Facility at Edwards AFB.

WILLIAM F. "BILL" ROBERTSON, CDR, born Oct. 16, 1929, Cumberland, MD. Earned BS, MA, MED degrees in supervision and administration.

He joined the service April 7, 1948, and served four years in USAF as a mechanic; active duty 1956-60, USN. Stationed at Lackland AFB; Lowry AFB; NR Warren AFB and Sheppard AFB. Units served with include VQ2 Navy-Electronic Countermeasures Sqdn. 2, Port Lyautey, Morocco; Italy; Turkey; and Rota, Spain.

Discharged 1952 from USAF and 1974 from Navy Reserve to inactive duty. He retired from the USN in 1989. Awards include Good Conduct, National Service and Naval Reserve.

Memorable experience was being held over for Korean War; ASW Operations in MED; monitoring Russian space shots in Turkey; first navy squadron in Rota, Spain. He really enjoyed service life.

Single, he was an education math teacher for 28 years and retired in May 1986.

ANTONIA J.G. "ROBBIE" ROBINSON, LCDR, born Jan. 6, 1907 in Holyoke, MA. Attended Holyoke High School; Farren Memorial Hospital, Montague City, MA, Nursing School degree, RN.

She joined the service April 1944. Military locations and stations: Chelsea Naval Hospital, Chelsea, MA; Portsmith Naval Hospital, Portsmith, VA; Yorktown Dispensary, Yorktown, VA. Served with Med Co. M-3 He achieved the rank of LCDR and was discharged May 1946.

His memorable experiences: Founder NRA, 1954, national vice-president, Women of the Naval Reserve, 1959. Very active member of NRA. Has attended Conference that was held from 1954 to 1992 except one in 1987. She also attended the Mid-Winter Conferences yearly. Employed for the Department of Hospitals from 1932-75. She is a widower and has no children.

DUNLAP ROBERT "ROBBIE" ROBINSON, CAPT, was born in Auburn, in northern California on April 19, 1917. He was commissioned as an ENS upon graduation from the University of California NROTC, at Berkeley, CA in May 1939. After duty on board USS *Shaw* in 1939 and USS *Saratoga* in 1940, he served as war plans officer on the staff of ComLDF, 12ND. When war came he received orders to USS *Caldwell* (DD-605), building at Bethlehem Steel, San Francisco, and upon commissioning, was her first lieutenant, Communications and CIC officer. Later, during his four years on board, he also served as the XO and navigator, for 18 months, until on Dec. 12, 1944, the CO was a casualty in a kamikaze attack on the Second Resupply Echelon at Ormoc Bay in the Philippines. Robbie, then a lieutenant commander, at age 27 fleeted up to

command *Caldwell,* got her safely back to Leyte, and some three months later, received permanent orders as commanding officer, one of the first Reserve officers to command a fleet destroyer. He remained in command until 1946 when the ship came back to San Diego on her way to post-war mothballs on the East Coast.

During this four year wartime period, the *Caldwell* and her crew first operated mostly in the Aleutians and North Pacific for 18 months (Task Forces 8 & 16), then participated in the 1943 and 1944 Carrier Task Force raids on Wake, Nauru, Truk, Panape and other Pacific Islands, the recapturing of the Gilbert, Marshall and Philippine Islands (mostly with Task Force 38/58) and the New Guinea, Borneo, Palau and Okinawa campaigns. Remaining in the USNR after WWII, he commanded USS *Laws* (DD-558) Reserve Crew for four years at Naval Reserve Destroyer Div., Twelfth Naval District for two more years, retiring in 1977.

He has been active in the Navy League of the US since 1956, serving as state president for Northern California, and also as an assistant national judge advocate and its National Board of Directors for many years. He currently remains a national director emeritus.

CAPT Robinson holds an AB degree from the University of California and an LL.D degree from Boalt Hall, its graduate law school, and has practiced law in California continuously for 50 years, the third generation of his family to do so.

His decorations include the Silver Star Medal, the Bronze Star Medal w/star, the American Defense Medal w/ star, the Asiatic-Pacific Campaign Medal w/8 stars, WWII Victory Medal and the Naval Reserve Medal. He and his wife, Dulcie divide their retirement time between their home in Auburn and their mountain home at Kingvale, elevation 6,000 feet, on the Donner Summit, near Lake Tahoe, in California.

JAMES F. ROBINSON, LCDR, born March 3, 1926, Sparta, MO. Earned BS degree, Southwestern Missouri State; MA degree, New Mexico Highlands Univ. and graduate school at U of Arizona.

Joined the service Feb. 24, 1944, Farragut, ID; NTS San Diego; NTC Bainbridge; SM Base New London; West Coast Sound, San Diego; and SS-362. Units served with electrical division, Santa Fe, NR Tractor Tucson, NRTC, Hollywood, CA. He spent several years as Blue & Gold with USNA.

Discharged March 3, 1991. Awards include American Area and WWII Victory Medal. Most memorable was association with RMCB-17 and belated community service projects.

Married Patricia (deceased 1963), second to Gladys (1965-73) and third to Etta (1975-present). His children are Rebecca, Jenita and Mary Ann; grandchildren are Trane, China Rose, Patisse, Wendy and William. Civilian employment as teacher, Bureau of Indian Affairs, US probation and parole officer, retiring 1977.

WILLIAM H. "BILL or ROBBIE" ROBINSON, CAPT, born May 21, 1930 in Boston, MA, received a BS and BA from Syracuse University.

He joined the service March 24, 1952. His military locations and stations: March 1952-July 1952, Officer Candidate School, Newport, RI; August 1952-January 1953, USS *Rhea* (MSCO-5L), minesweeping officer, 1LT; January 1953-August 1955, USS *Inflict* (MSO-456), XO; September-October 1955, Naval Reserve Surface Div., division officer.

From November 1955-July 1964, Military Sealift Div., various billets; August 1964-July 1968, Naval Control of Shipping Unit, XO; August 1968-July 1917, Naval Control of Shipping Unit, CO; July 1971-September 1972, Group Commander Staff; September 1972-March 1974, Naval Reserve Officer School Student and instructor; March 1974-August 1975, Naval Reserve Officer School, CO; August 1975-March 1977, Ready Reserve Unit 1101, CO; March 1977-March 1977, Volunteer Training Unit, CO; March 1977-September 1979, Naval Control of Shipping,

NE North 101, CO; October 1979-October 1981, Military Sealift Command, VTU 0115, CO; October 1981-July 1983, Military Sealift Command, VTU-0115, Special Projects Officer.

Married Rhoda and they have three children: Stephen, Carolyn and Julie; and grandchildren, Alison and Melani.

WILLIAM R. "BILL or ROBBIE" ROBINSON, CDR, born Dec. 23, 1943 in Warren, PA, received a BS in psychology, University of Pittsburgh; AA Liberal Art, Point Park College; AS Construct Technology; Metropolitan Community College.

His military locations and stations: NAS Pensacola; Lowry AFB; NAS Alameda; Hawaii; NAS Alanta; Offutt AFB; NAS Whidbey Island. Served with VA-152; FICPAC; IPAC; RIP NAS Atlanta; JSTPS; NEACP; COMMATVAQ WING PAC. He participated in Vietnam in 1970.

He was awarded the Defense Merit Service Medal, Joint Service Commendation Medal, NAM, NDSM and Vietnam Service Campaign. He achieved the rank of CDR, USNR-TAR and discharged Sept. 1, 1988. His memorable experiences: VA-152 (WESTPAC 70) FICPAC (Operation Homecoming) NEACP 1984 European Economic Summit, Presidential Support).

He married Phyllis and they have two children, Scott Andrew and Jennifer Marie. Employed as sales associate for a jewelry company and cabinet dealer, C.E. Smith Custom Cabinets.

STANLEY A. ROCK, CAPT, born April 22, 1937 in Pontiac, MI, received a BA in engineering, University of Michigan, 1959; MA in psychology and education, Univerity of Michigan, 1960; BD Western Theo Seminary, 1966; Ed.D Drake University, 1975.

He joined the service Dec. 1, 1966. His military locations and stations: NY, Chaplain Co. 3-1; NY, RMCB19; T&S, Des Moines, IA; T&S Muskegon, MI. Served with SURFLANT REL113, Battle Creek, MI; Readiness Command 13, Great Lakes, IL.

Rock was awarded the Naval Reserve Medal, National Defense Medal and the Meritorious Service Award. Achieved the rank of CAPT, Chaplain Corp, USNR. His memorable experiences: remarkable ecumenical spirit of the Navy Chaplain Corp.

He is married to Nancy Rupp Rock and they have children: Lisa, Amy and Julie and grandchildren: Terri Ann and Jack. Employed as professor, Western Theological Seminary, 1976-present; Inter-Varsity staff, Officer of American Bible Society; minister, Reformed Church, New Jersey, IA.

ROBERT GOODE "BOB" ROCKWELL, LT, born Aug. 20, 1922 in LaJunta, Otero County, CO. He received a BS(EE) from the University of Colorado, 1944; MS(EE) Stanford 1948; degree of engineer, Stanford, 1949.

He joined the service April 1, 1942. His military locations and stations: NROTC, U of C, 1942-44; USS *Lexington* (CV-16), 1944-46. Served with 3rd and 5th Fleets. His battles and campaigns: Marianas Turkey Shoot; Battle of Philippine Sea; Leyte Gulf; S. China Sea; Iwo Jima. He was awarded six Battle Stars and the Presidential Unit Citation. Achieved the rank of LT and discharged May 1, 1946.

His memorable experiences: On the morning the atom bomb fell on Hiroshima, at 500 miles away to the south the noise was heard aboard. Flying at 200' over the Japanese airfields to check that propellers had been removed in early September 1945.

He is married to Norma Jean Fosnaugh and they have children: Laura Jean Amundsen, Michael Dennis Rockwell, Melanie Ann Vizenor, Robyn Goode Elkins; and grandchildren: Mark Amundsen; Katy, Rachel, Travis and Matthew Rockwell and Alexander Elkins. Employed as manager, CRT Tube Technology, Zenith Electronics.

STEPHEN J. "DOC" RODGERS, CAPT, born July 10, 1943 in Philadelphia, PA, received a medical degree from Hahnemann Medical College, 1969 and law degree from Delaware Law School in 1989.

He joined the service June 5, 1965. His military locations and stations: Binh Thuy RVN, NADC Warminister, NAEC Lake Hurst; VAL-4; NRMC PHILA; NARU Lakehurst; NAVAIRES NFK; 4th MAW. His battles and campaigns: Operation Game Warden.

His awards and medals: six rows. He achieved the rank of CAPT and is presently still serving in the USNR. His memorable experiences: He recalls with terror still fresh, getting Air Medals flying with CDR Bob Porter-skipper of the Black Ponies. A day doesn't go by that he is not reminded of the negligent treatment he received at Pendleton Naval Hospital after 1985 injuries at 29 Palms.

He is married to Bobbi-Rhine and they have children Abigail and Rebecca. Employed as medical legal consultant. He is presently awaiting determination of a physical evaluation board.

MAX E. ROHE, CDR, born March 21, 1917 in Kenmare, ND, received a BS in Industrial Arts.

He joined the service July 1, 1941. His military locations and stations: NAS Dallas, Corpus, Atlantic City, Pasco, Otis. Served as instructor, VF-81 fighter pilot, WASP. Achieved the rank of CDR.

He is married to Marine and they have children: CPO Nancy Hubbell and Dan Rohe and grandchildren: Katherine and Philipp. Employed as pilot American Airlines for 25+ years. He retired March 20, 1977.

PAUL W. ROHRER, RADM, born Oct. 19, 1925 in Milwaukee, WI, received a BA from the University of North Carolina, Chapel Hill and BS from Georgetown University.

He joined the service March 23, 1943. His military locations and stations: Command, LSM 238, 1946; Eng., PC 1149, 1946-47; Command, PCS 1387, 1950-53; XO, *Pollux* (AK-54), 1964-66; Command, *Shields* (DD-596), 1966-68; Command, *Great Sitkin* (AE-17), 1970-72; Command, Destroyer Sqdn. 34, 1972-74; Commander, Cruiser/Destroyer, Rep Charleston, 1972-74.

Notable shore tours: NATO Southern Europe, Naples, Italy, 1953-55; Naval Intelligence School, 1956-57; Japanese Language School, 1957-58; Assistant Naval Attache, US Embassy, Tokyo, 1958-61; US Embassy, Tokyo, 1958-61; Head Promotions Branch, BUPERS, 1961-64; DCOS, Plans, NAV RES TravCom, Omaha 2, 1968-70; Asst CNP for NR, 1974-75; NR, OPNAV, 1975-78.

His battles and campaigns: Pacific Theater, WWII, 1943-46; Vietnam (four stars), 1966-68.

He was awarded the Good Conduct Medal and the Legion of Merit (two awards).

Rohrer achieved the rank of RADM (upper half 108) and retired Sept. 1, 1978. His memorable experiences: 40+ years of marriage to the most wonderful woman in the world/commands at sea; knowing their Naval Reservists; completed 36+ years of active duty.

His wife Sarah T. Rohrer is deceased. They had a son John Marshall Rohrer.

WILLIAM S. "BILL" ROMANO, LCDR, born Sept. 20, 1934, Huntington, NY. Earned BA degree at Brown University and MS degree, Columbia Graduate School of Business.

Joined Brown's NROTC program in September 1952 and commissioned in June 1956. Other stations include Naval Communications Station, Washington, DC, 1956; units served with Communications Facility, London, as communications officer; USS *O'Hare*, 1957-58. Active Reserve to Retired Reserve, Oct. 1, 1976.

Married Carol and they have three children: William, Philip and Lynn. Civilian employment as data processing, IBM, Honeywell, Marine Midland. He is still working.

ROBERT BLAIR "BOB" ROPER, LCDR, born Feb. 13, 1925 in Hazelhurst, Cophia County, MS. Attended the University of Alabama, September 1942-June 1943 and September 1946-December 1947, receiving a BS in marketing, member of Kappa Sigma Fraternity.

He joined the service Jan. 9, 1943 at Birmingham, AL. His military locations and stations: University of North Carolina (Chapel Hill), V-12, (July 1943-June 1944); USN Brooklyn Navy Yard, Bayonne, NJ Annex, (July 1943-August 1943); NSCS Harvard (Navy Supply Corps School), (September 1943-January 1945). Commissioned ENS USNASB (Amphibious Supply Base) Exeter (Devon) England, February 1945-January 1946; USNARB (Advanced Receiving Barracks) Exeter, England (February 1946-July 1946; (USN Washington Navy Yard, Anacostia (June 1951-July 1951); USN ASO (Aviation Supply Office) Philadelphia, PA (August 1951-June 1953).

Served with USNR Surface Div. in Tuscaloosa, AL and Jackson, MS. Achieved the rank of LCDR. Retired with pay February 1985.

His memorable experiences: In England, he served as assistant supply officer and collected additional responsibilities as officers were being transferred/returned to the US, culminating with surplus disposal officer for USN Europe. When the base was redesignated an advanced receiving base, he became its supply officer two weeks prior to his 21st birthday. Strong fellow officers and dedicated enlisted men made this one of his most memorable experiences. CAPT D.B. Hilton, fleet supply officer, COMNAVEUR (London) was the guiding force.

When recalled for the Korean Conflict, he received an abbreviated course in military procurement at NAS, Anacostia and assigned ASO Philadelphia as a contracting officer. CDR Otis Stafford, procurement officer, and CAPT Humes, XO, as well as the dozens of other recalled Reservists demonstrated that the USN is an organization of quality people, eager for opportunity and responsibility.

He married Helen Marie Cannon, Oct. 25, 1952, in Philadelphia, PA. They have six sons, one daughter and 13 grandchildren. Initially was with a family (farm supply/fresh produce) business; later with Univac sales, followed by computer operations manager for Vickers Div., Sperry Corp., then a computer systems analyst for the Army & Air Force Exchange Service, Dallas, TX, (which operates the PX and BX's for the Army and Air Force worldwide). Retired from AAFES July 1987.

PETER P. "PETE" ROSASCHI JR., CDR, born May 24, 1931 in Brooklyn, NY, and received BA and MS degrees.

He joined the service Feb. 6, 1951, USNR R, Fort Schuyler, NY. His military locations and stations: Floyd Bennett, NY; Elizabeth, NJ; NCSO, VTU. Served with USS *Heyliger* (DE-510), 1953-55.

He achieved the rank of CDR and was discharged August 1955. His memorable experiences: prior to the 1978 Army/Navy classic, his West Point cadet son-in-law to be and a few of his friends arrived at their home in Monroe, NY (14 miles from W.P.) with the Navy Goat to add insult to injury they requested that he cares for it until the weekend of the game. They got the goat, Navy got the socre 28-0.

He is married to Florence and has children: Ellen, Nora, Peter, Daniel, Thomas, John and Amy and grandchildren: Katie, Lindsey, Peter, Michael, Tom, Joe, Allison, Ehrin, Charles, Sarah. Employed as principal, Special Schools, NYC Board of Education.

JOHN F. ROSCOE, CDR, born Oct. 1, 1946 in Philadelphia, PA. Received a BA from St. Leo College, St. Leo, FL in 1969. Entered naval service under the Aviation Officer Candidate Program at Pensacola. He was one of the 880 that were riffed in September 1970. He joined the USNR at NAS Willow Grove, spent a year in an intermediate maintenance unit, and then joined Replacement Training Unit 64 as a student naval flight officer. They were flying SP-2H Neptunes, two turning two burning! He received his NFO wings in July 1972 with over 400 hours in the Neptune. He worked his way up from patrol plane navigator to mission commander with over 2,000 flight hours most operational, all in the Reserves. He flew P-3, A, B, C, and sub types of each. He spent almost 15 years in Patron Six-Four.

The most memorable and demanding flying was out of Thule, Greenland over the Ice while attached to PATWING at Willow Grove. He is married to Candace (Convertino of Endicott, NY) and has two young men.

He has owned and operated JFR Automotive Supplies since 1974, and has an import/export firm, Roscoe Associates LTD. He retired in 1995 from naval service.

A. DEAN ROSEN, CAPT, bornNov. 21, 1939, Mott, ND. Schools include Firefighting, Great Lakes; Medical Effects of CBR Warfare; SELRES Readiness Crs.; Medical Mgmt. of Combat Casualties (C-4); SMRCC; Landing Force Medical Staff Planning Crs.; and Fleet Hospital Operations Trng. Crs.

Active duty from 1967-70, Internship, National Naval Medical Ctr., Bethesda, MD; Vietnam General Medical Officer aboard hospital ship USS Sanctuary; general medical officer, Naval Hospital, Long Beach, CA.

Naval Reserve Duty, 1973-present: Naval Reserve Ctr., Fargo, ND; held billets in various medical reserve units; staff billet, medical officer REDCOM REG 16, Minneapolis, MN and Blue & Gold officer at USNA.

Voluntary Recall to Active Duty: 1979, TEMAC general medical officer aboard USS Independence, Med cruise; 1991, ADSW staff billet USCOMSOLANT WATC 91-UNITAS 32/91, served aboard USS Barnstable County; 1992, ADSW staff billet, USCOMSOLANT senior TF medical officer WATC 92, UNITAS 33/92, served aboard USS Manitowac; 1993, ADSW, 2nd Mar. Div., Camp Lejeune staff medical officer 2/8 Marines. Battalion participated and provided medical care during mountain warfare training exercise at Bridgeport, CA.

Awards and Medals: Navy Commendation, Navy Unit Commendation, Meritorious Unit Commendation, Navy E Ribbon, Fleet Marine Force Ribbon, National Defense Medal, Vietnam Service Medal, Sea Service Deployment Ribbon, Naval Reserve Sea Service Ribbon, Armed Forces Reserve Medal, RVN Campaign Medal and Gallantry Cross Unit Citation, Expert Rifle and Pistol Medals. He is still active in the Naval Reserve.

Civilian employment as family practice physician, Devils Lake, ND, 1971 until retirement in 1994. Married A. Jane and they have three children: Terrence, Lynnell and Janine; and grandchildren: Brandon, Chad, Dylan and Alexis.

BERYL G. ROSENBERGER, (CAPT), born Jan. 4, 1935 in Scotland County, MD, received BA (Georgia State University); M Div. (Columbia Theology Seminary); DMin (Fuller Theology Seminary).

He joined the service Nov. 29, 1955 (USA-active). Commissioned LTJG, June 4, 1964. His military and locations and stations: NS Charleston, Camp Lejuene, NC; NAS Atlanta. Served with 4th MARDIV; RSU Chaplain (Camp Lejeune); REDCOM Reg. 7 Chaplain.

Rosenberger he was awarded the Navy Commendation and Armed Forces Medal. He achieved the rank of CAPT and discharged Sept. 1, 1990.

His memorable experiences: many ACDUTRAS, across CONUS; aboard USS *Wasp;* USS *Ranger,* Navy Annex (Res. Affairs); Scotland; Puerto Rico (Roosevelt Rds); promotion boards.

He is married to Mary Rosenberger and has six children: John, Richard, Peter, James, Thomas and Mary Elizabeth and 13 grandchildren. He is pastor, Presbyterian Church (USA); director, Healing Ministries.

NOAH SCHANFIELD ROSENBLOOM, born Sept. 30, 1924 in Minneapolis, MN, received BA June 1948 and LLB, June 1952, University of Minnesota.

He enlisted V-1 USNR, August 1952. His military locations and stations: USNR V-12, University of Dubuque and University of Illinois, July 1, 1943-June 30, 1945; USNMS, Columbia University, USNTC, GLakes July 1, 1945-Nov. 30, 1945. Served with USS PC 1147 and USS ATA-182, Jan. 1, 1946-Aug. 1, 1946.

He was awarded the Good Conduct Medal, American Theater, Pacific Theater and Victory Medals. He achieved the rank at discharge from AcDu, QM3; later commissioned ENS USNR in 1951, attained LCDR prior retirement on Oct. 1, 1969. Retired from active duty Feb. 8, 1946.

His memorable experiences: began flight training at WIS, Dubuque, IA, July 1943, based on USNR V-12 status (at own expense) realizing a dream and following a lifelong interest; entering Manila harbor on PA 177 in early morning Dec. 20, 1945, and seeing the damage everywhere, and what remained of Corregidor; observing the difficulty of lowering a boat in significant sea when they put their longboat in the water offo Johnston Island to refuel ATA 182 from its ABSD. Section tow in open sea with substantial swells, in June 1945, enroute, Eniwetok to Pearl Harbor.

He married Janet M. (Biesterfeldt) Rosenbloom, August 1953 and they have children: David, Sara, Nan and Thomas Rosenbloom and grandchildren: Benjamin, Alec and Leah Peters and Isaac Rosenbloom.

Employed in general law practice, Redwood Falls, MN, January 1953 to October 1963; Judge District Court chambered at New Ulm, MN, October 1963-March 1990.

CARL J. ROSSETTI, LCDR, born in Hillsdale, MI on April 6, 1926. Attended Western Michigan University, Columbia University obtaining a BSME degree. Commissioned ENS in June 1945. He served in an advanced base communications unit and at the Pearl Harbor Naval Shipyard until released in September 1946. He was recalled to active duty in the Korean Conflict serving aboard the USS *Glynn* APA-239 as boat group commander January 1951 to December 1952.

His service medals are American and Pacific Theater, WWII Victory Medal, Naval Reserve Medal and Korean Service. Joined NRA in 1958. He served in several Reserve units while on inactive status and retired in 1969.

In civilian life he worked for the Naval Facilities Engineering Command Pacific Div. serving as its Director of Transportation from 1975 until his retirement in 1992.

He resides in Honolulu, HI with his wife Gloria Aguilar. They have four children and nine grandchildren.

OKEY C. ROUSH, LCDR, born Jan. 8, 1926, is a native of Pt. Pleasant, WV, and joined the USN after high school graduation in 1943. He was first assigned in the V-12 program, and in 1944 became a NavCad in V-5, graduating from Pensacola in 2-46. He remained on active duty through 1947, having been trained in the N2S, SNJ, SBD, and TBM, and then flew the SB2C and AD types while assigned to the USS *Midway* in VA-1B.

Returning to Ohio State University in Columbus in 1948, he flew more hours than any other single engine reservist all years through 1950 at NAS Columbus, serving as an "on-call" ferry and test pilot in the TBM, F4U, F6F and SNB types until college graduation.

After serving five plus years as an FBI agent in New York City, in 1956 he returned to Columbus to North American Aviation where he worked in industrial security as the aviation incident and accident investigator, during the period in which the FJ-4, T2J, A5A Vigilante, and OV-10 types were being developed.

He retired from the USNR in 1973 with 30 years combined active and reserve service. From 1971 through his 1992 retirement, he was in the Special Investigation Unit of the Prudential Insurance Co.

Always an avid proponent of a strong national defense, especially by naval air power, he is a life member of NRA, as well as the Reserve Officers Association and The Retired Officers Association, and is currently a functionary in the Helldiver Sqdn. of the Association of Naval Aviation in Columbus, and is a member of the USS *Midway* reunion association.

He and his wife, Kathleen, live in Reynoldsburg, OH, and have a combined family of six children and 11 grandchildren, all of whom are his pride and joy.

GERALD J. "JERRY" ROUSSELL, LT, born April 14, 1943, Long Branch, NJ. Attended numerous business and accounting courses at St. Petersburg Jr. College; completed the Dale Carnegie Sales and Leadership Course and numerous Personnel and Restaurant Mgmt. Courses sponsored by McDonald's Corp. at Hamburger University.

He joined the USNR May 21, 1961 and had many memorable experiences which really sums up why he stayed in the Reserves for 24 years. He was a division officer and maintained records of all unit personnel and held training sessions on Naval Mine Warfare for the officers and enlisted men.

Awards include Naval Reserve Meritorious Service (4th Awd.), National Defense Ribbon and Armed Forces Reserve Medal (2nd Awd.)

Began a career with McDonald's Corp. and held a variety of management positions including store management, supervisor of multiple locations and business liaison consultant between 14 owners and parent corporation. Married Cheryl and they have two children, Amber Joy and Marie.

JOSEPH T. "REDDOG" ROZIC, CAPT, born July 15, 1939 in Butler, PA, received and EE from Penn State. He joined the service Sept. 26, 1962.

His military locations and stations: NAS Alameda, NAS Glenview, NAS Whidbey; NAF, Andrews. His battles and campaigns: VR-48.

He was awarded the Vietnam Service Medal, Bronze Star in lien of second. Achieved the rank of CAPT and discharged May 1994.

His memorable experiences: CO VR-48; CO, NR CV-62 *Independence.*

He is married to Sharon and they have children: Jeffrey J. and Scott T. Employed as with United Airlines as pilot, chief pilot.

BEETHOVEN T. "BEN" RUEDAS, MD, CAPT, born Feb. 8, 1935 in Lilo-An, Cebu, Philippines. He received a bachelor of arts science major; doctor of medicine.

He joined the service June 26, 1975. His military locations and stations: Beaufort Naval Hospital, Branch Clinic, Parris Island, SC; Mayport BMC, FL until 1982. USNR FFSG, Fleet Hospital 14, Naval Hospital 108.

His battles and campaigns: Desert Shield/Desert Storm, Persian Gulf War, 3rd Marine Air Wing.

He was awarded the National Defense Service Medal, Meritorious Unit Commendation, Expert Rifle Pistol, Naval Reserve, Meritorious Service Medal. He achieved the rank of CAPT and retired Feb. 8, 1995.

His memorable experiences: becoming a doctor as a working student; emigrated to the USA having voluntarily serve the USN Medical Corps, having stayed in the USNR, having called to active duty during the Persian Gulf War, Desert Shield/Desert Storm.

He is married Pacita S. Ruedas and has children:

Bennett S. Reudas, civil engineer and Roel S. Reudas, pharmacy student. Employed in general and family practice against loyal and good intention to continue to service the US Navy Medical Corps.

LAWRENCE E. "LARRY" RUGGIERO, LCDR, born Sept. 14, 1961 in Schenectady, NY, received a BS ME (US Naval Academy); MBA (Rutgers University); JD (George Washington University).

He was commissioned May 25, 1988. His military locations and stations: Naval Ordnance Station, Louisville, KY; Naval Weapons Station, Earle, NJ; Naval Mobile Construction Bn. 23 (Ft. Belvoir, VA).

He achieved the rank of LCDR, CEC, USNR. Awarded the Navy Achievement (2), Pistol, Rifle, National Defense Service Medal, Best of type ribbon.

He is single and an attorney with Fried, Frank, Harris, Shriver & Jacobson.

RICHARD LEE RUMBAUGH, CAPT, born June 25, 1941 in Akron, OH, graduated from the University of Akron, BSBA and was commissioned from OCS in August 1964.

ENS Rumbaugh qualified as OOD-UW on USS *Midway* (CVA-41) before serving on board USS *Haleakala* (AE-25). He augmented into the regular Navy and completed 5 1/2 years of active duty on a tour as director of security at NTC San Diego.

Upon leaving active duty he became a full time law student and commenced his career as a SWOS in the reserve. Ten years and seven ACDUTRA's in the Persian Gulf were with COMIDEASTFOR 105 Youngstown and ended with his term as CO 1985-86. In the restructured reserve CAPT Rumbaugh served 2 1/2 years as CO of the Readiness Unit Youngstown. CAPT Rumbaugh retired as of April 1, 1991. In 1997 attorney Rumbaugh completed 25 years in the private practice of law in Carrollton, OH.

GAIL E. "SPIKE" RUMLEY, LCDR, born Oct. 29, 1946 in Ft. Collins, CO, received a bachelor of architecture, masters of business administration from the University of Colorado, Boulder.

He joined the service March 1966. His military locations and stations: NCTC Pensacola, FL; HSGA Bremerhaven Germany, NW Radio Sta, Norfolk, VA. Served with NRGC Misawa Japan; NSGATO, Washington, DC; NSGA, Ft. Meade, USS *Tarawa*, USS *New Orleans*.

His units include: CO, NAVSECGRU Unit, Albuquerque, NM; CO, NAVSECGRU Unit, Lincoln, NE; CO, NAVSECGRU Unit, Springfield, MO.

Rumley achieved the rank of LCDR and was awarded the Armed Forces Service Medal, Good Conduct, National Defense Medal, Naval Reserve, Expert Pistol Marksman, Expert Rifle, and retired July 1996.

His memorable experiences: first airplane ride to boot camp; promotion to ensign; first trip outside the US to Germany; first experience on a warship; first experience as unit CO, storm on transit from San Diego to Hawaii on Tarawa.

He is divorced and has children: Michael, Joseph, Marc Edward and Nathan Andrew Rumley. He is president, Land Development and Construction Consulting Ltd.

HOWARD F. RUNDELL JR., CAPT, born in Pittsburgh, PA, CAPT Rundell entered AOC training in 1969 upon graduation from Youngstown State University (Ohio). Following his "winging" in February 1971, he reported to VT-2, Whiting Field, as a SERGRAD flight instructor in the T-28. In 1973 he transferred to the "World Famous Woodpeckers" of VP-49 in Jacksonville. CAPT Rundell participated in two P-3 deployments, Iceland in 1974 and Sigonella in 1975 as patrol plane/mission commander and the squadron's NATOPS (Standardization) officer. He finished his active duty in September 1976 with the VP-5 "Mad Foxes".

CAPT Rundell immediately affiliated with the USNR joining VP-93 aboard NAF Detroit (Selfridge ANGB). He was among the squadron's original PPC's, participating as an instructor pilot in the unit's transition from S-2s to P-3s. Serving in a variety of billets over the next 11 years including command of VP-2473 (SAU), CAPT Rundell assumed command of the VP-93 "Executioners" in 1987.

Follow-on command assignments at NAF Detroit/ NARA Selfridge included VR-24 COMP 0573 flying UC-12's, NR NAS Sigonella 0573 and NR CVN 69 0373 USS *Eisenhower*. A strong proponent of direct support of the gaining command, he led two detachments to Sigonella, the first following the unit's recall during ODS, and three detachments aboard "IKE", including a 100 man det (seven different units/three different drill sites) participating in "IKE's historic 94-95 Med cruise, the first with women as part of the permanent ship's company.

His current assignment is as XO of NR DCNO N4 (LOGISTICS) COMP 106 at the Pentagon.

CAPT Rundell has flown more than 5,000 hours as a naval aviator. He attended the Naval War College Strategy and Policy Course as well as the Senior Naval Reserve Officer Orientation Course. A corporate pilot for the General Motors Air Transport Section, he lives in Michigan with his wife Jacquie and children Darcy and Bill. He plans to retire in Pensacola.

DOUGLAS K. "POSSUM" RUSH, CAPT, born Oct. 11, 1950 in Patterson, NJ, received a BS from the USN Academy in 1972 and a JD degree from St. Louis University in 1981.

He joined the service June 23, 1968. Served with USS *Decatur* (DDG-31), HS-3; HT-8; MIUWU 112, MIUWU 114. His battles and campaigns: Vietnam, Desert Storm.

He was awarded the Naval Aviator Medal, two Navy Commendation Medals, Navy Achievement Medal and achieved the rank of CAPT, USNR.

His memorable experience: also was elected as a member St. Louis City Board of Education, 1985-91. Served as president, 1990-91. USN active 1968-78; Reserve 1978-97. Recalled to active duty, 1990-91. Served in Saudi Arabia and Kuwaited as KO MIUWU 112 and served as CO, MIUWU 114.

He is married to Janette M. Lohman and employed as attorney at law.

ROGER M. "ROG" RUSS, CWO 4, born Nov. 5, 1913 in Lewistown, PA, attended Philadelphia College of Pharmacy and Science, 1932-37; St. Bonaventure University 1938-39.

He joined the service July 19, 1940. His military locations and stations: Philadelphia Naval Hospital; NAS, Lakehurst; USS *Leedstown;* Bethesda Naval Hospital. Served with FMF, 3rd Marine Brigade, NAS Whidbey Island; USNH Camp Pendleton, NAS, Grosse Ile, MI.

Russ participated in the invasion of North Africa, November 1942. He achieved the rank of CWO4 and retired Aug. 1, 1963. His memorable experiences: participated in the first assault wave on the coast of North Africa in November; the torpedoing and loss of his ship, the USS *Leedstown.*

He received the Naval Reserve Medal, American Defense Service Medal, American Campaign Medal, Asiatic-Pacific Campaign Medal, EAME Campaign Medal w/ Bronze Star, Good Conduct Medal, WWII Victory Medal and American Campaign Medal.

He is married to Grace Iona Russ and they have children, Carolyn Ann and Susan Elaine and grandchildren: Tamatha, Christine, Tracy, Sandra, Angela, Patricia, Michelle and Carlyle. Employed as senior pharmacist for Drug Fair, Inc. in Pennsylvania area. Retired from Drug Fair, Inc. on Nov. 1, 1978.

JOHN J. RUSSELL, LCDR, born March 29, 1921 in Plainfield, NJ, received a BS from Kean College (Farm Newark State); MA San Diego State University; graduate of SDSC and UCSD.

He joined the service Aug. 15, 1943. His military locations and stations: Midshipman School, Notre Dame. Served with YP's Mayport, FL; USS *Munro;* USS *Biven*, Korean Conflict, USS *Colahan.*

He received the American Defense Medal, Asiatic-Pacific Campaign Medal, WWII Victory, Navy Occupational Service and Philippine Liberation.

Russell achieved the rank of LCDR, USNR and discharged November 1953 to Reserves. Transferred to Retired Reserves July 1, 1969.

His memorable experiences: Thanks from the ground troops on the Korean bombline for interdiction fire and star shells at night; Served as navigator for one foreign tour and sailed through two February 29ths. Qualified OOD, CIC officer, gun director officer.

He is married to Gloria and they have children: Kathleen, Kenneth, Bradley and Mary Louise and grandchildren: Sarah Hardin, Alex and Trent Tieger. Employed with San Diego City Schools as teacher, math chairman; SD City College: Inst-math.

RAYMOND M. "RAY" RUSSO, CAPT, born May 27, 1934, Yonkers, NY. Earned AB at Brown Univ., 1955 and MD at Univ. of Bologna, Italy in 1960. Served in the USAF, 1966-74 and in the USNR until 1978.

Stationed at Rhein-Main AFB, Germany, 1966-68; NR Center, Elizabeth, NJ, 1978-83, MEDCRU 504; Navy Reserve Ctr., Clifton, NJ, 1983, MEDCRU 504, CO; NR Readiness Ctr., Kearny, NJ, 1984-86, Fleet Hosp. 4A, CO; NR Readiness Ctr., 1986-91, Fleet Hosp 20, XO and OIC; Fleet Hosp. 15, Jubail, Saudi Arabia, 1991; NR Ctr., Ft. Schuyler, Bronx, NY, 1991-93, Bethesda Navy Hosp. Unit 302, CO; USS *Ortolon*, ASR-22, 1993-95, physician; NR Ctr., Ft. Schuyler, 1995-96, Fleet Hosp. 7, physician.

Awards and medals include Small Arms Expert Ribbon, National Defense Service Medal (2), Armed Forces Reserve Medal, Meritorious Unit Commendation Ribbon, Sea Service Deployment Ribbon, Navy Overseas Service Ribbon, Southwest Asia Service Medal, Kuwait Liberation Medal, Liberation of Kuwait Award and Bronze M, Mobilization Device.

Last civilian employment as assistant dean and professor of pediatrics, Albert Einstein College of Medicine, 1994-present and executive VP, Medical Affairs, Catholic Medical Center, 1990-present.

Married Joanne and they have three children: Victoria, Raymond Jr. and Joseph; and one grandchild, Helen.

HAROLD G. "HAL" RUTHERFORD, CAPT, born July 9, 1933 in Piney Flats, TN, received a BS in agriculture from the University of Tennessee, one year graduate studies.

He joined the service August 1956. His military locations and stations: Naval OCS, Newport, RI; NAS Jacksonville, NS Bermuda; NAS Whiting Field, Milton FL. Served with: VP-45, VT-2, USS *Forrestal;* Reserve Sqdn. VP-671; VP-5B3 ROAC chairman, Naval Academy info officer.

Rutherford was awarded the National Defense Service Medal, Naval Reserve Medal and Air Med Forces Reserve Medal. He achieved the rank of CAPT.

He is married to Virginia Warren Rutherford and has children: Maria Wagner, Lisa Hayes, John Rutherford, Sara Loos and grandchildren: Derek Wagner, Alison Wagner, Cathrine Hayes, Madeline Hayes, Brittany Rutherford, Jacquelyn Rutherford.

Employed as national advertising manager, Bristol Newspapers; maintenance and engineer services supervisor, Southern Airways; quality engineer Sperry/Univac, industrial sales rep, air and hydraulic equipment.

CLARENCE E. "GENE" SAGE, CDR, born July 8, 1934, Missouri Valley, IA, earned a BS degree in engineering-business in 1956. He joined the service Oct. 5, 1956. His military locations and stations: OCS Newport, RI and USNAD Earle, NJ. He served with various reserve units in four states.

Sage was awarded the National Defense Service Medal and the Naval Reserve Medal. He achieved the rank of CDR and was discharged from active duty March 1, 1960, and from reserve duty July 1, 1983.

His memorable experiences: Nike Base on fire near waterfront area while he had the duty officer assignment. DD crashing into pier while he had duty. Docked in foul weather against his reconnaissance.

He and his wife, Lynda, have three children: Crystal, Cyle and Chad; and four grandchildren: Luke, Kate, Kimberly and Cassidy. He was employed with ACF Ind. and as project engineer with Martin Marietta. Sage retired Oct. 1, 1993.

DANIEL F. SALLOWS, LCDR, born March 9, 1919, Brooklyn, NY, earned his AB degree at the University of Notre Dame, 1940 and his MBA degree at New York University, 1956.

He joined the service and was commissioned ENS Oct. 15, 1942. His military locations and stations: WWII, port director, New York, NY; Advanced Base Unit 3B; staff, ComSoPac; port director, Mobile, AL; staff, CINCLANT. Sallows participated in the Korean War, MSTS/NCSO Whittier, AK and MSTS/NCSO Long Beach, CA. He also participated in WWII, Asiatic-Pacific and American Defense campaigns.

Sallows was awarded the WWII Letter of Commendation by VADM W.C. Calhoun, ComSoPac in April 1945; Letter of Commendation by Adm. Jonas Ingram in March 1946; and the National Defense Service Medal.

He achieved the rank of LCDR and was discharged Oct. 28, 1952. His memorable experiences: ADM Halsey's staff, 1943; Noumea, New Caledonia; visit to Sydney, Australia, 1943; and ADM Ingram's Cinclant staff aboard *Vixen*, 1946.

He and his wife, Joan Marie, have three children: Barry, Gerard and Tracy; and one granddaughter, Gabrielle Marie. Civilian employment/positions include Alcoa SS Co., operations assistant then assistant chartering manager; Continental Grain Co., assistant traffic manager then traffic manager; Miles Metal Corp, traffic manager; TTT Ship Agencies Mgr. Full Cargo Div., Gerald Metals Corp., general traffic manager; Mitsui & Co. (USA) deputy general traffic manager. Sallows retired July 1, 1987.

WAYNE S. "FISH" SALMON, CAPT, born May 24, 1936, Gilroy, CA, earned his BS degree in 1958 at Humboldt State College and his MS degree in 1968 at Butler University.

He joined the service Oct. 24, 1958, NAS Glenview, IL. His military locations and stations: Pensacola, FL and NAS Barber's Point, HI. While on active duty he served with AEW Barron, PA, VW-14 and in the reserves he served with VR-722, VR-513, VR-70V1 and VR-51.

Salmon was awarded the Vietnam Service Medal, Reserve Medal (30 year), Navy Commendation Medal and the National Defense Service Medal.

He achieved the rank of CAPT and was discharged Aug. 23, 1963.

His memorable experiences: Landing on Johnston Island with #3 caged and the other three dying during an AF security exercise and being the CO of VR-51 Dec. 6, 1980-March 6, 1982.

He and his wife, Mary Elizabeth, have two children, Kathleen Elizabeth and Jeffrey Wayne; and two grandchildren, Stephanie E. and Caroline D. He was employed as a 7-8-9th grade science teacher. Salmon retired June 10, 1995.

ALBERT P. "AL" SALTER, CDR, born Aug. 23, 1929, Buffalo, NY, earned his BS degree in physics in 1952 at Canisius College.

He joined the USNR in February 1954, OCS, Newport, RI. His military locations and stations: NTS Keyport, WA; Bureau Naval Research, Washington, D.C.; supervisor of shipbuilding office, Camden, NJ (New York Shipbuilding Corp.).

He achieved the rank of CDR and was discharged in June 1960.

He and his wife, Janice Roes, have two children, Timothy A. (LT USN) and Christopher R. (LTJG USN). He was employed by D.J. Martin Equip. Co. as senior sales engineer. Salter retired Dec. 31, 1996.

RONALD A. SAMUELSON, CAPT, born Oct. 17, 1931, graduated from Roosevelt University and earned his MA degree in political science at the University of Chicago.

Attended OCS, Newport, RI, commissioned ENS in December 1956. Various billets as line officer and, subsequently, cryptologic officer. Designated "TAR" officer in 1962. Sequential active duty tours: NAVRADSTA Adak, Underwater Demolition Team Eleven, USS *Allagash* (AO-97), Defense Language Institute (Russian language student), National Security Agency, NAVCOMMSTA Honolulu, US CINCPAC (operations directorate). Spent 12 months of the latter three year tour in Vietnam and Southeast Asia. Inactive reserve duty assignments were with Naval Security Gp. units in Washington, D.C. area; three as CO. Promoted to CAPT in April 1981; retired in October 1989.

Campaign and service awards include: Navy Unit Commendation Medal, National Defense Service Medal, Armed Forces Expeditionary Medal, Vietnam Service Medal w/Silver Service Star, Armed Forces Reserve Medal, Vietnam Civil Actions Unit Citation and RVN Campaign Medal.

Retired from the National Security Agency in 1996. During his 24 year career he worked closely with military commands in operations throughout the world. Received the Meritorious Civilian Service Award in 1989. Currently a consultant for PSC, Inc. of Reston, VA. Also a nautical artist with paintings in Navy Art Gallery, National Cryptologic Museum and elsewhere. CAPT Samuelson and his wife, Darlene, reside in Columbia, MD.

WILLIAM A. "BILL" SANCHEZ, CAPT, born Oct. 2, 1936, Brooklyn, NY, earned his BA degree in marine engineering at New York State Maritime College, Ft. Schuyler, NY, 1958 and his MA degree in management science at Stevens Institute of Technology, Hoboken, NJ, 1967.

He joined the service Aug. 6, 1958. His military locations and stations: USS *McClelland* (DE-750) Philadelphia, PA (duties: DCA, chief engineer). He served with NR MSTS/NCSO Div 3-4(M), Elizabeth, NJ and NRSD 3-95(L), Liverpool, NY.

Sanchez was awarded the 20 year Reserve Ribbon and star. He achieved the rank of CAPT and was discharged Aug. 5, 1960.

His memorable experiences: CO, Convoy Com, San Juan 105, McKeesport, PA (1976-79) CO, NR VTU 0515, Pittsburgh, PA (1980-84).

He and his wife, Mary, have two children, Keith and Caroline; and three grandchildren: Ryan, Jay and Sean. He was employed as vice president - sales, Hankison International, Canonsburg, PA. Retired Sept. 30, 1984 after 26-1/2 years satisfactory government service (all with the USN).

JOHN F. "JACK" SANDERS, CAPT, born Dec. 14, 1921, Buckeye, AZ, graduated from Phoenix Junior College. He joined the USN in September 1942. He was an aviation/cadet at Kingsville, TX. He received his wings/ENS commission in 1944, Corpus Christi, TX.

In WWII he flew the F6F (hellcat) with VF21, "Black Jack Sqdn." on the USS *Independence* in the Pacific. During the Korean War, Sanders flew the TBM in VS-871 aboard the carrier USS *Bataan*. From 1954-56 Sanders flew blimps (K-ship, squadron ZP1) LTA at Elizabeth City, NC. Then selected TAR, he went to Oakland NAS.

He next served in Korea from 1959-60 with COMNAVFORKOREA, Naval Air Intelligence. From 1961-71 he served TAR, Dallas NAS and New Orleans NAS; Glenview (CNARESTRA); Atlanta NAS; South Weymouth, MA NAS; CO-NARTU Lakehurst, NJ; and CO-NRTC Tucson, AZ. Sanders retired June 30, 1971, with 29-1/2 years service.

He and his wife, Jo, have three sons: John Frank Jr., Navy seal; and Steven D. and Scott E., Navy pilots; and eight grandchildren: Sarah, Matthew, Ashley, Kelly, Michael, Matt, Jack and Corry. He was self-employed until 1985.

JEROME T. SANDS, LCDR, began his naval career in 1955 as a SR in the USNR in Pittsburgh, PA.

He obtained a BBA degree from the University of Pittsburgh in 1959. While attending college, he progressed to personnelman second class. He was commissioned ENS, USNR at Officer Candidate School, Newport, RI in 1961.

ENS Sands was assigned to the USS *Eversole* (DD-789), Yokosuka, Japan, and served in operations billets. *Eversole* took part in the "Formosa Patrol" during the Quemoy and Mateu Island dispute and returned to Bremerton, WA for FRAM I overhaul in 1962.

He served temporary duty aboard USS *Fechteler* (DDR-870) on radar picket patrol in CIC and AIC billets, and was promoted to LTJG. He returned to *Eversole* for sea trials and operated out of Long Beach, CA in 1963 until her next Westpac deployment in 1964. His billets included operations officer, navigator, senior watch officer and OOD (fleet operations) as well as AIC/ASAC and DASH CIC controller. *Eversole* deployed to Westpac with DESRON 23 and was attached to the Ready HUK Gp. with the ASW carrier USS *Bennington*.

LTJG Sands departed *Eversole* at Yokosuka in 1964 and resumed his active USNR career serving in Surface Div. and Gp. Command units in New Jersey, Florida, Massachusetts and Rhode Island. He was promoted to LT and LCDR, served as EO of NRSD 6-52(L) in Riviera Beach, FL, completed the Naval Reserve Command and Staff and War Gaming Courses at the Naval War College and retired from active reserve duty in 1975. He was awarded the Armed Forces Expeditionary Medal, the Armed Forces Reserve Medal and the Naval Reserve Meritorious Service and National Defense Medals.

LCDR Sands' post active duty career has been in senior Human Resource management positions with Ford Motor Co., ITT, Motorola and with Human Resource consulting firms. He did graduate work in the MBA program at Bryant College in Rhode Island and has been president of Sands & Assoc., his own consulting firm, since 1991. The Sands family consists of wife, Mary; son, Sean; and daughter, Lisa.

NED R. SANDS, CAPT, born Jan. 27, 1947, Burbank, CA, graduated from the University of California at Los Angeles with a BA degree and the University of Southern California with a MA degree.

He began his naval career in the USNR in Santa Monica, CA in March 1967 as an E-1. After being advanced to third class quartermaster he was accepted into the Reserve Officer Candidate program from which he was commissioned an ENS Nov. 1, 1968, at the Naval Officer Candidate School, Newport, RI. Ordered to active duty during the Vietnam war, he served for a year aboard USS *Wedderburn* (DD-684). After its decommissioning, he was ordered to Japan to serve aboard USS *Whitfield County* (LST-1169) as a part of LANSHIPFLOT ONE, and subsequently served as assistant training officer at Fleet Training Gp., San Diego.

As a member of the selected reserve after 1971, he served in various billets in Texas, California and Oregon. He was selected for command on five different occasions, serving as: CO NR Wepsta Concord 2622, Medford; CO NR Wepsta Concord 3022, Eugene; CO NR Readiness Unit 2204, Eugene; CO NR Readiness Unit 2207, Salem; and CO NR MSCO PAC NW 122, Seattle. Training duty included three tours at the Pentagon, two tours at the Naval War College, selection board duty and staff duty with Commander Cruiser Destroyer Gp. 3. He was selected for advancement to the rank of CAPT in February 1989 and retired at that rank Oct. 1, 1993.

He is married to the former Durinda Booth and is the father of three children: Reed W., Blake Hunter and Amy Rebecca, all of Portland, OR.

JOAO MIGUEL SANTOS, LCDR, born Nov. 28, 1953, Lisboa, Portugal. He moved to Luanda, Angola in 1959 when his father was appointed Director of Luanda International Airport. In September 1974, four months after the revolution in Portugal, he moved to Seattle, WA. He is a 1978 graduate of the University of Washington with degrees in aeronautical and astronautical engineering and in geological sciences. He joined the engineering staff at Douglas Aircraft Co. in January 1979. He received his MBA from Pepperdine University in June 1983. In July 1984 he returned to the northwest to join the marketing department of the Boeing Commercial Airplane Gp. of The Boeing Co.

In December 1984, still before receiving his US citizenship, he enlisted in the USN. His goal was to serve his "new" country and to get his "hands dirty" by working on airplanes. He served at VR-61 as a DC-9 plane captain, at Whidbey Island NAS, WA. He is recognized as fleet logistics "1986 Sailor of the Year" and "1986 Sailor of the Quarter" and earns his enlisted Air Warfare Specialist Wings. On June 12, 1987, in a ceremony at VR-61, he was granted a direct commission and commissioned as an ENS in the USN.

On June 13, 1987, he joined and became a plankowner of a new reserve engineering unit, Naval Air Systems Command 1589 at NAS Whidbey Island, attached to NAVAIR, Washington, D.C. Within three months of commissioning and on his first assignment to NAVAIR, he assumed responsibility for the "Foreign Weapons Evaluation for the F/A-18 and other Navy Tactical Aircraft" program. He was a guest speaker at two ASP conferences on that subject. His leadership on that program earned him his first Navy Achievement Medal.

In January 1989, to enhance his naval officer career, he transferred on a TAD status to VAQ-309 for a period of one year. He was given the Line Div. and prepared the plane captains to transition from the EA-6A Electric Intruder to the EA-6B Prowler. Worked with VAQ-129, on the arrival of the first Prowler, VAQ-309 as a fully qualified Line Div. His performance at VAQ-309 kept him in the squadron until 1994. He successively held the Quality Assurance and CTPL officer, acting assistant maintenance officer and assistant training officer and PAO positions. He sets-up the maintenance training department and the EA-6B Aircrew Ground Training Program at -309. He participated in three carrier deployments, CVWR-30 wing Fallon and Red Flag exercises, attended Strike Leader School and participated in the first Harm missile firing by a reserve squadron. While still at -309, he returns to NAVAIR to develop a "Structural Data Recording Set installation program for the EA-6B," which was implemented a year later.

He returned to NAVAIR by joining NADEP 0187 Alameda. His team's work in the "NADEP Post-Closure Project" contributed to NADEP's closure six months earlier and a savings of $20.0M to NAVAIR - his contributions earned him his second Navy Achievement Medal. He is currently assigned to NASC 0187 and supporting NAVAIR AIR-3.0 Logistics department working on Affordable Readiness and Reliability Centered Maintenance.

Having joined the non-residency Naval War College program as an ENS, he graduated in 1990 with Distinction, the only LTJG in a class of 472 students. In 1994 he earned the Armed Forces Reserve Medal. In 1995, he was selected as the NRA 13th District Junior Officer of the Year and he was a finalist for both the ROA and the NRA 1995 Officer of the Year. He was invited to attend the DOD Junior Officer Professional Development Seminar. In September 1995 he participated in the Reserve Foreign Exchange Program to the UK and the Royal Navy. He is a life member of the NRA, ROA, ANA, Navy League, Naval Institute, Naval War College Foundation and the AIAA.

In his current civilian assignment he is the director, International Sales for Boeing in Southwest Asia and Portuguese and French speaking countries in Africa. He holds a private pilot license. He speaks five languages. He is married to Debra Lynn, a director, product analysis at Boeing and also a LT, USN. They have a six-year-old girl, named Desiree LaRee. As a family, they have traveled extensively, enjoy boating and water skiing during the summer and snow skiing during the winter. LCDR Santos enjoys reading, music and theater.

GEORGE K. SARGENT III, CAPT, born in New York City, graduated from Hamilton College, Clinton, NY. Enlisted in the USN in 1966; recruit training, Great Lakes; assigned to US Naval Facility, Antigua, West Indies for 18 months. Commissioned ENS in February 1968.

Assigned to the USS *Taconic* (LCC-17), Little Creek, VA. In 1969 received orders to Naval Inshore Operations Training Center, Vallejo, CA then to Coastal Div. 13, Sa Dec, Republic of Vietnam, PCF-71. He participated in over 39 combat patrols in the Mekong River Delta and into Cambodia.

CAPT Sargent completed active service in December 1970 and continued in the active reserves at San Jose and Santa Ana, CA, Pittsburgh, and Great Lakes, IL.

CAPT Sargent was awarded the Bronze Star Medal w/Combat D Device, Navy Commendation Medal (Gold Star in lieu of second award), Combat Action Ribbon, Meritorious Unit Commendation Medal, Meritorious Civil Action Medal and Vietnam Campaign Medal.

He has been an investment broker with Prudential Securities since 1983. Prior to 1983 he spent 13 years in various sales and management positions with Benjamin Moore and Company.

WILLIAM B. SARGENT, CDR, a native of Muncie, IN, enlisted in the USN in June 1943. He attended the Illinois Institute of Technology and graduated from the University of Michigan in the Navy V-12 Program.

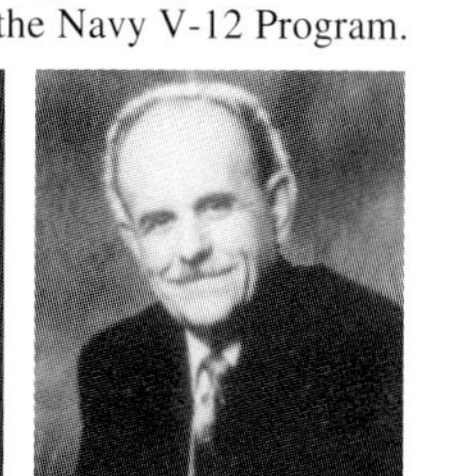

On July 1, 1946, he reported to the USS *Zellars* (DD-777). A month later he was transferred to USS *New* (DD-818). During the next 10 months he served as assistant gunnery and torpedo officer.

The *New*, a member of DesDiv B, was very active in the operations of a task group built around the *Franklin D. Roosevelt* (CV-42) and assigned to the Mediterranean Sea. Before returning to the US, in the spring of 1947, the *New* participated in the first post-war games in the Caribbean.

From 1954, after settling in the Detroit, MI metropolitan area, to 1964, he served in various billets in NRSD 9-104(L) Pontiac. In 1964 he associated with NRSD 9-90(L) Detroit where he served in several billets including CO.

CDR Sargent spent 30 years in various capacities of teaching and administration of vocational education in the Berkley, MI School District and Oakland County.

His family consists of wife, Rosemary; three grown children: William A., Marianne and Lorraine; and seven grandchildren.

JOHN H. "JACK" SARICKS, CAPT, born Jan. 5, 1917, Sandy Run, Luzerne County, PA, earned his EMA at Temple University.

He enlisted in the service Feb. 1, 1943, USNR Training Center. His military locations and stations: Cape May Frontier Base, USS *Rednour* (APD-202). He served with the Transport Div. 105 in the Pacific. Saricks participated in battle at Okinawa and American Theatre. Periods of active military service: WWII, Feb. 1, 1943-March 6, 1946, and Korean War, March 1, 1951-Jan. 27, 1954.

Saricks was awarded the American Theatre Medal, Asiatic-Pacific Campaign Medal, WWII Victory Medal and Naval Reserve Medal. He achieved the rank of CAPT and retired from the reserves Jan. 27, 1954.

His memorable experiences: USS *Rednour* suffered major battle damage from a suicide bomber hit during the battle of Okinawa.

He married the former Florence Toole (deceased) and they have one daughter, Marianne, and two grandchildren, Colleen and Christian. He was employed as a high school teacher for 36-1/2 years. Saricks retired Jan. 5, 1977.

ARTHUR H. "ART" SASS, CAPT, born Nov. 22, 1928, The Bronx, NY, earned his BS degree at the University of New York, Oswego and his EDM degree at Rutgers University.

He joined the service March 19, 1952, OCS Newport, Class VI, 6-2. His military locations and stations: USS LSMR 512 (Little Creek); Merchant Marines Reserve Officer Training Facility; NAVSEA. He served with Eng. Off. and OOD LSMR 512, BUPERS, Academic director NR EDO training program, NARTU Lakehurst, WEPTU-NAVWEPSTA-EARLE (CO).

Sass was awarded the National Defense Service Medal, Armed Forces Reserve Medal, Navy Commendation Medal and Meritorious Unit Commendation. He achieved the rank of CAPT and retired Nov. 26, 1985.

His memorable experiences: OOD at sea, developing and implementing program for placement and assignment officers at BUPERS; developing standards for evaluation of naval instructors; serving as academic director of reserve EDO training program; supervising complete overhaul of a naval vessel.

He and his wife, Eleanor, have four children: Nancy, Arlene, Susan, Eric; eight grandchildren; and one great-grandchild. Civilian employment and positions: Teacher, principal, *Who's Who in World*, *Who's Who in American Education*, superintendent of schools, director education and training, Naval Research Lab, president DEVPRO Inc. (training and conference consultant).

FRANCIS G. "GEORGE" SATTERTHWAITE, CAPT, born Nov. 26, 1936, San Francisco, CA, earned his BS degree in physics at Fordham University; MA degree in math at American University; and PhD in public administration at American University.

He joined the service July 22, 1958, New York City. His military locations and stations: USS *Benham*, USS *Hisi* EM, USS *Jaliao* USNA, USDAO. He served with Lima, USDAO Jakarta and USDAO New Delhi. Served in USDAO New Delhi during Gulf War.

Satterthwaite was awarded the Defense Superior Service Medal, Legion of Merit, JMUC and Navy and Marine Unit Commendation. He achieved the rank of CAPT.

His memorable experiences: Being recalled to active duty after 22 years. Swimming the Suuda with the Indonesian Marines.

He married the former Mary Jo McDonough and they have six children: Kim, Jory, Penny, Ted, Dan and Bart; and six grandchildren: Diana, Andrew, Todd, Bonnie, Gregory and Chris. He was employed as professor US Naval War College.

RICHARD SCOTT SAUNDERS, CDR, born June 13, 1955, Richmond, VA, earned his BS degree in psychology at North Georgia College and his MS in business management at National University, San Diego.

He joined the service Feb. 10, 1978. His military locations and stations: NAS Pensacola, 1978; VT-3, NAS Whiting Field, 1978; VT-19, NAS Meridian, MS, 1979-80; VT-7, NAS Meridian, MS, 1980; VS-41, NAS North Island, CA 1980-81; VS-33, NAS North Island, CA, 1981-83; VS-41, NAS North Island, CA, 1983-84; CNAVEURINT 0167, NAS Atlanta, GA, 1984-86; NAVAIRES San Diego, NAS North Island, CA, 1986-89; VR-56, NAS Norfolk, VA, 1989-91; VR-60, NAS Memphis, TN, 1991-93; NAVAIRES Norfolk, NAS Norfolk, VA, 1993-95; COMNAVAIRESFOR, New Orleans, LA, 1995-97; NAS New Orleans, LA, 1997-. He participated in Desert Shield.

Saunders was awarded the Meritorious Service Medal, Navy and Marine Corps Commendation Medal and Navy Achievement Medal.

He and his wife, Denise, have three children: Bo, Sam and Rachel. Still active in the military.

RICHARD P. "RICK" SAWYER JR., CAPT, born Dec. 15, 1941, New York City, NY, earned his AB and MBA degrees at Dartmouth College, Hanover, NH.

He joined the service July 19, 1963. His military locations and stations: OCS, Navy Supply Corps School, USS *Marshall* (DD-676), Twelfth Naval Dist. He served with MARDIV 911 at NAS So. Weymouth, DESC Dayton, DCS Columbus, DCMD Philadelphia. He participated in Vietnam and Operation Desert Storm.

Sawyer was awarded the Defense Superior, Defense Meritorious Service Medal w/Gold Star in lieu of second, Joint Service Commendation Medal, National Defense Service Medal w/Bronze Star and Armed Forces Reserve HG.

His memorable experiences: Recalled to active duty as CDR, Defense Contracts Management Area Office, Baltimore during Operation Desert Storm.

He and his wife, Carol, have two children, Dori and Kathy. He was employed as division sales manager, Avon Products and data processing manager, Ohio River Co. Sawyer retired Jan. 1, 1997.

PHILIP G. "PHIL" SAYLOR, CAPT, born May 21, 1918, Omaha, NE, attended V-7 OCS at New York and commissioned ENS in September 1941. Assigned to USS *Trever* (DMS-16) as assistant engineering officer, reporting in Pearl Harbor in October 1941. During Japanese attack on December 7, *Trever* was moored with division in Pearl City. Reassigned to USS *Harrison* (DD-573) as engineering officer from December 1943-March 1945.

During WWII participated in 11 battles in Asiatic-Pacific Campaign, including Guadalcanal-Tulagi, New Guinea, Marshall Islands, Guam-Saipan, Leyte Gulf and Iwo Jima.

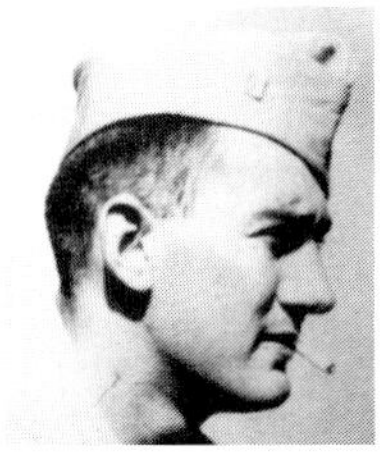

After WWII augmented to Regular Navy. Sea duty assignments included XO, USS *John R. Craig* (DD-885); CO, USS *Gatling* (DD-671); COMDESRON-18; AND CO, USS *Grand Canyon* (AD-28). Shore assignments included NOB Trinidad, Naval War College, OPNAV, JCS, BUPERS, office of SECDEF, and 3rd Naval Dist. Awards included Navy Unit Commendation (USS *Trever*) and Legion of Merit.

Retired as CAPT in July 1972. Employed as NRA HQ staff member from July 1975-January 1996 as assistant executive director and deputy executive director. Married the former Nila Ash in 1945 and had three daughters. Currently retired and residing with wife in Front Royal, VA.

NANCY C. SAYNER, CAPT, born June 3, 1935, Cleveland, OH, earned her doctorate, masters and baccalaureate degrees in nursing.

She joined the service in 1967. Her military locations and stations: Pensacola, FL; London, England (active duty); Raleigh, NC; Mountain View, CA; and Dallas NAS (reserves). She served in active duty during the Vietnam era.

Sayner was awarded the 20 Year Accommodation Medal and National Defense Service Medal.

She achieved the rank of CAPT and retired in September 1988.

RICHARD J. SCAPPINI, CDR, born Dec. 2, 1946, Waterbury, CT, earned his BA degree at the University of Connecticut in 1968 and his JD degree at the University of Connecticut School of Law in 1971.

He joined the USNR Aug. 9, 1983, NLSO NETC Newport, RI. His military locations and stations: NLSO Sub Base Groton, CA and Naval Justice School, Newport, RI. He served with NR LSO 101; NR LSO 201; and NR NIS 101.

Scappini was awarded the National Defense Service Medal, 10 Year Reserve Service, Expert Rifle, Expert Pistol, Navy Achievement Medal, Military Outstanding Volunteer Community Service.

His memorable experiences: Selected in 1995 as Junior Officer of the Year for Eastern Region USNR JAG Corps.

He and his wife, Lisa Kay, have two children, Jennie and Jeffrey. Civilian employment as senior partner in the law firm of St. John, Scappini, Lombard and Stevens.

CAROL ANN SCARLATA, LCDR, born in Manchester, NH, and received her registered nurse degree at Notre Dame School of Nursing.

She joined the service Feb. 4, 1981. Her military locations and stations: Tampa, FL; Manchester, NH; and Ft. Dix, NJ. She served with several hospital units; Fleet Hospital 7 and 15; and 4th FSSG MB. She participated in Operation Desert Shield/Storm.

Scarlata was awarded the Naval Reserve Medal, National Defense Service Medal, Southwest Asia Medal, Kuwait Medal and Unit Meritorious Medal.

Her memorable experiences: Shortly after arriving in country during Operation Desert Storm, she was reunited with her son in the laundry area. He also served in Saudi Arabia in the USMC.

She is divorced and has two children, Mark and Teresa. She is employed at the VA Hospital, Manchester, NH as assistant head nurse in the operating room. Scarlata is currently in the USNR.

DANIEL R. SCHAEFER, LCDR, born April 1, 1939, Hartford, CT, earned his AB and LLB degrees.

He joined the service Sept. 3, 1963. His military locations and stations: USNAS Paxutent River; USS *Shangri La* CVA-38; USN & MC Reserve Center, Hartford and New Haven.

Schaefer was awarded the National Defense Service Medal. He achieved the rank of LCDR and was discharged Dec. 10, 1967 (RAD).

His memorable experiences: Serving with some truly great people, true privilege of serving our country and our Navy.

Schaefer retired June 30, 1982.

MICHAEL A. SCHALL, CDR, born April 23, 1956, Granite City, IL. Commissioned through the NROTC program at Jacksonville University in 1978. His active duty tours included the USS *Schofield* (FFG-3) and the Fleet Anti-Submarine Training Center in San Diego, CA. Since affiliating with the Naval Reserve he has served with NR supervisor of shipbuilding, Conversion and Repair 306, NR OPNAV 04 Comp 106, NR Naval Surface Warfare Center Indian Head 106 (EO), NR Personnel Support Detachment 206 (CO), NR Naval Surface Gp. Long Beach, NR Naval Control of Eastern Pacific 119 (EO), and NR Personnel Mobilization Team 2619 (CO).

His awards include the Navy Commendation Medal, Navy Achievement Medal, Navy Expeditionary Medal, National Defense Medal and the Armed Forces Reserve Medal.

CDR Schall earned a MS in engineering management from George Washington University in 1992. He is a managing partner in an information and engineering firm in Bethesda, MD. He and his wife, Susan, and their two children, Michael and Karen, live together in Howard County, MD.

MYRON R. SCHALL JR., CAPT, enlisted in 1944. He qualified for naval flight training and was commissioned in 1947. He served with the Naval Air Transport Service (VR-3) flying the Continental Div. and then the Atlantic Div. routes in support of the Berlin Airlift. Still an ENS, he qualified and was designated an aircraft commander in R5D type aircraft.

After active duty release in late 1949, he graduated from the University of Washington in 1953 and was employed by North American Aviation (later Rockwell International) in Southern California. CAPT Schall remained active in Naval Air Reserve Transport Sqdns. serving in various capacities with VR-892, VR-772 and VR-773.

As CO of VR-772, his squadron operated C-118s, sometimes under fire, in Chu Lai and Da Nang, Vietnam (1966). In 1967, his squadron provided airlift support to the Mediterranean Fleet during the Israeli Missile Crisis. He was promoted to CAPT in 1968 and retired in 1971 with over 5200 flight hours.

JAMES P. SCHEAR, RADM, is an Ohio native and a 1967 graduate of the USNA. He was designated a naval aviator in 1968.

Active duty assignments: Patrol Sqdn. 50, with three WestPac deployments supporting operations in Vietnam and Patrol Sqdn. 31 as a P-3 pilot instructor.

Reserve assignments: Command of Patrol Sqdn. 91 in 1985-86; command of Patrol Sqdn. Master Augment Unit; battle CAPT in the PACFLT Command Center; and command of Reserve Patrol Wing Pacific's augment unit.

Flag officer assignments: Deputy/vice commander, Patrol Wings, Pacific; deputy commander, Maritime Defense Zone, Atlantic; CDR, Readiness Command Region Four; member - Reserve Forces Policy Board; and VADM in chief, US Atlantic Fleet.

ADM Schear is a captain with USAirways, and has served in several senior management positions. He resides in Pennsylvania with Sharon and their children: Ryan, Matthew, Adam and Caroline.

RENEE GREV SHEEHANM, CDR, born Sept. 3, 1948, in St. James, MN. Earned BS in nursing, College of St. Catherine, St. Paul, MN; MS, UCSF, 1981. She joined the Navy Nurse Corps Officer Candidate, Aug. 2, 1968.

Stationed at Naval Hospitals at Camp Lejeune, Guam, Great Lakes, Oakland and served with various Navy Medical Reserve Units in San Francisco Bay area including NHO 120 and recalled August 1990-May 1991 for Operation Desert Storm to Oakland Naval Hospital.

Discharged June 1, 1991. She received the National Defense and Naval Reserve Medals. She retired June 1, 1993, with almost 25 years of combined active and reserve Navy nursing duty.

Married William J. and they have three children:

Jonathan, Kelly and David. Retired in June 1993 from civilian employment as RN, Coronary Care Unit.

DANIEL J. SCHENKE, CDR, born in 1954, Cincinnati, OH, graduated with honors from the USNA in 1977, and earned his Navy wings of gold at NAS Corpus Christi, TX in 1978.

CDR Schenke served with VP-40 at NAS Moffett Field, CA from 1979-82, with VT-6 at NAS Whiting Field, FL from 1982-83, and VT-10 at NAS Pensacola, FL from 1983-85.

CDR Schenke separated from active duty in 1985, affiliated with VP-94 at NAS New Orleans, LA. A VP-94 highlight included 95 combat flight hours in Operation Sharp Guard during the Bosnia conflict.

CDR Schenke is presently attached to NR CARGU-0282, and has over 3400 pilot hours in various Naval aircraft.

CDR Schenke is employed as a client executive for the IBM Corp. He has served as president and chairman of the Republican Club of Greater Pensacola, committeeman to the Escambia County, FL Republican Executive Committee, chairman of the Northwest Florida Republican Presidents Council, and boardmember of Foundation for Excellence in Education in Escambia County, FL. He has an MBA from the University of West Florida.

CDR Schenke's family consists of wife Teri, son Jon and daughter Sarah.

CARL F. SCHETTER, CDR, born May 12, 1932, Minneapolis, MN, earned his BS degree in business administration in 1954 and his JD degree in 1959 at Marquette University.

He joined the service in June 1955, Marquette NROTC, commissioned ENS. Military locations and stations: San Diego, CA and Yokosuka, Japan. He served with LST 887 (USS *Lawrence County*) 1955-57 and JAG Reserve 1965-89.

Schetter was awarded the NSD and Naval Reserve Medal. He achieved the rank of CDR and was discharged from active service in July 1957 and retired from the reserves in June 1989.

His memorable experiences: VADM Dennisson attended change of command ceremony on their LST in June 1956.

He married the former Susan Mandeville, and they have four children: Fred, Mary, John and Tom; and three grandchildren: Nicole, Daniel and Elliott. He was a self-employed attorney for 30 years in Wisconsin and for eight years in California. He is still active in the practice of law.

ALBERT J. "BUD" SCHIFF, CAPT, born June 4, 1942, New York City, NY, earned his BSBA degree in 1964 at the University of North Carolina-Chapel Hill; his CLU degree in 1968 at American College; and his MSPA degree in 1974 at American Society of Pension Actuaries.

He was commissioned June 1, 1964. Active duty: Navy Supply Corps School (July 1964-October 1964) and USS *Berkeley* (DDG-15) (November 1964-July 1966). He served with Transportation, Traffic and Terminal Management (TT&TM 3-2), New York (1966-73); Advance Supply Base (ASB 102), New York (CO, 1974-77); Defense Fuel Supply Center (DFSC HQ 106), Washington, D.C. (XO, 1977-86); and Navy Resale Office (NAVRESO HQ 102), New York (1986-88).

Schiff was awarded the Armed Forces Service Medal, Vietnam Service Medal w/2 stars, Armed Forces Reserve Service Medal w/HG and Defense Meritorious Service Medal. He achieved the rank of CAPT and retired in 1988.

He and his wife, Jayne N., have two children, Matthew E. and Kara A., and one grandchild, Courtney F. He was employed as president and CEO, Insurance Alliances Gp., Inc.

DENNIS H. "DENNY" SCHIPKE, CAPT, born Feb. 29, 1944, San Francisco, CA, earned his BS degree at the University of Redlands and his DDS degree at the University of Southern California.

He joined the service in July 1969. His military locations and stations: NTC San Diego, dental officer. He served with various dental units including service with Marine Corp Unit.

Schipke was awarded the National Defense Medal, Armed Forces Reserve Medal and Meritorious Unit Commendation. He achieved the rank of CAPT and was discharged from active duty in July 1971.

His memorable experiences: CO, NR, NDCL San Francisco 320 (1990-92).

He has two sons, Steve (a member of a Seabee Unit in Denver, CO) and Brad. Schipke ran a family dental practice. He retired July 1, 1995.

DONALD R. SCHMIDT, CDR, born Feb. 20, 1934, Huron, SD, attended college for two years.

He joined the service in April 1955. His military locations and stations: Flt. training, Pensacola; VU5 Atsugi, Japan; and VA94 Alameoa, LA. He served with NAVCAD, VU5, VA94, VF727 and VA727.

He achieved the rank of CDR and was discharged in July 1959.

He and his wife, Beatrice M., have four children: Michael, Kathleen, Kristina and Deborah; and one grandchild, Margaret. He was employed by Ozark Airline and TWA Airline. Schmidt retired from civilian employment in November 1992 and from the USNR in February 1994.

JURGEN B. SCHMIDT, CAPT, born Feb. 3, 1923, Hamburg Germany, earned his BS degree at Illinois Institute of Technology.

He joined the service Dec. 29, 1942. His military locations and stations: Great Lakes, IL; Shoemaker, CA; Fleet City. He served aboard the USS *Porterfield* DD-682 and the USS *Grand Canyon* AD-28.

He achieved the rank of CAPT and was discharged April 16, 1946.

He and his wife, Adrienne, have three children: Karen, Diane and Kurt; and six grandchildren: Randy, Noah, Nicole, Sarah, Collin and Kate. He was treasurer-controller, Chicago Mold Eng. Co. Schmidt retired Feb. 3, 1983.

PETER ROBERT "PETE" SCHMIDT, CAPT, born Dec. 11, 1934, New York City, NY, earned his AB degree in 1956 at Duke University, Durham, NC.

He joined the service June 4, 1956. His military locations and stations: NAS Cecil Field, Jacksonville, FL, 1956-58 and USS *Randolph* CVA-15, 1958-60. From 1960-85 he served with various units; CO 4-51M; MOB officer COM 4; CO.

Schmidt was awarded the Naval Reserve Medal (second award). He achieved the rank of CAPT.

His memorable experiences: Deployment with Sixth Flt.; Task Force Alpha; deployment with reserves to Scotland in 1985.

He married Anne Parker and they have three children: Elizabeth Anne, Robert C and Susan; and three grandchildren: Todd, Jesse and Lindsey. He was president, Boydeu Consulting Corp. 1969-present.

RAYMOND P. SCHMIDT, CAPT, a Nebraskan, enlisted USNR and graduated from Great Lakes boot camp in 1955, commissioned in 1959, promoted CAPT in 1981, transferred to retired reserve in 1993, and retired in 1997.

Following Communication School, he reported to NSA as an analyst and instructor, accepting emergency orders to NAVSECGRUACT Key West, FL at the height of the Cuban Missile Crisis in October 1962. He rejoined the Ready Reserve in November 1964, recalled to active duty to conduct research and write naval history over the summer of 1967, and again mobilized in 1982 as the first Reserve Forces Advisor to Director NSA/Chief CSS through July 1, 1988.

During a 39-year career, he affiliated with 14 units in Nebraska, California, Wisconsin, Maryland, Virginia and Washington, D.C. He commanded two, and was EO of three, NAVRESSECGRU units. He was also a group command staff officer, supported BUPERS and CNA, and served as the executive assistant to commanding officer, Armed Forces/Naval and Marine Corps Reserve Center, Washington.

Degrees: Bachelor's in history from the University of Nebraska and master's in modern European history from the University of Wisconsin. He also completed part-time all courses for a master's in international relations from the University of Maryland, interrupted by the Cuban Missile Crisis, and for a PhD in government and politics from The American University, Washington, D.C., terminated by new civilian employment with ONI in 1981.

Major civilian positions: First naval cryptologic historian, and senior security reviewer for ONI in the Pentagon. He currently serves as head of the Dept. of the Navy information security program, and as editor of *Viewpoints* for the National Classification Management Society. Honored in Marquis *Who's Who in the East, Who's Who in America* and *Who's Who in the World*.

He helped organize the NRA ADM Yeager chapter and his units earned 100% NRA membership. Naval Academy information officer for 15 years. Life member of NRA, ROA, TROA, DAV and USNI.

Awards: Defense Meritorious Service Medal, Navy Meritorious Service Medal, Meritorious Unit Commendation, Naval Reserve Meritorious Service Medal, National Defense Service Medal and Armed Forces Reserve Medal w/2 HG devices.

CAPT Schmidt and his wife, Roberta (nee Schrom, Navy Junior) reside in Bethesda, MD. Their son, Dr. Douglas C. Schmidt, is professor of computer science, Washington University, St. Louis.

CHARLES FREDERICK "CHUCK" SCHNEIDER JR., CAPT, born Feb. 22, 1945, Charlottesville, VA, earned his BA degree at the University of Alabama in 1971 and his MA degree at San Diego State University in 1975.

He joined the USN June 13, 1963, and was discharged March 10, 1967. He joined the USNR March 10, 1967. His military locations and stations: 1963-64, Naval Aviation Technical Training Center, Memphis, TN, Anti-Submarine Warfare Technician's School; June-October 1964, Patrol Sqdn. 31, NAS, North Island, CA; October 1964-March 1967, Patrol Sqdn. 2, NAS Whidbey Island, WA, Iwakuni, Japan, Naha, Okinawa, Sangley Point, Philippines, Tan Son Nhut, Vietnam; March 1967, Naval Reserve Center, La Crosse, WI; January 1968, Patrol Sqdn. 10E4, NAS Twin Cities, MN; September 1969, Naval Air Facility, Atlanta, GA; September 1971, Group Command Staff, Naval Reserve Center, San Diego, CA; December 1978, Naval Reserve Public Affairs Center, San Diego, CA; October 1983, CO, NR Pacific Fleet, Audio Visual Command; October 1986, CO, NR Public Affairs Center, San Diego; October 1988, chief, Mobile Public Affairs, commander-in-chief, Pacific Fleet; September 1991, executive officer, NR VTU-PAO 1942; October 1992, CO, NR Public Affairs Center, San Diego (second tour); and October 1994-present, executive officer, NR VTU-PAO (second tour). He served in the Vietnam War.

Schneider was awarded the Naval Commendation Medal, Good Conduct Medal, National Defense Service Medal (2), Armed Forces Expeditionary Service Medal, Vietnam Service Medal, Naval Overseas Service Ribbon, Armed Forces Reserve Medal (2), Vietnam Campaign Medal, Expert Rifle - Gold Medal, and Expert Pistol - Gold Medal. He achieved the rank of CAPT (E1 thru E5/01 thru 06).

His memorable experiences: Aviation Anti-Submarine Warfare Technician (AX) 1963-72; commissioned ENS in 1972; promoted to CAPT in 1995; CO 1984-86 Pacific Fleet Audio-Visual Command.

He is married to the former Jana L. Jeske. Civilian employment as technical director, WCFT-TV, Tuscaloosa, AL.

OLIVER J. "OLLIE" SCHNEIDER, LCDR, born Jan. 31, 1926, Ravenna, OH, earned his BS degree in 1948 at Kent State University and his JD degree in 1951 at Western Reserve University.

He joined the service July 1, 1943, and served until Sept. 14, 1946. He re-entered the service Dec. 9, 1951, and served until Nov. 30, 1953. His military locations and stations: USS *Royal* (AMC-102), YMS-74 (1945-46) LCI(L) 404; USS *Jeffers* (DMS-27) (1951-52); 4th Naval District Legal Office (1952-53). He participated in the minesweeping in Inland Sea of Japan (1945-46). He achieved the rank of LCDR and was discharged in 1965.

He and his wife, the former Joann Pickens, have four children: Oliver J. Jr., Ruth M., Barbara A., and Susan. Civilian employment as special agent, Naval Intelligence (1953-57) and general law practice (1957-95). Schneider retired in 1995.

W. EDWARD "ED" SCHNITZ, CDR, born Sept. 22, 1905, St. Joseph, MO, earned his EE degree at Marquette University.

He joined the service Jan. 16, 1941. His military locations and stations: NOB Norfolk, VA, Public Works Dept. (June 9, 1941-1944), ROICC 4349 and 4365; PWO, Marine Corps Air Security, Walnut Ridge, AR (1944-May 1945); and NTC Gulfport, MS (May 1, 1945-September 1946).

He achieved the rank of CDR and retired from the USNR Oct. 1, 1965.

His memorable experiences: ROTC "Wolf Trap Magnetic Range," Cape Charles, VA; Fleet Supply Storehouse and Piers, Norfolk, VA NOB; Deperming Station and Oil Storage, Craney Island, VA; C&P Charles, VA; and Norfolk, VA.

He and his wife, Anna K., have three children: Nancy, Elisabeth and Janet; and three grandchildren: Jenifer, Edith and Erica. Civilian employment as chief engineer, Plant & Structures Facilities Branch, Federal Aviation Administration, Kansas City, MO. Schnitz retired Feb. 28, 1969.

ROBERT M. SCHOLES, CDR, enlisted at 17, May 5, 1955, never dreamed of spending 38 years with the USNR, retiring as a CDR. His enlisted service consisted of 14 years in surface units attaining the rate of first class radarman. Annual training schools and afloat in the amphibious fleet gave him a high degree of proficiency in his field. He was awarded the Meritorious Service Ribbon and selected as the Utah Naval Reserve Sailor of the Year in 1968.

He attended Brigham Young University, earning both his BS and MS degrees in industrial education. He received a direct appointment, Oct. 20, 1969, as a LTJG in the USNR for Special Duty (Intelligence), 1635. He attended intelligence schools and completed OJTs as required while drilling in Salt Lake City, UT. In 1978 the Salt Lake Intelligence unit was disbanded. The next seven years he flew the "airlift" to Alameda monthly, where he drilled with FIRSTPAC 1187. Having been selected as a CDR, he was instrumental in re-establishing an intelligence unit in Salt Lake City. He served as the first executive officer of IPAC 0220 receiving the Joint Meritorious Unit Award during this period.

As an IVTU member, he traveled to many parts of the world including Okinawa, Korea and England to participate in military exercises. In 1989 he became a NAIO (Blue & Gold Officer), serving as the assistant state coordinator.

CDR Scholes and his wife, Virginia (Brockbank), are the parents of five children: Kimberli, Wendi, Robert B., Cynamon and Tyson. They have five grandchildren and expect that number to grow. They reside in American Fork, UT enjoying his retirement from 32 years of teaching in the public schools.

JOHN P. "JACK" SCHOLL, CAPT, born May 15, 1941, Allentown, PA, attended San Diego State University where he earned his BS degree in zoology in June 1965 and his MS degree in biology in August 1972.

He joined the service in October 1963. His military locations and stations: NAV OCS; Diving School, Washington, D.C.; USS *Safeguard* ARS-25, Pearl Harbor. He served with HQ Co. 420 NAVSTA TI; NAVSEA 1006, Washington, D.C.; MASU 220 NAVSTA TI; NAVSEA 1006. He served in the Vietnam War.

Scholl was awarded the Humanitarian Service Medal, Armed Forces Reserve Medal, National Defense Service Medal, Naval Unit Commendation Medal, Vietnam Service Medal, and Armed Forces Expert Rifle Award. He achieved the rank of CAPT and was discharged in December 1968.

His memorable experiences: Salvage and diving jobs in Vietnam, Eniwetok Atoll and Palau.

He and his wife, Marilyn, have two children, Kaili and Damon. Civilian employment as biologist 1972-78 and general contractor/owner, Scholl Construction, 1978-present.

DOUGLAS D. "DOUG" SCHROEDER, CAPT, born July 30, 1936, Ridgeville Township, WI, earned his BS degree at the University of Wisconsin La Crosse and his MS degree at Winona State College, Minnesota.

He joined the service July 24, 1954. His military locations and stations: USS *Power* (DD-839); National Security Agency; Naval Security Gp.; and Naval Reserve.

Schroeder was awarded the National Defense Service Medal and Naval Reserve Ribbon w/2 HGs. He achieved the rank of CAPT and was discharged Oct. 1, 1989.

His memorable experiences: Served as a radioman from 1955-57; four years in Naval Reserve 1957-61; commissioned in 1961 and served mainly with Naval Reserve Security Gp. Units until 1989; commanded three different security groups for 10 years.

He and his wife, Mary, have three children: Marybeth, Julie and Susan; and three grandchildren: Emily, Tyler and Ashley. Civilian employment included 30 years as teacher/administrator in public schools and 19 years as superintendent of schools. Schroeder retired in 1990.

HENRY "JERRY" SCHROEDER JR., LCDR, born March 30, 1922, Carbondale, IL, earned his BA degree at Santa Barbara College.

He joined the service July 27, 1942. His military locations and stations: Preflight, Iowa City; NAS Olathe; NAS Corpus Christi; NAS Opa Locka; NAS Glenview (CV qualification USS *Wolverine*); NAS Norfolk; NAS Quonset Point; NAS Sand Point; NAS Klamath Falls; USS *Anzio*; NAS Barbers Point; NAS Pensacola; NAS North Island; and NAS Oceania. He served with VC-82, USS *Anzio*, Naval Flight School, Pensacola, VC-32 Office of Information, USS *Independence* CVA-62, VR-21, NAS Oceania. Battles and campaigns participated in were Iwo Jima, Okinawa and Philippines.

Schroeder was awarded the American Campaign Medal, Asiatic-Pacific Campaign Medal, Philippine Liberation Medal, Philippine Theater Medal, WWII Victory Medal, Navy Unit Commendation Ribbon, National Defense Service Medal, ADM Flatley Momoriel Award (February 1962), CNO Aviation Safety Award (1963 and 1964), Naval Reserve Medal, and Armed Forces Expeditionary Medal for Vietnam. He achieved the rank of LCDR and retired June 30, 1968.

His memorable experience was the Spring 1945 typhoon.

He and his wife, Gertrude M., have three children: Gary, Susan and Rebecca; one grandchild, Grace; and two great-grandchildren, Alex and Garyon. Civilian employment was as a teacher in California.

KENT L. SCHROEDER, CAPT, born May 29, 1935, Chicago, IL, earned his BS degree in arts and sciences at Florida State University in 1956.

He joined the service May 26, 1953, and was commissioned ENS (1956) via ROC program. His military locations and stations: NABTC Pensacola, FL (1956-57); USS *Des Moines* CA-134 (1957-59); CWO/CCWO-Staff COMSIXFLT; NROS 6-10 (1960-61); division officer, training officer, executive officer and CO of various surface divisions (1962-72); NRCG 1-10 (1972); NROS 1-9(M) (1973); NAVFAC HQ 101 (1974); VTU 0103 (1977); and NAIO for Naval Academy (Blue & Gold) (1971-87).

Schroeder was awarded the Armed Forces Reserve Medal (third award), National Defense Service Medal and Commandant's Award (1982) for Naval Academy, Region I. He achieved the rank of CAPT and retired reserve status July 1, 1987, and from the USN May 29, 1995.

His memorable experiences: Participated in the Lebanon crises of 1958-59; completed senior officer course at Naval War College, 1977; completed ADUTRA aboard USSD *Elokomin* (AO-55), USS *Telfair* (APA 210), USS *Liddle* (APD-60), USS *Spiegel Grove* (LSD-32) and USS *Yorktown* (CVS-10). Also at NAVFACENGCMD, Alexandria, VA; NAVCOMSTA NORVA; USNA; NAVCOMSYS HQ, Baileys Crossroads, VA; Amphibious War Indoctrination Course, Little Creek, VA; COT course, Brooklyn, NY; Naval Declassification Team, Washington, D.C.; and Senior Officers Indoctrination Course, Dam Neck, VA.

He is a widower with three children: David, Deborah Sheppard and Jennifer Poore; and three grandchildren: Jake, Samantha and Kenneth. He retired from C.R. Bard, Inc. in February 1995 after 33 years of service. Held various positions including director of marketing and vice-president of Regulatory Affairs in the USCI International and Bard Europe Divs. Completed employment at the corporate office as director, International Gp. Admin. Schroeder is a life member of NRA and a member of the Navy League.

CHESTER J. SCHULER, CAPT, born Feb. 15, 1933, Louisville, KY. After graduating from the University of Notre Dame in 1955 he began active duty as radio officer on the USS *Princeton* (CVS-37). After two years he was assigned as assistant communications officer on the staff of COMCARDIV-17 in the USS *Philippine Sea* (CVS-47), and one year later as nuclear physics instructor at the Nuclear Power School, then part of the Submarine School in New London.

Upon release from active duty in 1959, CAPT Schuler attended the Massachusetts Institute of Technology, obtaining a PhD in physics. During this time he was an instructor at Naval Reserve Officers' Schools at NAS So. Weymouth, MA and later in Boston. His civilian employment was as manager of the laser laboratory at the GCA Technology Div.

In 1971 CAPT Schuler began civilian employment with the Central Intelligence Agency in Washington, and began a 14 year reserve career with the Office of Naval Research (ONR). He served as CO of four ONR reserve units: the Naval Research Laboratory Unit, ONR's Tactical Development & Evaluation Unit, the ONR HQ Unit, and the ONR VTU. Just prior to his Navy retirement in 1985, CAPT Schuler was a member of the ONR Management Team where he coordinated the activities of 20 senior reserve scientific officers nationwide. He retired from the CIA in 1995 and resides in Fairfax Station, VA.

BLANCHE B. "DUTCH" SCHULTZ, CAPT, born Aug. 23, 1920, Palm, PA, earned her BS degree at Ursinus College and her MS degree at the University of Michigan.

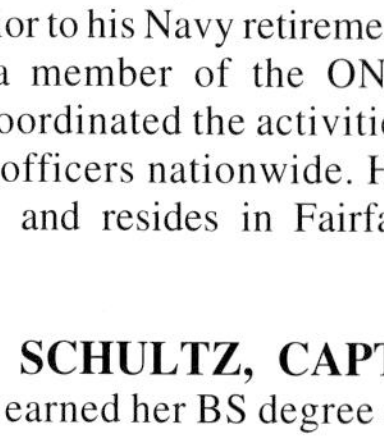

She joined the service in August 1942. Her military locations and stations: Philadelphia Navy Base, Pensacola NATC and NAS Willow Grove.

She achieved the rank of CAPT and was discharged in November 1978.

Her memorable experiences: Receiving her wings as an NAO aerial navigator.

Schultz' civilian employment was as professor of math/assistant academic dean at Ursinus College. She retired in 1986.

WARREN M. "KIP" SCHUR, CAPT, born May 4, 1947, earned his AB degree in philosophy at Harvard University in 1969; his AM degree in linguistics at the University of Michigan in 1974; and his JD degree at the University of Michigan in 1976.

He joined the service Sept. 19, 1969. His military locations and stations: Pearl Harbor and WestPac. He served aboard the USS *Ernest G. Small* (DD-838), USS *Epperson* (DD-719) and with various reserve units. He participated in the Vietnam War and Operation Earnest Will.

Schur was awarded the Meritorious Service Medal, Navy Achievement Medal, Vietnam Service Medal, Navy Expeditionary Medal, National Defense Service Medal and other miscellaneous awards.

His memorable experiences: Gunfire support and other operations in Vietnam conflict; acting as naval liaison officer on Kuwaiti tankers during Operation Earnest Will in Persian Gulf (1987-88).

He is a divorced father of three children: Emily B., Michael H. and Stephanie A. Schur is an attorney in private practice.

DOROTHY R. SCHWARTZ, CAPT, a native of Lawrenceburg, IN, received her BS degree in medical technology from the University of Cincinnati, OH in 1970. She began her military career as a laboratory officer at Naval Regional Medical Center, San Diego, CA in 1972. She worked in many billets during her tour at San Diego including assistant department head of microbiology, assistant blood procurement officer for Southeast Asia and Regional Dispensary liaison officer.

CAPT Schwartz continued her military association in the USNR serving in various billets such as staff medical service officer at Rediness Command Region Four, Philadelphia (also working with the Inspector General Inspection Team), CO, executive officer and administrative officer. CAPT Schwartz was reactivated in 1987 for four months as acting laboratory manager at Naval Hospital Oakland, CA.

She continued her education at Salisbury State University located at Salisbury, MD receiving her MBA in 1989. CAPT Schwartz was selected for CAPT in 1990 for FY 1991. Her service medals are: Navy Achievement Medal, two awards of the National Defense Medal, Reserve Meritorious Service Medal and the recent Military Outstanding Volunteer Service. The Schwartz family consists of son Samuel. They reside in Salisbury, MD.

STANLEY P. "STAN" SCHWARTZ, CAPT, born Oct. 28, 1930, Bronx, NY, earned his BS degree in education (math).

He joined the service Dec. 30, 1952. His military locations and stations: NAS Newport, RI; NAS Jacksonville; NAS Pensacola; NAS Glenview; NAS Quonset Point, RI. He served with OCS, AGO School, fitness training, CIC Officers School and VC-12.

Schwartz was awarded the National Defense Service Medal, European Occupation Service Medal and the Armed Forces Reserve Medal. He achieved the rank of CAPT USNR (Ret.) and was separated from service May 2, 1956.

His memorable experiences: Qualified as NAO(C), went to sea with detachments of ADs performing AEW. In the reserves worked in VS squadrons performing ASW and then transforming to patrol planes achieving NFO(N) and NFO(T) status. Progressed through ranks to command.

He and his wife, Edith H., have three children: Steven, Susan and Lauren; and two grandchildren, Brian and Evan. As a civilian he was employed as a sales representative and regional manager. Schwartz retired July 1, 1983.

EDWARD S. "ED" SCHWEIZER, CAPT, born May 6, 1938, Houston, TX, earned his DS degree at the University of Paris; his BA degree at Occidental College; and his MA degree at Pepperdine University.

He joined the service June 9, 1962. His military locations and stations: NRC, No. Hollywood, CA; NRC Santa Ana, CA; N&MCRC Phoenix, AZ; NS Long Beach; NRC Fall River, MA; NS Sasebo, Japan; New Orleans, LA; Scotia, NY; Norfolk, VA; and Washington, D.C. He served in three campaigns in the Vietnam War and served in Operation Frequent Wind, the evacuation of the Republic of Vietnam.

Schweizer was awarded the Meritorious Service Medal, Navy Commendation Medal w/Gold Star, Navy Unit Commendation Ribbon, Armed Forces Expeditionary Medal, Vietnam Service Medal, Humanitarian Service Medal, Sea Service Deployment Ribbon, Navy/Marine Corps Overseas Service Medal, Expert Pistol Shot Medal and the Navy Rifle Sharpshooter Marksmanship Ribbon. He achieved the rank of CAPT (O-6) and was discharged Aug. 31, 1988.

His memorable experiences: His military wedding to Suzan Lee Peterson, June 20, 1964; first military leadership role as section leader at OCS; coastal patrol North/South Vietnam demarkation line, observing the land battle, seeing corpses floating in the Saigon River; his first shore leave in Olongapo!

He and Suzan have three children: Ward Schweizer, Sally Depasquale and Elizabeth Scarbrough; and two grandchildren, Ricci and Joseph Depasquale. Civilian employment was as assistant center director, INS Western Service Center. He retired as deputy commander, NR REDCOM Six, Washington, D.C. Aug. 31, 1988.

HENRY C. "CHRIS" SCHWEMM JR., CDR, born Oct. 14, 1941, Baltimore, MD. CDR Schwemm is a 1964 graduate of the USNA. During his five years of active duty, he served on USS *Dewey* (DLG-14), served for a year as advisor to the Vietnamese Navy in Saigon and completed his active duty with a tour as counterinsurgency instructor at Little Creek, VA.

CDR Schwemm spent 16 years in the inactive reserves at both the Atlantic City, NJ and the Philadelphia Reserve Centers. He served for eight years on the staff of Readiness Command Region Four. During his active and inactive tours, CDR Schwemm earned the Navy Commendation Medal, the Navy Achievement Medal and the Purple Heart Medal. He retired from the USNR in 1985.

His civilian career has been with Atlantic Electric where he started employment as an associate engineer and advanced through several positions. His final position was as vice-president Power Generation and Fuels Management. He retired with over 26 years of service Oct. 1, 1996, and has now started his own corporation, Professional Services Inc., a management consulting business. The Schwemms reside in Egg Harbor Township in southern New Jersey.

The Schwemm family consists of wife, Mary Jane; daughter, Rebecca; and daughter, Robin, with her husband, Bob, and son, Christian.

JOHN G. SCOBLE, CAPT, born June 5, 1934, Nevada City, CA, graduated from California State University-Sacramento, degreed in life science, in their first graduating class June 1956.

Graduated, class 29 OCS, Newport, RI in December 1956. Served three years aboard the salvage ship USS *Grapple* (ARS-7) at Pearl Harbor where he served as stores officer, OPS/communications officer and executive officer. Made one NorthPac resupply trip and two WestPac deployments. One of these during the Quemoy-Matsu crisis. Earned the Armed Forces Expeditionary Medal (Taiwan). Released from active duty Dec. 2, 1959, and retired June 5, 1994, after 20 years in the Surface Reserve program where he held four command billets in the Los Angeles and Chicago areas.

Is married to Judith R. Scoble with seven children and six grandchildren. Civilian occupations include starting the first large liquid hydrogen plant and operating seven landfill gas recovery plants. He and Judy are now naturalists on whale watch boats.

ROBERT A. "ROB" SCOFIELD, CAPT, born April 27, 1940, Norwalk, CT, earned his BA degree at Bates College and his MBA degree at Rutgers University.

He joined the service in December 1963. He served on active duty 1964-67 in CIC Dept. on USS *Princeton* (LPH-5) off Vietnam, participated in numerous landings and with Assault Craft Div. 11 in San Diego and Japan. Served in the reserves 1967-91 in California, Connecticut and Colorado including various reserve units, particularly COMNAVFOR Korea and as CO of NAVCOMSTA Philippines.

Scofield was awarded the Navy Commendation Medal. He achieved the rank of CAPT USNR (Ret.) and was discharged in September 1991.

He married the former Jane Stubbs. Employed as investment broker and certified financial planner with A.G. Edwards in Denver since 1984.

DANIEL JOYNER SCOTT JR., CAPT, born in Humboldt, TN, graduated from Humboldt High School; undergraduate school at Marion Military Institute, Marion, AL 1948-49 and University of Tennessee-Knoxville, Knoxville, TN; Medical School at University of Tennessee-Memphis, Memphis, TN, January 1951-March 1954.

He served in the USN from April 1954-March 1965 and with the USNR from 1965-82 from which he retired as CAPT.

He and his wife, Sara Jane, have three children: Daniel III, Roy B. and Hugh B. He served as clinical instructor (1966) and clinical assistant professor (1971) at the University of Tennessee-Memphis. Hospital appointments: active staff Methodist Hospital System 1965-96; chief orthopaedics, Methodist Hospital System 1984-95; Baptist Memorial consultant 1965-96; St. Francis consultant 1965-96; VA Hospital consultant 1982; and US Naval Hospital-Memphis 1982. He also served in teaching positions with the USN, Portsmouth, VA, 1959-63; USN, Memphis, corpsman; consultant US Naval Hospital and VA Hospital 1965-82; Methodist Hospital System 1965-96; Campbell Clinic Rotation Methodist, 1970s-80; UT Medical Students Casts & Splints, 1965-70; and Methodist Myelo Clinic, 1965-80.

Scott has published several books, book chapters, journals and abstracts. He has served on many committees over the years.

HAROLD FRANK "SCOTTY" SCOTT, LCDR, born July 13, 1920, Madison Township, Clarion County, PA, earned his BS degree in civil engineering at Pennsylvania State University.

He enlisted in the NAVCAD program during summer 1942. Sworn in at Philadelphia, PA in October 1942 and called to active duty Feb. 8, 1943. Ordered to Fullerton Jr. College, Prescott, AZ for WTS, then to St. Mary's College for Preflight School. To Pasco, WA for primary flight training. To Corpus Christi for advanced flight training, receiving commission in USN as ENS with designation naval aviator April 5, 1944. Assigned to operational dive bomber training at Cecil Field, Jacksonville, FL, and subsequent carrier qualification at Glenview, IL (Lake Michigan). Assigned to Carrier Air Gp. 88 at Atlantic City, NJ, flying the SB2C 'Helldiver.'

With decrease in dive bombers aboard carriers, he transferred to VF88 flying the F6F 'hellcat.' This squadron was later split into two units - VB88 and VBF88. He was then in VBF88 flying the F4U corsair. They trained on the east coast until early 1945 staging to the Pacific Combat Zone via Hilo, HI; Saipan, Philippines, and boarded the *Yorktown* (CV-10) in Leyte Gulf. Departed Leyte Gulf in June 1945 headed for Japanese waters and carrier-based aircraft attacks on the Japanese mainland islands.

Following Japanese surrender, rest and recreation in Japan departing Tokyo Bay, Oct. 1, 1945, arriving in San Francisco October 20. Celebrating Navy Day at NAS Alameda, CA. Assigned to Navy Ferry Sqdn. VRF-1 Floyd Bennett, NY. Accepted into USN program assigned to Villanova University. Resigned USN commission in 1947 and left active duty. Participated in reserve activities joining the Weekend Warrior program (VA691) at Columbus, OH. This station closed in 1959. He opted out until going to Norfolk, VA in 1962 with ASW Sqdn. VS864. He retired from naval service in 1966 with the rank of LCDR.

His family consists of his wife, Eleanor; daughter, Eugenia (PhD); daughter, Carol (MEd); and son, Ray (attorney at law). Grandchildren include: Katy, Casey, Scotia, Corey, Molly and Neil. They live in the country on a farm. He enjoys deer and turkey hunting for recreation.

ROBERT BRUCE SCOTT, CDR, born Feb. 2, 1925, in Akron, OH. A two degrees NROTC graduate of Univ. of Michigan. Commissioned in 1946 and after ENS training cruise to Canada, he was assigned to USS *Fiske* in Boston. Transferred to OPDEVFOR Norfolk then assigned as CIC officer aboard USS *Winslow*.

Separated from active duty in 1948 and employed with B.F. Goodrich Rubber Company, supervising financial administration of Goodrich stores in the Dakotas and Minneapolis, MN.

Recalled to active duty during the "Korean Police Action" in 1953 as assistant navigator and helped sail the disabled Destroyer *Small* from Hawaii to Long Beach Naval Shipyard for repairs.

He completed Gunnery School in San Diego and reported at sea in the Western Pacific aboard USS *Hollister* (DD-788) as gunnery officer. CDR Scott completed recall as assistant operations and communications officer aboard troop flagship USS *Fremont* (AP-42) carrying Marines to and from Caribbean Sea training exercises.

His campaign and service medals are: ASM, WWIIM, China Service, NDSM, Naval Reserve Medal and KSM plus UNSM each w/2 Bronze Battle Stars.

Employed in southern California aerospace industry as an engineer in rocket engine propulsion with Rocketdyne Division of North American-Rockwell-Boeing, he joined the active Naval Reserve and qualified for designator change to Aeronautical Engineering Duty Only (AEDO). Completed his career as a supervisory electronic engineer with the Naval Air Warfare Center, Point Mugu, CA, before retirement in 1996.

The Scott family resides in Pacific Palisades and Malibu, CA, and consists of wife Lorle Anne, son Robert B. II, daughter-in-law Mary Devine and granddaughters, Catherine and Christina.

STEPHEN K. SCOTT, CAPT, a native of Mobile, AL, attended Auburn University and Columbus State University, obtaining MS and MBA degrees. He began his naval career at Newport, RI and was commissioned ENS in 1965.

During the Vietnam era he was stationed at Nuclear Weapons Training Center, Atlantic in Norfolk, VA as a training officer and then in USS *Damato* (DD-871) as 1st LT and Anti-Submarine Warfare officer. While aboard the *Damato* he served in Vietnam on various operations such as "Operation Sea Dragon." The *Damato* was also chosen to participate in Unitas IX, a six month deployment to South America. CAPT Scott was a fleet OOD and earned the Surface warfare officer designation.

Following several years of active duty he affiliated with the reserve and held command of units in Alabama, Georgia and Tennessee. His branch was with the Naval Control of Shipping Organization (NCSORG) and he served on the National NCSORG policy board. He attended the Naval War College and National Defense University for reserve policy and strategy courses. He was recalled to active duty during 1990 and served as commander, Military Sealift Command, Washington, D.C. and retired in September 1992.

CAPT Scott has been awarded the Meritorious Service Medal, the Navy Commendation Medal, and the Navy Unit Commendation Ribbon. His campaign and service medals include National Defense, Vietnam Service, Naval Reserve, Reserve Overseas and Expert Pistol and Rifle.

He is married to the former Adleyn Geiger of Columbus, GA and they reside in Phenix City, AL. He has four children from a previous marriage.

ROBERT N. "BOB" SCRUGGS, CDR, born Oct. 3, 1938, Nashville, TN, earned his BS degree at Texas A&I and his MS degree at the University of Arkansas.

He joined the service in January 1965. His military locations and stations: NAS Pensacola (NFO) 1965; August 1965-August 1967 VW1 (Guam); September 1967-September 1969 NAS Beeville, TX; February 1970-February 1973 VP-17 (Hawaii); March 1973-July 1975 NAS Millington; and 1975-84 VP60 NAS Glenview, IL. He served in Vietnam during various periods 1967-72.

Scruggs was awarded the Air Medal (3) and Unit Commendation Ribbon (2). He achieved the rank of CDR (O-5) and retired in October 1984.

His memorable experiences: First experience penetrating typhoons was awesome; first deployments to Chu Lai Vietnam were eerie and frightening; satisfying experience as AWACs for SAR.

He and his wife, Julia, have three children: Denice, Deborah and Jennifer; and six grandchildren: Jacob, Baylee, Kylee, Shelby, Remi and Keegan. Civilian employment as personnelist/personnel officer, Dyess AFB, Maxwell AFB and Aviano AFB.

CHARLES A. SCUDDER, CDR, born Aug. 8, 1916, Jefferson County, IN, received his Master's degree in education.

He joined the service Aug. 24, 1942. His military locations and stations: Northwestern University, North Africa, Sicily, Salerno, Italy, England, Omaha Beach, Newport, RI, Hawaiian Islands and Japan. He served with the USNR after WWII for 24-1/2 years. Participated in the invasions of North Africa, Sicily, Salerno and Normandy on Omaha Beach (Red).

Scudder was awarded four Battle Stars for North Africa, Sicily, Salerno and invasion of Normandy. He also earned the American Theater Medal, European Theater Medal, Asiatic-Pacific Campaign Medal, WWII Victory Medal, Faithful Service Ribbon and Good Conduct Medal. He achieved the rank of CMDR USNR and was discharged Dec. 31, 1945.

He and his wife, the former Mabel E. Stout, were married Aug. 4, 1945. She passed away Nov. 30, 1992. They have three children: Sharon Kay (married Richard Nehls), Gary Dean (married Martha Rogers) and Shirlene Diane (married Scott Burroughs); and seven grandchildren: Christopher, Jeffrey, Sarah, David, Clinton, James and Robert. Civilian employment at Harrison Township, Gusport, IN for three years; Dover School, Boone Co., IN for two years; Indianapolis Board of Education junior high teacher for five years; and principal K-8 for 30 years. Scudder retired in June 1981.

ROBERT H. "BOB" SCUDDER, CDR, born Feb. 9, 1938, Washington, IN, earned his BSME degree at Purdue University and his MBA degree at California Lutheran University.

He joined the service Jan. 31, 1962. His military locations and stations: NAV OCS, USS *Washburn*, Naval Missile Center, Point Mugu. He served with OPS Div. USS *Washburn*, Flight Test Div. Naval Missile Center. Participated in the Cuban Missile Crisis in 1962.

Scudder was awarded the Armed Forces Expeditionary Medal, American Defense Service Medal and Vietnam Service Medal. He achieved the rank of CDR and retired Oct. 31, 1989.

His memorable experiences: Aircrew designation in F-4 Phantom, A-3 Sky Warrior, E-2 Hawkeye; TEMAC in OPNAV/Pentagon (OP-90) as POM 84 coordinator. Various NAVAIR assignments as aeronautical engineering duty officer.

He is married to Charlene. Retired Jan. 2, 1994, after 34 years of Navy Civilian Service in engineering and management positions.

HAROLD "CHUCK" SEARCY, LCDR, born Jan. 24, 1923, Weston, MO, graduated from St. Joseph, MO Junior College (1948) and received his BA degree (1950) at Wm. Jewell College, Liberty, MO.

He enlisted in the USN V-5 program Nov. 16, 1942, Kansas City, MO. Active duty March 4, 1943. Commissioned ENS and naval aviator, NAS Corpus Christi, TX in September 1944. Military locations and stations: NAS Jacksonville, FL; NAS Glenview; COMAIRPAC San Diego; NAS Ream Field; NAS Kahuilui, Maui; NAS Marpi Point, Saipan; USS *Intrepid* CV-11, Tokyo Bay; NAS Yokosuka, Japan; NAS Kwajalein and NAS Yonabaru, Okinawa.

From 1947-57 was naval aviator, NAS Olathe, KS. Also attended school at Navy and Marine Corps Training Center, Kansas City, MO. In 1959-60 was educational officer, SURDIV 9-130M, St. Joseph, MO.

Searcy was awarded the American and Pacific Theatre Medals, WWII Victory Medal and Naval Reserve Medal. He achieved the rank of LCDR and was discharged May 26, 1947.

His memorable experiences: Cadet training; commissioning; assignment as pilot to VF-14; carrier qualifications; flying assigned FLTs over Japan from NAS Yokosuka, Japan. He accumulated 1500 hours flight time and 48 carrier landings on four carriers.

He married the former Beverly A. Farris Jan. 10, 1946. They have three children: John Patrick (married Cheryl Hunt), Terry Eugene (married Linda Evans) and Jody A. (married Greg Brown); and eight grandchildren: Jennifer L., Jason P., Ashley R., Kimberly R., Katie J., Trevor S., Brandon L. and Corey S. Was employed by TWA (30 years) and held positions in purchasing and in maintenance and scheduling. Retired Jan. 1, 1983, and lives on a farm. Enjoys traveling, golfing, grandchildren, fishing and flying.

JOHN M. "JACK" SEARLES, CDR, born July 25, 1915, Englewood, NJ, earned his AB degree in 1938 at Princeton University.

He joined the service Nov. 8, 1940. Military locations and stations: New York navy yard; Torredo and MTB, Newport, RI; SCTC Miami; Tulagi; Russell Island; Treasury; Peleliu; Leyte; and Okinawa. He served with MTBRONs 2,3,4,31, Shakedown SCTC Miami, CO MTBRON 3 and 31. Battles and campaigns participated in were: Guadalcanal, Palau, Leyte and Okinawa.

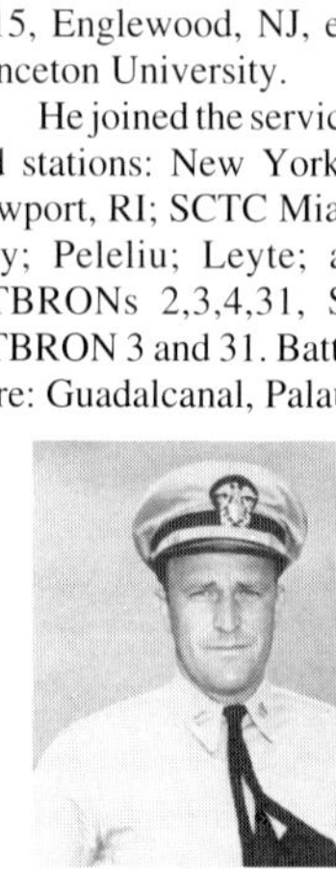
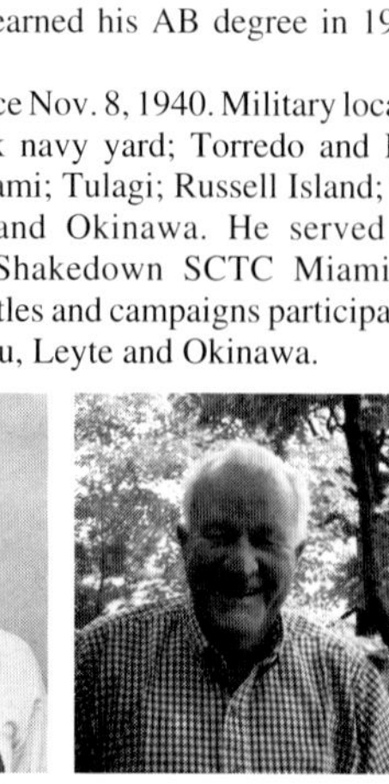

Searles was awarded the Navy Cross, Presidential Unit Citation and many others. He achieved the rank of CDR and was discharged Nov. 1, 1945.

His memorable experiences: Four months fighting

Jap navy at Guadalcanal. Published book in January 1993 *Tales of Tulagi*.

He and his wife, Barbara, have three children: Susan, John and Sarah; and seven grandchildren: Nicholas, Christopher, Tiana, Andrew, Abigail, Jessie and Kipp. Civilian employment at Pillsbury Mills in sales and at Martin Marietta as vice-president of sales. Searles retired Aug. 1, 1980.

MELVIN FRANK "MEL" SEIDENBERG, LCDR CEC, born June 1, 1927, Boston, MA, earned his BSCE degree at Purdue University in February 1949.

He joined the service in April 1945. Military locations and stations: CAPT Eddy RT program, Indianapolis, IN; boot camp, Great Lakes; NAS Philadelphia; Port Hueneme, CA; Subic Bay; Honolulu; Okinawa; Japan; Guam; Evanston, IL; and Glenview. He served with MCB-2.

Seidenberg earned various awards and medals. His memorable experiences: Their mission in Subic Bay where they obtained the topography and did the preliminary design, on site, of the NAS that was later built there.

He and his wife, Annette W., have three children: Steven, James and Sandra; and six grandchildren: Laura, Lindsay, Hannah, Adam, David and Jacob. He was public works officer of the Naval Armory in Chicago. Seidenberg retired upon completion of approximately 25 years of active and inactive duty.

SIGMUND L. "SIG" SEIGEL, CAPT, born Nov. 26, 1931, Bristol, PA, attended Undergraduate School at Gettysburg College and Dental School at the University of Pennsylvania.

He joined the service in June 1951. His military locations and stations: Bainbridge, Po Hang (K-3), Iwakuni, REDCOM 4. He served with FMF AirPac, 1st MAW, MWSG-17, NS-Key West, 4th PSSG. He participated in the Korean War.

Seigel was awarded the Expert Rifle Medal, Navy Commendation Medal, Meritorious Service Medal, Meritorious Unit Commendation Ribbon and Navy Unit Commendation Ribbon. He achieved the rank of CAPT O-6 and was discharged in July 1982. The whole 31 years of service was memorable.

He has two children, Mark and Kurt; and four grandchildren: Zach, Nick, Alexsis and Allison. Civilian employment includes dentist - CEO to janitor.

RICHARD W. SEILER, CAPT, a NRA life member, born and raised near St. Louis, MO. He enlisted in the USNR in February 1960 while attending Bradley University. Upon obtaining his BSEE degree in 1962 he reported to OCS where he was commissioned and ordered to active duty.

During almost 11 years of active duty, CAPT Seiler served in weapons and operations billets aboard USS *Hawkins* (DDR-873), USS *Harwood* (DD-861), USS *Preble* (DLG-15), and the Naval Ordnance Systems Command. He also received his MSEE from the Naval Postgraduate School and taught at the USNA. He received the Navy Commendation Medal while serving aboard *Preble* in Vietnam waters.

After returning to St. Louis and the USNR, CAPT Seiler performed missile, spacecraft, and aircraft engineering work with McDonnell Douglas, where he became chief flight simulation engineer. As a USNR engineering duty officer, he served in increasingly responsible billets, including command of three units before retirement in 1987.

WENDELL G. "WENDY" SELLERS, CDR, born Jan. 14, 1924. Enlisted in November 1942, entered active duty as a V-5 cadet at RPI, Troy, NY.

Designated naval aviator - multi-engine NAS Pensacola July 28, 1944. Completed operational in PVs in February 1945. Started as an instrument instructor Sqdn. VB2-2 NAS Lake City, FL. Entered Naval Hospital NAS Jacksonville in November 1945 for shoulder operations until released May 1946.

In July 1946 was hired by National Airlines as a co-pilot on the Lockheed Lodestar. Based in New York and Miami. In 1980 National merged with Pan American World Airways. He flew DC-3, DC-4, DC-6, DC-7, DC-8 and DC-10 aircraft. He retired in January 1983.

In July 1947 joined Reserve Sqdn. VP-ML-60, NAS Miami. Recalled to active duty in August 1948 at VR-22 NAS Norfolk. Released in August 1952, joined VR-801 NARTU Miami until base closed. Then joined VR-741 NARTU, Jacksonville. Finished as CO in May 1964. He enjoyed every facet of his Navy career!

He is a life member of NRA and TROA. Resides in Palm City, FL with his wife of 50 years, Cathy. They have two children, Sherry Marvin and Wendy Thomas; and four grandchildren: Wade, Will, Chris and Heather. He plays golf six days a week and is active in community affairs.

JAMES D. "JIM" SENCINDIVER, CDR, born June 23, 1955, Arlington, VA, earned his BS degree (physics) at the University of Virginia in 1977.

He joined the service May 20, 1977, NROTC, University of Virginia. Military locations and stations: Nuckar Power Pipeline, Orlando, FL; Idaho Falls, ID; and New London, CT. He served aboard the USS *James Monroe* (SSBN-622) (Gold), October 1978-July 1981; NR COMSUBLANT Staff, July 1981-December 1983; NR COMSUBLANT Det. 106, January 1984-September 1991; NR BUPERS 4 Comp. 606, October 1991-September 1994; NR DCNO OPS N3/N5, October 1994-present; post JO show tour, COMSUBLANT staff.

Sencindiver was awarded the Navy Achievement Medal, Meritorious Unit Commendation Ribbon (2), Expert Pistol Medal, Armed Forces Reserve Medal, National Defense Service Medal (2); "Battle E," and Sea Service Deployment. He achieved the rank of CDR and retired in August 1992.

His memorable experiences: Watching the movie "Das Boot" while submerged on SSBN-728 (Florida) initial sea trials (as COMSUBLANT material representative for the trials) and having his wife pin his Dolphins on after qualifying in submarines.

He married the former Sheila Hedrick. Civilian employment as ASW Targets project engineer in Unnamed Undersea Vehicles Program Management Office (PMS-403).

ARTHUR D. "SES" SESCLEIFER, CDR, born Dec. 13, 1922, Carlinville, IL, attended two years college at Notre Dame, IN and one year business college.

He enlisted in the USN Nov. 12, 1942, as Y3C for recruiting duty in Springfield, IL. Entered the Navy V-12 program at Notre Dame, IN for two years and went to OCS at Notre Dame.

Upon completion of OCS was commissioned an ENS and served on board the USS *Boyle* (DD-600) as communications officer. Ship was decommissioned and he joined the NRSD 9-30 in Springfield serving in several capacities.

He was recalled to active duty during the Korean Conflict in April 1952 and assigned to the USS *Knapp* (DD-653) in Charleston, SC as operations officer for two years. Following completion of this tour of duty returned to NRSD 9-30 serving in several more capacities including CO. Retired from the reserve formally Dec. 13, 1982. He received several medals for the services done.

Married to Marthann and has three children: JoAnn, Janet and Daniel; and five grandchildren: Amy, Daniel, Jane, Anne and David.

Was employed by the Chicago and Illinois Midland Ry Co., the United Way of Sangamon County, IL and the International Service Agencies from which he retired March 1, 1994.

PAUL J. SEVERSON, CAPT, born April 30, 1936, Northfield, MN, attended St. Olaf College, Northfield, MN and Navy Postgraduate School.

He reported for duty aboard the USS *Forester* (DER-334) Dec. 24, 1959. His military locations and stations: USS *Dahlgren* (DLG-12); USS *Forester* (DER-335); Navy support activity; Villefrance, France; and Navy Postgraduate School, Monterey, CA. He served with reserve units, Readiness Command, naval control of shipping, surface divisions and naval engineering division.

Severson was awarded the National Service Medal, Naval Reserve Medal and Armed Forces Expeditionary Medal. He achieved the rank of CAPT and was discharged in 1967.

His memorable experiences: Officer of the Deck for last nuclear tests Target Sled monitoring ship, Christmas Island, 1962 USS *Forester* (DER-334); assistant officer in charge, Naval Support Activity, Villefrance, France; special project officer for chief of naval training in naval studies for sea grant colleges.

He and his wife, Minny Katherine "Bitsy", have two children, Jonathan and Susan; and one grandson, Austin Ryan. He retired April 30, 1996.

VINCENT J. SHANAHAN, CAPT, born, Aug. 1, 1929, earned his BS degree at Massachusetts University.

He joined the USN in August 1952. He finished boot camp in Bainbridge, MD and then attended Officers Candidate School, Class 9, Newport, RI. He was commissioned ENS, USN in March 1953 to active duty.

Graduated salvage diver Bayonne, NJ. He reported aboard the USS *Opportune* ARS qualified for diving officer and command.

He left active duty and associated with the USNR in 1956. He held commands of Surface Unit, Naval Reserve Officers School, Harbor Clearance Units and finally, Group Command of Diving and Salvage 1006 Washington, D.C.

He raised the USS *Indiana*, sunk in 1847, from Lake Superior. His greatest achievement was receiving the Meritorious Service Medal for work in organizing diving and salvage. He retired from the USN in August 1989. He holds lifetime membership in both the Naval Reserve Assoc. and the Naval War College Foundation.

He and his wife, Irene, have three children: Vincent, Timothy and Chris; and four grandchildren: Michael, Meghan, Matthew and Sarah. Civilian employment as president, CEO, Kleen-Air Service Corp.

JOHN J. SHARER, LCDR, born Dec. 17, 1923, Washington, D.C., earned his BS degree (physics) at Maryland University.

He joined the USN in September 1942, as a V-5 cadet. Graduated as a naval aviator at NAS Corpus Christi, TX in April 1944. Trained as a fighter pilot with special training in ASW. Joined ASW Sqdn. VC-13 at Norfolk, VA in December 1944. The squadron was assigned to the jeep carrier CVE-57 Anzio in the Pacific and fought in the invasion of Okinawa.

Sharer was awarded the Air Medal, Unit Commendation Medal and Asiatic-Pacific Area Medal w/2 stars.

He served as flight instructor in Pensacola, FL at NAS Barin Field 1945-46. Recalled to active duty November 1952-June 1955. Pilot with VF-12 and later VC-4. Recalled a second time in October 1962-August 1963. Pilot with reserve squadron VS-751 at NAS Lakehurst, NJ. Remained with NARTU squadrons at Andrews AFB until retirement from the reserves in 1969.

Married the former Mary Haugh and has five children. Worked as an electronic engineer with Link Simulation Co. in Washington, D.C. area. Retired in February 1989.

PETER D. SHAY, CDR, born Aug. 30, 1942, New York City, NY, graduated from James Madison High School, Brooklyn, NY; attended the University of Florida and graduated from Adelphi University.

About to be drafted in 1965, he elected to enlist in the AOC Preflight training program, which he completed as an ENS in February 1967. He volunteered to fly with the newly formed HA(L)3, Seawolves in Vietnam, earning 26 Air Medals and Navy Commendation Medal w/Combat V and other unit medals and ribbons during his tour there.

On Feb. 1, 1968, he piloted one of the first Huey gunships to fly over Saigon, RVN during the commencement of the TET offensive. On the following day the flight team he was flying with, while under heavy enemy fire, provided escort for US Army aircraft while they rescued US Ambassador Ellsworth Bunker from the roof of the American embassy. During the balance of 1968 and during 1969 he flew H-46s on Vertrep missions in Norfolk and in the Mediterranean.

Upon completion of active duty CDR Shay joined NAIRU in Brooklyn, NY transferred to OPCON in Glenview, IL and then HS-75 in Lakehurst and Willow Grove, PA. While at HS-75 he was commended for founding Operation Santa Lift, a mission airlifting gifts to children in hospitals. He completed his service in VTU in Huntington, Long Island.

He is a member of the Seawolf Assoc., Naval Reserve Assoc. and the Naval Aviation Commandery. He is a founder and co-chairman of the Seawolf Memorial Fund involved with the raising of funds to restore a UH-1 on the Intrepid Museum.

He is an entrepreneur doing business in the graphics business in New York City. He collects art, is an avid runner and bodybuilder and enjoys reading about naval history and current military affairs.

JON "CHE" SHELLER, CAPT, born Aug. 29, 1950, a native of Santa Barbara, CA, graduated from the USNA in 1972, where he was awarded a BS degree in mechanical engineering and was commissioned an ENS, USN. Following graduation, he completed the Navy Nuclear Power Officer Training Program at Bainbridge, MD and Windsor, CT.

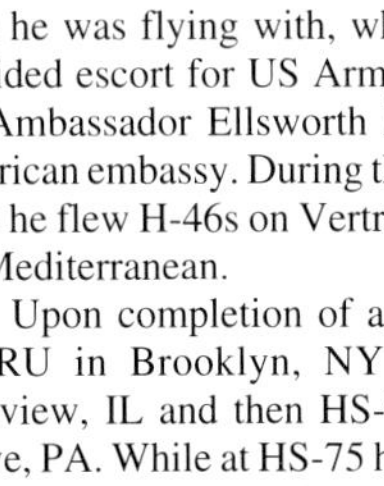

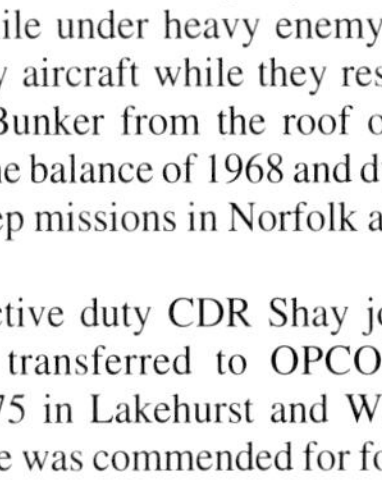

He subsequently served at sea aboard the attack submarine USS *Flying Fish* (SSN-673), earning the coveted "Gold Dolphins" for qualification in submarine warfare, and completing tours of duty as communications officer and electronic material officer while the *Flying Fish* conducted Cold War naval operations. He later "cross decked" to the USS *Sand Lance* (SSN-660) where he served as assistant engineer and weapons officer.

In 1977 CAPT Sheller left active duty and accepted a commission in the USNR, serving as staff readiness officer (ASW) in Commander-in-Chief US Atlantic Flt. Det. 206 until 1986. These duties included adjunct to the staff of the Center for War Gaming, Naval War College. During 1986-89 he served aboard the submarine tender USS *Ly Spear* (AS-36), where he completed surface warfare qualification while serving first as repair officer and later as executive officer of Reserve Det. 106. From 1989-91 he served as CO of this detachment.

During 1991-93, CAPT Sheller served as the first CO of Commander, Submarine Force US Atlantic Flt. C3 Det. 1. During 1993-95 he served as CO of International Armaments Cooperation Det. 206 on the staff of the chief of naval operations in the Pentagon. Since October 1995, he has been assigned as executive assistant to the CDR, Naval Base Norfolk and serves as the Navy's emergency preparedness liaison to the civil governments of the state of Maryland and the District of Columbia.

His personal awards include the Meritorious Service Medal (2), Navy Commendation Medal (5), and the Navy Achievement Medal. He holds a MS degree in management from National-Louis University, a MPA degree from George Washington University, and is a graduate cum laude of the Naval War College.

In his civilian career, CAPT Sheller leads the Program Team responsible for life cycle logistics planning and maintenance support for the high technology counter drug law enforcement equipment of the US Customs Service. He is married to the former Tink Deborah Cooper of Jonesboro, GA. They reside in Upper Marlboro, MD.

MARKE R. SHELLEY, CAPT, born Oct. 8, 1947, Moscow, ID, raised in Seattle, graduated from the University of Idaho, the University of Illinois and the Naval War College obtaining a bachelor of architecture, master of architecture and MA in national security and strategic studies. Following graduation he reported aboard USS *Wasp* (CVS-18) in August 1971 where he served as repair officer. He later served as DCA aboard USS *Dyess* (DD-880) until his release from active duty in July 1973.

A qualified architect in civilian life, he fashioned his USNR career upon that of an active duty afloat officer serving as division officer, department head, executive officer, CO, division commander and group commander. He commanded Mine Countermeasure Gp. Middle East in the Persian Gulf in 1989-90 and the Military Liaison Team in Bucharest, Romania in 1994-95. While commanding both at sea and ashore commissioned units, he served in a total of 10 command positions.

HENRY THOMAS "TOM" SHELLY, CDR, born Oct. 9, 1914, Roger Springs, TN, graduated high school and attended two years of college.

He joined the service Jan. 15, 1940. His military locations and stations: NTC Norfolk, VA and NAF French Morocco. He served with APA-6, APA-9, APA-59, IX-225 and APA-136. Battles and campaigns participated in include Atlantic, Pacific, European and Philippines.

Shelly was awarded the Theater Medal, Occupation Medal, Naval Reserve Medal and WWII Victory Medal. He achieved the rank of CDR and was discharged June 30, 1954. His memorable experiences: Tour of duty in French Morocco.

He and his wife, Roma, have three children: Carole, Tom Jr. and Elizabeth; and seven grandchildren: Jeffery, Trina, Paul, Louis, Janna, Sarah and Chrissy. Civilian employment as manager of mail processing, USPS. Shelly retired Oct. 9, 1974.

PHILIP D. SHENEFELT, CAPT, born July 31, 1943, Colfax, WA, graduated from the University of Wisconsin with a BS in zoology in 1966, MD in 1970, and MS in administrative medicine in 1984. In 1967 he was commissioned in the ENS 1915 program as a medical student.

He interned at Naval Hospital, Bethesda in 1970-71 and graduated from Naval Submarine and Diving Medical Schools in 1971. From 1972-74 he served as diving medical officer for COMNAVINSWARLANT with UDT 21, Seal Team 2 and EOD Gp. 2. After completing specialty training in dermatology at the University of Wisconsin Hospitals in 1978, he performed a TEMAC for three months as a dermatologist at Naval Hospital, Okinawa in 1979.

CAPT Shenefelt served as CO of NR MED CRU 1013 in 1980-81, as officer in charge of NR 4th MARDIV SUPDET D113 in 1982, and as CO of NR NAVHOSP GLAKES 1013 in 1985-87 at N&MCRC Madison, WI. In 1987 he moved to Tampa, FL, affiliating with NRC Tampa. He was promoted to CAPT, MC in 1987. In 1993 he retired after 26 years of service in the USNR. He is a life member of the NRA.

In civilian life, Dr. Shenefelt was an assistant professor of dermatology at the University of Wisconsin in Madison from 1978-87. He is currently an associate professor of dermatology at the University of South Florida in Tampa. He is a recognized expert in contact dermatitis and the epidemiology of skin diseases, having published articles on those topics in major dermatology journals.

CAPT Shenefelt's family includes his wife, Debrah, and three daughters: Elizabeth, Sara and Shaina. He enjoys flying as a private pilot, collecting books by Lafcadio Hearn on Japan, and studying Native American shamans and spirituality.

CHARLES WILLIAM "BILL" "SHEP" SHEPHERD, CAPT, born July 12, 1929, Pontiac, IL, earned his BS degree at the University of Illinois and his MS degree at New York University.

Joined US Merchant Marines in 1951. Enlisted in the USN in 1951 and served as CO. Graduated Class A Communications Technician School (both R and O branches), Bainbridge Island, WA in 1952. In 1953 he served at Yokosuka, Japan to Kami Seya and advanced to PO3 (CTR). He graduated Officer Candidate School, ENS, CO. Assigned Washington, D.C. OPNAV 202, Naval Security Gp. HQ, assistant department head. Was promoted to LTJG, department head in 1954 and transferred to Bremerhaven, Germany.

In 1955 was sent on 12-month TAD to Sixth Fleet as assistant officer-in-charge, NavSecGru Det. and served aboard the USS *Salem*, USS *Newport News*, USS *Northampton*, and USS *New Jersey*. Deployed to Carrier Gp. aboard USS *Coral Sea*. Was one of eight reserve officers fleetwide selected for Regular Navy Augmentation in 1956. Declined Regular Navy Augmentation and released from active duty in Bremerhaven in 1957. In 1958 he joined Clifton, NJ NavSecGru Reserve Unit NSGD 3-7 as training officer and promoted to LT, USNR. Clifton unit won national NavSecGru competition in 1959.

Joined New Rochelle, NY NavSecGru Reserve Unit NSGD 3-11 as executive officer in 1962. In 1965 affiliated with Chicago NR Public Affairs Co. 9-2. Joined Chicago NavSecGru Unit 9-2 as training officer. Maintained dual participation in both units until the Public Affairs Co. was decommissioned in 1979. Joined Illinois Naval Militia and Militia Officers Mess. Promoted to CO, Chicago NavSecGru Div. 9-2 in 1969. In 1971 was CO, Forest Park NavSecGruDiv 9-17. Served as public affairs officer, Gp. Command Staff and promoted to CDR, USNR in 1974.

In 1975 served as executive officer, NavSecGru Command and Control Unit 913, Chicago. Was CO, NavSecGru DIRNSA Unit 913, Chicago and Forest Park in 1977 and promoted to CAPT, USNR in 1979. Served with Readiness Command 13 as project officer in 1980, NavSecGru program officer in 1981 and project officer in 1983. In 1984 was CO, NavSecGru DIRNSA 1013. As CO, NavSecGru VTU 1311 in 1987 he was recalled to active duty for 90 days as acting reserve forces advisor to director, National Security Agency. In 1989 served as NavSecGru liaison and project officer, Readiness Command 13 VTU 1310. Retired from USN on 60th birthday at Great Lakes as senior cryptologic officer in midwest.

Shepherd was awarded the American Spirit of Honor Medal, Korean War Medal, UN Medal, European Occupation Medal, National Defense Medal, Armed Forces Reserve Medal w/2 HGs, Naval Reserve Meritorious Service Medal, Naval Security Group Meritorious Service Award, Readiness Command Region 13 Meritorious Service Award and numerous Reserve Unit citations and commendations.

He married the former Lorraine Hirvonen and has four children: John, Jennifer, Joan and Barbara; and one grandchild, Abigail. Civilian employment as field PR director with New York Life from which he retired Nov. 1, 1985.

KENNETH B. "KB" SHERMAN, CDR, was designated a naval flight officer in 1969 and then reported to VP-1 and served two tours as a P-3 PPTC at NAF Cam Rahn Bay, RVN, in support of Operation Market Time. Among Sherman's awards were the Air Medal, Armed Forces Expeditionary Medal, Navy Meritorious Unit Commendation and Vietnam Service Medal (5).

Leaving active duty, Sherman affiliated with VP-92 as a patrol plane mission CDR and department head. During this time he was awarded the Coast Guard Meritorious Unit Commendation and Special Operations Service Medal for duty involving drug interdiction operations. From 1991-95 Sherman was attached to VTU-9191, serving as NAS South Weymouth command duty officer until his retirement in 1995.

He is married to the former Karen Bellarosa. Sherman holds BA and MBA degrees from the University of Massachusetts at Amherst and works as a registered master electronics technician and flight instructor.

HENRY M. SHINE JR., CAPT, born June 7, 1921, Lincoln, ME, earned his AB degree in 1947 at Harvard and his JD degree in 1951 at the College of Law-Notre Dame.

He joined the service Oct. 3, 1942. His military locations and stations: Recruit training and QM School, San Diego and NASD Norfolk. He served aboard the USS *Hermitage* (AB-54) and USS *Pegasus* (IX-222).

Shine was awarded the WWII Victory Medal, American Campaign Medal and Asiatic-Pacific Campaign Medal. He achieved the rank of CAPT and was discharged from active duty July 26, 1946. He retired from the USN June 1, 1981.

His memorable experiences: Service on National Naval Reserve Policy Board; CO, JAG Reserve Command, Treasure Island; student, Defense Strategy Seminar, National Defense University (1977) and moderator 1978.

He and his wife Marguerite "Peggy" have two children, Mari Kotwicki and Christiana J.; and two grandchildren, Christopher and Cameron Kotwicki. Civilian employment as EVP California Bankers Assoc. and several federal commissions. He is still serving as professor, Troy State University.

STEPHEN P. SHIRIKIAN, CAPT, is a native of Schenectady, NY, and graduated from the New York State Maritime Academy. He began his naval career in 1942 as a cadet midshipman and was commissioned an ENS in 1944.

During WWII he served in the Pacific Theater as troop and cargo officer and deck watch officer on board the USS *President Monroe* (AP-104). He participated in the invasions of Lingayen Gulf, Iwo Jima, Okinawa and the occupation of Sasebo, Japan.

At the conclusion of WWII he was assigned officer in charge of the Naval Reserve Training Center at the US Naval Supply Depot in Scotia, NY. His next tour was that of assistant 1st LT and cargo officer on board the fleet oiler the USS *Waccamaw* (AP-109). While assigned to the Waccamaw the ship was deployed for seven months with the Sixth Fleet and won the famous "E" for its class.

Recalled for the Korean Conflict he was assigned to the USS *Whitley* (AKA-91) as 1st LT, command duty officer, cargo officer and gunnery officer.

His reserve billets were: CO Surface Div. 3-93, CO Surface Div. 3-13, battalion CO Surface Units in Albany, NY, then served as group commander at the Albany Naval Reserve Training Center. His last tour prior to being transferred to standby reserve was as naval liaison officer to adjutant general to the governor of New York state.

He was promoted to the rank of CAPT in 1965. His campaign and service medals are: American and Pacific Theater, Philippine Liberation Medal, Navy Expeditionary Medal, WWII Victory Medal, Occupation Medal, National Service Medal, Naval Reserve Armed Service Medal and Korean Service Medal.

WILLIAM ALBERT "AL" SHOW, LCDR, born Nov. 4, 1916, Alton, IL, earned his BS degree in chemistry and his MS degree in chemical engineering.

He joined the service in July 1942. His military locations and stations: BuOrd liaison with Army Chem. Corps., Edgewood ARS, Dugway Proving Grounds. He served with Army Chem. Corps; NOTS, China Lake; NOL, White Oak.

Show was awarded the American Theater Medal, etc. He achieved the rank of LCDR and was discharged from active duty Feb. 13, 1946, and then retired from USNR July 1, 1968.

His memorable experiences: CBR; explosives; weapons; Ammo-R&D, MFG; weapons tests and evaluation; Dugway Proving Grounds and NOTS (terminal ballistics officer).

He married the former Elizabeth Ellen Todd, and they have five children: James T., William T., Judith A., Michael J. and Thomas A.; and seven grandchildren: Laurence, Tammi, Kimberly, Adam, Katherine, Arah and Rebecca. Civilian employment: R&D, engineering, manufacturing, civil service-USAF PESO, Eglin AFB.

JAMES M. "JIM" SHULER, CPL (USMCR), CDR (USNR), born Nov. 18, 1939, Walla Walla, WA, earned his BA degree in business (accounting) at the University of Washington in 1963.

He joined the USMCR in June 1957; USNR in May 1965; and was commissioned in September 1965. His military locations and stations: Active - MCRD San Diego and Camp Pendleton; NAS Pensacola; and USS *Observation Island* (EAG-154). He served with 93rd Rifle Co. and MOB Unit; NR and various units NAS Seattle then Whidbey Island, VR-51 Det. and VR-61.

Shuler was awarded the Marine Corps Reserve Medal, Noel Davis Unit Award, National Defense Service Medal and Armed Forces Reserve Medal (2). He achieved the rank of CPL and was discharged from the USMCR in December 1957; and the rank of CDR, USNR from which he retired in October 1968.

He was married to Patsy J. but is presently divorced. Civilian employment with Blue Shield and Cross in marketing and accounting/property management. Shuler retired in July 1989.

RICHARD A. "DICK" SHULTZ, CAPT, born Oct. 26, 1942, Los Angeles, CA, earned his AA degree at Glendale College, his BS degree at San Diego State University and his MBA degree at Pepperdine University.

He attended Reserve Officer Candidate (ROC) School at Newport, RI during the summers of 1962 and 1964 and upon graduation from San Diego State University with a degree in biological sciences, was ordered to NFO preflight training at Pensacola, FL.

Prior to commissioning he served as QM aboard two USNR submarine units and was awarded his Silver Dolphins June 8, 1964, aboard SS-377, USS *Menhaden*.

He served in Patrol Sqdn. 22 (VP-22) at Barbers Point, HI after Flight School. VP-22 deployed to Naha, Okinawa and Sangley Point, Philippines during his tour duty in 1967-69.

Upon release from active duty, CAPT Shultz affiliated with VP-91 at Moffett Field, CA and was employed by Lockheed California Co. in foreign military aircraft sales. He was an aircrew tactical coordinator (TACCO) for seven years before he qualified as a special operations officer (Designator 1145) in ordnance management. He served aboard Navy Undersea Warfare Engineering Station, Keyport, WA, Reserve Det. Reno, NV until his retirement in 1987.

He is the managing partner of Shultz, Ferguson & Associates, an employee benefits marketing firm in Sacramento, CA that specializes in developing and marketing plans to the members of endorsing California trade associations. His wife, Nancy, works with him at the numerous conventions they attend each year. Their children, Stephen and Sherilyn, are independent and live in Sacramento as well.

DAVID M. "DAVE" SIEFKEN, CAPT, born June 9, 1942, Webster City, IA, earned his BA degree at Luther College, Decorah, IA.

His military locations and stations: USS *Everglades*, USS *Garcia*, and Naval Avionics Facility. He served with NSD, Subic Bay and Cinc Pac, Hawaii.

Siefken was awarded the Navy Commendation Medal. He achieved the rank of CAPT (O-6) and was discharged May 1, 1994.

He and his wife, Janice, have two children, Bryant and Kevin. Civilian employment as Winneshiek County clerk of court.

MERLIN D. SIMONSON, CAPT, born in 1943 in Powhatan Township, Pocahontas County, IA. He graduated from St. Olaf College in 1965; Naval OCS, commissioned March 18, 1966, followed by Naval Underwater Swimmers School and Explosive Ordnance Disposal (EOD) School.

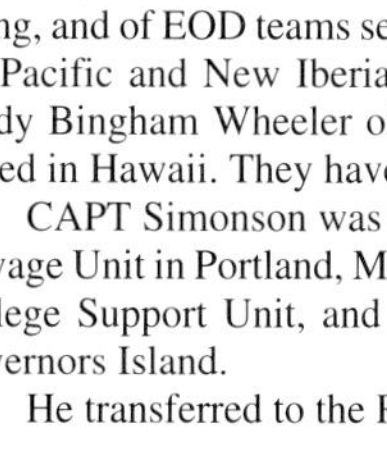

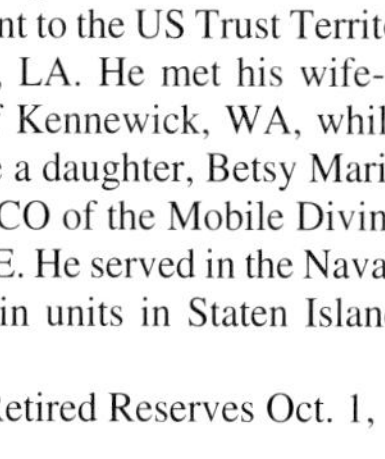

He completed a WestPac tour on the USS *Kearsarge* (CVS-33), was officer-in-charge of EOD Team #39 at Nha Trang, and of EOD teams sent to the US Trust Territory of the Pacific and New Iberia, LA. He met his wife-to-be, Trudy Bingham Wheeler of Kennewick, WA, while stationed in Hawaii. They have a daughter, Betsy Marie.

CAPT Simonson was CO of the Mobile Diving and Salvage Unit in Portland, ME. He served in the Naval War College Support Unit, and in units in Staten Island and Governors Island.

He transferred to the Retired Reserves Oct. 1, 1996.

MICHAEL V. "MIKE" SIMPSON, LCDR, born May 20, 1960, Colorado Springs, CO, earned his BA degree in political science at Southwest Texas State University.

He joined the service Jan. 9, 1985. His military locations and stations: VT-27/VT-28 NAS Corpus Christi; NAS Barbers Point; NAS Dallas, TX. He served with VT-27/VT-28, VQ-3, VT-27 and VR-59.

Simpson was awarded the National Defense Service Medal, Battle "E," Meritorious Unit Commendation Ribbon, Navy-Marine Corps Overseas Service Ribbon, Navy Reserve Service Medal and Expert Pistol Ribbon. He achieved the rank of LCDR and was discharged March 30, 1991.

He and his wife, Carmen, have three children: Stephen, Alexandra and Katarina. Civilian employment as pilot, Delta Air Lines MD 88/90 first officer.

FRANCIS J. SINCOX JR., CAPT, born May 11, 1932, Saginaw, MI, graduated Emory University, Atlanta, GA with a BA degree in 1954 and MD degree in 1958. Entered USNR PEBD June 11, 1954, became SELRES in 1955 at old NAS Atlanta (now PDK airport). First flight in a Navy aircraft was there, in 1955, the station aircraft being an SNJ! Also there were SNB, P2V and AD-5. Duty there was with AWS-67.

Active duty USNR 1957-58 at Emory University. Augmented to USN 1958-63. Duty at USNH St. Albans, NY, sea duty with CVSG-58 aboard USS *Randolph* (Essex Class) CVS-15 with HS-10, VS-26, VS-36 and VAW-12 aboard. Experiences there included recovery of Mercury Astronauts, Cuban Missile Crisis, fire at sea, and collision at sea. Shore duty NAS Norfolk until 1963. Designated naval flight surgeon in 1958.

Transferred to USNR inactive 1963-89, became SELRES again in 1989 at NMCTC Charlotte, NC, missed the air side, and transferred in June 1990 to NAS ATL HMA-773. Mobilized for Persian Gulf War November 1990-May 1991. Experiences there included mine strike aboard USS *Tripoli* LPH-10. Transferred to retired in May 1996.

Decorations include Combat Action, Navy Unit Commendation Ribbon, Meritorious Unit Commendation Ribbon, FMF, Naval Expeditionary Medal, Armed Forces Service Medal, Armed Forces Expeditionary Medal, Naval Reserve Medal, Southwest Asia, Sea Service, Kuwait Liberation (Saudi), Kuwait Liberation (Kuwait), Expert Rifle and Expert Pistol.

Continues to be a member of NRA and McROA. Resides in Gastonia, NC with wife Joyce. Still practices medicine, part-time.

ROBERT L. SINNOKRAK, CDR, born Aug. 17, 1946, Chicago, IL, graduated from George Washington High School in June 1963.

Enlisted in the USNR Oct. 26, 1965, Chicago Naval Armory. Completed QM "A" School at Fleet Training Center, San Diego in April 1967; advanced to QM3 in August 1967; and reported to Naval Station, Treasure Island for active duty in September 1967. After completing SERE Training at NAB Coronado in November 1967 reported for duty to Commander Naval Forces Vietnam, participating in seven campaign periods while attached variously to RIVSEC 543, NAVSUPPACT SAIGON, RIVDIV 595 and COMNAVFOR V.

Released from active duty in September 1969 and reaffiliated with the USNR. Promoted to QMC in February 1974 and assigned to USS *Crockett*, PG-88 with Coastal River Div. 21 in October 1974. Selected for Warrant Officer program in April 1975 and assigned as Reserve OinC of PTF 19 with COSRIVDIV 21. Selected for Limited Duty Officer program as LTJG in April 1981. Promoted to current rank (CDR) in July 1994.

Has held command of numerous reserve units, including MOMAG DET 1813 (three tours), MOTU 2 DET 113, WPNSTA CONCORD DET 813, ASSAULT CRAFT UNIT ONE DET 1813 and READINESS UNIT 1647.

While CO of Assault Craft Unit One Det. 1813, CDR Sinnokrak acted as both "CATF" and "CLF" while his unit and other afloat assets conducted a re-enactment of the "D" Day landings in front of over 100,000 spectators at Chicago's Montrose Beach, in commemoration of the 50th Anniversary of the original landing.

Has been awarded the Navy Commendation Medal w/Combat "V" and Gold Star, Navy Achievement Medal, Purple Heart Medal, Naval Reserve Meritorious Service Medal, National Defense Service Medal, Navy Pistol Marksman w/Silver "E," Navy Rifle Marksman, Vietnam Service Medal, Vietnam Campaign Medal, Armed Forces Reserve Medal, Combat Action Ribbon, Meritorious Unit Citation, Naval Unit Citation and Overseas Deployment Ribbon. Life member Naval Reserve Assoc.; past president NRA Great Lakes Chapter.

A self-employed Marine surveyor covering the Western Great Lakes and Illinois waterways.

The Sinnokrak family consists of wife Lynda and sons: Nicholas, Matthew and Christopher. They reside in Homewood, IL.

GEORGE P. SINNOTT, CDR, born July 5, 1934, obtained his BS degree from Villanova University. Advanced degrees, MPA and MA, were obtained from the City University of New York. He also attended the Naval War College and the Industrial College of the Armed Forces.

CDR Sinnott entered the USN OCS Program at Newport, RI, in August 1956 was commissioned and served on the USS *Pluck*. In the Ready Reserves, he served on the *Prescott, Lorikeet, Pierce, Huntington, Altair* and *LaSalle*. Other assignments included large surface divisions, staffs (including CINLANTFLT), Naval Recruiting Region V and VTUs.

In March 1970 he was involuntarily recalled to active duty. He retired in July 1983 from the USN and that same year from the New York City Police Dept. Thereafter, he retired from the New York State Naval Militia as a CAPT.

Presently employed with the Defense Logistics Agency, he provided support to Desert Shield and Desert Storm.

Over the years, many awards have been presented by the Navy, New York State, New York City Police Dept. and the Defense Logistics Agency.

The Sinnott's family consists of his wife, Ilean Mary (Beattie), and sons, George Patrick and James Andrew.

RAYMOND A. "RAY" SIUTA, CAPT., a native of New Kensington, PA, graduated from Indiana University of Pennsylvania with a bachelor's degree in physics in May 1968. He began his USN career at the Navy Officer Candidate School, Newport, RI and received his USNR commission in October 1968. Upon commissioning, he served over three and one-half years aboard the USS *Observation Island* (AG-154). He served as damage control assistant, weapons instrumentation officer and deck department head while aboard the *Observation Island*. CAPT Siuta also completed assignments as instructor at the Navy Officer Candidate School and as the BOOST Program evaluator at the Naval Education and Training Center, Pensacola, FL.

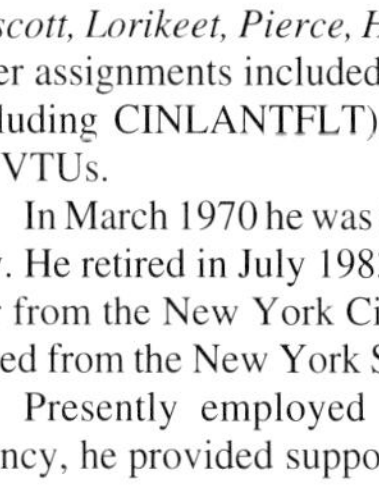
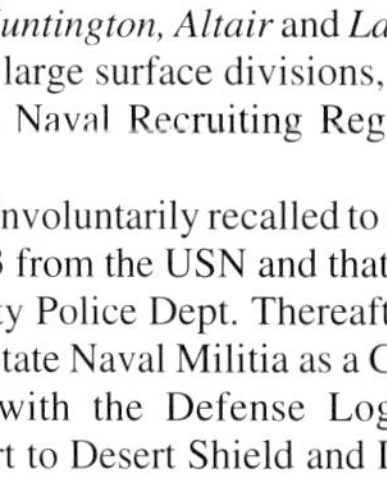

Following his release from active duty, he received a master's degree in Marine Affairs from the University of Rhode Island in 1973, and later, a MS in industrial administration from Carnegie-Mellon University in 1980.

CAPT Siuta affiliated with several reserve programs during his 25 year active duty and reserve career. He had over eight years with the Mobile Inshore Undersea Warfare (MIUW) Program and eight years with the Convoy Commodore Program. He served as CO, Personnel Mobilization Team 2922 in Eugene, OR and as CO, Volunteer Training Unit 2206 in Portland, OR. CAPT Siuta completed his USNR career as CO, Mobile Inshore Undersea Warfare Unit 110 in Portland, OR. He retired from the USNR in September 1993.

CAPT Siuta is married to the former Carolyn Fraze of St. Petersburg, FL. They have two sons, Gregory and Scott. He is the factory marketing manager for the Inkjet Supplies Business Unit of the Hewlett-Packard Co. CAPT Siuta and his family reside in Corvallis, OR.

WILLIAM JAMES "WJ" SKELLY, MD, CDR, born Jan. 29, 1959, Cleveland, OH, earned his BA degree at Williams College and his MD degree at Case Western Reserve.

He joined the service June 16, 1985. His military locations and stations: Buffalo, NY with Med Unit PO 513 A and Cleveland, OH with Med Unit PO 518 A.

He and his wife, Judy, have four children: William, Michael, George and Thomas. Civilian employment as an orthopaedic surgeon.

MARTA ISABEL COLEMAN SKELTON, LT/Q3, born Sept. 11, 1949, LaCeiba, Honduras, Central America, earned her diploma in nursing, BSN in nursing and MSN in nursing - trauma track.

She joined the service March 13, 1990. Her military locations and stations: Bessemer, AL for drills and Pensacola for OTS. She served with HMC11; Primus, IRR; and as reservist.

Skelton's awards and medals include Peace in the Gulf. She achieved the rank of LT/Q3 and is still serving. Her memorable experiences: The day she was sworn into the USNR.

She and her husband, Larry J. Skelton, MD, have one child, Isabel Cristino. Civilian employment as staff certified registered nurse anesthetist.

JOHN N. SKRIEF, CAPT, born July 22, 1946, Mandan, ND, earned his BA degree at the University of Minnesota and his MBA degree at the University of Portland.

He joined the service April 22, 1969. His military locations and stations: Officer Candidate School and basic flight training, Pensacola; NAS Corpus Christi where he received his navigator wings, was recognized as a Distinguished Naval Graduate and received a regular commission; VP training, Norfolk and NAS Pautuxent River; VP-10, NAS Brunswick with deployments to Sicily, Rota Spain, Lajes, Bermuda and Puerto Rico.

In November 1973 he left active duty and began a career in personnel in Vancouver, WA and joined VP-69 in Whidbey Island, WA from June 1974-July 1976. Relocated to the midwest and affiliated with VP-60, Glenview, IL from July 1976-October 1979. Transferred to VP-66, Willow Grove, PA from November 1979-June 1984. Assumed new duties with COMUSFORCESCARIB, Scranton, PA and Milwaukee, WI from July 1984-March 1987 with active duty in Key West. From April 1987-February 1989, served with CINCLANTFLT 106, Norfolk, VA.

While a member of OPNAV 09B in the Pentagon from March 1989 to being recalled to active duty with BUPERS in December 1990 reviewed exemptions and deferments from recall for Desert Shield and Storm for the SECNAV. His active duty was extended at the request of BUPERS from April 1991-August 1992, National Naval Medical Center, Bethesda, MD. Returned to active reserves until July 1993 when he retired. He still participates as a Blue and Gold officer for the USNA. Has received three USNA Commandants Award as outstanding Maryland Blue and Gold Officer.

He and his wife, Julie, have three sons, Bill, Matthew and Michael. Presently residing in Columbia, MD and enjoys golf and woodworking. Skrief is currently working for the Veterans Affairs in Washington, D.C. rating disabilities.

GREGORY J. SLAVONIC, CAPT, born May 6, 1949, Great Bend, KS, attended Oklahoma State University earning a BS degree in 1971, then enlisted in the USNR where he completed basic training and Signalman "A" School. Received orders to USS *Constellation* (CVA-64) and made two WestPac deployments to Vietnam.

In 1973 he was released from active duty and returned to graduate school where he earned an EDM from the University of Central Oklahoma. In 1974 he received a direct commission (public affairs) in the USNR.

From 1974-89 he served with Office of Information (OI) Det. 411 and held billets as administrative, training, operations and executive officer. In 1989, assigned staff public affairs officer for REDCOM ELEVEN. From 1992-98, CO of OI-411 and Navy Office of Information Southwest 111. 1994 unit received RADM Robert Ravitz Award for Public Affairs Excellence and 1996 unit was a finalist for REDCOM TEN's ADM John T. Natter Award (small unit).

In November 1990 he was recalled to active duty for Operation Desert Storm and Desert Shield. Assigned to GEN H. Norman Schwarzkopf's staff and dispatched to the Dhahran, Saudi Arabia and served as chief of Navy News Desk and as senior Combat Media Escort officer. He escorted several combat media pools during the Gulf War including embarks aboard USS *Curts* (FFG-38) where media documented the capture, processing and interrogation of over 35 Iraqi enemy prisoners of war and later he escorted media crew to USS *Tripoli* (LPH-10) when early in the morning of Feb. 18, 1991, *Tripoli* struck a submerged Iraqi mine in the Northern Persian Gulf.

His civilian jobs have included, general manager, advertising director and national account executive for several newspapers in the southwest and account executive for KFOR-TV (NBC).

CAPT Slavonic's civic positions have included president of Navy League-Oklahoma City Council, NRA (Central Oklahoma) and Reserve Officers Assoc. (VADM Marc Mitscher chapter), plus board of directors, Employer Support of Guard and Reserve. In 1984, selected as the Naval Reserve Assoc.'s first national Junior Officer of the Year recipient.

His awards include Bronze Star Medal, Meritorious Service Medal, Navy Commendation Medal (2), Navy Achievement Medal (3), Combat Action Ribbon, Presidential Unit Commendation, and other campaign service awards.

The Slavonic family consists of wife, Molly, daughters, Kara and Maggie, son Blake.

LEROY NEIL SLAYTON, LCDR, born Dec. 5, 1921, Self, TX. Enlisted in May 1942 at Dallas, TX for three months primary in N3Ns. From there to NAS Corpus graduating in seaplanes (OS2Us) thence sub patrol and combat tactics (gunnery, dive bombing, etc.). Then to SBDs, but was taken out due to a request for multi-engine pilots. NAS Atlanta, then to VRJ-1 on request by SECNAV and lower admirals. SECNAV said "door was open to regular" which did not accept requested forward area after seeing nuclear test, Cuba, etc. Covered several towns in China (moved Embassy from Chung King to Nan King). Seventeen house boys in Pieping covered all of the Pacific area. Checked out as PC in C-54s.

Flew all types in civilian life (P-51s, P-47s, C-54s, etc.). Recalled with unit to Korea, served the required two years then back to civilian life. In addition to BA earned an AE while on active duty and became flight test engineer on MacDonnell F-101. Applied at General Dynamics where he checked out in B-36, B-50 and several others, obtaining well over 11,000 hours (about 50% military).

Meanwhile obtained a Ranch Management degree and is still operating his own ranch.

Wife, Elizabeth Jean, has BS and BE degrees; three daughters and one son, all with one or more degrees; four grandchildren; and four great-grandchildren.

Has been through the inactive hobbies, but square dancing is active and always has new calls. They have danced in all 50 states and several foreign countries at various levels. All seem to be nice people - great fun!

ARTHUR C. "ART" SLOGGETT, LCDR, born Jan. 23, 1915, Hamakuapoko, Maui, HI, earned his AB degree at Dartmouth College.

Joined the service March 4, 1942 (date of rank). His military locations and stations: NAS Kaneohe, FWC Pearl and USS *Tolovana* (AO-64). He also served with the Pacific Fleet. Participated in the Philippine and Okinawa campaigns. He achieved the rank of LCDR and was released to inactive service Dec. 26, 1945.

He and his wife, Lydia E., have two children, Arthur W. and Anne S. Oxenberg; and three grandchildren: Kevin, Kristina and Ashley. Civilian employment with Hawaiian Electric Co. as assistant corporate secretary and manager Land Dept. Sloggett retired Aug. 1, 1975.

SIDNEY H. "SID" SMALL, CAPT, born Dec. 8, 1920, Wilkes-Barre, PA. Joined the service Dec. 8, 1941, and was sworn in on Dec. 17, 1941. His military locations and stations: Newport, RI; Bora Bora; USS *Saratoga*; and USS *D.H. Fox*.

He served with Bobcats, V-12, Colgate, and aboard USS *Saratoga* and USS *D.H. Fox*. Bobcats became part of first construction battalion which built oil tanks and air strip on Bora Bora. At the war's end participated in Magic Carpet operation which returned servicemen from the Pacific to San Francisco.

Small earned the Good Conduct Medal, Naval Reserve Medal, Armed Forces Reserve Medal, Atlantic Campaign Medal, Pacific Campaign Medal, Korean Service Medal, UN Service Medal, Korean Presidential Unit Citation, National Defense Service Medal and Victory Medal.

His memorable experiences: World cruise on USS *Fox*, Seventh Fleet, Korea. He achieved the rank of CAPT and was discharged Feb. 6, 1946. After being recalled in December 1950 Small was discharged again Dec. 31, 1952, and retired July 1, 1975.

He and his wife, Molly, have three children: Jay L., Charles H. and Faye P.; and two grandchildren, Adam A. and Peter C. Civilian employment as sales engineer.

EDWIN S. "DOC" SMISEK, CAPT, born Oct. 27, 1916, Lonsdale, MN, attended the University of Minnesota School of Dentistry.

He joined the service Feb. 28, 1942. His military locations and stations: California. He served with the USN and Seabees. Battles and campaigns: Central Pacific.

Smisek achieved the rank of CAPT and was discharged from active service Dec. 24, 1945, and from the reserves in July 1975.

His memorable experiences: Commissioned USS *Cascade*, destroyer tender in San Francisco in 1942. Spent two months in Pearl Harbor, 1942. Central Pacific 18 months. Discharged with Seabees in Shoemaker, CA in 1945.

He and his wife, Roslyn, have three children: Annette, Wayne and Joy. Civilian employment as doctor of dental surgery.

ALLEN H. SMITH III, LT, born Dec. 17, 1948, Philadelphia, PA, earned his BS degree, aerospace engineer, at Penn State and his MBA, law, at Drexel University.

He joined the service Oct. 10, 1966. His military locations and stations: NAS Pensacola, NAS Oceana, USS *Constellation* and *America*. He served with VT-1, VT-7, VT-9, VA-47, VA-165 and VA-35. Participated in the Vietnam War with TF-77.

Smith was awarded the Silver Star, Distinguished Flying Cross w/Bronze Star and Purple Heart (2). He achieved the rank of LT (03) and was discharged Feb. 21, 1973.

His memorable experiences: A-6A pilot, two consecutive tours in Vietnam with VA-165 and VA-35. Shot down over North Vietnam July 17, 1977.

He is divorced and has one child, Chris. Civilian employment with Wyeth-Ayerst International as vice-president of engineering.

BAKER A. SMITH, CDR, a native of Georgia, entered the USNA in 1965 and was commissioned in 1969. Anti-submarine warfare officer and nuclear warhead officer, USS *Myles C. Fox* (DD-829), he was the first officer selected for a Coast Guard-Navy exchange program, weapons officer, USCGC *Hamilton* (WHEC-715) and was certified engineer officer of the watch on the gas turbine-powered high endurance cutter. Qualifications included surface warfare officer, Helo landing signal officer and diving officer. As aide to the commandant, First Naval District, he was awarded the Navy Commendation Medal.

Following active duty assignments, CDR Smith completed several sea tours with Destroyer Sqdn. Twenty-Eight, USS *Damato* (DD-871) and USS *Steinaker* (DD-863). After executive officer, USS *Fortify* (MSO-446), he reported to the Pentagon, office of the Chief of Naval Operations, Surface Warfare (OP-03) as assistant for Mine Warfare where he supported the Avenger class minesweeper project. Next, as assistant for Strategy and Policy Planning (J5) to the commander in chief, Atlantic, he developed Caribbean contingency plans. He completed the Naval War College RO Strategy and Policy course.

Subsequently, CDR Smith became one of few reserve officers to wear both the Command-at-Sea Star and the Command-Ashore Trident. First, he assumed command as CDR, Mine Div. 123. He commanded at sea and operational forces as mine counter-measures CDR and was awarded the Meritorious Service Medal for development of Persian Gulf mine warfare tactics as well as a Gold Star in lieu of second Navy Commendation Medal. Finally, CDR Smith assumed command of the commissioned unit Mobile IUW Unit 206. He retired in 1989.

CDR Smith's civilian profession is corporate turnaround consulting and crisis management, serving as principal in the Atlanta office of Morris-Anderson & Assoc., Ltd. He holds the BS from the USNA, the MBA from Northeastern University, the JD from Suffolk University and the master in laws (Labor Law) from Georgetown University. He and his wife, Debby, have four children: Ellis, Elizabeth, Everett and Emery, and reside in Marietta, GA.

HAROLD P. "HAL" SMITH, LCDR, born March 14, 1929, New Haven, CT, earned his BCHE degree at Yale University in 1950.

He joined the service March 19, 1952. His military locations and stations: NAV Station, Newport, USS *Lewis*

Hancock (DD-675) and USS *Hyman* (DD-732). He served in the Korean War from 1952-55.

Smith was awarded the Korean Service Medal, UN Service Medal, Armed Services Medal, Naval Reserve Medal and European Occupation Medal. He achieved the rank of LCDR and was discharged May 25, 1955.

His memorable experiences: Recovery of downed British pilot in Korean waters. Shore bombardment of North Korean targets. As chief engineer of USS *Hyman* engineering department won DesLant full power competition in 1954.

He and his wife, the former Marceline Frey, have four children: Kevin, Marie, Brenda and Deborah; and eight grandchildren: James, Christine, Jennifer, Rebecca, Jessica, Michael, Jeremiah and Rachael. Civilian employment in various positions in rubber and chemical industries. Present position is senior tariff manager with ITM, Inc. Smith is a life member of the Naval Reserve Assoc. and US Naval Institute.

HORTON SMITH, RADM, born June 1, 1925, Seattle, WA. Enlisted in 1942 while still in high school and was a sailor in the V-12 program and NROTC at the University of Washington. He was in the Drum and Bugle Corps and was deputy CDR of the 3,100 students at his commissioning in 1945. As an ENS he was in the Pacific on the PC-1179 and then was CO of the LSM-288 and returned to graduate in 1947 (BA) from the University of Washington where he graduated from law school in 1950 (JD), admitted to the bar.

He was recalled for the Korean conflict and served on *Corsair* and *Grouper*, two submarines. In the summer of 1950 prior to his recall became the first West Coast officer to qualify in submarines earning his "Gold Dolphins" through the reserve program. He was an executive officer and qualified for command of submarines before his release from active duty in 1955.

RADM Smith was active in the submarine reserve program and in addition was CO of two minesweepers, *Waxbill* and *Ruff* and two reserve destroyers *Watts* and *Marshall*. He was CO of a MSC Div., a large group command of 1,500 and was the readiness CDR for Region 22 of 125 units and 6,000 reserves in the states of Washington, Oregon, Idaho, Montana and Alaska before his 1985 retirement.

RADM Smith was a judge of the Superior Court of the state of Washington for 20 years serving at the Juvenile Court, and in civil and criminal cases. Today he practices law in Seattle and has a master, Ocean license issued by the USCG and has been an adjunct faculty member at California and Texas Maritime Academies and the University of Washington Law School.

He has a wife, Lei-Llah, whom he married at the Submarine Base Chapel at Pearl Harbor and three adult children and resides in Seattle. On Labor Day 1996 he swam the Columbia River, a distance of two miles, for the sixth consecutive year for the Hoods River annual swim from Washington to Oregon. He is also a sailor and active in Rotary and a Silver Beaver recipient from the Boy Scouts in addition to being active in other civic groups.

JACKSON S. SMITH JR., CAPT, born May 23, 1943, Santa Cruz, CA, earned his BS degree in business at San Jose State College, San Jose, CA.

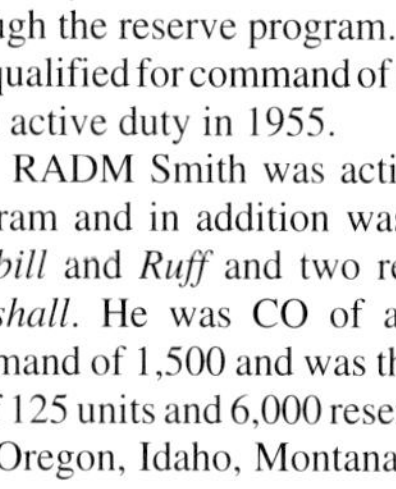

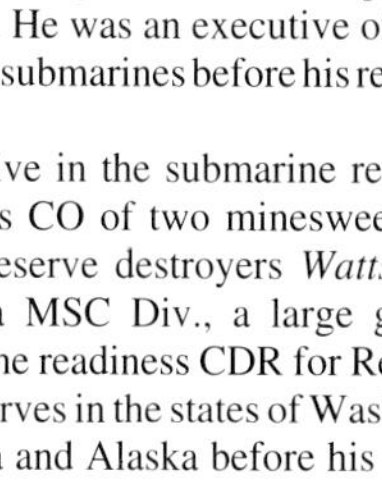

He joined the service Nov. 8, 1966. His military locations and stations: OCS, Newport, RI; DCA School, Philadelphia, PA; USS *Willard Keith* (DD-775), Norfolk, VA; ASW School, Key West, FL. Reserve duty: Fargo Building, Boston, MA; NRC Chicago, IL; NRC Alameda, CA; NRC Treasure Island, CA; N&MCRC Long Beach, CA; NRC Jacksonville, FL; and NRC, Nashville, TN. He served with the following units: Active - USS *Willard Keith*; reserve - USS *Compton*, Blue Crew; USS *Damato*, Blue Crew; USS *Holder*, Gold Crew; USS *Asheville*, Gold Crew; REDCOM 19 staff; Weapons Station, Concord, CA; Weapons Station, Seal Beach, CA; SIMA Long Beach, CA; and various VTUs.

Smith was awarded the Meritorious Service Medal, Navy Commendation Medal, National Defense Service Medal, Armed Forces Reserve Medal, Expert Rifle Medal, Expert Pistol Shot Medal and Naval Reserve Sea Service Ribbon. He achieved the rank of CAPT and was discharged May 12, 1970.

His memorable experiences: 30 years and seven command billets.

He and his wife, Susan Gale, have three children: Jackson III, Jennifer and Mark. Civilian employment as owner, auto dismantling yard. Smith retired May 12, 1997.

JUDSON L. "JUD" SMITH, RADM, born Oct. 3, 1917, Baltimore, MD, attended McDonogh School, Dickinson College, USNA and Johns Hopkins.

He joined the service June 15, 1939. His military locations and stations: USS *New York*; Fleet Mine Warfare - Yorktown; USS *Paducah*; USS *Chaffinch*; USS *Strive*; USS *Sustain*; USS *Dionysus*; USS *Roberts*; USS *Darby*; USS *Willard Keith*; USS *Henley*; and USS *Albany*.

Smith was awarded the Navy Meritorious Service Medal, Legion of Merit and Battle Star (6).

His memorable experiences: Sank German submarine and captured her crew off North Africa Coast, May 1944; Japanese surrender ceremony, Tokyo Bay; served as first NR readiness CDR; served as CDR destroyer division, three years, MOB billet, chief of staff 2nd Fleet.

He married the former Mary Honthardinge (deceased) and has two children, Judson Lord Jr. and Colby S. Hawks; and three grandchildren: Marshall Wells, Hardince I. and Comerys McK. Civilian employment as partner, Smith and Smith, estate planners. Smith retired June 30, 1975.

MILTON L. SMITH, CAPT, born Oct. 14, 1931, in Darien, CT. He is a graduate of Wesleyan University where he earned his BA and MAT degrees.

CAPT Smith joined the USN in July 1956 and was commissioned ENS, USNR in December 1956. After additional training at Naval Air Technical Training Center, NAS Jacksonville, FL, and the Naval Intelligence School in Anacostia, he was assigned to CDR, First Fleet, Det. "A" at NAS Ford Island, HI. In addition, he served with the Special Weapons Employment Gp., Pacific and Fleet Intelligence Center, Pacific.

Released to inactive duty in December 1960, he affiliated with Naval Air Intelligence Reserve Unit 831 at NAS Floyd Bennett Field, NY, and continued in intelligence production duties with increasing responsibilities, including department head. In July 1969 he joined Naval Reserve Intelligence Unit 3-1-11 in Syracuse, ultimately serving as assistant officer-in-charge and in August 1975 became executive officer of Contingency Gp. 503.

In October 1976 he assumed command of Contingency Support Unit 302 and in February 1978 became the original CO of Defense Documentation Unit 0102 and was selected for CAPT during this tour. He subsequently served on the staff of Reserve Intelligence Area Coordinator, Area 18, at NAS South Weymouth. On Oct. 1, 1982, he assumed the duties of Reserve Intelligence Area coordinator, Area 18, serving until 1984. CAPT Smith retired in 1985.

CAPT Smith and his wife, Gail, reside in Alder Creek, NY and Big Canoe, GA. They have four daughters: Lorna Montanaro, Westport, CT; Holly Smith, St. Paul, MN; Randi Smith, Denver, CO; and Julie Evans, Rome, NY.

PAUL E. "SMITTY" SMITH, CAPT, born Nov. 22, 1934, East Liverpool, OH, earned his BS degree and attended California State University, North Ridge College of Naval War and Naval War College.

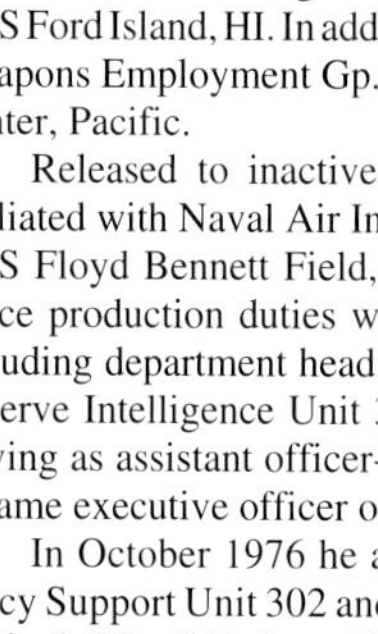

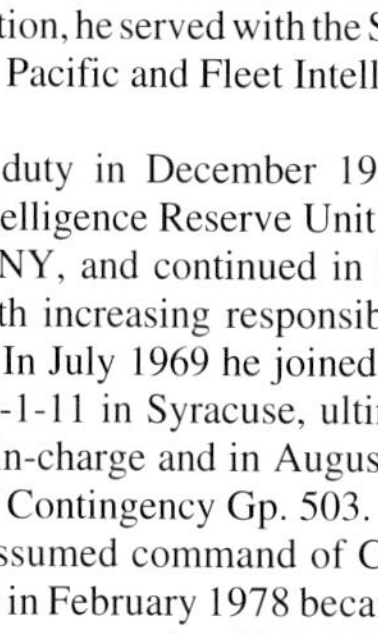

He joined the service Jan. 24, 1955. His military locations and stations: NAS Alameda, Pentagon, flight training, NAS North Island, NAS Quonset Point and NAS Lakehurst. He served with VT-2, VAW-11, VS-774, VS-32, VS-73, VS-30, OPNAV, DASO (reserve affairs) and DASN (reserve affairs).

Smith was awarded the Defense Meritorious Medal, Meritorious Service Medal and Navy Commendation Medal (3). He achieved the rank of CAPT and retired Oct. 1, 1984.

His memorable experiences: CO NAR Alameda (1981-83), 225 carrier landings, *Wasp, Hancock, Essex*, 4300 flight hours.

He is divorced and the father of three children: Jennifer, Scott and Karen.

PAUL W. SMITH, CAPT, born April 12, 1930, Albany, NY, attended the US Merchant Marine Academy, Kings Point, NY.

He joined the service in 1949 - midshipman, USNR at USMMA. Commissioned ENS March 2, 1953. His military locations and stations: USS *Zeal* (AM/MSF-131) 1953-55 (Mine Pac). He served as head, Maritime Affairs (OP-405E/OP-404) Strategic Sealift Div. office of the CNO, Washington, D.C. 1979-82.

Significant USNR duty: CO, USS *Reedbird* (MSC-51) NRF Long Beach, CA 1960-63; CO, USS *Cormorant* (MSC-122) NRF, Long Beach, CA 1963-65; CO NR Surface Div. 11-45, Santa Barbara, CA 1965-66; CO Inshore Undersea Warfare Div. 11-4, Long Beach, CA 1967-69; CDR Mine Div. 13 (NRF), Long Beach, CA 1969-70; CDR Mine Div. 51 (NRF), Long Beach, CA 1970-73; CO, NR Officer's School 11-2, San Diego, CA (1973-74); CO, Volunteer Training Unit (VTU) 1422, Tacoma, WA (1975-77); and CO YR-84 Ship Repair Base HQ 122, Tacoma, WA (1977-79).

Surface warfare officer (1115 designator). One of a select few naval reserve officers authorized to wear the Command-at-Sea Insignia. Manager, Marine Operations, Port of San Diego/director of transportation, Port of Tacoma. Former deck officer, US Flag Merchant Marine in dry cargo and tank vessels.

Smith was awarded the Navy Commendation Medal. He achieved the rank of CAPT and was discharged July 1, 1983.

He and his wife, Mildred, have four children: David, Michael, Stephen and Kenneth; and two grandchildren, Jessica and Matthew. Civilian employment as executive port director, Port of Pascagoula, MS. Smith retired in April 1993.

PHILIP W. SMITH, RADM, a native of Mt. Carmel, IL, attended the Universities of Eastern Illinois, Kansas City and Miami, FL obtaining a LLB and JD degree. He began his naval career in 1942 as an aviation cadet and was commissioned ENS in 1944.

During WWII he served in Pacific Campaign, Marianas and Iwo Jima with VB-116; Korean conflict (Pacific and Far East) in air transport operations with VR-5; Vietnam and Southeast Asia in C-118s with VR-743 in 1965-66. He served as transport plane CDR in the C-54 and C-118 aircraft, and as a plane CDR in PBY and PB4Y. He served as CO of VR-742, VR-50, VR-54, and as establishing director Naval Air Logistics Office (NALO), New Orleans, LA.

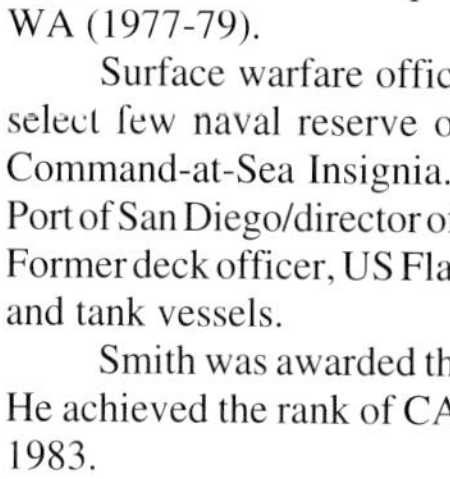

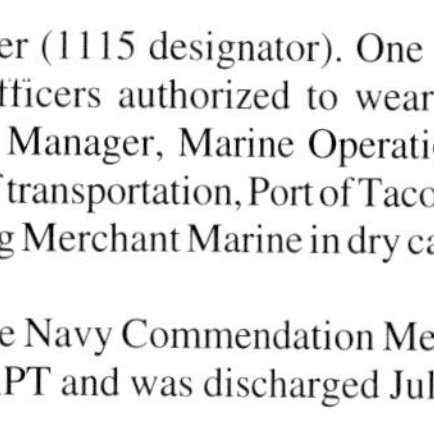

He was pilot of the 1952-53 Bob Hope Christmas Show to Alaska and the Far East. He advanced through the ranks and was promoted to RADM in September 1977. As a flag officer he served as DCOS and deputy COMNAVAIRLANT and was CDR Antilles Defense Command. He was an attorney with a financial and property practice in Miami, FL and simultaneously an airline captain with Delta Airlines, where he flew DC-3, DC-4, DC-6, DC-

7, DC-8 and DC-10 aircraft. RADM Smith was the national president of the Naval Reserve Assoc. in 1979-81.

ADM Smith served as assistant Federal Maritime Commissioner from January 1982-February 1987 during the Reagan Administration, and subsequently as the director of legislation for the Naval Reserve Assoc. RADM Smith has been awarded the Meritorious Service Medal and Meritorious Unit Commendation. His campaign and service medals are: American and Pacific Theatre, Navy Expeditionary Medal, WWII Victory Medal, and the Occupation, National Service, Naval Reserve, Armed Service, Korean Service and UN Medal.

RALPH W. SMITH JR., CAPT, born Feb. 16, 1933, Baltimore County, MD, earned his AA degree at Potomac State College in 1952 and his BS degree at WVU in 1963.

He joined the service Sept. 10, 1952, and was commissioned and designated naval aviator in May 1954. His military locations and stations: NAVTRACOM, NAS North Island, NAS Memphis, NAS Jacksonville, NAS Lemoore, NAS Willow Grove, BUPERS, NAS New Orleans, NAS Norfolk, and Pentagon. He served with VC-35, BTG-7, NATTC-Jacksonville AO "B" School, VA-25, PERS-D, OP-05R. He participated in the Vietnam War, Yankee Station, and two cruises aboard USS *Coral Sea* (CVA-43).

Smith was awarded the Distinguished Flying Cross, Air Medal (9), Navy Commendation Medal (2), China Service Medal, National Defense Service Medal, Armed Forces Expeditionary Medal, Vietnam Service Medal and RVN Campaign Medal Reserve Medal. He achieved the rank of CAPT and retired Jan. 1, 1983.

His memorable experiences: The phenomenal talent brought to active duty by their reserve members.

He and his wife, the former Dorothy Harper, have four children: Catherine, Jennifer, Sarah and Ralph W. III; and four grandchildren: Thomas, Benjamin, Evan and Madison. Civilian employment as Naval Reserve Assoc. consultant for officer record reviews.

ROBERT IRVING "BOB" SMITH, 0-6, born May 19, 1942, Washington, D.C., earned his BS degree at Holy Cross College.

He joined the service June 10, 1964. His military locations and stations: Patuxent River, MD and Pentagon. He served with Patrol Sqdn. 8, Patrol Sqdn. 68 and ACNO-Air Warfare.

Smith was awarded the Navy Commendation Medal. He achieved the rank of 0-6 and was discharged April 1, 1991. His memorable experiences: Being a VP-8 "Tiger" and a VP-68 "Blackhawk."

He and his wife, Patricia Marie, have two children, Christine Marie and Denise Marie; and two grandchildren, Robert Shawn and Michael Scott. Civilian employment as CAPT, USAir Shuttle.

ROY MICHAEL SMITH, CAPT, born May 17, 1949, Lake Charles, LA. His father was a career TAR naval aviator which resulted in the normal relocations of naval dependents. He graduated from the University of Texas at Arlington in 1971.

He received his Naval Flight Officer wings as a radar intercept officer at NAS Glynco, GA in June 1972. After completion of Replacement Air Gp. training in the F-4 aircraft, he was assigned to VF-161. In September 1973 VF-161 as part of the USS Midway Air Gp. was assigned to Yokosuka Japan as part of the Overseas Family Residency Program. Subsequent assignments as a TAR officer included VF-301, NARCEN Miramar, VF-202, NAS Dallas, Naval Postgraduate School, and two tours in the Pentagon.

CAPT Smith retired from active duty Oct. 1, 1996. He, his wife Shawn, a Selected Reserve CDR, and their three children now reside in Arlington, TX.

T.J. SMITH, CWO4, joined the USN Sept. 30, 1957. Graduated from Interior Communication Electrician "A" School, assigned to USS *Barton* (DD-722) (August 1958-December 1959). Transferred to USS *Wren* (DD-568) (December 1959-March 1960). After completing Basic Submarine/Nuclear Power Schools, attended prototype training at S1C Windsor Locks, CT. Assigned to USS *George Washington* (SSBN-598), Gold Crew as a reactor control operator making six FBM Deterrent Patrols (August 1961-August 1965), mainly standing IC watches forward. Transferred to USS *Angler* (AGSS-240) (August 1965-August 1966). Left active duty Aug. 13, 1996, as IC1 (SS) (DV).

Joined the USNR as company CDR for Recruit Training Program in Submarine Div. 11-6. ICC(SS) (DV) Smith was selected to become a warrant electrician (WO1) Dec. 15, 1970. He was the first reservist (SUBSUPPFAC 519) to qualify as command duty officer at Submarine Support Facility, San Diego. Served as CO of Naval Reserve Mobile Mine Assembly Gp. Unit 2419 from Dec. 16, 1976-Sept. 30, 1979. CWO4 Smith also served in the following reserve units: SURDIV 11-3(L), SUPDIV 11-1, SUBSUP 11-6 and SUBSUPPFAC 119. He retired June 1, 1988, as a chief engineering technician (CWO4).

A life member of the Naval Reserve Assoc. CWO4 Smith served as the 1983-84 president of the Terminal Island Chapter. Presently, he is active in the Sutter Chapter. He also served as the 1990 CDR of the Long Beach Commandery of the Naval Order of the US.

Married to the former Brenda Hanson from Santa Clara, CA. They have two children, Kari and Victor. On his retirement from the US Bureau of Prisons as a unit manager June 2, 1990, he changed his name to Hanson-Smith.

Currently, he is a volunteer serving as the Rio Del Oro district commissioner for the Golden Empire Council of the Boy Scouts.

WILLIE SMITH JR., CDR, born Feb. 19, 1945, Marquez, TX, earned his MA degree in management at Incarnate Word College, San Antonio, TX in 1989 after having earned a BS degree in law enforcement at Southwest Texas State University in 1974. Upon receipt of his undergraduate degree in 1974, the, then, 11 year naval reservist, was ordered to Naval Officer Candidate School in Newport, RI where he was commissioned as an ENS Nov. 1, 1974.

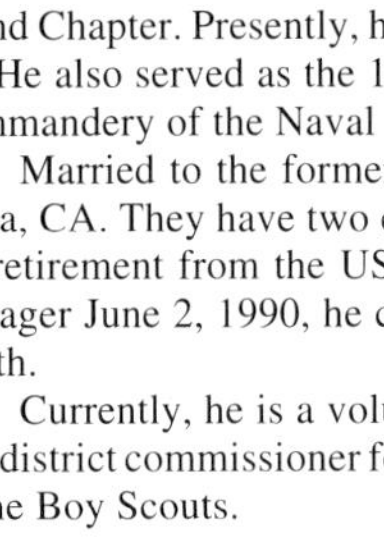

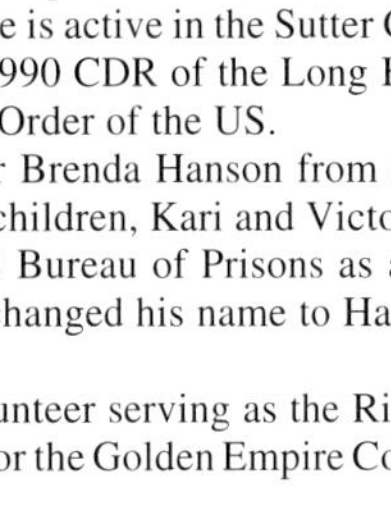

He served as a division officer in the Pacific Fleet for the ensuing three years and returned to the naval reserve attaining the rank of CDR (O-5) in 1990 and subsequently retiring Oct. 1, 1992, with 29 1/2 years of honorable service.

Additionally, he completed 22 1/2 years of honorable service with the San Antonio Police Dept. which culminated with a position as a planning officer in the department's Research and Planning Unit and retirement Aug. 1, 1992. CDR Smith attained additional graduate level education at Trinity University in San Antonio; Chapman College in Orange, CA; and Sam Houston State University in Huntsville.

Smith served as a police supervisor at University Hospital in San Antonio, TX from Sept. 19, 1994-Nov. 30, 1996, where he had direct supervisory responsibility for as many as 24 officers.

He is the founder of Smith CompuEnterprises and is presently developing services to be marketed via the Internet. He has four children: Steven, Sheldon, Sidney and Shareese.

CHARLES J. SMOLLEN, LCDR, born July 9, 1940, Vicksburg, MS, earned his BSCE degree at Louisiana Tech University, Ruston, LA.

He joined the USAR in December 1958 and later the USNR in August 1972. His military locations and stations: Various naval reserve VTU and CE units: NAVACTS UK 230 London; EFD PAC Dallas; HQ EUCOM 0130 Vaihingen, Germany; EFD CHES 106 Washington, D.C.; EFD 209 New Orleans, LA; EFD 108 Jacksonville, FL; RNMCB 24 Huntsville, AL; RNMCB 28 Shreveport, LA; and USAR 412th Engrs.

Smollen was awarded the National Defense Service Medal, Armed Forces Reserve Medal, Expert Rifle and Expert Pistol.

He is presently divorced and has six children: John, Susan, Brett, Joel, Emily and Luke; and two grandchildren, Jennifer and Patrick. Civilian employment as civil engineer, USN, Naples, Italy; Dallas, TX; and San Diego, CA; Corps of Engineers at Vicksburg, MS; Winchester, VA; Germany; and Egypt.

WINSTON R. "WIN" SMOYER, CAPT, born June 30, 1910, Akron, OH, earned his BA degree in 1932 at Dartmouth, his MA degree at the University of California, attended Claremont Graduate School, USC, and California State.

He joined the service in February 1943. His military locations and stations: NAS Quonset Point, NAS Corpus Christi, COM AIR 7th Fleet - Pacific and FLT AIR WING 10 - Philippines. He served in the Philippine Campaign 1945-46 and Admiralty Island to Leyte-Manila.

Smoyer was awarded the Meritorious Service Medal. He achieved the rank of CAPT and was discharged Oct. 10, 1946. His active Naval Air Reserve service was October 1946-July 1970.

His memorable experiences: Flag LT, aide to ADM O.B. Hardison, ComAir 7th Flt., NAS Sangley Point, Philippines, 1945 and Flag LT, aide to ADM Arthur Gavin, ComAir 7th Flt., Philippines, 1946.

He and his wife, Kathryn F., have two children, Anne and Susan; and five grandchildren: Winston, Lynn, Laura, Marsha and Matthew. Civilian employment with California Dept. of Vocational Rehabilitation for five years and Alhambra High School social science teacher for 20 years. Smoyer retired July 1, 1970.

ANNE BORK SNYDER, CDR, born Nov. 24, 1919, Buffalo, NY, earned her BS degree in public health nursing at the University of Buffalo in July 1955.

She joined the service Dec. 15, 1943, ENS, Navy Nurse Corps in USNR. Her military locations and stations: USNH, Sampson, NY; Brooklyn, NY; Oakland, CA; St. Albans, NY; and USS *Repose* (AH-16). On June 1, 1946, she was discharged to inactive duty subject to be recalled. She participated in WWII and Korean War while serving aboard the USS *Repose* (AH-16).

Snyder was awarded the WWII Victory Medal, American Campaign Medal, Naval Reserve Medal, Korean Service Medal and UN Service Medal. She achieved the following ranks: LTJG, April 1, 1945; LT, July 5, 1951; LCDR, Sept. 1, 1957; CDR, July 1, 1969. Snyder was discharged from active service June 29, 1954, and retired from the USNR Nov. 24, 1979.

Her memorable experiences: On Sept. 26, 1950, recalled to active duty in USNH Oakland, CA; Oct. 28, 1950, to USS *Repose* (AH-16) in Korea. In November 1950 ship sailed to Chinnampo, where they admitted patients from Pyeng-Yang, Korea. Returned to San Diego in February 1952. Next duty at St. Albans, NY. Discharged June 29, 1954, and joined USNR unit at Buffalo, NY in 1955.

She married Henry S. Snyder (USAF veteran) in 1967. Civilian employment with Erie County Health Dept. as public health nurse and Sup. public health nurse. Snyder retired from civilian employment in August 1976.

BRYAN SNYDER III, LTJG, born March 17, 1922, Dallas, TX, attended Dallas public schools and Texas A&M where he earned his BS degree.

He joined the V-7 program in college in 1941. His military locations and stations: Northwestern University Midshipman School; NAB Noumea; New Caledonia; and NAB Russell Islands. He served with COMSOPAC HQ,

3rd Fleet HQ and participated in shore based communications.

Snyder was awarded the Campaign Medal, US Defense Medal, Asiatic-Pacific Campaign Medal and WWII Victory Medal. He achieved the rank of LTJG and was discharged July 24, 1947.

His memorable experiences: Captain's orderly Midshipman School; CAPT Wegand officer in charge; on staff of ADM Halsey COMSOPAC; lived the story of South Pacific as portrayed by J.A. Michener.

He and his wife, Winna Gean, have three children: Bryan Chandler, Jane and Marguerite; and five grandchildren: Jenny, Lara, Sarah, Susan and Catherine. Civilian employment as purchaser, state of Texas; chicken farmer; printing salesman. Snyder retired Dec. 31, 1986.

JOSEPH E. "JOE" SOLAN, CDR, born June 3, 1918, Hammond, IN, received his BA degree from Upper Iowa University. Entered USN in August 1940. Selected for Midshipman School. Made cruise in USS *Arkansas* (BB-33), then to Northwestern University, Chicago campus. Commissioned ENS in March 1941.

Ordered to USS *Mount Baker* (AE-4), then to USS *Wyoming* (BB-32) for gunnery training. Assigned to CVE commissioning detail, Astoria, OR; commissioned 50 Kaiser class CVE carriers. Ordered to *Shamrock Bay*, then as navigator USS *Arcturus* (AKA-1), later as executive officer. Ended the war at Okinawa and occupation of Korea. Returned to Long Beach via Tokyo.

Transferred to USNR in 1946; commanded USNRTC Gary, IN and USS PC 1240; also served as squadron chief of staff in USS *Daniel Joy* (DE-585). Recalled for Korea emergency to staff COM Five Norfolk; then to USNRTC Chicago. Retired with rank of CDR; earned Battle Star (4), Commendation Medal and SWO Designator.

Civil service with US Dept. of HUD/FHA Chicago; retired as chief appraiser. Married 55 years to wife, Irene, and they live in Munster, IN. They have three children: Joseph E. Jr., David A. and Sara K.; and five grandchildren: Joseph III, Jason, Jerred, Jamie and Matthew. Attended convoy commodore course FTC San Diego, NBC & ASW courses FTC Key West.

ROLAND J. "SORNY" SORENSEN, LCDR, born March 22, 1923, Minneapolis, MN, graduated high school and attended trade school for A&E license.

He joined the service Feb. 2, 1942, and had flight training in November 1943. His military locations and stations: NATTC Norman, OK; Mustin Field, Philly Navy Yard; NAS Norman, OK; NAS Minneapolis, MN; NAAS Cuddihy, Corpus Christi, TX; NAAS Corry; NAAS Saufly, Pensacola; NAS Banana River; NAS North Island; NAS Miramar; NAS Sand Point; USS *Oriskany* CV-34 (Korea). He served with CVG-13, VA14A; CVG-102, VF-874 and participated in the Korean War.

Sorensen was awarded the Air Medal (2) and Commendation Medal w/Combat Distinguished Device. He achieved the rank of LCDR and was discharged Nov. 23, 1947, with retirement April 27, 1956.

His memorable experiences: All of the Carner operations.

He and his wife, Gladys, have six children: Terri, Cathie, Jim, Joan, Jeff and Kim; and 11 grandchildren: Michelle, Brett, Matt, Brian, Tony, Rhonda, Cory, Drew, Jay, Troy and Tori. Civilian employment as NWA mechanic and flight engineer. Sorensen retired March 22, 1983.

STANLEY A. "STAN" SOWINSKI, CDR, born May 5, 1952, Przyrow, Poland and immigrated to the US in 1963. He received his commission through the NROTC Scholarship Program at the Illinois Institute of Technology, Chicago, IL.

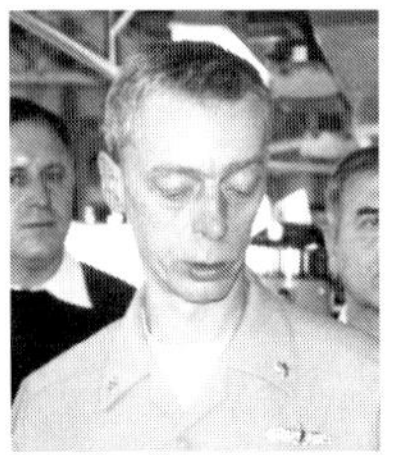

On active duty from 1976-80 he served aboard the USS *John F. Kennedy* (CV-67) in the engineering, legal and training departments. From 1980-81 he served aboard the USS *Sumpter* (LST-1181) as a gunnery and deck officer. From 1981-83 he served as the auxiliaries officer aboard the USS *Saipan* (LHA-2). From 1983-85 he served as the officer in charge of port services at the Philadelphia Naval Shipyard.

In the USNR he served as the training officer in a unit assigned to the USS *Perry* (FFG-7) then as the reserve administrator aboard the USS *Clark* (FFG-11) at Philadelphia, PA. Other USNR assignments included command of a SIMA unit, staff at CINCUSNAVEUR and two command tours of readiness units. In 1995 CDR Sowinski was voluntarily recalled for six months of active duty as the deputy chief, Military Liaison Team, Warsaw, Poland.

He is employed by the US Dept. of Labor as a regional safety and health manager. He and his wife, Sylvia, have two children, son, Anthony and daughter, Anastasia.

JAMES F. "JIM" SPAGNOLE, CAPT, born July 15, 1944, Seattle, WA, earned his AB degree at the University of San Francisco and his JD UOP degree at McGeorge Law School.

He joined the service Sept. 19, 1967. His military locations and stations: Norfolk, Philadelphia, Guantanamo and Pentagon. He served aboard the USS *Shakori* (ATR-102), USS *Preserver* (ARS-8), Naval War College and MDSU1, Det. 220.

Spagnole was awarded the Navy Commendation Medal, Unit Citation, Meritorious Service Medal, Naval Reserve Medal, Vietnam Era and National Defense Service Medal. He achieved the rank of CAPT USNR and was discharged June 1, 1991.

His memorable experiences: DV work, Guantanamo, USS *Enterprise*, NAS Alameda, NAS San Diego, Panama City, COMSERVRON 5 Honolulu, security duty, Pentagon.

He has one daughter, Lisa Marie. Civilian employment as an attorney.

STEPHEN A. SPANN, CAPT, born May 22, 1941, Albuquerque, NM, is a near lifelong resident of Colorado. Upon graduation from North High School, Denver, he enlisted for four years active duty in the USMC in 1959, he continued service in the USMCR for an additional four years with the 4th 155 Gun Btry. in Denver, and transferred to the Navy Seabee detachment in Cheyenne, WY as a plank owner. After earning a BS in civil engineering at Colorado State University in 1969, he applied for and received a direct commission as an ENS in the Civil Engineer Corps in May 1970.

CAPT Spann has held several billets in NMCB 15 which included "B" and "A" CO, training officer and Detachment OIC for both the Denver and Pueblo Detachments. He was selected CO of Reserve Div. Naval Facilities WEST and then as CO of NCFSU-2 at Fort Carson, CO. Currently he serves as command directorate J2 of MTU-CO-1 a joint officer Marine Corps command that provides exercise support to CINC PAC and CINC NORAD.

Spann earned a master's degree in public administration at the University of Denver in 1984. In civilian life he is the senior partner in Spann Engineering, LLC. Volunteer services include president of the Upper South Platte Water Conservancy District, executive secretary of the Como Civic Assoc., member of the US Committee on Large Dams, fellow in the Society of American Military Engineers. He is a registered professional engineer in Colorado.

CAPT Spann and his wife, the former Annemie Weinelt, have two sons, Dr. Stefan Spann and Christopher Spann, both married and with three children each. Colorado outdoors near Como absorbs all available spare time for the family.

JOHN PAUL SPATH, CAPT, born July 1, 1940, Great Barrington, MA, earned his BA degree in 1963 at Union College.

He joined the service in January 1963. His military locations and stations: January-May 1963, Officer Candidate School, Newport, RI; May 1963-October 1964, USS *Chilton* (APA-38), Norfolk, VA; October 1964-May 1966, USS *Calcaterra* (DER-390), Newport, RI; and May 1966-May 1968, commandant, Third Naval Dist., New York, NY.

Reserve duty background includes: 1968-70, NRSD 3-95(L), Liverpool, NY; 1970-71, NRSD 3-4(S), Liverpool, NY; 1971, NRGC 3/25(S), Liverpool, NY; 1972, NROS 5-11(S), Lexington, KY; 1972, NRGC 5-8(S), Lexington, KY; 1972-73, NRSD 5-7(S), Danville, KY; 1973-74, NROS 5-11(S), Lexington, KY; 1974-75, NRSD 5-30(M), Lexington, KY; 1975-76, IRU 123, Lexington, KY; 1976-77, VTU 123, Lexington, KY; 1977-78, NR Biddle CG-34, Lexington, KY; 1978-79, VTU 0908, Lexington, KY; 1979-81, VTU 0909, Louisville, KY; 1981-83, CONVOYCOM 209, Louisville, KY; 1983, VTU 0913, Nashville, KY; 1984, VTU 0504G, Cincinnati, OH; 1984, GEN IMA (D) 405, Cincinnati, OH; 1985-86, NCSO CARIB 209, Nashville, TN; 1986-88, VTU 0909, Louisville, KY; 1988-89, ABFC 109, Nashville, TN. He achieved the rank of CAPT (O-6) and retired July 1, 1984.

He and his wife, Patricia, have two children, John Jr. and James; and one grandchild, Katy. Civilian employment with Fluor Daniel Construction as project controller. Spath retired Oct. 1, 1991.

GEORGE E. SPEAKE, CDR, earned his BS degree in mechanical engineering at the University of Pennsylvania in 1936.

He joined the service June 15, 1936. His military locations and stations: Mostly carriers in the Pacific. Served with Old Ranger and several escort carriers in Pacific. Battles and campaigns participated in include Philippines, Iwo Jima, Okinawa, etc.

Speake was awarded the Air Medal. He achieved the rank of CDR and left active duty March 11, 1946. He retired from the USN June 6, 1975. His memorable experience: Typhoon in Wakayama Harbor in 1945.

He and his wife of 45 years, Jeanette H., have one son, George H. Jr.; and four grandchildren: George E. III, Wendy, Tina and Kim. Civilian employment for 30 years in Christian work with Moody Bible Institute.

FLOYD DAVIDSON SPENCE, CAPT, born April 9, 1928, Columbia, SC. He enlisted in the USNR May 27, 1947, and was honorably discharged May 23, 1952. Upon graduation from the University of South Carolina, he was commissioned as an ENS in the USNR June 6, 1952, and reported for active duty June 26, 1952. From June-July 1952 he received training at the Naval Amphibious Training Unit, Little Creek, Norfolk, VA.

He then served as an assistant navigator/legal officer

aboard the USS *Carter Hall* (LSD-3), from July 1952-July 1953. He was next assigned to the USS LSM-397, as a gunnery officer from July 1953-May 1954, during which time he was promoted to the rank of LTJG Dec. 6, 1953. He was honorably released from active duty May 14, 1954.

In the USNR he was promoted to LT July 1, 1956; LCDR Sept. 1, 1961; CDR July 1, 1966; and CAPT July 1, 1972. He is the former group commander of all Naval Reserve Units in Columbia, SC, and is the former CO of a Naval Reserve Surface Div. On Jan. 1, 1985, CAPT Spence transferred to the retired reserve and April 9, 1988, he retired from the USNR. He is the recipient of the Navy Occupation Service Medal w/Europe Clasp, National Defense Service Medal and the Armed Forces Reserve Medal.

CAPT Spence was elected to the US House of Representatives Nov. 3, 1970, and has been re-elected to each succeeding congress. He is currently the chairman of the House Committee on National Security, and a senior member of the House Committee on Veterans' Affairs.

He has four sons by his first wife, the late Lula Hancock Drake: David, Zack, Benjamin and Caldwell. He married the former Deborah Ellen Williams of Lexington, SC July 3, 1988.

GORDON D. SPENCE, CDR, born May 8, 1943, Springfield, MA, earned his BCE degree at Louisville and his MPA degree at Golden Gate University.

He joined the service in June 1966. His military locations and stations: Commissioned NROTC, University of Louisville; Civil Engineer School, Port Hueneme, CA; RVN with NMCB 74 (Da Nang); Naval Construction Regt. 21, Davisville, RI; Antarctic Support Force Det., Christ Church, NZ; Staff ComNavAirLant, Norfolk; assistant Public Works Offices, NAS Oceana; RNMCB 23 Det. 01; project manager, Atlantic Div., naval facility; OPS officer, ComCBLant, Little Creek, VA; staff, assistant chief of staff, CINCUSNAVEUR (ADP), Norfolk; staff, CNO Augment N44; Washington, D.C.

Spence was awarded the Navy Achievement Medal, Navy Unit Commendation Ribbon and Meritorious Unit Commendation Ribbon. He achieved the rank of CDR and was discharged in November 1989.

His memorable experiences: All of it.

He and his wife, Melinda, have four children: Scott, Adam, Aaron and Jonathan. He is employed as a civil engineer, DSCE and plans to retire in May 2003.

GEORGE OTIS SPENCER III, CDR, born Sept. 23, 1956, Houston, TX, earned his BS degree at the USNA, class of 1979 and his MA degree in chemistry at Rice University in 1989.

He joined the service May 30, 1979. His military locations and stations: USS *Texas* CGN-39 and USS *Brisco* DD-977.

He achieved the rank of LT on active duty and was discharged April 30, 1985. While still active in the reserves he has achieved the rank of CDR.

His wife's name is Colleen. Spencer is employed in chemical marketing.

PETER J. "PETER J" SPEROS, CAPT, born May 25, 1919, Salt Lake City, UT.

Voluntary enlistment March 13, 1942, DIO, 12ND, SF, Y2C. Early 1943 assigned to Naval Attaché, Cairo, Egypt and governments of Greece and Yugoslavia. While there, obtained and translated the operation report of Greek navy by ADM Alexandris, for which received letter of commendation from ADM of ONI, and shortly thereafter received spot commission to ENS.

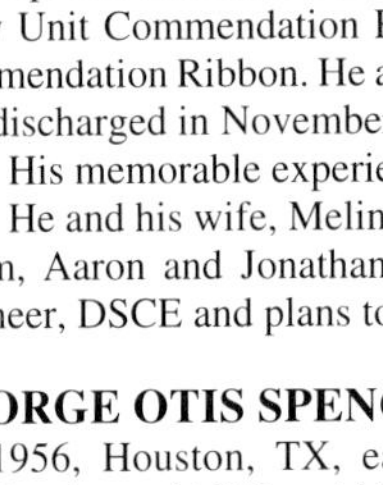
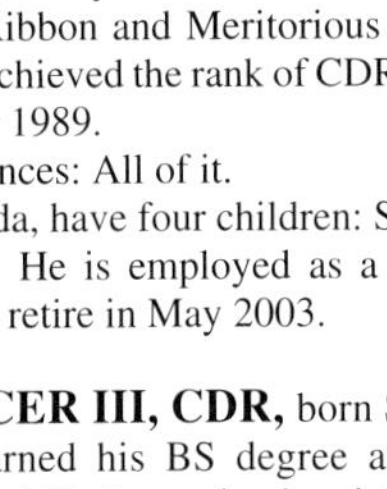

In the interim, volunteered to be parachuted in northern Greece, but Normandy invasion changed that to Japanese Language School in Boulder, CO, subsequently attending special schools in Platsburg, NY and Miami, FL, previous to his assignment as deck and gunnery officer, USS *Shipley Bay* (CXVE-85), also in charge of motor whale boat #2, picking up 130 miles south of Wake Island on January 19th, officers and crew of a Catalina patrol bomber, adrift in the Pacific for more than 12 hours—to this date he can still see the smiles of those 13 men delivered to Pearl Harbor.

Discharged in 1946 but remained in the reserves until called to duty for the Korean War, January 1951. Attended six months Naval Intelligence School, and later assigned to CNO, Pentagon, for release and receipt of information to foreign governments and classification control. In 1953 back to civilian life and retiring June 30, 1969, as CO NRID 12-1 having been appointed CAPT Sept. 1, 1966. Following his Dad's advice to all of them: "Your first duty is to your family, and second to your country"—well said for a Greek immigrant—thus his eldest son John is also retired as a full CDR.

He and his wife, Mary, have four children: John, Chriss, Peter Jr. and Tia Ann; and four grandchildren: Jeromie, Peter J., James and Benjamin. His civilian employment includes: Editing, *Business Digest*; executive secretary, CSPE; and president, Allen-Spero Corp.,etc.

PHILIP M. SPINA, LCDR, born Sept. 8, 1938, Brooklyn, NY, earned his BA degree at St. Michael's College, Vermont and his MBA at Long Island University.

He joined the service in January 1961 (OCS). His military locations and stations: USS *Lake Champlain* (CVS-39) Quonset Point, RI; COMASWFORLANT, Norfolk; and USS *Rockbridge* (APA-228), Norfolk.

Spina was awarded the National Defense Service Medal and Naval Reserve Medal. He achieved the rank of LCDR and was discharged in February 1967.

He and his wife, Marianne, have three children: Gregory, Ghia and Lea. He was self-employed as a tax grievance consultant until his retirement Sept. 1, 1981.

ROBERT H. "BOB" SPIRO JR., RADM, born Dec. 5, 1920, Biltmore, NC, earned his BS degree at Wheaton College; attended Harvard Graduate School, business administration (certificate Navy Supply Corps School); graduate studies, University of North Carolina; PhD at the University of Edinburgh; and post-doctoral studies at Duke and Columbia Universities.

He enlisted in the service Dec. 27, 1941, and was commissioned ENS March 15, 1943. His military locations and stations: District Intelligence Office, Norfolk, VA; Navy Supply Corps School at Harvard Business School; USS *Morris* (DD-417). He participated in the Marshall Island Operations, Asiatic-Pacific Raids, Western New Guinea Operations, Leyte Landings (Philippines), Lingayen Gulf and Okinawa Gunto Operations.

Spiro was awarded the Asiatic-Pacific Area Ribbon w/6 stars, WWII Victory Medal and Philippine Liberation Medal. He achieved the rank of RADM (upper half), SC, USNR and retired June 1, 1978.

His memorable experiences: He greatly enjoyed his four and one-half months of accelerated semester at the Harvard Graduate School of Business Administration in accounting, Navy Supply & Disbursing, 1943. He then served 19 months (without a night ashore) in combat aboard USS *Morris* (DD-417) in last six campaigns of Pacific War culminating in kamikaze destruction of *Morris* at Okinawa.

He is presently married to Terrie Gay. He has three children by a previous marriage: Robert Timothy, James Monroe and Elizabeth Susan; and five grandchildren: Robert, Keith, Lisa, Katie and Sarah. Civilian employment includes: University professor, dean, president; Under Secretary of Army; national executive director, Reserve Officers Assoc. of US; vice-president, American Security Council Foundation; chairman and CEO of RHS Imprinted Products, Inc.

LELAND BRYCE SPONHOLTZ, LT, born May 30, 1922, Utica, KS, graduated from the University of Florida in electrical engineering in May 1947.

He joined the service Sept. 7, 1943. His military locations and stations: Pensacola Naval AS, Cornell University, USS *Kidd* DD-661 and LSTs 1088 and 1090.

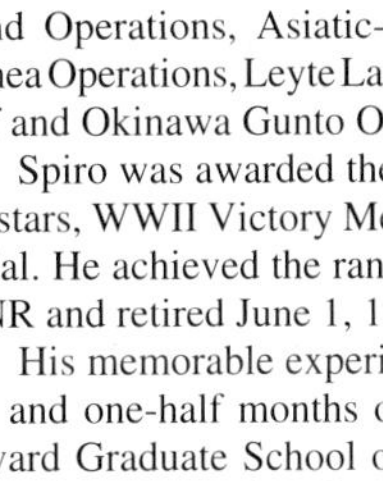
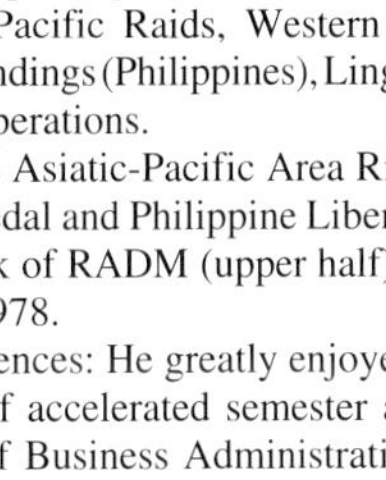

Sponholtz was awarded the American Theatre Medal, Asiatic-Pacific Theatre Medal and WWII Victory Medal. He achieved the rank of LT and was discharged from active duty July 16, 1946. (Retired with pay, USNR, May 30, 1982.)

His memorable experiences: Boot camp at Pensacola NAS; Midshipman School, Cornell University; graduated ENS Jan. 25, 1945; engineering officer, USS *Kidd* (DD-661), April 19, 1945-March 6, 1946; life member, Naval Reserve Assoc.

He married the former Virginia May Stiegelmeier Aug. 30, 1947. They have four children: Dennis, Carol, Warren and Allen; and three grandchildren: Benjamin, Karina and Victor Jr. He was employed for 35 years with Westinghouse Electric Corp. His last position was as manager of bids and proposals for the Oceanic Div. Sponholtz retired Aug. 31, 1982.

LOUIS E. "L.E." SPRADLIN, CAPT, born Jan. 13, 1940, Ropesville, TX, earned his BA degree in mathematics at the University of Texas at Austin and his MS in management at Pace University, New York.

He joined the Naval Reserve Officers Training Corp. (regular) in September 1958 and was commissioned ENS, USN, June 1, 1963. His military locations and stations: US Naval Submarine School, New London; USS *Harder* (SS-568), Charleston; Submarine Sqdn. Four, Charleston; Naval Reserve Surface Div. 6-32M, Charleston; Reserve Underway Training Unit, 6-15 which changed to Fleet Training Gp. 107; Naval Reserve Surface Unit 6-11; USS *Orion* (AS-18), Det. 107; Volunteer Training Unit 0705, Columbia, SC; Naval Reserve CDR Submarine Sqdn. 16, Det. 1208; and Voluntary Training Unit 0801G, Atlanta.

He achieved the rank of CAPT, USNR and was discharged Sept. 30, 1990.

His memorable experiences: As JO and diving officer was on *Harder* during the maneuvering recovery and depth charge trials for the *Harder's* class of submarine (the first post WWII class submarines). It was noisy and a little bumpy and the charges were not even close. The recovery trials included such exercises as holding the stern planes or bow planes on x degrees down at various speeds and then trying to recover before they got to test depth. One of the most interesting was when they could not use the motors to help recover. *Harder* also provided submarine services at the Guantanamo refresher training on a couple of occasions. Lots of all back full and all ahead flank maneuvers with the training destroyers with PDCs missing the mark and occasionally hitting the mark. The most interesting event at Guantanamo occurred in their *Harder* BOQ (a vacant house on the base) when a poker game was broken up by an attractive, "thought to be naive" Guantanamo school teacher bluffing their self proclaimed poker master XO out of the biggest pot of the evening.

He married the former Mary Helen Moore of Columbia, SC Aug. 26, 1966. They have three children: Helen Keels S. Jorn, Bernard Carl Jorn and Susan Elizabeth. Employed for 28 years with what started as the same family of companies: Southern Bell, AT&T, Southern Bell, BellSouth Corp., BellSouth Telecommunications, and BellSouth Business Systems. Retired in July 1995 from position of vice-president sales - The Carolinas for BellSouth Business Systems. Currently self-employed in consulting and networks.

JOHN STRICKLIN SPRATT, CAPT, born Jan. 3, 1929, San Angelo, TX, earned the following degrees: AS, BS, MSPH, MD, Diplomate American Board of Surgery and FACS.

He joined the service Sept. 19, 1952. Active duty stations: Chest Service, USNH, Oakland, CA, 1953; Field Medical Training School, Camp Pendleton, CA. 1953; Arctic Weather Training, Cold Water, CA, 1954; 1st Marine Div., Korea; USNH, Great Lakes, IL, 1955; NRMC, Guam, 1982; USNH, Portsmouth, VA, 1984; and National Naval Medical Center, Bethesda, MD, 1985-90.

Spratt attended many schools throughout his career at Bethesda, MD; Corpus Christi, Fort Sam Houston and Camp Bullis, TX; Great Lakes, IL; Albuquerque, NM; Rochester, MN; Newport, RI; Yokusuka, Japan; Minneapolis, MN; Pensacola and Jacksonville, FL; Colorado; Washington, DC; Gulfport, MS; Charleston, SC; San Diego and Coronado, CA; Alexandria, VA; New Orleans, LA; Mediterranean; and Millington, TN. He also completed numerous NR Officer Correspondence Courses from 1954-92.

Reserve Assignments at St Louis and Columbia, MO and in Louisville, KY: Medical officer, 3rd Inf. Bn, USMCR, 1956-59; MOB Team OIV 9-28, 1959-61; NRMC 9-1, 1961-62; representative of the commandant, 9th Naval

Dist., ENS 1915 Program, 1961-68; medical officer, 3rd Inf. Bn., USMCR, 1962-67; medical officer, USN Reserve Research Co., 9-28, 1967-72; naval representative of the commandant, 9th Naval Dist., Military Support Planning Office, HQ, MONG, 1972-74; executive officer, US Naval Regional Medical Center, 1975-77; CO, USNR Surgical Team 118, 1977; CO, USNR G-5 Station Hospital 118, 1977-79; CO, USNR Medical Contingency Response Unit 209, 1979-81; medical officer, MEDCRU 209, Louisville, KY, 1981-83; medical school liaison officer, University of Louisville, 1977-93; medical officer, NR USNH Camp Lejeune, Det. 309, 1983-84; officer in charge, PRIMUS (NR FH 500 COMMZ-11 DET P-6509), University of Louisville School of Medicine, 1984-92; and naval reserve volunteer, Training Unit 0909, NMRC, 1992-93.

Spratt was awarded the Navy Unit Citation Ribbon, National Defense Medal w/Bronze Star, Korean Service Medal w/Marine Corps Emblem, Armed Forces Reserve Medal w/3 HG Devices (for 40 years of uninterrupted satisfactory Federal service as a naval reserve officer), UN Medal, Navy Expert Rifleman Medal, Gold Wreath Award w/Letter of Commendation, Silver Star in lieu of second Gold Wreath, Conspicuous Service Medal, Letters of Appreciation and Commendation, medals for life membership in The Naval Order, The Association of Military Surgeons of the US, TROA and NRA. He was appointed "Admiral" for the Commonwealth of Kentucky in 1977 and "Military Advosor to the University of Louisville School of Medicine. He is a member and a past-president of Commodore Club of Navy League and in 1991 received a Letter of Commendation for seventh Gold Wreath Award from Naval Recruiting Command.

After 40 years, nine months and 21 days of uninterrupted federal service as a reserve officer his memorable experiences are too great to fit in this space. He transferred to the NR Retired List with pay July 1, 1993.

He and his wife, Beverly, have three children: John A. Spratt, CDR Shelley S. Young, and LCDR Robert Spratt; and six grandchildren: John Robert, Madelyn, Abigail, Austin, Molly and Thomas. Civilian employment as Clinical Professor of Surgery, F. Edward Hebert School of Medicine, Uniformed Services University of the Health Sciences, Louisville, KY, 1988-.

RANDALL E. "RANDY" SPRINKLE, born July 9, 1963, Winston-Salem, NC, earned his BA degree in political science at the University of North Carolina-Chapel Hill.

He joined the service in July 1986 and was commissioned ENS in December 1986. His military locations and stations: NAS Brunswick, ME; Keflawik Island, Sigonella Italy. He served with VP-11, Brunswick, ME and VP-94, New Orleans, LA. Participated in Operation Desert Storm/ Desert Shield; Sharp Guard/Maritime Guard (Adriatic).

Sprinkle was awarded the Navy Achievement Medal, National Defense Service Medal, Southwest Asian, NATO, Navy-Marine Corps Expeditionary Medal and Kuwait Liberation Medal. He achieved the rank of LCDR and was discharged from active duty in 1993.

His memorable experiences: Navigator on first US aircrew (P-3) to fly on Soviet Mike class submarine.

Civilian employment in property management, rental/ real estate agent.

MANFRED A. "FRED" STAAB, CDR, born May 3, 1941, Nuremberg, Germany, son of a U-boat officer, emigrated to the US in 1954 and was naturalized in 1959. Notwithstanding three years of Army ROTC, he graduated from Drexel Institute of Technology in 1965 with a BS in commerce and engineering. He entered the NAOC program in Pensacola and was commissioned in February 1966.

Active duty included three lively storm seasons with the Navy Hurricane Hunters (VW-4) in the Caribbean/ South Atlantic and AEW fleet support operations in the North Atlantic and Mediterranean.

In 1970 he returned to Michigan State University to earn an MBA in marketing and continued USNR duty with VP-66, VS-24, VR-54 and VP-90 squadrons. ACDUTRA deployments included service in the Atlantic, Mediterranean and WestPac. His official retirement date is March 1, 1994.

The years of Navy experiences helped balance a challenging career encompassing marketing management, environmental engineering and, most important, rearing a family with wife, Mary (Dudley). Their children are daughter Christina and son Peter.

Recreational diversions have included private flying, scuba diving, tennis and skiing. The future focuses on semi-retirement in the Caribbean area.

RICHARD K. "DICK" STACER, CAPT, born in 1921, Oregon City, OR, earned his BA and JD degrees.

He joined the service in 1943. His military locations and stations: Pacific Theater, Atlantic Theater, Mediterranean Sea and Alaska Sea Frontier. He served with COMPHIBPAC staff, JAG, COM 3, COM 11, COM 13, COMBATCRULANT, 6th Fleet. Battles and campaigns participated in were Saipan/Tinian, Leyte, Luzon and Iwo Jima.

Stacer was awarded the Meritorious Service Medal. He achieved the rank of CAPT (O-6) and retired in 1971. His memorable experiences: Battles of Leyte Gulf and Iwo Jima.

He and his wife, Evelyn, have three children: Jefferson, Richard T. and Tara; and two grandchildren, Paul and Savannah. Civilian employment as an attorney, National NRA JAG, National Naval Order JAG, and ABA Committee for Legal Assistance to Service Personnel.

DANIEL STACK, CAPT, born July 29, 1928, Brooklyn, NY, earned his BA degree at Brooklyn College in 1949; his LLB degree at Columbia University Law School in 1952; and his LLM degree at Georgetown University Law School in 1955.

He joined the service in July 1952. His military locations and stations: OCS, Newport, RI and Supply Corps School, Bayonne, NJ. He served with the USN purchasing office, Washington, DC, from April 1953-October 1955.

Stack was awarded the Joint Service Commendation Medal in 1980 and the Naval Order of US in 1984. He achieved the rank of CAPT and was discharged to inactive service Oct. 27, 1955.

His memorable experiences: Started tradition of Supply Corps School graduation trip (opening day, Yankee Stadium, 1953); bus and a championship softball team, 1953; two years as CO of Oscar 202 Reserve Unit, 1979-80. A lifetime serving with some of the finest people in the world.

He and his wife, Jane, married in 1953 and have two children, Joan and Gordon; and two grandchildren, Megan and Haley.

Civilian employment: Attorney, senior officer, bank, conglomerates, etc. He retired June 30, 1983, but continues recruiting for USNA.

ROBERT C. "BOB" STACK, RADM, a Raleigh, NC native, graduated from the University of North Carolina with an AB and NROTC commission in June 1965. After active duty from 1965-68 onboard USS *Fox* (DLG-33) and USS *Sproston* (DD-577), he took his MBA from the University of Virginia in June 1970.

As a reservist, he drilled onboard USS *R.K. Huntington* (DD-781) and USS *Fiske* (DD-842). He served as CO of Reserve Cargo Handling Bn. Seven, Navy Petroleum Office Fleet DET 211, NSC Charleston HQ A 107 and DCMD Atlanta B811. Promoted to RADM in June 1993, his mobilization billet was deputy commander, Aviation Supply Office (later-Naval Inventory Control Point). As president, Navy Supply Corps Foundation from 1987-96, he was responsible for awarding over $1 million in scholarships to dependents of Supply Corps officers, and supply enlisted ratings while increasing the investment account from $1.2 million to $2.6 million.

RADM Stack is executive vice-president of Texas Commerce Bank in Dallas, TX, where he lives with his wife Debbie. A NRA life member since June 1976, RADM Stack has been awarded the Legion of Merit, Defense Meritorious Service Medal, Meritorious Service Medal, Joint Service Commendation Medal, Navy Commendation Medal w/2 Gold Stars and Combat Action Ribbon. He is qualified as a Naval Aviation Supply Officer and Surface Warfare Supply Corps Officer.

RALPH J. STAEHLI JR., CAPT, born May 20, 1918, Portland, OR, earned his BA degree at the University of Washington. He enrolled at the University of Washington NROTC in September 1936 and was commissioned ENS USNR in June 1940.

His military locations and stations: USS *Spica*, assistant cargo officer, July-September 1940; USS *Crow* (AMc-20), engineering officer, executive officer, CO, September 1941-July 1942; USS *Liberator* (AMc-87), CO, August 1942-September 1943; Minewarfare School, Yorktown, VA, October-December 1943; USS YMS 140, CO, January 1944-April 1945; and Minecraft Training Center, Little Creek, VA, May 1945 until end of WWII. He participated in the Alaska Theater, 1942-43; Palau Islands Landings, 1944; Leyte Gulf Landing, 1944; and Lingayen Gulf Landings, 1945.

Staehli was awarded the Bronze Star Medal, American Theater Medal, Asiatic-Pacific Campaign Medal, and Philippine Liberation Medal. He achieved the rank of CAPT and was released from active duty in December 1945 and retired from the USNR in July 1972.

His memorable experiences: Clearing US laid minefields from entrances to St. Paul Harbor, Kodiak Island, Spring 1943; weathering a typhoon, Leyte Gulf, September 1944; being one-fourth mile from USS *Mt. Hood* when she blew up in Seeadler Harbor, Manus Island. Was alongside USS *Cebu* at the time which shielded them from direct force of blast.

He and his wife, Helen C., have two children, Ralene S. and Christopher P., LCDR USN (Ret.); and four grandchildren: Michael H., Tanya, Preston and Annie. Civilian employment: Ocean Transportation. Positions in sales, management, planning, estimating, terminal operations, etc. with various steamship companies and ocean towing companies. Staehli retired in March 1985.

HARLAN V. "TIPPY" STAGGS, CDR, born Sept. 19, 1916, Vanceburg, KY, earned his BSEE degree at the University of Kentucky in 1941. He accepted commission as ENS USNR Aug. 15, 1941. Entered active duty in October 1941, Noroton Heights, CT. From January-July 1942 was stationed at American Embassy, London, England as an observer with British navy in undersea warfare. Stationed at Bureau of Ordnance, Washington, D.C. from July 1942-December 1943.

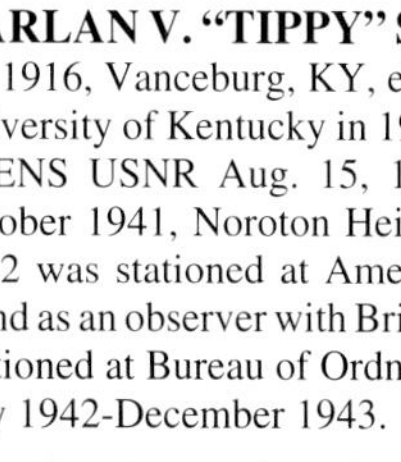

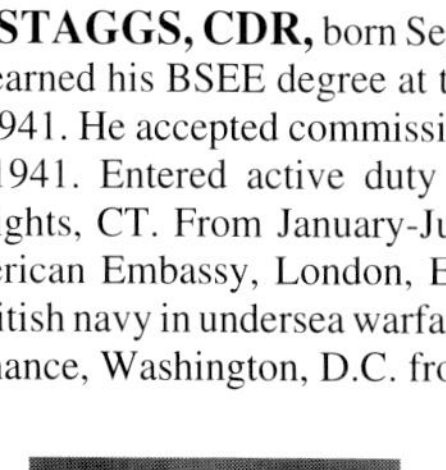

Duties involved development of undersea ordnance. From January-March 1944 he was stationed on US submarine R-5 at St. George, Bermuda. Attended Submarine School, New London, CT, April-June 1944. From July 1944-November 1945 assigned to Submarine Forces-Pacific. Assignments included two successful war patrols on USS *Mingo* (SS-261) and base assignments at Fremantle, Australia and Pearl Harbor, HI. In March 1946 was released to inactive duty with rank of LCDR. From February 1960-June 1965 served as CO, Naval Reserve Composite Co. 5-40, Lexington, KY.

Staggs' awards and medals: Pre-Pearl Harbor, three area ribbons and Victory Ribbon. Retired from USNR with the rank of CDR in February 1967. His memorable experiences: Operating with British navy in 1942 and two submarine war patrols in 1944.

He and his wife, Peggy, have one daughter, Katie Lou Howell and one grandchild, Clayton. Civilian employment as president of Staggs & Fisher Consulting Engineers. Staggs retired in February 1982.

RUDOLF SHEPHEARD "SHEP" STAHEL, CAPT, born in 1938, New Orleans, LA, was commissioned USN from the NROTC at Tulane University in 1959.

He served in USS *Mathews* (AKA-96) as administrative officer, then as navigator. In 1961 he became XO of USS *Duval County* (LST-758). In 1962 the ship and crew were used in filming the movie PT-109 about JFK's wartime naval service, then saw duty in the Cuban missile crisis. Upon return from Panama in 1963, Stahel resigned from active duty and joined the USNR.

His reserve assignments included tours as training officer for Surface Div. 8-22M in Baton Rouge, LA; CO of Surface Div. 4-65M in Trenton, NJ; CO of Convoy Commodore Unit 503 and CO of Reserve COMPHIBRON 6-605 in Buffalo, NY; and as the first CO of Readiness Unit Two in Dallas, TX responsible for seven reserve units. He is qualified as a Surface Warfare officer and his decorations include the Navy Expeditionary Medal (Cuba) and the Navy Commendation Medal.

Stahel joined the IBM Corp. in 1963 in Baton Rouge in marketing. He then taught in Princeton, NJ, managed sales and systems engineering teams in Buffalo and Rochester, NY, and has served in senior staff positions in Dallas since 1978. He currently handles business affairs for the IBM Legal Dept. responsible for 20 states.

He resides in Plano, TX where he has been involved in civic affairs as the community has rapidly grown to almost 200,000 people. He has served as chairman of the Planning and Zoning Commission, Horizon (master plan) Committee, Homeowners Council, Library Foundation, and Plano DART Advisory Board. He currently chairs the Parks and Recreation Board.

He is married to the former JoeAnn Womack of Baton Rouge. After their children entered high school, JoeAnn earned a master's degree and enjoyed a career as a market analyst in the personal computer industry, including tours with Future Computing, Inc., McGraw-Hill Co., and as president of StoreBoard, Inc., a subsidiary of Ziff-Davis Publishing Co. She retired from the business world in 1995 and now serves on the Plano Library Board and tends her horses.

Stahel and JoeAnn have three adult children: Anne and Shepheard reside in Plano, and Elizabeth lives near San Antonio. They have five young grandchildren.

DAVID M. STAHLEY, CWO4, born June 18, 1936, Cleveland, OH, joined the USNR in July 1953 and was discharged in October 1954; joined the USN in October 1954 and was discharged in August 1963; and the USNR in August 1963 and retired in January 1991.

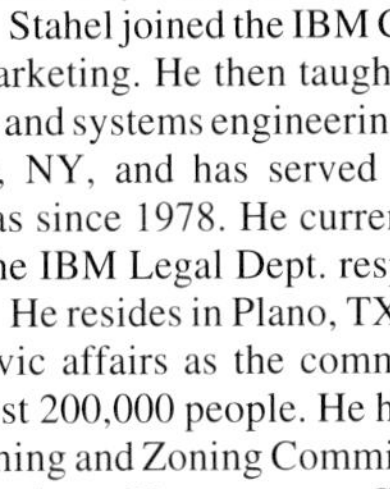

His military locations and stations: FSS San Diego and NavSeaCom Treasure Island. He served aboard the USS *Pluck* MSO-464, NavComSta Washington, D.C. and NavFac SNI Pt. Mugu, CA.

Stahley was awarded the Good Conduct Medal (2), National Defense Service Medal and Naval Reserve Medal.

He and his wife, the former Patricia J. Ryan, have four children: David, Michael, Mark and Sharon; and four grandchildren: Alicia, Michael, Katherine and Jeromy. Civilian employment as service engineer at Reliance Electric Co. and in sales at Allen-Bradley & Eurotherm Drives.

ROGER P. STAIGER JR., CAPT, enlisted in the USNR at Williamsport, PA in the Reserves Office Candidate (ROC) program in 1965. After graduating from Bucknell University with a BSEE, he was commissioned at the Naval Officer Candidate School, Newport, RI in October 1967.

His assignment was to the staff of the CDR in Chief Atlantic Fleet as Operations Control Center/Movement Report Center watch officer. In 1969 he was transferred to the USS *Mars* (AFS-1) homeported in Yokosuka, Japan, and made three deployments to the Gulf of Tonkin in support of Seventh Fleet Operations.

In 1971 CAPT Staiger was released from active duty and attended graduate school at the University of Pennsylvania where he earned a MS in engineering and a MBA from The Wharton School. Between 1971-78 he served as Selected Reserve engineering officer in USS *Douglas H. Fox* (DD-779) and USS *Robert L. Wilson* (DD-847) and as Selected Reserve operations officer in USS *Harold J. Ellison* (DD-864) homeported in Philadelphia, PA. Between 1974-77 he also served as part-time crew member and chief engineer on the 180 foot Barkentine *Gazela Primeiro* during her reconstruction and participation in the tall ship races from Bermuda to Newport, RI and Operation Sail '76.

CAPT Staiger has had five USNR commands including two SIMA (maintenance) detachments and a Replacement Unit supporting NRF FFG 7 class ships homeported in Philadelphia and the USNR unit which supported the assistant chief of Naval Operations for Surface Warfare. From 1986-88, CAPT Staiger was the Selected Reserve coordinator for USS *Clark* (FFG-11).

In 1979 CAPT Staiger was selected by the American Society of Mechanical Engineers and American Assoc. for the Advancement of Science as a congressional fellow. In 1983 CAPT Staiger was the first Selected Reserve to graduate from the Naval War College, Washington, D.C. campus. Between 1980-91 CAPT Staiger served as vice-president and member of the Board of Directors of the US Naval Sea Cadet Corps.

In his engineering career, CAPT Staiger has worked as oil refinery maintenance manager, and for 12 years as a professional staff member for the US House of Representatives, committee on Energy and Commerce. CAPT Staiger is currently the Washington representative for the Alyeska Pipeline Service Co., the operator of the Trans Alaska Oil Pipeline. CAPT Staiger is licensed as a registered professional engineer and a certified plant engineer.

At the end of 1997 CAPT Staiger continues to serve the USN as a duty CAPT in the Navy Command Center in the Pentagon and is in his 33rd year of USNR service.

GUSTAV J. STANGLINE, CDR, began his USN career in 1966 as an enlisted member of the USNR. In October 1978 YN1 Stangline was selected by the Naval Enlisted Reserve Assoc. and the chief of Naval Reserve as the 1978 National Sailor of the Year and meritoriously advanced to YNC. Received his commission as a LDO in April 1980. He has held billets in joint commands as ADP security officer and administrative officer and has served as CO of five USNR units and two command tours. He is currently serving as assistant training officer with COMTRALANT 106.

Stangline was awarded the Navy Commendation Medal, Navy Achievement Medal, National Defense Service Medal (2), Naval Reserve Meritorious Service Medal (3), Military Outstanding Volunteer Service Medal and Armed Forces Reserve Medal (2).

He holds a BS degree in accounting and management from the University of Scranton and a MS degree in computer information systems from Marywood College. He has additional advanced course work in educational technology from Lehigh University and has completed numerous courses in computer information systems, data network design and network administration.

CDR Stangline is a life member of the Naval Enlisted Reserve Assoc., member of the Naval Reserve Assoc., Fleet Reserve Assoc., Naval Recruiting District Assistance Council and several civic organizations.

He is married to the former Linda A. Steuben and has four sons: David, John, Daniel and Timothy; and one grandson, Adam. Presently resides in Springbrook, PA. Civilian employment at Lackawanna Jr. College, Scranton, PA as an associate professor from 1969-88. Since 1988 has been academic computing coordinator at Wilkes University, Wilkes-Barre, PA.

ARLEIGH C. "BILL" STANLEY, CDR, born Sept. 10, 1924, Pritchett, TX, attended Duke University (V-12), Northwestern University (midshipman) and University of Texas (post graduate).

He joined the service Nov. 17, 1942. His military locations and stations: Seabee Boot Camp, Norfolk, 23rd Regt., 2nd Div.; USMC, Camp Lejeune; USN, Atlantic Fleet. He served on anti-submarine duty on East Coast.

Stanley was awarded the WWII Victory Medal, American Theater Medal, Naval Reserve Medal and Secretary of Navy Commendation. He achieved the rank of CDR and was discharged June 30, 1978.

His memorable experience: Training Russian 5" 38 gun crews, participant in training of Chinese officers and commissioning of Nationalist Chinese Navy, USS *Philadelphia* (CL-41), USS *Houston* (CL-81), plankowner USS *Shreveport* (PCE-874), USS *Saratoga* (CU-60), and USS *Liberty* (AGTR-6).

He and his wife, Nell, have two sons, Mike and Dwayne; and five grandchildren: Ken, Carol, Adam, Aaron and Hunter. Civilian employment as district manager for The Sherwin-Williams Co.

FRANK S. STANLEY, CDR, born April 24, 1941, Chicago, IL, earned his MA degree in management at Central Michigan University.

He was commissioned ENS in 1963. Served as deck and gunnery officer, USS *Washburn* (AKA-107), 1963-65 with deployment to Vietnam; gunnery officer, USS *Herbert J. Thomas* (DD-833), 1965-66; instructor, director of Radarmen B and Missile Firecontrol B School, Fleet Anti-Air Warfare Training Center, San Diego, 1966-69; operations officer, USS *Passumpsic* (AO-108), 1969-70, deploying to Vietnam; CO, Naval and Marine Corps Training Center, Worcester, MA, 1970-74; executive officer and CO, Human Resource Management Center, Charleston, SC, 1974-77; promoted to CDR in San Diego, 1977-80; executive officer, office of the chief of Naval Reserve, Pacific, 1980-82; manpower officer and chief of staff, Naval Readiness Command, Region Nine, Memphis, TN, 1982-85; retired in July 1985 with 22 years of active duty.

Stanley was awarded the RVN Gallantry Cross w/ Palm Unit Citation, Armed Forces Reserve Medal, RVN Campaign Medal, National Defense Medal and Armed Forces Expeditionary Medal.

He and his wife, Sara Ellen, have four daughters: Kristina, Kara, Leah and Michelle; and one granddaughter, Renee. Civilian employment in public service including a tour as a county administrator. CDR Stanley is president of Old Ironsides Productions, Atlanta.

JAMES CHARLES "JIM" STANLEY, LCDR, born Sept. 23, 1931, Santa Barbara, CA, earned his BA degree at the University of Santa Barbara and is a graduate of the American Graduate School of International Management.

He joined the ROC program in 1952. His military locations and stations: ONI Pentagon, Washington, D.C., 1953-55; Dominican Republic; Mexico; Brazil; Panama; Pentagon; San Diego; and Port Hueneme.

Stanley was awarded the Defense Service Medal. He achieved the rank of LCDR and retired from the reserves in 1979.

His memorable experiences: CO, boarding group, Russian merchant vessels, Panama Canal, 1957 (Suez Crisis). Spent most of time on board during transit convincing Russian CAPT that they were not at war.

He and his wife, Clemencia, have three children: James, Sean and Mark; and three grandchildren: Daniela, Kristal and James. Civilian employment as general manager, Exxon in four Latin American countries, 1957-70 and president/owner, Fuertejioos, Costa Rica, 1970-96. Stanley retired Oct. 1, 1996.

EVA G. STAPLETON, CDR, born Aug. 6, 1932, Huntingdon, PA, attended nursing school and is an RN.

She joined the service in October 1958. Her military locations and stations: San Diego, CA; Subic Bay, Philippines; Yokosuka, Japan; Beaufort, SC; Patuxent River, MD (twice); Annapolis, MD; Guantanamo Bay, Cuba; and Da Nang, Vietnam. She served with the Navy Nurse Corps. CDR Stapleton served in Vietnam from September 1968-September 1969.

Stapleton was awarded the Navy Unit Commendation Medal, Navy Commendation w/Combat "V", National Defense Service Medal, RVN Campaign Medal w/device and Vietnam Service Medal w/2 Campaign Stars. She achieved the rank of LCDR (active) and CDR (reserves) and was discharged from active service in 1975 and from the reserves in 1984.

Her memorable experiences: All active years, especially Vietnam.

She is single and was employed at J.C. Blair Memorial Hospital, Huntingdon, PA until her retirement in 1992.

MAURICE Q. "JOE" STAPP JR., CDR CEC, born Sept. 21, 1925, Georgetown, TX, earned a BS degree in naval science and BS and MS degrees in civil engineering.

He joined the service Sept. 13, 1943. His military locations and stations: NAS Corpus Christi, TX; New Orleans, LA; Berea, KY; and Durham, NC. He served with V-12, NROTC, RMCB 28, RMCB 22, and 9th NCR.

Stapp achieved the rank of CDR CEC, USNR and retired Oct. 1, 1973. His memorable experiences: Served with dedicated and outstanding personnel in RMCB 28, RMCB 22 and 9th NCR.

He and his wife, Frances Paulyne, have three children: Sharlene Estelle, John Quinton and James Lee; and three grandchildren: Brandon, Kristen and Megan. Civilian employment as senior consultant with HBC Engineering, Inc. Stapp retired Sept. 21, 1985.

FRANK W. STARKS, LCDR, born March 11, 1948, Chicago, IL, earned his BS degree in electrical engineering at the Illinois Institute of Technology and his MBA degree at the University of Chicago.

He joined the service in October 1970. His military locations and stations: USS *Barney* (DDG-6), Norfolk; COSRIVDIV 21, Great Lakes, IL; and OIC Crane Det. 214, Crane, IN. He served with the Space and Naval Warfare Command, Washington, D.C.

Starks achieved the rank of LCDR and was released from active duty in August 1974. His memorable experiences: NATO cruise in North Atlantic in 1973.

He is the divorced father of two daughters, Nichele and Allison, and the grandfather of one granddaughter, Ashley. Civilian employment as technical supervisor, Johnson & Johnson and sales representative, R.R. Donnelley. Starks retired Nov. 1, 1992. He is presently chairman of Precision Printers.

WILBUR SNYDER "BILL" STAUFFER, LCDR, born Sept. 28, 1934, Lewistown, PA. He enlisted in the Army Security Agency in October 1951, after graduating from Montgomery Blair High School, Silver Spring, MD. He graduated from Army Language School, Presidio of Monterey, CA, in May 1952, and was designated translator and interpreter, Chinese-Mandarin. He served in Korea with Detachment Four, 303rd Communications Recon. Bn., 501st Communications Recon. Gp.

In October 1956 he was Outstanding Recruit of Plt. 235 at Marine Corps Recruit Depot, Parris Island, SC. Following assignments at Camp Pendleton, CA, and Naval Base Philadelphia, PA, he completed a Western Pacific tour with 3rd Bn., 9th Marines, 3rd Marine Div., as a SGT, intelligence assistant and interpreter.

He received a BA from University of Maryland, College Park, in June 1961. He was commissioned ENS, USNR, Unrestricted Line, aeronautical duties, specialty - intelligence, in February 1964, and retired in July 1984. His assignments included active duty in office of chief of Naval Operations and Patrol Sqdn. 65. His longest and most memorable reserve assignment was as air intelligence officer with Attack Sqdn. 305, Naval AS, Point Mugu, CA.

LCDR Stauffer worked more than 20 years in commercial airline cargo and passenger sales and service management, including 16 months as manager-military airlift for American Airlines at Norton AFB, CA, during the Vietnam conflict. That assignment included visits to Da Nang and Cam Ranh AB in March 1967. In June 1995, he retired from Defense Logistics Agency, Defense Distribution Region West. Currently, he is a substitute public school teacher in Northern California.

RODNEY C. "ROD" STEARNS, CAPT, born Oct. 17, 1943, Attleboro, MA, earned his BA degree in business administration at the University of Wyoming.

He joined the service Sept. 24, 1965. His military locations and stations: NAS Oceana, VA and NAS Brunswick, GA. He served with VF-41, 1967-70; NATTC, Brunswick, GA (VT-86), 1970-72; and VF-102, 1972-75. He participated in four Sixth Fleet cruises to the Mediterranean area.

Stearns was awarded the Pistol Sharpshooter, National Service Medal, Navy Reserve Medal and Navy Commendation. He achieved the rank of CAPT and retired Oct. 1, 1996.

His memorable experiences: 2,000 ARS and 600 carrier landings in F-4 Phantom, CO of three reserve units Air Systems 0891, Defense Plant Rep. Unit 0391, Lynn, MA, and Air Systems Unit 1391.

He and his wife, Connie, have two children, Randy and Julie; and one granddaughter, Jessica. Civilian employment as owner/manager, Stearns Stationery.

WILLIAM R. "BILL" STEBBINS JR., LCDR CEC, born Aug. 7, 1927, earned his BS degree in civil engineering and graduate certification in highway traffic.

He joined the service in January 1953 for active duty. His military locations and stations: K-3 and K-6, Korea and USNCBC, Gulfport, MS. He served with CBMU #1 (later #101) in Korea. Served two campaign periods in Korea.

Stebbins was awarded the WWII Victory Medal, Armed Forces Reserve Medal, Naval Reserve Medal, National Defense Service Medal, UN Korean Service Medal, Korean Service Medal w/2 Campaign Stars and Marine Corps Emblem. He achieved the rank of LCDR CEC and was discharged in December 1954.

Stebbins is single. Civilian employment as traffic engineer of highway departments in Illinois, Nebraska and Texas.

G. DONALD "DON" STEEL, CAPT, born March 11, 1953, Lansing, MI, earned his BS degree in 1975 at the USNA.

He joined the service June 30, 1971. His military locations and stations: Brunswick, ME; NAS Glenview, IL; NAF Detroit, MI; and NAS Willow Grove, PA. He served with VP-10, VP-11, VP-60, VP-93, VP-64, CONVOYCOM, NMCRC Columbus, OH. Battles and campaigns: Operations Sharp Guard, Maritime Monitor, Yugoslavia.

Steel was awarded the Navy Commendation Medal (2), Joint Meritorious Unit Award, Navy Unit Commendation, Navy Battle "E" (2), National Defense Service Medal (2), Sea Service Ribbon (3), Navy/Marine Corps Overseas Service Ribbon, Coast Guard Operations Ribbon, Armed Forces Reserve Medal, Navy Rifle Marksmanship Ribbon and Expert Pistol Medal.

His memorable experiences: CO, Patrol Sqdn. 64; executive officer, Patrol Sqdn. 93; CO, Readiness Unit One, Columbus, OH.

He and his wife, the former Alison A. Fisher, have five children: Shawn, Eric, Grace, Kathleen and Corinne. Civilian employment as president, Planned Investment Co.

SALVATOR M. "SAL" STEFULA, CDR, born Jan. 15, 1948, Spangler, PA, graduated from St. Francis College, Loretto, PA in May 1970. He attended St. Francis Seminary, Loretto, PA and was ordained to the Roman Catholic Priesthood May 11, 1974. He is a member of the Franciscan Order, Third Order Regular.

He was commissioned Dec. 8, 1981, and went to Chaplain School in January 1982. As a reservist he has been assigned to Marine Air Gp. 24, Kaneohe, HI; Personnel Mobilization Team, Honolulu, HI; NAS Barber's Point, HI; Fleet Hospital Det E, NAS Dallas, TX; Personnel Mobilization Team, NAS Dallas, TX; 14th Marines, NAS Dallas, TX; REDCOM 11, chaplain; and Fleet Hospital 21, HQ.

Stefula was awarded the 10 year Reserve Medal.

He has been pastor of St. Thomas the Apostle Church and Holy Trinity Church in Ft. Worth, TX for the past eight years.

DONALD L. STEGNER, CAPT, born March 1, 1933, Baltimore, MD, earned his BSEE degree at Duke University in 1955.

He joined the ROTC at Duke University in September 1951. His military locations and stations: Active duty, July 1955-June 1957, USS *Thuban* (AKA-19). Reserve duty: September 1958-June 1971, various MSTS and NCSO Divs., Baltimore; July 1971-June 1973, XO, NRNCSORG Div. 5-2, Baltimore; July 1973-June 1975, CO, NRNCSORG Div. 5-2, Baltimore; August 1975-September 1976, GVTU 1606, Adelphi; October 1976-September 1978, CO, NRNCSO Florida 506, Adelphi; October 1978-July 1979, CO, NR MSC HQWASHDET 106, Adelphi; October 1979-September 1981, XO, NR MSC HQWASHDET 106, Adelphi; October 1981-September 1983, CO, NR NCS/COMSC/MARAD 106; October 1983-June 1985, CO, NR GVTU 106, Adelphi; July 1985, retired reserve.

Stegner's awards include the Surface Warfare Officer, Armed Forces Reserve Medal w/HG and Marksman. He achieved the rank of CAPT (O-6) and was discharged June 30, 1985. His memorable experiences: DEW Line Resupply Mission and Suez War.

He and his wife, Edith B., have two daughters, Suzanne and Sandra; and three grandchildren: Patrick, Robert and Anne Morgan. Civilian employment as president, Eastern Electric Co.; forensic engineer and consultant, Hudson International and Potomac American Gp. Stegner retired from the USN in June 1985 and from Eastern Electric in January 1993.

DANIEL W. "DAN" STEHLY, LCDR, born Jan. 20, 1950, St. Louis, MO, earned his BSME degree at the University of Missouri-Rolla.

He joined the service Dec. 16, 1986. His military locations and stations: St. Louis, MO; Cape Girardeau, MO; Bahrain; Kansas City, MO; and Dallas, TX. He served with RNCFSU-4, CBHU-22, NMCB-15 and 9th NCR. He participated in Desert Shield/Storm.

Stehly was awarded the Navy Achievement Medal, Navy Unit Commendation, Sea Service Deployment, Southwest Asia Campaign, Liberation of Kuwait and Expert Rifle and Pistol.

He and his wife, Ellen, have two daughters, Jacqueline and Jennifer. Civilian employment as senior engineer, Union Electric Co.

EARL A. STEIN JR., CAPT, born Sept. 1, 1932, Fargo, ND, earned his BA degree in business education-economics.

He joined the service in March 1951, seaman recruit, USNR. His military locations and stations: USS *Gen. Breckenridge* (TAP-176). He served with Mobilization Unit, 9-57 Training Div., Selective Service Unit FLTT and FLTF.

Stein was awarded the China Service Medal, 10 year Naval Reserve Medal, Selective Service Meritorious Service Award and Naval Reserve Meritorious Service Award.

He achieved the rank of CAPT and was discharged June 31, 1985.

His memorable experiences: CO of MTD 9-57, SELSERVUNIT FLTT and SELSERVUNIT FLTF.

He and his wife, Kathryn, have three children: Craig, Kevin and Lisa; and seven grandchildren: Matthew, Benjamin, Elizabeth, Lindsay, Kelsey, Kendra and Zach.

ANNA M. STEINBERGER, CAPT, enlisted in the USN from McClusky, ND in 1967. She served as a photographer's mate after graduating from boot camp in Bainbridge, MD and PH "A" School in Pensacola, FL. She worked at the Naval Photographic Center in Washington, D.C. for 15 months and then finished her tour of duty as a public affairs photographer at NAS Alameda. The highlights of the tour were meeting GEN Jimmy Doolittle, serving as a photojournalist while on a carrier on board delivery CIA aircraft from VR-30 during an arrested gear landing and catapult take-off from the USS *Oriskany*, and being crowned Miss USO Armed Forces for the San Francisco Bay area in 1970.

After release from active duty, she attended Lansing Community College in East Lansing, MI. She was elected president of Phi Kappa Phi and graduated Summa Cum Laude with a full scholarship to Michigan State University. While there, she spearheaded a campaign to reform the GI Bill which ultimately led to legislation granting the same rights to both male and female veterans on dependency issues. She graduated with a master's degree in education and taught for six years in East Lansing before joining Hartland Consolidated Schools in 1979. In 1987 she was chosen as Hartland's Teacher of the Year and currently specializes in teaching math at the Village Elementary School and serving as the Outdoor Education director.

After enlisting in the USNR in 1974, she was soon advanced to PH2 and granted a direct commission. She has served in the following USNR units: NARDIV Y1/NAF 514, Investigative Operations Unit 2214, Fleet Intelligence Rapid Support Team 1773 (intelligence analyst, team leader, division officer, assistant production officer, administration officer and special assistant to training and production), Intelligence Volunteer Training Unit 0114 (Deputy Reserve Intelligence Area CDR for information and resources management), Atlantic Intelligence Command 1673 (XO and interim CO), Atlantic Intelligence Command 1773 (CO), and then back to AIC 1673 as XO after a unit consolidation. As of October 1996 she holds the position of deputy chief of staff for Administration for CDR, Naval Reserve Intelligence Command.

While on annual training, CAPT Steinberger served 14 tours at the Atlantic Intelligence Command, completed eight courses at training commands, participated in Ocean Venture '82 and '84, Solid Shield '83, Team Work '88 (Northwood in England), North Star '89 (Kolsas in Norway), and was CINCLANT intelligence analyst and deputy game administrator for Global War Games '89-'96 (Naval War College, Newport, RI).

CAPT Steinberger's decorations include the Navy Commendation Medal, Navy Achievement Medal, Joint Meritorious Unit Award, Meritorious Unit Commendation Award, National Defense Service Medal (2), and the Armed Forces Reserves Medal (20 years). She is a life member of the Naval Reserve Assoc. and Reserve Officers Assoc. She is an active member of the Naval Institute, Naval Intelligence for Professionals, WAVES National and State Organizations, and the American Legion. She serves on the steering committees for Greater Michigan Armed Forces Week (honors top ROTC graduates from seven colleges and universities) and International Night (social function for US and Canadian officers).

She and her two sons, Jeffrey and James, reside in Highland, MI. (Both young men have served nine months in the USNR and two weeks active duty since their mother never missed a drill or training duty!)

JOSEPH STEINER, CAPT, born Sept. 20, 1943, London, England, grew up in Brooklyn, NY, earned his BS degree in 1965 at The City College of New York and his DDS in 1969 at New York University College.

He joined the service May 12, 1965. His military locations and stations: Active duty, Dental Detachment MCRD Parris Island, SC, 1970-72. Reserve duty: 1972-74, Dental Co. 6-3, Charlotte, NC; 1974-77, NRDC 307, Charlotte, NC; 1977-80, OIC, NRDC Camp Lejeune Det. A 107, Charlotte, NC; 1980-82, NRDC Camp Lejeune HQ 107, Charlotte, NC; 1982-84, CO, NR Naval Hospital Camp Lejeune 107, Charlotte, NC; 1984-85, CO, NR Naval Hospital Camp Lejeune 307, Charlotte, NC; 1985-87, staff dental officer, NAVRESREDCOMREG SEVEN, Charleston, SC; 1987-90, CO, NR Naval Hospital Camp Lejeune 307, Charlotte, NC; 1990-91, CO, NR Naval Hospital Charleston 107, Charleston, SC; 1991-92, CO, NR VTU 0704, Charlotte, NC; 1992-93, CO, NR NDCL Parris Island 207, Charlotte, NC; 1993-94, CO, NR NDCL Camp Lejeune 207, Charlotte, NC; and 1994-95 CO, NR VTU 0704G, Charlotte, NC. He served on local and 2 REDCOM SEVEN policy boards.

Steiner was selected Outstanding Junior Officer by Reserve Officers Assoc. He was awarded the Meritorious Service Medal, National Defense Medal (2), Armed Forces Reserve Medal (2), Navy Marksmanship Pistol and Navy Recruiting Service Ribbon.

He has been a USNA information officer (Blue and Gold) since 1982.

His memorable experiences: Ability to lead units of excellent people and help many achieve greater rank and rate. Inspire people to achieve more in their military and civilian careers.

He and his wife, Renee Sherry, have two daughters, Michelle Jacqueline and Shara Deborah. Civilian employment as partner of Drs. Treat and Steiner, PA.

WALTER K. "WALT" STEINER, CAPT, born Dec. 20, 1944, Gardner, MA, earned his BS degree (history) at Tennessee Tech in 1967 and his MA degree (political science) at the University of Massachusetts in 1981.

He joined the service March 22, 1968. His military locations and stations: NAS Pensacola, NAS Brunswick, Tanson Nhut AB, Saigon, RVN. He served with Aircofat, HS-75, HM-18, USS *Independence*, Naval Inspector General, and US mission to NATO. He participated in the Vietnam War in 1972, Tran Hvong Dao 15, 16, 17.

Steiner was awarded the Defense Superior Service Medal, Defense Meritorious Service Medal, Meritorious Service Medal, Air Medal (9), Navy Commendation Medal, Navy Achievement Medal w/V, Navy Unit Commendation (3), Meritorious Unit Commendation, National Defense Service Medal (2), Philippine Presidential Unit Citation, RVN Cross of Gallantry and Civil Actions Service Medal. He achieved the rank of CAPT and is still on duty.

His memorable experiences: Flying in Vietnam; establishing a mission, introducing a type-model-series aircraft, and commissioning a new squadron in the USNR; serving US ambassador to NATO in post-Cold War Epoch.

He and his wife, the former Michelle Leblanc, have been married for 23 years and have three children: Brandon, D'Arcy and Erik. Civilian employment with US Dept. of Navy, DIA, NASA, and US Dept. of Labor.

EDWARD G. "ED" STEPHANY, CAPT, born March 13, 1931, New York, NY, earned BA, LLB and JD degrees.

He joined the service Sept. 17, 1952. His military locations and stations: Flight training, Weeksville, NC, and others. He served with ZP-1 and ZTU. Reserve duty 1958-83: VP Fasson 743 VS-743, VS-24, VP-743 and RTU-62. He achieved the rank of CAPT (O-6) and was discharged March 1, 1983.

His memorable experiences: Retirement March 13, 1961, 32 years of naval service - three commands, three types of aircraft, wide variety of aircraft flown, command of VS-743, VP-743 and RTU-62.

He and his wife, Dolores, have three children: Edward III, Pamela and Kurt. Civilian employment as a lawyer until his retirement March 13, 1991.

ALLAN W. STEPHENS, CDR, born Jan. 13, 1919, Rahway, NJ, earned his BA degree at Duke University.

He joined the service in June 1944. His military locations and stations: USS *Mattaponi, Titania, Penobscot, Paiute* and *Karin*.

Stephens was awarded the American Campaign Medal, European Theatre Campaign Medal, Pacific Theatre Campaign Medal and Korean Service Medal. He achieved the rank of CDR and was discharged in October 1953.

He and his wife, Muriel E., have three children: Ellen, Sue and John; eight grandchildren; and six great-grandchildren. Civilian employment as a manufacturing CEO. Stephens retired in 1979.

HORACE G. "STEVE" STEVENS, CDR, born Aug. 18, 1908, Los Angeles, CA, attended the University of California Law School.

He volunteered for duty Aug. 18, 1940. Sent to Great Lakes Training Station. Became assistant aide to ADM John Downes of 9th Naval Dist. Was then sent to Minelaying School, Yorktown, VA. Was given command of minesweep "Endurance."

His memorable experience: Rescuing 42 men off of torpedoed tanker. Leading the attack on beaches, Marianas-Palau, Leyte.

His wife is deceased. He wrote the commissioning speech for Nautilus Sub. Civilian employment as lobbyist in Marine Affairs. He retired in August 1964.

RICHARD G. STEVENS, CDR, born Dec. 29, 1925, Chicago, IL, served in the USNR 1943-54 and 1960-85 (active duty 1943-45, 1948-52, 1973, 1977-81). Life member, NRA. In combat as a fire controlman aboard *Intrepid* in the Pacific throughout 1944-45. Naval Reserve Intelligence Program 1960-85. "ACDUTRA" or "TEMAC" aboard *Kitty Hawk, Nimitz, Independence, Patwingone* and *Operation Homecoming* among others.

Earned his PhD in political science, University of Chicago, 1963. Carnegie Fellow, Harvard Law School, 1962-63. Fulbright Fellow, Hong Kong, 1986-87. Taught at Universities of Santa Clara, Washington and Waterloo (Canada) and Georgetown University, Defense Intelligence College, University of Hong Kong, National Defense University and the Institute of World Politics. Author of *American Political Thought, The Political Thought of American Statesmen, Frankfurter and Due Process, The American Constitution and its Provenance, The Sober Justices* and numerous articles in political science and law journals.

Stevens married the former Norma Jean Duncan in 1949. They have three children and seven grandchildren.

JOHN C. "JACK" STEVENSON, CAPT, born Nov. 23, 1929, Pittsburgh, PA, earned bachelor of architecture at Carnegie Institute of Technology.

He joined the service June 1, 1953. His military locations and stations: OCS, Newport, RI, July-November 1953; USS *Coggwell* (DD-651), November 1953-September 1957; COMPOSIT Co. 11-1, Sept. 11, 1957-April 5,

1960; Communications Div. 11-1, April 6, 1960-April 30, 1963; Harbor Defense Div. 11-1, March 1, 1963-May 22, 1964; Inshore Undersea Warfare Div. 11-1, May 22, 1964-June 30, 1969; Inshore Undersea Warfare Div. 11-3, July 1, 1969-June 30, 1970; CO, Military Training Div. 11-13, July 1, 1970-June 30, 1971; CO, Inshore Undersea Warfare Div. 11-3, July 1, 1971-Dec. 31, 1973; CO, Inshore Undersea Warfare Div. 3121, Jan. 1-June 30, 1974; Weapons Engineering Co. 121, July 1-Nov. 30, 1974; Military Sea Command 11-11, Dec. 1, 1974-June 30, 1975; General Volunteer Training Unit 321, July 1, 1975-Feb. 28, 1977; Naval Sea Det. 419, March 1-Aug. 30, 1977; General Volunteer Training Unit 1911, Aug. 31, 1977-Sept. 30, 1980; CO, Harbor Defense Coordination Component 119, Oct. 1, 1980-Sept. 30, 1982; General Volunteer Training Unit 1911, Oct. 1-Nov. 30, 1982; CO, Inshore Undersea Warfare Gp. One, Dec. 1, 1982-Sept. 30, 1984; Volunteer Training Unit 1511, Sept. 30, 1984-June 30, 1985; retired July 1, 1985.

Stevenson was awarded the China Service Medal, National Defense Service Medal, UN Service Medal, Korean Service Medal and Naval Reserve Medal.

His memorable experiences: Training younger officers to succeed and take command; counselling enlisted; and command tours (5).

He married the former Barbara Ann Wolfe and they have three children: John III, Kimberley Ann and Cameron Leigh; and six grandchildren: Candice Susann, John Charles IV, Joseph Michael, David Lee, Katelyn Nichole and Lauren Kimberley. Civilian employment as an architect - John C. Stevenson Architect, Inc.

JACK A. STEWARD, CDR, born Aug. 23, 1918, Moosejaw, Canada, earned his ME degree at Iowa State in 1941.

He joined the V-3 program Dec. 16, 1936 and the USNA V-7 program, Jan. 6, 1942. He served aboard the USS *Perry* DMS-17, USS *Butler* DMS-29 and USS *Jeffers* DMS-27 during WWII. He joined the USNR Feb. 1, 1946, Des Moines, IA. Served with Com 6 (Korean War) May 1, 1951. From May 1, 1953-Oct. 15, 1958 he served with the USNR Des Moines unit and Surface Div. 9-56.

Steward was awarded the Bronze Star Medal, Naval Unit Commendation, Naval Reserve Medal, American Area Medal, Asiatic-Pacific Campaign Medal w/8 stars, Victory Medal and Navy Occupation Service Medal.

His memorable experiences: USS *Perry* hit a mine at Palau and was sunk, eight men lost their lives. USS *Jeffers* swept the path for the USS *Missouri* into Tokyo Bay. He was young and enjoyed it all, but would not do it over again.

He and his wife, Marlyce G., have two children, Robert and Elaine, and four grandchildren: Robert Jr., Carl, Austin and Alison. Civilian employment as project manager, Firestone, 1946-81.

JOHN HAROLD STEWART, CAPT, born March 25, 1933, Kingston, NY, earned his BS degree at Sunymaritime College and his MBA degree at New York University.

He joined the service June 2, 1954. His military locations and stations: USS *Quail* (MSF-377) Naval Port Central, ComNav Base, New York City. He served with MineLant.

Stewart achieved the rank of CAPT and retired in September 1977. His memorable experiences: US Naval War College command, staff and senior officers courses and Armed Forces Staff College.

He and his wife, the former Jeanne Fennell, have five children: Alane, Michael, John, Sean and Keara; and three grandchildren: Joseph, Nickolas and Brianna. Civilian employment as IBM controller at the division, group and corporate level. Stewart retired March 31, 1996.

KENT A. STEWART, LT, born April 15, 1949, Crawfordsville, IN, earned his BS degree in 1975 at Purdue University and his MS degree in 1976 at Indiana University.

He joined the US Army in November 1968. His military locations and stations: Ft. Gordon, GA; Ft. Stewart, GA; NAS Jacksonville; and Cheatham Annex. He served with the 134th AHC, 365th Engr. Bn., 120th ARCOM, NMCB-14, and NRCHTB. He served in Vietnam from 1969-70.

Stewart was awarded the Army Commendation Medal, National Defense Service Medal, RVN Campaign Medal, Armed Forces Achievement Medal, Army Expert Rifle and Navy Expert Rifle and Pistol. He achieved the rank of LT and was released from active duty in September 1971.

His memorable experiences: Received direct commission as ENS, CEC, USNR in 1987.

He and his wife, Katherine, have two children, Jennifer and Robert. Civilian employment as safety engineer, Southern Div. NAVFACENG. Stewart plans to retire April 15, 2001.

RICHARD W. "DICK or STEW" STEWART, CDR, born Sept. 12, 1931, Tipton, IN, earned his BS degree in radio and television in 1957 at Indiana University.

He joined the USN in 1951 and served until 1955. Was an officer in the USNR from 1966-91. His military locations and stations: Great Lakes, boot camp/Electronics School. He served aboard the USS *Adria* PCS-1387, USS *Paiute*, experimental facility, CINCLANTFLT, US Forces, Caribbean.

Stewart was awarded the Good Conduct Medal, American Defense Medal, Expert Pistol, Sharpshooter Rifle and Naval Reserve w/4 clasps. He achieved the rank of CDR and was discharged in September 1991.

His memorable experiences: Serving with US forces Caribbean at Key West and on the island of Dominica. Serving alone as a local station operator on Grand Bahama Island for nine months.

He and his wife, Darlene, have two children, Brian and Lynne. Civilian employment as a broadcaster for 38 years (sportscaster) and owner of a Century 21 Real Estate Co. in Kokomo, IN. Stewart is an auctioneer.

RICHARD T. STIERER, CAPT, born April 21, 1929, New York, New York, earned his BS degree in 1969 at New York University and his JD degree in 1951 at the University of Florida.

He joined the service in November 1951. His military locations and stations: USS *F.M. Robinson* EDE-220, Korean War; Naval Reserve Center, West Palm Beach, FL (27 years). He participated in the Korean War.

Stierer was awarded the Navy Occupation Service Medal, Naval Reserve Medal and National Defense Service Medal. He achieved the rank of CAPT and transferred to retired reserve July 1, 1982.

He married the former Gloria Steffens, CAPT. Civilian employment as an attorney at law.

WILLIAM L. "BILL" STIFF, CAPT, born July 1, 1928, Oak Grove, VA, earned his BCE degree at the University of Virginia.

He joined the service in March 1953 (commissioned from NROTC at the University of Virginia). His military locations and stations: Active duty - USS *New Jersey*, March 1953-March 1956. Reserves - Inshore Undersea Warfare Div. 5-2 (CO from July 1968-July 1971). He served in the Korean War from April-November 1953.

Stiff was awarded the National Defense Service Medal, Korean Service Medal, Korean Presidential Unit Citation, UN Service Medal, European Occupation Medal, China Service Medal and Armed Forces Reserve Medal. He achieved the rank of CAPT and retired July 1, 1988.

His memorable experiences: Three years on the USS *New Jersey*.

He and his wife, Betty, have three children: Karen, W. Timothy and Charles L.; and two grandchildren, Kaitlin and Christina. After separation from active duty in 1956, he and his family settled in Richmond, VA. Until retirement in 1993 Bill worked for Richmond companies in various engineering/construction capacities: project manager and manager of construction administration for A/E design firms; director of construction on projects throughout the USA and in Nicaragua; director of facilities development for Best Products Co.; and president of Construction Management Assoc. Stiff retired Oct. 15, 1993.

MARK R. STILLWELL, CAPT, born Nov. 4, 1945, Springfield, MO, graduated in 1967 from Drury College, served four years as an enlisted reservist, and was commissioned through the Reserve Officer Candidate program in 1967.

He served on active duty from 1967-70 as deck division officer and public affairs officer aboard USS *Princeton* (LPH-5), acting 1LT aboard USS *Talladega* (LPA-208), and executive officer at Military Sealift Command Office, Inchon, Korea. Following active duty he served as executive officer of four units and as CO of four units at the Naval Reserve Center, Springfield, MO before retiring in 1991.

In civilian life he has worked since 1972 as Sports Information director, Southwest Missouri State University, Springfield, MO. His wife, Tina, is director of News Services at SMSU.

JOHN EASTWOOD STINCHFIELD, CDR, born Nov. 13, 1909, East Millinocket, ME, graduated from the University of Maine in 1934. He was a West Coast buyer for Montgomery Ward and was commissioned a LTJG in the Federal Building in San Francisco July 3, 1942. His first duty was at the Supply Dept. Navy Yard, Pearl Harbor, and eventually he became officer in charge of purchasing for Contractors Pacific Air Bases for all construction requirements for advanced bases. He was next ordered to the Aviation Supply Office, North Philadelphia in 1945, and became the West Coast field service representative for the ASO.

His first independent duty was supply and fiscal officer at NAS Moffet Field, CA, for two years before being ordered to Tokyo as the supply officer of the newly formed Fleet Air Wing Six in Iwakuni to supply air support operations during the Korean War. The next two assignments were S&FO of the Naval AS, Glenview, IL, the HQ for the Naval Air Reserve Command, for a three-year tour. Then Stinchfield went on to NAS Los Alamitos, CA, for a two-year tour.

The high point of his career was to have a three-year tour with MAAG (Military Assistance Advisory Gp.) Japan with his family. In Tokyo he was in logistics initially and then acted as protocol officer for the Joint Command under MGEN Rogers, USAF.

When it was time for his final tour, he obtained permission from CinPac to have concurrent travel with his family to continue travel on by air around the world to his final (and fourth) Naval AS as S&FO at Willow Grove, PA.

CDR Stinchfield ended his 21-year career in the USN by being asked by a Japanese VADM Orita to escort him and his aide to several USN Supply Commands in the US. He is entitled to eight campaign and service medals.

The Stinchfield family consisted of his wife, Pauline;

daughter, Shanna; son, John; daughter, Lida; two granddaughters, Shanna and Noel; and two grandsons, Alexis and Ryan.

JANICE C. STINSON, CAPT, born Jan. 14, 1952, and raised on Oahu, HI, began her naval career as an officer candidate hospitalman at the University of Hawaii in 1973 and received a BS in nursing in 1974. Following Officer Indoctrination School, she was assigned as charge nurse at Naval Hospital, Jacksonville, FL.

Following release from active duty CAPT Stinson worked on perinatal units in Yokosuka, Japan and Kailua, HI before returning to the University of Hawaii where she received a MS degree in maternal-child nursing in 1985. She has taught nursing at the Universities of Hawaii and San Francisco and California State University, Hayward. She currently works in labor and delivery at Alta Bates Medical Center in Berkeley and is a PhD candidate in nursing at the University of California, San Francisco.

Since affiliating with the Ready Reserve in 1979, CAPT Stinson performed as charge nurse, VTU 3005 (Yokosuka, Japan), assistant training officer, MEDCRU 120 (Newport, RI), training officer of ABFC 120 (Pearl Harbor, HI), administrative officer (Naval Hospital Oakland 120), executive officer (Naval Hospital Oakland 420, Mare Island) and has had three commands: Naval Hospital Oakland 420 and 220 (San Bruno) and 4th Marine Div., 2/23, E Co. She currently is REDCOM 22 Staff Nurse, Everett, WA. She was recalled to active duty for Desert Storm at Naval Hospital Oakland.

Her personal awards include Naval Commendation Medal, Naval Achievement Medal, Meritorious Unit Commendation Medal, National Defense Service Medal and the Armed Forces Reserve Medal.

CAPT Stinson and her husband, LCDR Ralf Stinson II, USN, RET, reside in Alameda, CA. They have two daughters, Akiko and Emiko.

SUSAN C. STOCKER, LCDR, born July 15, 1957, Elmhurst, IL, earned her BS degree in physical education/sports management in 1980 at Towson State University, Maryland.

She joined the service April 27, 1987, direct commission USNR. Her military locations and stations: April 1987-December 1988, RESPATWINGPAC 0179, NAS Point Mugu; January 1989-February 1992, RESPATWINGPAC 0180, NAS Moffett Field; March 1992-October 1993, ASWOC 0389, NAS Whidbey Island; October 1993-January 1996, RESPATWINGLANT 0186, NAS Norfolk; and February 1996-present, TACSUPPCEN 1079, NRRC Memphis.

Her memorable experiences: RIMPAC 1990, NAS Barbers Pt. and Northern Viking 1995, NAS Keflavik.

She and her husband, Bill (active duty USMC), were married in 1990. They have two sons, Michael and Robert. Civilian employment: 1980-84, youth director/outdoor pools manager, Annapolis Naval Complex; 1984-89, youth director, MWR NAS Point Mugu; 1989-90, computer operator, range instrumentation PMTC Point Mugu; 1990-93, computer operator, data processing, NAS Whidbey Island; 1993-present, mother/homemaker.

HOUSTON H. STOKES, CDR, currently professor of economics at the University of Illinois at Chicago, has written over 70 articles and four books in the areas of applied econometrics, time series, monetary economics, economic history and software design. He is the grandson of Richmond Pearson Hobson RADM and the son of LCDR W.E.D. Stokes Jr. Professor Stokes, former head of the Dept. of Economics (1982-87) and a member of the faculty since 1967, was educated at Cornell (BA 1962) and the University of Chicago (MA 1966 and PhD 1969).

He served in the USN for two years (1962-64) on the USS *Shasta* (AE-6) and for 20 years in the reserves in BUSHIPS 9-1 and NAVENG 9-1, retiring in 1984 with the rank of CDR. Professor Stokes has consulted widely in the financial area, the health economics area, the energy economics area and in software development.

Professor Stokes' main research agenda in recent years has been developing and applying diagnostic procedures for the specification of econometric models. To implement these procedures, he has developed the B34S® data analysis program, which has been used in a number of fields, including economics, business forecasting, health economics, labor economics and political science.

The Stokes family consists of wife, Diana, and sons, William and Houston, both currently at the University of Chicago Laboratory School.

LINUS M. "LINI" STOLL, CDR, born Jan. 30, 1926, LaCrosse, WI, earned his BS degree in chemical engineering at the University of Wisconsin, Madison, WI.

He joined the service Nov. 7, 1943. His military locations and stations: USNTS V-12, Madison, WI. He served aboard USS *Major* (DE-796), USS *Vogelgesang* (DD-862) and with CTCD 23 staff.

Stoll was awarded the American Theater Medal, WWII Victory Medal, Occupation Medal, Naval Reserve Medal, National Defense Service Medal and Armed Forces Reserve Medal. He achieved the rank of CDR and was discharged Dec. 7, 1952.

His memorable experiences: Mediterranean Sea cruise in 1951 (side trips: Rome [audience Pope Pius XII], Damascus, Syria, Venice, Athens and Volos, Greece); XO NRS Div. 9-224; CO NRS Div. 9-222.

He and his wife, Jeane, have four children: Michael, Barbara, Steven and Mary; and 11 grandchildren: Brian, Jenny, Becky, Jason, Ben, Matt, Samantha, Amanda, Zachary, Sam and Jacob. Civilian employment as chairman and CEO, Wisconsin Public Service Corp. Stoll retired Jan. 30, 1986.

GEORGE F. STOLTZ, CDR, born March 28, 1933, St. Paul, MN, earned his BA degree at the University of Minnesota, Duluth.

He joined the service Nov. 13, 1955. His military locations and stations: OCS Newport, RI; Bombardier Navigator School; NAS Sanford, FL. He served with VAH-3, VR-811, VP-811 and NARSV1. Stoltz was a Blue & Gold officer at the USNA. He achieved the rank of CDR and was discharged from active duty March 28, 1956. He transferred to retired reserves Aug. 5, 1982.

He and his wife, Kathryn, have two daughters, Julie and Jennifer. Civilian employment with Paine Webber Inc. for 30 years where he has served as vice-president.

EUGENE H. "GENE" STOMBLER, CAPT, born June 18, 1927, Philadelphia, PA, earned his bachelor's degree in mechanical engineering and certification in business and management.

He joined the service May 15, 1945. His military locations and stations: Great Lakes; Delmonte, CA; Corpus Christi, TX; Anacostia, D.C.; and Andrews, MD. He served with BARTU 661, WEPTU 661 and NASRU 661.

Stombler was awarded the WWII Victory Medal and Naval Reserve Medal. He achieved the rank of CAPT (O-6) and was discharged Aug. 9, 1946. He retired from the USNR in 1972.

His memorable experiences: Winning Noel Davis Trophy for best reserve unit of its type while being the CO of the unit (NASRU 661).

He and his wife, Charlotte, have four children: Ben, Steven, Fay and Miriam; and four grandchildren: Sophia, Hannah, Nadia and Mia. Civilian employment with the Dept. of the Army (8 years); Westinghouse (37 years), was manager of printed circuit operations and process control. Stombler retired from Westinghouse Dec. 30, 1992.

RAYMOND W. "RAY" STONE, TMC(SS), born Dec. 12, 1943, Seattle, WA, earned his AA degree in business management at Northwood Institute in 1987; his BBA w/major in management at Northwood in 1988; and his MS degree in management at Troy State University in 1990.

He joined the service May 28, 1963. His military locations and stations: NRED 6-49 (1963-64), SUBDIV 5-8 (1964), NR AS-18 ORION DET 406 (1975), NR AS-37 DIXON DET 3710 (1978), CNSRF CODE 32 (1985), and CNRF CODE 10 (1986-90). He served with the USS *Sea Cat* (SS-399), USS *Carp* (SS-338), USS *Grampus* (SS-523), USS *Runner* (SS-476), USS *Drum* (AGSS-228), USS *Torsk* (AGSS-423), USS *Trumpetfish* (SS-425), USS *Picuda* (SS-382), USS *Sirago* (SS-485), USS *Haddock* (SSN-621) and Trieste II (DSV-1).

Stone was awarded the Armed Forces Reserve Medal w/star, Good Conduct Award, Navy Achievement Medal, Expert Pistol Medal, Marksman Rifle Medal, National Defense Service Medal, Naval Reserve Meritorious Service Medal w/4 stars, Navy Commendation Medal, and Qualified SS Designator. He achieved the rank of TMC (SS) and retired from the USNR Oct. 1, 1990.

His memorable experiences: TEMAC recall 1985-90.

He is divorced and has one daughter, Jeanene Lynn. Civilian employment as program manager for VISTA Computer Services, Inc.

WILLIAM C. "BILL" STOOKEY, CDR CEC, entered the Navy V-12 program in July 1943 at that program's inception and left with a degree in civil engineering from CalTech in February 1946. He missed the benefit of a full midshipman's training because many war-time facilities were then being closed. He was assigned to the Civil Engrs.' Corp, and in July 1946, was one of 20 ENS sent to Port Hueneme to relieve many of the war-time active-duty officers then being discharged. This was an era of extreme activity at Port Hueneme which was still receiving large quantities of material, equipment, and men from the Pacific Islands. Leaving active duty in 1947 he remained in the reserve and spent the two years prior to retirement as CO of Reserve Mobile Construction Bn. 16.

He continued his career in civil engineering receiving an MCE from MIT and served nine years as city engineer for the city of Fullerton, CA. In 1964 he founded Willdan Assoc., a consulting firm specializing in municipal engineering where he continues as chairman of the board. Willdan today has 12 offices in three western states providing services to over 200 communities. He considers one of his major accomplishments the engineering to convert the agricultural community of Dairy Valley, CA, to what is now the model city of Cerritos, home of approximately 70,000 people.

Stookey's hobbies include traveling in Russia, particularly while it was still under communist rule, and aquaculture. He and his wife, Betty, have two children, Lori, a biologist and Chris an MD.

THEODORE S. "TED" STORCK, CDR, born Feb. 27, 1936, Morris, MN. In 1956 he joined the USN as an E-1 taking recruit training at Great Lakes. Then he was assigned to ComFAirWingOne in the Far East. He was discharged in 1958.

He received his bachelor's degree from the University of Minnesota in 1963; worked at *The Forum* newspaper in Fargo, ND. He received his commission as an ENS in 1965 and went on active duty with the Office of Information in the Pentagon in 1966. That fall he was sent to Syracuse University where he earned his master's degree in 1967.

He then went to Vietnam, serving with ComNavForVietnam as photo officer. There he was awarded the Bronze Star w/Combat V. Then it was to ComEleven in San Diego and back to the Pentagon where he left service in 1972.

Storck was given a reserve commission and reached the grade of CDR. He was awarded the Bronze Star, RVN Campaign Medal, Vietnam Service Medal, Combat Action Ribbon, Navy Unit Commendation, National Defense Medal and Armed Service Medal.

After he left the USN, he owned and managed radio stations in Wayne, NE; Ukiah and Red Bluff, CA; and Thief River Falls, MN. He now lives in Penn Valley, CA.

GEORGE A.W. "JUDGE" STOUFFER, CDR, born Oct. 11, 1911, Mechanicsburg, PA, earned his doctorate at the University of Pittsburgh (psychology).

He was commissioned Feb. 2, 1942. Active duty from March 21, 1942-Feb. 28, 1951. His military locations and stations: Great Lakes, Farragut USNTS, USN Disciplinary Barracks, Norfolk. He served with Post Psychologist, USNDB, USN Hospital, Yokosuka, Japan as senior clinical psychologist, 1st Marine Div., Korea.

Stouffer was awarded the American Campaign Medal, WWII Victory Medal, Navy Occupation Service Medal w/ Asia Clasp, Korean Service Medal, UN Service Medal, Naval Reserve Medal and Presidential Unit Citation. He achieved the rank of CDR, USNR and was discharged from active duty Aug. 4, 1946. He retired from the reserves July 27, 1952.

He and his wife, the former Anne Kerr, have four children: Mary Jean, Susan Anne, George "Rick" and Bruce. Civilian employment as dean, College of Education, Indiana University of Pennsylvania. Stouffer retired Sept. 1, 1976.

JOHN P. STREBEL, CAPT, born and raised in San Francisco, CA, graduated from San Francisco State University and commissioned at Newport, RI in February 1966. Following Weapons Officers School, he served on board the USS *James E. Kyes* (DD-787) as gunnery officer and 1LT, deploying twice to Vietnam.

Following active duty, tours included the USS *Cowell* (DD-547), USS *James C. Owens* (DD-776), USS *Hamner* (DD-718), CO, FTG 220 and FTG 120, SELRES coordinator, USS *Pyro* (AE-24), and various staff billets with CDR Third Fleet, US Maritime Zone Pacific, CinC Pac Fleet Det. 120, MARDEZPAC, sector Hawaii and CinC Pac Flt. Det. 920.

CAPT Strebel's awards include the Coast Guard Achievement Medal, Combat Action Ribbon, Naval Reserve Sea Service Ribbon, Armed Forces Expeditionary Medal, Vietnam Service Medal w/3 Bronze Stars, RVN Gallantry Cross and RVN Campaign Medal and is a designated surface warfare officer.

In civilian life CAPT Strebel is employed with the San Francisco Unified School Dist. in the Gifted and Talented Program and has two wonderful children, Scott and Shelly.

GEORGE F. STRINGER, CAPT, born Oct. 1, 1949, Kansas City, MO, earned his BS degree in naval architecture and took numerous business and military courses.

He joined the service June 28, 1968. His military locations and stations: USNA, USS *Nespelen* (AOG-55); USS *Pegasus* (PHM-1). He served with sup ships Pascagoula, MS; NAVSEADET 705; and ONRHQDET 106.

Stringer was awarded the Navy Achievement Medal, Navy Commendation Medal (2) and National Defense Service Medal (2).

His memorable experiences: Qualified as surface warfare officer and engineering duty officer with over 12 NOBC's assigned.

He and his wife, the former Suzanne E. Clarke, have three children: George C., James M. and David M. Civilian employment as general manager, Shenandoah Enterprises.

DAVID B. "DAVE" STRONG, CAPT, born Oct. 12, 1940, Vancouver B.C., Canada, earned his BA degree in 1963 and his JD degree in 1971, both at the University of Washington.

He entered the Naval Officer Candidate School, Newport, RI and received his commission as ENS, USNR in May 1963. ENS Strong attended the Naval Justice School course for non-lawyer legal assistance referral officers, and in August 1963, reported to USS *Mauna Kea* (AE-22) for duty as gunnery officer and assistant cargo officer. After service in the South China Sea following the Gulf of Tonkin Incident, he was detached to NAF El Centro, CA as legal assistance referral officer and assistant administrative officer, where he served until released from active duty in May 1966.

Returning to Seattle LT Strong affiliated with the USNR and served in NR Telecom Unit 13-1 and NROS Unit 13-1 while attending Law School. He was promoted to LCDR as a line officer. After receiving his JD degree he changed to a 2505 designator and affiliated with NR Law Co. 13-1, where he advanced to the position of CO. Subsequently, he served in VTU (Law) 2208, Naval Reserve Military Plans 2219, and NR LSO 122 in Seattle and Bremerton, holding various billets, culminating in the position of CO, NLSO Seattle 122. He was promoted to the rank of CAPT in June 1984.

In civilian life, CAPT Strong advanced to chief deputy prosecuting attorney for Island County, WA, before entering a 15 year career as a solo practitioner. He has served variously as president of the Skagit Valley College Board of Trustees; the Skagit Valley College Foundation; the Burlington Chamber of Commerce; the Skagit County Economic Development Council; and Burlington-Edison Kiwanis Club.

CAPT Strong is now semi-retired from the practice of law, managing various property and financial interests. He is divorced and has two children, Matthew and Erin.

FRED CLEGG STRONG, CDR, born Nov. 4, 1942, Baton Rouge, LA, earned his BS degree at Louisiana State University and his JD degree at Tulane University.

He joined the service Sept. 13, 1969. His military locations and stations: NR LSO Newport 310, New Orleans, LA and NR LSO Pensacola 310, New Orleans, LA.

Strong was awarded the National Defense Service Medal and Armed Forces Reserve Medal.

He and his wife, the former Isabel Bandy, have two children, Isabel Teissier and Emily Caldwell. Civilian employment as an attorney at law. Strong retired Jan. 1, 1990.

CHARLES L. "SKIP" STUART JR., born Sept. 5, 1938, Albany, NY, earned his BS degree in physics at Siena College, Loudonville, NY.

He joined the service in November 1961, attended Officer Candidate School, Newport, RI. Commissioned ENS USNR in April 1962. His military locations and stations: USS *Kennebeck* AO-36, San Francisco Naval Shipyard, July 1962-February 1964; USS *Interdictor* AGR-13, February-October 1964; Officer's Naval Nuclear Power School, Bainbridge, MD; Naval Nuclear Power Training Unit D1G, West Milton, NY; Officer's Submarine School, New London, CT, November 1964-June 1966; USS *Alexander Hamilton* SSBN-617, Blue Crew, assigned as communications and sonar officer.

Completed two Polaris patrols and was assigned Reactor Controls Div. during major shipyard overhaul at Electric Boat in Groton, CT. Released from active duty in June 1968 and immediately entered the active USNR program serving in increasing responsible positions, including CO of USS *Canopus* Submarine Support Unit in Liverpool, NY, as well as several tours as CO of the Volunteer Training Units until retirement Oct. 1, 1984, at the rank of CAPT. CAPT Stuart attended the Naval War College and the Armed Forces Staff College as a senior reserve officer.

Stuart was awarded the Polaris Patrol Pin w/star, National Defense Service Medal, Armed Services Medal and Naval Reserve Medal.

He and his wife, Anne, have four children: Scott, Brian (LT USNR), Sean and Tracy. Civilian employment: 23 years of commercial nuclear power plant experience at Nine Mile Point Nuclear Stations and Grand Gulf Nuclear Station. Six years as director of BWR Services with ABB Combustion Engineering Nuclear Operations, Windsor, CT.

DAVID G. STUBBEMAN, CAPT, born Aug. 28, 1938, Midland, TX, received a BA degree from Southern Methodist University and a JD degree from The University of Texas School of Law. He was commissioned Nov. 22, 1963, at the Officers Candidate School, Newport, RI and attended Naval Justice School at Newport.

He served as legal officer at NAS Adak and at NAAS Saufley Field, Pensacola, FL. In 1966 he left active duty and joined the Wagstaff Law Firm in Abilene, TX, where he practiced law for 30 years. He is a certified civil trial specialist. He was selected for CAPT in 1981 and retired from the USNR with 23 years of active and reserve service.

He is married to the former Sue Swenson and they have two children, Bill, a physician in Los Angeles, and Susan, an investment banker in New York City.

SHELDON D. STUCHELL, LCDR, born April 26, 1952, Buffalo, NY, earned his BS degree in business administration at Rochester Institute of Technology.

He joined the USN from March 1972-October 1980. From October 1980-present he has served in the USNR. His military locations and stations: 1973-77, USS *Kamehameha* (SSBN-642) Gold; 1977-80, Naval Submarine School, New London, CT; 1980-83, SIMA Chasn; 1983-1987, Fleet Training Gp.; 1987-1989, USS *Kalamazoo*; and 1989-present, DTINWU 214.

Stuchell was awarded the Navy Achievement Medal (2), Good Conduct Medal (2), Expert Pistol Medal, Expert Rifle Medal and US Coast Guard Joint Operations Medal.

He and his wife, the former Ria E. Hagen, have one daughter, Krista M. Civilian employment as North American customer marketing manager for Northern Telecom.

WILLIAM ERVIN "BILL" STUCKER, CAPT, born Feb. 15, 1923, Billings, MT, earned his AB degree at Hanover College and his DDS degree at Indiana University.

He joined the service Sept. 7, 1942, Louisville, KY. His military locations and stations: V-12 program, Wabash College; Midshipmen School, Columbia University, New York City, 1943, commissioned ENS; CO, Naval Training Station, Camp McDonough, Plattsburg, NY; Port Directors School, Port Hueneme, CA, 1943; port director, US Naval Operating Base, Argentia, Newfoundland 1944-46; traffic manager, US Naval Supply Depot, Argentia, Newfoundland 1946-50.

He transferred from Line to Dental Corps in 1957; Dental Co. 4-4, Cincinnati, OH, 1961-67; Dental Co. 5-5, Louisville, KY, 1967-75; executive officer, Dental Co. 5-5, 1969-72; CO, Dental Co. 5-5, 1972-75; liaison officer, University of Louisville Dental School, 1975-81; USNA information officer (Blue & Gold), 1977-date; dental recruiting officer, Louisville Recruiting Dist., 1977-81; VTU 0909, Louisville, 1976-81.

Stucker was awarded the Presidential Meritorious Service Medal. He achieved the rank of CAPT and retired Feb. 15, 1983.

He and his wife, the former Rose Mary Reddy, have three children: Joanne, Robert and Lisa. Civilian employment in private dental practice. He retired July 1, 1987.

JAMES WILLIAM "BILL" STUDDARD, CDR, born Feb. 6, 1950, Emory University, GA, earned his BBA degree at North Georgia College and his MBA degree at Georgia State University.

He joined the service March 5, 1971. His military locations and stations: COMIBERLANT (NATO) staff, Oeiras, Portugal. He served aboard the USS *Connole* (FF-1056), Norfolk, VA.

He and his wife, Diane, have two children, April and Travis. Civilian employment as senior special agent, US Customs Service.

PRESSLEY C. STUTTS JR., LCDR, born Feb. 3, 1957, Flint, MI, earned his BA degree in 1978 and his MDIV degree in 1981.

He joined the service May 9, 1980. His military locations and stations: Raleigh, NC; Knoxville, TN; and Charleston, SC. He served with MEF REL 107, MEF REL 309, and NR CINCLANFLT 0711A. He participated in Operation Desert Storm.

Stutts was awarded the Armed Forces Reserve Medal, National Defense Service Medal and FMF Ribbon.

His memorable experiences: Recalled for Operation Desert Storm (January-October 1991).

He is divorced and has two children, Pressley C. III and James Bartholomew. Civilian employment as president, Executive Communications, Inc.

WILLIAM DAVID "DAVID" SULLINS JR., RADM, born Aug. 3, 1942, Athens, TN, holds a BS and DO degree. He was commissioned a LTJG in the USN Medical Service Corps April 1, 1966. Upon his release from active duty in December 1969, he continued his USNR career as an optometrist providing primary eye care.

1982-85 Health Services support officer 4th FSSG; 1985-86 CO NR NAVHOSJAX-108; 1986-88 CO 4th Med. Bn., 4th FSSG, 4th Marine Div.; 1988-89 executive assistant OP-093R, NR OPNAV-093-106 the Pentagon; 1989-90 principal advisor BUMED code 07, Bureau of Medicine and Surgery. He was selected for Flag 1989. He was deputy director Medical Service Corps and assistant chief BUMED 04 (logistics).

RADM Sullins is the first medical representative to Marine Corps Reserve Officers Assoc. (MCROA); first MSC representative to the Navy Reserve Medical Dental Flag Council. Currently he is president of a professional corporation, Eyecare Clinics, PC of Athens, Madisonville and Etowah, and chief of Eye Services at Woods Memorial Hospital, Etowah, TN; past president, former chairman of the Board of Trustees of the American Optometric Assoc. (AOA).

RADM Sullins has two sons, William David III, an optometrist and Stuart Andrew, in optometry school. His wife's name is Leslie.

JOHN G. "JERRY" SULLIVAN, CAPT, born Sept. 28, 1942, El Paso, TX, earned his BS degree at the USNA in 1967 and his MSAE degree, USN Postgraduate School in 1968.

He joined the service in September 1960 as midshipman USNR at the University of California, Los Angeles. His military locations and stations: USNA; Naval Air Training Command (pilot); NAS Lemoore, CA; and NAS Point Mugu, CA. He served with VA-155, NAVAIRES Point Mugu and NAVAIRES San Diego. Battles and campaigns include two tours in Vietnam.

Sullivan was awarded the Meritorious Service Medal (2), Navy Commendation Medal (2), Vietnam medals and other individual and unit medals. He achieved the rank of LT on active duty and as CAPT in the USNR and was discharged in June 1973.

His memorable experiences: Seven years as midshipman at UCLA and USNA - probably a modern day record!

He and his wife, Barbara, have two children, Erin and Michael. Civilian employment in various aerospace engineering and executive positions. Sullivan plans to retire April 1, 1998, with 30 years commissioned.

LISA HEYDE SULLIVAN, LCDR, born Dec. 7, 1960, Honolulu, HI, earned her BSN degree.

She joined the service in September 1983. Her military locations and stations: NNMC Bethesda, Newport Naval Hospital. She served with Fleet Hospital, Providence, RI, Bethesda 206.

Sullivan was awarded the Meritorious Unit Commendation Ribbon and Navy Achievement Medal.

She and her husband, Timothy M. Sullivan, have two children, "Jack" John Patrick and Ryan Matthew.

JEFFERY H. SUNDAY, CDR, graduated with honors from the USCGA in 1975 with a BS degree where he was a company CDR and brigade protocol officer. Following training in naval communications and cryptography, he served as communications officer, electronics material officer, navigator and later as operations officer aboard USCGC *Firebush* (WLB-393).

CDR Sunday received a direct commission as a LT in the USNR in 1981 as an unrestricted line officer. His SELRES assignments have included tours as flag LT of Antilles Defense Command, XO then CO of NR WEPSTA Earle Det 1704 and CO of NR NAVSTA NORVA Det 104. He was staff communications officer for CDR, US Forces Azores and became CO of VTU 0402.

He is a graduate of the US Naval War College, with distinction.

CDR Sunday and his wife, Peggy, reside in Scranton, PA. They have three children: Andrew, James and Julie.

JOHN R. "JACK" SUNDINE, CAPT, born April 6, 1920, Moline, IL, earned his BS degree in journalism at Northwestern University, Evanston, IL.

He joined the service Oct. 6, 1942, USN Midshipman's School, Notre Dame University, Indiana. His military locations and stations: ATB, Ft. Pierce, FL; Pacific Theater; and San Francis, CA. He served with ComPhibLant, AdComPhibsPAC, ComTrans Ron 18, and ComTrans Ron 1.

His memorable experiences: Construction officer, 23,000-man base, Ft. Pierce, FL with garbage division and fire department; Navy motor pool O-in-C, San Francisco.

He has three children: Stephanie, Krista and John P.; and three grandchildren: Francesca, Katherine and Stephanie. Civilian employment as editor and co-editor, *Moline Daily Dispatch*, freelance writer.

DAVID ERNEST SWAFFORD, CAPT, born Jan. 26, 1943, Oklahoma City, OK, earned his BA degree in 1965 at the University of Oklahoma and his MBA degree in 1977 at the University of Central Oklahoma.

He joined the service in June 1965. His military locations and stations: Active duty - USS *Yorktown* (CVS-10). Reserve duty units: VTU 1108; NCSO Canal Zone 111; MSCO Korea 111; MSCO Northern Japan 211; DDG-20 GOLDSBOROUGH; FFGI BROOKE, Tulsa, OK; MAC G 1108, Oklahoma City, OK; FTG 3911, Oklahoma City, OK; GEN IRU 1111, Oklahoma City, OK; NROS 8-3 (M), Oklahoma City, OK; NROS 8-46 (M), Oklahoma City, OK; Training Support Unit 8-40; NRMTF 8-12, Oklahoma City, OK; NRSD 8-46 (M), Oklahoma City, OK; and NRG 8-10 (M), Oklahoma City, OK. He participated in the Vietnam War.

Swafford was awarded the National Defense Service Medal, Armed Forces Reserve Medal (2), Vietnam Service Medal w/1 Bronze Star and RVN Campaign Medal w/ device.

He and his wife, Cheryl M., have three daughters: Mandy, Micah and Sarah. Civilian employment as director of international sales, Star Building Systems. Swafford retired Sept. 21, 1991.

FRANKLIN A. "FRANK" SWANSON JR., CAPT, born Jan. 1, 1940, Chicago, IL, earned his BSEE and BNS degrees at the University of Wisconsin and his MSEE degree at Stanford University.

He joined the USNR in 1963 as ENS. His military locations and stations: USS *Arcadia* (AD-23) (communications officer). He served with various surface divisions, naval electronics commands and radiological commands.

Swanson was awarded the National Defense Service Medal, Meritorious Unit Commendation and Armed Forces Reserve Medal. He achieved the rank of CAPT (O-6) USNR and retired Oct. 1, 1985.

His memorable experiences: Acdutra's over period of 20 years; carriers, destroyers, mine sweepers, naval stations (Rota, Spain; Sasebo, Japan; London, England; Treasure Island, CA; and Mare Island, CA).

He married the former Beverly Rudholm and they have two children, Jeffrey C. and Jennifer A. Civilian employment in various engineering manager positions, currently I/O subsystems manager for Hal Computer Systems.

GARY K. SWEET, CWO3, born May 5, 1939, Detroit, MI. In 1954 his family moved to Memphis, TN where he earned his Boy Scout Eagle Award. He attended the University of Memphis. In 1958 he began his military career as an USMCR aviation ordnanceman with VMF-124 at NAS Memphis. While a member of this unit deployed at Yuma MCAS he was awarded the Navy/Marine Corp Medal. Also the Marine Corp Good Conduct Medal.

After three years of honorable service, he joined the USAF as a nuclear weapons specialist. While serving on a B-52 aircrew he earned his Air Crew Wings. Next assignment was at Vandenburg AFB where he earned the SAC Missileman Medal on the Minuteman ICBM project. In 1962 he was transferred to South Vietnam as a munitions maintenance specialist with the Mobile Riverine Forces. After a tour of duty he was transferred to the USAF SAC B-47 Nuclear Weapons Deactivation Team.

While serving in the USAF he was awarded the following decorations: Air Force Commendation Medal, Air Force Good Conduct Medal, National Defense Service Medal, Vietnam Service Medal, RVN Campaign Medal, Presidential Unit Citation, Air Force Longevity Service Award, Air Force Overseas Ribbon, RVN Gallantry Cross w/Palm Unit Citation.

After discharge from the USAF in 1965, he joined Boeing working on the NASA Saturn V Apollo Project. While working for United Technologies on the NASA Space Shuttle Solid Rocket Booster program he was selected NASA's Manned Space Honoree for his accomplishments.

In 1980 he joined the USNR as an APG aviation ordnanceman. He served in various units: VP-67 NAS, Memphis; NAS Jacksonville and currently as a weapons officer on the *JFK* (CV-67) Reserve Support Unit at NB Mayport, FL.

While serving with the USNR he has earned Aviation Aircrewman Medal and the Enlisted Aviation Warfare Specialist Medal. Also the Navy Reserve Meritorious Reserve Medal w/4 Bronze Stars, the second National Defense Service Medal and the Armed Forces Reserve Medal.

He currently resides in the Florence, AL area where he is the NIST/MEP regional director at the Bevill Center of Environmental Technology.

The Sweet family consists of wife, Rosemary Todd of Memphis, TN; son, Louis Todd, a computer programmer; daughter-in-law, Jennifer; two granddaughters, Savanah Lynn and Sydni Dawn; and son, Wesley Keith, a student at the University of New Orleans studying naval architecture.

EARL N. "DUTCH" SWILLUM, LT, born Nov. 15, 1919, Chicago, IL, earned his BS degree in business administration and candidate for MBA-CPA degree.

He joined the service March 9, 1943. His military locations and stations: Abbott Hall, Chicago; mostly Pacific Island. He served with the US Amphibious Forces. Battles and campaigns he participated in were Kwajalein, Saipan, Peleliu and Iwo Jima.

His memorable experiences: Fire explosion - Pearl Harbor (West Loch), watching Marines raise flag on Mt. Suribachi (Iwo) and mothballing LST 121(H).

He and his wife, Sally, have been married for 50 years and they have six children: Michael, Sarah, Kathy, Mark, David and Paul; and seven grandchildren: Jessica, John, Nikki, Erica, Justin, Sarah and David T. Civilian employment as financial staff analyst, Ford Motor Co., for 34 years with retirement Jan. 1, 1981, and director of financial affairs, John Hancock, for three years with retirement Jan. 1, 1984.

TERRY L.M. SWINNEY, LCDR, born Nov. 17, 1946, Olympia, WA, graduated from Olympia High School in June 1965, earned an associate degree from Centralia College in June 1972, his bachelor's degrees in public administration and business administration from the University Puget Sound in December 1978, and his master's degree in business administration at University Puget Sound in May 1980.

He joined the service in April 1966 on the delayed entry FAJC. Boot camp at San Diego, CA in August 1966; Submarine School in November 1966; USS *Henry Clay* (SSBN-625) (Blue) in April 1967; YN/PN "A" School, Bainbridge, MD in September 1967; USS *Cusk* (SS-348),

San Diego, CA in January 1968; COMSUBRON THREE, San Diego, CA in May 1978; COMSUBRON ONE, Pearl Harbor, HI in June 1978; USS *Halibut* (SSN-587) in July 1978; USS *Queenfish* (SSN-651) in August 1978; PAMIPACFLT School in December 1978; and USS *Sargo* (SSN-583) in January 1979.

Swinney was awarded the Enlisted Submarine Qualification, Polaris Patrol Pin, Navy Commendation Medal, Navy Achievement Medal, Navy Unit Commendation, Good Conduct Medal, Naval Reserve Meritorious Service Medal, Navy Expeditionary Medal, National Defense Medal (2), Armed Forces Expeditionary Medal, Vietnam Service Medal, Armed Forces Reserve Medal (2), RVN Gallantry Cross Unit Citation, RVN Campaign Medal, Expert Rifleman Medal (E), and Expert Pistol Shot Medal (E). He achieved the rank of LCDR in March 1992. He was released from active duty in May 1970; discharged in April 1972; entered the reserves in April 1974; and is still an active reserve.

His memorable experiences: Time spent aboard USS *Henry Clay* (SSBN-625) (Blue) in Scotland and WestPac tour with the USS *Sargo* (SSN-583) which included Japan, Singapore, Hong Kong and Philippines.

He married the former Gail E. Nicoletti. Civilian employment at Boeing Co., Philadelphia, PA, currently manager of office and food services, Helicopters Div.; Boeing Co., Seattle, WA, facilities, business management, operations management, contract administration, budget analyst and food service management; Hertz-Rent-A-Car, Seattle, WA, assistant city manager/airport operations; state of Washington, Chehalis, WA, Dept. of Social and Health Services, administrative assistant.

TORREY ALDEN "T.A." SYLVESTER, CAPT, born Nov. 10, 1936, Waterville, ME, earned his BA degree in 1959 at the University of Maine and his LLB degree in 1970 at the University of Maine School of Law.

He joined the service June 2, 1959. His military locations and stations: Reserves - Bangor, Augusta, Portland, ME; CO aboard USS *Hummingbird* (MSC-192). He served aboard the USS *Conway* (DD-507) and USS *Woodpecker* (MSC-209).

Sylvester was awarded the Navy Expeditionary Medal (2) and Vietnam Service Medal. He achieved the rank of CAPT (O-6) and was discharged June 2, 1986.

His memorable experiences: Mercury space shots, Bay of Pigs, Cuban Missile Crisis, mine sweeping, Vietnam bays Da Nang, Cam Ranh, Nha Trang, inspector GEN RESCOMREG ONE, member three selection boards (LT, CDR, CAPT).

He and his wife, Jennifer, have four children: Steven, Samuel, Susannah and Linda; and one grandchild, Jacob. Civilian employment as a lawyer and adult education teacher. Sylvester retired Nov. 10, 1996.

RICHARD JOHN SZATKIEWICZ, CAPT, born Jan. 4, 1939, Chicago, IL, earned his DDS degree at Loyola University, Chicago, IL in 1964.

He served in the 1925I program from 1961-64. Officer indoctrination at Newport, RI in 1963. Served with Naval Administrative Command, Great Lakes, IL, July-December 1964; 3rd Marine Div. FMF, Okinawa, Vietnam, December 1964-January 1966; Naval Administrative Command, Great Lakes, IL, January 1966-July 1967; Bureau of Medicine and Surgery, Washington, D.C., September 1988-September 1991.

Reserve units include: NRDC 9-2 student, 1961-64; NRGC 9-73, dental officer, 1967-74; NROS 9-17, student, 1972-74; COSRIVDIV21, dental officer, 1974-76; RECTRA 713, instructor, 1976-77; VTU 1314, 1977-79; NRDC HQ 113, XO, 1979-81; VTU 1314, CO, 1981-82; NAVRESREDCOM13 staff dental officer, 1982-84; NAVMEDCOMNEREG DET 413, CO, 1984-87; NAVMEDCOMNEREG DET 413 XO, 1987-88.

Szatkiewicz was awarded the Meritorious Service Medal, Navy Commendation Medal, Presidential Unit Citation, Navy Unit Citation, National Defense Medal, Vietnam Service Medal, Armed Forces Reserve Medal, Vietnam Campaign Medal, Vietnam Unit Citation and Navy Sharpshooters Medal. He achieved the rank of CAPT and retired in September 1991.

He married the former Antoinette Kathleen Willis and they have two children, Lisa Marie and Matthew Richard. Civilian employment as a dentist in private practice.

JOYCE TABBS, LT, has more than 24 years working in the Federal Government, specializing in the areas of information systems management, financial accounting and reporting, budgeting, marketing and customer service and contract management.

Earned BS degree in accounting, 1978; master's of public financial management, 1983; certified government financial manager, 1996; master's of general administration - Management Information Systems, 1997.

Employed with Federal Aviation Administration, 1974-75; Navy Dept., 1975-78; Naval Sea Systems Cmd., 1978-86; US Treasury, FMS, 1986-present and as Naval Reserve Officer, 1989-present, lieutenant: Project Manager (secret clearance with Top Secret in progress.

She is the recipient of the Navy and Marine Corps Achievement Medal awarded by the Dept. of the Navy, June 2, 1997.

EDWARD M. "SHREDLANT" TAICSICH, LCDR, born April 16, 1954 in Oak Park, IL, received a BS in biology from Marquette University, Milwaukee, WI, 1976.

He joined the service Dec. 19, 1976. His military locations and stations: active duty: USS *Caron* (DD-970); Norfolk, VA; Reserves: NAS Glenview, units: FIRSTEURLANT 1472; DEFINTEROG 0372; FIRST 1572/AIC1572; DIA HQ 0972; JAC MOLESWORTH.

He received the Battle "E", Sea Service, National Defense, JMUC, 10 years Reserves. Achieved the rank of LCDR and left active duty Dec. 23, 1980.

His memorable experiences: following a Russian task force in Orleans 80, from north of the British Isles to within 22 miles of Murmansk.

He married Marion and they have a daughter, Julia Elizabeth Taicsich. Employed with Abbott Laboratories as product specialist (biochemist).

W. ROBERT "BOB" TALBOT, 0-3, born July 20, 1950 in Manhatten, NY, received a BS in biology and oceanography; MS in geology and professional degree in hydrogeology and geological engineering.

He joined the service April 2, 1970. His military locations and stations: SUBRON 4, Charleston, SC; Sandpoint, NAS Seattle, WA. Served with USS *Trumpet Fish* (SS-425); USS *Pickeral* (SS-524); Reserves; Subgroup 7, Subbase Pearl.

Talbot participated in the Vietnam era. He was awarded the National Defense Medal. Achieved the rank of 0-3 and discharged April 2, 1991.

He is married to Susan Elizabeth and employed with US Geological Survey as hydrogeologist; US Bureau of Reclamation, geologist.

ALFRED F. "AL" TALLEY JR., CAPT, born July 10, 1932 in New Orleans, LA, received a BA from the University of California, Berkeley, 1950; LLB (JD) Boalt Hall, University of California, Berkeley, 1960; Coro Foundation, San Francisco, CA, Internship in Public Affairs.

His military history: Army ROTC, University of California, Berkeley, 1950-51; Air Force ROTC, University of California, Berkeley, 1951-54.

Commissioned 2nd LT, June 13, 1954. Served with USAF, Aug. 27, 1954 to Aug. 26, 1957: Flight School, Strategic Air Command, B-47s; MacDill AFB, Morocco; California Army National Guard, Stockton, CA, 1961-66; USNR, 1966-89. VA 876, NAS Los Alamitos, VA 776, VA 304, VA 303, VTU 8787, Cargru 0387, NAS Alameda.

His memorable experiences: CVWR 30 cruise and evaluation on USS *Ranger,* November 1976; CO VA 303, CO Cargru 0387, CO VTU 8787. Retired as CAPT (0-6).

He married Genevieve and they have four children: sons, Michael and John; daughters, Michelle Harris and Suzanna Shipman; grandchildren: Shelby and Cody Harris, Michael and Eugene Talley.

Employed as civilian lawyer, professional corporation and sole practitioner. Retired USNR August 1989.

THOMAS W. TANNER, LT, a retired Naval Air Reserve officer since July 1974, life member of the NRA, served 15 years on the National Nominating Committee, two years as chairman as well as on the National Advisory Committee and the Eleventh District vice president of budget and finance (12 years), and district chairman, Nominating Committee (13 years) and president of the South Bay Chapter (five years).

He enlisted in the USNR at NAS Los Alamitos, CA in 1950 assigned to Sqdn. VF-784. By March 1952, he became an naval aviation cadet at NAS Pensacola, Class 13-52. Many training assignments led to his commission as ENS and naval aviator in September 1953.

With the first "Special Weapons Sqdn." deployed in the fleet, his unit, VA 95, flying AD Skyraiders, served aboard the USS *Hornet* (CVA-12), rounding the world for service in Korea.

Beside the NRA, Tanner holds life memberships in the AFA; Association of Naval Aviation; Combat Pilots Association (founder); Marines Memorial Assoc.; Naval Order of the US; Reserve Officers Association and the Tailhook Assoc. He currently resides in Long Beach, CA.

ORA E. "GENE" TANNERY, LCDR, was born in the Shady Grove Community, Nacogdoches County, TX on Oct. 8, 1922. He was raised in San Augustine, TX. Tannery graduated from Stephen F. Austin State Teachers College, Nacogdoches, TX, in 1948, with a BS degree in general business.

Tannery joined the Navy's pre-flight training program, called V-5, shortly after Japan attacked the US at Pearl Harbor on Dec. 7, 1941. He was called to active duty on Oct. 1, 1943. On Aug. 4, 1944 he was among a fifty-percent (50%) quota restriction of Aviation Cadets being phased out of the flight program. Tannery transferred to the Surface Navy Program and on March 7, 1945 received a commission as ENS from the USNR Midshipmen's School, Columbia University, NY NY.

He was immediately called back into flight training as an officer trainee. Japan surrendered while he was in advanced single-engine training at NAS Corpus Christi, TX. While there Tannery received a down-check by a test pilot. Instead of asking for further training and another check flight he requested and was given orders to regular sea duty. After a year in Pearl Harbor in the port director's office he was ordered to inactive duty.

Tannery completed his education at SFA, now SFA State University. He then went to work in Beaumont, TX as manager of the appliance department of Gulf Hardware Co.

The Korean War broke out in 1950. Having stayed in the USNR, Tannery requested and was given orders back to active duty. He spent three years in Japan; two and a half of those years as CO of a northern naval outpost in Otaru, Hokkaido, Japan. It was called Military Sea Transportation Service/Naval Control of Shipping Office, MSTS/NCSO for short. His last year of active duty was spent at the HQ of the MSTS in Washington, DC. He continued in the USNR until Dec. 1, 1969 when he was transferred to the Retired Reserve.

The Tannery family includes his wife Doris; daughter Patricia and her husband Gary Wellborn; Sandra Bowden; son Greg and his wife Kelly. The Tannerys have six grandchildren. They live in Conroe, TX where Tannery retired from banking as vice president and lending officer. He enjoys Bass fishing and writing his family history.

ROBERT H. TANSOR, CDR, is a native of Chicago and a 1957 graduate of Northwestern University. In 1958 he was commissioned ENS, designated a special duty officer (Naval Intelligence), and assigned to the Office of Chief of Naval Operations, Washington, DC.

On release from active duty in 1961, he affiliated with the Naval Reserve Intelligence Program and served in various units at NAS Glenview, NAS New York and NAS Willow Grove. In 1972 he was promoted to CDR. He retired in 1980.

In civilian life, Tansor became a certified public accountant in 1961. In 1965 he entered private industry where he worked for several corporations before being elected a vice president and chief financial officer of a New York Stock Exchange listed company. In 1986 he became a private investor.

The Tansor family includes wife, Stephanie, and sons ,John and Adam. They reside in Berks County, PA near Reading.

MARK E. TARANGO, LCDR, born Feb. 29, 1960 in Phoenix, AZ, received a BS degree from the US Naval Academy in 1982.

He joined the service July 6, 1978. His military locations and stations: SWDS, Newport, RI; USS *Camden* (AOE-2), Bremerton, WA; USS *Mobile* (LKA-115), Long Beach, CA; Fleet Training Gp., San Diego, Naval Control of Shipping, Pacific Det 119.

Tarango is currently LCDR in Reserves. His memorable experiences: After five years of active duty he joined the Reserves and has been in the Reserves for 10 years now.

He married Sarah and they have a daughter, Kimberly, and son, Matthew. Employed with Microchip Technology, Inc., Chandler, AZ.

AMEL DEE TAUL, LT, born Sept. 2, 1911 in Kentucky, (Ohio County), received a BS degree from Western Kentucky University in 1938 and MA degree from the University of Missouri in 1948.

He joined the USNR, April 28, 1944. Commissioned as LTJG. His military locations and stations: US Naval Training Center, Service Schools Command, Great Lakes, IL.

Served with USS *Vogelgesang* (DD-862), Atlantic precommissioning fleet training. He was awarded the American Theater Ribbon and WWII Victory Ribbon. He achieved the rank of LT and discharged to inactive duty Aug. 30, 1946.

Activity in Reserve training programs: completed the remaining years for a total of 20 years with rank of LCDR.

He married Dorothy Helen Ducket and they have children: Gary Dee Taul, CDR/USN (Ret.); Patricia Ludene Taul Booth; grandchildren: Shannon and Sherri Taul, Kristin and Patrick Booth.

Employed as teacher in elementary schools in Kentucky and Missouri, 1931-34; high school teacher and principal in Missouri, 1935-42; electronics instructor in Army Air Force Technical Training Schools in Flordia, IL, 1942-44; Vocational advisor, Veterans Administration, September-November 1946; Education & Training with the Navy Department, Bureau of Naval Personnel, as a curriculum specialist, 1946-65; assistant director of Service Schools Training Division, 1965-72; 1972 Service Schools Division was transferred to Pensacola, FL; final position in Washington, assistant head of Undergraduate Education and Training, February 1972-March 1973, at this date elected to retire with a total of 31 years of military and civil service, March 30, 1973.

ALFRED R. TAYLOR, LT, born July 7, 1928 in North Carolina, received a BS in geology.

He joined the service July 27, 1945 and his military locations and stations: Atlantic and Pacific theaters. He served with 2nd Marine Div. (FMFLant), USMC Guam, Naval Gun Factory, Washington, DC, Camp Lejeune, NC; unit at Pensacola, Naval Shipyard, Ports, VA, Intell. Units.

He achieved the rank of LT USNR and SSGT Marine Corps (Reg. and Res). Retired from the USNR in 1972. His memorable experience was boot camp.

He married Eugenia Dare Eure and they have children: Patricia Dare and Teri Ann; grandchildren: Courtney, Cory, Dustin and Cameron. Employed as geologist, geologist senior, geologist supervisor.

ANDREW C. TAYLOR, CDR, received a BS in mechanical engineering in 1981 from the University of Oklahoma, and served aboard USS *Jacksonville* (SSN-699).

In 1986 he became a design engineer at Grand Gulf Nuclear Station in Mississippi. He served as training officer and XO for LSD-38 *Pensacola* 3809 and AS-18 *Orion* 918 in Jackson.

In 1992 he transferred to Arkansas Nuclear One, and became CO of LKA-114 *Durham* 1609 in Fort Smith. In 1993, he became TQL coordinator for REDCOM NINE. He became CO of WPNSTA CONCORD EOT 11 at Little Rock in 1995.

He holds a professional engineer's license and an MS in operations management. He was chairman of the Mississippi and Arkansas Sections of ASME, the ASME-International Nuclear Plant Systems Committee, and Arkansas RDAC. He was vice-president (Navy) of ROA-Arkansas.

CDR Taylor and his wife Carol are parents of Mandi, Jillian and Olivia.

MICHAEL F. "MIKE" TAYLOR, CWO-4 born April 24, 1932 in Flint, MI, received some college education.

He has been aboard many ships during his time in the USNR. He has seen duty on the USS *Sierra,* the *Lexington,* the *Orion,* the *Vulcan,* the *Piedmont,* the *Alamogordo* and the *Blakeley.* He also has been to numerous bases including Great Lakes, Norfolk, Charleston, Portsmouth, San Francisco, Long Beach, Newport, Subic Bay, Washington, DC, Pearl Harbor, Little Creek and Mayport.

When he entered the USNR, his pay entry base date was re-computed from April 21, 1952 to July 6, 1958. He retired April 23, 1992. His total naval service 33 years, nine months, 17 days. Total time since his first enlistment will span 40 years and three days. All of his 18 years with the US Naval Sea Cadets was a memorable experience.

He is entitled to wear the Pistol Marksmanship Ribbon w/Silver E, Armed Forces Reserve Medal w/Hourglass, National Defense Service Medal w/Bronze Star, Navy Occupation Service Medal (European), Reserve Meritorious Service Medal, Navy Good Conduct Medal and Navy Achievement Medal w/Gold Star.

Taylor is a life member of the Navy League of the US, a life member in that organizations Commandor Club, a life member of the NRA, a life member of the Reserve Officers Association of the United States and a life member of the VFW Fenton Post, transferring to Flint Post in 1980. He joined the Coast Guard Auxiliary after retiring from the USNR and has been active in teaching boating safety classes in the Flint area, he holds the position of FSO/PE flotilla staff office for public education) and FSO/MA (flotilla staff officer for materials). He is a member of the Flint Genealogical Society and holds the position of member at large on the board of directors.

He married Janet and has children: Tracy Jones, Tim Taylor, Todd Taylor and stepson Jon Nye; grandchildren: Michael and Hope Metivier; Lee Jones; Leah, Jenna, Sandi and Sara Taylor. Employed as supervisor AC Rochester, retired Aug. 30, 1991.

GUY J. TEATE, LT, born June 5, 1919 in Atlanta, GA, received a BS from a Georgia Tech.

December 3, 1936 while LT Teate was still in high school, he joined the USNR as an apprentice seaman. His first training cruise (1937) was on USS *Tattnall* (DD-125) and the next year his cruise was on the battleship USS *Arkansas.* In September 1939 while on a cruise aboard USS *Hopkins,* Hitler invaded Poland. *Hopkins* was ordered to return to Savannah.

His reserve unit was ordered to active duty in December 1940 aboard the USS *Lassen* (AE-3) at Mobile, AL. That year *Lassen* shuttled between east and west coast ports and Hawaii. It was on one of these trips that *Lassen* had passed through the Panama Canal and was 450 miles S.E. of Pearl Harbor on Dec. 7, 1941. *Lassen* immediately changed course for San Diego. In January 1942, *Lassen* left San Francisco with the task force that first shelled Wotje & Wolleii in the Marshalls, and continued on to Somoa alone.

In 1943, after several trips to the South Pacific, Teate was transferred to the new battleship USS *Iowa* (BB-61) in time for the commissioning ceremony on Feb. 22, 1943. He was a boatswain first and later made chief. He was aboard *Iowa* when it took President Roosevelt to the Teheran Conference in November 1943.

Iowa and *New Jersey* joined the Pacific Fleet immediatey for the Marshall and Gilbert bombardment and invasion. Next *Iowa* covered the invasion of New Guinea. On their way back to the Marshalls Task Force 38. Made strikes on Truk and sank several ships. *Iowa* etc. bombarded Ponapi and later was engaged in the battle for the Marianas Islands. *Iowa* was in the first battle of the Philippine Sea and the second battle. During this period *Iowa* was involved in the "Great Turkey Shoot" (338 Jap planes shot down in one day)!

Task Force 58 planes attacked Formosa and Japan and this was the beginning of the kamikazes attacks. The typhoon of late 1944, after the battle of Leyte Gulf bent one of *Iowa's* shafts and they returned to Hunters Point in California. Teate was transferred to Camp Elliott at San Diego and was discharged from active duty in October 1945.

He continued in the USNR and became a chief warrant boatswain. He later was promoted to lieutenant (03E). He was transferred to the Retired Reserve in 1965 after almost 42 years of service including five years active duty during WWII.

LT Teate graduated from Georgia Tech in 1951 and and was hired by Lockheed Aircraft. He retired from Lockheed as chief engineer - administration after 34 years. Married Maureen and they have one child, J. Michall, a 1992 naval academy graduate, and three grandchildren: Christopher, Kimberly and Jason.

WILLIAM G. TEATER SR., CAPT, born in Norwich, CT, July 2, 1946 to ENC Gilbert and Mary Teater, USN (Ret.). An engineering graduate of Texas A&M University; OCS Newport and DCA School Philadelphia.

CAPT Teater was assigned to the USS *Compton* (DD-705) and USS *William H. Standley* (DLG-32) as DCA and, as a reserve, engineering officer, NSD 6-28; training officer, CONVOYCOM 707 & NCSO JAXMEX East 108; CO, USS *Koelsch* (NRFF-1049); CO, COMNAVSTA Philippines 105; XO & admin officer, SPAWAR QA Det 1001; CO, NRRU Austin, TX; admin officer, NAVSEA Det 611, and CO, VTU 0410. He retired in September 1994.

An NRA member since 1974, as well as, a member of ROA and TROA. Teater held civilian positions as manager shop operations, General Electric Company; director manufacturing and plant manager, Cooper Industries; operations manager, Ingersoll-Rand Pumps; vice president operaions, Koch Engineering and president, W/M Group.

The Teater family consists of wife Maryann, son William Jr. and daughter, Catherine.

ELVAN D. "AL" TEEL, CDR, born in DuQuoin, IL May 14, 1924. Commissioned ENS and naval aviator NAS Pensacola, class of 1E-44. Graduated Instrument Flight Instructors School NAS Atlanta, PV-1 operational training at NAS Lake City.

Served as operational and maintenance test pilot as NAS Beaufort with the balance of WWII, HQ Sqdn as personal pilot to CDR Fleet Air Wing VI, NAS Whidbey Island. Recalled during Korean conflict, serving two years as line and test pilot in VR-3, Continental Div. MATS, NAS Moffett Field operating R4D-8, R5D and R6D aircraft.

Reserve duty and training accomplished at NAS Glenview and NAS Seattle in FASRON and VR Sqdn. utilizing PBYs and R4D and R5D aircraft. Served as CO VR 894, NAS Seattle, for three years ending July 1965, operating R5Ds. Retired April 1967.

Civilian career at United Air Lines, line flying DC-3, DC-4, DC-6, DC-7, DC-10, CV-340, B727 and B747 aircraft. Rated as piston and turbine flight engineer and airline transport rating in the foregoing aircraft. Served as DC-8 flight instructor, Denver during the introduction of that aircraft into the commercial fleet.

Al and his wife Perlle reside in Seattle, where they enjoy the boating and motor home life in retirement.

RICHARD A. TEMPLE, LCDR, a native of Williamsport, PA, he graduated from Northwestern State University. He was designated a naval aviator in February 1976.

Temple qualified in the A-7E Corsair aircraft while assigned to VA-87 and the T-2C aircraft while assigned to

VT-10. Following his assignment on board the USS *Constellation* (CV-64) he was released from active duty in November 1984.

Naval Reserve billets held include administrative officer and assistant training officer NR FAIRMED 1082, XO NR CV-60; USS *Saratoga* 0382 and XO NR CV-66 USS *America* 0193. Before retiring he was CO NR NAS JRB Willow Grove 1993.

LCDR Temple has been awarded the Navy Commendation Medal, Navy Achievement Medal, Meritorious Unit Commendation, Expert Pistol Medal, National Defense Service Medal and various unit awards.

He is employed by the Federal Aviation Administration in Washington, DC.

THOMAS M. TERPINAS, CAPT, born on July 31, 1922 in Frankfort, IN. After graduation from Frankfort High School, he entered Indiana University, completing his pre-medical education and receiving his AB degree in 1943. He enlisted in the naval service and upon completion of boot training was assigned to the Fleet Marine Force, Pacific Field Medical School, Camp Pendleton.

Once his training was completed, he was assigned to the 6th Marine Div. on Guadalcanal as a Marine Corpsman and served throughout the Okinawa Campaign. During the final phases of the fighting, he sustained a shrapnel wound but returned to duty once the division arrived on Guam. Shortly after Japan sued for peace, the division embarked for China to accept the surrender of all Japanese forces in North China and Manchuria. He was based in the Japanese barracks on the university grounds in Tsingtao during his China service. He returned to Conus in July 1946.

He re-entered Indiana University and completed his undergraduate medical and dental education. After an internship at Philadelphia General Hospital, he was accepted for a fellowship in surgery (maxillofacial and oral) at the Mayo Clinic, completing his training in 1955. He opened his practice in San Mateo, CA and continued his naval career as a reserve officer throughout the Korean and Vietnam conflicts. He was honorably discharged from the naval service in 1985 with 25 years of Meritorious Service.

He is a life member of the NRA, Navy League, 6th Marine Division Association, China Marine Association, and Marine Raiders Association. He is also a member of the American Legion (China Post #1) and the Forty & Eight (Indiana Voitore #145), Rotary Club International and the Commonwealth Club of California.

WILLIAM HARRISON "BILL" TEWELOW, CAPT, born May 25, 1935 in Wilmington, DE, received a masters (Ed Psy) in 1971 from Boston University.

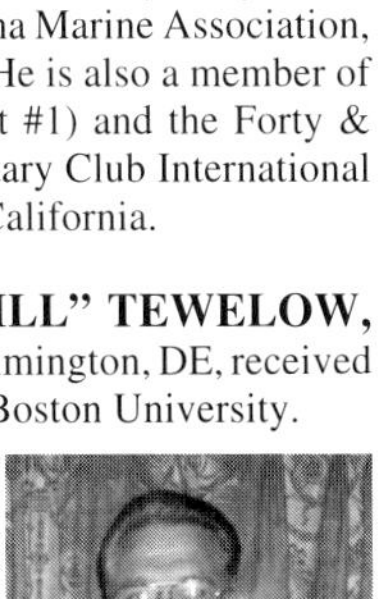

He entered the USN in 1954 as a seaman recruit, accepted a commission in 1963 and served aboard USS *Fremont* (APA-44), USS *Colleton* (APB-36) in country Vietnam, and NSA Naples, Italy. After affiliation with the USNR he served as Force Comptroller with the Sea Bees, ILO CENTCOM, CO ASB (M) 106, XO NSC JAX, and logistics officer VTU-6767. He retired in 1994 with the rank of CAPT

He has been awarded two Navy Commendation Medals and 23 other decorations. He has earned a BA from the University of Delaware, a masters in Educational Psychology (summa cum laude), AGS in in international relations both from Boston University. He is the president of the Tewelow Agency, Inc. in Mableton, GA where he resides with his wife, the former Ann Lamborn of West Grove, PA. Children are Beau, Valeri John; grandchildren, Trey and Devin.

TOMMY G. "TOM" THACKER, CDR, born May 17, 1939 in Poplar Bluff, MO, received a BA from Westminster College, Fulton, MO in 1960. He began his naval career in July 1961 in the USN School of Preflight. He was commissioned ENS in October 1961, designated NAO (N) in April 1962, air intelligence officer in October 1962, NAO (S) in August 1966, patrol plane tactical coordinator in March 1967, and mission commander in July 1975, NATOPS instructor in October 1976.

He served in Patrol Sqdn. 28 from 1962-65 including service in Vietnam following the Tolken Gulf incident. He served with VP-793, VP-882, VP-663, VP-661, VP-68A1, VP-68, and VP-67 accumulating about 3000 flight hours in SP2E/H & P3A/B aircraft in various capacities up to executive officer. He retired in October 1989 as a CDR.

He is currently an aeronautical information specialist with the National Imagery and Mapping Agency in St. Louis.

Divorced and has two children, Tommy II and Jonathan; and grandchildren, Stetson and Emerson. Employed at NIMA.

GARY F. THIELE, CAPT, born July 16, 1938 in Monona, IA, received a BS in math 1960 from Iowa State University, Ames, IA.

He joined the service September 1960. His military locations and stations: NAS Cubi Point; NAS Lemoore; NAS Alameda. Served with VU-5, VA-125, VA-212, VA-303. His battles and campaigns: SE Asia 1964-66. He was awarded the Distinguished Flying Cross, Air Medal (14 awards) and Navy Commendation w/Combat Distinguished Service.

He achieved the rank of CAPT and was discharged December 1966. His memorable experiences: triple Centurian USS *Hancock* (CVA-19); CO NR-CVN 68 *Nimitz;* CO NR NAS Norfolk.

Married Hanne Britt Slaata of Bergen, Norway and has two children: Kristian (USAF Academy cadet) and Gabriella. Employed with Pan American World Airways and Key Air Inc., pilot.

PAUL R. "DICK" THOMAS, CAPT, born in Fairfield Co., OH, on Oct. 1, 1932. He received a PhD in economics from the Ohio State University in 1967. He enlisted in the USN on April 24, 1951 and advanced to E-6 prior to the end of his enlistment on April 18, 1955. He served as communications technician with duty including the National Security Agency in Washington, DC and 19 months in the French Morocco, North Africa.

In February 1956, he enlisted in the USNR and drilled at NAS Columbus, OH and NAS Glenview, IL. On Sept. 5, 1961, he was commissioned as an ENS with a 1635 designator. He served with USNR Intelligence units, NRIU, 4-1-13, AAU 1815, NICHQ DET-0187, and IVTU 115. CAPT Thomas served as CO of NISCPROD 605, Dayton, OH and FIRSTEURLANT 1081, Columbus, OH, and as Reserve Intelligence area coordinator for AREA-12. He retired on July 1, 1992, concluding over 40 years of service to the USN. He is a life member of the NRA and TROA.

In 1988, Dr. Thomas retired as professor emeritus from The Ohio State University concluding over 27 years as a faculty member in teaching, research and administration. CAPT Thomas now serves as director, Advisory Services and Educational Programs for the Ohio Association of Chiefs of Police.

The Thomas family includes wife of 40 years, Ardis; sons: Scott, Jeffrey and David; daughter Cheryl; and grandchildren: Brian, Sarah, Jordon and Laura.

T. GILBERT THOMPSEN, CDR, born in Chicago, Dec. 11, 1916. Graduated from St. Olaf College, Northfield, MN and Luther Theological Seminary, St. Paul, MN.

He joined the Navy Chaplain Corps, July 5, 1944. Graduated from the Chaplains School, College of William and Mary, 1944.

On active duty, he served at NTC Farragut, ID, NT&DC Camp Elliott, San Diego, and aboard the USS *Montrose* (APA-212). On inactive duty, he served as CO NRCC 9-2 (Chaplain Company) Fargo, ND, NRTC Denver, CO 1960-67. He retired July 1, 1969.

His family consists of his wife Hedvig and five children: Loren, Paul, Harold, Naomi and Philip; and 12 grandchildren. In civilian life he served Lutheran congregations in North Dakota, Colorado and Minnesota. He retired from parish ministry in 1979.

His campaign and service medals are: American and Pacific Theatre, WWII Victory Medal and Naval Reserve. In retirement at Bemidji, MN, he enjoys hunting and fishing.

JAMES J. "JIM" THOMPSON, LT, born Oct. 22, 1940 in Waterbury, CT, received a BA in chemistry, 1962, University of New Mexico, Albuquerque, NM; MS Bionucleonics, 1970, Purdue University, W. Lafayette, IN; PhD Bionucleonics, 1972, Purdue University, W. Lafayette, IN.

He joined the service June 15, 1962. Commissioned at NROTC Unit, University of New Mexico. His military locations and stations: June 1962 to July 1964, USS *Staten Island* (AGB-5), homeport Seattle, WA; August 1964 to June 1966, USS *Caloosahatchee* (AO-98), homeport, Newport, RI; July 1966 to August 1968, NROTC Unit, Purdue University, W. Lafayette, IN.

Served with USS *Staten Island* (AGB-5); USS *Caloosahatchee* (AO-98); NROTC Unit Purdue University; Research Reserve Co., Lafayette, IN; Research Reserve Company, Albuquerque, NM; VTU-1102, Albuquerque, NM; USS *San Bernardino* (LST-1189) (Acdutra); USS *McKean* (DD-784) (Acdutra); Naval Research Laboratory, Washington, DC (Acdutra); Underwater Sound Reference Laboratory, Orlando, FL (Acdutra); Field Command Defense Nuclear Agency, Albuquerque, NM (Acdutra).

His campaigns: Operation Deepfreeze 63 in Antarctica (1962-63); Arctic deployment (1963); Mediterranean deployment (1964-65); Dominion Republic Intervention (1965).

Awarded the National Defense Service Medal, Antarctic Service Medal, Armed Forces Expeditionary Medal, Marksman Pistol and Marksman Rifle. He achieved the rank of LT and was discharged Aug. 31, 1968 from active duty. Achieved the rank of CAPT and retired Oct. 1, 1989 from the Reserves.

His memorable experiences: two months breaking ice in McMurdo Sound in 1962 and two months exploring the Palmer Peninsula in 1963. Getting their ship stuck in the ice north of Point Barrow in 1963. Many nice ports of call in Spain, Italy, Greece and Turkey in 1965. Serving as USN judge representing the Office of Naval Research at the International Science Fairs in Albuquerque and Pittsburgh.

Married April 20, 1963 in Santa Fe, NM to Jeanette Z. Gurulé. They have three children: James Tristan (LT, USNR) married with two children; Bruce Evan and Seth John; grandchildren are Chelsea Jessica and Paige Elizabeth.

Employed June 8, 1972-July 8, 1994, head, Health Protection Operations, Inhalation Toxicology Research Institute, Albuquerque, NM; Aug. 1, 1994 to present, director, Department of Radiological Health, radiation safety officer, Adjunct Associate Professor of Environmental Engineering, The University of Utah, Salt Lake City, UT.

Professional Certifications: American Board of Health Physics (1976); American Board of Industrial Hygiene (1978); Board of Certified Safety Professionals (1978).

MARK R. THOMPSON, RADM, born May 12, 1919 in Olean, NY, received a BA from Washington and Jefferson College; ThB Princeton Theological Seminary; DD Lafayette College; DD W&J College; Doctor Humane Letters, Scranton University.

He joined July 1, 1944. Served active duty April 1944-46; 1952-58. His military locations and stations: NAS Johnston Island; NAV MAG Bangor WS; NTC Bainbridge; NS San Juan; NAS Lakehurst. Served with COM DESLANT; Chaplain School, Newport, RI. Thompson achieved the rank of RADM, CHC, USNR Ret and discharged May 12, 1979.

Married Jean Williamson and has children: Mark Jr., John, Scott, Robert, Robin and Ross; grandchildren: Weslee, Nettie, Nathan, Matthew, Jessica, Trevor and Stacey. Employed as chaplain, Lafayette College (1946-52); pastor, Westminister Church (1958-81), Scranton, PA.

JOHN B. "THOR" THOREEN, LCDR, born April 11, 1964 in Phoenix, AZ, received a BS in Applied Mechanics from the University of Southern California.

He joined the service May 9, 1986. His military locations and stations: USS *Nassau* (LHA-4), Assault Craft Unit 2, (active). Served with USACOM, ACU 4, Naval Coastal Warfare Group 2 (Reserve).

Thoreen participated in Operations Desert Shield and Desert Storm. He was awarded the NAM (2), SWAS, Defense Medal, Kuwait Defense Medal, Kuwait Defense Medal (Saudia Arabia. He achieved the rank of LCDR and was discharged from active duty Sept. 30, 1992. Still drilling actively.

His memorable experiences: plankowner officer in charge of NR ACU4 DET 1; officer in charge of ACU 2 DET 206 for Desert Storm/Desert Shield.

He is married to Paige and is employed with Virginia Department of Environmental Quality, environmental engineer.

ARLIN ED "ED" THORLA, CAPT, born Aug. 17, 1929 in Zanesville, OH, received a Bsc Chemistry and Pharmacy, Ohio Northern University, ADA, OH.

He joined the service Dec. 13, 1951. His military locations and stations: OCS, Class IV, May 1952, Newport, RI; USN Amphibious Course, Coronado, CA; USN Deep Sea Diving School (October 1953); USN Special Devices School, Port Washington, LI; Industrial War College, Washington, DC, 1968; Sen. Reserve War College, Newport, RI, 1973; War gaming, Naval War College, Newport, RI, 1981.

Participated in the Korean War, 1952-53 and achieved the rank of CAPT, USNR, discharged RAD May 1955. His memorable experiences: grad. USN Amphibious School, Coronado, CA, June 1952; USN Diving School, Bayonne, NJ, October 1953; USNAVWAR College Senior Res. Off., August 1968.

He married Annette Eliz and they have children: Traci Ann, Jeffrey Arlin and Michelle Lynn. Employed at Park Davis and Co. (sen. prof. sales rep.)

WILLIAM B. "BILL" THURSTON, LCDR, born Aug. 6, 1919 in Baltimore, MD, received a BS from the University of Maryland.

He joined the service May 1943. His military locations and stations: OIC Princeton, Advanced Indoc. Ft. Schuyler, USATB, Ft. Pierce. Served with 5th and 7th Fleets. His battle and campaigns: Leyte Gulf and Lingayen Gulf. He achieved the rank of LCDR and discharged April 1946, inactive duty.

His memorable experiences: Japan Occupation battles. He is married to Virginia Lois Francis and they have children: William B. IV, Lee F., C. Grover and Todd D.; and seven grandchildren. Employed at TRW, Martin Aerospace, and MD Shipbuilding, Seagrams.

ALBERT W. TIEDEMANN JR., CAPT, born Nov. 7, 1924 in Baltimore, MD. Enlisted in USNR December 1942. Active duty, V-12 Program. Commissioned ENS at USNR Midshipman School, Notre Dame on Oct. 26, 1944. Ordered to 7th Amphibious Force. Assigned to USS LCI (L) 444 and participated in operations in the New Guinea and Philippine areas. Released to inactive duty as LTJG on Aug. 25, 1946.

USNR service included XO of Naval Reserve Officers School; CO of a Research Co., Surface Div., MobTeam Div. and System Analysis Div(SubLant). Served as OinC of a two week mobilization exercise.

Awarded Asiatic-Pacific, American, Philippine Liberation, Victory, Naval Reserve and Armed Forces Reserve Medals.

NRA service started as one of the 43 founders and one of 21 original steering group in 1954. Served as chairman, National Resolutions Committee (1954-56); National Naval Reserve Committee (1955-57); National Vice President for 5th District (1954-57); National Vice President, Naval Affairs (1962-63 and 1965-67); National Vice President, Education and Chapter Activities (1967-69). Originated the Navy Sabbath Program in 1965 and served as chairman (1965-75). Received the Association's National Meritorious Service and Twice a Citizen awards.

Received a PhD degree in chemistry, chief chemist, Emerson Drug Co., laboratory director with Hercules Inc. at Allegany Ballistics Laboratory and Radford Army Ammunition Plant, and director, VA Div. of Consolidated Laboratory Services. Resides in Richmond, VA.

JOHN G. TILGHMAN, LCDR, born Oct. 8, 1918 in Philadelphia, PA, received a BA from Baker University, Baldwin City, KS in 1947.

March 16, 1942-June 1944: enlisted service in Naval Hospital Corps. Various Navy, Seabee and Marine bases throughout the US. June 1944-October 1944: Midshipmen School, Northwestern Univ., Chicago and commissioned ENS, USNR. October 1944-November 1945: assigned to USS LCI(1) 523. Inactive from 1945-50.

Assigned in January 1950 to Organized Surface Div. 9-62, Topeka, KS; 1951-53, USN Communication Station, Pearl Harbor. Released from AcDu in 1953 and assigned April 12, 1954 to Naval Reserve Surface Div. 9-62, in Topeka. Received orders July 1, 1961 to NR Electronics Div. 9-14, Manhattan; Oct. 1, 1961, orders to NR Public Relations Co. 9-8, Topeka; Aug. 15, 1963, earmarked and ordered to appropriate duty with SVI US Army Corps Mobilization Designation Det 3, Topeka.

He was awarded the Naval Reserve Medal and Pacific Area Ribbon. Achieved the rank of LCDR, September 1956. His memorable experiences include inspecting various communication stations in Hawaii and Pacific areas as inspection officer at ComSta Pearl; also memorable was turning their ship over to the Russians in the Aleutian Islands.

Married Altha and they have children: Phillip, Douglas, Kenneth, Richard, Deborah (Packard); and grandchildren: Patrick, Susan, Andrea, Rebecca, Timmy, Bradley, Daniel, David, Michael, Benjamin and Elijah. Employed at Youth Center at Topeka as teacher 1945-51, business manager, 1953-83. He is active in his church and Boy Scouts (received his Beads in 1965 and the Silver Beaver Award. He is a member of the ROA and the NRA. He retired in 1983.

EDWARD TIMM, CAPT, entered the USNR while still in high school, as a seaman recruit in 1953 with Surface Div. 4-1-7 in Hazleton, PA. After graduation from high school in 1954, he enlisted in the regular Navy and served on active duty for four years. He completed Fleet Sonar School and several other electronics schools, and served as a sonarman second class aboard the destroyers USS *Haynesworth* and USS *Ault*. In 1961, after release from active duty, he graduated from Bloomsburg State University. While in college, he again affiliated with 4-1-7.

In 1966 he received a direct commission as an ENS and was assigned to USNR Intelligence Div. 3-1-5 at Scotia, NY. He later was assigned to Naval Air Intelligence Reserve Unit N-1, Area Analysis Unit 404, and Air Antisubmarine Sqdn. 71 (VS-71) at NAS, Lakehurst, NJ. At NAS Willow Grove, he served as administrative office and training officer with Fleet Intelligence Rapid Support Team Europe and Atlantic 0593, and as XO with NISRORU 0893.

He commanded the Naval Investigative Service Regional Office Reserve Unit 0893 from 1984 to 1986. In each of these years, the unit was awarded the prestigious Admiral John W. Cronin Trophy as the outstanding unit in Reserve Intelligence Area Sixteen. He was the first office to be awarded NIS Special Agent Credentials in Reserve Intelligence Area 16.

From 1986-89 he was assigned to the IVTU-0104 and served on the RIAC staff in various assignments. Timm commanded the Defense Intelligence Agency HQ Unit 0193 from October 1989 to October 1990. Again, his unit was awarded the Admiral Cronin Trophy as the outstanding unit in Reserve Intelligence Area Sixteen. He retired in 1990.

Retired from the Federal Aviation Administration Technical Center, Atlantic City, NJ, where he was the manager of Airport Operations and Emergency Services. He is now a consultant providing services on airport and security management. He lives in Northfield, NJ, with his wife Bettie. They have four children: Mark, Robert, Sharon and Edward IV.

ROBERT E. "BOB" TIPPING, CDR, born March 11, 1926 in Virginia, MN. He received a BBA from the University of Michigan and MA from the University of South Florida.

He served active duty March 1943-June 1946 and March 1951-March 1953. His military locations and stations: Great Lakes, Newport, Little Creek, NROTC University of Michigan. Served with 5th Amphib.-Pacific and Atlantic Fleet. His battles and campaigns: American Defense and Pacific.

Tipping was awarded the Good Conduct Medal, Pacific, American Defense, and Naval Reserve Medal. He achieved the rank of CDR, Supply Corps and released March 1953. His memorable experiences: Supply Corps School, Bayonne, NJ March 1951; Supply officer, USS *Kleinsmith* APD-134, 1951-53.

He married Dolores and they have two children, Barbara Armstrong and Melody Benoit. He is a retired teacher.

ALAN MITCHELL "AL" TODD, CAPT, born Dec. 1, 1944 in Long Beach, CA, received a BA from Cal State, Long Beach. He joined the service Jan. 18, 1963. Commissioned August 1966.

His military locations and stations: Coronado, CA (USN AB). Served with Underwater Demolition Team 12, Seal Team One, NAVAL Special Warfare Group One.

His battles and campaigns: Vietnam 1967, 1969, 1971. He was awarded the Bronze Star (combat V) and Navy Commendation with Combat V. He achieved the rank of CAPT (USNR) and discharged October 1975. His memorable experiences: basic underwater demolition/Seal training Class 40.

He is married to Sonja and they have two children, Cynthia and James; grandchildren, Danielle and Caleb. Employed as bank vice-president. He retired Nov. 1, 1996.

MARIAN ELAINE TOMS, CDR, born to missionary parents in Seoul, Korea on May 17, 1917. She graduated from Wilson College, Chambersburg, PA in June 1939. She enlisted in February 1943 in the WAVES as an officer candidate and was commissioned ENS on June 1, 1943.

She served eight months at NOB Norfolk. In February 1944 she was assigned to Pre-Radar School at Harvard and Radar School at MIT. She then was assigned to the Electronics Div. of BuShips for the remainder of her active duty to January 1946.

In August 1949 she earned her MS degree in physics at the University of Pennsylvania. After additional studies and research at Penn, she was employed as a physicist at NRL where she continued her research in nuclear physics until retirement. She participated in the NRL Research Reserve Co.

EARL J. TOOMBS, LCDR, born Jan. 14, 1921 in Springfield, MO, received a BS degree in business administration fromn Southwest Missouri State College. He served his country in WWII, the Korean War, the Cuban Crisis and the War in Vietnam.

After earning his commission and his wings at Corpus Christi, TX he served with Air Transport Sqdns. SEVEN and EIGHT, Hickam AFB, HI; HQ of the Military Air Transport Service, Scott AFB, IL; withthe Joint Target Planning Staff, Offutt AFB, NE; aboard NAS Agana, Guam; and with Fleet Tactical Support Sqdn. One, as pilot and C-131 training officer at VR-1.

He was awarded the American Theatre Medal, WWII Victory Medal, Asiatic-Pacific Campaign Medal, AF Commendation Medal, Vietnam Service Medal and National Defense Service Medal.

He married Rissie R. Hill and they had three children: Lou Ann, Thomas Earl and Mary Jane; grandchildren: Justin and Kenneth Toombs, Michele and Nicholas Ruppel. He died Nov. 17, 1980 in St. Petersburg, FL.

CAROLYN M. "TOP" TOPPING, CDR, born in Montclair, NJ, attended NJ State Teachers College, Newark, NJ. Graduated from Beaver College, Jenkintown, PA in 1941. Masters degree in education from University of Southern California, Los Angeles, CA. Completed all course work leading to Doctorates in School Administration and Psychology at USC.

Enlisted in WAVES 1943. Graduated from NAV Hospital (Chelsea, MA) as PH2. Assigned to NAV Hospital (Charleston, SC). Graduated as ENS (September 1944). Served at NTS (Bronx, NY) at: NAV Hospital (Charleston, SC); NOTS (Inyokern, CA); NAS (Camp Kearney, CA); NAS (El Centro, CA); NTS (San Diego, CA); NAV Hospital (Bremerton, WA); above billets in admin/personnel communications, educational training, WAVE representative.

Released from active duty in November 1947. Served in admin/personnel pay and non-pay billets at NAS Los Alamitos (CA) in VF-54, VF-781, VF-782, VF-784, US-776 and Air Wing Staff 77 from June 1948 until retirement from active reserve duty June 1969.

Taught 20 years in Los Angeles City Schools in 4th and 5th grades and special class for emotionally disturbed children. Retired from LA City Schools June 1969.

Active in cat shows in western states (CA, AZ, NE and UT) showing black Persian cats. Male Grand Champion was best long-haired cat in the US one year.

In January 1969 she entered the Episcopal religious order of the community of St. Mary, Peekskill, NY and as a sister has done volunteer work at a local nursing home for over 20 years. Also, calligraphy and ecclesiastical sewing. In 1994 she was selected volunteer of the year for all nursing homes in the Westchester District of Nursing Homes Association.

Her name in religion is Sister Mary Francis, CSM.

FELIX R. TORMES, CAPT, received a BS degree from Manhattan College (67) and an MD degree from Loyola University, Chicago in 1971.

He began his naval career as intern, NH San Diego 1971-72, then served as medical officer, USS *Truxtun* (DLGN-35) during the Vietnam Conflict. In 1973, he completed training as flight surgeon and was assigned to HS-15. He served a residency in orthopedic surgery at NRMC, Portsmouth, 1975-78 and was assigned to NH Camp Lejeune until 1980 when he transferred to the USNR.

As a reserve flight surgeon, he was assigned to NAS Memphis and VP-67 and served as CO, Chase Field Medical. He was activated in 1991 during Desert Storm and served with Fleet Hosp. 5, Al Jubail, Saudia Arabia as chief, ortho-neuro service. He is presently assigned as orthopaedic service. He is presently assigned as orthopaedic surgeon, PRIMUS, Vanderbilt.

Tormes holds a commercial, instrument pilots license with over 2,000 hours. He is board certified on orthopaedic surgery and is a member of the American College of Surgeons.

He has been awarded the Presidential Unit Citation, Unit Commendation Ribbon and Meritorious Unit Citation. Campaign and service medals are: National Defense Service (2), Vietnam Service, Southwest Asia Service, Kuwait Liberation Medals (2), Overseas Service and Armed Forces Reserve Medal.

THEODORE K. "TED or TK" TOUW, CDR, born Jan. 15, 1914 in Prospect Park Burough, NJ and received two year teacher certification Huron College.

He joined April 27, 1935. His military locations and stations: Naval Communications Reserve, South Dakota, Texas and Georgia. Served with COM 6, PG School Annapolis, COMSUBRON 10, Naval Base New London. His battles and campaigns: Atlantic and Pacific arenas.

His awards and medals: Naval Reserve Medal, American Theater and Asiatic-Pacific, Pre-Pearl Harbor and WWII Victory Ribbon.

Achieved the rank of CDR and was discharged Nov. 15, 1945. His memorable experiences: viewing the destruction at Kwajalein and Guam.

He married M. Iris Westbrook and has a son, Kenneth B., and grandchildren: Anduin, Kate and Adam. Last employer, Lockheed Aircraft, tech writer, retired Jan. 15, 1974.

SUSAN E. "SUSIE" TOWNSEND, CDR, born Sept. 5, 1946 in Philadelphia, PA, received a BS in Secondary Education/Biology West Chester State College (now West Chester University), West Chester, PA; 1978, MBA National University, San Diego, CA; and 1984 PhD Human Behavior, La Jolla University, San Diego, CA.

She was sworn in June 1970, OCS in 1970-71; commissioned ENS December 1970. Her military locations and stations: active duty: From OCS to Naval Electronics Laboratory Center-CO, RPS Officer and Scientific Technical Intelligence Liaison Officer (STILO) 1971-74; Defense Intelligence Agency, Intelligence Analyst/Soviet Navy 1974-76; Temporary active duty-CINCUSNAVEUR London, UK - terrorism analyst at Terrorism Fusion Analysis Cell 1986.

Reserve military locations and stations: area analysis unit, Intelligence analyst - 1976; FIRSTPAC 0894 - Intelligence Analyst - 1976-77; Naval Investigative Service Regional Office (NISRO) Reserve Unit 2794 (1977-84); intelligence officer; assistant training officer; training officer; assistant department head/training; department head/training; special projects officer (Fraud and Crisis Management); assistant operations officer and 1986-87, training department head.

NISRO Reserve Unit 2819 (1984-86) XO/intelligence officer; Naval Oceans Systems Center (NOSC) Reserve Unit 0194, 1987-89), XO. Battles and campaigns: Called to temporary active duty to CINCUSNAVEUR during the Gulf of Sidra Crisis during the summer of 1986.

Awards: Department of California, Reserve Officers Association Outstanding Naval Reserve Junior Officer of the Year 1982.

Her medals and ribbons: Joint Service Commendation Medal, Navy Commendation Medal, National Defense Medal, Overseas Reserve Ribbon, Expert Medal/Rifle and Expert Marksman/Pistol. Also has a Naval Reserve Association Medal. She achieved the rank of CDR, USNR-R and retired October 1993.

Her memorable experiences: active duty-testifying on "Soviet Naval Activities in Cuba" at a hearing of the Congressional Subcommittee on International Political and Military Affairs headed by Chairman Dante B. Fascell while stationed at DIA. Reserve-Serving as a terrorism analyst at the Terrorism Fusion Analysis Cell at CINCUSNAVEUR London during the Gulf of Sidra Crisis in 1986.

She married John H. Townsend. Current civilian employment and positions held: volunteer - GFWC California Federation of Women's Clubs San Diego District #25 - First vice-president/Dean of Chairmen; GFWC Parliamentary Law Club, president; GFWC Peninsula Woman's Club, recording secretary; San Diego County Jail Ministries Executive Board, scheduling coordinator and registrar of churches; Christ Fellowship Church, member of jail, convalescent home and rescue mission outreach ministries and Director of Music.

Past 1983-89 owner and consultant with Townsend Research Institute; 1986-89 Instructor with Fleet Aviation Specialized Operational Training Group Pacific (FASOTRAGRUPAC) Advanced SERE Course (Hostage survival); 1977-1980 La Jolla Research Corporation - business manager/member of the professional staff. Other jobs and volunteer positions held during that time. She retired October 1993.

JOSEPH F. "JOE" TRAINOR, CAPT, born Nov. 5, 1925 in NYC, NY, received a MPA from the University of Oklahoma, 1974. He enlisted in the USNR in 1943. He was assigned to officer training under the V-12 program at Hobart College, Geneva, NY, and received his commission in 1945. He served in USS *Walke* (DD-723) and later with the staff of the CDR, Cruiser-Destroyer Force, US Pacific Fleet, CAPT Trainor was released from active duty in 1947, and joined the Ready Reserve.

In 1950 CAPT Trainor volunteered for active duty in connection with the Korean Conflict and served in USS *Roan* (DD-853) and USS *Stockham* (DD-683). Released to inactive duty in January, 1953, he has held various billets in Ready Reserve units of the Third, Fifth, Sixth and Eleventh Naval Districts.

In October 1961, CAPT Trainor was XO of USS *Colahan* (DD-658) Reserve Crew, homeported to Long Beach, CA, when the ASW Selected Reserve Crews were called to active duty for the Berlin Crisis. USS *Colahan* was assigned to duties with the SEVENTH Fleet including patrols off the coast of Vietnam.

In May 1963, after returning to Washington, DC, CAPT Trainor assumed command of USS *R.F. Keller* (DE-419), the Navy's Reserve Ship in the Nation's Capital. Upon her deactivation he assumed command of USS *Loeser* (DE-680) and her reserve crew. From February 1967 to July 1968 he was CDR, Reserve Destroyer Div. 5ND.

In June 1973, upon his selection as CDR, NR Gp. 5-1(L), Baltimore, MD, he was relieved of duties with Reserve Operations Command Center Unit 5-2, Navy Command Support Center (CNO FLAG PLOT).

Following duties as CDR, Naval Reserve Group 5-1 (L) and 5-12 (L), CAPT Trainor returned in July 1975, as CO, to ROCCU 5-2, now designated as NCSC Staff 106. Concurrent with this change of command is CAPT Trainor's transfer to the Retired Reserve after 33 years as a dedicated Naval Reservist.

He is a graduate of Hobart College, and the Armed Forces Staff College, holds a master of Public Administration degree from the University of Oklahoma. He retired from the US Department of Justice in Washington, DC and is a security consultant.

CAPT Trainor and his wife, the former Miss Celestine Quinn of Lansford, PA, make their home in Potomac, MD. They have six children and five grandchildren.

RAYMOND W. TRAUB, CAPT, born Nov. 3, 1921 in East St. Louis, IL, received a BSC from St. Louis University.

He joined the service Aug. 1, 1943. His military locations and stations include: Samara Philippines and Guam. Served with VR-3, VR-5, VR-13, VR-6.

His battles and campaigns: Philippines. His awards include the usual WWII and South Pacific Medals and Navy Commendation. Achieved the rank of CAPT and was discharged Nov. 3, 1980.

He married Patricia and they have three children: Kevin, Brian and Mary Kathrine; grandchildren: Zach, Jeremy, Nick and Alex. Employed as captain and chief pilot, Ozark Airlines.

PAUL E. TREJO, CAPT, born Oct. 31, 1926 in San Jose, CA, received a BS from the University of Southern California; BS USNPG School Monterey; MA San Jose State University.

Service Schools: USN Submarine School, New London, CT, January 1951-June 1951; Senior Reserve Officers Course, Naval War College, Newport, RI, June 1974; Defense Strategy Seminar, National Defense University, Fort Myer, VA, July 1979.

Sea duty: USS *McCook* (DMS-36), USS *Floyd B. Parks* (DD-884), USS *Blenny* (SS-324), USS *Bashaw* (SSK-2411), USS *Barbero* (SSG-317).

Career milestones: qualified submarines May 1952; instructor guided missiles, NROS, San Jose, CA 1959-61; 1960 Redesignated EDO ordinance engineer (1455), 1961, redesignated aeronautical engineer (1515) on merger of BU ORD and BU AIR into BU WEPS; served Captain's Selection Board 1981; Significant periods of extended active duty: Sidewinder project officer, Air Systems Command, Washington, DC, staff commander Patrol Wings, Pacific Fleet, NAS Moffett.

Held commands: NAVAIRSYS Res Unit G-2; CAPTS Management Unit, AIRSYSCOM VTU 0380, AIRSYSCOM WEP RES Unit 1187; AIRSYSCOM WEP VTU 0187.

He was awarded the Navy Meritorious Service Medal 1980; Navy Commendation Medal 1983 and American Campaign WWII, WWII Victory, Navy Occupation (Asia Clasp), China Service, National Defense, Korean, Vietnam, Armed Forces Reserve Medal, (two Bronze Hourglasses, United Nations.

Trejo achieved the rank of CAPT and was discharged June 30, 1984. He married Kathleen Ruth and has son, Leslie Lee; daughter, Catherine Ann Trejo Collins; grandchildren: Sky Forrest, Oceana, Natasha, Ruth and Tammy. Employed as professor, engineering and astronomy, San Jose State University, Foothill College District.

ROBERT ALFRED "BOB" TREMANT, CAPT, born in New Orleans, May 3, 1930, graduate of Tulane University 1952, BS physics. He enlisted in the USNR 1948 and commissioned ENS (1105) from Class 8 (OCS) Newport December 1952. He was assigned to the Sixth Fleet flagship USS *Des Moines* (CA-134) from January 1953 to November 1955 as the main engine officer.

Upon release from active duty he reaffiliated with the Ready Reserve and subsequently held the CO billets of NRED 11-6, NRSD 8-93, MOB Team 8-15, NROS 8-4, GROUP COMMAND 8-26 and ONR/NRL TECH 410. He retired in 1983 with 35 years unbroken service.

CAPT Tremant entered his civilian career in January 1956 with the National Advisory Committee for Aeronautics (NACA), High Speed Flight Station, as an aeronautical research engineer working with the X-1, X-2, X-3, D558-2 and X-15 rocket aircraft, Century Series turbojets, Navy F8U, B-47 and flight certification of the Boeing 707 for first commercial use.

In April 1962 he transferred to the National Aeronautics and Space Administration (NASA) Manned Spacecraft Center (later renamed Johnson Space Center) into the Apollo Spacecraft Program Office. His work as subsystem manager included development of the stabilization and control system for the Apollo Command and Service Modules, development of the Extra Vehicular Communications System (EVCS) for the astronaut space suits on the lunar surface and the development of the Particles and Fields Subsatellites deployed into orbit around the moon on Apollo Missions 15 and 16 (for studies of the lunar magnetic fields and solar wind particles).

In the post Apollo period CAPT Tremant joined the Space Shuttle Engineering Office within the Engineering Directorate of the Johnson Space Center and served as the Contracts Officer for Communications and Tracking (C&T) Flight Hardware Systems procured as government furnished equipment. Additionally, he held the position of manager, Lockheed Engineering Support Contract for Operation and Maintenance of 18 C&T Avionics laboratories for development, fabrication and tests supporting Space Shuttle Orbiter, Orbiter payload systems, unique crew related devices and early space station studies. CAPT Tremant retired from the NASA Johnson Space Center in September 1994 with 38 years of service.

His family includes wife, Betsie Gillespie of New Orleans, son Robert Paul, daughter-in-law Tamara and granddaughter Cassandra Kay. He and his wife reside in Nassau Bay, TX and enjoy hunting, fishing and world travel.

GLENN E. TREWET, CDR, born May 25, 1925 in Fort Morgan, CO. He attended Colorado College, Colorado University and Southern Methodist University, obtaining a BBS degree in personnel management. He began his naval career in 1943 and received his ensign's commission and pilot's wings in January 1947.

He flew C-54s and C-47s in WESTPAC in 1947-48 with VR-6. The squadron was ordered to Germany in 1948 for the Berlin Airlift. He transferred to VR-5 in Seattle, flying C-54s to Alaska, then to VR-5 Detachment San Diego, flying C-47s and C-54s. LTJG Trewet's next tour was with GCA Unit #12 at NAS Norfolk, where he flew C-47s, C-54s, SNJs, F6Fs and TBMs. LT Trewet

reported to the ARTC at NAVSTA Midway in 1953 and flew HU-16s in SAR.

He next joined VP-19, NAS Alameda with deployment to NAVSTA Iwakuni in SP2Hs. He trained a Japanese crew in a SP2H for COMFAIR Alameda. Attended the 5-Term College Program at Colorado University, attended General Line School, Monterey, followed by an instructor tour at ATU-601, NAS Corpus Christi. He joined VP-702, NAS Dallas, then VP-872, NAS Alameda, under JTF-8, returning to VP-702 in 1962 for six more years finishing as commanding officer in 1968. CDR Trewet completed 30 years satisfactory service in 1973.

He flew 7,200 pilot hours in 11 naval aircraft types, holding plane commander designations in the C-47, C-54, HU-16 Albatross and the P2.

CDR Trewet's awards and medals are: Air Medal, American Campaign Service, WWII Victory, Naval Occupation w/Airlift Device, Medal for Humane Action, National Defense Service and Armed Forces Service Medal.

Among CDR Trewet's most memorable experiences are: a midair collision over the Gulf of Mexico, caused by his wingman. Both bailed out, spent 2 1/2 hours in rafts in 40 degree weather and survived a pickup and open-sea takeoff in a PBY. NAVCAD Trewet's first daughter was born that same day, Nov. 23, 1946; 132 round trips from Frankfurt to Berlin on the airlift (1948-49); 2 1/2 months flying daily patrols from Christmas & Johnston Islands with VP-872 under JTF-8, clearing atomic test areas. Not a single patrol was cancelled. The squadron received "well done" messages from all senior commanders, including CNO, SECDEF and President Kennedy; his unforgettable tour as commanding officer of VP-702 in 1967-68.

CDR Trewet served 24 years with FAA: four years as an airways flight inspection captain in the Convair 580; 18 years as a check airman in Convair 340s and 580s, Lockheed 188 Electra and Boeing 727; a year as an aviation safety inspector (Operations) on Pan American Airways in Africa; and a year as operations advisor to the flight standards staff chief and check pilot at the FAA Europe Africa and Middle East Region in Brussels. He retired from FAA in 1985, flew a Westwind 2 Jet to Hawaii for an Oklahoma millionaire, followed by three years as CAPT of piston and turboprop Grumman mallard flying boats for two airlines in Saint Croix, USVI and Puerto Rico.

CDR Trewet possesses an airline transport pilot certificate with 17 type ratings. He has flown 9,100 civilian pilot hours, for a total of 16,300 hours.

He was married to the former Barbara Gerow from 1945-66 and has been married to the former Jean Pugh Cain from 1968 to the present. He has four daughters: Glenda Owen and Candace Morse of Oklahoma City; Marilyn Trewet of Wichita Falls, TX; and Robin Lowrey of Tulsa. Also has three stepsons: Alan Cain of Grand Coulee, Washington; James A. (Andy) Cain of Atenas, Costa Rica; and Steven Cain of Oklahoma City. They have 10 granddaughters, five grandsons and three great-grandsons.

CDR Trewet has been an NRA member for 35 years, nearly all as a life member. He was the first president of the Oklahoma City Chapter in 1966-67 and the president of the Eighth District in 1975-77.

JOHN JOSEPH TROMBETTA, LCDR, born May 16, 1951 in New York City, received a PhD from SUNY Buffalo, 1981; MA, Hunter College, 1975; BA Lehman, 1973.

He joined the service Nov. 22, 1985. His military locations and stations include: NAS Pensacola, CHINFO, USEURCOM, NSA Naples, COMSUB PAC, (Pearl Harbor). His units include: Fleet SUP TRA NRC Syr. NY; OI DET 102, NY; NAVINFO East 102.

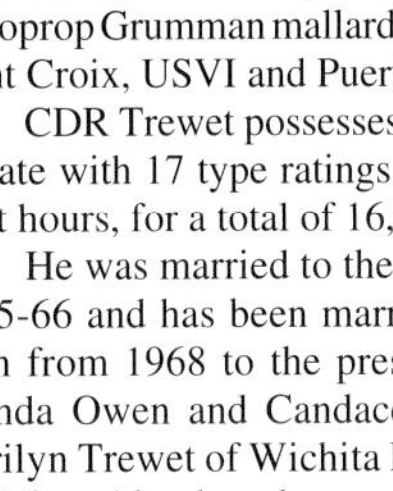

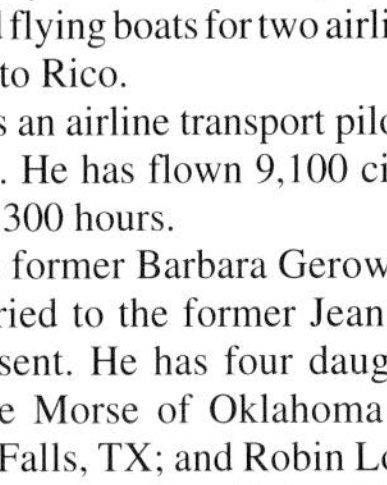

Trombetta received the Navy Achievement Medal, National Defense Service Medal, Naval Reserve Medal. He achieved the rank of LCDR and is currently USNR-R.

His memorable experiences: naval assignments aboard the USS *Alaska* (Pearl Harbor), P-3C (Orion) (Roosevelt Roads), Frigates assigned to the Great Lakes Cruise. Assignment: writing in the Pentagon for 50th anniversary of the end of WWII Fleet Week NYC 1986-95. AT at Naples, Italy, USEURCOM-Germany.

He is married to Joanna Gozzi and they have children, Jessica and Justin. Employed with Syracuse City School District (Nottingham High School), Department Special Education Chair/Speech-Language pathologist; owner, Executive Communications Co.

LEROY G. TROWBRIDGE, CDR, born in Chicago, IL, April 8, 1908. Grew up in Dubuque, IA. Graduated from University of Wisconsin in 1931; member of ROTC. Directed several emergency welfare programs in Wisconsin until time of commissioning in Navy as LTJG, August 1942. As CO of a USN Armed Guard Unit, he served on several merchant ships, in three Theatres of War. His last assignment in WWII was as flag secretary and personnel officer on the staff of Commodore A.H. Gray, CO of Service Sqdn. Eight, Pacific Fleet, Pearl Harbor.

His campaign and service medals include American Theater, EAME, Asiatic-Pacific, WWII Victory and Naval Reserve. He elected to remain in the USNR. While in Battle Creek, MI, he organized a Composite Training Unit, appointed CO and instructor. He fulfilled all requirements for transfer to the Retired Reserve as of June 30, 1966, with rank of CDR.

Beginning in 1946, his civilian positions included a Civil Service appointment with the Veterans Administration's rehabilitation division. Appointed the first director of Civil Defense for the city of Madison, WI; served 1952-54, resigned to accept a high level Civil Service administrative and training position with the Federal Civil Service Administration, headquarters in Battle Creek, MI. Selected as staff member of "Operation Cue" Yucca Flats, NV, 1955, the purpose being to study effects of atomic weapons on a pre-positioned community. Witnessed the blast and participated in civil defense training exercise. Retired from federal employment October 1969.

He is a life member of the NRA, TROA, American Legion and the ROA. He was ROA chapter president, Battle Creek, MI in 1961, and president of the Robert E. Lee Chapter, Northern Virginia, 1977.

He is now living with his wife Irene in a retirement center in Winter Park, FL, having enjoyed lots of recreational travel, church activities and volunteer work in the community. His children are Timothy Conklin and Thomas Jones Trowbridge; grandson is Timothy Jon Trowbridge (TJ).

BRUCE EVERETT TRUOG, CAPT, hails from Altoona, IA. He is 1971 graduate with honors of the US Merchant Marine Academy, Kings Point, NY. After teaching in Vermont public schools in the 1970s and earning a master's degree, he was called to active duty as an instructor of mathematics at the US Naval Academy from 1981 to 1986. Upon release from active duty, CAPT Truog moved to Florida and worked for the DOD as a engineer.

In the USNR, he is part of VTU 0809G in Orlando. He has been Charlie Co. CDR of RNMCB 27, CO of SIMA 508, CO of Mayport 508 and deputy commander of Readiness Unit Charlie.

His awards include the Navy Achievement Medal, Meritorious Unit Citation, Armed Forces Reserve Medal w/HG, the National Defense Service Medal, Expert Pistol Medal and Sharpshooter Ribbon.

Truog is married to the former Edythe Pfeiffer and they have three children, Jarrad, Corbet and Tiffany.

REUBEN T. "2-G" TSUJIMURA, CDR, born Oct. 12, 1938, St. Helena, CA. Earned BA in psychology at San Jose State Univ.; AA in personnel managment at Chabot College, Hayward, CA and attended numerous technical and specialized schools for employee enhancement. He joined the USAF, Jan. 28, 1957 and USNR March 3, 1961.

Active Duty: USAF 1957-61: Lackland AFB, TS; Chanute, IL; Tinker AFB, OK; Luke AFB, AZ; Nellis AFB, NV. Baisic training, Tech School, Bomber Wing, Aero Demonstration Team, tactical Fighter Wing. Crew chief, Flight Crew (B-52A), asst. dock chief. VA-122, 1961-63, TEMAC/TAD USS Forrestal, NAF Fallon, NV.

Aircraft type: B-52A; T-33; F-100A—D-F; F-94, F-104C, F-105, TV-2, A-4A-B-C-D-K; A-1H, F-2H-3; SH-3A; A-7A-B-C; F/A-18A and related engines for type aircraft. Served in USS *Forrestal* (CVA-59), USS *Ranger* (CVA/CV-61), USS *Enterprise* (CVN-65), USS *Nimitz* (CVN-68), USS *Dwight D. Eisenhower* (CVN-69), USS *Carl Vinson* (CVN-70), USS *Abraham Linoln* (CVN-72).

Awards include National Defense Service Medal (2), Navy Achievement Medal, Meritorious Unit Commendation, AF Good Conduct Medal (2), NR Medal (2), Sea Service Ribbon (Med-cruise), AF Longevity Ribbon, Armed Forces Reserve Medal (3), Battle E (6) Pistol (sharpshooter).

Reserve Affiliation: 1961, VF-876; 1961-70, VA-876; 1970-71, HS-85; 1971-72, RTU-303; 1972-83, VA-303; 1984-93, VFA-303; 1993-96, NR NADEPALA 0187; 1996 to present, NR Volunteer Training Unit 8787.

Civilian career 1962-67, L&B Welding Equip. Inc.; 1967, Filper Corp.; 1967-93, Pacific Gas & Electric Co.; 1996-present, Roberta Temporary Agency, Pinnacle Staffing, Inc. Married Janis Irene Echelberger and they have a daughter, Gina Kiku and son Jon Erik.

WALTER DUNN TUCKER, LTJG, born Oct. 5, 1931 in Richmond to Frederick Norton Tucker Sr. and Ethel Walker Dunn Tucker. Graduation from John Marshall High School in 1949 and University of Richmond in 1953 preceded OCS and Navy Supply Corps School.

ENS Tucker served in USS *Arneb* from June 1954 until May 1955 and as LTJG in USS *Betelgeuse* until released from active duty Oct. 31, 1956. Tucker spent 37 1/4 years with one bank, retiring from Crestar as vice president in 1994. He served in USNR Intelligence Unit 5-1-3 from 1957 until retirement as a CDR, June 30, 1973.

In November 1961 Tucker married Jacqueline Lee Randlette. Sons, Andrew Dunn and Bruce Edwin, daughter-in-law Margo and grandson Bryan David complete the family.

He is a board member of Richmond Council Navy League, serving as liaison with USS *Carter Hall*. He is a Sunday school teacher and a student of history.

THOMAS F. TURISCO, CAPT, a native of New London, CT, was commissioned in 1973 through the NROTC program at Rensselaer Polytechnic Institute, Troy, NY, where he earned both a BS and master of engineering degree in Nuclear Engineering. Upon completion, he reported to the Nuclear Power Directorate of the Naval Sea Systems Command (NAVSEA 08) for active duty on ADM Rickover's staff in the areas of submarine fluid systems design and reactor core manufacturing. This assignment included duty at the Bettis Reactor Engineering School in West Mifflin, PA, the S3G Reactor Prototype in West Milton, NY, and the Naval Reactors Representative's Office (NRRO) at electric boat in Groton, CT. He was released from active duty in 1977 and entered the Selected Reserve in 1979.

His initial Reserve assignment was with NR COMSERVRON 2-01, Hartford, CT, where he as a division officer in the Operations Department. In 1981, he transferred to NR NAVSHIPENGCTR Det 201, where he served as assistant administrative officer and was redesignated as an engineering duty officer. In October 1981, he joined NR NAVSEA Det 1101, Plainville, CT, where he served initially as projects officer and then the training officer until September 1990, when he was selected as CO of NR AS-11 *Fulton* DET 501, New Haven, CT.

With the inactivation of USS *Fulton* (AS-11) and the transfer of all reserve personnel to the Naval Submarine Support Facility, New London, CT, this unit was retitled NR SSF New London Det 801 with CAPT Turisco continuing as CO until Sept. 30, 1991. On Oct. 1, 1992, he became the CO of NR SUPSHIP Det 401. Upon selection to the rank of CAPT. CAPT Turisco was selected as CO of NR NAVSEA Det 1101, starting Oct. 1, 1993. Upon completion of this CO tour, he was selected for his current assignment as XO of NR NAVSEA Det 901, starting Oct. 1, 1995.

CAPT Turisco is authorized to wear the Navy Commendation Medal (two awards), Meritorious Unit Citation (two awards), National Defense Service Medal (two awards), Military Outstanding Volunteer Service Medal, Navy Recruiting Service Ribbon, Armed Forces Reserve Medal (two awards), Expert Rifle Medal, Expert Pistol Shot Medal and the Recruiting District Assistance Council Gold Wreath Award. He is a life member of the NRA.

In civilian life, CAPT Turisco is a project manager with ABB Combustion Engineering Nuclear Systems in Windsor, CT. He has been involved in a variety of nuclear power projects, including an extensive technology transfer project for the Republic of Korea. He resides in Simsbury, CT with his two children. His wife is deceased.

JAMES E. "JIM OR JET" TURNER JR., CAPT, born Aug. 8, 1942 in Phoenix, AZ, received a BA from UVA, 1964; JD, UVA, 1972; MBA University of Texas Arlington, 1983.

He joined the service June 6, 1964. His military locations and stations: Charleston, SC; DaNang, RVN; Washington, DC. Served with USS *R.H. McCard* (DD-822); CO SURVFOR/NAG I Corps; BUPERS.

His battles and campaigns: Vietnam (four campaigns). He was awarded the Combat Action Ribbon, Vietnam Service Medal, Meritorious Unit Citation. He achieved the rank of CAPT and was discharged Oct. 1, 1991.

His memorable experiences: The experience and pleasure of exotic travel throughout the Mediterranean, the Read Sea and Indian Ocean, to Southeast Asia and Australia, and the Caribbean, it was all memorable.

He is married to Lana and is employed as an attorney.

RAYMOND D. TUTTLE, CDR, born in Iowa (1909), raised in Minnesota, Illinois and Missouri. Received an AA degree from Missouri Junior College; also attended Missouri University. Lettered in college track, football and basketball (captain of the team). Joined Kappa Sigma Fraternity at Missouri University, (initiated later at UCLA (Past Grand Master of San Diego K.E. Alumni Association) and also attended San Diego State University.

Married Rose Marie Collins of San Diego in 1939. They have three sons: Dexter Daum of Brea, CA; Dennis Collins of La Mesa, CA and Jack Kirby of Descanso, CA; seven grandchildren, two step-grandchildren and two great-grandchildren.

CDR Tuttle joined the USN in 1942 and was commissioned a LTJG in 1943. (He resigned his position of general manager of the University Ford dealership to join the Navy); and served four years on active duty in WWII as project engineer in the Research & Invention's Dept. of the Special Devices Div. of the Navy's Bureau of Aeronautics in Washington, DC before duty in the American and Pacific theaters of operations as well as duty at various naval stations, including Quonset Point, RI, Pensacola, FL and the naval air station North Island, San Diego, CA where he was officer in charge of the Special Devices Div., operations duty officer and assistant training officer on the staff of the commander naval air bases in the 11th Naval District.

He remained active in the USNR after release from active duty as a LCDR in 1946, as a member of the USNR

Research Co. 11-5 in San Diego and was promoted to full commander status. He became commanding officer of this research company before retiring in 1969 with 27 years of credited service. He was awarded two campaign ribbons, the WWII Victory Medal the USNR ribbon with two stars and two commendations for his services.

During period of inactive reserve duty since WWII he joined General Dynamics/Convair as a material project administrator, and supervisor, and entered business for himself after leaving Convair (as a result of phase out of the atlas weapons system upon its completion), where he monitored the subcontractors installations at AF bases in Altus, Oklahoma, Walker AFB in Roswell, NM, and Plattsburgh, NY AFB in Plattsburgh, NY).

While in business in El Cajon dealing in government and industry surplus material and equipment, he was a charter member of the Kiwanis Club, vice-president in 1950 and elected president in 1951. He also was one of the originators and founders of the Grossmont Hospital Association in 1947, and served as its President for four years. This leadership effort resulted in the erection of the Grossmont Hospital in 1951 (now a 400 bed institution), sorely needed to serve the East San Diego County area.

Tuttle is also a "perpetual" member of the MOWW, a "life" member of TROA, NRA and DAV (with a WWII service connected injury which resulted in a 20% disability rating.)

He has been married 57 years to Rose Marie Collins and they have children: Dexter, Dennis and Jack; and grandchildren: Denise, Mark, Gregory, Alexandra, Jonathan, Cheyne and Jeremy and great-grandchildren: Christopher and Alyssa.

WALLACE D. TWEDEN, CAPT, born in Spokane and raised in Tacoma, WA. He began his naval career in 1957 as a NROTC midshipman at the University of Washington. Commissioned an ENS, USN, in 1961, he was assigned flight training in Pensacola, receiving his designation as naval aviator in January 1963.

He served aboard USS *Hornet* (CVS-12) with HS-2 in the Western Pacific in 1963 and 1964. After service as a flight instructor with VT-3 at Whiting Field, FL, he was released from active duty in 1966 and accepted a commission in the USNR. His reserve stations during a period of USNR transition and consolidation included NAS Seattle, NAS Glenview, and NAS Whidbey Island; then he served as standardization officer and safety officer with HS-85 at NAS Alameda, and then as XO and CO with HAL-5 at NAS Pt. Mugu.

In civilian life, CAPT Tweden is a captain with United Airlines, where his current assignment is as a flight operations manager at the San Francisco domicile. He has flown the B-727, B-737, B-747, (400), B-757, B-767, and, quite recently, worked to introduce the B-777 into commercial service.

CAPT Tweden has a married son, Michael, who lives in Waco, TX and his married daughter, Kristin, and granddaugther, Claire, live in Santa Rosa, CA.

ALBERT THEODORE "TED" TWESME, CAPT, born July 13, 1948 in LaCrosse, WI. He joined the service July 30, 1970.

His military locations and stations: Marine Corps Base, 29 Palms, CA. Served with H&S Bn., 1st Marine Div., 21st Dental Co.; NAV RES CTR, Oklahoma County, NAV RES Center, Las Vegas.

He was awarded the Unit Commendation, National Defense Medal (2) and Naval Reserve Medal (2). Achieved the rank of CAPT.

His memorable experiences: over 27 years of service he has enjoyed special active duty experiences such as Naval Dental Centers in Pearl Harbor, HI; Orlando, FL; Great Lakes Naval Station, San Diego, CA and aboard USS *Eisenhower.*

He is married to Mary Alice and they have children: Lee, Joe and Erika. Employed as oral and maxillofacial surgeon, Las Vegas, NV in private practice.

SUSAN PERRY TYE, CDR, born Nov. 13, 1957 in Troy, NY, received a BS in nursing; Boston University School of Nursing.

She joined the service May 31, 1979. His military locations and stations: Balboa Naval Hospital, Bethesda Naval Hospital. Served with NH Bethesda, NH Camp Lejeune, NH San Diego, NRFH500GB7222 DETG.

She was awarded the Navy Achievement Medal and Navy Commendation. She achieved the rank of CDR and was discharged April 1983 to Reserves.

Her memorable experiences: flying to Hawaii on a P-3 "nop", field exercises, Marines.

She is married to Charles M. Tye, LT COL/AD/USMC and has two children, Christopher and David. Employed as critical care nurse in various hospital settings over the past 18 years.

DAN M. TYSON, CAPT, born June 13, 1949 in Tokyo, Japan, received a MS in management from USC in 1978.

He joined the service Aug. 7, 1972. His military locations and stations: VA-97, VF-124, VF-114/USS *Enterprise,* USS *America* (active). He served with NAVAIRPAC 1094, VTU 3005, NAVAIR Support West (Reserve). His battles and campaigns: Vietnam, Frequent Wind.

He was awarded the Armed Forces Exp., Meritorious Unit Commendation, Battle E, National Defense and Sea Service Medal.

Tyson achieved the rank of CAPT and SEL RES. His memorable experiences include the Vietnam evacuation mission.

He is married to Harumi and they have children: Adriane, Ashley and Jasmine. Employed as general manager, East Asia, Varian Association.

THOMAS R. "TOM" ULRICH, CAPT, born Aug. 30, 1932, Louisville, KY, earned his BCHE degree at Villanova University in June 1956.

Commissioned ENS, NROTC regular scholarship to Villanova in September 1951. His military locations and stations: Flight training (Pensacola and Corpus Christi). He served with VP-42 (NAS North Island, CA); VT-31 (Corpus Christi, TX); VP-56 (NAS Norfolk, VA); COMFAIRMED/COMASW for Sixth Flt. (Naples, Italy); CINCLANTFLT; and OPNAV (Pentagon).

He achieved the rank of CAPT (O-6) and retired in June 1986. His memorable experiences: Naval aviator in P-5s, P-Zs and P-3 aircraft. Deployed to Far East, Iceland and Mediterranean against major Soviet submarine operations.

He and his wife, Barbara, have one daughter, Mary Anne, and three grandchildren: Brian, Michael and Megan. Civilian employment with Senior Executive Service (SES) in OPNAV (Pentagon); president and chief operating officer of Systems Technology and Applied Research (STAR) Corp. Ulrich retired in November 1994.

REX UMBENHAUR M.D., LT, Umby's Crumbies: the story of a musical organization aboard the *Tasker H. Bliss*, AP-131.

Dr.Clifford E. Ernst joined the *Tasker H. Bliss*, AP-131 in April 1945. Shortly after going aboard he met LT Rex Umbenhaur, the ship's dentist. Both of them were from the midwest. Umbenhaur was an amateur musician as was Ernst. Umbenhaur played drums, Ernst, woodwinds and piano. Their chaplain made instruments available to them, so they formed a small orchestra. A surprising number of men and officers could play instruments competently. They rehearsed daily after sick call. Since they ferried Army troops back and forth to Europe, they were soon invited to perform in the Army officers' mess, then entertain the Army troops aboard-on deck, weather permitting. They used some stock arrangements, some Ernst crafted, and some were "head" arrangements. The word "head" has a slightly different meaning than the usual Navy term.

Dr. Umbenhaur was a very affable and charismatic man, and was the personality of the little orchestra. The Bliss Sailors soon called them "Umbie's Crumbies," a name that stuck forevermore. For one of the deck shows, they composed a song which reflected the ship's activities:

Now Hear this, the voice will say;
Lay down to the galley at the break of day.
Joes and Gobs on *T.H. Bliss*, Lay down! lay down!
And now hear this! hear this!, hear this now!
Lay up to the bridge or to the bow,
When the Bosun pipes that sound,
All you're going to hear's: Lay down, Lay down!

Far out on the ocean you'll always find us ready, Lay up, lay down

Lay up, lay down, we're looking for Liberty Lady.

Sweepers, man your brooms, clean sweep down fore and aft!

Take all trash to the fantail aft, When the bosun pipes that sound,

All you're going to hear's, Lay down, lay down!

Dr. Umbenhaur was very popular with the Navy crew because of his great, good humor. He loved to argue politics. Their ward room was an area of lively discussion, because LT Tom Walker, son of a wartime US attorney general, was one of their officers. They also had aboard the late Phillip Wilkie, the son of Roosevelt opponent, Wendell Wilkie, who acted as an aid to the convoy admiral.

In their musical books, they had some slightly naughty songs, much enjoyed by the servicemen. One young lad could sing *Pistol Packing Mama* in five languages. Some of the troops had fiddles so they had some marvelous hoe-down music. The T.*H. Bliss* arrived in La Havre, France after V-E Day. Their first gig ashore was for a Navy installation near the port facility. "Umbie's Crumbies" played around the world after that until the day Dr. Umbenhaur was detached in late 1946.

In 1954 Ernst moved to Phoenix, AZ. He found Dr. Umbenhaur had also moved there. They resumed their musical association and formed a Dixieland Band playing gigs for local Chambers of Commerce and hospitals. Yes, they still called themselves "Umbies Crumbies."

LT Rex Umbenhaur died in 1996. His close friend sorely misses him. *Submitted by Dr. Clifford E. Ernst, LCDR aboard US Tasker H. Bliss AP-131.*

JOHN C. UMLAUF, LCDR, born July 24, 1918, Ashland, PA, earned his PhB degree from Muhlenberg College, Allentown, PA.

He joined the service June 6, 1943, Cornell University Naval Indoctrination School, deck officer. His military locations and stations: Little Creek, VA, Amphibious Training Base, Solomon Island Training Base, MD. He served with many different amphibious boats and groups, finally CO of LST. He participated in the Invasion of Normandy.

His memorable experiences: Served as executive officer of a British built gunboat at Invasion of Normandy. They were the first ship within 200 of beach on D-Day, sweeping mines and giving gunfire support. Stayed in area two weeks. End of first week picked out of water over 100 bodies for burial back in England as directed.

He married the former Madelon Frances Rufner and they have three children: Susan, Judith and Wendy; and two grandchildren, Kristina and Lisa. Civilian employment as looper at Bethlehem Steel; agent-manager, New York Life; and executive director, American Cancer Society Insurance Co., Lehigh County, PA. Umlauf retired in 1978.

LAWRENCE B. URBANO, CDR, is a graduate of Williams College and Harvard Law School. After private practice, he completed his civilian career as a judge of the Massachusetts Superior Court in December 1961.

He enlisted in the USN as a cadet and was designated a naval aviator in April 1944. He served in VF-3 and remained in that squadron until he was released from active duty Dec. 16, 1945.

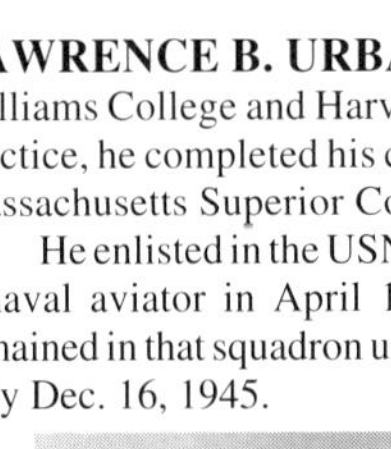

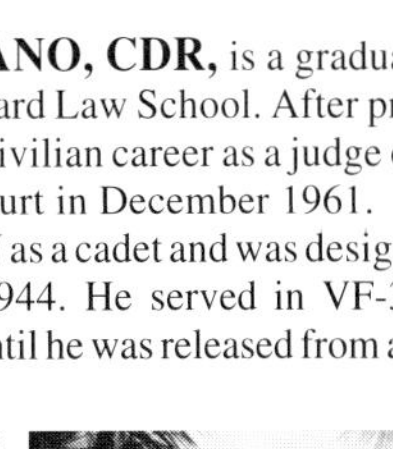

While in VF-3 he participated in combat strikes against various targets including Tokyo, the Philippines, Indo-China, Formosa, Hong Kong and Iwo Jima, and was awarded the following honors: Distinguished Flying Cross,

Air Medal w/Gold Star, Philippine Liberation Medal and Presidential Unit Citation (2).

On release from active duty he remained in the reserve, stationed at NAS Squantum and South Weymouth. In 1961 he became the CO of Anti-Submarine Sqdn. VS-915 which was activated Oct. 11, 1961, during the Berlin crisis. The unit was deactivated in August 1962 and at the same time, he terminated his association with the active reserve.

His most memorable moments were two: One was shooting down two Japanese fighters on one sortie Feb. 16, 1945. The second was earlier, Dec. 15, 1944, when he landed his damaged plane in the Philippine Sea and was picked up by a destroyer, put on a tanker and went through the typhoon of December 1944 aboard the tanker.

ANDREA JANE "CANDY" (FOREMAN) URSEM, LCDR, born Dec. 19, 1947, earned her LPN 91C20 (USA) and RN degrees at Elyria Memorial Hospital School of Nursing and her BSN degree at Bowling Green State University, Bowling Green, OH.

She joined the US Army June 30, 1969, and was discharged July 22, 1973. Her military locations and stations: Boot camp, Ft. McClellan, AL; 91A School, Ft. Sam Houston, TX; Ft. Hauchuca, AZ; 91C School, Ft. Bliss, TX; Baumont Gen. Hospital, Ft. Meade, MD. She joined the USAR Aug. 22, 1984, and was discharged Sept. 30, 1988, where she was attached to 256th General Hospital, Parma, OH.

On Oct. 1, 1988, she joined the USNR. Military locations and stations: Reserve Center, Cleveland Medical Unit, Oct. 1, 1988-May 1993; DS/DS Portsmouth General Hospital, Portsmouth, VA, Oct. 5, 1990-April 12, 1991; Selfridge ANGB, Mt. Clemens, MI, May 1993-April 1995; NRC Cleveland Medical Unit, April 1995-present.

Ursem was awarded the National Defense Medal, Good Conduct Medal, Army Commendation Medal, Army Service Medal, Meritorious Unit Citation, Marksmanship Ribbon and Coast Guard Bicentennial Ribbon.

She and her husband, Thomas, have four children: Thomas Jr., Joseph II, Zachary and Laurel. Civilian employment at Elyria Memorial Hospital, staff nurse; Elyria Methodist Home, supervisor; Allen Memorial Hospital, staff nurse; and Elyria City Health Dept., RN, BSN.

FREDERIC L. "TED" USHER, CAPT, born Oct. 5, 1920, Lynn, MA, graduated from high school and attended some college. He joined the service Oct. 7, 1943-Feb. 13, 1947 (active); 1947-51 (reserve); 1951-53 (active) and 1953-76 (reserve). He served in many military locations and stations.

His memorable experiences: Getting his wings; being selected a pilot in the 1st Jet Sqdn. VF-41; being CO of VA-912 and RS and Z1. He and his wife, Nancy S., have four children: Kathleen, Charles, Frederic Jr. and Douglas; 13 grandchildren; and two great-grandchildren. Civilian employment as system analyst, USN. Usher retired June 30, 1976.

VILIMANTAS S. "VIL" VAITAS, CDR, born Dec. 16, 1934, Lithuania, earned his BS degree in engineering at Tufts University and his MA degree in economics at Michigan State University.

He joined the ROTC in 1952 and was commissioned June 1, 1956. His military locations and stations: California, Atlantic and Mediterranean Flts., Philippine Islands and Cuba. He served with Net Depot, Tiburon; US *Boston* (CAG-1), Public Works Center, Subic Bay; and PWC, Guantanamo. Battles and campaigns participated in were Lebanon (1958), Vietnam Era service in Philippines (1965-68).

Vaitas was awarded the National Defense Medal, Armed Forces Expeditionary Medal (2), Armed Forces Reserve Medal (2) and Navy Meritorious Unit Citation.

He and his wife, Virginia E., have two children, Vilimantas S. Jr. and Virginia R., and four grandchildren: Vilimantas S. III, Timothy J., Amanda L. and Victoria E. Civilian employment in various engineering and management positions in federal government. Vaitas retired Dec. 1, 1983.

RICHARD A. "VAN" VAN CISE, CAPT, born June 20, 1934, Neah Bay, WA, earned his BS degree at Oregon State University. He joined the service June 26, 1957. His military locations and stations: USS *Marias* (AO-57), USS *Nitro* (AE-23) and 13th NO, Seattle. He served with the Sixth Flt., SERVLANT.

VanCise was awarded the National Defense Medal and Armed Forces Reserve Medal. He achieved the rank of CAPT and retired June 20, 1994.

His memorable experiences: On the USS *Marias* (AO-57) in January 1958, operating out of Barcelona with the Sixth Flt., he became ship's navigator. They were actively involved in the Lebanon crisis and the crisis in Iraq. Not having a workable loran onboard, nearly all of his posits (including Atlantic crossing) were by celestial.

In early '59 he was selected to be ship's navigator on the new ammunition ship *Nitro* (AE-23). Placing the ship in commission, organizing the Navigation Dept., and successfully getting the ship through underway training was a rewarding, but challenging experience. Enroute to the Mediterranean with a Second Flt. Task Force, COMSERVLANT brought his staff aboard their ship.

On the second day at sea he became the flag navigator as they led three other auxiliaries out of radar range to become an enemy combatant group with nuclear capability. Several days later, disguised as a holiday cruise ship with lights and dance music blaring, they steamed through the disposition in the dark of night without revealing their identity. After successfully pinpointing the carrier, the cruiser and most of the other combatants, they went over the horizon and launched their surface to surface attack.

Upon reaching Barcelona several weeks later they gathered charts of tracks and positions and compared them with those of their adversaries and discovered that they had disabled the carrier, the cruiser, and several DDs before they were put out of commission.

He and his wife, Marlies, have four children: Rick, Eric, Heike and Oliver; and six grandchildren: Lisa, Christopher, Taylor, Sam, Alex and Benjamin. Civilian employment as director, customer relations-Africa & Mideast, The Boeing Co., from which he retired Jan. 1, 1995.

JAMES L. "JIM" VAN DEN BIESEN, CAPT, born April 7, 1929, Hysham, MT, earned his BS degree at the University of Oregon in 1952.

He joined the service in June 1952 (Navcad class 26-52). His military locations and stations: Pensacola, Corpus Christi, Alameda, Iwakuni and Japan. He served with VP-50, ATU 501 (active), VS-892 and VS-83, RTU-83 (Reserves).

Van den Biesen was awarded the National Defense Service Medal, Navy Commendation Medal and Naval Reserve Medal. He achieved the rank of CAPT and was discharged from active duty Jan. 18, 1958, and retired May 1, 1978.

His memorable experiences: Flying patrols Yellow Sea/Sea of Japan operating out of Iwakuni Japan. Flight instructor. Requalifying aboard ship and keeping current in S2E while in VS-83 out of Whidbey Island, Oak Harbor, WA.

He and his wife, Marilyn, have two children, Vivian and Lisa. Civilian employment as captain for West Coast Airlines, Seattle, WA through several mergers - became Airwest, Hughes Airwest, Republic and North West Airlines. He retired Dec. 1, 1988.

RICHARD M. VANDERHOEVEN, CDR, born May 7, 1959, Montreal, Canada, earned his BS degree in electrical engineering.

He joined the service May 23, 1981. His military locations and stations: USS *Hammerhead* (SSN-663); Navy Recruiting Command. He served with SSN-663 from October 1982-October 1985 and CNRC from November 1985-December 1987.

Vanderhoeven was awarded the Navy Achievement Medal and Navy Commendation Medal. He achieved the rank of CDR and was discharged Dec. 30, 1987.

His memorable experiences: Reserves: Carrier Gp. Four, February 1988-October 1992 and COMSUBLANT, October 1992-October 1996.

He married the former Pamela Ann Kicak and they have two children, Andrew Richard and Erica Lynn. Civilian employment with US Government, Dept. of Navy, 1988-present.

PHILIP L. VANDER MYDE, USNR, born April 4, 1931, Sterling, IL. Graduated from the University of Minnesota (B.Arch) after four years of active duty in the USAF. Commissioned an ENS, USNR in 1959. Served as commanding officer of four Naval Reserve Intelligence units in the Washington DC area. Retired in 1991 with 36 years of service. Awarded the Meritorious Service Medal and Navy Commendation Medal among other.

In civilian life, he was a partner in three national architectural firms and is currently president/CEO of his own firm, Additions Inc. He is a past-president of the Alexandria Rotary Club and the Potomac Valley Chapter AIA and is on the board of the Navy Art Foundation. He has been listed in *Who's Who in the World* since 1984.

His family includes his wife Terry, son John, and daughters Jane and Missy. Plans are currently underway for a permanent, family waterfront home in the maritime forests of the Outer Banks of North Carolina.

JOEL K. VANDERWAAL, CDR, born in 1952, graduated with honors from the University of Illinois in January 1976 where he was commissioned through the regular NROTC.

He earned naval aviator wings in 1977 and until 1987 CDR VanderWaal served in various billets in VR-24, VP-50 and VP-90, earning designations of transport aircraft CDR (TAC) and patrol plane CDR (PPC). From 1987-95 he was plank owner and executive officer of NR CV-63 *Kitty Hawk* and PAO of NAVAIRESCEN Minneapolis.

CDR VanderWaal has been awarded the Meritorious Unit Citation and Battle "E,", Sea Service Deployment Medal, Overseas Deployment Medal, Naval Expeditionary Medal, National Defense Service Medal, Expert Pistol Medal and Expert Rifle Shot Medal.

In civilian life he is a CAPT for Northwest Airlines, life member of Naval Reserve Assoc. and Assoc. of Naval Aviation, volunteers for the National Ski Patrol and Bloomington Athletic Assoc. He resides in Bloomington, MN, with his wife and two daughters.

JOHN M. VANGORDON, CWO-4, born Oct. 23, 1915, North Kansas City, MO, attended school in the greater Kansas City area and joined the USN as a SA April 7, 1936, at a base pay of $21.00 per month.

He went by train to NTS San Diego for his boot camp training. His first duty at sea was aboard USS *Langley*, the Navy's first aircraft carrier. He was aboard *Langley* when it was converted to an aircraft tender servicing PBY-1 flying boats.

His next assignment was a 30 months tour of duty on the Asiatic Station. His first duty was on board USS *Isabel*, a patrol yacht that had previously served as flagship of the Yangtze River Patrol. China and Japan were then at war. While moored in the Whangpoo River he had a ring side seat to hand to hand combat between Chinese and Japanese soldiers and witnessed the burning of Shanghai. *Isabel* made several trips up the Yangtze River to rescue American citizens from beyond the front lines. On one of these trips he witnessed, what was then the capital of China, the burning of Nanking.

He later served on board USS *Black Hawk*, a destroyer tender and visited many China and Philippine Island places of interest.

Upon returning to the US he re-enlisted and was ordered to duty on board the new destroyer, USS *Niblack*. By then war had started in Europe. *Niblack* was assigned to North Atlantic patrol and escort duty. On one trip while rescuing survivors of a torpedoed Dutch freighter a U-boat was detected making preparations to attack *Niblack*. *Niblack* immediately got underway and dropped depth charges on the U-boat. It was the first time during WWII that the Navy fired a shot in anger.

After war was declared he was assigned to duty aboard USS *Essex*, a new aircraft carrier. She was immediately ordered to the Asiatic-Pacific Theater. On eight different occasions *Essex* entered into combat with the enemy while participating in the attacks on Marcos Island, Wake Island, Rabaul, Kwajalein Atoll, Truk, Saipan-Tinian and the occupation of Marshall Islands.

He was appointed to the rank of WO and thereafter ordered to duty with the newly formed Combat Aircraft Service Unit - 52. While in that unit he was in the occupation forces that landed on the island of Iwo Jima. He landed on D-plus three and was there until the signing of the Peace Treaty with Japan. One of the highlights of his career happened while on Iwo Jima. He listened to the radio broadcast of the Peace Treaty being signed on board USS *Missouri* in Tokyo Bay just a few hundred miles away.

After the war he returned to the US and was released from active duty. He later joined the USNR and became very active in the Reserve performing most of his drills at USNRTC Santa Barbara, CA. He is a life member of the Naval Reserve Assoc. He retired after 32 years of naval service.

The VanGordon family consists of wife, Genivra; daughters: Jeannine, Ginette and Claudette; son, Michel; and sons-in-law: John, Michael and Thomas.

ROBERT H. "VAN" VANNATTA, LCDR, born June 3, 1923, Toledo, OH, attended three years at Cleveland Marshall Law School.

He joined the service Oct. 31, 1942. His military locations and stations: NAS Pensacola, NAS Vero Beach, and NAS Barbers Point. He served with VF(N) 75 and VF-86 on USS *Wasp*. Battles and campaigns participated in were Okinawa and Japan.

Vannatta was awarded the Unit Citation. He achieved the rank of LCDR and retired Sept. 1, 1963. His memorable experiences: Flying from USS *Wasp*.

He is married to Mary G. Civilian employment as examiner with state of Ohio and state of Tennessee.

GRETCHEN ELIZABETH VAN ROY, LCDR, born Dec. 19, 1913, Crystal River, FL, earned her BS and MA degrees in music at Teachers College, Columbia University, New York City. She has done graduate study at the University of Southern California, Los Angeles and the University of Hawaii.

She joined the service in October 1942. Her military locations and stations: Smith College training, ENS, Jan. 8, 1943. She served in communications in Gulf Sea Frontier from January 1943-October 1944, transferred to 3rd Naval Dist., October 1944-August 1945, transferred to COMINCH August 1945 and released to inactive duty Dec. 1, 1945. During Jan. 6-Jan. 27, 1944, she was assigned to temporary duty in Washington, DC.

Van Roy achieved the rank of LCDR and retired July 1, 1971.

Civilian employment: Taught in Bloomington, IL at Illinois Wesleyan University; Long Beach State College, California; and Dade County Public Schools, Miami, FL.

PETER D. VAN SOEST, CAPT, born Aug. 10, 1937, Denver, CO, attended Central College, Pella, IA where he earned his BA degree and the University of Denver School of Law where he earned his JD degree.

Military service June 1959-December 1962. Active duty: training commands, naval flight officer VW-11 and VAW-12.

From 1963-91, drilling member of various Naval Reserve Intelligence units. Duties included service on Naval Reserve Intelligence staffs as CDR, command of intelligence units, and as CDR, Reserve Intelligence Area Five. USNA Blue and Gold officer and area coordinator.

CAPT Van Soest has actively practiced law in Denver, CO, since 1966-present. He is active as an officer and director of many professional and civic organizations. He is a life member of the Naval Reserve Assoc. and is a past president of the Mile High Chapter, NRA.

FRANK VAUGHAN, CAPT, born Oct. 24, 1920, Holloway, OH, graduated from the University of Cincinnati and Northwestern University where he earned his MBA. He also attended the Naval Postgraduate School, Naval War College and Oxford University.

CAPT Vaughan enlisted in the US Army in 1942 and was assigned to the 53rd Regt., Coast Arty., later transferred to the Army Air Corps. While at Maxwell Field he played on the base baseball team. In 1949 he received a direct commission in the USN as a LTJG. He volunteered for active duty again in 1950. CAPT Vaughan served with the 6th Flt. on the USS *Cambria* (APA-36) and in Korea as a department head on the USS *McNair* (DD-679). In 1953 the *McNair* completed an around the world cruise. Shore duty included CO, NADO Great Lakes. Reserve assignments included CO, NRSC 11-10 and SSOD 11-1.

In civilian life, CAPT Vaughan is a private investor.

MARK BASS VAUGHAN, CDR, born Oct. 9, 1958, Norfolk, VA, raised in Kentucky and attended the USNA. Upon graduation in 1980 he received a BS degree in aerospace engineering, was commissioned ENS and reported to NAS Pensacola, FL for flight training.

CDR Vaughan was awarded his wings as a naval aviator in September 1981 whereupon he reported to VT-3, NAS Whiting Field, Milton, FL as a T-34C flight instructor.

Following his flight instructor tour, CDR Vaughan taught leadership and management education and training for aviation division officers at Naval Technical Training Center, Treasure Island, San Francisco, CA.

CDR Vaughan affiliated with the USNR in 1986 upon his selection for the TAR (Training and Administration of Reserves) Officer program. His initial assignment was at Naval Air Facility, Washington, D.C. (on board Andrews AFB) as assistant operations officer flying the UC-12B.

In 1987 CDR Vaughan reported to "The Skyliners" of Fleet Logistics Support Sqdn. 48 (VR-48), onboard NAF *Washington*. He flew the C-131H (Convair 580) and served as assistant maintenance officer and assistant administration officer.

In 1989 he reported to VR-59, the world famous "Lonestar Express" at NAS Dallas, TX. He transitioned to the C-9B and served as the assistant training officer and assistant operations officer. In December 1990 CDR Vaughan mobilized and deployed with VR-59 in support of Operation Desert Shield/Storm where he flew logistics missions in support of combat operations throughout Europe and Southwest Asia.

In 1992 he reported to the staff of the CDR, Fleet Logistics Support Wing, NAS Dallas, TX as the C-9B NATOPS evaluator and model manager, training officer and assistant operations officer.

CDR Vaughan reported to "The Conquistadors" of VR-57 as the seventh officer in charge in 1993. Following his OIC tour, he reported to the staff of the CDR, Naval Air Reserve Force, New Orleans, LA as the VR program manager.

Selected for Aviation Command in 1995 CDR Vaughan reported to "The Conquistadors" as executive officer in 1996. He will relieve CDR Ray Farmer, the current CO, in October 1997.

CDR Vaughan is a 1994 graduate of the Naval War College, College of Continuing Education and has flown over 4200 hours in seven types of military aircraft. His personal decorations include the Meritorious Service Medal, the Navy and Marine Corps Commendation Medal (2) and the Navy and Marine Corps Achievement Medal (2).

He is married to the former Donna Ruth Hallberg of Orlando, FL. They reside in El Cajon, CA with their three children: Anna, Nathan and Rachel.

DAVID VAURIO, CDR, born Oct. 9, 1946, Pittsburgh, PA, earned his MA degree in criminal justice/computer science at Washington State University.

He joined the service Feb. 7, 1964. His military locations and stations: USS *Haynsworth* (DD-700) and USS *England* (DLG-22). He served with National Security Agency, National Reconnaissance Office, vice president's Counter Terrorism Task Force, National Security Council (counter terrorism), president's Foreign Intelligence Advisory Board and Advanced Research Projects Agency.

He achieved the rank of CDR and retired in September 1992. His memorable experiences: Completing theoretical design of secure computer hardware and software for the space BASFD element of the Strategic Defense Initiative (SDI). Design of the FBI Forensic Examination of Digital Data System (FEDDS).

He and his wife, Donna, have two children, Kimberly and John, and one grandchild, Michael. Civilian employment as senior research associate, science applications, International Corp.

CARL DEAN "DEAN" VAUSBINDER, CDR, born on a farm near Harveyville, Wabaunsee County, KS, Aug. 23, 1926. He graduated from North Hollywood High School (California) in 1944.

He entered the USN Jan. 12, 1945. His military locations and stations during WWII: NTC San Diego, boot camp; NTC Gulfport, MS, Quartermaster School; Navy 3864, Leyte Gulf, Philippine Islands; LCT #1301, Samar, Philippine Islands. He was honorably discharged as quartermaster, PO3, July 9, 1946, and signed up in the USNR July 10, 1946. He graduated from the University of Southern California (USC) in 1950 receiving a BA in psychology and a commission as ENS in USNR via NROTC program. In 1955 he became an active reserve in a Naval Air Intelligence unit at NAS Los Alamitos, CA; then to NAS Glenview, IL; NAS Grosse Ile, MI; NAS Mt. Clements, MI; and NAS Buffalo, NY. He was CO of NAIRU Y-1 until the squadron was decommissioned in 1974.

He retired as CDR USNR in 1975. His medals were: American Theatre Medal, Asiatic-Pacific Theatre Medal, WWII Victory Medal and Naval Reserve Medal.

His civilian employment included 36 years with Kimberly-Clark Corp., retiring March 1, 1991. Positions held were: manager of the marketing department, then national sales manager for Kimberly Clark of Canada; director, sales training for Kimberly Clark Corp. USA.

CDR Vausbinder's family consists of wife, Jean; three daughters: Jan, Julie and Jill; five grandchildren: Holly, Tara, Shane, Joshua and Karla; and one great-grandchild, Tabitha. He enjoys travel, genealogical research and reading.

CHARLES "CHUCK" VAVRIN M.D., O-6, born Nov. 15, 1936, Racine, WI, earned his MD degree at the University of Wisconsin.

He joined the service July 1, 1962. His military locations and stations: Ft. Sam Houston, San Antonio, TX.

Vavrin achieved the rank of O-5 from the US Army and O-6 from the USNR, 1972-90.

He and his wife, Karen, have two children, Louise and Susan. Self-employed as M.D.

GEORGE J. VELLELLA, CAPT, born April 4, 1931, Cleveland, OH, graduated from Illinois Institute of Technology, Chicago in 1954 with a BS degree in civil engineering and a commission through the Naval Reserve Officer Training Corps program.

He reported to USS *Richard B. Anderson* (DD-786) as main propulsion assistant. Aboard the *Anderson* he held the billets of damage control assistant, acting engineer officer, communications officer and navigator. His qualifications included OOD-U, fleet operations, ASW deck officer, and command duty officer.

Upon release from active duty in 1957 CAPT Vellella became active in the USNR. He served in many billets at eight different Naval Reserve Training Centers in five midwest states. He held nine commands including his first, Flt. Div. 9-20(D), Chicago and the Readiness Unit 1678-1, Sheboygan, WI, his last command. He retired April 4, 1991.

CAPT Vellella's active duty periods included: reporting officer in a major naval control of shipping exercise at Athens, Greece; naval control of shipping officer for a convoy exercise at Guam, Marianas; attended the O/P course at Naval War College, Newport; attended the Defense Resource Management course at Naval Post Graduate School, Monterey.

CAPT Vellella's involvement in the civilian community included: president of Marinette Rotary Club; Wisconsin Chapter American Public Works Assoc.; Northeast Wisconsin Post of Society of American Military Engineers; Retired Officers Club Northeast Wisconsin.

CAPT Vellella was presented the Distinguished Service award by the Wisconsin Section of the American Society of Civil Engineers, given in "Recognition of Distinguished Service to Community, Mankind and the Engineering Profession."

CAPT Vellella is authorized to wear the Surface Warfare Emblem, China Service (Extended) Medal, National Defense Medal, Naval Reserve Medal w/Hour Glass, Navy and Marine Corp Overseas Ribbon.

CAPT Vellella is married to the former Ann E. Westman of Chicago, IL. They have six sons: George Jr., Steven, Mark, Robert, Christopher and Paul.

PAUL C. "FOXHOLE" VESCOVO JR., MD, CAPT, born April 19, 1925, Memphis, TN. PreMed training at Memphis State University and graduated from University of Tennessee School of Medicine in March 1952. All post-graduate training at St. Joseph Hospital, Memphis, TN, training in internal medicine.

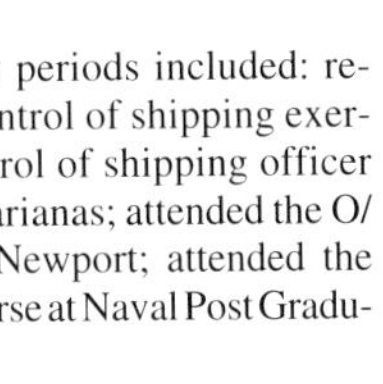

He was inducted into the US Army July 10, 1943. Basic training in Tank Destroyer Bn.; was transferred to 99th Inf. Div. and served in Europe with Co. F, 393rd Inf. Regt., 99th Inf. Div. Discharged from Army March 13, 1946. Battles and campaigns were the Ardennes, Rhineland and Central Europe. Wounded in battle once.

Awarded the American Theater Ribbon, EAME Theater Ribbon w/3 Bronze Stars, Good Conduct Medal, WWII Victory Medal, Occupation of Germany Medal, Bronze Star w/V, Purple Heart Medal, French Croix de Guerre w/ Bronze Star and Combat Infantry Badge. Highest Army rank was S/Sgt.

In the USN he was appointed as LT (Medical Corps) USNR, date of rank May 1, 1955, and date of commission June 27, 1960, promoted to CAPT, Medical Corps, USNR Aug. 1, 1975. Was retired from naval service April 19, 1985.

His memorable experiences: Battle of the Bulge in Europe and crossing of the Remagen bridge while serving with Co. F, 393rd Inf. Service with the Naval Readiness Command from March 2, 1977-Sept. 30, 1981. Service with the 4th Marine Div. as regimental surgeon. Received Letter of Commendation for service to the Readiness Command.

He and his wife, Paula Dianne, have seven children: Beth Marie, Paul C. III, Patrick, Joseph, Tyron, Tiffany and Karen Louise (killed July 1973); 10 grandchildren: Anne, Darren, Shannon, Amanda, Katherine, Laura Beth, Darren, Ryan, Tyron and Forrest; and one great-grandchild, Avianna. Civilian employment and positions held: Private practice of Internal Medicine from 1956-94. Now employed by Tri-Source, an HMO of Blue Cross Insurance. Was the medical examiner for Clay County, MO from February 1977-December 1992.

VERNON E. VICTORINE JR., LCDR, born Dec. 8, 1939, Chicago, IL, earned his BS degree at Culver-Stockton College, Canton, MO in 1962.

He joined the service in November 1962. His military locations and stations: OCS, Newport, RI; US NAVFAC, Barbados, WI; and US NAVFAC, Cape Hatteras, NC.

His memorable experiences: After leaving active duty spent 17 years in active reserve. Served in NRSD 9-1, Chicago; Gp. Command, Chicago and NRN CSO Chicago 413, Great Lakes, IL.

He married the former Linda Nelson and they have two children, Donald and Brian. Civilian employment at Vevco Inc., Naperville, IL as vice-president. Victorine retired July 1, 1983.

HAROLD EDOUARD "FRENCHY" VIEAU, CAPT, born Jan. 20, 1921, Minneapolis, MN, attended two years at Regis College, Denver, CO.

Enlisted as a V-5 cadet Wold-Chamberlain Field in August 1942. Graduated as naval aviator in March 1944 at NAS Corpus Christi. Reported for dive bombing at NAS Deland, FL. Completed car-quals on the USS *Sable*. In July 1944 reported to NAS North Island to CASU-5.

In August 1944 assigned to Sqdn. VC-91, NAS Los Alamitos. Squadron embarked on USS *Kitkin Bay* from Kaneohe to Admiralty Islands. At the Lingayan Gulf invasion Jan. 6, 1945, the carrier was kamikazed and knocked out of action. Transferred to USS *Savo Island*, CVE-78 at Tacloban. On to Okinawa for D-day, operating for 88 days in the area plus transferring at sea to USS *Makin Island* (CVE-93). Returned to States in July 1945 and reassigned to VT-20 NAS Norfolk.

Joined reserves at Los Alamitos from 1947-71, CO VS-776. Transferred to NAS Point Mugu, CO NARS-U1.

Vieau was awarded the Distinguished Flying Cross, Air Medal (3) and Philippine Liberation Medal. He achieved the rank of CAPT and retired in June 1976.

He married the former Beverly Brewster and they have three children: Jerome, Janet and Joan; and seven grandchildren: Jodi, Amanda, Diana, Jeffery, Lauren, Ryan and Nathan. Civilian employment as senior engineer, GTE of California. Vieau retired in December 1982.

WAYNE A. VINCENT, LT, born Oct. 17, 1962, Freeport, TX. Commissioned a Naval Reserve cryptologic officer in 1988, shortly after completion of a BS in government/history with the University of Maryland. Prior to his commissioning, he had served from 1982-88 as a US Marine. Assigned to NSGA Edzell, Scotland, Co. "B" Marine Support Bn. as a Cryptologic Manual Morse Intercept operator, he performed various mission duties including division watch coordinator, sergeant of the guard and platoon sergeant.

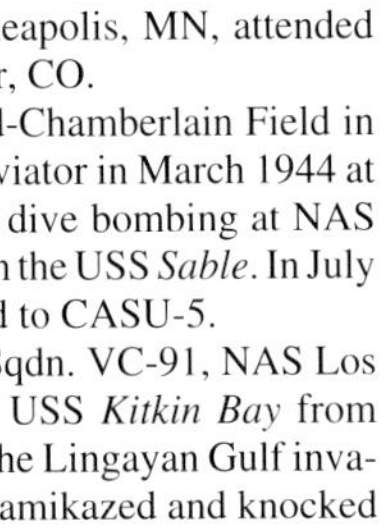

In the years following his commissioning, LT Vincent earned a BBA degree in management information systems and an EDM degree in administration and supervision. He has performed in various naval reserve security group areas including training, operations, and as AOIC of a Tactical Cryptologic Team. Additionally, he performed active duty training in support of missions aboard NSGA Naples, Italy; NSGA Edzell, Scotland; HQ US European Command, National Security Agency, NSGA Medina, San Antonio, TX; and Fleet Combat Training Center, Damneck, VA where he completed training as an electronic warfare officer.

LT Vincent's decorations include the Navy & Marine Corps Achievement Medal, Navy Unit Commendation, Marine Good Conduct Medal, National Defense Medal, Armed Forces Reserve Medal and Overseas Service Ribbon.

The Vincent family consists of wife, Theresa; daughters, Chelsea and Katelyn; and son, Stephen.

EUGENE J. "GENE" VITALE, CDR, born July 24, 1930, Chicago, IL, earned his BSME degree at Purdue University in 1952. He joined the service in December 1952. His military locations and stations: Quonset Pt., RI and Hutchinson, KS. He served with the Civil Engineering Corp., facilities management.

Vitale was awarded the usual Korean War medals. He achieved the rank of CDR and was discharged from active duty in May 1956 and from the reserves in June 1973.

His memorable experiences: Lots of great dedicated people and interesting places. Made many friends for life.

He and his wife, Lyda Rita, have two children, Alicia and Alison; and five grandchildren: Jessica, Jason, Jocelyn, Ryan and Adam. Civilian employment as operations manager, Hughes Aircraft Co. Vitale retired from the USN in 1990 and from Hughes in 1992.

WILLIAM OWEN "BOONDOC" VOELKER, LCDR, born June 15, 1953, Oakland, CA, earned his BS degree in biology and his MD degree, both from Loma Linda University. He spent three years in Danville, PA, completing a residency in emergency medicine, is a Fellow of the American Academy of Emergency Medicine and currently practices at Enloe Hospital in Chico, CA.

Dr. Voelker joined the USN April 1, 1993, serving PRIMUS 2063 in Sacramento. On Sept. 10, 1993, he graduated number one in the Direct Commission Officer Indoctrination Course Class 93230. While in PRIMUS he qualified for the National Defense Service Medal and Expert Rifleman award. He also served as 1/14 Arty. Bn. surgeon in CAX 8-94 at 29 Palms during record heat conditions in the summer of '94.

In December 1994 he was transferred to the 4th MAW and sent to NAS Pensacola for flight surgery training, graduating June 29, 1995, at the top of SFS Class 95002 to win the CO's Fox Flag Award. Downsizing resulted in transfer to M/D 0187 in October 1995 where he was billeted to the Branch Clinic at NAS Fallon. He served as a flight surgeon during Desert Rescue 5 and in Norway during Battle Griffin '96. He participated in four shakedown cruises (two each) aboard the USS *Constellation* and the USS *Lincoln* participating in both cats and traps with VRC-30 and VFA-125. In October 1996 LCDR Voelker was transferred to VFC-13 in Fallon where he currently serves as flight surgeon.

Dr. Voelker resides in Paradise, CA with his wife of 22 years, Betsy, and their two children, Erica and Christopher.

PAUL A. VOHS JR., CAPT, born Jan. 19, 1931, Kansas City, KS, earned his BS degree at Kansas State University, Manhattan in 1955; his MA degree at Southern Illinois University, Carbondale in 1958; and his PhD at Iowa State University in 1964.

He joined the service in 1949 and was called to active duty for Korean War as AN in 1950, discharged 1952 as AOUZ, direct commission 1955. His military locations and stations: VF-884 on USS *Boxer* (CV-21). He was affiliated with aviation squadrons and Office of Naval Research Reserve program.

Vohs was awarded the China Service Medal, Korean

War Medal, United Nations Medal, Naval Reserve Medal and Navy Commendation Medal (1984). He achieved the rank of CAPT (O-6) as 1355 designator and retired in 1991.

His memorable experiences: Served 39 years in Naval Reserve, 29 years were good (50+ points) included two years four days active duty.

He and his wife, Jeanette, have five children: Martha, Tori, Kathy, Alan and David; and seven grandchildren. Civilian employment: On faculty at Iowa State, Oregon State, South Dakota State, Oklahoma State unit leader; supervisor, Cooperative Wildlife Research Units; director, Denver Wildlife Research Center for US Fish and Wildlife Service.

JOHN L. VOLZ, CAPT, born Oct. 11, 1906, Phoenix, AZ. Was a graduate of the University of Southern California, magna cum laude, 1929 and did graduate work in Germany and Austria.

A petroleum engineer for Union Oil Co. he was commissioned LTJG in a BuShips program in March 1936. He was called to active duty Aug. 2, 1941, attached to the office of RINM, Los Angeles. His responsibilities included fueling Pacific and Asiatic Fleets (before Pearl Harbor), Priorities, Facilities (getting new companies started) and Terminations (settling the payoff of cancelled contracts).

From 1948-50 he was CO BuShips unit 11-2 in Los Angeles. Retired in 1966 after 30 years Reserve membership.

His wife of 65 years, the former Aleda King, died in 1994. They have two children, Edward and Margretta; and two grandchildren, Heather Ann and Micheal. Civilian employment as engineer, Union Oil and president, Volz Construction Co. He retired April 5, 1966.

WALTER D. VOLZ, CAPT, commissioned June 9, 1959, as an ENS in the USNR in St. Louis, MO. He received the degree of MDIV from Concordia Theological Seminary in St. Louis in 1961 and the degree of master of sacred theology from Luther Southern Seminary in Columbia, SC in 1965. He retired from the USN Sept. 1, 1989, after completing 30 years of naval service.

Significant billets which he held include: 4th Marine Div. chaplain, staff chaplain for Readiness Command - Region 7 in Charleston, SC, CO of 2MAFREL in Orlando, CO of 3MAFREL in Miami, CO of 2MAFREL in Raleigh, NC. He was the national chaplain for the Naval Reserve Assoc. from 1980-86 and is a life member of the association.

During the summer months, from 1972-76, he was on the staff of NETC at Newport, RI as an instructor for ROTC candidates. During October and November 1977 he was the command chaplain for the NATO Operation, "Bow Bell" in Norway, which was under the command of GEN Alexander Haig. He served as a member on three selection boards for captains, three boards for commanders and one board as recorder.

In civilian life he is the senior pastor of Our Savior Lutheran Church in Plantation, FL and active in numerous civic organizations. He married Gloria Kuhlmann, from St. Louis in 1961. They have two children, David and Kristin.

ROBERT D. "BOB" VON BERNUTH, born April 14, 1946, Del Norte, CO, began his naval career in February 1969 when he was sworn into the Naval Aviation Officer Candidate program at Naval AS, Seattle, WA. In July 1969 he entered Officer's Candidate School at Naval AS, Pensacola, FL and received his commission in November 1969 after graduating first in his class and having been designated a Distinguished Naval Graduate. He completed VT-10 in Pensacola in April 1970 and advanced jet navigation at Naval AS, Glynco, GA and was designated a naval flight officer in May 1970.

CAPT von Bernuth was assigned to the A-6 replacement air group (Attack Sqdn. 128) at Naval AS, Whidbey Island, WA for training as a fleet bombardier/navigator. He joined Attack Sqdn. 52 on Yankee Station in February 1971 and completed a Southeast Asia combat cruise with that unit on board the USS *Kitty Hawk* in July 1971. He received nuclear weapons training at Naval AS North Island, CA and again deployed with Attack Sqdn. 52 to Southeast Asia in January 1972. During the ensuing 11 months he served as nuclear safety officer and quality assurance officer and flew combat missions in the A-6A and A-6B aircraft. In total, he flew 275 combat missions and accumulated 315 carrier landings.

At the cessation of Vietnam conflict, CAPT von Bernuth was released from active duty, and he entered the ready reserve and later the inactive reserve. In July 1983 he affiliated with Naval Control of Shipping Organization-Med/Africa in Knoxville, TN and remained with that unit until affiliating with Patrol Sqdn. 60 in Naval AS, Glenview, IL in June 1986. In January 1994 CAPT von Bernuth assumed command of NR-CVN 69 Det 0272, USS *Dwight D. Eisenhower*, in January 1996 he assumed command of NR-CV 62 Det 0372, USS *Independence*, and in October 1966 he assumed command of NR-Battle Gp. Staff 8 in Forest Park, IL.

CAPT von Bernuth has earned the following awards: Distinguished Flying Cross (2), Meritorious Service Medal, Strike Flight Air Medal (20), Individual Air Medal (7), Navy Commendation Medal w/Combat "V" (12), Navy Commendation Medal, Navy Achievement Medal, Navy Unit Commendation (2), Battle "E," Vietnam Service, Navy and Marine Corps Overseas Service Ribbon, the Armed Forces Reserve Medal, the National Defense Medal (2), Vietnam Campaign (3), Vietnam Cross of Gallantry, Expert Rifle, and Expert Pistol.

In civilian life, Dr. von Bernuth is professor and chairman of Biosystems Engineering at Michigan State University. He is married to the former Judy Wehrman and they have two daughters, Jeanie, an engineering major at the University of Virginia; and Suzie, an allied health major at the University of Kentucky. The family resides in Eagle, MI.

JOSEPH F. VON SAUERS, CDR, a graduate of Officer Candidate School, CDR von Sauers initially served in the Engineering Dept. of the USS *Jouett* (CG-29), where he qualified as 1200lb steam engineering officer of the watch (EOOW) as an ENS. While in engineering, he served in the following billets: assistant MPA, A Div. officer, and R Div. officer. He then transferred to the Combat Systems Dept. where he served as gunnery and missile officer.

In 1984, CDR von Sauers was released from active duty and affiliated shortly thereafter with the Naval Reserve.

His first reserve tour was with the USS *Racine* (LST-1191), a ship of the Naval Reserve Force. While on board the *Racine*, CDR von Sauers served as the assistant deck department head and stood watch as officer of the deck (underway) and as command duty officer in (port).

In 1988 CDR von Sauers transferred to Mobile Inshore Undersea Warfare Unit (MIUWU) 105, stationed in Long Beach, CA. While with MIUWU 105, he served as the electronic material officer (EMO) and as the administration department head. In 1990 MIUWU 105 was mobilized in support of Operations Desert Shield and Desert Storm. While in Saudi Arabia, CDR von Sauers received the Navy Achievement Medal and qualified for the 9450 Inshore Undersea Warfare officer NOBC.

From 1991-92, CDR von Sauers served on the staff of CDR, Amphibious Gp. 3 in the operations and administration officer billets. In 1992 he relocated to the Dallas, TX area and affiliated with Naval Control of Shipping Southwest Asia Unit 111. While with NCSO SW Asia, CDR von Sauers was assigned to serve as liaison officer with the British High Commission, Singapore.

In 1994 CDR von Sauers relocated temporarily to New York City where he attended Columbia University School of Law. While in New York he served with Voluntary Training Unit 0203 and Military Sealift Command Force, Northern Persian Gulf 102. He was board-selected for his current assignment to NR CDR Naval Forces Japan 119 in 1995.

CDR von Sauers' significant annual training includes several naval reserve force cruises, several Mobile Inshore Undersea Warfare Unit exercises, Amphibious Warfare and Naval Control of Shipping schools, Bell-Buoy 94 (Singapore) and training with Military Sealift Command HQ.

CDR von Sauers' decorations include the Navy Achievement Medal, Meritorious Unit Commendation, Battle "E" Ribbon, National Defense Service Medal, Southwest Asia Service Medal (3), Sea Service Deployment (2), Naval Reserve Sea Service, Navy and Marine Corps Overseas Service, Kuwait Liberation Medal (Saudi Arabia), Kuwait Liberation Medal (KUWAIT), Armed Forces Reserve Medal, Expert Rifle Medal, and Marksman Pistol Ribbon.

CDR von Sauers holds BE (electrical engineering) (magna cum laude), MBA, JD (cum laude) and LLM degrees.

ROBERT W. "BOB" VOSS, LCDR, born Jan. 27, 1920, Pittsburg, KS, attended three and one-half years at Pittsburg State University (KSTC then).

He joined the service in May 1941. His military locations and stations: Pensacola NAS, San Diego, NAS Ford Island, New Hebrides (Espiritu Santos 1943). He served with VJ-2, VPB-100 and VP-12 (temporarily). Battles and campaigns included South Pacific and Marianas (Guam).

Voss was awarded area medals. He achieved the rank of LCDR and was separated in October 1945. His memorable experiences: Towing for Task Forces, various patrols. Graduated ENS NAS Pensacola Jan. 7, 1942. Night patrols in South Pacific 1943. Flew VJ & VPB, PB4Y2s. Left Navy in October 1945.

He and his wife, Mary Jane, have two children, Robert G. and Martha Jane; and five grandchildren: Jeff, Kim, Megan, Lindsay and Kate. Civilian employment as pilot with TWA from 1945-80. Voss retired from TWA in January 1980.

ALBERT P. "AL" VUCINOVICH, O-6, born Jan. 31, 1937, San Francisco, CA. Graduating from St. Patrick's Seminary, Menlo Park, CA, he was ordained a Catholic priest June 13, 1964. After five years of parochial ministry, he received a direct commission into the Naval Reserve Chaplain Corps April 28, 1969.

He was first assigned to USNR Surface Unit 12-37 for three years, and then began a typical, but highly interesting and enjoyable career. He spent eight years as the battalion chaplain of RNMCB-2, followed by two years as force chaplain. The same pattern followed with the USCG. Six years as the reserve chaplain for the Coast Guard's 12th CG Dist. was followed by becoming the CO of USCG RELSUP 106 which supplied reserve chaplains to the Coast Guard throughout the US.

Other assignments included time as chaplain of NAS 3087, several TEMACs working at NTC Great Lakes, as well as NTC and MCRD San Diego. It was mentioned above that Fr. Vucinovich's career was enjoyable and this was due in great part to the wonderful camaraderie within the Naval Reserve. His 23 year career passed as quickly as a busy drill week-end!

Vucinovich was awarded the Naval and Coast Guard

Commendation Medal and Naval Reserve Medal w/Hour Glass. His memorable experiences: Developing a lay leader program for Coast Guard cutters.

Civilian employment as pastor, St. Rita's Church, Fairfax, CA. Vucinovich retired Oct. 1, 1992.

DR. THOMAS L. "TOM" WACHTEL, CAPT, born Feb. 25, 1938, Mansfield, OH. Earned AB degree at Western Reserve Univ. and St. Louis Univ. CMM Tulane.

He joined the Medical Corps, USNR in 1961 and achieved the rank of CAPT, in 1981. Duty from 1969-77: served in USS Ticonderoga (CVA-14); Constellation (CVA-64); USA Avn. Med. Research Div., Ft. Rucker, AL; Training & Support NR Support Unit, Lexington, KY; NAVREGMEDCEN ORU 5218, Albuquerque, NM.

1978-88: 1st Mar. Div. MEDDET 111, Albuquerque; NR Surgical Team 119, San Diego, CA; NR Tarawa Hosp. Det. 219, San Diego; NR Naval Med. Cmd. SW Region, San Diego 119, San Diego; NR Naval Hospital 119, San Diego; NR REDCOM and Primus Progam, San Diego.

Duty from 1988-94, Res. Naval Const. Force, Lenexa, KS, and Gulfport, MS; NR Personnel Mobilization Team 3519, San Diego; NR Naval Hosp. Med. Readiness Unit 119, San Diego; Primus Program, NR Naval Hosp. San Diego P1942; and as CO, NR Naval Hosp. Med. Readiness Unit 119, San Diego, 1994-97.

Participated in Vietnam, 1969-70. His awards include the Vietnam Service Medal, Vietnam Campaign Medal, Navy Expert Pistor (E), Navy Commendation, Meritorious Unit Commendation w/star, National Defense w/star, Navy Reserve Medal w/HG and Armed Forces Expiditionary (Korea). He is still active.

Married Carolyn Coleman and they have children, John Matthew, David Earl-Martin and Julianne Marcia. Civilian employmnet as director, Trauma Services, Centura Health.

JOSEPH J. WACHTLER, LCDR, born Feb. 18, 1911 in Passaic, NJ.

He joined the service June 26, 1941. His military locations and stations: Ft. Schuyler, NY; Naval Gun Factory, Washington, DC; Bombsight operation and maintenance, Dahlgren, VA Sperry Gyro, Norfolk, VA, Bureau of Ordnance.

He was awarded the American Defense Service Ribbon and Award of Honor. Honorary retiree, USNR. He achieved the rank of LCDR, USNR and discharged Dec. 11, 1946.

He is married to Ethel and they have a daughter, June. Employed as president and owner, Dewitt Rubber Mfg. Co., Fairfield, NJ.

NELSYN E.B. WADE, CDR, born Nov. 4, 1922 in Nacogdoches, TX, received a BBA from Baylor University, Waco, TX.

He joined the service Sept. 19, 1942. His military locations and stations: USS *Alderamin* (AK-116), Pacific; Naval Reserve Units. Served with NRED 8-3, Nacogdoches, TX; NRSD 8-3, Lufkin, TX. His battles and campaigns: Asiatic-Pacific area; Okinawa, Iwo Jima. He was awarded the American Theater, Asiatic-Pacific (one star), and WWII Victory. He achieved the rank of CDR, USNR and discharged Nov. 4, 1982.

His memorable experiences: Losing LCVP on beach at Iwo Jima, rescue operations with their LCMs of a sinking ship at Okinawa. Typhoon at sea while enroute to Okinawa; bulldozer getting loose in #2 hole and banging up 2x6 trucks on each side, 43 degree rolls.

He married Julia Howard and they have children: Sylvia W. Drake, Lisa W. Drake, Wm. Alan Wade and David E. Wade; grandchildren: Alexandra and Wade Drake, Anjuli, Charles, Wm. Crouch, Sheridan and Caleb Wade. Employed as rancher, tree farmer, owner Nelsyn's Furniture Store.

RONALD J. WAGENBACH, CAPT, attended Loyola High School in Los Angeles where he entered the USNR prior to graduation in 1950. He was accepted in the ROC program and upon graduation from San Jose State College in 1954, he was commissioned an ENS.

Active duty assignments included harbor defense officer, Long Beach, CA and security officer, NAAS, Fallon, NV.

Wagenbach's USNR career included seven commands, a four-year assignment as inspector general of Readiness Command Region 19 in San Diego, service on three selection boards, attendance at the Armed Forces Staff College and the National War College, and special ACDUTRA assignments in Washington and Pearl Harbor.

Retiring in 1984, he was awarded the Meritorious Service Medal and the Navy Commendation Medal.

CAPT Wagenbach resides in Lake Arrowhead, CA with his wife Tana. They have three children and 10 grandchildren. He is a life member of NRA.

JAMES L. "JIM" WAITE, CDR 05, born Sept. 26, 1940 in Murray, UT, received a BA and MA from Brigham Young University 1968; 1969 PHD Southern Illinois University, Carbondale, 1971; Fulbright Scholar University of Stockholm, Sweden, 1970-71.

He served in the military 1957-63 and 1974-96. His military locations and stations: Treasure Island, CA; Home Port: Long Beach Naval Station, Kansas City Reserve Center, Ft. Leavenworth C&GSC, Washington, DC, DIA. Units served with: USS *Shelton* (DD-790); USS *Samuel Moore* (DD-747); DEFDOC 0218; NISCTRANS 0166; DEFATTACHE 0166.

Battles and campaigns: Formosa Patrol 1958-59. (WESTPAC). He was awarded the National Defense Medal, Armed Forces Expeditionary Medal (MATSU/Quemoy) Combat Action Ribbon.

Waite achieved the rank of CDR 05. His memorable experiences: many great assignments to DAO Stockholm, Sweden and also at DIA - attaches and operations during Desert Storm.

He married Susan and has children: Peter, Andrew, Phillip, Jaimie Sue and Paul; grandchildren: Jakob, Justin, Blake and Tiffany Sue. Employed as professor and chair of Department of Political Science and Geography, Central Missouri State University, Warrensburg, MO.

ZACHARY T. "ZACK" WAKEFIELD, CAPT, born Oct. 29, 1919 in Jarrolds Valley, WVA, received BS in chemistry from the University of Virginia.

He joined the service Feb. 12, 1943. His military locations and stations: NAS Seattle, Ovillayute, San Diego, Norfolk. Served with VC-72, aboard *Kasaan Bay* (CVE-69), April-August 1945 and received routine awards and medals.

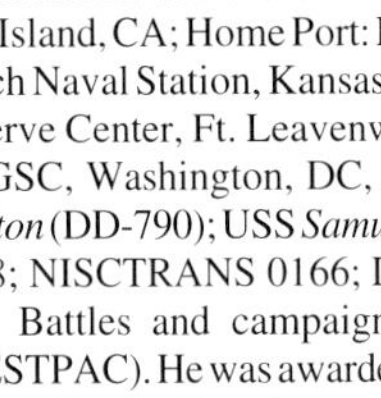

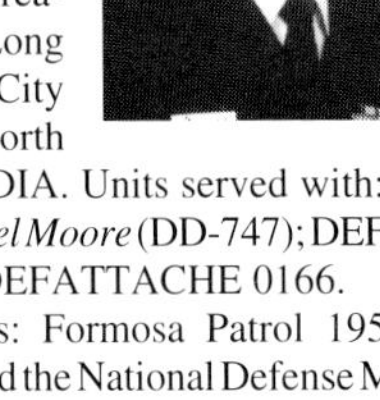

He achieved the rank of CAPT USNR-R and retired July 1, 1974. His memorable experiences: one on one dinner with ship's captain (discussing atom bombs).

He is married to Margaret and they have six children and 12 grandchildren. Employed as chief, Fundamental Research, TVA.

EDWARD B. "ED OR PETE" WALDMANN, CAPT, born Feb. 6, 1926, Council Bluffs, IA. Earned BS degree in 1946; MD in 1948 and MS, 1950; internal medicine, Mayo Clinic, 1952-55.

Student SV-12(S), Creighton Univ. School of Medicine, Omaha, NE, 1945; Student, School of Aviation Medicine, Pensacola, FL, 1950-51; Group Flight Surgeon, MAG 16, 1961-62. First Helicopter Air Group Medical Officer, US Military; Wing (Sr.) flight surgeon, WS-81, NAS, Minneapolis, MN, 1953-59; several billets at Phoenix, AZ, Navy and Marine Reserve Trng. Ctr., 1959-82; CO, Navy Regional Med. Ctr., Phoenix, 1974-77; CO, Med. Contingency Response Unit 1119, Phoenix, 1980-82; CAPT, Medical Corps, USNR and naval flight surgeon, Ready Reserve USN, 1945-82.

CAPT Waldman retired in July 1982. His awards include the WWII Victory Medal and Naval Reserve w/ cluster. He is life member of NRA, Naval Order of the US, Reserve Officers of Naval Service Club, Society of USN Flight Surgeons, TROA and American Legion.

Last civilian employment as employee healt director, St. Joseph's Hospital and Medical Ctr., Phoenix, AZ, 1968-90. He has served on numerous committees, board of directors, held various medical administrative positions, has 26 publications in medical literature and he is listed in several "Who's Who."

Married Arloa and they have children: Mary Kathleen Nelson, Mary Patricia, Stephen, Christine Tucker and Margaret Waldmann; grandchildren are Stephen and Geoffrey Nelson and Lena DeWald. He retired July 6, 1986.

CRAYTON C. WALKER, CDR, born Oct. 2, 1932 in Chickasha, OK, received BA (physics) 1954, Ricelli; PhD (psychology) 1965, Univerity of Illinois.

He joined the service June 6, 1954. His military locations and stations: USS *Isbell* (DD-869), 1954-55; USS *Albany* (CA-123), 1955-57; Research Reserve Units in Illinois, California and Maryland, 1962-84; Naval Hospital, Groton 101, 1984-85.

He was awarded the Naval Reserve Medal, National Service Medal and China Service. Walker achieved the rank of CDR (Medical Service Corps). His memorable experiences: XO Naval Reserves ONR Branch Office Boston 1983-84; CO Naval Reserve NH Groton 101 1984-85; OOD (IP/UW) USS *Albany* (1957-57).

He is married to Harriet and they have children: Scott, John, Jessica and Christopher. Employed as assistant professor, UCLA, 1965-70; Professor, University of Connecticut, 1970-89.

IRA GERARD "GERRY" WALKER, CDR, born Dec. 14, 1922 in Jeanerette, IA, received a DDS from Loyola University, New Orleans, LA, May 19, 1944.

He joined the service March 5, 1942. His military locations and stations: Loyola University, Great Lakes Center, Midway Island, Pearl Harbor, HI. Served with VMF Marine Air Wing (Midway Island-clinic Pearl Harbor. He received the Naval Reserve Medal, Armed Forces Reserve, WWII Freedom Medal, Asiatic Pacific Medal, American Campaign, WWII Victory Medal, Naval Reserve Medal and Meritorious Service.

He achieved the rank of CDR and was discharged Jan. 3, 1947. His memorable experiences: watching life cycle of "Gooney Birds" Midway Island; assembling of war ships at Pearl Harbor for atomic bombing at Bikini Atoll.

He is married to Mary Helen and they have children: Jennifer, Kathryn, James and Thomas and grandchildren: Scott, Melanie, Natalie, Stacy, Meridith, Whitney, Virginia, Therese and Luke. He is a self-employed dentist. Retired from the USNR in 1979.

JAMES R. WALKER, LT born in Houston, TX, July 20, 1960. A graduate of Tufts University, BA and Texas A&M University, MA, he began his naval career in 1987 as an Officer Sea and Mariner (OSAM) recruit. Upon completion of OCS in Newport, RI, he completed the Supply Corps' Basic Qualification Course (BQC) at Athens, GA. From 1989-1990 he served aboard USS *Mount*

Whitney (LCC-20) as food service/wardroom and disbursing officer.

Following active duty he drilled in Houston, TX with ABFC-FFTU 210 unit as a freight officer. His unit was recalled for Operation Desert Shield on Dec. 26, 1990. During this recall he served as air terminal OIC and AOIC at two forward logistics sites in the Persian Gulf.

Subsequent to Desert Storm he served as XO to NR J. Daniels 2708, supply officer to NR MDSU 2 DET 208, and Air Traffic to NR NAVMTO DET 108.

Walker was awarded the Navy Achievement Medal and the Meritorious Unit Commendation. His campaign and service medals are: National Defense, Navy and Marine Corps Overseas, Southwest Asia, Naval Reserve and the Kuwait Liberation Medal.

He resides in Key West, FL where he manages a horticulture program that provides meaningful employment for retarded citizens. He also provides contract management services to the Special Forces Underwater Warfare School Dining Facility. At home he enjoys cooking with his wife, Martha and his son, Mike.

RICHARD A. "DICK" WALKER, LT, got involved in the USNR by enlisting in March 1945 and was graduating from boot camp in San Diego at the time of VJ Day. He was reassigned to the amphibious training base in Coronado, CA. The first day at the base, a Chief called for volunteers to become firemen. No response! The second day, the Chief called out a list of names, including Walker and indicated they were going to be firemen! And that's how Walker became a member of the "Black Gang". However, he was assigned to a Standard Landing Craft Unit (#64) and spent his time driving a 36' LCVP around San Pedro Harbor as a liberty boat for ships in San Pedro Harbor, as well as performing routine maintenance work on the engine of the craft.

Upon discharge from active duty in August 1946, he remained in the USNR. Upon graduation from University of California, Berkeley, in 1950, it just so happened that Navy had a program wherein enlisted reservists who had completed four years of college could apply for a commission. And so he did and was commissioned in October 1950.

During his naval career, he drilled with Mobilization Teams 12-1 and 12-5. He was awarded the WWII Victory Medal, the Naval Reserve Medal and the Armed Forces Reserve Medal. And at the 43rd Convention of the NRA in Chandler, AZ in August-September 1996, he was presented with the NRA's National Award of Merit. He continues to serve as president of Sutter Chapter, Sacramento, CA.

In civilian life, he was a position classification specialist in McClellan AFB. And since there was not much call for salary and wage administrators, he was never called to active duty.

WM. BRUCE WALKER, CAPT, born July 4, 1925 in Chicago, IL. He enlisted while a senior in high school on Feb. 9, 1943 as a V-5 Naval Aviation Cadet (NAVCAD). The Navy sent him to Depauw University as a V-12A for two semesters, and 3/44 he started Navy Flight Prep School at Wooster, OH. The "Purge" (50% NAVCAD force reduction) interrupted cadet training and as a S2C he went through Aviation Machinist Mate School and as a S1c through aerial gunnery training. The V-5 program reopened and NAVCAD Walker went through St. Mary's preflight and on to primary flight training at NAS Glenview, IL, where on VJ-Day he stood the cadet duty watch. From there he was released to inactive duty in September 1945.

He returned to DePauw University, majored in Political Science and graduated in 1948.

With US involvement in Korea he requested and received a Reserve Commission, and ENS Walker affiliated with the USNR Intelligence Program which became a fascinating, challenging avocation/hobby that ended with retirement 32 years later (July 1, 1985). He commanded two intelligence reserve units and served as Reserve Intelligence area coordinator Area 9 (RIAC 9).

His education and specialized training enabled to him to become an effective "Cold War" warrior. He worked with the CNO Sea Power Presentation Team, other speaker teams including NAS Glenview, NTC Great Lakes and Education for American Security.

June 14, 1946 he married his high school sweetheart, Loraine McClelland, and their family consists of three sons: Bruce, Neil, Scott and three grandchildren.

In civil life CAPT Walker worked in the sales profession, retired in 1990, and currently lives in Wheaton, IL.

WILLIAM B. WALKER JR., CAPT, born April 13, 1947 in Pittsburgh, PA, received an MA from the Naval War College.

He enlisted as a seaman recruit in 1967. Upon commissioning in 1972, he served as security officer, Naval Communication, Greece. While on active duty, he earned a MA in Middle East History.

CAPT Walker's many assignments include three tours as XO and eight tours as CO. His awards include two Meritorious Service Medals, three Navy Commendation Medals and 24 other awards. Graduating from the Naval War College and the Armed Forces Staff College, he became the first naval reservist to complete all Joint Professional Military Education requirements.

As CO, CDR, US Naval Forces Central Command 108, his command was selected "Best Large Unit" in Readiness Command Region Eight two consecutive years, and was nominated for the "Best Naval Reserve Unit" in the nation. He resides in St. Petersburg, FL with his wife, the former Lynda Lynn of Fairfax, VA.

FRANCIS F. "FRANK" WALL, CDR, born Aug. 19, 1919 in Pisgah, IA, attended the University of Iowa for four years and received a BA degree and attended one year at the University of Southern California for MA.

He joined the service July 11, 1941, Northwestern Midshipmen's School. His military locations and stations: Northwestern Chicago 1941-January 1942, Minesweep School, San Francisco, CA 1942; USS *Sanderling,* San Diego 1942; USS YP422, 1943 Coral Sea and Solomon Islands; USS SC-521 Solomon Islands, US Naval Base Fiji Island 1943-44; USS *Pitt* (APA-223). His battles and campaigns: Solomons Campaign, Okinawa Campaign COM 3 New York, 1952-54.

He was awarded the National Defense, South Pacific Asiatic, Naval Reserve Medal, Air Force Medal, Japan Occupation Medal, Philippine Liberation Medal, China Sea Medal, Korean Victory Medal.

His memorable experiences: surviving grounding on coral reef off New Caledonia (1943), 77th Div. Troops on Kerama Retto, Okinawa, March 26, 1945, dodging kamikazes. Commanding USS *Twining* (DD-540) on an all reserve training cruise Long Beach to Ecuador, SA in June 1950 when Korean War started. Serving on CDR Selection Board in Washington in January and February 1964.

His wife Peggy died in 1991 and they have a son, Jim Wall and grandchildren: Christine, Cathleen, Justin and Shawn. Employed as president of Savings Bank, 10 years and chairman of same bank in Covina, CA for 12 years.

ROBERT C. "BOB" WALLACE, LT, born June 22, 1930 in Ephraim and received a BS in civil engineering.

He joined the service June 22, 1947 in USNR. His military locations and stations: Port Hueneme, CA; Guam, Chichi Jima; many bases USA. Served with MCB 9, MCB 10, RNMCB 18, Utah ANG 151CEF.

His battles and campaigns: Vietnam, made active duty on every campaign except Granada.

He achieved the rank of LT USNR and Chief MSGT USAFR, retired June 1990. He received the USMAC Cert. of Achievement, US National Defense Medal, Navy Unit Commendation, USAF Meritorious Service Medal and AF Achievement Medal. His memorable experiences: 1968 TET Vietnam.

He married June and is employed as civil engineer USBR Dept. Interior.

STANLEY "STAN" WALLACH, CAPT, born Dec. 10, 1928 in Brooklyn, NY, received an AB, Cornell 1948; MA Columbia 1949; MD, State of University of NY, Downstate Medical Center, 1953.

He joined the service in 1957. His military locations and stations: USNH, Chelsea, MA; Reserve: Reserve units at NAS, NY; Albany, NY; Scotia, NY; Burlington, VT; St. Petersburg, FL.

He achieved the rank of CAPT, MC, USNR, retired Dec. 10, 1988.

His memorable experiences: CO, NAV REGMEDCEN 503, Albany, NY, 1975-6; Medical Officer, REDCOM REG2, Scotia, NY 1977-79, 1982-83; CO, MEDCRU 405, Burlington, VT, 1980-82.

He married Pearl and they have children: Sara, Francine, Shellie, Rhonda, Allen, Corinne, Mara and Peter; and grandchildren: Lara, Steven, Andrea, Karen, Elizabeth, Carolyn, Gabrielle, Astin and Angelique. Employed as professor of medicine at State University of New York, Downstate Medical Center, Albany Medical College, University of Southern Florida College of Medicine, NYU School of Medicine; Chief, Medical Service, VA Medical Centers at Albany, NY and Bay Pines, FL.

HOLLIS W. "PETE' WALLS, CAPT enlisted in the USNR Nov. 28, 1949. Commissioned in 1953, he attended CICO School, NAS Glenview, and served as CICWO and all-weather air controller aboard the USS *Valley Forge* CVS-45. CAPT Walls was separated in 1955 but continued Reserve affiliation in Indianapolis.

His associaton with the Selective Service Reserve Forces Program began in 1969. Ten years later he returned to active duty on special work assignment with the Selective Service System, an independent executive agency. CAPT Walls served as region director and associate director for operations. At the time of his release from active duty and retirement, Nov. 20, 1991, he was executive director of the agency.

CAPT Walls received undergraduate and graduate degrees from Butler University. Most of his 24 years as an educator was in secondary school administration in Indiana. He is the recipient of the Defense Superior Service Medal, Legion of Merit, Defense Meritorious Service, Joint Service Commendation, Navy Commendation (w/second device), Navy Occupation (Europe), National Defense (w/second device), and Armed Forces Reserve (w/three devices). CAPT Walls' contributions to the operation of th Selective Service System were recognized by Meritorious and Exceptional Service Awards as well as the agency's highest, the Distinguished Service Award. Indiana Military Department awards were the Commendation and Distinguished Service Medals.

He and his wife Barbara have three children: Jeffrey, David and Holly. Having served nearly 42 years in the USNR the CAPT Walls family retired in Florida.

STEPHEN B. "STEVE" WALL-SMITH, LCDR, born Sept. 28, 1954 in Columbia, MO, received a BA from New York University; M. Div., St. Paul School of Theology (K.C. Missouri); MA and PhD (philosophy), University of Missouri.

He joined the service Sept. 26, 1986 (CHC, USNR). His military locations and stations: New York City, Midwest (Reserves). Units served with: NR SURFLATREL 202; NR MMEFREL 118; NR4 FFSG; NMCB-15.

He was awarded the Navy Commendation, USCG Meritorious Unit Commendation, AFR Medal, HUM Service Medal, Expert Rifle and Pistol Medal. He achieved the rank of LCDR. His memorable experiences: serving as station chaplain to a USCG SAR station on LI. Bored on punctuated by high tension!

He is married to Elisabeth Ellen Wall-Smith and they have a daughter, Mona. He is a pastor, Woodlandville United Methodist Church.

GERALD W. "JERRY" WALSH, CDR, born March 22, 1939, Brooklyn, NY. Earned AB, MBA, JD and MST degrees. Joined the military in July 1960 and stationed in Naples, ITaly; Norfolk; Boston; and Vietnam.

Units include USS Alameda; AVB-1; COMPHIRON 6 Staff; NAVMAI, Boston; CA-148. Participated in Cuban Missile Crisis, Vietnam, COMNAVSUPRACT, Da Nang and DCA5R, Boston.

Discharged from active duty June 30, 1971 and retired from Naval Reserve in 1990. Memorable experiences include being a 21-year-old chief engineman; a year in I Corps with CAPT Ron Hamby, USMC.

Married Nancy Harrington and they have three children: James, Michael and Andrew (LT, USN); and grandson Christopher. Civilian employment as an attorney.

PETER S. "PETE" WALTERS, LCDR, born April 18, 1934 in Detroit, MI, received a BS in foreign trade, University of Southern California; NROTC graduate, (1956). He joined the service Aug. 29, 1956, commissioned ENS, Supply Corps.

His military locations and stations: USS *Purdy* (DD-734), July-August 1956, Midshipman Cruise Charlie "S" Div. crew member; Navy Supply Corps School, October 1956-May 1957, student under instruction; USS *Carter Hall* (LSD-3), May 1957-August 1958, assistant supply officer, "S" Div. officer, disbursing officer.

Reserve Units served with: Member, various USNR Supply Companies, October 1958-May 1977. Retired, active reserve, June 1, 1977. He was awarded the UNSR Medal, National Defense Medal, and Armed Forces Reserve Medal. He achieved the rank of LCDR and retired April 18, 1994.

His memorable experiences: Panama Canal crossing on Midshipman Cruise, August 1956, aboard USS *Purdy;* WESTPAC cruise, January-August 1958, aboard USS *Carter Hall;* CO, Naval Reserve Supply Co. 11-6, Santa Monica, CA (1968-1970); Eleventh Naval District Reserve Policy Board, member (1970-71).

He married Sheila and they have children: Mrs. Heidi Bonn, Mrs. Aimee Barrere, and David Walters; and grandchildren: Ashley Barrere and Scott Barrere.

Retired as account executive, National Business Records Center, Inc. in Irvine, CA. His interest and hobbies include, master certified athletics official, USA Track & Field, since 1967; antique show exhibitor, since 1980.

PETER M. WANBAUGH, CAPT, born on April 8, 1931 and enlisted in the Naval Air Reserve at the age of 17 serving two years with FASRON-151 at NAF Anacostia, Washington, DC.

He attended the US Merchant Marine Academy, class of 1954. Commenced US naval flight training with class 8-54 and was designated September 1955.

His first tour assignment was VW-3, NAS Agana, Guam. On his second tour he was assigned to VP-19, NAS Alameda, CA which included a WESPAC deployment to MCAF Iwakuni, Japan.

CAPT Wanbaugh was released from active duty on February 1960. As a civilian he was employed as an aerospace administrator but maintained dual affiliation with NROS Long Beach and VP-776 NAS Los Alamitos, CA. He has served as CO of NARDIV-L1, NAS-119, NAS Cubi Point, and VTU-7676 and retired from naval service at NRTU Point Mugu, CA on October 1985.

CAPT Wanbaugh retired from this civilian occupation on April 1991 as director of procurement, G&H Technology Inc., Santa Monica, CA.

He and his wife Barbara have resided in Malibu, CA for over 20 years and have a son Jon, daughter Wendy and two grandchildren.

DIXON EARL WANSBURY, CAPT, born San Francisco, CA, Feb. 23, 1918, received an AB in economics from the University of California (Berkeley), 1941 Phi Beta Kappa; MS industrial management, MIT 1955, Sloan Fellow.

Commissioned as ENS (SC) USNR June 1941. On active duty June 1941-February 1946. Assigned Navy Supply Corps School Harvard; NSD Norfolk; USS *Sampson* (DD-394) (Pacific); NYD Charleston, SC (new construction); University of Colorado, Navy Language School (Russian) ONI Washington, DC.

Released from active duty February 1946. Remained in active reserve. Changed from supply designator 3105 to Line Intelligence 1635 in 1951. Attached to and CO of various intelligence units. CO of Res Intel Div. 12-1, 1969-74; RIAC Area 2 NAS Alameda at retirement 1978.

Employed with Bell System 1946-81. Various executive engineering and operational positions. Assistant vice president for labor relations at retirement.

He married Marjorie Aden at Williamsburg, VA October 1941. They have a son, Cory Aden, Wansbury, MD (served as LCDR (MC) USNR) and daughter, Lindsay Aden Wansbury, MFCC.

ARTHUR T. "ART" WARD, CDR, born June 15, 1927 in Cincinnati, OH, received a master of science, International Affairs, GWU 1970.

He enlisted June 14, 1945; USNA 1951; ENS, June 1, 1951. His military locations and stations: DIA and COMNAVFOR V. Served with USS *Missouri,* USS *Huse,* USS *Pinnacle,* and Min Div-43.

His battles and campaigns include Korea and Vietnam. He received the Bronze Star w/V, Meritorious Service Medal, Joint Service Commendation Medal and Combat Action Ribbon. He achieved the rank of CDR, USN and retired March 1, 1973.

His memorable experiences: one year in Vietnam, earned the Vietnamese Navy Distinguished Service Medal.

He is married to Esther and they have children: Cynthia, Jenny and Jessica; and grandchildren: Kirsten, John and Kelsey Bowen and Michael Caro. Employed as NRA consultant 1977 to date.

FREDERICK F. "FRED" WARD, CDR, born March 28, 1928 in Hartford, CT, received a BS from Hillyen College in 1952; MBA from the University of Hartford 1963.

Enlisted in the USN, June 1946. Assigned to boot camp, USNTC Bainbridge, MD. Assigned to USNATTC, Jacksonville, FL, Class P School, October 1946; USNAS, Lakehurst, NJ, Class A School, March 1947; NAS Norfolk, VA, ATFLT Weather Central; USNAAS Ocean, VA, aerographer, promoted to AERM 3/c, January 1948; USS *Mississippi* (EAG-128); weather forecasting and ballistic wind analysis, September 1948; RAD transfer to USNR, May 1948; enrolled Hillyer College, Hartford, CT, assigned to USNRTC, Hartford, CT, June 1948; recalled to active duty, Reserve Officer Candidate School, October 1949; ROCS Treasure Isle, CA, June 1950; ROCS, Long Beach, CA, commissioned ENS, USNR, June 1952; USNSC School, Bayonne, NJ, September 1952; USS *Purdy* (DD-734), supply officer, January 1953; *Purdy* assigned to 6th Fleet, October 1953; *Purdy* assigned to 7th Fleet, Task Force 77 in Korean Theater, April 1954; RAD promoted to LTJG, USNRTC, Cromwell, CT, May 1954; promoted to LCDR, 1955; USNRTC, Hartford, CT, 1956; promoted to CDR, 1958; retired CDR, USNR, 1988.

He received the WWII Victory Medal, American Theater of Operations, Korean Campaign and Singhman-Rhee Commendation Unit.

His memorable experiences: circumnavigation of the gale on the USS *Purdy* (DD-734) on way to Korea.

He married Janet and they have children: Frederick F. II; Terence, Timothy D., Esq.; and grandchildren: Rosemary and Thomas Jacob Ward. Employed as president & CEO Arrow Toll Co.

MICHAEL W. "SKIP" WARD, CAPT, born Oct. 9, 1941 in Norwich, NY, received a BS in marketing from the University of Arizona in 1963.

He joined the service Sept. 28, 1963. His military locations and stations: VT-2 Whiting Field, VR-21 NAS Barbers Point. Served with VR-USNR Gross Pointe, NAF Detroit, NAS Glenview, NAF Atlanta, NRF Washington, DC.

He achieved the rank of CAPT and was discharged Jan. 31, 1970. His memorable experiences: CO, VR2466; NAF Washington, DC 1982-84.

He is married to Cookie and they have a son, Horace. Employed as president of Atlantic Shores Distributors.

JOHN M. "JOHN" WARDEN, LT, born Nov. 3, 1922 in Woodbury, NY, received one year of college at the University of California at Davis.

He joined the service Oct. 15, 1942. His military locations and stations: Moffett Field, CA; Lakehurst, NJ; So Weymouth, MA and Richmond, FL. Served with ZP-11, ZP-21 and VR7. He participated in the American Area Campaign and was awarded the WWII Victory and Naval Reserve Medal. He achieved the rank of LT USNR and discharged Dec. 20, 1947.

Married Amy and they have children: Mary Villegas, Jeanne Rigali and Jim Warden; and grandchildren: Troy and Mark Watkins, John and Tina Rigali and Lindsay and Jenna Warden. Employed as president, CalWestern Sales Corp.

ROBERT W. "BOB" WARREN, CDR, born Oct. 11, 1931 in Rochester, NY, received a BA from the University of Rochester, 1954 and MBA from Cornell University in 1958.

He joined the service June 1954. His military locations and stations: USS *St. Paul* (CA-73), 1954-56; Batt 3-9 USNR, 1956-75, Rochester, NY.

He was awarded the National Defense, China Service and New York Naval Militia. Achieved the rank of CDR and retired June 1975.

His memorable experiences: had a great experience aboard the NR St. Paul. Their CAPT Charlie Ricketts, was a great inspiration to all who served aboard his ship. Also enjoyed the USNR after two years active.

Married Ann and they have children, Doug and Patricia; and grandchildren, Griffin and Haile. Employed at Marine Midland Bank, vice-president, retiring December 1991.

WALTER EMORY WARREN, CAPT, born in Palatka, FL, March 15, 1936, and a 1958 graduate of the University of Florida. He completed Naval Officer Candidate School in 1959 and was first assigned in the Naval Security Group to Pearl Harbor, HI. He also served on the USS *Hassayampa* (AO-145).

Released from active duty in 1962, he entered law school at Mercer University, Macon, GA, graduated in 1966 and commenced practice in Leesburg, FL. In 1969 he located in Virginia and split his civilian law career between private practice and US Government agencies, in Washington, DC. Among federal positions, he served as trial attorney in the Department of Justice, Deputy General Counsel of the Agency for International Development, and General Counsel for the Small Business Administration.

During his civilian career he remained active in the Reserve, holding billets ranging from litigation to operations law. He was CO of the legal assistance unit at the National Naval Medical Center in Bethesda, and of the CINCLANTFLT legal unit in Norfolk. He also served extended active duty periods at JAG HQ in Alexandria, and at the CNOs office in the Pentagon in 1982, 1983 and 1984. He was called up for the Gulf War from September 1990 through June 1991. He served in the US Central Command SJA Office as the head of the International Law Div., with cognizance over matters including rules of engagement and transnational issues. Transferred to the Retired Reserve in 1992. CAPT Warren and his wife, the former Mary Jane Detman, reside in Fairfax, VA. They have one son Walter Charles.

T. FRANK WATANABE, LTJG, born Feb. 3, 1968 in Indianapolis, IN, and received BA and MA degree from Georgetown University.

He received a direct commission as an intelligence officer in the USNR in February of 1994. He completed his initial intelligence training at NAF Washington, and joined the Anti-terrorism Watch at Naval Criminal Investigative Service. In August 1995, he transferred to NAS Fallon, NV, and assumed duties as an intelligence evaluator at the Naval Strike Warfare Center, now part of the Naval Strike and Air Warfare Center.

He has been a member of the NRA since shortly after joining the USNR, and served as vice president for Junior Officers and then president of the National Capital Region Chapter.

He received his education at Georgetown University, earning a bachelor's degree in International Relations in 1989, and a master's degree in National Security Studies in 1993.

In civilian life, he served briefly with the Departments of State and Army before becoming an intelligence analysts with the Central Intelligence Agency. In 1996 he was appointed the assistant director of the President's Foreign Intelligence Advisory Board. Later in 1996, he left federal service to become a strategic planner for Eli Lilly and Co. in Indianpolis, IN. He is married to Jaelee Watanbe.

ROBERT J. "BOB" WATSON, LCDR, born Oct. 24, 1920 in Blacksburg, VA, received BS, MA and PhD degrees.

He joined the service June 22, 1946. His military locations and stations: Washington, DC; South Pacific New Caledonia; Central Pacific (Guam). He served with supplementary radio activities (now Naval Security Group).

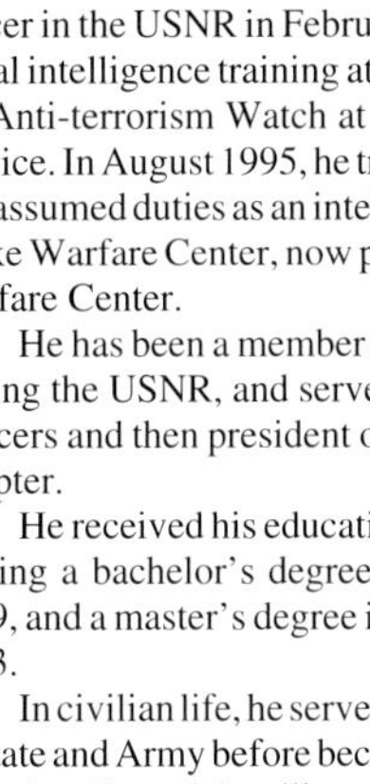

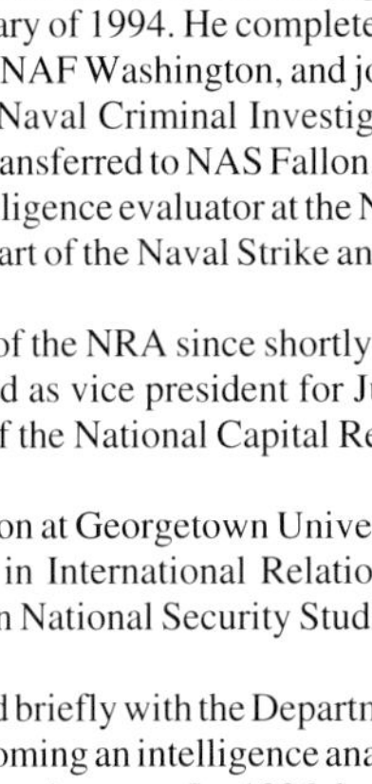

He was awarded the Navy Unit Citation, Pacific Theater, American Theater, Naval Reserve and Armed Forces Reserve Medals. He achieved the rank of LCDR and was discharged Oct. 24, 1980.

His memorable experiences: though he never heard a shot fired in anger, he participated in the "capture" of a Japanese soldier on Guam in WWII. While driving in a jeep with several fellow officers, several months after US troops recaptured the island, they came upon a Japanese soldier standing the the road, looking for a chance to surrender.

He married Laura and has children, Robert and Howard, and grandchildren: Anastasia, Taylor, Andrew and Victoria. Employed as historian with the Department of Defense.

WILLIAM HENLEY "BILL" WATSON, CAPT, born Jan. 28, 1923 in Rocky Mount, NC, received a BS in accounting (UNC-Chapel Hill); JD (LAW) (UNC-Chapel Hill).

He joined the service Sept. 28, 1941. His military locations and stations: Camp Bradford (1944) VA; (Univer-NROTC, 1941-44). Served with USS LST 705 (CO), Naval Reserve Units (Law) and (VTUs) (Composite Cos).

His battle and campaigns: Leyte Gulf invasion, Lingayen Gulf, Mindanao, Zamboango, Repatriation of POWS to Japan. He was awarded the American Theater, Pacific Theater, Japanese Occupation, Philippine Liberation, Naval Reserve and WWII Victory Medal. He achieved the rank of CAPT and was discharged Sept. 7, 1946.

Memorable experiences: invasion of Zamboango while beached, adjacent LST took a shell which exploded. The ship caught on fire, with a tank deck loaded with high octane gasoline and ammunition, but miraculously the broken water system put out the fire; kamikaze attacks.

He married Ruth (Nicky) and they have two children, William H. Jr., (now a commander in Reserves) and Margaret; and grandchildren: Caroline and William H. Watson III. Employed as an attorney, partner in firm since 1954.

EARL J. WATT, CAPT, born March 26, 1913 in Chelmsford, MA, graduating as valedictorian of the high school class of 1929. At Harvard he received the AB degree, cum laude, and the AM in 1934, becoming a teacher of French and athletic supervisor at his high school alma mater.

He volunteered for Navy service in 1943 and after training at Cornell and NTS in Norfolk had subsequent amphibious training at Coronado and San Bruno. He was assigned to beach party B4D312 aboard the USS Rochambeau and the USS *Henry T. Allen.*

His landings included Gamadodo, Ladava, Stringer Bay and Hollandia (Pancake Hill) in New Guinea and Luzon, Panay, Negros and Mindanao in the Philippines. At Subic Bay he was attached to Philippines Group 11 and participated in landings in Japan, Hiroshima and Baishinji, then heading home via Pearl Harbor in time for Christmas, 1945.

He remained in the USNR and was recalled for active duty from 1951 to 1954 at Little Creek and Norfolk, VA, the District of Columbia and Bethesda, MD. On inactive duty he served in Org Surf Divs 1-18 and 1-19 from 1946 to 1949. In 1951 he was again recalled to Navy Beach Group 2 at Little Creek and detached in 1953 when he reported to Recruiting Div. B6412 in Washington, DC. During that period he was hospitalized at Bethesda and was released from ACDUTRA Jan. 31, 1954, one day prior to his promotion to LCDR.

CAPT Watt then became CO of NR Surf Div 1-18 in Lowell, MA, which under his direction for three consecutive years was named best in New England in the First Naval District; in one of those years he was awarded a cruise to Greece and Italy, along with two enlisted men. The NR Mobilization Team 1-2 in Lawrence, MA, under his command as CO was named the 1960 outstanding First Naval District Team. In that period he was promoted to the rank of CDR. In 1963-64 he was group commander of Surf Batt 1-10, and he was appointed Commandant's representative for Public Relations, NRTC Lowell, 1964-65.

His professional life then took him to Europe where he was attached to and became CO of Naveur-2 in Paris, which included members from throughout France and Switzerland. Detached in 1967 he then was attached to NR Officers School 10-2 in San Juan, PR, in 1968, becoming CO from 1970-72.

He achieved the rank of CAPT, July 1, 1965 and was transferred to the NR Retired List March 26, 1973.

In civilian life he was coordinator of special services and assistant professor at Lowell Technological Institute (presently U. Mass-Lowell), then was affiliated with USO, Inc. in several capacities: as director of USO clubs in Paris, Malta, and San Juan and in executive positions in Atlanta, New York City, France, and Germany. His assignments included USO Caribbean and European area executive.

He retired from USO in 1979 and became active in several civic and community organization in Darien, CT, where he makes his home. He has two children: Claire (Mrs. Hugh Jenney) and Robert Watt, by his first wife, the late Clara Dexter who died in 1963, and two, Margaret and Michael, by his second wife, the former Barbara Browne. His special interests remain the cuisine and culture of other countries.

NORMAN E. "NORM" WAUGH, CAPT, born July 24, 1934 in Great Falls, MT. He has a BSBA (1956) and MBA (1974) from the University of Denver. He enlisted in the OCS program October 1956 and was commissioned an ENS February 1957 at the Officers Candidate School, Newport, RI.

He attended various schools at NAS Jacksonville and San Diego before reporting to the staff of Commander Barrier Pacific, NAS Barbers Point (1957-60) as assistant communications officer and Crypto security officer. He held the same billet at Naval Communications Station, San Diego (1960-61).

After active duty in 1961 he joined the Naval Air Reserve program in Denver, CO. Upon his retirement in 1984 he was CO USS *Midway* (Reserve).

CAPT Waugh is married to Lynn, whom he met while stationed in Hawaii. They have two daughters and live near Golden, CO. He is a broker/owner of Norman Waugh & Co., realtor.

WILLIAM JAMES "BILL" WAYLETT JR., CAPT, born Aug. 20, 1943 in Troy, NJ, received a MBA, masters of engineering (chemical), BS (engineering).

He joined the service Sept. 28, 1960 (as seaman recruit, USNR). His military locations and stations: Nuclear Power Training (Bainbridge, MD and WestMilton, NY). Served with USS *Sturgeon* (SSN637) February 1968-July 1970; USS *Tullibee* (SSN597); Submarine School, August 1970 to August 1971.

He was awarded the Navy Expeditionary Medal, Joint Services Expeditionary Medal, Navy Commendation w/Gold Star, Navy Achievement w/Gold Star and Navy Unit Commendation. He achieved the rank of CAPT (Reserve), LT (active) and was discharged Aug. 2, 1971.

His memorable experiences: four months Mediterranean cruise aboard USS *Tullibee,* when Lebanon crisis arose. There were three US SSNs competing for space in the ocean with three US carrier task forces and 50 Russian ships. Pucker time!

He married Eva Chloe Switzer and they have children: Katharine, William, Robert, Sherry, Fred and Danielle; and grandchildren: William (the ninth in succession to have the name) Madysin. Employed as senior vice-president, director of quality, Fleet Financial Group, Inc., retiring Feb. 1, 1993.

DONALD ALFORD WEADON JR., born in Brisbane, Australia on Sept. 15, 1945, he received his BA at Cornell University (where he spent a year in Japan and the Far East on a State Department sabbatical) in 1967, his JD from the University of California's Hastings College of the Law in 1975 and his MBA on a fellowship to the

Harvard Business School program at the Iran Center for Management Studies in Tehran (1976).

Commissioned an ENS in 1969 upon graduation from Naval OCS at Newport, RI, his first tour on DESDIV 112 Staff in Pearl Harbor saw participation in the Apollo 11 recovery and deployment to Vietnam aboard the USS *Saint Paul* (CL-73) and USS *Davidson* (DE-1045). He later served as aide and flag lieutenant to the CDR, Military Sealift Command, Far East in Yokohama Japan. For over a decade, he served in significant reserve intelligence billets, receiving numerous awards and commendations, including the Navy Commendation Medal.

His practice of international business, foreign trade and logistics law focuses on the international transfer on high technology goods and services and he was the first US lawyer to be authorized to practice in post-war Vietnam. He teaches at George Mason University, his articles have appeared in numerous journals and he has been a *Who's Who in the Law, Who's Who in America* and *Who's Who in the World* biography for over 20 years.

EDWIN F. WEAVER III, CAPT, born April 18, 1924 at Harrisburg, PA, attended public schools in Harrisburg and graduated Ursinus College in 1949. His naval career began when he enlisted in the V-1 Reserve Program in 1942, activated into the V-12 program at Ursinus College, transferred to Temple University School of Dentistry, was released to inactive duty in 1946. Dr. Weaver continued at the university getting his doctorate in 1948.

He was a dentist at the Milton Hershey School, following which he entered private practice in Hershey. In 1950 Dr. Weaver was recalled to active duty, a LT, to Parris Island Recruit Training Depot, thence to US NAS, Alameda, CA and COMFLOGWINGPAC, AIRSTRANSRON VR-2. In February 1953 the then LT Weaver was transferred to the Ready Reserve.

Upon return to civilian life, he practiced general dentistry, over 40 years, continuing his service in the Reserves, attending drills at the Reserve Center in Harrisburg, PA, along with summer two week ACDUTRA. In 1979 Dr. Weaver retired from the USN with the rank of CAPT.

He was active in organized dentistry, having gone through the chairs of the local dental society, belonging to Delta Sigma Delta Dental Fraternity, a fellow of The American and International Colleges of Dentists, Pierre Fauchard Academy, and Academy of General Dentistry. He is a life member of the Harrisburg Area Dental Society, the Pennsylvania Dental Association and the American Dental Association, founding member of the Dental Hygiene School at the Harrisburg Area Community College, an instructor in the Milton Hershey Medical Center and dental member of the hemophiliac team, 13 years on the State Board of Dentistry, (three as chairman) and a chief examiner for the Northeastern Board of Dental Examiners. CAPT Weaver is a Rotarian, served 12 years on Township Zoning Board, was member of vestry and Warden of his church and a member of various Masonic bodies.

He and his wife reside in Hershey, PA and they have three children.

JOSEPH WILLIAM WEAVER, CAPT, is the son of a career Navy flight surgeon, CAPT Joseph Walter Weaver (MC) USN Ret., and joined the USNR in the ENS 1915 Program after graduating from Duke University. He received his MD degree from Tulane University in 1966 after summer tours at NNMC Bethesda and in USS *Essex* (CVS-9), completing his senior year on active duty. He interned at USNH San Diego and after flight surgeon training at NAMI Pensacola, was assigned to CVW-7 embarked in USS *Independence* (CVA-62). He was transferred to Manned Spacecraft Center, NASA Houston for

Apollo 11, 12, and 13, receiving the Presidential Medal of Freedom as a member of the Apollo XIII Mission Operations Team, and several other NASA service awards.

CAPT Weaver then completed residency in Internal Medicine and fellowship in hematology/medical oncology at San Diego Naval Hospital, and was transferred to NARMC Pensacola as the assistant chief of medicine in 1974. He returned to San Diego Naval Hospital in 1978 for training in radiation oncology, and was sent to USNH Portsmouth as head of radiation oncology. He left active duty in 1983 and joined the USNR in the Tampa Bay area, drilling in St. Petersburg and Tampa. For the next nine summers, he was assigned to the aerospace medicine branch of BUMED, serving as its director in the absence of an active duty incumbent.

He served as XO and CO of his various units at St. Petersburg and as the medical officer, REDCOM 8 in Jacksonville. He retired Jan. 1, 1994 from the unit drilling with USF Medical School in Tampa completing over 31 years of active and reserve service. He now practices radiation oncology at Baptist Hospital in Pensacola where he is a consultant to NARMC. He is a consultant to the Armed Forces Medical Institute and a life member of AMSUS and the NRA.

He married the former Carol VanDenburg and has two sons, Jonathan, a securities trader in Chicago and Jeffrey, a law student at Washington and Lee University.

BRUCE C. "B.C." WEBB, CDR, born Dec. 6, 1939 in Bloomer, WI, received a BS from the US Naval Academy and attended US Naval Post Graduate School.

He joined the service July 7, 1959. His military locations and stations: USNA 59-63. Graduated 1963; active duty 1963-74; USNR, 1974-89. Units served with: Naval Air Training Command, VR-24, VP-2, VT-27, FASU-DaNang, VRF-31.

Participated in Vietnam and received Vietnam Service Medal w/Silver Star, Air Medal, MUC, NUC and Combat Action. Webb achieved the rank of CDR and was discharged Sept. 1, 1989.

His memorable experiences: served in USNR units: NAS Willow Grove, PA; VP-64; Staff SECDEF; Joint Staff JCS, J-3. Transferred to USNR from active duty in 1974.

He is divorced and has children, Eva and Charles; and grandchildren: Chris, Scott, Jessica and Brian. Employed as pilot with Trans American Airlines, 1974-76; US Department of Commerce, 1976-95.

JOHN O. WEBB JR., CAPT, a native of Metuchen, NJ, attended the University of Virginia and Northeastern University, obtaining a BS in nuclear engineering and MS in mechanical engineering. He was commissioned an ENS in 1973 through the NROTC program.

He served on active duty as communications officer, navigator and 3M coordinator on USS *Hammer* (DD-718). Following release from active duty in 1977, he affiliated with the USNR and was assigned to the USS *Damato* (DD-871) as SELRES operations officer. Upon Damato's decommissioning, he served in the Naval Control of Shipping Organization assigned to NCSO Boston 101.

In 1984, CAPT Webb entered the engineering duty qualification program, completing qualification as engineering duty officer in 1986. He was assigned to NSY Portsmouth NH 201 from 1985-93, where he served as administrative, training, production, executive and commanding officer. From 1993 to 1996, he was assigned as executive officer of NAVSEA Industrial Mobilization 101. In October 1996, CAPT Webb assumed command of NAVSEA DET 901.

From 1991 to 1994, CAPT Webb also served as the chairman of the Recruiting District Assistance Council (RDAC) supporting Navy Recruiting District Boston. He was selected as the Navy Recruiting Command's RDAC Chairman of the Year for 1993.

CAPT Webb has been awarded the Navy Commendation Medal, the Meritorious Unit Commendation, Battle E, Armed Forces Reserve Medal, National Defense Service Medal, Military Outstanding Volunteer Service Medal and Navy Recruiting Ribbon.

In his civilian occupation, he is the director of quality for ABB Service Inc. He resides in Wilmington, MA, with his wife, Ruth; daughters, Ruth and Amy; and son Jack.

SOLON D. WEBB, CAPT, born Feb. 5, 1947 in Navasota, TX, received a BS from the University of Houston (1969); MBA from Texas Tech (1974).

He joined the service March 21, 1969. His military locations and stations: NAVOCS November 1969-April 1970. Served with *Esteem* (MSO-438); *Persistent* (MSO-491); *Molala* (ATF-106); *Gallant* (MSO-489).

Webb received the Vietnam Service Medal w/two Bronze Stars, Vietnam Campaign Medal, Navy Commendation Medal, National Defense Service Medal, Armed Forces Reserve Medal (2nd) and Republic of Vietnam Meritorious Unit Citation. He achieved the rank of CAPT and released from active duty Nov. 26, 1972 and retired April 1, 1996.

He has a son, Solon D. Webb Jr., and is employed with American President Lines as vice president.

KENT SHEPARD "KENT" WEBBER, CAPT, born Jan. 9, 1940 in Belfast, ME, received a BS from Maine Maritime Academy, Castine, ME, 1960; The Naval War College, Newport, RI, diploma, 1989; MBA The George Washington University, Washington, DC, 1995.

He joined the service June 18, 1960. His military locations and stations: Navy Reserve Centers: Fargo Building, Boston, MA; Quincy, MA; New Orleans, LA; CDR, Military Sealift Command, Brooklyn, NY; CDR, Military Sealift Command, Gulf Sub-Area, New Orleans, LA; CDR, Military Sealift Command, Washington, DC, Naval War College, Newport, RI; USS *Iwo Jima* (LPH-2), Norfolk, VA; Naval School of Transportation Management, Oakland, CA; Fleet Anti-submarine Warfare Training Center, Norfolk, VA; Commander-in-chief, Western Atlantic, Norfolk, VA; Commander-in-chief, US Navy Europe, London, UK; National Defense University, Pensacola, FL; Navy Reserve Center, Adelphi, MD; Armed Forces Reserve Center, Anacostia, Washington, DC; Navy School of Transportation Management, Oakland, CA; Navy Command Center, Office of the CNO, Washington, DC.

Served with: Military Sealift Command Div 1-1, Merchant Marine Co. 1-1, Military Sealift Command HQ 101, Boston, MA; Naval Reserve Officers School 1-7; NAS, South Weymouth MA; Military Sealift Command Office 2001, Providence, RI; Military Sealift Command Boston/Eastern Mediterranean 101, Boston, MA; Naval Control of Shipping Office Venezuela 501, Quincy, MA; Naval Control of Shipping Office New Orleans 110, New Orleans, LA; CDR, Military Sealift Command HQ, Washington Det. 106; CDR, Military Sealift Command HQ, Washington Det. 206, Washington, DC; Naval Control of Shipping/Commander Military Sealift Command/Navy Command Center/Maritime Administration 106, Washington, DC; CDR, Military Sealift Command, Washington, DC.

He participated in battles and campaigns: Operation Desert Shield/Desert Storm/Provide Comfort. He was awarded the Meritorious Service Medal w/Gold Star, Navy Commendation Medal w/Gold Star, Navy Unit Commendation Ribbon, National Defense Medal w/ Bronze Star, Meritorious Unit Commendation Ribbon, Vietnam Service Medal, Vietnam Service Medal (Department of Transportation), Overseas Service Ribbon, Recruiting Ribbon, Armed Forces Reserve Medal w/2 HG devices and Bronze M, Expert Marksman (Pistol) and Expert Marksman (Rifle).

Webber achieved the rank of CAPT. His memorable experiences: Exercise director for "Rainbow Reach" which included 2500 reservists in the US in a CPX logistics exercise. This included Army, Navy, USCG and the Defense Logistics Agency.

He married Margaret Austin and they have children: Kristin, Stephen and Kendra.

Employment and positions held: Merchant Marine Officer 1960-66. Hold current license as chief engineer, Steam Vessels, Any HP and 3rd assistant engineer, motor vessels, Any HP, chief of marine engineering, General Dynamics/Quincy Shipbuilding Div., Quincy, MA, chief mechanical engineer, Avondale Shipyards, Inc. Avondale,

LA, manager technical liaison, Todd Pacific Shipyards Corporation, San Pedro, CA; senior project engineer, John J. McMullen Associates, Crystal City, VA, staff engineer, Technology Applications, Inc. Arlington, VA, General manager, Metropolitan Cash and Carry, Washington, DC, senior program manager, Designers and Planners, Inc., Arlington, VA.

WILLARD S. WEBBER, CAPT, born June 24, 1920 in Oakland, CA, received a BA from Stanford University, June 1942.

He joined the service Feb. 26, 1942 (AS-V7). His military locations and stations: Midshipmans School, Columbia University Dec. 1, 1942. Served with AO-21, AOG 8; AH 12; AH 15.

His battles and campaigns: Aleutian Islands, Philippines, Okinawa and two years, Korea. He was awarded five Battle Stars, 13 ribbons, Korean Presidential Unit Citation WWII-three Battle Stars. He achieved the rank of CAPT, USNR and retired July 1, 1965. His memorable experiences: over 100 consecutive days of kamikaze attacks at Okinawa; executive officer and navigator August 8; executive officer AH 12 and AH 15.

He married Arloine on July 25, 1943 and has on son Douglas (MD); daughter Candace A. Huston; and grandchildren, Matthew and Lauren Huston. Employed as investment manager, VP Loomis Sayles & Co.

ALBAN WEBER, RADM, a native of Chicago, IL, is a graduate of Harvard College and Harvard Law School, and has degrees from Northwestern University and John Marshall Law School as well.

During WWII, he commanded the USS *Chadron* (PC-564), USS *Dempsey* (DE-26) and USS *DeLong* (DE-684). On the USS *Dempsey,* he participated in most campaigns in the Pacific, including the Gilberts and Marshalls operations, in Guadalcanal and Bougainville in the Solomons, and finally at Guam and Palau and then Leyte in the Philippines. His final war command was the USS *DeLong*, a training ship for midshipmen. When the war ended, he was given command of DE Flotilla One, an administrative command for decommissioning.

After the war, he was appointed Chief of the International Law Section, JAG, where he participated in the War Crimes Trials and in the first meetings of the United Nations.

He then transferred to the State Department as counsel for the Foreign Liquidation Commission first in Italy and later as chief counsel for all Europe. On his return to the States in 1947, he was elected to the Chicago City Council where he served four years.

His subsequent civilian assignments were first, for 15 years as University Counsel for Northwestern University, and then as president of the Federation of Independent Illinois Colleges and Universities for another 15 years.

During this time he remained in the USNR and saw duty on several battleships and carriers and finally selected to command the Great Lakes Training Sqdn. consisting of seven ships which trained Reserves in the Great Lakes.

In 1970 he was one of three line officers from among 3000 eligible captains to be selected rear admiral. He was also selected by Governor Thompson as CDR of the Illinois Naval Militia.

He married Margaret Kenny and they have four adult children: Alban III, Peggy, Gloria and Brian; and six grandchildren. They presently live in Port St. Lucie, FL.

LINNEA J. (SOMMER) WEDDINGTON, LT, born Oct. 26, 1959 in Sierra Madre, CA, received a BS in 1981 from Slippery Rock State College and a MBA in 1996 from Meredith College.

She joined the USN, Feb. 22, 1982 and the USNR, June 1, 1989. Her military locations and stations: DLI Monterey, CA; NSGA Misawa, Japan; NRSECGRU Greensboro, NC.

She was awarded the NAM (2), Good Conduct, NDSM, Overseas Service and Marksman (Pistol). Achieved the rank of CTI2, USN and currently LT, USNR.

Her memorable experiences include living and working in Japan.

She married Anthony F. Weddington and is employed as business manager for Weddington Studio (Fine Arts).

WINSTON H. WEESE, RADM, is a native of Little Rock, AR. He graduated from Culver Military Academy Naval School and completed his pre-medical education at Cornell University, Ithaca, NY, where he finished two years of basic ROTC in field artillery. He volunteered for the USN, was assigned duty as an apprentice seaman in V-12 (medical), and was ordered to medical school at the University of Arkansas where he graduated with a BS and an MD degree, and a commission as LTJG (MC) USNR.

He then received orders to extended active duty at the NAS, North Island, San Diego, CA, followed by duty on Yap, Western Caroline Islands as a general medical officer, and later, a large native hospital on Koror, Palau Islands. He completed his tour of duty in the Pacific at the USNH, Guam, MI and remained in the Navy Medical Reserve, ultimately completing 38 years of satisfactory Navy service.

Upon return to civilian medical career he was selected for a four year residency at the Charity Hospital of Louisiana, New Orleans, followed by one year as a post-graduate house officer at the American Hospital of Paris in Paris, France.

Upon his return from France he joined Medical Reserve Co. 8-2 in New Orleans and served as CO from 1965-68, and as CO of the larger Group Command Hospital Replacement Unit 1970-72. He also served as the Navy medical liaison officer at Louisiana State University School of Medicine from 1955-85. This involved recruiting medical students to become probationary medical ensigns and overseeing their summer Navy training.

As a member of the NRA Dr. Weese served as president of the New Orleans Chapter and later, as NRA President of the Eighth Naval District. He was elected national surgeon and was active in the effort to change the title of this office to "Vice President for Health Affairs' and to certify any officer of the health corps as qualified to hold this elected office. Over a period of ten years he organized and chaired a "Health Corps Symposium" for officers of all the health corps, which was held the day before each annual NRA National Conference. He is a life member, number 159.

He was promoted to Flag Rank Sept. 1, 1974, served two years as consultant to the Bureau of Medicine and Surgery in OB/GYN and two additional years as chief of Naval Reserve Health Corps Training and Education. He was then selected for the BuMed Reserve Flag Billet as the officer assigned to operational medicine. Thereafter, he completed the US Marine Corps Amphibious Assault School at Camp Pendelton, CA and served as the senior medical officer on USMC Operation Solid Shield 1976. Among many other people, helped develop the fleet hospitals used in Desert Storm. His most recent mobilization billet was as CO of USNH Philadelphia. He was awarded the Legion of Merit, the Meritorious Service Medal and the Navy Commendation Medal. He retired from the USNR, Sept. 1, 1981 and since Aug. 1, 1981 has been on the faculty of LSU Medical School in New Orleans.

ROGER WEIGL, LCDR, born Jan. 24, 1921 in Collinsville, OK, attended University of Georgia and Oklahoma State University.

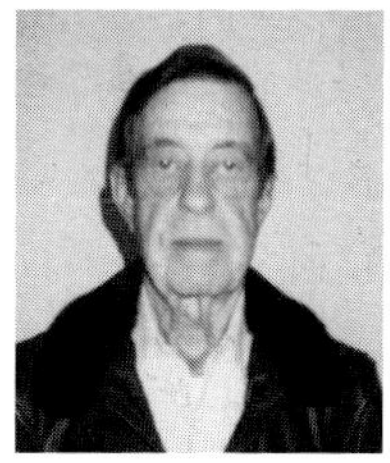

He joined the service Aug. 14, 1942. Began aviation cadet training Jan. 6, 1943. Commissioned ENS, Sept. 16, 1943. His military locations and stations: Pensacola, Jacksonville, Guantanamo, Norfolk, Alameda, Saipan and Dallas. Served with VPB-32 (flight officer), VP-7O3 (executive officer), AWS-70 (OpCon). His battles and campaigns: plane commander, PBY, PBM, P2V aircraft. About 3500 hours Navy flight time.

He was awarded the American Pacific Theater, WWII Victory Medal, Naval Reserve Medal and Armed Forces Reserve Medal. Weigle achieved the rank of LCDR and was discharged April 1, 1946. Retired from USNR April 1, 1965. His memorable experiences: In 1944, ferrying Russian aircraft PBN's from N. Carolina to Kodiak, AK; air sea rescue in Mariannas Islands.

He married Barbara Mary and they have children: Karen Lea Short, Julie Ann Beck and Rodney Alan Weigl; and grandchildren: Cindy Jarvis, Jenifer Short, Keith Short and David Beck. Employed at Phillips Petroleum (1946-82); manager excise taxes, retiring in 1982.

JOHN GEORGE "JACK" WEIHAUPT, CAPT, born March 5, 1930 in LaCrosse, WI, received a BS, MS, MS and PhD from the University of Wisconsin.

He joined the service January 1954. His military locations and stations: Ft. Belvoir, VA; Yangu, Korea; Eta Jima, Japan; Antarctica. Served with 3rd Engr. Bn. (Combat) and 24th Inf. Div. Participating in Korean campaign.

Weihaupt was awarded the National Defense Medal, United Nations, Antarctica, Rifle, Pistol and Reserve Medals. He achieved the rank of CAPT. His memorable experiences include (Korean winters), colleagues and friends.

He is married to Audrey and has been employed with AS&R; Anaconda; United Geophysical; Arctic Institute; DoD; Purdue University; Indiana University; San Jose State University and University of Colorado.

DOUGLAS B. WEILDING, LCDR, born in Attleboro, MA, Sept. 7, 1939. He received his BS from Boston University in 1961. LCDR Weilding enlisted in the USNR April 11, 1962 and was graduated from Officer Candidate School, Class Sixty, Aug. 16, 1962, in the same section, Echo Two, as Jeremy Michael Boorda.

From 1962-65 his active duty consisted of two years on board USS *Lake Champlain* (CVS-39) during which time he participated in the Cuban Blockade/Quarantine and was awarded the Armed Forces Expeditionary Medal; his next tour was on staff, Commander Fleet Air Quonset.

Upon discharge from active duty he attended law school on the GI Bill and earned the degrees of JD and LLM from Suffolk University. He was an assistant district attorney for four years.

He was affiliated with the mine warfare program until 1972, attached to the ResCrew of minesweepers USS *Turkey* (MSCO-56) and USS *Jacana* (MSC-193). During his tenure aboard USS *Turkey* (MSCO-56), the ship won the national award for command performance, received the Battle Efficiency "E" for two consecutive years and was awarded the Minesweeper "M" which was the first time the award was given to a USNR minesweeper. He became qualified for command at sea and was designated a surface warfare officer (1115), and also served aboard USS *Meadowlark* (MSC-196), USS *Falcon* (MSC-193), and USS *Frigatebird* (MSC-191) and USS *Dash* (MSO-428).

LCDR Weilding attended Jag School, NPT, basic and advanced courses; served aboard USS *Noxubee* (AOG-56); USS *Corry* (DD-817); USS *Newman K. Perry* (DD-883), and USS *Edson* (DD-946); other Acdutra was performed at N.S. Charleston, SC; NETC, NPT; DESRON 28 NRF, NPT; Keflavik, Iceland; and REDCOMONE. Reserve units included Iceland Defense Force 101, and Naval Control of Shipping Organization Boston 101. He transferred to the Retired Reserve July 1, 1983.

A life member of the NRA, Veterans of Foreign Wars of the US, American Legion, and all Masonic organizations, his law office is located in Attleboro, MA. He is a member of the Massachusetts Bar, Rhode Island Bar and the Supreme Court of the United States.

ROBERT P. WEINGART, CDR, born Oct. 3, 1937 in East Orange, NJ. Received two degrees from the University of Notre Dame, a BA in 1959 and a BSME in 1960. He received an MBA from the University of Michigan in 1970.

He enlisted in the USN at NAS Glenview, IL in March of 1962. He went through boot camp and electronic schools "A" and "B" at NAS Memphis, TN. Upon completion of his schooling, he served as an AT and crewmember in VP squadrons flying P2V Neptunes.

In July of 1967, he received a direct commission as an

ENS. He served in various intelligence billets at NAS Grosse Ile, Grosse Ile, MI, and later at NAF Detroit, Mount Clemens, MI. He was the intelligence officer for VS73, the "Blue Bandits" at the time of their decommissioning and also the intelligence officer and a plank owner in VP-93, the "Executioners". He retired from the USN in July of 1994. He was the intelligence officer for CINCUSNAVEUR DET 413 at the time of his retirement.

CDR Weingart worked as a sales engineer for General Motors Corporation from 1960 to 1970 and then as a district sales manager for the Xerox Corporation. In 1974, he started his own financial consulting firm and continues to this day as a financial consultant, with offices in Grosse Pointe Farms, MI.

He is married to the former Susan B. Lyman of Upper Montclair, NJ. They have four daughters: Nicole, Melinda, Deborah and Lindsey, and make their home in Troy, MI.

ROBERT W. "BOB" WEISE JR., CDR, born April 10, 1917 in Minneapolis, MN, received a BA from the University of Minnesota (1973); MSc (Econ), London School of Economics, London, England.

He enlisted in the service Dec. 5, 1943. His military locations and stations: NOB 153, Rio de Janeiro, Brazil 1943-44 (Feb); Great Lakes NTC, NSOTC Bayonne, NJ (April 1945-to Sept. 15, 1945) V-7 Supply Training, S1/c. Commissioned ENS Sept. 30, 1945.

Weise was awarded the American Theater Medal, Victory Medal and Naval Reserve Medal. He achieved the rank of CDR, SC, USNR, Sept. 30, 1945 and released from active duty Sept. 15, 1945.

His memorable experiences: served in Ready Reserve 1948-74; standby reserve, 1975-77; held assignments of pers. officer, admin. officer and executive officer for Politico-Military Affairs Unit 1-5, Washington, DC 1964-72; Acdutra, Norfolk, Bayonne, NJ, COMSEVLANT - 5th Fleet, 1952. Retired April 10, 1977 after 33 years in the USNR.

He married Ingrid and they have two children, Robert III and Carl Henrik; and grandchildren, Jennifer and Angela Weise. Employed as foreign service officer, Department of State, Washington, DC to 1966; American Consul and economic officer, Chile, Spain, Italy and Panama and Bolivia, 1941-66; Economist, Social Security Administration, International Staff, 1966-72.

CARTER B. "WIZ" WEISIGER, CAPT, born Dec. 26, 1917 in St. Louis, MO, received a BA from the University of Virginia in 1941.

He joined the NRAB, Anacostia, DC December 1941. His military locations and stations: LANT & PAC. Served with VP-208. His battles and campaigns include Saipan and Okinawa.

Weisiger was awarded the Air Medal and Distinguished Flying Cross. He achieved the rank of CAPT and was discharged October 1945. His memorable experiences: recalled in 1949 and retired in 1968.

His first wife Audrey Vivian died in October 1974. He married Ruth Loving Weed in September 1975. He has children: Jane, Louise, Ann and stepsons, Robert, Barry and Steve. Employed with Approach magazine for 12 years, retiring in 1980. Retired from Civilian Naval Aviation Safety Center in 1980.

CHARLES R. "CHARLIE" WEITHAUS, DDS, CDR, born Aug. 16, 1936 in Milwaukee, WI, attended University of Wisconsin for four years and received a DDS degree from Marquette University.

He served in enlisted service Jan. 27, 1955-Jan. 1, 1959; July 1967-July 1968, dentist (dental officer). His military locations and stations: Blimp Sqdn. ZW-1 Lakehurst, NJ (three years). Served with Dental Department, NAS Mayport, FL; Memphis; Norman, OK; Biloxi, MS. He is a Korean and Vietnam veteran.

Weithaus was awarded the Good Conduct Medal and various other medals. He achieved the rank of (AT1) enlisted and CDR (DC) 2205. Discharged from USNR in 1985.

His memorable experiences: the opportunity to serve as enlisted, and as an officer to be part of naval aviation, Electronics School Training, eight years of college (DDS degree), naval dental officer and 16 years USNR time, also dental private practice; crewman ZW-1 Sqdn.

He married Mary and they have children, Sara and Matthew. He has been employed in private dental practice for 26 years.

GEORGE WEITZ, CWO, born May 11, 1930 in Bronx, NY, enlisted in 1948 and completed boot camp at Great Lakes NTC. In January 1949 assigned to the Pacific Fleet he served on the USS *Fechteler* (DD-870) and then USS *Chandler* (DD-717) as a fire control striker. In October 1949 he was released from active duty as a FCSN and transferred to NRSD 3-71 at Ft. Schuyler, NY.

With the Korean War outbreak, he was activated and reported to the USS *Siboney* (CVE-112) in the Philadelphia Navy Yard. After 23 months of ASW operations with the Atlantic Fleet he was released from active duty as a FC3 and returned to NRSD 3-78. Advancing to FTM2 and NRSD 3-64 Brooklyn Navy Yard in 1953, he transferred to NRSD 8-81 El Paso, TX in 1954. He returned to Ft. Schuyler NRSD 3-78 in 1955 advancing to FT1, then FTC in 1964.

In 1968 he was selected as a warrant officer in the Ordnance Control field in the USNR and in 1970 promoted to CWO. Retired from Pitney Bowes Co. after 23 years in sales, as automation specialist in the Inserter Systems Div. His main interests are: naval history, US Naval Ship Covers, and travel. Weitz is divorced and has two sons, Adam and Paul.

DONALD S. "DON" WELCH, CDR (05), born Jan. 26, 1943 in Polk, OH, received a BS in business administration; MA in education and JD degree.

He joined the service Oct. 15, 1962. His military locations and stations: Navy Finance Center, Cleveland, Philadelphia Naval Shipyard. Served with USS *Mars* (AFS-1) and USS *Talbot* (DEG-4). He participated in the Vietnam era.

Welch received the Vietnam Campaign Medal, Vietnam Service Medal, United Nations Medal and Expeditionary Medal. He achieved the rank of CDR (0-5) and discharged Sept. 15, 1977.

His memorable experiences: flying aboard Sea Knight VH-46 Helicopter that must land on isolated beach in unfriendly area of Vietnam. Luckily repairs quickly made and no "Charley" in sight.

He married Cheryll Rae and they have a daughter, Kirsten Rae. Employed as real estate broker and attorney-at-law.

JACK L. WELLS, CDR, a native of Corry, PA, enlisted in the USN in June 1960, and served aboard USS *Observation Island* (EAG-154). Selected for the Navy Enlisted Scientific Education Program, he graduated from the Pennsylvania State University in electrical engineering in 1966. He attained the enlisted rate of IC1. After OCS he was commissioned ENS, USN in November 1966.

He served aboard USS *Gurke* (DD-783), USS *Banner* (AGER-1), and on the staff of COMINEFLOTONE embarked in USS *Catskill* (MCS-1), including combat operations in and in the waters surrounding the Republic of Vietnam and in the Tonkin Gulf. He served as Task Unit CDR for SEATO Mine Sweeping Exercises in the Republic of the Philippines.

In December 1970 he resigned his regular commission and returned to civilian life, affiliating with the USNR. He served in Service Div. 4-13(M), and as CO, LPD 203; CO, LST 1192-9205; CO, NAVTELCOMSTA ICELAND 205, CO, FLTRAGRU SANDIEGO Det 101 and on the staff of COMNAVSURFGRUONE, Newport, RI. He retired from the USNR in March 1983.

Surface Warfare qualified, he has been awarded the Navy Commendation Medal, Combat Action Ribbon, Navy Unit Commendation Ribbon, Enlisted Good Conduct Medal, and Battle Efficiency "E". Campaign medals include Vietnam Service (six stars), Armed Forces Expeditionary, Armed Forces Reserve, National Defense, Vietnam Campaign, Vietnam Gallantry Cross and Expert Rifle.

A life member of the NRA, he served as president of the USS *Niagara* Chapter for four years. CDR Wells is married to the former Catherine Marine Dowd and has one daughter, Laura. He is the technical director for the M&M/Mars Div. of Mars Inc. and resides in Stroudsburg, PA.

RANDOLPH S. "RANDY" WELLS, LCDR, born Nov. 6, 1929 in St. Louis, MO, received BS, LLB and JD degrees.

He joined the service Jan. 28, 1955. His military locations and stations: Iceland, Brunswick, ME; Argentia New Foundland. Served with VP-7 and VP-933. His participated in the Cuban Crisis, 1961-62.

Wells received the Foreign Service Medal and achieved the rank of LCDR. He was discharged July 1, 1979. His memorable experiences: as a naval aviator.

He is single and employed as estate administration attorney.

ALBERT G. "AL" WENDT, LT, born May 19, 1922 in Stanwood, IA, received a BA from the University of California at Berkeley.

He joined the service March 4, 1941. His military locations and stations: Great Lakes, USS *Cincinnati*, Northwestern University, St. Thomas, University of California at Berkeley.

His battles and campaigns: South Atlantic, two years, four months, October 1941-to January 1944. Awards and medals: Assoc. w/time and area. He achieved the rank of LT and discharged June 24, 1946. His memorable experiences: Chasing blockade runners in South Atlantic.

He married Margaret and they have two sons, Albert Guy and James Robert; daughter, Jeanette Wendt Hill; and grandchildren: Geordie, Katherine and Emily Wendt.

Employed by Cannon and Wendt Electric, March 1947-present.

DEAN F. WERNER, CAPT, born May 26, 1924 in St. Joseph, MO, received a BA from Westminster College and MD from the University of Oklahoma School of Medicine.

He enlisted in the service July 1942 and worked as a lab technician at the NH in Norman, OK. Attended medical school at the University of Oklahoma and was commissioned an ENS in June 1946. He was released from active duty in 1948 and recalled in May 1952 to serve at the NH, Memphis, TN. Subsequent tours included Field Medical School at Camp Lejeune; 2nd Bn., 90mm AA Gun Bn., Fleet Marine Force Atlantic and NAS, Olathe, KS. Released from active duty in June 1954.

He began his reserve career in 1954 in the Reserve Medical Co. 9-4. Subsequently served as CO of that unit. Other units he has served in are General IRU 317 and HQ 24th Marine, 4th Marine Div., Fleet Marine Force. CO of the 3rd Marine Div. Medical HQ 118.

His decorations include the Naval Reserve Medal, Armed Forces Reserve Medal, American Defense Medal, National Defense Medal and the Vietnam Service Medal.

He married Phyllis and they have children: Nancy Bybel, Sallie Brackman, Lynn Driggers and Barbara Douglas; and grandchildren: Matthew, Rebecca, Julie, Jill, Ashley, Adam, Michael and Amy. Employed in private practice of industrial medicine.

EDGAR FRANK WEST, CAPT, born March 31, 1923 in Long Branch, NJ. He earned a BS degree at Trenton State College and master's at Rutgers University.

He joined the service Oct. 24, 1942. His military locations and stations include Drew University V-12; Pacific Area; NAS Lakehurst, NJ. Served in USS LST-776 (Brodie), ZP Sqdns., VSs, AW Staff. His battles and campaigns: Iwo Jima, Okinawa, Philippine Occupation, Japan Occupation, America Theatre, Asiatic Theatre.

He was awarded the Armed Forces Reserve Medal (2) Naval Reserve Medal and WWII Victory Medal. He achieved the rank of CAPT and was discharged July 1, 1975. His memorable experiences: USS LST 776 (Brodie), first lieutenant to executive officer; Air Wing Staff 75, assistant chief of staff.

He married Jeanne and has three children: Deborah, Patricia and Thomas; and grandchildren: Bill, Katie, Michael and Peter. Employed as teacher, counselor, director pupil personnel services.

FREDERICK P. WEST JR., CAPT, born in New York on April 6, 1930. He graduated from New York State Maritime Academy in 1950 and sailed with the US Merchant Marine, earning his second mates license before going on active duty with the USN in 1952.

After two years with the Pacific Fleet aboard USS *Washburn* (AKA-108), he entered flight training, graduating in 1956 with both fixed and rotary wing qualifications. He was assigned to HU-2 Lakehurst, NJ from 1956-60, deploying with Helo Detachments to units of the Atlantic Fleet. His next duty was at NAS Port Isabel and NAS Kingsville, TX, as in transports and jets.

In 1963, he was transferred to VC-5 Det. NAHA, Okinawa, and took over the detachment as OinC. The Det. made two deployments to Vietnam during his tour. He was accepted into the TAR program and was transferred to NAS Alamitos, CA, as administration officer, was promoted to CAPT in 1973, and was XO until 1974, when the station was deactivated. CAPT West was then ordered to the NAVRESREDCOM, San Francisco, as chief of staff/deputy, and then to the NAVMARCORESCEN, Treasure Island as CO. He retired in October 1977.

STEPHEN F. "STEVE" WESTON, CDR, born April 15, 1945 in Colchester, VT, attended the University of Mississippi NROTC program, obtained a BA degree and commissioned, ENS, USN, in 1967. Assigned as electronics material officer in the USS *Mullinnix* (DD-944), providing NGFS in Southeast Asia in 1969, he qualified as SWO and operations officer in 1970. Next assigned as XO, and CO, of USS *Bulwark* (MSO-425), he later served as flag lieutenant to CDR, Mine Warfare Forces in 1971. From 1972 to 1974, he served as MCM Sqdn. staff officer to CDR, Mine Countermeasures, Royal Navy, in Scotland.

Later assignments, 1975-78, including weapons officer, USS *Richard L. Page* (FFG-5) and operations officer and detachment OIC for Fleet Composite Operations Readiness Group Two with COMCRUDESGRP EIGHT in the Mediterranean. Accepting a USNR commission in 1978, his selected reserve billets at NavResCtr, Burlington, VT, included USS *Barry* (DD-933); detachments with NavShipYard, Portsmouth, NH, and Maintenance Training Facility, Brooklyn, NY (as OIC). From 1988-91, he served as NavRes Military Plans Support Officer (Naval Liaison Officer) to the State of Vermont. Affiliation with the Reserve Center VolTraUnit, 1984-87 and 1991-93, included three years in command.

With 14 concurrent years in supervision of electronics manufacturing, he attained Certified Production & Inventory Control Manager (CPIM). In 1992, he became a state of Vermont Income Tax Examiner. CDR Weston has been awarded the Navy Achievement Medal, Navy Unit Commendation, Vermont National Guard Commendation, and the RVN Medal of Valor, 2nd Class. His campaign and service medals are: National Defense w/2 Stars, Vietnam Service w/2, RVN Campaign, Armed Forces Reserve and Sea Service Deployment.

He married Ruth and they have children: Glenn, Michele, Mathew and Nicholas; and grandchildren: Andrew, Alison, Kaitlyn and Devon.

FRANK W. WEYLER, CAPT, born April 21, 1937 in Cincinnati, OH, received a BS in physics from Xavier University in 1959.

He joined the service June 1959. His military locations and stations: OCS Newport, RI; NAS Jacksonville; NAS Memphis; NAS Oceana. Served with Attack Sqdn. 65 (1960-63), USS *Intrepid,* USS *Enterprise.*

His battles and campaigns: Cuban blockade 1962. He was awarded the Navy Expeditionary Medal, National Defense Medal, Armed Forces Reserve. Weyler achieved the rank of CAPT and discharged February 1963.

His memorable experiences: served 21 years in various USNR units in Ohio. Retired in April 1984.

He married Ann Sprague and they have five children: Katherine, Frank J., Steven, John and Carrie; and grandchildren: Austin and Grant Heminger, Joshua, Jared, and Olivia Weyler. Employed as Monsanto and E.G. & G., senior manager (engineering). Retired in August 1992.

WILLIAM P. WHALEN, CDR, born Dec. 29, 1920, Ashtabula, OH. Sworn into the USN on June 22, 1942. USNR Midshipmans School, Northwestern University. Commissioned ENS August 1943. Amphibian and beachmaster training at Little Creek, VA. USS *Custer* (APA-40), transferred to USS *Sheriden* (APA-51), Dec. 1, 1944.

Landed with assault troops at Roi-Namar Feb. 1, 1944, Saipan June 15, 1944; Guam July 22, 1944; Leyte Oct. 20, 1944, Lingayen Gulf Jan. 9, 1945; Okinawa April 1, 1945 and Tokyo Bay, Sept. 2, 1945.

Landed Leyte, Orange Beach TWO with 1st Bat., 383rd Regt., 96th Inf. Div. Hearing Japanese forces were moving through Surigao Straits and San Bernandino Straits, they set records in the most rapid ship to shore movement of troops and cargo in amphibious history.

Whalen awarded Bronze Star Medal, w/Combat V, China Service, American Campaign, Asiatic-Pacific w/5 Stars, WWII Victory, Navy Occupation, National Defense, USNR and Philippine Liberation. Retired as CDR USNR, Dec. 29, 1980.

Retired as vice-president, Joseph E. Seagram & Sons, Feb. 1, 1985. He married Margaret Riehl and they have nine children: Cathy, Mike, Pat, Billy, Mary, Rosemary, Sean, Tom and Kevin; and 20 grandchildren. He retired Feb. 1, 1985.

CLARENCE J. "CHAMP" WHEELER, CDR, born Sept. 25, 1917, Dallas, TX. He earned BS degree, 1941, in mathematics and his BA, 1946, psychology, both at Southern Methodist Univ., Dallas, TX; and his doctor of medicine, 1950, Johns Hopkins Univ. School of Medicine, Baltimore, MD.

He served in the USMCR, 1942-45, as CAPT, naval aviator and figher pilot. Retired Dec. 15, 1945 as MAJ. Recalled in 1968 and served as chief of surgery, Hue Central Hosp. in RVN; senior medical advisor, Thua Tien Province, South Vietnam; Commander, Medical Corps, USNR, 1968-70.

Awards include the Distinguished Flying Cross w/2 stars, Air Medal w/4 stars, Purple Heart, Presidential Unit Citation, Pacific Theater Combat Ribbon w/3 stars, American Theater Ribbon, Vietnamese Medal of Health, 1st Class, Vietnamese Medal of Social Welfare, Navy Commendation, Meritorious Bronze Star, National Defense, Vietnam Service and Campaign Medals, WWII Victory, Armed Forces Reserve, Marine Forces Reserve and Air Medal.

He has numerous professional memberships, honors, social and medical fraternities and cultural societies. Some of his civic memberships include MC League, MC and NR Officers Assocs., Navy League, MOWW, MC Avn. Assoc., Greater Dallas Reserve Officers Assoc and TROA.

Last civilian employment was from 1991-93 as medical director, Schick-Shadel Hospital, Dallas-Fort Worth, TX. Since 1994 he has been semi-retired. Married Patsy Lester and they have six children: Stephan, C.J. III, Robert, Thomas, David and Ritchey; nine grandchildren and two great-grandchildren.

ROSCOE O. WHIDDON, CDR, born April 10, 1935 in Thomasville, GA, attended Emory University 1953-54; Florida State 1954-56, University of Miami School of Medicine.

He joined the service Oct. 30, 1956. His military locations and stations: Mayport, FL. Served with ComDesDiv 162, 1961-63, 4th Anglico, FMF 1963-67, Miami, FL, 2nd Marine Div., Med. A 208, officer in charge 1977-79.

Whiddon achieved the rank of CDR, USN-R and was discharged July 9, 1963. He married Patricia and they have four children: David, Dana, Brian and Lisa; and a granddaughter, Lindsay Whiddon. He is a retired physician.

PHILIP A. "PHIL" WHITACRE, RADM, born May 17, 1932 in San Diego, CA, received an AB from San Diego State University and a MBA, Stanford University.

A veteran of over 37 years in the Navy and USNR, served on active duty during the Korean War and most recently served as deputy commander, Naval Supply Systems Command for Fleet Support, Logistics and Corporate Plans before his retirement on May 1, 1991. After having received his commission in May, 1955 at Officer Candidate School in Newport, RI, he attended Naval Supply Corps School in Athens, GA, for six months prior to his assignment aboard the USS *Sperry* (AS-12) in San Diego. Further active duty was as flag lieutenant and aide to commander, Naval Training Center in San Diego and OinC of Navy Exchanges, Barber's Point, HI.

His reserve affiliations were with units in San Diego, Muskegon, Mobile, Detroit, Milwaukee, Syracuse, Chicago and Washington, DC. He has served as CO, DCASR Chicago Unit 613, and also CO, SPCC Mechanicsburg, HQ 113.

Upon his release from active duty, he became associated with the Ryan Aeronautical Co. in San Diego and Continental Motors Corporation in Detroit. For the past 24 years, he has been prominent in the management of national performing arts organizations, having served as executive director of the San Diego Symphony Orchestra, associate manager and director of development of the Chicago Symphony Orchestra, and chief development officer for the National Symphony Orchestra at the John F. Kennedy Center for the Performing Arts in Washington, DC. He served as senior vice president for the National Foundation for Advancement in the Arts headquartered in Miami, FL and currently president and CEO of the Palm Beach Community Chest/United Way.

He is immediate past chairman and CEO of the Navy Supply Corps Foundation and past president of the US Navy Supply Corps Association and the Chicago Council, Navy League of the United States. He is a Chevalier of the Sovereign Order of St. John of Jerusalem Knights Hospitaller and is active on the boards of numerous national and international cultural, civic and philanthropic organizations and represents the US Information Agency as a special consultant in the International Bureau of Programs.

RADM Whitacre has been awarded the Legion of Merit, Defense Meritorious Service Medal, National Defense Medal, Navy and Marine Corps Overseas Service Medal and the Armed Forces Reserve Medal.

EDWARD FAITH WHITE JR., CAPT, born on Sept. 17, 1942 in Cincinnati, OH. He was graduated from Withrow High School, Cincinnati, OH in 1960.

During his senior year of high school he enlisted in the

USNR and his father (then CAPT Select) co-signed for him in February 1960. After boot camp at Great Lakes in June 1960, his first sea duty was a reserve cruise aboard the USS *Bears* (DD-654) in September 1960. He advanced from Seaman Recruit (E1) to Seaman (E3) while he drilled with Surface Div. 4-80 in Cincinnati, and attended the University of Cincinnati. Although selected for the Reserve Officer Candidate (ROC) program he entered active duty in June 1962 as a seaman.

His first ship was the USS *Antietam* (CVS-36) where he was in the Navigation Div. from July 1962 to May 1963. While the ship was originally out of Pensacola, FL, by January 1963 it was being decommissioned in the Philadelphia NSY. After that CAPT White reported, as ordered, to Submarine School at New London. The admin. office refused to enter him in class unless he extended his period of active duty. Therefore he served as a life guard and water safety instructor (WSI) at the base's two lakes (officer and enlisted) and swimming pool from May to December 1963.

While attached to the Base Special Services, he took the exam and advanced to 3rd Class Quartermaster. He reported to the Navigation Div. of his next ship, the USS *Franklin D. Roosevelt* (CVA-42), at the end of December 1963 and was released from active in April 1964 at Mayport, FL. He returned to Surface Div. 4-80, January to March 1965 as an instructor, while also attending the University of Cincinnati again.

After that he worked as a commercial diver in the oil fields offshore Louisiana, as a marine expediter on Dow Chemical's Texas tanker docks and as a Merchant Marine sailor on seagoing tugs and tankers. In 1971 he earned his original third mate's license in the Merchant Marine and in 1973 he earned a BBA from the University of Cincinnati. At that time he applied for and accepted an unrestricted line commission in the USNR as ENS, on Oct. 5, 1973.

He was married to Patricia Lynn Crawfis, of Milford, OH on Sept. 14, 1974 and spent the next 13 months in graduate school at the University of Wales Institute of Science and Technology, (UWIST) in Cardiff Wales. He sailed CAPT on sea going tugs and oilfield supply vessels from 1976 to 1986 and drilled with the Houston Naval Control Shipping Organization (NCSO). His seagoing schedule was such that he changed designator in 1983 to the Merchant Marine IRR (1625 deck) and has continued to make "satisfactory years" since 1979. From 1986 to 1994 he served repeatedly and on extended training duties as a shiphandling instructor on the sit-in scale models at the Naval Amphibious School, Little Creek, VA. In 1990 he earned his unlimited master's license in the Merchant Marine. He has served on numerous selection boards for BuPers since then. He received his "20 year" letter in 1994 and was promoted to CAPT, USNR-R (Oct. 1, 1995).

He currently holds the Naval Reserve Meritorious Service Medal, National Defense Medal, Armed Forces Reserve Medal, Merchant Marine Vietnam Service Ribbon, and the Rifle and Pistol Medals. CAPT White has been life member #4076 of the Naval Reserve Association since May 1984.

CAPT White, his wife Pat and their daughter Tara Marie live in Houston, TX. He has an older daughter, Lucinda Jane (White) Noble, who lives with her husband Dr. John W. Noble in Rock Hill, SC.

EDWARD FAITH WHITE SR., CAPT, born Aug. 30, 1910 in Mobile, AL. He spent the first nine years of his life in East Nashville, TN until his family moved to Cincinnati, OH, where he was graduated from Hughes High School in 1928. He then earned a degree in civil engineering through the University of Cincinnati's cooperative education program in 1933.

After several years as a highway supervisor he joined the University of Cincinnati's faculty as a math instructor and continued in this position until an ATO fraternity brother, ENS Guy G. Cummings D-V(g) USNR, wrote him about a new Navy program in May 1942. He accepted his appointment in the USNR June 29, 1942 as LTJG D-V(P), USNR, and reported on July 9, 1942 to USNR Midshipman's

School NY, NY. He subsequently reported on active duty, as a navigation instructor, Sept. 1, 1942 to USNR Midshipman's School, Abbott Hall, Northwestern University, Chicago, IL.

In June 1943, then LT White, D-V(S), USNR was ordered to 60 days TAD aboard the USS *McCook* (DD-496) as assistant navigator. The *McCook* was on escort duty for the "slow convoy UGS 11 to North Africa" (Casablanca), and escorting slow convoy GUS 10 in return". In June 1944 LT White performed three weeks TAD aboard the USS *Wilmette* (IX29) while continuing as navigation instructor at Abbott Hall three weeks TAD aboard the USS *Wilmette* (IX29) while continuing as navigation instructor at Abott Hall until March 1, 1945. from then until May 22, 1945 he was under instruction at the office of naval personnel in Washington, DC and then reported to director of training, 14th Naval District, Pearl Harbor for duty with the educational services office.

While on duty in Pearl Harbor he was promoted to LCDR effective Oct. 3, 1945 and was scheduled to be released from active duty the following month in Chicago (Great Lakes), but at his own request stayed on active duty as a "civilian readjustment interviewer" in Great Lakes until Aug. 31, 1946.

On July 2, 1947 he was assigned to Div. 9-159 at the Cincinnati, OH Naval Reserve Center in a pay billet. On Jan. 26, 1949 he was transferred from the Volunteer Reserve to the Organized Reserve, Surface Bn. 9-42, further modified to be "ONR SURF BATT 4-26" effective July 1, 1950. CAPT White's 10 year "Naval Reserve Medal" was awarded July 31, 1952 and he was selected for promotion to Commander in December 1952.

On July 1, 1954 he was assigned to NROS 4-4 Cincinnati, OH and was assigned as director (CO) on July 1, 1959. In February 1960 he was notified of his selection for promotion to CAPT, effective July 1, 1960. In August 1962 he received the Armed Servcies Reserve Medal (2nd 10 yr. award). CAPT White additionally served as CO of NR Research CO 4-11 (63 to 66) and on several 4th Naval District Reserve Poly Boards, lastly as president of said board.

He retired on Aug. 30, 1970 and was still a full professor of math and engineering at the University of Cincinnati, Engineering College, where he had returned at the end of WWII. He died July 23, 1971 from a brain tumor and is survived by his loving wife, the former Catharine Jane Bayles, their daughter Lucinda Edith White-Kelly, both currently residing in Freeport, ME; and their son Edward Faith White Jr., CAPT, USNR, from Houston, TX. He is also survived by four grandchildren and his brother George Guyger White III, a WWII Naval Reserve Officer in the Civil Engineering Corps, who left the USNR as a LCDR. *Submitted by his son, CAPT Edward Faith White Jr.*

RICHARD MAHAFFEY WHITE, CAPT, earned BS degree in systems analysis and MS in mathematical statistics, Miami Univ., Oxford, OH, 1972; National Security Agency, Astrodynamics Course, 1973 and Defense Intelligence College, Washington, DC, Indications and Warning Course, 1993.

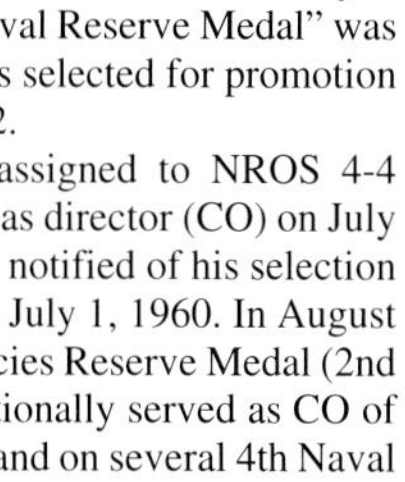

During the mid-80s, he served at CINCLANTFLT/ COMSECONDFLT; was OIC of the Navy Security Group DET in USS *Josephus Daniels* spanning two Mediterranean deployments; embarked in USS *Iowa* for a MED/NORLANT/Indian Ocean deployment; completed 16 month assignment as Command and Control Advisor to Commander Sixth Fleet embarked in the Flagship, USS *Belknap*; extended assignment afloat to COMSIXTHFLT (NATO DEPCOMSTRIKEFORSOUTH); served as sr. inspector for Reserve Inelligence Area 19; was FY94 XO of DIA Current Intelligence 0166; served as the Flagship Intelligence Support Rep.

His FY96-97 Reserve assignment in Pentagon as XO and Deputy Operations Coordinator in the Chairman Joint Chiefs of Staff Defense Intelligence Agency Joint Reserve Unit; extended active duty as exec. highlights managing editor and as DDI to the J2 Directorate for Intelligence, the Joint Staff and is now on active duty at DISA assigned to JWID '97 as the JDEF VIP briefer.

Military decorations include Joint Service Commendation Medal, Navy Commendation (2), JMUA (3), MUC, Armed Forces Expeditionary Medal, Battle E, Enlisted Good Conduct, Navy Expeditionary, National Defense (2), Navy Reserve Service, Sea Service (5), Foreign Service, Pistol (Marksman) and JCS Badge.

ROBERT L. WHITE, LCDR, born March 15, 1947 in Waynesville, MO, received a BS degree in Forest Science, Penn State University in 1969.

He joined the service May 1970. His military locations and stations: NAVOCS, Newport, RI; NISHQ, Washington, DC. Served with NISRA 1605 and DIAHQ 0705, Buffalo, NY.

He was awarded the COMNAVRESINTCOM Letter of Commendation, 1989; Navy Achievement Medal, 1991; DIA Joint Meritorious Unit Award, 1986.

White achieved the rank of LCDR and discharged December 1991. His memorable experiences: Operation Liberty, 1986; Reserve Weekends in Buffalo, NY; Hobbies: model railroading, jogging, camping, hiking, downhill skiing.

He is divorced and has a son, Robert L. White IV. Employed as Forester, program analyst, Forest Silviculturist, Allegheny National Forest.

ROBERT S. WHITE, LCDR, born Oct. 29, 1924 in Wakefield, MA, received a BS and ME, Tufts, 1945; MS in ME, Northeastern, 1955.

He enlisted in the service Dec. 7, 1942. His military locations and stations: V-12, Tufts, 1943-45; Midshipman School, Columbia University Class 26, 1945; APA 65 USS *Brisco*. Inactive duty NROS 1-1, Massachusetts; Korean Conflict. Active duty: February 1951: LST 603, engineering officer, six months, BU-2, salvage officer, repair officer, 2 1/2 years, inactive duty, on staff of NROS 4-8, Akron, OH 1958-71. He achieved the rank of LCDR and went on retired list Oct. 29, 1984.

White is a widower and has four children: Helen Stacy, Robert E. White, Donald White and Janet Smith; grandchildren: Patrick, Mary and Michelle White, and Gwendolyn and Jeremy Smith. Employed as assistant professor ME, Northeastern University 1946-51; 1953-55 Goodrich, Watertown, MA, engineer, 1955-57; BF Goodrich, Akron, OH, senior mechanical engineer, 1957-87, retired.

SCOTT WHITEHOUSE, RADM, born Jan. 29, 1914 in Louisville, KY, received a BA degree from Transylvania University, Lexington, KY, June 1936. Attended University of Kentucky Graduate School, 1936-38. Received an MD degree from University of Louisville School of Medicine, June 1942.

Joined the service June 1941, while a medical student, ENS (temporary) USNR, to keep the draft from sending him to the Army.

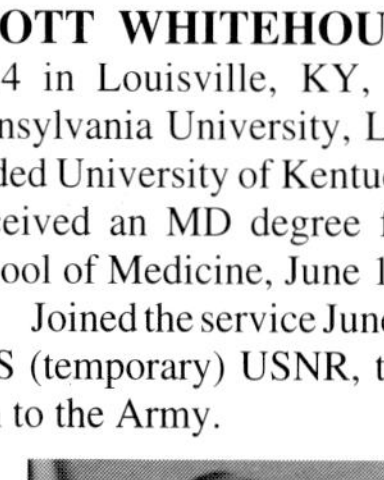 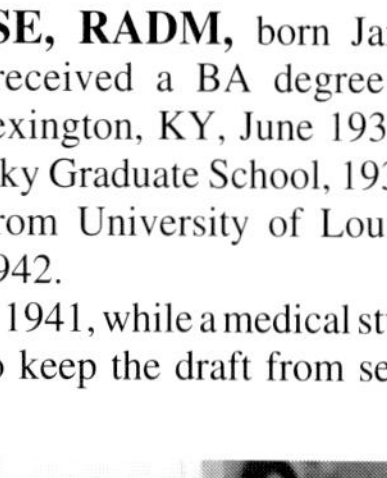

Reported for duty June 2, 1942, Norfolk Naval Hospital, Portsmouth, VA. June 1943, ordered to V-12 Unit, University of Richmond and Medical College of Medicine, Richmond, VA as medical officer. Ordered to pre-commissioning crew of USS *Boston* (DD-722), October 1943.

He had duty at Naval Hospitals: Oakland, CA; Long

Beach, CA; Mayo Foundation, Rochester, MN; St. Albans; Philadelphia (head anesthesiology section); Oakland.

Battles and campaigns: (on USS *Borton* (DD-722) (DesRon 60), Normandy Landings 1944. He then went to the Pacific and was screening the battleship *North Carolina* during the battle of the Philippine Sea. Participated in the troop landings of Ormoc Bay, Mindora Island, Lingayen Gulf, Iwo Jima and the Tokyo Bay carrier raid on Tokyo and Okinawa, and was part of the "screen" of the USS *Missouri* during the Japanese ceremonies.

He resigned his regular commission of CDR USN and joined the USNR as CDR. He then moved to Los Angeles and entered private practice with the McCuskey group of Anesthesiologists and was on the staffs of five major hospitals in Los Angeles. Selected for RADM, Medical Corps, USNR January 1970. Retired from USNR, Jan. 29, 1974 as RADM Medical Crops, USNR (upper half, retired). Retired from practice of medicine, Jan. 1, 1976.

He married Elizabeth Foley of Versailles, KY June 2, 1942 and they have three children: James (MD), Martha Jo (MD) and Elizabeth Jane (teacher); grandchildren: Gilliam, James and Martha

RALPH E. WHITEMAN, CAPT, born Feb. 12, 1928 in Monmouth, IL, received a BA from Monmouth College and MBA from Indiana University.

He enlisted August 1946 in the USN and Aviation Electronics, June 1952, commissioned. USNTC Great Lakes, NATTC Corpus Christi, NAS North Island, NSCS Bayonne, ESO Great Lakes.

Whiteman was awarded WWII Victory Medal, National Defense, Naval Reserve, American Theater. He achieved the rank of CAPT (03) and retired from the USNR February 1988.

His memorable experiences: served on board of directors, Navy Supply Corps Foundation; 40 years total service, Illinois coordinator (Blue & Gold) US Naval Academy.

He married Martha and has four children: Cynthia Ruszkowski, Laurie Seymour, Cheri and Robert Whiteman; grandchildren: Derek and Justin Ruszkowski, Kely and Eric Seymour. Employed Ppresident and CEO Security Savings Bank, retiring June 1992.

BROOKS L. WHITLOCK, LCDR, born June 3, 1939, Carl, WV. He earned BS degree in agriculture, West Virginia Univ. and master of agricultural management and resource development, Univ. of Florida.

Joined the service May 4, 1963 and was stationed at Pensacola, FL; Pearl Harbor; Washington, DC; Saigon; Vietnam (1966-67 and 1971-72); NAS Jacksonville, FL. Units served with include FICPAC, CVW-11, CVA-63, NRTSC, COMNAVFORV, CVWR-20, CINCPACFLT DET 1019.

Memorable experiences include debriefing pilots flying over north Vietnam; carrier qualification in a T-28 at VT-5 on USS Essex; standing on top of Ayres Rock in Australia while on R&R.

Awards and medals include NCM, NUC w/2 stars, VSM, VCM w/3 stars, VGM, MDSM w/star, Reserve Medal and Navy Pistol Medal. He achieved the rank of LCDR.

Married Karon and they have two children, Jonathan and Thomas. Civilian employment as county supervisor, FHA, USDA. He retired June 30, 1984.

JOHN B. WHITMORE JR., CAPT, born June 17, 1934 in Blacksburg, VA, received a BS, business administration, Virginia Tech. He joined the service Sept. 20, 1956, OCS Newport, RI. Commissioned as ENS USNR in March 1957.

His military locations and stations: USS *Hampden County* (LST-803); (1957), San Diego, CA, supply officer; DCA, JOOD; USS *Henrico* (APA-45), San Diego, CA (CONUS) and Yokosuka, Japan (WESTPAC), CO and officer of the deck (OOD), 1958-60; while aboard the ship conducted extensive operations in the Western Pacific as a participant in President Eisenhower's "People to People" Program.

During his tenure as an inactive USNR officer, he had the privilege of commanding five (5) individual USNR units, the most notable of which was Phib CB Two-Det 106. (2) In his last tour, he served as naval liaison officer to the Adj. Gen., VARNG while assigned to USNR Readiness Command Region Six, Washington, DC.

He was awarded Vietnam Service Medal, ARNG Commendation Medal, Armed Forces Reserve Medal, Pistol Expert Medal and Rifle Expert Medal.

His significant awards: Surface Warfare Officer Designation, Command Ashore Award, Region III MVP Award as a US Naval Academy Information (Blue and Gold) officer. He is a life member of NRA and past president of Southwest Virginia Chapter.

He achieved the rank of CAPT USNR and retired July 1, 1987.

ROBERT C. "BOB" WHITTEN, CDR, born Dec. 6, 1926 in Bristol, VA. Attended public schools in Baltimore County, MD 1933-44. Entered US Merchant Marine Academy 1944 and graduated May 1947 with license as third assistant engineer and commission as ENS, USNR. Sailed 1947 with American Export Lines as junior third assistant engineer.

Active duty: USN 1949-1953, serving as damage control assistant and liquid cargo officer in USS *Elokomin* (AO-55), main propulsion assistant and damage control assistant in USS *Ozbourn* (DD-846), and engineer officer in USS *Thomas F. Nickel* (DE-587). Qualified and stood watches as OOD in all three ships. Released from active duty in grade of lieutenant, August 1953. Most memorable service in USS *Ozbourn* on shore bombardment duty in Wonsan harbor, North Korea in January-March 1951 and June-July 1952. Division commander (then CAPT (later VADM) B.F. Roeder) in 1951 was the first "Mayor of Wonsan."

Inactive Reserve Service: After release from active duty in 1953, served in surface divisions in Buffalo, NY and Durham, NC. Later served as staff engineer of the Reserve DE Div. based in San Francisco Bay 1959-61. Attached to Research Reserve Co. based at Stanford University and Lockheed Missile and Space Co. from 1962 until retirement in 1971 in grade of commander. During the period 1960 to 1961 completed the Naval War College extension course in strategy and tactics.

Following release from active duty in 1953 completed work on BA in physics at SUNY Buffalo (New York). Enrolled as graduate student at Duke University in 1955, receiving PhD in theoretical nuclear physics in 1959.

Worked at Stanford Research Institute (now SRI International) as physicist and senior physicist (1959-67. Worked as research scientist at NASA-Ames Research Center 1967 (atmospheric science and space physics) to retirement in 1988. Published as author or co-author about 125 papers in scientific literature and authored, co-authored or edited six books during career. Co-investigator on Pioneer Venus spacecraft 1975-88.

He married Sally Marie Kriz, Aug. 2, 1953. He has two children, Robert Craig III and Lisa Marie (Marchese) and four grandchildren.

Civic and professional activities: assistant scoutmaster and scoutmaster, Boy Scouts of America (1970-72); president, Santa Clara Valley Council (1986), and later National Director (1988-present), Navy League of the United States; Adjutant-General, Scottish-American Military Society (1992-96); life member NRA; life member, ROA; president, USS *Ozbourn* (DD-846) Association (1993-present); Associate Fellow, American Institute of Aeronautics and Astronautics. Member of US Naval Institute since 1947.

Published commentaries and book reviews on military (mainly naval) topics (e.g. *Post Mortem,* a review of *The Generals' War: The Inside Story of the Conflict in the Gulf* by Michael Gordon and Bernard E. Traynor, Boston: Little Brown, 1994, published in Chronicles December 1995; review of *The Pentagon Wars: Reformers Challenge the Old Guard* by James G. Burton, Annapolis: Naval Institute Press, 1993, published in Social Science and Modern Society, May/June 1995; *Tailhook Part 2. Events at the Annual Tailhook Symposium,* assistant inspector general for investigations . . . , Washington, DC: US Government Printing Office, 1993, published in Journal of Political and Military Sociology, Summer 1996; comment on *Espionage or Negligence? A Sinking Mystery* (P. Huchthausen, *Naval History* (March/April 1996) published in Naval History, December 1996 deals with current Russian naval officers' opinion of ADM Gorshkov. A review of *Fall From Glory: The Men Who Sank The US Navy* by Gregory L. Vistica, New York: Simon and Schuster, 1995, in press in the *Journal of Political and Military Sociology.*

Participated in various conferences in Moscow on military reform of the armed forces of the Russian Federated Republic. Co-chairman of section 6 "Military and Social Work in Troops" at the 1994 *International Conference on the Social Dimensions of Military Reform* held in Moscow, May 23-25.

His campaign medals include the American Campaign, European Theater (both awarded for service as Midshipman, USNR in Atlantic convoys), WWII Victory Medal, Occupation Medal, China Service Medal, American Defense Medal, Korean Service Medal (four Battle Stars), UN Medal (Korea), Naval Reserve Medal, and Armed Forces Reserve Medal.

JOHN H. "WHITT" WHITTINGTON JR., CAPT, born Oct. 22, 1919 in Bloomington, IL, received a Dr. of Law from the University of Texas in 1942.

He joined the service Jan. 1, 1942. His military locations and stations: USS *Steele* (DE-8), SW Pacific, 1943-44-45. Served with the 2nd and 3rd Fleet. His battles and campaigns include the Pacific Theatre and Philippine Liberation.

Whittington achieved the rank of CAPT and discharged from inactive duty March 1946 and retired from the USNR Oct. 22, 1979. His memorable experiences: gunnery officer, Majuro, Guam, Saipan, Tinian invasions.

He married Margaret and they have four adult children and two grandchildren. Employed as justice, Court of Appeals in Dallas, TX.

JAMES P. "SCOTTY" WHYTE, CDR, born Aug. 25, 1921 in Columbus, MS, received an AB from Bucknell; MA from Syracuse University and a JD from the University of Colorado.

He joined the service June 16, 1943 (commissioned, Columbia Reserve Midshipmens School). Served with USS *Nucleus* (AM-268); USS *Steady* (AM-118). His battles and campaigns include the Atlantic, European and Asiatic-Theatres.

He achieved the rank of CDR and was retired April 1, 1967. His memorable experiences: mine sweeping, proposed invasion Japan; minesweeping, August-December 1945; Occupation Japan and Pacific areas.

He is married to Martha and they have three children: James, Ward and Scott; and grandchild, Jamie Melissa and Scott. Employed as professor and Dean of Law, W&M, practices law.

ROBERT E. "BOB" WICKMAN, CAPT, born March 21, 1933, received a BS from the University of California at Berkeley.

He joined the service June 1955. Active duty military locations and stations: Supply Corps School, Athens, GA; NAS North Island, CA. Reserve military locations and stations: various Supply Corps units, served as CO of POL 110 and NSC units.

He achieved the rank of CAPT. He is married to Allene and has children, William and Jenifer and grandchildren: Matthew, Lindsay, Sarah, Ryan and Will. Employed with Shell Oil Co., materials management and procurement management positions, worldwide.

LYNN H. WIDENER, CAPT, born in Bristol, VA and entered the US Naval Academy from Chattanooga, TN. He graduated with the class of 1970. He was the missile officer on USS *Mahan* (DLG-11), which deployed to the Western Pacific in 1973. As fire control officer on USS *Gridley* (CG-21), he was involved in the Vietnam evacuation and the Mayaguez incident in 1975. He then became an instructor in the Division Officer Course of SWOSCOLCOM in Newport, RI.

In 1977, he joined the USNR and completed his MBA from the University of Maryland in 1982. From 1978-87 he served in CONVOYCOM KEY WEST and NCSO TEXAS/SOUTH AMERICA at NRC Adelphi, MD. From 1987-89 he was in the SACLANT unit in Washington, DC until he became CO of CONVOYCOM SAN JUAN 106 in 1989.

He participated in the last live convoy exercise Rainbow Reef prior to Operation Desert Shield in 1990 at Diego Garcia. In 1991 he was selected for CAPT and CO of NR NCSO NORVA 306. After a tour as XO in the Washington, DC VTU, he became CO of NCSO Atlantic 206 in 1994. He became the XO of SACLANT DET 106 in 1996. In his civilian career, he is the deputy program manager for the Standard Missile contract of PEO (TAD) for LOGICON Syscon Corporation.

His memorable experiences: sea detail OOD for man overboard on April 1, 1974. He has been awarded the Meritorious Service Medal and two Navy Commendation Medals.

The Widener family consists of wife, Ann; and two sons, Christopher and Ryan. He enjoys personal investing, real estate restoration, and coin collecting.

CHESTER H. "CHET" WIDMANN, CDR, born Nov. 8, 1923 in Holding Twp., MN. Attended St. Cloud Tech High School. He received a BA from San Diego State University and post graduate of the University of Southern California. He joined the service Oct. 14, 1942. He served with VS-48, VT-5, VS-871, VS-773 and VS-775. His battles and campaigns include: Aleutian Islands Campaign. He was awarded the American Theater Medal, Aleutian Campaign and Marksman Medal.

He achieved the rank of CDR and retired Nov. 8, 1983 at the age of 60. His memorable experiences: survivor of mid-air collision, May 1944; fire in flight November 1945, fire in flight crash June 1960 and various close encounters in training and combat exercises.

Married Bernadine and they have seven children: Carmel, Dana, Jeff, Jamie, Beth, Steve and Leslie; and grandchildren: Shane, Amber, Brook, Brandon, Sean, Marina, Haley and Kelsey. Employed as president, Keenan Supply, CAP, California and Washington, retiring September 1986.

DELORIS JANE "LOIS or GORDIE" GORDON-WIEMERS, *See her bio in G Section.*

EUGENE H. "GENE" WIEMERS, ENS, born Oct. 22, 1955 in Alma, NE, spent 16 months V-12, Doane College and graduated from Hastings College in 1949.

He joined the service May 21, 1943. His military locations and stations: V-12, Doane College; PMS Asbury Park, NJ; MDM Columbia, NY; Adv. Line School, Miami, FL; Amphib. training, Camp Bradford, VA; USNH Norfolk, VA; USNH Sampson, NY. Wiemers was awarded the WWII Victory Medal. He achieved the rank of ENS, USNR, and was discharged Sept. 3, 1946. His memorable experiences: V-E Day, May 1945; V-J Day, Camp Bradford, VA; medical retirement, June 1, 1947.

Wiemers' wife, the former ENS Deloris J. Gordon, died July 14, 1996. They met as patients at USNH Sampson, NY. They were married by LTJG James O'Dea, Navy chaplain in St. Stephens Catholic Church, Geneva, NY. They celebrated their 50th anniversary June 14, 1996 and on July 14, 1996, her loving heart stopped while sitting in a chair at home. She is buried at Ft. McPherson National Cemetery, Maxwell, NE.

They had 11 children: Michael, Jeanne, Mary, Theresa, Joseph, Constance, Joanne, Annette, Christine and John and a son David died in infancy; grandchildren are Blaine Mathew, Melissa Theresa and Carolyn Kay Wiemers; Lynne Therese Van Volkenburg; Rebecca Jane and Stephanie Marie Green; Albert W. III and Daniel Eugene Kelly; Jerome Douglas Dolan; Aaron Neal, Jennifer Elaine and Kaitlin Jean Panowicz; Alexandra V. and Andrew John Gonzales; Michelle Marie Stumpf; Justin James and Forrest Gordon Eldred.

He was employed as parts manager, Kaiser-Frazer Dealer; salesman H.P. Lau (wholesale foods); insurance agency owner, January 1954. Retired August 1990.

JERRY A. WIENS, CAPT, born Jan. 22, 1941 in Newton, KS, received a MBA from the American University; BS from the University of Kansas.

He joined the service January 1964. His military locations and stations: Norfolk; Guantanamo Bay; Atlantic; Mediterranean; Washington, DC. Units served with: USS *DuPont;* Naval Scientific and Technical Intel. Center.

Wiens was awarded the National Defense and achieved the rank of CAPT and retired. His memorable experiences: sea duty.

He is married to Cynthia and has a daughter, Kristen. Self-employed and will retire w/pay Jan. 22, 2001.

WILIAM G. "BILL" WIEST, CAPT, born Feb. 5, 1917 in Cleburne, TX, received an AB, BM, MD from the University of Southern California.

He joined the service USN 1942, Intern Naval Hospital, San Diego 1943; 1946-47 OinC, Epidemiology Unit 11th N. District. His military locations and stations: Naval Hospital San Diego, Fleet Sonar School; Bead Bethlehem School. Served with USS *Mifflin* (APA-72). His battles and campaigns: Iwo Jima, 4th Marines, Okinawa 2nd Marines.

He was awarded the Purple Heart and Silver Star. He achieved the rank of CAPT and discharged from active duty July 1947.

His memorable experiences: battle of Iwo Jima 30 days 1945. USNR; Navy War College, RI; Medical Review Course, Bethesda, MD; Sr. Med. Officer Compton USNR and Los Alamitos Training Center.

His wife Patricia died July 22, 1996; he has children: Stephen C. and Kristine Wiest (deceased 1974); and grandchildren: Shane and Shalene Wiest. Employed as chief of staff and chairman, four Southern Californian hospitals. Retired Feb. 5, 1977.

JOHN P. "JACK" WILDER, LTSG, born Feb. 27, 1911 in South Pasadena, CA, received an AB in economics.

He joined the service April 1943. His military locations and stations: Quonset Point, USS *Prince William*. Served with Pacific and Atlantic. His battles and campaigns: transport, convoy, pilot training.

He achieved the rank of LTSG and was discharged December 1995. His memorable experiences: the miles covered at sea; as FDO in CIC the association with the ship and complement of aircraft. Also navigation SK & SG Radar.

He is a widower and has a son, Gary and grandchildren, Mark and Nicole. Employed as a life insurance underwriter.

ALBERT LEE WILEY JR., LCDR, born in Forest City, NC, June 9, 1936, and attended high school at Canton, NC, Phillips Academy, Andover, MA and Wilmington, NC. Although initially in the AFROTC, he joined the USNR in 1959. He received a bachelor of Nuclear Engineering (with honors) and did graduate stud-

ies in Nuclear Engineering on a US Atomic Energy Commission Fellowship at North Carolina State University. He worked for Lockheed Corp. as a nuclear engineer, but then entered the University of Rochester Medical School, receiving the MD degree in 1963. He interned in surgery and medicine at the University of Virginia Hospital and residency training at Stanford University Hospital, Palo Alto, CA and University of Wisconsin Hospitals, Madison, WI (later becoming board certified in radiation therapy, nuclear medicine and in Medical Health Physics.)

He was called to active duty in 1968, received nuclear and chemical defense training, and served as the medical director, safety officer and senior medical officer for a special radiation accident team at the US Naval Radiological Defense Laboratory, San Francisco. He was also periodic instructor on the medical management of radiation accidents at the Nuclear Weapons Training Center, Coronado, CA. In May 1969 he was ordered to Vienna, Austria, where he had some "Cold War" experience working with the US Embassy while representing the USN at International Atomic Energy Agency meetings on radiation accidents. Other memorable USNR experiences include active duties at the San Diego Naval Hospital, The San Francisco Hunter's Point Naval Shipyard Medical Dispensary, the Madison, Wisconsin, Navy-Marine Corps Reserve Unit and his last reserve active duty at the National Naval Medical Center, Bethesda, MD in November 1989.

He also had a PhD in Radiation Biology (major) and Nuclear Engineering (minor) from the University of Wisconsin-Madison and has served as assistant professor at the University of Texas, MD Anderson Cancer Center, Houston and as professor, associate director and clinical director of radiation oncology at the University of Wisconsin, Madison. While in Wisconsin he ran, unsuccessfully, as a candidate for US Congress (1982-1984) and as a candidate for Governor of Wisconsin (1986). In 1988 he became interim director of the East Carolina University Cancer Center, professor and chairman of the Department of Radiation Oncology and adjunct professor, Department of Physics. He was appointed by the US Secretary of Veterans Affairs as consultant to the US Department of Veterans Affairs and has served as consultant to the US Nuclear Regulatory Commission's Committee on Reactor Safeguards, the National Cancer Institute and was appointed by Gov. Thompson of Wisconsin to the Wisconsin Radioactive Waste Board and to the Governor's Council on Biotechnology.

He became a member of the USN Retired Reserve in 1989, and retired for almost 30 years of teaching/research in academic medicine in 1993, with over 150 articles, book chapters and abstracts on cancer causes and treatment, nuclear medicine, radiation accidents, and medical physics. He is now a Fellow of the American College of Radiology and a consultant in radiation medicine at the Watson Clinic, Lakeland, FL where he and his wife, Janet Lee, especially enjoy visits from their four daughters and four grandchildren.

KENNETH A. "KEN" WILLCOX, CAPT, born Sept. 12, 1945 in New York City, NY, received a BA from William College, Williamstown, MA and an MBA from the University of Minnesota, Minneapolis.

He joined the service June 24, 1968. His military locations and stations: Navy Supply Corps School, Athens, GA; OCS, Newport, RI.

Willcox served with USS *Carronade* (LFR-1) (homeported Yokosuka, Japan); USS *Shreveport* (LPD-12); NSD Subic Bay Reserve Units. His battles and campaigns: Vietnam.

He was awarded the Navy Commendation, Vietnam Service, Vietnam Campaign, Meritorious Unit Command, National Defense (2), Overseas Service, Armed Forces Reserve. Achieved the rank of CAPT 06, Supply Corps, USNR.

His memorable experiences include: gunfire support on inshore fire support rocket ship *Carronade* in Vietnam 1969-70; Nucleus crew USS *Shreveport* (LPD-12), built and commissioned in Seattle. CO of three NRA Subic Bay, NDS, NSC units in Minneapolis, MN; Fargo, ND; and Des Moines, IA. Chief of Logistics Readiness Team, CINCPAC.

He is married to Winnie and is executive vice-president of Delta-Waseca, Inc. and Tartan Transportation Systems, Inc.

DAVID GENE WILLE, CAPT, born Oct. 20, 1931 in Baltimore, MD, received a BA from Penn State and a MBA from American University.

He was commissioned an ENS, USNR, Jan. 28, 1954. His military locations and stations: Naval Security Station, Washington, DC. Served with Naval Security Group, National Security Agency.

Wille was awarded the Meritorious Service Medal and National Defense Service Medal and Armed Forces Reserve Medal. He achieved the rank of CAPT, USNR (naval cryptologic officer) and was discharged July 1987.

His memorable experiences: assistant commander for Reserve Plans and Readiness at the Naval Security Group Command on active duty July 1983-July 1987. Also participation in REDEX 2-84 aboard USS *Eisenhower,* USS *Mississippi* and USS *Mount Whitney.*

He is married to Joan Wille and they have three children: Robert, James and Cynthia; and grandchildren: Brittany, Kyle, Chad and Brooke. Employed at National Security Agency and in various management positions.

ARTHUR CLYDE WILLHELM JR., LCDR, born Oct. 10, 1921 in Galveston, TX, received a BS from Case Institute of Technology and spent four months as Midshipman at Annapolis (V-12).

He joined the service June 1942 (V-12). His military locations and stations: USS *Wolverine,* May 1944-November 1945; USS *Randolph,* December 1945-February 1946.

His awards and medals include the WWII and American Theater Medals. He achieved the rank of LCDR and was discharged February 1946. His memorable experiences: One millionth landing on USS *Wolverine.*

He has been married 53+ years to Mary Louise Coburn and they have three children: Linda Jean (Sparks), Bryan Clyde, Beverly Ann (Cromer); and eight grandchildren: Jeffrey Willhelm, Scott Hackstadt, Christopher Willhelm, Gary Chapin Jr., Amy (Chapin) Cantrell, Marc Willhelm, Andrew Cromer, Adam Cromer; and eight great-grandchildren. Employed as laboratory, plant and research metallurgist, then as engineer with US Pipe and Foundry Co. He retired in 1989.

GEORGE M. WILLIAMS, LCDR, born Aug. 24, 1933 in Moorfield, WV, received a BS from Hampden Sydney College, 1954 and PhD from West Virginia University, 1967.

He joined the service September 1957. His military locations and stations: Miramar NAS, CVA-38, CVA-19 and served with VA-113 Air Group 11. He participated in the Cold War and received the Good Conduct Medal for Vietnam.

Williams achieved the rank of LCDR and was discharged in 1961 and retired from USNR-R in 1977. His memorable experiences include the WestPac cruises, 1958-61.

He is married to Kay and they have three children: Sarah, Barbara and James. Employed at Hercules Inc. as hazards specialist, retiring in 1994.

JOHN C. "BIG JACK" WILLIAMS, CAPT, born June 17, 1928 in Denver, CO, received a BS in education from Notre Dame University (2) Regis College (2).

He enlisted in the service Oct. 10, 1950 (USS *Cardinal* MSC(O), CO; CO, USS *Phoebe* (MSC-199), Korea. Reserve stations: CO, Fleet Div. 5-31, Louisville, KY; CO, Surface Div. 5-30, Louisville, KY; staff, Red Com 4, Columbus, OH; Staff, Inspector General, Red Com Six, Memphis, TX; staff, Navy Liaison, Adj. General, Kentucky; Staff, Naval Reserve Center, Lexington, KY.

He participated in battles in Korea, Wonson Harbor, evac-1st Marine Div.; invasion Seoul, NK; Minesweeping OPS: Korea, Philippines, South China Sea, Republic of Vietnam, Quemoy Island, Pescadorees Island, China.

He was awarded the Vietnam Service Medal, Good Conduct Medal, Reserve Medal, Presidential Unit Citation, Korea, Congressional Commendation 1974, Reserve Operations State of Kentucky, Kentucky Meritorious Service Medal, Korean Service, United Nations, China Service, and Japan Operational Medal. Qualified Command at Sea, Surface Warfare, NBC Specialist. Graduate of US Naval War College (three summers).

His memorable experiences: minesweeping for Quemoy Island; landings; emergency evacuation; US Marine, 1st Div., Wonson, RNK; obtaining first command at sea after enlisted October 1950, E-1.

He married Beverly and they have a daughter Missy Schmidt and granddaughters Maggie and Emily Schmidt. Employed as sales manager, Alemite Co. and presently in ministry. He retired in June 1988.

MARK S. WILLIAMS, CDR born Feb. 13, 1956 in Olean, NY, received a BA from Alfred University in 1978 and a JD degree from the University of Richmond in 1981.

He joined the service June 1980. His military locations and stations: NLSO, Charleston, SC, 1981-83; NAVSUPFAC, Diego Garcia 1984; OJAG, Alexandria, VA, 1984-86.

He was awarded the Navy Achievement Medal, National Defense Medal and the Armed Forces Medal.

Williams achieved the rank of CDR. His memorable experiences: stationed at Diego Garcia, the foot print of freedom.

Married Patti and they have three children: Sarah, Jack and Kate. Employed as attorney and partner with Sharkey and Williams. Also a county legislator in New York.

RALPH AUGUSTINE "PINE APPLE" WILLIAMS, WO-4, born Feb. 12, 1909 in Richmond, VA. He joined the service May 13, 1943, USNR CB 5-1, 98 Bn. His military locations and stations: Pearl Harbor, Tarawa, Samar. Served with CB Div. 5-1 and Civil Engineer Corp.

Participated in the Tarawa campaign. He was awarded the WWII Victory Medal, Philippine Defense Medal, Philippine Liberation Medal and the American Theater Medal. He achieved the rank of 1-c petty, chief and WO-4 and retired Feb. 12, 1996.

His memorable experiences: they had orders to help the Marines 1943 on Tarawa, 2,000 Marines were marched down in front of them, before they could go ashore.

His wife Wilda E. Williams died Dec. 21, 1995. Employed as master plumber, design and inspection of plumbing, heating, air conditioning and electrical.

ROBERT E. "BOB" WILLIAMS, LT, born Nov. 24, 1919 in Logansport, IN, received his masters from IU.

He joined the service in 1940. His military locations and stations: Chicago, Pensacola, Memphis, TN. Served with the USNR and achieved the rank of LT and was discharged in 1945.

He is married to Marilyn and has children: Robert and Claudia; stepchildren: Glenna, Susan and Timmy; and grandchildren: Tommy, Stephen and Michael.

ROBERT W. "BOB" WILLIAMS, LCDR, born July 25, 1911 in Minneapolis, MN, received a BA from Carleton College 1933 and three years of graduate study in Psychology, University of Minnesota.

He joined the USN Aug. 28, 1943 as LTJG. Went to Ft. Schuyler, NY for basic training, then assigned (late 1943) as placement officer to Northwestern University, Midshipman's School and (in 1945) to Navy HQ, San Diego, CA. Rebased to inactive duty April 25, 1946; returned to DC; joined a Navy Reserve unit; retired in 1971.

He was awarded the American Defense Medal, Armed Forces Reserve, WWII Freedom, US Naval Reserve, and Military Order of Stars and Bars and other. He achieved the rank of LCDR in 1971. His memorable experiences: the excitement of VE and V-J days and handling the flood of returning Naval enlistees and officers, all eager to be home.

Married until his death on Jan. 15, 1997 at Woodbine Health Care Center, Alexandria, VA to Dorothy. They had two children, Susan Joyce Cluff and Sharon E. Williams; and grandchildren, Michael Sean Cluff and Kristin Michelle Shafer. Employed as personnel management specialist, Social Security Board, 1938-40; the Defense Department, 1940-43 and 1946-51; Office of the Secretary, HEW, 1951-71.

JOSEPH E. "JOE" WILLIAMSON, CAPT, born July 12, 1945 in Rochester, PA, received a BA from Wittenberg University and a MA from the University of Virginia; JD (Wayne State University).

He joined the service Dec. 19, 1968. His military locations and stations: Active, USS *Leary* (DD-879) Norva; Fleet Aviation Accounting Off. NAS NORVA; US Forces Japan; AIRPAC; NRMCB26; NRCC Long Beach; FISC San Diego; SUPSHIP.

Williamson was awarded the Navy Commendation Medal, National Defense, Armed Forces Reserve (w/Hourglass), Pistol Expert. He achieved the rank of CAPT, Supply Corps, USNR.

His memorable experiences: qualifying for "NASO" Wings on the *Kitty Hawk;* attending war college (twice); sea trials while in SUPSHIP C Bath Iron Wings and Stergeon Bay WI); exercises in Japan-Ulchi Focus Lens and Keen Sword.

He is married to Marsha and they have two children, Matthew and Stephen. He is employed at national logistics manager, Toyota USA, still in Drilling Reserve Status.

THEODORE M. "TED" WILLIAMSON, CDR, born May 8, 1957 in Ukiah, CA, received a BA in English Literature in 1979 from Lewis and Clark College, Portland, OR.

He joined the service Nov. 1, 1979, active and July 1, 1985 SEL RES. His military locations and stations: USS *Dalhgren* (DDG-43), active; USS *Leyte Gulf* (CG-55), Reserves, NCSO, Reserves.

He was awarded the Navy Achievement Award and two Sea Service Ribbons. Achieved the rank of CDR.

Serving as GCLO of Leyte Gulf unit was a rewarding experience (3 years).

He married Katherine and they have three children: Jennifer, Rebecca and Kimberly. He is employed as mail handler with the US Postal Service.

LESLIE A. WILLIG, CAPT, born and raised in Fort Wayne, IN, entered the USN as an apprentice seaman in February 1944. His sea duty was mostly in the Pacific on aircraft carriers on which he served, in operations, as part of Ship's Company. He was awarded several campaign medals, including the Korean Theater w/two Stars.

Willig became a member of NRA soon after it was founded and has served as chapter president, chairman of the Resolutions Committee, vice president for Naval Affairs (subsequently called executive vice president), and from 1971-73, as the National President.

He received a BS and BSNT degrees from Purdue University, an MA degree from the University of Louisville and a PhD from the University of Iowa. He worked on the administrative and teaching staffs of both Purdue and the University of Iowa, and for ten years, was the executive vice president of Tri-State University located in Angola, IN.

Since 1974, he has been chairman of the board, CEO and president of Photo Control Corporation, a publicly-held manufacturer of professional photographic equipment in Minneapolis, MN. His home is in Fremont, IN.

EDMUND P. WILLIS, CAPT, graduated from the University of Wisconsin, Madison, NROTC program in 1955. Ordered to duty aboard *Samuel B. Roberts* (DD-823), he served as damage control assistant and qualified as officer-of-the-deck underway. *Roberts* cruises in the Mediterranean as plane guard in 1956, and transited the Suez Canal, taking station with the Mid East Force in the Persian Gulf.

Willis returned to CONUS with orders to Rochester (CA-124) in Long Beach. As a gunnery division officer and OOD he sailed with *Rochester* in 1957 to WESTPAC where the ship served as Seventh Fleet flagship, making port calls in Sydney, Yokosuka, Hong Kong, Manila and Saigon.

Having returned to Long Beach in 1958, Willis took leave of *Rochester* as a Reserve LTJG and entered the graduate program at the University of California, Berkeley. He served in the inactive reserve aboard *Alvin C. Cockrell* (DE-366) stationed at Alameda, and was recalled to active duty in 1961, during the Berlin Crisis. *Cockrell* took station early in 1962 with the interdicting patrol off Vietnam. For six weeks, during that period, LT Willis served as naval liaison officer in DaNang, Vietnam.

When *Cockrell* returned to inactive status, in 1962, Willis returned to graduate studies at Berkeley and took the doctorate in history in 1967. As assistant professor of history he taught at Moravian College beginning in 1964. He went to York College, City University of New York in 1967. Promoted to associate professor, he taught history and served as department chair of a program for disadvantaged students.

In 1978 Willis turned to business, establishing Kinsale Mangers, Inc., in Oyster Bay, NY, a recruiting firm working primarily with Manhatten financial firms. He headed the firm until 1993 when he moved to the Washington, DC area to take a position as historian and political analyst with the Joint Warfare Analysis Center in Dahlgren, VA. He left JWAC in 1996.

While in New York, Willis served in the Reserve afloat program where he held various shipboard billets up through prospective commanding officer of *Moale* (DD-693). From there he moved to chief staff officer of Reserve Destroyer Div. 3ND. He next served as program coordinator with Readiness Commander Two. Upon promotion to CAPT he transferred into the Naval Control of Shipping Program as CO of NCSO NY 102, retiring from the inactive reserves in 1983.

CAPT Willis has two daughters and lives in Alexandria, VA.

ODETTE P. WILLIS, CAPT, born and raised in Nurnberg, Germany, attending nursing school at Texas Woman's University, Denton, TX and was commissioned as an ENS in November 1972. Duty stations included Naval Hospitals, Philadelphia, PA; Cherry Point, NC; Guam, where she cared for Vietnamese "Boat People" during "Operation New Life", and Oakland where she was recalled during ODS/S. Released from active duty in 1978, she received a MS in nursing from UCLA in 1981.

Naval Reserve duties included training, administrative, executive, commanding officer positions, REDCOM Staff, COMNAVSURFRESFOR ESC and National Policy Board membership. She also served as the assistant to the director of the Navy Nurse Corps (Reserve Matters) for three years.

CAPT Willis is a nurse educator and medical information systems analyst. She has an MBA in Health Services Administration from Golden Gate University.

Awards include the Meritorious Service Medal, Navy Commendation Medal w/Gold Star, Navy Achievement Medal, Meritorious Unit Commendation Ribbon, National Defense Service Medal w/Bronze Star, the Humanitarian Service Medal and Armed Forces Reserve Medal w/Silver HG and M Device.

WILLIAM H.R. WILLKENS, LT, born at New Britain, CT, Aug. 5, 1919 and received the AB, MEd and PhD degrees at the University of Pittsburgh and the BD degree at Colgate-Rochester Divinity School.

Following college graduation he was a German translator in the US Office of Censorship early in WWII until he enlisted in the USN July 1942. He served as a hospital corpsman until June 1943 when he was commissioned an ENS line officer. He served as XO and OinC on small anti-submarine patrol vessels and as communications officer and XO on small tankers.

In November 1945 he was released to inactive duty and in 1955 he was retired with the rank of LT USNR. His post-war career included service as an American Baptist pastor in New York and Pennsylvania followed by over 25 years in higher education as professor at Alderson-Broaddus College, Crozer Theological Seminary and Slippery Rock University of Pennsylvania where he was named professor emeritus in 1983.

In 1944 he was married to Rose Mary Gardner at Pittsburgh, PA where she was a nursing instructor. The Willkens have two sons, David, a former Navy ET on the USS *John F. Kennedy* and Daniel, a former Marine officer, now with the Department of Defense. In retirement the Willkens reside in Washington, PA.

JACK H. WILSON M.D., CAPT, born Sept. 5, 1928, Colorado Springs, CO, earned his BA degree at the University of Oklahoma and his MD degree at George Washington University.

He joined the OKNG April 10, 1947. His military locations and stations: Tripler, Hunter AFB, Walter Reed and Davis Monthan AFB. He served with the OKNG, USAF and USN.

Wilson's awards and medals include Flight Surgeon and Distinguished Military Graduate. He achieved the rank of CAPT USNR (MC) and retired Oct. 1, 1979.

He and his wife, Pat, have nine children: Kathy, Jim, Susan, Tom, Sharon, Stephanie, Laura, Mary and John; and nine grandchildren: Evan, Gary, Kyle, Beckie, Zachary, Ian, Sallie, Mitchell and Keats. Civilian employment as president of Phoenix Gynecologists LTD, Universal Laboratory, Inc.

JOHN A. "JACK" WILSON, LCDR, born Aug. 13, 1928, Parkersburg, WV, earned his BS degree at West Virginia University, his MDIV at San Francisco Theological Seminary, and his MA degree at San Jose State University.

He joined the USMC in May 1946 and the USN Chaplain Corps in July 1950. His military locations and stations: USS *Toledo*, 1st Mar. Div., Quantico, and Camp Pendleton while serving with the USMC; and Coastal Riv. Div. XI, 12th Destroyer Gp., and 2nd Mar. Div. while serving with the USN.

He was awarded the WWII Victory Medal, Navy Occupation Service Medal, Korean Medal (both), National Defense Service Medal and Reserve Medal.

His memorable experiences: Participating in prisoner swap following truce; North Korean Armistice at Pan Mon Von; commissioned USS *Toledo* took her on world cruise.

He married the former Arlene Kaiser and they have three children: John, Jesse and Melanie; and two grandchildren, John and Jesse Jr. Civilian employment as pastor of Newark Presbyterian Church and high school teacher for 25 years. Wilson retired in June 1979.

RAY L. WILSON, CWO, born Aug. 7, 1912, Phoenix, AZ. He was a graduate of Phoenix College and attended the University of Southern California. He is a registered pharmacist in Arizona, Nevada and New Mexico. His civilian employment was in various retail drug stores and in several Veteran Administration hospitals as director of the pharmacy service.

In 1940 he enlisted in the USNR as a chief pharmacists mate. He was ordered to active duty in March 1941 and his first station was at the Pearl Harbor Navy Dispensary. He was there at the onset of the Japanese attack Dec. 7, 1941. He was ordered aboard the USS *Platte*, a fleet oiler. His ship was very active through the Coral Sea Battle, the occupation of Guadalcanal, the Midway Battle and occupation of the Attu in the Aleutian Islands. The *Platte* received 11 Battle Stars for WWII service, six Battle Stars for the Korean War and five Battle Stars for Vietnam Service. He was privileged to be part of it.

His campaign and service awards are: Good Conduct Medal, American Defense Service Medal w/Base Clasp, Asiatic-Pacific Campaign Medal w/Bronze Star, American Campaign Medal, WWII Victory Medal, Armed Forces Reserve Medal, and Pearl Harbor Commemorative Medal. Wilson had several promotions while in the naval service and received retirement Aug. 7, 1972.

Other duty stations were San Diego and Great Lakes Training Stations. He was sanitation officer of San Bruno Advanced Base Personnel Depot and at the end of WWII he was the personnel officer of Pearl Harbor Hospital, Hawaii.

His wife, Bobbie, died in 1992, and his family now consists of daughter, Mary Lee; five grandchildren: Mary Jane, Robin, Michelle, Montiel and Rachel; and six great-grandchildren.

MICHAEL C. WIMPFHEIMER, CDR, born July 9, 1944, New York City, NY, earned his BA degree in 1964 at Columbia College and his JD degree in 1967 at Harvard Law School.

He joined the service in January 1968. His military locations and stations: OIS. He served with the Navy Justice School.

Wimpfheimer achieved the rank of CDR and was discharged in July 1970 and retired from the USNR-R in 1992.

He married the former Susanne Rabner and they have three children: Jan Steven, Barry Scott and Luba Rachel; and four grandchildren: Yair Etan, Ariel Dan, Avigayil Leah and Noam Shai. Civilian employment as an attorney, partner with Wimpfheimer & Wimpfheimer.

CARL D. WINDSOR, LCDR, born Oct. 2, 1942, Peoria, IL, earned his AA, BA, MA and PhD degrees at Ohio State in communications.

He joined the service Feb. 15, 1961, and was commissioned ENG in the USNR in January 1971. His military locations and stations: VR-6; VR-3 (NATWA, McGuire AFB, NJ); Cadillac/Lansing, MI; Fayetteville/Ft. Smith, AR; Columbus, OH; Roanoke/Norfolk/Arlington, VA (PAO/XO).

Windsor was awarded the Naval Reserve Medal, Defense Service Ribbon and Good Conduct Medal (3). He achieved the rank of LCDR and retired June 30, 1988.

His memorable experiences: PAO pre-com unit USS *JFK*, DoD presidential inaugural committee in January

1985. Acdutra in Bermuda; Roosevelt Roads; NTC San Diego; NTC Great Lakes; NAS Jacksonville, FL; NAS Pensacola, FL; and PAO Dallas, TX.

He and his wife, Beverly, have three children: Trent, Todd and Heather. Civilian employment with various radio/TV stations; professor of mass communications, John Brown University 1971-77; and president, Liberty University 1979-present.

RAYMOND L. "RAY" WINDT, CAPT, born May 21, 1919, Dallas, TX, graduated from SMU, Dallas in 1941 with a BS degree in business.

He joined the service in May 1942 and served until December 1945 and again from July 1950-December 1953. His military locations and stations: SCTC, Miami; ComServPac, Hawaii; FASRON 691, San Diego; and pre-flight, Pensacola. He served with VA-702, USS *Boxer*.

Windt achieved the rank of CAPT USNR and retired in 1979. His memorable experiences: Working with Brazilian naval mission in Miami; also while working in roll-up section of Advance Base Command, ComServPac, Hawaii.

He and his wife, Charline C., have three children: Dinah, Raymond L. Jr. "Ray" and Catherine Anne "Cathy"; and seven grandchildren: Jason R., Vicki, Kim, Nannette, Jason C., Joshua, and Jared. Civilian employment as sales representative, National Starch & Chem. from which he retired July 31, 1981; and owner/food broker, Southwest Food Service Assoc. from which he retired in 1989.

ROBERT DAVIS "DAVE" WINESETT JR., CAPT, born Dec. 14, 1941, Galax, VA, graduated from Virginia Polytechnic Institute with a BSA degree in 1965 and a master's degree in Urban and Regional Planning in 1966. He was commissioned ENS in January 1967 upon completion of Naval Officer Candidate School, Newport, RI.

After attending Civil Engineer Corps Officer School, Port Hueneme, CA, he was assigned to Chesapeake Div., Naval Facilities Engr. Command. He served as assistant resident officer in charge of construction at the Naval Ship Research and Development Center, Carderock, MD, and in the Planning Div. at CHESDIV in the Washington Navy Yard. He was released from active duty in October 1969.

His NR duty has included a variety of assignments in the Reserve Naval Construction Force, Defense Logistics Agency Depot support units, and Naval Transportation Office cargo terminal units. CAPT Winesett served in several positions on the staff of the NINTH Naval Construction Regt., including that of chief staff officer.

While serving as CO, NR officer in charge of construction, Middle Pacific, CAPT Winesett participated in mobilization studies in Okinawa and at Pearl Harbor, HI. He was then assigned as the administrative/personnel officer on the staff of the commander, Reserve Naval Construction Force. Having served for two years as CO, Naval Reserve Construction Bn. Center - Gulfport, he assumed command of Naval Reserve Engr. Field Div. South Aug. 1, 1991. CAPT Winesett has been awarded the Navy Commendation Medal, Meritorious Unit Commendation and National Defense Medal.

His memorable experiences: Training assignments in Okinawa and Hawaii; command of reserve units in the Reserve Naval Construction Force.

He is divorced and has two children, Jeffrey Paul and Melinda Evans. In his civilian career, CAPT Winesett has been a professional consultant in architecture and planning for over 27 years. He is registered as an architect in 11 states and holds certification from the American Institute of Certified Planners. He is a principal with The Onyx Group in the Washington, D.C. area. Active in a number of civic and professional organizations, he is a past president of the Oklahoma City Post, Society of American Military Engineers.

JOHN K. WINKLER, LCDR, entered the USNA in 1981 from the state of Missouri and graduated in 1985 with a BS degree in mechanical engineering. Upon selection of naval aviation, LCDR Winkler completed Flight School in Corpus Christi, TX and was designated a naval aviator Nov. 6, 1986. He completed E-2C Hawkeye training at NAS Miramar and was assigned to VAW-111 as the schedules officer.

After the decommissioning of VAW-111, LCDR Winkler reported to VAW-112 where he served for three years as an E-2C aircraft commander, ASO, NATOPS officer, and assistant operations officer. In December 1990 he transferred to VT-3 at NAS Whiting Field as a primary and intermediate flight instructor in the T-34C aircraft. During this period, LCDR Winkler served as the squadron ASO and schedules officer as well as a standardization check pilot.

Upon leaving active duty in 1994 he reported to NR DCMC/DPRO 0179 McDonnell Douglas where he served as the government flight representative and ASO. In January 1996, LCDR Winkler reported aboard VAW-77 at NAS Atlanta where he served as the assistant maintenance officer and is currently the squadron safety officer. LCDR Winkler has over 3700 accident free military flight hours and 194 carrier landings. In his civilian career he flies the Boeing 737 for Southwest Airlines.

MICHAEL W. "MIKE" WINKLER, CDR, born Oct. 30, 1946, Lakewood, OH, earned his BBA degree in management information systems at the University of Houston in 1976.

He joined the service June 24, 1968. His military locations and stations: Imperial Beach, CA; Tonkin Gulf; Beeville, TX; and New Orleans, LA (USNR). He served with HC-1 deployed in USS *Bon Homme Richard*; VT-24; and Reserve Intelligence units. He served in the Vietnam War in 1970.

Winkler was awarded the Naval Reserve Assoc. Junior Officer of the Year (SELRES) in 1985, Reserve Intelligence Area Seven, Junior Officer of the Year in 1985, Helicopter Aircrew Wings, Air Medal, Navy Achievement Medal, Vietnam Service Medal and various other medals and ribbons.

His memorable experiences: 1) Helicopter crewman, rescued downed A-4 pilot, Tonkin Gulf, 1970; 2) received USNR direct commission, Special Duty/Intelligence, 1970; 3) received credentials as Naval Criminal Investigative Service agent, 1986; and 4) general chairman, Naval Reserve Assoc. National Conference, 1987.

He married the former Shirley Heath and they have two children, Joy and Heath. Civilian employment with Brown & Root, Inc. as proposals specialist.

MARK J. WINTER, LT, born Feb. 9, 1966, Eglin AFB, FL, earned his BS degree in 1988 at Maine Maritime Academy and is expecting to earn his JD degree at the University of Maine in 1998.

He joined the service April 16, 1988. His military locations and stations: USS *Wabash* (AOR-5), Long Beach, CA (1988-92). He served with Mobile Inshore Undersea Warfare Unit 105 (1992-94) and COMNAVIUWGRU TWO (1994-).

He married the former Meredeth McCredie. Present civilian position as a law student.

PHILIP H. WINTER, CAPT, born Nov. 3, 1916, Chevy Chase, MD, graduated from Wilson Teachers College, Washington, D.C. in 1938.

In May 1943 received an ENS commission in the USNR. His military locations and stations: Navy Yard, New York, NTS(I), Ft. Schuyler, Bronx, NY, USS *Langley* (CVL-27), USNTC Camp Peary, VA, USNTC Bainbridge, MD. Battles and campaigns participated in were Philippine Sea, Leyte Gulf, Lingayen Gulf, Iwo Jima and Okinawa.

In 1949 he requested and received orders to the Naval Research Reserve Co. W-1, Washington, D.C. from which he retired in June 1973 with the rank of CAPT.

Following his release to inactive duty he was employed by the National Bureau of Standards as a physicist. Following the reorganization of the Bureau he was assigned to work in the Army's Mine Warfare program. In 1966 he was selected for assignment to the Defense Special Projects Gp. as a project officer. On termination of the Special Projects Gp. he retired from Federal Service after 31 years.

Winter was awarded the Pacific Theater Ribbon w/ 5 stars, American Theater Ribbon, Victory Medal, Navy Unit Commendation Ribbon, Philippine Liberation Ribbon w/2 stars, Naval Reserve Medal, and Armed Forces Reserve Medal.

In 1943 he married the former Catherine Groves of Charleston, SC and they had seven children: Catherine, Margaret, Philip E., Allen, Shirley Ann, Michjael and Warren; and 11 grandchildren: Katrina, Caroline, Davey, Jesse, Christine, Stephen, Laura, Tommy, Kate, Sara and Warren. This marriage ended in divorce in 1958. In 1961 he met and married Anne Dostert of Washington, D.C. and on retirement in 1972 they moved to Brookline, NH and involved themselves in pre-hospital medicine and volunteer ambulance work as nationally registered Emergency Medical Technicians until Anne died of cancer in 1988. CAPT Winter still resides in Brookline, NH.

DARES E. "DAY" WIRT, CAPT, attended Holyoke, MA public schools and earned his BS degree in IM at Georgia Institute of Technology.

He joined the service Dec. 13, 1942. His military locations and stations: NAS Ft. Lauderdale, NAS Chincoteague and NAS Atlanta. He served with VU-4, VT/VA-57 and VA/VF-672.

He achieved the rank of CAPT and was discharged Dec. 31, 1973.

His memorable experiences: VF-672 was awarded the Noel Davis Trophy as the best reserve fighter squadron during his tour as CO.

He and his wife Esther "Jo" have four children: Charles R., Peter E., Sarah Gale and Stephen D. Civilian employment with Lockheed GA, CO; administrative supervisor, customer service. He retired March 31, 1990.

IRENE N. WIRTSCHAFTER, CAPT, born in Elgin, IL, earned her BCS degree (accounting) at Columbus University.

She was commissioned ENS in 1944, USN (Supply Corps). Sea duty in 1956; CO, Supply Unit, 1973-75; honorary chief, USN, 1973; appointed by the Secretary of the Navy to the Advisory Committee for Retired Personnel, 1984-87; National Trustee, Association of Naval Aviation, 1984-date.

She is founder and comptroller, ANA Banana River Sqdn., 1985-date; appointed to Veterans Administration Advisory Committee on Women Veterans, 1987-89; Navy Liaison Officer, Commander's Retirees Council, Patrick AFB, 1985-88; national treasurer, NRA, 1975-77; National Advisory Committee, Naval Reserve Assoc., 1978-date; ROA, past chapter officer; TROA, Florida Conference Committee, 1987; Naval Order, past D.C. Commandry officer; Waves National, board member Space Coast Chapter, 1990; guest of honor - 1124th Recruit Revue, Orlando NTS; and Tailhook Assoc. member.

Over the years she has been a member of, and served on, numerous boards and civic committees. She was married to Burton Wirtschafter (deceased).

HAROLD R. "BOB" WISE, JR., CAPT., born July 20, 1942, Portsmouth, VA. He received his BS degree in education from West Chester University, West Chester, PA in May 1964 and is currently an MBA candidate at Kennesaw State University, Kennesaw, GA. He was commissioned an ENS in the USNR upon graduation from Officer Candidate School, Newport, RI in February 1965. After completing studies at Naval Damage Control Training Center, Philadelphia, he was assigned to USS *Intrepid* (CV-11) where he served as engineering watch officer and damage control assistant. He made deployments in the Atlantic, Caribbean and Western Pacific including duty in Vietnam waters.

In May 1967 CAPT Wise reported for duty at Naval Damage Control Training Center, Philadelphia as an instructor. He returned to the inactive Naval Reserve in May 1969.

As a drilling naval reservist, CAPT Wise served in various positions including CO, Auxiliary Floating Drydock Medium 104; assistant programs officer, Naval Reserve Readiness Command Region Four; plans officer, CINCEASTLANT 108; communications officer, SACLANT Det 205; director of training, Naval Reserve Center, Buffalo, NY; management information officer and retention officer, Naval Reserve Readiness Command Region Eight; CO, Naval Reserve Voluntary Training Unit 0801; assistant inspector general and assistant director of Plans for Readiness, Naval Reserve Readiness Command Region Eight; CO, Naval Reserve Readiness Unit C, Atlanta, GA; director of SELRES training, Naval Reserve Readiness Center, Atlanta, GA; and explosives safety inspector, Naval Ordnance Center, Naval Sea Systems Command, Washington, D.C.

CAPT Wise attended the Naval War College, served as a member of the Secretary of the Navy's National Naval Reserve Policy Board, completed the Reserve Components National Security Course presented by the National Defense University and served on several Naval Reserve officer line and staff selection boards.

CAPT Wise's civilian occupation is director, Credit Marketing Services for Equifax Inc., the leading provider of consumer and business information services in North America.

CAPT Wise is a life member and past national president of the National Naval Officers Assoc.; life member and past national vice president of the NRA; past national director for the Naval Sea Cadet Corps of the Navy League of the US; and served as the military appointee by the governor, state of Georgia to the Martin Luther King Jr. State Holiday Commission. His other affiliations include: Reserve Officers Assoc.; Naval Order of the US; US Naval Institute; Naval Enlisted Reserve Assoc. and life member Navy League of the US. He is also a performing member of the Atlanta Wind Symphony.

CAPT Wise was awarded the Navy Meritorious Service Medal, Navy "E", National Defense Medal, Vietnam Service Medal w/2 stars, Armed Forces Reserve Medal w/ Hour Glass, Republic of Vietnam Campaign Medal, Expert Rifle Medal and Sharpshooter Pistol Ribbon.

CAPT Wise has two children, Harold III and Stephanie. He is married to the former Dolores A. Peterson of Peekskill, NY and resides in Marietta, GA.

ROBERT J. WITTEBORT JR., CDR, born in Chicago, the son of CDR Robert J. Wittebort USNR and Marguerite Shaughnessy Wittebort. He attended Yale University, where he was commissioned through the NROTC program in September 1969.

After attending the CIC "O" course at NAS Glynco, GA, he was assigned to USS *Independence* (CVA-62), where he served as boat division officer, gunfire control officer, lifeboat officer, and court martial member during North Atlantic and Mediterranean deployments.

Released from active duty and graduated from Notre Dame Law School with a JD in 1974. In the Naval Reserve he has served in several amphibious units, commanding NR COMPHIBRON-3 in 1984-85. He was administrative officer of Second Naval Construction Regt. from 1986-92. He has performed seven active duty periods on the staff of the Chief of Naval Operations, Pentagon.

He was CDR of the Illinois Commandery of the Naval Order from 1986-88 and is a National VCDR. He has been awarded the Meritorious Unit Commendation, Navy "E," Foreign Shore Deployment, National Service Medal, Naval Reserve Medal, Expert Rifle Medal and Expert Pistol Medal. He lives with his wife, Nancy, in Chicago.

JOHN J. WOJNAR, CDR, born in Warrior Run, PA (near Wilkes-Barre) joined the USN in 1946 as an aviation midshipman in the V-5 Holloway program. After graduating as a naval aviator in 1949 at Corpus Christi, he was assigned to VP-21 in Patuxent River, MD, flying the Navy's newest patrol plane, P4M and on to Port Lyautey, French Morocco with the P4M-Q.

During his four years on active duty he (or as his fellow mates called him, "WOJ") flew the P4Y-2, R4D, SNB, F6F and F8F. He continued in the reserves for 19 more years, most of it flying the P2V. After getting his chemistry degree from Wilkes College, he worked for Hooker Chemical (now Occidental) 38 years holding positions in research, sales, marketing and purchasing.

He and wife, Lillian, reside in Amherst, NY. They have three children: John, Susan and Robert. CDR Wojnar is a member of the Naval Reserve and Flying Midshipmen Assocs.

ALFRED J. "FRED" WOLANIN JR., CAPT, born Nov. 18, 1951, Sharon, PA, graduated from Villanova University in 1973 and Temple University School of Dentistry in 1977.

He joined the service in June 1973; first active duty in July 1977. His military locations and stations: 2nd Dental Co., Camp Lejeune, NC; NAS Sigonella, Sicily; 14th Dental Co., 4th FSSG; Fleet Hospital #20; and Fleet Hospital #15. He participated in Operation Desert Storm and Gulf War.

Wolanin was awarded the Joint Meritorious Unit Award, Meritorious Unit Commendation, Fleet Marine Force Ribbon, National Defense Service Medal, Southwest Asia Medal, Overseas Deployment Ribbon, Sea Service Ribbon, Coast Guard Special Operations Ribbon, Naval Reserve Medal, Liberation of Kuwait Medal (Service Award), Liberation of Kuwait Medal (Kuwait), Navy Rifle Medal and Navy Pistol Medal.

He achieved the rank of CAPT (O-6) and was discharged in July 1982 followed by retirement in May 1991.

His memorable experiences: Recalled to active duty and deployed to Saudi Arabia for four months in Fleet Hospital #15 in Al Jabarl, Saudi Arabia during the Gulf War.

He and his wife, Mary Beth, have three children: Kristin, Alfred III and Alexander. Civilian employment as oral and maxillofacial surgeon in private practice.

JAMES L. WOLFSEN, CAPT, born Sept. 1, 1938, Fresno, CA, earned his BA in history at Loma Linda University in 1961 and his MD degree in 1966 at Loma Linda University School of Medicine.

In 1962 he was commissioned ENS USNR 1915 program. Commissioned USN in 1968. From 1969-70 was senior medical officer, USMCAS (H) Tustin, CA.

He was released from active naval service and transferred to the Reserves in 1970. Retired as CAPT USNR in 1992.

His memorable experiences: Surviving the rocket attack on Marble Mountain Air Facility Sept. 19, 1969.

He and his wife, Judith, have three children, Jay, Jon and Jennifer. Civilian employment as interventional radiologist, White Memorial Medical Center, California Hospital Medical Center, Los Angeles.

LOWELL A. WOMACK, CAPT, born March 25, 1938, Providence, KY, earned his BS degree at Auburn University in 1960 and his JD degree at the University of Alabama School of Law in 1966.

He joined the NROTC at Auburn in September 1956 and was commissioned in August 1960. His military locations and stations: USS *Talladega* (APA-208) at Long Beach, CA, September 1960-September 1961; US Submarine School, Groton, CT, September 1961-March 1962; USS *Tunny* (SSG-282), March 1962-September 1963, Pearl Harbor, HI.

Womack was awarded the Armed Forces Reserve Medal (3), Naval Reserve Medal, Navy Expeditionary Medal and Medal for Submarine Special Operations. He achieved the rank of CAPT and was discharged from active duty in September 1963 followed by retirement Sept. 1, 1990.

He married the former Betty C. Davis and they have four children: Bradley, Cyndy, Lowell Jr. and Haley; and four grandchildren: Carli, Courtney, Derek and Christy. Civilian employment with First National Bank of Tuskaloosa, September 1965-1988, rising from trust clerk (tax preparation) to senior vice president and senior trust officer; AmSouth Bank, N.A., September 1988-June 1993, serving as senior vice president and trust officer, head of institutional and corporate trust; and AmSouth Bank of Alabama, June 1993-present, as senior vice president and trust officer, head of trust administration.

DANIEL ANDREAE "DAN" WOOD, CAPT, born March 25, 1942, Norfolk, VA, graduated from the University of Michigan (BSIE) and Stanford University (MBA). He was commissioned ENS SC, USNR in 1964 from the NROTC program. Following his commissioning he reported to Navy Supply Corps School, Athens, GA. From 1965-67 CAPT Wood served aboard USS *Constellation* (CVA-64); deployed to the Western Pacific during 1966.

CAPT Wood's Naval Reserve career began at Treasure Island, San Francisco and continued with the NROS NAVEUR Reserve program in Paris, France and London England. In London he participated in the standup of CINCUSNAVEUR DET 130, London, England, the first SELRES drill-pay unit outside the US.

Returning to San Francisco in 1972, CAPT Wood served with the Naval Air Reserve program at NAS Alameda, including ACDUTRA aboard USS *Enterprise* CVN-65. Reassigned to London in 1976 he rejoined CNEDET130, performing management assistance projects for CINCUSNAVEUR, NAS Rota, Spain and aboard USS *Concord*.

In 1980 CAPT Wood returned to San Francisco where he served with CINCPACFLT DET420, reporting to inspector general, commander in chief Pacific Fleet. Serving with CINCPACFLT for seven years, he performed a series of management assistance assignments for CINCPACFLT, COMNAVAIRPAC, COMSUBPAC, COMFAIRMED, COMNAVLOGPAC and NAVSTA Seattle. Transferred to New York in 1986, CAPT Wood served with the NEXCOM HQ Unit, Naval Station New York, performing management assistance studies for NEXCOM.

From 1993-95 CAPT Wood served with Naval War College VTU 0119, Newport, RI. As director of the Advanced Concepts Dept., supporting The Center for Naval Warfare Studies, he participated in the design and management of the 1993-94 NWC "Global" War Games. In 1995 CAPT Wood returned to London, England, serving with CINCUSNAVEUR VTU 3001 until his retirement on June 30, 1995. CAPT Wood has been awarded the Navy Commendation Medal, Meritorious Unit Commendation (2),

Navy Overseas Service Medal, Vietnam Service Medal, Naval Reserve Medal (2), and National Defense Medal. He is a life member of the NRA, the Naval Order of the US, the Naval Institute and the Navy League.

CAPT Wood is currently vice president-Poland Country manager for Bank of America, responsible for BankAmerica's business in Poland. At Bank of America for the past 25 years, he has held corporate and investment banking positions in the US and Europe. CAPT Wood, his wife, Judy, and their children, Stuart, Gwendolyn and Graham, presently reside in Warsaw, Poland.

FRED M. WOOD, CAPT, born Sept. 24, 1927, Groveland, MA, earned his engineering degree from Tufts University.

He joined the USCG Sept. 22, 1945, and was commissioned ENS USNR Nov. 6, 1950. His military locations and stations: North Atlantic; Surface Div., Commander Naval Reserve Gp. Detroit; CO, 2 SACLANT Staff Support Units, Battle Creek, MI; and chairman, Recruiting District Assistance Council, Michigan.

He and his wife, Eleanor, had three sons: Barry, Don and Russell; and eight grandchildren: Nicole, Keith, Scott, Kevin, Joni, Kristen, Jenna and Joey. Wood (now deceased) retired Jan. 1, 1982.

DAVID H. "WOODY" WOODRUFF, CAPT, born April 12, 1930, Potsdam, NY, earned his AB degree at Bowdoin College in 1952 and his MBA degree at Boston University in 1958.

He joined the service Sept. 1, 1952 and graduated OCS, Newport, RI Jan. 1, 1953 (ENS USNR). His military locations and stations: Portsmouth Naval Shipyard, NH, 1953-55 and Boston, MA, HQ, 1955-56. He served with USS PCE-843, 1953-55 and US Naval Communications Station, Boston, MA, 1955-56.

Woodruff was awarded the American Defense Ribbon. He achieved the rank of CAPT and was discharged Sept. 1, 1956. Woodruff retired April 12, 1990.

He and his wife, Marlene Anne, have three children: Karen, Susan and Scott R.; and four grandchildren: Katie, Leyden, Jimmy and Richard M. Civilian employment as vice-president, Marine Midland Bank, 1975-80 and vice-president, First Fidelity Securities Gp., 1980-94.

GARRY L. WOODY, CAPT, born Nov. 6, 1946, Coeburn, VA, graduated from Virginia Tech in 1970 with a BS in industrial engineering. Entering the USN in September 1970 at NAS Norfolk, VA he began AOCS training in July 1971 at NAS Pensacola, FL. After completing basic flight training (VT-1) at Saufley Field, FL CAPT Woody moved to Whiting Field (VT-3) completing formation and instrument training.

In August 1972 he completed initial carrier quals aboard the USS *Lexington*. Following carrier training came multi-engine training at (VT-27) NAS Corpus Christi, TX and in February 1973 CAPT Woody was designed a naval aviator. Upon earning his wings CAPT Woody reported to NAS Norfolk, VA assigned to Fleet Tactical Support Sqdn. 40 (VRC-40). While attached to VRC-40 he became a carrier transport plane CDR providing logistics support for Atlantic Fleet aircraft carriers.

He accumulated over 1500 flight hours and 180 carrier landings flying the C1A. CAPT Woody's next tour of duty was again in Norfolk, VA assigned to the CINCLANTFLT Staff (Comm Div.). As a collateral duty he supported the ADM's staff flying C-131 aircraft which provided air transport for DOD personnel.

Upon release from active duty Sept. 30, 1978, CAPT Woody left Norfolk, VA for Denver, CO and thereafter affiliated with the USNR at NARCEN Denver. During his tenure with the USNR he served as avionics DO with ABFC H9B 0271, training and maintenance officer with NR NAS Fallon, NV, administrative officer and training officer with NR NAF Atsugi, Japan, XO of NR NAS Fallon, NV, and CO of NR NAF Atsugi, Japan. CAPT

Woody was awarded the Navy Commendation Medal, National Defense Medal w/Bronze Star, Expert Rifle Medal, Expert Pistol Medal, Reserve 10 Year Service Medal, and Reserve Overseas Service Ribbon. CAPT Woody retired from naval service Sept. 1, 1996.

Since his release from active duty he has been employed by Lockheed Martin Astronautics in Denver, CO as an industrial engineer. CAPT Woody has three children: Adam, Amie and Patrick.

CLARK E. WOOLDRIDGE, CDR, born July 26, 1926, Cross Cut, TX, earned his BA degree in psychology and his MS degree in social work at the University of Texas, Austin, TX.

He enlisted in the USN on a minority enlistment Oct. 6, 1943. Attended boot camp at San Diego in Co. 43-433. Attended Aviation Machinist Mate School, Norman, OK and Naval Air Gunners School, Purcell, OK. Received training as a combat air crewman at Miami, FL in the summer of 1944. Was attached to Torpedo Sqdn. #35 as a combat air crewman. The squadron was stationed intermittently at NAS Seattle, NAS Pasco, WA; Arlington, WA; Saipan and Guam.

Following WWII attended Engine Specialist School, Memphis, TN, in the Spring and Summer of 1946. Following graduation was attached to SB2C Bomber Sqdn. #5, San Diego.

Received honorable discharge from minority enlistment with rate of AMM2/C and signed up with the USNR in July 1947. Received a 1635 reserve commission as a LTJG in November 1959. Served as a member of Reserve Naval Intelligence Units in Houston, TX; Lubbock, TX; and Amarillo, TX.

Was OIC of the Amarillo Unit 1971-74. When the Amarillo Unit was phased out was inactive until affiliated with the Volunteer Training Unit as the mobilization officer of the Amarillo Naval Reserve Training Center. Retired with rank of CDR in 1984.

The Wooldridge family currently consists of wife, Barbara; son, Clark Jr.; daughters, Tracey and Christy; and five grandchildren. CDR Wooldridge is retired as professor emeritus from West Texas A&M University, Canyon, TX.

JOHN ROBERT "BOB" WRIGHT, CAPT, born Dec. 23, 1934, Quincy, MA, earned his AB degree at Northeastern University in 1957 and his MBA degree at Babson College in 1962.

He enlisted Sept. 27, 1952, Squantum Naval AS, Quincy, MA (Reserves). His military locations and stations: 1952-57, ZP-911 Lighter Than Air Sqdn. (specialty-meteorology); 1957-60, active duty, OCS, USS *Wasp* (CVS-18), deck officer (1105); 1960-1980, USNR Intelligence program, NRID 1-1, 3-1, CO-NR OPINTELL 0102 (1978-80).

Wright was awarded the Navy Expeditionary Medal and Naval Reserve Medal. He achieved the rank of CAPT - designator 1635 (Intell.) and retired in 1980.

His memorable experiences: Serving in Blimps (LTA), submarine, destroyer and aircraft carrier duty. ACDUTRA's at their gaining command NFOIO, Naval Field Operational Intelligence Office. Naval War College's Senior Reserve Officers course.

He and his wife, Marian, have three children: Susan, Betsy and John; and two grandchildren, Eric and Kristen. Civilian employment as executive vice president, A.C. Nielsen, a Dun & Bradstreet Co., from which he retired in 1988 after 21 years service. Presently manages his own executive recruiting company.

JOHNIE B. WRIGHT, CAPT, born Sept. 1, 1931, Marlow, OK, earned his BS degree in business at the University of Central Oklahoma, Edmond, OK.

He joined the service in May 1952. His military locations and stations: July-November 1953, OCS, Newport, RI; November 1953-July 1956, USS *Faribault* AK-179 (COMSERVPAC); July-November 1956, Texas Gp., Atlantic Reserve Flt., Orange, TX; November 1956-December 1977, Naval Reserve Training Center, Oklahoma City, OK.

Served with various Naval Reserve units at the NRTC, Oklahoma City including: command of Naval Reserve Surface Div. 8-16, command of the Naval Reserve Officer School, training officer of Surface Divs., executive officer of Surface Divs., and executive officer of Naval Reserve Security Gp.

Active duty for training included: destroyers, destroyer escort, aircraft carrier, ICAF National Security Seminar, Instructor School, NBC School and Damage Control and Firefighting School. He achieved the rank of CAPT and was released from active duty in November 1956.

His memorable experiences: Participated in the "Passage to Freedom" campaign in Indo-China when Vietnam was divided into North and South Vietnam. Was serving as a junior officer on board USS *Faribault*, AK-179. They issued supplies to the US ships that were carrying refugees that desired to relocate from North to South Vietnam. In addition to issue of regular food supplies, they issued rice, straw mats, chopsticks, "honey buckets" and human remains bags for the refugees.

He and his wife, Geraldine "Deanie," have two children, Shanna and Sholin, and four grandchildren: Michael, Kati, Cody and Alex.

Civilian employment as civil service employee 1959-87 as computer programmer, computer systems analyst, and supervisory computer systems manager. Retired from the Federal Aviation Administration in September 1986 with over 32 years of federal service.

From 1987 to date (February 1997) employed with FKW, Inc., Oklahoma City, OK presently as corporate manager, information systems. Wright retired from the USNR in December 1977 with over 25 years of service in the Naval Reserve.

RALPH G. WRIGHT JR., CDR, born in Sparta, IL. While attending St. Louis University, he enlisted in the USN in August 1940.

He attended Midshipman School at Northwestern University, Chicago, IL. He was commissioned ENS USNR Sept. 12, 1941, and was assigned to the tanker USS *Santee*. In December 1941 he completed the Armed Guard School, was CO Armed Guard on the merchant ship SS *Ponce de Leon*, with duty in the Pacific Theater until April 1943.

After completing studies at SCTC, Sound School and Tactical Radar School he spent the remainder of WWII on the USS *Kretchmer* (DE-329) where he served as anti-submarine warfare officer and executive officer.

After WWII he returned to St. Louis University and received his degree of BS in commerce and finance. He then moved to Cape Girardeau, MO to start an electric supply business and was CO, Naval Reserve Unit, Cape Girardeau, MO.

He was recalled to active duty in 1951 and was head

of Harbor Defense Logistics in the office of the Chief of Naval Operations, Washington, D.C. until May 1953.

His campaign and service medals are: American Defense Service Medal w/star, Asiatic-Pacific Campaign Medal, European Theatre of Operations Medal, American Theatre of Operations Medal, WWII Victory Medal and the Naval Reserve Medal.

The Wright family consists of wife, LaRose; daughters, Mary Louise and Katheryn Marie; son, Thomas Glenn; and granddaughters, Megan and Michelle.

WALTER G. WULFF, LCDR, born Aug. 30, 1923, East Rochester, NY, earned his BA degree in 1947 and his MA degree in 1948, both at the University of Michigan.

He joined the service in July 1943 and served until November 1950. His military locations and stations: Mediterranean and Pacific. He served with the Amphibious and Service Force, AIRPAC.

Wulff achieved the rank of LCDR and retired in 1969.

Civilian employment as vice-president international with McGraw-Hill.

FRANK L. WURST, CAPT, born June 2, 1948, Coral Gables, FL, earned his BSME degree in 1970 at the USNA, Annapolis, MD; his MSEM degree in 1971 at Georgia Tech, Atlanta, GA; and his MBA degree (finance) in 1986 at St. Joseph's, Philadelphia, PA.

He joined the service in June 1966, Midshipman USNA, Annapolis, MD. His military locations and stations: USNA, 1966-70; Nuclear Power School, 1972; USS *John Adams* SSIN-620, 1972-76; S1W Prototype, 1976-78; USS *Gilmore*, 1980; AMCC, 1981-82; USS *E.S. Land*, 1982-84; NSSF New London, 1985-91; VTU Atlanta, GA and Avoca, PA.

Wurst was awarded Submarine Qualification, the Patrol Pin, National Defense Service Medal, Expert Pistol Medal, Armed Forces Reserve Medal and Battle "E." He achieved the rank of CAPT and retired in October 1996.

His memorable experiences: Initial submarine sea trials, ship yard overhaul, Panama Canal crossing and other active duty submarine activities. CO of large (70 man) submarine support reserve unit offered numerous challenges and excitement.

He married the former Sharyn L. Cheshure and they have four children: Bethany, Frank, Nathan and Matthew. Civilian employment as supervisor maintenance engineering, Pennsylvania Power and Light.

ELMER STATEN WYNNE, CDR, born Oct. 23, 1917, El Paso, TX, received his BA, MA and PhD degrees from the University of Texas. He enlisted as a pharmacist's mate 2/c in 1942, was promoted to chief in 1944, and was commissioned ENS in 1944. He remained in the Reserve until August 1973.

He served with the 19th C.B. Special in New Guinea in 1944 and with an epidemic disease control group in the Philippines in 1945. Once in the Pacific he had the task of determining if a patient with meningitis had the epidemic type. Fortunately for the ship's company, he did not. The senior medical officer of an epidemic control group in Manila asked him if a bacterium detected there was the plague bacillus. Fortunately, it was not.

His medals are the American and Asiatic-Pacific Theater Medals, Philippine Liberation Medal, WWII Victory Medal and Good Conduct Medal.

His family consisted of wife, Regina (deceased); daughter, Edith Jane; and son, Frank Staten. His present wife is Opal Grace. They reside in San Antonio.

WIL YAMANOUCHI, RADM, graduated from the Washington University School of Dentistry with a BS and DDS degree in 1956. Served three years in the USN from September 1956-August 1959, at Navy 3912 and a year at CBC Port Hueneme.

Became a drilling reservist in February 1965 and held many command billets. First dental officer to command a medical unit; 11-2 Phoenix, AZ. In March 1977 became the officer-in-charge, 4th Dental Co., Det 2. December 1977 became the dental officer, 4th Marine Div. and in 1980 became the battalion CDR 4th Dental Bn., 4th Force Service Support Gp. during its inception.

He was a member of the USNR Flag Support Unit as assistant deputy director, surgeon general (OP-093D) and served as a member, Naval Reserve Medical Flag Council. He is a fellow in the American and International Colleges of Dentistry. He still functions as Blue and Gold officer responsible for many USNA graduates.

Awards and decorations include the Legion of Merit, Meritorious Service Medal, and Naval Recruiting Award w/2 Gold Clusters.

Married to the former Dee S. Lum of St. Louis, MO, they have one daughter, Kimberly J., M.D. who resides in Scottsdale, AZ; and one grandson, Barton Hill Sackett II.

Yamanouchi retired in March 1987.

CHRISTOPHER BARRETT "CHRIS" YATES, CAPT, born May 30, 1948, Toledo, OH, earned his BS degree (aerospace engineering) at the USNA in 1972; BS degree (mathematics) at the USNA in 1972; MA (international relations) in 1986 at Salve Regina College; MBA degree in 1987 at Southern Illinois University; MA degree (National Security and Strategic Studies) in 1991 at Naval War College.

He joined the service in June 1972. His military locations and stations: USS *New* (DD-818); USS *Ranger* (CVA-61); Fleet Training Gp., San Diego; SIMA, San Diego; Naval War College; OPNAV 095; and OPNAV 07.

Yates was awarded the Surface Warfare Medal, National Defense Service Medal, Sea Service Deployment Medal, Armed Forces Reserve Medal and NRA Junior Officer of the Year 1986 11th Dist. award.

His memorable experiences: Building and launching Tomahawk cruise missiles during the developmental phase.

He and his wife, Sally Marie, have three children: Monica, Christopher Jr. and Veronica. Civilian employment with Defense Intelligence Agency.

JAMES C. "JIM" YEARGIN, RADM, born Oct. 10, 1936, graduated from Stetson University, Deland, FL, with a BS degree in chemistry. He graduated from Emory University School of Dentistry in Atlanta, GA and later began private practice in Daytona Beach, FL.

RADM Yeargin's first naval assignment was as a general dental officer assigned to Naval AS Jacksonville, FL. He later became the training officer for Reserve Unit NDCL Jacksonville/Pensacola 1408. Following this he was assigned to the Marines with the 6th Dental Co. His next billet was as executive officer of NDCL Jacksonville/Pensacola 1408. He then became CO of NRDC Jacksonville 108. During his period of command, the unit was awarded the ADM Vaughn Award for the most outstanding dental unit in the Naval Reserve. His unit was also the recipient of the VFW National Defense Award for the most outstanding military reserve unit in the state of Florida during his tenure as CO.

RADM Yeargin's next assignment was as the staff dental officer for REDCOM Region Eight, where he later served as the assistant inspector general. He next became the CO of Naval Medical Command Southeast Region 208. Following this tour, he returned to REDCOM Region Eight as the director of Health Services.

He was next selected for the national billet of Force Dental Officer for the Reserve Naval Construction Force. From this position he was selected for the rank of RADM (lower half), USNR Dental Corps. In addition to this assignment, he served three years on the secretary of the Navy's Naval Reserve Policy Board.

RADM Yeargin's next assignments were as deputy flag officer to the assistant chief for Reserve Force Integration, BUMED, Washington, DC; reserve CINCPACFLT dental officer, San Diego, CA; and assistant deputy director for Reserve Medical Programs, chief of Naval Operations (NO95).

In July 1993 he accepted a collateral assignment as the Centralized Credentials Review and Privileging Activity (CCPA) Flag representative. On Feb. 3, 1994, he assumed additional responsibilities as the Reserve Flag coordinator for the Professional School Liaison Officer (PSLO) Program, which encompasses Navy Medical Dept. recruitment and scholarship programs at professional colleges and universities.

In November 1994 RADM Yeargin was selected to the rank of RADM (upper half). On Oct. 1, 1995, he assumed his current assignments as the reserve deputy assistant chief for dentistry (BUMED), assistant chief, Navy Dental Corps (BUMED) and force dental officer, COMNAVRESFOR.

RADM Yeargin has attended the Strategic Medical Readiness Course at Bethesda, MD and the Naval War College at Newport, RI. He also served as a Naval Academy Blue and Gold officer and is a member of the Naval Reserve Assoc. and Assoc. of Military Surgeons of the US (AMSUS).

RADM Yeargin's military awards include the Legion of Merit, Meritorious Service Medal w/Gold Star, Army Commendation Medal w/"V" for valor, Navy Commendation Medal w/3 Gold Stars, National Defense Service Medal w/Bronze Star, Medal for Humanitarian Service, Armed Forces Reserve Medal w/HG, Expert Rifle Medal and Expert Pistol Shot Medal. Additionally, RADM Yeargin has been designated as a Seabee Combat Warfare officer, and has received several unit commendations.

In civilian life RADM Yeargin is active in professional and civic organizations. He is a member of the American Dental Assoc. and the Florida Dental Assoc. and has served as a delegate. He is also a Fellow in both the International and American College of Dentists. For many years he served as an examiner for the Florida State Dental Boards. In addition, he has served on the Board of Directors of the American National Bank and the YMCA.

RADM Yeargin is an active sportsman who enjoys golf and horseback riding, and has a small ranch in central Florida. He is also an instrument rated multi-engine pilot who flies his own plane. He has a dental practice in Ormand Beach, FL and resides in Daytona Beach Shores, FL.

HERBERT C. "HERB" YIM, CAPT, born Feb. 1, 1930, Oakland, CA, earned his BS degree (engineering) in 1953 at the University of California-Berkeley and his MS degree (mechanical engineering) in 1962 at the University of Southern California.

He was a member of the USMCR from December 1947-December 1953. His military locations and stations: Marine Corps Reserve Center, Treasure Island, San Francisco, CA. He achieved the rank of CPL. He then served with the USNR from January 1955-September 1989. His military locations and stations: OCS Newport, Navy PG School, Monterey, CDR AEW Wing Pacific, Barbers Pt., BUAER, BUWEPS, NAVAIR, NAVMAT, Washington, D.C.

He achieved the rank of CAPT (O-6), USNR. His memorable experiences: 1) Served as CO of three Naval Reserve units. 2) Organized 12 Naval Reserve units in CONUS with AED/AMD officers to provide engineering and technical support to the NAVAIR P-3 Aircraft Project Office. 3) Developed and implemented plan to utilize engineering officers from NAVAIR, NAVSEA, NAVELEX and ONR Reserve Units to provide engineering and technical support to the OPNAV International Armaments Cooperation Program Office.

He married the former Pamela J. Nachman. He has five children: Debbie, Kenneth, Gregory, Theresa and Jade; four step-children: Susan, Carolyn, Robert and Lisa; two grandchildren, Derek and Riana; and two step-grandchildren, Adrienne and Britt.

Civilian employment and positions held: Hughes Aircraft Co., design supervisor, Air to Air Missiles and Airborne HF Radio Sets; North American Aviation, systems engineer, Space Vehicle Systems Integration, Apollo Program; Aerospace Corp., program manager, DoD Space Experiments Programs; Dept. of Energy, interchange executive, President's Executive Interchange Program; Executive Systems International, president, International technical consultant support to US agencies; AVTECH Marine International, president, Manufacturer's representative of power catamarans. Yim retired in September 1989.

DR. ELIHU "EL" YORK, CAPT, born July 7, 1927, Boston, MA, earned his BS degree at Tufts University in 1950, his MD degree at Tufts University in 1954 and his MPH degree at Harvard University in 1969.

He joined the US Army and served from Oct. 1, 1945-May 20, 1947. He then joined the USNR and served from Dec. 20, 1963-July 7, 1987. His military locations and stations: Nav Aero Med. Inst., Pensacola; Nav Aero Med. Research Lab, Pensacola. He served as senior medical officer aboard the USS *Essex* (CVS-9) from May 1966-August 1968.

York was awarded the WWII Victory Medal, Army Occupation of Japan Medal and Armed Forces Reserve Medal. He achieved the rank of CAPT (O-6), Medical Corps, USNR and retired July 7, 1987.

His memorable experiences: Helped to train both Gemini and Apollo NASA astronauts at USN Acceleration Lab, Johnsville, PA 1964-66. Solo Flight, T-34B Sauffley Field, Pensacola, May 1964. Ran Boston Marathon six times.

He married the former Audrey K. Spowart and they have three children: Peter, James and Sara; and one grandchild, Bailey. Civilian employment as director, Occupational Medicine, Hartford Hospital, Connecticut 1984-95.

ROBERT EDWIN "BOB" YORK, CAPT, born May 3, 1937, Auburn, KY, earned his BS degree in 1960 at Austin Peay State University, his MS degree in 1972 at Indiana University and 32 hours above master's at the University of Louisville.

He joined the service June 3, 1963 and was commissioned ENS from the Aviation Officer Candidate School at Pensacola, FL. His military locations and stations: Airborne Early Warning Sqdn. 13 (VW-13), Argentina, Newfoundland, Canada; July 1965-July 1966, Naval Advisory Gp., Vietnam, Saigon; August 1966-August 1969, air intelligence officer, Carrier Airborne Early Warning Sqdn. 11 (VAW-11 and RVAW-110).

After six years CAPT York was released from active duty in August 1969 and in September 1969 he started commuting by Naval Air Reserve Flights to Naval Air Facility, Washington, D.C. continuing for 23 years flying on C-54, C-118, C-131, C-130 and C-9 aircraft.

CAPT York's first NAF assignment was to VR-IA3 Sqdn. for three years. In 1972 after a change of designator from 1355 to 1635 he served in the following Naval Reserve Intelligence units: NAIRU A1, NIC Invest 506, IC 406, NICSPECSEC, NICFIN, NICRESOURCES 0166, OSD Tech Transfer 0166 and VTUI 0106.

CAPT York was awarded the Navy Commendation Medal, Joint Meritorious Unit Award, National Defense Medal, Vietnam Service Medal (3 campaigns), Naval Reserve Medal, Republic of Vietnam Campaign Medal and Navy Pistol Expert 45 Caliber. He achieved the rank of CAPT and retired in August 1992.

He married the former Reda Alice Goff and they have four children: Cecilia Lynn, Craig Alan, Connie Jo and Toni Marie; and five grandchildren: Morgan Leigh, Andrew York, Jenna Noell, Luke Daniel and David Alan. Civilian employment as mathematics teacher at Jeffersonville High School.

RONALD E. "RON" YORK, CDR, born Aug. 2, 1937, in Iowa, earned his BS degree (business administration) at Kansas State University.

He joined the service in October 1960 and served on active duty until February 1964. His military locations and stations: USS *Seminole* (AKA-104). He served with Samar 11-1(L), FITWING 321, FITWING 385 and VTU-8594.

York was awarded the National Defense Service Medal, Naval Reserve Medal and Pistol Marksman Medal. He achieved the rank of CDR and retired in April 1982 from the USNR.

He and his wife, Colette, have three children: Jennifer, Keith and Suzanne; and one grandchild, Conner. Civilian employment at Bank of America, Bank of Commerce and Wells Fargo Bank. York retired in August 1996.

LEHMAN H. YOUNG, LT, born Jan. 24, 1916, Fairfax, VA, attended George Washington University.

He joined the service Dec. 14, 1942. His military locations and stations: USNAS Corpus Christi; NAS Norman, OK; NATS, VRF-1, Floyd Bennett Field, NY. Battles and campaigns included Continental US.

Young was awarded an Honorable Discharge, Continental US Medal and US Sharp Shooter Medal.

His memorable experiences: Receiving instructor rating at Corpus and checking out in service aircraft at Willow Grove including instrument rating there.

He and his wife, Alice Marie, have three children: Deborah R., Pamela Ann and Lehman Jr. Civilian employment as printer, *Washington Post*; master printer, *The Virginia Press*, proprietor. Young retired Aug. 1, 1960.

PETER A. YOUNG, LT, born June 4, 1937, New York City, NY, attended Wheaton College in Illinois prior to beginning his naval career as an aviation cadet in 1961. While undergoing flight training at Pensacola he was also a member of the Naval Air Training Command Choir. He received his wings and commission in May 1963.

As a naval aviator, he served with HS-9 at NAS Quonset Point, RI from 1963-66 and with HS-832 at NAS New York from 1967-68. He had over 1300 hours of Navy flight time and more than 20,000 hours of civilian flight time as a pilot with TWA where he flew B-70, B-727, B-747, B-757, B-767 and L-1011 aircraft.

Young was awarded the National Defense Medal and Good Conduct Medal (Army). He achieved the rank of LT and was discharged Nov. 7, 1966.

He is married to the former Patricia Payette and they have two sons, Christopher and Jason. They reside in E. Greenwich, RI where they enjoy sailing and traveling. Young retired in 1997.

WILLIAM KEITH "BILL" YOUNG JR., RADM, born July 5, 1935, Utica, NY, earned his BS degree at the USNA and did graduate work at Georgetown University and Syracuse University.

He was commissioned in the service June 4, 1958. His military locations and stations: (Active duty) COMUSMACV, 1962-63; CNO OP941, Washington, D.C., 1963-65; Naval Oceanographic Office, 1967-69; CNO OP994, Washington, D.C. 1971-72;

(Reserve) NAVRESCEN, Charleston, SC, 1971; NAVRESCEN, Washington, DC, 1972-73; NAVRESCEN, Syracuse, NY - NRSD 3-4D 1973-74, CRUDESGRU 303 1974-77, VTU 0212 1977-80, FTG San Diego 1980-81, VTU 0212 1981-85; NAVMARCORRESCEN, Orlando, FL - MSC VTU 0809 1986-89 and VTU 0809 1989-91. He served with USS *Bache* (DDE-470), 1958-60; COMDESRON 4, 1960-62; and USS *Arneb* (AKA-56), 1965-67. He participated in the Vietnam War 1962-63.

Young was awarded the Air Force Meritorious Civilian Service Medal, Combat Action Ribbon, Joint Meritorious Unit Medal, AF Outstanding Unit Award (3), AF Organizational Excellence Award, AF Recognition Award (2), National Defense Medal (3), Vietnam Campaign Medal w/star, Armed Forces Expeditionary Medal, Armed Forces Reserve Medal w/HG, RVN Service Medal, RVN Cross of Gallantry Medal, RVN Civil Action Medal, Expert Pistol Medal, Expert Rifle Medal, New York State Military Commendation Medal, New York State Long and Faithful Service Medal and New York State Recruiting Medal.

He achieved the rank of CAPT, USNR, and retired Oct. 1, 1991; and RADM, New York Naval Militia, retiring Oct. 1, 1991. His memorable experiences: NRA Central New York Chapter president, two terms, while CO of CRUDESGRU 303, winning NAVRESCON Syracuse's first Outstanding Unit Award.

Civilian employment as high school math instructor, 1970-71; chief of the Communications-Electronics Plans and Programs Div., NORAD NE Region, 1973-85; communications manager, DOD Space Transportation Contingency Support Office, 1985-91; and computer lab education specialist, Brevard Community College, 1991-96. He is divorced.

STEPHEN G. "STEVE" YUSEM, RADM, born July 8, 1936, Cheltenham, PA, attended Cornell University, Ithaca, NY, as a NROTC midshipman, where he was commissioned in 1958. He served on the USS *Loyalty* (MSO-457) as deck officer and USS *Acme* (MSO-508) as operations officer in the western Pacific until released from active duty in 1960. He then attended the University of Pennsylvania Law School, Philadelphia, PA, where he received a JD degree in 1963.

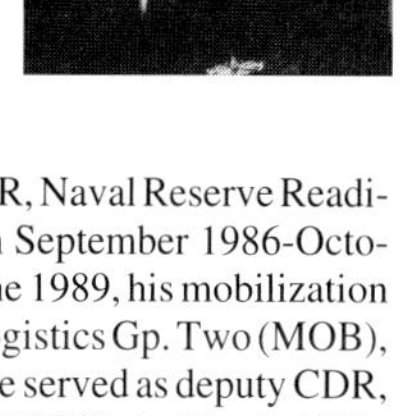

RADM Yusem served as CDR, Naval Reserve Readiness Command Region Four from September 1986-October 1988. From February 1986-June 1989, his mobilization billet was deputy CDR, Combat Logistics Gp. Two (MOB), and from July 1989-March 1992, he served as deputy CDR, Maritime Defense Zone Atlantic (MOB). In March 1992, he assumed duties as deputy chief of staff for Resources, US Naval Forces Europe (MOB). In October 1992 he assumed additional duty as CDR, Logistics Task Force Atlantic. He transferred to retired status in October 1993.

He commanded three Naval Reserve units: a surface division, an underway training unit and a special warfare task group. He has also served as chief of staff of Naval Reserve group commands and as inspector general, Naval Reserve Readiness Command Region Four.

On active duty for training, RADM Yusem has attended the Command and Staff Course, Defense Economics Course, Strategy and Policy Course and Naval Operations Course at the Naval War College. He attended the Senior Naval Reserve Officer Orientation Program at CINCLANTFLT, the Defense Strategy Seminar at the National Defense University and the Unconventional Warfare Course at the Air Force Special Operations School.

A Surface Warfare officer and Naval-Marine Corps parachutist, RADM Yusem holds the Legion of Merit (2) and the Meritorious Service Medal.

In civilian life, RADM Yusem is a senior partner of the law firm of High, Swartz, Roberts & Seidel, Montgomery County, PA, where he specializes in litigation and real estate law. He is past president of the Montgomery Bar Assoc. and the Montgomery Bar Foundation. He serves as director of the Montgomery County Community College Foundation.

As a member of the Naval Reserve Assoc., RADM Yusem is past president of the Fourth District and past president of the Philadelphia Chapter. He is also a member of the Reserve Officers Assoc., Naval War College Foundation and the Philadelphia Council of the Navy League.

He is a delegate to the Confederation of Interallied Officers of the Reserve.

He and his wife, Anita, who holds a doctorate in developmental psychology, reside in Gwynedd Valley, PA. They have two children, Caren Ann and Michael G.; and one grandchild, Maxwell. Yusem retired Oct. 31, 1993.

FRANK A. ZAVATTARO, CAPT, born March 8, 1931, Astoria, NY, is a 1953 graduate of St. Lawrence University.

Commissioned an ENS in 1953 he was promoted to CAPT in 1974 and retired in 1977.

In 1953 and 1954 CAPT Zavattaro served on the USS *Valcour* (AVP-55), a CDR Middle East Force flagship. In January 1955 he was ordered to the USS *Hopi* (ATF-71) and beginning in January 1956 until release from active duty to the USS *Lookout* (AER-2), an Atlantic Flt. Radar picketship.

Beginning in January 1957 CAPT Zavattaro served continually in Naval Reserve Surface and Staff organizations with assignments as a training officer, operations officer, executive officer and CO. He was CO of NRSD 3-13(M), Stamford, CT (1967-68), CO NCSORG 3-5, Stamford, CT (1970-73) and of MSCO 2001, Providence, RI (1974-76).

CAPT Zavattaro's ACDUTRA's included the Naval War College Command and Staff Course in 1968, and the Senior Officer Course and Global Strategy Conference in 1972.

In civilian life CAPT Zavattaro was employed by the Burndy Corp., a manufacturer of electrical and electronic connectors. He retired as director of sales, US Electronic Gp. in 1993. CAPT Zavattaro and his wife, Dorothy, reside in Cos Cob, CT. They have two children, Suzanne and Marcus N.

JOHN M. ZEB, CAPT, born Nov. 22, 1947, Cleveland, OH, graduated from Cleveland State University (BBA-accounting) and the University of Rhode Island (MBA). He enlisted in the USN in 1968 and was commissioned as an ENS in 1969. He served in both the Pacific and Atlantic Flts. during his active service, being assigned to destroyers and destroyer escorts.

During his Reserve career, he spent most of his assignments aboard ship or with shipboard related units. He served five command tours including service as CO or officer-in-charge of: Flt. Training Gp. Guantanamo Bay Cuba Det 201, USS *Exploit* (MSO 440), Reserve Construction Bn. Det 127 and 727 (as a special project for line officers and Seabees), and VTU 011.

Among the awards he received during his naval service are the Meritorious Unit Commendation, Coast Guard Meritorious Unit Commendation, National Defense Medal, Vietnam Service Medal, Overseas Service Medal, Active and Reserve Sea Service Medal, Battle E, Republic of Vietnam Campaign Medal and Expert Rifleman, Sharpshooter-Pistol, and Armed Forces Reserve Medal.

He was employed by the US Treasury Dept. as a bank examiner with a specialty in information systems. CAPT Zeb resides with his wife, Karen and son, John T., in East Freetown, MA. He enjoys gardening, scale modeling of ships, planes and armored vehicles, tropical fish and reading.

RICHARD J. "DICK" ZEH, LCDR, born Aug. 16, 1926, Jamaica, NY, earned his BSEE degree at Rensselaer Polytechnic Institute, Troy, NY in 1946 and his MBA degree at the University of Southern California, Los Angeles, CA in 1960.

He joined the service in May 1944. His military locations and stations: V-12 Program, RPI, Troy, NY; USS *Columbia*; NAVCOMSTA San Diego; NAVRADSTA Imperial Beach, CA. He served with MSTS as CO, XO, Military Dept. various USNS ships.

Zeh was awarded the Naval Reserve Medal, WWII Victory Medal, Occupation Medal, American Theater Medal, United Nations Medal, Korean Service Medal and National Defense Service Medal.

His memorable experiences: CO Military Dept. 1951 Bering Sea Resupply Expedition; XO Military Dept. Inchon Troop Transport; and daily "Red Ball" Pusan/Sasebo Troop Ferry.

He and his wife, Therese A., have two children, Anne M. and Kathrine L. and two grandchildren, David R. and Patrick K. Civilian employment as development engineer, Sylvania Electronics & Hughes Electronics; ground support program manager various Lockheed Missile Systems; manager, Systems Engineering, Lockheed Poseidon Missile Automatic Checkout System; and principal standards engineer, Ford Motor Co.

HELEN ZENKOVICH, LCDR, born April 16, 1910, Sioux City, IA, earned her BS degree at the University of Minnesota.

She was commissioned ENS May 29, 1944 with first active duty July 12, 1944. Her military locations and stations: USN, Farragut, ID; USNH, San Diego, CA; USNH, Sampson, NY; Great Lakes, IL; USS *Haven* AH-12, Far East to Inchon, Korea; Pusan, Korea; San Francisco, CA; St. Albans, NY; USNH Portsmouth, NH; Oak Knoll, Oakland, CA; USNH Hawthorne, NV; Treasure Island, CA; USNH Camp Pendleton; USNH San Diego; and Medical Center, Bethesda, MD.

Zenkovich was awarded the Naval Reserve Medal, Armed Forces Reserve Medal, American Campaign Medal, WWII Victory Medal, National Defense Service Medal, Korean Service Medal, United Nations Service Medal, and Korean Presidential Unit Citation.She retired from the Navy with the rank of LCDR in July 1970.

Civilian employment with USPHS Indian Medical Center from which she retired in 1972. Volunteer service with Maricopa County Health Dept., Phoenix, AZ in the Maternal, Child Health Clinic. She is single.

LOUIS P. "LOU" ZICARI, CAPT, born July 16, 1915, Rochester, NY, was one of nine children born of Italian immigrants who came to America at the turn of the century. He graduated from the New York State Merchant Marine Academy in 1937 and sailed in the Merchant Marine.

Commissioned in January 1991 he was ordered to active duty in May and assigned to USS *Chemung* (AO-30) as engineering officer. So urgent was the need for experienced engineering officers that in July 1941 he was at sea without the usual 90 day indoctrination period. He served as chief engineering officer of the *Chemung* and later in the same capacity of the USS *Lander* (APA-178).

During his service in the active reserves he served as CO, BuShips Co. 3-2, CO, Composite Co. 3-12 and CDR, Gp. Command 3-9(L).

His service medals are the American Defense Medal w/star, American Theatre Medal, European-African Theatre Medal, Asiatic-Pacific Theatre Medal and WWII Victory Medal.

He is married to the former Lena B. Bianchi. They have three children: Laura, Anne and Louis Jr.

PAUL L. "ZIP" ZIEMER JR., CAPT, enlisted in the Naval Reserve in 1965 as an aviation reserve officer candidate while attending San Diego State College. He drilled with VP-771 at NAS Los Alamitos, CA. Upon graduation in 1967, he was commissioned an ENS in the Naval Reserve.

After naval flight officer training, he received his wings in June 1968. During his naval service CAPT Ziemer served as Public Affairs/1LT, VAW-11, Det 34, 1969; training officer, VAW-111, Det 7, 1971; operations officer, VAW-88 and VAW-110 Det 4, 1972/75; OinC, VAW-88, 1979; student, DLI, Monterey, CA, and Peruvian Naval War College, La Punta, Peru, 1981/82; Director NFO training, Peruvian Naval AF, Callao, Peru, 1983; CO, VAW-78, 1985; XO, NAS Willow Grove, PA, 1986; CO, NMCRRC, Treasure Island, CA, 1989; CO, Naval Air Reserve Whidbey Island, WA, 1990; CO, NRRC, Seattle, WA, 1993.

CAPT Ziemer made five extended carrier deployments and flew over 4,200 flight hours, including 396 carrier landings and 218 combat missions in Vietnam. His personal awards include the Legion of Merit (2), Meritorious Service Medal (2), Air Medal (6), Navy Commendation Medal (3), Meritorious Unit Commendation (5), Battle "E" (4), the Vietnamese Air Gallantry Cross and other campaign and unit awards.

He resides with his family in Tucson, AZ, where he is a writing instructor for Pima Community College.

FREDERICK BEDELL "FRED" ZIESENHEIM, CDR, born Sept. 3, 1926, Erie, PA, earned his BS degree (industrial engineering) at Pennsylvania State University in 1947 and his LLB degree at the University of Pennsylvania Law School in 1957.

He enlisted as SA April 25, 1944, at age 17. Completed V-12 program at Pennsylvania State University and Midshipmen's School at Columbia University and commissioned ENS, USNR Nov. 2, 1945. Attached to USS LSM 381 as communications/supply officer, detached at Okinawa and returned to CONUS and separated Aug. 24, 1946.

Recalled to active duty as LTJG June 9, 1951. Served at CINCLANT/CINCLANTFLT HQ, Norfolk, VA until May 1953 and thereafter at COMSUBCOMNELM Naples, Italy in communications billet as Crypto watch officer, until returned to CONUS and separated June 10, 1954.

Reserve duty: 1954-57 - USNR Advanced Base CUB 4-1, Philadelphia, PA, communications officer; 1957-67 - USNR Surface Div. 4-12(M), McKeesport, PA, CO (1964-67); 1967-72 - USNR Law Co. 4-15, Pittsburgh, PA, CO (1969-72); and July 1, 1972 - transferred to USNR-Retired.

Ziesenheim was awarded the American Theatre Medal, WWII Victory Medal, National Service Medal and Naval Reserve Medal. His most memorable experience: The celebration of V-J Day in New York City, including marching in the Victory Parade with the midshipmen of Columbia University.

He married the former Martha Ann Midlam and they have one daughter, Cydney Ann. Civilian employment: Private Practice - Patent, Trademark & Copyright Law, Webb Ziesenheim et al., Pittsburgh, PA.

SAMUEL D. "SAM" ZIMTBAUM, CDR, born Feb. 1, 1927, Brooklyn, NY, earned his BS degree in 1948 at New York University and his MBA degree in 1949 at Indiana University.

He joined the service Jan. 8, 1945 and served on active duty until Aug. 3, 1946. Served with the reserves from Aug.

4, 1946-April 1, 1973. His military locations and stations: San Diego Naval Hospital, Brooklyn Naval Hospital. He served with NACTU Key West, FL, Jan. 10-Aug. 3, 1946.

Zimtbaum was awarded the WWII Victory Medal. Commissioned ENS Feb. 10, 1949 and CDR. July 1, 1964. In reserve served as supply officer Advanced Base Unit, Brooklyn Navy Yard July 1, 1949-Dec. 1, 1959. Mobilization Unit Phoenix NRTC Jan. 1, 1960-April 1, 1973. Had many training duties San Diego, CA as a supply officer from 1960-73.

He and his wife, Carol, have three children: Lorence, Ronald and Gary; and one grandchild, Amber. Civilian employment as president, Samuel D. Zimtbaum Co. Inc., a manufacturers representative firm. Zimtbaum retired April 30, 1996.

ROMAN GEORGE "DR. Z" ZIOLKOWSKI, RADM, born Aug. 30, 1917, Chicago, IL, attended Wright College, Loyola University and Chicago College of Dental Surgery. Loyola affiliate, graduated in June 1940.

December 22, 1941 he applied for commission USNR. Date of rank, February 1942, LTJG, USNR. Activated in April 1942 at Great Lakes NTC. Exam for USN, June 1942, to LTJG DC USN. August 1942, orders to NAS Moffet Field, CA, CUB 3 Unit. February 1943, shipped to Fiji awaiting further orders.

July - transferred to Guadalcanal via LST to Rendova, then to New Georgea, Munda. Built hospital on Munda (CUB 3) Base Hospital #11. Transferred to Dispensary Duty. New Caledonia (Noumea) January 1944. April 1944, orders to Naval Hospital, Corona, CA. September 1945, orders to USS *Columbus*, on station, Shanghai, China, on station Tsingtau, China, on station Yokasuka, Japan. Return to Oakland, CA. Returned to Long Beach, CA. Resigned USN commission, recommissioned USNR. Feb. 12, 1948, appointment USNR, Volunteer Dental Unit 11-6. July 1971, promoted to RADM L USNR. July 1975, retired RADM U USNR.

Ziolkowski was awarded the China Service Medal, American Pacific Medal, American Defense Medal, Asiatic-Pacific Victory Medal, National Defense Medal and Armed Forces Reserve Medal.

He and his wife, Dorothea Mae "Dolly", have five children: Patricia, Nancy, Tina, Stephen and Jeffrey; and four grandchildren: Emily, David, Daniel and Allison. Civilian employment as DDS in private practice, Pomona, CA. Ziolkowski retired in November 1980. Life member of NRA, served on several NRA committees.

BRUCE H. ZIRAN, LCDR, born Sept. 20, 1962, London, England, earned his BS degree in engineering in 1984 and his MD degree in 1988, both at Case Western Reserve University; orthopaedics 1994, trauma 1995.

He joined the service in July 1994 and served with PRIMUS.

Civilian employment as assistant professor, University of Pittsburgh, Dept. of Orthopedics. He is single.

VICTOR A. "ZUKE" ZUCARELLI, CDR, born Jan. 30, 1917, Buffalo, NY, earned his BS degree (chemistry) at Canisius College, Buffalo, NY in 1939.

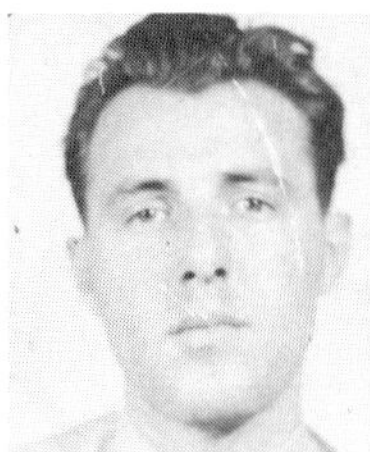

He joined the service in March 1941, aviation cadet, NAS Pensacola, FL through October 1941. His military locations and stations: Key West, Panama, Guantanamo Cuba, Bermuda and NAS Norfolk. He served with VP-32, VPB-201, Hedron 52; reserve duty NAS Niagara Falls. Anti-submarine and convoy duty in Atlantic, Caribbean and Pacific (Gallopolis Islands).

Zucarelli achieved the rank of CDR and retired April 1, 1965.

His memorable experiences: Met wife in 1942, Panama. Reserves, CO, Volunteer Unit 1947 resulting in the establishment of NAS Niagara Falls in 1948.

He and his wife, the former Isabel Gomez (native of Costa Rica) have four children: Judy Marie, James, Janice and Jon; and 11 grandchildren: Morgan, Matthew, Kristin, Paul, Michelle, Rebecca, Devin, Andrew, Jonathan, Abbey and Ashley. Civilian employment with Merck & Co. as executive sales representative. Zucarelli retired April 1, 1986.

ROBERT E. "BOB" ZUMWALT, CDR, born Oct. 1, 1939, Puyallup, WA, earned his BSA at the University of Washington, Seattle, WA.

He joined the service June 14, 1961 and was commissioned ENS, NROTC program. He served as XO/OPS officer, MSTS, Guam from 1961-63. From 1963-75, Naval Reserve Intelligence Unit, San Francisco; 1978-85, Naval Reserve Intelligence Unit, Poughkeepsie, NY (OPS officer).

Zumwalt was awarded the National Service Medal, Naval Reserve Medal, Expert Rifle Medal and Expert Pistol Medal. He achieved the rank of CDR and retired March 1, 1985.

His memorable experiences: Naval Reserve ACDUTRAs at CINCPAC, CINCPACFLT, Naval War College, and Div. Naval Intelligence.

He and his wife, Karen, have three children: Christina, Scott and Karin. Civilian employment as senior vice president Ellerbe Becket Architects & Engineers, 1969-92 and real estate broker, Coldwell Banker, 1993-97.

Head Table at the Awards Luncheon on Saturday, 4 October. First four figures, right to left: CAPT Robin Goodenough, Master of Ceremonies; CAPT Bolt; VADM William F. Raborn, USN, 1969 National Distinguished Service Award winner; and CAPT A. Winfield Chapin, USNR (Ret.).

Roster

This is the roster as provided by the Naval Reserve Association. The publisher regrets it cannot be responsible for any errors or omissions.

Aaby, Donn V, Capt
Aanstoos, Thomas K., CDR
Aardweg Jr, Henry T, Lt
Aaron, Gary W, CAPT
Aaron, Grace K., LCDR
Aaron, Harvey C, Capt
Aaron, James G, Cdr
Aarstad, Robert F, Capt
Abbey, Alfred E, CDR
Abbot Jr, J Lloyd, Radm
Abbott, Donald R, LCDR
Abbott, Howard W, LCDR
Abbott, Mark O, Capt
Abbott, Peter W., LT
Abbott, Rob, ENS
Abbott, Rodman P., LCDR
Abbott, William R, CDR
Abdelnour, George J, LCDR
Abel, E. Glynn, CDR
Abele, Richard S, Capt
Abele, Robert T, Capt
Abell, William J, LCDR
Abelson, Nathaniel O, CDR
Abendroth, Barbara V, Ens
Abernathey, Deana L., LCDR
Abernathy, Roger E, Capt
Abernathy, William J., CDR
Abernathy III, John M, Capt
Abert, James G, Capt
Ables, James A, Lt
Abraham, Reuben Israel, LT
Abrams, Joel S, Lt
Abrams, Mark M., CDR
Abrams, Robert C, Ltjg
Abramson, Harvey, Capt
Abrell, Gary A, Capt
Abriam, Ariel, CDR
Acebo, Roger L., Lt
Achenbach, John A., CAPT
Achilles, Henry R, Lt
Acker, Paul W., LCDR
Ackerman, Arthur P, LCDR
Ackerman, Ray, RADM
Ackermann, Peter G, Capt
Ackert, John P, LCDR
Ackert-Burr, Cheri N., LCDR
Acosta, Jose A., CDR
Acosta, Ronald, LCDR
Acosta, William, CAPT
Acquilano, Rocco D, CDR
Acquistapace, Tulio G, Capt
Acton, Daniel A, CAPT
Acuff Jr, John Edgar, LCDR
Adair, John W, Capt
Adair, Virgil K, CWO3
Adam, David J, Capt
Adam, Kenneth J, CAPT
Adamcin, Peter L, CDR
Adame, Juan J., CWO4
Adames, Ivan A., LCDR
Adamowicz, Anne Marie, CDR
Adams, Albert, Capt
Adams, Alfred Hugh, CDR
Adams, Alton Printess, CDR
Adams, Andrew R, CDR
Adams, Billy J, Capt
Adams, Bruce C, Capt
Adams, Charles E., LCDR
Adams, Charles F, CDR
Adams, David L., LT
Adams, Denis L, Capt
Adams, Dennis J., LCDR
Adams, Frank E, LCDR
Adams, Gregory S, CDR
Adams, Gregory S., LCDR
Adams, Gwendolyn D., LTJG
Adams, Henry R, Capt
Adams, Hugh T, LCDR
Adams, James D, CDR
Adams, James M, LCDR
Adams, Jane D., LT
Adams, Jesse J, CDR
Adams, Joan Fiske, Capt
Adams, John R., LTJG
Adams, Kandace, CDR
Adams, Melissa B., LT
Adams, Norris S, LCDR
Adams, Oscar W, CDR
Adams, Randall D., CDR
Adams, Ronald E, Capt
Adams, Ruth E, CDR
Adams, Sandra E, CDR
Adams, Thomas J, CDR
Adams, Tom W., CAPT
Adams, Wesley J, CDR
Adams, William H, Capt
Adams II, Charles J., LCDR
Adams Jr, Allan Blakely, Capt
Adams Jr, David Earl, Capt
Adams Jr, David J, Capt
Adams Jr, Edward A, Capt
Adams Jr, Lynn, LCDR
Adams Jr, Parks G, Capt
Adamson, Jeffrey D., CDR
Adamson, Mary A, LCDR
Adamson, Richard C., Capt
Adamson Jr, Nathaniel E, Capt
Adaschik, Anthony J, Capt
Adcook, Ann E., CAPT
Adell, James M, CDR
Adelmann Jr, Walter J., LCDR
Adema, Henry T, Capt
Adinolfi, Gregory, LCDR
Adkins, Roy E, CDR
Adkins, William E., Capt
Adkins Jr, Howard E, Capt
Adkisson, John F, Capt
Adler, Davis E, Capt
Adler, Lawrence E., CAPT
Adolphson, Nancy, CDR
Adolphson, Peter H., LCDR
Adrian, Shaun B., LCDR
Aduddell, Tommy D., LCDR
Advincula, Rodolfo Q, LCDR
Affuso, Chris D., CWO2
Afong, Danford S.K., CDR
Aga, John H, Lcdr
Agee, Kimberly R., Lcdr
Agger, James H, Capt
Aggson, Ronald B., CDR
Agnew, Patrick S., CDR
Agosti, Raymond E, Lt
Agra III, Joseph O, Lcdr
Aguila, Honorato M, Capt
Aguilar, Raymundo, CDR
Ahearn, A. Lawrence, CDR
Ahern, Robert P, CDR
Ahern, Thomas C, Capt
Ahern, Thomas E, CDR
Ahlberg, Arnold E, Capt
Ahlborn, Gale, Capt
Ahlenius, William M, CDR
Ahlers, Gordon K., Lcdr
Ahlfield, George S, Lcdr
Ahlin, Jeffrey H, CDR
Ahlsen-Girard, Edward F, Lcdr
Ahlstedt, Paul W., Lt
Ahlstrom, John D, Capt
Ahrens, William S, Lcdr
Aiello, Anthony R, CDR
Aiello, Virgil J, Capt
Aiello, William M, CDR
Aimers, Ralph F., Lt
Aires, James W, Lcdr
Akhdary, Andrew F., Lcdr
Akimoto, Nobuhiko J, Capt
Al-Saigh, Ayad Nafia, LCDR
Alaga Jr., Nick, LCDR
Alair, Gene L, Capt
Alaniva, Loyal V, CDR
Albani, James A., LCDR
Alberse Jr, Peter T, CDR
Albert, Charles E, Capt
Albert, James S, Lcdr
Albert, Margaret L, CDR
Albert, Timothy G., CDR
Albert, William E, CDR
Alberti, William T., CAPT
Albertin, Jean Christie, Lcdr
Alberts, Marion E, CDR
Alberts, Paul G., LCDR
Albertson-Ploucha, H L, LCDR
Albi, Lynn J, Capt
Albin, Frank I, Lcdr
Albin, Joseph S, CWO4
Albin, William M., LCDR
Albinder, Robert J, Lcdr
Albrecht, Robert A, Capt
Albright, Daphne G, CAPT
Albright, George E, CAPT
Albright, Michael J, CDR
Albright, Penrose L, Radm
Albright, Richard W, Lt
Albright Jr, Donald S, Radm
Albrigo, Vickie L., Lt
Albritton, Althea E., CDR
Albritton, John T, Capt
Albus, John P., Capt
Alcia, Russel E., LT
Alcivar, Michael L, Lt
Alda, John, CDR
Alden, Edward M., Capt
Alder, Kim L., CDR
Alderman, Lance R, CDR
Alderson, Jack B, Lcdr
Aldrich Jr, Arthur G, CDR
Aldunate, Carlos P, CDR
Alecxih, Peter C, Capt
Alexander, Angelo C, Capt
Alexander, Charles N, Lcdr
Alexander, David A., LTJG
Alexander, David B., LCDR
Alexander, Douglas E., LCDR
Alexander, Esther Kay, CDR
Alexander, George C., Lcdr
Alexander, Herbert M, CDR
Alexander, Jacqueline S, LCDR
Alexander, James R, CDR
Alexander, Marvin G, Capt
Alexander, Owens F, Capt
Alexander, Richard J, CDR
Alexander, William W, CDR
Alexander III, William P., CAPT
Alexander Jr, Al G., Capt
Alexander Jr, Robert C, Capt
Alf, Donald, Capt
Alferi, Sharon F., Lt
Alford Jr, W Vanmeter, Capt
Alfortish, Anthony C, Capt
Alfortish Jr, Lester A, Capt
Alfred, Larry R, Capt
Alger, John W, CDR
Aljoe, Daniel W, Capt
All, Kenneth O, Capt
Alla Jr, Francis V, CDR
Allan, James S., RADM
Allan-Martinez, Mary F., CDR
Allard, Civita M., CDR
Allard, Mary Catherine, LCDR
Allbrandt, Benjamin L., LCDR
Allen, Barbara B, Lt
Allen, Bruce L., Capt
Allen, Charles L, CDR
Allen, Clayton R., LCDR
Allen, Cynthia J., Lt
Allen, Dennis K, Capt
Allen, Elaine H, Capt
Allen, Francis F, Capt
Allen, Frank F, CDR
Allen, Gary Richard, Capt
Allen, Harry W., CDR
Allen, Henry W, CDR
Allen, Howard H., CDR
Allen, Hugh O, Lcdr
Allen, James D, Lt
Allen, Leroy D, CDR
Allen, Martyn A, Capt
Allen, Michael D, CDR
Allen, R Kendall, Capt
Allen, Richard E, Capt
Allen, Russell E., LCDR
Allen, Ruth M, LCDR
Allen, S. Summer, CAPT
Allen, Stephen R, Capt
Allen, Thomas E, Capt
Allen, Todd R., LCDR
Allen, Wayne L, Capt
Allen, William B, Capt
Allen, William C, Capt
Allen, William F., LCDR
Allen, William H, Capt
Allen, William L, Lt
Allen, Willis M, Lcdr
Allen Jr, Arthur E., CDR
Allen Jr, Frederick S, Capt
Allen Jr, Glenn L, Capt
Allen Jr, Robert G, Capt
Allenbach Jr, Charles E, Lcdr
Allenby, Douglas R., Lcdr
Aller, William L, Lcdr
Alley, James A, Capt
Alley, Robert G, Capt
Alley Jr, Thomas C., Capt
Allin, Stephen C., CDR
Allinder, Steven M., LT
Allington, Arthur K., CWO3
Allison, Christopher D, CAPT
Allison, Donna J., CDR
Allison, Gene L, Capt
Allison, George J, Lcdr
Allison, James L., CDR
Allison, Richard H, Capt
Allison II, Forrest H, CDR
Allman, David James, LT
Allman, Patrick H., Lcdr
Allman, Stephen B, Capt
Allston, Frank J, Radm
Alm, Robert R, Lcdr
Almeida, Gary S, CDR
Almeida, Sandra A., Cdr
Almon, Anthony K., LT
Alosi, James C, Capt
Alpaugh, Michael N., Lcdr
Alt, Kevin W., CDR
Althouse, Larry L, Lcdr
Althouse, Michael S., LCDR
Altizer Jr, Philip J., CDR
Altmann, Richard G, Radm
Altobelli, Charles G., LCDR
Altsman Jr, Robert J, CDR
Alvarez, Anthony R, CDR
Alvarez, David C, Capt
Alvarez, Pedro, CDR
Alvarez, Reinaldo G, Capt
Alvarez, William J., CDR
Alvis, Stephen G, Capt
Alwine Jr, Paul R, Capt
Amarel, James L., Lt
Amato, Peter E., CDR
Amberg Jr, Richard H, Capt
Ambos, John F, Lcdr
Ambrogi Jr, Joseph N, CAPT
Ambrose, John T, Capt
Ambrose, Robert C, Capt
Ambrose Jr, Harry D, CDR
Amdur, Gilbert N, CDR
Amerson, Thomas L, CDR
Ames, Oliver F, Capt
Ames, Robert S, CDR
Ames, Scott C., CDR
Ames Jr, Fred F, Capt
Amicangioli, Fred M., LCDR
Amidon, Donald A, Capt
Amidon, Robert B, CDR
Amisano, John, CDR
Ammann, Mark S., LCDR
Ammerman, Don J, CDR
Ammon, Peter C, Capt
Ammons, Dan L., CDR
Amon, Ronald J., Lt
Amos, James W, Lcdr
Amos, John D., LT
Amrhein, Alan G, Capt
Amundsen, Melvin S, Lcdr
Anagnostos, George, Capt
Anawalt, William K., CDR
Anciaux, Louis N, Capt
Andersen, Dale Jay, CWO4
Andersen, Folmer P, Capt
Andersen, Harold, Lcdr
Andersen, Margaret M, Lt
Andersen, Roland M, Lt
Anderson, Arthur H, Lt
Anderson, Billie P, Lcdr
Anderson, Blane K, CDR
Anderson, Burton M, Capt
Anderson, Charles O, Capt
Anderson, Charles L, Lcdr
Anderson, Charles S, Capt
Anderson, Curtis W, CDR
Anderson, Cyrus E, Capt
Anderson, David A., Lt
Anderson, David A., LT
Anderson, David J, Lcdr
Anderson, David K, CDR
Anderson, David W., CDR
Anderson, Dean T, Lcdr
Anderson, Don R, Capt
Anderson, Dorrance I, Capt
Anderson, Douglas J., CAPT
Anderson, Eddie G, CDR
Anderson, Edward E, Capt
Anderson, Emmet D, Lcdr
Anderson, Eric D., Lcdr
Anderson, Erik N., LCDR
Anderson, Frank A, CDR
Anderson, Frank H, Radm
Anderson, Gareld W., CAPT
Anderson, Gary B., CAPT
Anderson, Gary L, Capt
Anderson, George E, CDR
Anderson, Gerald A, Lcdr
Anderson, Gerald F., CDR
Anderson, Harriet L, CDR
Anderson, Harry D, Lcdr
Anderson, Jack W, Capt
Anderson, James B, Capt
Anderson, James D., Capt
Anderson, James Neil, CWO4
Anderson, James P, Lcdr
Anderson, Jerold F, Capt
Anderson, Joel M., CDR
Anderson, John G, CDR
Anderson, John M, CDR
Anderson, Kenneth O, CDR
Anderson, Lawrence R, Capt
Anderson, Lisa A., LCDR
Anderson, Mark D., CDR
Anderson, Mark E., CDR
Anderson, Mark J, CDR
Anderson, Oliver A, Capt
Anderson, Paul N., CDR
Anderson, Raymond D, CDR
Anderson, Richard A, Lcdr
Anderson, Richard A, Lcdr
Anderson, Robert L, CDR
Anderson, Robert L., CDR
Anderson, Robert W, CDR
Anderson, Roger L, Lcdr
Anderson, Rolph E, Capt
Anderson, Royce, CDR
Anderson, Russell F, Capt
Anderson, Russell L., Lcdr
Anderson, Samuel A, Capt
Anderson, Scott D., Lcdr
Anderson, Steven C., Lcdr
Anderson, Susan W., CDR
Anderson, Ted, LT
Anderson, Thomas Y, CDR
Anderson, Ward R, Capt
Anderson, William A, Capt
Anderson, William D, Capt
Anderson, William G, Lcdr
Anderson, William W, CDR
Anderson III, James O., Lt
Anderson III, Lane S, Capt
Anderson IV, John R., Capt
Anderson Jr, Albert C, Capt
Anderson Jr, Raymond V., CDR
Anderson Jr, Ross K, Capt
Anderson Jr, Walter E, Capt
Andes, Dean C, CDR
Andolina, Carrie A., Lt
Andolsek, Timothy G, LCDR
Andrade, John R, Lcdr
Andrea, Mario I, Lcdr
Andreoni Sr., David J., Lcdr
Andres, Gary M, Lcdr
Andres, Stephen M, Capt
Andresen, John R., LCDR
Andrew, John R, LT
Andrew, Michael J., Lcdr
Andrews, Dennis Allen, Lt
Andrews, Douglas M., CDR
Andrews, Harry W, CDR
Andrews, Howard F, Lcdr
Andrews, John B, Capt
Andrews, Marion L, Lcdr
Andrews, Randolph C., LCDR
Andrews, Robert L, CDR
Andrews, Robert S, Lcdr
Andrews, Thomas E, CDR
Andrews, Thomas H, CDR
Andrews, William R, Lcdr
Andrews III, Thomas L, CAPT
Andrews Jr, Robert D, Capt
Andrews Jr, Thomas L, Capt
Andriano Moore, Richard, CDR
Andrus, James R., CAPT
Andrus, Peter L., Capt
Andryuk, Nicholas A., Capt
Andy, Steven M, CAPT
Aneshansley, Charles W, LCDR
Ang, Elsa P., Capt
Angelo Jr., Joseph M., CDR
Anglada, Jose B R, Capt
Angle, Richard W., Lt
Anibal, Fred R, Capt
Annas, Cheri A., LCDR
Annin, Timothy E., Capt
Anschuetz, Herman G E, CWO3
Ansell Jr., Burness F., Capt
Anson, Harry C, Capt
Anthamatten, James W., CDR
Anthony, Lee S, Capt
Anthony, Rhory R, Lcdr
Anthony, Robert, Capt
Anthony Jr, John B., LT
Anthony Jr., James F., LCDR
Anthuis Jr, John W, Capt
Antle, Stephen B., CAPT
Antoff, Jerome W, Capt
Antonczyk, Raymond, CDR
Antonellis, James A., CDR
Antonio, Robert J, CAPT
Antoniou, Paul E, CDR
Antoniuk, Thomas H., LCDR
Anunson, M Gregory, Capt
Anyaehie, Frances D., CDR
Anzalone, Thomas A, CDR
Anzilotti, Vincent J, RADM
Anzinger, Emily M., LT
Apellaniz, Luis F., Lcdr
Apgar, Charles K., Lcdr
Apostol, John N, CDR
Apostolico, Frank E, CDR
Appel, Steven W., LT
Appicelli, Robert A, Capt
Aprahamian, George, CAPT
Apseloff, Roy I., CDR
Arader, Walter G, CDR
Arado, Joseph E, Capt
Aramino, Arthur D, LCDR
Aramino, Laura R.Z., ENS
Arasmith, Scot, Lt
Arbaugh, Linton D, CDR
Arbuthnot, Alice E, Lt
Arcano, Joseph T., CDR
Arceneaux, Aline M, CDR
Archambault, Gerard P, Capt
Archer, Douglas W, CDR
Archer, Gregg B, Capt
Archer, John T., LT
Archer, Wiley J, Capt
Archibald, R. Scott, Lt
Archibald III, D W, Ltjg
Archibald Jr, Robert E, CDR
Arcoleo, Richard K F, Lcdr
Ardell, Eric B, CDR
Arden, Daniel D., LCDR
Ardis, Robert B, Capt
Ardizzone, Vincent, Lcdr
Arenas Jr., Justo, Lcdr
Arens, Clarence E., CWO3
Arensmeyer Jr, Harold J, CDR
Argenzio-West, Jerry A, CDR
Arias, Angelito Mercado, Capt
Arildsen, Jesse A, CDR
Arkins, Joseph, Lcdr
Arledge, George E, CDR
Arlinghaus Jr, Frank H, Capt
Armen Jr, George K, Capt
Arment, Duane R, CAPT
Armento, William J, Ltjg
Armerding, Carl E, Capt
Armeson, Karen M., CDR
Armet, Harold R, Capt
Armond Jr, Ernest C, Lt
Armstrong, Edward B, Capt
Armstrong, Edward L, Capt
Armstrong, James F., LCDR
Armstrong, Jan V, Capt
Armstrong, Lloyd R, Capt
Armstrong, Margaret E, Capt
Armstrong, Melvin R, Capt
Armstrong, Patricia, LCDR
Armstrong, Robert J, Lcdr
Armstrong, Stevan M, CAPT
Armstrong, Thomas A, CDR
Armstrong, William A, Radm
Armstrong, William C, Capt
Armstrong Jr, George A, Capt
Arndt, Craig M., LT
Arndt, Edward K, Capt
Arnold, Alison M, Capt
Arnold, Billy R, LCDR
Arnold, Charles A., LCDR
Arnold, Kermit N, CDR
Arnold, Lance M, CAPT
Arnold, Ray W., LCDR
Arnold, Richard S., CDR
Arnold, Sharon K., LCDR
Arnsdorff, Samuel V., LCDR
Arny III, L. Wayne, Capt
Aron, Linwood M, Capt
Aronsohn, Jeff D., Lcdr
Arretteig III, Ulysses, CDR
Arrivee, Kim A, Lcdr
Arseneault Jr, Arthur J, LCDR
Arthur, James C, Lcdr
Arthur, Paul K, Radm
Artz, Samuel D, CWO3
Arveson, Donald F, Capt
Arvonen, George Mark, CDR
Asbjornsen, Douglas J., CDR
Asbury, Samuel W., LCDR
Aschan, Harry R, Capt
Aschenbrenner, Frank A, Capt
Aschwald, Howard B, CAPT
Ashbaker, Joseph W, CDR
Ashbaugh, Charles I, Capt
Ashbrook, Kay A, CDR
Ashbrook III, Heber C, CDR
Ashcraft, Dan D, Lcdr
Ashcraft, Jesse L, CDR
Ashcraft, Lee, Capt
Ashdown, Cecil S, Lcdr
Ashe, James H, CAPT
Ashe, Judith A., Lt
Asher, William M, Capt
Ashford, Russell P., LCDR
Ashley, Mark J., LCDR
Ashworth, Winthrop F, CDR
Askey, Charles B., CAPT
Asmus, Rodger W., Capt
Asmussen, Tod D., LCDR
Asper, Albert Leo, CDR
Asrael, Aaron R, Capt
Asseff, Carl F., Capt
Aster, George H, Capt
Asterino, William M, CDR
Astrella Jr, Trin J., CDR
Atalski, Walter P, Lt
Atcheson, Raymond W, Capt
Atchison, Keith E., CDR
Atchison, William R, CDR
Aten, Leo D, CWO4
Athas, David P., LCDR
Atherly, Harold E, Capt
Athey, Daniel D., CDR
Athmann, Ronald J., LCDR
Atienza, Fernando C, Capt
Atkins, Barry Kennedy, Radm
Atkins, George, Capt
Atkins, Harold T., CDR
Atkins, Kenneth P.,
Atkins, Thomas B, CDR
Atkins Jr, Francis A., Lcdr
Atkinson, E. Richard, Capt
Atkinson, John Russell, CDR
Atkinson, Nancy J, CDR
Atkinson III, Marion L, Lcdr
Atkinson Jr, Oliver T, Capt
Atkinson Jr, Thomas C, Capt
Atkisson, Robert D, Capt
Atterbury, George R., CDR
Atwater, Charles D, Capt
Atwater, Marcia R., ENS
Atwood, Alan F, CDR
Aubery, Ronald T., LCDR
Auborn, James J., Capt
Auer, Bart D, Capt
Auerbach, Richard C, Capt
Auestad, Craig A., Lcdr
Augelli, Robert A, CDR
Augonis, Frank E, CDR
Aukofer, James J, CWO4
Aulgur Jr, Robert T, CDR
Ausbrooks Jr, Erskine P, CDR
Auslander, Stanley E, Lcdr
Austen, Philip A, CDR
Austin, Clayton B, CDR
Austin, Edward U, Radm
Austin, George B., Capt
Austin, Gordon T., CAPT
Austin, James A, Radm
Austin, Tom A, CDR
Austin, William C., CDR
Austin Jr, Amory E., Capt
Austin Jr, Wayne R, CDR
Austin Jr., Paul O., CDR
Auten, Donald E., CAPT
Auth, Leslie D., CWO4
Auvil, William P., CDR
Avedian, Victor V, CDR
Avera Jr, Troy G, CDR
Avery, David J, CDR
Avery, Francis A, Capt
Avery, Kenneth, CDR
Avey Jr, Floyd G, Lcdr
Avignone III, Frank T., Capt
Avillan-Rosa, Ana M, CDR
Avison, David B, CDR
Axon, Theodore L., CWO3
Axtman, William H, CDR
Ayala II, Jose, Lt
Ayars, Charles O, Capt
Ayers, Gary R., LCDR
Ayers, James L, Capt
Ayers, James R, Capt
Ayers, Robert J, Capt
Aylward, Richard M, Lt
Aylward-Bingman, Lynn M, Capt
Aynesworth, James L, Capt
Ayotte, Brett D., CAPT
Ayres, Daniel M., CDR
Ayres, G Curtis, Capt
Ayres, John P., LT
Ayres, Joseph F, LCDR
Ayres, Keith D., LCDR
Azeris, Harry T, CDR
Azzarita, Frank C., LT
Azzarita, Frank X, Capt
Azzarone, Patricia A., LT
Azzolina, Joseph, Capt
Babb Jr, Robert M, Capt
Babcock, Donald W, Lcdr
Babcock, Edmund T, CWO4
Babcock, William F, Capt
Babij, Raymond, Capt
Babin, Henry J., CDR
Baca, Stephen P, LTJG
Bacastow, Arthur J, Capt
Baccanari, Patrick A., LCDR
Bacci, Raymond R, Lcdr
Bach, J. Jean, LT
Bach, William L., LCDR
Bacher, Anita C., LT
Bacher, Michael W., Lt
Bachman, Charles E, CDR
Bachman, Jennifer E., LT
Bachrach, Morton W, Lcdr
Bachtel, Mary A, CDR
Backer, John M, Capt
Backer Jr, Matthias H, Radm
Backman, John R, CDR
Backstrom, James A, CDR
Backus, John L, Capt
Bacus, William R, CAPT
Badders, Rebecca S., LCDR
Baddour, Barry A., LCDR
Baden, Edward C., CDR
Bader, Douglas A, Capt
Bader, Robert H, Capt
Badger, David H, Lcdr
Badger, Thomas K, CAPT
Badick, Joseph R, CAPT
Badley, Bruce P, Capt
Baer, Thomas H, Lt
Baer, Thomas S, CAPT
Baertlein, Richard D., CDR
Baerwald, Ross F, CDR
Baffa, Richard C., LCDR
Bagby, Calvin L., CAPT
Baggett Jr, Morris E., CDR
Baggs, Andrew H, CDR
Baggs, Cory E., ENS
Bagley, John N, Lcdr
Bagley II, David W, Capt
Bagshaw, Robert E, CWO2
Bagshaw, William, CWO2
Bagwill, Gary L, CDR
Baham, Paul M., LCDR
Bahill, S Larry, CDR
Bahry, Vera Jo, Capt
Baie, Warren E, CDR
Baier, Robert E, Capt
Bailey, Carlyle K, Capt
Bailey, David Z, Lcdr
Bailey, Don, ENS
Bailey, Donald F., Lcdr
Bailey, Elizabeth J., LCDR
Bailey, Franklin E., CDR
Bailey, Fred A, CDR
Bailey, Gary B, Capt
Bailey, Jack M, CDR
Bailey, Jack S, CDR
Bailey, James G, Capt

Bailey, John H, Capt
Bailey, Kevin R, Lcdr
Bailey, Mary P., CDR
Bailey, Michaelene, LCDR
Bailey, Robert C, CDR
Bailey, Robert J, Capt
Bailey, Robert N, Capt
Bailey, Ronald O., Capt
Bailey, Vernon Perry, Lcdr
Bailey, William A, Lcdr
Bailey, William C., Capt
Bailey Jr, Charles E, CWO3
Bailey Jr, Charles E., CDR
Bailey Jr, Henry G, Capt
Bailey Jr, William T, Capt
Bain, Edward D, CAPT
Bain, Elliot M, CDR
Bain, Stephen C., CDR
Baird, Donald R, CDR
Baird, James A, Capt
Bakalarski, Leo C, CDR
Baker, Beverley F., CDR
Baker, David E., LCDR
Baker, Donald L, CDR
Baker, Frank J, CDR
Baker, George M, Lcdr
Baker, Herman D, CDR
Baker, J Marshall, Lcdr
Baker, James G., CDR
Baker, James P, Capt
Baker, John C, Capt
Baker, Joseph W, Capt
Baker, Kevin R., LCDR
Baker, Kurt F, CDR
Baker, Larry O, CDR
Baker, Lynn D, Capt
Baker, Mark C, Capt
Baker, Michael C, Capt
Baker, Scott P., CDR
Baker, Sloan H, Ens
Baker, Vance, LCDR
Baker, William H, Capt
Baker, William J, Capt
Baker, Willie M, Lt
Baker III, Walter L, CDR
Baker Jr, Carlos P, Radm
Baker Jr, William R, CAPT
Baker Jr., Joe M., LT
Bakke, Marian M, CDR
Bakshis, John A, Lcdr
Balacki, Margaret F, Capt
Balano, Randy C., LCDR
Balaska, Mark A, CDR
Balboni, Raymond L, Lt
Balcerak, John W., CDR
Balch, Lloyd E, Lcdr
Baldauf, Joseph J., CDR
Balden, William H, CDR
Baldridge, Norma J., Lcdr
Baldwin, Dennis R., Lt
Baldwin, Donald R, CDR
Baldwin, Doreen L., LT
Baldwin, George H, Capt
Baldwin, John G, CDR
Baldwin, Paul G, CDR
Baldwin, Robert C, Capt
Baldwin, William C, Lcdr
Baldwin III, Simeon, CDR
Bale, Jens E, Lcdr
Bale, Mary, Lt
Balek, Ronald J., CWO4
Balent, Thomas R., LCDR
Bales, John A, Capt
Bales, Leonard D., CDR
Balestra Jr, Louis J, Capt
Balian, Harry, CDR
Balian, Robert S, Lt
Balint, Gregory G., Lt
Balk, Leo D., CDR
Ball, Allan C, Lt
Ball, Bryan K., CDR
Ball, Derwin R, CDR
Ball, Linda C., Lcdr
Ball, Richard L, Lcdr
Ball, Robert O, Capt
Ballance, George W, CDR
Ballance, Jerry, CDR
Ballew, Carol M, Capt
Ballew Jr, Paul K, Capt
Ballinger, Richard H, Capt
Ballister, Christopher, Lt
Ballou, Charles W., CDR
Ballou, Deborah S., LCDR
Ballou Jr., Thomas B., Capt
Ballow, Robert E., Capt
Ballwanz, Robert J., CWO2
Balnicky, Annette H, CDR
Balo, Harold A, Capt
Balog Jr, Julius, CDR
Balogh, William L, Lt
Baltierra, Richard C., ENS
Balunek, Adelbert A, CDR
Bambao, Marie A., LT
Bamrick, Edward E., CDR
Banas, Bryan, LT
Banda, Lionel A W, Capt
Bandler, Martin L, Lcdr
Bane Jr, Eugene M, CDR
Baney, Richard N, Lcdr
Bangma, John J, CDR
Bankerd, John P, Capt
Bankhead, Royce E, CDR
Banks, David E., CDR
Banks, Jane K., LT
Banks, Martin L, CDR
Banks, O. Gordon, Capt
Banks, Paul I, Lt
Banks Jr, Willie B, Capt
Banta, John M, Capt
Banta, Robert L, Lcdr
Baraldi, William J, Capt
Baralt, Juan R, Capt
Baranchulk, George, CDR
Baranowski, Arthur J, Capt
Baratta, Anthony J., CAPT
Barausky, Kenneth P., Radm
Baravik, Andrew W, Capt
Barber, Dean G, Lt
Barber, Francis A., Capt
Barber, John B, LT
Barber, Michael E, Capt
Barber, Teri J., LCDR
Barber Jr, Rufus R, Capt
Barber Sr, Richard L, Capt
Barbero, Eugene M, Capt
Barbero, Martin, CDR
Barbu, Walter P, Lcdr
Barcelo Jr, Peter, Lt
Barclay, William A, Lcdr
Barclay Jr, Charles O, CDR
Barclift, Steven C., LCDR
Barco, Robert O, Lt
Barden, William R., Lcdr
Bardenwerper, James E, Lcdr
Bare, Leland D, CDR
Barford, George, Capt
Barg, Michael D, Lcdr
Bargar, Robert S, Lcdr
Bargman, George R, Capt
Barham Jr, Charles D, Capt
Barile, Vincent L, CDR
Barker, Brian H, Capt
Barker, Corey B., ENS
Barker, Jimmy L, CDR
Barker, Thomas C, Capt
Barkhurst, Ralph, CWO4
Barkley, Joseph A, Lcdr
Barkman, Paul E, CDR
Barks, Horace B, Lcdr
Barling Jr, S Douglas, Capt
Barlow, Joseph Boyd, Lcdr
Barlow, Phyllis L, Capt
Barlow, Shirley, CWO4
Barlow Jr, Walter, Lt
Barlow Sr, Earl F., CWO3
Barnard, William F, Capt
Barnard Jr, John L, CDR
Barner, Mark F., LCDR
Barnes, Brian G., LCDR
Barnes, Carl M., ENS
Barnes, Donald J, CDR
Barnes, Forrest W, CDR
Barnes, Freddie I., LCDR
Barnes, John B, Capt
Barnes, Leonard E, CWO4
Barnes, Michael A, Capt
Barnes, Richard L, Capt
Barnes, Robert C., CAPT
Barnes, Robert W, Lcdr
Barnes, Roy H, Capt
Barnes, Thomas L, Capt
Barnes, Wendy L, CDR
Barnes, Wesley E, CWO4
Barnes, William C, CDR
Barnes, William J, Lcdr
Barnes Jr, Lawrence M, CDR
Barnes Jr, Maury E, CDR
Barnett, Charles B, Capt
Barnett, John S, Capt
Barnett, Pernel George, Ltjg
Barnett, Vyneta F., CDR
Barnett, William R, CDR
Barnette, Richard T, Capt
Barnhill, James O., Ltjg
Barnhill, Philip C, Capt
Barnum, Gaylene I., Lcdr
Baron, Rosalind, Lcdr
Baroudi, David G., Lt
Barr, Charles V, Capt
Barr, Edwin A, Lcdr
Barr, William S, Capt
Barrack, William S, CDR
Barrera, Manuel A., LCDR
Barret, Albert R, CDR
Barrett, Curtis D., CWO3
Barrett, Darwin S, Capt
Barrett, Michael J, Capt
Barrett, R Lloyd, Lcdr
Barrett, Stephen P, Capt
Barrett, William S, Capt
Barrett Jr, Edward L N, Capt
Barrett Jr, William H, Capt
Barrick III, Harry W, Capt
Barringer, Raymond E., CDR
Barron, Cameron A, Capt
Barron, Claude E, CDR
Barron, Gary R, LCDR
Barron, Gilbert, CDR
Barron, Leo E, CDR
Barron, Thomas J, Lcdr
Barron Jr, C Alex, CDR
Barrons, John C, Capt
Barrow, Charles W, Capt
Barrow, Robert W, CDR
Barrows, Frank B, Lcdr
Barry, Andrew M, Capt
Barry, George F, CAPT
Barry, Michael S., CDR
Barry, Paul L., CDR
Barry, Philip C, Lcdr
Barry, Stephen G, CDR
Barsness, John G, CDR
Bartee, Katherine, Lcdr
Barteluce, John J, CDR
Barth, Kristen L., Lt
Barth, Mark D., LCDR
Barth, Roger T, Capt
Bartholf, Robert, CDR
Bartholmey, Don S, Capt
Bartholomew, Charles W, CDR
Bartlett, John D, Capt
Bartlett, Michael B., Lcdr
Bartlett, Willard W, CDR
Bartlett, William H, Capt
Bartlett Jr, Draper, Lcdr
Bartley, Timothy J., Lt
Bartman, David F H, Ltjg
Barto, Stephen C., CDR
Bartolome, Scott, LCDR
Barton, Alexander J, Capt
Barton, Clark J, CDR
Barton, Edward B., Capt
Barton, Furman W, CDR
Barton, Gary F, CDR
Barton, Robert Wayne, Capt
Barton, Steven A., Cdr
Barton, Thomas E, CAPT
Barton, William B, Capt
Barton Jr, Douglas W, CDR
Bartoo, Rowland R, Lcdr
Bartosh, Stephen J, Lcdr
Bartsch, William, LCDR
Bartuska, Anthony J, Capt
Bartz, Jeffrey J., CDR
Bartz, Wayne, Capt
Bartz, William A, Capt
Bartz Jr, William G, CDR
Baruzzini, Joseph L, Lcdr
Basdekian, Edward A, Capt
Bash, Dennis M, CDR
Bashar, Peter N., LCDR
Basile, David, CDR
Basilio Jr, Anthony C, Capt
Basinger, Lorain A, Lt
Baskette Jr, Floyd K, Lcdr
Baskin, Walter H, Capt
Bass, Arthur E, CDR
Bass, Lunceford L., CDR
Bass, Malcolm E, CAPT
Bass, Orville D, LCDR
Bass II, John R, Capt
Bassanello, Adolph C, CDR
Bassett, Timothy S, CDR
Bassett III, Charles C, Ltjg
Bassford, Philip J, CDR
Bastien, Gill B, Capt
Bastob, Robert E, Lcdr
Batchelor, Kenneth F, Capt
Bateman, James A, CDR
Bateman, John W, Lcdr
Bateman, Mark, CDR
Bateman, Warren F, Lt
Bates, Donald C, Capt
Bates, Jack R, Capt
Bates, Nancy J., CDR
Bates, Roy L, Lcdr
Bates, Thomas R, Capt
Bates, Wayne L, Capt
Bath, David R, Capt
Batko, Anthony M, Capt
Batlle III, John A., CDR
Batten, James W, Lcdr
Battle, Steven S., LCDR
Battle, Virginia A, CDR
Battle, William E., CDR
Batton, Phil D, Capt
Battson, Vic, CDR
Baublitz, John E, Capt
Bauchspies Jr, Rollin L, Capt
Baudrau, Robert F., CDR
Bauer, Carl T, Capt
Bauer, Catherine M, Capt
Bauer, Kenneth H, Capt
Bauer, Lynton G, Capt
Bauer, Peter E, Lcdr
Baughman, Diane C., CDR
Baughman, Stephen D, CDR
Bauhan, Thomas L., Lt
Baum, Gerald R., Lcdr
Bauman, William E, Capt
Bauman, William L, CDR
Baumann, Dale N., Lt
Baumert, Frank T., CWO4
Baumgardner, Hugh W, Capt
Baumgardner, Loren L, Lt
Baumgardner, Walter L, CDR
Baumgartner, Wayne A, Lcdr
Baummer, Mark, ENS
Baumrucker, Marjorie L., CDR
Bautista, Manuel F, Capt
Baxter, Albert T, CWO4
Baxter, Valerie L., Lcdr
Baxter, Virginia R., CDR
Baxter, William A, CDR
Baxter, William L, Lcdr
Baxter Jr, James A, Lt
Bayard, Richard C, Lcdr
Bayer, David S, Lcdr
Bayer, Stephen R, Ens
Bayer III, Frederick B, CDR
Bayles, John J, Lcdr
Bayless, Charles S, CDR
Bayless Jr, Jon W, CDR
Bayley, John L, CDR
Baylies, Howard N, Lcdr
Baylor, Larry A, Capt
Baynes, Gerald T, Capt
Bayraktar, Ilker, Lt
Bazan, Robert S., LCDR
Bazemore, Russell, LCDR
Bazemore, Susan R., CDR
Bazemore, Webster C., CAPT
Beach, Richard L., LT
Beachler, Kenneth C, Capt
Beachley, Kenneth G, Lt
Beahan, Dennis J, Lcdr
Beahrs, John V, CDR
Beal, James A, Capt
Beal, John C, CDR
Beall III, L Clay, CDR
Beam, Bruce A, Capt
Beam, Gerald W, Lcdr
Beam, John M, Lcdr
Beam, Loudin J, Lt
Bean, Irving J, CDR
Bean, John L, Lcdr
Bean, Joseph L, Lt
Bear, Donald G, CWO2
Beard, James C, Lt
Beard, Joan M., CDR
Beard, Joseph J, Capt
Bearden III, James D, Capt
Beardman, Edward L, CDR
Beardslee, Jay W, CDR
Beardsley, James D., LT
Beardsley, Robert L, Lcdr
Beary III, John F., Capt
Beasley, John K., LT
Beasley, Katharine E., Lcdr
Beatty, Daniel A., CAPT
Beatty, Mark D., CDR
Beatty, William J, Lcdr
Beauchemin, A Ernest, CDR
Beaudrot, Lewis O., LCDR
Beaudry, Frederick H, Capt
Beaumont, William E., Lcdr
Beaver, Dennis T., CAPT
Beavers, Fred L., CDR
Beavins, Robert C, Capt
Bechard, Paul F, CDR
Bechtel, Melinda, Lcdr
Bechtelheimer, Thad C, Lcdr
Beck, Anthony W, Lt
Beck, Bryce L, Lcdr
Beck, Christopher V., LT
Beck, Donald A, LCDR
Beck, Donald G, Capt
Beck, Douglas A., ENS
Beck, George A, Capt
Beck, John A, Lcdr
Beck, L Brandt, Capt
Beck, Lemuel E, CWO4
Beck, Peter C., LCDR
Beck, Richard D., Lt
Beck, Robert C, Lcdr
Beck, Robert J, Capt
Beck, S Paul, Lt
Beck Jr, Robert F., CDR
Becka, James F., CDR
Becker, David E, Capt
Becker, Henry A,
Becker, Pierre R, Capt
Becker, Richard L, Radm
Becker, Robert E, CDR
Becker, William J, Capt
Becker Jr., Fred R., CAPT
Beckett, David A., LCDR
Beckham Jr, Walter H, Capt
Beckman, Ellen J, Lcdr
Beckman, Thomas G., CDR
Beckman Jr, Joseph C, Capt
Beckmann Jr, Paul A., CDR
Becknel, Gerald P, Capt
Beckum, Charles A., Lcdr
Beckwith, Peter H, Radm
Bedard, Albert, Lcdr
Bedell, Alfred C, Lcdr
Bedenbaugh, Garnett F, Capt
Bedenbaugh, William H, Capt
Bedford, John L, CDR
Bedinger, Robert W, CDR
Bedingfield, Hugh W, CDR
Bednar, Edward G, Capt
Bedoya, Francis Xavier, Lcdr
Bedsole, James T, CDR
Beeby, Gered H, Capt
Beecroft, James A, CDR
Beel, Todd L, Capt
Beem, Raymond M, Capt
Beeman, Thomas E, CDR
Beemer, Bruce A., Capt
Been, Robert J., Lt
Beers, Cloyd R, Lcdr
Beers, Donald W, Capt
Beers, Dorothy L, Lcdr
Beesburg, Robert Y, Capt
Beesley, Melvin H, CWO4
Beeson Jr, Edward L, CDR
Begley, Harry R, Lcdr
Behlendorf, Jack R., CDR
Behounek, Charles J, CDR
Beichley, Glenn L, Lcdr
Beigh, Clyde H, Lcdr
Beihl, William J, Lcdr
Beilke, Robert L., CDR
Beimer, Michael S, CDR
Belcher, James W, LCDR
Belden, Daniel C, CDR
Belfon, Robert F, Lcdr
Belinski, Edward J, Capt
Belisle, Kenneth C, RADM
Belko, Martin L., LT
Bell, Daniel B., CAPT
Bell, David P, CDR
Bell, Dennis D, Lcdr
Bell, Edmond V, Capt
Bell, Francis E, CDR
Bell, Garland E, CDR
Bell, George B, Capt
Bell, James J, Capt
Bell, Joel L., CWO3
Bell, John B, Capt
Bell, John H, Capt
Bell, Johnston B, CDR
Bell, King D, CDR
Bell, Lyndon R, CDR
Bell, Nick E., CWO4
Bell, Philip A, Lcdr
Bell, R Victor, Capt
Bell, Robert L, Capt
Bell, William F, Capt
Bell, William H, Capt
Bell, William Scott, Lcdr
Bellebaum, Donald E, CAPT
Bellenkes, Andrew H., CDR
Bellew, Francis H, Capt
Bellezza, Vincent A., CAPT
Bellino, Joseph M, Capt
Bellows, Douglas J, Capt
Bellringer, Edward A., LT
Belser, Richard E, Capt
Belser Jr, Townsend M, Capt
Belsher, Thaddeus R, Capt
Belt, Michael A., LT
Belton, Mark J., CDR
Beltrami, Albert P, CDR
Beltz, William R, Capt
Belyea, Eva M, Lt
Bembenek, Gail A.T., CDR
Benard, Gregory T., MIDN
Benator, Barry I, Capt
Benavage, Rae Eileen, CDR
Bencik, Charles A, Lcdr
Bendall, Michael L, CDR
Bender, John H, Capt
Bender, Robert D, Capt
Bender Jr, Edward J, Capt
Bendig, Judith J, CDR
Bendo, Allen J, CWO4
Benedetti, Aldo J, Capt
Benedict, Norman V, CDR
Benedikt, Cynan J. M., Lt
Benefield, Robert B, CAPT
Benevides Jr, Francis L, Capt
Benezette, Alyn L, CDR
Benharbit, Leila Jane, LT
Beninate, Douglas J, Lt
Benjamin, Analiza, LT
Benjamin, Charles H, Capt
Benjamin, Galen K, Capt
Benjamin, Julian R, Radm
Benjamin, Richard L, CDR
Benjamin, Stuart H, CDR
Benjes, William B, Capt
Benner, William B, CDR
Bennett, Alfred A, CDR
Bennett, Barry E, Capt
Bennett, Charles S., LCDR
Bennett, Daniel L, Lcdr
Bennett, David J., LCDR
Bennett, Francis V, Capt
Bennett, Gary D., LCDR
Bennett, Harry H, Capt
Bennett, J. Douglas, Capt
Bennett, James N, Lcdr
Bennett, James W., LCDR
Bennett, Jerry P, CDR
Bennett, John A, Lcdr
Bennett, John T, Lcdr
Bennett, Lawrence E, CDR
Bennett, Lincoln J, Lcdr
Bennett, Michael C, Lcdr
Bennett, Paul R, CDR
Bennett, R Lee, Lcdr
Bennett, Richard W, CWO4
Bennett, Robert A, CDR
Bennett, Robert L, CDR
Bennett, Scott A., LT
Bennett, William C, Capt
Bennett, William D., Lt
Bennett III, Earl C., Lcdr
Bennett Jr, Charles A, Capt
Bennion, Paul A, CDR
Bennion, Russell R., Lt
Benoist, J Hunt, CDR
Benoit, Richard P, Capt
Bensinger, Gregg A., CDR
Benson, Allen M, Capt
Benson, Arthur S., CDR
Benson, Dawn J., CDR
Benson, Donald D, CDR
Benson, H Curtis, Capt
Benson, Harold W, Capt
Benson, Horace P, Lcdr
Benson, James F, CDR
Benson, Jerome E, Capt
Benson, Joe N, CDR
Benson, John T, Lt
Benson, Joseph F, Lcdr
Benson, Larry Joe, Capt
Benson, Leo J., Lcdr
Benson, Michael J, Capt
Benson, Ray W, Capt
Benson, Richard Adam, CDR
Benson, Richard C, Lt
Benson, Robert B, Lcdr
Benson, Thomas H, Lcdr
Benson Jr, Charles W, Capt
Benson Jr., F. Delano R., LT
Bentivoglio Jr., Mario, ENS
Bentlage, Roger E., CDR
Bentle, Larry A, CAPT
Bentley, Charles R, Lcdr
Bentley, Donald R, Capt
Bentley, Geoffrey D, Capt
Bentley, Joe L., LCDR
Bentley, William R, CDR
Bentmann, Irwin F, Lcdr
Benton, Daniel A, Lcdr
Benton, John E, CDR
Benton, Robert R, CDR
Benton, Thomas H, CDR
Benton, W. Duane, Capt
Benton III, William C, Capt
Benton Jr, Robert C., CDR
Benton Sr., Mark G., Lt
Bentson, Gordon J, Capt
Bentz, Alan E, Capt
Benz, Stephen J, CDR
Berard, John D., CDR
Berard, Raymond W., Capt
Berardo, Joseph A., CDR
Berchielli, Gian L, Capt
Berck Jr, Henry Fred, Capt
Berdick, Marland L, Lcdr
Berens, James D, Capt
Beresford, Wilfred E., CDR
Berg, Donald G., LT
Berg, Donald M, CDR
Berg, Eric R, Capt
Berg, Martin W, CDR
Berg, Paul Doug, CDR
Berg, Peter E, Capt
Berg, Richard C., CDR
Berge, Eleanor H, Lcdr
Bergen, Blair A, CDR
Berger, Arthur W, Lcdr
Berger, Gregg M. S., Lcdr
Berger Jr, James A, CWO4
Bergeron, James H., Lcdr
Bergeron, Thomas W, CDR
Bergert, Frederick L, CDR
Berghult, Carl R, CDR
Bergquist, Carl W, Capt
Bergquist, Laurence C, CDR
Bergstrom, Arthur E., CDR
Berkaw, Bergen F, CDR
Berkeley Jr, Francis L, Capt
Berkey, Christopher Jon, ENS
Berkley, Leonard, CDR
Berlin, Bruce A, Capt
Berlin, Howard R, CDR
Berman, Lloyd A, Lcdr
Berman, Samuel I, Capt
Bernal, Ricarte A., CDR
Bernard, Francis X, Capt
Bernard, Franklin H, CAPT
Bernard, Robert J, Lt
Bernard, William N, CDR
Bernard, William V, Capt
Bernardon, Robert A., CDR
Bernays, Peter M, LCDR
Berner, Kenneth C, Capt
Berngard, Melvin, Capt
Bernhard, Paul G., Lcdr
Bernhardt, Paul A, Capt
Bernheim, Joseph B, Lt
Bernichon, Clay E, Lcdr
Bernie, Ronald J., CDR
Bernier, J Peter, Capt
Bernin, Victor M, Lt
Bernloehr, George R., LT
Berns Jr, Bernard A, Lcdr
Bernstein, Barbara E, CDR
Bernstein, Charles R, CDR
Bernstein, Robert S., Capt
Bernstein, Sidney S, Capt
Bernzen, Edward L., CDR
Berrett, M Victor, CDR
Berry, Alan R, CDR
Berry, Arthur F, CDR
Berry, Edward F, CWO4
Berry, John V, CDR
Berry, Patrisha L, Lcdr
Berry, R Crist, Capt
Berry, Robert W, CDR
Berry, Susan S, Lcdr
Berry III, Charles L., LT
Berry Jr., Vance D., CDR
Berryman, Edward B, CDR
Bersson, William T., CDR
Bertels Jr, Robert W, Lcdr
Bertish, Gerry J., LT
Bertoldi, John B, Capt
Bertolucci, Louie, Lcdr
Bertonneau, George A, Capt
Bertrand, John R, CDR
Bertsch, Fred S, Capt
Berube, Nathan E., LCDR
Beruvides, C. Marcelo, Lcdr
Berwick, Robert L, Capt
Berzins, Elwood J, CDR
Besch, Emerson L, Capt
Besheer, Roy P., CAPT
Besier, Rudolph F, Capt
Beskind, Robert L, Capt
Best, Mildred E, Capt
Best, Otto G, Lt
Best, Robert E, Capt
Best, Rosalind, Ens
Best, Scott A., Lcdr
Best, William V, Capt
Betancourt, Ronald, LCDR
Bethel, Edward L, CDR
Bethmann, James M., LCDR
Bethune, Graham E, CDR
Bethune, Walter F, Capt
Betit, Gregory C., CDR
Betleski, Adrian F, Lt
Betts, Fred R, Capt
Betts, Harry E, CDR
Betzold, John N, CDR
Beumer, Delbert H, Radm
Bevard, Bruce B., CDR
Beverage, Parker J, Capt
Bewick, James S, Capt
Beyer, Douglas C., LCDR
Beyer, Loraine M, CDR
Beyer, Nancy L, CDR
Beyer, Robert H, Capt
Beyer, Theodore H, Ltjg
Bezner, Hubert P, CDR
Bhattacharya, Arjun, CAPT
Biache Jr, Andrew, Capt
Biancalana, Silvio J, Capt
Bianchi, John R, Capt
Bibbs, Charles S., LCDR
Bick, Maurice, Lt
Bickford, Carol J, CDR
Bickle, Robert C., Lt
Biddle, Richard S, Lcdr
Biddulph, John, Capt
Bidewell, Len C., Lt
Bidwell, Ann E., CAPT
Biedebach, Herbert E, CDR
Biega, David P., CDR
Bielagus, Joseph F, Lcdr
Bielecki, Loretta F, CDR
Bielka, Robert P, CDR
Bien, Lyle G., VADM
Bierling-Slowey, Karin, LCDR
Bierman, Lawrence L, CDR
Biermann, Carlton A, CWO4
Biermanski, Lothar P, Capt
Bigelow, Brandon L., LT
Bigelow, Karen E., CDR
Bigelow, Richard L, CDR
Bigelow, Russell T, CDR
Bigenho, Roy M, CDR
Bigger, Charles H, Capt
Bigger, Darold F., Capt
Biggers, James C, Capt
Bigler, Glade S, Capt
Bigley Jr, Harry A, Capt
Bigsby, Charles F, Capt
Bik, John, Lcdr
Bilenker, Arthur J., Lcdr
Biles, Willis L., CDR
Billerbeck, Fred W, Capt
Billharz, David W, Capt
Billings, Thomas, CWO4
Billingsley, William M, Capt
Billington, Bryan J., ENS
Billman, John A, CDR
Billmyer III, George W, Lcdr
Bills, Eldon Dean, Capt
Billstone, Donald L, Capt
Bindon, Rodney B, Capt
Bingham, Elsie M, Lcdr
Bingham, Jane D, CDR
Bingham, Jerry D., LT
Bingham, Tad H, Capt
Bingham Jr, Harry P, CDR
Biolo, Mark W., Capt
Bipes, Mark E., LCDR
Birch, John W, CDR
Birch, William P., CDR
Bircher, Nicholas G, CDR
Birckbichler, Joseph E, Lcdr
Bird, Christopher W, CDR
Bird, James F., CDR
Bird, James S., CAPT
Bird, John A, Lcdr
Bird, Jon A, Capt
Bird, Kenneth D, Capt
Bird, Stephen A., CDR
Birdsall, John W, Lt
Birdwell, Thomas R, Capt
Birdwell Jr, James E, CDR
Birindelli, James B, Capt

Birk, William M, CDR
Birkner, Robert O, Capt
Birky, Zane F., CDR
Birmingham, William J, Capt
Birnbaum, Bernard I, CDR
Biron, Marshall R, Lcdr
Birosh, Dale A., Lcdr
Birr, Douglas R., CAPT
Birtcil Jr, Robert F, Radm
Bischoff, Craig E., CDR
Bisgard Jr., Carl V, Capt
Bisgrove, Michael E, Capt
Bish, Jerome D, CDR
Bishard, Pamela S., Lt
Bishop, Andrea C., LCDR
Bishop, Billy P, Lcdr
Bishop, Conrad C, CDR
Bishop, Deidre L., LCDR
Bishop, Donald C, CDR
Bishop, Edwin L, Capt
Bishop, Jerrold K, Capt
Bishop, John C., CDR
Bishop, John Robert, Capt
Bishop, Raymond E, CDR
Bishop, Robert J, CDR
Bishop, Timothy J, CDR
Bishop Jr, Harold E, CDR
Bishop Jr, Robert L, CDR
Biss, Joseph E, Capt
Bisset, Andrew E, Capt
Bitter, Marcus K, Lcdr
Bittick, Johnny M, Capt
Bittle, Lehner K, Capt
Biwer, Robert A, CDR
Bize III, David N., CAPT
Bizjak, William A, Lcdr
Bizzell, Dorothy E, CDR
Bjerke, Ardine L, Capt
Bjorge, Robert C, CDR
Bjork, Karl A., LCDR
Bjornsson, Gottskalk T, Capt
Black, Barbara S., CDR
Black, Bruce A, Radm
Black, Bruce M., Capt
Black, David L, CAPT
Black, Dock F, Lcdr
Black, Edward J, Lcdr
Black, Gerald D., LCDR
Black, James B, Capt
Black, Keith N, Capt
Black, Larry C., CDR
Black, Paul L, Capt
Black, R Powell, Capt
Black, Robert O, CWO4
Black, Robert R, CDR
Black, Ronald R, CWO4
Black, William C, CDR
Black, William D, Lcdr
Black, William H, Capt
Black Jr, Hugh, CDR
Black Jr, Robert G, Capt
Blackburn, Chris K., LCDR
Blackburn, Henry Lee, LCDR
Blackburn, James H., LCDR
Blackburn, Jeffrey E, CDR
Blackburn, John M., LCDR
Blackburn, Karen F, CDR
Blackburn, Robert S, Capt
Blackburn, Wendy S., CDR
Blackford, James D, Capt
Blackman, Geoffrey A, Lcdr
Blackman, Kay S, Lcdr
Blackman, Ralph E, CDR
Blackmon, Jack R, Capt
Blackshaw, Arthur D., CDR
Blackstock, James F, Capt
Blackwell, Edwin H, CDR
Blackwell, P. D., LT
Blackwood, Hugo G., CAPT
Blackwood, Martin S, Lcdr
Blackwood, Ricky L., CDR
Blackwood, Thomas R., CAPT
Blades, Judy G, Lcdr
Bladow, Craig W., LCDR
Blair, Brian E, CDR
Blair, Charles R, CDR
Blair, John E, CDR
Blair, Michael K., Lcdr
Blair, Richard J, Capt
Blair, Roger, CDR
Blair, William N., CAPT
Blaise, Francis J, CDR
Blake, David R, CDR
Blake, J. Cameron, CAPT
Blake, Kenneth W, CDR
Blake, Peter J, CDR
Blake, Raymond G, Capt
Blake, Robert W, Lcdr
Blake, Russell S, CDR
Blake, Thomas H., CAPT
Blake, William C, Capt
Blake Jr, Joseph C, CDR
Blakesley, Larry L, CDR
Blalock, Leonard Keith, Capt
Blanc, Robert D, Lcdr
Blanchard, Duane E, Capt
Blanco, Lucas J, Capt
Blaney, Walter A, CDR
Blanford, George T, CDR
Blank, Charles Anthony, Lcdr
Blanke, Lovetta R, CDR
Blankenship, Billy J, Capt
Blanton, Lindsay C, CDR
Blanton III, Sankey L, Lcdr
Blanz, C Dennis, CDR
Blaschke, James, CDR
Blasi, Don T, Capt
Blasko, Joseph A, Lcdr
Blatt, Norman W, Capt
Blatt Jr, Walter C, CDR
Blatus, Richard J, CDR
Blayney, Michael J., Lt
Blazek, James A, Capt
Blecken, William E, Lcdr
Bledsoe, Billy L, LCDR
Bleicher, Sheldon J, Capt
Blekicki, Christopher T., LT
Blemaster, Norman Lee, Capt
Blence Jr, F A, Capt
Blenkhorn, James M., Capt
Bleskey, Daniel M., CDR
Blethen, Lawrence P, Capt
Bletner, James K, LCDR
Bleveans, John, Capt
Blevins, Roger D., Lcdr
Bley, Dennis C, CDR
Bleyle Jr, George A, Capt
Bligh, Paul H., Lt
Blikken, Kathy L, LCDR
Blikken, Warren A., LCDR
Blish, Nelson A, Capt
Blitch, Thomas A, Lcdr
Block, Charles E, Capt
Block, Isaac, LTJG
Block, John A, Capt
Block, Julia A, Capt
Block, Leo, Capt
Block, Martin J, Capt
Block, Mervin, Lt
Blodgett, Emmett A, CDR
Blome Jr, George S, Lcdr
Blomquist, Roger N, Capt
Blondin, Carl A., Lcdr
Blood, Charles W, Lcdr
Blood Jr, Joseph B, Capt
Bloom, Elias, Capt
Bloom, Peter S., LCDR
Bloom, Robert F., CAPT
Bloomingdale, Lyman G, Lt
Blosberg, Gary K, CDR
Blose, Mary S., CDR
Blostan, Dell, Capt
Blot, Robert D., CDR
Blough, Herbert A, Capt
Blouin, Harvey L, Lcdr
Bloxom, Richard R, Capt
Blue, J. Guthrie, Lcdr
Blum, Felix E, CDR
Blum, John A, Capt
Blum, John R., LCDR
Blumberg, Herbert K, Capt
Blume, Russell E., Capt
Blume III, Arthur W, CDR
Blumenthal Jr, Philip L, Capt
Blunt, Kenneth J., LCDR
Blunt, Matthew R., LT
Blunt, Paul F, Capt
Blunt, Stephen R., LCDR
Bly, Raymond L, Capt
Blyckert, Warner A, Lcdr
Blythe, Homer D, Capt
Boardman, Keith A, CDR
Boasberg, Albert L, LT
Bobick, Catherine A., LT
Bobko, Allan D, CDR
Bobkowski, Sabrina J., LT
Bobrick, Edward A, Capt
Boccio, Dona V, LCDR
Boche, Harold M, Lcdr
Bochniarz, Thaddeus, Lcdr
Bock, Brett D., Lt
Bock, George R, CWO4
Bockeloh, James A., LTJG
Bockis Jr., Frederick J., LCDR
Boddington Jr, Edward M, Lt
Boddy, William G, CAPT
Bodenhorn, Gregory S, Capt
Bodenschatz, Robert B., CDR
Bodnar, George T, Capt
Bodnar, Louis A., LT
Bodour, Haig, Capt
Boeck, Larry J, Capt
Boehler III, Richard F, Lcdr
Boehme, Richard J., CDR
Boelens, John H, Capt
Boening, Andrew C., CDR
Boerbon, Floyd W, Capt
Boerke, Raymond H., Lcdr
Boerner, Gilbert W, Lt
Boerner, Peter R, Lt
Boerum, Thomas M., CDR
Boettcher, Thomas C, Capt
Bogan, John C, CDR
Bogart, James R, CDR
Bogart, S David, Capt
Bogden, Jim, LT
Bogel, Dorothy S, CDR
Bogert, Gary M, CDR
Bogert, John A, Capt
Boggeln, Laurence H., CAPT
Boggs, Harry W, Capt
Boggs, Lorraine C, Capt
Boggs, Marion W, Lcdr
Boggs, Robert E, CDR
Bogle, H Wesley, Lcdr
Bogle, Homer T, CDR
Bogle, Samuel L., CDR
Bogsted, Thomas, CDR
Bohan, Timothy W., LCDR
Bohannon, Edward Lee, Capt
Bohannon, Julie L., CDR
Bohannon Jr, M T, Capt
Bohensky, Donald F, CDR
Bohley, Michael Francis, Lcdr
Bohlke, Terry A., LCDR
Bohlmeijer, Janis L., Capt
Bohm, Dwight K, Capt
Bohmiller, Kevin D., Lcdr
Bohn, Robert W, Capt
Bohon, Charles R., LCDR
Boice, Kenneth C., Lcdr
Boies Jr, Lawrence R, Capt
Bois, Richard M., CDR
Boissonneault, Kevin A., LT
Bojarski, Eugene J., CAPT
Bokenkamp, Harry J, Lcdr
Boland, James P., LT
Boland, Kells M, Lcdr
Boles, James W, CWO4
Bolger, Francis T., CDR
Bolich Jr, Thomas F., LT
Boling, J Michael, Lcdr
Bolkan, Steve A., Lcdr
Bollam, Kenneth A, CDR
Bollard, Alice M, Lcdr
Boller III, George E, Lcdr
Bolles, Hebert W, Capt
Bollin, John C, CDR
Bollinger, Lawrence F, Capt
Bollman, Leslie F, Capt
Bollogh, Richard W, Lcdr
Bolser, William C, CDR
Bolstad, Richard J, CWO4
Bolt, Bernard L, CDR
Bolt, L Emet, Capt
Bolt, Robert B, Capt
Bolte Jr, Carl E, Capt
Bolton, Ronald G., Lt
Bolton Jr., Levi, LCDR
Boltz, Roger H, Capt
Boma, James R, CAPT
Boman, Kerry F., CDR
Bombara, Christopher M., LT
Bomberger, Charles R, Capt
Bomgaars, David A, Capt
Bomkamp, Gary W, Capt
Bonacker, Robert N, Lcdr
Bonaguidi, Robert, Lt
Boncek, Edward B., CAPT
Bond, Calhoun, CDR
Bond, Michael T., CDR
Bond, Ronald K, CDR
Bond, Victor P, Radm
Bond Jr, Lester E, Lcdr
Bonebreak Jr, Byron A, Capt
Bonelli, Garry J, Capt
Bonenberger, Bradley D., LT
Boney, Benjamin D, CDR
Boney, Fleeman H, CDR
Bong, Donald A, Lcdr
Bonich, Anthony P, CDR
Bonilla, Mary K., Lcdr
Bonn, Timothy A., LCDR
Bonnema, Norman G, CDR
Bonnen, Charles W, Lcdr
Bonner, Suzanne, CDR
Bonner, Walter J, Capt
Bonner Jr, James T, Lt
Bonwit Jr, Willard R., CDR
Bony, Guy A., Lcdr
Booher, Kermit R., CAPT
Booker, Charles W, CDR
Booker, John W, Capt
Bookhout Jr, George W, Lt
Boom, Glenn L, Capt
Boone, Harold K, CDR
Boone, Malcolm E, CDR
Boone, Paul R, Capt
Boone, Wilkerson, Capt
Boone, William B, CDR
Boos, Stuart L, Lcdr
Boostrom, Roger L, Capt
Booth, Carlen P, CDR
Booth, Celia A., CDR
Booth, Holly G., Lcdr
Booth, Perry J., Lcdr
Booth, Robert B., Lt
Booth Jr, E. G., Lt
Bopple, Stephen, LT
Borchardt, Paul D, Lcdr
Borches III, Joseph E, CDR
Borden, Richard C., Capt
Borden III, Frederick Y, CDR
Bordenave, Lee J, CDR
Bordley Jr, Madison B, Lcdr
Borer, Paul J, CDR
Borgard, Glenn E, Capt
Borges-Dubois, Linda L, CDR
Borgman Jr, Theodore J, Capt
Borgquist, Daryl, LT
Borgwardt, Loren C, Capt
Boring, Joseph F, Lcdr
Boris, Charles A, Lcdr
Borland, James A, CDR
Borlet, Brian R., LCDR
Bornet, Stephen F, Cdr
Bornholm, Earl G, CDR
Boro, Frank P, Capt
Boronkay, Peter H., LCDR
Borovicka, Stephen J, Lcdr
Borowski, Robert J., CDR
Borrebach, Bernard J, Lcdr
Borrmann, William B, CDR
Borrowdale, Kent, CDR
Bosnak, Dean A., Lcdr
Bossard, Frederick S., CAPT
Bosse, Michael J., Capt
Bosse Jr., John W., Lcdr
Bosserman, Elmer E, Capt
Bosshart, Arthur D., CWO3
Bosso, Daniel G., CDR
Bost, Floyd P, Capt
Bostick, Edgar J, Lt
Bostock, Craig W., ENS
Boston, James T., Lcdr
Boswell, Charles H, CDR
Boswell, James D., Lcdr
Boswell, Mitchell C., Capt
Boswell, Thomas E., LT
Boswell, William E, Lcdr
Bosworth, William P, Capt
Bosworth, William W, CDR
Bothwell III, Kenneth H, Lcdr
Bott, John F, Capt
Bottger, Ronald R, Capt
Botts, Susan Fay, CDR
Botwinick, Clyde, Lcdr
Bouchey, Stuart H, Capt
Boudreaux III, Luke S, CDR
Bouffard, Peter J., Lcdr
Bouldin, James A, Capt
Bouldry, John M, Capt
Boule Jr, Arthur E, Capt
Bouma, Kent G, Capt
Bouras, Johnny, Lcdr
Bourbon, Archie T, CDR
Bourcier, Robert S, CDR
Bourgeois, Brian S., CDR
Bourgeois, Gerald P, CDR
Bourn, Richard L, Capt
Bourne Jr, Kenneth B, Lt
Bourret, William L, Capt
Bouska, Allen G, CDR
Bouvet, J.L. Dominique, CDR
Bowar, Joseph J., LCDR
Bowden, L. Dan, Lcdr
Bowden, Ronald E, CDR
Bowden, Ronald R, CDR
Bowen, Buford L, Lcdr
Bowen, Cheryl P, CDR
Bowen, Daniel E, LCDR
Bowen, John F, CDR
Bowen, Ray H, Capt
Bowen Jr, Bradford T, CDR
Bower, Arlo R, CDR
Bower, Leslie A, Lcdr
Bower, Thomas E, Capt
Bowers, Ralph L, Lcdr
Bowers, Richard R, Capt
Bowers, Steven G, Capt
Bowers Jr, John F, Capt
Bowersox, Paul F., Lcdr
Bowes, David R, Capt
Bowie, Chester W, Capt
Bowie, David L, Lcdr
Bowles, Robert E, Capt
Bowles Jr, George A., Lcdr
Bowling, John M, Capt
Bowman, Anne E, CAPT
Bowman, David C., CDR
Bowman, Donald E, Capt
Bowman, Eugene W, Lcdr
Bowman, James, LT
Bowman, John L, Capt
Bowman, Kenneth W., LCDR
Bowman Jr, Gilbert P., Lcdr
Bowne, William L, Lcdr
Bowne Jr, Charles J, Capt
Bowser, John V, Capt
Bowser, Peter A, Capt
Box Jr, E O, Capt
Boxenbaum, Michael J., LT
Boxler, Thomas F, Lcdr
Boyajian, Charles A., LCDR
Boyce, Allen W, Capt
Boyce, Margaret L, Lt
Boyce, Paul A., LT
Boyce, Thomas H., CDR
Boyd, Ada A, Lcdr
Boyd, Gerald E, Capt
Boyd, Jack, CDR
Boyd, James A, Capt
Boyd, John A, Lcdr
Boyd, Joseph C, Capt
Boyd, Larry R., CAPT
Boyd, Laurance C., LT
Boyd, Lon V, Capt
Boyd, Ted W., ENS
Boyd Jr, John B, CDR
Boyd Jr, Marion S, CDR
Boyda, David J., CDR
Boyden, Fredric M., Capt
Boydston, Edward A, CAPT
Boyer, David A, Lcdr
Boyer, Michael F, Capt
Boyer, Michael F., CAPT
Boyer, Thomas R., LT
Boyer, Wesley E, CWO4
Boyett, Stephen N., Lcdr
Boykin, William S., LCDR
Boykin Jr, Wilburn T., Lcdr
Boylan, Christopher P, CDR
Boylan, James M, Lt
Boyland, Jack I, Capt
Boyle, Brian E, CDR
Boyle, Francis L, Lcdr
Boyle, George D, CWO4
Boyle, James R., CWO3
Boyle, Louis E., CAPT
Boyle, Marylou O, CDR
Boyles, Frederick H., LCDR
Boyles, Harry V, Capt
Boyles, Norman B, Lcdr
Boyles Jr, David Ray, Lt
Boyne, Philip J, Capt
Bozdech-Veater, Pauline, CDR
Bozek, Stanley M, CDR
Bozeman, John C., LCDR
Bozzelli, Robert R, Capt
Braatz, Matthew P, CDR
Brabant, Peter J, Capt
Brabston, William N, CDR
Brabyn, Philip M. H., Lcdr
Braccia, Anthony A, Radm
Bracken, Eugene E, CDR
Bracken, John O, Capt
Bracken, John P, Capt
Bracken, Robert M, CDR
Bracker, Heinrich W.R., Lt
Brackett, Sheila S., CAPT
Bradford, William L, Lcdr
Bradley, Clyde E, Capt
Bradley, David M, Lcdr
Bradley, Gary H., CWO4
Bradley, Katherine M W, CAPT
Bradley, Vincent D, Capt
Bradley, William J, CDR
Bradley Jr, Benjamin L, CDR
Bradshaw, Jerald S, CDR
Bradshaw, Robert D, Capt
Bradshaw Jr, John P, CDR
Bradshaw Jr., Charles T., LCDR
Brady, Brian J., ENS
Brady, Elizabeth W., CDR
Brady, James J, CDR
Brady, John E, CDR
Brady, John G., CAPT
Brady, Joseph F, Capt
Brady, Michael S., Lcdr
Brady, Paul S, Capt
Brady, Thomas F, CDR
Braeuninger, Albert D, Lt
Bragg, David G, CDR
Bragg, Kenneth N, Ltjg
Brahosky, Vance A., LCDR
Braid, Robert W, CDR
Braidwood, Douglas, Capt
Brainerd, Austin L, Capt
Brainerd, Donald L, Capt
Brainerd Jr, Robert P, Capt
Braisted, Lawrence E, Capt
Braithwaite Jr., Kenneth J., LCDR
Brake, Gary F, Lcdr
Bralliar, Robert B., CDR
Brallier Jr., Charles A, Lt
Bramble, Arthur W, Capt
Brame, John H, CDR
Brames, Thomas J, Lcdr
Brammer, David H, CDR
Bramson, Carl David, Lcdr
Branam, Stella P, Lcdr
Branch, Deborah F., LT
Branchflower, John L., Capt
Brandin, Alf E, CDR
Brandl, Richard J, Lcdr
Brandli, Allan E, CDR
Brands, Michael B, CAPT
Brandt, Chad, LT
Brandt, Keith D, Lcdr
Brandt, Norman, Lt
Brandt, Thomas E., LCDR
Branen, Dennis J., Lcdr
Brangaccio, Franklin D., Lcdr
Branham, William T, Capt
Brannan II, Tandy T., Capt
Bransfield, James J, Lcdr
Bransfield, William J, Capt
Branson, Donald C, Lt
Brant, James G, Lcdr
Brant, Joe Ed, Capt
Branum, James D, Capt
Braselman, Herbert P., CAPT
Brasfield, Randy B., CDR
Brasher, John D, CDR
Brasher, Samuel R, Capt
Brasseur, James W., Lcdr
Braswell, Joel H, Capt
Bratches, Harold F, CAPT
Bratiotis, Christos, Capt
Bratsven, Randell R, Lt
Bratton IV, George A., CDR
Brauchler, Debra S., CDR
Brauer, Carl F, CDR
Brauer Jr, Harrol A, Lt
Braun, Jack E., CDR
Braun, John C, Capt
Braun, John T., CAPT
Braun, Lisa Anne, LTJG
Braun, Michael G., CDR
Braun, Robert A., Capt
Braun, Robin R., CDR
Braun, Sharon K., LT
Braun, Steven E, CDR
Braun Jr, James H, CDR
Braverman, Allen S, Capt
Brawner, Thomas M, LCDR
Bray, John K, Capt
Bray, Kate, CDR
Bray, Stephen N., LT
Brayer, Michael N, Lcdr
Brazell, Stanley H, Capt
Brazelton, Michael D., LCDR
Brazier, Donald G, Capt
Brazier, Mark K, CDR
Brazier Jr, Emmett L, Lt
Breaux, Darren J., Lt
Breaux, Duane L., LT
Breaux, Frederick E., LCDR
Brechtel, Lynn Y, Capt
Brecken, David R, Lcdr
Breckner, Anthony J, CDR
Bredahl, Timothy D, Capt
Bredehoft, Joseph M., LCDR
Bredemeyer, Eric R, CAPT
Breeden Jr, Lester J, CDR
Breedlove, Tom S, Lt
Breen Sr, Vincent C, Capt
Breese, Thomas R, Capt
Breese, Todd R., CDR
Breeze, Harrel Ray, CDR
Bregel, Charles Emmett, CDR
Breingan, Peter L, Capt
Breinholt, Vance L, Capt
Breitzig, Robert W, Capt
Brendley, Debra A., LTJG
Brenkman, Robert L., LT
Brennan, Christopher P, Capt
Brennan, James P, Capt
Brennan, Lawrence B, Capt
Brennan, Lawrence L, CDR
Brennan, Nancy E., CDR
Brennan, Peter J, Capt
Brennan, Richard J., CDR
Brennan, Robert F, CDR
Brennan Jr., John W., CAPT
Brenneman, Thomas E., Lcdr
Brenner, Christopher D., CDR
Brentzel, James H, CDR
Bresee, William F, Capt
Bresee Jr, Miles H, Capt
Breslin, Fred T, Lcdr
Bresnahan, Joseph E, Capt
Bressette, Randal J., CWO3
Bressler, H T, CDR
Breuleux, Philip S, Capt
Brevik, Theodore J, Lcdr
Brewer, Deatri L., CDR
Brewer, Erin M., CDR
Brewer, Robert M, Capt
Brewer, Stephen A, Lcdr
Brewer Jr, Monroe F, Capt
Brewington, Rudolph W, CDR
Brewster, Harold M, Lcdr
Brewster, Thomas P, CDR
Brewton, Elmer C, Lt
Breyer, Thomas P., CAPT
Breznau, William G., Lcdr
Briand, Robert L, CDR
Brice, Karolla, Capt
Brice, Robert C, Lcdr
Brickey, Zane R, CWO4
Bridenhagen, Steven T., Lcdr
Bridenstine, Lorna E, CWO4
Bridge, Herbert M, Radm
Bridge, James B, CDR
Bridge, Jonathan J, Capt
Bridge, Robert L, CDR
Bridgeman, Harold E, CAPT
Bridgeman, Harry M, Capt
Bridgham, Alan M., LT
Brien, Robert E, Lcdr
Briere, Michael J, Lcdr
Briggs, Carroll G, Lcdr
Briggs, Fred M., CDR
Briggs Jr, Leon Royden, Capt
Brigham, John E., CDR
Bright, Frederick, CDR
Bright, J Barry, LTJG
Bright, Jeffery L., CDR
Bright, Kenneth M., CDR
Bright Jr, Richard G, Capt
Brightman, Peter S, Lcdr
Brigiani, William G, Lcdr
Briglio, Helen M., Lcdr
Brink, Benjamin M, CDR
Brink, Charles I, Capt
Brinkerhoff, Craig O., Lcdr
Brinkman, Daniel J, Capt
Brinkmann, Randy Steven, CDR
Brinks, Lawrence H, CDR
Brinson, Robert J, CDR
Brinson, Z Creighton, Capt
Brintzinghoffer, David, Capt
Brior, James E, Cwo4
Brisack, Philip R, CDR
Brisbin, Bradford A, Capt
Britt, Kenneth D., LCDR
Britt, Rodney D., Lcdr
Brittain, Patsy J, CDR
Britton, Melvin C, Capt
Broad, Lauriann M., LCDR
Broadbent, Charles D, Capt
Broadbent, David J., CDR
Broadbent, Steven W, CDR
Broaddus, William A, Lcdr
Broadhurst, Daniel E., MIDN
Broadway II, Blake, CWO4
Brock, Charles L, Capt
Brock, Don E, CDR
Brock, James E., MIDN
Brock, Michael T, CAPT
Brock Jr., Edward B., CDR
Brockett, Clyde Waring, Lcdr
Brockman, Ronald C., Capt
Brockman Jr, Homer S, CDR
Brockmeyer, Bradley S., LCDR
Broder, Joseph A, CDR
Broderick, Patrick J., LCDR
Brodigan, Charles J, CDR
Brodisch, Laurence W, CDR
Brodsky, Ronald D, Capt
Brody, Robert A., Capt
Broedel, Lloyd E, CDR
Brohm, Charles M, Capt
Brokob, Jeffrey F., LCDR
Bromwell, Kent S, Capt
Bronczyk, Henry V, CWO4
Bronson, Terrence P, Capt
Brook, Douglas A, Capt
Brooke III, James F, Capt
Brooker, Susan M, Capt
Brooks, Daniel A, Capt
Brooks, David L, CDR
Brooks, E Douglas, Lt
Brooks, Edward A, Lcdr
Brooks, Elbert D, Capt
Brooks, Judith A, CDR
Brooks, Kathleen A., Lcdr
Brooks, Kevin B, Capt
Brooks, Michael K, Capt
Brooks, R. C., CAPT
Brooks, Raymond O, CWO2
Brooks, Richard E, Capt
Brooks, Robert E, CDR
Brooks, Roman L, CDR
Brooks, Stanley P, Capt
Brooks, William M, Lt
Brooks, Windsor G, Lcdr
Brookstra, William R, Capt
Brose, Gregory W, Capt
Brose, Mark O, Capt
Broskus, Gerard, LCDR
Brossette, Shane R., LT
Brost, Ken R., CDR
Brostrom, Kent D., CDR
Brotherton, Matthew S., LCDR
Brough, Alexander S, CDR
Broussard, Warren J, Capt
Brovold, Christopher T., CDR
Brow, John J, CDR
Brow, Robert, Lcdr
Brower, Robert E, CDR
Brown, Allen K, Capt
Brown, Angela, LCDR
Brown, Arthur B, CDR
Brown, Barbara W., CDR
Brown, Barry L, LCDR
Brown, Beren I., MIDN
Brown, Bert S, Lcdr
Brown, Bruce H., Lcdr
Brown, Carol M., CDR
Brown, Charles E, CDR
Brown, Christopher M, Capt
Brown, Clyde M, CDR
Brown, David L., CDR
Brown, Dean R, Capt
Brown, Douglas W., CAPT
Brown, Earl F, Lt
Brown, Edward A, Capt
Brown, Ellis R, LCDR
Brown, Gary D, LCDR
Brown, George W, CDR
Brown, Glendon L, CDR
Brown, Glenn R, Capt
Brown, Gordon V, Capt
Brown, Honorene L., CDR
Brown, Jack R, Lcdr
Brown, Jeffrey E, Lcdr
Brown, Jennifer K., LCDR
Brown, Jimmy W, CWO4
Brown, John B, Capt
Brown, John N, Ltjg
Brown, Kevin D, CDR
Brown, Lawrence E., LCDR
Brown, Lawrence G, Capt
Brown, Lawrence O, Capt
Brown, Lawrence W, Capt
Brown, Lyle C, CDR
Brown, Lyman W, Ltjg
Brown, Malcolm C, Capt
Brown, Michael A., MIDN
Brown, Mykel, CDR
Brown, Owen C., Lcdr
Brown, Perry E., LT
Brown, Peter R, Radm
Brown, Philip B, CDR
Brown, Ralph H, CDR
Brown, Raymond, ENS
Brown, Rex, Lt
Brown, Richard A, CDR

Brown, Richard E, Lcdr
Brown, Richard R, Capt
Brown, Richard W, Lt
Brown, Richard W, Capt
Brown, Roy G, CDR
Brown, Sherry L., CDR
Brown, Stephen F., Lcdr
Brown, Stephen G, Capt
Brown, Stuart J, CAPT
Brown, Terence H, CDR
Brown, Thomas L, CDR
Brown, Tina L., LT
Brown, Vera Lee, LCDR
Brown, Victor H, CDR
Brown, Virginia M, Capt
Brown, Ward W, Capt
Brown, Wilbur A, Capt
Brown, William D, Capt
Brown, William E, CAPT
Brown, William G, CDR
Brown, William K, Capt
Brown, William S., LCDR
Brown II, Walter W., CDR
Brown III, John W, CDR
Brown Jr, Gordon L, Capt
Brown Jr, James T, CDR
Brown Jr, Stephen M, Capt
Brown Jr, Thomas J, Capt
Brown Jr, Thomas R, Capt
Browne, Aldis J, Capt
Browne, Christopher U., CDR
Browne, Michael J., CDR
Browning, Bernard S, Radm
Browning, James D, Capt
Browning, Philip L., LT
Broyles, Thomas M, Lt
Broz, Gordon A., LCDR
Broz, William R., CDR
Brubaker, David B, Capt
Brubaker, Richard C, CDR
Bruce, John W., Lcdr
Bruce, Robert G, Lcdr
Bruce, Robert L, CDR
Bruce, Sheron R., Lt
Bruce, Stephan R, Capt
Bruce, Thomas A, Lcdr
Bruce, Warren, Capt
Bruce Jr, Walter E, Capt
Bruckart, William L, CDR
Brucker, Robert E., LCDR
Bruckner Jr, Charles J., Capt
Bruder, Marion A, CWO2
Brueckmann, Michael A, CDR
Bruemmer, John L, Capt
Bruen, Walter P, Capt
Bruenner, David F, CDR
Brumage, Harold Richard, CDR
Brumfiel, Robert W., CDR
Brumley, David, LCDR
Brundage Jr, Donald J, Lcdr
Brune, David H, CAPT
Bruneau, Louis W, CWO4
Brunelli, John F., Radm
Bruner, Judith Neuffer, Capt
Bruner, Michael G., CAPT
Brunner, Robert L, Capt
Brunngraber, Rodric F, Capt
Bruno Jr, Anthony J, Lcdr
Bruno Jr, George I, Capt
Bruno Jr, John, Lt
Brunskill, Michael R, CDR
Brunsvold, Daniel P, CDR
Brush, John B, CDR
Brusoski, Jeffrey M., CDR
Bruttomesso Jr., Raymond I., LT
Bruyn, David L., Lcdr
Bryan, Douglas E, CDR
Bryan, Leslie E, Capt
Bryan, Shirley W, Lt
Bryan, Vivian B, Lt
Bryan, William A, CDR
Bryan Jr, George R, Capt
Bryant, Charles F, Capt
Bryant, Charles W., LCDR
Bryant, Christopher, ENS
Bryant, Dwane H., LTJG
Bryant, Maxey L., CDR
Bryant, Michael F, CDR
Bryant, Richard A, CWO4
Bryant, Robert L., CDR
Bryant, William G, Lt
Bryant-Wolf, Gary R., LCDR
Bryeans Jr, Henry P, CDR
Bryniak, Amy J., ENS
Bryson, Yvonne H., LT
Buaas, Marion Hugo, Capt
Bubb, Kevin W., CDR
Bubeck, Robert C, Capt
Bubendey, Paul F, Capt
Buch Jr, Walter Harding, CDR
Buchanan, Gail D, Capt
Buchanan, Richard S, Lcdr
Buchanan, Walter W, Lcdr
Buchanan III, Herbert L, Capt
Buchenkirch, Arthur J, Lcdr
Buchiarell, Philip A., CDR
Buchy, David C., LCDR
Buciak, James J, CDR
Buck, Earl F, Capt
Buck, Kenneth J, CDR
Buck, Robert M, CDR
Buckingham, William J., CAPT
Buckland, Susan E., Lcdr
Bucklew, Willard B, Capt
Buckley, Elizabeth A, Lcdr
Buckley, John R., CDR
Buckley, Robert H, Lcdr
Buckley, Robert L, Capt
Buckley, Thomas P, CDR
Buckley Jr, Ambrose W, Capt
Buckman, John M, CDR
Buckmaster, Laurence C, CDR
Buckner, Ernest Wesley, CDR
Buckner, Jerry L, Capt
Buckner, Paul H, CWO4
Bucko, Richard L, Capt
Buckta, David M., LCDR
Buda, Gary D, LCDR
Budai, Donald M, Capt
Buday, Anthony, Lt
Buday, Christopher Lee, LCDR
Budd, Richard M., Lcdr
Buddenbohn, Kennard M., CAPT
Budenz, George C, CDR
Budenz II, G Charles, CDR
Budway, John J, Capt
Budzik, Dennis M, Capt
Budzinski, Sylvester S, CWO4
Buechner, Barton D., CAPT
Buegler, Robert J, Lt
Buehler, David A., LTJG
Buehrig, Christopher J., CDR
Buehrle, Jeffrey D, CAPT
Buel Sr, Toby J, CDR
Buell, Byron W, Capt
Buffa, Robert J, Lcdr
Buffington, Edwin C, CDR
Bugden, C M, CDR
Bugg Jr, William A, Capt
Bulbin, Stanley D, Lt
Bulger, Ingrid A., LT
Bulger, Winfred H., Capt
Bull, Galen W, Lcdr
Bull, John S, Lt
Bullard II, William T, Lcdr
Bullions, Andrew C, CDR
Bullock, John A., CDR
Bulls, Leatrice A., LCDR
Buman, Charles F., Lcdr
Bumford, John S, CDR
Bunch Jr, William D, Lcdr
Buncher, Isabelle E, Lcdr
Bunde, Kermit A., ENS
Bundo, Neal G, CDR
Bundy, Paul F., CDR
Bundy III, Walter E, Capt
Bunje Jr, Ralph B, CDR
Bunke, Frederick B, Capt
Bunke, Frederick M., CDR
Bunker, Karl P., CDR
Bunker, Mark K., CDR
Bunkers, Melody F., CDR
Bunn, Dennis M., CAPT
Bunn, Gerald A, CDR
Bunten, David R, Capt
Buntin, David W., CDR
Bunting, Kathryn A., Lt
Bunting, Mark F, CDR
Buonanno, Aldo A, Capt
Burbank, John R, CDR
Burchill, George W, CDR
Burczyk, Maureen Anne, ENS
Burdoin, John F, Lcdr
Bures Jr, Frank, Capt
Burg, John L, Lt
Burger, Francis J, Capt
Burger, Frank Gregg, CDR
Burgess, Diane E., LCDR
Burgess, Henry C, CDR
Burgess, Paul C, Lt
Burgess, Robert E, Capt
Burgess, Ronald K., LCDR
Burgess III, Benjamin F, Capt
Burgess Jr, Robert L, Lt
Burgess Jr, Willard F, CDR
Burggraf, Don, Lcdr
Burgund, Gary L, CDR
Burich, John P, CDR
Burin, Ronald, CDR
Burjetka, Miro G, Capt
Burk, Bryce E., LT
Burk Jr, Jim L, Capt
Burkart, Hanes A, CDR
Burkart, Melvin G, Capt
Burke, Alfred D, Lcdr
Burke, Charles W, CDR
Burke, Donald B, Capt
Burke, John H., CDR
Burke, John P, Capt
Burke, Joseph P, CDR
Burke, Kenneth D., Lcdr
Burke, Kevin C., LTJG
Burke, Nicholas E, Capt
Burke, Patrick L, CDR
Burke, Paul A., LCDR
Burke, Paul V., CDR
Burke, Robert E, Lt
Burke, Robert E., LCDR
Burke, Thomas A., LTJG
Burke III, John J, Capt
Burke III, William G., Lcdr
Burke Jr, Edmund S, Lt
Burke Jr, Richard F, CDR
Burket, Copley, Lcdr
Burkett, Hugh Alan, Lcdr
Burkhardt, Michael T, Capt
Burkhart, Robert A, Capt
Burkhead Jr, Franklin, Capt
Burkholder, Thomas J., LT
Burkle, Frederick W, CWO4
Burkons, Hugh A, Capt
Burleigh, Brian R., LCDR
Burleson, Ira L, CDR
Burlingame, Esther E., LCDR
Burn, Gerald D., CWO4
Burn, Richard H, Capt
Burnett, Donathan, CDR
Burnett, Douglas R, Capt
Burnett, Gary G, CDR
Burnett, James V, CAPT
Burnett, Kathleen N., Lt
Burnette, P. Clifford, LCDR
Burnham Jr, Charles D, Capt
Burns, Barbara J, Capt
Burns, Charles W, CDR
Burns, Christopher S., CDR
Burns, Daniel, LT
Burns, Donald J., CDR
Burns, Donald R, LCDR
Burns, Joseph, LT
Burns, Kathleen A, CDR
Burns, Michael W, Lcdr
Burns, Patricia L, Capt
Burns, Patrick C., Lcdr
Burns, Robert A., Lcdr
Burns, Robert N, Capt
Burns, Thomas L, CDR
Burns, William A, Capt
Burns, William J, Lt
Burns Jr, Wesley R, CDR
Burnum, Steven T, CDR
Burque, Daniel J., Lcdr
Burr, David G, Lcdr
Burr, David S, Capt
Burr, Frederic W, CDR
Burr, Lawrence R, Capt
Burr, Michael E., Lcdr
Burr, Timothy A, CDR
Burrichter, Harry C, CDR
Burris, Gary M, Capt
Burris, Raymond M, Capt
Burris, Robert N, CDR
Burriston, Donald A, CDR
Burroughs, David S., CDR
Burroughs, Max A, CDR
Burroughs, Robert P, Capt
Burrow, Lamar, Capt
Burrow, Susan K., LCDR
Burrows, Drusilla, Lcdr
Burrud, Richard J, Capt
Burrus, Robert D., CDR
Burt, Laroy, Lcdr
Burt, William L, CAPT
Burthe II, Jack H, Lcdr
Burtner, Keith E., CDR
Burton, Daniel K., CDR
Burton, James D, Capt
Burton, John F, Capt
Burton, Lee C, CDR
Burton, Paul E, Lcdr
Burton, Robert C., LT
Burton, William E, Lcdr
Burwell, Clayton L, CDR
Burwell, William O, CDR
Burwell Jr, Armistead, CDR
Burzell, Linden R, CDR
Busby Sr, Glen A, CDR
Busch, Joseph H, Capt
Busch, Kenneth L, Capt
Busch, Raymond W., CDR
Busch Jr, William W, Lcdr
Buschbach, Thomas C, Lcdr
Buschbaum, Dennis E, CDR
Buschmann Jr, Robert O, Capt
Buser, Carol A, Capt
Bush, Gregory G, CDR
Bush, Richard S, CDR
Bush, Shane S., LT
Bush, Ward A, CDR
Bush II, Phillip B, Capt
Busharis, John G, CDR
Bushaw, William J, Capt
Bushika, Martin E., CAPT
Buska, Kurt B., CDR
Buskirk, William C., CAPT
Busma Jr, John H, Lcdr
Buss, Phillip T, CDR
Bussard, Clifford B, Capt
Busse, George K., Capt
Bussell, Robert B, Lcdr
Bussey, Michael D, CDR
Bustos, Marsell C., LCDR
Buswell, David H, Capt
Butcher, James H, CDR
Butcher, Michael G., CDR
Butcher, Richard E, Capt
Butemeyer, Kerri L., LT
Butler, Alley C., CDR
Butler, Carey R., CDR
Butler, Daniel S., CDR
Butler, David A, Lcdr
Butler, David J., LCDR
Butler, Eileen M, CDR
Butler, Frederick S, Lt
Butler, George D, Lcdr
Butler, Gladstone I, CDR
Butler, James L., CAPT
Butler, M Forest, Capt
Butler, Margaret J, Capt
Butler, Max E, Lcdr
Butler, Robert F, Lt
Butler, Scott R., CDR
Butler, Theodore H, Capt
Butler, William E., LCDR
Butler, William G, Capt
Butler, William H, Radm
Butler Jr, George Revis, CDR
Butler Jr, Lloyd M, Capt
Butler Jr, Vincent K., Capt
Buttner, Joseph R, CDR
Buttner, Patrick E, Capt
Button, John E, CDR
Butts, Edward L., LCDR
Butts Jr, Harold T, Lcdr
Butz, John Vernon, Capt
Buxton, Richard E, Lcdr
Buxton, Richard L, Capt
Buyse, George J., LCDR
Buzasi, Derek, ENS
Buzek, Mark C., CDR
Bybee Jr, Norbert A, Capt
Bye, Carlton P, CDR
Byerly, Dennis R, Capt
Byers, James W, CAPT
Byers, John E, CDR
Byington, Donald P, Capt
Bynum, Albert S, CDR
Byorth, John Patrick, Capt
Byrd, Darrell E, Lcdr
Byrd, David G, CDR
Byrd, John D, CWO4
Byrd, Larry D, CDR
Byrne, Charles M, CDR
Byrne, Robert A, Capt
Byrne IV, John W., CAPT
Byrne Jr, John E, Ens
Byrnes, Robert K, Capt
Byron, William H, Lcdr
Caballero, Daniel, LT
Cable, Christopher W, Lcdr
Cable, Robert J., LCDR
Cabrales, Steven X., Lcdr
Cacci, Ralph A., LT
Cadaing, G. L., Lcdr
Cadle, Richard M, Lcdr
Cadwallader, William L., Capt
Cady, Arlan Y, Lcdr
Cady, William D, Capt
Caesar, Julius S., CAPT
Caffey, Brian, ENS
Caggia Jr, Vincent P., CDR
Caglieri Jr., Guido E., Capt
Cahan, John M, Capt
Cahill, Brian W., ENS
Cahill, James M, Capt
Cahill, John J, Capt
Cahillane, John P, Lcdr
Cahoon, H Michael, CDR
Caille, Gary W., CAPT
Cain, James J, CDR
Cain, John C, LCDR
Cain, Robert C., LT
Caine, Patrick R, Capt
Caire, Albert D, Capt
Cairnes, George W, CDR
Cajski, Thomas A, Capt
Calabro, Art D, Capt
Caldbeck, Christopher W., LT
Calderone, James J, Capt
Calderone, Joseph D, CDR
Caldwell, Albert P, CDR
Caldwell, Carroll G, CWO4
Caldwell, Joseph W, CDR
Caldwell, Linda S., LCDR
Caldwell, Roland P, LCDR
Caldwell, William P, CAPT
Caldwell III, Mack D, CDR
Caldwell III, Robert E., Lt
Caldwell Jr, James W., CDR
Caldwell Jr, Robert K., Lcdr
Caletri, Joseph J, Capt
Calhoon Jr, William H, CDR
Calhoun, Frank B., CDR
Calhoun, John C, Lcdr
Calise, Kenneth J., CAPT
Calise, Nicholas J, Capt
Calise, Victor E, Capt
Calish, Sydney R, Lcdr
Calistro, Katherene I., Lt
Calkins, Dean L, Capt
Calkins, Leroy J, Lcdr
Call, Hughes, CDR
Callaghan, Nathan R, Capt
Callahan, A Joseph, Lcdr
Callahan, Frederic L, Lcdr
Callahan, Joseph F, CDR
Callahan, Kevin J, LCDR
Callahan, Kim F., CDR
Callahan, Robert M, CDR
Callahan, Robert V, Capt
Callahan, Victor L., LCDR
Callan, Daniel J, CDR
Callan, Leonard J., CAPT
Callas, Peter T., LCDR
Callaway, James G., Capt
Callender, Ruth E, Lt
Callicutt, Homer D., CWO4
Callies, Fred H, Capt
Callison, Harry R, CDR
Callison, Robert A., LT
Callo, Joseph F, Radm
Callow, Allan D, Radm
Calnek, Michael E., LCDR
Calomeni, Carolyn A., CDR
Calvert, Stanley D., LCDR
Calvin, William E, Lcdr
Camburn, William D, Lcdr
Camealy, Craig E., CDR
Camera, David A., CDR
Cameron, Jeffrey R., Lcdr
Cameron, John B, CDR
Cameron, John E, CWO4
Cameron, Ken, MIDN
Cameron, Richard F., Lcdr
Cameron II, Alan D, Capt
Camiel Jr, Edwin P, Capt
Cammack, Charles T, Lt
Camp, Kenneth N, CDR
Camp, S Lowry, Capt
Camp, Turner, Capt
Camp Jr, Worth, Capt
Campaigne, Robert J, Capt
Campbell, Ann M., Lcdr
Campbell, Bruce A., Capt
Campbell, Carroll J, Capt
Campbell, Charles I., CAPT
Campbell, Clarence H, CDR
Campbell, Clifton P, CDR
Campbell, Colin A., LTJG
Campbell, Courtney L, Lcdr
Campbell, Crom, Capt
Campbell, Dugald R, CDR
Campbell, Earl Wilson, CDR
Campbell, Fred H, CDR
Campbell, George E, CDR
Campbell, Harold J., LT
Campbell, Heidi L., LT
Campbell, James G, Capt
Campbell, Joseph C, Lcdr
Campbell, Joseph P, CWO4
Campbell, Kenneth W., CAPT
Campbell, Mary M., LTJG
Campbell, Norman L, Ltjg
Campbell, Phillip R, Capt
Campbell, Price F, CAPT
Campbell, Robert J, Capt
Campbell, Roy C, CDR
Campbell, Roy O, Lcdr
Campbell, Samuel S, CDR
Campbell, Scott E., CDR
Campbell, Susan E, LCDR
Campbell, Thomas H, Capt
Campbell, V. O., Lcdr
Campbell, William R, CDR
Campbell II, James E, CDR
Campbell Jr, Bowen, Capt
Campbell Jr, Boyce L, CDR
Campbell Jr, Homer R, Capt
Campbell Sr., Reginald C., CDR
Campen, Floyd G., CDR
Campfield, Michael, MIDN
Campillo, Armand V, CDR
Campillo, Fred G., LCDR
Campo Jr, Salvatore R., CDR
Campos, Mario A., LT
Camut, Kenneth J., Lcdr
Canada Jr, Paul M, CDR
Canavan, Kenneth, Capt
Canby, John C, Capt
Candelario, Edna Mae, LCDR
Candy, Gregory T., Capt
Canete, Danelo R, Capt
Canfield, Burt, ENS
Canfield, Margaret B., LT
Cann III, John P, Capt
Cannaday, Richard L, Lcdr
Cannan, Daniel E., CDR
Cannicott, Richard G, CDR
Cannon, Christopher L, Lcdr
Cannon, James D., CAPT
Cannon, James Thomas, LCDR
Cannon, Leo J, CDR
Cannon, Margaret Ann, CDR
Cannon, Michael A., LCDR
Cannon, Michael P, Lcdr
Cannon, Norman W, CDR
Cannon, Thomas M, Capt
Cannon IV, James M, Lcdr
Cannon Jr, Jesse D, CAPT
Cannon Jr., Harold F, CDR
Canon, Jack A., CAPT
Canon, Robert W, Capt
Cant, Geoffrey D, Capt
Canter, Thomas C, Capt
Canton, Julie I., LT
Cantrell, Kevin D., LT
Cantrell, Robert W, Capt
Cantrell Jr, Allen F., Cdr
Cantwell, Joseph E, CWO3
Canull, James A, CDR
Cape, Hays A, Lcdr
Capernaros, Peter S, Capt
Capirano, Medick M., CAPT
Capo, Louis J, Lcdr
Capra, Robert A, CDR
Capraro, Robert L., ENS
Capri, Dianne, CDR
Captor, Richard A, Capt
Caputo, Jay S, CDR
Caras, John A., CAPT
Carathers, Michael D, CDR
Card, Warren H, CDR
Cardell, Anthony F, CWO4
Carder, Kenneth H, Lcdr
Cardinal, James, Capt
Cardozo, Reginald H., CAPT
Cardwell, James N, CDR
Carey, Edwin W, CDR
Carey, F Joseph, Lcdr
Carey, Gerard M., LTJG
Carey, James J., RADM
Carey, John F., Lt
Carey, Merrick, Lt
Carey, Michael J, CDR
Carey, Omar J, CDR
Carey, Richard P., Capt
Carey, Sandra H., Capt
Carey Jr, Robert H, LCDR
Cariello, Francesca P., Lcdr
Carl, Charles T, Lcdr
Carl, Fred D, Capt
Carl, Louis Oscar, LT
Carlan, Kit, Lcdr
Carletti, James S, Capt
Carley, Steven M, Lcdr
Carlin, Ann Marie, LCDR
Carling, J Richard, Capt
Carlisle, Robert J, CDR
Carlo, James N, Capt
Carlone, Rodney A., Capt
Carlos, John M., LCDR
Carlsen, Charles R, Capt
Carlsen, Kenneth H, CDR
Carlson, Andrew A, Lt
Carlson, David S., CDR
Carlson, Eric A., LCDR
Carlson, F Paul, Capt
Carlson, Frank A, Lt
Carlson, James F., LCDR
Carlson, James W., CDR
Carlson, John Eric, Capt
Carlson, Lennart G, Lt
Carlson, Michael R., LCDR
Carlson, Peter T., LT
Carlson, Richard A, CDR
Carlson, Richard A, Capt
Carlson, Richard E, Capt
Carlson, Richard W, Lcdr
Carlson, Robert M, CDR
Carlson Sr, Robert O, Lcdr
Carlstrom, Glen R, CDR
Carlyon, Richard D, Lcdr
Carlyon, William H, Capt
Carman, Thomas P, Capt
Carmichael, Grant P., LT
Carmichael, Joseph R, Capt
Carmichael, Philip J., ENS
Carmichael Jr, David B, Radm
Carmody, Bert M., Capt
Carnegie, James D., CDR
Carneiro, Kenneth O, Capt
Carnes, Calland F., CDR
Carnes III, Clif B., ENS
Carney, Charles D, CDR
Carney, Gary, CDR
Carney, James E, CDR
Carney, James R, Capt
Carney, Kevin R., Lt
Carney, Marilyn L, Capt
Carof, Zelda, Lcdr
Carolus, Jay M, CDR
Caron, Joscelyn, LCDR
Carpenter, Anthony H., CAPT
Carpenter, Dante, Lcdr
Carpenter, David R., LCDR
Carpenter, Donald B, Lcdr
Carpenter, Elbert N, Lcdr
Carpenter, Galen D, CAPT
Carpenter, Harold F, Capt
Carpenter, James A., LT
Carpenter, Malcolm M, Lt
Carpenter, Marc F, CDR
Carpenter, Wesley J, LCDR
Carpenter, William S, Capt
Carpenter, Willis V, Capt
Carpenter III, George C, CDR
Carper, Alan J, LCDR
Carper, Eric, Lcdr
Carr, Charles R, Lcdr
Carr, Hiram T, Capt
Carr, James D., CDR
Carr, Julian, CDR
Carr, Lawrence P, Capt
Carr, Matthew A., CDR
Carr, Michael F, Lcdr
Carr, Rodney J, Lcdr
Carr, Thomas R, Capt
Carrano, Richard P, Lcdr
Carraway Jr, Robert H, CDR
Carrera, Richard N, Capt
Carrington, John P, Capt
Carrington, Richard J, CDR
Carroll, Anglique M., LT
Carroll, Cecil J., CDR
Carroll, George W., CDR
Carroll, John M., Lt
Carroll, Lon Day, Capt
Carroll, Michael A, Capt
Carroll, Robert, CDR
Carroll, Roy B, Capt
Carroll, Sean P., 2LT
Carroll, Thomas E., CDR
Carroll, William J, CDR
Carroll III, Phillip P, Capt
Carroll Jr, Charles M., CDR
Carroll Jr, Robert B, Lt
Carson, Cary D, CDR
Carson, Michael J, Capt
Carson, Mildred S, CDR
Carson, Phillip Roy, Capt
Carson III, Carl E., CDR
Carson Jr, Louis F, Capt
Cartagena, Miguel A, Capt
Carter, Barbara J., CAPT
Carter, Carroll L, CDR
Carter, Christopher D., ENS
Carter, Ernest V, Lcdr
Carter, Esther A., CDR
Carter, Frank H, Capt
Carter, Helen Pauline, Lt
Carter, Jack M, Capt
Carter, James E, Capt
Carter, Jerry M, Lcdr
Carter, John R, Lcdr
Carter, Juanita E, Lt
Carter, Michael A., Lcdr
Carter, Michael L., LT
Carter, Michael P., LCDR
Carter, Paul F, CWO4
Carter, Robert J, CDR
Carter, Robert T, CAPT
Carter, Rowland F, CAPT
Carter, Troy L., CDR
Carter, William L, Lt
Carter Jr, James H., LT
Carter Jr, Malcom K, Capt
Carter Jr, Norman M, Capt
Carterson, John S, Cdr
Cartier, David A, CDR
Carty, Christine E, CDR
Caruana, Robert, Lcdr
Cary, Freeman Hamilton, Radm
Casaburi, Daniel W., Lcdr
Casagrand, Robert F., LCDR
Casaus, Michael A, CDR
Case, Alex, Lt
Case, Edwin M., CWO4
Case, J. Fred, CAPT
Case, Jeffrey B, Capt
Case, John M, Lcdr
Casebeer, Harvey L, Capt
Casella, Jeremy T., MIDN
Casella, Leonard R, Capt
Caselli, John J, LCDR
Casey, Anita M, Lcdr
Casey, Danny G, Capt
Casey, Donnabelle A, Capt
Casey, John M, Capt
Casey, John S, Capt
Casey Jr, Harry J, Lcdr
Cashell, Alan W., Capt
Cashmere, J. Brian, LCDR
Cashmere, John J, Capt
Casler, William J, Lcdr
Caso, Peter J, Lcdr
Caspar, Bernard J, Lcdr
Casper, Thomas J, Capt
Cassalia, Peter T, Capt
Cassidy, Chuck E., CDR
Cassidy, Joseph M, Capt
Caster, Ernest M, CDR
Casterline, Kenneth R, CDR
Castinetti, Stephen L, Lcdr
Castle, Charles A., Capt
Castle, John P, CWO4
Castle, William M., Capt
Castner, Robert G, Capt
Castner II, Willis H, CAPT
Castor, John R, Capt
Castro, Ralph G, CDR
Caswell, Peter R, Lcdr
Caswell, Richard F, Capt
Catanese, Charles P, Lcdr
Catanzaro, Eugene W, Lcdr
Catena, Emanuela N M, CDR
Cates Jr, John Farley, Capt
Cathcart, David W, Lcdr
Cathey, Jesse L, Lcdr
Cathey, Nancy J, CDR
Caudill Jr, R. Paul, Radm
Caulfield, Thomas J., Capt
Causseaux, Orlis L, CDR
Cauthen, John M., Lcdr
Cauthen, Wiley M, Capt
Cauthers, Richard A, Lcdr
Cava, George L., Capt
Cavalieri, Ralph P, CDR
Cavallaro, Charles T., Lcdr
Cavanaugh, Kenneth C, Capt
Cavanaugh, Mary B, Capt
Cavanaugh, Thomas A., Capt
Cavanna, Cesar E, CDR
Cave, Alfred N., Lcdr
Cave, Alice R Severn, CDR
Cave, David B, Capt
Caveney, Yvonne M, CDR
Caveney Jr, Charles W, Capt
Caves, Roy D, Capt
Cavileer, Douglas R, CDR
Caviness, Claude P, Capt

Caviness, Jerry L, CWO3
Caviness, Stephen D, Capt
Cawthon, Franklin P, CDR
Cayangyang, Jerome R., LTJG
Caylor, Joseph J, CWO4
Cazares, David H., Capt
Cazier, James E., LCDR
Cebatoris, Joan M., CDR
Cecere, Anthony M, CDR
Cech, Albert J, Lcdr
Cecil, Mark A., Lcdr
Cecil Jr, Lawrence L, Lcdr
Cedarholm, Lloyd, Lcdr
Cederholm, Walter T, Capt
Ceglarski, John F., Lcdr
Celebrezze, James P, CDR
Celli, Marian L., LCDR
Celli, Robert R., CAPT
Cera, Mary Claire, LCDR
Cerda, James J, Radm
Cerine, Ferdinand C, Capt
Cerrone, Christopher, LT
Cervenka, Becky J, CDR
Cesear, Thomas M., Lt
Chabal III, Matthew, CDR
Chacey, Kenneth A, Capt
Chadbourn III, Charles C., Capt
Chadil, Robbins L, Lcdr
Chadwick, John B, Lt
Chadwick, Ralph G, CDR
Chadwick, Raymond S, Lcdr
Chaffee, Mark F., LCDR
Chakmakian, Carl, Capt
Chalfin, Steven, CAPT
Chamberlain, Aleta L., CDR
Chamberlain, Fred D., LCDR
Chamberlain, Mark S, CDR
Chamberlain, Robert M., CAPT
Chamberlain, Ross W., CDR
Chamberlin, Arthur M, Capt
Chamberlin, Willard L., Capt
Chambers, Brad K, Lcdr
Chambers, David M, Capt
Chambers, Richard E., LCDR
Chambers, Richard K, Radm
Chambers, Todd R., CDR
Chambers Jr, Robert H., Lcdr
Chamblee, Wayne M, CWO4
Chames, Nicholas A, CDR
Champagne, Joseph R.H., Lcdr
Champion, Lee, ENS
Chan, George C.S., LTJG
Chan, Rudolph J., LCDR
Chance, George R, CDR
Chanco, Albert B, Lt
Chandler, Eva-Marie H., CDR
Chandler, Floyd J, CDR
Chandler, Harold R, Lcdr
Chandler, James T, CDR
Chandler, Kathryn B, Capt
Chandler Jr, Alfred W, CAPT
Chaney, David A, Capt
Chaney, Marvin L, CDR
Chang, James C., Capt
Chang, Silvia K., CDR
Chapin, Joseph M, Lcdr
Chapin, Thomas H., LCDR
Chapin Jr, John E, Capt
Chaplin, Kenneth M, Lt
Chapman, Charles B, CDR
Chapman, David K., LCDR
Chapman, Donald R, Capt
Chapman, James G, Capt
Chapman, James G., LCDR
Chapman, James H, Capt
Chapman, James L, Capt
Chapman, John C, Capt
Chapman, Michael J., Lcdr
Chapman, Steven R, Lcdr
Chapman, Walter C., LCDR
Chapman, William R., LT
Chapman Jr, Carter S., Capt
Chappelka, Scott R., ENS
Charbonnet Jr, Pierre N, Vadm
Charleysale, Christopher H., ENS
Charnov, Bruce H, Capt
Charvat, William C, Capt
Chase, Gene R, Lcdr
Chase, Harry E., Cwo
Chase, John P, CDR
Chase, Jon G, CDR
Chase, Michael B., LT
Chase, Robert R, Capt
Chase, Ted E., Lcdr
Chase, William R, Capt
Chase Jr, Harlan C., Lt
Chassot Jr, Charles H., CAPT
Chastain, Edward M, Lcdr
Chastain, J D, Capt
Chastain, J Harold, Capt
Chastain, Jeffrey M, CAPT
Chastanet, Rachel I., CDR
Chaszeyka, Michael A., Lcdr
Chatelain, Jeanne M., LCDR
Chater, William T, Lt
Chatfield, Norman V, Lcdr
Chatowsky, Anthony P, Capt
Chatt, Rodrick H, CDR
Chatterton, Dalton K, CDR
Chauncey, Richard M, Lcdr
Chausse, Corinne, LTJG
Chavaree, Carl F, Lcdr
Chavez, Rudolph E., LCDR
Chazen, Bernard, Capt
Chazen, Milton, Capt
Cheatham, Kevin M., CAPT
Checkovich, John, Capt
Cheek, J Paul, CDR
Cheesman, Hugh C, Lcdr
Cheever, Francis S, Capt
Chelius, Richard E, Lcdr
Cheney, Charles E, CAPT
Chenis, John K, Capt
Chenoweth III, John P, Capt
Cherbak, Peter A, LT
Cherney, Thomas J, CDR
Chernish, Stanley M, CDR
Chernoff, Albert R., Capt
Chernoski, John J, CDR
Cherry, Ronald W., ENS
Chertok, Alan S., Lcdr
Cheslock, William B, Capt
Chesser, Wendell J, CDR
Chesser, Zane L, CDR
Chesson, Michael B, Capt
Chester, Mark D, Lcdr
Chesterman, Alan G, CDR
Chestnut, Julian S, Capt
Chetta, Serena A., CDR
Chew, Kenley Y., LCDR
Chewning, Thomas W, Lcdr
Chiang, Peter, CDR
Chiaro, David A., Lcdr
Chicas, Samuel A, Capt
Chicoine, Rene J, Capt
Chierico, Michael, LT
Chiffer, John M., Lcdr
Chigos, David K, Capt
Chikkappa, G, Capt
Child, E Tom, CDR
Childers Sr., David O, Capt
Childress, Carl W, CAPT
Childs III, John N., CAPT
Chilson, David W, Capt
Chin, Douglas, CAPT
Ching, Carl, CDR
Ching, Stephen T, Capt
Chinn, James A, CDR
Chinn, Ralph M., Capt
Chinque, Katherine M, Lcdr
Chiogioji, Melvin H, Radm
Chipchase, Calvert G, CDR
Chipley III, Charles L., CDR
Chisholm, Paul Grey, Lcdr
Chism, Ronald G, Capt
Chism Jr, Timothy K., CDR
Chismarich, Stephen R, Capt
Chisum, William R, Capt
Chiuminatto Jr, Martin, CDR
Chivers, Jeffrey W, LT
Chlosta, Robert J, Capt
Chmura Jr., John A., CDR
Chodak III, Paul, LCDR
Choppe, William J, CDR
Chounet, Frank Paul, CWO3
Chrisan, Paul E, Lcdr
Chrispens, John B, CDR
Christen, Samuel E, CDR
Christensen, Daniel W, Capt
Christensen, Donald C, Lcdr
Christensen, Harvey E, Capt
Christensen, James L, Capt
Christensen, John B, Capt
Christensen, Jon E, CDR
Christensen, Steven L., LCDR
Christensen, William L, CDR
Christensen Chc, C J, Lcdr
Christenson, Larry R, CDR
Christian, David S, Lcdr
Christian, John W., Lcdr
Christian, Robert B., LT
Christian, Ronald D., CDR
Christiansen, David W, CDR
Christiansen, John J, CWO4
Christiansen, Russell J, CDR
Christianson, Dean A., Lcdr
Christianson, Eric E., CDR
Christie, Alan P., Lt
Christie, Charles C., CDR
Christie, Neil M., CDR
Christman, Joseph G., LCDR
Christman, Patrick L, Capt
Christodoulou, Athena A., LCDR
Christofferson, John R, Capt
Christoph, Charles L, Capt
Christoph, Edward, Capt
Christopher, Allison L, Capt
Christopher, Michael D., CDR
Christopher, Philip D, Lt
Christopher, Steven C., CAPT
Christy, John W, CDR
Chriswell, Arlene E, CDR
Chron, Gustav N, Lcdr
Chruma, Joseph E, Capt
Chrzan, Joan T., CDR
Chubb, Dave M., Lcdr
Chuilli, Ernest J, Capt
Chumney, James R, Lcdr
Chung, Alice, Lcdr
Chung, Belinda W, Capt
Chung, Grace M, Lcdr
Chung, Mark M., CDR
Chung, Richard D, CDR
Churchill, Timothy J, CDR
Churchill, William B, Capt
Ciaffa, Philip R, LT
Cianci, Paul E., Capt
Ciccia-MacLean, Robert, LCDR
Cicero, Joseph S, Capt
Cienik, Kenneth C., LCDR
Cieslak, Randall C, CDR
Cieslinski, Peter K., LT
Ciggia, Donna M., CDR
Cilento, Bartley G., Capt
Cimino, Samuel P, Capt
Cindrich, Daniel E., Lt
Cindrich, Evan, CDR
Cinnamon, Jerome M, CDR
Cioe, Mary, CDR
Cipriano, Sam V, Lcdr
Cistriano, John J, Capt
Citrin, Marc D, Lcdr
Clabby, James A., CAPT
Cladek, Edward A, CDR
Clagett, Page B, Lcdr
Claiborn, Richard D, Lcdr
Clancy, Edward V, Capt
Clancy, George J, Capt
Clancy, Renee C, CDR
Clancy, Steven P., LCDR
Clancy Jr, Daniel H, CDR
Clapham, William T, Capt
Clarbour, Bertram N, Capt
Clardy, George L., CDR
Clare, David H, Lcdr
Clarendon, Colin C D, Capt
Clark, Anthony M., Lt
Clark, Arthur J., LCDR
Clark, Ben H, Capt
Clark, Benjamin C, Capt
Clark, Bonnie B., CDR
Clark, Bradley K, Lt
Clark, Bruce T., LCDR
Clark, Carl Richard, Lt
Clark, Carol S., Lcdr
Clark, Colin M, Capt
Clark, Daniel U, CDR
Clark, David A., ENS
Clark, David Alan, Capt
Clark, Donald B, CWO2
Clark, Donald H, CDR
Clark, Donald R, Lcdr
Clark, Douglas C, CDR
Clark, Ethel B, Capt
Clark, Hal A, Lcdr
Clark, Henry H, Lt
Clark, James Donald, Lcdr
Clark, James F, Capt
Clark, James L., CDR
Clark, James S., CDR
Clark, James Thaddeus, CDR
Clark, John Earle, Capt
Clark, Kenneth B, Capt
Clark, Neil A., CDR
Clark, Neill V, CDR
Clark, Philip C, Lt
Clark, Ray C, Lcdr
Clark, Raymond M, CDR
Clark, Richard D, Capt
Clark, Robert E, CDR
Clark, Scott M., Lt
Clark, Thomas A, Capt
Clark, Thomas H, CDR
Clark, Wallace H., Lt
Clark, Warren G, CDR
Clark, William W., LCDR
Clark Jr, Brozia H, CDR
Clarke, Frederick W, Capt
Clarke, Jon B, Capt
Clarke, Laurence B., Lt
Clarke, Richard W, Capt
Clarke, Robert R, Lt
Clarke Jr, Francis M, Capt
Clarke Jr, Raymond W, Capt
Clarkin Jr, Thomas R., Capt
Clarkson Jr, Andrew C, Capt
Clary, Irl C, Capt
Clary, John C, Lt
Clas, Richard J, CWO4
Classen, Donald C, Lcdr
Claudio, Michael A, CDR
Claughton, James C, Capt
Clause, Jr, Harry Paul, Capt
Clausen, Carl O, Capt
Clausen, Stephen R., Capt
Clausing, Richard B, Capt
Clauss, James T, CDR
Claxton, Norman L, Capt
Clay, Henry J., LCDR
Clay, Herbert D, Lt
Clay, Lyell B, CDR
Clays, Robert A, Lcdr
Clayton, George L, CDR
Clayton, John E, Capt
Clayton, John W, Capt
Clayton, Oramel I, Capt
Claytor, Ann Marie, LT
Cleaver, Lloyd J, Capt
Clebak Jr, Karl T, CAPT
Cleckler, Charles S., LCDR
Cleland, Gary L., CWO3
Clement, David E, CDR
Clement, James E., CDR
Clement, William S., Lcdr
Clements, Edward W, Capt
Clements, Joseph A., CDR
Clements, Robert F, Capt
Clements, Robert S., LCDR
Clemmer, Dayne E, CDR
Clemmons, Marvin C, Lcdr
Clemson, James F, CDR
Clendenin, G Gary, Capt
Clendenon, Claude E, Capt
Cleveland, Ronald L., LCDR
Cliff, David B, Lcdr
Cliff, Edmund Vail, CDR
Clifford, James W., CAPT
Clifford, Royce E., LCDR
Clifford, Verland J, CWO3
Clifford Jr, Charles F, Capt
Clifford Jr, Lawrence E, CAPT
Clift, James C, CDR
Clifton, George, LCDR
Clifton, Vincent P., LCDR
Cline, Eugene P, CDR
Cline, Rick, CDR
Cline, Walter L, Lt
Cline, William C, Capt
Clinkscales, Hulan F, Capt
Clinton, Rick P, Lcdr
Clinton, Robert M, CDR
Clites, Jeffrey L., Cdr
Cloak Jr., Stephen P., Lt
Clodfelter, Donald G, Lcdr
Cloninger, Kathryn A, CDR
Clontz, Stanford K, Lcdr
Closson, Spencer W, Capt
Cloud, Lutricia Lee, LCDR
Clow, Gordon H, Capt
Clum Jr, Woodworth B, CDR
Clusen, Gerald W, Lcdr
Cluster, Chris J., CAPT
Clyde, Payson J, Capt
Clyde, William W, Lcdr
Coaker, James W, Capt
Coan, Laurice O, Lcdr
Coane, Casey W, Radm
Coates, Edward E, Capt
Coates, Raymond W, CWO4
Coates Jr, John, Capt
Coates-Wamsley, Dianne, Lcdr
Cobb, Edgar H, Lcdr
Cobb, Edward M., CDR
Cobb, Glenn G, Capt
Cobb, Jeffrey D., LCDR
Cobb Jr., Nathan A., CDR
Cobbe, Stephen M., CDR
Cobble, Alex Y, CAPT
Cobern, James C, Capt
Cobry, Terrence N, Lcdr
Coburn, Dennis P., CDR
Coburn, George H, Capt
Coburn, Ralph G, Radm
Cocci, Barry J, Lt
Cocci, James A, CDR
Cochet, Alan Judson, Lcdr
Cochran, Carl E, Capt
Cochran, Jeffrey F., Lcdr
Cochrane, John C, Capt
Cochrane, John R., CDR
Cochrane, Michael H., CDR
Cochrane, Paul F, Capt
Cochrane, Paul Gordon, Lcdr
Cochrane Jr, John C., CDR
Cockelreas, Charles E, CDR
Cocker, Bruce J., LCDR
Cockrum Jr, Logan V., CDR
Coddington, Kenneth K, CDR
Coderre, David M, CDR
Coderre, Paul E, Lcdr
Cody, Art C., CDR
Cody, Joseph P, Capt
Coe, Douglas H, CAPT
Coe, John C, CDR
Coe, Priscilla B., CDR
Coelho, David F, Lcdr
Coelho, Joseph R, CDR
Cofer Jr, Joseph P, Capt
Coffel, Michael J., LT
Coffer, Alan B, Capt
Coffey, Edward J, Capt
Coffey, John P., CAPT
Coffey, Matthew J., Capt
Coffey, Richard W, Lcdr
Coffey, Thomas Edward, Capt
Coffman, Clarence L, CWO3
Coffman, Ralph B, CAPT
Coffman, Richard Leroy, Capt
Cogan, Phillip S, CDR
Cogan, Timothy C., Lcdr
Cohee III, Frank E., Capt
Cohen, Bernard S, Capt
Cohen, Howard C, Capt
Cohen, Jess B, Lcdr
Cohen, Joshua L., Lcdr
Cohen, Judith Ann, CAPT
Cohen, Martin F, CDR
Cohen, Paul R., LCDR
Cohen, Ronald A, Lcdr
Cohen, Samuel, Lcdr
Cohick, David G, Lcdr
Cohn, Joseph R, CDR
Cohn, Steven Lawre, CAPT
Coie Jr, John E, CDR
Coile, Wendy W., Lcdr
Coke, Byron W, CWO4
Cokely, Miriam W, Lcdr
Colaitis, Spiro Chris, Lcdr
Colan, Vincent J, Capt
Colasurdo, Michael M, CDR
Colbath, Robin A, Lcdr
Colbert, Marguerite D, Ltjg
Colbert Jr, Thomas E, CDR
Colbourn, Edward W, CDR
Colburn, Dwight W, Capt
Colby, Christopher J, Lcdr
Colby, Gage, Capt
Colby, Wendell S, Capt
Cole, Alonzo F, Lcdr
Cole, David J., LCDR
Cole, Ellerbe P, Capt
Cole, James N, Capt
Cole, Jeffrey P., LCDR
Cole, John E., CDR
Cole, Johnnie, CWO4
Cole, Joseph G, CDR
Cole, Lloyd A, CDR
Cole, Neal E, Lcdr
Cole, Robert J, Capt
Cole, Robert J., CDR
Cole, Robert R., CDR
Cole, Simon T., LT
Cole, Steven B, LCDR
Cole Jr., Charles W, Capt
Coleman, Clifford B, CDR
Coleman, George J, Radm
Coleman, James E M, Capt
Coleman, Lane P., Lt
Coleman, Michael C., LCDR
Coleman, Richard C., LCDR
Coleman, Robert L, Capt
Coleman II, Curtin R, Capt
Coleman Jr, Byron E, CDR
Coleman Jr, Joe T., CAPT
Coleman Jr, Robert F, Lcdr
Coles III, James R., Lcdr
Coletta, Paolo E, Capt
Colgan, Diane L, Capt
Colgan, John G, Capt
Collett, Joseph M, CDR
Collett Jr, Ronald D., LT
Colley, Craig C, Lcdr
Colley, Douglas J, CDR
Collier, Joe Barnard, CDR
Colligan, Robert P., CDR
Collin, Malcolm S, Lcdr
Collins, Alan W, Capt
Collins, Cherie B., Lcdr
Collins, Clyde, Capt
Collins, Frank A, Capt
Collins, James R, Capt
Collins, Jim F., CDR
Collins, Joe H, Lcdr
Collins, Joseph S, Capt
Collins, Max C, CDR
Collins, Michael J, CDR
Collins, Michael J., LCDR
Collins, Monte, CDR
Collins, Morris L, Capt
Collins, Patricia M., CAPT
Collins, Paul H, Capt
Collins, R Wayne, CAPT
Collins, Robert J, CDR
Collins, Robert R., LCDR
Collins, Robert W, Capt
Collins, Roy B, Capt
Collins, Thomas E, Capt
Collins, William B, Capt
Collins Jr, Charles L., Lcdr
Collins Jr, Howard D, Lt
Collins Jr, Hubert V., CAPT
Collinson, Arthur J, Capt
Collison Jr, Martin A, CDR
Collmann, Kathleen, ENS
Collyer, Robert O., CDR
Colman, Robert J, Lcdr
Colon, Steven W., CDR
Colona Jr, William H, Lcdr
Colp, Thomas A, CAPT
Colquhoun, R Bruce, CDR
Colquitt Jr, Richard E, Capt
Colt, Marshall, Capt
Colton, William S, CDR
Colvert Jr, Raymond G, CDR
Colvin, Donald R., LCDR
Colvin, Joseph T, Capt
Colvin, Robyn G, Lcdr
Colvin II, James G, Capt
Colwell, Lawrence S, Capt
Colwell, Robert N, Radm
Colyar, Ronald E, Lcdr
Combellick, Robert K, Lcdr
Combs, Edwin L, Capt
Combs, William A., CAPT
Comer, Sally A., CDR
Comerford, Gregory A, Capt
Comfort, Stephen A, CDR
Comins, John R, Capt
Commeree, Laurence M, Capt
Commons, Theodore J, CDR
Compardo, James Robert, Capt
Compton, Cecil C, Capt
Compton, Edward J, CDR
Conboy, Timothy Q., Lt
Concannon, T. Michael, Lcdr
Conder, James W, Capt
Cone, Theodore S, Capt
Confer, Donald D, CDR
Congdon, William C, Lcdr
Conger, Jon C, Capt
Conger, Michael D, Lcdr
Conkey, Becky, LCDR
Conkey, John A, Capt
Conklin, Dwight E, Capt
Conklin, George W, Lcdr
Conklin, Michael D, CDR
Conlan, Tye M., Ens
Conley, Don W, Capt
Conley, Gene R, Capt
Conley, James F, CDR
Conley, Tom M., LT
Conlin, Carter B, Capt
Conlin, Martin H, Lt
Conlin, Richard R, CDR
Conlon, Edward T, Capt
Conlon, James D, CDR
Conn, James E, Lcdr
Conn, Robert D, Capt
Connaughton, Sean T, Lcdr
Connell, James Joseph, Capt
Connell, Laurence M, Capt
Connell Jr, James G, Capt
Connell Jr, James J, CDR
Connelly, Floyd L., CDR
Connelly, Francis X, CDR
Connelly, Trulyn G, CDR
Conner, Frankie W, CDR
Conner, Steve E, CDR
Conner, Terry L, CAPT
Conners, Richard J, Lcdr
Conners, Robert T, CWO4
Connolly, James F, CWO4
Connolly, John F, Capt
Connolly, Joseph B, Lt
Connolly, Thomas P., CDR
Connor, Christopher D., LCDR
Connor, John J., LCDR
Connor, Robert T, Capt
Connor, Roger A, CDR
Connor Jr, G C, Lt
Connor Jr, John G, Lcdr
Connors, Tracy D, Capt
Connors Jr, William J, Capt
Conrad, Daniel L, Lcdr
Conrad, Ronald D., CDR
Conrad, Russell R, CDR
Conrad II, Gordon E W, CDR
Conradi, Genevieve E., CDR
Conroy, Frederick W, Capt
Conroy, James Michael, Capt
Conroy, John D, CDR
Conroy, John P, Capt
Conroy, Philip L, CDR
Conroy, Terry, Lt
Conry, Catherine E., Lt
Considine, Noreen, CDR
Conte, John V., CDR
Conter, Louis A, Lcdr
Conti, Anthony M, CAPT
Conti, Carmelo J., Capt
Conti, Emilio L, CDR
Conti, Francis A, Lcdr
Contino, Robert, CDR
Contreras, Bernard E., CDR
Contreras, William A., Lt
Converse, H. Peter, Lt
Conway, James R, Radm
Conway, Robert D., CAPT
Conway, Robert T, Capt
Conway, Thomas J, CWO4
Conway Jr, Curtis E., CWO3
Coogan Jr, James Joseph, CDR
Cook, Arthur G, CDR
Cook, Ben C, Capt
Cook, Charles L, CDR
Cook, David B., CWO3
Cook, David N, Capt
Cook, Dianne, Capt
Cook, Edward L, CDR
Cook, Frederick N., CDR
Cook, George Peter, Capt
Cook, Glen A, CDR
Cook, Harvey G, Capt
Cook, Hermon C., CDR
Cook, James L, Capt
Cook, John F, Capt
Cook, John W, CDR
Cook, Joseph K, Lt
Cook, Lynn K, CDR
Cook, Michael D, LCDR
Cook, Michael J, Lt
Cook, Michael L., CAPT
Cook, Nicholas A., CAPT
Cook, Patrick, LT
Cook, Rodney L, Capt
Cook, Shawn E., LCDR
Cook, Stephen M., LT
Cook, Theodore W, CDR
Cook, William C, CDR
Cook, William G., LT
Cook, William P, CDR
Cook Jr, Charles W, Capt
Cooke, Terrence A, Lcdr
Cooke, William Guy, LCDR
Cooke, William L, Lt
Cooke Jr., Vincent E., CDR
Cooksey, Mark G, Capt
Cooksey, Paul J, Lcdr
Cooksey, Susan G., CAPT
Cookson, John P, Capt
Coolbaugh Jr, George H, CDR
Cooley Jr, Lloyd E, Capt
Cooley Jr, Norman V, Capt
Coombs, Albert E., LT
Coomer Jr, Lesley B, CDR
Coomes, Edmund P, Lcdr
Coonan, Thomas R, CDR
Coons Jr, William R, CDR
Cooper, Alyson H., Lcdr
Cooper, Douglas E., LCDR
Cooper, Edward R, Capt
Cooper, George M, Lcdr
Cooper, Georgia S, Ltjg
Cooper, James H, Lcdr
Cooper, James R, Capt
Cooper, Jerry A, CDR
Cooper, Joan D., LT
Cooper, John B, Capt
Cooper, Laurie A., LT
Cooper, Lloyd F, Capt
Cooper, Philip P M, CDR
Cooper, Richard E., LCDR
Cooper, Robert L, Capt
Cooper, Roger M, Capt
Cooper, Scott L, CDR
Cooper, W. Scott, LT
Cooper, William B, Capt
Cooper, William David, Capt
Cooper, William F, Lcdr
Cooper Jr, David J, Capt
Cooper Jr, Tommy A, CDR
Cooper Jr, Wade T., CDR
Coopersmith, James M, Capt
Cope Jr, Stuart Samuel, Lcdr
Copeland, Mary T., CDR
Copeland, William N., CDR
Copeland III, Harrie E., CDR
Copley, David C, CDR
Copper, Bruce D, Capt
Copper Jr, Walter L, Capt
Coppins Jr, Cray J, Capt
Coppock Jr, Lawrence H., CDR
Corb, Bruce D, CDR
Corbett, James J., LCDR
Corbin, G Kimball, Lcdr
Corcoran, Brian, LCDR
Corcoran, Eugene F, Lcdr
Corcoran, John F, Capt
Corcoran, Peter J, CDR
Corcoran, Peter J., LTJG
Corcoran III, C Timothy, Lcdr
Corcoran III, George J, Cdr
Cordell, Donald E, CAPT
Cordwell, William R, Lcdr
Corgan, Randall J, Lcdr
Corkery, Jeffrey T, Lcdr
Corkery, Raymond M., Lt
Corley, Barton F, CDR
Corman, Cynthia D., CAPT
Cormier, Warren G, Capt
Cornelius, Walter, Capt
Cornell, Thomas B, Capt
Cornwell, Hubert R, Lcdr
Corpman, Daniel M., CAPT
Corrao, Eugene G., LT
Correa, John L., LCDR
Correnty, Frank A, Lcdr
Corrigan, Christopher M, CAPT
Corrigan, Donald T, Radm
Corriher, Cotesworth E, CDR
Corry, Daniel J, Capt
Corscadden, Henry J., CDR
Corsino, Marie F., LT
Corter Jr, Joseph, Cwo-4
Cortese, Charlene A., Lcdr
Cortez, Kimberly M., LT
Cortez, Rudolph M, Lt
Corum, Starling T, CDR
Cory, John A, Capt
Cos, Grant A, CDR
Cosgriff, Frederick W, Capt
Cosgriff, Lawrence E, Lt
Cosgrove, Joseph P, Capt
Cosgrove, Ronald P, Capt
Cosper, Lavelle W, CDR
Cosper, Thomas E, CDR
Cosper Jr, John C, CDR
Cossart, John W, Lcdr
Cossel, Stephen C, CDR
Cost, Richard W, Capt
Costa, Albert B, Lcdr
Costaldi, Mario E, CDR
Costantini, Vincent F., LT
Costantino, James, CDR
Costantino, Johm M., Capt
Costarakis, Dennis A., CDR
Costarakis, Matthew G, CDR
Costas, John N. H., RADM
Costello, Paul Keely, Capt
Costello, Peter L., CDR
Costen, Laurel M., CDR
Costenbader, David L, CDR
Costin, Leonard C, Lcdr
Coston, Wayland K, Lt
Cote, Kevin Jon, LCDR
Cote, Mark A., Lt
Cotey, Joseph E, Lt
Cothran, Garland R, Capt
Cothran, Walter W, CDR
Cotman, Lawrence, Capt

Cotner, Joseph T, CDR
Cotter, Joan E., CDR
Cotton, John B, Radm
Cotton, John G, RADM
Cotton, Michael J., LCDR
Cottone, James A, Capt
Cottrell, Davie, Capt
Cottrill, Christy W M, Lcdr
Cotugno, Paul J, CDR
Couch, Hugh R, Capt
Coughlan III, George R, Capt
Coughlen, Thomas D, Capt
Coughlin, Gerald W, Lt
Coughlin, John C, CDR
Coughlin, Richard J, Capt
Coughlin Jr, Joseph E, Capt
Coulombe, Serena M, Lcdr
Coulon Jr., Richard A., CDR
Coulter, James A, CDR
Countis, James T, CDR
Countryman, Roxie M, Lt
Counts, James L, CDR
Counts, John T., CDR
Couperthwait Jr, George, CDR
Courchesne, Penelope S., Lcdr
Courey, James M., Lcdr
Courneya, Thomas E, Radm
Coursey Jr, Bellford, Lcdr
Court, Donald B., Lcdr
Courtney, William A, Lcdr
Coutts, Alexander T, CWO4
Couture, Debra C., Lcdr
Couture, Raymond R, Radm
Covell, Peter E, CAPT
Covert Jr, Warren L, CDR
Coviello, Robert E, Lcdr
Covington, Michael B, Capt
Cowan, C Wayne, Lcdr
Cowan, Charles L, Capt
Cowan, David E, Capt
Cowan, Gary L, Lcdr
Cowan, Judy A, LT
Cowan, Marilyn, Lt
Cowan Jr, James H, CDR
Cowden, Anthony, Lcdr
Cowen, Joseph H, Capt
Cowen, Robert C, Lcdr
Cowing, Robert W., CDR
Cowley, Susan Kathleen, LT
Cox, Catherine W., CDR
Cox, Charles B, LCDR
Cox, David L., CAPT
Cox, Erik T., LT
Cox, Gordon L, Capt
Cox, Hamilton L, Lcdr
Cox, James D., Lt
Cox, James T, Lt
Cox, Jimmy Dean, LT
Cox, John Y, CAPT
Cox, Mack M, CAPT
Cox, Margaret M., Lcdr
Cox, Ray M, Lcdr
Cox, Rebecca Ann, Capt
Cox, Robert H, Capt
Cox, Ryan P., LCDR
Cox, Thomas L, CDR
Cox II, J Frederic, Capt
Cox Jr, James L, Capt
Cox Jr., Lyle A., CAPT
Coyle, Anthony L, Capt
Coyle, Francis X, Capt
Coyle, Richard H., CDR
Coyle, Robert J., LT
Cozad, Jerry E, Capt
Cozart, Thomas M., LCDR
Crabb, Joe F, CDR
Crabtree, Daniel B, Capt
Crabtree, Greg D., LCDR
Crabtree, Lionel C, CWO3
Craddock, Thomas A, CAPT
Craft, Keith C., LCDR
Craft, Lisa Ann, CDR
Craft Jr., Edward P., Lcdr
Cragin, Charles L, Capt
Cragin, Maureen P., Lcdr
Cragin, Naylor, Lcdr
Cragon, Stephen D, Capt
Craig, H. Allen, Capt
Craig, John E, Capt
Craig, Travis S., CDR
Craig, Wesley P, Capt
Craig, William H, Capt
Craig Jr, Robert M, CDR
Craige, Danny D, Capt
Craigie, James B, Capt
Crain, Joseph C, Capt
Crais, Henry R, CDR
Cramer, Mark, LCDR
Cramer, Richard S, CAPT
Crandall, David, LCDR
Crandall Jr, James W, CAPT
Crandell, William D, Capt
Crane, David D., Lt
Crane, Stephen N, Capt
Crane, William M., LCDR
Cranford, Reginald L, Capt
Cranmer, David J, Lcdr
Cranmer, Thomas F., Lcdr
Cranston, Craig H, CDR
Crates, Robert C, Radm
Cravaack Jr, Raymond J., CDR
Craven III, James B, Lcdr
Cravens, James O., LCDR
Crawford, Alvin H, Capt
Crawford, Charles W, Capt
Crawford, David C, CDR
Crawford, Duncan V, CAPT
Crawford, Edward C., CDR
Crawford, James K., LT
Crawford, Jerry E, LCDR
Crawford, John B., LCDR
Crawford, John D, Capt
Crawford, Wayne L, CDR
Crawford, William H, CDR
Crawford III, Tom H, CDR
Crawford Jr, B Hayden, Radm
Crawford Jr., Bertram M, Capt
Crawford Jr., John G., CAPT
Crawley, Carlton, Lt
Craycroft, Gary J, Lt
Creamer III, Frederic J, CDR
Crean, Sean F, CDR
Creed, James E, CDR
Creed, William J., CDR
Creeden, Dorsie G, CDR
Creekmore Jr, Thomas L, CDR
Creel, Loranne K., ENS
Creep Jr, Parker B, CDR
Creighton, Carolyn C., CDR
Creighton, Larry D, CDR
Crellin, Thomas E, Capt
Cresson, Ruth Ann, LCDR
Crews, Celinda R., CDR
Crews, John S, CAPT
Criez, Louis R, Capt
Crilly, Eugene R, CDR
Cripps, Larry D., CDR
Crismali, Michael C., CDR
Crisostomo, Danilo G, Lcdr
Crisp, Marvin H, CDR
Crispin, James L, Capt
Criss, Cecil M., LCDR
Crist, Jonathan L., Lcdr
Cristiani, Ralph D., CDR
Cristol, A Jay, Capt
Critchley, David W., LCDR
Critelli, JoAnn M., LCDR
Crites, Harold T., Capt
Crivelli, Albert F, Lcdr
Croce, John D., Lcdr
Crocker, Richard L, CDR
Crocker Jr, Walter, Capt
Crockett, Margaret E, Capt
Croker, Ben M, CDR
Crolick Jr, George R, Capt
Croll, Richard L, Capt
Crollard, Jerry, LCDR
Cromer, Howard E, CWO4
Crompton, David R, CDR
Cromwell, Robert D, Lcdr
Cronin, Hugh J, Capt
Cronin, John L, CAPT
Cronin, Judith Louise, CDR
Cronin Jr, C L, CDR
Cronin Jr, John W, Radm
Cronk, David J, Capt
Cronk, Elizabeth P., LT
Cronkite, Eugene P., Radm
Cronley, Maureen P, Lcdr
Crook, Delma L, Lcdr
Crook, Mary W., LCDR
Crook Jr, Richard H, Capt
Crooker, Peter E, CDR
Croom, Milton M, CDR
Crooms, Coletta B., CDR
Cropp, Ralph C, Lt
Crosbie III, James K, Lcdr
Crosland, William A, Capt
Cross, Daniel C, Lcdr
Cross, Jerry V, Capt
Cross, Raymond E, Lcdr
Cross, Robert A., Lcdr
Cross, Vilna A, Capt
Cross, William A, Capt
Crossland, Roger L, Capt
Crotty, Elizabeth A, CDR
Crouch, Mark B, CDR
Crouch, Myron K., CDR
Crouse, Carl N, CDR
Crouter, Brooke B., CDR
Crow, David L, Lcdr
Crow, Wayne H, CDR
Crowder, James M, Capt
Crowe, Richard C, Capt
Crowell, Robert L, Capt
Crowhurst, Thomas J, Capt
Crowley, James R, Capt
Crowley, Patricia Lynn, CDR
Crowson, James, MIDN
Crowther, Thomas B, CDR
Croy, Wallace W, Lt
Crozier, Marguerite M., LCDR
Crue Jr, Benjamin L, Capt
Cruikshank, Alan, Capt
Crum, James L., Capt
Crumbie, George J, Capt
Crump, Bankhead L, Lcdr
Crump, George W, Capt
Cruse, Patricia A., ENS
Crusoe, John A, Capt
Crutcher, John W, Capt
Crutchfield, James D, CAPT
Cruz Jr, A B, Capt
Cryan, Thomas M, Capt
Crysler, Melanie Beth, LCDR
Csencsits, James R., LCDR
Cubbage, Katie Marlene, Lcdr
Cuccia, Robert A, Capt
Cuccio, Michael V, Capt
Cuciti, Richard B, Capt
Cucullu, Irwin L, Capt
Cudal, Ted, LT
Cuddington, Michael E, CDR
Cuddy Jr, William F., Lcdr
Culbertson, Carol, Capt
Culbertson, Paul M., CDR
Culbertson III, William, Lcdr
Culhane, Edward C, CDR
Culhane, James R., Lt
Cullen, James G, CDR
Cullen, Raymond L, Capt
Cullers, Cleve, Lcdr
Culley, Joan M., CDR
Culley, Milton H, CDR
Cullum, Alcinda L., LTJG
Cully, Herbert L., CDR
Culotta, Anthony P, Capt
Culp, Melvin F, Capt
Culp, Robert V, Lcdr
Culpepper, Mary E, Lcdr
Culpepper, Michael L, Capt
Culpepper, Steven D, Lcdr
Culver, John A, Capt
Culverhouse, Elaine M., LCDR
Culverhouse Jr, Robert, Lcdr
Culwell Jr., C. William, CDR
Cumbie, Walter K., Lcdr
Cummer, John P, Capt
Cumming, Lisa A., LCDR
Cumming, Roy M, CDR
Cummings, Alan W, CDR
Cummings, Curtiss, Capt
Cummings, Francis, CDR
Cummings, Peter C., CAPT
Cummings, Richard L, LCDR
Cummings, Robert W, Lcdr
Cummings, William D., CDR
Cummins, Charles V, CDR
Cummins, Joshua W, CDR
Cummins, Ronald H, CDR
Cummins, Samuel A, Radm
Cummins, William E, Capt
Cuneo, James B, Capt
Cunningham, Charles R, Capt
Cunningham, Dale V, CDR
Cunningham, Hughes W, Capt
Cunningham, James A., Capt
Cunningham, John M., CAPT
Cunningham, Margaret A., CAPT
Cunningham, Melvin D, CDR
Cunningham, Paul J, Capt
Cunningham, Van H, Lcdr
Cunningham, Wallace R, Lcdr
Cunningham Jr, John E, Capt
Cuny Jr, Thomas B, Lcdr
Cuomo, Ralph J, Capt
Cupo, Douglas E, CDR
Cupples, Robert A, Capt
Curby, William A, Capt
Curcio, John B., Lt
Curley, Robert M, Capt
Curran, Daniel B., LCDR
Curran, Hugh D, CDR
Curran, James J, Capt
Curran, John P, Capt
Curran, Robert A., Capt
Curran, Thomas C, Lcdr
Curran Jr, John F, Capt
Current, Neal F, CDR
Current, Robert W, Lcdr
Curreri, Gino A, Capt
Currie, Duncan S., LCDR
Currie, Laurin C, CDR
Currie, Robert S, Capt
Currier, Stuart L, CDR
Curry, Eugene M, CWO4
Curry, Fred L., CDR
Curry, Irish O., CDR
Curry, Robert J, Capt
Curry, Thomas E., Lt
Curry III, Howard, Lcdr
Curtis, Craig, CDR
Curtis, Dawn L., CDR
Curtis, Edgar A, Capt
Curtis, Frederick G., CDR
Curtis, Gladys J., LTJG
Curtis, Harold E, CDR
Curtis, John F., CAPT
Curtis, Lawrence A, Capt
Curtis, Lee R, Capt
Curtis, R Dee, CDR
Curtis, Richard A, Capt
Curtis, Robert E, CDR
Curtis, William H, CDR
Curtis II, Leonard A, CDR
Curtis Jr, Dalton B, Capt
Curtis Jr, Nelson G, CDR
Curtiss, Dale R, CDR
Curtiss, Dean A, Capt
Curtze, Charles A, Radm
Cusano, Michael C, CDR
Cushing, Peter C., CAPT
Cushing Jr, Prentice, Capt
Cushing-Beyer, Eve A., CDR
Cusic, Marshall E., Capt
Cusick, Stephen K, Capt
Cusick Jr, Robert I, Capt
Cutchen, Bryan P., CDR
Cutillo, Richard T, Capt
Cutler, Robert J, Capt
Cutliffe, John N, CDR
Cutropia, James A, Capt
Cutting, Robert H., LT
Cutts, Leonard L, CDR
Cvrk, Stuart J., Capt
Cybulski, Jon V., LT
Cymerman, Z Adam, CDR
Cypen, Irving, Ltjg
Cyphers, Arthur B, CDR
Cyril, Jules, CDR
Cyrway, Christopher B., CDR
Czarnecki, Thomas R, Lcdr
Czarnecki-Cross, Johngy, CDR
Czarnomski, John E., LCDR
Czerewko, Dianna M., LT
Czerw, David J., Lcdr
Czerwinski, Linda S., Lcdr
Czerwonky, James H, Capt
D'Agostino, Joseph, Lt
D'Agostino, Robert, LCDR
D'Amico, Theodore A, Capt
D'Angelo, Edward W., CWO4
D'Aversa, Francis L., LT
Daack, Mary J, CAPT
Dabagh, Theodore D, Capt
Dabney, Roger L, CDR
Dabrowski, Jerome A, Capt
Dachille, Francis C., CDR
Dackert, Gerald G, Capt
Dacko, Joseph, Lcdr
Dadd, Alan C, CDR
Daeger, Edward A, Capt
Daggett, Steven B, Capt
Dagley, John C, CDR
Dagon, James E, Lcdr
Dahl, Kelley T., Lcdr
Dahl, Steven E., LCDR
Dahle, Leon A, CDR
Dahlheim, Harold D, CDR
Dahlinger, Henry O, Lcdr
Dahlquist, Arnold J, CDR
Daignault, Peter P, CDR
Dailey, Allen H, CDR
Dailey, Bruce R., Lcdr
Dailey, James A., LCDR
Dailey, William K., CDR
Dainer, Paul M, Capt
Dal Santo, John Q., CDR
Dalbey, Ralph J, CDR
Dale, Richard C., CDR
Dalessio, Robert M, Capt
Daley, John J, CDR
Daley, John T, Lcdr
Daley, Leonard T, CDR
Daley, Paul P, Capt
Dalkin, Joel M, Lt
Dallen, Robert D, Capt
Dalrymple, Jack, CDR
Dalton, Frederick E, LCDR
Dalton, Richard C, Capt
Dalton, Steven E, CDR
Dalton, Willis V, CDR
Daly, Brian F., CDR
Daly, Charles W, CDR
Daly, Christopher B, CAPT
Daly, John J, Capt
Daly Jr, Theodore M, Capt
Dam, Francis S, CDR
Dambaugh, John A, Capt
Damerow, Arno E, CDR
Damisch, Peter W., CDR
Damm, Charles A, CDR
Damore, Patrick R, Capt
Damron, Lawrence D, Capt
Damrow, Joseph W, Lt
Danaher, Gerald F., CDR
Danaher, James Wm, Capt
Dangler, Edward, Capt
Daniel, Douglas H., LCDR
Daniel, William, CDR
Daniel Jr, John C, LCDR
Daniele, Francis D, Lcdr
Daniels, A Raiford, Capt
Daniels, Brian L., CDR
Daniels, Franklin L, CDR
Daniels, Geoffrey, CDR
Daniels, Kirk J., Lcdr
Daniels, Michael A, Lcdr
Daniels, Thomas H., Capt
Daniels, William D, Lcdr
Daniels III, Hoyle H, CDR
Danielsen, Richard J, Lcdr
Danielson, Larry R, Capt
Danis, Karen T, Capt
Danis, Kurt D., LT
Danis-Litten, Carole, CAPT
Danisch, Herman J, Lcdr
Danks, Donald A, Lcdr
Dann, Alan O, CDR
Dann, Willard J, Capt
Danner, David L., CDR
Danner III, Paul K, CDR
Danni, Frank R, Capt
Dannov, Bernard A, Capt
Dansak, Daniel A, Capt
Dansker, Alfred S, Capt
Danskin, George A., CDR
Dantin, Karen C., CDR
Danzer, David B., CAPT
Dao, Huutri, LT
Dapser, Wayne R, CDR
Daquin, Frank J, CDR
Darby, Joseph J, Capt
Darcy, Thomas P, CDR
Darden Jr, David L, Lcdr
Darks, Ladonna L, CDR
Darley, Lee E, LCDR
Darling, Gerald T, Lcdr
Darling, John D, CDR
Darling, Robert F, Lcdr
Darling, Scott S., CDR
Darling, William E, Lcdr
Darosa, William A., Lt
Darr, Ruth E., CDR
Darrell, David E, LT
Darrow, John R, Lcdr
Darrow, Robert L, CWO4
Dartt, Darielle D, LCDR
Das, Adrian, LCDR
Dash, R Michael, Lcdr
Dashiell, Charles M, Capt
Dashiell, George F, CWO4
Dastugue, Fernand J, CDR
Dato, Leonard A., LCDR
Dattilo, Bruno F, Capt
Datzko, Stella, CDR
Daub, Wayne A., Lcdr
Daubenspeck, David G, Capt
Daugherty, Lawrence R., LCDR
Daughetee, Calvin C, LCDR
Daughtry, Douglas S, CDR
Daulong, Kevin N., CDR
Daum, Richard O, Capt
Daum, Robert S., ENS
Daus, Donald G, Capt
Dausin Jr., Lawrence R., LT
Daut, Pirro R, CWO4
Davan Jr, Clarence F, Capt
Davanzo, Eugene C., Lcdr
Davenport, Joseph D, Capt
Davenport, Shannon D., Lcdr
Davey, Lycurgus M, Capt
David, Kevin M., ENS
Davidenas, Leon M, Lcdr
Davidson, Bruce C., LCDR
Davidson, Gary R., CAPT
Davidson, Harry H, Capt
Davidson, Harry M, Lcdr
Davidson, James D, Capt
Davidson, John W, Capt
Davidson, Mark H., Mr
Davidson, Richard J, Capt
Davidson, Robert A, Lcdr
Davidson, Robin I, Capt
Davidson Jr, Harlan W, Lcdr
Davies, Anton F, Lcdr
Davies, Buck, CDR
Davies, Donna C, Lcdr
Davies, Francis M, CDR
Davies, John G, Capt
Davies, Scott R., Lcdr
Davis, Allen E., LCDR
Davis, Beatrice Kimsey, Ens
Davis, Billy F, LT
Davis, Charles G, Lcdr
Davis, Charles M, Lcdr
Davis, Charles S, Capt
Davis, Charles W, Lcdr
Davis, Charles W, Lcdr
Davis, Christopher R, LCDR
Davis, Craig M., CDR
Davis, Cynthia G., CAPT
Davis, David H., Capt
Davis, David H., Lt
Davis, Deirdre, LT
Davis, Dennis W, Lcdr
Davis, Donald A, CDR
Davis, Donald A, Capt
Davis, Donald L, CAPT
Davis, Donna P, Capt
Davis, Frederick C., Lcdr
Davis, Gerald L, Capt
Davis, Graham B, Lt
Davis, Grant, CDR
Davis, Homer B, CDR
Davis, Howland S, CDR
Davis, James E., Lt
Davis, James H, Capt
Davis, James M, Capt
Davis, Jerrie W, Lcdr
Davis, John B, CDR
Davis, John D, CDR
Davis, John E., LTJG
Davis, John M, Capt
Davis, John P, Capt
Davis, Jonathan L, CAPT
Davis, Joseph R, CDR
Davis, L Dow, Capt
Davis, Lloyd J, Capt
Davis, Lynn C, CWO4
Davis, Marc R, CDR
Davis, Norman C, CDR
Davis, Norman L, CDR
Davis, Patricia M., Lcdr
Davis, Paul L E, CDR
Davis, Peter W, CDR
Davis, Richard D, Capt
Davis, Richard S., CDR
Davis, Robert N., Lcdr
Davis, Robert W, Capt
Davis, Ronald P, Lcdr
Davis, Ross E., LCDR
Davis, Samuel A, Lcdr
Davis, Scott O., CDR
Davis, Stephen P, Lcdr
Davis, Teresa, Capt
Davis, Theodore E, Capt
Davis, Thomas S, Capt
Davis, William, CWO3
Davis, William E, Capt
Davis, William E, CDR
Davis, William H, Lcdr
Davis, William J, Capt
Davis, William J, Lcdr
Davis, William M, CDR
Davis, Wooldridge W, CDR
Davis II, David B, Capt
Davis III, Thomas E, CWO3
Davis IV, Dewitt, CDR
Davis Jr, Charles F, Capt
Davis Jr, Chester R, Capt
Davis Jr, Ernest J, Capt
Davis Jr, Frank L, Capt
Davis Jr, Norman O, LCDR
Davis Jr, Ozro L, Capt
Davis Jr, Raymond J., Lt
Davish, Joseph C, Lt
Davisson, Jeffrey D., LCDR
Davitt, William H, Lcdr
Davitt Jr, James, Capt
Dawber, Gordon L, Lcdr
Dawson, Charles F., CDR
Dawson, John H, Capt
Dawson, Kyle J, Capt
Dawson, Piers L., CDR
Dawson Jr, Northrop, Lcdr
Dawson Jr, S Cooper, Capt
Dawson Jr., Howard W., RADM
Dawydiak, Eugene Mario, Lcdr
Day, John E, Capt
Day, John G, Capt
Day, Stuart M., ENS
Day, Thomas E., Lcdr
Day Jr, Charles P, CDR
Day Jr, James S, Capt
Day Sr, Doyle K, CDR
Dayley, Alan J, Capt
Dayton, Jack O, Lcdr
Dayton, Regina L, Lcdr
Dayton, Timothy J, CAPT
Daywalt, Theodore L, Capt
Daza, Iuminada H., Capt
Daze, William Rock, CAPT
Dazet Jr, Clement S, Lcdr
De Atkine, Laura C., Lt
De Bois Jr, Oran P., Capt
De Bord, Timothy, Capt
De Buhr, Wallace E., Lt
De Burgos, Vicki A, LT
De Castro Jr, J Edmund, Capt
de Golian, William D, Capt
De Jean, Gregory O., CDR
de la Cruz, Julio S., CWO2
de La Cruz, Lorena M R, LCDR
de La Cruz Jr., L. Victor P., ENS
De La Torre, Jose L., CDR
De Lapo, Nicholas M., LT
de Leon, Fortunato S., LCDR
De Loach, Jay A., CAPT
De Lorenzo, Ronald G., CDR
De Luca, Susan Ellen, LTJG
De Maggio, John M, CDR
De Masi, Nicholas J., LCDR
De Paoli, Thomas M, CDR
De Polo, Rudolph M, CDR
De Socio, John A., LT
De Stefano, Mark, Lt
De Tata, James, Lcdr
De Tonnancourt, Linda I, Lcdr
De Villers, Rebecca E., CAPT
De Vito, Joseph P., CDR
De Vito, Joseph Peter, LT
De Vries, Donald C, Lcdr
De Wet, Martin C, CDR
Deacon Jr, Benjamin H, Capt
Deal, Donn E, Lcdr
Deal, Pamala K., CDR
Dean, Jeffrey S, Capt
Dean, John W., CDR
Dean, Kevin V, Capt
Dean, Marvin E, Capt
Dean, Robert A, CDR
Dean III, Robert M, Capt
Deaner, Florence E., CDR
Deangelis Jr, Joseph T, Capt
Dear, James M, CDR
Deare, Joel, LCDR
Dearie, David B., CDR
Dearth, Andrew H, Capt
Deas, William J, Capt
Deas III, Harry L, Capt
Deaugustinis, William C, CDR
Debarrows, John H, CWO2
Debaun, Gerard B, Capt
Debbink, Dirk J, CAPT
Debbout, John P., Capt
Debeck, Thomas W., CAPT
Debenedetto, Thomas J, Capt
Deberardinis, Franco, CDR
Deberardinis, John T, CDR
Deboer, William J, Lcdr
Debona, Donald J, Capt
Debona, Gerald J, Capt
Debons, Francis E, Capt
Debus, Peter J, Lcdr
Decarlo, Robert C., Lt
Decesari, Robert J., Lcdr
Decicco, Mary V., CAPT
Decicco, Virginia A, CDR
Decker, Andrew S., LT
Decker, Carolyn M., LTJG
Decker, Christian F, Capt
Decker, David R, Capt
Decker, Elanore P, LCDR
Decker, Geoffrey F., CDR
Decker, John T., Lt
Decker, Joseph W., Lt
Decker, Luis C, CDR
Decker, Petersen N., LCDR
Deckert, Warren P, Capt
Decordova Jr, Noel, Capt
Dedonatis, Samuel J, CWO4
Deem, Andrew J., Lcdr
Deemer, Leonard H, CDR
Deetz, Ralph A, Lcdr
Defiebre Jr, Bruce Karl, Capt
Deforth, Peter W, Capt
Degenfelder, Susan L, LT
Degenhardt, James C., CDR
Degeorge, Edwin W., CDR
Degermajian, Katherine R., Lcdr
Degiglio, Michael A, CAPT
Degiorgio, Jerry Robert, Capt
Deglandon, Ira G, Lcdr
Degnan, James E, CDR
Degrassie, John S, Capt
Dehaan, Jacob D, CDR
Dehner, Bruce E, Capt
Deiparine, Michael K., Lcdr
Deitle, Joy F., LT
Deitz, Terrence J., CDR
Dekay Jr, Rodman D, Lt
Dekleva, John F, Lcdr
Del Campo, John F, CDR
Del Mar, Alberto S, LCDR
Del Pizzo Jr, Charles W, Capt
Del Rosario, Juanito B, Lcdr
Del Tredici, Gene V, CAPT
Del Vecchio, Francis X, CDR
Delahoussaye, Deblanc A, Capt
Delair Jr., Louis, CDR
Delaney, Brian F., Capt
Delaney, Henry R, CDR
Delaney, Leo T, CDR
Delaney, Stephanie, LT
Delannoy, William C, Lcdr
Delaquis, Richard S., CDR
DeLaurentis, Robert G., LT
Delauter, Wayne E, CDR
Delcambre, John Bruce, CDR
Deldotto Jr, Edward F., Lcdr
Deleon, Leo M, Capt
Deleot III, Charles F, Capt
Delfin, Phil C., CDR
Delfs, Hugh N., CAPT
Delgado Jr, Rene, CDR
Dell, Robert E, Capt
Dell'Aquila, James V, Lcdr
Delloro, Andy, LCDR
Deloach, R Jeb, CDR
DeLoache, Timothy P., LCDR
Delong, David M, CDR
Delong, Howard J, CDR
Delpizzo, Richard D., Lcdr
Delrose, John M., Lcdr
Delwiche, David E, Capt
Delzell, William W., CDR
Delzer, Robert H, CDR
Demaggio, Bernard R, Capt
Deman, Anthony R.W., Lcdr
Demarche, Linda P., CDR
Demaree, Martha H, CDR
Demarinis Jr, Robert A, CDR
Demario, Ida M, Lcdr
Demarre, Carol S, Lcdr
Demartin, Rolando R, CDR
Dematta, Elliott K, CDR
Dembek, Stephen, Lcdr
Dembinski, Elsa B, Lcdr
Dembinski II, Thomas H, CDR
Demers, William A, CDR
Demetrio, Lianne S., LTJG
Deming, Richard W, CDR
Demontbreun, Wilbur R., CDR
Demoss, Jerry V, Cdr
Dempsey, Brian, Capt
Dempsey, Edward J, Capt
Dempsey, John M, CDR
Dempsey, Marie B, Capt
Dempsey, William J, Capt
Den Bleyker, James, Lt
den Dulk, Donald E., CAPT
Denardo, Gregory K., CDR
DeNardo, Janet E., CAPT
Denari, Mark E., CAPT
Dendinger, Joan E, CAPT
Denham, Charles R, CDR
Denholm, David Y, Lcdr
Denicola, Barbara Ann, Lt
Denman, Kenneth R, Lcdr

Dennard, Robert L, CDR
Denney, J Ronald, Radm
Denney Jr, John X, CDR
Denning, Richard G, Capt
Denninghoff, James P, Capt
Dennis, Creighton E., Ltjg
Dennis, Pamela R., LCDR
Dennis, Robert W, Capt
Dennis, Scott D., Lt
Dennis, Timothy A., CDR
Dennis III, John C, Capt
Dennis Jr, John P, Capt
Dennison, Robert L., CAPT
Denny, Douglas R, Capt
Denny, James H, Capt
Denson, Harry S., CDR
Denson, James D, CWO4
Denton Jr, Robert, Capt
Denz, Rudolph C., Lcdr
DePaio, Thomas C., ENS
Depascal Jr., Neil J, CDR
Depew, Clark B, Capt
DePizzol, Samuel N., LT
Deputy, Robert J, CDR
Deraps, Gordon D, Capt
Derby, William E, CDR
Derby Jr, Lawrence H, Capt
Dermyer, William D, Capt
Derr, Frederick M, Capt
Derr, John F, Capt
Derrick, G Ralph, Lcdr
Derryberry, Don M, Lcdr
Derthick, John W, CDR
Dervay, Joseph P., CDR
Des Jardins, Paul R, Capt
Desamper, Ernesto J, Capt
Desibour, Jay R, CDR
Desiderio, John J, CDR
Desjardins, Charles R, Capt
Despres, Joseph A, Capt
Desruisseaux, George L, CDR
Desserich Jr, Russell W, Capt
Dethier, Sandra, LCDR
Dethmann, Wernes W, Lcdr
Detrick, Mark, Capt
Detring, Linda M, CDR
Detterline, Carl L, Capt
Dettis, Johnna L., CAPT
Dettmer III, William H, Lcdr
Detweiler, John H, Capt
Detweiler, Joseph H, Lt
Detwiler, George D, Lcdr
Detwiler, Jerry F, Capt
Detwiler, Melvin L, Lcdr
Detzky, Michael L, CDR
Deuser, James C, CDR
Deussing, Eric C., ENS
Devaney, Charles W, CDR
Devaney, Patrick O., LCDR
Devault, Richard H., Capt
Devault Jr, George E., CAPT
deVeer, Donald H., CDR
Dever, Craig B, Capt
Dever, Kathryn I, Lt
Devereaux, William P, CAPT
Devers, Ann M., Lcdr
Devincenzi, Ronald D, Capt
Devine, Michael J, Lcdr
Devine, Vivian M., LCDR
Deviney, James R., CDR
DeVinney, David J., LT
Devito, George M, Capt
Devlin, Duke P, CDR
Devlin, John F, Capt
Devlin, Patrick F, Lt
Devoe, Carlyle J, Capt
Devoe, Sherman W, Capt
Devoe Jr, Harry L, CDR
Devoll, Marvin W, Lcdr
Devon, Thomas J, CDR
Devoss, Hugh A, CAPT
Devries, Larry G, Capt
Dewald, David A., Lcdr
DeWalt, Brook, LTJG
Dewalt, Ronald P, CDR
Dewar, John M., Lt
DeWeese, Kathryn J., ENS
Dewey, Glenn R., ENS
Dewey, Mary L., Lcdr
Dewey, William E, CDR
Dewey II, Earl F, CAPT
DeWoody, Wesley E., CDR
Dews Jr, Malachi, CWO4
Dewulf, Dennis D, CDR
Dewyer III, William L, CDR
Dexter, Stephen P, Capt
Deyak, John J, CDR
Deyoe, Harry B, Lcdr
Di Antonio, Mark J., LCDR
Di Bella, Richard M., Capt
Di Biase, Thomas L., Lcdr
Di Cairano, Canio, CDR
Di Carlo, Peter H., LCDR
Di Cola, Cynthia Ann, CDR
Di Maio, James J., CDR
Di Marcantonio, Albert, Capt
Di Marcantonio, Josette C., LCDR
Di Maria, James A., LT
Diacsuk, John S., CDR
Diamond, Michael W, Capt
Diamond, Stephen P, LCDR
Diamond, Thomas E, Lcdr
Diamond Jr, James L., Lt
Diaz, Richard, Lcdr
Dibble, Benjamin T, Capt
Dibella, Dominick J., Lcdr
Dibona, Lawrence B, Capt
Dicampli, James K., Lcdr
Dichov, Maria E, Lcdr
Dick, Ramon S, Lcdr
Dick, Robert D, Capt
Dickason, Royal V, Lcdr
Dickens, John W, CDR
Dickerson, Aubrey W., CDR
Dickerson, Daniel J, Capt
Dickerson, Jerauld L, Capt
Dickerson, Kenneth T., CDR
Dickey, Gene P, Radm
Dickey, Glen L, CDR
Dickinson, Arthur F, CDR
Dickinson, George E, Capt
Dickinson, William J, CDR
Dickson, David W., Lcdr
Dickson, Deborah S, CDR
Dickson, Edward H, Lcdr
Dickson, Gary W., Capt
Dickson, Hiram B, CDR
Diebold, Marilyn R., Lcdr
Diederich, Karl, LCDR
Diehl, James M, Capt
Diehl, Steven F., Lcdr
Diehl Jr, George F, Capt
Diehl Jr, Russell C, CDR
Dierckman, Thomas E, Capt
Dierks, Gordon R, CDR
Diesner, Harvey H, Capt
Dieter, Kenneth A., Capt
Dietrich, Donald P, Capt
Dietrich, Gerard L, Lcdr
Dietrich, Ronald L, Capt
Dietz, Clyde P., CDR
Diffendal, John W, Capt
Digan, Thomas E., CDR
Digiannantonio, Edmond, CDR
DiGiovanni, Augustine, CDR
Digiuseppe, John A, Capt
Dillan Jr, Johan A., Capt
Dillemuth, Thomas C., LT
Diller, David C., LCDR
Dillon, Gregory B., CDR
Dillon, Kevin, Capt
Dillon, Richard W, CAPT
Dillon, Ronald L, Capt
Dillon III, Lucius Polk, Capt
Dillon Jr, Frank T, Capt
Dillow, Richard M., Capt
Dilorenzo, Leonard A, Capt
Dilorenzo, Michael J., Lcdr
Dilworth, J Richardson, Lcdr
Dilworth, Robert Daniel, Capt
DiMauro, Nicholas J., CDR
Dimitro, Gregory J., LCDR
Dimmick, Lester W, CDR
Dimock IV, Joseph J., Lcdr
Dinegar, Robert H, CDR
Dingeldey, Peter E, Capt
Dinges, Thomas L, CDR
Dinkel, James R, LT
Dinkelspiel, Robert, Capt
Dinneen, Joseph C., CDR
Dinsmore, Bruce J, CDR
Dinsmore, Marvin B, Lcdr
Dinwiddie Jr, Parker H, CDR
Diodati, Ronald Ralph, Capt
Dipace, Gino, Capt
Dipalma, Elaine M, Capt
Dipiazza, Nicholas C, Capt
Dipierro, Gary S., Lcdr
DiPuccio, Andrew A., LCDR
Dipuppo, Michael T, Capt
DiRaimo, Edgar F, CDR
Dirga, Peter M, CDR
Dirks, Richard A, Capt
Dischner, Catherine U., CDR
Disharoon, Joseph L., ENS
Dishart, Daniel W, Lcdr
Dishman, David R, Capt
Dismukes, Newton B., Capt
Disney, Anna Charlene, CWO4
Disney, Lawrence A, CDR
Dissault, Jerry M, CDR
Distin, Sherman M., Lcdr
Ditmyer, Arthur T, Capt
Dittmann, Harry G, Capt
Dittmeier, David A, CDR
Dittmer, John C, CDR
Ditzler, David P, CDR
Divine II, Robert B, CDR
Dix, Richard C, Cwo
Dixon, Boyd R, Capt
Dixon, David Lynn, CDR
Dixon, David W., Lcdr
Dixon, Donald L, Capt
Dixon, Ella Jean, CDR
Dixon, Judy Ann, CAPT
Dixon, Kelvin N., Lcdr
Dixon, Philip E, CDR
Dixon, Thomas L, CDR
Dixon, Willard W, Lt
Dlwgosh, William R, CDR
Dobberteen, Kurt J, CDR
Dobbins, William L, CDR
Dobbins III, Robert N, LT
Dobeck, Richard H, CDR
Dobey, Patrick L, CDR
Dobrochowski, Michael A, CWO4
Docekal, William L, CAPT
Dockendorff, Robert D, Capt
Dockery, J Randel, CDR
Doclar Jr, Ernest P, CDR
Dodd, Joseph L, CDR
Dodd Jr, W Stanley, CDR
Dodd Jr., Bravell, LCDR
Dodds, John T, CDR
Dodge, David K, CDR
Dodge, Douglas D, Capt
Dodge, Elliott G, CDR
Dodge, Lawrence W, Lt
Dodge, Lee A, CDR
Dodge, Lloyd P, Capt
Dodge, Stephen A, Capt
Dodgen, Joey B., LT
Dodrill, David E, Capt
Dodson, Albertus F, Capt
Dodson, Calvin E, CWO3
Dodson, John R, CDR
Dodson, John R, CDR
Dodson, Max H, Capt
Dodson, Robert L, Lcdr
Dodson, Robert W, Lcdr
Doe, Paul C., LTJG
Doege, Arthur G, Lcdr
Doehnert, Mark P., CDR
Doellinger, John L, CDR
Doerr, Donald F, Capt
Doggett, William K, CDR
Doherty, Francis J, Lcdr
Doherty, Mark G, CDR
Doherty, Robert O, Capt
Doherty, Thomas J., Lcdr
Doherty, Warren K, CDR
Doherty Jr, Francis J, Capt
Dohner, David E, Capt
Dokurno, Anthony D., CDR
Dolan, Charles R., Lcdr
Dolan, John F, Ltjg
Dolan, Michael P, Capt
Dolan, Robert F, Capt
Dolan, Robert W, Lcdr
Dolan Jr, Michael B, CDR
Dolby, Leon M, Lcdr
Dolby, William R, Lcdr
Dolezal, Robert L., CDR
Doll, Bruce A., CAPT
Dollinger, Armand L, Capt
Dollins, Toney R., CDR
Dolman, Robert D, Lcdr
Domaguing Jr, Macario L, Capt
Domanski, Frank P, CDR
Dombrowski, Stanley J., Lcdr
Domeracki, Henry S, CDR
Domingo, Nestorio L., CWO3
Dominick, Jeffery A., Lt
Dommes, Ralph, LCDR
Donaghy, Francis D, Capt
Donahoe, David F, Capt
Donahue, Dennis M, CDR
Donahue, John G, CDR
Donahue, Michael, CDR
Donahue, Michael J, Capt
Donahue, William J, Lcdr
Donahue Jr, Daniel L., LT
Donahue Jr., Thomas M., LCDR
Donaldson, Theodore, Capt
Donaldson, Theodore E, Lcdr
Donalson, Robert C, Capt
Dondero, Robert T, CDR
Donegan, Roger P., Lcdr
Doneker, Thomas G, Capt
Donelson Jr, Martin, Capt
Donevant, Jerry A, Capt
Donges, William H., CAPT
Donley, Carlton R, CDR
Donley, William B, Capt
Donnefeld, Oscar W, Capt
Donnell, Robert E, Capt
Donnelly, Kathleen M, Lcdr
Donnelly, Mark E., LCDR
Donoho, Cecil C, Capt
Donohoo, Stanley F, Capt
Donohue, Edwin A, Lcdr
Donohue, Mike, LCDR
Donohue, Robert E., CDR
Donovan, Dendy D., CDR
Donovan, John Kenneth, Lt
Donovan, Michael R, Capt
Donovan, Patrick G, CDR
Donovan, Robert L, CDR
Donovan, Stephen J, CDR
Doolin, Don, Lcdr
Doolittle, Paul H., Lcdr
Doolittle, Theus Lee, Capt
Doose, Paul R, CDR
Doppes, Neil J, Lcdr
Doran, David J, Capt
Dorbin, James J., Lcdr
Doris, Karen E., Capt
Dorman, Alexander H, Capt
Dorman, Brent A., CDR
Dorman Jr, Marshall L, CDR
Dorn, Denise G., LT
Dornbier, Annette M., Lcdr
Doroff, David P., LCDR
Dorosky, John M., Lt
Dorratcague, Pierre E, Capt
Dorrell III, Russell E., LCDR
Dorris, Michael S., LT
Dorsch, Anthony J, CDR
Dorsett, Ralph A, CDR
Dorsey, James E, CDR
Dorsey, Joseph C, Capt
Dorsey Jr, Earl, CDR
Doscher Jr, J Henry, Capt
Dosh, Raymond C, Lt
Doskocil, William C., CDR
Dosta, John C, CDR
Dostal, Donald L, Capt
Dotseth, James H, CDR
Dotson, Bennie F, Capt
Dotson, Leslie A, CAPT
Dotson, Troy L., LT
Doty, Phillip E., LCDR
Doub, Thomas A, CDR
Doud, Howard R, CDR
Doudiet, Samuel B, CDR
Dougherty, Andrew P., Lcdr
Dougherty, Barry L, Capt
Dougherty, Edward N, Lcdr
Dougherty, Judith E., CDR
Dougherty, Ken R, CDR
Dougherty, Norris R, Capt
Dougherty III, Francis E., LCDR
Douglas, Michael L, Capt
Douglas, Peter H, Capt
Douglas, Scott K., Lcdr
Douglass, Gordon E, Lcdr
Douglass, Raymond B, CWO4
Douthat, Joel C., Lcdr
Dove, Larry E, CDR
Dow, David R, Lt
Dow, Jesse L, Lcdr
Dowbak, Gregory, CDR
Dowd, John F, Capt
Dowds, James W., LCDR
Dowdy III, William L, CDR
Dowie, Jon W, Capt
Dowis, James F, Capt
Dowley, Peter W, Capt
Dowling, George E, CAPT
Dowling, Michael L., CAPT
Downes, Arnold E., Lcdr
Downey, John B, Capt
Downey, Robert P., Lcdr
Downey Jr, William E, Capt
Downey Jr., Gerald J, CDR
Downie, Robert H, CDR
Downing, Jan S., CDR
Downing, John E, Capt
Downing, Mary Lou, Lcdr
Downing, Robert E, Capt
Downs, Harold L, Lcdr
Downs, James L, Lcdr
Downs, John R, Capt
Downton III, Charles E, Capt
Dowst, B Roger, Lcdr
Doxey, David, LCDR
Doyle, Craig P., LCDR
Doyle, Howard C, Lcdr
Doyle, Joseph S., Lcdr
Doyle, Michael H, CDR
Doyle, Michael T, CDR
Doyle, Michael T, Capt
Doyle, Norman, Lcdr
Doyle, Thomas E, Capt
Doyle, William J, Lcdr
Dozier, Stevan R, LCDR
Drabek, Richard J, CDR
Drabouski, John J, Lcdr
Draffone, A. Peter, CWO4
Drago Jr, Anthony J, Lcdr
Dragonir, George N, CWO4
Dragovich, Joseph D, Capt
Drake, Daniel J, Capt
Drake, David M, CDR
Drake, Joseph P, Ltjg
Drake, Michael J., CDR
Drake, Stephen Ralph, CDR
Drakes, William A., LCDR
Drance, Christopher P., CDR
Draney, Jerald L, CDR
Draper, Charles R, Lcdr
Draper, S Garland, CDR
Draper, Terry M., LT
Drasal, James E, Lcdr
Draut, Arthur K, CDR
Draut, Kenneth L, Lcdr
Drefahl, Steven P, Capt
Dreher, Richard W, Capt
Drehoff, John J, Capt
Dreiling, Bernard J, Lcdr
Drennen, Donald R., CDR
Drennen, Grant A, Capt
Dresser, George B, Capt
Dresser, William T, CDR
Dreswick, John A, CDR
Drew, Cynthia B., ENS
Drew, Daniel J., LCDR
Drew, Donald A., Lcdr
Drew, Marianne B, Capt
Drew III, Ernest W, Capt
Drexler, Joseph S, CDR
Dreyer, Douglas J., CDR
Dreyer, Irving, Lcdr
Dreyer, Jack K, CDR
Driessnack, Richard H, Lcdr
Drinkhouse, Joseph R, CDR
Drinkwater, Myron D, Capt
Driscoll, Charles C, Capt
Driscoll, Charles E, Capt
Driscoll, Edmund F., CDR
Driscoll, George C, LCDR
Driscoll, Joseph A, CDR
Driscoll, Margaret, Lt
Driscoll, Sondra L., CAPT
Driscoll, William P., CDR
Driscoll Jr, Wm Thomas, Capt
Driver, Douglas B., CDR
Driver, John J., Capt
Driver, R Temple, CDR
Driver, Russell B, Capt
Drobotij, Erin B., ENS
Droll, Michael A, CDR
Drone, Donald D, Capt
Droski, Stanley T, Lcdr
Drowota III, Francis F, CDR
Droy, Mark D., CDR
Droz, Cynthia Maravich, Capt
Drudik, Donald J, Capt
Drum, Edward A., CAPT
Drumbor, George A, CDR
Drumm, Donald K, Capt
Drumm, Thomas F, Capt
Drummond, Richard C, Capt
Drumwright, Henry E, CAPT
Drury, Jeffrey Paul, CDR
Dryhurst, George M., CDR
Dryman, Talmage L, Lcdr
Du Frayne, Francis J., CDR
Dubar, Pearson E, Capt
Dubay, Roland C, Capt
Dube, Daniel N, CDR
Dube, George, Capt
Dube, Laurent F., LCDR
Dube, William I., CDR
Duberstein, Norman, CDR
Dubois, H Keith, CDR
Dubois, Michel L, CDR
Dubois, William I, Lcdr
Dubose, Carlye T, CDR
Dubs, Theodore B, Capt
Dubuar, James D, Capt
Dubuc, Carroll E, Capt
Duby Jr, Charles W, Capt
Ducharme, Richard L, CWO4
Duchesne, Sandra M, Lcdr
Duck Jr, John R, CAPT
Duckett, Robert L, Lcdr
Duckworth, Edward W, Lcdr
Duckworth, Lowell C, Lcdr
Duclos, Michael E, CDR
Ducommun, Dale J, Capt
Dudley, Jack C, Lcdr
Dudley, James C, Capt
Dudley, Malcolm H, CAPT
Dudziak, Donald J, Capt
Duell, David P, Lt
Duerr, Edwin C, Capt
Duetsch, Robert A, Capt
Duff, Ann H., Lt
Duff, Joseph W, Capt
Duff III, John, CDR
Duffert, Charles M., LT
Dufficy, Thomas J, Capt
Duffy, Charles J, Capt
Duffy, Edward J P, Capt
Duffy, Laurence S, CDR
Duffy, Thomas M., LCDR
Duffy Jr, Thomas E, Capt
Dufresne, Dean A, CDR
Dugan, John J, CDR
Dugan Jr, Joseph J, CDR
Duggan, Maurice R, Ltjg
Duggar, Thomas E, CDR
Dugger, Roy W, Lcdr
Duguay, Robert F., CDR
Duhamel, David P, CDR
Duhamel, Robert R, Capt
Duke, Harvey L, Capt
Duke, James C, CDR
Duke, Roger M, Lcdr
Duke, William M., LT
Dukes, William E, Capt
Dulaney, Greg T., LCDR
Dulin, Robert A., MIDN
Dullea, Cynthia A., CDR
Dumbauld, Dennis B., CDR
Dunavent, Rebecca L, Lcdr
Dunbar, John D, Capt
Dunbar, Richard M, Capt
Dunbar, William M, Capt
Duncan, Alan B, LCDR
Duncan, Aldean, LCDR
Duncan, Craig L., LCDR
Duncan, Gregory L, CDR
Duncan, James M, CDR
Duncan, James W, Capt
Duncan, Jon A, CDR
Duncan, Keith T., Lcdr
Duncan, L. G. Earl, Capt
Duncan, Marvin E, Capt
Duncan, Randy E., CDR
Duncan, Richard G, Capt
Duncan Jr, William F, CDR
Duncavage, Josephine A, Lcdr
Duncheskie, Cristina E., CDR
Dundorf, James D., Lcdr
Dunegan, Linda C., Lt
Dunham, Richard S., Lt
Dunham, Robert M, CDR
Dunk III, Andrew P., Lcdr
Dunlap, Gene H, Capt
Dunlap, James A, Capt
Dunlap, Leonard G., Lcdr
Dunlap, Leslie H., CDR
Dunlap, Robert M, CDR
Dunlap, Roy L, Radm
Dunlap III, J Allen, CDR
Dunlop, A. Scott, LCDR
Dunn, Arthur K, CDR
Dunn, David P, Capt
Dunn, Duane A, CDR
Dunn, Edwin B, Lcdr
Dunn, Harold L, Capt
Dunn, Jon I, CDR
Dunn, Linda L., Lcdr
Dunn, Linda M, CAPT
Dunn, Martin C, Lcdr
Dunn, Michael L., CDR
Dunn, Nancy E, CAPT
Dunn, Paul H, Capt
Dunn, Paula, LT
Dunn, Peter G., Lt
Dunn, Peter H., Capt
Dunn, Robert F, Capt
Dunn, Robert F, Vadm
Dunn, Seymour B, Capt
Dunn, Stephen M, CDR
Dunn, Thomas P., LCDR
Dunn, Vance P, Ltjg
Dunn Jr, Edward I, CDR
Dunnahoo, Chad S., LCDR
Dunnavant, David E, CDR
Dunnavant, Ezra L, Lcdr
Dunne, Ellen T, Capt
Dunne, James M, Capt
Dunne, Michael J, CDR
Dunnenberger III, Henry, CAPT
Dunton, Wilton M., CDR
Dupuis, Louis P., CDR
Dupuis, Robert G., CDR
Durand, John E, CDR
Durando, Virgil G, Lt
Durant, Charles W., Lcdr
Durbin III, Frank, CWO4
Durfee, Frederick G, CDR
Durflinger, Don A, CDR
Durham III, John L, Capt
Durham Jr, Joseph H, Lcdr
Durkee, Kendall G, Capt
Durkin, Cheryl J., LCDR
Durkin, James J., LCDR
Durkin III, Timothy A, Lcdr
Durrance Jr, Frank M, CDR
Durst, George M F, Lcdr
Dusenberry, Harry A, CAPT
Dutour, William J., CDR
Dutton, Anthony L, CDR
Dutton, Arthur C, CDR
Dutton, Robert R., CDR
Dutton Jr, George R, Capt
Duvall, Ellen K., CDR
Duvall, George C, Ltjg
Duvall, S E, CDR
Duveen, Judi A, CAPT
Duxbury, Richard B, Capt
Duxbury, William H., Cdr
Dvorscak, Bernard J, Capt
Dwigans, Joseph L, Lcdr
Dwyer, Kathleen M, Lcdr
Dwyer, Martin J, CDR
Dwyer, Phillip W, Lcdr
Dwyer, Rodney C., Capt
Dwyer Jr., Louis E., Lcdr
Dyckman, Julia Y, Capt
Dye, Judy Lynn, LT
Dyer, Dwight D, Capt
Dyer, J Russell, Capt
Dyer, James E, CAPT
Dyer, Ted E, Capt
Dyer, William W, Lcdr
Dyer III, Hugh N, Capt
Dyke, James R, Capt
Dykema, Owen Wood, Capt
Dymond, John W, Lcdr
Dysart, Jack L, CDR
Dyson, Brenda S., ENS
Dyson, J. B., LCDR
Dzieciolowski, Lisa R., CDR
Dziedzic, Stanley F., CDR
Eade, Robert M, Capt
Eadie, John W, Lcdr
Eadie, Ronald E, Capt
Eagan, Joseph M., CAPT
Eagle, Milton, Ltjg
Eagleton, Alison P., LCDR
Eagleton, Richard E, Capt
Eakin Jr, Melton O, Lt
Eamigh, John R, CDR
Eardley, Daniel P., CDR
Earhart, Linda A., CDR
Earl, Lemuel R, Lcdr
Earl, Richard L, Capt
Earle, Otis K, Capt
Earles, Steven L., LCDR
Earley, James J, CDR
Earnshaw, John, CDR
Earwaker Jr, John S, CDR
East, Danny G., CDR
East, Richard H, CDR
East Jr, Edwin L, Lcdr
Eastham Jr, C Stacey, CDR
Eastman, Thomas W, CDR
Eastman Jr, Kenneth O, Capt
Easton Jr, Glenn H, CDR
Eastwick, Joyce M., CDR
Eastwood, James W, Radm
Eastwood, Myles E, Capt
Eatinger Jr., Robert J., CDR
Eatman, Genevieve B, Lcdr
Eaton, Gregory M., ENS
Eaton, Marvin A., LT
Eaton, William E, Lcdr
Eatough, Gregory G., LCDR
Eaves, Gregory E., LCDR
Eaves Jr, O Fred, CDR
Ebbert, Donald G, Capt
Ebe, Hubert J, Capt
Ebel, Frederick W, Capt
Eberhart, Andrew W., CDR
Eberling, George G., Lcdr
Eberly, Michael D., LT
Eberly, William G, Capt
Eberly III, George J., Lt
Ebert, Charles H., CDR
Eberth, Robert W, Capt
Eble, Frederick D., CDR
Eble, J. Thomas, Capt
Echevarrio, Philip, LCDR
Eck, Michael A, Capt
Eck, Scott W., CDR
Eckberg, Theodore J, Capt
Eckel, Mark Eugene, CDR
Ecker, John A, Capt
Ecker, Madeleine A., Lcdr
Ecker, Paul G, Capt
Ecker III, Charles T, Capt
Eckerson, Peter L., CDR
Eckfeld, Frederick J, CDR
Eckhardt, John H., Lcdr
Eckhoff, Harold E, CDR
Eckhoff, Melvin L, CWO4
Eckmann, Rolff W, Capt
Eckrote, Calvin C, Lcdr
Eckstein, Tim E., Capt
Eckstrom, Daniel R, CDR
Eckstrom, Richard A, Capt
Eddy, Dana G, CDR
Eddy, Douglas B., CWO3
Eddy, Lockwood R, CDR
Eddy, Malcolm L., Lcdr
Eddy Jr, James G, Lt
Ede, George, CWO3
Edelman, Leighton M, Lcdr
Edelman, Martin J., CAPT
Edelmann, Gregg D., CDR
Edelstein, Ronald E, CWO4
Eden Jr, Thomas M, CDR
Edens, Michael O W, Capt
Eder, Michael D, CDR
Edgar, A. Michael, CDR
Edgar, J Kenneth, Capt
Edgar, John B, CDR
Edgar, John W, Lcdr
Edgar, Vicki L., Lcdr
Edge, Mary F., CDR
Edgerly, John C., LCDR
Edinger, John S, Capt
Edman, Arnold J, CDR
Edmisson, Ken W., LTJG
Edmiston, Scott, LCDR
Edmonds II, Leland C, Capt
Edmondson, Locke F, Lcdr
Edmunds, Basil L, Capt
Edmunds, Elizabeth H, CDR
Edmunds, John H., CAPT
Edmunds, Stephen S, CDR
Edone, Joseph C, Lt
Edquist, Paul E, CDR
Edson, Mark W., LCDR
Edwards, Charles R., CAPT
Edwards, Christopher J, Capt
Edwards, Chuck N., CDR
Edwards, Doris V, CDR
Edwards, H Banks, CDR
Edwards, J Donald, Lcdr
Edwards, Jack H, Lcdr
Edwards, James I, CDR
Edwards, Jane W, Lcdr
Edwards, Jerry L., CDR
Edwards, Jerry M., Lt
Edwards, Kathleen D., Lcdr
Edwards, Luther H, LCDR
Edwards, Marcus S., LCDR
Edwards, Norman M., CDR
Edwards, Paul B, Lcdr
Edwards, Rustin, LT
Edwards, Tamara J., Lcdr
Edwards, Thomas H, Capt
Edwards, Thomas L, CDR
Edwards, Warren E, CDR
Edwards, Willarda V., CDR
Edwards, William H, Lcdr
Edwards, Willie M., CDR
Edwards III, James D, Capt
Edwards III, Olin M, Capt
Edwards Jr, James T, CDR
Edwards-Small, Amy C., LT
Edwin, James R, CDR
Eells, Gerdin D, CDR
Egan, Edward J, Lt

Egenberger, Thomas A, CDR
Egger, Robert A., LT
Egger, Teri Ann, Lcdr
Eggert, Kyle J., 2LT
Eggleston, Ervin L, Capt
Eggleston, J. Marvin, Lcdr
Eggleston, Rollin D, Lcdr
Eggleston Jr, Donald J, Lcdr
Egnotovich, Michael M, Capt
Ehlers Jr, Victor J., CDR
Ehmcke, Lance D, Capt
Ehr, Richard L, Capt
Ehrhardt III, Robert E., CDR
Ehrhart, Richard B, Lt
Ehrlich, Clarence L, Capt
Ehrlich, George E, CDR
Eiben, Donald T, CDR
Eich, John R, CDR
Eich, Richard Frank, Lcdr
Eich, Ritch K, Capt
Eichenbaum Jr, Harold, CDR
Eichenberger, Wm E, CDR
Eichenlaub, Valerie L., LCDR
Eicher, John D, Capt
Eichert III, Edwin S, Lt
Eichholz, Gerhard C, Lcdr
Eichold II, Bernard H, Capt
Eickenhorst, Julius W, CDR
Eickhoff, Roger C., Capt
Eickhoff Jr., Gregory, Lcdr
Eid, Norman V, CDR
Eide, Fred A, CDR
Eidson, Robert A, Capt
Eikel, Robert, CDR
Eiland, Gary L., CDR
Eiland, Guinn L, Capt
Eilberg, James S, Capt
Eilers, Barbara G., CAPT
Eilert, Mark A., LCDR
Eimas, Richard, CDR
Einbider, Morgan K., LT
Eisberg Jr, Arthur C, CDR
Eisenberg, Robert A., Lt
Eisenhart Jr, Don W., CDR
Eisenhauer, Robert G, CDR
Eisenhuth, Joseph P, Lcdr
Eisenman, Lawrence S, Capt
Eissler, Frederick, Lcdr
Eiting, Joseph P, Capt
Ekern, Ronald J, CDR
Eklem, John A, Lcdr
Eklund, Michael S., LCDR
Ekman, Glenn R, CDR
Ekola, Timothy, LTJG
Ekrut, Billy G, LCDR
Ekvall, Richard W., CDR
Elaine, Sharon, CDR
Elam, Thomas M., LCDR
Elam III, Harper J, CDR
Elarde-McCraig, Karen, LCDR
Elbert, Theodore F, CDR
Elbinger, Steven Z, CAPT
Eld, Doreen T, CDR
Elden, Thomas K, CDR
Elder, Anne M, CDR
Elder, Chris D., Lcdr
Elder, James L, Capt
Elder, Philip I, Lcdr
Elder, Robert M., CWO4
Elder II, W Robert, Lcdr
Elder Jr, George R, CDR
Eldred, Norman O., Ltjg
Eldred, Randy M., Lcdr
Eldredge, Ernest P., Lt
Eliason, Howard J., CDR
Elie, Thomas E., LCDR
Elkins, Danny W, Lcdr
Elkins Jr, John T, Capt
Ellabarger, Lowell G, CDR
Elledge, Elliott F., Capt
Elleman, Peter A, CWO4
Eller Jr., Floyd K., LT
Ellexson Jr, Stanley E, Capt
Ellingboe, James D, CDR
Ellingson, Luther J, Capt
Elliott, C Gerald, CDR
Elliott, Charles M, CDR
Elliott, David S., CAPT
Elliott, John S., CDR
Elliott, Kendal R, Lt
Elliott, Paul B, Capt
Elliott, Richard G, Capt
Elliott, Robert M, Capt
Elliott, Wade B., CDR
Elliott-Blakeslee, J K, CDR
Ellis, Brent L, Lcdr
Ellis, Donald A, Capt
Ellis, Douglas D, Capt
Ellis, Eugene D, Capt
Ellis, Frank D, Capt
Ellis, James T, Lcdr
Ellis, Jimmy L., Lcdr
Ellis, John C, CDR
Ellis, John F, Lt
Ellis, John R, CDR
Ellis, Mark S., CDR
Ellis, Michael V, Capt
Ellis, Osborne N, Capt
Ellis, Richard C, CDR
Ellis, Stephen L., CAPT
Ellis III, Howard B, Capt
Ellis Jr., George G, CDR
Ellison, James E, CDR
Ellison, Jeannine M., LCDR
Ellison, Leo C, CDR
Ellrich, Chris, CDR
Ellwood, Paul M, Capt
Elmore, Allan F, Capt
Elmore, Lowell E, Capt
Elrod, Scott, MIDN
Elrod, Stephen E., CDR
Elsen, Paul D, Lcdr
Elsesser, Brian A., ENS
Elsner, Chad, Lcdr
Elsner, Norbert B, Capt
Elsner, Robyn F., CDR
Elson, Peter H, Lt
Elster, Richard M, LT
Elstner Jr, Earl T, Capt
Elston, Sheryll L, Capt
Elston II, Robert G, Capt
Elwell, Robert G, CDR
Elwood, Albert A, Capt
Ely, Charles C, Lcdr
Ely, Jack V., CDR
Emberg, Jack S., LT
Embery, Vernon K, CDR
Embree, Frank G, Capt
Embry, Lloyd B, Capt
Emerson, Jack M, CDR
Emerson, John R, Lt
Emerson, Richard E, CDR
Emerson, Sidney, CDR
Emery, Austin H, CDR
Emery, Lee E, Capt
Emery, Todd B., LCDR
Emmanuele, Roland C, Capt
Emmel, Robert S, LT
Emory, Emerson, CAPT
Emory Jr, W Charles, CDR
Emrich, Allen H, Lt
Emry, Richard D, CDR
Emslie, William A, CDR
Enchelmayer, Peter A., CDR
Ende, Norman, Capt
Enders Jr, Lloyd W, Lt
Enderson, Laurence W, Capt
Endless, Margarita, CDR
Endres, John C, CDR
Eng, Benedict A, CDR
Eng, Wellington R L, Capt
Eng, Weylin G, Capt
Engbritson, Arthur C, Lcdr
Engebretson, Gary L, CDR
Engel, John H, Capt
Engel, Kirk E., Lcdr
Engel Jr, Wilson F, Capt
Engelhart, Darin A., LT
Engelke, Harold A, Capt
Engelman, Melvin A, Lcdr
Engen, Judith K, CDR
Engh, Byron R, CDR
Engh, D. Sandra, Lt
England, Negley F, CDR
England Jr, Carl D, Capt
Engle, Calvin R, CDR
Engle, Harold G, CDR
Engle, Paul M, Lcdr
Engle, William R, Lcdr
Englehart, David R., LCDR
Englehart, Thomas A., LT
Engler, Raymond V, Lcdr
Engler, Royce A, Capt
English, Glenn A, Capt
English, Joseph G., Capt
English, Raymond P, CAPT
English, Teresa J., LT
Engstrom, John T, Capt
Engvall, William O., CDR
Eninger, Larry A, Capt
Enloe, Daniel T., CDR
Ennis, Richard K, CDR
Eno Jr, Joseph E., CDR
Enos, Charles E., CDR
Enright, James R, Capt
Enright, Marc E, CDR
Ensz, Richard C., CDR
Entas, Leon J, Capt
Entenmann, Thomas E, CDR
Entriken, Robert K, Capt
Episcopo Jr, Nicholas J, CDR
Epp, Donald C, Lcdr
Epperson, Julien D., LCDR
Eppert III, John W., CDR
Epple, Erwin L., CDR
Epple, James E., LCDR
Epps, Elwood W, CDR
Epps, Richard C, CDR
Epstein, Charles S, Capt
Epstein, David M, Capt
Epstein, Michael J., Capt
Erb, Barry C., CDR
Erdenberger, Michael J., LT
Erdin, Forrest D, CAPT
Erdman Jr, Robert F., LCDR
Erhardt, Richard A, CDR
Erhardt, Richard O, Capt
Ericksen, Nancy F., CAPT
Erickson, Ann C, Ens
Erickson, Bernard J, Lcdr
Erickson, Carl A, Lcdr
Erickson, Charles A., Capt
Erickson, Clifford M, Capt
Erickson, Ernest G, Lcdr
Erickson, Gaylord L, CDR
Erickson, Hilmer E, CWO4
Erickson, James B, Lcdr
Erickson, James H, CDR
Erickson, John A, CDR
Erickson, Michael E., Lt
Erickson, Michael F, CDR
Erickson, Newell L, CDR
Erickson, Richard C, Capt
Erickson, Richard P, Capt
Erickson, William B., LCDR
Ericson, Paul J, Capt
Erie, Robert T, CDR
Eriksen, Einar N, CDR
Eriksen, Erik A., CDR
Erikson, John C, Capt
Erikson, Michael Scott, LCDR
Erland Jr, Robert J., CDR
Erlandson, Charles W, CDR
Erlandson, John C, CDR
Ermentrout, Bree A., Lcdr
Ernest, F L, Capt
Erno, Ruth L, CDR
Ernsberger, Deborah A., LCDR
Ernst, Clifford E, CDR
Ernst, Heidi M., CAPT
Ernst, Kevin M., CDR
Erpelding, Edward E, Capt
Errigo, John A., CAPT
Ertman, Stephen C, CDR
Ervin, Franklin A., Capt
Ervin, Harry C, Capt
Ervin, James B., CDR
Erwin, Jerry W., Capt
Erwin, Robert, Ltjg
Erz, Arthur O, Lcdr
Esaki, Evan H., LCDR
Escamilla, Elena L., LCDR
Eschen, Hermann J, Lcdr
Eschenbaum, John M, CDR
Escher Jr, Rudolph N., CDR
Escobar Jr., Pat, LT
Escudero, Ronald J., CAPT
Esfandiari, Mary Ann, Lt
Eskew, John B, Capt
Eslinger, Phillip W, CDR
Esperanza, Bonifacio C., Capt
Espeseth, Robert D, Capt
Espeseth Jr, Robert D, Lt
Esposito, Paul W, Capt
Essi, Vernon F, CDR
Esslinger, Jon H, Capt
Estabrooks, Edward G, Lt
Estefan, Jeffrey Allen, Lt
Estep, Eldon M, Capt
Esterbrooks, Robert C, Radm
Estes, Mark G., CAPT
Estes, Richard T., CDR
Estlund, Gregory J., Capt
Estrada, Debra D., Lcdr
Estrada, Richard P, Capt
Estrada, Stephen J., Lt
Estrin, S. Sharkey, Capt
Etheredge, James B, CDR
Etheridge, Tammy H, Radm
Ethridge, Stephen A., CDR
Etienne, Fleurette S, Lcdr
Etzkorn, John R, Lcdr
Eubank, Paul J, CDR
Eubanks, Ernest H, Capt
Eubanks, Marlene N, LCDR
Eubanks, Mary Ella, Lcdr
Eucare, Alexander J, CDR
Evanoff, John, Capt
Evanoff, Michael J., Lcdr
Evans, Albert C, Capt
Evans, Arnold A, Capt
Evans, Carletta Y., CDR
Evans, David J., LT
Evans, Dawn, LTJG
Evans, Douglas J., CDR
Evans, F Emmett, Capt
Evans, Frank C, Capt
Evans, John T, Capt
Evans, John W, Capt
Evans, M Dwight, CDR
Evans, Michael E., Lcdr
Evans, Miles J, CWO4
Evans, Nancy A., LCDR
Evans, Norman S, Lcdr
Evans, Robert B, Lcdr
Evans, Thomas W., Lcdr
Evans, Todd W., LCDR
Evans, William M, Lt
Evans, Winthrop S, Lcdr
Evans III, Henry J, CDR
Evans Jr, Earl F, Capt
Evans Jr, Evan G., Lcdr
Even, Sondra A., LCDR
Everett, David R., CDR
Everett, Dennis L., Lcdr
Everett, Edwin A, CDR
Everett, Terrell A, CDR
Everette, Robert B, Lcdr
Everman, Jerry D, Capt
Everson, Donald B, Capt
Everts, Robert A, Capt
Evora, Victor F., ENS
Ewan, Thomas K, CAPT
Ewan, William Kenneth, Capt
Ewing, Michael R., CDR
Ewing, William David, CAPT
Ewing Jr, William H, Capt
Eyer, Mark S., CDR
Eyerly, Robert C, Capt
Eynatten, Robert J, Capt
Ezell, Gerry D., CDR
Faas, Vicki L, CDR
Fabre Jr, Frank J, Capt
Fabrizio, Francis J, Radm
Facas, Charles M, CDR
Fackler, Nancy A, Radm
Fackner, Jean E, Lt
Fafinski, John A., Capt
Fagan, William A, Capt
Fagan, William J, Lcdr
Fagen, Raymond, CDR
Fagg, James C, LT
Faherty, Denis M., CAPT
Faherty, Genevieve M., CDR
Fahs, Philip James, CAPT
Fahy Jr, Richard, Lcdr
Faibisoff, Burt I., CAPT
Failor, Mark J., LT
Fain, John G, CDR
Fain, Robert D, Lcdr
Faina, Dean L., Ens
Fainberg, Bertrand M, CDR
Fair, Harlan W, Capt
Fair Jr, Louis, CDR
Fairbanks, Dale J, Capt
Fairbrother, Joseph T., CDR
Fairhurst, Franklin E, Capt
Fairley, David R., LCDR
Faist, John A, CAPT
Fajer, Vincent J, Capt
Fajnor, Craig R., CDR
Fajnor, John P, Lcdr
Fakess, John A., CDR
Falcon, Paul, CAPT
Faler, Dale, Capt
Fales, Willard C, CDR
Falkenstein, Robert F., Capt
Fall, Harry C, Capt
Fallon, Patrick J., CDR
Falsetti, Donna J., Lcdr
Falt Jr, Gordon H, Capt
Falvey, David J, CAPT
Fancher, Christopher S., CDR
Fanning, Frank L, CAPT
Fanning Jr, Timothy O, Radm
Fanseen, James F, Capt
Faraone, Annette L., CDR
Faria, Robert E, Lcdr
Farina, Anthony J., LCDR
Farkas, Daniel P, Capt
Farley, Evan T., LCDR
Farley, James C, CDR
Farley, Michael H, Capt
Farley, Randal D., CDR
Farley, Robert A., CDR
Farley, Russell A, Lt
Farling, Alan J, Lcdr
Farlow, Lynne A., CDR
Farmer, Alan C., CDR
Farmer, Michael K, CDR
Farmer, Ronald A., LCDR
Farnsworth, Harry N, Capt
Farnum, Leon V, Lcdr
Farr Jr, John F, Capt
Farrar, David R, Capt
Farrar, Ed, Capt
Farrell, George R, Capt
Farrell, John F, Lcdr
Farrell, Paul A., LT
Farrell, Vincent D, Capt
Farrell, Walter W., CDR
Farrell, William L, Capt
Farrell Jr, Richard L, Capt
Farris, Marc, CDR
Farris, Ryan, MIDN
Fasano, Vincent P, Capt
Faschini, Cynthia M., LCDR
Fashbaugh, Robert C, Capt
Fath, Perry M., Lt
Fauber, Stuart C, Capt
Faubus, Ikey R, Lcdr
Faul, Anthony, CDR
Faul Jr., John L, Lcdr
Faulders, J David, CAPT
Faulkner, Betty L, Lt
Faulkner, Harold H, CDR
Faulkner, Thomas B., Lcdr
Faus, Norton T, Capt
Fauser, Steven J., Lt
Faust, H Patrick, Capt
Faust, Robert B, Capt
Fava, Mark C., Lcdr
Favero, Val R, Capt
Favre, John A, Lcdr
Fawthrop, Roland P, Capt
Fawx, William, Lcdr
Fay, Robert A., CAPT
Fazio, Nicholas A, Lcdr
Fearnow, F Clay, CDR
Featherstone, Robert J, Capt
Feaver, Peter D., LT
Febach, M. Colleen, Lcdr
Fedarko, William, Capt
Feder, Joel M, Capt
Federoff, Jack A., CDR
Fedison, Dennis P., CAPT
Fedor, Louis P, CDR
Fedyski, Robert J, Lcdr
Fee, James W, Capt
Fee, Jason, LCDR
Fee, Richard A, CDR
Feeney, James F, Capt
Feeney, Joseph J, Capt
Feeney, Kevin J, Capt
Feeney, Saundra M., Lcdr
Feerick, John P, Capt
Fehrenbacher, Randall L., LCDR
Fehrle, Thomas L, CAPT
Feichtinger, Mark R, Capt
Feiden, Bart J, CDR
Feist, Robert W, CDR
Feist Jr, Raymond F, CDR
Felber, James A, Lcdr
Felchlin, James C, CDR
Felcyn, John J, Capt
Feld, Gerald A, Capt
Felder II, Wilson N., CDR
Feldhaus, Bernard R, CDR
Feldman, Joel M, Capt
Feldman, Lee R., LCDR
Feldman, Mervyn I, Lcdr
Feldman, Peter M, Capt
Felgate, George, Capt
Felger, Thomas R, CAPT
Felix, John L, CDR
Felix, Robin, CAPT
Felix, William V, CDR
Felker, John J, Capt
Fell, James R, CDR
Fell Jr, Bert H, CDR
Feller, Cynthia M, CDR
Feller, Gary F, Lcdr
Feller, Rainer, Capt
Fellers, George O, CDR
Fellers, William O, Lcdr
Fellows, Fred J, CDR
Felsenthal Jr, Harry, CDR
Felt, Carl A, Capt
Feltes, James E, CDR
Feltham, Francis M, Capt
Feltham, Marshall E, CDR
Felthous, Alan R., Capt
Felthous, Roger A., CWO4
Felton, Kenneth D, CDR
Fenimore, Julie A., Lcdr
Fenn, Dan E., CAPT
Fennessy, Eamon T, CDR
Fenning, Scott M., CDR
Fenstermacher, John W, CDR
Fenstermaker, Paul M, CWO4
Fenton, James R., CDR
Fenzl, Donald, Lcdr
Feraco, Frank C, CDR
Ferber, Arthur H, Capt
Ferchau, David Otto, LCDR
Ferdin, Peter H., Lt
Ferenchick, Richard J, CDR
Ferenczy, William J., CAPT
Fergione, John A, Capt
Fergus, David W., CDR
Ferguson, David W, CAPT
Ferguson, Earl A, Capt
Ferguson, Eugene M, CDR
Ferguson, Floyd A, CWO2
Ferguson, Gary W, CDR
Ferguson, Hugh M, CDR
Ferguson, James D, Capt
Ferguson, James L, CDR
Ferguson, Jere C, Capt
Ferguson, John C, LCDR
Ferguson, John David, CDR
Ferguson, John F, Capt
Ferguson, Larry W, Capt
Ferguson, Michael P., CDR
Ferguson, Oscar L, CWO4
Ferguson, Ray W, CDR
Ferguson, Robert K, Capt
Ferguson, Robert L, Capt
Ferguson, Walter V, Lcdr
Ferguson Jr, Emmet F, Capt
Ferguson Jr, Ernest F, CAPT
Ferguson Sr, Robert, Lt
Ferketich, Michael L., LCDR
Fern, George M, CDR
Fernandes, Ricardo A, Lcdr
Fernelius, Byrne C, Lcdr
Ferrante, Michael P, Lt
Ferrara, Alfred J, Capt
Ferrari, Daniel E, Capt
Ferre, Earl B, Lcdr
Ferrell, Kenneth J, Lcdr
Ferrell Jr, Dewitt T, CDR
Ferrer, Edward B., CDR
Ferrera, John J, Capt
Ferris, Edward, Capt
Ferris, Joyce M, CDR
Ferris III, William T, Capt
Ferry, Edward T, Capt
Ferry, Francis R, CDR
Ferry, John J, Capt
Fertig, Steven A, Capt
Fesler, Diane M., LT
Fessel Jr., T. F., CDR
Fetchko, Edward J, Lcdr
Fetrow, Blake M., LCDR
Fetter, Kenneth A, Capt
Fetter, Mark A., LT
Fetterman, Henry H, CDR
Fettes, Barbara E., ENS
Fewster, Timothy B., LCDR
Fey, Norman J, Capt
Fezza, Alessandro, Lcdr
Fiacco, Robert D., LCDR
Ficca, Gary P, Lcdr
Fichtner, Ernest, CDR
Fichtner, Michael L., Lcdr
Fick, J Howard, Lcdr
Fidler, Charles R., Lcdr
Fidler, William K, CDR
Fiebelkorn Jr., George A., CAPT
Field, Carl B, Lcdr
Field, Cynthia M, CDR
Field, David R, Capt
Field, Edward J, CDR
Field, Lee, Lcdr
Field, Mark W, Lt
Field, Richard H, Capt
Field, Richard P, Radm
Fielding, George M, Capt
Fieldman, Barbara, CAPT
Fields, Jay A, CWO4
Fields, Walter S., Capt
Fields, William J, Lcdr
Fiesel, Christopher A., Lt
Fietta, Peter M, Capt
Figueroa, Richard, CWO4
Figuras, R. J., CAPT
Filardi Jr., Charles J., CDR
Filep, Ernest J, CDR
Filipic, Allen J., Capt
Fillmore, Lila J, CDR
Fillo III, Stephen F., CDR
Filostrat II, John, LT
Filson, Steve N., CDR
Finagin, William B, Radm
Finan, Laurence G., LCDR
Finby, Mark A., LCDR
Finch, Walter D, Lcdr
Findley, Kenneth W, CDR
Fine, Charles B, CDR
Fine, Gregory R., CDR
Fingado, Brian H., LCDR
Finger, Ronnie R., Lt
Fingerhut, A. F., Lcdr
Fingerhut, Judy M., LT
Fink, Allan W, Lt
Fink, Daniel C., CDR
Fink, Daniel L., Lcdr
Fink, Edward R, Capt
Fink, Rodney J, CDR
Fink, Signor J L, Lcdr
Fink, Thomas E., LCDR
Finkbiner, Paul W, CDR
Finkel, Howard A, CDR
Finkenhagen, William L, Lcdr
Finlay Jr., John David, Capt
Finley, Michael S, CDR
Finley, Patricia M., LCDR
Finley, Robert H, Capt
Finnegan, Terrence E., LCDR
Finneran, John B, Capt
Finnerty, John B, Lt
Finnes, Roger G, CDR
Finney, Louis A, Capt
Finney, Peter T., Lcdr
Finney, Robert G, Capt
Fiorello, James R, CDR
Fiorino, Michael, Lcdr
Fireoved, Stephen J., CAPT
Firestone, Alan M., Capt
Firestone, Karen J., ENS
Firman, Gregory J., Capt
Firth, Richard E, Lcdr
Fischbach, Reynolds J, Capt
Fischbeck, Vicki Lynn, LCDR
Fischer, Douglas R., Capt
Fischer, G Jack, CAPT
Fischer, Gerald I., LT
Fischer, Richard W., Lcdr
Fischer, Walter R, CDR
Fischer Jr, John N, Capt
Fischetti, John A, Capt
Fischley, James H, CDR
Fish, Cameron H., Lt
Fish, David A, CDR
Fisher, Aimee F, CDR
Fisher, Charles J, Lcdr
Fisher, Charles K, Capt
Fisher, Helen C., LCDR
Fisher, Howard R., Capt
Fisher, J Lyndal, CDR
Fisher, Jack G, Capt
Fisher, Jere D., Lcdr
Fisher, Katherine M, LCDR
Fisher, Kenneth L, Radm
Fisher, Kymberly L., LCDR
Fisher, Ollie C, CAPT
Fisher, Paul E, Lcdr
Fisher, Paul R., CDR
Fisher, R. Stewart, CAPT
Fisher, Richard T, Lcdr
Fisher, Ronald A., Lcdr
Fisher, Ted D., CDR
Fisher, Thomas C, Lcdr
Fisher Jr, E. Byron, Capt
Fisher Jr, James F, Lcdr
Fisherkeller, Carl R, CDR
Fisk, Marcus J., CDR
Fiske, Gregory A., CDR
Fiske, Ronald N, CDR
Fitch Jr, Edward M, Capt
Fitch Jr, Morgan L, Lt
Fitzell, Joann Walker, LT
Fitzgerald, Amy L., LCDR
Fitzgerald, David M, Capt
Fitzgerald, Douglas P, Capt
Fitzgerald, Francis A, CDR
Fitzgerald, James A, Capt
Fitzgerald, James G, Capt
Fitzgerald, John B, CDR
Fitzgerald, John W, Capt
Fitzgerald, Malorie L., CDR
Fitzgerald, Marsha A, Capt
Fitzgerald, Michael J., CDR
Fitzgerald, Michael P., Capt
Fitzgerald, Patrick J., CDR
Fitzgerald, Peter L, Lt
Fitzgerald, Rebecca L, CDR
Fitzgerald, Richard S, Radm
Fitzgerald, Robert L, CDR
Fitzgerald, Stephen G, Lcdr
Fitzgerel, Betsy A., CDR
Fitzgibbons, Thomas B, CDR
Fitzmaurice Jr., Patrick J., CDR
Fitzpatrick, Garland M, CDR
Fitzpatrick, Norman G., LCDR
Fitzpatrick, Richard S, Lt
Fitzpatrick, Scott L., CDR
Fitzsimmons, Brian R., LT
Flack, Fred P, Capt
Flacke, Robert F, Capt
Flagg, Wilson F, Radm
Flagler, Norris J, Capt
Flaherty, David E., Lcdr
Flaherty, Dawn Mary, LCDR
Flaherty, James F., LCDR
Flaherty, Keith L, CDR
Flaherty, Michael B, Lcdr
Flaherty, Michael T., Capt
Flaherty, Richard L, Capt
Flanagan, Deborah A., Lcdr
Flanagan, Hugh M, Capt
Flanagan, John E, CDR
Flanagan, Michael S, Lcdr
Flanagan, Robert A, CDR
Flanders, Niles D, CWO4
Flandreau, Mark E, Capt
Flanigan, Joseph D, CDR
Flanigan, William M., LT
Flannery, James P., LCDR
Flaum, Joel M, Lcdr
Fleetwood Jr, Joseph A, Capt
Fleig, Otto J, CDR
Fleigel, Jeffrey D, Lcdr
Fleischmann, Dennis J., Lcdr
Fleming, Francis M, CDR
Fleming, Garwin K, CDR
Fleming, John M., CDR
Fleming, Michael B, Capt
Fleming, Robert W, CDR
Fleming, Timothy P., LCDR
Fleming Jr, John B, Capt
Fleming Jr, Myron T, Capt
Flemming III, John E, Capt
Flenner, James A, Lcdr
Fletcher, Gary R., Lcdr
Fletcher, George J, Lcdr
Fletcher, Glenn A., CDR
Fletcher, James P., LCDR
Fletcher, John M, CDR
Fletcher, Richard K, CDR
Fletcher, Robert, Capt
Fletcher, Robert M, Lcdr
Fletcher Jr, Albert E, CDR
Fletcher Jr, Roy W, CDR
Flinn, Thomas R, Lcdr
Flint, Michael, ENS
Flippin III, Walter N., Capt
Floberg, John F, Lcdr
Flodman, William C, CDR
Flomenbaum, Hyman S, Lt
Flood, Kevin P, Capt
Flor, Patrick J., LCDR
Flor Jr, Frederick H, Capt
Flora, Mark D, CDR
Florence, Richard A, Capt
Flores, Joseph A, Lcdr
Flores Jr, Enrique, Lcdr
Florimonte, Thomas S, CAPT
Florio, John A., CDR
Floss Jr, Walter J, Capt
Flournoy, Stephen P., LCDR
Flournoy, Thomas H, CDR
Flowers, Joel R., Lcdr
Flowers, Neal S., Capt
Floyd, David S, CDR
Floyd, Debra S., CDR
Floyd, Edward T, Capt
Floyd, Emmett O, Radm
Floyd, Thomas F., CDR
Fluaitt, David B, CDR
Flude, John W, CDR
Fluharty, Allan R., Lcdr
Flynn, Daniel L., CDR
Flynn, F Patrick, CAPT
Flynn, Gordon L, Capt
Flynn, J. William, LCDR
Flynn, James K., CDR
Flynn, John E, Capt
Flynn, John P, Ltjg

Flynn, Lawrence P, CDR
Flynn, Michael B, Capt
Flynn, Phillip W, CDR
Flynn, R. Patrick, Lcdr
Flynn, Timothy C, Capt
Flynn, Timothy T, CAPT
Flynn, William D, Capt
Flynn, William F, CDR
Flynn Jr, Henry M, Capt
Flynn Jr, Richard, Lcdr
Fobes, Duncan K., CDR
Fobes, W Stewart, CDR
Foege, Vernon F, CDR
Foerster, Bruce S, Capt
Fogarty, Thomas V, Capt
Fogerty, William K, Capt
Fogle, G Dean, Capt
Fogle, Philip R, Capt
Fogt, Norman W, Capt
Foil Jr, Horace O, Lt
Foleen, Carl R, CDR
Foley, Carl O, CDR
Foley, Christopher C., Lt
Foley, Daniel M, CDR
Foley, Daniel V, CDR
Foley, Dennis J, Capt
Foley, James K, CDR
Foley, James P., CDR
Foley, Louis T, CDR
Foley, Michael T, Capt
Foley, Robert B., ENS
Foley, Thomas F., LCDR
Folgate, Julius M, CAPT
Folger Jr, Ralph E., Lcdr
Folkes, Howard D., LCDR
Folmer, George W, CDR
Folsom, Edward R, CDR
Folsom, Walter C, Lcdr
Folsom Jr, Robert B, Lcdr
Foltz, Howard P, LCDR
Foltz, Peter C, Capt
Foltz, Teresa B, CDR
Folven, Jon G, CDR
Fong, Arthur C., ENS
Fonseca, Anthony, Lcdr
Fontaine, Scott A., CAPT
Fontanelli, Enio, CDR
Fontanier, Charles E, CDR
Fontecchio, Kenneth L, Capt
Fontenot, Paul R, CDR
Foody, Richard R, CDR
Foor III, Rolland E, Capt
Foote, David C., Capt
Foote, David L., CDR
Foote, Gerald L, CDR
Foote, Herbert W., Capt
Foote, Kim M., LT
Forbes, Albert, Ltjg
Forbes, Billy D, Lcdr
Forbes, Francis M, CDR
Forbes, Gregory A, Capt
Forbes, Jon E, CAPT
Forbes, Laurie E., CAPT
Force, Michael P., LT
Ford, Alan R., CDR
Ford, Barbara G, CDR
Ford, Janet M, CDR
Ford, Morris H., Lcdr
Ford, Richard R, LT
Ford, Thomas C, CDR
Ford, William E, Lcdr
Ford, William R, CDR
Ford Jr, Francis J, Capt
Fordham, Christopher P., LT
Fordham, Robert J, Capt
Forest, Edward F., LCDR
Forkner, Donald F, Capt
Forma, Steven A., Lcdr
Forman, Richard E, CDR
Formulak, Colin, LT
Forrest, Benjamin F., Capt
Forrest, James E., RADM
Forrest, L Albert, Capt
Forrest, Thomas Allen, LCDR
Forrester, James E, Lcdr
Forrester II, Joseph G, CDR
Fors, Roy E., CWO4
Forseth, James M, CDR
Forsgren, Philip, MIDN
Forsha, Anne V., Capt
Forshay, Bruce E., CDR
Forslund, Robert A, Capt
Forster, Barbara A., LCDR
Forstner, Gerald, CDR
Forsyth, Timothy, CDR
Forsyth, William G, Capt
Forsyth Jr, William D, Capt
Fort, C Margret, Lcdr
Fort, Thomas H, Capt
Fortanas, Jean C., Lcdr
Forte, Raymond, Lcdr
Fortenberry, Jack, Lcdr
Fortin, Ernest, LCDR
Fortin, Louis A, CDR
Fortna, Nancy M., Lcdr
Fortunati, Richard A, CDR
Fortunato, Dominic V, Capt
Fortune, Ben F, CDR
Forward, Richard G, CDR
Foshay, Wayne E., CAPT
Foshee, Amy J., LCDR
Foss III, George E, CDR
Fossa, James F., LCDR
Foster, Alfred F, Lcdr
Foster, Claudia J., LCDR
Foster, D. Brian, LCDR
Foster, Dirk L., LCDR
Foster, Douglas F., LCDR
Foster, George O, CDR
Foster, J. Michael, CAPT
Foster, Jerald D, CDR
Foster, Karen W., CDR
Foster, Kenneth E, Lt
Foster, Lloyd B, CDR
Foster, Mark W, Lcdr
Foster, Michael S, Lcdr
Foster, P Lyman, CDR
Foster, Raymond F, CDR
Foster, Raymond O, Capt
Foster, Rex D., LCDR
Foster, Robert L, Lcdr
Foster, Robert L., CDR
Foster, Steven S, CAPT
Foster, Timothy A., LT
Foster, Vincent E, Capt
Foster, Walter N, Capt
Foster Jr, Charles K, CDR
Foster Jr, John W, Capt
Foster Jr, John W, Lcdr
Fountain, Freeman P, Capt
Fourcade Jr, Emile J, Capt
Fournier, Donna J, CDR
Fournier, Leroy J, Capt
Foust, John T, Lcdr
Fowle, Dennis R, CDR
Fowle, William C, Lt
Fowler, Christopher B, LCDR
Fowler, James E., Lt
Fowler, James R, Radm
Fowler, Jeffrey B, Capt
Fowler, Rodney S, Lcdr
Fowler, Sara A., Lcdr
Fowler, Sarah M., LT
Fowler, Timothy N., LCDR
Fowler, William A, Lcdr
Fowler Jr, Eugene F, CAPT
Fowler Jr, Louis B., CDR
Fowlkes, Hobart V, Capt
Fox, Albert Daniel, Capt
Fox, Cecil C, Capt
Fox, Daniel W, Capt
Fox, Donna, CDR
Fox, E Inman, Capt
Fox, Edward M, CDR
Fox, Francis H, Capt
Fox, J Frank, Capt
Fox, James C, Capt
Fox, John F, Lcdr
Fox, Leslie S, CDR
Fox, Norman W, Lt
Fox, Patrick W., CWO3
Fox, Richard A, CDR
Fox, Stephen C, CAPT
Fox, Thomas F, CDR
Fox, Timothy R., Lcdr
Fox III, Charles D, Capt
Fox Jr, Milden J, CDR
Foy, Brian O., LCDR
Foy, David D, CDR
Foy, Patrick D, Lcdr
Frainier, Dennis A., CDR
Frainier, Richard J, Capt
Fraley Jr, Morris A, Lt
Fralin, Jay M, LT
Frame, Scott B., CDR
France, Barbara B., LT
France, Peter J, LCDR
France, Thomas D, Capt
France II, Ralph H, Capt
Francis, David L, Capt
Francis, Edward J., CDR
Francis, John W, Capt
Francis, Kelley L., ENS
Francis, Michael D., Capt
Francis, Perry G, Capt
Francis, William D, Capt
Francis Sr, Paul M., CDR
Franck, Ernest P, Capt
Franco, Edith L., CDR
Francois, Emmett W, Lcdr
Francom, Krista A., CDR
Franczyk, David P, CDR
Frank, Carl J., CDR
Frank, Joel S, CDR
Frank, Kenneth D, CDR
Frank, Matthew S., LT
Frank, Robert L, CDR
Frank, Roger Alan, Lcdr
Franke, Arnold G, Capt
Frankel, Andrew V., LT
Frankel, Donald D, Capt
Frankel, Stuart L., CAPT
Frankenfield, William H, Capt
Frankfurt, Alan I., CDR
Franklin, Douglas S, Lcdr
Franklin, Larry B, Radm
Franklin, Thomas G., LT
Franklin Jr, H Bland, Capt
Franks, David B., CDR
Franssen, Roger A, CAPT
Frantz, Richard L., Capt
Franz, Carmella A., CDR
Franz, Donald R, Capt
Franz, Harold C., Capt
Franz, Jean P, CDR
Franz, Richard L, Capt
Franzman, Rodney C, Capt
Frary, Lynn R, Capt
Fraser, David L, CDR
Fraser, Joseph L, Lcdr
Fraser, Nancy H., LCDR
Fraser, Richard R, CDR
Fraser III, Hugh E., CDR
Frates, Steven B, Capt
Fravel, Glen E, CDR
Frazer, James A, Capt
Frazer, Orie, LT
Frazier, Jeanne E., Lcdr
Frazier, Paul H, CAPT
Frazier, Stephen R, Capt
Freckleton Jr., W. C., CDR
Frederick, Lance A., Capt
Fredette, Michael E., Lcdr
Fredrickson, Robert B, Capt
Fredrickson, Ronald H, Capt
Free Jr, William T, Capt
Freeburn, Steven, CDR
Freed, Clarke S., CDR
Freeden, Harold R, Capt
Freedline, Yale, CDR
Freeman, Berkley I, Capt
Freeman, Emery T, Capt
Freeman, John A, Capt
Freeman, John F, Ltjg
Freeman, Julian D, Capt
Freeman, Manuel J, Lcdr
Freeman, Pamela J., CDR
Freeman, Robert M, Capt
Freeman, Rufus A., CDR
Freeman, William T, Capt
Freemire, Richard, CWO4
Freemyers, Michael C., CDR
Freese, Donald R, Capt
Freested, Robert C, Lcdr
Freiberger, Howard, Capt
Freidberg, Michael L., LT
Frein, James O, Capt
French, Charles G, Ens
French, Harold S, CDR
French, Henry M, CDR
French Jr, Frederic R, Lcdr
Frenkel, Robert L, Capt
Frerich, William R, CDR
Frerichs, Cletus T, CDR
Frerichs, Edward D, Lcdr
Fresher, Brian L., CDR
Fressilli, Thomas F, CDR
Fretz, Eric B., LT
Freund, Carole K, CDR
Freundlich, Richard, Radm
Frevola, Michael J., Lt
Frey, David A., CDR
Frey, Lawrence N., Lt
Frey, Robert H,
Frey, Thomas E, CDR
Freye, Daniel L, CAPT
Frick, Bill Lee, CDR
Frick, Edward D, CDR
Fricke, Dorothy E, Lcdr
Fricke, Joseph D., LCDR
Frickenschmidt, Dan, LT
Friedeberg, William J, CDR
Friedemann, Conrad, Capt
Friedenbach, Kenneth J, Capt
Friedenreich, John, LT
Friedlander, Todd, Lcdr
Friedman, Edward D, Capt
Friedman, Harry, CAPT
Friedman, Howard, Capt
Friedman, Leslie H, Capt
Friedman, Robert J, Ltjg
Friedman, Robert S, Lcdr
Friedt, Gail L., LCDR
Friel, John J, Capt
Friend Jr, Paul E, LCDR
Frier Jr, John M, CDR
Fries, Carsten H., LT
Friesen, Lindalou, Lt
Friske, Warren H, Lcdr
Fritch, Joanne E, CAPT
Fritsche, James J., CDR
Fritsvold, John D, Lcdr
Fritz, Raymond A., CDR
Fritz, Wayne R, Capt
Fritz, Wilbert, Capt
Fritzemeier, Ronald R., LCDR
Fritzler, Roger L., CAPT
Frizzell, George E, Capt
Froberg Jr., Kenneth J., ENS
Frohlich, Andrew S, Capt
Fromuth, Robert C, CDR
Froncillo Jr, Robert E., Lcdr
Frondorf, John R, Capt
Frost, Charles W., Capt
Frost, David, CDR
Frost, James H, Capt
Frost, Jeffrey E., CDR
Frost, Laurance A, CWO4
Frost, Stephen S., CDR
Fruchtmann, Jerome J, Lcdr
Fruner, Fred A., LT
Fry, Charles R, CDR
Fry, Lon A, LT
Fry, Vance H, Radm
Fryauf, Mark T., CDR
Frye, David B, Capt
Frye, Luther V, Lcdr
Fuderer, George A, CDR
Fugal, Boyd L, Lcdr
Fugina, John R, Lt
Fuhr, John C., CDR
Fujikawa, Janet, CDR
Fujimoto, Akira F, Capt
Fulcher, Clay W G, Radm
Fulcher, John W., Lcdr
Fulgham, Michael O, CWO4
Fulkerson, Glenn M, Capt
Fuller, Albert B, Lcdr
Fuller, Brian A., LCDR
Fuller, Dean Anthony, Lcdr
Fuller, George R, Capt
Fuller, John G C, Capt
Fuller, Joseph B, CDR
Fuller, Margaret L, Lcdr
Fuller, Richard C, Capt
Fuller, Robert S, Lt
Fuller, Thomas F., CDR
Fuller, William G, CDR
Fuller, William M., CDR
Fuller Jr, Chester E, Lcdr
Fuller Jr, Frederic J, CDR
Fuller Jr, Lee M, Capt
Fuller Jr, Robert G, Capt
Fullerton, William R, CDR
Fullmer, Lee D, Capt
Fulton, Lyman A, Capt
Fulton, Norman F, CWO4
Fulton, William C, LT
Fulton Jr, J Wayte, Capt
Funaro, Vincent M, CDR
Funchess, Dennis V, CDR
Funk, James C, CDR
Funk, Walter F, Lt
Funk, William H, Capt
Funken, Lee K, CDR
Fuoco, Frank J, Ltjg
Furches Jr, Stephen L, CDR
Furchner, David K, CDR
Furco, Joseph J, Capt
Furdak, Edward J, Capt
Furgason, David W, Capt
Furia, Joseph M., Lcdr
Furlan, Deborah M., CDR
Furlong, Francis J, Lt
Furlow III, Charles M, Capt
Furlow Jr, Lawrence F, Capt
Furniss Jr, Henry D, CDR
Furst, Barbara S, Capt
Furst, William A., Lcdr
Furukawa, Robert J., LCDR
Furuya, Lincoln A, Capt
Fusco, Edmund E, Cwo-4
Fussell, Bradley C., CDR
Fussell, M Eugene, Radm
Futch, Edward I, Capt
Fye, Robert F, CDR
Fyten, Richard A, CDR
Gabay, Mark R., Lcdr
Gabele, Richard E, LCDR
Gabriel, David C., CDR
Gabriel, Robert, Lcdr
Gabrielli, Louis A., Lcdr
Gabrielson, Robert L, Lcdr
Gacser, George R, LCDR
Gad, Shayne C., CAPT
Gadow, John R, Capt
Gadzala, Jack, CDR
Gaffga, Leonard H, CDR
Gaffney, William T, Capt
Gage II, Reginald P, Lcdr
Gagen, Michael H, Capt
Gahagan, Michael D., Lcdr
Gahr, Daniel J., CDR
Gahran, Brian H, Capt
Gainer Jr, James V, Capt
Gaines, Howard W, CDR
Gaines, Joyce E., LT
Gaines Jr, Ben P, CDR
Gaines Jr, McMurray, CDR
Gairing, Donald A, Capt
Gaither, Thomas A, Capt
Gaitley IV, Julian L., Lt
Gala, Gary J., LCDR
Galamaga, Donald P, Capt
Galamaga, Peter C., LT
Galang, Keith A., Lt
Galang, Remedios N., CAPT
Galayda-Bacon, Nancy A., Lcdr
Galbraith, John M, CDR
Galbraith, Robert E., CAPT
Galbreath, Richard N, CDR
Galbut, Hyman P, Capt
Galdun, Michael J, CDR
Gale, Steven A., LCDR
Galenian, Gregory J, CDR
Galeno, William A, Lcdr
Galkin, Warren B, Lcdr
Gall, Ralph C., LT
Gall, Robert A, Capt
Gallagher, Frank X, Lcdr
Gallagher, Gerald M, CDR
Gallagher, Jean E M, Lt
Gallagher, Kevin J., CDR
Gallagher, Richard J, Capt
Gallagher, Thomas E, Capt
Gallagher Jr, John E, Capt
Gallaway, Nada R., CDR
Gallaway Jr, Richard M., Lcdr
Gallegos, Louis A, CDR
Galligan, Donald J., CDR
Gallo, James J., Capt
Gallo, John, CDR
Gallogly, William W, Lcdr
Galloway, Darrell R., CAPT
Galloway, Karen L, LCDR
Galman, David P, Lt
Galt, William K., CDR
Galuppo, Lawrence J, CDR
Galusha, David E, Capt
Gamache, Gabriel J, Lcdr
Gamber, Marlan E, Capt
Gamble, F Trevor, Capt
Gamble, Paula E., Lcdr
Gamble, Peter A., CDR
Gamblin, Kenneth, CWO4
Gamboe, Thomas E, CDR
Gambone, Joseph C, Capt
Gambrell Jr, Millard E, CDR
Gammage, Robert A, Capt
Gamwell, Wayne R., LCDR
Gandrud, Gairel L, CDR
Gangoy, Leo C., Capt
Gangsaas, Catherine, CDR
Gann, Charles T, CDR
Gannon, Joseph A, CDR
Gannon, Warren C, Cwo
Gansa, Alexander N, Capt
Ganse, Terrence L, CDR
Ganskopp, Daryl S, Lcdr
Gant, Gregon L., CAPT
Gant, James L, CDR
Gantzer, William E, Lcdr
Gapinski, Raymond R, Lt
Garber, Paul W, Capt
Garber, Vera, CDR
Garbuzinski, Terrence J., LCDR
Garcia, Adrian A., LCDR
Garcia, Benjamin A, CDR
Garcia, Daniel, LCDR
Garcia, Felice A, Capt
Garcia, Frank C, Capt
Garcia, Joel T., Lcdr
Garcia, Michael C, CDR
Garcia, Philip A, CDR
Garcia, Raul E, CDR
Garcia III, Jose F., LCDR
Garcia Jr, Vincente C., CDR
Garcia-Paguio, Charlene, LT
Gard, Daniel L., Lcdr
Gardea Jr, David L, Capt
Gardenier II, John S, CDR
Gardiner, Margaret O, Lt
Gardiner, Steven P, Lt
Gardner, Charles R., Capt
Gardner, David K., LCDR
Gardner, Donna M., CDR
Gardner, H. Joseph, Capt
Gardner, James Knoll, CAPT
Gardner, Joseph T, Capt
Gardner, Morrison F., CDR
Gardner, Robert D, CDR
Gardner, Rodney L., Lcdr
Gardner, William V, CDR
Gardner Jr, George P, Lt
Gareffa, Joseph J, Capt
Garland, Catherine L, Ltjg
Garlick Jr, Arnold L, Lcdr
Garlinghouse, David M., Ens
Garlington, Benny V, Capt
Garman, Quentin P, Capt
Garmer, Stanley A, CDR
Garner, John, CDR
Garnett, John R, Lcdr
Garney, Geoffrey G., CDR
Garot, Otto Leon, CDR
Garrett, Aaron D, CDR
Garrett, Alan B, Lcdr
Garrett, Alex R, CDR
Garrett, Broox C, Ltjg
Garrett, Carlton Lee, CDR
Garrett, Henry F M, Capt
Garrett, James J, Lcdr
Garrett, Robert L, CDR
Garrett, Theodore E, Capt
Garrett Jr, Albert E, CDR
Garrett Jr, Tom W, CAPT
Garrick, Robert M, Radm
Garrigues, Ann L., CDR
Garrison Jr, Charles P, Capt
Garrison Jr, William H, Capt
Garrison-Backer, Georgia L., Lcdr
Garrison-Chaffee, R. E., LCDR
Garrity, Patricia A, CDR
Gartmann, Russell J., CDR
Gartner, Anne E, Capt
Garvin, Paul M., LCDR
Garvin, Robert G, Capt
Gascoigne, Bruce M, CDR
Gasinski, Christine I., LCDR
Gasper, Charles A, CDR
Gast, Ronald P, CDR
Gaston, Cynthia A., Lcdr
Gaston, Hartley R, CDR
Gates, Fred Hall, CDR
Gates, James C, Lcdr
Gates, James F., CDR
Gates, John W, Radm
Gates, Peter D., CDR
Gates, Timothy W., Lcdr
Gates Jr., Stephen, CAPT
Gatlin, Beth A., CDR
Gatlin, Larry L, CDR
Gatlin Jr, Carl E, Capt
Gato, David T, CDR
Gatton, Thomas H, Lcdr
Gattone, Jane L., LT
Gaucher, John H, Capt
Gaudette, Roland E, Capt
Gaudio, Joseph A, Capt
Gaul, William M, CDR
Gauld, Edwin S, CDR
Gauntlett, James F, CDR
Gauntt, Royce E, LCDR
Gaussiran, Charles D, Lcdr
Gautille, Joseph C, CAPT
Gauvin, Albert L, CDR
Gauvin, Raymond E, CWO4
Gavan, John D, Radm
Gawrys, Joseph A, Capt
Gawryszewski, Nancy N., LCDR
Gay, David T, Lcdr
Gay, Thomas A, Capt
Gay, William A., CDR
Gay III, Daniel J., CDR
Gayle, James A, CDR
Gayle, S Michael, CDR
Gaynor, Jeffrey W, Lt
Gayton, John J., LT
Gayton, Linda L, Lcdr
Gazda, Peter F., CAPT
Gean, Bret C., Lcdr
Gearhart, Donald C, Capt
Geary, William J, Capt
Gebhard, Karl T E, CDR
Gedeon Jr, Robert L., CDR
Gee, Darrell, Lcdr
Geeslin Jr, James M., Capt
Gehrig, Neil E, CAPT
Gehring, Linda L., Lcdr
Geib, Arthur J, Lt
Geib, Douglas J., CDR
Geier, Edward A, Capt
Geier-Bernhagen, Theresa, LT
Geil, Kenneth P., CAPT
Geilfuss, Steven K., Lcdr
Geis, Albert H, CAPT
Geis III, John Hammond, LCDR
Geiselman, Sandra L., CAPT
Geisinger, Bruce, CDR
Geist, John M, Capt
Gelesko, Nick, Lt
Gelineau, Ronald G, Capt
Geller, Jacob, Capt
Geller, Janet S., Lcdr
Gelman, Stephen R, Lcdr
Gemma, Lawrence E, Capt
Gemoets, Gary V., Capt
General, John A., Capt
Genereux, Bruce T, CDR
Genest, Armand R., CDR
Genever, Douglas, Lcdr
Genevro, Michael R, LCDR
Genna, Michael R, CDR
Genninger, Edward J, CDR
Genova, F Donald, Capt
Genrich, Gloria K, CDR
Gensel, Richard L, Capt
Genstil, Stephen M., Capt
Gentile, James H, CAPT
Gentile, Vincent J, CDR
Gentilella, John A, CDR
Gentle, Fred D, Lt
Gentle Jr, Alfred F., CDR
Gentry, J. Thomas, Capt
Genualdi, Frederick, CDR
Genz, Ross C, Lcdr
Geoffrey, Leon S, Lcdr
Geoffrion, David G., CDR
Geoghegan, James C, CDR
George, Bonita A., LCDR
George, Caroline W, Capt
George, George P, Capt
George, Jack G, Lcdr
George, James F, Capt
George, James F, Capt
George, John Reid, Lcdr
George, Joseph H, Lcdr
George, Thomas C, CDR
George, Thomas P., LCDR
George, William A, Capt
George Jr, Walter N, Lcdr
Georgeson, Gary V., CDR
Georgevits, Kenneth L., LT
Georgius, David R., Capt
Geraci, Robert P., Capt
Geraghty, John A, CDR
Geraghty, Patrick C., LCDR
Gerard, Maurice W, CAPT
Gerber, Gregory F., ENS
Gerell, John D, CDR
Gerety, David W, CDR
Gerety, Richard A, Lcdr
Gerfen, Thomas W, Lcdr
Gerhard, Robert C., Ltjg
Gerhardt, Emerson D, Capt
Gerlach, Henry O, CDR
Gerlach, James M., Lt
Gerlach Jr, Carl C, Lcdr
Germain, Eric Francis, CDR
Gernand, Bradley E., Ens
Gerosa, Vincent C, Capt
Gerow, George E, CDR
Gerrie Jr, Marshall J, Capt
Gerringer, Karen F, CDR
Gerson, Joe R, Lt
Gerth, Gary J, CDR
Gerth, Stephen B, Capt
Gertz, Dwight L, Capt
Gervais, Michael J., CDR
Gervason, Richard E, Lcdr
Gesell, Thomas R., Lt
Gettys Jr, Richard H., CAPT
Getz, Edwin S., CDR
Getz, Jesse L, Lcdr
Getz, Marshall S., CDR
Geyer, Bradley N., Lcdr
Ghaffari, Reza, LCDR
Ghiradella, Ronald J, CDR
Ghirardi, Mark S., LCDR
Giacin, Kenneth J, Capt
Giacomo, Charles L., Lcdr
Giaimo, Joseph A., CDR
Giancatarino, Anthony D, Capt
Giangiobbe, Frank T, Lcdr
Giannini, Richard J, CDR
Giardina, Joseph A, CDR
Giardini, Giovanni T, Capt
Giarrusso, Carl F, WO1
Gibaldi, Joseph G., CDR
Gibb, Bryce M., LT
Gibbons, Hugh E, Capt
Gibbons, James L, Lcdr
Gibbons, Juanita H., LT
Gibbons, Mark, Lt
Gibbons Jr, Hubert B, CDR
Gibbons Jr, Thomas, CWO3
Gibbs, Anthony W, Capt
Gibbs, David M., LCDR
Gibbs, Fredric, Radm
Gibbs, Janet M., CDR
Gibbs, John E., Lcdr
Gibbs Jr, Clarence J, Capt
Gibler, Kenneth C., LCDR
Giblett, David R, Capt
Gibney, Scott C, Lcdr
Gibson, Charles R, Lt
Gibson, Debra J., LT
Gibson, Dudley, LT
Gibson, Edward P., CDR
Gibson, Frank L, CDR
Gibson, Guy P, Lt
Gibson, Thomas Eugene, Capt
Gibson, William C, CDR
Gibson III, William, CDR
Giddings, John P., CAPT
Gideon, David H., Lt
Gideon, Jerry, Lcdr
Giegerich, Alfred J, CDR
Giel Jr, A Michael, Lcdr
Gieser, Frank A, Capt
Giesey, Sharon V., Capt
Giessing, George C, Capt
Giesting Jr, W E, CDR
Gift, Lawrence S, Lt
Giglio, Anthony G., LT
Gilbert, Gerald E, Radm
Gilbert, John R, CDR
Gilbert, Patricia M, Capt
Gilbert III, John Henry, CDR
Gilbertson, Dennis L, CDR
Gilbertson, Lawrence A., CDR
Gilbertson, Roger, Radm
Gilboe, David D, CDR
Gilbride, Ann DeBaets, CDR
Gilbride, Daniel T., LT
Gilchrist, Lionel F, Capt
Gilchrist, Norman S, Capt
Gilcrest, Andrew S, CDR
Gildea, Patrick F., CDR
Gildred, George L, Lt
Gile, Robert H, Capt
Giles, Delfred L, CWO4
Giles, Henry G, CWO4
Giles III, Alexander F., Lt
Giles Jr, Ira A, CDR
Gilio, Douglas A, Capt
Gilker, William M, CDR
Gill, Gerald W, Capt
Gill, James Joseph, CDR
Gill, Mary Rebecca, LT
Gill, Robert M, CAPT
Gill, William L, Capt
Gill III, Burgo D., LT
Gillease, Dennis B, Capt
Gillen, Bruce R, CDR
Gillen, Howard W, Capt
Gillen, William F, Lcdr
Gillespie, John C, Capt
Gillespie, Laurel C, Capt
Gillespie, Richard I, Capt
Gillett, Edward R., CDR
Gillette Jr, Thomas A, Lcdr
Gilley, Frank P, CDR
Gillies, Donald C, Capt
Gillies, John A., Capt
Gillies, Sue A, Lcdr
Gilligan, R. Brian, LT
Gilligan, Thomas F, CWO2
Gilliland, Charles H, Capt
Gilliland, Charles L, CDR

Gilliland, David E., LCDR
Gilliland, Greg A, CAPT
Gilliland, Jane G, CDR
Gilliland, Keith D, Lt
Gilliland Jr, C Herbert, CAPT
Gillin, J Chris, Capt
Gillin, Robert Arthur, Lcdr
Gillingham, David R, CDR
Gillingham, David R., CDR
Gillio, David A, CDR
Gillis, Dana G, Capt
Gillis, John W, CDR
Gillis, Lyman R, Capt
Gillis, Roderick C, CDR
Gillow, Stacy A., CDR
Gilman, Frank M, Lt
Gilman, George C., CDR
Gilman, John W., LCDR
Gilman, Joseph, Capt
Gilman, Thane, LCDR
Gilmartin, Paul A., LT
Gilmer, David S, Capt
Gilmer Jr., David S., LT
Gilmore, John, Capt
Gilmore, Sidney B, Lcdr
Gilpatric, James F, Lt
Gilsdorf, Mark T, Capt
Gilson, Benjamin J, Capt
Gilson, Carroll P, CDR
Gilson Jr., Paul J., Lcdr
Gimbel, Charles R, CDR
Gimmel, Daniel G, Lcdr
Gin, Steven, Capt
Ginn, Robert L, Lt
Ginn, Robert W, CDR
Ginsburg, Jay E, Lt
Gintowt, Charles C, CDR
Giordano, Frank J, CDR
Giovannoni, John J, CDR
Giovia, Virginia M., Lcdr
Gipe, Charles W, Lcdr
Giragosian, Newman H, Capt
Girardin, David W, Lcdr
Girolami, John J, Capt
Girouard, Marvin J, CDR
Girtman, Ben E, Capt
Gish, Joseph B., LT
Giuda, John B., CDR
Giuffrida, Sebastian J, Capt
Giuliano, Carmine F, Lcdr
Giusini Usnr (Ret), A V, Capt
Giza, Mary C, CDR
Gjerde, Kristine J, Lcdr
Gjerde, Kurt A, CDR
Gjovig, Allan J, CDR
Glace, Dean A, CDR
Gladbach, John J, CDR
Gladden, Riley Joe, CAPT
Gladstone, Kenneth M, Capt
Gladwin Jr, William J, Capt
Glahn, Gerald L, Capt
Glandt, Paul S., CDR
Glasby, Stuart F, CDR
Glaser, Kathleen M., LCDR
Glaser, William C, Capt
Glasnapp, Randy E, CDR
Glass, F. Stephen, Radm
Glass, Juandell D, CAPT
Glass, Kenneth M, Capt
Glass, Laurie K, CDR
Glass, Norman R, Capt
Glass, William O, Capt
Glass II, Harry John, CDR
Glass Jr, Park R, CDR
Glass Jr, William O, Capt
Glass Sr, Gary A., Lcdr
Glassberg, Arnold M, Capt
Glasscock, Douglas D., CAPT
Glasser, Robert A, Lcdr
Glassmaker, Jeffrey L., LT
Glassmire, David R, LT
Glathe, William Q, Capt
Glawson Sr, Stephen J., CDR
Glazer, J. Henry, Capt
Glazeski, Michael A, CDR
Glazeski, Richard J., CWO4
Glazier, John J, CDR
Gleason, Mary F, CAPT
Gleason, Ronald G, Capt
Gledhill, Virginia R., LCDR
Gleiter, Werner C L, Capt
Glenewinkel, Gloria S., CDR
Glenn, Robert C, Capt
Glenn-Scheer, Margaret, Lcdr
Glentzer, George A, Lt
Glesmann, Douglas C, CAPT
Glick, Maurie, Lcdr
Glidden, Richard L, Capt
Glisson, Harold K, Ltjg
Glisson, Patrick C, Capt
Glomb, James D, Lcdr
Glore, James W, Capt
Gloss, Gregory C, CDR
Glover, Anthony L., Lcdr
Glover, Elizabeth B, CDR
Glover, Lanny, CDR
Gluck, Scott R., LCDR
Glynn, Kevin C., CDR
Glynn, William C., Capt
Glynn, William J, CDR
Gnam, James H, CDR
Gnerlich, Janet L Berg, Lcdr
Goad, Samuel T, Capt
Goddard, Boyd T, Capt
Godfrey, Richard E, Capt
Godfrey, Robert B, Capt
Godfrey, Timothy D., LT
Godfrey Jr, Lawrence M, Capt
Godley, John B, Capt
Godshall, Douglas R, Capt
Godwin, Albert D, Capt
Godwin, John T, CDR
Godwin, Wallace G, CDR
Goebel, Glenn H., LCDR
Goebel, James B, CDR
Goebel, Ken C, Ltjg
Goeders, Calvin N, CDR
Goeks, Greg J., CAPT
Goeters, Eleanor, Lcdr
Goetz, John R, Capt
Goff, Robert A, Capt
Goff III, Ralph D, Lcdr
Goforth Jr, Harold W, Capt
Gogel, Bernard G, CDR
Goggin, Walter W, Lcdr
Gohde, Charles T., Lcdr
Golda, Eugene M, CAPT
Goldbeck, Larry O, Capt
Goldberg, Kenneth I, CDR
Goldberg, Melvin J, CDR
Goldberg, Robert J., CAPT
Goldberg, Victor Y, Capt
Goldcamp, Mary E, Lt
Golden, Edward R, Lt
Golden, Gerald D, CDR
Golden, James L., Capt
Golden, James R., ENS
Golden, Martin F, CWO4
Golden, Michael J, Lcdr
Golden, Patrick S., CDR
Golden, Robert H, Lt
Golden, Robert M, CDR
Goldenberg, Thomas A., CDR
Golding, John M., CWO4
Golding, Joseph F., LCDR
Goldman, Jeffery B, Lcdr
Goldman, Neil J, Lt
Goldman, Sherri M, CDR
Goldman NYNM, Harold J., Radm
Goldner, Richard S., CDR
Goldschmidt, Michael, LCDR
Goldschmitt, Marc L, CDR
Goldsmith, Kenneth M., LCDR
Goldsmith, Richard A, CDR
Goldsmith, Van E, CDR
Goldsmith Jr, James E, Capt
Goldstein, Charles, CDR
Goldstein, Kathleen M., CAPT
Goldszer, James F., CDR
Goldy, Donald C., Capt
Goll, Edward C, CDR
Golomb, Gary W, Lcdr
Gomes, Francine M, CDR
Gomes, Joseph P, Lcdr
Gomez, Edward, Capt
Gomez, Irasema, LT
Gomez, Wayne D, CDR
Gonion, Gordon E, CDR
Gonsalves, Paul G., Lt
Gonzales, Pedro J, Lcdr
Gonzales III, Roman, Capt
Gonzalez, James, Lt
Gonzalez, Michael A, Lt
Gonzalez, Philip J., CWO3
Good, Beverly F, Capt
Good, Edwin M, Capt
Good, James L, CDR
Good, Todd A, CDR
Goodale, James E, Lcdr
Goode, Quenneth Cameron, Lt
Goode, Ronald A, CDR
Gooden, Mark E., LCDR
Goodenough, Robin W, Capt
Goodey, Wallace W., Lcdr
Goodfellow, James P, CDR
Goodfried, Herbert J, CDR
Goodhart, Frederick W, Cwo
Goodin, Robert H, Lcdr
Gooding, Thomas C., LCDR
Gooding III, Harry G, Capt
Goodlatte, Peter R, CDR
Goodloe, Wallace C., CDR
Goodman, Craig W., CDR
Goodman, James B, CDR
Goodman, Martin B, CDR
Goodrich, Donald D., LCDR
Goodrich, Freeman R, Capt
Goodrich, Vincent N, CDR
Goodroe, H Wayne, Capt
Goodson, Ray A., CAPT
Goodspeed, Manley J, Lt
Goodwin, Bonnie S, CDR
Goodwin, Cecil A, Lcdr
Goodwin, Jack C, Capt
Goodwin, James W, Lcdr
Goodwin, William A, Capt
Googins, June E., LT
Goorgian, Richard, CDR
Goorjian, Paul Mike, Capt
Gordon, Allan P, CDR
Gordon, James J, CDR
Gordon, James P., CDR
Gordon, James W, Capt
Gordon, Lance B, CDR
Gordon, Michael D, CDR
Gordon, Sandra E., LT
Gordon, Timothy W., Lcdr
Gore, Diane D, Capt
Gore, Richmon C, Lt
Gorecki, Richard T., Lcdr
Gorgas, Chester W, CDR
Gorham, Harvey E, Lcdr
Gorka, Carolyn Coe, LT
Gorman, Daniel J, Capt
Gorman, Irving X, Capt
Gorman, James Lou, CAPT
Gorman, Jennifer Lynn, CAPT
Gorman, Mark K., CDR
Gorman, Russell W, Radm
Gorman, Thomas M, CAPT
Gornick, George F, CWO4
Gorrell, Robert P, Lt
Gorring, Ronald A., Lcdr
Gosciniak, Albert R., CDR
Gosha, Michael J, Capt
Goshow, John P, Capt
Gosler, James R, Capt
Gosnell, Charles E, Capt
Gosnell, F Laurence, Lcdr
Gosnell, Stephen G., CDR
Goss, Gordon W., CDR
Goss, Jerry L., Ltjg
Goss, Michael A, Capt
Gosse, Clinton G, Capt
Gossett, Gilbert Barton, Lcdr
Gotautas, Vito A, CDR
Gotch, Merry Ann, CDR
Gott, Carl F, Capt
Gottschalk, Robert G, Capt
Gottschall, Alvin G, CDR
Gotzinger, Cheryl A., LTJG
Goudy, Ronald C, Capt
Gouge, Michael J., CAPT
Gough, Richard J, Capt
Gould, Mike K, CDR
Gould, Richard H, CDR
Gould, Robert L., CAPT
Goulding, Bruce R, Capt
Goward, Eleanor L, CAPT
Goward Jr, Richard F., CAPT
Gowen, William R, CDR
Gowling, Thomas A, Lt
Gowrie, Christopher M, CDR
Goynes, Freddie A, Lcdr
Graaff, Paul D., CDR
Grabarz, Theodore L., LT
Grabe, James R, CDR
Grabfield, Philip R, Lcdr
Grabow, Wilbur G, CDR
Grabulis, Dennis D, CDR
Grace, John C, CDR
Grace, Marcellus N, Capt
Grace, Perry R, Capt
Grace, William J, Capt
Graczyk, Robert J, Lcdr
Grade, Richard E, CDR
Grady, Brian M., LT
Grady, J Martin, Lcdr
Graessle II, Grant, LCDR
Graf, Donald V., CAPT
Graf, Holly A., LCDR
Graf, John, Lcdr
Graf, Robin L, CDR
Graf, Walter C, CDR
Graf, William D., LCDR
Graff, Rebecca, LT
Graff, Thomas P., CDR
Grage, Walter A, Capt
Gragg, Jody M., LT
Graham, Bruce M, CDR
Graham, Cam L, Lcdr
Graham, Charles R, Capt
Graham, Francis J, Capt
Graham, Gerold W, CDR
Graham, J. Randolph, LCDR
Graham, James T, Capt
Graham, Kendall P, CDR
Graham, Michael A, LCDR
Graham, Stanley S, CDR
Graham, William O, Lcdr
Graham Jr, Clive, Lt
Graham Jr, James E, Capt
Grahs, Paula J., LCDR
Gralla Jr, Arthur R, Capt
Gramling, Alfred S, Lt
Gramount Jr, Victor H, Lt
Gramza, Allen E, CDR
Grandin, Edward S, Capt
Grandy, Olander G, Lcdr
Graner, Richard E, Capt
Granger, Allan E, Capt
Granger, Arthur J, CWO4
Granger, David L, CDR
Granger Jr, Charles F, CDR
Granier, Russell J., CDR
Granlund, Harold, Lt
Grannis, William G, CDR
Grant, Albert O, CDR
Grant, Francis B, CDR
Grant, James W., CDR
Grant, John T, Capt
Grant, Margaret E, Capt
Grant, Robert T, Lcdr
Grant Jr, Ernest E, Capt
Grant Jr, William A, Lt
Grantham, Armon G, Lcdr
Graske, Theodore W, Capt
Grasseschi, Gary L, CDR
Gratonik, Edwin, Lt
Grauer, John Allen, CDR
Graupner, John G, Capt
Gravallese, Albert J, Capt
Gravely Jr, Samuel L, Vadm
Gravem, Otis E, Lcdr
Graves, Dale V, Capt
Graves, Michael L., CDR
Gravitt, Patrick D, Capt
Gray, Anna E, Capt
Gray, Daniel D., LCDR
Gray, David J, Capt
Gray, David M, CDR
Gray, David P., CAPT
Gray, Donald M, CDR
Gray, Garold G, CDR
Gray, Glen B, Capt
Gray, Harrison D, CDR
Gray, Kathleen S, Lcdr
Gray, Kathryn T, CDR
Gray, Mark D., CAPT
Gray, Patrick T, Lcdr
Gray, Peter T, Lt
Gray, Richard L, CDR
Gray, Robert E, CDR
Gray, Robert E, Lt
Gray, William M, Lcdr
Gray Jr, James W, Radm
Gray Jr, John W, Capt
Gray Jr, Paul A, CAPT
Graybeal, J Patrick, Lcdr
Grayson, Charles W., CDR
Grayson, Ellison C, Capt
Grayson Jr, William J, CDR
Grazel, Christopher P, Capt
Greabell, Dean A., Lcdr
Grealish, James V, Radm
Grebasch, Virginia S., CDR
Greco, Frank A., CDR
Greco, Michael A, Capt
Greeley, Catherine S, Capt
Green, Arthur F., CWO4
Green, Carolyn J., CDR
Green, Christine A., LT
Green, Don H, CDR
Green, F Lee, CAPT
Green, George H., CDR
Green, George R, CDR
Green, Harold C, CDR
Green, James T., ENS
Green, Jeanne M, CDR
Green, John M, Capt
Green, Joseph B, Lcdr
Green, Kenneth E, CAPT
Green, Kenneth J., LCDR
Green, Peter D., CDR
Green, Philip D., Lt
Green, Raymond B, CDR
Green, Rex R, Capt
Green, Robert A., Lt
Green, Robert L., Capt
Green, Robert W, CDR
Green, William C., LCDR
Green, William L, CDR
Green, Winthrop, Capt
Green Jr, Clyde L, Capt
Green Jr, Gerald E., LCDR
Greenan, Edward J, Capt
Greenaway, James F, CDR
Greenberg, Robert N., CDR
Greenberg, Stephan M, CAPT
Greene, Andrew F., Capt
Greene, Bennett A, Lcdr
Greene, Dennis, CDR
Greene, Elvin, Lcdr
Greene, Eric K., Lcdr
Greene, Ira M, Capt
Greene, Joseph T., LT
Greene, Kenneth C, Capt
Greene, Raymond A, CDR
Greene, Walter L, Lcdr
Greene Jr, Robert E, Capt
Greenfield, Gordon A, CDR
Greenfield, Sandy R, Lt
Greenfield, Theodore A, Lcdr
Greenleaf, George A, Capt
Greenleaf, Wayne K, Capt
Greenlee, Ralph R, Ltjg
Greenlees, James N, Capt
Greenspan, Alan E, Capt
Greenwald, Marvin, Capt
Greenway, Bradley K, CDR
Greenway, V. William H., CDR
Greenway, William J, CDR
Greenwood, Burniel K, Lcdr
Greenwood, Richard L, Lcdr
Greenwood, Richard M, CDR
Greer, Howard E, Vadm
Greer, Robert D, CDR
Greer, William E, Capt
Greer IV, Robert C, CDR
Greer Jr, G. Bruce, Lcdr
Greer Jr, William H, Capt
Greever III, Charles F, CDR
Greeves, Robert E, Capt
Gregersen, Scott, Capt
Gregg, Frederick P., Lcdr
Gregg, Joseph, CWO4
Gregg, William, CDR
Greggerson Jr, Vernon L, CDR
Gregoire, Robert E., CDR
Gregory, Bion M, Lt
Gregory, H Jackson, Capt
Gregory, Irma L., Lcdr
Gregory, James S., CDR
Gregory, Kenneth, ENS
Gregory, Lowell C., CDR
Gregory, Martha J, Capt
Gregory, Robert S., Lcdr
Gregory, Thomas B, CDR
Gregov, John R., LCDR
Greiner, Harold F, Capt
Gremer, Dennis J, CDR
Grenier, Donna Marie, LCDR
Gresco, William J, CDR
Gresham, Tyler R, CDR
Gresham Jr, Henry W, CDR
Gribben, Walter J, Capt
Grice, Paul F, Capt
Grieco, Anthony M., Lcdr
Grieco, Joseph P, CDR
Grier, Charles P, CAPT
Griesbach, Philip J, Lcdr
Griessel, Rodger F, CDR
Griffin, Charles E., LCDR
Griffin, Chris E., LT
Griffin, Edward T, Lt
Griffin, Elizabeth B., LCDR
Griffin, Gary L, Capt
Griffin, Gerald B, CDR
Griffin, John, Lcdr
Griffin, John J, Capt
Griffin, Larry P, Capt
Griffin, Martin K., Cdr
Griffin, Richard B., CDR
Griffin, Thomas J., Lcdr
Griffin II, Dorsey W., CAPT
Griffin Jr, Gerald L, Capt
Griffin Jr, Robert E, Capt
Griffin Jr., Joseph G., Lcdr
Griffis, Nancy E, CDR
Griffith, Andrew R., Lcdr
Griffith, James B, Capt
Griffith, Lora J., LCDR
Griffith, Robert C, CDR
Griffiths, Kay A., LCDR
Griggs, Austin C, Capt
Grigsby Jr, Gilbert R, Capt
Grim, David J.A., Lcdr
Grimball, William H, CDR
Grimes, Clinton A., LCDR
Grimes, Fred M, CAPT
Grimes, James A, Capt
Grimes, Michael E., Lcdr
Grimes, Roy T, Capt
Grimsley, Elizabeth B, CDR
Grimson, James A, Capt
Griner, Roland F, Lcdr
Grinley, Richard A, CDR
Grinnell, Eltinge, Lcdr
Grinspoon, Alan M, Capt
Grishman, Lee H., CAPT
Grisnik III, Francis J, Lt
Griswold, Edson, Capt
Griswold, Henry C, Capt
Griswold, Ivan C, Capt
Griswold, James E., CDR
Griswold, Lynn C., Capt
Gritsavage, Clem E., Lcdr
Grittner, Gary J, CDR
Groce, Thomas C, Capt
Grocki, Alma M.O.L., CDR
Grocki, Chet J., CAPT
Grodecki, Edward F, Capt
Groden, Joseph A, Lcdr
Groene, Sandra H, CDR
Groenewold, Carol A., Lcdr
Groesbeck, Deborah, Capt
Groff, David A., CAPT
Groff, Peter F, CDR
Grogan, Robert F, Capt
Grogan, Steven M, Lcdr
Grogg Jr, William, Lcdr
Gronell, Mary L, Lcdr
Groneman, Allen J, Lcdr
Gronkiewicz, Bruce V., CDR
Groom, Claire R., CDR
Groom, Craig C, Capt
Groom, John C, Capt
Groome, Ralph B., CDR
Grooters, Ronald D, CDR
Grose, Frederick J, Capt
Groselle, Francis X, Lcdr
Grosheider, Arthur R, Capt
Gross, Charles A, Lcdr
Gross, Gail, Lcdr
Gross, Kenneth J., Lcdr
Gross, Larry, MIDN
Gross, Mose, Capt
Gross, Philip J, CDR
Gross, Thomas J, Radm
Gross Jr, Omer S, Capt
Grossbaum, John J, Capt
Grossheim, David F, Lt
Grosspietsch, Carl W, CDR
Groszer, Todd A., CDR
Grounds, Sylvester Q, Lcdr
Grove, C Samuel, Capt
Grove, Pearce S, Lcdr
Grover, David H, CDR
Grover, Gretchen G, Capt
Grover, Steven F, Lt
Grover, Woodbury L, Capt
Groves, Chris, ENS
Groves, Jeremy, ENS
Grubb, J Rodney, Radm
Grubb, William S, CDR
Grubbs, Paul E., CDR
Grube, Edward M, CDR
Gruber, Karen N., CDR
Grumbling Jr, Hudson V, Capt
Grun, William A, Lcdr
Grunert, John L, CDR
Grunke, James M, CDR
Grupe, David A., Capt
Grushinski, Jeanne E, Capt
Grussendorf, Mark J., Capt
Grutzius, Charles R, Capt
Guarnieri, Kristen G., Lcdr
Guba, Richard D., LCDR
Guderian, William, Capt
Guebert, David R, Capt
Gueco Jr, Vicente A., Capt
Guerci, Thomas C, CDR
Guerin, James W, CDR
Guerrero, Joe, CWO4
Guerrero, Trent, LT
Guertin, Nickolas H., LT
Guess, Gary Paul, LCDR
Guest, Lawrence F., Lcdr
Guest, Robert W, Lcdr
Guetter, Paul M, CDR
Guevremont, Marc R., Lcdr
Guffey, Elton Eugene, CDR
Gugisberg, Donald, Lcdr
Gugisberg, Mark R., CAPT
Guhl, M John, Capt
Guice Jr, William T., CAPT
Guido, Anthony S., CAPT
Guidoboni, Brian T., CDR
Guidoboni, Cheryl A., CDR
Guidoboni, Mark R., CDR
Guidry Jr, Norris J., CDR
Guilbert Jr, Edward H, CDR
Guiler, Thomas A, CDR
Guinn III, John P, Capt
Guinn Jr, John A, CDR
Guise, Loretta M, Capt
Gukeisen, Terry R, Lcdr
Gulbrandson, Charles R.,
Gulck, Frederick K, Capt
Gulko, Edward, Lcdr
Gulledge, Hal A, Capt
Gulley, Teresa, LT
Gulley II, Robert B., CDR
Gulyas, Paul J, CDR
Gum, Charles L, Capt
Gum III, Carl D., Lcdr
Gummer III, E Garrett, CAPT
Gunderson, Don H, CDR
Gunderson, Edward C, Capt
Gunion, Edward R, Lcdr
Gunn, Alan L., CDR
Gunn, Raymond G, Lcdr
Gunning, Jason B., ENS
Gunter Jr, Malcolm L, Lt
Gunzel, Arthur L, Lt
Gurekian, Patti R., CDR
Gurioli, Lawrence V., Lcdr
Gurney, Peter L., LCDR
Gurry, Michael J., Lcdr
Gurson, John J, Capt
Gurtowsky, Ernest J., LCDR
Gus Jr, Carl J, CDR
Gussis, Michael N, CDR
Gustafson, Charles R, Capt
Gustafson, Chris O., LCDR
Gustafson, Emil, CDR
Gustafson, Jack E, Capt
Gustafson, Timothy M., CDR
Gustavson, Arne G., Lcdr
Gustavson, Donald R, Capt
Gustin, John M, Capt
Gustin, Kurt L, Lcdr
Gustin, Stephen A., CDR
Gutheim, Robert J, Capt
Guthridge, David L, CDR
Guthridge, Donald J, Capt
Guthrie, Bruce B., CDR
Guthrie, Gene E, Capt
Guthrie, Richard G, CDR
Guthrie Jr, Wallace N, Radm
Gutierrez, Phillip A., ENS
Gutman, Alan J., LT
Gutzwiller, Kenneth J, Capt
Guyer, John O, Lcdr
Guza, Anne L, CDR
Guzauskis, Steven A., CDR
Guzman, Andrew R, Lt
Guzzo, Louis M, Lcdr
Gwiazda, Stanley J, Lcdr
Gworek, Mary Ann, LCDR
Gwyn, Blanche H, Lcdr
Gwyn, James Richard, CDR
Gwynn, Robert H, CDR
Haag, Philip G, CAPT
Haakon, Christopher P., Capt
Haas, Christopher, LCDR
Haas, Donald J, CDR
Haas, James E, Capt
Haas, Jon M., Capt
Haas, Robert, CDR
Haas, Sara F, Lt
Haas Jr, Bernard C., CDR
Habeger, David J, CDR
Haben, Gary J., LCDR
Haberlach, William P, CDR
Haby Jr, Hayden G., CDR
Hacala, Martin J, Capt
Hackbart, Robert W, CDR
Hacke, Charles M, Lcdr
Hacker, Richard B, Lcdr
Hackett, Charles J, CDR
Hackett, Vincent T, Capt
Hackmann, David C., CAPT
Hadden, Eddie R, CDR
Haddock, Carol J., LCDR
Haddock Jr, Paul E., LT
Haden, John R, Capt
Hadjis, Thomas J, Lcdr
Hadley, John R., Lt
Hadley, Theresa I., LT
Haertling, Lambert L, Lcdr
Hafer, Dale V, CDR
Hafer, S Charles, CAPT
Haffey, James J., Lcdr
Haffner Jr, John G, Capt
Hafner, John K., LCDR
Haftel, Michael E., LCDR
Hagan, Kenneth J, Capt
Hagele Jr, Herbert C, Capt
Hagelin, John Andrew, CDR
Hagen, David C., Lt
Hagen, Mark D, CDR
Hagen, Merle E, Capt
Hagen, Patricia S., LT
Hagen, Patrick K., LT
Hagen, Thomas B, Capt
Hager, Dennis L, Capt
Haggard, Phil E, CDR
Haggerty, Joseph P., CAPT
Hagin, Gerald D, Capt
Hagins Jr., Sylvester R., CDR
Hagist, Warren M, CDR
Hagood Jr, Thomas A., Lt
Hagopian, Alan M, CDR
Hague, Frank H, Lcdr
Hahn, Carl G, Capt
Hahn, Herman W, Lt
Hahn, Howard Dwight, Lt
Hahn, Peter W, Capt
Hahn, Richard C, Capt
Hahn, Richard E, Capt
Hahn, Thomas A, Capt
Haidet, Walter R, Capt
Hailey, Rufus B, CDR
Haimes, Arthur S, CWO4
Haines, David D., CDR
Haines, John P., CDR
Hains, John P, CDR
Hair, Max A, Capt
Haizlip, Arthur L, CAPT
Hajduk, Daniel J., CDR
Halberg, Paul V, Capt
Halbert, Lynette M., Lcdr
Halbert, Robert H, Lt
Hale, Bradford Allen, Lcdr
Hale, Douglas R, CAPT
Hale, John H, CDR
Hale, Joseph H., Lcdr
Hale, Kevin T., CDR
Hales, Stephen R., Lcdr
Hales, Thomas S., Lt
Haley, Dennis F, Lcdr
Haley, Francis L, CDR
Haley, James R, CDR
Haley, John D, Lcdr
Haley, Kathleen G., CDR
Haley, Larry J., CDR
Haley, Mark M., Lcdr
Haley, Paul L, Capt
Haley, Paul R., CAPT
Haley, Thomas O, Capt
Halfon, Solomon, Ltjg
Halicks, William G., LCDR
Hall, Aaron L, Capt
Hall, Arthur L, Capt
Hall, Brian J., ENS
Hall, C. Ernest, CDR
Hall, Charles M, Lt
Hall, Charles R, CDR
Hall, Charles W, Capt
Hall, D. B., CDR
Hall, David E, LCDR
Hall, David H, Capt
Hall, David L., Lcdr
Hall, Donald M, CDR
Hall, Gerald A, Capt
Hall, James C, Capt
Hall, Jerry A, CDR
Hall, Kenneth R, Lcdr
Hall, Lawrence J, Capt
Hall, Lyle Ross, Radm
Hall, May B, Capt
Hall, Ralph N, CDR
Hall, Ricardo M, CDR
Hall, Robert N, CDR
Hall, Robert R, CDR
Hall, Robert T, Capt
Hall, Stephen F, Lt
Hall, Thomas F., Radm
Hall, Vaughn E, Capt

Hall, Wallace H, CDR
Hall, Wilford H, CDR
Hall, William A, CAPT
Hall Jr, Daniel B, CDR
Hall Jr, Kenneth T, CDR
Hallahan, Timothy V, CDR
Hallas, Donald J, CDR
Hallberg, Earle G, CDR
Hallberg, Robert K, CDR
Haller, Brian Q, Capt
Haller, Dale H, Lcdr
Haller, Robert K, Capt
Hallissey, Ronald J, CDR
Halloran, Michael R., CDR
Hallquist, John I, Capt
Hallstead, Richard, CDR
Hallwachs, Donald A, CDR
Halpern, Harry S, CDR
Halprin, Arthur E, Capt
Halsey, Maurice E, CDR
Halsey, William G., Lt
Halstead III, Arthur L, Capt
Halter, Ellen F., CAPT
Halter II, Leroy W, Capt
Halupka, Clement A, Lcdr
Halverson, Arthur G, CDR
Halverson, Charles R., LCDR
Halverson, Guy D, Capt
Halverson, Kenneth P., Lcdr
Halzel, Isadore, Capt
Ham, Robert J., CDR
Ham Jr, John E, Capt
Hamaday, Ronald A., LCDR
Hamblen, Don P., CAPT
Hambleton, Duane T, CDR
Hamblin, Marilyn, Lcdr
Hamby, Dale L, Lcdr
Hamelberg, Kim S., Capt
Hamelin, Gregory R, Capt
Hamilton, Bruce, CDR
Hamilton, Charles E, Capt
Hamilton, Frank W, Lcdr
Hamilton, Harry W., LTJG
Hamilton, J Mark, Lt
Hamilton, James A., CDR
Hamilton, John W, CDR
Hamilton, Lynne L., Capt
Hamilton, Maxine J., Lt
Hamilton, Michael P, Capt
Hamilton, Richard L., CDR
Hamilton, Robert L., CAPT
Hamilton, Rolland M, Capt
Hamilton, Samuel D, CDR
Hamilton, Steven W., CDR
Hamilton, William R, Capt
Hamilton Jr, George K., Capt
Hamlin, Harvey E, CWO2
Hamlin, Kent W, Capt
Hamm, Horace A, Capt
Hamm, Marvin J., CDR
Hamm, Myron G, Capt
Hamman, Larry V., CDR
Hammann, Arthur H, Lcdr
Hammar, Kevin D., Capt
Hammel, I E, Lcdr
Hammel Jr, John C, Lcdr
Hammer, David C, CDR
Hammer, Eric J., LT
Hammer, Leon, Capt
Hammer, Robert S, CDR
Hammergren, Carl K, CDR
Hammes, Carl H, CDR
Hammes, Peter L, Capt
Hammett, Peter H, LCDR
Hammett, Robert A., LCDR
Hammond, Fred W., CDR
Hammond, Julie A, LCDR
Hammond, Robert H, Capt
Hammond, Weldon W, Capt
Hammond, William J., CDR
Hammond, Wilton N, CDR
Hammond III, Edward P, Capt
Hammons, Virginia A., Lcdr
Hamner, Craig A., Lcdr
Hamner, Richard T., CDR
Hampton, James F, Capt
Hampton, John D., LCDR
Hampton, Lana D., LCDR
Hampton, Rebecca C., CDR
Hampton, W. Christopher, LCDR
Hampton, William R, Capt
Hampton, Woodward M, Lcdr
Hampton III, Hood C, Capt
Hamrick, William H, Lt
Hance, Vernell M, CDR
Hancock, Deborah A., LCDR
Hancock, William E, Lt
Hancox, Jack D, CDR
Hand, Bonnie L., Lcdr
Hand, Christopher M., LCDR
Hand, Gregory F., LCDR
Hand Jr., Thomas E, Capt
Handler, Scott S., LCDR
Handley, Chester A, CDR
Handrop, Ronald S, CDR
Handy Jr, John R., CDR
Haneghan, William J, Capt
Haney, Cheryl M., LCDR
Haney, Clarion B, Lcdr
Haney, James M, CDR
Hanf, William F., CDR
Hanft, George J, Lt
Hanisee, John J., CWO3
Hanke, Harold W, CAPT
Hankins, Charles G, CDR
Hankins, Paul W, Capt
Hanks, Gale S, Lcdr
Hanley, Frank A, CDR
Hanley, Gerald S, CDR
Hanley, William B, Capt
Hanley Jr, Arthur B., LCDR
Hanlin, John P, Capt
Hanlock, Arthur W., CDR
Hanna, C Vernon, CDR
Hanna, Cody A, CDR
Hanna, Doris V, CDR
Hanna, Harry C., LT
Hanna, Mary M, Capt
Hanna, Robert N, CDR
Hanna Jr, Alvin T, Capt
Hannan, Christina A., CDR
Hannan, William T, Capt
Hannigan, Louis F, CDR
Hannigan Jr., John T., LT
Hannon, Gerard K, CDR
Hannon, Timothy Patrick, CAPT
Hannum, Kenneth J, Lt
Hansbrough, Bruce A., CDR
Hanscom, John F, CDR
Hansen, Allen C., CAPT
Hansen, Arthur L, CDR
Hansen, Donald H, Capt
Hansen, Gary L., Lcdr
Hansen, Henry P, Lcdr
Hansen, John K, Capt
Hansen, Julian R, Capt
Hansen, Kurt F., Capt
Hansen, Leo E, Lcdr
Hansen, Maris D, CDR
Hansen, Peter C, Capt
Hansen, Ralph J, Lcdr
Hansen, Richard A, Capt
Hansen, Robert C, Capt
Hansen, Robert H, Capt
Hansen, Torleif S, Lcdr
Hansen Jr, Oluf M, CDR
Hanson, Alan L., Lcdr
Hanson, Charles P, Capt
Hanson, Edward D, CDR
Hanson, Elizabeth C., LT
Hanson, Erik A, CDR
Hanson, Hugh G, Lt
Hanson, Marshall A, Capt
Hanson, Martin P, Capt
Hanson, Michael D., LCDR
Hanson, Monte A, CDR
Hanson, Stephen L, Capt
Hanson III, Paul G, CAPT
Hapeman, Michael H., Lcdr
Haplea, George E, CDR
Happ, John T, CDR
Happ, Nellie L, CDR
Haraby, William A., Lcdr
Harada, Kikuo, Capt
Harap, Sean, LT
Harbord, Chris, LCDR
Hard, Douglas A, Capt
Hardas, Billy R., Lcdr
Hardee, Robert M, CDR
Hardeman Jr, Frank, Capt
Harden, Richard W., CDR
Harden-Lozier, Maureen, CDR
Harder, David F, Capt
Harder, Henry L, Capt
Hardesty, Stella M, Lcdr
Hardesty III, William H, Capt
Hardiman, Joseph M, CDR
Hardin, Catalina M, LCDR
Hardin, James T, Capt
Hardin, Patricia M, Lcdr
Harding, Joseph C., LCDR
Harding, Mary M., CDR
Harding, Muriel D, Lcdr
Harding, Ralph L, Lcdr
Harding III, T Peter, Capt
Hardnett, Alphonso, LCDR
Hardt, Dan A, CDR
Hardy, Howard O, CDR
Hardy, Randall W, Capt
Hardy III, G. Mark, CDR
Hare, A Craig, Capt
Hare, Jack A, CDR
Hare, Joseph C, Radm
Hare, Thomas Truxtun, CDR
Hargadon, Joseph M., Lcdr
Hargrave, Charles W, Capt
Hargrove, Robert J, Capt
Hargrove III, Harry L, Capt
Haring, Michael H., CDR
Harker, Donald P., Cdr
Harker, Robert W, Lcdr
Harkey, John Paul, CAPT
Harkin, Edward C, CDR
Harkins, James C, CDR
Harkins, Richard E, Capt
Harkins, Steve R., Capt
Harkleroad, John F, CDR
Harkness, Gregg E, CDR
Harkness Jr, Daniel H, Capt
Harlan, Wayne A, Capt
Harley, David A., Lcdr
Harley, Larico, ENS
Harlin, Floyd, Lcdr
Harman, Joe H, Capt
Harmeling, R Kent, CDR
Harmeyer, Karen A, RADM
Harmon, David L, CWO3
Harmon, George P W, Capt
Harmon, Harold W, Capt
Harmon, Joseph R, Lcdr
Harmon, Stephen C, CAPT
Harms, Harold B, CDR
Harmuth, William B, CDR
Harnar, James A., CAPT
Harnden, Paul F, CDR
Harned, Charles C, Capt
Harned, Stephen W, Capt
Harner, H. Allen, CDR
Harness, Francis W, Radm
Harnett, Jeffrey H, Capt
Harold Jr, Elliotte M, Capt
Harper, Bruce G, Capt
Harper, Conrad R, Ltjg
Harper, Earnest F., Lt
Harper, James A, CDR
Harper, James R, CDR
Harper, John L, Lcdr
Harper, M. Bruce, LCDR
Harper, Mark D., CDR
Harper, Norman W, Capt
Harper, Pamela J., Lt
Harper, Robert L, Capt
Harper, Vernon M, Lcdr
Harpold, Gary J., Capt
Harr, Charles D., CDR
Harrell, Albert C, CWO4
Harrell, Altha, LCDR
Harrell, Edward E, Lt
Harrell, Edward K., Lcdr
Harrell, John M., Lcdr
Harrell, Margaret R., LCDR
Harrer, Max W, Capt
Harrer, Wilbur J, Lt
Harrigan, Daniel J., CDR
Harrington, Douglas M., LCDR
Harrington, James J, CAPT
Harrington, James R, Capt
Harrington, Michael G, Capt
Harrington, Michael J., CAPT
Harrington, Patrick H, Capt
Harrington, Wayne B, Capt
Harris, Benjamin K, CDR
Harris, Benny J, Lcdr
Harris, Bill J, Capt
Harris, Charles R, Lcdr
Harris, Clarence M., CAPT
Harris, David A, Capt
Harris, David J, LT
Harris, David L, CDR
Harris, David M., LCDR
Harris, Dennis J, CDR
Harris, Doyle V, CDR
Harris, Edward L, CDR
Harris, Eric R., Lt
Harris, Frederick B., LCDR
Harris, George C, Lt
Harris, Gregory C, Lcdr
Harris, Jack D., LCDR
Harris, James E, Lcdr
Harris, James V., LT
Harris, John A., Lcdr
Harris, John W., CDR
Harris, Kenny D, CDR
Harris, Lee B, Lcdr
Harris, Lisa A., CDR
Harris, Norman G, Capt
Harris, Paul E., CDR
Harris, Peter W., Capt
Harris, Richard N, Capt
Harris, Robert L, Capt
Harris, Robin L., LCDR
Harris, Ronald S, Capt
Harris, Sharon E., LT
Harris, William A, Capt
Harris Jr, Henry E, Capt
Harrison, Cathy A., CDR
Harrison, Claiborne J., Lt
Harrison, Curtis B, Capt
Harrison, Daniel E, Capt
Harrison, Edwin D., CAPT
Harrison, Horace, Capt
Harrison, James B, CDR
Harrison, Jeffrey A, CDR
Harrison, Leo J., Lcdr
Harrison, Robert W, Capt
Harrison, V. Paul, Radm
Harrison Jr, Edward H., Lcdr
Harrison Jr, Ralph H, Capt
Harrison Jr, Reginald W., CDR
Harrold, Beatrice E, CAPT
Harrow, Thomas L, Capt
Harrower, Eugene, Capt
Harryman, Carrel R, CAPT
Harshbarger, Arthur A, Lcdr
Harshfield, James B., Capt
Hart, Charles A, CDR
Hart, Charles B, Capt
Hart, Dwight D, LCDR
Hart, E Thomas, Capt
Hart, John C, Capt
Hart, John M, Lcdr
Hart, John T, Lcdr
Hart, Kenneth R., Capt
Hart, Martin J, CDR
Hart, Max F, CWO4
Hart, Michael David, CDR
Hart, Randey R., Lcdr
Hart, Richard S, Capt
Hart, Royce N, Lcdr
Hart, Terrel L., LCDR
Hart, Thomas L, CDR
Hart, William L, CWO4
Hart Jr, Harold J, Capt
Hartbower, Carl E, CDR
Harter, Philip J, Lcdr
Hartik, Lorraine, CDR
Hartley, Melinda DeAngelis, LT
Hartman, Clayton A., LCDR
Hartman, James L, Capt
Hartman, Mileva M, Capt
Hartman Jr., William L, CDR
Hartmann, Richard P, Capt
Hartowicz, Patricia A, Lcdr
Harts, Justin, ENS
Hartsook, Burnett E, Lcdr
Hartstirn, Stephany L., LT
Hartung, Stacy, ENS
Hartwell, Kenneth W, Capt
Hartwig, Edward J., CDR
Harty, George F, CDR
Harty, James D., CDR
Harvey, Craig T., CDR
Harvey, Edward P, CDR
Harvey, Grace Ellen, Lcdr
Harvey, Harlon N, Capt
Harvey, Harry, Lt
Harvey, Norman C, CDR
Harvey Jr, Addison J, Lcdr
Harvey Jr, Daniel G, Capt
Harvey Jr, Pierce J., Lcdr
Harvey Jr, Ray W, Capt
Harvey-Smith, Derek, Lcdr
Harvin Jr, Edwin L, Lcdr
Harwayne, Thomas W, Capt
Harwell, Isabella, LCDR
Harwell, Layne H, CAPT
Harwell, Thomas W, CDR
Harwood II, Theodore L, Capt
Hashbarger, Charles T, Lcdr
Haskew, Eugene S, LCDR
Haskins, Edward B., CDR
Haslup, Allen L, Capt
Haspedis, Walter L, Lt
Hass, Kathleen A., CDR
Hass, Rodney C, CDR
Hassall Jr, George C, Lcdr
Hasselbach, Gerard, CAPT
Hasselquist, Maynard B, CDR
Hassenplug, John K, Capt
Hasson, James J, CWO4
Hastings, Edmund S, CDR
Hastings, Harold W, Lcdr
Hastings, Michael S., Lcdr
Hastings II, Robert W, CDR
Hatch, Robert A., Lcdr
Hatcher, Fredric F, Capt
Hatfield, Don A, CDR
Hathaway, Carmrid G, Capt
Hathaway, Robert Don, CDR
Hathaway, Robert R, Capt
Hatloy, David S, CWO4
Hatton, Bruce N, Capt
Hatton, Leonard, Lcdr
Hauber, Mahlon E., CWO4
Haubner, Michael J, CDR
Hauck, David J, CDR
Hauck, Fred P, CDR
Hauck, Fredrick M., CDR
Haug, Richard J, Lt
Haug, Shain B, Capt
Haugh, John D., Capt
Haughey, Kevin J., CDR
Haughtelin, Craig D., LCDR
Haunss, Charles E., Lt
Haupt, Eric Robert, Capt
Haupt SR., Robert R., Capt
Hauschild, Daniel C, CDR
Hauser, Kenneth L, Lcdr
Hausmann, Albert C, CDR
Haussmann, Norman J, Capt
Haven Jr, Robert C, CDR
Havens, Justis E, CDR
Haverty, John L, Capt
Haviland, Timothy J., LCDR
Havlick, Ronald G., CDR
Havlik, John R., LCDR
Hawes, George R, Lcdr
Hawes, Stephen P, Capt
Hawkins, Charles F., Lcdr
Hawkins, Charles J, Ltjg
Hawkins, Deborah C., CDR
Hawkins, Frederick C, Capt
Hawkins, Isaac K, CDR
Hawkins, Norman G., CDR
Hawkins, Robert W., Lcdr
Hawkins, William F, CDR
Hawkins III, Chas W, CDR
Hawkinson, Lawrence D, CDR
Hawks, Charles T, CDR
Hawks, Donald Ray, CDR
Hawks, Susan L., Lcdr
Hawley, David H., LCDR
Haworth, Charles T, Capt
Hawthorne, Mary E, CDR
Hawthorne, Thomas S, CDR
Hayase, Peter J., Lcdr
Hayashida, Alan K, CDR
Haycraft, Glenn W, Lcdr
Hayden, Jeffrey R., Lcdr
Hayden, John W, Capt
Haydu, James A, CDR
Hayek, Charles P, Capt
Hayes, Charles R, Capt
Hayes, Donald P, Lt
Hayes, Everett A, CDR
Hayes, Francis W, CDR
Hayes, Fred E., Capt
Hayes, Gerard B, LT
Hayes, Gregory K., LCDR
Hayes, Holland W, Lcdr
Hayes, James E, Lcdr
Hayes, John B., Lcdr
Hayes, John J, CDR
Hayes, Kenneth S., LCDR
Hayes, Larry E., CAPT
Hayes, Richard A., CDR
Hayes, Robert T, Capt
Hayes, Rufus D, Capt
Hayes III, Richard D., CAPT
Hayes Jr, David J A, Capt
Hayes Jr, Don L., CDR
Hayes Jr, Walter A, CDR
Hayn Jr, Frank A, CDR
Haynes, John S, CDR
Haynes, Roy G., Lcdr
Haynes, Willard R., CDR
Hays, James F, Capt
Hays, Jeff V., CDR
Hays, Michael E., CWO4
Hays, Richard L, Lcdr
Hays, Robert L, Lcdr
Hayslett, Roderick J, Capt
Haythorn, Gary W, Lcdr
Hayward, Thomas B, Adm
Hayward, Winchell T, Lcdr
Haywood, Herbert H, Capt
Hazam, Stephen T, CDR
Hazelrig, John P, Capt
Hazelwood, Dorothy B, CDR
Hazen, Donald S, Lcdr
Hazley, George J, Capt
Heacock, Louis W, CDR
Head, Peter S, Lcdr
Head, Rufus W, Capt
Head, William I, Capt
Headley, G Robert, CDR
Headley, William A, Capt
Healey, Dennis T, CWO4
Healy, Edward R., Capt
Healy, James B, Ltjg
Healy, James V, Capt
Healy, John G, Lcdr
Heard Jr, William H, Capt
Hearl, Alan F, Capt
Hearl, Joe, Ens
Hearn, Christine E., LT
Hearn, Neal E, CDR
Heath, Jerry L, CDR
Heath, Stanley P, CWO4
Heatley, Richard T, Capt
Heavey, James L, Capt
Heavey Jr, Martin R, CDR
Hebb, Ralph K, Capt
Heberer, Eugene O, CDR
Hebert, George J., CDR
Hebert, Veronica D., Capt
Hebert III, Joseph N, Capt
Heburn, George W, Capt
Hecht, George E, CDR
Hecker, Robert Wyatt, Capt
Heckman, Henry S, CDR
Hecox, Ronald E, CDR
Hedeen, Paul M, Lcdr
Hedeen, Richard L, Lcdr
Hedgepeth, Beverly D., CDR
Hedman, William D, CDR
Hedrick, Donald E, Lcdr
Hedrick, Lyle E, CDR
Hedrick Jr, Charles W, CDR
Heebner, Charles, CDR
Hefestay Jr, Frank J, CDR
Heffernan, Darrlyn M., CWO4
Hegarty Sr, Daniel P., Lcdr
Hegele, Thomas R, CDR
Heget, Harry S, CAPT
Heggie Jr, Alfred D, Capt
Hegi Jr, Ernest A, CAPT
Hegidio, Joe L., CDR
Heideman, Kenneth K, Lcdr
Heidrick, Thomas C, Lcdr
Heidt, Edward A, Capt
Heidt, Vincent E, CWO4
Heidt Jr, Arthur J, Capt
Heiken, Frederick C, Capt
Heil, H William, Lcdr
Heil, Louis L, CAPT
Heilenday, Richard M., CAPT
Heim, David L, Capt
Heim, David T., CDR
Heimark, John J, CDR
Heimerdinger, Marie C., CDR
Heimerdinger, Theresa M., LTJG
Heimgartner, Jean L, Lt
Hein, Richard A, Capt
Heindel, John H, CDR
Heine, William W., CDR
Heinkel, Robert L, Lcdr
Heinrich, Steven P, Capt
Heinrichs, George R, Lcdr
Heins, Edward W., CDR
Heintzman, John S, Capt
Heinze, John R, Lcdr
Heise, Robert E., ENS
Heisey, Walter F, Capt
Heiss, Jeffrey L, CDR
Heister Jr, Robert J., CAPT
Heiter, Kathy A., CAPT
Heithaus, Roger, Lcdr
Held, Tim D, Lcdr
Helfstein, Michael L., Lcdr
Helgeson, James D, CDR
Helland, Richard J., LCDR
Hellawell Jr, G Alan, Capt
Helle, James A, CDR
Helleis, John G, Capt
Heller, Deborah D., CDR
Heller, Susan M., LT
Hellewell, Martin S, Capt
Hellickson, James A, Capt
Hellman, Allen D, CDR
Hellman, Todd T., CDR
Hellweg, Henry S, CDR
Helm, Robert A, Lcdr
Helms, L. Lynn, Lcdr
Helms, Maureen Martin, Lcdr
Helpap, John Frederick, CDR
Helsel, Charles E, CDR
Helson, Ronald B, Capt
Helsper, James T, Capt
Helton, Ronald L, Capt
Helweg, Otto Jennings, Capt
Hemberger, Jay M, CDR
Heminger, Rex A., LCDR
Hemings Jr, Robert M, Lcdr
Hemmelgarn, Stephen A., LCDR
Hemming, Raymond, Radm
Hemmy Jr, Victor H, Capt
Hempel, Robert E, Capt
Hemphill, James E, CDR
Hendel, Donna M, CDR
Hendel, Richard W, Capt
Hendel, Thomas H, Lcdr
Henderson, Barbara J., CDR
Henderson, Carl L., CWO4
Henderson, Dale L, CDR
Henderson, David L, CDR
Henderson, Earl D, Ltjg
Henderson, Earl W, CDR
Henderson, Edwin C, Lcdr
Henderson, Finlay C, Lcdr
Henderson, Jack D, CDR
Henderson, Kenneth P, Lcdr
Henderson, Marvin D, CDR
Henderson, Richard A., Lcdr
Henderson, Thomas J., Capt
Henderson, Todd A., CDR
Henderson, Tomas G, Lcdr
Hendren, Cyrus E, Capt
Hendricks, Frederick B, Capt
Hendricks, Nelson P., Lcdr
Hendricks, Robert L., LCDR
Hendricks, William C., LCDR
Hendricks, William R., Capt
Hendricks Jr, A. T., Capt
Hendrickson, Paul B, Capt
Hendrickson, Robert C, Capt
Hendrickson, William C, CDR
Hendrix, David G, Capt
Hendrix, Ralph M, CDR
Heney, William P, Lcdr
Henk, David R., CDR
Henke, Robert J., LCDR
Henkel, Charles A., CDR
Henline, Wilbur P, CAPT
Hennebury, Paul E., LT
Hennessey, Michael, Capt
Hennessey Jr, Thomas V, Capt
Henning, Richard A, CDR
Henning Jr, Fred L, Capt
Hennings, Craig A., Lt
Hennings III, Louis W, CDR
Henry, Dan O, Lt
Henry, David R, Lcdr
Henry, Dennis L., CDR
Henry, Edwin S., CDR
Henry, James R, Capt
Henry, John B, Capt
Henry, John D., Lcdr
Henry, John M, CDR
Henry, Jon, LCDR
Henry, Lee L, Capt
Henry, Richard J, Capt
Henry, Steven D, CDR
Henry, Thomas E, Capt
Henry, Thomas S., LCDR
Henry, Wayne O, Capt
Henry, William O E, CDR
Henry, Zachary A, LCDR
Henry III, Curtis R, Lcdr
Henry Jr, Robert A., Capt
Hensgen, Oscar Eugene, Capt
Henshaw Jr, Paul C, CDR
Hensley, Frank M, Capt
Hensley, Larry D, Capt
Hensley Jr, John C, CDR
Henson, Larry D, Capt
Henson, Leonard J, CAPT
Henson, Mark L., ENS
Henson, William H, Lcdr
Henson, William N, Capt
Henze, Robert J, Lcdr
Heppe, Frances Z., Lt
Herb, Martha E.G., CDR
Herbein, John G, Capt
Herberg, James L., Lcdr
Herbert, Bentley D, Lcdr
Herbert, Donald V, Capt
Herbert, George M, Lcdr
Herbert, Hansel K, LCDR
Herbert, Thomas J, CAPT
Herbert, William G, Capt
Herbig IV, Henry F., CDR
Herbol, Wesley C, CDR
Herbst Jr, Albert J, Lcdr
Herbst Jr, John C, CDR
Herciga, William C, Lcdr
Hereford Jr, Robert E, Capt
Herfel, Bruce A., LCDR
Herge, Henry C, Lcdr
Herkner, Richard T, Capt
Herlevi, Michael D., Lt
Herlihy, Charles E, Capt
Herlson, Mark C, Capt
Herman, Harry A, Capt
Herman, Robert L, CDR
Herman, William N., LT
Herman Jr, Laurence T, CDR
Hermann, Edward P., Capt
Hern, Herbert R, Capt
Hernandez, Alice E, Lcdr
Hernandez, John R, Capt
Hernandez, Rudy, Capt
Hernandez, Suzan P., LCDR
Herndon, Lucia A., Lcdr
Herndon, William A, CAPT
Herold, James L, CDR
Herold, John L, CDR
Herold, Merlin C, Capt
Heron, Donald J, Lcdr
Heron, Paul J, CDR
Heronen, Daniel J, CDR
Heroux Jr, Pierre J, CDR
Herr Jr, William M, CDR
Herrick, Peter S, Lcdr
Herrick, Richard E, CDR
Herriges, Michael J., CDR
Herring III, Robert A, CDR
Herrington, Alfred H, CDR
Herrington, Robert M., LCDR
Herrman, Roger W, CDR
Herrmann, Lacy B., Lt
Herrmann, Thomas J., CDR
Herron, Jennifer L., ENS
Herron, John R., LCDR
Herron, Michael K., LT
Herron, William E, Radm
Hersch, Marvin H, Capt
Hersch, Stephen D., LCDR
Herschel, David M., LT
Herscher, Herbert L, CDR
Herschler, Daniel A., LT
Hersh, David W, CDR
Hersh Jr, Jerre F, Ltjg
Hershberger, Donald A, Capt
Hershberger, John David, LT
Hershman, Jerald B., Capt
Hersman, Norman R., Lt
Herspring, Dale R, Capt
Hertz, Robert D., LCDR
Hertzberg, Hutz H., LCDR
Hervey Jr, John C, Capt
Herzbrun, Herman, Lt
Herzel, Albert M, CWO4
Herzog, Daryl R, Lcdr
Herzog, Stuart K, CDR
Herzog, William L, CDR
Heselton, Leon W, Lt
Hess, B Anton, Lcdr
Hess, Brenda K., LT
Hess, Curt, Capt
Hess, David W, CDR
Hess, Glenn C, Lt
Hess, James B, Capt
Hesse, Richard S., CWO3
Hesse, Robert L, Lcdr
Hesser, E. Grant, Lcdr
Hessert, Richard S, Capt
Hessey V, John H, Capt
Hessinger, Kevin A, CDR
Hester, Gerald G, Capt
Hetherington, Donald L, Capt
Hetherington, John W, Capt
Hetrick, Richard B, CAPT
Hetzel, Gary N., CDR
Hetzel, Noel J, Lcdr
Hever, John J, Radm
Hevia, Jorge, Lcdr
Hewell, Charles L., CAPT
Hewett, Paul N, CDR
Hewitt, Billy L, Capt
Hewitt, Cullen, Capt
Hewitt, Dean J, Capt
Hewitt, Jerry P, CDR
Hewitt, Winton Edward, Lcdr
Hewitt 3rd, John D, Ltjg
Hexdall, Eric J., LCDR
Heyck Jr, Joseph G, Capt
Heyer, Nelson O, Radm
Heyer, Nora F, Ltjg
Heymann, Herbert, Lcdr

Heyne, Delmar H, CDR
Heywood, Richard, ENS
Hiatt, Charles M, Lt
Hiatt, Keith S, Lcdr
Hibbard Jr, Curt W, CDR
Hibben, Quentin L., LCDR
Hibberd, Charles P, CDR
Hibl, David J, Capt
Hibner, Larry A, CDR
Hickam Jr, A Wayne, Capt
Hickey, Christopher M., Lcdr
Hickey, Dennis J, Capt
Hickey, John P, Capt
Hickey, Kevin S., Lcdr
Hickey Jr, Edward J, CDR
Hickman, John W., LT
Hickman, Warren J, Capt
Hickman, William H, CAPT
Hickman Jr, William R, Ltjg
Hickmann, Charles A, CDR
Hicks, Charles G., Lcdr
Hicks, Fred M., CDR
Hicks, Gary L, CDR
Hicks, Gerald F, Capt
Hicks, Gregory L., LT
Hicks, Jimmie D, CWO2
Hicks, Weldon B, LCDR
Hicks II, Ronald B., LCDR
Hicks Jr, Coy S, CWO3
Hicks Jr., George R., CAPT
Hide, Donald L, CAPT
Hiel, Richard J., CDR
Hier, Gerald V, Lt
Hierlwimmer, Ulf R, Capt
Hietbrink, Russell A, CWO2
Higdon, James N, Capt
Higer, Ralph W, Capt
Higginbotham, Allen B, Capt
Higginbotham, Kent R, CAPT
Higgins, Christina J., Lcdr
Higgins, David L., Lcdr
Higgins, Eugene P, Capt
Higgins, John H, CDR
Higgins, John J, CDR
Higgins, Linda W, CDR
Higgins, Michael S., MIDN
Higgins, Raymond J, Lcdr
Higgins, Robert W, CDR
Higgins, Vincent Paul, LCDR
Higgins, Walter M, Capt
Higgins, William J., Lcdr
Higgins III, Thomas J, CDR
Higgins Jr, Edward B, Lcdr
Higgins Jr, William E., Lcdr
Higgins Jr, William F., Capt
Higgs, Gerry E, Capt
Higgs, William T., Capt
Hildebrand, Robert A., Lcdr
Hilditch, John, CDR
Hildreth, Jon P, CWO4
Hildum, Jack E, Lcdr
Hile, Carol L, CDR
Hilf, Paul T, Capt
Hilke, James L, Capt
Hill, A Jackson, CAPT
Hill, Bryce E., LT
Hill, Clarence D, LT
Hill, Dale A, Lcdr
Hill, David H, CDR
Hill, Ernest H, Lcdr
Hill, Gary W., Lcdr
Hill, Gordon R, CAPT
Hill, James D, Capt
Hill, James K., LCDR
Hill, James M, CDR
Hill, Jay W, Capt
Hill, Jeffrey A, CDR
Hill, Jerry W, Capt
Hill, Jon J., Lcdr
Hill, Joseph B, CDR
Hill, Karen D., CDR
Hill, Lawrence R, CDR
Hill, Leo B, CDR
Hill, Mark, LCDR
Hill, Ralph A, CWO4
Hill, Ralph P, Lt
Hill, Richard A, CAPT
Hill, Richard A, Capt
Hill, Richard A., Capt
Hill, Ross A, Capt
Hill, Terrance G, Capt
Hill, Thomas J, Radm
Hill, Wade Lee, CDR
Hill II, George J, Capt
Hill Jr, Donald C., CAPT
Hill Jr, Edgar Andrew, Lcdr
Hill Jr, Maurice B, Capt
Hill Jr., Luther, LCDR
Hiller, John L, Capt
Hillerman, Neal H, Capt
Hilliard, Robert M., CAPT
Hilliard, William N, Capt
Hillig, Kenneth J, CDR
Hillman, Albert S, Lcdr
Hillman, James F., Lcdr
Hillman, Mark J, Capt
Hillman, Tatnall L, Capt
Hills, Gary Lee, CDR
Hills Jr, Chester I, CDR
Hilt, Thomas G., Capt
Hilton, Helen J., CDR
Hilton, Jackson M, CDR
Hilton, Mark D., LCDR
Hilton, Thomas F., CAPT
Hilton Jr, Charles E, CDR
Hiltz, Thomas G., CDR
Himchak, David P, CDR
Himschoot, Robert A, CWO4
Himstreet, Thomas R, CDR
Hinck, John A., Lcdr
Hinckley, George A., LT
Hinckley Sr., Edward S, CDR
Hind-Marsh, B Douglas, CDR
Hinderleider, Thomas A., Lcdr
Hine, Francis C, CDR
Hine, Patricia S, CDR
Hine, Roger C., CDR
Hine SR, Gilbert C, CWO3
Hinebaugh, Mark A., LCDR
Hines, Hubert O, CDR

Hines, Jerome H., Capt
Hines Jr, John R, Capt
Hines Jr., Clifford G, Lt
Hinkle, Daniel B, Capt
Hinman, Katherine J, Capt
Hinman, Leah Rae, Lcdr
Hinrichs, Tamra L., LT
Hinson, Bruce H, CAPT
Hinson, David E, Capt
Hinson Jr, William H, Capt
Hinsvark, Don G, Capt
Hintermister, Frederick, Lcdr
Hinton, David O, Capt
Hintze, Douglas E., CDR
Hinz, Donald E, CDR
Hinz, Eric S., CDR
Hinz, Paul T., LCDR
Hipol, Manuel A, Capt
Hippe, Kenneth G., Capt
Hippeli, Fredrick, LTJG
Hire, Kathryn P., CDR
Hirman, Joyce I., LT
Hirsch, Erwin F, Capt
Hirsch, Glenn A., LCDR
Hirsh, Louis M., Capt
Hirshon, Sidney, Lcdr
Hiscock, William W., LCDR
Hitchcock, Donald J, Lt
Hitchcock, George H, CDR
Hitchcock, Paul D., LCDR
Hite, John P, Lcdr
Hite, Patrick L., Lcdr
Hitt, Curtis L, Capt
Hittner, George C, CDR
Hjerpe, Erik D., CDR
Hoak III, Frank M, Capt
Hobbes, Paul P., CDR
Hobbs, William E, CDR
Hobbs, William H, Capt
Hobby, Lawrence J, CWO3
Hobcroft, Bryan G, Lcdr
Hobokan, Andrew, Capt
Hobson, Gary T., LCDR
Hobson, Robert L, Capt
Hocevar, Ronald B., LT
Hoch Jr, Robert G, CAPT
Hochberg, Peter J, Lcdr
Hochreiter, Herbert, CDR
Hochschild III, Arthur, Lcdr
Hochstetler, James E, Lcdr
Hock, Steven R., CDR
Hock Jr, W Fletcher, Capt
Hodak, Gary W., CDR
Hodder, Kenneth L, Capt
Hodella, Fred H, CDR
Hodge, Danny B., CDR
Hodge, G Lowrance, CAPT
Hodge, Gerald L., LCDR
Hodge, Rowland S, Capt
Hodge, Susan F., LT
Hodge, Walter O, Capt
Hodgell, Tolin W, Capt
Hodges, Alan S, Capt
Hodges, Charles S., LCDR
Hodges, Elizabeth C, CDR
Hodges, J Norman, Capt
Hodges, Michael P., LCDR
Hodges Jr, James P, CDR
Hodgkins, Richard E., LCDR
Hodgson, Brian D., Lcdr
Hodson, Robert W, Capt
Hoeck, Jeanne F, Capt
Hoefer, John H, Radm
Hoefer, Richard A, CDR
Hoefer III, Albert, Lcdr
Hoeg, Walter W, Lcdr
Hoel, Willard R, CDR
Hoelscher, John H., CDR
Hoelting, Alan, Capt
Hoener, James H, Capt
Hoerr, Robert C, Lcdr
Hoertkorn, Walter R, CDR
Hoertz, Paul G., CDR
Hoey, Lee Danny, LCDR
Hofer, John P, CDR
Hoff, James L, Capt
Hoff, Matthew J, Lcdr
Hoff, Monica B., Lcdr
Hoff, Richard W, Capt
Hoff, Robert A, CDR
Hoff, Walter M, CDR
Hoffecker, William A, Lcdr
Hoffer, Louis J, Lcdr
Hofferbert, Louis E, CAPT
Hoffman, Charles H, Lt
Hoffman, James D., CAPT
Hoffman, Joel J, Capt
Hoffman, Joseph, CWO4
Hoffman, Joseph P., LT
Hoffman, Katherine D, CDR
Hoffman, Kay E, LCDR
Hoffman, Michael E, LT
Hoffman Jr, Charles E, Capt
Hoffman Jr, Raymond H, CWO4
Hoffmann, Brent A., Capt
Hoffmann, Thomas M., LT
Hoffmann, Wayne A., CAPT
Hoffmann Jr, Arthur D, Capt
Hoffmeyer, John T, Capt
Hofhine, Duane J, Capt
Hofmann, Joseph P, Lcdr
Hofmann, Richard J, CDR
Hogan, Donald A, Capt
Hogan, Eddie Walsh, Capt
Hogan, James R, Lcdr
Hogan, John David, CAPT
Hogan, Richard R, Capt
Hogan, Richard W., CWO4
Hogan, Robert L., CAPT
Hogan III, Thomas F, CDR
Hogden, Raymond L, Capt
Hoge Jr, William R., Capt
Hogg, Elaine M., Lcdr
Hogg, Kevin R., LT
Hoglund, Eric P., LT
Hogue, James E, CWO4
Hohenadl, Eike H, CDR
Hoisington, Richard R, CDR
Hoke, Howard H, CDR
Hokokian, Paul S, Capt

Holcomb, Harold L, Lcdr
Holcomb, Wade G, Lcdr
Holden Jr, William H, Capt
Holder, Lawrence R, CDR
Holdren, Larry R, CDR
Holdridge, Michael J., Lcdr
Hole, Cheryl D., CDR
Holewinski, Joseph A, Lt
Holford, Thomas W, Capt
Holihen, Edward M, CDR
Holland, Bill E, Lcdr
Holland, Christopher S., CAPT
Holland, Jeffrey W, CDR
Holland, Joann S., Lcdr
Holland, Zack D, Ltjg
Holle, Karlton, CDR
Hollenbach, Paul D., Capt
Hollenbeck, Orville S, Lcdr
Holler, Ann F, Capt
Hollerbach, Robert R, Capt
Hollett Jr, Grant T, Radm
Holley Jr, Paul E, CDR
Holliday, Dorothy V, Capt
Hollinberger, David, CDR
Hollinger, Frederick R., Capt
Hollingsworth, Mary L, CDR
Hollis, Kathleen S, Lcdr
Hollis, Randall H, Lt
Hollis, Robert H, CDR
Hollister, Jack D, Capt
Hollman, William J., CDR
Hollmeier, Eric J., LT
Hollmer, Alvin, CDR
Hollow, John L, Capt
Holloway, Joseph F., LCDR
Holloway, Richard E, Capt
Holloway, Sallie B, CAPT
Holloway Jr, Richard L, CDR
Hollyfield Jr, Wallace, Capt
Holman, Bud G, Capt
Holman, David, Lt
Holman, Donald R, Capt
Holman, Harland E, Radm
Holman, Larry D, Lcdr
Holman, William A., Capt
Holmbeck, Robert E, Ltjg
Holmberg, Holly, ENS
Holmes, Carl F, Lcdr
Holmes, Edward A, CDR
Holmes, Gary L, Lcdr
Holmes, George E, CDR
Holmes, George W, CDR
Holmes, J Frederick, Capt
Holmes, James E, Capt
Holmes, John B, Radm
Holmes, John M, Capt
Holmes, John M., Lt
Holmes, Melanie A, Capt
Holmes, Milburn J, Capt
Holmes, Peggy R, Capt
Holmes, Peter A., Lcdr
Holmes, Robert C., LCDR
Holmes, Robert O., LCDR
Holmes, William F, Capt
Holmes, William W., CDR
Holmes II, M. Dennis, CDR
Holmes Jr, Jack I, CDR
Holmes Jr, Robert E, CWO4
Holmes Jr, Sidney L, Capt
Holmquist, Eric A, Capt
Holmstrom, Robert D, Capt
Holoubek, Leonard, CDR
Holpuch, Michael O, CDR
Holsenbeck, Daniel C, Capt
Holshey, Michael L, CDR
Holstead, Roland E, Capt
Holston, James R., Lcdr
Holstrom, Robert W P, Capt
Holt, David L, CDR
Holt, James B., Lcdr
Holt, Michael S, CDR
Holt, Waldo S, Capt
Holt Jr, Maltbie L, Lcdr
Holtzman, Maureen D., Capt
Holwell, Kevin, CDR
Holyfield, George W, Capt
Holz, Jane R, CDR
Holzhausen, Joseph E, Lcdr
Holzworth, Gail W., LT
Hom, John J, CAPT
Hom, Tony C., Lt
Homan, Russell W., CAPT
Homan, William G.H., CDR
Homer Jr, Joseph J, Capt
Homolash, Cheryl R., LT
Honan, Michael P, Capt
Hone, Lawrence R, Capt
Honer, Harold R, CDR
Honer, James A, CDR
Honey, James L., CDR
Honigschmidt, James O, Capt
Hood, David D., CAPT
Hood, Michael D., CDR
Hood, Raleigh M, Capt
Hood III, Horace, Capt
Hook, Byron J, Capt
Hook, Stephen H, CDR
Hooker, James D., Capt
Hooker, John E, Lcdr
Hooker, Joseph H., LTJG
Hooks, James H B, Capt
Hooks Jr, J Thel, Capt
Hooks Jr., C Dock, LCDR
Hoon, Harold D, CDR
Hooper, Charles I., LCDR
Hooper, George R, CDR
Hooper, James E, Capt
Hooper, Richard B, Capt
Hooper, Robert M, Capt
Hooper, Robert W, Capt
Hooper Jr, John R, Capt
Hoornbeek, Louis A, Lcdr
Hoos, E Lou, Capt
Hooton, Philip L., CDR
Hoover, Frederick A., CDR
Hoover, Harwood, Capt
Hoover, Joan M.J., LT
Hoover, Kerwin E, CDR
Hope, John C, CAPT
Hopfenspirger, Thomas M, CDR

Hopkins, Billy G, Lt
Hopkins, David H, Capt
Hopkins, Donald D, Capt
Hopkins, Donna L., CAPT
Hopkins, Fred E, CDR
Hopkins, James D, Lt
Hopkins, John D, Capt
Hopkins, Leonard E, Lt
Hopkins, Mariane S, CDR
Hopkins, William H P, CDR
Hopkins III, Elwood W, CDR
Hopp, Carol A, LCDR
Hoppe, Harold W, Capt
Hopper, Hugh R, Capt
Hopper, J Alan, Lt
Hopper, Richard E, Capt
Hopper, Sally Jo H., LCDR
Hopper, Veronica Anna, Lt
Hopwood Jr, Herbert G, Capt
Horan, Helen M, Ltjg
Horan, James J, Capt
Horbiak, Stephen C., CDR
Hord, Phillip G, Lcdr
Hormozi, James, LTJG
Horn, Dennis L, Capt
Horn, Dion E, CDR
Horn, Elford Wade, Capt
Hornbacker, Kurt, LTJG
Hornburg, Kirk D., LCDR
Horne, Gregory, Capt
Horne, Ricky L, CDR
Horne, William A, CDR
Horne III, Charles E, Capt
Horneman, Douglas J., ENS
Horner, John B, Capt
Horner, Monte L, Lcdr
Hornung, Francis C, Lcdr
Horowitz, Emanuel, Lt
Horowitz, Gary R., CDR
Horsell, Dennis D, CAPT
Horsley, James M, Lcdr
Horsman, Douglas E., CDR
Horsmon, Frederick M, Capt
Horst Jr, Julian L, Capt
Horton, Douglas J, Capt
Horton, Jay S, Lcdr
Horton, Laura M, CDR
Horton, William E, Capt
Horvath, Joseph J., LCDR
Horvath, Julius O, CWO4
Horwitz, Stephen H, Capt
Hosannah II, Hilton O., CDR
Hosch, Joseph W., CWO4
Hosea, Andrew J, CDR
Hosek, Timothy J, Lt
Hoshko, Christopher J., Lcdr
Hosier, Charles W, Lcdr
Hoskins, Iffath A., Capt
Hoskins, Michael H, CDR
Hoskins, Richard C, Capt
Hoskins III, Thomas T, CDR
Hoskins Jr, Richard C, CDR
Hotaling, Eric J., Capt
Hotchkiss, Lois E, Lt
Hotes, Frederick L, Capt
Hotham, James A, CDR
Hotter, Ann N, CDR
Hotton, Randy, Capt
Houg, Martin M, Lcdr
Hough, Devon L, CWO4
Hough, Marvin F, Capt
Hough, Phillip G., CAPT
Houghten, Donald, Lcdr
Hougland, Charles A, Capt
Houk Jr, Thomas L, Capt
Houlihan, Edmund M, CDR
Hourihan, Richard T, Lcdr
Hourihan, Thomas J, Lcdr
House, Karl T, Capt
House, Ronald R, Lcdr
Houseknecht, Peter, Capt
Housepian, Edgar M, Lcdr
Houser, David A, CDR
Houser, Raymond J, CWO4
Houser, William D, Vadm
Houser Jr, Donald F, CDR
Housiaux, John E, Capt
Houston, Hugh E, CDR
Houston, W. Brock, Capt
Houting, Peter R, CDR
Hov, David Thor, CDR
Hov, Mary Ruth, CDR
Hove, Peter C., Lcdr
Howard, Andrew C, Lcdr
Howard, Bruce T., CDR
Howard, Charlton T., CDR
Howard, Craig M, LCDR
Howard, Earlyn Perry, Capt
Howard, Herbert G, CDR
Howard, Jackson B, Capt
Howard, John C., CDR
Howard, Joseph D, Lcdr
Howard, Linda G, Lcdr
Howard, Lowell Bennett, Capt
Howard, Macauley, Capt
Howard, Mark B., Lcdr
Howard, Paul L, CWO4
Howard, Robert G, CDR
Howard, Ronald E., Lcdr
Howard, Thomas K, Capt
Howard, William L, Capt
Howard Jr, Horace W, Capt
Howard Jr, William J, Capt
Howarth, Peter T., Lcdr
Howatt, Franklyn J, CAPT
Howe, David B, Capt
Howe, Frederick W, CDR
Howe, John W, Capt
Howe, Linnie A, Lcdr
Howe, Paul T, Lcdr
Howe Jr, Wesley W, Capt
Howe USPHS, Patrick K., LT
Howell, Christian F., ENS
Howell, George E, Lcdr
Howell, James B., MIDN
Howell, Jerry G., CDR
Howell, John G, Capt
Howell, Jones V, CDR
Howell, Lembhard G, CDR
Howell, Thomas W, CAPT

Howell, W Robert, CDR
Howells, Gary N, CDR
Howerter, Craig E., Capt
Howerton, Dean R., Lcdr
Howes, Sandra L, Capt
Howie, Bruce C, Capt
Howison, Philip W, CDR
Howlett, Fred J, Capt
Howlett, James W, Capt
Howlett, Leonard G, Ltjg
Howren Jr, Harry H, CDR
Howse, William S., CAPT
Howson, George O, Capt
Hoy, Jeffrey C, CAPT
Hoy, Thomas L, CDR
Hoyem, Mary K, Capt
Hoyer, Carl I, Capt
Hoyle, William R, Capt
Hoyne, Susan D, CDR
Hoyt, Ramon W, CWO4
Hoyt, William C, Lcdr
Hrbacek, Mary Jean T, CDR
Hruska, Harvey J, CDR
Hryshchyshyn, Michael, CDR
Hsia, Lawrence, CAPT
Hubbard, George T, CWO4
Hubbard, Harry H, Lt
Hubbard, John R, Radm
Hubbard, Richard P., CAPT
Hubbell, James H., Capt
Hubbert, Frederick D., Lt
Hubbs, John M, CDR
Huber, Eileen M, Lcdr
Huber, Gerald P, Capt
Huber, Judy E., LCDR
Hubert, John F, Capt
Hubley, Edward S., Lcdr
Huddleston, Leo N., CDR
Hudgens, Robert K., CAPT
Hudgins, Michael J., Capt
Hudnall, Arnold, Capt
Hudock, Joseph M., LCDR
Hudson, Agnes S, CDR
Hudson, Charles J, Capt
Hudson, George M., LCDR
Hudson, Harold G, Lt
Hudson, John I, CDR
Hudson, Michael A, CDR
Hudson, Stanley P, CDR
Hudson Jr., Bobby, CDR
Hudson Sr., Hannis H., CAPT
Hudspeth, Barry E., Cdr
Hudspeth, John R, LCDR
Huebner, Allan C., MIDN
Huebner, Richard H., CAPT
Huegel, Valerie K., CDR
Huerter, Cyril B., Lcdr
Huessy, Alan, LTJG
Huey, Eric G, Lcdr
Huey, Joel W., Lcdr
Huey, Pamela R., LT
Huey Jr, Arthur S, CDR
Huey Jr, H David, Lcdr
Huf, William L, CDR
Hufeld, William J., Lt
Huff, Howard W, Capt
Huff, James A, Capt
Huff, Maxwell T, Lcdr
Huff, Raymond J, CWO4
Huffman, James C, Capt
Huffman, Margaret C., Lcdr
Huggard, Ernest D, Capt
Huggard, John P, Capt
Huggins, Bobby W., CDR
Huggins, Harry L, CDR
Huggins, Hermon, LT
Huggins Jr, Judson D, CDR
Hughes, A L, Capt
Hughes, Arthur C, Capt
Hughes, Bruce N, Capt
Hughes, Charles S., CDR
Hughes, Dennis E., CDR
Hughes, Eric H., LCDR
Hughes, Frederick P., CAPT
Hughes, Gayle D., LT
Hughes, Herbert H, Capt
Hughes, Herman W, Capt
Hughes, I Harding, Lcdr
Hughes, James B, CDR
Hughes, Jason A., LCDR
Hughes, John H, Capt
Hughes, Mark A., MIDN
Hughes, Michael F., CDR
Hughes, Richard W, CDR
Hughes, Russell K., LCDR
Hughes, Samuel P, Lcdr
Hughes, Stephen D, CDR
Hughes, William E, Lcdr
Hughes, William G, CDR
Hughes, William R, CDR
Hughes III, Richard C., CDR
Hughes Jr, Charlie B, Capt
Hughes Jr, F Francis, Capt
Hughes Jr, William J, Capt
Hughey, Michael J., Lcdr
Hughey, Thomas M, CDR
Hugman, Kevin H., CDR
Huish, Alexander P., CAPT
Hujarski, Steve R, CDR
Hulett, Charles E, Capt
Hulick, Timothy P, Capt
Huling, Gary D., ENS
Hull, Herbert D, CDR
Hull, Robert E, Ltjg
Hull, Steven L, Capt
Hulme II, Woodrow W, Capt
Hults, Thomas P, CDR
Hults III, Arthur S, Capt
Hulun, Ralph J, CDR
Humber, John L, Capt
Humenuck, Richard J, CAPT
Hummel, Paul L, Capt
Humphreville, Thomas R, CDR
Humphrey, Barry E, CDR
Humphrey, Herbert, Lcdr
Humphrey, Michael J., Lcdr
Humphrey Jr, Thomas F., Lt
Humphreys, Charles E., Capt
Humphreys, Keith C, Capt
Humphries, Charles S, Capt

Humphries, Howard L, CDR
Hundley, Herbert A., LCDR
Hungerford, James M, CDR
Hungerford, William P, CDR
Hungerman, Gregory P, Lcdr
Hunley, Burt H, Lcdr
Hunn, Mark G., CDR
Hunt, Brian R., Lt
Hunt, Christine M., LCDR
Hunt, David R., Lcdr
Hunt, Dennis J, CDR
Hunt, James V, Capt
Hunt, John R, Capt
Hunt, Julia Agnes, Lcdr
Hunt, Kirby P, CDR
Hunt, Robert A., CDR
Hunt, William R, Lcdr
Hunt Jr, Charles J, CDR
Hunt Jr, Warren C, Lt
Hunter, David P., LCDR
Hunter, Elisa F., CDR
Hunter, Robert L, Lcdr
Hunter, Ronald E, Capt
Hunter, Tom W, Lcdr
Hunter, William G, CDR
Hunter III, Malcolm K, Capt
Hunter Jr., Harold F, CDR
Huntley, Larry T, CDR
Hupp, Charles J, CDR
Hupp, James G, CDR
Hupp, Thomas H, CDR
Hurd, John W., LT
Hurd, Steven R, Lcdr
Hurlbut, Wendell P, Capt
Hurley, Edward T, Lt
Hurley, John W, Capt
Hurley, Rodney G, Capt
Hurley II, Thomas F., LCDR
Hurley Jr, Daniel C, Capt
Hurst, Charles J, Lcdr
Hurst, John M, CAPT
Hurst, Richard D, Capt
Hurst, Warren L, LCDR
Hurt, Clarence, Capt
Hurt, Marc A, Capt
Hurt Jr, Alfred B, Capt
Hurtt, David J., CDR
Husband, Willard R, CDR
Huse, James A, Lcdr
Hushaw, Walter D, Capt
Husmann, Donald, Capt
Huss, Marvin C., CDR
Huss, Stephen L, CDR
Huss, William W, Capt
Husta, Peter A., CDR
Hustvedt, Erling H, Capt
Hutchings, Charles W., LT
Hutchinson, David B., ENS
Hutchinson, Derrick, LCDR
Hutchinson, Otis H, CDR
Hutchinson, Ralph W, Lcdr
Hutchinson, Thomas H., CAPT
Hutchinson, Thomas W, Lcdr
Hutchinson, Thurlow Q, Capt
Hutchison, James H, CDR
Hutchison, Mark D., LCDR
Hutchison, Thomas K., Lcdr
Hutchison Jr, Claude B, Capt
Hutchko, Alvan J, Capt
Hutsell, Riddell C, CDR
Hutton, Renny M., Lcdr
Huus, Helen, CDR
Huyong, Douglas G., CDR
Hyatt, Wynne T., CDR
Hybiske, Thomas H, Capt
Hyde, George W, Capt
Hyde Jr, Robert T, Capt
Hyka, Edwin W, CDR
Hyland, Patrick P., CDR
Hyman, Mark E., CDR
Hyman, Ronald B., Lcdr
Hymel, Gene J, Lcdr
Hyndman, David S, Lcdr
Hynes, Alan P., CDR
Hynes, Michael W, Capt
Hynes II, Joseph E, CDR
Hyslop, David J, CWO4
Ianno, John K., CAPT
Ibach, Maryanne T, Radm
Iborg, Eugene L., CDR
Icenhour, Alan S., LCDR
Icsman, James A, CDR
Iffert, Herbert C, Capt
Ignaut, James E, Mr
Ihrig, Stephen D, Capt
Ijams, Charles C, Capt
Ikeler, Alfred E., CDR
Ikerd, Edward E, Lcdr
Iler Jr, William B, LCDR
Ilk, Joseph L., Lcdr
Imboden, Lester E, CDR
Imel, John M, Capt
Imeson Jr, H Louis, CDR
Imlay-Spangler, Ann, LT
Immik, Raymond, Lt
Indall, Norm E, CDR
Inderwiesen, Charles E, Capt
Indiviglia, Salvatore J, CDR
Infranco, Michael P., LCDR
Ingersoll, F Thomas, Capt
Ingham, John E., Lt
Inghram, Richard C, Lcdr
Ingison, Malcolm S, CDR
Ingle Jr, Hugh C, Capt
Ingley III, Fred, Lcdr
Inglis, James H, Capt
Ingold, Jack D, Capt
Ingraham, Donald S, Capt
Ingraham, Marshall T, Capt
Ingraham, Murray J, Capt
Ingram, Charles M, Capt
Ingram, John G, Capt
Ingram, Mark W., ENS
Ingram Jr, Earl T, CWO4
Inguanzo, Osvaldo M, Capt
Ingwersen, William F, Capt
Inman, Bradley M, Capt
Inman, William H, LT
Inouye, Michael J., CDR
Iovino, Pamela M., CDR

Irby, Frank R, CDR
Ireland, Wilfred C, CWO2
Irigoyen, Erwig R., ENS
Irish, Jermain S, CDR
Irland, Richard L, CDR
Irmisch, Louis W, CAPT
Irons, Roy L, Capt
Irrgang, Otto R, Lcdr
Irving, Buster L, Capt
Irwin, Hugh S, Lcdr
Irwin, Thomas C, Radm
Isaacks Jr, Stanley A, Lcdr
Isaacson, Isaac Dale, CDR
Isaacson, Jeffrey A., LT
Ise, Nancy P., Capt
Isenburg, Charles D, Lcdr
Ishihara, James, Capt
Ishol, Lyle M, CDR
Iskowitz, Robert, Capt
Islam, Mohammed S, Capt
Isleib, Charles C., Lcdr
Isom, Charles E, Cdr
Isquith, David A, Capt
Israel, David O, Capt
Israel, Jacob S., CDR
Israel, Stephen S, RADM
Isselhard, David E, Lt
Istre, Verna J, Lcdr
Iverson, Gary W, Capt
Iverson, Selmer O, Capt
Iverson, Thomas R., LCDR
Ives, Barbara A., CDR
Ives, Christopher L., CAPT
Ives, George Stoughton, CDR
Ivey, Louis A, Capt
Ivory III, Thomas M, Capt
Iyonmahan, Karla A., LT
Izard, Robert F, CDR
Izzo Jr, Wallace J, CDR
Jablonski, Ian, CDR
Jablonski, Joseph R., CDR
Jablonski, Raymond G, Lcdr
Jablonski-Hodsdon, Rhonda K., LT
Jack, John L, CDR
Jack, Robert W, Capt
Jackson, Bruce K., CDR
Jackson, Byron C, CDR
Jackson, Charles U, CDR
Jackson, David S, CWO4
Jackson, Eldon L, Lcdr
Jackson, Frederick P, Capt
Jackson, Gerald S, CAPT
Jackson, Harry Bonneau, Capt
Jackson, Howard E, CWO4
Jackson, Jack E, Lt
Jackson, Joann T, Lcdr
Jackson, John A, Capt
Jackson, John J, CWO4
Jackson, Kenneth A., Lcdr
Jackson, Robert C., CDR
Jackson, Robert E, CDR
Jackson, Robert J, CDR
Jackson, Robert W, Capt
Jackson, Roger W, Lcdr
Jackson, Rose M., LT
Jackson, Samuel C, Capt
Jackson, Saundra C., Lcdr
Jackson, Susan S, Capt
Jackson, Warren E, Capt
Jackson III, Alexander, Radm
Jackson Jr, Alexander, Radm
Jackson Jr, Charles A, Capt
Jackson Jr, Harry K, CDR
Jackson Jr, John R, Capt
Jackson Jr, William, Capt
Jaco Jr, Allen A, CDR
Jacob, Cynthia R., LCDR
Jacob, Thomas R, CDR
Jacob, W. Hannan, LCDR
Jacobs, David W, Capt
Jacobs, Francis T, Capt
Jacobs, Jan C, CDR
Jacobs, Kay L, Capt
Jacobs, Louis S, Lt
Jacobs, Morton S, Capt
Jacobs, Roberta L, CDR
Jacobs, Roger P, Capt
Jacobs, Sheila Denise, Lcdr
Jacobs, Stuart K, Capt
Jacobs, William C, LCDR
Jacobsen, Adolf M B, Capt
Jacobsen, Craig A, Capt
Jacobsen, Douglas A, Capt
Jacobsen, Mary K., LCDR
Jacobsen III, Charles P, Lcdr
Jacobson, Albert P, Capt
Jacobson, Carl E, Capt
Jacobson, F. Russell, Lcdr
Jacobson, John R., CDR
Jacobson, Michael H, Capt
Jacobson, Raymond G, CDR
Jacobson, Richard P, CDR
Jacobson, Stephen E, CDR
Jacoby, Benedicta R., LCDR
Jacot, Jay F, Capt
Jacques, Philippe A, Capt
Jacques, Richard D, Lcdr
Jacroux, Kenneth R, Capt
Jaeger, Arthur F, Lcdr
Jaeger, Boi J, Capt
Jaeger, Edward A, Capt
Jaeger, James Michael, CDR
Jaeger, Richard N, Capt
Jaehne, Madeleine A., Capt
Jaehnig, James L, CDR
Jaffe, Jack W, Lcdr
Jagel, William R, CDR
Jagger, Frank A, CDR
Jaggers, James D, Capt
Jagoda Jr, John S, CWO4
Jahn, David E, CDR
Jahnke, Fred R, Capt
Jahns, Arthur W, Capt
Jakaboski, Theodore P, Capt
Jaklitch, Patricia A., Lcdr
Jakovic, Louis G, Lcdr
Jaksha, Steven F, CDR
Jakuboski, Jeff, LT
Jakubowski, Wayne, CAPT
James, Earle F, Lcdr
James, Fred M., CDR
James, Kenda C., Lcdr
James, Laura A, CAPT
James, Robert G, Radm
James, William F, Lcdr
James, William L, CDR
James Jr, Thomas C, Lcdr
Jameson, David W, CAPT
Jameson, Dwight C, Ltjg
Jameson, Robert Q, CDR
Jameson, William M, CDR
Jamharian, Reuben A, CAPT
Jamieson, J W, CWO4
Jamison, Earl J, CDR
Jamison, Lyle C, Capt
Jancuski, John, Capt
Janczak, Martin E, Radm
Janes, Charles E, Capt
Janes, David A, Radm
Janetos, Dion S, CDR
Janiec, Jan D, Capt
Janigian, Bruce J, Capt
Janikowsky, Linda C., CDR
Janke, John R, CDR
Jankiewicz, Joseph J., CDR
Jankowski, Leonard, Lcdr
Janning, Timothy J, LT
Janosco, Kathryn M., LCDR
Jansen, Ralph F, Lcdr
Janssen, Erwin T, Capt
Jarchow, John H, CAPT
Jarrett, David B, Lcdr
Jarvi, Reed K., Cdr
Jarvis, Arthur R W, CDR
Jarvis, Ernest V, Lcdr
Jasitt, Donna W., CDR
Jaskowiak, Robert B., Lt
Jasparro, Christopher R., LTJG
Javaras, Eftyhia, Capt
Jaworowski, Ronald O., LCDR
Jay, John, CDR
Jayne, Hilton, CDR
Jaynes, John E., CAPT
Jayson, Herman L, Lt
Jeffers, Gene E, Capt
Jeffers Jr, Joseph B, Capt
Jefferson, James F, Capt
Jeffords, Douglass C, Capt
Jeffrey, Richard W., CDR
Jeffries, Henry C., Lt
Jeffries, Margaret L., Lcdr
Jelinek, Stephen J, Lcdr
Jeng, James C., LCDR
Jenik Jr, Edward C, CDR
Jenista, George B, Lcdr
Jenkins, Arthur R, CAPT
Jenkins, Bonnie D., LT
Jenkins, Dwight, Lcdr
Jenkins, Eddie L, Capt
Jenkins, Elizabeth R, CDR
Jenkins, George P, Lt
Jenkins, Kenneth R, Lcdr
Jenkins, Mark R., Lcdr
Jenkins, Terry L., ENS
Jenkins, Tim M, Radm
Jenkins, Walter W, Capt
Jenkins, William, Capt
Jenks, David M, CDR
Jennings, Donald R, Capt
Jennings, Jody L., LCDR
Jennings, John M, Capt
Jennings, Mark E, CDR
Jennings, R. C., LCDR
Jennings, Robert H, CDR
Jennings, Sandra L., Lcdr
Jennings, Tappan H, Capt
Jennings, William E, CDR
Jennings II, Jay R, CDR
Jennings III, Belton E, Capt
Jennings Jr, Robert W, Capt
Jennison, Christopher W, CDR
Jensen, Alfred H, CWO4
Jensen, Arthur M, Capt
Jensen, Arthur S, Capt
Jensen, Bruce E., LT
Jensen, Carl O, Lcdr
Jensen, Christine E, Lcdr
Jensen, Christopher H., CDR
Jensen, Eric A, Capt
Jensen, Eric C., Lcdr
Jensen, Harry W., Lcdr
Jensen, John T, Capt
Jensen, Kathleen, CDR
Jensen, Kenneth R, CDR
Jensen, Richard A, Capt
Jensen, Robert G, CDR
Jensen, Thomas A, Lcdr
Jensen Jr, Joseph E, Capt
Jeppesen Jr, Lee J, CDR
Jepsen, Edward P, Lcdr
Jepson, William B, Capt
Jerman, Joseph L., LTJG
Jernigan, Jeffery R., CDR
Jerome, Jeffrey B., CDR
Jerome, Michael S, Capt
Jervis, Jeannie R, Lcdr
Jessup, David H., Capt
Jestus, Roberta L., Lt
Jett-Parmer, Jonathan J., ENS
Jetter, Deborah A, CDR
Jew, Kelvin Y, Lcdr
Jewell, Edwin David, CDR
Jewell, Michael Boyd, Lcdr
Jewett, Ann E, Capt
Jewett, David M, Capt
Jewett, Ellen M., CDR
Jezukewicz, Joseph A, CDR
Jimenez, Arthur, Lt
Jindra, Frank E., LT
Jindra, Lawrence F., CDR
Jindrich, Charles A., Capt
Joerger, John M, CDR
Jofko, John, Capt
Jofre, Sandra S., Lt
Joh, Leopold F., CDR
Johansen, Marvin W, Lcdr
Johansen, Peter C, CDR
Johansen, Roy Dean, Lcdr
Johansmeier, Ray, Cdr
Johanson, Walter J, CDR
Johansson, William A., MIDN
John, David F, Capt
Johnescu, Robert J, Capt
Johns, Marianne S, LCDR
Johnsen, Arvin W., CDR
Johnson, Allan D, CDR
Johnson, Alphonse I, Lcdr
Johnson, Arnold C., CDR
Johnson, Ben H, Lt
Johnson, Bennett H., CDR
Johnson, Bjarne E, CDR
Johnson, Bruce A, Capt
Johnson, Bryan L., Lcdr
Johnson, Carl E, CWO4
Johnson, Charles C, CDR
Johnson, Charles E, Lt
Johnson, Charles E, Capt
Johnson, Charles L, CDR
Johnson, Clarence A E, Radm
Johnson, Clarence S, Lcdr
Johnson, Clinton B, Capt
Johnson, Clinton V, Lcdr
Johnson, David G, Capt
Johnson, David K, Capt
Johnson, David S, CDR
Johnson, Dean M., LT
Johnson, Don N, Capt
Johnson, Donald A, Lcdr
Johnson, Donald C, Capt
Johnson, Donald Leroy, Capt
Johnson, Douglas A, Capt
Johnson, Douglas A., Lcdr
Johnson, Douglas L, Radm
Johnson, Douglas S, CDR
Johnson, Edward W, Lcdr
Johnson, Ellsworth L, Lcdr
Johnson, Francis C, Capt
Johnson, Francis C., LT
Johnson, Fredrick, Capt
Johnson, George E, Lcdr
Johnson, George I, CDR
Johnson, George W, Lt
Johnson, Glenn E, Lt
Johnson, Gregory C., CAPT
Johnson, Henry O., CDR
Johnson, James C, CDR
Johnson, James L, CDR
Johnson, James M, Capt
Johnson, James M, CDR
Johnson, James R., CDR
Johnson, Jean Willard, Lcdr
Johnson, Jeff W, CDR
Johnson, John A, Lcdr
Johnson, John B, Radm
Johnson, John H, Lcdr
Johnson, John W, Capt
Johnson, Jolene T, Lcdr
Johnson, Jordan E, CDR
Johnson, Joseph R., LT
Johnson, Keith R, Lt
Johnson, Kenneth A, CDR
Johnson, Kenneth A., Capt
Johnson, Larry C, Capt
Johnson, Larry E, Capt
Johnson, Leonard C, Lcdr
Johnson, Marilyn D., LT
Johnson, Matthew L., LTJG
Johnson, Melanie M., CDR
Johnson, Michael D., CDR
Johnson, Minh T., LCDR
Johnson, Nancy R, Capt
Johnson, Nicholas R., Capt
Johnson, Norman L, CDR
Johnson, Norman L, Lcdr
Johnson, Patrick D., Lcdr
Johnson, Paul M., CDR
Johnson, Perry E., LT
Johnson, Pierce J, Radm
Johnson, Robert A, Capt
Johnson, Robert F, CDR
Johnson, Robert Louis, Capt
Johnson, Robert S, Capt
Johnson, Roger D, CDR
Johnson, Ronald E., CAPT
Johnson, Ronald K., LCDR
Johnson, Roxan J, LCDR
Johnson, Roy C, CDR
Johnson, Roy W, Capt
Johnson, Russell C, Capt
Johnson, Ryan E., CDR
Johnson, Seth H, Lcdr
Johnson, Stanley H., CDR
Johnson, Stanwood W, Capt
Johnson, Steven J., LT
Johnson, Steven K, Capt
Johnson, Susan V., Lcdr
Johnson, Terry D, CDR
Johnson, Terry L., LCDR
Johnson, Theodore J, Capt
Johnson, Thomas H, CDR
Johnson, Thomas T, Capt
Johnson, Walter E, Lcdr
Johnson, Walter F, Capt
Johnson, Walter R., LCDR
Johnson, Warren B, CDR
Johnson, Wayne E, CDR
Johnson, Wesley H., CDR
Johnson, William A, Capt
Johnson, William F, Capt
Johnson, William G, CDR
Johnson, William H, CDR
Johnson, Winton H, Capt
Johnson II, Weston M, Capt
Johnson III, Charles M, Capt
Johnson III, Clayton H, Capt
Johnson Jr, Charles B, CDR
Johnson Jr, Clarence E, Capt
Johnson Jr, Dale C, Capt
Johnson Jr, Earl W, Capt
Johnson Jr, George L, Capt
Johnson Jr, Harry I, Capt
Johnson Jr, Mathews M, CAPT
Johnson Jr, Melvin F, CDR
Johnson Jr, Milton E, Capt
Johnson Jr, Philip, Capt
Johnson Jr, Raymond A, CDR
Johnson Jr, Stephen A, Lcdr
Johnson Jr, Sterling, Capt
Johnson Jr, Theodore W, CDR
Johnson Jr., Sigvard B., CDR
Johnston, Bruce T, Capt
Johnston, Burrell D, Capt
Johnston, C. E., Capt
Johnston, Charles D., CDR
Johnston, Charles F, CAPT
Johnston, Charles R, Capt
Johnston, David L, Capt
Johnston, David R, CDR
Johnston, Douglas A, CDR
Johnston, Ervin F, Capt
Johnston, Greg R., Capt
Johnston, Harvey D, CDR
Johnston, James R, Capt
Johnston, James R, CDR
Johnston, John W, Capt
Johnston, Mary Ann, Capt
Johnston, Norman W, Lcdr
Johnston, Patricia, CDR
Johnston, Peter M., Lcdr
Johnston, Richard H, Capt
Johnston, Richard W, CDR
Johnston, Robert G, Capt
Johnston, Roy W, CDR
Johnston, Steven A., CAPT
Johnston, William E, Capt
Johnston, William R, Capt
Johnston III, John L., LT
Johnston Jr, William C, CDR
Johnstone, Charles P, Capt
Johnstone, Denver T, Capt
Johnstone, James T, Lcdr
Joia, Michael F., LT
Jokela, Albert, Lcdr
Jolivette, Mark E, CAPT
Jolliffe, John E., CDR
Jolly, John E, CAPT
Jolly, Jon B, Lt
Jolly, Sandra H, CDR
Jonassen, Robert N, Lcdr
Jones, Arthur V, Capt
Jones, Bluett E, Capt
Jones, Bradley W., CAPT
Jones, Brian G, CDR
Jones, Carl D, Capt
Jones, Charles L, Lcdr
Jones, Charles W, CDR
Jones, Charlie C, CAPT
Jones, Clayton D, CDR
Jones, Coral S, CDR
Jones, Corliss L, CDR
Jones, Cyril D, Capt
Jones, Daniel M, Capt
Jones, Dean E, Capt
Jones, Donald L, Capt
Jones, Dorothy, Lcdr
Jones, Douglas E, CDR
Jones, Doyle E L, CDR
Jones, Dwayne C., LT
Jones, Edward G, Lcdr
Jones, Edward M, Capt
Jones, Emmett L., CDR
Jones, Fred C, Capt
Jones, Frederick L., Lcdr
Jones, George C, Capt
Jones, George W, CDR
Jones, Gordon J., LCDR
Jones, James B., CDR
Jones, James E, CDR
Jones, James M, Capt
Jones, Jeffrey B, Capt
Jones, Jimmy E, Capt
Jones, John A Drew, CDR
Jones, John G, Capt
Jones, John H, CDR
Jones, John H., CAPT
Jones, John H., CWO3
Jones, John L, Capt
Jones, Johnnie K., LT
Jones, Johnny R., LCDR
Jones, Joseph N., CWO2
Jones, Judith N, CDR
Jones, Keith M., Lcdr
Jones, Kenneth L., CDR
Jones, Kenneth M, CDR
Jones, Louis E, Capt
Jones, Mary P, Lcdr
Jones, Michael D, CDR
Jones, Norris E., Capt
Jones, P W, CDR
Jones, Paul, CDR
Jones, Randall E., CDR
Jones, Randy K., CDR
Jones, Raymond Paschal, WO1
Jones, Rebecca, MIDN
Jones, Richard A, Lcdr
Jones, Richard L., CDR
Jones, Richard Milton, CDR
Jones, Richard W., CDR
Jones, Robert G., CDR
Jones, Robert P, CDR
Jones, Roger F., CAPT
Jones, Roger John, Capt
Jones, Scott D., LT
Jones, Shari L., LT
Jones, Stephen E, CDR
Jones, Stephen J, Capt
Jones, Stephen P., Lcdr
Jones, Steven C, Lcdr
Jones, Steven M, CDR
Jones, T. David, LCDR
Jones, Talmadge C, Capt
Jones, Thomas E., CAPT
Jones, Thomas E., CWO4
Jones, Thomas E., LCDR
Jones, Thomas L, Capt
Jones, Thomas L, Capt
Jones, Walter M, LCDR
Jones, Wendell E., LT
Jones, William J, Capt
Jones, William L, Lt
Jones, William N, Lt
Jones, William R, Capt
Jones, William R, Lt
Jones Jr, Everett J, Capt
Jones Jr, George B, Capt
Jones Jr, Gordon W., CDR
Jones Jr, Leonard D., LCDR
Jones Jr, Paul H, Lcdr
Jones Jr, Royden E., CDR
Jones Jr, T Embury, Capt
Jones Jr, Thomas L, Capt
Jones Jr., M Addison, Capt
Joransen, William S, Capt
Jorczak, Paul J, Capt
Jordan, Dwight S, CDR
Jordan, Edward C, Lcdr
Jordan, James H, CDR
Jordan, James R, Capt
Jordan, Jeremy D, Lcdr
Jordan, John H, CDR
Jordan, John S., Lcdr
Jordan, Larry H, CAPT
Jordan, William S, Lcdr
Jordan, Zenobia, ENS
Jordan Jr, Alexander J, Capt
Jordan Jr., Robert G., LCDR
Jorgensen Jr, R E, CDR
Jorgenson, Cecilia, LCDR
Jory, Jerrold G, Capt
Joseph, Elaine L, Lcdr
Joseph, John B, Capt
Joseph, Mark R, Capt
Joseph, Michael A., CWO4
Joseph II, Byron J., CDR
Joshi, Raj L., Lcdr
Joslin, Ivan L, Capt
Joslin, John F, Capt
Joslin, Robert S, Capt
Jouannet, Peter R., Capt
Jouannet, Richard F, Capt
Joy, John A, Ltjg
Joyce, Michael P., LCDR
Joyce III, Thomas H, Capt
Joyce Jr, Harrison T, Capt
Joyner, George W D, CDR
Joynson, Mary Ellen, LCDR
Joynt, Monica M., LCDR
Joynt, Thomas R, Capt
Juarez Jr, Robert N, CDR
Juda, Richard J, Capt
Judd, Harlan B, Capt
Judd, Steven M., CDR
Judge, John W., CDR
Judge, Thomas J, Capt
Judge, Timothy M, Lcdr
Judice, Carroll M, CDR
Judkins Jr, Lawrence W, Capt
Judson, David E, Lt
Judson, Lee A., LCDR
Judson II, Arthur, CDR
Juergens, Kenneth C, Capt
Juett, John M, CDR
Jugar, Richard F, CWO4
Juhl, Gordon E, Capt
Julian, Ronald H, Capt
Juliano, Paul J., CAPT
Jungst, Robert A, CWO3
Junkin, Anthony Bruce, Capt
Junkins, Steven M., CDR
Juraschek, Robert F, Lcdr
Jurkonis, Albinas K, CDR
Jurkowski, Nancy J, Lcdr
Justesen Jr, Norman S, Capt
Justice, Lisa M., ENS
Justus, James S, Capt
Justus, Norman E, Capt
Kabis, Walter L, Lcdr
Kabzinski, Dolores J., LT
Kachmarik, George, CDR
Kack, Kenneth E, Lcdr
Kaczmarczyk, Francis D, Capt
Kaczmarek, Jean A, Lt
Kaczorowski, James P, Capt
Kaegebein, Russ E, Capt
Kaelin, Joseph, Capt
Kaessner, Maryanne P., CDR
Kahl, Bruce W., CDR
Kahl, Thomas R, Capt
Kahle, Jerold E, Lcdr
Kahler, Daniel E., Lcdr
Kahler, Stephen R, Lt
Kahler III, John H, Capt
Kahn, Bruce E, Capt
Kahn, Donna L., CDR
Kahn, Lester A, Lcdr
Kahn, Rolf, Lcdr
Kahoe, John F, CWO3
Kail, Robert P, Lcdr
Kaino, Kerry L., LCDR
Kainz, Vicki L, CDR
Kaiser, Clifford Y., CDR
Kaiser, Donald G, Lcdr
Kaiser, James M, Lt
Kaiser, John P., CDR
Kaiser, Ralph L, CDR
Kaiser, Theresa L., CDR
Kaiser, Thomas A., CDR
Kaiser, Thomas L, Lcdr
Kaiser III, Theodore S., LTJG
Kaish, Terri, LCDR
Kaker, Kyle F., CDR
Kalal, Ralph, Lcdr
Kalathas, Nicholas T, CDR
Kaleel, David W., Lcdr
Kalien, Orman E, Lcdr
Kalinoski, Edward F, Capt
Kalinowski, Robert D., CDR
Kalland, Paul L, CDR
Kallberg, Keith T, Capt
Kallio, Peter J., Lcdr
Kallman, Raymond G, Capt
Kallsen, Homer D, Lcdr
Kallus, Ernest R, Capt
Kallus, Lynn M., Capt
Kalman, Thomas J, Capt
Kalmus, William Bernard, Lcdr
Kaltenbach, John L, Capt
Kaltenbach, Terry M., CAPT
Kamberos, Louis S., CDR
Kaminski, Mark W., CDR
Kaminski Jr, Joseph, CDR
Kaminsky, Allan L., Capt
Kamm, Thomas A, Radm
Kammer, William N, Capt
Kampe Jr., Delmar F., CDR
Kamps, William J, Capt
Kanakaris, Peter N, Lcdr
Kanaley, Thomas M, Capt
Kandzer, Michael E., LT
Kane, Donna L., LTJG
Kane, Edward J, CDR
Kane, Francis M., LCDR
Kane, John J, Capt
Kane, Michael P., LCDR
Kane, Thomas E, Capt
Kane III, John V, Capt
Kanninen, Michael L, Capt
Kansteiner, Beau K, Lcdr
Kantner, James L, CAPT
Kantor, Robert J, Lcdr
Kantrowitz, Steven B, Capt
Kanuck, Mark J., CAPT
Kanwisher Jr, Walter C, CDR
Kaphingst, Scott T., CWO2
Kapigian, John M, CDR
Kaplan, Allen S, Capt
Kaplan, Daniel R, Lt
Kaplan, Julian, Lcdr
Kaplan, Sanford S, CDR
Kaplon, Ronald K, Lcdr
Kapurch, Thomas J, LCDR
Karagianis, Michael G, Capt
Karagiannis, Harriet, Lcdr
Karagounis, John D, CDR
Karalekas, S Steven, Capt
Karas, Robert E, CDR
Kardulias, George D, CDR
Karell, Eric J., CDR
Karlsson, Carl R, Capt
Karnbach, Christopher R., LCDR
Karnes, Lori V., LCDR
Karnes, Richard D, Capt
Karnes, Robert D, Capt
Karnowski, Thomas J, CDR
Karns, William E, Ens
Karol, Ralph L, Lcdr
Karpstein, Arick R., Lt
Karson, John J., CDR
Karsten, Jackie Lee, CWO4
Kary, D. Jean, LT
Kary, James A, CDR
Kasch, Randall B, Lcdr
Kaseno, Timothy A., LCDR
Kaseote, George, Capt
Kasin, Lynn R, Capt
Kasinac, Joseph F., Lcdr
Kaske, Herbert M, Capt
Kaskin, Jonathan D, Capt
Kaslavage, Leo J., CDR
Kasper Jr, Robert J, Capt
Kasperski, Daniel C, Capt
Kass, Stephen B., LCDR
Kass Jr, Frederick J, Capt
Kast, Daniel K, Lt
Kastner, William J, CAPT
Katana, Thomas E., CDR
Kataoka, Y Glenn, Capt
Kataryniak, William, CDR
Kathan, Robert H, Lcdr
Kathka, Timothy L, Capt
Katilus Jr, George J, Capt
Katin, Marjorie R, Capt
Katin, Robert A, CAPT
Katz, Arnold E, Capt
Katz, Ira R, Lt
Katz, Marvin A, CDR
Katz, Robert I, Capt
Kau, Melvin E M K, Capt
Kauchak, Martin P, CAPT
Kaucher, James E., CDR
Kauderer, Howard Todd, CDR
Kauffman, Daniel G, Capt
Kauffman, Theodore G, CDR
Kauffman, William F., CDR
Kaufhold, Edmund Earl, LT
Kaufman, Boris R., Ltjg
Kaufmann, Anders J, CDR
Kaufmann, Erwin L, Lcdr
Kausak, David A., Lt
Kavulia, John M, Capt
Kawamoto, Owen N, CDR
Kawamura, Randall M, Capt
Kay, Richard O, Capt
Kay Jr, James, Capt
Kaysak, John M., Lcdr
Kayye, Paul T, Radm
Kazanjian, Phillip C, Capt
Kazarian, Kirk K, Capt
Keane, Francis W, Radm
Keane, Thomas P, Lt
Kear, Harold J, Capt
Kearney, Channing L., LT
Kearney, Gary P, Capt
Kearney, Kevin, Lt
Kearns, Edward L, Lcdr
Keasler, Warren Kent, Capt
Keathley, Bryan O, CDR
Keathley, Jerry W, Capt
Keating, Donald F, CDR
Keating, Michael T, CDR
Keating, Pius F, CDR
Keating, Steven G., LCDR
Keating, Thomas J, CDR
Keating, Walter G, Capt
Keaveny, Patrick J, Capt
Keck, Steven L., CDR
Kee, Kenneth R, CDR
Keefe, Daniel D., LT
Keefer, Charles J, Capt
Keefer, Kenneth M., Lcdr
Keegan, John C, CDR
Keele, Earl Eugene, Lcdr
Keeler, James M, CDR
Keeley, Patrick F., Lcdr
Keeling, Robert A, CDR
Keely, Wesley M, Capt
Keen, Cecil B, Lcdr
Keenan, Shawn D., MIDN
Keenan Jr, John J., Capt
Keene, Charles R, CDR
Keeney, Timothy R E, Capt
Keep, Philip R, CDR
Keepers, B. J., CDR
Kees, William R, CDR
Keffer, Walter N, CDR
Kehoe, Donald J, Lcdr
Kehoe, Eleanor F, Capt
Kehoe, Francis P, CWO4
Keicher, Charles, CDR
Keigher, Donald J, Lt

Keigler, George A, Capt
Keihn, Leroy H, WO1
Keil, Allen W, CDR
Keiller, Stuart J, Capt
Keiran, Joseph P, CDR
Keisling, Marion B, CDR
Keister, Richard R, Capt
Keitel, Gunther P, Capt
Keith, Edward L., CDR
Keith, Phillip A., Capt
Keith, Stephen T., Radm
Keithline, Harold J., CWO4
Keitz, Daniel M., LCDR
Kelch, Gary M., ENS
Kelchner, Jonathan L., Lcdr
Keleher, Peter D, Capt
Kelker, John S, Lcdr
Kellard, Michael G., CAPT
Kelleher, Arthur J., Capt
Kelleher, Marie B, Capt
Kelleher Jr., Kenneth S., CAPT
Keller, Donald G, Capt
Keller, Joyce P., Cdr
Keller, Paul D, CDR
Keller, Thomas M, CDR
Keller, Virginia N., CDR
Keller Jr, Arthur V, Lcdr
Keller Jr, Charles H, CDR
Keller Jr, William R, Capt
Kellerhals, Thomas M, Capt
Kellerman, George, Lcdr
Kelley, Daniel J., LT
Kelley, Eugene E, CDR
Kelley, H. C., CDR
Kelley, Jack D., CAPT
Kelley, James H, CWO4
Kelley, Jeffrey B., LCDR
Kelley, Jerry R., CAPT
Kelley, John D., CDR
Kelley, John P., LT
Kelley, Kathleen, Capt
Kelley, Peter E., CDR
Kelley Jr, James W., CDR
Kellick, William L, Capt
Kellner, Timothy L., LCDR
Kellogg, Huston G, Capt
Kellogg, John S, Capt
Kellogg, Steven C, Capt
Kells, Keith E, CDR
Kellstrand, Robert E, CDR
Kellum Jr, Albert F, CDR
Kelly, Albert B, Lcdr
Kelly, Andy, Capt
Kelly, Daniel J, CAPT
Kelly, Edward J, CDR
Kelly, Gerald T, CDR
Kelly, James C, Capt
Kelly, Kenneth D, Lcdr
Kelly, Kevin T, CDR
Kelly, Larry T., LT
Kelly, Michael J., LT
Kelly, Richard E, Capt
Kelly, Thomas R, Lcdr
Kelly, Thomas V, CDR
Kelly, Wayne, Lt
Kelly, William C, Capt
Kelly, Wilson C, Lcdr
Kelly Jr, James Patrick, Capt
Kelly Jr, William M, CDR
Kelly Jr., William A., CDR
Kelly Sr, Stephen S., Lcdr
Kelman, Edward M, CDR
Kelner, Richard W, CDR
Kelsey, Donna G, CDR
Kelso, Albert P, Capt
Kelso, Melvin F, Capt
Keltz, Mildred M, Lcdr
Kemery Jr, Frank X, Lcdr
Kemmerer, Clayton M., LT
Kemp, Angela R., Lcdr
Kemp, Marion I., LCDR
Kempf, Cecil J, Vadm
Kempi, Greg D., ENS
Kempter, Rudolph H, Lcdr
Kendall, Alf R, CDR
Kendall, Julius, Lcdr
Kendle, Louise R., Capt
Kendra, Robert J, Capt
Kendrick, James A, CDR
Kendrick, William D, Lcdr
Kendrick Jr, Thomas F, CDR
Kendziorski, Thomas F, CDR
Kenlon, Frank D, Lcdr
Kenna Jr, Mark J., LCDR
Kennard, Deborah E, LCDR
Kennedy, Bobby J, Capt
Kennedy, Bruce A, Capt
Kennedy, Cain J, Capt
Kennedy, David N, Lcdr
Kennedy, Edwin C, CDR
Kennedy, Gordon L, CDR
Kennedy, James D, Capt
Kennedy, James R., CDR
Kennedy, John M., CDR
Kennedy, Kriss M., CDR
Kennedy, Lawrence D, Lcdr
Kennedy, Michael E., CDR
Kennedy, Peter L, CDR
Kennedy, Robert F, CDR
Kennedy, Robert H., CDR
Kennedy, Troy W, Capt
Kennedy, Victoria A., CWO2
Kennedy, William G., CAPT
Kennedy Jr, Guy H, CWO2
Kennedy Jr, James J, CDR
Kennedy Jr, John W., CDR
Kennerly, John C, CAPT
Kenney, Brian P., CDR
Kenney, Charles F., Capt
Kenney, David J., Lt
Kenney, James M, Capt
Kenney, James R, Capt
Kenney, Michael J., Lcdr
Kennicutt, Roger T., CDR
Kenniff, Robert L, Capt
Kennon, William, CDR
Kenny, Dwight Allen, LCDR
Kenny, L Allen, Capt
Kenny, Michael F, Capt
Kenny Jr, Frank J., CDR
Kenny Jr, John R., LT
Kenoff, Jay S, CDR
Kensrue, Milo M, Lcdr
Kent, Jamon H, Capt
Kent, John O, Capt
Kent, Kenneth M, CDR
Kent, Leonard J, Lcdr
Kent, Michael G, LTJG
Kenyon, A Prentice, Capt
Kenyon III, Robert E, CDR
Kenyon Jr, Walter C, Capt
Keogh, Brian P., LCDR
Keough, John J, Lcdr
Keough, Steven J, CAPT
Keown, Rhoda L, CDR
Keppel, Julie E., Lcdr
Keppel, Robert E, Capt
Kerchner Jr, Charles F., CDR
Kerekes, Michael J., CDR
Kerfoot, Branch P, Lcdr
Kermoian, Ralph, Capt
Kerning, Daniel P., LCDR
Kerns, Cameron E, CDR
Kerns, James T, Lcdr
Kerpa, Fritz E, CWO4
Kerr, Gerald N, Capt
Kerr, John E, Radm
Kerr, Lawrence E, CDR
Kerr, Leslie H, CDR
Kerr, Lynne A., LTJG
Kerr, Ronald A., LCDR
Kerr, Thomas C, Lcdr
Kerrebrook, Robert A, Capt
Kerrigan, John F, Capt
Kersh, Jack R, Lt
Kersh, Kristin R., ENS
Kerstein, Morris D, Radm
Kerwick, Richard P, CDR
Kesner, Kathi D, CDR
Kessel, Glenn R, Capt
Kessel, Jason L., CDR
Kessler, David B, Capt
Kessler, George A., CDR
Kessler, James M, Capt
Kestner, Charles P, CDR
Keswick, Curtis W, CDR
Ketchum, Daniel C, Capt
Keter, Yaron A., LT
Ketner, Bruce A, Capt
Kett, John W, Lcdr
Kettenhofen, Billy W, Capt
Keville, Franklin J, Capt
Kevlin, Steven G, Capt
Key, John M, CDR
Key, Tony H, Lcdr
Keyes, Fenton G, Ens
Keyes, Samuel W, Capt
Keyserling, Michael S, CDR
Kice, Charles J, CDR
Kidd III, Isaac C, Capt
Kidd Jr, Isaac C, Adm
Kidd Jr, James H, CAPT
Kiely, Barbara A, CAPT
Kienitz, Douglas A, CDR
Kiergan, Christopher D., Lcdr
Kierl, T Phil, CAPT
Kiess, Sally, LCDR
Kiker, William Bruce, CAPT
Kilcoyne, Kevin, CWO2
Kilday, Jeremy, ENS
Kildea, James T, CDR
Kilduff, Thomas M, Lcdr
Kiley, Joseph P, Ltjg
Kilger Jr, Peter F., CDR
Kilgore, Andrew L, CDR
Kilgore, James C, CDR
Kilgore, Sidney J, Capt
Kilhoffer, Carl D, Capt
Kilian, Walter D, Capt
Kilkenny, Brian J., Ltjg
Kilkenny, Robert T, CDR
Killea, William R., CDR
Killey, John M, CAPT
Killian, Durmot E,
Killian, Michael J, Capt
Killian III, Samuel P, CDR
Killingsworth, Joel E, CDR
Killingsworth, Steven M, LCDR
Killingsworth, Thomas A., LCDR
Killion, Steven B., ENS
Kilmurray, Robert B, CDR
Kilpatrick, James A, CDR
Kilpatrick, Joseph W., Capt
Kilpatrick, Steve K., LT
Kilroy, John Rhys, Lcdr
Kilty, John J, Lcdr
Kim, John, CDR
Kim, Ronald H.Y., CDR
Kim, Thomas H, Capt
Kim, Thomas J., LCDR
Kim, Thomas P., LCDR
Kimball, Martin A, CDR
Kimberling, Stephen M, CDR
Kimbrough, Hugh H., Lt
Kimbrough, Warren O, Capt
Kimpel, William A, Capt
Kinane, Keith W, CAPT
Kinard, Richard D, Capt
Kincannon, Thomas B, CDR
Kinczel, John Rogers, Capt
Kindelan, James J, Lcdr
Kinder, Clifford L, CDR
Kindig, David Allan, Lcdr
Kindsvatter, Anne L, CDR
Kindwall, Eric P., CDR
Kiner, Richard A, Capt
King, Alex, Lt
King, Benjamin F, Capt
King, Benny G, CAPT
King, Charles G, Lcdr
King, David W, Lt
King, Delores D, Lcdr
King, Donald L, Lt
King, Douglas W., LT
King, Gary E, Capt
King, George, CDR
King, Jeffrey D., MIDN
King, John B, Lt
King, John M, Capt
King, John S, Lcdr
King, John W., Capt
King, Kathy Ann, Lcdr
King, Larry A., Capt
King, Marvin E., CAPT
King, Michael J, CDR
King, Nelson R., CDR
King, Pat, CAPT
King, Philip Alan, CDR
King, Philip V, Radm
King, Phillip E, Capt
King, Richard J., LCDR
King, Richard W., CDR
King, Robert C, Lcdr
King, Robert J, CDR
King, Robert K, Capt
King, Sara A, Ltjg
King, Susan G., LT
King, Thomas Joseph, CWO4
King, Timothy A., CWO3
King, William A, CDR
King, William Bruce, Capt
King, William C, Lcdr
King, William C, CDR
King, William H, Lcdr
King III, Stephen J., Lt
King IV, Benjamin F., CDR
King Jr, Archibald E, Lcdr
King Jr, Charles J, Lcdr
King Jr, Franklin L, CDR
King Jr, John B, CWO4
King Jr., Arthur T, CDR
Kingen, Thomas F, Capt
Kingery, Michael E., CDR
Kingsbury, Robert L, Capt
Kingsbury, William S, CDR
Kinkade, Kenneth, CDR
Kinkin, Frederic M., CDR
Kinnett, Gary E, Capt
Kinney, Bruce H, Capt
Kinney, David L, Lcdr
Kinney, Fred H., CDR
Kinney, James R, Capt
Kinney, Michael J., LCDR
Kinney, Susan C, Lcdr
Kinports, Kevin J, CAPT
Kinsella Jr, James J., Capt
Kinsey, Patrick M., CDR
Kintzinger, Paul R, Capt
Kinzer, Charles M, Lcdr
Kinzie, Larry L, CDR
Kipp, David C, CDR
Kipp, John L., CDR
Kipple, Dwayne W., LT
Kipps, Harry J, Lcdr
Kiraly, Joseph, Lcdr
Kiraly, Ruth A, CDR
Kiraly, Susan A., LCDR
Kirby, Brendan T., LT
Kirby, H Ralph, Lcdr
Kirby, James E, Capt
Kirchberg, Daniel J, Lcdr
Kirchgessner, David E, CDR
Kirchmann, Steven R, CDR
Kirchner, Diane, Lcdr
Kirchner, Gary P., LCDR
Kirk, David Q, CAPT
Kirk, Edward, Lcdr
Kirk, Keith T., LCDR
Kirk, Mark S., LCDR
Kirk, Richard E, CWO4
Kirk, Samuel D., CDR
Kirkeeng, Alf E, Lcdr
Kirkland, Douglas I., CDR
Kirkland, Faris A., CDR
Kirkland, James R, CDR
Kirkland, Otis D, CDR
Kirkley, Mary Eliz, Capt
Kirkpatrick Jr, Donald, Capt
Kirkwood, Kenneth S, Capt
Kirman, Irving A, Lt
Kirn, Peter C, Capt
Kirshenbaum, Arnold S., CDR
Kirstein, Philip L, Capt
Kirsten, Jerald, Capt
Kirtland, John C., CDR
Kirts, Russell L., CDR
Kirwan, Katharyn G, Lcdr
Kirwan, Lawrence P., CAPT
Kisela Jr, Andrew J, Capt
Kiser, Colin L, Lcdr
Kiser, Donald L, Lcdr
Kiser, Mitchell R., CWO4
Kishel, Joan C., CAPT
Kishida, Earl K., CDR
Kislak, Jay I, Ltjg
Kislan, Thomas A, CDR
Kisor, William K, Lcdr
Kiss, Joseph T, Capt
Kissinger, George M, Capt
Kissinger Jr, Donald J, Capt
Kissling Jr, John R, CDR
Kistler, John S., Capt
Kistler, Michael R., Lt
Kitchings Jr, Atley A, Capt
Kite, Marcia, Lcdr
Kite, Robert M, CDR
Kittell, John F., CAPT
Kittle Jr., William W., LCDR
Kittler, John L, CAPT
Kitts Jr, Earle L, Capt
Klaas, John E, Capt
Klaben, Donald E, CDR
Klarman, Edward J, Capt
Klassen, Alvin H, CAPT
Klatt, Richard W., Capt
Klauber, Richard A, Capt
Klauer, Patricia A, CAPT
Klee, Carl R, Capt
Klee, Matthew C., Lt
Klehr, Charles W., LCDR
Klein, Anthony J, CDR
Klein, David S., CDR
Klein, Eugene, CDR
Klein, James P, CWO4
Klein, John C, CAPT
Klein, Joshua B., LCDR
Klein, Melvin W, Capt
Klein, Murray J, CDR
Klein, William E, Lcdr
Klein Jr, George F, CDR
Kleinbauer, Horst, Capt
Kleingartner, Elmer F, Lcdr
Kleinke, Dave A., LT
Kleinknecht, Kenneth, Lt
Kleinmaier, Guenter, CWO4
Kleinschmidt, Sharon K, CDR
Kleitsch, Edward C., CDR
Klementik, David C, Capt
Klempnauer, Richard G, CDR
Klenk, Eugene L, Capt
Klentzman, Chris A., CDR
Klepac, William H., LTJG
Kletke, Dale B, CDR
Kletz, Charles R., CWO3
Klicar, Frank M, CDR
Kliefoth, A B, CAPT
Kligis, John J, Lcdr
Kliman, Douglas H., Lt
Klimkowski, Eileen A., Lcdr
Klimkowski, Johnnie F, Capt
Kline, Nelson H, CDR
Kline, Russell L, CAPT
Kling, Frank, CDR
Klinger, Brian K., LT
Klinger, Joseph G, Lcdr
Klinges, David H, Lcdr
Klinges, Edward F, Lt
Klingler, Dan R, CDR
Klingler, Peter J, CDR
Klink, Patricia A., LCDR
Klink, Stephen Colby, Capt
Klobukowski, Jerome J, Lcdr
Klocek, Kenneth D, Capt
Klocek, Matthew W, Capt
Klock, Robert B, CDR
Kloeppel, Daniel L, Radm
Kloner, William, Capt
Klooster, Alan T., Lcdr
Kloppenburg, Robert D, Lt
Klos, James J., LCDR
Klos, Warren W, CDR
Klose, John P., CDR
Klose, Ronald M, CDR
Klosowski Jr., Bernard S., LCDR
Kloss IV, David S, Lt
Klote, Paul J, Capt
Klotz, Steven Irving, Capt
Klotz Jr, Alfred, CAPT
Klump, Robert W, Lcdr
Klun, John S, CDR
Klusza, Michael C., LCDR
Klutts, William A, Lcdr
Kmosko, Bruce C, CDR
Knape, Raymond E, Capt
Knappenberger, Margaret, CAPT
Knauer II, William W, CDR
Kneizys, Vincent C, CDR
Knepper, Eugene M, CWO3
Kniff, Stacey L., LT
Knight, Barbara A, Capt
Knight, Fred B, CDR
Knight, Henry M, CWO4
Knight, James R, Lcdr
Knight, Jeffery R., LCDR
Knight, Linda K, Lcdr
Knight, Peter L, Capt
Knight, Robert Herman, CDR
Knight, Robert Kendal, Capt
Knilans, Raymond T, CDR
Knipe, Francis G, Capt
Knipe, Norman L, CDR
Knippa Jr, Paul H, CDR
Kniskern, Kenneth L, Capt
Knisley III, Charles R., Lcdr
Knisley III, Harold L, CDR
Knobloch, William F, CDR
Knoesel, Donald E, Capt
Knoetgen, Peter M., CAPT
Knoll, Anna, Lcdr
Knollmeyer, Peter M., LCDR
Knott, Keith R, Lt
Knouse, Melford G, CDR
Knowles, Edward R, Capt
Knowles, James B, Capt
Knowles, Jonathan R, CDR
Knowles, Kent N, Capt
Knowles, Philip D., CDR
Knowlton, Chadbourne R, CDR
Knox, David J., CDR
Knox, H Donald, Capt
Knox, Jerral F, Lcdr
Knox, Sherryl L, Capt
Knudsen, Orlando S, CDR
Knudsen, Russell D, Capt
Knudson, Peter W., LCDR
Knudson, Richard T, Capt
Knudson, Theodore E, CDR
Knuth, Alan E, CDR
Knuth, Dean L., Capt
Knuth, Jeffrey, LT
Knuth, Joan K, CAPT
Knutsen, Edward W, Capt
Knutson, Donal L, CDR
Koban, George A., LCDR
Kobayashi, Daizo, LTJG
Kobet, Louis S, CDR
Kobosko, Timothy S., CDR
Kobryn, Joyce D, Lt
Koch, David H., ENS
Koch, David L, CDR
Koch, Dennis P, CAPT
Koch, Mark R, Capt
Koch, Walter P, CDR
Kochanik, Randall W., CDR
Kochanowski, Andrew P., LT
Kochey, John G, Capt
Koeber Jr, John F, CDR
Koehler, George J, Lcdr
Koehler, Henry C, Capt
Koelsch, James P, Capt
Koenig, Daniel J., Capt
Koenig, Freddie R, CAPT
Koenig, John C, Lt
Koenig, Rodney C, Capt
Koenigseder, Lauretta A, CDR
Koerner, Douglas G, CDR
Koerner, Osborne S, Capt
Kofsky, Richard J, Lcdr
Kohanek, Curtis W, Lt
Kohl, E James, Capt
Kohlweiss, Greg L., LT
Kohman, Victor E, Lcdr
Kohring, Mark W., CAPT
Kohut, John G., Capt
Kokosz, Paul R., LT
Kokron, Colette E, LCDR
Kolaczyk, Kenneth S., LCDR
Kolar, Michael J, Capt
Kolar Jr, Thomas R, Lcdr
Kolb, Roger D, Capt
Kolb Jr, Stanley D., CDR
Kolbe, Harold W, Lt
Kolchak, George M, CDR
Kole, Betty L., CDR
Kolgen, Peter J, Capt
Kolko, Richard J, Lcdr
Kollar, Albert J, CWO4
Kollath, Newell E, CDR
Kolleda, Richard E., CDR
Kolostov, Linda S., LCDR
Kolpacke, Alvin F., CDR
Kolshak, Ann M., CDR
Komara, Stephen R, LCDR
Kometas, Athas N., LCDR
Komlosy, Leslee S., LT
Kommer, Dennis D, Capt
Kompelien, Verlin B, Capt
Komraus, David K., CDR
Kondrack, Charles J, Capt
Konefes, William A., LCDR
Konetchy, Howard A, Capt
Kong, Chin Ho, LCDR
Konka, Paul V., CAPT
Konopinski, Virgil J, Lt
Konopka, Anthony P, CDR
Konsugar, Brian E., CAPT
Koopmann, Charles H, CDR
Kooy, H Peter, Capt
Kopp, James E, Capt
Kopp, Richard S., CDR
Koppel Jr, Peter J., CDR
Koppernolle, Mary N., Lcdr
Kopplin, John F, Capt
Kopstain, C Craig, Capt
Koral, Eugene A, Capt
Korb, Lawrence J, Capt
Korchowsky, George W., CDR
Koren, John P, Capt
Korn, George B, CDR
Kornblue, Edwin B, Capt
Kornblut, Alan D, CDR
Kornegay, Sidney W, Lcdr
Korschun, Sanford L, Ltjg
Korth, Bernard, LCDR
Korthals, William C, CAPT
Kortum, Robert W, Capt
Kosaveach, Bruce E, Capt
Koschak III, John Karl, CAPT
Koscielny, Justine F.G., CDR
Kosek, Ernest, Lt
Koski, Arthur E, Lcdr
Koski, Cheryl L., CAPT
Koski, David R, Capt
Koslov, John, Capt
Koss, Richard L, Lcdr
Kossovsky, Nir, CDR
Kossowan, Laurie S., CDR
Kosteczko, Joseph, ENS
Kostelnik, Mark S, CDR
Koster, William, CAPT
Kostera, Thomas A, Lcdr
Kothe, Bruce W, Lcdr
Kough, Thomas J, Lcdr
Koury, Chris R, CAPT
Koussa, Harold A, Lcdr
Kouyoumjian, Joel N., Lcdr
Kovac, John N., CDR
Kovacinski, Bolek A, CDR
Koval, Jeffrey R., CDR
Kovarik, John F., Capt
Kovch, George N, CDR
Kowal, Norbert W, CDR
Kowalczyk, Eleanor M, CDR
Kowalczyk-Beckworth, J, Capt
Kowalewski, Dennis J., Lt
Kowalick, David J., LCDR
Kowalski, Daniel J., LCDR
Kowalsky, Bradley D., Lt
Kozicz, Mark S., CDR
Koziewicz, Michael Roy, Capt
Kozischek, Albert J, Capt
Kozol, Solomon M, CDR
Kozub, Victoria M., CDR
Kracke, William F, Capt
Kraft, Edwin E, Capt
Kraft, Leonard H, CDR
Kraft, Robert J., LCDR
Krahe, Joseph A, CDR
Krajc, Joseph P., Capt
Krajewski, Walter M, Capt
Krakosky, John A., LCDR
Krakowka, James L, CDR
Krall, Francis Michael, LT
Kram, Barry W, CDR
Kramer, Duane M, Lcdr
Kramer, Joanne L., CAPT
Kramer, Kevin Bear, CAPT
Kramer, Mary L, CDR
Kramer, Ralph G, CDR
Kramer, Robert, Lt
Kramer, Robert M, CDR
Kramer, Scott, Lcdr
Kranda, William R, Capt
Kranz, Helen, Capt
Krapin, Lloyd J., CDR
Kraska, Jon C., Lt
Kratt, Theodore D, Capt
Krattiger, John T, Capt
Kratzer, Jeffrey J., CDR
Kraus, William A., LCDR
Kraus, William L, Capt
Krause, Brian, ENS
Krause, Kerry J, Capt
Krawitz, Michael F, Lcdr
Krayer, Peter H, CDR
Krayer, William J, Lcdr
Kraynak Jr, Andrew C., CWO4
Kraynak Jr, Michael, Capt
Krcha, Clifford W, CAPT
Krebs, Richard F, Capt
Krehl, Donald G, Capt
Kreidler, Frank A, CDR
Kreimier, Alan R, CDR
Kreiner Jr, Charles F, Capt
Kreiser, Alfred J, Capt
Kreitner, Richard E., Lcdr
Krejci, Emil J, CDR
Krejci, James F, Lcdr
Krejci, Robert W., ENS
Krekelberg, Robert J., LT
Krembs, Carla M., CDR
Krennerich, Penelope A, Lcdr
Krepp Jr, Martin W, Capt
Kreutzberg, Karen M., CDR
Kreutzer, John L, CDR
Kriebel Jr, James M, Capt
Krieg, E. Michael, Capt
Krienke, Nadyne, CDR
Krier, Eugene M, CDR
Krigel, Jack Z, Capt
Kriger, Milton N, CDR
Krimbill, Norman B, Capt
Krimmer, Lee Albert, LT
Kriner, John F., Lcdr
Krinsky Usms, Paul L., Radm
Krisiak, Joseph Allan, CDR
Kriss, Charles L, Lcdr
Kristiansen, Walter K, Capt
Kritselis-Gaume, A J, LCDR
Krivan, William R, Capt
Kriz, Paul J, CDR
Krizov, Karen L., Lcdr
Kroeger, Robert E., Lt
Kroell Jr, Hugh S, CDR
Kroetch, Mary A., LCDR
Krogen, Richard A, Capt
Kroll, C Douglas, CDR
Krommenhock, Jacob A, Lt
Kroncke, William W, Capt
Kronzer, Richard L, Capt
Kroon, George D, CDR
Kroon, M. Thomas, Lcdr
Kropp, Theron L, Capt
Kross, James M, CDR
Krost, Richard F, CDR
Krouch, Carl W, Capt
Krouse, Philip C, Lt
Krucke, Hans H, Capt
Krueger, Grant E., CDR
Krueger, Roger D, CDR
Krueger Jr, Paul L, CDR
Krug, Peter J., CDR
Kruger, Fred E, Capt
Kruger, Martin R., Lcdr
Kruger, Richard J, Capt
Kruger, Susan S., LT
Kruger, Wayne K., CAPT
Krumholtz, Charles R, CDR
Krumlauf, David J, CWO4
Krumm Jr., Harold T., CDR
Krup, Leonard G, Capt
Krupski, Darla L., LCDR
Kruse, Dean H, Capt
Kruse, Thomas E, Lt
Kruse, William D, CDR
Kryschtal, Jaroslaw J, Lcdr
Krysiak, Stanley F, CWO4
Kuasi-Gassaway, Ase O., LCDR
Kubiniec, Michael T., LCDR
Kuch, Vicki L., CDR
Kucharski, Michael J, Capt
Kucharz, Kenneth M, Capt
Kuchinski, John G, Capt
Kuchinski, John J, CDR
Kucinski, Henry J, Capt
Kuegler, Alexander C, Capt
Kuehl, Winston C, Capt
Kuehn, John L, Capt
Kuehnert, Robert G, CDR
Kuemper, Albert J, CDR
Kuerbitz, Brian A., LCDR
Kuester, Janet A, CDR
Kuester, William G, Capt
Kuffrey, C Eugene, Lcdr
Kufus, John R, CDR
Kuhl, Frederick W, Lcdr
Kuhlenkamp, John F, Lcdr
Kuhls, Christopher Earl, CWO4
Kuhn, F Stuart, Capt
Kuhn, Heinz W G, Lcdr
Kuhn, John G, Lt
Kuhn, John S, Capt
Kuhn, Rik D, Capt
Kulavich, Roger L., Lcdr
Kuldell, Randall B., CDR
Kulick, Francis J, CDR
Kull, Robert J, Capt
Kullberg, Wayne R, CDR
Kulp, Alan K., CDR
Kulp, Lisa T. Davis, CAPT
Kumagai, Robert H, Lcdr
Kumamoto, Steven, Capt
Kumpf, Dwight R., CAPT
Kuncik, Thomas M., CDR
Kunkel, Paul H., Lcdr
Kunst, Arthur F, Capt
Kupec, Anthony P., Lt
Kupresin, Sam H, Capt
Kuraitis, Kestutis V., CDR
Kuras, Anthony T, CDR
Kurek, Raymond H, Capt
Kurihara, Russell A., CAPT
Kurilec, Robert E, CDR
Kurowski, John M., CDR
Kurtis, James T, Capt
Kurtz, Abram G, Capt
Kurtz, Kenneth J, Capt
Kurtz, Robert, Lcdr
Kurtzke, John F, Radm
Kuruzovich, George, CDR
Kurz, Connie S., Lcdr
Kurz, Michael A., CDR
Kurz, William C, Capt
Kurzenhauser, Peter, CDR
Kusaba, Jack S, Lt
Kushner, Christopher M, Lcdr
Kuskis, Robert A., LCDR
Kussman, Robert N, CDR
Kutas, Leslie, CDR
Kutch, Fred, Capt

Kutch, Raymond M., CDR
Kuthy, Eugene W, Capt
Kuziela, Thaddeus J., LCDR
Kvidera, Patrick L, CWO4
Kyle, Alan M, Capt
Kyle, Deborah M., Lt
Kynett, Michael Smith, CDR
Kynett Jr, Gerald P, CDR
Kyser, Russell H, CDR
Kyto, David J., Lcdr
La Brecque Jr, A E, Lcdr
La France, Daniel R., Lcdr
La Mica, Arlington F., Lcdr
La Nicca, Sandra L., CDR
La Palme, A. F., CDR
Laabs Jr, Walter W, CDR
Laakso, Jeffrey W., LCDR
Labak, Stanley J., CDR
Labarge Jr, Vincent A, Lcdr
LaBella, Nancy A., CDR
Labelle, Francis F, CDR
Labeouf, Alan A, CDR
Labhard, Susan Cooper, CDR
Labonte, William W, CDR
Labouisse III, John P, Capt
Labuda, Gary L, CDR
Lacey, Howard W, Capt
Lachenmayer, Greg T., CDR
Lackey Jr, Marvin L, Capt
Lackey Jr, Vaden M, CDR
Lackie, John H., CDR
Lackowski, Donald H, Capt
Lacomb, Henry B, CWO4
Lacoste, Steven G., Lcdr
Lacour Jr, Joseph A, CDR
Lacroix, Mary R., LT
Lacy, Newton T., Lcdr
Ladas, John P, Capt
Ladd, Paul V, Capt
Ladd, Richard A, Lcdr
Laderas, Rodney T, CWO4
Ladley, Leah M., CAPT
Lady, Andrew J., Lt
Laedlein, Charles E, CDR
Laegen, Robert F, CDR
LaFleur-Gershator, Judith L., Lcdr
Lafond, Lou L., Capt
LaFontaine, Pamela S., Lt
Laforet, Eugene G, Capt
Lagdon Jr, Richard H., Lcdr
Lagergren, Theodore F., CAPT
Lagow, Atwell C, CDR
Lagrant, George R, Lt
Laguardia, Thomas L., CDR
Lahr, Charles H., Capt
Lai, Eddy William, CDR
Lai, Emily H., LT
Laible, John C, Capt
Lain, Horton, Capt
Lain, Katherine L., LCDR
Laing, Bruce C, Capt
Laing, John Wailon, Capt
Laingen, Leah F., Lt
Lainhart IV, John W, Capt
Lair, George L, CWO4
Laird, Marshall R., Lcdr
Lajiness, Michelle J., Lcdr
Lake, Miguel Angel, Lt
Lakefield, Bruce R, CDR
Lakhavani, Shankar T, Lt
Lakis, Nicholas P., CDR
Laliberte, Lauryl L., LT
Lalicker, Robert G, Capt
Lalonde, Richard C. A., LT
Lamar, Sheila L., LT
Lamar III, Robert O, Capt
Lamarco, A J, Lcdr
Lamas, Judith A, Lcdr
Lamb, Charles H, CDR
Lamb, Dennis W, CDR
Lamb, Frederic P, Lcdr
Lamb, George E, Capt
Lamb, John A., LCDR
Lamb, Theodore H, CDR
Lamb, William M, Capt
Lambden, Jerry D, Capt
Lambden, William J, Capt
Lamberson, Arvin L, Lcdr
Lambert, Gregory M J, Capt
Lambert, Jason A., LCDR
Lambert, Robert D, CWO4
Lambert Jr, Ray B, Capt
Lamberth, Gretchen D, Capt
Lambeth, John D, CDR
Lambrechts, Robert J., LT
Lambright, Korotha C., Lcdr
Lambright, Stephen K, CDR
Lambroff Jr, Stephen, LCDR
Lamermayer, Richard N, Capt
Lamesa, Salvatore F, Lcdr
Lamm, Ross D., ENS
Lamm, William D, Capt
Lammers, Charles M, Capt
Lamont, Terrence J., Lt
Lamont Jr, William J, CDR
Lampton, Richard E, CDR
Lamson, John M, Lt
Lancaster, Dale M, Capt
Lancaster, Herman L, CDR
Lancaster, Joel R, Capt
Lancaster Jr, George B, Capt
Lancaster Jr, James O, CDR
Lance, Brenda R, Capt
Lance, Jeanette A., Lcdr
Lance, Jeffrey J, LCDR
Lance, Leon L, CWO4
Lance, Theodore D, CDR
Lance Jr, Lee Roy, CDR
Lancia, Robert B., Lt
Lanctot Jr, Roy C, Capt
Landau, Joseph W, CDR
Landauer, Edwin G, CAPT
Landerholm, Robert W., CDR
Landers, David T, Capt
Landers, Gregory P, Capt
Landers, John D, Capt
Landers, Rhonda S., CDR
Landes, Aaron, Radm
Landes, Lee E, Radm
Landherr Jr, William J, Capt
Landis, David H, CDR
Landis, Scott J., CDR
Landis, Sky Terrence D, Lcdr
Landkammer, Kristen D, Capt
Lando, Robert E, CDR
Landon Jr, Lyndon H, CDR
Landrigan, Philip J., CDR
Landry, Leo R, Capt
Landry, Philip K, CWO4
Landry, Robert A, Capt
Landsman, Jeffrey M., Lcdr
Lane, C Robert, Lcdr
Lane, Charles H, Lcdr
Lane, Delford L, Ltjg
Lane, Homer P, Lt
Lane, James A, Lt
Lane, Sandra L., LCDR
Lane, William J., CDR
Lang, Gwyn E, CDR
Lang, James R, Capt
Lang, Michael D., LT
Lang, Paul R, CDR
Lang, Robert H., Capt
Lang, William A, Capt
Lang, William W, Capt
Langan, William J, Lcdr
Langdale, Emory L, CDR
Langdon, Arthur H, CDR
Lange, James B, CDR
Lange, Robert J., CDR
Lange, Walden A, Capt
Langen Jr, William A., CDR
Langenbach III, Ross K, Capt
Langenberg, William H, Radm
Langer, Ludy E, Lt
Langevin, Adele M, Lcdr
Langford, Bruce, CDR
Langford, John A, Capt
Langford, Marvin E, LT
Langland, Rodger A, Capt
Langley, Llewllyn R, Lcdr
Langley Jr, Rolland A, Capt
Langlois, Paul M, CDR
Langmaid, Ed P, Lt
Langness, Jay C, Capt
Langston, Randall A, Capt
Langwith, Dennis K, CDR
Langworthy, Robert B, Capt
Lanham, Jane E., LT
Lanier, Henrietta R, CDR
Lanier, Lou Ann, LCDR
Lankin, Eric M, LT
Lanning, Gail A, CWO4
Lansing, James C., Lt
Lant, James H., Capt
Lanterman, Beryl A, CDR
Lantz, Eric B., LCDR
Lanza, Carl F, Capt
Lapan, Donald P, CDR
Lapham, Arthur E, CDR
Lapp, Milton C, Capt
Lappan, George J., CDR
Lappi, Ronald V, Lcdr
Lara, James C., LCDR
Laracy, Martin F, CDR
Laraway, George A, CDR
Laray, William R., LCDR
Lardis, Alexander E, Capt
Large, Nelson D, Capt
Largent II, William D, CDR
Larger, Chris J, CDR
Larick, William A., CDR
Larkin, Samuel Patrick, CDR
Larkin, Terrance E, Lcdr
Laroche, Brian L., LCDR
Laroche, Harold I., CDR
Larocque Jr, Edward A, Lcdr
Larrabee, Robert K, Capt
Larrabee Jr, Robert K, Lcdr
Larrew, Duane J, CWO3
Larsen, John S, Capt
Larsen, John T, Capt
Larsen, Peter M., CDR
Larson, Arthur D, CDR
Larson, Brian T., LCDR
Larson, David L., Capt
Larson, Erwin R, Lt
Larson, Glenn A, CDR
Larson, Gregory R., LCDR
Larson, Harold B R, CDR
Larson, Herbert W, CDR
Larson, Norman L, Lt
Larson, Reuben R, Capt
Larson, Richard D, Capt
Larson, Roland H, CWO4
Larson, Thomas L, Capt
Larson Jr, E Arthur, CDR
Larue, Charles F, Lcdr
Larue, James Parker, Lt
Lasham, Gary A, CAPT
Lasher, James M, CDR
Lashley, James E, Capt
Laskey, Robert L, CDR
Laskowski, Edward J, Capt
Laskowski, Leo S, Capt
Lasseter, Carolyn J, CDR
Lasseter II, Granville D., CAPT
Lassetter, Clarence R, CDR
Latham, Elizabeth P., Lcdr
Latham, Henry J, Capt
Lathbury, Vincent T, CDR
Lathrop, Loyt L, Lcdr
Lathrop, Steven J., LT
Lathrop JAGC USNR, M L, Capt
Latimer, Charles L, CDR
Latini, David J., Lt
Latka, Robert A, CDR
LaTour III, George, ENS
Latsch, Lois F, Lcdr
Latshaw, K Michael, CDR
Lattimore Jr, B Gresh, Capt
Latus, Charles K, CDR
Lau, Bernard S K, CDR
Lau, Gayle J, CAPT
Laubach, William S, Capt
Laubscher, Priscilla J., CAPT
Laudenslager, Wilbur M, Capt
Lauder, John J, CDR
Lauer, Michael J, CDR
Lauer, William J, CDR
Laughead, Robert R, CDR
Laughlin, John R, CDR
Laughon, James H, Capt
Laughter, David W., CDR
Laulor, Richard J, Capt
Laurel, A Ernest, CDR
Laury, Frank B, CDR
Lava, Irving M, CDR
Lavenau, Arthur A, Capt
Laverty, James H, CWO4
Lavezzoli, Peter J, Lcdr
LaVigne, Barry A., CAPT
Lavoie, Mark A., LCDR
Law, Frank G, Capt
Law, Kenneth M, CAPT
Law, Stephen P, LT
Lawing, Kenneth L., CAPT
Lawless, Douglas P, CDR
Lawless, Peter N., LT
Lawless, Richard J., Lcdr
Lawman, Thomas K., CDR
Lawn, Robert M, Capt
Lawrence, David W, Capt
Lawrence, Gerald S, Capt
Lawrence, Gregory Edson, CDR
Lawrence, Harold W, CDR
Lawrence, Jack L, CDR
Lawrence, Paul B, Capt
Lawrence, Roseann F., CAPT
Lawrence, William H., CAPT
Lawrence II, Fred P, Capt
Lawrence Jr, Robert E, Capt
Lawrence Jr, Vernie C, Capt
Lawrie, Peter H, Capt
Lawrie, William T, Capt
Lawson, Donald O, CDR
Lawson, Donald R, CDR
Lawson, Jeffrey E., Capt
Lawson, Lexie W, LT
Lawson, Shirley A., CDR
Lawson, Thomas F., LT
Lawson Jr, David S, Capt
Lawton, Dale A., LCDR
Lawton, Elizabeth N., LCDR
Lawton, Richard W., Lcdr
Lawton, Thomas J, Lt
Lawver, Philip J., LCDR
Laxson, Lon M, CDR
Layman, David K., Lt
Layman Jr, Harvey, CAPT
Layman Jr, James H., Capt
Laynor Jr, William G, Capt
Layton, John R, Capt
Lazar Jr, James T, Capt
Lazarre, Julie S., CDR
Lazo, Leo M, Capt
Lazzarevich, Richard E, Capt
Lazzaro, John D., LCDR
Lea, Jonathan D, CDR
Leach, Charles H, CAPT
Leach, James R., LCDR
League, Jan J., LTJG
Leahy, John H, Capt
Leahy, Richard A, CDR
Leahy, Terence P, Capt
Leahy III, Joseph P., Lcdr
Leak, Mildred B, Capt
Leake, Eugene H, CDR
Leaman, Harold L, Capt
Leanos, Stanley, Lt
Lear, Daniel B, Capt
Learned, Charles E, Capt
Learson, Harold W, Capt
Leary, Charles C, Capt
Leary, Donald W, Capt
Leary, Doreen V., Lcdr
Leary, Robert D, CAPT
Lease, Morris D, Lcdr
Leathem, Douglas B, Capt
Leatherman, Edward H, Capt
Leatherman, Jerry F, Lcdr
Leavell, Daniel A, CDR
Leavenworth, Robert A, Capt
Leavitt, Morton, Capt
Leavitt Jr, Charles H, CDR
Leban, Frank A., LCDR
Lebar, Tony, CDR
Leberman, Donald G, CDR
Leblanc, Marvin L, LCDR
Leblanc, Sandra M, Lcdr
Leblanc, Stephen A, Lcdr
Leblang, Theodore R, Lcdr
Leboeuf, Mark F., Lcdr
Leboeuf III, Herve J, Capt
Lebold, Robert C, Capt
Lebovich, David A., LT
Lecain Jr, Lloyd G, CDR
Leckie Jr, James N, LCDR
Leckrone, Robert E, CDR
Lecocq, Rhoda P, Lt
Lecompte, Warren J, Lcdr
Ledbetter, Charles S., CDR
Ledbetter, Donald L, CDR
Ledbetter, Jack W, Capt
Ledbetter, Logan Scott, Lcdr
Lederer, Edward B., LT
Lederhaas, Joseph, Capt
Ledford, John G, Lcdr
Ledford, Kevin J, Lt
Ledin, Donald H, CDR
Lednicky, John A, Ltjg
Ledoux, Larry, Capt
Lee, Benjamin H K, Capt
Lee, Craig W., CDR
Lee, Daniel S, Lcdr
Lee, Debra S., LCDR
Lee, Edward M, CDR
Lee, James C, Capt
Lee, James H., LT
Lee, Jeffrey A., CDR
Lee, Raymond J, Capt
Lee, Robert E, Lcdr
Lee, Robert J., CDR
Lee, Sara R., CDR
Lee, Stephen R., CDR
Lee, Thomas A, Capt
Lee, William A, CDR
Lee, William E, Capt
Lee II, Earl E., CDR
Lee Jr, James A, CDR
Lee Jr., Phillip E., LT
Leech, James N, Capt
Leech, Richard C., CAPT
Leeds, Ronald P E, Lt
Leef, Carolyn D, CAPT
Leemhuis, John P, CDR
Leer, Hans J., CWO4
Lees, Frederick J, CDR
Lees, Frederick L., CDR
Lees, Jody L, Capt
Lees, Joseph F, CAPT
Lees, Robert W, CDR
Leeson Jr., Joel W, CDR
Leetz, Daniel M, Lcdr
Leever, George R, Capt
Lefevre, George F, CDR
Legard Jr, Edwin N., CDR
Legg, William E, CDR
Leggett, Gregory K, CDR
Lehman, Benjamin J, Radm
Lehman, David H., CAPT
Lehman, David K., CDR
Lehman, Douglas C, CAPT
Lehman, James J., CDR
Lehman, John F., CDR
Lehmann, James P., CDR
Lehner, James L, Lcdr
Lehre, Edward J, Capt
Lehwalder, Laurus W, Capt
Leibell III, Vincent L, CAPT
Leichtfuss, Fred H, CDR
Leigh, Carl G, Lcdr
Leigh, Michael A, CDR
Leighty, William E, Capt
Leininger, Evelyn, Capt
Leininger, G Frank, Lcdr
Leis, Robert P., Lt
Leisch, Donald G, CDR
Leiser Jr., Fred C., CAPT
Leist, Frederick O, Ltjg
Leisz, George W, Capt
Leite, Michael J, Capt
Leithold, Kenneth K, Capt
Leitz, Edward M., LCDR
Lejeune, Robert L, CDR
Leland, Edward L, Lcdr
Leland, Gerald A., Ltjg
Lell, Virgil G, Capt
Lemay Jr, J W, Capt
Lembo, Joseph, CDR
Lemezis, James, Lcdr
Lemley, Edwin P, CDR
Lemme, David R, LT
Lemmen, Mark E, Lcdr
Lemmons, Jeffrey A., CDR
Lemoine, Freman, CWO2
Lemoine, Jerry Dean, Lcdr
Lemon, James R., CDR
Lemons, Edward P., CDR
Lenahan, Linda M., CDR
Lenderink, Donald R, Lcdr
Lenfest, Harold F, Capt
Lennon, George W, Capt
Lennon, James, LCDR
Lenoir, Christine B., Lcdr
Lent, Norman F, Lt
Lenz, George F, Lt
Leo, Paul P, Capt
Leoffler, Kathleen C., CDR
Leon, Anne C, Capt
Leon, Peter F, Capt
Leonard, Arthur E, LT
Leonard, Guy J, Capt
Leonard, Janice W., CDR
Leonard, Mark H., Lt
Leonard, Michael D, Lcdr
Leonard, Michael W., CDR
Leonard, Ralph V, CDR
Leonard, Richard R., CDR
Leonard, Ronald M, Lcdr
Leonardy, Donald B, Capt
Leone, Lawrence T., CDR
Leopard, Patricia A., Lcdr
Lepage, Jerome P., LT
Leperi, Karin A., Lcdr
Lepkowski, Steven E, CDR
Lepore, John A., Lcdr
Lepore, John J, CDR
Lequeux, Malcolm F, CDR
Lerch, Russel T, Capt
Lerette, Kevin S, Lcdr
Lerner, Julian, Capt
Leslie, William H, Capt
Lesperance, Richard D., Lcdr
Less, Tony, VADM
Less Jr, Frank L, Capt
Lessard, Louis W., CAPT
Lesslie, William H, Capt
Lester, James P., CDR
Lester, John W, Capt
Lester, William J, CDR
Lester Jr, Walter C, Lcdr
Letcher, Jerry L, CDR
Leto, David D, CDR
Lett, Celia E., LT
Lett, Thomas W., CDR
Lett Jr, Royal A, CDR
Letten, James B, Lcdr
Leukhardt, Laura J., LCDR
Leukhardt, Martin W, Radm
Leuning, Ross L., CDR
Levasseur, George E, CDR
Level, Howard R., Lt
Levenson, Saul, Capt
Leverett, Gary L, Capt
Leverich, Richard W., LT
Leveson, James M, Capt
Levesque, Louis P, Lcdr
Levesque, Steven P., LT
Levi, Jeffery W., CDR
Levin-Homme, Jo-Anne T, Capt
Levine, Lowell J, Capt
Levine, Michael N, Lcdr
Levine, Robert B., CDR
Levine, Stephen M., Capt
Levinge, Claude, Capt
Levinson, Henry G, Capt
Levis, William, CDR
Levy, James M, Lcdr
Levy, Joseph H, LT
Levy, Milton L, CDR
Levy, Ted, Radm
Levy III, Edmund B, Capt
Lewandowski, Anthony, Capt
Lewandowski, Melvin A, Capt
Lewelling, Dale A., CAPT
Lewers, Freddie R, LCDR
Lewis, Cynthia A, Lcdr
Lewis, Erwin C, LCDR
Lewis, Eugene R, CDR
Lewis, Gary E., Lcdr
Lewis, George E, CAPT
Lewis, Gregory A, Lt
Lewis, James H, Capt
Lewis, John B, Capt
Lewis, John M., CDR
Lewis, John M., ENS
Lewis, Kenneth A, Capt
Lewis, Kenneth D, Lcdr
Lewis, Leon D, CDR
Lewis, Maxwell L, Capt
Lewis, Quinn E., Lt
Lewis, Richard F, Lcdr
Lewis, Robert E, Capt
Lewis, Robert H, Capt
Lewis, Ronald A, Lcdr
Lewis, Stanton C, Lt
Lewis, Stephen A., CAPT
Lewis, Thomas M, Capt
Lewis, Virgil E, CWO4
Lewis II, Benjamin E, CDR
Lewis Jr, Millard E., LCDR
Lewis Jr, Paul S, Lcdr
Lewis Jr, Wallace L, Capt
Lewko, Ken, CDR
Ley, Robert J, Lcdr
Leyson, Albert T., LTJG
Liashek Jr, Peter, CAPT
Libbey, Louis E, Lt
Libby, Frederic C, CDR
Libby, Lewis S, Capt
Libhart, Wayne P, Capt
Libnoch, Joseph A, Capt
Lichtenbarger, Luana C, Lcdr
Liddle, Boyd N, Lcdr
Lieb, Drew F, Lcdr
Lieberman, Jack P, Capt
Liebman, Marc E, CAPT
Liebman, William M, Capt
Liechty, Robert B., LCDR
Liefer, Maureen C., LCDR
Lier, Douglas W, Capt
Lierman, Roy J, Capt
Liersemann, Frank P., CDR
Liggett Jr, Hiram S, Capt
Lightstone, William H, CAPT
Liguori, Joseph A., CAPT
Liipfert Jr, Benjamin B, Capt
Lile, Ronald D., LCDR
Liles, Gary B, CDR
Liles, Janet S., CDR
Lilie, James C, Lcdr
Liljestrand, Roger P, Lcdr
Liljestrand, Walter E, Capt
Lill, Joseph, CDR
Lillard, Michael L, CDR
Lilliendahl, Alfred S, CDR
Lilljedahl, Lee Blythe, Capt
Lillquist, David B, CDR
Lilly, Daniel, ENS
Lilly, Gerald E, Capt
Lilly, Josiah K., Capt
Lilly, Michael A, Capt
Lilly, Richard B., Lt
Lilly, Thomas G, Radm
Lilly III, Sale T., CDR
Lim, Arnold S., LCDR
Limbaugh, Thomas L., Lcdr
Liming, James K., CDR
Limjoco, Uriel R, Capt
Lin, Willington, LCDR
Linard, Andrew Lee, ENS
Linch, Joan E., Lcdr
Lincoln, Walter B, CAPT
Lind, Harold E, Capt
Lind, James J, Capt
Lind, Robert R, Capt
Lind, Thomas O, CDR
Lindberg, Amy D., Lcdr
Lindblad, Karl-Albert M, LT
Lindborg, Leonard D., CDR
Lindell, John Eric, Capt
Lindell, Peter R, Capt
Lindeman, Bruce, Lcdr
Lindeman, Jon B, Capt
Lindeman, Roger D, Lcdr
Lindenauer, Richard M, Capt
Lindenmayer, Kathleen M., LCDR
Linderman, James S., CDR
Linderman Jr, Ralph D, CDR
Lindgren, J. Stanley, LCDR
Lindholm, Lori A., CDR
Lindler, Leslie S., CDR
Lindner, Orlin R, Ltjg
Lindner, Robert W, Capt
Lindquist, John R, Capt
Lindquist, Karl M., LT
Lindsay, G Hunter, Capt
Lindsay, John W, Capt
Lindsay, Robert L, CDR
Lindsay III, Joseph L, Capt
Lindsey, Abutan, Lcdr
Lindsey, Andrew E, CAPT
Lindsey, Fredrick J, Lt
Lindsey, Gary J, CDR
Lindstrom, John D, Capt
Lingerfelt, Luther H, Lt
Lingerfelt, Michael L, Capt
Lingley, Gordon S, CDR
Lininger, Murt K, Capt
Link, Wendell R, CDR
Link, William R, Capt
Linker, Edward M, Lt
Linn, Daniel A, Capt
Linn, Robin A., CDR
Linn, Roger David, CDR
Linnemeyer, Robert F, Ltjg
Linow Jr, William H, Lcdr
Linscott, Roger H, Capt
Linsenmeyer, Charles M, CDR
Linton, George H, Capt
Lintzenich, Gary M, Lcdr
Linz, Andrew M, Capt
Lion, Robert J., Lt
Lipinski, David R, Capt
Lippitt, Alvin W, CDR
Lippman, Ronald V., LCDR
Lipscher, Allen S, CDR
Lipscomb, Jack C, Capt
Lipscomb, Roland D, CDR
Lipscomb II, David, Capt
Lipsey, E. Spivey, CDR
Lipson, Merek E, Capt
Liptak, Joseph R, Lcdr
Lirio, Philip A, Lt
Liska, Mitchell C, Lcdr
Liska, R. Bruce, Capt
Lisle, Paul C, Capt
Lisowski, Christian, MIDN
Liss, Jim J., LT
Lister, Robert H, CDR
Lister, Toney J, Capt
Liston, Thomas L, Lcdr
Liszewski, Martin A, CDR
Litchfield, Rodger A, Capt
Litka, Adam L, CDR
Littell, Linda R., CDR
Litten, Robert D, CWO4
Littig, Gerald V, CDR
Littig, Ronald C., LT
Little, Bradley J, CDR
Little, Charles O, CDR
Little, Daniel K., Lcdr
Little, James S., LCDR
Little, Lothrop S, CDR
Little, Munson W, Lt
Little, Palmer O, Lt
Little, Robert H, Lcdr
Little, Steven M, CDR
Little Jr, O Morgan, Capt
Little Jr, Sherman C, Lcdr
Littlejohn, Robert G., CDR
Littleton, Marguerite, CDR
Littman, C Arthur, Lcdr
Littman, David L, Lt
Littrell, James L, CAPT
Litzenberger, Ronald W., Capt
Liu, Montgomery P., Lcdr
Livengood, Rex L, CDR
Livermore, Leroy C, Lt
Livesay II, James R, CDR
Livingston, Arno R, CDR
Livingston, Gordon, CDR
Livingston, James M, CDR
Livingston, Paul J, Capt
Livingston III, Henry H, Capt
Livingston III, Robert D., CDR
Llewellyn, Sandra L., LT
Lloyd, Douglas L, CDR
Lloyd, Harold F, Lcdr
Lloyd, Marjorie A, Lcdr
Lloyd, Morgan P, Capt
Lloyd, Richard D, Capt
Lloyd, William V, Lcdr
Lloyd III, Wallace H, CDR
Lloyd Jr, John J, LT
Lloyd Jr, Wallace H, Capt
Lob, Leo C, Lt
Lobaugh, Cheryl L, CDR
Lobecker, Robert N, CDR
Lobell, Wayne T., CDR
Lobue, Thomas J, Capt
Lochead, Stuart G, Lcdr
Lockard, James C., Lcdr
Lockard, Noff S., Capt
Locke, Richard E, Capt
Locke, Richard M, Lcdr
Locke, Robert L, Capt
Lockett, Geoffrey K., Lcdr
Lockett III, Robert J, Lt
Lockhart, Charles W, Lcdr
Lockhart, Jacqueline J., CDR
Lockhart, Timothy J, CAPT
Lockie, Thomas F, Lt
Locks, Salamah, LCDR
Lockwood, George W, Capt
Lockwood, John J, Lcdr
Lockwood, William L, Capt
Lockwood Jr, William H, Capt
Loeb, Jesse B, Lt
Loeben Jr, Arthur F, CAPT
Loeblein, Robert T, Lcdr
Loeffler, Roland T, CDR
Loeffler, William P., CAPT
Loes, Philip A, Capt
Loesel, Richard O, Capt
Loftis, John F, CDR
Logan, A Scott, Radm
Logan, Cedryc A, LCDR
Logan, Edward W, Capt
Logan, Scott Brian, Lt
Logan, Thomas D., LT
Logan, William J, Radm
Logue, Thomas A., Lcdr
Lohman, J William, Capt
Lohr, Peter James, Lcdr
Lohr, Steven E., LCDR
Lohse Jr, Alfred C, Lcdr
Loigman, Harold, Capt
Lokey, Charles M, Capt
Lokken, Larry J, Lt
Lokker, Anthony P, Lt
Loktu, Alfred M, CDR
Loll, Leslie, LTJG
Lollar, Benjamin D., Lcdr
Lomacchio, Thomas D, Capt
Loman Jr., Richard F., Lcdr
Lomax, James R, CDR
Lombardi, Charles G, CDR
Lonergan Jr, Harold R, Capt
Lones, John E, CDR
Loney, Timothy J, Lcdr
Long, Arthur L, Lcdr
Long, Barry, Lcdr
Long, Byron D, Capt
Long, Charles E., Capt
Long, Clayton L, CDR
Long, George C, CDR
Long, George E, Lcdr
Long, James F, Capt

Long, James T, Capt
Long, Kermit K, CDR
Long, Louis J, Lcdr
Long, Neville S, CDR
Long, Polly Hatch, CDR
Long, Raymond A, Capt
Long, Robert E, Lcdr
Long, Walter J, Lcdr
Long Jr, Noah H, Radm
Long Jr, William A, Lcdr
Longacre, Daniel C, LCDR
Longshore III, William, CDR
Longwell, Charles R, Lt
Longworth, David W, Lcdr
Looby Jr, John A, Capt
Loockerman, William D, Capt
Lookabill, Robert L, Capt
Loomans, Bernard D.A., Lcdr
Loomer, Arthur R, CAPT
Loomis, Karl F, Capt
Loose, Ronald R, Capt
Lootens, Thomas E, Lt
Loparco, Paul R., Ltjg
Lopez, Adriel, CDR
Lopez, Bernard T, Lcdr
Lopez, Daniel, LT
Lopez, George I., LCDR
Lopez, Gordon K, CDR
Lopez, J. Adam, CDR
Lopez, Jose L., ENS
Lopez, Victor C, Capt
Lopez II, Thomas J., Lt
Lopez III, John, CDR
Lopez Jr, Gilberto J, Capt
Lopez-Cepero, Henry, Capt
Lord, Frank E, Capt
Lord, James R, CDR
Lord, Rose Marie, Lcdr
Lorentson, Adrian D., CAPT
Lorentzsen, Norman M, Lcdr
Lorenzen, Susan R., LTJG
Lorenzo, Michael, Radm
Loring, Lewis D, CDR
Loring, Morton A, CDR
Loring Jr, Caleb, Lt
Loschinkohl, Victor J., LT
Loser II, Robert W, Capt
Losi, Francis P., Lcdr
Losquadro, John L, Capt
Lostutter, John M, Lcdr
Lota, Victor J, CDR
Loth, Richard W., CDR
Lotterhos, Joseph E, Capt
Lotto, Paul A., Ltjg
Lotuaco, Gamaliel G, Capt
Lotz, J Wendell, CDR
Lotz, William Gregory, CDR
Loucks, Daniel P, Capt
Loucks, Gregory, LCDR
Louderback, Frederic C, CDR
Loughary, Blaine S, CDR
Loughlin, Patrick J, Capt
Loughran, Joseph L, Radm
Louie, Cheryl C., CDR
Louis, Richard A, Capt
Loustaunau, Paul J, Capt
Louwsma, David L., Capt
Love, Craig R., CDR
Love, Cynthia J., ENS
Love, David L, LCDR
Love, Dennis L., CDR
Love, James M, Capt
Love, John E, Radm
Love, John J, CDR
Love, Peter W, Capt
Love, Robert J, Lcdr
Love IV, Milas S., Lcdr
Love Jr, Robert W, Capt
Lovecchio, Peter H, Lcdr
Lovejoy, Douglas A, CDR
Lovejoy, Owen B, CDR
Lovelace, Mozart, Lcdr
Lovell, Rilla C, Capt
Lovell Jr, Franklin A, Capt
Lovering, James R, Capt
Lovett, John F, Lcdr
Loving, Roger G., LTJG
Loving Jr, John I, Capt
Lovisone, Robert D., Lt
Lovitt Jr, Lewis D, Lcdr
Low, Milton K., CAPT
Lowder II, James F., CDR
Lowe, B. James, Radm
Lowe, George E., Capt
Lowe, Richard E, Lcdr
Lowe, Russell K., Lcdr
Lowe III, Thomas M., Capt
Lowell, James R., CDR
Lowell, Vincent M., LT
Lowell, Walter S, CDR
Lowell II, William A, Capt
Lowenstein, Duane C., CDR
Lowery, Beverly C., LCDR
Lowes, Glenn S, Capt
Lowndes Jr, Henry H, CDR
Lowrance Jr, Eugene H, Capt
Lowry, Kurt S., ENS
Lowry, Robert C, Capt
Lowry Jr, Edward R, CDR
Lowry Jr, Harold J, CDR
Lowther, Jack E, CWO4
Lowther, William L, CDR
Loy, Franklin E, Lt
Loy, Raymond E, Ltjg
Loyola, Benito, Lcdr
Lozier, Thomas G., CAPT
Lubach, Donald W., Capt
Luban, James R, CWO4
Lubbers, Phillip R., Capt
Lubrano, Joseph A., Capt
Lucas, Bryan D, Capt
Lucas, Paul, LCDR
Lucas, Richard C., Capt
Lucas, Robert, LCDR
Lucas, W. Max, Capt
Lucas, Wilmuth C, CDR
Lucas Jr, Daniel B, Lcdr
Lucci, Pasquale R, Capt
Luce, Edgar A, Capt
Luchs, George H., CDR
Lucia, Russell J., CDR
Lucier Sr, John E, CDR
Luckam, Thomas G, Capt
Luckard, Joseph J, Capt
Luckhardt, Alexandra R., LCDR
Lucks, Steven J, CAPT
Lucky, George W, CDR
Lucy, Robert W., Lcdr
Luczak, Charles T, CDR
Ludlow, Lester A, CDR
Ludwick, Joan F., Lcdr
Ludwick Jr, Ralph W, Capt
Ludwig, Karl D, Capt
Ludwig, Leonard C., LCDR
Ludwig II, Frederick E, Capt
Ludwigsen Jr., Lawrence R., Capt
Ludwikowski, Joseph F., CAPT
Luetschwager, Bruce E., LT
Luevano, Rafael V., CDR
Luffman, Jerry W, Lcdr
Lufkin Jr, Fitz O, Capt
Lugash, Marc S, Lcdr
Lugo, Jose A., LCDR
Lui, Kevin M., LCDR
Lujan, Jason P., LT
Lujan, Oscar E., CAPT
Lukacz Jr, John, CDR
Luke, Norman J, Lcdr
Luke, Randall D, Capt
Luke, Richmond M., Lt
Lukens, Reeves A, Capt
Lukinbeal, Donald Lee, Capt
Lumetta, John P., CDR
Lumpkin III, Claude C, Capt
Lun, Ben M., MIDN
Luna, Antonio X., LCDR
Luna, Paul, E-6
Lunan, Ralph G, Capt
Lund, James L, Lcdr
Lund, John N., LCDR
Lund, Robert L, Capt
Lund, Rolf G., CDR
Lund Jr, Warner E, Capt
Lundbeck, Paul J, CDR
Lundberg, Jon C., LCDR
Lunde, Tarrol A, Lcdr
Lundeen, Lyle A, CDR
Lundell, Douglas J, Capt
Lundell, Glenn H, Lcdr
Lunden, Francis G, Capt
Lundgaard, Stanley H, Capt
Lundin, Carl D, Capt
Lundin, Thomas C., LCDR
Lundquist, Carol Linda, CDR
Lundstrom, Robert A., CAPT
Lunney, J Robert, Capt
Lunny, Thomas J, Lt
Lunt, Martin A, Lcdr
Lupo, Thomas J, Como
Lupone, Donald F, Capt
Luppino Jr., Rocco, Capt
Lupton, William James, Capt
Lush Jr, Richard M, Lcdr
Lusk, William E, CDR
Lussier, Normand V, Capt
Lutche, Michael W, CDR
Lutes, L. Dan, LCDR
Luther, John A, Lcdr
Luther, Marty G, CDR
Luthringer, John D, CDR
Lutkenhouse, John P, CWO4
Lutsch, Nicholas T, CDR
Lutton, Andrew E., LCDR
Luttrell III, Robert T, CAPT
Lutz, Larry G., Capt
Lutz Jr, Harold G, Capt
Lydic III, Frederick W, Capt
Lydon, Robert H, Lt
Lykins, Lidle L, CDR
Lyle, Deborah A., CDR
Lyle, John W, Lcdr
Lyle Jr, John D, CDR
Lyles, David G., LT
Lyles, Kevin S., CDR
Lyman, Clark E, CDR
Lyman, Frank L, Capt
Lynch, Harvey J, Capt
Lynch, James P., Lcdr
Lynch, John J, Capt
Lynch, John M., CDR
Lynch, Karen Ann, LCDR
Lynch, Maurice P, Capt
Lynch, Paul L, Capt
Lynch, Philip H, Capt
Lynch, Richard C, Capt
Lynch, Robert K, Capt
Lynch, William C, Capt
Lynch, William J., RADM
Lynch Jr, Charles, Lcdr
Lynn, Bradley D., CDR
Lynt, Richard K, Lt
Lyon, James P, CDR
Lyon, Michael E. U., Capt
Lyon, Paul S., CDR
Lyon, Stephen J., LCDR
Lyons, Alvin J., Lcdr
Lyons, Daniel K., Lcdr
Lyons, Edward A, Capt
Lyons, Frank H, CDR
Lyons, M Adelaide, Lcdr
Lyons, Mary E, Capt
Lyons, Mary Jo, CDR
Lyons, Richard B, Capt
Lyons, Robert A, Lcdr
Lyons, Thomas F, Capt
Lyons, William M, CDR
Lyons Jr, John M, CDR
Lyons Jr., Edward E., ENS
Lysko, Paul G., CDR
Lytle, Charles A, CDR
Lytle Jr, Donald L, CWO4
Maass, Peggy A., Lcdr
Mabry, Alanson H, Capt
Macario, Glenn, LT
Macbain, Richard D, CDR
Maccune, Gregory M., Lcdr
MacDonald, Dennis A, Capt
MacDonald, John F, Capt
MacDonald, Paul C., LCDR
MacDonald, Roger, CDR
Macdougall, Donald D, Lt
MacDougall, Robert W., CDR
Mace, Jack R, Lcdr
Mace, Nancy A., LCDR
Maceluch, Robert W, Capt
Macewicz, Edna May, Capt
Macgillivray, Colin A, Capt
MacGinnis, John A., CAPT
Macgregor, Maclaren H, CDR
Mach, George R, Capt
Mach, L Frank, Capt
Machina, Henry M, Lt
Machol, Katherine H, Lcdr
Macie, Joseph C, Capt
Macintyre Jr, Dual A, Capt
Maciolek, Ronald J, CDR
Mack, Bruce B., CDR
Mack, Caprice M., LCDR
Mack Jr, John A, Capt
Mackay, Melissa J., Lcdr
Mackay, Stanley R, CDR
Mackebon, Barry L., CWO3
MacKenzie, Andrea B., LTJG
MacKenzie, Bradford W., LT
Mackenzie, Curtis M.S., CDR
Mackenzie, Glenn W, Capt
Mackenzie, Nicol Ian, CDR
Mackenzie, Violet M, Lcdr
Mackey, Edward T, CDR
Mackey, Jeffrey A, Capt
Mackey, Robert G, Lt
Mackinnon, Richard M, CDR
Mackley Jr, Robert W., LCDR
Maclafferty, Paul S, Lcdr
Maclaren, Ron J., CDR
Maclear, Frank R, Lt
MacMullen, Clark, Capt
Macmullen, John H, Lcdr
Macneil Sr, John F, LT
Macnish, Stephen M, CDR
Macon, Christopher, Lcdr
Macpherson, John, Capt
Mactarnaghan, John T, Lcdr
Macvaugh, Edwin S., Lt
Macvaugh III, Horace, Radm
MacWilkinson, Nigel J., LCDR
Macwilliams, Michael L, Capt
Madary, Paul R, Lcdr
Maddalena, Michael A, Capt
Madden, Barry E, Capt
Madden, Daniel M, Lt
Madden, George T, Lcdr
Madden, John M, Lcdr
Maddock, Mickey T., LCDR
Maddock, Thomas S, Radm
Maddox, Don C, CDR
Maddy, Janet M, CDR
Madge, Norman W, Capt
Madigan, Christopher H, Capt
Madill, Wm. Robertson, LCDR
Madison, Steven J., LT
Madison, Thomas D, CDR
Madrigal, Pete, LCDR
Maduell, Louis M, Lcdr
Maduro, Henry B., CAPT
Maffeo, Steven E, CDR
Magee, Jeffrey D., LT
Magee, Marshall, CDR
Magee, Michael H., LT
Magee, Thomas H, Capt
Magee Jr, Alfred W, CDR
Mager Jr, George L, Capt
Magers, F Deane, Capt
Magette, Susan H, CDR
Maggi Jr, Charles G, Lt
Maggiore, Joseph C., Lcdr
Magill, Alfred Andrew, Capt
Magill, Sandra L, Lcdr
Magnuson, Enar V, Capt
Magnuson, Peter M., LCDR
Magruder, David B., LT
Maguire, Jean F, Capt
Maguire, Kevin, CDR
Maguire, Virgil D, CDR
Mahaffey, John F, Capt
Mahaffey, Willard A, Capt
Mahala, Terrie Ann, LCDR
Mahan, Daniel R., CDR
Mahan, Michael L., LCDR
Mahan, Richard J, Capt
Mahan, Stan R., CAPT
Mahany, Terry M, Lcdr
Maher, James J, Capt
Maher, Jay A, Capt
Maher, Lawrence M, CDR
Maher, Thomas W, CDR
Mahnken, Thomas G., LT
Mahon, Martin J., CAPT
Mahon, Richard C., CDR
Mahon, William M, Lcdr
Mahon Jr., Roy O., CAPT
Mahoney, Carol A, CDR
Mahoney, David J., CDR
Mahoney, John, CDR
Mahoney, John W, Capt
Mahoney, Leo T, CDR
Mahoney, Richard P, Capt
Mahoney Jr, Harry L, Lcdr
Maibaum, David W., CDR
Maienschein, Joseph E, Lcdr
Maier, Paul V, Capt
Maier, Richard L, CDR
Maihoff, Nancy A., CDR
Mailander, Gary D, Capt
Main, Donald J, Capt
Main, James H, Capt
Main Jr, Arthur W, Lcdr
Main Jr, James S, CDR
Mainart, Stephen J., LCDR
Maine, William L, CWO3
Mainland, Michael L., CDR
Maiorana, Mario, CDR
Maisey, William A., Capt
Maissian, Edward D, CDR
Maitland, James R., CAPT
Majewski, Kenneth E., LCDR
Majka, Stanley J, CDR
Majors, Donald R., CWO3
Majors, Mary Jo, Capt
Majors, Oliver W, Capt
Makarainen, Dennis C, Capt
Maki, Bernard A, Lcdr
Makofske, James P, CDR
Malamut, Lewis, Capt
Malarney, Jeff J., LT
Malcolmson, Fredrk C, Lcdr
Malcomson, Richard A, Lcdr
Maldonado, Frank R., ENS
Maldonado, Virginia, Lcdr
Malec, Vernon M, Capt
Maleska, Andrea C, Lcdr
Maley, Michael J., Lt
Malick, George, CDR
Malick, Jewell E., CAPT
Malicki, Richard C., CDR
Malinic, Robert J, Lt
Malinick, John D., LCDR
Malins, Chester W, Capt
Mallery, Glenn O, Lcdr
Mallin, Bruce A, Capt
Mallman, Thomas L, Capt
Mallone, Joseph P, Lcdr
Mallonee, Jack L, Lcdr
Malloy, Henry L, Lt
Malloy Jr, William F., CDR
Mally, Patrick J., Lcdr
Malmberg, Norman R, Capt
Maloian, Richard D, Lcdr
Malone, David G, CDR
Malone, Frances A, Lt
Malone, James O, CDR
Malone, John H, Capt
Malone, Kathleen E, LCDR
Malone, Richard D, Capt
Malone, William F, Lcdr
Maloney, Joel N., LT
Maloney, John J, CDR
Maloney, Timothy, CDR
Maloney, William A., Lcdr
Maloney, William E, Lcdr
Maloney Jr., Thomas C., LCDR
Maloy, David F, CDR
Maloy Jr, Edward J, Lt
Malters, Edward H, Capt
Malvin, David W, Lcdr
Malyevac, David S, Capt
Mamaril, Myrna E., Lcdr
Mamber, Judith S, CAPT
Man Jr, Robert M, Capt
Manas, Todd M., CDR
Manchura, Tony, Lt
Mancias Jr, Joe, Capt
Mancini, Shawn David, LT
Mancini Jr, John A, Lcdr
Mancino, Richard A, Capt
Mandato, Kieran G, Lcdr
Mandel, Philip N, CAPT
Mandell, Alvin, Lcdr
Mandell Jr., Samuel J., LCDR
Mandich, Joseph, Capt
Maneatis, George A, Capt
Maneatis, James A, Capt
Mangan, John L, CDR
Manganaro, William F., CAPT
Mangiapane, Joseph S, CDR
Mangimeli, James W, CDR
Mangin, Albert M, Capt
Mangino, Joseph M, CDR
Manglicmot, Romeo L, CDR
Mangum, Wallace S, CDR
Manheim, Robert, LT
Maniatis, Charlynn C., CAPT
Manion, Charles F, CDR
Maniscalco, Horris G, Lcdr
Mankey, Gregory A., CAPT
Mankus, Roman T, Capt
Manley, Clovis E, CDR
Manlove, Mark W, Capt
Mann, Burton L, Lcdr
Mann, Charles K, CDR
Mann, Gregory J, CDR
Mann, Harlan W, CDR
Mann, John B, Lcdr
Mann, Steve L, Lcdr
Mann, William P, Capt
Manning, Dennis E, Capt
Manning, James H, Ltjg
Manning, John C, Capt
Manning, John E., CDR
Manning, John J., Lt
Manning, Kenneth P, Radm
Manning, Michael P., Capt
Manning, Robert W, Capt
Manning, Timothy M, Lcdr
Mannion, Robert J, CDR
Manocheo, D Nicholas, Lt
Manolescu, Timothy A, CDR
Mansel, Wendell B., LT
Mansell, William R, Lt
Manser, Philip B., LCDR
Mansfield, George F, CDR
Mansfield, James D, Capt
Manship, James Renwick, Lcdr
Manske, Edgar J, CDR
Mantegna, Carl J, CDR
Mantel, James E, Radm
Manter, James C., Lcdr
Manthei, B Michael, CDR
Manuia, Marjorie M H, CDR
Manus Jr, Louis J, CDR
Manuzak, Hubert F, Lcdr
Manzella, Vincent J, Capt
Mapes, John L, CDR
Mapes, Kenneth W, Capt
Mapes, William M, Capt
Maple, James A, CDR
Maple, William Robert, Capt
Maplethorpe, William S, CDR
Mara, Gary N, CDR
Maran, Janet L., CDR
Marasco, David F., LT
Marberger, Martin R., LCDR
Marble, David C., CDR
Marcello, Craig M, Capt
March, Richard C, CDR
Marchant, William M, CDR
Marchese, Rodolfo, CDR
Marconyak, Louis J, Capt
Marcotte, Timothy J., CDR
Marcoux, J Paul, CDR
Marcus, William W, Lt
Marcussen, William J, Capt
Marean, Martha M., CDR
Marek, Gene A, Capt
Marett, Michael J., Lcdr
Margolin, Robert S, Capt
Margolis, Jack H, CDR
Marguth, Jerry W., Capt
Mariam, Ethel, Capt
Marich, Milutin, Capt
Marilao, Joseph M, Lcdr
Marin, Rene M., Lcdr
Marin, Richard L., CDR
Marin, Robert G., LCDR
Marine, Dewey R, Lcdr
Marinelli, Lawrence, Capt
Marino, David J., LTJG
Marion, Robert D, Capt
Maris, Alan H, CDR
Marjenhoff, August J, Capt
Mark, Charles J., CDR
Marker, Howard W., CAPT
Markert Jr., Philip J., CDR
Markey, Patrick J, CDR
Markfort, Gary Leo, CDR
Markham, Douglas G., LT
Markiewicz, Eugene C, Lt
Marking-Camuto, Richard, Lcdr
Markley, James T, Lcdr
Markloff, James D., Lcdr
Marko, Darlene S., CDR
Markov, David R., Lt
Markovich, John Joseph, CDR
Markowicz, John C, Capt
Markowitz, Stephen A, Capt
Marks, Edward D, Lcdr
Markunas, Dennis J, CDR
Markussen, Dean B, CDR
Markwell Jr, Jim F, CDR
Marland III, Lloyd W., CAPT
Marlay, Robert C, Radm
Marlin, David H., LCDR
Marlow, Edmond O, CDR
Marlowe, Frank I, Capt
Marlowe, Harold C., Capt
Marlowe, John W, CDR
Marlowe, Kevin L., LCDR
Marne, Elvin A, Capt
Maronick, Steve E., LCDR
Maroon, Samuel A., LT
Maroun, Paul A., LCDR
Marozsan, George M., CAPT
Marr, Gary G., CWO4
Marr, William R, Capt
Marra, Bruce L, CDR
Marra, James P, Capt
Marrero, Michael A., ENS
Marrese, Marylynn, CAPT
Marryott, Ronald F., RADM
Marsalek, Daniel E, Capt
Marsalis, Forrest S, Capt
Marschall, Charles S, LCDR
Marsh, Charles C., Lcdr
Marsh, Myrle Frederick, Capt
Marsh, Pamela M., CDR
Marshall, Albert P., Lcdr
Marshall, Barbara A., Lcdr
Marshall, Cheryl L., LCDR
Marshall, Eric M, Capt
Marshall, Guy S, CDR
Marshall, Jack A, Lcdr
Marshall, John S, CAPT
Marshall, Karin E., CDR
Marshall, Linn G., LCDR
Marshall, Matthew, CDR
Marshall, Nathaniel M, Lcdr
Marshall, Oren W, CDR
Marshall, Paul G., Lcdr
Marshall, Robert L, Capt
Marshall, Robert L., CAPT
Marshall, Stanmore B, Lcdr
Marshall Jr, Henry A., CDR
Marsolek, Paul N, CDR
Martel, Leon C, Capt
Martella Jr, Alex A., Capt
Martens, Kathleen, LT
Martenson, Craig S, CDR
Martin, Allen D, Lcdr
Martin, Arthur H, Capt
Martin, Bobby J, CWO4
Martin, Brad Allen, LT
Martin, Burton L, Lt
Martin, C. Allen, LT
Martin, Carlton S, Capt
Martin, Charles A, Capt
Martin, Christopher D, Lcdr
Martin, Cletus O, Capt
Martin, David P, Capt
Martin, David P, Capt
Martin, Donald W, Capt
Martin, Edgar H, Capt
Martin, Edward F, Lcdr
Martin, Edwin P, Lt
Martin, Gary L, Capt
Martin, George A, CDR
Martin, George E, Lcdr
Martin, George R, CDR
Martin, Harry J, CDR
Martin, Henry D, Capt
Martin, James F, Lt
Martin, James M, Capt
Martin, James M, Capt
Martin, John Darnall, CDR
Martin, John W, Capt
Martin, John W, Capt
Martin, Joseph C., CDR
Martin, Kimber D, CDR
Martin, Larry D, Capt
Martin, Len A., LCDR
Martin, Luke S., Lcdr
Martin, Melba A, Lcdr
Martin, Michael J, Capt
Martin, Norman F, Capt
Martin, Paul E., LCDR
Martin, Peter S, Capt
Martin, Richard G, Capt
Martin, Richard G., CDR
Martin, Richard V, Lcdr
Martin, Robert E, Capt
Martin, Robert J.T., LCDR
Martin, Robert S, Capt
Martin, Robert V, Capt
Martin, Roman G, Capt
Martin, Ronald C, Capt
Martin, Ronald R, Capt
Martin, Theodore I., Lt
Martin, Thomas J, CDR
Martin, William B, Capt
Martin, William C., CDR
Martin, William H, Capt
Martin, William M, Lcdr
Martin, William R., Lcdr
Martin, William R., CDR
Martin III, William H, Capt
Martin Jr, Alfred J, Capt
Martin Jr, Irwin R, CAPT
Martin Jr, Marshall L, CDR
Martin Jr, Paul B., CDR
Martin Sr, Charles W., CDR
Martindale, Julie F, CDR
Martinez, Armando M., CDR
Martinez, Brian M., Lcdr
Martinez, Eduardo V., Lcdr
Martinez, Gilbert H, Capt
Martinez, James Richard, CDR
Martinez, John R, Lcdr
Martinez, Jose E, Capt
Martinez, Miguel, MIDN
Martinez, Modesto T, CDR
Martinez, Nancy J, Capt
Martinez, Richard P, CDR
Martinez Jr, Jack, Capt
Martini, Joseph C, CDR
Martinka, Steven C, CAPT
Martino, Richard C, CAPT
Martinuzzi, Eric, LCDR
Martone, Douglas A, CDR
Martonosi Jr, Anthony N, LCDR
Marts, Gordon A, CDR
Martucci Jr, Joseph A, Capt
Maruszewski Jr, R F, Capt
Marvel, William C., LCDR
Marvin, Diane G., CDR
Marx, Barbara J, CDR
Marx, David C, CDR
Marx, Douglas W., CDR
Marx-Cunitz, Wendy L., LCDR
Marzullo, Raymond L, Capt
Masarachia, Sam C, CDR
Masclee, Kent I, Lt
Masdon, Steven R., CDR
Masengil, William H, Lcdr
Mashburn, Telfair J, CDR
Masica, Eugene M, Capt
Masoero, Robert E, CDR
Mason, Charles C, Capt
Mason, David O, Capt
Mason, Donald G, Lcdr
Mason, Edward H, Capt
Mason, James N, Capt
Mason, John A, Capt
Mason, Lawrence J, CDR
Mason, Leonard R, CDR
Mason, Richard L, CDR
Mason, Robert, Lcdr
Mason, Robert B., LCDR
Mason, Stephen A., CDR
Mason, Theodore T, Capt
Mason, Walter A, CDR
Mason, William F, CDR
Mason Jr, A O, CDR
Mason Jr, Marvin H, Capt
Mason Jr., George B, Capt
Massaro, John A., Lt
Massenburg, Walter B., MIDN
Massengale, Robert G, CDR
Massey, Anthony G., LCDR
Massey, Charles D., CDR
Massey III, Walter B, Lcdr
Massion, Dennis D, Capt
Masslofsky, Anthony P, CDR
Mast, Harold J, Capt
Mast, Ross W., Capt
Masten, Keevan D., LT
Masters, Gregory R, Capt
Masters, Tracy D., Lt
Masters Jr, William D., Capt
Masterson, Russell W, Capt
Masterson Jr, Robert P, Capt
Mastrocola, John R., LT
Matarazzo, Joseph D, Capt
Matasic, Charles S, CDR
Matchette, Eric E, Capt
Matera, Joan E, Lcdr
Matetich, Michael G, CAPT
Mather, David B., CAPT
Mather II, Samuel S, CDR
Matheson, Cheryl, LCDR
Mathews, Gerald K, CDR
Mathews, James M., LCDR
Mathews, William M, CDR
Mathias, Joseph M, CDR
Mathiesen, Martin L., CDR
Matia, Thomas E, Lt
Matjasko, Louis S, Capt
Matlock, James A, Capt
Matlock, Richard E, CDR
Matos, Luis E., CDR
Matregrano, Raymond J., Lcdr
Matsey, James A., CDR
Matson, Jack B, Capt
Matson, James C, Lcdr
Matson, Thomas W, CDR
Matt, Michael R, Capt
Matt Jr, Harold A, Capt
Matta, Robert M, Capt
Matta, Todd H, CDR
Matteoli, Richard L., CDR
Mattern, Allan L, Capt
Mattern, Diane K., LCDR
Mattern, Randall M, CDR
Matthew, Tyler R., Capt
Matthews, Billy C, Capt
Matthews, Carey E., LCDR
Matthews, Daphne S., LCDR
Matthews, David T., LT
Matthews, Frederic E., CDR
Matthews, John E, Lt
Matthews, Joseph D., Lcdr
Matthews, Keith B, Capt

Matthews, Robert S, Capt
Matthews, Thomas C, CDR
Matthias Jr, William, CDR
Matthiesen, Fred H, Lcdr
Mattingly, John M, CDR
Mattingly, William H, Capt
Mattison, Frank G, CDR
Mattson, Kenneth R, Capt
Maturi, Harold J, CDR
Matzie, Regis A, Capt
Mauck, Sam M, Capt
Mauck, William L, Capt
Mauck III, Charles J, Capt
Maudlin, Paul J, Lcdr
Mauer, William F, Capt
Maughan, Richard V, Capt
Maughlin, Richard K, Radm
Mauldin, P. Christopher, Lt
Maultsby, James A, Capt
Maultsby III, J. Alex, CDR
Maurer, Kathryn L., LCDR
Maurer, Roy J, Capt
Maurer, Thomas E, CDR
Maurer, Valerie A., CDR
Mauro, Lance Roger, Lcdr
Mauro, Robert D, Capt
Mauroner, Norman L, Capt
Maurovich, Thomas J, Capt
Mawn, Paul E., CAPT
Maxey, James R, CDR
Maxin, Christopher J., Lcdr
Maxwell, Byron R., Capt
Maxwell, David E.J., CDR
Maxwell, Gary L., Lcdr
Maxwell, Hugh M, Lcdr
Maxwell, John M, Capt
Maxwell, Kenneth A, CDR
Maxwell, Neil L, Capt
Maxwell, Stephen L, Capt
May, Chester A, Lt
May, Donald L, Capt
May, Donald R, CDR
May, John D, CDR
May, Kenneth C, Lcdr
May, Lisa A., LT
May, Patricia Lynn, CDR
May, Peter J., LCDR
May, Robert F, Lcdr
May, Robert R, Lt
May, Ronald P., LT
May, Thomas E., Lt
May Jr, Clifford J, CDR
Mayberry, Clarence D, CDR
Maye III, Peter R, Lcdr
Mayer, Conrad J., Capt
Mayer, Donald W, Lt
Mayer, Matthew L, CDR
Mayer, Philip E, CDR
Mayer Jr, Earl H, CDR
Mayer Jr, Joseph P, Capt
Mayers, Jean, Lcdr
Mayes, Richard F, CDR
Mayfield, Carl, Capt
Mayfield, David M, Capt
Mayfield, Richard A., LCDR
Mayfield, Vernon C, CDR
Mayhew, Richard H, Lcdr
Mayhew, Thomas P., Lt
Maymi, Carlos A., LT
Maynard, Elden C, CDR
Maynard, Gary A., CDR
Maynard, Michael J., CDR
Maynard, Theodore W, Capt
Maynor, Pamela A., CDR
Mayo, Donald S, CWO4
Mayo, George D, Capt
Mayo, John R, CWO4
Mayo, Thomas J, CDR
Mayo, William R, Lcdr
Mayo Jr, William M, Lcdr
Mays, Donald R., LCDR
Mays, John P, Lcdr
Mays Jr, Lewis V, Capt
Mazan, Walter A, Lcdr
Mazer, Stephen B, Capt
Mazurczak II, Michael, Capt
Mazza, Richard C., CDR
Mazza Jr, John A, Capt
Mazzarella, Victoria E., LCDR
Mazzeo, Michael D, Capt
Mazzeo, Russell A, CDR
Mc Carthy, William D, Capt
Mc Pherran, Archie L, Lcdr
McAdams, Stephen A., Capt
McAdoo, Robert D, CDR
McAfee, David R., Lt
McAfee, Frank M, Capt
McAleer, Irene M., CDR
McAlick, John, CDR
McAliley III, John R., LT
McAlister, Elvin W, Lt
McAllister, Albert L, CDR
McAllister, Kendall H., CWO4
McAllister, Lee F., LT
McAlpine, William F., CDR
McAnaney, Edward G, CDR
McAndrew III, Robert G, Capt
McAndrews, Mark L, CDR
McAneny, Colin C, Lcdr
McAniff, Edward J, Capt
McAninch, Robert A., Lcdr
McAnney, Harold B, Cwo
McArdle, Patrick F, CDR
McArthur, Lawrence B, Capt
McAtee, Thomas L, Capt
McAteer, Christopher E., CDR
McAuley, Dennis J, CDR
McAvey, Dorothy P., CDR
McAvoy, Mary, MIDN
McAvoy, Maureen F., CDR
McBeth, Daniel P, CDR
McBeth, Lloyd D, Capt
McBrayer, Robert L, Capt
McBride, Robert A., Lcdr
McBride, Walter H, Lcdr
McBride Jr, Frank W, CDR
McBroom, Dennis B., CDR
McBroom, Virgil R, Lcdr
McCabe, Donald F, Lcdr
McCabe, Erin P., Lt
McCabe, Paul F, CDR
McCabe, Theresa A, LCDR
McCabe II, Francis J., Lcdr
McCabe Jr, Robert L., CDR
McCabe Jr, Thomas J, CAPT
McCaffrey, James D, CWO4
McCaffrey, Stanley E, Capt
McCall, Gary A, CDR
McCall, James A, Capt
McCall, Julius C., LCDR
McCall, Kenneth D, CDR
McCalla, John H, CDR
McCalley, Royce A, Capt
McCallum Jr., R. E., Lcdr
McCamant, Robert R, CDR
McCambridge, Donald W, CDR
McCameron, James D, Capt
McCamy, John D, CDR
McCance, David M., Capt
McCandless, Edward R W, CDR
McCann, David D., Lt
McCann, David T, Capt
McCann, Donald R, Lt
McCann, Jett C, CDR
McCann, John F., Capt
McCann Jr, Charles J, Capt
McCann Jr., William E., Capt
McCanne, Robert H, CDR
McCarel, Dan C, CAPT
McCarel, Stephan C, CDR
McCarroll, Michael T., CWO3
McCartan, Daniel P, LT
McCarthey, Nancy M, CDR
McCarthy, Brian N, Radm
McCarthy, Campbell J., Lcdr
McCarthy, Helen M, Capt
McCarthy, John J, CDR
McCarthy, John J., Lcdr
McCarthy, John P, Lt
McCarthy, Joseph B., Capt
McCarthy, Kevin M, Capt
McCarthy, Lynn A., CDR
McCarthy, Susan S., LT
McCarthy, Terrence F, Capt
McCarthy, Thomas, Capt
McCarthy III, George E., LCDR
McCarthy Jr, Joseph E., CAPT
McCartin, Dennis W., Lcdr
McCartney, Stephen F., Capt
McCartney, William R, Lcdr
McCarty, Anne P., LCDR
McCarty, Charles R, Lt
McCarver, Sammy C., CDR
McCaslin, Joe C, Capt
McCauley, Andrew J., LT
McCauley, Joan E, CDR
McCausland, Agnes A, CDR
McCaw III, John H., CDR
McChesney, Lee D, CDR
McClain, Charles M, Capt
McClain, Rhonda J., LT
McClanahan, Michael W, Capt
McClatchey, William M, Lcdr
McClave, Ross R, Capt
McClean, Edward J, CDR
McClellan, David S., CDR
McClellan, Francis I., CDR
McClellan, Richard J, CDR
McClelland, Richard L, Capt
McClelland Jr, W. R., Capt
McClendon, John D, Capt
McCleskey, Charles M., CDR
McCleskey, Kevin B, Capt
McClintock, Raymond D, Lcdr
McClintock, William R, Capt
McClish, Lee C., Lcdr
McCloskey, Jhea M, CDR
McCloskey, Roy J, CDR
McCloskey Jr, John F, Capt
McClune, Pamela, CDR
McClure, Charles E, CDR
McClure, George F, CDR
McClure, John C., LT
McClure, Robert E, CDR
McClure, William D, Capt
McClure Jr, James J, Lt
McCole Jr, George C., Lcdr
McCollom, Kyle L., Lcdr
McCollum, Alfred J, Capt
McCollum, John W., CAPT
McCollum, Luke M., Lcdr
McCollum, Malcolm C., CDR
McCollum Jr, I William, CDR
McComb, Dennis N, Capt
McConaghy, Richard, Lcdr
McConnell, Larry A, Capt
McConnell, Lawrence G, Capt
McConnell, Warren H, Lcdr
McConville, John R., Lcdr
McCooey Jr., Alfred E., CDR
McCord, Martin A., Lt
McCord, Michael T., Lcdr
McCorkle, Dianna E, Lcdr
McCorkle, Richard K, Capt
McCorkle, Timothy R, Lcdr
McCormack, Edward J, Lcdr
McCormack, Kevin S., CDR
McCormack Jr, John J., CDR
McCormick, Donald G, CDR
McCormick, Edna E, Lcdr
McCormick, Leighton E, CDR
McCormick, Mary R., LCDR
McCormick, Matthew A., Capt
McCormick, Richard B, CDR
McCormick, Robert C, CDR
McCormick, Shawn M., LCDR
McCoskey, James L, Lcdr
McCoy, Alan J., LCDR
McCoy, Bruce E., LT
McCoy, Charles L, CDR
McCoy, James M, Capt
McCoy, John Harold, Capt
McCoy, Patrick J., CWO4
McCoy, Raymond H, Lt
McCoy, Roberta M., CAPT
McCoy Jr, Raymond, Lt
McCracken, Ann B., CDR
McCracken, Billie R, Lcdr
McCracken, David J, Capt
McCracken Jr, Wm D, CDR
McCrea, William E, Capt
McCreery, Frank D, CWO2
McCrillis, William G., LT
McCroan, Bruce, LT
McCrory, Richard C, CDR
McCue, Thomas P, CDR
McCue III, Edward O, CDR
McCullar, Bruce H., CAPT
McCulloch, Andrew C, Lt
McCulloch, Thomas O, Capt
McCullough, Daniel V, Lcdr
McCullough, John A, Capt
McCullough, Stuart W, Capt
McCullough III, F H, Lcdr
McCurdy, Wells B, Capt
McCuskey, William H., CDR
McCutchen, Jean M., CDR
McCutchin, Robert C., Lcdr
McDaniel, Charles C, Capt
McDaniel, David T, CDR
McDaniel, John C, CDR
McDaniel, R Pat, CDR
McDaniel, Robert Bruce, Capt
McDaniel, William L, Capt
McDaniel II, Hugh H, CDR
McDaniel Jr, Joseph W, CDR
McDaniel Jr, Samuel D, Lcdr
McDavid, Harry F., CAPT
McDavid Jr., Elijah L., CDR
McDermott, Barbara A., LCDR
McDermott, Cindy L., LCDR
McDermott, Daniel J, Capt
McDermott, John J, CDR
McDevitt, Carol J, Capt
McDivitt, John D., CDR
McDonald, Bruce M, Capt
McDonald, Charles R, Ltjg
McDonald, Craig O., CAPT
McDonald, Douglas A., LT
McDonald, Douglas H, Capt
McDonald, Frances C., Capt
McDonald, Harry C., CDR
McDonald, James C, CAPT
McDonald, James M, Capt
McDonald, James P., Lcdr
McDonald, Jerre P., CDR
McDonald, John E, CDR
McDonald, John J, Lcdr
McDonald, Oscar D, CDR
McDonald, Robert A, Capt
McDonald, Susan C., Lcdr
McDonald Jr, John B, CDR
McDonald Jr, Paul C, Lcdr
McDonnell, James J, Capt
McDonnell, Joseph L, CDR
McDonnell, Patrick J., LCDR
McDonnold, Robert O, Lt
McDonough, James, LCDR
McDonough, Mark J, CDR
McDonough, William L, Capt
McDougal, Robert H, CDR
McDougall, Thomas D., LCDR
McDougle, John D, CDR
McDougle, Sarah Lynne, ENS
McDowall, Mark T, Capt
McDowell, Mark A., Lcdr
McDowell, Robert G., LCDR
McDowell, William M, Capt
McElhenny Jr, James P, Capt
McEligot, Donald M, Capt
McElligatt, Edward, Capt
McElligott, Michael J., LCDR
McElroy, D. W., CAPT
McElroy, Michael L, Lcdr
McElwaine, Laurence P, Lcdr
McElwreath, Sally C, Capt
McEnery, Thomas A, Capt
McEvers, James E, LT
McEvoy, Charles P., Lcdr
McEwan Jr, William T, Lcdr
McEwen, Randall J, Capt
McFadden, Gunnar, MIDN
McFadden, Patricia A, CDR
McFadden, Paul M, Capt
McFadden Jr, R Lee, Capt
McFarlan Jr, Edward, Lt
McFarland, Richard L, Capt
McFedries, George W, CDR
McGaha Jr, John R, Capt
McGalliard, Bruce E, Capt
McGarrah, James M, Capt
McGarrigle, Donald R, Lcdr
McGarry, Edward R, Capt
McGarry, James V., CAPT
McGaughey, James D, Capt
McGavick, Leo J, Lt
McGavran, Frank J, Capt
McGee, David L, CDR
McGee, James M., LCDR
McGee, Kathryn B., CAPT
McGee, Michael B., CDR
McGee II, Francis P, LT
McGee III, Harry S, Capt
McGee Jr, Willie E, Capt
McGettrick, Garvin, CDR
McGhee, B L Paul, Lcdr
McGiffin III, John G, Capt
McGiffney, Gregory J., Lcdr
McGill, Charles F, LCDR
McGill, Jerry, LCDR
McGillicuddy, Thomas F, Capt
McGinley, James L., CDR
McGinn, Edward J., Capt
McGinn, Frank L, Capt
McGinn, Neil F., CDR
McGinn Jr., John P, CAPT
McGinnis, Audrey L, Lcdr
McGinnis, Michael W, CDR
McGinnis, Paul B., LT
McGinnis, Robert S, Capt
McGinnis, Terence A., Capt
McGlew, John J, CDR
McGlone, Hugh F, Lcdr
McGlynn, William F, Capt
McGoey Jr, William B, CDR
McGonagle, E Richard, CDR
McGovern, James J., LCDR
McGovern, Peter J, CDR
McGowan, John P, Capt
McGowan, Neil R, CDR
McGrath, James A., LCDR
McGrath, James J, CDR
McGrath, John A., LCDR
McGrath, Mark J., 2LT
McGrath, Martin E, Capt
McGrath, Michael F., LCDR
McGrath, Michael S, CDR
McGrath, Patrick E., CDR
McGrath, Richard S., LCDR
McGrath, Thomas B., Capt
McGrath, William H, CDR
McGraw, Doonan D, Capt
McGraw, James P., LCDR
McGraw, Jeanette L, CDR
McGraw, Michael M, Lcdr
McGree, Mark P, CAPT
McGregor, Bruce M, CDR
McGregor, Thomas P, CDR
McGrew, Edwin W, CDR
McGroarty, James E, Capt
McGuinn, Phillip, CDR
McGuire, Daniel J., Lcdr
McGuire, James, ENS
McGuire, John Charles, Lcdr
McGuire, Thomas P, Lcdr
McGuire, Vincent, Capt
McGuirk, George S, Lcdr
McGurk, William J, Capt
McGurr, Thomas J, CWO4
McGurty, William H, Capt
McGurty Jr, John A., Capt
McHale, James W., CDR
McHargue, Gary R, CDR
McHenry, Rebecca W, Lcdr
McHoes, Lewis N, Lcdr
McHugh, Donald E, Capt
McHugh, Jon W, CDR
McHugh Jr., Terence J, Capt
McInnis, John G, Lcdr
McInnis, Malcolm A, Capt
McIntosh, William J, CDR
McIntyre, George W, CDR
McIntyre, John D, Capt
McIntyre, Loren A, Capt
McIntyre, Moni, CDR
McIntyre, Richard T, Capt
McIntyre, Robert B, Lcdr
McIntyre, Stewart Laird, CAPT
McIsaac Jr, John M., Lcdr
McIver, Renwick S, Capt
McIvor, Bruce H, CDR
McKallagat, Peter B, Capt
McKaughan, Robert R, Lt
McKavitt, Margaret O., Lcdr
McKay, Angus, LT
McKay, Hunter A, Capt
McKay, Ronald V., CDR
McKay, William S., Lcdr
McKee, Joanne F., CDR
McKee, Roger C, Capt
McKeehan, Robert D., Capt
McKeen, Harold R, CDR
McKeever, Michael S, Capt
McKenna, John F, CDR
McKenna, Michael P, Capt
McKenna, Patrick H., CAPT
McKenna, Robert S., LCDR
McKenney, Peter J., LCDR
McKenzie, Alfred J., CDR
McKenzie, Michael J, CDR
McKeon, John E., LCDR
McKeon, John P, Lcdr
McKeown, Michael J, Lcdr
McKesson, Richard W, CDR
McKewon, Ray W, Capt
McKibben, Barry L., LCDR
McKillop, Marjorie J., Lcdr
McKim, Patrick R., CAPT
McKinlay, Robert T, Capt
McKinley, Samuel H, CDR
McKinley Jr, John H, Radm
McKinney, Jerome Lee, CDR
McKinney, Kendall C., CDR
McKinney, Kenneth W, Capt
McKinney, Russell R, CDR
McKinney, William L, CDR
McKinney Jr, James E, Ltjg
McKinnon, Henry J, CDR
McKinzie, Dorothy A, Capt
McLachlan, John D, Capt
McLain Jr, James A, CDR
McLaughlin, J Robert, CWO3
McLaughlin, John E., Lcdr
McLaughlin, John P., Capt
McLaughlin, Mary C., LT
McLaughlin, Matthew W, Lcdr
McLaughlin, Peter A., LT
McLaughlin, Richard F, Capt
McLaughlin, William J, Capt
McLaughlin Jr, Francis, LCDR
McLaughlin Jr, Robert H, CDR
McLean, David S, Capt
McLean, Mark N., LT
McLean, Walter Leo, Capt
McLellan, Laureston H, CDR
McLemore, Albert S., CAPT
McLendon Jr, Dana C, Capt
McLeod, John P., CDR
McLeod, Robert O, Lcdr
McLeod, Roderick D, Capt
McLeod III, Walton J, Capt
McLernon, Thomas A., CDR
McLester, Paul H., LCDR
McLoone, Hugh E., CAPT
McLoskey, Robert D, Capt
McLoughlin, Peter D, Capt
McMackin, Donald E, CDR
McMackin, Mary Anne, LT
McMahon, Edward J, CDR
McMahon, James E, Capt
McMahon, James M, Lcdr
McMahon, John M, Lcdr
McMahon, Michael O., LT
McMahon, Thomas J, Capt
McMahon, Thomas T, CDR
McMaken, Glenn J., CDR
McManaman, Kenneth C, Capt
McManus, Arthur T, Capt
McManus, Charles E, CDR
McManus, Rory A., Lcdr
McManus, William M, Capt
McManus Jr, Charles H, CDR
McMaster, Joni M., Lcdr
McMaster, Marla Jill, Capt
McMaster Jr, Spencer R, Lcdr
McMasters, Clyde V, Capt
McMasters, Ronald D, Lcdr
McMasters Jr, Donald H, Capt
McMenamin, James P, CDR
McMichael, George P, Lcdr
McMichael, J Richard, CDR
McMichael Jr, E D, Capt
McMicken, Marion W, CDR
McMikle, Steven W, CDR
McMillan, Charles M, Lcdr
McMillan, Donald J., Lcdr
McMillan, Matthew P., LT
McMillan, Richard A, CDR
McMillan, Thurmond D, LCDR
McMillan Jr, Louis K, Capt
McMinn, William E., CDR
McMonagle, Kevin P, Lcdr
McMullin, Jeffrey M., Lcdr
McMurray, Kenneth L., LCDR
McMurry, William S, Capt
McMurtray, Austin T, Lt
McNair, Jeffry L, CAPT
McNair, Martin B, Capt
McNallen, James B, Capt
McNally, Brian J., Lt
McNally, Gerald E, CDR
McNally, John F, Lcdr
McNamar, David F, Capt
McNamara, Eugene T, CDR
McNamara, Richard M., LT
McNamara, Robert H, Capt
McNamara, Rose Marie, Capt
McNamee, Davis P, Lcdr
McNary, William M, CDR
McNaughton, Drummond, Capt
McNaughton, William H, Lcdr
McNeely, Donald, Lcdr
McNeely, Richard I, Capt
McNeil, Richard J, CDR
McNeill, Michael F, Capt
McNeilly, Dennis, CDR
McNerney, William, Lcdr
McNicholas, William E, Capt
McNichols, Chris, Ltjg
McNichols, Lynn Phillip, Capt
McNulty, James A, Capt
McNulty, James A, Capt
McNulty, Richard P, CDR
McPeek Jr, Harley I, CDR
McPhee Jr, Malcolm J., CDR
McPherson, Raymond G, Capt
McPherson III, L F, Capt
McPhillips, Robin R., CDR
McPoland, Patric R., CAPT
McQuarrie, Irvine G., Capt
McQueen, Eric G, CDR
McQueen III, Ralph L., ENS
McQuerry, Wayne H, CDR
McQuillan III, Robert A., AT1
McQuinn, Jack H, Capt
McQuoid III, Frank, Lt
McRae, David G., CDR
McRae, Donald A, Capt
McRae, Donald J, Lcdr
McReaken, David W, Capt
McRoberts, J William, Capt
McSwain III, Wade H., LCDR
McSweeny, Robert H, Capt
McTague, Robert H, CDR
McTighe, Kathleen P., CAPT
McVeigh, John J, Capt
McVeigh, Rose A, CDR
McVey, Francis B, CDR
McVicar, Douglas J., CAPT
McVicar, John W, CWO4
McVicker, Dennis F., LCDR
McVicker, William, Lcdr
McWhirter, Maurice J, CDR
McWhite, Peter B, Capt
McWhorter, William C, Capt
McWilliams, Alan R, Lt
McWilliams, Samuel E, Radm
Meacham, Larry R, CAPT
Mead, Carsten J, CDR
Mead, David R, Capt
Meador, Darrell R, Lcdr
Meador, Malvin E, Capt
Meador Jr, James C, Capt
Meadowcroft, Robert A, CDR
Meadows, Brian K., LT
Meagher Jr, Maurice F, Capt
Meakins, Keith D, Lcdr
Mealey Jr, Thomas H, Capt
Meals, Kimberly A., LCDR
Meaney, Patrick J, Capt
Means, Charles L., Capt
Means, Howard L, Lcdr
Mease, Frank B, CDR
Meatte, Luke L., Lcdr
Mecca, Joseph N, CDR
Meck, Bret E, Lcdr
Mecleary, Read B, Capt
Medd, Charles H, Capt
Medd Jr, John S., CAPT
Medina, Christopher L., LCDR
Medlin, Delyle B, CDR
Meehan, Karen A, Capt
Meehan, MaryAnn, LT
Meehan, Richard T., CAPT
Meek, Gordon E, CDR
Meek, Jon R., LCDR
Meeker, Natividad F., ENS
Meeker, Theodore G, Capt
Meeker Jr, Harmon S, Capt
Meeks, Harman Taylor, Capt
Meese, Harold, Capt
Meetze, James J., Lt
Megna, Anthony J., Capt
Mehail, James J., Lt
Mehas, Andrew G, Lt
Mehldau, Alan P, Capt
Mehner, Orman Vaughn, LCDR
Mehrhoff, William R., Capt
Mehringer, Ralph G, Capt
Meier, Gregory L., LCDR
Meier, Michael, CDR
Meierdierks, John P, Lcdr
Meighan, Jacob W., Capt
Meighan, MaryAnn, CDR
Meighen, Thomas J, CDR
Meilert, John J, Capt
Meinhold, Arthur J, Capt
Meinz, Daniel W., Lcdr
Meisel, Robert H, CDR
Meiselas, Robert W, CAPT
Meisenbach, Edward W, CDR
Meister, Bradley C., LCDR
Meixner, Charles F., CDR
Mejico, Antonio, Lcdr
Meldahl, Craig D, CDR
Melinosky, Theodore F, CDR
Melis, Mairead Holly, CDR
Mellard, David H., CWO3
Mellblom, Frank V., CDR
Mellblom, Mark H., LT
Melley, Peter C., ENS
Mellin, Richard W, Lcdr
Mellish, Lucille W, Capt
Mellish, Ronald W., Lcdr
Mellish, William J, CDR
Mello, Gilbert D, CDR
Melnicoff, Mark A, CDR
Melton, Evan L, Lcdr
Meltz, Fred, Capt
Meltzer, Larry A, Lcdr
Melucci, Robert J, CDR
Melville, Thomas F, CDR
Melvin, Albert A, Capt
Melvin, Charles S, Capt
Melvin, Stephen R., ENS
Melyan, Wesley R C, Lt
Memon, Abdul H., LTJG
Menard, Richard D, LCDR
Mendel, John E., LCDR
Mendelson, Burton G, Lt
Mendez, Manuel A., Lcdr
Mendez, Marcos O, Ltjg
Mendralski, Frank T, Capt
Menefee, Curtis H P, CAPT
Menez, Martin C, Capt
Mengel, William L, CDR
Mengler, Alan, CWO4
Menker, Donald F, CDR
Menkes, Richard J, Lcdr
Mensch, John E, Capt
Menser, Daniel H, Capt
Menzuber, Mark T., CDR
Meo, Joseph Michael, CDR
Mercante, James E, CDR
Mercer, Buck, Lt
Mercer, David C, CDR
Mercer, William E, CAPT
Merchlewitz, Blaise L., Lcdr
Merckel, Charles G, Capt
Mercury, James, CWO4
Merdes, Daniel W, Capt
Meredith, Frank R, CDR
Meredith, Harlan C, Lcdr
Merenda, Peter F, Capt
Merg, Ralph G., CDR
Mergele, Edwin W, CDR
Merino, Michael R., LT
Meriwether, William P., LT
Meriwether III, Gordon, Capt
Merkel, Daniel G, Capt
Merkel, Jule F., CAPT
Merkowsky Jr, Andrew J, CDR
Merola, Lawrence J., LTJG
Merolla, Louis B, Lt
Merriam, Dwight H, Capt
Merrick, Robert G, Lcdr
Merrifield, Robert G, Capt
Merrill, Patrick H, CAPT
Merrill, Philip R, Capt
Merrill, Ray A., Lcdr
Merrill, Ronald E., Lcdr
Merrill, Thomas L., LCDR
Merrill Jr, Charles B, Capt
Merrill Jr, George D, Capt
Merriman, Frank A, CDR
Merring, Robert L, Capt
Merring III, William H., Lt
Merritt, Clarence H, Lcdr
Merritt, Ernest A, Capt
Merritt, Hardy L, Capt
Merritt, Stanley Y, CDR
Merritt, Susan J, CDR
Merritt, William E, CWO4
Merritt, William S., Lcdr
Merritt Jr, William A, Capt
Merselis, Stephen W, CDR
Merten, Mark J, Lcdr
Mertz, Wayne W, Capt
Merz, E William, Lt
Meserve, Richard O, Lcdr
Messerschmidt, John G, CDR
Messick, Charles E., LT
Messier, Jeffrey M., Lcdr
Messikomer, Edwin E, Capt
Messina, Donald M, CDR
Messner, Karl R, CDR
Metcalf, Michael D, Capt
Metcalf, Sidney M, LT
Metcalf, Stephen, CAPT
Metcalfe, James A., CAPT
Metcalfe III, Robert D, CDR
Meth, Bruce M., CAPT
Metres Jr, Philip J, Capt
Metro, Richard C., CDR
Metskas, Michael A, CAPT
Mettille, Frank C, LT
Mettke, Winston D, CDR
Metz, Hermann A, Capt
Metz, Thomas R, Capt
Metz Jr, John Grattan, CAPT
Metzger, Alan W, CDR
Metzger, John M, Capt
Metzner, Robert J, Lcdr
Meves, Tim, LT
Meyer, Arthur E., CWO4
Meyer, Charles A, CDR
Meyer, Christian J, Lcdr
Meyer, Douglas R., CDR
Meyer, Glenn W, CDR
Meyer, Harvey K, CDR

Meyer, Jack A, Capt
Meyer, James F, Capt
Meyer, Lawrence W, Capt
Meyer, Mary J, LCDR
Meyer, Melvin E., Capt
Meyer, Robert A, CDR
Meyer, Robert H, Capt
Meyer, Stephanie P., Lcdr
Meyer, Stephen D, CDR
Meyer, Stephen W., LT
Meyer, Victor A, Capt
Meyer, Walter C, CDR
Meyer, William F, Capt
Meyers, Edward W, Capt
Meyers, Gerson E, Capt
Meyers, Mae E., LCDR
Meyers, Robert H., CDR
Meylan, Charles L, Capt
Meys, Patricia J., LCDR
Mezger, Erik B, CDR
Mezzacappa, Carmine J, CWO4
Micciche, Frank V, Lcdr
Michael, Dennis J, Lcdr
Michael, Kirk B., Capt
Michael, Melinda Kaye, Lt
Michael, Peter R., Lt
Michaels, Gregory, Lt
Michaels Jr, John F, Capt
Michals, Steven L, CDR
Michalski, Ronald A, Lcdr
Michaud, Brian K, Capt
Michaud Jr, Wilfred P, Lcdr
Micheels, Scott R, CDR
Michel, Richard C, CAPT
Michels, Bruce E, Capt
Michie, John C, CWO2
Mickelberry, William C, Capt
Mickelson, Charles A, Capt
Mickelson, Hugh L, Lcdr
Mickelson, Keith H, CDR
Micken, Charles M, Capt
Mickunas, Thomas A, Lcdr
Middleton, Blake, Capt
Middleton, Caryl A, Lt
Middleton, Clyde W, Capt
Middleton, Greg K., Lt
Middour, Dean L, CDR
Mieczynski, Philip M., Lt
Miele, Joel A, Capt
Mielke, Russell N., LCDR
Miener, Clyde J, Capt
Mierzejewski Jr, C V, Lt
Miesbauer, Donald T, CDR
Miesowitz, John J, Lcdr
Mieszczanski, Paul P., CDR
Mighell, Kenneth J, CAPT
Migliazzo, Salvatore C, Lcdr
Mihail, R. Clifford, CDR
Mihailovic, Jevrem, LTJG
Mihalcik, Joseph A., Capt
Mihaley, Ronald G, CDR
Mihalik, Mark S., Lcdr
Mihan, T Richard, Lcdr
Mihelich, Jack H, Capt
Mihocik, Robert A, CDR
Mikac, Joseph S., CDR
Mikell Jr, Bernard J, Lcdr
Mikesell, Willard L, CDR
Miklas, George T, Lcdr
Mikrut, Fred R, CDR
Mikulka, Robert M, Lcdr
Mikwold, Harold R, Capt
Milam, Benjamin T, CDR
Milam, Jerry M, Lcdr
Milam, Roger A, Capt
Milan, Roberto, ENS
Milazzo, Charles F, Lcdr
Milbrath Jr, A Gordon, Capt
Milburn Jr, Robert H., CDR
Miles, John P, Capt
Miles, Lewis N, Capt
Miles, Paul A, CDR
Miles, Richard N, Capt
Miles Jr, Ernest P., Lcdr
Miles Jr, James L., Lcdr
Milford, Robert F, CDR
Milham, Russell O, CDR
Millen, Joseph A, Capt
Miller, A Scott, Capt
Miller, Allen L, Lcdr
Miller, Allen W, LCDR
Miller, Barry G, Capt
Miller, Ben W, CDR
Miller, Brett C., Capt
Miller, Brian H., Lcdr
Miller, Brian Knox, CDR
Miller, Bryan D., LCDR
Miller, Carl, Capt
Miller, Charles D, Capt
Miller, Charles E, CDR
Miller, Christine M, CDR
Miller, Clarence E, Lcdr
Miller, D. F., Capt
Miller, David L, Capt
Miller, David L, Capt
Miller, Donald Milton, CDR
Miller, Donna M, CDR
Miller, Elizabeth S, Lcdr
Miller, Floyd H, Radm
Miller, Forrest A, CDR
Miller, Forrest A, CDR
Miller, Francis F, Lcdr
Miller, Frank Richard, CAPT
Miller, G Geoffrey, Capt
Miller, George E, Lcdr
Miller, George H, Lcdr
Miller, George H, CDR
Miller, George R, Lcdr
Miller, Gerald A, CDR
Miller, Gloria, LT
Miller, Gregory A., LCDR
Miller, Grover C, Lcdr
Miller, Henry R, Lcdr
Miller, James A, CDR
Miller, James C, CWO4
Miller, James E, Radm
Miller, James L, Capt
Miller, James R, CDR
Miller, James R., LCDR
Miller, James S, Lcdr
Miller, Jay M, Capt
Miller, John J, Capt
Miller, John K, Capt
Miller, John S, Capt
Miller, John U, CDR
Miller, Joseph D, Capt
Miller, Joseph G, Capt
Miller, Judith J, Lcdr
Miller, Kathleen, Capt
Miller, Kerwin E., CDR
Miller, Kevin Stacy, LT
Miller, Lawrence M., LT
Miller, Louis L, CWO4
Miller, Malcolm A, Lcdr
Miller, Marc D., Lcdr
Miller, Mark T., CAPT
Miller, Mary H., CDR
Miller, Maynard M, Capt
Miller, Neal D, Lcdr
Miller, Norman L, Capt
Miller, R Warburton, Capt
Miller, Raymond C, Capt
Miller, Richard E, CDR
Miller, Richard H, Capt
Miller, Robert B, Capt
Miller, Robert C, CAPT
Miller, Robert L, Capt
Miller, Roger L., CWO2
Miller, Roger M, Lcdr
Miller, Roland Drew, Capt
Miller, Roy E, CAPT
Miller, Samuel D, Lt
Miller, Stephen A, CDR
Miller, Stephen J., CDR
Miller, Susan S, Capt
Miller, Theresa M., Lt
Miller, Thomas R, Capt
Miller, Timothy T., CDR
Miller, Warren J., LCDR
Miller, William J, Capt
Miller, William O, Radm
Miller, William S, CAPT
Miller II, John Howell, CDR
Miller III, Albert E, CDR
Miller Jr, Albert Ray, Capt
Miller Jr, Andrew J, Capt
Miller Jr, George B., CDR
Miller Jr, Harry E, Capt
Miller Jr, Jay Ralph, Radm
Miller Jr, Joseph J., Lt
Miller Jr, Raleigh, CAPT
Miller Jr, William D, Lcdr
Miller Jr., William, LCDR
Milliard, Roland L, Lt
Milligan, Allie W., Lcdr
Milliken, Teresa E., CDR
Milliken, William J, Capt
Millikin III, Charles W, Capt
Millikin Jr, Raymond W, Capt
Millot, Thomas S, Lt
Mills, C. Robert, CDR
Mills, Dyann M., Capt
Mills, Jim F, Lcdr
Mills, Michael T, Capt
Mills, Nile D., Capt
Mills, Robert V., CDR
Mills, Stanley, Lt
Mills, Thomas E, Lcdr
Mills, William J., LCDR
Mills III, Ladson F., CDR
Mills Jr, Stuart K, Capt
Mills Jr, William J, Radm
Millsap, Byron B., CDR
Millsaps, Martha Jean, Lcdr
Milne, Robert S, Lcdr
Milner, Belinda C., LCDR
Milner, Ernest Charles, Capt
Milner, Larry D., CDR
Milroy, Rose A, Capt
Milstead, Andrew H, CDR
Milstein, Eric B., LT
Mims Jr, James R, Capt
Mina, Virginia A., Lcdr
Minamyer, William Eric, CAPT
Minck, Gary R., CAPT
Minehart, R Eric, Lcdr
Miner, Cameron B, Lcdr
Miner, George D, Capt
Miner, Robert G, CDR
Minette, David M, Lcdr
Minette, Dennis J, Lcdr
Mingle, Clifford E, Capt
Mingo, Jerome J., LCDR
Minich, Frank A, CAPT
Mink, Charles E, CDR
Minneci, William H, Capt
Minning, Steven E., Lcdr
Minor, Craig M, Lt
Mintline, Douglas D, Lcdr
Minton, Adam W., Lcdr
Minton, Rebecca H, CDR
Minton, Ronald Lee, CAPT
Minton Jr, Worley F, Lcdr
Mintun, Michael G, CWO4
Minturn Jr, Robert B, CDR
Mintz, Alan K., LTJG
Mirabile, Joseph M., LCDR
Miralia, Lauren M, Capt
Miralles, Vincent E., LTJG
Miranda, Llewelyn N., LCDR
Miranda, Mario C, Capt
Miranne Jr, E John, Lcdr
Mire, Kenneth T., CDR
Mirkes, Melanie, ENS
Miroff, Andrew B., ENS
Mischke, Thomas, CDR
Miskill, Diana S, CDR
Miskosky, Robert E, CDR
Misner, Arthur J, Lcdr
Missik, William J, Lt
Missildine, Dale S, Capt
Misurek Jr, John T, Capt
Mitcham, John B, Capt
Mitchell, Anne E., CDR
Mitchell, Charles J., CAPT
Mitchell, Charles R, Lcdr
Mitchell, David M., CDR
Mitchell, Donald L., CDR
Mitchell, E Harold, CDR
Mitchell, Gary C, Capt
Mitchell, Gary E, CAPT
Mitchell, Gregg F, Capt
Mitchell, J Pat, Capt
Mitchell, Mark E, CDR
Mitchell, Paul M, CDR
Mitchell, Paul M, CDR
Mitchell, Richard J, CDR
Mitchell, Robert L., CAPT
Mitchell, Russell H, Capt
Mitchell, Russell J, Capt
Mitchell, Stephen A, Capt
Mitchell, Theodore C, Lt
Mitchell, Thomas C, Capt
Mitchell, William F, Lcdr
Mitchell, William H, CDR
Mitchell Jr, Hugh M, Lcdr
Mitchell Jr, John P, CDR
Mitchell Jr, Michael G, Capt
Mitchell Jr, Robert E, Capt
Mitchell Jr., Archie, CDR
Mitkowski, Wendy G., MIDN
Mitskevich, Geoffrey S., ENS
Mitsoff, Gregory L., LCDR
Mitts, Murray G, Capt
Mixer, Bruce W, Lcdr
Mize, Robert S., Capt
Mizeras, Vincent F, CDR
Mjelde, Carroll V, Capt
Moak, Donald L., CDR
Moak II, Robert G., LCDR
Mobbs, Evelyne O., CDR
Mock Jr, Kermit T, Capt
Moczarny, James J, CDR
Modell, Robert L, Capt
Modrak, George P, Capt
Moebius, Arthur P, Lt
Moeller, Jo Anne Rioli, Lcdr
Moeller, Roy P, Capt
Moens, Mary Alison, LTJG
Moffett, Sulinda, CDR
Moffett, Virgil V, CDR
Moffitt, Richard E, Capt
Moffitt, Richard L, CDR
Mogabgab, William J, CDR
Mohan Jr, Herbert, CDR
Mohning, David D, Lt
Mohr, Ronald E, CDR
Moir, Edwin L, Capt
Moir, Van E., CDR
Moisenco, Alexander S, CDR
Moitoret, Victor A, Capt
Mojica, Virginia I., LT
Molans, James A, Capt
Molder, Glenn D., Lcdr
Molenda, Edward J, CDR
Moll, Alan L, CDR
Mollere, Gail E, CDR
Mollerus, Robert J., CAPT
Molligo, Anthony J, Lcdr
Molloy, Paul M, CDR
Molochko, Anthony, CWO4
Molthop, Richard W, Lcdr
Moltzau Jr, Ralph H, CDR
Momano, Steven F, Lcdr
Monaco, Theodore J., CAPT
Monaghan, Michael J., ENS
Monahan, James E, CDR
Monahan, Lawrence K, Capt
Monahan, Martha A., LT
Monahan, Michael J., ENS
Monahan, Michael T., CDR
Monahan, William J., LTJG
Monahan Jr, Edward J, Capt
Monahan Jr, Robert E, Capt
Monahan Jr, Thomas P, Capt
Moncada, Richard L., Lcdr
Moncilovich, Milan, LT
Moncrief, Lloyd E, CDR
Moncure, John, Capt
Moneghan, John E, Capt
Monell, Gary Elwyn, Capt
Monge, Humberto L., LT
Monile, Albert J, Capt
Monk, Ronald P, Capt
Monkman, David M., Lcdr
Monks Sr, James W., Lcdr
Monroe, Alexander G, Capt
Monroe, Alphonso, LCDR
Monroe, Donald W, Capt
Monroe, Howard M, CDR
Monroe, John E, Capt
Monroe, Mary J, CDR
Monroe, Maxwell D, Capt
Monroe, Nick L., CDR
Monsell, Marshall A, Capt
Monson, Arthur C, Capt
Monsour, Edward, Capt
Montagna Jr, Anthony L, Capt
Montague, John E, CDR
Montague Jr, Edgar B, Lcdr
Montanio, Richard A., CDR
Monteau, Levi, CDR
Monteiro, Donna D, Lcdr
Monteleone, Anthony J, Capt
Montemurro, Frank, CDR
Montes, Daniel, Lt
Montgomery, David L., CDR
Montgomery, Donald G, Capt
Montgomery, Douglas W., LCDR
Montgomery, Howard G, Capt
Montgomery, John D, CAPT
Montgomery, Michael D, Lcdr
Montgomery Jr, Clarence P, Capt
Montilla, Jonathan G., LT
Montoya, Samuel, Capt
Monuro, Mark, ENS
Mooberry, Marion W, Lcdr
Moody, Daniel L, Lcdr
Moody, James R, Capt
Moomy, David H, CDR
Moon, Carl F, Capt
Moon, Mary V, CDR
Moon, Timothy D, CDR
Mooney, Frank E, Capt
Mooney, Joseph, CDR
Mooney, Kevin E, CDR
Mooney Jr, William M, Capt
Moonier Jr., James E., LCDR
Moor, Marion M, CDR
Moore, Alvy P, CDR
Moore, Andrew F., ENS
Moore, Barton A., CDR
Moore, Charles E, CDR
Moore, Christopher S., LCDR
Moore, Donald B, CDR
Moore, Donald R., Lcdr
Moore, Douglas W, Capt
Moore, E Earl, CDR
Moore, Forrest G, Capt
Moore, Fred M, CWO4
Moore, Frederick W, Lcdr
Moore, Gary L, CDR
Moore, Gary L, CDR
Moore, George M, Capt
Moore, Henry N, Capt
Moore, John H, Lcdr
Moore, Joseph P., CDR
Moore, Judith L., LT
Moore, Kathy R, CDR
Moore, Layne D., LCDR
Moore, Mark A., LCDR
Moore, Melinda S., LT
Moore, Merlyn D, Capt
Moore, Michael D, Capt
Moore, Michael Kevin, Capt
Moore, Micheal A., Lcdr
Moore, Patrice M., CDR
Moore, Ray H, CDR
Moore, Robert H, CDR
Moore, Robert W, Capt
Moore, Robert W, Lcdr
Moore, Robin H S, CDR
Moore, Roger D, Capt
Moore, Sarah E, Lt
Moore, Stephen E, LCDR
Moore, Stephen P., LCDR
Moore, Terry A, CAPT
Moore, Thomas A, CDR
Moore, Wayland W, CDR
Moore, William A, Capt
Moore, William R., CAPT
Moore, William T, CDR
Moore II, John H, CDR
Moore III, James G, Capt
Moore III, Lawrence E, Capt
Moore Jr, Charles L, CAPT
Moore Jr, Ernest V, CDR
Moore Jr, Moore, Radm
Moore Jr, Richard T, Lt
Moore Jr, Robert H, Ltjg
Moore Jr, Saul F, Capt
Moore Jr, Vernon J, Capt
Moore Jr, Walter J, Capt
Moore Jr, William A, Capt
Moorehead, Josef D, Capt
Moorer, Thomas H, Adm
Moorhead, Robert G C, CDR
Moorman, George R, Capt
Moorman, Mickey C, CDR
Moorman Jr, William L, Capt
Moose Jr, James M, Capt
Moquin, William N., Lcdr
Morales, Maurice, LT
Moran, Basilio, Lt
Moran, Bryan P, Lcdr
Moran, Douglas R, Capt
Moran, James A., Lt
Moran, John P., LT
Moran, John W, CDR
Moran, Michael D., Lcdr
Moran, Richard, Capt
Moran, Robert P, CDR
Moran, Thomas M, Lcdr
Moran, William T, Capt
Morant, Victor A, Capt
Moravec, Daniel J, LCDR
Morehead, John A, Capt
Moreland, James R., CDR
Moreland, Maynard W, Lcdr
Morell, Allen N, Capt
Morency, Donald C, Capt
Moreno, Cathy Ann, CDR
Moret, Douglas H., CDR
Moretti, Joseph G, Lt
Moreus, Richard W, Capt
Morey, Benjamin W, Capt
Morey, James H, Lcdr
Morey Jr, William C, CDR
Morgan, Arthur F, Capt
Morgan, Barbara P, CDR
Morgan, Charles D, CAPT
Morgan, Charles V, CDR
Morgan, Christopher L, Capt
Morgan, David C, Capt
Morgan, David R, Lcdr
Morgan, Donald A, Capt
Morgan, Donald T, CDR
Morgan, Edward A, Capt
Morgan, Elizabeth T, Capt
Morgan, Floyd T, CDR
Morgan, Hilary S, CDR
Morgan, J Kirk, Capt
Morgan, James B., LT
Morgan, James L, Capt
Morgan, Kathleen M., Lt
Morgan, Kenneth J, Capt
Morgan, Lesley C, CDR
Morgan, Michael D, Capt
Morgan, Patricia J., CWO2
Morgan, Richard A., CDR
Morgan, Richard S., Lcdr
Morgan, Ronald R, Radm
Morgan, Russell E, Lcdr
Morgan, Steven R., Radm
Morgan, Thomas K., CDR
Morgan, Thomas L, Lt
Morgan, Tony F., LCDR
Morgan, Uley T, CDR
Morgan, William J, LT
Morgan Jr, Frank J, Capt
Morgan Jr, James F., Capt
Morgan Jr, James W, Capt
Morganelli, Patrick D., CDR
Mori, David K., CDR
Moriarty, Christopher P, Capt
Moriarty, John T, Lcdr
Moriarty, Thomas H, Capt
Moriarty, Vincent C., CDR
Morien, John F, CDR
Morin, Paul E, CDR
Morin, Roland L,
Morissette, Garry G, CDR
Morissette, Robert G., CDR
Moritz, Robert F, Capt
Morledge, George A, CDR
Morley, Helen F, Lcdr
Moroney, Patrick M., LT
Morr, Arthur J, CDR
Morrell, James M, Capt
Morrell, Verena Hart, Lcdr
Morrill, Cheri C., Lcdr
Morrill, Philip J, Capt
Morrill, William C., CDR
Morrione, James W., CDR
Morris, Dennis P., LCDR
Morris, Drewry H, Capt
Morris, Edward N, CDR
Morris, Elizabeth M, Capt
Morris, Francis M., CDR
Morris, Glenn P., CDR
Morris, John F, Capt
Morris, John G, Lcdr
Morris, John J, Capt
Morris, John L., Lt
Morris, John Lloyd, CAPT
Morris, Linda L., Lcdr
Morris, Mark F., CDR
Morris, Michael C, CDR
Morris, Robert H, Capt
Morris, Thomas E, CDR
Morris Jr, Henry E, Lcdr
Morris Jr, William L, Capt
Morris Sr, Jeremy G, Capt
Morrison, Charles D, CAPT
Morrison, Christine I., Lt
Morrison, Donald C., CDR
Morrison, Evelyn M, Capt
Morrison, Hugh E, Capt
Morrison, J. Hagood, CDR
Morrison, John K, CDR
Morrison, Michael K, CDR
Morrison, Robert L, Capt
Morrison, Ronald W, CAPT
Morrison, W. Lane, CAPT
Morrison, William R, CWO4
Morrison III, William J, Lcdr
Morriss, George A, CDR
Morrissey, Michael F., LCDR
Morrissey, Stephen J, Lcdr
Morrow, Bennett A, CAPT
Morrow, George E, Capt
Morrow, James P, Lcdr
Morrow, Michael S., MIDN
Morrow, Richard H, CWO4
Morrow, Sharon R., LT
Morrow, Stephen G, Capt
Morse, David A., CAPT
Morse, Gary A, Capt
Morse, Jack H, CDR
Morse, Martin A., CDR
Morse, Russell P, Capt
Morse, Stanley C, Capt
Morse III, Joshua M, Capt
Mort, Linda L, Lcdr
Mortensen, David R, CDR
Mortensen, Lloyd P, CDR
Mortimer, Louis R, Capt
Morton, Barry V, CAPT
Morton, Calvin T, Capt
Morton, Catherine J, CDR
Morton III, Anselm H, CDR
Morton Jr, Thomas G, Lcdr
Morton Jr, W Wallace, CDR
Mortsolf, Larry A, Capt
Mosbacher, Michael H., Lcdr
Mosby Sr, Ronald F., Lcdr
Moscardini, James J, CDR
Moscariello, Savino N., CDR
Moschella, Christopher, Capt
Mosebach, Allen F, Lcdr
Moseley, Donald L, CDR
Moser, June M., LCDR
Moser, Kerrie S, Capt
Moses, Kenneth V, CDR
Moses, Sandra Ganus, CDR
Mosher, Richard G, CWO4
Moshier, Margaret T, CDR
Mosier, Russell B, Capt
Moskowitz, Michael, Lcdr
Mosley, Mike H., LCDR
Moss, Mel D, Capt
Moss, Robert B, Capt
Moss, Todd M., LCDR
Moss Jr, George M, Capt
Mossberg, Carl L., CDR
Mosteller III, James W, Capt
Motl, Gerald P, Capt
Motley Jr., Charles E., CDR
Mott, Ann M., Lcdr
Mott, Robert A., ENS
Motter, Jeffry C., CDR
Mottet, Elizabeth A, Capt
Mottet, Mary K, CDR
Motz, Scott W., CDR
Moulin, Gerald R, Lcdr
Moulketis, James C., Capt
Moulton, Carl T., LT
Moulton, David W, Capt
Moulton, Jeffrey S., LT
Moultrie-Lizana, Angelyn K., LCDR
Mounsdon, Thomas A., CDR
Mount, Howard A., Lcdr
Mounts, William E, CDR
Mourad, Amin, LT
Mouritsen, T Edgar, CAPT
Moutos, Minelaos, Capt
Moxon, Timothy S., CDR
Moy, Evelyn L, LCDR
Moyer, Harold E, Lt
Moynahan, Dennis M, Capt
Moynihan Jr, Daniel J, Capt
Moynihan Jr, Garrett J., LCDR
Mrgudich, John, Lcdr
Mudge, Daniel D., Lcdr
Mudrock, John, Lcdr
Muehlbach Jr, Donald S, Lcdr
Mueller, Donald C, Capt
Mueller, John P, CDR
Mueller, Louis P., LCDR
Mueller, Ron, Capt
Muenkel, David R., CDR
Mues, Christopher J., LCDR
Muhitch, Joseph M., Lcdr
Muhm, William J, LT
Muir, Robert J, CDR
Muir, Roger W, Capt
Mulcahy, Bridget A, CDR
Mulcahy, Patrick R., LCDR
Mulcahy, Roger J., Capt
Mulhern, Edward M, Capt
Mulhern, John J, Capt
Mulieri, C. Chris, LT
Mulkeen, Martin J., Capt
Mull Jr, Robert S., Capt
Mullaney, James F, CDR
Mullaney, Michael John, Capt
Mullarkey, Christopher J., Lcdr
Mullen, James M, CDR
Mullen, John T., LCDR
Mullen, Merritt D, CDR
Mullen, Michael, LT
Mullen, Peter L, Capt
Mullen-Green, Mary J., LCDR
Mullenhour, Donald E, CWO4
Muller, Donna Sue, CAPT
Muller, George W, Lcdr
Muller, Robert J, CDR
Muller Jr, Richard F, Capt
Mullett, John F, CDR
Mulligan, John J, CDR
Mulligan, John R, Capt
Mullin, John G, CDR
Mullin Jr, William E, CDR
Mullins, Friel, Capt
Mullins, Michael L., LT
Mullins Jr., Alden F, Capt
Mullis, Dana D., CDR
Mulqueeney, John J, CDR
Mulquin, James J, Lt
Mulski, John H, Capt
Mulvaney, Robert M, CDR
Mulvany, George M, Capt
Mulvey, David R, CDR
Mulvey, Kevin J., Lcdr
Mulvihill, Paul, Capt
Muma, Jack W, CDR
Mumaw, John J, Radm
Mumpower, Kevin N., LCDR
Mundell Jr, John A., Capt
Mundt, Michael R, CDR
Munger, Bruce E, CDR
Munn, John A, Capt
Munoz, Edwardo T., LCDR
Munro, Donna L, Capt
Munro, Edna E., Lcdr
Munsell, Rick W., Capt
Munson, George J, CDR
Munson, Mark A, CAPT
Munsterman, James I., CAPT
Muoth, Mary L., CDR
Mur, Raphael, Capt
Muraczewski, Matthew, ENS
Murdock, John S., CAPT
Murdock, Robert V, Capt
Murin, Thomas, Lt
Muro, Robert D., CDR
Murphey, Cassius L, Lcdr
Murphey, Clifford W, CDR
Murphey, Kathryne O, CDR
Murphree, Edwin K, Capt
Murphy, Arthur R, CDR
Murphy, Christopher L, Lcdr
Murphy, Christopher P., LCDR
Murphy, David B, Lcdr
Murphy, Denise E, CDR
Murphy, Elaine D, Lcdr
Murphy, George A., Lcdr
Murphy, Harry B, CDR
Murphy, Hugh R, Capt
Murphy, James L, CDR
Murphy, James T, Lcdr
Murphy, Jo Anne, LCDR
Murphy, John F, Capt
Murphy, John K, Lt
Murphy, John L, CDR
Murphy, John R, CDR
Murphy, Joseph M., CDR
Murphy, Kenneth J, CDR
Murphy, Lawrence J, Capt
Murphy, Mark A., LT
Murphy, Mary Margaret, CDR
Murphy, Michael D, CDR
Murphy, Michael J., LCDR
Murphy, Michael O, LCDR
Murphy, Ned D, Lcdr
Murphy, Patrick J, CDR
Murphy, Patrick J., Lcdr
Murphy, Richard M., CDR
Murphy, Richard T, Capt
Murphy, Robert B, Lcdr
Murphy, Robert L, CDR
Murphy, Robert O, CDR
Murphy, Robin S., LTJG
Murphy, Stephen J, Lcdr
Murphy, Stephen P, Capt
Murphy, Thomas F, Capt
Murphy, William A, CDR
Murphy, William A, CDR
Murphy III, Edmund D., Capt
Murphy III, James J, CDR
Murphy Jr, Alexander J, Lcdr
Murphy Jr, Benjamin F, CDR
Murphy Jr, Joseph P, LCDR
Murphy Jr, Joseph P., CWO4
Murphy Jr., P. Lawrence, CDR
Murray, Douglas P, CDR
Murray, James W, Capt
Murray, John J, Lt
Murray, Kenneth D., LCDR
Murray, Laurel A, CDR
Murray, Michael A, Capt
Murray, Michael J, CDR
Murray, Michael R., Lcdr
Murray, Richard J, Capt
Murray, Robert C, Capt
Murray Jr, Anthony H, Radm
Murray Jr, Thomas H, Capt
Muscher, Arthur R., CDR
Musgrove, Robert W., LT
Musick, James E, CDR

Musselman, D Eugene, CDR
Musselman, Warren E, Capt
Musser, James R, Capt
Musser, Steven J., Lcdr
Musser Jr., John R., Capt
Mussler, Hans K, Capt
Musso, Carrie L, Lcdr
Mussoni, Frank L, Lt
Musto, Pasquale A, CDR
Muszynski, Larry, Lcdr
Mutton, James O, Capt
Mutty, Lawrence B, Capt
Myatt, Kenneth E, Radm
Myers, Allen L, Lcdr
Myers, Dale P, Capt
Myers, Frank H., LCDR
Myers, Gayle E, Lcdr
Myers, James C, CAPT
Myers, Jimmy L., Lcdr
Myers, John A, CDR
Myers, John E., CDR
Myers, Joseph, LT
Myers, Joseph H, Capt
Myers, Martine R., CDR
Myers, Michael Fraser,
Myers, Michael J., LCDR
Myers, Peter S, Lcdr
Myers, Richard A, Capt
Myers, Richard A, Capt
Myers, Robert H, CDR
Myers, Ronald G, Capt
Myers, Steven G., CDR
Myers, Thomas F, Capt
Myers Jr, Edwin L, Capt
Myers Jr, Randolph, Capt
Myhre, Charles A, Lcdr
Myhre, John J, CDR
Myhre Jr, John S, Capt
Mykityshyn, Mark G., CDR
Myles, Wendell K, Capt
Myllenbeck, Richard W., LCDR
Mynard, C Randall, Capt
Mynderse, Lance A., CDR
Myrick, David G, Capt
Myrick, James S, CDR
Mytinger, Helen A, Capt
Naab, Earl F, Capt
Naas, David D, Capt
Nabors Jr, Thomas D, Capt
Nacion, Karla J, CDR
Nadarajah, Vish, Ens
Nadeau, Roger, CDR
Nadoolman, Milton, Lt
Nagel, Norman J, Capt
Nagelin Jr, Thomas F, CAPT
Naggatz, Jeffrey D, CDR
Nagle, Willard E, Lcdr
Nagle III, Fred S, Capt
Nagy, Bettye G, Capt
Nahr Jr, Ernest V, CAPT
Naill, Timothy P., LT
Nakahata, Duane T., CDR
Nakamura, Donald F, CDR
Nakashima, Kenneth H., LT
Nakasone, Harriet H, Capt
Nanartowich, Edward J., CAPT
Nancarrow, Clifford A., Capt
Nance, Granval J, LCDR
Nankervis, John T, CDR
Nankervis Jr, John T., Lcdr
Nannini, Albert A, Capt
Nanny, Jackie D, CAPT
Nants Jr, J Stanley, Capt
Nardella, Anthony L., CDR
Nardi, Francesco P, Capt
Narloch, Joseph A., LCDR
Nash, Bernard E, Radm
Nash, Clarice A., CAPT
Nash, Jennifer S., CDR
Nash, John F, CAPT
Nash, Madeline A., LCDR
Nash, Max E, LCDR
Nash, Spencer J, CDR
Nash Jr., Charles R., ENS
Naslund, Charles W, LCDR
Nason III, Merton D, Capt
Nasson, Scott L., LCDR
Nastasi, John C, Lt
Nastos, Alvin E, Lcdr
Nathanson, Andrew J, CDR
Nathanson, Irwin D, Capt
Nathe, Al H, CDR
Nation, James T, CDR
Nation, Margaret J, Capt
Natter, James A., Capt
Natter, John T, Radm
Naugher Jr, Loran D., CAPT
Naughten, Kevin S., LTJG
Naujokas, Joseph M, Lcdr
Nault, Bonnie A, CDR
Nauman Jr, St. Elmo, Cdr
Nauta, Pieter F., CDR
Nava, Edward J, LCDR
Navard, Lear Ansel, CDR
Navaretta, Stephen, CDR
Navarre Jr, John E., CDR
Navas, George E., Lcdr
Naville, Timothy J., Capt
Navin Jr, John J, Capt
Nawrocki, James T, CDR
Nawrocki, Timothy J, CAPT
Nawrocki, Tina L., LCDR
Nazarenus, Dorothy M, Capt
Neal, John H, CDR
Neal, Joseph F H, Capt
Neal, Robert P., Lt
Neal Jr, John S, Capt
Neary, Gregory D., CDR
Neary, Thomas E, Capt
Neathery, James W, Capt
Neaton, Ronald A, CAPT
Nedwicki, Edward G, Capt
Neece, David W, CWO4
Needham, Donald M., Lcdr
Needham, John J, CAPT
Needham, Richard, Lcdr
Needham Jr, John H, CDR
Neel, David L, Lcdr
Neel, Jerald C, CDR
Neely, Robert A, Lcdr
Nees, Randy E., Capt
Neeson, Marcus K, CDR
Neff, Robert C, CDR
Neff, Sherry D, CDR
Negaard, Galen, LCDR
Negron, Bill, CDR
Neibauer, James E., CDR
Neighbors, Charles W., CAPT
Neil, George, Lcdr
Neill Jr, James S, Capt
Neiman, Richard J, CDR
Neiterman, Bertram, CWO4
Nelsen, Norman W, Capt
Nelson, Bonnie M., LT
Nelson, Bradley K., Lcdr
Nelson, Carroll T, Ltjg
Nelson, Clinton A, CDR
Nelson, Dean C, Capt
Nelson, Deborah M, Capt
Nelson, Dennis G, CDR
Nelson, Donald L, Capt
Nelson, Drew D., CDR
Nelson, Gregg C, CDR
Nelson, Gregory F., Capt
Nelson, Harry A., Lcdr
Nelson, Harvey G, Capt
Nelson, Homer B, Lt
Nelson, Jeffrey A., CDR
Nelson, Joe L, Capt
Nelson, John D., Lt
Nelson, John E, Lcdr
Nelson, John H, CDR
Nelson, Kevin W., LT
Nelson, Marc H, CDR
Nelson, Mary Jane, CDR
Nelson, Michael S., LCDR
Nelson, Ralph A., CAPT
Nelson, Reginald, LCDR
Nelson, Richard A, CWO4
Nelson, Richard H, CDR
Nelson, Richard N, CAPT
Nelson, Richard T, Capt
Nelson, Robert D., LCDR
Nelson, Robert E, Lcdr
Nelson, Robert H, CAPT
Nelson, Robert H, Capt
Nelson, Roger B, Capt
Nelson, Rolland G, Capt
Nelson, Ronald D, CDR
Nelson, Roy E, Capt
Nelson, Stephen C., CAPT
Nelson, Tiffany T, CDR
Nelson, Travis E, Lcdr
Nelson, William H, Capt
Nelson II, Albert L C, CDR
Nelson Jr, George G, Lcdr
Nelson Sr., Charlie S, Capt
Neman, David C, CDR
Nemanic, Joseph J, Capt
Nemchick, Michael P, Radm
Nemechek, Paul M., CDR
Nemmers, Raymond F, CDR
Nemmers, Robert William, Lt
Nenneau, Thomas D, CDR
Nerrie, George K, Capt
Neruda, Michael E, CAPT
Nesbitt, Charles D, Capt
Nesbitt, Robert S, CDR
Ness, Harry H, Capt
Ness, Kent M, LCDR
Neth, Samuel L., CDR
Nettles, Willard H, Capt
Nettum, Mary C., CAPT
Netzlof, William H, Lt
Neu, Stephen E., LT
Neubauer, Eric H., LCDR
Neubeck, Bruce P, CDR
Neudecker Jr, Joseph W, Lt
Neuhaus, Joan T., LT
Neuman, Carl F, CDR
Neuman, Edwin C, Capt
Neuman, James W, Lcdr
Neumeister, Alvin F, CDR
Neupert, John A., CAPT
Neureither, George R., LCDR
Neve, James K, Capt
Nevel, William G, Capt
Nevels, Jeffrey M, CDR
Neven, George, CWO4
Neville, James A, Lcdr
Neville, Thomas F, Capt
Neville Jr, William J, Capt
Nevins, Freeman J, CDR
Nevitt, James E, CDR
Newberg Jr, Eric G., CDR
Newberry III, Arthur N., LCDR
Newburn, Robert, Capt
Newburn Jr., C. L., LT
Newby, J Cameron, Capt
Newby, John G., Capt
Newcomb, Gene W., LCDR
Newcomb, Joseph Scott, Capt
Newcomb, Karen Reidy, LCDR
Newcomb, Robert M, Lcdr
Newell, Dwight J., LCDR
Newell, Robert T, Capt
Newill, Vaun A, CDR
Newlan, Bruce M, Capt
Newman, Benjamin G, Capt
Newman, Carol E., Lt
Newman, Cyril, Capt
Newman, Everett D, Lt
Newman, James L, Capt
Newman, John T., LCDR
Newman, Nancy L, Lcdr
Newman, Rick, Lcdr
Newman, Stuart T., CAPT
Newman Jr, James B., Capt
Newsom, Ernest B, Capt
Newton, Clinton B, CDR
Newton, Hood G, CDR
Newton, James C., CAPT
Newton, Jerry A., CAPT
Newton, Kenneth R, CAPT
Newton, Wayne T, CDR
Newton Jr, Alexander A, Capt
Ng, Clinton H., LCDR
Nguyen, Chung, Lcdr
Nguyen, Huan T., LT
Nguyen, Tuan Anh, LT
Nguyen, Van T., Lcdr
Niccolai, Marino J, CDR
Niccoli, Dominic J, Lcdr
Nicely, Kip W, Capt
Nichamoff, Seth, ENS
Nicholas, Jeffrey C., CDR
Nicholas, Robert R, Capt
Nicholas Jr, Jack R., CAPT
Nichols, Charles E, CDR
Nichols, David W., LCDR
Nichols, Herman A., CDR
Nichols, John David, Lcdr
Nichols, Kenneth R, Lcdr
Nichols, Roger J, Capt
Nichols, Stephen D., CDR
Nichols, Walter F., Lt
Nichols, William B, CDR
Nichols, William C., CDR
Nichols Jr, C Philip, Capt
Nichols Jr, Sterling, Capt
Nicholson, Daniel A, Capt
Nicholson, Leland W, CDR
Nicholson Jr, Robert J, Capt
Nicholson Jr., Charles E., LCDR
Nickell, Darryl C, CDR
Nickelsen, Dale C, Capt
Nickens, Anthony D, Lt
Nickoli, Wally Ray, CDR
Nickson, Douglas C, CDR
Nicol, Roland N., CDR
Nicolai, Lynn M., CDR
Nicolai, Robert L, CDR
Nicolaus, Nelson A, Lt
Nicolls, Keith K, Capt
Nicoloff, Michael J., CDR
Nicols Jr, Joseph J, Capt
Niebruegge, Marvin W, CDR
Niedermaier, David G., Capt
Niedling, Ivan M, Lcdr
Nielsen, Charles A, CDR
Nielsen, George I T, Capt
Nielsen, Jack S, Capt
Nielsen, Kelley L., LCDR
Nielsen, Orville F, Capt
Nielsen, Robert G, Capt
Nielsen, Robert L., CDR
Niemtzow, Michael A., ENS
Niemyer, Andrew S., Capt
Nienstedt, John F, Capt
Niesen, Jeffrey S, Capt
Niesse, John E, Capt
Nieto Jr, William, Capt
Nieusma, F Luehrs, CDR
Nieves, Rafael F., Lcdr
Nieznay, Walter M, Capt
Nifong, Robert L, Capt
Nigrelli, Joan Mary, CDR
Nigro, Charles D, Lt
Nikischer Sr, Frank W, Lcdr
Niklor, Matthew, 2LT
Niles, Christopher P., ENS
Niles, James H, Capt
Nilsen, Anders M., Lcdr
Nilsson, Jeffrey S, CDR
Ninteman, Vincent J, CWO4
Nischwitz, Jack E, CWO4
Nissen, Peter F, CDR
Nitchman, Mark E, Lt
Nitschke, Eric R, Lt
Nix, Carleton D, Capt
Nixon, Edward C, Capt
Nixon, Walter B, Capt
Nixon Jr, James J, CDR
Nizinski, William, Capt
Noble, Gary D., CDR
Noble, Janis Lynn, LCDR
Noble, Michael A., LCDR
Noble, R. Scott, CAPT
Noble, Russell E, CDR
Noble, Vernon H, CDR
Noble Jr, Herbert G, Capt
Noble Jr, J Kendrick, Capt
Nocton, Michael E, Capt
Noe, Scott S., LCDR
Noel, Katherine, ENS
Noelsch, Robert L, Lcdr
Nohrden, Daniel C., CWO3
Nolan, James C, Capt
Nolan, Janiece S, Capt
Nolan, Lawrence J, Lcdr
Nolan, Mary Etta, CDR
Nolan, Michael E, Capt
Nolan, Philip O., CDR
Nolan, Robert L, Capt
Nolan, Roger T., Capt
Nolan, Thomas E, Capt
Nolan, Thomas R, Lt
Nolan, Wendy C., LT
Noland, J Lynn, Capt
Noland, Robert T, Lcdr
Nolen, Stanley L., LT
Nolen, Ulysses L, Capt
Nolf, Jacob S, CDR
Nolin, Lasalle L, Lcdr
Noll, Gerald O, Lcdr
Nolstad, Arnold R., Lcdr
Nolte, John N., CDR
Nontelle, Donald R, CDR
Noonan, James F, CDR
Noonan, Patrick A, Capt
Noone, James A, Capt
Nooney, William L., Lt
Norby, Karl G, CDR
Norcross Jr, Philip F, CDR
Nord, Frederic J, CAPT
Nordgren, Lee A, CDR
Nordhougen, Gregory P., CDR
Nordin, Michael T, Capt
Nordmeyer, Philip J, Capt
Nordquist, Thomas G, CDR
Nordstrom, G Ray, Lcdr
Noreen, Frederick B, Lcdr
Norman, Michael T., LCDR
Normand Jr, Louis L., Capt
Normile, Michael M., LCDR
Norona, Robert N, Lcdr
Norris, David M, Capt
Norris, E Wynn, CDR
Norris, Hugh B, Lt
Norris, James C, Capt
Norris, John C, Capt
Norris, Richard G., CDR
Norris, Stuart E, CDR
Norris, Thomas J, Capt
Norris, William B, BTCM
Norris III, Morgan E., Lcdr
North, Carl E., LCDR
North, Henry R, CDR
Northcraft, Julian O, Capt
Northcutt, William B, CDR
Northrop, Joseph A, CDR
Northrup, Richard E., LCDR
Norton, David K, Lcdr
Norton, E Dennis, Capt
Norton, Howard C., Capt
Norton, Terry L., CDR
Nortonik, Amy, ENS
Norvik, Eiliv M, CDR
Norwood, Bradley K., Lcdr
Norwood, Linda A., Lcdr
Norwood, Sandra J., LCDR
Nossaman, Jerry, Capt
Nossen, Robert P, CDR
Nottingham Jr, George M, Lcdr
Nottingham Jr., Milton G., CDR
Notz, Philip L., ENS
Nourse, Richard D, CDR
Novak, James M., LCDR
Novak, Joseph F., CWO3
Novak, Michael J., CDR
Novak Jr, Edward A, Capt
Novak-Tibbet, Susan M., LCDR
Novas, Kathleen F., Lcdr
Novelli, James R, CWO4
Novotny, Kenneth W., CDR
Nowack, Joseph M., CDR
Nowaczyk, Mary Lenore, Lcdr
Nowaczyk, Timothy I, Lcdr
Nowak, Michael S, CDR
Nowak, Stephen F, CDR
Noyes, Charles H, Lcdr
Nuar, Frank L, Capt
Nuckles, Robert W, Lcdr
Nugent, James M., CDR
Nummi, Lisa Horton, ENS
Nunally, Merrill L., LCDR
Nunan, Thomas R, Capt
Nunez, Charles A., CAPT
Nunez, Juan R., Ens
Nunez, Richard D, Lcdr
Nunnenkamp, John M, Capt
Nunns, Scott E, Capt
Nusbaum, William N., Lcdr
Nusinson, Louis, Capt
Nuss, Robert C, Radm
Nutt, Andrew T, Capt
Nutt Jr, Thomas A, Capt
Nuttall Jr, Arthur, CDR
Nutter, John M, Capt
Nutting Jr, Perley G, Capt
Nutty, John E, Lcdr
Nyboer, Jan H., Radm
Nye, Kerry L., CDR
Nyilis, Michael J., LCDR
Nystedt, Charles M., Capt
O'Brien, Andrew J, Capt
O'Brien, Ann B, Capt
O'Brien, Anthony T., Lcdr
O'Brien, Argyle Q, CDR
O'Brien, Charles F., Lcdr
O'Brien, Cornelius J, Lt
O'Brien, David A, CDR
O'Brien, Francis D, Capt
O'Brien, Henry J, Capt
O'Brien, James J, Lt
O'Brien, John K., LCDR
O'Brien, Mark A., CDR
O'Brien, Martin J, Capt
O'Brien, Michael J.J., Capt
O'Brien, Neal J., CDR
O'Brien, Paula J., Lcdr
O'Brien, Raymond D, CDR
O'Brien, Sean P., LT
O'Brien, Thomas J., CDR
O'Brien, Thomas P., LCDR
O'Bryant, Deborah A., LCDR
O'Bryon II, Tom W, CDR
O'Connell, John R., Capt
O'Connell, Matthew P, Capt
O'Connor, Francis E, Lt
O'Connor, Gary S, Lcdr
O'Connor, Hugh M, Capt
O'Connor, James F, CDR
O'Connor, James J, Capt
O'Connor, Joan E., CDR
O'Connor, John J., LCDR
O'Connor, Michael S., CDR
O'Connor, Sean P., LT
O'Connor, Sheryl A., CDR
O'Connor Jr., James J., Lcdr
O'Dea Jr., David P., LCDR
O'Donnell, Jeffrey R., LT
O'Donnell, John J., Capt
O'Donnell, Kim Marie, CDR
O'Donnell, Michael D, CDR
O'Donnell III, John H., CAPT
O'Farrell Jr, William J, CDR
O'Hanlan, Edward V, CDR
O'Hanlon, Michael D, CDR
O'Hara, Joseph P, CDR
O'Hara II, Joseph M., Lt
O'Hora, Thomas M., Lcdr
O'Keefe, Francis J., CAPT
O'Leary, Joseph E, CWO4
O'Malley, Charles A, Capt
O'Malley, Susan M, CDR
O'Malley, William E., LT
O'Malley Jr., Thomas J., LT
O'Neal, Michael A., Capt
O'Neil, Edward C., CDR
O'Neill, Ellen J, CDR
O'Neill, John James, CAPT
O'Neill, Kathleen, LCDR
O'Neill, Mary F., CAPT
O'Rourke, Andrew P, Capt
O'Rourke, Ronald E., LCDR
O'Rourke Jr, James F, CDR
O'Shaughnessy, John L., CAPT
O'Sullivan, John P., LCDR
O'Sullivan, Marybeth A, Lcdr
O'Toole Jr., Charles S., Capt
O'Toole Jr., Raymond D., CDR
Oakes, George P., Lcdr
Oakes, Montford E., Lcdr
Oaks, Michael D, CDR
Oaks, Steven D., CDR
Oard, Douglas W, CDR
Oaster Jr, Leo James, CDR
Obanion, Lois, CDR
Obenhaus, Richard R, CDR
Obenour, Brent J., CAPT
Oberacker, Martin F, CDR
Oberdorfer, Anthony H, CDR
Oberg, David A, Lt
Oberg, Russell C, CDR
Oberheu, David M, CDR
Oberholzer, David J, Capt
Oberle, George A, Capt
Oberle, William J, Capt
Oberley, William J, Lcdr
Oberman, M Dean, Lcdr
Obermeier, Carl J., LT
Obermiller, William J, Lt
Obersat, Thomas F., LT
Oberstein, Howard B., CDR
Obley, Loren L, Capt
Obrien, George W, CWO4
Obrien, Robert E, Lcdr
Obrien, Ward J, Capt
Obrien Jr, Brian, Lt
Obusek, Joseph B, Capt
Obyrne, Carole A., Capt
Ochs, Eugene T, CDR
Ochsner, Judy R., CDR
Ocker, Jerold R, Capt
Ocon, Henry F, Lcdr
Oconnell, Charles F, Capt
Oconnell, Daniel W, Capt
Oconnell, John R, CDR
Oconnell, Kevin J, Capt
Oconnell, Margaret M, Lt
Oconnell 3rd, Charles A, Lcdr
Oconnell III, Paul J, Capt
Oconnell Jr, Edward J, Capt
Oconnor, David, CDR
Oconnor, Thomas P, Capt
Ocorr, Jerome H., LCDR
Odegaard, Alf T, CDR
Odegard, Ronald S, Capt
Odell, Jean M, Capt
Odell, Michael L, Capt
Odenthal, Joseph T., Cwo
Odo, Clarence S, Lcdr
Odom, William C, LCDR
Odonnell, James K, Capt
Odonnell, Mary L, CDR
Odonnell, Raymond V, CDR
Odonnell, Thomas M., Lcdr
Odonnell Jr, Joseph T, CDR
Odonnell Jr, William P, Radm
Odor, Louis, CDR
Oechslin, Peter E, Capt
Oefelein, John J, Capt
Oekerman, Victor E, CWO4
Oeser, David J., CDR
Oestreicher, Edward, LT
Offen, Lillian L, Lcdr
Offner, Dennis J., Cdr
Ogan, Richard W, Lcdr
Ogawa, Mitsuo K, CDR
Ogden, Marion W, CDR
Ogilvie, Gregory G., CDR
Ogilvy, Lester E, CDR
Oglesbee, Jack L, Capt
Ogletree, Darlice S, CDR
Ogletree, Maurice, LCDR
Oh, William Y., CAPT
Ohalloran, John R, Capt
Ohalloran, John W, CDR
Ohara, Ramon D, Capt
Ohara Jr, John F, Lcdr
Ohare, Robert E, Capt
Ohearn, Lawrence S, CDR
Ohlemeier, David W., LCDR
Ohler, Robert L, Capt
Ohm, Wang S., CDR
Ohme Jr, Rudolph M, CDR
Ohnstad, Dianne M., CDR
Ohnstad, Donald A, CDR
Oien, Gregory L, Lcdr
Okada, Floyd, Lcdr
Okeefe III, John J, CDR
Okelley, Robert V, Capt
Oki, George M, Capt
Okihiro, Glenn M., CAPT
Okimura, Raymond, CDR
Okonek, Lois Marie, Lcdr
Oksol, Richard F., CDR
Olah Sr., Marcus, LT
Olaughlin, Terrence J., Lcdr
Old, Philip E., Lcdr
Old Jr, Harold E, Capt
Oldach Jr., Raymond P., ENS
Oldershaw, John H., CDR
Oldfield, John C, Capt
Olds Jr, Hugh W, Capt
Oletzky, Sheldon, Lcdr
Olig, Denice Ellen, LCDR
Olinder, Everett J, Capt
Olivarez, Edward J, Lcdr
Olivas, Lee M., LCDR
Oliver, Robert G, Capt
Oliver, Ronald Leigh, CWO4
Oliver, Vincent C, CDR
Oliver, William F., CDR
Oliver Jr., Robert A., LT
Oliveras, Angel I., Lcdr
Olivier, John L, CDR
Olkhovsky, Paul, Lcdr
Olmstead Jr, Richard G., CDR
Olsen, Eugene A, Lt
Olsen, Jack A, Capt
Olsen, Karen M., LCDR
Olsen, Lawrence E., CDR
Olsen, Lloyd G, Capt
Olsen, Raymond, CWO4
Olsen, Roger P., CWO4
Olsen, Steve A, LCDR
Olsen, William, LCDR
Olsen Jr., Charles W, Lt
Olsgaard, Richard B, CDR
Olshinski, John A, Capt
Olson, Carl L, Lcdr
Olson, Dennis D, CDR
Olson, Donald C, CDR
Olson, Earle Z, CDR
Olson, Harold O, Lt
Olson, Harold R., Lcdr
Olson, Jeffrey K., CDR
Olson, John R., LT
Olson, Kenneth C, Lcdr
Olson, Lloyd W, CDR
Olson, Maureen C., Lcdr
Olson, Richard A., Lcdr
Olson, Richard R., LTJG
Olson, Robert D, CDR
Olson, Ronald W, Capt
Olson, Stephen R, Capt
Olson, Wayne R., Capt
Olson II, James D., Radm
Olson Jr, David O, Capt
Olvera, Carlos N, CDR
Omalley, George C, Capt
Oman, Christopher S., CDR
Oman, Ray C., CDR
Oman, Thomas J., Lt
Omara, William M, Capt
Omark, Warren R, CDR
Omeara, James E, CWO4
Omeilia, Charles R, Capt
Omelia, Michele J, LCDR
Omundson, Jerrold W, Lcdr
Ona, Ryan J., MIDN
Onan, Thomas M, Capt
Ondrus, Sandra L., Lcdr
Oneal, Patrick W., Lcdr
Oneal, Robert J, Capt
Oneil, Danny T., CDR
Oneil, Sean P., CAPT
Oneil, Timothy L, CDR
Oneil, William A, Capt
Oneil Jr, John J, Capt
Oneill, John M, CDR
Oneill Jr, Samuel J, Capt
Opacki, Dennis G, CDR
Opalinshi, John, MIDN
Opalka, William B, CWO4
Opitz, Martin E, CDR
Opitz, Paul F, Lcdr
Opkins, Stephen A, Lcdr
Oplinus, Ronald A, Lcdr
Opon, Stanislaus, Lcdr
Oppenheim, David S, Capt
Oppenheimer, Douglas L, Capt
Oppenheimer, Philip J., CAPT
Orange, E Neil, Capt
Ordich, Scott A., LCDR
Orecchio, John J, CDR
Oreck, Steven L., Capt
Oreilly, Clarke F, Capt
Oreilly, Robert W, Capt
Oreilly Jr, James A, Capt
Oriatti, Mario D, Capt
Orjales, Rodolfo, LCDR
Orlandi-Alvarez, R, CAPT
Orlando, Ralph P., Capt
Orlando Jr, Pompei L, CDR
Orlaskey, William H, Lcdr
Orley, William J, Capt
Ornelas, Ernesto, CDR
Ornelas, Ernesto E, CDR
Orologas, Gus N., CDR
Oropeza, Antonio, LT
Orourke, John F, Lcdr
Orourke, John G, Capt
Orr, Alvin G, CDR
Orr, John D, Capt
Orr, Nancy A, LCDR
Orr, Thomas R., Lt
Orr, William T., LT
Orski, Gary A, CAPT
Orson, William E., LCDR
Ortloff, Fredrick P., CDR
Ortman, Kay E., CDR
Orvis, John E, Lt
Orwig, Charles E, Lcdr
Orzech, James Kenneth, Capt
Osborn, Donald R., CDR
Osborn, Johnny R., CDR
Osborn, Pamela A, CDR
Osborn II, Otis B., Lcdr
Osborn Jr, William W, CDR
Osborne, Donald V, Capt
Osborne, K. Michael, CDR
Osborne, Lloyd B, Capt
Osborne, Oliver K, Capt
Osborne, Thomas M, Capt
Osbourn, James Morgan, Lt
Oscarson, Carl H., CDR
Osgood, Robert P., LT
Oshea, Edward J, Radm
Oshea, John P, CDR
Oshea, Michael G, CDR
Oshields, Donald B, CDR
Osier, Christopher, Capt
Oslund, Joellen D, Capt
Osseck Jr, Thomas A., CDR
Ostach, Jim, Lt
Ostapski, S Andrew, CAPT
Osteen, Sandra K, CDR
Oster, Ted A, Capt
Oster, William H, CDR
Osterhout, Royal H, CDR
Osterlund, Richard L., Capt
Ostertag, Larry G, Capt
Osthaus, John G, Capt
Ostien Jr, Paul H, Capt
Ostow, S Michael, CDR
Ostrander, John M, CDR
Ostrander, Peter H, Capt
Ostrom, Paula L, CDR
Ostronic, F. Patrick, CDR
Ostrowski, Gregory R., Capt
Oswald, Jamie A., LCDR
Oswald, Kim Dotson, CDR
Oswald, Roderick D, Lcdr
Oswald, Stephen S, Capt
Oswald Jr, John C, CWO4
Oswald Jr., Richard E., CDR
Ota, Roy T, Capt

Otey, Glen R, Capt
Otis, Linda L., CDR
Ott, Luther M., CDR
Ott, William A, CAPT
Ott Jr, Norman K., Capt
Otterberg, Carl O, Capt
Ottinger, Rodney L., Lcdr
Otto, Berton R, Capt
Otto, James P, LCDR
Oudshoorn, John P, CAPT
Outcalt, Dudley M, Capt
Outland, George F, Capt
Ove, Peter N., CAPT
Overby, Duane L, Lcdr
Overman, Thomas F, Lcdr
Overstreet, Edwin D, Capt
Overton, Elvin E, CDR
Overton, Eugene W, Capt
Overton, John C, Capt
Overvold, Paul M., CDR
Oviatt, Gary A, CDR
Owen, Bruce W, Capt
Owen, Donald F., LCDR
Owen, Duncan E, Capt
Owen, Graydon S, Capt
Owen, James G, CDR
Owen, John W, CDR
Owen, K. Kip, CDR
Owen, Steven E., LTJG
Owens, Anthony J, Lt
Owens, Brian M, Capt
Owens, Bruce H, Capt
Owens, James L, CDR
Owens, James W, Capt
Owens, Kathleen P., LCDR
Owens, Larry B., Lcdr
Owens, Robert P, Radm
Owens, Thomas D, Lcdr
Owens, Vickie L., Lcdr
Owens Jr, James K, Capt
Owens Jr, John R, CAPT
Owings, Derrick W, CDR
Oxenford, Robin A, Capt
Oxenstierna, Maria Teresa, LT
Oxford, Gregory E., CDR
Oxford, Lewis A, CDR
Ozab, Philip W, Capt
Ozanne, William, Lcdr
Ozawa, Kenneth H, Capt
Paaske, James H, CWO4
Pabian, Charles T, Ltjg
Pabinquit, David B., CDR
Pabst, Julius F, Lcdr
Pabst, Richard J, Capt
Pacala, Theodore, Capt
Pace, Annette, Capt
Pace, Charles E., LCDR
Pacey, Charles W, Lcdr
Pache Jr, Eugene P, Capt
Pachuta, Mark T., Capt
Pacilio, Jerry A, Capt
Pack, Richard, Lcdr
Packard, Michael J, Capt
Packard II, Charles E, Lt
Padama, Arthur D., LTJG
Paddock, Charles G., CDR
Paddock, John F, Radm
Paddock, Michael J, LCDR
Paddock Jr, Peter B., LCDR
Paden, Jack G, Capt
Padfield, Boyd F., LTJG
Padilla, Amilcar, LT
Padon, Thomas G, CDR
Padovani, Bruno S., CDR
Pagan, Victor M., Lt
Pagano, William J, Capt
Page, Alfred E., CDR
Page, Arnold S, CDR
Page, Bradley A., LCDR
Page, Daniel R., LT
Page, Jack W, Capt
Page, James F, CDR
Page, Loren H, Capt
Page, Lynda J., CDR
Page, Paul, Lcdr
Page, Ronald L., LCDR
Page, William R., Capt
Pagels Jr, George, Lcdr
Paget Jr, James S, Capt
Pagter, Lawrence B, Lcdr
Paige, Charles J, CDR
Paige, Kevin A., LCDR
Painter, Allen C, Capt
Painter, Chad L., Lcdr
Painter, Donald J, CDR
Painter, Steven S., CDR
Pair, Claude H, Lcdr
Pala, Anthony J., CDR
Paladini, Victor C, Lcdr
Palazzo Jr, Anthony J, Capt
Palcich, James A, Lt
Palermo, Robert S, Lcdr
Palia, Richard G, Lt
Pallick, Harry D, Capt
Pallme, Robert C, Lcdr
Pallot, E Albert, Lcdr
Palm, Steven B, Capt
Palmatier, Harry C, Capt
Palmer, Candace L, Lcdr
Palmer, Cary R, Capt
Palmer, J David, Capt
Palmer, Jean K., CDR
Palmer, Keith F, Lt
Palmer, Randall W, Lcdr
Palmer, Richard L, Capt
Palmer, Rodney L, CDR
Palmer, Steve F., CDR
Palmer, Timothy G, Capt
Palmer, William R., Capt
Palmer, William S., LCDR
Palmer, William W, Capt
Palmer-Willis, Harriet, CDR
Palmore, Carlyle C., Lcdr
Palmucci, Victor J, Capt
Palombo, Richard A., LCDR
Palos, Miguel Martin, CDR
Pals, Larry A, CDR
Pandolfo, Edward P., Lcdr
Pandzik, Michael L, Capt
Pane, Karen W., LT
Panganiban, Rudy S., Capt
Pangborn Jr, Willard H, Lcdr
Panger, Duane J, LCDR
Panhorst, Donald L, CDR
Panico, Earl A., LCDR
Panko, Susan M., CDR
Pankovics, James, CDR
Pankratz, John M, CAPT
Pankratz, Walter L, Capt
Panthen, Donald C., Ens
Pantoja, Anthony, LCDR
Pantone, Donald S., LCDR
Panuncialman, Robert I, Lcdr
Papadopoulos, Loukas, LTJG
Papalski, Joseph, CDR
Papas, Charles Reedy, Ltjg
Papas, Eugenia, Lcdr
Pape, Susan S., CDR
Papenthien, Michael R., CDR
Papineau, Daniel A, CDR
Papke, Craig A., Lt
Papke, Mary Beth, Ens
Papoulias Jr, Anthony L, CDR
Pappalardo, Salvatore J, Capt
Pappas, Gregory T, Lcdr
Pappas, James P, Capt
Pappas, Sophocles G, Capt
Papsdorf, F Warren, Lcdr
Paradiso, Angelo E, Capt
Parado, John, ENS
Paras, Mark J, CDR
Parchman, George W, Lcdr
Pardee, Charles A., Lt
Pardo, Stanley T, CDR
Pardue, Oscar W, Capt
Pare, John A, Capt
Parenti, Thomas M, CDR
Paretti Jr, Andrew, Capt
Parham, Louis, Capt
Parham, William N., CDR
Parins, David, ENS
Paris, Bret, Lcdr
Parisi, Frank, Capt
Park, Patrick J, Capt
Parke, John S, CDR
Parke, Thomas L., Capt
Parker, Alan C, Capt
Parker, Arthur H, Lcdr
Parker, Bradford Wayne, CDR
Parker, David A., LCDR
Parker, Debra D., CDR
Parker, Gary D, CDR
Parker, Gilbert H., Lcdr
Parker, James J, Lcdr
Parker, James W, Capt
Parker, Jeffrey K, Lt
Parker, John C, Capt
Parker, Joseph M, CDR
Parker, Kevin E., CDR
Parker, Matthew, LCDR
Parker, Michael J., LT
Parker, Randall A., ENS
Parker, Robert C., Capt
Parker, Stephen L., LCDR
Parker III, Charles W., LT
Parker Jr, Leonard A., Capt
Parker Jr, Robert C, Capt
Parker Jr., Robert H., LT
Parker-Falzoi, Judith, LT
Parkerson, George C, CDR
Parkes, James A, CDR
Parkes II, James C, CDR
Parkinson, James W, CDR
Parkinson, Leonard S, Capt
Parkis, Clyde, CDR
Parks, Braddock L., Lcdr
Parks, Brent G., Lcdr
Parks, James C, CDR
Parks, John B., LT
Parks, Leroy C, Lcdr
Parks, Patricia A., CDR
Parks, Thomas B, Lcdr
Parlee, Kenneth R, CDR
Parmenter, Lester L, Capt
Parmentier, Christine M, Lcdr
Parmentier, Timothy J, CAPT
Parmley, Philip L., MIDN
Parnell, John E, Lt
Parr, Jack R, CAPT
Parr, James F, Capt
Parr, Jerry Deane, CAPT
Parrino, Jack J, CDR
Parrino, Sam, Capt
Parris, Charles C, CDR
Parrish, Bruce A, Lt
Parrish, David O., Capt
Parrish, William G, CDR
Parrott, Michael T., CDR
Parrr, William Lee, ENS
Parry, William E, CDR
Parry Jr, Howell J, CDR
Parry Jr, John C, Capt
Parsons, Clara D, CDR
Parsons, David W, Capt
Parsons, Frederick L, CWO4
Parsons, George T, CDR
Parsons, Kent, Capt
Parsons, Lee, Capt
Parsons, Paul I, CDR
Parsons, Richard H, Lcdr
Parsons, William C, Lt
Partnoy, Ronald A, Capt
Partrick, Robert E, CDR
Paschal, Brent E., LCDR
Paschall, Robert L, Capt
Pasciucco Jr, John J, Capt
Pascual, Francis M, CDR
Pasinski, Walter V, Lcdr
Paskauskas, Denis P, Lcdr
Pasley, Robin F., CDR
Pasqualucci, Joseph A., CDR
Passaglia, Charles T., Lcdr
Passen, Christine Ann, Lcdr
Passman, Frederick J, Capt
Passmore, Robert O, Capt
Pasternak, Carolyn M., LT
Pastore, Michael J, Lt
Pastula, Dave, LCDR
Patafio Jr, John J, CDR
Patane, Louis L, Lcdr
Patch, Allan K., CDR
Patchen, Stephen, CDR
Pate, Allen S, Capt
Pate, Mary E, CAPT
Pate, Peter S, Lcdr
Pate, Ronald M, Capt
Paternoster, Stephen J, CDR
Paternostro, C. David, CDR
Paterson, Dale R, Capt
Paterson, Robert L, LCDR
Paterson, Thomas M, CDR
Pathman, Ronald C, Capt
Patience USAFR, William J., 1LT
Patin, Dennis J., CDR
Patras Jr, William N J, LT
Patriquin, David A, Capt
Pattavina, Vincent, Capt
Patten, Lynden D, CDR
Patterson, Claire J, Capt
Patterson, Daniel J, Capt
Patterson, David C, Capt
Patterson, Edward J, CWO4
Patterson, Harold W, Ltjg
Patterson, Hiram M., LCDR
Patterson, Homer E, LT
Patterson, James R, Lcdr
Patterson, Joan L, CDR
Patterson, John Paul, Lcdr
Patterson, Kenneth F., CDR
Patterson, Mark C., LCDR
Patterson, Michael J., ENS
Patterson, Robert F., Capt
Patterson, Timothy J., CWO3
Patterson, Walter D, CDR
Patterson, William H, Capt
Patterson, William M, Capt
Patterson, Wyndell S., LCDR
Patterson II, John L, CDR
Patterson Jr, Donald J, Capt
Patterson Jr, George, Capt
Patterson Jr, Thomas J, Radm
Patton, Christopher A., CDR
Patton, Edward L, LCDR
Patton, Richard C., Lcdr
Patton, Thomas C., CDR
Patton, William B, Capt
Patton Jr, James H, CDR
Paty, Charles R, Capt
Patz, Theodore J, CWO4
Paul, Christopher J., CDR
Paul, David L, Capt
Paul, Harold R, CDR
Paul, R. John, CAPT
Paul, Richard L, Capt
Paul, Thomas R, CDR
Paule, Philip A, Capt
Paules III, Granville E, Capt
Paulette, Stephen W., LCDR
Paulino, Richard G, Capt
Paulk, Robert R., LCDR
Paulsen, Albert G, Radm
Paulsen, M. Melody, LCDR
Paulson, David W, Capt
Paulus, Richard A, Capt
Pausa, Clements E, Capt
Pausche, Michael R., Lcdr
Pauze, John A, Capt
Pauze, Philip C, Capt
Pavelle, James R, Capt
Pavlicek, Otto, CDR
Pavlick, Donna L, Capt
Pavlick, Michael J., CDR
Pavlick Jr, Charles T, Capt
Pavlisin, Robert G, Capt
Pavur, Robert J., LCDR
Pawlak, Mark J, CDR
Pawlak, William J, Capt
Paxson, James T, Capt
Paxton, Charles N, Radm
Paylor Jr, Frank R, Lcdr
Payne, Charles A, Capt
Payne, David R, Lcdr
Payne, George H, CDR
Payne, Harvey M, Capt
Payne, John T, Capt
Payne, Louis K, Capt
Payne, Michael A, Capt
Payne, Michael D, Capt
Payne, Philip M., CDR
Payne, Randolph A., CDR
Payne, Stephen J., Lcdr
Payne, William H, Capt
Payne, William J., CDR
Payne, William M, Lcdr
Payson, Albion L, CDR
Payton, Gladstone A, Capt
Payton, James E., Lt
Payton, Roland W, Lcdr
Peace II, H W, Capt
Peacher, Robert W, Capt
Peacock, Charles A, Capt
Peacock, Donald J, Capt
Peacock, James L, CWO4
Peak, Martin A, Lcdr
Peake, Drew, CDR
Peaker, Robert H, CWO4
Peaks, Pamela V., LCDR
Pearce, Jeffrey R., CDR
Pearce, Robert W, Capt
Pearce, Scott A., LT
Pearce, Ted, ENS
Pearcey, Richard C, CDR
Pearcy, Donald W, CDR
Pearse, Jack F, Radm
Pearson, Conrad F, Capt
Pearson, John B, CDR
Pearson, Julie A., CDR
Pearson, Kenneth F, CDR
Pearson, Raymond H, Lcdr
Pearson Jr, C Thomas, CDR
Pease, Frank V, CDR
Pease, Rodney D, Capt
Peavey, Michael P, CDR
Pebbles Jr, Harold A, Capt
Pebley, Charles F, CDR
Pecha, Brian S., LCDR
Pechulis, John J, Capt
Pecic, John W., CAPT
Peck, Joseph J., Radm
Peck, Larry A., CDR
Peck, Michael David, Lcdr
Peck, Roger R, Capt
Peck, Thomas L, CDR
Peck, Willis I, Lcdr
Peckham, Edward W, CDR
Peckham, Gail F, Lcdr
Peckham, William R, CWO4
Pecorelli, Joseph P., CDR
Peden, Ann M, CDR
Peden III, William J, CDR
Pedersen, Calvin J, Capt
Pedersen, Dean W, Capt
Pedersen, John B, CDR
Pedersen, John H, Radm
Pedersen, Mark W., CDR
Pedersen, Paul Michael, Lt
Pederson, Donald H, CDR
Pederson, George W, CDR
Pedmo, Richard P, CDR
Pedrick IV, Daniel W, Capt
Peebles, Leroy, Lcdr
Peecook, Keith M., CDR
Peek, Michael J., LCDR
Peery, Larry J, Capt
Peet, John R, CDR
Peeters, Gerald W, CAPT
Peeters, Richard Thomas, Capt
Peickert, Marlin W, CDR
Peifer, Mark Allen, CDR
Peirano, Douglas R, LCDR
Pelaez, Wayne R, Capt
Pelczarski, Danielle L., LT
Pell, John H, CDR
Pella, Reynaldo L., LT
Pellegrini, Arthur E., Capt
Pellegrino, Anthony, CDR
Pello, Richard A, Lt
Pellosie, Carmine J, CDR
Pelloth, Michael J., CAPT
Peloquin, Brett A., Lcdr
Peloquin, Peter J., CDR
Peloquin, Robert A, Capt
Peloquin, Robert D, CDR
Pelosi, Richard P, CDR
Pelowski, Gregg R., CDR
Pelton, Montross G, CDR
Peltzer, Thomas J., LT
Pena, Teresita D, Lcdr
Pena III, Fernando, LCDR
Penalver, Juan R., Lt
Pence, Donald E, Lcdr
Pence, Douglas E., CDR
Pence, John R, Lt
Pendergast, Debra K., CDR
Pendergraph, Frankie, LT
Pendleton, James E, CDR
Pendleton, Robert L, Capt
Pendleton, William C, Capt
Penfield Jr, Scott R, Capt
Penn, William H, CDR
Penn, William L, Capt
Penn III, Washington K, CDR
Penne, James J, CDR
Penney, William M, Capt
Pennington, Bennie J, Lcdr
Pennington, Shelley, CAPT
Pennisi, Francis E., Lcdr
Penny, Ralph E, Capt
Peoples, Denton L, Capt
Peoples, Philip M, Capt
Peplinski, Richard P, CDR
Pepple, William S, CDR
Percer, William P, Lt
Perch, Robert L, CDR
Percy III, Robert R., RADM
Perdigon III, Gus J, Capt
Perdue, Peter L, Capt
Perduto, James A., Lt
Pereira, Awilda, LT
Pereira, James F., CDR
Pereira, Ronald M, Lcdr
Peretti, Angela, ENS
Perez, Armando, CDR
Perez, Joseph M, Capt
Perez, Luis F., Lcdr
Perez-Maiy, Osvaldo, LT
Perez-Vergara, Eliezer, CDR
Perfetto, Chester, Capt
Perkerson, T. Robert, CAPT
Perkins, Curt G., LCDR
Perkins, Katherine M., Lcdr
Perkins, Kevin C., LCDR
Perkins, Martin C, Capt
Perkins, Richard A., Capt
Perkins, William Scott, CAPT
Perkins, William T, CDR
Perkison, William K, CDR
Perlick, Christopher F., Lcdr
Perlmutter, William J., Capt
Pernokas, Louis N, Capt
Perotti, Joseph G, Capt
Perreault, Mark D., Capt
Perron, Dean R, CDR
Perroni, Joseph R, Capt
Perry, Alden T., CDR
Perry, Earl A., CAPT
Perry, Gilbert B, Capt
Perry, Herbert S, Capt
Perry, Jack Robb, Lt
Perry, James A, CDR
Perry, Joel W, CDR
Perry, Kempton J, CDR
Perry, Leroy W, CDR
Perry, Linda A., Lcdr
Perry, Robert Edwin, Lcdr
Perry, Roland J, Lcdr
Perry Jr, Robert J., LCDR
Persing, Kenneth F, CDR
Persinger, William C, CDR
Persson, Vernon H, Capt
Pertle, Gregory J., Lcdr
Pertsch, John J, CDR
Pervez, Nilofer, LCDR
Perwien, James D, Lcdr
Perz, Lucille L., Lcdr
Pestrue, Darell W, CDR
Petering, Roland V, Capt
Peters, Charles B, CDR
Peters, David E, Capt
Peters, David W, Lcdr
Peters, Donald W, Lcdr
Peters, Frederich J, Capt
Peters, Geoffrey W, Capt
Peters, Wayne E, CDR
Peters, William E, CDR
Peters Jr, Clarence A, Lt
Petersen, Dean L, Capt
Petersen, Donald B, CDR
Petersen, Gary L, Capt
Petersen, Harry J, Capt
Petersen, James A, Capt
Petersen, John C, Lcdr
Petersen, Kenneth J., LT
Petersen, Paul C, CDR
Petersen, Robert A, Capt
Petersen, Roy David, Capt
Petersen, Russell J, CDR
Petersen, Woodrow T, Capt
Peterson, A Harold, Capt
Peterson, Bradley A., CDR
Peterson, Bradley J., CDR
Peterson, Dan L, CDR
Peterson, Dana Mark, CDR
Peterson, David A, CDR
Peterson, David D, Lcdr
Peterson, David I, Capt
Peterson, Diana C., LCDR
Peterson, Donald D, Capt
Peterson, Donald Olaf, Capt
Peterson, Douglas D, Capt
Peterson, Everett M., CDR
Peterson, Frederick W, Capt
Peterson, Harley H., Capt
Peterson, Harry E., CDR
Peterson, James R, Capt
Peterson, Jean C, Lcdr
Peterson, Jon P, Capt
Peterson, Kenneth L, Capt
Peterson, Lowell H, Capt
Peterson, Neil T., CAPT
Peterson, Norman G, Capt
Peterson, Omas L, CAPT
Peterson, Patrick B., Capt
Peterson, Richard A, Lcdr
Peterson, Richard M., Capt
Peterson, Robert C, Capt
Peterson, William M, Capt
Peterson Jr, Aldor C, Lt
Peterson Jr, Robert E, Capt
Pethick, Donald R, Capt
Petit, Noel J, Capt
Petitfils, Preston C, Lcdr
Petitt, Thomas D, Lcdr
Petrella, Albert D, Lt
Petrella, Karen M, CAPT
Petrick, Ben J, LCDR
Petrie, George C, CDR
Petrie, Vernor H, Lcdr
Petrosky, Joseph C, Capt
Petrus, Robert M, CWO4
Pettengill Jr, Howard W, Capt
Pettersen, Erik G, Capt
Pettett, Lawrence W., LTJG
Petteys, Nelson B, Lcdr
Pettigrew, Kenneth W., Radm
Pettit, William E, Lcdr
Pettway, Annie C., LCDR
Petty, David T, CDR
Petzoldt, Ronald W., CDR
Pevovar, Michael D, Lcdr
Pezzella, Carmen C, Lt
Pezzopane, Bernard, Capt
Pfefferle, Warren, CDR
Pfeifer Jr, H Arthur, Capt
Pfeiffer, Steven B, CDR
Pfeiffer, William G., Capt
Pfeil, Don H, Capt
Pfeil, Richard A, Capt
Pfleger, James W, CDR
Pflieger, James E, CDR
Pfluger, Werner M, CDR
Pflugheber, Edward G, Capt
Pflum, Frederick W, Lcdr
Pfotzer, William, Capt
Pfrimmer Jr, John G., Capt
Pham, Chau H., LCDR
Phan, Phan, Lt
Phan, Sydney Hungquoc T., ENS
Phares, H William, Lcdr
Pharmakis, Rusty W., LCDR
Pharo SR, Charles W, CWO4
Phelan, James R., CDR
Phelan, John T, CDR
Phelps, Cheri D., LT
Phelps, James C, Capt
Phelps III, Edwin R, Capt
Phelps Jr, Lloyd L, CWO4
Phelps Sr, Norman T, CDR
Phenicie, Charles K, CDR
Phifer, Fred T, Lt
Philactos, George A, Lcdr
Philbin, Peter E, Capt
Philbrick, Bruce W, Lcdr
Philbrick Jr, Charles R, Lcdr
Philipps, Charles J, CWO4
Philipson, Willard D, Capt
Phillip, John W, Capt
Phillippe, Angela L., LCDR
Phillips, Benedict I., LT
Phillips, Bradford G, CDR
Phillips, Carol Ann, CAPT
Phillips, Charles L, CDR
Phillips, Ewart E, CDR
Phillips, Frank H, Capt
Phillips, Geoffrey L., Lcdr
Phillips, George B, Capt
Phillips, Grant E., Lcdr
Phillips, Harold W, CDR
Phillips, Harry R, Capt
Phillips, Henry, CDR
Phillips, Howard E, Lt
Phillips, John F, CWO4
Phillips, Kenneth R., CDR
Phillips, Linus L, CDR
Phillips, Lynn Ann, LTJG
Phillips, Mark A., Lcdr
Phillips, Michael J., LCDR
Phillips, Neal C, Capt
Phillips, Patricia A., Lcdr
Phillips, Reed M, Capt
Phillips, Robert G., CDR
Phillips, Ronald E., CDR
Phillips, Roy S, CDR
Phillips, Vicki Lee, Lcdr
Phillips, Wilkerson P., LT
Phillips, William M, Lcdr
Phillips, William R., CDR
Phillips, Wilson H, Lt
Phillips Jr, Earl W, CDR
Phillips Jr, Richard J, CAPT
Phillips Jr, W Lee, Capt
Phillipson Jr, William, Capt
Philpitt, James B, CDR
Phinizy, Robert B, Capt
Phinney, Sean C., Lcdr
Phipps, James B, Lcdr
Pianko, Theodore, CDR
Piazza, Robert W, Capt
Piazza Jr, Ben J, CAPT
Picard, Barbara, LT
Piccirillo, Darren W., ENS
Picco, John J, Capt
Piccolo, John A., CDR
Piccone, David B., Lcdr
Pichler, Donald E., CAPT
Pick, Samuel E, CDR
Pickard, Martin D., CDR
Pickel, Diane M, Lcdr
Pickering, James C, CDR
Pickering, Joel C., CDR
Pickett, Charles E, Capt
Pickett, Ellis B, Lcdr
Pickett, Wayne J., LCDR
Pickett Jr, Robert D, Capt
Pickle, Franklin P, CDR
Pickrell, John H, Capt
Pickup, Jack H, CDR
Picogna, Joseph L, CDR
Picou, David S., CDR
Picquet, Roger, CDR
Piela, Michael W., LCDR
Pieper, Don B, Lcdr
Pieper, Gregory J., CDR
Pierce, Allyn M, Capt
Pierce, Chester M, CDR
Pierce, Glenn I, CDR
Pierce, Gregory D., Capt
Pierce, Jerry S, CAPT
Pierce, Judith L, CDR
Pierce, Julie A., LT
Pierce, Keith A, CDR
Pierce, Scott F., LCDR
Pierce, Vernon L, CAPT
Pierce, William L., LCDR
Pierotti Jr, Julius V., Lcdr
Pierpan, Henry J., LT
Piersee, Charles R., Capt
Pierson, George M, CDR
Pierson III, Robert C., CDR
Pietrangelo, Gary P, Capt
Pifel, David B., LCDR
Piianaia, Abraham, CDR
Piianaia, Gordon Y., Lcdr
Pike, William Y., Ltjg
Pilafian, Harry, Lt
Pilapil, Caroline E., Lt
Pilapil, Virgilio R, Capt
Pilch, Edward D, CDR
Pilcher, Jan K., LCDR
Pilcher, Stuart T, CWO4
Pilcher, Von B., LT
Pilcher, Willa J, Lt
Pimentel, Charles M., LT
Pinard, Thomas C, Capt
Pincus, Irwin D, CDR
Pine, Judith M., Lcdr
Pine II, Norman J., Lt
Pinetti, Randall L., CAPT
Pingel, Earl G, Lt
Pinkard, Dennis M, Capt
Pinkerton, Danny C, Capt
Pinkerton, Roy H, Capt
Pinkney, D Timothy, CDR
Pinkowski, Francis S, Capt
Pinkozie, Dennis P, Lcdr
Pinkston, Jerry D., ENS
Pinney, Virginia R., CDR
Pino, Richard M., Lcdr
Pinsker, Clive D, Capt
Pinskey, Howard S, CDR
Pinson III, Thomas J., CDR
Pinsoneault, Michael J., LCDR
Pintar, John F, CDR
Pintelon, Lyn M, Lcdr
Pinto, Edmund, Capt
Pionk, Daniel R, LCDR
Piontek, Ray, Capt
Piotrowski, Roman E, Capt
Piotrowski, Walter E, CDR
Piotrowski, William C, Capt
Piotter, Glen W, Capt
Piper, Edwin E, Lcdr
Piper Jr, Charles E, CDR
Pipes, Larry S., CAPT
Pipiringos, Margaret A., LCDR
Pipkin, Cyril M, Capt
Pipkin, Richard M, CDR
Piranian, Alfred G, CDR
Pirie, Robert B, Capt
Pirotte, Steven J., Lt
Piscitelli, Peter A, Capt
Pishdad Jr, Amir M, Lcdr
Pistole, Carl, LT
Pitchford, Gayel A, Capt
Pitchford, James E., LCDR
Pitman, Peter F., LT
Pitman, Shaunee K., CDR
Pitney, John C, Capt
Pitocco Jr, Philip J, Capt
Pitt, Donald A, CDR
Pitt, Donald F, CDR
Pitt Jr, Woodrow W, CDR
Pittard, Michael J, CDR
Pittis, Albert D, Lt
Pittman, Alan R., Capt
Pittman, Galen W, Capt
Pittner, Douglas E., ENS
Pitts, Staley L, Lcdr
Pitts, Theodore R, Capt
Pittsey, Tony E, Capt
Pitz, Rosemarie Ruth, Lcdr

Pitzer, Paul D, CAPT
Pizzeck, Eugene X, CDR
Pizzuti, Glenn A, Lcdr
Plank, Dick A, CDR
Plante, Normand E, Capt
Plasko, Michael G., LT
Platt, Marcor B., LT
Platten, Diane M., CDR
Platz, George M., CAPT
Plaud, Joseph J., LT
Pleas, Robert B, CDR
Pleasanton, David Frank, Lcdr
Plecha, Zigmund, CDR
Plehal, James B, Capt
Pletkovich, Terry L., LCDR
Plettner, Richard M, CDR
Pletzke, Carl J., CWO3
Plisco, Loren R., CAPT
Pluhar, Robert E, CDR
Plumb, Joseph C, Capt
Plumb, Ralph H, Lcdr
Plumlee, Richard A., CDR
Plummer, Harold C, CDR
Plunkett, Carl A, Ltjg
Plunkett, Kenneth J., LCDR
Plyer, Bruce A, Capt
Po, Taina Villanueva, LCDR
Pocaro, Victor J., Lcdr
Pock, Arnold, Capt
Poe, Larry L., Radm
Poe, Oliver Conrad, Capt
Poehler, Allen, Capt
Poelker, George F., CAPT
Poff, Philip W, CDR
Pogue, Joe H, Capt
Pohl, Gregory R., Capt
Pohlman, Robert, CDR
Pohto, Daniel A, Lcdr
Poimiroo, John R, Capt
Points, David Stephen, CDR
Poirier, Gerard R, Lcdr
Poirier Jr., Phillip L., CAPT
Polak, Barbara S, Capt
Poland Jr., Robert L., CWO3
Polanski, Charles M, CDR
Polechetti, John S., LCDR
Polich, Timothy J, Lcdr
Poling, John D, Capt
Poling, P. Adam, Lcdr
Polk, Frederick E, Lt
Poll, Harry F, Capt
Pollak, Kenneth D., Lcdr
Pollak, Sanford, CDR
Pollard, Alexandra G, Capt
Pollock, Earl M, Lcdr
Pollock, Edwin E, Lcdr
Pollock, Louis W., Capt
Pollock, Robert E, Capt
Pollock, Wendell B, CDR
Pollock, William E, CDR
Polson, Herbert E, CDR
Polunsky, Robert I, CDR
Polutnik, Francis L, Capt
Pomeroy, Edward E, Lcdr
Pond, James N., CDR
Pond, Karl J E, Capt
Pong, George Y, Lcdr
Ponseigo, Andrew C., Lt
Ponta Jr, Andrew J, Capt
Pontus Jr., Stephen P., CDR
Ponzeline, Dante C, Capt
Pool, Robert A F, CDR
Poole, Bradley C., Radm
Poole, Danny G, Lcdr
Poole, Peter, Capt
Pooler, Kim D, LCDR
Pooley, David W, CDR
Pope, Bruce M, Capt
Pope, F Patrick, Lcdr
Pope, Gary S, Lcdr
Pope, Larry G, CDR
Pope, Rose H., LCDR
Pope Jr, Lawrence L, CDR
Popham, Neal R, Capt
Popielarz, Stephen J, LT
Poplin, Woodrow M., CDR
Popovich, Daniel E, Capt
Popp, Joseph M, Capt
Popp, Ralph W, Lcdr
Poppe, Christopher R, LT
Porch, Harriett E, Capt
Porpora, Gregory A, Lcdr
Porras, Oscar J., CDR
Portell, Michael S., Lcdr
Porter, Lawrence H, Lcdr
Porter, Sally Mae, Ltjg
Porter, Stuart Thomas, CDR
Porter, Tana K, Lcdr
Porter, Terrance M, Capt
Porter, William C, CDR
Porter Jr, Nathan F.S., Capt
Porter Jr, William R, CDR
Porterfield, Troy M, Lcdr
Portman, A Fritz, Lcdr
Porto, Joseph A., Lcdr
Porzio, Arthur, Capt
Posada, Luis E, CDR
Posadas, Bradley, LCDR
Posadas, John G., LCDR
Possumato, Roberto N., CDR
Post, Ephraim R., LCDR
Post, Jeffrey M., LCDR
Poston, James M., LCDR
Potash, Matthew E, CDR
Poteet, Billy J, Capt
Potempa, Ronald E, Lcdr
Potenza, Bruce M., CDR
Potochney, Peter J, CDR
Potochnik, Ventzel J, CDR
Potosnak, Richard E, LCDR
Pottenger, Carol M., CAPT
Potter, Charles B., Lcdr
Potter, Charles R, Capt
Potter, Joel M, Lt
Potter, Kevin J., Lt
Potter, Logan H, Lcdr
Potter, Michael C, Capt
Potter, Thomas A, Lcdr
Pottinger, Beverley, CDR
Pottinger, Scott M., CDR
Potts, Dale G, Capt
Potts, Edward A, Capt
Potts, John A, CAPT
Potvin, Louis E, CDR
Pou, Robert L, CDR
Poulton, William L, CDR
Pound Jr, Ralston M, CDR
Pounds, David S, CDR
Pounds, Truman E., Lt
Pourron Jr, Louis E, CDR
Powel III, Samuel F, Capt
Powell, A. Edward, LCDR
Powell, David G, Capt
Powell, Elmer C, CDR
Powell, Gary E., LT
Powell, George E, Lcdr
Powell, Henry F., CDR
Powell, Hurley J T, Capt
Powell, John C, LCDR
Powell, Matthew J, Lt
Powell, Richard R, Capt
Powell, Robert H W, Capt
Powell, Ronald L, Lcdr
Powell, Thomas B., LCDR
Powell, William Eugene, Lcdr
Powell, William M, Lcdr
Powell III, Teneyck B., CDR
Power, Edward J, Capt
Power, John F, Lcdr
Power, Timothy H, Capt
Powers, Anne T, CAPT
Powers, Donald C, Capt
Powers, Elizabeth Anne, LT
Powers, Gary S., Lcdr
Powers, Kendall G, Capt
Powers, Michael A., Lcdr
Powers, Michael C, Capt
Powers, Michael E, Capt
Powers, Rhoda Amadio, LCDR
Powers, Thomas J, CDR
Powers Jr, William J., Lcdr
Poyer, David C., Capt
Poynter, Drexel E, Capt
Pozek, Linda M., LT
Pozsgai, Timothy G., Lcdr
Prada, George B, CDR
Prall, Gary W., LT
Prather, Constance M, Lcdr
Prather, Royace H., Capt
Pratt, Alma P., CAPT
Pratt, John B, CDR
Pratt, Loren M, CDR
Pratt, Raymond L, Lcdr
Pratt, Steven G., Capt
Pratt, William L, CDR
Pratt, Wyatt B, CDR
Pratt II, Robert W., CAPT
Preble, John D., LCDR
Prechtl, Jerome S, Capt
Pred, Gordon D, Capt
Preddy, Raymond R, Capt
Prehn Jr, W Lawrence, CDR
Premeske, James W., CWO4
Prendergast Jr, Finis H, Capt
Preovolos, James P, CDR
Prescott Jr, D Clinton, CDR
Press, William A., CDR
Pressley, Stevie, Capt
Preston, Irving S, Lcdr
Preston, Noel G, Capt
Preston, Robert E., CDR
Preston, Zelldrick L., LCDR
Pretor, Ernest P, Lt
Prettyman, George D, Capt
Prevatt, Bruce C, Capt
Previ, Ronald W, Capt
Prey, John H, Capt
Preziose, Paul W, CDR
Pribnow Jr, Harry W, CDR
Pribyl, Silvin, Lcdr
Price, Alan H, Capt
Price, Charles M, CWO4
Price, Daniel M, CDR
Price, David A, Capt
Price, David G, CDR
Price, Donald S, CDR
Price, Frank L, Capt
Price, J Thomas, Lcdr
Price, Jacob M, Capt
Price, James O, CAPT
Price, Joel A., CDR
Price, Mary E, CDR
Price, Michael, LCDR
Price, Michael J., CDR
Price, Rebecca M., ENS
Price, Richard H, Capt
Price, Robert L, Lcdr
Price, Ronald J, Lcdr
Price, Sam T, CAPT
Price Jr, Edward W, CWO4
Price Jr, Raymond F, Lcdr
Price Jr, William J, Lcdr
Price Jr., John P., LT
Pricer, Mark A., LT
Prichard, Barry, Capt
Prickett, Jerry L, Capt
Prickett Jr, David C, Capt
Pride Jr, Jack G, Lcdr
Pries, William E, Capt
Priest, John E, Capt
Priest III, Fenton F, Capt
Priest Jr, William G, Capt
Priester Jr, Hugh W, Capt
Priewe, Raymon D., CDR
Prigge, Roger L, CDR
Prijic Jr., Daniel, LCDR
Primich, Frederic F, Capt
Primm, Charles E, LCDR
Primm, W. Rogers, Lt
Prina, L Edgar, Capt
Prince, Andrew S, Capt
Prince, Eric T, CDR
Prince, Milton S, Capt
Prince, Prescott L, CDR
Prince, Suzanne V., CDR
Prince, Warren E, CDR
Prince Jr, Aaron E, CDR
Prindle, Regina M., LCDR
Prindle, Ronald W, CDR
Prindle, William J, Lcdr
Pring, Brian L., Capt
Printer, Norman G, CDR
Pripp, Alwin F, Capt
Prise, Kevin M., LCDR
Prisk, Archibald S, CDR
Pristernik, Edward A, CAPT
Pritchard, Gary L., CDR
Probst, Theodore G, Capt
Prochnow, James E, CDR
Proctor, Christopher A, CDR
Proctor, Donald J, Capt
Proctor, Robert H, CAPT
Proctor, Robert W, Lt
Profilet, William J., Lcdr
Prohaska, Gerald F, CDR
Prose, Steven E., Lt
Proskey, Aloysius J, Capt
Prosser, Gwynneth M, Lcdr
Protacio, Barbara J., Lcdr
Proteau, Thomas J, Capt
Proulx, Gregory M., Lt
Prouty, Charles S, Capt
Provost, Catherine A., LCDR
Prozzillo, Frederick J., Lt
Prucha, James T, Capt
Pruden, Ethel E., CDR
Pruessner, Robert D, Capt
Pruet, Charles W, Capt
Pruett, Richard K., CAPT
Pruitt, W Bruce, Capt
Prukop, Ralph A, Lcdr
Pruneda Jr, Augustin C, CDR
Prusiecki, David E, CDR
Pruske, James E, Capt
Pryor, Roger W, Capt
Pryor III, Julius, CDR
Pryor Jr, Charles A, Capt
Psihas, Andrew P, Capt
Publicover, Francis G, CDR
Puckett, John R, Capt
Puckett, Philip L., CDR
Puckett, Ronald G, CDR
Pugh, Marion L, CWO4
Pugliese, Paul A., CDR
Puhrmann, Dale L., Capt
Pullara, James L., LT
Pullen, Mark C, CDR
Pulliam, Morris W, Capt
Pulliam, Zane W, LT
Pully, Maria T., LT
Puls Jr, George, CDR
Pulsinelli, John A, Capt
Pulst, Marvin L, Capt
Punzelt, David L, Capt
Pupa, John J., CDR
Purcell, Kieran J., Lcdr
Purcell, Michael C., CAPT
Purcell, Robert C, Capt
Purdy, Allan W, Lcdr
Purdy, Claire Haynes, Capt
Purdy, Thomas J, Lcdr
Puripongs-Aziz, Naboone J., Lcdr
Purpus, Harold E, Lcdr
Pursley, Charles R, CDR
Pursley, Scott J., CDR
Purves, William, Lcdr
Purvis, J. Scott, LCDR
Puryear, Harry H, CDR
Pusch, Robert Andrew, Capt
Puterbaugh, Jackie L., LT
Putman Jr, Thomas E., CDR
Putnam, Alan G., Lcdr
Putnam, David A., Lt
Putnam, Donald S, CDR
Putnam, Hollis T, Capt
Putnam, Jack M, Lcdr
Putnam, James K, CDR
Putnam, Keith L, CDR
Putnam, Randy B., Lcdr
Putnam Jr, Joe B, Capt
Putnam Jr, John G, Capt
Putt Jr, Kenneth F, Capt
Putterman, Donald E, CDR
Puttre, John M., CDR
Puzon, Daniel I., Capt
Pyeatte, Jesse E, Capt
Pykosh, Henry P, Capt
Pyle, Richard E, Lcdr
Pyle III, Howard, Capt
Pyne, Gerard W, CDR
Pyron, John E, Capt
Quaglino Jr, Joseph, CDR
Quaintance, Carl F, CDR
Qualey Jr., Michael P., LT
Quam, Paul E, CDR
Quandahl, Jerald R, CDR
Quaranta, Edward A., Lt
Quartulli, O Thomas, Capt
Queenan, John M., CDR
Query, William C, CDR
Quidatano Jr, Anthony F, CDR
Quier, John R, Cwo-4
Quigley, Kenneth L, Capt
Quigley, Stephen T, Radm
Quijada, Edward J, Capt
Quinlan, Gary J., LCDR
Quinlan, Matthew M, Lcdr
Quinlan, Timothy J., LCDR
Quinn, Anthony D., CDR
Quinn, Frances M, CDR
Quinn, James J, Capt
Quinn, Janet J.L., CDR
Quinn, John J., Lcdr
Quinn, Kenneth A, Lcdr
Quinn, Paul V, Radm
Quinn, Randy L., Lcdr
Quinn Jr, Edward B, CDR
Quinones, Gregg A., ENS
Quinones, Rafael, CDR
Quintana, Josephine F., Lt
Quintero, Donald B, CDR
Quirk, Richard G, CDR
Quisenberry, A Doyle, CDR
Raab, Frank, Radm
Raack, James D, Capt
Rabenhorst, Karl A., LCDR
Rabideau, Clarence J, Capt
Rabideau, Joseph, CDR
Rabith, Michael C., CDR
Rabun, James R, Capt
Rachko, Barbara G., LCDR
Rachupka, Craig, LT
Radcliffe, Robert L, CDR
Radcliffe, William M, Lt
Radel Jr, John L, Lt
Rader, Mark T., CDR
Radford, Francis M, Capt
Radford, Robert J, Capt
Radvansky, Joseph R, Lcdr
Radway, Robert W, Capt
Rae, Harry G., CAPT
Rae, James W, Lcdr
Raetz, Greg C, CDR
Rafanelli, Gregory D, CDR
Raff, John B., CAPT
Raff, Loretta, Lcdr
Raffe, Lance W, CDR
Raffetto, Shackley F, CDR
Raftshol, Rudolph W, CDR
Ragan, Alan K., CDR
Ragan, Andrew J, CWO4
Ragan, Helen Frances, Lcdr
Rager, Terrence R, Lcdr
Ragland, John R, Capt
Ragland, Robert C., Lt
Ragusa, Christopher M., Lt
Raher, Steven J., Lcdr
Rahmes, Mark D., Lcdr
Raichle, Henry F, Capt
Rainey, Barry E, CAPT
Rainey, Charles A., LCDR
Rainey, John C, CAPT
Rainey, William M, CDR
Rainone, Henry M., LCDR
Rains, Zane Bruce, LT
Raisbeck Jr, Clifford C, Capt
Rake, Richard L., LCDR
Raley Jr, John W, CAPT
Ralston, James S, CDR
Ralston, Robert R, CWO4
Ram Jr, Edward J, CDR
Ramage, James D, Radm
Rambo, Rex H, CDR
Ramian, Randall J, LCDR
Ramige, Melvin D, CWO4
Ramirez, Adolf A., CDR
Ramlo, John H, Capt
Ramm, William K, Lcdr
Ramos, Enrique R., LT
Rampey, James W., CDR
Ramsay, Selwyn P, Capt
Ramsdell, Richard G, Capt
Ramsden, Mary C., Lcdr
Ramsen, John C, Lcdr
Ramsey, Allan A, CDR
Ramsey, Brenton L, Capt
Ramsey, George N, CDR
Ramsey, Philip D., Lcdr
Ramsey, Robert Bruce, Lcdr
Ramsey, Wiley F, Capt
Ramsey, William P., CDR
Ramsland, Russell J, Lcdr
Ramstad, Marshall T, Capt
Ramstead, Gunard, CDR
Ranalli, Michael P, Capt
Rancourt, Norman G, CDR
Rand, Michael M, Capt
Randall, Daniel J., LT
Randall, Debra Louise, Lcdr
Randall, Ellsworth J, Capt
Randall, Frank R, Lcdr
Randall, George T, CDR
Randall, John M, Capt
Randall Jr, Dwight T, Lcdr
Randinelli, J. Scott, Lcdr
Ranftle, Michael A., Capt
Range, Robert W, CDR
Ranken, William B, Lt
Rankin, Eric D., Lt
Rankin, Loren A, Lcdr
Rankin, Robert A, LCDR
Ranney, Bruce W, Capt
Ransbottom, Catherine E, Lcdr
Rant, Michael J., Lt
Ranta, Silius C, Capt
Ranum, Kenneth G, CDR
Rao, Bolar R, Capt
Raper Jr, Hal S, Capt
Rapp, W Keith, CDR
Rappana, Duane, CDR
Rappeport, Myron H, Capt
Rapps, Richard E, Capt
Rappuhn, Ralph F, Capt
Rappuhn Jr, Alfred A, Capt
Rarig, Scott L., CAPT
Ras, Lee R, CDR
Rash, Tom J, Lcdr
Rask, Stephen J, Capt
Rasmus, Ronald C, Capt
Rasmussen, Eric A, Capt
Rasmussen, Nicholas R, Capt
Rasmussen, Richard A, Capt
Rast, T. Paul, CAPT
Rataczak, Mark H., Cdr
Ratajczyk, Craig Alan, LT
Rath, Jack A, Capt
Rath, Katherine E, CDR
Rathbone, Thomas M., Capt
Rathbun, Howard J., LTJG
Rathbun, Lawrence A, Capt
Ratkovich, John, Lcdr
Ratliff, Russell T., CDR
Ratner, William D, Capt
Rattan, Donald P., LCDR
Ratte, Charles L., CDR
Rau, Edwin M., CDR
Raudabaugh, Richard L, Capt
Rausa, James M, CDR
Rausch, Charles K, Lcdr
Rausch, Harry A, Capt
Rausch, Sheila B., CDR
Ravanpey, Ramin, ENS
Rawden, Roger G, Lcdr
Rawding, Henry C, CDR
Rawhouser, Marjorie, Lcdr
Rawlings Jr, John W, Lcdr
Rawlins, James A, CAPT
Rawls, Douglas S, Capt
Ray, Tracy L., LTJG
Rayburn, Ruth Ann, ENS
Rayder, Warren T., Capt
Rayermann, Richard E., LCDR
Raymond, Douglas R, CDR
Raymond, Edward J, Ltjg
Raymond, Lawrence W, Capt
Rayner, Ralph M, Lt
Razorsek, Nancy C., CDR
Read Jr, Dale, Capt
Read Jr, Frederick W, Capt
Readdy, William, CAPT
Reade Jr, Richard S, Capt
Ready, George E, Capt
Ready, Scott A., Lcdr
Reagan, John H, Ltjg
Reagan, Scott D., CAPT
Reale, Dennis M, CDR
Reamer, Mark G., LT
Reap, James B, Radm
Reaves, Daniel L., LCDR
Reavis, Karen J., LT
Reber, Karl W, CDR
Reberger, J Philip, CDR
Rechson, Mitchell H, Lcdr
Rechsteiner, Roy O, CDR
Reckert, Richard A, Lcdr
Record, Lloyd C., LT
Recto, Efren E., CDR
Rector, Robert W, Capt
Redd, J C, Capt
Reddan, Mark J, CAPT
Redding, Caleb Richard, CDR
Redding, Marshall S, Capt
Redfearn, Richard W, Capt
Redfield, Alva L, Lcdr
Redfield, James W, CDR
Redford Jr, Thomas G, Lcdr
Reding, Curtis C., Capt
Redinger, Rosalinda M., LT
Redington III, Joseph J, Lcdr
Redman, David B., CDR
Redmon, Alan R., LT
Redmond, James W, Capt
Redmond, Mary J, Lt
Redmond, Robert F, Lt
Redmond Jr., Donald D., CDR
Redpath, Sharon H., CDR
Reed, Bruce W., CWO4
Reed, Carl, Capt
Reed, David D, CDR
Reed, Dorothy J., LCDR
Reed, Garland G., LCDR
Reed, Howard I, CDR
Reed, James David, CDR
Reed, John G, Capt
Reed, Louis J, Capt
Reed, Margaret M.L., Lt
Reed, Nannette J, Lcdr
Reed, Ralph E, Capt
Reed, Robert M., CAPT
Reed, Wallace S, Capt
Reed III, Emile L, CDR
Reeder, Jonathan D., CDR
Reedy, Kathleen A., LT
Reem, Glen M, Capt
Reemsnyder, Harding E, CDR
Rees, Gayle H, Capt
Reese, Dennis F, Capt
Reese, Donald A, CDR
Reese, James C, CDR
Reese, Joseph W, Capt
Reese, Noel R., Lcdr
Reeser, John D., Lcdr
Reetz, Harvey F, CDR
Reeve, Kirk W, CDR
Reever, John S, CAPT
Reeves, Billy D, CDR
Reeves, Charles M, CDR
Reeves, Donald J, Capt
Reeves, George W, CAPT
Reeves, Gevan D., LT
Reeves, Warren E, CDR
Reeves, William B, Capt
Regadanz, Bruce E, CWO4
Regal, Roger William, Capt
Regan, James D, Capt
Regan, James F, Capt
Regan, John W, CDR
Regan, Rodney M., CDR
Register, Stephen T, Capt
Reglin, Gary L., CDR
Rehkemper, Alan C, Capt
Rehme Jr, John A, CDR
Reich, Richard J, Capt
Reich, Robert W., Capt
Reich Jr, Charles A, CDR
Reichard, Edmund C, Capt
Reichel, Richard L, CDR
Reichenbach, Donald D, Capt
Reichert, Harold H, CDR
Reichle, Garland E, Lcdr
Reid, Andrea J., LT
Reid, Doris A., CAPT
Reid, Gary B, CAPT
Reid, Kenneth J., Lt
Reid, Louis A., LT
Reid, Priscilla C, Lcdr
Reid, Rudolph, CWO4
Reid, Sean F., LCDR
Reid, Stephen P., LCDR
Reid, Van Cleave, Lcdr
Reid, Whitelaw, Lt
Reid, William S, Capt
Reid Jr, Owen P, CDR
Reid Jr, William T, CDR
Reidell, Christine E, Lcdr
Reidenbach, Robert P, Ltjg
Reider, George, Radm
Reidy, Brian P., LCDR
Reidy, James G, CWO4
Reidy, Patrick J., LCDR
Reidy III, Edward T, Capt
Reierson, Paul D, Capt
Reierson, Ronald M, CDR
Reifenstein, George H, Radm
Reiger, Edward J., LT
Reilley, John B, CDR
Reilly, Curtis G, CDR
Reilly, Edward C, Capt
Reilly, Michael F, CDR
Reilly, Michael J, Capt
Reilly, Thomas J, CDR
Reilly, William J., CDR
Reim, Kenneth R, CDR
Reimer, Ronald F., CDR
Rein, Douglas J., Capt
Reinauer, James R., CDR
Reinbold, Robert E., Lt
Reinbold, Thomas G, Lcdr
Reinhard, Michelle, Lcdr
Reinhardt, Laurent C.G., CDR
Reinheimer, Richard, CDR
Reininger, Robert J, CDR
Reinke, Henry S, Capt
Reinke, Sharon F, Capt
Reintjes, Robert J, CDR
Reis, Frederick S, Capt
Reis, John C, Lt
Reis, Peter S, CDR
Reisdorf, Stephanie, LT
Reisenauer, H Michael, CDR
Reiser, Leroy F, Capt
Reiser, Phyllis E., Lcdr
Reisert Jr, Joseph H, CWO4
Reising, Joseph A, Lt
Reit, Ronald O, Capt
Reite, Charles D, Capt
Reiter, David R., LCDR
Reiter, Gary Michael, CDR
Reiter, Glenn M, Lcdr
Reiter, Joshua, Lt
Relph, Virginia A., Lcdr
Remar, John M, CWO4
Remick, William C, CDR
Remington, Daniel W., LCDR
Remus, Fred E, CDR
Remy, Sandra L., LT
Renella, Marco Salvador, Capt
Renfree, Edward W., Lt
Renick, Charles M, Capt
Reninger Jr, Charles W, Capt
Renne, Merlin M, Capt
Renner Jr, Conrad J, Lcdr
Renton, William J, Capt
Rentschler, Donald R, Capt
Reott Jr., Donald E., CDR
Repchak, Richard N, Lcdr
Repp, Frank J, Lcdr
Reppard, David B, CDR
Reppert, Jerry L, Lcdr
Reppert, Joseph R, Lt
Resch Jr, Philip R, Capt
Resedean, Peter R, CDR
Resendez, Reynaldo, CAPT
Retson, Philip N, CDR
Rettinger, Donald H, Capt
Retzke, David E, Capt
Retzlaff, John E., CDR
Reuss, Mark R., LCDR
Reuter Jr., William C, CAPT
Reutter, Walter E, Capt
Reveal, Kenneth H, CDR
Revercomb, Jeffrey P, Lcdr
Rexford, Robert S, Lcdr
Reynierse, Peter J, CDR
Reynolds, Arlington, CWO2
Reynolds, Charles L., Lt
Reynolds, Courtenay B, Lcdr
Reynolds, David A, Capt
Reynolds, Don R, Capt
Reynolds, Edward L, CDR
Reynolds, Frank P, Capt
Reynolds, Frank R, CDR
Reynolds, Gary L, CDR
Reynolds, John A S, Capt
Reynolds, Mary M., LTJG
Reynolds, Paul F, Capt
Reynolds, Paul T, Lcdr
Reynolds, Robert L., LCDR
Reynolds, Robert W, Capt
Reynolds, Robert W, Capt
Reynolds, Roger W, Capt
Reynolds, Vincent G, Lcdr
Reynolds, William H, Capt
Rheker III, Joseph F., MIDN
Rhett, William, CDR
Rhoades, Gael A, Lcdr
Rhoades, William W, CAPT
Rhoads, Dewain, Lcdr
Rhoads, Donald R., Capt
Rhodeman Jr, Thomas J., CDR
Rhodenbaugh, Ralph M, CAPT
Rhodes, Dan C., CDR
Rhodes, Henry V., CDR
Rhodes, J William, CAPT
Rhodes, John M, Capt
Rhodes, Raymond, CDR
Rhodes, Robin, CDR
Rhodes, Thomas W, Capt
Rhodes, William G, CDR
Rhodes II, Richard A, CDR
Rhyne, Dennis A, Capt
Ricca, Arthur, LCDR
Riccio, Cole A, LT
Ricco, Nicholas D, CDR
Rice, C Erwin, Lcdr
Rice, Charles L., Capt
Rice, Charles W, Capt
Rice, David M, Lcdr
Rice, David P, Capt
Rice, Edward W, CDR
Rice, Gilbert T, CWO3
Rice, Kenneth K, Capt
Rice, Samuel Stone, Lcdr
Rice, Steven T, CDR
Rice, William T., Capt
Rice Jr, Emory C, Capt
Rice Jr, John C, Capt
Rich, Brian C, CDR
Rich, Harold G., RADM
Rich, James C, CDR
Rich, Robert T, Capt
Rich, William L, Capt
Rich Jr, Charles H, Capt
Rich Jr, Howard L, Capt
Richard, D. Eugene, CDR
Richard, Patricia A., Lcdr
Richards, Anthony C, Capt
Richards, Catherine R, CDR
Richards, Charles L, CDR
Richards, Gregory W, Capt

Richards, James H, Capt
Richards, Jeffrey, LCDR
Richards, John R, Capt
Richards, Richard L, Capt
Richards, Ricky V., CDR
Richards, Robert J., CDR
Richards, Stuart A., Capt
Richards III, Davis A, CDR
Richards Jr, Bernard N, Lcdr
Richards Jr., J. Permar, LCDR
Richardson, Charles C, Capt
Richardson, Charles W, Lt
Richardson, Dale A., LCDR
Richardson, Diane C., LT
Richardson, Emmett W, Capt
Richardson, Frank A, Capt
Richardson, J D, CDR
Richardson, James D, CDR
Richardson, Joyce F., LT
Richardson, Robert L, CDR
Richardson, Silvana F., CDR
Richardson, Weinert, Lt
Richardson, Zachary M., Lcdr
Richardson III, F. W., CDR
Richardson Jr, John M, Lcdr
Richey, Robert C, CDR
Richman, Michael Scott, ENS
Richmond, Lyle L, Capt
Richter, Gary P., LT
Ricker Jr, Charles T, Capt
Rickerman, William J, Lt
Rickert, John M, Lcdr
Rickert, Walter K., Lt
Ricketson Jr., Charles, LCDR
Rickett, Joseph W, Capt
Ricketts, Donald B, Capt
Rickords, Ralph E., CDR
Ricks, Henry C, Capt
Riday Jr., Raymond S., CDR
Riddell, Gerald D, Capt
Riddiough, Peggy L, CDR
Riddle, Glenden Paul, Lcdr
Riddle, Peter E, Capt
Ridgway, David W, Lcdr
Ridler, Richard L., LT
Riebe, Donald E., CDR
Riebeling, Herbert A, CDR
Rieber, Christopher, Capt
Rieder, Albert E, Radm
Rieger, Jane E, Lcdr
Rieger, Robert V., LCDR
Rieke, Edward T, Lcdr
Rielly, Thomas C, Lcdr
Rieper, Alan G, Capt
Riess, Daniel M, Capt
Riester, John E., CDR
Riffel, Robert G, CWO4
Riffle, Ronald K, Lt
Rigazio, Richard C., Capt
Rigg, Michael D, CDR
Riggin, Scott A., CDR
Riggins, Timothy L., CDR
Riggins Jr, Freddie, CDR
Riggins Jr, Tracy, Capt
Riggs, Joseph Ronald, CDR
Riggs, Mary C., Lcdr
Riggs, Wayne L, CDR
Riggs Jr, Carl O, CDR
Righter, Donald L, CWO4
Rigoli, William L, Lcdr
Rigone, John L, Capt
Rigoulot II, Kenneth G., CDR
Rigsbee, Clifford M, Capt
Riisma, Richard A., MIDN
Riley, Edward F, CDR
Riley, Frederick L, Capt
Riley, James E, CDR
Riley, John H., Capt
Riley, John J, Capt
Riley, John T, Capt
Riley, Keith P., Lt
Riley, Michael R, Capt
Riley, Terrence L, Capt
Riley, Thomas F, Capt
Riley, Timothy C, CDR
Rima Jr, William H, Lcdr
Riman, Bernard B, Lcdr
Rimas, Generoso S., Lcdr
Rimbach Jr., Francis L., Lcdr
Rinaldi, Karyn M., LT
Rinando, Alan R., Capt
Rinard, Tommie F, Radm
Rinderer Jr, William P, LCDR
Rindin, James A, Capt
Ring, Paul D., CDR
Ringelberg, John M, CAPT
Ringer, James H, Capt
Ringstad, Robert C, Capt
Ringstrom, Philip J., LCDR
Rio Jr, Manuel, Capt
Riordan, John J, Capt
Rioux, Michael P, CDR
Ripley Jr, James M, CAPT
Rippee, Earl F, Capt
Rippel, George J, Capt
Rippey, Joseph S, Lcdr
Ripsom, George A, Capt
Rish, Berkley L, Capt
Risher, John F, Capt
Rising, Richard P, Capt
Risko, Annabelle B, CDR
Rissel, Martin C, CDR
Risser, Thomas A, Capt
Ritchie Jr, George P, Lcdr
Ritter, Charles E, Lt
Ritter, James P, CDR
Ritter, Laurence B, Capt
Ritter, Robert A, Capt
Ritter Jr, Jack F, Capt
Rittinger, Thomas E, Lcdr
Ritz, Mary L., CDR
Ritzenberg, Phillip, CDR
Rivas, Juan A., LT
Rivel, Robert B, CDR
Rivenbark, Michael D., LCDR
Rivenbark, Terry G, LCDR
Rivera, Carmen M., CDR
Rivera, Hector J., LCDR
Rivera, Homar, LCDR
Rivera, Sean P., Lcdr
Rivera-Alsina, Manuel E, Capt
Rivers, Jere W, Capt
Rivers, Michael Thomas, CAPT
Rivers, Sophia, LCDR
Rix, William H., Capt
Rizack, Martin A, Capt
Rizzi, Louis L, Capt
Rizzo, Anthony J., CDR
Roach, George E., Capt
Roach, James P, Lcdr
Roach, Mary V., Lcdr
Roach, Thomas D., LT
Roach, William J., CDR
Roan, Leo C, CDR
Roark, Cecil A, Lcdr
Roark III, George J, Capt
Robards III, Frank B., CDR
Robb Jr, David M, CDR
Robbins, Charles R, CDR
Robbins, Dayton D, Lcdr
Robbins, Duane E, Capt
Robbins, Edward M, Lcdr
Robbins, Joseph, Capt
Robbins, Kenneth M, CDR
Robbins, Richard C., Lcdr
Robbins, Richard R, Capt
Robbins, Robert J, CDR
Robbins Jr, Joseph A, CDR
Robbins Jr., David Lee, LCDR
Roberson, Carlton Lewis, Lcdr
Robert, James R, Capt
Robertie, Frederick T, Capt
Roberts, Bayard H, Lt
Roberts, Brian N., CDR
Roberts, Charles B., CDR
Roberts, Charles E, CDR
Roberts, Charles G, CDR
Roberts, Charles H., CDR
Roberts, Charles Terry, CDR
Roberts, Derek G, CDR
Roberts, Dexter F, Lcdr
Roberts, Donald J, LCDR
Roberts, Duke E., Lcdr
Roberts, Edward R, CDR
Roberts, Frank D, CDR
Roberts, Gordon L, Lcdr
Roberts, James B, Capt
Roberts, James L, CDR
Roberts, James R, Capt
Roberts, Jared I, CDR
Roberts, Jean E, Capt
Roberts, John S, Capt
Roberts, Joseph A., LCDR
Roberts, Paul G, Capt
Roberts, Rex G, CAPT
Roberts, Samuel R, Capt
Roberts, Steven W., CAPT
Roberts, Thomas L, CDR
Roberts, Tim Hal, Capt
Roberts, Vincent B., ENS
Roberts, Wayne L, Lcdr
Roberts, Wilbur E, Radm
Roberts, Willard G, LCDR
Roberts, William A, Capt
Roberts, William B, Capt
Roberts Jr, Clarence L, Capt
Roberts Jr, James E, Capt
Roberts Jr, Thomas C, Capt
Robertson, Alexander A, CAPT
Robertson, Alton F, Capt
Robertson, Daniel S, CAPT
Robertson, Francis G, Capt
Robertson, Harry L, Lcdr
Robertson, John Sherrod, CAPT
Robertson, Larry Allen, CDR
Robertson, Morgan M, CDR
Robertson, Richard A, Capt
Robertson, Stanley M, Capt
Robertson, Steven M., CDR
Robertson, William F, CDR
Robertson Jr, Lawrence, Capt
Robey, Dana P., LCDR
Robey, Richard E., CAPT
Robichaud, Eugene J, CDR
Robichaud, Paul G., CDR
Robillard, Edward L, CDR
Robin, Clinton G, Capt
Robins, Derek A., Lcdr
Robinson, Antonia J G, Lcdr
Robinson, Arthur L., LT
Robinson, Barry H, CDR
Robinson, Brant F., LT
Robinson, Darryl W., LT
Robinson, David, CDR
Robinson, Don E, Capt
Robinson, Dunlap R, Capt
Robinson, Eugene I, Lcdr
Robinson, Gary L, Capt
Robinson, Greg, LCDR
Robinson, Harry C, CDR
Robinson, Henry G., LT
Robinson, Isaiah N., Lcdr
Robinson, James F, Lcdr
Robinson, John A., ENS
Robinson, John C, Lt
Robinson, John E., LCDR
Robinson, Kenneth W, CDR
Robinson, Lawrence H, Capt
Robinson, Mark H., Lcdr
Robinson, Michael F, Capt
Robinson, Milton A, CDR
Robinson, Richard H, Capt
Robinson, S Jesse, CDR
Robinson, Suzette E., Lcdr
Robinson, Ted G., Lcdr
Robinson, Teressa H., LCDR
Robinson, Terry L, Capt
Robinson, Wilbert C, Lcdr
Robinson, William H, Capt
Robinson, William R, CDR
Robinson Jr, Paul H., Lt
Robinson Jr, William G., Lcdr
Robinson Jr., Ronald A., LCDR
Robison, Edward C., LT
Robles-Coombs, Judith M, LCDR
Robson, Gloria S., CDR
Robson, John H, Capt
Roby, David D, CDR
Roby, Jon Lee, Lcdr
Rocco, David A., CWO3
Roche, Clinton E, CDR
Roche, Jerome J., Capt
Roche, John R, CDR
Rochon, Anthony C., ENS
Rock, Calvin C, Capt
Rock, Robert R, Lcdr
Rock, Stanley Arthur, Capt
Rocker, James T, CDR
Rockey Jr, Robert D, CAPT
Rockwell, Charles D, Capt
Rockwell, Lewis A, CDR
Rockwell, Robert G, Lt
Rockwell, Susan L, CDR
Rockwood, John S., CDR
Rockwood, Richard J., CDR
Rocque, Paul F, Capt
Roddenberry, Devin S., LCDR
Roddy, Beverly A., LCDR
Rodefeld, Kristine M., ENS
Rodelli, Melvyn A, Capt
Roderick, Sandra A, CDR
Rodes Jr, James F, LCDR
Rodgaard, John A, CDR
Rodgers, James, CWO3
Rodgers, James A, CAPT
Rodgers, Stephen F, Capt
Rodgers, Stephen J, Capt
Rodgers, Vernon P, Capt
Rodi, Louis A., LT
Rodman, M Robert, CDR
Rodman, Stanley R, CDR
Rodman Jr, James H, Lcdr
Rodriguez, Antonio J, Capt
Rodriguez, Carol A, Capt
Rodriguez, David B., LT
Rodriguez, James A., LCDR
Rodriguez, James T, Capt
Rodriguez, Libertad, CDR
Rodriguez, Pedro, LCDR
Rodriguez, Pedro P., Lt
Rodriguez, Robert J., CDR
Rodriguez, Ruben A., LCDR
Rodriguez, Ruben S., Lt
Roe, Jack W, Capt
Roe, John E, Capt
Roeder, Erich R., LCDR
Roedersheimer, Charles, Capt
Roegner, Craig W., LT
Roehrig Jr, Richard W, CDR
Roelant, James G., CDR
Roemer, Eileen J., CDR
Roemer, John J, CDR
Roemer Jr, Ernest, Capt
Roenbeck, Richard G, CDR
Roese, William E, CWO4
Roettger Jr., Norman C., Capt
Roffey Jr, Robert C, Capt
Rogers, Anne K., ENS
Rogers, Betty A., LT
Rogers, Charles E, Capt
Rogers, Dorothy M, Lt
Rogers, Fred B, Capt
Rogers, John C., LCDR
Rogers, John H., CDR
Rogers, John W., CDR
Rogers, Jonathan D, Lcdr
Rogers, Mary R., Lcdr
Rogers, Michael H, CDR
Rogers, Paul C, Lcdr
Rogers, Philip J., CDR
Rogers, Richard M, CDR
Rogers, Roy R, CDR
Rogers, Sarah F.R., CDR
Rogers, Stacey A., LT
Rogers, Steven L., LT
Rogers, Theodore E., CDR
Rogers, Thomas G, Lcdr
Rogers, William F, Lcdr
Rogers, William J., Lcdr
Rogers, William W, CDR
Rogers Jr, Walter J, CDR
Rohay, Charles D, CDR
Rohde, Dennis M, CWO4
Rohde, Francis E, Capt
Rohe, Max E, CDR
Rohleder, John R, Radm
Rohlman, Mary R, LCDR
Rohr, John A, CDR
Rohrer, Paul W, Radm
Rohrssen, Steven W., CDR
Roig, William A., CAPT
Roland, Dennis A, Lcdr
Roland, Elizabeth J. D., CAPT
Roland, Jane M., Lcdr
Roland, Michael J, Capt
Rolf, Robert P, CDR
Rolfes, David H, Capt
Rolfs, James D, CDR
Rollandi, Victor L., Capt
Rollberg, Charles A, Lcdr
Roller, George S, Capt
Roller, Harry E, Lcdr
Roller, James P, Lcdr
Rollins, Clyde T, Capt
Rollins, James A, CDR
Rollins, Laurence R, Capt
Rollo Jr., Lawrence D., LT
Rollow Jr., Thomas A., CDR
Roloff, Marston V, Capt
Rolph, H Renton, Capt
Roman, Daniel R., LCDR
Roman, Elmer L., ENS
Roman, Richard J., LT
Romanchek, Walter R, Capt
Romano, James J, CDR
Romano, William S, Lcdr
Romanoli, Peter J, Lcdr
Romanski, Peter J, Lt
Romary, Dennis C, Capt
Romero, Francisco J., CDR
Romero, Jose L., CDR
Romine, David E., LCDR
Ronan, Terrence, CDR
Roney Jr, Ernest E, Capt
Rongone, Mark E., CDR
Ronzio, Margaret J., CAPT
Rood, Larry Norman, Capt
Roof, William H, CDR
Roomian, Richard S, CDR
Roon, Eugene R, Capt
Rooney, Daniel J, Capt
Rooney, James T., Lcdr
Rooney, William E, CDR
Roop, Howard, Radm
Roots, Charles R., Lcdr
Roots, David W., Lcdr
Roper, Daniel A., Lt
Roper, Leah Fe, ENS
Roper, Robert B, LCDR
Roquemore, Michael B., Capt
Rosa, A. K., Lcdr
Rosales, Roland, LT
Rosanelli, Gabriel D, CDR
Rosar, Honor A., Lcdr
Rosaschi Jr, Peter P, CDR
Rosborough, Bruce R., LT
Rosburg, William D, Lcdr
Roscoe, John F, CDR
Rosdahl, Robert F, CDR
Rose, Barbara A., LCDR
Rose, Charles S, CDR
Rose, Edward W, CWO4
Rose, Guy P, Capt
Rose, Harold D, CDR
Rose, Julian F., CDR
Rose, Richard J, CDR
Rose, Robert E., LCDR
Rose, Ronn S., Lcdr
Rose, Victus F., Lcdr
Rose, William N, CDR
Rosemond Jr, Julian B, Lcdr
Rosen, Arvin D, Capt
Rosen, Julius J, CDR
Rosen, Paul, LCDR
Rosen, Robert S, Lcdr
Rosen NYNM, Robert A., Radm
Rosenbaum, Lester R, CWO3
Rosenberg, B Robert, Lcdr
Rosenberg, Charles, CDR
Rosenberg, David Alan, CDR
Rosenberg, Garrett H, Lcdr
Rosenberg, Marvin A., CDR
Rosenberg, Richard M, CDR
Rosenberger, Beryl G, Capt
Rosenberger, Carol A, CDR
Rosenberger, Jay N, Capt
Rosenberger, John, Capt
Rosenbloom, Noah S., Lcdr
Rosenbloom, Richard M, Lt
Rosenblum, David A, Capt
Rosenblum, Everett A, Capt
Rosene, Philip G, Capt
Rosene, Robert W, CDR
Rosenkranz, Lois M, Lcdr
Rosenthal, Joseph M, Capt
Rosenthal, Lawrence E, CDR
Rosenthal, Stephen A, Capt
Rosholt, Gary W., CDR
Rosnes, Bruce H, Capt
Ross, Ben A, CDR
Ross, Chet S., CDR
Ross, Christopher T W, CDR
Ross, Donald H, CWO4
Ross, Douglas A., LCDR
Ross, Edward M, Capt
Ross, George, CDR
Ross, Harold M, Capt
Ross, Kenneth, CDR
Ross, Leslie J, CWO4
Ross, Paul F, Capt
Ross, Paul P., CAPT
Ross, Ralph M, Capt
Ross, Robert F, Lcdr
Ross III, Raphael, Capt
Ross Jr, William C, CDR
Rossell, Harold J., Lcdr
Rossen, Matthew, Lcdr
Rosser, William T., Lt
Rossetti, Carl J, Lcdr
Rossetti, Lee V., CDR
Rossi, Albert J, CDR
Rossi, Louis A, Capt
Rossi, Robert J, Capt
Rossio, Michael, LCDR
Rossland, Robin L., Lt
Rosson, Frank W, LCDR
Rostad, Jerald A., Lt
Roth, Charles W, Lcdr
Roth, Lewis F, CDR
Rothenberg II, H J, Capt
Roths, John M, Capt
Rothschild, Bernard B, Lt
Rothschild, Joel S., Lcdr
Rothstein, Jerome P, Capt
Rothstein, Robert, Lt
Rothwell, Peter S, Capt
Roti, Susan J., LCDR
Rottmann, Gary E., Lcdr
Rotton, Sietze Jeff, CDR
Rouch, Sidney A., LCDR
Rouen, Douglas Michael, Capt
Rouillard, Irene C, CDR
Rounds, Horace P, Lcdr
Rouse, Daniel P., CDR
Rouse, David A., LT
Rouse, Luonne A., LCDR
Rouse, Robert M, Lcdr
Rouse Jr, Robert G, Capt
Roush, Okey C, Lcdr
Roussakies, John C., MIDN
Rousseau, Charles E, Capt
Rousseau, Kevin, LTJG
Rousseau, Larry J, Capt
Roussell, Gerald J, Lt
Routson, Janet L., CDR
Routzahn, Karl D., CDR
Roux Jr, Leo G, Capt
Rouzer, William W, Capt
Rowan, Robert L., Lt
Rowe, Daniel R., LCDR
Rowe, Earl L, CAPT
Rowe, Henry C, Lcdr
Rowe, John E, Lcdr
Rowe, Mary Ann F, CAPT
Rowe, William H, Capt
Rowe Jr, George, Lt
Rowe Jr, Joseph V, Lcdr
Rowland, Dean C, Lcdr
Rowland, John K., LCDR
Rowlett, Caius M, Lcdr
Rowley, Charles W, Capt
Roy, Donald E, Radm
Roy, Larry W, Capt
Roy, Regina C., Lcdr
Roy, Robert L, Capt
Roy Jr., William Glenn, CDR
Royal, Clifford H, Capt
Royal, Raymond, CDR
Royal, Thomas A, CDR
Roybal, Sandra A., CDR
Royce, John C., CDR
Royce, Philip A, Lt
Royder, James O, Capt
Royer, Thomas E, CDR
Roys, Michael S., CAPT
Roys, Perry F, CDR
Royston, Clyde E., LCDR
Rozic, Joseph T, Capt
Rozsypal, Edward J, Capt
Rozul, Alex V., CDR
Rubenstein, Richard B, Capt
Rubiano, Bolivar, Lcdr
Rubin, David L, Capt
Rubino, James F, Capt
Rubino, Simon, Lcdr
Ruch, Roger E, LCDR
Rucker, Brian R, CDR
Rucker, William K., CDR
Rudd Jr, Edwin I, Capt
Rude, LeRoy D, LCDR
Rudell, Frederick L, Capt
Rudinger, Edwin A, Capt
Rudman, Joseph W, CWO4
Rudy, John C, CDR
Ruedas, Beethoven T, Capt
Rufe, Robert W, Capt
Ruff, Richard K, Capt
Ruffin, Lynn N, Capt
Rufo, Paul A., Lt
Rugaber, Christopher R., LCDR
Rugg, William R, Capt
Ruggiero, Lawrence E., CDR
Ruggiero, Ralph L., CDR
Ruiz, Arid Sepulveda, LTJG
Ruiz, Eleazar M, Lcdr
Ruiz, Fernando A, CDR
Ruiz, Lawrence D., LT
Ruiz, Marisol Garcia, ENS
Ruiz III, Adalberto, CDR
Ruland, Thomas C, Capt
Rule III, Adrian O, Capt
Rumans, Larry W., Capt
Rumbaugh, Richard Lee, Capt
Rumley Jr., Gail E, Lcdr
Rummele, John E, Capt
Rump, Michael U, Lcdr
Rumrill, James K, Capt
Rumsch, Bruce A, CDR
Runals, Michael L, CDR
Rundell Jr, Howard F, Capt
Rungo, Ralph J, Capt
Runkle Jr, William A, Capt
Runnels, John B, Capt
Running, Kristi A., LCDR
Running, Richard B, Capt
Runyan, Joshua B., CDR
Rupert, M Kimberly, Capt
Rupinski Jr., Daniel T., LCDR
Ruple Jr, Frank W, Lcdr
Ruppel, David E., LCDR
Ruppel, John J, CWO4
Ruppel, Robert W, Capt
Rupprecht, George J, Capt
Ruseling, Gary L, CDR
Rush, Douglas K, Capt
Rush, John D, CDR
Rusher, Albert H, CDR
Rushing, Jack, Lcdr
Rusnak, Martha A, Capt
Russ, Carol A, CDR
Russ, Henry H., CWO3
Russ, Myron H, CDR
Russ, Roger M, CWO4
Russe II, Laurence P., CDR
Russell, Andrea J., Lcdr
Russell, Ann J., CDR
Russell, Charlie S, Capt
Russell, David G, CDR
Russell, David L, CDR
Russell, David T, Capt
Russell, Harold J, Capt
Russell, Henry C, CDR
Russell, Howard L., CAPT
Russell, John J, Lcdr
Russell, John T, Capt
Russell, Kenneth L., LT
Russell, Richard Lee, CDR
Russell, Robert J, Capt
Russell, Sanford H, Capt
Russell, Steven L, CAPT
Russell, Walter J, CDR
Russell, William H, Capt
Russell III, Lee C., Lt
Russert, Jamie, ENS
Russo, Raymond M, Capt
Russo, Ronald F, Capt
Russo Jr, Frank R., LCDR
Rust, Charles A., CDR
Rust, Robert D, Capt
Rust Jr, Paul D, Lcdr
Ruter, Charles M, Lcdr
Ruth, Edward J., CDR
Ruth, Philip M, CAPT
Rutherford, Bruce J, CDR
Rutherford, Harold G, Capt
Rutledge, Belinda J., CDR
Rutledge, Delbert L, Lcdr
Rutledge, John W, Lcdr
Rutledge, Larry L., CDR
Ruttenberg, Deborah K., LT
Ruwwe, George R, Capt
Ryals, Paula A., Capt
Ryals, Willard G, Capt
Ryan, Alan S, Lcdr
Ryan, Burns S, Lcdr
Ryan, Charles T, Capt
Ryan, Christopher A., LCDR
Ryan, Edward J, CDR
Ryan, Gary T, CDR
Ryan, Harriet R, CDR
Ryan, James A, CAPT
Ryan, James P, Capt
Ryan, Jeremiah P, CDR
Ryan, Margaret Anne, LCDR
Ryan, Michael G., Lcdr
Ryan, Michael J, LCDR
Ryan, Paul J, Capt
Ryan, Raymond M, Lcdr
Ryan, Richard S., Capt
Ryan, William R., CDR
Ryan Jr, Calvin C, CDR
Ryberg, William J, CDR
Rydland, Robert H, Capt
Rye, Scott C., Lt
Ryer, Thomas A., LCDR
Ryerson, David R, CDR
Ryerson, Joe F, Lcdr
Rykken, Edward W, LCDR
Rykowski, Margaret A, CDR
Rypien, David V., Lt
Rytell Jr, Frank E, Capt
Rzeszotko, Joseph A., CDR
Saare, Eugene D, CDR
Sabelli, Anthony J, Lcdr
Saberson, Lee A, CDR
Sabol, Alexander J, Capt
Sabree, George K, Lcdr
Sachuk, Robert J., CDR
Sacra, James W, Lcdr
Sadler, Dudley C, Capt
Sadler, John C., CDR
Sadler, Linville R, CWO4
Sadler, Mary Jean, LCDR
Sadler, Michael J., LCDR
Sadlowski, Michael E., LCDR
Safina, Michael J., Lcdr
Sage, Clarence E, CDR
Sagen, Paul G, CDR
Sager, Edward R, CDR
Sager, Harlan R, Capt
Sager Jr, Paul H, CDR
Sahaida, John, Capt
Sahrbeck, Douglas J., CAPT
Saiki, Kenneth T, Capt
Sailer, Joseph R., Capt
Sain, Frank C, CDR
Saint Jr, William K, CDR
Sajna, Michael, CDR
Sakel, Daniel E., Lcdr
Salamanca, Raymond J, CDR
Salamoni, Vincent A., LCDR
Salander, Samuel, CDR
Salazar, Guillermo J, CAPT
Salazar, Pedro E, LCDR
Salcedo, Sylvester L, Lcdr
Salisbury, Roger E, CAPT
Salkeld, Stephen A, CDR
Sallade, Frederick E, Capt
Sallah, Stephen J., LCDR
Salley, Leonard B, Capt
Sallows, Daniel F., LCDR
Salman, Kenneth A., Lcdr
Salmen, Mark J, Capt
Salmon, Edward P, Capt
Salmon, Wayne S, Capt
Salter, Albert P, CDR
Salvatore, Anthony D., MIDN
Salyer-Caldwell, Ann L., LCDR
Samchuck, Gerald M., LT
Sammons, Lucy N, Capt
Samms Jr, Floyd T, Capt
Sample, Harvey C, Lcdr
Sample Jr, William B, Capt
Sampson, George C., LT
Sams, Marion A, Capt
Samtmann, David G, Capt
Samuelson, Howard B., LT
Samuelson, Ronald A, Capt
Sanborn, James H, Capt
Sanchez, Daniel M., CWO4
Sanchez, Lorlee, Lt
Sanchez, Robert P, Lcdr
Sanchez, William A, Capt
Sand, Wayne R, Capt
Sandahl, Lee F, Lt
Sandberg, David E., Lcdr
Sandberg, Nils S., Capt
Sandelli, Leonard C, CDR
Sander, Karl, ENS
Sanders, David M.S., LT
Sanders, George H, Capt
Sanders, Herold C, Lcdr
Sanders, Jerrell L, CDR
Sanders, John F, Capt
Sanders, Kenneth R., Lcdr
Sanders, Max F, Capt
Sanders, Norman R., CDR
Sanders, Robert E, Lcdr
Sanders, Scott E, CDR
Sanders, Shawn G., Sgt
Sanders, Thomas E, Capt
Sanders, Wade R, Capt
Sanders Jr, Jesse A, Lcdr
Sanderson, Owen J, CDR
Sanderson, Tom L., Capt
Sanderson III, B S, Lcdr
Sanderson Jr, Joseph T., Lcdr
Sandhoo, Gurpartap S., LT
Sandmann, Robert E, Capt
Sandor, Mark, LCDR
Sands, Jerome T, Lcdr
Sands, Mark R., LT
Sands II, Ned R., Capt
Sandstrom, John E, CWO4
Sandvig, Dennis L, CDR
Sandweg, Robert F., CAPT
Sandy, Grant Ferrell, Lcdr
Sanford, Cecil L, CDR
Sanford, Frederic G., RADM
Sanford, Katherine L., CAPT
Sanford, Robert L, CDR
Sanford III, Edward W., LCDR
Sang, Gary E., CDR
Sanks, Harold L., LT
Sanmiguel, Ruben D., Lcdr
Sanstead, Peter N., LT
Santeiu, Elizabeth M, Capt
Santelli, Takae Francis, Lcdr
Santiago, Jose F., ENS
Santini, Basil A, CDR

Santisteban, Jackie, CDR
Santo, Daniel C, Lt
Santopietro Jr, Thos V, Capt
Santos, Joao Miguel, LCDR
Santos Jr, William, CDR
Santucci, Steven J, Capt
Sanwick Jr, Paul B., Capt
Sapp, Dennis J, Capt
Sapp, Michael E., Lcdr
Sapp, Virginia Lynn, Lcdr
Sappenfield, James A., LCDR
Saracino Jr., Joseph E., LCDR
Sarber Jr, John D, Capt
Sargeant, Alan D., CAPT
Sargent, Joel A, CDR
Sargent, William B, CDR
Sargent III, George K, Capt
Sargent Jr, Edward C, Capt
Saricks, John H, Capt
Sarkisian, Haig M, Lcdr
Sarnie, Robert W, Capt
Sarnoski, Stephen R., CDR
Sartin, T R, Capt
Sartori, Howard J, Capt
Sarwacinski, Belen M, Lcdr
Sasmor, James C., Lt
Sass, Arthur H, Capt
Sassaman, Robert W, CDR
Sassone, Charles H., CDR
Sather, Andrew O, CDR
Satterfield, Gary T, Capt
Satterfield, Grey W, CAPT
Satterfield, John R., LT
Satterthwaite, F G, Capt
Sauer, Gerald L, Capt
Sauer, Karl F, LCDR
Sauer, William F, Capt
Saul, Bill H, Capt
Saul, Gregory D., CDR
Sauls, Charles B., LT
Saunders, Glenn M, CAPT
Saunders, Hobie, Lt
Saunders, James R, CDR
Saunders, Laura S., Lt
Saunders, Richard S., CAPT
Saunders Jr., Charles L, Lt
Sava, Samuel G, Capt
Savage, John E, Capt
Savage, Robert L, CDR
Savage, Robert M., CAPT
Savage, Ronald William, Lcdr
Savage, Stephen A, CWO4
Savage, Warren H, CWO4
Savard, Arthur E, CDR
Savarese Jr, Charles J, Capt
Savides, George, CDR
Savidge, Thomas W., Lcdr
Savio, Leo J, Capt
Saviola, Carl P, Lcdr
Savory, William A, CDR
Sawada, Kathleen Y, CAPT
Sawczuk, Stanley R, CDR
Sawyer, Brooke E, CDR
Sawyer, Christopher T.J., LT
Sawyer, Dennis L, CDR
Sawyer, Diana L., Lcdr
Sawyer, Gwendolyn Anne, CDR
Sawyer, Robert L., LCDR
Sawyer, Ronald G, CDR
Sawyer, Russell D, CDR
Sawyer, Thomas E, Capt
Sawyer, William A, Capt
Sawyer Jr, Richard P, Capt
Sawyer Sr., Edward G., Lcdr
Sax, Samuel W, Capt
Saxberg, Joel T, CDR
Saxey II, Edward, CDR
Sayer, George C, Radm
Saylor, Charles M., CDR
Saylor, David S, Lcdr
Saylor, Philip G, Capt
Sayne, Patrick J, CDR
Sayner, Nancy C, Capt
Sayre, Edwin D, CDR
Scaboo, George N, Capt
Scacchi, Michael P, CAPT
Scagos, George A, Lcdr
Scalenghe, Steven W, CDR
Scanlan, Emmett A, CDR
Scanlon, Jerome B, Capt
Scanlon, John J, Lcdr
Scanlon, Robert M, Lcdr
Scaperotta, Ralph M, CDR
Scappini, Richard J, CDR
Scaring, William A, Capt
Scaring, William E, LCDR
Scarlata, Carol A., Lcdr
Scarlata-Naatz, Lisa J., Lcdr
Scarpulla, Norman K, Lcdr
Scattergood, J. Henry, Capt
Scattolini, Joan Marie, CDR
Scavullo, Joseph J, CDR
Scee, Joseph E, CDR
Schaan, Daniel J., LT
Schaap, Steven G, CDR
Schade Jr, Walter B, Capt
Schaeberle, Robert M, Lcdr
Schaefer, Dale G, Capt
Schaefer, Daniel R., Lcdr
Schaefer, James A, Capt
Schaefer, Lisa A, CDR
Schaefer, Mark G., Lcdr
Schaefer, Melvin C, Lcdr
Schaefer, Rick A., Lt
Schaefer, Thomas R, CDR
Schaefer Jr, William A, CDR
Schaeffer, Paul B, CDR
Schafer, Cora Marie, CDR
Schafer, Daniel L., LCDR
Schafer, Donald R., CAPT
Schafer, Michael W., LT
Schaffer, Walter W, CDR
Schalk, Thomas E, CAPT
Schall, Michael A, CAPT
Schall Jr, Myron R, Capt
Schaller, Clement H, CDR
Schaller, Lora S., LT
Schanen, Patrick A., CDR
Schantz Jr, John P, CDR
Schapley II, Ramon F., Lt
Scharf, James W, CDR
Scharff Jr, Aaron R, Capt
Schauf, Michael C, CDR
Schaughency, Richard M, Lcdr
Schaumann, Robert C, CDR
Schaus, Nicholas J, Capt
Schear, James P, Radm
Schechinger, Terry D, Capt
Schedler, William H, Lcdr
Scheer, Andrew J, CDR
Scheffer, James H, Capt
Scheiderich, William A, Lcdr
Scheiner, Robert, CDR
Schell, Norman B, Capt
Schell, William W, CAPT
Schellhorn, Charles W, Capt
Schellinger, Douglas W, Lt
Schemenauer, Loyd G, Capt
Schempp, Theodore L, CDR
Schenck, Alexander W, Lcdr
Schenck, Francis Roland, Capt
Schenke, Daniel J., CDR
Scherbel, Paul N, CDR
Scherer, Daffney L., LCDR
Scherf Jr., Paul H., Capt
Scherman, Craig, CDR
Scherrer, John W., Ltjg
Schetter, Carl F, CDR
Scheurer, Jeanne, CAPT
Scheurich, Norman C., CDR
Schexnayder, Mark R., LCDR
Schiada, James J., Capt
Schiano, Richard J, Capt
Schiavo III, Louis J, CDR
Schick, Herbert A, Capt
Schick, Philip F, Capt
Schiefen, Richard E., CAPT
Schiff, Albert J, Capt
Schillinger, Dana R., CDR
Schimmelpfenig, Fred J, Lcdr
Schindler, Dan S., LT
Schinzing, Michael R., Lcdr
Schipke, Dennis H, Capt
Schipper, Jack W, CWO4
Schisler, Robert A, CDR
Schjaastad, Dolores F, Lt
Schjolberg, Marsha A., CAPT
Schlaefer, Douglas C, CAPT
Schlagheck, David R., CDR
Schlameus, Alfred B, CDR
Schlang, Lawrence H, CDR
Schlegel, E Bernard, Capt
Schlegel, Richard C., LCDR
Schleif Jr., Norman W, CAPT
Schlenk Jr, Peter P., CDR
Schless, James M, Lt
Schlichter, Stephen L, Lcdr
Schlies, Edward W., Lt
Schlinker, Barbara Ann, Lt
Schloetzer, Frederick F, Lcdr
Schloss, Philip William, CDR
Schlote Jr, Russell C, CAPT
Schlotterer, John S, Lcdr
Schloz, John M, Capt
Schlundt, Howard A, Capt
Schmahl, Donald V., CDR
Schmehl, Laurene, LCDR
Schmid, Gordon E, Lcdr
Schmidheiser, Charles J., LT
Schmidt, Donald L, Capt
Schmidt, Donald R, CDR
Schmidt, Ernest L, CDR
Schmidt, Henry E, Lcdr
Schmidt, John, Lcdr
Schmidt, John E, Capt
Schmidt, Jurgen B, Capt
Schmidt, Margaret M, Capt
Schmidt, Melbourne M, Lcdr
Schmidt, Peter J., CDR
Schmidt, Peter R, Capt
Schmidt, Raymond P, Capt
Schmidt, Richard J, CDR
Schmidt, Steven A., CDR
Schmidt, Walter A, CDR
Schmidt Jr, C Miles, Capt
Schmiedel, Peter S, Lcdr
Schmitt, Paul S, CDR
Schmitt, Randall K., CDR
Schmitt, Roger R, Lcdr
Schmitt, William E., CDR
Schmitz, Dennis, Capt
Schmitz, Fidel, CDR
Schmitz, Richard S, Lt
Schmitz, William E, CDR
Schnabel, Walter F, CDR
Schnaedelbach, Gerald, CDR
Schnatz, Paul E, CDR
Schneckloth, Sue E, Lt
Schneider, Alan K, CDR
Schneider, Christine L., Lcdr
Schneider, James G, Lt
Schneider, John P., CDR
Schneider, John R., CDR
Schneider, Joseph C, Capt
Schneider, Michael W, Capt
Schneider, Oliver J, Lcdr
Schneider, Paul H, Lt
Schneider, Robert C., LCDR
Schneider, Roger L, Capt
Schneider, Ronald H, CDR
Schneider, Thomas H, Capt
Schneider, William B, CDR
Schneider, William J, Capt
Schneier, Ezra D, Lt
Schnell, Richard A, CDR
Schnell, Stanley W., LCDR
Schoem, Scott R., CDR
Schoen Jr, Robert M, Capt
Schoenagel Jr, Fred C, CDR
Schoenborn, Terrence, Lcdr
Schoendienst, James A., LCDR
Schoendorf, Judson R, Capt
Schoenig, Arthur F, Capt
Schofield, Calvin O, Capt
Schofield, R Scott, CDR
Scholes, Robert J, CAPT
Scholes, Robert M, CDR
Scholl, John P, Capt
Scholle, Harold C, Lcdr
Schoman Jr, Charles M, Capt
Schomer, Don C, CDR
Schoppy, Joseph G., Lcdr
Schorn, David M., Capt
Schott, Charles L., LCDR
Schott, Patricia T. F., LCDR
Schowalter, Mark W, Capt
Schowe, Ronald R, CAPT
Schrade, Donald E, Capt
Schrader, Dennis R., CDR
Schrader, Terri L., LCDR
Schrader, William H, Capt
Schrage, Warren A, LCDR
Schram, William R, Lcdr
Schraut, Gary E., Capt
Schrobo, Bernice R., Capt
Schroeder, Daniel L., CDR
Schroeder, Douglas D, Capt
Schroeder, Duane E., LCDR
Schroeder, George J., Lt
Schroeder, Kent L, Capt
Schroeder, Mark A, CDR
Schroeder, Thomas E, CDR
Schroeder Jr, Henry, Lcdr
Schroeder Jr, Oliver C, Capt
Schryver, Chandler, Lt
Schuber III, Fred J., Radm
Schubert, Albert E, CDR
Schubert, Frederick W, Capt
Schubert, Ron, CDR
Schuchard, Alfred S, Capt
Schuelke, Jeffrey D., CDR
Schuerman, John E, Lcdr
Schuermann, David C, Lt
Schuetz, Jeffrey D., Lcdr
Schuh, Marlin D, Capt
Schuhle, John E, Capt
Schuldt Jr, T J C, CDR
Schulenburg, Robert W, Capt
Schuler, Chester J, Capt
Schuler, Frederick T, Lcdr
Schuler, John J., ENS
Schuler, Lawrence E, Capt
Schuler, Richard J, Lcdr
Schuler, Timothy D., CDR
Schuller, David D, Capt
Schulman, Dale C., LCDR
Schulte, James T., LT
Schulte, Robert H, Lcdr
Schulthess, Bruce, LCDR
Schulting, Eugene O, CDR
Schultz, Blanche B, Capt
Schultz, David P., CDR
Schultz, Edward R, CDR
Schultz, James B, CDR
Schultz, Louis D, Capt
Schultz, Melvin R, Capt
Schultz, Stephen C., ENS
Schultz, Thomas A, Capt
Schultz, William W, LCDR
Schultz Jr, Arthur V, Capt
Schulz, Roy S, Capt
Schulze, Curtis P, Ltjg
Schulze Jr, Henry A, CDR
Schumacher, H Richard, Lt
Schumacher, James J, Capt
Schumann, Alfred W, Capt
Schumann, O Douglas, Capt
Schunk, Edward A., CDR
Schupbach, Neal A, Lcdr
Schur, Warren M., Capt
Schust, Robert L, Lcdr
Schuster, John R., CDR
Schuster, Victor K, Capt
Schuster, William F, CDR
Schutt, Robert C., CDR
Schutte, Robert J, CDR
Schwab, Donald E, CDR
Schwanke, Louis C, CDR
Schwarb, Frank J, Lcdr
Schwartz, Dorothy R, Capt
Schwartz, James S., CDR
Schwartz, Martha E, Lcdr
Schwartz, Martin D, Lcdr
Schwartz, Norbert V, Capt
Schwartz, Pamela A., LT
Schwartz, Peter A, CDR
Schwartz, Phyllis, LCDR
Schwartz, Robert H, Lt
Schwartz, Robert P., CDR
Schwartz, Robert R, Lcdr
Schwartz, Ronald L, Capt
Schwartz, Stanley P, Capt
Schwartz, Susan K, CAPT
Schwartzkopf, Gary P, Capt
Schwarzenberg, Jan, CDR
Schwarzkopf, Albert B, CDR
Schwedhelm, Martin H, CDR
Schweigart, Joseph A, CDR
Schweitz, David T, Lcdr
Schweitzer, Ann C, CDR
Schweitzer, Robert C, Capt
Schweizer, Edward S, Capt
Schweizer, Jack R, Capt
Schwemm Jr, Henry C, CDR
Schwichtenberg, Linda, CDR
Schwikert, Richard S, Lcdr
Schwoebel Jr, Richard, Lcdr
Scichili, Carl J, Radm
Scobee, Mitchell O, Lt
Scoble, John G, Capt
Scoffield, William T., Capt
Scofield, Robert A, Capt
Scoggin, Steven W., CDR
Scoggins, Robert G, Lcdr
Scolpino, Anthony J., CDR
Scotece-Armstrong, C E, Ltjg
Scott, Arnold D, Ltjg
Scott, Augustus B, Capt
Scott, Barbara-Anne V, Lcdr
Scott, Brian E, Capt
Scott, Carven A., Lcdr
Scott, David J, CDR
Scott, Donald B., CAPT
Scott, George J, CDR
Scott, Harold Frank, Lcdr
Scott, Hugh P, Lcdr
Scott, James H, Radm
Scott, John D., CAPT
Scott, Joseph R, CAPT
Scott, Kenneth T, CDR
Scott, Michael J., CDR
Scott, Michael L, CAPT
Scott, Michael R, Radm
Scott, Neil D., Lcdr
Scott, Owen F, CDR
Scott, Raymond T., Lt
Scott, Richard L, Lt
Scott, Robert B, CDR
Scott, Robert G, Capt
Scott, Ronald C, Capt
Scott, Stephen K, Capt
Scott, Steven J., CDR
Scott, Thomas E, Capt
Scott, Warren C, Lcdr
Scott, William D., LCDR
Scott II, Ralph H, Capt
Scott III, Palmer Mars, Lcdr
Scott Jr, George C, Lcdr
Scott Jr, Gerald J, RADM
Scott Jr, Thomas J, Capt
Scott Jr, Wade H, Lt
Scott Jr, Walter J, CDR
Scott-McKnight, C. V., CDR
Screnci, Giuseppe A., LTJG
Scribner, David A., CAPT
Scruggs, Robbie D., CAPT
Scruggs, Robert N, CDR
Scruton, Robert J, Lt
Scudder, Charles A, CDR
Scudder, Robert H, CDR
Scudder, Ronald L., Capt
Scullion, William R., Lcdr
Scully, Richard R, CDR
Scyoc, Paul G., Lcdr
Seabold, Charles E, Lcdr
Seabrook, Archibald F, Lcdr
Seafeldt, Albert C, CDR
Seagren, Dwight S., LTJG
Seaman, Billy M, Lcdr
Seaman, Kevin G., CDR
Seaman, Stephen D., Lt
Seanor, James E, CDR
Search, Beth M., Capt
Searcy, Harold, Lcdr
Searcy, William C, Lcdr
Searles, John M, CDR
Searls Jr, Henry H, Lcdr
Sears, Dale D, CDR
Sears, John W, Capt
Sears, Michael E., Lcdr
Sears Jr, Richard I, CDR
Seashols, Stuart L., CWO4
Seastrom, George R, CDR
Seastrom, James R, CDR
Seate, John H, CDR
Seaton, Jerome M, CDR
Seaward, Randall Allen, CDR
Seawright, Lori A, Lcdr
Seawright, Robert E, CDR
Sebastian, Mark C., LCDR
Sebelius, Peter W, CDR
Sebrell, J Emmett, Capt
Sechrist, Ronald E., CWO4
Secker, Martin D, CDR
Secord, William J, Capt
Secrest, Glenn J, Capt
Sedgley, Ronald M, CDR
Sedita, Josephine R, CDR
Sedlack, Richard E, Capt
Sedlack, Robert E, LT
Sedlack, Thomas R., LCDR
Sedlar, Frank J, CWO4
Sednek, Carol Frances, CDR
Sedor, Leonard T, CWO4
Sedor, Valerie Ann, CDR
Seeley, Jimmie W, Radm
Seeley, Stephen M., Capt
Seelig, Stephen W, CDR
Seelig, William C, Lcdr
Seely, Thomas K., CDR
Seemann, Sue Ann, CAPT
Seff, James M, Capt
Seffel, Joel S, Capt
Seftas, G. Randall, Capt
Segal, Daniel D, Lcdr
Segal, Joshua E., LT
Segert, Randall L., CAPT
Seghers, Philip E, Capt
Seiberlich, Carl J, Radm
Seibert, Moody G., Capt
Seibert, Peter K, Capt
Seibold, Frederick W, CDR
Seidel, James A., CDR
Seidenberg, Mel F, Lcdr
Seidenstein, William R, Lcdr
Seiders, George D, CDR
Seidler, Edward R, CWO4
Seidman, Gilbert, Lcdr
Seigel, Sigmund L, Capt
Seiler, Dean B, Radm
Seiler, Richard W, Capt
Seiple, Herbert G, Lt
Seixas, Kim G, CDR
Sekula, Paul, CDR
Sekula, Theodore, Capt
Selby, John H., Lt
Selden, Thomas L., CDR
Self, Dennis D., LT
Self, John F, Capt
Selkirk, James M, Lt
Selle, Robert W., CDR
Selleck, Zeno E, CDR
Sellers, Charles W, CDR
Sellers, Wendell G, CDR
Sellery, George F., LT
Sellery Jr., Robert A., LT
Sellner, James M, Lcdr
Sells, John E, Lcdr
Selvidge, Ross S, CDR
Semcken Jr, John H, Capt
Semerad, James M., Lcdr
Semet, Robert J, Capt
Semmes, Oliver J, Capt
Sempek, Marie L., LCDR
Sen, Cameron A., CDR
Sen, Jeffrey C., LCDR
Sencindiver, James D., CDR
Seneff, Michael G., CAPT
Senf, Randall L, CDR
Senger Jr, James M, Capt
Senich, Robert J, CDR
Senior, John R, Radm
Sensenbach, Wilson E, Lcdr
Sensoli, Joseph A, Capt
Sentz Jr, John D, Capt
Senykoff, Ronald S., LCDR
Sepulveda, Martin C., LCDR
Sequeira, F E, Capt
Sera, Enrique J., LCDR
Serafini, Charles P, LCDR
Serafini Jr, Attilio, Capt
Serber, David, LCDR
Serbin, Kathryn M., CDR
Serfustini, Anthony B., Capt
Sergio, Frederick A, Capt
Serif, Michael L, LT
Serota, Cornelia C, CDR
Serpente, James J, CDR
Serrano, Ronaldo A.C., Lcdr
Servis, Ronald W, Capt
Sescleifer, Arthur D, CDR
Sessions, Catherine P., CDR
Seton, Charles B, Lcdr
Sette, Karl R, CAPT
Settle, James B, Lcdr
Settle III, Halsey M, CAPT
Severance, David L, Lt
Severino, John A., CDR
Severson, Andrew J, Capt
Severson, Paul J, Capt
Severson Jr, Paul M., CWO3
Sevitski, Linda M., Lcdr
Sewall, Jean C., CDR
Seward, Diane M., Lcdr
Sexsmith, Gerald T, Capt
Sexton, Charles E, Capt
Sexton, David L, Lcdr
Sexton, Donald R., CDR
Sexton, James R., Lcdr
Sexton, Scott Wayne, Lcdr
Seybel, Roger B, CWO4
Seyfert, Harry G, CWO4
Seymour, Christopher R., LCDR
Seymour, Curtis A, CWO4
Shackelford III, Tom F., CAPT
Shadbolt, Craig, CDR
Shaddock, James M., CDR
Shaefer Jr., Charles F., CDR
Shafer, Martha J., LCDR
Shaffer, J Nevin, Radm
Shaffer, Jesse S, Capt
Shaffer, Keith, Lcdr
Shaffer, Robert F, Capt
Shaffer, William O., Capt
Shaffer Jr, John Nevin, CDR
Shaft, Jearald J, Capt
Shanahan SR, Vincent J, Capt
Shaneman, Stephen C, LT
Shank, Nancy E., CDR
Shankar, T. S., Lt
Shanley, Patrick M, CDR
Shannon, Gwyn A., Lcdr
Shannon, James F., LCDR
Shannon, James O, CDR
Shannon, John G, Capt
Shannon, Sean M., LCDR
Shannon, Terry L., LCDR
Shannon, Thomas R, CDR
Shannon Jr, Edgar F, Capt
Shannon Jr, George M., Lcdr
Shapbell Jr, William B, Capt
Shapiro, John L, LCDR
Shappell, Joseph R., Lcdr
Sharer, John J, Lcdr
Sharkey, Charles W., LCDR
Sharkey, Steven M., CDR
Sharkey, Thomas P, CAPT
Sharp, Alexander V.L., CDR
Sharp, James T, CDR
Sharp, Keith A., LCDR
Sharp, Robert A, CDR
Sharp, Winfred H, Capt
Sharp Jr, Charles E, CAPT
Sharratt, Charles R, CAPT
Sharts, Clay M, Capt
Shaske, Jeff, LCDR
Shatford II, Walter T, CDR
Shatynski, Michael M., CDR
Shatynski, Stephan J., Lcdr
Shaughnessy, John, CDR
Shaughnessy, William K, CDR
Shaver, James C, CDR
Shaver, Philip J., CAPT
Shaver Jr, Henry M, Capt
Shaw, Burton A, Capt
Shaw, Elizabeth M, Ens
Shaw, Frederick G, Capt
Shaw, Harold M, Capt
Shaw, James D., LCDR
Shaw, John C, Lcdr
Shaw, Joseph T, Capt
Shaw, Robert R, CDR
Shaw, Stephen D, CDR
Shaw II, James E., Lcdr
Shaw Jr, George G, CDR
Shaw Jr, Joseph W, Capt
Shaw Jr, Thomas R., Lcdr
Shay, Dennis R, CDR
Shay, Peter D, CDR
Shayne, George R, Capt
Shea, John L., LCDR
Shea Jr, Walter E, Lcdr
Shealy Jr, Wilson O., CAPT
Shean-Craig, Monica M, CDR
Shearer, Jerry R, Lcdr
Shearer III, Oliver V., CDR
Shears Jr, Howard F, CWO4
Shebalin, Paul V, Capt
Sheddan, Mark A., Lcdr
Shedlock, David B., LCDR
Sheeder, Elwood B, Lcdr
Sheedy, Eugene J, Lcdr
Sheehan, Brian T, Radm
Sheehan, Renee Grev, CDR
Sheehan Jr, William H, Capt
Sheehe, Lawrence G, CWO4
Sheehy, John J, CDR
Sheen Jr, Robert V, CDR
Sheets, Charles I., Lcdr
Sheets, Paul M., LCDR
Sheff, Gerald W, Lcdr
Sheffer, William G, CDR
Sheffield, Brian E, Capt
Sheffield, Glenn A, CDR
Sheibels, Russell A, Lcdr
Shein, Paul D, Lcdr
Shelar Jr, Eugene, Capt
Shelby, Daniel D, Lt
Shelby, John H., CAPT
Shelesny, Stephen G, Lt
Shelin, Randal G., CDR
Shell, Barbara A, CAPT
Shell, Mark D., CDR
Shellenbarger, Tedd, LT
Shellenberger, J. S., CDR
Sheller, Jon, Capt
Shelley, Marke R, RADM
Shelley, Ronald N, Capt
Shelley Jr, Henry C, CDR
Shelly, Henry T, CDR
Shelnut, Charles A, Capt
Shelor, Jean G., Lcdr
Shelter, Andrew P., CDR
Shelton, B Ron, Lcdr
Shelton, Christina L., CDR
Shelton, James E., CAPT
Shelton, Lloyd C., Lcdr
Shelton, Michael C, Capt
Shelton, Robert D, Lcdr
Shelton, Robert R, Lt
Shelton, Tamara L., Lt
Shenefelt, Philip D, Capt
Shepherd, Arthur E, CWO4
Shepherd, C. William, Capt
Shepherd, Orrin W., Capt
Sheppard, Gerald R, Capt
Sheppard, Laurie C., CDR
Sheppard, Robert H, Lcdr
Sheppard Jr, Harry L, Capt
Shepperd, James W, Lcdr
Sherbo, Paul S., CAPT
Sherer, Richard S, Lcdr
Sheridan, Earl A, CDR
Sheridan, James J, CDR
Sheridan, Robert J, CDR
Sherlock, Robert S., CAPT
Sherman, Anthony M., Lcdr
Sherman, Christopher M, Capt
Sherman, Paul M, Lcdr
Sherrard Jr., Frank C, CDR
Sherrill, William D., CDR
Sherry, Henry W, Lcdr
Sherwin, George R, Capt
Sherwood, Bruce M, Lcdr
Sherwood, John R, CDR
Shetterly, Andrew H., CDR
Shiber, Eugene J, Cwo
Shields, Bryan W., Lcdr
Shields, John D, Capt
Shields, Woodrow R., CAPT
Shields Jr, Daniel L, CDR
Shiflett, Edward E, Capt
Shiflett, Shirley B., CDR
Shigekawa, David G, CDR
Shimberg, Edmund, Capt
Shimizu, Wayne S., CAPT
Shimkus, Elizabeth A., Lcdr
Shindler, Carlton H, CDR
Shine III, Francis E, CDR
Shine Jr, Henry M, Capt
Shine Jr, Joseph L, LCDR
Shiner, Alfred L, CDR
Shingler, John J, CDR
Shintani, Calvin M., CDR
Shipley Jr, Raymond H, Capt
Shippee, Michael J., LCDR
Shiraki, Clyde Y, Capt
Shiraki, Ross K, Lcdr
Shirikian, Stephen P, Capt
Shiring Jr, Paul B, Lcdr
Shirley, James Arthur, Lcdr
Shirley, Robert L, CAPT
Shirley, Thomas W, Capt
Shirley Jr, J. C., Lcdr
Shively, Albert E, CDR
Shively, Isaac G, Lcdr
Shively, Robert W, Lcdr
Shiver, William P, CWO4
Shiverick, Asa, Ens
Shkapsky, Jean M., CDR
Shock Jr, Glenn H, CDR
Shockley, Keith I., LCDR
Shockley, Steven H, CDR
Shoemake, Harold R, CDR
Sholes, William H, Lcdr
Sholl, Theodore G, Capt
Shoop, Thomas V, Capt
Shores, Hubert A, CDR
Short II, Dwight H, CDR
Shorten, David R, CDR
Shorter, Robert E., CDR
Shorter Jr, James R, Capt
Show, James T, CDR
Show, William A, Lcdr
Showalter, Amos T, CDR
Showalter, Merle R, CDR
Showalter, Nelson, Lcdr
Showers, Arthur V, Lcdr
Shown, Ted G, CDR
Shreeve, Daniel F., CAPT
Shreve, Lloyd H, CDR
Shrewsbury, Walter B, Capt
Shrieves, John R., LCDR
Shuart, Esler L, Capt
Shudy Jr, John, CDR
Shuford, Charles R, CDR
Shuler, James M, CDR
Shull, Debra Kay, CDR
Shull Jr, James M, CDR
Shulman, Nisson E, Capt
Shulock, Charles, Ltjg
Shultz, Richard Allard, Capt
Shultz, S. Peter, CDR
Shultz, Theodore B, Capt
Shumate, Sandra Ann, LTJG
Shumlas, John A, Capt
Shunk, Kimberly, CDR
Shur, Nicholas M., Lt
Shurling, James R, CWO4
Shuttleworth, Joe A., LCDR

Sibick, Eugene M, CDR
Sibold, Robert D, Capt
Siddall, Luke D., CDR
Sidebottom, Kristi L, LCDR
Sidebottom, Stephen W, Lcdr
Sides, Curtis D, Lcdr
Sidler Jr, Norman F, Capt
Sidlow, John W, Lcdr
Sidoni, James A, Capt
Sidrow, Michael R., Cdr
Sie, Anthony K, Capt
Sieberns, David O, Capt
Siebert, Douglas A, Capt
Siefken, David M, Capt
Siegel, F Carlton, Lcdr
Siegel, Gary, Capt
Siegel, Gerald, CDR
Siegel, Herbert E, Lcdr
Siegel, Jay M., Capt
Siegel, Leonard, CDR
Siegel, Melvin A, Lcdr
Siegel, Samuel L, Capt
Siemon, Elwin F, Capt
Siemsen, Clara J, Lt
Siftar, Frank C, Capt
Sigel, Steven C, Capt
Sigg, Joseph A, CDR
Sightler, Carl L, Capt
Sigismund, Mark, Lcdr
Sigler, Titus S, CAPT
Signell, Warren I, Lt
Signor, Peter L., Lt
Sigvartsen, Herbert C, Lcdr
Siken, James P, Capt
Sikes, James K, CDR
Sikes, Stanley C., Lcdr
Silaghi, Robert A, CDR
Silakoski, Anthony F., CAPT
Siler Jr, Eugene E, CDR
Silinsky, Paul S, CDR
Silliman Jr, Edwin L, Lcdr
Silloway, Richard F., Capt
Sills, James R., LCDR
Silva II, Charles E., CDR
Silva Jr, Edward J, Capt
Silver, Phillip W, Lcdr
Silverman, Norman R., CAPT
Silverman, Preston Z, CDR
Silverman, Robert G., CDR
Silverman, Ronald S., LT
Silvert Jr, Robert M, CDR
Silverthorne, Craig W, CDR
Silverwood, Thomas E, Capt
Silvester, Lindsey M., CDR
Silvey, James H., LCDR
Silvey, Larry D, Capt
Sime, Margaret, LCDR
Simenstad, Archibald M, CDR
Simeone, John B, CDR
Simmonds, Thomas L., Capt
Simmons, Glen Dale, Lcdr
Simmons, Guy J, Capt
Simmons, Harold Blair, CDR
Simmons, Joseph L, Capt
Simmons, Mason S., Lcdr
Simmons, Robert M, CDR
Simmons, Robert T, Capt
Simmons, William G, CDR
Simmons Jr, Glenn B, Capt
Simms, Ernest Lee, CAPT
Simon, Matthew H, Capt
Simon, Ronald J, CDR
Simon, Steven J., LT
Simon, Theodore G, CDR
Simon, William J, Capt
Simon, William S., LCDR
Simonds, Charles R, Lt
Simonetti, Teresa M., LCDR
Simonitsch, Mark A, CDR
Simons, Henry M, CAPT
Simons, Richard L, CAPT
Simonson, Merlin D, Capt
Simonson, Todd J., LT
Simontacchi, Alexander, Capt
Simpkins, Becky J, CDR
Simpson, Branden D., LCDR
Simpson, Duncan A, LCDR
Simpson, Michael K, CDR
Simpson, Michael V., LCDR
Simpson, Robert M, Capt
Simpson, Thomas E., LT
Simpson, Virginia Renee, CDR
Simpson, William M, Capt
Simpson, William P, CDR
Simpson III, Charles C, CDR
Simpson Jr, John P, Lcdr
Simpson Jr, Robert M, Capt
Sims, James H, Capt
Sims, Joseph R, CDR
Sims, Philip D, Capt
Sims, Stephen A, CDR
Sims, T Stanley, Capt
Sims Jr William M, CAPT
Sincerbeaux, Robert, Lcdr
Sinclair, Robert E, Capt
Sinclair, Robert R, Lt
Sinclair, Todd M., ENS
Sincox Jr., Francis J., CAPT
Sindall Jr, Albert P, Capt
Sindelar, James H, Capt
Singer, Donald R, CDR
Singer, Lawrence E, CDR
Singer, Richard B, Capt
Singer, Scott F., LT
Singletary, Deanna, LCDR
Singleton, Milton J., Lt
Singley, John H., CDR
Singzon, Jaime M., Capt
Sinibaldi Jr., Robert A., CDR
Sink, James M, Capt
Sink, John D, Capt
Sink, William F, Lcdr
Sinnett, Kevin P., CDR
Sinnokrak, Robert L, CDR
Sinnott Jr, George P, CDR
Sinoply, Ann E., LCDR
Sipe, Gary H, CDR
Sipe, Thomas R, Capt
Sipes, Max E, Lcdr
Sipos, Kathleen I., Lcdr
Sisak, Michele T, Lcdr
Sisco, August C, Capt
Sisco, Barbara A., CDR
Sisco, Kenneth L, Capt
Sisk, Arnold J, CDR
Sisk Jr., Darrel B., CDR
Sisk Jr., Fred L., LCDR
Sisley, Dale L, Capt
Sison, Carlito D., LT
Sisson, Kurt D, CAPT
Sites, Bruce L, Capt
Sitko, Rodger A., CWO2
Sittig, Raymond L, Lcdr
Sitz III, Joe H, CDR
Siuta, Raymond A, Capt
Siver, Chester A, CDR
Sivers, Mark R, CAPT
Sivills, Glenn R., CAPT
Siwik, Joseph C, CDR
Six, Steve, Capt
Skaar, Harvey E, Capt
Skaggs, Charles R, CDR
Skalko, Deborah A., LCDR
Skarupa, Thomas E, CDR
Skelly, John F, Lcdr
Skelly, William J., CDR
Skelton, Marta Coleman, LT
Skelton, Robert T, Capt
Skerchock, Peter, Capt
Skiber, Carol A., CDR
Skidmore, Rosemary S., LT
Skillin, James H, CDR
Skillman, Don K, Capt
Skinner, David O, Lcdr
Skinner, Robert L, CDR
Skinner, Susan M., CDR
Skinner, Victoria G, CDR
Skinner, William Edward, Capt
Skjei, Kevin R., CAPT
Sklarin, Michael L, Lcdr
Skokan, Bryan A, Capt
Skolfield Jr, John T, CDR
Skomra, Richard, LT
Skone, Matthew R., LT
Skora, Joyce A., LCDR
Skoropowski, Ernest P, Capt
Skotty, Donald R, Capt
Skrabak, Cletus, LT
Skrief, John N, Capt
Skrzat, Richard P., CAPT
Skucy, Andrew S, CDR
Skwaryk, Robert F., LCDR
Skye, George E., CAPT
Slabe, Barbara J, Capt
Slack, Theodore C, CDR
Slagle, Michael L., Capt
Slaton, Malcolm W, CWO3
Slaton, Randolph R, Lcdr
Slattery, Thomas J., CDR
Slattery, William E, Lcdr
Slaughter, Charles N, CDR
Slaughter, Vernon D, Lcdr
Slavonic, Greg J, CAPT
Slavonic, Martha J, Capt
Slay, Robert H, Lcdr
Slayton, Leroy N, LCDR
Sleavin Jr, Frank R, CAPT
Slebos, Christian G, CDR
Sledd, Melvin F, Capt
Sleeper, Donald C, Capt
Sleeper Jr, Thomas E, Capt
Slehofer, Gerald A, CDR
Slichko, Matthew J., Lcdr
Slick III, Charles E., LCDR
Slight, Stephen I, Capt
Slingerland, Gregory D., LT
Sliwa, Norman W, CDR
Sloan, Charles E, Capt
Sloan, Charles R, Capt
Sloan, Louis R, Lcdr
Sloan, Robert L, Capt
Sloan, Scot G., LCDR
Sloan, Stephen J., Lcdr
Slocum, Russell L., LCDR
Slone, Martin A., Lt
Slonosky, Nicholas V., Lcdr
Sloop, Phillip A., CDR
Slusher, Gerald M., LT
Sluyter, Vern C, CDR
Slye, George E, CDR
Slyfield, James T., CDR
Smack, Michael B., CDR
Small, Douglas E, Capt
Small, Gilbert S, CDR
Small, James C, CDR
Small, Sidney H, Capt
Small Jr, Kenneth E, CDR
Smalley, Robert A, CDR
Smaltz, N Michael, Capt
Smark, Alan, LCDR
Smart, Bruce Allen, CDR
Smart, John A, Lt
Smart IV, David O, Radm
Smathers, Edmond W., CDR
Smedstad, Eric R., LTJG
Smeltzer, Ronald J., CAPT
Smiley, Leonard V, Lcdr
Smisek, Edwin S, Capt
Smith, Alan D, LCDR
Smith, Allen R, Lt
Smith, August B, CWO4
Smith, B A, CDR
Smith, Barry G, CDR
Smith, Brian W, Capt
Smith, Canie B, Capt
Smith, Carey A, CDR
Smith, Carl L, Capt
Smith, Cecil N, CDR
Smith, Charles M, Lcdr
Smith, Charles R, Radm
Smith, Charles R., CDR
Smith, Charles R., LCDR
Smith, Christopher J., Lt
Smith, Craig P, CDR
Smith, Dan, CDR
Smith, Daniel C, LT
Smith, David A, CDR
Smith, David Albert, CAPT
Smith, David H, Lcdr
Smith, David L, Capt
Smith, Denise L., LT
Smith, Dennis E., LTJG
Smith, Dixon J, Capt
Smith, Dolores M., LCDR
Smith, Donald E, CWO4
Smith, Donald F, CDR
Smith, Donald H, Capt
Smith, Dorothie S, Lcdr
Smith, Douglas E, CDR
Smith, Douglas G, CDR
Smith, Douglas G., ENS
Smith, Douglas W., CDR
Smith, Duret S, Capt
Smith, E Brian, Capt
Smith, Edgar L, Capt
Smith, Edward G., CDR
Smith, Edward S, Lcdr
Smith, Eliza I, CDR
Smith, Emmett C, Lcdr
Smith, Eric W, LCDR
Smith, Frank L, Capt
Smith, Frankland, CDR
Smith, Franklyn P., Capt
Smith, Frederick M, CDR
Smith, Frederick S, Lcdr
Smith, Gail A., CDR
Smith, Gary D, Capt
Smith, Gary D., LCDR
Smith, Gene D, CDR
Smith, Gene S, CDR
Smith, George A, CDR
Smith, Gerald, Capt
Smith, Gerald A, Capt
Smith, Gerard A, CDR
Smith, Gregory A., LCDR
Smith, Gregory F, Capt
Smith, Harold P, Lcdr
Smith, Harriet Lee, LCDR
Smith, Harry L, Lcdr
Smith, Harry Lee, Capt
Smith, Henry A., Lcdr
Smith, Henry W, Capt
Smith, Horton, Radm
Smith, Howard W, Capt
Smith, Hugh Virden, CDR
Smith, J R, CWO4
Smith, J. Kirk, LCDR
Smith, J. Michael, LT
Smith, Jack S, Radm
Smith, James F, Capt
Smith, James F, CAPT
Smith, James H, Lcdr
Smith, James H, Capt
Smith, James L, Capt
Smith, James Michael, CAPT
Smith, James P, CDR
Smith, James Robert, Capt
Smith, Jamie L., CDR
Smith, Janice Foss, LCDR
Smith, John P, CDR
Smith, Jonathan C., Lt
Smith, Joseph C, Capt
Smith, Judson L, Radm
Smith, Kathleen A., LT
Smith, Kathleen K., CDR
Smith, Kenneth A, Capt
Smith, Kenneth W., CDR
Smith, Kerschiel D., CAPT
Smith, Lankford T, CDR
Smith, Larry D., LCDR
Smith, Lee M, Capt
Smith, Lester R, RADM
Smith, Lewis M, Capt
Smith, Lincoln S., LCDR
Smith, Loena W, Lcdr
Smith, Lorena A., CDR
Smith, Marilyn A., Capt
Smith, Mark A, CDR
Smith, Martin J, Capt
Smith, Mary K, CAPT
Smith, Maurice R, Capt
Smith, Michael E., LCDR
Smith, Michael L., LT
Smith, Michael P, Capt
Smith, Mildred E, Lcdr
Smith, Milton L, Capt
Smith, Nathan B., LT
Smith, Norman A, Capt
Smith, Norman D, CDR
Smith, Orville A, CDR
Smith, Paul C, CDR
Smith, Paul E, Capt
Smith, Paul J., Lcdr
Smith, Paul W, Capt
Smith, Peter M, Lcdr
Smith, Philip W, Radm
Smith, Phillip J., CDR
Smith, Pierre F, Capt
Smith, Rachel E., LCDR
Smith, Ralph B, LCDR
Smith, Ralph W, Capt
Smith, Randall D, CDR
Smith, Ray E, Lcdr
Smith, Raymond J, CDR
Smith, Rex L, Lcdr
Smith, Richard A, CDR
Smith, Richard A., Capt
Smith, Richard B, LCDR
Smith, Richard C, Capt
Smith, Richard C, Capt
Smith, Richard C., CDR
Smith, Richard D, Capt
Smith, Richard Franklin, Capt
Smith, Richard L, CDR
Smith, Richard L, CDR
Smith, Richard L, Capt
Smith, Robert A., Lcdr
Smith, Robert B, Lcdr
Smith, Robert B., Lt
Smith, Robert E, CDR
Smith, Robert E., CDR
Smith, Robert F, Lcdr
Smith, Robert G, CDR
Smith, Robert I, Capt
Smith, Robert J, Ltjg
Smith, Robert M, CDR
Smith, Robert M., Lcdr
Smith, Robert Vincent, CDR
Smith, Rodney K., Lcdr
Smith, Roger D., LCDR
Smith, Roger M., Lcdr
Smith, Ronald E, Capt
Smith, Ronald L, Capt
Smith, Ronald W, Capt
Smith, Roy M, CAPT
Smith, Ruth Harris, Lt
Smith, Sanfield B, Lt
Smith, Scott R., LCDR
Smith, Sherrill Edwin, Capt
Smith, Stephen L, Lt
Smith, Stephen R, Capt
Smith, Sterling C., LCDR
Smith, Steven D, CDR
Smith, Stewart E, CDR
Smith, Sue K, Lcdr
Smith, Teresa L, LCDR
Smith, Theodore A, Capt
Smith, Thomas G., CDR
Smith, Thomas Joseph, CWO4
Smith, Thomas Manly, CDR
Smith, Thomas P, Capt
Smith, Urban Eugene, Capt
Smith, Vernon D, Capt
Smith, Victor S, Lcdr
Smith, Wallace D, CDR
Smith, Walter D, Lcdr
Smith, Walter S, Capt
Smith, Ward B, Capt
Smith, Warren T., CDR
Smith, Wendell O, Lcdr
Smith, Willard G, CDR
Smith, William A, Capt
Smith, William A, Lt
Smith, William B., LCDR
Smith, William E, Capt
Smith, William E, Lcdr
Smith, William O, Capt
Smith III, Allen H, Lt
Smith III, Arthur L, Lcdr
Smith III, Lee, CDR
Smith III, Philip J, CDR
Smith III, Robert, RADM
Smith IV, William A., LT
Smith Jr, Chester C, LCDR
Smith Jr, Clyde R, Capt
Smith Jr, Columbus L, CDR
Smith Jr, Frank A, Lcdr
Smith Jr, Fred B., Capt
Smith Jr, George H, Lcdr
Smith Jr, Hugh R, Radm
Smith Jr, Jackson S, Capt
Smith Jr, John E, CDR
Smith Jr, Kenneth A, Capt
Smith Jr, Myron L., Lcdr
Smith Jr, Ralph E, Lt
Smith Jr, Randell C, CDR
Smith Jr, Warren J, Capt
Smith Jr, William H, Lt
Smith Jr, William J, CDR
Smith Jr, William M, Capt
Smith Jr, Willie, CDR
Smith-Yeoman, Tracie M, Lt
Smithen, Charles M, Lcdr
Smithers, Samuel J., CDR
Smithhart, Richard A, CAPT
Smits, Walter I, CDR
Smollen, Charles J, Lcdr
Smolsky, M. Alison, LCDR
Smouse, David P., CAPT
Smoyer, Winston R, Capt
Smyers, Tracy D., Lcdr
Smyre, Melford E., CDR
Smyth, Denis C, CDR
Smyth, Theodore H, Lcdr
Smythe, Robert B, Capt
Snead, Harry J, Lt
Snead, Ronald W, Capt
Snedeker Jr, E Kenneth, Capt
Sneden, Kathleen M., Capt
Sneden II, Roger P, CDR
Sneyd, William G, Lcdr
Snider, Lance A., LCDR
Snider, Ted L, Capt
Snow, Richard V., Lcdr
Snow, Sharon Kay, LT
Snowden, Ernest M, Capt
Snowdon, David E., Lcdr
Snyder, Anne B, CDR
Snyder, Bruce E, Capt
Snyder, Clinton C, CDR
Snyder, Daniel R., Capt
Snyder, David M, Capt
Snyder, David R, LT
Snyder, Douglass G, CDR
Snyder, Edward P., Lcdr
Snyder, Eric, LT
Snyder, George A, CDR
Snyder, Harry D, Capt
Snyder, John E, Capt
Snyder, Larry R, Lcdr
Snyder, Matthew J., LT
Snyder, Randall C., CDR
Snyder, Richard Charles, Capt
Snyder, Robert C, CDR
Snyder, Robert M, CDR
Snyder, Thomas L, Capt
Snyder, Willard M, Lcdr
Snyder, William J., LCDR
Snyder, William P, CDR
Snyder III, Bryan, Ltjg
Snyder Jr, Orville H, Capt
Soares, Joseph O, WO1
Soares, Michael L, Capt
Sobke, Gene E, Capt
Sodd, Timothy T.A., Lcdr
Soden Jr., Ralph G., LTJG
Soderman, Arne P., Capt
Soine, Jon C, Capt
Sokel, William D, Capt
Sokolowski, Joseph W, Capt
Sokolski, P Gerard, CDR
Sola, Gary G., LCDR
Solan, Joseph E, CDR
Soldano, Daniel A, CDR
Soldwedel, Fred D., CDR
Solgaard, Albert L, Capt
Solkshinitz, Stephen E., CDR
Solli, Richard E, Capt
Solomon, Brenda J., LCDR
Solomon, Jonathan G, Capt
Solum, James M, CDR
Somerndike, John M, Capt
Somers, James W, Capt
Somers, John A, Lcdr
Somerville, Peter M, CDR
Sommer, Robert W, CDR
Sommer-Weddington, Linnea J., Lt
Sommerhalder, John O, CDR
Sommers, Chester F, CDR
Sommerville Jr, Robert, Lt
Sondag, Mark Richard, Lcdr
Sonnemaker Jr, Harry H, Capt
Sonnenberg, Raymond W, Capt
Sooy, Evelyn N, CDR
Soran, Patrick D, Capt
Soran Jr, James J, CDR
Soranno, Michael A, CDR
Sorem-Hughlett, Barbara, Lcdr
Sorensen, Charles R, Ltjg
Sorensen, Harvey T., Lcdr
Sorensen, Marc L, Capt
Sorensen, Michael G, Lcdr
Sorensen, Raymond E, CDR
Sorensen, Roland J, Lcdr
Sorensen, Scott D, CDR
Sorensen, Vale H, Capt
Sorenson, Edmund H, Lt
Sorenson, Helmer E, Lt
Sorenson, John G., Lt
Sorenson, Loren Ora, Lt
Sorg, W Stuart, Capt
Soriano, Paul, CDR
Sorrells, Elizabeth, Lcdr
Sorrentino, Wayne P, Lcdr
Sosnowski, Robert G., CDR
Soth, Forrest C, CWO4
Souder, Charles E, CDR
Soulant, Kenneth A, CDR
Soule, John G, Capt
Soule, Rolland L, CDR
Sours, Charles L, CDR
Sours, Kenneth P., LCDR
Sours, Martin H, Capt
Southard Jr, Richard B., CDR
Southby, Richard F., Capt
Southerland, Arnold J, CDR
Southerland, Louis F, CDR
Southern, Samuel O, Capt
Southworth, Thomas W, Capt
Soward, William W, CWO3
Sowinski, Stanley A, CDR
Soza, Joel Raymond, LTJG
Spacek, Calvin E., CDR
Spane, William T, Capt
Spangler Jr, Henry J, Capt
Spann, Stephen A, Capt
Sparger, James H, CDR
Sparks, Robert B, CDR
Sparks, Terry L., Lcdr
Sparks, Timmy J, Lcdr
Sparling, Mark J, Lcdr
Sparling, Ronald A, Capt
Sparrow Jr, William C, Capt
Spatafore, Joseph L, CDR
Spath, James A, CDR
Spath, John P, Capt
Spatz, Robert R, Lcdr
Spaulding, Dana C, CDR
Spaulding, Peter E, CDR
Spaulding, R. Frank, CDR
Spaulding, Vance, Lcdr
Spawton, James M., Lcdr
Speake, George E, CDR
Spear, Theodore E., CDR
Spears, David V., LCDR
Spears, Margo, Lcdr
Spears, Morton F, CDR
Spears, Terrel J., Lcdr
Spears Jr, Franklin S., CDR
Specht, Malcolm R, Capt
Speer, Randall M, Capt
Speicher, James W., LCDR
Speights, William D, Capt
Speise, William D, Lcdr
Spell, David J, Lt
Speller, Bruce H, Capt
Spellman, John L, LCDR
Spellman, John P, Capt
Spellman, Kelley J., Lcdr
Spence, Charles A, LCDR
Spence, Gordon D, CDR
Spence, Karl D, CDR
Spencer, Barbara H., LCDR
Spencer, Billie H, Capt
Spencer, Craig W., CDR
Spencer, Duane E, CDR
Spencer, Glenna L., Lcdr
Spencer, Leland G, Lt
Spencer, Lennie W., CDR
Spencer, Robert G, Lcdr
Spencer, Thaxter P, Capt
Spencer III, George O., CDR
Spencer Jr, Robert E, CDR
Spenser, Kenneth V, Capt
Spero, Robert C., CDR
Speros, John P, CDR
Speros, Peter J, Capt
Speth Jr, Harry F, Capt
Spethman, Lana M., CAPT
Spiegel, Irving, Capt
Spielberger, Charles D, CDR
Spierling Jr, Paul R, CDR
Spies, Roderic A, Capt
Spiher, William Bruce, CDR
Spikes, George A, Capt
Spilios, George B, Capt
Spillane, Paul J, CDR
Spillis, James P, Capt
Spina, Philip M, Lcdr
Spindler, Charles R, Lcdr
Spinelli, Robert B., CDR
Spinelli, Thomas M., Lcdr
Spink Jr, G Thomas, Capt
Spinks, Brian M, CDR
Spiritosanto, John M, Lcdr
Spiro, Robert H, Radm
Spitek, Joseph C., CDR
Spitt, Wallace G, Lcdr
Spitzer, Jan W, CDR
Spitznagel, Mark W, Lcdr
Spivey, Elbert G, Lt
Spivey IV, Joseph M., Lcdr
Splan, Francis Xavier, Ltjg
Sponholtz, Leland B, Lt
Spooner, Daniel J., CAPT
Spradlin, Louis E, Capt
Sprague Jr, Gilbert G, CWO2
Spratt Jr, John S, Capt
Sprenger, Michael A, Lcdr
Spriggs, David A, Capt
Spriggs, Donald K, Capt
Spriggs, Gregory D., Lcdr
Springer, Allen L, Capt
Springer, Clinton H, Lcdr
Springfels, Carl T., CDR
Springfels, Lura A., CDR
Sprinkle, Christian E., CDR
Sprott, Robert E, Lcdr
Sproull, James E, Lcdr
Sprowls, Paul Alan, LCDR
Sprowls, Paul H, Capt
Sprowls, Sharon L, CAPT
Spruill, John P, Capt
Sprung, Stewart R., CDR
Spurck, Danial N, Capt
Spurlin, Leanne E., Lcdr
Spurling, Nelson E, CDR
Spurr, Richard P., CDR
Squatrito, Jerome J, CDR
Srnka, David M., LCDR
St Angelo, Donald G, Radm
St Martin, Roy J, CDR
St. Aubin, Robert J, CWO4
St. Clair, Harry, Capt
Staab, Fred A., CDR
Staats, Grant S., LT
Stacer, Richard K, Capt
Stacey, Jerusha L., LCDR
Stacey, Lindley C, Capt
Stack, Daniel, Capt
Stack, Robert C, Radm
Stack Jr, Jerome D., CDR
Stackhouse, William A., CDR
Stacy, Janice M, CDR
Stader Jr., L David, CDR
Staeheli, Bruce W., CAPT
Staehle, John H, Capt
Staehli Jr, Ralph J, Capt
Staes, James P, Capt
Stafford, Richard M, Capt
Stafford, William C, Lcdr
Stagg, Frances I, Capt
Stagg, Nancy L, CDR
Stagg, Robert G, Lcdr
Staggs, Harlon V, CDR
Staggs, Robert P, Capt
Stahel, Rudolf S, CAPT
Stahl, Archie A, Capt
Stahl, Gary, CDR
Stahl Jr, Robert J, Capt
Stahley, David M, CWO4
Stahlman, James E., CDR
Stahura, Edwin, CDR
Staiger, Donna D, CAPT
Staiger, Roger P, Capt
Staiger Jr., John A., CAPT
Stainbrook, Michael B, CAPT
Stainton, Samuel D., CDR
Stallman III, George L, Capt
Staman, William T, CWO4
Stamberg, Allan J, Capt
Stamer, Paul Joseph, CDR
Stamm, C William, Lcdr
Stampler, Glenn L., Lcdr
Stampley Jr, O K, CDR
Stamps, Peter D., Lcdr
Stanczak, Allan M, CDR
Standard, Sarah M., LCDR
Stanfield, Wayne S., CAPT
Stanford, Harold M, Capt
Stanford, James L, Capt
Stanford, L Bradley, Capt
Stanford, Roy S, Lcdr
Stang, Alan V, Capt
Stange, Mark E., CAPT
Stangline, Gustav J, CDR
Stankiewicz, Zbigniew S, CDR
Stanley, Arleigh C, CDR
Stanley, Frank S, CDR
Stanley, H C, Capt
Stanley, Harold G, CDR
Stanley, Harry C, CDR
Stanley, Irving C., Lt
Stanley, James C, Lcdr
Stanley, Timothy D., Capt
Stansbury, John W., CDR
Stansbury, Thomas A, Radm
Stanton, David A, CDR
Stanton, Rodney A, Capt
Stapf, Michael J, Capt
Staples Jr, Fitzhugh D, Capt
Stapleton, Eva G, CDR
Stapleton, Robert J, Capt
Stapleton, Stephen H, Capt
Stapp Jr, Maurice Q, CDR
Stark, Alvin, CDR
Stark, David A, Lcdr
Stark, Joan F., LCDR
Stark, Lawrence W, Capt
Stark, Michael G., CDR
Stark, Timothy J, CDR
Stark, Willard A, CDR
Starks, Frank W, Lcdr
Starkweather, Robert L, Lcdr
Starowski, Robert M., LT
Starr, Betty W, Capt
Starr, Dennis M, CDR
Starr, Larry Waylan, CAPT
Starr, Nikki, LCDR
Starr, William H., LCDR
Starsiak Jr, William J., Capt
Stasco, Gary W., CDR
Staser, David R., CDR
Stasiak, Roger S, CDR
Stasney Jr, Joseph, Capt
Staszewski, Robert M., Lcdr
States, Marilin B., CDR
Staub, Bernard F, CDR
Staudmeister, A Larry, Capt
Staudt Jr, Calvin C., CDR

Stauff, Arthur W., LCDR
Stauffer, Dale E, Capt
Stauffer, Wilbur S, Lcdr
Stauffer, William H, Capt
Stauning, Herbert G, Lcdr
Stavros, Gus G, Capt
Stayman, Kathleen M., CDR
Steagall Jr, William F, CAPT
Stearns, Christopher R., LT
Stearns, Rodney C, Capt
Stebbins, Joanne L., Lcdr
Stebbins Jr, William R., Lcdr
Stechmann, Erik D., LCDR
Steckel, John, LCDR
Steedly, Wendell Melvin, Lcdr
Steel, Francis P, CDR
Steel, G Donald, Capt
Steel, Paul W., CDR
Steel, Robert W, Lcdr
Steele, Francis Xavier, CDR
Steele, John Luther, LCDR
Steele, William A, Capt
Steelman, Mark B, CDR
Steen, Robert P, CDR
Steensma, Robert C., CAPT
Steenwerth, James K, CDR
Stees, Joseph S, Lt
Stefan, William F, CWO4
Stefanak, Matthew A., Lcdr
Stefancic, Robert L., Ltjg
Stefani, Rob, ENS
Steffens, Theodore C, Capt
Steffenson, Michael R, Capt
Steffes Jr, Arnold M, Capt
Steffy, John S., CAPT
Stefula, Salvator M, CAPT
Steger, William G, CDR
Stegman, Royce R., LCDR
Stegman, Walter Carl, Lt
Stegner, Donald L, Capt
Stehlik, Michael V., Lcdr
Stehly, Daniel W, LCDR
Steiger, Keith E, Lcdr
Steigerwalt Jr, Richard, Lt
Stein, David A, Capt
Stein, Kenneth J., Lcdr
Stein, Marvin L, CDR
Stein, Thomas F, Capt
Stein, William B, Capt
Stein Jr, Earl A, Capt
Steinam Jr, Edwin J, Capt
Steinauer, Richard A, Lcdr
Steinbeck, John R., Lcdr
Steinberger, Anna M, Capt
Steinbrecher, Alan K., Capt
Steinbronn, Jeffrey P., LCDR
Steiner, Frank G, CDR
Steiner, Howard J., CDR
Steiner, Joseph, Capt
Steiner, Raymond E., Lcdr
Steiner Jr, Walter K, Capt
Steinhauer, Jules V, Lcdr
Steinke, Jay F., LT
Stella, Joseph W, Capt
Stellman, William J, Lcdr
Stemper, Marion F, Lt
Stemplewicz, John T., Capt
Stender, Jeffrey D., Lcdr
Stensland, John N., CAPT
Stentz, Terry L, Capt
Stephan, C. Alan, Capt
Stephan, Howard W, Lcdr
Stephan, Paul E, Lcdr
Stephany, Edward G, Capt
Stephens, Allan W, CDR
Stephens, Arthur L., CAPT
Stephens, David A., CDR
Stephens, David P., LT
Stephens, Gene M, CWO3
Stephens, Gregory F., Lcdr
Stephens, John R, CAPT
Stephens, Novis G, LCDR
Stephens, Paul A, LT
Stephens, Richard G, CDR
Stephens, Ronald C, Capt
Stephens, Vicki S., Lcdr
Stephenson, Daniel K, CAPT
Stephenson, Edward V, Capt
Stephenson, Henry R, Lt
Stephenson, Horace L, CAPT
Stephenson, Keith R., CDR
Stephenson, Michael W., Lt
Stephenson, Richard L, Lcdr
Stephenson, Samuel S., LT
Stepler, Kay B, CDR
Sterchi Jr, Charles F, LCDR
Sterett, Barry L, CDR
Stern, Philip H, Capt
Sternitzke, Vincnt Leo, CDR
Stetson, Jeane H, Capt
Stetz Jr, Joseph A, Capt
Steuart, Allen W, Lcdr
Steuble, Roger H, CDR
Steuer Jr, Rudolph R, Capt
Steuernagel, Kathryn A, Lcdr
Stevens, Alfred E, Lcdr
Stevens, George F., Lcdr
Stevens, John A, Capt
Stevens, Lenhard A, Lcdr
Stevens, Paul Mark, Lcdr
Stevens, Renee Minella, CDR
Stevens, Richard G, CDR
Stevens, Rollin M., CAPT
Stevens, Rom A., CDR
Stevens, Ronald W, Capt
Stevens, Timothy F, Capt
Stevens, Ward W, Capt
Stevens, Weaver L, Lt
Stevens, William L., LCDR
Stevens, William R, CDR
Stevenson, John C, Capt
Stevenson, Kenneth A, CAPT
Stevenson, Robert E, CDR
Stevenson, Wallace J, CDR
Stevenson III, Richard G., LT
Stevenson Jr, William R, CDR
Stevenson Sr, Eugene D., Capt
Stevinson, Donald, CDR
Steward, Jack A, CDR
Steward, Michael R., ENS
Stewart, Billy A, Capt
Stewart, Carl J, CDR
Stewart, Edith A, Lcdr
Stewart, Frank C., CAPT
Stewart, George A, CDR
Stewart, Gilbert L, CWO4
Stewart, James A, Capt
Stewart, James F, CDR
Stewart, James W., ENS
Stewart, John H, Capt
Stewart, Kent A., LT
Stewart, Lewis E., CDR
Stewart, Otis F, Lt
Stewart, Richard W, CDR
Stewart, Robert E, Capt
Stewart, Shirley J, Lcdr
Stewart, Stephen D., LCDR
Stewart, Tabitha J.T., LT
Stewart, Thomas C, CDR
Stewart Jr, Charles E., CDR
Stewart Jr, John M, Capt
Stice, Theresa M., LCDR
Sticinski, Ron J., CDR
Stickles, Duane H., CDR
Stickney, Douglas L, Lcdr
Stickney, Michele V, CAPT
Stiebel, Victor G, CDR
Stieber, Conrad H, CDR
Stiegelmeyer, Gary L, CDR
Stiehl, William D, CDR
Stierer, Gloria Steffens, Capt
Stierer, Richard T, Capt
Stiff, William L, Capt
Stiglitz, Michael J., LT
Stihler, Robert E, CDR
Stiles, Steven D., LT
Still, Rodney A, Capt
Stillings, David G, Capt
Stillwell, Mark R, Capt
Stilwell, Eric L, CDR
Stimac, Michael V, Capt
Stimpson, John W, Lcdr
Stimpson, Mark A, Lt
Stinchfield, John E, CDR
Stinson, Janice C, CAPT
Stinson, Jonathan F., LCDR
Stinson, Joseph D., CAPT
Stitzlein, David R., CDR
Stiver, Charles E, Capt
Stock, Fred J. T., Lt
Stockdale, Martin H, CDR
Stockdill, Pat W, Capt
Stocker, Susan C., LCDR
Stockham, John K, CDR
Stocking, Howard B, Lt
Stocking, Richard R, Capt
Stockmaster, Brian K., LT
Stockreiter, Francis M, Capt
Stoddard, Dennis M., LT
Stoddard, Richard C, Capt
Stoehs, William F, Capt
Stoeltzing, Richard H, Capt
Stoffel, Daniel M, CDR
Stoffel, Mark A., CDR
Stoffels, Gary B., Capt
Stohlmann, John C, CAPT
Stohr, Richard J, Capt
Stokes, Daniel A., MIDN
Stokes, Robert P, Lt
Stokes Phd, Houston H, CDR
Stokich Jr, Theodore M, Capt
Stokke-Nelson, Cathy M, CDR
Stoll, Alice M, CDR
Stoll, Linus M, CDR
Stolle, Robert C., CDR
Stoltz, George F, CDR
Stoltz, Timothy G., LT
Stombler, Eugene H, Capt
Stonage, Charles L, CDR
Stone, David L, Lcdr
Stone, Edna M, Lcdr
Stone, John R, CDR
Stone, Joseph L, Capt
Stone, Juan Girard, CDR
Stone, Kent N, CAPT
Stone, Robert E, CDR
Stone, Scott C S, CDR
Stone, Thomas E, Lcdr
Stone Jr, Charles E, Capt
Stonelake, Henry T, Capt
Stoner, William G, Capt
Stonesifer, Frederic S., Capt
Stoney, Kurt A, CDR
Stookey, William C, CDR
Stopkey, Waldemar D, CDR
Storck, Theodore S, CDR
Storey, Harold R, Lt
Storey, Kurt D., LCDR
Storey, Robert W, CDR
Storey, Sam M, Capt
Storjohann, Gordon, Lcdr
Storm, Dorothy B., Lcdr
Story, John D, Lcdr
Stouffer, George A, CDR
Stouffer, Jill W., CDR
Stough, Richard H, CAPT
Stoughton, Robert A, CDR
Stouse, Robert, LT
Stout, Daniel P., LCDR
Stout, James S, LTJG
Stout, Peter C, Capt
Stoutamore, James B, Capt
Stovall, Eula M, Capt
Stovall III, J. Henry, CDR
Stover, Dan M, Capt
Stover, Gregory D, CDR
Stowe, Charles R B, Capt
Stowe, Martin D., CDR
Strachan, Leland B, Lcdr
Strachan Jr, Alexander, CWO4
Strain, William E, Capt
Strand, Donald F, Capt
Strand, Michael G, Capt
Strand, Susan M., LCDR
Strand, William R., CDR
Strandberg, Gregory R., CDR
Strandberg, Josiah R W, Capt
Strande, Edward J, Lcdr
Stranz, Helen S, Lcdr
Strassheim, Thomas J., Capt
Stratbucker, Gerald L, Capt
Stratemeier, John S., CDR
Stratford, Milton W., LCDR
Stratton, Richard A, Capt
Straugh, Thomas W, Capt
Straughn, William R, Capt
Strausberg, Haskel, CDR
Strauss, David S, CDR
Strauss, Michael B, Capt
Strauss, Stephen J., LT
Straws, John A., LT
Strebel, John P, Capt
Strebel, Joseph F, Capt
Street, Jimmy C, CAPT
Streeter, Bruce A., Capt
Streets, Richard R, Capt
Strehle, Hugh A E, Capt
Streit Jr, Peter L, Capt
Strenfel, Joseph M., LT
Stretch, Robert M, Lt
Stribling, William R, CDR
Strickland, Karen, CDR
Strickland, Stacy S, LCDR
Strickler, David W, CDR
Stride Jr, John L., CAPT
Stringer, George F, Capt
Stringer, Robert D., Lcdr
Stringer-Gowdy, Amelia F., LCDR
Strittmatter, Lawrence, CDR
Strobl, Karl J, Capt
Strock, Sylvia S, Capt
Strohman, James R., LCDR
Stromberg Sr, David P, Lcdr
Strong, Clarence E, CDR
Strong, Craig S, CDR
Strong, David B, Capt
Strong, Dorothy O, CDR
Strong, Douglas S., LCDR
Strong, Fred Clegg, CDR
Strong, Robert C., CDR
Stroschein, Thomas M., Lcdr
Strother, Jennifer M., CDR
Strother, Robert B, Capt
Stroubakis, Demetri S., LT
Stroud Jr, Lisle A, Capt
Stroup, Eugene R, Lcdr
Strubel, Thomas, LT
Struble, Dan, LCDR
Strudas, Alan J, CDR
Strunk, Harold Kenneth, Capt
Strunk II, William M., Capt
Strutner, Richard E, Capt
Stuart, Dana Catheen, LCDR
Stuart, Dean H, CDR
Stuart, Douglas A, Capt
Stuart, Wayne S., Capt
Stuart, William L, Lt
Stuart Jr, Charles L, Capt
Stubbeman, David G, Capt
Stubblefield, Roderick W, Capt
Stubbs, Campbell L, CDR
Stubbs, Laura D., CAPT
Stubbs, Willard B, Lcdr
Stuber, Edward J, Lcdr
Stuber, John E., Lcdr
Stuchell, Sheldon D, Lcdr
Stuckel, John E, CDR
Stucker, William E, Capt
Stuckert Jr, George A, Capt
Studdard, James W, CDR
Studley, Michael H, Capt
Studwell, Theodore D., CDR
Studzinski, Edward A, Capt
Stueber, Henry B., CDR
Stuhlmann, Allen H, CDR
Stuhr, Bernard C., Lcdr
Stuhr, William S, CDR
Stull, Joseph D, CDR
Stull, Judson T, Capt
Stundon, Dorothy A., CDR
Stunz, Emil G, Capt
Sturek, Clifford J, Radm
Sturgell, Robert A., CDR
Sturges, David K, Lcdr
Sturla, George H, CWO2
Sturm, Mabelle K, CAPT
Sturm Jr, Harold F, Capt
Sturtevant, Brewster, CDR
Sturtz, Barbara Q., CAPT
Sturtz, David H., CDR
Stutts Jr., Pressley C, Lcdr
Styles, Philip C., Lcdr
Suarez, Angel M., Lcdr
Subin, David K, CDR
Suda, George A, Capt
Sudbeck, Gerald F, CDR
Sudler Jr, Louis C, Capt
Sudnick, Daniel R, Capt
Sudol, Walter E, Capt
Sudranski, Herbert F, CDR
Suffridge, Buford J, Capt
Sugars, George P, CDR
Suggs, Robert C, Capt
Sugimoto, Cynthia R., LCDR
Sugino, Gary S., CDR
Suhrke, Elton H, Lcdr
Suida, Eleanor F., Lcdr
Suiter, William C., Lcdr
Suko, Todd A., Lt
Sullins Jr, W David, Radm
Sullivan, Andrew M, CDR
Sullivan, Brien D., CDR
Sullivan, Charles F., LCDR
Sullivan, Christopher, CDR
Sullivan, David J, Capt
Sullivan, David L, CDR
Sullivan, Elizabeth J., LT
Sullivan, Eugene T, Lcdr
Sullivan, Gerald C, Radm
Sullivan, J Hugh, CDR
Sullivan, James, Lcdr
Sullivan, James E, CDR
Sullivan, John G, Capt
Sullivan, Kathryn D, CDR
Sullivan, Kevin J., CAPT
Sullivan, Lisa A., CDR
Sullivan, Mark F, Capt
Sullivan, Mary R, Capt
Sullivan, Patrick J., LTJG
Sullivan, Paul V., LCDR
Sullivan, Philip L, CWO3
Sullivan, Raymond C, Capt
Sullivan, Richard J., LCDR
Sullivan, Ruth M, Lcdr
Sullivan, Timothy B, CDR
Sullivan Jr, John J, Capt
Sullo, Joseph G, Ltjg
Sulzer, Arthur H, Capt
Sumigray, Alan, Lcdr
Summers, Anthony T., Lcdr
Summers, Bruce G, CDR
Summers, Stephen C, CDR
Summitt, Robert L, Radm
Sumner, Johnnie L, Lcdr
Sumner, William Jay, CAPT
Sumwalt, Samuel G., Lcdr
Sun, Harry H., Lcdr
Sunday, Jeffery H, CAPT
Sundberg, Edward A, Capt
Sunderland, Charlotte A., CDR
Sundgren, Helen H, Capt
Sundine, John R, Capt
Sundstrom Jr, Carl W, Capt
Sunley, John F, Capt
Sunseri, Richard M, Capt
Sunseri, Stephen M., CDR
Suozzo, Joseph M, CDR
Supple Jr, George C, CDR
Surdel, Melvin M, CWO4
Surles, Oliver S, Capt
Surratt, Randal L., Capt
Sus, George M, Lcdr
Susick, Nancy A., Lcdr
Susik, Michael B, Capt
Suszan, Melinda A., Lcdr
Suter, Dale D, Lcdr
Suter, Vane E, Capt
Sutherland, Sandra F, Capt
Sutphen, Bob L, Capt
Sutter, John L, Capt
Suttle, Catherine, CDR
Sutton, David P, Capt
Sutton, John D, Lcdr
Sutton, Terence M, Capt
Sutton, Thomas L, Capt
Sutton, Thomas W., LCDR
Svazas, Brian C., CDR
Svendsgaard, David J., LT
Svenningsen, Mark E, Lcdr
Svenson, William J, CDR
Svoboda, Paul J, Lcdr
Swafford, David, CAPT
Swailes, Don W, Capt
Swailes, John H, Capt
Swain Jr., John W., CDR
Swalchick, Jeffrey M., CAPT
Swall, Don T, CDR
Swan, Arthur D, Lcdr
Swan, Benjamin D, Capt
Swanbeck Jr, Robert R., CDR
Swanberg, J Marvick, Capt
Swanberg, John W, Capt
Swanberg, Paul M, Capt
Swancutt, Mark A., LCDR
Swanson, Aleda J, Lcdr
Swanson, Alvin J, Ltjg
Swanson, David H, Capt
Swanson, Dorothy F, Lcdr
Swanson, Douglas W., Lcdr
Swanson, Erik R., LCDR
Swanson, James R, Capt
Swanson, Karen A., CDR
Swanson, Keith S, Capt
Swanson, Michael R, CAPT
Swanson, Nels H., LT
Swanson, Robert J., CDR
Swanson, Stephen R, Capt
Swanson Jr, Franklin A, Capt
Swarbrick, Kirk B., CDR
Swartley, Steven W, Capt
Swartz, Philip J, Capt
Swartz, Thomas J, Capt
Swartz, Warren L, Capt
Swarzkopf, James F, LT
Swazee, Stephen D., CAPT
Sweat, James P, Lcdr
Sween, David N., LT
Sweeney, Bernard D, Lcdr
Sweeney, Eugene W, Capt
Sweeney, James D, CDR
Sweeney, James M, Capt
Sweeney, John J, Radm
Sweeney, Laurie, ENS
Sweeney, Michael R., CDR
Sweeney, Robert E., CDR
Sweet, Gary K., CWO4
Sweet, James E., LT
Sweigert, Cloyd S, CDR
Swensen Jr, Charles C, CAPT
Swenson, Carl E, Capt
Swenson, Clarence D, Lcdr
Swenson, Craig J, Capt
Swenson, David P., Lcdr
Swenson, Donald L, CDR
Swenson, J Bruce, Capt
Swenson, Lyle H, Lcdr
Swenson, Richard F, Lcdr
Swenson Jr, Eric H, Capt
Swepston, Thomas D, CDR
Swetnam, Michael S., CDR
Swickard, Ben R, WO1
Swiebel, Robert M., CDR
Swift, Frank R., Capt
Swift, William D, CDR
Swigart, Curtis D, Lcdr
Swillum, Earl N, Lt
Swindal Jr, Frederick L, Capt
Swindle, David W., Lcdr
Swindler, Steven B., LCDR
Swineford, Michael C, CDR
Swink, Harley H, Capt
Swinney, Terry L M, CDR
Swinton, William F., CDR
Swinton III, Charles W, CAPT
Swisshelm, Daniel M., LT
Swistak, Anthony J, CDR
Switzer, Harry A, Capt
Switzer, Robert E, Radm
Swoboda, Sandra M., LT
Swofford, James H., Lcdr
Swords, Claude E, LCDR
Sylvester, Torrey A, Capt
Symasko, Peter W, Capt
Symeonides, Joseph H., LT
Symolon, Walter C, Lcdr
Symonds, Ronald D, Capt
Symons, Irving J, Lcdr
Synk, Frederick A, Lcdr
Syring, Robert E., LCDR
Syversen, Carl E., CDR
Sywassink, Craig S, CDR
Szabo, Alex J, CDR
Szabo, George J, Capt
Szakaly, John S, CDR
Szalwinski, Duane E., CDR
Szarleta, M Anthony, Capt
Szatkiewicz, Richard J, Capt
Szekretar, Allen R., CDR
Szwez, Peter J, Lcdr
Szymanski, Rita M, Lcdr
Szymczak, Lucian Henry, Lcdr
Tabbert, Gary D, Capt
Tabbs, Joyce, Lt
Tabbut, Spencer, LCDR
Tabenken, Lisa A., LCDR
Taborsky, Lawrence E., LCDR
Tachovsky, Thomas J, Capt
Tackett, Theodore T, Capt
Tackitt, Patricia H, Capt
Tacoronte, Linda B., Lcdr
Tadlock, Timothy A., Lcdr
Taff, Dillan W, CDR
Taflin, David E., Lcdr
Taft, David N, CDR
Tagney, Patrick J., Lcdr
Tague, Brian C., LCDR
Tague, Charles L, CDR
Tahler, Graham, Radm
Tahler, James W., CDR
Tait, John I, Capt
Tajima, Luke R., LT
Takahashi-Niotta, Jae K., LCDR
Takenaka, Harold H, Capt
Talbot, Edward B, Capt
Talbot, Russell W., CDR
Talbot, William R, Lt
Talbot Jr, John H, Capt
Talbott, William U, Capt
Talley, Darrell L, Lcdr
Talley Jr, Alfred F, Capt
Talley Jr, James M, Capt
Talley Jr, William L, Capt
Talmadge, Bruce A, CDR
Talson, Steven M., CAPT
Tambini, Robert J, Capt
Tamisiea, Donald E, CDR
Tamny, Peter D, CDR
Tan, Rosemarie C., LCDR
Tanck, Calvin E., Lt
Tandberg, Sharon P, Capt
Tandy, David J, Capt
Tandy, Jeffrey S., Lt
Tanguay, Gerard E, CDR
Taninbaum, Warren, Lcdr
Tanner, Lewis W, CDR
Tanner, Linda D, CDR
Tanner, Thomas W, Lt
Tannery, Ora E, Lcdr
Tanowitz, Herbert B, Capt
Tansor, Robert H, CDR
Tanzar, Paul D., CDR
Tapley, John T, CAPT
Taquino, Maurice A, Capt
Taramasso, Daryl, CDR
Tarango, Mark E., CDR
Taranow, Myron J., CDR
Tarantino, Francis, Lt
Tardino Jr, Victor J, CDR
Tarr, Robert J, Capt
Tarrant, Stanley R, Capt
Tarvin, Cleveland R, Lt
Taschner, Bruce, Capt
Tassell, Peter C, CDR
Tassinari, James P, Lt
Tassinari, Silvio J, Capt
Tata, John B, CDR
Tata, Robert M., CDR
Tate, Gene, LT
Tate, John T, Capt
Tate, Johnnie D, Lt
Tate, William H., CDR
Tate, Willie Harold, Capt
Tate Jr, Lee, Lt
Taubman, Arthur I, CDR
Taufer, Thomas J, CDR
Taul, Amel D, Lcdr
Taurman, Keith L., CDR
Taylor, Alfred R, Lt
Taylor, Andrew C, CDR
Taylor, Bruce K, Lcdr
Taylor, Charles A, Capt
Taylor, Creigh W, CDR
Taylor, Daniel M, Lcdr
Taylor, Deborah Gleason, CAPT
Taylor, Donald A, Capt
Taylor, Donald D, Lcdr
Taylor, Edward J, CDR
Taylor, Geo Pete, Capt
Taylor, Gregory M., CDR
Taylor, Homer C, Lcdr
Taylor, Horace H, Capt
Taylor, James A, CDR
Taylor, James E., Radm
Taylor, James M., CDR
Taylor, Jayne A, CDR
Taylor, Judith, CDR
Taylor, Jules Blake, Lcdr
Taylor, Kent D, Capt
Taylor, Lawrence Robert, CDR
Taylor, Loren L, Capt
Taylor, Louise A, CDR
Taylor, Maurice E, CDR
Taylor, Michael F, CWO4
Taylor, Michael P., CDR
Taylor, Norman H, Capt
Taylor, Norman W, Capt
Taylor, Ray P, CDR
Taylor, Raymond F, Capt
Taylor, Ronnie Lee, Lcdr
Taylor, Sherry A, Capt
Taylor, Steven J, Capt
Taylor, Steven L., CDR
Taylor, Stuart M., CDR
Taylor, Thomas A, Capt
Taylor, Thomas J, Capt
Taylor, Timothy M, Capt
Taylor, Timothy O., CDR
Taylor, Ward J, CDR
Taylor, Warren H, Lcdr
Taylor, William D, Capt
Taylor, William F, Lcdr
Taylor, William H, Capt
Taylor Jr, Dean, Capt
Taylor Jr, E Lee, Capt
Taylor Jr, Edwin J, CDR
Taylor Jr, Joseph Z, CDR
Taylor Jr, Samuel K, CDR
Taylor Jr., Ben E., LT
Tchinski Jr, John J, Capt
Teace, Charles E., LT
Teague, Michael M., CDR
Tear, Philip F, Lt
Teare, Paul L, CDR
Teate, Guy J, Lt
Teater, William G, Capt
Tebbetts, Stanley R, Lt
Teborek, James S, Capt
Tedder, Virgil R, CDR
Teegarden, Susann, LCDR
Teel, Elvan D, CDR
Teel, Robert L, Lt
Teeter, David M., CDR
Teitel, Kevin, LTJG
Teller, Donald E, CDR
Tello, Gil, LCDR
Tempest, Mark J, Capt
Temple, K. P., CDR
Templeton, Robert E, CDR
Templeton Jr, John M., Lcdr
Templin, Lee F., Lcdr
Tenaro, Leslie J., CDR
Tenbrink, Robert E, Lt
Tennant, Betty R, CDR
Tennant, Harry W, CDR
Tennant, Juanita D., LCDR
Tenney, Charles I, Lt
Tennyson, Richard B., Lcdr
Ter Kuile, Roger M, Lcdr
Terchek, Joseph J, Lcdr
Terhune, Paul H., Lcdr
Terhune, Robert J, Capt
Terpinas, Thomas M, Capt
Terrell, Richard A., Lcdr
Terry, J Timothy, Capt
Terry, Robert M., Capt
Tesar, Charles B, Capt
Tesar, Susan C, CDR
Tetreault, Malcolm E, Lcdr
Tewelow, William H, Capt
Teynor, Joseph W, CDR
Tezza, David A., CDR
Thacker, Tommy G, CDR
Thackery, Donald N., CAPT
Thalheimer Jr, John C, Capt
Tharps, Morris L, Lcdr
Thayer, Christopher D, Lcdr
Thayer, James L, CDR
Thayer Jr, Wendall M, Lcdr
Thedford, Ronald Bryan, Lcdr
Theiss, Clifford A, Capt
Themmen, Arne R, Capt
Theodoratos, Lya Nicole, LT
Theodore, George, Lcdr
Theodore, Michael G, CDR
Theodore, Pierre R., LT
Theriot, Glenn J, Capt
Therrell, Tim S., CDR
Thesenvitz, Victor W., LCDR
Thetford, David D., CDR
Thetford, Thomas C, Lcdr
Theus, Eugene P., Capt
Thews Jr, Albert W, CDR
Thibadeau, Barbara M., CDR
Thiede, Theodore D., LCDR
Thiel, Douglas J, Capt
Thiele, Gary F, Capt
Thiele, John M, CWO4
Thier, Fred W., LTJG
Thiessen, Richard I, CWO4
Thiringer, Sheridan A., CAPT
Thissell, Charles W, CDR
Thode, Thomas A, Capt
Thoma, Margaret F, Lcdr
Thoman, Mark E., Capt
Thomas, Alexander B, CDR
Thomas, Arthur J, Capt
Thomas, Charles B, Lcdr
Thomas, Craig H., CDR
Thomas, Creed C, Capt
Thomas, Dennis G, Lcdr
Thomas, Donald E., LCDR
Thomas, Faith E., CDR
Thomas, Gene A, CDR
Thomas, Gregory A., Lcdr
Thomas, Harvey F, CDR
Thomas, James E, Capt
Thomas, James E, CDR
Thomas, James W, Lcdr
Thomas, Jeffrey H, CAPT
Thomas, Jerold J, Capt
Thomas, John K, CDR
Thomas, John P, Capt
Thomas, John R, Capt
Thomas, John R, Capt
Thomas, Kenneth J., Lcdr
Thomas, L., LCDR
Thomas, Larry W, Capt
Thomas, Louis N, CDR
Thomas, Lynn Anne, LCDR
Thomas, Marilyn M, Capt
Thomas, Mark A., MIDN
Thomas, Maureen B., LT
Thomas, Michael D, CDR
Thomas, Michael D., CDR
Thomas, Michael E, Lcdr
Thomas, Michael O., Lcdr
Thomas, Orville Gordon, Capt
Thomas, Paul R, Lt
Thomas, Paul R, Capt
Thomas, Phillip A, Capt

Thomas, Quinn D, Capt
Thomas, Robert J, Capt
Thomas, Robert J, Capt
Thomas, Robert J., LT
Thomas, Robert L, Capt
Thomas, Roger W., CDR
Thomas, Steven E, CDR
Thomas, Walter D, CDR
Thomas III, George J, Lcdr
Thomas IV, John F., CDR
Thomas Jr, Joseph B, CDR
Thomas Jr, Walter E, CWO4
Thomason, Francis J, Lcdr
Thomason, George W, Capt
Thomasson, John A., LT
Thomforde, James H, Capt
Thompkins, Gerald O., LCDR
Thompsen, T. Gilbert, CDR
Thompson, Alfred H, Lcdr
Thompson, Alfred S, CDR
Thompson, Allan R, CDR
Thompson, Alonso O., ENS
Thompson, Ann E, LT
Thompson, Azariah G, Capt
Thompson, Bobby C, Capt
Thompson, Bower Patrick, Capt
Thompson, Brandon J., CDR
Thompson, Charles E, Lcdr
Thompson, Clarence C, CWO2
Thompson, Daryl S., CDR
Thompson, David A, Capt
Thompson, Don L, CDR
Thompson, Donald H, Capt
Thompson, Douglas A., CDR
Thompson, Edmonstone, CDR
Thompson, Elton N, Capt
Thompson, Fred W, CDR
Thompson, Gary B., LCDR
Thompson, George, CDR
Thompson, Guy R, CWO4
Thompson, J. Stanton, CAPT
Thompson, James B, Lt
Thompson, James J, Capt
Thompson, James J., Lcdr
Thompson, James T., LT
Thompson, Jeffrey L, Capt
Thompson, John E, Lcdr
Thompson, John Hart, Capt
Thompson, John M., CAPT
Thompson, Joseph R, Capt
Thompson, Joseph S, Lt
Thompson, Julian B, Capt
Thompson, Kenneth A., Lcdr
Thompson, Kenneth E, Capt
Thompson, Kermit D, Capt
Thompson, Lloyda L, CDR
Thompson, Mark C., CDR
Thompson, Mark R, Radm
Thompson, Melvin R, Lcdr
Thompson, Nancy D, CDR
Thompson, Oscar M, Lt
Thompson, Pamela L.M., CDR
Thompson, Peter M, CDR
Thompson, Richard A, Capt
Thompson, Robert B, Lcdr
Thompson, Robert W, CDR
Thompson, Ronald L, CDR
Thompson, Scott G, Lcdr
Thompson, Sumner L, CDR
Thompson, Thomas R, Lcdr
Thompson, William D., CDR
Thompson, William R, CDR
Thompson III, Charles S., LT
Thompson Jr, Frederic, CDR
Thompson Jr, James E, Capt
Thompson Jr, Reuel, Lcdr
Thompson Jr, Wade H, Lt
Thompson Jr, William H, Capt
Thomsen, Jeffrey E, Lcdr
Thomson, Don Cameron, CDR
Thomson, Frank W, CDR
Thomson, Ian C, VADM
Thoreen, John B., Lcdr
Thoresen, Stephen J., LCDR
Thorfinnson, Stanley T, CDR
Thorla, Arlin E, Capt
Thorn, James D, CDR
Thorne, Douglass E, Capt
Thorne, E Dayton, Lcdr
Thorne, Kevin B., CDR
Thorne, Lewis G, Capt
Thornhill, Daniel B, CDR
Thornton, Carolyn, ENS
Thornton, Donald A., Lt
Thornton, Jack W, CDR
Thornton, Rebecca C, Lcdr
Thornton, Steven J., LCDR
Thornton, Thomas R, Lcdr
Thornton, William E, CDR
Thorp, Burt D., Capt
Thorp, John E, CAPT
Thorp, M J, CDR
Thorpe, Bert D, Capt
Thorson, Donald P, CDR
Thorstad, Roger S., LCDR
Thow, Kimberly A., Lcdr
Thrasher, Marvin P, Lt
Threlkeld Jr, Walter F, Capt
Thrush, George F, Lcdr
Thuot Sr., Clifford G, Capt
Thur, James A, Capt
Thurman Jr, Jack E, Capt
Thurmond, Thomas G, Capt
Thurmond Jr, James E., Capt
Thurston, Donald R, Capt
Thurston, William B, Lcdr
Thurwachter, Loron F, Lcdr
Thweatt Jr, Frank M, Capt
Tiber, Anne, ENS
Tichy, John M, CDR
Tidball, Jack P, Lcdr
Tidd, Emmett H, Vadm
Tidstrom, Fred L, CDR
Tieaskie, Linda J., CDR
Tiedeman, Mary E, Capt
Tiedemann Jr, Albert W, Capt
Tiernan, Robert P, Radm
Tiernan, Robert R., Lcdr
Tierney, Francis J, Lcdr
Tierney, Ralph C, Capt
Tierney III, John J., LCDR
Tiffin, David S, Lcdr
Tigue, Joseph B, Lt
Tikkanen, Thomas W., LT
Tilghman, John G, Lcdr
Till, John E, Radm
Tillerson Jr, William E, CDR
Tillman, Danny J., Lcdr
Tillman, Henry C., LCDR
Tillotson, Drew R, Capt
Tilock, Fred H, CDR
Tilton, Charles H, Capt
Tilton, Tanya Tylene, Lcdr
Timberlake, Gregory A., Capt
Timer, Albert V, CDR
Timm, David R., CDR
Timm, Edward, Capt
Timm, Leonard O, Lt
Timmerman Jr., Claude E., CAPT
Timmons, John P, Lt
Tindall, Ray C, Radm
Tiner, Billy D, Capt
Tinguely, John E., LT
Tinker, Roy B, Lcdr
Tinley, John H, Capt
Tinlin, Ronald G, LCDR
Tinsman, Stewart H W, Capt
Tipper, Paul R, Lcdr
Tippett Jr, James R, CDR
Tipping, Robert E, CDR
Tipton, Harry B, Capt
Tisch, Alfred C, CDR
Tisher, Jim W., CDR
Tittle Jr, Robert R, Lcdr
Titunik, Ira R, CDR
Titus, John A, Lcdr
Titus, Robert G, Capt
Titus, Ronnie G., CDR
Tivnan, Thomas P., LCDR
Tober, William A, Lcdr
Tobi Jr, George B, Capt
Tobin, Daniel J, Capt
Tobin, John L, CDR
Tobin, William E, CWO4
Todd, Alan M, Capt
Todd, David N, Capt
Todd, Gordon L., CWO3
Todd, Harold C, Capt
Todd, Larry J, CDR
Todd, Robert W., Capt
Todd, Vincent W, Capt
Toedt, Elizabeth M., CDR
Toellner, David A., LCDR
Toleno, Robert P., CDR
Tolentino, Robert P., CWO4
Toler, Melissa A, CDR
Toler, William G, Capt
Tolkoff, Edward, Capt
Toll, Giles D, Capt
Tollefson, Jo E., LCDR
Tolleris, Ralph, Lcdr
Tollison, Thomas M, Capt
Tollison Jr, Alfred C, Capt
Tom, Raymond, CDR
Tomaselli, John J, Capt
Tomaszewski, Thomas N., LCDR
Tombasco, Peter A., LCDR
Tomlin, Ronald D., LT
Tomlin III, Henry B, Capt
Tomlinson, Craig S, Capt
Tomlinson III, Allen R, Capt
Tompkins, Albert E, Capt
Tompkins, John F., CDR
Toms, Marian E, CDR
Tomski, Sharon J., Lcdr
Tomzik, John M., Lt
Ton, Gary M., CDR
Tonahill, Joe H, LT
Toncray, George W, Capt
Tonkin, Russell J, Capt
Toole II, Theron C., CDR
Toomey, John B, Capt
Toothman, Douglas R, Lcdr
Tootle, Glenn A., LT
Topp, Walter S., CDR
Topping, Carolyn M, CDR
Torelli Jr, Nicholas M, CDR
Torgerson, Robert, CDR
Torma, David P., CDR
Tormes, Felix R, Capt
Tormey Jr, Edward N, Capt
Torok, Helmut A., Capt
Torp, Eric C., CDR
Torp, Raymond J., Lcdr
Torreon, Ben E., LT
Torres, James D, Capt
Torris, Donna M., LCDR
Torsch, Virginia M., CDR
Torsiello, John B, CDR
Toth, Christopher A., Lcdr
Toth, Susan M., Lcdr
Toth Jr, Joseph, Lcdr
Totushek, John B., Radm
Tougas, Lee A, CDR
Toups Jr, James E, Capt
Tourigny, Clement G, Capt
Tousignant, Peter M., CDR
Touw, Theodore K, CDR
Towne, Joseph H, CDR
Townend, Sally S., Lcdr
Townsend, Bobby N, Lcdr
Townsend, J Russell, Lcdr
Townsend, Susan E, CDR
Townsend III, John H, Lcdr
Toye, James C, CDR
Toyias, James P, Capt
Trabue, Turner, Ltjg
Tracey, Marsha L., CDR
Trafas, Peter C, Ltjg
Trafton, Rose M., LCDR
Trainor, Joseph F, Capt
Trame, Lawrence C., Lcdr
Trammel, James A, CDR
Trammell, Calvin D, CAPT
Trammell, William L., LT
Trandal, David W, CDR
Trapp, Frederick P, CDR
Trapp, Harry D, Lcdr
Tratnyek, Joseph P., LT
Traub, Daniel J., CDR
Traub, Raymond W, Capt
Traupe, Robert L, Capt
Traut, Arthur J, Capt
Trautwein, Robert T, Capt
Trautwein, Tom A, Radm
Traver, Stephen A, CDR
Travers, Paul J., LT
Travers, Verne, LT
Travis, James E, Capt
Traweek, Billy B, Capt
Treacy, Stephen T., CDR
Treadwell Jr, Kenneth, Capt
Treat, Charles W, Capt
Tredennick, Harry L, Capt
Tredway, Donald R, Capt
Treece, Guy E, CDR
Trefny, Thomas W., LT
Treitz, Robert G., Capt
Trejo, Paul E, Capt
Trelease, Joan W, CAPT
Treloggen, Johnny N., Lcdr
Tremant, Robert A, Capt
Tremblay, Harold G, Lcdr
Tremblay, Janet L., CDR
Tremblay, Victor J, Capt
Trenholm, Beverly A., LCDR
Trently, William M., CDR
Trepagnier, Albert J, Capt
Trethric, Richard A, Lt
Trevathan, Larry E, Capt
Trevett, Robert E, Capt
Trevino, Robert C, Capt
Trewet, Glenn E, CDR
Triandafilou, Andrew N, Lt
Triantafyllides, George, Lt
Tribelhorn, Ronald W, CDR
Trice, R. D., CDR
Triebold, Louis H, Lcdr
Trierweiler, James R, CDR
Triftshauser, Roger W, Radm
Trigilio, John J., Lcdr
Trimmer, Terry H, Capt
Tringale, Michael, Lcdr
Trinh, Nam K., Lt
Trinkwalder, Janet C, Capt
Triplett, William H, CDR
Tripodi, Raymond A, Lcdr
Tripp, Daniel B, Lt
Tripp, James S, CDR
Tripp, Jeffrey J., LCDR
Trippe Jr, Richard M, Capt
Tritchler, James A, Capt
Troiani, Joseph E, CDR
Trojack, John E, Capt
Trollope, Richard G, Lcdr
Trombetta, John J., Lcdr
Trotter, Thomas A, Capt
Troup, Tony E., LT
Trout, B. Thomas, CAPT
Troutman, David L, CDR
Troutman, Stephen B, Capt
Trowbridge, Leroy G, CDR
Trowse, Michael J, Capt
Troxel, Charles F, Capt
Troxell, Lewis M, CDR
Troy, Dennis L, Lcdr
Troyan, Patricia J., CDR
Trucco Jr, John A, LT
Trucker, Don, Lt
Trudeau, Cathy L, Lcdr
Trudeau, Michael A., LCDR
Trudel, Stephen J., Lcdr
True, Jay W, Lcdr
True, Robert E., LT
Truitt, Benjamin, Lcdr
Trujillo, Lionel G, Capt
Trujillo, Patricia, CDR
Trulock, Ronald E, Capt
Truman, Ramon R, Lcdr
Trumbull, Edward Roy, Lt
Trumbull, William C, CDR
Truog, Bruce E, Capt
Truslow, Alfred R, Lt
Tryon, Stephen H S, CDR
Tsai, Pedro T.H., Lcdr
Tschaepe, Kurt O E, Capt
Tscheschlok, Debra M., LCDR
Tschirhart, John A, CDR
Tschirhart, Paul M, Capt
Tse, Kwok B., LCDR
Tshontikidis, Thomas W., Lcdr
Tsujimura, Reuben T, CDR
Tubbs, Nelson C., LCDR
Tubbs, Richard H, CDR
Tubbs, William B, Capt
Tucci, Donald J, CDR
Tuck, Steven Glenn, LT
Tucker, David J., CDR
Tucker, Eugene F, Capt
Tucker, Richard M, CDR
Tucker, Samuel S, Capt
Tucker, Walter D, CDR
Tucker Jr., Thomas R., CAPT
Tuckerman, Bruce A., CDR
Tudisco, Ellen M, Lcdr
Tudor Jr, Frederic, Capt
Tuebner, David L, Lt
Tufts, Arthur W, Capt
Tuleya, Raymond R, Capt
Tulip, S. Craig, LCDR
Tullai, Jason, LCDR
Tulleners, Frederic A, CDR
Tulley, Edward M, Lcdr
Tulley Jr., James H., CAPT
Tullis, Joe M, Lcdr
Tulloh, John C, Lt
Tumbleson, William C, CDR
Tumminia, Richard L, CWO4
Tunison, Colin R, Lt
Tunnicliffe, William W, Capt
Tuohig, Gail M., CDR
Tupper II, Russell L, CDR
Turbeville, John F, Capt
Turbeville III, Daniel, CDR
Turchi, Charles R, CDR
Turco, Robert I, Capt
Turek, Gabriella, Lt
Turets, Thomas F., Lt
Turiello, Ronald J, CDR
Turisco, Thomas F, Capt
Turk, Alan R, CDR
Turk, James D, Capt
Turk, Raymond J, Lt
Turk Jr, Edward J, Capt
Turley, Ella M., Lt
Turnage, W. Sherman, Lcdr
Turnbow, Kristofer P., Lcdr
Turnbull, Fred D, Capt
Turnbull, Harold M, Lcdr
Turnbull Jr, Wade W, Lcdr
Turner, Brian E., CDR
Turner, Bruce E, CDR
Turner, David E., CDR
Turner, Ellis L, Lcdr
Turner, Gerry D, Capt
Turner, Gloria J., CDR
Turner, James M, CDR
Turner, John S, Capt
Turner, Karen A, Lcdr
Turner, Lawrence P, LCDR
Turner, Otis L, Capt
Turner, Philip L., LT
Turner, Russell S., Lt
Turner, Thomas A, Lt
Turner, Thomas S., LCDR
Turner, Thorn, LT
Turner, Wade S., LT
Turner, William D, Capt
Turner Jr, Benjamin B, Capt
Turner Jr, James E, Capt
Turner Jr, James E, Capt
Turner Jr, Ralph W., CDR
Turner Jr, Roger W., Lcdr
Turner Jr, Russell G, Capt
Turpin, Dennis T, Lcdr
Turpin, Robert, Capt
Turrentine, William E, Lcdr
Tuthill, John E, Capt
Tuthill, W Lee, Capt
Tutt, Billy D, Capt
Tuttle, George T, Lcdr
Tuttle, Jackson C, Capt
Tuttle, Raymond D, CDR
Tuttle Jr, Lloyd P., LCDR
Tutunjian, Stephen J, CDR
Tutwiler Jr, M N, CDR
Twardos, Helen K, Lcdr
Tweden, Wallace D, Capt
Tweedy, Roxanne D., Lcdr
Twesme, Albert T, Capt
Twigg, David W, CDR
Twilla Jr, William M, Capt
Twining, Joseph P., CAPT
Twiss, Ralph B, CDR
Twist, Gerald A, Capt
Twite, Gaylord M, Capt
Twombly, Arthur L, Capt
Twomey, Thomas J, Capt
Tyahla, Stephen F., CDR
Tyau, Wesley Y.E., Lcdr
Tychsen Jr, Paul E, CDR
Tydings III, William E, CWO4
Tye, Clifford F, CDR
Tye, Susan Perry, CDR
Tyler, Daniel I, CDR
Tyler, Devin R., MIDN
Tyler, Ina June, LT
Tyler, Robert L, CWO4
Tylicki, Robert A, CAPT
Tyma, John H, CDR
Tyner, Michael J., LCDR
Tyner Jr, Weldon H., CDR
Tyre, Timothy E., CDR
Tyree, Wilbur S, CWO4
Tyson, Charles C., Lcdr
Tyson, Dan M., Capt
Tyson, George J, Capt
Tyson, Karen S, CDR
Udd, Elmer, Capt
Uddenberg Jr, Bert, Capt
Ude, Charles S., CDR
Udicious, Richard A, Capt
Uecker, John M., LCDR
Uehling, Donald E, CDR
Uehlinger, Scott C., LCDR
Ueland, Christopher, Lcdr
Ueland, Ole M, CDR
Ugelow, Stephen A., CDR
Ugorcak, Elizabeth Ann, Lcdr
Uhl, Steven L, Lcdr
Uhl, Thomas W, Capt
Ulery, William H, Lcdr
Ullberg, Annette L., LCDR
Ulmer, Edward W, CDR
Ulmer, Garold G., Lcdr
Ulmer, Michael F., LCDR
Ulmer II, Frederick J., Lcdr
Ulrey, Richard G, Lcdr
Ulrich, Theresa Ann, Lt
Ulrich, Thomas R, Capt
Ulrich Jr, V Kenneth, Capt
Ulsh, Merle F, CDR
Ulstein, Lawrence, Lcdr
Umbach, John C, CDR
Umbel, Richard P, CDR
Umberg, Linda R., LCDR
Umberger III, Paul R., LT
Umlauf, John C, Lcdr
Umphenour, Jess H., LCDR
Underdahl, Patricia Jo, CDR
Underhill, Burton, Capt
Underhill, Douglas V., CAPT
Underhill, Thurlow R, Capt
Underwood, Brian C, Capt
Underwood, Glenn R, Capt
Underwood, Horace H, Capt
Underwood Jr, George D, Capt
Unger, Eloise J., LT
Unger, Verner B, Capt
Unger Jr, Norbert S, Lcdr
Unger Jr, Paul R, Lt
Unkle, David W., Lcdr
Unruh Jr, Howard Kirk, Capt
Unser, Philip J, CDR
Upchurch, Douglas B., LCDR
Upton, Cheryl A., CDR
Upton, Dana L, CDR
Upton, Douglas R., LT
Urban, Henry J, CDR
Urban, Jack, Capt
Urbana, Robert D, Capt
Urbano, Lawrence B, CDR
Urbon, Victor A, CDR
Urewicz, Charles J., Lcdr
Urinyi, John R, Capt
Urquhart III, Charles F, CDR
Urquidi, Teresa, LT
Ursem, Andrea J., Lcdr
Usher, Frederic L, Capt
Usher, Robert J, Lt
Utley, Gregory L., CDR
Utley, Wallace R, Capt
Utne, James T, CDR
Utterback, Wiley V., CDR
Uyeda, Dale, LT
Uzarski, Donald R., Capt
Uznanski, Kenneth M, Capt
Vacca, Angelo W, Capt
Vacca, Fred W, Capt
Vaccarino, Frank, Lcdr
Vacek, Deborah A., Lcdr
Vacura, Jean M, Lcdr
Vadeboncoeur, E M, CDR
Vail, Richard M, Lcdr
Vaitas, Vil S, CDR
Valace, James J, CDR
Valcis, Andrew M, CWO4
Valenta, Joseph R, Capt
Valenti, Gabriel M, Lcdr
Valentine, Deberall A, CAPT
Valentine, Judith D., CDR
Valentino, Charles R., CDR
Valiquet, Albert W, Lcdr
Valle, Paul M, CDR
Vallely, Edward M, Lt
Valley, Raymond L, CDR
Valliere, Michael L., Lcdr
Valori, Mark F., LTJG
Valvo, Gary M, CDR
Van Abel, Robert J., CDR
Van Almelo, Frederick K, CDR
van Avery, Christopher E., LT
Van Benthem, Paul J., Lt
van Benthuysen, Baldwin, Lcdr
Van Cise, Richard A., Capt
Van De Mark, John S, CDR
Van Deavender, Ben, CDR
Van Den Biesen, James L, Capt
Van Der Kamp, Kathryn J, Lcdr
Van Der Veen, James A., CDR
Van Dette, John J, CDR
Van Deusen, William B, Lcdr
Van Dop, Richard K., Lcdr
Van Duinen, Neil G., Lcdr
Van Dyke, William H, CAPT
Van Etten, Jack A, Lt
Van Gorden, Charles, CDR
Van Hala, Mark D., Lcdr
Van Heest, Victor J, CDR
Van Horn, Bradley R., LCDR
Van Horn, Michael A, CDR
Van Kopp, Eddy, Lt
Van Lier, Herman R, CDR
Van Male, Nicholas R, Capt
Van Ness, Peter H, CDR
Van Nest, James A., Lcdr
Van Norde, Peter J, Capt
Van Ooyen, William P, CWO4
Van Oss, David J, Capt
Van Otegham, James H, Lcdr
Van Petten, David J., CAPT
Van Reeth, Philip J, Lcdr
Van Rensler, Barry W., CDR
Van Rossem, Henry N, Lcdr
Van Roy, Gretchen E, Lcdr
Van Sickle, James L, CDR
Van Sittert, Thomas P., CDR
Van Skike, Jeffrey B, CDR
Van Soest, Peter D, Capt
Van Strydonck, Thomas, CDR
Van Vactor, Gary R., CDR
Van Vleck, Michael F, CDR
Van Voast, Herbert, CDR
Van Voorhees, F. David, LCDR
Van Voorhis, Lorraine, Lt
Van Winkle, Robert L, Capt
Vanande, Cheryl L., LCDR
Vanburg, Herbert, Lcdr
Vance, Apostle P, CDR
Vance, George F, Capt
Vance, John, CDR
Vance, William A, Capt
Vandale, John J., Capt
Vandam, Irwin K, CDR
Vande Kerkhoff, Henry T, Lcdr
Vandegriff, John Henry, Capt
Vanden Bosch, Martin L., Lcdr
Vanden Heuvel, Robert E, Capt
Vandenberg, James N., ENS
Vander Lugt, Robert D., LT
Vander Myde, Paul A., Capt
Vander Myde, Philip L., Capt
Vander Yacht, Brian K., CDR
Vanderbloemen, Lisa A., Lt
Vanderbrook, Robert J., ENS
Vandercook, Jon R, Lcdr
Vanderhoeven, Richard M, CDR
Vanderplaats, Roger A., LCDR
Vanderschoot, David A., Lcdr
Vanderveen, Gordon H, Lcdr
Vandervoort, Albert F, CDR
Vanderwaal, Joel K., CDR
Vanderweele, John T, Lcdr
Vandyke, James A, CDR
Vangordon, John M, CWO4
Vanhoomissen, John E, Lcdr
Vannatta, Dennis Roy, Lcdr
Vannatta, John O, Capt
Vannatta, Robert H, Lcdr
Vanni, G. Steven, Lt
Vannortwick, William A, CDR
Vanover, Albert D, CDR
Vanover, Thomas D., LT
Vanschaick, Otto E, Lcdr
Vansickel, Harold C, Lcdr
Vant Hull, Marietta P, Lcdr
Vantrigt, Peter C, Capt
Vanwye, Harlan E, CDR
Vanzandt, W Robert, Capt
Vanzwalenburg, Paul R, CDR
Varboncouer, Anthony P., CAPT
Varenholt, Susan M., Capt
Varga, Dennis J., LCDR
Vargas, Patric J., Lt
Vargas, Philip J., CDR
Varnadore, D. Craig, CWO2
Varney, Philip R, CDR
Varricchio, Philip T, CDR
Varsanyi, Peter S., CDR
Vartan, Karen S, Capt
Vastardis, John E., Lcdr
Vaughan, Bobby F, Capt
Vaughan, Charles F., CDR
Vaughan, Daniel E., Lcdr
Vaughan, Francis, Capt
Vaughan, G Dennis, Radm
Vaughan, Mark B, CDR
Vaughan, Mary E., CDR
Vaughan, Matthew G., LT
Vaughan, Michael W, Capt
Vaughan, Raymond A, CWO4
Vaughn, Donald W., CDR
Vaughn, Hershey B., LCDR
Vaughn, Kenneth G, Lcdr
Vaughn, Thomas K, CAPT
Vaughn, William A., CDR
Vaupel, George B, Capt
Vaurio, David G, CDR
Vausbinder, Carl D, CDR
Vavrin, Charles R, CAPT
Vawter Jr, Wallace, Capt
Veary, Michael, ENS
Veatch Jr, Walter D, CDR
Veilleux, Joseph E., Lcdr
Veit, Walter A, CDR
Veith, Joshua P., ENS
Velazquez, Edward, Lcdr
Vellella, George J, Capt
Vellenga, Daniel R, Capt
Vellenga, David B, Capt
Velling, Joseph J., CDR
Vellucci, Frank R, CDR
Velotta, John C, Capt
Velte, Michael J, CAPT
Venable, Bruce S, CDR
Venable, Laura L, Lcdr
Vendrzyk, Judith M., LCDR
Venes Jr, John W, CDR
Venezia, Maribeth, Lcdr
Vengrow, Michael I., CAPT
Veninger, Frank, CDR
Venter, Josiah B, CDR
Ventura, Gustavo M., Lcdr
Ver Steeg, Carl, CDR
Vera, Albert J, CDR
Verbrycke IV, John R., CAPT
Verburg, Robert L, Lcdr
Vercher, John R, Lt
Verdel, Thomas H, CAPT
Vere, Harry W, CDR
Veremakis, Dean S, Capt
Verlin, James A, Capt
Verna, Bernard J, CDR
Vernet, Frank G., Lcdr
Vernon, Melinda H., CDR
Vernon, Robert L., CDR
Verrengia, Thomas, Capt
Verrill, John A, Lt
Verschuyl, Ronald, CWO4
Verstandig, Robert J, Capt
Verwohlt, Hermann R, Lcdr
Verzino Jr, William J, Capt
Vescovo Jr, Paul C, Capt
Vessels, Virginia A., Lcdr
Vestal, Ellis B, Capt
Vester, Stephen J., CDR
Vetrano, Anita J, CDR
Vetterick, Rolland C, Capt
Vezeridis, Michael P, Capt
Vicens, Jose J, CDR
Vicinanzo, Paul C, Lcdr
Vickerman, M. John, Capt
Vickers, Timothy C., CDR
Vickers III, Thomas A, Lcdr
Vickery, Thomas E, Capt
Victorian, Inez M., Lcdr
Victorine Jr., Vernon E, Lcdr
Vidacovich, Richard P, Capt
Vidinhar, Lamond R., LCDR
Vidoni, Thomas J., LTJG
Vieau, Harold E, Capt
Viele, Jack T, Capt
Viering, Edward W., CDR
Vierzba, Edmund A, CDR
Vietmeier, Ronald E, CWO4
Vigil, Samuel A., CDR
Vikingson, Vernon F, CDR
Vilano, Samuel H, Lt
Villanova, Margo M., LCDR
Villard, Anna Y, Lcdr
Villarin, Nolan D., Lcdr
Vincent, Brian S., LCDR
Vincent, Steven F., Lcdr
Vincent, Wayne A., Lt
Vincent II, Harold T., LCDR
Vine, Gary L., Capt
Vineyard, Michael H., CDR
Vinovich, Gary K., CDR
Vinton Jr, Benjamin, Capt
Violante, Lawrence J, Lt
Virgens, Victor E, Lt
Virgilio Jr, Joseph, CDR
Visher, John C., Lcdr
Viskup, Steven A, Capt
Vitale, Alexander F, Lcdr
Vitale, Eugene J, CDR
Vitale, Joseph S, Capt
Vitali Jr., Henry R., CDR
Vitanza, Joseph T, Capt
Vite, Jean H., Lcdr
Vitek, Randall W, Lcdr
Vitolo, Robert A, Lt
Vivadelli, John H., CDR
Vivar, Juan C., LCDR
Vivona, Richard B, Capt
Voegeli, Nicholas J, Lcdr
Voelker, William, CDR
Voepel Jr, William C, CDR
Vogel, Frederick Q, Capt
Vogel, Martin, Lcdr

Vogel, Philip L., Lcdr
Vogel, Robert R., CAPT
Vogeley, John C, Capt
Vogelsang, Carl E, CDR
Vogler, Bernard E, CDR
Vogt, Brian J., CDR
Vogt, John H, Capt
Vohs Jr, Paul A, Capt
Voiss, Daniel V, Capt
Voldeng Jr, Karl E, CDR
Volk, Austin N, Capt
Volk, Charles F, Lt
Volker, Kenneth L., CDR
Vollaire, Robert L, CDR
Vollbracht, Joe, LCDR
Vollkommer, Frederick C., LCDR
Volpe, David M., ENS
Volpe, Joseph R, Capt
Volstorf, Michael P., LT
Volz, Gary M., CDR
Volz, John L, Capt
Volz, Vincent J, Capt
Volz, Walter D, Capt
Von Almen, Adelbert M, Capt
Von Berg, Donald R., CDR
Von Bernuth, Robert D., Capt
Von Bodungen, Joseph P., Capt
Von Buelow, Carl Eric, CDR
Von Saunder, Raymond O, CDR
Von Tungelin, Lloyd L, CDR
Vonkempf, Paul, Capt
Vonrichter, Frank C, CDR
Voracek, Thomas J, Lcdr
Vorhies, Lynn R, CDR
Vorse, Lawrence B, Capt
Vortmann, Thomas A, Capt
Vorwerk, Arno Carl F, CDR
Voss, David A., Lcdr
Voss, Gregory E., LCDR
Voss, Joseph G, Lcdr
Voss, Kenneth L, CDR
Voss, Robert W, Lcdr
Voth, Harold M, Radm
Voughs, Tyrone Y., MIDN
Voulgarakis, George R., CDR
Vowell, Joe L, CAPT
Vowinkel, Russell C, Capt
Vrablec, John J, Capt
Vraney, Robert E, CDR
Vranicar, Russell J., CDR
Vroom, Oliver E, Lcdr
Vucinovich, Albert P, Capt
Vukovitz, John C, Capt
Vuksta, Michael J, Capt
Wachowicz, Fred S, CAPT
Wachtel, Thomas L, Capt
Wachtler, Joseph J., LCDR
Wacker, David C., CWO4
Wada, Douglas T, CDR
Waddell, Daniel W, CDR
Waddell, Ray K, Capt
Waddell, Robert A, Capt
Waddell, Ronald M., CDR
Waddell, Watson W, CDR
Wade, Dorothy J, CDR
Wade, Lillian F, Lt
Wade, Michael S, Capt
Wade, Nelsyn E, CDR
Wade, Patrick J., LCDR
Wadley, William J, Lcdr
Wadnola, Frederick J., LCDR
Wagaman, Michael J, CDR
Wagenbach, Ronald J, Capt
Wages Jr, Ward, CDR
Waggener, Anna T., CDR
Wagner, Charles S, Capt
Wagner, Francis H T, Lcdr
Wagner, John David, CDR
Wagner, John R., LT
Wagner, Robert C, CDR
Wagner, Ronald P., Lcdr
Wagnon, William G., Capt
Wagoner, Glen, CDR
Wagoner, Jon D, Capt
Wagstaffe, Desmond H, Capt
Wahler, Christopher C., Lcdr
Wahler, Joseph L, Capt
Wainwright, Dale W, CDR
Wainwright, Norman, CDR
Waite, James L, CDR
Waite, Michael P, Lcdr
Waite Jr, John J, CDR
Wakefield, Mannsell C, Capt
Wakefield, Zachary T, Capt
Walaitis, William R, Capt
Walden, James C, CDR
Waldmann, Edward B, Capt
Waldron, James F, CAPT
Waldron, Jonathan C., CAPT
Wales, Leonard P, Capt
Wales, William Douglas, Capt
Walge, Ernst K., LT
Walish, John P., LCDR
Walker, Andrew A, Capt
Walker, Archie C, CDR
Walker, Chris G., LT
Walker, Christopher L., LT
Walker, Clemons F, Lcdr
Walker, Crayton C, CDR
Walker, Dan L., Capt
Walker, David R, Capt
Walker, Donald R, CDR
Walker, Douglas B, Capt
Walker, Frederick D, CDR
Walker, Gary R., CWO2
Walker, George K, Capt
Walker, Henry M, Capt
Walker, Ira G, CDR
Walker, J Mark, Capt
Walker, James R., LCDR
Walker, Jeffrey K, CDR
Walker, John P., CDR
Walker, Joseph E, Lt
Walker, Joseph M, Capt
Walker, Judith M, Capt
Walker, Larry V., LT
Walker, Lisa M., LCDR
Walker, Lynn W, Capt
Walker, Mary E., CDR
Walker, Michael L., Capt
Walker, Norman L, Capt
Walker, Richard A., Lt
Walker, Steven K., Lcdr
Walker, Stuart J., LCDR
Walker, W Bruce, Capt
Walker Jr, William B., Capt
Walko, Joseph J, CWO4
Wall, Francis F, Capt
Wall, James A, Capt
Wall, Jeffrey, LCDR
Wall, Marc D., Capt
Wall, Richard L, CDR
Wall, Richard L., LCDR
Wall, Woodrow L, CWO4
Wall III, Thomas P., Lcdr
Wall Jr, Henry J, Lcdr
Wall-Smith, Stephen B., Lcdr
Wallace, Charles M, Lcdr
Wallace, Hugh H, Lt
Wallace, John C, Capt
Wallace, Kenneth F, CDR
Wallace, Nathan G, CWO4
Wallace, Prentice L, CDR
Wallace, R Eugene, CDR
Wallace, Richard J, CDR
Wallace, Robert C, Lt
Wallace, William A., Lt
Wallach, Andrew J, Capt
Wallach, Stanley, Capt
Waller, John C H, Capt
Walley, James E, Capt
Walley Jr., James M., CAPT
Wallin, Gene A, Capt
Wallin, William R, Capt
Wallingford, Marvin C, CDR
Wallington, George L, Capt
Wallio, Richard B., CDR
Walls, Hollis W, Capt
Walls, Lynda S., LCDR
Walls, Robert J, CDR
Walluk, Annette Fullmer, Lcdr
Walsh, Austin M, Capt
Walsh, Bernard, Capt
Walsh, Daniel K., LCDR
Walsh, David E, Lcdr
Walsh, Donald J, CDR
Walsh, George J, Lcdr
Walsh, Gerald W, CDR
Walsh, James A, CDR
Walsh, Jeanne L, Lcdr
Walsh, John M, CDR
Walsh, Jon C., CDR
Walsh, Joseph D, Capt
Walsh, Kathleen E, Lcdr
Walsh, Leo A, Capt
Walsh, Neil R, Lcdr
Walsh, Owen B, Capt
Walsh, Patrick J., LCDR
Walsh, Peter E., CAPT
Walsh, Stephen A., CDR
Walsh, William J, CDR
Walsh, William J, Capt
Walsh II, Joseph A, LCDR
Walsh Jr, Joseph F, Capt
Walsh Jr, William J, CDR
Walsh-Lund, Carol, CDR
Walstad, John O, CDR
Walston, David E, CDR
Walter, Georgette C., Lt
Walter, Katherine A., Lcdr
Walters, Joycelyn B., CDR
Walters, Julian Phillip, Lcdr
Walters, Peter S, Lcdr
Walthall, Frank G, Lt
Walton, Fredrick H, CDR
Walton, Michael L, Lt
Walton, Stephan B., LCDR
Walton, Steven D, CDR
Walton III, John W, Capt
Walton Jr, Harry B, Lt
Waltzer, Jack R., CDR
Walus, Walter E, CDR
Walvoord, David R, Lcdr
Walz, Deborah J., CDR
Wamsley, Brian W., CAPT
Wamsley, Colleen M., LT
Wanamaker, Warren A, Capt
Wanbaugh, Peter M, Capt
Wanders, Joseph A., Capt
Wang, Po H., CDR
Wanger, William H., Capt
Wansbury, Dixon E, Capt
Wanzenberg, Frederick W, Lcdr
Ward, Arthur D, Capt
Ward, Arthur J, Capt
Ward, Arthur T, CDR
Ward, David Francis, Capt
Ward, Donovan Frederick, CDR
Ward, Dorothy B, Lcdr
Ward, Emmet G, Capt
Ward, Frederick F, CDR
Ward, Gary J., Lt
Ward, George H, Capt
Ward, Hugh P, Lcdr
Ward, Jack G, Capt
Ward, Jack H, CDR
Ward, James T, Capt
Ward, John M., CDR
Ward, Michael J., LCDR
Ward, Michael W, Capt
Ward, Ralph C, CWO4
Ward, Robert C, Capt
Ward, Robert D., LCDR
Ward, Roderic C, Capt
Ward, Stacey B., LCDR
Ward, Warren E, Lcdr
Ward Jr, Arthur E, CDR
Ward Jr, Henry H, Lt
Ward Jr, Norman J, CDR
Warden, James M, CAPT
Warden, John M, Lt
Ware, Alan F., CAPT
Ware, Dale K., CDR
Ware Jr, Lewis L., Capt
Warfel, Suzanne Lee, Lcdr
Warfield, Colin, MIDN
Warfield, Lemuel O, Radm
Wargelin, David W, CDR
Warger, Gregory P, LTJG
Wargo, Linda E., Capt
Warhola, Joseph G, Capt
Waring, Jerome E, Lcdr
Warlick, Joseph Dee, CDR
Warmouth, Nicholas, ENS
Warner, Bertram B, Lcdr
Warner, John R, Capt
Warner, Joseph E., CDR
Warner, Kristine D., CAPT
Warner, Michael E., CDR
Warner, Suzan P., LCDR
Warner, Terry O, CAPT
Warner, William T, CAPT
Warner Jr, James E, CDR
Warnes, Philip G, Capt
Warnock, Jeffrey J, Capt
Warren, Dean, Lcdr
Warren, John A, Capt
Warren, Robert W, CDR
Warren, Walter E, Capt
Warren, William C, Ltjg
Warren, William H, CAPT
Warren, William L, Lcdr
Warren, William R, CDR
Warren Jr, Donald C, CAPT
Warrick, James C, Capt
Warrick Jr, Woodward A, Lcdr
Warrington, Robert G, CDR
Warthen, Nowell John, Lcdr
Wasalaski, Robert G., Capt
Wasek Jr, Joseph, CAPT
Waselchuck, John M., LCDR
Washburn, Kent L, Capt
Washburn, Rolla G, Capt
Washington, Robert J, CDR
Wasikowski, Lawrence J., Lcdr
Wasileski, Jeffrey M., LT
Wasko, John J, LT
Wasley, James C, Capt
Wass, Leonard R, Capt
Wassel, Mary L, CAPT
Wasserman, Richard A, Lcdr
Wasserman, William L, Capt
Wassman, E. Robert, LCDR
Wasson, Eugene F, CDR
Wasylyk, Sylvia H, Lt
Wasz, Edwin J., CDR
Waszak, Louise C., CDR
Watanabe, Frank, LT
Waterman, Larry W, CDR
Waters, Bradford, CDR
Waters, Douglas E, Lt
Waters, James P, Lt
Waters III, Marshall P, Capt
Waters Jr, Albert A, CDR
Waters Jr, Harry, CDR
Watkins, Cliff P, CDR
Watkins, Gary E., CDR
Watkins, James D, Capt
Watkins, Phillip L, Lt
Watkins, Roger D, Capt
Watkins, Steven D., Lcdr
Watkins Jr., Kenneth S., Capt
Watry, Margaret M, CAPT
Watson, Ann Olivia, CDR
Watson, Becky J., LCDR
Watson, Clyde M, CWO4
Watson, Douglass C, Capt
Watson, Edward B, CDR
Watson, Gary E, Capt
Watson, Jay W, Lcdr
Watson, John, Capt
Watson, John E., LTJG
Watson, Louise, Lt
Watson, Michael, CAPT
Watson, Michael G., LTJG
Watson, Michele, Lt
Watson, Robert J, Lcdr
Watson, Scott M, CDR
Watson, Thomas H, CDR
Watson, Timothy B., CAPT
Watson, William H, Capt
Watson III, Earl E, CDR
Watson Jr, Noel A, CWO4
Watson Jr, William H, CAPT
Watt, Earl J, Capt
Watt, John W, Capt
Watt, Thomas P., Lcdr
Watt Jr, George P, Capt
Wattay, Leslie J, Capt
Wattay Sr, Alexander E, CDR
Watters, Hubert, CDR
Watters, William J., CDR
Watters III, Jesse D., LT
Wattez, Richard N, CDR
Watts, Aaron D, CDR
Watts, Costello, LT
Watts, David J, CDR
Watts, David R, Capt
Watts, Franklin S, Lcdr
Watts, James T, CAPT
Watts, Patrick R, Capt
Watts, William E, Lcdr
Watts Jr, Donald L, Capt
Waugh, Norman E, Capt
Waylett, Nancy E, Lcdr
Waylett, Stephen F, CDR
Waylett Jr, William J, Capt
Wayne, Anthony, Capt
Wayne, Robert S, Lcdr
Wayne, Thomas P, CDR
Wayt, Sims M., LT
Wazenegger, Charles J, CDR
Weadon Jr, Donald A, Lcdr
Wear, Terrance J, Capt
Wear Jr, Lawrence E, Capt
Wears, Millie R., Lcdr
Weatherbee, Artemus E, CDR
Weatherford Jr, Wade S, CDR
Weatherley, Kent W, Lcdr
Weatherly, William G, Lcdr
Weatherly Jr, Harley O, Lcdr
Weathers, Dale, LCDR
Weathers III, Lawrence G., CDR
Weaver, David E., LCDR
Weaver, David G, CDR
Weaver, David G., CDR
Weaver, Douglas W., Lcdr
Weaver, George A, Radm
Weaver, Gordon F, Capt
Weaver, Harry B, Capt
Weaver, Irene K, Lcdr
Weaver, Joel T., LT
Weaver, Joseph W, Capt
Weaver, Kenneth M, Lcdr
Weaver, Lester G., CDR
Weaver, Peter Mark, Capt
Weaver, Robert H, CDR
Weaver, Ronald S, CDR
Weaver, Salvatore J., CDR
Weaver, Steve E., LCDR
Weaver III, Edwin F, Capt
Webb, Bruce C, CDR
Webb, David R., Capt
Webb, Michael K., CDR
Webb, Newton H, CDR
Webb, Raymond E, CDR
Webb, Robert E, Capt
Webb, Solon D, Capt
Webb III, Robert W, CDR
Webb Jr, John O, Capt
Webb Sr, Aubury R, CDR
Webber, Edward G, Lcdr
Webber, James R., LT
Webber, Kent S, Capt
Webber, Willard S, Capt
Weber, Alban, Radm
Weber, Burrell U, CDR
Weber, Craig T., Lcdr
Weber, John F, CDR
Weber, Karen V., LT
Weber, Paul J., ENS
Weber, Robert L, Capt
Weber, Ronald G., CWO3
Weber Jr, John W., Lcdr
Weber Jr, Max G, CDR
Webster, Bobby R, Capt
Webster, Bruce D., CDR
Webster, Guy N, CDR
Webster, J Mark, Lcdr
Webster, Jeffrey L., Lcdr
Webster, Reg, LCDR
Webster, Richard A., CDR
Webster Jr, Richard M, Capt
Wechselberger, Jacob F, Capt
Weckman, Paul F., Lcdr
Wecksell, Alan, Lcdr
Weckworth, Thomas V, Capt
Weddington, Michael E, CDR
Weed, Peter B, Capt
Weed Jr, John C, Radm
Weekley, Brett H, Lcdr
Weekley, Gregory, Lt
Weeks, Peter, LCDR
Weeks, Randall W, Capt
Weeks, William E, CDR
Weems, Bill W, CDR
Weems, Thomas M., Lcdr
Weenink, Robert A., CDR
Weese, Winston H, Radm
Weesner, James E, CDR
Weeth, John B, Capt
Wefald, Robert O, Capt
Wegner, Waldo W, Capt
Wegryn, Robert, CDR
Wehe, Donna J, CAPT
Weickert, Paul M, Lcdr
Weidenkopf, David W, Capt
Weidman, Catherine M., LCDR
Weidman, Clare David, CDR
Weidman Jr, Robert H, Radm
Weidner, Arthur H, CDR
Weidner, S Edward, Lt
Weigand, Peter J., Lcdr
Weigandt Jr, Carl F, Capt
Weigel, John M., Lcdr
Weigl, Roger, Lcdr
Weigmann, Douglas A., LCDR
Weigold, John F., Lcdr
Weihaupt, John G, Capt
Weikel, Charles P, Lcdr
Weilding, Douglas B, Lcdr
Weiler, Maurice D, CDR
Weiler Jr, Wayne B, Lt
Weill, Lawrence E., CDR
Weinberger, Robert G, Capt
Weiner, Ronald A, Capt
Weiner, Steven A, Capt
Weinert, Bruce R, CDR
Weinert Jr., William E., LCDR
Weingart, Clifford A, Capt
Weingart, Robert P, CDR
Weinman Jr, William V., Lcdr
Weinschenk Jr, Alfred J, Capt
Weinstein, Martin, CDR
Weirick, Nils T., CDR
Weise Jr, Robert W, CDR
Weisenburger, Philip J, Capt
Weishaupt, Paul J, Lt
Weisiger, Carter B, Capt
Weiss, Anne C., CDR
Weiss, Ernest E, Capt
Weiss, Joseph, Lcdr
Weiss, Michael E, Lcdr
Weiss, Mitchell L, Capt
Weiss, Richard A, Lcdr
Weiss Jr, Robert F, CAPT
Weissbach Jr, Albert F, Capt
Weistroffer, Gary G., Lcdr
Weitfle Jr, Paul L, Capt
Weitgenant, Harold R, Capt
Weithaus, Charles R, CDR
Weithman, Don F., LCDR
Weitz, George, CWO2
Welch, Betty Ann, LCDR
Welch, David D, Capt
Welch, Donald S, CDR
Welch, Jeffrey W., CAPT
Welch, John A., Lcdr
Welch, Kenneth J, Capt
Welch, Patrick E, Lt
Welch, Peter B., Capt
Welch, Richard J, CDR
Welch, Stine M., Lcdr
Welch, William G., CDR
Welch Jr, Henry T, Lcdr
Welcome, Allan T, Lcdr
Weld, Louis M., Lcdr
Weldon, William K, Lcdr
Wellens, Charles H, CDR
Wellington, Graham P, Capt
Wellish, Carl S, Capt
Wells, Carl S, Lt
Wells, Clifford P, Capt
Wells, Connie W., Lcdr
Wells, Donald C, Lt
Wells, Elliott L, Lt
Wells, Jack L, CDR
Wells, John F, CDR
Wells, John G, Capt
Wells, Judson L, CDR
Wells, M Claudette, CDR
Wells, Maurice I, Lcdr
Wells, Michael E., CDR
Wells, Randolph S, Lcdr
Wells, Richard, Lcdr
Welnick, Robert N., Lcdr
Welsh, Charles J, LT
Welsh, Susan, LCDR
Welter, David D., CDR
Welter, Gerard R, Capt
Welton, John W, Wo
Welton, Robert G, Capt
Welty, Christopher J., CDR
Weltzien, Brian D, CDR
Wendells, Arthur T, Capt
Wendelstadt, Vaughn A, LCDR
Wendling, Catherine M, Lcdr
Wendt, Albert G, Lt
Wendt, Christopher D., MIDN
Wendt, Robert W, Capt
Wenger, Max, Capt
Wenner, Margaret S., CDR
Wennet, Richard I, Capt
Wenschlag, Roger E., CDR
Wentland, Norman H, Capt
Wentz, Joan D., CDR
Wentz, Timothy A., Lcdr
Weny, Bruce M, CAPT
Wenze, Gloria Tansits, CAPT
Wenzel, Robert M., Lcdr
Wenzler, Robert T., Capt
Werner, Dean F, Capt
Werner, John, CDR
Werner, Mark R., CDR
Werner IV, George A., Lt
Wernes, John Erik, Lcdr
Wernli, Daniel E., CAPT
Werp, Peter, LCDR
Wertalik, Charles S, CDR
Wertz, Douglas E., CDR
Wesely, Laura J, Lcdr
Wesenberg, William W, CDR
Wesendorf, Leonard E, CDR
Wessman, Richard H, Capt
Wessner, Steven C., MIDN
Wesson, Mike T., LT
West, Charles A, CDR
West, E Frank, Capt
West, James B, Capt
West, Jerry D, Capt
West, Kerry A., LCDR
West, Kirby P, Ltjg
West, Maxwell C, Lcdr
West, Raymond L, CWO4
West, Robert P, Lcdr
West, William L, Lcdr
West Jr, Frederick P, Capt
Westall, Charles L, CDR
Westerfield, Jack, Capt
Westergren, Richard C, CDR
Westerhof, Henry A, Capt
Western, John, LT
Westfall, Daniel E, CWO4
Westfall, Susan L., CDR
Westhaus, Randy, CDR
Weston, Stephen F, CDR
Weston III, Joseph C., CDR
Westover, Bruce E, CDR
Westover, Stanley R, CDR
Westrich, Charles L, Capt
Westura, Warren S, CDR
Westwood, George E., CDR
Wetherill, Christopher, Lcdr
Wetlesen, Einar A, Capt
Wetmore, Ronald W, CDR
Wettach, George E, Capt
Wetzel, Blair H, Capt
Wetzel, James A., LCDR
Wetzel, James C., CDR
Wetzel, Steven L., Capt
Wetzel Jr, James L, Lcdr
Wexler, Donald L., CDR
Weyler, Frank W, Capt
Weyrich, Jack G, CDR
Whale, William T., CDR
Whalen, Donald C., LCDR
Whalen, James, Lcdr
Whalen, William P, CDR
Whalley, Christopher J., Lcdr
Whalley, Joseph S., LCDR
Wharton, James M., LCDR
Whatley, Lawrence A., LCDR
Whealy, Robert A, Capt
Wheatley, Chester H, Lcdr
Wheeler, Charles E, Lcdr
Wheeler, Fred A, Capt
Wheeler, Gerald E., CDR
Wheeler, John R, Capt
Wheeler, Mary Case, Capt
Wheeler, Philip C., Lcdr
Wheeler, Richard A, Capt
Wheeler, Sean M., LT
Wheeler, William F, Lcdr
Wheeler Jr, Clarence J, CDR
Wheelock, Franklin K, Lcdr
Wheelwright, Joan R, Capt
Whelan, Edward W, Lt
Whelan, Joseph R, Lcdr
Whelan, Thomas J, CDR
Whelan, Thomas W., Lcdr
Whiddon, Roscoe O, CDR
Whigham, J Earl, Lcdr
Whipple, Walter, Capt
Whisenhunt, Ronald A., CDR
Whisenhunt, Warren, LTJG
Whiston, Edward D., LT
Whitacre, Jan M., CAPT
Whitacre, Philip A, Radm
Whitaker, Robert H, Capt
Whitaker, Thomas N, Lcdr
Whitaker, Thomas P, Capt
Whitcomb, James A, Lcdr
Whitcomb, Lee E, Capt
White, Brents R, Capt
White, C David, Capt
White, Charles D., CDR
White, Charles E, CAPT
White, David, Lt
White, David A, CDR
White, Donald B, CDR
White, Douglas G, CDR
White, Edward L, LT
White, Edward P, CDR
White, Gerald A, CAPT
White, James F, Capt
White, James Lee, CDR
White, John C, Capt
White, John P, Lcdr
White, Kenneth R, CDR
White, Larry R., Capt
White, Nancy A., CDR
White, Nicholas H., LT
White, Patricia M., Lcdr
White, Patricia Naomi, CDR
White, Paul F., Lcdr
White, Paul M, CDR
White, Reuben C, CWO3
White, Richard M., Capt
White, Richard R, CDR
White, Robert F, Capt
White, Robert S, Lcdr
White, Rolfe K., CDR
White, Ron T., CAPT
White, Stewart S, Lcdr
White, Todd, LT
White, Tracy M., Ens
White, Vernon E, CDR
White, Vincent B, CDR
White, William D, Capt
White, William J, CDR
White, Willis A, CDR
White II, James G, CAPT
White III, Willis S., CDR
White Jr, Edward Faith, Capt
White Jr, Henry F, Radm
Whited, Brian M., CDR
Whitehead, Robert M, Lt
Whitehead Jr, Daniel L., LCDR
Whitehouse, Scott, Radm
Whitehurst, Brian R., CDR
Whiteman, James L, Lcdr
Whiteman, Ralph E, Capt
Whitesides, James D, Capt
Whitfield, Richard W, Capt
Whitford, Peggy R., Lcdr
Whiting, Jeffrey B., Lcdr
Whiting, Teresa S, Lcdr
Whiting, William A, Capt
Whitley, Dewey L., CDR
Whitlock, Brooks L, Lcdr
Whitlock, Lloyd M, CWO4
Whitlock, Robert B, Capt
Whitlock, Roy E, CDR
Whitman, Deborah J., CDR
Whitman, Jeffrey D., Lcdr
Whitman, Michael W., LT
Whitman Jr, Gerald, CDR
Whitmore, Robert C, Capt
Whitmore Jr, John B, Capt
Whitney, Frank E., CDR
Whitney, Robert C, Capt
Whitney, Steven K, Capt
Whitson, William F, Capt
Whitt, James W, Lcdr
Whittaker, Ronald G, Capt
Whittemore, Laurence, CDR
Whitten, James P, CDR
Whitten, Jimi D., Lt
Whitten Jr, Robert C, CDR
Whittenberg, William R, CDR
Whittington, Joel L., LT
Whittington, Stephen J., LT
Whittington Jr, John H, CAPT
Whittles, Thomas E, Lcdr
Whitty, Richard J., LT
Whyte Jr, James P, CDR
Wiberg, John L, Capt
Wick, Clayton P., CDR
Wickerham, Donna L, Lcdr
Wickerson, Stephen F., Lcdr
Wickliffe, John R, Lcdr
Wicklund, John D, CDR
Wickman, Robert E, Capt
Wickman, Victor P, Capt
Wicks, Charlene D., CDR
Wicks, Franklin D, CDR
Widener, Lynn H., Capt
Widis, Marvin L, Lcdr
Widmann, Chester H, CDR
Widner, William R, Capt
Wiebke, Eric A., CDR
Wiebush, Joseph R, Capt
Wiedmaier, Harry O, Capt
Wiehage, James W, Capt
Wiemers, Eugene H, Ens
Wieneke, Steven J., CDR
Wiens, Jerry A, Capt
Wierzba, Dwayne, LTJG
Wiese, Richard A, Capt
Wiesnet, Donald R, Capt
Wiest, William G, Capt
Wieting, William F, Capt
Wietman, Randy G., Lcdr
Wiggins, Charlotte L, LCDR
Wiggins, Glenn R, Capt
Wikstrom, Michael B., Lcdr
Wikstrom, R. B., Lcdr
Wilborn, Christopher S, Lcdr
Wilbur, Barry N., CAPT
Wilbur, Daniel E., CDR
Wilbur, Edward S, Lcdr
Wilbur, Gayle S., CDR
Wilbur, Richard M, Lcdr
Wilbur, Roger F., Lt
Wilbur Jr, Warren A, Capt
Wilburn, Christopher M, Lt
Wilburn, E Gordon, CDR
Wilcox, Bruce, Lcdr
Wilcox, Charles E., CDR
Wilcox, John D, Capt

Wilcox, John W, Capt
Wilcox, Walter G, LCDR
Wilcox, William M, Capt
Wilcox Jr, John R., Capt
Wilcox Jr, Leonard A, Capt
Wild, William L, CDR
Wilde, David W, CDR
Wilde, William T., CDR
Wilde Jr, E Andrew, CDR
Wildenradt, Theodore H, Lt
Wilder, Charles H, CWO4
Wilder, Stephen F, CDR
Wilder, W. Michael, Capt
Wilder III, David S., Capt
Wilder Jr, Calvin R, CDR
Wildhack III, William A., LCDR
Wildrick, William S, Capt
Wiley, Matthew D., Lt
Wiley, Richard A, LCDR
Wiley Jr., Albert L, Lcdr
Wilhelm, Paul J, Capt
Wilhoit, Charles V, CAPT
Wilhoit, James P., LCDR
Wilken Jr, Leon O, Capt
Wilker, John F., CDR
Wilkes, John M., LCDR
Wilkes, Laudius, CDR
Wilkie, James E., Lcdr
Wilkie, W Michael, Lt
Wilkins, Loring P, Lcdr
Wilkins, Steven L., MIDN
Wilkinson, Edward E, Lt
Wilkinson, Gerald E, Capt
Wilkinson, Harvey E, CAPT
Wilkinson, James Edward, LT
Wilkinson, Jim, ENS
Wilkinson, Timothy C., LCDR
Wilkinson, Wilfrid P., LT
Wilkinson Jr, John F, Capt
Wilkinson Jr, Joseph B, CAPT
Wilks, Gerald L., Capt
Wilks, Jack A, CDR
Willard, Daniel S, Capt
Willard, Gerald W, CDR
Willcox, Kenneth A, Capt
Wille, D Gene, Capt
Willeford, Raymond F, Capt
Willert, Roger F, Lcdr
Willett, Katherine J., Lcdr
Willey, Dale D, CDR
Willey, H Fletcher, Capt
Willey, Michel Frederic, Capt
Willey, Richard D, Lcdr
Willgress, Edwin R, Capt
Willhelm Jr, Arthur C, Lcdr
Williams, Alan L, Capt
Williams, Alvin R, Capt
Williams, Beverly J., CDR
Williams, Charles A., CAPT
Williams, Claude M, CDR
Williams, Danny C., Lt
Williams, David A, Capt
Williams, David M, CDR
Williams, Dennis S, CAPT
Williams, Don L, Lcdr
Williams, E Cameron, Capt
Williams, E Nash, Capt
Williams, Eddie R, CDR
Williams, Edward D, Capt
Williams, Edward R, CDR
Williams, Eileen J, CDR
Williams, Frederick L, CDR
Williams, George F, Capt
Williams, George M, Lcdr
Williams, Harmon M, Capt
Williams, Harris E., CAPT
Williams, Herbert B, Capt
Williams, J Bruce, Lcdr
Williams, James E, Capt
Williams, James L, Capt
Williams, James M, Capt
Williams, James Michael, Lt
Williams, James S, Capt
Williams, James T, Lcdr
Williams, James T., Lcdr
Williams, Jeffrey C., Capt
Williams, Jeffrey G., CDR
Williams, John A, Capt
Williams, John C, Capt
Williams, John C, CDR
Williams, John C., LT
Williams, John L, Capt
Williams, John S, CDR
Williams, Kathryn M, CDR
Williams, Kenneth M, CDR
Williams, Kenneth T, CDR
Williams, Louis R, Capt
Williams, Mark S., CDR
Williams, Michael D., LT
Williams, Michael G, Capt
Williams, Morton D, Capt
Williams, Nancy L, LCDR
Williams, Patricia G., LCDR
Williams, Priscilla A., CDR
Williams, Reg A, CAPT
Williams, Richard L, CDR
Williams, Robert Barclay, CDR
Williams, Robert C, CWO4
Williams, Robert E, Lt
Williams, Robert L, CWO4
Williams, Robert L, CDR
Williams, Ronald D, Lcdr
Williams, Roy E, Capt
Williams, Ruth A., Lcdr
Williams, Scott K, CAPT
Williams, Scott R., ENS
Williams, Sharon M., CDR
Williams, Terry C, Capt
Williams, Thomas M, CDR
Williams, Ulysses J C, Capt
Williams, Walter K, CDR
Williams, Warren Allan, CDR
Williams, Wesley E, Ltjg
Williams III, A H, Capt
Williams III, Dudley D, Capt
Williams III, Josh T., Lcdr
Williams Jr, Charles B, LCDR
Williams Jr, James C, CDR
Williams Jr, Tyler E, Capt
Williams Jr, Walter A, CAPT
Williams Jr, William F, Capt
Williams Jr., Elwyn C., LT
Williamson, Issac R, CAPT
Williamson, James E, CAPT
Williamson, James K, CDR
Williamson, Joseph, Capt
Williamson, Larry A, Capt
Williamson, Marcus N., CWO3
Williamson, Ruth I, Lt
Williamson, Theodore M, CDR
Williamson, Thomas D J, CDR
Williamson, Walter D., ENS
Williamson-Sutton, C. A., LT
Willich, Nicholas A, Capt
Willig, Leslie A, Capt
Willis, Barry S, Capt
Willis, Beauford N, Capt
Willis, Clifton F, CDR
Willis, Edmund P, Capt
Willis, Jerome A, CDR
Willis, John R., Capt
Willis, Karen D., CDR
Willis, Odette P, Capt
Willis, Ron L, LCDR
Willis, Roy, Lcdr
Willis III, Park W, Radm
Willis Jr, Guion H., LCDR
Willkens, William H, Lt
Willman, Thomas D, LT
Willoughby, J Hamil, Capt
Willoughby, Kenneth R, Capt
Willoughby, Kevin, LT
Wills, Douglas L, CAPT
Willson, Harry T, Lcdr
Wilmer, William B, CDR
Wilpitz Jr, L Wayne, CAPT
Wilson, Andrew H., Cdr
Wilson, Andy H., Lcdr
Wilson, Charles G, CDR
Wilson, Charles L, Lcdr
Wilson, Charles R, CDR
Wilson, David G., Capt
Wilson, David S., CDR
Wilson, Donald H, Capt
Wilson, Douglas M, CDR
Wilson, Eldon J, CAPT
Wilson, G Richard, Capt
Wilson, Gabriel H, CDR
Wilson, George A, CDR
Wilson, George R., Capt
Wilson, Gerald W, CDR
Wilson, Gerald Wayne, CDR
Wilson, Gregory C., CDR
Wilson, Harold E, Capt
Wilson, Hugh P, Capt
Wilson, Jack H, Capt
Wilson, James G, Capt
Wilson, James S, Capt
Wilson, James W, CDR
Wilson, Joe H, Capt
Wilson, John A, Lcdr
Wilson, John H, CDR
Wilson, Joseph A., Lcdr
Wilson, Paul C., LCDR
Wilson, Ray L, CWO4
Wilson, Renay, LCDR
Wilson, Richard A, CDR
Wilson, Richard H, CDR
Wilson, Robert W, CDR
Wilson, Ryan M., CDR
Wilson, Stanton R, Capt
Wilson, Steven D, CAPT
Wilson, Thomas A, CDR
Wilson, W Graham, Capt
Wilson, Walter D, Lcdr
Wilson, Ward T., CDR
Wilson, Wilbur A, CDR
Wilson III, William J, Lcdr
Wilson Jr, Edwin M, Radm
Wilton, James P, Lcdr
Wiltrant, James C., ENS
Wiltse, James C., CDR
Wiltse, Leslie K, Lcdr
Wiltse, Ronald A, Lcdr
Wiltshire, James A, CDR
Wimmer, David L, CWO4
Wimmer, Leith S., Lt
Wimmer, Ronald W, CDR
Wimmer, William A., LCDR
Wimpfheimer, Michael C, CDR
Winbush II, N. Wyman, CDR
Wincek, Arthur T, Lt
Winchester, Nancy A, CDR
Windsor, Carl D, Lcdr
Windt, Raymond L, Capt
Wines, Richard L, Capt
Winesett Jr, Robert D, Capt
Wing, Thomas M., LCDR
Wing, Timothy D., CDR
Wingert, Neil S., CDR
Winget, Frank W., Lcdr
Wingett, Wayne A., LCDR
Wingfield, Thomas C., LCDR
Winglewich, Eugene E, Lcdr
Winkelman, Herman A, Lcdr
Winkler, Christie W, Capt
Winkler, David F., CDR
Winkler, James P., Lcdr
Winkler, John K., Lcdr
Winkler, Michael W, CDR
Winkler, Paul A, CWO3
Winkles, Paul F, Capt
Winkowski, David J., CDR
Winnard, David E., Capt
Winokur, James L, Lt
Winowicz Jr, Stanley J, Capt
Winsor, Mark R, CDR
Winston, Delores L., LT
Winter, Eugene J, CDR
Winter, Gerald N, Lcdr
Winter, Jeffrey D., CDR
Winter, Mark J., LCDR
Winter, Philip H, Capt
Winter, Steven M, Lcdr
Winter Jr, Wilburn J, Capt
Winterroth, James M., CDR
Winters, Earl T, Lcdr
Winters, Harvey A., LCDR
Winters, Kenneth M., CWO4
Winters, Richard D, Capt
Winterton, Boyd W, CDR
Winton, Calhoun, Capt
Wireman, Ray B, Capt
Wirt, Charles R, Lcdr
Wirt, Dares E, Capt
Wirt, Morgan E, CDR
Wirth, Milton J, CDR
Wirtschafter, Irene N, Capt
Wirtz, Charles J, Lcdr
Wise, John E, CDR
Wise, Lawrence A., Lcdr
Wise, Timothy A., Lcdr
Wise Jr, Harold R, Capt
Wiseman, Robert J., CDR
Wisenbaker, John D, LCDR
Wiser, Gary A, Capt
Wiser Jr, Charles F, CDR
Wisnewski, Paul S, Lcdr
Wisnom, Jeffrey R., CDR
Wistl, Albert R, Lcdr
Witamyer III, John H, CDR
Witbeck, Robert L, Lcdr
Witcher, Murray H, Capt
Witcher, Robert C, Capt
Wither, William A, Lcdr
Withers, Eleanor L., LCDR
Withers III, Jerome B, CDR
Witt, Dean E, Capt
Witt, Florence M, Capt
Wittcoff, Mark, LCDR
Witte, Barry R., Lcdr
Witte, David A, Capt
Witte, Donald M, Capt
Wittebort, Robert J, CDR
Wittebort Jr, Robert J, CDR
Witten, John J., Lcdr
Wittenberg, Charles F., CDR
Wittler, Leroy F, Lcdr
Wittman, Joseph J, CDR
Wittmann, Bertrand R, Lcdr
Wix, Donald M, CDR
Woerfel, George A, CDR
Woerner, Mark H., CDR
Woerz, Gary F., CDR
Wohlert, Earl Ross, LT
Woikey, Douglas R, CDR
Wojcicki III, Peter J, CDR
Wojciechowicz, Michael, CDR
Wojnar, John J, CDR
Wolanin Jr, Alfred J., Capt
Wold, Richard L, Lcdr
Wolensky, Irene, Capt
Wolf, Charles R, CAPT
Wolf, Clinton E, Capt
Wolf, Elwood J, Ltjg
Wolf, James P, CDR
Wolf, Lee E, Capt
Wolf, Richard A, Lcdr
Wolf, William F, CDR
Wolf Sr, Horace L, Capt
Wolfe, Alan M, CDR
Wolfe, Charles F., Lt
Wolfe, David E, Lt
Wolfe, Gilbert, LCDR
Wolfe, John C, Capt
Wolfe, John D., Lcdr
Wolfe, Keith N, CDR
Wolfe, Michael J, Capt
Wolfe, Patricia E., CDR
Wolfer, Marvin F, Lt
Wolff, Gregg, LCDR
Wolff, Lippman W, Lcdr
Wolff, Peter U, Capt
Wolff Jr, Stephen J, Capt
Wolford, Fred T, CDR
Wolford, Helen G, CDR
Wolfram, Michael C., Lcdr
Wolfsen, James L, Capt
Wolfson, Louann I, ENS
Wolters, Timothy S., LCDR
Wolven, Donald L., CDR
Womack, Lowell A, Capt
Womack Jr, William L., CDR
Wombwell, James A., CDR
Womer, J Robert, Lcdr
Won, Jeffrey C., CDR
Wong, Chester W., CDR
Wong, Dennis E., CDR
Wong, Michael F., Lcdr
Wong, R. D., CDR
Wong, Randal S, LCDR
Wong, Richard, CDR
Wong, Terrence C., CDR
Wonsala, Stanley W, CDR
Wood, Cleo B, Lt
Wood, Daniel A, Capt
Wood, Dennis Patrick, CAPT
Wood, Gregory A., LT
Wood, J. Jeffrey, CDR
Wood, John M, Capt
Wood, John P., LCDR
Wood, Joseph R, Lcdr
Wood, Kenneth L, Capt
Wood, Kenneth P, CDR
Wood, Kent A.L., LCDR
Wood, Maynard A, Capt
Wood, Randel W., CDR
Wood, Terray E, CAPT
Wood, Thomas D, Capt
Wood, Thomas H., CDR
Wood, Will O, CDR
Wood, William M, Lt
Wood Jr, John W, Ltjg
Wood Jr, Joseph H, Capt
Wood Jr, Leon G, Capt
Wood Jr, Thornley B, Lcdr
Wood Jr., James A.E., LCDR
Woodall, Jonathan H, CDR
Woodall, Mark E., LCDR
Woodard, David R, CAPT
Woodard, James Anderson, Capt
Woodard, Richard L, Capt
Woodard, Timothy D., LT
Woodard Jr, Luther E., Lt
Woodbury, Jason, ENS
Woodcock, David A., CDR
Wooddell, John G, Capt
Woode, Bradley S., LCDR
Woodell, Marshall E, CDR
Woodfin, Robert H, CDR
Woodford, Robert G, Capt
Woodings, Matthew H., CDR
Woodman, Charles B, CDR
Woodman, William J, CWO3
Woodruff, David H, Capt
Woodruff, James J, Capt
Woodruff, Susan L, CDR
Woodruff, William M, Lcdr
Woodruff III, Berryman, CDR
Woods, Charles J, Capt
Woods, David L, Capt
Woods, Howard R, Lcdr
Woods, Joseph G, Lcdr
Woods, Mary Ellen, LTJG
Woods, Michael D, Lcdr
Woods, Raymond W, CAPT
Woods, Ruel H, CDR
Woods Jr, Samuel N, Lt
Woodson, Helen M., LTJG
Woodson, Riley D, Capt
Woodward, Dennis N., Lt
Woodward, George R., CDR
Woodward, Michael B., CDR
Woodward, S Lee, LT
Woodward Jr, George P, Capt
Woodward Jr, James F, CDR
Woodworth, Scott A., LCDR
Woody, Garry L, Capt
Woody, Johnny W, Capt
Woody Jr, John H, Capt
Woofter, Dennis D, Capt
Wooldridge, Clark E, CDR
Woollard, Wilma J., LT
Woolley, Richard B, CDR
Wools, Ronald J, Capt
Woolston, David E., LT
Woomer, Jerry J., CDR
Woomer, John R, CDR
Wooten, William M, CDR
Worachek, John E, CDR
Workinger, Thomas G, Capt
Workman, William T, Capt
Workman Jr, Ellery C, CDR
Worley, Mary Ellen, Ltjg
Worley, Regis C, CDR
Wormwood, Vance A, CDR
Worrall, Chris, Lcdr
Worrall Jr, Joseph H, Lt
Worrilow, Mark D., CDR
Worringer, Robert, LCDR
Worsham, Bertrand R, CAPT
Worth III, Joseph C, Capt
Worthey, Michael K, LCDR
Worthington, Arthur E., CWO3
Worthy, James R., LT
Worthy, Paul F., Ltjg
Wotring, Mary Louise, Lcdr
Wouk, Herman, Lt
Wrate, Christopher A., LCDR
Wright, Bruce C, Capt
Wright, Burton G, CDR
Wright, Clayton H, CDR
Wright, Dewey S, Lcdr
Wright, Edwin S, CAPT
Wright, Francis D, Capt
Wright, Frederick M, Capt
Wright, Garland P, CAPT
Wright, George William, Lcdr
Wright, James B., CDR
Wright, James C, CDR
Wright, James W., CAPT
Wright, John C, Capt
Wright, John R, Capt
Wright, John S, Capt
Wright, Johnie B, CAPT
Wright, Kathleen R., LCDR
Wright, Kenneth T, Ltjg
Wright, Loraine E., LCDR
Wright, Mitchell, LT
Wright, Russell A, Capt
Wright, Samuel F, Capt
Wright, Thomas D, Capt
Wright, W Scott, Lcdr
Wright, William H, Lcdr
Wright IV, Donald A., LCDR
Wright Jr, Harry T, Capt
Wright Jr, M Blair, Lcdr
Wright Jr, Ralph G, CDR
Wright Jr, William D, CDR
Wright Jr, William G, Lt
Wright Jr, William H, CDR
Wright Jr, Willie F, Lcdr
Wright Jr., Thomas E., Capt
Wrightnour, G Brooks, CDR
Wrigley, Francis J, CWO4
Wuestner, Robert J., CDR
Wuichet II, John W, Lcdr
Wulff, Alden T, Capt
Wulff, John T, Radm
Wulff, Robert A, CDR
Wulff, Walter G, Lcdr
Wunsch, William F, Capt
Wurgler, Patricia M., CDR
Wurz, Kenneth F, CDR
Wusstig, Stefan M., Lt
Wyant, Richard A., CDR
Wyatt, David R, Capt
Wyatt, Philip W, CDR
Wyble, Jerome B, Capt
Wyker, John J, CDR
Wyld Jr, Charles W, Cwo-4
Wylie, Donald G, Capt
Wyllys, Ronald E, Capt
Wyma, Bruce H, Capt
Wyman, Kathleen, Lcdr
Wynne, Elmer S, CDR
Wyre Jr., Harry Wilmer, Capt
Wysocki, Joseph, Lcdr
Wyvill II, Samuel A., CDR
Wyzywany, Frank T, CDR
Xenos, Nicholas C., CDR
Xides, Gust P., LT
Yabut, Bernardo R, CDR
Yackley, Mark, 2LT
Yagi, Edward L., Lcdr
Yalon, Jerome M, CDR
Yamada, Ted Y., Lcdr
Yamanouchi, Haruto W, Radm
Yamashiro, Robert Y., CDR
Yamashita, Craig T., CDR
Yancey, Billy L., CDR
Yancey, Michael M., LCDR
Yanders, Armon F, Capt
Yandoh, Judith A., Capt
Yap, Melvyn M L, Capt
Yarber, William J., Capt
Yarborough, Jerry L., LCDR
Yardley Jr, John L M, Capt
Yarger, Albert A., LCDR
Yarger, Peter D., LCDR
Yarling, Richard W, Lt
Yasui, Homer, Capt
Yates, Benjamin S., CDR
Yates, Christopher B, Capt
Yates, David E, CAPT
Yates, Jerry M, LCDR
Yates, Joe W., ENS
Yates, R Douglas, Capt
Yates, Rebekah J., LCDR
Yates, Ronald L, LCDR
Yatsko, George J, Capt
Ybanez, Vicky L.T., Capt
Yeager, Thomas A., CAPT
Yeargin, James C, Radm
Yeaste, Edward A., LCDR
Yedinak, Thomas A., LTJG
Yee, Daniel O, Lt
Yelinek, Leonard J, Lcdr
Yerick, Martin R, CDR
Yerkes, Margaret E., CDR
Yesh, David M., Lcdr
Yetsko, Edward P, CAPT
Yett, John D., LCDR
Yiannias, N J, Lt
Yim, Herbert C, Capt
Yip, David W., CDR
Yoder, Gregory J., CDR
Yoder, John T., Lcdr
Yoder, William E, Capt
Yoho, Robert K., LCDR
Yonemoto, Ronald M, Capt
Yonker, Robert H, Capt
Yopp, George M., Capt
York, Elihu, Capt
York, Gary A, Capt
York, Robert E, Capt
York, Ronald E, CDR
York, Ronald F, Capt
York, Sam A, CDR
Yost, Lori A., LCDR
Yost II, Floyd G, CDR
Young, Allen C., CDR
Young, Bernerd E, CDR
Young, Bruce A., LT
Young, Charles F, Capt
Young, D Michael, CDR
Young, Donald A, CDR
Young, Dykes M., LCDR
Young, Edwin Harold, Capt
Young, Eric Coy, Lcdr
Young, Fred J, CDR
Young, Gary A, Capt
Young, Guy E, CDR
Young, Harold H, CDR
Young, Jacqueline T E, Capt
Young, James F, Capt
Young, Jeffrey A, Lcdr
Young, Jeffrey A., LT
Young, Johanna L, Capt
Young, John H., CDR
Young, John W, Lt
Young, John W, Capt
Young, Karen Gail, LT
Young, Karl S., CDR
Young, Larry R, Capt
Young, Laura M, CAPT
Young, Nancy Z., CDR
Young, Newton L, CWO4
Young, Randall W, Capt
Young, Richard E, Radm
Young, Richard E., LT
Young, Robert L., CDR
Young, Roger A, CDR
Young, Roland S, CDR
Young, Rose M., CDR
Young, Samuel R., Lcdr
Young, Sanford N, Capt
Young, Stephen G, Capt
Young, Steven L., Lcdr
Young, William F., Lcdr
Young, Woodrow W, Lcdr
Young, Worchester D., Cdr
Young Jr, Charles, CDR
Young Jr, George L, Capt
Young Jr, Kenneth L, CDR
Young Jr, William K, Capt
Young Jr., Peter A., Lcdr
Young Sr, Lehman H, Lt
Young VII, William O., CDR
Youngberg, Kevin R., Lcdr
Youngblood Jr, James C, Lcdr
Younger, Michael T., LT
Younger, Robert D, CDR
Younts, Austin R, CDR
Yoviene, Ronald J, CDR
Yraceburn, Michael J, CDR
Yuchasz, Joseph W, CDR
Yue, William M, Capt
Yuen, Charlotte N., CDR
Yuhas, Michael A, CDR
Yuknes, David F., LT
Yungfleisch, Lee, Lt
Yusem, Stephen G, Radm
Yuziuk, Jeffrey A., LCDR
Zabel, Michael L., Lcdr
Zaborowski, James J, CDR
Zaborsky, Mary Jane, LCDR
Zaccone, Michael J., CDR
Zachary, John D, Lcdr
Zaeh, William D, CDR
Zagano, Phyllis, CDR
Zagorites, Jerry A, Lt
Zagranis, James A., CDR
Zagurskie, Leonard, Lcdr
Zahn, Howard E, CDR
Zahner, Carl J, Capt
Zahornacky, James M, CAPT
Zaiger, Kimo K, Lcdr
Zaino, John C., LCDR
Zaino, Teresa A., Lcdr
Zaiser Jr, Ray W, CDR
Zajicek, Roger A, CDR
Zakriski, Timothy G, CDR
Zakula, Michael N, Lcdr
Zaloga, Leo T., CDR
Zander, Andrew T, Capt
Zander, Warren E, Capt
Zander Jr, Richard F, Lcdr
Zandstra, Steven P, Capt
Zaneski, Robert L., LCDR
Zanetti Jr, Joseph R, CDR
Zanin, Eric F., LCDR
Zannis, Thomas N, CDR
Zapata, Michael, Lcdr
Zappone, Julie A., LCDR
Zarem, John, CDR
Zaremba, James W, Capt
Zaroulis, Charles G, Capt
Zarrillo, Eugene M., LT
Zauper, Robert, CDR
Zavatsky, Patricia A., Lcdr
Zavattaro, Frank A, Capt
Zavisza, Daniel M, Capt
Zeb Jr, John M, Capt
Zeh, Richard J, Lcdr
Zeh Jr, Thurber G, Capt
Zehner, Kenneth W, Capt
Zehrt, William Harold, Capt
Zeigler, Robert W, Capt
Zekos, Nicholas V, Capt
Zelensky, Eugene D, Lcdr
Zelko, James J, Capt
Zeller III, William F., CDR
Zeman Jr, Charles D, Lcdr
Zemnickas, William V., Capt
Zemp Jr., John W., LCDR
Zenkovich, Helen, Lcdr
Zeno, Allan S, Capt
Zepp, Eugene J, Lcdr
Zequeira Jr, Marcos A., Capt
Zerbo, Joseph R., CDR
Zerda, Kenneth V, Capt
Zeswitz, Sandra L., Lcdr
Zetterberg, Forrest L, Capt
Zetterstrom, Karen O., LT
Zettler, John W, CDR
Zgolinski, Albert G, CDR
Ziamandanis, John J, CDR
Zicari, Louis P, Capt
Zickos, Theodore R, Capt
Ziebelman, Michael S., CAPT
Zieger, Herbert F, Lcdr
Ziehm, Linda J., Lt
Ziemer Jr, Paul L, Capt
Ziesenheim, Frederick B, CDR
Zignego, Christine P, CDR
Zimble, James A, Vadm
Zimmer, Ithan B., LCDR
Zimmer, James A, CDR
Zimmer, Paul A., Lcdr
Zimmerman, Elaine E, LCDR
Zimmerman, H Stephen, CDR
Zimmerman, Herbert, CWO4
Zimmerman, John D., CDR
Zimmerman, John L, CDR
Zimmerman, Joseph, Lcdr
Zimmerman, Keith M, CDR
Zimmerman, Richard C, Capt
Zimmerman, Ronald L, Capt
Zimmerman, Stephen A, Capt
Zimmerman, Sylvia S, CDR
Zimmerman, Wayne L, Capt
Zimmermann Jr, Herman T, CWO4
Zimtbaum, Samuel D, CDR
Zin, John, Lcdr
Zinchak, William J, Lt
Zingarelli, Michael A., Lt
Zinman, Robert M, Capt
Zinn, Clyde D, CDR
Zinn, Daniel W., Lcdr
Ziolkowski, Roman G, Radm
Zipp, Eugene G, Capt
Ziran, Bruce H., Lcdr
Zirkelbach, Harry W, Capt
Zirkle, Matthew A., Lcdr
Zitka, Glenn L., CDR
Zivovic, Sreten, CDR
Zmuda, Marcia L, CDR
Zmuda Jr, Raymond A., Capt
Zoch, Donald G, CDR
Zock, James L, Capt
Zody, Mary, Lcdr
Zohuri, Bahman, Lcdr
Zolkowski, John A., Lcdr
Zoller, Ernest K, Capt
Zoller II, John B, CDR
Zongrone, Joyce J., Lcdr
Zotter, Edward J, CDR
Zounek, Daniela, LT
Zralek, Robert L, Radm
Zschau, Julius J, Capt
Zuberano, Henry A, Lcdr
Zucarelli, Victor A, CDR
Zucchetto, Nelson R., Lt
Zuhr, Kenneth Christian, CDR
Zuidema, Peter Brian, CDR
Zuino, Angelo R., CWO4
Zukowski, Chester M, Lcdr
Zulauf, Earl C, Lcdr
Zuley, Richard P., LTJG
Zulich, Michael J, CDR
Zumkehr, Wendell F, Lcdr
Zummo, John J, Lcdr
Zumwalt, Crane E, Capt
Zumwalt, Eric R, CAPT
Zumwalt, Robert E, CDR
Zundel, David W., MIDN
Zuniga, Guillermo, Capt
Zupko III, George M, Capt
Zurcher, James L, Capt
Zurnieden Jr, Ludwig A, Capt
Zveare, Robert W, Lt
Zvetina, James R, CDR
Zwall, Richard, Capt
Zwanzig, Peter E, CDR
Zwemke, Roger E, CDR
Zwick, Michael E., LT
Zwicker, William E, CDR

Index

Editor's note: The index does not include entries from the biography and roster sections due to those entries appearing in alphabetical order in their respective sections.

– D –

– E –

– F –

– G –

– H –

– O –

– P –

– Q –

– R –

– S –

FRANKLIN D ROOSEVELT